PROCEEDINGS

of the

2003 IEEE International Symposium on Circuits and Systems

VOLUME IV of V

-Digital Signal Processing-
-Computer Aided Network Design-
-Advanced Technology-

Sunday, May 25 – Wednesday, May 28, 2003

Imperial Queen's Park Hotel
Bangkok, Thailand

Sponsored by the IEEE Circuits and Systems Society and
the Mahanakorn University of Technology

IEEE

Mahanakorn University of Technology

For the latest information please visit the Symposium website:
www.iscas2003.org

Proceedings of the 2003 IEEE International Symposium on Circuits and Systems

IEEE Catalog Number: 03CH37430
ISBN: 0-7803-7761-3
Library of Congress: 80-646530

Additional Copies of this Publication are Available From:

IEEE Service Center
445 Hoes Lane
P.O. Box 1331
Piscataway, NJ 08855-1331
1-800-678-IEEE

Table of Contents

Message
from
His Excellency Thaksin Shinawatra
Prime Minister of Thailand

It is my great pleasure to welcome you to Thailand on the occasion of the 2003 ISCAS Conference. For many, it is probably a first opportunity to visit our country, and I hope it will be a memorable one.

Thailand is presently moving from being a manufacturing based economy to playing a global role in both telecommunications and silicon chip technology. It is indeed a great honour and pleasure that the IEEE Circuits and Systems Society has recognised this role by choosing Thailand to host this timely conference.

The theme of the ISCAS Conference, healthcare and lifestyle technology, is most important, and we are very pleased that Thai engineers will have an opportunity to learn from this international symposium, attended by experts and researchers who are foremost in their field. I am immensely pleased to note that Thai engineers are also submitting papers in the conference. The progress of our country depends on our intellectual resources. Your presence here draws attention to the value we in Thailand place on new technologies and our potential to participate in their creation.

May I wish you all a very stimulating conference, and that you will enjoy the opportunity to experience Thai culture.

(Thaksin Shinawatra)
Prime Minister of Thailand

General Chairs Message

Sawasdee! (a friendly Thai greeting)

As we enter a new wave of technology inspired by lifestyle, healthcare and the environment, Bangkok offers an ideal setting for the 2003 ISCAS: "Life-style Technologies: A New Wave of Circuits and Systems".

In May 1999 the CAS Society, under the presidency of Professor George S. Moschytz, agreed to host ISCAS 2003 in Thailand. This was a momentous result. The foresight of the CAS Society coupled with the dedication and enthusiasm shown by the host, Mahanakorn University of Technology, ensures that ISCAS 2003 will stimulate one of the most rapidly growing technological countries in Far East Asia.

Having successfully established a CAS Society chapter in 1998, the Thai CAS enthusiasts have been pivotal in helping to establish several new initiatives for the CAS Society, including the development of advanced continuing education courses.

ISCAS 2003 is an event not to miss. The technical and social programme will provide outstanding networking opportunities, globally placed where east meets west. The Keynote speakers Sir Richard Sykes, Sir Ara Darzi and Dr Arika Matsuzawa will focus on healthcare from the point of view of technology, cybersurgery and the role of advanced mixed signal design.

Inspired by ISCAS 1994 in London, the technical programme promises to be of the highest quality, reintroducing Forum and Invited sessions. The dedication of our technical program chair, Professor Tor Sverre Lande (Bassen), and his team of technical track chairs, representing all tracks of CAS, has been flawless. Bassen's attention to detail, his strictness and integrity during the review process has been exceptional. The management of the web-based submission system provided by Tom and Barbara Wehner has been sterling. The sheer professionalism in which the review process was conducted was far more than managing software, but an incredible personal touch offered through years of handling the administration of the CAS Society.

Unfortunately, ISCAS 2003 has fallen at a time when the CAS Society is experiencing financial difficulties which would normally have compromised the financial flexibility of this conference. If it were not for the financial assistance graciously provided by Mahanakorn University of Technology, then the entire quality of the conference would have been at stake. However, with head and shoulders up high the ISCAS technical committee can be proud of its record breaking successes.

Last but not least, we thank Mrs. Pornpan Pookaiyaudom (Local Arrangement Chair) together with her colleagues at Mahanakorn University of Technology, for their hard work and creativeness in putting together a local program which is even keeping us in suspense! Also many thanks to the Tourism Authority of Thailand for all the organization and information provided to make your stay in Bangkok most memorable.

Sitthichai Pookaiyaudom
General Co-Chair

Chris Toumazou
General Co-Chair

The centre of Bangkok provides a spectacular setting for ISCAS 2003, a tranquil, thought provoking environment for ISCASers to continue discussions after technical programmes or enjoy the beauty of Thailand.

Temples, museums, people, street markets, cuisines, bars and resorts coupled with the best technical and local arrangement programs will make this an ISCAS to remember.

Welcome to Bangkok!

Technical Program Chair Message

ISCAS 2003 in Bangkok is bound to be successful. New records have been set in whatever way you put it: record number of papers submitted, record number of tutorial proposals, record number of invited (special) sessions. Hopefully you will find the ISCAS schedule familiar with a major emphasis on the technical program. We have framed in the technical sessions with plenary sessions. Every day starts with a keynote session and closes with a forum session.

Our theme *Life-style Technologies: A New Wave of Circuits and Systems* is reflected in the topics for our keynote speakers. On Monday, Sir Richard Sykes, the past Chairman of the worldwide medical company GlaxoSmithKleine will talk about *The Fusion of Biology and Engineering: The Dawn of a New Generation of Medicine.* On Tuesday morning Sir Ara Darzi, a pioneer in Keyhole- and Telesurgery, will follow up with more specific technical issues on engineering in his talk on *Cybersurgery.* Our last keynote speaker on Wednesday morning will be Dr. Akira Matsuzawa from Matsushita Electric Industrial Company talking about advanced mixed-mode integrated circuits and the future impact of lifestyle consumer technologies. His talk is titled: *Mixed Signal SoC: A New Technology Driver in LSI Industry.*

The afternoon forum sessions will hopefully wind up an exciting day for you with interesting panel discussions in a social setting with complementary refreshments. Monday's forum is titled *"CASS, Quo Vadis?"* debating the future of our society, which is important for all of us. Tuesday afternoon's forum session titled *"How Technology Drives Both Globalisation and Helps Anti-globalisers"* is controversial enough to get most people going. The forum session titled *"Schroedinger vs. Kirchhoff: Is the EE Department (and Circuits and Systems) Obsolete?"* is provoking enough for rounding off the conference on Wednesday night.

With a new record of 1812 submitted papers, a little less than 1200 papers were selected through a rigorous review process:

- A technical committee of 18 members assigned papers to 236 different review committee members.
- The members of the different review committees then assigned papers to selected reviewers adding up to 2506 reviewers producing 6074 reviews.
- The acceptance decisions were taken by the review committee members and checked by the technical committee members before our TC meeting in London in December finalizing the technical program.

ISCAS is a true international conference with papers from 50 different countries. Most papers are as expected from the US with 253 contributions, but we also have contributions from Algeria, Bangladesh, Tanzania and Macau. We accepted a little more than 63% of the regular submissions.

We decided to reduce the number of parallel sessions to only 12 leaving a reasonable audience for all sessions. With few parallel sessions and many papers we had to schedule 4 lecture sessions each day with 5 papers in each session. To make the schedule come together, we removed the traditional coffee break making coffee available whenever you need a break. In this way we are able to trade long waiting lines for coffee, for only 10 minutes breaks between sessions. The interactive sessions will go in parallel with the lecture sessions. We have tried to organize the interactive contributions in 13 topical groups of 8 papers. We wanted to give plenty of time for discussions (more than 3 hours) and made only two daily sessions. We received a record number of 30 proposals for invited sessions and accepted 22. We decided to merge these sessions with the other regular technical sessions to make topical tracks. In this way you should not have to run around to find your favorite sessions.

As usual we will have our tutorial day on Sunday. Again we received a record number of 45 tutorial proposals. 17 carefully selected tutorials were accepted. It is a mixture of full-day and half-day tutorials with interesting topics for everyone.

As you can imagine, the apparatus for getting through a smooth review process and setting up a technically sound and good program is demanding in many ways. Tom Wehner and his team have provided a remarkable technical solution for electronic treatment of papers and reviews. For the first time the complete program composition was done electronically during our technical meeting. Tom's assistance is not only technical, but also practical in many ways. His knowledge of the ISCAS review process and procedures is the only record we have and as a resource person he is a major asset for the CASS community hardly understood by the CASS leadership.

The enormous efforts done by technical committee members, review committee members and reviewers together with the large number of great contributions is the real soul of ISCAS. The compelling enthusiasm and efforts of the ISCAS community to contribute in different ways is the strength of ISCAS. It is important, however, to emphasize the tremendous support ISCAS 2003 received from Prof. Sitthichai Pookaiyaudom and Mahanakorn University of Technology. Without the financial and administrative support they have offered, ISCAS 2003 would never have happened.

ISCAS 2003 in Bangkok will be an experience from a technical point of view, but also socially. The facility is great and the friendliness and beauty of the Thai people will make ISCAS 2003 unforgettable for all of you. There is only one record left, let us have a record number of attendees as well!

See you all in Bangkok.

Tor Sverre Lande
Technical Program Chair

Organizing Committee

General Co-Chairs
Sitthichai Pookaiyaudom, *Mahanakorn University of Technology, Thailand*
Chris Toumazou, *Imperial College, London*

Vice Chair
Jitkasame Ngarmnil, *Mahanakorn University of Technology, Thailand*

Technical Program Chair
Tor Sverre Lande, *University of Oslo, Norway*

Technical Advisory Committee
Martin Hasler, *Swiss Federal Institute of Technology (EPFL), Lausanne, Switzerland*
Nobuo Fujii, *Tokyo Institute of Technology, Japan*
Georges Gielen, *Katholieke Universiteit Leuven, Belgium*
Thanos Stouraitis, *University of Patras, Greece*
Wanlop Surakampontorn, *King Mongkut's Inst. of Tech. Ladkrabang, Thailand*
Sawasd Tantaratana, *Sirindhorn International Institute of Technology, Thailand*

Finance Chair
Sujate Jantarang, *Mahanakorn University of Technology, Thailand*

Keynote Session Chair
Magdy A. Bayoumi, *University of Louisiana, USA*

Forum Sessions Chairs
George S. Moschytz, *Swiss Federal Institute of Technology (ETH), Switzerland*
Hanspeter Schmid, *Bernafon AG, Switzerland*

Tutorial Sessions Chairs
Andreas G. Andreou, *Johns Hopkins University, USA*
Wouter A. Serdijn, *Delft University of Technology, The Netherlands*

Invited Sessions Chairs
Ljiljana Trajkovic, *Simon Fraser University, Canada*
Gert Cauwenberghs, *Johns Hopkins University, USA*

Industry Liaisons
Graham Hellestrand, *VaST Systems Technology, USA*
Hua Su, *KLA-Tencor Corporation, USA*

ISCAS Steering Committee Chair
Geert De Veirman, *Agere Systems, USA*

Local Arrangements Chair
Pornpan Pookaiyaudom, *Mahanakorn University of Technology, Thailand*

Local Arrangements Committee
Jitkasame Ngarmnil, *MUT, Thailand*
Narissara Intrachan, *MUT, Thailand*
Apinunt Thanachayanunt, *King Mongkut's Institute of Technology Ladkrabang, Thailand*
Sompop Purivikaipong, *MUT, Thailand*
Satida Kailad, *MUT, Thailand*
Jirayut Mahattanakul, *MUT, Thailand*
Thanwa Sripramong, *MUT, Thailand*
Phaophak Sirisuk, *MUT, Thailand*
Manop Orphimai, *MUT, Thailand*
Mitchai Chongcheawchamnan, *MUT, Thailand*
Peerapol Yuvapusitanont, *MUT, Thailand*
Athikom Roeksabutr, *MUT, Thailand*
Nuanrat Padungkul, *MUT, Thailand*
Jatoron Teeratchaichayuti, *MUT, Thailand*
Kosin Jumnongthai, *The University of the Thai Chamber of Commerce, Thailand*
Nunthana Noygert, *MUT, Thailand*

IEEE CAS Society Liaison
S. Panchanathan, *Arizona State University, USA*

IEEE CAS Society Administrator
Barbara Wehner, *IEEE CAS Society*

International Coordinators
USA & Canada
Edgar Sanchez-Sinencio, *Texas A&M University, USA*
Europe, Middle East & Africa
Anthony C. Davies, *Kingston University, UK*
Latin America
Sergio L. Netto, *Federal University of Rio de Janeiro, Brazil*
Asia
Nobuo Fujii, *Tokyo Institute of Technology, Japan*
Australia & Pacific
Branko Celler, *University of New South Wales, Australia*

Professional Conference Organizer
CM Organizer Co., Ltd., 1417 Town in Town, Bangkok, 10310, Thailand

Tour Host
Tour of Siam Co., Ltd., 61/20 Rama 9, Bangkok, 10310, Thailand

Technical Program Committee

Analog Circuits and Signal Processing
Robert Fox, *University of Florida*
Tuna Tarim, *Texas Instruments, Inc.*

Bio-medical Circuits and Systems
Esther Olivia Rodriguez Villegas, *Instituto de Microelectronica de Sevilla*

General and Nonlinear Circuits and Systems
Joos Vandewalle, *Katholieke Universiteit Leuven*
Rueywen Liu, *University of Notre Dame*
(Blind Signal Processing)
K. Thulasiraman, *University of Oklahoma*
(Graph Theory and Computing)
Marian Kazimierczuk, *Wright State University*
(Power Systems)

Digital Signal Processing
Tapio Saramaki, *Tampere University of Technology*

Communications
Huifung Sun, *Mitsubishi Electric Research Laboratories*
Magdy Bayoumi, *University of Louisiana*
(Circuits and Systems for Communications)

Multimedia Systems & Applications
Liang-Gee Chen, *National Taiwan University*

VLSI Systems & Applications
Keshab Parhi, *University of Minnesota*
Chein-Wei Jen, *National Chiao Tung University*

Computer Aided Network Design
Trond Ytterdal, *Norwegian Institute of Science and Technology*

Neural Networks and Systems
Wai-Chi Fang, *California Institute of Technology*
Luigi Fortuna, *Universita'degli Studi di Catania*
(Cellular Neural Networks)
Ralph Etienne-Cummings, *University of Maryland*
(Biologically Inspired Systems)

Advanced Technology
Ralph Etienne-Cummings, *University of Maryland*
Peter (Chung-Yu) Wu, *National Chiao Tung University*
(Nanoelectronics & Gigascale)

Invited Sessions
Gert Cauwenberghs, *Johns Hopkins University*
Ljiljana Trajkovic, *Simon Fraser University*

Review Committee Members

We would like to thank these individuals for contributing to the success of ISCAS 2003 through their efforts in coordinating the review process:

Abshire, Pamela
Afghahi, Cyrus
Ahmad, Ishfaq
Ahmadi, Majid
Aizawa, Kiyoharu
Alarcon, Eduard
Allen, Phil
Andreou, Andreas
Arena, Paolo
Aronhime, Peter
Arslan, Tughrul
Asai, Hideki
Au, Oscar C.
Badawy, Wael
Baglio, Salvatore
Barrettino, Diego
Barrows, Geoffrey
Barry, Mark
Baschirotto, Andrea
Basu, Sankar
Bayoumi, Magdy
Binkley, David
Branciforte, Marco
Burdett, Alison
Cabbouj, Moncef
Cauwenberghs, Gert
Cetin, A. Enis
Chan, Shing-Chow
Chau, Lap-Pui
Chen, ChangW
Chen, Jie
Chen, Liang-Gee
Chen, Oscal T.C.
Chen, Xiang
Chiang, Tihao
Chrzanowska-Jeske, M.
Chung, Henry
Czarkowski, Dariusz
Degertekin, Levent
Delbruck, Tobi
Deng, Tian-Bo
Duque, Francisco
Egiazarian, Karen
Einwich, Karsten
Eisenstadt, Bill
ElGamal, Mourad
Elwakil, Ahmed
Endo, Tetsuro
Etienne-Cummings, R.
Fiez, Terri
Filanovsky, Igor
Fliege, Norbert
Fortuna, Luigi
Foty, Daniel
Fox, Rob
Frasca, Mattia
Freely, Orla
Friedman, Eby
Galkowski, Krysztof
Gao, Wen
Geiger, Randy
Genov, Roman
Georgiou, Julio
Ghodssi, Reza
Gielen, Georges
Gilli, Marco
Green, Michael
Hang, Hsueh-Ming
Harris, John
Hasan, Anwar
He, Zhihai
Helfenstein, Markus
Hernandez, Luis
Higgins, Charles
Hinamoto, Takao
Hodge-Miller, Angela
Hong, Xianlong
Horiuchi, Timothy
Hu, Bo
Huang, Jiwu
Ibrahim, Mohammad
Indiveri, Giacomo
Inouye, Yujiro
Ioinovici, Adrian
Ismail, Yehea
Ivanov, Vadim
Jamali, Mohsin
Johansson, Hakan
Julian, Pedro
Karam, Lina
Karhunen, Juha
Katti, Raj
Kawamata, Masayuki
Kiaei, Sayfe
Kolumban, Geza
Korotkov, Alexander
Kung, David
Kuo, C. C. Jay
Kuo, Chin-Hwa
Kuo, Chung J.
Kwan, H.K.
Laakso, Timo
Lai, Y.K.
Lee, Edward
Lehmann, Torsten
Lie, WenNung
Lim, Sung Kyu
Lim, Yong Ching
Lin, Chia-Wen
Lin, ChinTeng
Lin, D. W.
Ling, Nam
Liou, Ming
Liu, Bin-Da
Liu, Derong
Liu, Shih-Chii
Loui, Alexander
Lu, Wu-Sheng
Luo, Hui
Lustenberger, Felix
Maggio, Gian Mario
Malcovati, Piero
Maloberti, Franco
Manganaro, Gabriele
Markov, Igor
Mason, Andrew
Massie, Mark
Mathis, Wolfgang
Mehta, Dinesh
Milanovic, Jovica
Minch, Bradley
Miyanaga, Yoshikazu
Mok, Philip
Moon, Un-Ku
Mukund, P.R.
Ndjountche, Tertulien
Newcomb, Robert
Ng, Tony T.
Ngi, NganKing
Nguyen, Truong
Niamat, M.
Nils, Matthias
Nilsson, Peter
Nishio, Yoshifumi
Nowrouzian, Behrouz
Nwankpa, Chika
Occhipinti, Luigi
Ogorzalek, Maciej
Palumbo, Gaetano
Pastore, Stefano
Pei, Soo-Chang
Phoong, See-May
Pirsch, Peter
Pouliquen, Philippe
Rajan, P.K.
Ramachandran, Ravi
Ramirez-Angulo, Jaime
Reddy, Hari
Renfors, Markku
Roberts, Gordon
Rodriguez-Vazquez, A.
Rodriguez-Villegas, E.
Roy, Kaushik
Saito, Toshimichi
Saluja, Kewal
Sanchez-Sinencio, E.
Sapatnekar, Sachin
Sarmiento-Reyes, A.
Schmid, Hanspeter
Sengor, Neslihan
Serdijn, Wouter
Setti, Gianluca
Shanbhag, Naresh
Sheu, Bing
Shi, Bertram
Shi, Weiping
Shi, Yun-Qing
Shibata, Tadashi
Silva, Jose
Sobelman, Gerald
Song, Xiaoyu
Soudris, Dimitros
Sridhar, Ramalingam
Stan, Mircea
Stocker, Alan
Stouraitis, Thanos
Suetsugu, Tadashi
Sun, MingTing
Sun, Yichuang
Sunwoo, Myung
Sylvester, Dennis
Szolgay, Peter
Tabus, Ioan
Tarim, Tuna
Temes, Gabor
Tetzlaff, Ronald
Theodoridis, Sergios
Thulasiraman, K.
Tong, Lang
Trajkovic, Ljiljana
Tsai, Tsunghan
Vainio, Olli
Van der Speigel, Jan
van Schaik, Andre
Vandevoorde, Glenn
Vandewalle, Joos
Wang, C.C.
Wang, Eric
Wang, Jun
Wang, Xiao Fan
Wang, Yuke
Wanhammar, Lars
Weeks, Michael
Worapishet, Apisak
Wu, Andy
Wu, Chung Yu
Wu, Ja-Ling
Yadid Pecht, Orly
Yamamura, Kiyotaka
Yang, Jar-Ferr
Yong, Lian
Ytterdal, Trond
Zaghloul, Mona
Zarandy, Akos
Zeng, Bing
Zheng, Wei Xing
Zhiping, Lin
Zhuang, Xinhua

Reviewers

We would like to thank the following individuals for their efforts in the review process of ISCAS 2003:

Abbasi, S.
Abdul-Aziz, A
Abed, Khalid
Abeysekera, Saman
Aboulhamid, M.
Aboulnasr, Tyseer
Abshire, Pamela
Adams, K. J.
Adams, Michael
Adde, Patrick
Adeniran, Olujide
Adil, Farhan
Adiseno, A
Adlerstein, Miriam
Adut, Jozef
Aeag, Aeag
Afghahi, Cyrus
Aga, Arshan
Agarwal, Amit
Aggoun, Amar
Agrawal, Dakshi
Ahmad, M.O.
Ahmad Azli, N.
Ahmadi, M
Ahmadi, Shahrokh
Ahmed, Ayman
Ahn, Gil-Cho
Ahn, Hong Jo
Aihara, Kazuyuki
Aizawa, Kiyoharu
Akers, Lex
Akhtar, M.
Akkarakaran, Sony
Akramullah, S.
Aksen, Ahmet
Alarcon, Eduard
Albinet, Xavier
Al-Hashimi, Bashir
Alioto, Massimo
Allam, Mohamed
Allen, Phillip
Allstot, David
Almaini, A. E. A.
Alonso, Marcos
Alpert, Charles
Al-Rabadi, Anas
Al-Sarawi, Said
Al-Tawil, Khalid
Altun, Gulsah
Amin, Chirayu
Amira, A
Amourah, Mezyad
Amuso, Vincent
Andrea, Baschirotto
Andreani, Piero
Andreani, Pietro
Andreev, Boris
Andreou, Andreas
Annunziato, Mauro
Antonio, John
Antoniou, Andreas
Anvar, Ali
Apsel, Aliyssa
Apsel, Alyssa
Aramvith, S.
Ardalan, Sasan
Arena, Paolo
Arik, Sabri
Arnold, Mark
Aronhime, Peter
Arora, Nimarta
Arslan, Tughrul
Arthur, John
Asai, Hideki
Åström, Pontus
Atwell, Bob
Au, Oscar
Aunet, Snorre
Ausin, Jose L.
Auvergne, Daniel
Awrejcewicz, Jan
Ayazi, Farrokh
Aydin, N
Bacciotti, Andrea
Badawy, Wael
Bae, Hyeon-Min
Baghaie, Ramin
Baglio, Salvatore
Bai, Junfeng
Baier, Steven
Baier, Steve
Bajard, J.
Bakhshai, Alireza
Baki, Rola
Bakkaloglu, Bertan
Balamurugan, G.
Balsara, Poras
Balteanu, Florinel
Balya, David
Bandela, Chaitanya
Barmada, Sami
Barrettino, Diego
Barrows, Geoffrey
Barry, Mark
Barthelemy, Hervé
Bartlett, Viv
Bashagha, Akil
Batalama, S. N.
Batten, Robert
Beccari, Claudio
Belenky, Alex
Belkasim, Saeid
Bell, Kristin
Bellar, Maria
Bellini, Armando
Ben Dhaou, Imed
Benaissa, M.
Bendix, Peter
Benesty, Jacob
Ben-Yaakov, Sam
Berberidis, C
Berkeman, Anders
Berket, Karlo
Bermak, Amine
Bernhard, Bernhard
Berrou, Claude
Bertoni, G
Bettayeb, Maamar
Bharath, Anil
Bharath, K. T.
Bhushan, Manjul
Bi, Guoan
Bialek, Janusz
Bian, Jinian
Bibyk, Steven
Bickerstaff, Mark
Biere, Armin
Biernacki, Janusz
Biey, Mario
Binkley, David
Bisdounis, Labros
Bjørnsen, Johnny
Blaian, Jennifer
Blalock, Travis
Blalock, Benjamin
Bomar, Bruce
Bonani, Fabrizio
Boni, Andrea
Boonchoo, B.
Boonyaroonate, I.
Borghetti, Fausto
Bose, N
Bose, Tamal
Bouldin, Donald
Bouridane, Ahmed
Boussaid, Farid
Bouzerdoum, A.
Bowman, Robert
Boylston, Eddie
Brajovic, Vladimir
Branciforte, Marco
Brandolini, M.
Bregovic, Robert
Brehler, Matthias
Brennan, Paul
Britton, Chuck
Brooke, Martin
Brorsson, Mats
Brown, Lyndon
Bruce, Jay
Bruce, Thompson
Bruckner, T.
Brueske, Dan
Bruton, Leonard
Bruun, Erik
Bui, Hung
Burdett, Alison
Burger, Thomas
Burian, Adrian
Burleson, Jeffrey
Burt, Rodney
Cabeza Laguna, R.
Cai, Jianfei
Cai, Xiaodong
Cai, Hua
Cakici, Tamer
Calabro, Antonino
Calvo-Lopez, Belen
Cao, Lei
Cao, Yong-Yan
Cao, Kevin
Cao, Jun
Cao, Jinde
Carlosena, Alfonso
Carmona, Ricardo
Carro, Luigi
Carroll, Thomas
Cartaxo, Adolfo
Carusone, Anthony
Carvajal, Ramon
Castagnetti, R.
Castilla, Miguel
Castorina, S.
Cauwenberghs, G.
Caverly, Robert
Caviglia, Daniele
Celasun, Isil
Celma, Santiago
Cervantes, Ilse
Cetin, Enis
Cevik, Kulmiz
Chai, D.
Chaing, Jen-Shiun
Chakrabarti, C.
Chakrabartty, S.
Cham, Wai-Kuen
Champac, Victor
Champneys, Alan
Chan, Thomas
Chan, Wai Sum
Chan, Yui-Lam
Chan, Yuk-Hee
Chan, S.C.
Chandra, Naveen
Chandramouli, R.
Chang, Pao-Chi
Chang, L.H.
Chang, Shih-Cheng
Chang, Shih-Fu
Chang, Te-Hao
Chang, Wen Tsung
Chang, Yuyu
Chang, Minkuan
Chang, Yu-Lin
Chang, Hao-Chieh
Chang, Feng-Cheng
Chang, D.
Chang, Chip Hong
Chang, Chin-Chen
Chang, Chen-Hao
Chang, C.
Chang, Ben-Jye
Chang, Ruey-Feng
Chang, Hsuan Ting
Chao, Su
Chao, Wei-Min
Charkraborty, S.
Chau, Lap-Pui
Chaudhry, Irfan
Chaudhuri, Bikram
Cheely, Matthew
Chen, Wen-Cheng
Chen, Sao Jie
Chen, Sau-Gee
Chen, Tao
Chen, Tongwen

Chen, Trista
Chen, Tsuhan
Chen, Oscal T.-C
Chen, Xiang
Chen, Xiaofan
Chen, Xilin
Chen, Yen-Fu
Chen, Ying-Cheng
Chen, Yiqin
Chen, Yirng-An
Chen, Guanrong
Chen, Yuh-Shyan
Chen, Charlie
Chen, Mely Chi
Chen, Guozhu
Chen, Chang Wen
Chen, Ching-Yeh
Chen, Chuen-Yau
Chen, Chung-Ho
Chen, Chunhong
Chen, Degang
Chen, Guoqing
Chen, Hua-Mei
Chen, Huiting
Chen, Jiann-Jone
Chen, Jie
Chen, Lih-Shyang
Chen, Kevin
Chen, Lauren Hui
Chen, Bor-Sen
Cheng, Ming-Yang
Cheng, C.
Cheng, Sheu-Chih
Cheng, Hui
Cheng, D.
Cheong, K.-W.
Cheong, Cha-Keon
Chern, Ming-Yang
Chi, Chong-Yung
Chiacchiarini, H.
Chiang, Jen-Shiun
Chiang, Yu-Ming
Chico, Jorge Juan
Chien, Shao-Yi
Chik, Raymond
Ching, P. C.
Chiu, Chin-Fong
Chiueh, Tzi Dar
Cho, Changhyuk
Cho, Choongeol
Cho, Choonsk
Cho, Kyeongsoon
Cho, Seonghwan
Choi, Byungcho
Choi, Seungjin
Choi, Jun Rim
Choi, Seung Hoon
Chong, Philip
Chongcheawchum-narn, Mitchai
Choo, Hunsoo
Chou, Chung-Yun
Chou, Cheng-Fu
Chou, Chun-Hsien
Chow, Tommy
Chowdhury, Sazzadur
Christopoulos, C.
Chrzanowska-Jeske, Malgorzata
Chu, Chong-Nuen
Chuang, T.
Chuang, Chiaoyao
Chueh, Juan-Ying
Chung, Jen-Feng
Chung, Jin-Gyun
Chung, Pau-Choo
Chung, Ronald
Chung, Wonzoo
Chung, Henry
Chuvychin, V.
Cimen, Ebru
Cioffi, John M.
Ciriani, Valentina
Civalleri, P.
Claesson, Ingvar
Clapp, Matthew
Clara, Martin
Clark, John
Clauss, Christoph
Cockborn, Bruce
Cohen, Marc
Coleman, Jeffrey
Concalves, Jorge
Cong, Hong-Ih
Cong, Yonghua
Connelly, J Alvin
Considine, Peter
Constandinou, T.
Conti, Massimo
Copeland, M
Courbage, M.
Cox, David
Cqco, Cqco
Crozier, S
Cruz Martins, D.
Culurciello, E.
Curzan, Jon Paul
Cusinato, Paolo
Czarkowski, D.
Czarkowski, D.
Dallago, Enrico
D'Amico, Belen
D'Amore, Dario
Daneshbeh, Amir
Dante, Andreas
Darabi, Jeff
Darwish, Tarek
Das, Manohar
Dasgupta, Pallab
Datta, Animesh
Daumas, Marc
Davoodi, Azadeh
De Caro, D.
De Feo, Oscar
De La Cruz-Blas, A
De La Vega, A.
De Lima, Jader
De Queiroz, A.
De Silva, Liyanage
De Wilde, Wim
Debrunner, Linda
Deen, Jamal
Deever, Aaron
Degertekin, Levent
Deiss, Armin
Dejhan, Kobchai
Del Rio, Rocio
Delbruck, Tobias
Delgado Restitut, Manuel
Demosthenous, A.
Dempster, Andrew
Deng, Yunbin
Deng, Irene
Deng, Tian Bo
Denk, Tracy
Derryberry, R.T.
Deshmukh, R.G.
Dessouky, M.
Deval, Yann
Dewilkins, William
Di Bernardo, Mario
Di Garbo, Angelo
Di Giandomenico, Antonio
Diamante, Olga
Diaz-Mendez, A.
Diaz-Sanchez, A.
Dimitrakopoulos, G.
Ding, Jian-Jiun
Diniz, Paulo
Dipaola, Nunzio
Dittmar, Andre
Do, Minh
Doboli, Alex
Doddo, Francesco
Doerrer, Lukas
Dogaru, Radu
Domanski, Marek
Donato, Cafagna
Dong, Sheqin
Dong, Zhiwei
Dongming, Xu
Doorenbos, Jerry
Dores Costa, J. M.
Doyle, Johnn
Draou, A.
Du, Limin
Dudek, Piotr
Dulger, Fikret
Duman, Tolga
Dumitrescu, B.
Dung, Lanrong
Dupret, Antoine
Duque-Carrillo, F.
Dutta, Santanu
Dys, Grzegorz
Ebeid, N
Effelsberg, W.
Eguchi, Kei
Eguiazarian, Karen
Einwich, Karsten
Eisenstadt, Bill
Ekelund, Jan-Erik
El Leithy, Nevine
Elakkumanan, P.
El-Erian, Heba
Eles, Petru
El-Gamal, Mourad
Elgharbawy, Walid
Elhak, Hany
Elliott, Duncan
El-Moursy, Magdy
Elrabaa, M.
Elwakil, Ahmed
Elwan, Hassan
Emira, Ahmed
Endo, Tetsuro
Enright, Michael
Enz, Christian
Eo, Yungseon
Erdogan, Ahmet
Ericson, M Nance
Ernst, Rolf
Espejo, Servando
Etienne-Cummings, Ralph
Evangelista, G.
Ewing, Robert
Fabre, Alain
Fakhraie, Sied
Falkowski, Bogdan
Fan, Chih-Peng
Fan, Yi
Fang, Wai-Chi
Fang, Hung-Chi
Fang, Jing-Jing
Farshbaf, Hamed
Faulkner, Mike
Fayed, Ayman
Fei, Haibo
Femia, Nicola
Ferguson, Paul
Fernandes, Jorge
Ferrari, Vittorio
Ferrer, Enrique
Fiez, Terri
Figueras, Johan
Fijalkowski, B.
Filanovsky, Igor
Fischer, Godi
Fish, Alex
Fliege, Norbert
Flueck, Alexander
Flügel, Sebastian
Fnaiech, Farhat
Foldesy, Peter
Foo, Say-Wei
Forgues, Scott
Forti, M.
Fortuna, Luigi
Fossas, Enric
Foty, Daniel
Fox, Rob
Francesconi, F.
Franchi, Eleonora
Francken, Kenneth
Franken, Dietrich
Frasca, Mattia
Freimann, Achim
Freitas, L.C.
Frey, Doug
Frey, Matthias
Friebe, Lars
Friedman, Eby
Fu, Dongdong
Fu, Ming
Fujishima, Minoru
Fujita, Masahiro
Fukui, Yutaka
Fuller, Arthur T.G.
Funada, Tetsuo
Fung, Chun Yu
Furber, S.
Furth, Paul
Gabbouj, Moncef
Gabrea, Marcel
Gacsadi, Alexandru
Galias, Zbigniew

Galijasevic, Enisa
Galkowski, K.
Galkowski, K.
Gamulkiewicz, B.
Gan, Woon-Seng
Gandelli, A.
Gandhi, Bhavan
Garcera, Gabriel
Garcia, Antonio
Garcia Ortiz, A.
Garcia-Andrade, M.
García-Armada, A.
Gardini, Laura
Garello, R.
Garg, Adesh
Garg, Nitin
Garrett, David
Gatica-Perez, D.
Gatti, Umberto
Gehrke, Winfried
Genov, Roman
Georgiou, Julius
Georgoulas, Nikos
Gerek, Omer
Gerosa, Andrea
Gevorkian, David
Ghallab, Yehya
Gharavi, Hamid
Ghassemi, Forooz
Ghodssi, Reza
Ghoneima, Maged
Ghosh, Donna
Giannakis, G.
Gierkink, Sander
Gildenblat, G.
Gilli, Marco
Giral, Roberto
Girlando, Giovanni
Giurcaneanu, C.
Giustolisi, Gianluca
Glentis, George
Goh, Say Song
Gokcay, Didem
Gola, Alberto
Goldberg, David
Goldberg, Rick
Goldgeisser, L.
Gomez, Gabriel
Gong, Xiaoxiang
Gonzalez Jimenez, Jose Luis
Goodman, Jim
Gopalarathnam, T.
Goras, Liviu
Goren, Leyla
Gotchev, Atanas
Gramacki, Artur
Gramacki, Jaroslaw
Grant, Doug
Grassi, Giuseppe
Graves, Zaminah
Graziani, Salvatore
Grimaldi, D.
Grimm, Christoph
Grueger, Klaus
Gruev, Viktor
Gu, Yongru
Gu, Jeffrey
Gu, Ming
Guan, Ling
Guan, Yong Liang
Gubina, Ferdinand
Gui, Xiang
Guinjoan, Francesc
Guo, Cheng-An
Gura, Nils
Gustafsson, Oscar
Gutierrez De Anda, Miguel A.
Guzelis, Cuneyt
Habili, Nariman
Haddad, Sandro
Hadjicostis, C.
Haehnle, Reiner
Haenggi, Martin
Hafed, Mohamed
Hagleitner, C.
Halonen, Kari
Ham, Donhee
Hamilton, Alister
Hamoui, Anas
Hamzaoglu, Fatih
Han, Dong Seog
Hang, Hsueh-Ming
Hänggi, Martin
Hanjalic, Alan
Hanumolu, Pavan
Hao, Shu-Sheng
Haroun, Baher
Harris, John
Harrison, Reid
Hartenstein, Reiner
Harvey, Tim
Hasan, M
Hasegawa, Akio
Hashim, Ahmed
Hasler, Martin
Hasler, Paul
Hata, Yukata
Hatrack, Paul
He, Chen
He, Chenming
He, Ming
He, Shousheng
He, Xinhua
He, Yong
He, Yun
He, Yuwen
He, Zhenya
He, Zhihai
He, Zhongli
Hegazy, Hazem
Heijmans, Henk
Heinzelman, W.
Helfenstein, M.
Hellberg, Richard
Henkel, Frank
Hernandez, David
Hernandez-Martinez, Luis
Hernes, Bjørnar
Herrmann, Klaus
Hespanha, Joao
Heydari, Payam
Higashimura, M.
Higgins, Charles
Hikihara, Takashi
Hill, Jim
Himsoon, T.
Hinamoto, Takao
Hiser, Douglas
Hiskens, Ian
Hjalmarsson, Emil
Hjorungnes, Are
Ho, Chia-Chiang
Ho, Jungle
Ho, K. C.
Ho, Yo-Sung
Ho, Antony T. S.
Hodge, Tiffany
Hodge, Angela
Hodge, Robin
Hoekstra, Jaap
Hogervorst, Ron
Holland, Rod
Holmberg, Johnny
Hong, David
Hong, Li
Hong, Yeguang
Honig, Michael L.
Horio, Yoshihiko
Horita, Eisuke
Horiuchi, Timothy
Hoshuyama, O.
Hou, Yanfeng
Hou, Zeng-Guang
Hristu, Dimitris
Hsia, Shih-Chang
Hsiang-Cheh, H.
Hsieh, Pi-Fuei
Hsieh, Ming-Ta
Hsieh, Junwei
Hsieh, Chaur-Heh
Hsieh, Ching-Tang
Hsu, Pochin
Hsu, Huai-Yi
Hsu, Chun-Lung
Hsu, Chiou-Ting
Hsu, Chih-Wei
Hsu, Terng-Yin
Hsung, Tai-Chiu
Hu, Yongjian
Hu, Bo
Hu, Haitian
Hu, Jiang
Hu, Ming
Hua, Yingbo
Huang, Yueh-Min
Huang, Shih-Hsu
Huang, Weimin
Huang, Feng-Jung
Huang, Tiejun
Huang, Shih-Way
Huang, Yu-Wen
Huang, Jiwu
Huang, Jiun-Lang
Huang, Fred
Huang, Chunglin
Huang, C.
Huang, Han
Huang, Kuan-Hsun
Hueper, Knut
Huerta, Ramon
Hughes, John
Hung, William
Hung, Chih-Ming
Hung, Hsien-Sen
Huo, Q.
Hurst, Paul
Hutter, Andreas
Hwang, Tingting
Hwang, Yin-Tsung
Hwang, Wen-Liang
Hwang, Shin-Jai
Hwang, Shih-Arn
Hwang, Chien-Hwa
Hwang, Ren-Hung
Hwee, Gwee Bah
Hyvarinen, Aapo
Iannazzo, Mario
Ichiro, Kuroda
Iinatti, Jari
Ikehara, Masaaki
Indiveri, Giacomo
Inki, Mika
Inoue, Yasuaki
Inouye, Yujiro
Ioinovici, Adrian
Irie, Hisaichi
Isao, Nakanishi
Ishibash, Koichiro
Ishiguro, Makio
Islam, Syed
Ismail, Yehea
Itoh, Yoshio
Ivanov, Vadim
Ivascu, Alex
Ivlev, Ruslan
Izquierdo, Ebroul
Jachalsky, Jörn
Jain, Faquir
Janda, Zarko
Jang, Hyuk-Jae
Jang, Jyh-Shing
Jank, Wolfgang
Janke, W.
Jankovic, Sasa
Janssen, Geert
Jayadeva, M
Jeon, Okjune
Jeon, Woo Chul
Jeong, Deog-Kyoon
Jeong, Kyong-Pil
Jeong, Woopyo
Jeppson, Kjell
Jepsson, Kjell
Jiang, Hanjun
Jiang, Yingtao
Jiang, Yu
Jin, Craig
Jin, Le
Jin'no, Kenya
Jinzenji, Kumi
Jitrangsri, Tasapon
Johannessen, Erik
Johansson, Hakau
Johansson, Stefan
Johansson, Hakan
Johnson, Gregory
Jørgensen, Ivan
Jose, De La Rosa
Joseph, Joseph
Joshi, Rajan
Joshi, Rajiv
Jou, Jing-Yang
Jovcic, Dragan
Juffer, Lance
Julian, Pedro
Jullien, Graham
Jung, M

Jung, Kooho
Kage, Tetsuro
Kaiser, Andreas
Kajitani, Yoji
Kale, Izzet
Kale, Kaustubh
Kaler, Karan
Kamal, Abu-Hena
Kamata, Hiroyuki
Kamendje, G
Kammula, A.
Kan, Kou-Sou
Kanade, Takeo
Kaneko, Mineo
Kankanhalli, M.
Kanyogoro, Ndiritu
Kapur, Ateet
Karp, Tanja
Karsilayan, Aydin
Karthikeyan, S.
Karvanen, Juha
Kastner, Ryan
Kato, Toshiji
Kaul, Richard
Kawamoto, Mitsuru
Kay, Douglas
Kay, Kay
Kearney, Ian
Keller, Bob
Kempel, Leo
Kennedy, Michael
Kennings, Andrew
Ker, Ming Dou
Kerness, Nicole
Keskin, Mustafa
Khasawneh, M.
Khawam, Sami
Khoury, John
Khumsat, P.
Ki, Wing-Hung
Kilchoer, José
Kim, Yong-Bin
Kim, Jae Joon
Kim, Kuenwoo
Kim, Pansop
Kim, Seokjin
Kim, Wonjong
Kim, Yong Kwan
Kim, Bruce
Kimura, Tomohisa
King, Irwin
Kirkham, Tony
Kis, Gabor
Kiss, Peter
Kiya, Hitoshi
Kleine, Ulrich
Kleinfelder, Stuart
Kleismit, Rick
Kliewer, Joerg
Kloos, Helge
Klumperink, E.
Ko, Hyung-Jong
Ko, Yusuck
Kocak, Taskin
Kocarev, Ljupco
Kofidis, E
Kohda, Tohru
Kok, Ted
Kolokotronis, N
Kolumban, Geza
Konishi, Keiji
Koon, Lee
Kopmann, Heiko
Kopsinis, Y
Korman, Can
Kormanyos, Brian
Korotkov, A.
Kou, Yajun
Koufopavlou, O.
Koukoulas, P
Kourouma, M.
Koutroumpezis, G.
Kozak, Muchit
Kraig, Olejniczak
Kravets, Victor
Krishnamohan, S.
Krishnan, Shoba
Kristensen, Fredrik
Krizhanovsk, V.
Krukowski, Artur
Kubota, Toshiro
Kuchcinski, K.
Kucic, Matt
Kugi, Andreas
Kuhn, Bill
Kuhn, Jay
Kukimoto, Yuji
Kumar, Kapil
Kummaranguntla, Ravi
Kummert, Anton
Kumwilaisak, W
Kundur, Deepa
Kuo, Tien-Ying
Kuo, Chung J.
Kuo, C.-C. Jay
Kuo, Peggy
Kuo, Sen-Maw
Kuroda, Ichiro
Kursun, Volkan
Kwan, H. K.
Kwok, Chee Yee
Kwon, S.
Kyaw, Kyaw
Laakso, Timo
Laberge, Sebastien
Lagudu, Sateesh
Lai, Yen-Tai
Lai, Yeong-Kang
Lai, Justin
Lai, Xiaoping
Lai, Shang-Hong
Lai, Y. K.
Lai, Yuk Ming
Laih, Chi-Sung
Lalithambika, V.
Lam, Wai
Lampinen, Harri
Landolt, Oliver
Lang, Markus
Langemeyer, Stefan
Langlois, Pierre
Langmann, Ulrich
Larson, Lawrence
Larson, Tony
Larsson, Eric
Larsson-Edefors, P.
Lasanen, Kimmo
Lau, F.C.M.
Lawrance, Anthony
Lawson, SS
Lazarus, Kenneth
Lazic, Dejan
Leblebici, Yusuf
Ledesma, Francisco
Lee, Kong-Aik
Lee, Yvette
Lee, Yim-Shu
Lee, Tong-Yee
Lee, Seung-Hoon
Lee, Seok-Jun
Lee, Kyoung Keun
Lee, Chen-Yi
Lee, Hae-Seung
Lee, Edward
Lee, Duan-Shin
Lee, Dan
Lee, Chong-Ho
Lee, Chiou-Yng
Lee, Cheung Fai
Lee, Hoi
Lee, Pei-Jun
Leelavattananon, K.
Leenaertz, Domine
Leger, Gildas
Lehmann, Torsten
Lei, Jing
Leman, Karianto
Lenart, Thomas
Leou, Jin-Jang
Leppanen, Tuomas
Leung, Ka
Leung, Ka Nang
Leung, Pak-Keung
Leung, S. H.
Leung, C. H.
Levi, Viktor
Levin, Peter
Li, Xi
Li, Zhuo
Li, Zhiwu
Li, Zhimin
Li, Zeyu
Li, Yue
Li, Xiaoli
Li, Gang
Li, Li
Li, John
Li, Jipeng
Li, Jinhua
Li, Hua
Li, Haiyu
Li, Xu
Lian, Yong
Liang, Victor
Liang, Xue-Bin
Liang, Yi
Liang, Bor-Sung
Liao, Teh-Lu
Liao, Yuan-Fu
Liavas, A
Liberali, Valentino
Liberzon, Daniel
Lie, Wen-Nung
Lilius, Johan
Lillis, John
Lim, Yong-Ching
Limpaphayom, K.
Lin, James
Lin, Zhi-Ping
Lin, Yuan-Pei
Lin, Yao-Chung
Lin, Shyh-Feng
Lin, Shih-Chun
Lin, Pin-Hsun
Lin, Li Ju
Lin, Bor-Ren
Lin, Hsin-Lei
Lin, Guo-Shiang
Lin, Fil
Lin, Daw-Tung
Lin, Chin-Teng
Lin, Ching-Yung
Lin, Chia-Wen
Lin, Huei-Yung
Lin, Phone
Linan, Gustavo
Linares-Barranco, Bernabe
Lindberg, Erik
Lindmark, Björn
Ling, Nam
Liu, B
Liu, Dake
Liu, Chun-Nan
Liu, Derong
Liu, Jianhua
Liu, Jin
Liu, Jun
Liu, Shen-Iuan
Liu, Shih-Chii
Liu, Shing-Min
Liu, Wei
Liu, Xiaoping
Liu, Yongliang
Liu, Yuan-Chen
Liu, Zhongmin
Liu, Bin-Da
Livermore, Carol
Lo, Alan
Lo Presti, Matteo
Long, Howard
Long, John
Loose, Markus
Lopez, Antonio
Lopez, Jose
Lopez Valcarce, R.
Loubaton, Philippe
Louis, Jean-Paul
Loulou, Mourad
Lowenborg, Per
Lowther, Rex
Lu, Jinhu
Lu, Yan
Lu, Xiang
Lu, Haiping
Lu, Chun-Shien
Lu, Chia-Hsiung
Lu, Baoliang
Lu, Albert
Lu, Wu-Sheng
Lucyszyn, Stepan
Luo, Fa-Long
Luo, Qiang
Luo, Hui
Luo, Daniel
Luo, Jiebo
Lustenberger, Felix
Lutovac, Miroslav
Lutz, Jonathan

Lyshevski, Sergey
Ma, Kai-Kuang
Ma, K.-K.
Ma, Fan
Madden, Patrick
Maggio, Gian M.
Mahadevan, Raj
Mahapatra, Nihar
Mahattanakul, J.
Mahmoodi, Hamid
Maillard, Xavier
Maio, Ivan
Majstorovic, M.
Mak, Brian
Makarov, Valeri
Makowski, Marek
Makur, Anamitra
Malcovati, Piero
Malik, Saqib
Malisuwan, S.
Maloberti, Franco
Mammela, Aarne
Mämmelä, Aarne
Man, K. F.
Mandal, Mrinal
Mandolesi, Pablo
Manganaro, G.
Manolakos, Elias
Mansour, M.
Mantooth, Alan
Mao, Xun
Maranesi, Piero
Marcianesi, Andrea
Margala, Martin
Marsolais, A.
Martens, Ewout
Martin, Mark
Martin Langerwerf, Javier
Martinez-Peiro, M.
Martinez-Salamer,L
Martinez-Smith, A.
Mason, Andrew
Masselos, K.
Massie, Mark
Masuda, Naoki
Mathiopoulos, T
Mathis, Wolfgang
Matzke, W.
Mauer, Volker
Maundy, Brent
Mauris, Gilles
Maxim, Adrian
Mayaram, K.
Mazumder, Pinaki
Mazzini, Gianluca
Mcandrew, Colin
Mccanny, John
Mcconnell, Ross
Mccune, Earl
Mcdonald, Eric
Mcgregor, J
Mcguinness, P.
Mcneill, John
Medeiro, Fernando
Mehta, Swati
Meitzler, Richard
Mekhallalati, M.
Menon, P
Merakos, P.
Merkli, Patrick
Mertins, Alfred
Messer, Hagit
Mezhiba, Andrey
Miaou, S.
Michelakis, Kostis
Mikhael, Wasfy
Milanovic, Miro
Milanovic, Veljko
Milic, Ljiljana
Miller, Michael
Miller, William
Milosevic, Dusan
Milovanovic, B.
Min, Yinghua
Minaei, S.
Minato, Shin-Ichi
Minch, Bradley
Minima, Minima
Minz, Jacob
Mirabbasi, Shahriar
Miranda, Vladimiro
Mircean, Cristian
Mirhassani, Mitra
Miri Lavasani, H.
Mishra, Manu
Mitra, Subhasish
Mitros, Ania
Mitsubori, K.
Miu, Karen
Miyanaga, Y.
Miyasato, Y.
Mizuno, Masayuki
Mladenov, Valeri
Moghaddam, Saeed
Mohieldin, Ahmed
Moiola, Jorge
Mojarradi, M.
Momtaz, Afshin
Montuschi, Paolo
Moon, U.
Moon, James
Moon, Sung
Morgan, D. R.
Morgan, Matthew
Morgul, Omer
Mori, Kazuyoshi
Moritz, James
Morling, Richard C
Moro, Seiichirou
Morris, Brad
Moshnyaga, Vasily
Mourad, Samiha
Mow, Wai Ho
Mu, Karen
Mudra, Regina
Mueller, P.C.
Mueller, Paul
Mukherjee, D.
Mukherjee, Souvik
Mukhopadhyay, S.
Mukund, P.R.
Mulgrew, B
Muneyasu, Mitsuji
Munoz, Felipe
Murao, Kenji
Murata, Tadao
Murch, Ross
Mutale, Joseph
Muto, Virginia
Myers, Chris
Myers, Brent
Nabky, Frederic
Nadeem, Ahmed
Nader, Ahmed
Nadine, A.
Næss, Øivind
Nagai, Takayuki
Naik, Sagar
Nakagaki, Atsushi
Nakagawa, Akira
Nakajima, Hiryouki
Nakamoto, M.
Nakamura, T.
Nakanishi, Isao
Nakaoka, Mutsuo
Nakashizuka, M.
Nakaya, Yuichiro
Nakhla, Michel
Nakkar, Mouna
Nallanathan, A
Nancy, Nancy
Narasimhan, Ashok
Narayan, Prasat
Naroska, Edwin
Nassif, Sani
Nastov, Ognen
Nauta, Bram
Naware, Vidyut
Nayak, Ghanshyam
Ndjountche, T.
Nebel, Wolfgang
Nedic, Dusko
Netto, Sergio
Neves, Jose
Newcomb, Robert
Ng, Tung-Sung
Ng, Chiu Wah
Ngamroo, I.
Ngarmnil, J.
Nguyen, Truong Q
Nguyen, Truong
Ni, Zhi-Cheng
Ni, Yang
Niamat, M.
Niebur, Dagmar
Nielsen, Jannik H.
Niemisto, Riitta
Nijmeijer, Hendrik
Niknejad, Ali
Nikolaidis, Spyros
Nilsson, Peter
Nimrihter, Miroslav
Ninomiya, Kei
Nishi, Tetsuo
Nishihara, Akinori
Nishikawa, Kiyoshi
Nishimura, Shotaro
Nishio, Yoshifumi
Nizzoli, Gabriele
Njoelsatd, Tormod
Nolte, Norman
Nongpiur, R. C.
Nordebo, Sven
Nowrouzian, B.
Nucci, Carlo.
Nwankpa, C
Nygard, Tony
Nys, Olivier
Öberg, Johnny
Obote, Shigeki
Occhipinti, Luigi
Ochi, Hiroshi
Oconnor, William
Oelmann, Bengt
Ogawa, Makoto
Ogorzalek, Maciej
Ogunfunmi, T.
Ohki, Makoto
Ohlsson, Henrik
Ohmacht, Martin
Ohnakado, T.
Ohno, Shuichi
Okello, James
Oklobdzija, Vojin
Okumura, Makiko
Olejniczak, Kraig
Oliaei, Omid
Olivar, Gerard
Olleta, Beatriz
Olsson, Thomas
Omeni, Okundu
O'Nils, Mattias
Onodera, Hidetoshi
Oota, Ichirou
Opal, Ajoy
Oraintara, Soontorn
Oralkan, Omer
Orcioni, Simone
Orla, Feely
Orozco, Javier
Orre, Roland
Ortega, Juan
Ossieur, Peter
Ouhyoung, Ming
Oulmane, Mourad
Outhong, N.
Ovaska, Seppo
Ovod, Vladimir
Owall, Viktor
Ozalevli, Erhan
Ozer, Ibrahim
Ozoguz, Serdar
Paasio, Ari
Pal, Nikhil R.
Palaniappan, Kannappan
Palaskas, Yorgos
Paliouras, Vassilis
Palmisano, G.
Palmkvist, Kent
Palomaki, Kalle I.
Palumbo, G.
Pamunuwa, Dinesh
Pan, Zhengang
Pan, Su
Panagiotopoulos, D
Pandey, Pramod
Pandharipande, A
Pang, Ai-Chun
Pant, Vivek
Paoli, Gerhard
Papa, David
Papageorgiou, D.
Pappalardo, D.
Parameswaran, Ash
Parhi, Keshab
Parikh, Jigesh
Park, In-Cheol
Park, Junghun
Parlitz, Ulrich

Parthasarathy, K.
Partio, Mari
Paserba, John
Passeraub, Philippe
Pastore, Stefano
Patane, Luca
Paton, Susana
Patru, Dorin
Paul, Bipul
Pavan, Shanthi
Paviol, Jim
Pei, Soo-Chang
Pemmaraju, A.
Peng, Zebo
Pengwu, C.
Pennisi, Salvatore
Peralias, Eduardo
Perez, Lance
Perkins, Dmitri
Perkowski, Marek
Perreault, David
Perrott, Michael
Persson, Per
Pesquet, J-C
Pesquet-Popescu, Beatrice
Peterson, Lena
Petilli, Gene
Petitcolas, Fabien
Petraglia, Antonio
Petrakis, E.
Petrou, Maria
Phadoongsidhi, M.
Phang, Khoman
Philipp, Markus
Philipp, Ralf
Phillips, Joel
Phoong, See-Mat
Pialis, Tony
Piazza, Francesco
Piguet, Christian
Pikovsky, Arkadi
Pillai, S. U.
Pineda De Gyvez, Jose
Piriyapoksombut, Pramote
Pirsch, Peter
Pishdad, Bardia
Plataniotis, K.
Plett, Calvin
Plosila, Juha
Pollock, Lori
Poncino, Massimo
Pontikakis, Bill
Poorfard, Ramin
Popel, Denis
Popovic, Radivoje
Popovich, Mikhail
Porta, Sonia
Porto, Domenico
Poveda, Alberto
Premaratne, Kamal
Prentice, John
Proakis, John
Profunser, Dieter
Prokop, Thomas
Psychalinos, Costas
Pu, Chiang
Pulincherry, A.
Pureswaran, Veena
Puri, Ruchir
Qi, Xin
Qian, Tongfeng
Qiao, Changge
Quintana, Jose
Ra, Jong Beom
Radhakrishnan, D.
Radvanyi, Andras
Raffo, Luigi
Ragaie, H
Ragonese, Edigio
Raheli, Riccardo
Rahkonen, Timo
Rajan, Periasamy
Ramachandran, R.
Ramirez, Jorge
Ramirez-Angulo, J.
Rao, K.R.
Rao, Raghuveer
Rategh, Hamid
Rathfelder, Peter
Raut, Rabin
Ravinuthula, V.
Raychowdhury, A.
Reddy, Hari
Regalia, P. A.
Reggia, James
Reggiani, Luca
Reiss, Josh
Rekeczky, Csaba
Renfors, Markku
Rentala, Vijay
Reuter, Carsten
Reyhani-Masoleh, Arash
Reynaldo, Pinto
Richard, Johnson
Ridler, Oliver
Rincon-Mora, G.
Rintamaki, Matti
Riskin, Eve
Ritcey, J. A.
Rivoir, Roberto
Rizzo, Alessandro
Rjoub, Abdoul
Ro, Yoonhyuk
Robertson, Ian
Rocha-Perez, M.
Rochelle, Jim
Rodrigues, J.
Rodriguez-Vazque, Angel
Rodriguez-Villegas, Esther
Rogers, John
Rogers, Eric
Rokhsaz, Shahriar
Rombouts, Pieter
Rontogiannis, A.
Roo, Pierte
Roos, Jane
Roose, Dirk
Rosas, Juventino
Rosenbaum, Linnea
Rout, Saroj
Rovatti, Riccardo
Rowley, Matt
Rudiakova, Anna
Rueda, Adoracion
Rulkov, Nikolai
Rundel, Bernd
Saari, Ville
Sabadell, Justo
Sadler, Brian
Saeedifard, M.
Safi-Harb, Mona
Sahoo, Bibu
Saito, Toshimichi
Sakai, H
Salama, Khaled
Salama, C A
Salama, Aly
Salib, Philippe
Sallese, J.
Salo, Teemu
Sanchez, Stephen
Sandner, Christoph
Sangster, Alan
Sankur, Bulent
Sanubari, Junibakti
Sanz-Pascual, M.
Sarcinelli-Filho, M.
Sarkar, Sandip
Sarpeshkar, Rahul
Sarwate, Dilip
Sathe, Visvesh
Satoh, Akashi
Saubhayana, M.
Saul, Peter
Savaria, Yvon
Sawan, Mohamad
Sawasd, T.
Sayyed, A
Scarlet, Scarlet
Schaik, Andre
Schaumann, Rolf
Schimming, T.
Schlarmann, Mark
Schmid, Hanspeter
Schneider, Marcio
Schniter, Phil
Schobben, D. W. E.
Schreiber, Thomas
Schreier, Richard
Schupke, Dominic
Sculley, Terry
Sekiya, Hiroo
Selesnick, I.
Sengoku, Masakazu
Seng-Pan, Ben
Seng-Pan U, Ben
Sengupta, Susanta
Seo, Inchang
Serdijn, Wouter
Serra, Paco
Sethuraman, R
Setti, Gianluca
Sewell, John
Shadaydeh, Maha
Shah, Rahul
Shaheen, Mohamed
Shakiba, M.
Shalash, Ahmed
Shan, Shiguang
Shan, Ying
Shand, Mark
Shanfeng, Cheng
Shang, Weijia
Shapiro, Linda
Sharif-Bakhtiar, M.
Shaw, Ruey-Shiang
Shcherbakov, Pavel
Shea, John
Sheikh, H.
Shen, Jacky
Shen, Day-Fann
Shepherd, Bruce
Sheu, Ming-Hwa
Sheu, S.
Shi, Bertram
Shi, Pengcheng
Shi, Rock
Shi, Yun-Qing
Shibata, Tadashi
Shieh, L.
Shieh, Ming-Der
Shih, Ishiang
Shih, Timothy K.
Shih, Yu Chuan
Shim, Byonghyo
Shin, Kyung Wook
Shiu, Pun Hang
Shiue, Wen-Tsong
Shivalingaiah, H.
Shmerko, Vlad
Shmilovitz, Doron
Shoyama, Masahito
Shrivastava, Yash
Shu, Keliu
Signell, Svante
Silva, Jose'
Silva, Christopher
Silva, Fernando
Silva-Martinez, J.
Sim, Calvin
Simon, Marvin
Simoni, Andrea
Simula, Olli
Sina, Balkir
Sing, Choy Chiu
Sinha, Amit
Siozios, K.
Sirisantana, Naran
Siriwongpairat, W.
Sirna, Guglielmo
Siu, K W
Sjöland, Henrik
Sklavos, N.
Skowronski, Mark
Skytta, Jorma
Smedley, Keyue
Smekal, Zdenek
Smirnov, A.
Smith, Malcolm
So, Hing-Cheung
Soenen, Eric
Sohn, Kwanghoon
Soliman, Ahmed
Solsona, Jorge
Sommen, P. C. W.
Sommer, Ralf
Song, Xiaoyu
Song, Kee-Bong
Sonkusale, Sameer
Sotiriadis, Paul
Soudan, Bassel
Soudris, Dimitrios
Sparsö, Jens
Sridhara, Srinivasa

Sridharan, Sharmila
Sridharan, Ranjani
Srikanthan, T.
Srirattana, N.
Staber, Michael
Stan, Mircea
Stanacevic, Milutin
Stankovic, A
Stankovic, Radomir
Starzyk, Janusz
Stathaki, Tania
Steensgaard, Jesper
Stefanoiu, Dan
Steiner, Ian
Stengel, Bob
Steyaert, Michiel
Stocker, Stocker
Stocker, Alan
Storace, Marco
Stouraitis, A.
Stoyanov, Georgi
Strandberg, Roland
Streiff, Matthias
Strollo, A. G. M.
Stroud, Chuck
Styczynski, Z.
Su, Po-Chyi
Su, Mu-Chun
Su, Jonathan
Su, Jianbo
Su, Hsiao Wei
Su, Chi-Hsi
Su, Wen-Yu
Su, Gui-Jia
Suematsu, Noriharu
Suetsugu, Tadashi
Sun, Yu
Sun, Changyin
Sun, Hung-Ming
Sun, Ming-Chang
Sun, Qibin
Sun, Yichuang
Sunar, Berk
Sung, W
Sung, Kihyuk
Suykens, Johan
Svelto, Francesco
Svensson, Christer
Swann, Brian
Swetharanyan, L.
Szczepanski, S.
Szirányi, Tamas
Szolgay, Peter
Szymanek, R.
Tabus, Ioan
Tadokoro, Yoshiaki
Taillefer, Chris
Takagi, Naofumi
Takahashi, N.
Takala, Jarmo
Talley, Mike
Tam, Clarence
Tamasevicius, A.
Tamura, Yoshiyasu
Tan, Ying
Tan, Siang Ton
Tan, Yap-Peng
Tanaka, Tetsuro
Tanaka, Hisa-Aki
Tanaka, Mamoru
Tang, Ao
Tang, Kit-Sang
Tang, Pushan
Tang, S.
Tang, Ying
Tang, Yiyan
Tang, Yonghui
Tang, Zhilong
Tang, Zhiwei
Tanitteerapan, T.
Tanji, Yuichi
Tao, Liang
Tarim, Nil
Tarim, Tuna B.
Tatas, Konstantinos
Tavares, Vitor
Tay, D.
Tay, D.B.H.
Tayebi, A.
Taylor, Desmond P.
Taylor, Clark
Tchamov, Nikolay
Tchobanou, M.
Teich, Juergen
Tejada, Francesco
Tekalp, Murat
Temel, Turgay
Temes, Gabor
Tenca, A
Tenore, Francesco
Terzija, Vladimir
Terzioglu, Esin
Tetzlaff, Ronald
Thamvichai, R.
Thanachayanont, A.
Theodoridis, G.
Theuwissen, Albert
Thiran, Patrick
Thoidis, Ioannis
Thoka, Sreenath
Thomas, Charles
Thompson, Charlie
Thörnberg, Benny
Thulasiraman, K.
Tian, Qi
Tian, Lianfang
Ticli, Lucio
Tiebout, Marc
Tiew, Kee-Chee
Timmermann, Dirk
Tlelo-Cautle, E.
Tlili, Fethi
Toker, Ali
Tokuda, Isao
Tom C.-I., Lin
Tombras, G
Tommy, Tommy
Tomsovic, Kevin
Ton, Coenen
Tor Lande, Bassen
Torelli, Guido
Torikai, Hiroyuki
Torkelson, Mats
Torralba, Antonio
Torralba Silgado, A
Touba, Nur
Trajkovic, Ljiljana
Tran, To
Tran, Trac
Trappe, Peter
Trask, Chris
Trautmann, Steffen
Trifiletti, A.
Tsai, Chien-Wu
Tsai, Chun-Jen
Tsai, Pei-Yun
Tsai, Tsung-Han
Tsakalides, P.
Tsang, Tommy
Tsay, Frank
Tschanz, James
Tse, Chi Kong
Tseng, C.
Tseng, Chia-Jeng
Tseng, C.
Tsimring, Lev
Tsividis, Yannis
Tsubone, Tadasi
Tsui, C. Y.
Tsukahara, Tsuneo
Tsuneda, Akio
Tu, Yu-Kuang
Tucker, Steve
Tung, Yi-Shin
Tuqan, Jamal
Turchetti, Claudio
Tureli, Uf
Turner, Laurence
Turney, Robert
Typpo, Jukka Tapio
Ubdesai, U. B.
Ueta, Tetsushi
Um, Junhyung
Urbanek, Wolfram
Ushida, Akio
Ushio, Toshimitsu
Uyttenhove, Koen
Vachoux, Alain
Vaidyanathan, PP
Vainio, Olli
Valabrega, Paolo
Valdes-Garcia, A.
Valero-Lopez, Ari
Valimaki, Vesa
Valkama, Mikko
Valle, Maurizio
Van, Lan-Da
Van Der Tang, J.
Vanassche, Piet
Vandewalle, Joos
Vankayala, Vijay
Vary, Peter
Vasconselos, J.A.
Vasilache, Adriana
Vaucher, Cicero
Vazquez-Leal, H.
Velenis, Dimitrios
Vemuru, Srinivas
Venchi, Giuseppe
Venkatasubramanian, Mani
Verghese, George
Vergos, H
Verhoeven, C.
Vesterbacka, Mark
Vetro, Anthony
Vidal-Lopez, Eva
Vidal-Verdu, F.
Vidojkovic, Vojkan
Vignoli, V.
Viholainen, Ari
Vinci, Chira
Vinson, Vinson
Vipul, Katyal
Visweswariah, C.
Vitelli, Massimo
Vlach, Jiri
Vlassis, S.
Volkovskii, Alex
Vora, Janki
Vournas, Costas
Vucic, Mladen
Wackersreuther, G.
Waltari, Mikko
Walter, Colin
Wambacq, Piet
Wang, Sheng-Jyh
Wang, Yuke
Wang, Yuzhen
Wang, Yongtao
Wang, Yi-Chuu
Wang, Xufang
Wang, Xingguo
Wang, Xiaofeng
Wang, Wen-Chieh
Wang, Tu-Chih
Wang, Tu-Chi
Wang, Zhou
Wang, Lipo
Wang, Chua-Chin
Wang, Chung-Neng
Wang, Chunyan
Wang, Hsin-Min
Wang, Hwei-Tseng
Wang, Jinn-Shyan
Wang, Jun
Wang, Li
Wang, Lige
Wang, Benyi
Wanhammar, Lars
Wassal, Amr
Watanabe, Takayuki
Watanabe, Toshimasa
Weber, Robert
Weeks, Michael
Wei, Yung-Chiang
Wei, Xiaohui
Wei, Che-Ho
Wei, Gu-Yeon
Wei, Shugang
Weigel, Robert
Weinstein, Eugene
Weldon, Tom
Wells, Kevin
Wen, Johnny
Wentzloff, David
Werner, Stefan
Westall, Fred
Wichlund, Sverre
Wiesbauer, A.
Wilcock, R.
Wilde, Andreas
Wilson, Denise
Wing, Omar
Winkler, Stefan
Winograd, Gil
Winter, Matthias
Winzker, Marco

Wittenburg, J.
Wlshen, Wlshen
Wong, Tan F.
Wong, Hon-Sum
Wong, Kit
Wong, Mike
Wong, Ngai
Wong, P.
Wongkomet, N.
Wongsirimunkong, Anchalee
Woods, R
Worapishet, Apisak
Wu, Jianhong
Wu, Y. C.
Wu, Wei-Chung
Wu, Chung-Bin
Wu, Shinq-Jen
Wu, Yung-Gi
Wu, Sauhsuan
Wu, Rui-Cheng
Wu, Lifeng
Wu, Ja-Ling
Wu, Hsiao-Kuang
Wu, Hongyi
Wu, H.
Wu, Feng
Wu, Chung-Hsien
Wu, Chai
Wu, An-Yeu
Wu, Chung-Yu
Wulff, Carsten
Wyers, Eric
Xia, X.G.
Xia, Xiang-Gen
Xia, Bin HKU
Xian, Feng
Xiao, Chengshan
Xiao, Yegui
Xiao, Yequi
Xie, Lihua
Xin, Jun
Xiong, Xiaoxu
Xiong, Zixiang
Xu, Hai
Xu, Li
Xu, Wei Ze
Xue, Guoliang
Xue, Ping
Yadid-Pecht, Orly
Yalcin, Mustak
Yamada, Isao
Yamagishi, Akihiro
Yamamoto, Toru
Yamamura, K.
Yamasaki, Toshi
Yamazato, Takaya
Yan, Hong
Yan, Zhiyuan
Yan, Shouli
Yan, Xianlang
Yang, Shiqiang
Yang, Xiaojian
Yang, Zaohong
Yang, Xiaokang
Yang, Simon X.
Yang, Jin
Yang, Congguang
Yang, Jar-Ferr
Yang, Xin-Xing
Yang, Huazhong
Yang, Meng-Da
Yanovsky, G.
Yanushkevich, S.
Yao, Kung
Yap, Kim-Hui
Yardim, Anush
Yarman, B. S.
Yasushi, Yuminaka
Yasuura, Hiroto
Yau, Elson
Yazdi, Navid
Ye, Song
Yeatman, Eric
Yeh, Chia-Hung
Yen, Hsiu-Huei
Yen, Shwu-Huey
Yeo, Hypeogoo
Yim, Seong-Mo
Yin, Qizhang
Ying, Changsheng
Yip, Kun-Wah
Yoon, Jang Sup
Yoon, Kwang Sub
Yoshikawa, T.
Youn, Jeongnam
Youn, Yong-Sik
Young, F.Y.
Ytterdal, Trond
Yu, Yajun
Yu, Chu
Yu, Ya -Jun
Yu, Zhiping
Yu, June
Yu, Guoyao
Yu, Meng-Lin
Yu, Heather
Yuan, Baozong
Yuan, Fei
Yuan, Jiren
Yuan, Lujun
Yuan, Xiaoming
Yufera, Alberto
Zaghloul, Mona
Zanchi, Alfio
Zarandy, Akos
Zebulum, Ricardo
Zelikson, Michael
Zencir, Ertan
Zeng, Bing
Zeng, Xuan
Zeng, Y. H.
Zeng, Zhigang
Zerbe, Jared
Zhang, Tao
Zhang, Yueping
Zhang, Yuan
Zhang, Youming
Zhang, Yi
Zhang, Ximin
Zhang, Yan
Zhang, Zengyan
Zhang, C.
Zhang, Qian
Zhang, Yanqing
Zhang, Gong
Zhang, Hanzen
Zhang, Hongjiang
Zhang, Jiye
Zhang, Junmou
Zhang, Ming
Zhao, Debin
Zhao, Qing
Zhao, Yinyin
Zhao, Dan
Zhaonian, Zhang
Zheng, Wei Xing
Zheng, Weiguo
Zheng, Jinghong
Zheng, Yuanjin
Zheng, Li-Rong
Zhou, Hai
Zhou, Sean
Zhou, Yiqing
Zhu, Yunshan
Zhu, Zhigang
Zhu, Wei-Ping
Zhu, Ce
Ziehe, Andreas
Zierhofer, Clemens
Zimmerman, Reto
Zou, Dekun
Zou, Mouyan
Zou, Qiyue
Zourntos, Takis
Zwolinski, Mark

Invited Sessions

The invited sessions complement the regular program with topics of special interest to the circuits and systems community. They cut across and beyond disciplines traditionally represented at ISCAS. Several invited sessions revolve around the conference theme of "Life-style Technologies".

We thank all invited session organizers for their initiative and enthusiasm in compiling an exciting program. Their dedicated effort and the valuable support of Tom Wehner, made it possible to collect invited contributions and coordinate the peer review process on a very tight schedule.

Gert Cauwenberghs
Ljiljana Trajkovic
ISCAS 2003 Invited Sessions Chairs

New Trends in Switching Power Converters Towards Integration
Co-Chair - Eduard Alarcon, *Universitat Politecnica de Catalunya*
Co-Chair - Luigi Fortuna, *University of Catania*

High-speed Circuits and Interconnects
Chair - Hideki Asai, *Shizuoka University*
Organizer - Michel Nakhla, *Carleton University*

IC Interfaces for Sensors/MEMS
Chair - Franco Maloberti, *University of Texas at Dallas*
Co-Chair - Andrea Baschirotto, *University of Lecce*
Co-Chair - Piero Malcovati, *University of Pavia*

Variable Digital Filters
Co-Chair - Tian-Bo Deng, *Toho University*
Co-Chair - Masayuki Kawamata, *Tohoku University*

Theoretical Aspects in Cellular Neural Networks
Chair - Marco Gilli, *Polytechnic University of Torino*

EDGE is Cutting Edge
Chair - Markus Helfenstein, *Philips Semiconductor, Switzerland*

Biomedical Devices, Sensors and Related Systems
Chair - Angela Hodge-Miller, *Naval Research Laboratory*
Co-Chair - Robert W. Newcomb, *University of Maryland*

Piecewise Linear Circuits and Systems
Chair - Pedro Julian, *Johns Hopkins University*
Co-Chair - Marco Storace, *University of Genoa*

Multidimensional Circuits, Systems and Signal Processing
Chair - Sankar Basu, *National Science Foundation*
Co-Chair - Hari Reddy, *California State University Long Beach*
Chair - Krzysztof Galkowski, *University of Zielona Gora*
Co-Chair - Zhiping Lin, *Nanyang Technological University*

Advances in Recurrent Neural Networks
Chair - Derong Liu, *University of Illinois*
Co-Chair - H. K. Kwan, *University of Windsor*

Speech and Language Processing
Chair - Ravi P. Ramachandran, *Rowan University*
Co-Chair - H. K. Kwan, *University of Windsor*

Nonlinear Dynamics for Coding Theory and Network Traffic
Chair - Gianluca Setti, *University of Ferrari*
Co-Chair - Gian Mario Maggio, *University of California, San Diego*
Co-Chair - Gialuca Mazzini, *University of Ferrari*
Co-Chair - Riccardo Rovatti *University of Bologna*

Sensor Arrays for Visual Tracking and Navigation
Chair - Bertram Shi, *University of Science and Technology*
Co-Chair - Csaba Rekeczky, *Hungarian Academy of Science*

Security and Data Hiding
Chair - Yun Q. Shi, *New Jersey Institute of Technology*
Co-Chair - Jiwu Huang, *Zhongshan University*
Co-Chair - Oscar Au, *Hong Kong University of Science and Technology*

Bionics
Chair - Ronald Tetzlaff, *University of Frankfurt*
Co-Chair - Paolo Arena, *Universita degli Studi di Catania*

Fault Detection and Tolerance in Distributed Networks
Chair - Krishnaiya Thulasiraman, *University of Oklahoma*

Frequency Response Masking Techniques
Chair - Lian Yong, *National University of Singapore*
Co-Chair - Hakan Johansson, *Linkoping University*

Behavioral Modeling and Simulation of Mixed-Signal Systems
Chair - Trond Ytterdal, *Norwegian University of Science and Technology*

MEMS and Applications
Chair - Mona Zaghloul, *George Washington University*
Co-Chair - Majid Ahmadi, *University of Windsor*
Co-Chair - Veljko Milanovic, *University of California at Berkeley*

Spatiotemporal Cellular Vision Systems
Chair - Akos Zarandy, *Hungarian Academy of Science*
Co-Chair - Angel Rodriguez-Vazquez, *University of Sevilla*

VOLUME I OF V: -ANALOG CIRCUITS & SIGNAL PROCESSING-

Track 1: ANALOG CIRCUITS & SIGNAL PROCESSING

ALGORITHMS FOR NONUNIFORM BANDPASS SAMPLING IN RADIO RECEIVER I-1

Yi-Ran Sun and Svante Signell, *Royal Institute of Technology*

MODELING OF ACCUMULATION MOS CAPACITORS FOR HIGH PERFORMANCE ANALOG CIRCUITS I-5

Aránzazu Otín and Santiago Celma, *University of Zaragoza*, and Concepción Aldea, *Electronic Design Group*

A HIGHLY LINEAR FRONT-END BASED ON A LOGARITHMIC MULTIPLIER-FILTER I-9

Ganesh Kathiresan, *Toumaz Technology Ltd.*, Emmanuel Drakakis and Chris Toumazou, *Imperial College of Science, Technology & Medicine*

A NOVEL DESIGN TECHNIQUE FOR VERY LOW VOLTAGE MOS TRANSLINEAR CIRCUITS I-13

Antonio Lopez-Martin and Alfonso Carlosena, *Public University of Navarra, Pamplona*, and Jaime Ramirez-Angulo, *New Mexico State University*

A SMALL ANALOG VLSI INNER HAIR CELL MODEL I-17

Andre Van Schaik, *The University of Sydney*

SOME STRUCTURAL CONDITIONS UNDER WHICH AN RLC NETWORK IS CONTROLLABLE OVER F(Z) I-21

Kai-Sheng Lu, *Wuhan University of Technology*

A MULTI-LEVEL STATIC MEMORY CELL I-25

Philipp Häfliger and HåVard Kolle Riis, *University of Oslo*

CMOS OPTICAL RECEIVER CHIPSET FOR GIGABIT ETHERNET APPLICATIONS I-29

Sung-Eun Kim, Seong-Jun Song, Sung Min Park, and Hoi-Jun Yoo, *KAIST*

A BIASED LOW-VOLTAGE BICMOS MIXER FOR DIRECT UP-CONVERSION I-33

Esa Tiiliharju and Kari Halonen, *Helsinki University of Technology*

A HIGH-RESOLUTION AND FAST-CONVERSION TIME-TO-DIGITAL CONVERTER I-37

Chorng-Sii Hwang, *National Taiwan University*, and Poki Chen, *National Taiwan University of Science & Technology*, and Hen-Wai Tsao, *National Taiwan University*

A HIGH PERFORMANCE CMOS CURRENT-MODE PRECISION FULL-WAVE RECTIFIER (PFWR) I-41

Varakorn Kasemsuwan and Surachet Khucharoensin, *King Mongkut's Institute of Technology Ladkrabang*

A TIME-CONTINUOUS OPTIMIZATION METHOD FOR AUTOMATIC ADJUSTMENT OF GAIN AND PHASE IMBALANCES IN FEEDFORWARD AND LINC TRANSMITTERS I-45

Bo Shi, *Institute for Communications Research*, and Lars Sundstrom, *Ericsson Mobile Platforms AB*

A BICMOS 10GB/S ADAPTIVE CABLE EQUALIZER I-49

Guangyu Zhang, Pruthvi Chaudhari, and Michael Green, *University of California*

A 1.8 V 10-BIT 80 MS/S LOW POWER TRACK-AND-HOLD CIRCUIT IN A 0.18 UM CMOS PROCESS I-53

Erik Säll, *Linköping University*

INTEGRATED INDUCTORS OVER MOSFETS – EXPERIMENTAL RESULTS OF A THREE DIMENSIONAL INTEGRATED STRUCTURE I-57

Nikolaos Nastos and Yannis Papananos, *National Technical University of Athens*

A RESONANT PAD FOR ESD PROTECTED NARROWBAND CMOS RF APPLICATIONS I-61

Jaynie Shorb, Xiaoyong Li, and David Allstot, *University of Washington*

ANALYSIS OF SHIELDED PLANAR CIRCUITSBY A MIXED VARIATIONAL-SPECTRAL METHOD I-65

Mohamed Lamine Tounsi and Houda Halheit, *USTHB University*, Mustapha Cherif Yagoub, *University of Ottawa*, and Abdelhamid Khodja,. *USTHB University*

A 7MW 1GBPS CMOS OPTICAL RECEIVER FOR THROUGH WAFER COMMUNICATION I-69

Alyssa Apsel, *Cornell University*, and Andreas Andreou, *Johns Hopkins University*

A HIGH-BANDWIDTH WIRELESS INFRARED RECEIVER WITH FEEDFORWARD OFFSET EXTRACTOR I-73

Chin-Shan Hsieh and Hong-Yi Huang, *Fu-Jen Catholic University*

A 10 MILLIWATT 2 GBPS CMOS OPTICAL RECEIVER FOR OPTOELECTRONIC INTERCONNECT I-77

Alyssa Apsel, *Cornell University*, and Andreas Andreou, *Johns Hopkins University*

AN ADAPTIVE SILICON SYNAPSE I-81

Elisabetta Chicca, Giacomo Indiveri, and Rodney Douglas, *ETH Zurich*

ANALOG-DECODER EXPERIMENTS WITH SUBTHRESHOLD CMOS SOFT-GATES I-85

Matthias Frey and Hans-Andrea Loeliger, *ETH Zuerich*, Felix Lustenberger, *CSEM, Zuerich*, Patrick Merkli and Patrik Strebel, *ETH Zuerich*

A SYMMETRIC MINIATURE 3D INDUCTOR I-89

Srinivas Kodali, and David Allstot, *University of Washington*

ON-CHIP INDUCTOR STRUCTURES: A COMPARATIVE STUDY I-93

Srinivas Kodali, Taeik Kim, and David Allstot, *University of Washington*

RF TRANSFORMER AS A DIRECTIONAL COUPLER WITH ARBITRARY LOAD I-97

Janusz Biernacki and Dariusz Czarkowski, *Polytechnic University*

A DC CURRENT MEASUREMENT CIRCUIT FOR ON-CHIP APPLICATIONS I-101

Clarence Tam and Gordon W. Roberts, *McGill University*

A DIGITALLY SKEW CORRECTABLE MULTI-PHASE CLOCK GENERATOR USING A MASTER-SLAVE DLL I-105

Atsushi Suzuki, Shoji Kawahito, Daisuke Miyazaki, and Masanori Furuta, *Shizuoka University*

A SELF-TESTING METHOD FOR THE PIPELINED A/D CONVERTER I-109

Jaeki Yoo, Edward Lee, and Earl Swartzlander, *Univeristy of Texas at Austin*

SYNTHESIS OF A PULSE-FORMING REACTANCE NETWORK TO SHAPE A DELAYED QUASI-RECTANGULAR PULSE I-113

Igor Filanovsky, *University of Alberta*, and Platon Matkhanov, *St. Petersburg University of Electrical Engineering*

MODELING ALL-MOS LOG FILTERS AND ITS APPLICATION TO SIGMA-DELTA MODULATORS I-117

Jofre Pallarès, *Institut de Microelectrònica de Bercelona - CNM*, Justo Sabadell, *Barcelona Design Center - Epson*, and Francisco Serra-Graells, *Institut De MicroelectròNica De Bercelona - CNM*

ANALOG WAVELET TRANSFORM EMPLOYING DYNAMIC TRANSLINEAR CIRCUITS FOR CARDIAC SIGNAL CHARACTERIZATION I-121

Sandro Haddad, *Delft University of Technology*, Richard Houben, *Medtronic*, and Wouter A. Serdijn, *TU Delft*

A LOW VOLTAGE SECOND ORDER BIQUAD USING PSEUDO FLOATING-GATE TRANSISTORS I-125

ØIvind Næss, Espen A. Olsen, Yngvar Berg, and Tor Sverre Lande, *University of Oslo*

TIMING-MISMATCH ANALYSIS IN HIGH-SPEED ANALOG FRONT-END WITH NONUNIFORMLY HOLDING OUTPUT I-129

Sai-Weng Sin, Seng-Pan U, and R.P. Martins, *University of Macau*, and J.E. Franca, *Instituto Superior Tecnico*

DESIGN OF A DIGITALLY PROGRAMMABLE DELAY-LOCKED-LOOP FOR A LOW-COST ULTRA WIDE BAND RADAR RECEIVER I-133

Nuno Paulino, Marco Serrazina, Joao Goes, *UNINOVA*, and Adolfo Steiger-Garcao, *UNINOVA*

RESISTIVE FET IQ VECTOR MODULATOR USING MULTILAYER PHOTOIMAGEABLE THICK-FILM TECHNOLOGY I-137

Choon Ng, *University of Surrey*, Mitchai Chongcheawchamnan, *Mahanakorn University of Technology*, Ian Robertson, *University of Surrey*, and K Cho, *RF Hitec Inc*

HIGH-SPEED LOW INPUT IMPEDANCE CMOS CURRENT COMPARATOR I-141

Varakorn Kasemsuwan and Surachet Khucharoensin, *King Mongkut's Institute of Technology Ladkrabang*

LOW-VOLTAGE LOW-POWER CMOS ANALOGUE CIRCUITS FOR GAUSSIAN AND UNIFORM NOISE GENERATION I-145

Guiomar Evans, Joao Goes, Adolfo Steiger-Garcao, Manuel Ortigueira, and Nuno Paulino, *UNINOVA*, and Joao Lopes, *Universidade de Lisboa*

ON THE FEASIBILITY OF APPLICATION OF CLASS E RF POWER AMPLIFIERS IN UMTS I-149

Dusan Milosevic, Johan Van Der Tang, and Arthur Van Roermund, *Eindhoven University of Technology*

Track 1.1: Amplifiers

A NOVEL 1-V CLASS-AB TRANSCONDUCTOR FOR IMPROVING SPEED PERFORMANCE IN SC APPLICATIONS I-153

Gianluca Giustolisi and Gaetano Palumbo, *Università di Catania*

A NEW 1.5V LINEAR TRANSCONDUCTOR WITH HIGH OUTPUT IMPEDANCE IN A LARGE BANDWIDTH I-157

Juana Martínez-Heredia, Antonio Torralba, and Ramón Carvajal, *Universidad de Sevilla*, and Jaime Ramírez-Angulo, *New Mexico State University*

A WIDE LINEAR RANGE LOW-VOLTAGE TRANSCONDUCTOR I-161

Adrian Leuciuc, *State University of New York At Stony Brook*

CONSTANT-GM CONSTANT-SLEW-RATE HIGH BANDWIDTH LOW-VOLTAGE RAIL-TO-RAIL CMOS INPUT STAGE FOR VLSI CELL LIBRARIES I-165

Juan Carrillo, and J. Francisco Duque-Carrillo, *University of Extremadura*, and Guido Torelli, *University of Pavia*, and José L. Ausín, *University of Extremadura*

AN AUTO-INPUT-OFFSET REMOVING FLOATING GATE PSEUDO-DIFFERENTIAL TRANSCONDUCTOR I-169

Timothy Constandinou, and Julius Georgiou, and Chris Toumazou, *Imperial College*

A TECHNIQUE FOR DC-OFFSET REMOVAL AND CARRIER PHASE ERROR COMPENSATION IN INTEGRATED WIRELESS RECEIVERS I-173

Stephen Shang, and Shahriar Mirabbasi, and Resve Saleh, *University of British Columbia*

A GENERAL THEORY OF THIRD-ORDER INTERMODULATION DISTORTION IN COMMON-EMITTER RADIO FREQUENCY CIRCUITS I-177

Liwei Sheng and Lawrence Larson, *University of California, San Diego*

CAPACITIVELY COUPLED MULTIPLE RESONANCE NETWORKS I-181

Antonio de Queiroz, *Federal University of Rio De Janeiro*

DIGITALLY TUNEABLE ON-CHIP RESISTOR IN CMOS FOR HIGH-SPEED DATA TRANSMISSION I-185

Kyoung-Hoi Koo, *Samsung Electronics*

SIMPLE NOISE FORMULAS FOR MOS ANALOG DESIGN I-189

Alfredo Arnaud, *Facultad De Ingenieria - Urou*, and Carlos Galup-Montoro, *Universidade Federal de Santa Catarina*

AN HEURISTIC CIRCUIT-GENERATION TECHNIQUE FOR THE DESIGN-AUTOMATION OF ANALOG CIRCUITS I-193

Esteban Tlelo-Cuautle and Alejandro Diaz-Sanchez, *INAOE*

DESIGN AND OPTIMIZATION OF LOW-VOLTAGE TWO-STAGE CMOS AMPLIFIERS WITH ENHANCED PERFORMANCE I-197

Rui Tavares, Bruno Vaz, João Goes, Nuno Paulino, and Adolfo Steiger-Garcao, *UNINOVA*

COMPACT INTEGRATED TRANSCONDUCTANCE AMPLIFIER CIRCUIT FOR TEMPORAL DIFFERENTIATION I-201

Alan Stocker, *Institute of Neuroinformatics*

LNA DESIGN OPTIMIZATION WITH REFERENCE TO ESD PROTECTION CIRCUITRY I-205

Sharmila Sridharan, Ghanshyam Nayak, and P.R. Mukund, *Rochester Institute of Technology*

A NOVEL NOISE OPTIMIZATION DESIGN TECHNIQUE FOR RADIO FREQUENCY LOW NOISE AMPLIFIERS I-209

Byunghoo Jung, Anand Gopinath, and Ramesh Harjani, *University of Minnesota, Minneapolis*

BIPOLAR LNA DESIGN AT DIFFERENT OPERATING FREQUENCIES I-213

Giovanni Girlando, *STMicroelectrionics*, Egidio Ragonese, Alessandro Italia, and Giuseppe Palmisano, *Università di Catania*

DUAL-BAND SUB-1 V CMOS LNA FOR 802.11A/B WLAN APPLICATIONS I-217

Tommy Tsang and Mourad El-Gamal, *McGill University*

LOW NOISE AMPLIFIER DESIGN FOR ULTRA-WIDEBAND RADIOS I-221

Won Namgoong and Jongrit Lerdworatawee, *University of Southern California*

A 1-V CMOS OUTPUT STAGE WITH HIGH LINEARITY I-225

Walter Aloisi, Gianluca Giustolisi, and Gaetano Palumbo, *Università di Catania*

A NEW RAIL-TO-RAIL DRIVING SCHEME AND A LOW-POWER HIGH-SPEED OUTPUT BUFFER AMPLIFIER FOR AMLCD COLUMN DRIVER APPLICATION I-229

Chih-Wen Lu, *National Chi Nan University*

A NEW METHOD FOR EVALUATING HARMONIC DISTORTION IN PUSH-PULL OUTPUT STAGES I-233

Gianluca Giustolisi and Gaetano Palumbo, *Università di Catania*

A NEW COMPACT LOW-POWER HIGH SLEW SATE CLASS AB CMOS BUFFER I-237

Antonio Torralba, Ramon Carvajal, and Juan Galan, *Universidad de Sevilla*, and Jaime Ramirez-Angulo, *New Mexico State University*

SELF-REGULATED FOUR-PHASED CHARGE PUMP WITH BOOSTED WELLS I-241

Joseph Shor, Yan Polansky, Yair Sofer, and Eduardo Maayan, *Saifun Semiconductors*

A 1V FULLY DIFFERENTIAL CMOS LNA FOR 2.4GHZ APPLICATION I-245

Chih-Lung Hsiao, Ro-Min Weng, and Kun-Yi Lin, *National Dong Hwa University*

A PARAMETRIC APPROACH TO DESCRIBE DISTRIBUTED TWO-PORTS WITH LUMPED DISCONTINUITIES FOR THE DESIGN OF BROADBAND MMIC'S I-249

Ahmet Aksen and B. Siddik Yarman, *Isik University*

HIGH BANDWIDTH TRANSIMPEDANCE AMPLIFIER DESIGN USING ACTIVE TRANSMISSION LINES I-253

Hyeon-Min Bae and Naresh Shanbhag, *University of Illinois at Urbana-Champaign*

A DIFFERENTIAL DIFFERENCE COMPARATOR FOR MULTI-STEP AD CONVERTERS I-257

Gang Xu and Jiren Yuan, *Lund University*

MODELLING DIFFERENTIAL PAIRS FOR LOW-DISTORTION AMPLIFIER DESIGN I-261

Enno de Lange and Oscar de Feo, *Swiss Federal Institute of Technology, Lausanne* Arie Van Staveren, *Delft University of Technology*

A FAST SETTLING CMOS OPERATIONAL AMPLIFIER I-265

Haibin Huang and Ezz El-Masry, *Dalhousie University*

NEW OUTPUT STAGE FOR LOW SUPPLY VOLTAGE, HIGH-PERFORMANCE CMOS CURRENT MIRROR I-269

Antonio Torralba, Ramon Carvajal, and Fernando Muñoz, *Universidad de Sevilla*, and Jaime Ramirez-Angulo, *New Mexico State University*

A 1.5 V HIGH-SPEED CLASS AB OPERATIONAL AMPLIFIER FOR HIGH-RESOLUTION HIGH-SPEED PIPELINED A/D CONVERTERS I-273

Saeid Mehrmanesh, Hesam Amir Aslanzadeh, Mohammad Bagher Vahidfar, and Mojtaba Atarodi, *Sharif University of Technology*

1-V QUASI CONSTANT-GM INPUT/OUTPUT RAIL-TO-RAIL CMOS OP-AMP I-277

Juan Carrillo and J. Francisco Duque-Carrillo, *University of Extremadura*, Guido Torelli, *University of Pavia*, and José L. Ausín, *University of Extremadura*

LOW POWER DEMODULATORS WITH PHASE QUANTIZATION FOR A ZERO-IF BLUETOOTH RECEIVER I-281

Sohrab Samadian, *University of California, Los Angeles*, Ryoji Hayashi, *Mitsubishi Electric Company, Japan*, and Asad Abidi, *University of California, Los Angeles*

A PRECISE TEMPERATURE-INSENSITIVE AND LINEAR-IN-DB VARIABLE GAIN AMPLIFIER I-285

Sungho Beck, *FCI, Inc.*, Myung-Woon Hwang, *KAIST*, Sang-Hoon Lee, *Future Communications IC Inc.*, Gyu-Hyeong Cho, *KAIST*, and Jong-Ryul Lee, *Future Communications IC Inc.*

HIGH-DYNAMIC-RANGE DECIBEL-LINEAR IF VARIABLE-GAIN AMPLIFIER WITH TEMPERATURE COMPENSATION FOR WCDMA APPLICATIONS I-289

Francesco Carrara, *Universita' di Catania - Facolta' di Ingegneria – DIEES*, Pietro Filoramo, *STMicroelectronics*, and Giuseppe Palmisano, *Università di Catania*

A LOW-POWER LOW-NOISE AMPLIFIER IN 0.35-ìm SOI CMOS TECHNOLOGY I-293

Ertan Zencir and Numan S. Dogan, *North Carolina A&T State University*, Ercument Arvas, *Syracuse University*, and Mohammed Ketel, *North Carolina A&T State University*

INTERFERENCE OF ESD PROTECTION DIODES ON RF PERFORMANCE IN GIGA-HZ RF CIRCUITS I-297

Ming-Dou Ker and Chien-Ming Lee, *National Chiao Tung University*

AN 8-BIT, 1MW SUCCESSIVE APPROXIMATION ADC IN SILICON-ON-SAPPHIRE CMOS I-301

Eugenio Culurciello and Andreas Andreou, *Johns Hopkins University*

APPLICATION OF ACM MODEL TO THE DESIGN OF CMOS OTA THROUGH A GRAPHICAL APPROACH I-305

Henrique Santos and Ana Isabela Cunha, *Universidade Federal da Bahia*

RESOLUTION PREDICTION FOR BANDPASS-SD-MODULATORS USING SIMULINK BEHAVIOR SIMULATION I-309

Dirk Weiler, Thomas Van Den Boom, and Bedrich Hosticka, *Fraunhofer IMS*

A SYMMETRIC QUADRATURE-LESS IMAGE REJECTION ARCHITECTURE FOR RF RECEIVERS I-313

Rami Salem and Mohamed Tawfik, *Mentor Graphics*, and Hani Ragaie, *Ain Shams University*

A LOW-POWER LOW-NOISE 600MHZ CMOS IF DEMODULATOR FOR SUPERHETERODYNE RECEIVERS I-317

Jihua Zheng, Yongming Li, and Hongyi Chen, *Tsinghua University*

MATCHING OF LOW-NOISE AMPLIFIERS AT HIGH FREQUENCIES I-321

Aleksandar Tasic, Wouter A. Serdijn, and John R. Long, *TU Delft*

DESIGN AND OPTIMIZATION OF CMOS CLASS-E POWER AMPLIFIER I-325

Zhan Xu and Ezz El-Masry, *Dalhousie University*

A 2.4 GHZ CMOS IMAGE-REJECT LOW NOISE AMPLIFIER I-329

Ming-Chang Sun, Shing Tenqchen, Ying-Haw Shu, and Wu-Shiung Feng, *National Taiwan University*

DESIGN OF A 4.4 TO 5 GHZ LNA IN 0.25 UM SIGE BICMOS TECHNOLOGY I-333

Paolo Crippa, Simone Orcioni, Francesco Ricciardi, and Claudio Turchetti, *University of Ancona*

A CMOS INFRARED OPTICAL SIGNAL PROCESSOR FOR REMOTE CONTROL I-337

Joon-Jea Sung, *Korea University*, Guen-Soon Kang, *Silicomtech*, and Suki Kim, *Korea University*

A NEW DIFFERENTIAL CMOS CURRENT PRE-AMPLIFIER FOR OPTICAL COMMUNICATIONS I-341

Fei Yuan and Bendong Sun, *Ryerson University*

EXPLOITING HYPERBOLIC FUNCTIONS TO INCREASE LINEARITY IN LOW-VOLTAGE FLOATING-GATE TRANSCONDUCTANCE AMPLIFIERS I-345

Yngvar Berg, *University of Oslo*, Snorre Aunet, *Norwegian University of Science and Technology*, Ølvind Næss, Johannes Lomsdalen, and Mats Høvin, *University of Oslo*

A WIDE-BAND CURRENT-MODE OTA-BASED ANALOG MULTIPLIER-DIVIDER I-349

Khanittha Kaewdang, Chalermpan Fongsamut, and Wanlop Surakampontorn, *King Mongkut's Institute of Technology Ladkrabang*

A LOW-VOLTAGE COMPATIBLE TWO-STAGE AMPLIFIER WITH >=120DB GAIN IN STANDARD DIGITAL CMOS I-353

Chengming He, Degang Chen, and Randall Geiger, *Iowa State University*

A FULLY-INTEGRATED SELF-TUNED TRANSFORMER BASED STEP-UP CONVERTER I-357

A. Pretelli, Anna Richelli, Alessandro Savio, Luigi Colalongo, and Zsolt Kovacs-Vagna, *University of Brescia*

EFFECTS OF RESISTIVE LOADING ON UNITY GAIN FREQUENCY OF TWO-STAGE CMOS OPERATIONAL AMPLIFIERS I-361

Uday Dasgupta, *Institute of Microelectronics*, and Yong-Ping Xu, *National University of Singapore*

CASCODE TRANSCONDUCTANCE AMPLIFIERS FOR HF SWITCHED-CAPACITOR APPLICATIONS I-365

Joseph Adut, *Texas Instruments, Inc.*, and Jose Silva-Martinez, *Texas A&M University*

THE ELECTROTHERMAL MODEL OF THE LINEAR POWER SUPPLIES I-369

Janusz Zarêbski and Krzysztof Górecki, *Gdynia Maritime University*

A 900 MV 25 UW HIGH PSRR CMOS VOLTAGE REFERENCE DEDICATED TO IMPLANTABLE MICRO-DEVICES I-373

Yamu Hu and Mohamad Sawan, *Ecole Polytechnique de Montreal*

INCREASING THE IMMUNITY TO ELECTROMAGNETIC INTERFERENCES IN A BANDGAP VOLTAGE REFERENCE I-377

A. Pretelli, Anna Richelli, Luigi Colalongo, and Zsolt Kovacs, *University of Brescia*

A 1-VOLT, HIGH PSRR, CMOS BANDGAP VOLTAGE REFERENCE I-381

Saeid Mehrmanesh, Mohammad Bagher Vahidfar, Hesam Amir Aslanzadeh, and Mojtaba Atarodi, *Sharif University of Technology*

NEW CMOS BALANCED OUTPUT TRANSCONDUCTOR AND APPLICATION TO GM-C BIQUAD FILTER I-385

Soliman Mahmoud and Inas Awad, *Cairo University*

ONE CLASS OF TRANSFER FUNCTIONS WITH MONOTONIC STEP RESPONSE I-389

Igor Filanovsky, *University of Alberta*

ANALOG IMPLEMENTATION OF MOS-TRANSLINEAR MORLET WAVELETS I-393

Carlos Sanchez-Lopez, *INAOE*, Alejandro Diaz-Sanchez, *National Institute for Astrophysics, Optics and Electronics*, and Esteban Tlelo-Cuautle, *INAOE*

A FULLY DIFFERENTIAL LOW-VOLTAGE CMOS HIGH-SPEED TRACK-AND-HOLD CIRCUIT I-397

Tsung-Sum Lee and Chi-Chang Lu, *National Yunlin University of Science & Technology*

ADAPTIVITY FIGURES OF MERIT AND K-RAIL DIAGRAMS -COMPREHENSIVE PERFORMANCE CHARACTERIZATION OF LOW-NOISE AMPLIFIERS AND VOLTAGE-CONTROLLED OSCILLATORS I-401

Aleksandar Tasic, Wouter A. Serdijn, and John R. Long, *TU Delft*

A CMOS CURRENT-MODE SQUARER/RECTIFIER CIRCUIT I-405

Boonchai Boonchu and Wanlop Surakampontorn, *Mahanakorn University of Technology*

BROAD-BAND TRANSIMPEDANCE AMPLIFIER FOR MULTIGIGABIT-PER-SECOND (40GBPS) OPTICAL COMMUNICATION SYSTEMS IN 0.135UM PHEMT TECHNOLOGY I-409

Miguel Madureira, *Instituto De Telecomunicações*, Paulo Monteiro, Rui Aguiar, and Manuel Violas, *Universidade de Aveiro*, Maurice Gloanec, Eric Leclerc, and Benoit Lefebvre, *UMS S.A.S. - United Monolithic Semiconductors*

A NEW NON-ITERATIVE, ADAPTIVE BASEBAND PREDISTORTION METHOD FOR HIGH POWER RF AMPLIFIERS I-413

Nikos Naskas and Yannis Papananos, *National Technical University of Athens*

A 22-MW 435MHZ SILICON ON INSULATOR CMOS HIGH-GAIN LNA FOR SUBSAMPLING RECEIVERS I-417

Te-Hsin Huang, Ertan Zencir, Mehmet R. Yuce, Numan S. Dogan, and Wen-Tai Liu, *North Carolina State University*, and Ercument Arvas, *Syracuse University*

CONCEPT OF TRANSFORMER-FEEDBACK DEGENERATION OF LOW-NOISE AMPLIFIERS I-421

Aleksandar Tasic, Wouter A. Serdijn, and John R. Long, *TU Delft*

A HIGH COMPLIANCE CMOS CURRENT SOURCE FOR LOW VOLTAGE APPLICATIONS I-425

Michele Quarantelli, *University of Brescia*, Marco Poles, Marco Pasotti, and PierLuigi Rolandi, *ST Microelectronics*

DESIGN OF ACCURATE ANALOG CIRCUITS FOR LOW VOLTAGE LOW POWER CMOS SYSTEMS I-429

Christian Falconi and Arnaldo D'Amico, *University of Tor Vergata, Roma*, and Marco Faccio, *University of L'aquila, Monteluco di Roio*

A HIGH SPEED LOW INPUT CURRENT LOW VOLTAGE CMOS CURRENT COMPARATOR I-433

Kornika Moolpho, Jitkasame Ngarmnil, and Suchada Sitjongsataporn, *Mahanakorn University of Technology*

SIGE HBT POWER AMPLIFIER FOR IS-95 CDMA USING A NOVEL PROCESS, VOLTAGE, AND TEMPERATURE INSENSITIVE BIASING SCHEME I-437

Nuttapong Srirattana and Muhammad Shakeel Qureshi, *Georgia Institute of Technology*, Arlo Aude and Vikram Krishnamurthy, *National Semiconductor Corporation*, Deukhyoun Heo, Phillip Allen, and Joy Laskar, *Georgia Institute of Technology*

LARGE-SIGNAL S-PARAMETERS CAD TECHNIQUE APPLIED TO POWER AMPLIFIER DESIGN I-441

Eric Kerhervé, Mathieu Hazouard, and Laurent Courcelle, and Pierre Jarry, *IXL Laboratory*

LUMPED ELEMENT BASED DOHERTY POWER AMPLIFIER TOPOLOGY IN CMOS PROCESS I-445

Chaiwat Tongchoi, Mitchai Chongcheawchamnan, and Apisak Worapishet, *Mahanakorn University of Technology*

CLASS-E CMOS POWER AMPLIFIERS FOR RF APPLICATIONS I-449

Tang Hung and Mourad El-Gamal, *McGill University*

Track 1.2: Analog Filters

POWER - AREA - DR - FREQUENCY - SELECTIVITY TRADEOFFS IN WEAKLY NONLINEAR ACTIVE FILTERS I-453

Yorgos Palaskas and Yannis Tsividis, *Columbia University*

3.3-V BASEBAND GM-C FILTERS FOR WIRELESS TRANSCEIVER APPLICATIONS I-457

Jaeyoung Shin, *Samsung Electronics Co., Ltd.*, Sunki Min, Soosun Kim, and Joongho Choi, *University of Seoul*, Soohyoung Lee, Hojin Park, and Jaewhui Kim, *Samsung Electronics Co., Ltd.*

ANALYTICAL SYNTHESIS OF VOLTAGE MODE OTA-C ALL-PASS FILTERS FOR HIGH FREQUENCY OPERATION I-461

Chun-Ming Chang and Bashir M. Al-Hashimi, *University of Southampton*

A NEW RESISTORLESS ELECTRONICALLY TUNABLE VOLTAGE-MODE FIRST-ORDER PHASE EQUALIZER I-465

Shahram Minaei, *Dogus University*, and Oguzhan Cicekoglu, *Bogazici University*

LOW-NOISE LOW-POWER ALLPOLE ACTIVE-RC FILTERS MINIMIZING RESISTOR LEVEL I-469

Drazen Jurisic, *University of Zagreb, Croatia*, George Moschytz, *Swiss Federal Institute of Technology, Zurich*, and Neven Mijat, *University of Zagreb*

NOVEL INTEGRABLE NOTCH FILTER IMPLEMENTATION FOR 100 DB IMAGE REJECTION. I-473

Jayant Parthasarathy and Ramesh Harjani, *University of Minnesota, Minneapolis*

TUNABLE ANALOG LOUDSPEAKER CROSSOVER NETWORK I-477

Antonio Petraglia, Fernando Baruqui, and Eduardo Rapoport, *Federal University of Rio de Janeiro*

A 1V 0.54μW FOURTH ORDER SWITCHED CAPACITOR FILTER WITH SWITCHED OPAMP TECHNIQUE FOR CARDIAC PACEMAKER SENSING CHANNEL I-481

Jen-Shiun Chiang and Cheng-Ming Ying, *Tamkang University*

A CMOS 2 MHZ SELF-CALIBRATING BANDPASS FILTER FOR PERSONAL AREA NETWORKS I-485

Craig Frost, Gary Levy, and Bryan Allison, *Cadence Design Systems Inc.*

A LOW-POWER CMOS COMPLEX FILTER FOR BLUETOOTH WITH FREQUENCY TUNING I-489

Ahmed Emira and Edgar Sanchez-Sinencio, *Texas A&M University*

0.18μ CMOS GM-C DIGITALLY TUNED FILTER FOR TELECOM RECEIVERS I-493

Stefano D'Amico and Andrea Baschirotto, *University of Lecce*

A 0.7-μm CMOS ANTI-ALIASING FILTER FOR NON-OVERSAMPLED VIDEO SIGNAL APPLICATION I-497

Pairote Sirinamaratana and Naiyavudhi Wongkomet, *Chulalongkorn University*

A COMPACT BIQUADRATIC GM-C FILTER STRUCTURE FOR LOW-VOLTAGE AND HIGH FREQUENCY APPLICATIONS I-501

Armin Tajalli and Mojtaba Atarodi, *Sharif University of Technology*

ON THE INVARIANCE OF THE SECOND-ORDER MODES OF CONTINUOUS-TIME SYSTEMS UNDER GENERAL FREQUENCY TRANSFORMATION I-505

Masayuki Kawamata, *Tohoku University*

PROPERTIES OF ANALOG SYSTEMS WITH VARYING PARAMETERS I-509

Roman Kaszynski, *Technical University of Szczecin*

A NOVEL APPROACH TO THE DESIGN AND SYNTHESIS OF GENERAL-ORDER BODE-TYPE VARIABLE-AMPLITUDE ACTIVE-RC EQUALIZERS I-513

Behrouz Nowrouzian, *University of Alberta*, and Arthur T. G. Fuller, *Nortel Networks*

CMOS TRANSCONDUCTOR DESIGN FOR VHF FILTERING APPLICATIONS I-517

Luo Zhenying, Ming Fu Li, and Yong Lian, *National University of Singapore* and S.C.Rustagi, *Institute of Microelectronics*

DESIGN CONSIDERATIONS FOR A 1.5-V, 10.7-MHZ BANDPASS GM-C FILTER IN A 0.6-UM STANDARD CMOS TECHNOLOGY I-521

Armin Tajalli and Mojtaba Atarodi, *Sharif University of Technology*

PERFORMACE COMPARISON OF TOW-THOMAS BIQUAD FILTERS BASED ON VOAS AND CFOAS I-525

Rosario Mita, Gaetano Palumbo, and Salvatore Pennisi, *University of Catania*

THE DESIGN OF AN ACTIVE BAND PASS FILTER USING UNIFORMLY DISTRIBUTED RC LINE I-529

Prakit Tangtisanon, Anchana Khempila, Kanok Janchitrapongvej, Nouanchang Panyanouvong, and Sorapong Saetia, *King Monkut's Institute of Technology Ladkrabang*, Shirou Sudo and Mitsuo Teramoto, *Toki University*

NOVEL HIGH PERFORMANCE SINGLE AMPLIFIER BIQUADS I-533

Brent Maundy, *University of Calgary*, Ezz El-Masry, *Dalhousie University*, and Peter Aronhime, *University of Louisville*

RC POLYPHASE FILTERS WITH FLAT GAIN CHARACTERISTICS I-537

Kazuyuki Wada and Yoshiaki Tadokoro, *Toyohashi University of Technology*

REALIZATION OF ELECTRONICALLY TUNABLE LADDER FILTERS USING MULTI-OUTPUT CURRENT CONTROLLED CONVEYORS I-541

Amorn Jiraseree-Amornkun, *King Mongkut's Institute of Technology Ladkrabang*, Nobuo Fujii and Wanlop Surakampontorn, *King Mongkut's Institute of Technology Ladkrabang*

LOG-DOMAIN COMPLEX FILTER DESIGN WITH XFILTER I-545

Mykhaylo Teplechuk and John Sewell, *University of Glasgow*

COMPARISON OF CURRENTS IN DIFFERENTIAL LOG-DOMAIN FILTERS WITH COMMON-MODE FEEDBACK I-549

Hyung-Jong Ko and Robert Fox, *University of Florida*

FULLY-DIFFERENTIAL LOG-DOMAIN INTEGRATOR WITH ORTHOGONAL COMMON-MODE AND DIFFERENTIAL-MODE RESPONSES I-553

Jirayut Mahattanakul and Y.Piputtawutchai, *Mahanakorn University of Technology*

A CMOS DIGITALLY TUNABLE TRANSCONDUCTOR FOR VIDEO FREQUENCY OPERATION I-557

Belen Calvo, Maria Teresa Sanz, Santiago Celma, and Pedro A. Martinez, *University of Zaragoza*

AN ULTRA LOW-VOLTAGE GM-C FILTER FOR VIDEO APPLICATIONS I-561

Saeid Mehrmanesh, Mohammad Bagher Vahidfar, Hesam Amir Aslanzadeh, and Mojtaba Atarodi, *Sharif University of Technology*

PERFORMANCE OPTIMIZATION IN MICRO-POWER, LOW-VOLTAGE LOG-DOMAIN FILTERS IN PURE CMOS TECHNOLOGY I-565

Andrea Maniero, Andrea Gerosa, and Andrea Neviani, *University of Padova*

THE TAU-CELL: A NEW METHOD FOR THE IMPLEMENTATION OF ARBITRARY DIFFERENTIAL EQUATIONS I-569

Andre Van Schaik and Craig Jin, *The University of Sydney*

INDUCTORLESS RF AMPLIFIER WITH TUNEABLE BAND-SELECTION AND IMAGE-REJECTION I-573

Apinunt Thanachayanont and Somkid Sae-Ngow, *King Mongkut's Institute of Technology Ladkrabang*

RF, Q-ENHANCED BANDPASS FILTERS IN STANDARD 0.18UM CMOS WITH DIRECT DIGITAL TUNING I-577

Hesham Ahmed, Chris Devries, and Ralph Mason, *Carleton University*

LUMPED ELEMENT MODEL APPROACH FOR THE BANDWIDTH ENHANCEMENT OF COUPLED MICROSTRIP ANTENNA I-581

Wiset Saksiri, *Mahanakorn University of Technology*, and Monai Krairiksh, *King Mongkut's Institute of Technology Ladkrabang*

Track 1.3: Low-Power Analog Circuits

A TIME DOMAIN TECHNIQUE FOR COMPUTATION OF NOISE SPECTRAL DENSITY IN SWITCHED CAPACITOR CIRCUITS I-585

Vinita Vasudevan, *Indian Institute of Technology Madras*

NEW LOW-POWER LOW-VOLTAGE DIFFERENTIAL CLASS-AB OTA FOR SC CIRCUITS I-589

Ramon Carvajal and Juan Antonio Galan, *Universidad de Sevilla*, Jaime Ramirez-Angulo, *New Mexico State University*, and Antonio Torralba, *Universidad de Sevilla*

A 0.35UM CMOS VOLTAGE DERIVATIVE SENSOR WITH SIGN AND INFLECTION OUTPUTS I-593

Gerry Quilligan and Phil Burton, *University of Limerick*

NOISE-SHAPING MODULATION IN HIGH-Q SC FILTERS I-597

Jose Ausin, M.A. Dominguez, and J. Francisco Duque-Carrillo, and G. Torelli, *University of Extremadura*

NON-UNIFORM SAMPLING SC CIRCUITS BASED ON NOISE-SHAPING FEEDBACK CODING I-601

Jose Ausin and J. Francisco Duque-Carrillo, *University of Extremadura*, G. Torelli, *University of Pavia*, and J. V. Valverde, *University of Extremadura*

DESIGN OF LOW-VOLTAGE LOW-POWER SC FILTERS FOR HIGH-FREQUENCY APPLICATIONS I-605

Walter Aloisi, Gianluca Giustolisi, and Gaetano Palumbo, *Università di Catania*

LOW-POWER SWITCHED-CAPACITOR FILTERS USING CHARGE-TRANSFER INTEGRATORS I-609

Inchang Seo and Robert Fox, *University of Florida*

A 2.5V SWITCHED-CURRENT SIGMA-DELTA MODULATOR WITH A NOVEL CLASS AB MEMORY CELL I-613

Shuenn-Yuh Lee, *National Chung Cheng University*, and Yueh-Lun Tsai Wei-Zen Su, and Po-Hui Yang, *Southern Taiwan University of Technology*

Track 1.4: Oscillators and Reference Circuits

ANALYSIS OF PHASE NOISE DUE TO BANG-BANG PHASE DETECTOR IN PLL BASED CLOCK AND DATA RECOVERY CIRCUITS I-617

Kasin Vichienchom and Wentai Liu, *North Carolina State University*

ANALOGUE INTERPOLATION BASED DIRECT DIGITAL FREQUENCY SYNTHESIS I-621

Alistair McEwan and Steve Collins, *Oxford University*

A SINE-OUTPUT ROM-LESS DIRECT DIGITAL FREQUENCY SYNTHESISER USING A POLYNOMIAL APPROXIMATION I-625

Charan Meenakarn, *National Electronics and Comupter Technology Center, Thailand*, and Apinunt Thanachayanont, *King Mongkut's Institute of Technology Ladkrabang*

NONLINEAR ANALYSIS OF A COLPITTS INJECTION-LOCKED FREQUENCY DIVIDER I-629

Uroschanit Yodprasit, *Swiss Federal Institute of Technology*, and Christian Enz, *Centre Suisse D'electronique Et de Microtechnique (CSEM)*

LOW-POWER CMOS PLL FOR CLOCK GENERATOR I-633

Wen-Chi Wu and Chih-Chien Huang, *Industrial Technology Research Institute*, Chih-Hsiung Chang, *National Chiayi University*, and Nai-Heng Tseng, *Industrial Technology Research Institute*

PHYSCICAL SCALING OF INTEGRATED INDUCTOR LAYOUT AND MODEL AND ITS APPLICATION TO WLAN VCO DESIGN AT 11GHZ AND 17GHZ I-637

Marc Tiebout, *Infineon Technologies AG*

A LOW POWER 2.2-2.6GHZ CMOS VCO WITH A SYMMETRICAL SPIRAL INDUCTOR I-641

Tser-Yu Lin, Ying-Zong Juang, Hung-Yu Wang, and Chin-Fong Chiu, *CIC, National Science Council, Taiwan*

AN ARBITRARILY SKEWABLE MULTIPHASE CLOCK GENERATOR COMBINING DIRECT INTERPOLATION WITH PHASE ERROR AVERAGE I-645

Lixin Yang and Jiren Yuan, *Lund University*

INFLECTION POINT CORRECTION FOR VOLTAGE REFERENCES I-649

Kee-Chee Tiew, *Iowa State University*, Jim Cusey, *Dallas Semiconductor*, and Randall Geiger, *Iowa State University*

HIGH-CURRENT CLAMP FOR FAST-RESPONSE LOAD TRANSITIONS OF DC-DC CONVERTER I-653

Thilak Senanayake and Tamotsu Ninomiya, *Kyushu University*

A NEW CMOS CHARGE PUMP FOR LOW-VOLTAGE (1V) HIGH-SPEED PLL APPLICATIONS I-657

Rola Baki and Mourad El-Gamal, *McGill University*

IMPACT OF MUTUAL INDUCTANCE AND PARASITIC CAPACITANCE ON THE PHASE-ERROR PERFORMANCE OF CMOS QUADRATURE VCOS I-661

Xiaoyan Wang and Pietro Andreani, *Technical University of Denmark*

STUDY AND SIMULATION OF CMOS LC OSCILLATOR PHASE NOISE AND JITTER I-665

Michael McCorquodale, Mei Kim Ding, and Richard Brown, *University of Michigan*

ANALYSIS OF EMITTER DEGENERATED LC OSCILLATORS USING BIPOLAR TECHNOLOGIES I-669

Jing-Hong Zhan, Kyle Maurice, Jon Duster, and Kevin Kornegay, *Cornell University*

LOW-NOISE BIASING OF VOLTAGE-CONTROLLED OSCILLATORS BY MEANS OF RESONANT INDUCTIVE DEGENERATION I-673

Aleksandar Tasic, Wouter A. Serdijn, and John R. Long, *TU Delft*

A LOW-POWER LOW-VOLTAGE OTA-C SINUSOIDAL OSCILLATOR WITH MORE THAN TWO DECADES OF LINEAR TUNING RANGE I-677

Juan Galan, Ramon Carvajal, Fernando Muñoz, and Antonio Torralba, *Universidad de Sevilla*, and Jaime Ramirez-Angulo, *New Mexico State University*

ANALYSIS AND DESIGN OF A DOUBLE TUNED CLAPP OSCILLATOR FOR MULTI-BAND MULTI-STANDARD RADIO I-681

Wim Michielsen, *Royal Institute of Stockholm*, Li-Rong Zheng, *Royal Institute of Technology*, and Hannu Tenhunen, *Royal Institute of Technology Stockholm*

ANALYSIS OF TIMING JITTER IN RING OSCILLATORS DUE TO POWER SUPPLY NOISE I-685

Tony Pialis and Khoman Phang, *University of Toronto*

A QUADRATURE RELAXATION OSCILLATOR-MIXER IN CMOS I-689

Luís B. Oliveira and Jorge Fernandes, *IST/INESC-ID Lisboa*, Michiel H.L. Kouwenhoven, *National Semiconductors B.V. Delft*, Chris Van den Bos and Chris J. M. Verhoeven, *TU Delft*

A 7-MHZ PROCESS AND TEMPERATURE COMPENSATED CLOCK OSCILLATOR IN 0.25UM CMOS I-693

Krishnakumar Sundaresan, *Georgia Institute of Technology*, Keith Brouse, *Texas Instruments Inc.*, Kongpop U-Yen, Farrokh Ayazi, and Phillip Allen, *Georgia Institute of Technology*

PERFORMANCE CHARACTERISTICS OF AN ULTRA-LOW POWER VCO I-697

Jamal Deen, Mehdi Kazemeini, and Sasan Naseh, *McMaster University*

PHASE NOISE IN A BACK-GATE BIASED LOW VOLTAGE VCO I-701

Mehdi Kazemeini, *McMaster University*, Jamal Deen, *Université Montpellier Ii*, and Sasan Naseh, *McMaster University*

LARGE-SIGNAL AND PHASE NOISE PERFORMANCE ANALYSIS OF ACTIVE INDUCTOR TUNABLE OSCILLATORS I-705

Xibo Zhang, Philip Mok, Mansun Chan, and Ping Ko, *Hong Kong University of Science and Technology*

AN INDUCTIVELY-TUNED QUADRATURE OSCILLATOR WITH EXTENDED FREQUENCY CONTROL RANGE I-709

Raymond Koo and John Long, *Delft University of Technology*

SIDEBAND NOISE REDUCTION IN TRANSPOSED GAIN OSCILLATORS I-713

Sawat Bunnjaweht, Michael Underhill, and Ian Robertson, *University of Surrey*

A 40 GHZ MODIFIED-COLPITTS VOLTAGE CONTROLLED OSCILLATOR WITH INCREASED TUNING RANGE I-717

Chao Su, Sreenath Thoka, Tiew Kee-Chee, and Randall Geiger, *Iowa State University*

A DUAL EDGE-TRIGGERED PHASE-FREQUENCY DETECTOR ARCHITECTURE I-721

Syed Ahmed and Ralph Mason, *Carleton University*

Track 1.5: Mixing and Synthesis

A 1.6 GHZ DOWNCONVERSION SAMPLING MIXER IN CMOS I-725

Darius Jakonis and Christer Svensson, *Linkoping University*

EXTRACTION OF TIMING JITTER FROM PHASE NOISE I-729

Omid Oliaei, *Motorola Labs*

PHASE NOISE IMPROVEMENT IN FRACTIONAL-N SYNTHESIZER WITH 90 DEGREE PHASE SHIFT LOCK I-733

Joohwan Park, *University of Texas at Dallas*, and Franco Maloberti, *University of Texas at Dallas & University of Pavia*

ON THE EFFECTS OF TIMING JITTER IN CHARGE SAMPLING I-737

Sami Karvonen, Thomas Riley, and Juha Kostamovaara, *University of Oulu*

BONDING-PAD-ORIENTED ON-CHIP ESD PROTECTION STRUCTURES FOR ICS I-741

Haigang Feng, Rouying Zhan, Guang Chen Qiong Wu, Xiaokang Guang, H. Xie, and Albert Wang, *Illinois Institute of Technology*

PERFORMANCE ANALYSIS OF GENERAL CHARGE SAMPLING I-745

Gang Xu and Jiren Yuan, *Lund University*

PULSE EXTRACTION: A DIGITAL POWER SPECTRUM ESTIMATION METHOD FOR ADAPTATION OF GBPS EQUALIZERS I-749

Xiaofeng Lin, Guangbin Zhang, and Jin Liu, *University of Texas at Dallas*

OPTIMIZATION OF SHIELD STRUCTURES IN ANALOG INTEGRATED CIRCUITS I-753

Ken Yamamoto, Minoru Fujishima, and Koichiro Hoh, *The University of Tokyo*

DESIGN OF A FREQUENCY SYNTHESIZER FOR WCDMA IN 0.18-UM CMOS PROCESS I-757

Young-Mi Lee, Ju-Sang Lee, Sang Jin Lee, and Ri-A Ju, *Kyungpook National University*

DESIGN OF ENHANCEMENT CURRENT-BALANCED LOGIC FOR MIXED-SIGNAL ICS I-761

Li Yang and J.S. Yuan, *University of Central Florida*

AN EFFICIENT MIXED-SIGNAL ARCHITECTURE FOR MINIMUM OUTPUT ENERGY BLIND MULTIUSER DETECTION I-765

Phaophak Sirisuk, Apisak Worapishet, and Saifon Tanoi, *Mahanakorn University of Technology*

ALGORITHMIC PARTIAL ANALOG-TO-DIGITAL CONVERSION IN MIXED-SIGNAL ARRAY PROCESSORS I-769

Roman Genov, *University of Toronto*, and Gert Cauwenberghs, *Johns Hopkins University*

AN ANALOG SIGNAL INTERFACE WITH CONSTANT PERFORMANCE FOR SOCS I-773

Eric Fabris, Luigi Carro, and Sergio Bampi, *Universidade Federal do Rio Grande do Sul*

MIXED-SIGNAL GRADIENT FLOW BEARING ESTIMATION I-777

Milutin Stanacevic and Gert Cauwenberghs, *The Johns Hopkins University*

SUB-VOLT SUPPLY ANALOG CIRCUITS BASED ON QUASI-FLOATING GATE TRANSISTORS I-781

Jaime Ramirez-Angulo, *New Mexico State University*, Ramon G. Carvajal, *Universidad de Sevilla*, Carlos Urquidi, *Delphi Automotive*, and Antonio Torralba, *Universidad de Sevilla*

DIRECT DIGITAL SYNTHESIZER WITH TUNABLE PHASE AND AMPLITUDE ERROR FEEDBACK STRUCTURES I-785

Jouko Vankka, Jonne Lindeberg, and Kari Halonen, *Helsinki University of Technology*

EQUIVALENT CIRCUIT MODELS FOR STACKED SPIRAL INDUCTORS IN DEEP SUBMICRON CMOS TECHNOLOGY I-789

Subhash Rustagi and Chun-Geik Tan, *Institute of Microelectronics*

A CMOS CURRENT-CONTROLLED OSCILLATOR AND ITS APPLICATIONS I-793

Chunyan Wang, M. Omair Ahmad, and M.N.S. Swamy, *Concordia University*

A 5-GHZ SELF-CALIBRATED I/Q CLOCK GENERATOR USING A QUADRATURE LC-VCO I-797

Hyung Ahn, In-Cheol Park, and Beomsup Kim, *KAIST*

LOW-VOLTAGE LOW-POWER WIDEBAND CMOS CURRENT CONVEYORS BASED ON THE FLIPPED VOLTAGE FOLLOWER I-801

Antonio Lopez-Martin, *Public University of Navarra*, Jaime Ramirez-Angulo, *New Mexico State University*, and Ramon G. Carvajal, *Universidad de Sevilla*

NEW VERY COMPACT CMOS CONTINUOUS-TIME LOW-VOLTAGE ANALOG RANK-ORDER FILTER ARCHITECTURE I-805

Jaime Ramirez-Angulo, *New Mexico State University*, Ramon Carvajal, *Universidad de Sevilla*, and Gladys Ducoudray, *New Mexico State University*

AN OPTIMIZED CMOS PSEUDO-ACTIVE-PIXEL-SENSOR STRUCTURE FOR LOW-DARK-CURRENT IMAGER APPLICATIONS I-809

Yu Shih and Chung-Yu Wu, *National Chiao Tung University*

LOW-VOLTAGE CLOSED-LOOP AMPLIFIER CIRCUITS BASED ON QUASI-FLOATING GATE TRANSISTORS I-813

Jaime Ramirez-Angulo, *New Mexico State University*, Antonio Lopez-Martin, *Public University of Navarra*, Ramon G. Carvajal, *Universidad de Sevilla*, and Chad Lackey, *New Mexico State University*

EXTREMELY LOW SUPPLY VOLTAGE CIRCUITS BASED ON QUASI-FLOATING GATE SUPPLY VOLTAGE BOOSTING I-817

Fernando Muñoz, *Universidad de Sevilla*, Antonio Lopez-Martin, *Universidad Publica de Navarra*, Ramon Carvajal, *Universidad de Sevilla*, Jaime Ramirez-Angulo, *New Mexico State University*, Antonio Torralba, *Universidad de Sevilla*, Meghraj Kachare, *New Mexico State University*, and Bernardo Palomo, *Universidad de Sevilla*

Track 1.6: Signal Conversion

DESIGN TECHNIQUES FOR A FULLY DIFFERENTIAL LOW-VOLTAGE LOW-POWER FLASH ANALOG-TO-DIGITAL CONVERTER I-821

Tsung-Sum Lee, Li-Dyi Luo, and Chin-Sheng Lin, *National Yunlin University of Science & Technology*

A SELF-CALIBRATION TECHNIQUE FOR TIME-INTERLEAVED PIPELINE ADCS I-825

Välnö Hakkarainen, Lauri Sumanen, and Mikko Aho, *Helsinki University of Technology*, Mikko Waltari, *Conexant Systems Inc.*, and Kari Halonen, *Helsinki University of Technology*

AN EXTENDED RADIX-BASED DIGITAL CALIBRATION TECHNIQUE FOR MULTI-STAGE ADC I-829

Jipeng Li and Un-Ku Moon, *Oregon State University*

NEW SAMPLING METHOD TO IMPROVE THE SFDR OF TIME-INTERLEAVED ADCS I-833

Kamal El-Sankary, Ali Assi, and Mohamad Sawan, *Ecole Polytechnique de Montreal*

IMPROVED CURRENT-SOURCE SIZING FOR HIGH-SPEED HIGH-ACCURACY CURRENT STEERING D/A CONVERTERS I-837

Miquel Albiol, José Luis González, and Eduard Alarcon, *Technical University of Catalunya*

DESIGN AND REALIZATION OF A MODULAR 200 MSAMPLE/S 12-BIT PIPELINED A/D CONVERTER BLOCK USING DEEP-SUBMICRON DIGITAL CMOS TECHNOLOGY I-841

Zeynep Toprak, *ST Microelectronics*, and Yusuf Leblebici, *Swiss Federal Institute of Technology*

A METHOD TO REDUCE POWER CONSUMPTION IN PIPELINED A/D CONVERTERS I-845

Liviu Chiaburu and Svante Signell, *Royal Institute of Technology*

ANALYSIS OF THE AVERAGING TECHNIQUE IN FLASH ADCS I-849

Pedro Figueiredo and João Vital, *Chipidea Microelectrónica, S.A*

AN 8-BIT, 1.8V, 20 MSAMPLES/S ANALOG-TO-DIGITAL CONVERTER USING LOW GAIN OPAMPS I-853

Douglas Beck and David Allstot, *University of Washington*, and Douglas Garrity, *Motorola, Inc.*

FREQUENCY-INTERLEAVING TECHNIQUE FOR HIGH-SPEED A/D CONVERSION I-857

Guillaume Ding, *Swiss Federal Institute of Technology*, Kamran Azadet, *Agere Systems*, Catherine Dehollain and Michel Declercq, *Swiss Federal Institute of Technology, Lausanne*

OUTPUT IMPEDANCE REQUIREMENTS FOR DACS I-861

Susan Luschas and Hae-Seung Lee, *Massachusetts Institute of Technology*

AN IMPROVED BINARY ALGORITHMIC A/D CONVERTER ARCHITECTURE I-865

Liviu Chiaburu and Svante Signell, *Royal Institute of Technology*

A 3 V 12B 100 MS/S CMOS D/A CONVERTER FOR HIGH-SPEED SYSTEM APPLICATIONS I-869

Hyuen-Hee Bae, *Sogang University*, Jin-Sik Yoon, *TLI Inc.*, Myung-Jin Lee, Eun-Seok Shin, and Seung-Hoon Lee, *Sogang University*

A MONOTONIC DIGITAL CALIBRATION TECHNIQUE FOR PIPELINED DATA CONVERTERS I-873

Waisiu Law, Jianjun Guo, Charles Peach, Ward Helms, and David Allstot, *University of Washington*

EFFICIENT DIGITAL SELF-CALIBRATION OF VIDEO-RATE PIPELINE ADCS USING WHITE GAUSSIAN NOISE I-877

Martin Unterweissacher, *Technische Universitaet Graz*, Joao Goes, Nuno Paulino, Guiomar Evans, and Manuel Ortigueira, *UNINOVA*

A DIGITAL BACKGROUND CALIBRATION TECHNIQUE FOR PIPELINED ANALOG-TO-DIGITAL CONVERTERS I-881

Hung-Chih Liu, *Silicon Integrated Systems Corporation*, Zwei-Mei Lee, and Jieh-Tsorng Wu, *National Chiao Tung University*

A 1.8-V HIGH-SPEED 13-BIT PIPELINED ANALOG TO DIGITAL CONVERTER FOR DIGITAL IF APPLICATIONS I-885

Hesam Amir Aslanzadeh, Saeid Mehrmanesh, Mohamad Bagher Vahidfar, and Mojtaba Atarodi, *Sharif University of Technology*

A MIXED-SIGNAL CALIBRATION TECHNIQUE FOR LOW-VOLTAGE CMOS 1.5-BIT/STAGE PIPELINED DATA CONVERTERS I-889

Jianjun Guo, Waisiu Law, Charles Peach, Ward Helms, and David Allstot, *University of Washington*

AN 8-BIT 2-GSAMPLE/S ANALOG-TO-DIGITAL CONVERTER IN 0.5-UM SIGE TECHNOLOGY I-893

Farhang Vessal and C. Andre T. Salama, *University of Toronto*

A POWER-EFFICIENT ARCHITECTURE FOR HIGH-SPEED D/A CONVERTERS I-897

Kamran Farzan and David Johns, *University of Toronto*

1-GS/S, 12-BIT SIGE BICMOS D/A CONVERTER FOR HIGH-SPEED DDFS I-901

Kwang-Hyun Baek, Myung-Jun Choe, and Edward Merlo, *Rockwell Scientific Company*, and Sung-Mo Kang, *University of California, Santa Cruz*

POWER EFFICIENT SCALABLE PRECISION RATIONAL DIGITAL-TO-ANALOGUE CONVERTERS I-905

Calvin Sim and Chris Toumazou, *Imperial College*

1-D AND 2-D SWITCHING STRATEGIES ACHIEVING NEAR OPTIMAL INL FOR THERMOMETER-CODED CURRENT STEERING DACS I-909

Zhongjun Yu, Degang Chen, and Randall Geiger, *Iowa State University*

HIGH ORDER 1-BIT DIGITAL SIGMA DELTA MODULATION FOR ON CHIP ANALOGUE SIGNAL SOURCES I-913

Chiheb Rebai, Dominique Dallet, and Philippe Marchegay, *IXL Laboratory*

DIRECT DIGITAL SYNTHESIZER WITH TUNABLE DELTA SIGMA MODULATOR I-917

Jouko Vankka, Jonne Lindeberg, and Kari Halonen, *Helsinki University of Technology*

A 1V, 12-BIT WIDEBAND CONTINUOUS-TIME SIGMA-DELTA MODULATOR FOR UMTS APPLICATIONS I-921

Friedel Gerfers, Maurits Ortmanns, and Yiannos Manoli, *Albert Ludwig University, Freiburg*

INFLUENCE OF FINITE INTEGRATOR GAIN BANDWIDTH ON CONTINUOUS-TIME SIGMA DELTA MODULATORS I-925

Maurits Ortmanns, Friedel Gerfers, and Yiannos Manoli, *University of Freiburg*

A SIGMA-DELTA BASED OPEN-LOOP FREQUENCY MODULATOR I-929

Denis Daly and Anthony Chan Carusone, *University of Toronto*

LOW VOLTAGE 2-PATH SC BANDPASS DELTA-SIGMA MODULATOR WITHOUT BOOTSTRAPPER I-933

Lei Wang, *Maxim Integrated Products*, and Sherif Embabi, *Texas A&M University*

A LOW-COMPLEXITY LOW-DISTORTION TOPOLOGY FOR WIDEBAND DELTA-SIGMA ADCS I-937

Zhenghong Wang, *Princeton University*, Xieting Ling, and Bo Hu, *Fudan University*

A TIME-INTERLEAVED SWITCHED-CAPACITOR BAND-PASS DELTA-SIGMA MODULATOR I-941

Minho Kwon, Jungyoon Lee, and Gunhee Han, *Yonsei University*

DESIGN OF NOISE SHAPING FIR FILTERS BY MINIMIZING IN-BAND PEAK AMPLITUDE FOR STABLE SINGLE- AND MULTI-BIT DATA CONVERTERS I-945

Mitsuhiko Yagyu, *Tokyo University of Agriculture and Technology*

A LOW-POWER CMOS FOLDING AND INTERPOLATION A/D CONVERTER WITH ERROR CORRECTION I-949

Renato T. Silva, *Portugal Telecom*, and Jorge Fernandes, *IST/INESC-ID Lisboa*

GAIN AND OFFSET MISMATCH CALIBRATION IN MULTI-PATH SIGMA-DELTA MODULATORS I-953

Vincenzo Ferragina and Andrea Fornasari, *University of Pavia*, Umberto Gatti, *Siemens Mobile Communications S.p.A.*, Piero Malcovati, *University of Pavia*, and Franco Maloberti, *University of Texas at Dallas*

CONTINUOUS-TIME SIGMA-DELTA MODULATOR INCORPORATING SEMI-DIGITAL FIR FILTERS I-957

Omid Oliaei, *Motorola Labs*

STABILITY ANALYSIS OF A SIGMA DELTA MODULATOR I-961

Jianxin Zhang, Paul Brennan, and D. Jiang *University College London*, E. Vinogradova and P.D. Smith, *University of Dundee*

COST-ORIENTED DESIGN OF A 14-BIT CURRENT STEERING DAC MACROCELL I-965

Janusz Starzyk, *Ohio University*, and Russell Mohn, *Sarnoff Corporation*

DNL AND INL YIELD MODELS FOR A CURRENT-STEERING D/A CONVERTER I-969

Marko Kosunen, Jouko Vankka, Ilari Teikari, and Kari Halonen, *Helsinki University of Technology*

ANALYSIS OF THE DYNAMIC SFDR PROPERTY OF HIGH-ACCURACY CURRENT-STEERING D/A CONVERTERS I-973

Tao Chen and Georges Gielen, *K.U.Leuven, ESAT-MICAS*

MISMATCH-BASED TIMING ERRORS IN CURRENT-STEERING DACS I-977

Konstantinos Doris, *Technical University Eindhoven*, Arthur Van Roermund, *Eindhoven University of Technology*, and Domine Leenaerts, *Philips Research Labs*

AN 8-BIT CURRENT MODE RIPPLE FOLDING A/D CONVERTER I-981

Huseyin Dinc, *Texas A&M University*, and Franco Maloberti, *University of Texas at Dallas*

STABLE HIGH-ORDER DELTA-SIGMA DACS I-985

Peter Kiss, *Agere Systems*, Jesus Arias, *University of Valladolid*, and Dandan Li, *Agere Systems*

CONTINUOUS TIME SIGMA-DELTA MODULATORS WITH TRANSMISSION LINE RESONATORS AND IMPROVED JITTER AND EXCESS LOOP DELAY PERFORMANCE I-989

Luis Hernandez and Susana Paton, *Universidad Carlos III*

HIGH-SPEED DACS WITH RANDOM MULTIPLE DATA-WEIGHTED AVERAGING ALGORITHM I-993

Yu-Hong Lin, *National Cheng Kung University*, Da-Huei Lee, *Advanic Technologies, Inc.*, Cheng-Chung Yang, *Advanic Technologies, Inc.*, and Tai-Haur Kuo, *National Cheng Kung University*

DUAL-MODE SIGMA-DELTA MODULATOR FOR WIDEBAND RECEIVER APPLICATIONS I-997

Jen-Shiun Chiang , Pao-Chu Chou, and Teng-Hung Chang *Tamkang University*

AN ULTRA-LOW POWER DOUBLE-SAMPLED A/D MASH SIGMA-DELTA MODULATOR I-1001

Remi Le Reverend, *Zarlink Semiconductor*, Izzet Kale, *University of Westminster*, Guy Delight, *Zarlink Semiconductor*, Dik Morling, *University of Westminster*, and Steve Morris, *Zarlink Semiconductor*

QUANTIZATION NOISE IN THE FIRST-ORDER NON-FEEDBACK DELTA-SIGMA MODULATOR WITH DC-INPUT I-1005

Dag Wisland, Mats E. Høvin, and Tor Sverre Lande, *University of Oslo*

A 1.4G SAMPLES/SEC COMB FILTER DESIGN FOR DECIMATION OF SIGMA-DELTA MODULATOR OUTPUT I-1009

Daeik Kim and Martin Brooke, *Georgia Institute of Technology*

CONTINUOUS-TIME, FREQUENCY TRANSLATING BANDPASS DELTA-SIGMA MODULATOR I-1013

Anurag Pulincherry, *Oregon State University*, Michael Hufford and Eric Naviasky, *Cadence Design Systems*, and Un-Ku Moon, *Oregon State University*

BP DECIMATION FILTER FOR IF-SAMPLING MERGED WITH BP DELTA-SIGMA MODULATOR I-1017

Teemu Salo, *Helsinki University of Technology*, Saska Lindfors, *Aalborg University*, and Kari Halonen, *Helsinki University of Technology*

0.65V SIGMA-DELTA MODULATORS I-1021

Jens Sauerbrey and Martin Wittig, *Infineon Technologies*, Doris Schmitt-Landsiedel, *TU Munich*, and Roland Thewes, *Infineon Technologies*

A HYBRID DELTA-SIGMA MODULATOR WITH ADAPTIVE CALIBRATION I-1025

Jae Hoon Shim, In-Cheol Park, and Beomsup Kim, *Korea Advanced Institute of Science and Technology*

EXCESS LOOP DELAY EFFECTS IN CONTINUOUS-TIME QUADRATURE BANDPASS SIGMA-DELTA MODULATORS I-1029

Frank Henkel and Ulrich Langmann, *Ruhr-University Bochum*

A NEW LOW-POWER SIGMA-DELTA MODULATOR WITH THE REDUCED NUMBER OF OP-AMPS FOR SPEECH BAND APPLICATIONS I-1033

Amin Safarian, *Sharif University of Technology*, Farzad Sahandi, *Emadsemicon*, and Mojtaba Atarodi, *Sharif University of Technology*

FUNDAMENTAL LIMITS OF JITTER INSENSITIVITY IN DISCRETE AND CONTINUOUS-TIME SIGMA DELTA MODULATORS I-1037

Maurits Ortmanns, Friedel Gerfers, and Yiannos Manoli, *University of Freiburg*

A BANDPASS SIGMA-DELTA MODULATOR EMPLOYING MICRO-MECHANICAL RESONATOR I-1041

Xiaofeng Wang and Yong Ping Xu, *National University of Singapore*, Zhe Wang and Saxon Liw, *Institute of Microelectronics*, Wai Hoong Sun and Leng Seow Tan, *National University of Singapore*

A VERY LOW-VOLTAGE, LOW-POWER AND HIGH-RESOLUTION SIGMA-DELTA MODULATOR FOR DIGITAL AUDIO IN 0.25-UM CMOS I-1045

Mohammad Yavari, Omid Shoaei, and Ali Afzali-Kusha, *University of Tehran*

AN 82 DB CMOS CONTINUOUS-TIME COMPLEX BANDPASS SIGMA-DELTA ADC FOR GSM/EDGE I-1049

Farzad Esfahani, Philipp Basedau, Roland Ryter, and Rolf Becker, *Philips Semiconductors AG*

NEW DUAL-QUANTIZATION MULTIBIT SIGMA-DELTA MODULATORS WITH DIGITAL NOISE-SHAPING I-1053

Francisco Colodro and Antonio J. Torralba Silgado, *Universidad de Sevilla*

A 10-BIT, 4 MW CONTINUOUS-TIME SIGMA-DELTA ADC FOR UMTS IN A 0.12 ìM CMOS PROCESS I-1057

Lukas Doerrer, Antonio Di Giandomenico, and Andreas Wiesbaur, *Infineon*

A 10.7 MHZ BANDPASS SIGMA-DELTA MODULATOR USING DOUBLE-DELAY SINGLE OPAMP SC RESONATOR WITH DOUBLE-SAMPLING **I-1061**

Kuai Fok Au, Kuok Hang Mok, Ho Ieng Leong, and Chon In Lao, *University of Macau*

A SPUR-FREE FRACTIONAL-N SIGMA-DELTA PLL FOR GSM APPLICATIONS: LINEAR MODEL AND SIMULATIONS **I-1065**

Marco Cassia, *Orsted DTU*, Peter Shah, *Qualcomm Inc.*, and Erik Bruun, *Technical University of Denmark*

A DESIGN METHODOLOGY FOR POWER-EFFICIENT CONTINUOUS-TIME SIGMA-DELTA A/D CONVERTERS **I-1069**

Jannik Nielsen and Erik Bruun, *Technical University of Denmark*

A LOW-VOLTAGE 38μW SIGMA-DELTA MODULATOR DEDICATED TO WIRELESS SIGNAL RECORDING APPLICATIONS **I-1073**

Yamu Hu, Zhijun Lu Lu, and Mohamad Sawan, *Ecole Polytechnique de Montreal*

VOLUME II OF V: -COMMUNICATIONS- -MULTIMEDIA SYSTEMS & APPLICATIONS-

Track 5: COMMUNICATIONS

A METHOD FOR GENERATING NON-GAUSSIAN NOISE SERIES WITH SPECIFIED PROBABILITY DISTRIBUTION AND POWER SPECTRUM II-1

Shen Minfen, *Shantou University*, Francis H. Y. Chan, *Hong Kong University,* and Patch J. Beadle, *Portsmouth University*

PERFORMANCE EVALUATION OF OPTIMAL DMT TRANSCEIVERS FOR ADSL APPLICATION II-5

Shang-Ho Tsai, *University of Southern California*, and Yuan-Pei Lin, *National Chiao Tung University*

A NOVEL CHANNEL IDENTIFICATION METHOD FOR FAST WIRELESS COMMUNICATION SYSTEMS WITH TRANSMITTER AND RECEIVER DIVERSITY II-9

Honghui Xu and Soura Dasgupta, *The University of Iowa*, and Zhi Ding, *University of California, Davis*

A NEW TIMING RECOVERY ARCHITECTURE FOR FAST CONVERGENCE II-13

Piya Kovintavewat, *Georgia Institute of Technology*, M. Fatih Erden and Erozan Kurtas, *Seagate Technology*, and John Barry, *Georgia Institute of Technology*

A NOVEL CHANNEL ESTIMATION AND TRACKING METHOD FOR WIRELESS OFDM SYSTEMS BASED ON PILOTS AND KALMAN FILTERING II-17

Yuanjin Zheng, *Institute of Microelectronics*

A FEEDFORWARD TIMING RECOVERY SCHEME USING TWO SAMPLES PER SYMBOL: ALGORITHM, PERFORMANCE AND IMPLEMENTATION ISSUES II-21

Wei-Ping Zhu, Yupeng Yan, M.O. Ahmad, and M.N.S. Swamy, *Concordia University*

JOINT CARRIER AND FRAME SYNCHRONIZATION FOR MPSK DEMODULATION II-25

Wei-Ping Zhu, M.O. Ahmad and M.N.S. Swamy, *Concordia University*

ON THE DEVELOPMENT OF A MODEM FOR DATA TRNASMISSION AND CONTROL OF ELECTRICAL HOUSEHOLD APPLIANCES USING THE LOW-VOLTAGE POWER-LINE. II-29

Sara Escalera, J. Manuel Garcia-Gonzalez, Carlos M. Dominguez-Matas, Oscar Guerra, and Angel Rodriguez-Vazquez, *IMSE-CNM-CSIC*

Track 5.1: Systems and Architectures

RECONFIGURABLE MEMORY BUS SYSTEMS USING MULTI-GBPS/PIN CDMA I/O TRANSCEIVERS II-33

Jongsun Kim, Zhiwei Xu, and M. Frank Chang, *University of California at Los Angeles*

IP BASED RECONFIGURABLE DIGITAL PLATFORM FOR SATELLITE COMMUNICATIONS II-37

Marco Re, Andrea Del Re, and Gian Carlo Cardarilli, *University of Rome Tor Vergata*

A RECONFIGURABLE CHANNEL CODEC COPROCESSOR FOR SOFTWARE RADIO MULTIMEDIA APPLICATIONS II-41

Alessandro Pacifici, Caterina Vendetti, Fabrizio Frescura, and Saverio Cacopardi, *University of Perugia*

A HIGHLY EFFICIENT RECONFIGURABLE ARCHITECTURE FOR AN UTRA-TDD MOBILE STATION RECEIVER II-45

Ronny Veljanovski, Aleksandar Stojcevski, Jugdutt Singh, Mike Faulkner, and Aladin Zayegh, *Victoria University*

CONFIGURABLE PREAMBLE SYNCHRONIZER FOR SLOTTED RANDOM ACCESS IN W-CDMA APPLICATIONS II-49

Chi-Fang Li, Wern-Ho Sheen, Fu-Chang Chuang, and Yuan-Sun Chu, *National Chiao Tung University*

LOW HARDWARE COMPLEXITY PARALLEL TURBO DECODER ARCHITECTURE II-53

Zhongfeng Wang, *National Semiconductor Co.*, Yiyan Tang, and Yuke Wang, *University of Texas at Dallas*

ARCHITECTURE-AWARE LOW-DENSITY PARITY-CHECK CODES II-57

Mohammad Mansour and Naresh Shanbhag, *University of Illinois at Urbana-Champaign*

A MASSIVELY SCALEABLE DECODER ARCHITECTURE FOR LOW-DENSITY PARITY-CHECK CODES II-61

Anand Selvarathinam, Gwan Choi, Krishna Narayanan, Abhiram Prabhakar, and Euncheol Kim, *Texas A&M University*

IMPLEMENTATION OF A PARALLEL TURBO DECODER WITH DIVIDABLE INTERLEAVER II-65

Jaeyoung Kwak, Sook Min Park, Sang-Sic Yoon, and Kwyro Lee, *Korea Advanced Institute of Science and Technology*

DESIGN OF SOFT-OUTPUT VITERBI DECODERS WITH HYBRID TRACE-BACK PROCESSING II-69

Yun-Nan Chang, *National Sun Yat-Sen University*

ALTERNATIVE DIRECT DIGITAL FREQUENCY SYNTHESIZER ARCHITECTURES WITH REDUCED MEMORY SIZE II-73

Dimitrios Soudris, Marios Kesoulis, and Christos Koukourlis, and A. Thanailakis, *Democritus University of Thrace,* and S. Blionas, INTRACOM

A LOW-POWER, MEMORYLESS DIRECT DIGITAL FREQUENCY SYNTHESIZE R ARCHITECTURE II-77

Kalle Palomaki and Jarkko Niittylahti, *Tampere University of Technology*

QUADRATURE DIRECT DIGITAL FREQUENCY SYNTHESIZER USING AN ANGLE ROTATION ALGORITHM II-81

Florean Curticapean and Kalle I. Palomäki, *Tampere University of Technology*, and Jarkko Niittylahti, *Atostek Ltd., Tampere, Finland*

A NOVEL FREQUENCY SYNTHESIZER CONCEPT FOR WIRELESS COMMUNICATIONS II-85

Cesar Caballero Gaudes, *Universidad de Zaragoza*, Mikko Valkama and Markku Renfors, *Tampere University of Technology*

SYNCHRONIZATION OF FRACTIONAL INTERVAL COUNTER IN NON-INTEGER RATIO SAMPLE RATE CONVERTERS II-89

Jaakko Ketola, Jouko Vankka, and Kari Halonen, *Helsinki University of Technology*

AN INNOVATIVE SCHEDULING SCHEME FOR HIGH SPEED NETWORK PROCESSORS II-93

Ioannis Papaefstathiou, *FORTH*, Nelly Leligou, *National Technical University of Athens*, Fanis Orphanoudakis, George Kornaros, and Nick Zervos, *Ellemedia Technologies*, and George Konstantoulakis, *Inaccess Networks*

ACTIVE FLOW IDENTIFIERS FOR SCALABLE, QOS SCHEDULING IN 10-GBPS NETWORK PROCESSORS II-97

George Kornaros, *Ellemedia Technologies*, Yannis Papaefstathiou, *ICS-FORTH*, and Fanis Orphanoudakis, *Ellemedia Technologies*

AN IMPROVED LINK PRICE ALGORITHM FOR INTERNET FLOW CONTROL II-101

Dongliang Guan and Songyu Yu, *Institute of Image Communication & Information Processing*

WIRE-SPEED TRAFFIC MANAGEMENT IN ETHERNET SWITCHES II-105

Shridhar Mubaraq Mishra, Pramod Pandey, Ardhanari Guruprasad, Chun Feng Hu, and Ming Hung, *Infineon Technologies (AP)*

PRECODED OFDM FOR POWER LINE BROADBAND COMMUNICATION II-109

Fethi Tlili, Fatma Rouissi, and Adel Ghazel, *Ecole Supérieure des Communication de Tunis*

BUFFER IMPLEMENTATION FOR PROTEO NETWORKS-ON-CHIP II-113

Ilkka Saastamoinen, Mikko Alho, and Jari Nurmi, *Tampere University of Technology*

A VLSI DESIGN OF TURBO DECODER CORE FOR INTEGRATED COMMUNICATION SYSTEM-ON-CHIP APPLICATIONS II-117

Wai-Chi Fang, *NASA's JPL, California Institute of Technology*

IMPLEMENTATION OF A PROGRAMMABLE 64~2048-POINT FFT/IFFT PROCESSOR FOR OFDM-BASED COMMUNICATION SYSTEMS II-121

Jen-Chih Kuo, Ching-Hua Wen, and An-Yeu Wu, *National Taiwan University*

A COMPUTATIONAL TECHNIQUE AND A VLSI ARCHITECTURE FOR DIGITAL PULSE SHAPING IN OFDM MODEMS II-125

Eleni Fotopoulou, Vassilis Paliouras, and Thanos Stouraitis, *University of Patras*

AN EFFICIENT MEMORY-BASED FFT ARCHITECTURE II-129

Chao-Kai Chang, Chung-Ping Hung, and Sau-Gee Chen, *National Chiao Tung University*

A CONTINUOUS FLOW MIXED-RADIX FFT ARCHITECTURE WITH AN IN-PLACE ALGORITHM II-133

Jae H. Baek and Byung Soo Son, *Ajou University*, Byung Gak Jo, *Agency for Defence Development*, Myung Hoon Sunwoo and Seung K. Oh, *Ajou University*

IMPLEMENTATION OF CHANNEL DEMODULATOR FOR DAB SYSTEM II-137

Chien-Ming Wu and Ming-Der Shieh, *National Yunlin University of Science & Technology*, Hsin-Fu Lo, *National Science Council Chip Implementation Center*, and Min-Hsiung Hu, *National Yunlin University of Science & Technology*

ENHANCEMENT OF DATA TRANSMISSION IN OFDM-WLAN SYSTEM USING TRANSMIT DIVERSITY METHOD II-141

Tasapon Jitrangsri, *Mahanakorn University of Technology*, and Suvepon Sittichivapak, *King Mongkut's Institue of Technology Ladkrabang*

ML FRAME SYNCHRONIZATION FOR IEEE 802.11A WLANS ON MULTIPATH RAYLEIGH FADING CHANNELS II-145

Yik-Chung Wu, *Texas A&M University*, Kun-Wah Yip and Tung-Sang Ng, *The University of Hong Kong*

COMBINING ADAPTIVE SMOOTHING AND DECISION-DIRECTED CHANNEL ESTIMATION SCHEMES FOR OFDM WLAN SYSTEMS II-149

Hsuan-Yu Liu, Yi-Hsin Yu, Chien Jen Hunh, Terng-Yin Hsu, and Chen-Yi Lee, *National Chiao Tung University*

A FLEXIBLE DESIGN OF A DECISION FEEDBACK EQUALIZER AND A NOVEL CCK TECHNIQUE FOR WIRELESS LAN SYSTEMS II-153

Hsin-Lei Lin, Robert C. Chang, Chih-Hao Huang and Hongchin Lin, *National Chung Hsing University*

A HIGH SPEED COMPLEX ADAPTIVE FILTER FOR AN ASYMMETRIC WIRELESS LAN USING A NEW QUANTIZED POLYNOMIAL REPRESENTATION II-157

Adesh Garg, *ATIPS Laboratory*, Ian Steiner, *University of Calgary*, Graham Jullien and Jim Haslett, *ATIPS Laboratory*, and Grant Mcgibney, *TRLabs*

A 2GHZ IMAGE-REJECT RECEIVER IN A LOW IF ARCHITECTURE FABRICATED IN A 0.1UM CMOS TECHNOLOGY II-161

Magnus Wiklund, *Ericsson Technology Licensing Ab*, Stefan Nilsson, *Ericsson Mobile Platforms AB*, Christian Bjork, *VIA Networking Sweden AB*, and Sven Mattisson, *Ericsson Mobile Platforms AB*

SELF TUNED FULLY INTEGRATED HIGH IMAGE REJECTION LOW IF RECEIVERS: ARCHITECTURE AND PERFORMANCE II-165

Yuanjin Zheng and Chin Boon Terry Tear, *Institute of Microelectronics*

A COMPLEX CHARGE SAMPLING SCHEME FOR COMPLEX IF RECEIVERS II-169

Sami Karvonen, Thomas Riley, and Juha Kostamovaara, *University of Oulu*

HIGH GAIN GAAS 10GBPS TRANSIMPEDANCE AMPLIFIER WITH INTEGRATED BONDWIRE EFFECTS II-173

Miguel Madureira, *Instituto De Telecomunicações*, Paulo Monteiro, Rui Aguiar, and Manuel Violas, *Universidade de Aveiro*, Maurice Gloanec, Eric Leclerc, and Benoit Lefebvre, *UMS S.A.S. - United Monolithic Semiconductors*

A BROADBAND UPCONVERTER UNIT FOR DOUBLE-CONVERSION RECEIVERS II-177

Kari Stadius, Arto Malinen, Petteri Paatsila, and Kari Halonen, *Helsinki University of Technology*

A 10GB/S CDR WITH A HALF-RATE BANG-BANG PHASE DETECTOR II-181

Mehrdad Ramezani and C. Andre T. Salama, *University of Toronto*

CLOCK RECOVERY CIRCUIT WITH ADIABATIC TECHNOLOGY (QUASI-STATIC CMOS LOGIC) II-185

Wing Ki Yeung, Cheong Fat Chan, Chiu Sing Choy, and Kong Pang Pun, *The Chinese University of Hong Kong*

A 1.8V 2.4GHZ CMOS ON-CHIP IMPEDANCE MATCHING LOW NOISE AMPLIFIER FOR WLAN APPLICATIONS II-188

Baoyong Chi and Bingxue Shi, *Tsinghua University*

AN ACCURATE CURRENT SOURCE WITH ON-CHIP SELF-CALIBRATION CIRCUITS FOR LOW-VOLTAGE DIFFERENTIAL TRANSMITTER DRIVERS II-192

Guangbin Zhang and Jin Liu, *University of Texas at Dallas*, and Sungyong Jung, *University of Texas at Arlington*

A MIXED-MODE DELAY-LOCKED LOOP FOR WIDE-RANGE OPERATION AND MULTIPHASE OUTPUTS II-196

Kuo-Hsing Cheng, Yu-Lung Lo, and Wen-Fang Yu, *Tamkang University*

A LINEARIZED 2-GHZ SIGE LOW NOISE AMPLIFIER FOR DIRECT CONVERSION RECEIVER II-200

Jouni Kaukovuori, Mikko Hotti, Jussi Ryynänen, Jarkko Jussila , and Kari Halonen, *Helsinki University of Technology*

DESIGN OF CMOS CML CIRCUITS FOR HIGH-SPEED BROADBAND COMMUNICATIONS II-204

Ullas Singh and Michael Green, *University of California*

DESIGN OF ULTRA HIGH-SPEED CMOS CML BUFFERS AND LATCHES II-208

Payam Heydari and Ravi Mohavavelu, *University of California, Irvine*

A NOVEL ESTIMATOR FOR THE VELOCITY OF A MOBILE STATION IN A MICRO-CELLULAR SYSTEM II-212

Ghasem Azemi, Bouchra Senadji, and Boualem Boashash, *Queensland University of Technology*

HIGH LOOP-FILTER-ORDER SIGMA-DELTA-FRACTIONAL-N FREQUENCY SYNTHESIZERS FOR USE IN FREQUENCY-HOPPING SPREAD-SPECTRUM COMMUNICATION SYSTEMS II-216

Niels Christoffers, Rainer Kokozinski, Stephan Kolnsberg, and Bedrich Hosticka, *Fraunhofer Institute for Microelectronic Circuits and Systems*

CHANNEL ESTIMATION FOR WIRELESS COMMUNICATIONS USING SPACE-TIME BLOCK CODING TECHNIQUES II-220

J. Yang, Yichuang Sun, John Senior, and Nandini Pem, *University of Hertfordshire*

FAST-POLARIZATION-HOPPING TRANSMISSION-DIVERSITY TO MITIGATE PROLONGED DEEP FADES II-224

Kainam Wong and So Leung Alex Chan, *University of Waterloo*, and Rafael P. Torres, *Universidad de Cantabria*

SPACE-TIME EQUALIZER FOR ADVANCED 3GPP WCDMA MOBILE TERMINAL: EXPERIMENTAL RESULTS II-228

As Madhukumar, Z Gou Ping, T Kian Seng, Y Kuck Jong, Tz Minqian, Ty Hong, and Francois Chin, *Institute for Communications Research*

A NOVEL PILOT STRUCTURE FOR THE DOWNLINK TRANSMISSION OF CYCLIC PREFIX ASSISTED SINGLE CARRIER CDMA SYSTEM WITH FREQUENCY DOMAIN EQUALISATION II-232

As Madhukumar, Francois Chin, and Ying-Chang Liang, *Institute for Communications Research*

PEAK-TO-AVERAGE POWER-RATIO REDUCTION VIA CHANNEL HOPPING FOR DOWNLINK CDMA SYSTEMS II-236

Yajun Kou, Wu-Sheng Lu, and Andreas Antoniou, *University of Victoria*

PARALLEL BUS SYSTEMS USING CODE-DIVISION MULTIPLE ACCESS TECHNIQUE **II-240**

Shinsaku Shimizu, Toshimasa Matsuoka, and Kenji Taniguchi, *Osaka University*

HIGH PERFORMANCE DS-CDMA SYSTEM USING NOVEL SC-I CHIP SHAPING **II-244**

Triratana Metkarunchit and Prasit Prapinmongkolkarn, *Chulalongkorn University*

PREDICTING CDMA SPECTRAL REGROWTH USING A GENERAL STATISTICAL BEHAVIOURAL MODEL FOR POWER AMPLIFIERS WITH MEMORY EFFECTS **II-248**

Tracey Goh and Roger Pollard, *University of Leeds*

EFFICIENT BIT-SERIAL SYSTOLIC ARRAY FOR DIVISION OVER GF(2^M) **II-252**

Chun Pyo Hong, Chang Hoon Kim, Soonhak Kwon, and In Gil Nam, *Daegu University*

FPGA REALIZATION OF AN OFDM FRAME SYNCHRONIZATION DESIGN FOR DISPERSIVE CHANNELS **II-256**

Yin-Tsung Hwang and Kuo-Wei Liao, *National Yunlin University of Science & Technology*, and Chien-Hsin Wu, *National Chung Cheng University*

QUADRATURE DIRECT DIGITAL FREQUENCY SYNTHESIZERS: AREA-OPTIMIZED DESIGN MAP FOR LUT-BASED FPGAS **II-260**

Francisco Cardells-Tormo, *Hewlett-Packard Company*, and Javier Valls-Coquillat, *Polytechnic University of Valencia*

A NOVEL ACS SCHEME FOR AREA-EFFICIENT VITERBI DECODERS **II-264**

Yiqun Zhu and Mohammed Benaissa, *The University of Sheffield*

FPGA DESIGNS OF PARALLEL HIGH PERFORMANCE GF(2^233) MULTIPLIERS **II-268**

Marcus Bednara, Cornelia Grabbe, Jamshid Shokrollahi, Jürgen Teich, and Joachim von zur Gathen, *Paderborn University*

PROVIDING FLEXIBILITY IN A CONVOLUTIONAL ENCODER **II-272**

Matthias Kamuf, John Anderson, and Viktor Öwall, *Lund University*

A LOW-POWER SYSTOLIC ARRAY-BASED VITERBI DECODER AND ITS FPGA IMPLEMENTATION **II-276**

Man Guo, M. Omair Ahmad, M.N.S. Swamy, and Chunyan Wang, *Concordia University*

REVERSE TRACING OF FORWARD STATE METRIC IN LOG-MAP AND MAX-LOG-MAP DECODERS **II-280**

Jaeyoung Kwak Sook Min Park, and Kwyro Lee, *Korea Advanced Institute of Science and Technology*

STRUCTURED DESIGN OF AN INTEGRATED SUBSCRIBER LINE INTERFACE SYSTEM AND CIRCUIT **II-284**

Armin Tajalli and Mojtaba Atarodi, *Sharif University of Technology*

A LOW POWER 4.3 GHZ PHASE-LOCKED LOOP WITH ADVANCED DUAL-MODE TUNING TECHNIQUE INCLUDING I/Q-SIGNAL GENERATION IN 0.12μM STANDARD CMOS **II-288**

Georg Konstanznig and Andreas Springer, *Korea Advanced Institute of Science and Technology*, and Robert Weigel, *University of Erlangen-Nürnberg*

A 1.35 GHZ CMOS WIDEBAND FREQUENCY SYNTHESIZER FOR MOBILE COMMUNICATIONS **II-292**

Esdras Juarez-Hernandez, Alejandro Diaz-Sanchez, and Esteban Tlelo-Cuautle, *National Institute for Astrophysics, Optics and Electronics*

A CMOS ANALOG CONTINUOUS-TIME FIR FILTER FOR 1GBPS CABLE EQUALIZER **II-296**

Xiaofeng Lin and Jin Liu, *University of Texas at Dallas*

MIXER TOPOLOGY SELECTION FOR A 1.8-2.5 GHZ MULTI-STANDARD FRONT-END IN 0.18 UM CMOS **II-300**

Vojkan Vidojkovic and Johan Van Der Tang, *Eindhoven University of Technology*, Arjan Leeuwenburgh, *National Semiconductor BV*, and Arthur Van Roermund, *Eindhoven University of Technology*

ALGORITHM AND ARCHITECTURE DESIGN FOR A LOW-COMPLEXITY ADAPTIVE EQUALIZER II-304

Chun-Nan Chen, Kuan-Hung Chen, and Tzi-Dar Chiueh, *National Taiwan University*

DESIGN AND IMPLEMENTATION OF HIGH-SPEED ARBITER FOR LARGE SCALE VOQ CROSSBAR SWITCHES II-308

Chun Kit Hung, Chi Ying Tsui, and Mounir Hamdi, *Hong Kong University of Science and Technology*

A VIRTUALLY JITTER-FREE FRACTIONAL-N DIVIDER FOR A BLUETOOTH RADIO II-312

Zhenhua Wang, *Philips Semiconductors*

A NOVEL C-BAND CMOS PHASE SHIFTER FOR COMMUNICATION SYSTEMS II-316

Sotoudeh Hamedi-Hagh and C. Andre T. Salama, *University of Toronto*

HIGH-SPEED VLSI ARCHITECTURE FOR PARALLEL REED-SOLOMON DECODER II-320

Hanho Lee, *University of Connecticut*

MODIFIED REDUCED DIMENSIONALITY CHANNEL MODEL II-324

Kitti Lertsirimit, *Asian University of Science and Technology*

ADAPTIVE DIGITAL COMPENSATION IN DSP BASED FM MODULATORS II-328

Anthony G Lim, Guo Qing Wang, and Victor Sreeram, *University of Western Australia*

A WINDOW-GATE FOR THE ECHO CANCELLER OF DMT ADSL II-332

Hsin-Yung Wang and Shyue-Win Wei, *National Chi-Nan University*

DESIGN AND SIMULATION OF MINIATURIZED COMMUNICATION SYSTEMS EMPLOYING SYMMETRICAL LOSSLESS TWO-PORTS CONSTRUCTED WITH TWO KINDS OF ELEMENTS II-336

Siddik Yarman, Ahmet Aksen, and Ebru Cimen, *Isik University*

A MODELLING OF IONOSPHERIC DELAY OVER CHIANG MAI PROVINCE II-340

Suporn Suwantragul, Pong-In Rakariyatham, Tharadol Komolmis, and Akachai Sang-In, *Chiang Mai Unvie rsity*

A NEW MULTIPLIERLESS CORRELATOR FOR TIMING SYNCHRONIZATION IN IEEE 802.11A WLANS II-344

Kun-Wah Yip and Tung-Sang Ng, *The University of Hong Kong*, and Yik-Chung Wu, *Texas A&M University*

CONTENT-BASED LOAD BALANCING WITH MULTICAST AND TCP-HANDOFF II-348

Daorat Kerdlapanan and Akharin Khunkitti, *King Mongkut's Institute of Technology Ladkrabang*

COOPERATIVE BIT-LOADING AND FAIRNESS BANDWIDTH ALLOCATION IN ADSL SYSTEMS II-352

Nikos Papandreou and Theodore Antonakopoulos, *University of Patras*

Track 5.2: Video Transmission

A CONTINUOUS TRACKING ALGORITHM FOR LONG-TERM MEMORY MOTION ESTIMATION II-356

C.J. Duanmu, M.Omair Ahmad, and M.N.S. Swamy, *Concordia University*

OPTIMAL JOINT SOURCE-CHANNEL BIT ALLOCATION FOR MPEG-4 FINE GRANULARITY II-360

Jizheng Xu, Qian Zhang, and Wenwu Zhu, *Microsoft Research Asia*, Xiang-Gen Xia, *Univ. of Delaware,* and Ya-Qin Zhang, *Microsoft Research Asia*

FGS-BASED VIDEO STREAMING TEST BED FOR MPEG-21 UNIVERSAL MULTIMEDIA ACCESS WITH DIGITAL ITEM ADAPTATION II-364

Chung-Neng Wang, Chia-Yang Tsai, Han-Chung Lin, Hsiao-Chiang Chuang, Yao-Chung Lin, Jin-He Chen, Kin Lam Tong, Feng-Chen Chang, Chun-Jen Tsai, Shuh-Ying Lee, Tihao Chiang, and Hsueh-Ming Hang, *National Chiao Tung University*

CONTENT BASED ERROR DETECTION AND CONCEALMENT FOR IMAGE TRANSMISSION OVER WIRELESS CHANNEL II-368

Shuiming Ye, *Laboratories for Information Technology*, Xinggang Lin, *Tsinghua University*, and Qibin Sun, *Laboratories for Information Technology*

DYNAMIC BIT-RATE REDUCTION BASED ON FRAME-SKIPPING AND REQUANTIZATION FOR MPEG-1 TO MPEG-4 TRANSCODER II-372

Kwang-Deok Seo, *LG Electronics Inc.*, Soon-Kak Kwon, Sug Ky Hong, and Jae-Kyoon Kim, *Dongeui University*

INTEGRATED VISION SENSOR FOR DETECTING BOUNDARY CROSSINGS II-376

Samuel Zahnd, Patrick Lichtsteiner, and Tobi Delbruck, *ETH Zuerich*

MULTIPLE SINGLE PIXEL DIM TARGET DETECTION IN INFRARED IMAGE SEQUENCE II-380

Mukesh Zaveri, Uday Desai, and S.N. Merchant, *Indian Institute of Technology, Bombay*

AUTOMATIC FACE COLOR SEGMENTATION BASED RATE CONTROL FOR LOW BIT-RATE VIDEO CODING II-384

Datchakorn Tancharoen, Hatairat Kortrakulki, Sak Khemachai, Supavadee Aramvith, and Somchai Jitapunkul, *Chulalongkorn University*

Track 5.3: Image Coding and Processing

MULTIPLIERLESS PREDICTOR FOR DPCM IMAGES II-388

Ratchaneekorn Thamvichai, *St. Cloud State University*, Tamal Bose, *Utah State University*, and Miloje Radenkovic, *University of Colorado, Denver*

AN ENCHANCED HEXAGONAL SEARCH ALGORITHM FOR BLOCK MOTION ESTIMATION II-392

Ce Zhu, *Nanyang Technological University*, Xiao Lin, *Laboratories for Information Technology*, and Lap-Pui Chau, *Nanyang Technological University*

ADAPTIVE VECTOR MEDIAN FILTER FOR REMOVAL IMPULSES FROM COLOR IMAGES II-396

Khumanthem Singh, *Centre for Electronics Design and Technology of India*, and Prabin K. Bora, *IITG, India*

OPTIMUM WORD LENGTH ALLOCATION FOR MULTIPLIERS OF INTEGER DCT II-400

Masahiro Iwahashi and Osamu Nishida, *Nagaoka University of Technology*, Somchart Chokchaitam, *Thammasat University*, and Noriyoshi Kambayashi, *Nagaoka University of Technology*

A NOVEL ALGORITHM FOR COLOR QUANTIZATION BY 3D DIFFUSION II-404

K. Lo, Y.H. Chan, and M.P. Yu, *The Hong Kong Polytechnic University*

NEW PARTITION-BASED FILTERS FOR SUPPRESSING MIXED HIGH PROBABILITY IMPULSE AND GAUSSIAN NOISES IN IMAGES II-408

Noritaka Yamashita, Hiroo Sekiya, Jianming Lu, and Takashi Yahagi, *Chiba University*

FAST CONTROL GRID POINT ESTIMATION FOR MESH BASED MOTION ESTIMATION II-412

Kai Tsang and Oscar Au, *The Hong Kong University of Science and Technology*

ROI-BASED SCALABILITY FOR PROGRESSIVE TRANSMISSION IN JPEG2000 CODING II-416

Osamu Watanabe and Hitoshi Kiya, *Tokyo Metropolitan University*

AN ROI IMAGE CODING BASED ON SWITCHING WAVELET TRANSFORM II-420

Shinij Fukuma, S. Ikuta, M. Ito, S. Nishimura, and M. Nawate, *Shimane University*

ADAPTIVE DIRECTIONAL ZEROTREE IMAGE CODING II-424

Vutipong Areekul, *Kasetsart University*, and Suksan Jirachawang, *NECTEC*

AN EFFICIENT WAVELET-VQ METHOD FOR IMAGE CODING II-428

Momotaz Begum, Nurun Nahar, Kaneez Fatimah, and Md. Hasan, *Bangladesh University of Engineering & Technology*

ERROR PROTECTION FOR JPEG2000 ENCODED IMAGES AND ITS EVALUATION OVER OFDM CHANNEL II-432

Khairul Munadi, Masayuki Kurosaki, Kiyoshi Nishikawa, and Hitoshi Kiya, *Tokyo Metropolitan University*

A POST-PROCESSING METHOD FOR VECTOR QUANTIZATION TO ACHIEVE HIGHER PSNR AND NEARLY CONSTANT BIT RATE II-436

Zhibin Pan, Koji Kotani, and Tadahiro Ohmi, *Tohoku University*

SPATIAL VARYING FILTERING FOR COLOR FILTER ARRAY INTERPOLATION IN DIGITAL STILL CAMERAS II-440

Oscar Au and Ming Sun Fu, *Hong Kong University of Science and Technology*

SCALABLE GIGA-PIXELS/S BINARY IMAGE MORPHOLOGICAL OPERATIONS II-444

Songpol Ongwattanakul, Phaisit Chewputtanagul, David J. Jackson, Kenneth G. Ricks and *The University of Alabama*

FPGA IMPLEMENTATION OF BLOCKTRUNCATION CODING ALGORITHM FOR GRAY SCALE IMAGES II-448

Sherif Saif, *ERI*, Hazem Abbas, *Mentor Graphics*, and Salwa Nassar, *ERI*

FPGA IMPLEMENTATION OF A FREQUENCY ADAPTIVE LEARNING SOFM FOR DIGITAL COLOR STILL IMAGING II-452

Menon Shibu, Chip Hong Chang, and Rui Xiao, *Nanyang Technological University*

UNSUPERVISED IMAGE SEGMENTATION USING LOCAL HOMOGENEITY ANALYSIS II-456

Feng Jing, *Tsinghua University*, Mingjing Li and Hong-Jiang Zhang, *Microsoft Research Asia*, and Bo Zhang, *Tsinghua University*

MULTIPLE CONTOUR SEGMENTATION WITH AUTOMATIC THRESHOLDING II-460

Y.B. Chen and Oscal Chen, *National Chung Cheng University*

A BAYESIAN SKIN/NON-SKIN COLOR CLASSIFIER USING NON-PARAMETRIC DENSITY ESTIMATION II-464

Douglas Chai, Son Lam Phung, and Abdesselam Bouzerdoum, *Edith Cowan University*

VEHICLE IMAGE CLASSIFICATION VIA EXPECTATION-MAXIMIZATION ALGORITHM II-468

Suree Pumrin, *Chulalongkorn University*, and Daniel Dailey, *University of Washington*

A LOW BIT-RATE HYBRID DWT-SVD IMAGE-CODING SYSTEM (HDWTSVD) FOR MONOCHROMATIC IMAGES II-472

Humberto Ochoa and Kamisetty Rao, *University of Texas at Arlington*

FAST CODEWORD SEARCH ALGORITHM FOR ECVQ USING HYPERPLANE DECISION RULE II-476

Kousuke Imamura, Ahmed Swilem, and Hideo Hashimoto, *Kanazawa University*

AN EFFIEIENT ALGORITHM FOR FRACTAL IMAGE CODING USING KICK-OUT AND ZERO CONTRAST CONDITIONS II-480

Cheung-Ming Lai, Kin-Man Lam, and Wan-Chi Siu, *The Hong Kong Polytechnic University*

Track 6: MULTIMEDIA SYSTEMS & APPLICATIONS

TCP-FRIENDLY ASSURED FORWARDING (AF) VIDEO SERVICE IN DIFFSERV NETWORKS II-484

Young-Gook Kim and C.-C. Jay Kuo, *University of Southern California*

A GA-BASED ROUTING METHOD WITH AN UPPER BOUND CONSTRAINT II-488

Jun Inagaki, Miki Haseyama, and Hideo Kitajima, *Hokkaido University*

A COST-EFFECTIVE 2-D DISCRETE COSINE TRANSFORM PROCESSOR WITH RECONFIGURABLE DATAPATH II-492

Yeong-Kang Lai and Han-Jen Hsu, *National Chung Hsing University*

AN EFFICIENT BINARY MOTION ESTIMATION ALGORITHM AND ITS ARCHITECTURE FOR MEPG-4 SHAPE CODING II-496

Tsung-Han Tsai and Chia-Pin Chen, *National Central University*

A NEW LINEAR PREDICTOR EMPLOYING VECTOR QUANTIZATION IN NONORTHOGONAL DOMAINS FOR HIGH QUALITY SPEECH CODING I-500

Wasfy Mikhael, *University of Central Florida*, and Venkatesh Krishnan, *Georgia Institute of Technology*

DISK I/O MIXED SCHEDULING STRATEGY FOR VOD SERVERS II-504

Hai Jin, Jie Xu, Bibo Tu, and Shengli Li, *Huazhong University of Science and Technology*

ENCHANCED CHAOTIC IMAGE ENCRYPTION ALGORITHM BASED ON BAKER'S MAP II-508

Mazleena Salleh, Subariah Ibrahim, and Ismail Fauzi Isnin, *University Technology Malaysia*

A HARDWARE-LIKE HIGH-LEVEL LANGUAGE BASED ENVIRONMENT FOR 3D GRAPHICS ARCHITECTURE EXPLORATION II-512

Inho Lee, Joung-Youn Kim, Yeon-Ho Im, Yunseok Choi, Hyunchul Shin, Changyoung Han, Donghyun Kim, Hyoungjoon Park, Young-Il Seo, Kyusik Chung, Chang-Hyo Yu, Kanghyup Chun, and Lee-Sup Kim, *KAIST*

LIMITS IN FIR SUBBAND BEAMFORMING FOR SPATIALLY SPREAD NEAR-FIELD SPEECH SOURCES II-516

Nedelko Grbic, Sven Nordholm, and Antonio Cantoni, *University of Western Australia*

INTERACTIVE INTERFACE OF REALTIME 3D SOUND MOVEMENT FOR EMBEDDED APPLICATIONS II-520

Shinya Komata, Ablonczy Pal, Noriaki Sakamoto, Wataru Kobayashi, Takao Onoye, and Isao Shirakawa, *Osaka University*

AUTOMATIC MOVING OBJECT EXTRACTION IN MPEG VIDEO II-524

Wei Zeng, *Harbin Institute of Technology*, Wen Gao and Debin Zhao, *Chinese Academy of Sciences*

RECURSIVE BLOCK MATCHING PRINCIPLE FOR ERROR CONCEALMENT ALGORITHM II-528

Mei-Juan Chen, Che-Hsing Chen, and Ming-Chieh Chi, *National Dong Hwa University*

A NEW ERROR RESILIENT CODING SCHEME FOR JPEG IMAGE TRANSMISSION BASED ON DATA EMBEDDING AND VECTOR QUANTIZATION II-532

Li-Wei Kang and Jin-Jang Leou, *National Chung Cheng University*

RESOURCE ADAPTATION BASED ON MPEG-21 USAGE ENVIRONMENT DESCRIPTIONS II-536

Huifang Sun and Anthony Vetro, *Mitsubishi Electric Research Laboratories*, and Kohtaro Asai, *Mitsubishi Electric Corp.*

Track 6.1: Speech Processing and Coding

CHANNEL COMPENSATION OF MODULATION SPECTRAL FEATURES II-540

Somsak Sukittanon and Les Atlas, *Univerisity of Washington*

TWO MICROPHONES SPEECH ENHANCEMENT SYSTEM BASED ON A DOUBLE AFFINE PROJECTION ALGORITHM II-544

Marcel Gabrea, *École de Technologie Supérieure*

TWO-CHANNEL MICROPHONE ARRAY PROCESSING FOR SPEECH ENHANCEMENT II-548

Zhaoli Yan, Limin Du, Jianqiang Wei, and Hui Zeng, *Chinese Academy of Sciences*

DESIGN OF A LOW POWER PSYCHO-ACOUSTIC MODEL CO-PROCESSOR FOR MPEG-2/4 AAC LC STEREO ENCODER II-552

Tsung-Han Tsai, *National Taiwan University* Shih-Way Huang, *National Central University*, and Liang-Gee Chen, *National Taiwan University*

DETERMINATION OF PITCH OF NOISY SPEECH USING DOMINANT HARMONIC FREQUENCY II-556

Celia Shahnaz, S. Fattah, and Md. Hasan, *Bangladesh University of Engineering & Technology*

CONSTRAINED OPTIMIZATION FOR A SPEECH-DRIVEN TALKING HEAD II-560

Kyoungho Choi and Jonghoon Lee, *ETRI*

SINGLE GAUSS MODEL SET-BASED DATA IMPUTATION METHOD FOR COMPLEX ASR TASK II-564

Yu Luo and Limin Du, *Chinese Academy of Sciences*

AUTOMATIC SYNCHRONIZATION OF SPEECH TRANSCRIPT AND SLIDES IN PRESENTATION II-568

Yu Chen and Wei Jyh Heng, *Laboratories for Information Technology*

A TWO-CHANNEL TRAINING ALGORITHM FOR HIDDEN MARKOV MODEL TO IDENTIFY VISUAL SPEECH ELEMENTS II-572

Say Wei Foo, *Nanyang Technological University*, Liang Dong, and Yong Lian, *National University of Singapore*

THE NEW NADPCMB^MLT CODING SCHEME: FROM NON LINEAR FULLBAND TOWARD NON LINEAR SUBBAND PREDICTION CODING OF SPEECH SIGNAL II-576

Guido D'alessandro, *University of Ancona*, Marcos Faúndez Zanuy, *Escola Universitària Politècnica de Mataró, Barcelona*, and Francesco Piazza, *University of Ancona*

SLAP: A SYSTEM FOR THE DETECTION AND CORRECTION OF PRONUNCIATION FOR SECOND LANGUAGE ACQUISITION II-580

Lingyun Gu and John Harris, *University of Florida*

AN IMPROVED SPEAKER IDENTIFICATION TECHNIQUE EMPLOYING MULTIPLE REPRESENTATIONS OF THE LINEAR PREDICTION COEFFICIENTS II-584

Wasfy Mikhael and Pravinkumar Premakanthan, *University of Central Florida*

A NEW ALGORITHM FOR VOICE ACTIVITY DETECTION II-588

Jianqiang Wei, Limin Du, Zhaoli Yan, and Hui Zeng, *Chinese Academy of Sciences*

Track 6.2: Video Coding and Compression

EVALUATE LOSS PROBABILITIES OF PACKET-AWARENESS CODER INTO QUEUING NETWORKS II-592

Jie Chen, *Brown University*

SECURE THE IMAGE-BASED SIMULATED TELESURGERY SYSTEM II-596

Yanjiang Yang, Zhenlan Wang, Feng Bao, and Robert H. Deng, *Laboratory for Information Technology*

DESIGN OF A CMOS IMAGE SENSOR WITH PIXEL-LEVEL ADC IN 0.35 μM PROCCESS II-600

F. Hashemi, KH. Hadidi, and A. Khoei, *Urmia University*

A NOVEL DE-INTERLACING TECHNIQUE BASED ON PHASE PLANE CORRELATION MOTION ESTIMATION II-604

Mainak Biswas and Truong Nguyen, *University of California, San Diego*

A ZEROTREE STEREO VIDEO ENCODER II-608

Anil Fernando and Suthinee Thanapirom, *Asian Institute of Technology*, and Eran Edirisinghe, *Loughborough University*

A MODIFIED METHOD FOR CODEBOOK DESIGN WITH NEURAL NETWORK IN VQBASED IMAGE COMPRESSION II-612

Safar Hatami, Mohammad Javad Yazdanpanah, Omid Fatemi, and Behjat Frozandeh, *University of Tehran*

PERFORMANCE OPTIMIZATION FOR MOTION COMPENSATED 2D WAVELET VIDEO COMPRESSION TECHNIQUES II-616

Zhen Li, *Purdue University*, Feng Wu and Shipeng Li, *Microsoft Research Asia*, and Edward Delp, *Purdue University*

EFFICIENT AND FULLY SCALABLE ENCRYPTION FOR MPEG-4 FGS II-620

Chun Yuan 1, *Tsinghua University*, Bin B. Zhu, *Microsoft Research Asia*, Yidong Wang, *Beijing University*, Shipeng Li, *Microsoft Research Asia*, and Yuzhuo Zhong *Tsinghua University*

LOW-COMPLEXITY GLOBAL MOTION ESTIMATION BASED ON CONTENT ANALYSIS II-624

Fang Zhu and Ping Xue, *Nanyang Technological University*, and Eeping Ong, Laboratories for Information Technology

GLOBAL MOTION ESTIMATION FROM COARSELY SAMPLED MOTION VECTOR FIELD AND THE APPLICATIONS II-628

Yeping Su and Ming-Ting Sun, *University of Washington*, and Vincent Hsu, *CCL/ITRI, Taiwan*

A ROBUST GLOBAL MOTION ESTIMATION SCHEME FOR SPRITE CODING II-632

Hoi-Kok Cheung and Wan-Chi Siu, *The Hong Kong Polytechnic University*

FAST BLOCK MATCHING BASED ON MULTI RESOLUTION MOTION ESTIMATION FOR DISCONTINUITY BLOCKS II-636

Hwangsik Bae and Jongwha Chong, *Hanyang University*

VIDEO-ON-DEMAND SERVER SYSTEM DESIGN WITH RANDOM EARLY MIGRATION II-640

Yinqing Zhao and C.-C. Jay Kuo, *University of Southern California*

CANFIND - A SEMANTIC IMAGE INDEXING AND RETRIEVAL SYSTEM II-644

Chin-Hwa Kuo, Tzu-Chuan Chou, Nai-Lung Tsao, and Yung-Hsiao Lan, *Tamkang University*

CONTENT BASED PHOTOGRAPH SLIDE SHOW WITH INCIDENTAL MUSIC II-648

Xian-Sheng Hua, Lie Lu, and Hong-Jiang Zhang, *Microsoft Research Asia*

MOTION FIELD INTERPOLATION FOR FRAME RATE CONVERSION II-652

Mohammed Al-Mualla, *Emirates Telecommunications Corporation*

DCT-BASED VIDEO FRAME-SKIPPING TRANSCODER II-656

Kai-Tat Fung and Wan-Chi Siu, *The Hong Kong Polytechnic University*

LOW DELAY RATE-CONTROL IN VIDEO TRANSCODING II-660

Yu Sun, Xiaohui Wei, and Ishfaq Ahmad, *The University of Texas at Arlington*

ROBUST SEMI-REGULAR MESH REPRESENTATION OF 3D DYNAMIC OBJECTS II-664

Jeong-Hyu Yang, Chang-Su Kim, and Sang-Uk Lee, *Seoul National University*

OBJECT TRACKING METHOD USING BACK-PROJECTION OF MULTIPLE COLOR HISTOGRAM MODELS II-668

Jung-Ho Lee, Woong-Hee Lee, and Dong-Seok Jeong, *Inha University*

A MUTUAL INFORMATION APPROACH TO ARTICULATED OBJECT TRACKING II-672

Evangelos Loutas, Nikos Nikolaidis, and Ioannis Pitas, *Aristotle University of Thessaloniki*

AN EFFICIENT COLOR COMPENSATION SCHEME FOR SKIN COLOR SEGMENTATION II-676

Kwok-Wai Wong, Kin-Man Lam, and Wan-Chi Siu, *The Hong Kong Polytechnic University*

AN UNSUPERVISED APPROACH TO DOMINANT VIDEO SCENE CLUSTERING II-680

Hong Lu and Yap-Peng Tan, *Nanyang Technological University*

EFFECTIVE HARDWARE-ORIENTED TECHNIQUE FOR THE RATE CONTROL OF JPEG2000 ENCODING II-684

Te-Hao Chang, Chung-Jr Lian, Hong-Hui Chen, Jing-Ying Chang, and Liang-Gee Chen, *National Taiwan University*

OBJECTIVE QUALITY ASSESSMENT FOR COMPRESSED VIDEO II-688

Susu Yao, Weisi Lin, Zhongkang Lu, EePing Ong, Laboratories for Information Technology, and Minoru Etoh, Multimedia Signal Processing Lab

ADAPTATION OF VIDEO ENCODERS FOR IMPROVEMENT IN QUALITY II-692

Ramkishor Korada, P S S B K Gupta Pallapothu, Raghu Tippuru Srikantharao, and Suman Kopparapu, *Emuzed India Private Limited*

MOTION ADAPTIVE DE-INTERLACING BY HORIZONTAL MOTION DETECTION AND ENHANCED ELA PROCESSING II-696

Shyh-Feng Lin, Yu-Lin Chang, and Liang-Gee Chen, *National Taiwan University*

DIRECT MODE CODING FOR BI-PREDICTIVE PICTURES IN THE JVT STANDARD II-700

Alexis Tourapis, Feng Wu, and Shipeng Li, *Microsoft Research Asia*

A NOVEL APPROACH TO FAST MULTI-FRAME SELECTION FOR H.26L VIDEO CODING II-704

Andy Chang, Oscar Au, and Yeung Yick Ming, *HKUST*

AN IMPROVED ADAPTIVE ROOD PATTERN SEARCH FOR FAST BLOCK-MATCHING MOTION ESTIMATION IN JVT/H.26L II-708

Kai-Kuang Ma and Gang Qiu, *Nanyang Technological University*

VERY LOW BIT RATE WATERCOLOR VIDEO II-712

Jiang Li, Jizheng Xu, Shipeng Li, and Keman Yu, *Microsoft Research Asia*

Track 6.3: VLSI Architectures

A HIGH DATA-REUSE ARCHITECTURE WITH DOUBLE-SLICE PROCESSING FOR FULL SEARCH BLOCK-MATCHING ALGORITHM II-716

Yeong-Kang Lai and Lien-Fei Chen, *National Chung Hsing University*

ANALYSIS AND HARDWARE ARCHITECTURE FOR GLOBAL MOTION ESTIMATION IN MPEG-4 ADVANCED SIMPLE PROFILE II-720

Shao-Yi Chien, Ching-Yeh Chen, Wei-Min Chao, Yu-Wen Huang, and Liang-Gee Chen, *National Taiwan University*

A HIERARCHICAL DEPTH BUFFER FOR MINIMIZING MEMORY BANDWIDTH IN 3D RENDERING ENGINE: DEPTH FILTER II-724

Chang-Hyo Yu and Lee-Sup Kim, *KAIST*

A PN TRIANGLE GENERATION UNIT FOR FAST AND SIMPLE TESSELLATION HARDWARE II-728

Kyusik Chung and Lee-Sup Kim, *KAIST*

A RESCHEDULING AND FAST PIPELINE VLSI ARCHITECTURE FOR LIFTING-BASED DISCRETE WAVELET TRANSFORM II-732

Bing-Fei Wu and Chung Fu Lin, *National Chiao Tung University*

HIGH SPEED MEMORY EFFICIENT EBCOT ARCHITECTURE FOR JPEG2000 II-736

Hung-Chi Fang, Tu-Chih Wang, Chung-Jr Lian, Te-Hao Chang, and Liang-Gee Chen, *National Taiwan University*

DESIGN FRAMEWORK FOR JPEG2000 ENCODING SYSTEM ARCHITECTURE II-740

Yoshiteru Hayashi, Hiroshi Tsutsui, Takahiko Masuzaki, Tomonori Izumi, Takao Onoye, and Yukihiro Nakamura, *Kyoto University*

AN HMM-BASED SPEECH RECONGITION IC II-744

Wei HAN, Kwok-Wai HON, Cheong-Fat CHAN, Tan LEE, Chiu-Sing CHOY, Kong-Pang PUN, and P. C. CHING, *The Chinese University of Hong Kong*

A 24-BIT FLOATING-POINT AUDIO DSP CONTROLLER SUPPORTING FAST EXPONENTIATION II-748

Sung-Won Lee, Hyeong-Ju Kang, and In-Cheol Park, *KAIST*

A NEW 2-D 8X8 DCT/IDCT CORE DESIGN USING GROUP DISTRIBUTED ARITHMETIC II-752

Jiun-In Guo, *National Chung Cheng University*, Jia-Wei Chen, and Han-Chen Chen, *National Lien-Ho Institute of Technology*

SUPERSYSTOLIC ARRAYS ON LARGE-SCALE FPGA STRUCTURES II-756

Surin Kittitornkun, *King Mongkut's Institute of Technology Ladkrabang*, and Yu Hen Hu, *University of Wisconsin, Madison*

EMBEDDED RECONFIGURABLE ARRAY TARGETING MOTION ESTIMATION APPLICATIONS II-760

Sami Khawam and Tughrul Arslan, *The University of Edinburgh*, and Fred Westall, *Epson Scotland Design Centre*

IMPLEMENTATION OF PSK DEMODULATOR FOR DIGITAL BS/CS BROADCASTING SYSTEM II-764

Masahide Hatanaka and Toshihiro Masaki, *Osaka University*, Minoru Okada, *Nara Institute of Science and Technology*, and Koso Murakami, *Osaka University*

A HIGH-EFFICIENCY RECONFIGURABLE DIGITAL SIGNAL PROCESSOR FOR MULTIMEDIA COMPUTING II-768

Li-Hsun Chen and Oscal T-C. Chen, *National Chung Cheng University*, and Ruey-Liang Ma, *Industrial Technology Research Institute*

VLSI IMPLEMENTATION OF A REAL-TIME VIDEO WATERMARK EMBEDDER AND DETECTOR II-772

Nebu John Mathai, Ali Sheikholeslami, and Deepa Kundur, *University of Toronto*

MPEG-4 VIDEO CODEC IP DESIGN WITH A CONFIGURABLE EMBEDDED PROCESSOR II-776

Jin-Gyeong Kim and C.-C. Jay Kuo, *University of Southern California*

A SOC FEATURING VARIABLE BUS ARCHITECTURE AND ENHANCED VIDEO COPROCESSORS FOR MPEG-4 MULTIMEDIA APPLICATIONS II-780

Min Yong Jeon, Hyunil Byun, Jooho Ha, Kitaek Lee, Joohyung Kim, Jiyoung Seo, Kyungwoo Lee, and Seungho Lee, *C&S Technology Inc*

HARDWARE-ORIENTED OPTIMIZATION AND BLOCK-LEVEL ARCHITECTURE DESIGN FOR MPEG-4 FGS ENCODER II-784

Chih-Wei Hsu, Yung-Chi Chang, Wei-Min Chao, and Liang-Gee Chen, *National Taiwan University*

COMPUTATIONALLY CONTROLLABLE INTEGER, HALF, AND QUARTER-PEL MOTION ESTIMATOR FOR MPEG-4 ADVANCED SIMPLE PROFILE II-788

Wei-Min Chao, Tung-Chien Chen, Yung-Chi Chang, Chih-Wei Hsu, and Liang-Gee Chen, *National Taiwan University*

A HALF-PEL MOTION ESTIMATION ARCHITECTURE FOR MPEG-4 APPLICATIONS II-792

Mohammed Sayed and Wael Badawy, *University of Calgary*

HARDWARE ARCHITECTURE DESIGN FOR VARIABLE BLOCK SIZE MOTION ESTIMATION IN MPEG-4 AVC/JVT/ITU-T H.264 II-796

Yu-Wen Huang, Tu-Chih Wang, Bing-Yu Hsieh, and Liang-Gee Chen, *National Taiwan University*

PARALLEL 4X4 2D TRANSFORM AND INVERSE TRANSFORM ARCHITECTURE FOR MPEG-4 AVC/H.264 II-800

Tu-Chih Wang, Yu-Wen Huang, Hung-Chi Fang, and Liang-Gee Chen, *National Taiwan University*

ARCHITECTURES FOR FUNCTION EVALUATION ON FPGAS II-804

Nalin Sidahao, *Mahanakorn University of Technology*, George Constantinides, and Peter Cheung, *Imperial College*

SONICmole: A DEBUGGING ENVIRONMENT FOR THE ULTRASONIC RECONFIGURABLE COMPUTER II-808

Theerayod Wiangtong, *Mahanakorn University of Technology*, Chun Tee Ewe and Peter Cheung, *Imperial College*

Track 6.4: Multimedia Transmission

MULTI-SERVER OPTIMAL BANDWIDTH MONITORING FOR QOS BASED MULTIMEDIA DELIVERY II-812

Anup Basu, Irene Cheng, and Yinzhe Yu, *University of Alberta*

RATE CONTROL FOR REPLICATED VIDEO STREAMS II-816

Jiangchuan Liu, Kin-Man Cheung, and Bo Li, *Hong Kong University of Science and Technology*, and Ya-Qin Zhang, *Microsoft Research, Asia*

MULTICAST VIDEO SYNCHRONIZATION VIA MPEG-4 FGS/XML REPRESENTATION II-820

Xiaoming Sun and C.-C. Jay Kuo, *University of Southern California*

REAL-TIME SCHEDULING ON SCALABLE MEDIA STREAM DELIVERY II-824

Kui Gao, Wen Gao, Simin He, Peng Gao, and Yuan Zhang, *Chinese Academy of Sciences*

CACHE REPLACEMENT AND SERVER SELECTION FOR VIDEO PROXY ACROSS WIRELESS INTERNET II-828

Zhe Xiang, Qian Zhang, and Wenwu Zhu, *Microsoft Research Asia*

UNEQUAL PACKET LOSS RESILIENCE FOR MPEG-4 VIDEO OVER THE INTERNET II-832

Xiaokang Yang, *Institute for Infocomm Research*, Ce Zhu, *Nanyang Technological University*, Zhengguo Li and Xiao Lin, *Institute for Infocomm Research, Singapore*, and Nam Ling, *Santa Clara University*

AN END-TO-END RATE CONTROL PROTOCOL FOR MULTIMEDIA STREAMING IN WIRED-CUM-WIRELESS ENVIRONMENTS II-836

Kun Tan, Qian Zhang, and Wenwu Zhu, *Microsoft Research Asia*

USE OF CONCATENATED FEC CODING FOR REAL-TIME PACKET VIDEO OVER HETEROGENEOUS WIRED-TO-WIRELESS IP NETWORKS II-840

Yong Pei and James Modestino, *University of Miami*

VIDEO BROADCASTING OVER MIMO-OFDM SYSTEMS II-844

Zhu Ji, *Tsinghua University*, Qian Zhang and Wenwu Zhu, *Microsoft Research Asia,* and Jianhua Lu, *Tsinghua University*

FAST PREDICTIVE INTEGER- AND HALF-PEL MOTION SEARCH FOR INTERLACED VIDEO CODING II-848

Lifeng Zhao and C.-C.Jay Kuo, *University of Southern California*

IMPROVING THE VIDEO QUALITY AT SCENE TRANSITIONS FOR MPEG-1/2 VIDEO ENCODERS II-852

Anil Fernando, Hemantha Kodikara Arachchi, *Asian Institute of Technolo gy*

FAST MOTION VECTOR AND BITRATE RE-ESTIMATION IN ARBITRARY DOWNSIZING VIDEO TRANSCODING II-856

Haiwei Sun, Yap-Peng Tan, and Yongqing Liang, *Nanyang Technological University*

COMPUTATION REDUCTION IN CASCADED DCT-DOMAIN VIDEO DOWNSCALING TRANSCODING II-860

Yuh-Reuy Lee, Chia-Wen Lin, and Yen-Wen Chen, *Industrial Technology Research Institute*

VIEW-DEPENDENT TRANSMISSION OF 3-D NORMAL MESHES II-864

Jae-Young Sim and Chang-Su Kim, *Seoul National University*, C.-C. Jay Kuo, *University of Southern California*, and Sang-Uk Lee, *Seoul National University*

REGION OF INTEREST DETERMINED BY PICTURE CONTENTS IN JPEG 2000 II-868

Chih Chang Chen, Oscal Chen, and Horng Hsinn Wu, *National Chung Cheng University*

ADAPTIVE SCHEME FOR INTERNET VIDEO TRANSMISSION II-872

Osama Lotfallah and Sethuraman Panchanathan, *Arizona State University*

BIT ALLOCATION FOR PROGRESSIVE FINE GRANULARITY SCALABLE VIDEO CODING WITH TEMPORAL-SNR SCALABILITIES II-876

Jizheng Xu, Feng Wu, and Shipeng Li, *Microsoft Research Asia*

CONSTANT QUALITY RATE ALLOCATION FOR SPECTRAL FINE GRANULAR SCALABLE (SFGS) VIDEO CODING II-880

Wen-Nung Lie and Ming-Yang Tseng, *National Chung Cheng University*, and I-Cheng Ting, *Institute of Information Industry*

RATE DISTORTION OPTIMIZATION IN THE SCALABLE VIDEO CODING II-884

Zhijie Yang, *Chinese Academy of Science*, Feng Wu and Shipeng Li, *Microsoft Research Asia*

PERFORMANCE AND COMPLEXITY JOINT OPTIMIZATION FOR H.264 VIDEO CODING II-888

Jianning Zhang, Yuwen He, Shiqiang Yang, and Yuzhuo Zhong, *Tsinghua University*

RATE CONTROL FOR ADVANCE VIDEO CODING(AVC) STANDARD II-892

Ma Siwei, *Institutue of Computing Technology*, Gao Wen and Gao Peng, *Chinese Academy of Science*, and Lu Yan, *Harbin Institute of Technology*

CHANNEL CODING FOR H.264 VIDEO IN CONSTANT BIT RATE TRANSMISSION CONTEXT OVER 3G MOBILE SYSTEMS II-896

Ngoc Dung Dao and W.A.C Fernando, *Asian Institute of Technology*

REAL-TIME IMPLEMENTATION OF H.263+ USING TI TMS320C6201 DIGITAL SIGNAL PROCESSOR II-900

Kuie-Tsong Shih, Chia-Yang Tsai, and Hsueh-Ming Hang, *National Chiao Tung University*

Track 6.5: Watermarking and Encryption

SEMANTIC IMAGE CLUSTERING USING RELEVANCE FEEDBACK II-904

Xiaoxin Yin, *University of Illinois at Urbana-Champaign*, Mingjing Li, Lei Zhang, and Hongjiang Zhang, *Microsoft Research Asia*

DOMINANT COLOR IMAGE RETRIEVAL USING MERGED HISTOGRAM II-908

Ka-Man Wong, Chun-Ho Cheung, Tak-Shing Liu, and Lai-Man Po, *City University of Hong Kong*

REVERSIBLE DATA HIDING II-912

Zhicheng Ni, Yun Q. Shi, Nirwan Ansari, and Wei Su, *New Jersey Institute of Technology*

A HIGH CAPACITY DISTORTION-FREE DATA HIDING ALGORITHM FOR PALETTE IMAGE II-916

Hongmei Liu, Zhefeng Zhang, Jiwu Huang, and Xialing Huang, *Zhongshan University*, and Yun Q. Shi, *New Jersey Institute of Technology*

DATA HIDING IN HALFTONE IMAGES BY CONJUGATE ERROR DIFFUSION II-920

Ming Sun Fu and Oscar Au, *Hong Kong University of Science and Technology*

2-D AND 3-D SUCCESSIVE PACKING INTERLEAVING TECHNIQUES AND THEIR APPLICATION TO DATA HIDING II-924

Zhicheng Ni, Yun Q. Shi, Nirwan Ansari and Jiwu Haung, *New Jersey Institute of Technology*

WATERMARK RE-SYNCHRONIZATION USING LOG-POLAR MAPPING OF IMAGE AUTOCORRELATION II-928

Adnan Alattar and JoelMeyer, *Digimarc Corporation*

GEOMETRIC MOMENT IN IMAGE WATERMARKING II-932

Li Zhang, *South China University of Technology*, Sam Kwong, *City University of Hong Kong*, and Gang Wei, *South China University of Technology*

A BLIND WATERMARKING TECHNIQUE FOR MULTIPLE WATERMARKS II-936

Peter H.W. Wong and Oscar Au, *The Hong Kong University of Science and Technology*

A NOVEL FREQUENCY DOMAIN WATERMARKING ALGORITHM WITH RESISTANCE TO GEOMETRIC DISTORTIONS AND COPY ATTACK II-940

Shao Yafei, Zhang Li, Wu Guowei, and Lin Xinggang, *Tsinghua University*

ROBUST MULTIBIT AUDIO WATERMARKING IN THE TEMPORAL DOMAIN II-944

Charalampos Laftsidis, Anastasios Tefas, Nikolaos Nikolaidis, and Ioannis Pitas, *Aristotle University of Thessaloniki*

IMAGE WATERMARKING ALGORITHM APPLYING CDMA **II-948**

Yangmei Fang and Jiwu Huang, *Zhongshan University*, and Yun Q. Shi, *New Jersey Institute of Technology*

A HYBRID SYSTEM FOR AUTOMATIC FINGERPRINT IDENTIFICATION **II-952**

Sanpachai Huvanandana, *Chulachomklao Royal Military Academy*, Settapong Malisuwan and Jakkapol Santiyanon, *Rangsit University*, and Jenq-Neng Hwang, *University of Washington*

A FRAGILE WATERMARK ERROR DETECTION SCHEME FOR JVT **II-956**

Peng Zhou and Yun He, *Tsinghua University*

TOWARDS TAMPER DETECTION AND CLASSIFICATION WITH ROBUST WATERMARKS **II-959**

Henry Knowles, Dominque Winne, Nishan Canagarajah, and David Bull, *University of Bristol*

VOLUME III OF V:

-GENERAL & NONLINEAR CIRCUITS & SYSTEMS-

-INVITED SESSIONS-

Track 3: GENERAL & NONLINEAR CIRCUITS & SYSTEMS-

REAL-TIME PROCESSING OF NOISY RF PULSES III-1

Mohamed I. Sobhy, Khaled H. Moustafa. and Mostafa Y. Makkey, *The University of Kent at Canterbury*

POWER FACTOR CORRECTION USING FRACTIONAL CAPACITORS III-5

Wajdi Ahmad, *University of Sharjah*

A NOVEL HOMOTOPY-BASED ALGORITHM FOR THE CLOSEST UNSTABLE EQUILIBRIUM POINT METHOD IN NONLINEAR STABILITY ANALYSIS III-8

Jaewook Lee, *Pohang University of Science and Technology*

MULTIPLE OUTCOMES OF IDEALIZED SWITCHING III-12

Paul Dan Cristea and Rodica Tuduce, *Polytechnical University of Bucharest*

SHARPER BOUNDS FOR THE ZEROS OF POLYNOMIALS III-16

Mohammed Hasan, *University of Minnesota, Duluth*

SUBGRIDDING METHOD FOR SPEEDING UP FD-TLM CIRCUIT SIMULATION III-20

Baohua Wang and Pinaki Mazumder, *University of Michigan*

SYSTEM OBSERVABILITY AND NONLINEAR PARAMETER IDENTIFICATION OF NONYLPHENOL BIODEGRADATION KINETICS III-24

Vishal Shah, *California Institute of Technology*, Aditya Chaubal, *Virginia Polytechnic Institute*, Ravi Ramachandran, *Rowan University*, Raul Ordonez, *University of Dayton*, and Kauser Jahan, *Rowan University*

PUBLIC-KEY ENCRYPTION BASED ON CHEBYSHEV MAPS III-28

Ljupco Kocarev and Zarko Tasev, *University of California, San Diego*

Track 3.1: Blind Signal Processing

NEW HYPERBOLIC SOURCE DENSITY MODELS FOR BLIND SOURCE RECOVERY SCORE FUNCTIONS III-32

Khurram Waheed and Fathi M. Salem, *Michigan State University*

BLIND ADAPTIVE EQUALIZER BASED ON LOCALLY GENERATED SUB-CARRIER III-36

James Okello and Masashi Mizuno, *Kyushu Institute of Technology*, Yoshio Itoh, *Tottori University*, and Hiroshi Ochi, *Kyushu Institute of Technology*

CONVERGENCE BEHAVIOUR OF AN ADAPTIVE STEP-SIZE CONSTANT MODULUS ALGORITHM FOR DS-CDMA RECEIVERS III-40

Peerapol Yuvapoositanon, *Mahanakorn University of Technology*, and Jonathon Chambers, *King's College London*

BLIND SIGNAL SEPARATION USING FIXED OVERCOMPLETE BASIS FUNCTION DICTIONARIES III-44

Paul Sugden and Nishan Canagarajah, *University of Bristol*

BLIND DECONVOLUTION OF MIMO-FIR CHANNELS DRIVEN BY COLORED INPUTS USING SECOND-ORDER STATISTICS III-48

Mitsuru Kawamoto and Yujiro Inouye, *Shimane University*

A SUBSPACE METHOD FOR CHANNEL ESTIMATION OF MULTI-USER MULTI-ANTENNA OFDM SYSTEM III-52

Yonghong Zeng and Tung Sang Ng, *The University of Hong Kong*

STRUCTURAL ANALYSIS OF A ONE-PORT FORMED BY A RESISTIVELY COUPLED NONLINEAR RING III-56

Stefano Pastore, *University of Trieste*

Track 3.2: Chaotic Circuits and Systems

A NEW APPROACH TO GENERATE HYPERCHAOTIC 3D-SCROLL ATTRACTORS IN A CLOSED CHAIN OF CHUA'S CIRCUITS III-60

Donato Cafagna and Giuseppe Grassi, *University of Lecce*

STABILIZING UNSTABLE EQUILIBRIUM POINTS OF CHAOTIC SYSTEMS USING PI REGULATOR III-64

Guo-Ping Jiang and Suo-Ping Wang, *Nanjing University of Posts & Telecommunications*

CHAOS IN CROSS-COUPLED BVP OSCILLATORS III-68

Tetsushi Ueta and Hiroshi Kawakami, *Tokushima University*

A HYPERCHAOTIC CIRCUIT FAMILY INCLUDING A DEPENDENT SWITCHED CAPACITOR III-72

Yusuke Takahashi, Hidehiro Nakano, and Toshimichi Saito, *Hosei University*

NORMAL FORMS OF BORDER-COLLISIONS IN HIGH-DIMENSIONAL NONSMOOTH MAPS III-76

Mario Di Bernardo, *University of Sannio*

MOVEMENT OF SMALL AMPLITUDE PARTS IN A COUPLED CHAOTIC SYSTEM III-80

Yasuteru Hosokawa, *Shikoku University*, and Yoshifumi Nishio, *Tokushima University*

GENERATION OF MULTI-SCROLL CHAOS USING SECOND-ORDER LINEAR SYSTEMS WITH HYSTERESIS III-84

Fengling Han, *Central Queensland University*, Xinghuo Yu, *Royal Melbourne Institute of Technology*, Guanrong Chen, *City University of Hong Kong*, Wenbo Liu, *Nanjing University of Aeronautics & Astronautics*, and Yong Feng, *Harbin Institute of Technology*

RECONSTRUCTION OF PIECEWISE CHAOTIC DYNAMICS USING A MULTIPLE MODEL APPROACH III-88

Nan Xie and Henry Leung, *University of Calgary*

DELAYS IN PWM CONTROL LOOPS IMPLY DISCONTINUITY IN SAMPLED DATA MODELS OF POWER ELECTRONIC CIRCUITS III-92

Soumitro Banerjee, *Indian Institute of Technology*, and Sukanya Parui, *Indian Institute of Technology, Kharagpur*

TRANSIENT DYNAMICS AND CHAOS OBSERVED IN STRONGLY NONLINEAR MUTUALLY-COUPLED OSCILLATORS III-96

Tetsuro Endo, Yuhki Aruga, and Kentaro Yamauchi, *Meiji University*

CONTROL AND SYNCHRONIZATION OF A 4-SCROLL CHAOTIC SYSTEM III-100

Wen bo Liu, *Nanjing University of Aeronautics & Astronautics*, and Guanrong Chen, *City University of Hong Kong*

DESIGN OF CHAOTIC SPREAD-SPECTRUM SEQUENCES WITH GOOD CORRELATION PROPERTIES FOR DS/CDMA III-104

Chengquan An and Tingxian Zhou, *Harbin Institute of Technology*

A SIMPLE NONAUTONOMOUS CHAOTIC CIRCUIT WITH A PERIODIC PULSE-TRAIN INPUT III-108

Hidehiro Nakano, Keita Miyachi, and Toshimichi Saito, *Hosei University*

COMPLEX BEHAVIOURS IN TWO BI-DIRECTIONALLY COUPLED LORENZ SYSTEMS III-112

Silvano Cincotti and Simona Di Stefano, *University of Genoa*

ADVANCED CHAOS-BASED FREQUENCY MODULATIONS FOR CLOCK SIGNALS EMC TUNING III-116

Stefano Santi, Riccardo Rovatti, and Gianluca Setti, *Di Università di Bologna*

BIFURCATIONS AND CHAOS IN THE TURBO DECODING ALGORITHM III-120

Zarko Tasev, Ljupco Kocarev, and Gian Mario Maggio, *University of California, San Diego*

OPTIMIZATION OF 3 PHASE SPREADING SEQUENCES OF MARKOV CHAINS III-124

Hiroshi Fujisaki, *Kanazawa University*

GIBBS-LIKE PHENOMENA IN CHAOS-BASED FREQUENCY MODULATED SIGNALS III-128

Stefano Santi, Riccardo Rovatti, and Gianluca Setti, *Università di Bologna*

CONTROLLING BIFURCATION AND CHAOS IN INTERNET CONGESTION CONTROL MODEL III-132

Liang Chen, Xiaofan Wang, and Z.Z. Hahn *Shanghai Jiao Tong University*

Track 3.3: Nonlinear Circuits and Systems

NONAUTONOMOUS PULSE-DRIVEN CHAOTIC OSCILLATOR BASED ON CHUA'S CIRCUIT III-136

Ahmed Elwakil, *University of Sharjah*

GENERATION OF HOMOCLINIC OSCILLATION IN COUPLED CHUA'S OSCILLATORS III-140

Syamal Dana, *Indian Institute of Chemical Biology*, Prodyot Kumar Roy, *Presidency College*, and Satyabrata Chakraborty, *Indian Institute of Chemical Biology*

DETERMINING THE OSCILLATION OF DIFFERENTIAL VCOS III-144

Antonio Buonomo and Alessandro Lo Schiavo, *Seconda Università di Napoli*

A CMOS THREE-STATE FREQUENCY DETECTOR COMPLEMENTARY TO AN ENHANCED LINEAR PHASE DETECTOR FOR PLL, DLL OR HIGH FREQUENCY CLOCK SKEW MEASUREMENT III-148

Mathieu Renaud and Yvon Savaria, *Ecole Polytechnique de Montreal*

A CMOS NEURAL OSCILLATOR USING NEGATIVE RESISTANCE III-152

Hanjung Song, *Chungcheong College*, John Harris, *University of Florida*

CONTROLLING CHAOS VIA SECOND-ORDER SLIDING MODES III-156

Barbara Cannas, *University of Cagliari*, Silvano Cincotti, *University of Genoa*, Alessandro Pisano, *University of Cagliari*, and Elio Usai, *University of Cagliari*

A NOVEL LOW PHASE NOISE 1.8V 900MHZ CMOS VOLTAGE CONTROLLED RING OSCILLATOR III-160

Dean Badillo, *Motorola/Arizona State University*, and Sayfe Kiaei, *Arizona State University*

SOME CHARACTERIZATIONS OF INTERVAL SYSTEMS: FURTHER EXTENSIONS AND APPLICATIONS III-164

Long Wang, *Peking University*

AN ERROR DISTRIBUTION BASED NONLINEAR COMPANDING METHOD FOR ANALOG BEHAVIORAL MODELING VIA WAVELET APPROXIMATION III-168

Xuan Zeng, Jun Tao, Yang Feng Su, and WenBing Chen, *Fudan University*, and Dian Zhou, *Univeristy of Texas at Dallas*

AN ALGORITHM GMST FOR EXTRACTING MINIMAL SIPHON-TRAPS AND ITS APPLICATION TO EFFICIENT COMPUTATION OF PETRI NET INVARIANTS III-172

Akihiro Taguchi, Satoshi Taoka, and Toshimasa Watanabe, *Hiroshima University*

GENERATION OF N-SCROLL CHAOS USING NONLINEAR TRANSDUCTORS III-176

Ahmed Elwakil, *University of Sharjah*, Khaled Salama, *Stanford University*, and Serdar Ozoguz, *Istanbul Technical University*

AN INTEGRATED MULTI-SCROLL CIRCUIT WITH FLOATING-GATE MOSFETS III-180

Tetsuya Fujiwara and Yoshihiko Horio, *Tokyo Denki University* and Kazuyuki Aihara, *University of Tokyo*

TOPOLOGICAL CRITERIA FOR SWITCHED MODE DC-DC CONVERTERS III-184

Masato Ogata and Tetsuo Nishi, *Kyushu University*

A HETEROCLINIC POINT AND BASIN BOUNDARIES IN A PIECEWISE LINEAR CHAOTIC NEURON MODEL III-188

Hiroto Tanaka and Ushio Toshimitsu, *Osaka University*

AN INTERVAL ALGORITHM FOR FINDING ALL SOLUTIONS OF NONLINEAR RESISTIVE CIRCUITS III-192

Kiyotaka Yamamura and Naoya Igarashi, *Chuo University*, and Yasuaki Inoue, *University of East Asia*

AN EFFECTIVE INITIAL SOLUTION ALGORITHM FOR GLOBALLY CONVERGENT HOMOTOPY METHODS III-196

Yasuaki Inoue, *Waseda University*, Saeko Kusanobu, *University of East Asia*, Kiyotaka Yamamura and Makoto Ando, *Chuo University*

CREATING IMPLICIT HOMOTOPY METHODS USING HARDWARE DESCRIPTION LANGUAGES III-200

Leonid Goldgeisser, *Synopsys Inc*

CO-EXISTENCE OF CHAOS-BASED AND CONVENTIONAL DIGITAL COMMUNICATION SYSTEMS III-204

Francis Lau and Chi Tse, *Hong Kong Polytechnic University*

PERFORMANCE ANALYSIS OF MULTIPLE ACCESS CHAOTIC-SEQUENCE SPREAD-SPECTRUM COMMUNICATION SYSTEMS EMPLOYING PARALLEL INTERFERENCE CANCELLATION DETECTORS III-208

Wai Tam, Francis C.M. Lau, and Chi K. Tse, *Hong Kong Polytechnic University*

SYNCHRONIZATION OF A NETWORK OF INTERMITTENTLY COUPLED CAPACITOR CIRCUITS III-212

Junya Shimakawa and Toshimichi Saito, *Hosei University*

PARTIAL STABILITY OF DISCONTINUOUS DYNAMICAL SYSTEMS UNDER ARIBITRARY INITIAL Z-PERTURBATIONS III-216

Ye Sun and Anthony N. Michel, *University of Notre Dame*

L2 GAIN ANALYSIS FOR SWITCHED SYMMETRIC SYSTEMS WITH TIME DELAY UNDER ARBITRARY SWITCHING III-220

Guisheng Zhai, *Wakayama University*, Xinkai Chen, *Kinki University*, Ye Sun and Anthony N. Michel, *University of Notre Dame*

Track 3.4: Graph Models and Algorithms

ARCHITECTURAL DESIGN AND ANALYSIS TOOLBOX TO IMPLEMENT SHORTEST PATH ALGORITHMS IN HARDWARE III-224

Kai-Hock Quek and Siew-Kei Lam, *Nanyang Technological University*, Neelkamal Agrawal, *IIT Guwahati*, and Thambipillai Srikanthan, *Nanyang Technological University*

USING FPGAS TO SOLVE THE HAMILTONIAN CYCLE PROBLEM III-228

Micaela Serra, *University of Victoria*, and Ken Kent, *University of New Brunswick*

GLOBAL SCHEDULING AND REGISTER ALLOCATION BASED ON PREDICATED EXECUTION III-232

Rogerio de Azambuja, *ULBRA Santa Maria*, and Luiz C. V. Dos Santos, *Universidade Federal de Santa Catarina*

A 2-APPROXIMATION ALGORITHM FSA+1 TO (L+1)-EDGE-CONNECT A SPECIFIED SET OF VERTICES IN A L-EDGE-CONNECTED GRAPH III-236

Satoshi Taoka, *Hiroshima University*, Toshiya Mashima, *Hiroshima International University*, and Toshimasa Watanabe, *Hiroshima University*

GENERATION OF WAVE DIGITAL STRUCTURES FOR CONNECTION NETWORKS CONTAINING IDEAL TRANSFORMERS III-240

Dietrich Fränken and Jörg Ochs, *University of Paderborn*, and Karlheinz Ochs, *Ruhr-University Bochum*

INSUFFICIENTLY MARKED SIPHON OF PETRI NETS - EXTENSION OF TOKEN-FREE SIPHON - III-244

Atsushi Ohta and Kohkichi Tsuji, *Aichi Prefectural University*

CONVERGENT TRANSFER SUBGRAPH CHARACTERIZATION AND COMPUTATION III-248

Wing Li, *University of Arkansas*

OPTIMUM REGISTER ASSIGNMENT FOR HETEROGENEOUS REGISTER-SET ARCHITECTURES III-252

Thomas Zeitlhofer and Bernhard Wess, *Vienna University of Technology*

Track 3.5: Power Electronics and Systems

THREE-PHASE HIGH-POWER-FACTOR RECTIFIER WITH THREE AC POWER SWITCHES III-256

Bor-Ren Lin, T.Y.Yang, and Y.C. Lee, *National Yunlin University of Science & Technology*

A SIMPLE CRITERION TO JUDGE PFC CONVERTER STABILITY III-260

Mohamed Orabi and Tamotsu Ninomiya, *Kyushu University*

SINGLE-SWITCH FLYBACK POWER-FACTOR-CORRECTED AC/DC CONVERTER WITH LOOSELY REGULATED INTERMEDIATE STORAGE CAPACITOR VOLTAGE III-264

Dylan Dah-Chuan Lu, David Ki-Wai Cheng, and Yim-Shu Lee, *Hong Kong Polytechnic University*

A NEW ADAPTIVE HARMONIC EXTRACTION SCHEME FOR SINGLE-PHASE ACTIVE POWER FILTERS III-268

Alireza Bakhshai, *Isfahan University of Technology*, Houshang Karimi, *University of Toronto*, and Maryam Saeedifard, *Isfahan University of Technology*

SOFT-SWITCHING BOOST POWER FACTOR CORRECTION CONVERTER WITH NEGATIVE SLOPE RAMP CARRIER CONTROL III-272

Tanes Tanitteerapan, *King Mongkut's University of Technology Thonburi*, and Shinsaku Mori, *Nippon Institute of Technology*

LOSSLESS VOLTAGE-CLAMPING OF A CLASS E AMPLIFIER WITH A TRANSFORMER AND A DIODE III-276

Tadashi Suetsugu and Marian Kazimierczuk, *Wright State University*

DESIGN AND ANALYSIS OF CLASS DE AMPLIFIER WITH ANY OUTPUT Q, ANY DUTY RATIO AND SWITCH ON REISISTANCE III-280

Hiroo Sekiya, Satoki Oshikawa, Jianming Lu, and Takashi Yahagi, *Chiba University*

THE LOW STRESS VOLTAGE BALANCE CHARGING CIRCUIT FOR SERIES CONNECTED BATTERIES BASED ON BUCK-BOOST TOPOLOGY III-284

Yuttasak Rungruengphalanggul, Itsda Boonyaroonate, and Charnyut Karnjanapiboon, *King Mongkut's University of Technology Thonburi*

STEADY-STATE ANALYSIS OF PWM DC-TO-DC REGULATORS III-288

Luigi Egiziano, Nicola Femia, and Giovanni Spagnuolo, *University of Salerno*, and Massimo Vitelli, *Second University of Naples*

DERIVATION OF THE CUK PWM DC-DC CONVERTER CIRCUIT TOPOLOGY III-292

Brad Bryant and Marian Kazimierczuk, *Wright State University*

BOOST CONVERTER WITH HIGH VOLTAGE GAIN USING A SWITCHED CAPACITOR CIRCUIT III-296

Oded Abutbul, Amir Gherlitz, Adrian Ioinovici, and Yefim Berkovich, *Holon Academic Institute of Technology*

DESIGN OF A STEP-UP/STEP-DOWN SC DC-DC CONVERTER WITH SERIES-CONNECTED CAPACITORS III-300

Kei Eguchi, *Kumamoto National College of Technology*, Zhu Hongbing, *Hiroshima Kokusai Gakuin University*, Fumio Ueno, *Sojo University*, and Toru Tabata, *Kumamoto National College of Technology*

A DESIGN SPACE EXPLORATION FOR INTEGRATED SWITCHING POWER CONVERTERS III-304

Gerard Villar, Eduard Alarcon, Francesc Guinjoan, and Alberto Poveda, *Technical University of Catalunya*

MONOLITHIC DISTRIBUTED POWER SUPPLY FOR A MIXED-SIGNAL INTEGRATED CIRCUIT III-308

Siamak Abedinpour and Sayfe Kiaei, *Arizona State University*

BIFURCATION ANALYSIS OF A POWER-FACTOR-CORRECTION BOOST CONVERTER: UNCOVERING FAST-SCALE INSTABILITY III-312

Chi Tse and Octavian Dranga, *Hong Kong Polytechnic University*, and Herbert Iu, *University of Western Australia*

TIME-DELAY MODELLING FOR MULTI-LAYER POWER SYSTEMS III-316

Ian Hiskens, *University of Wisconsin, Madison*

STEADY STATE AND DYNAMIC SECURITY ASSESSMENT IN COMPOSITE POWER SYSTEMS III-320

Chanan Singh and Hyungchul Kim, *Texas A&M University*

AN EVOLUTIONARY GAME APPROACH TO ENERGY MARKETS III-324

Leontina Pinto and Jacques Szczupak, *Engenho*

ESTIMATING LOADS IN DISTRIBUTION FEEDERS USING A STATE ESTIMATOR ALGORITHM WITH ADDITIONAL ADJUSTMENT OF TRANSFORMERS LOADING FACTORS III-328

Manoel Medeiros, Marcos Almeida, and Daniel Silveira, *Federal University of Rio Grande do Norte*

"INTERMITTENT" CHAOS AND SUBHARMONICS IN SWITCHING POWER SUPPLIES III-332

Chi Tse, Yufei Zhou, and Francis Lau, *Hong Kong Polytechnic University*, and Shui-Sheng Qiu, *South China University of Technology*

APPLICATION OF WAVELET TRANSFORM TO STEADY-STATE APPROXIMATION OF POWER ELECTRONICS WAVEFORMS III-336

Pik Wan Michelle Ho, Ming Liu, and Chi Tse, *Hong Kong Polytechnic University,* and Jie Wu, *South China University of Technology*

SINGLE-PHASE AC/AC CONVERTER BASED ON HALF-BRIDGE NPC TOPOLOGY III-340

Bor-Ren Lin, T. Y. Yang, and T. C. Wei, *National Yunlin University of Science & Technology*

V2 CONTROL OF INTERLEAVED BUCK CONVERTERS III-344

Veerachary Mummadi, *Indian Institute of Technology Delhi*

MODELING OF CASCADE BUCK CONVERTERS III-347

Veerachary Mummadi, *Indian Institute of Technology Delhi*

ROBUST POLE LOCATION FOR A DC-DC CONVERTER THROUGH PARAMETER DEPENDENT CONTROL III-351

Vinícius Montagner and Pedro L. D. Peres, *University of Campinas*

HIGH-EFFICIENCY ELECTRONIC TRANSFORMER FOR LOW-VOLTAGE HALOGEN LAMP III-355

Kamon Jirasereeamornkul, Itsda Boonyaroonate, and Kosin Chamnongthai, *King Mongkut's University of Technology Thonburi*

PUSH-PULL DC/DC INVERTER FOR LARGE ELECTROLUMINESCENT LAMP III-359

Surachet Bumrungkeeree and Itsda Boonyaroonate, *King Mongkut's University of Technology Thonburi*

ALLOCATING USAGES OF VOLTAGE SECURITY MARGIN IN DEREGULATED ELECTRIC MARKETS III-363

Garng Huang and Nirmal-Kumar Nair, *Texas A&M University*

OSCILLATION MODE ANALYSIS IN POWER SYSTEMS BASED ON DATA ACQUIRED BY DISTRIBUTED PHASOR MEASURMENT UNITS III-367

Takuhei Hashiguchi, Masamichi Yoshimoto, Yasunori Mitani, Osamu Saeki, and Kiichiro Tsuji, *Osaka University*

SUPPORT VECTOR REGRESSION BASED ADAPTIVE POWER SYSTEM STABILIZER III-371

Udomsak Boonprasert, Nipon Theera-Umpon, and Chewasak Rakpenthai, *Chiang-Mai University*

OPTIMAL PLACEMENT OF MULTI-TYPE FACTS DEVICES BY HYBRID TS/SA APPROACH III-375

Pornrapeepat Bhasaputra and Weerakorn Ongsakul, *Asian Institute of Technolgy*

AN HVDC-BASED CONTROLLER DESIGN FOR STABILIZATION OF FREQUENCY OSCILLATION III-379

Sanchai Dechanupaprittha, *Sirindhorn International Institute of Technology*, Adual Patanapakdee, *Sripatum University*, and Issarachai Ngamroo, *Sirindhorn International Institute of Technology*

FLUX HARMONIC SPECTRUM PROCCESSING OF DIRECT TORQUE CONTROLLED INDUCTION MOTOR III-383

Shahriyar Kaboli and Mohammad Reza Zolghadri, *Sharif University of Technology*

ELECTRICAL PROTECTION SELECTIVITY IN DC AUXILIARY INSTALLATIONS IN POWER PLANTS AND SUBSTATIONS III-387

Srdjan Skok, Ante Marusic, and Sejid Tesnjak, *Faculty of Electrical Engineering and Computing*

SUBBAND DECOMPOSITION ORIENTED MULTIRATE ELECTRICAL POWER NETWORK DIGITAL SIMULATION III-391

Jacques Szczupak, *Engenho*, Silvana Faceroli, *Univ. Federal de Juiz de Fora*, and Karla Silva, *Univ. Federal Fluminense*

EXAMINING CHARACTERISTICS OF AN OBSERVABILITY FORMULATION FOR NONLINEAR POWER SYSTEMS III-395

Chris Dafis and Chika Nwankpa, *Drexel University*

PHOTO-VOLTAIC POWER CONVERTER WITH A SIMPLE MAXIMUM-POWER-POINT-TRACKER III-399

Subbaraya Yuvarajan, *North Dakota State University*, and Shanguang Xu, *Tyco Electronics*

CONJECTURAL VARIATION BASED LEARNING OF GENERATOR'S BEHAVIOR IN ELECTRICITY MARKET III-403

Yiqun Song, *The University of Hong Kong*, Zhijian Hou, *Shanghai Jiao Tong University*, Fushuan Wen, Yixin Ni, and Felix F. Wu, *The University of Hong Kong*

A HEURISTIC APPROACH FOR POWER SYSTEM MEASUREMENT PLACEMENT DESIGN III-407

Garng Huang, Jiansheng Lei, and Ali Abur, *Texas A&M University*

DECENTRALIZED H-INFINITY LOAD FREQUENCY CONTROL USING LMI CONTROL TOOLBOX III-411

Dulpichet Rerkpreedapong and Ali Feliachi, *West Virginia University*

ANALYTICAL COMPUTATION OF MULTIPATH COMPONENTS IN THE INDOOR POWER GRID III-415

Despina Anastasiadou and Theodore Antonakopoulos, *University of Patras*

A POWER DIVIDER USING LINEAR ELECTRIC PROBES COUPLING INSIDE CONDUCTING CYLINDRICAL CAVITY III-419

Sanya Amnartpluk, Chuwong Phongcharoenpanich, Sompol Kosulvit, and Monai Krairiksh, *King Mongkut's Institute of Technology Ladkrabang*

CLASSIFICATION OF POWER SYSTEM FAULTS USING WAVELET TRANSFORMS AND PROBABILISTIC NEURAL NETWORKS III-423

Jayachandra Shenoy, *Indian Institute of Science*, and Harish Kashyap, *The National Institute of Engineering*

RANDOM NOISE IN SWITCHING DC-DC CONVERTER: VERIFICATION AND ANALYSIS III-427

Anawach Sangswang and Chika Nwankpa, *Drexel University*

PRACTICAL RESULTS AND FINITE DIFFERENCE METHOD TO ANALYZE THE ELECTRIC AND MAGNETIC FIELD COUPLING BETWEEN POWER TRANSMISSION LINE AND PIPELINE III-431

Mahmoud Elhirbawy and Les Jennings, *University of Western Australia*, Sulaiman Dhalaan, *General Organization for Technical Education and Vocational Training, Riyadh, Saudi Arabia*, and W.W. L. Keerthipala, *Curtin University of Technology*

INVITED SESSIONS

Track 11.1: Switching Power Converters Towards Integration and Recurrent Neural Networks

TRANSFORMERLESS DC-DC CONVERTERS WITH A VERY HIGH DC LINE-TO-LOAD VOLTAGE RATIO III-435

Boris Axelrod, Adrian Ioinovici, and Yefim Berkovich, *Holon Academic Institute of Technology*

USE OF STATE TRAJECTORY PREDICTION IN HYSTERESIS CONTROL FOR ACHIEVING FAST TRANSIENT RESPONSE OF THE BUCK CONVERTER III-439

Henry Chung, Kelvin Leung, and Ron Hui, *City University of Hong Kong*

FEASIBILITY STUDY OF ON-CHIP CLASS E DC-DC CONVERTER III-443

Tadashi Suetsugu, S. Kiryu, and Marian Kazimierczuk, *Wright State University*

SINGLE-INDUCTOR DUAL-INPUT DUAL-OUTPUT SWITCHING CONVERTER FOR INTEGRATED BATTERY CHARGING AND POWER REGULATION III-447

Yat-Hei Lam, Wing-Hung Ki, Chi-Ying Tsui, and Philip K. T. Mok, *The Hong Kong University of Science and Technology*

OPTIMIZED DESIGN OF MOS CAPACITORS IN STANDARD CMOS TECHNOLOGY AND EVALUATION OF THEIR EQUIVALENT SERIES RESISTANCE FOR POWER APPLICATIONS III-451

Gerard Villar, Eduard Alarcon, Francesc Guinjoan, and Alberto Poveda, *Technical University of Catalunya*

CMOS CURRENT-MODE ANALOG CIRCUIT BUILDING BLOCKS FOR RF DC-DC CONVERTER CONTROLLERS III-455

James Masciotti and Lessing Luu, *Columbia University*, and Dariusz Czarkowski, *Polytechnic University*

RECENT TRENDS IN FUZZY CONTROL OF ELECTRICAL DRIVES: AN INDUSTRY POINT OF VIEW III-459

Chira Vinci, *ST Microelectronics*, Luigi Fortuna and Antonino Cucuccio, and Matteo Lo Presti, *Università degli Studi di Catania*

DESIGN AND REALISATION OF A NANO-INDUCTANCE FOR INTEGRATED POWER CONVERTERS III-462

Bruno Estibals, Corinne Alonso, Franck Carcenac, Alain Salles, Laurent Malaquin, and Christophe Vieu, *LAAS/CNRS*

OUTPUT CONVERGENCE ANALYSIS OF CONTINUOUS-TIME RECURRENT NEURAL NETWORKS III-466

Derong Liu and Sanqing Hu, *University of Illinois*

A RECURRENT NEURAL NETWORK FOR NONLINEAR CONVEX PROGRAMMING III-470

Youshen Xia and Jun Wang, *Chinese University of Hong Kong*

RECURRENT NEURAL NETWORKS: OVERVIEW AND PERSPECTIVES III-474

Anthony Michel, *University of Notre Dame*

ON GLOBAL STABILITY OF HOPFIELD NEURAL NETWORKS WITH DISCONTINUOUS NEURON ACTIVATIONS III-478

Mauro Forti and Paolo Nistri, *University of Siena*

THREE-LAYER BIDIRECTIONAL ASYMMETRICAL ASSOCIATIVE MEMORY III-482

H. K. Kwan, *University of Windsor*

Track 11.2: Cellular and High-Speed Circuits & Systems

MODELING OF DIFFUSION PROCESS IN P-N JUNCTION DIODE USING MATRIX RATIONAL APPROXIMATION III-486

Yuichi Tanji, *Kagawa University*

A REDUCTION TECHNIQUE OF LARGE SCALE RCG INTERCONNECTS IN COMPLEX FREQUENCY DOMAIN III-490

Yoshihiro Yamagami, Yoshifumi Nishio, Atsumi Hattori, and Akio Ushida, *Tokushima University*

REALIZABLE REDUCTION OF RLC CIRCUITS USING NODE ELIMINATION III-494

Masud Chowdhury, Chirayu Amin, and Yehea Ismail, *Northwestern University*, Chandramouli Kashyap, *IBM Austin Research Lab.*, and Byron Krauter, *IBM Microelectronics*

ANALYSIS OF PCB INTERCONNECTS USING ELECTROMAGNETIC REDUCTION TECHNIQUE III-498

Takayuki Watanabe, *University of Shizuoka*, and Hideki Asai, *Shizuoka University*

PASSIVE MACROMODELING OF SUBNETWORKS CHARACTERIZED BY MEASURED DATA III-502

Dharmendra Saraswat, Ramachandra Achar, and Michel Nakhla, *Carleton University*

A 32X32 CELLULAR TEST CHIP TARGETING NEW FUNCTIONALITIES III-506

Ari Paasio and Mika Laiho, *Helsinki University of Technology*, Jonne Poikonen, *University of Turku*, Asko Kananen and Kari Halonen, *Helsinki University of Technology*

A CNN-BASED CHIP FOR ROBOT LOCOMOTION CONTROL III-510

Paolo Arena, Salvatore Castorina, Luigi Fortuna, Mattia Frasca, and Marco Ruta, *Università degli Studi di Catania*

ON-CHIP TEMPLATE TRAINING FOR PATTERN MATCHING BY CELLULAR NEURAL NETWORK UNIVERSAL MACHINES (CNN-UM) III-514

Ralf Schoenmeyer, Dirk Feiden, and Ronald Tetzlaff, *Johann Wolfgang Goethe-University*

VISION SYSTEMS BASED ON THE 128X128 FOCAL PLANE CELLLAR VISUAL MICROPROCESSOR CHIP III-518

Akos Zarandy, Csaba Rekeczky, and Istvan Szatmari, *MTA - SZTAKI*

ANALOG WEIGHT BUFFERING STRATEGY FOR CNN CHIPS III-522

Angel Rodriguez-Vazquez, Gustavo Liñan-Cembrano, Ricardo Carmona, Servando Espejo, and Rafael Dominguez-Castro, *Instituto de Microelectronica de Sevilla*

Track 11.3: IC Interfaces for Sensors/MEMS

A PARTICLE DETECTOR FULLY-PROGRAMMABLE INTERFACE CIRCUIT FOR SATELLITE APPLICATIONS III-526

Fausto Borghetti, *University of Pavia*, Massimiliano Gobbi, *MISARC*, Andrea Fornasari and Piero Malcovati, *University of Pavia*, Franco Maloberti, *University of Texas at Dallas*, and Marco Pagano, *LABEN*

ELECTROSTATICAL COUPLING-SPRING FOR MICRO-MECHANICAL FILTERING III-530

Dimitri Galayko, Andreas Kaiser, and Lionel Buchaillot, *ISEN*, Dominique Collard, *CIRMM*, and Chantal Combi, *ST Microelectronics*

HIGH-ACCURACY INSTRUMENTATION AMPLIFIER FOR LOW VOLTAGE LOW POWER CMOS SMART SENSORS III-534

Christian Falconi, *University of Tor Vergata, Roma*, Marco Faccio, *University of L'aquila, Monteluco di Roio*, Arnaldo D'amico and Corrado Di Natale, *University of Tor Vergata, Roma*

DESIGN CONSIDERATIONS FOR AN AUTOMOTIVE SENSOR INTERFACE SIGMA-DELTA MODULATOR III-538

Fernando Medeiro, Jose Manuel de la Rosa, Rocio del Rio, Belen Perez-Verdu, and Angel Rodriguez-Vazquez, *Instituto de Microelectronica de Sevilla IMSE-CNM-CSIC*

INTERFACE CIRCUITRY FOR CMOS-BASED MONOLITHIC GAS SENSOR ARRAYS III-542

Christoph Hagleitner, Diego Barrettino, Andreas Hierlemann, Oliver Brand, and Henry Baltes, *ETH Zuerich*

Track 11.4: Variable Digital Filters

DESIGN OF SINUSOID-BASED VARIABLE FRACTIONAL DELAY FIR FILTER USING WEIGHTED LEAST SQUARES METHOD III-546

Chien-Cheng Tseng, *National Kaohsiung First University of Science and Technology*

VECTOR-ARRAY DECOMPOSITION-BASED DESIGN OF VARIABLE DIGITAL FILTERS III-550

Tian-Bo Deng, *Toho University*

LINEAR PROGRAMMING DESIGN OF LINEAR-PHASE FIR FILTERS WITH VARIABLE BANDWIDTH III-554

Per Lowenborg and Hakan Johansson, *Linkoping University*

AUDIO SIGNAL PROCESSING VIA HARMONIC SEPARATION USING VARIABLE LAGUERRE FILTERS III-558

David Tay, *Latrobe University*, Saman Abeysekera and Arjuna Balasuriya, *Nanyang Technological University*

ON THE APPLICATION OF VARIABLE DIGITAL FILTERS (VDF) TO THE REALIZATION OF SOFTWARE RADIO RECEIVERS III-562

Shing Chow Chan and Kim Sang Yeung, *The University of Hong Kong*

VARIABLE BIQUADRATIC DIGITAL FILTER SECTION WITH SIMULTANEOUS TUNING OF THE POLE AND ZERO FREQUENCIES BY A SINGLE PARAMETER III-566

Georgi Stoyanov, *Technical University of Sofia*, and Masayuki Kawamata, *Tohoku University*

EFFICIENT SYMBOL SYNCHRONIZATION TECHNIQUES USING VARIABLE FIR OR IIR INTERPOLATION FILTERS III-570

Martin Makundi and Timo Laakso, *Helsinki University of Technology*

Track 11.5: Theoretical Aspects in Cellular Neural Networks

ON THE PREDICTION OF PERIOD-DOUBLING BIFURCATIONS IN ALMOST RECIPROCAL CELLULAR NEURAL NETWORKS III-574

Mauro Di Marco and Mauro Forti, *Universita' di Siena*, and Alberto Tesi, *Universita' di Firenze*

RELATIONS BETWEEN SPATIO-TEMPORAL PHENOMENA AND EIGENVALUES IN MUTUALLY COUPLED CNNS III-578

Zonghuang Yang, Masayuki Yamauchi, Yoshifumi Nishio, and Akio Ushida, *Tokushima University*

COUPLED CHAOTIC SIMULATED ANNEALING PROCESSES III-582

Johan Suykens, Mustak E. Yalcin, and Joos Vandewalle, *Katholieke Universiteit Leuven*

CLASSIFICATION OF SYNCHRONIZED STATES IN CNNS WITH HIGHER ORDER CELLS III-586

Antonio Andreescu, Zbigniew Galias, and Maciej Ogorzalek, *University of Mining and Metallurgy*

ON THE EFFECT OF BOUNDARY CONDITIONS ON CNN DYNAMICS:STABILITY AND INSTABILITY, BIFURCATION PROCESSES AND CHAOTIC PHENOMENA III-590

Istvan Petras, *Hungarian Academy of Science*, Paolo Checco and Marco Gilli, *Politecnico Di Torino*, Tamas Roska, *The Hungarian Academy of Sciences*, and Mario Biey, *Politecnico di Torino*

Track 11.6: EDGE is Cutting Edge

EDGE TRANSMITTER ALTERNATIVE USING POLAR MODULATION III-594

Earl Mccune and Wendell Sander, *Tropian Inc.*

GSM/EDGE EVOLUTION, BASED ON 8-PSK III-598

Ilya Gonorovsky, Niels Andersen, and Mark Pecen, *Motorola*

RF POWER CONTROL IN GSM SYSTEMS FOR CONSTANT AND NON CONSTANT ENVELOPE MODULATION SCHEMES III-602

Rolf Becker, Ralf Burdenski, and Willem Groeneweg, *Philips Semiconductors*

EDGE DATA RECEIVER DESIGN III-606

Doug Grant, Marko Kocic, Lidwine Martinot, and Zoran Zvonar, *Analog Devices*

TRANSMIT ARCHITECTURES AND POWER CONTROL SCHEMES FOR LOW COST HIGHLY INTEGRATED TRANSCEIVERS FOR GSM/EDGE APPLICATIONS III-610

Aristotele Hadjichristos, *Ericsson Mobile Platforms*

COMBINED GMSK AND 8PSK MODULATOR FOR GSM AND EDGE III-614

Markus Helfenstein, Peter Bode, and Alexander Lampe, *Philips Semiconductors*

Track 11.7: Biomedical Devices, Sensors, and Related Systems

CELL CLINICS FOR BIOELECTRONIC INTERFACE WITH SINGLE CELLS III-618

Pamela Abshire, Jean-Marie Lauenstein, Yingkai Liu, and Elisabeth Smela, *University of Maryland*

ADVANCED BIOCHIP: PRINCIPLE AND APPLICATIONS IN MEDICAL DIAGNOSTICS AND PATHOGEN DETECTION III-622

Tuan Vo-Dinh, G.D. Griffin, A.L. Witenberg, D.L. Stokes, J. Mobley, M. Askari, and R. Maples, *Oak Ridge National Laboratory*

ANALYSIS OF A SIMPLE A/D CONVERTER WITH A TRAPPING WINDOW III-626

Toshimichi Saito and Hiroshi Imamura, *Hosei University*

AUTOMATIC LOCALIZATION OF CRANIOFACIAL LANDMARKS FOR ASSISTED CEPHALOMETRY III-630

Idris El-Feghi, Maher Sid-Ahmed, and Majid Ahmadi, *University of Windsor*

PORTABLE ARRAY BIOSENSORS III-634

Chris Taitt and Joel P. Golden, *Naval Research Laboratory*, Yura Shubin, *Geo-Centers, Inc.* and Lisa C. Shriver-Lake, *Naval Research Laboratory*, Kim E. Sapsford, *George Mason University*, James B. Delehanty and Frances S. Ligler, *Naval Research Laboratory*

MEMS MODULES FOR LIFE-ON-A-CHIP III-638

Deirdre Meldrum, Mark Holl, Pahnit Seriburi, Stephen Phillips, Joseph Chao, Ling-Sheng Jang, and Fettah Kosar, *University of Washington*

Track 11.8: Piecewise Linear Circuits and Systems

FINDING ALL SOLUTIONS OF PIECEWISE-LINEAR RESISTIVE CIRCUITS USING THE SIMPLEX METHOD III-642

Kiyotaka Yamamura and Takehisa Kitakawa, *Chuo University*

ON MODELING, ANALYSIS AND DESIGN OF PIECEWISE LINEAR CONTROL SYSTEMS III-646

Mikael Johansson, *Royal Institute of Technology*

EXPLOITING PIECEWISE LINEAR FEATURES: MULTINESTED AND SIMPLICIAL CELLULAR NEURAL/NONLINEAR NETWORKS III-650

Pedro Julian, *The Johns Hopkins University*, Radu Dogaru, *Polytechnic University of Bucharest*, and Leon O. Chua, *University of California, Berkeley*

PWL APPROXIMATION OF DYNAMICAL SYSTEMS: AN EXAMPLE III-654

Marco Storace, *University of Genova*, and Oscar de Feo, *Swiss Federal Institute of Technology, Lausanne*

DC ANALYSIS OF PWL ELECTRIC NETWORKS AND SUB-NETWORKS BY MEANS OF SET THEORY III-658

Stefano Pastore, *University of Trieste*, and Amedeo Premoli, *Politecnico di Milano*

Track 11.9: Multidimensional Circuits and Systems

SKEW-SYMMETRY IN THE EQUIVALENT REPRESENTATION PROBLEM OF A TIME-VARYING MULTIPORT INDUCTOR III-662

Nirmal Bose, *The Pennsylvania State University*, and Alfred Fettweis, *Ruhr-University Bochum*

PRIME FACTORIZATION OF N-D POLYNOMIAL MATRICES III-666

Hyungju Park, *Oakland University*

STABILITY AND STABILISATION OF 2D DISCRETE LINEAR SYSTEMS WITH MULTIPLE DELAYS III-670

Wojciech Paszke and Krzysztof Galkowski, *University of Zielona Gora* and James Lam, *University of Hong Kong*, Shengyuan Xu, *University of Hong Kong*, Eric Rogers, *University of Southampton*, and David Owens, *University of Sheffield*

AN LMI APPROACH TO THE DESIGN OF 2D FIR MULTIRATE FILTER BANKS III-674

Ran Yang, *The Univeristy of Melbourne*, Cishen Zhang and Lihua Xie, *Nanyang Technological University*

A NEW PARAMETERIZATION METHOD FOR ALL STABILIZING CONTROLLERS OF ND SYSTEMS WITHOUT COPRIME FACTORIZABILITY III-678

Kazuyoshi Mori, *The University of Aizu*

A STABILITY TEST FOR CONTINUOUS-DISCRETE BIVARIATE POLYNOMIALS III-682

Yuval Bistritz, *Tel Aviv University*

MULTDIMENSIONAL CONVOLUTIONAL CODE: PROGRESSES AND BOTTLENECKS III-686

Chalie Charoenlarpnopparut and Sawasd Tantaratana, *Sirindhorn International Institute of Technology*

DELTA OPERATOR BASED 2-D FILTERS: SYMMETRY, STABILITY, AND DESIGN III-690

I-Hung Khoo, *University of California, Irvine*, Hari Reddy, *California State University, Long Beach*, and P.K. Rajan, *Tennessee Tech University*

NEW METHOD FOR WEIGHTED LOW-RANK APPROXIMATION OF COMPLEX-VALUED MATRICES AND ITS APPLICATION FOR THE DESIGN OF 2-D DIGITAL FILTERS III-694

Wu-Sheng Lu and Andreas Antoniou, *University of Victoria*

A SPLIT-RADIX ALGORITHM FOR 2-D DFT III-698

Saad Bouguezel, M. Omair Ahmad, and M.N.S. Swamy, *Concordia University*

REDUCING THE COMPUTATIONAL COMPLEXITY OF NARROWBAND 2D FAN FILTERS USING SHAPED 2D WINDOW FUNCTIONS III-702

Leila Khademi and Leonard Bruton, *University of Calgary*

HIGH-SPEED FPGA-IMPLEMENTATION OF MULTIDIMENSIONAL BINARY MORPHOLOGICAL OPERATIONS III-706

Joerg Velten and Anton Kummert, *University of Wuppertal*

TRANSFORMING 1-D EVEN-LENGTH FILTERBANKS INTO 2-D FILTERBANKS III-710

David Tay, *Latrobe University*

THE FISHER INFORMATION MATRIX FOR TWO-DIMENSIONAL CONTINUOUS SEPARABLE-DENOMINATOR SYSTEMS III-714

Qiyue Zou and Zhiping Lin, *Nanyang Technological University*, and Raimund Ober, *University of Texas at Dallas*

Track 11.10: Speech and Language Processing

VLSI ARCHITECTURE FOR THE EFFICIENT COMPUTATION OF LINE SPECTRAL FREQUENCIES III-718

David Reynolds, Linda Head, and Ravi Ramachandran, *Rowan University*

JOINT TIME DELAY AND PITCH ESTIMATION FOR SPEAKER LOCALIZATION III-722

L.Y. Ngan, *The Chinese University of Hong Kong*, Y. Wu, *Xidian University*, H.C. So, *City University of Hong Kong*, P.C. Ching and S.W. Lee, *The Chinese University of Hong Kong*

FEATURE EXTRACTION BASED ON PERCEPTUALLY NON-UNIFORM SPECTRAL COMPRESSION FOR SPEECH RECOGNITION III-726

Kam-Keung Chu and Shu-Hung Leung, *City University of Hong Kong*

SHORT SEGMENT AUTOMATIC LANGUAGE IDENTIFICATION USING A MULTIFEATURE-TRANSITION MATRIX APPROACH III-730

John Grieco, *AFRL*, and Elis Pomales, *RADC*

USABLE SPEECH MEASURES AND THEIR FUSION III-734

Robert Yantorno, Brett Smolenski, and Nishant Chandra, *Temple University*

Track 11.11: Nonlinear Dynamics for Coding Theory and Network Traffic

CHARACTERIZATION OF A SIMPLE COMMUNICATION NETWORK USING LEGENDRE TRANSFORM III-738

Takashi Hisakado and Kohshi Okumura, *Kyoto University*, Vladimir Vukadinovic and Ljiljana Trajkovic , *Simon Fraser University*

APPLICATIONS OF NONLINEAR DYNAMICS TO THE TURBO DECODING ALGORITHM III-742

Ljupco Kocarev and Zarko Tasev, *University of California, San Diego*, and Gian Mario Maggio, *CWC/UCSD (STMicroelectronics)*

INTERNET PACKET TRAFFIC CONGESTION III-746

David Arrowsmith, Raul Mondragon-C, Jonathan Pitts, and Matthew Woolf, *Queen Mary, University of London*

SPREAD-SPECTRUM MARKOVIAN-CODE ACQUISITION IN ASYNCHRONOUS DS/CDMA SYSTEMS III-750

Tohru Kohda, Yutaka Jitsumatsu, and Tahir Khan, *Kyushu University*

A NONLINEAR COMPETITIVE MODEL OF TRAFFIC FLOWS AT CONGESTED LINKS IN NETWORKS III-754

Mario Di Bernardo, *University of Sannio*, Franco Garofalo and Sabato Manfredi, *University of Naples Federico II*

ACTIVE SUBSET SELECTION APPROACH TO NONLINEAR MODELING OF ECG DATA III-758

Christian Merkwirth, *Max-Planck-Institut Für Informatik*, Joerg Wichard and Maciej Ogorzalek, *University of Mining and Metallurgy*

CODED MODULATIONS BASED ON CONTROLLED 1-D AND 2-D PIECEWISE LINEAR CHAOTIC MAPS III-762

Thomas Schimming and Martin Hasler, *Swiss Federal Institute of Technology, Lausanne*

ON THE AVERAGE CODEWORD-CAPACITY OPTIMALITY OF CHAOS-BASED ASYNCHRONOUS DS-CDMA III-766

Gianluca Setti, *University of Ferrara*, Riccardo Rovatti, *University of Bologna*, and Gianluca Mazzini, *University of Ferrara*

Track 11.12: Sensor Arrays for Visual Tracking and Navigation

NEUROMORPHIC SELECTIVE ATTENTION SYSTEMS III-770

Giacomo Indiveri, *Univeristy of Zurich - ETH Zurich*

BIO-INSPIRED FLIGHT CONTROL AND VISUAL SEARCH WITH CNN TECHNOLOGY III-774

Csaba Rekeczky, David Balya, Gergely Timar, and Istvan Szatmari, *MTA -SZTAKI*

AN ADAPTIVE CENTER OF MASS DETECTION SYSTEM EMPLOYING A 2-D DYNAMIC ELEMENT MATCHING ALGORITHM FOR OBJECT TRACKING III-778

Alexander Fish, Dmitry Akselrod, and Orly Yadid-Pecht, *Ben-Gurion University*

A FLEXIBLE GLOBAL READOUT ARCHITECTURE FOR AN ANALOGUE SIMD VISION CHIP III-782

Piotr Dudek, *University of Manchester Institute of Science and Technology*

VISION CHIP FOR NAVIGATING AND CONTROLLING MICRO UNMANNED AERIAL VEHICLES III-786

Mark Massie, Chris Baxter, and J.P. Curzan, *Novabiomimetics, Inc.*, Paul McCarley, *Eglin Air Force Base*, and Ralph Etienne-Cummings, *Johns Hopkins University*

Track 11.13: Security and Data Hiding

A NOVEL SELF-CONJUGATE HALFTONE IMAGE WATERMARKING TECHNIQUE III-790

Ming Sun Fu and Oscar Au, *Hong Kong University of Science and Technology*

AN IMAGE FUSION BASED VISIBLE WATERMARKING ALGORITHM III-794

Yongjian Hu, *South China University of Technology*, and Sam Kwong, *City University of Hong Kong*

ANALYSIS OF THE ROLE PLAYED BY ERROR CORRECTING CODING IN ROBUST WATERMARKING III-798

Limin Gu and Jiwu Huang, *Zhongshan University*, and Yun Q. Shi, *New Jersey Institute of Technology*

INFORMATION EMBEDDING IN JPEG-2000 COMPRESSED IMAGES III-802

Po-Chyi Su and C.-C. Jay Kuo, *University of Southern California*

SECURE DATA HIDING IN BINARY DOCUMENT IMAGES FOR AUTHENTICATION III-806

Haiping Lu, Alex C. Kot, and Jun Cheng, *Nanyang Technological University*

MOTION TRAJECTORY BASED VIDEO AUTHENTICATION III-810

Weiqi Yan and Mohan Kankanhalli, *National University of Singapore*

A SEMI-FRAGILE OBJECT BASED VIDEO AUTHENTICATION SYSTEM III-814

Dajun He, Qibin Sun, and Qi Tian, *Labs for Information Technology*

ENHANCEMENT OF BLIND WATERMARK RETRIEVAL IN DRIFT-COMPENSATED MPEG VIDEO III-818

Yong Guan, Sugiri Pranata, Viktor Wahadaniah, Hock Chuan Chua, and Habib Mir Hosseini, *Nanyang Technological University*

A DNA-BASED, BIOMOLECULAR CRYPTOGRAPHY DESIGN III-822

Jie Chen, *Brown University*

ROBUST DIGITAL IMAGE-IN-IMAGE WATERMARKING ALGORITHM USING THE FAST HADAMARD TRANSFORM III-826

Anthony Ho and Jun Shen, *Nanyang Technological University*, Andrew Chow and Jerry Woon, *Datamark Technologies*

ACTIVE STEGANALYSIS OF SPREAD SPECTRUM IMAGE STEGANOGRAPHY III-830

Rajarathnam Chandramouli, and Koduvayur Subbalakshmi, *Stevens Inst. of Technology*

Track 11.14: Bionics

A MICROPOWER COCHLEAR PROSTHESIS SYSTEM III-834

Julius Georgiou and Christopher Toumazou, *Imperial College*

DETECTING GLOBAL SPATIAL-TEMPORAL EVENTS: SACCADIC SUPPRESSION III-838

David Balya, *MTA-Sztaki*

IMPLEMENTATION OF TURING PATTERNS FOR BIO-INSPIRED MOTION CONTROL III-842

Paolo Arena, Adriano Basile, Luigi Fortuna, Mattia Frasca, and Luca Patane', *Università degli Studi di Catania*

BIO-INSPIRED OPTICAL FLOW CIRCUITS FOR THE VISUAL GUIDANCE OF MICRO-AIR VEHICLES III-846

Franck Ruffier, StéPhane Viollet, StéPhane Amic, and Nicolas Franceschini, *CNRS/Univ. de la Méditerranée*

FEATURE EXTRACTION IN EPILEPSY USING A CELLULAR NEURAL NETWORK BASED DEVICE - FIRST RESULTS III-850

Christian Niederhoefer, Philipp Fischer, and Ronald Tetzlaff, *University of Frankfurt*

Track 11.15: Fault Detection and Tolerance in Distributed Networks

OPTIMAL ADAPTIVE PARALLEL DIAGNOSIS FOR ARRAYS III-854

Toshinori Yamada, Kumiko Nomura, and Shuichi Ueno, *Tokyo Institute of Technology*

ENCODED FINITE-STATE MACHINES FOR NON-CONCURRENT ERROR DETECTION AND IDENTIFICATION III-858

Christoforos Hadjicostis, *University of Illinois*

THE MULTI-LEVEL PARADIGM FOR DISTRIBUTED FAULT DETECTION IN NETWORKS WITH UNRELIABLE PROCESSORS III-862

Krishnaiyan Thulasiraman, *University of Oklahoma*, Ming-Shan Su, *Southeastern Oklahoma State University*, and Vakul Goel, *University of Oklahoma*

MULTIPLE FAILURE SURVIVABILITY IN WDM NETWORKS WITH P-CYCLES III-866

Dominic Schupke, *Munich University of Technology*

WORMHOLE ROUTING IN DE BRUIJN NETWORKS AND HYPER-DEBRUIJN NETWORKS III-870

Elango Ganesan, *Foreminds Corp*, and Dhiraj Pradhan, *University of Bristol*

Track 11.16: Frequency Response Masking Techniques

AN ITERATIVE METHOD FOR OPTIMIZING FIR FILTERS SYNTHESIZED USING THE TWO-STAGE FREQUENCY-RESPONSE MASKING TECHNIQUE III-874

Ya Jun Yu, *National University of Singapore*, Tapio Saramäki, *Tampere University of Technology*, and Yong Ching Lim, *National University of Singapore*

OPTIMAL DESIGN OF FIR FREQUENCY-RESPONSE-MASKING FILTERS USING SECOND-ORDER CONE PROGRAMMING III-878

Wu-Sheng Lu, *University of Victoria*, and Takao Hinamoto, *Hiroshima University*

COSINE AND SINE MODULATED FIR FILTER BANKS UTILIZING THE FREQUENCY-RESPONSE MASKING APPROACH III-882

Linnéa Rosenbaum, Per Löwenborg, and HåKan Johansson, *Linkoping University*

AN ALTERNATING VARIABLE APPROACH TO FIR FILTER DESIGN WITH POWER-OF-TWO COEFFICIENTS USING THE FREQUENCY-RESPONSE MASKING TECHNIQUE III-886

Wei Rong Lee and Volker Rehbock, *Curtin University of Technology*, K. L. Teo, *The Hong Kong Polytechnic University*, and Lou Caccetta, *Urtin University of Technology*

OPTIMIZATION TECHNIQUES FOR COSINE-MODULATED FILTER BANKS BASED ON THE FREQUENCY-RESPONSE MASKING APPROACH III-890

Miguel Furtado, Paulo Diniz, and Sergio Netto, *Federal University of Rio De Janeiro*

Track 11.17: Behavioral Modeling and Simulation of Mixed-Signal Systems

THE CONTINUOUS-DISCRETE INTERFACE - WHAT DOES THIS REALLY MEAN? MODELLING AND SIMULATION ISSUES III-894

Andrew Brown, *University of Southampton*, and Mark Zwolinski, *Electronic Systems Design Group*

A SYSTEMC EXTENSION FOR BEHAVIORAL LEVEL QUANTIFICATION OF NOISE COUPLING IN MIXED-SIGNAL SYSTEMS III-898

Mattias O'Nils, Jan Lundgren, and Bengt Oelmann, *Mid Sweden University*

MODELING AND VERIFICATION OF A PROGRAMMABLE MIXED-SIGNAL DEVICE USING VERILOG III-902

Monte Mar, *Orora Design Technologies*, and Bert Sullam, *Cypress Microsystems*

BEHAVIORAL MODELING AND SIMULATION OF HIGH-SPEED ANALOG-TO-DIGITAL CONVERTERS USING SYSTEMC III-906

Johnny Bjoernsen and Trond Ytterdal, *Norwegian University of Science and Technology*

A SURVEY OF BOTTOM-UP BEHAVIORAL MODELING METHODS FOR ANALOG CIRCUITS III-910

Alan Mantooth, Liming Ren, Xiaoling Huang, Yongfeng Feng, and Wei Zheng, *University of Arkansas*

ANALOG AND MIXED SIGNAL MODELING WITH SYSTEMC-AMS III-914

Karsten Einwich, *Fraunhofer IIS/EAS*, Alain Vachoux, *EPFL Lausanne*, and Christoph Grimm, *University Frankfurt*

Track 11.18: MEMS and Applications

GATELESS DEPLETION MODE FIELD EFFECT TRANSISTOR FOR MACROMOLECULE SENSING III-918

Angela Hodge Miller, F. Keith Perkins, Martin Peckerar, and Stephanie Fertig, *Nova Research*, and Leonard Tender, *Naval Research Laboratory*

DESIGN, FABRICATION, AND MODELING OF MICROBEAM STRUCTURES FOR GAS SENSOR APPLICATIONS IN CMOS TECHNOLOGY III-922

Ioana Voiculescu and Mona Zaghloul, *George Washington University*, and R. Andrew Mcgill, *Naval Research Laboratory*

A MODULAR SENSOR MICROSYSTEM UTILIZING A UNIVERSAL INTERFACE CIRCUIT III-926

Andrew Mason, *Michigan State University,* Navid Yazdi, *Corning Intellisense,* J. Zhang, *Michigan State University*, and Z. Sainudeen, *Corning Intellisense*

TOWARD A WIRELESS OPTICAL COMMUNICATION LINK BETWEEN TWO SMALL UNMANNED AERIAL VEHICLES III-930

Matthew Last, *UC Berkeley*, Brian Leibowitz, Baris Cagdaser, Anand Jog, Lixia Zhou, Bernhard Boser, and Kristofer Pister, *Berkeley Sensor and Actuator Center*

APPLICATION OF MEMS TECHNOLOGIES TO NANODEVICES III-934

Lance Doherty, *UC Berkeley*, Hongbing Liu, *Adriatic Research Institute*, and Veljko Milanovic, *Adriatic Research Institute*

A MEMS CUSTOM MICROPACKAGING SOLUTION III-938

Sazzadur Chowdhury, Majid Ahmadi, and William Miller, *University of Windsor*

VOLUME IV OF V:
-DIGITAL SIGNAL PROCESSING-
-COMPUTER AIDED NETWORK DESIGN-
-ADVANCED TECHNOLOGY-

Track 4: DIGITAL SIGNAL PROCESSING

CLASSIFICATION OF BPSK AND QPSK SIGNALS WITH UNKNOWN SIGNAL LEVEL USING BAYES TECHNIQUE IV-1

Liang Hong and Dominic Ho, *University of Missouri, Columbia*

JOINTLY MINIMUM SYMBOL ERROR RATE FIR MIMO TRANSMITTER AND RECEIVER FILTERS FOR PAM SIGNAL VECTORS IV-5

Are Hjørungnes, *Helsinki University of Technology*, and Paulo S. R. Diniz, *Universidade Federal do Rio de Janeiro*

ANALOG REPRESENTATION AND DIGITAL IMPLEMENTATION OF OFDM SYSTEMS IV-9

Yuan-Pei Lin, *National Chiao Tung University*, and See-May Phoong, *National Taiwan University*

JOINTLY MINIMUM MSE TRANSMITTER AND RECEIVER FIR MIMO FILTERS IN THE PRESENCE OF NEAR-END CROSSTALK AND ADDITIVE NOISE IV-13

Are Hjørungnes, *Helsinki University of Technology*, Paulo S. R. Diniz and Marcello L. R. Campos, *Universidade Federal do Rio de Janeiro*

LOCALIZATION OF A MOVING SOURCE USING TDOA AND FDOA MEASUREMENTS IV-17

K.C. Ho and Wenwei Xu, *University of Missouri, Columbia*

IMPLEMENTING OTSU'S THRESHOLDING PROCESS USING AREA-TIME EFFICIENT LOGARITHMIC APPROXIMATION UNIT IV-21

Hui Tian, Siew Kei Lam, and Thambipillai Srikanthan, *Nanyang Technological University*

STEADY-STATE PROPERTIES OF THE SIGN ALGORITHM FOR THE CONSTRAINED ADAPTIVE IIR NOTCH FILTER IV-25

Yegui Xiao and Rabab Ward, *University of British Columbia*, and Akira Ikuta, *Hiroshima Prefectural Women's University*

ROBUST BEAMFORMER DESIGN BY POWER MINIMIZATION AND ITS UNCONSTRAINED PARTITIONED IMPLEMENTATION IV-29

Zhu Liang Yu and Meng Hwa Er, *Nanyang Technological University*

A MEMORY EFFICIENT REALIZATION OF CYCLIC CONVOLUTION AND ITS APPLICATION TO DISCRETE COSINE TRANSFORM IV-33

Hun-Chen Chen, Jiun-In Guo, and Chein-Wei Jen, *National Chiao Tung University*

BIT RATE OPTIMIZED TIME-DOMAIN EQUALIZERS FOR DMT SYSTEMS IV-37

Chun-Yang Chen and See-May Phoong, *National Taiwan University*

FAST ALGORITHMS FOR COMPUTING FULL AND REDUCED RANK WIENER FILTERS IV-41

Mohammed Hasan, *University of Minnesota, Duluth*, and Mahmood Azimi-Sadjadi, *Colorado State University*

A 2048 COMPLEX POINT FFT PROCESSOR USING A NOVEL DATA SCALING APPROACH IV-45

Thomas Lenart, *Lund Institute of Technology*, and Viktor Öwall, *Lund University*

ON THE RELEVANCY OF THE PERFECT RECONSTRUCTION PROPERTY WHEN MINIMIZING THE MEAN SQUARE ERROR IN FIR MIMO FILTER SYSTEMS IV-49

Are Hjørungnes, *Helsinki University of Technology*, and Paulo S. R. Diniz, *Universidade Federal do Rio de Janeiro*

USE OF NOISE SHAPING FOR RATE CONVERSION IN OVERSAMPLED VIDEO SIGNALS IV-53

Ivan Ryan and Oliver Mccarthy, *University of Limerick*

GENERATION OF EMBEDDING WATERMARK SIGNALS FROM REFERENCE WATERMARK IV-57

Tae Young Kim and Taejeong Kim, *Seoul National University*, and Kiryung Lee, *Electronics and Telecommunications Research Institute*

EFFICIENT SIGNAL PROCESSING IN EMBEDDED JAVA SYSTEMS IV-61

Rafael Krapf, *UFRGS - Federal University of Rio Grande Do Sul*, and Luigi Carro, *Universidade Federal do Rio Grande do Sul*

AN EFFICIENT SPLIT-RADIX FFT ALGORITHM IV-65

Saad Bouguezel, M. Omair Ahmad, and M.N.S. Swamy, *Concordia University*

PERFORMANCE OF AN ADAPTIVE HOMODYNE RECEIVER IN THE PRESENCE OF MULTIPATH, RAYLEIGH-FADING AND TIME-VARYING QUADRATURE ERRORS IV-69

Ediz Cetin, Izzet Kale, and Richard Morling, *University of Westminster*

ON-LINE SIGNATURE VERIFICATION METHOD UTILIZING FEATURE EXTRACTION BASED ON DWT IV-73

Isao Nakanishi, Naoto Nishiguchi, Yoshio Itoh, and Yutaka Fukui, *Tottori University*

DESIGN OF A DIGITAL REACTION-DIFFUSION SYSTEM FOR RESTORING BLURRED FINGERPRINT IMAGES IV-77

Koichi Ito Takafumi Aoki, and Tatsuo Higuchi, *Tohoku University*

A DISCRETE FRACTIONAL FOURIER TRANSFORM BASED ON ORTHONORMALIZED MCCLELLAN-PARKS EIGENVECTORS IV-81

Magdy Hanna, *Cairo University*

FAST INTEGER FOURIER TRANSFORM (FIFT) BASED ON LIFTING MATRICES IV-85

Ratchaneekorn Thamvichai, *St. Cloud State University*, Tamal Bose, *Utah State University*, and Miloje Radenkovic, *University of Colorado, Denver*

SAVING THE BANDWIDTH IN THE FRACTIONAL DOMAIN BY GENERALIZED HILBERT TRANSFORM PAIR RELATIONS IV-89

Soo-Chang Pei and Jian-Jiun Ding, *National Taiwan University*

ANGULAR DECOMPOSITION FOR THE DISCRETE FRACTIONAL SIGNAL TRANSFORMS IV-93

Min-Hung Yeh, *National I-LAN Institute of Technology*

EFFICIENT PRUNING ALGORITHMS FOR THE DFT COMPUTATION FOR A SUBSET OF OUTPUT SAMPLES IV-97

Saad Bouguezel, M.Omair Ahmad, and M.N.S. Swamy, *Concordia University*

A ROBUST WIDEBAND ARRAY BEAMFORMER USING FAN FILTER IV-101

Zhu Liang Yu, Qiyue Zou, and Meng Hwa Er, *Nanyang Technological University*

A REGULARIZED SIMULTANEOUS AUTOREGRESSIVE MODEL FOR TEXTURE CLASSIFICATION IV-105

Yaowei Wang, Yanfei Wang, and Wen Gao, *Chinese Academy of Sciences*, and Yong Xue, *Institute of Remote Sensing Applications*

EFFICIENT FREQUENCY ESTIMATION USING THE PULSE-PAIR METHOD AT DIFFERENT LAGS IV-109

Saman Abeysekera, *Nanyang Technological University*

PERFORMANCE ANALYSIS OF ALGORITHMIC NOISE-TOLERANCE TECHNIQUES IV-113

Byonghyo Shim and Naresh Shanbhag, *University of Illinois at Urbana-Champaign*

A MINIMUM VARIANCE FILTER FOR DISCRETE-TIME LINEAR SYSTEMS PERTURBED BY UNKNOWN NONLINEARITIES IV-117

Alfredo Germani, *University of L'aquila*, Costanzo Manes, *Universita degli Studi dell 'Aquila*, and Pasquale Palumbo, *Istituto di Analisi dei Sistemi Ed Informatica del CNR*

ON-LINE HIGH-RADIX EXPONENTIAL WITH SELECTION BY ROUNDING IV-121

Jose-Alejandro Pineiro and Javier Bruguera, *Univ. Santiago de Compostela*, and Milos Ercegovac, *University of California, Los Angeles*

FAST IMPLEMENTATONS OF MONTGOMERY'S MODULAR MULTIPLICATION ALGORITHM IV-125

Ananda Mohan Pemmaraju Venkata, *I.T.I. Limited*

GDFT TYPES MAPPING ALGORITHMS AND STRUCTURED REGULAR FPGA IMPLEMENTATION IV-129

Hassan Saleh and Mahmoud Ashour, *NCRRT, Aea, Egypt*, and Aly Salama, *Cairo University*

A HIGH-SPEED, LOW LATENCY RSA DECRYPTION SILICON CORE IV-133

Ciaran Mcivor, Máire Mcloone, and John Mccanny, *Queen's University Belfast*

EUTDSP: A DESIGN STUDY OF A NEW VLIW-BASED DSP ARCHITECTURE IV-137

G. Chaji, Reza Pourrad, and Mehdi Fakhraie, *University of Tehran*, and Mohammad Tehranipour, *University of Texs at Dallas*

Track 4.1: Digital Filters

EFFICIENT DESIGN OF PR BIORTHOGONAL COSINE-MODULATED FILTER BANKS USING CONVEX LAGRANGIAN RELAXATION AND ALTERNATING NULL-SPACE PROJECTIONS IV-141

Wu-Sheng Lu, *University of Victoria*, Robert Bregovic and Tapio Saramäki, *Tampere University of Technology*

THREE CLASSES OF IIR COMPLEMENTARY FILTER PAIRS WITH AN ADJUSTABLE CROSSOVER FREQUENCY IV-145

Ljiljana Milic, *M. Pupin Institute, University of Belgrade*, and Tapio Saramäki, *Tampere University of Tecnology*

A NEW CLASS OF EVEN LENGTH WAVELET FILTERS IV-149

David Tay, *Latrobe University*

SDP FOR MULTI-CRITERION QMF BANK DESIGN IV-153

Hoang Tuan and Le Hai Nam, *Toyota Technological Institute*, Hoang Tuy, *Institute of Mathematics, Hanoi*, and Truong Q. Nguyen, *University of California, San Diego*

EFFICIENT IMPLEMENTATION OF COMPLEX EXPONENTIALLY-MODULATED FILTER BANKS IV-157

Juuso Alhava, Markku Renfors, and Ari Viholainen, *Tampere University of Technology*

FUZZY FILTERS FOR NOISY IMAGE FILTERING IV-161

H. K. Kwan, *University of Windsor*

HIGH-SPEED TUNABLE FRACTIONAL-DELAY ALLPASS FILTER STRUCTURE IV-165

Ji-Suk Park, Byeong-Kuk Kim, and Jin-Gyun Chung, *Chonbuk National University*, and Keshab Parhi, *University of Minnesota*

DESIGN OF ULTRASPHERICAL WINDOWS WITH PRESCRIBED SPECTRAL CHARACTERISTICS IV-169

Stuart Bergen and Andreas Antoniou, *University of Victoria*

ANALYTICAL DESIGN OF FRACTIONAL HILBERT TRANSFORMER USING FRACTIONAL DIFFERENCING IV-173

Chien-Cheng Tseng, *National Kaohsiung First University of Science and Technology*

EFFICIENT DESIGN OF SVD-BASED 2-D DIGITAL FILTERS USING SPECIFICATION SYMMETRY AND ORDER-SELECTING CRITERION IV-177

Tian-Bo Deng, *Toho University*

REDUCE THE COMPLEXITY OF FREQUENCY-RESPONSE MASKING FILTER USING MULTIPLICATION FREE FILTER IV-181

Chun Zhu Yang and Yong Lian, *National University of Singapore*

NEW DESIGNS OF FREQUENCY-SELECTIVE FIR DIGITAL FILTERS IV-185

Ishtiaq Khan and Masahiro Okuda, *The University of Kitakyushu*, and Ryoji Ohba, *Hokkaido University, Japan*

DESIGN OF 2-D VARIABLE FRACTIONAL DELAY FIR FILTER USING 2-D DIFFERENTIATORS IV-189

Chien-Cheng Tseng, *National Kaohsiung First University of Science and Technology*

SOME OBSERVATIONS ON MULTIPLIERLESS IMPLEMENTATION OF LINEAR PHASE FIR FILTERS IV-193

Mrinmoy Bhattacharya and Tapio Saramäki, *Tampere University of Technology*

A SYSTEMATIC TECHNIQUE FOR OPTIMIZING ONE-STAGE TWO-FILTER LINEAR-PHASE FIR FILTERS FOR SAMPLING RATE CONVERSION IV-197

Peyman Arian and Tapio Saramäki, *Tampere University of Technology*

CONCISE REPRESENTATION AND CELLULAR STRUCTURE FOR UNIVERSAL MAXIMALLY FLAT FIR FILTERS IV-201

Saed Samadi, *Concordia University*, and Akinori Nishihara, *Tokyo Institute of Technology*

DESIGN AND IMPLEMENTATION OF A RECONFIGURABLE FIR FILTER IV-205

Kuan-Hung Chen and Tzi-Dar Chiueh, *National Taiwan University*

DESIGN OF IIR DIGITAL ALLPASS FILTERS USING LEAST PTH PHASE ERROR CRITERION IV-209

Chien-Cheng Tseng, *National Kaohsiung First University of Science and Technology*

DESIGN OF IIR DIGITAL FILTERS IN THE COMPLEX DOMAIN BY TRANSFORMING THE DESIRED RESPONSE IV-213

Tatsuya Matsunaga, Masahiro Yoshida, and Masaaki Ikehara, *Keio University*

A GENERALIZED DIRECT-FORM II TRANSPOSED STRUCTURE FOR IIR FILTER IMPLEMENTATION WITH MINIMAL ROUNDOFF NOISE GAIN IV-217

Gang Li, Zixue Zhao, and Jinxin Hao, *Nanyang Technological University*

M-CHANNEL LIFTING-BASED DESIGN OF PARAUNITARY AND BIORTHOGONAL FILTER BANKS WITH STRUCTURAL REGULARITY IV-221

Ying-Jui Chen, *Massachusetts Institute of Technology*, Soontorn Oraintara, *University of Texas at Arlington*, and Kevin Amaratunga, *Massachusetts Institute of Technology*

FILTER STRUCTURES FOR DECIMATION: A COMPARISON IV-225

Artur Wroblewski and Josef A. Nossek, *Munich University of Technology*

AN OPTIMAL ENTROPY CODING SCHEME FOR EFFICIENT IMPLEMENTATION OF PULSE SHAPING FIR FILTERS IN DIGITAL RECEIVERS IV-229

Vinod Prasad, Annamalai Benjamin Premkumar, and Edmund Lai Ming-Kit, *Nanyang Technological University*

EXPONENTIALLY-MODULATED FILTER BANK-BASED TRANSMULTIPLEXER IV-233

Juuso Alhava and Markku Renfors, *Tampere University of Technology*

ALLPASS STRUCTURES FOR MULTIPLIERLESS REALIZATION OF RECURSIVE DIGITAL FILTERS IV-237

Mrinmoy Bhattacharya and Tapio Saramäki, *Tampere University of Technology*

JOINT OPTIMIZATION OF ERROR FEEDBACK AND COORDINATE TRANSFORMATION FOR ROUNDOFF NOISE MINIMIZATION IN STATE-SPACE DIGITAL FILTERS IV-241

Takao Hinamoto and Hiroaki Ohnishi, *Hiroshima University*, and Wu-Sheng Lu, *University of Victoria*

IIR EQUALIZER DESIGN BASED ON THE IMPULSE RESPONSE SYMMETRY CRITERION IV-245

Mladen Vucic and Hrvoje Babic, *Faculty of Electrical Engineering and Computing*

MULTIPLIERLESS IMPLEMENTATION OF BANDPASS AND BANDSTOP RECURSIVE DIGITAL FILTERS USING ALLPASS STRUCTURES IV-249

Mrinmoy Bhattacharya and Tapio Saramäki, *Tampere University of Technology*

THE DESIGN OF TWO-CHANNEL PERFECT RECONSTRUCTION FIR TRIPLET WAVELET FILTER BANKS USING SEMIDEFINITE PROGRAMMING IV-253

K. S. Yeung and Shing Chow Chan, *The University of Hong Kong*

MULTIPLIER-LESS REAL-VALUED FFT-LIKE TRANSFORMATION (ML-RFFT) AND RELATED REAL-VALUED TRANSFORMATIONS IV-257

S. C. Chan and K. M. Tsui, *The University of Hong Kong*

ON UNBIASED PARAMETER ESTIMATION OF AUTOREGRESSIVE SIGNALS OBSERVED IN NOISE IV-261

Wei Xing Zheng, *University of Western Sydney*

PAIRING AND ORDERING TO REDUCE HARDWARE COMPLEXITY IN CASCADE FORM FILTER DESIGN IV-265

Hyeong-Ju Kang and In-Cheol Park, *KAIST*

DIRECT RECURSIVE STRUCTURES FOR COMPUTING RADIX-R TWO-DIMENSIONAL DCT IV-269

Che-Hong Chen, Bin-Da Liu, and Jar-Ferr Yang, *National Cheng Kung University*

A SIMPLIFIED LATTICE FACTORIZATION FOR LINEAR-PHASE PARAUNITARY FILTER BANKS WITH PAIRWISE MIRROR IMAGE FREQUENCY RESPONSES IV-273

Lu Gan and Kai-Kuang Ma, *Nanyang Technological University*

A DESIGN FLOW FOR LINEAR-PHASE FIXED-POINT FIR FILTERS: FROM THE NPRM SPECIFICATIONS TO A VHDL CODE IV-277

Chia-Yu Yao, Chin-Chih Yeh, Tsuan-Fan Lin, Hsin-Horng Chen, and Chiang-Ju Chien, *Huafan University*

IMPROVING THE FILTER BANK OF A CLASSIC SPEECH FEATURE EXTRACTION ALGORITHM IV-281

Mark Skowronski and John Harris, *University of Florida*

A NEW HEURSTIC SIGNED-POWER OF TWO TERM ALLOCATION APPROACH FOR DESIGNING OF FIR FILTERS IV-285

Tetsuya Fujie, *Kobe University of Commerce*, Rika Ito, *Tokyo University of Science*, Kenji Suyama, *Toyko DENKI University*, and Ryuichi Hirabayashi, *Tokyo University of Science*

RECONFIGURABLE IMPLEMENTATION OF RECURSIVE DCT KERNELS FOR REDUCED QUANTIZATION NOISE IV-289

Suleyman Demirsoy, Robert Beck, Andrew Demspter, and Izzet Kale, *University of Westminster*

DESIGN GUIDELINES FOR RECONFIGURABLE MULTIPLIER BLOCKS IV-293

Suleyman Demirsoy, Andrew Dempster, and Izzet Kale, *University of Westminster*

AN EFFICIENT INTERLEAVED TREE-STRUCTURED PERFECT RCONSTRUCTION FILTERBANK IV-297

Nan Li and Behrouz Nowrouzian, *University of Alberta*

THE INTEGER MDCT AND ITS APPLICATION IN THE MPEG LAYER III AUDIO IV-301

Soontorn Oraintara and Tharakram Krishnan, *University of Texas at Arlington*

MINIMAL ARMA LATTICE DIGITAL FILTER REALIZATION IV-305

H. K. Kwan, *University of Windsor*

FAST CHARACTERIZATION OF THE NOISE BOUNDS DERIVED FROM COEFFICIENT AND SIGNAL QUANTIZATION IV-309

Juan Lopez, Carlos Carreras, Gabriel Caffarena, and Octavio Nieto-Taladriz, *Universidad Politecnica Madrid*

Track 4.2: Wavelets and Multirate Signal Processing

N STAGE NON-SEPARABLE TWO DIMENSIONAL WAVELET TRANSFORM FOR REDUCTION OF ROUNDING ERRORS IV-313

Masahiro Iwahashi, Munkhbaatar Delgermaa, Koji Ueno, and Noriyoshi Kambayashi, *Nagaoka University of Technology*

PROLONGED TRANSPOSED POLYNOMIAL-BASED FILTERS FOR DECIMATION IV-317

Djordje Babic, Tapio Saramäki, and Markku Renfors, *Tampere University of Technology*

DISCRETE-TIME MODELING OF POLYNOMIAL-BASED INTERPOLATION FILTERS IN RATIONAL SAMPLING RATE CONVERSION IV-321

Djordje Babic, Vesa Lehtinen, and Markku Renfors, *Tampere University of Technology*

ANALYSIS OF RATIONAL SAMPLE RATE CONVERSION USING IMAGE RESPONSE COMBINING IV-325

Vesa Lehtinen and Markku Renfors, *Tampere University of Technology*

OPTIMALLY WEIGHTED LOCAL DISCRIMINANT BASES IV-329

Kamyar Hazaveh and Kaamran Raahemifar, *Ryerson University*

REDUCED ORDER RLS POLYNOMIAL PREDISTORTION IV-333

Minglu JIN, Sooyoung Kim Shin, Deockgil. Oh, and Jaemoung Kim, *Electronics and Telecommunications Research Institute*

WAVELET-TRANSFORM-BASED STRATEGY FOR GENERATING NEW CHINESE FONTS IV-337

Jiu-Chao Feng, *Southwest China Normal University,* Chi K. Tse, *Hong Kong Polytechnic University*, and Yuhui Qiu, *Southwest China Normal University*

LOCAL DISCRIMINANT BASIS ALGORITHM – A REVIEW OF THEORY AND APPLICATION IN SIGNAL PROCESSING IV-341

Kamyar Hazaveh and Kaamran Raahemifar, *Ryerson University*

Track 4.3: Adaptive Signal Processing

NEXT CANCELLATION IN XDSL SYSTEMS USING VARIABLE-LENGTH CANCELLERS IV-345

Rajeev Nongpiur, Dale Shpak, and Andreas Antoniou, *University of Victoria*

NORM AND COEFFICIENT CONSTRAINTS FOR ROBUST ADAPTIVE BEAMFORMING IV-349

Qiyue Zou, Zhu Liang Yu, and Zhiping Lin, *Nanyang Technological University*

INTEGRATED ACTIVE NOISE CONTROL COMMUNICATION HEADSETS IV-353

Woon-Seng Gan, *Nanyang Technological University*, and Sen M Kuo, *Northern Illinois University*

MINIMUM SELECTION GSC AND ADAPTIVE LOW-POWER RAKE COMBINING SCHEME IV-357

Suk Kim, *Samsung Electronics Co., Ltd.*, Dong Ha, and Jeffrey Reed, *Virginia Tech*

ADAPTIVE IIR NOTCH FILTER WITH CONTROLLED BANDWIDTH FOR NARROW-BAND INTERFERENCE SUPPRESSION IN DS CDMA SYSTEM IV-361

Aloys Mvuma, *Hiroshima University*, Shotaro Nishimura, *Shimane University*, and Takao Hinamoto, *Hiroshima University*

A FINITE PRECISION LMS ALGORITHM FOR INCREASED QUANTIZATION ROBUSTNESS IV-365

Fredric Lindstrom, *Konftel Technology AB*, Mattias Dahl and Ingvar Claesson, *Blekinge Institute of Technology*

A NEW LMS-BASED FOURIER ANALYZER IN THE PRESENCE OF FREQUENCY MISMATCH IV-369

Yegui Xiao and Rabab Ward, *University of British Columbia*, and Li Xu, *Akita Prefectural University*

TRANSFORM DOMAIN APPROXIMATE QR-LS ADAPTIVE FILTERING ALGORITHM IV-373

Xinxing Yang and S. C. Chan, *The University of Hong Kong*

PERFORMANCE ANALYSIS OF ADAPTIVE IIR NOTCH FILTERS BASED ON LEAST MEAN P-POWER ERROR CRITERION IV-377

Maha Shadaydeh and Masayuki Kawamata, *Tohoku University*

RAMP: AN ADAPTIVE FILTER WITH LINKS TO MATCHING PURSUITS AND ITERATIVE LINEAR EQUATION SOLVERS IV-381

John Håkon Husøy, *Stavanger University College*

EXTENDED RLS LATTICE ADAPTIVE FILTERS IV-385

Ricardo Merched, *Federal University of Rio De Janeiro*

ON USE OF AVERAGING IN FXLMS ALGORITHM FOR SINGLE-CHANNEL FEEDFORWARD ANC SYSTEMS IV-389

Muhammad Akhtar, Masahide Abe, and Masayuki Kawamata, *Tohoku University*

HARDWARE IMPLEMENTATION OF EVOLUTIONARY DIGITAL FILTERS IV-393

Masahide Abe and Masayuki Kawamata, *Tohoku University*

TURBO CODED MULTIPLE SYMBOL DIFFERENTIAL DETECTION FOR CORRELATED RAYLEIGH FADING CHANNEL IV-397

Pisit Vanichchanunt, Chantima Sritiapetch, Suvit Nakpeerayuth, and Lunchakorn Wuttisittikulkij, *Chulalongkorn University*

A STUDY ON THE STEP SIZE OF CASCADED ADAPTIVE NOTCH FILTER UTILIZING ALLPASS FILTER IV-401

Yasutomo Kinugasa, *Matsue National College of Technology*, Yoshio Itoh, *Tottori University*, Masaki Kobayashi, *Chubu University*, Yutaka Fukui, *Tottori University*, and James Okello, *Kyushu Institute of Technology*

POLYPHASE IIR FILTER BANKS FOR SUBBAND ADAPTIVE ECHO CANCELLATION APPLICATIONS IV-405

Artur Krukowski and Izzet Kale, *University of Westminster*

A NEW ALGORITHM FOR HOWLING DETECTION IV-409

Jianqiang Wei and Limin Du, *Chinese Academy of Sciences*, Zhe Chen and Fuliang Yin, *Dalian University of Technology*

INTEGRATED NEAR-END ACOUSTIC ECHO AND NOISE REDUCTION SYSTEMS IV-412

Sen Kuo, *Northern Illinois University*, Dianwei Sun, *Analog Devices*, and Woon Gan, *Nanyang Technological University*

SIMPLIFIED STRUCTURES FOR TWO-DIMENSIONAL ADAPTIVE NOTCH FILTERS IV-416

Soo-Chang Pei and Chang-Long Wu, and Jian-Jiun Ding, *National Taiwan University*

INPUT BALANCED STATE-SPACE REALIZATION BASED ADAPTIVE RECURSIVE FILTERS IV-420

Jiong Zhou and Gang Li, *Nanyang Technological University*

ROBUST RECURSIVE BI-ITERATION SINGULAR VALUE DECOMPOSITION (SVD) FOR SUBSPACE TRACKING AND ADAPTIVE FILTERING IV-424

Yu Wen, S.C. Chan, and K.L. Ho, *The University of Hong Kong*

CASCADED-PARALLEL ADAPTIVE NOTCH FILTER BASED ON ORTHOGONAL DECOMPOSITION IV-428

Yaohui Liu, *Helsinki University of Technology*, Paulo Diniz, *Universidade Federal do Rio de Janeiro*, and Timo Laakso, *Helsinki University of Technology*

ON PRE-WHITENED SIGN ALGORITHMS IV-432

Sofia Ben Jebara, *Sup'com*, and Hichem Besbes, *Celite Systems Inc.*

OPTIMUM MIMO TRANSMIT-RECEIVER DESIGN IN PRESENCE OF INTERFERENCE IV-436

Unnikrishna Pillai, *Polytechnic University*, and Hyun S. Oh, *Samsung Electronics Co., Ltd.*

ESTIMATION OF TRANSMISSION LINE PARAMETERS BY ADAPTIVE INVERSE SCATTERING IV-440

Akihiro Yonemoto, Takashi Hisakado, and Kohshi Okumura, *Kyoto University*

A LEAST-SQUARES BASED ALGORITHM FOR FIR FILTERING WITH NOISY DATA IV-444

Wei Xing Zheng, *University of Western Sydney*

A SPACE-TIME DECORRELATING RAKE RECEIVER FOR DS-CDMA COMMUNICATIONS OVER FAST FADING CHANNELS IV-448

Chun Yu Fung, Shing Chow Chan, and Kai Wing Tse, *The University of Hong Kong*

REALIZATION OF THE NLMS BASED TRANSVERSAL ADAPTIVE FILTER USING BLOCK FLOATING POINT ARITHMETIC IV-452

Abhijit Mitra and Mrityunjoy Chakraborty, *Indian Institute of Technology, Kharagpur*

CONVERGENCE ANALYSIS OF A CORDIC-BASED GRADIENT ADAPTIVE LATTICE FILTER IV-456

Shinichi Shiraishi, Miki Haseyama, and Hideo Kitajima, *Hokkaido University*

ON THE IMPROVEMENT OF BLIND MOE DETECTOR VIA A POSTERIORI ADAPTATION AND ADAPTIVE STEP-SIZE IV-460

Sampan Pampichai and Phaophak Sirisuk, *Mahanakorn University of Technology*

A HIGH-SPEED BLIND DFE EQUALIZER USING AN ERROR FEEDBACK FILTER FOR QAM MODEMS IV-464

Jung Hoo Lee, Weon Heum Park, Ju Hyung Hong, and Myung Hoon Sunwoo, *Ajou University,* and Kyung Ho Kim *Telecommunications R&D Center,*

Track 4.4: Multidimensional DSP Systems

STABILITY, CONTROLLABILITY AND OBSERVABILITY OF 2-D CONTINUOUS-DISCRETE SYSTEMS IV-468

Yang Xiao, *Northern Jiaotong University*

ROUNDOFF NOISE MINIMIZATION IN TWO-DIMENSIONAL STATE-SPACE DIGITAL FILTERS USING ERROR FEEDBACK IV-472

Takao Hinamoto and Keisuke Higashi, *Hiroshima University*, and Wu-Sheng Lu, *University of Victoria*

A 4D FREQUENCY-PLANAR IIR FILTER AND ITS APPLICATION TO LIGHT FIELD PROCESSING IV-476

Donald Dansereau and Leonard Bruton, *University of Calgary*

REALIZATION OF HIGH ACCURACY 2-D VARIABLE IIR DIGITAL FILTERS BASED ON THE REDUCED-DIMENSIONAL DECOMPOSITION FORM IV-480

Hyuk-Jae Jang and Masayuki Kawamata, *Tohoku University*

A NEW RECURSIVE FORMULATION FOR 2-D WHT IV-484

Ayman Elnaggar, *Sultan Qaboos University*, and Mokhtar Aboelaze, *York University*

Track 4.5: Parallel and Real-Time Signal Processing

PARALLEL ITERATIONS FOR RECURSIVE MEDIAN FILTER IV-488

Adrian Burian and Jarmo Takala, *Tampere University of Technology*, and Marina Dana Topa, *Technical University of Cluj-Napoca*

AN IMPLEMENTATION OF NUMERICAL INVERSION OF LAPLACE TRANSFORMS ON FPGA IV-492

Akihiro Yonemoto, Takashi Hisakado, and Kohshi Okumura, *Kyoto University*

A NEW MEMORY REFERENCE REDUCTION METHOD FOR FFT IMPLEMENTATION ON DSP IV-496

Yiyan Tang, Lie Qian, and Yuke Wang, *University of Texas at Dallas*, and Yvon Savaria, *Ecole Polytechnique de Montreal*

REAL-TIME ACQUISITION AND TRACKING FOR GPS RECEIVERS IV-500

Abdulqadir Alaqeeli, Janusz Starzyk, and Frank Van Graas, *Ohio University*

A NOVEL SAMPLING PROCESS AND PULSE GENERATOR FOR A LOW DISTORTION DIGITAL PULSE-WIDTH MODULATOR FOR DIGITAL CLASS D AMPLIFIERS IV-504

Bah-Hwee Gwee, Joseph Chang, Victor Adrian, and Haryanto Amir, *Nanyang Technological University*

A COMPARISON OF ALGORITHMS FOR SOUND LOCALIZATION IV-508

Pedro Julian and Andreas Andreou, *Johns Hopkins University*, Larry Riddle, *Systems Signals Corporation*, Shihab Shamma, *University of Maryland*, and Gert Cauwenberghs, *Johns Hopkins University*

MODULI SELECTION IN RNS FOR EFFFICIENT VLSI IMPLEMENTATION IV-512

Wei Wang, *University of Western Ontario*, M.N.S. Swamy and M. Omair Ahmad, *Concordia University*

NONUNIFORM SAMPLING DRIVER DESIGN FOR OPTIMAL ADC UTILIZATION IV-516

Frank Papenfuß, *University of Rostock,* Y. Artyukh and E. Boole, *Institute of Applied Microelectronics and Computer Science,* and D. Timmermann, *University of Rostock*

DESIGN OF A HIGH SPEED REVERSE CONVERTER FOR A NEW 4-MODULI SET RESIDUE NUMBER SYSTEM IV-520

Bin Cao, Thambipillai Srikanthan, and Chip Hong Chang, *Nanyang Technological University*

CONFLICT-FREE PARALLEL MEMORY ACCESS SCHEME FOR FFT PROCESSORS IV-524

Jarmo Takala, Tuomas Järvinen, and Harri Sorokin, *Tampere University of Technology*

PARALLEL SUB-CONVOLUTION FILTER BANK ARCHITECTURES IV-528

Andrew Gray, *JPL/Caltech*

DUAL CLOCK RATE BLOCK DATA PARALLEL ARCHITECTURE IV-532

Winser Alexander and An-Te Deng, *North Carolina State University*

NEW EFFICIENT RESIDUE-TO-BINARY CONVERTERS FOR 4-MODULI SET {2^n-1, 2^n, 2^n+1, 2^(n+1)-1} IV-536

Bin Cao, Chip Hong Chang, and Thambipillai Srikanthan, *Nanyang Technological University*

A PARALLEL/PIPELINED ALGORITHM FOR THE COMPUTATION OF MDCT AND IMDCT IV-540

N. Rama Murthy, *Center for Artificial Intelligence & Robotics*, and M.N.S. Swamy, *Concordia University*

ARCHITECTURE FOR CORDIC ALGORITHM REALIZATION WITHOUT ROM LOOKUP TABLES IV-544

Chuen-Yau Chen and Wen-Chih Liu, *I-Shou University*

Track 8: COMPUTER AIDED NETWORK DESIGN

RELATIONSHIP BETWEEN HAAR WAVELET AND REED-MULLER SPECTRA IV-548

Bogdan Falkowski, *Nanyang Technological University*

IMPROVEMENTS ON LAYOUT OF GARMENT PATTERNS FOR EFFICIENT FABRIC CONSUMPTION IV-552

Sophon Vorasitchai and Suthep Madarasmi, *King Mongkut's University of Technology Thonburi*

PROPERTIES OF FASTEST LIA TRANSFORM MATRICES AND THEIR SPECTRA IV-556

Bogdan Falkowski and Cicilia Lozano, *Nanyang Technological University*

FAST LINEARLY INDEPENDENT TERNARY ARITHMETIC TRANSFORMS IV-560

Bogdan Falkowski and Cheng Fu, *Nanyang Technological University*

TERNARY ARITHMETIC POLYNOMIAL EXPANSIONS BASED ON NEW TRANSFORMS IV-564

Bogdan Falkowski and Cheng Fu, *Nanyang Technological University*

STABILITY ROBUSTNESS OF INTERCONNECTED DISCRETE TIME SYSTEMS WITH SYNCHRONIZATION ERRORS IV-568

Peter H. Bauer and Cedric Lorand, *University of Notre Dame*, and Kamal Premaratne, *University of Miami*

Track 8.1: Modeling Algorithms

MICROWAVE AMPLIFIER DESIGN FOR MOBILE COMMUNICATION VIA IMMITTANCE DATA MODELLING IV-572

Ali Kilinc, Haci Pinarbasi, Siddik Yarman, and Akmet Aksen, *Isik University*

WORST-CASE TOLERANCE ANALYSIS OF NON-LINEAR SYSTEMS USING EVOLUTIONARY ALGORITHMS IV-576

Biagio de Vivo and Giovanni Spagnuolo, *University of Salerno*, and Massimo Vitelli, *Second University of Naples*

COMPUTATION OF PROJECTIONS AND EIGEN DISTRIBUTION IN HALF-PLANES AND DISKS IV-580

Mohammed Hasan, *University of Minnesota, Duluth*, and Ali Hasan, *University of Nasser*

Track 8.2: Analog Modeling and Simulation

FITTING CONSIDERATIONS OF POLYNOMIAL DEVICE MODELS IV-584

Timo Rahkonen and Antti Heiskanen, *University of Oulu*

MODELING SKIN EFFECT WITH REDUCED DECOUPLED R-L CIRCUITS IV-588

Shizhong Mei and Yehea Ismail, *Northwestern University*

EFFICIENT INTERCONNECT MODELING BY FINITE DIFFERENCE QUADRATURE METHODS IV-592

Qinwei Xu and Pinaki Mazumder, *University of Michigan*

1/F NOISE MODELING USING DISCRETE-TIME SELF-SIMILAR SYSTEMS IV-596

Rajesh Narasimha, Sripriya Bandi Rachaiah, Raghuveer Rao, and P.R. Mukund, *Rochester Institute of Technology*

HIGH LEVEL ACCURACY LOSS ESTIMATES FOR A CLASS OF ANALOG/DIGITAL SYSTEMS IV-600

Cesare Alippi and Marco Stellini, *Politecnico di Milano*

CLOSED FORM METRICS TO ACCURATELY MODEL THE RESPONSE IN GENERAL ARBITRARILY-COUPLED RC TREES IV-604

Dinesh Pamunuwa, *Royal Institute of Technology*, and Shauki Elassaad, *Cadence Berkeley Labs*

PODEA: POWER DELIVERY EFFICIENT ANALYSIS WITH REALIZABLE MODEL REDUCTION IV-608

Rong Jiang and Tsung-Hao Chen, *University of Wisconsin, Madison*, and Charlie Chung-Ping Chen, *National Taiwan University*

5TH ORDER ELECTRO-THERMAL MULTI-TONE VOLTERRA SIMULATOR WITH COMPONENT-LEVEL OUTPUT IV-612

Antti Heiskanen and Timo Rahkonen, *University of Oulu*

A 5TH ORDER VOLTERRA STUDY OF A 30W LDMOS POWER AMPLIFIER IV-616

Antti Heiskanen, *University of Oulu, Finland*, Janne Aikio, *Elektrobit*, and Timo Rahkonen, *University of Oulu*

A SIMULINK-BASED APPROACH FOR FAST AND PRECISE SIMULATION OF SWITCHED-CAPACITOR, SWITCHED-CURRENT AND CONTINUOUS-TIME SIGMA-DELTA MODULATORS IV-620

Javier Moreno-Reina, José M. de la Rosa, Fernando Medeiro, and Rafael Romay, *Instituto de Microelectronica de Sevilla IMSE-CNM-CSIC*, Rocio del Rio, *Universidad de Sevilla*, Belen Perez-Verdu and Angel Rodriguez-Vazquez, *Instituto de Microelectronica de Sevilla IMSE-CNM-CSIC*

DISCRETE-TIME MODELING AND SIMULATION OF VEHICLE AUDIO SYSTEMS IV-624

Francesco Piazza, *University of Ancona*, Ferruccio Bettarelli and Ariano Lattanzi, *Leaff Engineering Srl*, Romolo Toppi, Massimo Navarri, and Michele Pontillo, *FAITAL Spa*, and Stefano Bartoloni, *Università degli Studi di Ancona*

AN EFFICIENT REDUCED-ORDER INTERCONNECT MACROMODEL FOR TIME-DOMAIN SIMULATION IV-628

Timo Palenius and Janne Roos, *Helsinki University of Technology*

ACCURATE VHDL-BASED SIMULATION OF SIGMA-DELTA MODULATORS IV-632

Rafael Castro-López, Francisco V. Fernández, Fernando Medeiro, and Angel Rodriguez-Vazquez, *IMSE-CNM-CSIC*

AN ACCURATE BEHAVIORAL MODEL OF PHASE DETECTORS FOR CLOCK RECOVERY CIRCUITS IV-636

Marco Balsi, Francesco Centurelli, Giuseppe Scotti, Pasquale Tommasino, and Alessandro Trifiletti, *Universita' la Sapienza di Roma*

SYMBOLIC ANALYSIS: A FORMULATION APPROACH BY MANIPULATING DATA STRUCTURES IV-640

Esteban Tlelo-Cuautle, *INAOE-ITP*, Carlos Sánchez-López, *INAOE*, and Federico Sandoval-Ibarra, *CINVESTAV*

ACCURATE COMPACT MODEL EXTRACTION FOR ON-CHIP COPLANAR WAVEGUIDES IV-644

Taeik Kim, Xiaoyong Li, and David Allstot, *University of Washington*

INTERCONNECT MODELING AND SENSITIVITY ANALYSIS USING ADJOINT NETWORKS REDUCTION TECHNIQUE IV-648

Herng-Jer Lee, Chia-Chi Chu, and Wu-Shiung Feng, *Chang Gung University*

MIXED-MODE ESD PROTECTION CIRCUIT SIMULATION-DESIGN METHODOLOGY IV-652

Haigang Feng, Rouying Zhan, Qiong Wu, Guang Chen, Xiaokang Guan, and Albert Wang, *Illinois Institute of Technology*

A TECHNIQUE FOR REDUCTION OF UNCERTAIN FIR FILTERS IV-656

Anthony G Lim and Victor Sreeram, *University of Western Australia*, and Ezra Zeheb, *Technion-Israel Institute of Technology*

CONCURRENT LOGIC AND INTERCONNECT DELAY ESTIMATION OF MOS CIRCUITS BY MIXED ALGEBRAIC AND BOOLEAN SYMBOLIC ANALYSIS IV-660

Sambuddha Bhattacharya and Richard Shi, *University of Washington*

AN EFFICIENT SYLVESTER EQUATION SOLVER FOR TIME DOMAIN CIRCUIT SIMULATION BY WAVELET COLLOCATION METHOD IV-664

Xuan Zeng, Sheng Huang, and Yang Feng Su, *Fudan University*, and Dian Zhou, *Univeristy of Texas at Dallas*

TABLE LOOK-UP BASED COMPACT MODELING FOR ON-CHIP INTERCONNECT TIMING AND NOISE ANALYSIS IV-668

Haitian Hu, *Motorola Inc.*, David Blaauw, *University of Michigan*, Vladimir Zolotov, Kaushik Gala, Min Zhao, and Rajendran Panda, *Motorola Inc.*, and Sachin Sapatnekar, *University of Minnesota*

PROCESS VARIATION DIMENSION REDUCTION BASED ON SVD IV-672

Zhuo Li, Xiang Lu, and Weiping Shi, *Texas A&M University*

ANALOG IP DESIGN FLOW FOR SOC APPLICATIONS IV-676

Marwa Hamour, Res Saleh, Shahriar Mirabbasi, and Andre Ivanov, *University of British Columbia*

Track 8.3: CAD Algorithms

SINGLE VARIABLE SYMMETRY CONDITIONS IN BOOLEAN FUNCTIONS THROUGH REED-MULLER TRANSFORM IV-680

Sudha Kannurao, *Temasek Polytechnic*, and Bogdan Falkowski, *Nanyang Technological University*

A BOOLEAN EXTRACTION TECHNIQUE FOR MULTIPLE-LEVEL LOGIC OPTIMIZATION IV-684

Oh-Hyeong Kwon, *Uiduk University*

OPTIMAL USE OF 2-PHASE TRANSPARENT LATCHES IN BUFFERED MAZE ROUTING IV-688

Soha Hassoun, *Tufts University*

CONCURRENT OPTIMIZATION OF PROCESS DEPENDENT VARIATIONS IN DIFFERENT CIRCUIT PERFORMANCE MEASURES IV-692

Ayhan Mutlu, *Santa Clara University*, Norman Gunther, *Electron Devices Lab*, and Mahmud Rahman, *Electron Devices Lab*

TIME DOMAIN RESPONSE AND SENSITIVITY OF PERIODICALLY SWITCHED NONLINEAR CIRCUITS IV-696

Fei Yuan and Quan Li, *Ryerson University*

CAD SYSTEM FOR DESIGN AND SIMULATION OF DATA CONVERTERS IV-700

Pedro Estrada, *Texas A&M University*, and Franco Maloberti, *University of Texas at Dallas*

AUTOMATIC ANALOG LAYOUT RETARGETING FOR NEW PROCESSES AND DEVICE SIZES IV-704

Nuttorn Jangkrajarng, Sambuddha Bhattacharya, Roy Hartono, and Richard Shi, *University of Washington*

EVALUATING A BOUNDED SLICE-LINE GRID ASSIGNMENT IN O(NLOGN) TIME IV-708

Song Chen, Xianlong Hong, Sheqin Dong, Yuchun Ma, and Yici Cai, *Tsinghua University*, Chung-Kuan Cheng, *University of California,San Diego,* and Jun Gu, *Science & Technology University of Hong Kong*

NOISE CONSTRAINT DRIVEN PLACEMENT FOR MIXED SIGNAL DESIGNS IV-712

William H. Kao and Wen Kung Chu, *Cadence Design Systems, Inc.*

ALGORITHMS FOR ANALOG VLSI 2D STACK GENERATION AND BLOCK MERGING IV-716

Rui Liu, *Chinese Academy of Sciences*, Sheqin Dong, Xianlong Hong, and Di Long, *Tsinghua University*, and Jun Gu, *Science & Technology University of Hong Kong*

COMBINING CLUSTERING AND PARTITIONING IN QUADRATIC PLACEMENT IV-720

Yongqiang Lu, Xianlong Hong, Wenting Hou, Weimin Wu, and Yici Cai, *Tsinghua University*

FLOORPLANNING WITH PERFORMANCE-BASED CLUSTERING IV-724

Malgorzata Chrzanowska-Jeske, Benyi Wang, and Garrison Greenwood, *Portland State University*

THE STRUCTURE DETERMINATION FOR THE TIME-OPTIMAL SYSTEM DESIGN ALGORITHM IV-728

Alexander Zemliak, *Puebla Autonomous University*

AN EFFICIENT APPROACH FOR ERROR DIAGNOSIS IN HDL DESIGN IV-732

Che-Hua Shih and Jing-Yang Jou, *National Chiao Tung University*

YIELD OPTIMIZATION WITH CORRELATED DESIGN PARAMETERS AND NON-SYMMETRICAL MARGINAL DISTRIBUTIONS IV-736

Kumaraswamy Ponnambalam, *University of Waterloo*, Abbas Seifi, *Amirkabir University of Technology*, and Jiri Vlach, *University of Waterloo*

GENERATION AND PROPERTIES OF FASTEST TRANSFORM MATRICES OVER GF(2) IV-740

Bogdan Falkowski and Cicilia Lozano, *Nanyang Technological University*

SOC DESIGN INTEGRATION BY USING AUTOMATIC INTERCONNECTION RECTIFICATION IV-744

Chun-Yao Wang, Shing-Wu Tung, and Jing-Yang Jou, *National Chiao Tung University*

SYNTHESIZING CHECKERS FOR ON-LINE VERIFICATION OF SYSTEM-ON-CHIP DESIGNS IV-748

Rolf Drechsler, *University of Bremen*

Track 10: ADVANCED TECHNOLOGY

LOAD CELL RESPONSE CORRECTION USING ANALOG ADAPTIVE TECHNIQUES IV-752

Mehdi Jafaripanah, Bashir Al-Hashimi, and Neil White, *University of Southampton*

BIAS-ADAPTIVE CROSS-COUPLED CMOS MAGFET PAIR FOR BIPOLAR MAGNETIC FIELD DETECTION IV-756

Zhiqing Li and Xiaowei Sun, *Nanyang Technological University*, Wei Fan and Guojun Qi, *Singapore Institute of Manufacturing Technology*

PHOTO-THERMOELECTRIC POWER GENERATION FOR AUTONOMOUS MICROSYSTEMS IV-760

Salvatore Baglio, Salvatore Castorina, Luigi Fortuna, and Nicolò Savalli, *Università degli Studi di Catania*

DEVELOPMENT OF A CMOS LOW-NOISE ANALOG FRONT-END ASIC FOR X-RAY IMAGING APPLICATIONS IV-764

Emmanuel Zervakis, *National Technical University of Athens*, Dimitris Loukas, *National Center of Scientific Research*, Nikos Haralabidis and A. Pavlidis, *Athena Semiconductors S.A.*

FLUXGATE MAGNETIC SENSORS WITH A READOUT STRATEGY BASED ON RESIDENCE TIMES MEASUREMENTS IV-768

Bruno Andò and Salvatore Baglio, *University of Catania*, Adi Bulsara, *Space and Naval Warfare Systems Center*, and Luca Gammaitoni, *University of Perugia*

A 96 X 64 INTELLIGENT DIGITAL PIXEL ARRAY WITH EXTENDED BINARY STOCHASTIC ARITHMETIC IV-772

Tarik Hammadou and Magnus Nilson, *Motorola Labs*, Amine Bermak, *Hong Kong University of Science and Technology*, and Philip Ogunbona, *Motorola Labs*

A CMOS IMAGER WITH PIXEL PREDICTION FOR IMAGE COMPRESSION IV-776

Daniel Leon, Sina Balkir, Khalid Sayood, and Michael Hoffman, *University of Nebraska, Lincoln*

ON CHIP GAUSSIAN PROCESSING FOR HIGH RESOLUTION CMOS IMAGE SENSORS IV-780

Sanjayan Vinayagamoorthy, *University of Waterloo*, and Richard Hornsey, *York Universtiy*

NORMAL OPTICAL FLOW CHIP IV-784

Swati Mehta and Ralph Etienne-Cummings, *Johns Hopkins University*

HIGH-SPEED POSITION DETECTOR USING NEW ROW-PARALLEL ARCHITECTURE FOR FAST COLLISION PREVENTION SYSTEM IV-788

Yusuke Oike, Makoto Ikeda, and Kunihiro Asada, *University of Tokyo*

A SILICON RETINA SYSTEM THAT CALCULATES DIRECTION OF MOTION IV-792

Seiji Kameda and Tetsuya Yagi, *Osaka University*

A FAST AND LOW POWER CMOS SENSOR FOR OPTICAL TRACKING IV-796

Nicola Viarani, Nicola Massari, Lorenzo Gonzo, Massimo Gottardi, David Stoppa, and Andrea Simoni, *ITC - IRST*

AN ORIENTATION SELECTIVE 2D AER TRANSCEIVER IV-800

Thomas Choi and Bertram Shi, *Hong Kong University of Science and Technology*, and Kwabena Boahen, *University of Pennsylvania*

A SILICON RETINA WITH CONTROLLABLE WINNER-TAKE-ALL PROPERTIES IV-804

Shih-Chii Liu, *University and ETH Zurich*

SINGLE CHIP STEREO IMAGER IV-808

Ralf Philipp and Ralph Etienne-Cummings, *Johns Hopkins University*

A SCANNING THERMAL MICROSCOPY SYSTEM WITH A TEMPERATURE DITHERING, SERVO-CONTROLLED INTERFACE CIRCUIT IV-812

Joohyung Lee, *University of Wisconsin, Madison*, and Yogesh Gianchandani, *University of Michigan, Ann Arbor*

CHARACTERIZATION OF IPMC STRIP'S SENSORIAL PROPERTIES: PRELIMINARY RESULTS IV-816

Luigi Fortuna, Salvatore Graziani, and Claudia Bonomo, *Università Degli Studi di Catania*, and Ciro Del Negro, *INGV-Catania*

A LOW-POWER ADAPTIVE INTEGRATE-AND-FIRE NEURON CIRCUIT IV-820

Giacomo Indiveri, *Univeristy of Zurich - ETH Zurich*

EVIDENCE UPDATING IN A HETEROGENEOUS SENSOR ENVIRONMENT IV-824

Kamal Premaratne and Duminda A. Dewasurendra, *University of Miami*, and Peter H. Bauer, *University of Notre Dame*

ENERGY-BALANCING STRATEGIES FOR WIRELESS SENSOR NETWORKS IV-828

Martin Haenggi, *University of Notre Dame*

POWER DISSIPATION LIMITS AND LARGE MARGIN IN WIRELESS SENSORS IV-832

Shantanu Chakrabartty and Gert Cauwenberghs, *Johns Hopkins University*

GNOMES: A TESTBED FOR LOW POWER HETEROGENEOUS WIRELESS SENSOR NETWORKS IV-836

Erik Welsh, Walt Fish, and Jeremy Frantz, *Rice University*

ANALYSIS OF SHORT DISTANCE OPTOELECTRONIC LINK ARCHITECTURES IV-840

Alyssa Apsel, *Cornell University*, and Andreas Andreou, *Johns Hopkins University*

BULK CARBON NANOTUBE AS THERMAL SENSING AND ELECTRONIC CIRCUIT ELEMENTS IV-844

Wong Tak Sing and Wen Jung Li, *The Chinese University of Hong Kong*

CMOS INTEGRATED GAS SENSOR CHIP USING SAW TECHNOLOGY IV-848

Shahrokh Ahmadi, Mona Zaghloul, Can Korman, and Kuan-Hsun Huang, *The George Washington University*

A MICRO-HOTPLATE-BASED MONOLITHIC CMOS GAS SENSOR ARRAY IV-852

Diego Barrettino, Markus Graf, Martin Zimmermann, Christoph Hagleitner, Andreas Hierlemann, and Henry Baltes, *ETH Zuerich*

MICROMACHINED PIEZORESISTIVE TACTILE SENSOR ARRAY FABRICATED BY BULK-ETCHED MUMPS PROCESS IV-856

Tanom Lomas, *King Mongkut's Institute of Technology Ladkrabang*, Adisorn Tuantranont, *National Electronics and Computer Technology Center*, and Fusak Cheevasuvit, *King Mongkut's Institute of Technology Ladkrabang*

FABRICATION PROCESS FOR A MICROFLUIDIC VALVE IV-860

Antonio Luque and Jose Manuel Quero, *Universidad de Sevilla*, Cyrille Hibert and Philippe Flückiger, *Swiss Federal Institute of Technology (EPFL)*

A NEUROMORPHIC SOUND LOCALIZER FOR A SMART MEMS SYSTEM IV-864

Andre Van Schaik, *The University of Sydney*, and Shihab Shamma, *University of Maryland*

A TIME-SERIES PROCESSOR FOR SONAR MAPPING AND NOVELTY DETECTION IV-868

Timothy Horiuchi and Ralph Etienne-Cummings, *University of Maryland*

A VLSI MODEL OF RANGE-TUNED NEURONS IN THE BAT ECHOLOCATION SYSTEM IV-872

Matthew Cheely and Timothy Horiuchi, *University of Maryland*

DEVELOPMENT OF AN AA SIZE ENERGY TRANSDUCER WITH MICRO RESONATORS IV-876

Ming Ho Lee, Chi Lap Yuen, Wen Jung Li, and Heng Wai Leong, *Chineses University of Hong Kong*

A DISPLACEMENT-TO-VOLTAGE CONVERTER CIRCUIT USING A SWITCHED-CAPACITOR TECHNIQUE IV-880

Wee Liang Lien, Boon Seah Quek, and Rajan Walia, *Institute of Microelectronics*

HIGH-SPEED PROCESSOR FOR QUANTUM-COMPUTING EMULATION AND ITS APPLICATIONS IV-884

Minoru Fujishima, Kosuke Saito, Masafumi Onouchi, and Koichiro Hoh, *The University of Tokyo*

INFORMATION EXCHANGES IN QUANTUM ARRAYS DUE TO SPATIAL DIVERSITY IV-888

Maide Bucolo, Luigi Fortuna, and Manuela la Rosa, *Universita' Degli Studi di Catania*, Donata Nicolosi and Domenico Porto, *ST Microelectronics*

A GLOBAL WIRE PLANNING SCHEME FOR NETWORK-ON-CHIP IV-892

Jian Liu, Lirong Zheng, Dinesh Pamunuwa, and Hannu Tenhunen, *Royal Institute of Technology Stockholm*

QUANTUM DOT NETWORKS WITH WEIGHTED COUPLING IV-896

Koray Karahaliloglu and Sina Balkir, *University of Nebraska, Lincoln*

PERFORMANCE MODELING OF RESONANT TUNNELING BASED RANDOM-ACCESS MEMORIES IV-900

Hui Zhang, Pinaki Mazumder, and Li Ding, *The University of Michigan*, and Kyounghoon Yang, *KAIST*

THE CMOS/NANO INTERFACE FROM A CIRCUITS PERSPECTIVE IV-904

Matthew Ziegler and Mircea Stan, *University of Virginia*

THIN FILM PIN PHOTODIODES FOR OPTOELECTRONIC SILICON ON SAPPHIRE CMOS IV-908

Alyssa Apsel, *Cornell University*, Eugenio Culurciello and Andreas Andreou, *Johns Hopkins University*, and Keith Aliberti, *Army Research Laboratory*

THE IMD CANCELLATION CHARACTERISTICS OF PREDISTORTION LINEARIZER IN MICROWAVE TRANSISTOR IV-912

Unghee Park and Kyung Hee Lee, *ETRI*

VOLUME V OF V:

-BIO-MEDICAL CIRCUITS & SYSTEMS-

-VLSI SYSTEMS & APPLICATIONS-

-NEURAL NETWORKS & SYSTEMS-

Track 2: BIOMEDICAL CIRCUITS & SYSTEMS

A MICROPOWER ENVELOPE DETECTOR FOR AUDIO APPLICATIONS V-1

Michael Baker, Serhii Zhak, and Rahul Sarpeshkar, *Massachusetts Institute of Technology*

A SMART BI-DIRECTIONAL TELEMETRY UNIT FOR RETINAL PROSTHETIC DEVICE V-5

Rizwan Bashirullah, Wentai Liu, Ying Ji, Alper Kendir, Mohanasankar Sivaprakasam, Guoxing Wang, and Bharat Pundi, *North Carolina State University*

DESIGN OF AN INTEGRATED POTENTIOSTAT CIRCUIT FOR CMOS BIO SENSOR CHIPS V-9

Alexander Frey, Martin Jenkner, Meinrad Schienle, Christian Paulus, Birgit Holzapfl, Petra Schindler-Bauer, and Franz Hofmann, *Infineon Technologies AG*, Dirk Kuhlmeier and Jürgen Krause, *november AG* , Jörg Albers, *ISiT*, Walter Gumbrecht, *Siemens AG* , Doris Schmitt-Landsiedel, *Technical University of Munich*, and Roland Thewes, *Infineon Technologies AG*

DISTRIBUTED NEUROCHEMICAL SENSING: IN VITRO EXPERIMENTS V-13

Grant Mulliken, Mihir Naware, Abhishek Bandyopadhyay, Gert Cauwenberghs, and Nitish Thakor, *Johns Hopkins University*

VISUAL SIGNAL PROCESSING AND IMAGE UNDERSTANDING IN BIOMEDICAL SYSTEMS V-17

Marek Ogiela and Ryszard Tadeusiewicz, *Univeristy of Mining and Metallurgy*

AN EFFICIENT BLOOD VESSEL DETECTION ALGORITHM FOR RETINAL IMAGES USING LOCAL ENTROPY THRESHOLDING V-21

Thitiporn Chanwimaluang and Guoliang Fan, *Oklahoma State University*

PHASE ANALYSIS OF DNA GENOMIC SIGNALS V-25

Paul Dan Cristea, *Polytechnical University of Bucharest*

ENHANCED TIME-FREQUENCY FEATURES FOR NEONATAL EEG SEIZURE DETECTION V-29

Hamid Hassanpour, *Signal Processing Research Centre*, Mostefa Mesbah and Boualem Boashash, *Queensland University of Technology*

AN OPTIMAL FEATURE SET FOR SEIZURE DETECTION SYSTEMS FOR NEWBORN EEG SIGNALS V-33

Pega Zarjam, Mostefa Mesbah and Boualem Boashash, *Queensland University of Technology*

Track 2.1: Implantable Electronics

AN ULTRA LOW-POWER DYNAMIC TRANSLINEAR CARDIAC SENSE AMPLIFIER FOR PACEMAKERS V-37

Sandro Haddad and Sebastian Gieltjes, *Delft University of Technology*, Richard Houben, *Medtronic*, and Wouter A. Serdijn, *TU Delft*

A MICROPOWER ANALOG VLSI PROCESSING CHANNEL FOR BIONIC EARS AND SPEECH-RECOGNITION FRONT ENDS V-41

Timothy Lu, Michael Baker, Christopher Salthouse, Ji-Jon Sit, Serhii Zhak, and Rahul Sarpeshkar, *Massachusetts Institute of Technology*

A HIGH-RATE FREQUENCY SHIFT KEYING DEMODULATOR CHIP FOR WIRELESS BIOMEDICAL IMPLANTS V-45

Maysam Ghovanloo and Khalil Najafi, *University of Michigan*

A VERY LOW-POWER 8-BIT SIGMA-DELTA CONVERTER IN A 0.8UM CMOS TECHNOLOGY FOR THE SENSING CHAIN OF A CARDIAC PACEMAKER, OPERATING DOWN TO 1.8V V-49

Andrea Gerosa and Andrea Neviani, *University of Padova*

A CORTICAL STIMULATOR WITH MONITORING CAPABILITIES USING A NOVEL 1 MBPS ASK DATA LINK V-53

Jonathan Coulombe, Jean-Francois Gervais, and Mohamad Sawan, *Ecole Polytechnique de Montreal*

VLSI IMPLEMENTATION OF WIRELESS BI-DIRECTIONAL COMMUNICATION CIRCUITS FOR MICRO-STIMULATOR V-57

Shuenn-Yuh Lee, Shyu-Chyang Lee, and Jia-Jin Jason Chen, *National Chung Cheng University*

A CMOS MICRO-POWER WIDEBAND DATA/POWER TRANSFER SYSTEM FOR BIOMEDICAL IMPLANTS. V-61

Okundu Omeni and Chris Toumazou, *Imperial College*

A CMOS CURRENT-TO-LCD INTERFACE FOR PORTABLE AMPEROMETRIC SENSING SYSTEMS V-65

Karn Opasjumruskit and Naiyavudhi Wongkomet, *Chulalongkorn University*

Track 7: VLSI SYSTEMS & APPLICATIONS

XLIW - A SCALEABLE LONG INSTRUCTION WORD V-69

Christian Panis, *Carinthian Tech Institut*, Raimund Leitner, *Infineon Technologies*, Herbert Grünbacher, *Carinthian Tech Institut*, and Jari Nurmi, *Tampere University of Technology*

FIELD PROGRAMMABLE GATES ARRAYS AND ANALOG IMPLEMENTATION OF BRIN FOR OPTIMIZATION PROBLEMS V-73

Sui Tung Mak, Hoi Shing Raymond Ng, and Kai Pui Lam, *Chinese University of Hong Kong*

A LOW-COMPLEXITY POWER-EFFICIENT SIGNALING SCHEME FOR CHIP-TO-CHIP COMMUNICATION V-77

Kamran Farzan and David Johns, *University of Toronto*

EFFICIENT MODELING AND SYNTHESIS OF ON-CHIP COMMUNICATION PROTOCOLS FOR NETWORK-ON-CHIP DESIGN V-81

Robert Siegmund and Dietmar Müller, *TU Chemnitz*

SIGNALING CAPACITY OF FR4 PCB TRACES FOR CHIP-TO-CHIP COMMUNICATION V-85

Marcus Van Ierssel, Tooraj Esmailian, Ali Sheikholeslami, and Pas Pasupathy, *University of Toronto*

A CMOS CHARGE PUMP FOR SUB-2.0V OPERATION V-89

Kuo-Hsing Cheng, Chung-Yu Chang, and Chia-Hung Wei, *Tamkang University*

NEW DYNAMIC LOGIC-LEVEL CONVERTERS FOR HIGH PERFORMANCE APPLICATION V-93

Nam-Seog Kim, Yong-Jin Yoon, Uk-Rae Cho, and Hyun-Geun Byun, *Samsung Electronics, Korea*

DESIGN OF 2.5V/5V MIXED-VOLTAGE CMOS I/O BUFFER WITH ONLY THIN OXIDE DEVICE AND DYNAMIC N-WELL BIAS CIRCUIT V-97

Ming-Dou Ker and Chia-Sheng Tsai, *National Chiao Tung University*

AREA-EFFICIENT MEMORY-BASED ARCHITECTURE FOR FFT PROCESSING V-101

Sang-Chul Moon and In-Cheol Park, *KAIST*

OPTIMIZATION OF POWER CONSUMPTION FOR AN ARM7-BASED MULTIMEDIA HANDHELD DEVICE V-105

Hoseok Chang, Wonchul Lee, and Wonyong Sung, *Seoul National University*

LOOP SCHEDULING FOR MINIMIZING SCHEDULE LENGTH AND SWITCHING ACTIVITIES V-109

Zili Shao, Qingfeng Zhuge, and Edwin H.-M. Sha, *University of Texas At Dallas*, and Chantana Chantrapornchai, *Silpakorn University*

VARIABLE DELAY RIPPLE CARRY ADDER WITH CARRY CHAIN INTERRUPT DETECTION V-113

Andreas Burg, Frank Gürkaynak, Hubert Kaeslin, and Wolfgang Fichtner, *ETH Zuerich*

5 GHZ PIPELINED MULTIPLIER AND MAC IN 0.18UM COMPLEMENTARY STATIC CMOS V-117

Jos Sulistyo and Dong Ha, *Virginia Tech*

A 3.3V 1GHZ LOW-LATENCY PIPELINED BOOTH MULTIPLIER WITH NEW MANCHESTER CARRY-BYPASS ADDER V-121

Hwang-Cherng Chow and I-Chyn Wey, *Chang Gung University*

LOW COST LOGARITHMIC TECHNIQUES FOR HIGH-PRECISION COMPUTATIONS V-125

Siew Kei Lam, *Nanyang Technological University*, Devendra Kumar Chaudhary, *Indian Institute of Technology Kanpur*, and Thambipillai Srikanthan, *Nanyang Technological University*

A METHODOLOGY FOR IMPLEMENTING FIR FILTERS AND CAD TOOL DEVELOPMENT FOR DESIGNING RNS-BASED SYSTEMS V-129

Dimitrios Soudris, Kyriakos Sgouropoulos, and Kostantinos Tatas, *Democritus University of Thrace*, Vasilis Pavlidis, *University of Rochester*, and Antonios Thanailakis, *Democritus University of Thrace*

A FLEXIBLE LUT-BASED CARRY CHAIN FOR FPGAS V-133

Andrea Lodi, Carlo Chiesa, Fabio Campi, and Mario Toma, *University of Bologna*

LOW-ERROR FIXED-WIDTH SQUARER DESIGN V-137

Kyung-Ju Cho, Eyn-Min Choi, Jin-Gyun Chung, and Myung-Sub Lim, *Chonbuk National University,* and J.W.Kim, *ETRI*

AREA-TIME OPTIMAL ADDER WITH RELATIVE PLACEMENT GENERATOR V-141

Aamir Farooqui, *Synopsys*, Vojin Oklobdzija, *University of California, Davis*, and Sadiq Sait, *King Fahd University of Petroleum & Minerals*

ABOUT THE PERFORMANCES OF THE ADVANCED ENCRYPTION STANDARD IN EMBEDDED SYSTEMS WITH CACHE MEMORY V-145

Guido Bertoni, *Politecnico Di Milano*, Aril Bircan, *ALaRI*, Luca Breveglieri, *Politecnico di Milano*, Pasqualina Fragneto, *ST Microelectronics*, Marco Macchetti, *ALaRI*, and Vittorio Zaccaria, *Politecnico di Milano*

AN EFFICIENT INVERSE MULTIPLIER/DIVIDER ARCHITECTURE FOR CRYPTOGRAPY SYSTEMS V-149

Junhyung Um, *Korea Advanced Institute of Science and Technology*, Sangwoo Lee, Youngsoo Park, and Sungik Jun, *ETRI*, and Taewhan Kim U, *KAIST*

ON THE HARDWARE IMPLEMENTATIONS OF THE SHA-2 (256, 384, 512) HASH FUNCTIONS V-153

Nicolas Sklavos and Odysseas Koufopavlou, *University of Patras*

A FAST-SERIAL FINITE FIELD MULTIPLIER WITHOUT INCREASING THE NUMBER OF REGISTERS V-157

Wonjong Kim, Seungchul Kim, and Hanjin Cho, *ETRI*, and Kwangyoub Lee, *Seokyeong University*

SHIFT-ACCUMULATOR ALU CENTRIC JPEG2000 5/3 LIFTING BASED DISCRETE WAVELET TRANSFORM ARCHITECTURE V-161

Kay-Chuan Tan and Tughrul Arslan, *The University of Edinburgh*

DESIGN OF A SELF-TIMED ASYNCHRONOUS PARALLEL FIR FILTER USING CSCD V-165

Harri Lampinen, Pauli PeräLä, and Olli Vainio, *Tampere University of Technology*

ACCURATE DELAY MODEL AND EXPERIMENTAL VERIFICATION FOR CURRENT/VOLTAGE MODE ON-CHIP INTERCONNECTS V-169

Rizwan Bashirullah and Wentai Liu, *North Carolina State University*, and Ralph Cavin, *Semiconductor Research Corporation*

AREA-EFFECTIVE FIR FILTER DESIGN FOR MULTIPLIER-LESS IMPLEMENTATION V-173

Tay-Jyi Lin, Tsung-Hsun Yang, and Chein-Wei Jen, *National Chiao Tung University*

AN APPROACH FOR IMPROVING THE SPEED OF CONTENT ADDRESSABLE MEMORIES V-177

Jihad Hyjazie and Chunyan Wang, *Concordia University*

WATERMARKING BASED IP CORE PROTECTION V-181

Yu-Cheng Fan and Hen-Wai Tsao, *National Taiwan University*

HISTORY-BASED MEMORY MODE PREDICTION FOR IMPROVING MEMORY PERFORMANCE V-185

Seong-Il Park and In-Cheol Park, *KAIST*

A CODING METHOD FOR 123 DECISION DIAGRAM PASS TRANSISTOR LOGIC CIRCUIT SYNTHESIS V-189

Mutlu Avci and Tulay Yildirim, *Yildiz Technical University*

NOVEL RECHARGE SEMI-FLOATING-GATE CMOS LOGIC FOR MULTIPLE-VALUED SYSTEMS V-193

Yngvar Berg, *University of Oslo*, Snorre Aunet, *Norwegian University of Science and Technology*, Omid Mirmotahari and Mats Høvin, *University of Oslo*

AN EFFICIENT TRANSISTOR OPTIMIZER FOR CUSTOM CIRCUITS V-197

Xiao Yan Yu and Vojin Oklobdzija, *University of California, Davis*, and William Walker, *Fujitsu Laboratory of America*

A FRAMEWORK OF EVOLUTIONARY GRAPH GENERATION SYSTEM AND ITS APPLICATION TO CIRCUIT SYNTHESIS V-201

Naofumi Homma, Takatumi Aoki, Makoto Motegi, and Tatsuo Higuchi, *Tohoku University*

A ZERO-TIME-OVERHEAD ASYNCHRONOUS FOUR-PHASE CONTROLLER V-205

Nattha Sretasereekul, Hiroshi Saito, Masashi Imai, Euiseok Kim, and Metehan Ozcan, *The University of Tokyo*, Krerkchai Thongnoo, *Prince of Songkla University*, Hiroshi Nakamura and Takashi Nanya, The University of Tokyo

A NEW ROBUST HANDSHAKE FOR ASYMMETRIC ASYNCHRONOUS MICRO-PIPELINES V-209

Kuo-Hsing Cheng, Yang-Han Lee, and Wei-Chun Chang, *Tamkang University*

A FAST ALGORITHM TO REDUCE 2-DIMENSIONAL ASSIGNMENT PROBLEMS TO 1-DIMENSIONAL ASSIGNMENT PROBLEMS FOR FPGA-BASED FAULT SIMULATION V-213

Reza Sedaghat, *Ryerson University*

DESIGN OF A SWITCH FOR NETWORK ON CHIP APPLICATIONS V-217

Partha Pande, Cristian Grecu, Andre Ivanov, and Res Saleh, *University of British Columbia*

MODULO (2p+/-1) MULTIPLIERS USING A THREE-OPERAND MODULAR ADDITION AND BOOTH RECODING BASED ON SIGNED-DIGIT NUMBER ARITHMETIC V-221

Shugang Wei and Kensuke Shimizu, *Gunma University*

A SYSTEMATIC METHODOLOGY FOR DESIGNING AREA-TIME EFFICIENT PARALLEL-PREFIX MODULO 2^N-1 ADDERS V-225

Giorgos Dimitrakopoulos, Haridimos. T. Vergos, Dimitrios Nikolos, and Costas Efstathiou, *University of Patras*

HIGH-RADIX REDUNDANT CIRCUITS FOR RNS MODULO R^N - 1, R^N , OR R^N + 1 V-229

Ioannis Kouretas and Vassilis Paliouras, *University of Patras*

AREA EFFICIENT, HIGH SPEED PARALLEL COUNTER CIRCUITS USING CHARGE RECYCLING THRESHOLD LOGIC V-233

Peter Celinski, *The University of Adelaide*, Sorin D. Cotofana, *Delft University of Technology*, and Derek Abbott, *The University of Adelaide*

VIRTUAL-SCAN : A NOVEL APPROACH FOR SOFTWARE-BASED SELF-TESTING OF MICROPROCESSORS V-237

Giorgos Dimitrakopoulos and Xrisovalantis Kavousianos, *University of Patras*, and Dimitris Nikolos, *University of Patras, and Computer Technology Institute*

DESIGN OF PROGRAMMABLE EMBEDDED IF SOURCE FOR DESIGN SELF-TEST V-241

Sanghoon Choi, William Eisenstadt, and Robert Fox, *University of Florida*

FORMAL VERIFICATION OF LTL FORMULAS FOR SYSTEMC DESIGNS V-245

Daniel Große and Rolf Drechsler, *University of Bremen*

OPEN COMPUTATION TREE LOGIC WITH FAIRNESS V-249

Ansuman Banerjee, Pallab Dasgupta, and Partha Chakrabarti, *Indian Institute of Technology Kharagpur*

DESIGN AND EXPERIMENTAL RESULTS FOR A CMOS FLIP-FLOP FEATURING EMBEDDED THRESHOLD LOGIC V-253

Marius Padure, Sorin Cotofana, and Stamatis Vassiliadis, *Delft University of Technology*

VARIABLE SAMPLING WINDOW FLIP-FLOP FOR LOW-POWER APPLICATION V-257

Sang Dae Shin, Hun Choi, and Bae Sun Kong, *Hankuk Aviation University*

DESIGN OF MUX, XOR AND D-LATCH SCL GATES V-261

Massimo Alioto, *University of Siena*, and Gaetano Palumbo, *Università di Catania*

Track 7.1: Memory Circuits

PARAMETERIZED AND LOW POWER DSP CORE FOR EMBEDDED SYSTEMS V-265

Ya-Lan Tsao, Ming Hsuan Tan, Jun-Xian Teng, and Shyh-Jye Jou, *National Central University, Taiwan*

MACROMODEL FOR SHORT CIRCUIT POWER DISSIPATION OF SUBMICRON CMOS INVERTERS AND ITS APPLICATION TO DESIGN CMOS BUFFERS V-269

Shrutin Ulman, *Goa University*

INDUCTIVE INTERCONNECT WIDTH OPTIMIZATION FOR LOW POWER V-273

Magdy El-Moursy and Eby Friedman, *University of Rochester*

ON OPTIMIZING POWER AND CROSSTALK FOR BUS COUPLING CAPACITANCE USING GENETIC ALGORITHMS V-277

Edwin Naroska, *University of Dortmund*, Shanq-Jang Ruan, *Synopsys Taiwan Inc.*, Feipei Lai, *National Taiwan University*, Uwe Schwiegelshohn, *University of Dortmund*, and Len-Chin Liu, *Synopsys Taiwan Inc.*

LOW OPERATING VOLTAGE AND SHORT SETTLING TIME CMOS CHARGE PUMP FOR MEMS APPLICATIONS V-281

David Hong and Mourad El-Gamal, *McGill University*

ANALYSIS OF OUTPUT RIPPLE IN MULTI-PHASE CLOCKED CHARGE PUMPS V-285

Khoman Phang and Louie Pylarinos, *University of Toronto*

NO-RACE CHARGE RECYCLING COMPLEMENTARY PASS TRANSISTOR LOGIC (NCRCPL) FOR LOW POWER APPLICATIONS V-289

Ali Abbasian, Seid Hadi Rasouli, and Ali Afzali-Kusha, *University of Tehran*, and Mehrdad Nourani, *University of Texas at Dallas*

A PARAMETERIZED LOW POWER DESIGN FOR THE VARIABLE-LENGTH DISCRETE FOURIER TRANSFORM USING DYNAMIC PIPELINING V-293

Jiun-In Guo, Chih-Da Chien, and Chien-Chang Lin, *National Chung Cheng University*

MINIMIZING SWITCHING ACTIVITY IN INPUT WORD BY OFFSET AND ITS LOW POWER APPLICATIONS FOR FIR FILTERS V-297

Hojun Kim, *Globespanvirata*, and Jin-Gyun Chung, *Chonbuk National University*

A LOW-COMPLEXITY CORRELATION ALGORITHM V-301

Sau-Gee Chen and Kuang-Fu Cheng, *National Chiao Tung University*

A LOW-POWER CMOS INTEGRATED CIRCUIT FOR BEARING ESTIMATION V-305

Pedro Julian, Andreas Andreou, Pablo Mandolesi, and David Goldberg, *Johns Hopkins University*

SIMULTANEOUS TASK ALLOCATION, SCHEDULING AND VOLTAGE ASSIGNMENT FOR MULTIPLE-PROCESSORS-CORE SYSTEMS USING MIXED INTEGER NONLINEAR PROGRAMMING V-309

Lap-Fai Leung, Chi-Ying Tsui, and Wing-Hung Ki, *Hong Kong University of Science and Technology*

AN ILP-BASED SCHEDULING SCHEME FOR ENERGY EFFICIENT HIGH PERFORMANCE DATAPATH SYNTHESIS V-313

Saraju Mohanty, N. Ranganathan, and Sunil Chappidi, *University of South Florida*

A NOVEL HYBRID PASS LOGIC WITH STATIC CMOS OUTPUT DRIVE FULL-ADDER CELL V-317

Mingyan Zhang, Jiangmin Gu, and Chip Hong Chang, *Nanyang Technological University*

ULTRA LOW VOLTAGE, LOW POWER 4-2 COMPRESSOR FOR HIGH SPEED MULTIPLICATIONS V-321

Jiangmin Gu and Chip Hong Chang, *Nanyang Technological University*

DESIGN OF A 32-BIT SQUARER EXPLOITING ADDITION REDUNDANCY V-325

Asim Al-Khalili and Aiping Hu, *Concordia University*

AREA AND POWER OPTIMIZATION OF FPRM FUNCTION BASED CIRCUITS V-329

Yinshui Xia, B. Ali, and A.E.A. Almaini, *Napier University*

A CONFIGURABLE DIVIDER USING DIGIT RECURRENCE V-333

Anders Berkeman and Viktor Öwall, *Lund University*

A LOW POWER ASYNCHRONOUS GF(2^173) ALU FOR ELLIPTIC CURVE CRYPTO-PROCESSOR V-337

Pak Keung Leung, Chiu Sing Choy, Cheong Fat Chan, and Kong Pang Pun, *The Chinese University of Hong Kong*

LOW POWER BLOCK BASED FIR FILTERING CORES V-341

Ahmet Erdogan and Tughrul Arslan, *The University of Edinburgh*

DYNAMIC OPERAND TRANSFORMATION FOR LOW-POWER MULTIPLIER-ACCUMULATOR DESIGN V-345

Masayoshi Fujino and Vasily Moshnyaga, *Fukuoka University*

VOLTAGE SCALING AND REPEATER INSERTION FOR HIGH-THROUGHPUT LOW-POWER INTERCONNECT V-349

Vinita Deodhar and Jeffrey Davis, *Georgia Institute of Technology*

A TRIPLE PORT RAM BASED LOW POWER COMMUTATOR ARCHITECTURE FOR A PIPELINED FFT PROCESSOR V-353

Mohammad Hasan and Tughrul Arslan, *The University of Edinburgh*

DSP ENGINE FOR ULTRA-LOW-POWER AUDIO APPLICATIONS V-357

Dik Morling and Izzet Kale, *University of Westminster*, Steve Morris and Frank Custode, *Zarlink Semiconductor*

AN ASYNCHRONOUS PIPELINE COMPARISONS WITH APPLICATION TO DCT MATRIX-VECTOR MULTIPLICATION V-361

Sunan Tugsinavisut and Suwicha Jirayucharoensak, *University of Southern California*, and Peter Beerel, *Fulcrum Microsystems*

A MULTI-PHASE CHARGE-SHARING TECHNIQUE WITHOUT EXTERNAL CAPACITOR FOR LOW-POWER TFT-LCD COLUMN DRIVERS V-365

Shao-Sheng Yang, *National Tsing Hua University*, Pao-Lin Guo, *Da-Yeh University*, Tsin-Yuan Chang, *National Tsin-Hua University*, and Jin-Hua Hong, *Da-Yeh University*

AN ACTIVE LEAKAGE-INJECTION SCHEME APPLIED TO LOW-VOLTAGE SRAMS V-369

Jader de Lima, *Universidade Estadual Paulista*

LOW-POWER AND LOW-VOLTAGE FULLY PARALLEL CONTENT-ADDRESSABLE MEMORY V-373

Chi-Sheng Lin, Kuan-Hua Chen, and Bin-Da Liu, *National Cheng Kung University*

A LOW POWER CHARGE SHARING ROM USING DUMMY BIT LINES V-377

Byung-Do Yang and Lee-Sup Kim, *KAIST*

A LOW-VOLTAGE MICROPOWER ASYNCHRONOUS MULTIPLIER FOR A MULTIPLIERLESS FIR FILTER V-381

Chien-Chung Chua, Bah-Hwee Gwee, and Joseph Chang, *Nanyang Technological University*

INCREASING THE LOCALITY OF MEMORY ACCESS PATTERNS BY LOW-OVERHEAD HARDWARE ADDRESS RELOCATION V-385

Alberto Macii and Enrico Macii, *Politecnico Di Torino*, and Massimo Poncino, *Universit`a di Verona*

A SEMI-GRAY ENCODING ALGORITHM FOR LOW-POWER STATE ASSIGNMENT V-389

Chunhong Chen, Jiang Zhao, and Majid Ahmadi, *University of Windsor*

A POWER EFFICIENT REGISTER FILE ARCHITECTURE USING MASTER LATCH SHARING V-393

Marek Wroblewski, *Munich University of Technology*, Matthias Mueller, Sven Simon, Andreas Wortmann, and Wolfgang Pieper, *Hochschule Bremen*, and Josef A. Nossek, *Munich University of Technology*

A NEW SYNCHRONOUS MIRROR DELAY WITH AN AUTO-SKEW-GENERATION CIRCUIT V-397

Sei Hyung Jang, *Hynix Semiconductor*

AN SOI 4 TRANSISTORS SELF-REFRESH ULTRA-LOW-VOLTAGE MEMORY CELL V-401

Amara Amara and Olivier Thomas, *ISEP*

HIGH PERFORMANCE AND LOW POWER COMPLETION DETECTION CIRCUIT V-405

Hing Mo Lam and Chi Ying Tsui, *Hong Kong University of Science and Technology*

LOW-POWER CMOS CIRCUIT TECHNIQUES FOR MOTION ESTIMATORS V-409

Tadayoshi Enomoto and Tomomi Ei, *Chuo University*

POWER-DELAY TRADEOFFS IN RESIDUE NUMBER SYSTEM V-413

Alberto Nannarelli, Gian Carlo Cardarilli, and Marco Re, *University of Rome Tor Vergata*

UPPER AND LOWER BOUNDS ON FSM SWITCHING ACTIVITY V-417

Eleptheria Athanasopoulou, and Christoforos Hadjicostis, *University of Illinois*

EXPLOITING RECONFIGURABILITY FOR LOW-POWER CONTROL OF EMBEDDED PROCESSORS V-421

Luigi Carro, Edgard Corrêa, and Rodrigo Cardozo, *Universidade Federal do Rio Grande do Sul*, Fernando Moares, *PUCRS - Faculdade de Informática*, and Sergio Bampi, *Universidade Federal do Rio Grande do Sul*

A DUAL-PULSE-CLOCK DOUBLE EDGE TRIGGERED FLIP-FLOP FOR LOW VOLTAGE AND HIGH SPEED APPLICATION V-425

Kuo-Hsing Cheng and Yung-Hsiang Lin, *Tamkang University*

MTCMOS WITH OUTER FEEDBACK (MTOF) FLIP-FLOPS V-429

Mircea Stan and Marco Barcella , *University of Virginia*

Track 7.2: VLSI Design

COMPARISON OF SYNTHESIZED BUS AND CROSSBAR INTERCONNECTION ARCHITECTURES V-433

Vesa Lahtinen, Erno Salminen, Kimmo Kuusilinna, and Timo HämäläInen, *Tampere University of Technology*

A DIGITALLY CONTROLLED PLL FOR DIGITAL SOCS V-437

Thomas Olsson and Peter Nilsson, *Lund University*

NOVEL SUB-1V CMOS DOMINO DYNAMIC LOGIC CIRCUIT USING A DIRECT BOOTSTRAP (DB) TECHNIQUE FOR LOW-VOLTAGE CMOS VLSI V-441

Po-Chen, *National Taiwan University*, and James Kuo, *University of Waterloo*

A LOW POWER DELAYED-CLOCKS GENERATION AND DISTRIBUTION SYSTEM V-445

Su Kio, Kian Chong, and Carl Sechen, *University of Washington*

CLOCK RECOVERY IN HIGH-SPEED MULTILEVEL SERIAL LINKS V-449

FaisalMusa and Anthony Carusone, *University of Toronto*

FULL-CUSTOM CMOS REALIZATION OF A HIGH-PERFORMANCE BINARY SORTING ENGINE WITH LINEAR AREA-TIME COMPLEXITY V-453

Tugba Demirci, *Sabanci University*, Ilhan Hatirnaz and Yusuf Leblebici, *Swiss Federal Institute of Technology*

SIMULATED EVOLUTION ALGORITHM FOR MULTIOBJECTIVE VLSI NETLIST BI-PARTITIONING V-457

Sadiq Sait, Aiman H. El-Maleh, and Raslan H. Al-Abaji, *King Fahd University of Petroleum & Minerals*

PLIC-PLAC: A NOVEL CONSTRUCTIVE ALGORITHM FOR PLACEMENT V-461

Renato Hentschke and Ricardo Reis, *Universidade Federal do Rio Grande do Sul*

GRAPH-BASED APPROACH TO EVALUATE NET ROUTABILITY OF A FLOORPLAN V-465

Niwat Waropas, Rajendar Koltur, and Malgorzata Chrzanowska-Jeske, *Portland State University*

CSD MULTIPLIERS FOR FPGA DSP APPLICATIONS V-469

Michael Soderstrand, *Oklahoma State University*

ELECTRICAL CHARACTERISTICS OF MULTI-LAYER POWER DISTRIBUTION GRIDS V-473

Andrey Mezhiba and Eby Friedman, *University of Rochester*

ACCURATE RISE TIME AND OVERSHOOT ESTIMATION IN RLC INTERCONNECTS V-477

Noha Mahmoud and Yehea Ismail, *Northwestern University*

NOISE-CONSTRAINED INTERCONNECT OPTIMIZATION FOR NANOMETER TECHNOLOGIES V-481

Mohamed Elgamel, Kannan Tharmalingam, and Magdy Bayoumi, *University of Louisiana at Lafayette*

A CROSSTALK AWARE TWO-PIN NET ROUTER V-485

Ming-Fu Hsiao, *National Taiwan University*, Malgorzata Marek-Sadowska, *University of California, Santa Barbara*, and Sao-Jie Chen, *National Taiwan University*

PLACEMENT WITH SYMMETRY CONSTRAINTS FOR ANALOG LAYOUT USING RED-BLACK TREES V-489

Sarat C. Maruvada, Karthik Krishnamoorthy, Subodh Annojvala, and Florin Balasa, *University of Illinois at Chicago*

ARBITRARY CONVEX AND CONCAVE RECTILINEAR BLOCK PACKING BASED ON CORNER BLOCK LIST V-493

Yuchun Ma, Xianlong Hong, Sheqin Dong, Song Chen, and Yici Cai, *Tsinghua University*, Chung-Kuan Cheng, *University of California, San Diego*, and Jun Gu, *Science & Technology University of Hong Kong*

GENERAL ITERATIVE HEURISTICS FOR VLSI MULTIOBJECTIVE PARTITIONING V-497

Sadiq Sait, Aiman H. El-Maleh, and Raslan H. Al-Abaji, *King Fahd University of Petroleum & Minerals*

TILE-GRAPH-BASED POWER PLANNING V-501

Jyh Perng Fang and Sao Jie Chen, *National Taiwan University*

HIGH PERFORMANCE ASYNCHRONOUS BUS FOR SOC V-505

Eun-Gu Jung, Byung-Soo Choi, and Dong-Ik Lee, *Kwang-Ju Institute of Science and Technology*

MINIMIZING COUPLING JITTER BY BUFFER RESIZING FOR COUPLED CLOCK NETWORKS V-509

Ming-Fu Hsiao, *National Taiwan University*, Malgorzata Marek-Sadowska, *University of California, Santa Barbara*, and Sao-Jie Chen, *National Taiwan University*

STATISTICAL MODELING OF GATE-DELAY VARIATION WITH CONSIDERATION OF INTRA-GATE VARIABILITY V-513

Kenichi Okada, *Tokyo Institute of Technology*, Kento Yamaoka and Hidetoshi Onodera, *Kyoto University*

EFFICIENT CORE DESIGNS BASED ON PARAMETERIZED MACROCELLS WITH ACCURATE DELAY MODELS V-517

Makram Mansour, Mohammad Mansour, and Amit Mehrotra, *University of Illinois at Urbana-Champaign*

Track 7.3: VLSI Testing

CONFIGURABLE TWO-DIMENSIONAL LINEAR FEEDBACK SHIFTER REGISTERS FOR DETERMINISTIC AND RANDOM PATTERNS V-521

Chien-In Chen and Kiran George, *Wright State University*

A RESOURCE BALANCING APPROACH TO SOC TEST SCHEDULING V-525

Dan Zhao and Shambhu Upadhyaya, *The State University of New York at Buffalo*

ALIASING PROBABILITY CALCULATIONS IN NONLINEAR COMPACTORS V-529

Christoforos Hadjicostis, *University of Illinois*

A DETERMINISTIC DYNAMIC ELEMENT MATCHING APPROACH TO ADC TESTING V-533

Beatriz Olleta, Lance Juffer, Degang Chen, and Randall Geiger, *Iowa State University*

EXPERIMENTAL EVALUATION AND VALIDATION OF A BIST ALGORITHM FOR CHARACTERIZATION OF A/D CONVERTER PERFORMANCE V-537

Kumar Parthasarathy and Le Jin, *Iowa State University*, Turker Kuyel, *Texas Instruments Inc,* Dana Price, *Motorola Inc*, Degang Chen and Randall Geiger, *Iowa State University*

SYSTEMATIC TEST PROGRAM GENERATION FOR SOC TESTING USING EMBEDDED PROCESSOR V-541

Mohammad Tehranipour and Mehrdad Nourani, *University of Texas at Dallas*, Mehdi Fakhraie and Ali Afzali-Kusha, *University of Tehran*

ON EFFICIENT EXTRACTION OF PARTIALLY SPECIFIED TEST SETS FOR SYNCHRONOUS SEQUENTIAL CIRCUITS V-545

Aiman El-Maleh and Khaled Al-Utaibi, *King Fahd University of Petroleum & Minerals*

COMBINATIONAL CIRCUIT FAULT DIAGNOSIS USING LOGIC EMULATION V-549

Shyue-Kung Lu and Jian-Long Chen, *Fu-Jen Catholic University,* Cheng-Wen Wu, *National Tsin Hua University* , Wen-Feng Chang, *Wan-Leng Institute of Technology*, and Shi-Yu Huang, *National Tsin Hua University*

THE FAULT DETECTION OF CROSS-CHECK TEST SCHEME FOR INFRARED FPA V-553

Meng-Lieh Sheu, Tai-Ping Sun, Chi-Wen Lu, and Mon-Chau Shie, *National Taiwan University of Science & Technology*

MEASUREMENT AND SPICE PREDICTION OF SUB-PICOSECOND CLOCK JITTER IN A/D CONVERTERS V-557

Alfio Zanchi, Ioannis Papantonopoulos, and Frank Tsay, *Texas Instruments, Inc.*

A TESTER-ON-CHIP IMPLEMENTATION IN 0.18U CMOS UTILIZING A MEMS INTERFACE V-561

Rashid Rashidzadeh, William C. Miller, and Majid Ahmadi, *University of Windsor*

ON-CHIP DEBUG SUPPORT FOR EMBEDDED SYSTEMS-ON-CHIP V-565

Klaus Maier, *University of Kent*

A MODULAR TEST STRUCTURE FOR CMOS MISMATCH CHARACTERIZATION V-569

Massimo Conti, Paolo Crippa, Francesco Fedecostante, Simone Orcioni, Francesco Ricciardi, Claudio Turchetti, and Loris Vendrame, *University of Ancona*

EFFICIENT BIST SCHEMES FOR RNS DATAPATHS V-573

Dimitrios Nikolos, Dimitris Nikolos, Haridimos Vergos, *University of Patras*, and Constantinos Efstathiou, *TEI of Athens*

BIST FOR CLOCK JITTER MEASUREMENT V-577

Kuo-Hsing Cheng, Shu-Yu Jiang, and Zong-Shen Chen, *Tamkang University*

A NEW INITIALIZATION TECHNIQUE FOR ASYNCHRONOUS CIRCUITS V-581

Kaamran Raahemifar, *Ryerson University*, and Majid Ahmadi, *University of Windsor*

A CASE STUDY OF COST AND PERFORMANCE TRADE-OFF ANALYSIS FOR MIXED-SIGNAL INTEGRATION IN SYSTEM-ON-CHIP V-585

Meigen Shen, Zheng Li-Rong, and Hannu Tenhunen, *Royal Institute of Technology Stockholm*

Track 7.4: VLSI Implementations

FAST PROTOTYPING OF RECONFIGURABLE ARCHITECTURES FROM A C PROGRAM V-589

Sebastien Bilavarn, *EPFL*, Guy Gogniat, *Universite de Bretagne Sud*, Jean Luc Philippe and Lilian Bossuet, *LESTER - South Britany University*

INTERFACE DESIGN APPROACH FOR SYSTEM ON CHIP BASED ON CONFIGURATION V-593

Issam Maalej, *ENIS* and Guy Gogniat, *Universite de Bretagne Sud*, Mohamed Abid, *ENIS*, and Jean-Luc Philippe, *LESTER - South Britany University*

CIRCUIT DESIGN FROM OPTIMAL WAVELET PACKET SERIES EXPRESSIONS V-597

Radomir Stankovic, *Faculty of Electronics, Yugoslavia,* Karen Egiazarian and Jaakko Astola, *Tampere University of Technology*, and Milena Stankovic, *Faculty of Electronics, Yugoslavia*

AN INTEGRATED FRAMEWORK OF DESIGN OPTIMIZATION AND SPACE MINIMIZATION FOR DSP APPLICATIONS V-601

Qingfeng Zhuge and Edwin H.-M. Sha, *University of Texas At Dallas*, and Chantana Chantrapornchai, *Silpakorn University*

REDUCING THE NUMBER OF VARIABLE MOVEMENTS IN EXACT BDD MINIMIZATION V-605

Ruediger Ebendt, *University of Kaiserslautern*

A NOVEL IMPROVEMENT TECHNIQUE FOR HIGH-LEVEL TEST SYNTHESIS V-609

Saeed Safari, *Sharif University of Technology*, Hadi Esmaeilzadeh, *University of Tehran*, and Amir Hossein Jahangir, *Sharif University of Technology*

SYNTHESIS AND OPTIMIZATION OF INTERFACES BETWEEN HARDWARE MODULES WITH INCOMPATIBLE PROTOCOLS V-613

Vassilis Androutsopoulos, Thomas Clarke, and Mike Brookes, *Imperial College*

CONTROL SIGNAL SHARING OF ASYNCHRONOUS CIRCUITS USING DATAPATH DELAY INFORMATION V-617

Hiroshi Saito, Euiseok Kim, Masashi Imai, Nattha Sretasereekul, Hiroshi Nakamura, and Takashi Nanya, *The University of Tokyo*

MULTITASKING IN HARDWARE-SOFTWARE CODESIGN FOR RECONFIGURABLE COMPUTER V-621

Theerayod Wiangtong, *Mahanakorn University of Technology*, Peter Cheung and Wayne Luk, *Imperial College*

SYSTEM-ON-CHIP DESIGN USING INTELLECTUAL PROPERTIES WITH IMPRECISE DESIGN COSTS V-625

Byoung-Woon Kim, *Hynix Semiconductor*, and Chong-Min Kyung, *KAIST*

AN EXPLORATION-BASED BINDING AND SCHEDULING TECHNIQUE FOR SYNTHESIS OF DIGITAL BLOCKS FOR MIXED-SIGNAL APPLICATIONS V-629

Natt Thepaysuwan, Hua Tang, and Alex Doboli, *State University of New York at Stony Brook*

Track 7.5: Fault Detection and Fault Tolerance

A SYSTOLIC MULTIPLIER WITH LSB FIRST ALGORITHM OVER GF(2^M) WHICH IS AS EFFICIENT AS THE ONE WITH MSB FIRST ALGORITHM V-633

Soonhak Kwon, *Sungkyunkwan University*, Chang Hoon Kim and Chun Pyo Hong, *Daegu University*

A LOGIC-AWARE LAYOUT METHODOLOGY TO ENHANCE THE NOISE IMMUNITY OF DOMINO CIRCUITS V-637

Yonghee Im and Kaushik Roy, *Purdue University*

PARTIAL REROUTING ALGORITHM FOR RECONFIGURABLE VLSI ARRAYS V-641

Wu Jigang and Srikanthan Thambipillai, *Nanyang Technological University*

FAULT TOLERANT DATAPATH BASED ON ALGORITHM REDUNDANCY AND VOTE-WRITEBACK MECHANISM V-645

Mineo Kaneko and Kazuaki Oshio, *Japan Advanced Institute of Science and Technology*

A FAULT TOLERANT HARDWARE BASED FILE SYSTEM MANAGER FOR SOLID STATE MASS MEMORY V-649

Giancarlo Cardarilli, Marco Ottavi, Salvatore Pontarelli, Marco Re, and Adelio Salsano, *University of Rome Tor Vergata*

Track 9: NEURAL NETWORKS & SYSTEMS

MIXED-SIGNAL VLSI IMPLEMENTATION OF THE PRODUCTS OF EXPERTS' CONTRASTIVE DIVERGENCE LEARNING SCHEME V-653

Patrice Fleury and Alan Murray, *The University of Edinburgh*

ANALOG CONTINUOUS-TIME RECURRENT DECISION CIRCUIT WITH HIGH SIGNAL-VOLTAGE SYMMETRY AND DELAY-TIME EQUALITY V-657

Akira Hirose and Kazuhiko Nakazawa, *The University of Tokyo*

HYBRID NEURAL NETWORK ARCHITECTURE FOR AGE IDENTIFICATION OF ANCIENT KANNADA SCRIPTS V-661

Harish Kashyap, Bansilal P H, and Arun Koushik, *The National Institute of Engineering*

PLASTIC NNS FOR BIOCHEMICAL DETECTION V-665

Hoda Abdel-Aty-Zohdy and Jacob Allen, *Oakland University*, and Robert Ewing, *Air Force Research Laboratory*

ALGORITHMS FOR COMPUTATING PRINCIPAL AND MINOR INVARIANT SUBSPACES OF LARGE MATRICES V-669

Mohammed Hasan, *University of Minnesota, Duluth*

ANALYSIS OF GLOBAL EXPONENTIAL STABILITY FOR A CLASS OF BI-DIRECTIONAL ASSOCIATIVE MEMORY NETWORKS V-673

Hongxia Wang and Chen He, *Shanghai Jiao Tong University*, and Juebang Yu, *University of Electronic Science and Technology of China*

SYNTHESIS FOR SYMMETRIC WEIGHT MATRICES OF NEURAL NETWORKS V-677

Montien Saubhayana and Robert Newcomb, *University of Maryland, College Park*

EXPONENTIAL PERIODICITY OF NEURAL NETWORKS WITH DELAYS V-681

Changyin Sun, *Southeast University*, Changgui Sun, *NJUST*, and Chun-Bo Feng, *Southeast University, China*

Track 9.1: Neural Systems and Applications

A HIGHLY SCALABLE 3D CHIP FOR BINARY NEURAL NETWORK CLASSIFICATION APPLICATIONS V-685

Amine Bermak, *Hong Kong University of Science and Technology*

COMPACT IMAGE COMPRESSION USING SIMPLICIAL AND ART NEURAL SYSTEMS WITH MIXED SIGNAL IMPLEMENTATIONS V-689

Radu Dogaru, *Polytechnic University of Bucharest*, Manfred Glesner, *Technical University of Darmstadt*, and Ioana Dogaru, *B&C Microsystems SRL, Romania*

RBF NETWORK BASED ON GENETIC ALGORITHM OPTIMIZATION FOR NONLINEAR TIME SERIES PREDICTION V-693

Qingnian Zhang, Xiang Yang He, and Jianqi Liu, *Wuhan University of Technology, P.R. China,*

NEUROSEARCH: A PROGRAM LIBRARY FOR NEURAL NETWORK DRIVEN SEARCH META-HEURISTICS V-697

Ewin Mardhana and Tohru Ikeguchi, *Saitama University*

ARTIFICIAL NEURAL NETWORKS AS RAIN ATTENUATION PREDICTORS IN EARTH-SPACE PATHS V-701

Gilson Alencar and Luiz CalôBa, *Coppe - UFRJ*, and Mauro Assis, *IME*

STEPSIZE CONTROL IN NLMS ACOUSTIC ECHO CANCELLATION USING A NEURAL NETWORK APPROACH V-705

Giovanni Tummarello, Fabio Nardini, and Francesco Piazza, *University of Ancona*

DEVELOPMENT OF A DECISION SUPPORT SYSTEM FOR HEART DISEASE DIAGNOSIS USING MULTILAYER PERCEPTRON V-709

Hongmei Yan *Chongqing University*, Jun Zheng and Yingtao Jiang, *University of Nevada*, Chenglin Peng, *Chongqing University*, and Qinghui Li, *Southwest Hospital*

PREPROCESSING BASED SOLUTION FOR THE VANISHING GRADIENT PROBLEM IN RECURRENT NEURAL NETWORKS V-713

Francesco Piazza, *University of Ancona*, Stefano Squartini, *Università degli Studi di Ancona*, and Amir Hussain, *University of Stirling*

MEDICAL DATA MINING MODEL FOR ORIENTAL MEDICINE VIA BYY BINARY INDEPENDENT FACTOR ANALYSIS V-717

Jeong Shim, *Yongin Songdam College*, and Lei Xu, *Chineses University of Hong Kong*

GLOBAL ASYMPTOTIC STABILITY OF A LARGER CLASS OF DELAYED NEURAL NETWORKS V-721

Sabri Arik, *Istanbul University*

GLOBAL OPTIMIZATION OF NEURAL NETWORK WEIGHTS USING SUBENERGY TUNNELING FUNCTION AND RIPPLE SEARCH V-725

Hong Ye and Zhiping Lin, *Nanyang Technological University*

A RECURRENT NEURAL NETWORK FOR 1-D PHASE RETRIEVAL V-729

Adrian Burian and Jarmo Takala, *Tampere University of Technology*

CLASS-BASED NEURAL NETWORK METHOD FOR FAULT LOCATION OF LARGE-SCALE ANALOGUE CIRCUITS V-733

Yigang He and Tan Yanghong, *Hunan University*, and Sun Yichuang, *University of Hertfordshire*

DYNAMICAL HYSTERESIS NEURAL NETWORK FOR GRAPH COLORING PROBLEM V-737

Kenya Jin'no, Hiroshi Taguchi, Takao Yamamoto, and Haruo Hirose, *Nippon Institute of Technology*

APPLICATION OF CRITICAL SUPPORT VECTOR MACHINE TO TIME SERIES PREDICTION V-741

Thanapant Raicharoen and Chidchanok Lursinsap, *Chulalongkorn University*, and Paron Sanguanbhokai, *Air Force Acedemy*

A NEURAL NETWORK TO RETRIEVE ATMOSPHERIC PARAMETERS FROM INFRARED HIGH RESOLUTION SENSOR SPECTRA V-745

Antonio Luchetta, *University of Florence*, Carmine Serio, *University of Basilicata*, and Mariassunta Viggiano, *Istituto di Metodologie per l'Analisi Ambientale - CNR*

A CONVOLUTIONAL NEURAL ARCHITECTURE: AN APPLICATION FOR DEFECTS DETECTION IN CONTINUOUS MANUFACTURING SYSTEMS V-749

Jose Calderon-Martinez, *Instituto Tecnologico De Aguascalientes*, and Pascual Campoy-Cervera, *Universidad Politecnica de Madrid*

Track 9.2: New and Emerging Techniques

NEW CRITERIA FOR THE EXISTENCE OF STABLE EQUILIBRIUM POINTS IN NONSYMMETRIC CELLULAR NEURAL NETWORKS V-753

Neyir Ozcan and Sabri Arik, *Istanbul University*, and Vedat Tavsanoglu, *Isik University*

AN AREA-EFFICIENT FULL-WAVE CURRENT RECTIFIER FOR ANALOG ARRAY PROCESSING V-757

Jonne Poikonen, *University of Turku*, and Ari Paasio, *Helsinki University of Technology*

CMOS IMPLEMENTATION OF AN EXTENDED CNN CELL TO DEAL WITH COMPLEX DYNAMICS V-761

Gianluca Giustolisi, *Università di Catania*, and Alessandro Rizzo, *Politecnico di Bari*

ON DYNAMIC BEHAVIOR OF FULL RANGE CNNS V-765

Fernando Corinto, Marco Gilli, and Civalleri Pier Paolo, *Politecnico di Torino*

HIGH SPEED ROAD BOUNDARY DETECTION FOR AUTONOMOUS VEHICLE WITH THE MULTI-LAYER CNN V-769

Hyongsuk Kim, Seungwan Hong, and Hongrak Son, *Chonbuk National University*, Tamas Roska, *The Hungarian Academy of Sciences*, and Frank Werblin, *University of California, Berkeley*

Track 9.3: Fuzzy Logic Circuits

CELLULAR NEURAL NETWORKS WITH SECOND-ORDER CELLS AND THEIR PATTERN FORMING PROPERTIES V-773

Radu Matei, *Technical University of Iasi*

A NEW METHOD FOR IMPLEMENTING GATE OPERATIONS IN A QUANTUM FACTORING ALGORITHM V-777

Domenico Porto, *STMicroelectrionics*, Luigi Fortuna, Paola Barrera, and Antonino Calabro, *Università degli Studi di Catania*

PERIODIC ORBITS AND BIFURCATIONS IN ONE-DIMENSIONAL ARRAYS OF CHUA'S CIRCUITS V-781

Marco Gilli, Paolo Checco, and Fernando Corinto, *Politecnico di Torino*

FUZZY POWER CONTROL WITH WEIGHTING FUNCTION IN DS-CDMA CELLULAR MOBILE COMMUNICATION SYSTEM V-785

Watit Benjapolakul and Wasimon Panichpattanakul, *Chulalongkorn University*

FUZZY TEMPORAL REPRESENTATION AND REASONING V-789

Phayung Meesad, *King Mongkut's Institute of Technology*, and Gary Yen, *Oklahoma State University*

SYNCHRONIZATION IN ARRAYS OF FUZZY CHAOTIC OSCILLATORS V-793

Maide Bucolo, Luigi Fortuna, and Manuela la Rosa, *Universita' Degli Studi di Catania*

A HIGH SPEED SCALABLE AND RECONFIGURABLE FUZZY CONTROLLER V-797

Rogelio Palomera Garcia, *Univ. of Puerto Rico- Mayaguez*, and Felix Homburg, *University of Dortmund*

Track 9.4: Neural Learning and Intelligent Systems

DESIGN OF A SELF-ORGANIZING LEARNING ARRAY SYSTEM V-801

Janusz Starzyk and Tsun-Ho Liu, *Ohio University*

FUZZY ASSOCIATIVE DATABASE FOR MULTIPLE PLANAR OBJECT RECOGNITION V-805

Shahed Shahir, Xiang Chen, and Majid Ahmadi, *University of Windsor*

TIGHT BOUNDED LOCALIZATION OF FACIAL FEATURES WITH COLOR AND ROTATIONAL INDEPENDENCE V-809

Suphakant Phimoltares and Chidchanok Lursinsap, *Chulalongkorn University*, and Kosin Chamnongthai, *King Mongkut's University of Technology Thonburi*

ANALOG CMOS IMPLEMENTATION OF GALLAGER'S ITERATIVE DECODING ALGORITHM APPLIED TO A BLOCK TURBO CODE V-813

Matteo Perenzoni, Andrea Gerosa, and Andrea Neviani, *University of Padova*

LEARNING TEMPORAL CORRELATIONS IN BIOLOGICALLY-INSPIRED AVLSI V-817

Adria Bofill-i-Petit and Alan Murray, *The University of Edinburgh*

DIGITAL PULSE MODE NEURON WITH LEAKAGE INTEGRATOR AND ADDITIVE RANDOM NOISE V-821

Hiroomi Hikawa, *Oita University*

LOSSLESS NON-ARBITRATED ADDRESS-EVENT CODING V-825

Vladimir Brajovic, *Carnegie Mellon University*

A WIDE-FIELD DIRECTION-SELECTIVE AVLSI SPIKING NEURON V-829

Shih-Chii Liu, *University and ETH Zurich*

FOCAL PLANE IMAGE SEGMENTATION USING LOCALLY INTERCONNECTED SPIKING PIXEL ARCHITECTURE V-833

Amine Bermak, *Hong Kong University of Science and Technology*, and Matthias Hofinger, *The University of Ulm*

DESIGN OF A REDUCED KII SET AND NETWORK IN ANALOG VLSI V-837

Dongming Xu, Liping Deng, John Harris, and Jose Principe, *University of Florida*

CLASSIFICATION OF BPSK AND QPSK SIGNALS WITH UNKNOWN SIGNAL LEVEL USING THE BAYES TECHNIQUE

Liang Hong *K.C. Ho*

Department of Electrical and Computer Engineering
University of Missouri-Columbia
Columbia, MO 65211, USA

ABSTRACT

Automatic modulation classification through digital signal processing has found many applications in wireless communication systems, including interference identification, spectrum management and electronic warfare. In practice, wireless communication environments suffer from ill-characterized interference, fading and multipath. As a result, the level of the received signal is always time-varying and not known. This paper studies the use of Bayes method to distinguish BPSK signal and QPSK signal without *a priori* knowledge of the received signal level. The maximum *a posteriori* classifier is derived with Rayleigh distributed signal amplitude. Simulations show that the classifier using the Bayes technique is superior to the classifier without considering the variation of signal level.

1. INTRODUCTION

Automatic modulation classification is a digital signal processing technique that automatically reports the modulation type of a received signal under the situation where the signal exists and some modulation parameters such as carrier frequency, symbol time, are available. It is a rapidly evolving area in various digital signaling systems currently deployed or planned in a variety of environments. Once the correct modulation type is identified, data demodulation, information extraction and signal exploitation can be followed. Modulation classification has found wide applications in both non-cooperative and cooperative scenario, including electronic warfare, reconnaissance, surveillance, threat analysis, as well as interference identification, adaptive channel modulation and spectrum management [1]-[7].

A variety of techniques have been proposed to distinguish binary phase shift keying (BPSK) signal and quadrature phase shift keying (QPSK) signal. Liedtke [1] proposed to use the difference between the consecutive information-bearing phase for classification. Wei and Mendel [2] transformed a general pattern classification problem into one of function approximation and applied a fuzzy logic method for classification. Beidas and Weber [3] proposed a classifier based on the time domain higher-order correlations. Hsue and Soliman [4] introduced a modulation recognizer based on the zero-crossing characteristics of the received signals. Ho *et al.* [5] applied the histogram of the Haar wavelet transform magnitude peaks for identification. These methods require high signal-to-noise ratio (SNR).

One promising approach for BPSK and QPSK modulation type identification is the decision-theoretic approach. In this approach, modulation classification is considered as a composite hypothesis statistical testing problem and probabilistic arguments are employed for its solution. The resulting classifier is optimum in the sense that it minimizes the average cost function of misclassification probability. Polydoros and Kim [6] derived an approximate likelihood ratio classifier, which works well at low SNR, by assuming the signal level is constant and known to a receiver. However, in practice, wireless communication environments suffer from ill-characterized interference, fading and multipath. The level of the received signal is always varying and not known. Hong and Ho [7] tackled the problem by using generalized likelihood ratio test. The unknown signal level is first estimated using the maximum likelihood method and is then used in the likelihood ratio test for classification. In this paper, a modulation classifier based on the Bayes test is investigated to discriminate BPSK signal and QPSK signal with unknown signal level. The maximum *a posteriori* classifier is derived by averaging over the unknown signal level with the prior knowledge of the probability distribution of the signal level.

The paper is organized as follows. The next section introduces the signal model and assumptions. Section 3 elaborates in detail the classification test statistics. Section 4 presents the simulation results. Finally, conclusion is made in Section 5.

2. SIGNAL MODEL AND ASSUMPTIONS

The received signal consists of two uncorrelated components, a signal $s(t)$ and an additive white Gaussian noise (AWGN) $n(t)$, that is $r(t) = s(t) + n(t)$, $0 \leq t \leq NT_s$, where $s(t)$ is a BPSK signal or a QPSK signal. The noise $n(t)$ is zero-

0-7803-7761-3/03/\$17.00 ©2003 IEEE

mean and has two-sided power spectral density (PSD) equal to $N_0/2$ W/Hz. T_s is the symbol duration and N is the number of observed symbols. The signal $s(t)$ can be described in quadrature form as

$$s(t) = s_I(t)cos(2\pi f_c t + \theta_c) - s_Q(t) sin(2\pi f_c t + \theta_c), \quad (1)$$

where f_c is the carrier frequency, θ_c is the carrier phase, $s_I(t)$ and $s_Q(t)$ are the in-phase and quadrature components respectively. They are given by

$$s_I(t) = \sum_{n=1}^{N} \sqrt{2S}\alpha cos\theta_n p(t - nT_s),$$
$$s_Q(t) = \sum_{n=1}^{N} \sqrt{2S}\alpha sin\theta_n p(t - nT_s), \quad (2)$$

where θ_n is the modulation phase sequence. It is regarded as a random variable which varies from symbol to symbol. For QPSK signal, $\theta_n \in \{i\pi/4 : i = 1, 3, 5, 7\}$. For BPSK signal, $\theta_n \in \{0, \pi\}$. $p(t)$ is the standard unit pulse of duration T_s seconds. S is the nominal signal power which is a constant. $\alpha \geq 0$ is the signal level factor. It is a random variable and indicates the variations of signal level along time. The probability density function of α is assumed to follow the Rayleigh distribution:

$$P_r(\alpha) = \frac{\alpha}{\sigma_\alpha^2} e^{-\frac{\alpha^2}{2\sigma_\alpha^2}}, \quad (3)$$

where σ_α^2 is the variance of α.

In this study, the symbol time T_s, the carrier frequency f_c, the carrier phase θ_c, the PSD of AWGN N_0, the nominal power level S and the variance of α are presumed known to a receiver. Without loss of generality, θ_c will be set to zero. Furthermore, to simplify illustration, the variation of signal level is assumed to be slow. That is, the signal level factor α is relatively constant but unknown within an observation period, and it varies from one observation period to another.

3. BPSK AND QPSK MODULATION CLASSIFIER USING THE BAYES TECHNIQUE

The problem of classifying BPSK signal and QPSK signal without prior knowledge of signal level can be stated as the composite binary hypothesis testing problem:

$$H_1: \quad r(t) = s(t; BPSK, \theta_n, \alpha) + n(t)$$
$$H_0: \quad r(t) = s(t; QPSK, \theta_n, \alpha) + n(t) \quad (4)$$

The likelihood function of the received waveform $r(t)$ in AWGN, given the signal $s(t)$ and signal level factor α, is defined as [8]

$$\Lambda[r(t)|s(t;\alpha)] = E_{\theta_n}\left\{ exp\left[\frac{2}{N_0}\int_0^{NT_s} r(t)s(t;\alpha)dt - \frac{1}{N_0}\int_0^{NT_s} s^2(t;\alpha)dt\right]\right\}, \quad (5)$$

where E_{θ_n} denotes expectations with respect to the modulation phase sequence θ_n.
Denote

$$r_{I,n} = \int_{(n-1)T_s}^{nT_s} r(t)cos(2\pi f_c t)dt,$$
$$r_{Q,n} = \int_{(n-1)T_s}^{nT_s} r(t)sin(2\pi f_c t)dt, \quad (6)$$

as the in-phase and quadrature matched filter outputs during the nth symbol interval. It has been shown in [7] that, substituting the signal model equations (1) and (2) into (5), and taking into account that the phase sequence θ_n is independent yields

$$\Lambda[r(t)|s(t;\alpha)] = \prod_{n=1}^{N} E_{\theta_n}\left\{exp\left[\frac{2}{N_0}\left(\sqrt{2S}\alpha cos\theta_n r_{In} - \sqrt{2S}\alpha sin\theta_n r_{Qn}\right) - \frac{ST_s}{N_0}\alpha^2\right]\right\}. \quad (7)$$

Upon applying the expectation over θ_n, (7) can be simplified further. For BPSK signal, $\theta_n \in \{0, \pi\}$ and has equal probability, so that

$$\Lambda[r(t)|s(t;\alpha)]_{BPSK} = \prod_{n=1}^{N}\left\{\frac{1}{2}\left[exp\left(\frac{2\sqrt{2S}}{N_0}\alpha r_{In}\right) + exp\left(-\frac{2\sqrt{2S}}{N_0}\alpha r_{Qn}\right)\right]\cdot exp\left(-\frac{ST_s}{N_0}\alpha^2\right)\right\}. \quad (8)$$

For QPSK signal, $\theta_n \in \{i\pi/4 : i = 1, 3, 5, 7\}$ and has equal probability. Applying the expectation over θ_n gives

$$\Lambda[r(t)|s(t;\alpha)]_{QPSK} = \prod_{n=1}^{N}\left\{\frac{1}{4}\left[exp\left(\frac{2\sqrt{S}}{N_0}\alpha r_{In}\right) + exp\left(-\frac{2\sqrt{S}}{N_0}\alpha r_{In}\right)\right]\cdot\left[exp\left(\frac{2\sqrt{S}}{N_0}\alpha r_{Qn}\right) + exp\left(-\frac{2\sqrt{S}}{N_0}\alpha r_{Qn}\right)\right]\cdot exp\left(-\frac{ST_s}{N_0}\alpha^2\right)\right\}. \quad (9)$$

Since the signal level factor α is a random variable with known Rayleigh distribution, applying the expectation over α gives the Bayes likelihood function

$$\Lambda'[r(t)|s(t)] = E_\alpha\{\Lambda[r(t)|s(t;\alpha)]\}. \quad (10)$$

Upon using (3), taking expectation of (8) over α gives the Bayes likelihood function for BPSK signal

$$\Lambda'[r(t)|s(t)]_{BPSK} = \int_0^\infty \Lambda[r(t)|s(t;\alpha)]_{BPSK} P_r(\alpha)d\alpha$$

$$= \left(\frac{1}{2}\right)^{N-1} \sum_{i_2=0}^{1} \sum_{i_3=0}^{1} \cdots \sum_{i_N=0}^{1} \left[\frac{1}{2\sigma_\alpha^2 A} + \frac{1}{\sigma_\alpha^2 A N_0} \sqrt{\frac{\pi S}{2A}} B_{i_2,i_3\cdots,i_N} \cdot exp\left(\frac{2S}{N_0^2 A} B^2_{i_2,i_3\cdots,i_N} \right) \cdot erf\left(\frac{1}{N_0} \sqrt{\frac{S}{2A}} B_{i_2,i_3\cdots,i_N} \right) \right], \quad (11)$$

where

$$A = \frac{NST_s}{N_0} + \frac{1}{2\sigma_\alpha^2} \quad (12)$$

and

$$B_{i_2,i_3,\cdots,i_N} = \left| r_{I1} + \sum_{w=2}^{N} (-1)^{i_w} r_{Iw} \right|. \quad (13)$$

$erf(x) = \frac{2}{\sqrt{\pi}} \int_0^x e^{-t^2} dt$ is the error function and there are 2^{N-1} terms for summation in (11).

The expectation of (9) over α gives the Bayes likelihood function for QPSK signal

$$\Lambda'[r(t)|s(t)]_{QPSK} = \int_0^\infty \Lambda[r(t)|s(t;\alpha)]_{QPSK} P_r(\alpha) d\alpha$$

$$= \left(\frac{1}{2}\right)^{2N-2} \sum_{j_2=0}^{1} \sum_{j_3=0}^{1} \cdots \sum_{j_N=0}^{1} \sum_{k_2=0}^{1} \sum_{k_3=0}^{1} \cdots \sum_{k_N=0}^{1} \left[\frac{1}{2\sigma_\alpha^2 A} + \frac{1}{4\sigma_\alpha^2 A N_0} \sqrt{\frac{\pi S}{A}} C_{j_2,j_3\cdots,j_N,k_2,k_3\cdots,k_N} \cdot exp\left(\frac{S}{N_0^2 A} C^2_{j_2,j_3\cdots,j_N,k_2,k_3\cdots,k_N} \right) \cdot erf\left(\frac{1}{N_0} \sqrt{\frac{S}{A}} C_{j_2,j_3\cdots,j_N,k_2,k_3\cdots,k_N} \right) + \frac{1}{4\sigma_\alpha^2 A N_0} \sqrt{\frac{\pi S}{A}} D_{j_2,j_3\cdots,j_N,k_2,k_3\cdots,k_N} \cdot exp\left(\frac{S}{N_0^2 A} D^2_{j_2,j_3\cdots,j_N,k_2,k_3\cdots,k_N} \right) \cdot erf\left(\frac{1}{N_0} \sqrt{\frac{S}{A}} D_{j_2,j_3\cdots,j_N,k_2,k_3\cdots,k_N} \right) \right], \quad (14)$$

where

$$C_{j_2,j_3\cdots,j_N,k_2,k_3\cdots,k_N} = \left| r_{I1} + r_{Q1} + \sum_{u=2}^{N} (-1)^{j_u} r_{Iu} + \sum_{v=2}^{N} (-1)^{k_v} r_{Qv} \right|,$$

$$D_{j_2,j_3\cdots,j_N,k_2,k_3\cdots,k_N} = \left| r_{I1} - r_{Q1} + \sum_{u=2}^{N} (-1)^{j_u} r_{Iu} - \sum_{v=2}^{N} (-1)^{k_v} r_{Qv} \right|, \quad (15)$$

Note that there are 2^{2N-2} terms for summation in (14).

Now taking the ratio of the two Bayes likelihood functions and comparing it with a certain threshold gives the Bayes test for classifying a BPSK signal and a QPSK signal when the signal level is unknown. It is equal to

$$l(r) = \frac{\Lambda'[r(t)|s(t]_{BPSK}}{\Lambda'[r(t)|s(t]_{QPSK}} \quad (16)$$

If $l(r)$ is bigger than a certain threshold, the input is classified as a BPSK signal. Otherwise it is classified as a QPSK signal. When the probability of error is minimized, the threshold should be set to $\gamma = P_{QPSK}/P_{BPSK}$, where $P_M; M = BPSK, QPSK$ is the prior probabilities of the two hypothesis. It is equal to one for equal prior probabilities of the two hypothesis.

Because the large number of the terms for the summation in (11) and (14), the computational loads in obtaining the exact solution for the Bayes likelihood functions of BPSK signal and QPSK signal increase exponentially with the increase of the number of observation symbols. Numerical integration techniques for the expectation over α may be preferred to find the Bayes likelihood functions of BPSK signal and QPSK signal.

4. SIMULATIONS

In this section, we use simulations to compare the performance of the classifier using the Bayes technique with that of the classifier assuming known and constant signal level. The results are illustrated in terms of the average probability of correct classification (PCC), which is defined as

$$PCC = \frac{P_r(BPSK|BPSK) + P_r(QPSK|QPSK)}{2}, \quad (17)$$

where $P_r(M|M); M = BPSK, QPSK$ is the probability that the classifier gives the correct modulation type of an input signal. The number of ensemble runs for all simulations is 1000. The number of the observed symbols is 100 and the variance of the signal level factor is 8.5. Left Riemann sums method is used to find the numerical solution for the expectation over α in the likelihood functions of BPSK signal and QPSK signal.

Figure 1 shows the PCC versus $E[\alpha^2]ST_s/N_0$, the average SNR. The solid line is for the performance of the classifier in [7] using the true signal level factor. This curve represents the best performance and serves as a reference. The dashed line represents the classification accuracy of the classifier using the Bayes technique. The dotted line shows the performance of the classifier assuming known and constant signal power that was set to the average signal power $E[\alpha^2]S$. The classification thresholds in both cases are set to one because the prior probabilities of BPSK signal and QPSK signal are equal. It is clear that when the average SNR is larger than -4 dB, the classifier using the Bayes technique has much better performance than the classifier without considering the variation of signal level.

When the best classification threshold is determined before hand according to the prior knowledge of the average SNR, the performance of the classifier without considering amplitude variation can be improved. Figure 2 shows the PCC versus the average SNR. The classification algorithms are the same as those used in Figure 1. The thresholds of the classifier using the Bayes technique was set to one. The threshold of the classifier assuming constant and known signal level was adjusted using the histogram of the likelihood values under the two hypotheses so that the PCC shown was the largest at the given average SNR. There is slight performance improvement for the classifier using the Bayes technique in this particular simulation. If the best threshold can be found in the case assuming constant signal amplitude, it may be possible to obtain a comparable performance as the Bayes method.

5. CONCLUSIONS

This paper proposed and studied the use of Bayes method to distinguish BPSK and QPSK signal with unknown signal level. The Bayes test is derived by averaging over the random signal level that has Rayleigh distribution. By setting the threshold to one for equally likely prior probabilities, the classifier using Bayes method has much improvement than the classifier without considering amplitude variation when the average SNR is larger than -4 dB. When the threshold of the classifier assuming constant known signal level is adjusted based on prior knowledge and SNR to give maximum PCC, there is slight improvement in performance of the classifier using Bayes method.

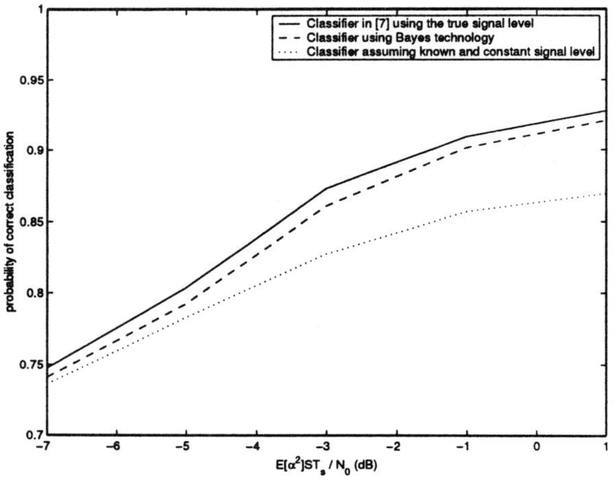

Figure 1: PCC versus average SNR.

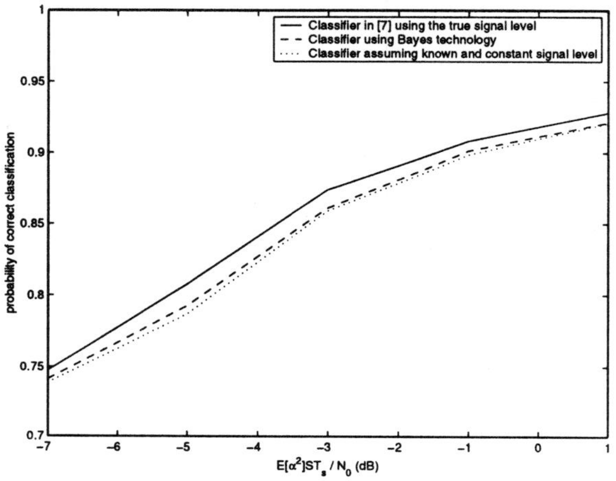

Figure 2: PCC versus average SNR.

6. REFERENCES

[1] F. F. Liedtkd, "Computer simulation of an automatic classification procedure for digitally modulated communication signals with unknown parameters," in *Signal Processing*, vol. 6, pp. 311-323, 1984.

[2] W. Wei and J. M. Mendel, "A fuzzy logic method for modulation classification in nonideal environments," in *IEEE Trans. Fuzzy Systems*, vol. 7, no. 3, pp. 333-344, June 1999.

[3] B.F. Beidas and C.L. Weber, "Higher-order correlation-based approach to modulation classification of digitally modulated signals," in *IEEE J. Select. Areas Commun*, vol. 13, no.1, pp. 89-101, January 1995.

[4] S.-Z. Hsue and S.S. Soliman, "Automatic modulation classification using zero crossing," in *IEE Proceedings F (Radar and Signal Processing)*, vol. 137, no. 6, pp. 459-464, December 1990.

[5] K.C. Ho, W. Prokopiw and Y.T. Chan, "Identification of M-ary PSK and FSK signals by the wavelet transform," in *Proc. IEEE MILCOM*, pp. 886-890, California, November 1995.

[6] A. Polydoros and K. Kim, "On the detection and classification of quadrature digital modulations in broad-band noise," in *IEEE Trans. Commun.*, vol. 38, no. 8, pp. 1199-1211, August 1990.

[7] L. Hong and K.C. Ho, "BPSK and QPSK modulation classification with unknown signal level," in *Proc. IEEE MILCOM.*, Los Angeles, California, October 2000.

[8] H. L. Van Trees, '*Detection, Estimation, and Modulation Theory*, New York: Wiley, 1971.

Jointly Minimum Symbol Error Rate FIR MIMO Transmitter and Receiver Filters for PAM Signal Vectors

Are Hjørungnes
Signal Processing Laboratory
Helsinki University of Technology
P. O. Box 3000, FIN-02015 HUT, Finland
Tel. +358 9 451 5386, Fax +358 9 452 3614
Email: arehj@wooster.hut.fi

Paulo S. R. Diniz
Signal Processing Laboratory
Universidade Federal do Rio de Janeiro
P. O. Box 68504, CEP: 21945-970 RJ, Brazil
Tel. +55 21 2562 8211, Fax +55 21 2562 8205
Email: diniz@lps.ufrj.br

***Abstract*— A theory is developed for jointly minimizing the symbol error rate (SER) between the desired and decoded signals with respect to the coefficients of the transmitter and receiver FIR multiple-input multiple-output (MIMO) filters. The original source signal is assumed to be a vector source with equally likely memoryless PAM vector components. The channel model constitutes of a known FIR MIMO transfer function and a Gaussian additive channel noise independent of the original signal. The channel input signal is assumed to be power constrained. When compared to the methods available in the literature, the proposed method yields better results due to the joint optimization of the transmitter-receiver pair.**

I. Introduction

The problem of transmitting a vector time series with memoryless equally likely PAM vector components over a vector channel is investigated. It is assumed that the channel is discrete in time and corrupted by additive signal-independent noise which is Gaussian complex circularly symmetric with zero mean and with known second-order statistics. Furthermore, it is assumed that the channel input vector time series has a maximum allowable average power and that the channel transfer function is known. Figure 1 indicates that the dimensions of some of the vectors and matrices that are used in this article. The dimension N of the vectors in the original time series $\boldsymbol{x}(n)$ and the dimension M_t and M_r of the input $\boldsymbol{y}(n)$ and output $\hat{\boldsymbol{y}}(n)$ time series of the channels may in general be different. It is considered that the $M_t \times N$ transmitter FIR MIMO $\boldsymbol{E}(z)$ and the $N \times M_r$ receiver FIR MIMO equalizer filter $\boldsymbol{R}(z)$ are FIR, i.e., they have the following model:

$$\boldsymbol{E}(z) = \sum_{k=0}^{m} \boldsymbol{e}(k) z^{-k} \text{ and } \boldsymbol{R}(z) = \sum_{k=0}^{l} \boldsymbol{r}(k) z^{-k}, \tag{1}$$

where the nonnegative integers m and l are assumed to be known. The matrices $\boldsymbol{e}(k)$ and $\boldsymbol{r}(k)$, have dimensions $M_t \times N$ and $N \times M_r$, respectively. At the output of the FIR MIMO receiver filter $\boldsymbol{R}(z)$, the real part of the vector $\hat{\boldsymbol{x}}(n)$ is kept. Then, the resulting vector is fed to a hard limiter which outputs one of the allowed values of the original source signal. In Figure 1, the combination of the real part and hardlimiting operations is denoted DEC$(\cdot)$. It is remarked that the used memoryless decisions are suboptimal and that better performance can be obtained if more advanced decoding techniques, such as decision feedback equalization, maximum likelihood sequence estimation, or maximum a posteriori sequence estimation [1], are employed.

The channel transfer function is given by an FIR polynomial matrix of order q, and is denoted by $\boldsymbol{C}(z)$, and expressed as:

$$\boldsymbol{C}(z) = \sum_{k=0}^{q} \boldsymbol{c}(k) z^{-k}, \tag{2}$$

where the $M_r \times M_t$ matrices $\boldsymbol{c}(k)$ and the order q are assumed to be known. The matrix $\boldsymbol{C}(z)$ can be used to model a single-input single-output (SISO) channel with transfer function $C(z)$ as in [2], when

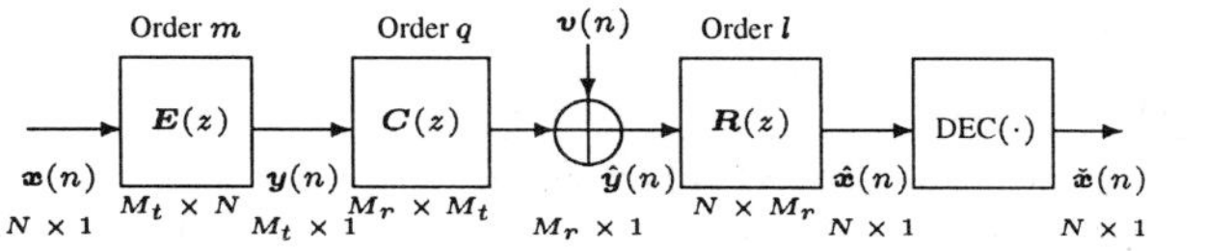

Fig. 1. The power-constrained FIR MIMO block system with given transfer function and additive channel noise.

$M_t = M_r > 1$. In this case, $\boldsymbol{C}(z)$ is a pseudo-circulant matrix whose elements can be found from the impulse response $C(z)$, see [2] for details.

A generalization of the considered system was optimized with respect to the minimum mean square error (MSE) in [3]. Analytical expressions for optimal transform coders that minimizes the MSE were derived in [4] for the identity channel transfer function. This result was extended to include an FIR channel transfer function for redundant filter banks in [2], under restrict assumptions on the transform filter length and the order of the scalar channel transfer function. In the performance results presented in [2], the bit error rate (BER) versus channel quality is used as a performance measure even though the optimization criterion used was the MSE. In the present work, the SER, which is a very important criterion in communication theory, is employed as optimization criterion. Various works [5], [6], [7] have minimized the BER with respect to the receiver parameters when it is modeled by a memoryless transform, and only optimization of the receiver transform is considered.

There are several contributions in this paper. The receiver is treated as an FIR MIMO filter of order l, unlike earlier publications, where only the case of $l = 0$ has been considered when minimum BER is the optimization criterion. However, for the minimum MSE criterion, all values of l have been considered in the literature. The additive noise on the channel can be colored and is complex valued. The channel transfer function is modeled as a known MIMO channel. It is assumed that the filter coefficients are complex. For further improvements in the system performance, the transmitter and receiver FIR MIMO filters are jointly optimized to minimize the SER between the desired output and the actual output vector time series of the receiver, subject to a channel input power constraint. Therefore, a complete optimization of the communication system is proposed.

II. Problem Formulation

A. Operator Definitions

The *row-expanded* matrix $\boldsymbol{R}_{-}$ of the matrix polynomial $\boldsymbol{R}(z)$, see Equation (1), is the $N \times (l+1)M_r$ matrix given by

$$\boldsymbol{R}_{-} = [\boldsymbol{r}(0) \;\; \boldsymbol{r}(1) \;\; \cdots \;\; \boldsymbol{r}(l)]. \tag{3}$$

The *column-expanded* matrix $\boldsymbol{R}_{|}$ of the matrix polynomial $\boldsymbol{R}(z)$ is the $(l+1)N \times M_r$ matrix given by

0-7803-7761-3/03/$17.00 ©2003 IEEE

$$\boldsymbol{R}_{|} = \left[\boldsymbol{r}^T(l)\ \boldsymbol{r}^T(l-1)\ \cdots\ \boldsymbol{r}^T(1)\ \boldsymbol{r}^T(0)\right]^T. \tag{4}$$

The *row-diagonal-expanded* matrix $\boldsymbol{E}_{\urcorner}^{(q)}$ of the matrix polynomial $\boldsymbol{E}(z)$, see Equation (1), is the $(q+1)M_t \times (m+q+1)N$ matrix given by:

$$\boldsymbol{E}_{\urcorner}^{(q)} = \begin{bmatrix} \boldsymbol{e}(0) & \cdots & \boldsymbol{e}(m) & \cdots & \boldsymbol{0} \\ \vdots & \ddots & \ddots & \ddots & \vdots \\ \boldsymbol{0} & \boldsymbol{e}(0) & \cdots & \cdots & \boldsymbol{e}(m) \end{bmatrix}. \tag{5}$$

Let ν be a positive integer. The column-expansion of a vector time series $\boldsymbol{x}(n)$ of dimension $(\nu+1)N \times 1$ is defined as:

$$\boldsymbol{x}(n)_{|}^{(\nu)} = \left[\boldsymbol{x}^T(n), \boldsymbol{x}^T(n-1), \ldots, \boldsymbol{x}^T(n-\nu)\right]^T. \tag{6}$$

Let k and ν be positive integers and let the operator denoted by $\mathcal{T}_{\nu}^{(k)}$ be the *reshape operator*. $\mathcal{T}_{\nu}^{(k)} : \mathbb{C}^{\nu \times (m+k+1)N} \rightarrow \mathbb{C}^{(k+1)\nu \times (m+1)N}$ produces a $(k+1)\nu \times (m+1)N$ block Toeplitz matrix from a $\nu \times (m+k+1)N$ matrix, where ν is a positive integer. Let $\boldsymbol{W}_{-}$ be a $\nu \times (m+k+1)N$ matrix, where the ith $\nu \times N$ block denoted by $\boldsymbol{w}(i)$, $i \in \{0, 1, \ldots, m+k\}$. Then, the operator $\mathcal{T}_{\nu}^{(k)}$ applied to the matrix $\boldsymbol{W}_{-}$ is defined as:

$$\mathcal{T}_{\nu}^{(k)}\{\boldsymbol{W}_{-}\} = \begin{bmatrix} \boldsymbol{w}(k) & \boldsymbol{w}(k+1) & \cdots & \boldsymbol{w}(m+k) \\ \vdots & \vdots & \ddots & \vdots \\ \boldsymbol{w}(1) & \boldsymbol{w}(2) & \cdots & \boldsymbol{w}(m+1) \\ \boldsymbol{w}(0) & \boldsymbol{w}(1) & \cdots & \boldsymbol{w}(m) \end{bmatrix}. \tag{7}$$

B. Input-Output Relationship

The total row-expanded transfer polynomial matrix from the original vector time series $\boldsymbol{x}(n)$ to the output $\hat{\boldsymbol{x}}(n)$ of the FIR MIMO receiver is given by $\boldsymbol{R}_{-}\boldsymbol{C}_{\urcorner}^{(l)}\boldsymbol{E}_{\urcorner}^{(q+l)}$. This matrix has dimension $N \times (m+q+l+1)N$. The overall expression for the output of the FIR MIMO receiver filter can be written as:

$$\hat{\boldsymbol{x}}(n) = \boldsymbol{R}_{-}\boldsymbol{C}_{\urcorner}^{(l)}\boldsymbol{E}_{\urcorner}^{(q+l)}\boldsymbol{x}(n)_{|}^{(m+q+l)} + \boldsymbol{R}_{-}\boldsymbol{v}(n)_{|}^{(l)}, \tag{8}$$

where the vector $\boldsymbol{x}(n)_{|}^{(m+q+l)}$ has dimension $(m+q+l+1)N \times 1$, the vector $\boldsymbol{v}(n)_{|}^{(l)}$ has dimension $(l+1)M_r \times 1$, and $\boldsymbol{R}_{-}$ is given by Equation (3). Each vector component of $\boldsymbol{x}(n)_{|}^{(m+q+l)}$ contains the constellation symbols with equal probability.

The FIR MIMO receiver filter $\boldsymbol{R}(z)$ can be interpreted as consisting of N FIR receiver filters where the impulse response of FIR receiver filter number i, where $i \in \{0, 1, \ldots, N-1\}$, is given by $(\boldsymbol{R}_{-})_{i,:}$. The notation $(\boldsymbol{R}_{-})_{i,:}$ denotes row number i of the matrix $\boldsymbol{R}_{-}$. In this article, the numbering of the rows and columns begins with 0.

C. Input Signal Models

It is assumed that the input vector time series to the system are jointly wide sense stationary and uncorrelated with each other.

The channel is assumed to be corrupted by additive Gaussian complex circularly symmetric noise which is independent of the transmitted signal. The noise is assumed to have zero mean and known second-order statistics. The Hermitian positive definite autocorrelation matrix of dimension $(l+1)M_r \times (l+1)M_r$ of the $(l+1)M_r \times 1$ vector $\boldsymbol{v}(n)_{|}^{(l)}$ is defined as:

$$\boldsymbol{\Phi}_{\boldsymbol{v}}^{(l,M_r)} = E\left[\boldsymbol{v}(n)_{|}^{(l)}\left(\boldsymbol{v}(n)_{|}^{(l)}\right)^H\right], \tag{9}$$

where the operator ^{H} denotes complex conjugated transposed. The average variance of the components of the complex, Gaussian, circularly symmetric, additive channel noise $\boldsymbol{v}(n)$ is denoted N_0.

Let the ith component of the $N \times 1$ vector $\boldsymbol{x}(n)$ be drawn from a PAM constellation with L_i symbols, such that $x_i(n)$ is taken from the set: $\mathcal{A}_i \triangleq \{2k+1-L_i \mid k \in \{0, 1, \ldots, L_i-1\}\}$. It is assumed that the components of the vector $\boldsymbol{x}(n)$ is uncorrelated with each other, white and each symbol of $x_i(n)$ is assumed to be equally likely such that $x_i(n)$ has zero mean and variance $\sigma_{x_i}^2 \triangleq E\left[x_i^2(n)\right] = \frac{L_i^2-1}{3}$. For the assumed source, $\boldsymbol{\Phi}_{\boldsymbol{x}}^{(m,N)} = \boldsymbol{I}_{m+1} \otimes \operatorname{diag}\left(\sigma_{x_0}^2, \sigma_{x_1}^2, \ldots, \sigma_{x_{N-1}}^2\right)$.

When the component $x_i(n)$ is taken from the set $\mathcal{A}_i$, there exist $\prod_{i=0}^{N-1} L_i^{m+q+l+1}$ different possibilities for $\boldsymbol{x}(n)_{|}^{(m+q+l)}$. Let these vectors be arbitrarily indexed by k and denote the indexed symbols as $\boldsymbol{x}_k(n)_{|}^{(m+q+l)}$. The $(m+q+l+1)N \times 1$ vector $\boldsymbol{x}_k^{(i)}(n)_{|}^{(m+q+l)}$ is defined to be equal to $\boldsymbol{x}_k(n)_{|}^{(m+q+l)}$ except that component number $i+\delta_v N$ of the vector $\boldsymbol{x}_k^{(i)}(n)_{|}^{(m+q+l)}$ is equal to $+1$, where the integer $\delta_v \in \{0, 1, \ldots, m+q+l\}$ is called *vector delay*, and it should be chosen carefully depending on the transfer function of the channel $\boldsymbol{C}(z)$ and the orders of the transmitter and receiver filters. The desired output vector $\boldsymbol{d}(n)$ is given by $\boldsymbol{d}(n) = \boldsymbol{x}(n-\delta_v)$. The number of possible vectors $\boldsymbol{x}_k^{(i)}(n)_{|}^{(m+q+l)}$ is denoted by K_i and this number is given by:

$$K_i = \prod_{\nu=0}^{N-1} L_{\nu}^{m+q+l+1}/L_i. \tag{10}$$

D. Inner Products

The ordinary Euclidean inner product for complex vectors will be useful in the transmitter optimization: Let $\boldsymbol{a}_i \in \mathbb{C}^{(m+1)NM_t \times 1}$ be a complex-valued *column vector*, and the Euclidean inner product be denoted $\langle \boldsymbol{a}_0, \boldsymbol{a}_1 \rangle = \boldsymbol{a}_0^H \boldsymbol{a}_1$.

For the receiver optimization, the following inner product will be used: Let $\boldsymbol{b}_i \in \mathbb{C}^{1 \times (l+1)M_r}$ be a complex-valued *row vector*, and let the *receiver inner product* be defined as:

$$\langle \boldsymbol{b}_0, \boldsymbol{b}_1 \rangle_{\boldsymbol{\Phi}_{\boldsymbol{v}}^{(l,M_r)}} = \boldsymbol{b}_0 \boldsymbol{\Phi}_{\boldsymbol{v}}^{(l,M_r)} \boldsymbol{b}_1^H. \tag{11}$$

E. Power Constraint Formulation

The constraint on the power used by the channel input vector $\boldsymbol{y}(n)$ per *source symbol* can be expressed as:

$$\begin{aligned} \frac{1}{N} E\left[\|\boldsymbol{y}(n)\|^2\right] &= \frac{1}{N} \operatorname{Tr}\left\{\boldsymbol{E}_{-}\boldsymbol{\Phi}_{\boldsymbol{x}}^{(m,N)}\boldsymbol{E}_{-}^H\right\} \\ &= \frac{1}{N} \operatorname{vec}^H(\boldsymbol{E}_{-})\left[\left(\boldsymbol{\Phi}_{\boldsymbol{x}}^{(m,N)}\right)^* \otimes \boldsymbol{I}_{M_t}\right] \operatorname{vec}(\boldsymbol{E}_{-}) = E_s, \end{aligned} \tag{12}$$

where it is assumed that the original signal is a memoryless PAM source where the ith component has values taken from a PAM constellation with L_i levels with equal probability, the operator $\otimes$ is the Krönecker product, E_s denotes the energy per source symbol. The vec operator stacks the columns of the matrix it is applied to into a long column vector, placing the first column on the top, then the second column, and so on [8]. The power constraint means that the vector $\operatorname{vec}(\boldsymbol{E}_{-})$ lies on a hyper-ellipsoid with radius $\sqrt{E_s N}$ with the center at the origin.

F. SER Expression

Let the symbol $\boldsymbol{s}_k^{(i)}(n)_{|}^{(l)}$ denote the kth vector of dimension $(l+1)M_r \times 1$, being defined by:

$$\boldsymbol{s}_k^{(i)}(n)_{|}^{(l)} = \boldsymbol{C}_{\urcorner}^{(l)}\boldsymbol{E}_{\urcorner}^{(q+l)}\boldsymbol{x}_k^{(i)}(n)_{|}^{(m+q+l)}. \tag{13}$$

Whenever the index k is not required, the notation can be simplified, that is, the symbols $\boldsymbol{x}^{(i)}(n)_{|}^{(m+q+l)}$ and $\boldsymbol{s}^{(i)}(n)_{|}^{(l)}$ will be used to denote one of the vectors $\boldsymbol{x}_k^{(i)}(n)_{|}^{(m+q+l)}$ and $\boldsymbol{s}_k^{(i)}(n)_{|}^{(l)}$, respectively. The following relation holds: $\boldsymbol{s}^{(i)}(n)_{|}^{(l)} = \boldsymbol{C}_{\urcorner}^{(l)}\boldsymbol{E}_{\urcorner}^{(q+l)}\boldsymbol{x}^{(i)}(n)_{|}^{(m+q+l)}$.

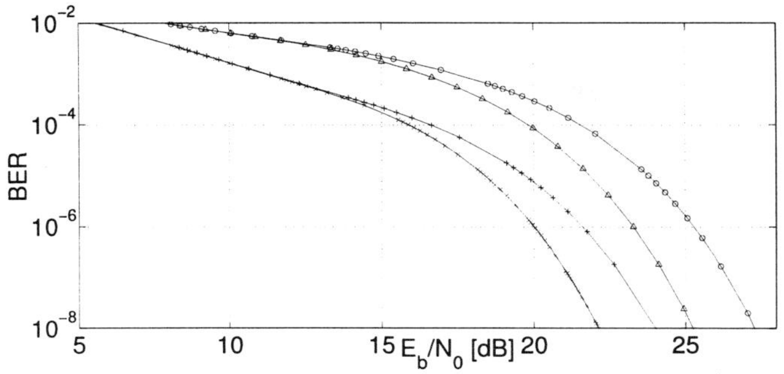

Fig. 2. BER versus E_b/N_0 for five different systems: minimum MSE with $m = 0$ and $l = 4$: $-\circ-$, minimum BER with $m = 0$ and $l = 4$: $-\triangle-$, minimum MSE with $m = l = 4$: $-+-$, and minimum BER with $m = l = 4$: $-\times-$. The vector delay is chosen according to $\delta_v = \left\lfloor \frac{m+q+l}{2} \right\rfloor + 1$.

Let $\boldsymbol{H}(z) = \boldsymbol{R}(z)\boldsymbol{C}(z)$ denote the z-domain representation of the convolution of the receiver and channel polynomial matrices. This matrix has order $q + l$ and dimension $N \times M_t$. In the transmitter optimization, the $(m+1)M_tN \times 1$ vector $\boldsymbol{t}_k^{(i)}(n)$ will be required, being defined as:

$$\boldsymbol{t}_k^{(i)}(n) = \text{vec}\left(\left((\boldsymbol{H}_{\mid})_{i:N:(q+l)N+i,:}\right)^H \cdot \mathcal{T}_1^{(q+l)} \left\{ \left(\boldsymbol{x}_k^{(i)}(n)_{\mid}^{(m+q+l)}\right)^H \right\} \right), \quad (14)$$

where the notation $\boldsymbol{a}_{k_0:N:k_1,:}$ means the sub-matrix of $\boldsymbol{a}$ that consists of the rows whose indices are in the set $\{k_0, k_0 + N, k_0 + 2N, \ldots, k_1\}$.

It can be shown that the kth branch in the ith eye diagram when $+1$ was transmitted can be written as:

$$\text{Re}\left\{ (\boldsymbol{R}_-)_{i,:} \boldsymbol{C}_{\urcorner}^{(l)} \boldsymbol{E}_{\urcorner}^{(q+l)} \boldsymbol{x}_k^{(i)}(n)_{\mid}^{(m+q+l)} \right\} =$$

$$\text{Re}\left\{ \left\langle (\boldsymbol{R}_-)_{i,:}, \left(\boldsymbol{s}_k^{(i)}(n)_{\mid}^{(l)}\right)^H \left[\boldsymbol{\Phi}_{\boldsymbol{v}}^{(l,M_r)}\right]^{-1} \right\rangle_{\boldsymbol{\Phi}_{\boldsymbol{v}}^{(l,M_r)}} \right\}$$

$$= \text{Re}\left\{ \left\langle \boldsymbol{t}_k^{(i)}(n), \text{vec}\left(\boldsymbol{E}_-\right) \right\rangle \right\}, \quad (15)$$

for all $i \in \{0, 1, \ldots, N-1\}$ and $k \in \{0, 1, \ldots, K_i - 1\}$ where the last equality can be shown by utilizing the structure of the matrices $\boldsymbol{C}_{\urcorner}^{(l)}$, $\boldsymbol{E}_{\urcorner}^{(q+l)}$, the vector $(\boldsymbol{R}_-)_{i,:}$, and the formulae in [8]. If the system has an open eye diagram in the absence of channel noise at the output of the ith receiver filter, the expressions in Equation (15) must be positive for all $k \in \{0, 1, \ldots, K_i - 1\}$.

It can be shown that the SER for the system can be expressed as:

$$\text{SER} = \frac{1}{N} \sum_{i=0}^{N-1} \frac{2L_i - 2}{L_i} E\left[Q\left(\frac{\sqrt{2}\,\text{Re}\left\{ (\boldsymbol{R}_-)_{i,:} \boldsymbol{s}^{(i)}(n)_{\mid}^{(l)} \right\}}{\left\| (\boldsymbol{R}_-)_{i,:} \right\|_{\boldsymbol{\Phi}_{\boldsymbol{v}}^{(l,M_r)}}} \right) \right], \quad (16)$$

where the expectation operator $E[\cdot]$ is performed over the original signal vectors $\boldsymbol{x}_k^{(i)}(n)_{\mid}^{(m+q+l)}$. From this expression, it is seen that there is no loss in optimality by scaling the receiver filter vectors $(\boldsymbol{R}_-)_{i,:}$ according to:

$$\left\| (\boldsymbol{R}_-)_{i,:} \right\|^2_{\boldsymbol{\Phi}_{\boldsymbol{v}}^{(l,M_r)}} = (\boldsymbol{R}_-)_{i,:} \boldsymbol{\Phi}_{\boldsymbol{v}}^{(l,M_r)} (\boldsymbol{R}_-)_{i,:}^H = 1. \quad (17)$$

G. Joint Optimization

Assume that the normalization of the receiver filters are used. For the optimization of the ith receiver filter, the following result is needed in the steepest decent method:

$$\frac{\partial}{\partial (\boldsymbol{R}_-)_{i,:}^*} \text{SER} = -\frac{L_i - 1}{\sqrt{\pi} K_i L_i N} \sum_{k=0}^{K_i - 1} e^{-\text{Re}^2\left\{ (\boldsymbol{R}_-)_{i,:} \boldsymbol{s}_k^{(i)}(n)_{\mid}^{(l)} \right\}}$$

$$\left\{ \left(\boldsymbol{s}_k^{(i)}(n)_{\mid}^{(l)} \right)^H - \text{Re}\left\{ (\boldsymbol{R}_-)_{i,:} \boldsymbol{s}_k^{(i)}(n)_{\mid}^{(l)} \right\} (\boldsymbol{R}_-)_{i,:} \boldsymbol{\Phi}_{\boldsymbol{v}}^{(l,M_r)} \right\}. \quad (18)$$

And for the transmitter optimization, the following result is used in the steepest decent method:

$$\frac{\partial}{\partial \text{vec}\left(\boldsymbol{E}^*{}_-\right)} \zeta = \frac{\mu}{N} \left[\left[\boldsymbol{\Phi}_{\boldsymbol{x}}^{(m,N)} \right]^* \otimes \boldsymbol{I}_{M_t} \right] \text{vec}\left(\boldsymbol{E}_-\right)$$

$$- \frac{1}{\sqrt{\pi} N} \sum_{i=0}^{N-1} \frac{L_i - 1}{K_i L_i} \sum_{k=0}^{K_i - 1} e^{-\text{Re}^2\left\{ \left\langle \boldsymbol{t}_k^{(i)}(n), \text{vec}(\boldsymbol{E}_-) \right\rangle \right\}} \boldsymbol{t}_k^{(i)}(n), \quad (19)$$

where $\zeta = \text{SER} + \frac{\mu}{N} E\left[\|\boldsymbol{y}(n)\|^2 \right]$ is the Lagrange objective function and μ is the positive Lagrange multiplier for the power constraint.

The optimization of the FIR MIMO filters is performed as follows: first initial values of the transmitter FIR MIMO filter $\boldsymbol{E}(z)$, vector delay δ_v, μ, N, M_t, M_r, q, $\boldsymbol{C}(z)$, m, and l are chosen according to the application of interest. Second, an iterative search is performed where the receiver FIR MIMO filter $\boldsymbol{R}(z)$ is found by using the steepest decent method and the result from Equation (18) with the current value of the transmitter filter, and then the transmitter filter is found by using the steepest decent algorithm and the result in Equation (19) with the current value of the receiver filter. This iteration is performed until convergence occurs. The algorithm is guaranteed to converge at least to a local minimum since at each step the objective function is decreased and the objective function is lower bounded by zero. Convergence can also be shown by the Global Convergence Theorem [9]. If the filters do not give average transmit energy E_s, then μ should be adjusted and the procedure should be done all over again.

III. Results

In the two examples given here, the original signal is bits such that $L_i = 2$, SER = BER (bit error rate), and $E_s = E_b$.

A. Channel A Used in [10]

Set $N = 1$, and let the channel transfer function be given by:

$$C(z) = 1.2 + 1.1z^{-1} - 0.2z^{-2}, \quad (20)$$

with $M_t = M_r = 1$ and $q = 2$. δ_v was chosen optimally, leading to $\delta_v = 2$ when $l = 2$ and $\delta_v = 4$ when $l = 4$. This is the channel called Channel A in [10].

Figure 2 shows the BER versus E_b/N_0 performance of the proposed minimum BER and the minimum MSE systems when using $m = 0$ or $m = 4$ and $l = 4$. The results reported in [10] are in agreement with the results given in the figure. In [10], the transmitter was not optimized, and this corresponds to the case where $m = 0$. It can be seen from Figure 2, that the performance can be improved by including the transmitter in the optimization, since the performance when using $m = 4$ is significantly better than when $m = 0$. For example, for BER $= 10^{-3}$ a gain in the value of E_b/N_0 of approximately 5.1 dB or 6.2 dB can be achieved by allowing the transmitter to have order $m = 4$ as compared to the identity transmitter of order $m = 0$, when minimum SER or minimum MSE is the optimization criterion, respectively. The complexity of the transmitter will increase in this way, whereas, the complexity of the receiver is unchanged.

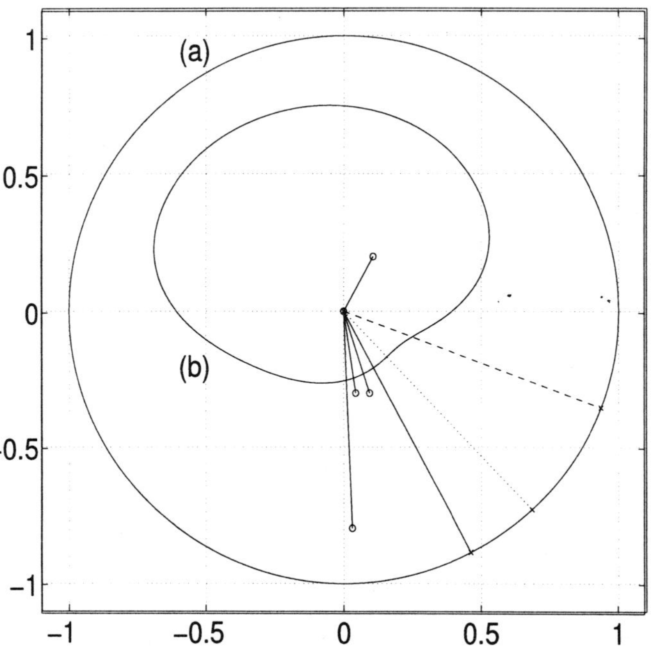

Fig. 3. (a) shows the unit circle. (b) shows the polar plot of the BER versus angle θ for normalized receiver filter $\boldsymbol{R}_{-} = [\cos(\theta), \sin(\theta)]$. The lines ended with circles show *scaled* versions of the $K = 4$ vectors in Equation (22). The unit-norm minimum MSE receiver is shown by the solid line ($\theta = -62.5°$), the unit-norm minimum BER receiver vector is shown by the dotted line ($\theta = -46.9°$), and one open eye unit-norm receiver is shown by the dashed line ($\theta = -20.7°$).

B. Optimal FIR MIMO Filter with Closed Eye

Assume that the channel transfer function is given by:

$$\boldsymbol{C}(z) = \begin{bmatrix} -0.0125 \\ -0.5 \end{bmatrix} + \begin{bmatrix} 0.1375 \\ -0.6 \end{bmatrix} z^{-1} + \begin{bmatrix} 0.0625 \\ -0.5 \end{bmatrix} z^{-2}, \tag{21}$$

such that $q = 2$, $M_t = 1$ and $M_r = 2$, furthermore, let $N = 1$, $\delta_v = 1$, $m = l = 0$, $\boldsymbol{E}(z) = 1$, and $\boldsymbol{\Phi}_v^{(l,M_r)} = 0.5\boldsymbol{I}_2$. The vectors $\boldsymbol{s}_k^{(0)}(n)_{\mathsf{I}}^{(0)} = \boldsymbol{C}_{-}\boldsymbol{x}_k^{(0)}(n)_{\mathsf{I}}^{(2)}$ for this example are given by

$$\left[\boldsymbol{s}_0^{(0)}(n)_{\mathsf{I}}^{(0)} \;\; \boldsymbol{s}_1^{(0)}(n)_{\mathsf{I}}^{(0)} \;\; \boldsymbol{s}_2^{(0)}(n)_{\mathsf{I}}^{(0)} \;\; \boldsymbol{s}_3^{(0)}(n)_{\mathsf{I}}^{(0)}\right] = \begin{bmatrix} 0.2125 & 0.0875 & 0.1875 & 0.0625 \\ 0.4 & -0.6 & -0.6 & -1.6 \end{bmatrix}. \tag{22}$$

Since the receiver in this example has dimension 1×2 and since the receiver norm of the receiver can be chosen arbitrarily, the receiver filter can be set equal to $\boldsymbol{R}_{-} = [\cos(\theta), \sin(\theta)]$, where $\theta \in (-180°, 180°]$ is the angle that the receiver vector $\boldsymbol{R}_{-}$, lying on the unit circle, makes with the first axis $[1, 0]$. Figure 3 shows the unit circle, the BER as a function of the angle θ, the scaled vectors in Equation (22), and three different unit-norm receiver filters.

In Figure 4, an artificial noiseless eye diagram, employing a triangular pulse shape function, is presented for this example for the minimum MSE, the minimum BER, and an arbitrary chosen open eye, with $\theta = -20.7°$. All three receiver filters are normalized according to Equation (17). This eye diagram was constructed by drawing all the $K = 4$ possible values of the eye opening, given in Equation (15). For $\left\{\boldsymbol{s}_1^{(0)}(n)_{\mathsf{I}}^{(0)}, \boldsymbol{s}_2^{(0)}(n)_{\mathsf{I}}^{(0)}, \boldsymbol{s}_3^{(0)}(n)_{\mathsf{I}}^{(0)}\right\}$ it is seen that that the eye opening is positive for all three receiver filters, but for $\left\{\boldsymbol{s}_0^{(0)}(n)_{\mathsf{I}}^{(0)}\right\}$, the eye diagram is open only for the open eye receiver and not for the other three receiver solutions. In this example, the

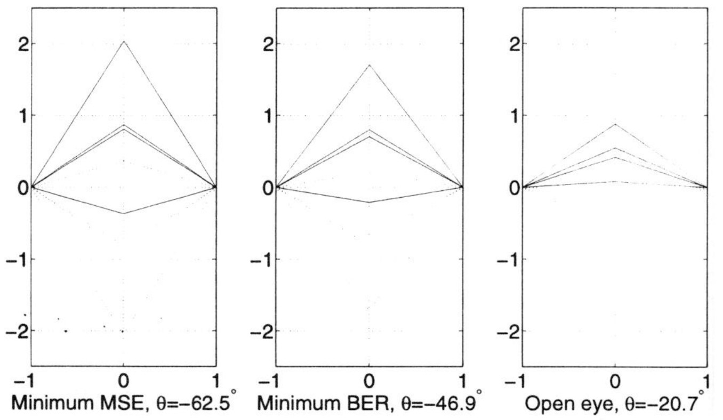

Fig. 4. Noiseless eye diagrams for minimum MSE, minimum BER, and an open eye example when all the receiver filters are normalized according to Equation (17). The solid lines correspond to $+1$ and the dotted lines corresponds to -1 sent.

optimal minimum BER receiver is given by $\theta = -46.9°$ resulting in a BER $= 0.2279$. For the minimum MSE solution, $\theta = -62.5°$ leading to BER $= 0.2330$. If $\theta = -20.7°$, the eye diagram is open but BER $= 0.2643$. This shows that when there is high variance noise present on the channel, it is not in general true that the eye is open, as was assumed in [5].

IV. Conclusions

A theory was derived which enables the minimization of the SER by jointly optimizing the transmitter and receiver FIR MIMO filters for communication of a vector time series with uncorrelated PAM vector components over a power-constrained FIR MIMO channel. Based on the theory, an iterative solution was developed which is able to converge to a locally minimum SER solution. An example showed that the eye can be closed for the minimum SER solution when the channel noise is large.

References

[1] J. G. Proakis, *Digital Communications*, McGraw-Hill, 3rd edition, 1995.

[2] A. Scaglione, G. B. Giannakis, and S. Barbarossa, "Redundant filterbank precoders and equalizers Part I: Unification and optimal designs," *IEEE Trans. Signal Processing*, vol. 47, no. 7, pp. 1988–2006, July 1999.

[3] A. Hjørungnes, P. S. R. Diniz, and M. L. R. de Campos, "Jointly minimum MSE transmitter and receiver FIR MIMO filters in the presence of near-end crosstalk and additive noise," in *Proc. Int. Symp. on Circuits and Systems*, Bangkok, Thailand, May 2003, IEEE, accepted for publication.

[4] K.-H. Lee and D. P. Petersen, "Optimal linear coding for vector channels," *IEEE Trans. Commun.*, vol. COM-24, no. 12, pp. 1283–1290, Dec. 1976.

[5] X. Wang, W.-S. Lu, and A. Antoniou, "Constrained minimum-BER multiuser detection," *IEEE Trans. Signal Processing*, vol. 48, no. 10, pp. 2903–2909, Oct. 2000.

[6] C.-C. Yeh, R. R. Lopes, and J. R. Barry, "Approximate minimum bit-error rate multiuser detection," in *Proc. IEEE GLOBECOM*, Sidney, Australia, Nov. 1998, vol. 6, pp. 3590–3595.

[7] I. N. Psaromiligkos, S. N. Batalama, and D. A. Pados, "On adaptive minimum probability of error linear filter receivers for DS-CDMA channels," *IEEE Trans. Commun.*, vol. 47, no. 7, pp. 1092–1102, July 1999.

[8] R. A. Horn and C. R. Johnsen, *Topics in Matrix Analysis*, Cambridge University Press Cambridge, UK, 1991.

[9] D. G. Luenberger, *Linear and Nonlinear Programming*, Addison–Wesley Publishing Company, Reading, Massachusetts, USA, 2nd edition, 1984.

[10] C.-C. Yeh and J. R. Barry, "Adaptive minimum bit-error rate equalization for binary signaling," *IEEE Trans. Commun.*, vol. 48, no. 7, pp. 1226–1235, July 2000.

ANALOG REPRESENTATION AND DIGITAL IMPLEMENTATION OF OFDM SYSTEMS

Yuan-Pei Lin
Dept. Elect. and Control Engr.,
National Chiao Tung Univ.,
Hsinchu, Taiwan

See-May Phoong
Dept. of EE & Grad. Inst. of Comm Engr.,
National Taiwan Univ.,
Taipei, Taiwan

ABSTRACT

Many existing results on the analysis of OFDM systems are based on an analog representation. The actual implementation of OFDM transmitters typically consists of a discrete DFT matrix and a digital-to-analog (DAC) converter. In this paper, we show that the analog representation admits the implementation of a DFT matrix followed by a DAC converter only in special cases. Necessary and sufficient conditions for such cases will be given. Analyses based on the analog representation, e.g., spectral roll-off of transmitter outputs and carrier frequency offset, may not be appropriate when digital implementation is employed.

1. INTRODUCTION

The OFDM (orthogonal frequency division multiplexing) systems [2] are well-known for applications in wireless LAN (local area network) and broadcast of digital audio and digital video. Fig. 1 shows the schematic of an analog OFDM transmitter with M subcarriers. Assuming that the subcarrier spacing $= \Omega_0$, the output of the transmitter is given by

$$x(t) = \sum_{k=0}^{M-1} x_k g(t) e^{jk\Omega_0 t}. \quad (1)$$

The pulse shaping filter $g(t)$ is usually the rectangular pulse,

$$g(t) = \begin{cases} 1, & 0 \le t \le T_0, \\ 0, & \text{otherwise}, \end{cases} \quad \text{where } T_0 = 2\pi/\Omega_0 \quad (2)$$

Many studies on OFDM systems are carried out using the expression in (1), e.g., the spectral roll-off of the outputs of OFDM transmitters [3][4], the effect of carrier frequency offset [5], and crest factors of the transmitter outputs [6]. A number of non-rectangular pulse shaping $g(t)$ has been proposed to improve the spectral roll-off of the transmitted signal $x(t)$, e.g., [4][6]. Optimal pulses for minimizing interference is considered in [7] and in [8] parameters of Gaussian pulses are optimized to minimize bit error rates. Although the representation in Fig. 1 is convenient for analysis, the modulation of subcarriers is typically done in the discrete time. Such a transmitter (Fig. 2) consists of two parts [3]: a DAC (digital to analog converter) and the part performing digital modulation of subcarriers, which can be efficiently implemented using an M by M IDFT matrix. The sampling period is $T_s = T_0/M$ and the discrete sequence $w[n]$ shown in Fig. 2 is typically the rectangular window,

$$w[n] = \begin{cases} 1, & 0 \le n \le M-1, \\ 0, & \text{otherwise}, \end{cases} \quad (3)$$

In Chapter 5 of [3], it is mentioned that when we use the digital implementation with an ideal lowpass reconstruction filter $h(t)$ in the DAC converter, the shaping filter $g(t)$ is no longer the rectangular pulse. A precise connection between the analog representation and the digital implementation has not been stated earlier in the literature. In this paper we consider the equivalence of the analog representation and the digital implementation of OFDM transmitters in Fig. 1 and Fig. 2. For the commonly considered case of a rectangular pulse $g(t)$, we show that the two transmitters are not equivalent regardless of the choices of the window $w[n]$ and the reconstruction filter $h(t)$ in the DAC. Also, the digital implementation with a rectangular window $w[n]$ and an ideal lowpass $h(t)$ does not have an equivalent analog representa-

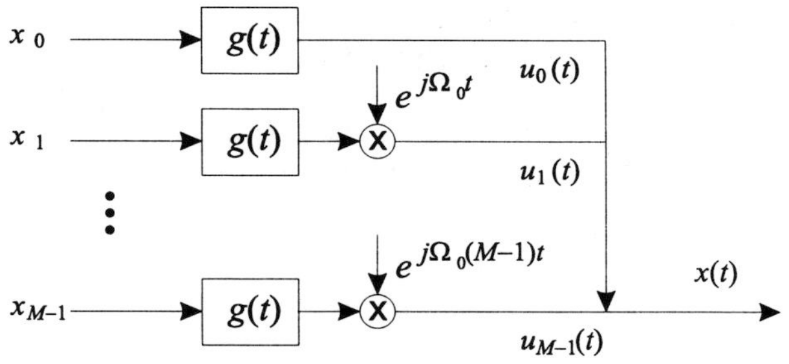

Figure 1: The baseband analog representation of the OFDM transmitter with M subcarriers and pulse shaping filter $g(t)$.

This work was supported in parts by National Science Council, Taiwan, R. O. C., under NSC 90-2213-E-002-097 and 90-2213-E-009-108, Ministry of Education, Taiwan, R. O. C, under Grant # 89E-FA06-2-4, and the Lee and MTI Center for Networking Research.

0-7803-7761-3/03/$17.00 ©2003 IEEE

tion in Fig. 1 for any $g(t)$. Given an arbitrary pulse shaping filter $g(t)$, there does not exist an equivalent digital implementation in general. The analysis of OFDM systems based on the analog representation may not be appropriate if the underlying pulse shaping filter does not allow an equivalent digital implementation. The two systems in Fig. 1 and Fig. 2 can be made equivalent in some special cases. A necessary and sufficient condition for such cases will be derived. An example of a set of $g(t)$, $w[n]$ and $h(t)$ that satisfies the condition will be given.

2. OFDM TRANSMITTERS WITH RECTANGULAR PULSE SHAPING

Suppose the reconstruction filter of the DAC is $h(t)$ as indicated in Fig. 2. The output of the DAC with sampling period T_s is given by

$$y(t) = \sum_{n=0}^{M-1} y[n]h(t - nT_s), \tag{4}$$

where $y[n]$ is the input of the DAC as indicated in Fig. 2.

Theorem 1 *Let the OFDM transmitter in Fig. 1 have a rectangular pulse shaping filter $g(t)$ as given in (2) and the transmitter in Fig. 2 have a discrete rectangular pulse $w[n]$ as given in (3). The outputs of the two systems, respectively $x(t)$ and $y(t)$, are not the same for any choice of reconstruction filter $h(t)$.*

Proof: The signal $y[n]$ is given by

$$y[n] = w[n] \sum_{k=0}^{M-1} x_k e^{j\frac{2\pi}{M}kn}.$$

Substituting the above expression of $y[n]$ into (4), we arrive at

$$y(t) = \sum_{k=0}^{M-1} x_k \sum_{n=0}^{M-1} e^{j\frac{2\pi}{M}kn} h(t - nT_s).$$

Comparing this expression with (1), we conclude that $x(t)$ and $y(t)$ are equal for an arbitrary sequence x_k if and only if

$$\sum_{n=0}^{M-1} e^{j\frac{2\pi}{M}kn} h(t - nT_s) = g(t)e^{jk\Omega_0 t}, \tag{5}$$

for $k = 0, 1, \cdots, M-1$. In particular, the above equation is true for $k = 0$ and $k = 1$. When $k = 0$, we have $\sum_{n=0}^{M-1} h(t - nT_s) = g(t)$. Applying Fourier transform on the both sides of the equation and using $G(j\Omega) = e^{-jT_0\Omega/2}\sin(T_0\Omega/2)/\Omega$, we can verify that $h(t)$ is the rectangular pulse of duration T_s, i.e.,

$$h(t) = \begin{cases} 1, & 0 \le t \le T_s, \\ 0, & \text{otherwise.} \end{cases} \tag{6}$$

When $k = 1$, we have

$$\sum_{n=0}^{M-1} e^{j2n\pi/M} h(t - nT_s) = g(t)e^{j\Omega_0 t}.$$

Let $f(t) = e^{-j2\pi t/T_0}h(t)$. Then the above equation can be written as $\sum_{n=0}^{M-1} f(t-nT_s) = g(t)$. Similarly, this requires $f(t)$ be the rectangular pulse of duration T_s, which contradicts the solution of $h(t)$ obtained for $k = 0$. Therefore (5) can not be satisfied for any reconstruction filter $h(t)$; hence $x(t)$ and $y(t)$ are not the same. △△△

Example 1. Let us consider the typical choice of $g(t)$, $h(t)$ and $w[n]$ described in [3]. The pulse $g(t)$ is chosen to be the rectangular pulse of duration T_0 given in (2), and $w[n]$ the discrete rectangular window of length M given in (3). The reconstruction filter is an ideal lowpass filter

$$H(j\Omega) = \begin{cases} 1, & |\Omega| < \pi/T_s, \\ 0, & \text{otherwise.} \end{cases} \tag{7}$$

The corresponding impulse response is

$$h(t) = \frac{\sin \frac{\pi t}{T_s}}{\frac{\pi t}{T_s}}. \tag{8}$$

We can easily see that the two systems shown in Fig. 1 and Fig. 2 are not equivalent by observing that $x(t)$ has a finite duration T_0 and the duration of $y(t)$ is not finite. Also $y(t)$ is bandlimited while $x(t)$ is not. Notice that $X(j\Omega)$ has most of its energy in the frequency band $(-\Omega_0/2, (M-0.5)\Omega_0)$ as the mainlobe of $G(j\Omega)$ stretches from $-\Omega_0/2$ to $\Omega_0/2$. On the other hand, the energy of $Y(j\Omega)$ is in the frequency band $(-M\Omega_0/2, M\Omega_0/2)$.

With the above choices of $g(t)$, $h(t)$ and $w[n]$, the two transmitters in Fig. 1 and Fig. 2 are not the same. However, the two transmitters has the following connection. When $x(t)$ and $y(t)$ are sampled with sampling period T_s, the samples are the identical. That is, $x(nT_s) = y(nT_s)$ for $n = 0, 1, \cdots, M-1$. To see that, we can verify $x(nT_s) = \sum_{k=0}^{M-1} x_k e^{jkn2\pi/M}$, for $n = 0, 1, \cdots M-1$, are the M-point DFT of x_k. Therefore

$$x(nT_s) = y[n], n = 0, 1, \cdots, M-1.$$

On the other hand, (4) yields $y(nT_s) = \sum_{m=0}^{M-1} y[m]h((n-m)T_s)$. As $h(t)$ is the ideal lowpass given in (8), it is a Nyquist filter with $h(nT_s) = \delta[n]$, we have

$$y(nT_s) = y[n], n = 0, 1, \cdots, M-1.$$

Such a connection means that if the channel $C(j\Omega)$ is deal ($C(j\Omega) = 1$) and the received signals are sampled with sampling period T_s, the received samples are the same using either one of the two transmitters. The relationship may

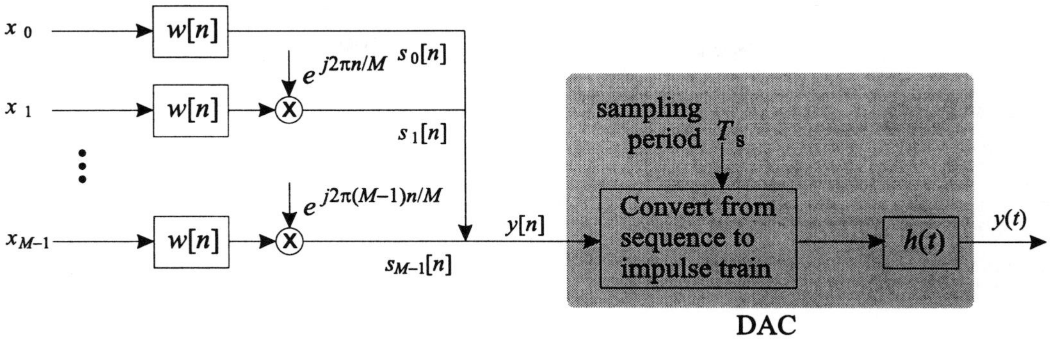

Figure 2: Commonly used digital implementation of the OFDM transmitter.

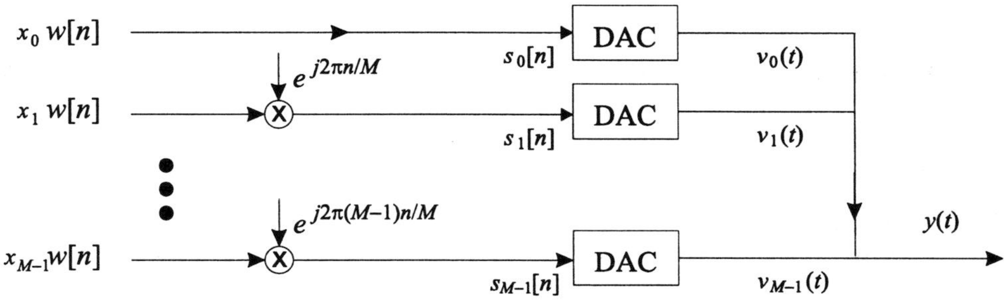

Figure 3: Equivalent block diagram of system in Fig. 2.

not hold for general $C(j\Omega)$. Consider, for example, the case that the channel is a delay, $c(t) = \delta(t - \Delta)$. We can verify that the samples of the received signals for the two transmitters in Fig. 1 and Fig. 2 are different.

3. CONDITIONS FOR EQUIVALENCE OF ANALOG REPRESENTATION AND DIGITAL IMPLEMENTATION

The equivalence of the two systems in Fig. 1 and Fig. 2 can be established in certain cases. For the convenience of derivation, we redraw the system in Fig. 2 as Fig. 3, in which the DAC block is as in Fig. 2. The output due to the k-subcarrier is given by $V_k(j\Omega) = S_k(e^{jT_s\Omega})H(j\Omega)$, [9]. Notice that $S_k(e^{j\omega})$ is a frequency-shifted and scaled version of $W(e^{j\omega})$, i.e.,

$$S_k(e^{j\omega}) = x_k W(e^{j(\omega-2\pi k/M)}), k = 0, 1, \cdots, M-1.$$

Therefore, we have

$$\begin{aligned} V_k(j\Omega) &= x_k W(e^{j(T_s\Omega-2\pi k/M)})H(j\Omega) \\ &= x_k W(e^{jT_s(\Omega-k\Omega_0)})H(j\Omega), \end{aligned} \quad (9)$$

where we have used the facts that $\Omega_0 = 2\pi/T_0$ and $T_0 = MT_s$. On the other hand, the output of the analog representation in Fig. 1 due to the k-th subcarrier is given by

$$U_k(\Omega) = x_k G(j(\Omega - k\Omega_0)).$$

The equivalence of the two systems in Fig. 1 and Fig. 2 means $V_k(\Omega) = U_k(\Omega)$ and therefore

$$W(e^{jT_s(\Omega-k\Omega_0)})H(j\Omega) = G(j(\Omega - k\Omega_0)),$$

for $k = 0, 1, \cdots, M-1$. Summarizing, we have the following theorem.

Theorem 2 *The OFDM transmitter in Fig. 1 can be implemented as in Fig. 2, namely, the two systems are equivalent, if and only if the pulse shaping filter $g(t)$, the digital window $w[n]$ and the reconstruction filter $h(t)$ satisfy:*

$$W(e^{j\Omega T_s})H(j(\Omega + k\Omega_0)) = G(j\Omega), k = 0, 1, \cdots, M-1. \quad (10)$$

In other words, if we are to use a shaping filter $g(t)$ that allows a digital implementation as in Fig. 2, the pulse $g(t)$ should be such that we can find $h(t)$ and $w[n]$ that satisfy (10). Given a discrete window $w[n]$ and a reconstruction filter $h(t)$, there does not exist an equivalent analog representation in general. For example consider the choices of

$h(t)$ and $w[n]$ in Example 1. Using Theorem 2, we can verify that $W(e^{j\Omega T_s})H(j\Omega) \neq W(e^{j\Omega T_s})H(j(\Omega+\Omega_0))$; there does not exist a pulse shaping filter $g(t)$ such that Fig. 1 and Fig. 2 are equivalent. Notice that we did not place any constraint on the duration of $g(t)$, $h(t)$ and $w[n]$ in the derivation; the condition in (10) is valid for non-finite pulses as well.

Corollary 1 *The analog OFDM transmitter with a rectangular pulse $g(t)$ in Fig. 1 does not admit the digital implementation in Fig. 2.*

A proof can be found in [10]. An example of $g(t)$, $w[n]$ and $h(t)$ that meet the requirement in (10) is given in the next example.

Example 2. Consider the case that $g(t)$ is an ideal filter bandlimited to $0 < \Omega < \Omega_0$, as shown in Fig. 4(a),

$$G(j\Omega) = \begin{cases} 1, & 0 < \Omega < \Omega_0, \\ 0, & \text{otherwise.} \end{cases}$$

We choose $h(t)$ to be an ideal filter of the following frequency characteristics (Fig. 4(b)),

$$H(j\Omega) = \begin{cases} 1, & 0 < \Omega < M\Omega_0, \\ 0, & \text{otherwise.} \end{cases}$$

The discrete window $w[n]$ is an ideal filter bandlimited to $0 < \omega < 2\pi/M$ in the period of $0 \leq \omega < 2\pi$. A plot of $W(e^{j\Omega T_s})$ is given in Fig. 4(c). Then $W(e^{j\Omega T_s})$ is periodic with period $2\pi/T_s = M\Omega_0$. In this case we can verify that the condition given in (10) is satisfied and the analog representation and the digital implementation are equivalent.

4. CONCLUSION

The OFDM transmitter used in practice is based on a DFT matrix followed by a DAC converter. On the other hand the analysis of OFDM systems is usually based on a convenient analog representation with a continuous-time pulse $g(t)$. In this paper, we show that the DFT based OFDM transmitter has an analog representation only in restricted cases. Conversely, given an arbitrary pulse $g(t)$ for the analog transmitter, there is usually no digital DFT based implementation. Therefore designs or analyses of OFDM systems based directly on the digital schematic will be more useful than those based on the analog schematic.

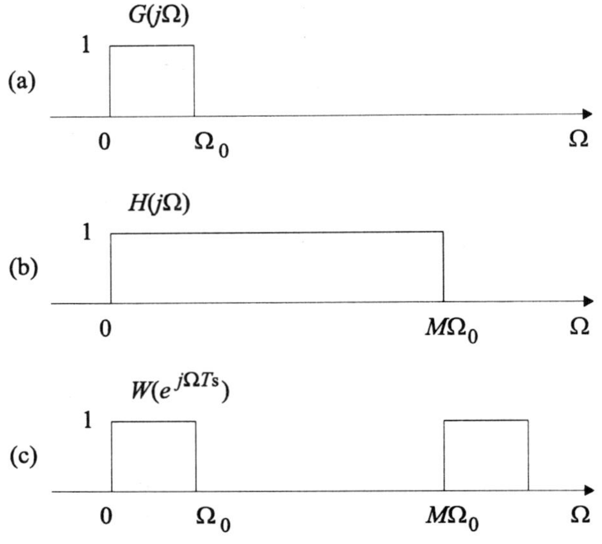

Figure 4: *Example 2.* Illustration of $G(j\Omega)$, $H(j\Omega)$ and $W(e^{j\Omega T_s})$.

5. REFERENCES

[1] S. B. Weinstein and P. M. Ebert, "Data Transmission by Frequency-Division Multiplexing Using the Discrete Fourier Transform," *IEEE Trans. Communication Technology*, vol. 19, no. 5, Oct. 1971.

[2] L. J. Cimini, "Analysis and Simulation of a Digital Mobile Channel Using Orthogonal Frequency Division Multiple Access," *IEEE Trans. Comm.*, 1985.

[3] G. L. Stuber, *Principles of mobile communication*, 2nd. ed., Kluwer Academic Publishers, 2001.

[4] A. Vahlin and N. Holte, "Optimal Finite Duration Pulses for OFDM," *IEEE Trans. Comm.*, 1996.

[5] T. Pollet, M. Van Bladel, and M. Moeneclaey, "BER Sensitivity of OFDM Systems to Carrier Frequency Offset and Wiener Phase Noise ," *IEEE Trans. Comm.*, 1995.

[6] M. Pauli and P. Kuchenbecker, "On the reduction of the out-of-band radiation of OFDM-signals," Proc. ICC, 1998.

[7] H. Nikookar and R. Prasad, "Optimal waveform design for multicarrier transmission through a multipath channel," Proc. IEEE Vehicular Tech. Conf., 1997.

[8] K. Matheus, and K.-D. Kammeyer, "Optimal design of a multicarrier systems with soft impulse shaping including equalization in time or frequency direction," Proc. IEEE Globalcom, 1997.

[9] A. V. Oppenheim, R. W. Schafer, with J. R. Buck, *Discrete-Time Signal Processing*, 2nd. ed., Prentice-Hall, 1999.

[10] Yuan-Pei Lin and See-May Phoong, "OFDM Transmitters: Analog Representation and DFT Based Implementation," Submitted to IEEE Trans. Signal Processing.

Jointly Minimum MSE Transmitter and Receiver FIR MIMO Filters in the Presence of Near-End Crosstalk and Additive Noise

Are Hjørungnes
Signal Processing Laboratory
Helsinki University of Technology
P. O. Box 3000, FIN-02015 HUT, Finland
Tel. +358 9 451 5386, Fax +358 9 452 3614
Email: arehj@wooster.hut.fi

Paulo S. R. Diniz, and Marcello L. R. de Campos
Signal Processing Laboratory
Universidade Federal do Rio de Janeiro
P. O. Box 68504, CEP: 21945-970 RJ, Brazil
Tel. +55 21 2562 8211, Fax +55 21 2562 8205
Email: {diniz,campos}@lps.ufrj.br

***Abstract*—A theory for jointly minimizing the mean square error (MSE) between the desired and decoded signal with respect to the transmitter and receiver FIR multiple-input multiple-output (MIMO) filters is developed in the presence of near-end crosstalk and two additive noise sources independent of the original signal. The additive noise sources account for noise at the input signal and on the channel. The transfer functions are known FIR MIMO transfer functions that can model the desired communication link and the undesired near-end crosstalk channel. The channel input vector time series has an average power constrained. An iterative numerical optimization algorithm is proposed. When compared to the methods available in the literature, the proposed method yields better results due to the joint optimization of the transmitter-receiver pair, and is applicable to a more general scenario that may include correlated sources and near-end crosstalk.**

I. Introduction

The problem of transmitting the original discrete-time vector time series $\boldsymbol{x}(n)$ corrupted by signal-independent additive noise $\boldsymbol{u}(n)$ over a vector channel is investigated. It is assumed that the channel is discrete in time and corrupted by additive signal-independent noise $\boldsymbol{v}(n)$. Furthermore, it is assumed that the channel input vectors have constrained power, and that the vector channel transfer functions are known and can model the desired channel $\boldsymbol{C}(z)$ and the undesired near-end crosstalk channel $\boldsymbol{G}(z)$. The transmitter and receiver FIR MIMO filters are jointly optimized with respect to the block MSE between the desired output and the actual output vector time series of the receiver filter, subject to a channel power constraint. An iterative numerical solution is proposed based on formulae for finding the optimal transmitter FIR MIMO filter for a given receiver FIR MIMO filter, and vice versa.

Analytical expressions for optimal transform coders were derived in [1] for the identity channel transfer function, and this was extended to include an FIR channel transfer function for redundant filter banks in [2] under restrict assumptions on the transform filter length and the order of the scalar channel transfer function. In the field of power-constrained FIR MIMO filters, the transmitter and receiver filter are jointly optimized by random search in [3]. Formulae for finding the optimal receiver FIR MIMO filter for a given transmitter FIR MIMO filter and vice versa are proposed in [4]. In [5], an algorithm is given for finding jointly optimized, linear phase, single input single output (SISO), FIR pre- and post-filters with decimator and expander for communication over a channel with additive noise. For a given transmitter FIR MIMO filter, the optimal FIR receiver filter has been derived in different ways [4], [6], [7]. In [8], a multirate Kalman receiver filtering approach was proposed for solving the FIR problem for a fixed transmitter FIR MIMO filter.

The system considered is shown in Figure 1. The original vector time series $\boldsymbol{x}(n)$ is corrupted by an uncorrelated additive vector noise series $\boldsymbol{u}(n)$. The reason for including this additive noise is to make the developed theory as general as possible and at the same time there are practical situations where this occur. Two examples are when images or audio are sampled in order to store them in a computer. All the input signals to the system are assumed to have zero mean and to be uncorrelated with each other, but all of them can be correlated in itself with known second order statistics.

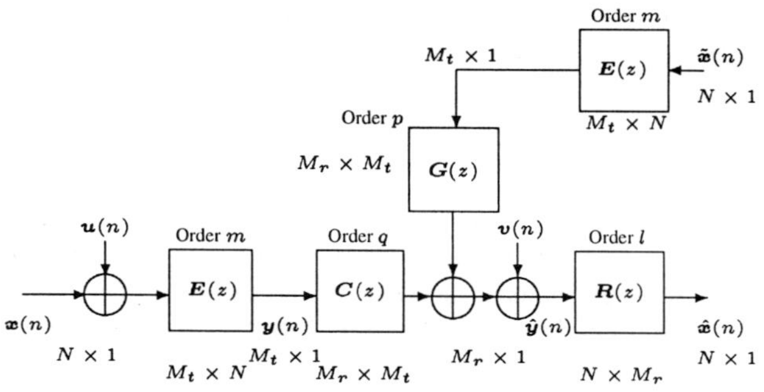

Fig. 1. The power-constrained MIMO block system with near-end crosstalk and additive input and channel noise.

Near-end crosstalk is modeled by a feedback from the receiver to itself by the known $M_r \times M_t$ transfer matrix $\boldsymbol{G}(z)$. Figure 1 indicates that the dimensions of the times series that is used in the article. N, M_t, and M_r are positive integers that can be different. It is assumed that the $M_t \times N$ transmitter FIR MIMO $\boldsymbol{E}(z)$ and the $N \times M_r$ receiver FIR MIMO equalizer filter $\boldsymbol{R}(z)$ are FIR, i.e., they have the following model:

$$\boldsymbol{E}(z) = \sum_{k=0}^{m} \boldsymbol{e}(k) z^{-k} \quad \text{and} \quad \boldsymbol{R}(z) = \sum_{k=0}^{l} \boldsymbol{r}(k) z^{-k}, \tag{1}$$

where the numbers m and l are assumed to be known. This means that the matrices $\boldsymbol{e}(k)$ and $\boldsymbol{r}(k)$, have dimensions $M_t \times N$ and $N \times M_r$, respectively. It is assumed that all the vector time series are jointly wide sense stationary (WSS).

II. Expansion Operators

For FIR polynomials matrices, four expansion operators will be used and one expansion of a vector time series will be needed as well. Let $\boldsymbol{A}(z) = \sum_{i=0}^{k} \boldsymbol{a}(i) z^{-k}$ be an FIR polynomial of order k and dimension $N_0 \times N_1$. The *row-expanded* matrix $\boldsymbol{A}_{-}$ of the matrix polynomial $\boldsymbol{A}(z)$ is an $N_0 \times (k+1)N_1$ matrix given by

$$\boldsymbol{A}_{-} = [\boldsymbol{a}(0) \;\; \boldsymbol{a}(1) \;\; \cdots \;\; \boldsymbol{a}(k)]. \tag{2}$$

The *column-expanded* matrix $\boldsymbol{A}_{|}$ of the matrix polynomial $\boldsymbol{A}(z)$ is a $(k+1)N_0 \times N_1$ matrix given by

$$\boldsymbol{A}_{|} = \left[\boldsymbol{a}^T(k) \;\; \boldsymbol{a}^T(k-1) \;\; \ldots \;\; \boldsymbol{a}^T(1) \;\; \boldsymbol{a}^T(0)\right]^T. \tag{3}$$

The *row-diagonal-expanded* matrix $\boldsymbol{A}_{\urcorner}^{(q)}$ of the matrix polynomial $\boldsymbol{A}(z)$ is a $(q+1)N_0 \times (k+q+1)N_1$ matrix given by:

0-7803-7761-3/03/$17.00 ©2003 IEEE

$$A_{\lnot}^{(q)} = \begin{bmatrix} a(0) & \cdots & a(m) & \cdots & 0 \\ \vdots & \ddots & \ddots & \ddots & \vdots \\ 0 & \cdots & a(0) & \cdots & a(m) \end{bmatrix}. \quad (4)$$

The *column-diagonal-expanded* matrix $A_{\llcorner}^{(q)}$ of the matrix polynomial $A(z)$ is a $(k+q+1)N_0 \times (q+1)N_1$ matrix given by

$$A_{\llcorner}^{(q)} = \begin{bmatrix} a(k) & \cdots & 0 \\ \vdots & \ddots & \vdots \\ a(0) & \ddots & a(k) \\ \vdots & \ddots & \vdots \\ 0 & \cdots & a(0) \end{bmatrix}. \quad (5)$$

Let ν be a positive integer. The column-expansion of a vector time series $x(n)$ of dimension $(\nu+1)N \times 1$ is defined as:

$$x(n)_{\shortmid}^{(\nu)} = \left[x^T(n) \; x^T(n-1) \; \cdots \; x^T(n-\nu)\right]^T. \quad (6)$$

III. Problem Formulation

Let the desired channel and undesired crosstalk transfer function be modeled by FIR polynomial matrices of order q and p, respectively, and let these polynomials be denoted $C(z)$ and $G(z)$. These matrices have dimension $M_r \times M_t$. Expressed as matrix polynomials these matrices are expressed as:

$$C(z) = \sum_{k=0}^{q} c(k) z^{-k} \quad \text{and} \quad G(z) = \sum_{k=0}^{p} g(k) z^{-k}, \quad (7)$$

where the $M_r \times M_t$ matrices $c(k)$ and $g(k)$ are assumed to be known.

Let $\mathcal{T}^{(k)} : \mathbb{C}^{N\times(m+k+1)N} \to \mathbb{C}^{(k+1)N\times(m+1)N}$ be called the *reshape operator* producing a $(k+1)N \times (m+1)N$ block Toeplitz matrix from an $N \times (m+k+1)N$ matrix. Let W_- be an $N \times (m+k+1)N$ matrix, where the ith $N \times N$ block is given by $w(i)$, $i \in \{0, 1, \ldots, m+k\}$. Then, the operator $\mathcal{T}^{(k)}$ acting on the matrix W_- yields:

$$\mathcal{T}^{(k)}\{W_-\} = \begin{bmatrix} w(k) & w(k+1) & \cdots & w(m+k) \\ \vdots & \vdots & \ddots & \vdots \\ w(1) & w(2) & \cdots & w(m+1) \\ w(0) & w(1) & \cdots & w(m) \end{bmatrix}. \quad (8)$$

The autocorrelation matrix of dimension $(\nu+1)N \times (\nu+1)N$ of the $(\nu+1)N \times 1$ vector $x(n)_{\shortmid}^{(\nu)}$ is defined as:

$$\Phi_{x}^{(\nu,N)} = E\left[x(n)_{\shortmid}^{(\nu)} \left(x(n)_{\shortmid}^{(\nu)}\right)^H\right], \quad (9)$$

where the operator ^{H} denotes complex conjugated transposed. For other column-expanded vectors the autocorrelation matrices are defined in a similar way.

Let the $(m+1)N \times (m+1)N$ matrix $\Psi_{x}^{(m,N)}(i)$ be defined as follows

$$\Psi_{x}^{(m,N)}(i) = E\left[\left(x(n)_{\shortmid}^{(m)}\right)^* \left(x(n+i)_{\shortmid}^{(m)}\right)^T\right], \quad (10)$$

where $i \in \{-q-l, -q-l+1, \ldots, q+l\}$, the operator * means complex conjugation of the components, and the operator ^{T} represents matrix transposition.

The desired receiver output signal $d(n)$ can be chosen arbitrarily, therefore, the results are a generalization of the result as found in [4]. Let the ith component of the desired vector signal $d(n)$ be denoted $d_i(n)$, where $i \in \{0, 1, \ldots, N-1\}$. Let $x_i(n)$ and $\hat{x}_i(n)$ denote the ith vector component of the vectors $x(n)$ and $\hat{x}(n)$, respectively. The scalar time series $x(n)$ can be used to produce the N input vector $x(n)$ to the FIR MIMO transmitter filter through the following equation

$$x(n) = [x(nN) \; x(nN-1) \; \cdots \; x(nN-(N-1))]^T. \quad (11)$$

The relationship between the $N \times 1$ output vector $\hat{x}(n)$ and the scalar time series $\hat{x}(n)$ can be given by:

$$\hat{x}(n) = [\hat{x}(nN+N-1) \; \hat{x}(nN+N-2) \; \cdots \; \hat{x}(nN)]^T. \quad (12)$$

The most general way to choose the components of the desired vector when using the blocking and unblocking structures given in Equations (11) and (12) is:

$$d_i(n) = x((n-\delta_v)N - \delta_s - i). \quad (13)$$

The terms δ_v and δ_s are the *vector* and *scalar delay* through the system, respectively. These delays are introduced such that the developed theory is as general as possible, and they add the extra parameters δ_v and δ_s to the optimization of the FIR MIMO filter system.

The cross-covariance matrix $\phi_{x,d}^{(\nu,N)}$ of dimension $(\nu+1)N \times N$ is defined as:

$$\phi_{x,d}^{(\nu,N)} = E\left[x(n)_{\shortmid}^{(\nu)} d^H(n)\right], \quad (14)$$

where the vector $x(n)_{\shortmid}^{(\nu)}$ has dimension $(\nu+1)N \times 1$ and where $d(n)$ is the desired $N \times 1$ vector time series used in the calculation of the block MSE of the system. The block MSE, denoted $\mathcal{E}_{N,M_t,M_r}$, is defined as:

$$\mathcal{E}_{N,M_t,M_r} = E\left[\|\hat{x}(n) - d(n)\|^2\right]. \quad (15)$$

Among all the input signals to the system, see Figure 1, it is assumed that the desired vector time series $d(n)$ can only be correlated to the original vector time series $x(n)$.

The output of the transmitter FIR MIMO filter $E(z)$ is represented by the $M_t \times 1$ vector $y(n)$ given by

$$y(n) = E_-\left(x(n)_{\shortmid}^{(m)} + u(n)_{\shortmid}^{(m)}\right), \quad (16)$$

where the notation in Section II is used and the dimension of the column-expended vectors $x(n)_{\shortmid}^{(m)}$ and $u(n)_{\shortmid}^{(m)}$ both have dimension $(m+1)N \times 1$.

The row-expanded matrix which is used to express the total transfer function from the original signal $x(n)$ to the output signal $\hat{x}(n)$ is denoted $W(z)$ and it has order $m+q+l$. It can be shown that $W_- = R_- C_{\lnot}^{(l)} E_{\lnot}^{(q+l)}$, and the dimension of this matrix product is $N \times (m+q+l+1)N$. The matrix $C_{\lnot}^{(l)}$ has dimension $(l+1)M_r \times (q+l+1)M_t$ and the matrix $E_{\lnot}^{(q+l)}$ has dimension $(q+l+1)M_t \times (m+q+l+1)N$.

By rewriting the convolution sum with the notation introduced in Section II, it is possible to express the $N \times 1$ output vector $\hat{x}(n)$ of the receiver filter as follows:

$$\begin{aligned} \hat{x}(n) = & R_- C_{\lnot}^{(l)} E_{\lnot}^{(q+l)} \left(x(n)_{\shortmid}^{(m+q+l)} + u(n)_{\shortmid}^{(m+q+l)}\right) \\ & + R_- G_{\lnot}^{(l)} E_{\lnot}^{(p+l)} \tilde{x}(n)_{\shortmid}^{(m+p+l)} + R_- v(n)_{\shortmid}^{(l)}, \end{aligned} \quad (17)$$

where R_- is an $N \times (l+1)M_r$ matrix, the vectors $x(n)_{\shortmid}^{(m+q+l)}$ and $u(n)_{\shortmid}^{(m+q+l)}$ have dimension $(m+q+l+1)N \times 1$, $v(n)_{\shortmid}^{(l)}$ is an $(l+1)M_r \times 1$ vector, and $\tilde{x}(n)_{\shortmid}^{(m+p+l)}$ is an $(m+p+l+1)N \times 1$ vector. Note that in Equation (17), the matrix $E_{\lnot}^{(q+l)}$ has dimension $(q+l+1)M_t \times (m+q+l+1)N$ and the matrix $E_{\lnot}^{(p+l)}$ has dimension $(p+l+1)M_t \times (m+p+l+1)N$. The matrix $G_{\lnot}^{(l)}$ has dimension $(l+1)M_r \times (p+l+1)M_t$.

It can be shown that the block MSE $\mathcal{E}_{N,M_t,M_r}$ for the system is given by Equation (18), and the average power constraint for the channel input vector $y(n)$ can be expressed as:

$$E\left[\|y(n)\|^2\right] = \operatorname{Tr}\left\{E_- \Phi_{x+u}^{(m,N)} E_-^H\right\} = P. \quad (19)$$

Problem 1: Minimize the block MSE given by Equation (18) with respect to the transmitter FIR MIMO filter $E(z)$ and the receiver FIR MIMO filter $R(z)$, subject to the constraint in Equation (19).

IV. Proposed Solution

The constrained optimization problem stated in Section III can be converted to an unconstrained optimization problem by using a positive Lagrange multiplier μ. The unconstrained objective function ζ_{N,M_t,M_r} can be expressed as

$$\zeta_{N,M_t,M_r} = \mathcal{E}_{N,M_t,M_r} + \mu \operatorname{Tr}\left\{ \boldsymbol{E}_{-} \boldsymbol{\Phi}_{\boldsymbol{x}+\boldsymbol{u}}^{(m,N)} \boldsymbol{E}_{-}^{H} \right\}. \tag{20}$$

It can be shown [9], that the equation for the optimal FIR MIMO transmitter filter, for a given FIR MIMO receiver filter, can be written as:

$$\boldsymbol{A} \cdot \operatorname{vec}(\boldsymbol{E}_{-}) = \boldsymbol{b}, \tag{21}$$

where matrix $\boldsymbol{A}$ is an $(m+1)M_tN \times (m+1)M_tN$ matrix given by Equation (22) and vector $\boldsymbol{b}$ is of dimension $(m+1)M_tN \times 1$ given by

$$\boldsymbol{b} = \operatorname{vec}\left(\boldsymbol{C}_{\urcorner}^{H} \left(\boldsymbol{R}_{\llcorner}^{(q)}\right)^{H} \mathcal{T}^{(q+l)} \left\{ \left(\boldsymbol{\phi}_{\boldsymbol{x},\boldsymbol{d}}^{(m+q+l,N)}\right)^{H} \right\} \right). \tag{23}$$

In Equation (22), the operator $\otimes$ is the Krönecker product and the matrix $\boldsymbol{I}_{M_t}$ is the identity matrix of dimension $M_t \times M_t$.

It is important to mention that Equation (21) is not equivalent to the corresponding equation found in [4]. In Subsection V-A, it will be shown by a design example that the formula presented in [4] does not lead to the same results as the proposed formula for a correlated source.

It can be shown [9], that the optimized receiver FIR MIMO filter for a given transmitter FIR MIMO filter is given by the following equation:

$$\begin{aligned} \boldsymbol{R}_{-} = {} & \left(\boldsymbol{\phi}_{\boldsymbol{x},\boldsymbol{d}}^{(m+q+l,N)}\right)^{H} \left(\boldsymbol{E}_{\urcorner}^{(q+l)}\right)^{H} \left(\boldsymbol{C}_{\urcorner}^{(l)}\right)^{H} \\ & \cdot \left[\boldsymbol{G}_{\urcorner}^{(l)} \boldsymbol{E}_{\urcorner}^{(p+l)} \boldsymbol{\Phi}_{\tilde{\boldsymbol{x}}}^{(m+p+l,N)} \left(\boldsymbol{E}_{\urcorner}^{(p+l)}\right)^{H} \left(\boldsymbol{G}_{\urcorner}^{(l)}\right)^{H} + \boldsymbol{\Phi}_{\boldsymbol{v}}^{(l,M_r)} \right. \\ & \left. + \boldsymbol{C}_{\urcorner}^{(l)} \boldsymbol{E}_{\urcorner}^{(q+l)} \boldsymbol{\Phi}_{\boldsymbol{x}+\boldsymbol{u}}^{(m+q+l,N)} \left(\boldsymbol{E}_{\urcorner}^{(q+l)}\right)^{H} \left(\boldsymbol{C}_{\urcorner}^{(l)}\right)^{H} \right]^{-1}. \end{aligned} \tag{24}$$

The problem of jointly optimizing the overall system performance is performed by the following iterative approach: For a fixed transmitter FIR MIMO filter, the receiver FIR MIMO filter is optimized by solving Equation (24), then the transmitter FIR MIMO filter is optimized by using the previously optimized value of the receiver filter in Equation (21), and this procedure is repeated until convergence is reached.

V. Results, Critique, and Comparisons

A. Comparison with the Method Proposed by Honig et al. [4]

In [4], the desired signal was chosen as in Equation (13) with $\delta_s = 0$. It can be shown that if the performance of the proposed system is compared to the results found by the formulae in [4], the same results are obtained if $x(n)$ is uncorrelated, where $x(n)$ is the scalar time series used to produce the input vector time series $\boldsymbol{x}(n)$ in Equation (11), and $\boldsymbol{u}(n) = \boldsymbol{0}$.

The signal to noise ratio (SNR) is given by $\sigma_x^2 N/\mathcal{E}_{N,M_t,M_r}$, where σ_x^2 is the variance of the original input time series $x(n)$ and $\mathcal{E}_{N,M_t,M_r}/N$ is the MSE per source sample. In this example, it is assumed that the input vectors $\boldsymbol{x}(n)$ to the transmitter FIR MIMO filter are produced by a blocking operation of a WSS time series $x(n)$, see Equation (11). The channel signal to noise ratio (CSNR) is given by $(P/M_t)/\sigma_v^2$, where P/M_t is the output energy from the transmitter per input channel sample, see Equation (19), and σ_v^2 is the variance of the white additive channel noise $\boldsymbol{v}(n)$ with equal variance for each vector component.

In Figure 2, the performance of the proposed system using the parameters that is given in the caption of the figure. The best possible performance of all systems is given by the optimal performance theoretically attainable (OPTA) curve. Because of Shannon's separation theorem [10], the OPTA curve can be found by evaluating the distortion rate function of a source with a given metric at the channel capacity function [11], and at the same time it is taken into consideration that N source samples are transmitted per M_t channel input and M_r channel output samples. In Figure 2, the OPTA curve is found for a memoryless Gaussian source and channel and the metric used is the squared error distortion. The curve with the circles is obtained by using the receiver filters of the proposed system with the transmitter FIR MIMO filter found by Equation (4.14) in [4]. The reason for doing this is that there is an inconsistency in the equations for the transmitter filters, but the receiver equations are the same in [4] and Equation (24). The rightmost $\times$-mark and circle correspond to the same value of Lagrange multiplier μ.

From Figure 2, it is seen that the performance obtained by Equation (4.14) in [4] gives negative values of SNR. Therefore, this formula cannot be correct for the correlated input source used in the example. The correct formula is given in Equation (21).

In [12], which is an earlier article by the same authors as [4], it was assumed that the input signal was uncorrelated, and it was mentioned that the results could be easily generalized to correlated sources. This is what has been attempted in [4], but the equation for an optimal transmitter FIR MIMO filter for a given receiver filter cannot be used for correlated sources. For a detailed study of the formulae for the optimal FIR MIMO transmitter filter and a comparison with the formulae in [4], see [9].

B. Comparison with the Method Proposed by Malvar [5]

In [5], jointly optimal transmitter and receiver FIR SISO filters with linear phase were proposed for $M_t = M_r = 1$. It was assumed that signal-independent noise was added both at the original signal and at the channel. No near-end crosstalk was included, i.e., $\boldsymbol{G}(z) = \boldsymbol{0}$.

Let the input signal to noise ratio (ISNR) be defined as the ratio between the input signal variance and the input noise variance σ_x^2/σ_u^2, where σ_u^2 is the variance of the additive input noise. In this definition, it is assumed that all vector components of $\boldsymbol{x}(n)$ and $\boldsymbol{u}(n)$ have zero mean and variance equal to σ_x^2 and σ_u^2, respectively.

$$\begin{aligned} \mathcal{E}_{N,M_t,M_r} = \operatorname{Tr}\Big\{ & \boldsymbol{\Phi}_{\boldsymbol{d}}^{(0,N)} + \boldsymbol{R}_{-}\boldsymbol{\Phi}_{\boldsymbol{v}}^{(l,M_r)}\boldsymbol{R}_{-}^{H} - \boldsymbol{R}_{-}\boldsymbol{C}_{\urcorner}^{(l)}\boldsymbol{E}_{\urcorner}^{(q+l)}\boldsymbol{\phi}_{\boldsymbol{x},\boldsymbol{d}}^{(m+q+l,N)} - \left(\boldsymbol{\phi}_{\boldsymbol{x},\boldsymbol{d}}^{(m+q+l,N)}\right)^{H}\left(\boldsymbol{E}_{\urcorner}^{(q+l)}\right)^{H}\left(\boldsymbol{C}_{\urcorner}^{(l)}\right)^{H}\boldsymbol{R}_{-}^{H} \\ & + \boldsymbol{R}_{-}\boldsymbol{C}_{\urcorner}^{(l)}\boldsymbol{E}_{\urcorner}^{(q+l)}\boldsymbol{\Phi}_{\boldsymbol{x}+\boldsymbol{u}}^{(m+q+l,N)}\left(\boldsymbol{E}_{\urcorner}^{(q+l)}\right)^{H}\left(\boldsymbol{C}_{\urcorner}^{(l)}\right)^{H}\boldsymbol{R}_{-}^{H} + \boldsymbol{R}_{-}\boldsymbol{G}_{\urcorner}^{(l)}\boldsymbol{E}_{\urcorner}^{(p+l)}\boldsymbol{\Phi}_{\tilde{\boldsymbol{x}}}^{(m+p+l,N)}\left(\boldsymbol{E}_{\urcorner}^{(p+l)}\right)^{H}\left(\boldsymbol{G}_{\urcorner}^{(l)}\right)^{H}\boldsymbol{R}_{-}^{H} \Big\}. \end{aligned} \tag{18}$$

$$\begin{aligned} \boldsymbol{A} = {} & \sum_{i_0=0}^{q}\sum_{i_1=0}^{l}\sum_{i_2=0}^{l}\sum_{i_3=0}^{q} \boldsymbol{\Psi}_{\boldsymbol{x}+\boldsymbol{u}}^{(m,N)}(i_0+i_1-i_2-i_3) \otimes \left(\boldsymbol{c}^{H}(i_0)\boldsymbol{r}^{H}(i_1)\boldsymbol{r}(i_2)\boldsymbol{c}(i_3)\right) + \boldsymbol{\Psi}_{\boldsymbol{x}+\boldsymbol{u}}^{(m,N)}(0) \otimes \mu\boldsymbol{I}_{M_t} \\ & + \sum_{i_0=0}^{p}\sum_{i_1=0}^{l}\sum_{i_2=0}^{l}\sum_{i_3=0}^{p} \boldsymbol{\Psi}_{\tilde{\boldsymbol{x}}}^{(m,N)}(i_0+i_1-i_2-i_3) \otimes \left(\boldsymbol{g}^{H}(i_0)\boldsymbol{r}^{H}(i_1)\boldsymbol{r}(i_2)\boldsymbol{g}(i_3)\right). \end{aligned} \tag{22}$$

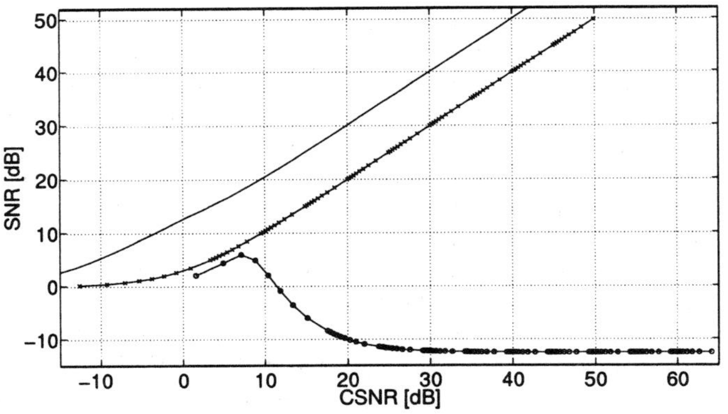

Fig. 2. SNR versus CSNR performance of proposed power-constrained FIR MIMO filters are shown by the ×-marks while the system proposed in [4] is shown by the circles. The input time series is a Gaussian AR(1) time series with correlation coefficient 0.95, Gaussian white noise is added on the channel, with: $N = M_t = M_r = m = l = \delta_v = 2$, $q = p = \delta_s = 0$, $\boldsymbol{u}(n) = \boldsymbol{0}$, $\boldsymbol{C}(z) = \boldsymbol{I}$, and $\boldsymbol{G}(z) = \boldsymbol{0}$. The upper curve is OPTA.

In Subsection 3.4.3 in [5], an example is given using the parameters that are given in Figure 3. Furthermore, it was assumed that the subband samples were uncorrelated with the white additive channel noise. Linear phase is assumed in [5], and the filter lengths are 13 in the transmitter and 25 in the receiver filter. The same parameters were used in the example with the proposed theory. Figure 3 shows the results achieved with the proposed method. The OPTA curve for Figure 3 can be found from Subsection 4.5.4 in [11], where real noise on the input signal is taken into consideration when finding the rate distortion function.

Figure 3 shows that the performance of the proposed system reaches a certain limit when the CSNR values are increased. The reason for this is that when $\min\{M_t, M_r\} < N$, it is impossible to achieve perfect reconstruction. In the example in [5], the noise level in both the input signal and the channel was 30.0 dB, that is ISNR = CSNR = 30.0 dB, and the overall performance of the system was SNR = 11.87 dB in [5]. This is 0.02 dB worse than the results obtained by the proposed system. If $\min\{M_t, M_r\}$ is increased, the performance of the system will improve at the expense of increasing the bandwidth used on the channel. As a reference a system where only the receiver filter where optimized is also included in Figure 3. In this system, the transmitter filter is given by $\boldsymbol{E}(z) = z^{-6}$ and the other system parameters were the same as above. From Figure 3, it is seen that 1.9 dB can be gained in SNR by jointly optimizing the transmitter and receiver compared to only optimizing the receiver when CSNR=30 dB.

VI. Conclusions

Equations were found for finding jointly minimized MSE transmitter and receiver FIR MIMO filters for a power-constrained channel. An iterative algorithm was proposed that is able to converge to a locally optimal solution.

An inconsistency was found in the literature in the case of correlated sources for finding the optimal transmitter FIR MIMO filter. A new equation was proposed in this article, which was validated by simulations. The theory was extended to include also the case of uncorrelated additive noise to the original signal.

The filters proposed in this article are a generalization of the method proposed in [5], where an algorithm for finding jointly optimal transmitter and receiver FIR SISO filters having one transmitter and one receiver filter, i.e., $M_t = M_r = 1$, was given.

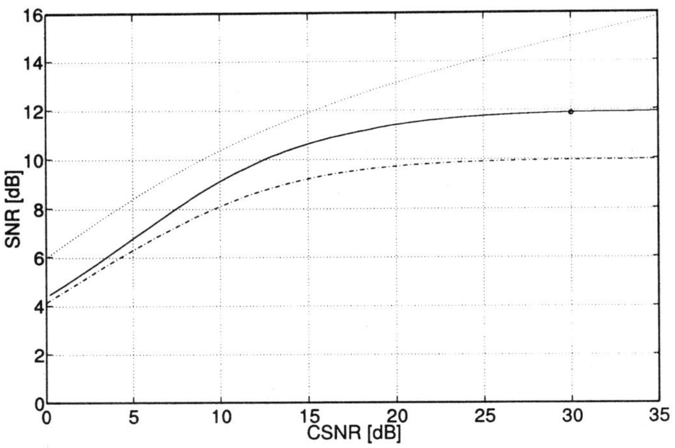

Fig. 3. The solid curve shows the SNR versus CSNR performance using the proposed theory with the following parameters: $N = 6$, $M_t = M_r = 1$, $m = 2$, $l = 4$, $q = p = 0$, $\delta_v = 2$, $\delta_s = 1$, $\boldsymbol{C}(z) = \boldsymbol{I}$, $\boldsymbol{G}(z) = \boldsymbol{0}$, and the input PSD is an AR(1) source with correlation coefficient 0.95. The circle shows the performance of the system proposed in [5], and OPTA is given by the dotted curve. ISNR = 30.0 dB in all systems. The dash-dotted line show the system where the receiver is given by $\boldsymbol{E}(z) = z^{-6}$ and only the receiver is optimized.

The proposed iteratively optimized joint transmitter-receiver filter pair was confronted with methods found in the literature in simulations for different application scenarios. In all of them, the proposed method performed better, setting new benchmark values towards the optimal performance.

References

[1] K.-H. Lee and D. P. Petersen, "Optimal linear coding for vector channels," *IEEE Trans. Commun.*, vol. COM-24, no. 12, pp. 1283–1290, Dec. 1976.

[2] A. Scaglione, G. B. Giannakis, and S. Barbarossa, "Redundant filterbank precoders and equalizers Part I: Unification and optimal designs," *IEEE Trans. Signal Processing*, vol. 47, no. 7, pp. 1988–2006, July 1999.

[3] B.-G. Song and J. A. Ritcey, "Joint pre and postfilter design for spatial diversity equalization," *IEEE Trans. Signal Processing*, vol. 45, no. 1, pp. 276–280, Jan. 1997.

[4] M. L. Honig, P. Crespo, and K. Steiglitz, "Suppression of near- and far-end crosstalk by linear pre- and post-filtering," *IEEE J. Select. Areas Commun.*, vol. 10, no. 3, pp. 614–629, Apr. 1992.

[5] H. S. Malvar, *Optimal Pre- and Post-Filtering in Noisy Sampled-Data Systems*. Ph.D. thesis, Massachusetts Institute of Thechnology, Cambridge, MA, USA, Sept. 1986.

[6] K. Gosse and P. Duhamel, "Perfect reconstruction versus MMSE filter banks in source coding," *IEEE Trans. Signal Processing*, vol. 45, no. 9, pp. 2188–2202, Sept. 1997.

[7] A. N. Delopoulos and S. D. Kollais, "Optimal filter banks for signal reconstruction from noisy subband components," *IEEE Trans. Signal Processing*, vol. 44, no. 2, pp. 212–224, Feb. 1996.

[8] B.-S. Chen, C.-W. Lin, and Y.-L. Chen, "Optimal signal reconstruction in noisy filter bank systems: Multirate Kalman synthesis filtering approach," *IEEE Trans. Signal Processing*, vol. 43, no. 11, pp. 2496–2504, Nov. 1995.

[9] A. Hjørungnes, M. L. R. de Campos, and P. S. R. Diniz, "Jointly optimized transmitter and receiver FIR MIMO filters in the presence of near-end crosstalk," *IEEE Trans. Signal Processing*, submitted November 2002.

[10] R. E. Blahut, *Principles and Practice of Information Theory*. Addison Wesley, Reading, Massachusetts, USA, 1987.

[11] T. Berger, *Rate Distortion Theory*. Prentice-Hall, Inc, Englewood Cliffs, New Jersey, USA, 1971.

[12] P. Crespo, M. L. Honig, and K. Steiglitz, "Optimization of pre- and post-filters in the presence of near and far-end crosstalk," in *Proc. Int. Conf. on Communications*, Boston, USA, June 1989, vol. 1, pp. 541–547.

LOCALIZATION OF A MOVING SOURCE USING TDOA AND FDOA MEASUREMENTS

K. C. Ho , Wenwei Xu

Dept. of Electrical and Computer Engineering
University of Missouri - Columbia, Columbia, M0 65211

ABSTRACT

Closed-form solution that estimates the position and velocity of a moving source using time difference of arrival (TDOA) and frequency difference of arrival (FDOA) measurements of a signal received by a number of sensors are developed. The method requires only two least-squares minimization stages and is not iterative. The local convergence and large computation problems associated with the traditional linear iterative solutions can thus be eliminated. The estimation accuracy of the proposed method achieves the Cramer-Rao lower bound. Simulation results are included to examine the algorithm's performance and compare it with the Taylor-series technique.

1. INTRODUCTION

The problem of locating a source passively has been of considerable interest for many years. For a stationary source, one common technique is to measure the TDOA's of the source signal to a number of spatially separated sensors. Each TDOA defines a hyperbola in which the emitter must lie. The intersection of the hyperbolae gives the source location estimate [1] - [3]. When the source is moving, FDOA measurements should be used in addition to TDOA's to estimate the source position and velocity accurately.

Obtaining the source location and velocity from TDOA and FDOA measurement is known to be a non-trivial task. This is because the relationship between TDOA and FDOA measurements and the source location is nonlinear. One straightforward method to the localization problem is exhaustive search in the solution space. This is very computationally intensive, inefficient and prohibits real-time processing. A possible alternative is to linearize the measurement equations through Taylor-series expansion [4]. The Taylor-series linearization method requires a proper initial position and velocity guess close to the true solution that may not be easy to obtain in practice. In addition, it also suffers from convergence problem and is computationally expensive because the method is iterative. We shall present in this paper an attractive closed-form solution to the source position and velocity through TDOA and FDOA measurements. The proposed solution attains Cramer-Rao lower bound (CRLB) for Gaussian noise in small error region.

The proposed solution method follows the idea first introduced in [5]. It transforms the measurement equations to a set of linear equations by introducing nuisance parameters. It then solves the source location, velocity and the nuisance parameters by linear least-squares. Next, the nuisance parameters are eliminated through the use of another linear least-squares minimization to further improve the position and velocity estimates. The proposed solution is computationally efficient and does not suffer from local convergence problem.

2. THE CLOSED-FORM SOLUTION

Consider a two-dimension scenario where an array of M moving sensors is used to determine the position $\mathbf{u} = [x,\, y]^T$ and velocity $\dot{\mathbf{u}} = [\,\dot{x},\, \dot{y}\,]^T$ of an unknown moving source as shown in Figure 1. Without loss of generality through coordinate transformation, let sensor 1 be stationary and located at the origin so that $x_1 = y_1 = \dot{x}_1 = \dot{y}_1 = 0$. The other sensor positions $\mathbf{s}_i = [\,x_i,\, y_i\,]^T$ and velocities $\dot{\mathbf{s}}_i = [\,\dot{x}_i,\, \dot{y}_i\,]^T$ are assumed known. We wish to estimate the source position and velocity using TDOA and FDOA measurements.

Let the distance between the source and the sensor i as r_i. Then

$$r_i = |\,\mathbf{s}_i - \mathbf{u}\,| = \sqrt{(x_i - x)^2 + (y_i - y)^2}. \qquad (1)$$

If the TDOA of a signal received by the sensor pair i and 1 is t_{i1}, and if c is the signal propogation speed, then the set of TDOA measurement equations are:

$$r_{i1} = ct_{i1} = r_i - r_1 \qquad (2)$$

where r_{i1} is the range difference. Upon rewriting (2) as $r_{i1} + r_1 = r_i$, squaring both sides and substituting (1), the TDOA measurement equations become

$$r_{i1}^2 + 2r_{i1}r_1 = R_i^2 - R_1^2 - 2x_i x - 2y_i y \qquad (3)$$

where $R_i = \sqrt{x_i^2 + y_i^2}$, and $i = 2, 3, \cdots, M$. The intersection of (3) gives an estimate of the source location $\mathbf{u}$.

The TDOA equations can only identify the source position and not the velocity. The estimation of velocity requires

0-7803-7761-3/03/$17.00 ©2003 IEEE

FDOA measurements. To devise the equations that make use of FDOA measurments, we take the time derivative of (3):

$$r_{i1}\dot{r}_{i1} + \dot{r}_{i1}r_1 + r_{i1}\dot{r}_1 = x_i\dot{x}_i + y_i\dot{y}_i - \dot{x}_i x - x_i\dot{x} - \dot{y}_i y - y_i\dot{y} \quad (4)$$

for $i = 2, 3, \cdots, M$, where $\dot{r}_{i1}$ is the range rate differences that are derived from the FDOA measurements. Note that (4) is a set of equations that relates both TDOA and FDOA with the source position and velocity. The task is to estimate the source position $\mathbf{u}$ and velocity $\dot{\mathbf{u}}$ from (3) and (4), when a set of TDOA and FDOA measurements, or equivalently range difference and range rate difference measurements r_{i1} and $\dot{r}_{i1}$, $i = 2, 3, \cdots, M$ are available.

Let the vector $\mathbf{u}_1 = [\, x, y, r_1, \dot{x}, \dot{y}, \dot{r}_1 \,]^T$. It contains the unknown source parameters and two nuisance variables r_1 and $\dot{r}_1$. In the presence of TDOA and FDOA measurement noise, the equation error vector from (3) and (4) is

$$\varepsilon_1 = \begin{bmatrix} \varepsilon_t \\ \varepsilon_f \end{bmatrix} = \mathbf{h}_1 - \mathbf{G}_1\mathbf{u}_1 \quad (5)$$

where

$$\mathbf{h}_1 = \begin{bmatrix} r_{21}^2 - R_2^2 \\ \vdots \\ r_{M1}^2 - R_M^2 \\ 2\,(r_{21}\dot{r}_{21} - x_2\dot{x}_2 - y_2\dot{y}_2) \\ \vdots \\ 2\,(r_{M1}\dot{r}_{M1} - x_M\dot{x}_M - y_M\dot{y}_M) \end{bmatrix} \quad (6)$$

$$\mathbf{G}_1 = -2\begin{bmatrix} x_2 & y_2 & r_{21} & 0 & 0 & 0 \\ & & & \vdots & & \\ x_M & y_M & r_{M1} & 0 & 0 & 0 \\ \dot{x}_2 & \dot{y}_2 & \dot{r}_{21} & x_2 & y_2 & r_{21} \\ & & & \vdots & & \\ \dot{x}_M & \dot{y}_M & \dot{r}_{M1} & x_M & y_M & r_{M1} \end{bmatrix} \quad (7)$$

Equation (5) is a set of linear equations with respect to $\mathbf{u}_1$. The weighted LS solution of $\mathbf{u}_1$ that minimizes $\varepsilon_1^T\mathbf{W}_1\varepsilon_1$ is

$$\mathbf{u}_1 = (\mathbf{G}_1^T\mathbf{W}_1\mathbf{G}_1)^{-1}\mathbf{G}_1^T\mathbf{W}_1\mathbf{h}_1 \quad (8)$$

where $\mathbf{W}_1 = E[\varepsilon_1\varepsilon_1^T]^{-1}|_{\mathbf{u}_1=\mathbf{u}_1^o}$ is the weighting matrix and $\mathbf{u}_1^o$ is the true solution of $\mathbf{u}_1$. To find $\mathbf{W}_1$, we shall evaluate the TDOA equation error vector ε_t and FDOA equation error vector ε_f separately. Upon putting the true solution $\mathbf{u}_1^o$ into (3) and expressing r_{i1} as $r_{i1}^o + c\Delta t_{i1}$, where Δt_{i1} denotes the TDOA noise, we have

$$\varepsilon_t = c\mathbf{B}\Delta\mathbf{t} + c^2\Delta\mathbf{t}\otimes\Delta\mathbf{t} \approx c\mathbf{B}\Delta\mathbf{t} \quad (9)$$

where the second order error term has been ignored. The matrix $\mathbf{B} = 2\,diag\{\, r_2^o, r_3^o, \cdots, r_M^o\}$ is a diagonal matrix whose elements are the true ranges between the source and receivers, $\Delta\mathbf{t} = [\,\Delta t_{21}, \cdots, \Delta t_{M1}\,]^T$ is a vector of TDOA measurement error and $\otimes$ represents the element by element multiplication. Similarly, expressing r_{i1} as $r_{i1}^o + c\,\Delta t_{i1}$, $\dot{r}_{i1}$ as $\dot{r}_{i1}^o + c\Delta\dot{t}_{i1}$, evaluating at the true solution $\mathbf{u}_1^o$ and ignoring the second order error term, we have

$$\varepsilon_f \approx c\,(\dot{\mathbf{B}}\Delta\mathbf{t} + \mathbf{B}\Delta\dot{\mathbf{t}}) \quad (10)$$

where $\dot{\mathbf{B}} = 2\,diag\{\dot{r}_2^o, \dot{r}_3^o, \cdots, \dot{r}_M^o\}$ is a diagonal matrix whose elements are the range rates between source and receivers, and $\Delta\dot{\mathbf{t}}$ is a vector of FDOA measurement error. Hence, we have

$$\varepsilon_1 = c\mathbf{B}_1\begin{bmatrix} \Delta\mathbf{t} \\ \Delta\dot{\mathbf{t}} \end{bmatrix} \quad (11)$$

so that

$$\mathbf{W}_1 = \mathbf{B}_1^{-T}\mathbf{Q}^{-1}\mathbf{B}_1^{-1} \quad (12)$$

where

$$\mathbf{B}_1 = \begin{bmatrix} \mathbf{B} & 0 \\ \dot{\mathbf{B}} & \mathbf{B} \end{bmatrix}, \quad \mathbf{Q}_1 = c^2\begin{bmatrix} \mathbf{Q}_t & \mathbf{Q}_{tf} \\ \mathbf{Q}_{tf}^T & \mathbf{Q}_f \end{bmatrix} \quad (13)$$

$\mathbf{Q}_t$ and $\mathbf{Q}_f$ are known covariance matrices for TDOA noise and FDOA noise, respectively. $\mathbf{Q}_{tf}$ is the cross-covariance matrix of TDOA and FDOA noise. The weighting matrix is not known in practice because $\mathbf{B}$ and $\dot{\mathbf{B}}$ contain the true emitter position and velocity. However, if $\dot{\mathbf{B}} \ll \mathbf{B}$, i.e. if the source is moving relatively slowly compared to the range, $\mathbf{B}_1$ becomes diagonal. Furthermore, if the source is far away from the sensor array, we have $r_1 \approx r_2 \approx \cdots \approx r_M$ so that $\mathbf{B}_1$ reduces to a scalar multiple of an identity matrix. Since the effect of weighting matrix is invariant under scaling, $\mathbf{W}_1$ can be approximated by

$$\mathbf{W}_1 = \mathbf{Q}^{-1}. \quad (14)$$

On the contrary, in the situation where $\dot{\mathbf{B}}$ is close to $\mathbf{B}$ or the source is not far from sensors, (14) can first be used in (8) to obtain an initial solution $\mathbf{u}_1$ to estimate $\mathbf{B}$ and $\dot{\mathbf{B}}$. Now $\mathbf{B}_1$ is available and a better weighting matrix can be formed from (12) to produce a more accurate solution. This process can be iterated several times to improve the accuracy of $\mathbf{u}_1$. From weighted LS theory, the estimated $\mathbf{u}_1$ has a covariance matrix given by

$$cov(\mathbf{u}_1) = (\mathbf{G}_1^T\mathbf{W}_1\mathbf{G}_1)^{-1}. \quad (15)$$

Although the first two elements of $\mathbf{u}_1$ give the emitter position and the 4th to 5th elements of $\mathbf{u}_1$ give the source velocity, this answer is not accurate. This is because in solving for $\mathbf{u}_1$, it is assumed that the nuisance variables r_1 and $\dot{r}_1$ are independent of the source position and velocity. They are, in fact from (1), related to the source location parameters by (note that $x_1 = y_1 = \dot{x}_1 = \dot{y}_1 = 0$):

$$r_1^2 = x^2 + y^2 \tag{16}$$

$$\dot{r}_1 = \frac{x\dot{x} + y\dot{y}}{r_1} \tag{17}$$

We shall apply a second stage minimization to refine the location estimate through the relationship (16) and (17). The final source position and velocity estimate should minimize the equation errors in (16) and (17) while maintaining as close as possible to the source location values contained in $\mathbf{u}_1$. The second stage process constructs another set of equations

$$\boldsymbol{\varepsilon}_2 = \mathbf{h}_2 - \mathbf{G}_2\mathbf{u}_2 \tag{18}$$

$$\mathbf{h}_2 = \begin{bmatrix} u_1(1)^2 \\ u_1(2)^2 \\ u_1(3)^2 \\ u_1(1)u_1(4) \\ u_1(2)u_1(5) \\ u_1(3)u_1(6) \end{bmatrix}, \quad \mathbf{G}_2 = \begin{bmatrix} 1 & 0 & 0 & 0 \\ 0 & 1 & 0 & 0 \\ 1 & 1 & 0 & 0 \\ 0 & 0 & 1 & 0 \\ 0 & 0 & 0 & 1 \\ 0 & 0 & 1 & 1 \end{bmatrix},$$

$$\mathbf{u}_2 = \begin{bmatrix} x^2 & y^2 & x\dot{x} & y\dot{y} \end{bmatrix}^T \tag{19}$$

The error vector $\boldsymbol{\varepsilon}_2$ is resulted from non-zero covariance of $\mathbf{u}_1$. Minimizing the weighted second norm of $\boldsymbol{\varepsilon}_2$ yields

$$\mathbf{u}_2 = (\mathbf{G}_2^T\mathbf{W}_2\mathbf{G}_2)^{-1}\mathbf{G}_2^T\mathbf{W}_2\mathbf{h}_2 \tag{20}$$

where $\mathbf{W}_2$ is a weighting matrix given by $E[\boldsymbol{\varepsilon}_2\boldsymbol{\varepsilon}_2^T]^{-1}|_{\mathbf{u}_2=\mathbf{u}_2^o}$ and $\mathbf{u}_2^o$ is the true solution of $\mathbf{u}_2$. When putting $\mathbf{u}_1 = \mathbf{u}_1^o + \Delta\mathbf{u}_1$, where $\mathbf{u}_1^o$ is the true value of $\mathbf{u}_1$ and $\Delta\mathbf{u}_1$ is the estimation noise of $\mathbf{u}_1$, it is straightforward to deduce from (18) that

$$\boldsymbol{\varepsilon}_2|_{\mathbf{u}_2=\mathbf{u}_2^o} = \Delta\mathbf{h}_2 \approx \mathbf{B}_2\Delta\mathbf{u}_1 \tag{21}$$

where

$$\mathbf{B}_2 = \begin{bmatrix} 2x & 0 & 0 & 0 & 0 & 0 \\ 0 & 2y & 0 & 0 & 0 & 0 \\ 0 & 0 & 2r_1 & 0 & 0 & 0 \\ \dot{x} & 0 & 0 & x & 0 & 0 \\ 0 & \dot{y} & 0 & 0 & y & 0 \\ 0 & 0 & \dot{r}_1 & 0 & 0 & r_1 \end{bmatrix} \tag{22}$$

and second order error terms have been ignored. Hence

$$\mathbf{W}_2 = \mathbf{B}_2^{-T}\text{cov}(\mathbf{u}_1)^{-1}\mathbf{B}_2^{-1} = \mathbf{B}_2^{-T}(\mathbf{G}_1^T\mathbf{W}_1\mathbf{G}_1)\mathbf{B}_2^{-1}. \tag{23}$$

Since $\mathbf{B}_2$ depends on the true source location and velocity, they will be replaced by their estimates from $\mathbf{u}_1$ for implementation. Simulation results show that there is no significant loss in accuracy. Since $\mathbf{G}_2$ is constant, subtracting (20) by its true value yields

$$\Delta\mathbf{u}_2 = (\mathbf{G}_2^T\mathbf{W}_2\mathbf{G}_2)^{-1}\mathbf{G}_2^T\mathbf{W}_2\Delta\mathbf{h}_2 \tag{24}$$

Table 1: Positions and Velocities of Sensors

sensor no. i	x_i	y_i	$\dot{x}_i$	$\dot{y}_i$
1	0	0	0	0
2	400	0	-40	-10
3	0	500	0	-30
4	100	200	20	-40
5	-100	-100	-30	-10

The covariance matrix of $\mathbf{u}_2$ is therefore, upon using the definition of $\mathbf{W}_2$ below (20),

$$\text{cov}(\mathbf{u}_2) = (\mathbf{G}_2^T\mathbf{W}_2\mathbf{G}_2)^{-1} \tag{25}$$

The source position and velocity are contained in $\mathbf{u}_2$, as shown in its definition in (19). The position estimate $\mathbf{u} = [\,x,\, y\,]^T$ is obtained from the first two elements of $\mathbf{u}_2$:

$$\mathbf{u} = \mathbf{U}\begin{bmatrix} \sqrt{u_2(1)} & \sqrt{u_2(2)} \end{bmatrix} \tag{26}$$

where $\mathbf{U} = diag\{sgn\ (u_1(1)), sgn\ (u_1(2))\}$, and is used to remove the sign ambiguity of the sqaure root operations in (26). Furthermore, the final source velocity estimate $\dot{\mathbf{u}} = [\,\dot{x},\, \dot{y}\,]^T$ is given by

$$\dot{\mathbf{u}} = \mathbf{U}\begin{bmatrix} \frac{u_2(3)}{\sqrt{u_2(1)}} & \frac{u_2(4)}{\sqrt{u_2(2)}} \end{bmatrix}^T \tag{27}$$

3. SIMULATIONS

This section compares the performance of the proposed solution method with that of the Taylor-series method and the CRLB. The positions and velocities of sensors are listed in Table 1. TDOA and FDOA estimates are generated by adding to the true values Gaussian noises having a correlation matrix $\mathbf{R}$, where $\mathbf{R}$ is set to $c^2\sigma^2$ in the diagonal elements and $0.5c^2\sigma^2$ otherwise. That is, $\mathbf{Q}_t = \mathbf{Q}_f = \mathbf{R}$. The noises added to TDOA and FDOA are independent so that $\mathbf{Q}_{tf}$ is zero.

The position and velocity of the emitter were set to be $(12000, 5000)$, and $(-20, 5)$, respectively. Since the source is distant away from the receivers and is moving relatively slowly, the required assumptions are satisfied for (14) to hold. Figure 2 shows the accuracy of the position and velocity estimate of the proposed method in terms of root mean-square error (RMSE). Also given is the results from the Taylor-series method [4]. The method uses truncated Taylor-series expansion to linearize the TDOA and FDOA equations and solves for the solution iteratively. The source position and velocity estimate was initialized to have 10% relative mean-square error by adding Gaussian white noise to the true solution. It is clear from Figure 2 that the Taylor-series method fails to converge, while the proposed method achieves the Cramer Rao Lower Bound (CRLB) when the noise power is below 0.1. Above this noise level, the thresholding effect from the non-linarity of the estimation problem occurs and the estimation accuracy deviates from the CRLB. Fgiure 3 is the result when the Taylor-series method is initialized to the true solution added with Gaussian white noise that has a standard deviation twice the CRLB. The Taylor-series method is able to converge in this case. However, it deviates from the CRLB earlier than the proposed method.

4. CONCLUSIONS

We have developed a closed-form solution for estimating the position and velocity of an emitter based on TDOA and FDOA measurements from an arbitrary array of sensors. The proposed method is computationally attractive and does not suffer from convergence and initialization problems. The proposed solution requires less computation than the Taylor-series method and has higher noise threshold. Simulation shows that estimation accuracy of the proposed source location method achieves the CRLB for Gaussian noise around small error region.

5. REFERENCES

[1] D. J. Terrieri," Statistical theory of passive location systems," *IEEE Transactions on Aerospace and Electronic Systems*, vol.AES-20, no. 2, pp.183-198, May 1984.

[2] J. O. Smith, and J. S. Abel," Closed-form least-squares source location estimation from range-difference measurements," *IEEE Transactions on Acoustic, Speech, Signal Processing*, vol.ASSP-35, no.12, pp.1661-1669, Dec. 1987.

[3] W. W. Smith Jr., and P. G. Steffes, " Time delay techniques for satellite interface location system," *IEEE Transactions on Aerospace and Electronic Systems*, vol.25, no.2, pp.224-230, Mar. 1989.

[4] W. H. Foy," Position-location solution by Taylor-series estimations," *IEEE Transactions on Aerospace and Electronic Systems*, vol.AES-12, pp.187-194, Mar. 1976.

[5] K. C. Ho, and Y. T. Chan, "Solutions and performance analysis of geolocation by TDOA," *IEEE Transactions on Aerospace and Electronic Systems*, vol.29, no.4, pp.1311-1322, Oct. 1993.

[6] S. M. Kay, *Fundamentals of Statistical Signal Processing, Estimation Theory.* Prentice-Hall, NJ: Engle-Wood Cliffs, 1993.

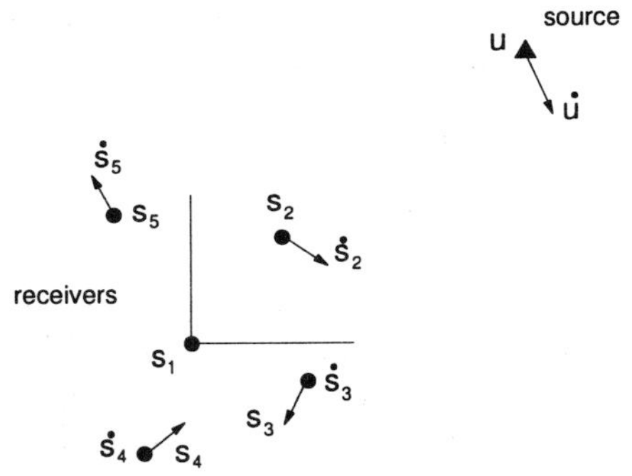

Figure 1: Source Location Scenario.

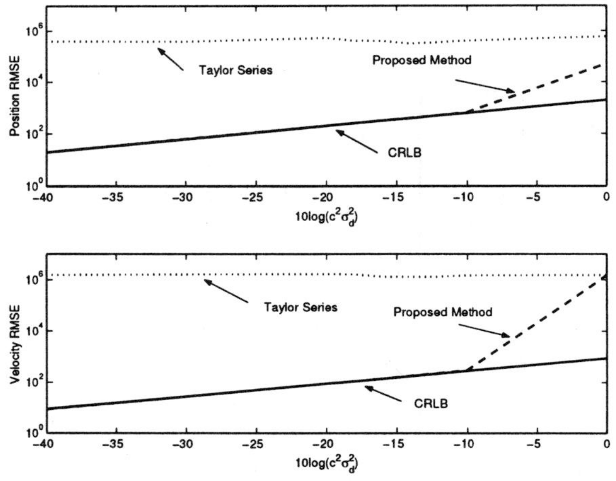

Figure 2: Location accuracy of the proposed method, the Taylor-series method is initialized to have 10% relative mean-square error.

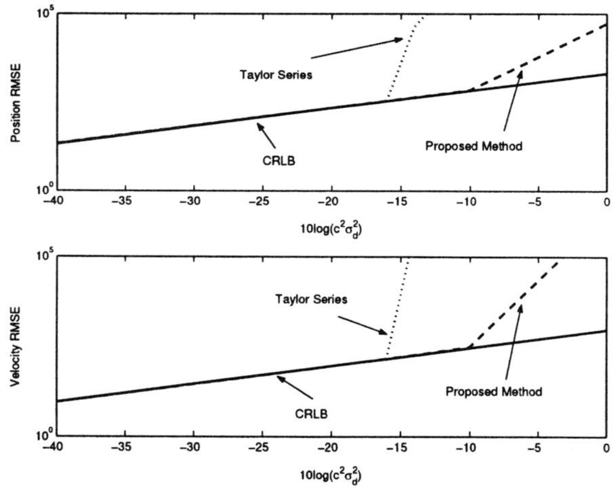

Figure 3: Location accuracy of the proposed method, the Taylor-series method is initialized to have a standard deviation twice the CRLB.

IMPLEMENTING OTSU'S THRESHOLDING PROCESS USING AREA-TIME EFFICIENT LOGARITHMIC APPROXIMATION UNIT

H. Tian, S. K. Lam, T. Srikanthan

Centre for High Performance Embedded Systems,
Nanyang Technological University,
Nanyang Avenue, SINGAPORE 639798

ABSTRACT

Otsu's global automatic image thresholding method has been widely employed in various real-time applications. In this paper, a novel architecture for the BCVC (Between Class Variance Computation) of Otsu's method is presented to meet these high-speed requirements. The proposed implementation employs a binary Logarithmic Conversion Unit (LCU) to eliminate the complex divisions and multiplications in the Otsu's procedure. Implementations on the FPGA (Field Programmable Gate Array) platform show that our method achieves a computation speed-up of about 2.75 times by occupying only 1/6th of the FPGA slices required by one that relies on the direct implementation.

1. INTRODUCTION

Otsu's method [6][8] is a very popular global automatic thresholding technique, which can be applied to a wide range of applications such as noise reduction for human action recognition [1], adaptive progressive thresholding to segment lumen regions from endoscopic images [2][4][10], document segmentation [3], pre-processing of a neural-network classifier for hardwood log inspection using CT images [5], low cost in-process gauging system in removing illumination dependencies well [7], real-time segmentation of images with complex background environment [9] and segmentation of moving lips for speech recognition [11].

These applications demand real-time performance and a hardware implementation is essential to increase the computational efficiency of the Otsu's procedure. However, the basic Otsu's thresholding computation involves many iterative complex arithmetic operations such as multiplications and divisions, which does not lend well to a high-speed and low cost implementation.

In this paper, a novel architecture for the BCVC of Otsu's method is developed to remove the architecture's bottleneck through simplification of the complex operations by converting them to less complex binary logarithmic operations. Minimal LUT (Look-Up Table) size along with only adders and shifters are employed in the logarithmic conversion to reduce the hardware cost. We have implemented both the direct implementation of the Otsu's method and the proposed method using binary logarithmic operations on the FPGA platform and the results shows that the latter has substantially reduced area-time complexity.

In the next section, we provide the expressions of the Otsu's computation with emphasis on the BCVC portion, which is the bottleneck of the entire procedure. Next, we introduce the LCU and describe how it improves the hardware efficiency of the BCVC module. Lastly, we provide the results analysis of the FPGA implementations on the direct implementation and proposed architecture.

2. BOTTLENECK OF OTSU'S METHOD

Otsu's method maximizes the a posteriori between-class variance $\sigma_B^2(t)$, given by (1), which is reduced to as shown in (2) with the terms defined in (3), where n_i represents the number of pixels with gray level i, L is the number of gray levels and N is the total number of pixels in the image.

$$\sigma_B^2(t) = w_0(t) w_1(t)[\mu_1(t) - \mu_0(t)]^2 \quad (1)$$

$$\sigma_B^2(t) = w_0(t)[1 - w_0(t)][\frac{\mu_T - \mu_t(t)}{1 - w_0(t)} - \frac{\mu_t(t)}{w_0(t)}]^2 \quad (2)$$

$$w_0(t) = \sum_{i=0}^{t} \frac{n_i}{N},\ w_1(t) = 1 - w_0(t),\ \mu_t(t) = \sum_{i=0}^{t} i\frac{n_i}{N} \ \ and \ \ \mu_T = \sum_{i=0}^{L-1} i\frac{n_i}{N} \quad (3)$$

Optimal threshold t^* is found through a sequential search for the maximum of $\sigma_B^2(t)$ for the values of t where $0 \le t < L$. The architecture of Otsu's procedure can be divided into 3 computation modules shown in Fig. 1.

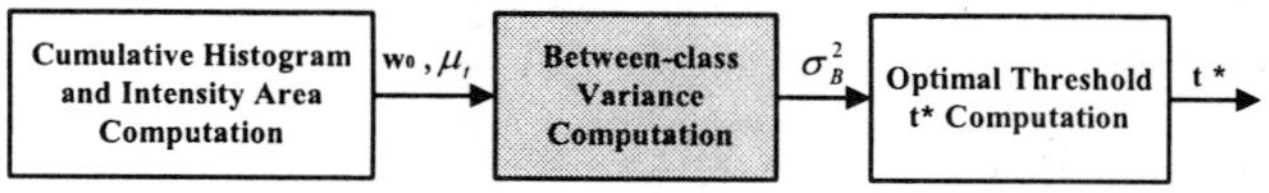

Figure 1. Computational modules of Otsu's architecture

The Cumulative Histogram (CH) and Cumulative Intensity Area (CIA) computation modules are used to compute and store $w_0(t)$ and $\mu_t(t)$ for L gray levels. The Optimal Threshold (OT) computation module is a comparison process to choose the maximum $\sigma_B^2(t^*)$ and its corresponding optimal threshold t^*. Both of these modules are the simple pre and post-processing of

0-7803-7761-3/03/$17.00 ©2003 IEEE

the BCVC and they are reasonably straightforward operations in comparison to the BCVC process.

From (2), we observe that the direct implementation of the computation of between-class variance requires a large number of compute intensive operations of squaring, multiplication and division. A direct implementation of the BCVC module is illustrated in Fig. 2, whereby the CIA and CH arrays contain the values computed from the CH and CIA computation modules.

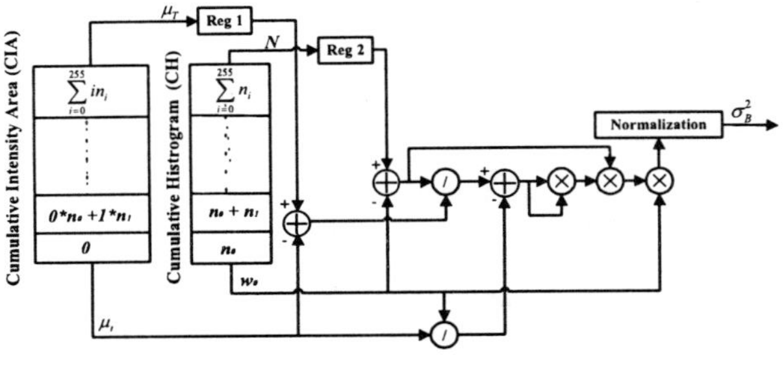

Figure 2. Direct implementation for BCVC module

In a typical Otsu's operation, the computation of the between-class variance has to be iterated for L gray levels, and hence the bottleneck in the Otsu's architecture lies in the computation of the between-class variance after the CH and CIA arrays has been calculated. This necessitates a development of a novel architecture to improve the efficiency of this module.

3. INCORPORATING BINARY LOGARITHM FOR BCVC

In this section, we propose the use of an efficient binary LCU to convert the compute intensive multiplication and division operations of the BCVC into simplified fixed-point circuitry through the use of a small LUT and simple computation unit.

3.1 Binary Logarithm Conversion Process

Let $z_{n-1}z_{n-2}\cdots z_0$ be an n-bit binary representation of Z and z_c is the first non-zero bit of Z. The value of this number can be written as shown in (4) where $0 \leq x < 1$. Hence the binary logarithmic value of Z is shown in (5), where c is the characteristic of the logarithm and $\log_2(1+x)$ constitutes the mantissa.

$$Z = 2^c + \sum_{i=0}^{c-1} 2^i z_i = 2^c(1 + \sum_{i=0}^{c-1} 2^{i-c} z_i) = 2^c(1+x) \quad (4)$$

$$\log_2 Z = c + \log_2(1+x) \quad (5)$$

To obtain a reasonable conversion accuracy, a LUT can be used to represent the mantissa, i.e. $\log_2(1+x)$. However, the depth of the LUT will have a complexity of $O(2^n-1)$. As the input data width increases, this method becomes costly in view of the large VLSI area required. To reduce the size of the LUT, we identify the mantissa values based on the combinations of the u-bit binary word formed by $z_{c-1}z_{c-2}\cdots z_{c-u}$, which is the next u bits following bit position c. These bits are used to address a small LUT that stores the minimum mantissa L_{min} of each group. The remaining m bits of Z ($z_{c-u-1}z_{c-u-2}\cdots z_{c-u-m}$) are used to compute a more accurate approximated value of the mantissa. These bits are divided into $\lceil m/2 \rceil$ multiplier groups as shown in (6), whereby $k = 0,1,\ldots,\lceil m/2 \rceil - 1$.

$$f_k = z_{c-u-(2k+1)} z_{c-u-(2k+2)} \quad (6)$$

The multiplicands are calculated for each combination of $z_{c-1}z_{c-2}\cdots z_{c-u}$ and they form the content D of another small LUT addressable by $z_{c-1}z_{c-2}\cdots z_{c-u}$. The multiplicands are then shifted accordingly for computation with each multiplier groups. The complete approximation algorithm can be mathematically expressed as follows:

$$\log_2 Z = c + L_{\min} + \sum_{k=0}^{\lceil m/2 \rceil -1} \left(\left(D / 2^{2k} \right) \times f_k \right) \quad (7)$$

It is noteworthy that the accuracy of the approximation can be varied through the factors u and m. A larger u and m implies larger LUT and more computation units, which in return provides a higher approximated accuracy and vice versa. The binary logarithmic conversion method mentioned is employed to calculate the value of $\sigma_B^2(t)$ as shown in (8).

$$\begin{aligned}\log_2 \sigma_B^2(t) &= \log_2 \frac{[w_0(t)\mu_T - \mu_t(t)]^2}{[1-w_0(t)]w_0(t)} \\ &= 2\log_2[w_0(t)\mu_T - \mu_t(t)] - \log_2[1-w_0(t)] - \log_2 w_0(t)\end{aligned} \quad (8)$$

The anti-logarithm function is not required in this case as we are only interested in finding out at which threshold the logarithm of $\sigma_B^2(t)$ has attained its maximal value.

3.2 Proposed BCVC Module

Fig. 3 shows the proposed architecture for the BCVC of Otsu's method.

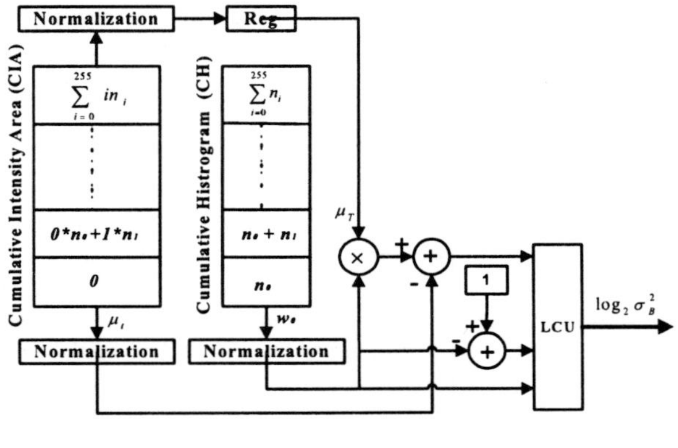

Figure 3. Proposed architecture for the computation of between-class variance

The contents of *Reg* store the value shifted out from the top of the CIA register array for computing the value of μ_T. The block of CIA and CH register arrays are circularly shifted and the values of (1-w_0) and ($w_0\mu_T - \mu_t$) are calculated, after which the value of $\log_2 \sigma_B^2(t)$ could be obtained from (8) using the LCU. The normalization unit consists of several right shifters to adjust $w_0(t)$ and $\mu_t(t)$ to meet the accuracy requirements of various applications. Comparing to the direct implementation in Fig. 2, we can immediately see that the proposed architecture eliminates the required dividers in the architecture. Moreover, the number of multipliers has been significantly reduced.

4. PERFORMANCE ANALYSIS

The proposed architecture employs a LCU unit, which includes a multiplexer, two small LUTs and a simple computation unit. The multiplexer is used to select one out of the three operands for computation at any one time. The other trivial resources within the LCU consist of shifters and adders to manipulate the contents retrieved from the LUT for the logarithmic conversion. For a 16-bits input binary number, the maximum error incurred by the logarithmic conversion for various values of *u* is shown in Table 1. The choice of *u* for the LCU can be finalized by referring to Table 1 based on the accuracy requirements of the Otsu's application.

u	LUT size (bits)	Maximum error
2	80	0.013636
3	224	0.003620
4	416	0.001774
5	1024	0.000398
6	2048	0.000115

Table 1. Conversion precision with various LUT size

In order to verify the efficiency of our method, both the direct implementation and the proposed architecture for the BCVC have been modeled using VHDL and synthesized with Synplicity Synplify Pro 7.0.3 targeted at the Xilinx Virtex XCV800 HQ240-4 FPGA device. The LCU used in the implementation is configured with $u = 4$ and $m = 8$. The critical path delay and the area for a one-time BCVC by both methods are listed below.

	Direct implementation	Proposed implementation
Latency (ns)	362.4	132.0
Area (slices)	622	109
Number of IOBs	113	49

Table 2. FPGA comparison results

It can be observed that the latency, area and number of external IO blocks have been substantially reduced in the proposed architecture. For example, a speed-up of 2.75 times is achieved along with a reduction in area and IO pins by a factor of 6 and 2.3 respectively. Based on the result analysis in Table 2, we can conclude that the proposed architecture provides a hardware efficient approach to the Otsu's method when compared to the direct implementation.

5. SUMMARY

An area-time efficient re-configurable architecture for BCVC of Otsu's method has been proposed to realize high-speed image thresholding, which can be employed in various real-time image-processing methods. The design exploits a logarithmic conversion technique, consisting of small LUT and computation units to overcome the bottleneck of finding the maximum between-class variance in the Otsu's method. Notable reduction in area was made possible as a direct result of eliminating the complex dividers and simplifying multiplication circuitry using logarithmic operations. FPGA based implementations show that our method achieves a computation speed-up of about 2.75 times by occupying only 1/6th of the number of slices required by one that relies on direct implementation.

6. REFERENCES

[1] Arseneau S. and Cooperstock J.R., "Real-Time Image Segmentation for Action Recognition". *Proc. IEEE Pacific Rim Conference on Communications, Computers and Signal Processing*, Victoria, B. C., Canada, pages 86-89, 1999.

[2] Asari K.V., Srikhanthan T., Kumar S. and Radhakrishnan D., "A Pipelined Architecture for Image Segmentation by Adaptive Progressive Thresholding". *Journal of Microprocessors and Microsystems*, 23(8-9): 493-499, 1999.

[3] Eikvil L., Taxt T. and Moen K., "A Fast Method for Adaptive Binarization". *Proc. 1st Int. Conf. Document Analysis and Recognition (ICDAR)*, St. Malo, France, 1991.

[4] Kumar S., Asari K. V. and Radhakrishnan D., "Real-Time Automatic Extraction of Lumen Region and Boundary from Endoscopic Images". *Med. Biol. Eng. Comput.*, 37(5): 600-604, 1999.

[5] Li P., Abbot A.L. and Schmoldt D.L., "Automated Analysis of CT Images for the Inspection of Hardwood Logs". *Proc. IEEE Int. Conf. on Neural Networks*, pages 1744-1749, June 1996.

[6] Liao P.S., Chen T.S. and Chung P.C., "A Fast Algorithm for Multilevel Thresholding". *Journal of Information Science and Engineering*, 17: 713-727.

[7] Miller J.W.V, Shridhar V., Wicker E. and Griffth C., "Very Low-Cost In-Process Gauging System", *Proc. IEEE Pacific Rim Conf. on Communications*, Computers and Signal Processing, Victoria, B. C., Canada, pages 86-89, 1999.

[8] Otsu N., "A Threshold Selection Method from Gray-Level Histogram". *IEEE Trans. System Man Cybernetics*, SMC-9(1): 62-66, 1979.

[9] Seow M. J. and Asari K. V, "A Parallel VLSI Architecture for Real-Time Segmentation of Images with Complex Background Environment". *Proc. 10th NASA Sym. on VLSI*

Design, Albuquerque, New Mexico, USA, pages 1031-1036, 2002.

[10] Tian H., Srikanthan T. and Asari K.V., "Automatic Segmentation Algorithm for the Extraction of Lumen Region and Boundary from Endoscopic Images". *Med. Biol. Eng. Comput.*, 39(1): 8-14, 2001.

[11] Zhang X.Z., "Automatic Speechreading for Improved Speech Recognition and Speaker Verification", PhD thesis, Georgia Institute of Technology, USA, 2002.

STEADY-STATE PROPERTIES OF THE SIGN ALGORITHM FOR THE CONSTRAINED ADAPTIVE IIR NOTCH FILTER

Yegui Xiao† [1], *Rabab Kreidieh Ward*†, *Akira Ikuta*††

† Institute for Computing, Information and Cognitive Systems (ICICS)
University of British Columbia (UBC)
2356 Main Mall, Vancouver, BC, Canada V6T 1Z4
E-mail: xiao@ece.ubc.ca

†† Hiroshima Prefectural Women's University (HPWU)
1-1-71, Ujina-Higashi, Minami-ku, Hiroshima, Japan 734-8558

ABSTRACT

Many algorithms have been proposed for the constrained adaptive IIR notch filter for frequency estimation. The sign algorithm (SA) is a good option in terms of low computational cost and robustness against additive noise of impulsive nature. However, unlike most of the other algorithms, the performance of the SA has not been reported on. This is because of the difficulty due to the presence of the sign function. To overcome this difficulty, we, here, present an effective approach where relatively slow adaptation and Gaussianity of the notch filter output are assumed. Two difference equations are first established for the convergences in the mean and in the mean square, respectively. Steady-state estimation error and mean square error (MSE) of the SA are then derived in closed forms. Theory-based comparison between the SA and the plain gradient (PG) algorithm is done in some detail. Extensive simulations demonstrate the validity of our analytical results not only for the slow adaptation cases but also for cases of relatively fast adaptation.

1. INTRODUCTION

Adaptive IIR notch filtering for frequency estimation has attracted much research in the signal processing community: this is because adaptive IIR notch filters require considerably fewer filter coefficients compared with their FIR-type counterparts for the same notch bandwidth and similar performance.

Many adaptive algorithms have been developed to adjust the filter coefficients of the IIR notch filters [1]-[6]. These include the sign algorithm (SA) [2], the plain gradient (PG) algorithm [4], the normalized gradient (NG) algorithm [2], the recursive prediction error (RPE) algorithm [1, 6], the lattice algorithm (LA) [3], the p-power (PP) algorithm [5], and the memoryless nonlinear gradient (MNG) algorithm [6].

The SA has the advantage of being the simplest one among them, and hence has greatest merit in hardware implementation. Its problem is the slow convergence. However, in some applications such as ECG sinusoidal interference cancellation [5] and sinusoidal engine noise reduction [11], the SA forms a very good option in terms of both computational cost and performance. This is because a coarse estimate for the frequency is available in advance and the notch filter just needs to adapt to the signal to reduce the small mismatch between the estimate and the real frequency. Moreover, when the additive noise is of impulsive nature, the SA is more stable than the PG and PP ($p = 3, 4$) [5].

So far, several investigations have been made to analyze the performances of some of the above-mentioned gradient-based adaptive algorithms. Nishimura et al. [7] proposed a technique to analyze the performance of the SA and the PG for an IIR notch filter that is similar to the lattice notch filter. Their technique is based on gradient linearization. Petraglia et al. presented similar analysis of the PG for the bilinear IIR notch filter [8]. However, it has been found that the gradient linearization technique used by Nishimura and Petraglia can not be applied to the SA and the PG for the constrained IIR notch filter. Recently, the PG and the MNG for the constrained IIR notch filter have been analyzed in [9, 10], where gradient non-linearization is used. *Unfortunately, the SA for the constrained IIR notch filter can not be analyzed by all the above-mentioned techniques because of the difficulty posed by the existence of the sign function in the algorithm.*

This paper presents a detailed performance analysis of the SA for the constrained IIR notch filter. To overcome the difficulty introduced by the sign function,

[1] The 1st author is on leave from the HPWU, Hiroshima, Japan 734-8558.

we assume 1) Gaussianity of the filter output and 2) relatively slow adaptation, all for the sake of analytical tractability. The 1st assumption holds as long as the additive noise is white, and not necessarily Gaussian. The 2nd assumption is not so restrictive, because the analytical results obtained also explains well the simulations for relatively fast adaptation. The steady-state bias and MSE expressions are derived in closed forms. Theory-based comparison between the SA and the PG is done in some detail. Extensive simulations are provided to confirm the analytical results.

2. CONSTRAINED IIR NOTCH FILTER AND THE SIGN ALGORITHM

The constrained adaptive IIR notch filter [1] is expressed by

$$H_N(z) = \frac{1 + az^{-1} + z^{-2}}{1 + \rho a z^{-1} + \rho^2 z^{-2}} \tag{1}$$

where ρ is a pole contraction factor (pole radius) over $(0, 1)$ which controls the notch bandwidth of the filter. a is the filter coefficient whose true value is calculated by $a_0 = -2\cos\omega_0$. ω_0 is the unknown frequency of a noisy input sinusoidal signal

$$x(t) = A\cos(\omega_0 t + \theta) + v(t) \tag{2}$$

where $v(t)$ is an additive white Gaussian noise. θ is the phase of the signal that is uniformly distributed over $[0, 2\pi)$. The adaptive IIR notch filter (1) is used to estimate the frequency of the sinusoid. The SA that updates the filter coefficient is given by

$$\hat{a}(t+1) = \hat{a}(t) - \mu\ sgn(e(t))s(t) \tag{3}$$

where $sgn(\cdot)$ is the sign function. $e(t)$ is the notch filter output, also referred to as error signal. μ is a step size parameter. $\hat{a}(t)$ is the estimate of the filter coefficient a. $s(t) = -\rho e(t-1) + x(t-1)$ is the gradient signal.

3. PERFORMANCE ANALYSIS

A: Steady-state error and gradient signals

According to [9], at steady-state, the error and gradient signals in (3) may be expressed by

$$\begin{aligned} e(t) &= AB\delta_a(t)\cos(\omega_0 t + \theta - \phi) \\ &\quad - \rho AB^2\delta_a^2(t)\cos(\omega_0 t + \theta - 2\phi) + v_1(t), \end{aligned} \tag{4}$$

$$\begin{aligned} s(t) &= A\cos(\omega_0 t + \theta - \omega_0) \\ &\quad - \rho AB\delta_a(t)\cos(\omega_0 t + \theta - \omega_0 - \phi) \\ &\quad + \rho^2 AB^2\delta_a^2(t)\cos(\omega_0 t + \theta - \omega_0 - 2\phi) + v_2(t) \end{aligned} \tag{5}$$

where $\delta_a(t) = \hat{a}(t) - a_0$. See [9] for B and ϕ that are constants determined by ω_0 and ρ. $v_1(t)$ is a zero-mean noise signal at the output of the notch filter whose variance is $\sigma_{v_1}^2$. $v_2(t)$ is also a zero-mean noise signal in the gradient signal whose variance is $\sigma_{v_2}^2$. The correlation between these two noise signals is indicated by $R_{1,2}$. See [9] for their explicit expressions.

B: Estimation bias

Using (4) and (5) in (3), the difference equation of convergence in the mean for the estimation error $(\delta_a(t))$ can be expressed as

$$\begin{aligned} &E[\delta_a(t+1)] \\ &= E[\delta_a(t)] - \mu A\cos(\omega_0 t + \theta - \omega_0)\underbrace{E[sgn(e(t))]}_{=I_1(t)} \\ &\quad + \mu\rho AB\cos(\omega_0 t + \theta - \omega_0 - \phi)\underbrace{E[sgn(e(t))\ \delta_a(t)]}_{=I_2(t)} \\ &\quad - \mu\rho^2 AB^2\cos(\omega_0 t + \theta - \omega_0 - 2\phi) \\ &\qquad \times\underbrace{E[sgn(e(t))\ \delta_a^2(t)]}_{=I_3(t)} \\ &\quad - \mu\underbrace{E[sgn(e(t))\ v_2(t)]}_{=I_4(t)}. \end{aligned} \tag{6}$$

To calculate $I_1(t)$, $I_2(t)$, $I_3(t)$, and $I_4(t)$, we need to have further information on the probability distribution $p(e(t))$ of $e(t)$, probabilistic relations between $e(t)$ and $\delta_a(t)$, $e(t)$ and $v_2(t)$. When $v(t)$ is white and Gaussian, we have found that $e(t)$ follows a Gaussian distribution, and $e(t)$ and $\delta_a(t)$, $e(t)$ and $v_2(t)$, are jointly Gaussian distributed. Mathematical details are omitted due to space limitation. Following these findings, we have, after tedious and technical calculations,

$$\begin{aligned} I_1(t) &= \int_{-\infty}^{\infty} sgn(e(t))p(e(t))de(t) \\ &= -\int_{-\infty}^{0} p(e(t))de(t) + \int_{0}^{\infty} p(e(t))de(t) \\ &= 2sgn\left(\frac{\mu_e}{\sigma_e}\right) erf\left(\left|\frac{\mu_e}{\sigma_e}\right|\right), \end{aligned} \tag{7}$$

$$\begin{aligned} I_2(t) &= 2E[\delta_a(t)]sgn\left(\frac{\mu_e}{\sigma_e}\right) erf\left(\left|\frac{\mu_e}{\sigma_e}\right|\right) \\ &\quad + \sqrt{\frac{2}{\pi}}\frac{Q_{e\delta_a}}{\sigma_e}\exp\left\{-\frac{1}{2}\left(\frac{\mu_e}{\sigma_e}\right)^2\right\} \end{aligned} \tag{8}$$

Expressions for $I_3(t)$ and $I_4(t)$ are omitted here. In the derivations of the above equations, simulation-based facts are used, that $\delta_a(t)$ and $v_1(t)$, $\delta_a(t)$ and $v_2(t)$ are uncorrelated to each other. Furthermore, the terms of $\delta_a(t)$ with orders equal to or higher than 3 are very small and are ignored for analytical simplicity. Substituting these equations back in (6) and ignoring many insignificant terms, one ultimately reaches the difference equation for $\delta_a(t)$

$$\begin{aligned} E[\delta_a(t+1)] &= (1 - \mu\psi_{11})\,E[\delta_a(t)] \\ &\quad + \mu\psi_{12}E[\delta_a^2(t)] + \mu\eta_1 \end{aligned} \tag{9}$$

where

$$\psi_{11} = \frac{A^2 B}{\sqrt{2\pi}\sigma_{v_1}} \cos(\omega_0 - \phi), \tag{10}$$

$$\psi_{12} = \frac{\rho A^2 B^2}{\sqrt{2\pi}\sigma_{v_1}} \{\cos(\omega_0 - 2\phi) + \cos\omega_0\}, \tag{11}$$

$$\eta_1 = -\sqrt{\frac{2}{\pi}} \frac{R_{1,2}}{\sigma_{v_1}} . \tag{12}$$

C: Estimation MSE

From (3), we have

$$E[\delta_a^2(t+1)] = E[\delta_a^2(t)] - 2\mu \underbrace{E[\delta_a(t) sgn(e(t)) s(t)]}_{=M_1(t)} + \mu^2 \underbrace{E[s^2(t)]}_{=M_2(t)} . \tag{13}$$

After very complicated and technical calculations, $M_1(t)$ and $M_2(t)$ can be derived in a similar way as $I_i(t)$ $(i = 1,2,3,4)$. Putting the results in (13) and removing the insignificant terms, we get the difference equation for the convergence in the mean square,

$$E[\delta_a^2(t+1)] = -\mu\psi_{21} E[\delta_a(t)] + (1 - \mu\psi_{22}) E[\delta_a^2(t)] + \mu^2 \eta_2 \tag{14}$$

where

$$\psi_{21} = \sqrt{\frac{2}{\pi}} \frac{2R_{1,2}}{\sigma_{v_1}}, \tag{15}$$

$$\psi_{22} = \sqrt{\frac{2}{\pi}} \frac{A^2 B}{\sigma_{v_1}} \cos(\omega_0 - \phi), \tag{16}$$

$$\eta_2 = \frac{1}{2} A^2 + \sigma_{v_2}^2 . \tag{17}$$

Linear difference equations (9) and (14) govern the dynamics of the SA in the vicinity of its steady state. A coarse stability bound for the step size parameter can be derived from these linear equations with ease.

D: Steady-state estimation bias and MSE

Next, we derive the steady-state estimation bias and MSE. At steady-state, using

$$E[\delta_a(t+1)]|_{t\to\infty} = E[\delta_a(t)]|_{t\to\infty} = E[\delta_a(\infty)],$$
$$E[\delta_a^2(t+1)]|_{t\to\infty} = E[\delta_a^2(t)]|_{t\to\infty} = E[\delta_a^2(\infty)]$$

in (9) and (14), and solving the resultant simultaneous equations, we obtain the estimation bias and the MSE:

$$E[\delta_a(\infty)] = \frac{\mu\eta_2\psi_{12} + \eta_1\psi_{22}}{\psi_{11}\psi_{22} + \psi_{12}\psi_{21}}, \tag{18}$$

$$E[\delta_a^2(\infty)] = \frac{\mu\eta_2\psi_{11} - \eta_1\psi_{21}}{\psi_{11}\psi_{22} + \psi_{12}\psi_{21}} . \tag{19}$$

From the above explicit expressions (18) and (19), we have the following interesting and important results:

C1 The SA is inherently biased, no matter how small the positive step size parameter μ is.

C2 The estimation bias and the MSE are all proportional to the step size value when the step size is relatively large ($> 10^{-4}$).

C3 When μ is very small ($\leq 10^{-4}$), both the steady-state bias and the MSE become independent of the step size.

E: Comparison between the SA and the PG

To make a fair comparison, we first equalize the dynamics of both algorithms and then compare their steady-state estimation biases and MSEs. Set

$$\mu_{sa} = \left(\sqrt{\frac{\pi}{2}}\sigma_{v_1}\right) \mu_{pg}. \tag{20}$$

where μ_{sa} and μ_{pg} indicate the step size parameters for the SA and the PG, respectively. Then it is easy to find that the SA and the PG [9] have the same dynamics around their steady states, because the two difference equation sets for the convergences in the mean and mean square for the SA and the PG have the same coefficient matrices, as long as (20) holds. It is also found that under the above condition, both algorithms produce very similar bias for both slow and relatively fast adaptations. They also yield similar MSEs for slow adaptation. But they present different MSEs for relatively fast adaptation; i.e., the PG gives smaller MSE, which fits our common observation that the PG usually works better than the SA for the same convergence rate. See simulations (Fig.3) for their comparisons.

4. SIMULATION RESULTS

Here, we show some typical simulation results. A pdf of the error signal at $t = 1500$ ($\mu = 0.001$, 1000 *runs*) is depicted in Fig.1. Fig.2 shows comparisons between theory and simulation of the estimation bias and the MSE, versus the pole radius, for relatively fast and slow adaptations, respectively. It can be noticed that the analytical estimation bias indicates excellent fit to the simulated values, and the analytical estimation MSE agrees on the whole with the simulation reasonably well. Therefore, one of the two assumptions, i.e., small step size, is made for the sake of analysis tractability and is actually not so restrictive. In Fig. 3, the estimation biases of the SA and the PG are compared where their step sizes satisfy (20). It is seen that the theory for the SA and the PG agrees with the simulation well, and both algorithms present similar convergence rate and similar steady-state bias.

5. CONCLUSIONS

In this paper, we first derive closed form expressions for the steady-state estimation bias and MSE of the SA. From these expressions, the following interesting results are analytically concluded: 1) the SA is inherently biased; 2) both the estimation bias and the MSE are proportional to the step size value in relatively fast adaptation; 3) both the estimation bias and the MSE are independent of the step size when the step size is very small; 4) the SA produces the largest absolute bias and MSE around frequencies 0.15π and 0.85π, and presents zero bias and the least MSE at frequency 0.5π. Theory-based comparison between the SA and the PG is also presented. Simulation results indicate the validity of the analytical expressions. Using the analysis technique used in this work, other adaptive algorithms that contain the sign function may now be tackled.

6. REFERENCES

[1] A. Nehorai, *"A minimal parameter adaptive notch filter with constrained poles and zeros,"* IEEE Trans. Acoust., Speech, Signal Processing, vol.ASSP-33, no.4, pp.983-996(1985).

[2] K. Martin and M. T. Sun, *"Adaptive filters suitable for real-time spectral analysis,"* IEEE Trans. Circuits, Syst., vol.CAS-33, no.2, pp.218-229(1986).

[3] N. I. Cho and S. U. Lee, *"On the adaptive lattice notch filter for the detection of sinusoids,"* IEEE Trans. Circuits, Syst., vol.40, no.7, pp.405-416(1993).

[4] J. F. Chicharo and T. S. Ng, *"Gradient-based adaptive IIR notch filtering for frequency estimation,"* IEEE Trans. Acoust., Speech, Signal Processing, vol.ASSP-38, no.5, pp.769-777(1990).

[5] S.-C. Pei, and C.-C. Tseng, *"Adaptive IIR notch filter based on least mean p-power error criterion,"* IEEE Trans. on Circuits, Syst., vol.40, no.8, pp.525-529(1993).

[6] Y. Xiao, Y. Kobayashi, and Y. Tadokoro, *"A new memoryless nonlinear gradient algorithm for a second-order adaptive IIR notch filter and its performance analysis,"* IEEE Trans. on Circuits, Syst.-II, vol.45, no.4, pp.462-472(1998).

[7] S. Nishimura, J. K. Kim, and K. Hirano, *"Mean-squared error analysis of an adaptive notch filter,"* in Proc. IEEE Int. Symp. Circuits, Syst. Portland, pp.732-735(1989).

[8] M. R. Petraglia, J. J. Shynk, and S. K. Mitra *"Stability bounds and steady-state coefficient variance for a second-order adaptive IIR notch filter,"* IEEE Trans. Signal Processing., vol.42, no.7, pp.1841-1845(1994).

[9] Y. Xiao, Y. Takeshita, and K. Shida, *"Steady-state analysis of a plain gradient algorithm for a second-order adaptive IIR notch filter with constrained poles and zeros,"* IEEE Trans. on Circuits, Syst.-II, vol.48, no.4, pp.733-740(2001).

[10] Y. Xiao and N. Tani, *" Statistical properties of a memoryless nonlinear gradient algorithm for an adaptive constrained IIR notch filter,"* in Proc. IEEE Int. Symp. Circuits, Syst., Sydney, May 2001.

[11] S. M. Kuo, and D. R. Morgan, *"Active noise control systems — algorithms and DSP implementation,"* John Wiley & Sons INC, 1996.

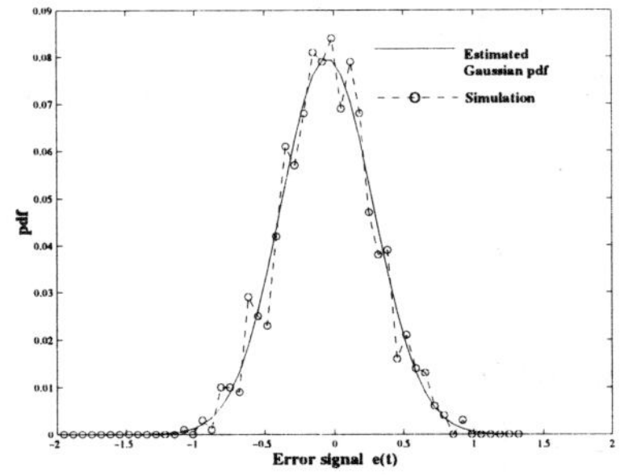

Fig. 1 A pdf of the error signal.

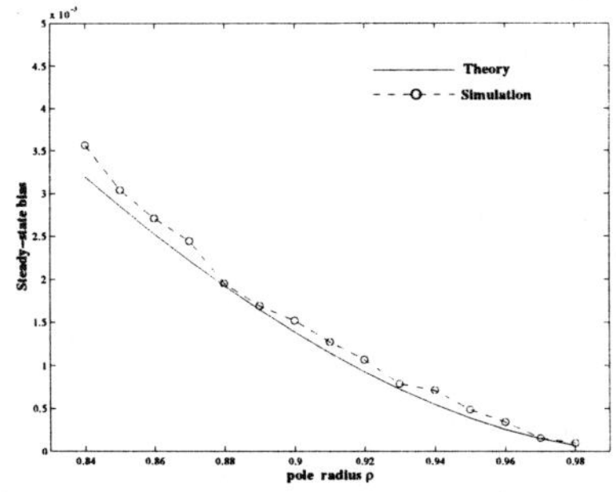

(a) Estimation bias (fast adaptation, $\mu = 10^{-3}$)

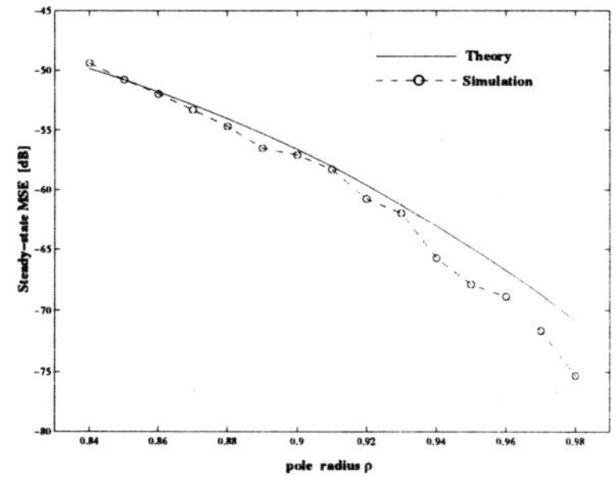

(b) Estimation MSE (slow adaptation, $\mu = 10^{-5}$)

Fig. 2 Comparison between theory and simulation for each of the estimation bias and the MSE, versus the pole radius ($\omega_0 = 0.2\pi$, $A = \sqrt{2}$, SNR=10 [dB], 40 runs).

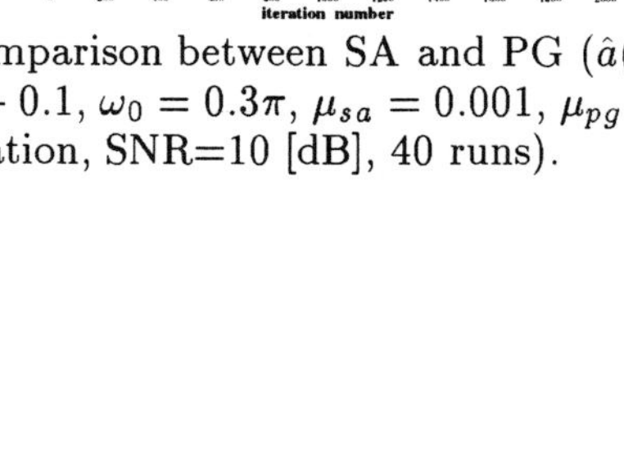

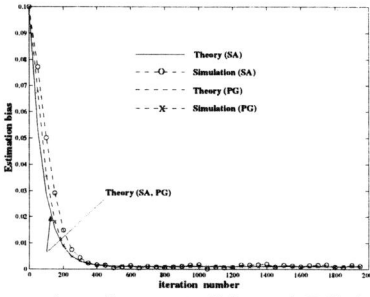

Fig. 3 Comparison between SA and PG ($\hat{a}(0) = -2\cos\omega_0 + 0.1$, $\omega_0 = 0.3\pi$, $\mu_{sa} = 0.001$, $\mu_{pg} = 0.0024$, fast adaptation, SNR=10 [dB], 40 runs).

ROBUST BEAMFORMER DESIGN BY POWER MINIMIZATION AND ITS UNCONSTRAINED PARTITIONED IMPLEMENTATION

Zhu Liang Yu

Center for Signal Processing,
School of EEE,
Nanyang Technological University,
Singapore, 639798
Email: ezlyu@ntu.edu.sg

Meng Hwa Er

School of EEE,
Nanyang Technological University,
Singapore, 639798
Email: emher@ntu.edu.sg

ABSTRACT

A robust array beamformer design method by power minimization and its unconstrained partition implementation is proposed in this paper. An orthogonal blocking matrix is obtained through the proposed method. The new blocking matrix ensures that a desired look-direction response of the processor over a frequency band of interest can be closely approximated. Furthermore, if the spectrum of the target signal is known a priori, such as in microphone array application where the statistical property of speech is known, the new method may use fewer/less degree of freedom. This new method also provide a theoretical support for the derivation of the blocking matrix through array calibration. Simulation results show that the effectiveness of the proposed method.

1. INTRODUCTION

Adaptive array processing has received considerable attention over decades due to its wide applications in the fields of wireless communication, speech acquisition, sonar, Radar [1, 2], etc. An adaptive array has the capacity to retrieve the weak target signal from strong interferences and background noises.

The well studied broadband adaptive array processor is the linearly constrained minimum variance (LCMV) beamformer [3], also named Frost processor. The output beam is formed as a weighted sum of the real signal on the taps of multichannel tapped delay lines connected to the sensors. A set of linear constraints is used to ensure that a desired frequency response characteristic in the look direction is achieved provided that the array signals are presteered with broadband delays. The tapped delay weights are chosen to minimize the total mean output power as an indirect way of rejecting the interference and noise. Other broadband array implementations were proposed in [4], etc. One difficulty of these approaches is that a set of presteering delays is required. Another difficulty is that the performance of these approaches will degrade when there exist array imperfections.

A set of linear constraints was derived in [5] by minimizing the power response over frequency band of interest as well as other parameters. The array processor with this set of constraints can handle a variety of steering situations, namely, no presteering, coarse/quantized presteering, and exact presteering. Furthermore, the approach enables various type of errors and mismatches between signal model and actual scenario to be incorporated in the problem formulation.

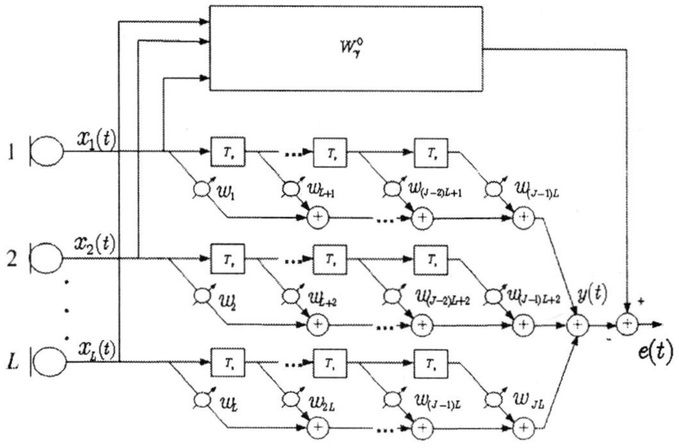

Figure 1: The wideband array processor

In this paper, we derive another set of linear constraints which is based on minimization of the power instead of the power response over interested parameters. In addition to its similar characteristics as the approach in [5], this new method has other advantages. For examples, it is possible to use less constraints when the spectrum distribution of target signal is known. Furthermore, this new approach can be used to design the robust array processing by array calibration.

This paper is organized as follows. In Section 2, the new method is deduced in detail. Some numerical examples are shown in Section 3 to illustrate the effectiveness of the new method. In the last section, a brief conclusion is given.

2. THE PROPOSED METHOD

For the wideband partitioned array processor shown in Fig. 1, the array processor has L sensors and $J-1$ tapped delay sections. The intertap delay is T_s. The processor's weight vector W is defined as a LJ dimensional real vector

$$\mathrm{W}^T = [\mathrm{w}_1^T \ \cdots \ \mathrm{w}_J^T] \tag{1}$$

where T denotes the transpose operator. w_k is the L dimensional real vector defined by

$$[\mathrm{w}_k]_i = w_{(k-1)L+i} \quad k = 1, 2, \cdots, J, \ i = 1, 2, \cdots, L \quad (2)$$

The frequency response of the processor can be expressed as

$$H(f, \theta) = \mathrm{B}(f, \theta)\mathrm{W} \quad (3)$$

where $\mathrm{B}(f, \theta)$ is expressed as

$$\mathrm{B}(f, \theta) = E^T(f) \otimes (\mathrm{s}^T(f, \theta)\mathbf{T}(f)) \quad (4)$$

where $\otimes$ denotes the Kronecker product operator. $E(f)$ is a L-dimensional vector defined by

$$E^T(f) \triangleq [1 \ exp(-j2\pi f T_s) \ \cdots \ exp(-j2\pi f(J-1)T_s)], . \quad (5)$$

where f is the frequency of the plane wave signal $x(t)$ and T_s is the sampling period. $\mathrm{s}(f, \theta)$ is the L dimensional vector defined by

$$\mathrm{s}^T(f, \theta) \triangleq [exp(j2\pi f \tau_1) \ \cdots \ exp(j2\pi f \tau_L)], \quad (6)$$

where $\{\tau_i\}$ are the spatial delays which are given by

$$\tau_i = \frac{\vec{\mathrm{u}}(\theta) \cdot \vec{\mathrm{r}}_i}{v} \quad i = 1, 2, \cdots, L, \quad (7)$$

where v is the speed of propagation of the wave front in the medium in which the array is immersed, $\vec{\mathrm{u}}(\theta)$ is the unit vector in the direction θ. $\vec{\mathrm{r}}_i$ is the position vector of the ith array element.

The matrix $\mathbf{T}(f)$ is the $L \times L$ dimensional matrix defined by

$$\mathbf{T}(f) \triangleq diag[exp(-j2\pi f T_1), \ \cdots, \ exp(-j2\pi f T_L)] \quad (8)$$

where $\{T_i\}$ are the presteering delays. Presteering is not required in the proposed method. Therefore, all these delays are set as zeros.

The output signal $y(t)$ of the lower part array processor can be expressed in frequency domain as

$$Y(f) = X(f)H(f, \theta). \quad (9)$$

where $X(f)$ and $Y(f)$ are the Fourier transform of the signal $x(t)$ and $y(t)$, respectively. The power of signal $y(t)$ can be obtained as

$$P_y = E\{\int_{f_l}^{f_u} |X(f)H(f, \theta)|^2 df\} = \int_{f_l}^{f_u} P(f)\rho(f, \theta)df \quad (10)$$

where $E\{\cdot\}$ denotes the expectation operator. Values f_l and f_u are the lower bound and upper bound of the interested frequency band respectively. $P(f)$ is the power spectrum of signal $x(t)$, which can be expressed as

$$P(f) = E\{X(f)X^*(f)\} \quad (11)$$

where the asterisk denotes complex conjugate operator. The power response function $\rho(f, \theta)$ of the lower part array processor can be expressed as

$$\rho(f, \theta) = H^*(f, \theta)H(f, \theta) \quad (12)$$

The power of signal $y(t)$ in (10) can be simplified as

$$P_y = \int_{f_l}^{f_u} P(f)\rho(f, \theta)df = \mathrm{W}^T\mathbf{Q}_1\mathrm{W} \quad (13)$$

where $\mathbf{Q}_1$ is a $LJ \times LJ$ dimensional positive semidefinite symmetric matrix. It is given by

$$\mathbf{Q}_1 = \mathrm{Re}\{\int_{f_l}^{f_u} P(f)\mathrm{B}^H(f, \theta)\mathrm{B}(f, \theta)df\} \quad (14)$$

where H denotes the complex conjugate transpose operator and $\mathrm{Re}\{\cdot\}$ denotes the real part.

In the discussion above, we assume that the directions-of-arrival (DOA) of the target signals are known exactly and the array is ideal. However, in the real application, the array has some steering error as well as other imperfections. The array imperfections include array geometry error, array receiver phase error, etc. In order to achieve robust performance in presence of these errors, we make some modification to the matrix Q_1 [6, 5]. Suppose that the array response is a function of independent parameters p_1 to p_m, these parameters can be the steering error, phase error, geometry error, etc. We can obtain the integral of the power of $y(t)$ over the frequency band of interest $[f_l, f_u]$ as well as over variation in parameters p_1 to p_m

$$\int_{p_m-\Delta p_m/2}^{p_m+\Delta p_m/2} \cdots \int_{p_1-\Delta p_1/2}^{p_1+\Delta p_1/2} \int_{f_l}^{f_u} P(f)\rho(f, \theta) df dp_1 \cdots dp_m = \mathrm{W}^T\mathbf{Q}_\gamma\mathrm{W} \quad (15)$$

where $\mathbf{Q}_\gamma$ is a $LJ \times LJ$ dimensional symmetric matrix. It is given by

$$\mathbf{Q}_\gamma = \mathrm{Re}\{\int_{p_m-\Delta p_m/2}^{p_m+\Delta p_m/2} \cdots \int_{p_1-\Delta p_1/2}^{p_1+\Delta p_1/2} \int_{f_l}^{f_u} P(f)\mathrm{B}^H(f, \theta, p_1, \cdots, p_m)\mathrm{B}(f, \theta, p_1, \cdots, p_m) df dp_1 \cdots dp_m\} \quad (16)$$

Therefore, as a general form, the matrix $\mathbf{Q}_\gamma$ is used in the deduction of the new method.

Here we assume that the upper part of array processor is designed, which means W_γ^0 is given. To prevent desired signal from cancellation, the lower part of the array processor should be configured to output zero target signal power in direction θ. According to (15), we have

$$\mathrm{W}^T\mathbf{Q}_\gamma\mathrm{W} = 0 \quad (17)$$

Note that $\mathbf{Q}_\gamma$ is symmetric, it can be factorized as

$$\mathbf{Q}_\gamma = \mathbf{U}_\gamma\mathbf{\Lambda}_\gamma\mathbf{U}_\gamma^T \quad (18)$$

where $\mathbf{U}_\gamma$ is the $LJ \times LJ$ dimensional orthogonal matrix given by

$$\mathbf{U}_\gamma = [U_{\gamma,1}, \cdots, U_{\gamma,LJ}] \quad (19)$$

where $\{U_{\gamma,i}, i = 1, 2, \cdots, LJ\}$ are the LJ orthogonal eigenvectors of Q_γ and have property that

$$U_{\gamma,i}^T U_{\gamma,j} = \delta_{ij} = \begin{cases} 1, & \text{if } i = j \\ 0, & \text{if } i \neq j \end{cases} . \quad (20)$$

The matrix Λ_γ is the $LJ \times LJ$ dimensional diagonal matrix given by

$$\mathbf{\Lambda}_\gamma = diag\{\lambda_{\gamma,1}, \cdots, \lambda_{\gamma,LJ}\} \quad (21)$$

where $\lambda_{\gamma,i}, i = 1, 2, \cdots, LJ$ are the LJ eigenvalues of Q and

$$\lambda_{\gamma,1} \geq \lambda_{\gamma,2} \geq \cdots \lambda_{\gamma,LJ} \geq 0 \quad (22)$$

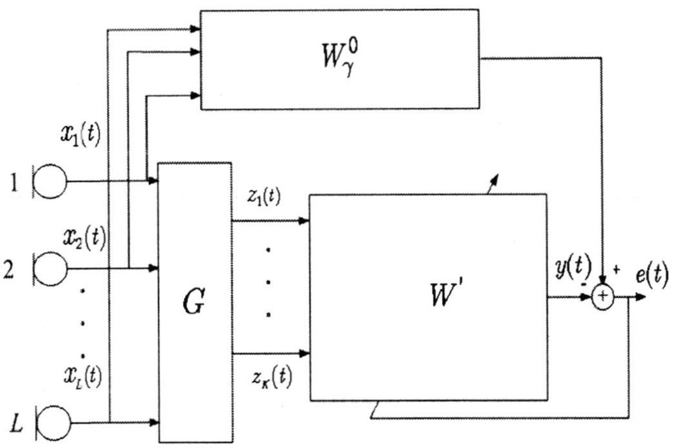

Figure 2: The unconstrained partitioned wideband array processor

Substitute (18) into (17), and if we assume that Q_γ has a rank of n_O, the necessary and sufficient condition for (17) to be satisfied is

$$U_{\gamma,i}^T \mathrm{W} = 0, \qquad i = 1, 2, \cdots, n_O \tag{23}$$

A set of linear constraints of the form given by (23) can be used to ensure that the array has zero power response to the direction θ. The number of constraints n_O can be obtained by finding the smallest integer such that the percentage trace ξ of Q_γ matrix defined by

$$\xi \triangleq \frac{100 \sum_{i=1}^{n_O} \lambda_{\gamma,i}}{\sum_{i=1}^{LJ} \lambda_{\gamma,i}} \tag{24}$$

is greater than or equal to some threshold.

According to (23), there are n_O constraints applied to the array weight vector W. The degree of freedom of the weight vector for adaptive processing is $K = LJ - n_O$. In the new unconstrained partitioned implementation as shown in Fig. 2, the weight vector W is decomposed as

$$\mathrm{W} = \mathbf{G}^T \mathrm{W}' \tag{25}$$

where $\mathbf{G}$ is the blocking matrix of size $K \times LJ$, and W′ is K dimensional unconstrained real vector which is used as the adaptive array weight in new implementation. According to (23), the weight vector W should be orthogonal to the eigenvectors $U_{\gamma,i}$, $i = 1, \cdots, n_O$. That requires

$$U_{\gamma,i}^T \mathbf{G}^T \mathrm{W}' = 0, \qquad i = 1, 2, \cdots, n_O \tag{26}$$

Hence, the blocking matrix $\mathbf{G}$ should be designed to satisfy

$$\mathbf{G} U_{\gamma,i} = 0. \tag{27}$$

A simple way to design this blocking matrix $\mathbf{G}$ is to use the eigenvector $U_{\gamma,i}$, $i = n_O + 1, \cdots, LJ$. The blocking matrix $\mathbf{G}$ is formed as

$$\mathbf{G} = [U_{\gamma,n_O+1}, \cdots, U_{\gamma,LJ}]^T \tag{28}$$

It can be proved that the unconstrained partitioned implementation of the array processor shown in Fig. 2 with the designed blocking matrix $\mathbf{G}$ has the same output as the processor shown in Fig. 1 [6].

In the discussion above, it is assumed that the upper part of array processor W_γ^0 is given. In the real application, W_γ^0 should be designed according to some criteria. A method to obtain W_γ^0 of the upper filter which ensures the closest approximation to the desired look direction response $\mathrm{A}(f, \theta_O)$ over a frequency band of interest $[f_l, f_u]$ and for variation of system parameters are given in [5]. Since W_γ^0 does not affect the lower part of the array processor, hence, in this paper, we still use the method proposed in [5].

When the fixed beamformer W_γ^0 and the blocking matrix $\mathbf{G}$ are obtained, the optimization problem of the array processor can be expressed as

$$\min_{W'} (\mathrm{W}_\gamma^0 - \mathbf{G}W')^T \mathbf{R}_{xx} (\mathrm{W}_\gamma^0 - \mathbf{G}W'), \tag{29}$$

where $\mathbf{R}_{xx}$ is the covariance matrix of the array received signal. The optimal weight vector W'_{opt} can be obtained as

$$W'_{opt} = (\mathbf{G}\mathbf{R}_{xx}\mathbf{G}^T)^{-1} \mathbf{G}\mathbf{R}_{xx}\mathrm{W}_\gamma^0. \tag{30}$$

3. NUMERICAL STUDY

To demonstrate the performance of the proposed method, computer simulations involving a uniformly linear microphone array processor with 10 sensors and 32 taps has been carried out. A speech signal is used as the target signal whose frequency band is $250 - 3500Hz$. The inter-element spacing of sensors is $4cm$. The sampling rate is $8KHz$ to avoid temporal aliasing. In the simulation, the target signal with 6dB power and white elements noise with $-10d$B power are used. In this paper, only the simulation results for $\gamma = 1$ are given. The upper part beamformer W_1^0 is designed using method proposed in [5].

The statistical speech power spectrum distribution [7] has been used as the $P(f)$ to derive the $\mathbf{Q}_\gamma$ matrix in (14). It was concluded by experiments that the statistical speech power spectrum is peaked at about 250-500Hz and above this frequency, the spectrum falls off at about $8 - 10d$B/octave. This property was used in [**?**] to obtain the weighted optimal design solution for filter design. It is also adopted in this paper. The results are presented for three processors, including the processor with $P(f)$ using statistical speech spectrum, the processor using flat power spectrum [5] and the Griffiths-Jim beamformer [4].

The threshold for percent trace of $\mathbf{Q}_1$ matrix is selected as $\xi = 99.9\%$. The corresponding number of constraints is $n_1 = 10$ for the proposed processor and $n_2 = 28$ for the processor with flat spectrum.

Fig. 3 shows the optimal power estimate outputs for Griffiths-Jim beamformer, the new proposed beamformer and the processor with constraints obtained using flat spectrum assumption. In this simulation, the target signal is assumed to impinge on the array from direction 0°. It shows that the proposed processor has slightly better performance than the processor with flat spectrum assumption. Their performances are both better than the Griffiths-Jim beamformer's. This result also shows that the proposed method with less constraints than other methods can still work well for speech application.

Fig. 4 shows the optimal power estimate plot for three processors. In this simulation, the target signal is assumed to impinge on the array from the direction 30°. Exact presteering delay is used for Griffiths-Jim processor. No presteering delay is used for the other two processors. It shows that the proposed method can work without presteering delay. This is an attractive characteristic of the proposed method.

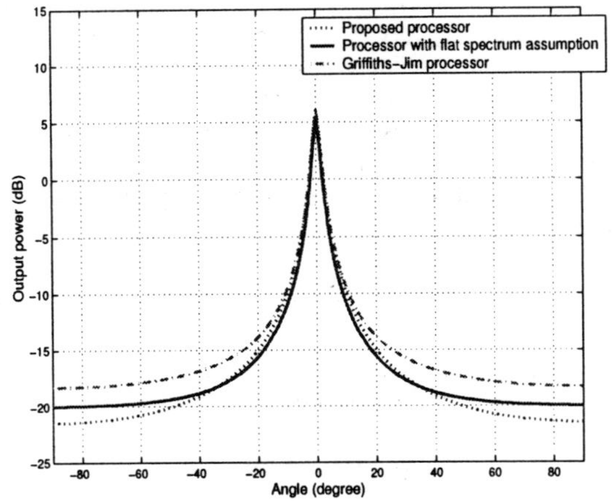

Figure 3: Optimal power estimate plots of array processors

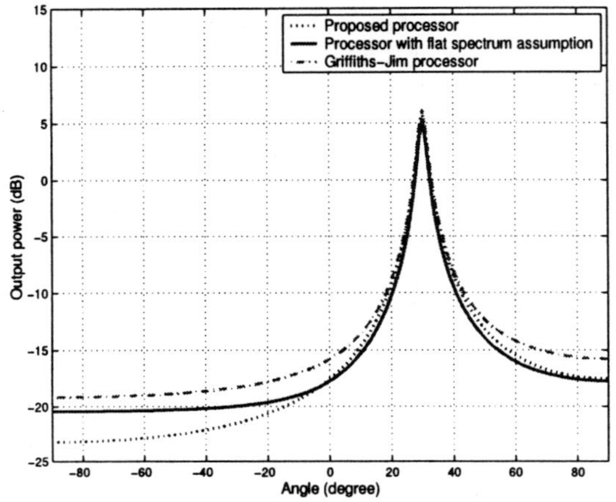

Figure 4: Optimal power estimate plots of array processors

4. CONCLUSION AND DISCUSSION

In this paper, a new robust array design method based on power minimization as well as its unconstrained partitioned implementation is proposed. This method provides some attractive properties. It can be used to design a robust wideband array with its response closely approximates the desired one. When the statistical property of the target signal spectrum is known, it is possible to use less constraints. It can also be shown that this proposed method can handel various array imperfections as well as three types of presteering, coarse presteering, no presteering and exact presteering. Furthermore, this new method provides the theoretical support for the design of the array through array calibration.

5. REFERENCES

[1] G. V. Tsoulos, *Adaptive Antennas for Wireless Communications*, New York: IEEE Press, 2001.

[2] H. Krim and M. Viberg, "Two decades of array signal processing research: the parametric approach," *IEEE Signal Processing Mag.*, vol. 13, no. 4, pp. 67–94, Jul. 1996.

[3] O. L. Frost III, "An algorithm for linearly constrained adaptive array processing," in *Proc. IEEE*, Aug. 1972, pp. 926–935.

[4] L. J. Griffiths and C. W. Jim, "An alternative approach to linearly constrained adaptive beamforming," *IEEE Trans. Antennas Propagat.*, vol. AP-30, no. 1, pp. 27–34, Jan. 1982.

[5] M. H. Er and A. Cantoni, "A new set of linear constraints for broad-band time domain element space processor," *IEEE Trans. Antennas Propagat.*, vol. AP-34, no. 3, pp. 320–329, Mar. 1986.

[6] M. H. Er, *Optimum Antenna Array Processors with Linear and Quadratic Constraints*, Ph.D. thesis, The University of Newcastle, 1985.

[7] L. R. Rabiner and R. W. Schafer, *Digital processing of speech signals*, Prentice-Hall, Englewood Cliffs, N.J., 1978.

A Memory Efficient Realization of Cyclic Convolution and its Application to Discrete Cosine Transform

Hun-Chen Chen[1] *Jiun-In Guo*[2] *Chein-Wei Jen*[1]

[1]Department of Electronics Engineering and Institute of Electronics, National Chiao Tung University, Hsin Chu, Taiwan, ROC
Email:hcchen@swallow.ee.nctu.edu.tw

[2]Department of Computer Science and Information Engineering, National Chung Cheng University, Chia Yi 621, Taiwan, ROC
Email:jiguo@cs.ccu.edu.tw

ABSTRACT

This paper presents a memory efficient design for realizing the cyclic convolution and its application to the discrete cosine transform (DCT). We adopt the way of distributed arithmetic computation, and exploit the symmetry property of DCT coefficients to merge the elements in the matrix of DCT kernel and then separate the kernel to be two perfect cyclic forms to facilitate an efficient realization of 1-D N-point DCT using (N-1)/2 adders or substractors, one small ROM module, a barrel shifter, and $\frac{N-1}{2}+1$ accumulators. The comparison results with the existing designs show that the proposed design can reduce delay-area product significantly.

1. INTRODUCTION

The efficient hardware implementation of DCT is still a challenging problem, which plays a key function in image and signal processing, especially for the demanding multi-media and portable applications. To achieve efficient hardware realization, many researches have been done on realizing the multiplications needed in the DCT through ROM [1-5]. One is the memory-based systolic array design [1] that proposed cyclic convolution based architecture with the features of simple I/O behavior and removing data redundancy in the DCT coefficients. The other is called distributed arithmetic (DA) based design that is an efficient method for computing inner products [2-5] by using ROM tables and accumulators. The DA technique has been widely adopted in many DSP applications such as DFT, DCT, convolution, and digital filters. Therefore, there has been great interest in reducing the ROM size required in the implementations of the DA-based DCT architectures [6-7]. Most of the DA-based DCT designs exploit some memory reduction techniques such as the partial sum techniques and the offset binary coding (OBC) techniques. However, their designs are still not efficient enough since they only exploit the constant property of the transform coefficients without considering further optimization in reducing hardware.

In this paper, we appropriately combine the features of cyclic convolution and DA technique to propose a memory efficient architecture in realizing the cyclic convolution and apply it to the 1-D DCT. By exploiting the cyclic convolution, we find that different DCT outputs can be computed using the same coefficients and the same input data samples in a rotated order. If we directly realize the DCT using conventional DA technique [2], we find that N ROM modules are used and only one word in a ROM module is accessed at a time in computing the DCT outputs. This reveals a message that the ROM utilization is not good enough. To increase the ROM utilization, we re-arrange the contents of ROM in different way. That is, we first group the candidates of DA inputs with rotated order as the same candidate, and then arrange the ROM contents in such a manner that the partial products for accumulating different DCT outputs according to the candidate are grouped together and accessed simultaneously. The partial products arranged in a group should be rotated suitably before accumulating. In this way, the ROM module will contain only a few groups of contents and only one ROM module, instead of N identical ROM modules, is needed to compute the 1-D N-point DCT. Unfortunately, the rotated input samples in the input-data matrix of DCT possess different signs so that it is not easy to apply the proposed approach directly to DCT realization. According to the symmetry property of DCT coefficients, we can merge the elements in the matrix of DCT kernel, and separate the matrix to two perfect cyclic forms. Then these two smaller cyclic convolutions can be realized with the proposed design approach efficiently. This realization facilitates reducing the ROM size exponentially. As compared with the existing memory-based and DA-based designs, the proposed design can reduce the delay-area product exponentially.

2. PROPOSED DA REALIZATION OF CYCLIC CONVOLUTION

2.1 The proposed approach

Let us first consider a cyclic convolution example:

$$U=\begin{bmatrix}u1\\u2\\u3\\u4\end{bmatrix}=\begin{bmatrix}a&b&c&d\\d&a&b&c\\c&d&a&b\\b&c&d&a\end{bmatrix}\cdot\begin{bmatrix}v1\\v2\\v3\\v4\end{bmatrix},\tag{1}$$

Using the commutative property of convolution, we can rewrite (1) as follow

$$U=\begin{bmatrix}u1\\u2\\u3\\u4\end{bmatrix}=\begin{bmatrix}v1&v2&v3&v4\\v2&v3&v4&v1\\v3&v4&v1&v2\\v4&v1&v2&v3\end{bmatrix}\cdot\begin{bmatrix}a\\b\\c\\d\end{bmatrix},\tag{2}$$

where {v1, v2, v3, v4} are input data, {a, b, c, d} are coefficients, and {u1, u2, u3, u4} are output data. Observing (2), we find that different outputs in vector *U* can be computed using the same input data with rotated order and the same set of coefficients {a, b, c, d}. According to the DA technique, using the same set of coefficients implies that identical ROM modules are used to compute all the different outputs. And using the same input data samples with rotated order implies that we can arrange the partial products generated from the rotated combinations of the input data samples as a group and these partial products can be accessed simultaneously in accumulating all the outputs. Fig. 1 shows the proposed DA architecture for computing the vector *U*. We use a group ROM to store all the shared ROM content that contains only 24 words instead of 64 words needed in the DA-based architecture.

0-7803-7761-3/03/$17.00 ©2003 IEEE

While, we need an additional barrier shifter and address decoder. But, this part of overhead is minor as compared with the saving we get in the ROM size, as we analyzed in the previous works [8].

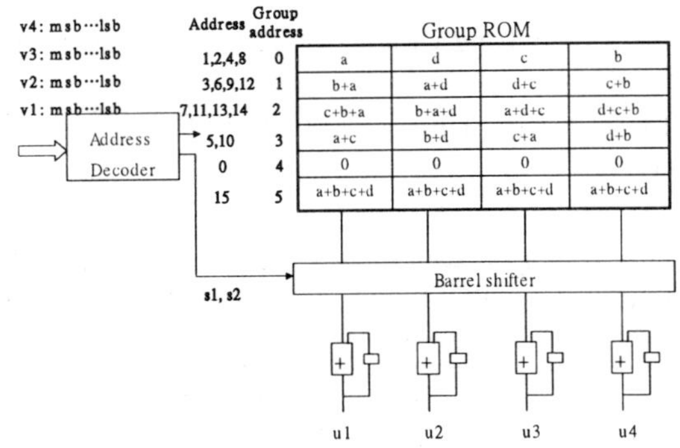

Fig. 1: The architecture of the design example formulated in cyclic convolution.

2.2 The advantages of the proposed approach

In the following, we focus on evaluating the delay time and hardware cost of the circuits in the proposed design approach and the conventional DA approach. For a fair comparison, we adopt the TSMC 0.35um CMOS data-path cell-library [9] in the performance evaluation in terms of delay-area product shown in Fig. 2. We find that the delay-area product of the design with the proposed design approach is much lower than that of the conventional DA design as N increases, which illustrates that the proposed approach possesses better performance than the conventional DA-based designs.

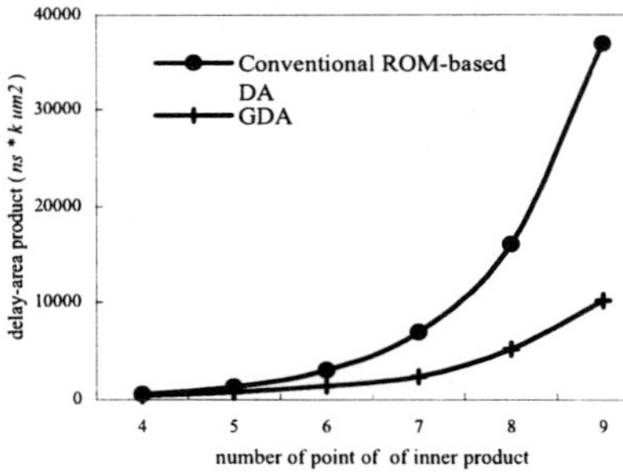

Fig. 2: The delay-area product comparison of the designs with the proposed approach and the conventional DA-based approach using 16-bit data word length.

3. THE PROPOSED DESIGN FOR 1-D DCT

3.1 Algorithm derivation

If transform length N is prime, we can write the 1-D N-point DCT of an input sequence $\{y(n),\ n = 0,\ 1,\ \ldots,\ N-1\}$ in cyclic convolution form by exploiting the property of I/O data permutation as

$$Y(0) = \sum_{n=0}^{N-1} y(n)$$

$$Y((g^k)_N) = [2 \cdot T((g^k)_N) + x(0)] \cdot \cos(\tfrac{\pi}{2N} \cdot ((g^k)_N)); k = 1, \ldots, N-1 \qquad (3)$$

$$T((g^k)_N) = \sum_{n=1}^{N-1} x((g^{n-k})_N) \cdot (-1)^m \cdot \cos(\tfrac{\pi}{N} \cdot (g^n)_N)$$

, where $(g^k)_N$ denotes the result of "g^k modulo N" for short, g is a primitive element, and the sequence $\{x(n)\}$ is defined as

$$\begin{Bmatrix} x(N-1) = y(N-1) \\ x(n = y(n) - x(n+1)); x = 0, \ldots N-2 \end{Bmatrix}$$

By using the symmetry property of cosine kernel as

$$\cos(\tfrac{\pi}{N} \cdot (g^k)_N) = \cos(\tfrac{\pi}{N} \cdot (N - (g^k)_N)) = -\cos(\tfrac{\pi}{N} \cdot (g^{n+\frac{N-1}{2}})_N),$$

we can re-write the $T((g^k)_N)$ in (3) as

$$T((g^k)_N) = \sum_{n=1}^{(N-1)/2} [x((g^{n-k+1})_N) \cdot (-1)^m + x((g^{(n-k+1+\frac{N-1}{2})})_N) \cdot (-1)^{m+\frac{N-1}{2}}] \qquad (4)$$

$$\times \cos(\tfrac{\pi}{N} \cdot (g^{n+1})_N; k = 1, \ldots, N-1$$

Exploiting the symmetry property of the DCT coefficient, we can merge the elements in the matrix of DCT kernel and separate the kernel into the two perfect cyclic forms, which facilitates the efficient realization of the DCT through the proposed design approach. Fig. 3 shows the area reduction of the ROM cost when applying the symmetry property of the DCT coefficients, which reduces the ROM size greatly.

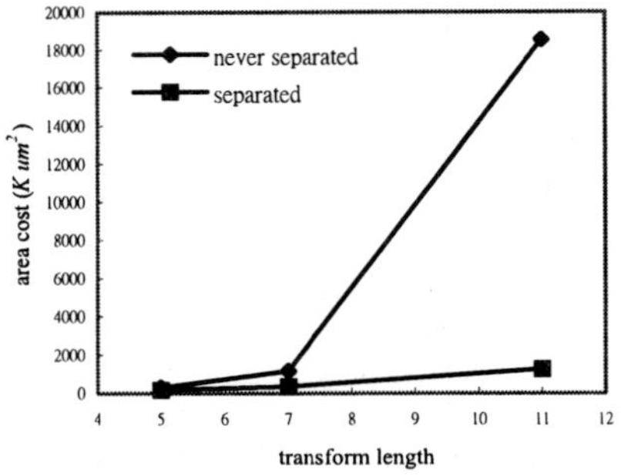

Fig. 3: The area reduction of the ROM cost when applying the symmetry property of the DCT coefficients.

For facilitating the proposed approach, we can formulate the $T((g^k)_N)$ specified in (4) as the DA formulation as

$$T((g^{k+1})_N) = \begin{cases} \bullet \sum_{j=1}^{L} (\sum_{n=1}^{(N-1)/2} G_1(x_j((g^{n-k})_N)) \cdot \cos(\tfrac{\pi}{N} \cdot (g^{n+1})_N)) \cdot 2^{-j}; k = 1, \ldots, \tfrac{N-1}{2} \\ \sum_{j=1}^{L} (\sum_{n=1}^{(N-1)/2} G_2(x_j((g^{n-k})_N)) \cdot \cos(\tfrac{\pi}{N} \cdot (g^{n+1})_N)) \cdot 2^{-j}; k = \tfrac{N-1}{2}+1, \ldots, N-1 \end{cases} \qquad (5)$$

where L denotes the data word length of the variable x, N denotes the transform length, the variable $G(x_j((g^{n-k})_N))$ denotes the jth-bit group address of the ROM access operations, and the preprocessed input sequence $\{x(n)\}$ is defined as

$$x((g^{n-k+1})_N) = \begin{cases} x((g^{n-k+1})_N); if\ n-k+1 \geq 0 \\ x((g^{(N-1)+(n-k+1)})_N); if\ n-k+1 < 0 \end{cases} \qquad (6)$$

$$x((g^{n-k+1+\frac{N-1}{2}})_N) = \begin{cases} x((g^{n-k+1+\frac{N-1}{2}})_N); if\ n-k+1+\frac{N-1}{2} \geq 0 \\ x((g^{(N-1)+(n-k+1+\frac{N-1}{2})})_N); if\ n-k+1+\frac{N-1}{2} < 0 \end{cases} \qquad (7)$$

The value of m is determined by

$$(g^{n+1})_N + m \cdot N = (g^{n-k+1})_N \cdot (g^k)_N; n, k = 1, \ldots, N-1 . \qquad (8)$$

3.2 Architecture design

Fig. 4 shows the proposed architecture design that realizes

the 1-D 7-point DCT. It consists of a pre-processing stage, a group distributed arithmetic unit (GDAU), and a post-processing stage. The input buffer and pre-processing in the preprocessing stage shown as Fig. 5 are designed by using the bidirectional shift registers and an accumulator, which is used to generate the data sequence $x(n)$ from input sequence $y(n)$. The GDAU shown in Fig. 6 is used to carry out the computation of $T((3^k)_7)$ based on the proposed design approach. Due to the same content of group ROM the only one group ROM in the GDAU is required to compute the outputs of the separated cyclic operation. In Fig. 6, the combined input vector Xj=[$x1(j)$, $x2(j)$, $x3(j)$] is first fed into an address decoder to determine which group it should belong to. The address decoder will compute the seed-value Xj'=[$x1'(j)$, $x2'(j)$, $x3'(j)$], group address Gj=[$g2(j)$, $g1(j)$], and the rotating factor Sj=[$s2(j)$, $s1(j)$] by decoding the input vector according to Table 1. Table 2 shows the ROM content arrangement in the proposed design. It is noted that we need one small group ROM modules of size (N-1)/2 x G_{num} words for computing $T((3^k)_7)$. In above, G_{num} denotes the number of groups in the group ROM modules, which is dependent on the transform length N. The serial multiplier of post-processing stage shown in Fig. 5(b) is used to perform the final output $Y((3^k)_7)$.

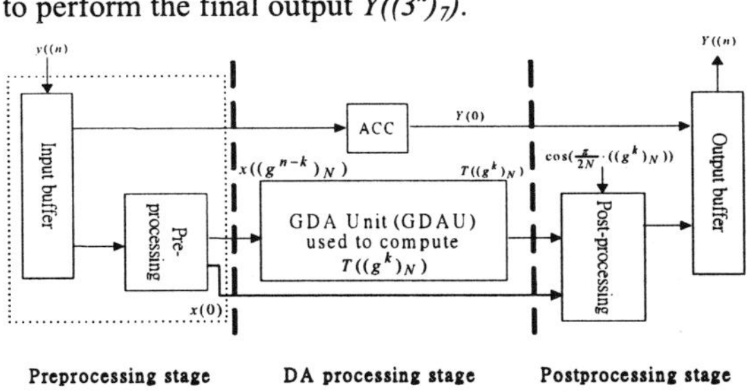

Fig. 4: The block diagram of the proposed architecture for computing the 1-D N-point DCT.

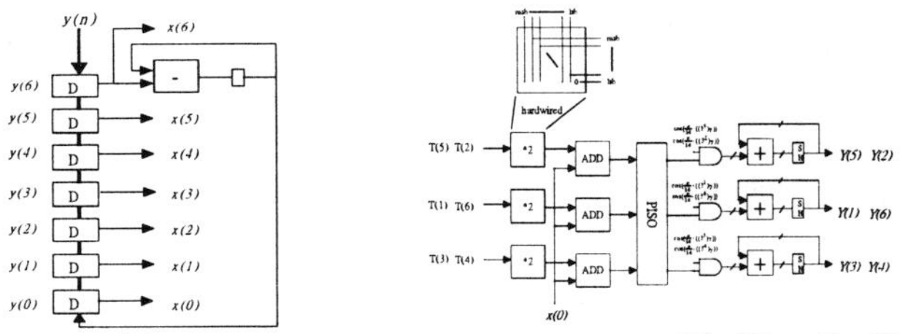

(a) (b) Fig. 5: The design of the (a) pre-processing stage; the (b) constant serial-multiplier design of post-processing stage.

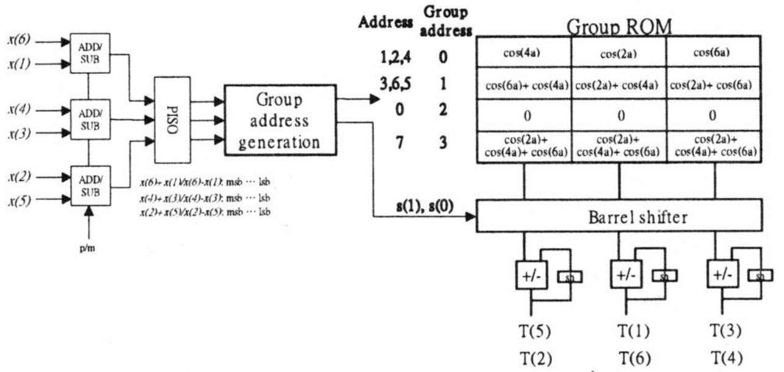

Fig. 6: The GDAU designs used to compute $T((3^k)_7)$ in the 1-D 7-point DCT.

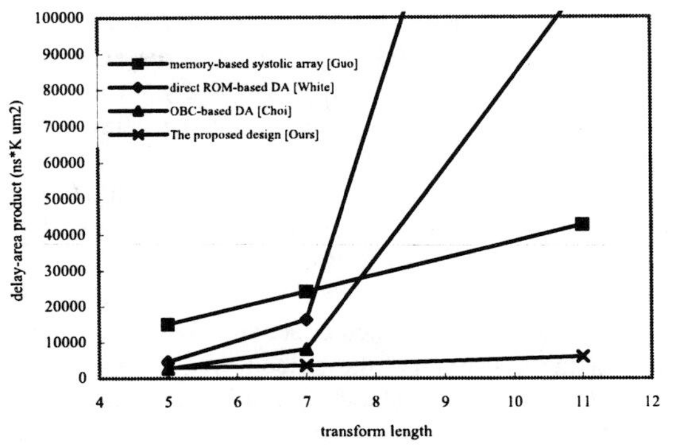

Fig. 7: The delay-area product of the proposed design and the existing DCT designs [1,2,6] in realizing the 1-D DCT.

Table 1: The seed value, group address, and rotating factor used for the design of group address generator of 1-D 7-point DCT

Input value (Xj)			*Seed value (Xj')*			*Group address (Gj)*		*Rotating factor (Sj)*	
x3(j)	*x2(j)*	*x1(j)*	*x3'(j)*	*x2'(j)*	*x1'(j)*	*g2(j)*	*g1(j)*	*s2(j)*	*s1(j)*
0	0	0	0	0	0	1	0	0	0
0	0	1	0	0	1	0	0	0	0
0	1	0	0	0	1	0	0	0	1
0	1	1	0	1	1	0	1	0	0
1	0	0	0	0	1	0	0	1	0
1	0	1	1	1	1	0	1	1	0
1	1	0	0	1	1	0	1	0	1
1	1	1	1	1	1	1	1	0	0

Table 2: 8-word ROM contents arranged into groups

address	Group address			
1, 2, 4	*0*	*cos(4a)*	*cos(2a)*	*cos(6a)*
3, 5, 6	*1*	*cos(6a)+ cos(4a)*	*cos(2a)+ cos(4a)*	*cos(2a)+ cos(6a)*
0	*2*	*0*	*0*	*0*
7	*3*	*cos(2a)+ cos(4a)+ cos(6a)*	*cos(2a)+ cos(4a)+ cos(6a)*	*cos(2a)+ cos(4a)+ cos(6a)*

Table 3: The hardware cost comparisons of the proposed design and some existing DCT designs in realizing the 1-D N-point DCT.

	Adder (word)	FF (word)	ROM (word)	Barrel shifter (word)	RAM (word)
Guo [1] (Memory-based systolic array)	*N*	*N-1*	$(N-1)\cdot 2^{(L/2)}$	*0*	*(N-1)*
White [2] (directly DA)	*N*	*N*	$2^N \cdot N$	*0*	*0*
Choi [6] (OBC-based DA)	*2N*	*N*	$2^{(N-2)} \cdot N$	*0*	*0*
The proposed design	*2N*	*N-1*	$2^{\frac{6(N-1)}{7\ 2}}\ \frac{(N-1)}{2})$	*0.75L* [-0.072 + 0.435 * (N-1) + 0.053 * (N-1)^2]*	*0*

Note: L denotes word length, and N denotes the transform length.

4. PERFORMANCE EVALUATION

In this section, we will illustrate the performance evaluation of the design using the proposed design approach and some existing DCT designs. The existing DCT designs used in this evaluation include memory-based systolic array designs [1], direct DA design [2], OBC DA design [6]. Table 3 shows the hardware cost comparisons of the above-mentioned designs. For a fair comparison, we adopt the Passport 0.35 μm, 3.3-volt CMOS cell-library [9] in the evaluation in terms of the delay time and area cost. According to the two measures, we can accurately evaluate these designs in delay-area product with respect to different values of N. Table 4 shows the comparisons of these designs. The design in [1] is a ROM-based systolic array design. It needs about N adders, $(N\text{-}1)\cdot 2^{(L/2)}$ words of ROM if the ROM tables in the design are partitioned once, and N-1 words of RAM. According to the cell-library, the silicon area of this design is equal to 1237N-1217 Kum^2. The design in [2] is the conventional DA-based design, it requires about N 16-bit adders and $2^N\cdot N$ words of ROM, and the silicon area of this design is equal to $29.3N+4.75\cdot 2^N\cdot N$ Kum^2. The design in [6] is the other DA-based design using the reduction technique of OBC, it requires about 2N 16-bit adders and $2^{(N-2)}\cdot N$ words of ROM, and the silicon area of this design is equal to $35.2N+4.75\cdot 2^{(N-2)}\cdot N$ Kum^2. Fig. 7 shows the delay-area product of the proposed design and the existing DCT designs [1,2,6] in realizing the 1-D DCT. As shown in Fig. 7, in the case of 16-bit data word-length, the delay-area product of the proposed design is much lower than the ROM-based DCT designs [1,2,6]. As a result, regarding the long length transform, we suggest to use the methodology of realizing the long length DCT with short ones, like the prime-factor decomposition algorithm, and etc. Then, it is suggested to realize the short length DCT with the proposed design approach for achieving better delay-area product.

Table 4: The performance comparisons of the proposed design and the existing DCT designs [1,2,6] in realizing the 1-D N-point DCT in terms of delay time and silicon area. (Word length = 16 bits)

	Cycle time *(T)*	Adder *(16-bit)*	FF *(16-bit)*	ROM *(16-bit)*	Barrel shifter *(16-bit)*	RAM *(16-bit)*	Delay*Area *(ns * Kum^2)*
Guo [1] (Memory-based systolic array)	*T=tmux+trom +tadd+tadd*	*5.9N*	*7.8(N-1)*	*1216·(N-1)*		*7.4·(N-1)*	*[(N-1)T/N] (1237N-1217)*
White [2] (Direct DA)	*T = trom+ tadd*	*5.9N*	*23.4N*	$4.75\cdot 2^N\cdot N$			$[(16T)/N]*(29.3N+4.75\cdot 2^N\cdot N)$
Choi [6] (OBC-based DA)	*T = trom+2tadd*	*11.8N*	*23.4N*	$4.75\cdot 2^{(N-2)}\cdot N$			$[(16T)/N]*(35.2N+4.75\cdot 2^{(N-2)}\cdot N)$
The proposed design	*T= trom+tbr+tadd*	*11.8N*	*23.4(N-1)*	$4.75\cdot 2^{\frac{6}{7}\frac{(N-1)}{2}}\cdot\frac{(N-1)}{2}$	$12*[-0.072+0.435*(N-1)+0.053*(N-1)^2]$		$[(32T)/N]*[-28.8+39.1N+0.64N^2+4.75\cdot 2^{\frac{6}{7}\frac{(N-1)}{2}}\cdot\frac{(N-1)}{2}]$

5. CONCLUSION

This paper presents a memory efficient design for realizing the cyclic convolution and its application to the DCT. We combine the advantages of DA computation and cyclic convolution, and exploit the symmetry property of DCT coefficients to facilitate an memory efficient realization of 1-D N-point DCT using (N-1)/2 adders or substractors, one small ROM module, a barrel shifter, and $\frac{N-1}{2}+1$ accumulators. To increase the ROM utilization, we re-arrange the content of ROM into several groups in which all the elements in a group will be accessed simultaneously in computing all the DCT outputs. Comparing with the existing designs, the proposed design can reduce delay-area product significantly. Since we adopt the algorithm for prime length DCT in formulating the cyclic convolution, we have a limitation on the transform length that N should be a prime number in the proposed design. This is not a restriction on the proposed design approach since there are other algorithms proposed in the literature that any length sinusoidal transform can be formulated into cyclic convolution with little overhead [10-11]. The proposed design approach can be easily applied in the transform problems formulated into cyclic convolution.

ACKNOWLEDGMENT

This work is supported by the National Science Council, Taiwan, Republic of China, under the grant NCS-91-2215-E009-033.

REFERENCES

[1] J. I. Guo, C-M. Liu, and C-W Jen, "The efficient memory-based VLSI array designs for DFT and DCT," *IEEE Trans. Circuits Syst. II.* vol. 39, pp. 723-733, Oct. 1992.

[2] S. A. White, "Applications of distributed arithmetic to digital sequence processing: a tutorial review," *IEEE ASSP Mag.*, vol. 6, pp. 4-19, 1989.

[3] W.P. Burleson and L.L. Scharf, "A VLSI Design Methodology for Distributed Arithmetic," *J. VLSI Signal Processing,* vol. 2, pp. 235-252, 1991.

[4] S. Wolter, A. Scubert, H. Matz, and R. Laur, "On the Comparison between Architectures for the Implementation of Distributed Arithmetic," ISCAS1993, vol. 3, pp. 1829-1832, 1993.

[5] K. Nourji and N. Demassieux, "Optimal VLSI Architecture for Distributed Arithmetic-based Algorithm," ICASSP1994, vol. 2, pp. 509-512, 1994.

[6] J. P. Choi, S. C. Shin, and J. G. Chung, "Efficient ROM size reduction for distributed arithmetic," *ISCAS 2000,* pp. II61-II64, May 2000.

[7] M. Sheu, J. Lee, J. Wang, A. Suen, and L. Liu, "A High Throughput-Rate Architecture for 8*8 2-D DCT," ISCAS1993, vol. 3, pp. 1587-1590, 1993.

[8] H. C. Chen, J. I. Guo, and C. W. Jen, "A New Group Distributed Arithmetic Design for The One Dimensional Discrete Fourier Transform, " ISCAS2002, pp. I-421-I-424, 2002.

[9] Compass, PASSPORT library, 0.35 micron 3.3-volt high performance standard cell library, 1996.

[10] J. I. Guo, "An Efficient Parallel Adder Based Design for One Dimensional Discrete Fourier Transform," *Proceedings of the National Science Council, ROC, Part A*, vol.24, no.3, pp.195-204, May 2000.

[11] J. I. Guo, "A New Distributed Arithmetic Algorithm and its Hardware Architecture for the Discrete Hartley Transform," *Pattern Recognition and Image Analysis,* vol.10, no.3, pp.368-378, 2000.

BIT RATE OPTIMIZED TIME-DOMAIN EQUALIZERS FOR DMT SYSTEMS

Chun-Yang Chen, See-May Phoong

Dept. of EE and Grad. Inst. of Comm. Engr., National Taiwan Univ., Taiwan, R.O.C.

ABSTRACT

The discrete multitone (DMT) transceivers have enjoyed great success in high speed data transmission. It is known that when the cyclic prefix is no shorter than the channel impulse response (CIR), the DMT system is ISI free. For channels with very long CIR such as DSL loops, a time-domain equalizer (TEQ) is typically added at the receiver to shorten the effective impulse response. This paper proposes a filterbank approach to the design of TEQ for maximizing the bit rate. Moreover we introduce a structure of DMT system with multiple TEQs. The optimal solution for multiple TEQs is given in closed form and it can serve as a theoretical upper bound for all other TEQs. From the multiple TEQ structure, we propose a DMT system with a pair of complex conjugating TEQs. Simulation examples are given to verify the merit of the proposed TEQ.

1. INTRODUCTION

Discrete Multitone modulation (DMT) has been successfully employed for high speed data transmission over frequency selective channels such as DSL. Fig. 1 shows a DMT system. In a DMT scheme, the input vector $\mathbf{s}$ consisting of modulation symbols is passed through an M-point IDFT matrix. For every block of M data samples, the transmitter adds a cyclic prefix of length L. At the receiver, the L samples corresponding to the cyclic prefix are first removed before the DFT operation. It is known that when cyclic prefix is no shorter than the channel impulse response (CIR), we can obtain ISI free by multiplying the DFT output with a set of scalars known as the frequency domain equalizers (FEQ). In a DMT scheme, the longer the CIR is, the longer the cyclic prefix is needed to avoid ISI. For applications such as DSLs where the CIR can be very long, a time-domain equalizer (TEQ) is used to shorten the effective CIR.

In the past, many methods have been proposed for the design of the TEQ [1]-[8]. These methods can be categorized into two types. The first approach is to design the TEQ by optimizing objective functions depending on the TEQ output. In [1] [2] [3], TEQ is designed to shorten the effective CIR or delay spread by maximizing the energy (or weighted) of the effective CIR within a certain window. In [2] [4] [5], the authors design the TEQ so that SNR at the TEQ output is maximized. TEQs designed using these methods are not optimal in the sense that the resulting TEQs do not maximize the bit rate. The second approach optimizes the geometrical mean (GM) of SNRs of all tones [6] [7] [8]. This approach involves highly nonlinear optimization though it is optimal. Suboptimal solution by replacing the GM with arithmetic mean has been given in [8]. However SNR values estimated using formulations in [6] [8] can deviate from the actual values by several decibels.

In this paper, we propose a filterbank (FB) approach to the TEQ design problem. Using this approach, the expressions for the ISI error and noise error at the FEQ output can be obtained. A TEQ minimizing the GM of these error variances can be designed. Moreover the FB approach gives rise to a DMT scheme with multiple TEQs that can be viewed as a generalization of the dual-tone DMT scheme in [9]. The optimal multiple TEQs are given in closed form. Though having a very high implementation cost, DMT scheme with multiple TEQs can serve as a valuable theoretical bound on the performance. Moreover we propose a DMT scheme with a pair of complex conjugating TEQs. Simulations are carried out to demonstrate the usefulness of the proposed scheme.

Boldfaced upper-case and lower-case letters denote matrices and vectors respectively. The symbols *, ^{T}, and ^{H} represent respectively complex conjugate, transpose and complex conjugate followed by transpose.

2. FILTERBANK FORMULATION OF DMT SCHEMES

Fig. 1 shows a DMT system. In this paper, M denotes the size of the DFT matrix and L represents the cyclic prefix length. The channel is modelled as an LTI real FIR filter $c(n)$ with a real additive WSS noise $\nu(n)$ whose power spectrum is $S_\nu(e^{j\omega})$. Let N_c and N_t be respectively the order of the channel $c(n)$ and the time-domain equalizer $t(n)$. Their z-domain expressions are

$$C(z)=\sum_{n=0}^{N_c} c(n)z^{-n}, \text{ and } T(z)=\sum_{n=0}^{N_t} t(n)z^{-n}.$$

The effective channel becomes $P(z) = C(z)T(z)$. The scalar multipliers $1/P_k$ are known as the frequency-domain equalizers (FEQ), where P_k are equal to the product $C(e^{j2\pi k/M})T(e^{j2\pi k/M})$.

In the following derivation, we will employ the FB interpretation of DMT transceiver. Using multirate identities, one can verify that operations of cyclic removal, serial-to-parallel conversion and the DFT matrix of Fig. 1 can be redrawn as Fig. 2. The symbol $\downarrow N$ denotes subsampling by a factor of N, where $N = M + L$. The receiving filters $H_k(z)$ are the DFT filters

$$H_k(z)=\sum_{i=L}^{M+L-1} e^{-j2\pi ki/M} z^i.$$

In many transmission environments, the channel and noise characteristics vary in different frequency regions. We would like to exploit these characteristics to design a good TEQ. For example, we can use a specific TEQ, say $T_a(z)$ for low-frequency tones and another TEQ, say $T_b(z)$ for high-frequency tones. By doing so, we are able to optimize $T_a(z)$ (or $T_b(z)$) so that its noise gain at

This work was supported by National Science Council under contract #NSC91-2219-E-002-047 and Ministry of Education under grant #89-E-FA06-2-4, Taiwan R.O.C.

0-7803-7761-3/03/$17.00 ©2003 IEEE

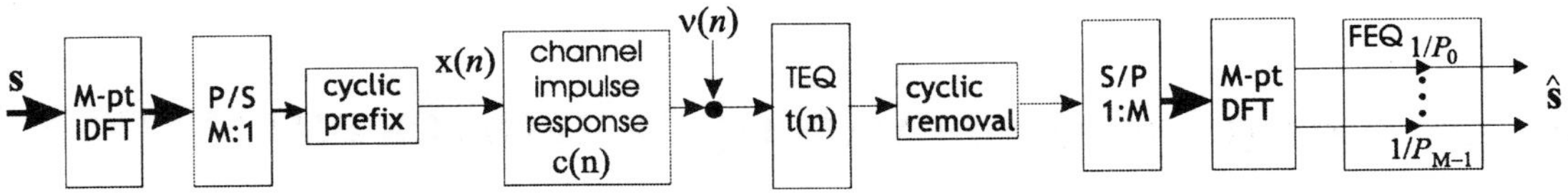

Figure 1: DMT scheme with the time-domain equalizer $t(n)$.

low-frequency (correspondingly high-frequency) region is small. To achieve this flexibility, we introduce a DMT receiver with multiple TEQs as shown in Fig. 3. In this case, the scalars P_k are given by

$$P_k = C(e^{j2\pi k/M})T_k(e^{j2\pi k/M}). \tag{1}$$

It is not difficult to verify that if the cyclic prefix is no shorter than the impulse responses of shortened channels $C(z)T_k(z)$ all k, then the DMT scheme with the receiver given in Fig. 3 continues to enjoy the ISI free property. By setting the TEQs $T_k(z) = T(z)$ for $k = 0, \cdots, M-1$, it is straightforward to verify that Fig. 3 reduces to the conventional case in Fig. 2. In the design of $T_k(z)$, we can exploit the extra freedom of the proposed receiver so that the ISI error and noise error are minimized. When only two TEQs are used, then the multiple-TEQ scheme reduces to to the dual-tone DMT system in [9]. In Fig. 3, each tone uses a different TEQ and this results in a costly receiver. Though its implementation cost is very high, this multiple-TEQ scheme can serve as a theoretical bound. Later we will see that by carefully designing a pair of complex-conjugating TEQs, one can obtain a very satisfactory performance.

Formulation of ISI Errors and Noise Errors: One of the objectives of TEQ design is that the convolution $c(n) * t_k(n)$ will have most of its energy within a specific window of length L. Impulse responses outside the window will generate interblock ISI. Define the sequence

$$d(n) = \begin{cases} 0 & \text{for } n_w < n \le n_w + L, \\ 1 & \text{for } 0 \le n \le n_w \text{ or } n_w + L < n \le N_c + N_t, \end{cases}$$

where n_w is the starting location of the desired window. Then we can describe the ISI term of the kth tone as

$$p_{isi,k}(n) = d(n)\big(c(n) * t_k(n)\big). \tag{2}$$

From Fig. 3, we see that the output error at the kth tone is given by $e_k(n) = [e_{isi,k}(n) + e_{\nu,k}(n)]_{\downarrow N}$, where

$$e_{isi,k}(n) = h_k(n) * p_{isi,k}(n) * x(n)/P_k \tag{3}$$
$$e_{\nu,k}(n) = h_k(n) * t_k(n) * \nu(n)/P_k. \tag{4}$$

As the downsampler $[\bullet]_{\downarrow N}$ does not change the variance, we have

$$\sigma^2_{e_k} = \sigma^2_{isi,k} + \sigma^2_{\nu,k},$$

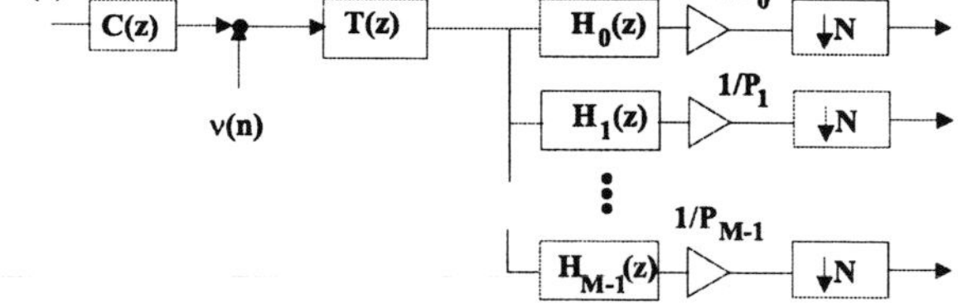

Figure 2: DMT receiver redrawn using the FB structure.

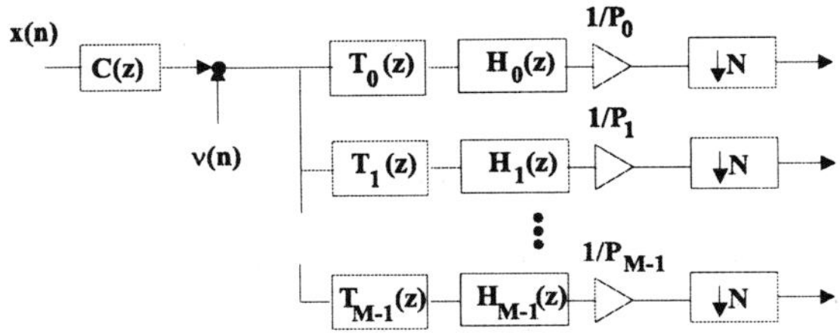

Figure 3: DMT receiver with multiple TEQs $T_k(z)$.

where we have assumed that the signal and noise are uncorrelated.

One can express the error variances using a matrix formulation. Define the vectors

$$\mathbf{t}_k = (t_k(0)\ t_k(1)\ \cdots\ t_k(N_t))^T$$
$$\mathbf{w}_k = (1\ e^{j2\pi k/M}\ \cdots\ e^{j2\pi k N_t/M})^T.$$

Let $\mathbf{C}$ and $\mathbf{H}_k$ be respectively $(N_c + N_t + 1) \times (N_t + 1)$ and $(M+N_c+N_t) \times (N_c+N_t+1)$ lower triangular Toeplitz matrices whose first columns are given by

$$(c(0)\ c(1)\ \cdots\ c(N_c)\ 0\ \cdots\ 0)^T$$
$$(e^{j2\pi k(M-1)/M}\ \cdots\ e^{j2\pi k/M}\ 1\ 0\ \cdots\ 0)^T.$$

Let $\mathbf{D}$ be an $(N_c + N_t + 1) \times (N_c + N_t + 1)$ diagonal matrix with entries $d_{ii} = d(i)$. Using the above definitions, the error variances can be rewritten as[1]:

$$\sigma^2_{isi,k} = \frac{\sigma_x^2 \mathbf{t}_k^H \mathbf{C}^H \mathbf{D}^H \mathbf{H}_k^H \mathbf{H}_k \mathbf{D} \mathbf{C} \mathbf{t}_k}{|C(e^{j2\pi k/M})|^2 \mathbf{t}_k^H \mathbf{w}_k \mathbf{w}_k^H \mathbf{t}_k}, \tag{5}$$

$$\sigma^2_{\nu,k} = \frac{\mathbf{t}_k^H \tilde{\mathbf{H}}_k^H \mathbf{R}_\nu \tilde{\mathbf{H}}_k \mathbf{t}_k}{|C(e^{j2\pi k/M})|^2 \mathbf{t}_k^H \mathbf{w}_k \mathbf{w}_k^H \mathbf{t}_k}, \tag{6}$$

where $\mathbf{R}_\nu$ is the $(M + N_t) \times (M + N_t)$ autocorrelation matrix of $\nu(n)$, and $\tilde{\mathbf{H}}_k$ is a lower triangular Toeplitz matrix having the same form as $\mathbf{H}_k$ but with dimensions of $(M + N_t) \times (N_t + 1)$.

3. OPTIMIZATION OF TEQ

We have formulated the error variances due to ISI and channel noise. To simplify the notations, we define two $(N_t+1) \times (N_t+1)$ Hermitian matrices:

$$\mathbf{Q}_{ISI,k} = \frac{\mathcal{E}_x \mathbf{C}^H \mathbf{D}^H \mathbf{H}_k^H \mathbf{H}_k \mathbf{D} \mathbf{C}}{|C(e^{j2\pi k/M})|^2},\quad \mathbf{Q}_{\nu,k} = \frac{\tilde{\mathbf{H}}_k^H \mathbf{R}_\nu \tilde{\mathbf{H}}_k}{|C(e^{j2\pi k/M})|^2}.$$

[1]To simplify the analysis, we assume that $x(n)$ is a white WSS process. This assumption is usually quite accurate when optimal bit and power loading are employed. Numerical simulations show that except for the first tone, all the variances of ISI estimated under this assumption are within 0.5 dB of the actual values.

Note that the matrix $\mathbf{Q}_{ISI,k}$ is semi positive definite and the matrix $\mathbf{Q}_{\nu,k}$ is positive definite for all k. Moreover these matrices satisfy

$$\mathbf{Q}_{ISI,M-k} = \mathbf{Q}^*_{ISI,k}, \quad \mathbf{Q}_{\nu,M-k} = \mathbf{Q}^*_{\nu,k}, \tag{7}$$

for $k = 1, \cdots, M/2-1$. For $k = 0$ and $k = M/2$, these matrices are real. In the following, we consider the optimization of the TEQs with different criteria:

A. Single TEQ Minimizing mse at the FEQ output (**mmse-f**)*:* In this case, $\mathbf{t}_k = \mathbf{t}$ for all k. The mmse-f TEQ can be obtained by solving the following optimization problem:

$$\arg\min_{\mathbf{t}} \sum_{k=0}^{M-1} \frac{\mathbf{t}^H(\mathbf{Q}_{ISI,k} + \mathbf{Q}_{\nu,k})\mathbf{t}}{\mathbf{t}^H \mathbf{w}_k \mathbf{w}_k^H \mathbf{t}}.$$

Note that using the complex conjugate relations in (7) and the fact that $\mathbf{w}_k = \mathbf{w}^*_{M-k}$, one can verify that the mmse-f TEQ has real coefficients. The above optimization problem is highly nonlinear.

B. Single TEQ Minimizing Geometrical-Mean of $\sigma^2_{e_k}$ (**1-real-opt**)*:* It is known [6] [8] that the MMSE-f TEQ is not optimal in terms of bit rate maximization or transmission power minimization. Under bit and power loading, the optimal TEQ is the $\mathbf{t}$ that minimizes the following geometrical mean:

$$\arg\min_{\mathbf{t}} \prod_{k=0}^{M-1} \frac{\mathbf{t}^H(\mathbf{Q}_{ISI,k} + \mathbf{Q}_{\nu,k})\mathbf{t}}{\mathbf{t}^H \mathbf{w}_k \mathbf{w}_k^H \mathbf{t}}.$$

Note that this is the optimal solution for the single TEQ case. Using the same reasoning as above, this optimal TEQ has real coefficients. Such an optimization is also highly nonlinear.

C. Optimal Multiple TEQs (**multi-opt**)*:* Note that $\mathbf{t}_k$ affects only $\sigma^2_{e_k}$. Therefore when each tone has its own TEQ, the global optimal solution can be obtained by solving

$$\arg\min_{\mathbf{t}_k} \frac{\mathbf{t}_k^H(\mathbf{Q}_{ISI,k} + \mathbf{Q}_{\nu,k})\mathbf{t}_k}{\mathbf{t}_k^H \mathbf{w}_k \mathbf{w}_k^H \mathbf{t}_k},$$

for $k = 0, \cdots, M-1$. Note that $(\mathbf{Q}_{ISI,k} + \mathbf{Q}_{\nu,k})$ is positive definite. Let $\mathbf{Q}_k^{1/2}$ be the unique positive definite matrix such that $\mathbf{Q}_k^{1/2}\mathbf{Q}_k^{1/2} = (\mathbf{Q}_{ISI,k} + \mathbf{Q}_{\nu,k})$. Then by letting $\mathbf{u}_k = \mathbf{Q}_k^{1/2}\mathbf{t}_k$, the optimal $\mathbf{t}_k$ can be obtained by solving

$$\arg\max_{\mathbf{u}_k} \frac{\mathbf{u}_k^H(\mathbf{Q}_k^{-1/2})\mathbf{w}_k \mathbf{w}_k^H \mathbf{Q}_k^{-1/2}\mathbf{u}_k}{\mathbf{u}_k^H \mathbf{u}_k}.$$

As the matrix $\mathbf{Q}_k^{-1/2}\mathbf{w}_k\mathbf{w}_k^H\mathbf{Q}_k^{-1/2}$ has rank one, it has only one nonzero eigenvalue and the $\mathbf{u}_k$ that maximizes the above function is given by $\mathbf{u}_{k,opt} = \mathbf{Q}_k^{-1/2}\mathbf{w}_k$. Therefore we have the closed form solution (no nonlinear optimization is needed)

$$\mathbf{t}_{k,opt} = \mathbf{Q}_k^{-1}\mathbf{w}_k. \tag{8}$$

Note that this is the optimal solution that minimizes the average as well as GM of error variances. The performance of all linear TEQs with the same number of coefficients will be bounded by this solution.

D. Complex-Conjugate Pair of TEQs (**2-complex**)*:* Though the multi-opt TEQ is globally optimal, its implementation cost is too high. One way to reduce the complexity is to use a small number of TEQs and each TEQ equalizes a number of adjacent tones. The TEQ in each group can be designed separately. From the simulation results on typical CSA loops, we found that it gives a very satisfactory performance if we partition the M tones into 2 groups. Group 1 contains Tones $0, \ldots, M/2-1$ whereas Group 2 contains Tones $M/2, \ldots, M-1$. In Group 1, we choose the tone with the highest SNR, say Tone J. The $\mathbf{t}_{J,opt}$ defined in (8) is used as the TEQ for those tones in Group 1. For Group 2, the best tone will be Tone (M-J) due to complex conjugate property and we have $\mathbf{t}_{M-J,opt} = \mathbf{t}^*_{J,opt}$. As $t_{J,opt}$ has the closed form solution (8), the design cost of this 2-complex TEQ is very low. Even though there are 2 TEQs, we need only to design and implement one TEQ. The reason is as follows. The outputs of the tones in Group 2 are simply complex conjugates of those in Group 1. At the receiver, we need to implement only those tones in Group 1 and hence only 1 complex TEQ is implemented. Note that this 2-TEQ structure is different from the dual path DMT in [8]. In [8], two receivers, each with a different real TEQ for all tones, are implemented and a tone selector is employed to select outputs with higher SNRs from the two receivers.

4. SIMULATION EXAMPLES

The transmission channels considered in the simulation are the 8 typical CSA loops as in [7] [8]. The channel noise consists of a additive white Gaussian noise (AWGN) with -140 dBm/Hz and a near-end crosstalk (NEXT) whose power spectral density is

$$S_{next}(f) = \gamma\, k_{NEXT}\, f^{3/2} \times \frac{[\sin(\pi f/f_0)\sin(\pi f/2f_0)]^2}{[1 + (f/f_{1,3dB})^6](f^2 + f^2_{2,3dB})},$$

where $f_0 = 1.455$ MHz, $f_{1,3dB} = 3$ MHz, $f_{2,3dB} = 40$ kHz and $k_{NEXT} = 2.1581 \times 10^{-9}$. The parameter $\gamma = 0.0282$ (-15.5 dB) represents the adjacent binder effect. The simulation assumes that all the receivers have a perfect estimation of the channel response. The sampling rate T_s is 2.208 MHz. The DFT size is $M = 512$ and the cyclic prefix length is $L = 32$. The bits and power are optimally allocated using a water-filling type algorithm. The modulation scheme used is QAM. The order of the TEQ is $N_t = 4$ (i.e., 5 taps).

We compare the performance of the 4 TEQs in Sec. 3.A–3.D, the TEQ that maximizes the signal-to-interference at the TEQ output (maxsir) [1], and the TEQ that minimizes mse at the TEQ output (mmse-t) [2]. In Table 1, we list the maximum achievable bit rate when the transmission power is $\sigma_x^2 = 14$ dBm. The maximum achievable bit rate is given by

$$\sum_{k=0}^{255} b_k = \sum_{k=0}^{255} \lfloor \log_2\Big(1 + \frac{\sigma_x^2/\sigma_{e_k}^2}{10}\Big)\rfloor.$$

For uncoded QAM constellations, the above formula corresponds to a bit error probability of 10^{-7}. It should be emphasized that the values of $\sigma^2_{e_k}$ are the true error variances obtained from the actual implementation of the DMT systems.

From Table 1, we see that though having smaller error variances, both mmse TEQs do not necessarily give a better performance than the maxsir TEQ. As expected, the 1-real-opt TEQ outperforms the maxsir, mmse-t and mmse-f TEQs. Despite having a very low design cost, the 2-complex TEQ has an excellent performance and in some cases it outperforms the 1-real-opt TEQ,

	Loop1	Loop2	Loop3	Loop4	Loop5	Loop6	Loop7	Loop8
maxsir	4.93	4.14	4.83	3.98	4.42	4.49	3.93	3.30
mmse-t	4.67	5.42	4.72	4.09	4.83	4.50	3.86	3.25
mmse-f	4.88	5.36	4.70	4.19	4.79	4.37	4.40	4.24
1-real-opt	4.96	5.46	4.89	4.24	4.84	4.59	4.38	4.36
2-complex	4.95	5.50	4.74	4.11	4.87	4.64	4.61	4.40
multi-opt	5.30	5.93	5.20	4.78	5.34	4.88	4.92	4.57

Table 1: Maximum achievable bit rate (Mbps). The transmission power is 14 dBm.

whose design procedure involves a highly nonlinear optimization problem. The multi-opt TEQ has the best performance.

In [6] [8], the SNR at the kth tone is estimated as follows:

$$SNR_k = \frac{\sigma_x^2 |P_{sig}(e^{j2\pi k/M})|^2}{\sigma_x^2 |P_{isi}(e^{j2\pi k/M})|^2 + S_\nu(e^{j2\pi k/M})|T(e^{j2\pi k/M})|^2}, \tag{9}$$

where $p_{sig}(n) = c(n) - p_{isi}(n)$ represents the impulse response of the signal path. When the noise at the TEQ output is nonflat, this can result in a large error. In Fig. 4, we plot the actual SNRs, SNRs estimated using the FB formulation and SNRs given in (9) versus the subband (or tone) index. The transmission channel in this case is CSA Loop 1 and the TEQ is maxsir TEQ, whose design is independent of the SNR formula. From the figures, we see that our estimates match nicely with the actual values, whereas SNRs estimated using (9) can deviate from the actual values by as much as 10 dB.

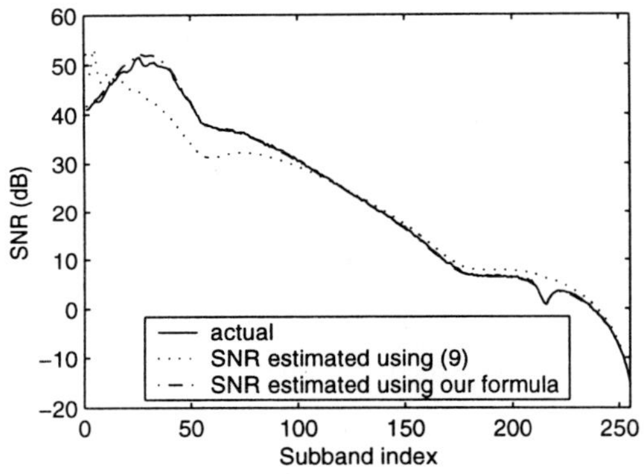

Figure 4: SNR estimated using different formulas.

In Fig. 5, we plot the bit error rate versus transmission power. The transmission channel is CSA Loop 7 and the transmission rate is 6 Mpbs. From the figure, we see that multi-opt TEQ has the best performance. The 2-complex TEQ though having a relatively low design and implementation cost is only slightly worse than the multi-opt TEQ. As expected, the 1-real-opt TEQ outperforms the other non optimal 1-real TEQs (maxsir, mmse-t and mmse-f).

5. CONCLUSIONS

In this paper, we proposed a FB formulation for the TEQ design problem. Such a formulation gives rises to a DMT structure with multiple TEQs. The optimal solution for the multiple TEQs is given in closed form and its performance can serve as a valuable upper bound. A suboptimal solution with a pair of complex conjugating TEQs is given. Simulations show that the suboptimal solution though having a relatively low design and implementation cost is only slightly worse than the multi-opt TEQ.

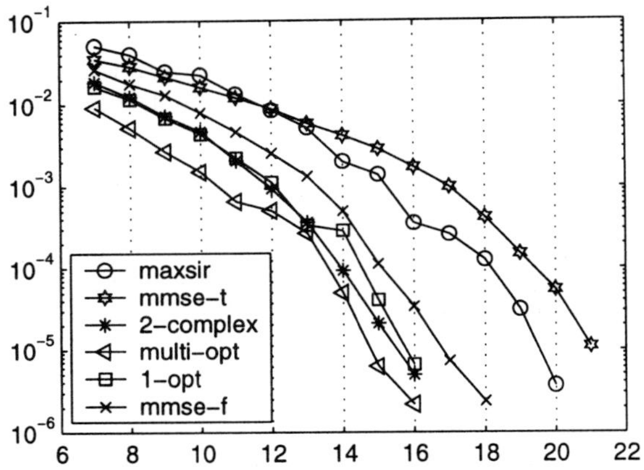

Figure 5: Bit error rate versus transmission power in dBm.

6. REFERENCES

[1] P. J. W. Melsa, R. C. Younce, and C. E. Rohrs, "Impulse response shortening for discrete multitone transceivers," *IEEE Trans. Commun.*, Dec. 1996.

[2] I. Djokovic, "MMSE equalizers for DMT systems with and without crosstalk," *31st Asilomar Conf.*, pp. 545-549, 1997.

[3] R. Schur and J. Speidel, "An efficient equalization method to minimize delay spread in OFDM/DMT systems," *Proc. IEEE Int. Conf. Commun.*, June 2001.

[4] M. Nafie and A. Gatherer, "Time-domain equalizer training for ADSL," *Proc. IEEE Int. Conf. Commun.*, June 1997.

[5] A. Tkacenko and P. P. Vaidyanathan, "Noise optimized eigenfilter design of time-domain equalizers for DMT systems Communications," *Proc. IEEE Int. Conf. Commun.*, 2002.

[6] N. Al-Dhahir and J. M. Cioffi, "Optimum finite-length equalization for multicarrier transceivers," *IEEE Trans. Commun.*, Jan. 1996.

[7] B. Farhang-Bouroujeny and M. Ding, "Design methods for time-domain equalizers in DMT transceivers," *IEEE Trans. Commun.*, Mar. 2001.

[8] G. Arslan, B. L. Evans, and S. Kiaei, "Equalization for discrete multitone transceivers to maximize bit rate," *IEEE Trans. Signal Proc*, Dec. 2001.

[9] Ming Ding, A. J. Redfern, B. Evans, "dual-path TEQ structure for DMT-ADSL systems," *Proc. IEEE Int. Conf. Acoutics, Speech, and Signal Proc*, 2002.

Fast Algorithms for Computing Full and Reduced Rank Wiener Filters

Mohammed A. Hasan(1) and M.R. Azimi-Sadjadi(2),
(1) Department of Electrical & Computer Engineering, University of Minnesota Duluth
(2) Department of Electrical & Computer Engineering, Colorado State University
E.mail:mhasan@d.umn.edu

Abstract

Wiener filter is an important tool in many signal processing applications. This paper proposes a fast inverse free method based on mean-square-error (MSE) criterion for the design of Wiener filters when both the given signal and the desired signal are wide-sense stationary random processes. Specifically, we present several approaches for tracking Wiener filter using gradient descent, line search, and adaptive methods. A new framework for computing the full rank Wiener filter and order update for reduced rank Wiener filter serially is presented. Emphasis is placed on methods that require least amount of matrix inversion including Rayleigh Quotient like methods. Simulations are also presented to examine the performance of the proposed methods.

1 Introduction

The Wiener filter was developed by Nobert Wiener in the early 1900's. It can be used to reconstruct a noisy signal by minimizing the Mean-Square Error (MSE) between the original and received signal. In typical applications of Wiener filters, the input signal $y(n)$ consists of the reference signal $x(n)$ embedded in additive noise $v(n)$. This can be expressed as: $y(n) = x(n) + v(n)$. The main assumptions are that both the signal and the noise are zero-mean wide-sense stationary processes, the signal and noise $x(n)$, $v(n)$, are statistically independent, and the autocovariance functions $R_{yy}(k)$ and the cross-covariance $R_{xy}(k)$ of the signals $x(n)$, $y(n)$, are known.

Wiener filter (WF) [1,2,3] has widely been used in various correlation-based statistical signal processing areas such as system identification, predictive deconvolution, channel equalization, noise cancellation and suppression, echo cancellation and time delay estimation.

The full and reduced rank Wiener filter (RRWF) can be utilized wherever a desired signal needs to be extracted from random background noise or deterministic interference. Common applications are echo cancellation, equalization, neural network learning, and spectral line enhancement. Computation of Wiener filter involves $R_{xy}R_{yy}^{-1}$ which is unsuitable if the dimension of the vector y is very large. To overcome this difficulty several algorithms are developed. In [4] a multi-stage Wiener filter was developed which can be derived using only matrix vector multiplications. The reduced rank Wiener filter was first developed in [5,6]. In this paper we present several approaches for on-line computation of the full rank and reduced rank Wiener filter using criteria that involve minimization of trace or determinants of some cost functions. Simulations show that the proposed methods provide computational saving in addition to having good estimation accuracy and fast convergence.

2 Basic Definitions and Notation

The classical wiener filtering problem can be described as follows. Given a desired signal $x(n) \in \mathbb{R}^m$ and observed signal $y(n) \in \mathbb{R}^n$, find $W \in \mathbb{R}^{m\times n}$ so that

$$E(||x(n)-Wy(n)||^2) = trace\{R_{xx}-WR_{yx}-R_{xy}W^T+WR_{yy}W^T\}$$

is minimum, where E denotes statistical expectation and $||.||$ is the standard Euclidean norm. It will be assumed that both signals are zero mean wide sense stationary processes.

It is well known that if $x(n)$ and $y(n)$ are complex stationary processes, then the exact solution is given by $W = R_{xy}R_{yy}^{-1}$ and therefore $x(n)$ is estimated by $\hat{x}(n) = Wy(n)$. The error covariance matrix Q_{xx} is defined by

$$Q_{xx} = E(\hat{x}(n) - x(n))(\hat{x}(n) - x(n))^T = R_{xx} - R_{xy}R_{yy}^{-1}R_{yx}.$$

This matrix can be rewritten as

$$Q_{xx}(W) = Q_{xx} + (R_{xy} - WR_{yy})R_{yy}^{-1}(R_{xy} - WR_{yy})^T.$$

Throughout this paper, the notation *trace*(.) and det(.) will stand for the matrix trace and determinant operators, respectively.

3 Reduced Rank Wiener Filtering

The main goal in computing reduced rank Wiener filter is to find the rank r minimizer W_r that minimizes

$$Q_{xx}(W_r) = E(||x(n) - W_r y(n)||^2) \quad W_r \text{ has rank r.}$$

After simplification,

$$Q_{xx}(W_r) = Q_{xx} + (R_{xy} - W_rR_{yy})R_{yy}^{-1}(R_{xy} - W_rR_{yy})^T.$$

This formulation is useful if a reduced rank W_r is to be computed. Assume that W_r is of rank r, then W_r can be written as $W_r = ab^T$ where a and b are two matrices of full rank r. This approach was considered in [7]. In this paper, we will assume that $W_r = U\Sigma V^H$ where U and V are orthogonal and Σ is diagonal and positive definite. The matrices U, Σ, and V are solutions of the minimization problem:

$$\text{Minimize } \frac{1}{2}trace\{Q_{xx} + (R_{xy} - U\Sigma V^T R_{yy})R_{yy}^{-1} \times (R_{xy} - U\Sigma V^T R_{yy})\}$$

subject to $U^TU = I$, $V^TV = I$ and Σ is diagonal with positive entries.

For simplicity assume now that W_r is of rank 1, then $U \in \mathbb{R}^m$, $V \in \mathbb{R}^n$ and $\Sigma = \sigma > 0$.

Let $F(u, v, \sigma)$ be the lagrangian of the above problem given by

$$F(u, v, \sigma) =$$
$$\frac{1}{2}trace\{Q_{xx} + (R_{xy} - u\sigma v^T R_{yy})R_{yy}^{-1}(R_{xy} - u\sigma v^T R_{yy})\}$$
$$- (u^T u - 1)\frac{\lambda_1}{2} - (v^T v - 1)\frac{\lambda_2}{2},$$

where λ_1 and λ_2 are the Lagrange multipliers. Clearly,

$$\frac{\partial F}{\partial u} = -\sigma R_{xy} v + \sigma^2 u v^T R_{yy} v - \lambda_1 u,$$
$$\frac{\partial F}{\partial v} = -\sigma R_{yx} u + 2\sigma^2 R_{yy} v u^T u - \lambda_2 v,$$
$$\frac{\partial F}{\partial a} = trace\{-uv^T R_{yx} - R_{xy} v u^T + 2\sigma u v^T R_{yy} v u^T\},$$
$$= -2v^T R_{yx} u + 2\sigma v^T R_{yy} v u^T u.$$

At optimality, we have

$$\sigma = \frac{v^T R_{yx} u + u^T R_{xy} v}{2v^T R_{yy} v}.$$

From the above derivation, an alternative (but equivalent) formulation can be described as follows:

$$\text{Maximize } \{\frac{u^T R_{xy} v}{v^T R_{yy} v} : u^T u = 1, \ v^T v = 1.\}$$

Then if this problem achieves its maximum σ for some u and v, then $w = u\sigma v^T$ is a rank 1 Wiener filter.

4 Wiener Filter Computation Using a Line Search

One can also use a line search to compute the full rank Wiener filter. Assume that

$$F(w) = \frac{1}{2}trace\{Q_{xx} + (R_{xy} - wR_{yy})R_{yy}^{-1}(R_{yx} - R_{yy}w^T)\},$$

then

$$\nabla_w F = R_{xy} - wR_{yy}.$$

Therefore, given w and a direction h, we are looking for minimizing $F(w + h\alpha)$ with respect to α. This implies that

$$\frac{\partial F(w + h\alpha)}{\partial \alpha} = 0.$$

Assuming that $h = \nabla_w F$, we obtain after some algebraic simplifications

$$\alpha = (h^T R_{yy} h)^{-1}(h^T R_{yx} - h^T R_{yy} w^T)$$
$$= (h^T R_{yy} h)^{-1}(h^T h).$$

The updated Wiener filter will be $w' = w + h\alpha$.

There are two approaches for updating w. If $m >> n$, then it is more efficient to use the updating rule $w' = w + h\alpha$ for some $n \times n$ matrix. However, if $m << n$, it is easier to search for an $m \times m$ matrix α and then w is updated using the rule $w' = w + \alpha h$.

4.1 Adaptive Estimation of Wiener Filter

Now assume that the more recent data is used so that R_{xy} and R_{yy} are updated as $\hat{R}_{xy} = \beta R_{xy} + (1-\beta)xy^T$ and $\hat{R}_{yy} = \beta R_{yy} + (1 - \beta)yy^T$, where $x = x(n)$, $y = y(n)$ are the new input and output vectors, and $0 \leq \beta < 1$. Then α can be computed in similar way as follows: Let W_r and $h_0 = R_{xy} - wR_{yy}$, our goal is to compute a new w which minimizes

$$F(w) = \frac{1}{2}trace\{Q_{xx} + R_{xy} - wR_{yy})R_{yy}^{-1}(R_{yx} - R_{yy}w^T)\}.$$

Then,

$$\alpha = [\beta h^T R_{yy} h + (1 - \beta)h^T y^T y h]^{-1}[\beta h^T R_{yx} + (1 - \beta)h^T y x^T -$$
$$\beta h^T R_{yy} w - (1 - \beta)h^T y y^T w^T]$$
$$= [\beta h^T R_{yy} h + (1 - \beta)h^T y y^T h]^{-1}[\beta h^T h_0 + (1 - \beta)h^T y(x^T - y^T w^T)]$$

where h_0 is the increment vector of the previous step. This can be summarized in the following algorithm.

Algorithm 1:

Given an initial guess W_0. For $k = 1, \cdots$, do

1. compute $h = R_{xy} - W_k R_{yy}$
2. set $\alpha = (hh^T)(hR_{yy}h^T)^{-1}$
3. Update W_k so that $w_{k+1} = W_k + \alpha h$.
4. given new input and output vectors $x(k), y(k)$, update the matrices R_{yy} and R_{xy} as $\hat{R}_{xy} = \beta R_{xy} + (1 - \beta)x(k)y(k)^T$ and $\hat{R}_{yy} = \beta R_{yy} + (1 - \beta)y(k)y(k)^T$.
5. set $W_r = w$ and go to Step 1.

If the dimension of $hR_{yy}h^T$ is large, matrix inversion in Step 2 can be avoided by using the approximation

$$\alpha \approx trace(hh^T)/trace(hR_{yy}h^T). \qquad (1)$$

An alternative but more efficient formulation is described in the following algorithm which is based on the least squares method of solving the equation $R_{xy} - WR_{yy} = 0$. .

Algorithm 2:

Given an initial guess $w_0(:, 1 : r)$ of rank n.

1. set $r_0 = R_{yy}w_0(:, r) - R_{xy}(:, k)$ and compute $h = R_{yy}r_0$
2. compute $h_1 = R_{yy}h$
3. set $\alpha = -(h^T h)(h_1^T h_1)^{-1}$
4. update w_0 so that $w(:, r) = w_0(:, r) + \alpha h$.
5. k=k+1, $w_0(:, r) = w(:, r)$, go to step 1.
6. repeat Steps 1-4 until convergence.

4.2 A Rayleigh Quotient Like Approach

The derivation of the last section suggests that u, v can be computed by solving the maximization problem:

$$\text{Maximize } \{\frac{u^T R_{xy} v}{v^T R_{yy} v} : u^T u = 1, \ v^T v = 1.\} \qquad (2)$$

This problem represents an optimization problem over a compact set and thus it achieves its maximum and minimum in that set. i.e., there exist σ, the maximum of (2), and unit vectors u and v which maximize (2) such that $W_1 = u\sigma v^T$ is the rank 1 Wiener filter. Let $\mathcal{L} = \frac{u^T R_{xy} v}{v^T R_{yy} v} - (u^T u - 1)\frac{\lambda_1}{2} - (v^T v - 1)\frac{\lambda_2}{2}$ be the Lagrangian, then necessary conditions of optimality are

$$\frac{R_{xy} v}{v^T R_{yy} v} - \lambda_1 u = 0,$$
$$\frac{(v^T R_{yy} v)R_{yx} u - 2(u^T R_{xy} v)R_{yy} v}{(v^T R_{yy} v)^2} - \lambda_2 v = 0, \qquad (3a)$$

for some Lagrange multipliers λ_1 and λ_2. If u, v is an optimal solution of (2), then

$$\lambda_1 = \frac{u^T R_{xy} v}{v^T R_{yy} v}$$
$$\lambda_2 = -\frac{u^T R_{xy} v}{v^T R_{yy} v} = -\lambda_1 \tag{3b}$$

A gradient descent method can be applied to solve the two equations in (3a), however a faster method would be to use a line serach as follows. Let h_1, h_2 be two directions and let α be a scalar which minimizes (2) in the directions of h_1 and h_2. An optimal value of the scalar α may be determined by solving

$$\frac{\partial}{\partial \alpha}\{\frac{(v+\alpha h_2)^T R_{yx}(u+\alpha h_1)}{(v+\alpha h_2)^T R_{yy}(v+\alpha h_2)} - \frac{\lambda_1}{2}(u+\alpha h_1)^T(u+\alpha h_1) - \frac{\lambda_2}{2}(v+\alpha h_2)^T(v+\alpha h_2)\} = 0, \tag{3c}$$

where λ_1 and λ_2 are as in (3b) and

$$h_1 = \frac{R_{xy} v}{v^T R_{yy} v} - \lambda_1 u,$$
$$h_2 = \frac{(v^T R_{yy} v) R_{yx} u - 2(u^T R_{xy} v) R_{yy} v)}{(v^T R_{yy} v)^2} - \lambda_2 v. \tag{3d}$$

Using the fact that $u^T h_1 = 0$ and $v^T h_2 = 0$, (3c) yields the following fifth order polynomial equation:

$$d_5\alpha^5 + d_4\alpha^4 + d_3\alpha^3 + d_2\alpha^2 + 2d_1\alpha + d_0 = 0, \tag{4}$$

where

$$d_5 = -(\lambda_1 h_1^T h_1 + \lambda_2 h_2^T h_2) b_2^2$$
$$d_4 = -2(\lambda_1 h_1^T h_1 + \lambda_2 h_2^T h_2) b_1 b_2$$
$$d_3 = -(\lambda_1 h_1^T h_1 + \lambda_2 h_2^T h_2)(b_1^2 + 2b_0 b_2)$$
$$d_2 = c_2 - 2b_0 b_1(\lambda_1 h_1^T h_1 + \lambda_2 h_2^T h_2)$$
$$d_1 = c_1 - b_0^2(\lambda_1 h_1^T h_1 + \lambda_2 h_2^T h_2)$$
$$d_0 = c_0,$$

and where

$$a_0 = u^T R_{xy} v$$
$$a_1 = h_1^T R_{xy} v + u^T R_{xy} h_2$$
$$a_2 = h_1^T R_{xy} h_2$$
$$b_0 = v^T R_{yy} v$$
$$b_1 = h_2^T R_{yy} v + v^T R_{yy} h_2$$
$$b_2 = h_2^T R_{yy} h_2.$$
$$c_0 = a_1 b_0 - a_0 b_1$$
$$c_1 = 2(b_0 a_2 - a_0 b_2)$$
$$c_2 = b_1 a_2 - b_2 a_1.$$

The fifth order equation (4) has five solutions at least one of which is real. The formulas

$$u' = u + \alpha h_1$$
$$v' = v + \alpha h_2$$

can be used to update u and v, where α is a nonzero real root of (4).

5 Reduced Rank Wiener Filter: Alternative Formulation

In this section, we consider a determinantal cost function for computing a reduced rank Wiener filter. The error covariance matrix can be shown to be

$$Q_{xx}(w_r) = Q_{xx}(w) + (w - w_r) R_{yy} (w - w_r)^T.$$

Thus consider

$$F(w_r) = \det(Q_{xx}(w_r)).$$

It can be shown that

$$\frac{\partial F}{\partial w_r} = 2Q_{xx}(w_r)^{-1}(w - w_r) R_{yy}.$$

Assume that w_r is of rank r, then $w_r = ab^T$, where a, b are two matrices of full rank r. The optimal matrices a and b can be determined by solving the minimization problem:

$$\text{Minimize} \quad F(a,b) = \det(Q_{xx} + (w - ab^T) R_{yy} (w - ab^T)^T)$$

over all matrices a, b of full rank r. The partial derivatives of F with respect to a and b are given in the next result.

Theorem 1. *Let $F(a,b)$ as given above, then*

$$\frac{\partial F}{\partial a} = -2[Q_{xx} + (w - ab^T) R_{yy} (w - ab^T)^T]^{-1} (w - ab^T) R_{yy} b,$$
$$\frac{\partial F}{\partial b} = -2R_{yy}(w - ab^T)^T (Q_{xx} + (w - ab^T) R_{yy} (w - ab^T)^T)^{-1} a.$$

Proof: Let $H(a,b) = Q_{xx} + (w - ab^T) R_{yy} (w - ab^T)^T$, then

$$\frac{\partial F(a + \epsilon E_{ij}, b)}{\partial \epsilon} = trace(H(a + \epsilon E_{ij}, b)^{-1} \frac{\partial H(a + \epsilon E_{ij}, b)}{\partial \epsilon}).$$

Here $E_{ij} = e_i e_j^T$, where e_i denotes the ith column of an identity matrix so that a and E_{ij} have same size. Clearly,

$$\frac{\partial H}{\partial \epsilon} = -E_{ij} b^T R (w - ab^T)^T - (w - ab^T) R_{yy} b E_{ji}.$$

Therefore,

$$\frac{\partial F}{\partial a} = (-b^T R (w - ab^T)^T H^{-1})^T - H^{-1}(w - ab^T) R_{yy} b$$
$$= -H^{-1}(w - ab^T) R_{yy} b - H^{-1}(w - ab^T) R_{yy} b$$
$$= -2(Q_{xx} + (w - ab^T) R_{yy} (w - ab^T)^T)^{-1} \{(w - ab^T) R_{yy} b\}.$$

Similarly,

$$\frac{\partial F(a, b + \epsilon E_{ij})}{\partial \epsilon} = trace(-H^{-1} a E_{ji} R_{yy} (w - ab^T)^T - H^{-1}(w - ab^T) R_{yy} E_{ij} a^T).$$

Therefore,

$$\frac{\partial F}{\partial b} = -R_{yy}(w - ab^T)^T H^{-1} a - (a^T H^{-1} (w - ab^T) R_{yy})^T$$
$$= -2R_{yy}(w - ab^T)^T H^{-1} a.$$

Remark: If A and P are invertible matrices, one can show that the following relation holds:

$$Px^T(A + xPx^T)^{-1} = (I + Px^T A^{-1} x)^{-1} Px^T$$
$$= (P^{-1} + x^T A^{-1} x)^{-1} x^T.$$

From this remark, $\frac{\partial F}{\partial b}$ simplifies to

$$\frac{\partial F}{\partial b} = -2R_{yy}(w - ab^T)^T (Q_{xx} + (w - ab^T) R_{yy} (w - ab^T)^T)^{-1} a$$
$$= (R_{yy}^{\frac{1}{2}} + (w - ab^T)^T Q_{xx} (w - ab^T))^{-1} (w - ab^T)^T Q_{xx}^{-1} a.$$

Thus

$$(w - ab^T)^T Q_{xx}^{-1} a = 0,$$

or

$$R_{yy}^{\frac{1}{2}} (w - ab^T)^T Q_{xx}^{-1} a = 0.$$

To solve these two equations, let us make the change of variables:

$$Q_{xx}^{-\frac{1}{2}} w R_{yy}^{-\frac{1}{2}} = G, \; b_1 = R_{yy}^{-\frac{1}{2}} b, \; a_1 = Q_{xx}^{-\frac{1}{2}} a.$$

This leads to:

$$F = \det(Q_{xx})\ det\{I + (Q_{xx}^{-\frac{1}{2}} w R_{yy}^{-\frac{1}{2}} - Q_{xx}^{-\frac{1}{2}} b^T R_{yy}^{-\frac{1}{2}}) \times (R_{yy}^{-\frac{1}{2}} H^T Q_{xx}^{-\frac{1}{2}} - R_{yy}^{-\frac{1}{2}} b a^T Q_{xx}^{-\frac{1}{2}})\}$$
$$= \det(Q_{xx}) \det\{I + (G - a_1 b_1^T)(G - a_1 b_1^T)^T\}.$$

This implies that $\frac{\partial F}{\partial a_1} = 0$ yields $(G - a_1 b_1^T) b_1 = 0$ and hence $G b_1 = a_1 b_1^T b_1$. Similarly, $\frac{\partial F}{\partial b_1} = 0$ yields $(G - a_1 b_1^T)^T a_1 = 0$ or $G^T a_1 = b_1 a_1^T a_1$. Theoretically, if G is known, then a_1, b_1 can be computed using an alternating power method [7] as follows:

Remark: Let $G_r = a_1 b_1^T$, then the above equations can be rewritten as

$$(G - G_r) G_r^T = 0$$
$$(G^T - G_r^T) G_r = 0 \tag{5}$$

Clearly, if $G = U_r \Sigma_r V_r + U_n \Sigma_n V_n$, $\Sigma_r > \Sigma_n$ is the SVD of G, then one can set $G_r = U_r \Sigma_r V_r$. Thus $G - G_r = U_n \Sigma_n V_n^T$. It is easy to verify that G_r is a solution of (5) and G_r is of rank r. Hence, $F_{opt} = \det(Q_{xx}) \det(I + G_n G_n^T) = \det(Q_{xx}) \det(I + \Sigma_n^2)$.

Algorithm 3:

Let $b_1(0)$ be an initial matrix of full rank r. For $k = 1, 2, \cdots$, let

$$a_1(k) = G b_1(k),$$
$$b_1(k+1) = G^T a_1(k). \tag{6}$$

After several steps, b_1 can be approximated as $b_1 \approx G^T a_1(K)(a_1(K)^T a_1(K))^{-1}$, and $a_1 \approx G b_1(K)(b_1(K)^T b_1(K))^{-1}$, where K is sufficiently large integer.

6 Simulations

In this section we present two sets of simulations that demonstrate the numerical performance of the proposed algorithms. Figure 1 shows the convergence for the full rank Wiener filter using exact and approximate line search when Algorithm 3 is applied. The exact line search is accomplished by applying Algorithm 1, while the inexact one is accomplished by applying Equation (1). The signals x and y are generated using the Matlab function rand. Clearly, convergence occurs within about 30 iterations. One stopping criterion would be that $||\alpha h|| \leq \epsilon$. Figure 2 is generated using Algorithm 2 and it shows the convergence for the full rank Wiener filter when the covariance matrices R_{xy} and R_{yy} are known in advance.

7 Conclusion

We have presented an online based MSE criterion for the design of full and reduced rank Wiener filters. The proposed methods are adaptive and can be used to order update reduced rank Wiener filter serially. Matrix calculus of determinants and traces are applied to derive and compute reduced rank Wiener filters, where we have shown that the proposed trace and determinantal MSE criterion for deriving reduced rank Wiener filters are equivalent.

References

[1] S. Haykin, Modern Filters, Macmillan, New York, 1989.

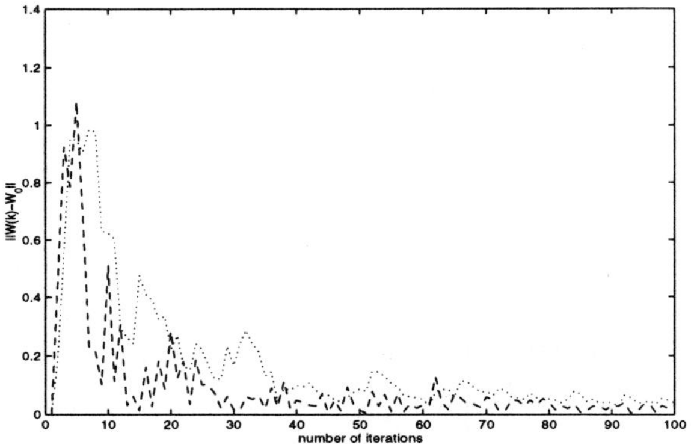

Figure 1: *Convergence behavior of Algorithm 1. The dotted line represents $||W(k) - W_0||$ versus the number of iterations. The dashed line is generated using Algorithm 1 but α is estimated using Equation (1). Here W_0 represents the exact Wiener filter.*

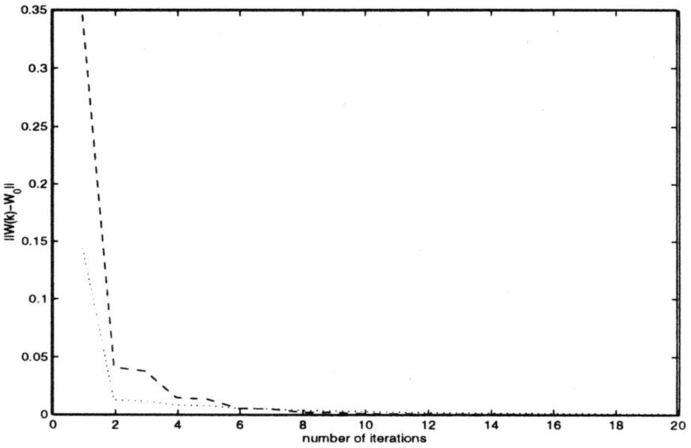

Figure 2: *Convergence behavior of Algorithm 2 in batch mode. The dotted line represents $||W(k) - W_0||$ versus the number of iterations. The dashed line is generated using Algorithm 2 but α is estimated using Equation (1). Here W_0 represents the exact Wiener filter.*

[2] S.M. Kay, Modern Spectral Estimation, Prentice-Hall, Englewood Cliffs, NJ, 1988.

[3] C.W. Therrien, Discrete Random Signals and Statistical Signal Processing, Prentice-Hall, Englewood Cliffs, NJ, 1992.

[4] Goldstein, J.S.; Reed, I.S.; Scharf, L.L., "A multistage representation of the Wiener filter based on orthogonal projections," IEEE Transactions on Information Theory, Volume: 44 Issue: 7 , Nov. 1998 Page(s): 2943-2959.

[5] L. L. Scharf, Statistical Signal Processing, Detection, Estimation, and Time Series Analysis, Addison-Wesley, 1991.

[6] L. L. Scharf. "The SVD and reduced-rank signal processing", Signal Processing, Vol. 25, pp. 113-133, 1991.

[7] Y. Hua, and M. Nikpour, "Computing the Reduced Rank Wiener Filter by IQMD", IEEE Signal Processing Letters, Vol. 6, No. 9, pp. 240-242, September 1999.

A 2048 COMPLEX POINT FFT PROCESSOR USING A NOVEL DATA SCALING APPROACH

Thomas Lenart and Viktor Öwall

CCCD, Department of Electroscience, Lund University
Box 118, SE-221 00 Lund, Sweden
Phone: +46 (0)46 222 91 05
Email: {thomas.lenart, viktor.owall}@es.lth.se

ABSTRACT

In this paper, a novel data scaling method for pipelined FFT processors is proposed. By using data scaling, the FFT processor can operate on a wide range of input signals without performance loss. Compared to existing block scaling methods, like implementations of Convergent Block Floating Point (CBFP), the memory requirements can be reduced while preserving the SNR. The FFT processor has been synthesized and sent for fabrication in a 0.35μm standard CMOS technology. In netlist simulations, the FFT processor is capable of calculating a 2048 complex point FFT or IFFT in 27μs with a maximum clock frequency of 76MHz.

1. INTRODUCTION

The Fast Fourier Transform (FFT) has a wide range of applications in digital signal processing [1]. For instance, FFT's are used in communication systems like DAB, DVB and IEEE 802.11a. The FFT is also used for analysing sound, images and video when removing undesired or perceptual irrelevant information, in radar applications as well as in different instrumentations. The N-point Discrete Fourier Transform (DFT) is defined as

$$X(n) = \sum_{k=0}^{N-1} x(k) W_N^{nk} \qquad n = 0,1\ldots,N-1 \qquad (1)$$

where $W_N = e^{-j(2\pi/N)}$. The direct implementation of the DFT have a complexity of $O(N^2)$. Using the FFT, the complexity can be reduced to $O(N \cdot \log_2(N))$. The FFT is also more suitable for hardware implementation due to the physical regularity of the algorithm, but requires memory buffers to store parts of the sequence during calculation. The memory requirements are a crucial parameter since memories are rather expensive in both terms of area and power consumption. Reducing the memory requirement will therefore significantly affect the total size of the design. In this paper, the implementation of a high performance pipelined FFT processor with low memory requirements using a novel scaling approach is presented.

2. FFT ARCHITECTURES

There are many different ways to implement an FFT processor. The computations can be done in a number of iterations by time multiplexing a single memory and arithmetic unit, Fig. 1.a, or by using a pipelined architecture, Fig. 1.b. A pipelined radix-2 architecture requires $\log_2(N)$ arithmetic units, one for each butterfly stage, and is therefore more area expensive than using one single radix-2 unit. In return, the calculations will be $\log_2(N)$ times faster when using pipelining.

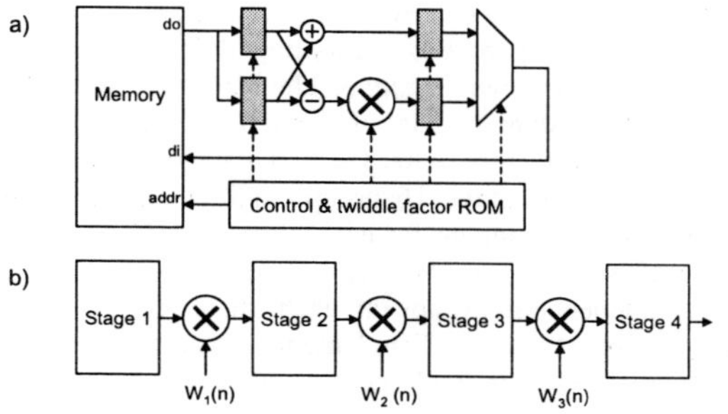

Fig. 1. a) Time multiplexed FFT processor using a single memory and butterfly unit. b) Pipelined FFT processor.

The problem with a fixed point FFT is to maintain the accuracy and preserve the dynamic range at the same time. One way to achieve a high signal-to-noise ratio is to increase the internal wordlength for every stage in the pipelined FFT (variable datapath), i.e. the wordlength will be wider at the output than at the input. Another way to improve the signal-to-noise ratio without increasing the internal wordlength is to use data scaling. One example of data scaling is *block floating point* that uses exponents, or scaling factors, for internal representation to improve the SNR. The exponents are usually shared between the real and imaginary part of a complex value, or even shared among a set of complex values [2-4], unlike normal floating point representation. Fig. 2 shows the almost constant internal wordlength when using data scaling and also the increasing internal wordlength when using a variable datapath. The width of the internal scale factor representation can be optimised for each stage. In this paper, focus will be on different scaling approaches.

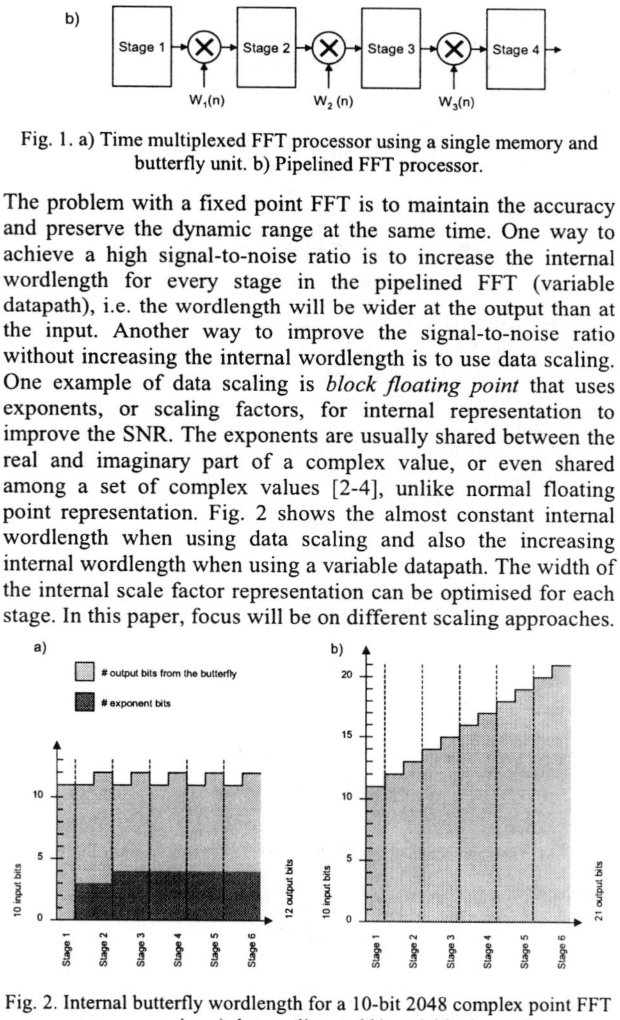

Fig. 2. Internal butterfly wordlength for a 10-bit 2048 complex point FFT processor using a) data scaling and b) variable datapath.

0-7803-7761-3/03/$17.00 ©2003 IEEE

2.1 Block floating point

Time multiplexed FFT processors can use a data scaling method called Block Floating Point (BFP). After calculating all outputs from stage N, the largest output value can be detected and the intermediate result is scaled to improve the precision. When using BFP, all values share one single scale factor. BFP requires that the scale factor for stage N can be determined before starting the calculations of stage N+1. This approach cannot be applied to pipelined architectures due to the continuous dataflow.

2.2 Convergent block floating point

When a pipelined architecture is used, it is not efficient to wait until stage N has finished to determine the scaling factor. Instead a method called Convergent Block Floating Point (CBFP) has been proposed [2-4] as shown in Fig. 3. The basic idea is that the output from a radix-4 stage is a set of 4 independent groups that can use different scale factors. After the first stage there will be 4 groups, after the second stage 16 groups and so on. This will converge towards one exponent for each output sample from the FFT. The same scheme can be applied for a radix-2 stage, generating 2 independent groups at each stage. If the initial butterfly is of radix-2 type, most implementations omit the CBFP logic in the first stage due to the large memory overhead.

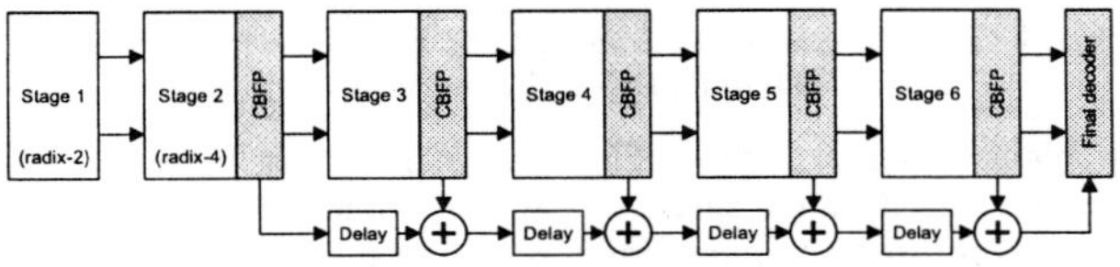

Fig. 3. A pipelined 2048-points FFT using CBFP.

The drawback with CBFP is that it requires a lot of memory. For the 2048 point pipelined FFT in Fig. 3, the input values are split into 2 groups of size 1024 in the initial radix-2 stage. The second stage produces 4 new groups containing 256 values each. These 256 values from the complex multiplier have to be stored in a buffer, as illustrated in Fig. 4, before the scaling factor for the group can be determined. Furthermore, it has to be saved in full precision because normalization cannot be done until the scaling factor is known. The length of the delay buffer after stage k is

$$l = 4^{\lceil \log_4(N) \rceil - k} \qquad (2)$$

Another drawback is that the latency will increase, caused by the delay in the intermediate buffer.

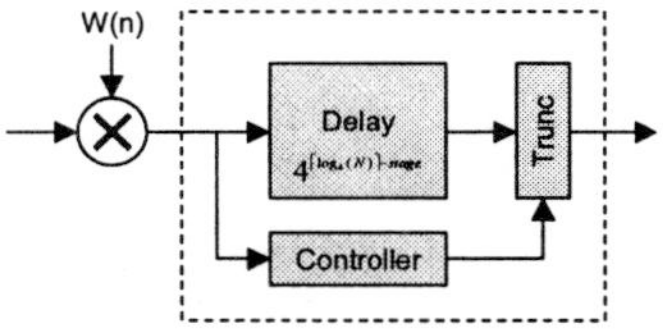

Fig. 4. CBFP logic between the consecutive stages.

2.3 Presented approach

Current CBFP techniques require a buffer to store N/4 outputs from a radix-4 stage with N inputs. The proposed method in this paper is to remove this buffer and rescale the data on the fly. After the complex multiplier, the result is normalized without delay and sent to the next butterfly stage. Therefore, each butterfly must be able to rescale one of the input values, if they are represented with different exponents. Starting from the second stage, the wordlength in the delay feedback has to be widened to hold both the complex value and the scale factor. This approach towards a hybrid floating-point processor still has much in common with the block scaling technique since only negative scale factors are allowed, leading to a reduced exponent representation and simple scaling logic. Furthermore, there is no input exponent, which reduces the memory overhead in the first (and largest) memory stage.

For further memory reduction, recall that each block in CBFP is represented with the same exponent. Accordingly, if the exponent in a block is only allowed to increase, the number of possible changes is limited. This will lead to minor performance degradation, but it will be shown that this limited number of changes can be stored in a special way to lower the memory requirements.

3. IMPLEMENTATION

The presented FFT processor is based on the radix-2^2 decimation-in-frequency algorithm [5]. The radix-2^2 algorithm is well suited for the presented approach because of the simple single-path delay feedback structure. Fig. 5 shows the modified FFT radix-2^2 processor, based on three different kinds of building blocks. The IBF unit is a normal butterfly that is only used in the first stage. The MUL unit contains a complex multiplier and a normalizing unit with proper rounding. The output value, $z(x)$, from a MUL block is represented by a complex value, $a(x)+jb(x)$, and a positive scale factor $s(x)$ as

$$z(x) = 2^{-s(x)}\left(a(x) + jb(x)\right) \qquad (3)$$

The MBF unit is a radix-2^2 butterfly with a wider delay feedback to hold scale factors. To avoid problems with data alignment, equalizing units are used in conjunction with the butterfly units. Unlike CBFP, the large intermediate buffers between the FFT stages are not needed and replaced with the extra logic required to implement the equalizing units.

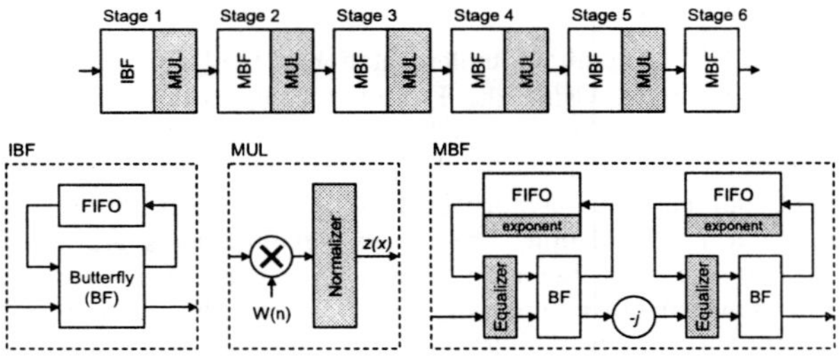

Fig. 5. The 2048 complex point FFT mainly consists of three different kinds of building blocks.

Compared to a floating-point implementation, no changes to the butterfly and complex multiplier units are required. The equalizer aligns the information to the butterfly, while the input to the complex multiplier is aligned by default.

3.1 Butterfly and complex multiplier

There are two butterfly stages in each radix-2^2 stage, calculating the sum and the difference between the input values and the output from the single-path delay feedback. When scale factors are used, it must be possible to align the inputs if they do not share the same exponent. An equalizer unit, only activated when the butterfly is not filling or draining the delay feedback, performs the alignment of input values to the butterfly stage. The equalizer compares the exponents of the two inputs to detect if the values are aligned or not. If there is a difference, the smallest input value is right shifted with the same number of bits as the difference between the two exponents. The aligned values are propagated to the butterfly unit. The output from the complex multipliers is normalized and sent to the next FFT stage. The normalizing unit is based on a number of compare and shift units connected in series. At the same time as the value is shifted by the normalizing unit, the exponent is incremented accordingly.

3.2 Delay feedback

The reordering method in a radix-2^2 FFT is the single-path delay feedback [5]. For shorter delays, several flip-flops can be connected in series. However, when the length of the FIFO increases, this approach is no longer area efficient. The large flip-flop cells can then be replaced with a single or dual port memory together with logic for control and address generation. Three different approaches to memory-based delays has been synthesized to a 0.35μm cell library for the Alcatel Microelectronics CMOS process and compared. The most straightforward approach is to use a dual port memory connected to an address generator. This allows simultaneous reads and writes to any memory location. One drawback with dual port memories is the required area, which is considerably larger than for single port memories. It is also possible that dual port memories are not always available, which makes the implementation process dependent.

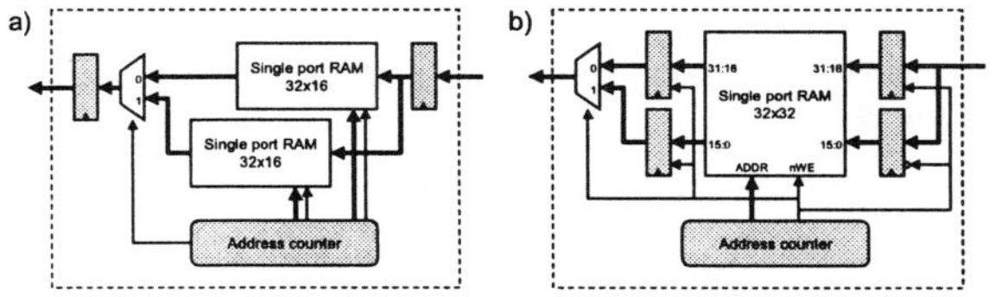

Fig. 6. a) Two single port memories. b) Single port memory with double wordlength.

One solution is to use two single port memories, alternating between reading and writing every clock cycle, Fig. 6.a. The drawback is the duplication of the address logic when using two memories instead of one. A third approach uses only one single port memory but with double wordlength, Fig. 6.b. This is possible, due to the consecutive addressing scheme used in the delay feedback. In addition to removing the duplicated address logic, the total number of memories for placement will be reduced. An area comparison of four different ways of building a delay feedback, or FIFO, is presented in Fig. 7.

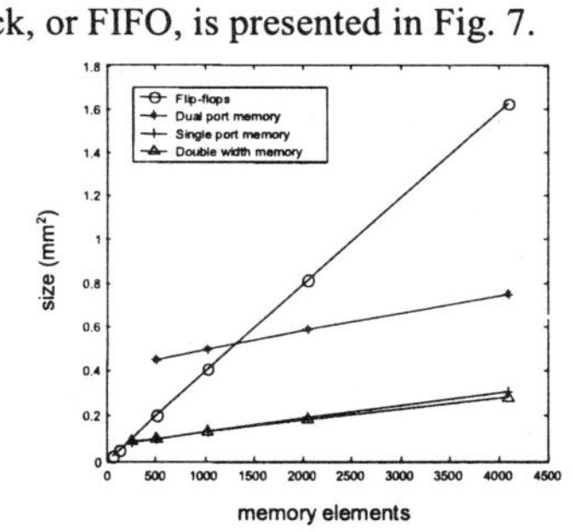

Fig. 7. Area requirements for different FIFO implementations.

Considerations have been taken to compare the actual size on the chip by adding an additional 50μm space for the power ring around the memories. When less than approximately 250 bits are required, flip-flops is the preferred method. In the presented design, single port memories with double wordlength have been used for the largest delay feedbacks. Flip-flops have only been used for the four shortest delays.

In order to remove the intermediate buffers between the stages, the exponents have to be saved in the delay feedback. When using floating point, two separate exponents are required for the real and imaginary parts. However, both CBFP and the presented approach represent a complex value with one shared exponent. Allocating space for only one exponent for each complex value reduces the memory requirements for the delay feedbacks.

3.3 Further memory reductions

The memory requirements can be reduced even further, by analysing the characteristics of block scaling. In the CBFP approach, the output from a radix-4 stage is a set of 4 independent groups, where each group share a single scale factor. The same principle can be applied to the current approach by introducing a Restricted Delay Line (RDL).

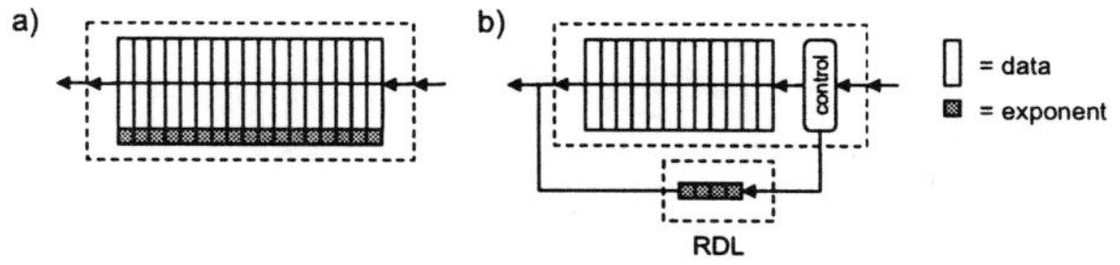

Fig. 8. a) FIFO with shared exponent. b) FIFO using RDL.

The basic idea is to only update the RDL with important scale factors, with a small performance penalty, instead of storing all scale factors in the delay feedback. A scale factor can be considered important if the current value is larger than the previous value and thus has to be stored, otherwise data is shifted corresponding to the scale factor most recently saved in the RDL. The maximum number of changes for a delay feedback of size N using a scale factor representation of K bits is 2^K, which is the minimal length of the RDL. The RDL is restarted periodically, initialised by storing the current scale factor. The stored value appears on the output after N cycles and is used for consecutive output values until a new scale factor is present. In addition to the performance penalty, the drawback with using a RDL is the additional logic required. Therefore it is only useful when building large sized FFT's. The RDL will however affect the size in a greater extent if the FFT processor is implemented so that it supports a scale factor input. In this case, the largest delay feedback can take advantage of the RDL, instead of storing scale factors the traditional way. For the current implementation, the largest delay feedback could be reduced by 15%, assuming a 4-bit input scale factor.

4. COMPLEXITY ANALYSIS

In this section, the presented approach will be compared with other implementations in terms of memory requirements, chip area and accuracy. An early version of the design has been presented at NORCHIP [6], and sent for fabrication in a standard 0.35μm CMOS process. The presented FFT processor is capable of performing a 2048 complex point FFT or IFFT in approximately 27μs, running at a maximum clock frequency of 76MHz. The FFT sent for fabrication is limited to 50MHz.

4.1 Memory and area requirements

For a 2048 complex point FFT processor, the memory occupies approximately 55% of the chip area, as can be seen in Diagram 1. Reducing the memory requirement will therefore significantly affect the total size of the design. A comparison between an FFT using variable data path, CBFP and the presented approach, all with 10 bit inputs has been made. Fig. 9 shows the number of memory elements required and it can be seen that the CBFP implementation requires substantially more memory than the other two, despite that CBFP logic has not been used in the initial stage. Fig. 10 shows the size of the total design, which follows the same trend. According to Diagram 1, the logic overhead in the presented approach, i.e. the equalizing units, does not have a large impact on the total chip area. Consequently, the area expensive intermediate buffers that are used in CBFP can be replaced with logic that requires less space.

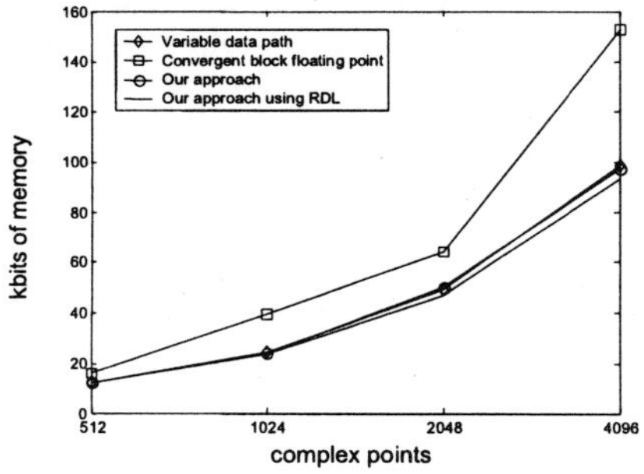

Fig. 9. Total number of memory elements required in the delay feedbacks for the different implementations.

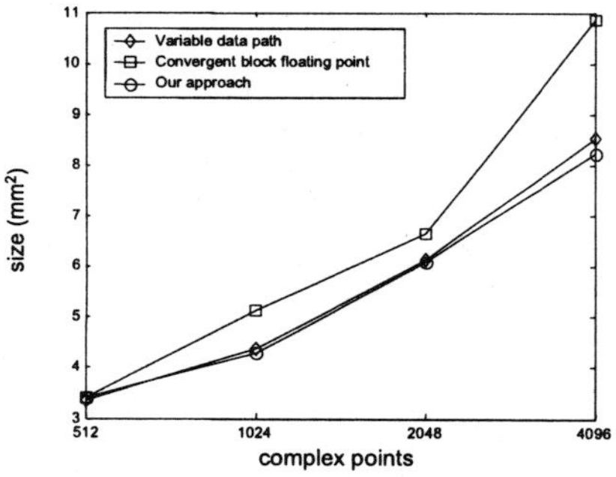

Fig. 10. Total chip size for the different implementations.

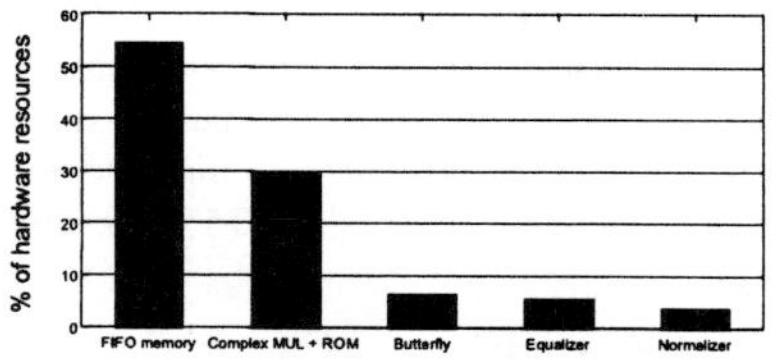

Diagram 1. Allocation of hardware resources.

4.2 Precision

When data scaling is used, the FFT processor can operate on a wide range of input signals. Even when the input signal has low amplitude, the signal will be scaled to full amplitude in the first stage, preserving the accuracy. The architectures described in section 4.1 have been simulated with various input signals including random noise, sine waves, OFDM and step response, all resulting in a higher SNR than for the CBFP. Fig. 11 shows the SNR for input signals starting with full dynamic range and then with down scaled input signal. The FFT with variable datapath produces a higher SNR when utilizing the full dynamic range. However, for arbitrary input signals, scaling is preferred.

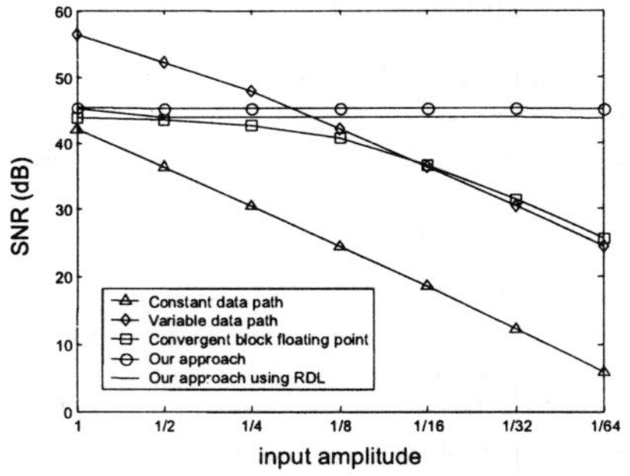

Fig. 11. Signal to noise ratio versus the amplitude of the input signal for different FFT implementations, all using 10-bit input wordlength.

As expected, the presented approach can maintain a high SNR even for down scaled input values. One of the reasons that CBFP is not capable of keeping a constant SNR is that there is usually no CBFP logic in the first radix-2 stage as shown in Fig. 3. The cost of adding CBFP logic to the first radix-2 stage is very high due to the large amount of intermediate buffer memory required. In the presented approach, the scaling part takes place in the subsequent pipeline stage. Hence, scaling is applied in all stages. If CBFP logic were added to the first stage, the corresponding SNR curve in Fig. 11 would remain constant as in our approach, but the memory requirements will be even higher than what is shown in Fig. 9.

5. CONCLUSION

An FFT processor using a novel data scaling approach that can operate on a wide range of input signals, keeping the SNR at a constant level, has been presented. Compared to block scaling approaches, such as CBFP, the proposed design requires significantly smaller chip area due to the reduced memory requirements. At the same time it is capable of producing a higher SNR since scaling is applied in all pipeline stages. The FFT processor has been sent for fabrication in a 0.35µm CMOS process.

6. REFERENCES

[1] E. Oran Brigham, *The fast Fourier transform and its applications*, Prentice-Hall, 1988.

[2] Se Ho Park *et.al.* "A 2048 complex point FFT architecture for digital audio broadcasting system", In *Proc. ISCAS* 2000.

[3] Se Ho Park *et.al.* "Sequential design of a 8192 complex point FFT in OFDM receiver", In *Proc. AP-ASIC*, 1999.

[4] E. Bidet, D. Castelain, C. Joanblanq and P. Senn, "A Fast Single-Chip Implementation of 8192 Complex Point FFT", *IEEE J. of Solid-State Circuits*, Vol. 30, NO. 3, 1995.

[5] Shousheng He, *Concurrent VLSI Architectures for DFT Computing and Algorithms for Multi-output Logic Decomposition*, PhD Thesis, Lund University, 1995.

[6] Thomas Lenart and Viktor Öwall, "A Pipelined FFT Processor using Data Scaling with Reduced Memory Requirements", In *Proc. NORCHIP*, 2002.

On the Relevancy of the Perfect Reconstruction Property when Minimizing the Mean Square Error in FIR MIMO Filter Systems

Are Hjørungnes
Signal Processing Laboratory
Helsinki University of Technology
P. O. Box 3000, FIN-02015 HUT, Finland
Tel. +358 9 451 5386, Fax +358 9 452 3614
Email: arehj@wooster.hut.fi

Paulo S. R. Diniz
Signal Processing Laboratory
Universidade Federal do Rio de Janeiro
P. O. Box 68504, CEP: 21945-970 RJ, Brazil
Tel. +55 21 2562 8211, Fax +55 21 2562 8205
Email: diniz@lps.ufrj.br

***Abstract*— Under the assumption that the transmitter FIR multiple-input multiple-output (MIMO) filter is given and FIR left invertible, conditions are derived for when the minimum minimum mean square error (MSE) FIR MIMO filter system does possess the perfect reconstruction (PR) or zero forcing (ZF) property. In most cases, examples using common models for both source coding and communication systems show that constraining the system to have the PR or ZF property is suboptimal in the minimum MSE sense.**

I. Introduction

For a given left invertible transmitter FIR MIMO filter, when is it optimal in the minimum MSE sense to use a system possessing the PR or ZF property? This is the main question answered in this article. The system treated can model for example filter banks used in source coding and communication systems that use linear FIR MIMO filters in both the transmitter and receiver.

In source coding systems using analysis and synthesis filter banks, the PR property is often enforced to the system [1], [2], [3], [4]. In communication systems using transmitter and receiver FIR MIMO systems, the filters can be designed under the so-called zero forcing (ZF) constraint [5], [6], [7], [8]. Here, the PR or ZF constraint is questioned, and it is shown under which conditions there is no loss in optimality, in the minimum MSE sense, by enforcing the PR or ZF constraint. It is shown that in almost all source coding and channel models considered, it is suboptimal to enforce the PR or ZF constraint in the system design.

The dimension N of the vectors in the original time series and the dimension M of the input time series to the channel may in general be different, and the system is assumed to be discrete in time. The transmitter and receiver are represented by polyphase matrices. In Figure 1, the $M \times N$ matrix $\boldsymbol{E}(z)$ and the $N \times M$ matrix $\boldsymbol{R}(z)$ are causal FIR MIMO filters of order m and l, respectively, which represent linear time invariant signal processing units, and they can be expressed as:

$$\boldsymbol{E}(z) = \sum_{k=0}^{m} \boldsymbol{e}(k) z^{-k} \quad \text{and} \quad \boldsymbol{R}(z) = \sum_{k=0}^{l} \boldsymbol{r}(k) z^{-k}. \tag{1}$$

This means that the matrices $\boldsymbol{e}(k)$ and $\boldsymbol{r}(k)$, have dimensions $M \times N$ and $N \times M$, respectively. It is assumed that all the vector time series are jointly wide sense stationary (WSS). The PR or ZF property is present if it is possible to recover a delayed version of the $N \times 1$ vector $\boldsymbol{x}(n)$ from $\hat{\boldsymbol{y}}(n)$, in the absence of noise, i.e., $\boldsymbol{q}(n) = \boldsymbol{0}$. This is only possible if $M \geq N$, because if $M < N$, then the matrix $\boldsymbol{E}(z)$ has normal rank [3] less than or equal to M, and then this matrix cannot be left invertible and PR is impossible. Therefore, it is assumed that $M \geq N$ in this article.

In [9], the transform and non-causal infinite order cases were considered for the case where $N = M$. This article is an extension of the results found in [9] to the FIR case and to the case where $M \neq N$. In [9], it was shown that the condition for PR being optimal in the non-causal infinite order case is:

$$E\left[\boldsymbol{q}(n)\hat{\boldsymbol{y}}^H(m)\right] = \boldsymbol{0}, \quad \forall n, m, \tag{2}$$

where the operator ^{H} means complex conjugated transposed, and the condition for the transform case is:

$$E\left[\boldsymbol{q}(n)\hat{\boldsymbol{y}}^H(n)\right] = \boldsymbol{0} \ \ \forall n. \tag{3}$$

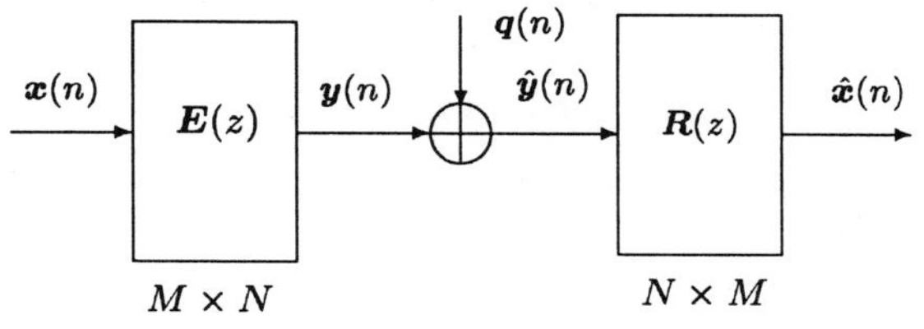

Fig. 1. System model.

It will be shown that, when $N = M$ and non-causal infinite order or transform systems are considered, the conditions proposed in the current article reduces to the results in Equations (2) or (3), respectively.

In [9], the transform and the non-causal infinite-order cases were treated for zero block delay through the MIMO filter system. The FIR case has to be treated in a different manner, because in [9], it was assumed that the receiver filter consists of the PR receiver filter followed by a Wiener filter. Then, conditions for when the Wiener polyphase matrix is equal to the identity matrix were given for the transform case and the non-causal infinite-order case. In both these cases, it can be assumed that the Wiener filter has the same order as the first part of the receiver filter, which is either infinite or zero. In the FIR case, this is not possible because, if the first part of the receiver filter is the FIR inverse of the transmitter filter, the following Wiener polyphase filter can only be a memoryless matrix if the order of the total receiver filter is to be kept constant. By using this method, the optimization does not include a search over all polyphase matrices of a given order. Therefore, a different method must be used for the FIR case.

The related problems for conditions when the PR solution exists are investigated in [10], [11], and the jointly minimum MSE transmitter and receiver FIR MIMO filters are found in [12].

The rest of this article is organized as follows: In Section II, some of the notation and definitions needed in the article is introduced and in Section III, the equation for the optimal FIR MIMO receiver filter is derived. The conditions for when the PR property is optimal are derived in Section IV. Section V gives examples showing that these conditions for PR being optimal are not in general satisfied for many

0-7803-7761-3/03/$17.00 ©2003 IEEE

source coding and communication systems. Finally, conclusions and discussions are given in Section VI.

II. Notation and Definitions

A common choice for the desired receiver output signal $\boldsymbol{d}(n)$ is the following:

$$\boldsymbol{d}(n) = \boldsymbol{x}(n - \delta_v), \tag{4}$$

where the integer δ_v denotes the *vector delay* through the overall system. In this article, the theory developed is valid for any choice of the desired vector signal $\boldsymbol{d}(n)$, but in the numerical examples presented, the choice in Equation (4) is made.

In this article, some special expansions of matrix polynomials will be needed, and they are defined next. A *row-expanded* matrix $\boldsymbol{R}_{-}$ is an $N \times (l+1)M$ matrix given by

$$\boldsymbol{R}_{-} = [\boldsymbol{r}(0)\ \ \boldsymbol{r}(1)\ \ \cdots\ \ \boldsymbol{r}(l)]. \tag{5}$$

The *row-diagonal-expanded* matrix $\boldsymbol{E}_{\urcorner}^{(l)}$ is an $(l+1)M \times (m+l+1)N$ matrix given by:

$$\boldsymbol{E}_{\urcorner}^{(l)} = \begin{bmatrix} \boldsymbol{e}(0) & \boldsymbol{e}(1) & \cdots & \boldsymbol{e}(m) & \cdots & \boldsymbol{0} \\ \vdots & \ddots & \ddots & \ddots & \ddots & \vdots \\ \boldsymbol{0} & \cdots & \boldsymbol{e}(0) & \boldsymbol{e}(1) & \cdots & \boldsymbol{e}(m) \end{bmatrix}. \tag{6}$$

Let ν be a positive integer. The column-expansion of a vector time series $\hat{\boldsymbol{y}}(n)$ of dimension $(\nu+1)N \times 1$ is defined as:

$$\hat{\boldsymbol{y}}(n)_{|}^{(\nu)} = \begin{bmatrix} \hat{\boldsymbol{y}}(n) \\ \hat{\boldsymbol{y}}(n-1) \\ \vdots \\ \hat{\boldsymbol{y}}(n-\nu) \end{bmatrix}. \tag{7}$$

The column-expansion of other vector time series are defined in the same manner as shown by Equation (7).

The autocorrelation matrix of dimension $(\nu+1)N \times (\nu+1)N$ of the vector $\hat{\boldsymbol{y}}(n)_{|}^{(\nu)}$ is defined as:

$$\boldsymbol{\Phi}_{\hat{\boldsymbol{y}}}^{(\nu,N)} = E\left[\hat{\boldsymbol{y}}(n)_{|}^{(\nu)} \left(\hat{\boldsymbol{y}}(n)_{|}^{(\nu)}\right)^{H}\right]. \tag{8}$$

For other column-expanded vector time series the autocorrelation matrices are defined in an analog fashion as shown in Equation (8).

Let the $N \times (l+1)M$ matrix $\boldsymbol{\phi}_{\boldsymbol{d},\hat{\boldsymbol{y}}}^{(l,N,M)}$ be defined as

$$\boldsymbol{\phi}_{\boldsymbol{d},\hat{\boldsymbol{y}}}^{(l,N,M)} = E\left[\boldsymbol{d}(n) \left(\hat{\boldsymbol{y}}(n)_{|}^{(l)}\right)^{H}\right]. \tag{9}$$

III. FIR MIMO Wiener Receiver Filter

The FIR MIMO Wiener receiver filter is used in the derivation of the conditions for optimality of PR in Section IV, and it is derived in this section.

The following relationship between the input and the output of the causal FIR MIMO receiver filter $\boldsymbol{R}(z)$ is valid:

$$\hat{\boldsymbol{x}}(n) = \sum_{k=0}^{l} \boldsymbol{r}(k)\hat{\boldsymbol{y}}(n-k), \tag{10}$$

where the order l of the FIR MIMO receiver filter is known.

By means of the orthogonality principle [13], [14] for vector time series, the FIR MIMO Wiener receiver filter will now be derived. The error vector for the FIR MIMO filter system is given by $\hat{\boldsymbol{x}}(n) - \boldsymbol{d}(n)$. According to the orthogonality principle, the error vector has to be orthogonal to all the available observations at the input of the FIR MIMO receiver filter. This can be expressed mathematically as

$$E\left[\left(\hat{\boldsymbol{x}}(n) - \boldsymbol{d}(n)\right)\hat{\boldsymbol{y}}^{H}(n-p)\right] = \boldsymbol{0}, \tag{11}$$

for all $p \in \{0, 1, \ldots, l\}$. By using the result of Equation (10), Equation (11) can be rewritten as

$$\sum_{k=0}^{l} \boldsymbol{r}(k)\boldsymbol{\kappa}_{\hat{\boldsymbol{y}}}(p-k) = E\left[\boldsymbol{d}(n)\hat{\boldsymbol{y}}^{H}(n-p)\right], \tag{12}$$

for all $p \in \{0, 1, \ldots, l\}$, where the $M \times M$ matrix $\boldsymbol{\kappa}_{\hat{\boldsymbol{y}}}(n)$ is defined as $\boldsymbol{\kappa}_{\hat{\boldsymbol{y}}}(n) = E\left[\hat{\boldsymbol{y}}(n+k)\hat{\boldsymbol{y}}^{H}(k)\right]$. By putting the $l+1$ matrix equations in Equation (12) together, it is possible to rewrite these equations as:

$$\boldsymbol{R}_{-} = \boldsymbol{\phi}_{\boldsymbol{d},\hat{\boldsymbol{y}}}^{(l,N,M)} \left(\boldsymbol{\Phi}_{\hat{\boldsymbol{y}}}^{(l,M)}\right)^{-1}, \tag{13}$$

where the correlation matrices defined through Equations (8) and (9) are used. When the optimal FIR MIMO receiver filter is expressed as in Equation (13), it has a form related to the Wiener receiver filter in the non-causal infinite-order and the transform cases, see [9] for details. If any of the coefficients of the matrix $\boldsymbol{R}_{-}$ are set to zero, a procedure for optimizing the remaining coefficients was developed in [15].

IV. Conditions for Optimality of PR or ZF

In this section, the conditions for when PR or ZF FIR MIMO filter are optimal, for a given left invertible FIR MIMO transmitter polyphase filter, are derived by means of Wiener filter theory.

Assume that the transmitter FIR MIMO filter $\boldsymbol{E}(z)$ is FIR left invertible, and let $\hat{\boldsymbol{R}}(z)$ be the FIR MIMO receiver filter for which the filter system possesses the PR property, that is:

$$\hat{\boldsymbol{R}}_{-}\boldsymbol{E}_{\urcorner}^{(l)} = [\boldsymbol{0}\ \cdots\ \boldsymbol{0}\ \boldsymbol{I}\ \boldsymbol{0}\ \cdots\ \boldsymbol{0}], \tag{14}$$

where the right hand side of the equation is an $N \times (m+l+1)N$ matrix, and where the non-zero element in the first row of the $N \times N$ identity matrix $\boldsymbol{I}$ is placed in column number $N\delta_v$. The numbering of the columns starts with 0.

The output of the FIR MIMO receiver filter is equal to $\hat{\boldsymbol{x}}(n) = \boldsymbol{d}(n) + \boldsymbol{u}(n)$, where $\boldsymbol{d}(n)$ is the desired output signal and $\boldsymbol{u}(n)$ is the perturbation output vector signal. In general the following is valid:

$$\boldsymbol{d}(n) = \boldsymbol{R}_{-}\hat{\boldsymbol{y}}(n)_{|}^{(l)} - \boldsymbol{u}(n). \tag{15}$$

The optimal FIR MIMO row-expanded polyphase matrix $\boldsymbol{R}_{-}$ of dimension $N \times (l+1)M$ can be expressed as

$$\begin{aligned} \boldsymbol{R}_{-} &= \boldsymbol{\phi}_{\boldsymbol{d},\hat{\boldsymbol{y}}}^{(l,N,M)} \left(\boldsymbol{\Phi}_{\hat{\boldsymbol{y}}}^{(l,N)}\right)^{-1} \\ &= E\left[\boldsymbol{d}(n)\left(\hat{\boldsymbol{y}}(n)_{|}^{(l)}\right)^{H}\right]\left(\boldsymbol{\Phi}_{\hat{\boldsymbol{y}}}^{(l,N)}\right)^{-1} \\ &= E\left[\left(\boldsymbol{R}_{-}\hat{\boldsymbol{y}}(n)_{|}^{(l)} - \boldsymbol{u}(n)\right)\left(\hat{\boldsymbol{y}}(n)_{|}^{(l)}\right)^{H}\right]\left(\boldsymbol{\Phi}_{\hat{\boldsymbol{y}}}^{(l,N)}\right)^{-1} \\ &= \boldsymbol{R}_{-}\boldsymbol{\Phi}_{\hat{\boldsymbol{y}}}^{(l,N)}\left(\boldsymbol{\Phi}_{\hat{\boldsymbol{y}}}^{(l,N)}\right)^{-1} - E\left[\boldsymbol{u}(n)\left(\hat{\boldsymbol{y}}(n)_{|}^{(l)}\right)^{H}\right]\left(\boldsymbol{\Phi}_{\hat{\boldsymbol{y}}}^{(l,N)}\right)^{-1} \\ &= \boldsymbol{R}_{-} - E\left[\boldsymbol{u}(n)\left(\hat{\boldsymbol{y}}(n)_{|}^{(l)}\right)^{H}\right]\left(\boldsymbol{\Phi}_{\hat{\boldsymbol{y}}}^{(l,N)}\right)^{-1}, \end{aligned} \tag{16}$$

where the results from Equations (8), (9), and (15) have been used. Since the matrix $\left(\boldsymbol{\Phi}_{\hat{\boldsymbol{y}}}^{(l,N)}\right)^{-1}$ is assumed to be invertible, Equation (16) can be rewritten as:

$$E\left[\boldsymbol{u}(n)\left(\hat{\boldsymbol{y}}(n)_{|}^{(l)}\right)^{H}\right] = \boldsymbol{0}, \tag{17}$$

where the zero matrix on the right side has dimension $N \times (l+1)M$. Equation (17) is valid for all systems that use the optimal FIR MIMO receiver filter.

Assume now that the system possesses the PR or ZF property, then $\boldsymbol{u}(n) = \hat{\boldsymbol{R}}_{-}\boldsymbol{q}(n)_{|}^{(l)}$, which is the additive noise filtered through the PR FIR MIMO receiver filter $\hat{\boldsymbol{R}}(z)$, in Equation (14). If the expression of $\boldsymbol{u}(n)$ is inserted in Equation (17), it is seen that PR is optimal if, and only if, the following holds:

$$\hat{\boldsymbol{R}}_{-} E\left[\boldsymbol{q}(n)_{l}^{(l)}\left(\hat{\boldsymbol{y}}(n)_{l}^{(l)}\right)^{H}\right]=\mathbf{0}, \quad \forall n, \tag{18}$$

where the matrix $\hat{\boldsymbol{R}}_{-}$ has dimension $N \times (l+1)M$, the cross-correlation matrix $E\left[\boldsymbol{q}(n)_{l}^{(l)}\left(\hat{\boldsymbol{y}}(n)_{l}^{(l)}\right)^{H}\right]$ has dimension $(l+1)M \times (l+1)M$, and the zero matrix on the right hand side is of dimension $N \times (l+1)M$.

If the conditions found in Equation (18) are compared to the Equation (3) for the case when $N = M$, which gives the conditions for the transform case, it is seen that they are equivalent when $m = l = 0$ is used in the FIR case. In this case, the $N \times (l+1)M$ matrix $\hat{\boldsymbol{R}}_{-}$ has rank N and, therefore, it can be dropped from the equation when $N = M$.

The conditions in Equation (18) can be generalized to non-causal FIR filters where filter index goes from $-l$ up to l. By letting $l \to \infty$, these conditions can be rewritten by means of the z-transform to the equivalent from:

$$\hat{\boldsymbol{R}}(z)\boldsymbol{S}_{\boldsymbol{q},\hat{\boldsymbol{y}}}(z)=\mathbf{0}, \tag{19}$$

where the $M \times M$ cross power spectral density matrix $\boldsymbol{S}_{\boldsymbol{q},\hat{\boldsymbol{y}}}(z)$ is the z-transform of $E\left[\boldsymbol{q}(k+n)\hat{\boldsymbol{y}}^{H}(n)\right]$. Since the normal rank of the $N \times M$ matrix $\hat{\boldsymbol{R}}(z)$ is N, this matrix can be dropped from this equation in the case when $N = M$. This means that in this case, the conditions for the FIR case agree with the non-causal infinite order case, see Equation (2).

If the MIMO transmitter filter is left invertible and no noise is present, it follows from the conditions in Equation (18) that PR or ZF property is optimal. In this case, PR or ZF filter systems achieve zero MSE. Since MSE is always non-negative, this is optimal.

Given the conditions for PR being optimal when $N = M$ in the non-causal infinite-order case, see Equation (2), and in the transform case, see Equation (3), it is intuitively surprising that the matrix $\hat{\boldsymbol{R}}_{-}$ is part of the conditions in the FIR PR case. It might be more intuitive to guess that the condition in the FIR PR case would be that the matrix $E\left[\boldsymbol{q}(n)_{l}^{(l)}\left(\hat{\boldsymbol{y}}(n)_{l}^{(l)}\right)^{H}\right]$ should be the zero matrix. However, the conditions in the FIR case are *not* that strict because, in the following example, it will be shown that the matrix $E\left[\boldsymbol{q}(n)_{l}^{(l)}\left(\hat{\boldsymbol{y}}(n)_{l}^{(l)}\right)^{H}\right]$ is non-zero, but the conditions in Equation (18) are satisfied.

Let $N = M = 2$, $m = l = \delta_v = 1$, and the matrix $\hat{\boldsymbol{R}}_{-}$ be given by

$$\hat{\boldsymbol{R}}_{-}=\begin{bmatrix} 0 & 0 & 1 & 0 \\ 0 & 1 & 0 & 0 \end{bmatrix}. \tag{20}$$

With this choice the following non-zero block Toeplitz matrix will satisfy the condition in Equation (18):

$$E\left[\boldsymbol{q}(n)_{l}^{(l)}\left(\hat{\boldsymbol{y}}(n)_{l}^{(l)}\right)^{H}\right]=\begin{bmatrix} 0 & 0 & x_2 & x_3 \\ 0 & 0 & 0 & 0 \\ 0 & 0 & 0 & 0 \\ x_0 & x_1 & 0 & 0 \end{bmatrix}, \tag{21}$$

which is different from the zero matrix, because x_i is an arbitrary number for all $i \in \{0,1,2,3\}$.

An alternative method is now given for showing that when $N = M$ and $l \geq 1$ the cross-correlation matrix in Equation (18) is different from the zero matrix. Since rank $\left(\hat{\boldsymbol{R}}_{-}\right) = N$, the dimension of the null-space of $\hat{\boldsymbol{R}}_{-}$ is equal to $(l+1)M - N$, which is greater than zero when $M = N$ and $l \geq 1$. Therefore, the null-space of the matrix $\hat{\boldsymbol{R}}_{-}$ is non-empty and the matrix $E\left[\boldsymbol{q}(n)_{l}^{(l)}\left(\hat{\boldsymbol{y}}(n)_{l}^{(l)}\right)^{H}\right]$ in Equation (18) can be chosen different from the zero matrix.

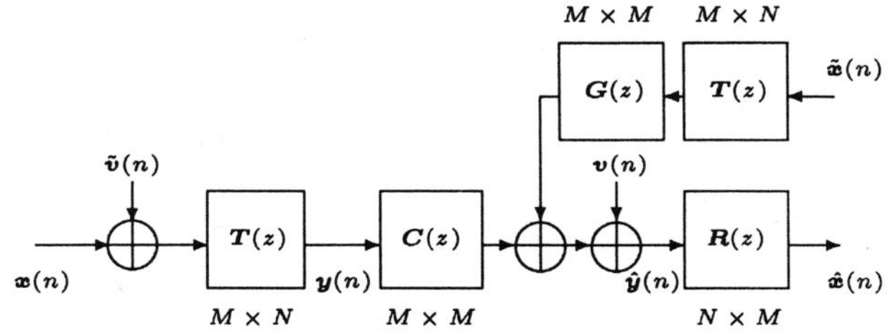

Fig. 2. FIR MIMO block system with near-end crosstalk and additive noise in the input signal and at the channel.

V. Examples

A. Source Coding Systems

1) High Rate Quantization Model: With the high rate quantization model [16], the additive quantization noise and the unquantized subband signals are assumed to be uncorrelated. In this case, the cross-correlation matrix in Equation (18) is given by $E\left[\boldsymbol{q}(n)_{l}^{(l)}\left(\hat{\boldsymbol{y}}(n)_{l}^{(l)}\right)^{H}\right] = \boldsymbol{\Phi}_{\boldsymbol{q}}^{(l,M)}$, which is an invertible matrix for finite rates. The condition in Equation (18) cannot be satisfied because the only solution is $\hat{\boldsymbol{R}}_{-} = \mathbf{0}$, which is certainly not a PR filter system. Therefore, with the high rate quantization model, it is never optimal to use a PR filters for finite rates. However, as the rate approach infinite, the quantization noise will approach zero, and the cross-correlation matrix $E\left[\boldsymbol{q}(n)_{l}^{(l)}\left(\hat{\boldsymbol{y}}(n)_{l}^{(l)}\right)^{H}\right]$ will approach zero as well. For infinitely high rates, the condition of Equation (18) is satisfied, and the PR filter system is optimal.

2) Centroid Representation in Scalar Quantizers: If centroid is used in scalar quantizers designed for each of the vector components of $\boldsymbol{y}(n)$, all the *diagonal* elements of the matrix $E\left[\boldsymbol{q}(n)_{l}^{(l)}\left(\hat{\boldsymbol{y}}(n)_{l}^{(l)}\right)^{H}\right]$ will be zero. PR filters are optimal if $N = M = 1$ and $m = l = 0$. If the signal dependent quantization noise model developed in [17] is used, the condition in Equation (18) is in general not satisfied for other cases than $N = M = 1$ and $m = l = 0$. The reason for this, is that for all other values for N, M, m, and l, the off-diagonal elements of the cross-correlation matrix $E\left[\boldsymbol{q}(n)_{l}^{(l)}\left(\hat{\boldsymbol{y}}(n)_{l}^{(l)}\right)^{H}\right]$ are non-zero, and examples can be constructed showing that in general the PR property will be suboptimal. Several examples were given in [17].

3) Centroid Representation in Vector Quantizers: As was observed in [9], if centroid is used as representation levels in vector quantizers that is optimized for the subband vector $\boldsymbol{y}(n)$, the relation in Equation (3) holds [18], which is the same condition for when PR is optimal using transforms. In the FIR case, where $l \geq 1$, the condition in Equation (18) is *not* in general satisfied when using centroids as representation levels in the vector quantizer.

B. Communication System

Consider the communication system shown in Figure 2, where all the input vector time series in the system are assumed to be uncorrelated with each other. The $M \times M$ polynomials $\boldsymbol{C}(z)$ of order q and $\boldsymbol{G}(z)$ of order p are modeling the desired and undesired near-end crosstalk communication channels, respectively. The transmitter FIR MIMO filter is assumed to be given by the $M \times N$ polynomial $\boldsymbol{T}(z)$ having order m.

Calculation the $(l+1)M \times (l+1)M$ cross-correlation matrix $E\left[\boldsymbol{q}(n)_{|}^{(l)} \left(\hat{\boldsymbol{y}}(n)_{|}^{(l)}\right)^H\right]$ that is part of Equation (18) results in:

$$E\left[\boldsymbol{q}(n)_{|}^{(l)} \left(\hat{\boldsymbol{y}}(n)_{|}^{(l)}\right)^H\right] = \boldsymbol{\Phi}_{\boldsymbol{v}}^{(l,M)} + \boldsymbol{C}_{\neg}^{(l)} \boldsymbol{T}_{\neg}^{(q+l)} \boldsymbol{\Phi}_{\tilde{\boldsymbol{v}}}^{(m+q+l,N)} \left(\boldsymbol{T}_{\neg}^{(q+l)}\right)^H \left(\boldsymbol{C}_{\neg}^{(l)}\right)^H + \boldsymbol{G}_{\neg}^{(l)} \boldsymbol{T}_{\neg}^{(p+l)} \boldsymbol{\Phi}_{\tilde{\boldsymbol{x}}}^{(m+p+l,N)} \left(\boldsymbol{T}_{\neg}^{(p+l)}\right)^H \left(\boldsymbol{G}_{\neg}^{(l)}\right)^H, \quad (22)$$

which is an invertible and positive definite matrix. If this matrix is inserted and eliminated from Equation (18), it is seen that $\hat{\boldsymbol{R}}_{-} = \boldsymbol{0}$, which shows that the system shown in Figure 2 is not optimal in the minimum MSE sense when the ZF property is enforced. The ZF property is enforced for simplifications of this model in, for example, the following references [6], [10], [11].

VI. Conclusions and Discussions

Conditions were derived for when the PR or ZF property is optimal in the minimum MSE sense when using FIR MIMO filter systems. Different source and communication systems were considered and it was shown that the PR or ZF condition is seldom satisfied. This means that if minimum MSE is the optimization criterion, it is almost always suboptimal to constrain the system to have the PR or ZF property. For low bit rates or poor quality channels, the transmitter FIR MIMO filter will not be left invertible, so it is not surprising that the PR or ZF property is a questionable property to impose to the system.

Figure 3 shows the bit error rate (BER) versus channel signal to noise ration (CSNR) performance of two transform systems using $m = l = \delta_v = 0$ without interblock interference proposed in [11] which minimize the MSE. The original time series is uncorrelated and produce only -1 or 1 with equal probability and the scalar channel noise $v(n)$ is complex white circulant Gaussian.

From the figure, it is seen that the system which possesses the PR property outperforms the non-PR system for good channel conditions. This shows that if another performance measure is used than the system is designed for, in this example BER, the PR property can be valuable. Whether PR is beneficial or does not depend on how different the nature of the two performance measures are. In [19], a system that jointly minimizes the symbol error rate with respect to the transmitter and receiver FIR MIMO filters under the average channel input power constraint is proposed.

In general, all conditions which are imposed to a system will reduce the system performance if the performance criterion is the same as the design criterion.

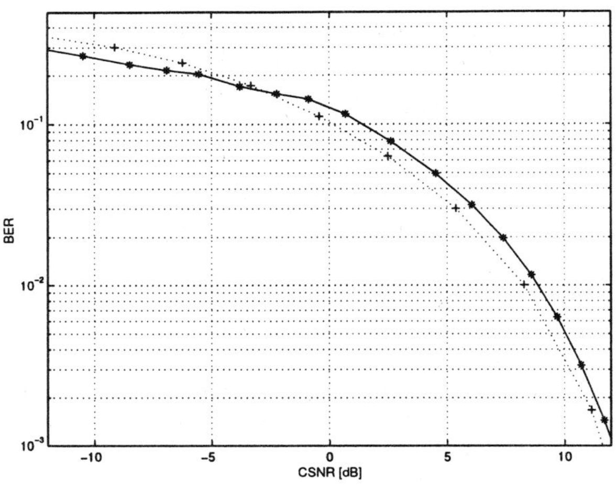

Fig. 3. BER versus CSNR for two transform systems $m = l = \delta_v = 0$ optimized for the same coefficients minimizing the MSE having PR: $-+-$ and non-PR: $-*-$. $N = 8$, $M = 11$. The scalar channel used have transfer function $\left(1 - 0.9z^{-1}\right)\left(1 - 0.7e^{j2\pi 0.256}z^{-1}\right)\left(1 - 0.4e^{j2\pi 0.141}z^{-1}\right)$, furthermore, $\boldsymbol{G}(z) = \boldsymbol{0}$ and $\tilde{\boldsymbol{v}}(n) = \boldsymbol{0}$.

References

[1] Didier Le Gall and Ali Tabatabai, "Subband coding of digital images using symmetric short kernel filters and arithmetic coding techniques," in *Proc. Int. Conf. on Acoustics, Speech, and Signal Proc.*, Apr. 1988, pp. 761–764.

[2] Marc Antonini, Michel Barlaud, Pierre Mathieu, and Ingrid Daubechies, "Image coding using wavelet transform," *IEEE Trans. Image Processing*, vol. 1, no. 2, pp. 205–220, Apr. 1992.

[3] P. P. Vaidyanathan, *Multirate Systems and Filter Banks*. Prentice Hall, Englewood Cliffs, New Jersey, USA, 1993.

[4] Henrique S. Malvar, "Lapped biorthogonal transforms for transform coding with reduced blocking and ringing artifacts.," in *Proc. Int. Conf. on Acoustics, Speech, and Signal Proc.*, Munich, Germany, Apr. 1997, vol. 3, pp. 2421–2424.

[5] A. N. Akansu, P. Duhamel, X. Lin, and M. de Courville, "Orthogonal transmultiplexers in communication: A review," *IEEE Trans. Signal Processing*, vol. 46, no. 4, pp. 979–995, Apr. 1998.

[6] Y.-P. Lin and S.-M. Phoong, "ISI-free FIR filterbank transceivers for frequency-selective channels," *IEEE Trans. Signal Processing*, vol. 49, no. 11, pp. 2648–2658, Nov. 2001.

[7] Y.-P. Lin and S.-M. Phoong, "Optimal ISI-free DMT transceivers for distorted channels with colored noise," *IEEE Trans. Signal Processing*, vol. 49, no. 11, pp. 2702–2712, Nov. 2001.

[8] Y.-P. Lin and S.-M. Phoong, "Minimum redundancy for ISI free FIR filterbank transceivers," *IEEE Trans. Signal Processing*, vol. 50, no. 4, pp. 842–853, Apr. 2002.

[9] P. P. Vaidyanathan and T. Chen, "Statistically optimal synthesis banks for subband coders," in *Proc. for Twenty-Eighth Asilomar Conference on Signals, Systems and Computers*, October–November 1994, pp. 986 –990.

[10] X.-G. Xia, "New precoding for intersymbol interference cancellation using nonmaximally decimated multirate filterbanks with ideal FIR equalizers," *IEEE Trans. Signal Processing*, vol. 45, no. 10, pp. 2431–2441, Oct. 1997.

[11] A. Scaglione, G. B. Giannakis, and S. Barbarossa, "Redundant filterbank precoders and equalizers Part I: Unification and optimal designs," *IEEE Trans. Signal Processing*, vol. 47, no. 7, pp. 1988–2006, July 1999.

[12] A. Hjørungnes, P. S. R. Diniz, and M. L. R. de Campos, "Jointly minimum MSE transmitter and receiver FIR MIMO filters in the presence of near-end crosstalk and additive noise," in *Proc. Int. Symp. on Circuits and Systems*, Bangkok, Thailand, May 2003, IEEE.

[13] L. L. Scharf, *Statistical Signal Processing: Detection, Estimation, and Time Series Analysis*. Addison-Wesley Publishing Company, 1990.

[14] C. W. Therrien, *Discrete Random Signals and Statistical Signal Processing*. Prentice – Hall Inc., Englewood Cliffs, New Jersey, USA, 1992.

[15] A. Hjørungnes, H. Coward, and T. A. Ramstad, "Minimum mean square error FIR filter banks with arbitrary filter lengths," in *Proc. Int. Conf. on Image Processing*, Kobe, Japan, Oct. 1999, vol. 1, pp. 619–623.

[16] N. S. Jayant and P. Noll, *Digital Coding of Waveforms, Principles and Applications to Speech and Video*. Prentice-Hall, Inc., Englewood Cliffs, New Jersey, USA, 1984.

[17] A. Hjørungnes, *Optimal Bit and Power Constrained Filter Banks*. Ph.D. thesis, Norwegian University of Science and Technology (NTNU), Trondheim, Norway, 2000.

[18] A. Gersho and R. M. Gray, *Vector Quantization and Signal Compression*. Kluwer Academic Publishers, Boston, MA, USA, 1992.

[19] A. Hjørungnes and P. S. R. Diniz, "Jointly minimum symbol error rate FIR MIMO transmitter and receiver filters for PAM signal vectors," in *Proc. Int. Symp. on Circuits and Systems*, Bangkok, Thailand, May 2003, IEEE.

Use of Noise Shaping for Rate Conversion in Oversampled Video Signals

I. Ryan, O. McCarthy

PEI

University of Limerick

ABSTRACT

This paper deals with the use of sigma delta techniques to allow a low resolution interpolator to achieve the same noise performance as a higher resolution interpolator when processing oversampled signals[1].

1. INTRODUCTION

In our previous paper [1], we dealt with the use of a sigma delta system for the reduction of the noise caused by rate conversion, when the rate conversion is performed on a signal while the signal has not yet been decimated and is thus still oversampled. This signal is the output of an oversampled video rate ADC. Rate conversion at video bandwidth is becoming necessary due to the new video standards [9,10,11,13]. This paper will show the results of imposing more realistic and stricter design requirements. In the original paper the requirements was that the noise in the inband part of the output spectrum caused by any inband signal in the input must be 60dB's lower than the input sine wave that caused it. The paper ignored the problem of output inband noise caused by out of band input signals. Out of band noise in the output signal can be ignored asthe rate converter is followed by a decimating filter, which eliminates this high frequency noise.

2. BASIC SUMMARY OF THE SYSTEM

Sigma delta modulators are becoming quite popular in the design of DAC's and ADC's. [4] They allow DAC's and ADC's with a low number of output levels to obtain the same performance as ones with a higher number of levels. The gain is not without a cost. They are required to operate at a higher rate than normal ADC's and DAC's. This means that the sample rate is not just limited by theNyquist requirement of being double the highest frequency in the input signal. In sigma delta systems, the sample rate is generally considerably higher than this minimum requirement. The gain for using sigma delta noise shaping improves with the number of doublings of the sampling frequency. This means that the higher the sampling rate the better. In some cases the sampling rate can be made so high that only one bit is required in the output. Also, they generate high frequency noise, this means thatsigma delta systems must be followed by low pass filters. ADC's and DAC's are basically quantisers and the basic point is that sigma delta systems allow low resolution quantisers to match the performance of high resolutionquantisers but produces high frequency noise.

The system that this paper deals with is a rate converter that takes as its input the output from a sigma delta ADC. The ADC operates at video bandwidth. This means that the bandwidth of the input signal is 6Mhz. Due to the high sample rate required for this signal, it is not possible to massively oversample it. The processing would have to happen in the high hundreds ofMhz. This would not be feasible for either the analogue or digital components. The signal from the ADC is 8 bits wide at 108Mhz, but it contains the accuracy of 12.5bits of resolution due to the sigma delta process. This signal then undergoes rate conversion and is decimated down to the 13.5 - 27Mhz range. The represents an output sample rate from the rate converter ranging from 108Mhz to 216Mhz. These data rates for the rate converter mean that the selected architecture must be capable of operating at high speed. The 60dB SNR requirement equates to the 10 bit accuracy buses in the blocks that follow the rate converter. The out of signal band noise in the input signal due to the sigma delta ADC means that there is out of band noise generated at 30 dB's below the input signal power in the input signal. The basic block diagram of the sigma delta interpolator is shown in fig 1. The sigma delta block is fed with the desired phase point. The accuracy of the accumulator that chooses this phase point would be higher than the interpolation accuracy of the interpolator. This interpolator would use one of the standard interpolation methods [2,3]. The sigma delta block then takes this high accuracy phase point and converts it to a lower resolution one. An added benefit of this design is that it allows the rate conversion ratio to be pretty arbitrary as the accuracy of the accumulator rather than the accuracy of the interpolator is the limiting factor in the accuracy of the conversion ratio. Other papers dealing withhigh resolution conversion ratios are given in the references [5,6,7,8,12]. The output of the sigma delta system is then fed into the interpolator. The timing jitter due to the low resolution of the interpolator is thus noise shaped. We showed in [1] that this results in the noise in the output signal due to the timing jitter also being noise shaped. However, unlike in a sigma delta ADC or DAC, increasing the order of the sigma delta past a certain point resulted in no further in band noise attenuation. It was

[1] The author wishes to acknowledge the support of Analog Devices, Limerick in this project

0-7803-7761-3/03/$17.00 ©2003 IEEE

also shown that if the interpolator did not interpolate and could only pass through input samples, then the required noise performance was not possible. This would be the equivalent of a 1-bit sigma delta modulator and given the relatively low oversampling ratio, that was not a surprising result.
We showed in [1] that the required interpolation resolution of the interpolator was 16 phase points per input sample. Also, the sigma delta block would have to perform 2nd order sigma delta noise shaping on the ramp function from the phase accumulator.

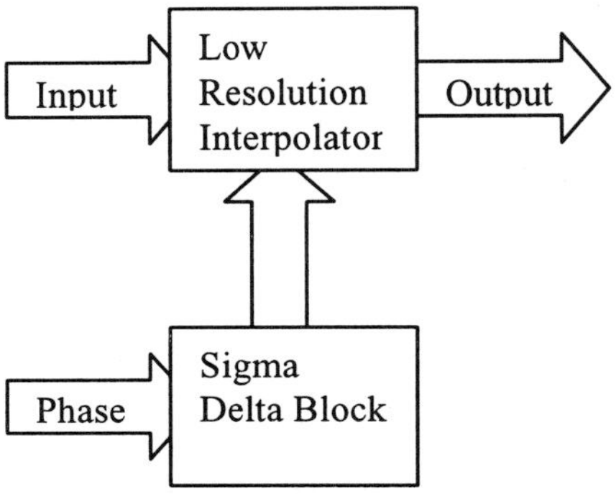

Figure 1. Block Diagram of Sigma Delta System

3. Results of Realistic Circuit Requirements

In [1], realistic circuit requirements were not imposed on the design. The two main requirements are that real circuits cannot operate at arbitrary high speeds and also that multipliers must be kept to a minimum area. Obviously, restriction of the size of the multipliers also helps the speed issue.

3.1 Output Data Rate

The rate converter has a range of allowable rate conversion ratios ranging from 1 to 2. This means that the maximum sample rate at the output of the interpolator is 216Mhz. This is the input into the following decimating filter. This is quite a high sample rate and a clock operating at this frequency is not available. The 108Mhz clock used for clocking the sigma delta ADC is the highest frequency that is available. However, even if a 216Mhz clock were available, operating a block at that frequency would be difficult. The clock period would be slightly less than 5ns and 1ns could easily be used up for clock jitter and flip flop set up and hold times. This leaves virtually no time for calculation.
Tests indicated that if the rate converter was made to operate using a range of rate conversion ratios from 0.5 to 1, meaning that the range of the output sample rates would be 54Mhz to 108Mhz and modifying the decimating filter to decimate by 4 instead of 8, then the required SNR could not be achieved. The results of the simulations are shown in Fig 2.
Due to the fact that we are testing a rate converter, it is not possible to merely give a single transfer function. Fig. 2 gives the superimposed results for a number of simulation runs with a range of rate conversion ratios covering the 0.5 to 1 rate conversion ratio range.
Another point to note is that the noise transfer function refers to the total signal band noise generated by a sine wave at the given frequency in the input. Noise generated in the out of signal band frequency range is not included.
The simulations showed a drop in the worst case noise generated by a signal band signal from 63dB's lower to only 57dB's lower. This is lower than the required 60dB's. The reason for this drop in performance is that the output frequency range is now only half as wide but the total noise power is the same resulting in twice as much power ending up in the signal band.

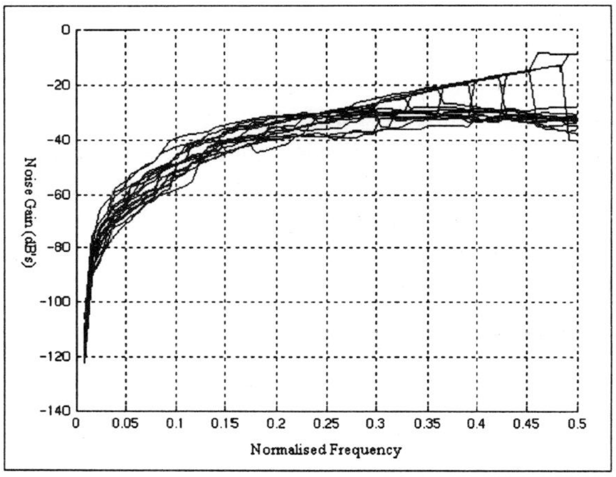

Figure 2. Noise Gain for 54Mhz – 108Mhz system

A second problem with the response is that the noise attenuation for out of band signals is in the worst case only 10dB's. In the previous case this would have been around 30dB's. This noise occurs when the output rate is near 54Mhz and it represents the aliasing of the high frequency component into signal band. There is noise in the input signal from the ADC at a power level 30dB's below the signal power due to the sigma delta action in the ADC. This noise is only decimated by a further 10dB's before it is aliased into the signal band. Thus noise is generated in the signal band of the output of the interpolator at a power which is only 40dB's lower than the input signal. This represents unacceptably low performance. Both the noise generated due to signal band signals in the input and also noise generated due to high frequency noise in the input are both unacceptably high.

The solution that will be used in implementing the block is to allow the interpolator to interpolate two output samples per clock cycle and also to design the decimator to be able to accept up to two samples per clock cycle. This will allow them to both to be operated using a 108Mhz clock while allowing a 216Mhz-output sample rate. The disadvantage is that some components in both blocks will have to be duplicated to allow this.

3.2 Noise Gain for High Frequency Input

The interpolator presented in [1] has a worst case noise gain of –30dB's. The desired noise gain for out of signal band signals would be in the region of –40dB's. This represents anti-aliasing requirements for the overall system. The filter coefficients used in the interpolator in previous sections were merely three 16 tap averaging filters convolved with each other. Since the interpolator interpolates to 16 phases, this filter would only require three registers even though it has 46 filter coefficients.

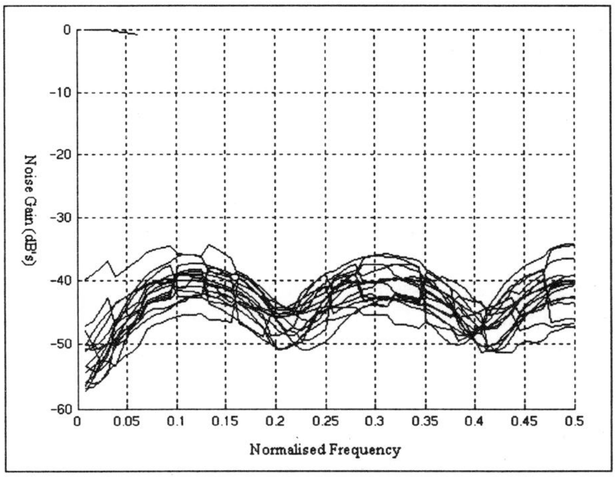

Figure 3. Noise Gain non Prioritised Coefficients

In order to improve the high frequency noise performance a Matlab function was created that would estimate the worst case noise performance for a given interpolating filter. Actual testing of a given filter would take too long as it would require testing over the full range of frequencies and also over the full range of rate conversion ratios. This was then used to select filter coefficients by usingMatlab to search for local maxima of this function. This would be run many times and the best maxima chosen. Fig. 3 shows the first attempt at using this system. This filter is an 80-tap filter, which would require 5 registers if implemented. The overall worst case noise has been improved from 30dB's to 35dB's. The same simulation run with the best 64-tap filter was a noise gain of 30dB's. However, in both cases the noise due to signal band signals has increased. Whereas before it was 63dB's below the signal that caused it, it is now only 35dB's lower. This result is due to the fact that the function does not prioritise the signal band noise performance.

The function was then modified so that the signal band frequencies were prioritised before then max function was called. Table 1 gives the priorities that result in the best performance. This resulted in the functionprioritising the signal band so that it had to be 60dB's better than the high frequency noise gain before the high frequency noise gain would be taken into account. What is surprising is that even though the high frequency band would appear to be ignored, these settings resulted in the best noise gain performance.

Fig 4 shows the results of the best 80-tap filter that uses those coefficients. This gives a signal band noise gain of –62.5dB's. The 64-tap version of this filter had a signal band noise gain of –61dB's. In both cases the required signal band noise performance is attained. The high frequency noise gain of the two filters is where the differences occur. As can be seen from the diagram, the worst case noise gain for this filter is better than 40dB's. However, for the 64-tap filter the worst case gain is around 37dB's. This means that the 80-tap filter will have to be used.

Frequency Band (MHz)	Priority (dB)
0 – 6.75	60
6.75 – 16	10
13.5 – 27	00

Table 1. Priority Values

3.3 Quantising Coefficients

The final problem that I will address here is the problem of multiplier size. An interpolator is normally implemented using a polyphase structure. In this structure each register tap is multiplied by a specific coefficient. The set of coefficients for each output phase point is stored in a ROM of some kind. It can be implemented in logic as opposed to being an actual ROM block. This requires that each register has a multiplier connected to it that can handle whatever coefficient the ROM requests for the current phase point. However, implementation of a 10-bit multiplier would be difficult at 108Mhz. The complexity of the multiplier can be reduced if the set of coefficients is converted to CSD values[14]. CSD is a number system where the digits can be [-1,0,1]. This allows then number of non-zero digits to be reduced, i.e. 7 can be represented as (0,1,1,1)=4+2+1 or (1,0,0,-1)=8-1. In a multiplier, each non-zero digit in the coefficient requires a shift of the input andan input into the adder tree. Thus, reducing the number of non-zero terms reduces the size of the multiplier and also the delay through the adder tree. What is important is the total number of non-zero terms in the coefficients for each phase point. This is because the total number of inputs into the adder tree must be as large as the number of non zero digits in the worst case phase point.

The matlab function that chooses the filter coefficients was modified so that it would only usequantised coefficients. Each phase point was evaluated and the total number of non-zero digits was calculated for each phase point. Then a penalty depending on the worst case phase point was imposed. This meant that the set of filter coefficients that was selected had a low worst case number of non-zero digits. However, it did not require that any specific coefficient to be restricted. This was important as previous tests showed that if each coefficient was restricted to 2 non-zero terms, then it was not possible to attain the required performance, 3 non-zero digits were required.

The result of this process was a set of coefficients where no phase required more than 13 non-zero digits, this represents 2.6 digits per register. It will be necessary to usemuxes to swap between the taps depending on which tap requires the non-zero digit, but this would have been necessary anyway as the coefficients change depending on phase point.

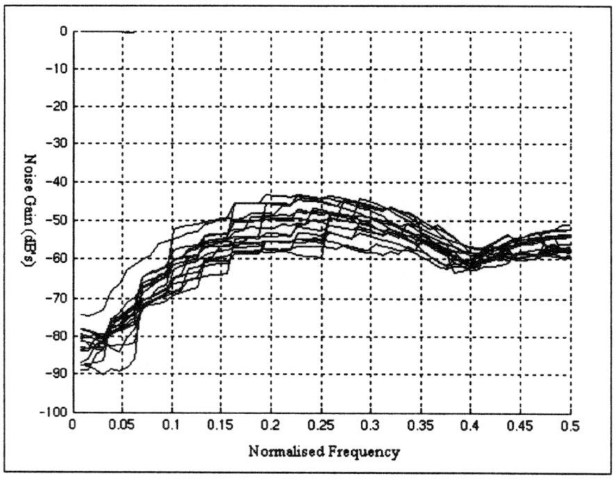

Figure 4. Noise Gain Prioritised Coefficients

4. Conclusion

The more detailed analysis shown in this paper shows that the proposed architecture can achieve improved noise performance by noise shaping the noise in ahigh speed rate converter. The two problems associated with operation at that speed can be solved. The requirement forhigh speed multipliers is solved by the use ofcsd filter coefficients. This also has the added benefit of reducing the overall area of block. The second method for allowing the required frequency to be achieved is to use two parallel interpolators to allow a data rate of 216Mhz while still operating at 108Mhz.

5. REFERENCES

[1] I. Ryan, O. McCarthy, "using noise shaping to allow arbitrary rate conversion in high speedoversampled signals", Irish Signals and Systems Conference 2001, pp. 213-217.

[2] R. E. Crochiere, L.R. Rabiner. "Multirate Digital Signal Processing", Prentice-Hall, Englewood Cliffs, NJ, 1983.

[3] P. P. Vaidyanathan. "Multirate systems and Filter Banks", Prentice-Hall, Englewood Cliffs, NJ, 1993.

[4] "Delta-Sigma Data Converters: Theory, Design, and Simulation" By Stephen R. Norsworthy, Richard Schreier and Gabor C. Temes, IEEE Press.

[5] R. Adams, T. Kwan. "A Stereo Asynchronous Digital Sample-Rate Converter for Digital Audio". IEEE Journal of Solid-state Circuits, Vol 29, No 4, April 1994, pp 481ff.

[6] F. Ling. "Digital Rate Conversion with a Non-rational ratio for High Speed Echo-Cancellation Modem", IEEE International Conference on Acoustics, Speech, and Signal Processing, Vol. 3, 1993, p 13 -16

[7] Y. Medan, U. Shvadron. "Asynchronous Rate Conversion". IEEE First Workshop on Multimedia Signal Processing, 1997, p 107 -112

[8]. R. Fitzgerald, W. Anderson. "Spectral Distortion in Sampling Rate Conversion by Zero-Order Polynomial Interpolation", IEEE Transations on Signal Processing, Vol. 40, No. 6, June 1992, p 1576 -1579

[9]. J. G. Janssen, J. H. Stessen, P. H. de With. "An Advanced Sampling Rate Conversion Technique For Video And Graphics Signals", IPA97, 15-17 July 1997, Conf Publ No. 443, IEE, 1997.

[10]. K. Nakamura, M. Kurokawa, A. Hashiguchi, M. Kanou, K. Aoyama, H. Okuda, S. Iwase, T. Yamazaki. "Video DSP architecture and its application design methodology for sampling rate conversion", Workshop on VLSI Signal Processing, IX, 1996, p418 - 427

[11] D. Gillies, J. Doty, A. Rothermel, R. Schweer. "Combined TV Format Control and Sampling Rate Conversion IC", IEEE International Conference on Consumer Electronics, 1994. Digest of Technical Papers.

[12] J. M. de Carvalho, J. V. Hanson. "Efficient sampling rate conversion with cubicsplines", SBT/IEEE International Telecommunications Symposium, 1990. Symposium Record, p 439 - 442

[13] D.-H. Lee, J.-S. Park, Y.G. Kim. "Video format conversions between HDTV systems", IEEE Transactions on Consumer Electronics, Aug. 1993, Vol. 39, No. 3, p 219-224

[14] H. Samueli, "*An improved search algorithm for the design of multiplierless FIR filters with powers-of-two coefficients*", IEEE Trans. on Circuits and Systems, vol.36, pp.1044--1047, July 1989.

GENERATION OF EMBEDDING WATERMARK SIGNALS FROM REFERENCE WATERMARK OF THE DETECTOR

Tae Young Kim, Taejeong Kim* and Kiryung Lee***

* School of Electrical Engineering and Institute of New Media and Communications
Seoul National University, Seoul, 151-742, ROK
kty@infolab.snu.ac.kr, tkim@snu.ac.kr
** Electronics and Telecommunications Research Institute, Daejeon, 305-350, ROK
kiryung@etri.re.kr

ABSTRACT

In this paper, we study asymmetric watermarking schemes which use different keys for the embedder and the detector. In an additive asymmetric marking system, the reference watermark of the detector and the embedded watermark in the host signals should be different but have some correlation value above a threshold if the detector does correlation test. Watermark signals can be decomposed into the magnitude and the phase components in the frequency domain. It is shown that the latter plays a more critical role in the correlation detector than the former does. It motivates us to propose a method to generate a set of embedding signals by randomly shifting the phase of the reference watermark. We also analyze the security of our method when the reference watermark is open to the attackers.

1. INTRODUCTION

A watermark is a secret signal embedded in digital media, which carries some information related to the owner, the distributor, and the copy control, etc. Most existing additive watermarking schemes are symmetric, meaning that an embedding signal is the same as the reference signal of the detector. If the detector is publicly available in this symmetric system, we have a security problem. Once an attacker has a full knowledge of the reference watermark, he can easily remove all the watermarks from the marked data by subtraction attack.

A solution to the problem is the asymmetric scheme whose embedding and reference watermark signals are different [1][2][3][4]. A general asymmetric watermarking system is described in Figure 1. It is also possible to do detection test using the embedding watermark itself. The detection using the reference watermark is called the *public* detection and the detection using the embedding watermark is called the *private* detection. If we use a correlation-based detector, the embedding watermark and the reference one should have some correlation value above a threshold. In this paper, we present a method to generate a set of embedding watermark signals from a reference watermark in the frequency domain. We also address the merit of our algorithm and the security against subtraction attack.

This work was supported by the Brain Korea 21 project.

2. CORRELATION DETECTION

Most watermark detectors correlate input signals with a reference watermark and compares the correlation (coefficient) values with a threshold. It is easily proved the correlation is not varied under unitary transform $\mathbf{U}$: $\mathbf{U}\mathbf{U}^H = \mathbf{I}$, where $(\cdot)^H$ is the hermitian transpose and $\mathbf{I}$ is the identity matrix.

$$\underline{x}^H \underline{y} = \underline{x}^H \mathbf{U}^H \mathbf{U} \underline{y} = (\mathbf{U}\underline{x})^H (\mathbf{U}\underline{y}) = \underline{X}^H \underline{Y},$$

where $\underline{x} = (x(0), x(1), \cdots .x(N-1))^T$, and $\underline{X} = \mathbf{U}\underline{x}$. One of unitary transform is the Discrete Fourier Transform (DFT) whose definition is $X(k) = \frac{1}{\sqrt{N}} \sum_{n=0}^{N-1} x(n) e^{-j\frac{2\pi kn}{N}}$ $(k = 0, 1, \cdots, N-1)$ and it can be expressed in polar form, $X(k) = |X(k)| e^{j\theta_{X(k)}}$. Using the conjugate-symmetry or $X(k) = X^*(N-k)$ for real valued $x(n)$, the correlation between two real sequences $x(n)$ and $y(n)$ can be represented as follows:

$$\sum_{k=0}^{N-1} X^*(k) Y(k) = \sum_{k=0}^{N-1} |X(k)||Y(k)| \cos(\theta_{X(k)} - \theta_{Y(k)}).$$

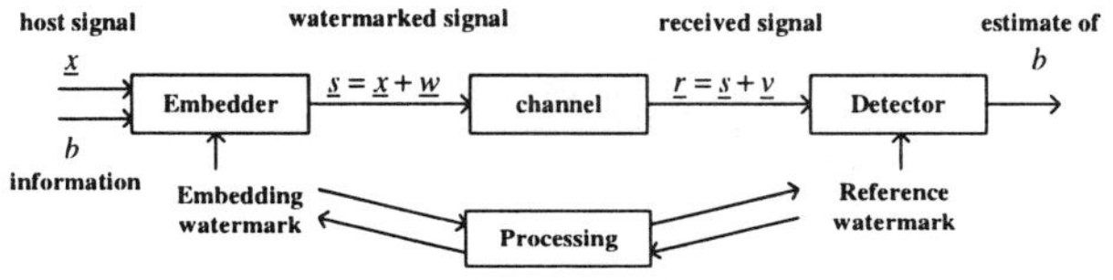

Fig. 1. A general asymmetric watermarking system

In the above equation, the component to determine the polarity of the correlation is the phase difference. The magnitude component plays just the role of weighting.

If the host signal is not white, prewhitening such as *linear prediction coding* (LPC) filtering may be used to improve the detection reliability [5][6]. Let us consider the case $x(n)$ and $y(n)$ are filtered by a filter $h(n)$ before the correlation test. If the filtering is implemented by circular convolution, the DFT coefficients of the filtered sequences $x_h(n)$ and $y_h(n)$ are given by $H(k)X(k)$ and $H(k)Y(k)$, respectively. In this case, therefore, the correlation is

$$\sum_{k=0}^{N-1} |H(k)|^2 |X(k)||Y(k)| \cos(\theta_{X(k)} - \theta_{Y(k)}).$$

The whitening filtering is, therefore, considered as a method to optimize the weighting factors in the DFT domain correlation test.

3. THE PROPOSED METHOD

In this section we propose a method to generate embedding watermark signals from a reference watermark $w_r(n)$ using random-phase-shifting as follows:

1. Calculate the DFT coefficients of $w_r(n)$:
   $W_r(k) = \frac{1}{\sqrt{N}} \sum_{n=0}^{N-1} w_r(n) e^{-j\frac{2\pi kn}{N}}$.
2. Randomly shift the phase of $W_r(k)$ to make an embedding watermark:
   $W_e(k) = W_r(k)e^{j\phi(k)}$, $W_e(N-k) = W_r(N-k)e^{-j\phi(k)}$, $0 < k < \frac{N}{2}$, where $\phi(k)$ is i.i.d. random variable whose probability density function (pdf) is uniform in $[-\phi_0, +\phi_0]$, and we assume $W_r(0) = 0$, and $W_r(\frac{N}{2}) = 0$ if N is even.
3. Get an embedding watermark signal by the IDFT:
   $w_e(n) = \frac{1}{\sqrt{N}} \sum_{k=0}^{N-1} W_e(k) e^{j\frac{2\pi kn}{N}}$.

If we define β as the expected correlation coefficient between a given reference watermark and an embedding watermark, then we have

$$\begin{aligned}\beta &= \frac{E[\sum_{k=0}^{N-1} W_e^*(k) W_r(k)]}{\sqrt{\sum_{k=0}^{N-1} |W_e(k)|^2}\sqrt{\sum_{k=0}^{N-1} |W_r(k)|^2}} \\ &= E[\cos(\phi(k))] \\ &= \frac{\sin(\phi_0)}{\phi_0}. \qquad (1)\end{aligned}$$

If we set $\phi_0 = 108.6°$, $\beta \approx 0.5$. The expected correlation coefficient between two different reference watermarks $w_{e1}(n)$, $w_{e2}(n)$ generated by the proposed method is given by β^2.

For a cover signal $x(n)$, the embedder performs the insertion of the watermark by $s(n) = x(n) + w_e(n)$. Since the watermarked signal $s(n)$ may be changed by some channel (attack) noise, the input $r(n)$ to the detector is modelled by $r(n) = s(n) + v(n) = x_v(n) + w_e(n)$, where $v(n)$ is an additive channel noise and $x_v(n) = x(n) + v(n)$. For the public detection, we find the linear correlation between $r(n)$ and $w_r(n)$.

$$\rho_{\text{public}} = \frac{1}{N}\sum_{n=0}^{N-1} x_v(n) w_r(n) + \frac{1}{N}\sum_{n=0}^{N-1} w_e(n) w_r(n). \quad (2)$$

If we assume that $x_v(n)$ is independent of $w_r(n)$, the expected correlation of the watermarked case (H_1) is given by

$$E[\rho_{\text{public}}|H_1] = \beta E[\rho_{\text{private}}|H_1], \quad (3)$$

where ρ_{private} is the private detection test statistic or the linear correlation between $r(n)$ and $w_e(n)$.

If the cover signal is not white, pre-whitening is done before correlation. We use the autoregressive (AR) model for $x(n)$ and whiten $x(n)$ using the inverse filter $h(n)$. In fact, we use $r(n)$ instead of $x(n)$ since $x(n)$ is not available in the blind detector. To decide the presence of the watermark, the public detector compares $|\rho^h_{\text{public}}|$ with a properly chosen decision threshold η. Since the filtering plays just the role of frequency-weighting in the correlation test, the expected public-to-private detection test statistic ratio $\frac{E[\rho^h_{\text{public}}|H_1]}{E[\rho^h_{\text{private}}|H_1]}$ is still β.

4. SECURITY ANALYSIS OF THE PROPOSED ALGORITHM

In this section, we analyze the security of our algorithm when an adversary has a thorough knowledge of the reference watermark somehow.

4.1. Domain of Possible Embedding Watermarks

Firstly, we consider the case that the adversary tries to reproduce the really embedded watermark in the host signal. Since $w_e(n)$ is a randomly phase-shifted version of $w_r(n)$ in the frequency domain, the domain of possible embedding watermark signals is order of the bandwidth of $w_r(n)$. For example, if $w_r(n)$ is white (flat magnitude-spectrum), the domain is $(2\phi_0)^{\frac{N}{2}-1}$ for even N. If $w_r(n)$ is not very narrow-banded, therefore, we can avoid the security threat for a sufficiently large N.

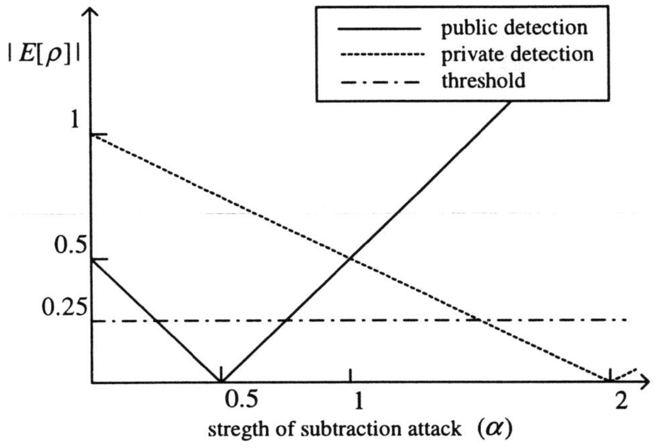

Fig. 2. The expected correlations of the private and the public detection $E[\rho_{\text{private}}|H_1]$, $E[\rho_{\text{public}}|H_1]$ under subtraction attack when $\beta = 0.5$

4.2. Subtraction Attack

Now, we consider the subtraction attack achieved by $r'(n) = r(n) - \alpha w_r(n) = x_v(n) + w_e(n) - \alpha w_r(n)$, where α is the strength of the subtraction attack. To make the public detection impossible, we have $E[(w_e(n) - \alpha w_r(n))w_r(n)] = 0$. This equation is the same as the orthogonality condition in the *linear minimum mean square error* estimation problem and leads to $\alpha = \beta$, when the distortion by the watermark signal terms or $w_e(n) - \alpha w_r(n)$ is $(1-\beta^2)E[w_e^2(n)]$ [7]. It implies that the subtraction attack to disable the public detection always reduces the power of the watermark, so the attacked signal has a better objective quality than the originally marked signal.

The adversary may want to disable the private detection. To make the private detection impossible, we have $E[(w_e(n) - \alpha w_r(n))w_e(n)] = 0$. This equation leads to $\alpha = \frac{1}{\beta}$, when the distortion by the watermark signal terms is $(\frac{1}{\beta^2} - 1)E[w_e^2(n)]$. If $\beta = 0.5$, the distortion is increased by about 4.77dB.

In the case that the adversary has one of the embedding watermarks, he can also disable the public or the private detection in the work embedded by different embedding watermark. When the public detection is disabled by the subtraction attack ($E[(w_{e1}(n) - \alpha w_{e2}(n))w_r(n)] = 0$), the distortion is $2(1-\beta^2)E[w_e^2(n)]$. When the private detection is disabled ($E[(w_{e1}(n) - \alpha w_{e2}(n))w_{e1}(n)] = 0$), the distortion is $(\frac{1}{\beta^4} - 1)E[w_e^2(n)]$. If β is less than 0.5, the subtraction attack may severely decrease the quality of the work. If he has several embedding watermarks, he may estimate the reference watermark by averaging them, so this case becomes the same case that the reference one is open.

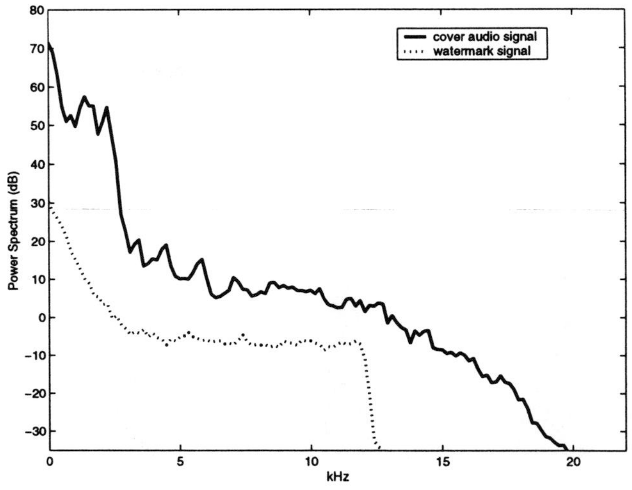

Fig. 3. Power spectrum of a jazz audio signal and a reference watermark signals

To secure the watermarks, at least, we must select the design parameters so that the subtraction attack using the reference watermark cannot disable the private detection and the public detection simultaneously. Now we consider how to select the design parameters for the case that an identical threshold is used in both detections. We assume $E[w_e^2(n)] = 1$ and $0 < \beta < 1$ without loss of generality. The expected correlations between $r'(n)$ and $w_e(n)$ or $w_r(n)$ under the subtraction attack is $E[\rho_{\text{private}}|H_1] = 1 - \alpha\beta$ or $E[\rho_{\text{public}}|H_1] = \beta - \alpha$. Not to disable both the detection at the same time, $\{\alpha : |1 - \beta\alpha| \leq \eta \cap |\beta - \alpha| \leq \eta\}$ must be null, which leads to $\eta < 1 - \beta$. Since it is reasonable that η in the public detection should be less than β, we, finally, obtain $\eta < \min(1-\beta, \beta) = 0.5$. A proper choice of the parameter set is $\beta = 0.5$ and $\eta = 0.25$. Figure 2 describes how the expected correlations are changed by the subtraction attack in this parameter set.

5. IMPLEMENTATION

We applied our scheme for audio watermarking in the time domain. We prepared five jazz audio clips sampled at 44.1 kHz with 16 bits resolution for experiments. Using some listening test, we designed the power spectrum of the embedding watermark signals to make no audible distortion. Figure 3 illustrates the power spectrums of a jazz audio signal and a reference watermark signal using the Welch averaged periodogram method [7]. We segmented the audio sequences into the frames of the length $N = 2^{10}$. For each frame, we embedded different embedding watermarks by $s(n) = x(n) + \gamma w_e(n)$, where γ is a scaling factor to ensure no perceptible distortion. After the embedding process, we attacked the marked audio signals by white noise addition

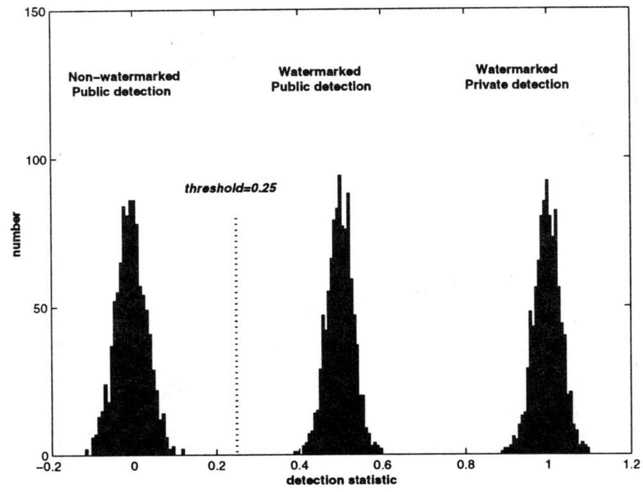

Fig. 4. Distribution of the detection statistics

and 96kbps MP3 compression/decompression.

To whiten the input frames of the detector, we used the fifth order AR model for them: $r(n) + \sum_{i=1}^{5} a_i r(n-i) = e(n)$. We found $\hat{a}_i$'s that minimize the power of $e(n)$ by the autocorrelation method [7]. After filtering $r(n)$ and $w_r(n)$ using $h(n) = \delta(n) + \sum_{i=1}^{5} \hat{a}_i \delta(n-i)$, we calculated the correlation coefficient of the two filtered sequences. We averaged each set of successive 64 coefficients to make a reliable public detection statistic. The private detection was also performed in the same manner. The distributions of the detection statistics are plotted in Figure 4, where horizontal scaling is done to make the expected correlation of the private detection normalized to 1. With a threshold of 0.25, we had no detection error in the experiment. Empirically obtained expected public-to-private detection test statistic ratio, *i.e.*, $\frac{E[\rho_{\text{public}}|H_1]}{E[\rho_{\text{private}}|H_1]}$ was $0.4998 \approx \cos(60°)$.

The implemented asymmetric audio watermarking system is vulnerable to the desynchronization attacks such as cropping or time scale modification since the time domain is very sensitive to the attacks. To tackle the problems, we maybe have to resort to the self-synchronization by the content analysis or the DFT magnitude domain watermarking [8][9].

6. CONCLUSIONS

We proposed a method to make embedding watermark signals from a reference watermark of the detector. First, we decompose reference watermark to the magnitude and the phase components by the DFT. Then, we modify only the phase components to generate a set of embedding watermark signals. Whether the detector uses pre-filtering or not, the expected correlation coefficient between the reference watermark and the embedding watermarks will not be changed. We also addressed how to select the design parameters for the security against subtraction attack.

7. REFERENCES

[1] J. J. Eggers, J. K. Su, and B. Girod, "Asymmetric watermarking schemes", Tagungsband des GI Workshops "Sicherheit in Mediendaten", Berlin, Germany, September 2000, Springer Verlag, ISBN 3-540-67926-X, pp. 107-123.

[2] F. Hartung and B. Girod, "Fast public-key watermarking of compressed video", In *Proc. of the IEEE Intl. Conf. on Image Processing 1997*, vol. 1, pp. 528-531, October 1997.

[3] J. Picard and A. Robert, "On the Public Key Watermarking Issue", In *Proc. of SPIE Vol. 4314 : Security and Watermarking of Multimedia Contents III*, pp. 290-299, San Jose, Jan. 2001.

[4] H. Choi, K. Lee, and T. Kim, "Transformed-key asymmetric watermarking system", In *Proc. of SPIE Vol. 4314 : Security and Watermarking of Multimedia Contents III*, pp. 280-289, San Jose, Jan. 2001.

[5] G. Depovere, T. Kalker, and J. P. Linnartz, "Improved watermark detection reliability using filtering before correlation", In *Proc. of the IEEE Intl. Conf. On Image Proc.*, pp. 430-434, 1998.

[6] J.W. Seok and J. W. Hong, "Audio watermarking for copyright protection of digital audio data", *IEE Electroncis Letters*, vol. 37, No. 1, pp. 60-61, Jan. 2001.

[7] Monson H. Hayes, *Statistical Digital Signal Processing and Modeling*, , Wiley, 1996.

[8] Chung-Ping Wu, Po-Chyi Su and C.-C. Jay Kuo, "Robust Audio Watermarking for Copyright Protection," in *Proc. SPIE* Vol. 3807, July 1999, pp. 387-397.

[9] Darko Kirovski and Henrique Malvar, "Robust Spread-Spectrum Audio Watermarking," in *Proc. ICASSP* Vol. 3, 2001, pp. 1345-1348.

EFFICIENT SIGNAL PROCESSING IN EMBEDDED JAVA SYSTEMS

Rafael Krapf & Luigi Carro

Instituto de Informática - Universidade Federal do Rio Grande do Sul
Av. Bento Gonçalves, 9500 - Campus do Vale - Bloco IV – Caixa Postal 15064
91501-970 - Porto Alegre - RS - Brasil

ABSTRACT

Digital Signal Processing is a necessary capability of the new portable devices, as multimedia contents become common place for embedded digital systems. However, because of the stream-like behavior of multimedia applications, specialized processors and compilers are required to manipulate contents such as audio and video. This paper presents a set of architecture modifications that interact with the compiler, generating an optimized code so that one can use a single language (Java) for the specification of complex System-on-Chips with signal processing characteristics. Besides reaching a higher efficiency for DSP, the designer maintains the portability of the Java code for general purpose computing. Experimental results show that with a 5% of area overhead one can reach from 23% to 50% performance improvement in FemtoJava microcontroller.

1. INTRODUCTION

Digital Signal Processing(DSP) is the manipulation of data by specific algorithms that take advantage of the high integration scale, robustness and flexibility of the digital circuits to process data streams [1]. The stream-like behavior of this class of applications requires the use of specialized processors with architectural characteristics specifically developed for continuous processing of data. These specialized processors offer processing options that only a compiler made for that processor is able to explore at the maximum, and, in some cases, even assembly programming is needed, because the DSP instruction set is highly unregular, making the creation of an efficient compiler difficult [2].

In highly integrated systems, where hybrid parts with different behaviors are combined, originating the concept of System-on-Chip (SOC), the complexity of the design task makes necessary the use of an object oriented programming language to develop the target application. The advantages of such languages are the reusability of the code, and the modularity to deal with the increasing complexity.

The use of a single, object-oriented language to describe all the behavior to be mapped to cores on a SOC would reduce the design time by the reuse of software components [3]. Java [4] is a suitable language for this need, not only because of the object-orientation paradigm, but also because of its portability. The designer has the option to develop the application, debug and test it in his preferred environment for later transfer to the target device, and the Java compatibility will ensure that the application will work properly. Java compilers also generate small code, that implies in less memory necessary to store the application, and also in less dissipated power to access this memory.

However, all Java processors found in the literature are not tuned for stream processing. As an example, the Java machines proposed by Sun Microsystems [5]-[6] are simply execution engines for Java bytecodes. For the family of DSP applications, the performance of Java processors is generally poor. The reasons for the low performance of Java machines running stream-like algorithms are the low level of specialization of the Java processors to deal with stream-like applications, and the poor resources in the Java language to describe streaming behavior.

When developing a stream application in Java (e.g.: a FIR filter) the resources to describe such application are conditional loops. These conditional loops after compilation are transformed in a set of jump instructions, that have no resemblance with the stream-like behavior required to maintain high throughput.

The optimization approach proposed in this paper focus on two different levels at the synthesis workflow: architecture modification and compiler interaction. These modifications together can boost the performance of native Java Processors for DSP applications, with minimum area overhead. The reason for the speed-up is the fact that the use of customizable classes make the application execution more efficient, once the program is able to explore all the available hardware of the processor without penalty on design time. Experimental results have shown that typical DSP applications, when compiled with a standard Java compiler, spend a lot of instructions and time controlling the program flow, and executing functions that are more suited to be resolved by hardware. These paper will focus on the specific hardware modification developed for the processor, that will maintain peak performance with low cost.

This paper is organized as follows: in the next section, previous efforts in the direction of making general purpose processors support DSP applications are presented. Section 3 discusses case studies, and how these case studies have been compared. Section 4 presents the architecture enhancements for the FemtoJava microcontroller targeting DSP applications, and the class identification needed to address such new architecture features. The results of these architecture modifications in terms of speed-up improvement for the case studies are resumed in section 5, and the conclusions of this work are presented in section 6.

0-7803-7761-3/03/$17.00 ©2003 IEEE

2. RELATED WORK

Kang's et al. research on RISC-based DSP processors [7] has shown that with few architectural modifications, their RISC processor achieved execution time speedups up to 3.5 times. Their strategy was based on the insertion of DSP processor characteristics in a RISC processor, and the comparison of the resulting processor with a superscalar one. The modifications proposed in [7], however, showed that they were only effective for very specific pieces of code in real applications. Real world applications, composed of a mix of behaviors, presented a much smaller speed-up, as an example, the 50 taps FIR filter written in Java has additions and multiplications as the main task of the algorithm. This filter processing 84 inputs spent only 64761 cycles in FemtoJava microcontroller, that is only 7.24% of the 755914 cycles of all computation. The reason for the [7] authors' results in QCELP algorithm was the small portion of the program that was actually executing DSP functions. In fact, most of real life applications have a DSP part for data manipulation and general purpose computation models for all other functions.

In the development of a multimedia extension to ARM7 processor, Huang et al. [8] found that hardware modifications to make that processor more efficient on executing DSP applications have a small cost (11% of the processor) and the processing time of such applications could be reduced up to 79%. It is important to notice that in the work of [8] the algorithms used as benchmark (mostly video applications) are very adaptable to the optimizations proposed. In real life applications the same pattern of smaller speed up reported in [7] is expected.

Other studies have been made in the direction of hybrid DSP-RISC architectures, and Smith [9] discussed about the common characteristics among RISC processors and DSP ones. In fact, there are several characteristics on both processors types that perform similar functions but are described with different names and intend to solve peculiarities on each type of processor.

All the mentioned previous work was based on the idea that the speed up of the RISC could be developed only at the hardware level. This is not the real case, since the compiler must be aware of the available resources, so that it can generate efficient code for the microprocessor. In [7], the way achieved to interact with the compiler was a code converter that adapts the original assembly code to the optimized one. But, the poor results for complete applications leads one to investigate better ways to describe signal processing applications, and also general computing functions that may take advantage from the hardware included to optimize DSP applications.

In this work we present the combination of hardware and software resources that can effectively tackle the optimization of a general purpose microprocessor to the stream-like behavior of DSP applications, without losing performance in other computational models. The target processor is the FemtoJava machine, and SASHIMI is the software environment that supports its development [10].

3. CASE STUDIES

To better understand the case studies and their behavior, first we address the typical characteristics of the DSP applications.

3.1 Signal Processing Computational Needs

DSP applications deal with streams. Signal Processing, as the name suggests, involves continuously evaluation and processing of inputs that come from a given source. The processing of streams means that the algorithm for DSP applications will be executed while the stream is being received, or while the user does not interrupts its execution.

DSP algorithms are computational intensive. Much computation is necessary to transform a bit stream in an image or in an audio output. Any filtering or compression algorithm involves a large amount of numerical transformations.

Another important feature is that a common characteristic of DSP algorithms is that most of the processing is placed in small, much repeated loops. Also, a big amount of DSP functions would be better implemented in hardware, because most of these processing functions stay in the critical execution path of the application and, also, because they are simpler in hardware. The best example of this is the bit reverse function, often used in FFT computations. For each bit reverse addressing it is necessary to spend 4 java machine instructions. The same operation, while performed in hardware, is just the connection of some wires in a certain order.

The need for data manipulation on signal processing algorithms leads to the final important characteristic: the need to load from memory more than a value at a time and the need to deal with continuous processing with finite memory. To deal with this, it is common to find in DSP processors more than one memory bank, memory banks with multiple possible accesses and complex memory management functions.

3.2 Signal Processing Evaluation

For the evaluation of the Java processor on executing DSP algorithms, the FIR (Finite Impulse Response) filter, the FFT (Fast Fourier Transform) and the DCT (Discrete Cosine Transform) functions were implemented in Java. These three applications were chosen because of their presence in most DSP systems (including the MP3 decoder). To analyze the speed-up provided by the technique, the number and the type of the instructions executed by the running application were mapped for each of the methods in the applications. Thus, one can count the number of machine cycles necessary to execute the application. The total number of cycles is the metric to evaluate the programs in the Java processor.

In order to count the instructions executed on the Java programs we used a modified version of the Bytecode Instrumentation Tool (BIT) [11]. This set of Java classes allows the user to insert code into a compiled `.class` file. The modifications made in BIT include the reference to the machine cycles in the FemtoJava microcontroller. With proper code insertion one can compute the calls to specific methods, basic blocks, and even instructions, so it is possible to count the total number of instructions executed by the application. The number of cycles needed to perform an instruction is a known characteristic of the processor, and combining these two information one can measure how many cycles will be necessary to run an specific application, an specific application method or part of an algotithm.

Different versions of the FIR, FFT and DCT algorithms were implemented in order to make their code suitable to the new hardware available in the processor, so the algorithms could take advantage of the new DSP structures. For all the versions implemented, the FIR filter was made with 50 taps processing 84 inputs, the FFT was implemented for 16 points une group of 16 points presented for FFT processing, and the DCT was implemented processing 64 input data points.

4. SUGGESTED MODIFICATIONS

Observing common DSP applications one can see that most of them usually needs one or more of the following DSP characteristics:

- Multiply-accumulate instruction
- Efficient looping with minimum (or zero) overhead
- Circular buffer
- Bit reverse addressing

By providing classes with the software behavior of circular buffer, multiply-accumulate, bit reverse addressing and zero-overhead loop, one can easily create optimized DSP applications to the processor, while at the same time extending the Java ability to describe such applications.

To correctly model the behavior of circular buffers in Java one must provide to the DSP programmer tools that increase the Java ability to describe such structures. This was implemented through a set of classes. Each one of these classes has one of the circular buffer behaviors (size configuration, data input and data output).With these classes the Java language gained a higher representativity for DSP applications that use the circular buffer.

The same occurs to other DSP characteristics studied and presented in this work. The SASHIMI tool is able to detect the call to those specific methods and change these calls for special DSP instructions that activate the specific DSP hardware. This way, a call to a specific method is substituted by a special instruction in the optimized processor.

4.1 The signal processing structures proposed

As the FemtoJava processor has in its structure a bus arbiter to control the I/O ports and other structures mapped in memory, it is convenient to extend the functionality of this arbiter so that it can be configured by the user to address a circular buffer. Once the end of the buffer is reached, the modified bus arbiter points to the begin of the buffer, performing, this way, a circular buffer.

The original arithmetic and logic unit (ALU) of the FemtoJava processor already contains a multiplier and an adder/subtractor, so, in order to implement the integer multiply and accumulate (IMAC) instruction the only need is to feed the adder with the result of the multiplier, and store the temporary value in an ALU internal register. It is important to notice that this internal register has the length of the multiplier result (double of the word length of the processor) and all multiply-accumulate operations are performed with this word length. This leads to a better precision when using a series of *imac* instructions instead of isolated multiplications and additions.

An efficient looping was implemented based on the fact that most of inner DSP loops are of fixed size – typically in a FOR loop. On these inner loops, it is possible to define the loop size and configure an automated jump on the program counter (PC). Once the end of the loop is reached, a certain counter is decremented and the PC is fed with the address relative to the begin of the loop. The cost of the structure is three registers (to store the end and begin of the loops and the counter), an adder, a simple AND comparator, and a small logic to control the output. To configure the efficient loop registers, their location is memory mapped.

Bit reverse addressing is an important feature in DSP processors and it has zero cost for a fixed number of bits to be reversed. It is a simple reorder of the output wires in the desired manner from the input wires. But, in a configurable bit reverse, one can create a series of simple bit reverse addressing modes, and select the desired bit reverse output with another parameter. In the FemtoJava microcontroller this was implemented in the ALU, using one control bit to select between the adder or the bit reverse block.

5. RESULTS

5.1 The speedup provided by DSP structures

One can observe in Table 1 that the speedup provided by the DSP structures is significant in the FIR filter. Also the DCT algorithm was heavily improved with the insertion of the DSP instructions in the FemtoJava microcontroller as can be seen on table 2.

Table 1 - Speedup provided by DSP structures in FIR filter

Application	Cycles	Speedup
Standard FIR filter	755914	-
FIR filter with imac	665741	11.93 %
FIR filter with circular buffer	482113	36.22 %
FIR filter with circular buffer and imac	379091	49.85 %
FIR filter with circular buffer, imac and efficient looping	289127	61.75 %

On table 3 is the optimization results for the FFT using the DSP structures proposed.

Table 2 - Speedup provided by DSP structures in DCT

Application	Cycles	Speedup
DCT	230695	-
DCT with imac	189095	18.03 %
DCT with imac and efficient looping	170279	26.19 %

At the first moment it was not expected any reduction in the number of memory access, since none of the DSP hardware structures was targeting memory accesses reduction. But in the DCT and FIR filter one can see in tables 4 and 5 that there is a significant memory access reduction, since using the imac DSP function, the temporary value of the computation does not need to be stored in the memory. Instead, this temporary value is stored in the imac register inside the modified FemtoJava ALU. Tables 4 and 5 take into account the number of cycles spent on memory access instructions, in the FemtoJava microcontroller that corresponds to instructions *iaload, iastore, getstatic, putstatic, store_idx* and *load_idx*. The reason behind counting number of cycles instead of simple instruction counting is that the circular buffer implemented access the memory through *store_idx* and *load_idx* instructions, that are more efficient in cycles than *getstatic* and *putstatic*. So, this metric for memory access is more adequate.

Table 3 – Speedup provided by DSP structures in FFT

Application	Cycles	Speedup
Standard FFT	78215	-
FFT with bit reverse function	62727	19.80 %
FFT with bit reverse function and efficient looping	62370	20.26 %

Table 4 – Cycles spent in memory access instructions on FIR filter

Application	Cycles	Speedup
FIR filter	566559	-
FIR filter with imac	274099	51.62 %
FIR filter with circular buffer	267050	52.86 %
FIR filter with circular buffer and imac	176505	68.85 %

Table 5 – Cycles spent in memory access instructions on DCT

Application	Cycles	Speedup
Integer DCT	184437	-
Integer DCT with imac	152195	17.48 %

5.2 The cost of DSP structures

All DSP instructions and structures presented in this work were implemented targeting minimum area overhead, so their implementation reused most of system structures already available while introducing as less additional hardware as necessary. The cost of these structures are displayed in table 6, where "logic cells" is relative to a Maxplus II compilation of those structures in a EPF10K20RC240-4 (Flex 10K20) device from Altera.

Table 6 – Size of DSP structures inserted in FemtoJava microcontroller

Application	Logic Cells	Relative Size
Complete FemtoJava	1492	100 %
Bit reverse function	22	1.47 %
Imac instruction	34	2.3 %
Efficient looping	81	5.4 %
Circular buffer	94	6.3 %

6. CONCLUSIONS

With few cost in hardware area one can achieve a high performance improvement in the FemtoJava microcontroller, the best result is for the FIR filter that could have its performance improved by 61% with a total cost in hardware of 15% of the processor.

The main advantage of the approach proposed is that the designer does not need to program to these special structures in low level language. There is a high level access to these structures, so the SASHIMI environment optimizes all the application with the special structures when they are required via high level structures by the programmer.

7. REFERENCES

[1] Proakis J. G. and Manolakis D. G. "Digital signal processing : Principles, algorithms, and applications". Third Edition. Upper Saddle River, Prentice Hall: 1996. 968 pp.

[2] Eyre J. and Bier J. "Dsp processors hit the mainstream". *IEEE Computer Magazine*, 17 (2). August, 1998. pp 51-59.

[3] Booch G. "Object oriented design: With applications". Redwood City, The Benjamin/Cummings: 1991. 580 pp.

[4] Gosling J., Joy B. and Steele G. L. "The java^{tm} language specification". Java series.Second Edition. Reading, Mass., Addison-Wesley: 2000. 825 pp.

[5] Mcghan H. and O'connor M. "Picojava: A direct execution engine for java bytecodes". *IEEE Computer*, 31 (10). October 1998. pp 22-30.

[6] Sun M. I.;"Picojava-ii programmer's reference manual".Sun Microsystems: 1999. 512 pp.

[7] Kang J., Lee J. and Sung W. "A performance evaluation of a risc-based digital signal processing architecture". *IEEE Workshop on Signal Processing Systems (SiPS): Design and Implementation.* 1998. pp pp 538-547.

[8] Huang I.-J., Huang W.-K., Gu R.-T. and Kao C.-F. "A cost effective multimedia extension to arm7 microprocessors". *ISCAS 2002.* IEEE: 2002. pp pp II.304 - II.307.

[9] Smith M. R. "How riscy is dsp?" *IEEE Micro*, 12 (6). November - December, 1992. pp 10-23.

[10] Ito S. A., Carro L. and Jacobi R. "System design based on single language and single-chip java asip microcontroller". *Design, Automation and Test in Europe Conference and Exibition.* Paris, IEEE Computer: 2000. pp

[11] Lee H. B. and Zorn B. G. "Bit: A tool for instrumenting java bytecodes". *USENIX Symposium on Internet Technologies and Systems.* Monterey, California, USENIX Association: 1997. pp pp 73-82.

An Efficient Split-Radix FFT Algorithm

Saad Bouguezel, M. Omair Ahmad, Fellow, IEEE, and M.N.S. Swamy, Fellow, IEEE

Department of Electrical and Computer Engineering
Concordia University
1455 de Maisonneuve Blvd. West
Montreal, P.Q., Canada H3G 1M8
E-mail: {b_saad, omair, swamy}@ece.concordia.ca

ABSTRACT

In this paper, an efficient split-radix FFT algorithm is proposed for computing the length-2^r DFT that reduces significantly the number of data transfers, index generations, and twiddle factor evaluations or accesses to the lookup table. It is shown that the arithmetic complexity of the proposed algorithm is no more than that of the existing split-radix algorithm. The basic idea behind the proposed algorithm is that a radix-2 and a radix-8 index maps are used instead of a radix-2 and a radix-4 index maps as in the classical split-radix FFT. In addition, since the algorithm is expressed in a simple matrix form using the Kronecker product, it facilitates an easy implementation of the algorithm, and allows for an extension to the multidimensional case.

I. INTRODUCTION

In 1984, Duhamel and Hollmann [1] introduced the split-radix FFT algorithm which is the simplest algorithm having the lowest number of arithmetic operations. They claim that the use of the radix-8 index map in the split-radix approach does not bring any reduction in the number of arithmetic operations. However, in this paper, we show that by suitably combining the twiddle factors and forming a special butterfly, and using the radix-8 index map in the split-radix approach, a substantial reduction in the number of data transfers, and twiddle factor evaluations or accesses to the lookup table can be achieved. Further, this is achieved with the arithmetic complexity being no more than that of the split-radix algorithm [1],[2]. The basic idea behind the proposed algorithm is that a radix-2 and a radix-8 index maps are used instead of a radix-2 and a radix-4 index maps as in the classical split-radix FFT. Although the butterfly size is larger, the implementation does not need any additional complexity, since modern processors, both general and DSP, posses sufficient number of internal fast registers to store the intermediate results.

II. PROPOSED ALGORITHM

The DFT is defined by

$$X(n) = \sum_{k=0}^{N-1} x(k) W_N^{nk}, \quad n = 0, 1, ..., N-1 \qquad (1)$$

where $W_N = exp(-j2\pi/N)$, $N = 2^r$. The decomposition of (1) into even-indexed and odd-indexed terms provides

$$X(2n) = \sum_{k=0}^{N/2-1} y_e(k) W_{N/2}^{nk}, \quad n = 0, 1, ..., N/2-1 \qquad (2)$$

$$X(2n+1) = \sum_{k=0}^{N/2-1} y_o(k) W_N^k W_{N/2}^{nk}, \quad n = 0, 1, ..., N/2-1 \qquad (3)$$

where the input sequences $y_e(k)$ and $y_o(k)$ of (2) and (3) can be written as

$$\begin{bmatrix} y_e(k) \\ y_o(k) \end{bmatrix} = H_2 \begin{bmatrix} x(k) \\ x(k+N/2) \end{bmatrix}, H_2 = \begin{bmatrix} 1 & 1 \\ 1 & -1 \end{bmatrix} \qquad (4)$$

This first step of the decomposition is processed by applying a second order Hadamard matrix H_2 to the input sequence $x(k)$ without requiring the twiddle factors. A significant improvement, especially in the reduction of the number of data transfers, can be achieved when a radix-8 index map is used in the decomposition of the odd-indexed terms. Let us consider $k = \frac{N}{8}k_0 + k_1$, where $k_1 = 0, 1, ..., (N/8)-1$ and $k_0 = 0, 1, 2, 3$. Then, the odd-indexed terms given by (3) become

0-7803-7761-3/03/$17.00

$$X(2n+1) = \sum_{k_1=0}^{N/8-1} \left(\sum_{k_0=0}^{3} y_o(k_1 + k_0 \frac{N}{8}) e^{-j\frac{\pi}{4}k_0} W_4^{nk_0} \right) W_N^{k_1} W_{N/2}^{nk_1} \quad (5)$$

Since the length-4 DFT does not require multiplications, its appearance in (5) is very advantageous when the following decimation-in-frequency is used: $n = 4n_1 + n_0$, where $n_1 = 0, 1, ..., (N/8) - 1$ and $n_0 = 0, 1, 2, 3$. Then, (5) can be expressed in a closed form as,

$$X(8n + \beta) = \sum_{k_1=0}^{N/8-1} y_o^{\beta}(k_1) W_{N/8}^{n_1 k_1}, \quad \beta = 1, 3, 5, 7 \quad (6)$$

where the input sequences $y_o^{\beta}(k_1)$, $\beta = 1, 3, 5, 7$, are the components of a vector that can be expressed, after certain mathematical transformations, in a simple matrix form given by

$$\widetilde{Y}_o = \widetilde{B}_4^o Y_o. \quad (7)$$

The components of the column vectors $\widetilde{Y}_o$ and Y_o in (7) are $\widetilde{Y}_o((\beta - 1)/2) = y_o^{\beta}(k_1)$ and $Y_o((\beta - 1)/2) = y_o(k_1 + N(\beta - 1)/16)$ respectively, and the matrix $\widetilde{B}_4^o$ is given by

$$\widetilde{B}_4^o = F_4^o B_4 \quad (8)$$

where the twiddle factor matrix F_4^o is a 4x4 diagonal matrix, whose non-zero elements are $F_4^o((\beta-1)/2, (\beta-1)/2) = W_N^{\beta k_1}$, β=1, 3, 5, 7. The constant matrix B_4 can be further factored as

$$B_4 = W_4 S_4, \quad (9)$$

W_4 being a length-4 DFT matrix operator, which may factored as

$$W_4 = Z_4 D_4 R_4 \quad (10)$$

where $Z_4 = H_2 \otimes I_2$, I_2 is a 2x2 identity matrix, $j = \sqrt{-1}$, $D_4 = \begin{bmatrix} I_2 & 0 \\ 0 & J_2 \end{bmatrix}$, $J_2 = \begin{bmatrix} 1 & 0 \\ 0 & -j \end{bmatrix}$, $R_4 = \begin{bmatrix} 1 & 0 & 1 & 0 \\ 1 & 0 & -1 & 0 \\ 0 & 1 & 0 & 1 \\ 0 & 1 & 0 & -1 \end{bmatrix}$ and $\otimes$ denotes the Kronecker product [3]. The diagonal matrix S_4 can be factored using the Kronecker product as follows:

$$S_4 = (I_2 \otimes C_2) J_g^o \quad (11)$$

where $C_2 = \begin{bmatrix} 1 & 0 \\ 0 & e^{-j\pi/4} \end{bmatrix}$ and $J_g^o = J_2 \otimes I_2$. Since the multiplication operation $-j(a + jb)$ renders a complex number $(a + jb)$ to become $(b - ja)$, the matrix J_g^o can be implemented within index generation process, and hence, J_g^o does not require extra processing time.

Using (8)-(11) in (7) and the fact that $R_4(I_2 \otimes C_2) = (C_2 \otimes I_2)R_4$, after some manipulations, the general sub-butterfly of the proposed algorithm can be expressed in a simple matrix form as

$$\begin{bmatrix} y_o^1(k_1) \\ y_o^3(k_1) \\ y_o^5(k_1) \\ y_o^7(k_1) \end{bmatrix} = F_4^o Z_4 T_4 R_4 J_g^o \begin{bmatrix} y_o(k_1) \\ y_o(k_1 + N/8) \\ y_o(k_1 + N/4) \\ y_o(k_1 + 3N/8) \end{bmatrix} \quad (12)$$

where $k_1 = 0, 1, ..., (N/8) - 1$, and the constant twiddle factor matrix $T_4 = D_4\,(C_2 \otimes I_2)$. For a given value of k_1, the four inputs of (12) can be obtained from (4) as follows:

$$\begin{bmatrix} y_e(k_1) \\ y_e(k_1 + N/8) \\ y_e(k_1 + N/4) \\ y_e(k_1 + 3N/8) \\ y_o(k_1) \\ y_o(k_1 + N/8) \\ y_o(k_1 + N/4) \\ y_o(k_1 + 3N/8) \end{bmatrix} = (Z_4 \otimes I_2) \begin{bmatrix} x(k_1) \\ x(k_1 + N/8) \\ x(k_1 + N/4) \\ x(k_1 + 3N/8) \\ x(k_1 + N/2) \\ x(k_1 + 5N/8) \\ x(k_1 + 3N/4) \\ x(k_1 + 7N/8) \end{bmatrix} \quad (13)$$

Finally, the general butterfly of the proposed FFT algorithm is constructed using (13) and the general sub-butterfly (12). It is clear that for a sequence of length N, the required number of butterflies to perform the first stage of the decomposition is $N/8$. These butterflies are indexed by k_1 from 0 to $(N/8) - 1$ as indicated in (12) and (13). It can also be seen that the sub-butterfly for $k_1 = 0$ can be considered as a special sub-butterfly, since the twiddle factor matrix F_4^o becomes an identity matrix, and hence, the number of multiplications can be reduced. Let us consider the sub-butterfly when $k_1 = N/16$. By using the relations that $W_N^{5k_1} = W_N^{4k_1} W_N^{k_1}$ and $W_N^{7k_1} = W_N^{4k_1} W_N^{3k_1}$, the twiddle factor matrix F_4^o becomes

$$F_4^o\,(k_1 = N/16) = J_g^o\,(I_2 \otimes F_2^o) \quad (14)$$

where $F_2^o = \begin{bmatrix} e^{-j\pi/8} & 0 \\ 0 & e^{-j3\pi/8} \end{bmatrix}$. By substituting (14) in (12), using the fact that $(I_2 \otimes F_2^o)\,Z_4 = Z_4\,(I_2 \otimes F_2^o)$, and by combining the twiddle factor matrices $(I_2 \otimes F_2^o)$ and T_4, the matrix $\widetilde{B}_4^o$ is expressed as

$$\widetilde{B}_4^o\,(k_1 = N/16) = J_g^o Z_4 T_8 R_4 J_g^o \quad (15)$$

where $T_8 = (I_2 \otimes F_2^o)\,T_4$. The number of operations to process this sub-butterfly is less than that needed to process the general sub-butterfly, (see Section IIIA) and hence, the sub-butterfly corresponding to $k_1 = N/16$ can also be considered as a special sub-butterfly. A general butterfly which uses the special sub-butterfly for $k = 0$ or $N/16$ is called a special butterfly.

We can now summarize the concept of the proposed split-radix FFT algorithm for efficiently computing the length-N DFT, where N is an integral power of two. Using the butterfly constructed by (12) and (13) and by performing $(N/8) - 2$ general butterflies and two special butterflies corresponding to $k = 0$ and $N/16$, the initial sequence $x(k)$ of length N is decomposed into five sub-sequences. The first sub-sequence of length $N/2$ is constructed by the first four outputs of (13) and the other four sub-sequences each of length $N/8$ represent the four outputs of the sub-butterfly (12). This process is repeated successively for each of the new resulting sub-sequences, until the size is reduced to a 4 or 2-point DFT without twiddle factors.

III. COMPUTATION COMPLEXITY OF THE PROPOSED ALGORITHM

In this section we consider the efficiency of the proposed FFT algorithm by analyzing its computational complexity and comparing it with that of the existing split-radix FFT algorithm [2]. The analysis and comparison will not only include arithmetic operations, but also operations such as data transfers, and twiddle factor evaluations or accesses to the lookup table, since they contribute significantly to the execution time of the algorithms. Since the new algorithm is based on a split-radix approach, important properties such as the use of an arbitrary length $N = 2^r$ and in-place computation are preserved.

A. Arithmetic Complexity

We first consider the case, where the complex multiplication is performed by three real multiplications and three real additions (3mult-3add scheme). Then, the general butterfly constructed using (12) and (13) requires 16 real multiplications and 48 real additions. The special butterfly when $k = 0$ requires 4 real multiplications and 36 real additions. The special butterfly when $k = N/16$ requires 12 real multiplications and 44 real additions. The proposed decomposition consists of dividing a length-N DFT to one length-$N/2$ DFT and four length-$N/8$ DFTs in the first stage that requires $(N/8) - 2$ general butterflies and two special butterflies corresponding to $k = 0$ and $N/16$, and repeating successively the process until the size is reduced to a 4 or 2-point DFT without twiddle factors. Therefore, it is seen that the expressions for the number of real multiplications and real additions of the proposed FFT algorithm are, respectively,
$m_N^{33} = 2N - 16 + m_{N/2}^{33} + 4m_{N/8}^{33} \quad (N \geq 16)$
and $a_N^{33} = 6N - 16 + a_{N/2}^{33} + 4a_{N/8}^{33} \quad (N \geq 16)$,
with $m_8^{33} = 4$, $m_4^{33} = m_2^{33} = 0$,
and $a_8^{33} = 52$, $a_4^{33} = 16$, $a_2^{33} = 4$.
Similarly, when a complex multiplication is performed using 4 real multiplications and 2 real additions (4mult-2add scheme), it can be shown that the corresponding numbers are
$m_N^{42} = \frac{5}{2}N - 20 + m_{N/2}^{42} + 4m_{N/8}^{42} \quad (N \geq 16)$
and $a_N^{42} = \frac{11}{2}N - 12 + a_{N/2}^{42} + 4a_{N/8}^{42} \quad (N \geq 16)$,
with $m_8^{42} = 4$, $m_4^{42} = m_2^{42} = 0$,
and $a_8^{42} = 52$, $a_4^{42} = 16$, $a_2^{42} = 4$.

The arithmetic complexities of the 3-butterfly implementation of the proposed FFT algorithm and the 3-butterfly implementation of the split-radix FFT algorithm [2] for complex data for various values of N are given in Tables I and II. In the 3mult-3add scheme, both the algorithms have exactly the same number of arithmetic operations, whereas in the 4mult-2add scheme, the proposed algorithm has a reduced number of multiplications at the cost of a slight increase in the number of additions. However, it is interesting to note that the total number of operations (i.e., multiplications+additions) is exactly the same for the two algorithms.

TABLE I.
Comparison of arithmetic complexities using the 3mult-3add scheme

Size	Algorithm in[2]		Proposed algorithm	
N	Mult	Add	Mult	Add
8	4	52	4	52
16	20	148	20	148
32	68	388	68	388
64	196	964	196	964
128	516	2308	516	2308
256	1284	5380	1284	5380
512	3076	12292	3076	12292
1024	7172	27652	7172	27652
2048	16388	61444	16388	61444
4096	36868	135172	36868	135172

TABLE II.
Comparison of arithmetic complexities using the 4mult-2add scheme

Size	Algorithm in [2]		Proposed Algorithm	
N	Mult	Add	Mult	Add
8	4	52	4	52
16	24	144	24	144
32	84	372	84	372
64	248	912	240	920
128	660	2164	636	2188
256	1656	5008	1592	5072
512	3988	11380	3812	11556
1024	9336	25488	8896	25928
2048	21396	56436	20364	57468
4096	48248	123792	45832	126208

B. Data Transfers

The implementation of the butterfly for a given value

of k_1, constructed using (12) and (13), of the proposed FFT algorithm, consists of reading two points from the external memory of the processor and performing the operations of addition and subtraction using these two points as the operands. The result of the addition is returned to the external memory, whereas that of the subtraction is kept in an internal register of the processor. This process is repeated four times to compute (13) of the butterfly. The four points kept in the processor are used to process the sub-butterfly (12). The results of the additions for $k_1 = 0, 1, ..., (N/8) - 1$ returned to the external memory are grouped to form the first sub-sequence of length $N/2$, and the four subsequences of length $N/8$ are formed by grouping separately the results of each line of the output vector of (12). This scheme of implementation reduces significantly the number of data transfers and index generations. To compare the number of data transfers and index generations of the proposed and existing split-radix FFT algorithms, we assume for a given length, that the respective repeated butterfly operation is applied in the same way to both the algorithms. Then, it can be shown that the expressions for the number of data transfers (real and imaginary parts) of the proposed and existing split-radix algorithms are respectively $D_N^p = 2N + D_{N/2}^p + 4D_{N/8}^p$ for $N \geq 8$, and $D_N^e = 2N + D_{N/2}^e + 2D_{N/4}^e$ for $N \geq 4$, where $D_4^p = 8$, $D_2^p = D_2^e = 4$, $D_1^p = D_1^e = 0$. The number of data transfers for both the algorithms are compared in Table III. The existing split-radix algorithm requires 20% more data transfer operations than the proposed algorithm. Thus, the same savings are also obtained in the index generation. The use of radix-8 Cooley-Tukey FFT can bring about even more savings; however, it will have a higher arithmetic complexity compared to that of the proposed algorithm and imposes more restrictions on the choice of the transform length.

C. Twiddle Factors

In counting the number of twiddle factor (sine and cosine) evaluations or accesses to the lookup table required by each of the split-radix algorithms, it is assumed that the constant factors $1/\sqrt{2}$, $\cos(\pi/8)$ and $\sin(\pi/8)$ are initialized and kept in the internal registers of the processor during the processing time of the corresponding algorithm. However, for the proposed and existing split-radix algorithms, 8 and 4 real factors are respectively required to be evaluated or read from the lookup table by the general butterfly. Therefore, the number of twiddle factor evaluations or accesses to the lookup table can be obtained by counting the required number of the general butterflies in each algorithm. They are respectively $T_N^p = N - 16 + T_{N/2}^p + 4T_{N/8}^p$ for $N \geq 16$, $T_N^e = N - 8 + T_{N/2}^e + 2T_{N/4}^e$ for $N \geq 8$, where $T_8^p = T_4^p = T_2^p = 0$, and $T_4^e = T_2^e = 0$. It is seen from Table III that saving of over 30% in the evaluation of twiddle factors or in the access to the lookup table can easily be achieved by the proposed algorithm. Notice that, when the lookup table is used, identical savings are obtained by the proposed algorithm in the address generation for reading the twiddle factors.

Table III.
Comparison of the number of twiddle factor evaluations or accesses to the lookup table, and data transfers.

Size	Twiddle factors			Data transfers		
N	T_N^e	T_N^p	%	D_N^e	D_N^p	%
8	0	0	00.00	36	24	33.33
16	8	0	100	92	72	21.73
32	32	16	50	228	168	26.31
64	104	64	38.46	540	392	27.40
128	288	176	38.88	1252	936	25.23
256	744	480	35.48	2844	2120	25.45
512	1824	1232	32.45	6372	4712	26.05
1024	4328	2944	31.97	14108	10504	25.54
2048	10016	6896	31.15	30948	23080	25.42
4096	22760	15904	30.12	67356	50120	25.58

IV. CONCLUSION

In this paper we have proposed a split-radix FFT algorithm for computing the DFT of an arbitrary length $N = 2^r$. It is based on the use of a radix-2 and a radix-8 index maps instead of a radix-2 and a radix-4 index maps as in the existing split-radix FFT. It has been shown that the proposed algorithm outperforms the existing split-radix algorithm in terms of the numbers of data transfers, index generations, and twiddle factor evaluations or accesses to the lookup table, without an increase in the arithmetic complexity.

ACKNOWLEDGMENT

This work was supported by NSERC (Canada) and FCAR (Quebec) grants.

REFERENCES

[1] P. Duhamel and H. Hollmann, "Split radix FFT algorithm," Electron. Lett., vol. 20, pp. 14-16, Jan. 5, 1984.

[2] H. V. Sorensen, M. T. Heideman and C. S. Burrus, "On computing the split-radix FFT," IEEE Trans. Acoust., Speech, Signal Processing, vol. ASSP-34, pp. 152-156, Feb. 1986.

[3] J. Granata, M. Conner and R. Tolimieri, "The tensor product: A Mathematical programming language for FFT's and other fast DSP operations," IEEE SP Magazine, pp. 40-48, Jan. 1992.

PERFORMANCE OF AN ADAPTIVE HOMODYNE RECEIVER IN THE PRESENCE OF MULTIPATH, RAYLEIGH-FADING AND TIME-VARYING QUADRATURE ERRORS

Ediz Çetin, Izzet Kale and Richard C. S. Morling

University of Westminster, Department of Electronic Systems,
Applied DSP and VLSI Research Group,
London W1W 6UW, United Kingdom

ABSTRACT

In this paper, we carry out a detailed performance analysis of the blind source separation based I/Q corrector operating at the baseband. Performance of the digital I/Q corrector is evaluated not only under time-varying phase and gain errors but also in the presence of multipath and Rayleigh fading channels. Performance under low-SNR and different modulation formats and constellation sizes is also evaluated. What is more, BER improvement after correction is illustrated. The results indicate that the adaptive algorithm offers adequate performance for most communication applications hence, reducing the matching requirements of the analog front-end enabling higher levels of integration.

1. INTRODUCTION

The homodyne/Zero-IF receivers provide high levels of integration. With this architecture, I/Q signal processing is used to downconvert the RF signal to baseband. However, this architecture, in common with all I/Q architectures, is vulnerable to mismatches between the I and Q channels.

Both analog and digital methods for correcting the I/Q mismatches of homodyne receivers have been proposed in the literature [1]-[6]. This paper explores the performance capability of a non-pilot aided adaptive DSP technique developed for the quadrature receivers in [5], [6]. The ability of the algorithm to work under time-varying phase and gain errors as well as under multipath and Rayleigh fading environments has been explored. Its performance under low-SNR and different modulation formats and constellation sizes is also evaluated.

This paper is organized as follows: Section 2 defines the model of the source separation based zero-IF receiver. Section 3 describes the performance measures and simulation results. Concluding remarks are given in Section 4.

2. ADAPTIVE HOMODYNE RECEIVER

2.1 Architecture

A block diagram and the equivalent system for the source separation based homodyne receiver is given in Figure 1. The incoming signal $s(t)$ consists of the desired/wanted signal $s(t)$ at f_{RF} and can be expressed as:

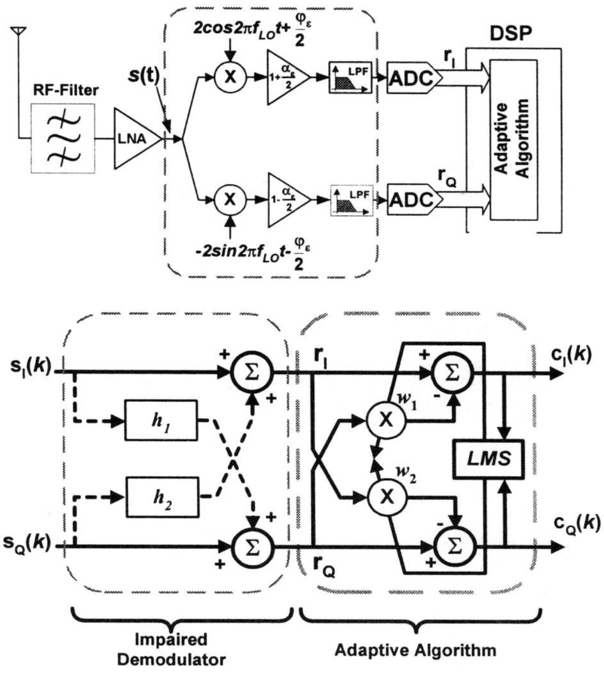

Figure 1 Blind source separation based receiver

$$s(t)=\tfrac{1}{2}[u(t)e^{j2\pi f_{RF}t}+u^{*}(t)e^{-j2\pi f_{RF}t}] \quad (1)$$

where $u(t)$ is the complex envelope of the wanted signal. To simplify the analysis, the whole phase and gain imbalances between the I and Q channels are modelled as an unbalanced quadrature downconverter [5] and can be expressed as:

$$x_{LO}=e^{j2\pi f_{LO}t}(g_1e^{j\frac{\varphi_\varepsilon}{2}}-g_2e^{-j\frac{\varphi_\varepsilon}{2}})+e^{-j2\pi f_{LO}t}(g_1e^{-j\frac{\varphi_\varepsilon}{2}}+g_2e^{j\frac{\varphi_\varepsilon}{2}}) \quad (2)$$

where $g_1=(1+0.5\alpha_\varepsilon)$, $g_2=(1-0.5\alpha_\varepsilon)$ and φ_ε is the phase and α_ε is the gain mismatch between the I and Q channels and $f_{LO}=f_{RF}$. As shown in Figure 1, the received signal $s(t)$ is quadrature mixed with the non-ideal LO signal, x_{LO}, and low-pass filtered resulting in the baseband signal with in-phase and quadrature components:

$$\begin{aligned} r_I(k)&=\overbrace{(1+0.5\alpha_\varepsilon)\cos(\varphi_\varepsilon/2)}^{\psi}s_I(k)+\overbrace{(1+0.5\alpha_\varepsilon)\sin(\varphi_\varepsilon/2)}^{h_2}s_Q(k)\\ r_Q(k)&=\underbrace{(1-0.5\alpha_\varepsilon)\sin(\varphi_\varepsilon/2)}_{h_1}s_I(k)+\underbrace{(1-0.5\alpha_\varepsilon)\cos(\varphi_\varepsilon/2)}_{\gamma}s_Q(k)\end{aligned} \quad (3)$$

0-7803-7761-3/03/$17.00 ©2003 IEEE

where ψ and γ are ≈ 1 and can be safely ignored. The in-phase signal $r_I(k)$ is corrupted by the quadrature signal $r_Q(k)$ leaked due to phase and gain mismatches. A leakage from the quadrature signal into the in-phase signal also exists. Ideally the I and Q channels are not correlated with each other. However, in the presence of the quadrature phase and gain errors this relationship no longer exists and they are correlated. The proposed algorithm, depicted in Figure 1, acts as a decorrelator and tries to de-correlate the I and Q channels hence eliminating phase and gain errors. The source estimates, $c_I(z)$ and $c_Q(z)$ can be expressed as:

$$\begin{aligned} c_I(z) &= (1 - w_1 h_2) s_I(z) + (h_1 - w_1) s_Q(z) \\ c_Q(z) &= (h_2 - w_2) s_I(z) + (1 - w_2 h_1) s_Q(z) \end{aligned} \qquad (4)$$

When the coefficients converge, i.e. $w_1 = h_1$ and $w_2 = h_2$ then the source estimates become:

$$\begin{aligned} c_I(z) &= (1 - h_1 h_2) s_I(z) \\ c_Q(z) &= (1 - h_1 h_2) s_Q(z) \end{aligned} \qquad (5)$$

As it can be seen from (5) the sources have been separated. Furthermore, $(1 - h_1 h_2) \approx 1$ and can be safely ignored. The description blind or unsupervised implies that we do not know the mixing coefficients h_1, h_2, nor the probability distribution of the sources except that they are not correlated.

2.2 BER for QPSK

The effect of gain and phase mismatches on a QPSK constellation is illustrated in Figure 2.

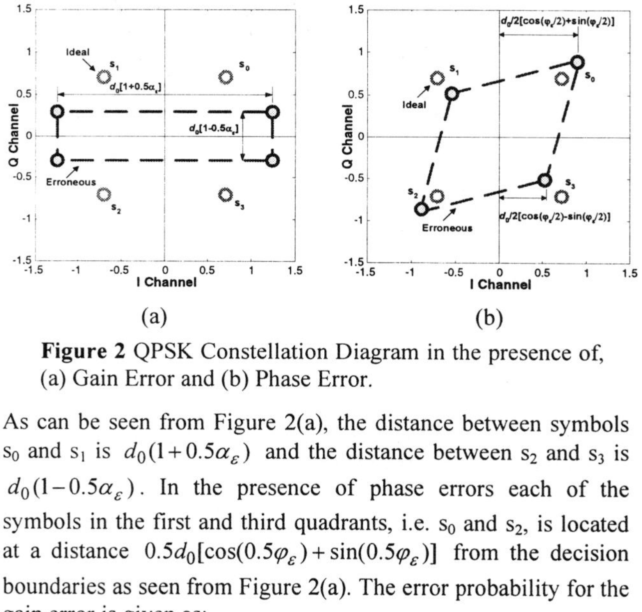

Figure 2 QPSK Constellation Diagram in the presence of, (a) Gain Error and (b) Phase Error.

As can be seen from Figure 2(a), the distance between symbols s_0 and s_1 is $d_0(1 + 0.5\alpha_\varepsilon)$ and the distance between s_2 and s_3 is $d_0(1 - 0.5\alpha_\varepsilon)$. In the presence of phase errors each of the symbols in the first and third quadrants, i.e. s_0 and s_2, is located at a distance $0.5 d_0[\cos(0.5\varphi_\varepsilon) + \sin(0.5\varphi_\varepsilon)]$ from the decision boundaries as seen from Figure 2(a). The error probability for the gain error is given as:

$$P_b \approx \frac{1}{2}\left\{ Q\left[(1 + 0.5\alpha_\varepsilon)\sqrt{\frac{2E_b}{N_0}} \right] + Q\left[(1 - 0.5\alpha_\varepsilon)\sqrt{\frac{2E_b}{N_0}} \right] \right\} \qquad (6)$$

and for the phase error:

$$P_b \approx \frac{1}{2}\left\{ Q\left(\cos\left(\frac{\varphi_\varepsilon}{2}\right) + \sin\left(\frac{\varphi_\varepsilon}{2}\right)\right)\left[\sqrt{\frac{2E_b}{N_0}}\right] + Q\left[\left(\cos\left(\frac{\varphi_\varepsilon}{2}\right) - \sin\left(\frac{\varphi_\varepsilon}{2}\right)\right)\sqrt{\frac{2E_b}{N_0}}\right] \right\} \qquad (7)$$

Equations (6) and (7) express the BER as a function of phase and gain errors for QPSK modulated signals.

3. PERFORMANCE ANALYSIS

3.1 Simulation Setup

The performance of the proposed structure is analysed considering QPSK and 16-QAM signals with ideal symbol rate sampling. AWGN and Multipath Rayleigh Fading channels were assumed with phase and gain error values varying from 30° to 7.5° and 3dB to 1dB respectively are investigated.

The performance of the adaptive algorithm is characterized by the modelling-error [6]. This gives a global figure for the quality of the identification of the coupling coefficients h_1 and h_2 by w_1 and w_2. It is defined as the squared norm of the difference of the values between the original coefficients used in the mixture and the estimated coefficients, relative to the squared norm of the mixture coefficients. Another performance measure used is the *Image Rejection Ratio* (IRR) [5] that can be achieved. This can be interpreted as the *Signal-to-noise-Ratio* (SNR) in the desired channel.

3.2 Tracking Capabilities

Another performance measure is the capability of the adaptive algorithm in tracking non-stationary environments i.e. time varying phase and gain errors. In order to show the robustness of the proposed approach we start by adapting the coefficients to 15° phase and 0.34 dB of gain error. After 1750 frames, the amplitude imbalance is changed linearly from 0.34 dB to 0.87 dB. After 750 frames, an abrupt change from 0.87 dB to 3 dB is made and the phase error is abruptly changed to 30°. After 1500 frames, the phase error is changed linearly from 30° to reach 40° for the next 500 frames. Figure 3 depicts the tracking capability of the proposed algorithm for (a) QPSK and (b) 16-QAM modulation schemes in the presence of an AWGN channel with an SNR of 20 dB.

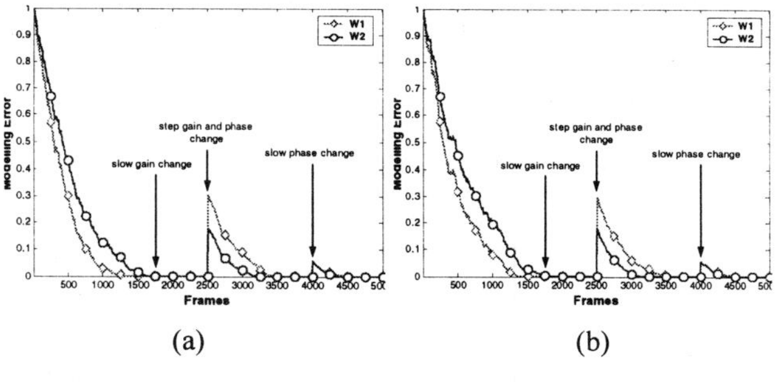

Figure 3 Tracking capability of the proposed algorithm, (a) QPSK and (b) 16-QAM case.

As can be seen from Figure 3, a sudden change in the mixture coefficients, phase and gain errors, does not cause the algorithm to diverge and the algorithm tracks the changes rapidly and the modelling error is zeroed. In addition, the compensator performance is not affected by time-variant phase and gain errors. This indicates that the proposed method is also capable of tracking time-varying imbalances.

3.3 Multi-path and Fading Channels

Another performance measure is the capability of the adaptive algorithm in fading and multi-path environments. The robustness of the proposed approach in a more realistic environment than the AWGN channel is demonstrated using a Rayleigh Fading channel with multipath. Figure 4 depicts the channel profiles, received signal power over time for (a) slow fading and (b) fast fading with a multipath Rayleigh channel.

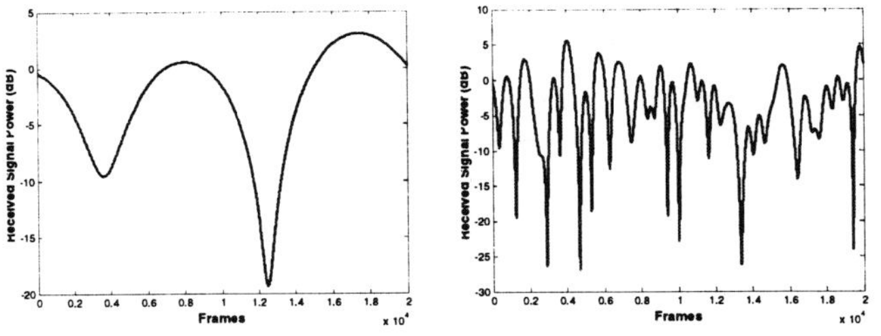

(a) (b)

Figure 4 Channel profiles for (a) slow fading and (b) fast fading, multipath Rayleigh Channel.

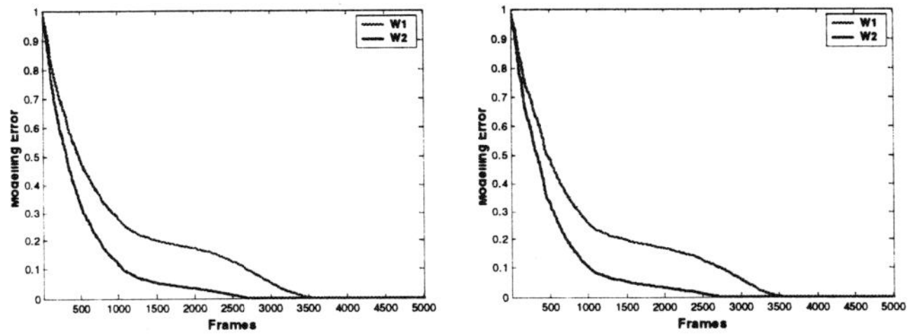

(a) (b)

Figure 5 Modelling error for first (a) QPSK and (b) 16-QAM for phase error of 30° and amplitude imbalance of 3 dB, for slow fading.

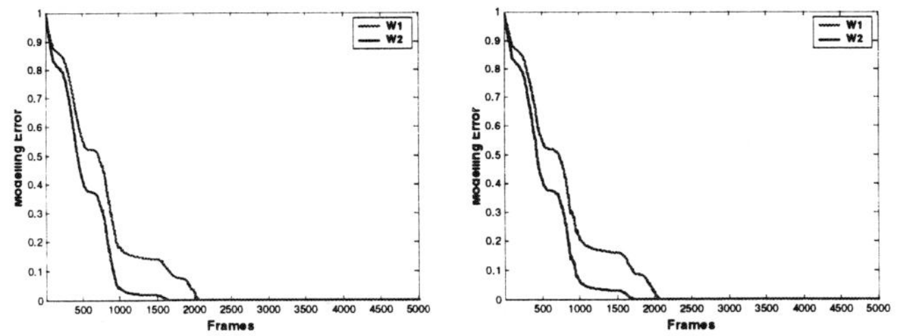

(a) (b)

Figure 6 Modelling error for first (a) QPSK and (b) 16-QAM for phase error of 30 ° and amplitude imbalance of 3 dB, for fast fading.

As can be seen from Figures 5 and 6, the proposed algorithm is able to work under both slow fading and fast fading multipath channels and the modelling error is effectively zeroed. Table 1 depicts the resulting tap estimates w_1 and w_2, residual gain and phase errors and the steady-state IRR for QPSK and 16-QAM modulated signals in slow and fast fading multipath environments.

3.4 BER Improvement for QPSK

In Section 2.2 the BER degradation for a QPSK signal in the presence of phase and gain errors was derived. The resulting BER after the application of the proposed algorithm is shown in Figure 7 where the ideal BER, BER in the presence of phase error of 30° and an amplitude imbalance of 3dB and the BER after the application of the proposed correction scheme are depicted.

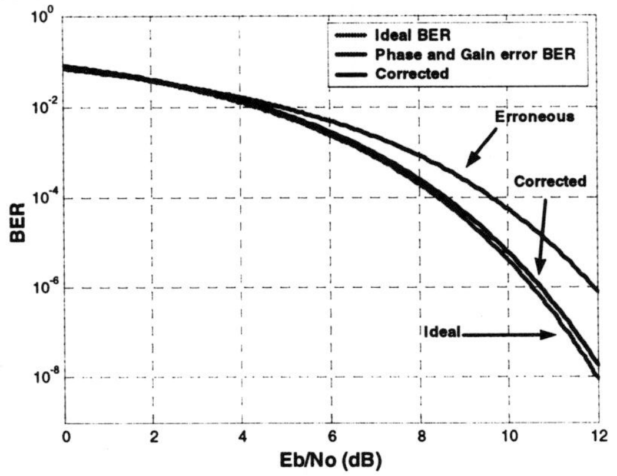

Figure 7 BER improvement.

3.5 Different Modulation Formats and Low SNR

The performance of the proposed algorithm under low SNR values is shown in Table 2, where the SNR required to achieve 10^{-1} BER is shown. For 8-PSK and 32-PSK as well as 32-QAM and 256-QAM cases the algorithm is able to identify the mixing matrix hence eliminating the phase and gain errors. IRR in the order of 58–93.3 dB after imbalance compensation was shown to be achievable. This IRR is much more than the required amount.

4. CONCLUDING REMARKS

In this paper we have reported on the performance of a source separation based compensator for homodyne receivers. The properties of the algorithm can be summarized as follows:

- The algorithm enables fast and very accurate I/Q imbalance compensation at low cost. Simulation results show IRR values from 58-94 dB after compensation.
- The algorithm compensates for phase and gain imbalances in the whole receiver chain not only those introduced by the quadrature downconverter.
- The algorithm is able to work under multipath and Rayleigh fading environments.
- The algorithm is able to work under low SNR.
- The algorithm works on the fly and is able to track time-varying errors.
- The algorithm works with any type of modulation formats and constellation sizes, since compensation takes place before any modulation specific operation.
- The algorithm is very simple to implement consisting of two, single-tap adaptive filters with LMS coefficient update hardware.

Modulation type		Before Correction					After Correction				
		Gain Error (dB)	Phase Error (deg)	h_1	h_2	IRR (dB)	Gain Error (dB)	Phase Error (deg)	w_1	w_2	IRR (dB)
Slow Fading	QPSK	3	30	0.3018	0.2158	10.0	1.0e-4	4.3e-4	0.3016	0.2156	73.2
	16-QAM						0.0018	4.4e-4	0.3016	0.2156	72.3
Fast Fading	QPSK						0.0121	1.8e-5	0.3020	0.2156	63.1
	16-QAM						0.0059	1.4e-4	0.3020	0.2158	69.2

Table 1 Parameter values for QPSK and 16-QAM for slow and fast fading multipath environments.

Modulation type	Before Correction					After Correction				
	Gain Error (dB)	Phase Error (deg)	h_1	h_2	IRR (dB)	Gain Error (dB)	Phase Error (deg)	w_1	w_2	IRR (dB)
8-PSK SNR= 5.6 Eb/No=0.87	3	30	0.3018	0.2158	10.0	1.9e-4	3.7e-5	0.3018	0.2158	93.3
	2	15	0.1454	0.1155	15.2	0.0145	5.0e-5	0.1453	0.1157	61.5
	1	7.5	0.0691	0.0616	21.2	0.0073	1.7e-6	0.0692	0.0616	67.5
32-PSK SNR=13.7 Eb/No=6.7	3	30	0.3018	0.2158	10.0	0.0012	2.4e-4	0.3017	0.2157	77.3
	2	15	0.1454	0.1155	15.2	0.0034	8.7e-5	0.1454	0.1156	74.1
	1	7.5	0.0691	0.0616	21.2	0.0218	1.9e-4	0.0691	0.0615	57.9
32-QAM SNR=15.5 Eb/No=8.5	3	30	0.3018	0.2158	10.0	9.1e-4	2.4e-4	0.3017	0.2157	77.7
	2	15	0.1454	0.1155	15.2	0.0120	6.8e-5	0.1454	0.1156	63.2
	1	7.5	0.0691	0.0616	21.2	0.0032	7.2e-5	0.0691	0.0616	74.6
256-QAM SNR=24.9 Eb/No=15.9	3	30	0.3018	0.2158	10.0	0.0068	7.1e-5	0.3017	0.2160	68.1
	2	15	0.1454	0.1155	15.2	0.0108	3.9e-5	0.1454	0.1157	64.1
	1	7.5	0.0691	0.0616	21.2	0.0045	4.7e-5	0.0692	0.0617	71.7

Table 2 Parameter values for BER of 10^{-1}.

5. REFERENCES

[1] Crols, J. and M.S.J. Steyaert, "Low-IF Topologies for High-Performance Analog Front Ends of Fully Integrated Receivers", *IEEE Transactions on Circuits and Systems II: Analog and Digital Signal Processing*, vol. 45, issue 3, pp. 269–282, March 1998.

[2] Lohtia, A., P. Goud and C. Englefield, "An Adaptive Digital Technique for Compensating for Analog Quadrature Modulator/Demodulator Impairments", *IEEE Pacific Rim Conference on Communications, Computers and Signal Processing*,vol. 2, pp. 447–450, May 1993.

[3] Churchill, F.E., G.W. Ogar and B.J. Thompson, "The Correction of I and Q Errors in a Coherent Processor", *IEEE Transactions on Aerospace and Electronic Systems*, vol. AES-17, no.1, pp. 131–137, January 1981.

[4] McLeod, M.D., "Fast Calibration of IQ Digitiser Systems", *IEE Colloquium on System Aspects and Applications of ADCs for Radar, Sonar and Communications*, pp. 1–4, November 1987.

[5] Cetin, E., I. Kale and R. C. S. Morling, "Adaptive Compensation of Analog Front-end I/Q Mismatches in Digital Receivers", *International Symposium on Circuits and Systems, (ISCAS 2001)*, vol. 4, pp. 370-373, May 2001.

[6] Cetin, E., I. Kale and R. C. S. Morling, "On The Structure, Convergence And Performance of An Adaptive I/Q Mismatch Corrector", *IEEE Vehicular Technology Conference (VTC 2002 Fall) Pathway to Ubiquitous Wireless Communications*, vol. 4, pp. 2288-2292, September 2002.

ON-LINE SIGNATURE VERIFICATION METHOD UTILIZING FEATURE EXTRACTION BASED ON DWT

I. Nakanishi†, *N. Nishiguchi*‡, *Y. Itoh*‡, *Y. Fukui*‡

†Faculty of Education and Regional Sciences, Tottori University
4-101 Koyama-minami, Tottori-shi, 680-8551 Japan
‡Faculty of Engineering, Tottori University
4-101 Koyama-minami, Tottori-shi, 680-8552 Japan

ABSTRACT

In this paper, we propose a new signature verification method. Parameters of the on-line signature are decomposed into multi-level signals by utilizing the DWT (discrete wavelet transform). Personal features are emphasized by the DWT sub-band decomposition and extracted. The extracted features are verified using the adaptive algorithm. On-line signature is subtly various comparing with other biometrics, therefore, there is fluctuation in the number of strokes. Then, we also propose a method which is robust to fluctuation of the number of strokes by using the DP (dynamic programming) matching. Through computer simulations, the effectiveness of these proposed methods is confirmed.

1. INTRODUCTION

Information services over the internet such as the electronic commerce and the information resource management have been developed. However, there is a possibility that the wiretapping and the data alteration in a network or database, and the deception of a genuineness using non-meeting may arise. Then, it is important to verify the user from its forgery, in especially a non-meeting commercial transaction. Conventionally, as the user authentication method, the static biometrics such as fingerprint scan, hand geometry, iris patterns and face have been utilized. However, they require special detective equipments and advanced processing software, and restrict the surroundings which can be used. To the contrary, on-line signature which utilizes the difference of personal dynamic features such as the stroke order, the pen-pressure and the pen-inclination brings high security by simple operation in the PDA (personal digital assistants) system with the pen input method. It is superior in user receptiveness[1],[2].

In this paper, we propose a new signature verification method. Time-varying parameters of text dependent on-line signatures are decomposed into multi-level detail signals, high-frequency components and an approximation, low-frequency component by using the DWT sub-band decompositions. In the filtering process, the extracted details are the signals which emphasize the nuances in each frequency band, and become personal features. They are then verified using the adaptive algorithm. On-line signature subtly varies unlike other biometrics; therefore, there is fluctuation in the number of strokes. We also propose a method which is independent of the fluctuation to number of strokes utilizing DP matching[3],[4]. Through computer simulations, the effectiveness of the proposed method is examined.

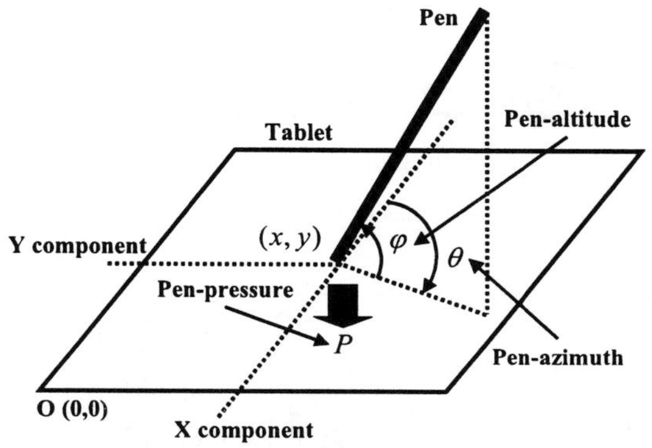

Fig.1 On-line signature parameters

2. FEATURE EXTRACTION

2.1 Extraction of Signature Parameters

We digitize on-line signature with a pen-tablet as shown in Fig.1 and extract the time-varying parameters such as x and y components concerning the pen-position, the pen-pressure P, the pen-altitude φ and pen-azimuth θ. The pen-position parameter is normalized to decrease the personal fluctuation. Therefore, we define the normalized pen-position as

$$x^*(t) = \frac{x(t) - x_{\min}}{x_{\max} - x_{\min}} \cdot \alpha_x \quad \left(x_{\min} \le x(t) \le x_{\max}\right) \tag{1}$$

$$y^*(t) = \frac{y(t) - y_{\min}}{y_{\max} - y_{\min}} \cdot \alpha_y \quad \left(y_{\min} \le y(t) \le y_{\max}\right) \tag{2}$$

where $x(t)$ and $y(t)$ are the original pen-position parameters. α_x and α_y are coefficients for scaling each parameter, and these are 100 in this paper. Subscripts max and min correspond to each maximum and minimum value, respectively.

2.2 Signature Data

Before signing, the writer is required to do some practices so as to get accustomed to using the pen-tablet. Also, when the writers sign, the genuine authors are not able to refer their original signatures. On the other hand, the forgers are made to admit signing over referring character shapes of genuine signatures. An example of the genuine signature and its forgery is shown in Fig.2. On-line parameters about x and y components concerning the pen-position are shown in Fig.3. It is clear that x and y components concerning the pen-position of the forgery are

0-7803-7761-3/03/$17.00 ©2003 IEEE

similar to those of its genuine signature in the time-domain signal. That is to say, it is difficult to discriminate signature in the time-domain signal, especially for x component. Also, the pen-position data is easy to forge since it remains in the written signature. As space is limited, pen-pressure, pen-altitude and pen-azimuth are not listed. However, inversely, it is obvious that these parameters have personal features, so that it is easy to discriminate writers.

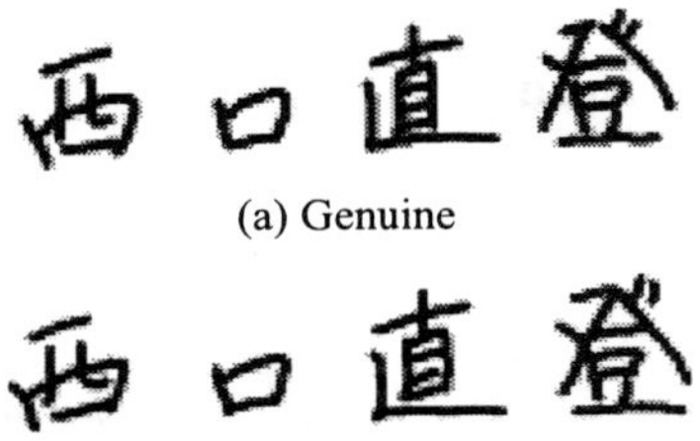

(a) Genuine

(b) Forgery

Fig.2 Examples of the signature

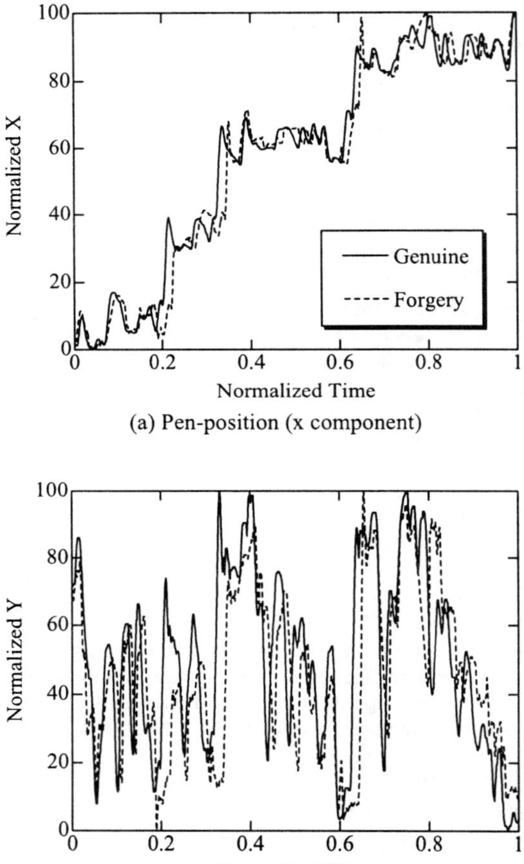

(a) Pen-position (x component)

(b) Pen-position (y component)

Fig.3 On-line parameters

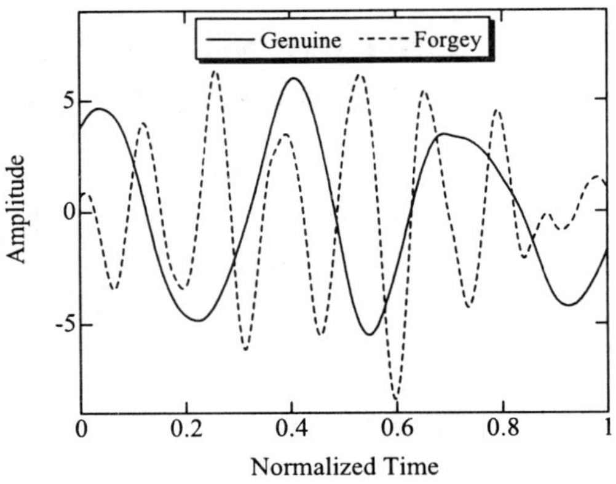

Fig.4 DWT detail signals at level 8

Also, they are not visible features; therefore, their imitation is impossible even if the writing motion is watched. As a result, these on-line parameters are advantageous to signature verification. However, particular pen-input terminal is required to detect them, so that they are inferior in portability. In this paper, we try to make it easy to discriminate signatures by utilizing the pen-position parameters as the verification targets which require no additional pen-input terminal.

2.3 Feature Extraction by DWT

Time series pen-position parameter is decomposed into multi-level *details* and an *approximation* by utilizing the DWT. At one-stage DWT sub-band decomposition, high-frequency component is the detail and low-frequency component is the approximation. In the filtering process, the details are the signals which make the nuances emphasize in each frequency band, and become personal features. For example, x components at the time-domain of Fig.3 are decomposed into 8 levels by using Daubechies8 filters. The details at frequency level 8 are shown in Fig.4. These results show that the DWT analysis makes it easy to extract the difference between the genuine signature and its forgery while it is difficult in the time-domain. By using the DWT analysis, we require no particular functions such as pen-pressure and pen-inclination with the pen-input terminal. Moreover, by considering all level results, the number of comparative targets is increased, so that we can obtain higher accuracy of verification than that in each level.

3. SIGNATURE VERIFICATION

In this section, a signature verification method utilizing adaptive algorithm is proposed. A block diagram of the proposed signature verification system is illustrated in Fig.5. At verification phase, the on-line parameter concerning the pen-position is decomposed by using the DWT, and the details are extracted as the personal features. In the verification algorithm, matching of time series between template and verification signal are needed. Then, both number of strokes are matched by DP matching using the pen-pressure parameter. Where the template is the average of past five details of the genuine signature. Moreover, the adaptive processing is applied to each level detail. The converged value of the updated adaptive weight is outputted at each level. The total verification is achieved by considering all level results. In this paper, the pen-pressure parameter is formed into 2 values given by Eq.(3).

$$P^*(t)=\begin{cases}1 & (P(t)>0)\\ 0 & (P(t)=0)\end{cases} \quad (3)$$

3.1 Stroke Corresponding by DP Matching

We discuss the method which makes number of strokes both input and template signature match by using the DP matching. In this paper, the difference of number of strokes γ between template and input signature are extended within ±2 strokes. That is to say, if $\gamma>2$, it is decided that the input signature is a forgery. First, γ is calculated. If $0\le\gamma\le 2$, the number of strokes are reduced by coupling strokes for the signature which has larger number of strokes.

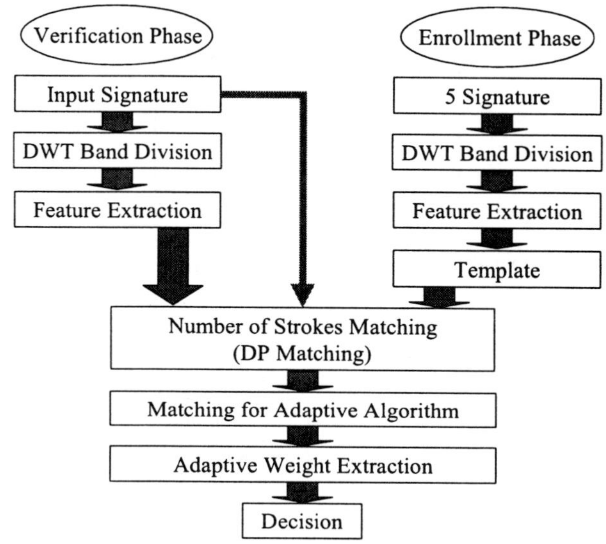

Fig.5 Block diagram of proposed signature verification system

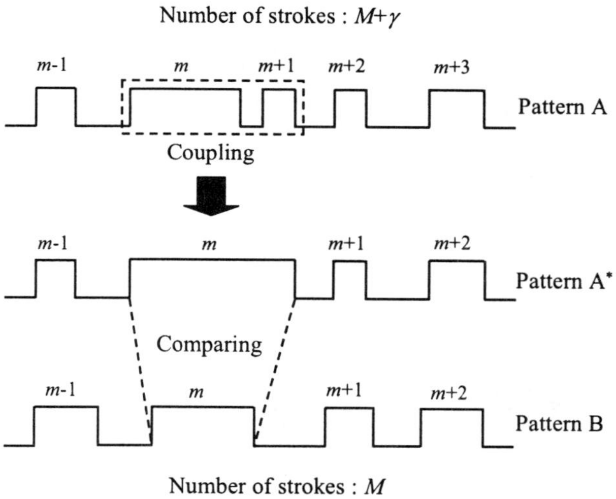

Fig.6 Stroke corresponding by using pen-pressure parameter

As shown in Fig.6, for example, in the case of $\gamma=1$, one set of case coupled two strokes is formed for a pen-pressure parameter which has $M+\gamma$ strokes as pattern A, and by this process, number of strokes become M as pattern A^*. Next, the DP matching is implemented for both pattern A^* and pattern B. These processes are performed to all cases, and the coupled stroke which DP distance is the minimum value, becomes a coupling target. Also, in the case of $\gamma=2$, two sets of case coupled two strokes or one set of case coupled three strokes are formed for a pen-pressure parameter which has $M+\gamma$ strokes. Next process is the same as that of $\gamma=1$. In the case of $\gamma=0$, anything is not implemented. When number of strokes for both input pattern ψ and template one Φ are matched, the DP distance between input pattern ψ_m and template one Φ_m in *m-th* stroke is defined as follows;

$$D(\Psi_m,\Phi_m)=\frac{g_m(I_m-1,J_m-1)}{I_m+J_m} \quad (4)$$

$$g_m(i,j)=\min\begin{bmatrix} g_m(i,j-1)+d_m(i,j)\\ g_m(i-1,j-1)+2d_m(i,j)\\ g_m(i-1,j)+d_m(i,j)\end{bmatrix} \quad (5)$$

$$(i=1,2,\cdots,I_m-1;\quad j=1,2,\cdots,J_m-1)$$

$$g_m(0,0)=2d_m(0,0) \quad (6)$$

where I_m, J_m are number of samples contained in *m-th* stroke of input and template, respectively. $d_m(i,j)$ is the distance between data when *i-th* sample of input and *j-th* sample of template contained in *m-th* stroke correspond, and given by Eq.(7)

$$d_m(i,j)=\left|t_\Psi^m(i)-t_\Phi^m(j)\right| \quad (7)$$

where $t_\Psi^m(i)$, $t_\Phi^m(j)$ are normalized time at *i-th* sample of input and at *j-th* sample of template contained in *m-th* stroke, respectively. The normalized time is the discrete time to all writing time and value which divided all number of samples into each sample.

3.2 Feature Selection from the Detail

Through the preparatory experiments, we found that only the intra-stroke of the detail was suitable for the matching. The inter-stroke was greatly influenced by the personal fluctuation, so that it was unsuitable for the personal verification. The intra-stroke and the inter-stroke are pen tip movement from pen-down to pen-up and from pen-up to pen-down, respectively. In this paper, to extract the intra-stroke, we utilized the pen-pressure parameter for convenience. However, the intra-stroke can be extracted by using whether the pen tip is on the tablet or not instead of the pen-pressure parameter. In such a case, no particular equipment is necessary.

3.3 Adaptive Process for Signature Verification

An adaptive processing for signature verification at level k is shown in Fig.7, where $x_k(n)$ is the input detail and $d_k(n)$ is the template at level k. $e_k(n)$ is the error signal. Matching in each time series within a stroke is implemented as follows. First, the writing time both the input and the template are normalized at each stroke unit. In this process, the normalized time is the discrete time and value which divided number of samples into each sample at each stroke unit. Next, based on the template, each sample which the distance between normalized time both the input and the template is the nearest, is matched. Then matching of 1 to 1 is implemented. The coefficient $w_k(n)$ is adaptively updated based on the LMS (least-mean-square) algorithm given by[5]

$$w_k(n+1) = w_k(n) + \mu_k e_k(n) x_k(n) \tag{8}$$

$$e_k(n) = d_k(n) - w_k(n) x_k(n) \tag{9}$$

where μ_k is the step size parameter. Ideally, the adaptive coefficient converges on 1 since the similarity is strong if the input signal $x_k(n)$ is of the genuine signature. Otherwise, it converges on smaller value than 1. The signature verification is achieved by whether the converged value of the adaptive coefficient is nearly 1 or not.

4. SIMULATION RESULTS

The number of iterations of the adaptive algorithm in Fig.7 was set as 100 thousand in order to obtain fully convergence of adaptive coefficient. For using only one adaptive weight, the computational complexity for convergence is reduced. The levels of the details used for the matching were from 5 to 8. The details at level 1 to 4 have large personal fluctuations since they correspond to small fluctuation of pen movements. The step size parameters μ_k are shown in Table 1.

Fig.8 shows the convergence characteristics of the adaptive coefficient $w_k(n)$ in x component at level 8. When the input signal was of the genuine signature, the adaptive coefficient converged on 1 closer than that of its forgery. Consequently, it was confirmed that the proposed method was capable of distinguishing the genuine signature from its forgery.

5. SUMMARY

We presented a new on-line signature verification method utilizing feature extraction based on DWT. In the proposed method, time-varying parameters concerning the pen-position of on-line signature are decomposed into the multi-level details by using the DWT. Then, the details become the features which make the nuances emphasize in each frequency band. They were verified using the adaptive algorithm. Moreover, we also proposed a method which is independent of the fluctuation to number of strokes utilizing the DP matching. Consequently, discrimination whether the writer was the genuine or not could be performed effectively. Signature verification method considering multi-level results or introduction of the nonlinear processing like the neural networks which is superior in the pattern recognition is studied further.

6. REFERENCES

[1] M. Yoshimura and I. Yoshimura, "Writer recognition the state-of-the-art and issues to be addressed (in Japanese)," IEICE Technical Report, PRMU96-48, Jun. 1996.

[2] Y. Yamazaki and N. Komatsu, "Extraction of Personal Features from On-Line Handwriting Information in Context-Independent Characters," IEICE Trans. Fundamentals, vol.E83-A, no.10, pp.1955-1962, Oct. 2000.

[3] C.J. Jin, M. Watanabe, T. Kawashima and Y. Aoki, "On-Line Signature Verification by Non-public Parameters (in Japanese)," IEICE Trans., vol.J75-D-2, no.1, pp.121-127, Jan. 1992.

[4] Y. Komiya, T. Ohishi and T. Matsumoto, "A Pen Input On-Line Signature Verifier Integrating Position, Pressure and Inclination Trajectories," IEICE Trans. Inf. & Syst., vol.E84-D, no.7, pp.833-838, Jul. 2001.

[5] S. Haykin, Introduction to Adaptive Filters, Macmillan publishing Company, New York, 1984.

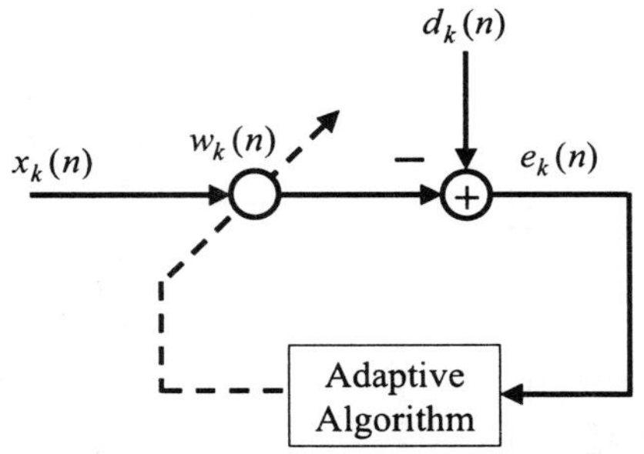

Fig.7 Adaptive process for signature verification

Table 1 Step size parameters

level	step size parameter	
	x component	y component
5	0.0001	0.00001
6	0.00003	0.000003
7	0.00001	0.000001
8	0.00001	0.000001

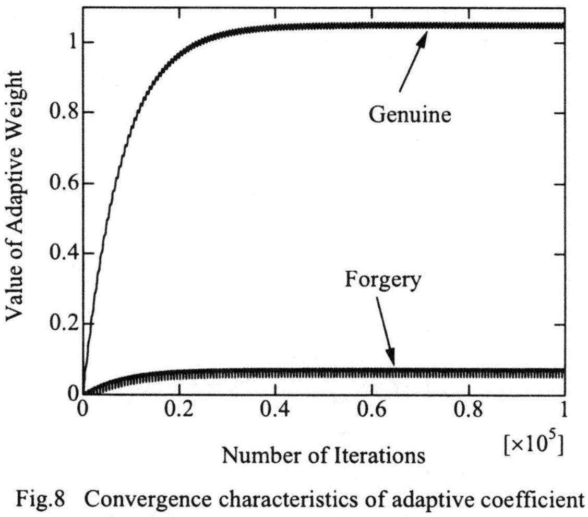

Fig.8 Convergence characteristics of adaptive coefficient at level 8

DESIGN OF A DIGITAL REACTION-DIFFUSION SYSTEM FOR RESTORING BLURRED FINGERPRINT IMAGES

Koichi Ito, Takafumi Aoki and Tatsuo Higuchi

Graduate School of Information Sciences, Tohoku University
Aoba-yama 05, Sendai, 980-8579 JAPAN
Phone: +81-22-217-7169, Fax: +81-22-263-9308,
E-mail: ito@aoki.ecei.tohoku.ac.jp

ABSTRACT

This paper presents an algorithm for fingerprint image restoration using a Digital Reaction-Diffusion System (DRDS). The DRDS is a model of a discrete-time discrete-space nonlinear reaction-diffusion dynamical system, which is useful for generating biological textures, patterns and structures. This paper focuses on design and evaluation of a special DRDS having a capability of restoring incomplete fingerprint images. The phase-only image matching technique is employed to evaluate the similarity between the original fingerprint images and the restored images. The proposed algorithm is useful for person identification applications using fingerprint images.

1. INTRODUCTION

Living organisms can create a remarkable variety of patterns and forms from genetic information. In embryology, the development of patterns and forms is sometimes called *Morphogenesis*. In 1952, Alan Turing suggested that a system of chemical substances, called *morphogens*, reacting together and diffusing through a tissue, is adequate to account for the main phenomena of morphogenesis [1]. From an engineering viewpoint, the insights into morphogenesis provide important concepts for devising a new class of intelligent signal processing algorithms inspired by biological pattern formation phenomena [2].

Recently, we have proposed a framework of *Digital Reaction-Diffusion System* (DRDS) – a discrete-time discrete-space reaction-diffusion dynamical system – for designing signal processing models exhibiting active pattern/texture formation capability [3]. This paper describes an application of the DRDS to fingerprint image restoration. The problem considered here is to restore the original fingerprint patterns from blurred fingerprint images. We design a special reaction-diffusion system to generate the most likely fingerprint pattern for a given incomplete fingerprint image. The proposed system is useful for identifying a person even from a blurred fingerprint image and could enhance the performance of conventional fingerprint identification systems. The restoration capability is evaluated by using the phase-only matching technique [4] for fingerprint identification, which has already been applied to practical fingerprint identification systems by the authors' group [5].

2. DIGITAL REACTION-DIFFUSION SYSTEM

A Digital Reaction-Diffusion System (DRDS) – a model of a discrete-time discrete-space reaction-diffusion dynamical system – can be naturally derived from the original reaction-diffusion system defined in continuous space and time. The general M-morphogen reaction-diffusion system with two-dimensional (2-D) space indices (r_1, r_2) is written as

$$\frac{\partial \tilde{\boldsymbol{x}}(t, r_1, r_2)}{\partial t} = \tilde{\boldsymbol{R}}(\tilde{\boldsymbol{x}}(t, r_1, r_2)) + \tilde{\boldsymbol{D}}\nabla^2 \tilde{\boldsymbol{x}}(t, r_1, r_2), \quad (1)$$

where

$\tilde{\boldsymbol{x}} = [\tilde{x}_1, \tilde{x}_2, \cdots, \tilde{x}_M]^T$,
$\tilde{x}_i$: concentration of the i-th morphogen,
$\tilde{\boldsymbol{R}}(\tilde{\boldsymbol{x}}) = [\tilde{R}_1(\tilde{\boldsymbol{x}}), \tilde{R}_2(\tilde{\boldsymbol{x}}), \cdots, \tilde{R}_M(\tilde{\boldsymbol{x}})]^T$,
$\tilde{R}_i(\tilde{\boldsymbol{x}})$: reaction kinetics for the i-th morphogen,
$\tilde{\boldsymbol{D}} = diag[\tilde{D}_1, \tilde{D}_2, \cdots, \tilde{D}_M]$,
$diag$: diagonal matrix,
$\tilde{D}_i$: diffusion coefficient of the i-th morphogen.

We now sample a continuous variable $\tilde{\boldsymbol{x}}$ in (1) at the time sampling interval T_0, and at the space sampling intervals T_1 and T_2. Assuming discrete time-index to be given by n_0 and discrete space indices to be given by (n_1, n_2), we have

$$\boldsymbol{x}(n_0, n_1, n_2) = \tilde{\boldsymbol{x}}(n_0T_0, n_1T_1, n_2T_2). \quad (2)$$

Using this discritization, the general DRDS can be obtained as

$$\begin{aligned}\boldsymbol{x}(n_0+1, n_1, n_2) &= \boldsymbol{x}(n_0, n_1, n_2) \\ &+\boldsymbol{R}(\boldsymbol{x}(n_0, n_1, n_2)) + \boldsymbol{D}(l * \boldsymbol{x})(n_0, n_1, n_2), \quad (3)\end{aligned}$$

where

$$\begin{aligned}\boldsymbol{x} &= [x_1, x_2, \cdots, x_M]^T, \\ \boldsymbol{R} &= T_0\tilde{\boldsymbol{R}} = [R_1(\boldsymbol{x}), R_2(\boldsymbol{x}), \cdots, R_M(\boldsymbol{x})]^T, \\ \boldsymbol{D} &= T_0\tilde{\boldsymbol{D}} = diag[D_1, D_2, \cdots, D_M],\end{aligned}$$

$$l(n_1, n_2) = \begin{cases} \frac{1}{T_1^2} & (n_1, n_2) = (-1, 0), (1, 0) \\ \frac{1}{T_2^2} & (n_1, n_2) = (0, -1), (0, 1) \\ -2(\frac{1}{T_1^2} + \frac{1}{T_2^2}) & (n_1, n_2) = (0, 0) \\ 0 & \text{otherwise,} \end{cases}$$

0-7803-7761-3/03/$17.00 ©2003 IEEE

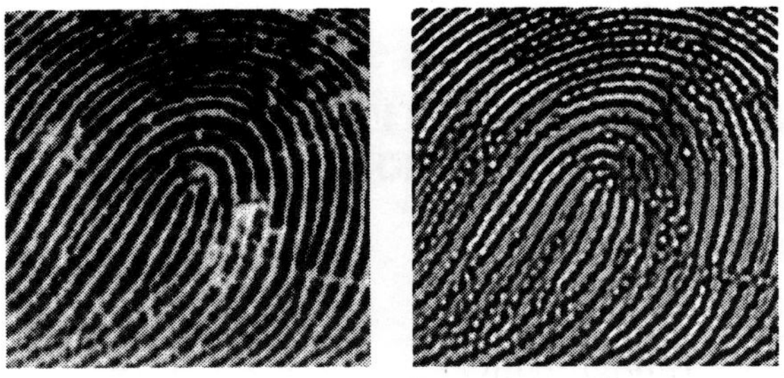

(a) (b)

Figure 1: Enhancement of a fingerprint image: (a) original image, (b) enhanced image.

and $*$ is the spatial convolution operator defined as

$$
\begin{aligned}
(l * \boldsymbol{x})(n_0, n_1, n_2)
&= \begin{bmatrix} (l * x_1)(n_0, n_1, n_2) \\ (l * x_2)(n_0, n_1, n_2) \\ \vdots \\ (l * x_M)(n_0, n_1, n_2) \end{bmatrix} \\
&= \begin{bmatrix} \sum_{p_1=-1}^{1} \sum_{p_2=-1}^{1} l(p_1, p_2) x_1(n_0, n_1 - p_1, n_2 - p_2) \\ \sum_{p_1=-1}^{1} \sum_{p_2=-1}^{1} l(p_1, p_2) x_2(n_0, n_1 - p_1, n_2 - p_2) \\ \vdots \\ \sum_{p_1=-1}^{1} \sum_{p_2=-1}^{1} l(p_1, p_2) x_M(n_0, n_1 - p_1, n_2 - p_2) \end{bmatrix}.
\end{aligned}
$$

In this paper, we use the two-morphogen DRDS ($M = 2$) with the Brusselator reaction kinetics, which is one of the most widely studied chemical oscillators [6]. The two-morphogen Brusselator-based DRDS is defined as follows:

$$
\begin{bmatrix} x_1(n_0{+}1, n_1, n_2) \\ x_2(n_0{+}1, n_1, n_2) \end{bmatrix} = \begin{bmatrix} x_1(n_0, n_1, n_2) \\ x_2(n_0, n_1, n_2) \end{bmatrix}
+ \begin{bmatrix} R_1(x_1(n_0, n_1, n_2), x_2(n_0, n_1, n_2)) \\ R_2(x_1(n_0, n_1, n_2), x_2(n_0, n_1, n_2)) \end{bmatrix}
+ \begin{bmatrix} D_1(l * x_1)(n_0, n_1, n_2) \\ D_2(l * x_2)(n_0, n_1, n_2) \end{bmatrix}, \tag{4}
$$

where

$$
\begin{aligned}
R_1(x_1, x_2) &= T_0 \left\{ k_1 - (k_2 + 1) x_1 + x_1^2 x_2 \right\}, \\
R_2(x_1, x_2) &= T_0 \left(k_2 x_1 - x_1^2 x_2 \right).
\end{aligned}
$$

In this paper, we employ the parameter set: $k_1 = 2$, $k_2 = 4$, $T_0 = 0.01$, $D_1 = T_0$ and $D_2 = 5T_0$.

The DRDS thus defined can be used to enhance fingerprint patterns [3]. To do this, we first set the initial fingerprint image in $x_1(0, n_1, n_2)$, at time 0. Note that spatial sampling parameters T_1 and T_2 should be adjusted according to the inherent spatial frequency of the given fingerprint image. The dynamics (4) has the equilibrium $(x_1, x_2) = (2, 2)$, and the variation ranges of variables (x_1, x_2) are bounded around the equilibrium point as $1 \le x_1 \le 3$ and $1 \le x_2 \le 3$ in the case of given parameter set. Hence, we first scale the [0,255] gray-scale fingerprint image into [1,3] range. The scaled image becomes the initial input $x_1(0, n_1, n_2)$, while the initial condition of the second morphogen is given by $x_2(0, n_1, n_2) = 2$ (equilibrium). The zero-flux Neumann boundary condition is employed for computing the dynamics. After n_0 steps of the DRDS computation, we obtain $x_1(n_0, n_1, n_2)$ as the output image, which is scaled back into the [0,255] gray-scale image to produce the final output. Figure 1 shows the enhancement of a fingerprint image using the DRDS.

Our initial observation, however, shows that the DRDS with a spatially isotropic diffusion term (4) often produces some broken ridge lines in processing fingerprint images as shown in Fig. 1(b), since it does not take account of the local orientation of ridge flow. In order to solve this problem, the next section defines an *adaptive DRDS* model, in which we can use the local orientation of the ridge flow in a fingerprint image to guide the action of DRDS. This can be realized by introducing orientation masks to be convolved with the diffusion terms in DRDS (4).

3. ADAPTIVE DRDS FOR FINGERPRINT RESTORATION

In this section, we modify the definition of the simple two-morphogen DRDS (4) to have an adaptive DRDS dedicated to fingerprint restoration tasks. The two-morphogen adaptive DRDS with the Brusselator reaction kinetics can be written as

$$
\begin{bmatrix} x_1(n_0{+}1, n_1, n_2) \\ x_2(n_0{+}1, n_1, n_2) \end{bmatrix} = \begin{bmatrix} x_1(n_0, n_1, n_2) \\ x_2(n_0, n_1, n_2) \end{bmatrix}
+ \begin{bmatrix} R_1(x_1(n_0, n_1, n_2), x_2(n_0, n_1, n_2)) \\ R_2(x_1(n_0, n_1, n_2), x_2(n_0, n_1, n_2)) \end{bmatrix}
+ \begin{bmatrix} D_1(h_1^{n_1 n_2} * l * x_1)(n_0, n_1, n_2) \\ D_2(h_2^{n_1 n_2} * l * x_2)(n_0, n_1, n_2) \end{bmatrix}, \tag{5}
$$

where

$h_i^{m_1 m_2}(n_1, n_2)$: orientation mask at the pixel (m_1, m_2) for the i-th morphogen,

$R_1(x_1, x_2)$, $R_2(x_1, x_2)$: the Brusselator reaction kinetics.

In the above equation, we define the orientation mask $h_1^{m_1 m_2}(n_1, n_2)$ at the pixel (m_1, m_2) as a 32×32 matrix of real coefficients within the window $(n_1, n_2) = (-16, -16) \sim (15, 15)$. The mask $h_1^{m_1 m_2}(n_1, n_2)$ controls the dominant orientation of the generated pattern at every pixel (m_1, m_2) according to the local ridge flow in the given fingerprint image. The orientation mask can be automatically derived as follows (Fig. 2): (i) take the Fourier transform of the local image around the pixel (m_1, m_2), (ii) extract the dominant ridge orientation θ from the transformed image, (iii) generate a mask pattern $H_1^{m_1 m_2}(j\omega_1, j\omega_2)$ having the orientation θ in frequency domain as

$$
H_1^{m_1 m_2}(j\omega_1, j\omega_2) = \begin{cases} 1 & \text{for unstable frequency band (black pixels in Fig. 2(iii))} \\ 2 & \text{otherwise,} \end{cases}
$$

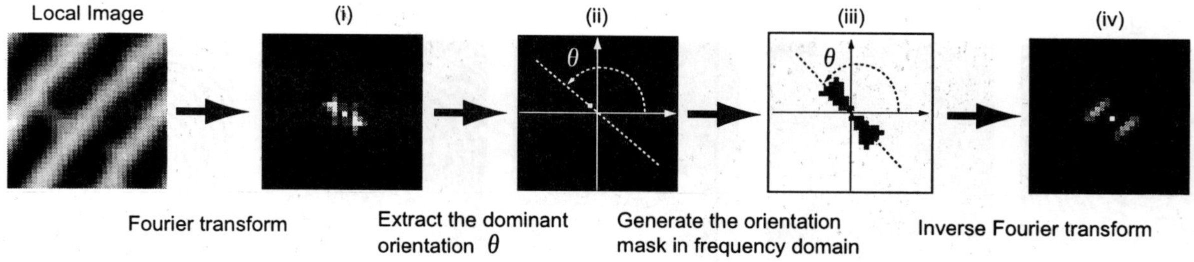

Figure 2: Generation of the orientation mask.

procedure *Adaptive DRDS with Hierarchical Orientation Estimation*

1. **begin**
2. $p := 2$; { initialize the image partitioing factor}
3. **while** time step n_O equals to 500 **do**
4. **begin**
5. **if** p is less than 9 **then**
6. **begin**
7. partition the input image into p^2 sub-images;
8. generate independent orientation masks for p^2 sub-images;
9. run the adaptive DRDS (Eq. (5)) for 10 time steps;
10. $p := p + 1$
11. **end**
12. **else**
13. **begin**
14. generate independent orientation masks for all pixels;
15. run the adaptive DRDS (Eq. (5)) for 10 time steps
16. **end**
17. **end**
18. **end**.

Figure 3: Algorithm for the adaptive DRDS with hierarchical orientation estimation.

and (iv) take the inverse Fourier transform to obtain the orientation mask $h_1^{m_1 m_2}(n_1, n_2)$. The orientation mask $h_2^{m_1 m_2}(n_1, n_2)$ for the second morphogen, on the other hand, has the value 1 at the center $(n_1, n_2) = (0, 0)$, and equals to 0 for the other coordinates (n_1, n_2). Thus, the dynamics for the morphogen $x_2(n_O, n_1, n_2)$ does not take account of the local orientation.

In practical situation, it is difficult to obtain the exact orientation masks from blurred fingerprints directly. Addressing this problem, we estimates local orientation masks recursively using a coarse-to-fine approach as shown in Fig. 3. This restoration algorithm starts with rough estimation of local orientation for four sub-images ($p = 2$). The image partitioning factor p gradually increases as restoration step n_O increases. We can obtain pixel-wise orientation masks $\boldsymbol{h}^{m_1 m_2}(n_1, n_2)$ after 80 time steps. This simple strategy makes possible significant improvement in the precision of orientation estimation.

4. EXPERIMENT

This section describes a set of experiments for evaluating restoration performance of the proposed algorithm. The problem considered here is to restore the original fingerprint image from its "subsampled" image. For this purpose, we generate a subsampled fingerprint image from the original image as follows: (i) partition the original image into $R \times S$-pixel rectangular blocks, and (ii) select one pixel randomly from every block and eliminate all the other pixels (set 127, middle gray-level, to the pixels). The image thus obtained has the same size as the original image, but the number of effective pixels is reduced to $1/(R \times S)$.

The restoration capability of the proposed algorithm is evaluated by calculating the similarity between the original fingerprint image and the restored image. To measure the similarity, we employ the phase-only image matching technique [4], which has been proved to have an efficient discrimination capability in practical fingerprint identification tasks [5]. In this experiment, we use 15 distinct fingerprint images (Finger01–Finger15). Restoration experiments are carried out for various subsampling rates $1/(3 \times 3)$, $1/(3 \times 4)$, $1/(4 \times 4)$, $1/(4 \times 5)$, $1/(5 \times 5)$, $1/(5 \times 6)$, $1/(6 \times 6)$, $1/(6 \times 7)$, $1/(7 \times 7)$, $1/(7 \times 8)$ and $1/(8 \times 8)$.

For example, Fig. 4 shows the original image, the subsampled image ($n_O = 0$) and restored images at $n_O = 100$, 200 and 400, respectively, for the case of $1/(6 \times 6)$ subsampling. We can observe that the fingerprint pattern is reconstructed from the subsampled image gradually as time step n_O increases. Figure 5 shows the variation of matching scores calculated between the original image of Finger01 and the restored images of Finger01–Finger15 for the case of subsampling rate $1/(6 \times 6)$. The matching score of the restored image of Finger01 increases selectively as the number of steps n_O increases. For every experimental trial, the optimal discrimination capability could be obtained at around $n_O = 400$ steps, which is indicated with a vertical dashed line in Fig. 5. The horizontal dashed line indicates the threshold for discrimination.

Table 1 shows the success rate of fingerprint identification for various subsampling rates. In the case where subsampling rates are from $1/(3 \times 3)$ to $1/(6 \times 6)$, we can restore the subsampled images completely. This experiment demonstrates a potential ca-

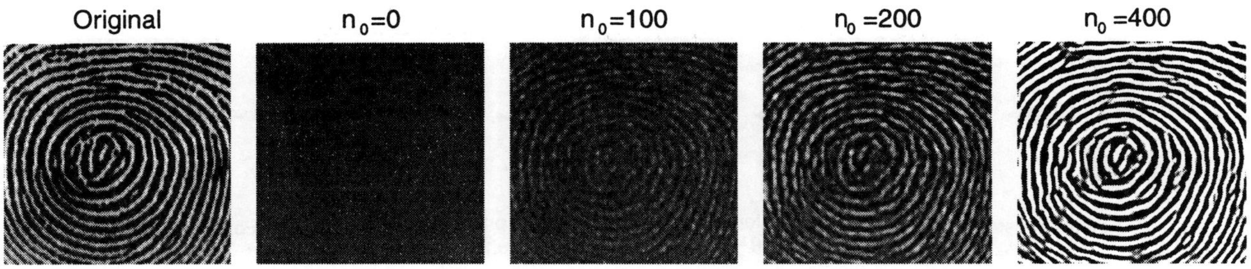

Figure 4: Fingerprint restoration from a subsampled image of *Finger01* with subsampling rate $1/(6 \times 6)$.

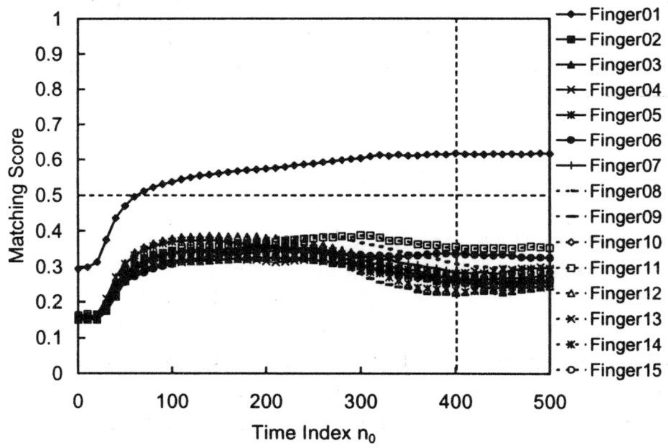

Figure 5: Matching scores between the original image of *Finger01* and the restored images of Finger01–Finger15 (restoration from $1/(6 \times 6)$ subsampled images).

Table 1: Identification rate.

Subsampling Rate	Number of Identified Samples	Identification Rate
$1/(3 \times 3)$	15	100%
$1/(3 \times 4)$	15	100%
$1/(4 \times 4)$	15	100%
$1/(4 \times 5)$	15	100%
$1/(5 \times 5)$	15	100%
$1/(5 \times 6)$	15	100%
$1/(6 \times 6)$	15	100%
$1/(6 \times 7)$	14	93%
$1/(7 \times 7)$	12	80%
$1/(7 \times 8)$	5	33%
$1/(8 \times 8)$	1	7%

pability of adaptive DRDS to enhance the performance of matching algorithms for blurred fingerprint images. For subsampling rates higher than $1/(6 \times 6)$, it becomes increasingly difficult to find correct orientation masks. In this region, dedicated fingerprint models (such as deformable templates) may be required for further improvement of restoration performance.

5. CONCLUSION

This paper presents an application of the DRDS to fingerprint image restoration. The adaptive DRDS combined with a coarse-to-fine orientation estimation technique can reconstruct complete fingerprint patterns even from $1/(6 \times 6)$-subsampled images. The proposed algorithm may be useful in many person identification applications based on fingerprint images.

6. REFERENCES

[1] A. M. Turing, "The chemical basis of morphogenesis," Phil. Trans. R. Soc. London, Vol. B237, pp. 37–72, Aug. 1952.

[2] A. S. Sherstinsky and R. W. Picard,"M-Lattice: From morphogenesis to image processing," IEEE Trans. Image Processing, vol. 5, no. 7, pp. 1137–1150, July 1996.

[3] K. Ito, T. Aoki, and T. Higuchi, "Digital reaction-diffusion system — A foundation of bio-inspired texture image processing —," IEICE Trans. Fundamentals, Vol. E84-A, No. 8, pp. 1909–1918, Aug. 2001.

[4] K. Takita, T. Aoki, Y. Sasaki, T. Higuchi and K. Kobayashi, "High-accuracy image registration based on phase-only correlation and its experimental evaluation," Proc. of 2002 IEEE International Symposium on Intelligent Signal Processing and Communication Systems, pp. 86–90, Nov. 2002.

[5] http://www.higuchi.ecei.tohoku.ac.jp/poc/

[6] J. D. Murray, "Mathematical Biology," Springer-Verlag, Berlin, 1993.

A DISCRETE FRACTIONAL FOURIER TRANSFORM BASED ON ORTHONORMALIZED McCLELLAN-PARKS EIGENVECTORS

Magdy Tawfik Hanna
Department of Engineering Mathematics and Physics
Cairo University / Fayoum Branch
Fayoum, Egypt.

ABSTRACT

A version of the discrete fractional Fourier transform (DFRFT) is developed with the objective of approximating the continuous fractional Fourier transform (FRFT). First the McClellan-Parks nonorthogonal eigenvectors of the discrete Fourier transform (DFT) matrix are generated analytically after deriving explicit expressions for the elements of those vectors. Second the Gram-Schmidt technique is applied to orthonormalize the eigenvectors in each eigensubspace individually. Third Hermite-like approximate eigenvectors are generated. Finally exact orthonormal eigenvectors as close as possible to the Hermite-like approximate eigenvectors are obtained by the orthogonal procrustes algorithm.

I. INTRODUCTION

A legitimate definition of the discrete fractional Fourier transform (DFRFT) should approximate its analog counterpart namely the fractional Fourier transform (FRFT). Guided by this goal, a recent definition emerged for the DFRFT that will have eigenvectors that resemble samples of the Hermite-Gaussian functions which are the eigenfunctions of the FRFT [1-2].

A fundamental step in the definition of a DFRFT is the eigendecomposition of the DFT matrix F. McClellan and Parks proved that matrix F has only the 4 distinct eigenvalues $\{1,-j,-1,j\}$ and they determined their multiplicities [3]. Moreover they constructed a complete set of linearly independent nonorthogonal eigenvectors. Dickinson and Steiglitz presented a technique for computing orthonormal eigenvectors of matrix F by a detailed analysis of a special matrix S that commutes with F [4]. They proved that if λ is a distinct eigenvalue of S then its corresponding eigenvector will also be an eigenvector of F but with a different eigenvalue. Dickinson and Steiglitz conjectured that the eigenvalues of S are distinct except when N is divisible by 4.

In their recent development, Pei et. al. [1] adopted the eigenvectors of the special matrix S. In the present paper the elegant technique of McClellan and Parks for the analytical generation of independent nonorthogonal eigenvectors of F will be adopted. Next the Gram-Schmidt technique will be applied to orthonormalize these eigenvectors. Finally the Hermite-like approximate eigenvectors will be projected on the last generated orthonormal eigensubspaces using the orthogonal procrustes algorithm [5] in order to get Hermite-like eigenvectors to be used as a basis for defining a DFRFT. This proposed technique has the advantage of avoiding the numerical evaluation of the eigenvectors of matrix S. It has the merit of not needing to revert to a classification technique for assigning the eigenvectors of S to the 4 eigensubspaces of F. It has the extra merit of avoiding the difficulties that arise when S has a repeated eigenvalue. Actually in the latter case an ordinary eigenvector evaluation program will not generate the circularly even and odd eigenvectors corresponding to a double eigenvalue of S and one has to take into account some computational considerations in order to generate the right circularly symmetric eigenvectors of F [4].

II. THE McCLELLAN-PARKS EIGENDECOMPOSITION

The elements of the DFT matrix F are given by:

$$F_{m,n} = \frac{1}{\sqrt{N}} W^{(m-1)(n-1)} \qquad , m, n = 1, \cdots, N \tag{1}$$

where $W = \exp(-j2\pi / N)$.

McClellan and Parks expressed the eigenvectors of F as $z_r = Fu_r \pm u_r$ for $\lambda = \pm 1$ and as $z_r = jFv_r \pm v_r$ for $\lambda = \mp j$ where u_r and v_r are simple column vectors to be given shortly. The values of the index r are given in Table 1 where the integer v is defined by:

$$v = \lfloor 0.5N \rfloor + 1 . \tag{2}$$

where $\lfloor b \rfloor$ is the largest integer not exceeding b.

The simple N-dimensional column vectors u_r for $1 \le r \le v$ are defined by:

$$u_1 = [1 \;\; 0 \;\; \cdots \;\; 0]^T ,$$
$$u_2 = [0 \;\; 1 \;\; 0 \;\; \cdots \;\; 0 \;\; 1]^T , \tag{3}$$
$$u_3 = [0 \;\; 0 \;\; 1 \;\; 0 \;\; \cdots \;\; 0 \;\; 1 \;\; 0]^T , \cdots$$

For $2 \le r \le v-1$, the elements $(u_r)_k$, $k = 1, \cdots, N$ are all zero except for $k = r, N+2-r$ which are unities. The same applies to $(u_v)_k$ when N is odd. When N is even, all elements $(u_v)_k$, $k = 1, \cdots, N$ will be zero except for the element $(u_v)_v$ which will be unity.

On the other hand, the simple N-dimensional column vectors v_r for $1 \le r \le N - v$ are defined by:

$$v_1 = [0 \;\; 1 \;\; 0 \;\; \cdots \;\; 0 \;\; -1]^T ,$$
$$v_2 = [0 \;\; 0 \;\; 1 \;\; 0 \;\; \cdots \;\; 0 \;\; -1 \;\; 0]^T , \cdots \tag{4}$$

All elements $(v_r)_k$, $k = 1, \cdots, N$ are zero except $(v_r)_{r+1} = 1$ and $(v_r)_{N+1-r} = -1$.

0-7803-7761-3/03/$17.00 ©2003 IEEE

In preparation for deriving explicit expressions for the elements of the McClellan-Parks eigenvectors, one starts by defining the vectors:

$$x_r = Fu_r \qquad , 1 \le r \le v \tag{5}$$

$$y_r = jFv_r \qquad , 1 \le r \le N - v . \tag{6}$$

From definition (1) of matrix F and definition (3) of vector u_1, one gets the elements of vectors x_1 as:

$$(x_1)_k = 1/\sqrt{N} \qquad , k = 1, \cdots, N . \tag{7}$$

For $2 \le r \le v-1$, one gets:

$$\begin{aligned}(x_r)_k &= \frac{1}{\sqrt{N}}\left(W^{(k-1)(r-1)} + W^{(k-1)(N+1-r)}\right) \\ &= \frac{2}{\sqrt{N}}\cos\left(\frac{2\pi}{N}(k-1)(r-1)\right)\end{aligned} \quad ,1 \le k \le N \tag{8}$$

For odd N, the elements of x_v are also given by (8). For even N, vector u_v has only one nonzero element, and consequently the elements of x_v reduce to:

$$\begin{aligned}(x_v)_k &= \frac{1}{\sqrt{N}} W^{(k-1)(v-1)} = \frac{1}{\sqrt{N}} W^{0.5N(k-1)} \\ &= \frac{1}{\sqrt{N}}(-1)^{(k-1)} \quad ,1 \le k \le N \text{ , N even}\end{aligned} \tag{9}$$

On the other hand, the elements of vector y_r of (6) can be derived as:

$$\begin{aligned}(y_r)_k &= j\left(F_{k,r+1} - F_{k,N+1-r}\right) \\ &= j\frac{1}{\sqrt{N}}\left(W^{(k-1)r} - W^{(k-1)(N-r)}\right) \quad , 1 \le k \le N . \\ &= \frac{2}{\sqrt{N}}\sin\left(\frac{2\pi}{N}(k-1)r\right)\end{aligned} \tag{10}$$

By utilizing (7)-(10), one can derive explicit expressions for the elements of the McClellan-Parks real eigenvectors of Table 1. Those expressions are given in Tables 2 and 3 corresponding to the eigenvalues $\lambda = \mp j$ and $\lambda = \pm 1$ respectively.

III. ORTHONORMAL HERMITE-LIKE EIGENVECTORS

1. Orthonormalizing the McClellan-Parks Eigenvectors

Since matrix F is unitary, eigenvectors corresponding to distinct eigenvalues are orthogonal. Since F has only 4 distinct eigenvalues, the N-dimensional real space will be divided into 4 eigensubspaces E_k corresponding to the 4 distinct eigenvalues $\lambda_k = (-j)^k$, $k = 0,1,2,3$. The McClellan-Parks technique of last section generates real eigenvectors. For each eigenvalue λ_k, the corresponding nonorthogonal eigenvectors span the eigensubspace E_k. Since eigenvectors lying in different spaces E_k are orthogonal, it remains to apply the Gram-Schmidt technique to the McClellan-Parks eigenvectors in each subspace E_k individually in order to get orthonormal eigenvectors. The resulting orthonormal eigenvectors of E_k will be arranged to form the columns of a matrix to be denoted by V_k.

2. Generation of Hermite-Like Approximate Eigenvectors

Samples of the Hermite-Gaussian functions will be used for generating approximate eigenvectors u_n of matrix F [1,2]:

$$Fu_n \approx (-j)^n u_n . \tag{11}$$

where $n \in \Psi$ and $\Psi = \{n_1, n_2, \cdots, n_N\}$ is the set of indices suggested in [2]. Actually for odd N, the set Ψ is $\{0,1,\cdots,N-1\}$; and for even N, the set Ψ is $\{0,1,\cdots,N-2,N\}$. The Hermite-like approximate eigenvectors u_{n_k}, $k = 1,\cdots,N$ will be classified into 4 sets corresponding to the 4 distinct eigenvalues of F. The vectors corresponding to λ_k will form the columns of a matrix to be denoted by U_k.

3. Hermite-Like Orthonormal Exact Eigenvectors

The matrix $\hat{U}_k$ of eigenvectors of F corresponding to the eigenvalue λ_k that is as close as possible to matrix U_k will be expressed as $\hat{U}_k = V_k Q_k$ where Q_k is a unitary matrix to be evaluated by the orthogonal procrustes algorithm [5]. One should mention that although the orthogonal procrustes algorithm was employed in [1], it was erroneously applied since $\hat{U}_k$ was wrongly taken as $\hat{U}_k = Q_k V_k$ rather than as $\hat{U}_k = V_k Q_k$.

IV. A DISCRETE FRACTIONAL FOURIER TRANSFORM

The Hermite-like orthonormal eigenvectors of matrix F obtained in the last section will be arranged to form the unitary matrix $\hat{U}$ of order N as follows:

$$\hat{U} = \begin{bmatrix} \hat{u}_{n_1} & \hat{u}_{n_2} & \cdots & \hat{u}_{n_N} \end{bmatrix} . \tag{12}$$

The transform kernel of the discrete fractional Fourier transform (DFRFT) is defined by [1]:

$$F^{\frac{2}{\pi}\alpha} = \hat{U} D^{\frac{2}{\pi}\alpha} \hat{U}^{+} \tag{13}$$

where the diagonal matrix $D^{\frac{2}{\pi}\alpha}$ is defined by:

$$D^{\frac{2}{\pi}\alpha} = diag\{\exp(-j\alpha n_1),\cdots,\exp(-j\alpha n_N)\}. \quad (14)$$

Let the column vector x represent the time-domain signal. The discrete fractional Fourier transform X_α of x is defined by:

$$X_\alpha = F^{\frac{2}{\pi}\alpha} x . \quad (15)$$

V. CONCLUSION

A version of the discrete fractional Fourier transform (DFRFT) has been developed to approximate its continuous counterpart, namely the fractional Fourier transform (FRFT).

REFERENCES

[1] S.-C. Pei, M.-H. Yeh and C.-C. Tseng, "Discrete fractional Fourier transform based on orthogonal projections," *IEEE transactions on Signal Processing*, vol. SP-47, pp. 1335-1348, May 1999.

[2] S.-C. Pei, C.-C. Tseng and M.-H. Yeh, "A new discrete fractional Fourier transform based on constrained eigendecomposition of DFT matrix by Lagrange multiplier method," *IEEE Transactions on Circuits and Systems, Part II*, vol. 46, pp. 1240-1245, September 1999.

[3] J.H. McClellan and T.W. Parks, "Eigenvalue and eigenvector decomposition of the discrete Fourier transform," *IEEE Transactions on Audio and Electroacoustics*, vol. AU-20, pp. 66-74, March 1972.

[4] B.W. Dickinson and K. Steiglitz, "Eigenvectors and functions of the discrete Fourier transform," *IEEE Transactions on Acoustics, Speech and Signal Processing*, vol. ASSP-30, pp. 25-31, February 1982.

[5] G.H. Golub and C.F. Van Loan, *Matrix Computations*, Johns Hopkins University Press, Baltimore, M.D., 1989.

Table 1: The McClellan-Parks eigenvectors of matrix F

N	λ	1	-j	-1	j
	Eigenvectors	$Fu_r + u_r$	$jFv_r + v_r$	$Fu_r - u_r$	$jFv_r - v_r$
4m	r	$1,2,\cdots,m,v$	$1,\cdots,m-1,v-2$	$1,2,\cdots,m$	$1,2,\cdots,m-1$
4m+1	r	$1,2,\cdots,m,v$	$1,2,\cdots,m$	$1,2,\cdots,m$	$1,2,\cdots,m$
4m+2	r	$1,2,\cdots,m+1$	$1,2,\cdots,m$	$1,2,\cdots,m+1$	$1,2,\cdots,m$
4m+3	r	$1,2,\cdots,m+1$	$1,2,\cdots m,v-1$	$1,2,\cdots,m+1$	$1,2,\cdots,m$

Table 2: The elements $(z_r)_k$, $k = 1,\cdots,N$ of the eigenvectors $z_r = jFv_r \pm v_r$

λ	-j	j
Eigenvectors	$z_r = jFv_r + v_r$	$z_r = jFv_r - v_r$
$1 \le r \le N-v$	$(z_r)_k = \begin{cases} \frac{2}{\sqrt{N}}\sin\left(\frac{2\pi}{N}(k-1)r\right)+1 & ,k=r+1 \\ \frac{2}{\sqrt{N}}\sin\left(\frac{2\pi}{N}(k-1)r\right)-1 & ,k=N+1-r \\ \frac{2}{\sqrt{N}}\sin\left(\frac{2\pi}{N}(k-1)r\right) & \text{otherwise} \end{cases}$	$(z_r)_k = \begin{cases} \frac{2}{\sqrt{N}}\sin\left(\frac{2\pi}{N}(k-1)r\right)-1 & ,k=r+1 \\ \frac{2}{\sqrt{N}}\sin\left(\frac{2\pi}{N}(k-1)r\right)+1 & ,k=N+1-r \\ \frac{2}{\sqrt{N}}\sin\left(\frac{2\pi}{N}(k-1)r\right) & \text{otherwise} \end{cases}$

Table 3: The elements $(z_r)_k$, k = 1, ⋯, N of the eigenvectors $z_r = Fu_r \pm u_r$

λ	1	-1
Eigenvectors	$z_r = Fu_r + u_r$	$z_r = Fu_r - u_r$
$r = 1$	$(z_1)_k = \begin{cases} \frac{1}{\sqrt{N}} + 1 & \text{for } k = 1 \\ \frac{1}{\sqrt{N}} & \text{otherwise} \end{cases}$	$(z_1)_k = \begin{cases} \frac{1}{\sqrt{N}} - 1 & \text{for } k = 1 \\ \frac{1}{\sqrt{N}} & \text{otherwise} \end{cases}$
$2 \le r \le \nu - 1$	$(z_r)_k = \begin{cases} \frac{2}{\sqrt{N}} \cos\left(\frac{2\pi}{N}(k-1)(r-1)\right) + 1 & \text{for } k = r, N+2-r \\ \frac{2}{\sqrt{N}} \cos\left(\frac{2\pi}{N}(k-1)(r-1)\right) & \text{otherwise} \end{cases}$	$(z_r)_k = \begin{cases} \frac{2}{\sqrt{N}} \cos\left(\frac{2\pi}{N}(k-1)(r-1)\right) - 1 & \text{for } k = r, N+2-r \\ \frac{2}{\sqrt{N}} \cos\left(\frac{2\pi}{N}(k-1)(r-1)\right) & \text{otherwise} \end{cases}$
$r = \nu$, N odd	$(z_\nu)_k = \begin{cases} \frac{2}{\sqrt{N}} \cos\left(\frac{2\pi}{N}(k-1)(\nu-1)\right) + 1 & \text{for } k = \nu, N+2-\nu \\ \frac{2}{\sqrt{N}} \cos\left(\frac{2\pi}{N}(k-1)(\nu-1)\right) & \text{otherwise} \end{cases}$	$(z_\nu)_k = \begin{cases} \frac{2}{\sqrt{N}} \cos\left(\frac{2\pi}{N}(k-1)(\nu-1)\right) - 1 & \text{for } k = \nu, N+2-\nu \\ \frac{2}{\sqrt{N}} \cos\left(\frac{2\pi}{N}(k-1)(\nu-1)\right) & \text{otherwise} \end{cases}$
$r = \nu$, N even	$(z_\nu)_k = \begin{cases} \frac{1}{\sqrt{N}}(-1)^{(k-1)} + 1 & \text{for } k = \nu \\ \frac{1}{\sqrt{N}}(-1)^{(k-1)} & \text{otherwise} \end{cases}$	$(z_\nu)_k = \begin{cases} \frac{1}{\sqrt{N}}(-1)^{(k-1)} - 1 & \text{for } k = \nu \\ \frac{1}{\sqrt{N}}(-1)^{(k-1)} & \text{otherwise} \end{cases}$

FAST INTEGER FOURIER TRANSFORM (FIFT) BASED ON LIFTING MATRICES

R. Thamvichai[1], *Tamal Bose*[2] *and Miloje Radenkovic*[3]

[1]Electrical & Computer Engineering, St. Cloud University
St. Cloud, MN 56301-4498
[2]Center for High-speed Information Processing (CHIP)
Electrical & Computer Engineering, Utah State University, Logan, UT 84322-4120
[3]Dept. of Electrical Engineering, University of Colorado, Denver, CO 80217-3364

ABSTRACT

This paper proposes a fast algorithm for computing the approximated DFT, called the Fast Integer Fourier Transform (FIFT). The new transform is based on factorization of the DFT matrix into a product of some specified matrices and lifting matrices. The elements of the lifting matrices are quantized to the nearest binary-number representation. Therefore, the proposed algorithm can be implemented in fixed-point arithmetic using only shifting operations and additions. Any length-2^l DFT sequence for $l \geq 1$ can be computed using this algorithm.

1. INTRODUCTION

Orthogonal transforms, such as Discrete Cosine Transform (DCT) and Discrete Fourier Transform (DFT), are important tools in digital signal processing. Floating-point multiplications naturally occur in these transforms. During the past few years, there has been significant effort to find algorithms such that no floating-point multiplications are needed, [1]-[4] and reference therein. The motivation is based on the fact that algorithms implemented in fixed-point arithmetic result in reducing cost, low power and smaller chip area in VLSI implementation. Often, the DFT is implemented in hardware using the efficient Fast Fourier Transform (FFT) algorithms [6] and fixed-point arithmetic. However, if coefficients of DFT are directly quantized in the conventional implementation (for fixed-point arithmetic), its invertible property is no longer valid. A few publications [3], [4] proposed algorithms to approximate the DFT such that no floating-point multiplications are needed and its invertible property is guaranteed. In [3], elements of the transform matrix, only integer numbers, are obtained by solving a system of nonlinear equations. In [4], each twiddle factor, $W_N^k = e^{-\frac{j2\pi k}{N}}$, of the FFT algorithm is approximated by a finite-length binary number and is implemented using the lifting scheme which ensures that the transform is invertible even though the twiddle factors are quantized.

This paper proposes an alternate algorithm, called Fast Integer Fourier Transform (FIFT), for approximating the DFT with finite-length binary number coefficients such that the transform is invertible with no error. The idea is to factor the DFT matrix based on the decimation-in-frequency FFT algorithm into a product of matrices. Then, the factored matrix, whose elements are twiddle factors, is factored into products of lifting matrices whose elements are quantized to the nearest finite-length binary number. This algorithm yields perfect reconstruction of an original sequence and can be implemented using only shifting and addition operations. The paper is organized as follows. In Section 2, definitions and lemmas used to derive the FIFT algorithm are stated. The algorithm is presented in Section 3 and its computational complexity is discussed in Section 4. Illustrative examples are given in Section 5 and Section 6 is the conclusion.

2. DEFINITIONS AND LEMMAS

Definition 1: A lifting matrix has all its diagonal elements equal to 1 and only one nonzero off-diagonal element [1]-[5]. Let $M_{k,l}(s)$ be a $N \times N$ lifting matrix whose nonzero off-diagonal element, called a lifting coefficient s, is at the (k,l) position given by

$$M_{k,l}(s) = \begin{bmatrix} 1 & 0 & \cdots & 0 \\ 0 & \ddots & & \\ & s & \cdots & 0 \\ 0 & \cdots & & 1 \end{bmatrix}. \quad (1)$$

It is easy to show that an inverse of a lifting matrix is also a lifting matrix whose nonzero off-diagonal element has the same value with the opposite sign, i.e. $M_{k,l}^{-1}(s) = M_{k,l}(-s)$. This implies that the inverse transform can be found by simply subtracting out what was added in at the

This work was supported in part by NASA Grant NAG5-10716 and the State of Utah Centers of Excellence Program

0-7803-7761-3/03/$17.00 ©2003 IEEE

forward transform. This property has a powerful significance such that even if the lifting coefficient, s, is quantized to a nearest number representation, i.e. rounding to an integer or to numbers that are of the form $\pm\frac{p}{2^m}$ where p, m are integers, a perfect reconstruction of an original sequence can be obtained. The advantage of multiplication with numbers of the form $\pm\frac{p}{2^m}$ is that no floating-point multiplication is needed. The above provides the motivation to factor the DFT matrix into products of some specified matrices and lifting matrices whose lifting coefficients will be quantized to the nearest number of the form $\pm\frac{p}{2^m}$. The quantized version of DFT matrix will be called the FIFT matrix. The application of lifting matrices for DFT matrix factorization ensures that an inverse FIFT will yield a perfect reconstruction of an original sequence.

Definition 2: Let $s = e^{j\theta}$ for $\theta \in R$, $\beta(s)$ denote an approximation of s which is written as a number of the form $\pm\frac{p}{2^m} \pm j\frac{k}{2^m}$ where p, m, k are integers and $\pm\frac{p}{2^m}$ and $\pm\frac{k}{2^m}$ are used to approximate real and imaginary parts of s, respectively. Since $\frac{1}{s}$ is the complex conjugate of s, $\beta(\frac{1}{s})$ is also a complex conjugate of $\beta(s)$.

Observe that if s is approximated by $\beta(s)$, matrix $M_{k,l}(\beta(s))$ is an approximation of matrix $M_{k,l}(s)$. Therefore, given any sequence $\mathbf{x}$, a transformed sequence $\mathbf{y} = M_{k,l}(s)\mathbf{x}$ can be approximated as $\overline{\mathbf{y}} = M_{k,l}(\beta(s))\mathbf{x}$. Since $M_{k,l}^{-1}(\beta(s)) = M_{k,l}(-\beta(s))$, the original sequence $\mathbf{x}$ can be perfectly reconstructed, i.e. $M_{k,l}(-\beta(s))\overline{\mathbf{y}} = \mathbf{x}$.

The following two lemmas will be used in a factorization of the DFT matrix in the next section. These lemmas were established in [2], and therefore their proofs are omitted here.

Lemma 1 *Let D be a $L\times L$ diagonal matrix whose determinant is 1, i.e. $D = diag(b_0, b_1, ..., b_{L-1})$ where $b_0b_1...b_{L-1}=1$. Let $\alpha_0 = b_0$, $\alpha_k = \alpha_{k-1}b_k$, $k = 1, 2, ..., L-1$. Then $D = B_1B_2 = B_2B_1$ where B_1 and B_2 are as follows. If L is even,*

$$B_1 = diag(\alpha_0, \frac{1}{\alpha_0}, \alpha_2, \frac{1}{\alpha_2}, ..., \alpha_{L-2}, \frac{1}{\alpha_{L-2}})$$
$$B_2 = diag(1, \alpha_1, \frac{1}{\alpha_1}, \alpha_3, \frac{1}{\alpha_3}, ..., 1).$$

If L is odd,

$$B_1 = diag(\alpha_0, \frac{1}{\alpha_0}, \alpha_2, \frac{1}{\alpha_2}, ..., 1)$$
$$B_2 = diag(1, \alpha_1, \frac{1}{\alpha_1}, \alpha_3, \frac{1}{\alpha_3}, ..., \alpha_{L-2}, \frac{1}{\alpha_{L-2}}).$$

Lemma 2 *For even L, matrix D can be factored into a product of lifting matrices as*

$$D = \left\{ \prod_{k=0}^{L/2-1} \begin{bmatrix} M_{2k,2k+1}(\alpha_{2k} - 1)\cdot \\ M_{2k+1,2k}(1)\cdot \\ M_{2k,2k+1}(\frac{1}{\alpha_{2k}} - 1)\cdot \\ M_{2k+1,2k}(-\alpha_{2k}) \end{bmatrix} \right\} \cdot$$
$$\left\{ \prod_{k=1}^{L/2-1} \begin{bmatrix} M_{2k-1,2k}(\alpha_{2k-1} - 1)\cdot \\ M_{2k,2k-1}(1)\cdot \\ M_{2k-1,2k}(\frac{1}{\alpha_{2k-1}} - 1)\cdot \\ M_{2k,2k-1}(-\alpha_{2k-1}) \end{bmatrix} \right\} \quad (2)$$

where $M_{k,l}(s)$ is a $L \times L$ lifting matrix defined in (1).

3. FAST INTEGER FOURIER TRANSFORM (FIFT) ALGORITHM

The DFT sequence of a length-N sequence, $x(n)$, is defined as $X(k) = \sum_{n=0}^{N-1} x(n)W_N^{nk}$, $k = 0, 1, ..., N$ where $W_N = e^{-\frac{j2\pi}{N}}$. Note that both $x(n)$ and $X(k)$ are in general complex numbers. The DFT can be written in a matrix form as $\mathbf{X} = F_N\mathbf{x}$ where $\mathbf{X} = [X(0), X(1), ..., X(N-1)]^T$, $\mathbf{x} = [x(0), x(1), ..., x(N-1)]^T$ and F_N, an N-point DFT matrix given as

$$F_N = [W_N^{nk}]_{n,k=0,1,...,N-1}. \quad (3)$$

For $N = 2^l$, $l \geq 1$, the decimation-in-frequency FFT algorithm [6], can be employed to compute a DFT sequence. Based on this algorithm, the following is obtained.

Lemma 3 *The DFT matrix for length-N sequence, F_N, can be factored as*

$$F_N = P_N \begin{bmatrix} F_{\frac{N}{2}} & \mathbf{0}_{\frac{N}{2}} \\ \mathbf{0}_{\frac{N}{2}} & F_{\frac{N}{2}} \end{bmatrix} \begin{bmatrix} I_{\frac{N}{2}} & \mathbf{0}_{\frac{N}{2}} \\ \mathbf{0}_{\frac{N}{2}} & \mathbf{W}_{\frac{N}{2}} \end{bmatrix} \begin{bmatrix} I_{\frac{N}{2}} & I_{\frac{N}{2}} \\ I_{\frac{N}{2}} & -I_{\frac{N}{2}} \end{bmatrix} \quad (4)$$

where $\mathbf{0}_{\frac{N}{2}}$ is a $\frac{N}{2} \times \frac{N}{2}$ zero matrix, $I_{\frac{N}{2}}$ is a $\frac{N}{2} \times \frac{N}{2}$ identity matrix, and $\mathbf{W}_{\frac{N}{2}}$ is a $\frac{N}{2} \times \frac{N}{2}$ diagonal matrix given as

$$\mathbf{W}_{\frac{N}{2}} = diag(W_N^0, W_N^1, ..., W_N^{\frac{N}{2}-1}),$$

and P_N is a $N \times N$ permutation matrix given by

$$P_N = \begin{bmatrix} 1 & 0 & & & \cdots & & 0 \\ 0 & 0 & \cdots & 0 & 1 & 0 & \cdots \\ 0 & 1 & 0 & & \cdots & & 0 \\ 0 & & \cdots & & 0 & 1 & \cdots \\ & & & \vdots & & & \\ 0 & & & 1 & \cdots & & 0 \\ 0 & & & \cdots & & & 1 \end{bmatrix}.$$

Observe that the transform needs no multiplication for $N = 2$ and 4 since elements of the DFT matrix, (3), are only ± 1 and $\pm j$. For $N = 2^l$ where $l \geq 3$, floating-point multiplications are needed for multiplying any vector with matrix $\mathbf{W}_{\frac{N}{2}}$. To obtain an algorithm which needs no floating-point multiplication, we propose a process of factorizing matrix $\mathbf{W}_{\frac{N}{2}}$ into a product of one speci ed matrix and lifting matrices, and then approximating the lifting coef cient of each lifting matrix by numbers of the form $\pm\frac{p}{2^m}$ where p, m are integers. To use Lemmas 1 and 2, matrix $\mathbf{W}_{\frac{N}{2}}$ is rst factored as follows.

Lemma 4 *For $N = 2^l$ and $l \geq 3$, matrix $\mathbf{W}_{\frac{N}{2}} = diag(W_N^0, W_N^1, ..., W_N^{\frac{N}{2}-1})$ can be written as*

$$\mathbf{W}_{\frac{N}{2}} = diag(j, 1, ..., 1) \cdot diag(-j, W_N^1, ..., W_N^{\frac{N}{2}-1})$$

where $\det(diag(-j, W_N^1, ..., W_N^{\frac{N}{2}-1})) = 1$*;* $\det(\mathbf{X})$ *denotes a determinant of matrix* $\mathbf{X}$.

Proof. The idea is to factor the matrix $\mathbf{W}_{\frac{N}{2}}$ into a product of two matrices such that one of them has determinant equal to 1. We rst nd a determinant of $\mathbf{W}_{\frac{N}{2}}$.

$$\det(\mathbf{W}_{\frac{N}{2}}) = W_N^{(0+1+...+\frac{N}{2}-1)}$$
$$= W_N^{\frac{1}{2}\times(\frac{N}{2}-1)\times\frac{N}{2}} = W_N^{(\frac{N^2}{8}-\frac{N}{4})}.$$

Let B and K be integers such that $W_N^B W_N^{(\frac{N^2}{8}-\frac{N}{4})} = W_N^{NK} = 1$. Therefore $B = NK - \frac{N^2}{8} + \frac{N}{4}$, which yields

$$W_N^B = W_N^{(NK-\frac{N^2}{8}+\frac{N}{4})} = W_N^{-\frac{N^2}{8}} \cdot W_N^{\frac{N}{4}}$$
$$= e^{\frac{-j2\pi(-\frac{N^2}{8})}{N}} \cdot e^{\frac{-j2\pi(\frac{N}{4})}{N}}$$
$$= e^{\frac{j\pi N}{4}} \cdot e^{\frac{-j\pi}{2}} = -j.$$

The last equation results from the fact that for $N = 2^l$ and $l \geq 3$, $e^{\frac{j\pi N}{4}}$ always equal to 1. Thus, matrix $\mathbf{W}_{\frac{N}{2}}$ can be factored into 2 matrices, one of which has determinant equal to 1 as follows:

$$\mathbf{W}_{\frac{N}{2}} = diag(W_N^{-B}, 1, ..., 1) \cdot diag(W_N^B, W_N^1, ..., W_N^{\frac{N}{2}-1})$$
$$= diag(j, 1, ..., 1) \cdot diag(-j, W_N^1, ..., W_N^{\frac{N}{2}-1}).$$

■

Since $\det(diag(-j, W_N^1, ..., W_N^{\frac{N}{2}-1})) = 1$, Lemmas 1 and 2 can be used to factor this matrix into a product of lifting matrices as in (2) where $L = \frac{N}{2}$, $\alpha_0 = -j$, and $\alpha_k = \alpha_{k-1} W_N^k$ for $k = 1, 2, ..., \frac{N}{2} - 1$. Note that α_k is an angle rotation, i.e. $e^{j\theta}$ for $\theta \in R$, whose real and imaginary parts always have magnitude less than or equal to 1. To avoid floating-point multiplication, the lifting coef cient of each lifting matrix of $\mathbf{W}_{\frac{N}{2}}$ is approximated by a number of the form $\pm\frac{p}{2^m}$. Based on *Definition 2*, let $\beta(\alpha_k)$ and $\beta(\frac{1}{\alpha_k})$ be an approximation of α_k and $\frac{1}{\alpha_k}$, respectively, $\mathbf{W}_{\frac{N}{2}}$ can then be approximated as

$$\widetilde{\mathbf{W}}_{\frac{N}{2}} = diag(j, 1, ..., 1) \cdot \left\{\prod_{k=0}^{N/4-1} \begin{bmatrix} M_{2k,2k+1}(\beta(\alpha_{2k}) - 1) \cdot \\ M_{2k+1,2k}(1) \cdot \\ M_{2k,2k+1}(\beta(\frac{1}{\alpha_{2k}}) - 1) \cdot \\ M_{2k+1,2k}(-\beta(\alpha_{2k})) \end{bmatrix}\right\} \cdot \left\{\prod_{k=1}^{N/4-1} \begin{bmatrix} M_{2k-1,2k}(\beta(\alpha_{2k-1}) - 1) \cdot \\ M_{2k,2k-1}(1) \cdot \\ M_{2k-1,2k}(\beta(\frac{1}{\alpha_{2k-1}}) - 1) \cdot \\ M_{2k,2k-1}(-\beta(\alpha_{2k-1})) \end{bmatrix}\right\} \quad (5)$$

From the above, (4) can be approximated as

$$\widetilde{F}_N = \underbrace{P_N}_{\text{Step 4}} \underbrace{\begin{bmatrix} \widetilde{F}_{\frac{N}{2}} & 0_{\frac{N}{2}} \\ 0_{\frac{N}{2}} & \widetilde{F}_{\frac{N}{2}} \end{bmatrix}}_{\text{Step 3}} \underbrace{\begin{bmatrix} I_{\frac{N}{2}} & 0_{\frac{N}{2}} \\ 0_{\frac{N}{2}} & \widetilde{W}_{\frac{N}{2}} \end{bmatrix}}_{\text{Step 2}} \underbrace{\begin{bmatrix} I_{\frac{N}{2}} & I_{\frac{N}{2}} \\ I_{\frac{N}{2}} & -I_{\frac{N}{2}} \end{bmatrix}}_{\text{Step 1}} \quad (6)$$

Note that a complete factorization of matrix $\widetilde{F}_N$ can be obtained by factoring $\widetilde{F}_{\frac{N}{2}}$ into matrices composed of $\widetilde{F}_{\frac{N}{4}}$ and then recursively factoring $\widetilde{F}_{\frac{N}{4}}$ into matrices composed of $\widetilde{F}_{\frac{N}{8}}$, and so on. This yields a new fast and computationally ef cient transform which employs only additions and shifting operations, called the FIFT algorithm. From (6), the algorithm can be detailed as follows:

Step 1) Compute $u(n) = x(n) + x(n + N/2)$ and $l(n) = x(n) - x(n + N/2)$, for $n = 0, 1, ..., \frac{N}{2} - 1$.

Step 2) Compute $\widetilde{\mathbf{l}} = \widetilde{\mathbf{W}}_{\frac{N}{2}}\mathbf{l}$ where $\mathbf{l} = [l(0), l(1), ..., l(\frac{N}{2} - 1)]^T$ and $\widetilde{\mathbf{l}} = [\widetilde{l}(0), \widetilde{l}(1), ..., \widetilde{l}(\frac{N}{2} - 1)]^T$. This step is performed by multiplying each matrix of $\widetilde{\mathbf{W}}_{\frac{N}{2}}$, (5), with the sequence $\mathbf{l}$ until $\widetilde{\mathbf{l}}$ is obtained. Note that given $\mathbf{x} = [x_0, ..., x_{N-1}]^T$, $y = M_{k,l}(\beta(\alpha)) \cdot \mathbf{x} = [x_0, ..., x_k + \beta(\alpha)x_l, x_{N-1}]$. Only one element of $\mathbf{x}$ is changed when multiplied with a lifting matrix. Since real and imaginary parts of $\beta(\alpha)$ are in term of $\frac{p}{2^m}$, only shifts and additions are employed in a computation of $x_k + \beta(\alpha)x_l$.

Step 3) Apply a $\frac{N}{2}$-point FIFT matrix, $\widetilde{F}_{\frac{N}{2}}$, to both sequences $u(n)$ and $\widetilde{l}(n)$ and let the outputs be $U(k)$ and $\widetilde{L}(k)$, respectively. Note that $\frac{N}{2}$-point FIFT sequence can be recursively calculated using a $\frac{N}{4}$-point FIFT which in turn is calculated using $\frac{N}{8}$-point FIFT, and so on. This implies that Step 1) and Step 2) will be used recursively for the computation of Step 3).

Step 4) $X(2k) = U(k)$ and $X(2k + 1) = \widetilde{L}(k)$, $k = 0, 1., ..., N/2 - 1$.

For reconstruction, an inverse of FIFT matrix is given as

$$(\widetilde{F}_N)^{-1} = \frac{1}{2}\begin{bmatrix} I_{\frac{N}{2}} & I_{\frac{N}{2}} \\ I_{\frac{N}{2}} & -I_{\frac{N}{2}} \end{bmatrix}\begin{bmatrix} I_{\frac{N}{2}} & \mathbf{0}_{\frac{N}{2}} \\ \mathbf{0}_{\frac{N}{2}} & (\widetilde{\mathbf{W}}_{\frac{N}{2}})^{-1} \end{bmatrix}$$

$$\begin{bmatrix} (\widetilde{F}_{\frac{N}{2}})^{-1} & \mathbf{0}_{\frac{N}{2}} \\ \mathbf{0}_{\frac{N}{2}} & (\widetilde{F}_{\frac{N}{2}})^{-1} \end{bmatrix} P_N^{-1}$$

where $P_N^{-1} = P_N^T$, and

$$(\widetilde{\mathbf{W}}_{\frac{N}{2}})^{-1} = diag(-j, 1, ..., 1)\cdot$$

$$\left\{ \prod_{k=1}^{N/4-1} \begin{bmatrix} M_{2k,2k-1}(\beta(\alpha_{2k-1}))\cdot \\ M_{2k-1,2k}(-\beta(\frac{1}{\alpha_{2k-1}})+1)\cdot \\ M_{2k,2k-1}(-1)\cdot \\ M_{2k-1,2k}(-\beta(\alpha_{2k-1})+1) \end{bmatrix} \right\}\cdot$$

$$\left\{ \prod_{k=0}^{N/4-1} \begin{bmatrix} M_{2k+1,2k}(\beta(\alpha_{2k}))\cdot \\ M_{2k,2k+1}(-\beta(\frac{1}{\alpha_{2k}})+1)\cdot \\ M_{2k+1,2k}(-1)\cdot \\ M_{2k,2k+1}(-\beta(\alpha_{2k})+1) \end{bmatrix} \right\}$$

Note that matrix $\widetilde{F}_N$ is not in general orthogonal, that is $\widetilde{F}_N^T\widetilde{F}_N \neq I_N$. Therefore, matrix $(\widetilde{F}_N)^{-1}$is employed for reconstruction, i.e. $(\widetilde{F}_N)^{-1}\widetilde{F}_N = I_N$, and may be very different from $\widetilde{F}_N^T$.

4. COMPUTATIONAL COMPLEXITY

A lifting step is defined as a multiplication of a lifting matrix with a vector. Since a vector and lifting coefficient are in general complex numbers, a lifting step is composed of one complex addition and one complex multiplication (four real additions and four real multiplications). If the lifting coefficients are of the form $\pm\frac{p}{2^m}$ where p, m are integers, multiplying a number with this coefficient can be implemented using only shifting operations and additions. The number of lifting steps and complex additions used in the FIFT algorithm are as follows.

There is no computation in Step 4). Step 1) uses N complex additions. Step 2) employs $3(\frac{N}{2}-1)$ lifting steps, $(\frac{N}{2}-1)$ complex additions, and one multiplication (multiply with j). The proof is shown in [2] and is omitted. The computation in Step 3) results from multiplying a vector from Step 2) with matrix $\widetilde{F}_{\frac{N}{2}}$ which is also composed of lifting steps and complex additions. Let $Lift(N)$ and $Add(N)$ denote the number of lifting steps and the number of complex additions, respectively, used in length-N FIFT algorithm. The computations can then be written as

$$Lift(N) = 2Lift(\frac{N}{2}) + 3(\frac{N}{2} - 1)$$

$$Add(N) = 2Add(\frac{N}{2}) + (\frac{N}{2} - 1) + N$$

$$= 2Add(\frac{N}{2}) + \frac{3N}{2} - 1.$$

Applying this method recursively to $Lift(\frac{N}{2})$ and $Add(\frac{N}{2})$ and so on, we then obtain $Lift(N) = \frac{3}{2}N\log_2 N - 3N + 3$ and $Add(N) = \frac{3}{2}N\log_2 N - N + 1$. In addition, there are $N-1$ multiplications (multiply with j) needed in the algorithm. If we ignore these computations, the FIFT algorithm employs $3N\log_2 N - 4N + 4$ complex additions and $\frac{3}{2}N\log_2 N - 3N + 3$ complex multiplications which results in $9N\log_2 N - 14N + 14$ real additions and $6N\log_2 N - 12N + 12$ real multiplications. Recall that multiplication with $\pm\frac{p}{2^m}$ is implemented using only shifting and adding operations. For an inverse transform, its computation is similar to that of a forward transform except that extra shifting operations are needed for multiplying with the extra $\frac{1}{2}$.

5. CONCLUSION

A FIFT algorithm is proposed to compute the DFT for any length-N sequence, where $N = 2^l$ for integer $l \geq 1$, in fixed-point arithmetic. The proposed algorithm is implemented using only shifting operations and additions. In addition, the invertible property of the transform is preserved even though the transform coefficients are quantized and the computational complexity of the inverse transform is comparable to that of the forward transform. The error due to quantization of the lifting coefficients can be reduced by increasing the number of bits in the quantizer. However, this increases the computational complexity due to an increase in the number of additions. Therefore, there is a trade-off between its accuracy and its computational complexity.

6. REFERENCES

[1] J. Liang and T. D. Tran, "*Fast multiplierless approximations of the DCT with the lifting scheme,*" IEEE. Trans. Sig. Proc., vol. 49, no. 12, pp. 3032-3044, Dec 2001.

[2] Y. Zeng, L. Cheng, G. Bi, and A. C. Kot, "*Integer DCTs and fast algorithms,*" IEEE Trans. Sig. Proc., vol. 49, no. 11, pp. 2774-2782, Nov. 2001.

[3] S. C. Pei and J. J. Ding, "*Integer Discrete Fourier Transform and its extension to integer trigonometric transforms,*" ISCAS 2000, pp. 513-516, May 2000.

[4] S. Oraintara, Y. J. Chen, and T. Nguyen, "*Integer Fast Fourier Transform (INTFFT),*" ICASSP 2001, pp. 3485-3488, May 2001.

[5] W. Sweldens, "*The lifting scheme: A construction of second generation wavelets,*" SIAM J. Math. Anal., vol. 29, no. 2, pp. 511-546, 1998.

[6] A. V. Oppenhiem and R. W. Schafer, *Discrete-time Signal Processing*, Prentice Hall, 1989.

SAVING THE BANDWIDTH IN THE FRACTIONAL DOMAIN BY GENERALIZED HILBERT TRANSFORM PAIR RELATIONS

Soo-Chang Pei, Jian-Jiun Ding

Department of Electrical Engineering, National Taiwan University,
Taipei, Taiwan, R.O.C
Email address: pei@cc.ee.ntu.edu.tw

ABSTRACT

In this paper, we develop some methods to save the bandwidth required in the fractional domain. The fractional domain is the transformed domain of the fractional Fourier transform (FRFT). It is the intermediate of the time domain and the frequency domain. We find that, with the aid of the fractional Hilbert transform and other techniques, we can save 1/2 or 3/4 of the bandwidth in the fractional domain if the signal is causal, real, a real signal multiplied by chirp, a fractal, or a finite duration signal. The efficiency of the FRFT can hence be improved.

1. INTRODUCTION

In communication, we usually use the Hilbert transform and other techniques to save the bandwidth in the frequency domain. The Hilbert Transform (HLT) [1] is defined as:

$$O_{HI}(x(t)) = IFT(-j\,\mathrm{sgn}(\omega)\cdot FT(x(t))) \tag{1}$$

where FT, IFT are forward / inverse Fourier transforms. If $x(t)$ is causal (i.e., $x(t) = 0$ for $t < 0$), the real part and imaginary part of $X(\omega)$ ($X(\omega) = FT(x(t))$) form a Hilbert transform pair:

$$\mathrm{Im}(X(\omega)) = O_{HI}(\mathrm{Re}(X(\omega))) \qquad \text{if } x(t) \text{ is causal.} \tag{2}$$

Since $\mathrm{Im}(X(\omega))$ can be recover from $\mathrm{Re}(X(\omega))$, in the frequency domain we can store $\mathrm{Re}(X(\omega))$ instead of $X(\omega)$, and half of the bandwidth can be saved (because the imaginary part of the spectrum is not required). Besides, the even part and odd part of $X(\omega)$ also form a Hilbert transform pair:

$$X_o(\omega) = O_{HI}(X_e(\omega)) \qquad \text{if } x(t) \text{ is causal.} \tag{3}$$

So, in the frequency domain we can also store $X_e(\omega)s(\omega)$ ($s(\omega) = 0$ for $\omega < 0$, $s(0) = 1/2$, $s(\omega) = 1$ for $\omega > 0$) instead of $X(\omega)$ to save half of the bandwidth (since $X_e(\omega)s(\omega) = 0$ for $\omega < 0$).

Besides, if $x(t)$ is real, $X(\omega) = FT(x(t))$ is conjugate-symmetric:

$$X(\omega) = \overline{X(-\omega)} \qquad \text{if } x(t) \text{ is real} \tag{4}$$

Since $X(-\omega)$ can be recovered from $X(\omega)$ by the above relation, in the frequency domain, we can store $X(\omega)s(\omega)$ instead of $X(\omega)$, and half of the bandwidth can be saved.

Thus, in the frequency domain, we can use the above techniques to save half of the bandwidth if $x(t)$ is causal or real. Then, we may ask whether the same things can be done in the fractional domain. The **fractional domain** is the **transformed domain of the fractional Fourier transform (FRFT)** (see equation (5)). Since the FRFT becomes more and more important in signal processing, how to process signal efficiently in the fractional domain also becomes an important topic.

For example, the FRFT have been used for fractional modulation and fractional multiplexing [6], i.e., store and transmit the signal in the fractional domain instead of the frequency domain. Besides, the FRFT can be used for fractional filter and fractional system design [2][6]. There are also many other operations related to the FRFT [2]. They process signals in the fractional domain instead of the frequency domain. If we can develop some techniques to save the bandwidth required in the fractional domain, the efficiency of those operations can all be improved.

In this paper, we develop some methods to save the bandwidth in the fractional domain. We find that, if $x(t)$ is causal, real, a real function multiplied by chirp, a finite duration signal, or a fractal, **half** of the bandwidth in the fractional domain can be **saved**. If $x(t)$ is causal-real, a real signal with finite duration, or a real fractal, **3/4** of the bandwidth in the fractional domain can be **saved**.

First, in Sec. 2, we introduce the FRFT and the fractional Hilbert transform (FRHLT). Then, in Secs. 3~7, we develop some methods to save the bandwidth in the fractional domain. In Sec. 8, we give some examples. In Sec. 9, we give a conclusion.

2. FRACTIONAL FOURIER / HILBERT TRANSFORMS

The **fractional Fourier transform (FRFT)** [2], which is the generalization of the Fourier transform (FT), is defined as

$$O_F^{\alpha}(x(t)) = \sqrt{\frac{1-j\cot\alpha}{2\pi}}\cdot e^{j\cdot\cot\alpha\cdot u^2/2}\int_{-\infty}^{\infty} e^{-j\csc\alpha\cdot u\cdot t}\cdot e^{j\cot\alpha\cdot t^2/2}\,x(t)\cdot dt \qquad \text{when } \alpha \neq N\pi, \tag{5}$$

$$O_F^{2N\pi}(x(t)) = x(t), \quad O_F^{2(N+1)\pi}(x(t)) = x(-t). \tag{6}$$

When $\alpha = \pi/2$, it becomes the FT. It is reversible and additive:

$$O_F^{-\alpha}\left(O_F^{\alpha}(x(t))\right) = x(t), \quad O_F^{\beta}\left(O_F^{\alpha}(x(t))\right) = O_F^{\alpha+\beta}(x(t)). \tag{7}$$

In this paper, we use $X_\alpha(u)$ to denote the FRFT of $x(t)$

$$X_\alpha(u) = O_F^{\alpha}(x(t)). \tag{8}$$

The FRFT can extend the utilities of the FT, and is useful for filter design, optical system analysis, pattern recognition, communication, etc. Many signal processing problems that can't be solved well by the original FT will be solved by the FRFT. The FRFT becomes more and more important in signal processing.

The **fractional Hilbert transform (FRHLT)** [3] is defined based on the FRFT. Its formula is as follows (we generalize the formula in [3] a little):

0-7803-7761-3/03/$17.00 ©2003 IEEE

$$O_{HI}^{(a,b),\alpha}(x(t)) = O_F^{-\alpha}\left(H_{(a,b)}(u)\cdot O_F^{\alpha}(x(t))\right) \tag{9}$$

where $H_{(a,b)}(u) = a - j\,\mathrm{sgn}(u)b$. (10)

The original HLT, the analytic signal, and the fractional analytic signal [4] are all the special cases of the FRHLT. We will show that the FRHLT is very helpful for saving the bandwidth in the fractional domain (the transformed domain of the FRFT).

3. REDUCING THE BANDWIDTH OF CAUSAL SIGNALS

For the case of the FRFT, when $x(t)$ is causal, there is no obvious relation between the real part and odd part of $X_\alpha(u)$ (which is the FRFT of $x(t)$), and it is hard to find the counterpart of (2). However, we can find the counterpart of (3). If $x(t)$ is **causal**:

$$x(t) = 0 \qquad \text{for } t < 0, \tag{11}$$

in the fractional domain, the **even part** and the **odd part** of $X_\alpha(u)$ form a **fractional Hilbert transform pair (FRHLT pair)**:

$$X_{\alpha,o}(u) = O_H^{(0,-j),\pi-\alpha}\left(X_{\alpha,e}(u)\right) \tag{12}$$

where $X_{\alpha,e}(u) = \left(X_\alpha(u) + X_\alpha(-u)\right)/2$,

$$X_{\alpha,o}(u) = \left(X_\alpha(u) - X_\alpha(-u)\right)/2. \tag{13}$$

It is a generalization of (3). Its proof is shown in Appendix.

Thus, if $x(t)$ is causal and in the fractional domain we have known the even part of $X_\alpha(u)$, we can use the FRHLT to recover the odd part of $X_\alpha(u)$ from (12), and hence the whole value of $X_\alpha(u)$. Therefore, in the **fractional domain**, we only have to store the **positive even part** of $X_\alpha(u)$:

$$\tilde{X}_{\alpha,e}(u) = X_{\alpha,e}(u)\cdot s(u) \tag{14}$$

where $s(u)$ is the step function:

$$s(u) = 0 \text{ for } u < 0, \quad s(0) = 1/2, \quad s(u) = 1 \text{ for } u > 0. \tag{15}$$

Since $\tilde{X}_{\alpha,e}(u) = 0$ for $u < 0$, if we store $\tilde{X}_{\alpha,e}(u)$ instead of $X_\alpha(u)$, **half** of the bandwidth can be saved. We can recover $X_\alpha(u)$ from $\tilde{X}_{\alpha,e}(u)$ by the following process:

① $X_{\alpha,e}(u) = \tilde{X}_{\alpha,e}(u) + \tilde{X}_{\alpha,e}(-u)$, (16)

② $X_\alpha(u) = X_{\alpha,e}(u) + O_H^{(0,-j),\pi-\alpha}\left(X_{\alpha,e}(u)\right)$

$$= O_H^{(1,-j),\pi-\alpha}\left(X_{\alpha,e}(u)\right). \tag{17}$$

Similarly, if $x(t)$ is **anti-causal**:

$$x(t) = 0 \qquad \text{for } t > 0, \tag{18}$$

the **even part** and **odd part** of $X_\alpha(u)$ also form a **FRHLT pair**:

$$X_{\alpha,o}(u) = O_H^{(0,j),\pi-\alpha}\left(X_{\alpha,e}(u)\right). \tag{19}$$

Notice that the parameters $(0, -j)$ in (12) is changed into $(0, j)$. Thus, in this case we can also store $X_{\alpha,e}(u)\cdot s(u)$ instead of $X_\alpha(u)$, and **half** of the bandwidth in the fractional domain can be **saved**.

4. REDUCING THE BANDWIDTH OF REAL SIGNALS

The FRFT of a real signal does not have conjugate-symmetry property. Thus, the relation as in (4) is no longer satisfied in the fractional domain. However, we can still reduce the bandwidth in the fractional domain with a more complicated method. If $x(t)$ is **real**, in the **fractional domain** we can store $A_\alpha(u)$ defined as follows instead of $X_\alpha(u)$

$$A_\alpha(u) = \mathrm{Re}\left[\sqrt{je^{-j\alpha}}\, e^{-j\cot\alpha\cdot u^2/2}\cdot \tilde{X}_{\alpha,e}(u)\right] + j\cdot\mathrm{Im}\left[\sqrt{je^{-j\alpha}}\, e^{-j\cot\alpha\cdot u^2/2}\cdot \tilde{X}_{\alpha,o}(u)\right] \tag{20}$$

where

$$\tilde{X}_{\alpha,e}(u) = X_{\alpha,e}(u)s(u), \quad \tilde{X}_{\alpha,o}(u) = X_{\alpha,o}(u)s(u). \tag{21}$$

$X_{\alpha,e}(u)$, $X_{\alpha,o}(u)$ are the even part and odd part of $X_\alpha(u)$, and $s(u)$ is defined as in (15). Because

$$A_\alpha(u) = 0 \qquad \text{for } u < 0, \tag{22}$$

half of the bandwidth in the fractional domain can be saved. We can recover $X_\alpha(u)$ from $A_\alpha(u)$ by

① $Z_\alpha(u) = A_\alpha(u) + \overline{A_\alpha(-u)}$, (23)

② $X_\alpha(u) = \sqrt{-je^{j\alpha}}\cdot e^{j\cot\alpha\cdot u^2/2}\cdot\{Z_\alpha(u) +$

$$FT\left[j\tan\left(\sin(2\alpha)t^2/4\right)\cdot IFT\left(Z_\alpha(u)\right)\right]\}. \tag{24}$$

In fact, in (23), $Z_\alpha(u)$ is equal to

$$Z_\alpha(u) = \mathrm{Re}\left(\sqrt{je^{-j\alpha}}\cdot e^{-j\cot\alpha\cdot u^2/2}\, X_{\alpha,e}(u)\right) + j\cdot\mathrm{Im}\left(\sqrt{je^{-j\alpha}}\cdot e^{-j\cot\alpha\cdot u^2/2}\, X_{\alpha,o}(u)\right). \tag{25}$$

The proof of (24) is rather complicated, and can be seen from our manuscript.

Similarly, if $x(t)$ is **pure imaginary**, we also only have to store the value of $A_\alpha(u)$, and **half** of the bandwidth in the fractional domain can be saved. We can recover $X_\alpha(u)$ from (23) and

② $X_\alpha(u) = \sqrt{-je^{j\alpha}}\cdot e^{j\cot\alpha\cdot u^2/2}\cdot\{Z_\alpha(u) -$

$$FT\left[j\cot\left(\sin(2\alpha)t^2/4\right)\cdot IFT\left(Z_\alpha(u)\right)\right]\}. \tag{26}$$

5. REDUCING THE BANDWIDTH OF REAL SIGNALS MULTIPLIED BY CHIRP

The results in Sec. 4 can be further generalized. If $x(t)$ is a **real signal multiplied by chirp**:

$$x(t) = e^{j\cdot\tau\cdot t^2}\, y(t) \qquad \text{where } y(t) \text{ is a real function,} \tag{27}$$

and τ is any real number, then in the fractional domain, we also only have to store the value of $A_\alpha(u)$ defined as in (20), and **half** of the bandwidth in the fractional domain can be saved. We can follow the process similar to (23) and (24) to recover $X_\alpha(u)$ from $A_\alpha(u)$, except for that (24) is modified as

② $X_\alpha(u) = \sqrt{-je^{j\alpha}}\cdot e^{j\cot\alpha\cdot u^2/2}\cdot\{Z_\alpha(u) +$

$$FT\left[j\tan\left((\sin(2\alpha)+\eta)t^2/4\right)\cdot IFT\left(Z_\alpha(u)\right)\right]\}. \tag{28}$$

That is, $\sin(2\alpha)$ is changed into $\sin(2\alpha)+\eta$.

6. REDUCING THE BANDWIDTH OF THE SIGNALS WITH FINITE DURATION

In the case where $x(t)$ is a **finite duration** signal:

$$x(t) \neq 0 \qquad \text{only when } t \in [t_1, t_2], \tag{29}$$

we can use the FRHLT pair relation to save half of the bandwidth required in the fractional domain. In this case, we only have to store the value of $B_\alpha(u)$ the fractional domain

$$B_\alpha(u) = \left[e^{jt_1 \sin\alpha \cdot u} X_\alpha(u + t_1 \cos\alpha)\right]_{even} s(u) \tag{30}$$

where $[f(u)]_{even} = [f(u) + f(-u)]/2$. Since $B_\alpha(u) = 0$ for $u < 0$, so **half** of the bandwidth in the fractional domain can be saved. We can recover $X_\alpha(u)$ from $B_\alpha(u)$ by the following process:

① $B_{1,\alpha}(u) = (B_\alpha(u) + B_\alpha(-u))/2$. (31)

② $B_{2,\alpha}(u) = O_H^{(0,-j),\pi-\alpha}(B_{1,\alpha}(u)) + B_{1,\alpha}(u)$. (32)

③ $X_\alpha(u) = e^{-jt_1 \sin\alpha(u - t_1\cos\alpha)} B_{2,\alpha}(u - t_1 \cos\alpha)$. (33)

The above results can be proved from the fact that the even and odd parts of $\exp(j \cdot t_1 \sin\alpha \cdot u) \cdot X_\alpha(u + t_1 \cos\alpha)$ form a **FRHLT pair**:

$$O_H^{(0,-j),\pi-\alpha}\left\{\left[e^{jt_1 \sin\alpha \cdot u} X_\alpha(u + t_1 \cos\alpha)\right]_{even}\right\}$$

$$= \left[e^{jt_1 \sin\alpha \cdot u} X_\alpha(u + t_1 \cos\alpha)\right]_{odd}. \quad \text{if } x(t) = 0 \text{ for } t \notin [t_1, t_2] \tag{34}$$

In signal processing, the signal we deal with is usually a finite duration signal. Using the above method, we can save half of the bandwidth required in the fractional domain.

Some signals are not finite duration signals, but, with a little modification, we can also use the above method to save the bandwidth required in the fractional domain. For example, for a scaling invariant signal (i.e., a **fractal**) which satisfies:

$$x(\sigma \cdot t) = \lambda \cdot x(t) \qquad \text{where } t \in (-\infty, \infty), \tag{35}$$

if we know the value of $y(t)$ where

$$y(t) = x(t) \quad \text{for } t \in [t_1, t_2],\ t_1 < 0 < t_2,$$

$$y(t) = 0 \quad \text{otherwise}, \tag{36}$$

we can find all the values of $x(t)$ for $t \in (-\infty, \infty)$ by iterative scaling. Thus, we can first convert the fractal $x(t)$ into the finite duration signal $y(t)$, then apply the method described in (30)~(33) to save **half** of the bandwidth in the fractional domain.

7. THE CASES WHERE WE CAN SAVE 3/4 OF THE BANDWIDTH

In Secs. 3~6, we have described several conditions where we can save half of the bandwidth in the fractional domain. In fact, sometimes we can further reduce the bandwidth.

If $x(t)$ is **real and causal**, in the fractional domain, we only have to store the value of

$$C_\alpha(u) = \text{Re}\left(\sqrt{je^{-j\alpha}} \cdot e^{-j\cot\alpha \cdot u^2/2}\ \tilde{X}_{\alpha,e}(u)\right) \tag{37}$$

where $\tilde{X}_{\alpha,e}(u)$ is defined as in (21). Since $\tilde{X}_{\alpha,e}(u) = 0$ for $u < 0$, so $C_\alpha(u)$ is a positive real function. If we store $C_\alpha(u)$ instead of $X_\alpha(u)$, **3/4** of the bandwidth in the fractional domain can be **saved** (since $C_\alpha(u)$ has no negative part and positive-imaginary part). We can recover $X_\alpha(u)$ from $C_\alpha(u)$ by the following process, which is in fact the combination of (23, (24), and (17):

① $Z_\alpha(u) = C_\alpha(u) + C_\alpha(-u)$, (38)

② $X_{\alpha,e}(u) = \sqrt{-je^{j\alpha}} \cdot e^{j\cot\alpha \cdot u^2/2} \cdot \{Z_\alpha(u) + FT\left[j\tan(\sin(2\alpha)t^2/4) \cdot IFT(Z_\alpha(u))\right]\}$, (39)

③ $X_\alpha(u) = O_H^{(1,-j),\pi-\alpha}(X_{\alpha,e}(u))$. (40)

Similarly, if $x(t)$ is **real and anti-causal**, or **pure imaginary and causal** (or **anti-causal**), we can also store $C_\alpha(u)$ defined as in (37) instead of $X_\alpha(u)$, and **save 3/4** of the bandwidth in the fractional domain. We can also use the above process with a little modification to recover $X_\alpha(u)$ from $C_\alpha(u)$. If $x(t)$ is pure imaginary and causal (or anti-causal), the 2nd step is modified as:

② $X_{\alpha,e}(u) = \sqrt{-je^{j\alpha}} \cdot e^{j\cot\alpha \cdot u^2/2} \cdot \{Z_\alpha(u) - FT\left[j\cot(\sin(2\alpha)t^2/4) \cdot IFT(Z_\alpha(u))\right]\}$. (41)

If $x(t)$ is real (or pure imaginary) and anti-causal, the 3rd step is modified as

③ $X_\alpha(u) = O_H^{(1,j),\pi-\alpha}(X_{\alpha,e}(u))$. (42)

Moreover, if $x(t)$ is a **real *(*or pure imaginary*)*** signal with **finite duration**, we only have to store the value of $D_\alpha(u)$ defined as follows to **save 3/4** of the bandwidth in the fractional domain.

$$D_\alpha(u) = \text{Re}\left[\sqrt{je^{-j\alpha}}\, e^{j\sin(2\alpha)t_1^2/4} e^{-j\cot\alpha \cdot u^2/2} B_\alpha(u)\right] s(u) \tag{43}$$

where $B_\alpha(u)$ is defined as in (30).

8. EXPERIMENTS

Wee do some experiments to illustrate the concepts introduced in this paper. Fig. 1(a) is a real signal. Its FRFT is shown in Fig. 1(b). It is not band-limited, but we can use (20)~(25) to save half of the bandwidth in the fractional domain. Figs. 1(c) and 1(d) are

Fig. 1(c): $\exp(-j\cot\alpha \cdot u^2/2) X_{\alpha,e}(u)$

Fig. 1(d): $\exp(-j\cot\alpha \cdot u^2/2) X_{\alpha,o}(u)$. (44)

Fig. 1(e) is the combination the real part of Fig. 1(c) and the imaginary part of Fig. 1(d). It is equal to $Z_\alpha(u)$ defined in (25). Since $Z_\alpha(u) = conj(Z_\alpha(-u))$, the negative part of $Z_\alpha(u)$ is redundancy. Thus, we can take its positive part, as in Fig. 1(f). It is in fact equal to $A_\alpha(u)$ defined in (20). Since $A_\alpha(u) = 0$ for $u < 0$, if we store $A_\alpha(u)$ instead of $X_\alpha(u)$, half of the spectrum in the fractional domain is saved. We can recover $Z_\alpha(u)$ from $A_\alpha(u)$ by (23), and use (24) to recover $X_\alpha(u)$ from $Z_\alpha(u)$.

In Fig. 2, we give another example. The signal in Fig. 2(a) is a real-causal signal. It is not band-limited in the fractional domain, as in Fig. 2(b). However, we can use (37)~(40) to save 3/4 of the bandwidth in the fractional domain. We can take the even part of $X_\alpha(u)$ (as in Fig. 2(c)), then multiply it by $(je^{-j\alpha})^{1/2} e^{-j\cot\alpha \cdot u^2/2}$ and take the real part (as in Fig. 2(d)), then take the positive part (as in Fig. 2(f)). The result in Fig. 2(f) is equal to $C_\alpha(u)$ defined in (37). Since

$$C_\alpha(u) = 0 \text{ for } u < 0, \qquad \text{Im}(C_\alpha(u)) = 0, \tag{45}$$

3/4 of the spectrum in the fractional domain is saved. We can recover $X_\alpha(u)$ from $C_\alpha(u)$ by (38), (39), and (40). The outputs of Steps 1, 2, 3 are just Figs. 2(d), 2(c), and 2(b), respectively.

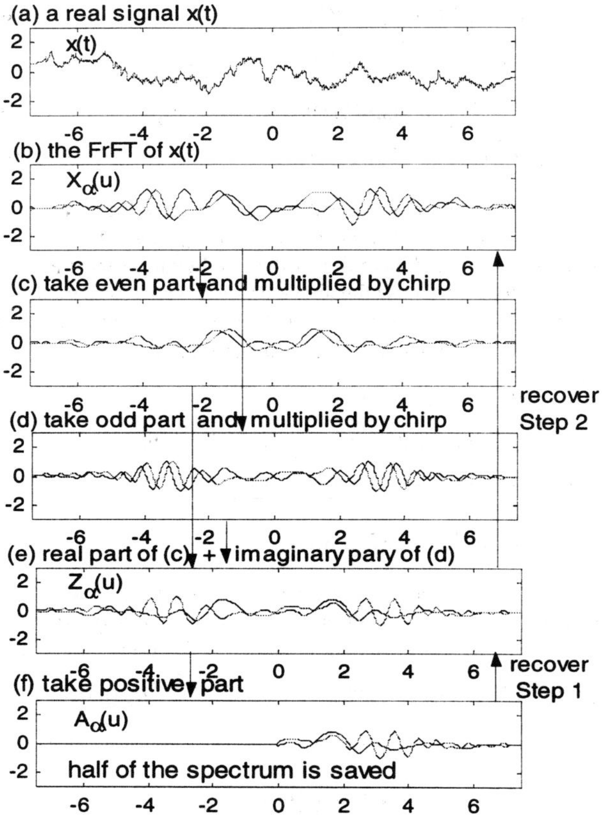

Fig. 1 Saving bandwidth in the fractional domain for a real signal. We use two lines to represent real and imaginary parts.

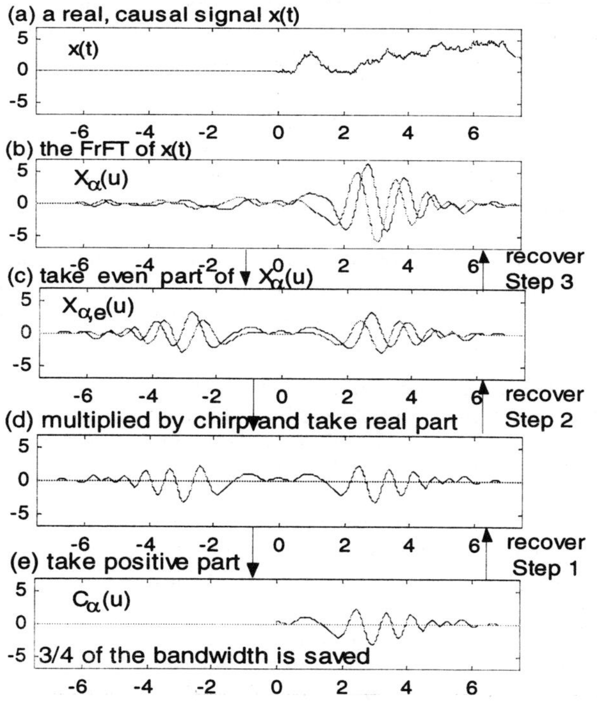

Fig. 2 Saving bandwidth in the fractional domain for a real and causal signal.

9. CONCLUSION

In this paper, we have introduced some methods to reduce the bandwidth required in the fractional domain (the transformed domain of the fractional Fourier transform (FRFT)). We find that:

- In following conditions, we can **save half** of the bandwidth in the fractional domain:
  (1) real (or pure imaginary), (2) causal (or anti-causal),
  (3) real function × chirp, (4) finite duration, (5) fractal.
- In following conditions, we can **save 3/4** of the bandwidth in the fractional domain:
  (1) real (or pure imaginary) & causal (or anti-causal),
  (2) real (or pure imaginary) & finite duration (or a fractal).

Thus, by the methods introduced in this paper, we can improve the efficiency of the FRFT in signal processing.

10. REFERENCES

[1] S. C. Hahn, "*Hilbert Transforms in Signal Processing*", Boston: Artect House, 1996.

[2] H. M. Ozaktas, M. A. Kutay, and Z. Zalevsky, "*The Fractional Fourier Transform with Applications in Optics and Signal Processing*", John Wiley & Sons, 2000.

[3] A. W. Lohmann, D. Mendlovic, and Z. Zalevsky, 'Fractional Hilbert transform', *Opt. Lett.*, vol. 21, no. 5, p 281-283, Feb. 1996.

[4] A. I. Zayed, 'Hilbert transform associated with the fractional Fourier transform', *IEEE Signal Processing Lett.*, vol. 5, no. 8, p 206-208, Aug. 1998.

[5] S. C. Pei and M. H. Yeh, 'Discrete fractional Hilbert transform', *Proc. of IEEE Int'l Symp. on Circuits and Systems*, vol. 4, p. 506-509, Jun. 1998.

[6] H. M. Ozaktas, B. Barshan, D. Mendlovic, L. Onural, 'Convolution, filtering, and multiplexing in fractional Fourier domains and their rotation to chirp and wavelet transform', *J. Opt. Soc. Am. A*, vol. 11, no. 2, p. 547-559, Feb. 1994.

[7] S. C. Pei and J. J. Ding, 'Bandwidth saving and edge detection by fractional Hilbert transforms', submitted to *IEEE Trans. Signal Processing.*

11. APPENDIX

(Proof of (12)):

If $x_e(t)$, $x_o(t)$ are the even part and the odd part of $x(t)$, then

$$X_{\alpha,e}(u)=O_F^{\alpha}(x_e(t)),\quad X_{\alpha,o}(u)=O_F^{\alpha}(x_o(t)). \qquad (\text{'}15)$$

Then, from the additivity property of the FRFT (see (7)), and the fact that $O_F^{\pi}(x(t)) = x(-t)$ (see (6)), we obtain

$$O_F^{\pi-\alpha}(X_{\alpha,e}(u))=O_F^{\pi}(x_e(t))=x_e(-t)=(x(-t)+x(t))/2,$$

$$O_F^{\pi-\alpha}(X_{\alpha,o}(u))=O_F^{\pi}(x_o(t))=x_o(-t)=(x(-t)-x(t))/2. \qquad (\text{'}16)$$

Since $x(t)$ is causal, we conclude that $x_o(-t) = -\mathrm{sgn}(t)\cdot x_e(-t)$. So

$$O_F^{\pi-\alpha}(X_{\alpha,o}(u))=-\mathrm{sgn}(t)\cdot O_F^{\pi-\alpha}(X_{\alpha,e}(u)).$$

Since $-\mathrm{sgn}(t) = H_{0,-j}(t)$,

$$X_{\alpha,o}(u)=O_F^{\alpha-\pi}\left(H_{0,-j}(t)O_F^{\pi-\alpha}(X_{\alpha,e}(u))\right)=O_H^{(0,-j),\pi-\alpha}(X_{\alpha,e}(u)).$$

□

ANGULAR DECOMPOSITIONS FOR THE DISCRETE FRACTIONAL SIGNAL TRANSFORMS

Min-Hung Yeh

Department of Electronic Engineering, National I-LAN Institute of Technology, I-Lan, Taiwan, R. O. C.

ABSTRACT

In this paper, a generalized angular decomposition algorithm for the discrete fractional signal transforms is proposed. By the angular decomposition algorithm, the discrete fractional signal transforms can be computed by a weighted summation of the transform evaluated at special angles. And the weighting coefficients are just equal to the inverse discrete Fourier transform of the eigenvalues of the fractional signal transform kernels.

1. INTRODUCTION

The fractional Fourier transform(FRFT) is interpreted as a rotation of signal in the time-frequency plane, and it has mixed time and frequency characteristics of signals[1][2].. Because of the importance of the FRFT, finding the discrete fractional Fourier transforms (DFRFT) are discussed and developed in the recently years. In [3] and [4], the DFRFT based upon the eigen decomposition method can obey the rotation property and have similar outputs as the continuous cases.

Because the DFRFT is developed based up the eigen decomposition, and its needs the product of transform kernel and input signal to compute the transform output. Such a computation requires $\mathcal{O}(N^2)$ operations. Many researchers have tried to develop other methods for computing and defining the DFRFT. In [5], a method for digital computation of fractional Fourier transform with order $\mathcal{O}(N\log N)$ is proposed. In [6], a novel method for computing the DFRFT was proposed. By this method, any angles of the DFRFT can be computed by a weighted summation for the DFRFTs. This angular decomposition algorithm provides an optional computational method , and it is useful for the signal detection by the FRFT.

Besides the discrete Fourier transform(DFT), , several other fractional signal transforms are proposed and discussed in the recent year[7]-[12]. Many fractional unitary transforms have also been developed by the eigen decomposition method, such as the discrete fractional Hartley transform(DFRHT)[8], the discrete fractional cosine transform (DFRCT)[9][10], the discrete fractional sine transform (DFRST)[9], and the discrete fractional Hadamard transform (DFRHaT)[12].

2. PRELIMINARY

2.1. Discrete fractional Fourier transform

The transform kernel of DFRFT is defined as[3][4]:

$$\mathbf{F}_\alpha = \sum_{k=0}^{N-1} e^{-jk\alpha}\mathbf{v}_k\mathbf{v}_k^T \quad \text{(N is odd)} \quad (1)$$

$$\mathbf{F}_\alpha = \sum_{k=0}^{N-2} e^{-jk\alpha}\mathbf{v}_k\mathbf{v}_k^T + e^{-jN\alpha}\mathbf{v}_N\mathbf{v}_N^T \quad \text{(N is even)} \quad (2)$$

where α indicates the rotation order of DFRFT. $\mathbf{v}_k$ is the k-th order DFT Hermite eigenvector. The methods for finding the DFT Hermite eigenvectors $\mathbf{v}_k$ are presented in [3][4]. It must be noted that there exists a jump in the last DFRFT eigenvalue for the even N case.

2.2. Other discrete fractional signal transforms

2.2.1. DFRHT

It is known that the eigenvectors of DHT are the same as those of DFT, but the eigenvalues λ_k are only 1 and -1[8][13]. The DFRHT is written as[8]:

$$\mathbf{H}_\alpha = \sum_{k=0}^{N-1} \lambda_k\mathbf{v}_k\mathbf{v}_k^T \quad (3)$$

where λ_k is the eigenvalue of the DFRHT, and it is as follows:

$$\lambda_k = e^{-j\lfloor \frac{k}{2} \rfloor 2\alpha} \quad (4)$$

where the $\lfloor n \rfloor$ operator indicates the largest integer which is less than or equal to n.

2.2.2. DFRCT and DFRST

The fractional versions of DCT-I/DST-I, DCT-II, DCT-IV/DST-IV are proposed in [9], [10] and [11], respectively. All of these fractional versions of DCTs and DSTs are developed based up on the eigen decomposition. But the DCT-II can not have closed form for the eigenvectors and eigenvalues[10]. Here we only use the fractional DCT-I/DST-I and DCT-IV/DST-IV in this paper. The DFRCT and DFRST kernels are written as:

$$\mathbf{C}_\alpha = \sum_{k=0}^{N-1} e^{-j2k\alpha}\hat{\mathbf{v}}_{2k}\hat{\mathbf{v}}_{2k}^T \quad (5)$$

0-7803-7761-3/03/$17.00 ©2003 IEEE

$$\mathbf{S}_\alpha = \sum_{k=0}^{N-1} e^{-j2k\alpha} \tilde{\mathbf{v}}_{2k+1} \tilde{\mathbf{v}}_{2k+1}^T \tag{6}$$

where $\hat{\mathbf{v}}_k$ and $\tilde{\mathbf{v}}_k$ are the DCT and DST eigenvector that is obtained from the k-th order DFT Hermite eigenvector[9].

2.2.3. DFRHaT

The eigenvectors of Hadamard transform with length 2^n can be generated iteratively from the eigenvector $[\sqrt{2}-1 \;\; 1]^T$ of Hadamard transform with length 2[12]. Similar to the development of DFRFT, the normalized Hadamard eigenvectors obtained by and with k sign-changes are assigned to the eigenvalue $e^{-jk\alpha}$.

$$\mathbf{H}_{n,\alpha} = \sum_{k=0}^{2^n-1} e^{-jk\alpha} \bar{\mathbf{v}}_k \bar{\mathbf{v}}_k^T \tag{7}$$

where $\bar{\mathbf{v}}_k$ is the k-th order normalized Hadamard eigenvector.

2.3. Previous angular decomposition for the DFRFT

An angular decomposition for the DFRFT is developed in [6]. Here we quote the result for the further development. In [6], the DFRFT for angle α can be written as a linear combination of DFRFT.

$$\mathbf{F}_\alpha = \begin{cases} \sum_{n=0}^{N-1} B_n \mathbf{F}_{n\beta} & N \text{ is odd} \\ \sum_{n=0}^{N} B_n \mathbf{F}_{n\beta} & N \text{ is even} \end{cases} \tag{8}$$

where $\beta = \frac{2\pi}{N}$ for N is odd, and $\beta = \frac{2\pi}{N+1}$ for N is even. The weighting coefficients B_n are computed as:

$$B_n = \begin{cases} IDFT\{e^{-jk\alpha}\}_{k=0,1,2,\ldots,N-1} & N \text{ is odd} \\ IDFT\{e^{-jk\alpha}\}_{k=0,1,2,\ldots,N} & N \text{ is even} \end{cases} \tag{9}$$

Two implementation methods for the above algorithm are developed[6]: One is **parallel method**, the other is **cascade method**. The block diagrams for these two implementation methods are shown in [6].

3. THE GENERAL ANGULAR DECOMPOSITION FOR THE DFRFT

To begin with, we will consider a general form for the angular decomposition by the DFRFT. The $N \times N$ matrix $\mathbf{G}$ is any matrix which can be written as the eigen decomposition form with the DFT Hermite eigenvectors as its eigenvectors.

$$\mathbf{G} = \mathbf{V}\mathbf{D}\mathbf{V}^* \tag{10}$$

where $\mathbf{D}$ is a diagonal matrix with entries $d_0, d_1, \ldots, d_{N-1}$ in the diagonal. $\mathbf{V}$ is the $N \times N$ matrix which is composed by the DFT Hermite eigenvector.

In developing the general angular decomposition algorithm, M terms for the weighted summation will be considered.

Proposition 1 *For any odd $N \times N$ matrix $\mathbf{G}$ with the DFT Hermite eigenvectors can be written as the weighed summation for the DFRFT evaluated at the angles $\frac{2\pi}{M}$, where $M \geq N$. The weighting coefficients are equal to the M-point IDFT of the sequence*

$$\mathbf{d} = [d_0, d_1, \ldots, d_{N-1}, \underbrace{0, \ldots, 0}_{(M-N)\text{'s zeros}}] \tag{11}$$

Proof :

The inverse DFT (IDFT) for the sequence $\mathbf{d}$ is computed to get a new sequence $\mathbf{c}$. It means that the sequences $\mathbf{c}$ and $\mathbf{d}$ are just the DFT pairs.

$$c_m = \frac{1}{M} \sum_{k=0}^{M-1} d_k e^{j\frac{2\pi}{M}mk} \tag{12}$$

for $m = 0, 1, 2, \ldots, M-1$. It is well-known that d_n can also be written as a weighted summation of c_k by the forward DFT.

$$d_k = \sum_{m=0}^{M-1} c_m e^{-j\frac{2\pi}{M}km} \tag{13}$$

for $k = 0, 1, 2, \ldots, M-1$. Thus the matrix $\mathbf{G}$ can be decomposed as:

$$\begin{aligned} \mathbf{G} &= \sum_{k=0}^{N-1} d_k \mathbf{v}_k \mathbf{v}_k^T \\ &= \sum_{k=0}^{N-1} \sum_{m=0}^{M-1} c_m e^{-j\frac{2\pi}{M}mk} \mathbf{v}_k \mathbf{v}_k^T \\ &= \sum_{m=0}^{M-1} c_m \sum_{k=0}^{N-1} e^{-j\frac{2\pi}{M}mk} \mathbf{v}_k \mathbf{v}_k^T \\ &= \sum_{m=0}^{M-1} c_m \mathbf{F}_{m\frac{2\pi}{M}} \end{aligned} \tag{14}$$

□

Clearly, the above proposition can not be suitable to the even N case. For the even N, we can generate the sequence $\mathbf{d}$ in the other way : a zero is inserted between the last two diagonal elements d_{N-2} and d_{N-1}.

Proposition 2 *For any even $N \times N$ matrix $\mathbf{G}$ with the DFT Hermite eigenvectors can be written as the weighted summation for the DFRFT evaluated at the angles $\frac{2\pi}{M}$, where $M > N$. The weighting coefficients are equal to the M-point IDFT of the sequence*

$$\mathbf{d} = [d_0, d_1, \ldots, d_{N-2}, 0, d_{N-1}, \underbrace{0, \ldots, 0}_{(M-N-1)\text{'s zeros}}] \tag{15}$$

Proof :

Similar to the odd N case, the sequence $\mathbf{d}$ can be rewritten as a linear combination of the IDFT coefficients c_k, for $k = 0, 1, \cdots, N-2$, and the eq.(13) can also be used to compute d_k. The particular last entry d_{N-1} in the diagonal matrix **D** also can be obtained by the DFT computation, and it is written as follows:

$$d_{N-1} = \sum_{m=0}^{M-1} c_m e^{-j\frac{2\pi}{M}mN} \tag{16}$$

$$\begin{aligned}
\mathbf{G} &= \sum_{k=0}^{N-2} d_k \mathbf{v}_k \mathbf{v}_k^T + d_{N-1}\mathbf{v}_N \mathbf{v}_N^T \\
&= \sum_{k=0}^{N-2}\sum_{m=0}^{M-1} c_m e^{-j\frac{2\pi}{M}km} \mathbf{v}_k \mathbf{v}_k^T + \sum_{m=0}^{M-1} c_m e^{-j\frac{2\pi}{M}Nm} \mathbf{v}_N \mathbf{v}_N \\
&= \sum_{m=0}^{M-1} c_m (\sum_{k=0}^{N-2} e^{-j\frac{2\pi}{M}km} \mathbf{v}_k \mathbf{v}_k^T + e^{-j\frac{2\pi}{M}Nm} \mathbf{v}_N \mathbf{v}_N^T) \\
&= \sum_{m=0}^{M-1} c_m \mathbf{F}_{m\frac{2\pi}{M}} \qquad (17)
\end{aligned}$$

□

From the above discussion, we know that any operator **G** with the DFT Hermite eigenvectors as its eigenvectors can be written as a linear combination of the DFRFTs. The weighting DFRFTs are evaluated at the angles $m\frac{2\pi}{M}$ (for $m = 0, 1, \ldots, M-1$). And the weighting coefficients are just equal to the inverse DFT of the diagonal entries of fractional signal transform kernel and the padding zeros.

3.1. Specific operators by the weighted summation of DFRFT

The weighted summation of DFRFT discussed in the previous subsection can be used for several specific operations.

- **Discrete fractional Fourier transform ($\mathbf{G} = \mathbf{F}_\alpha$)**
  It is obvious that DFRFT is a special case of the **G** matrix. So the DFRFT with angle can be rewritten as a weighted summation of the DFRFT evaluated at the angles $m \times \frac{2\pi}{M}$, for $m = 0, 1, \cdots, M-1$.
- **Discrete fractional Hartley transform ($\mathbf{G} = \mathbf{H}_\alpha$)**
  It is known that the DFRHT is with the same eigenvectors as those of the DFRFT[8]. So the DFRHT kernel is also a special case of the matrix **G**. And the eigenvalues of the DFRHT are equal to $e^{-j\lfloor \frac{k}{2} \rfloor 2\alpha}$[8], for $k = 0, 1, \cdots, N-1$
- **Outer product of the discrete Hermite eigenvectors ($\mathbf{G} = \mathbf{v}_k \mathbf{v}_k^T$)**
  The outer product of the k-th order Hermite eigenvectors can be trivially obtained by assigning d_k to be 1 and the other diagonal d_k to be zero.

$$d_k = \delta(k) \tag{18}$$

- **DFT commuting matrix S ($\mathbf{G} = \mathbf{S}$)**
  In [14], a commuting matrix **S** is introduced to find real, symmetric and orthogonal eigenvectors of DFT matrix. Moreover, the eigenvectors of **S** are used to define the DFRFT in [4]. The computing matrix **S** is also a special form of **G**, so it can be written as a weighting summation of the DFRFTs. The weighting coefficients are the IDFT of the eigenvalues of **S**.

3.2. Discussion

In the above subsection, we have developed a generalized version of the angular decomposition for the DFRFT. By the method in [6], the number of weighting terms is N when N is odd, and it is $(N+1)$ when N is even. In our new decomposition, the DFRFT can be decomposed into M terms, where M can be any integer which is greater than N. The parallel and cascade implementation methods in [6] can also be applied to the new general angular decomposition form. Moreover, If M is power of 2, the weighting coefficients for the angular decomposition can be computed by the FFT which is the fast algorithm of DFT.

4. ANGULAR DECOMPOSITION FOR THE OTHER DISCRETE FRACTIONAL SIGNAL TRANSFORMS

Many discrete fractional signal transforms are similar to the DFRFT, and are developed based up on the eigen decomposition form. In the following, we will discuss the angular decomposition forms for these transforms.

4.1. Decomposition of the DFRCT and DFRST

Proposition 3 *For the DFRCT/DFRST kernel matrix $\mathbf{C}_\alpha/\mathbf{S}_\alpha$ can be written as the weighted summation for the DFRCT/DFRST evaluated at the angles $\frac{\pi}{M}$, where $M \geq N$. The weighting coefficients are equal to the M-point IDFT of the sequence $\hat{\mathbf{d}}$.*

$$\hat{\mathbf{d}} = [1, e^{-j2\alpha}, \ldots, e^{-j2(N-2)\alpha}, e^{-j2(N-1)\alpha}, \underbrace{0, \ldots, 0}_{(M-N)\text{'s zeros}}] \tag{19}$$

Proof :

Similar to the method for the DFRFT angular decomposition, we can use the sequence $\hat{\mathbf{d}}$ to compute its inverse DFT $\hat{\mathbf{c}}$. Thus the definition of DFRCT kernel in eq.(5) can be written as:

$$\begin{aligned}
\mathbf{C}_\alpha &= \sum_{k=0}^{N-1}\sum_{m=0}^{M-1} \hat{c}_m e^{j\frac{2\pi}{M}km} \hat{\mathbf{v}}_{2k} \hat{\mathbf{v}}_{2k}^T \\
&= \sum_{m=0}^{M-1} \hat{c}_m \sum_{k=0}^{N-1} e^{-j\frac{2\pi}{M}km} \hat{\mathbf{v}}_{2k} \hat{\mathbf{v}}_{2k}^T \\
&= \sum_{m=0}^{M-1} \hat{c}_m \mathbf{C}_{m\frac{\pi}{M}} \qquad (20)
\end{aligned}$$

From eq.(20), we can know that the DFRCT can be computed by a weighted summation of DFRCTs. And the weighting DFRCTs are evaluated at the angles $m\frac{\pi}{M}$, for $m = 0, 1, \ldots, (M-1)$.

By comparing the definitions of DFRST and DFRCT in eqs.(5) and (6), we can observe that their definitions are similar and both of them are with the same eigenvalues $e^{-j2k\alpha}$. So the angular decomposition of DFRST can be developed similar to the case in DFRCT, and it can be written as follows:

$$\mathbf{S}_\alpha = \sum_{m=0}^{M-1} \hat{c}_m \mathbf{S}_{m\frac{\pi}{M}} \tag{21}$$

□

4.2. Decomposition of the DFRHT

In [8], we know that the DFRHT can be written as the summation of the two auxiliary matrices, $\mathbf{Fr}_\alpha$ and $\mathbf{Fi}_\alpha$.

$$\mathbf{H}_\alpha = \mathbf{Fr}_\alpha + \mathbf{Fi}_\alpha \tag{22}$$

$$= \sum_k e^{-j2k\alpha}\mathbf{v}_{2k}\mathbf{v}_{2k}^T + \sum_k e^{-j2k\alpha}\mathbf{v}_{2k+1}\mathbf{v}_{2k+1}^T \tag{23}$$

The two auxiliary matrices are with the similar form as the DFRCT and DFRST kernel, so they can be written as the summation of transforms evaluated at the angle $\frac{\pi}{M}$.

$$\mathbf{H}_\alpha = \sum_{m=0}^{M-1} c_m \mathbf{Fr}_\alpha + \sum_{m=0}^{M-1} c_m \mathbf{Fi}_\alpha \tag{24}$$

$$= \sum_{m=0}^{M-1} c_m \mathbf{H}_{m\frac{2\pi}{M}} \tag{25}$$

where c_m is the IDFT of the sequence (19).

4.3. Decomposition of the DFRHaT

From eq.(7), we can observe that the DFRHaT has the same eigen decomposition to the DFRFT with odd length. So we can use the inverse DFT to compute the weighting coefficients. And the DFRHaT with any angle can be written as a weighted summations for the DFRHaT evaluated at the angle $m\frac{2\pi}{M}$, for $m = 0, 1, \ldots, M-1$.

5. CONCLUSION

In this paper, a generalized angular decomposition for the DFRFT is developed. Besides the DFRFT decomposition, other discrete fractional signal transforms can also be decomposed into the weighting terms of transforms. Table 1 gives us a summary of the angular decomposition for the discrete fractional signal transforms.

6. REFERENCES

[1] L. B. Almeida, "The fractional Fourier transform and time-frequency representation," *IEEE Trans. Signal Process.*, vol. 42, pp. 3084–3091, Nov. 1994.

[2] H. M. Ozaktas, Z. Zalevsky, and M. A. Kutay, *The Fractional Fourier Transform with applications in optics and signal processing.* John Wiley & Sons, 2000.

[3] S. C. Pei, M. H. Yeh, and C. C. Tseng, "Discrete fractional Fourier transform based on orthogonal projections," *IEEE Trans. Signal Process.*, vol. 47, pp. 1335–1348, May 1999.

[4] C. Candan, M. A. Kutay, and H. M. Ozaktas, "The discrete fractional Fourier transform," *IEEE Trans. Signal Process.*, vol. 48, pp. 1329–1337, May 2000.

[5] H. M. Ozaktas, O. Arikan, M. A. Kutay, and G. Bozdagi, "Digital computation of the fractional Fourier transform," *IEEE Trans. on Signal Process.*, vol. 44, pp. 2141–2150, Sept. 1996.

[6] S. C. Pei and M. H. Yeh, "A novel method for discrete fractional Fourier transform computation," in *Proceedings of IEEE International Symposium on Circuits and Systems*, vol. II, pp. 585–588, May 2001.

[7] A. W. Lohmann, D. Mendlovic, Z. Zalevsky, and R. G. Dorsch, "Some important fractional transforms for signal processing," *Optics Communications*, vol. 126, pp. 18–20, 4 1996.

[8] S. C. Pei, C. C. Tseng, M. H. Yeh, and J. J. Shyu, "Discrete fractional Hartley and Fourier transforms," *IEEE Trans. Circuit and System, Part II*, vol. 45, pp. 665–675, 1998.

[9] S. C. Pei and M. H. Yeh, "The discrete fractional cosine and sine transforms," *IEEE Trans. Signal Process.*, vol. 49, no. 6, pp. 1198–1207, 2001.

[10] F. Cariolaro, T. Erseghe, and P. Kraniauskas, "The fractional discrete cosine transform," *IEEE Trans. Signal Process.*, vol. 50, no. 4, pp. 902–911, 2002.

[11] C. C. Tseng, "Eigenvalues and eigenvectors of generalized DFT, generalized DHT, DCT-IV and DST-IV matrices," *IEEE Trans. Signal Process.*, vol. 50, no. 4, pp. 866–877, 2002.

[12] S. C. Pei and M. H. Yeh, "Discrete fractional Hadamard transform," in *Proceedings of IEEE International Symposium on Circuits and Systems*, pp. 179–182, June 1999.

[13] S. C. Pei, C. C. Tseng, M. H. Yeh, and J. J. Ding, "A new definition of continuous fractional Hartley transform," in *Proceedings of IEEE International Conference on Acoustics, Speech, and Signal Processing*, pp. 1485–1488, May 1998.

[14] B. W. Dickinson and K. Steiglitz, "Eigenvectors and functions of the discrete Fourier transform," *IEEE Trans. Acoust., Speech, and Signal Process.*, vol. ASSP-30, pp. 25–31, Feb. 1982.

transform	sequence $k = 0, 1, \ldots, N-1$	weighted terms	evaluated angles $m = 0, 1, \ldots, M-1$
DFRFT	$e^{-jk\alpha}$	DFRFT	$m\frac{2\pi}{M}$
DFRHT	$e^{-j\lfloor\frac{k}{2}\rfloor 2\alpha}$	DFRFT	$m\frac{2\pi}{M}$
	$e^{-j2k\alpha}$	DFRHT	$m\frac{\pi}{M}$
DFRCT	$e^{-j2k\alpha}$	DFRCT	$m\frac{\pi}{M}$
DFRST	$e^{-j2k\alpha}$	DFRST	$m\frac{\pi}{M}$
DFRHaT	$e^{-jk\alpha}$	DFRHaT	$m\frac{2\pi}{M}$

Table 1: Angular decompositions for the discrete fractional signal transforms

Efficient Pruning Algorithms for the DFT Computation for a Subset of Output Samples

Saad Bouguezel, M. Omair Ahmad, Fellow, IEEE, and M.N.S. Swamy, Fellow, IEEE

Department of Electrical and Computer Engineering
Concordia University
1455 de Maisonneuve Blvd. West
Montreal, P.Q., Canada H3G 1M8
E-mail: {b_saad, omair, swamy}@ece.concordia.ca

ABSTRACT

This paper presents efficient pruning algorithms for computing the DFT for a subset of output samples based on radix-2 decimation-in-time and decimation-in-frequency FFTs. They provide efficient implementations with a minimum number of stages. Comparisons are made with previously reported algorithms in terms of the computational complexity. The proposed algorithms are shown to provide a substantial reduction in the number of arithmetic operations, data transfers, address computations, and twiddle factor evaluations or accesses to the lookup table. The proposed algorithms retain all the features and characteristics, such as the simplicity and regularity, of the well-known Cooley-Tukey radix-2 FFT algorithms.

I. INTRODUCTION

The discrete Fourier transform (DFT) finds applications in almost every field of science and engineering. A major reason for its widespread use is the existence of efficient techniques for its computation. The related algorithms for the computation of the length-N DFT are generally known as the fast Fourier transforms (FFT). Since the discovery of the radix-2 FFT by Cooley-Tukey in 1965, considerable research has been carried out resulting in a number of algorithms [1], which are usually implemented in such a way that the input and output sequences have the same lengths. However, there are many applications where a significant part of the output samples are not needed, for example, where only a narrow spectrum band is desired.

In order to speed up the FFT algorithm in computing the DFT for only a subset of output samples, Markel has proposed the well-known pruning algorithm [2] that is based on a modification of the Cooley-Tukey standard radix-2 (CTR2) decimation-in-time (DIT) FFT algorithm. Assuming that only the first L, ($L < N = 2^r$), output points are needed and is restricted to be a power of two, $L = 2^d$, Markel's algorithm retains the first d stages of the standard radix-2 DIT FFT algorithm and alters the remaining $(r-d)$ stages by avoiding the computations of the undesired $(N - L)$ points. The ith stage, ($i = d+1$ to r), contains $2^{r-i}L$ half butterflies each requiring 4 real multiplications (M_r), 4 real additions (A_r), and 2 RAM read (R_r) and 1 RAM write (R_w) operations of complex numbers. The different requirements of the Markel's algorithm are given in Table I. Skinner proposed a method to prune the input samples by modifying the CTR2 DIT FFT algorithm [3]. Sorensen and Burrus [4] showed how this idea can be adapted to prune the output samples by modifying the CTR2 DIF FFT algorithm and called it as the Skinner algorithm. This Skinner's algorithm retains the first d stages of the radix-2 DIF FFT algorithm and alters the remaining $(r-d)$ stages by avoiding the twiddle factors, and using only additions. The ith stage, ($i = d+1$ to r), requires $2 \times 2^{r-i}L$ real additions, and $2 \times 2^{r-i}L$ RAM read and $2^{r-i}L$ RAM write operations of complex numbers. The requirements of the Skinner's algorithm are also given in Table I. Sorensen and Burrus [4] proposed another method, that seems to be better than the algorithm of Markel or that of Skinner in terms of arithmetic complexity. This is due to the use of the split-radix FFT in the computation of the sub-transforms. However, the use of the split-radix FFT increases some what the structural complexity of the algorithm in [4]

0-7803-7761-3/03/$17.00 ©2003 IEEE

in comparison with those of Markel and Skinner, which are based on the simple and regular structure of the CTR2 FFT. These three pruning algorithms decrease the number of arithmetic operations relative to the full length FFT. However, no attempt has been made to minimize the number of data transfers, address computations, and twiddle factor evaluations or accesses of a lookup table, which also contribute significantly to the execution time of the algorithm.

In this paper, we propose efficient algorithms for the pruning of the output samples based on both the DIT and DIF versions of the CTR2 FFT. They provide efficient implementations with a minimum number of stages and a significantly reduced structural and computational complexities, compared to those of the previously reported pruning algorithms that are based on the CTR2 FFT algorithm. The main difference between the proposed pruning algorithms and those mentioned above is in the decimation process. In the decomposition of the proposed method, the decimation is carried out only on a few stages, all the other stages being grouped by an appropriate recursive process so as to minimize the number of arithmetic operations, data transfers, address computations and twiddle factor evaluations or accesses to the lookup table.

II. PRUNING BASED ON THE RADIX-2 DIT FFT

The proposed method is similar to that of Markel in the sense that both the methods are based on the CTR2 DIT FFT algorithm. The main difference, however, is that in the proposed method only d stages are needed in pruning the output samples. Further, the remaining $(r-d)$ stages are grouped and incorporated in the dth stage of the radix-2 DIT FFT. This grouping provides a significant reduction in data transfers, address computations and control operations, and certain other advantages. Assuming $L = 2^d < N = 2^r$, the DFT becomes

$$X(m) = \sum_{k=0}^{N-1} x(k) W_N^{mk}, \qquad m = 0, 1, 2, ..., L-1 \qquad (1)$$

where $W_N = exp(-j2\pi/N)$. Therefore, the index k can be expressed by r binary bits, whereas for the index m, d bits are sufficient. Thus, the product mk can be written as

$$\begin{aligned} mk = {} & (2^{d-1}m_{d-1} + 2^{d-2}m_{d-2} + ... + 2m_1 + m_0)(2^{r-1}k_{r-1} \\ & +2^{r-2}k_{r-2} + ... + 2^{r-d+1}k_{r-d+1} + 2^{r-d}k_{r-d}) \\ & +(2^{d-1}m_{d-1} + 2^{d-2}m_{d-2} + ... + 2m_1 + m_0). \\ & .(2^{r-d-1}k_{r-d-1} + ... + 2k_1 + k_0) \qquad (2) \end{aligned}$$

It is obvious that the DIT of the second term of the product mk does not provide any simplification for the corresponding term of the right side of (1). It means that all the corresponding stages cannot provide the advantage of redundant operations. Consequently, in order to reduce the number of stages, we apply the DIT only to the first term in the right side of (2). After substituting (2) in (1), we can perform the pruning operation with only the following d stages:

$$X_i(m_0...m_{i-1}k_{r-i-1}...k_0) = \sum_{k_{r-i}=0}^{1} X_{i-1}(m_0...m_{i-2}k_{r-i}...k_0) W_N^{(2^{i-1}m_{i-1}+...+m_0)2^{r-i}k_{r-i}} \qquad (3)$$

and

$$X_d(m_0...m_{d-1}0...0) = \sum_{k_0=0}^{1} ... \sum_{k_{r-d-1}=0}^{1} \sum_{k_{r-d}=0}^{1} X_{d-1}(m_0...m_{d-2}k_{r-d}...k_0) W_N^{m2^{r-d}k_{r-d}} W_N^{m(2^{r-d-1}k_{r-d-1}+...+k_0)} \qquad (4)$$

where $i = 1, 2, ..., d-1$, $X_0(k) = x(k)$ and $m_i \in [0, 1]$. The input is taken in the natural order so as to obtain the output in the bit-reversed sequence. Thus, the desired narrow spectrum is obtained by the bit-reversal of only L points of the dth stage. The set of equations given by (3) represents the first $(d-1)$ stages of the radix-2 DIT FFT [5], and (4) represents the dth stage obtained by a grouping of the last $(r-d+1)$ stages. Therefore, the problem of pruning the output samples using a radix-2 DIT FFT can be divided into two distinct parts. After expressing the indices by a decimal notation, the first part that contains the $(d-1)$ stages corresponding to (3) can be efficiently calculated by the following recursive relation

$$X_{m,k,i}^{ou} = H_2 diag\begin{bmatrix} 1 & W_N^{m2^{r-i}} \end{bmatrix} X_{m,k,i}^{in}, \; i = 1, 2, ..., d-1 \qquad (5)$$

where $c_{m,i}^0(k) = 2^{r-i+1}\overleftarrow{m} + k$, $c_{m,i}^1(k) = C_{m,i}^0(k) + 2^{r-i}$, $X_{m,k,i}^{ou} = \begin{bmatrix} X_i\left(c_{m,i}^0(k)\right) & , & X_i\left((c_{m,i}^1(k)\right) \end{bmatrix}^t$, $X_{m,k,i}^{in} = \begin{bmatrix} X_{i-1}\left(c_{m,i}^0(k)\right) & , & X_{i-1}\left((c_{m,i}^1(k)\right) \end{bmatrix}^t$, t denotes the transpose of a matrix, $H_2 = \begin{bmatrix} 1 & 1 \\ 1 & -1 \end{bmatrix}$, $diag\begin{bmatrix} a & b \end{bmatrix} = \begin{bmatrix} a & 0 \\ 0 & b \end{bmatrix}$ and $\overleftarrow{m}$ denotes the bit-reversed value of m. For each value of i, (5) corresponds to a stage. However, for a given stage and for a specific value of m and of k, (5) represents a butterfly. For each stage there are $N/2$ such butterflies. Each butterfly requires

2 RAM address computations of complex numbers, 4 real multiplications, 6 real additions, and 2 RAM read and 2 RAM write operations of complex numbers. In order to simplify the procedure for the address generation, we do the bit-reversal of the exponents of the twiddle factors, which are generally precomputed and stored in a lookup table. Recall that the first stage of the radix-2 DIT FFT does not contain any twiddle factor and also the second stage requires only multiplications by $-j$. Therefore, the first $(d-1)$ stages require $(Nlog_2L - N)$ RAM address computations (R_{ac}) of complex numbers, $(2Nlog_2L - 6N)$ real multiplications, $(3Nlog_2L-5N)$ real additions, and $(Nlog_2L-N)$ RAM read and $(Nlog_2L - N)$ RAM write operations of complex numbers.

It is seen from (5) that the indexing process introduced in the above development not only simplifies the address generation, but has two other advantages. First, it allows the processing of all the butterflies of the ith stage having the same twiddle factors, and this can be easily done by fixing m and varying k from 0 to $(2^{r-i} - 1)$, $i = 1, 2, ..., d-1$. This technique provides a significant reduction in the number of accesses of the lookup table and address computations. The second advantage is that, by this indexing process, it is easy to avoid trivial multiplications in (5) by starting each stage with $m = 0$ and varying k. Additional savings are possible if the implementation is designed to take advantage of the twiddle factors at the third stage.

The second part of the proposed algorithm is concerned with the last stage, represented by the second equation of (4). The $(r - d + 1)$ summations in (4) are replaced by a single summation, and is expressed, after using the decimal notation of the indices, by the recursive equation:

$$X^{ou}_{m,0,d} = \sum_{k=0}^{(N/L)-1} \left(D_{m,k} H_2 D_m X^{in}_{m,0,d}\right) \quad (6)$$

where $D_{m,k} = diag\left[\begin{array}{cc} W_N^{mk} & W_N^{((L/2)+m)k} \end{array}\right]$, $D_m = diag\left[\begin{array}{cc} 1 & W_L^m \end{array}\right]$.

Since modern processors posses sufficient number of internal fast registers to store the intermediate results, (6) can be efficiently implemented without requiring intermediate storing in and loading from the RAM. This reduction in the number of RAM address generations, and read and write operations is one of the advantages of the recursive equation (6). There are only L points to be computed using (6), hence this part of pruning requires only L write operations, and can be implemented as follows. For each value of m, $m = 0, 1, ..., (L/2) - 1$, vary k in decreasing order from $(N/L) - 1$ to 0, generate the two addresses for reading, read two points from the RAM, process the general butterfly of the radix-2 DIT FFT, multiply the butterfly outputs by the twiddle factors, store the results in registers, repeat the process for the second value of k, add these results to the previous ones and store in the same registers, and repeat the whole process until $k = 0$. Write the final results in RAM using the last two address generated for reading. It is clear from (6) that a large number of trivial multiplications can be easily avoided when $m = k = 0$. Thus, it can be easily deduced that the second part of our decomposition requires N RAM address computations of complex numbers, $6N - (2N/L) - 4L$ real multiplications, $7N - (N/L) - 4L$ real additions, and N RAM read and L RAM write operations of complex numbers. If we combine the computational complexities of the both the parts, we find that our decomposition presents a significant improvement in comparison with Markel's algorithm, especially in the address generation and data transfer; however, the number of real multiplications and additions is only slightly reduced. In order to further decrease this number, we rearrange (6) as

$$X^{ou}_m = \sum_{k=0}^{(N/4L)-1} D_{m,k} \overline{X}^{in}_{m,k} \quad (7)$$

where

$$\overline{X}^{in}_{m,k} = H_2 D_m X^{in}_{m,k} + J_4 H_2 D^2_m X^{in}_{m,k+\frac{N}{2L}} + J_8 \left(H_2 D^1_m X^{in}_{m,k+\frac{N}{4L}} + J_4 H_2 D^3_m X^{in}_{m,k+\frac{3N}{4L}}\right) \quad (8)$$

$J_8 = diag\left[\begin{array}{cc} 1 & (1-j)/\sqrt{2} \end{array}\right]$, $D^1_m = diag\left[\begin{array}{cc} W^m_{4L} & W^{5m}_{4L} \end{array}\right]$, $J_4 = diag\left[\begin{array}{cc} 1 & -j \end{array}\right]$, $D^2_m = W^m_{4L} D^1_m$, and $D^3_m = W^{2m}_{4L} D^1_m$.

A close examination of the computational complexity of (7), taking into account the number of operations introduced by each matrix in the calculation of (8) and the number of trivial multiplications that can be avoided when $m = k = 0$, shows that the second part of pruning requires only $\{(19/4)N - (7N/2L) - 4L\}$ real multiplications and $\{(13/2)N - (7N/4L) - 4L\}$ real additions.

The overall computational complexity of the proposed algorithm is summarized in Table I. The comparison of this complexity is made only with that of Markel's pruning algorithm, since the both these are based on the CTR2 DIT FFT algorithm. It is clear that the proposed algorithm is far better than Markel's, since all the operations that contribute significantly to the execution time are reduced.

There remains another point, which needs to be discussed. It concerns the number of read operations for the twiddle factors from the lookup table. It can be shown that the number of read operations for twiddle factors from the lookup table needed to compute (7) is $(9N/4) - (3N/2L) - 2L - 1$. To reduce this number, we introduce a technique to implement (7) with a minimum number of read operations from the lookup table. Since $D_{m,k}$ is a diagonal matrix and $W_N^{m(k+1)} = W_N^{mk} W_N^m$, we get $D_{m,k+1} = D_{m,k} D_{m,1}$. Hence, by performing an appropriate factorization, (7) can be efficiently implemented by expressing it as

$$X_m^{ou} = \overline{X}_{m,0}^{in} + D_{m,1}(\overline{X}_{m,1}^{in} + D_{m,1}(\overline{X}_{m,2}^{in} + \dots$$
$$\dots + D_{m,1}(\overline{X}_{m,\frac{N}{4L}-2}^{in} + D_{m,1}(\overline{X}_{m,\frac{N}{4L}-1}^{in}))\dots)) \quad (9)$$

It is seen from (9) that for a given value of m, $m = 0, 1, \dots, (L/2) - 1$, it requires only 9 read operations for the twiddle factors. Note that for $m = 0$, one read operation from the lookup table is enough. Thus, the number of read operations for the twiddle factors is only $(9L/2) - 8$.

III. PRUNING BASED ON THE RADIX-2 DIF FFT

In this section, a second efficient algorithm for the pruning of the output samples is proposed. It is similar to that of Skinner in the sense that both the algorithms are based on the CTR2 DIF FFT algorithm. This algorithm is derived by using techniques similar to those in Section II. It can be shown that it consists of two parts. The first part can be efficiently calculated by the recursive equation.

$$X_m^{ou} = diag\left[\ 1 \quad W_N^{2^{i-1}k}\ \right].H_2 X_{m,k}^{in} \quad (10)$$

where $k = 0, 1, \dots, 2^{r-i} - 1$, $m = 0, 1, \dots, (L/2) - 1$ and $i = 1, 2, \dots, d - 1$. The second part is given by the expression

$$X_m^{ou} = \overline{X}_{m,0}^{in} + D(\overline{X}_{m,1}^{in} + D(\overline{X}_{m,2}^{in} + \dots$$
$$\dots + D(\overline{X}_{m,\frac{N}{4L}-2}^{in} + D(\overline{X}_{m,\frac{N}{4L}-1}^{in}))\dots)) \quad (11)$$

where $D = D_{0,1}$. Since the entire second part needs only one access to the lookup table, this decomposition not only reduces the number of accesses to the lookup table, but also is suitable for the lookup table reduction techniques. Note that the first d stages of Skinner's algorithm can be computed by any radix-2 DIF FFT. Assuming now that they are computed by (10), the dth stage requires $(N/L) - 1$ accesses to the lookup table, since in general $N \gg L$. Other advantages of the proposed algorithm compared to that of Skinner are included in Table I.

IV. CONCLUSION

In this paper we have proposed two pruning algorithms to compute the DFT for a subset of output samples, one based on the radix-2 DIT FFT and the other on the radix-2 DIF FFT. These two algorithms have been compared respectively with those of Markel [2] and Skinner [3] with regard to computational complexity. It has been shown that these algorithms outperforms those of Markel and Skinner in terms of the number of arithmetic operations, data transfers, address computations, and twiddle factor evaluations or accesses to the lookup table, yet retaining all the features, such as simplicity and regularity, of the CTR2 FFT algorithms.

TABLE I
Computational complexities of the different algorithms in the computation a subset of output samples

	Pruning based on the radix-2 DIT FFT	
	Markel's algorithm	Proposed algorithm
M_r	$2Nlog_2L - 4L$	$2Nlog_2L - \frac{5}{4}N - \frac{7N}{2L} - 4L$
A_r	$3Nlog_2L + 2N - 4L$	$3Nlog_2L + \frac{3}{2}N - \frac{7N}{4L} - 4L$
R_{ac}	$Nlog_2L + 2N - 2L$	$Nlog_2L$
R_r	$Nlog_2L + 2N - 2L$	$Nlog_2L$
R_w	$Nlog_2L + N - L$	$Nlog_2L - (N - L)$
	Pruning based on the radix-2 DIF FFT	
	Skinner's algorithm	Proposed algorithm
M_r	$2Nlog_2L$	$2Nlog_2L - \frac{5}{4}N - 2L$
A_r	$3Nlog_2L + 2N - 2L$	$3Nlog_2L + \frac{3}{2}N - 3L$
R_{ac}	$Nlog_2L + 2N - 2L$	$Nlog_2L$
R_r	$Nlog_2L + 2N - 2L$	$Nlog_2L$
R_w	$Nlog_2L + N - L$	$Nlog_2L - (N - L)$

ACKNOWLEDGMENT

This work was supported by NSERC (Canada) and FCAR (Quebec) grants.

REFERENCES

[1] P. Duhamel and M. Vetterli, "Fast Fourier transforms: A tutorial review and a state of the art," Signal Processing, vol. 19, pp. 259-299, 1990.

[2] J. D. Markel, "FFT Pruning," IEEE Trans. Audio Electroacoust., vol. AU-19, pp. 305-311, Dec. 1971.

[3] D. P. Skinner, "Pruning the decimation-in-time FFT algorithm," IEEE Tans. Acoust., Speech, Signal Processing, vol. ASSP-24, pp. 193-194, Apr. 1976.

[4] H. V. Sorensen and C. S. Burrus, "Efficient computation of the DFT with only a subset of input/output points," IEEE Trans. Signal Processing, vol. 41, pp.1184-1200, Mar. 1993.

[5] S. Bouguezal, D. Chikouche and A Khellaf, " An efficient algorithm for the computation of the multidimensional discrete Fourier transform," Multidimensional Systems and Signal processing, vol. 10, pp. 275-304, July 1999.

A ROBUST WIDEBAND ARRAY BEAMFORMER USING FAN FILTER

Zhu Liang Yu, Qiyue Zou

Center for Signal Processing,
School of EEE,
Nanyang Technological University,
Singapore, 639798
Email: ezlyu@ntu.edu.sg
equzou@ntu.edu.sg

Meng Hwa Er

School of EEE,
Nanyang Technological University,
Singapore, 639798
Email: emher@ntu.edu.sg

ABSTRACT

A new method on designing robust wideband generalized sidelobe canceller (GSC) using fan filter as blocking matrix is proposed in this paper. The idea presented is to exploit the line shape spectrum of the propagating wave and design a fan filter which suppresses the signal from a specific direction as well as over interested frequency band. Robust performance to steering error can be achieved by designing the fan filter with protected region around the assumed target direction. Moreover, the fixed beamformer in GSC can use fan filter also. Simulation results show that the proposed array processor has robust performance to large steering error, and it has high interference cancellation performance.

1. INTRODUCTION

The wideband adaptive array processors such as the linearly constrained minimum variance (LCMV) beamformer [1] and its alternative generalized sidelobe canceller (GSC) implementation [2] have been well studied. It is well known that the adaptive array processors can highly degrade its performance when there are array imperfections in practical scenarios. The well-discussed severe imperfection is the mismatch between array steering direction and target signal direction. This is also called steering error.

Some methods were proposed in various literatures [3–8] to overcome this problem. However, for large steering error, these methods have poor performance or even fail to work.

An important robust beamforming method [9, 10] uses spatial filter to replace the blocking matrix in GSC structure and can achieve robust beamforming performance to steering error. In this method, the spatial filter is designed as 1-D highpass filter through standard filter design approach. The stopband of the spatial filter is obtained by a simple mapping approach [9, 10]. This method does not fully exploit the temporal-spatial property of the propagating signal wave. This design method has some drawbacks. It can only work well when the desired signal impinges on the array from broadside. Moreover, it cannot perform well in the case of larger steering error. Otherwise, the spatial filter will suppress not only the target signal but also the interference and noise. This array processor only works well for suppressing the high frequency interferences. These drawbacks are discussed in detail in [11].

In this paper, a 2-D filter is introduced to replace the blocking matrix in GSC. This 2-D filter is called fan filter which is already well applied in seismic signal processing and array signal processing [12–14]. The propagating wave signal has a line-shape temporal-spatial spectrum as the thick line which corresponds to the direction of arrival (DOA) θ in Fig. 1. Moreover, this spectrum line crosses the origin in the spatial-temporal plane and its slope is determined by the DOA of the propagating signal. Therefore, the incident target signal from the direction in $[\theta_l, \theta_u]$ has the spectrum which distributes in the gray area as shown in Fig. 1. If an filter is designed to have null in this gray area, it can block the target signal from direction θ where $\theta \in [\theta_l, \theta_u]$. This filter can be used as the block matrix of the GSC system shown in Fig. 2. It is obvious that the fan filter is a suitable choice of this 2-D filter.

This paper is organized as following. In Section 2, the 2-D spectrum of the propagating wave signal is discussed. In Section 3, the proposed blocking filter is explained in detail. A simple fan filter design method is also presented. Some numerical examples are shown in Section 4 to illustrate the effectiveness of the new proposed beamformer. In the last section, a brief conclusion and some comments on future improvement are given.

2. SPECTRUM OF PROPAGATING WAVE

An wideband array processor has N equally spaced isotropic sensors along the x-axis and L tap delay sections. Let θ denote the direction-of-arrival (DOA) of the wave signal, and d denote the inter-sensor spacing. If the signal propagating speed is v, it can be obtained [13, 14] that

$$\omega_1 = \omega T_s, \quad \omega_2 = \omega \frac{d \sin\theta}{v}, \tag{1}$$

where T_s is the sampling period and ω is the signal frequency, ω_1 and ω_2 are normalized temporal and spatial frequencies, respectively.

It is easy to show that the 2-D spectrum of the target signal $s(t)$ is distributed on the line corresponding to θ as shown in Fig. 1. From (1), it is straightforward to obtain that

$$\omega_2 = \frac{d \sin\theta}{v T_s} \omega_1. \tag{2}$$

If d, v, T_s are all given, the slope of the spectrum line is determined by θ only. This means, if two signals have the same temporal spectrum but impinge on the array from different directions, the spectrums of these signals distribute around two different lines in 2-D spectrum plot.

0-7803-7761-3/03/$17.00 ©2003 IEEE

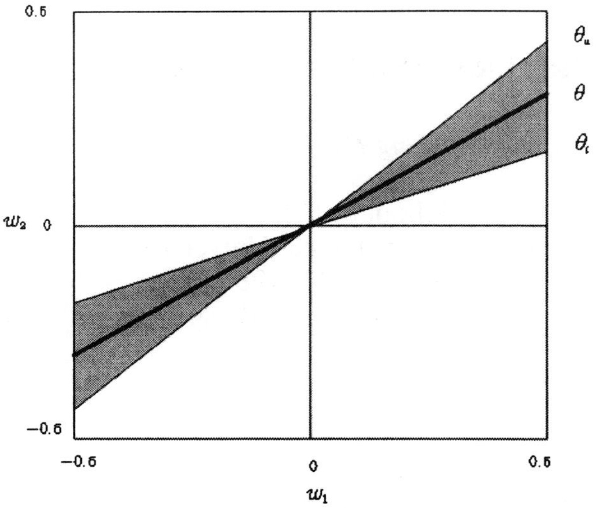

Figure 1: 2-D spectrum of propagating wave signal

This is an important property of the travelling signal wave in array processing. Consequently, the desired signal from direction θ can be passed by using a filter which has peak response along the spectrum line of the desired signal and very small gain in other area. This filter suppresses all the interferences from other directions. It is also possible to design opposite filter that has null along the spectrum line and high gain in other area. Thus, the target signal is blocked by this filter and other signals are passed. In the practical array design, the direction of the target signal varies in a range $[\theta_l, \theta_u]$, where θ_l and θ_u are the boundaries of the signal direction. From the discussion above, it can be concluded that the spectrum of the target signal distributes in the gray area as shown in Fig. 1. To design the array processor which has protected response in the gray region between lines corresponding to θ_l and θ_u in Fig. 1 is important in designing the robust array processor against the steering error. The angle region $[\theta_l, \theta_u]$ can be determined by designer according to the requirements of the tolerance to steering error.

3. PROPOSED BEAMFORMER

An adaptive GSC is shown in Fig. 2, where G and B are fixed beamformer and blocking matrix respectively. H_1 to H_K are adaptive filters. The general discussion on GSC can be found in [2].

In [9, 10], the blocking matrix is designed as spatial filter, which has wider null in the direction of broadside to prevent desired signal from cancellation where there exists steering error. This method can be considered as an extension of the blocking matrix proposed in [2].

In this paper, the fan filter is used to replace the blocking matrix B as well as the fixed beamformer G in Fig. 2. The special structure of the proposed blocking matrix B is that each row of matrix B is replaced by an individual blocking filter which is designed as fan filter. Each blocking filter is of size ML and operates on M sensor signals and L taps. The new blocking matrix can be

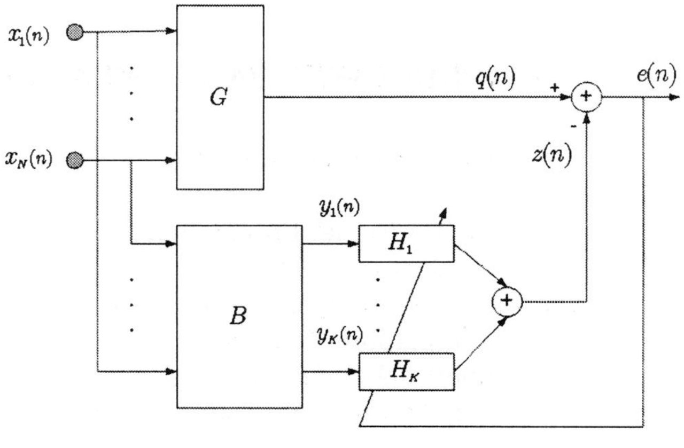

Figure 2: Broadband GSC beamformer

express as

$$B^T = \begin{bmatrix} \mathbf{b}_1^T & \cdots & \mathbf{b}_M^T & \mathbf{0}^T & \cdots & \mathbf{0}^T \\ \mathbf{0}^T & \mathbf{b}_1^T & \cdots & \mathbf{b}_M^T & \cdots & \mathbf{0}^T \\ \vdots & & \ddots & & \ddots & \vdots \\ \mathbf{0}^T & \mathbf{0}^T & \cdots & \mathbf{b}_1^T & \cdots & \mathbf{b}_M^T \end{bmatrix}, \tag{3}$$

where $\mathbf{b}_m^T = [\beta_{m1} \cdots \beta_{mL}]$ and $\mathbf{0}^T$ is a $1 \times L$ vector whose entries are all zero. The number of adaptive filters is $K = N - M + 1$. We assume that the array is linear and the propagating wave is plane. The blocking filter response can be expressed as

$$\mathbf{B}(\omega, \theta) = \sum_{m=0}^{M-1} \sum_{l=0}^{L-1} \beta_{ml} e^{-j\omega l T_s} e^{j\omega m \frac{d \sin\theta}{v}}. \tag{4}$$

According to (1), (4) can be simplified to

$$\mathbf{B}(\omega_1, \omega_2) = \sum_{m=0}^{M-1} \sum_{l=0}^{L-1} \beta_{ml} e^{-j\omega_1 l} e^{j\omega_2 m}. \tag{5}$$

The kth spatial filter $\mathbf{B}_k(\omega_1, \omega_2)$ is given as

$$\mathbf{B}_k(\omega_1, \omega_2) = e^{-j(k-1)\omega_2} \mathbf{B}(\omega_1, \omega_2). \tag{6}$$

Since the task of the fixed beamformer G in GSC shown in Fig. 2 is to favour the protected region, which can be designed according to the specification of quiescent response of the array processor. The task of blocking filter B is opposite. Some constraints, such as orthogonality [2] can be imposed on the design of these two filters. A very simple method is to design the fixed beamformer G with the assumption that its response in the protected region is one and zero outside, and vice versa for the lower beamformer B. However, in the practical applications, since the filter can not possess ideal response characteristics, we suggest to construct B through unconstrained filter design method and design G with the constraints which ensure a zero cross-correlation between the quiescent response and the blocking matrix output signal [10]. This method will be discussed in detail in the future work.

Once the beamformer B and G are designed, the adaptive filter H_k can be found by solving the Wiener equation, when all incoming signals are band-limited and correctly sampled. A discrete-time Wiener solution can be found and approximated as FIR filters.

The Wiener solution of the adaptive filter can be obtained by

$$\begin{aligned} R_{YY}(\omega)H(\omega) &= R_{qY}(\omega), \\ Y &= [y_1(n) \cdots y_K(n)]^T, \end{aligned} \tag{7}$$

where $R_{xy}(\omega)$ represents the cross-covariance function between signal x and y (auto-covariance function when $x = y$), and the signal $q(n)$ is the output of fixed beamformer G. The adaptive method to find the optimal filter $H(\omega)$ in time domain can use some well known algorithms such as Least Mean Square (LMS) [15]. To obtain a robust performance when there is small desired signal leakage to the adaptive filter, the norm constrained LMS [7, 16] can be applied to prevent the desired signal from cancellation. A tractable way to obtain a real-time adaptive beamformer with prescribed resolution is using the leaky LMS algorithm [9, 10] to update the adaptive weights. Some conclusions on the specification of blocking filter were also given there to prevent desired signal from cancellation.

The fan filter used in the proposed beamformer can be designed by various 2-D filter design methods [17–19]. Here we introduce a simple method based on McClellan transform [17]. The design method is summarized as:

Step 1 Design a 1-D prototype FIR filter. This filter can be lowpass filter (for fixed beamformer design) or highpass filter (for blocking filter design) according to the requirements on the fan filter. It's response function can be expressed as

$$P(\omega) = \sum_{m=1}^{M} p(m)e^{-j\omega n}, \tag{8}$$

where $p(m)$ is the impulse response.

Step 2 Apply transformation

$$\omega = T(\omega_1, \omega_2) = (\frac{vT_s\omega_2}{d\omega_1} - \sin\theta_0)/2 \tag{9}$$

to (8). The response of 2-D filter is obtained as

$$Q(\omega_1, \omega_2) = \sum_{m=1}^{M} p(m)e^{-jT(\omega_1,\omega_2)n}. \tag{10}$$

In the region where the 2-D spectrum of the incident signal does not appear, the value of response function can be set arbitrarily.

Step 3 Apply inverse Fourier transform to get system response function $q(n_1, n_2)$. In order to obtain the filter coefficient $\beta_{m,l}$ according to the number of elements M and the number of taps L, $q(n_1, n_2)$ is then truncated as

$$\begin{aligned} \beta_{m,l} &= q(l - \frac{L-1}{2}, -m + \frac{M-1}{2}) \\ (m &= 0, 1, \cdots, M-1; l = 1, 2, \cdots, L-1) \end{aligned} \tag{11}$$

The designed fan filter coefficients $\beta_{m,l}$ are used in matrix B to form the proposed blocking filter. The fixed beamformer can also be replaced by the filter designed by this method.

4. NUMERICAL STUDY

In this section, some simulation results are shown to illustrate the effectiveness of proposed method. It can handle the array without presteering delay sections. In the simulation, the desired direction of the target signal is assumed as $\theta_0 = -30°$ with protected region $[-20° \quad -40°]$. There are no presteering section in the array processor.

In the following simulations, the inter-sensor spacing is $d = vT_s$. The number of sensors is $N = 20$. The sensors used to design blocking filter is $M = 15$. The filter has $L = 15$ taps. The 1-D filter prototype is designed as Kaiser filter. The filter prototypes used for fixed beamformer and blocking filter are lowpass filter and highpass filter respectively.

The array is steered to direction $\theta_0 = -30°$. The responses of designed blocking filter B and fixed beamformer G are shown in Fig. 3 and Fig. 4 respectively.

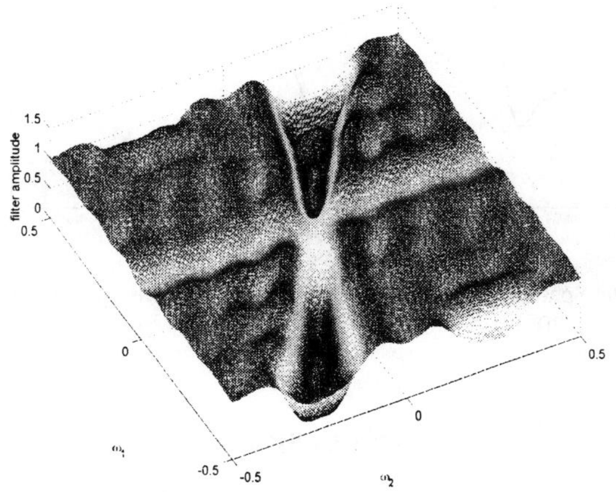

Figure 3: Blocking filter designed using 15 sensors and 15 taps ($\theta_0 = -30°$)

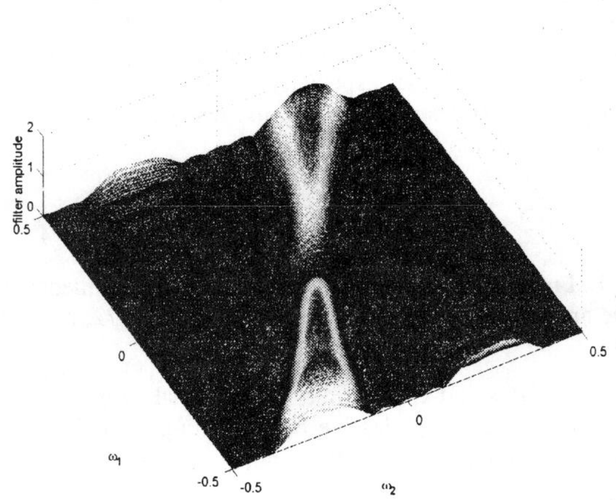

Figure 4: Fixed beamformer filter designed using by 15 sensors and 15 taps ($\theta_0 = -30°$)

To prevent desired signal from cancellation, the norm constrained LMS is used to update the weights of the adaptive array processor. The simulation signal is a band-limited white signal, whose bandwidth is $0.3 - 3.7KHz$. The input signal-to-noise ration (SNR) is $20d$B, where the noise is spatially and temporally white. In order to obtain robust performance against the steering error, we design the array processor with beamwidth about $20°$ around its looking direction. Fig. 5 shows the spatial power response of the adaptive array processor whose steering direction is $\theta_0 = -30°$.

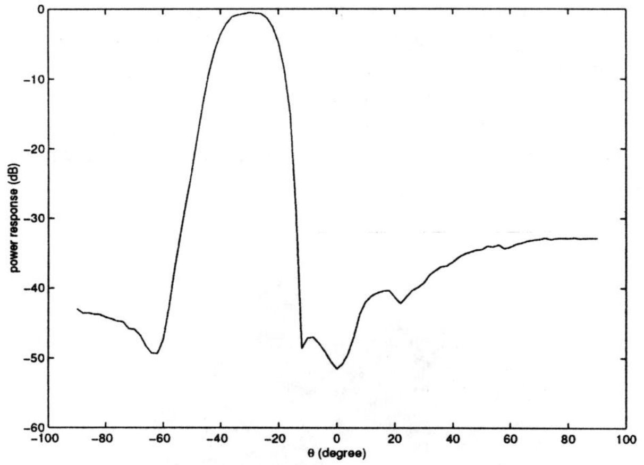

Figure 5: Spatial response of the proposed adaptive array processor ($\theta_0 = -30°$)

5. CONCLUSIONS

A new method on designing robust wideband GSC processor is proposed based on fan filter design. The advantage of this new method is that it works properly for wideband signal with robust performance. Meanwhile, the beamwidth of the array processor can be determined by 2-D filter design. Furthermore, this method can work for array processor with/without presteering sections. Simulation results show the effectiveness of the proposed method. However, the fan filter needs relatively more sensors to achieve the desired spatial response, especially for low frequency signal. The optimal fan filter design method to use less sensors and taps is under our investigation.

6. REFERENCES

[1] O. L. Frost III, "An algorithm for linearly constrained adaptive array processing," in *Proc. IEEE*, Aug. 1972, pp. 926–935.

[2] L. J. Griffiths and C. W. Jim, "An alternative approach to linearly constrained adaptive beamforming," *IEEE Trans. Antennas Propagat.*, vol. AP-30, no. 1, pp. 27–34, Jan. 1982.

[3] J. E. Hudson, *Adaptive Array Principles*, Peter Peregrinus Ltd., 1981.

[4] D. Nunn, "Performance assessments of a time-domain adaptive antenna processor in a broadband environment," in *Proc. Inst. Elect. Eng. F, H*, 1983.

[5] M. H. Er and A. Cantoni, "Derivative constraints for broadband element space antenna array processors," *IEEE Trans. Acoust., Speech, Signal Processing*, vol. ASSP-31, no. 6, pp. 1378–1393, Dec. 1983.

[6] M. H. Er and A. Cantoni, "A new set of linear constraints for broad-band time domain element space processor," *IEEE Trans. Antennas Propagat.*, vol. AP-34, no. 3, pp. 320–329, Mar. 1986.

[7] H. Cox, R. M. Zeskind, and M. M. Owen, "Robust adpative beamforming," *IEEE Trans. Acoust., Speech, Signal Processing*, vol. ASSP-35, no. 10, pp. 1365–1376, Oct. 1987.

[8] N. K. Jablon, "Adaptive beamforming with the generalized sidelobe canceller in the presence of array imperfection," *IEEE Trans. Antennas Propagat.*, vol. 34, no. 8, pp. 996–1012, Aug. 1986.

[9] I. Claesson and S. Nordholm, "A spatial filtering approach to robust adaptive beamforming," *IEEE Trans. Antennas Propagat.*, 1992.

[10] S. Nordebo, I. Claesson, and S. Nordholm, "Adaptive beamforming: spatial filter designed blocking matrix," *IEEE J. Ocean. Eng.*, vol. 19, no. 4, pp. 583–590, Oct. 1994.

[11] Z. L. Yu, "Robust algorithm for microphone array system," Technical Report EEE3/022/2002, School of EEE, Nanyang Technological University, Singapore, Aug. 2002.

[12] S. Treitel, J. L. Shanks, and C. W. Frasier, "Some aspects of fan filtering," *Geophysics*, vol. 32, no. 5, pp. 789–800, Oct. 1967.

[13] K. Nishikawa, T. Yamamoto, K. Oto, and T. Kanamori, "Wideband beamforming using fan filter," in *Proc. ISCAS'92*, 1992.

[14] T. Sekiguchi and Y. Karasawa, "Wideband beamspace adaptive array utilizing FIR fan filters for multibeam forming," *IEEE Trans. Signal Processing*, vol. 48, no. 1, pp. 277–284, Jan. 2000.

[15] B. Widrow, J. R. Glover, J. M. McCool, J. Kaunitz, C. S. Williams, R. H. Hearn, J. R. Zeidler, and R. C. Goodlin, "Adaptive noise cancellating: Principles and applications," in *Proc. IEEE*, 1975.

[16] B. Widrow and S. Stearns, *Adaptive Signal Processing*, Prentice-Hall, Englewood Cliffs, NJ, 1985.

[17] W. S. Lu and A. Antoniou, *Two-Dimenstional Digital Filters*, Marcel Dekker Inc., 1992.

[18] E. Z. Psarakis and G. P. Mertzios, V. G. Alexiou, "Design of two-dimensional zero phase FIR fan filters via the McClellan transform," *IEEE Trans. Circuits Syst.*, vol. 37, no. 1, pp. 10–16, Jan. 1990.

[19] S. S. Kidambi, "Design of two-dimenstional nonrecursive filters based on frequency transformations," *IEEE Trans. Signal Processing*, vol. 42, no. 12, pp. 3025–3029, Dec. 1995.

A REGULARIZED SIMULTANEOUS AUTOREGRESSIVE MODEL FOR TEXTURE CLASSIFICATION*

Yao-wei WANG[1] *Yan-fei WANG*[2] *Wen GAO*[3] *Yong XUE*[4]
1; 3 Institute of Computing Technology,
Graduate School of Chinese Academy of Science,
Chinese Academy of Sciences, POBox 2704 Beijing 100080, P.R.China
Email: ywwang@jdl.ac.cn, wgao@jdl.ac.cn
2; 4 Laboratory of Remote Sensing Information Sciences,
Institute of Remote Sensing Applications, Chinese Academy of Sciences,
POBox 9718, Beijing 100101, P.R.China
Email: wyf@lsec.cc.ac.cn, xuey@unl.ac.uk

ABSTRACT

In this paper, we present a new method for texture classification which we call the regularized simultaneous autoregressive method (RSAR). The regularization technique is introduced. With the technique, the new algorithm RSAR outperforms the traditional algorithm in texture classification. Particularly, our new algorithm is useful for extracting texture from the image which is coarse or contains too much noise.

1. INTRODUCTION

Texture plays an important role in image processing. It provides information for recognition and interpretation for human beings. There are many research works on texture analysis, such as texture classification and texture segmentation (see [2] [5] [9]). The algebraic methods are mainly the least square error (LSE) and the maximum likelihood estimation (MLE) method. It is well known that the texture model is in fact a kind of discrete operator equations. The coefficient matrix is very ill-conditioned. To get a stable texture analysis, LSE is infeasible. This method is unstable even in the noiseless case. MLE is a little better than LSE, but it cannot give us any estimation even in the simplest case (see [8]). Furthermore, MLE is much more time-consuming. Therefore, we take LSE as the basic model and resort to other stable methods. Regularization is such a technique introduced by Tikhonov and Arsenin in their famous book [7]. Originally this method is for linear and nonlinear integral equations of the first kind. These equations are known to be ill posed in the sense of Hadamard [1]. Numerically, the computer can only deal with discrete data. Hence we have to transform the continuous problem into discrete problem, i.e., the discrete ill-posed problems. The SAR model is an example of discrete model of ill-posed problems. Let $g(s)$ be the gray level of a pixel at the position $s=(s_1,s_2)$ in an $m \times m$ image $(s_1,s_2=1,2,\cdots,m)$. The basic SAR model for image texture is usually in the form

$$\sum_{r\in I}\theta(r)g(s+r)+\mu+\varepsilon(s)=g(s) \tag{1}$$

where I is all the neighboring pixels of the pixel s, $\varepsilon(s)$ is an independent Gaussian random variable with zero mean and variance σ^2. $\theta(r)$, $r \in I$, are the model parameters characterizing the dependence of a pixel to its neighbors, and μ is the bias which is dependent on the mean gray value of the image. All parameters μ, σ, and $\theta(r)$ can be estimated from a given window (sub image) by the least square estimation technique or the maximum likelihood estimation method. As we have noted that this kind of estimation is unstable. Other researchers have also made some research on SAR model. J Mao and A K Jain proposed the multi-resolution simultaneous auto-regressive model to enable multi-scale texture analysis [5]. In [3] [6] [4], the comparison of the MRSAR features with other features using Brodatz texture images had been made. However they did nothing to improve the quality of the SAR estimation model.

In this paper, we will use the regularization technique to tackle the parameter estimation in texture classification problems. We demonstrate that solution of the regularized simultaneous auto regressive model (RSAR) is better than which from the least squares error method. The paper is organized as follows: in section 2, we describe that how the RSAR works and the related problems. In section 3, numerical simulation is performed based on the texture image from Brodatz album. Finally, in section 4, we give some discussions and future research.

2. RSAR TEXTURE CLASSIFICATION

We usually use the spatial relations among neighboring pixels to characterize the texture. A main class of model for specifying the underlining interaction among the given observation is the SAR model in the form (1), where all parameters μ, σ, and $\theta(r)$ can be estimated by LSE or

0-7803-7761-3/03/$17.00 ©2003 IEEE

MLE. The parameters $\theta(r)$ are usually used for texture classification and segmentation. But as we have said, the above two methods are not stable. Therefore, we prefer the regularization method, which transfer the ill-conditioned system into a well-conditioned system.

What we are interested is to estimate the parameter $\theta(r)$, which characterizes the texture of the given image. Clearly, (1) can be written as

$$\sum_{r\in I}\theta(r)g(s+r)=g_{\varepsilon}(s)-\mu \qquad (2)$$

where $g_{\varepsilon}(s)=g(s)-\varepsilon(s)$. We assume that ε is random variable and $\varepsilon \sim N(0,\sigma^2)$.

Notice that $s=(s_1,s_2),\ s_1,s_2=1,2,\cdots,m,\ r\in I,\ I$ is the neighboring pixels at the position s. Then (2) can be written as

$$\sum_{r\in I}\theta(r)g(s_t+u)=g_{\varepsilon}(s_t)-\mu \qquad (3)$$

and each s_t is in the form of s. For different r_i and s_t, (3) form a linear equation

$$G\theta^u = g_{\varepsilon}^{\mu} \qquad (4)$$

with $G=(g_{ij})$ the matrix and θ^u, g_{ε}^{μ} the corresponding vectors.

In theory, we can estimate the parameter θ^u by solving the above equations. But we must keep in mind that, the direct methods should be avoided. (4) can be considered as a discrete linear operator equation. The matrix G is very ill-conditioned. We must resort to new stable technique.
For (4), we consider the minimization problem

$$\min J(\theta^u)=\left\|G\theta^u-g_{\varepsilon}^{\mu}\right\|_{l_2}^2 \qquad (5)$$

However this formulation is still unstable. Because it is equivalent to the normal equation

$$G^tG\theta^u = G^tg_{\varepsilon}^{\mu} \qquad (6)$$

Now it is self-evident that $cond(G^tG)>cond(G)$, hence the problem is much more ill-conditioned. To overcome this problem, we introduce the regularization technique, i.e., instead of (5), we solve the minimization problem

$$\min J(\theta^u)=\frac{1}{2}\left\|G\theta^u-g_{\varepsilon}^{\mu}\right\|_{l_2}^2+\frac{\alpha}{2}\left\|\theta^u\right\|_{l_2}^2 \qquad (7)$$

where $\left\|\theta^u\right\|_{l_2}^2$ serves as the stabilizer, α is the parameter to balance the bias between the original and the new problem. α can also be considered as the Lagragian parameter for the constrained problem

$$\min \tilde{J}(\theta^u)=\left\|G\theta^u-g_{\varepsilon}^{\mu}\right\|_{l_2}^2 \qquad (8)$$

$$\text{s.t.} \left\|\theta^u\right\|\le b \qquad (9)$$

where b is the upper bound for the solution θ^u.
If we denote the singular system of G as $\{\lambda_i;x_i,y_i\}$, i.e.,

$$Gy_i=\lambda_i x_i \qquad (10)$$

$$G^tx_i=\lambda_i y_i \qquad (11)$$

and the singular values satisfy $\lambda 1\ge\lambda 2\ge\cdots\ge 0$. Thus the solution can be expressed as

$$G^{+}\theta^u=\sum_i\frac{(g_{\varepsilon}^{\mu},x_i)}{\lambda_i}y_i \qquad (12)$$

Now it is clear that $(g_{\varepsilon}^{\mu},x_i)/\lambda_i$ grows very quickly for small singular values λ_i and the instability occurs. This is the reason that we introduce regularization in this context. Solving (7) leads to the following equation:

$$G^tG\theta^u+\alpha\theta^u=G^tg_{\varepsilon}^{\mu} \qquad (13)$$

Notice that the parameter $\alpha>0$ can be chosen by user, hence the coefficient matrix $G^tG+\alpha E$ (E is the identity matrix) can be positive definite. Therefore the Cholesky decomposition can be employed to get the solution. In the following, we will describe the method for solving equation (13).
Assume that the Cholesky decomposition of the matrix $G^tG+\alpha E$ as

$$G^tG+\alpha E=LDL^t$$

where L is the lower triangular matrix with the diagonal elements all ones and D is the diagonal matrix. With such configuration, the parameters θ^u are solved though the following two systems:

$$v = L^{-1}G^t g_\varepsilon^\mu \qquad (14)$$

$$\theta^u = L^{-t}D^{-1}v \qquad (15)$$

Note that L is the triangular matrix, the cost of computation of the above tow linear system is very small. Overall, the cost of Cholesky decomposition is about $O(n^3/6)$. For current computer, the amount of computation is reasonable. The algorithm is described as follows;

Algorithm 2.1 (Regularized SAR)

STEP 1 Factor $G^tG + \alpha E = LDL^t$;

STEP 2 Solve $Lv = G^t g_\varepsilon^\mu$;

STEP 3 Solve $DL\theta^u = v$.

3. NUMERICAL SIMULATION

(-1,1)	(0,1)	(1,1)
(-1,0)	(0,0)	(1,0)
(-1,-1)	(0,-1)	(1,-1)

Figure 1: The second-order neighborhood for pixel at site (0, 0)

We use the basic SAR model for textured images, see equation (1). For our problem, the parameters are determined by the choice of the neighborhood I and the choice of the window size. Here, for convenience, we choose a simple second-order neighbor-hood as shown in Figure 1. The parameter θ is estimated using 25×25 overlapping windows.

To be simple, the SAR model used here is rotation-variant, which means that when image rotates, the model parameters also change. Clearly, our algorithm is also fit for rotation-invariant model.

We tested all images in Brodatz album, here we give example for a group of images. Each image belongs to different class (see Figure 2-Figure 11). In each figure, the left is the original image, the middle is the texture classified by LSE, and the right is the texture classified by RSAR. For the algorithm RSAR, the choice of the parameter α is apriori. Here we chose $\alpha = 0.01$. α can not be too large or too small. A very large α leads to a well-conditioned system, but it is a poor approximation to the original problem; a very small α leads to a well approximation, but the perturbation caused by noise is still activated. Usually α is chosen between 0 and 1. In our image, the horizon axes represents the eight textures coordinates; the vertical axes represents the corresponding texture whose value at a pixel is the model parameter $\theta(r)$ and is scaled to 0-255 for the purpose of display.

Figure 2-Figure 11 list the plot of the textured images from different class. In table 1 and 2, we compute the norms of the results from LSE and RSAR respectively. Very large norm values mean that the results are far away from the true value and the corresponding algorithm is unstable, which also means that the texture of the image may be very coarse or contains too much noise. Small norm values mean that the results are better and the corresponding algorithm is stable. Clearly, we can see from the table and these figures that the textures classified by RSAR are better than that from LSE. In figures 2, 3, 6, 8 and 9, almost no texture is classified by LSE. The norm values are quite large. But all of these can be overcome by RSAR. The texture classified by RSAR is satisfactory; the norm values are also small. In figures 4, 5, 7, 10 and 11, both algorithms can give us satisfactory results. The texture classified by LSE is a little clearer than RSAR, we think this is due to these images are "good" themselves. But we have noted that the parameter α =0.01 is not optimal. If we adjust α appropriately, the results obtained by RSAR will be better.

4. CONCLUSION

The new algorithm RSAR presented in this paper is useful for texture classification, especially for the images with coarse texture or containing too much noise. In our experiments, the window we choose is 25×25. Clearly, the size can be larger. But the cost of computation will increase enormously. We consider other optimization technique may be used there. The choice of the parameter α is a delicate matter. In our tests, the choice of α as 0.01 is not optimal. The best value of α should be matched with the error due to noise and machine truncation. But this is difficult to do. All of these are our future work.

Table 1: The norm of the data from LSE and RSAR (d001-d005)

	d001	d002	d003	d004	d005
LSE	1.18×10^6	2.8×10^5	40.71	41.85	1.33×10^3
RSAR	115.12	49.41	136.30	136.27	99.23

Table 2: The norm of the data from LSE and RSAR (d006-d010)

	d006	d007	d008	d009	d010
LSE	56.41	625.76	2.28×10^3	28.51	34.32
RSAR	107.91	167.38	25.52	111.64	122.72

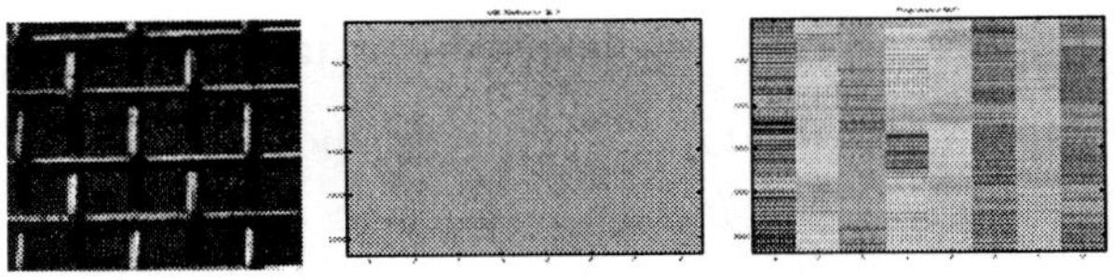

Figure 2: Comparison of the LSE and RSAR for d001

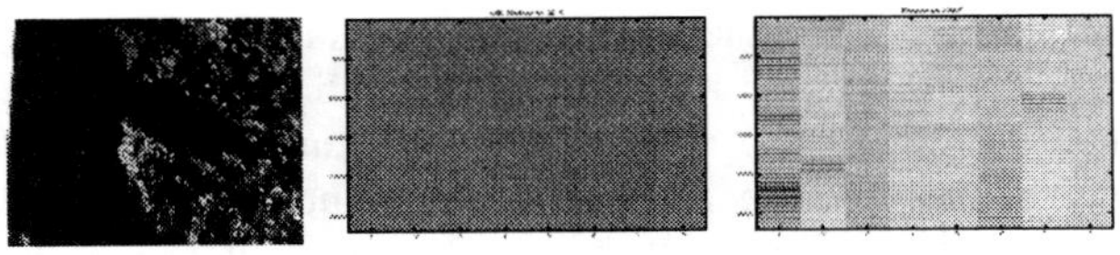

Figure 3: Comparison of the LSE and RSAR for d002

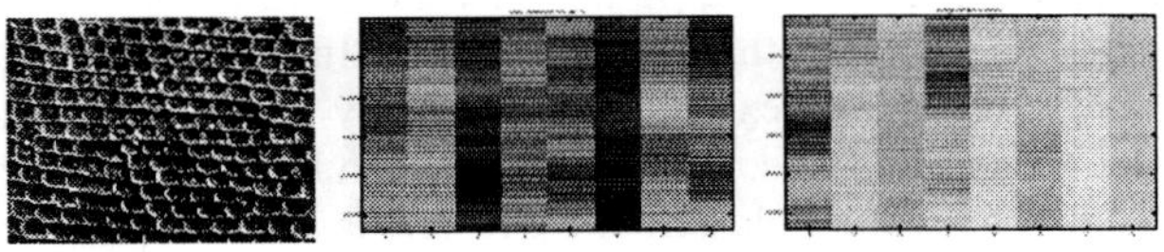

Figure 4: Comparison of the LSE and RSAR for d003

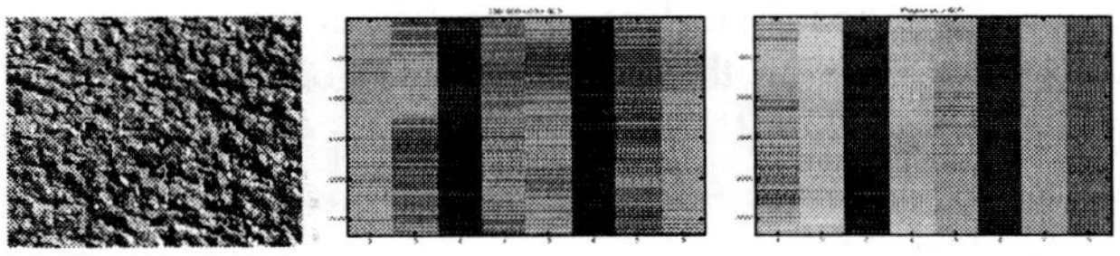

Figure 5: Comparison of the LSE and RSAR for d004

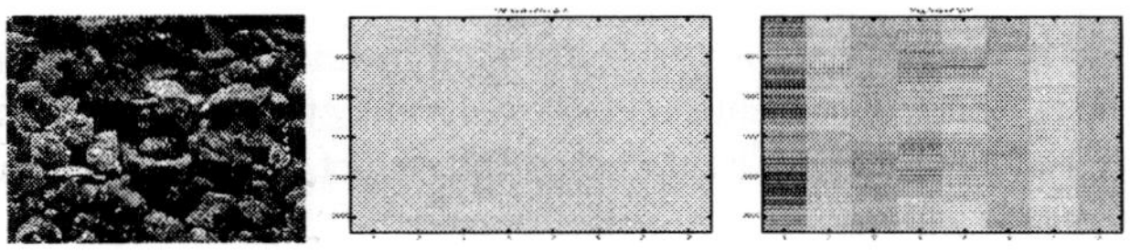

Figure 6: Comparison of the LSE and RSAR for d005

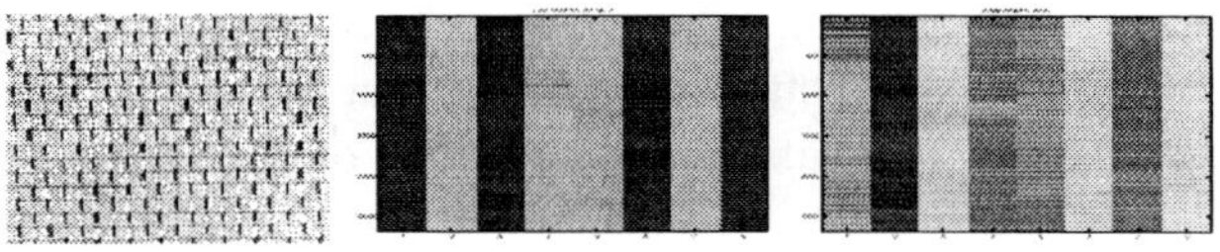

Figure 7: Comparison of the LSE and RSAR for d006

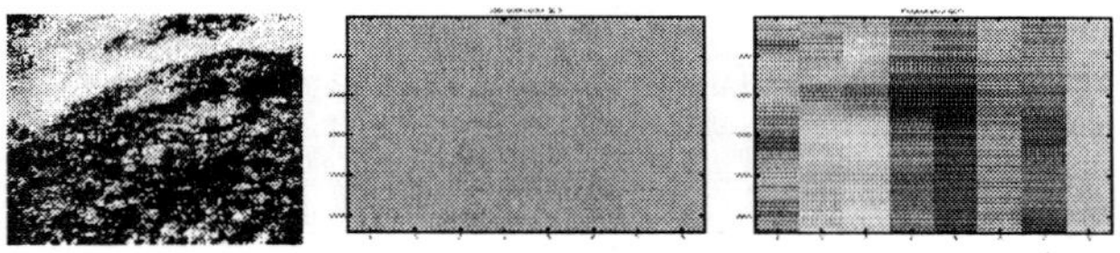

Figure 8: Comparison of the LSE and RSAR for d007

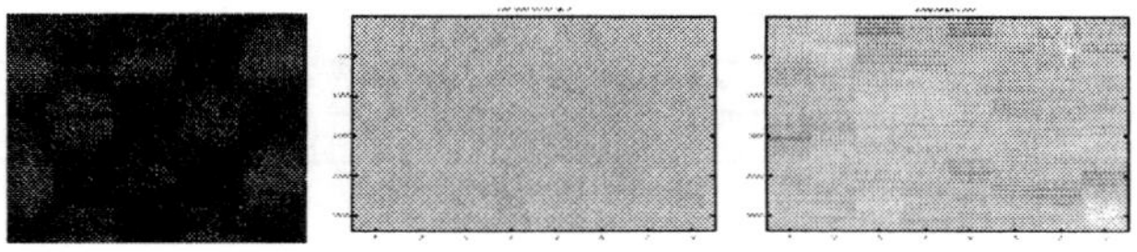

Figure 9: Comparison of the LSE and RSAR for d008

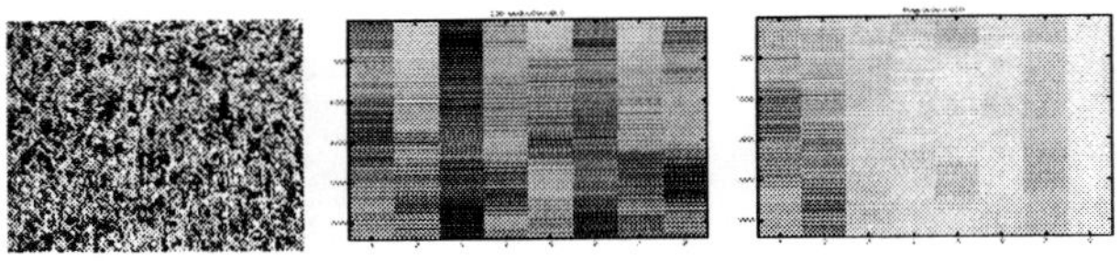

Figure 10: Comparison of the LSE and RSAR for d009

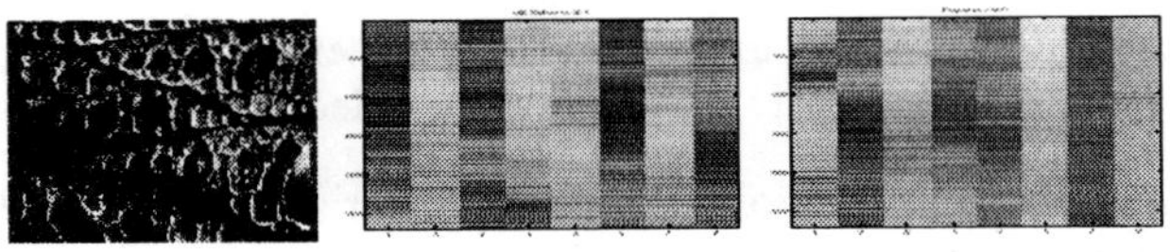

Figure 11: Comparison of the LSE and RSAR for d010

5. REFERENCES

[1] J Hadmard, *Lectures on the Cauchy Problems in Linear Partial Differential Equations,* Yale University Press New Haven, 1923.

[2] R M Haralick, *Statistical and structural approaches to texture*, Proc. IEEE 67, pp.786-840, 1979.

[3] F Liu and R W Picard, *Periodicity, directionality and randomness: World features for image modeling and retrieval*, MIT Media Lab Technical Report, No. 320.

[4] B S Manjunath and W Y Ma, *Texture features for browsing and retrieval of image data*, IEEE Trans. on Pattern Analysis and Machine Intelligence, Vol. 18, No. 8, pp.837-842, Aug 1996.

[5] J Mao and A K Jain, *Texture classification and segmentation using multi-resolution simultaneous autoregressive models*, Pattern Recognition, Vol. 25, No. 2, pp.173-188, 1992.

[6] R W Picard, T Kabir and F Liu, *Real-time recognition with the entire Brodatz texture database*, Proc. IEEE Int. Conf. on Computer Vision and Pattern Recognition, pp.638-639, New York, June 1993.

[7] A N Tikhonov and V Y Arsenin, *Solutions of Ill-posed Problems*, Winston/Wiley, 1977.

[8] V N Vapnik, *Inductive principles of statistics and learning theory*, Springer-Verlag New York, 2000.

[9] H Wechsler, *Texture analysis-a survey*, Signal Process. 2, pp.271-282, 1980.

* Yao-wei WANG and Wen GAO are supported by the research projects "Broad-Band Network Mutual Multimedia System for 4C Amalgamation (KGCXZ103)" "Initial Phase of the Knowledge Innovation Program"; Yan-fei WANG and Yong XUE are supported by the research projects "CAS Hundred Talents Program" and "Digital Earth (KZCX2-312)".

Efficient Frequency Estimation using the Pulse-Pair Method at Different Lags

Saman S. Abeysekera

School of Electrical and Electronic Engineering,
Nanyang Technological University, Nanyang Avenue, SINGAPORE 639798.
E-mail: esabeysekera@ntu.edu.sg

***Abstract**: Pulse-Pair (PP) method is an efficient technique of frequency estimation, which requires a very low number of multiplications. Therefore, it is ideally suitable for FPGA type hardware implementations, especially at high frequencies, as the multiplications are relatively expensive for Silicon area optimization. The optimum frequency estimation using the PP method needs a lag value equal to the two-third of the signal length. The optimal PP method, however, requires phase unwrapping that usually makes the technique computationally un-attractive. In this paper, we propose a new efficient technique for evaluating the optimum PP method without resorting to phase unwrapping. Detailed discussion of the performance of the optimum PP method is provided in the paper, together with a comparison with the technique of DFT based maximum likelihood frequency estimation.*

1. Introduction

Accurate frequency estimation of a mono-tone (single sinusoid) signal, in the presence of noise, is one of the most commonly encountered problems in signal processing applications. This problem has been intensively studied under the condition when the noise present can be represented using an additive model. In such cases, the peak estimate of the Discrete Fourier Transform (DFT) of the signal would results in a maximum likelihood estimation for the frequency [1]. The DFT estimate would also yield optimum results as they achieve Cramer-Rao bound (CRB) for the frequency estimate [2]. However, the DFT peak estimate is computationally exhaustive, especially for implementation in hardware at very high frequencies and thus simpler, computationally efficient, algorithms are often sought for practical applications [3]. A simple algorithm for frequency estimation of a sinusoid is the Pulse-Pair (PP) algorithm, originally proposed by Benham et. al. ,in an application to Doppler spectrum estimation [4]. Since then the PP method has been quite widely used in many radar and sonar applications. We have investigated, previously, the statistical performance of the PP method as reported in [5]. This analysis is comprehensive and accounts for all forms of noise models. In an additive noise environment, we have clearly shown the need of using a parabolic window in the analysis. The use of a parabolic window achieves CRB at relatively high signal-to-noise ratios.

In certain applications it is necessary to include a multiplicative noise model. For example, multiplicative noise models are encountered in modeling the fading process in mobile communications, coherent radar processing, array processing of spatially distributed signals, and back scatter acoustic signals etc. [6]. In the presence of additive noise, at low signal-to-noise ratios (SNR), the PP method does not perform as well as the DFT based frequency estimation methods. However, recent literature has shown that, in the presence of a multiplicative noise process the PP method shows very good performance even at low SNRs, and it achieves CRB for the frequency estimate [7]. Thus, being a computationally simple technique, the PP method provides an efficient means of frequency estimation for a wide range of applications.

One of the important parameters need to be selected in using the PP method is the lag value for computation. Usually, a lag value of 1 is used. The optimum value for the lag, which gives the best performance is given by $2N/3$ [3][4]. However, estimates having lag values greater than 1 requires phase unwrapping, which is computationally intensive and could result in erroneous results if improperly implemented. A recent publication has discussed how the phase unwrapping could be achieved efficiently for a Field Programmable Gate Array (FPGA) implementation [3]. This method identified as the 'Crozier Technique', however, requires successive approximations and iterations for the phase unwrapping, which is non-trivial. In this paper, we propose a very simple and efficient technique to evaluate the frequency estimates using the PP method at large lag values, without resorting to complicated phase unwrapping techniques. Based on our work on [5], details of the statistical properties of the frequency estimates derived via the PP method are also provided. Although the work presented in this paper addresses only the additive noise models, the work could be easily extended for situations encountering multiplicative noise as well.

2. Pulse-Pair Method

The Pulse-Pair (PP) method, first proposed in [4] and analyzed in [8] by Lank et. al. is briefly described here. Consider a mon-tone signal $z(n)$ in the presence of an additive complex white noise sequence $\{\theta_1(n) + j\theta_2(n)\}$ i.e.,

$$z(n) = \{A(n) + jB(n)\}e^{j2\pi f_0 n} + \theta_1(n) + j\theta_2(n) \ , \qquad (1)$$

where f_0 is the signal frequency and n the sampling index. The term $\{A(n) + jB(n)\}$ accounts for the signal amplitude. Having $\{A(n) + jB(n)\}$ as a constant means that the signal amplitude is non-time varying and thus is free from multiplicative noise. The PP method of frequency estimation using N data samples at a lag value of m samples is given by

$$\hat{f}_0 = \left(\frac{1}{2\pi m}\right)\tan^{-1}\left(\frac{\mathrm{Im}(S_m)}{\mathrm{Re}(S_m)}\right) , \qquad (2)$$

where

$$S_m = \left(\frac{1}{N-m}\right)\sum_{n=0}^{N-1-m} w(n)z(n+m)z^*(n) \ , \qquad (3)$$

0-7803-7761-3/03/$17.00 ©2003 IEEE

and $w(n)$ is a data windowing function. Note that in general, if $f_0 \geq (2\pi m)^{-1}$, the estimate in equation (2) needs phase unwrapping. Therefore, having $m=1$ needs no phase unwrapping. We have comprehensively studied the statistics of the estimate of equation (2), and its variants, in [5].

The estimate $\hat{f}_0$ is unbiased, and for a constant amplitude sinusoidal signal the estimate variance is given by (equation 33, ref 5)

$$E\left[\left(\hat{f}_0 - f_0\right)^2\right] = \left(\frac{1}{2\pi m}\right)^2 \left(\frac{\min(m, N-m)}{(N-m)^2\, SNR} + \frac{SNR^{-2}}{2(N-m)}\right) \quad (4)$$

where $min(x,y)$ denotes either x or y, whichever is the minimum value. The result in equation (4) is achieved when a rectangular window is used in the estimation of equation (3). For a given N, it can be easily shown that the variance in (4) achieves a minimum value when $m=2N/3$ and a quasi-minimum when $m=N/3$. For $m=2N/3$, the minimum variance is given by,

$$E\left[\left(\hat{f}_0 - f_0\right)^2\right] = \frac{27}{16\pi^2 N^3}\left(SNR^{-1} + 0.5 SNR^{-2}\right) \quad (5)$$

It is also shown in [5] that using a parabolic window in equation (3), with $m=1$, the variance of the estimate is obtained as,

$$E\left[\left(\hat{f}_0 - f_0\right)^2\right] = \frac{3SNR^{-1}}{2\pi^2 N(N^2-1)} + \frac{SNR^{-2}(6N^3)}{40\pi^2 (N^2-1)^2} \quad (6)$$

The parabolic window is defined as

$$w(n) = \frac{6\left[(N-1) + (N-2)n - n^2\right]}{N(N^2-1)} \qquad n = 0\ldots(N-2) \quad (7)$$

At large SNR, equation (6) results in

$$E\left[\left(\hat{f}_0 - f_0\right)^2\right] = \frac{3SNR^{-1}}{2\pi^2 N(N^2-1)} \quad , \quad (8)$$

which is the CRB of the frequency estimate [2]. From equation (6) it can also be shown that if

$$SNR < N^2/10 \quad , \quad (9)$$

the estimate in equation (6), obtained using a parabolic window, deviates from the CRB. On the other hand, the estimate in equation (5), obtained using rectangular window, do not achieve CRB even at high SNR. The excess mean square error from the CRB is given by, $9/8 = 0.51dB$, and the SNR threshold value for deviation from the (CRB + $0.51dB$) is given by

$$SNR < 0.5 \quad . \quad (10)$$

Thus we have demonstrated the following two properties for the frequency estimate using the PP method.

- Using a parabolic window with $m=1$ in the PP method CRB can be achieved. However, this is only at a high SNR, i.e. when $SNR > N^2/10$. Larger the N, a higher SNR is required to achieve the CRB.
- A rectangular window with $m=2N/3$ results in an estimate error variance of $0.51dB$ above the CRB. The SNR threshold is given by $SNR > 0.5$.

Figure 1 shows a simulation plot obtained using *1000* iterations. Since the estimate in equation (2) is unbiased, a normalized frequency of *0.0* is used for the simulation. (This avoids the need of phase unwrapping). The number of samples is $N = 40$. Both a rectangular and a parabolic window have been used in the frequency estimation using different lag value for m. For comparison CRB is also shown in this Figure. It can be seen from Figure 1 that at low SNR values it is advantageous to use a rectangular window with $m=2N/3$. (The use of $m=N/3$ as the lag value would produce similar results.) However, in comparison with PP method having $m=1$, the use of $m=2N/3$ requires phase unwrapping, which is not trivial and often is computationally expensive.

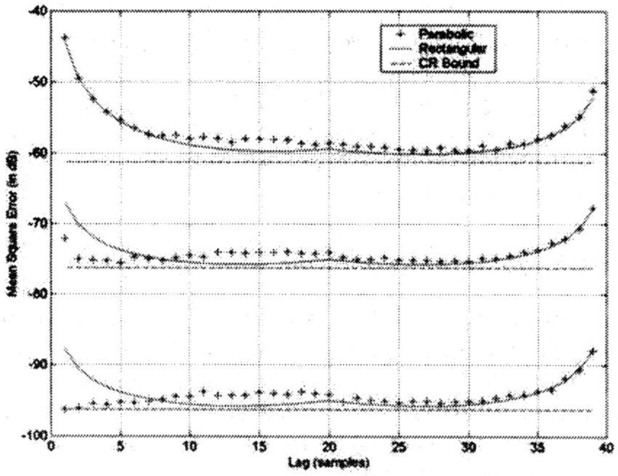

Figure 1: *Mean square error of frequency estimation when Parabolic (+) and Rectangular (solid) windows are used in the Pulse-Pair Method with different lag values, m. Three different SNRs are used: 5dB (top), 20dB (middle), 40dB (bottom). The CR bound is also shown in dashed lines.*

Figure 2 shows another set of simulations obtained for a parabolic windowed PP method with $m=1$. The number of iterations is 1000 and the normalized signal frequency was selected as *0.45*. The signal length has been increased form $N=8$ to $N=512$ in the simulations. The CRB is also shown in Figure 2 and at low SNR the deviation from the CRB is evident which is as noted in equation (9).

Figure 3 is similar to Figure 2, but a rectangular window has been used with the optimum lag value of $m = 2N/3$. In Figure 3, the deviation of estimate error variance from the ($CRB+0.51dB$) is evident, as noted in equation (10). Note that for the estimates in Figure 3, care should be taken due to the phase wrapping around $\pm\pi$. How this can be achieved efficiently, is discussed in the following section.

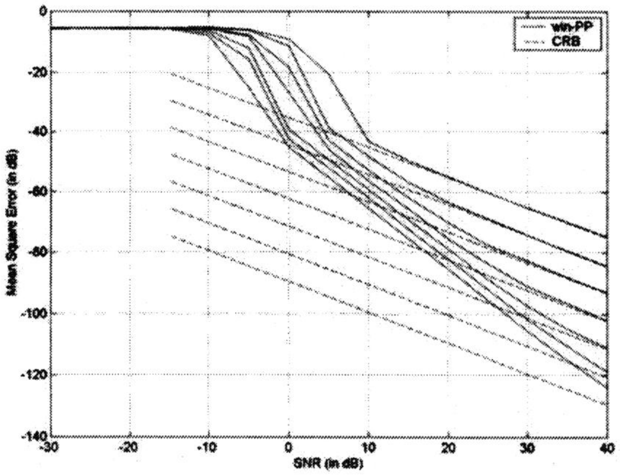

Figure 2: *Mean square error of frequency estimation versus SNR, when the Parabolic (solid) window is used in the Pulse-Pair Method with a lag value, m=1. The signal length varies from 8, 16, 32, 64, 128, 256 and 512 from top plot to bottom plot. The CR bound is also shown in dashed lines.*

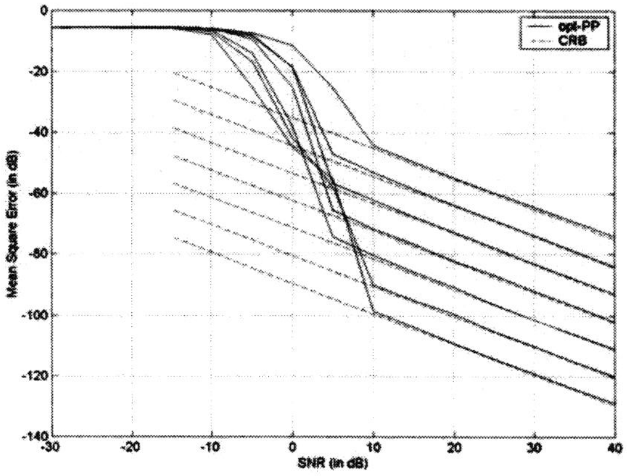

Figure 3: *Mean square error of frequency estimation versus SNR, when the Rectangular (solid) window is used in the Pulse-Pair Method with an optimum lag value of m=2N/3. The signal length varies from 8, 16, 32, 64, 128, 256 and 512 from top plot to bottom plot. The CR bound is also shown in dashed lines.*

3. Efficient Frequency Estimation for *m > 1*

It is clear from the discussion in the previous section that a rectangular window used at an optimal lag value of *m=2N/3* is the suitable choice for frequency estimation using the PP method, especially at low *SNRs*. However, using the optimal lag value of *m=2N/3*, requires phase unwrapping for correct frequency estimation. In the following we describe an efficient algorithm to estimate the frequency without resorting to complicated phase unwrapping.

Suppose at a lag value of *m*, the frequency error of the estimate in equation (3) is given by ε_m, then equation (2) can be expressed as,

$$\tan^{-1}\left(\frac{\mathrm{Im}(S_m)}{\mathrm{Re}(S_m)}\right) = 2\pi m\hat{f}_0 = 2\pi m(f_0 + \varepsilon_m) , \qquad (11)$$

Similarly for *m=1*,

$$\tan^{-1}\left(\frac{\mathrm{Im}(S_1)}{\mathrm{Re}(S_1)}\right) = 2\pi\hat{f}_0 = 2\pi(f_0 + \varepsilon_1) , \qquad (12)$$

Therefore,

$$\tan^{-1}\left(\frac{\mathrm{Im}(S_m)}{\mathrm{Re}(S_m)}\right) - m\tan^{-1}\left(\frac{\mathrm{Im}(S_1)}{\mathrm{Re}(S_1)}\right) = 2\pi m(\varepsilon_m - \varepsilon_1) , \qquad (13)$$

Noting that the errors are usually small, i.e. $\varepsilon_1, \varepsilon_m \ll 1$ we get $2\pi m(\varepsilon_m - \varepsilon_1) < 1$, and thus dividing equation (13) by *m* we get,

$$\frac{1}{m}\left[\tan^{-1}\left(\frac{\mathrm{Im}(S_m)}{\mathrm{Re}(S_m)}\right) - m\tan^{-1}\left(\frac{\mathrm{Im}(S_1)}{\mathrm{Re}(S_1)}\right)\right]_{pv} = 2\pi(\varepsilon_m - \varepsilon_1) \qquad (14)$$

where *pv* denote the principal value of the angle in range $\pm\pi$. Evaluation of equation (14) need no phase unwrapping. Combining equations (12) and (14) we get,

$$\frac{1}{2\pi m}\left[\tan^{-1}\left(\frac{\mathrm{Im}(S_m)}{\mathrm{Re}(S_m)}\right) - m\tan^{-1}\left(\frac{\mathrm{Im}(S_1)}{\mathrm{Re}(S_1)}\right)\right]_{pv} + \frac{1}{2\pi}\tan^{-1}\left(\frac{\mathrm{Im}(S_1)}{\mathrm{Re}(S_1)}\right) = (f_0 + \varepsilon_m) \qquad (15)$$

Thus, we have obtained an estimate for the signal frequency with an error given by ε_m, which is the error obtained using the optimal lag value. That is, the estimate in equation (15) is optimum in the mean square sense, and needs no phase unwrapping in the computation. Figure 3 was obtained using the above method with a normalized signal frequency very close to the Nyquist frequency ($f_0 = 0.45$).

It can also be argued that, a combination of the PP estimates (instead of a single lag estimate) would improve its performance. However, the use of combined *N/3* estimates in the range *m=N/3* to *m=2N/3* did not show any improvement in simulation studies.

4. Computational Efficiency of the PP Method

We will compare the computational efficiency of the PP method with the DFT based maximum likelihood method. The comparison is based on the number of multiplications required for the estimation. This is because in most hardware implementations (e.g. in FPGA) the number of multiplications is the most expensive for Silicon area optimization [3][9].

Frequency estimation using the DFT peak search can be implemented by the evaluation of successive DFTs to increase the accuracy of the estimated frequency. Figure 4 shows the error variance of the maximum likelihood frequency estimate, obtained via simulations using 1000 iterations. The DFT peak search technique has been used in obtaining the plot in Figure 4. The CRB is also shown in Figure 4. (In comparing Figure 4 with Figure 3, it can be stated the maximum likelihood estimates have an advantage over the PP based methods when the *SNR* is very low.)

Using the CRB in equation (8) as the required frequency accuracy, the number of multiplication operations required in the DFT peak search can be expressed as

$$M_{DFT} = \frac{N}{2}\log_2\left(\frac{2\pi^2 N(N^2-1)}{3SNR^{-1}}\right) , \qquad (16)$$

As can be seen from equation (15), the PP method with optimum lag requires the evaluation of estimates for $m=1$ and $m=2N/3$. Therefore, multiplication operations required in the PP method is given by,

$$M_{PP} = (N-1) + \frac{N}{3} . \qquad (17)$$

Thus at large *SNR* and at large *N*, $M_{PP} << M_{DFT}$, and since the optimum lag PP method achieves an error variance marginally higher than the CRB (i.e. *CRB + 0.51dB*), the PP method is the suitable choice for a low computational load hardware implementation.

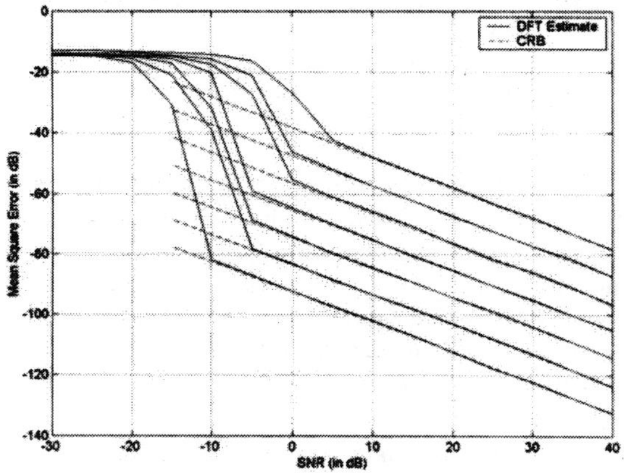

Figure 4: *Mean square error of frequency estimation versus SNR, when the maximum likelihood DFT peak search technique is used. The signal length varies from 8, 16, 32, 64, 128, 256 and 512 from top plot to bottom plot. The CR bound is also shown in dashed lines.*

5. Conclusions

In this paper, the computation of the Pulse-Pair (PP) method, which is an efficient technique of frequency estimation, is discussed. It is noted that the PP method has a very low number of multiplications, and is ideally suitable for FPGA type hardware implementations, especially at high frequencies. It is noted that the multiplications are relatively expensive for Silicon area optimization in FPGA implementations. The disadvantage of optimum frequency estimation using the PP method is that the method requires a lag value equal to the two-third of the signal length. This needs phase unwrapping that usually makes the technique computationally un-attractive. In this paper, we have proposed a new efficient technique for optimal frequency estimation via the PP method, without resorting to phase unwrapping. Detailed discussion of the performance of the optimum PP method is provided in the paper. A comparison with the performance of the DFT based maximum likelihood frequency estimation technique is also provided.

6. References

[1] Rife, D. C. and Boorstyn, R. R. "Single-tone Parameter Estimation from Discrete-Time Observations," *IEEE Transactions on Information Theory*, vol. 20, pp. 591-598, 1974.

[2] Kay, S. M., "A Fast and Accurate Single Frequency Estimation by Linear Prediction," *IEEE Transactions on Signal Processing*, vol. 37, pp. 1987-1990, 1989.

[3] Lafrance, L-P., Cantin, M-A., Savaria, Y., Sung, S. H. and Lavoie, P. , "Architecture and Performance Characterization of Hardware and Software Implementations of the Crozier Frequency Estimation Algorithm," *IEEE International Symposium on Circuits and Systems – ISCAS 2002*, pp. 823-826, Pheonix AZ, USA, 2002.

[4] Benham, F. C., Groginsky, H. L. Soltes, A. S. and Works, G., "Pulse Pair Estimation of Doppler Spectrum Parameters," *Final Report, Contract F-19628-71-C-0126, Journal of Circuit Theory Applications*, Raytheon Co., Wayland, MA USA, 1972.

[5] Abeysekera, S. S., "Performance of Pulse-Pair Method of Doppler Estimation", *IEEE Transactions on Aerospace and Electronic Systems*, vol. 34, pp. 520-531, 1998.

[6] Coulon, M., Tourneret, J-Y. and Swami, A., "Detection of Multiplicative Noise in Stationary Random Processes using Second-and Higher Order Statistics," *IEEE Transactions on Signal Processing*, vol. 48, pp. 2566-2575, 2000.

[7] Ghogho, M., Swami, A. and Durrani, T. S., "Frequency Estimation in the Presence of Doppler Spread: Performance Analysis," *IEEE Transactions on Signal Processing*, vol. 49, pp. 777-789, 2001.

[8] Lank, G. W., Reed, I. S. and Pollon, G. E., "A Semicoherent Detection and Doppler Estimation Statistics", *IEEE Transactions on Aerospace and Electronic Systems*, vol. 9, pp. 151-165, 1973.

[9] Abeysekera, S. S. and Charoensak, C., "Optimum Sigma-Delta De-Modulator Filter Implementation Via FPGA," 14th IEEE International ASIC/SOC Conference, Washington DC, USA, September 2001.

PERFORMANCE ANALYSIS OF ALGORITHMIC NOISE-TOLERANCE TECHNIQUES

Byonghyo Shim and Naresh R. Shanbhag

Coordinated Science Laboratory, ECE Dept.
University of Illinois at Urbana-Champaign
1308 West Main Street, Urbana, IL 61801
Email: [bshim,shanbhag]@mail.icims.csl.uiuc.edu

ABSTRACT

In this paper, we present performance analysis of algorithmic noise-tolerance (ANT) techniques. First, we analyze the predictor and RPR based ANT schemes. Next, we present a hybrid ANT scheme which is resilient to burst errors usually occurring in a high soft-error rate (P_{er}) region. For a frequency selective FIR filtering, it is shown that simulation results match well with the analytic bounds while providing about 40 dB improvement in the mean square error over a conventional DSP system at a $P_{er} = 10^{-4}$. It is also shown that the proposed hybrid ANT scheme maintains its robustness in noise mitigation even in the high soft-error rate region of upto $P_{er} = 10^{-2}$.

1. INTRODUCTION

We have shown in the past that algorithmic noise-tolerance (ANT) technique [3]-[5] are very effective in combating deep submicron (DSM) noise [1]-[2]. However, past work has focused on simulations to prove the effectiveness of ANT. This paper has two contributions: 1) performance analysis of two existing ANT techniques, the prediction based [3] and reduced-precision redundancy (RPR) [5], and 2) a new ANT technique that is referred to as hybrid ANT and its performance analysis. This paper is organized as follows. After discussing ANT scheme in Section II, we present its performance analysis and propose hybrid ANT scheme in Section III. Simulation results and discussion are given in Section IV.

2. PRELIMINARIES

We first present the ANT based digital signal processing (DSP) system that ensures high reliability in the presence of DSM noise and briefly discuss previous ANT schemes.

2.1. ANT based DSP system

The output of a DSP system is represented by

$$y_{o,n} = d_n + Q_n \tag{1}$$

where d_n is desired output signal and Q_n is the noise due to channel effects, ADC quantization noise, etc.. The output SNR is

$$SNR_{out,org} = 10\log_{10}(\frac{\sigma_d^2}{\sigma_Q^2}) \tag{2}$$

This work is supported by NSF grant CCR 99-79381, ITR 00-85929 and DARPA MSP grant

The output in the presence of soft errors is

$$y_{a,n} = d_n + Q_n + \eta_n \tag{3}$$

where $\eta[n]$ is the noise due to the soft error. In this case, the output SNR is

$$SNR_{out,err} = 10\log_{10}(\frac{\sigma_d^2}{\sigma_Q^2 + \sigma_\eta^2}) \tag{4}$$

where $Q[n]$ and $\eta[n]$ are assumed to be uncorrelated. Equation (4) can be rewritten as $SNR_{out,err} = SNR_{out,org} - \Delta$ where $\Delta = 10\log_{10}(1 + \frac{\sigma_\eta^2}{\sigma_Q^2})$. By employing ANT, we reduce the noise power due to soft error σ_η^2 and thereby guaranteeing that Δ is sufficiently small.

The decision rule for an ANT-based DSP system is given by

$$\hat{y}_n = \begin{cases} y_{a,n} & \text{if } |y_{a,n} - y_{ANT,n}| \le T_h \\ y_{ANT,n} & \text{if } |y_{a,n} - y_{ANT,n}| > T_h. \end{cases} \tag{5}$$

where T_h is a precomputed threshold and $y_{ANT,n}$ is the corrected output. In order to guarantee that $y_{a,n} = y_{o,n}$ in the absence of errors, the threshold becomes

$$T_h = \max_{\forall y_{o,n}} |y_{o,n} - y_{ANT,n}|. \tag{6}$$

In doing so, no false alarms can occur and therefore only three possible cases exist; (1) no error, (2) undetected error, (3) detected error, with P_{ner}, P_{uer}, and P_{der}, being the corresponding probability of occurrence, respectively. The estimation error power is given by

$$\begin{aligned} \sigma_{y_o-\hat{y}}^2 &= E[\,|y_o - \hat{y}|^2\,] \\ &= P_{der}E[\,|y_o - y_{ANT}|^2\,] \\ &+ \; P_{uer}E[\,|y_o - y_a|^2 \mid (y_a - y_{ANT} \le T_h)\,] \\ &+ \; P_{ner}E[\,|y_o - y_a|^2\,] \end{aligned} \tag{7}$$

Note that the third term in (7) is zero. Furthermore, we denote the noise power due to ANT σ_{ANT}^2, and soft error power σ_η^2, as

$$\sigma_{ANT}^2 = E[\,|\,y_o - y_{ANT}\,|^2] \tag{8}$$

$$\sigma_\eta^2 = E[\,|\,y_o - y_a\,|^2 \mid (y_a - y_{ANT} \le T_h)\,] \tag{9}$$

Substituting (8) and (9) in (7), we get

$$\sigma_{y_o-\hat{y}}^2 = P_{der}\cdot\sigma_{ANT}^2 + P_{uer}\cdot\sigma_\eta^2 \tag{10}$$

0-7803-7761-3/03/$17.00 ©2003 IEEE

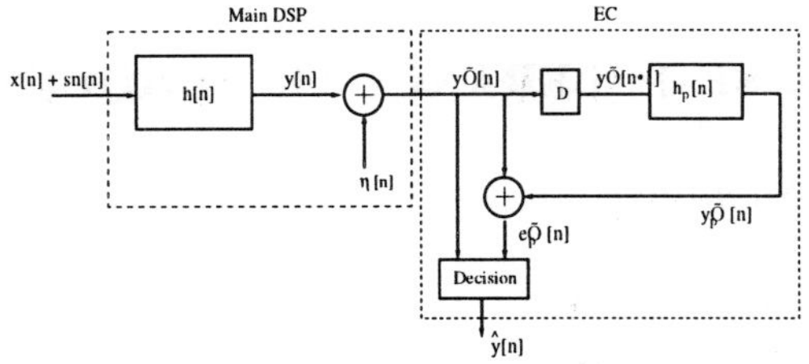

(a)

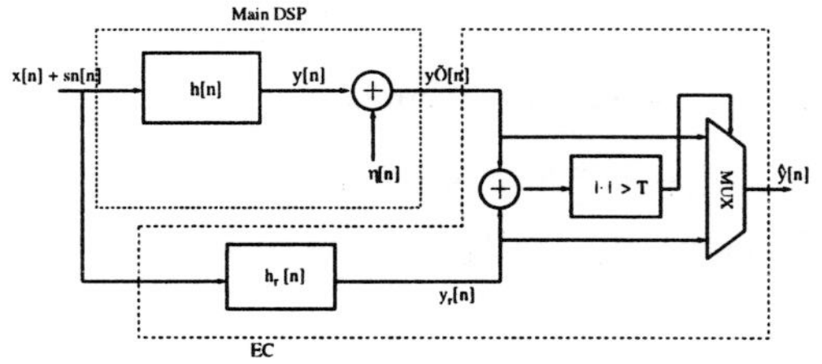

(b)

Figure 1: Algorithmic noise-tolerance schemes: (a) prediction based ANT scheme, and (b) RPR based ANT.

2.2. Prediction based ANT

The block diagram of predictor based ANT is shown in Fig. 1(a), where a forward predictor is added to the original system for error detection and correction [3]. Since the adjacent output y_n, of MDSP is highly correlated for a sufÞciently narrowband system, the predictor generates an estimate to be used as an output when an error occurs.

2.3. RPR based ANT

Figure 1(b) shows the block diagram of an RPR based ANT scheme. The RPR is a replica of the original system with small precision operands [5]. Though RPR output $y_{r,n}$ is generally not equal to the original one $y_{o,n}$ due to the LSB quantization noise, it is a good estimate in case of an error event. When an error is detected using (5), $y_{r,n}$ is employed as the Þnal output.

3. PERFORMANCE ANALYSIS OF ANT

In this section, we present an analysis of the predictor, RPR and the hybrid ANT technique which is composed of both techniques. Here, we assume that the error-control block is designed to be error free. Indeed, this assumption is reasonable since the complexity of the error control block is much lesser than that of MDSP and hence the likelihood of soft-error is signiÞcantly smaller.

3.1. Predictor based ANT

Let the output vector of an original system is $\boldsymbol{Y} = [y_{n-1}, \ldots, y_{n-N}]^T$ and the predictor coefÞcient vector is $\boldsymbol{W} = [w_1, \ldots, w_N]^T$. Then, the forward predictor output is

$$y_{p,n} = \sum_{k=0}^{N} w_k y_{n-k-1} = \boldsymbol{W^T Y} \qquad (11)$$

where the prediction error $e_{p,n} = y_n - \boldsymbol{W^T Y}$.

By using the derivative of $E[e_{p,n}^2]$, we can show the minimum MSE which corresponds to noise power of forward predictor $\sigma^2_{ANT,pre}$, is given by

$$\sigma^2_{ANT,pre} = \min E[e^2_{p,n}] = E[y_n^2] - \boldsymbol{P^T R^{-1} P} \qquad (12)$$

where

$$\begin{aligned} \boldsymbol{P^T} &= E[y_n \boldsymbol{Y^T}] = E[(\, y_n y_{n-1} \cdots y_n y_{n-N}\,)^T] \\ \boldsymbol{R} &= E[\boldsymbol{YY^T}] = E[\,(y_{n-1} \cdots y_{n-N})^T (y_{n-1} \cdots y_{n-N})\,] \end{aligned}$$

In addition, by applying the Triangular inequality, we can obtain the upper bound on σ^2_η as

$$\begin{aligned} \sigma^2_\eta &= E[\,|y_a - y_o|^2 \mid (y_a - y_{pre} \le T_{h,pre})\,] \\ &= E[\,|y_a - y_{pre} + y_{pre} - y_o|^2 \mid (y_a - y_{pre} \le T_{h,pre})\,] \\ &\le T^2_{h,pre} + \sigma^2_{ANT,pre} \qquad (13) \end{aligned}$$

Inserting (12) and (13) into (10), we get the upper bound on the minimum noise power for the prediction based ANT system as

$$\sigma^2_{y_o - \hat{y}} \le P_{der}\, \sigma^2_{ANT,pre} + P_{uer}(T^2_{h,pre} + \sigma^2_{ANT,pre}) \qquad (14)$$

Predictor based ANT works well in narrowband systems (typically $\omega_b < 0.3\pi$). However, its prediction performance reduces when the Þlter bandwidth increases. In addition, when a burst error occurs in the system, as would be the case in high P_{er} region, the predictor cannot provide reliable detection thereby degrading the system performance severely.

3.2. RPR based ANT

Assume that the operands precision in a reference original system is $B_1 + 1$ bits and that of a RPR system is $B_2 + 1$ bits, where $B_1 > B_2$. In addition, we denote the quantization step size of an original DSP and that of RPR as $\Delta_o = \frac{1}{2^{B_1}}$ and $\Delta_r = \frac{1}{2^{B_2}}$. The quantization noise N_x and N_h between the original value and that of RPR x_r and h_r is deÞned as $N_x = x - x_r$ and $N_h = h - h_r$, where x and h are the input and Þlter coefÞcients, respectively. With this assumption, one can show that the noise power of the RPR ANT scheme $\sigma^2_{ANT,rpr}$, is given by

$$\sigma^2_{ANT,rpr} = \sum_i [N^2_{h_i} \sigma^2_x + \frac{h^2_{r,i}}{6}(2\Delta^2_r - 3\Delta_r\Delta_o + \Delta^2_o)] \qquad (15)$$

where σ^2_x is the input signal power. By applying an analysis similar to the one in (13), we get the noise power σ^2_η as

$$\sigma^2_\eta \le T^2_{h,rpr} + \sigma^2_{ANT,rpr} \qquad (16)$$

Employing (15), (16), and (10) we can obtain the upper bound on the noise power for the RPR based ANT as

$$\begin{aligned} \sigma^2_{y_o - \hat{y}} &= P_{der} \cdot \sigma^2_{ANT,rpr} + P_{uer} \cdot \sigma^2_\eta \qquad (17) \\ &\le P_{der} \cdot \sigma^2_{ANT,rpr} + P_{uer} \cdot (T^2_{h,rpr} + \sigma^2_{ANT,rpr}) \end{aligned}$$

Note that the noise power term in (17) depends only on precision and quantization noise but not on bandwidth. While the RPR based ANT provides good performance for a wide range of bandwidths, it consumes more power than the predictor based ANT system.

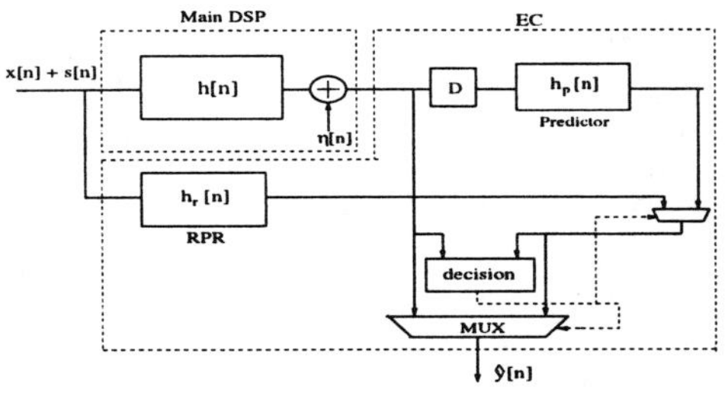

(a)

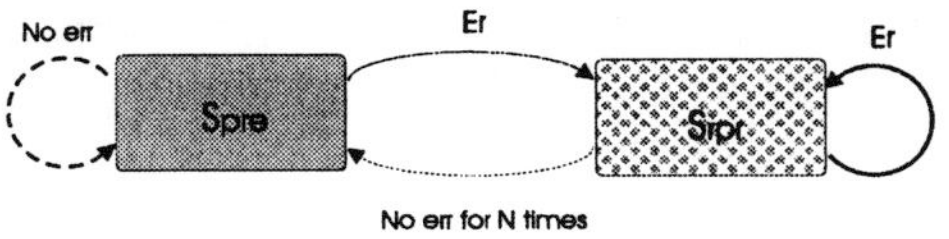

(b)

Figure 2: The hybrid ANT technique: (a) block diagram, and (b) the state machine.

3.3. Hybrid ANT

The predictor based ANT, which produces good performance in low soft-error rate P_{er}, has a problem as P_{er} increases. Once error is made, an erroneous output is fed back into the predictor, leading to an incorrect decision until the erroneous output is purged. On the other hand, the RPR based ANT has relatively high power consumption than the predictor based scheme.

Figure 2(a) shows the block diagram of the proposed hybrid ANT scheme that overcomes this problem. The key idea in the hybrid ANT is to use a predictor when there are no errors (i.e., for error detection) and RPR when an error occurs (i.e., error correction). As illustrated in Fig. 2(b), the error control is transferred to an RPR state (S_{rpr}) from a predictor state (S_{pre}) after error detection. If RPR does not detect an error for N consecutive cycles, i.e., the error propagation is terminated, then the error control is passed back to the predictor.

Assuming that the error event is independent, the probability of being in the RPR and predictor states are, respectively,

$$\begin{aligned} P_{rpr} &= P_{er} + (1 - P_{er})P_{er} + \cdots (1 - P_{er})^{N_p - 1} P_{er} \\ P_{pre} &= (1 - P_{er})^{N_p} \end{aligned} \tag{18}$$

where N_p is the number of taps in the predictor. Clearly, $P_{rpr} + P_{pre} = 1$. Then, the noise power of the hybrid ANT system is

$$\sigma^2_{ANT,hybrid} = P_{rpr}\, \sigma^2_{ANT,rpr} + P_{pre}\, \sigma^2_{ANT,pre} \tag{19}$$

Using (18), (19), and (14), we can obtain the upper bound on the noise power of the hybrid ANT as

$$\begin{aligned} \sigma^2_{y_o - \hat{y}} \leq\; & P_{der} P_{pre} \cdot \sigma^2_{ANT,pred} + P_{der} P_{rpr} \cdot \sigma^2_{ANT,rpr} \\ +\; & P_{uer} P_{pre} \cdot (T^2_{h,pre} + \sigma^2_{ANT,pre}) \\ +\; & P_{uer} P_{rpr} \cdot (T^2_{h,rpr} + \sigma^2_{ANT,rpr}) \end{aligned} \tag{20}$$

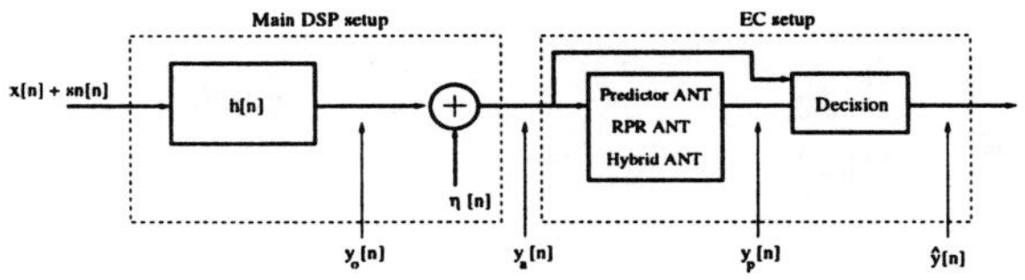

Figure 3: Simulation setup.

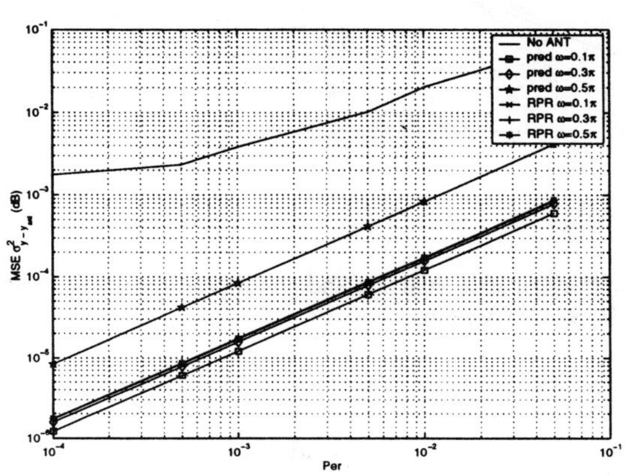

Figure 4: Analytic results of prediction and RPR ANT scheme for bandwidth variation.

Since RPR is used only when error is detected, we can turn this block off in the predictor state S_{pre} and thereby save power while maintaining robustness when P_{er} is high.

4. SIMULATIONS AND DISCUSSION

The setup used to measure the performance of the proposed scheme is shown in Fig. 3, where a frequency selective Þlter is used to generate a bandlimited signal $y_o[n]$ from a wideband input with noise $x[n] + s[n]$. The SNR without ANT is given by

$$SNR_{in} = 10 \log_{10} \frac{\sigma^2_{y_o}}{\sigma^2_{y_a - y_o}} \tag{21}$$

whereas, the SNR at the ANT output is given by

$$SNR_{out} = 10 \log_{10} \frac{\sigma^2_{y_o}}{\sigma^2_{y_o - \hat{y}}} \tag{22}$$

where $\sigma^2_{y_o}$, is the signal power of original Þlter output. In this simulation, we employ a 31-tap Þlter with a 12×12 multipliers and a 26-bit accumulator. For a DSM noise model, we employ random ßipping of an output bit in the digital Þlter with a speciÞed soft-error probability P_{er}. Figure 4 shows the analysis results of predictor and RPR based ANT, where the main Þlter is a low-pass Þlter having a bandwidth from $\omega_b = 0.1\pi$ to 0.5π. For the RPR MAC, we used the half precision of the original Þlter (i.e., 6×6 multiplier) and a three tap of predictor is employed in the prediction based ANT. While the performance of RPR is quite similar

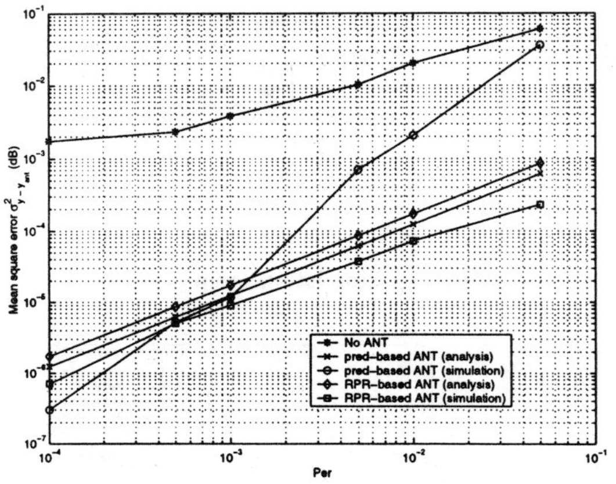

Figure 5: Performance analysis and simulation results of prediction and RPR ANT scheme.

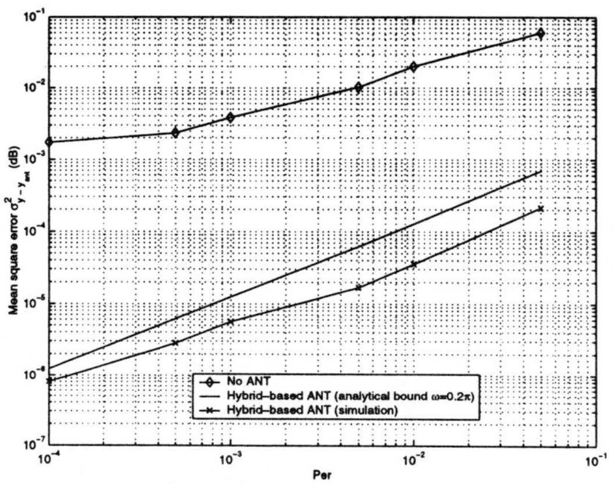

Figure 6: Performance analysis and simulation results of hybrid ANT scheme.

over the entire frequency range, the performance of the predictor deteriorates as the bandwidth increases.

Figure 5 shows the analysis and simulation results of the predictor and RPR based ANT, where the bandwidth of main Þlter is now set to $\omega_b = 0.2\pi$. While the conventional system has high mean square error regardless of P_{er}, predictor ANT has MSE less than 10^{-6} at $P_{er} = 10^{-4}$ resulting in 40 dB gain over conventional MDSP. As discussed in Section II.A, the performance of predictor degrades severely as P_{er} increases due to the error propagation. On the contrary, the performance of RPR is close to the analytic results even when P_{er} is high.

Next, we considered the hybrid ANT scheme under the same simulation setup. As shown in Fig. 6, the performance of hybrid ANT is similar to the prediction scheme in low P_{er} region. However, unlike to prediction scheme, we observe that hybrid scheme maintains its performance even when P_{er} becomes high.

Finally, in order to observe the comprehensive picture includ-

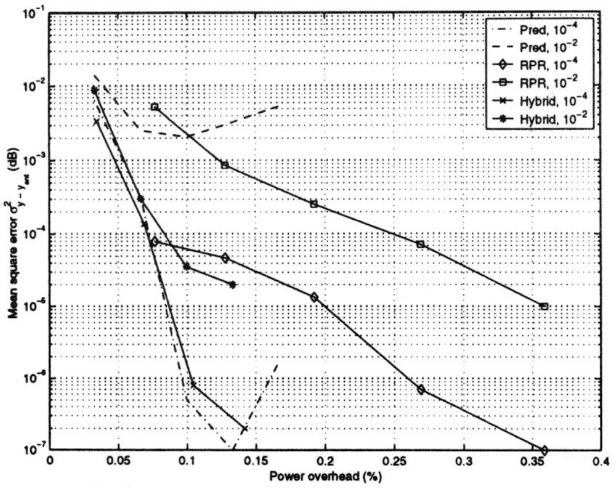

Figure 7: Contour plot for P_{er} and power overhead variation.

ing additional power consumption, we show the contour plot for the power overhead as well as P_{er}, which is given by

$$P_{overhead}(\%) = \frac{P_{ANT} - P_{MDSP}}{P_{MDSP}} \times 100\%. \quad (23)$$

where P_{MDSP} and P_{ANT} are the power of original DSP and ANT based DSP, respectively. As shown in Fig. 7, the power overhead of hybrid ANT is about 15 $\sim$ 20 % lesser than that of RPR in a similar MSE region. Thus, the hybrid ANT scheme is not only robust to soft-errors when P_{err} is high, it is also energy-efÞcient.

5. REFERENCES

[1] T. Karnik and S. Vangal, ÒSelective node engineering for chip-level soft error rate improvement,Ó *Proc. of Symp. on VLSI circuits*, vol. 4, pp. 132-135, 2002.

[2] N. Shanbhag, K. Soumyanath, and S. Martin, ÒReliable low-power design in the presence of deep submicron noise,Ó *Proc. of Intl. Symp. on Low-Power Electronics and Design*, pp 295-302, 2000.

[3] R. Hedge and N. Shanbhag, ÒSoft digital signal processing,Ó *IEEE Trans. VLSI*, vol. 9, pp. 813-823, Dec. 2001.

[4] L. Wang and N. Shanbhag, ÒLow-power signal processing via error-cancellation,Ó *Proc. of IEEE workshop on Signal Processing Systems*, Lafayette Oct. 2000.

[5] B. Shim, and N. R. Shanbhag, ÒLow-power digital Þltering via reduced-precision redundancy,Ó *Proc. of Asilomar conf.*, Vol. 1, pp. 148-152, Nov. 2001.

A Minimum Variance Filter for Discrete-Time Linear Systems Perturbed by Unknown Nonlinearities

A. Germani[1,2]   C. Manes[1,2]   P. Palumbo[2]

[1]Dipartimento di Ingegneria Elettrica, Università degli Studi dell'Aquila, Poggio di Roio, 67040 L'Aquila - Italy, {germani,manes}@ing.univaq.it

[2]Istituto di Analisi dei Sistemi e Informatica del CNR, IASI-CNR Viale Manzoni 30, 00185 Roma - Italy, palumbo@iasi.rm.cnr.it

Abstract

This paper investigates the problem of state estimation for discrete-time stochastic systems with linear dynamics perturbed by unknown nonlinearities. The Extended Kalman Filter (EKF) can not be applied in this framework, because the lack of knowledge on the nonlinear terms forbids a reliable linear approximation of the perturbed system. Following the idea to compensate this lack of knowledge suitably exploiting the information brought by the measured output, a recursive linear filter is developed according to the minimum error variance criterion. Differently from what happens for the EKF, the gain of the proposed filter can be computed off-line. Numerical simulations show the effectiveness of the proposed filter.

1 Introduction

This paper is concerned with the filtering problem for systems that have a nominal linear behavior perturbed by nonlinear terms that are unknown to the filter designer. In addition, random noises affect both the state and output equations. Many works in literature deal with the problem of state estimation when the system model and/or the noise model are uncertain. The H_∞ approach provides a solution for the filtering problem of linear systems when the noise energy is bounded [7], or when the uncertainties on the system matrices are bounded [8]. In [9] an H_∞ filter is presented for uncertain linear systems perturbed by known nonlinear terms satisfying a Lipschitz condition. A classical filtering approach in the presence of known nonlinearities is the so called Extended Kalman Filter (EKF), whose performance is not guaranteed, but is easy to implement and in many cases gives satisfactory results. A robust version of the EKF filter, in the case of known nonlinear terms, is presented in [2]. In [3] a robust H_∞ filter is presented for continuous time linear systems in the presence of uncertain nonlinear perturbations that satisfy a known bound on the norm. All the cited papers do not handle a stochastic noise model. Moreover, a partial knowledge on the system nonlinearities (the norm bound or the Lipschitz constant) is required and appears in the filter equations or plays a role in the filter design.

This paper develops a filter that does not require any knowledge on the form of the nonlinear perturbation terms, and that is optimal, according to the minimum error variance criterion, in a specific class of estimators, following the approach in [4], [6], [5]. The key-point in the filter derivation is a clever use of the measured output vector in such a way to compensate the lack of knowledge on the nonlinear terms.

This work is supported by *CNR* (Italian National Research Council) and by *ASI* (Italian Aerospace Agency).

2 Linear Systems with Nonlinear Perturbations

This paper considers stochastic systems of the type

$$\begin{aligned} x(k+1) &= A(k)x(k)+B(k)u(k) \\ &\quad + R(k)h(k,x(k))+\mathcal{N}_f(k), \\ y(k) &= C(k)x(k)+\mathcal{N}_g(k), \quad x(0)=x_0, \quad k\geq 0, \end{aligned} \tag{1}$$

where $x(k) \in I\!R^n$ is the system state, $u(k) \in I\!R^p$ is a known input, $y(k) \in I\!R^q$ is the measured output, $h(\cdot,\cdot): I\!R^+ \times I\!R^n \mapsto I\!R^m$ is a nonlinear map, and is assumed to be unknown. $\{\mathcal{N}_f(k)\}, \{\mathcal{N}_g(k)\}$ are the state and output noise sequences, and are assumed white, mutually uncorrelated and not necessarily Gaussian, with known covariances $\{Q_f(k)\}$ and $\{Q_g(k)\}$. The initial state x_0 is a random variable, with mean $\bar{x}_0$ and covariance Ψ_0, uncorrelated with the noise sequences. $A(k), B(k), R(k), C(k)$ are matrices of suitable dimensions.

We are able to show that the lack of any knowledge on the nonlinear map $h(\cdot,\cdot)$ can be compensated by the output measurements, provided that the output vector $y(k)$ has a sufficiently large dimension. The following Lemma illustrates how this can be done.

0-7803-7761-3/03/$17.00 ©2003 IEEE

Lemma 1. *Define the extended state $X_e(k) \in I\!R^{n+m}$ as follows:*

$$X_e(0) = \begin{bmatrix} x_0 \\ 0 \end{bmatrix}, \quad X_e(k) = \begin{bmatrix} x(k) \\ h(k-1, x(k-1)) \end{bmatrix}, \quad k > 0. \tag{2}$$

Suppose that for all $k > 0$ the matrix sequence

$$H(k) = \begin{bmatrix} I_n & -R(k-1) \\ C(k) & O_{q\times m} \end{bmatrix}, \qquad k > 0 \tag{3}$$

has full column rank $n+m$, and define a sequence of left-inverses $\{H^\dagger(k)\}$ (i.e., matrices such that $H^\dagger(k)H(k) = I_{n+m}$). Then the evolution of $X_e(k)$ obeys the following (non-strictly causal) model:

$$\begin{aligned} X_e(k+1) &= A_e(k)X_e(k) + B_e(k)u(k) \\ &\quad + D_e(k)y(k+1) + \mathcal{F}_e(k), \\ y(k) &= C_e(k)X_e(k) + \mathcal{G}_e(k), \quad k \geq 0, \end{aligned} \tag{4}$$

where

$$\begin{aligned} A_e(k) &= H^\dagger(k+1)\begin{bmatrix} A(k) & O_{n\times m} \\ O_{q\times n} & O_{q\times m} \end{bmatrix}, \\ D_e(k) &= H^\dagger(k+1)\begin{bmatrix} O_{n\times q} \\ I_q \end{bmatrix}, \\ B_e(k) &= H^\dagger(k+1)\begin{bmatrix} B(k) \\ O_{q\times p} \end{bmatrix}, \\ C_e(k) &= \begin{bmatrix} C(k) & O_{q\times m} \end{bmatrix}, \end{aligned} \tag{5}$$

and with $\mathcal{F}_e(k)$ and $\mathcal{G}_e(k)$ extended noise sequences defined as:

$$\mathcal{F}_e(k) = H^\dagger(k+1)\begin{bmatrix} \mathcal{N}_f(k) \\ -\mathcal{N}_g(k+1) \end{bmatrix}, \quad \mathcal{G}_e(k) = \mathcal{N}_g(k). \tag{6}$$

The proof of this Lemma is simply obtained through direct substitution and simple computations.

Remark 2. Since all matrices of the sequence $\{H(k)\}$ have dimensions $(n+q) \times (n+m)$, the full column-rank condition assumed in Lemma 1 requires that $q \geq m$: this means that the number of output components must be at least equal to the number of independent nonlinear perturbation terms. This is a reasonable necessary condition, in that the measurements have been used to *replace* the nonlinear terms in a state-space equation. The important result is that in equation (4) the unknown nonlinear term $h(\cdot,\cdot)$ does not appear. On the other hand, the strict causality of the original system is lost, because the linear recursive state equation requires $y(k+1)$ to compute $X_e(k+1)$.

Remark 3. The extended state noise $\{\mathcal{F}_e(k)\}$ is still a white sequence, but is no more uncorrelated with the output noise $\{\mathcal{G}_e(k)\}$: from definition (6) it is evident a one-step correlation.

3 The Filtering Algorithm

The class of estimators considered in this paper is defined below:

Definition 4. *A state estimator for the class of linear stochastic systems perturbed by unknown nonlinear terms of the type (1) is said to be insensitive to the Nonlinear Perturbation (shortly, NLP-insensitive) if its structure does not depend explicitly on the nonlinear terms.*

The filtering algorithm here proposed is based on the estimation of the extended state of system (4). The same ideas developed in [6] are used here to deal with the non-strict causality of system (4). The following decomposition is suggested:

Proposition 5. *The extended state (2) and the output of system (4) can be decomposed as:*

$$\begin{aligned} X_e(k) &= X_d(k) + X_s(k), \\ y(k) &= y_s(k) + C_e X_d(k), \end{aligned} \tag{7}$$

where the dynamics of $X_d(k)$ and $X_s(k)$ are described by the two systems

$$\begin{aligned} X_d(k+1) &= A_e(k)X_d(k) + B_e(k)u(k) + D_e(k)y(k+1), \\ X_d(0) &= I\!E\big[X_e(0)\big] = \big(\bar{x}_0^T \quad 0^T\big)^T, \end{aligned} \tag{8}$$

$$\begin{aligned} X_s(k+1) &= A_e(k)X_s(k) + \mathcal{F}_e(k), \\ X_s(0) &= X_e(0) - I\!E\big[X_e(0)\big] = \big(x_0^T - \bar{x}_0^T \quad 0^T\big)^T, \\ y_s(k) &= C_e(k)X_s(k) + \mathcal{G}_e(k). \end{aligned} \tag{9}$$

According to Proposition 5, the extended state is split into two components: one, named $X_d(k)$, is not directly affected by the noises and can be exactly computed as a linear function of the measured output up to the current time k; the other, named $X_s(k)$, is generated by a strictly causal stochastic system, whose output is $y_s(k)$. Hence, when facing the problem to estimate $X_e(k)$, it is clear that only the component $X_s(k)$ needs to be estimated. When an estimate of $X_e(k)$ is available, the estimate of $x(k)$, the state of the original system, is obtained by picking up the first n components.

Definition 6. *The class $\mathcal{P}$ of estimators of $x(k)$ in (1) is defined as the set of all the NLP-insensitive estimators of the form:*

$$\begin{aligned} \widetilde{X}_e(k) &= X_d(k) + \widetilde{X}_s(k), \\ \tilde{x}(k) &= S\widetilde{X}_e(k), \end{aligned} \tag{10}$$

where $S = \begin{bmatrix} I_n & O_{n\times m}\end{bmatrix}$ *and* $\widetilde{X}_s(k)$ *is any estimate of the process* $X_s(k)$ *among all the Borel functions of the measurements* $\{y_s(\tau),\ \tau \le k\}$.

It is well known that the optimal choice for $\widetilde{X}_s(k)$ is given by the conditional expectation $\widehat{X}_s(k) = \mathbb{E}\big[X_s(k)|\mathcal{Y}_s^k\big]$, whose computation in general can not be obtained through algorithms of finite dimensions. From an applicative point of view, it is useful to develop finite-dimensional approximations of the optimal filter, for instance the optimal linear filter:

$$\widetilde{X}_s(k) = \Pi\big[X_s(k)|L_t(Y_s^k)\big], \tag{11}$$

that is the projection of $X_s(k)$ onto the Hilbert space of all linear transformations of the measurements $\{y_s(\tau),\ \tau \le k\}$. The $\mathcal{P}$-estimator (10) with $\widetilde{X}_s(k)$ computed as in (11), will be denoted the *optimal* $\mathcal{P}$*-linear estimator of* $x(k)$.

Remark 7. Since the component $X_d(k)$ can be exactly computed, the error covariance matrix for a $\mathcal{P}$-linear estimate is:

$$\begin{aligned}\mathrm{Cov}\big(x(k) - \tilde{x}(k)\big) &= S\,\mathrm{Cov}\big(X_e(k) - \widetilde{X}_e(k)\big)\,S^T \\ &= S\mathrm{Cov}\big(X_s(k) - \widetilde{X}_s(k)\big)S^T\end{aligned} \tag{12}$$

and therefore it depends only on the filtering error on the component $X_s(k)$.

Theorem 8. *The optimal* $\mathcal{P}$*-linear estimate* $\tilde{x}(k)$ *of the state of system* (1) *is achieved by the following filter:*

$$\begin{aligned}\tilde{x}(0|-1) &= \bar{x}_0,\\ \tilde{x}(0) &= \tilde{x}(0|-1) + K_G(0)\big[y(0) - C(0)\tilde{x}(0|-1)\big],\\ \tilde{x}(k{+}1|k) &= \mathcal{A}(k)\tilde{x}(k) + \mathcal{B}(k)u(k),\\ \tilde{x}(k+1) &= \tilde{x}(k+1|k) + \mathcal{D}(k)y(k+1) + K_G(k+1)\\ &\quad\cdot\big[y(k+1) - C(k+1)\mathcal{D}(k+1)y(k+1)\\ &\quad - C(k+1)\tilde{x}(k+1|k)\big],\end{aligned} \tag{13}$$

$$\textit{where}: \quad \begin{aligned}\mathcal{A}(k) &= SH^\dagger(k+1)\begin{bmatrix} I_n \\ O_{q\times n}\end{bmatrix},\\ \mathcal{B}(k) &= SB_e(k), \quad \mathcal{D}(k) = SD_e(k),\end{aligned} \tag{14}$$

with $S = \begin{bmatrix} I_n & O_{n\times m}\end{bmatrix}$ *and the filter gain* $K_G(k)$ *recursively computed as:*

$$\begin{aligned}M_P(k+1) &= \begin{bmatrix} \mathcal{A}(k)P(k)\mathcal{A}(k)^T + Q_f(k) & O_{n\times q}\\ O_{q\times n} & Q_g(k+1)\end{bmatrix},\\ K_G(k+1) &= -SH^\dagger(k)M_P(k+1)L(k)^T\\ &\quad\cdot\big[L(k)M_P(k+1)L(k)^T\big]^\dagger,\\ P(k+1) &= \big[SH^\dagger(k) + K_G(k+1)L(k)\big]\\ &\quad\cdot M_P(k+1)H(k)^{\dagger T}S^T,\end{aligned} \tag{15}$$

$$\textit{with} \quad L(k) = \begin{bmatrix} O_{q\times n} & I_q\end{bmatrix}\big(I_{n+q} - H(k)H^\dagger(k)\big), \tag{16}$$

$$\begin{aligned}K_G(0) &= \Psi_0 C(0)^T\big[C(0)\Psi_0 C(0)^T + Q_g(0)\big]^\dagger,\\ P(0) &= \big[I_n - K_G(0)C(0)\big]\Psi_0,\end{aligned} \tag{17}$$

as initial conditions. $P(k)$ *is the sequence of covariance matrices of the error* $\tilde{x}(k) - x(k)$.

Proof. The Proof, although not difficult, is too long to be reported in this conference paper. The idea is to develop the Kalman filter equations for the component $X_s(k)$ of the extended state, because it is well known that the Kalman filter provides the best linear state estimate. The component $X_d(k)$ is exactly computed through the linear equation (8) and added to the best linear estimate of $X_s(k)$, thus providing a linear estimate of $X_e(k)$. Finally, from this estimate the subvector $\tilde{x}(k)$ is extracted. All computations involved can be put in the form of the algorithm presented in the Theorem. ■

Remark 9. The performance of the proposed filter could be improved by extending the class of the estimator to the wider class of polynomial transformations of the output, following the approach described in [1].

Remark 10. Note that the *correction* gain $K_G(k)$ used in the proposed filtering algorithm can be computed off-line, differently from what happens for the Extended Kalman Filter, where the correction gain is a function of the current state estimate, and needs to be computed on-line.

4 Simulation Results

As it has been said in the introduction, the proposed filtering algorithm is useful in all cases in which the linear part of the system dynamics is well-known, while the nonlinear part is completely unknown. In these circumstances the Extended Kalman Filter can not use the linear approximation (around the previous estimate) of the nonlinear terms. An interesting point is that the gain of the proposed filter is not forced by the measurements, and can be computed off-line.

Let system (1) be described by the time-invariant matrices and the nonlinear term reported below

$$\begin{aligned}&A = \begin{bmatrix} 0.3 & 0.2 & 0\\ 0 & 0.7 & 0\\ 0.2 & 0 & 0.3\end{bmatrix}, \quad B = \begin{bmatrix} 2\\ 0.4\\ 0.8\end{bmatrix}, \quad R = \begin{bmatrix} -2\\ 2\\ 0\end{bmatrix},\\ &C = \begin{bmatrix} -1 & 1 & 1\\ 1 & 2 & -1\end{bmatrix}, \quad h\big(x(k)\big) = x_2(k)\cos\big(x_1(k)\big).\end{aligned} \tag{18}$$

The state and output noises are discrete zero-mean asymmetric distribution, whose covariance matrices are:

$$Q_f = \begin{bmatrix} 0.024 & 0.024 & 0 \\ 0.024 & 0.064 & -0.080 \\ 0 & -0.080 & 0.160 \end{bmatrix}, \qquad (19)$$
$$Q_g = \begin{bmatrix} 0.4532 & -0.5636 \\ -0.5636 & 6.2228 \end{bmatrix}.$$

There is a deterministic drift given by the unitary step $u(k) = \delta_{-1}(k)$. The initial state is given by $x_0 = (8 \ \ -6.5 \ \ 10)^T$, far from the initialization of the filter: without any information concerning the statistics of x_0, the *a priori* initial estimate $\tilde{x}(0|-1)$ is chosen null with the identity as covariance matrix.

As it can be easily verified, matrix A is an asymptotically stable matrix, so that the convergence of the filter is guaranteed.

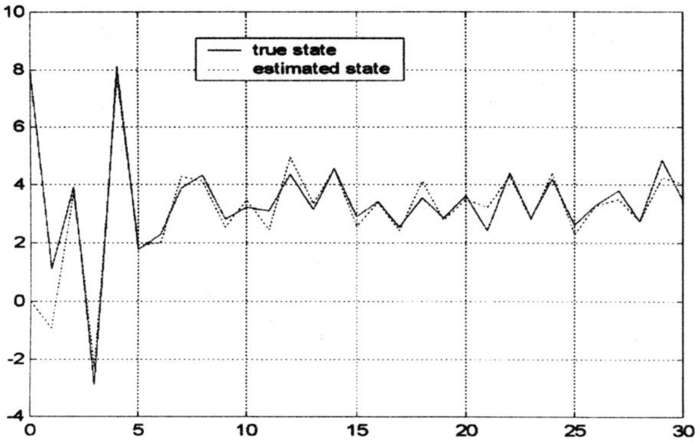

Fig. 1. True and estimated state: first component.

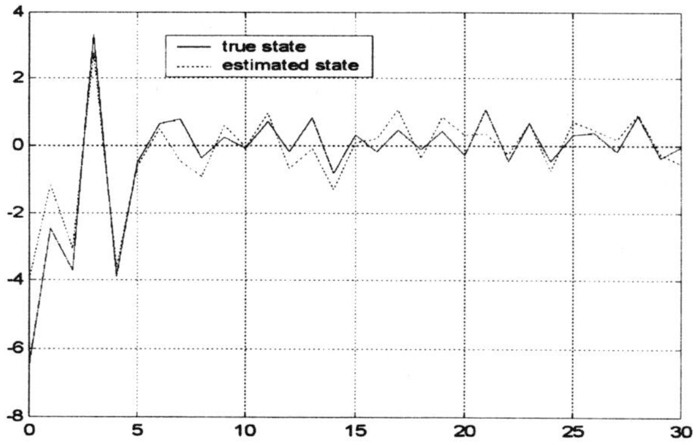

Fig. 2. True and estimated state: second component.

5 Conclusions

This work presents a new approach in filtering a stochastic linear system perturbed by uncertain nonlinearities. A clever use of the measurements allow the construction of a recursive filter that does not need any knowledge on the nonlinear perturbation terms.

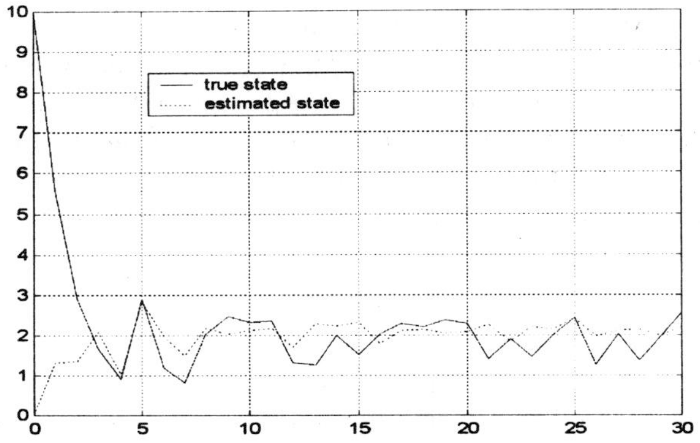

Fig. 3. True and estimated state: third component.

References

[1] F. Carravetta, A. Germani and M. Raimondi, "Polynomial filtering for linear discrete-time non-Gaussian systems," *SIAM J. Control and Optim.*, Vol. 34, No. 5, pp. 1666–1690, 1996.

[2] G.A. Einicke and L.B. White, "Robust Extended Kalman Filetring," *IEEE Trans. on Signal Processing*, Vol. 47, No. 9, pp. 2596–2599, 1999.

[3] M.R. Filho and C.J. Munaro, "Robust H_∞ filtering for continuous-time linear systems with norm-bounded nonlinear uncertainties," *Proc. of the 2001 IEEE Int. Conf. on Control and Applications*, pp 686–690, Mexico, City, 2001.

[4] A. Germani, C. Manes and P. Palumbo, "Optimal linear filtering for stochastic non-Gaussian descriptor systems," *Proc. 40th Conference on Decision and Control*, Orlando, Florida, 2001.

[5] A. Germani, C. Manes and P. Palumbo, "Kalman Bucy filtering for linear stochastic differential systems with unknown inputs," *Proc. 15th IFAC World Conference on Automatic Control*, Barcellona, Spain, 2002.

[6] A. Germani, C. Manes and P. Palumbo, "Optimal linear filtering for bilinear stochastic differential systems with unknown inputs," *to appear on IEEE Trans. Autom. Contr.*, Vol. 47, October, 2002.

[7] K.M. Nagpal and P.P. Khargonekar, "Filtering and Smoothing in an H_∞ setting," *IEEE Trans. Autom. Contr.*, Vol. 36, pp. 152–166, 1991.

[8] L. Xie, C.E. de Souza and M. Fu, "H_∞ estimation for discrete-time linear uncertain systems," *Int. J. on Robust and Nonlinear Control*, Vol. 1, pp. 111-123, 1991.

[9] L. Xie, C.E. de Souza and Y. Wang, "H_∞ filter design for discrete-time uncertain nonlinear systems," *Proc. of 33rd IEEE Conf. on Decision and Control*, Lake Buena Vista, Fl, pp. 3937-3942, 1994.

ON–LINE HIGH–RADIX EXPONENTIAL WITH SELECTION BY ROUNDING

J.-A. Piñeiro, J. D. Bruguera

Dept. of Electronic and Computer Eng.
Univ. Santiago de Compostela, Spain
alex,bruguera@dec.usc.es

M. D. Ercegovac

Computer Science Dept.
Univ. California, Los Angeles (UCLA)
milos@cs.ucla.edu

ABSTRACT

An on-line high-radix algorithm for computing the exponential function (e^X) with arbitrary precision n is presented. Selection by rounding and a redundant digit-set for the digits e_j are used, with selection by table in the first iteration to guarantee the convergence of the algorithm, and the on-line delay is $\delta = 2$ cycles. A sequential architecture implementing the algorithm is proposed, and the execution times and hardware requirements are estimated for 32-bit and 64-bit computations for several radix values. An analysis of the tradeoff between area and speed shows that the most efficient implementations are obtained for radix values from $r = 32$ to 256, depending on the precision.

1. INTRODUCTION

On-chip and off-chip communications are commonly the bottleneck of today's applications, having a negative impact in the overall performance of a system. Therefore, there is a need to develop algorithms which allow the computations of series of operations with a reduced communication bandwidth. On-line algorithms are a good solution for this problem [2, 7], since they perform computations in a *digit–serial, left-to-right* manner allowing the overlap between successive operations to reduce the execution time.

An iteration consists of computing one radix-r digit of the result and a new value of an internal state function (residual), but unlike conventional digit-recurrence methods, online algorithms begin producing the result digits on the basis of partial information. That is, once the $\delta + 1$ most significant digits of the input operand(s) have been received (δ is called the *on-line delay*), the most significant digit of the result can be generated, as shown in Figure 1. The use of a redundant number system allows updating of the partial result based on the input digits received in later iterations.

The main advantages of on–line arithmetic are (i) a reduced interconnection bandwidth, (ii) variable precision capability, and (iii) overlapping dependent operations, which results in reduced execution time.

On–line algorithms have already been proposed for the basic arithmetic operations and a set of elementary functions, important in scientific computing, artificial neural networks, and signal processing applications [1, 2, 3, 4].

In this paper we present an on–line algorithm for the computation of the exponential function with arbitrary precision n. We propose the use of a high-radix scheme ($r = 2^b$), potentially reducing the latency regarding a radix-2 implementation by a factor of b, and the execution time proportionally. However, the use of high-radix results in a increased area, and therefore an accurate analysis is necessary to determine which radix values lead to a good tradeoff between area and speed, thus resulting in an efficient implementation of the algorithm.

The on–line algorithm for exponential computation is explained in Section 2. In Section 3, the sequential architecture implementing our algorithm is proposed. Estimates of the execution time and area requirements for all radices are presented in Section 4. Finally, the main contributions made in this paper are summarized in Section 5.

2. ALGORITHM

In this paper we present an on–line algorithm for the computation of the exponential function (He^X), with the use of a high-radix and selection by rounding (details in [6]). The exponential is computed by using an *additive normalization*, with a set of constants $\ln(1 + e_j r^{-j})$, where $r = 2^b$ is the radix and e_j is a radix-r digit. The computation begins on the basis of partial information:

$$X[0] = \sum_{j=1}^{\delta} x_j r^{-j}, \tag{1}$$

with δ the *on-line delay* and x_j the radix-r digits of the input operand X, as shown in Figure 1.

This work was developed while J.-A. Piñeiro was with the University of California, Los Angeles (UCLA), USA. J.-A. Piñeiro and J. D. Bruguera were partially supported by the Ministry of Science and Technology (MCYT) under contract TIC2001-3694-C02.

0-7803-7761-3/03/$17.00 ©2003 IEEE

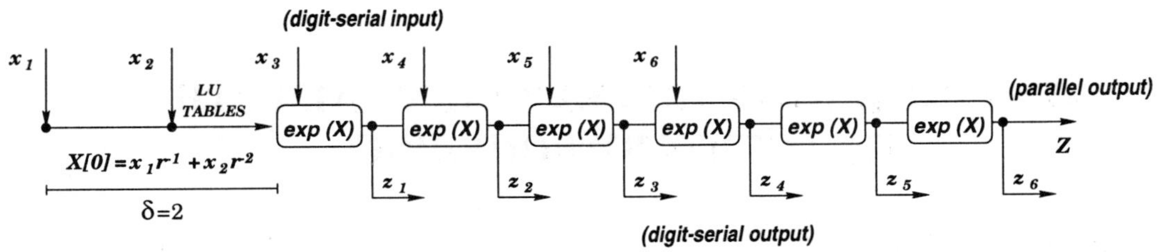

Figure 1: Timing diagram of our algorithm (for $N = 6$; parallel output produced via on-the-fly conversion)

The recurrences involved in the computation of the exponential are the following:

$$\begin{aligned} W[j+1] &= rW[j] - B[j] + x_{j+\delta}r^{-\delta+1} \\ E[j+1] &= E[j](1 + e_j r^{-j}) \end{aligned} \tag{2}$$

with $B[j] = r^{j+1}\ln(1 + e_j r^{-j})$, $j \geq 1$, $W[1] = rX_0$ and $E[1] = H$ being an input constant. For a precision of n bits, a total number of $N = \lceil n/b \rceil$ iterations is necessary to obtain the final result, $E[N+1] \approx H\exp(X)$. An approximation $-e_j r$ to the logarithm constants is used in iterations $j \geq \lceil N/2 \rceil + 1$, reducing the overall hardware requirements of the algorithm.

The selection of digits e_j for $j \geq 2$ is done by rounding an estimate of the residual:

$$e_j = round(\hat{W}[j]_t), \tag{3}$$

with a digit set for e_j of $\{-(r-1), \ldots, -1, 0, 1, \ldots, (r-1)\}$. The estimate is obtained by truncating $W[j]$ to t fractional bits and adding a 1 in position with weight 2^{-t}.

The convergence conditions [6] require selection by table look-up in the first iteration. This table is addressed by the b most significant bits of the input operand, and the selection is performed so that $-(r-3) \leq e_2 \leq (r-2)$. The convergence conditions also determine a minimum value of $t = 2$, an on-line delay $\delta = 2$, and $r \geq 8$.

The use of an on-line scheme requires a redundant number system, which also results in a faster execution of the recurrences, since the additions are independent of the precision. The use of redundancy, however, results in a more complex implementation.

We now summarize the algorithm for He^X with $\delta = 2$:

- First iteration ($j = 1$):

$$\begin{aligned} W[2] &= r^2 X[0] - B[1] + x_3 r^{-1} \\ E[2] &= H + e_1 r^{-1} H \end{aligned} \tag{4}$$

  with $B[1] = r^2 \ln(1 + e_1 r^{-1})$, and e_1 selected by table look-up so that $-(r-3) \leq e_2 \leq (r-2)$. The b most significant bits of the input operand $X[0]$ are used to address the tables storing both $e_1 r^{-1}$ and $B[1]$. This table look-up can be performed one cycle in advance.

- Iterations $j = 2$ to $\lceil \frac{N}{2} \rceil$:

$$\begin{aligned} W[j+1] &= rW[j] - B[j] + x_{j+2}r^{-1} \\ E[j+1] &= E[j] + e_j E[j] r^{-j} \end{aligned} \tag{5}$$

  with $B[j] = r^{j+1}\ln(1+e_j r^{-j})$, e_j selected by rounding the truncated residual $\hat{W}[j]$, the constants $B[j]$ read from a look-up table addressed by the digits e_j and the iteration index j, and x_{j+2} the digit of the input operand X received in iteration j.

- Iterations $j = \lceil \frac{N}{2} \rceil + 1$ to N:

$$\begin{aligned} W[j+1] &= rW[j] - e_j r + x_{j+2}r^{-1} \\ E[j+1] &= E[j] + e_j E[j] r^{-j} \end{aligned} \tag{6}$$

  with e_j selected by rounding the truncated residual and x_{j+2} the digit of the input operand X received in iteration j.

3. ARCHITECTURE

A block diagram of a sequential architecture implementing our algorithm is shown in Figure 2. The main features of this architecture are:

- All variables are in signed-digit (SD) representation[1] to allow a faster execution of iterations, since the addition operations become independent of the required precision n.
- All products of the type $E[j]r^{-j}$ are performed as shifts, since $r = 2^b$. Barrel shifters are employed to carry out these computations.
- SDAα is a *signed-digit binary adder* with α input bit-vectors. The addition of two SD operands requires a SDA4 adder, while SDA3 adders can be used for accumulating an operand in SD representation and an operand in two's complement ($2C$) representation.
- The constants $-r^{j+1}\ln(1+e_j r^{-j})$ are approximated by $-e_j r$ for iterations $j \geq \lceil \frac{N}{2} \rceil + 1$. Therefore, no look-up table is needed in those iterations.

[1] An equivalent architecture could be obtained by using *carry–save* (CS) representation of the variables.

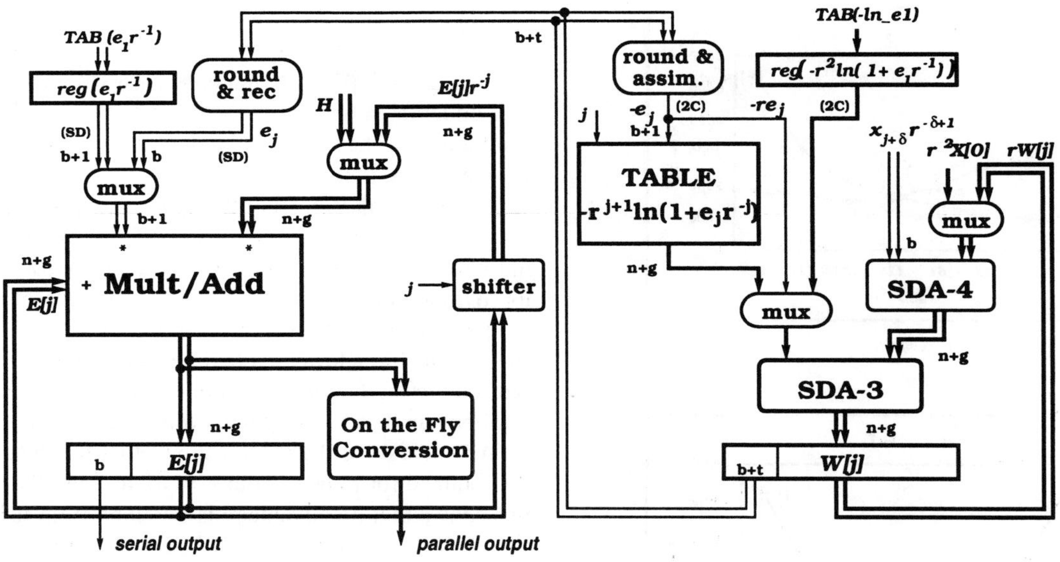

Figure 2: Block diagram of the proposed architecture

- g guard bits are used in the look-up tables, registers and other main logic blocks to prevent the truncation error from affecting the final result. Since N is the total number of iterations, $g = \lceil log_2(N) \rceil + 1$ bits.
- An SD multiply/add unit is used in the $E[j]$ recurrence, instead of separated SD multiplier and adder. The multiplier operand is in SD radix-4 representation to reduce by half the number of partial products to be accumulated. However, since the multiplicand is represented in SD radix-2, the size of this unit is roughly double of a regular multiplier with non-redundant multiplicand, and one extra level in the accumulation tree is required.
- The recoding to SD-4 of the multiplier operand in the $E[j]$ recurrence is performed outside the multiply/add unit (*round&rec* unit, and directly storing of e_1 in SD-4 representation in the initial look-up table).
- The first iteration of the exponential can be performed once the first $\delta + 1$ digits of the input operand X are known, as shown in Figure 1. δ previous cycles are necessary to obtain $X[0]$. The look-up tables used in the first iteration can be addressed one cycle in advance, while $x_{1+\delta}$ is being computed, and the looked-up values stored in registers, as shown in Figure 2. Therefore, these look-up tables are not included in the critical path, reducing the cycle time of the on-line scheme regarding its conventional counterpart.
- The selection by rounding is performed in two different units. The first one, the *round&rec* unit, takes the $(b + t)$ most significant bits of $W[j]$ $(t = 2)$ and produces a SD representation of the digit e_j, to be used as multiplier in the SD multiply/add unit. On the other hand, the *round&assim* unit takes the same $(b + t)$ input word, but produces a 2C representation of the digit with opposite sign $(-e_j)$, by rounding the input word and performing its assimilation to non-redundant representation. The output of this unit is used for addressing the look-up table storing the logarithm constants and for generating the approximation $-e_j r$ to the logarithm, both in the residual recurrence.

The latency of our algorithm is $(\delta + 1 + N)$ cycles, since N iterations are necessary in the computation of the exponential, but the first one cannot be computed until $x_{1+\delta}$ has been obtained ($\delta + 1$ cycles). The execution time is therefore $(\delta + 1 + N) \times t_c$, with t_c the cycle time of the on-line scheme (assuming that the digits of the input operand x_j can be obtained within that cycle time).

4. EVALUATION AND ANALYSIS

An analysis of the tradeoffs between area and speed in the implementation of this architecture is presented, based on estimates of the execution time for the parallel output and the total area, obtained for precisions $n = 32$ and 64-bits, and for radix values from $r = 8$ to $r = 1024$, as shown in Figure 3. These estimates have been obtained [6] using an approximate technology-independent model [5, 8]. The time and area units are τ and fa, the delay and area of a 1-bit full-adder.

The most remarkable points in this analysis are:

- The cycle time depends on r and not on the required precision, due to the use of redundant arithmetic.

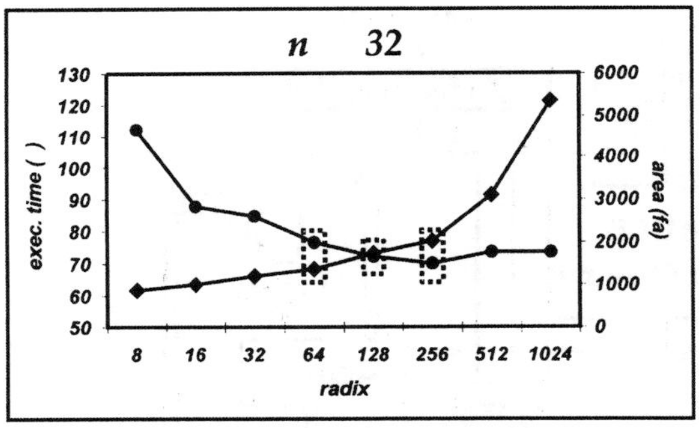

(a)

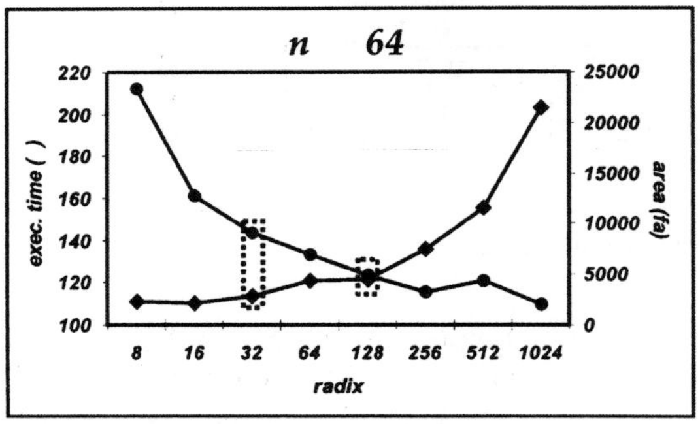

(b)

Figure 3: Execution time and area cost estimates for the proposed architecture

- The cycle time increases with the radix, but the latency decreases. A good tradeoff between latency and cycle time can be achieved for specific values of r, leading to low execution times. The area requirements must also be considered. The best tradeoffs for the proposed architecture are highlighted in Figure 3: $r = 64, 128$ and 256 for 32-bit computations, and $r = 32$ and 128 for $n = 64$-bits.

- There is no advantage in using higher radix values like $r = 512$ and 1024), because the execution times are similar to those achieved with $r = 128$ and 256, while the area requirements are prohibitive, due to the exponential growth in the table sizes with the radix.

5. SUMMARY AND CONCLUSIONS

A high-radix on–line scheme for computing the exponential function with arbitrary precision n, its implementation, and tradeoffs in the selection of the radix have been presented in this paper. Selection is done by rounding for $j \geq 2$ and by a table lookup in the first iteration. The use of high radix significantly reduces the latency of the algorithm, leading to low execution times. On the other hand, the area increases with higher values of the radix.

The exponential is computed by an additive normalization, using the values of elementary logarithms stored in a look-up table addressed by the coefficient e_j and the iteration index j. The use of redundant representation reduces the cycle time, making it independent of the precision.

A sequential architecture implementing our algorithm has been proposed, and estimates of the execution times and hardware requirements are obtained, according to an approximate model for the delay and the area of the main components, for radix values from $r = 8$ to 1024. The on–line scheme has a lower cycle time than its parallel counterpart, due to the fact that the look-up tables used in the first iteration can be addressed one cycle in advance.

An analysis of the results shows that the best tradeoffs are obtained for radices from $r = 32$ to 256. This analysis also shows that no advantage is obtained from using very-high radices, such as $r = 512$ or 1024, because of their prohibitive hardware requirements, due to the exponential growth in the size of the tables with the radix.

6. REFERENCES

[1] J. D. Bruguera and T. Lang. 2D DCT using On-Line Arithmetic. In *Intl. Conference on Acoustics, Speech, and Signal Processing (ICASSP)*, volume 5, pages 3275–3278, 1995.

[2] M.D. Ercegovac and T. Lang. On-Line Arithmetic: A Design Methodology and Applications. In R.W. Brodersen and H.S. Moscovitz, editors, *VLSI Signal Processing, III*, chapter 24. IEEE Press, 1988.

[3] B. Girau and A. Tisserand. On-Line Arithmetic-based Reprogrammable Hardware Implementation of Multilayer Perceptron Back-Propagation. In *5th International Conference on Microelectronics for Neural Networks and Fuzzy Systems (MicroNeuro '96)*, volume 5, pages 168–174, 1996.

[4] S. Kla, C. Mazenc, X. Merrheim, and J. M. Muller. New Algorithms for On-Line Computation of Elementary Functions. In *Proc. SPIE: Advanced Signal Processing Algorithms, Architectures, and Implementations II*, 1991.

[5] J.-A. Piñeiro, M. D. Ercegovac, and J. D. Bruguera. High-Radix Logarithm with Selection by Rounding. In *Proc. IEEE 13th Intl. Conference on Application-specific Systems, Architectures and Processors (ASAP'02)*, pages 101–110, 2002.

[6] J.-A. Piñeiro, M. D. Ercegovac, and J. D. Bruguera. On-Line High-Radix Exponential with Selection by Rounding. Technical report, Univ. Santiago de Compostela, 2002. At http://www.ac.usc.es.

[7] S. Rajagopal and J. R. Cavallaro. On-Line Arithmetic for Detection in Digital Communication Receivers. In *Proc. IEEE 15th Intl. Symp. Computer Arithmetic (ARITH15)*, pages 257–265, 2001.

[8] W. F. Wong and E. Goto. Fast Hardware-Based Algorithms for Elementary Function Computations. *IEEE Trans. on Computers*, 43(3):278–294, March 1994.

FAST IMPLEMENTATIONS OF MONTGOMERY'S MODULAR MULTIPLICATION ALGORITHM

P.V.Ananda Mohan

Research and Development,
I.T.I.Limited,
Bangalore 560016 India.

ABSTRACT

In this paper, techniques for fast implementations of Montgomery's modular multiplication algorithm are described. These use non-redundant multi-bit recoding. These are compared regarding the area and computation requirements and compared with conventional implementation.

1. INTRODUCTION

Modular multiplication using Brickell's algorithm [1-4] has received considerable attention in literature. Techniques for fast implementation using two or more bits at a time have been described [5-7]. In these techniques, the operations start from the MSBs of the multiplicand and proceed towards LSB and requires fewer steps (<n) for n-bit operands. Recently, Montgomery's modular multiplication algorithm has received considerable attention [8-13]. This interestingly starts from the LSB of the multiplicand and proceeds towards the MSB. This algorithm needs n steps considering one bit at a time. However, this technique yields only scaled result $(A.B.2^{-n})$ mod M which can be modified to yield (A.B) mod M by pre-multiplication of both A and B by 2^n and post multiplication of the result by 2^n. In this letter, we explore algorithms considering two or more bits at a time so as to speed up the modulo multiplication operation.

The Montgomery algorithm can be applied in two ways basically. In the first of these denoted as "*Binary to RNS conversion after pre-multiplication*", the multiplication of the two n bit operands A and B is carried out first to yield a $2n$ bit number X=A.B. The scaled residue corresponding to the n LSBs of the $2n$ bit product obtained in n steps is next added mod M to the n MSBs to obtain the final result [10,11]. In the second technique denoted as "*Integrated modulo multiplication with scaling*", the operations of multiplication and modulo reduction with scaling are integrated and this computation also needs n steps. We describe in this letter, new techniques for fast implementation for both these requirements by considering l bits (l = 2, 3 and 4) at a time.

2. MONTGOMERY'S MODULO MULTIPLICATION ALGORITHM

Considering that (A.B) mod M is to be obtained, first the values of A and B are scaled by 2^n so that $A'=A.2^n$ and $B'=B.2^n$. Next, we estimate $X=(A'.B'.2^{-n})$ mod M
$=(A.B.2^n)$ mod M. In other words, this operation has yielded a scaled product corresponding to scaled inputs A and B. Thus, the repeated squarings and multiplication of these products appropriately as needed in exponentiation operations are on scaled numbers. In other words, if B=A in RSA algorithm, A^K mod M evaluation yields $(A^K.2^n)$ mod M. A post scaling step by 2^{-n} of this result yields the desired result A^K mod M. The overheads that are needed are the front end operation of scaling A by 2^n mod M to obtain A' and the rear end operation of scaling again by 2^{-n} mod M.

3. BINARY TO RNS CONVERSION AFTER PRE-MULTIPLICATION

Considering that the product A.B is the 2n bit word denoted as $x_{2n}\, x_{2n-1} \ldots\ldots x_n\, x_{n-1} \ldots\ldots x_2 x_1$, we need to find $(A.B.2^{-n})$ mod M = $(W_1+W_o.2^{-n})$ mod M where W_1 is the n bit word $x_{2n}\, x_{2n-1} \ldots\ldots x_n$ and W_o is the n bit word $x_{n-1} \ldots\ldots x_2 x_1$. We first determine $W_o.2^{-n}$ mod M and add it W_1 to obtain $(A.B.2^{-n})$ mod M. We will first consider the case of taking l bits of W_o at a time (with l = 2). The basic operation in each step is given by

$$E_i = (\alpha/4) \bmod M = [(E_{i-1} + x_{2i}\, x_{2i-1})/4] \bmod M \quad \text{for } i = 1, \ldots, (n-1)/2 \qquad (1)$$

where $E_o=0$ and $\alpha = E_{i-1}+x_{2i}\, x_{2i-1}$.
The mechanization of this basic step needs two cases to be considered based on the two LSBs of the modulus M viz., $01_{(2)}$ and $11_{(2)}$. For these respective cases, (1) becomes

$$E_i=[(E_{i-1} + x_{2i}\, x_{2i-1} + 3M.\, \alpha_2\alpha_1)/4] \bmod M \quad \text{for } i = 1,\ldots,(n-1)/2 \qquad (2a)$$

and

$$E_i = [(E_{i-1} + x_{2i}\, x_{2i-1} + M.\, \alpha_2\alpha_1)/4] \bmod M \quad \text{for } i = 1,\ldots,(n-1)/2 \qquad (2b)$$

where α_2 and α_1 are the two LSBs of α and $E_o = 0$. The computation starts from LSBs $x_2 x_1$. Evidently, since $\alpha_2\, \alpha_1$ can be at most

0-7803-7761-3/03/$17.00 ©2003 IEEE

3, the value of the term being divided by 4 in (2a) and (2b) can be at most 4M+2, considering that the term 3 $\alpha_2\alpha_1$.M can be reduced mod 4M due to the division by 4 in (2a) and (2b).In other words, for all the four possibilities of $\alpha_2\alpha_1$ viz., $11_{(2)}$, $10_{(2)}$, $01_{(2)}$ and $00_{(2)}$, we have $3\alpha_2\alpha_1$ M as M, 2M, 3M and 0 respectively. The object of the computation in (2) is to add a quantity *(3M.* $\alpha_2\alpha_1$*)* or *(M.* $\alpha_2\alpha_1$*)* or *zero* to $(E_{i-1} + x_{2i+1}x_{2i})$ so that the sum is exactly divisible by 4. Undoubtedly, there is no modulo reduction needed and only four operand addition *(i.e* E_{i-1}, $x_{2i}x_{2i-1}$, *M, 2M)* is needed for hardware implementations. The weighting of M by 0,1,2 and 3 can be by adding conditionally 0 or M or one bit left shifted M i.e. 2M. All the terms in (2) can be added by carry- save-adders followed by a carry look-ahead adder or Brent-Kung adder using regular layout [15]. It may be noted that the complete calculation of α is not needed to decide the value of $(3M\alpha_2\alpha_1)$ in (2) since only the two bit word $x_{2i+1}x_{2i}$ and the two LSBs of E_{i-1} need to be added to decide the $\alpha_2\alpha_1$ term in (2b).

The extension to the case of $l = 3$ i.e. taking three bits at a time is slightly more involved. Depending on the three LSBs of M, the expressions (1) and (2) change as

$$E_i = (\beta/8) \bmod M = [(E_{i-1} + x_{3i}x_{3i-1}x_{3i-2})/8] \bmod M \quad for\ i = 1,\ldots,(n-1)/3 \qquad (3a)$$

and

$$E_i = [(E_{i-1} + x_{3i}x_{3i-1}x_{3i-2} + kM.\ \beta_3\beta_2\beta_1)/8] \bmod M \quad for\ i = 1,\ldots,(n-1)/3 \qquad (3b)$$

where k = 7, 5, 3 and 1 respectively for the cases corresponding to LSBs of M being $001_{(2)}$,$011_{(2)}$,$101_{(2)}$ and $111_{(2)}$ and $E_o = 0$. It can be seen that the third term in (3b) is at most 7M and thus the numerator is less than 8M+6. There is no need for modulo reduction as the three LSBs will be zero. Note once again that the three LSBs of E_{i-1} need to be added to $x_{3i}x_{3i-1}x_{3i-2}$ to obtain the word $\beta_3\beta_2\beta_1$ and in this case a five operand addition is needed.

The results for four bit scanning ($l = 4$) can be derived in a similar manner depending on the four LSBs of M:

$$E_i = (\gamma/16) \bmod M = [(E_{i-1} + x_{4i}x_{4i-1}x_{4i-2}x_{4i-3})/16] \bmod M \quad for\ i = 1,\ldots,(n-1)/4 \qquad (4a)$$

and

$$E_i = [(E_{i-1} + x_{4i}x_{4i-1}x_{4i-2}x_{4i-3} + k'M.\ \gamma_4\gamma_3\gamma_2\gamma_1)/16] \bmod M \quad for\ i = 1,\ldots,(n-1)/4 \qquad (4b)$$

where k' = 1,3,5,7,9,11,13,15 corresponding to the four LSBs of M viz., $1111_{(2)}$, $0101_{(2)}$, $0011_{(2)}$,$1001_{(2)}$,$0111_{(2)}$,$1101_{(2)}$,$1011_{(2)}$ and $0001_{(2)}$ and $E_o = 0$. In this case as well, there is no need for modulo reduction since the four LSBs of the numerator in (4b) will be zero. Note however that six operand addition is needed. In a similar manner, this technique can be applied to the cases of considering more bits at a time so as to reduce the number of steps.

4. INTEGRATED MODULO MULTIPLICATION WITH SCALING

We next consider fast algorithms for the integrated modulo multiplication *(A.B.2^{-n}) mod M.* In this case the multiplicand A weighted by l bits of B at a time, first needs to be added to the result in the previous step E_{i-1} and then the scaling by 2^k needs to be done. Considering that *B* is represented as $b_n\, b_{n-1}\ldots b_2\, b_1$, we have for the case considering two bits at a time (i.e. k=2),

$$E'_i = (\alpha'/4) \bmod M = [(E'_{i-1} + A.(b_{2i}\,b_{2i-1}))/4] \bmod M \quad for\ i = 1,\ldots,(n-1)/2 \qquad (5)$$

As illustrated in the previous technique, in this case also, next, based on the two LSBs of the modulus M viz., $01_{(2)}$ and $11_{(2)}$, (3M. $\alpha'_2\alpha'_1$) or (M. $\alpha'_2\alpha'_1$) will be added respectively to the numerator in (5), so that E'_i can be obtained by ignoring the two zeroed LSBs. Note that α'_2 and α'_1 are the LSBs of α'. It may be noted that the value of the numerator in (5) after addition of 3M. $\alpha'_2\alpha'_1$ or M. $\alpha'_2\alpha'_1$ will be at most 7M thus needing one step of modulo reduction. At most five operands need to be added considering that A, 2A, M, 2M and E_{i-1} need to be added at most. The extension to the cases for l = 3 and 4 is straight forward with appropriate changes in (3) and (4) and use of the same choices of k and k' based on the LSBs of M.

5. EXAMPLES

We will consider the evaluation of ($18.13.2^{-4}$) mod 37. Direct multiplication yields 18.13 = 234 = 1110 1010 $_{(2)}$. We use two bits at a time. Thus W_1=14 and W_0=10. The two steps are as follows:

$E_1 = [(0 + 10_{(2)})/4] \bmod 37 = [(0+10_{(2)} + (3.37).\underline{10}_{(2)})/4] \bmod 37 = 19$.

$E_2 = [(19 + 10_{(2)})/4] \bmod 37 = (19 + 10_{(2)} + (3.37).\underline{01}_{(2)})/4] \bmod 37 = 33$.

The result is (14 +33) mod 37 = 10 where 14 is the value of the four bit MSB word of the product. Note that the underlined words are obtained after adding the two LSBs of the first two terms and 3.37. (Note that 3M is added since LSBs of this chosen M are $01_{(2)}$).

The case of integrated multiplication for the same example considering two bits at a time yields the following:

$E_1 = [(0 + 01_{(2)}.18)/4] \bmod 37 = [(0 + 01_{(2)}.18 + \underline{10}_{(2)}.(3.37))/4] \bmod 37 = 23$.

$E_2 = [(23 + 11_{(2)}.18)/4] \bmod 37 = [(23 + 11_{(2)}.18 + \underline{01}_{(2)}.(3.37))/4] \bmod 37 = 10$

6. AREA AND COMPUTATION TIME EVALUATION

The area and time needed for the implementation of both the above techniques using more than one bit in each step (l > 1) are next considered and compared with the case using only one bit multiplication in one step. In the first method, the initial multiplication time of the first stage is same irrespective of the number of bits chosen in each step in the Montgomery reduction. In the second stage, using one bit at a time, the architecture employed in each step comprises of one three operand addition of E_i, b_i and conditionally M. Thus, a two input adder with a carry input can be employed followed by a latch. In the two bit case, four operands need to be added needing a $(n+2)$ bit data path. Thus, two carry save adder stages preceding a two-input adder are needed. In a similar manner, for the three and four bit cases, five and six operands need to be added. Assuming that adders with regular VLSI layout [14] are employed, the area needed and addition time are known to be $n(\log_2 n+1)A_{FA}$ and $(\log_2 n+1)D_{FA}$ where A_{FA} and D_{FA} are the area and delay of a full –adder respectively. We also assume that the same hardware is used in several steps in order to conserve area since in cryptographic applications demand the operands lengths of the order of 1024 bits typically. The area and time requirements for the above four cases are presented in Table I for the general n case. As an illustration for the n=1024 bit case, the area and time requirements in the case n=1,2,3 and 4 are respectively as follows:

$$Area=1024\ A_{AND} + 12289\ A_{FA}+2048\ A_{Latch}$$

$$Area=2048\ A_{AND} + 13315\ A_{FA}+2048\ A_{Latch}$$

$$Area=3072\ A_{AND} + 14345\ A_{FA}+2048\ A_{Latch}$$

$$Area=4096\ A_{AND} + 15376\ A_{FA}+2048\ A_{Latch}$$

$$Delay = 1024[12\ D_{FA} + D_{Latch}+ D_{AND}] = = [12,288\ D_{FA} +1024\ D_{Latch}+1024\ D_{AND}]$$

$$Delay = 512[13\ D_{FA} + D_{Latch}+ D_{AND}] = = [6656\ D_{FA} +512\ D_{Latch}+512\ D_{AND}]$$

$$Delay = 342[14\ D_{FA} + D_{Latch}+ D_{AND}] = = [4808\ D_{FA} +3424\ D_{Latch}+342\ D_{AND}]$$

$$Delay = 256[15\ D_{FA} + D_{Latch}+ D_{AND}] = = [3840\ D_{FA} +256\ D_{Latch}+256\ D_{AND}]$$

It can be seen that the computation time reduction achieved is significant while the increase in area is less than 36.5% upto the four bit case. Note that the delay needed for the front-end n x n multiplier using modified Booth's algorithm is about (512 D_{FA}+11D_{FA}) = 523.D_{FA}. Unfortunately, the area needed for this multiplier is exorbitant even if modified Booth's algorithm is employed.

In the second technique, the pre-multiplication hardware is not needed, however, needing slight increase in area and computation time over that of the second stage in the previous technique. The increase in area is because of the two reasons: (a) due to the need for (k-1) additional binary weighted inputs of A based on the number of bits k being considered in each step and (b) due to the modulo M reduction needed for estimating E_i in (5). The modulo reduction needs an additional adder stage to parallelly compute (x+y-M) so that using a 2:1 multiplexer, the result (x+y) or (x+y-M) can be selected. Note that –M is realized as the two's complement of M. The inputs to this auxiliary adder can be tapped from the main carry save adder adding E_{i-1} and b_i.A. It can be noted that the modulo M addition need to be done excluding the last bit. It can be seen that the increase in area is n.($\log_2$n+2)A_{FA} + n.$A_{2:1MUX}$ + n.A_{Inv} whereas the increase in delay per step is D_{FA} +$D_{2:1Mux}$.

Table.I. Area and delay requirements for the second stage for evaluating $(A.B.2^{-n}) \bmod M$

Bits in each step	*Area*	*Time*
1	$n.A_{AND}+(n+1).A_{FA} +n(\log_2 n+1).A_{FA}+2n.A_{latch}$	$n.[(\log_2 n+2) D_{FA}+D_{latch}+D_{AND}]$
2	$2n.A_{AND}+2(n+2).A_{FA} +n(\log_2 n+1).A_{FA}+2n.A_{latch}$	$(n/2).[(\log_2 n+3) D_{FA}+D_{latch}+D_{AND}]$
3	$3n.A_{AND}+3(n+3).A_{FA} +n(\log_2 n+1).A_{FA}+2n.A_{latch}$	$(n/3).[(\log_2 n+4) D_{FA}+D_{latch}+D_{AND}]$
4	$4n.A_{AND}+4(n+4).A_{FA} +n(\log_2 n+1).A_{FA}+2n.A_{latch}$	$(n/4).[(\log_2 n+5) D_{FA}+D_{latch}+D_{AND}]$

In the architecture for the two bit case, five operands need to be added E_{i-1}, 2A.b_{2i}, A.b_{2i-1}, $(\alpha'_2\alpha'_1 .M)$, $(2\alpha'_2\alpha'_1 .M)$ before the carry and sum vectors are fed to the Brent-Kung adder. First E_{i-1}, 2A.b_{2i} and A.b_{2i-1} need to be added first in a carry-save adder. The last two sum and carry bits together with the two LSBs of M decide whether M or 2M or both need to be added in the next two carry save adder stages. The result of this summation will be to make the two LSBs zero. It can be noted that the modulo M addition need to be done excluding the last k bits (e.g. last 2 bits 2 in the 2 bit case), since the carry-save-adder gives these bits as zeroes in any case in order to facilitate division by 2^k. The additional area needed is n.($\log_2$n+2)A_{FA} and n.$A_{2:1MUX}$ + n.A_{Inv} , for the modulo adder and due to the two additional inputs to be added is 2(n+2) A_{FA}. The increase in delay for each step is (D_{FA}+D_{MUX}) due to the modulo adder and 2D_{FA} for the additional carry-save-adder stage. Thus, the total increase in delay is (3.(n/2).D_{FA} +(n/2)D_{MUX} for k=2. The results for the three and four bit case are respectively (n/3). (5D_{FA}+D_{MUX}) and (n/4).(7D_{FA}+D_{MUX}) regarding delay and the areas are respectively (4n+12)A_{FA}+ n.$A_{2:1MUX}$ and (6n+24)A_{FA} + n.$A_{2:1MUX}$ in addition to the increase in area mentioned above.

7. IMPLEMENTATION OF PRE-SCALING AND POST-SCALING STAGES

It has been mentioned before that the inputs need to be scaled by 2^n mod M for cryptographic applications and similarly the output of the final Montgomery step should be scaled by 2^{-n} mod M to restore the original result. The pre-scaling is the well-known binary to RNS conversion problem. The conversion starts from the n th bit till LSB as explained next. As an illustration, P.2^n mod M where P and M are n bit words yields with the definition of P as $b_{n-1}b_{n-2}\ldots b_1b_0$,

$$P.2^n = b_{n-1}b_{n-2}\ldots b_1b_0\ 00\ldots.00$$

First, the residue of the word P is determined mod N. Next, it is doubled and its residue mod N is determined. This operation is performed N times to obtain $(P.2^n)$ mod M. An example will be in order to determine 11.2^4 mod 13:11 mod 13 =11, 22 mod 13=9, 18 mod 13 =5, 10 mod 13 =10, 20 mod 13 =7.
Thus the architecture needs a modulo adder and latches as shown since doubling is achieved by appending a LSB of zero. Thus, the area needed is $2n.(\log_2 n+1)\ A_{FA}\ +n.A_{\ 2:1Mux}.+\ n.A_{Inv}$. and the conversion time is $n.[(\log_2 n+1)\ D_{FA}\ +\ D_{2:1Mux}]$. Evidently, the conversion time is quite comparable to the Montgomery multiplication step but, however, in the case of RSA implementations, this pre-scaling is done only once in the beginning. The post scaling step is similar to the Montgomery step already described.

8. COMPARISON WITH OTHER TECHNIQUES

It is relevant to compare the techniques described in this paper with conventional higher radix implementations for (A.B) mod N. We follow the approach of Lu Harn [4] and Prasanna and Ananda Mohan [6,7] for one and two and higher bit cases respectively. The reader is referred to [6,7] for details about these architectures. We first consider the case of taking one bit at a time. It may be noted that the previous result is scaled by 2^b where b is the number of bits taken in each step, and added to $b_i.A$. The result can be atmost 3M and hence using two additional parallel subtractors, the residue of this is found. The area increase is hence largely due to these parallel adders and the multiplexer following these adders. It can be seen that the area is 50% larger than that of the second techniques described earlier using Montgomery algorithm with integrated multiplication and modulo reduction. This is because the dominant term $k.n(\log_2 n+1)\ A_{FA}$ is fixed i.e k=2 in the latter case whereas it is 2^b-1 for the former. Note that the area can be reduced in the scheme of Prasanna and Ananda Mohan by using only one hardware subtractor but at the expense of computation time.

9. CONCLUSION

It has been shown that faster implementation of Montgomery's multiplication is possible considering two or more bits at a time. The only overhead is the carry-save addition needed to add various operands. The word length increase is marginal by 2,3,4 bits etc compared to the word lengths of M which for authentication schemes using RSA are about 512 to 1024 bits. It may be noted that we have also described a faster implementation of Meehan et al [13] binary to RNS conversion algorithm taking more bits in a step.

10. REFERENCES

[1] G.R. Blakley, A computer algorithm for calculating the product AB modulo M, IEEE Transactions on Computers, Vol. 32, pp 497-500, May 1983.

[2] E.F. Brickell, A fast modular multiplication algorithm with application to two-key cryptography, Advances in Cryptography, Proceedings Crypto '82, Plenum, New York, pp 51-60,1983.

[3] K.R. Sloan Jr., Comments on " A computer Algorithm for calculating the product A.B mod M", IEEE Transactions on Computers, Vol. C-34, pp 290-292, March 1985.

[4] E. Lu Lein Harn, J. Lee and W. Hwang, A programmable VLSI architecture for computing multiplication and polynomial evaluation modulo a positive integer, IEEE Journal of Solid-State Circuits, Vol. SC-23, pp 204-207, 1988.

[5] T. Beth and D. Gollmann, Algorithm engineering for public key algorithms, IEEE Journal of Selected Areas in Communications, Vol. SAC-7, pp 458-466, 1989.

[6] B.S. Prasanna and P.V. Ananda Mohan, Fast VLSI architectures using non-redundant multi-bit recoding for computing AY mod N, Proc. IEE, Part G , Vol. 141, pp 345-349, 1994.

[7] B.S. Prasanna and P.V. Ananda Mohan, Fast VLSI architectures using non-redundant multi-bit recoding for computing exponentiation modulo a positive integer, Proc. ISCAS, Singapore, pp 3054-3057, 1991.

[8] P.L. Montgomery, Modular multiplication without trial division, Math. Comput., Vol. 44, pp 519-521, 1985.

[9] C.K. Koc, T. Aca and B.S. Kaliski. Jr. Analyzing and comparing Montgomery Multiplication Algorithms, IEEE Micro, Chip, Systems, Software and Applications, pp 26-33, 1996.

[10] C.C. Yang, T.S. Chang and C.W. Jen, A new RSA cryptosystem hardware design based on Montgomery's algorithm, IEEE Transactions on Circuits and Systems, Part-II, Analog and Digital Signal Processing Vol. 45, pp 908-913, 1998.

[11] C.Y. Su, S.A. Wang, P.S. Chen and C.W. Vu, An improved Montgomery's algorithm for high speed RSA Public key Cryptosystem, IEEE Transactions on VLSI Systems, Vol. 7, pp 280-283, 1999.

[12] S.E. Eldridge and C.D. Walter, Hardware implementation of Montgomery's modular multiplication algorithm, IEEE Transactions on Computers, Vol. 42, pp 693-699, 1993.

[13] S.J. Meehan, S.D. O'Neil and J.J. Vaccaro, An Universal Input and output RNS Converter, IEEE Transactions on Circuits and Systems, Vol. 37, pp 799-803, June 1990.

[14] R.P. Brent and H.T. Kung, A regular layout for parallel adders, IEEE Trans. on Computers, Vol. 31, pp 260-264, March 1982.

GDFT TYPES MAPPING ALGORITHMS AND STRUCTURED REGULAR FPGA IMPLEMENTATION

*H. I. Saleh, M. A. Ashour, A. E. Salama**

NCRRT, AEA, Nasr City, P.O.B. 29, Cairo, Egypt

*Faculty of Engineering, Cairo University, Giza, Egypt

ABSTRACT

In this paper, regular and non-multiplicative mapping algorithms between different types of Generalized Discrete Fourier Transform (GDFT) are proposed. The proposed mapping algorithms are used to build regular and real twiddle factors FFT algorithms. It presents a more regular FFT than that recently presented GDFT type-1, which in addition requires log (N/2) stages of permutation [3]. Hardware realization of 16-point FFT, based on the proposed mapping algorithms, with real twiddle factors butterfly rather than complex twiddle factors in traditional FFT algorithms, is implemented in Xilinx XC4000 and Vertix series Field Programmable Gate Array (FPGA). Cost comparisons with alternative approaches are presented. Our proposed algorithms achieve a significant improvement in the FPGA-based designs.

1. INTRODUCTION

The Fast Generalized DFT (GDFT) was studied in recent publications [1-3]. There are four types of GDFT according to the shifts in either or both indices. The shift by 1/2 in the time index, frequency index, or both gives the odd-time, odd-frequency, or odd-squared DFTs. The traditional, odd-time, odd-frequency, odd-squared DFTs are referred to as types I, II, III, and IV, respectively. The traditional DFT has no shifts in the indices. The length and the type of the DFT will be given in the subscript and superscript of the DFT symbol, respectively.

The four version of the GDFT are given in (1), (2), (3), and (4).

Traditional (Type I) N-point DFT ($\mathbf{DFT}_N^{I}$):

$$\mathbf{X}(k) = \sum_{n=0}^{N-1} x(n) e^{-j\frac{2\pi}{N}nk}, k = 0,.., N-1 \tag{1}$$

Odd-time (Type II) N-point DFT ($\mathbf{DFT}_N^{II}$):

$$\mathbf{X}(k) = \sum_{n=0}^{N-1} x(n) e^{-j\frac{2\pi}{N}(n+0.5)k}, k = 0,.., N-1 \tag{2}$$

Odd-frequency (Type III) N-point DFT ($\mathbf{DFT}_N^{III}$):

$$\mathbf{X}(k) = \sum_{n=0}^{N-1} x(n) e^{-j\frac{2\pi}{N}n(k+0.5)}, k = 0,.., N-1 \tag{3}$$

Odd-squared (Type IV) N-point DFT ($\mathbf{DFT}_N^{IV}$):

$$\mathbf{X}(k) = \sum_{n=0}^{N-1} x(n) e^{-j\frac{2\pi}{N}(n+0.5)(k+0.5)}, k = 0,.., N-1 \tag{4}$$

The real and complex split-radix Generalized Fast Fourier Transform (SRGFFT) was developed in [1], and simplified in [2]. Special forms of the GDFT matrices were investigated and their sparse matrix factorizations were presented to complete Wang's [4] set of real sparse matrix factorizations for the family of discrete sinusoidal transforms [3].

Wang's algorithm utilized the conjugate symmetry of twiddle factors (W_N^{-kn} = conj. ($W_N^{-k(N-n)}$) = cos ($\frac{2\pi kn}{N}$) - j sin ($\frac{2\pi kn}{N}$)) in order to get a real sparse factorization of DFT. A real sparse factorization of GDFT matrices leads to simple fast algorithms for their computation, where only real arithmetic is involved. The resulting generalized signal flow graphs for the computation of different versions of the GDFT represent simple and compact unified approach to the fast discrete sinusoidal transform computation [3]. All algorithms in [3] are based on the universal DCT-II/DST-II (DCT-III/DST-III) computational structure, which is used as the basic processing component. The main advantage of real arithmetic GFFT [3] is the real twiddle factors rather than complex twiddle factors in traditional FFT algorithms [1,2], [4-9]. Wang's real sparse factorization of DFT is illustrated for 16-point in Fig. 1. The real sparse factorizations of other three types of GDFT were presented [3].

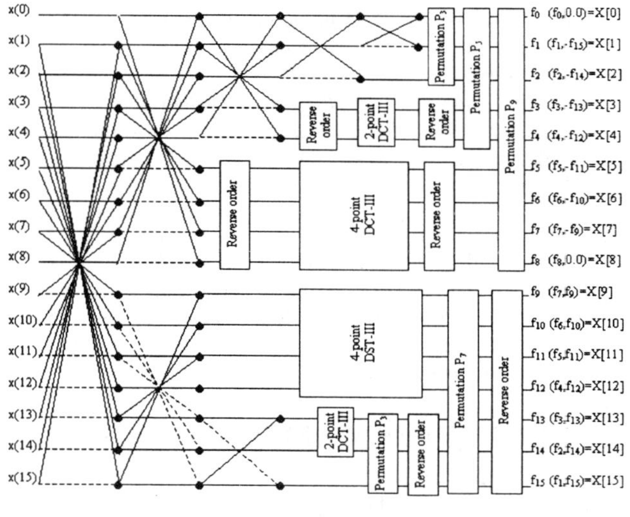

Fig. 1. Signal flow graph of DFT-I for N=16 as in reference [3]. *Solid lines represent unity transfer factors while dashed lines represent transfer factors –1. • represents addition. Permutation P_i is an ordering of the even indexed elements in normal order followed by the odd indexed in a reverse order, i.e. P(x[0], x[1], x[2], x[3], ..., x[N-2], x[N-1])=(x[0], x[2], ..., x[N-2], x[N-1], ..., x[3], x[N-1]).*

Fig. 2 shows the 2, 4, 8-point DCT^{II}/DST^{II} (DCT-III/DST-III) processing components. The real sparse factorization of GDFT

0-7803-7761-3/03/$17.00 ©2003 IEEE

algorithm offers the most efficient FFT algorithm in terms of computation complexity, where only real arithmetic is involved. For the first type of GDFT ($GDFT^{I}$), the real sparse factorization which was proposed in [3], shown in Fig. 1 for 16-point, has a lack of regularity and requires $\log_2$ (N/2) stages of permutation. Therefore, a more regular real sparse factorization of $GDFT^{I}$ is needed. On the contrary, computing the GDFT by the traditional fast algorithms, such as split-radix [1] and CT-FFT [5], is more regular and simpler in case of the $\mathbf{DFT^{I}}$ type than the other types but required a complex arithmetic. Therefore, regular and simple mapping from type to another where the computing is more regular or simpler is required.

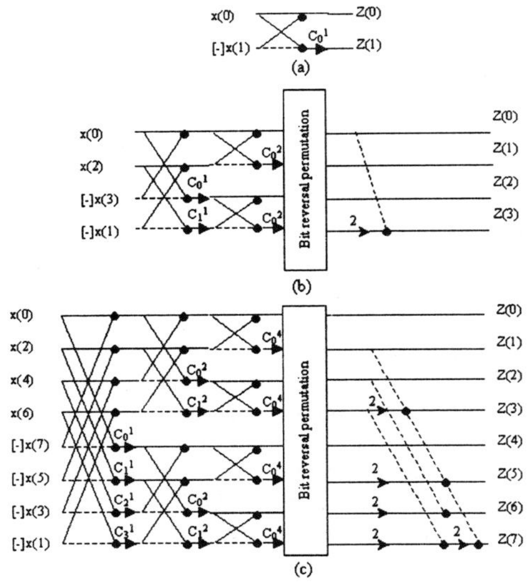

Fig. 2. DCT-II [DST-II] universal block for (a) N=2, (b) N=4, and (c) N=8. *Solid lines represent unity transfer factors while dashed lines represent transfer factors –1. • represents addition, and → represents multiplication.* C_n^k = *cos(k(4n+1)* π */2N). The DCT output set is in normal order and the DST output set is in reverse order, i.e.* $DCT_{out}(n)=Z(n)$, *and* $DST_{out}(n)= Z(N-n-1)$.

This paper is organized as follows. In Section 2 we propose mapping algorithms between different types of GDFT. In Section 3 we then compare our proposed structure to that given in [3] and compare the FPGA-based implementation of the proposed structure with recently published implementations. Finally, we summarize our work and results.

2. GDFT TYPES MAPPING ALGORITHMS

In order to map one type of GDFT into another type, we developed mapping algorithms. Mapping a GDFT with odd-time, odd-frequency, or odd-squared type into traditional type is easily derived and tends to weighting process of input vector, or output vector, or both, respectively. In mapping type III onto type I, the input to the traditional type after the weighting factors is $x(n)e^{-j\frac{\pi}{N}n}$, where $x(n)$ is the input vector of the type III. As well as, in mapping type II onto type I, the output of the type II after the weighting factors is $\mathbf{X}(k)e^{-j\frac{\pi}{N}k}$, where $\mathbf{X}(k)$ is the output vector of the traditional type. Mapping in the reverse direction, i.e. from non-shift to shift in index, requires a regular and simple arithmetic algorithm, which can be employed to map between different types in real arithmetic GDFT as in [3]. In the following, we provide these types mappings.

2.1 Mapping $\mathbf{DFT_N^{I}}$ ($\mathbf{DFT_N^{III}}$) into shorter-length $\mathbf{DFT^{II}}$ ($\mathbf{DFT_N^{IV}}$) blocks:

Separating the summation of $\mathbf{DFT_N^{I}}$ in equation (1) into two summations, one for the odd-index input data and the other for the even-index, gives the equation:

$$\mathbf{X}(k) = \sum_{\substack{n=1\\ n=odd}}^{N-1} x(n)e^{-j\frac{2\pi}{N}nk} + \sum_{\substack{n=0\\ n=even}}^{N-2} x(n)e^{-j\frac{2\pi}{N}nk}, \quad k = 0,..,N-1.$$

$$= \sum_{n=0}^{N/2-1} x(2n+1)e^{-j\frac{2\pi}{N}(2n+1)k} + \sum_{n=0}^{N/2-1} x(2n)e^{-j\frac{2\pi}{N}(2n)k}, \quad k = 0,...,N-1. \qquad (5)$$

Putting $\boldsymbol{x''(n) = x(2n+1)}$ and $\boldsymbol{x'(n) = x(2n)}$

$$\mathbf{X}(k) = \sum_{n=0}^{N/2-1} x''(n)e^{-j\frac{2\pi}{N}k(2n+1)} + \sum_{n=0}^{N/2-1} x'(n)e^{-j\frac{2\pi}{N}k(2n)}, \quad k = 0,...,N-1. \qquad (6)$$

Let
$$\mathbf{X''}(k) = \sum_{n=0}^{N/2-1} x''(n)e^{-j\frac{2\pi}{N}k(2n+1)}, \qquad (7)$$

and
$$\mathbf{X'}(k) = \sum_{n=0}^{N/2-1} x'(n)e^{-j\frac{2\pi}{N}k(2n)}. \qquad (8)$$

Hence,

$$\boldsymbol{X(k) = X''(k) + X'(k)}, \ k = 0, \ldots, N\text{-}1. \qquad (9)$$

Putting $k = k + N/2$ in equations (7), and (8):

$$\mathbf{X''}(k+\mathbf{N/2}) = \sum_{n=0}^{N/2-1} x''(n)e^{-j\frac{2\pi}{N}(k+N/2)(2n+1)} = -\,\mathbf{X''}(k). \qquad (10)$$

$$\mathbf{X'}(k+\mathbf{N/2}) = \sum_{n=0}^{N/2-1} x'(n)e^{-j\frac{2\pi}{N}(k+N/2)(2n)} = \mathbf{X'}(k). \qquad (11)$$

Substitute (10), and (11) in (9):

$$\mathbf{X}(k+N/2) = -\,\mathbf{X''}(k) + \mathbf{X'}(k), \ k = 0, \ldots, N/2\text{-}1. \qquad (12)$$

Equations (7), and (8) can be reformed as:

$$\mathbf{X''}(k) = \sum_{n=0}^{N/2-1} x''(n)e^{-j\frac{2\pi}{N/2}(n+0.5)k}, \ k = 0, \ldots, N/2\text{-}1, \qquad (13)$$

and

$$\mathbf{X'}(k) = \sum_{n=0}^{N/2-1} x'(n)e^{-j\frac{2\pi}{N/2}nk}, \ k = 0, \ldots, N/2\text{-}1. \qquad (14)$$

Equations (13), and (14) represent an N/2-point DFT^{II}, and N/2-point DFT^{I}, respectively. By recursive decomposing of (14), the $\mathbf{DFT_{N/2}^{I}}$ can be calculated in terms of shorter-length DFT^{II} blocks. Repeating the same decomposition criteria on the DFT^{III} in equation (3); by replacing k by (k+0.5) in equations (5) to (14); the DFT^{III} is mapped into shorter-length DFT^{IV} blocks. Hence, the mapping algorithm of DFT^{III} into shorter-length (DFT^{IV}) blocks has the same criteria of mapping DFT^{I} into shorter-length DFT^{II} blocks. Fig. 3 shows the mapping criteria for 16-point DFT^{I} (DFT^{III}) into shorter-length DFT^{II} (DFT^{IV}) blocks.

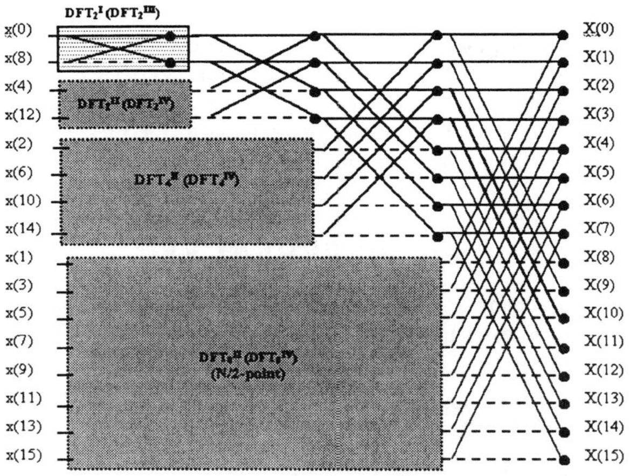

Fig. 3. Mapping 16-point DFT^{I} (DFT^{III}) into shorter-length DFT^{II} (DFT^{IV}) blocks (decimation in time). *Solid lines represent unity transfer factors while dashed lines represent transfer factors –1. Small solid circles represent addition.*

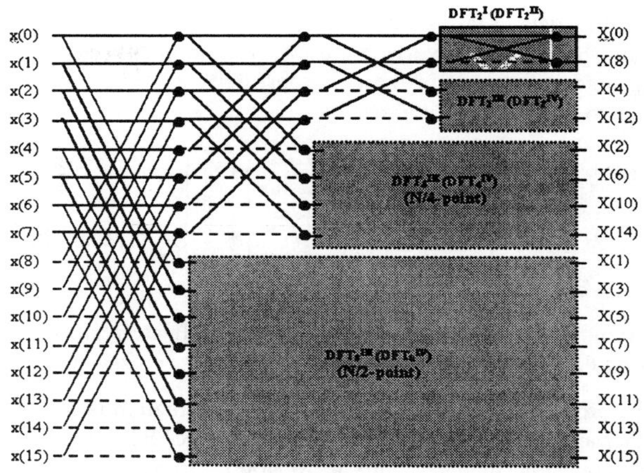

Fig. 4. Mapping 16-point DFT^{I} (DFT^{II}) into shorter-length DFT^{III} (DFT^{IV}) blocks (decimation in frequency). *Solid lines represent unity transfer factors while dashed lines represent transfer factors –1. Small solid circles represent addition.*

2.2 Mapping DFT^{I} (DFT^{II}) into shorter-length DFT^{III} (DFT^{IV}) blocks:

Separating the output of the $\mathbf{DFT}_N^I$ (X(k)) in equation (1) into two equations, one for the odd-index output transformed data (X(2k+1)) and the other for the even-index (X(2k)), gives the next equations:

$$\mathbf{X}(2k+1) = \sum_{n=0}^{N-1} x(n) e^{-j\frac{2\pi}{N}n(2k+1)}, \quad k = 0,..,N/2-1. \tag{15}$$

$$\mathbf{X}(2k) = \sum_{n=0}^{N-1} x(n) e^{-j\frac{2\pi}{N}n(2k)}, \quad k = 0,..,N/2-1. \tag{16}$$

Putting $X'(k) = X(2k+1)$ in (15), and $X''(k) = X(2k)$ in (16):

$$X'(k) = \sum_{n=0}^{N-1} x(n) e^{-j\frac{2\pi}{N}n(2k+1)}, \quad k = 0,..,N/2-1. \tag{17}$$

$$X''(k) = \sum_{n=0}^{N-1} x(n) e^{-j\frac{2\pi}{N}n(2k)}, \quad k = 0,..,N/2-1. \tag{18}$$

Following similar mathematical manipulation as in Section 2.1, we arrive at:

$$\mathbf{X}'(k) = \sum_{n=0}^{N/2-1} (x(n) - x(n+N/2)) e^{-j\frac{2\pi}{N/2}(k+0.5)n}, \quad k = 0,..,N/2-1. \tag{19}$$

and

$$X''(k) = \sum_{n=0}^{N/2-1} (x(n) + x(n+N/2)) e^{-j\frac{2\pi}{N/2}kn}, \quad k = 0,..,N/2-1. \tag{20}$$

As, equations (19), and (20) represent N/2-point DFT^{III}, and N/2-point DFT^{I}, respectively. Hence, by recursive decomposing of equation (20), the DFT^{I} can be calculated in terms of shorter-length DFT^{III} blocks. Repeating the same decomposition criteria on the $\mathbf{DFT}_N^{II}$ in equation (2); by replacing n by (n+0.5) in equations (15) to (20), the DFT^{II} is mapped into shorter-length DFT^{IV} blocks. Hence, the mapping algorithm of DFT^{II} into shorter-length DFT^{IV} blocks has the same criteria of mapping DFT^{I} into shorter-length DFT^{III} blocks. Fig. 4 shows the mapping criteria for 16-point DFT^{I} (DFT^{II}) into shorter-length DFT^{III} (DFT^{IV}) blocks.

3. COMPARISONS AND REALIZATIONS

3.1 Computation Complexity

Using Fig. 2, which has the 2, 4, 8-point DCT^{II}/DST^{II} components, Fig. 4 is expanded in Fig. 5, in terms of DCT/DST blocks instead of DFT^{III} blocks.

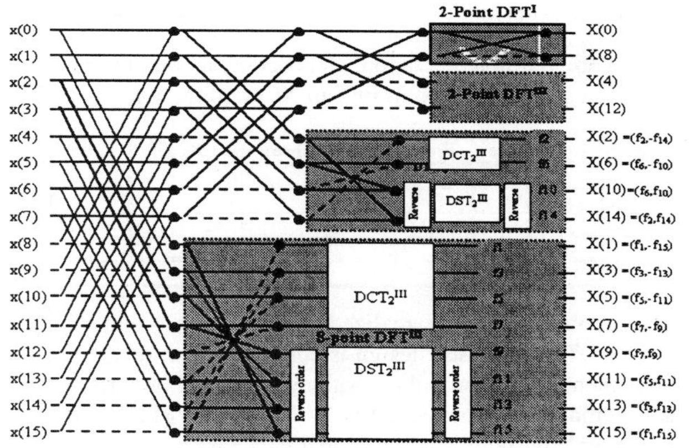

Fig. 5. Mapping 16-point DFT^{I} into shorter-length DFT^{III} blocks. *Solid lines represent unity transfer factors while dashed lines represent transfer factors –1. Small solid circles represent addition.*

From Figs. 3, and 4, it is clear that mapping one type of GDFT into another type, which has shift by 1/2 in time, follows the pattern given in Fig. 3. Wherever, mapping one type into another with 1/2 shift in frequency follows the pattern given in Fig. 4. Figures 3 and 4 are more regular and simpler than Fig. 1 [3], which calculates the DFT^{I}. Figs. 3, and 4 give the same complexity of DCT, and DST blocks as in Fig. 1, but with higher degree of regularity and without need of permutation blocks.

3.2 Hardware Realization

An 8-point FFT network with 8-bit complex data and 8-bit coefficient is realized on FPGA Xilinx XC4000. A serial pipelined real data 8-point FFT consumes half the area of that of the complex data case. In the complex data 8-point FFT network, two real 8-point FFT networks are used, see Fig. 6, processing the real and imaginary parts of the data separately, and extra last stage is used to compute the final results.

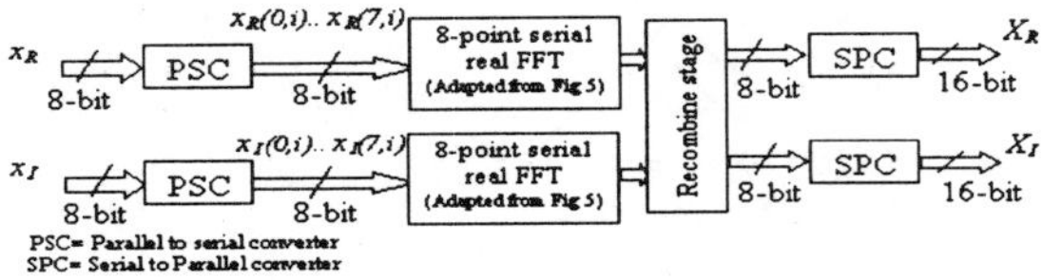

Fig. 6. Block Diagram of the proposed hardware realization of 8-point FFT network in XC4000.

The proposed 8-point FFT network implementation is compared to other approaches for the design of an 8-point FFT network with 8-bit precision mapped onto a XC4000 series. In [10], radix-2 twiddle factors for a butterfly are precomputed and contained within lookup tables. In [12], FFT distributed arithmetic (DA) sub-equations are precomputed and contained within lookup tables. In [11], a DA approach using 4-operand multiply-accumulate modules is presented. In [13], an on-line FFT network implementation, using complex on-line arithmetic, based on adopting a redundant complex number system to represent complex operands as a single unified number. Table 1 shows the implementation results of the compared realizations, input vector feeding rate (throughput) in clock cycles, and the latency of computing the first output vector in cycles. The proposed realization has the fastest execution time, and the lowest cost in terms of CLB's, half of that of the alternative approaches.

Table 1 Implementation results comparison of complex data 8-point FFT in XC4000 family.

Method	CLB's	Freq. /MHz	Throughput/Cycle	Latency/Cycle
[10]	684	N. A.	N. A.	N. A.
[11]	596	118	32	40
[12]	432	N. A.	N. A.	N. A.
[13]	206	42	16	25
Proposed	127	91	16	24

The proposed design is realized using a radix-4 butterfly process element. The proposed design is compared with recently proposed implementation of an online-arithmetic radix-4 butterfly processor, using embedded Block RAM (BRAM) in XCV812E, which is used to calculate 64-point FFT and satisfies the 4 microseconds required calculation time for Orthogonal Frequency Division Multiplexing in HIPERLAN2 [14]. The proposed realization has half the area of that in [14], as shown in Table 2. Since the power consumption is proportional to the design area and frequency, hence the two designs have similar power consumption.

Table 2 Implementation of 64-point FFT in XCV812E

Method	Slices	Freq. /MHz	BRAM	64-point execution time
[14]	443	150	3	2.88 μ sec.
Proposed	242	210	3	3.657 μ sec.

4. SUMMARY

Regular and simple arithmetic mapping algorithms for transforming the different types of GDFT are presented. The proposed mapping algorithms are used to build regular and real twiddle factors FFT algorithms, based on the real arithmetic GDFT presented in [3]. The FPGA-based hardware realizations are implemented on Xilinx XC4000 and Vertix series. Cost comparisons with alternative approaches are presented. Due to regularity of the structure and real factors of the FFT algorithm, a significant improvement in hardware resources for FPGA realization is achieved.

5. REFERENCES

[1] Soo-chang P., and tzyy-Liang L., "Split-radix generalized fast Fourier transform", Signal Processing 54, 1996, pp.137-151.

[2] Guoan B., and Yanqui C., "Fast generalized DFT and DHT algorithms", Signal Processing 65, 1998, pp 383-390.

[3] Britanak, V., and Rao, K. "The fast generalized discrete Fourier transforms: A unified approach to the discrete sinusoidal transforms computation", Signal Processing 79, 1999, pp. 135-150.

[4] Wang Z., "Fast Algorithms for the Discrete W Transform and for the Discrete Fourier Transform", IEEE Trans. On Acoustics, Speech, and Signal Processing, Vol. ASSP-32, No. 4, August 1984, pp. 803-816.

[5] Cooley J. W., and Tukey J., "An Algorithm for the Machine Calculation of Complex Fourier Series", Math. Comput., Vol. 10, April 1965, pp. 297-301.

[6] Vetterli M., and Duhamel P., "Split Radix Algorithms for Length-P^m DFTs", IEEE Trans. On Acoustics, Speech, and Signal Processing, Vol. 37, No. 1, January 1989, pp. 57-64.

[7] Jia L., Gao Y., and Tenhunen H., "A New VLSI-Oriented FFT Algorithm and Implementation", in Proc. of IEEE International ASIC Conference, Rochester, USA, September 1998, pp.337-341.

[8] Suzuki Y., Sone T., and Kido K., "A New FFT Algorithm of Radix 3, 6, and 12", IEEE Trans. On Acoustics, Speech, and Signal Processing, Vol. 34, No. 2, April 1986, pp. 380-383.

[9] Martens J., "Recursive Cyclotomic Factorization- A New Algorithm for Calculating the Discrete Fourier Transform", IEEE Trans. On Acoustics, Speech, and Signal Processing, Vol. 32, No. 4, August 1984, pp.750-761.

[10] Mintzer L., "The XILINX FPGA as an FFT processor", Electronic Engineering, 69, May 1997, pp. 81-84.

[11] Lau D., Schneider A., Ercegovac M., and Villasenor J., "FPGA-based structure for on-line signal processing". Journal of VLSI Signal Processing, 1999, pp. 310-311.

[12] Helal H., Mashali S., and Salama A., "A new implementation of FFT on FPGA using distributed arithmetic", proceedings of the 4th IEEE international Conference on Electronics, Circuits and Systems, Egypt, Dec. 1997, pp. 1437-1443.

[13] McIlhenny R., and Ercegovac M., "On the Design of an On-line FFT Network for FPGA's", proceedings of the 33rd Asilomar Conference on signals, systems, and computers, USA, vol. 2, Oct. 1999, pp. 1484 -1488.

[14] Perez-Pascual A., Sansolani T., Valls J., "FPGA-BASED RADIX-4 BUTTERFLIES FOR HIPERLAN/2", ISCAS 2002, proceedings III, Arizona, USA, pp. 277-280.

A HIGH-SPEED, LOW LATENCY RSA DECRYPTION SILICON CORE

Ciaran McIvor, Máire McLoone, John V McCanny

DSiP Laboratories, School of Electrical and Electronic Engineering,
The Queen's University of Belfast, Northern Ireland.
E-Mail: c.mcivor@ee.qub.ac.uk, maire.mcloone@ee.qub.ac.uk, j.mccanny@ee.qub.ac.uk

Abstract

This paper introduces a novel and generic approach to the hardware implementation of the RSA decryption function, which may be used to create digital signatures in an RSA based signature scheme. The algorithm used for modular multiplication is Montgomery's multiplication algorithm. The design is speed optimised and as such employs the R-L binary method as a means for modular exponentiation. An RSA decryption can be performed in only $(k/2 + 3)^2$ clock cycles, where k is the size of the modulus, by employing carry save adders in order to achieve fast parallel addition and the Chinese Remainder Theorem to speed up exponentiation. To the authors' knowledge, this is the lowest number of clock cycles required for any radix 2 based RSA decryption system reported in the literature. As such the design can achieve a data throughput rate of 234.47 kb/s for a 512-bit modulus and a rate of 90.58 kb/s for a 1024-bit modulus when implemented onto a Xilinx Virtex2 XC2V8000 chip.

1. Introduction

The increasing use of e-commerce and the increase in demand for secure communications over the Internet, for example, has led to an ever-greater demand for reliable high-speed security products, which can carry out data encryption in real time. In particular, public-key cryptosystems provide a mechanism for secure symmetric key exchange, as well as being a useful method for creating and verifying digital signatures.

The RSA system [1] is one of the most commonly used public-key cryptosystems and is the subject of this paper. An RSA operation is essentially a modular exponentiation, which in turn requires modular multiplication. Also, in order for an RSA cryptosystem to be considered secure, it is widely recognised that key sizes should be in excess of 1024-bits [2], [3] which, in many cases, can only be achieved through direct hardware implementation. Thus, it follows that RSA cryptography is extremely computationally intensive.

This paper therefore introduces a novel hardware design aimed at tackling this problem, with a view to achieving a high RSA exponentiation throughput rate when implemented in silicon. The focus of the paper is on RSA decryption and all results are given in relation to the decryption function. The RSA private exponent d may be as large as $[(p-1)(q-1)-1]$, where p and q are two large prime numbers and the k-bit modulus n is equal to $p*q$. Thus, it is assumed that the private exponent d is a k-bit number. This is in contrast to the RSA public exponent e, which can be chosen arbitrarily (usually 3, 17 or $2^{16}+1$) as long as it satisfies the conditions stated in section 2 of this paper. In general, therefore, the RSA decryption function is more computationally intensive than the RSA encryption function. Also, RSA decryption is usually the most common operation carried out in applications such as smart cards. However, as both functions essentially perform identical arithmetic operations, the approach described in this paper can also be utilised to carry out encryption.

Previous RSA designs have mainly focused on two particular strategies. Firstly, the Montgomery multiplication algorithm [4], [5], [6] has been used alongside a redundant number representation [7], [8] in order to avoid long carry propagation. Secondly, a number of 2-D systolic array designs have been proposed [9], [10] for use alongside the Montgomery multiplication algorithm and typically consist of a k×k matrix of 1-bit processing elements. However, these arrays have proved costly to implement due to the amount of resources required since RSA cryptosystems typically need a modulus of around 1024-bits in length.

The approach described here utilises the Montgomery multiplication algorithm [4], [5], [6] for carrying out the modular multiplication operation. The corresponding modular exponentiation algorithm is the well-known R-L binary method [11], [12]. The design also makes extensive use of carry save adders as a means of fast, parallel addition. By combining these along with the Chinese Remainder Theorem (CRT) [13], an architecture has been developed, which has a latency of only $(k/2+3)^2$ clock cycles per decryption. To the authors' knowledge this is the lowest number of clock cycles required for any radix 2 based RSA decryption system reported in the literature. Carry propagation is avoided by using carry save adders and as such the corresponding combinational logic between clock cycles means that a good clock speed can still be achieved. Thus, since an RSA decryption requires a very low number of clock cycles, a high data throughput rate is achievable.

The concepts developed have been verified and demonstrated through an FPGA implementation, in this case using a Xilinx Virtex2 XC2V8000 chip. However, as will be discussed, the design is completely technology independent and can therefore just as easily be implemented onto ASIC technology.

The paper is arranged as follows. The RSA cryptosystem and modular exponentiation are summarised in section 2. The silicon implementation is discussed in section 3. Performance results are provided in section 4 and some concluding remarks are given in section 5.

2. Background Information

2.1 RSA Cryptography

The RSA cryptosystem as proposed by Rivest, Shamir and Adleman in 1978 [1] is summarised as follows:

0-7803-7761-3/03/$17.00 ©2003 IEEE

Step 1: Two prime numbers p and q are generated and their product is calculated, denoted by, $n = p*q$. n is called the modulus.
Step 2: A positive integer e is then determined, satisfying,
$3 \leq e < (p-1)(q-1)$;
e must also be relatively prime to $(p-1)(q-1)$.
Step 3: e is then used to determine another positive integer d for which, $ed = 1 \ (mod\ (p-1)(q-1))$. i.e. d is the multiplicative inverse of e, modulo $(p-1)(q-1)$.
Step 4: Thus, the public key is represented as the pair (e, n) and the private key is represented as the pair (d, n).
Step 5: If M represents the plaintext and C represents the ciphertext, then encryption is carried out using the encryption function $C = M^e$ (mod n). Decryption is carried out in an almost identical manner using the function $M = C^d$ (mod n).

The underlying mathematics demonstrating the correctness of the RSA algorithm is given in [1]. It should also be noted that no public-key cryptosystem has ever been proven to be secure. The security of public-key systems is based on the presumed difficulty of a small set of number theory problems. The security of the RSA system lies in the fact that it would be computationally infeasible to calculate the value of d given only the public key pair (e, n), assuming that large enough prime numbers p and q are used in determining e, d and n. This would be equivalent to trying to factorise a 1024-bit number n, which theoretically has only two factors p and q. This will eventually be achieved but at present with current technology it is sufficient that n should be between 1024-bits and 2048-bits in length depending on the level of security required in the system.

2.2 Montgomery Multiplication

This section describes the Montgomery multiplication algorithm, which originated from Montgomery's paper published in 1985 [4]. In order for the Montgomery multiplication algorithm to operate correctly it is necessary for the radix to be relatively prime to n. This condition is satisfied here as n is clearly the product of two large prime numbers and the radix is equal to 2.

Now, given an integer $a < n$, α is said to be its n-residue with respect to r if, $\alpha = a*r \ (mod\ n)$, where $r = 2^k$. Likewise, given an integer $b < n$, β is said to be its n-residue with respect to r if, $\beta = b*r \ (mod\ n)$. The Montgomery product of α and β can then be defined as, $z = \alpha*\beta*r^{-1} \ (mod\ n)$, where r^{-1} is the inverse of r, modulo n. A variant of Montgomery's multiplication algorithm [5], which computes the Montgomery product of α and β, is presented in pseudo code below. All numbers are represented in binary form where n is the k-bit modulus, α and β are the k-bit operands and $\alpha, \beta < n$.

Algorithm 1: *Montgomery Multiplication* (α, β, n)
S[0] <= (others => '0');
for i in 0 **to** k-1 **loop**
q_i <= $(S[i]_0 + \alpha_i * \beta_0)$ **mod** 2;
$S[i+1]$ <= $(S[i] + \alpha_i * \beta + q_i * n)$ **div** 2;
end loop;
return S;

2.3 The R-L Binary Method

The R-L binary method utilizes the Montgomery multiplication algorithm, which determines its intermediate results. Some on-chip pre-computation is required in order to convert the ordinary residue inputs to the chip into their n-residue form for use with algorithm 1, as described in section 2.2. In order to calculate the correct n-residues, it is necessary to pre-compute the value 2^{2k} $(mod\ n)$ externally, which can then be stored alongside the relevant private or public keys. This value is relatively inexpensive to compute as once it is determined it need only be changed if the value of the modulus n changes. The value 2^{2k} $(mod\ n)$ is used to convert (on-chip) the ordinary residue operands a and b into their n-residue values α and β respectively as follows:

$$\text{MontMult}(a, 2^{2k} \ (\text{mod } n), n) = a*2^{2k}*r^{-1} \ (\text{mod } n) = a*r \ (\text{mod } n) = \alpha \qquad \text{(i)}$$

b is converted to β in an identical manner. Also, it can be shown that the Montgomery product of two n-residues is indeed an n-residue itself as follows:

$$z = \text{MontMult}(\alpha, \beta, n) = a*r*b*r*r^{-1} \ (\text{mod } n) = a*b*r \ (\text{mod } n) \qquad \text{(ii)}$$

Therefore, as modular exponentiation here is simply a series of Montgomery multiplications, then the result from the loop statement of algorithm 2 below will also be in n-residue form (i.e. $R_k = M*r \ (mod\ n)$). Thus, a further Montgomery multiplication is required in order to convert this result back into its ordinary residue form as follows:

$$M = R_k*r^{-1} \ (\text{mod } n) = 1*R_k*r^{-1} \ (\text{mod } n) = \text{MontMult}(1, R_k, n) \qquad \text{(iii)}$$

Now, if the computation to be carried out is $M = C^d$ (mod n), where d is the k-bit private exponent (as d may be as large as $[(p-1)(q-1)-1]$), then the following modular exponentiation algorithm may be used.

Algorithm 2: *Modular Exponentiation (C, d, n)*
$K = 2^{2k}$ (mod n); (computed externally)
P_0 <= MontMult (K, C, n);
R_0 <= MontMult (K, 1, n);
for i in 0 to k-1 **loop**
if $d_i = 1$ **then** R_{i+1} <= MontMult (R_i, P_i, n) **end if**;
P_{i+1} <= MontMult (P_i, P_i, n);
end loop;
M <= MontMult (1, R_k, n);
return M;

2.4 The Chinese Remainder Theorem

The basic application of the CRT to achieve faster RSA decryption is outlined in this section. A more detailed explanation of the background is given in [13]. As outlined in section 2, the RSA decryption function is given by $M = C^d$ (mod n), which in turn can be written as, $M = C^d$ (mod p*q), since $n = p*q$. Fundamentally, the CRT allows this computation to be broken down into two smaller, less complex computations given as follows,

$$M_p = C_p^{Dp} \ (\text{mod } p)$$
$$M_q = C_q^{Dq} \ (\text{mod } q)$$

where

$$C_p = C \pmod p,\ D_p = d \pmod{p-1}$$
$$C_q = C \pmod q,\ D_q = d \pmod{q-1}$$

The final value M is then obtained using the equation,

$$M = [(M_q - M_p)(p^{-1} \pmod q)) \pmod q]p + M_p \qquad \text{(iv)}$$

3. Novel Hardware Implementation

The approach used to implement the RSA system in hardware is based on algorithm 2 in section 2.3. This incorporates algorithm 1 in section 2.2. The CRT is also applied to significantly speed up the RSA decryption computation. A modulus size of 512-bits has been chosen here for illustrative purposes. However, by capturing this generically (in VHDL) the modulus size is completely variable and is easily changed by altering two parameters.

3.1 Montgomery Exponentiation Overview

Figure 3.1 provides an overview of the Modular exponentiation design approach used. Even though the modulus size is 512-bits, all major inputs to the exponentiation core are only 256-bits due to the use of the CRT. This clearly illustrates the speed advantage that CRT based designs hold over alternative structures. The signals p and q represent the two large prime numbers that form the modulus $n = p*q$. The signals C_p and C_q correspond to the ciphertext, which is to be decrypted. The symbols D_p and D_q refer to the corresponding values outlined in section 2.4. The signal *p_constant* is the externally computed value $2^{2(k/2)} \pmod p$ and corresponds to the value $2^{2k} \pmod n$ as discussed in section 2.3. Likewise, the signal *q_constant* signifies the value $2^{2(k/2)} \pmod q$. When the *ready* signal is set to high this indicates that the correct values of the decrypted ciphertext M_p and M_q should now be registered. The component *COUNTER* essentially implements the iterative loop statement of algorithm 2 in section 2.3.

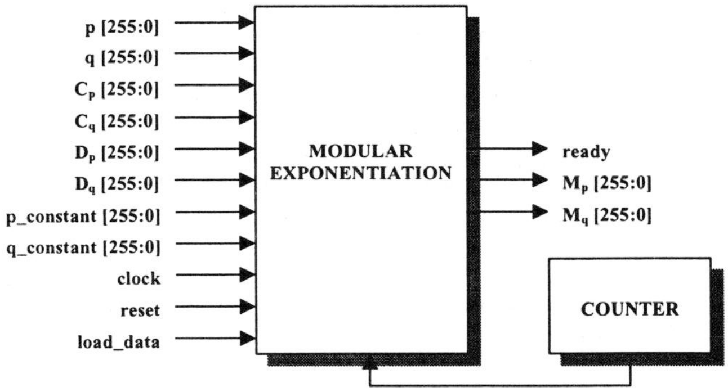

Figure 3.1: Overview of Modular Exponentiation Core

3.2 Montgomery Multiplication Overview

Figure 3.2 gives an outline of the Montgomery multiplier architecture developed. Three rounds of carry save logic are utilised to construct a 5-input carry save adder, which is used to calculate the intermediate results S from algorithm 1 in section 2.2. These intermediate results are kept in redundant form until the loop statement of algorithm 2 in section 2.3 has terminated, at which point the redundant representation of the output from the exponentiation algorithm is added together using a 1-D array of full adders. The 5-input carry save adder requires only one clock cycle to produce its output and so each iteration of the loop statement of algorithm 1 requires only one clock cycle to compute.

The inputs *a1*, *a2*, *b1* and *b2* correspond to the α and β inputs of algorithm 1. Again the iterative 'for loop' is represented by the component *COUNTER* in figure 3.2. The function of the array of *FA* (full adder) cells are to perform a summation of the carry save representation of the α input to algorithm 1. This is to ensure that the correct values of the α_i are used at each iteration of algorithm 1. These values are registered in the *FA REGISTER* but are readily available when they are required at each iteration of the loop in algorithm 1.

The main purpose of the *Q_ADDER* component is to determine the q_i values of algorithm 1. These, in turn, are fed into a multiplexer to determine whether or not the modulus should be fed into the *CSA* component along with other relevant input values needed to calculate the intermediate values S at each iteration of the loop statement in algorithm 1.

When algorithm 2 has terminated the correct outputs from the *CSA REGISTER* are then added together to form the correct output value M_p as shown in figure 3.1. The output value M_q is computed using a separate Montgomery multiplier that operates in parallel with the M_p calculations.

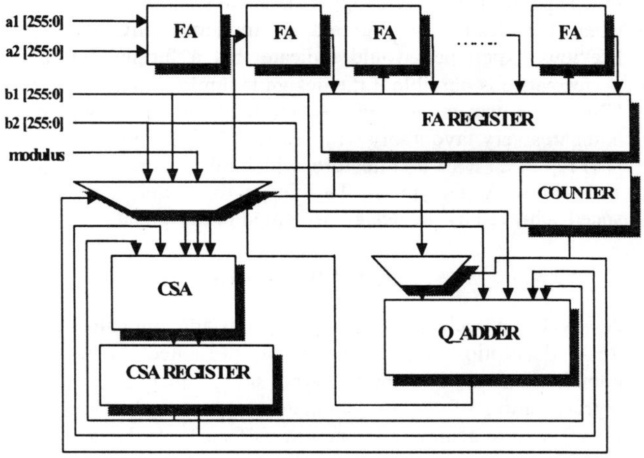

Figure 3.2: Overview of Montgomery Multiplication Component

4. Performance Results

The system described has been captured generically using VHDL and used to create a silicon demonstrator using a Xilinx Virtex2 XC2V8000BF957-5 device. Table 1 provides performance and area values obtained for varying bit lengths k.

The performance results can be compared with those recently reported in the literature. Daly and Marnane [11] recently reported a rate of 115.5 kb/s for a 480-bit modulus and 45.8 kb/s for a 1080-bit modulus, which are only half the speed of the equivalent throughput values reported in Table 1. Their design was targeted for a Xilinx Virtex V1000FG680-6 device.

Bit Length (k)	No. of CLB Slices	Data Throughput Rate	Time For One RSA Decryption (ms)
128	3110	1.51 Mb/s	0.08
256	6147	537.63 kb/s	0.48
512	12041	234.47 kb/s	2.18
1024	24174	90.58 kb/s	11.30

Table 1: Performance and Area Results Obtained

Blum and Paar [2] presented an FPGA systolic array implementation and were able to process a 1024-bit RSA decryption in 10.18ms, 12.41ms and 12.52ms dependent on the size of the processing element u=4, u=8 and u=16 respectively. These times are comparable to the 1024-bit decryption time reported in Table 1. Their design was implemented onto the Xilinx XC4000 series of FPGAs. It should be noted that these implementations may yield better performance results if implemented onto the latest Xilinx FPGAs such as the Virtex2 family of devices. However, a major drawback of their implementations is that they are technology dependent and targeted to exploit certain features of specific Xilinx devices. For example, Daly and Marnane [11] exploit the fast carry chains of their chosen Xilinx Virtex device. Similarly, Blum and Paar [2] rely on the dedicated hardware within their target device, which accelerates the carry path of adders and counters. This is in contrast to the design presented, which is completely technology independent. Thus, even higher data rates should be achievable when the design is migrated to modern ASIC technology. Previous experience would indicate that a further doubling in performance is possible if the design is implemented in a 0.18μm CMOS. However, the FPGA implementation presented still compares very favourably with recent VLSI designs. Kwon *et al.* [14] report an RSA architecture, which takes 22ms to complete a 1024-bit RSA decryption. This is approximately only half the speed achieved by the design presented in this paper.

5. Conclusion

A novel approach to the silicon design and implementation of RSA decryption cores has been presented. This uses a combination of a 5-input carry save adder and redundant representation of the intermediate results of algorithm 1 and algorithm 2 to achieve a latency of $(k/2 + 3)^2$ clock cycles, where k is the length of the modulus. We believe that this latency is the smallest reported to date. This combined with a shallow logic depth, achieved using the fast, parallel carry save approach, produces a circuit with very impressive throughput rates. This has been demonstrated through a generic implementation on a Xilinx Virtex2 XC2V8000 FPGA. The design is also completely technology independent and can be migrated to ASIC or other FPGA / PLD technologies.

6. Acknowledgements

This research has been funded by Amphion Semiconductor Ltd. and by a Northern Ireland Department of Learning postgraduate studentship in the form of a CAST award.

7. References

[1] R.L. Rivest, A. Shamir, L. Adleman: "A Method for Obtaining Digital Signatures and Public-Key Cryptosystems". Communications of the ACM, 21(2): 120–126, February 1978.

[2] T. Blum, C. Paar: "Montgomery Modular Exponentiation on Reconfigurable Hardware". Proceedings 14th Symposium on Computer Arithmetic, pp. 70–77, 1999.

[3] A. J. Elbirt, C. Paar: "Towards an FPGA Architecture Optimized for Public-Key Algorithms". SPIE Symposium on Voice, Video and Communications, Sept 1999.

[4] P.L. Montgomery: "Modular Multiplication without Trial Division". Math. Computation, Vol. 44, 1985, pp. 519–521.

[5] C.D. Walter: "Montgomery Exponentiation Needs No Final Subtractions". Electronics Letters, 35(21): 1831–1832, October 1999.

[6] C.K. Koc, Tolga Acar, Burton S. Kaliski Jr.: "Analyzing and Comparing Montgomery Multiplication Algorithms". IEEE Micro, 16(3): 26–33, June 1996.

[7] C.D. Walter: "Fast Modular Multiplication using 2-Power Radix". International Journal of Computer Mathematics, Vol. 3, pp. 21–28, 1991.

[8] S.E. Eldridge, C.D. Walter: "Hardware Implementation of Montgomery's Modular Multiplication Algorithm". IEEE Transactions on Computers, Vol. 42, pp. 693–699, July 1993.

[9] A. Tiountchik: "Systolic Modular Exponentiation via Montgomery Algorithm". Electronics Letters, Vol. 34, pp. 874–875, April 1998.

[10] C.D. Walter: "Systolic Modular Multiplication". IEEE Transactions on Computers, Vol. 42, pp. 376–378, Mar 93.

[11] A. Daly, W. Marnane: "Efficient Architectures for implementing Montgomery Modular Multiplication and RSA Modular Exponentiation on Reconfigurable Logic". FPGA 2002, February 2002.

[12] C.K. Koc: "High-Speed RSA Implementation". Technical Report, RSA Laboratories, RSA Data Security, Inc., Redwood City, CA, 1994.

[13] J. Quisquater, C. Couvreur: "Fast Decipherment Algorithm for RSA Public-Key Cryptosystem". Electronics Letters, Vol. 18, pp. 905–907, October 1982.

[14] Taek-Won Kwon, Chang-Seok You, Won-Seok Heo, Yong-Kyu Kang, Jun-Rim Choi: "Two Implementation Methods of a 1024-bit RSA Cryptoprocessor Based on Modified Montgomery Algorithm". IEEE International Symposium on Circuits and Systems, pp. 650–653, May 2001.

eUTDSP: A Design Study of a New VLIW-Based DSP Architecture

G. R. Chaji, R. M. Pourrad, S. M. Fakhraie
VLSI Circuits and Systems Laboratory
The Univ. of Tehran
Tehran 14399, Iran
{rzchaji,pourrad}@yahoo.com, fakhraii@ut.ac.ir

M. H. Tehranipour
Center for Integrated Circuits & Systems
The Univ. of Texas at Dallas
Richardson, TX 75083
mht021000@utdallas.edu

Abstract

This paper presents a new DSP architecture called eUTDSP that is based on a traditional VLIW architecture. It is able to perform maximum of 4 instructions per cycle with a 128-bit instruction word size. VLIW systems usually suffer from the disadvantage of larger program memories due to longer instructions. As well, branches usually have some added delay slots and this also causes more drawbacks. In order to solve such problems, some architectural solutions are presented in this paper. Benchmarking results obtained from VHDL and C++ models show the efficiency of eUTDSP for better utilization of functional units in each cycle, code size shirinking and reducing branch/loop overheads.

1 Introduction

Very Long Instruction Word (VLIW) architecture is one of the highest performance architectures available for DSP processors. VLIW architecture provides predictable, efficient, higher performance, and benefit from a mature compiler technology. A VLIW processor can execute multiple instructions simultaneously. However, it is relying depends on the compiler and programmer to find the parallelism in the instructions. In VLIW architecture, the instruction scheduling is performed by software at the compilation time prior to the execution of the code [1] [2]. This leads to simpler VLSI implementations since most of the complexity has been moved from hardware to software, while allowing executation of more instructions in parallel.

There are several characteristics in the VLIW that cause some performance drawbacks. VLIW usually suffers from the disadvantage of larger program memory due to longer instructions [3]. This happens because every computation elements has to get a command in every cycle, even if it should do nothing. However, this can be overcome by inserting NOP instructions directly by the hardware. This problem also decreases the performance of a VLIW DSP. As well, branches usually have some delay slots and this also causes more drawbacks. By increasing the number of data path elements and the length of the VLIW instructions in order to enhance performance, the existing optimized DSP software needs to be rewritten. It impacts not only the time-to-market of a new DSP product, but also the lead time in enhancing custom-made DSP applications. In particular, some critical hand-written assembly code must be rewritten. As a results, it presents a big challenge to applications programmers and DSP compiler designers [4].

Recently, several high-performance DSP processors have been introduced based on VLIW architecture such as TI's C6x [5] and ADI's Tiger SHARC [6]. Although today's silicon chips are dense enough to allow us build cost effective single-chip VLIW-based DSP processors, there are still some technical limitations. Several research groups have worked on VLIW architecture [7] [8]. In [7], a low-cost VLIW DSP architecture is presented. It is able to perform maximum of 8 instructions per cycle with only a 64-bit instruction word by fetching the address of the previously stored instructions instead of the entire instruction word. Since loops are in the critical paths in most of DSP applications, some optimization approaches have been reported in [4] and [9].

In order to solve such problems, some architectural solutions are presented in this paper. These are implemented on a VLIW DSP that has been designed in VLSI Circuits and Systems Laboratory, University of Tehran to be code-compatible with the instruction set of TMS320C62x of TI. It executes maximum of 4 instructions per cycle with a 128-bit instruction word. Some features are added to increase the performance for using of this DSP in data communication applications. Our proposed architectural improvements are implemented by adding several instructions to, by redesigning the pipeline flow and adding data and instruction caches of the original architecture to overcome the above mentioned performance drawbacks of a VLIW architecture. The benchmarking results show the efficiency of the using of functional units in the proposed DSP when compared to the published performance of C62x [2].

The rest of this paper is organized as follows. Section 2 shows a VLIW-based architecture called UTDSP (University of Tehran DSP). In Section 3, several enhancments are presented and implemented on UTDSP. Simulation results are shown in Section 4. Conclusion ends the paper in Section 5.

2 UTDSP: A VLIW-Based DSP

UTDSP is a 32-bit fixed-point DSP based on a VLIW architecture. Using a VLIW architecture for UTDSP makes it capable of multifunctional applications. UTDSP supports the instruction set of TMS320C62x from TI [5]. A major difference is the width of memory access bus that is 256 bits in C62x and 128 bits in UTDSP. It has a four-parallel-issue data path with four parallel functional units and its computational power

0-7803-7761-3/03/$17.00 ©2003 IEEE

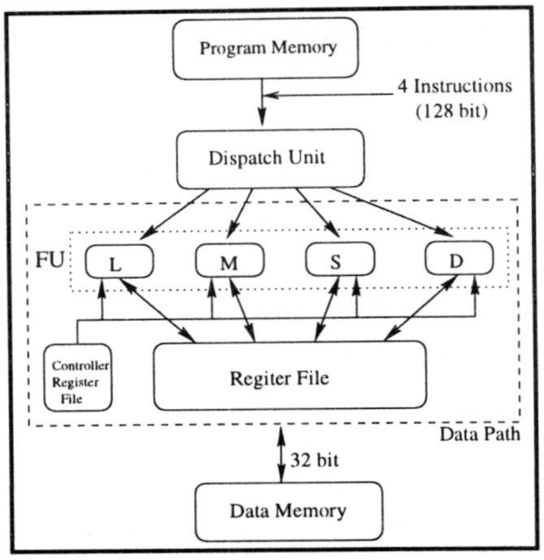

Figure 1. UTDSP Architecture with 4 functional units.

has been tuned for 500-800 MIPS portable DSP applications when implemented in a 0.25 mm CMOS or compatible technology. As well, its functional units (FUs) are more likely to be used efficiently as our results will reveal. For its relative simplicity, it is also a good starting point for applying different modifications to enhance the performance. Figure 1 shows the data path of this processor.

Four parallel 32-bit instructions are fetched in the fetch phase of UTDSP. Fetch phase is performed in four clock cycles. Decode phase of the pipeline consists of two cycles. In the first cycle that is called dispatch, the job of assigning instructions to FUs is performed and major control signals accompanying the generated execute packets are produced. After dispatch phase, instructions of the execute packets are processed in the decode phase of FUs, in which the control signals for each FU are generated with respect to the instruction.

After decode phase, instructions enter into the FUs. L unit is responsible for arithmetic and logic operations. M unit does the multiply operations and S unit does the branch, shift and bit manipulation and some arithmetic and logic operations. D unit performs memory accesses and address calculations. All data transfers between the register files and the memory are performed by D unit. Multiply takes 2 cycles, store takes 3 cycles, and load instructions takes 5 cycles. However, the FUs of these instructions are pipelined and can accept a new instruction at each cycle. All other instructions are one-cycle instructions.

The processor has a 32x32 bit register file that is accessed by the FUs. There is also a control register file including the addressing mode register, status register, and interrupt registers. Modeling of this processor has been performed at RT level and several test benches have been written for verifying the design. After passing initial tests, some more complex test benches have been written to check the clock cycles and some of the performance parameters.

3 eUTDSP: Enhanced UTDSP

A Modified VLIW Architecture based on UTDSP with Enhanced Functionality has been designed. In VLIW architectures, memory latency is the main reason for data conflicts. Compiler or the programmer has to consider fixed memory latency for instruction scheduling. In order to provide real time processing, most DSP processors have the ability to perform the multiply-and-accumulate (MAC) and Add-Compare-Select (ACS) operations in a single instruction cycle [8]. MAC and ACS operations, probably the most often cited features of a DSP processor, are highly useful in many DSP applications such as digital filters, correlation, fourier transform, and viterbi operations. General purpose processors that lack hardware MAC and ACS take several tens of clock cycles to compute them. Simultaneous multi threading in each FU is another way to improve the performance. Complex functional units can employ this feature. For example, when there are two separate blocks for MAC and ACS, we can easily use them together.

A vectored architecture is also useful to remove misleading MIPS numbers and data conflicts. In addition, it reduces the code size and increases the performance. This architecture uses a vectored dual-operation multiplying unit and parallel simultaneous threads for critical operations, such as ACS, MAC and ADD are provided. In addition, it can support two simultaneous threads. Figure 2 shows the block diagram of this architecture that has been called $eUTDSP$. As shown, there are *Trace Cache*, *Loop Table* and *dynamic pipeline* in this architecture for enhancing UTDSP. Also, the functional units have been modified and some operations such as MAC and ACS have been added to them. This architecture will be described in the next subsections.

3.1 Enhancements for UTDSP

Here, we describe the enhancements implemented on UTDSP.

Dynamic pipeline: In a VLIW processor, the length of instructions execution phases is fixed. The reason is that the compiler should consider a previously agreed cycle number for each instruction, and then, schedule a set of instructions. eUTDSP has a *Dcache* and employs a dynamic data pipeline. The compiler considers the number of phases that are needed in the case of a hit access. Processor checks the data conflicts between load instructions and other instructions. In the case of a conflict, processor stops the execution and waits until data is prepared. loop instructions are checked in Loop Checking block. After that some information related to loop instruction are issued to Loop Table block.

Trace Cache: The processor has a cache for instructions; in which a trace of program is stored. Noteworthy, in trace cache only those instructions that pass execution stage are saved, not those whose might be rejected after a loop or branch operation. Therefore, the cache performance increases. This unit is used for loop and branch and provides very good benefits for many DSP algorithms such as FFT, FIR, and Viterbi.

Loop-Table: This block supports the loop structure with minimum delay overhead. The loop instructions are checked by this block. Some information used in Loop Table come from pipeline controller. The information are beginning loop address, ending address, and loop number. For example, C62x uses conditional branches for loops and this instruction needs five-delay slots. These slots should be filled with useful instructions or NOPs, and filling the slots with useful instructions is not always possible. Therefore, these slots are often filled with NOPs. As a result, the time and memory overhead increases. The loop Table is very effective in executing the loops and reduces loop and branch overheads.

3.2 eUTDSP Major Units

The major units of eUTDSP are described in the following:

Instruction Unit: This unit is composed of fetch unit, pipeline controller unit and decode unit. Fetch unit fetches four 32-bit instructions at each time. The first stage generates the address of the next fetch packet. If the address that is generated exists in the Loop Table, next address will be obtained from Loop Table. At the second stage, if the address exists in the cache, fetch packet is fetched from cache and is sent to the next stage. However, if a cache miss occurs, the fetch packet is fetched from instruction memory. The third stage manipulates variable length memory latencies, and at the fourth stage fetch packet becomes ready.

Pipeline control unit determines the members of an execution packet that can be executed in parallel together. Few special instructions are executed in this unit, such as NOP and Loop. If NOP has been detected, the fetch operations are stalled. If loop instruction has been detected, some information is calculated from the instruction and is sent to the Loop Table. At the end of each cycle, instructions of an execute packet are sent to decode unit. Decode unit decodes instructions of execution packet and detects data conflicts between previous load instructions and instructions of the execution packet. If the operands of the instructions are from register files, it reads them.

Execution Unit and File Registers: The instructions that satisfy their conditions are executed in one of the L, M, S, D units. Arithmetic, Logical, and ACS instructions are executed at L unit. The ACS is used for Viterbi decoding or encoding. This instruction returns the superior metrics and its path. The returned path can be used for viterbi encoding. M unit supports vectored dual-multiply operation feature for multiply and MAC instructions. It is very useful for complex multiplications and for two parallel vector MAC operations. The dual MAC provides additional efficiency for performing different kinds of multiply/accumulate operations that is very useful in DSP algorithms especially in Communication applications. This unit has pipeline architecture and supports SIMD operations. There are many useful applications such as FFT for complex multiplying instruction. D unit generates the data address and performs memory accesses. S unit performs arithmetic, logical, branch and bit manipulation instructions. This unit supports the SIMD operations (Single Instruction Multiple Data) for 16 bit addition and subtraction. The FFT butterfly can use this feature.

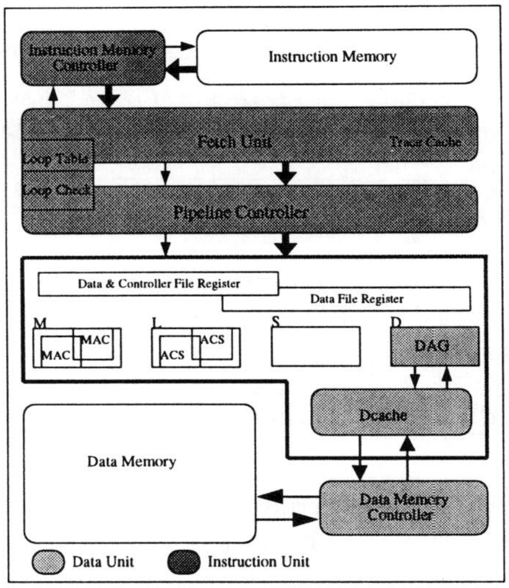

Figure 2. eUTDSP block diagram.

The processor has two 16x32 independent file registers for data and one control register for control and status information. It is useful for multi threading. In addition, the delay of this structure is less than a 32x32 register file.

Data unit: This unit is composed of Dcache, memory controller and data address generation unit (DAG). It can read or write 32, 16 or 8 bits per each memory access. Dcache is a 2-way cache and uses the Least Recently Used (LRU) method for replacing data. The data cache is faster than the data memory, so using it reduces the data conflicts and program cycles. A FIFO has been used in this unit to keep the trace of consecutive misses. So if misses occur serially for byte and half-word memory operations, storage of the existing half of the word in this FIFO makes it instantly available for when the second part of the word is read from memory (to be written into the cache on a single word). This feature is very useful for applications such as computation of 16-bit FFT.

4 Simulation Results

C62x is a fixed-point 32-bit VLIW architecture, which is able to execute maximum of 8-instructions in parallel. There are two identical data paths in C62x and each one has four independent functional units. This parallelism is obtained by using eight parallel FUs [5]. Memory accesses in C62x are performed in 5 cycles for load and 3 cycles for store. Branch instruction has five delay slots. Most of DSP algorithms have several loops. Therefore, this 5-cycle delay slot of the branch instruction causes a significant performance drawback. Some special functions such as ACS and MAC are not directly supported by C62x and must be implemented by software, and this may increase the delay and code size.

After preparing a software model (with C++) and a hardware model (VHDL based [10]) of the UTDSP and eUTDSP,

Table 1. Code size of the programs on DSPs.

	$C62x$	$UTDSP$	$eUTDSP$	*Reduction*
RFIR	80	76	28	65%
CFIR	160	160	32	80%
DCT	350	350	63	83%

Table 2. Number of clock cycles.

	$C62x$	$UTDSP$	$eUTDSP$	*Reduction*
RFIR	3705	3820	1910	48%
CFIR	8787	12852	3516	60%
DCT	15623	21076	5792	63%

the models have been verified by several test benches. Some benchmarks have been chosen to compute the performance parameters for the considered three DSP processors (eUTDSP, UTDSP and C62x). Codes of the benchmarks have been initially written in C62x code composer and then optimized as much as possible [5]. These codes have been directly executed on a C6211 simulator due to compatibility of the instructions of UTDSP and eUTDSP with C62x. Then, some modifications have been applied to the codes so that they could be executed on the UTDSP model. In order to have optimized codes for eUTDSP, further modifications must be performed to use the added features of eUTDSP. The results of benchmarking are presented for two FIR filters such as a Real FIR (RFIR) and a Complex FIR (CFIR) with different complexities and also Discrete Cosine Transform (DCT). The used C++ model also is very flexible and different architectural ideas can be applied to it.

Table 1 shows the code sizes of the three programs on three DSPs. As seen, with increasing the complexity of the programs, code size of C62x and UTDSP has intensively increased in comparison to eUTDSP. Table 1 also shows the code size reduction percentage for eUTDSP in comparison to C62x. As shown, with increasing the complexity of the programs, the number of code size for eUTDSP has been reduced by 83% in comparison to the case of C62x. Table 2 shows the number of clock cycles required for execution of each program on each DSP. The clock cycle for eUTDSP is significantly lower when compared to the other two DSPs. The fifth column of the table shows the reduction percentage of the required clock cycle for each program in the case of eUTDSP compared to C62x. It shows 63% of clock cycle reduction for executing the DCT by eUTDSP.

Our benchmarking results reveal that usually some of the FUs are empty even if the code intended to be optimized. The usage of FUs by each DSP has been shown in Table 3. The number beside of DSPs in the second row shows the number of available FUs in each DSP. As shown, for example for DCT, C62x uses 3.85 FUs of its eight FUs in each cycle, while UTDSP and eUTDSP use 1.81 and 2.83 FUs of their four FUs, respectively. Table 3 also shows the percentage of using FUs by the three DSPs. As seen, with increasing the complexity, eUTDSP uses a higher percentage of FUs in comparison to UTDSP and C62x.

Table 3. Number of utilized FUs by each DSP.

	$Number\ of\ utilized\ FUs$		
	$C62x$ (8)	$UTDSP$ (4)	$eUTDSP$ (4)
RFIR	5.60[70%]	2.44[61%]	2.44[61%]
CFIR	4.06[51%]	1.92[48%]	2.48[62%]
DCT	3.85[48%]	1.81[45%]	2.83[71%]

5 Conclusion

The eUTDSP core is designed of individual modules using VHDL [10]. By connecting all of the modules and controlling them in parallel, it is possible to have a flexible architecture with a high performance. In order to control and support the parallel behavior of these modules, the VLIW concept have been utilized in the core. Some new architectural enhancements for eUTDSP have been presented in this paper that improve the relative performance and have several benefits in the context studied. As the benchmarking results show, by using these architectural enhancements in eUTDSP, the performance is significantly increased. The final point to remark is that although we have reduced the hardware size and total number of FUs in eUTDSP, the overall system performance have been improved for the considered communication algorithms by added utilization of the remaining FUs and some hardware blocks such as Loop Table and Trace Cache..

References

[1] J. L. Hennesy and D. A. Patterson, *Computer Architecture: A Quantitive Approach*, Morgan Kaufmann, San Mateo, California, 1996.

[2] *VLIW Architecture for DSP*, Berkeley Design Technology Inc., 2000, available online at www.bdti.com.

[3] O. Shalvi, N. Sommer, "New Generation DSPs: Concepts and Architecture," in Proc. *IEEE Convention of Electrical and Electronic Engineers*, pp. 233-237, 2000.

[4] J. Wang, B. Su, and E. W. Hu, "A Scalable Loop Optimization Approach for Scalable DSP Processors," in Proc. *Int. Conf. on Acoustics, Speech, and Signal Processing*, pp. 3646-3649, 2000.

[5] *TMS320C6000 CPU and Instruction Set Reference Guid*, Texas Instrument Inc., 1999.

[6] O. Wolf and J. Bier, *Tiger SHARC Sinks Teeth Into VLIW*, Microprocessor Report, 1998

[7] L. Petit and J. D. Legat, "A Low-Cost VLIW DSP Architecture for Communication Equipment," in Proc. *Int. Symp. on Signals, Systems and Electronics*, pp. 278-281, 1998.

[8] L. Lee, et. al, "DSP Design Using VLIW Architecture," in Proc. *Int. Conf. on Semiconductor Electronics*, pp. 160-167, 2001.

[9] Z. Tang, et. al, "A New Architecture for Branch Intensive Loops," in Proc. *Advanced in Parallel and Distributed Computing*, pp. 241-246, 1997.

[10] Z. Navabi, *VHDL Analysis and Modeling Digital Systems*, MC Graw-Hill, 1999.

Efficient Design of Perfect-Reconstruction Biorthogonal Cosine-Modulated Filter Banks Using Convex Lagrangian Relaxation and Alternating Null-Space Projections

Wu-Sheng Lu[1], *Robert Bregovic*[2], *and Tapio Saramäki*[2]

1. Dept. of Electrical and Computer Eng.
University of Victoria
Victoria, BC, Canada V8W 3P6

2. Institute of Signal Processing
Tampere University of Technology
P. O. Box 553, FIN-33101 Tampere, Finland

ABSTRACT

In essence, designing a perfect-reconstruction (PR) biorthogonal cosine-modulated filter bank (BCM) is a non-convex constrained optimization problem that can be solved in principle using general optimization solvers. However, when the number of channels is large and the order of the prototype filter (PF) is high, numerical difficulties in using those optimization solvers often occur, and the computational efficiency also becomes a concern. This paper proposes an algorithm that carries out the design in two stages. In the first stage, a convex Lagrangian relaxation technique is used to obtain a near PR (NPR) filter bank and, in the second stage, the coefficient vector of the PF obtained is alternately projected onto the null-spaces that are associated with the PR constraints, which turns the NPR filter bank into a PR filter bank. Simulation results are included to demonstrate the robustness of the proposed algorithm for designing BCM filter banks with a large number of channels and high-order PF as well as satisfactory design efficiency.

1. INTRODUCTION

Biorthogonal cosine-modulated filter banks (BCM) have played an increasingly important role in multirate signal processing because they offer reduced system delays compared to what linear-phase cosine-modulated filter banks can offer and their efficient implementation can be readily substantiated through the polyphase decomposition. In addition, the optimal synthesis of a BCM-based multirate system can be focused on the prototype filter (PF) alone.

Recent progress in the analysis and design of BCM filter banks has been reported by several authors, see, for example, [1] – [13]. Available design techniques include the quadratic-constrained least-squares (QCLS) method [4], [9], [10] that minimizes the stopband energy of the PF subject to the time-domain PR constraints; the factorization-based method [8], [11] that yields a parameterized realization in which the PR property is ensured while minimizing the stopband energy of the PF; and the sequential design method [13] that is carried out by first designing a filter bank with small number of channels and a relatively short filter length and then gradually increasing the number of channels as well as the filter length using a technique initiated in [3].

The optimization problem formulated in the time-domain is nonconvex. Although, in principle, general optimization solvers can be applied to find a solution, when the channel number is large and the order of the prototype filter (PF) is high, numerical difficulties in using those optimization solvers often occur, and the computational efficiency also becomes a concern. This paper proposes an algorithm that carries out the design in two stages. In the first stage, a convex relaxation technique is used to obtain a near PR (NPR) filter bank. The relaxation is carried out by a sequential convex approximation of the Lagrangian associated with the original (nonconvex) optimization problem, and can be viewed as an enhanced version of sequential quadratic programming (SQP) [14]. In the second stage, the coefficient vector of the PF obtained from the first stage is alternately projected onto the null-spaces that are associated with the PR constraints. The projections turn the NPR filter bank into a nearby PR filter bank with a fairly moderate increase of the stopband energy for the PF. Simulation results are included to demonstrate the robustness of the proposed algorithm for designing BCM filter banks with a large number of channels and high-order PF as well as satisfactory design efficiency.

2. DESIGN PROBLEM

2.1 BCM Filter Banks

An M-channel maximally decimated BCM filter bank is characterized by the coefficients of its analysis and synthesis filters that are given by

$$h_k(n) = 2h(n)\cos\left[\frac{\pi}{M}(k+\frac{1}{2})(n-\frac{D}{2}) + (-1)^k\frac{\pi}{4}\right] \qquad (1a)$$

and

$$f_k(n) = 2h(n)\cos\left[\frac{\pi}{M}(k+\frac{1}{2})(n-\frac{D}{2}) - (-1)^k\frac{\pi}{4}\right] \qquad (1b)$$

for $1 \le k \le M-1$ and $0 \le n \le N-1$, respectively, where $\{h(n)\}$ is the impulse response of the finite-impulse-response (FIR) PF, and D denotes the system delay. BCM filter bank structures other than that of (1) can also be obtained using different DCT modulations [10]. In this paper, however, we shall concentrate on the DCT-IV BCM filter banks as specified by (1) along with the following assumptions: (i) the channel number M is even, (ii) the filter length N assumes the form $N = 2mM$ for some positive integer m, and (iii) the system delay assumes the form $D = 2sM + d$ where s is an integer and $d = 2M - 1$. The rationale of these assumptions have been addressed in the literature [10] – [12]. The input-output relation of the system in the z-domain is given by

0-7803-7761-3/03/$17.00 ©2003 IEEE

$$Y(z) = T_0(z)X(z) + \sum_{l=0}^{M-1} T_l(z)X(ze^{-j2\pi l/M}) \quad (2a)$$

where

$$T_0(z) = \frac{1}{M}\sum_{k=0}^{M-1} F_k(z)H_k(z) \quad (2b)$$

$$T_l(z) = \frac{1}{M}\sum_{k=0}^{M-1} F_k(z)H_k(ze^{-j2\pi l/M}) \text{ for } l = 1,2,\ldots,M-1 \quad (2c)$$

It follows that the filter bank holds the PR property if and only if

$$T_0(e^{j\omega}) = e^{-jD\omega} \text{ for } \omega \in [0,\pi]$$

$$T_l(e^{j\omega}) = 0 \text{ for } \omega \in [0,\pi] \text{ and } 1 \le l \le M-1$$

In the time-domain, the PR condition can be described by the following set of quadratic equations [10]:

$$\boldsymbol{h}^T \boldsymbol{Q}_{l,n}\boldsymbol{h} = c_{l,n} \text{ for } 0 \le n \le 2m-2 \text{ and } 1 \le l \le M-1 \quad (3a)$$

where $\boldsymbol{h} = [h_0\ h_1 \cdots h_{N-1}]^T$ collects the coefficients of the PF, and

$$\boldsymbol{Q}_{l,n} = \boldsymbol{V}_{d-l}\boldsymbol{D}_n\boldsymbol{V}_l^T + \boldsymbol{V}_{d-M-l}\boldsymbol{D}_n\boldsymbol{V}_{M+l}^T \quad (3b)$$

$$\boldsymbol{D}_n(i,j) = \begin{cases} 1 & \text{if } i+j=n \\ 0 & \text{otherwise} \end{cases} \quad (3c)$$

$$\boldsymbol{V}_l(i,j) = \begin{cases} 1 & \text{if } i = l+2jM \\ 0 & \text{otherwise} \end{cases} \quad (3d)$$

and $c_{l,n} = \delta(n-s)/2M$. The performance of a BCM filter bank is typically measured by:

- Amplitude distortion: $e_m(\omega) = 1 - |T_0(e^{j\omega})|$
- Group-delay distortion: $e_{gd}(\omega) = D - \arg|T_0(e^{j\omega})|$
- Worst-case aliasing error: $e_a(\omega) = \max_{1\le l\le M-1} |T_l(e^{j\omega})|$

where $\omega \in [0,\pi]$. A filter bank is said to be NPR if the above measures are uniformally small in magnitude for all frequencies. Concerning the PF, it is often desirable to construct a PR or NPR filter bank with the PF's stopband energy

$$e_2(\boldsymbol{h}) = \int_{\omega_s}^{\pi} |H(e^{j\omega})|^2 \, d\omega \quad (4a)$$

minimized, where $H(e^{j\omega}) = \sum_{k=0}^{N-1} h_k e^{-jk\omega}$ and

$$\omega_s = \frac{(1+\rho)\pi}{2M} \text{ with } \rho > 0 \quad (4b)$$

It can easily be verified that $e_2(\boldsymbol{h}) = \boldsymbol{h}^T\boldsymbol{P}\boldsymbol{h}$ where $\boldsymbol{P}$ is a symmetric positive definite Toeplitz matrix determined by its first row given by $[\pi - \omega_s\ \ -\sin\omega_s\ \cdots -\sin(N-1)\omega_s/(N-1)]$.

2.2 PR Constraints

It can be readily verified that with $d = 2M - 1$, the constraints in (3a) for $0 \le n \le 2m-2$ and $M/2 \le l \le M-1$ are identical to those for $0 \le n \le 2m-2$ and $0 \le l \le M/2-1$. Therefore, the PR constraints to be considered in this paper are given by

$$\boldsymbol{h}^T\boldsymbol{Q}_{l,n}\boldsymbol{h} = c_{l,n} \text{ for } 0 \le n \le 2m-2 \text{ and } 0 \le l \le M/2-1 \quad (5)$$

where $\boldsymbol{Q}_{l,n} = \boldsymbol{V}_{2M-l-1}\boldsymbol{D}_n\boldsymbol{V}_l^T + \boldsymbol{V}_{M-l-1}\boldsymbol{D}_n\boldsymbol{V}_{M+l}^T$.

2.3 Problem Formulation

The design problem can be stated in the time-domain as

$$\text{minimize} \quad e_2(\boldsymbol{h}) = \boldsymbol{h}^T\boldsymbol{P}\boldsymbol{h} \quad (6a)$$

$$\text{subject to: constraints in (5)} \quad (6b)$$

A difference between (6) and the one in [10] is that the number of constraints involved in (6b) is a half of that in Eq. (65) of [10].

3. DESIGN METHOD

3.1 Basic Sequential Quadratic Programming

The Lagrangian of the constrained problem (6) is given by [14]

$$L(\boldsymbol{h},\boldsymbol{\lambda}) = \boldsymbol{h}^T\boldsymbol{P}\boldsymbol{h} - \sum_{i=1}^{K}\lambda_i a_i(\boldsymbol{h}) \quad (7)$$

where $K = M(2m-1)/2 = (N-M)/2$ is the number of constraints in (6b), and $a_i(\boldsymbol{h}) = \boldsymbol{h}^T\boldsymbol{Q}_{l,n}\boldsymbol{h} - c_{l,n}$ with $i = nM/2 + l + 1$. It is well known that a solution of problem (6) must satisfy the following Karush-Kuhn-Tucker (KKT) condition [14]:

$$\nabla L(\boldsymbol{h},\boldsymbol{\lambda}) = \begin{bmatrix} \nabla_h L(\boldsymbol{h},\boldsymbol{\lambda}) \\ \nabla_\lambda L(\boldsymbol{h},\boldsymbol{\lambda}) \end{bmatrix} = \boldsymbol{0} \quad (8)$$

Suppose we start with a reasonable initial PF coefficient vector $\boldsymbol{h}_0$ and an initial Lagrange multiplier vector $\boldsymbol{\lambda}_0 = \boldsymbol{0}$. In the kth iteration, $\{\boldsymbol{h}_k, \boldsymbol{\lambda}_k\}$ is updated to $\{\boldsymbol{h}_{k+1}, \boldsymbol{\lambda}_{k+1}\} = \{\boldsymbol{h}_k, \boldsymbol{\lambda}_k\} + \{\boldsymbol{\delta}_h, \boldsymbol{\delta}_\lambda\}$ such that

$$\nabla L(\boldsymbol{h}_{k+1}, \boldsymbol{\lambda}_{k+1}) \approx \nabla L(\boldsymbol{h}_k, \boldsymbol{\lambda}_k) + \nabla^2 L(\boldsymbol{h}_k, \boldsymbol{\lambda}_k)\begin{bmatrix} \boldsymbol{\delta}_h \\ \boldsymbol{\delta}_\lambda \end{bmatrix} = \boldsymbol{0} \quad (9)$$

which leads to the following linear system of equations:

$$\begin{bmatrix} \boldsymbol{W}_k & -\boldsymbol{A}_k^T \\ -\boldsymbol{A}_k & \boldsymbol{0} \end{bmatrix}\begin{bmatrix} \boldsymbol{\delta}_h \\ \boldsymbol{\delta}_\lambda \end{bmatrix} = \begin{bmatrix} \boldsymbol{A}_k^T\boldsymbol{\lambda}_k - \boldsymbol{g}_k \\ \boldsymbol{f}_k \end{bmatrix} \quad (10)$$

where $\boldsymbol{W}_k = 2(\boldsymbol{P} - \sum_{i=1}^{K}\lambda_i\boldsymbol{Q}_i)$, $\boldsymbol{g}_k = 2\boldsymbol{P}\boldsymbol{h}_k$, $\boldsymbol{A}_k = 2[\boldsymbol{Q}_1\boldsymbol{h}_k \cdots \boldsymbol{Q}_K\boldsymbol{h}_k]^T$, and $\boldsymbol{f}_k = [a_1(\boldsymbol{h}_k) \cdots a_K(\boldsymbol{h}_k)]^T$. Equation (10) can be written as

$$\boldsymbol{W}_k\boldsymbol{\delta}_h + \boldsymbol{g}_k = \boldsymbol{A}_k^T\boldsymbol{\lambda}_{k+1} \quad (11a)$$

$$\boldsymbol{A}_k\boldsymbol{\delta}_h = -\boldsymbol{f}_k \quad (11b)$$

Note that (11a) and (11b) are the *exact* KKT conditions for the following quadratic programming (QP) problem:

$$\text{minimize} \quad \frac{1}{2}\boldsymbol{\delta}^T \boldsymbol{W}_k \boldsymbol{\delta} + \boldsymbol{\delta}^T \boldsymbol{g}_k \qquad (12a)$$

$$\text{subject to:} \quad \boldsymbol{A}_k \boldsymbol{\delta} = -\boldsymbol{f}_k \qquad (12b)$$

Once a solution of (12) is obtained, based on (11a) the Lagrange multiplier vector can be computed as

$$\boldsymbol{\lambda}_{k+1} = (\boldsymbol{A}_k \boldsymbol{A}_k^T)^{-1} \boldsymbol{A}_k (\boldsymbol{W}_k \boldsymbol{\delta}_h + \boldsymbol{g}_k) \qquad (13)$$

and $\boldsymbol{W}_k$, $\boldsymbol{g}_k$, and $\boldsymbol{A}_k$ can be updated to $\boldsymbol{W}_{k+1}$, $\boldsymbol{g}_{k+1}$, and $\boldsymbol{A}_{k+1}$ accordingly. The iteration continues until certain criterion, such as the norm of $\boldsymbol{\delta}_h$ is less than a prescribed tolerance or the number of iterations reaches a given bound, is satisfied.

3.2 Convex Relaxation of Problem (12)

In general, the objective function in problem (12) is not convex. To obtain a meaningful iterate from the approximate KKT condition in (9), a convex relaxation of (12) is desirable. This can be accomplished in two ways. Perhaps the simplest way is to replace matrix $\boldsymbol{W}_k$ with constant matrix $2\boldsymbol{P}$. As a result, the modified problem in (12) is a *convex* QP problem that possesses a unique global minimizer. Also note that the modified Hessian matrix requires no update during the iteration process. However, because of the modification, the Lagrange multiplier $\boldsymbol{\lambda}_k$ is no longer able to influence the Hessian and the modified algorithm usually cannot enjoy a fast convergence rate. Another way to relax the problem in (12) into a convex QP is to use a quasi-Newton update, such as the Broyden-Fletcher-Goldfarb-Shanno formula [14], [15] that replaces $\boldsymbol{W}_k$ by $\boldsymbol{Y}_k$ where $\boldsymbol{Y}_k$ is updated as follows:

$$\boldsymbol{Y}_{k+1} = \boldsymbol{Y}_k + \frac{\boldsymbol{\eta}_k \boldsymbol{\eta}_k^T}{\boldsymbol{\delta}_h^T \boldsymbol{\eta}_k} - \frac{\boldsymbol{Y}_k \boldsymbol{\delta}_h \boldsymbol{\delta}_h^T \boldsymbol{Y}_k}{\boldsymbol{\delta}_h^T \boldsymbol{Y}_k \boldsymbol{\delta}_h} \qquad (14a)$$

where $\boldsymbol{Y}_0 = \boldsymbol{I}$, $\boldsymbol{\delta}_h = \boldsymbol{h}_{k+1} - \boldsymbol{h}_k$, $\boldsymbol{\eta}_k = \theta \boldsymbol{\gamma}_k + (1-\theta)\boldsymbol{Y}_k \boldsymbol{\delta}_h$,

$$\boldsymbol{\gamma}_k = (\boldsymbol{g}_{k+1} - \boldsymbol{g}_k) - (\boldsymbol{A}_{k+1} - \boldsymbol{A}_k)^T \boldsymbol{\lambda}_{k+1} \qquad (14b)$$

$$\theta = \begin{cases} 1 & \text{if } \boldsymbol{\delta}_h^T \boldsymbol{\gamma}_k \ge 0.2 \boldsymbol{\delta}_h^T \boldsymbol{Y}_k \boldsymbol{\delta}_h \\ \dfrac{0.8 \boldsymbol{\delta}_h^T \boldsymbol{Y}_k \boldsymbol{\delta}_h}{\boldsymbol{\delta}_h^T \boldsymbol{Y}_k \boldsymbol{\delta}_h - \boldsymbol{\delta}_h^T \boldsymbol{\gamma}_k} & \text{otherwise} \end{cases} \qquad (14c)$$

3.3 Further Enhancements

The algorithm can be further enhanced by including a norm constraint on vector $\boldsymbol{\delta}_h$ and a line search step. The norm constraint is of importance because it validates the approximation (9). In doing so, the convex relaxation of problem (12) becomes

$$\text{minimize} \quad \frac{1}{2}\boldsymbol{\delta}^T \boldsymbol{Y}_k \boldsymbol{\delta} + \boldsymbol{\delta}^T \boldsymbol{g}_k \qquad (15c)$$

$$\text{subject to:} \quad \boldsymbol{A}_k \boldsymbol{\delta} = -\boldsymbol{f}_k \qquad (15b)$$

$$\|\boldsymbol{\delta}\| \le \beta \qquad (15c)$$

where β is a small positive scalar. The problem in (15) is a second-order cone programming problem [16] that can be solved using, for example, SeDuMi [17]. Having obtained the solution $\boldsymbol{\delta}$, a line search is carried out by finding a positive scalar α_k that minimizes the following merit function

$$\psi(\boldsymbol{h}_k + \alpha \boldsymbol{\delta}) = e_2(\boldsymbol{h}_k + \alpha \boldsymbol{\delta}) + \mu \sum_{i=1}^{K} a_i^2 (\boldsymbol{h}_k + \alpha \boldsymbol{\delta}) \qquad (16)$$

where $\mu > 0$ weighs the importance of the constraints in (6b) in relative to the stopband energy. Having done this, the PF coefficient vector is updated from $\boldsymbol{h}_k$ to $\boldsymbol{h}_{k+1} = \boldsymbol{h}_k + \alpha_k \boldsymbol{\delta}$.

3.4 Alternating Null-Space Projections

The above method can be used to obtained a practically PR BCM filter bank when a sufficient number of iterations are carried out. Below we sketch a method that can be used to turn an NPR into a PR filter bank quickly provided that the NPR filter bank is *sufficiently* "close" to its PR counterpart.

A careful examination of the constraints in (5) shows that these equations can be expressed as either $\boldsymbol{C}_o \boldsymbol{h}_{ek} = \boldsymbol{b}_k$ or $\boldsymbol{C}_e \boldsymbol{h}_{ok} = \boldsymbol{b}_k$, where $\boldsymbol{h}_{ek}$ and $\boldsymbol{h}_{ok}$ are $N/2$-dimensional vectors formed by the even-indexed and odd-indexed components of $\boldsymbol{h}_k$, respectively, $\boldsymbol{C}_o$ and $\boldsymbol{C}_e$ are $(N - M)/2$ by $N/2$ matrices that are linearly determined by $\boldsymbol{h}_{ok}$ and $\boldsymbol{h}_{ek}$, respectively, and $\boldsymbol{b}_k$ is a constant vector of dimension $(N - M)/2$. Matrices $\boldsymbol{C}_o$ and $\boldsymbol{C}_e$ are in general of full row-rank. Consequently, for a fixed $\boldsymbol{h}_{ok}$ (or $\boldsymbol{h}_{ek}$), the null-spaces of linear operators $\boldsymbol{C}_o$ (or $\boldsymbol{C}_e$) are $M/2$-dimension subspaces in space $R^{N/2}$. Therefore, for a *fixed* $\boldsymbol{h}_{ok}$, if we denote a special solution of the *linear* system $\boldsymbol{C}_o \boldsymbol{h}_{ek} = \boldsymbol{b}_k$ by $\boldsymbol{h}_{es}$, then all solutions of the system can be expressed as $\boldsymbol{h}_{ek} = \boldsymbol{h}_{es} + \boldsymbol{V}_e \boldsymbol{\xi}_e$ where $\boldsymbol{V}_e$ is a $N/2$ by $M/2$ matrix whose columns are a set of basis vectors in the null space of $\boldsymbol{C}_o$, and $\boldsymbol{\xi}_e$ is an $M/2$-dimensional "free" vector that can be determined by minimizing the stopband energy of the PF. The above process can be viewed as projecting vector $\boldsymbol{h}_k$ onto the null space so as to force the resulting coefficient vector to be PR. As such, it is expected that the change in the resulting coefficient vector will remain moderate if vector $\boldsymbol{h}_k$ is already close enough to its PR counterpart. Next, a similar projection is performed by fixing an $\boldsymbol{h}_{ek}$ and expressing the solutions of $\boldsymbol{C}_e \boldsymbol{h}_{ok} = \boldsymbol{b}_k$ as $\boldsymbol{h}_{ok} = \boldsymbol{h}_{os} + \boldsymbol{V}_o \boldsymbol{\xi}_o$ where $\boldsymbol{V}_o$ is formed by the basis vectors of the null space of $\boldsymbol{C}_e$, and $\boldsymbol{\xi}_o$ is an $M/2$-dimensional free vector that can be determined by minimizing the stopband energy of the PF. The projection continues several times until the difference between the PF coefficient vectors before and after the projection becomes insignificant.

4. DESIGN EXAMPLES

The proposed algorithm was applied to design several BCM filter banks. In each design $\rho = 1$ and $\mu = 100$ were assumed. The algorithm was implemented using MATLAB on a Pentium III 1GHz PC. The design parameters and performance evaluation results are shown in Table I, where K_i denotes the number of iterations carried out in the first stage of the design, and Proj. # denotes the number of projections performed. As a representative of the designs, the amplitude responses of the PF and those analysis filters in the frequency range $0 \leq \omega \leq \pi/16$ for the 256-channel filter bank are shown in Figs. 1a and b, respectively.

Concerning the computational efficiency, note that solving the problem in (15) takes most of the CPU time in each iteration of the first design stage. The average CPU time for solving (15) in the four designs listed in Table I was 6.46, 40.60, 81.45, and 402.71 seconds, respectively. The CPU time required to carry out the second stage of the design was found insignificant in relative to that of the first stage.

Table I: Design Parameters and Performance Evaluation Results

M	32	64	128	256
N	320	640	1280	2560
D	255	511	1023	1535
$e_2(\boldsymbol{h})$	$1.04 \cdot 10^{-6}$	$5.77 \cdot 10^{-7}$	$3.01 \cdot 10^{-7}$	$2.65 \cdot 10^{-7}$
$\max\lvert e_m\rvert$	$2.68 \cdot 10^{-14}$	$4.34 \cdot 10^{-14}$	$1.24 \cdot 10^{-13}$	$2.05 \cdot 10^{-13}$
$\max\lvert e_{gd}\rvert$	$4.71 \cdot 10^{-11}$	$1.17 \cdot 10^{-10}$	$1.29 \cdot 10^{-11}$	$1.44 \cdot 10^{-11}$
$\max\lvert e_a\rvert$	$2.99 \cdot 10^{-14}$	$5.87 \cdot 10^{-14}$	$1.26 \cdot 10^{-13}$	$2.68 \cdot 10^{-13}$
K_i	100	200	550	590
Proj. #	10	10	0	0

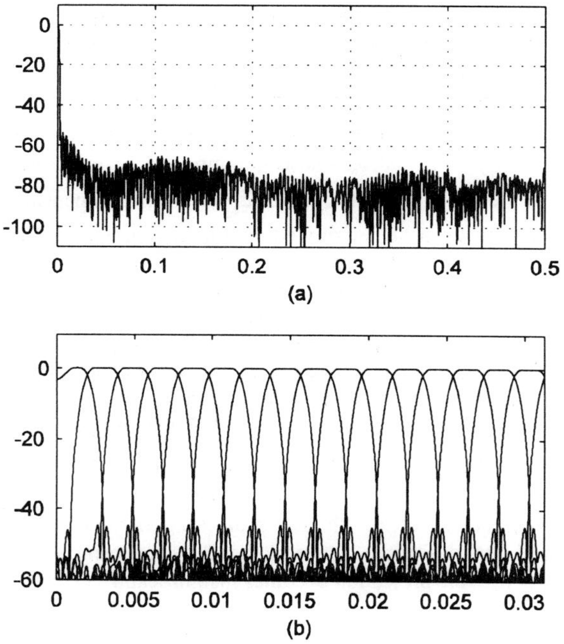

Figure 1. Amplitude responses of (a) the PF for the BCM filter bank with $M = 256$, $N = 2560$ and $D = 1535$; and (b) its analysis filter bank in the frequency range $0 \leq \omega \leq \pi/16$.

Acknowledgement: This work was supported by the Tampere International Centre for Signal Processing, Tampere University of Technology, Tampere, Finland, and the Academy of Finland, project No. 44876 (Finnish centre of Excellence program (2000-2005)).

5. REFERENCES

[1] Malvar S. H. *Signal processing with lapped transforms*, Artech House, 1992.

[2] Koilpillai R. D. and Vaidyanathan P. P. "Cosine-modulated FIR filter banks satisfying perfect reconstruction", *IEEE Trans. Signal Processing,* vol. 40, pp. 770-783, April 1992.

[3] Saramäki T. "Designing prototype filters for perfect-reconstruction cosine-modulated filter banks", *IEEE Int. Symp. Circuits Syst.*, vol. 3, pp. 1605-1608, San Diego, CA., May 1992.

[4] Nguyen T. Q. "A quadratic-constrained least-squares approach to the design of digital filter banks", *IEEE Int. Symp. Circuits Syst.*, San Diego, CA., May 1992.

[5] Lin Y.-P. and Vaidyanathan P. P. "Linear phase cosine modulated maximally decimated filter banks with perfect reconstruction", *IEEE Trans. Signal Processing,* vol. 43, pp. 2525-2539, Nov. 1995.

[6] Nguyen T. Q. and Koilpillai R. D. "The theory and design of arbitrary-length cosine-modulated filter banks and wavelets, satisfying perfect reconstruction", *IEEE Trans. Signal Processing,* vol. 44, pp. 473-483, Mar. 1996.

[7] Xu H., Lu W.-S. and Antoniou A. "Efficient iterative design method for cosine-modulated QMF banks", *IEEE Trans. Signal Processing,* vol. 44, pp. 1657-1668, July 1996.

[8] Schuller G. "A new factorization and structure for cosine modulated filter banks with variable system delay", *Asilomar Conf. Signals, Systems, and Computers*, Pacific Glove, CA, vol. 2, pp. 1310-1314, Nov. 1996.

[9] Nguyen T. Q. and Heller P. N. "Biorthogonal cosine-modulated filter banks", *IEEE Int. Conf. Acoustics, Speech, and Signal Processing*, vol. 3, pp. 1471-1474, Atlanta, GA, May 1996.

[10] Heller P. N., Karp T. and Nguyen T. Q. "A general formulation of modulated filter banks", *IEEE Trans. Signal Processing,* vol. 47, pp. 986-1002, Apr. 1999.

[11] Karp T., Mertins A. and Schuller G. "Efficient biorthogonal cosine-modulated filter banks", *Signal Processing*, vol. 81, pp. 997-1016, May 2001.

[12] Saramäki T. and Bregovic R. *Multirate systems and filter banks*, Chap. 2 in *Multirate Systems: Design and Applications*, ed. By Jovanovic-Dolecek, Idea Group Publishing, Hershey PA., 2002.

[13] Bregovic R. and Saramäki T. "An efficient approach for designing nearly perfect-reconstruction low-delay cosine-modulated filter banks", *IEEE Int. Symp. Circuits Syst.*, Scottsdale AZ., vol. 1, pp. 825-828, May 2002.

[14] Fletcher R. *Practical methods of optimization*, 2nd ed., Wiley, New York, 1987.

[15] Powell M. J. D. *"Algorithms for nonlinear constraints that use Lagrangian functions", Math. Programming*, vol. 14, pp. 224-248, 1978.

[16] Ben-Tal A. and Nemirovski A. *Lectures on modern convex optimization*, SIAM, Philadelphia, 2001.

[17] Sturm, J. F. "Using SeDuMi1.02, a MATLAB toolbox for optimization over symmetric cones", *Optimization Methods and Software*, vol. 11-12, pp. 625-653, 1999.

THREE CLASSES OF IIR COMPLEMENTARY FILTER PAIRS WITH AN ADJUSTABLE CROSSOVER FREQENCY

Ljiljana Milic[1] *and Tapio Saramäki*[2]

[1]Tampere International Center for Signal Processing (TICSP)
[2]Institute for Signal Processing, Tampere University of Technology, P. O. Box 553, FIN-33101 Tampere, Finland
[1] Permanent affiliation: University of Belgrade, M. Pupin Institute, P. O. Box 15, 11001 Belgrade, Yugoslavia

ABSTRACT

Three classes of complementary recursive low-pass/high-pass filter pairs are introduced. For each class, the crossover frequency can be arbitrarily selected without changing the predetermined stop-band attenuation being the same for both filters and the filter pair is constructed using two all-pass sub-filters as building blocks. Based on the properties of elliptic minimal Q-factors transfer functions, simple expressions are derived for evaluating the coefficients for the all-pass sections to arbitrarily change the crossover frequency. Design procedures are developed for synthesizing filter pairs implemented as a parallel connection of two all-pass sub-filters and for two classes of filter pairs constructed as tapped cascaded interconnections of two identical all-pass sub-filters. The first of them together with the direct parallel connection of two all-pass filters provide the power-complementary property, whereas the second one provides the magnitude-complementary property.

1. INTRODUCTION

Complementary infinite-impulse response (IIR) filter pairs can be synthesized for generating power-complementary, all-pass complementary, or magnitude-complementary filter pairs (see, e.g., [1]). A very attractive alternative to generate a power-complementary low-pass/high-pass IIR filter pair is to use lattice wave digital filters (parallel connections of two all-pass filters) [2]–[4]. Usually, the crossover frequency for this filter pair occurs in the middle of the base-band, that is, it is located exactly at 1/4 in terms of the normalized frequency. If the stop-band attenuation of both filters is the same, then the low-pass and high-pass filters are both half-band IIR filters (see, e.g., [4], [5]).

This paper proposes a technique to change, by means of simple formulae, the location of the crossover frequency of the half-band filter pair to an arbitrary location while still retaining the attenuation properties of the initial half-band filters. For this purpose, elliptic minimal Q-factors (EMQF) transfer functions introduced in [6], [7] provide directly the desired solution.

In addition to the direct parallel connection of two all-pass filters, two structures constructed as tapped cascaded interconnections of two identical all-pass filters are considered based on the use of the synthesis schemes described in [8]–[10]. The first (second) structure allows one to generate power-complementary (magnitude-complementary) filter pairs with an adjustable crossover frequency. The advantage of using tapped cascaded interconnections of two identical all-pass sub-filters is fact that that the all-pass sub-filters are of very low orders.

2. USE OF EMQF FILTERS

This section shows how the properties of EMQF filters [6], [7] can be exploited in a very straightforward manner for generating complementary IIR filter pairs with an adjustable crossover frequency such that the stop-band attenuation of both filters remains the same. Due to the fact that EMQF filters comprise a half-band filter as a special case, we start with a half-band filter pair to generate the desired complementary filter pair whose crossover frequency can be arbitrarily chosen.

2.1 Properties of EMQF filters for Generating Power-Complementary Filter Pairs

A complementary IIR filter pair constructed as a parallel connection of two all-pass sub-filters is shown in Fig. 1. For half-band IIR filters, this filter pair is given by (see, e.g., [4], [5])

$$\mathbf{G}^{HB}(z)=\frac{1}{2}\left[A_0^{HB}(z)\pm A_1^{HB}(z)\right]=\frac{1}{2}\left[\prod_{i=2,4,\ldots}^{(N+1)/2}\frac{\beta_i^{HB}+z^{-2}}{1+\beta_i^{HB}z^{-2}}\right.$$
$$\left.\pm z^{-1}\prod_{i=3,5,\ldots}^{(N+1)/2}\frac{\beta_i^{HB}+z^{-2}}{1+\beta_i^{HB}z^{-2}}\right],\ \beta_i^{HB}<\beta_{i+1}^{HB}. \quad (1)$$

Here, N, the filter orders, is restricted to be odd and $\mathbf{G}^{HB}(z)=[\,G_{LP}^{HB}(z)\ \ G_{HP}^{HB}(z)\,]^T$ is a vector containing the low-pass (with the plus sign) and high-pass (with the minus sign) half-band filter transfer functions $G_{LP}^{HB}(z)$ and $G_{HP}^{HB}(z)$. The subscript 'HP' is used to emphasize that the filters are half-band filters. One of poles of these transfer functions is located at the origin and the remaining poles are complex-conjugate pairs being located on the imaginary axis at $z=\pm jr_i$ for $i=2,3,\ \cdots,(N+1)/2$, giving $\beta_i^{HB}=(r_i)^2$. The notation $\beta_i^{HB}<\beta_{i+1}^{HB}$ in Eq. (1) indicates how the poles are shared between the all-pass sections.

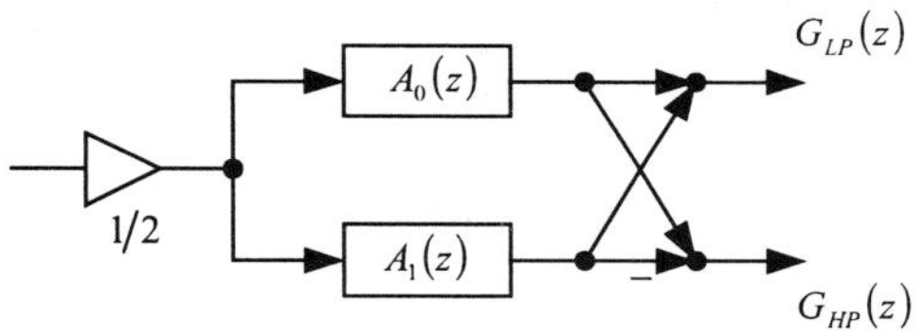

Figure 1. Double-complementary IIR filter pair constructed as a parallel connection of two all-pass filters.

This filter pair when generated by means of an odd-order elliptic filter is characterized by the following attractive properties. First, the sum of the squared-magnitude responses of the two filters is identically equal to unity. Hence, the filter pair has the power-complementary property. Second, the 3-dB crossover frequency (the frequency where the squared-magnitude responses of both $G_{LP}^{HB}(z)$ and $G_{HP}^{HB}(z)$ achieve the value of 1/2) is located at $f_{3dB}=1/4$ in terms of the normalized frequency. Third, the pass-band and stop-band edges, denoted by f_p and f_s in terms of the normalized frequency, satisfy $f_s=1/2-f_p$. Fourth, the attenuation of both the low-pass and high-pass filters is the same. Since the sum of $G_{LP}^{HB}(z)$ and $G_{HP}^{HB}(z)$ is $A_0^{HB}(z)$, this filter pair is also an all-pass complementary filter pair, thereby called as a double-complementary filter pair [1].

0-7803-7761-3/03/$17.00 ©2003 IEEE

The low-pass/high-pass EMQF filter pair is generated with the aid of the above half-band filter pair as follows [6], [7]:

$$\mathbf{G}(z)=\frac{1}{2}\left[A_0(z)\pm A_1(z)\right]=\frac{1}{2}\left[\prod_{i=2,4,\ldots}^{(N+1)/2}\frac{\beta_i+\alpha(1+\beta_i)z^{-1}+z^{-2}}{1+\alpha(1+\beta_i)z^{-1}+\beta_i z^{-2}}\right.$$
$$\left.\pm\frac{\alpha_1+z^{-1}}{1+\alpha_1 z^{-1}}\prod_{i=3,5,\ldots}^{(N+1)/2}\frac{\beta_i+\alpha(1+\beta_i)z^{-1}+z^{-2}}{1+\alpha(1+\beta_i)z^{-1}+\beta_i z^{-2}}\right],\ \beta_i<\beta_{i+1}, \quad (2)$$

where $\mathbf{G}^{HB}(z)=[\,G_{LP}^{HB}(z)\ G_{HP}^{HB}(z)\,]^T$ is again a vector containing the resulting low-pass and high-pass filter transfer functions $G_{LP}(z)$ and $G_{HP}(z)$ that can be still implemented as shown in Fig. 1. The very simple formulae for converting the delay term z^{-1} and the second-order all-pass sections in Eq. (1) to the first-order and the second-order all-pass sections in Eq. (2) will be developed in the following subsection.

The properties of EMQF filters guarantee that for the EMQF filter pair, as given by Eq. (2), the crossover frequency can be changed while keeping the attenuation of both the low-pass filter and high-pass filter the same as for the start-up half-band filter pair. Second, for these filter pairs, the z-plane poles are on the circle that is orthogonal with the unit circle and centered on the real axis [6], [7]. When the center of the circle approaches infinity, the circle degenerates into the imaginary axis, and the half-band filter with the poles on the imaginary axis is obtained. This property provides a very straightforward approach to transforming a half-band filter pair to an EMQF filter pair whose crossover frequency can be arbitrarily chosen, as will be described in the following subsection.

2.2 Frequency transformations

This subsection shows how an EMQF filter with an arbitrary 3-dB cutoff frequency f_{3dB} can be generated with the aid of a prototype half-band filter and the following bilinear transform:

$$s=k(z-1)/(z+1)\,, \quad (3)$$

where k is a constant to be adjusted.

Figure 2 illustrates the overall transformation procedure. The start-up filter is an odd-order half-band filter with the selectivity factor ξ defined by

$$\xi=\tan(\pi f_s)/\tan(\pi f_p), \quad (4)$$

where $f_s=1/2-f_p$. The z-plane poles of the half-band filter poles are located on the imaginary axis and the 3-dB cutoff frequency is located at $z_{3dB}^{HB}=e^{j\pi/2}$, as illustrated in Fig. 2(a) for a third-order half-band filter. Then, the inverse bilinear transform with $k=\sqrt{\xi}$ is used to map this half-band filter from the z-plane to the s-plane to generate the corresponding analogue prototype filter, as shown in Fig. 2(b). The resulting s-plane poles of the analogue prototype are located on the circle centered at the origin with the radius equal to $k=\sqrt{\xi}$ and the 3-dB cutoff frequency being located at $s_{3dB}=j\sqrt{\xi}$ [6]. The analogue prototype is then mapped from the s-plane the to the z-plane by using $k=\sqrt{\xi}/\tan(\pi f_{3dB})$ in Eq. (3) resulting in the EMQF filter whose 3-dB cutoff frequency is placed at the desired location $z_{3dB}=e^{j2\pi f_{3dB}}$, as shown in Fig. 2(c). Since the start-up half-band filter and resulting EMQF filter have the common analogue prototype, the double-complementary property of the half-band filter is preserved when using these transformations.

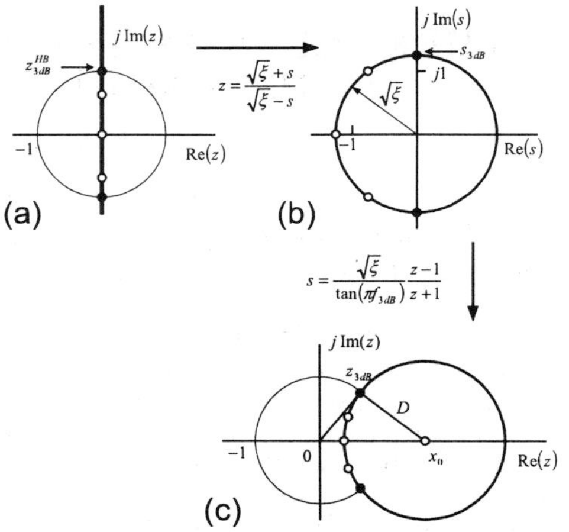

Figure 2. Frequency transformations. (a) Start-up half-band filter. (b) Corresponding analogue prototype filter. (c) EMQF filter with the desired 3-dB cutoff frequency at $z_{3dB}=e^{j2\pi f_{3dB}}$.

The key role in the mappings of Fig. 2 is the bilinear transform, as given by Eq. (3). The value of the constant k is first adjusted to map the digital half-band filter to the analogue prototype, and adjusted again to map the analogue prototype to the EMQF filter with the desired f_{3dB}. The goal is to express the constants α_1, α, and β_i of the EMQF filter [see Eq. (2)] in terms of the constants β_i^{HB} of the start-up half-band filter [see Eq. (1)], and the desired value of f_{3dB}. This is achieved when the transformations of Fig. 2 are directly applied to the individual all-pass sections in Eqs. (1) and (2). Table 1 summarizes the formulae for converting the start-up half-band filter to the EMQF filter with the given 3-dB cutoff frequency [6], [7]. The normalized pass-band and stop-band edges for the resulting EMQF filter are given by

$$f_p=\frac{1}{\pi}\tan^{-1}\!\left(\tan(\pi f_{3dB})/\sqrt{\xi}\right)\ \text{and}\ f_s=\frac{1}{\pi}\tan^{-1}\!\left(\sqrt{\xi}\tan(\pi f_{3dB})\right). \quad (5)$$

It should be pointed out that the 3-dB cutoff frequency of a single filter is equivalent to the 3-dB crossover frequency of a complementary filter pair.

Table 1: Parameters for a power-complementary filter pair with the given 3-dB cutoff frequency f_{3dB}

First-order Section	$\alpha_1=(1-\tan(\pi f_{3dB}))/(1+\tan(\pi f_{3dB}))$	
Second-order Section	$\alpha=\dfrac{1-(\tan(\pi f_{3dB}))^2}{1+(\tan(\pi f_{3dB}))^2}$	$\beta_i=\dfrac{\beta_i^{HB}+\alpha_1^2}{\beta_i^{HB}\alpha_1^2+1}$

3. SYNTHESIS OF FILTER PAIRS

This section shows how exploit the properties of the EMQF filters for synthesizing complementary IIR low-pass/high-pass filter pairs with an adjustable crossover frequency for the three filter pair classes mentioned in Introduction.

3.1 Parallel connection of two all-pass sub-filters

Consider an IIR power-complementary filter pair as depicted in Fig. 1. Given the odd order N, the 3-dB crossover frequency f_{3dB}, and the minimum stop-band attenuation A_s in decibels, the synthesis can be carried out as follows:

1) Design the start-up power-complementary IIR half-band filter pair, as given by Eq. (1), such that the minimum stop-band attenuation is exactly A_s.[1] Let the resulting pass-band and stop-band edges be f_p and $f_s = 1/2 - f_p$.
2) Design the power-complementary IIR filter pair, as given by Eq. (2), having the desired 3-dB crossover frequency f_{3dB} by determining its parameters using the formulae of Table 1.
3) Determine the pass-band and stop-band edges of the resulting filter pair according to Eqs. (4) and (5).

Figure 3 shows responses of some filter pairs for $N = 7$ and $A_s = 60$ dB. The thick line shows the response of the half-band filter pair. To obtain a new filter pair, only 5 constant values have to be computed using the formulae of Table 1.

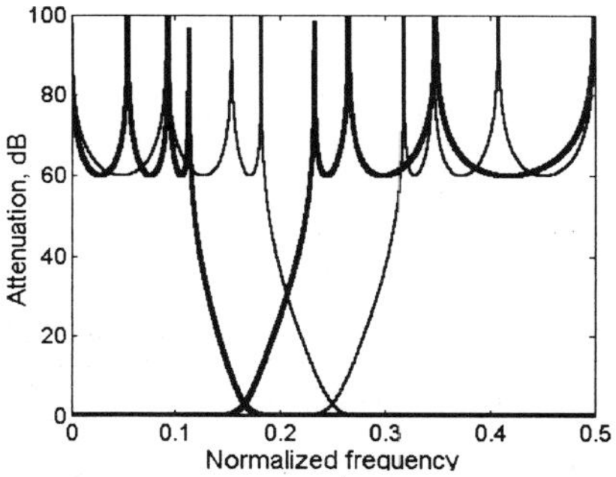

Figure 3. Responses for power-complementary seventh-order filter pairs. Thick-line: f_{3dB} =1/4. Thin line: f_{3dB} =1/6.

3.2 Tapped cascaded interconnection of two all-pass filters having the power-complementary property

For a power-complementary filter pair constructed as a tapped cascaded interconnection of two identical all-pass filters, the low-pass and high-pass transfer functions are given by [8], [10]

$$H_{LP}(z) = \sum_{m=0}^{M} a[m][A_0(z)]^m [A_1(z)]^{M-m} \tag{7a}$$

$$H_{HP}(z) = \sum_{m=0}^{M} (-1)^m a[M-m][A_0(z)]^m [A_1(z)]^{M-m}, \tag{7b}$$

where M is an odd integer and $A_0(z)$ and $A_1(z)$ are the all-pass filters in Eq. (1) or Eq. (2). An efficient implementation form for the above filter pair is show in Fig. 4 [10] [2]. The details on how to convert the $a[m]$'s to the k_l's can be found in [10].

Given the odd orders M and N, the 3-dB crossover frequency f_{3dB}, and the minimum stop-band attenuation A_s, the optimized filter pair can be generated as follows:

1) Determine $\Delta = 10^{-A_s/10}$.
2) Optimize the $a[m]$'s to maximize θ_p such that the amplitude response of a nonlinear-phase transfer function $F(z) = \sum_{m=0}^{M} a[m] z^{-m}$ stays within the limits 1 and $\sqrt{1-\Delta}$ in the normalized pass-band region [0, θ_p] and the maximum amplitude value is $\sqrt{\Delta}$ in the normalized stop-band region [1/2−θ_p, 1/2].
3) Perform the synthesis scheme of Subsection 3.1 with the main exception that now the desired stop-band attenuation for the start-up half-band filter pair, as given by Eq. (1) and determining the all-pass sections in Fig. 4, is given by

$$\hat{A}_s = -10\log_{10}\{\cos[\pi(1/2-\theta)]\}. \tag{8}$$

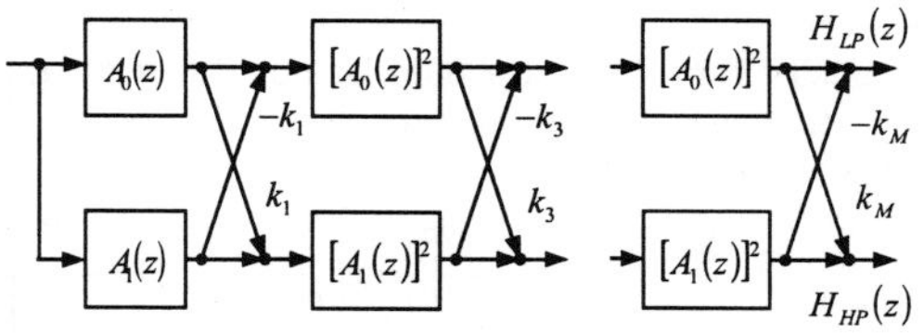

Figure 4. Lattice structure for the proposed power-complementary filter pair.

As shown [8]–[10], the basic idea of using tapped cascaded interconnections of two identical all-pass sub-filters is based on following two facts. First, the pass-band and stop-band regions for the direct parallel connections and the overall structures are the same. Second, due to use of several copies of two all-pass sections, the requirements for the parallel connection of the all-pass sections become significantly milder compared with the overall filter. This enables one to use only first-order and second-order all-pass sections, thereby making the tuning of the crossover frequency of the filter pair extremely simple without changing the tap coefficients combining the two all-pass sections. Step 2 in the above procedure can be performed by slightly modifying the design scheme described in [8].

Figure 5 shows responses of the some power-complementary filter pairs in the case where A_s = 60 dB and M = 5 identical copies of two all-pass filters of orders 2 and 1 are used, that is, N = 3. For this design, $\hat{A}_s$ = 19.37 dB and the new filter pair is generated by evaluating only 3 constant values according to the formulae of Table 1.

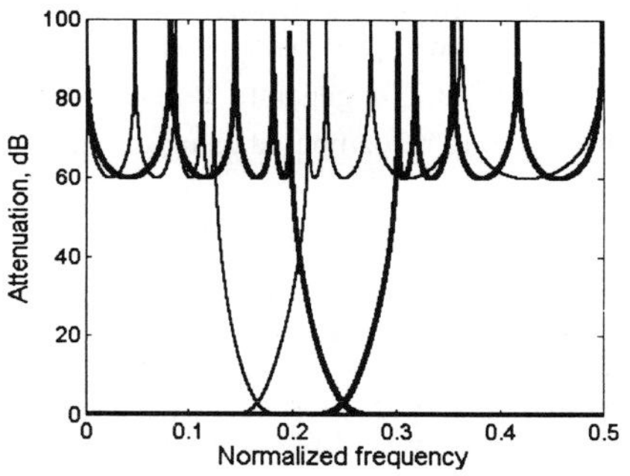

Figure 5. Power-complementary filter pairs constructed using 5 identical copies of the first-order and the second-order all-pass filters. Thick-line: f_{3dB} =1/4. Thin line: f_{3dB} =1/6.

3.3 Tapped cascaded interconnection of two all-pass filters having the magnitude-complementary property

For a magnitude-complementary filter pair, the low-pass and high-pass transfer functions are given by [9]

$$H_{LP}(z) = \sum_{m=0}^{M} a[m][A_0(z)]^m [A_1(z)]^{M-m} \tag{9a}$$

[1] At this stage, $A_0(z) \equiv A_0^{HP}(z)$ and $A_1(z) \equiv A_1^{HP}(z)$ in Fig. 1 (see Eq. 1).

[2] It should be pointed out that this implementation form is valid only for the power-complementary filter pairs under consideration in this subsection. The structure of Fig. 4 has been derived based on Fig. 11 in [10] by removing the decimation by a factor of two. Note that the subscripts and signs of the k's are different.

$$H_{HP}(z) = \sum_{m=0}^{M} b[m][A_0(z)]^m [A_1(z)]^{M-m} \qquad (9b)$$

Here, M is an integer being two times an odd integer, $a[M-m] = a[m]$, $b[M-m] = b[m]$, and $b[m] = -a[m]$ for $m = 0, 1, \cdots, M/2-1$. Furthermore, $a[M/2] = b[M/2] = 1/2$ and $a[m] = b[m] = 0$ for the remaining odd values of m. An efficient implementation for the above magnitude-complementary filter pair is shown in Fig. 6 [9]. The sum of the above two transfer functions is $[A_0(z)]^{M/2}[A_1(z)]^{M/2}$ so that they form also an all-pass complementary filter pair. It should be noted that for the above filter pair, the magnitude responses achieve the value of 1/2 (–6 dB) at the crossover frequency.

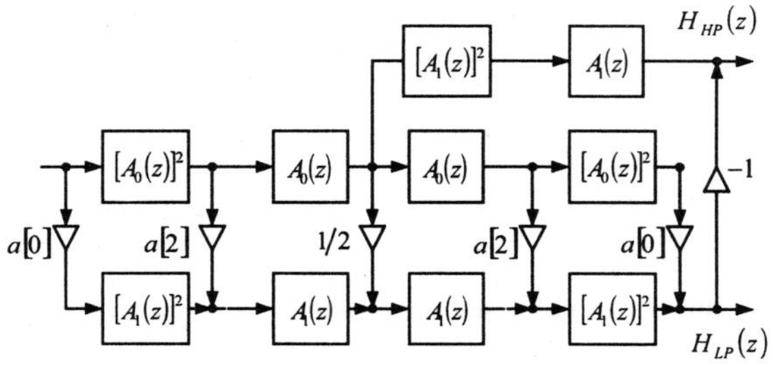

Figure 6. Structure for the proposed magnitude-complementary (all-pass complementary) filter pair. $M=6$.

In this case, the overall synthesis can be performed as in Subsection 3.2 by replacing Steps 1 and 2 by

1) Determine $\Delta = 10^{-A_s/20}$.
2) Optimize the $a[m]$'s to maximize θ_p such that the zero-phase frequency response of the FIR filter transfer function [3] $F(z) = z^{-M/2}\{1/2 + \sum_{n=1}^{(M+2)/4} a[M/2-(2n-1)][z^{2n-1} + z^{-(2n-1)}]\}$ stays within the limits 1 and $1-\Delta$ in the normalized pass-band region $[0, \theta_p]$ and within the limits 0 and Δ in the normalized stop-band region $[1/2-\theta_p, 1/2]$.

Figure 7 shows responses of the some filter pairs in the case where A_s = 60 dB and M = 6 identical copies of two all-pass filters of orders 2 and 1 are used, that is, N = 3. For this design, $\hat{A}_s$ = 14.53 dB. To generate a new filter pair, only 3 constant values have to be computed using the formulae of Table 1.

4. CONCLUSION

This paper has introduced three design approaches for generating complementary low-pass/high-pass IIR filter pairs with an adjustable crossover frequency. The start-up complementary filters are low-pass/high-pass half-band filter pairs constructed as a parallel connection of two all-pass filter sections (a lattice wave digital filter). Exploiting the properties of elliptic minimal Q-factors transfer functions, simple formulae for the direct calculation of the coefficients have been derived. This allows one to implement programmable complementary filter pairs with an adjustable crossover frequency in a very simple and straightforward manner. For this purpose, future work is devoted to generate simplified forms for the formulae of Table 1. The errors caused by these new formulas can be compensated by slightly over-designing the start-up filter pair so that the filter pair with the desired crossover frequency still meets the given criteria.

Particular benefits have been obtained for the realization structures based on the use of the tapped cascaded interconnection of two identical all-pass sub-filters leading to power-complementary or magnitude-complementary filter pairs. For these structures, the all-pass sub-filters are of a very low order, thereby making the adjustment of the crossover frequency significantly easier.

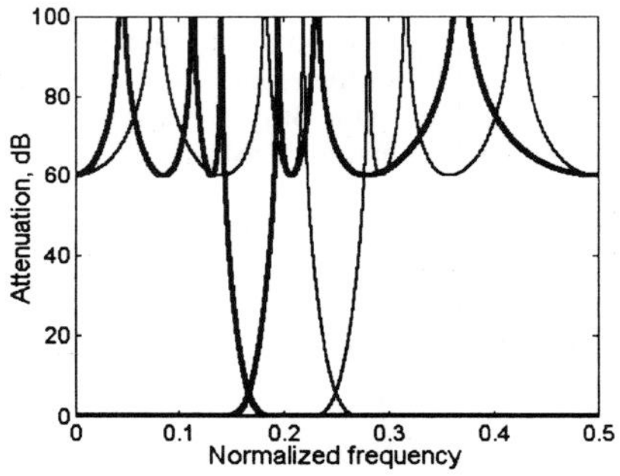

Figure 7. Magnitude-complementary filter pairs constructed using 6 identical copies of the first-order and the second order all-pass filters. Thick-line: f_{3dB} =1/4. Thin line: f_{3dB} =1/6.

5. ACKNOWLEDGEMENTS

This work was supported by the Academy of Finland, project No. 44876 (Finnish Centre of Excellence Program, 2000–2005). Authors greatly appreciate the inspiring discussions with Prof. Markku Renfors, Tampere University of Technology.

REFERENCES

[1] S. K. Mitra, Digital Signal Processing: A Computer-Based Approach. McGraw-Hill, 2001.

[2] R. Nouta, "The Jaumann structure in wave-digital filters," *Int. J. Circuit Theory Applicat.*, vol. 2, pp. 163–174, June 1974.

[3] A. Fettweis and H. Levin, and A. Sedlmeyer, "Wave digital lattice filters," *Int. J. Circuit Theory Applicat.*, vol. 2, pp. 203–211, June 1974.

[4] L. Gazsi, "Explicit formulas for lattice wave digital filters," *IEEE Trans. Circuits and Syst.*, vol. CAS-32, pp. 203–211, Jan. 1985.

[5] M. Renfors and T. Saramäki, "Recursive Nth-band digital filters – Part I: Design and properties, Part II: Design of multistage decimators and interpolators," *IEEE Trans. Circuits and Syst.*, vol. CAS-34, pp. 24–51, Jan. 1987.

[6] M. D. Lutovac, D. V. Tošić, B. L. Evans, *Filter Design for Signal Processing Using MATLAB and Mathematica.* Upperside River, New Jersey: Prentice Hall, 2000.

[7] M. D. Lutovac and Lj. D. Milić, "Design of computationally efficient elliptic IIR filters with a reduced number of shift-and-add operations in multipliers," *IEEE Trans. Signal Processing*, vol. 45, no. 10, pp. 2422 –2430, Oct.1997.

[8] T. Saramäki and M. Renfors, "A novel approach for the design of IIR filters as a tapped cascaded interconnection of identical allpass subfilters," in *Proc. 1987 IEEE Int. Symp. Circuits Syst.* (Philadelphia, Pennsylvania), May 1987, vol. 2, pp. 629–632.

[9] H. Johansson and T. Saramäki, "A class of complementary IIR filters," in *Proc. 1999 IEEE Int. Symp. Circuits Syst.* (Orlando, Florida), July 1999, vol. 3, pp. 299–302.

[10] H. Johansson and L. Wanhammar, "High-speed recursive filter structures composed of identical all-pass subfilters for interpolation, decimation, and QMF banks with perfect magnitude reconstruction," *IEEE Trans. Circuits Syst. I*, vol. 46, no. 1, pp. 16–28, January 1999.

---

[3] This $F(z)$ is a transfer function of a linear-phase FIR half-band filter. The zero-phase frequency response is obtained by omitting the phase term and is expressible as $F(\omega) = 1/2 + 2\sum_{n=1}^{(M+2)/4} a[M/2-(2n-1)]\cos[(2n-1)\omega]$.

A New Class of Even Length Wavelet Filters

David B. H. Tay
Department of Electronic Engineering,
LaTrobe University,
Bundoora, Victoria 3086, Australia.
Fax: +61 3 9471 0524; Tel: +61 3 9479 2529; E-mail: d.tay@ee.latrobe.edu.au

Abstract— **A new class of biorthogonal wavelet filters and its design is presented in this work. The filters are even in length and have linear phase response. The new filter class is a modification of the Halfband Pair Filter Bank (HPFB, which only yields odd length filters) and is constructed using the Parametric Bernstein Polynomial. Perfect reconstruction is inherent in the structure of the filters and the desired number of vanishing moments can be easily achieved by setting the appropriate parameters of the Bernstein Polynomial to zero. The design of the non-zero parameters is achieved through a least squares method which is non-iterative. The design techniques allows filters with different characteristics to be designed with ease.**

I. Introduction

The DWT (Discrete Wavelet Transform) is a versatile tool for many signal processing applications [1],[2],[3]; most notably in image compression. The design of wavelet filters is equivalent to the design of perfect reconstruction (PR) multirate filter banks. The main difference between traditional filter bank design and wavelet filter design is the imposition of vanishing moments in the latter case. Vanishing moments is achieved by imposing zeros at the aliasing frequency (for the 2-channel case the zeros are at $z = -1$) and is the main mechanism for regularity in the corresponding wavelet function. The PR and regularity conditions impose constraints on the filter coefficients which can complicate the design process. The design process basically amounts to the optimization of the coefficients subject to the PR and vanishing moment constraints.

In the previous work [4] the design of the class of HPFB (Halfband Pair Filter Bank) [5] was considered. The parametric Bernstein polynomial was used in the design and this lead to a structure in which perfect reconstruction and vanishing moments constraints were automatically satisfied. These structural properties help to reduce the complexity of the design process and a simple least squares design process was presented in [4]. However the filters in [4] always have odd length. It has been observed by some researchers [6] that even length filters might be better in some image coding applications as it overcomes some of the problems associated with the shift variant nature of multirate systems. Even length filters are fundamentally different from odd length filters as the former yields antisymmetric wavelets while the latter yields symmetric wavelets. Even length filters are able to provide half sample delay and this property is crucial in the construction of the dual-tree complex wavelet transform [7].

In this paper a new class of filters is introduced by modifying the HPFB. The filters have even length and will be referred to as EBFB (Even-length Bernstein Filter Bank). The EBFB has the same advantages as the HBFB as described above. An efficient two-stage, least squares design technique is employed for the design of the EBFB. The technique is non-iterative and involves solving linear equations and filters with different characteristics can be designed with ease.

II. Even Length Bernstein Filter Bank

We begin by describing the Parametric Bernstein Polynomial (first introduced in [8]) which is used in the construction of the EBFB. The Parametric Bernstein Polynomial is defined as:

$$B_N(x;\alpha) \equiv \sum_{i=0}^{N} f(i) \binom{N}{i} x^i (1-x)^{N-i} \quad (1)$$

where N is odd, $\alpha = [\ \alpha_0 \ \ \ldots \ \ \alpha_{(N-1)/2}\ ]^T$ and

$$f(i) \equiv \begin{cases} 1-\alpha_i & 0 \le i \le \frac{1}{2}(N-1) \\ \alpha_{N-i} & \frac{1}{2}(N+1) \le i \le N \end{cases}$$

The polynomial can be transformed into a z-transform filter function by the following substitution:

$$x = \frac{1}{4} z (1-z^{-1})^2 = \sin^2\left(\frac{\omega}{2}\right)$$

and satisfies the halfband filter condition:

$$B(x) + B(1-x) = 1 \quad (2)$$

where for brevity $B(x) = B_N(x;\alpha)$. If we set

$$\alpha_i = 0 \quad \text{for} \quad i = 0, \ldots, L.$$

then $B(x)$ can be factorized as:

$$B(x) = (1-x)^{L+1} R(x)$$

where $R(x)$ is the remainder polynomial, ie. $B(x)$ has a $(L+1)$ order zero at $x = 1$. The equivalent z-transform filter function has a $2(L+1)$ order zero at $z = -1$.

We now define two functions $H(x)$ and $F(x)$ as:

$$H(x) \equiv (1-x)^{-1/2} B_1(x) \quad (3)$$

$$F(x) \equiv (1-x)^{1/2} [\, B_1(x) + 2 B_2(x) \quad (4)$$

$$- 2 B_1(x) B_2(x) \,] \quad (5)$$

0-7803-7761-3/03/$17.00 ©2003 IEEE

where

$$B_1(x) = B_N(x; \alpha/\alpha_0 = 0) \quad (6)$$

$$B_2(x) = B_M(x; \beta/\beta_0 = 0) \quad (7)$$

In (6) and (7), the first parameter is set to zero; ie. $\alpha_0 = \beta_0 = 0$. This means that both $B_1(x)$ and $B_2(x)$ have *at least one* $(1-x)$ factor each. Using the functions defined above the low-pass filters of the EBFB are given by

$$H_0(z) \equiv z^{1/2} H(\tfrac{1}{4}z(1-z^{-1})^2) \quad (8)$$

$$F_0(z) \equiv z^{-1/2} F(\tfrac{1}{4}z(1-z^{-1})^2) \quad (9)$$

and the high-pass filters are given by:

$$H_1(z) = z^{-1}H_0(-z) \qquad F_1(z) = zF_0(-z) \quad (10)$$

Note that although there appears to be irrational factors (eg. $z^{1/2}$ and $(1-x)^{1/2}$) in the expressions above, it can be shown that the resulting filter functions $H_0(z)$ and $F_0(z)$ are valid FIR filter functions, ie. Laurent polynomial in z. Furthermore the filters are of even length and possess linear phase response. Using (2), (3), (4), (8) and (9), it can be easily verified that:

$$H_0(z)F_0(z) + H_0(-z)F_0(-z) = 1$$

so that perfect reconstruction is satisfied.

The desired degree of vanishing moments for $H_0(z)$, $(2L_H + 1)$, can be achieved by setting $\alpha_i = 0$ for $i = 0, \ldots, L_H$. The degree of vanishing moments for $F_0(z)$ is $(2\min(L_H, L_F) + 3)$ where $\beta_i = 0$ for $i = 0, \ldots, L_F$.

The advantages of the EBFB are that:
1. perfect reconstruction and
2. vanishing moments

are structurally imposed. Hence, the optimization that needs to be carried out on the remaining non-zero free parameters is an unconstrained one. In this paper, the objective function that is considered in the optimization is the energy of the ripples in the frequency domain.

III. Least Squares Design of Free Parameters

The filter function $H(x)$ in (3) and $F(x)$ in (4) are irrational function of the variable x. A least squares formulation in the variable x requires numerical integration for its solution, ie. non-analytical solution. To overcome this problem we propose the following change of variable:

$$x = 1 - y^2 \iff y = (1-x)^{1/2} = \cos\left(\frac{\omega}{2}\right) \quad (11)$$

Over the range $0 \le \omega \le \pi$, y monotonically decreases with ω. Note in particular that when $\omega = 0$ (DC), $y = 1$ and when $\omega = \pi$ (aliasing frequency), $y = 0$.

Using the change of variable (11), it can be shown that $B_1(x)$ and $B_2(x)$ in (6) and (7) can be expressed as:

$$B_1(y) = K^0(y) - \sum_{l=L_H+1}^{(N-1)/2} k_l^0(y)\,\alpha_l \quad (12)$$

$$B_2(y) = K^1(y) - \sum_{l=L_F+1}^{(M-1)/2} k_l^1(y)\,\beta_l \quad (13)$$

where

$$K^0(y) = \sum_{i=0}^{(N-1)/2} \binom{N}{i} (1-y^2)^i y^{2N-2i}$$

$$K^1(y) = \sum_{i=0}^{(M-1)/2} \binom{M}{i} (1-y^2)^i y^{2M-2i}$$

$$k_l^0(y) = \binom{N}{l} [(1-y^2)^{N-l} y^{2l} - (1-y^2)^l y^{2N-2l}]$$

$$k_l^1(y) = \binom{M}{l} [(1-y^2)^{M-l} y^{2l} - (1-y^2)^l y^{2M-2l}]$$

Using the expressions above, it can be shown that $H(x)$ and $F(x)$ in (3) and (4) can be expressed as:

$$H(y) = y^{-1}K^0(y) - \sum_{l=L_H+1}^{(N-1)/2} y^{-1}k_l^0(y)\,\alpha_l \quad (14)$$

$$F(y) = y\,K^2(y) - \sum_{l=L_F+1}^{(M-1)/2} y\,k_l^2(y)\,\beta_l \quad (15)$$

where

$$K^2(y) \equiv B_1(y) + 2(1 - B_1(y))K^1(y)$$

$$k_l^2(y) \equiv 2(1 - B_1(y))k_l^1(y)$$

Note the slight abuse of notation in (12), (13), (14) and (15) where the same symbols for the functions B_1, B_2, H and F are used after the change of variable. With the change of variable, the filter functions H and F are now polynomials in y (note that the factor y^{-1} in (14) will be cancelled by the factors in $K^0(y)$ and $k_l^0(y)$). It can also be readily seen that the *filter function in (14) is a linear function of the free parameter α_i and the filter function in (15) is a linear function of the free parameters β_i (when α_i is fixed).*

There are two sets of free parameters (α and β) to be determined in the design process. Filter H_0 depends only on α, but filter F_0 depends on both α and β. A *two-stage* approach is adopted in the design process:
1. The α parameters are first determined by optimizing H_0.
2. The β parameters are next determined by optimizing F_0 with fixed values of α that was determined from the first stage.

For convenience, we denote the generic form of (14) and (15) by:

$$G(y) = K(y) - \sum_{l=L_1}^{L_2} k_l(y)\,\theta_l$$

where $G(y)$ can be either $H(y)$ or $F(y)$ and θ_l can be either α_l or β_l. Note that the factor y^{-1} in (14) or y in (15) is absorbed into $K(y)$ and $k_l(y)$. Consider now the following objective function

$$E \equiv (1-\gamma)\int_{y_p}^{1} (G-1)^2\,dy + \gamma\int_0^{y_s} G^2\,dy \quad (16)$$

$$= (1-\gamma)\int_{y_p}^{1}\left(K(y) - 1 - \sum_{l=L_1}^{L_2} k_l(y)\,\theta_l\right)^2 dy$$

$$+ \gamma\int_{0}^{y_s}\left(K(y) - \sum_{l=L_1}^{L_2} k_l(y)\,\theta_l\right)^2 dy \qquad (17)$$

where $0 \le \gamma \le 1$; $y_s = \cos\left(\frac{\omega_s}{2}\right)$ and $y_p = \cos\left(\frac{\omega_p}{2}\right)$ represents the stopband and passband edges respectively. Now E is a quadratic function of the θ_l's and represents the weighted sum of the passband-ripple energy and stopband energy. The optimum values of θ_l (by minimizing E) can be obtained by solving the set of simultaneous equations (similar to the Wiener-Hopf or normal equations found in adaptive filter theory):

$$\mathbf{R}\,\theta = \mathbf{p} \qquad (18)$$

where

$$\theta = [\ \theta_{L_1} \ \ \ldots \ \ \theta_{L_2}\ ]^T$$

and the elements of $\mathbf{R} = [r_{lm}]$ and $\mathbf{p} = [p_l]$ ($0 \le l, m \le (L_2 - L_1)$) are respectively

$$r_{lm} \equiv \gamma\int_{0}^{y_s} k_l(y)\,k_m(y)\,dy + (1-\gamma)\int_{y_p}^{1} k_l(y)\,k_m(y)\,dy$$

and

$$p_l \equiv \gamma\int_{0}^{y_s} K(y)\,k_l(y)\,dy + (1-\gamma)\int_{y_p}^{1} (K(y)-1)\,k_l(y)dy$$

Note that all integrations can be performed analytically as the functions involved are polynomials in y.

IV. Design Examples

For all the examples in this section, $N = M = 5$ and this gives filters H_0 and F_0 with lengths 10 and 22 respectively. In the plots, the frequency response of filter H_1 (which is the high-pass version of F_0 - see (10)) will be shown instead of F_0.

Example 1: We choose $L_H = L_F = 1$ ($\alpha_{0,1} = \beta_{0,1} = 0$) and this gives H_0 and F_0 with 3 zeros and 5 zeros, respectively, at $z = -1$. In the optimization $\omega_p = 1.26$, $\omega_s = 2.12$, $\gamma = 0.5$ for H_0 and $\omega_p = 1.26$, $\omega_s = 2.04$, $\gamma = 0.5$ for F_0. The optimized free parameter values are: $\alpha_2 = -0.2031$, $\beta_2 = -0.6243$. Figure 1 shows the frequency response of the filter pair. Also shown in Figure 1 are the responses of the maximally flat pair (all free parameters set to zero). We readily see the increased sharpness of this example over the maximally flat equivalent. Figure 2 shows the corresponding analysis scaling functions of this example. Also shown are the scaling functions of Daubechies [9] (minimum phase and least symmetric). All scaling functions shown have comparable degree of smoothness.

Example 2: We choose $L_H = L_F = 0$ ($\alpha_0 = \beta_0 = 0$) and this gives H_0 and F_0 with 1 zeros and 3 zeros, respectively, at $z = -1$. In the optimization $\omega_p = 1.26$, $\omega_s = 2.12$, $\gamma = 0.5$ for H_0 and $\omega_p = 1.26$, $\omega_s = 2.04$, $\gamma = 0.5$ for F_0. The optimized free parameter values are: $\alpha_1 = 0.0217$, $\alpha_2 = -0.2773$, $\beta_1 = 0.2998$, $\beta_2 = -1.4159$. Figure 3 shows the frequency response of the filter pair. Also shown in Figure 3 are the responses from example 1. Comparing the two examples, we see that with a reduction in the degree of flatness (number of zeros at $z = -1$), a higher stopband attenuation is attainable. Note also that filter H_1 has a sharper roll-off in example 2.

Example 3: We choose $L_H = L_F = 0$ ($\alpha_0 = \beta_0 = 0$) and this gives H_0 and F_0 with 1 zeros and 3 zeros, respectively, at $z = -1$. In the optimization $\omega_s = 2.12$, $\gamma = 1$ for H_0 and $\omega_s = 1.88$, $\gamma = 1$ for F_0. The optimized free parameter values are: $\alpha_1 = 0.0423$, $\alpha_2 = -0.4430$, $\beta_1 = 0.4920$, $\beta_2 = -1.9982$. Figure 4 shows the frequency response of the filter pair. Also shown in Figure 4 are the responses from example 3. Compared with example 2, the filters in example 3 has a sharper roll-off but at the expense of increased ripple magnitude.

The examples above illustrate the flexibility of being able to design wavelet filters with different characteristics.

V. Conclusion

A new class of wavelet filters, called the EBFB (Even-length Bernstein Filter Bank), has been presented in this work. The filters have even length, linear phase and are constructed from the parametric Bernstein polynomial. The advantages of the EBFB are that perfect reconstruction and vanishing moments constraints are structurally imposed. A two-stage approach for the design of the filters has been presented. At each stage of the design, the least squares criterion was applied and this led a simple design process which involves solving a set of linear normal equations. Design examples were presented to illustrate the effectiveness and flexibility of the design technique. A possible extension of this work is to use the minimax criterion instead of the least squares for filter design.

References

[1] P. P. Vaidyanathan. *Multirate Systems and Filter Banks.* Prentice-Hall, 1993.

[2] M. Vetterli and J. Kovacevic. *Wavelets and Subband Coding.* Prentice-Hall, 1995.

[3] G. Strang and T. Nguyen. *Wavelets and Filter Banks.* Prentice-Hall, 1996.

[4] D. B. H. Tay. Two-Stage, Least Squares Design of Biorthogonal Filter Banks. In *Proc. IEEE Symp. Circuits Sys.*, 2000.

[5] S. M. Phoong, C. W. Kim, P. P. Vaidyanathan, and R. Ansari. A New Class of Two-Channel Biorthogonal Filter Banks and

Wavelet Bases . *IEEE Trans. Signal Proc.*, 43(3):649, March 1995.

[6] J. D. Villasenor, B. Belzer, and J. Liao. Wavelet Filter Evaluation for Image Compression . *IEEE Trans. Image Proc.*, 4(8):1053, August 1995.

[7] N. G. Kingsbury. Shift Invariant Properties of the Dual-Tree Complex Wavelet Transform. In *Proc. ICASSP*, 1999.

[8] H. Caglar and A. N. Akansu. A generalized parametric PR-QMF design technique based on Bernstein polynomial approximation. *IEEE Trans. Signal Proc.*, 41(7):2314, July 1993.

[9] I. Daubechies. *Ten Lectures on Wavelets.* Society for Industrial and Applied Maths., 1992.

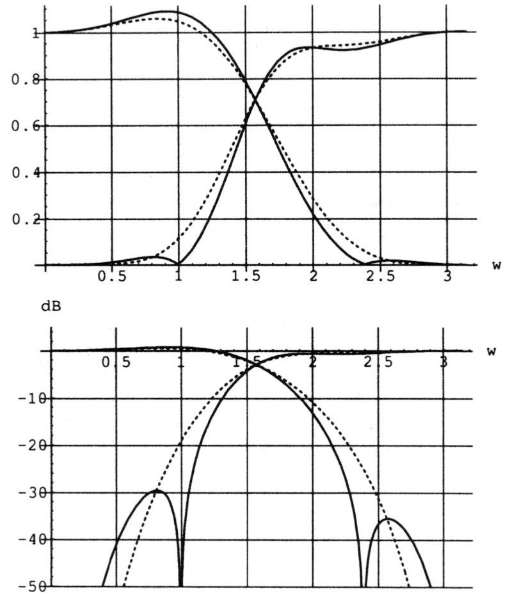

Fig. 1. Frequency response of filter pair (H_0, H_1) in example 1. Top diagram - linear, bottom diagram - logarithmic. Solid line: least-squares optimized. Dashed line: maximally flat equivalent.

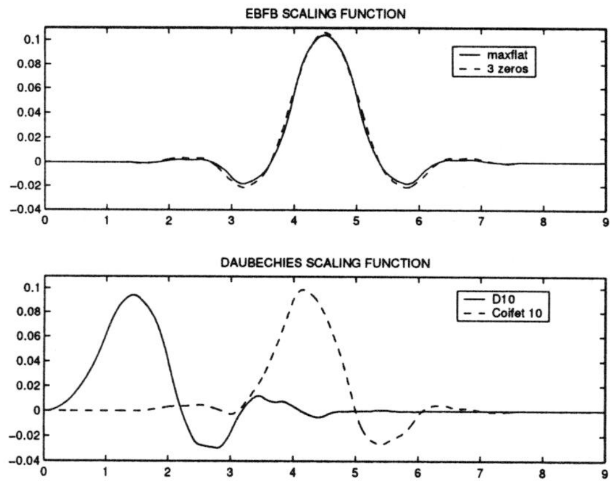

Fig. 2. Scaling function of EBFB and Daubechies. Top diagram: analysis filter H_0 from example 1. Bottom diagram: orthonormal filters of Daubechies.

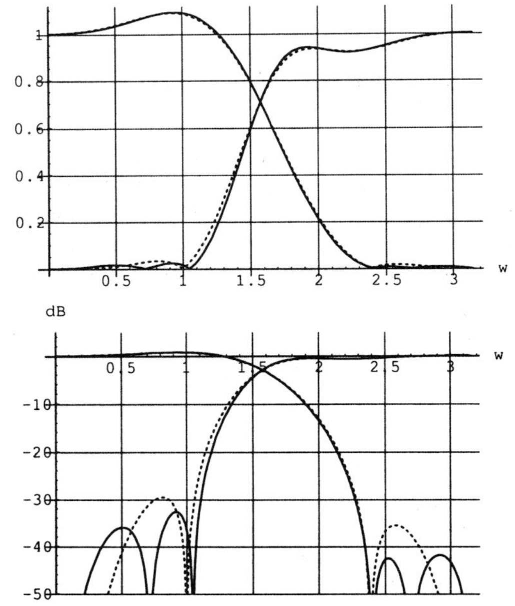

Fig. 3. Frequency response of filter pair (H_0, H_1) Top diagram - linear, bottom diagram - logarithmic. Solid line: example 2. Dotted line: example 1.

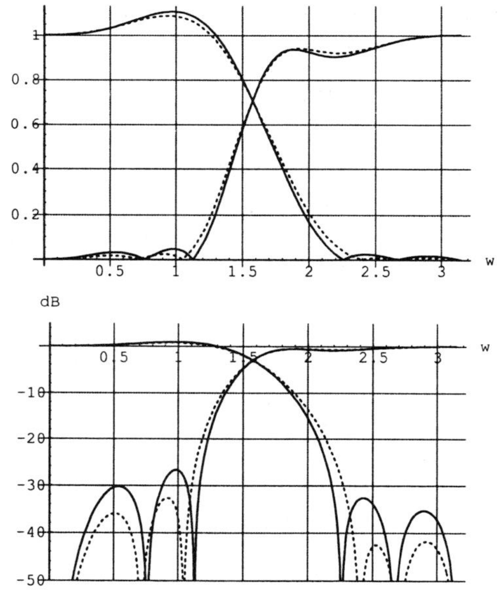

Fig. 4. Frequency response of filter pair (H_0, H_1) Top diagram - linear, bottom diagram - logarithmic. Solid line: example 3. Dotted line: example 2

SDP FOR MULTI-CRITERION QMF BANK DESIGN

H.D. Tuan, L.H. Nam

Dept. of Electrical and Computer Eng.,
Toyota Technological Institute,
2-1-12 Hisakata, Tenpaku, Nagoya, Japan
Email: {tuan,nam}@toyota-ti.ac.jp

H. Tuy

Institute of Mathematics,
P.O. Box 631, Bo Ho,
Hanoi, Vietnam
Email: htuy@hn.vnn.vn

T. Q. Nguyen

Dept. of Electrical and Computer Eng.,
University of California in San Diego,
9500 Gilman Dr., La Jolla CA 92093-0407, USA
Email: nguyent@ece.ucsd.edu

ABSTRACT

Quadrature mirror filter (QMF) bank with multi-criterion constraints such as minimal aliasing and/or minimal error coding is among the most important problems in filter bank design, for solving which linear algebra-based methods are still heuristic and do not always work, especially for large filter length. It is shown in this paper that when filters are of nonlinear phase this problem can be reduced to convex linear matrix inequality (LMI) optimization, which can be very efficiently solved either by the standard LMI solvers or our previously developed solver. The proposed computationally tractable optimization formulations are confirmed by several simulations.

1. INTRODUCTION

Quadrature mirror filter (QMF) bank is a very important class in digital filter bank because of its potential applications in subband coding [1, 7]. The starting point is the two-channel QMF bank design problem, which can be formulated as follows (see Fig. 1).
Given two analysis filters $H_0(z)$ and $H_1(z)$, design two (synthesis) filters $G_0(z)$ and $G_1(z)$ such that

$$H_0(z)G_0(z) + H_1(z)G_1(z) = cz^{-n_0}, \quad n_0 > 0, c \neq 0, \quad (1)$$

$$H_0(-z)G_0(z) + H_1(-z)G_1(z) = 0. \quad (2)$$

It is well known that (1)-(2) deal with the perfect reconstruction of the filter bank in the absence of the noises q_0, q_1: (1) is the distortion-free condition while (2) is the aliasing-free one. To accomplish this common performance measure, the standard design procedure (see e.g. [1, p. 138]) is first to make a particular choice $G_0(z) = -H_1(-z), G_1(z) = H_0(-z)$ for the synthesis filters to satisfy the aliasing-free condition (2) and then H_i are designed to complement the perfect reconstruction condition (1). With this procedure, there may not be much freedom left for the designed analysis filters H_i to satisfy other performance measures to make QMF bank reliable for practical applications.
By changing G_i to G_i/c, $i = 0, 1$, if necessary, without loss of generality we can set $c = 1$. Thus, a natural optimization formulation for handling constraints (1), (2) is as follows

$$\min_{G_0(z), G_1(z)} ||z^{-n_0} - [H_0(z)G_0(z) + H_1(z)G_1(z)]||_\infty : (2), \quad (3)$$

where $||.||_\infty$ denotes the standard $\mathcal{H}_\infty$-norm, i.e. we try to minimize the distortion effect while keeping the aliasing-free condition.
Note that constraint (2) implies that

$$\begin{bmatrix} G_0(z) \\ G_1(z) \end{bmatrix} = G(z) \begin{bmatrix} H_1(-z) \\ -H_0(-z) \end{bmatrix} \quad (4)$$

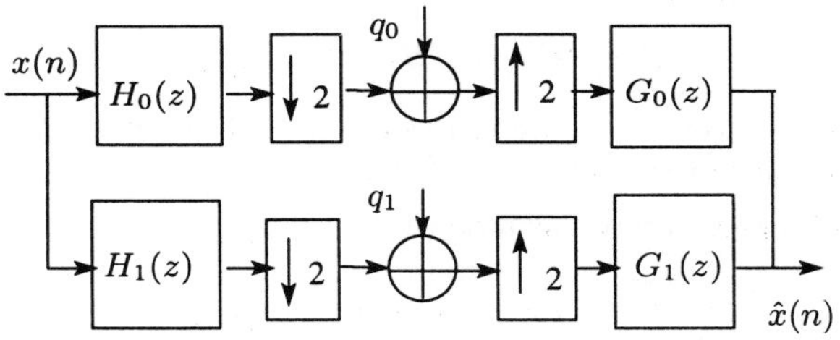

Figure 1: QMF filter bank with 2 analysis and 2 synthesis channels.

with some $G(z)$. Substituting this value into (3), one gets the equivalent optimization formulation of problem (3)

$$\min_{G(z)} ||z^{-n_0} - H(z)G(z)||_\infty, \quad (5)$$

$$\text{where } H(z) = H_0(z)H_1(-z) - H_1(z)H_0(-z). \quad (6)$$

When $G(z)$ is assumed stable proper rational (i.e. $G(z) \in \mathcal{RH}^\infty$), problem (5) is just a model-matching or Nehari problem, which has been well studied in the classical $\mathcal{H}_\infty$ control and thus the optimal solution $G_{opt}(z)$ of (3) can be easily found [2]. However, the order of this optimal solution $G_{opt}(z)$ is too high and generally the resulting synthesis filters $G_i(z)$ are required to be of infinite impulse response (IIR) even when the analysis filters H_i are of finite impulse response (FIR). The classical frequency-domain based techniques of $\mathcal{H}_\infty$ control are unable to address the case of optimal fixed order of G. Therefore, for computational implementation one has to use the heuristic balanced truncation technique to reduce the order of $\mathcal{H}_\infty$ solution [4], which does not always work. Moreover, it is impossible within this approach to impose additional constraints for multi-performances.
In this paper, we propose a most natural approach to problem (5) by considering it just as a filtering problem, as it is ! The rationale is that the filtering problem is simpler than the general $\mathcal{H}_\infty$ control problem and therefore can be more properly and efficiently handled than the latter. Actually, our purpose is twofold:

- To show that all the above mentioned difficulties can be overcome in the $\mathcal{H}_\infty$ filtering approach.
- More importantly, to show that a variety of additional constraints to give the designed filter bank desired characteristics such as no distortion, aliasing energy minimization, constrained nonlinearity minimization and coding error minimization, which cannot be handled by the above mentioned existing approaches, can be dealt with by our approach relatively easily.

0-7803-7761-3/03/$17.00 ©2003 IEEE

The structure of the paper is as follows. Section 2 gives different optimization formulations for the QMF bank design problems. The semi-definite programming based solution method for nonlinear phase filters is presented in Section 3. Section 4 is specially devoted to the development of an optimization algorithm for solving the error coding minimization of QMF. Some simulation examples described in Section 5 confirm the viability of our theoretical development in the previous sections.

2. OPTIMIZATION-BASED DESIGN FORMULATIONS

First, let us mention that the optimal solution $G_0(z), G_1(z)$ of optimization problem (3) is not unique and actually there is a set of such optimal solutions. Thus, assuming that the optimal value of (3) is γ, its optimal solution can be fully characterized by constraint (2) and the inequality

$$||z^{-n_0} - [H_0(z)G_0(z) + H_1(z)G_1(z)]||_\infty \leq \gamma. \quad (7)$$

Under these constraints, we can tackle more challenging QMF bank design problems such as the following:

- Aliasing energy minimization [1, pp.295]

$$\min_{G_0(z),G_1(z)} \frac{1}{2\pi}\int_{\pi}^{\pi} |G_0(e^{\jmath\omega})|^2 S_{xx}(e^{\jmath(\omega+\pi)})|H_0(e^{\jmath(\omega+\pi)})|^2 d\omega$$
$$\text{subject to } (2), (7), G_0(z) = G_1(-z), \quad (8)$$

where $R_{xx}(k)$ is the autocorrelation function of the input signal $x(n)$, and S_{xx} is the frequency domain representation of R_{xx}.

- Constrained nonlinearity minimization [1, pp. 296]

$$\min_{G_0(z),G_1(z)} \sum_{i=0}^{1}\sum_{\ell=0}^{[(n+m)/2]-1} [\tilde{h}_{i\ell} - \tilde{h}_{i(n+m-\ell)}]^2 \quad , \quad (9)$$
$$\text{subject to } (2), (7)$$

where $\tilde{h}_{i\ell}$ are the coefficients of $H_i(z)G_i(z)$, $i = 0, 1$, which can be easily seen as linear functions in the coefficients of the synthesis filters G_i

It is clear that the objectives in (8), (9) are convex quadratic in the coefficients of synthesis filters $G_0(z), G_1(z)$, while in subsequent sections, the constraints (2), (7) will be shown to be convex in these coefficients as well. Consequently, both (8) and (7) are convex optimization problems, which can be solved efficiently by existing optimization codes. We now formulate another important problem of coding error minimization, which is nonconvex but nevertheless can be very efficiently solved by the method to be developed in Section 4.

Let $q_0(n), q_1(n)$ in Fig. 1 be the noise source by quantization, which are assumed wide stationary (WSS), white, uncorrelated and uniformly distributed with zero means. Let b_i denote the number of bits per sample of the $i-$subband signal $X_i(z) = \frac{1}{2}H_i(z)[X(z^{1/2}) + X(-z^{1/2})]$ with variance $\sigma^2_{x_i}$. Then the variance $\sigma^2_{q_i}$ of $q_i(n)$ is given by $\sigma^2_{q_i} = c2^{-b_i}\sigma^2_{x_i}$ with some constant c [5]. Thus, the distortion caused by quantization is

$$e = ||G_0||_2^2\sigma^2_{q_0} + ||G_1||_2^2\sigma^2_{q_1} = c(2^{-b_0}||G_0||_2^2\sigma^2_{x_0} + 2^{-b_1}||G_1||_2^2\sigma^2_{x_1}),$$

where $||.||_2$ is the standard $\mathcal{H}_2$-norm. Now, under the fixed bit rate assumption $b_0 + b_1 = b =$ const, the coding error minimization problem can be formulated as

$$\min_{G_0(z),G_1(z),b_0,b_1} \left\{ c\sum_{i=0}^{1} 2^{-b_i}||G_i||_2^2\sigma^2_{x_i} : (2), (7), b_1 + b_2 = b \right\}$$
$$= c2^{1-b/2}\sigma^2_{x_0}\sigma^2_{x_1} \times \min_{G_0(z),G_1(z)} \{||G_0||_2^2||G_1||_2^2 : (2), (7),$$
$$2^{-b}||G_1||_2 \leq ||G_0||_2 \leq 2^b||G_1||_2\}. \quad (10)$$

Now, the objective function (10) is nonconvex so the optimization problem (10) seems to be a very difficult one. Nevertheless, as it will be shown in Section 4, (10) belongs to the realm of concave and monotonic optimization [10, 11] and based on this fact, we can develop an extremely efficient algorithm which requires just a few convex or linear programs to be solved.

3. NONLINEAR PHASE FIR FILTER DESIGN

Suppose that the FIR analysis filters $H_0(z)$ and $H_1(z)$ are of m-order, i.e.

$$\begin{bmatrix} H_0(z) \\ H_1(z) \end{bmatrix} = \begin{bmatrix} h_{0m}z^{-m} + ... + h_{01}z^{-1} + h_{00} \\ h_{1m}z^{-m} + ... + h_{11}z^{-1} + h_{10} \end{bmatrix}$$
$$= \begin{bmatrix} h^0 \\ h^1 \end{bmatrix} [z^{-m} \quad ... \quad z^{-1} \quad 1]^T, \quad (11)$$
$$h^i = [h_{im} \quad ....h_{i1} \quad h_{i0}], \; i = 0, 1.$$

Then we can express the FIR filter $H(z)$ defined by (6) as

$$H(z) = \sum_{k=0}^{2m} h_k z^{-k}, \quad (12)$$

where

$$h_k = \sum_{\ell=\max\{0,k-m\}}^{\min\{k,m\}} (-1)^{k-\ell}(h_{0\ell}h_{1(k-\ell)} - h_{1\ell}h_{0(k-\ell)}),$$
$$0 \leq k \leq 2m. \quad (13)$$

Accordingly, the synthesis filters $G_0(z)$ and $G_1(z)$ are designed among the classes of FIR with some order n:

$$\begin{bmatrix} G_0(z) \\ G_1(z) \end{bmatrix} = \begin{bmatrix} g_{0n}z^{-n} + ... + g_{01}z^{-1} + g_{00} \\ g_{1n}z^{-n} + ... + g_{11}z^{-1} + g_{10} \end{bmatrix}$$
$$= \begin{bmatrix} g^0 \\ g^1 \end{bmatrix} [z^{-n} \quad ... \quad z^{-1} \quad 1]^T, \quad (14)$$
$$g^i = [g_{in} \quad .... \quad g_{i1} \quad g_{i0}], \; i = 0, 1.$$

First, we can write

$$H_0(-z)G_0(z) + H_1(-z)G_1(z) =$$
$$\sum_{k=0}^{n+m}[\sum_{\ell=\max\{0,k-m\}}^{\min\{k,n\}} (-1)^{k-\ell}(g_{0\ell}h_{0(k-\ell)} + g_{1\ell}h_{1(k-\ell)})]z^{-k} \quad (15)$$

Hence, (2) is equivalent to the following linear constraints on $g = [g^{0T} \quad g^{1T}]^T$:

$$\sum_{\ell=\max\{0,k-m\}}^{\min\{k,n\}} (-1)^{k-\ell}(g_{0\ell}h_{0(k-\ell)} + g_{1\ell}h_{1(k-\ell)}) = 0, \quad (16)$$
$$0 \leq k \leq n+m.$$

On the other handwe can write

$$H_0(z)G_0(z) + H_1(z)G_1(z) - z^{-n_0} = \sum_{k=0}^{n+m} L_k(g)z^{-k} - z^{-n_0}, \quad (17)$$

where $L_k(g)$ are affine functions on g and defined by

$$L_k(g) = \sum_{\ell=\max\{0,k-m\}}^{\min\{k,n\}} (g_{0\ell}h_{0(k-\ell)} + g_{1\ell}h_{1(k-\ell)}), \quad (18)$$

$$0 \leq k \leq n+m.$$

The following state space representation is well known

$$H_0(z)G_0(z) + H_1(z)G_1(z) - z^{-n_0} = \left[\begin{array}{c|c} A_{cl} & B_{cl} \\ \hline C_{cl}(g) & L_0(g) \end{array}\right], \quad (19)$$

with matrices $A_{cl}, B_{cl}, C_{cl}(g)$ of dimensions $(n+m) \times (n+m), (n+m) \times 1, (m+n) \times 1$ given by

$$A_{cl} = \begin{bmatrix} 0 & 1 & 0 & .... & 0 & 0 \\ 0 & 0 & 1 & .... & 0 & 0 \\ ... & ... & ... & ... & ... & .... \\ 0 & 0 & 0 & .... & 0 & 1 \\ 0 & 0 & 0 & ... & 0 & 0 \end{bmatrix}, \; B_{cl} = \begin{bmatrix} 0 \\ 0 \\ ... \\ 0 \\ 1 \end{bmatrix}, \quad (20)$$

$$C_{cl}(g) = [\, L_{m+n}(g) \;\; \ldots \;\; L_{n_0}(g) - 1 \;\; \ldots \;\; L_1(g)\,]. \quad (21)$$

Note that A_{cl}, B_{cl} are independent of the design filter coefficient g while $C_{cl}(g)$ is affine dependent on g.
The following result is well known in control theory (see e.g. [12]).

Lemma 1 *One has $||G_0(z)H_0(z)+G_1(z)H_1(z)-z^{-n_0}||_\infty < \gamma$ if and only if the following linear matrix inequality (LMI) is feasible in the symmetric matrix variable P of size $(m+n) \times (m+n)$ and the vector coefficient variable g,*

$$\begin{bmatrix} -P & * & * & * \\ 0 & -\gamma I & * & * \\ PA_{cl} & PB_{cl} & -P & * \\ C_{cl}(g) & L_0(g) & 0 & -\gamma I \end{bmatrix} < 0. \quad (22)$$

Using this leads to the following LMI optimization formulation for the optimization problem (3), (8), (9).

Theorem 1 *Suppose that A_{cl}, B_{cl}, C_{cl} are defined by (20) and (21). Then problem (3) is equivalent to the following LMI optimization problem*

$$\min_{P,\gamma,g} \gamma \text{ subject to } (19), (21), (16), (22). \quad (23)$$

Analogously, the optimization problems (8), (9) are equivalent to the following LMI optimization problems

$$\min_{P,g} \frac{1}{2\pi}\int_{\pi}^{\pi} |G_0(e^{\jmath\omega})|^2 S_{xx}(e^{\jmath(\omega+\pi)})|H_0(e^{\jmath(\omega+\pi)})|^2 d\omega \quad (24)$$

$$\text{subject to } (19), (21), (16), (22),\; G_0(z) = G_1(-z),$$

$$\min_{P,g} \sum_{i=0}^{1} \sum_{\ell=0}^{[(n+m)/2]-1} [\tilde{h}_{i\ell} - \tilde{h}_{i(n+m-\ell)}]^2$$

$$\text{subject to } (19), (21), (16), (22), \quad (25)$$

$$\tilde{h}_{ik} = \sum_{\ell=\max\{0,k-m\}}^{\min\{k,n\}} g_{i\ell}h_{i(k-\ell)}.$$

Before closing this section, let us mention some efficient LMI codes in Matlab such as [3, 8], which will be used in our simulation.

4. GLOBAL OPTIMIZATION FOR CODING ERROR MINIMIZATION

Back to the coding error minimization problem (10), we observe that its objective function $||G_0||_2^2||G_1||_2^2$ can be equivalently replaced by $\log(||G_0||_2^2||G_1||_2^2)$, i.e. (10) is equivalent to following optimization problem

$$\min_{G_0(z),G_1(z),\nu_0,\nu_1} [\log \nu_0 + \log \nu_1] : \qquad (2), (7) \quad (26)$$

$$||G_i||_2^2 < \nu_i,\; i = 0, 1. \quad (27)$$

Note that all constraints (2), (7) and (27) are convex LMI or convex quadratic while the objective function $f(\nu_0, \nu_1) = \log \nu_0 + \log \nu_1$ in (26) is nonconvex but separable concave [10] and monotonic in (ν_0, ν_1) [11]. These useful structures will be exploited later to develop an efficient algorithm for solving (26).
Denote by Ω the projection of the feasible set onto the (ν_0, ν_1)-space:

$$\Omega = \{(\nu_0, \nu_1)|\; \exists(G_0(z), G_1(z)) \text{ s.t. } (2), (7), (27)\}. \quad (28)$$

This is easily seen to be a convex set and the problem now is to minimize the concave function $f(\nu_0, \nu_1) = \log \nu_0 + \log \nu_1$ over this convex set.

We will solve this problem by a branch and bound (BB) algorithm which requires two basic operations [10]:
1. *Bounding*: Given any rectangle $M = [p, q] := \{(\nu_0, \nu_1)|\; p_i \leq \nu_i \leq q_i,\; i = 0, 1\}$, compute a number $\beta(M)$ (lower bound) by the following SDP

$$\min_{G_0(z),G_1(z),\nu_0,\nu_1} \varphi(\nu_0) + \psi(\nu_1) : \quad p \leq \nu_0 \leq q,\; r \leq \nu_1 \leq s,$$

$$(2), (7), (27), \quad (29)$$

with $\varphi(\nu_0) = \log p + \frac{\log q - \log p}{q-p}(\nu_0 - p)$ and $\psi(\nu_1) = \log r + \frac{\log s - \log r}{s-r}(\nu_1 - r)$.

2. *Branching*: Given any rectangle $M = [p, q]$, divide it into subrectangles (subdivision rule).

The algorithm starts from an initial partition consisting of the single rectangle $[p^0, q^0] \times [r^0, s^0]$. At any iteration κ, each rectangle M in the current partition is assigned the lower bound $\beta(M)$. Also a feasible solution is known which is currently the best solution so far obtained (the value of the objective function associated with this current best solution is the current best value). Obviously, all M with $\beta(M)$ larger than this currently best solution can be deleted from further consideration since there will be no better solution inside M. The partition member M_κ with smallest lower bound among the current partition is then selected and further subdivided according to the subdivision rule, before the algorithm goes to the next iteration. The algorithm stops when $\beta(M_\kappa)$ differs from the current best value less than a prescribed tolerance ε : the current best solution is then an approximate optimal solution within tolerance ε (ε-optimal solution).

5. SIMULATION

In this section, the simulation is set up for demonstrating the viability of results developed in the previous Sections. The analysis filters are the same as in Example 1 of [4], i.e. H_0 is a 19th-order low-pass filter designed using the Remez algorithm with transition band $[0.45\pi, 0.55\pi]$, and $H_1(z) = H_0(-z)$, i.e. $h_{1m} = (-1)^m h_{0m},\; m = 0, 1, ..., 19$. Some details of simulation for this example are following:

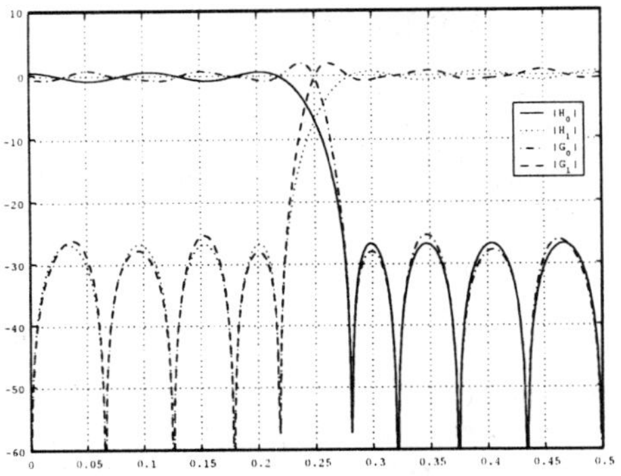

Figure 2: Frequency response of nonlinear phase filters in $\mathcal{H}_\infty$-norm optimization

- $\mathcal{H}_\infty$-norm optimization: the order of the synthesis filters G_i is $n = 75$, and the delay $n_0 = 39$ is chosen like in [4]. Solving using Matlab LMI toolbox gives around 2% reconstruction error over all frequency range (the optimal value of the corresponding optimization problem (23) is $\gamma = 0.0212346443801$), which is as good as the result reported in [4] though our synthesis filters are of FIR while their counterpart are of IIR with the same order in [4]. This is really a surprising result. The resulting frequency responses for these filters are plotted in Fig.2. Frequency responses of H_i are also shown in the same figure.
- Nonlinearity optimization: The same data is used for simulation of direct case with nonlinearity minimization constraint by solving problem (25). The reconstruction error is set to a fixed value at $\gamma = 0.03$. Then the nonlinearity is improving 5.1% compared with the no constraint case of $\mathcal{H}_\infty$-norm only optimization (23). In particular, the values of $\sum_{i=0}^{1}\sum_{\ell=0}^{[(n+m)/2]-1}[\tilde{h}_{i\ell} - \tilde{h}_{i(n+m-\ell)}]^2$ in (9) are 0.92419943 and 0.97367140 for (25) and (24), respectively. Magnitude frequency responses for H_i and G_i are shown in Fig. 3.
- Error coding optimization: As for the example of $\mathcal{H}_\infty$-norm optimization, the coding error $||G_0||_2 * ||G_1||_2 = 0.5316$. Using the same data and setting $\gamma = 0.025$ then the coding error after optimazation procedure is 0.5167, nearly 3% improvement compared to the above case. The frequency responses for H_i and G_i are presented in Fig. 4.

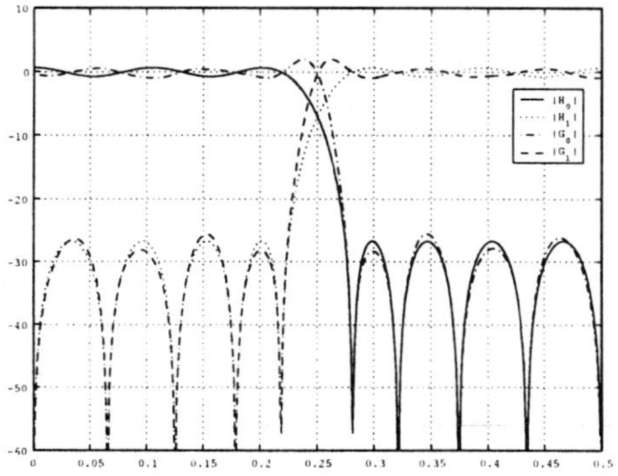

Figure 3: Frequency response of nonlinear phase filters in nonlinearity minimization

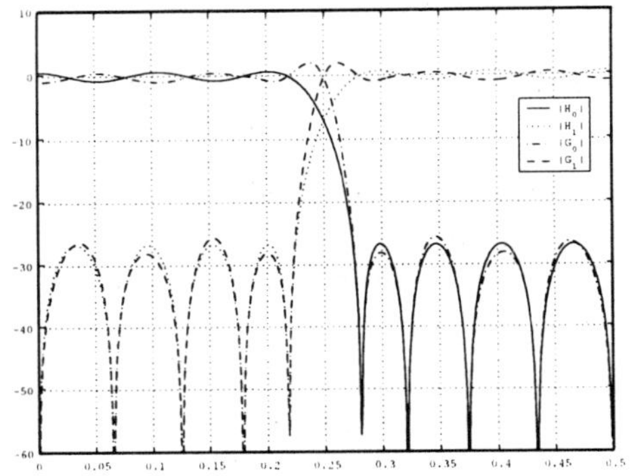

Figure 4: Frequency response of nonlinear phase filters in coding error minimization

6. CONCLUDING REMARKS

In this paper, we have shown that several multicriterion QMF bank design problems can be formulated as computationally tractable optimization problems and thus are effectively solved by the existing softwares in PC. Because of space limitation, we have presented only results related to some of them such as aliasing energy, constrained nonlinearity and error coding minimization for nonlinear phase FIR filter. For other results including that for linear phase FIR we refer the interested reader to [9] and some our subsequent papers.

7. REFERENCES

[1] A.N. Akansu, R.A. Haddad, *Mutiresolution signal decomposition: transforms, subbands, wavelets*, Academic Press, 2001.

[2] T. Chen, B.A. Francis, Design of multirate filter banks by $\mathcal{H}_\infty$ optimization, *IEEE Trans. Signal Processing* 43(1995), 2822-2829.

[3] P. Gahinet, A. Nemirovski, A. Laub, M. Chilali, *LMI control toolbox*, The Math. Works Inc., 1995.

[4] J. Huang, G. Gu, A direct approach to the design of QMF banks via frequency domain optimization, *IEEE Trans. Signal Processing* 46(1998), 2131-2138.

[5] N.S. Jayant, P. Noll, *Digital coding of waveforms*, Prentice-Hall, 1984.

[6] T.Q. Nguyen, Digital filter bank design quadratic constrained formulation, *IEEE Trans. Signal Processing* 43(1995),

[7] G. Strang, T.Q. Nguyen, *Wavelet and filter banks*, Wellesley-Cambridge Press, 1996.

[8] J.F. Sturm, SeDuMi: a Matlab toolbox for optimization over symmetric cones, *http://www.unimaas.nl/ sturm/software/sedumi.html.*

[9] H.D. Tuan, L.H. Nam , H. Tuy, T.Q. Nguyen, Multi-criterion optimized QMF bank design, To appear in *IEEE Trans. on Signal Processing*.

[10] H. Tuy, *Convex analysis and global optimization*, Kluwer Academic, 1998.

[11] H. Tuy, Monotonic optimization: problems and solution approach, *SIAM J. Optimization* 11(2001), No 2, 464-494.

[12] K. Zhou, J.C. Doyle, K. Glover, *Robust and optimal control*, Prentice-Hall, 1996.

EFFICIENT IMPLEMENTATION OF COMPLEX EXPONENTIALLY-MODULATED FILTER BANKS

Juuso Alhava, Ari Viholainen, and Markku Renfors

Tampere University of Technology
Institute of Communications Engineering
P.O. Box 553, FIN-33101 Tampere, FINLAND
e-mail: jualhava@cc.jyu.fi, avi@cs.tut.fi, markku.renfors@tut.fi

ABSTRACT

In Exponentially-Modulated Filter Bank (EMFB) the complex subfilters are generated from a real-valued prototype filter by multiplying the filter impulse response with complex exponential sequences. This gives the possibility to use block transforms with computationally fast algorithms. Our definition for the EMFB is based on Extended Lapped Transform (ELT). Then we may apply Fast ELT based algorithms for cosine-modulated and sine-modulated filter banks (CMFB, SMFB) as basic building blocks in the implementation of this complex filter bank. Alternatively, a modified polyphase structure with 2 branches (being the sampling rate conversion factor) can be used. The latter structure can be used also with non-perfect reconstruction filter bank designs that allow some additional degrees of freedom in the optimization. In this paper we explore these alternative realization structures for CMFBs and SMFBs.

1. INTRODUCTION

The complex exponentially-modulated filter bank we proposed in [1][2] is a tool for subband signal processing of complex signals. The EMFB definition grounds on cosine-modulated filter banks, where the phase of the modulating cosines are as in the extended lapped transform. Then we can use fast algorithms derived for the ELT as a component in EMFB implementations. In addition, we need the "sister filter bank" of the CMFB, namely the sine-modulated filter bank, that can also be implemented using ELT type of structure.

The -band ELT implementation can be divided into prototype filter and modulation parts. The cosine-modulation is done with block transform, Discrete Cosine Transform IV (DCT-IV). If the prototype filter coefficients satisfy the Perfect Reconstruction (PR) condition [3], then we can apply Fast ELT structure with cascaded butterfly sections [4]. When we use nearly-perfect designs [5], the prototype filter can be implemented with 2 polyphase filters, but not with the Fast ELT. These CMFB algorithms require only small changes for SMFB implementations.

This research was carried out in the project "Advanced Multicarrier Techniques for Wireless Communications" funded by the Academy of Finland.

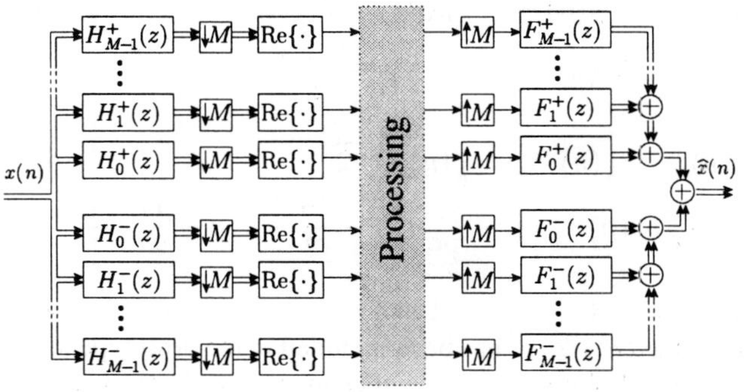

Figure 1: Critically sampled complex filter bank.

The relevant parts of the EMFB topic are considered in Section 2. We define the complex subfilters with CMFB and SMFB subfilter equations and represent the subband signal processing system with Lapped Transform (LT) notations. As the EMFB can be computed with two real filter banks, we give the necessary modifications to convert Fast ELT algorithm for SMFB in Section 3. Then in Section 4 we derive the 2 polyphase structure for the SMFB. The same structure for the CMFB is also given for comparison purposes.

2. EXPONENTIALLY-MODULATED FILTER BANK

The cosine- and sine-modulated synthesis filter equations, based on the ELT definitions, are

$$_k(n) = \sqrt{\frac{2}{}}\, h\ (n) \cos\ \ n + \frac{+1}{2}\ \ k + \frac{1}{2}\ \ \frac{}{} \qquad (1)$$

$$_k(n) = \sqrt{\frac{2}{}}\, h\ (n) \sin\ \ n + \frac{+1}{2}\ \ k + \frac{1}{2}\ \ \frac{}{}\ . \qquad (2)$$

The subscript k is the subchannel index. From these real synthesis filters we can construct the exponentially-mod-

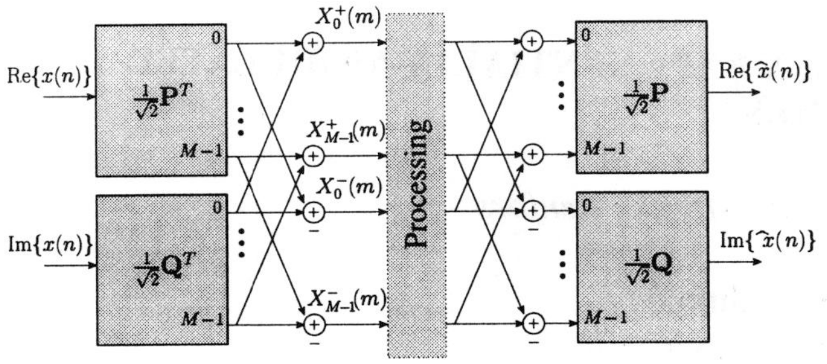

Figure 2: Implementation of the EMFB with LTs.

ulated filter bank whose synthesis filters are

$$_k(n) = \ _k(n) + \cdot \ _k(n) \quad \overline{2} \tag{3}$$
$$= \frac{1}{\overline{2}} h\ (n) \qquad n + \frac{+1}{2} \quad k + \frac{1}{2} \quad —$$

$$_k^-(n) = \ _k(n) - \cdot \ _k(n) \quad \overline{2} \tag{4}$$
$$= \frac{1}{\overline{2}} h\ (n) \qquad - \quad n + \frac{+1}{2} \quad k + \frac{1}{2} \quad — \quad ,$$

where $k = 0, 1, \ldots, \quad - 1$ and $n = 0, 1, \ldots, N - 1$. The scaling factor $\overline{2}$ has been included in the prototype filter. Figure 1 shows the 2 subchannel filter bank, where the analysis filters are time-reversed and complex conjugated versions of the synthesis filters. It is critically sampled, as we take only the real parts of the analysis filter outputs [6]. This system satisfies the PR condition [1].

We collect the CMFB and SMFB subfilters in real N lapped transform matrices and [1][3]

$$[\ ]_{nk} = \ _k(n) \tag{5}$$
$$[\ ]_{nk} = \ _k(n). \tag{6}$$

Then we can represent the computational structure of the complex filter bank with lapped transforms as shown in Figure 2. (Note: The subscipt notations in matrix equations (5) and (6) refer to element on nth row and kth column.) The forward transform matrix ^{T} is equivalent to analysis CMFB and is the inverse transform matrix (synthesis CMFB). ^{T} and are connected similarly with SMFB. We find computationally efficient structure for the EMFB when we implement the LTs with fast algorithms.

3. FAST EXTENDED LAPPED TRANSFORM

3.1. Butterfly Matrices

Fast ELT was presented by Malvar in [4]. The algorithm consists of DCT-IV modulation matrix and cascaded butterfly matrix section. Between each butterfly half of the outputs have delays (forward Fast ELT). The elements of the th butterfly matrix are cosines and sines of the ELT angles θ_i. The number of butterfly matrices depends on the overlapping factor as $= 0, 1, \ldots, \quad - 1$. The design parameter defines the length of the ELT basis functions: $N = 2$. When we write

$$= \text{i} \quad \{\cos\theta_0, \cos\theta_1, \ldots, \cos\theta \quad _{2-1}\} \tag{7}$$

$$= \text{i} \quad \{\sin\theta_0, \sin\theta_1, \ldots, \sin\theta \quad _{2-1}\} \tag{8}$$

the butterfly matrix is

$$= \quad - \quad . \tag{9}$$

Matrix is the reversing block matrix (dimensions 2 2) with ones on its antidiagonal and the other elements are zero

$$= \begin{matrix} & & 1 \\ & 1 & \\ \cdot^{\cdot^{\cdot}} & & \\ 1 & & \end{matrix} \quad . \tag{10}$$

3.2. Fast ELT for Sine-Modulated Filter Bank

The same ELT angles of the CMFB provide fast algorithm for the SMFB. Figures 3 and 4 show the modified Fast ELT flowgraphs. The th butterfly matrix for the SMFB is

$$= \quad - \quad - \quad - \tag{11}$$

only the signs of the sine-terms have been reversed. The other necessary change is to replace DCT-IV with Discrete Sine Transform IV (DST-IV). The elements of the DST-IV matrix are

$$[\ ]_{nk} = \frac{\overline{2}}{} \sin \quad n + \frac{1}{2} \quad k + \frac{1}{2} \quad — \quad . \tag{12}$$

There is a simple connection between and . The basis functions of the DST-IV can be obtained from the DCT-IV by reversing the columns and changing the sign of odd columns:

$$[\ ]_{nk} = (-1)^k [\ ] \ _{-1-n\ k}. \tag{13}$$

This means that fast DST-IV transform can be computed in three steps: I) Change the signs of odd elements in input data vector II) Compute the DCT-IV transform with the fast algorithm [3] III) Reverse the elements in DCT-IV transformed data vector.

4. $2M$ POLYPHASE FILTERS FOR MODULATED FILTER BANKS

4.1. Matrix Decomposition of

The modulating sines for the SMFB can be generated from the DST-IV transform matrix. To show this we decompose the basis functions of the lapped transform in the following manner:

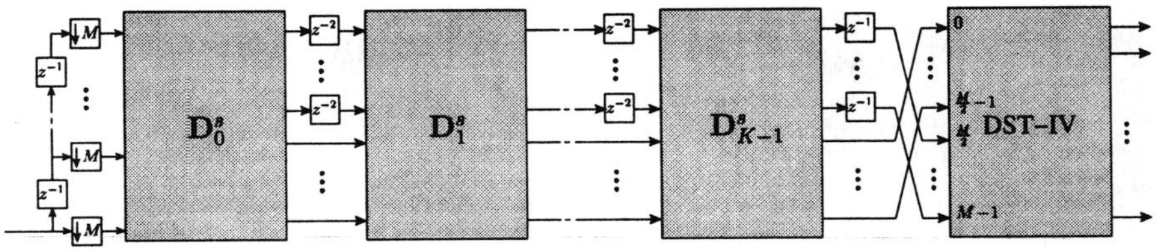

Figure 3: SMFB (analysis) implementation with Fast Direct ELT.

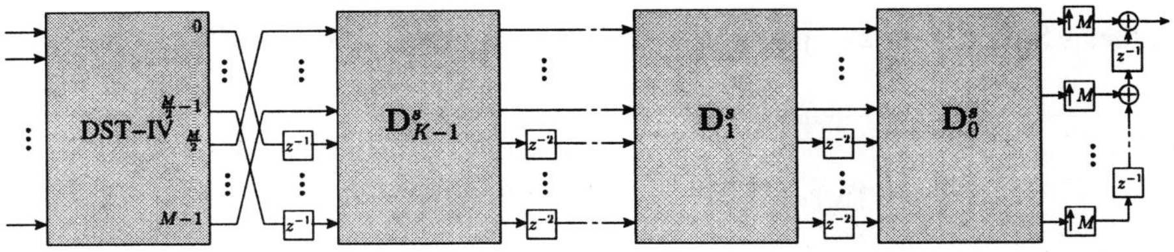

Figure 4: SMFB (synthesis) implementation with Fast Inverse ELT.

First we separate the prototype filter from . The diagonal matrix

$$= \mathrm{i} \quad \{h\ (0), h\ (1), \ldots, h\ (2 \qquad - 1)\} \tag{14}$$

contains the prototype filter coefficients and the modulating sines are grouped in . The elements of the modulating matrix are

$$[\quad]_{nk} = \sin \quad n + \frac{\quad + 1}{2} \quad k + \frac{1}{2} \quad — \tag{15}$$

where $n = 0, 1, \ldots, 2 \qquad - 1$ and $k = 0, 1, \ldots, \qquad - 1$. Furthermore we partition the matrices as follows

$$= \qquad = \qquad {}_{0} \quad {}_{1} \quad \ddots \quad {}_{-1} \qquad {}_{0} \quad {}_{1} \quad \vdots \quad {}_{-1} \quad . \tag{16}$$

Here $_i$ is a 2 2 matrix and $_i = \mathrm{i} \quad \{h\ (2i \quad), h\ (1 + 2i \quad), \ldots, h\ (2 \quad - 1 + 2i \quad)\}$. The submatrices $_i$ of the modulating matrix are

$$[\quad_i]_{nk} = \sin \quad n + 2i \quad + \frac{\quad + 1}{2} \quad k + \frac{1}{2} \quad — \quad . \tag{17}$$

It can be shown that $_i = - \quad _{i\,1}$ and $_i = \quad _{i\,2}$.

Now we define a 2 matrix , which consist of 2 2 square submatrices:

$$= \quad -\mathbf{I} \quad - \quad - \quad \mathbf{I} \quad . \tag{18}$$

Now the DST-IV comes into play: We can verify that

$$_0 = - \tag{19}$$

Then we write as a decomposition of the prototype filter, matrices and DST-IVs:

$$= \quad - \ {}_0 \quad {}_1 \quad - \ {}_2 \quad \ddots \quad \ddots \quad \vdots \quad \frac{}{2} \quad . \tag{20}$$

This system of matrices describes the synthesis filter bank shown in Figure 5. The subfilters $_i(\ )$ are defined as

$$_i(\ ) = \sum_{=0}^{-1} (-1) \quad ^{1}\ h\ (i+2 \qquad) \quad ^{-2} = \sum_{=0}^{-1} \ _i(\ ) \quad ^{-2} \tag{21}$$

This is not a pure polyphase realization of the prototype filter, because there are 2 subfilters and the interpolation factor is .

The analysis bank is constructed similarly. Only now we transpose matrix and multiply it by $(-1) \quad ^{-1}$ to map 2 subsignals from the polyphase filters to the DST-IV. The analysis bank is shown in Figure 6.

4.2. Cosine-Modulated Filter Bank with $2M$ Polyphase Filters

The same decomposition for matrix gives CMFB implementation where polyphase filters $_i(\ )$ are identical with the SMFB case in Figure 5 and 6. But the modulation matrix is changed to DCT-IV and is replaced by

$$= \quad -\mathbf{I} \quad \mathbf{I} \quad . \tag{22}$$

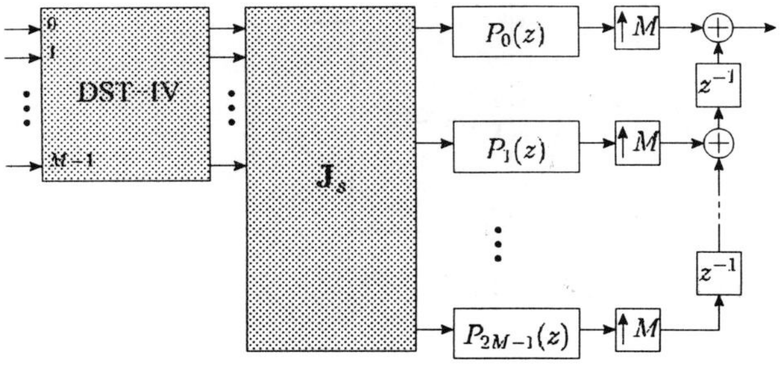

Figure 5: ELT-based synthesis SMFB with 2 polyphase filters.

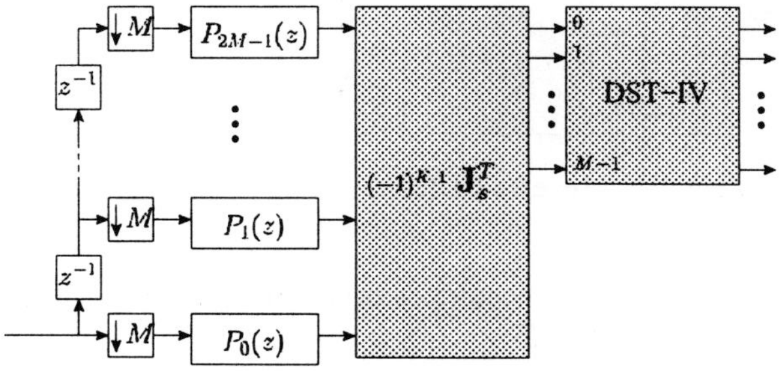

Figure 6: ELT-based analysis SMFB.

The analysis CMFB with 2 polyphase filters is shown in Figure 7.

5. CONCLUSIONS

In this paper the implementation structure of the critically sampled complex exponentially-modulated filter bank uses CMFB and SMFB as building blocks. We have presented a fast computation structure for SMFBs by modifying the extended lapped transform algorithms. When we implement these real transforms with ELT-based algorithms, the computational complexity is of the same order with conventional block transforms. By applying the computational complexity formulas of the Fast ELT [3], we find the number of arithmetic operations per complex input samples for the EMFB. The number of real multiplications and additions (respectively) is

$$(\ ) = (2 \ + \ + \mathrm{o}\ _2\ ) \quad (23)$$
$$\alpha(\ ) = (2 \ + \ + \ \mathrm{o}\ _2\ ). \quad (24)$$

The complexity is dependent on the decimation factor and the overlapping factor . (Note: There are 2 subchannels in the EMFB.) The number of multiplications have been reduced by scaling some butterfly coefficients to 1 [3]. The complexity of the EMFB with 2 polyphase filters is

$$(\ ) = (4 \ + 2 + \mathrm{o}\ _2\ ) \quad (25)$$
$$\alpha(\ ) = (4 \ + \ \mathrm{o}\ _2\ ). \quad (26)$$

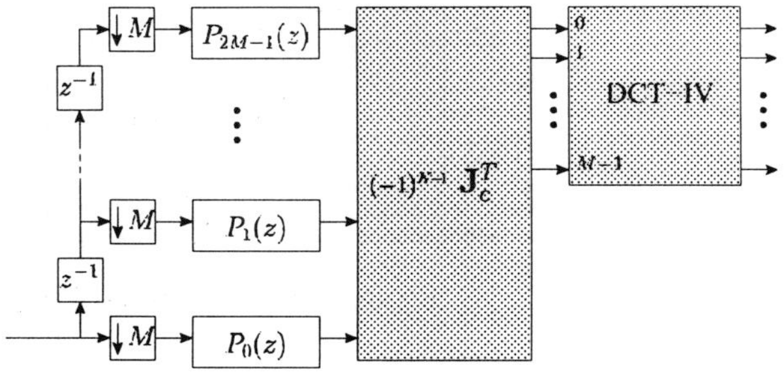

Figure 7: Forward ELT (analysis CMFB).

Thus the EMFB computations can be done with high efficiency.

In the continuation we shall consider efficient implementation of oversampled filter banks that are obtained by taking the complex (instead of real) subchannel output signals from the analysis bank. Such oversampled filter banks are very useful in many applications, like channel equalization in filter bank based transmultiplexer systems.

6. REFERENCES

[1] J. Alhava and M. Renfors, "Complex lapped transforms and modulated filter banks," in *Proc. Int. TICSP Workshop on Spectral Methods and Multirate Signal Processing*, Tolouse, France, Sept. 7–8, 2002, pp. 87–94.

[2] J. Alhava and M. Renfors, "Exponentially-modulated filter bank-based transmultiplexer," in *Proc. IEEE Int. Symp. on Circuits and Systems*, Bangkok, Thailand, May 25–28, 2003.

[3] H. S. Malvar, *Signal Processing with Lapped Transforms*. Norwood, MA: Artech House, 1992.

[4] H. S. Malvar, "Extended lapped transforms: Properties, applications, and fast algorithms," *IEEE Trans. Signal Processing*, vol. 40, pp. 2703–2714, Nov. 1992.

[5] A. Viholainen, T. Saramäki, and M. Renfors, "Nearly perfect-reconstruction cosine-modulated filter bank design for VDSL modems," in *Proc. Int. Conf. on Electronics, Circuits and Systems*, Pafos, Cyprus, Sept. 5–8, 1999, pp. 373–376.

[6] A. Viholainen, T. Hidalgo Stitz, J. Alhava, T. Ihalainen, and M. Renfors, "Complex modulated critically sampled filter banks based on cosine and sine modulation," in *Proc. IEEE Int. Symp. on Circuits and Systems*, Scottsdale, Arizona, USA, May 26–29, 2002, vol. 1, pp. 833–836.

FUZZY FILTERS FOR NOISY IMAGE FILTERING

H. K. Kwan

Department of Electrical and Computer Engineering
University of Windsor
401 Sunset Avenue, Windsor, Ontario, Canada N9B 3P4
kwan1@uwindsor.ca

ABSTRACT

In this paper, seven fuzzy filters for noise reduction in images are introduced. Each of these fuzzy filters, applies a weighted membership function to an image within a window to determine the center pixel, is easy and fast to implement. Simulation results on the filtering performance of these seven fuzzy filters, the standard median filter (MED), and the standard moving average filter (MAV) on images contaminated with low, medium, high impulse and random noises are presented. Results indicate that these seven fuzzy filters achieve varying successes in noise reduction in images as compared to the MED and MAV filters.

1. INTRODUCTION

The two common types of noise in images are impulse (or salt and pepper) noise, and random (or Gaussian) noise. Impulse noise can be expressed by noise density. Random noise can be expressed in terms of its mean and variance values. Noise can be generated during image capture, transmission, storage, as well as during image copying, scanning, and display. For examples, impulse noise can be generated through TV broadcasting and due to information losses; and random noise can be generated during film exposure and development. Noise reduction in images has been one of the common tasks in image processing. For the case of impulse noise, most part of an original image is unaltered, and the image is characterized by some corrupted samples that vary drastically. Compared to impulse noise, random noise is a more challenging type of noise, it is important to be able to reduce random noise effectively in images. In image processing, various linear and nonlinear filtering methods have been proposed. Linear filtering techniques used for noise reduction in images are characterized by mathematical simplicity and can effectively reduce noise with spectral components that do not overlap with those of an image. However, linear filters cannot effectively reduce impulse noise and have a tendency to blur the edges of an image. In such situations, median filters [1-3], which are nonlinear filters, provide an effective solution. Median filters have good edge preserving ability, can eliminate impulse noise, and have moderate noise attenuation ability in the flat regions of an image. The operations of a classical median (MED) filter involve the application of a window to move over an image and to replace the value at the center pixel with the median of all the pixel values within the window. In so doing, a pixel with a distinct intensity (in the case of an impulse) as compared to those of its predefined neighbors will be eliminated. The implementation of a standard median filter is simple and the filter can process an image in a fast manner. The performance of a median filter is average for filtering random noise in an image. This difficulty can be overcome with some success by another nonlinear filtering technique using moving average (MAV) filters [1-3]. Moving average filters can smooth random noise, but they cannot suppress impulse noise and cannot preserve sharp edges of an image. The idea of a standard moving average filter is to replace its center pixel by the average value of its predefined neighboring pixels, which can be easily implemented. The ability to filter unwanted impulse noise and random noise while preserving the edges and details of an image is a non-trivial task. Various nonlinear filters based on classical and/or fuzzy techniques [1-8] have emerged in the past few years for this challenging task. Review on fuzzy-type of filters can be found in [5-7] and a comparison study has been reported in [8]. Depending on their filtering strategies, these filters can be classified as classical filters, classical-fuzzy filters, and fuzzy filters. In [9], we have described median filtering using fuzzy concept. Symmetrical and asymmetrical triangular membership functions with median center are used for filtering impulse, random, and mixed noises of a periodic rectangular pulse. Similar fuzzy filters consisting of symmetrical and asymmetrical triangular membership functions with median center and moving average center have been applied to filtering of images contaminated with impulse, random and mixed noises. In this paper, we present a summary of our earlier study on 2-dimensional fuzzy filters for noise reduction in images. Seven fuzzy filters are defined and their filtering performance on impulse noise and random noise are presented.

2. DEFINITIONS OF FUZZY FILTERS

Let x(i, j) be the input of a 2-dimensional fuzzy filter, the output of the fuzzy filter is defined as:

$$y(i,j)=\frac{\sum_{(r,s)\in A} F[x(i+r),j+s)]\cdot x(i+r,j+s)}{\sum_{(r,s)\in A} F[x(i+r,j+s)]} \quad (1)$$

F[x(i, j)] is the general window function and A is the area of the window. For a square window of dimensions NxN, the range of r and s are: $-R \le r \le R$ and $-S \le s \le S$, where $N = 2R+1 = 2S+1$. With the definitions of different window functions, seven fuzzy filters are obtained, which we shall call the Gaussian fuzzy filter with median center (GMED), the symmetrical triangular fuzzy filter with median center (TMED), the asymmetrical triangular fuzzy filter with median center (ATMED), the Gaussian fuzzy filter with moving average center (GMAV), the symmetrical triangular fuzzy filter with moving average center (TMAV), the asymmetrical triangular fuzzy filter with moving average center

0-7803-7761-3/03/$17.00 ©2003 IEEE

(ATMAV), and the decreasing weight fuzzy filter with moving average center (DWMAV).

2.1 MED

In the case of a standard median filter, the window function is defined as:

$$F[x(i+r,j+s)]=\begin{cases}1 & for\, x(i+r,j+s)=x_{med}(i,j)\\ 0 & otherwise\end{cases} \quad (2)$$

such that the output value y(i, j) at the center of a window A is replaced by the median value $x_{med}(i, j)$ among all the input values x(i+r, j+s) for r, s ∈ A at discrete indexes (i, j).

2.2 MAV

In a standard moving average filter, the window function is defined as:

$$F[x(i+r,j+s)]=1 \quad for\; r,s\in A \quad (3)$$

The moving average filter is equivalent to a fuzzy filter with a rectangular window covering all the input values x(i+r, j+s) for r, s ∈ A in the window A.

2.3 GMED

The Gaussian fuzzy filter with the median value within a window chosen as its center value is defined as:

$$F\,[x(i+r,j+s)] = e^{-\frac{1}{2}[\frac{x(i+r,j+s)-x_{med}(i,j)}{\sigma(i,j)}]^2} \quad for\; r,s \in A \quad (4)$$

$x_{med}(i, j)$ and σ(i, j) represent, respectively, the median value and the variance value of all the input values x(i+r, j+s) for r, s ∈ A in the window A at discrete indexes (i, j).

2.4 TMED

The symmetrical triangular fuzzy filter with the median value within a window chosen as the center value is defined as:

$$F[x(i+r,j+s)] = \begin{cases}1-|x(i+r,j+s)-x_{med}(i,j)|/x_{mm}(i,j)\\ for\,|x(i+r,j+s)-x_{med}(i,j)|\le x_{mm}(i,j)\\ 1 \quad for\, x_{mm}=0\end{cases} \quad (5a)$$

$$x_{mm}(i, j) = \max [\, x_{max}(i, j) - x_{med}(i, j),\, x_{med}(i, j) - x_{min}(i, j)\,] \quad (5b)$$

$x_{max}(i, j)$, $x_{min}(i, j)$ and $x_{med}(i, j)$ are, respectively, the maximum value, the minimum value, and the median value of all the input values x(i+r, j+s) for r, s ∈ A within the window A at discrete indexes (i, j).

2.5 ATMED

The asymmetrical triangular fuzzy filter with the median value within a window chosen as the center value is defined as:

$$F[x(i+r,j+s)] = \begin{cases}1-[x_{med}(i,j)-x(i+r,j+s)]/[x_{med}(i,j)-x_{min}(i,j)]\\ for\, x_{min}(i,j)\le x(i+r,j+s)\le x_{med}(i,j)\\ 1-[x(i+r,j+s)-x_{med}(i,j)]/[x_{max}(i,j)-x_{med}(i,j)]\\ for\, x_{med}(i,j)\le x(i+r,j+s)\le x_{max}(i,j)\\ 1\; for\, x_{med}(i,j)-x_{min}(i,j)=0\; or\; x_{max}(i,j)-x_{med}(i,j)=0\end{cases} \quad (6)$$

In (6), the degree of asymmetry depends on the difference between $x_{med}(i, j)-x_{min}(i, j)$ and $x_{max}(i, j)-x_{med}(i, j)$. $x_{max}(i, j)$, $x_{min}(i, j)$ and $x_{med}(i, j)$ are, respectively, the maximum value, the minimum value, and the median value among all the input values x(i+r, j+s) for r, s ∈ A within the window A at discrete indexes (i, j).

2.6 GMAV

The Gaussian fuzzy filter with the moving average value within a window chosen as its center value is defined as:

$$F\,[x(i+r,j+s)] = e^{-\frac{1}{2}[\frac{x(i+r,j+s)-x_{mav}(i,j)}{\sigma(i,j)}]^2} \quad for\; r,s \in A \quad (7)$$

$x_{mav}(i, j)$ and σ(i, j) represent, respectively, the moving average value and the variance value of all the input values x(i+r, j+s) for r, s ∈ A in the window A at discrete indexes (i, j).

2.7 TMAV

The triangular fuzzy filter with the moving average value within a window chosen as its center value is defined as:

$$F[x(i+r,j+s)] = \begin{cases}1-[|x(i+r,j+s)-x_{mav}(i,j)|]/x_{mv}(i,j)\\ for\,|x(i+r,j+s)-x_{mav}(i,j)|\le x_{mv}(i,j)\\ 1 \quad for\, x_{mv}=0\end{cases} \quad (8a)$$

$$x_{mv}(i, j) = \max [\, x_{max}(i, j) - x_{mav}(i, j),\, x_{mav}(i, j) - x_{min}(i, j)\,] \quad (8b)$$

$x_{max}(i, j)$, $x_{min}(i, j)$ and $x_{mav}(i, j)$ represent, respectively, the maximum value, the minimum value, and the moving average value of x(i+r, j+s) within the window A at discrete indexes (i, j).

2.8 ATMAV

The asymmetrical triangular fuzzy filter with the moving average value within a window chosen as its center value is defined as:

$$F[x(i+r,j+s)] = \begin{cases}1-[x_{mav}(i,j)-x(i+r,j+s)]/[x_{mav}(i,j)-x_{min}(i,j)]\\ for\, x_{min}(i,j)\le x(i+r,j+s)\le x_{mav}(i,j)\\ 1-[x(i+r,j+s)-x_{mav}(i,j)]/[x_{max}(i,j)-x_{mav}(i,j)]\\ for\, x_{mav}(i,j)\le x(i+r,j+s)\le x_{max}(i,j)\\ 1\; for\, x_{mav}(i,j)-x_{min}(i,j)=0\; or\; x_{max}(i,j)-x_{mav}(i,j)=0\end{cases} \quad (9)$$

The degree of asymmetry depends of the difference between $x_{mav}(i, j)$-$x_{min}(i, j)$ and $x_{max}(i, j)$-$x_{mav}(i, j)$. $x_{max}(i, j)$, $x_{min}(i, j)$ and $x_{mav}(i, j)$ represent, respectively, the maximum value, the minimum value, and the moving average value of x(i+r, j+s) within the window A at discrete indexes (i, j).

2.9 DWMAV

The decreasing weight fuzzy filter with the moving average value within a window chosen as its center value is defined as:

$$F[x(i+r, j+s)] = 1 - \frac{\max(|r|, |s|)}{\max(|R|, |S|) + t} \quad (10a)$$

$-R \le r \le R$ and $-S \le s \le S$, and $2R+1 = 2S+1 = N$ (10b)

N is the width of a square window of dimensions NxN. t is the threshold value that determines the height of the decreasing triangular-shape weighted function at |r|=R and/or |s|=S. In general, t = 1, 2, and 3 gives a varying degree of FIR filtering performance. For ease of explanation, we shall call the DWMAV filters with t = 1, 2, and 3 as DWMAV1, DWMAV2, and DWMAV3 respectively.

3. SIMULATIONS

The 8-bit mono Lena image of dimensions M1xM2 (= 256x256) pixels is used for simulations. The pixels s(i, j) for $1 \le i \le M1$ and $1 \le j \le M2$, of the image is corrupted by adding two kinds of noise, namely, impulse (or salt and pepper noise) noise $n_i(i, j)$, and random (or Gaussian) noise $n_g(i, j)$. Low, medium, and high levels of impulse noise, each with respective density values of 0.03, 0.15, and 0.3 is added to the image. Also, low, medium, and high levels of random noise, each has a mean value of 0.0 and a respective variance value of 0.0052, 0.021, and 0.106 is added to the image.

In all the simulations, square windows of dimensions NxN pixels and with different values of width N (= 3, 5, 7) are used. The mean squared error (MSE) is used to compare the relative filtering performance of various filters. The MSE between the filtered output image y(i, j) and the original image s(i, j) of dimensions M1xM2 pixels is defined as:

$$MSE = \frac{\sum_{i}^{M1}\sum_{j}^{M2}[y(i,j) - s(i,j)]^2}{M1xM2} \quad (11)$$

The MSE of the original and filtered noisy Lena image for the 3 levels of impulse noise and the 3 levels of random noise for N = 3, 5, 7 are respectively summarized in Tables 1 and 2. As seen from Tables 1-2, the MSE values of the impulse and random noise filtered images share some similar properties. As the window width N increases, nearly all the MSE values increase for low-level noises while majority of the MSE values decrease for high-level noises, and there is a combination of MSE values increase and decrease for medium-level noises. In general, for reduced MSE performance, a narrower window width is appropriate for low-level noises, and a wider window width is appropriate for high-level noises. It should be noted that the edges and details of an image become blur as the window width N increases. From the filtered images, it is observed that edges and details are well preserved for N = 3 in all the seven filters. The top 3 out of all the seven filters (in which the DWMAV has three sub-filters) for low, medium, and high levels of impulse and random noises are ranked according to their MSE values for N = 3. The filter with the minimum MSE value is ranked first and so on as summarized in Tables 3-4.

4. SUMMARY

In this paper, a study of seven fuzzy filters and their filtering performance has been presented. Each of these fuzzy filters applies a weighted membership function to an image within a window to compute the value of the center pixel, is easy and fast to implement and can suppress low, medium, and high levels of impulse noise and random noise with a varying degree of success. Depend on the features of an image, the performance of each of these seven filters varies slightly. In general, the filtering performance of each of these fuzzy filters is quite consistent among images of similar characteristics. In practice, the edges and details of an image can be preserved when the window width is small (for N = 3). As the window width increases (for N = 5 or 7), filtered images become blur, but under a high-level noise, the filtering capability of the majority of these fuzzy filters increases. As a general guideline, a small window width appears to be appropriate for a low level of noise, and a larger window width may be considered for a higher level of noise.

5. REFERENCES

[1] Pitas I. and Venetsanopoulos A. N., *Nonlinear Digital Filters*, Kluwer Academic Publishers, Boston, 1990.

[2] Agaian S., Astola J., and Egiazarian K., *Binary Polynomial Transformations And Nonlinear Digital Filters*, Marcel Dekker, Inc., 1995.

[3] Mitra S. K. and Sicuranza G., Editors, *Nonlinear Image Processing*, Academic Press, 2000.

[4] Kerre E. E. and Nachtegael M., Editors, *Fuzzy Techniques In Image Processing*, Series on Studies in Fuzziness and Soft Computing, Vol. 52, Springer-Verlag, 2000.

[5] Russo F., "Recent advances in fuzzy techniques for image enhancement", IEEE Transactions on Instrumentation and Measurement, 47(6):1428-1434, 1998.

[6] Nachtegael M., Van der Weken D., Van De Ville A., Kerre E., Philips W., Lemahieu I., "An overview of classical and fuzzy-classical filters", Proceedings of IEEE International Conference on Fuzzy Systems, pp. 3-6, 2001.

[7] Nachtegael M., Van der Weken D., Van De Ville A., Kerre E., Philips W., Lemahieu I., "An overview of fuzzy filters for noise reduction", Proceedings of IEEE International Conference on Fuzzy Systems, pp. 7-10, 2001.

[8] Nachtegael M., Van der Weken D., Van De Ville A., Kerre E., Philips W., Lemahieu I., "A comparative study of classical and fuzzy filters for noise reduction", Proceedings of IEEE International Conference on Fuzzy Systems, pp. 11-14, 2001.

[9] Kwan H. K. and Cai Y., "Median filtering using fuzzy concept", Proceedings of 36th Midwest Symposium on Circuits and Systems, Detroit, vol. 2, pp. 824-827, August 1993.

Filters	N	Density of Impulse Noise		
		Low 0.03	Medium 0.15	High 0.3
Noisy Image		578.40	2894.60	5841.80
MED	3	55.68	102.92	387.44
	5	122.62	159.92	254.00
	7	190.35	235.65	337.69
GMED	3	53.69	104.50	367.21
	5	118.85	150.75	230.05
	7	184.42	214.00	277.41
TMED	3	55.60	105.02	386.14
	5	124.21	158.22	245.93
	7	193.37	220.91	284.32
ATMED	3	69.17	109.12	319.19
	5	150.48	161.97	190.27
	7	226.52	228.68	253.53
MAV	3	167.54	486.44	1025.60
	5	242.14	406.55	745.69
	7	339.47	461.70	742.63
GMAV	3	64.46	157.20	520.40
	5	138.85	170.63	280.98
	7	214.73	240.04	306.87
TMAV	3	57.57	108.73	382.35
	5	135.52	169.61	251.08
	7	216.21	241.85	289.91
ATMAV	3	96.58	128.39	142.49
	5	202.80	208.50	214.30
	7	293.10	285.62	307.75
DWMAV1 (t=1)	3	138.88	484.09	1054.00
	5	169.27	354.52	720.39
	7	225.45	365.24	671.28
DWMAV2 (t=2)	3	141.93	472.10	1023.10
	5	183.01	358.48	711.53
	7	246.27	379.78	676.75
DWMAV3 (t=3)	3	143.71	471.14	1018.60
	5	189.37	362.20	711.34
	7	256.78	388.08	681.97

Table 1. MSE values of original and filtered images.

Filters	N	Variance of Random Noise		
		Low - 0.0052	Medium - 0.021	High - 0.106
Noisy Image		324.97	1248.80	4782.40
MED	3	120.90	312.97	1174.50
	5	157.38	258.12	673.65
	7	223.36	304.38	613.52
GMED	3	111.29	284.18	1052.70
	5	144.15	225.47	575.41
	7	203.91	259.65	492.34
TMED	3	121.61	321.53	1222.60
	5	146.04	227.68	591.19
	7	205.87	258.15	489.32
ATMED	3	111.81	263.74	958.55
	5	166.04	261.22	790.65
	7	238.10	334.26	884.61
MAV	3	133.05	244.98	751.83
	5	225.00	271.95	546.10
	7	327.27	356.43	567.96
GMAV	3	104.14	237.63	820.96
	5	157.04	217.54	477.50
	7	228.25	270.04	440.88
TMAV	3	106.38	257.58	916.52
	5	154.54	218.87	490.08
	7	226.01	270.10	450.10
ATMAV	3	116.40	253.71	956.70
	5	189.38	280.84	919.77
	7	273.14	367.86	1039.40
DWMAV1 (t=1)	3	102.06	223.87	770.79
	5	149.95	204.16	509.79
	7	211.37	245.91	483.01
DWMAV2 (t=2)	3	106.63	222.31	746.01
	5	164.79	214.70	506.21
	7	232.99	264.87	492.57
DWMAV3 (t=3)	3	108.62	223.20	742.33
	5	171.55	220.23	507.85
	7	243.76	274.69	499.34

Table 2. MSE values of original and filtered images.

Filters	Low	Medium	High
MED	3	1	6
GMED	1	2	3
TMED	2	3	5
ATMED	6	5	2
MAV	11	11	10
GMAV	5	7	7
TMAV	4	4	4
ATMAV	7	6	1
DWMAV1	8	10	11
DWMAV2	9	9	9
DWMAV3	10	8	8

Table 3. MSE ranking of impulse noise filtered images.

Filters	Low	Medium	High
MED	9	10	10
GMED	6	9	9
TMED	10	11	11
ATMED	7	8	8
MAV	11	5	3
GMAV	2	4	5
TMAV	3	7	6
ATMAV	8	6	7
DWMAV1	1	3	4
DWMAV2	4	1	2
DWMAV3	5	2	1

Table 4. MSE ranking of random noise filtered images.

HIGH-SPEED TUNABLE FRACTIONAL-DELAY ALLPASS FILTER STRUCTURE

Ji-Suk Park[1], *Byeong-Kuk Kim*[1], *Jin-Gyun Chung*[1], *Keshab K. Parhi*[2]

[1]Div. of Electronic & Information Engr., Chonbuk National University, Chonju, Korea
[2]Dept. of Electrical & Computer Engr., University of Minnesota, Minneapolis, MN, USA
[1]{jspark, bkkim, jgchung}@vlsidsp.chonbuk.ac.kr, [2]parhi@ece.umn.edu

ABSTRACT

By using a fractional delay (FD), a filter can perform various operations such as asynchronous sample rate conversion, synchronization of digital modems and sound synthesis of musical instruments. An IIR tunable FD filter can be designed using Thiran-based FD allpass filter structure. However, this IIR structure has the disadvantage of long critical path and large hardware overhead.

In this paper, a modified Thiran-based FD allpass filter structure is proposed. Compared with the existing structure, the proposed structure requires shorter critical path, less hardware complexity and improved frequency characteristics.

1. INTRODUCTION

FD filter is an interpolation device for continuously varying time delay. FD filters can be used in many signal processing applications such as synchronization of digital modems, asynchronous sample rate conversion (ASRC) and sound synthesis of musical instruments.

FD filters can be designed using either FIR or IIR structures. The Farrow structure is one of the most widely known FIR-based FD filter and can be derived using the Horner's rule [1][2]. The delay of the Farrow structure is controlled by one single parameter. In low frequency applications, the Farrow structures exhibit sufficient accuracy with low-order filters (e.g., third or fourth order). However, in high frequency applications, higher-order filters should be used to obtain sufficient accuracy.

Thiran-based FD allpass filters are IIR-based FD filters [3] and generally require much less hardware than the FIR-based FD filters. The coefficients of Thiran-based FD allpass filters are obtained using Thiran formulas [4]. However, this IIR structure has the drawback of long critical path. In addition, large coefficient word-lengths are required for VLSI implementation of the Thiran-based FD allpass filters.

In this paper, a modified IIR-based FD filter structure is proposed. Although this new structure is based on Thiran formulas, it has shorter critical path than the conventional Thiran-based FD allpass filters. In addition, the coefficients of the new structure require shorter word-lengths than the conventional Thiran-based FD allpass structure.

In Section 2, FIR and IIR FD filter structures are reviewed. In Section 3, the modified FD allpass filter structure is proposed. Simulation results are presented in Section 4 and brief conclusions are given in Section 5.

2. FD FILTER STRUCTURES [3]

2.1 FIR FD Filter Structure

Generally, the transfer function of an N-th order FIR filter is expressed as

$$H(z) = \sum_{n=0}^{N} h_n z^{-n} . \quad (1)$$

In order to control the delay of FIR filter with a single delay parameter d, each filter coefficient h_n can be approximated using L-th order polynomial as

$$h_n(d) = \sum_{l=0}^{L} c_{ln} d^l . \quad (2)$$

In Farrow structure, the variable d is extracted from the transfer function as follows:

$$\begin{aligned} H(d,z) &= \sum_{n=0}^{N} \left[\sum_{l=0}^{L} c_{ln} d^l \right] z^{-n}, \\ &= \sum_{l=0}^{L} \left[\sum_{n=0}^{N} c_{ln} z^{-n} \right] d^l . \end{aligned} \quad (3)$$

To reduce the processing needed at the high output sample rate, we can canonically express the variable delay parameter d and delay elements z^{-1} by applying the Horner's rule. Then we can get the output as a function of a single delay parameter d as follows:

$$\begin{aligned} y(k) &= \sum_{n=0}^{N} x(k-n) h_n, \\ &= \sum_{l=0}^{L} \left[\sum_{n=0}^{N} c_{ln} x(k-n) \right] d^l . \end{aligned} \quad (4)$$

Implementation of (4) leads to the Farrow structure.

2.2 Thiran-based FD Allpass Filter

The general transfer function of an allpass filter can be expressed as

$$H(z) = \frac{B(z)}{A(z)} = \frac{z^{-N} A(z^{-1})}{A(z)}, \quad (5)$$

0-7803-7761-3/03/$17.00 ©2003 IEEE

where $A(z)=\sum_{n=0}^{N} a_n z^{-n}$ and $a_0 = 1$.

If the same method as in FIR structure is applied to each coefficient, a_n can be expressed as

$$a_n = \sum_{l=0}^{L} e_{ln} d^l, \qquad n = 0,1,\ldots,N. \tag{6}$$

Then, $A(z)$ can be obtained as

$$\begin{aligned} A(z) &= \sum_{n=0}^{N} a_n z^{-n} = 1 + \sum_{n=1}^{N} a_n z^{-n}, \\ &= 1 + \sum_{n=1}^{N}\left[\sum_{l=0}^{L} e_{ln} d^l\right] z^{-n}, \\ &= 1 + \sum_{l=0}^{L}\left[\sum_{n=1}^{N} e_{ln} z^{-n}\right] d^l. \end{aligned} \tag{7}$$

However, it can be shown that the implementation of the transfer function $H(z)$ obtained by (7) requires delayfree loops.

To avoid this problem, the output can be expressed by difference equation in time domain as

$$y(k) = x(k-N) + \sum_{l=0}^{L}\left\{\sum_{n=1}^{N}\left[x(n+k-N) - y(k-n)\right] e_{ln}\right\} d^l . \tag{8}$$

The coefficients $e_{ln}(0 \le l \le L)$ for each coefficient a_n can be determined by Thiran formulas [4][5] as

$$\begin{aligned} a_k &= (-1)^k \binom{N}{k} \prod_{n=0}^{N} \frac{d+n}{d+k+n}, \\ &= (-1)^k \binom{N}{k} \prod_{n=0}^{k-1} \frac{d+n}{d+N+n+1}, \end{aligned} \tag{9}$$

where $\binom{N}{k} = \frac{N!}{k!(N-k)!}$.

The coefficient a_k is expressed as a form of rational polynomial of the fractional delay d. Since it is difficult to implement the division including a time-varying factor, we need to transform the coefficients to a polynomial form.

The transfer function of an allpass filter can be expressed as

$$H(z) = \frac{g(d)[\hat{a}_N + \cdots + \hat{a}_1 z^{-(N-1)}] + z^{-N}}{1 + g(d)[\hat{a}_1 z^{-1} + \cdots + \hat{a}_N z^{-N}]}. \tag{10}$$

The modified coefficient $\hat{a}_k$ is obtained by multiplying a common factor to each coefficient a_k as

$$\begin{aligned} \hat{a}_k &= (-1)^k \binom{N}{k} \frac{\prod_{n=0}^{k-1}(d+n)}{\prod_{n=1}^{k}(d+N+n)} \prod_{n=1}^{N}(d+N+n), \\ &= (-1)^k \binom{N}{k} \prod_{n=0}^{k-1}(d+n) \prod_{n=k+1 \le N}^{N}(d+N+n), \\ &\cong \sum_{l=1}^{N} \hat{e}_{lk} d^l, \qquad \text{for } k = 1,2,\cdots,N, \end{aligned} \tag{11}$$

where $a_0 = 1$.

The canceling division $g(d)$ can be approximated by a polynomial in fractional delay d:

$$g(d) = \frac{1}{\prod_{n=1}^{N}(d+N+n)} \cong \sum_{i=0}^{I} g_i d^i . \tag{12}$$

The coefficient g_i is approximated by the truncated Maclaurin series or Talyor series expansion as

$$g(d) \cong \frac{N!}{(2N)!} \prod_{n=1}^{N}\left[1 + \sum_{k=1}^{I}(-1)^k \left(\frac{d}{N+n}\right)^k\right], \tag{13}$$

where I is the order of the approximating polynomial. Then the general IIR FD filter structure can be obtained as shown in Fig. 1.

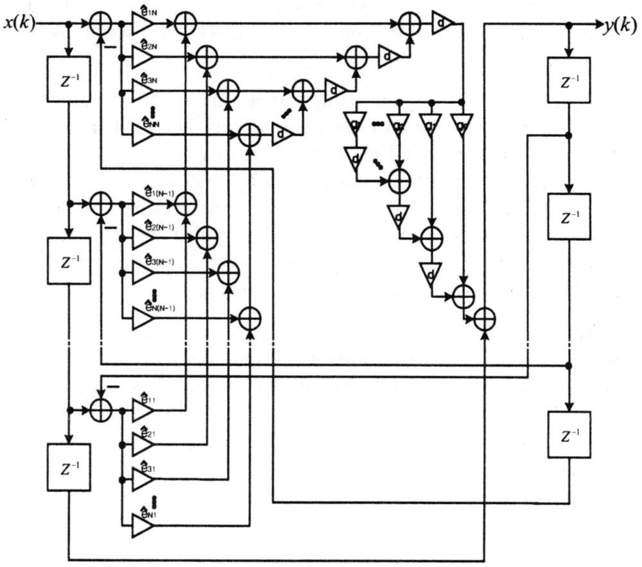

Figure 1. General Thiran-based FD allpass filter.

3. PROPOSED STRUCTURE

The critical path is defined as the longest path between any two storage elements or between inputs and outputs. The critical path computation time determines the minimum feasible clock period of a DSP system [6]. As can be seen in Fig. 1, the Thiran-based FD allpass filter has long critical path. In addition, the difference between the values of $\hat{a}_k$ and the values of $g(d)$ is about 10^2 times in second order case and about 10^4 times in third order case. Thus large word-lengths are needed for VLSI implementation of the filter structure. To solve these problems, modified Thiran-based FD allpass filter structure is proposed in this section.

From (11) and (12), each coefficient a_k of the IIR filter is

$$a_k = g(d) \cdot \hat{a}_k, \\ = \left[\sum_{i=0}^{I} g_i d^i\right] \cdot \left[\sum_{l=1}^{N} \hat{e}_{lk} d^l\right]. \tag{14}$$

It can be noticed that d^l is close to zero when l is large since the range of the delay parameter d is from -0.5 to 0.5. Thus a_k can be approximated as

$$a_k = g(d) \cdot \hat{a}_k, \\ \cong \sum_{m=1}^{M} \left[\sum_{l=1}^{m} g_{m-l} \hat{e}_{lk}\right] d^m, \tag{15}$$

where $a_0 = 1$.

Using the coefficient approximation of (15), $A(z)$ can be expressed as

$$A(z) = 1 + \sum_{n=1}^{N} a_n z^{-n}, \\ = 1 + \sum_{n=1}^{N}\left[\sum_{m=1}^{M}\left(\sum_{l=1}^{m} g_{m-l}\hat{e}_{\text{ln}}\right) d^m\right] z^{-n}, \\ = 1 + \sum_{m=1}^{M}\left[\sum_{n=1}^{N}\left(\sum_{l=1}^{m} g_{m-l}\hat{e}_{\text{ln}}\right) z^{-n}\right] d^m, \\ = 1 + \sum_{m=1}^{M}\left[\sum_{n=1}^{N} c_{mn} z^{-n}\right] d^m. \tag{16}$$

In (16), $c_{mn} = \sum_{l=1}^{m} g_{m-l}\hat{e}_{ln}$ is a constant value.

We can get the corresponding difference equation in the time domain directly as

$$y(k) = x(k-N) + \sum_{m=1}^{M}\left\{\sum_{n=1}^{N}[x(n+k-N) - y(k-n)]c_{mn}\right\} d^m. \tag{17}$$

Conventional third order Thiran-baed FD filter structure is shown in Fig. 2. Corresponding third order structure obtained by the proposed method is shown in Fig. 3. Notice that the proposed structure has much shorter critical path than the conventional structure. Several simulation results are given in detail in section 4.

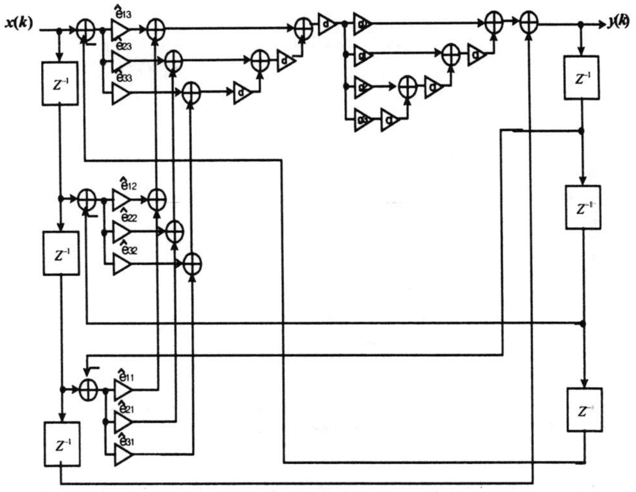

Figure 2. Thiran-based FD allpass filter (N=3, I=3).

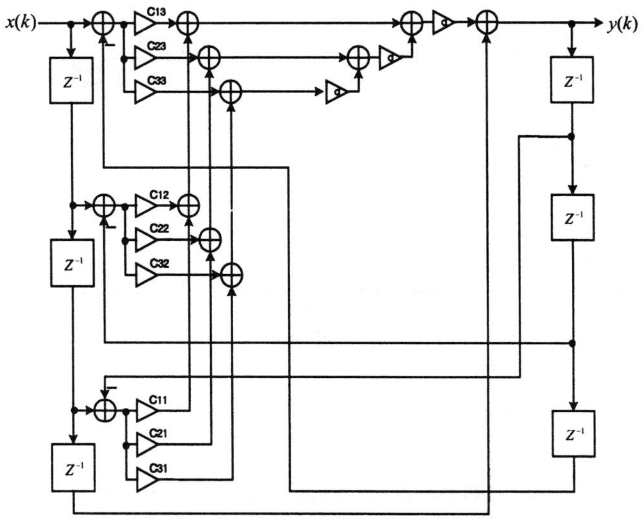

Figure 3. Proposed FD allpass filter (N=3, M=3).

4. SIMULATION RESULTS

In general, One would expect that the filter performance is deteriorated by removing some of the terms, but the performance of the proposed structure is improved. Fig. 4 compares the phase delay of each second order FD filter for worst-case d=0.5. In conventional FD allpass filter, maximum phase delay error is 0.0203 samples. However, the maximum phase delay error of the proposed FD allpass filter is only 0.0084 samples.

Table 1 compares critical path computation time and required hardware for second order conventional Thiran-based FD filter and proposed FD filter. It can be seen that the critical path computation time is reduced by 39% and hardware complexity is reduced by 30% by the proposed method.

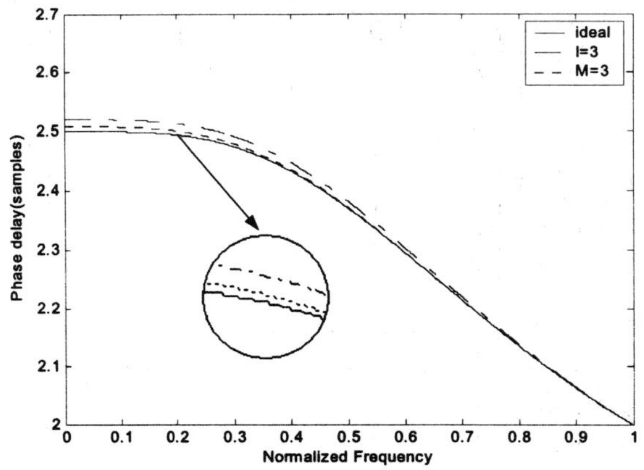

Figure 4. Phase delay comparison in second order case (d=0.5): ideal (solid line), conventional Thiran-based FD filter (N=2, I=3) (dotted line) and proposed FD filter (M=3) (dash-dot line).

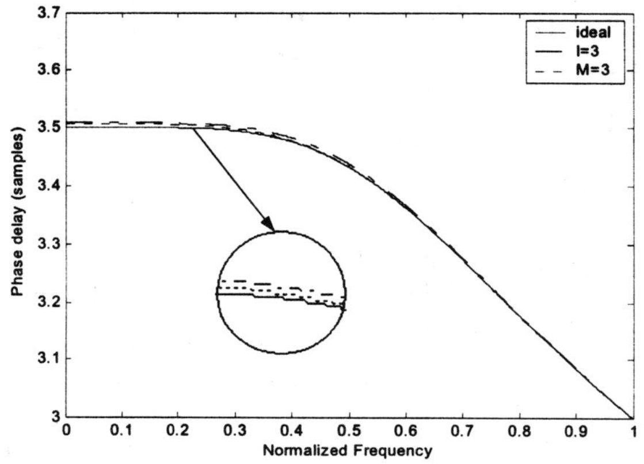

Figure 5. Phase delay comparison in third order case (d=0.5): ideal (solid line), conventional Thiran-based FD filter (N=3, I=3) (dotted line) and proposed FD filter (M=3) (dash-dot line).

Table 1. H/W comparison of second order conventional Thiran FD filter and proposed FD filter (Tm = multiplication time, Ta = addition time)

		Thiran FD filter			Proposed FD filter		
		I=2	I=3	I=4	M=2	M=3	M=4
Critical Path Comp. Time		6Tm +5Ta	7Tm +6Ta	8Tm +7Ta	3Tm +4Ta	4Tm +5Ta	5Tm +6Ta
Total H/W	Multiplier	11	13	15	6	9	12
	Adder	8	9	10	6	8	10

Table 2. H/W comparison of third order conventional Thiran FD filter and proposed FD filter

		Thiran FD filter			Proposed FD filter		
		I=2	I=3	I=4	M=2	M=3	M=4
Critical Path Comp. Time		7Tm +8Ta	8Tm +9Ta	9Tm +10Ta	3Tm +5Ta	4Tm +6Ta	5Tm +7Ta
Total H/W	Multiplier	17	19	21	8	12	16
	Adder	14	15	16	9	12	15

In third order FD filter, the phase delay for worst-case d=0.5 is shown in Fig. 5. The maximum phase delay error of conventional FD allpass filter is 0.0086 samples. However, the maximum phase delay error of the proposed FD allpass filter is only 0.0076 samples.

In third order case, from Table 2, it can be shown that critical path is reduced by 45% and hardware complexity is reduced by 37% by the proposed method.

5. CONCLUSIONS

In this paper, a modified design method for IIR FD filters has been presented. By simulations, it was shown that the proposed FD allpass filter structure requires much less critical path computation time and hardware complexity. Also, it was shown that the proposed FD allpass filter structure exhibits better frequency domain characteristics and requires shorter coefficient word-lengths.

6. REFERENCES

[1] C. W. Farrow, "A continuously variable digital delay element." in Proc. 1988 IEEE Int. Symp. Circuits and Systems, Espoo, Finland, vol. 3, pp. 2641-2645, June, 1988.

[2] L. Erup, F. M. Gardner and R. A. Harris, "Interpolation in digital modems - Part II: Implementation and performance." IEEE Trans. Comm., vol. 41, no. 6, pp. 998-1008, June, 1993.

[3] M. Makundi, V. Valimaki, and T. I. Laakso, "Closed-form design of tunable fractional-delay allpass filter structures." in Proc. 2001 IEEE Int. Symp. Circuits and Systems, ISCAS 2001, vol. 4, pp. 434 -437, April, 2001.

[4] T. I. Laakso, V. Valimaki, M. Karjalainen and U. K. Laine, "Splitting the unit delay - tools for fractional delay design." IEEE Signal Processing Mag, vol. 13, no. 1, pp. 30-60, Jan, 1996.

[5] J. P. Thiran, "Recursive digital filters with maximally flat group delay." IEEE Trans. Circuit Theory, vol. 18, no. 6, pp. 659-664, Nov, 1971.

[6] K. K. Parhi, *VLSI digital signal processing systems : design and implementation*, A Wiley-Interscience Publication, 1999.

DESIGN OF ULTRASPHERICAL WINDOWS WITH PRESCRIBED SPECTRAL CHARACTERISTICS

Stuart W. A. Bergen and Andreas Antoniou

Dept. of Elec. & Comp. Engineering, University of Victoria,
P.O. Box 3055 STN CSC, Victoria, B.C. V8W 3P6, Canada.
Tel: +1 (250) 721-8781; Fax: +1 (250) 721-6052
e-mail: sbergen@ece.uvic.ca, aantoniou@ece.uvic.ca

ABSTRACT

A method for designing ultraspherical windows with prescribed spectral characteristics is proposed. The characteristics considered include the ripple ratio, main-lobe width, null-to-null width, and a user defined side-lobe pattern. The method employs a variety of short algorithms that calculate the ultraspherical window's independent parameters based on the prescribed spectral characteristics.

1. INTRODUCTION

Window functions are used to reduce Gibbs' oscillations and are employed in a variety of applications including power spectral estimation, beamforming, and digital filter design. With a large number of applications available, window flexibility becomes a key concern. The ultraspherical window is a flexible window that has three independent parameters for controlling its properties and can be generated by the use of a closed-form equation [1].

Adjustable windows such as the Dolph-Chebyshev [2], Kaiser [3], and Saramäki [4] windows have two independent parameters for controlling their properties, namely, the window length which alters the main-lobe width and a parameter which alters the relative side-lobe amplitude. The use of ultraspherical polynomials (also known as Gegenbauer polynomials) for window designs introduces another degree of freedom relative to adjustable windows which allows one to achieve a variety of side-lobe patterns in the window's spectral representation.

In this paper, a method is proposed which enables one to design ultraspherical windows with prescribed spectral characteristics such as ripple ratio, main-lobe width, null-to-null width, and a user defined side-lobe pattern. The method employs a variety of short algorithms that calculate the ultraspherical window's independent parameters based on the prescribed spectral characteristics. Two different equations are available for the generation of the window coefficients. A comparison shows that the two equations differ significantly with respect to computational complexity.

2. ULTRASPHERICAL WINDOW COEFFICIENTS

The coefficients of the ultraspherical window can be calculated as [5]

$$w(nT) = \frac{\mu x_\mu^{2M}}{M+|n|}\binom{\mu+M+|n|-1}{M+|n|-1} \cdot \sum_{m=0}^{M-|n|}\binom{\mu+M-|n|-1}{M-|n|-m} \cdot \binom{M+|n|}{m} A^m \quad \text{for } |n| \le M, \tag{1}$$

where μ, x_μ, and M are independent parameters and $A = 1 - x_\mu^{-2}$. The window length $N = 2M+1$ is assumed to be odd throughout, $w(nT) = w(-nT)$, and the normalized ultraspherical window is given by $\widehat{w}(nT) = w(nT)/w(0)$. Alternatively, one may generate the ultraspherical window coefficients as

$$w(nT) = \frac{1}{2M+1}\left[C_{2M}^{\mu}(x_\mu) + 2\sum_{s=1}^{M} C_{2M}^{\mu}\left(x_\mu \cos\frac{\pi s}{2M+1}\right) \cdot \cos\frac{2\pi s n}{2M+1}\right] \quad \text{for } |n| \le M. \tag{2}$$

The values $C_n^\mu(x)$ can be calculated using the recurrence relationship

$$C_r^\mu(x) = \frac{1}{r}\left[2x(r+\mu-1)C_{r-1}^\mu(x) - (r+2\mu-2)C_{r-2}^\mu(x)\right] \tag{3}$$

for $r = 2, 3, \ldots, n$, where $C_0^\mu(x) = 1$ and $C_1^\mu(x) = 2\mu x$. When $\mu = 0$ in Eqs. (1) and (2), the Dolph-Chebyshev window is obtained and can be calculated as described in [6].

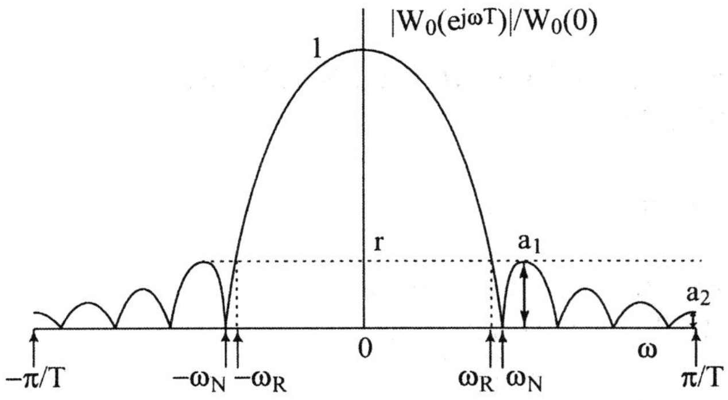

Fig. 1. Normalized spectral representation of a typical window function.

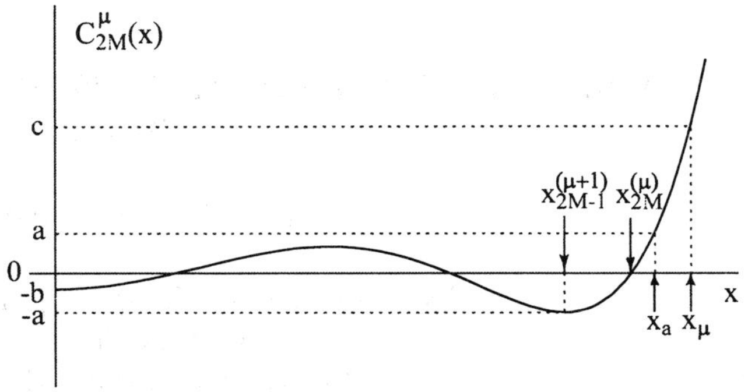

Fig. 2. Some important quantities of ultraspherical polynomial $C_{2M}^{\mu}(x)$ for the values $\mu = 2$ and $M = 3$.

3. PRESCRIBED SPECTRAL CHARACTERISTICS

As shown in Section 2, the ultraspherical window function entails three independent parameters which influence the shape of its frequency spectrum. Typically, a window function is specified in terms of certain spectral characteristics and, therefore, a method is needed by which one can determine the window parameters that would yield prescribed spectral characteristics. Two basic spectral characteristics are the null-to-null width B_N and the main-lobe width B_R which, in this paper, are defined as $B_N = 2\omega_N$ and $B_R = 2\omega_R$, respectively, where ω_N and ω_R are identified in Fig. 1.[1] Another common characteristic of window functions is the ripple ratio r, which is defined as

$$r = \frac{\text{maximum side-lobe amplitude}}{\text{main-lobe amplitude}}. \tag{4}$$

If R is the ripple ratio in dB, then r is given by

$$r = 10^{-R/20}. \tag{5}$$

A useful way of comparing the side-lobe pattern of two windows is in terms of the side-lobe roll-off ratio s defined by

$$s = a_1/a_2 \tag{6}$$

where a_1 is the amplitude of the side lobe nearest to the main lobe and a_2 is the amplitude of the side lobe furthest from the main lobe. If S is the side-lobe roll-off ratio in dB, then s is given by

$$s = 10^{S/20}. \tag{7}$$

To satisfy the above specifications, consider the ultraspherical window's zero-phase spectral representation

$$W_0(e^{j\omega T}) = C_{2M}^{\mu}(x_\mu \cos \omega T/2). \tag{8}$$

[1]Note that the null-to-null and main-lobe widths have been used interchangeably to refer to the bandwidth between $-\omega_N$ and ω_N in the past.

Figure 2 shows $C_{2M}^{\mu}(x)$ for the values $\mu = 2$ and $M = 3$ and identifies some important parameters to be used in the succeeding sections.

3.1. Side-lobe roll-off ratio

From Fig. 2, the side-lobe roll-off ratio is defined as

$$s = \frac{a}{b} = \left| \frac{C_{2M}^{\mu}(x_{2M-1}^{(\mu+1)})}{C_{2M}^{\mu}(0)} \right| \tag{9}$$

where the values of $C_n^{\mu}(x)$ can be calulated using Eq. (3), $C_{2M}^{\mu}(0) = \frac{\Gamma(M+\mu)}{\Gamma(\mu)\Gamma(M+1)}$, $\Gamma(\cdot)$ is the gamma function, and $x_{2M-1}^{(\mu+1)}$ is the largest zero of the derivative of $C_{2M}^{\mu}(x)$, namely, $2\mu C_{2M-1}^{\mu+1}(x)$. Unfortunately, closed-form expressions for the zeros of the ultraspherical polynomial do not exist but they can be found iteratively. Using the Newton-Raphson method, the iteration to find $x_{2M-1}^{(\mu+1)}$ becomes

$$y_{k+1} = y_k - \frac{C_{2M-1}^{\mu+1}(y_k)}{2(\mu+1)C_{2M-2}^{\mu+2}(y_k)} \quad \text{for } k = 1, 2, \ldots \tag{10}$$

where the starting point is set to the upper bound $y_1 = x_{2M-1}^{(0)} = \cos[\pi/(4M-2)] < 1$.

Achieving a prescribed side-lobe roll-off ratio s is possible by selecting the independent parameter μ when M is known in advance. In accordance with Eq. (9), μ can be found through minimization of the function

$$f(\mu) = \left(s - \left| \frac{C_{2M}^{\mu}(x_{2M-1}^{(\mu+1)})}{C_{2M}^{\mu}(0)} \right| \right)^2. \tag{11}$$

Simple algorithms such as dichotomous, Fibonacci, or golden section line searches, as outlined in [7], may be used to perform the minimization.

M	min S (dB)	max S (dB)
2	−6.0193	4.9557
3	−10.1985	12.7875
4	−13.0503	20.8269
5	−15.2041	28.5520
6	−16.9323	35.8374
7	−18.3748	42.6709
8	−19.6123	49.0768
9	−20.6958	55.0899

Table 1. Limiting side-lobe roll-off ratio's for small M.

Upper and lower bounds for the optimization algorithm should be set to

$$\begin{array}{ll} \mu_L = 0 \text{ and } \mu_U = 10 & \text{for } s > 1 \\ \mu_L = -0.9999 \text{ and } \mu_U = 0 & \text{for } 0 < s < 1. \end{array} \tag{12}$$

If $s = 1$, then minimization is unnecessary and $\mu = 0$. The bound $\mu_L = -0.9999$ was chosen because $C^{\mu}_{2M}(x)$ has a singularity at the value $\mu = -1$. Also, for values of $\mu \leq -1.5$ the second null of the window crosses the first null rendering the result useless for our purposes. The bound $\mu_U = 10$ was chosen as values of $\mu > 10$ yield minimal gains in s for low values of M.

An important consequence of selecting the bounds $\mu_U = 10$ and $\mu_L = -0.9999$ is for lower values of M some side-lobe roll-off ratios are not attainable. For example, if $M = 3$, the values $S = 20\log_{10} s = 50$ dB and -18 dB are not attainable for any value of μ. For this reason, we limit the possible design range of S to that produced using $\mu_L = -0.9999$ and $\mu_U = 10$ for small values of M as shown in Table 1. The values of M given in Table 1 include only those which do not span the side-lobe roll-off ratio range, $-20 \leq S \leq 60$, given in dB.

3.2. Null-to-null width

From Fig. 2, the null-to-null width is defined as

$$\omega_N = 2\cos^{-1}(x^{(\mu)}_{2M}/x_\mu) \tag{13}$$

where $x^{(\mu)}_{2M}$ is the largest zero of $C^{\mu}_{2M}(x)$. As in the previous section, we can find $x^{(\mu)}_{2M}$ with the Newton-Raphon iteration

$$y_{k+1} = y_k - \frac{C^{\mu}_{2M}(y_k)}{2\mu C^{\mu+1}_{2M-1}(y_k)} \quad \text{for } k = 1, 2, \ldots \tag{14}$$

where the starting point is set to the upper bound $y_1 = x^{(0)}_{2M} = \cos(\pi/4M) < 1$.

With μ and M known, a desired null-to-null width can be achieved by selecting the independent parameter x_μ. Rearranging Eq. (13), an expression for x_μ which satisfies a prescribed null-to-null width ω_N is given by

$$x_\mu = x^{(\mu)}_{2M}/\cos(\omega_N/2). \tag{15}$$

3.3. Main-lobe width

From Fig. 2, the main-lobe width is defined as

$$\omega_R = 2\cos^{-1}(x_a/x_\mu) \tag{16}$$

where x_a is given by $C^{\mu}_{2M}(x_a) = a$. The value of x_a can be found through a two-step process for a given μ and M. First we find $x^{(\mu+1)}_{2M-1}$ by applying the iteration in Eq. (10). Next, since $a = |C^{\mu}_{2M}(x^{(\mu+1)}_{2M-1})| = C^{\mu}_{2M}(x_a)$, we find x_a through the second Newton-Raphson iteration

$$y_{k+1} = y_k - \frac{C^{\mu}_{2M}(y_k) - a}{2\mu C^{\mu+1}_{2M-1}(y_k)} \quad \text{for } k = 1, 2, \ldots \tag{17}$$

where the starting point is set to the upper bound $y_1 = x^{(0)}_{2M} = \cos(\pi/4M) < 1$.

With μ and M known, a desired main-lobe width can be achieved by selecting the independent parameter x_μ. Rearranging Eq. (16), an expression for x_μ that yields a prescribed main-lobe width ω_R is given by

$$x_\mu = x_a/\cos(\omega_R/2). \tag{18}$$

3.4. Ripple ratio

From Fig. 2, the ripple ratio is defined as

$$r = \frac{a}{c} = \left|\frac{C^{\mu}_{2M}(x_a)}{C^{\mu}_{2M}(x_\mu)}\right|. \tag{19}$$

With μ and M known, a desired ripple ratio can be achieved by selecting the independent parameter x_μ. This can be accomplished through the Newton-Raphson iteration

$$y_{k+1} = y_k - \frac{C^{\mu}_{2M}(y_k) - a/r}{2\mu C^{\mu+1}_{2M-1}(y_k)} \quad \text{for } k = 1, 2, \ldots \tag{20}$$

where $a = |C^{\mu}_{2M}(x^{(\mu+1)}_{2M-1})| = C^{\mu}_{2M}(x_a)$. The starting point can be set to $y_1 = x_0 = \cosh\left[\frac{1}{2M}\cosh^{-1}\left(\frac{1}{r}\right)\right]$.

4. COMPUTATIONAL COMPLEXITY

Figure 3 shows a comparison of the computational complexity associated with Eqs. (1) and (2) for increasing values of M. The high computational complexity in Eq. (2) is primarily due to the repeated calculation of $C^{\mu}_{n}(x)$ by the recurrence relationship given in Eq. (3). Evidently, Eq. (1) offers reduced computational complexity.

5. EXAMPLES

Example 1: For $M = 25$ and $S = 20$ dB, generate the ultraspherical window coefficients that yield (a) $\omega_R = 0.25$ rads/s and (b) $\omega_N = 0.25$ rads/s.

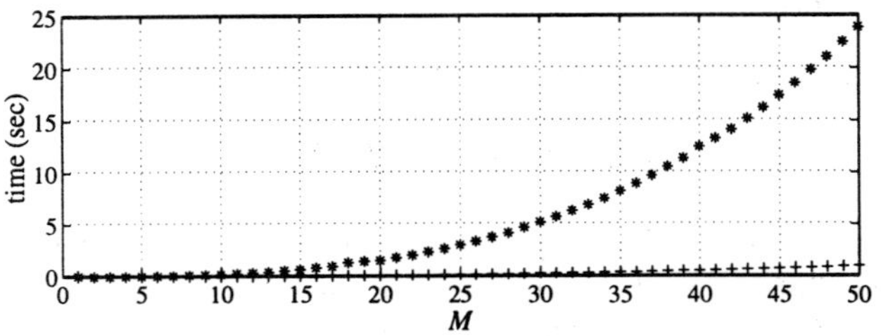

Fig. 3. Computational complexity for generating the ultraspherical window coefficients using Eqs. (1) and (2) represented by '+' and '*', respectively.

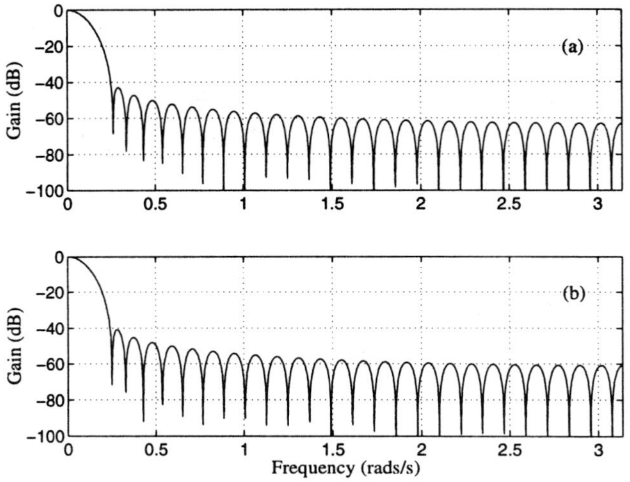

Fig. 4. Ultraspherical window spectral representations for $M = 25$, $S = 20$ dB and (a) $\omega_R = 0.25$ rads/s, (b) $\omega_N = 0.25$ rads/s.

Figure 4 shows spectral representations obtained for the windows. For both cases, the minimization of Eq. (11) resulted in $\mu = 0.9517$ and Eqs. (18) and (15) gave $x_\mu = 1.0067$ and 1.0060 for $\omega_R = 0.25$ and $\omega_N = 0.25$ rads/s, respectively.

Example 2: For $M = 25$ and $R = 50$ dB, generate the ultraspherical window coefficients that yield (a) $S = -10$ dB and (b) $S = 30$ dB.

Figure 5 shows spectral representations obtained for the windows. Minimizing Eq. (11) resulted in $\mu = -0.3914$ and 1.5151 for $S = -10$ and 30 dB, respectively, and Eq. (20) gave $x_\mu = 1.0107$ and 1.0091, respectively.

6. CONCLUSIONS

A method was proposed for calculating the ultraspherical window's independent parameters to achieve window designs that have prescribed spectral characteristics. The char-

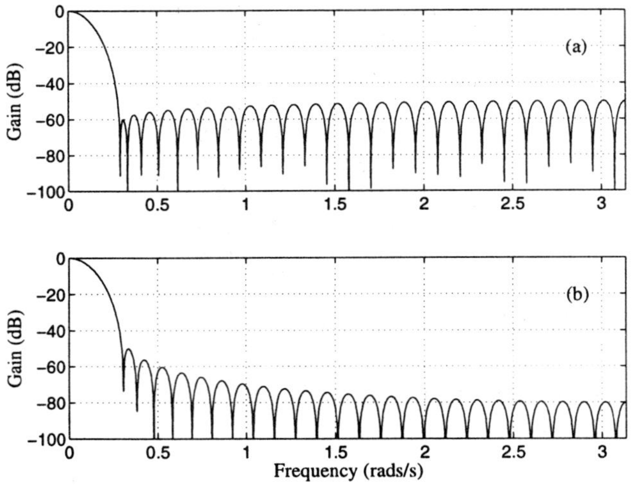

Fig. 5. Ultraspherical window spectral representations for $M = 25$, $R = 50$ dB, and (a) $S = -10$ dB, (b) $S = 30$ dB.

acteristics considered included the ripple ratio, main-lobe width, null-to-null width, and side-lobe roll-off ratio. The method was found to yield the desired characteristics to a high degree of precision as shown through examples.

The computational complexities of two different equations for generating the ultraspherical window coefficients, namely, Eqs. (1) and (2), were also compared. The former was found to be significantly more efficient.

7. REFERENCES

[1] R. L. Streit, "A two-parameter family of weights for nonrecursive digital filters and antennas," *IEEE Trans. on Acoustics, Speech, and Signal Processing*, vol. 32, no. 1, pp. 108-118, February 1984.

[2] C. L. Dolph, "A current distribution for broadside arrays which optimizes the relationship between beamwidth and side-lobe level," *Proc. IRE*, vol. 34, pp. 335-348, June 1946.

[3] J. F. Kaiser, "Nonrecursive digital filter design using I_0-sinh window function," *IEEE Int. Symp. on Circuits and Systems*, pp. 20-23, April 1974.

[4] T. Saramäki, "A class of window functions with nearly minimum sidelobe energy for designing FIR filters," *IEEE Int. Symp. on Circuits and Systems*, Portland, Oregon, pp. 359-362, May 1989.

[5] S. W. A. Bergen and A. Antoniou, "Generation of ultraspherical window functions," XI European Signal Processing Conference, Toulouse, France, vol. 2, pp. 607-610, Sept. 2002.

[6] A. Antoniou, *Digital Filters*, McGraw-Hill, 1993.

[7] R. Fletcher, *Practical Methods of Optimization*, Wiley, 1987.

ANALYTICAL DESIGN OF FRACTIONAL HILBERT TRANSFORMER USING FRACTIONAL DIFFERENCING

Chien-Cheng Tseng

Department of Computer and Communication Engineering, National Kaohsiung First University of Science and Technology, Kaohsiung, Taiwan
E-mail: tcc@ccms.nkfust.edu.tw

ABSTRACT

Conventionally, fractional differencing (FD) has been successfully used to generate fractal process called fractional Brownian motion, and fractional Hilbert transformer (FHT) has been also applied to the edge detection of images and the construction of secure single-side band (SSB) communication system. In this paper, the relation between FD and FHT will be investigated such that FD can be applied to design FHT directly. The proposed design method not only provides the closed-form transfer function but also ensures that the designed fractional Hilbert transformer is stable. One design example is demonstrated to illustrate the effectiveness of the new design method.

1. INTRODUCTION

The Hilbert transformer has been widely used in communication applications to generate single sideband (SSB) signals by splitting the modulating signal into two components which are 90 degrees out of phase. This approach reduces the required bandwidth for transmission of the signal by half [1]. In 1996, Lohmann et al. [2] generalized the classical Hilbert transform by introducing two different definitions of what they called the fractional Hilbert transform (FHT). One definition is a modification of the spatial filter with a fractional parameter, the other is based on the fractional Fourier transform. In [3], Pei and Yeh developed the discrete version of fractional Hilbert transform and applied it to the edge detection of images. In [4], Zayed introduced another generalization of the Hilbert transform to obtain the signal's analytic part by suppressing the negative frequencies of the signal's fractional Fourier transform. In [5], Tseng and Pei proposed a secure SSB communication in which the fractional order of Hilbert transformer is used as a secrete key for demodulation. Recently, the designs of maximally flat FIR and allpass fractional Hilbert transformers are also investigated in details [6][7].

On the other hand, the fractional differencing (FD) have been shown to be a good model of long-correlation signal [8]. The long-correlation signals are often referred to as 1/f noise or random fractal and have been reported in various applications including hydrology [9], infrared remote sensing [10], texture image [11] and network traffic [12] etc. The definition and property of fractional differencing operator have been described in [13]. And, the algorithms to estimate the parameters of discrete fractionally differenced Gaussian noise process have been studied in [14][15]. In this paper, the fractional differencing operator will be used to design fractional Hilbert transformer. The new method not only provides the closed-form transfer function but also ensures that the designed Hilbert transformer is stable. The paper is organized as follows: First, fractional Hilbert transformer and fractional differencing are reviewed briefly. Then, the details that uses fractional differencing to design fractional Hilbert transformer are described. Finally, one design example is demonstrated to illustrate the effectiveness of the proposed approach.

2. FRACTIONAL HILBERT TRANSFORMER

The ideal frequency response of fractional Hilbert transformer (FHT) with order ν and delay D is defined by

$$H_\nu(\omega) = \begin{cases} e^{-j\phi}e^{-j\omega D} & 0 < \omega < \pi \\ e^{j\phi}e^{j\omega D} & -\pi < \omega < 0 \end{cases} \tag{1}$$

where $\phi = \frac{\nu\pi}{2}$. When we choose $\nu = 1$ and $D = 0$, FHT becomes the conventional Hilbert transformer [1]. The details that use FHT to construct single-side band signal for saving communication bandwidth are described below: Given a real signal $x(n)$, its complex analytic signal $\hat{x}_\nu(n)$ is defined by

$$\hat{x}_\nu(n) = x(n-D) - e^{-j\phi}\bar{x}_\nu(n) \tag{2}$$

where $\bar{x}_\nu(n)$ is the output of the fractional Hilbert transformer with order ν and delay D by feeding signal $x(n)$. Taking the Fourier transform at both sides in eq(2), we have the expression in the frequency domain as

$$\hat{X}_\nu(\omega) = X(\omega)e^{-j\omega D} - e^{-j\phi}\bar{X}_\nu(\omega) \tag{3}$$

where $\bar{X}_\nu(\omega) = H_\nu(\omega)X(\omega)$. Substituting eq(1) into eq(3), we obtain

$$\hat{X}_\nu(\omega) = \begin{cases} 2je^{-j\phi}\sin(\phi)e^{-j\omega D}X(\omega) & 0 < \omega < \pi \\ 0 & -\pi < \omega < 0 \end{cases} \tag{4}$$

Thus, the negative frequency content of $\hat{x}_\nu(n)$ is all suppressed. Because $2je^{-j\phi}\sin(\phi)$ is a constant and $e^{-j\omega D}$ corresponds to a pure delay, the signal $\hat{x}_\nu(n)$ contains the same information as $x(n)$. Now, let us use $x(n) = \cos(n\omega_0)$ to demonstrate the above fact. It is easy to show that

$$\bar{x}_\nu(n) = \cos(\omega_0(n-D) - \phi) \tag{5}$$

and the analytic signal $\hat{x}_\nu(n)$ is given by

$$\begin{aligned} \hat{x}_\nu(n) &= \cos(\omega_0(n-D)) - e^{-j\phi}\cos(\omega_0(n-D) - \phi) \\ &= je^{-j\phi}\sin(\phi)e^{j\omega_0(n-D)} \end{aligned} \tag{6}$$

When we choose $\phi = \frac{\pi}{2}$ and delay $D = 0$, this result is reduced to $\hat{x}_\nu(n) = e^{j\omega_0 n}$. So far, fractional Hilbert transformer has been reviewed briefly. In next section, let us study the fractional differencing.

0-7803-7761-3/03/$17.00 ©2003 IEEE

3. FRACTIONAL DIFFERENCING

The fractional differencing has been successfully used to generate a fractal process called Brownian motion (fBm) [14][15]. The spectrum of fBm is of the form ω^{-2d}, where d is a fractional number and ω is the frequency. Usually, the fBm can be modeled as the output $y(n)$ of a dth order fractional differencing filter $G(z)$ driven by a Gaussian white noise $v(n)$ with zero mean and unit variance. The parameter d plays the role of the Hurst parameter (H) in fBm. In [14][15], the transfer function of $G(z)$ is chosen as

$$\begin{aligned} G(z) &= \frac{1}{(1-z^{-1})^d} \\ &= \sum_{k=0}^{\infty} \binom{-d}{k} (-z)^{-k} \\ &= \sum_{k=0}^{\infty} \frac{(k+d-1)!}{k!(d-1)!} z^{-k} \\ &= \sum_{k=0}^{\infty} \frac{d(1+d)\cdots(k-1+d)}{k!} z^{-k} \end{aligned} \tag{7}$$

where d is a fractional number. The frequency response of $G(z)$ is given by

$$\begin{aligned} G(e^{j\omega}) &= \frac{1}{(1-e^{-j\omega})^d} \\ &= \left(2j \sin\frac{\omega}{2}\right)^{-d} e^{j\frac{\omega}{2}d} \end{aligned} \tag{8}$$

Since $\sin\frac{\omega}{2}$ can be approximated as $\frac{\omega}{2}$ when ω tends to zero, we have

$$G(e^{j\omega}) \approx (j\omega)^{-d} e^{j\frac{\omega}{2}d} \tag{9}$$

for small value of frequency ω. This means that the power spectrum $S_y(\omega)$ of fractal process $y(n)$ is

$$S_y(\omega) = |G(e^{j\omega})|^2 \approx \omega^{-2d} \tag{10}$$

Thus, the spectrum behaves as ω^{-2d} as ω tends to zero. That is, at low frequency the spectrum of the process $y(n)$ has a behavior similar to that of the fBm. In this paper, we will use $G(z)$ to design fractional Hilbert transformer rather than to generate fBm. The details are described in next section.

4. DESIGN METHOD

The main idea that uses fractional differencing to design fractional Hilbert transformer is given below: Replacing z by z^{-1}, the transfer function $G(z^{-1})$ becomes:

$$G(z^{-1}) = \frac{1}{(1-z)^d} \tag{11}$$

From eq(8), the frequency response of $G(z^{-1})$ is given by

$$G(e^{-j\omega}) = \left(2j \sin\frac{-\omega}{2}\right)^{-d} e^{-j\frac{\omega}{2}d} \tag{12}$$

Now, let us define the allpass filter $A(z)$ below:

$$A(z) = \frac{G(z^{-1})}{G(z)} \tag{13}$$

then, from equations (8) and (12), we have the frequency response of $A(z)$ as

$$\begin{aligned} A(e^{j\omega}) &= \frac{G(e^{-j\omega})}{G(e^{j\omega})} \\ &= (-1)^{-d} e^{-j\omega d} \\ &= (e^{-j\pi})^{-d} e^{-j\omega d} \\ &= e^{j\pi d} e^{-j\omega d} \end{aligned} \tag{14}$$

Compare eq(14) with eq(1), it is clear that $A(e^{j\omega}) = H_\nu(\omega)$ if we choose $d = \frac{-\nu}{2}$ and $D = d$. Thus, allpass filter $A(z)$ is an ideal fractional Hilbert transformer. However, $G(z)$ and $G(z^{-1})$ have the infinite terms, so they can not be implemented in practical hardware. For implementation, $G(z)$ and $G(z^{-1})$ must be truncated into the finite terms. Now, let us define the truncated fractional differencing filter $\hat{G}(z)$ below:

$$\hat{G}(z) = \sum_{k=0}^{N} \binom{-d}{k} (-1)^{-k} z^{-k} \tag{15}$$

then the truncated allpass filter

$$\hat{A}(z) = \frac{\hat{G}(z^{-1})}{\hat{G}(z)} \tag{16}$$

will approximate $e^{j\pi d}e^{-j\omega d}$ well. When order N approaches ∞, allpass filter $\hat{A}(z)$ becomes the ideal fractional Hilbert transformer. However, $\hat{G}(z^{-1})$ is a noncausal filter. For implementation, we multiply $\hat{A}(z)$ by pure delay z^{-N} to obtain the causal transfer function:

$$H(z) = z^{-N}\hat{A}(z) = \frac{z^{-N}\hat{G}(z^{-1})}{\hat{G}(z)} \tag{17}$$

The $H(z)$ will approximate the response $e^{j\pi d}e^{-j\omega(N+d)}$ well, so $H(z)$ is the final designed fractional Hilbert transformer in which order $\nu = -2d$ and delay $D = N + d$. The entire design algorithm is now summarized as follows: Given fractional parameter ν and order N:
Step 1: Compute the parameter $d = \frac{-\nu}{2}$.
Step 2: Use eq(15) to compute filter $\hat{G}(z)$.
Step 3: The designed fractional Hilbert transformer is given by $H(z) = \frac{z^{-N}\hat{G}(z^{-1})}{\hat{G}(z)}$. The filter $H(z)$ will approximate $e^{-j\frac{\nu}{2}\pi}e^{-j\omega(N-\frac{\nu}{2})}$ well.

Finally, two remarks are made. First, because eq(15) is a closed-form formula, the proposed method provides a analytical design of fractional Hilbert transformer. Second, because $H(z)$ is an IIR allpass filter, $H(z)$ needs to be stable for implementation. However, the above derivation does not guarantee the stability of filter $H(z)$. In next section, stability issue will be discussed.

5. STABILITY ISSUE

In this section, we first show that the zeros of the polynomial $G(z) = 0$ are all inside the unit circle, i.e., $G(z)$ is a minimum phase polynomial. Then, we show that the zeros of the truncated polynomial $\hat{G}(z) = 0$ are also inside unit circle. Thus, the designed fractional Hilbert transformer is stable. In order to show $G(z)$ is minimum phase, the following theorem is reviewed.

Fact 1: The zeros of polynomial $P(z) = 0$ are all inside the unit circle if

$$Re[P(e^{j\omega})] > 0 \quad \text{for } \omega \in [0, \pi] \tag{18}$$

where $Re[P(e^{j\omega})] > 0$ denotes the real part of $P(e^{j\omega})$.
Pf: See [16].
From the eq(8), the real part of $G(e^{j\omega})$ can be written as

$$Re[G(e^{j\omega})] = \left(2\sin(\frac{\omega}{2})\right)^{-d} \cos\left(\frac{d}{2}(\omega - \pi)\right) \tag{19}$$

When $\omega \in [0, \pi]$ and $d \in (-1, 1)$, it can be shown that

$$\sin(\frac{\omega}{2}) > 0$$

$$\cos\left(\frac{d}{2}(\omega - \pi)\right) > 0$$

These two equations imply that $Re[G(e^{j\omega})] > 0$ for $d \in (-1, 1)$. Thus, $G(z)$ is a minimum phase polynomial when $d \in (-1, 1)$. Now, two remarks are made as follows:
Remark 1: Because $G(z)$ is a minimum-phase polynomial, the magnitude $|G(e^{j\omega})|$ and phase $arg(G(e^{j\omega}))$ will satisfy the following relation [17]:

$$ln|G(e^{j\omega})| = C + \frac{1}{2\pi} P \int_{-\pi}^{\pi} arg[G(e^{j\theta})] \cot\left(\frac{\omega - \theta}{2}\right) d\theta \tag{20}$$

where P denotes the Cauchy principal value of integral and C is a constant given by

$$C = \frac{1}{2\pi} \int_{-\pi}^{\pi} ln|G(e^{j\omega})| d\omega \tag{21}$$

Remark 2: Although $G(z)$ is stable, the degree of $G(z)$ is infinite. Thus, the fact that $G(z)$ is stable does not imply that the truncated polynomial $\hat{G}(z)$ is stable. In the following, the Enestrom-Kakeya theorem in [7][18] will be applied to show $\hat{G}(z)$ is stable if $d \in (0, 1)$.
Fact 2: Let $P(z) = \sum_{k=0}^{N} a_k z^{-k}$ be a polynomial with $a_k > 0$ and ratio $r_k = \frac{a_{k+1}}{a_k}$. Then all the zeros of $P(z) = 0$ are contained in the annulus

$$\min_k r_k \le |z| \le \max_k r_k \tag{22}$$

Pf: See [18].
When we choose $P(z) = \hat{G}(z)$, we have coefficients:

$$a_k = \frac{d(1 + d) \cdots (k - 1 + d)}{k!} \tag{23}$$

If the parameter d is in the range $(0, 1)$, the coefficients $a_k > 0$. Thus, we can use Fact 2 to test the stability of $\hat{G}(z)$. In this case, the ratio r_k can be calculated by

$$r_k = \frac{a_{k+1}}{a_k} = \frac{k + d}{k + 1} \tag{24}$$

Clearly, when $d \in (0, 1)$, we have $r_k < 1$ for all $0 \le k \le N$. This implies that $\max r_k < 1$. According to Enestrom-Kakeya theorem, all zeros of $\hat{G}(z) = 0$ are contained in the unit circle. Thus, $\hat{G}(z)$ is stable, i.e., the designed FHT $H(z) = \frac{z^{-N}\hat{G}(z^{-1})}{\hat{G}(z)}$ is stable for $d \in (0, 1)$. Finally, One remark is made as follows:
Remark 3: In the above, Enestrom-Kakeya theorem can not show that $\hat{G}(z)$ is stable for $d \in (-1, 0]$. However, in the lattice-form realization of fractional difference [19], the partial correlation coefficients are known to be

$$\kappa_k = \frac{d}{k - d} \tag{25}$$

Clearly, $|\kappa_k| < 1$ for $d \in (-1, 0]$. This implies that $\hat{G}(z)$ is stable, i.e., the designed FHT $H(z) = \frac{z^{-N}\hat{G}(z^{-1})}{\hat{G}(z)}$ is also stable for $d \in (-1, 0]$.

6. DESIGN EXAMPLE

In this section, one example performed with MATLAB language is presented to illustrate the effectiveness of the proposed design method. Because the designed fractional Hilbert transformer approximates the response $e^{-j\frac{\nu}{2}\pi} e^{-j\omega(N-\frac{\nu}{2})}$ and the term $e^{-j\omega(N-\frac{\nu}{2})}$ corresponds to a pure delay, the normalized phase response

$$PH(\omega) = \frac{arg(H(e^{j\omega})) + (N - \frac{\nu}{2})\omega}{-\frac{\pi}{2}} \tag{26}$$

is used to evaluate the performance. The more $PH(\omega)$ close to ν is, the better designed results we have. When order N is chosen as 20, the $PH(\omega)$ of the designed FHT for $\nu = 0.4$ is shown in Fig.1. Clearly, the $PH(\omega)$ approximates the ideal response $\nu = 0.4$ well except around frequency $\omega = 0$. Fig.2 also shows the pole diagram of the designed FHT. It is obvious that all poles are inside the unit circle. The maximum pole radius is 0.8990. Thus, the designed FHT is stable. Finally, three remarks are made as follows:
Remark 1: For a stable allpass filter $H(z)$, the phase response $arg(H(e^{j\omega}))$ is always equal to zero at $\omega = 0$. This implies that $PH(0)$ is always equal to zero, as shown in Fig.1.
Remark 2: For a stable allpass filter $H(z)$, the phase response $arg(H(e^{j\omega}))$ is always equal to $-N\pi$ at $\omega = \pi$. This implies that $PH(\pi)$ is always equal to ν, as shown in Fig.1.
Remark 3: Because FHT is designed directly from the FD, FHT and FD can be implemented together in one structure. Fig.3 shows the joint implementation of fractional differencing and fractional Hilbert transformer. When input $x(n)$ is a white noise, the fractional differencing output is fBm with order $-d$. And, the order ν of fractional Hilbert transformer is $-2d$.

7. CONCLUSION

In this paper, the relation between FD and FHT has been studied such that FD can be applied to design FHT directly. The proposed design method not only provides the closed-form transfer function but also ensures that the designed fractional Hilbert transformer is stable. However, only 1-D filter is considered here. Thus, it is interesting to extend the proposed method to design 2-D fractional Hilbert transformer in future work.

REFERENCES

[1] S.L. Hahn, *Hilbert Transforms in Signal Processing*, Artech House, 1996.

[2] A.W. Lohmann, D. Mendlovic and Z. Zalevsky, "Fractional Hilbert transform," *Optics Letters*, vol.21, pp.281-283, Feb. 1996.

[3] S.C. Pei and M.H. Yeh, "Discrete fractional Hilbert transform," *IEEE Trans. Circuits Syst. II*, vol.47, pp.1307-1311, Nov. 2000.

[4] A.I. Zayed, "Hilbert transform associated with the fractional Fourier transform," *IEEE Signal Processing Letters*, vol.5, pp.206-208, Aug. 1998.

[5] C.C. Tseng and S.C. Pei, "Design and application of discrete-time fractional Hilbert transformer," *IEEE Trans. Circuits Syst. II*, vol.47, pp.1529-1533, Dec. 2000.

[6] S.C. Pei and P.H. Wang, "Analytical design of maximally flat FIR fractional Hilbert transformers," *Signal Processing*, vol.81, pp.643-661, 2001.

[7] S.C. Pei and P.H. Wang, "Maximally flat allpass fractional Hilbert transformers," *IEEE Int. Symp. on Circuits and Systems*, vol.4, pp.701-704, May 2002.

[8] D.G. Manolakis, V.K. Ingle and S.M. Kogon, *Statistical and Adaptive Signal Processing*, McGraw-Hill, New York 2000.

[9] J.R.M. Hosking, "Modeling persistence in hydrological time series using fractional differencing," *Water Resources Research*, vol.20, pp.1898-1908, Dec. 1984.

[10] S.M. Kogon and D.G. Manolakis, "Signal modeling with self-similar α-stable processes: The fractional Levy stable motion model," *IEEE Trans. Signal Processing*, vol.44, pp.1006-1010, Apr. 1996.

[11] K.B. Eom, "Long-correlation image models for textures with circular and elliptical correlation structures," *IEEE Trans. Image Processing*, vol.10, pp.1047-1055, July, 2001.

[12] J. Ilow, "Forecasting network traffic using FARIMA models with heavy tailed innovations," in *Proc. ICASSP*, vol.6, pp.3814-3817, 2000.

[13] J.R.M. Hosking, "Fractional differencing," *Biometrika*, vol.68, pp.165-176, 1981.

[14] M. Deriche and A.H. Tewfik, "Signal modeling with filtered discrete fractional noise processes," *IEEE Trans. Signal Processing*, vol.41, pp.2839-2849, Sep. 1993.

[15] M. Deriche and A.H. Tewfik, "Maximum likelihood estimation of the parameters of discrete fractionally differenced Gaussian noise process," *IEEE Trans. Signal Processing*, vol.41, pp.2977-2989, Oct. 1993.

[16] A.T. Chottera and G.A. Jullien, " A linear programming approach to recursive filter design with linear phase", *IEEE Trans. Circuits and Systems*, vol.29, pp.139-149, Mar. 1982.

[17] A.V. Oppenheim, R.W. Schafer and J.R. Buck, *Discrete-time Signal Processing*, 2nd Edition, Prentice-Hall, 1999.

[18] N. Anderson, E.B. Saff and R.S. Varga, "On the Enestrom-Kakeya theorem and its sharpness," *Linear Algebra and Its Applications*, vol.28, pp.5-16, 1979.

[19] S.M. Kogon and D.G. Manolakis, "Efficient generation of long-memory signals using lattice structures," *Asilomar Conference on Signals, Systems, and Computers*, vol.2 pp.990-994, 1996.

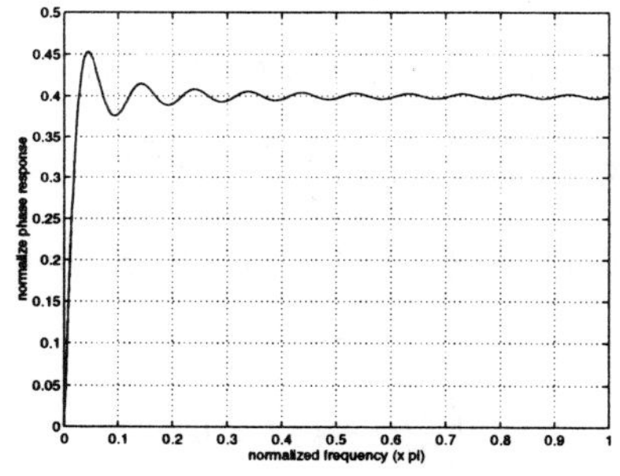

Figure 1. The normalized phase response $PH(\omega)$ of the designed fractional Hilbert transformer with $\nu = 0.4$.

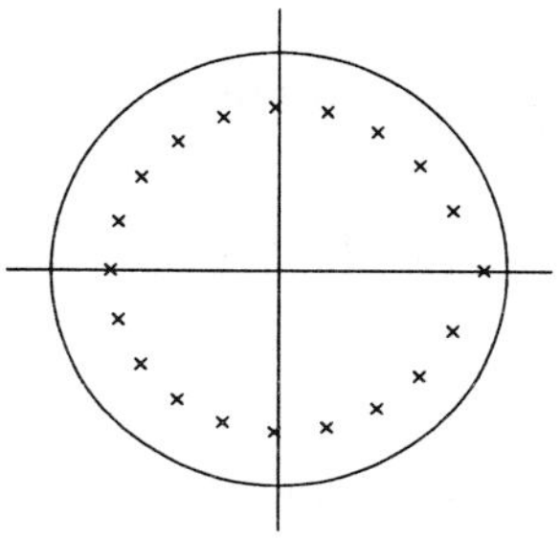

Figure 2. The pole diagram of the designed fractional Hilbert transformer with $\nu = 0.4$.

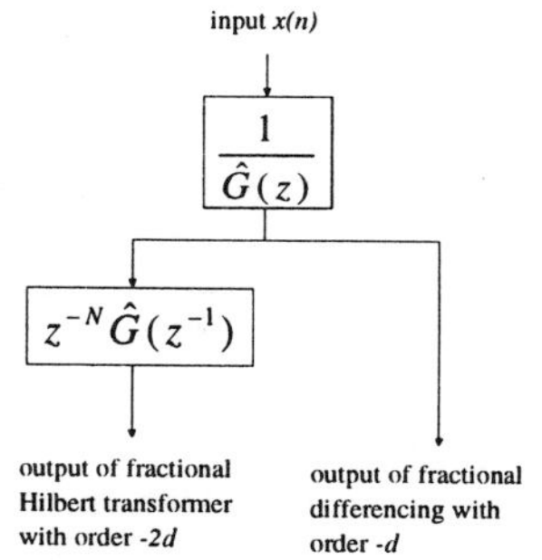

Figure 3. The joint implementation of fractional differencing and fractional Hilbert transformer.

EFFICIENT DESIGN OF SVD-BASED 2-D DIGITAL FILTERS USING SPECIFICATION SYMMETRY AND ORDER-SELECTING CRITERION

Tian-Bo Deng

Department of Information Science
Faculty of Science, Toho University
Miyama 2-2-1, Funabashi, Chiba 274-8510, Japan

E-mail: deng@is.sci.toho-u.ac.jp

ABSTRACT

Two-dimensional (2-D) digital filters are widely useful in image processing and other 2-D digital signal processing fields, but designing 2-D digital filters is much more difficult than designing one-dimensional (1-D) ones. This paper provides a new insight into the existing singular value decomposition (SVD)-based design approach in the sense that the SVD-based design can be performed more efficiently by exploiting the new symmetry of the 2-D magnitude specifications. By using the specification symmetry, only half of the 1-D filters (sub-filters) need to be designed, which significantly simplifies the design process and reduces the computer storage required for 1-D sub-filter coefficients. Another novel point of this paper is that an objective criterion is proposed for selecting appropriate sub-filter orders in order to reduce the hardware implementation cost. A design example is given to illustrate the effectiveness of the SVD-based design approach by exploiting specification symmetry and new order-selecting criterion.

1. INTRODUCTION

Two-dimensional (2-D) digital filtering is one of the most fundamental and most important processing techniques in digital image processing and other 2-D digital signal processing fields. Up to this point, many methods have been developed for implementing and designing 2-D digital filters, among the developed techniques, the indirect approaches that decompose the original 2-D problems into 1-D ones have received considerable attention. The reason is that the 2-D problem can be easily attacked by solving a set of easier 1-D problems through using the accumulated 1-D techniques, thus the original 2-D problems can be indirectly solved in an elegant way. The SVD-based approaches have been developed in the frequency-domain by a few researchers in an increasingly improved manner as follows:

⟨1⟩ Separable 2-D filter with only one section [1].
⟨2⟩ SVD-based 2-D filters with biased circuits [2].
⟨3⟩ ISVD-based 2-D filters without biased circuits [3].
⟨4⟩ Nonnegative decomposition-based design [4, 5].
⟨5⟩ SVD-based non-quadrantal symmetric design [6].
⟨6⟩ SVD-based design with different sub-filter orders [7].

This paper is aimed to further advance the SVD-based design methods ⟨5⟩ and ⟨6⟩ in the following aspects:

- By exploiting the symmetry of the desired 2-D magnitude responses, we show that the design of 1-D sub-filters can be significantly simplified such that only one 1-D sub-filter in each parallel section needs to be designed, and the other one has identically the same filter coefficients as the designed one.
- An objective error criterion is proposed for selecting appropriate orders for different 1-D sub-filters such that each sub-filter contributes to the final 2-D filter design accuracy at the same extent.

A design example is given to illustrate the above two points.

2. DESIGN USING SYMMETRIES

In this section, we briefly review the mirror-image symmetry and mirror-image anti-symmetry existing in the SVD of the desired 2-D zero-phase frequency response [6], and then exploit a new symmetry that can be efficiently utilized in the SVD-based design for simplifying the 2-D filter design process.

2.1. Symmetry and Anti-symmetry

As proved in [6], both quadrantally symmetric and non-quadrantally symmetric 2-D zero-phase frequency responses can be approximated by using the SVD method, which decomposes the original zero-phase 2-D filter design problem into the problems of designing zero-phase or $\pi/2$-phase 1-D sub-filters. The design approach can be outlined follows.

Assume that $H_d(\omega_1, \omega_2)$ is the desired zero-phase 2-D frequency response, where $\omega_1, \omega_2 \in [-\pi, \pi]$ are normalized angular frequencies. By using the equally-spaced samples of $H_d(\omega_1, \omega_2)$, we can form a specification matrix

$$A = [H_d(\omega_{1l}, \omega_{2m})] = [a_{l,m}] \in R^{L \times M} \tag{1}$$

where

$$\omega_{1l} = -\pi + \frac{2\pi(l-1)}{L-1}, \qquad l = 1, 2, \cdots, L$$
$$\omega_{2m} = -\pi + \frac{2\pi(m-1)}{M-1}, \qquad m = 1, 2, \cdots, M.$$

The SVD of the matrix A results in

$$A = \sum_{i=1}^{r} \sigma_i u_i v_i^T = \sum_{i=1}^{r} \tilde{u}_i \tilde{v}_i^T \tag{2}$$

0-7803-7761-3/03/$17.00

where r is the rank of the matrix $\boldsymbol{A}$, and $\sigma_1, \sigma_2, \cdots, \sigma_r$ are the singular values, $\sigma_1 \geq \sigma_2 \cdots \geq \sigma_r > 0$,

$$\tilde{\boldsymbol{u}}_i = \sqrt{\sigma_i}\boldsymbol{u}_i, \quad \tilde{\boldsymbol{v}}_i = \sqrt{\sigma_i}\boldsymbol{v}_i. \tag{3}$$

The column vectors $\tilde{\boldsymbol{u}}_i$ and $\tilde{\boldsymbol{v}}_i$ are *either* mirror-image symmetric *or* mirror-image anti-symmetric simultaneously [6]. In the SVD-based design of 2-D filters, only the first K pairs of $\tilde{\boldsymbol{u}}_i, \tilde{\boldsymbol{v}}_i$ are used, and the others are neglected as

$$\boldsymbol{A} \approx \hat{\boldsymbol{A}} = \sum_{i=1}^{K} \tilde{\boldsymbol{u}}_i \tilde{\boldsymbol{v}}_i^T. \tag{4}$$

The normalized root-mean-squared (RMS) error is

$$\begin{aligned} E_K &= \frac{\|\boldsymbol{A}-\hat{\boldsymbol{A}}\|}{\|\boldsymbol{A}\|} \\ &= \frac{\left\| \sum_{i=K+1}^{r} \tilde{\boldsymbol{u}}_i \tilde{\boldsymbol{v}}_i^T \right\|}{\left\| \sum_{i=1}^{r} \tilde{\boldsymbol{u}}_i \tilde{\boldsymbol{v}}_i^t \right\|} = \frac{\left(\sum_{i=K+1}^{r} \sigma_i^2 \right)^{1/2}}{\left(\sum_{i=1}^{r} \sigma_i^2 \right)^{1/2}} \end{aligned} \tag{5}$$

where $\|\cdot\|$ denotes the Euclidean norm.

After truncating the last several $\tilde{\boldsymbol{u}}_i, \tilde{\boldsymbol{v}}_i$, the remaining $\tilde{\boldsymbol{u}}_i, \tilde{\boldsymbol{v}}_i$ can be regarded as the desired frequency responses of zero-phase or $\pi/2$-phase 1-D filters $F_i(z_1)$ and $G_i(z_2)$ respectively, and the sub-filters $F_i(z_1)$ and $G_i(z_2)$ are separately designed by using the existing 1-D design techniques.

2.2. New Symmetry

If $H_d(\omega_1, \omega_2)$ is symmetric with respect to the straight lines $\omega_1 = \omega_2$ and $\omega_1 = -\omega_2$, then the SVD in (4) generates the vectors $\tilde{\boldsymbol{u}}_i, \tilde{\boldsymbol{v}}_i$ that satisfy either $\tilde{\boldsymbol{u}}_i = \tilde{\boldsymbol{v}}_i$ or $\tilde{\boldsymbol{u}}_i = -\tilde{\boldsymbol{v}}_i$.

Proof: Without loss of generality, we let $L = M = 2N$ in (1) and thus the matrix $\boldsymbol{A}$ is a real 2N-by-2N matrix whose entries satisfy

$$a_{l,m} = a_{2N+1-l,2N+1-m}.$$

The matrix $\boldsymbol{A}$ can be partitioned as

$$\boldsymbol{A} = \begin{bmatrix} \boldsymbol{A}_1 & \boldsymbol{A}_2 \\ \boldsymbol{A}_3 & \boldsymbol{A}_4 \end{bmatrix} \tag{6}$$

and we assume that it has distinct singular values. If matrices $\hat{\boldsymbol{I}}_N, \boldsymbol{I}_N$ are the N-by-N backward permutation matrix and N-by-N identity matrix defined by

$$\hat{\boldsymbol{I}}_N = \begin{bmatrix} & \boldsymbol{0} & & 1 \\ & & 1 & \\ & \cdot^{\cdot^{\cdot}} & & \\ 1 & & & \boldsymbol{0} \\ 1 & & & \end{bmatrix} \tag{7}$$

$$\boldsymbol{I}_N = \begin{bmatrix} 1 & & & \\ & 1 & \boldsymbol{0} & \\ & & \ddots & \\ & \boldsymbol{0} & 1 & \\ & & & 1 \end{bmatrix} \tag{8}$$

and the matrix $\tilde{\boldsymbol{I}}$ is formed by using $\hat{\boldsymbol{I}}_N$ and $\boldsymbol{I}_N$ as

$$\tilde{\boldsymbol{I}} = \begin{bmatrix} \boldsymbol{I}_N & \boldsymbol{0} \\ \boldsymbol{0} & \hat{\boldsymbol{I}}_N \end{bmatrix} \tag{9}$$

then we can form the matrix

$$\boldsymbol{H} = \tilde{\boldsymbol{I}}\boldsymbol{A}\tilde{\boldsymbol{I}} = \begin{bmatrix} \boldsymbol{A}_1 & \boldsymbol{A}_2\hat{\boldsymbol{I}}_N \\ \hat{\boldsymbol{I}}_N\boldsymbol{A}_3 & \hat{\boldsymbol{I}}_N\boldsymbol{A}_4\hat{\boldsymbol{I}}_N \end{bmatrix} = \begin{bmatrix} \boldsymbol{H}_1 & \boldsymbol{H}_2 \\ \boldsymbol{H}_2 & \boldsymbol{H}_1 \end{bmatrix}$$

where $\boldsymbol{H}_1$, $\boldsymbol{H}_2$ are N-by-N matrices,

$$\begin{aligned} \boldsymbol{H}_1 &= \boldsymbol{A}_1 = \hat{\boldsymbol{I}}_N\boldsymbol{A}_4\hat{\boldsymbol{I}}_N \\ \boldsymbol{H}_2 &= \boldsymbol{A}_2\hat{\boldsymbol{I}}_N = \hat{\boldsymbol{I}}_N\boldsymbol{A}_3. \end{aligned} \tag{10}$$

The SVD of $\boldsymbol{A}$ results in

$$\begin{aligned} \boldsymbol{A} &= \tilde{\boldsymbol{I}}\boldsymbol{H}\tilde{\boldsymbol{I}} \\ &= \boldsymbol{U}\boldsymbol{\Sigma}\boldsymbol{V}^T \\ &= \begin{bmatrix} \boldsymbol{u}_1 & \boldsymbol{u}_2 & \cdots & \boldsymbol{u}_{2N} \end{bmatrix} \boldsymbol{\Sigma} \begin{bmatrix} \boldsymbol{v}_1 & \boldsymbol{v}_2 & \cdots & \boldsymbol{v}_{2N} \end{bmatrix}^T \end{aligned} \tag{11}$$

where $\boldsymbol{u}_i, \boldsymbol{v}_i$ are the normalized eigenvectors of $\boldsymbol{H}\boldsymbol{H}^T$ and $\boldsymbol{H}^T\boldsymbol{H}$ respectively, and $\boldsymbol{\Sigma}$ is the diagonal matrix

$$\boldsymbol{\Sigma} = \mathrm{diag}(\sigma_1 \quad \sigma_2 \quad \cdots \quad \sigma_{2N}).$$

Thus

$$\begin{aligned} \boldsymbol{H} &= \tilde{\boldsymbol{I}}\boldsymbol{U}\boldsymbol{\Sigma}\boldsymbol{V}^T\tilde{\boldsymbol{I}} \\ &= (\tilde{\boldsymbol{I}}\boldsymbol{U})\boldsymbol{\Sigma}(\tilde{\boldsymbol{I}}\boldsymbol{V})^T \end{aligned} \tag{12}$$

where

$$\begin{aligned} \tilde{\boldsymbol{I}}\boldsymbol{U} &= \tilde{\boldsymbol{I}}\begin{bmatrix} \boldsymbol{u}_1 & \boldsymbol{u}_2 & \cdots & \boldsymbol{u}_{2N} \end{bmatrix} \\ \tilde{\boldsymbol{I}}\boldsymbol{V} &= \tilde{\boldsymbol{I}}\begin{bmatrix} \boldsymbol{v}_1 & \boldsymbol{v}_2 & \cdots & \boldsymbol{v}_{2N} \end{bmatrix}. \end{aligned} \tag{13}$$

If $\boldsymbol{u}_i$ and $\boldsymbol{v}_i$ are simultaneously mirror-image symmetric, then

$$\tilde{\boldsymbol{I}}\boldsymbol{u}_i = \begin{bmatrix} \boldsymbol{x}_i \\ \boldsymbol{x}_i \end{bmatrix}, \quad \tilde{\boldsymbol{I}}\boldsymbol{v}_i = \begin{bmatrix} \boldsymbol{y}_i \\ \boldsymbol{y}_i \end{bmatrix}. \tag{14}$$

Otherwise, $\boldsymbol{u}_i$ and $\boldsymbol{v}_i$ are simultaneously anti-symmetric as

$$\tilde{\boldsymbol{I}}\boldsymbol{u}_i = \begin{bmatrix} \boldsymbol{x}_i \\ -\boldsymbol{x}_i \end{bmatrix}, \quad \tilde{\boldsymbol{I}}\boldsymbol{v}_i = \begin{bmatrix} \boldsymbol{y}_i \\ -\boldsymbol{y}_i \end{bmatrix}. \tag{15}$$

If $H_d(\omega_1, \omega_2)$ is symmetric with respect to the straight line $\omega_1 = \omega_2$, then we can verify that $\boldsymbol{A}_1$ is symmetric, i.e.,

$$\boldsymbol{A}_1 = \boldsymbol{A}_1^T. \tag{16}$$

Similarly, if $H_d(\omega_1, \omega_2)$ is also symmetric with respect to the straight line $\omega_1 = -\omega_2$, then $\boldsymbol{A}_2\hat{\boldsymbol{I}}_N$ is symmetric, i.e.,

$$\boldsymbol{A}_2\hat{\boldsymbol{I}}_N = (\boldsymbol{A}_2\hat{\boldsymbol{I}}_N)^T. \tag{17}$$

The symmetries (16) and (17) together with (10) lead to

$$H_1 = H_1^T, \quad H_2 = H_2^T. \tag{18}$$

Since

$$HH^T = \begin{bmatrix} H_1H_1^T + H_2H_2^T & H_1H_2^T + H_2H_1^T \\ H_2H_1^T + H_1H_2^T & H_2H_2^T + H_1H_1^T \end{bmatrix}$$

and

$$H^TH = \begin{bmatrix} H_1^TH_1 + H_2^TH_2 & H_1^TH_2 + H_2^TH_1 \\ H_2^TH_1 + H_1^TH_2 & H_2^TH_2 + H_1^TH_1 \end{bmatrix}$$

it is evident that H is a normal matrix, i.e.,

$$HH^T = H^TH.$$

Substituting (12) into above equations obtains

$$\begin{aligned} HH^T &= (\tilde{I}U)\Sigma(\tilde{I}V)^T\left[(\tilde{I}U)\Sigma(\tilde{I}V)^T\right]^T \\ &= (\tilde{I}U)\Sigma(\tilde{I}V)^T(\tilde{I}V)\Sigma(\tilde{I}U)^T \\ &= (\tilde{I}U)\Sigma^2(\tilde{I}U)^T \end{aligned} \tag{19}$$

and

$$\begin{aligned} H^TH &= \left[(\tilde{I}U)\Sigma(\tilde{I}V)^T\right]^T(\tilde{I}U)\Sigma(\tilde{I}V)^T \\ &= (\tilde{I}V)\Sigma(\tilde{I}U)^T(\tilde{I}U)\Sigma(\tilde{I}V)^T \\ &= (\tilde{I}V)\Sigma^2(\tilde{I}V)^T \end{aligned} \tag{20}$$

thus

$$(\tilde{I}U)\Sigma^2(\tilde{I}U)^T = (\tilde{I}V)\Sigma^2(\tilde{I}V)^T \tag{21}$$

which implies

$$U = V \quad \text{or} \quad U = -V. \tag{22}$$

Consequently, we can conclude that

$$u_i = v_i \quad \text{or} \quad u_i = -v_i \tag{23}$$

where u_i and v_i are *either* mirror-image symmetric *or* mirror-image anti-symmetric as shown in (14) and (15). The new symmetry (23) can be utilized to design 1-D sub-filters $F_i(z_1)$ and $G_i(z_2)$ efficiently. If we use $F_i(z_1)$ to approximate $\tilde{u}_i$, and set the coefficients of another sub-filter $G_i(z_2)$ identically the same as those of $F_i(z_1)$, i.e.,

$$G_i(z_2) = F_i(z_2)$$

then $T_iF_i(z_2)$ approximates $\tilde{v}_i$ just as $F_i(z_1)$ approximates $\tilde{u}_i$, where

$$T_i = \begin{cases} 1 & \text{if } \tilde{u}_i = \tilde{v}_i \\ -1 & \text{if } \tilde{u}_i = -\tilde{v}_i. \end{cases} \tag{24}$$

As a result, only sub-filters $F_1(z_1), F_2(z_1), \cdots, F_K(z_1)$ need to be designed, and $T_iF_i(z_2)$ can be readily obtained. This symmetry exploitation can

- reduce the design work by 50%,
- save the computer storage for sub-filter coefficients by 50%.

The resulting 2-D digital filter is shown in Fig. 1, where

$$S_i = \begin{cases} 1 & \text{for zero-phase sub-filters} \\ -1 & \text{for for } \pi/2\text{-phase sub-filters.} \end{cases} \tag{25}$$

2.3. Order-Selecting Criterion

An important step in SVD-based 2-D filter design is how to select the orders of 1-D sub-filters for approximating different vectors $\tilde{u}_i$ and $\tilde{v}_i$. Most existing SVD-based designs use the same order for different 1-D sub-filters [6], but the only one exception proposed in [7] utilizes different orders for different $\tilde{u}_i$ and $\tilde{v}_i$. That is, low-order sub-filters are used for low-energy vectors, and high-order filters are for high-energy vectors. This paper will show that this order-selecting policy is not appropriate since lower order sub-filters cannot achieve good approximations to the last several vectors whose elements become more and more irregular (zigzag) as the number of parallel sections increases. Instead, we propose a new objective criterion for selecting appropriate sub-filter orders not only based on the vector energy but also based on the irregularity of vector elements.

First, let us define a set of approximation errors. Assume that f_i and g_i are the actual vectors for approximating $\tilde{u}_i$ and $\tilde{v}_i$, respectively, and that the approximation error vectors are

$$\begin{aligned} \Delta\tilde{u}_i &= \tilde{u}_i - f_i \\ \Delta\tilde{v}_i &= \tilde{v}_i - g_i. \end{aligned} \tag{26}$$

Clearly, the normalized RMS errors are

$$e_{\tilde{u}_i} = \frac{\|\tilde{u}_i - f_i\|}{\|\tilde{u}_i\|} = \frac{\|\Delta\tilde{u}_i\|}{\sqrt{\sigma_i}} \tag{27}$$

$$e_{\tilde{v}_i} = \frac{\|\tilde{v}_i - g_i\|}{\|\tilde{v}_i\|} = \frac{\|\Delta\tilde{v}_i\|}{\sqrt{\sigma_i}}. \tag{28}$$

It should be noted here that relatively large approximation errors $e_{\tilde{u}_i}$ and $e_{\tilde{v}_i}$ do not necessarily affect the final 2-D design accuracy significantly. On the other hand, too large errors $e_{\tilde{u}_i}$ and $e_{\tilde{v}_i}$ imply that f_i and g_i do not contribute to the improvement of the final 2-D design accuracy anymore, and those extra f_i and g_i should be completely removed. Based on this philosophy, we should select the orders of 1-D sub-filters $F_i(z_1)$ and $G_i(z_2)$ by considering how the individual errors $e_{\tilde{u}_i}$ and $e_{\tilde{v}_i}$ affect the whole design accuracy. In this paper, we define the following normalized RMS approximation error

$$E_{f_j} = \frac{\|A - A_{f_j}\|}{\|A\|} \times 100\% \tag{29}$$

$$E_{g_j} = \frac{\|A - A_{g_j}\|}{\|A\|} \times 100\% \tag{30}$$

where

$$A_{f_j} = \sum_{i=1(i\neq j)}^{K} \tilde{u}_i\tilde{v}_i^T + f_j\tilde{v}_j^T \tag{31}$$

$$A_{g_j} = \sum_{i=1(i\neq j)}^{K} \tilde{u}_i\tilde{v}_i^T + \tilde{u}_jg_j^T. \tag{32}$$

The orders of 1-D sub-filters $F_i(z_1)$ and $G_i(z_2)$ are selected such that the approximation errors E_{f_j}, E_{g_j} are almost the same for all the vectors f_j and g_j, where $j = 1, 2, \cdots, K$.

3. DESIGN EXAMPLE AND CONCLUSION

[**Elliptical Filter**]: The desired zero-phase frequency response is

$$H_d(\omega_1, \omega_2) = \begin{cases} 1 & 0 \le \omega_g \le \omega_p \\ \dfrac{(\omega_a - \omega_g)}{(\omega_a - \omega_p)} & \omega_p \le \omega_g \le \omega_a \\ 0 & \text{otherwise} \end{cases} \quad (33)$$

where

$$\omega_g = \sqrt{\tilde{\omega_1}^2 + \frac{\tilde{\omega_2}^2}{2}} \quad (34)$$

$$\omega_p = 0.35\pi, \quad \omega_a = 0.50\pi$$

$$\begin{bmatrix} \tilde{\omega_1} \\ \tilde{\omega_2} \end{bmatrix} = \begin{bmatrix} \cos\alpha & \sin\alpha \\ -\sin\alpha & \cos\alpha \end{bmatrix} \begin{bmatrix} \omega_1 \\ \omega_2 \end{bmatrix}, \quad \alpha = -\frac{\pi}{4}. \quad (35)$$

Since the 2-D frequency response specification satisfies the new symmetry, thus the 2-D filter can be efficiently designed.

To form the specification matrix $\boldsymbol{A}$, the frequencies ω_1, ω_2 are equally sampled at the step size $\pi/40$, and then the corresponding samples $H_d(\omega_{1l}, \omega_{2m})$ are used to construct $\boldsymbol{A} \in \boldsymbol{R}^{81 \times 81}$. In [7], the first 12 sections are approximated, i.e., K=12, the normalized RMS decomposition error is 1.0983%. When the sub-filter orders in [7] are used, the normalized magnitude response error of the designed 2-D filter is 2.0777%. In our design, we just use the first 8 channels, the decomposition error E_8 is 1.7094%, and the new order-selecting criterion is applied to the selection of 1-D sub-filter orders such that the errors E_{f_j} and E_{g_j} defined in (29) and (30) are below 1.7200%. The magnitude response of our designed 2-D filter is plotted in Fig. 2, whose normalized RMS error is 1.8373%. By comparing our design results with those in [7], we can make the following conclusions:

- Exploting the new symmetry in the SVD-based 2-D filter design enables us to accomplish the design by designing only 8 sub-filters, but 24 sub-filters need to be designed in [7].
- The number of our total multiplier coefficients is 142, which is less than 50% of the total multiplier coefficients (292) used in [7].
- Our design error (1.8373%) is smaller than that (2.0777%) by the method [7].

That is, the new SVD-based technique can achieve higher design accuracy with significantly reduced design complexity and much less hardware implementation cost than the design approach [7].

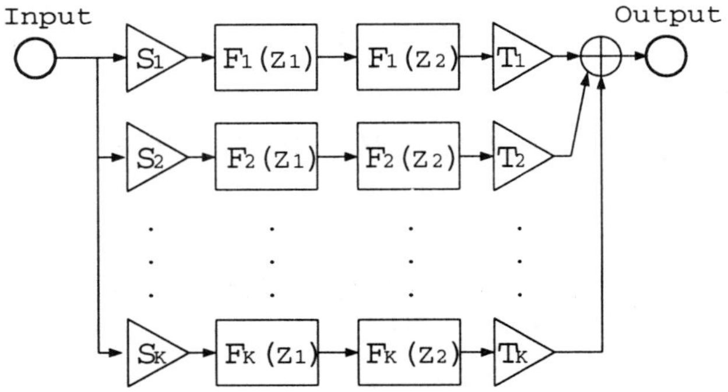

Fig. 1. SVD-based 2-D filter using new symmetry.

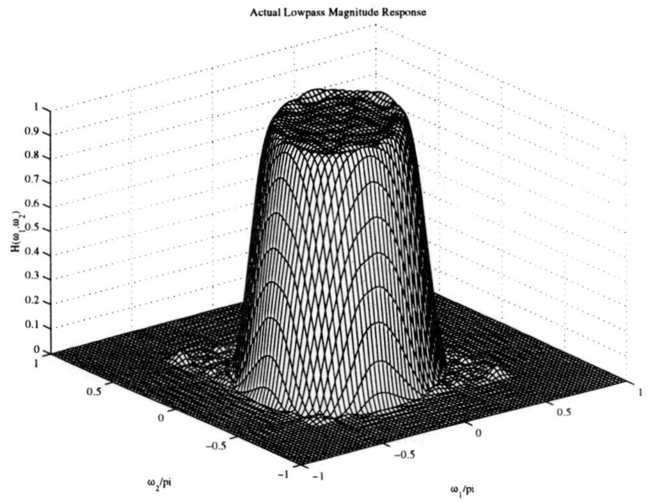

Fig. 2. Actual 2-D magnitude response.

4. REFERENCES

[1] R. E. Twogood and S. K. Mitra, "Computer-aided design of separable two-dimensional digital filters", *IEEE Trans. Acoust., Speech, Signal Processing*, vol. ASSP-25, no. 2, pp. 165-169, April 1977.

[2] A. Antoniou and W.-S. Lu, "Design of two-dimensional digital filters by using the singular value decomposition", *IEEE Trans. Circuits Syst.*, vol. CAS-34, pp. 1191-1198, Oct. 1987.

[3] T.-B. Deng and M. Kawamata, "Frequency-domain design of 2-D digital filters using the iterative singular value decomposition", *IEEE Trans. Circuits Syst.*, vol. 38, pp. 1225-1228, Oct. 1991.

[4] T.-B. Deng and T. Soma, "Successively linearized non-negative decomposition of 2-D filter magnitude design specifications", *Digital Signal Processing*, vol. 3, no. 2, pp.125-138, April 1993.

[5] T.-B. Deng, M. Kawamata, and T. Higuchi, "Design of two-dimensional recursive digital filters based on the optimal decomposition of magnitude specifications", *Circuits, Systems, and Computers*, vol. 3, no. 3, pp. 733-756, Sept. 1993.

[6] W.-S. Lu, H.-P. Wang, and A. Antoniou, "Design of two-dimensional digital filters using singular value decomposition and balanced approximation method", *IEEE Trans. Signal Processing*, vol. 39, pp. 2253-2262, Oct. 1991.

[7] N. Yazdi, M. Ahmadi, and J. J. Soltis, "Efficient design of 2-D linear phase FIR filters using singular value decomposition (SVD)" *Electronics Letters*, vol. 28, no. 24, pp. 2256-2258, Nov. 1992.

REDUCE THE COMPLEXITY OF FREQUENCY-RESPONSE MASKING FILTER USING MULTIPLICATION FREE FILTER

Chun Zhu Yang and *Yong Lian*
ECE Dept., National University of Singapore
Singapore 119260
Email : eleliany@nus.edu.sg

ABSTRACT

A modified frequency-response masking (FRM) approach is presented for the design of arbitrary bandwidth sharp FIR filters. The new filter achieves significant savings in the number of arithmetic operations by replacing the band-edge shaping filter in the FRM technique with a modified interpolated finite impulse response (IFIR) filter. Our example shows that the proposed one-stage FRM filter outperforms an original two-stage FRM filter in terms of the number of multipliers and group delay.

1. INTRODUCTION

For the arbitrary bandwidth sharp FIR filter design, the frequency-response masking (FRM) technique [1]-[6] provides one of the most computationally efficient realizations. The FRM technique utilizes a delay-complementary concept to realize the sharp transition-band by a band-edge shaping filter with a sparse coefficient vector. The sparse coefficient vector brings some desirable features required by application specific integrated circuit (ASIC) implementation, such as reduction of arithmetic operations and low power consumption. These features make the FRM technique very attractive in the modern filter design.

Let $F_s(z)$ be an odd length N_{F_s} linear phase FIR filter. Its complement can be formed as

$$F_c(z) = z^{-(N_{F_s}-1)/2} - F_s(z) \tag{1}$$

If we interpolate $F_s(z)$ by a factor of M, a band-edge shaping filter $F_s(z^M)$ is formed, whose transition bandwidth is M times narrower than that of $F_s(z)$. Two masking filters $F_1(z)$ and $F_2(z)$ are employed to remove the undesired frequency components from $F_s(z^M)$ and its complement. The transfer function of the overall filter is given by:

$$H(z) = F_s(z^M)F_1(z) + [z^{-M(N_{Fs}-1)/2} - F_s(z^M)]F_2(z) \tag{2}$$

One of the possible realizations for the FRM approach is shown in Fig. 1. It is noted that the length of each subfilter trades off with the interpolation factor M, i.e. the larger is M, the longer are the lengths of the masking filters. To distribute the computational load evenly among all subfilters and to achieve maximum savings in the arithmetic operations, a multistage FRM [2] has to be considered. The drawbacks for a multistage FRM approach are the increased number of group delay and complex design procedures associated with the design of $2N+1$ subfilters for a N-stage FRM design. To address these problems, several modified single-stage FRM structures were introduced recently in [3]-[6]. One of the interesting approaches [5] is to replace the band-edge shaping filter with an interpolated finite impulse response (IFIR) filter [7] by adopting a large M. This approach yields additional savings in the number of multipliers and adders compared with the original FRM technique [1]. In this paper, we will show that further savings in the number of arithmetic operations can be achieved by replacing the interpolator in [5] with a multiplication free filter.

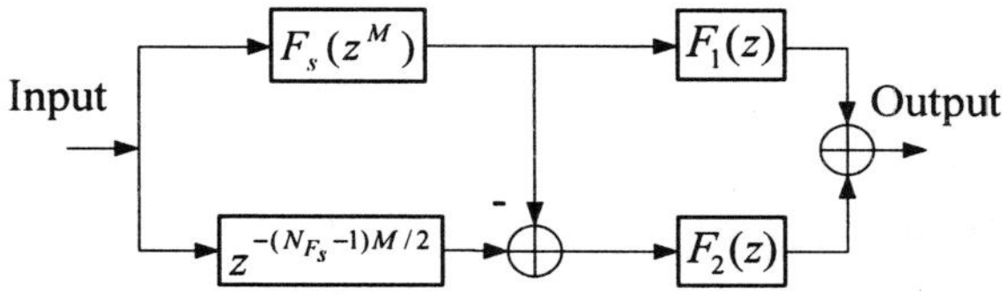

Fig. 1. A realization of FRM approach

The organization of the rest of this paper is as follows: in Section 2, we present a new multiplication free filter and apply it to the FRM approach. The design equations are given in Section 3. Design procedure is discussed in Section 4. Section 5 is dedicated to a design example.

2. A MODIFIED FRM STRUCTURE BASED ON MULTIPLICATON FREE FILTER

In the FRM approach, we may replace the band-edge shaping filter with an IFIR filter if a proper M is selected as shown in Fig. 2, where GD is the total number of group delay of $F_s(z^M)H_r(z)$. It is a well-known fact that the ripple of each subfilter compensates each other in the FRM approach. In other words, it is not necessary for the $F_s(z^M)H_r(z)$ to satisfy the passband specification of the overall filter. The passband error caused by $F_s(z^M)H_r(z)$ can be corrected by $F_1(z)$ or $F_2(z)$. This prompts us to relax the passband requirement for $F_s(z^M)H_r(z)$. As a result, $H_r(z)$ can be replaced by a simple filter that provides sufficient stopband attenuation while has large ripple in the passband. It is highly desirable that $H_r(z)$ is a multiplication free filter. To find such a filter, let us consider a 4th-order FIR filter $H_N(z)$ introduced in [8] whose z-transform transfer function is given by:

$$H_N(z) = 0.125(1 + 2z^{-1} + 2z^{-2} + 2z^{-3} + z^{-4}) \tag{3}$$

The stopband attenuation of $H_N(z)$ can be improved considerably if the "response sharpening" technique [9] is employed. A new filter $H_r(N, L, z)$ can be formed with the z-transform transfer function and the zero-phase frequency response given by

$$H_r(N, L, z) = 0.125^L[(1 + z^{-2N})(1 + z^{-N})^2]^L \tag{4}$$

$$H_r(\omega) = 0.5^L[\cos(N\omega) + \cos^2(N\omega)]^L \tag{5}$$

where N and L are positive integers. As the coefficients of $H_r(N, L, z)$ only involve 1 and 2, $H_r(N, L, z)$ is basically a multiplication free filter. It can be seen from Eq. (5) that N and L determine the nulls and stopband attenuation of $H_r(N, L, z)$, respectively. $H_r(N, L, z)$ has periodic frequency response with

0-7803-7761-3/03/$17.00 ©2003 IEEE

lobes centered at $2\pi k/N$, where $k = 0, 1, \ldots, N\text{-}1$, and provides good stopband attenuation. Fig. 3 shows the frequency response of $H_r(8, 3, z)$.

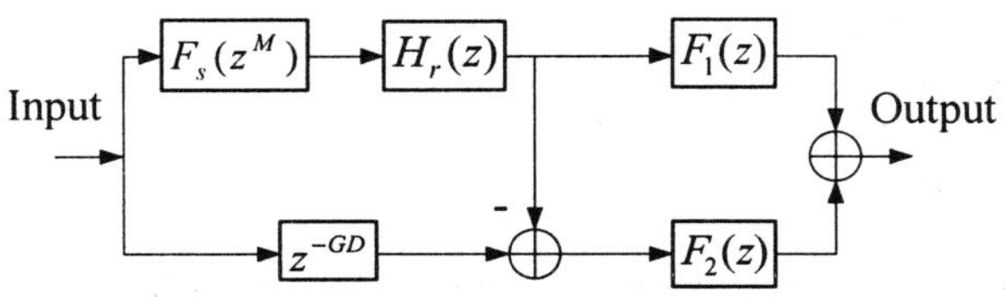

Fig. 2. A modified FRM structure

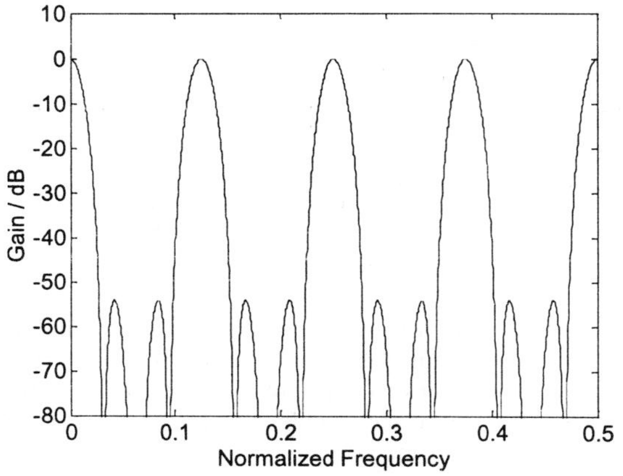

Fig. 3. A $H_r(z)$ with N=8, L=3.

Replacing $H_r(z)$ in Fig. 2 with $H_r(N, L, z)$, a modified FRM approach is created. The frequency responses of various subfilters used in such an approach are sketched in Fig. 4. In Fig. 4(a), $F_s(z)$ is a model filter with cutoff frequencies at θ_{F_s} and ϕ_{F_s}, and its interpolated version $F_s(z^M)$ is shown in Fig. 4(b). Cascading $F_s(z^M)$ with $H_r(N, L, z)$, a pair of complementary band-edge shaping filters $A(z)$ and $C(z)$ are formed, as shown in Figs. 4(c) and 4(d). It can been seen from Fig. 4(c) that $A(z)$ has much wider stopband width than that of the masking filter in the original FRM technique. As a result, the transition-band of $F_1(z)$ is relaxed significantly. The corresponding frequency responses of $F_1(z)$, $F_2(z)$ and the overall filter $H(z)$ are shown in Figs. 4(c)-4(e). Figs. 4(f)-4(h) shows the case in which the transition-band of $H(z)$ is determined by $C(z)$. We denote the case of Fig. 4(e) as Case A and the case of Fig. 4(h) as Case B.

3. DESIGN EQUATIONS

In this section, we derive the design equations for lowpass filters. The same procedure can be employed to obtain the design equations for highpass filters. Let us consider the synthesis of a lowpass filter with passband and stopband edges at ω_p and ω_s, respectively. Let θ and ϕ denote the passband and stopband edges of any subfilter, respectively. Therefore, the band-edges of $F_1(z)$ are θ_{F_1} and ϕ_{F_1}, respectively. Similar notations are applied to $F_2(z)$.

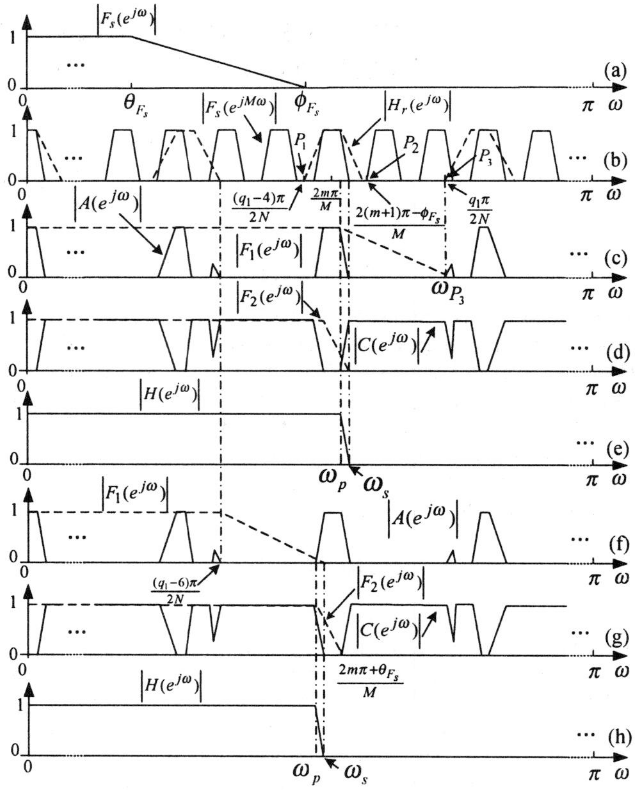

Fig. 4. Frequency responses of subfilters in Fig. 2.

To determine the band-edges of $F_s(z)$, it can be shown that

$$\omega_s = \begin{cases} (2m\pi + \phi_{F_s})/M & \text{For Case A} \\ (2m\pi - \theta_{F_s})/M & \text{For Case B} \end{cases} \tag{6}$$

where m is an integer as shown in Fig. 4(b). To ensure that (6) yield a solution with $0 < \theta_{F_s} < \phi_{F_s} < \pi$, we find two sets of values for Case A and B.

For Case A:

$$m = \lfloor \omega_p M / 2\pi \rfloor \tag{7}$$

$$\theta_{F_s} = \omega_p M - 2m\pi \tag{8}$$

$$\phi_{F_s} = \omega_s M - 2m\pi \tag{9}$$

where $\lfloor x \rfloor$ denotes the largest integer less than or equal to x.

For Case B:

$$m = \lceil \omega_s M / 2\pi \rceil \tag{10}$$

$$\theta_{F_s} = 2m\pi - \omega_s M \tag{11}$$

$$\phi_{F_s} = 2m\pi - \omega_p M \tag{12}$$

where $\lceil x \rceil$ denotes the smallest integer larger than or equal to x. For the masking filter $F_2(z)$, its band-edges can be found from Figs. 4(d) and 4(g). For Case A, we have

$$\theta_{F_2} = (2m\pi - \theta_{F_s})/M \tag{13}$$

$$\phi_{F_2} = \omega_s \tag{14}$$

For Case B, we have

$$\theta_{F_2} = \omega_p \tag{15}$$

$$\phi_{F_2} = (2m\pi + \theta_{F_s}) / M \tag{16}$$

For the masking filter $F_1(z)$, the passband edge for Case A can be found from Fig. 4(c),

$$\theta_{F_1} = \omega_p \tag{17}$$

The stopband edge ϕ_{F_1} depends on the nulls of $H_r(z)$. As shown in Fig. 4(b), one of the lobes of $H_r(z)$ is used to mask the "tooth" of $F_s(z^M)$ to form the band-edges of the overall filter. Let us denote such a lobe as band-edge shaping lobe. It is clear that the center frequency ω_o of the band-edge shaping lobe should locate inside "tooth" of $F_s(z^M)$. We have:

$$(2m\pi - \theta_{F_s}) / M < \omega_o = 2k\pi / N < (2m\pi + \theta_{F_s}) / M \tag{18}$$

where k is an integer less than N. From (5), the nulls of $H_r(z)$ occur at

$$\omega_{null} = q\pi / 2N \tag{19}$$

where q is an odd integer. For Case A, the stopband edge is determined by the null of the neighboring lobe at the right side of bandedge shaping lobe, therefore, we have

$$\phi_{F_1} \approx q_1\pi / 2N \tag{20}$$

where q_1 is an odd integer and $4k < q_1 < 4(k+1)$.

For Case B, we have

$$\theta_{F_1} \approx (q_1 - 6)\pi / 2N \tag{21}$$

$$\phi_{F_1} = \omega_s \tag{22}$$

To use the design equations (7)–(22), we need to find M, N, and L. The determination of optimal M, N, and L will be discussed in the next section.

4. DESIGN PROCEDURE

To synthesize the overall filter $H(z)$, we shall first decide the optimal interpolation factor M as well as a set of values of N and L for $H_r(N, L, z)$. Unfortunately, there is no known closed-form expression for M, N and L. However, N and L should meet some conditions for given specifications. These conditions will help us select a set of N's and L's. As shown in Fig. 4(b), for Case A, the first null P_1 to the left side of ω_o should meet the condition:

$$\begin{cases} \omega_{P_1} < [2m\pi - \phi_{F_s}] / M \\ \omega_{P_1} = (q_1 - 4)\pi / 2N \end{cases} \tag{23}$$

For Case B, P_1 is the first null to the right side of ω_o. We have

$$\begin{cases} \omega_{P_1} > [2m\pi + \phi_{F_s}] / M \\ \omega_{P_1} = (q_1 - 2)\pi / 2N \end{cases} \tag{24}$$

We know that each $H_N(z)$ produces about 18 dB stopband attenuation [8]. Therefore, the minimum stopband attenuation of $H_r(N, L, z)$ are about $18L$ dB. To provide sufficient stopband attenuation, we have

$$\begin{cases} L > -(20\log\delta_s)/18 & \text{(For Case A)} \\ L > -(20\log\delta_p)/18 & \text{(For Case B)} \end{cases} \tag{25}$$

There is a salient point P_2 to help to adjust the value of L, as shown Fig. 4(b) and 4(g), the attenuation provided by $H_r(N, L, z)$ at point P_2 should meet the stopband ripple requirement for Case A and passband ripple requirement for Case B, respectively. Further compensation can be achieved by $F_1(z)$. Therefore, for Case A, we have,

$$\begin{cases} 0.5^L \,|[\cos(N\omega_{P_2}) + \cos^2(N\omega_{P_2})]^L| \approx \delta_s \\ \omega_{P_2} = [2(m+1)\pi - \phi_{F_s}] / M \end{cases} \tag{26}$$

For Case B, we have

$$\begin{cases} 0.5^L \,|[\cos(N\omega_{P_2}) + \cos^2(N\omega_{P_2})]^L| \approx \delta_p \\ \omega_{P_2} = [2(m-1)\pi + \phi_{F_s}] / M \end{cases} \tag{27}$$

where δ_p and δ_s are the required passband and stopband ripples, respectively. For a given M, using (18), (23)–(28), we can find a set of N's and L's. It should be noted that smaller N and L will lead to less group delay.

So far we have developed some guidelines for selecting N and L, it is difficult to find a closed-form expression for the optimal M. We use a simple searching program to obtain a set of optimal values for M, N and L. It is a well-known fact that the filter length is mainly determined by the transition bandwidth of the filter and is inversely proportional to the transition band-width. We shall define a cost function $C(\Delta)$ to measure the overall complexity of the filter as:

$$C(\Delta) = 1/\Delta_{F_s} + 1/\Delta_{F_1} + 1/\Delta_{F_2} \tag{28}$$

where $\Delta_{F_s} = \phi_{F_s} - \theta_{F_s}$ is the transition bandwidth of $F_s(z)$. Similar definitions are applied to other subfilters. For a given set of M, N and L, using the design equations derived in section 3, $C(\Delta)$ can be easily calculated. Therefore, the searching program is to find the values of M, N and L that minimize the cost function $C(\Delta)$.

To meet the overall specifications, when $H_r(z)$ is determined, 3 subfilters have to be designed simultaneously in order to achieve global optimization. However, this is a non-linear problem and is difficult to solve. We have adopted a sub-optimized technique to overcome the difficulties. The proposed design procedure is based on an iterated design approach that optimizes one subfilter at a time while uses the rest subfilters as prefilters. The following steps are recommended.

Step 1. Find a suitable M and the set of values of N and L for $H_r(N, L, z)$.

Step 2. Design $F_1(z)$ and $F_2(z)$ with the band-edges given by equations discussed in Section 3 for Case A or for Case B, respectively. Set the ripple of each filter to 85% of $H(z)$.

Step 3. Determine the band-edges of $F_s(z)$ according to (7)–(9) for Case A and (10)–(11) for Case B. Let the zero-phase frequency response of $F_s(z^M)$ be written as

$$F_s(\omega) = \sum_i f_s(n)\mathrm{trig}(M\omega, i) \tag{29}$$

where $\mathrm{trig}(M\omega, i)$ is a proper trigonometric function depending on the type of the filter under design and $f_S(n)$ is the impulse response of $F_S(z)$. Similar notations can be applied to other subfilters. In the passband, $F_s(z)$ needs to satisfy the following,

$$1-\delta_p(\omega)-F_2(\omega)\le[F_1(\omega)-F_2(\omega)]H_r(\omega)\cdot \left[\sum_i f_s(n)\text{trig}(M\omega,i)\right]\le 1+\delta_p(\omega)-F_2(\omega) \quad (30)$$

In the stopband, we have

$$-\delta_s(\omega)-F_2(\omega)\le[F_1(\omega)-F_2(\omega)]H_r(\omega)\cdot \left[\sum_i f_s(n)\text{trig}(M\omega,i)\right]\le \delta_s(\omega)-F_2(\omega) \quad (31)$$

Linear programming can be used to solve (30) and (31) by evaluating these two equations on a dense grid of frequencies.

Step 4. Design $F_2(z)$ again by using the rest subfilters, i.e. $F_s(z)$, $F_1(z)$ and $H_r(z)$ as prefilters. The design equations can be derived as in Step 3.

Step 5. Repeat Step 4 by replacing $F_2(z)$ with one subfilter from $F_1(z)$ or $F_s(z)$ and use the rest subfilters as prefilters, until there is no further improvement being observed.

5. AN EXAMPLE

To illustrate the new design approach, we design a lowpass filter with passband edge at $\omega_p = 0.3\times 2\pi$ and stopband edge at $\omega_s = 0.301\times 2\pi$. The passband ripple is at most 0.01 and the stopband attenuation is at least 40 dB, respectively. To design such a filter by conventional design method, the estimated filter length is 1851. If designed by single stage FRM technique, the optimum lengths for $F_s(z)$, $F_1(z)$ and $F_2(z)$ are 139, 49 and 69, respectively, and M is 14. 130 multipliers are needed. If the method in [5] is adopted, we need 104 multipliers with two interpolation factors of 7 and 3 for the IFIR pair. Using our proposed method with an interpolation factor of 21, we get $N = 7$ and $L = 3$ for $H_r(z)$. The lengths of $F_s(z)$, $F_1(z)$ and $F_2(z)$ are 93, 17 and 69, respectively, 91 multipliers are needed which yields more than 29% and 27% savings in the number of multipliers and adders, respectively, compared with the original FRM approach. Table 1 lists the comparisons among different design approaches. As indicated in table 1, our approach even outperforms the 2-stage FRM technique with fewer multiplications and group delay. The frequency response of the overall filter is shown in Fig. 5.

Table 1 Comparison of different design methods

Design	Multipliers	Adders	Group delay
Minimax	925	1850	925
1-stage FRM	130	256	1000
2-stage FRM	92	177	1105
Method in [5]	104	202	1084
Proposed	91	185	1050

6. CONCLUSION

In this paper, a modified structure to reduce the complexity of the design of sharp FIR filter using frequency-response masking technique was introduced. The success of the proposed method is based on a multiplication free filter. This method can achieve considerable savings in terms of number of multipliers and adders while does not need additional steps in designing the overall filter.

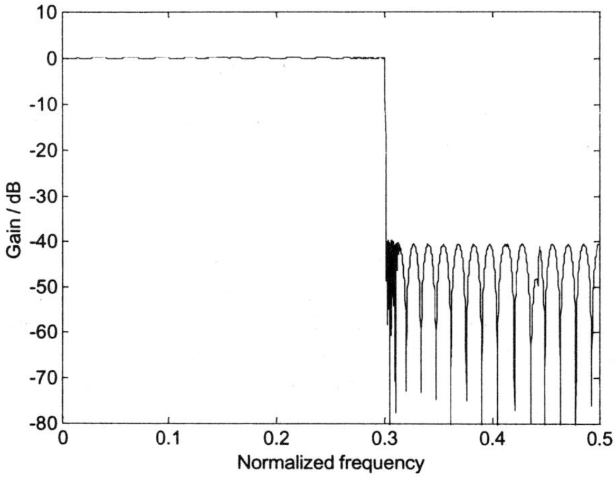

Fig. 5. Frequency response of the example

7. ACKNOWLEDGEMENT

This work was supported by the National University of Singapore Academic Research Fund R263-000-160-112.

8. REFERENCES

[1] Y.C. Lim, "Frequency -response masking approach for the synthesis of sharp linear phase digital filters," *IEEE Trans. Circuits Syst.*, vol. CAS-33, pp. 357-364, April 1986.

[2] Y.C. Lim and Y. Lian, "The optimal design of one - and two- dimensional FIR filters using the frequency response masking technique", *IEEE Trans. Circuits Syst., Part 2.*, vol. 40, pp. 88-95, February 1993.

[3] Y.C. Lim and Y. Lian, "Frequency response masking approach for digital filter design: complexity reduction via masking filter factorization", *IEEE Trans. Circuits Syst.*, part2, vol. 41, pp. 518-525, August 1994.

[4] Y. Lian, "A new frequency response masking structure with reduced complexity for FIR filter design," *Proc. IEEE Int. Symp. Circuits Syst., ISCAS 2001*, vol. II, pp. 609-612, May 2001.

[5] Y. Lian, L. Zhang, and C.C. Ko, "An improved frequency response masking approach for designing sharp FIR filters," *Signal Processing*, 81, pp. 2573-2581. 2001.

[6] Chun Zhu Yang and Yong Lian, "A modified structure for the design of sharp FIR filters using frequency-response masking technique," *Proc. IEEE Int. Symp. Circuits Syst., ISCAS 2002*, vol. III, pp. 237-240, May 2002.

[7] Y. Neuvo, C. Y. Dong and S.K. Mitra, "Interpolated finite impulse response filters," *IEEE Trans Acoust. Speech, Signal Processing*, vol. ASSP-32, pp. 563-570, Jun. 1984.

[8] Y. Lian and Y. C. Lim, "New prefilter structure for designing FIR filters," *IEE Electron. Letters*, vol. 29, pp. 1034-1036, May 1993.

[9] J. F. Kaiser and R. W. Hamming, "Sharpening the response of a symmetric non-recursive filter by multiple use of the same filter," *IEEE Trans Acoust. Speech, Signal Processing*, vol. ASSP-25, pp. 415-422, Otc. 1977.

New designs of frequency selective FIR digital filters

Ishtiaq R. Khan[1, 2], *M. Okuda*[2], *R. Ohba*[3]

[1]Foundation for Advancement of Industry and Science, Kitakyushu, Japan

[2]Department of Environmental Engineering, The University of Kitakyushu, Japan

[3]Division of Applied Physics, Graduate School of Engineering, Hokkaido University, Japan

Abstract

Fourier series expansion is used to explore the relationship between the coefficients of halfband lowpass and general bandpass FIR digital filters. This relationship is used to derive new designs of band pass/stop and low/high pass MAXFLAT FIR digital filters, from an available design of halfband lowpass MAXFLAT FIR digital filters. The presented designs have explicit formulas for the filter coefficients, and can give infinite number of distinct filters of any finite length. The formulas are modified to obtain a new class of filters, which allow a reduction in the transition bandwidth at the expense of smoothness of the magnitude response, and have a performance comparable to that of the minimax filters.

1. Introduction

Maximally flat (MAXFLAT) digital filters (DFs) are one of the most important types of filters [1], which find their applications when time domain properties are of more importance or smooth frequency responses and higher stopband attenuations are desirable. A merit of the MAXFLAT DFs over others is the simplicity of their design, as their frequency responses can be expressed in closed forms. The filter coefficients are calculated by taking inverse discrete Fourier transforms (IDFT) of the closed form expressions. Some of the design techniques, including those based on Hermite [2], Krawtchouk [3] and Bernstein polynomials [4] can be found in [2-8].

Recently we used Taylor series to further simplify some of the MAXFLAT designs by eliminating the need of IDFT for finding their coefficients [9-13]. It was shown [9] that central difference approximations based on the Taylor series were in fact the same as the maximally linear type III digital differentiators [14]. A new type of approximations, also based on the Taylor series, was presented and implemented as more accurate and wide-band type IV digital differentiators [10]. Inter-relationships between the coefficients of different types of DFs [15] were used to design different types of MAXFLAT DFs including low/high pass DFs [11-12], mid pass/stop DFs [12] and discrete and differentiating Hilbert transformers [13]. All of these designs give explicit formulas for the filter coefficients and therefore are very simple as compared to the classical design procedures.

In this paper, we establish a relationship between the coefficients of a halfband lowpass DF and a general bandpass DF by comparing their ideal impulse responses (coefficients) obtained by Fourier series expansion of their frequency responses. This relationship is used with the formulas of the coefficients of halfband lowpass MAXFLAT DFs [11] to obtain similar formulas for general band pass/stop and low/high pass MAXFLAT DFs. The presented designs not only eliminate the need of IDFT for finding the coefficients, but also meet the desired cutoff frequencies more closely, because they can give infinite number of distinct filters of any finite length. Most of the classical MAXFLAT designs give only finite number of DFs for a finite length and cannot exactly meet the desired cutoff frequencies.

It is well known that the MAXFLAT DFs have highest accuracy (smoothness) among the available designs, but at the same time their transition bands are relatively wider, and the only way to reduce the transition bandwidth is to increase their length. On the other hand, transition bandwidth of the equiripple (minimax) filters [16-17] can be narrowed for a fixed length, at the expense of an increase deviation from the ideal frequency response. Due to uniformly distributed error (ripple), these filters can meet the desired specifications (transition width and maximum error) with a minimum length. However, unlike for MAXFLAT DFs, finding the coefficients of these filters is not straightforward. Famous MP algorithm generally used for their design is an iterative procedure and performs intensive search over a dense frequency grid in each iteration [18]. In this paper, we modify the formulas presented for MAXFLAT DFs, such that, the resultant filters have variable transition bandwidths for fixed lengths, like the minimax DFs. Though the design complexity of presented filters is far less than that of minimax filters, design examples show that performance of the both is comparable.

2. Interrelationships between coefficients of halfband lowpass and general bandpass DFs

Consider a halfband lowpass DF with a passband $0 \le \omega \le \pi/2$ and a general bandpass DF with a passband $\omega_{c1} \le \omega \le \omega_{c2}$. A periodic frequency response can be expanded as a Fourier series, whose coefficients give the ideal impulse response (coefficients) of the corresponding FIR digital filter. Denoting the ideal impulse responses of halfband lowpass and general bandpass DFs by **h** and **g** respectively, we can write

$$h_n = \frac{1}{2\pi}\int_{-\pi/2}^{\pi/2} e^{j\omega n} d\omega = \frac{\sin(n\pi/2)}{n\pi},$$

and

$$g_n = \frac{1}{2\pi}\left[\int_{-\omega_{c2}}^{-\omega_{c1}} e^{j\omega n} d\omega + \int_{\omega_{c1}}^{\omega_{c2}} e^{j\omega n} d\omega\right] = \frac{\sin(n\omega_{c2}) - \sin(n\omega_{c1})}{n\pi},$$

where n can take any integer values. The coefficients can be simplified as

0-7803-7761-3/03/$17.00 ©2003 IEEE

$$h_n = \begin{cases} 0.5 & n = 0 \\ 0 & n = \text{even} \\ \dfrac{(-1)^{(n+1)/2}}{n\pi} & n = \text{odd,} \end{cases}$$

and

$$g_n = \begin{cases} \dfrac{\omega_{c2} - \omega_{c1}}{\pi} & n = 0 \\ \dfrac{\sin(n\omega_{c2}) - \sin(n\omega_{c1})}{n\pi} & \text{otherwise.} \end{cases}$$

From the above coefficients of ideal DFs, we can see that odd indexed h_n and g_n differ only in the numerator terms. Even indexed h_n are zero, whereas even indexed g_n are given by the same formulas as for odd indexed g_n. Similarly the difference of central coefficients h_0 and g_0 can be observed and the following four-step procedure to transform the coefficients of a halfband lowpass DF to those of a general bandpass DF can be obtained:

1. Remove the sign term of the odd indexed h_n.
2. Use the same formula to replace even indexed $h_n = 0$.
3. Multiply all h_n by $\sin(n\omega_{c2}) - \sin(n\omega_{c1})$.
4. Set $h_0 = (\omega_{c2} - \omega_{c1})/\pi$.

3. Frequency selective MAXFLAT FIR DFs

In [11], we presented explicit formulas for the MAXFLAT halfband lowpass DFs, and here we transform them to those of the MAXFLAT bandpass DFs, using the procedure described in the previous section. Similar formulas are obtained for bandstop, lowpass and highpass DFs as well.

3.1 Bandpass DFs

Coefficients of a MAXFLAT halfband lowpass FIR DF of length $4N$-1 are given as [11]

$$h_0 = 1/2,$$

$$h_{\pm(2k-1)} = \frac{(-1)^{k+1}(2N-1)!!^2}{2^{2N}(N+k-1)!(N-k)!(2k-1)}, 1 \le k \le N,$$

$$h_{\pm 2k} = 0, 1 \le k \le N-1, \tag{1}$$

where double factorial of an integer n is given as $n(n$-2)(n-4)…(>0).

Step 1 of the transformation procedure simply requires removing the sign term of $h_{\pm(2k-1)}$ in Eq. (1). In the resultant equation, we make a substitution of m = 2k-1 to write it in a suitable form for implementation of step 2, and obtain

$$h_{\pm m} = \frac{(2N-1)!!^2}{2^{2N}\left(\dfrac{2N+m-1}{2}\right)!\left(\dfrac{2N-m-1}{2}\right)!m}, \; 0 < m < 2N, m = \text{odd.} \tag{2}$$

To apply step 2 of the transformation procedure, we replace factorials in the denominator of Eq. (2) by more general gamma functions, which are defined for a complex value z as

$$\Gamma(z) = \int_0^\infty t^{z-1} e^{-t} dt.$$

For integer values of z, $\Gamma(z) = (z$ -1)!.

Using gamma function, Eq. (2) can now be used both for odd and even values of m as

$$h_{\pm m} = \frac{(2N-1)!!^2}{2^{2N}\Gamma\left(\dfrac{2N+m-1}{2}+1\right)\Gamma\left(\dfrac{2N-m-1}{2}+1\right)m}, 0 < m < 2N. \tag{3}$$

It can be shown mathematically that for an integer n

$$\Gamma\left(\frac{n}{2}+1\right) = \begin{cases} \dfrac{n!!}{2^{n/2}} & n = \text{even} \\ \dfrac{n!!}{2^{n/2}}\sqrt{\dfrac{\pi}{2}} & n = \text{odd,} \end{cases}$$

and therefore Eq. (3) can be simplified as

$$h_{\pm m} = \begin{cases} \dfrac{(2N-1)!!^2}{2m(2N+m-1)!!(2N-m-1)!!} & 0 < m < 2N, m = \text{odd} \\ \dfrac{(2N-1)!!^2}{\pi m(2N+m-1)!!(2N-m-1)!!} & 0 < m < 2N, m = \text{even,} \end{cases} \tag{4}$$

and this completes the step 2 of transformation procedure.

Now applying steps 3-4, we obtain the following complete set of coefficients of a MAXFLAT bandpass FIR DF:

$$g_{\pm n} = \begin{cases} \dfrac{\omega_{c2} - \omega_{c1}}{\pi} & n = 0 \\ \dfrac{(2N-1)!!^2 \pi^{p-1}[\sin(n\omega_{c2}) - \sin(n\omega_{c1})]}{2^p n(2N+n-1)!!(2N-n-1)!!} & 0 < n < N, \end{cases} \tag{5}$$

where p is the remainder of n/2.

3.2 Bandstop DFs

A bandstop frequency response can be obtained by subtracting a bandpass response of the same cutoff frequencies from an allpass response. This is equivalent to subtracting the coefficients of the bandpass DF from a unit impulse sequence, which has all zeros except a one at the center. Therefore, the coefficients of a MAXFLAT bandstop DF with lower and upper cutoff frequencies ω_{c1} and ω_{c2} respectively, can be written from Eq. (5) as

$$g_{\pm n} = \begin{cases} 1 - \dfrac{\omega_{c2} - \omega_{c1}}{\pi} & n = 0 \\ \dfrac{(2N-1)!!^2 \pi^{p-1}[\sin(n\omega_{c1}) - \sin(n\omega_{c2})]}{2^p n(2N+n-1)!!(2N-n-1)!!} & 0 < n < N. \end{cases} \tag{6}$$

3.3 Lowpass DFs

A lowpass response of cutoff frequency ω_c can be obtained from Eq. (5) by setting $\omega_{c1} = 0$ and $\omega_{c2} = \omega_c$ as

$$g_{\pm n} = \begin{cases} \dfrac{\omega_c}{\pi} & n = 0 \\ \dfrac{(2N-1)!!^2 \pi^{p-1} \sin(n\omega_c)}{2^p n(2N+n-1)!!(2N-n-1)!!} & 0 < n < N. \end{cases} \quad (7)$$

3.4 Highpass DFs

A highpass response of cutoff frequency ω_c can be obtained from Eq. (6) by setting $\omega_{c1} = \omega_c$ and $\omega_{c2} = \pi$ as

$$g_{\pm n} = \begin{cases} 1 - \dfrac{\omega_c}{\pi} & n = 0 \\ \dfrac{-(2N-1)!!^2 \pi^{p-1} \sin(n\omega_c)}{2^p n(2N+n-1)!!(2N-n-1)!!} & 0 < n < N. \end{cases} \quad (8)$$

Apparently Eqs. (5-8) contain large factorials terms, however in practice, we can find simple relationships between g_k and g_{k-1}, and all the coefficients can be calculated iteratively without need of calculating any factorials.

Typical magnitude responses of DFs designed with Eqs. (5-8) are shown in Figs. 1-2. It should be noted that these designs can give infinite number of distinct filters of any finite length and therefore can exactly meet the desired cutoff frequencies. This is not possible with most of the classical designs like given in [2-4], which give only N distinct filters for a length of $2N+1$.

4. Reducing the transition bandwidth

4.1 Method

The MAXFLAT filters have relatively wider transition bands, as can be seen from Figs. 1-2, and the only way to reduce the transition bandwidths is to increase their lengths. It can be noted from Eqs. (5-8) that the filter coefficients decrease sharply as we move away from the center. For example, the ratio between the last and the center coefficient for the low/high pass filters shown in Fig. 2 is of the order of 10^{-20}. If these small coefficients are ignored while implementing the filter, the effect is almost negligible on the transition bandwidth but smoothness of the filter deteriorates by appearance of a small ripple on the magnitude response. This effect can be used for designing the ripple filters of narrow transition bands with smaller lengths.

If we design a filter of length $4M$-1, and implement it with a smaller number of central $4N$ -1 ($N < M$) coefficients, the resultant filter of a smaller length $4N$ -1 will have a narrower transition band corresponding to a larger length $4M$ -1. To do this, we do not need to calculate all of the $4M$-1 coefficients; rather it can be done simply by multiplying N in Eqs. (5-8) by $\alpha = M/N \geq 1$. For example, the coefficients of a lowpass ripple filter can be written as

$$g_{\pm n} = \begin{cases} \dfrac{\omega_c}{\pi} & n = 0 \\ \dfrac{(2\alpha N-1)!!^2 \pi^{p-1} \sin(n\omega_c)}{2^p n(2\alpha N+n-1)!!(2\alpha N-n-1)!!} & 0 < n < N. \end{cases} \quad (9)$$

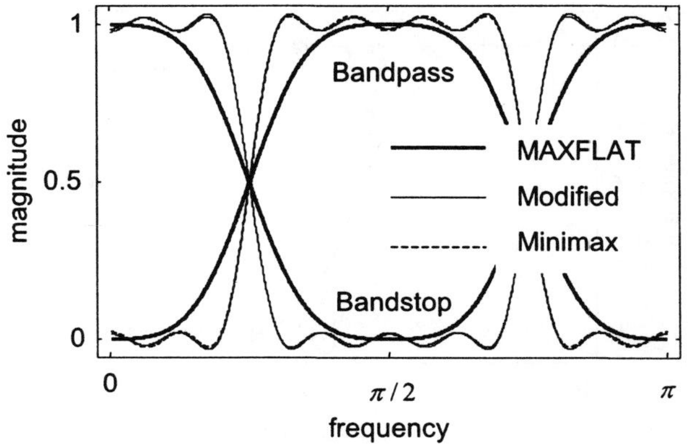

Fig. 1: Magnitude responses of MAXFLAT band pass/stop filters of length 31 (N = 8), designed with Eqs. (5-6) are plotted for cutoff frequencies $\omega_{c1} = 0.25\pi$ and $\omega_{c2} = 0.75\pi$. Transition bandwidth of the responses is reduced using modified procedure for $\alpha = 10$, and the resultant responses are compared with minimax responses of the same specifications.

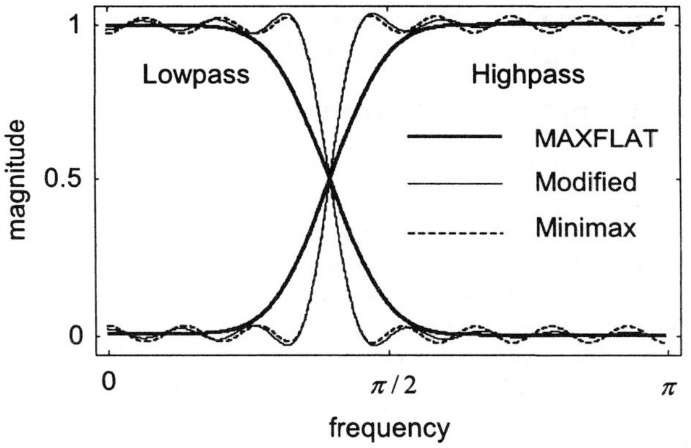

Fig. 2: Magnitude responses of MAXFLAT low/high pass digital filters of length 31 (N = 8), designed with Eqs. (7-8) are plotted for a cutoff frequency $\omega_c = 0.4\pi$. Transition bandwidth of the responses is reduced using modified procedure for $\alpha = 10$, and the resultant responses are compared with minimax responses of the same specifications.

For $\alpha = 1$, these formulas give MAXFLAT designs, while higher values of α narrow the transition bands but a small ripple appears on the magnitude responses. This can be seen in Fig. 3, where magnitude responses of DFs designed by using Eq. (9) for $\omega_c = 0.4$, and N = 8 are shown for different values of α . Clearly, the effect of increasing the value of α on the transition bandwidth is the same as that of increasing the order for the MAXFLAT filters. It should however be noted that an increase in α beyond a certain value, which increases with order of the filter, has very

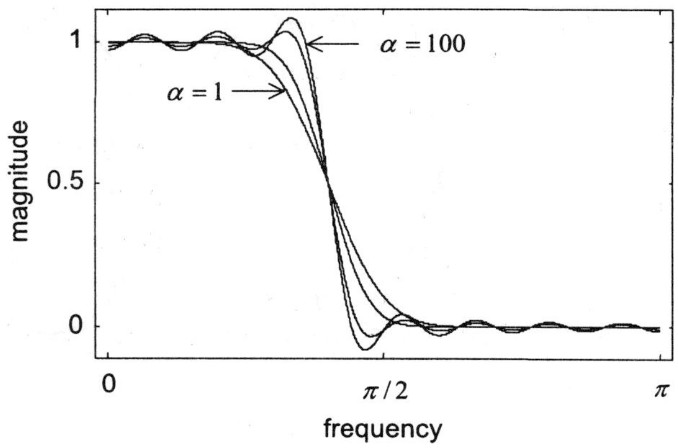

Fig. 3: Effect of increasing the value of α on transition bandwidth is shown by plotting the magnitude responses of filters of length 31 (N = 8) and cutoff frequency $\omega_c = 0.4\pi$ for α = 1, 2, 10 and 100.

small effect on transition bandwidth. Based on our experience, we suggest that the value of α should be kept smaller than N.

In Figs. 1-2, the thin lines show the magnitude responses of the DFs designed with the modified formulas for $\alpha = 10$. Reduction in the transition bandwidths at the expense of the smoothness of the magnitude responses is clearly visible.

4.2 Comparison with the minimax DFs

DFs designed with modified formulas become comparable to the minimax DFs in the sense that for both types, transition bandwidth can be modified for a fixed length. Minmax designs have been explored by many researchers and their behavior can be well anticipated, whereas relationship of α with the transition bandwidth or the ripple size is not exactly known for our modified designs. Moreover, unlike minimax DFs, the size of ripple in our designs is not uniform, and therefore the maximum value of error is smaller in the minimax DFs. As far as the design complexity is concerned, the minimax designs use iterative procedures and are far more complicated than our presented designs, which give the explicit formulas for the coefficients. This simplicity can make our modified designs attractive in certain real time applications.

The dotted lines in Figs. 1-2 show the magnitude responses of the minimax DFs of the same specification as for the other filters shown in the figures. It can be seen that our modified designs have performances comparable to the minimax designs.

5. Conclusions

Interrelationships between the coefficients of halfband lowpass and general bandpass digital filters are explored and used to modify the existing designs of MAXFLAT halfband lowpass FIR filters to new designs of general low/high pass and band pass/stop MAXFLAT FIR digital filters. The presented designs are not only simpler but also meet the desired cutoff frequencies more closely.

The presented designs are modified such that a design parameter controls the transition bandwidth of the resultant filters by changing the size of ripple on the magnitude response. The resultant filters have magnitude responses comparable to those of the minimax filters.

6. References

[1] P. P. Vidydnathan, Design and implementation of digital FIR filters, in Handbook of digital signal processing engineering applications, D. F. Elliott, Ed., Academic Press, Inc., 1987.

[2] O. Herrmann, "On the approximation problem in nonrecursive digital filter design," IEEE Trans. Circuit Theory, 18:411-423, May 1971.

[3] M. U. A. Bromba and H. Ziegler, "Explicit formula for filter function of maximally flat nonrecursive digital filters," Electronics Letters, 16(24):905-906, 1980.

[4] L. R. Rajagopal and S. C. Dutta Roy, "Design of maximally-flat FIR filters using the Bernstein polynomial," IEEE Trans. Circuits Syst., 34(12):1587-1590, 1987.

[5] P. Thajchayapong, M. Puangpool, S. Banjongjit, "Maximally flat F.I.R. filter with prescribed cutoff frequency," Electronics Letters, 16(13), 1980.

[6] M. T. Hanna, "Design of linear phase FIR filters with a maximally flat passband," IEEE Trans. Circuits Syst.-II, 43(2):142-147, 1996.

[7] T. Cooklev and A. Nishihara, "Maximally flat FIR filters," Proc. IEEE Int. Symp. Circuits Syst., Chicago, IL, May 1993, pages 96-99.

[8] C. Gumacos, "Weighting coefficients for certain maximally flat nonrecursive digital filters," IEEE Trans. Circuits Syst., CAS-25(4): 234-235, 1978.

[9] I. R. Khan, and R. Ohba, "Digital Differentiators based on Taylor series," IEICE Trans. Fundamentals, E82-A(12): 2822-2824, 1999.

[10] I. R. Khan, and R. Ohba, " New design of full band differentiators based on Taylor series," IEE Proc.-Vision, Image and Signal Processing, 146(4):185-189, 1999.

[11] I. R. Khan, and R. Ohba, "Efficient design of halfband low/high pass FIR filters using explicit formulas for tap coefficients," IEICE Trans. Fundamentals, E83-A(11): 2370-2373, 2000.

[12] I. R. Khan, and R. Ohba, "New designs of low/high pass and mid pass/stop FIR digital filters, Int. J. Circuit Theory and Applications," 29: 423-431, 2001.

[13] I. R. Khan, and R. Ohba, "New efficient designs of discrete and differentiating FIR Hilbert transformers," IEICE Trans. Fundamentals, E83-A(12): 2736-2738, 2000.

[14] B. Carlsson, " Maximum flat digital differentiators", Electronic Letters, 27(8):675-677, 1991.

[15] S. C. Pei and J. J. Shyu, "Relationships Among Digital One/Half Band Filters, Low/High Order Differentiators, and Discrete/Differentiating Hilbert Transformers," IEEE Trans. Signal Processing, 40(3): 694-700, 1992.

[16] T. W. Parks and J. H. McClellan, " Chebyshev approximation for non recursive digital filters with linear phase," IEEE Tran. Circuit Theory, 19: 189-194, 1972.

[17] A. E. Cetin, O. N. Gerek, and Y. Yardimci, "Equiripple FIR filter design by the FFT algorithm," IEEE Signal Processing Magazine, Mar. 1997, pages 60-64.

[18] IEEE Digital Signal Processing Committee, Programs for Digital Signal Processing, IEEE Press, New York 1979.

DESIGN OF 2-D VARIABLE FRACTIONAL DELAY FIR FILTER USING 2-D DIFFERENTIATORS

Chien-Cheng Tseng

Department of Computer and Communication Engineering, National Kaohsiung First University of Science and Technology, Kaohsiung, Taiwan
E-mail: tcc@ccms.nkfust.edu.tw

ABSTRACT

In this paper, the design of two-dimensional (2-D) variable fractional delay FIR filter using 2-D differentiators is presented. First, the Taylor series expansion is used to transform the specification of 2-D variable fractional delay filter into the one of 2-D differentiator. This transform makes the design problem of 2-D fractional delay filter reduce to the design of 2-D differentiators with different orders. Then, a simple window method is proposed to design the 2-D differentiators. And, the analysis of design error is also provided. Finally, the design examples are demonstrated to illustrate the effectiveness of this new design approach.

1. INTRODUCTION

In many applications of signal processing, there is a need for a delay which is a fraction of the sampling period. These applications include time adjustment in digital receivers, beam steering of antenna array, speech coding and synthesis, modeling of music instruments, sampling rate conversion, time delay estimation, comb filter design and AD conversion etc [1]-[10]. An excellent survey of the fractional delay filter design is presented in tutorial paper [3][4]. Generally speaking, the design methods can be classified into two categories. One is the fixed fractional delay (FFD) filter design, the other is variable fractional delay (VFD) filter design. In FFD case, the delay value is fixed, so conventional FIR and IIR filter design techniques can be utilized to design FFD filter directly. In VFD case, the delay is adjustable or tunable, so variable filter design method must be developed to design this type filters. Until now, there have been several methods to design VFD FIR filters such as Farrow method [5], weighted least squares method [6]-[9], and constrained minimax optimization [10] etc.

On the other hand, the digital differentiator is a very useful tool to determine and estimate the time derivatives of a given signal. So far, several methods have been developed to design IIR and FIR digital differentiators such as eigenfilter method [11], weighted least squares method [12], and quadratic programming [13] etc. In [14], the relation between one-dimensional (1-D) VFD filter design problem and the differentiator design problem has been established. As a result, the comprehensive design tools of the differentiator in the literature can be applied to design 1-D VDF filter. In this paper, the differentiator bank technique in [14] will be extended to design 2-D variable fractional delay filter. The design procedure can be divided into the following two steps. First, the Taylor series expansion is used to transform the specification of 2-D variable fractional delay filter into the one of 2-D differentiator. This transform makes the design problem of 2-D fractional delay filter reduce to the design of 2-D differentiators with different orders. Second, a simple window method is proposed to design the 2-D differentiator. Now, the details are described in next section.

2. DESIGN OF 2-D VARIABLE FRACTIONAL DELAY FIR FILTER

2.1 Problem Statement

For the design of 2-D variable fractional delay filter, the desired frequency response is given by

$$H_d(\omega_1,\omega_2) = e^{-j(\omega_1(D_1+p_1)+\omega_2(D_2+p_2))} \qquad (1)$$

where delay (D_1, D_2) are integers and (p_1, p_2) are variable fractional numbers in the range $[-0.5, 0.5]$. The transfer function of the 2-D FIR filter is chosen as follows:

$$H(z_1,z_2,p_1,p_2) = \sum_{n_1=0}^{N_1}\sum_{n_2=0}^{N_2} a_{n_1n_2}(p_1,p_2) z_1^{-n_1} z_2^{-n_2} \qquad (2)$$

where $a_{n_1n_2}(p_1,p_2)$ are the polynomial functions in p_1 and p_2, i.e.,

$$a_{n_1n_2}(p_1,p_2) = \sum_{k_1=0}^{M_1}\sum_{k_2=0}^{M_2} a_{n_1n_2k_1k_2} p_1^{k_1} p_2^{k_2} \qquad (3)$$

Substituting eq(3) into eq(2), the transfer function can be expressed as

$$\begin{aligned} &H(z_1,z_2,p_1,p_2) \\ &= \sum_{k_1=0}^{M_1}\sum_{k_2=0}^{M_2}\sum_{n_1=0}^{N_1}\sum_{n_2=0}^{N_2} a_{n_1n_2k_1k_2} z_1^{-n_1} z_2^{-n_2} p_1^{k_1} p_2^{k_2} \\ &= \sum_{k_1=0}^{M_1}\sum_{k_2=0}^{M_2} G_{k_1k_2}(z_1,z_2) p_1^{k_1} p_2^{k_2} \qquad (4) \end{aligned}$$

where $G_{k_1k_2}(z_1,z_2) = \sum_{n_1=0}^{N_1}\sum_{n_2=0}^{N_2} a_{n_1n_2k_1k_2} z_1^{-n_1} z_2^{-n_2}$. Now, the design problem is how to design 2-D subfilters $G_{k_1k_2}(z_1,z_2)$ such that the filter $H(z_1,z_2,p_1,p_2)$ approximates the desired response $H_d(\omega_1,\omega_2)$ as well as possible. In the following, a method based on differentiator will be presented to solve this problem.

2.2 Decomposition of 2-D Fractional Delay Filter

Using the Taylor series expansion, 2-D fractional delay response $H_d(\omega_1,\omega_2)$ can be decomposed into the combination of differentiators. The main result can be summarized

0-7803-7761-3/03/$17.00

as the following fact:
Fact: If the frequency response of (k_1, k_2)th order differentiator is defined by

$$F_{k_1k_2}(\omega_1, \omega_2) = (j\omega_1)^{k_1}(j\omega_2)^{k_2}e^{-j(\omega_1 D_1+\omega_2 D_2)} \quad (5)$$

then it can be shown that the 2-D fractional delay filter $H_d(\omega_1, \omega_2)$ can be decomposed into the form:

$$H_d(\omega_1, \omega_2) = \sum_{k_1=0}^{\infty}\sum_{k_2=0}^{\infty}\frac{(-1)^{k_1+k_2}}{k_1!k_2!}F_{k_1k_2}(\omega_1, \omega_2)p_1^{k_1}p_2^{k_2} \quad (6)$$

Pf: Using the Taylor series expansion, the exponential function can be expressed as a polynomial as follows:

$$\begin{aligned} e^{-j\omega_1 p_1} &= \sum_{k_1=0}^{\infty}\frac{(-p_1)^{k_1}}{k_1!}(j\omega_1)^{k_1} \\ e^{-j\omega_2 p_2} &= \sum_{k_2=0}^{\infty}\frac{(-p_2)^{k_2}}{k_2!}(j\omega_2)^{k_2} \end{aligned} \quad (7)$$

The product of the above both terms is given by

$$e^{-j(\omega_1 p_1+\omega_2 p_2)} = \sum_{k_1=0}^{\infty}\sum_{k_2=0}^{\infty}\frac{(-p_1)^{k_1}(-p_2)^{k_2}}{k_1!k_2!}(j\omega_1)^{k_1}(j\omega_2)^{k_2} \quad (8)$$

Multiply both side by the factor $e^{-j(\omega_1 D_1+\omega_2 D_2)}$, we get

$$\begin{aligned} & e^{-j(\omega_1(D_1+p_1)+\omega_2(D_2+p_2))} \\ &= \sum_{k_1=0}^{\infty}\sum_{k_2=0}^{\infty}\frac{(-p_1)^{k_1}(-p_2)^{k_2}}{k_1!k_2!}(j\omega_1)^{k_1}(j\omega_2)^{k_2}e^{-j(\omega_1 D_1+\omega_2 D_2)} \\ &= \sum_{k_1=0}^{\infty}\sum_{k_2=0}^{\infty}\frac{(-1)^{k_1+k_2}}{k_1!k_2!}F_{k_1k_2}(\omega_1, \omega_2)p_1^{k_1}p_2^{k_2} \end{aligned} \quad (9)$$

The proof is completed.

Because p_1 and p_2 are both in the range $[-0.5, 0.5]$, the power functions $p_1^{k_1}$ and $p_2^{k_2}$ approach to zero when k_1 and k_2 are very large. Thus, by truncating the high order differentiators, the frequency response of 2-D fractional delay filter can be approximated by

$$\hat{H}_d(\omega_1, \omega_2) = \sum_{k_1=0}^{M_1}\sum_{k_2=0}^{M_2}\frac{(-1)^{k_1+k_2}}{k_1!k_2!}F_{k_1k_2}(\omega_1, \omega_2)p_1^{k_1}p_2^{k_2} \quad (10)$$

Because $p_i^{k_i}$ is an increasing function of p_i $(i = 1, 2)$, the smaller p_1 and p_2 are, the better approximation eq(10) has. To evaluate the performance that $\hat{H}_d(\omega_1, \omega_2)$ in eq(10) approximates ideal response $H_d(\omega_1, \omega_2)$, the normalized root-mean-squares error is defined by

$$NRMS = \frac{\left[\int_0^{\alpha\pi}\int_0^{\alpha\pi}\int_{-0.5}^{0.5}\int_{-0.5}^{0.5}|E(\omega_1, \omega_2)|^2 dp_1 dp_2 d\omega_1 d\omega_2\right]^{1/2}}{\left[\int_0^{\alpha\pi}\int_0^{\alpha\pi}\int_{-0.5}^{0.5}\int_{-0.5}^{0.5}|H_d(\omega_1, \omega_2)|^2 dp_1 dp_2 d\omega_1 d\omega_2\right]^{1/2}} \times 100\% \quad (11)$$

where $E(\omega_1, \omega_2) = \hat{H}_d(\omega_1, \omega_2) - H_d(\omega_1, \omega_2)$. By using the eq(1)(10) and equality $|H_d(\omega_1, \omega_2)| = 1$, we obtain

$$NRMS = \frac{\left[\int_0^{\alpha\pi}\int_0^{\alpha\pi}\int_{-0.5}^{0.5}\int_{-0.5}^{0.5}|v(\omega_1, \omega_2)|^2 dp_1 dp_2 d\omega_1 d\omega_2\right]^{1/2}}{\alpha\pi} \times 100\% \quad (12)$$

where

$$v(\omega_1, \omega_2) = \sum_{k_1=0}^{M_1}\sum_{k_2=0}^{M_2}\frac{(-j)^{k_1+k_2}(\omega_1 p_1)^{k_1}(\omega_2 p_2)^{k_2}}{k_1!k_2!} - e^{-j(\omega_1 p_1+\omega_2 p_2)}$$

Clearly, the $NRMS$ only depends on the choice of α, M_1 and M_2. Choosing $\alpha = 0.9$, Table 1 lists the $NRMS$ for various M_1 and M_2. From this result, we see that $NRMS$ is a monotonically decreasing function of M_1 and M_2. When M_1 and M_2 are greater than 5, the $NRMS$ is less than 0.2%. Finally, compare eq(4) and eq(10), we see that if the filters $G_{k_1k_2}(z_1, z_2)$ are designed to approximate $\frac{(-1)^{k_1+k_2}}{k_1!k_2!}F_{k_1k_2}(\omega_1, \omega_2)$, then the filter $H(z_1, z_2, p_1, p_2)$ will approximate $H_d(\omega_1, \omega_2)$. Thus, the design problem of 2-D fractional delay filter becomes the designs of $(M_1+1)(M_2+1)$ 2-D differentiators. This design problem will be discussed in next subsection.

2.3 Design Method

In the following, the 2-D window method will be presented to design 2-D differentiators. If the ideal impulse response of 2-D (k_1, k_2)th order differentiator $\frac{(-1)^{k_1+k_2}}{k_1!k_2!}F_{k_1k_2}(\omega_1, \omega_2)$ is $f_{k_1k_2}(n_1, n_2)$, then the coefficients $a_{n_1n_2k_1k_2}$ of the filter $G_{k_1k_2}(z_1, z_2)$ can be obtained by multiplying the $f_{k_1k_2}(n_1, n_2)$ by a 2-D window function $w(n_1, n_2)$,i.e.,

$$a_{n_1n_2k_1k_2} = f_{k_1k_{k_2}}(n_1, n_2)w(n_1, n_2) \quad (13)$$

Thus, the design problem reduces to find the impulse response $f_{k_1k_2}(n_1, n_2)$ and choose a suitable window function $w(n_1, n_2)$. In the following, these two problems will be investigated. Using the 2-D Fourier transform, we have

$$\begin{aligned} & f_{k_1k_2}(n_1, n_2) \\ &= \frac{1}{4\pi^2}\int_{-\pi}^{\pi}\int_{-\pi}^{\pi}\frac{(-1)^{k_1+k_2}}{k_1!k_2!}F_{k_1k_2}(\omega_1, \omega_2)e^{j(\omega_1 n_1+\omega_2 n_2)} d\omega_1 d\omega_2 \\ &= \left(\frac{1}{2\pi}\int_{-\pi}^{\pi}\frac{(-1)^{k_1}}{k_1!}(j\omega_1)^{k_1}e^{-j\omega_1 D_1}e^{j\omega_1 n_1}d\omega_1\right) \\ & \quad \left(\frac{1}{2\pi}\int_{-\pi}^{\pi}\frac{(-1)^{k_2}}{k_2!}(j\omega_2)^{k_2}e^{-j\omega_2 D_2}e^{j\omega_2 n_2}d\omega_2\right) \\ &= f_{k_1}(n_1)f_{k_2}(n_2) \end{aligned} \quad (14)$$

where $f_{k_i}(n_i)$ $(i = 1, 2)$ can be computed by

$$f_{k_i}(n_i) = \frac{1}{\pi}\sum_{m=0}^{k_i}\frac{(-1)^{m+k_i}\pi^{k_i-m}}{\Gamma(k_i-m+1)}\frac{\sin\left((n_i-D_i)\pi + \frac{(k_i-m)\pi}{2}\right)}{(n_i-D_i)^{m+1}} \quad (15)$$

for $n_i \neq D_i$ and

$$f_{k_i}(D_i) = \begin{cases} 0 & k_i \text{ is odd} \\ \frac{(-1)^{1.5k_i}\pi^{k_i}}{\Gamma(k_i+2)} & k_i \text{ is even} \end{cases} \quad (16)$$

for $n_i = D_i$. As to the choice of 2-D window function, there exist several 2-D window functions in the literature [15]. If we choose the separable window function $w(n_1, n_2) = w(n_1)w(n_2)$, then from eq(13) (14), we have

$$a_{n_1n_2k_1k_2} = [f_{k_1}(n_1)w(n_1)][f_{k_2}(n_2)w(n_2)] \quad (17)$$

where $w(n)$ is a 1-D window function. Thus, 2-D fractional delay filter design can be decomposed into two 1-D fractional delay filter designs for this choice. Moreover, if we choose a unseparable 2-D window function:

$$w(n_1, n_2) = w(n)|_{n=\sqrt{n_1^2+n_2^2}} \quad (18)$$

,then filter coefficients can be computed by eq(13) and the designed 2-D fractional delay filter is unseparable.

2.4 Design Example

Now, we describe a design example to illustrate the proposed method. To evaluate the performance, the maximum absolute error $\epsilon_{max}(p_1, p_2)$ is defined by

$$\epsilon_{max}(p_1, p_2) = \max\{|e(\omega_1, \omega_2, p_1, p_2)| \mid \omega_1 \in [0, \alpha\pi], \omega_2 \in [0, \alpha\pi]\}$$

where error

$$e(\omega_1, \omega_2, p_1, p_2) = H_d(\omega_1, \omega_2) - H(e^{j\omega_1}, e^{j\omega_2}, p_1, p_2) \quad (19)$$

The parameters chosen are $\alpha = 0.9$, $N_1 = 45$, $N_2 = 49$, $D_1 = 22$, $D_2 = 25$, $M_1 = M_2 = 5$, and window function $w(n_1, n_2)$ is the separable Hamming window function $w(n_1)w(n_2)$. Table 2 lists the $\epsilon_{max}(p_1, p_2)$ for some fixed (p_1, p_2) value. It is clear that the errors are very small. Moreover, we can find that ϵ_{max} for $(p_1, p_2) = (0, 0)$ are the smallest among all combination of p_1 and p_2. Fig.1 shows the 2-D magnitude response for $(p_1, p_2) = (0, 0)$. We see that the magnitude response is almost equal to one for all frequencies. If the phase response of filter $H(z_1, z_2, p_1, p_2)$ are denoted by $\theta(\omega_1, \omega_2, p_1, p_2)$, then two group delay responses are defined by

$$\tau_1(\omega_1, \omega_2, p_1, p_2) = -\frac{\partial\theta(\omega_1, \omega_2, p_1, p_2)}{\partial\omega_1}$$
$$\tau_2(\omega_1, \omega_2, p_1, p_2) = -\frac{\partial\theta(\omega_1, \omega_2, p_1, p_2)}{\partial\omega_2} \quad (20)$$

Fig.2 and 3 show the group delay responses $\tau_1(\omega_1, 0, p_1, 0)$ and $\tau_2(0, \omega_2, 0, p_2)$. Clearly, the specification is well satisfied. Finally, the analysis of design error is provided as follows: Using eq(4) and eq(6), the error defined in eq(19) can be written as

$$\begin{aligned} & e(\omega_1, \omega_2, p_1, p_2) \\ &= H_d(\omega_1, \omega_2) - H(e^{j\omega_1}, e^{j\omega_2}, p_1, p_2) \\ &= \sum_{k_1=0}^{\infty}\sum_{k_2=0}^{\infty} \frac{(-1)^{k_1+k_2}}{k_1!k_2!} F_{k_1k_2}(\omega_1, \omega_2) p_1^{k_1} p_2^{k_2} \\ &\quad - \sum_{k_1=0}^{M_1}\sum_{k_2=0}^{M_2} G_{k_1k_2}(e^{j\omega_1}, e^{j\omega_2}) p_1^{k_1} p_2^{k_2} \\ &= \sum_{k_1=0}^{M_1}\sum_{k_2=0}^{M_2} e_{k_1k_2}(\omega_1, \omega_2) p_1^{k_1} p_2^{k_2} + e_t(\omega_1, \omega_2, p_1, p_2) \end{aligned}$$

where error $e_{k_1k_2}(\omega_1, \omega_2)$ is given by

$$\begin{aligned} & e_{k_1k_2}(\omega_1, \omega_2) \\ &= \frac{(-1)^{k_1+k_2}}{k_1!k_2!} F_{k_1k_2}(\omega_1, \omega_2) - G_{k_1k_2}(e^{j\omega_1}, e^{j\omega_2}) \end{aligned}$$

and $e_t(\omega_1, \omega_2, p_1, p_2)$ is

$$\begin{aligned} & e_t(\omega_1, \omega_2, p_1, p_2) \\ &= \sum_{k_1=M_1+1}^{\infty}\sum_{k_2=M_2+1}^{\infty} \frac{(-1)^{k_1+k_2}}{k_1!k_2!} F_{k_1k_2}(\omega_1, \omega_2) p_1^{k_1} p_2^{k_2} \end{aligned}$$

Note that $e_t(\omega_1, \omega_2, p_1, p_2)$ is the truncation error due to the removal of the high order differentiators. And, $e_{k_1k_2}(\omega_1, \omega_2)$ is the approximation error that filter $G_{k_1k_2}(z_1, z_2)$ fits ideal response $\frac{(-1)^{k_1+k_2}}{k_1!k_2!} F_{k_1k_2}(\omega_1, \omega_2)$ in the window method. The truncation error can be reduced by increasing number of differentiators M_1 and M_2, and the approximation error can be reduced by increasing filter orders N_1 and N_2. When M_1, M_2, N_1 and N_2 approach infinity, the designed filter $H(z_1, z_2, p_1, p_2)$ will approach the ideal response $H_d(\omega_1, \omega_2)$.

3. CONCLUSION

In this paper, a new 2-D variable fractional delay FIR filter design method using 2-D differentiators has been presented. The design examples are demonstrated to illustrate the effectiveness of this new design approach.

REFERENCES

[1] K. Rajamani, Y.S. Lai and C.W. Farrow, "An efficient algorithm for sample rate conversion from CD to DAT", *IEEE Signal Processing Lett.*, vol. 7, pp. 288-290, Oct. 2000.

[2] S. Basu and Y. Bresler, "An empirical study of minimax-optimal fractional delays for low-pass signals", *IEEE Trans. Circuits Syst. II*, vol. 49, pp. 288-292, Apr. 2002.

[3] T.I. Laakso, V. Valimaki, M. Karjalainen and U.K. Laine, "Splitting the unit delay: Tools for fractional delay filter design", *IEEE Signal Processing Mag.*, vol. 13, pp. 30-60, Jan. 1996.

[4] V. Valimaki and T.I. Laakso, "Principle of fractional delay filters," *Int. Conf. Acoust. Speech Signal Processing*, pp. 3870-3873, May 2000.

[5] C.W. Farrow, "A continuously variable digital delay element", *Int. Symp. Circuits and Systems*, pp. 2641-2645, 1988.

[6] A. Tarczynski, G.D. Cain, E. Hermanowicz and M. Rojewski, "WLS design of variable frequency response FIR filters", *Int. Symp. Circuits and Systems*, pp. 2244-2247, 1997.

[7] W.S. Lu and T.B. Deng, "An improved weighted least-squares design for variable fractional delay FIR filters", *IEEE Trans. Circuits Syst. II*, vol. 46, pp. 1035-1040, Aug. 1999.

[8] T.B. Deng and W.S. Lu, "Weighted least-squares method for designing variable fractional delay 2-D FIR digital filters", *IEEE Trans. Circuits Syst. II*, vol. 47, pp. 114-124, Feb. 2000.

[9] T.B. Deng, "Discretization-free design of variable fractional-delay FIR filters", *IEEE Trans. Circuits Syst. II* vol. 48, pp. 637-644, June 2001.

[10] J. Vesma and T. Saramaki, "Optimization and efficient implementation of FIR filters with adjustable fractional delay", *Int. Symp. Circuits and Systems*, pp.2256-2259, 1997.

[11] S.C. Pei and J.J. Shyu, "Eigenfilter design of higher-order digital differentiators," *IEEE Trans. Acoust. Speech Signal Process.* vol. 37, pp. 505-511, 1989.

[12] S. Sunder and V. Ramachandra, "Design of equiripple nonrecursive digital differentiators and Hilbert transformers using a weighted least-squares technique," *IEEE Trans. Signal Processing*, vol. 42, No. 9, pp. 2504-2509, Sep. 1994.

[13] C.C. Tseng, "Stable IIR digital differentiator design using iterative quadratic programming approach," *Signal Processing*, vol. 80, pp. 857-866, 2000.

[14] C.C. Tseng, "Design of variable fractional delay FIR filter using differentiator bank", *Int. Symp. Circuits and Systems*, vol. 4, pp. 421-424, 2002.

[15] W.S. Lu and A. Antoniou, *Two-Dimensional Digital Filters*, Marcel Dekker, 1992.

Table 1 The $NRMS$ error for various M_1 and M_2.

M_1	M_2	$NRMS_2$ (%)
1	1	36.9392
1	2	24.8454
1	3	21.8010
1	4	21.7055
1	5	21.5683
2	2	11.7798
2	3	7.9984
2	4	7.8506
2	5	7.8156
3	3	3.3257
3	4	2.3956
3	5	2.2990
4	4	0.8217
4	5	0.5741
5	5	0.1710

Table 2 The maximum absolute error ϵ_{max} for various (p_1, p_2).

p_1	p_2	ϵ_{max}
-0.5	-0.25	0.0118
-0.5	0	0.0108
-0.5	0.25	0.0131
-0.25	-0.25	0.0056
-0.25	0	0.0030
-0.25	0.25	0.0033
0	-0.25	0.0045
0	0	0.0021
0	0.25	0.0029
0.25	-0.25	0.0064
0.25	0	0.0043
0.25	0.25	0.0054
0.5	-0.25	0.0108
0.5	0	0.0089
0.5	0.25	0.0103

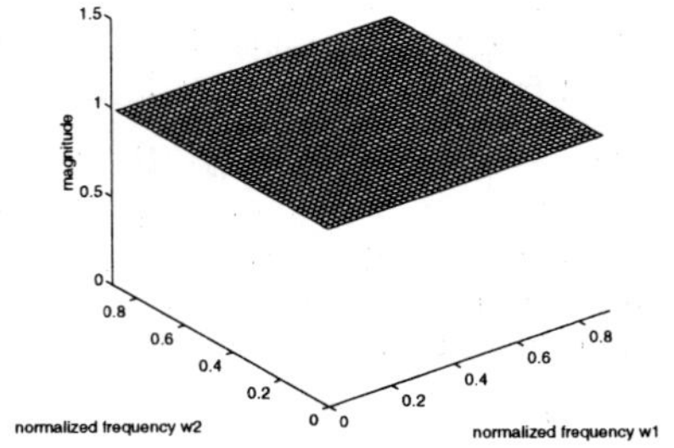

Figure 1. The magnitude response of the designed 2-D variable fractional delay filter with $(p_1, p_2) = (0, 0)$.

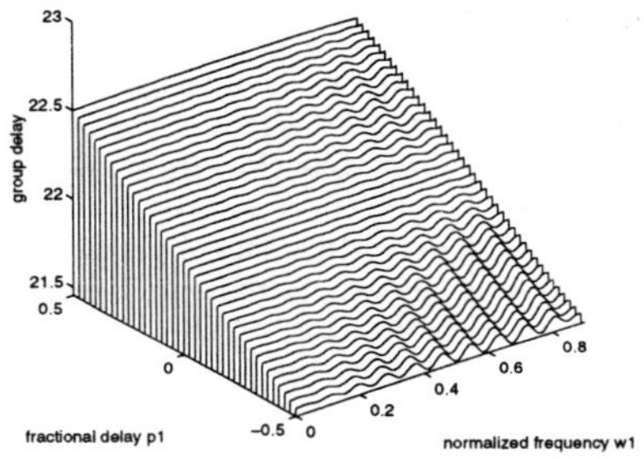

Figure 2. The group delay response $\tau_1(\omega_1, 0, p_1, 0)$ of the designed 2-D variable fractional delay filter.

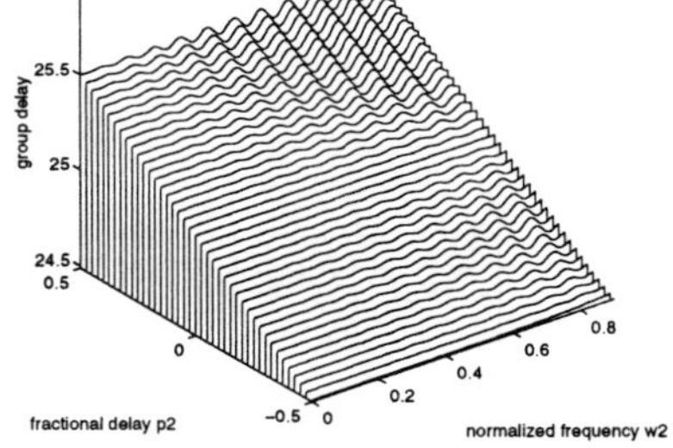

Figure 3. The group delay response $\tau_2(0, \omega_2, 0, p_2)$ of the designed 2-D variable fractional delay filter.

SOME OBSERVATIONS ON MULTIPLIERLESS IMPLEMENTATION OF LINEAR PHASE FIR FILTERS

Mrinmoy Bhattacharya and Tapio Saramäki

Institute of Signal Processing
Tampere University of Technology
P. O. Box 553, Tampere, FIN 33101, Finland
E-mail: {mrinmoy,ts}@cs.tut.fi

ABSTRACT

This paper investigates the case of multiplierless implementation of linear phase FIR filters by converting the multiplier coefficients to minimum signed powers-of-two (MNSPT) or canonic signed digit (CSD) forms. It was observed that if one is willing to accept some deviation from the given specifications, the required number of nonzero bits becomes quite low, making multiplierless implementation feasible. Alternatively, one can start with a filter that exceeds the given criteria that may involve acceptable level of increase in the filter order, but with much lesser total number of nonzero bits than the initial design. Then, the coefficient values are quantized into the desired representation forms such that the given overall criteria are still met. Fairly exhaustive investigation suggests that less than three nonzero bits are quite sufficient, along with reduction in number of arithmetic operations with attendant increase in rate of data throughput.

1. INTRODUCTION

In a multiplierless implementation of a digital filter, generally minimum number of signed powers-of-two (MNSPT) or canonic signed digits (CSD) representations of binary digits are extensively used for representing the multiplier coefficients. An MNSPT representation of a coefficient value is given by $\sum_i a_i 2^{-t_i}$, where each a_i is either 1 or –1 and t_i is a positive or negative integer. For instance, 1.93359375 can be realised as $2-2^{-4}-2^{-8}$. In this example case, the multiplication is achieved not by a nine-bit multiplier, but with aid of three bit shifts and two subtracts.

For the cases of FIR filters, the major approach for multiplierless implementations comprises of optimizing [5–8, 11] the filter coefficient values such that the resulting filter meets the given criteria with its coefficients values being expressible in MNSPT or CSD forms. For filter design being basically a problem of approximation due to the tolerances in specifications, optimization methods are used to find the optimal transfer functions under the given constraints. While optimization methods are considered to be quite satisfactory, one may not assure or guarantee that the optimal solution will always be found under the given constraints. The solution can be unsatisfactory, for example, in terms of the filter order, the given wordlength of the multipliers, or the specified number of shifts and adds (in the case of multiplierless implementation), or some combination of them. In such cases, some parameters or characteristics of the filter have to be relaxed to obtain an acceptable design.

Another approach is based on combining simple sub-filters [1, 2, 12] that can be implemented using only a few shifts and adds and/or subtracts. Although quite attractive, to make this approach as a viable one, a large database of such filters will have to be generated and some optimization method will have to be evolved in order to combine some of them to meet the desired specifications.

In [3,4] the feasibility of implementing multiplierless recursive digital filters based on low-sensitivity structures has been demonstrated. Allowing a marginally insignificant deviation in the specifications a gross reduction in number of nonzero bits (effectively the number of shifts and adds and/or subtracts required) was seen to be feasible without any increase in the filter order.

Here, we investigate the case of feasibility of multiplierless implementation of linear phase FIR filters in direct form structure by allowing deviation in the specifications that would involve some increase in filter length. We observe that a similar approach (as in the case of IIR Filters) for multiplierless implementation is feasible where the increase in the filter length is offset by a gross reduction in the total number of nonzero bits and number of arithmetic operations with attendant increase in data throughput.

At this stage we digress to point out that a filter being implemented is a sub-system of some system that the system designer wants to implement and his requirement is not exactly rigid but generally flexible within limits. His considerations are that of performance and cost (or performance vis-a-vis cost), and not necessarily the strict adherence to the initial specification he has issued; most likely, he has included a good amount of design margin in his goal system. As such, the filter designer may find out all the options in consultation with the system designer.

2. IMPLEMENTATION

We consider the typical direct form structure [9,10] for the linear phase FIR system with transfer function as

0-7803-7761-3/03/$17.00 ©2003 IEEE

$$H(z) = \sum_{n=0}^{L-1} h(n) z^{-n} \quad (1)$$

where

$$h(n) = \pm h(L-1-n) \quad (2)$$

and both symmetry and antisymmetry conditions are incorporated in (2). The number of multiplier coefficients are $L/2$ or $(L+1)/2$ depending on the filter length being even or odd, respectively.

The steps leading to multiplierless implementation are outlined as follows:

(a) Initially, the filter is designed as per the specifications provided using the Parks-McClellan algorithm [9,10] as implemented in the Signal Processing Toolbox of Matlab.

(b) Performance in terms of degradation of passband ripple A_p dB and stopband attenuation A_s dB with gradual reduction (upto a certain level) in number of bits for quantization is noted. Also, the total number of nonzero bits after converting the quantized multiplier coefficients to MNSPT/CSD forms are noted.

(c) As mentioned earlier, we have parameters like filter length L, passband ripple A_p, stopband attenuation A_s, passband edge ω_p , and stopband edge ω_s, where we may allow certain deviations. Noting the pattern of degradation as in sub-para (b) above we allow certain deviation in the specification and design a fresh filter with the same algorithm. For example, our observation shows that the rate of of degradation of stopband attenuation in direct form realization is much more than that of passband ripple, the specification of the fresh filter may be that the stopband attenuation is much higher than the initial specification.

(d) As in sub-para (b) earlier the number of nonzero bits to meet the initial specification is noted.

We note that that the total number in nonzero bits is much lower than that in initial design and the results are better explained in the next section.

3. RESULTS AND DISCUSSIONS

Table 1 illustrates the results of the implementation over a broad representative of linear phase FIR filters, where N_m, N_b, and N_{bm} indicate the number of multiplier coefficients, the total number of nonzero bits for the multipliers, and the average number of nonzero bits per multiplier coefficient, respectively. Figures 1, 2, 3, and 4 show the amplitude characteristics of some of the filters of Table 1.

For elaborating the contents of Table 1, we consider the case of Filter 1 with three stopband attenuation levels of 40 dB, 39 dB, and 33 dB. In the case of 40 dB (first row of Filter 1 performance results), we find that the filter designed with revised specifications is better in requirements of N_b as 29 only vis-a-vis 47 for the filter designed from initial specification, with an increase in length of three only (i.e., 32 vis-a-vis 29). In contrast to this, in the cases of stopband attenuation levels of 39 dB and 37 dB, the opposite is true, i.e., the requirements of N_b are much less in the case of degraded performance of the initial design case than in the case of revised specification that is associated with marginal decrease in filter length. The choices are obvious (the values of L, N_m, and N_b are shown in bold and italicized letters for the chosen ones) and confirm our earlier statements. Similar observations are made in respect of all the filter examples.

Further observation is made in respect of a substantial reduction in arithmetic operations like number of bit shifts and number of additions. The reduction can be computed as $(L_1 - L_2 + N_{b1} - N_{b2} - N_{m1} + N_{m2})$ additions and $(N_{b1} - N_{b2})$ bit shifts, where we have considered a case similar to the case of 40 dB stopband attenuation of Filter 1 (with subscript 1 refers to the initial design case and subscript 2 refers to the revised specification design case; we also mention that N nonzero bits imply $(N-1)$ additions). This reduction in arithmetic operations is quite substantial in most cases and would provide an increase in throughput rate of data. We also observe that with increasing filter length, the reduction in arithmetic operations (for example, see the cases pertaining Filter 4) increases.

4. CONCLUSIONS

In this paper, we have shown that the multiplierless implementation of FIR filters using the approach outlined above and evidenced over a large spectrum of FIR filters is a feasible and attractive proposition. Either, we can accept deviations in passband and stopband tolerance specifications compared with the initial infinite-precision design, or one can start with a design with stricter specification followed by the coefficient values being quantized to a level such that the given overall criteria are met. In both the cases some increase in filter length is involved that is offset by a gross reduction of total number of nonzero bits. In addition, there is substantial reduction of total number of arithmetic operations like number of additions and bit shifts, leading to an increase in data throughput. Our analysis indicates that utilizing this approach, it is possible to achieve a multiplierless realization with less than three nonzero bits per multiplier with about ten percent increase in length. Further, it is seen that there is some reduction of leakage of energy through stopband. Future work is devoted to applying optimization techniques towards further reduction in the number of nonzero bits.

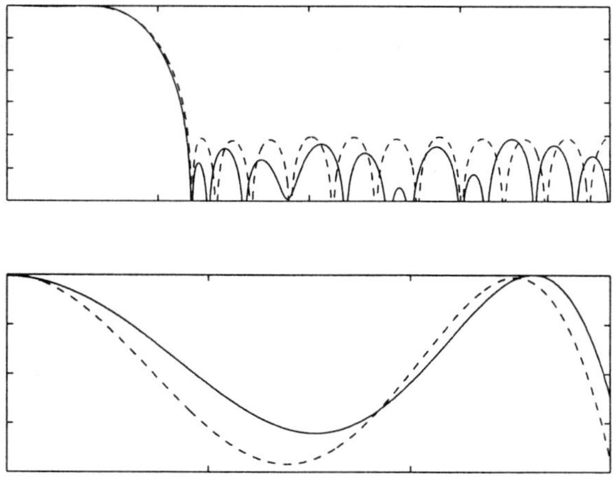

Fig. 1. Amplitude responses in the case of Filter 1 for initial design (dashed line) and the alternative feasible (solid line) that requires twenty-nine nonzero bits.

Table 1. Results of some representative FIR filters indicating the options.

Filter specifications	Feasibility from initial design						Alternatives feasible from revised design						Revised design values of	
	L	N_m	N_b	N_{bm}	A_p dB	A_s dB	L	N_m	N_b	N_{bm}	A_p dB	A_s dB	A_p dB	A_s dB
Filter 1: Lowpass ω_p=0.15π, ω_s=0.3π, A_p=0.2 dB,and A_s=40 dB, 39 dB, and 37 dB	29	15	47	3.13	0.19	40.35	***32***	***16***	***29***	1.81	0.16	41.2	0.15	45.0
	"	"	***37***	2.46	0.205	39.9	28	14	49	3.5	0.2	39.32	0.21	39.0
	"	"	***31***	2.07	0.195	37.0	27	14	51	3.64	0.192	37.3	0.2	37.0
Filter 2: Lowpass ω_p=0.2π, ω_s=0.3π, A_p=0.5 dB, and A_s=50 dB, 39 dB, and 35 dB	42	21	78	3.71	0.42	51.0	***44***	***22***	***59***	2.68	0.49	52.1	0.5	55.0
	"	"	***41***	1.95	0.485	39.5	36	18	65	3.61	0.456	39.4	0.5	39.0
	"	"	***34***	1.62	0.48	35.82	31	16	41	2.56	0.49	35.5	0.5	35.0
Filter 3: Lowpass ω_p=0.1π, ω_s=0.2π, A_p=0.3 dB, and A_s=60 dB, 50 dB, and 49 dB	54	27	107	3.96	0.275	60.3	***62***	***31***	***83***	2.68	0.272	62.25	0.3	75.0
	"	"	***72***	2.67	0.275	53.4	47	24	103	4.29	0.297	50.2	0.3	50.0
	"	"	***63***	2.33	0.288	49.88	"	"	"	"	"	"	"	"
Filter 4: Lowpass ω_p=0.1π, ω_s=0.15π, A_p=0.1 dB, and A_s=50 dB and 45 dB	104	52	177	3.40	0.093	50.5	***113***	***57***	***135***	2.36	0.095	52.01	0.09	60.0
	"	"	***108***	2.07	0.098	45.2	92	46	176	3.82	0.099	45.1	0.1	50.0
Filter 5: Bandpass ω_{p_1}=0.2π, ω_{p_2}=0.3π, ω_{s_1}=0.1π, ω_{s_2}=0.4π, A_p=0.5 dB, and A_s=50 dB, 48 dB and 45 dB	42	21	71	3.38	0.496	50.7	***49***	***25***	***53***	2.12	0.45	50.9	0.5	60.0
	"	"	***55***	2.62	0.498	48.78	41	22	70	3.18	0.5	48.1	0.5	48.0
	"	"	***44***	2.09	0.499	45.03	40	20	72	3.6	0.5	45.2	0.5	45.0
					Absolute error less than						Absolute error less than		Absolute error (Design value)	
Filter 6: Hilbert Transformer ω_{p_1}=0.1π, ω_{p_2}=0.9π, ω_{s_1}=0, ω_{s_2}=π,	31	8	36	4.5	0.00278		***39***	***10***	***25***	2.5	0.00265		0.000683	
	"	"	***21***	2.62	0.006		29	7	27	3.86	0.0055		0.00545	
Filter 7: Differentiator (full band)	32	16	66	4.12	0.00604		***36***	***18***	***38***	2.11	0.00601		0.005065	
Filter 8: Differentiator (partial band) ω_p=0.4π, ω_s=0.45π, (magnitude at 0.4π is 0.4π)	51	26	116	4.46	0.0234		***57***	***29***	***77***	2.66	0.019		0.0173	
	"	"	***53***	2.04	0.024		...	...	...	...	....		......	

Note:- For odd length Hilbert transformer alternate multipliers are zero-valued.

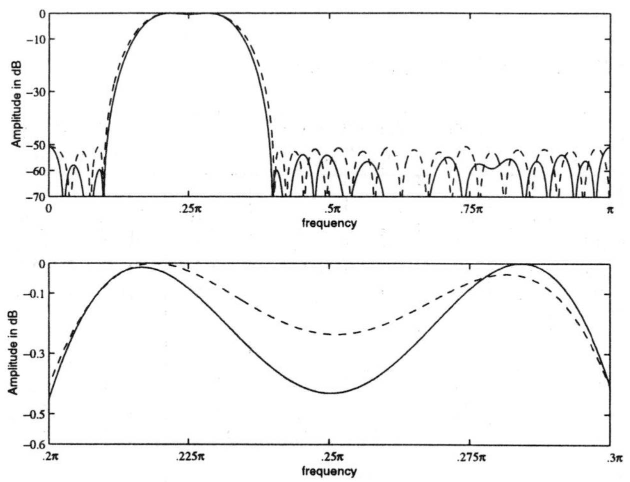

Fig. 2. Amplitude responses in the case of Filter 5 (bandpass) for initial design (dashed line) and the alternative feasible (solid line) that requires fifty-three nonzero bits.

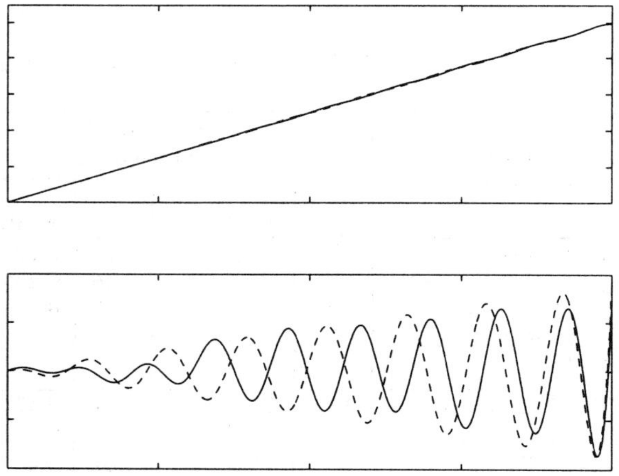

Fig. 4. Amplitude responses in the case of the Filter 7 (full band differentiator) for initial design (dashed line) and the alternative feasible (solid line) that requires thirty-eight bits.

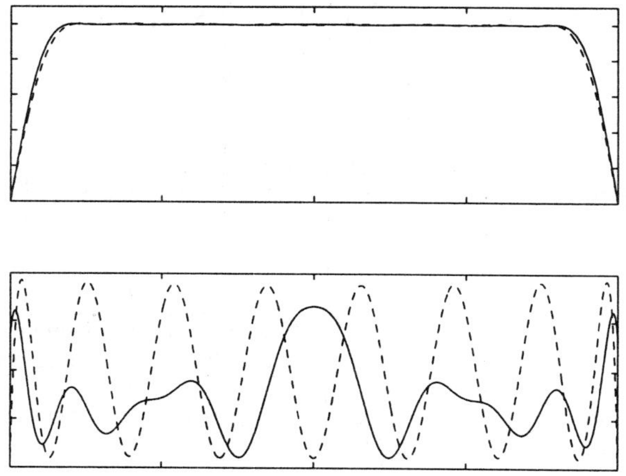

Fig. 3. Amplitude responses in the case of the Filter 6 (Hilbert transformer) for initial design (dashed line) and the alternative feasible (solid line) that requires twenty-five nonzero bits.

5. ACKNOWLEDGEMENT

This work was supported by the Academy of Finland, project No. 44876 (Finnish Centre of Excellence program (2000-2005)).

6. REFERENCES

[1] John A. Adams and Alan N. Williamson, Jr., "A New Approach to FIR Digital Filters with Fewer Multipliers and Reduced Sensitivity," *IEEE Trans. Circuits Syst.*, vol. CAS-30, p. 277-283, May 1983.

[2] ....".............., "Some efficient Efficient Digital Prefilter Structures," *IEEE Trans. Circuits Syst.*, vol. CAS-31, pp. 260-266, March 1984.

[3] M. Bhattacharya and J. Astola, "Multiplierless Implementation of Recursive Digital Filters based on Coefficient Translation Methods in Low Sensitivity Structures," *Proc. IEEE Intl. Symp. Circuits Syst.,(ISCAS 2001)*, Sydney, Australia, vol.- II, pp. 697-700, 2001.

[4] M. Bhattacharya, T. Saramäki, and J.Astola, "Multiplierless Realization of Recursive Digital Filters," *Proc. 2nd Intl. Symp. Image and Signal Processing and Analysis, (ISPA 2001)*, Pula, Croatia, pp. 469-474, 2001.

[5] Y. C. Lim and S.R. Parker, "FIR Filter Design over a Discrete Powers-of-Two Coefficient Space," *IEEE Trans. Acoust. Speech, Signal Processing,* vol. ASSP-31, pp. 583-591 June 1983.

[6] Y. C. Lim, "Design of Discrete-Coefficient-Value Linear phase FIR Filters with optimum Normalized Peak Ripple Magnitude," *IEEE Trans. Circuits Syst.,* vol. CAS-37, pp. 1480-1486, Dec. 1990.

[7] Y. C. Lim and Bede Liu, "Design of Cascade Form FIR Filters with Discrete Valued Coefficients," *IEEE Trans. Acoust. Speech, Signal Processing,* vol. ASSP-36, pp. 1735-1739, Nov. 1988.

[8] Y. C. Lim, J. B. Evans, and B. Liu, "Decomposition of Binary Integers into Signed Powers-of-Two Terms," *IEEE Trans. Circuits Syst.,* vol. CAS-38, pp. 667-672,June 1991.

[9] S. K. Mitra, *Digital Signal Processing : A Computer-Based Approach.* (Second Edition), McGraw-Hill, 2001.

[10] J. G. Proakis and D. G. Manolakis, *Digital Signal Processing: Principles, Algorithms and Applications.* Prentice-Hall, Inc., Upper Saddle River, NJ, 1996.

[11] H. Samueli, "An Improved Search Algorithm for the Design of Multiplierless FIR Filters with Powers-of-Two Coefficients," *IEEE Trans. Circuits Syst.,* vol. CAS-36, pp. 1044-1047, July 1989.

[12] P. P. Vaidyanathan and G. Beitman, " On Prefilters for Digital Filter Design," *IEEE Trans. Circuits Syst.,* vol. CAS-32, pp. 494-499, May 1985.

A SYSTEMATIC TECHNIQUE FOR OPTIMIZING ONE-STAGE TWO-FILTER LINEAR-PHASE FIR FILTERS FOR SAMPLING RATE CONVERSION

Peyman Arian and Tapio Saramäki

Institute of Signal Processing, Tampere University of Technology
P. O. Box 553, FIN-33101 Tampere, Finland
e-mail: {peyman, ts}@cs.tut.fi

ABSTRACT

It is well-known that the computational complexity of a one-stage linear-phase finite-impulse response (FIR) decimator (interpolator) can be drastically reduced by using an additional linear-phase FIR filter at the output sampling rate (at the input sampling rate). The main difficulty in synthesizing these decimators and interpolators is to find the orders for these filters as well as their frequency-response shaping responsibilities for minimizing the overall number of multipliers per input (output) sample in the decimation (interpolation) case. This paper proposes a systematic approach for solving this problem.

1. INTRODUCTION

It has been observed by several authors [1]–[5] that the computational complexity of a one-stage linear-phase impulse response (FIR) decimator (interpolator) can be drastically reduced by using an addition filter stage at the output sampling rate (at the input sampling rate). Figures 1 and 2 show the resulting structures and their single-stage equivalents used for the analysis and synthesis purposes. This observation has been first made by Martinez and Parks in [6]. In their design scheme for decimators [1], $A(z)$ is a transfer function of a linear-phase FIR filter and $B(z)$ is an all-pole filter. The role of $A(z)$ is to shape the stopband in the desired manner, whereas $B(z)$ gives the desired response for the overall passband. This results in a significant reduction in the overall number of multiplications per input sample at the expense of a nonlinear-phase performance in the passband. In order to achieve a linear-phase performance, Saramäki modified this approach by using a linear-phase FIR transfer function for $B(z)$ [4]. The resulting filters require a slightly higher number of multipliers per input sample. In [4], $B(z)$ has been designed to provide one zero at $z = -1$ in order to reduce the multiplication rate even further in the case where the stopband edge of the decimator is located at $\omega = \pi/D$.

In the above-mentioned two approaches, the role of $A(z)$ is mainly to take care of the stopband shaping, whereas the role of $B(z)$ is to generate the desired passband response. In [1], [2], Chu and Burrus have proposed an opposite strategy in designing linear-phase FIR decimators. In their design scheme, the goal is to meet the given overall criteria such that $A(z)$ has the minimum complexity. As has been observed by Saramäki in [5], the best results in terms of the multiplication rate are obtained between the above two extreme cases.

This work was supported by the Academy of Finland, project No. 44876 (Finnish Centre of Excellence Program (2000–2005)).

[1] In this contribution, we shall concentrate on the design of decimators. The results can be applied directly to the interpolation case since decimators and interpolators are dual structures.

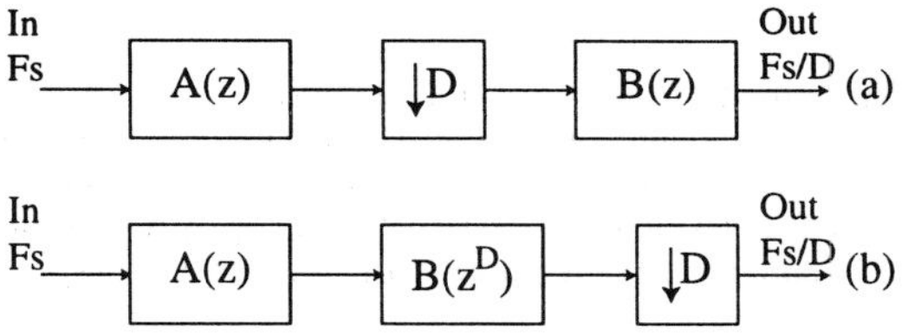

Figure 1: One-stage two-filter structure for a D-to-1 decimator. (a) Actual implementation. (b) Single-stage equivalent.

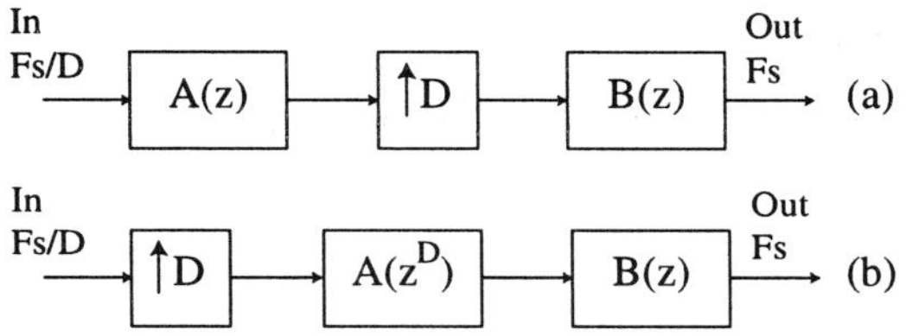

Figure 2: One-stage two-filter structure for a 1-to-D interpolator. (a) Actual implementation. (b) Single-stage equivalent.

The main purpose of this paper is to give a systematic approach for designing the decimator structure of Figure 1(a) in such a manner that the overall number of multipliers is reduced.[2] We concentrate on the case where both $A(z)$ and $B(z)$ are linear-phase FIR filters. In [5], the optimum solution has been found by trying various frequency-response-shaping responsibilities between $A(z)$ and $B(z)$ and then minimizing their orders for each selection. Since there are a large number of alternatives, this approach is very time consuming. Furthermore, it is very difficult to generate a systematic design scheme for automatically finding the optimum solution based on this approach.

2. PROPOSED CLASS OF DECIMATORS

This section introduces the proposed class of decimators. First, the transfer function is considered. Then, the zero-phase impulse response for it is expressed for the optimization purposes.

[2] It should be pointed out that in the case of multistage decimator, the best results are usually obtained in the case where the filter stage working at the output sampling rate takes care of the passband shaping only or has a single zero at $z = -1$ in the case where the stopband edge of the overall decimator filter is located at $\omega = \pi/D$. Therefore, these cases are not worth studying.

2.1. Decimator Transfer Function

The transfer function of the proposed linear-phase FIR decimators is of the form

$$H(z) = A(z)B(z^D), \quad (1)$$

where

$$A(z) = \sum_{n=0}^{N_A} a[n]z^{-n} \quad (2)$$

with $a[N_A - n] = a[n]$ for $n = 0, 1, \cdots, N_A$ and

$$B(z) = \sum_{n=0}^{N_B} b[n]z^{-n} \quad (3)$$

with $b[N_B - n] = b[n]$ for $n = 0, 1, \cdots, N_B$.

Here, D is the sampling rate conversion ratio. When this filter is used for decimation, $B(z^D)$ is realized as $B(z)$ at the lower output sampling rate as shown in Figure 1(a). This reduces significantly the number of delay elements required in the implementation.

2.2. Zero-Phase Frequency for the Decimator Transfer Function

The zero-phase frequency response for the above transfer function is expressible as (the phase term $e^{-j\omega(N_A+DN_B)/2}$ is omitted from the frequency response)

$$H(\omega) = A(\omega)B(D\omega), \quad (4)$$

where

$$A(\omega) = \begin{cases} \left[\frac{N_A}{2}\right] + 2\sum_{n=1}^{N_A/2} a[N_A - n]\cos(n\omega) & \text{for } N_A \text{ even} \\ 2\sum_{n=0}^{\frac{N_A-1}{2}} \left[\frac{N_A-1}{2} - n\right]\cos\left(\frac{2n\omega+1}{2}\right) & \text{for } N_A \text{ odd} \end{cases} \quad (5)$$

and

$$B(\omega) = \begin{cases} \left[\frac{N_B}{2}\right] + 2\sum_{n=1}^{N_B/2} b[N_B - n]\cos(n\omega) & \text{for } N_B \text{ even} \\ 2\sum_{n=0}^{\frac{N_B-1}{2}} \left[\frac{N_B-1}{2} - n\right]\cos\left(\frac{2n\omega+1}{2}\right) & \text{for } N_B \text{ odd.} \end{cases} \quad (6)$$

3. STATEMENT OF OPTIMIZATION PROBLEM FOR THE PROPOSED CLASS OF DECIMATORS

This section states the optimization problem for the proposed decimators.

When exploiting the coefficient symmetries of $A(z)$ and $B(z)$ and utilizing the fact that only every Dth output sample of $A(z)$ has to be evaluated, the overall number of multipliers per input sample becomes

$$R_M = \lfloor (N_A + 2)/2 \rfloor + \lfloor (N_B + 2)/2 \rfloor, \quad (7)$$

where $\lfloor x \rfloor$ stands for the integer part of x.

We consider the following criteria for $H(\omega)$, as given by Equations (4), (5), and (6):

$$1 - \delta_p \le H(\omega) \le 1 + \delta_p \text{ for } \omega \in [0, \alpha\pi/D] \quad (8)$$

$$-\delta_s \le H(\omega) \le \delta_s \text{ for } \omega \in [\pi/D, \pi], \quad (9)$$

where D is the decimation ratio and $\alpha < 1$ specifies the passband region. Alternatively, these criteria can be expressed as

$$|E(\omega)| \le \delta_p \text{ for } \omega \in [0, \alpha\pi/D] \cup [\pi/D, \pi], \quad (10)$$

where

$$E(\omega) = W(\omega)[H(\omega) - D(\omega)] \quad (11)$$

with

$$D(\omega) = \begin{cases} 1 & \text{for } \omega \in [0, \alpha\pi/D] \\ 0 & \text{for } \omega \in [\pi/D, \pi] \end{cases} \quad (12)$$

and

$$W(\omega) = \begin{cases} 1 & \text{for } \omega \in [0, \alpha\pi/D] \\ \delta_p/\delta_s & \text{for } \omega \in [\pi/D, \pi]. \end{cases} \quad (13)$$

The optimization problem under consideration is the following: *Optimization Problem:* Given D, α, δ_p, and δ_s, find the orders and coefficients of $A(z)$ and $B(z)$, as given by Equations (2) and (3), to meet the criteria given by Equations (8) and (9) such that, first, R_M, as given by Equation (7), is minimized and, second,

$$\epsilon = \max_{\omega \in [0, \alpha\pi/D] \cup [\pi/D, \pi]} |E(\omega)| \quad (14)$$

is minimized.

4. DESCRIPTION OF THE PROPOSED OPTIMIZATION ALGORITHM

This section describes the proposed algorithm for finding the optimum solution to the problem stated in the previous section.

4.1. Sub-algorithm Used in the Main Algorithm

Before describing the overall algorithm, a sub-algorithm is introduced. Given the decimator criteria as well as N_A and N_B, the orders of $A(z)$ and $B(z)$, this algorithm is carried out using the following three steps:

- Step 1: Use the Remez algorithm or linear programming to determine the coefficients of $B(z)$ of order N_B to minimize

$$\tilde{\delta}_p = \max_{\omega \in [0, \alpha\pi]} |B(\omega) - 1| \quad (15)$$

  subject to the condition $B(\omega) = 0$ at $\omega = \pi$.

- Step 2: Use the Remez algorithm or linear programming to determine the coefficients of $A(z)$ of order N_A to minimize

$$\tilde{\delta}_s = \max_{\omega \in [\pi/D, \pi]} |A(\omega)B(D\omega)| \quad (16)$$

  subject to the condition $A(\omega) = 1$ at $\omega = 0$.

- Step 3: Use sequential quadratic programming to determine simultaneously the coefficients of $A(z)$ and $B(z)$ to minimize ϵ, as given by Equation (14).

4.2. Main Algorithm

Based on the use of the above sub-algorithm, the overall procedure is performed using the following steps:

- Step 1: Use the Remez algorithm to find the minimum-order linear-phase FIR filter to meet the criteria of Equations (8) and (9). Let this order be N_{min}. Determine an initial guess for N_B as $N_B^{(1)} = \lceil N_{min}/D \rceil$ [3].
- Step 2: Use the above sub-algorithm with a small number of grid points for designing various overall transfer functions $H(z)$, as given by the Equation (1), for the fixed $N_B = N_B^{(1)}$. First use $N_A = D$ and increment it by D until the overall stopband ripple becomes less than or equal to $2\delta_s$. Save this order as $N_A^{(2)}$.
- Step 3: Use the sub-algorithm with a small number of grid points in such a way that $N_A = N_A^{(2)}$ is fixed and $N_B = N_B^{(1)}$ or is incremented by 1 until the passband ripple becomes less than or equal to $1.2\delta_p$. Save this order as $N_B^{(3)}$.
- Step 4: Use the sub-algorithm with a small number of grid points in such a way that $N_B = N_B^{(3)}$ is fixed and $N_A = N_A^{(2)}$ or is incremented by 1 until the stopband ripple becomes less than or equal to $1.15\delta_s$. Save this order as $N_A^{(4)}$.
- Step 5: Use the sub-algorithm with a small number of grid points in such a way that $N_A = N_A^{(4)}$ is fixed and $N_B = N_B^{(3)}$ or is incremented by 1 until the passband ripple becomes less than or equal to $1.1\delta_p$. Save this order as $N_B^{(5)}$.
- Step 6: Form candidate order pairs (N_A, N_B) for the values of N_B in the range $N_B^{(5)} \leq N_B \leq N_B^{(5)} + J$ with J being an integer. To do this, the sub-algorithm with a small number of grid points is used for each N_B in the above range in order to find the minimum value of N_A to meet the given criteria. A good initial guess for N_A is $N_A = N_A^{(4)}$. Save the resulting pairs (N_A, N_B).
- Step 7: Apply the sub-algorithm with a high number of grid points in order to check whether the given criteria are met by each candidate pair (N_A, N_B). If this is not true for some order pairs, then increment N_A until the criteria are met. Save the resulting pairs (N_A, N_B).
- Step 8: Select among the order pairs the one minimizing R_M as given by Equation (7). If there are more than one pair giving the same minimum value, then select the pair having the smallest value of N_B.

In the above algorithm, fewer grid points are used at Steps 2, 3, 4, 5, and 6 in order to make them faster. At these steps there is no need to use a large number of grid points since we are looking for the potential candidates and not for the final optimum solution. The choice of the number of grid points depends strongly on the specifications, particularly on the width of the transition band. Our simulations suggest that in the few-grid-point stages of the algorithm, the proper number of grid points in the passband and stopband are

$$M_p = \left\lfloor \frac{\pi}{\omega_s - \omega_p} \right\rfloor \tag{17}$$

and

$$M_s = \left\lfloor \frac{\pi(\pi - \omega_s)}{\omega_p(\omega_s - \omega_p)} \right\rfloor, \tag{18}$$

respectively. For the steps with a high number of grid points, M_p and M_s are multiplied by four.

[3] $\lceil x \rceil$ is the smallest integer larger than or equal to x.

5. NUMERICAL EXAMPLES

This section illustrates, by means of two examples taken from the literature, the efficiency and flexibility of the proposed algorithm.

5.1. Example 1

Consider the decimator specifications [5]: $D = 10$, $\alpha = 0.5$, $\delta_p = 0.01$, and $\delta_s = 0.001$. This means that the passband and stopband edges are located at $\omega_p = 0.5\pi/D$ and at $\omega_s = \pi/D$ respectively.

The PC on which the simulation was carried out was a Pentium 1.6 GHz with 512 Mb of memory. For simplicity, for all the steps of the sub-algorithm of Subsection 4.1 a function `fminimax` from the optimization toolbox provided by MathWorks Inc. has been used [7].

First the algorithm described in the previous section starts with the order of $B(z)$ being equal to 11 (end of Step 1) and that of $A(z)$ being equal to 10. After 5 seconds, the orders of $A(z)$ and $B(z)$ become 40 and 11, respectively (end of Step 2). At this stage, the stopband ripple is less than or equal to $2\delta_s$.

It turns out that the same orders of $A(z)$ and $B(z)$ make the passband ripple less than $1.2\delta_p$. Therefore, Step 3 is skipped. At the end of Step 4, which takes only 3 seconds, the orders of $A(z)$ and $B(z)$ become 39 and 11, respectively. After 2 seconds, Step 5 is finished, resulting in $A(z)$ and $B(z)$ of orders 39 and 12, respectively. Step 6 takes 12 seconds and at the end of this step, we have the following pairs of candidates: (39,12), (37,13), and (38,14). The first number of each pair represents the order of $A(z)$ and the second one represents the order of $B(z)$. In our simulation, we used $J = 2$.

At the end of Step 7, we have the candidate pairs (40,12), (38,13), and (38,14). The last stage is the selection of the best result corresponding to Step 8. The pair (38,13) is the solution resulting in the smallest R_M, as given by Equation (7), and is, therefore, the optimal solution. Various responses for the optimum overall filter with $N_A = 38$ and $N_B = 13$ are depicted in Figures 3 and 4.

The optimized filter requires 27 multipliers, 51 delay elements, and 2.7 multiplications per input sample. The minimum order for the optimized direct-form linear-phase FIR filter to meet the given criteria is 108. When exploiting the coefficient symmetry and using the fact that only every tenth output sample has to be evaluated, then the resulting decimator requires 55 multipliers, 108 delay elements, and 5.5 multiplications per input sample. Based on the above values, the proposed two-stage two-filter decimator provides an excellent performance when compared with its direct-form counterpart.

5.2. Example 2

If we apply the same procedure for the following specifications [1], [3]–[6]: $D = 20$, $\delta_p = 0.05$, $\delta_s = 0.005$, $\omega_p = 0.045\pi$, and $\omega_s = 0.05\pi$, then the algorithm of Section 4 shows that the optimimum solution is achieved by $N_A = 105$ and $N_B = 40$. The corresponding decimator requires 74 multipliers, 145 delay elements, and 3.7 multiplications per input sample. The corresponding figures for the direct-form linear-phase FIR filter of order 652 exploiting the coefficient symmetry and the fact that only every twentieth output sample has to be evaluated are 327, 652, and 16.35. Once again the superiority of the proposed design is clearly evident. Figures 5 and 6 show some responses for the optimized design.

5.3. Comparison of the Proposed Algorithm with Other Existing Algorithms

The results obtained using the proposed algorithm are the optimum ones and better than those resulting when using the design schemes

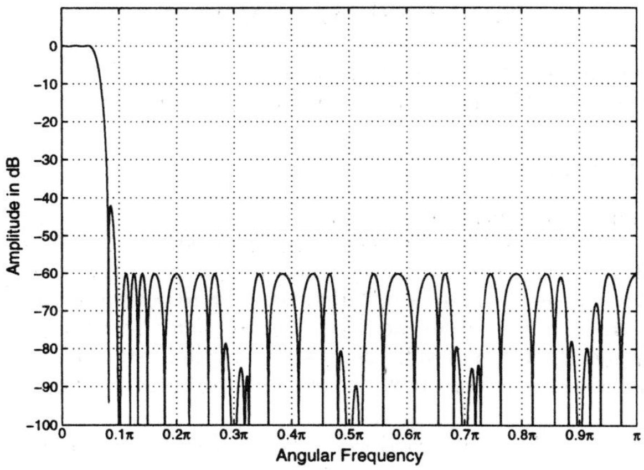

Figure 3: Amplitude response of the optimal one-stage two-filter decimator of Example 1.

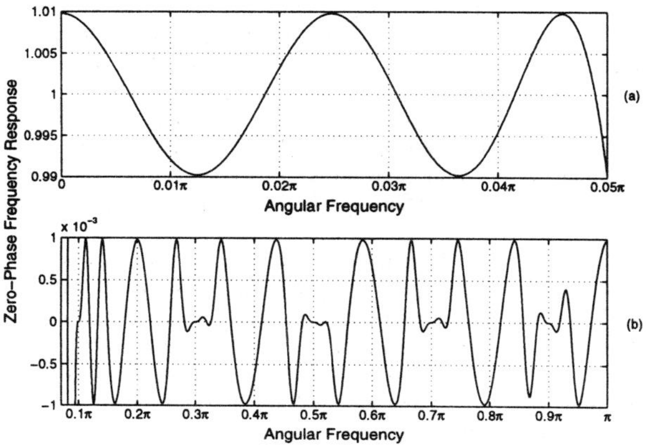

Figure 4: Zero-phase frequency response of the optimal one-stage two-filter decimator of Example 1. (a) Passband details. (b) Stopband details.

considered in Introduction. Although they do not excel those provided by Saramäki [5] in terms of filter orders, the simplicity of the proposed algorithm makes it an attractive alternative for designing one-stage two-filter linear-phase FIR filters structures for sampling rate alteration purposes.

6. CONCLUSION

This paper has introduced a systematic approach for designing one-stage two-filter linear-phase FIR filters for sampling rate conversion. The earlier design techniques are either sub-optimal or lack systematism. The proposed algorithm has turned out to find the optimal solution for various criteria. Furthermore, as long as the transition band is not too sharp, the optimal solution can be found quite fast, as has been shown in Section 5. Future work is devoted to making the proposed algorithm faster. For this purpose, it is crucial to find rather accurate estimates for the orders of the subfilters in Figure 1(a) that minimize the number of multiplications per input sample.

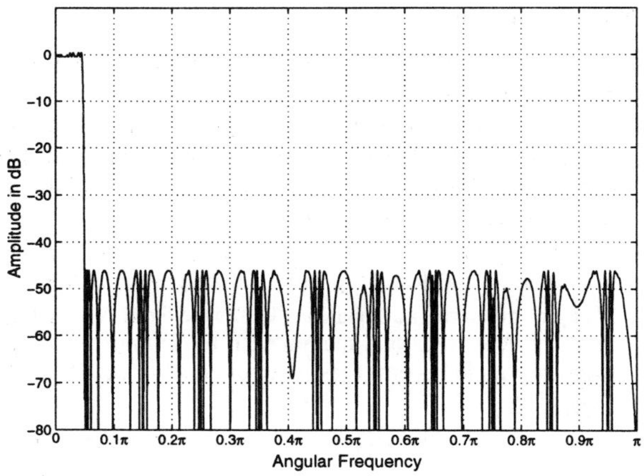

Figure 5: Amplitude response of the optimal one-stage two-filter decimator of Example 2.

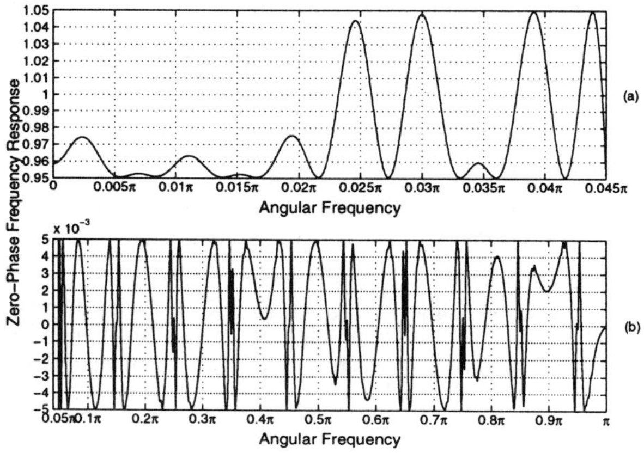

Figure 6: Zero-phase frequency response of the optimal one-stage two-filter decimator of Example 2. (a) Passband details. (b) Stopband details.

7. REFERENCES

[1] S. Chu and C. S. Burrus, "Optimum FIR and IIR multistage multirate filter design," *Circuits, Systems, and Signal Processing,* vol. 2, pp. 361–386, No. 3, 1983.

[2] S. Chu and C. S. Burrus, "Multirate filter designs using comb filters," *IEEE Trans. Circuits Syst.,* vol. CAS-31, pp. 913–924, Nov. 1984.

[3] J.-K. Liang and R. J. P. de Figueiredo, "A new class of nonlinear phase FIR digital filters and its application to efficient design of multirate digital filters," *IEEE Trans. Circuits Syst,* vol. CAS-32, pp. 944–948, Sept. 1985.

[4] T. Saramäki, "A class of linear-phase FIR filters for decimation, interpolation, and narrow-band filtering," *IEEE Trans. Acoust., Speech, Signal Processing,* vol. ASSP-32, pp. 1023–1036, Oct. 1984.

[5] T. Saramäki, "Design of optimal multistage IIR and FIR filters for sampling rate alteration," in *Proc. IEEE Int. Symp. Circuits Syst.* (San Jose, CA), pp. 227–230, May 1986.

[6] H. G. Martinez and T. W. Parks, "A class of infinite-duration impulse response digital filters for sampling rate reduction," *IEEE Trans. Acoust., Speech, Signal Processing,* vol. ASSP-27, pp. 154–162, Apr. 1979.

[7] T. Coleman, M. A. Branch, and A. Grace, *Optimization Toolbox User's Guide,* The MathWorks, Inc., Version 2, 1999.

CONCISE REPRESENTATION AND CELLULAR STRUCTURE FOR UNIVERSAL MAXIMALLY FLAT FIR FILTERS

Saed Samadi

Concordia University

Montreal, QC, Canada

Akinori Nishihara

The Center for Research and Development of Educational Technology

Tokyo Institute of Technology
Tokyo, 152-8552 Japan

ABSTRACT

The universal maximally flat lowpass FIR filters of Baher possess closed-form expressions for their transfer function. The expressions involve binomial coefficients and nested sums. We show that there exists a concise formula for the universal maximally flat lowpass filter $H_{N,K,d}(z)$ in the form of the Nth power of a linear algebraic operator acting on a finite-length sequence. The linear operator involves the forward shift operator, widely used in numerical analysis and the theory of finite differences. We also employ the formula to develop a hierarchical cellular structure for realization of arbitrary universal maximally flat filters. The structure is comprised of identical cells that are interconnected regularly. The structure is especially suitable for a variable realization where the values of the parameters can be varied by adding or deleting extra cells or changing the value of a single multiplier coefficient.

1. INTRODUCTION

Two of the most outstanding contributions to the theory of lowpass maximally flat FIR filters have been due to O. Herrmann [1] and H. Baher [2]. In his 1970 article, Herrmann gives the first closed-form formula for the transfer function of linear phase maximally flat FIR filters. Baher's 1982 results on the design of filters with simultaneous conditions on the amplitude and group-delay response, on the other hand, generalize Herrmann's results and form a broad class of maximally flat filters [3]. This class, encompassing various types of FIR systems, is parameterized by N, the order, K, the number of zeros at $z = -1$, and a real-valued parameter denoted by d. The role of d is to control the value of the group delay at $\omega = 0$. The authors have simplified Baher's formula for the transfer function of the class and showed that lowpass filters of even or odd lengths, with linear or nonlinear phase response characteristics, as well as fractional delay and half-band filters may be readily obtained by setting N, K, and d in an appropriate manner. The class of filters is referred to as the universal maximally flat digital filters in appreciation of the breadth of its members. Other results show that Baher's universal maximally flat filters are exact and optimal for delaying and upsampling of polynomial signals [4].

The formulas and expressions presented in [3], facilitate the design of the filters provided that a computer algebra system or mathematical routines for handling polynomials and binomial coefficients are available. The filters may then be realized using any one of the techniques available for implementation of FIR filters. In [3], it has also been shown that for certain values of the parameters, the filters may be realized using a dedicated multiplier-free structure. The structure is an array of simple cells that are interconnected in a regular manner. Existence of such multiplier-free structures was first discovered in [5] for the linear-phase filters of Herrmann. Besides the virtue of being multiplier-free, these dedicated cellular structures are hierarchical and lend themselves to scalable realizations. A scalable realization allows the parameters, N and K for this structure, to be changed actively by adding or deleting extra cells. Thus comes the term hierarchical, meaning that for fixed values of N and K, outputs taken from certain cells at the upper levels belong to filters with lower values of N or K.

In view of the current knowledge on the universal maximally flat filters, we are still faced with two open questions. The first question arises from the recognition of the fact that the existing simplified formulas for the transfer function are still unwieldy. Thus, we are interested in knowing whether other forms of closed-form expressions are available. Among the desirable properties of such alternative formulas are conciseness and freedom from binomial coefficients. The second question is concerned with the existence of dedicated cellular structures, with desirable properties such as scalability, for arbitrary members of the class. We naturally predict that for an arbitrary member of the class, the parameter d takes on any real value and, thus, a general multiplier-free structure is a fictitious entity. However, an investigation for a cellular structure that is controlled by real-valued multiplier coefficients is a realistic cause. The objective of this paper is to derive a concise closed-form formula for the universal maximally flat filters that is free from the cumbersome binomial coefficients. We start by deriving a recurrence relation for generation of a sequence that forms the coefficients of the Bernstein-form representation of the transfer function. The Bernstein-form representation of the transfer function has been fully developed in [3] and we build on the results obtained therein. Then, a simple method for converting the Bernstein-form polynomials, expressed over the bases z and $1-z$, to the power form is introduced. Subsequently, a variance of this method is derived to convert the representation of the transfer function at hand to a concise formula. This formula is in the form of the Nth power of a first-degree linear operator. An accompanying cellular structure to the formula is then developed. The structure is a triangular interconnection of simple cells and is scalable and variable with respect to K and d, respectively. A detailed description of the main results is given in Section 2. A cellular structure is developed in Section 3, and conclusions are drawn in Section 4.

0-7803-7761-3/03/$17.00 ©2003 IEEE

2. MAIN RESULTS

The transfer function of the universal maximally flat digital filters is given by [3]

$$H_{N,K,d}(z) = \sum_{j=0}^{N-K} b_j \left(\frac{1-z^{-1}}{2}\right)^j \left(\frac{1+z^{-1}}{2}\right)^{N-j}. \quad (1)$$

The three-term recurrence

$$j\, b_j + 2d\, b_{j-1} - (j - N - 2)\, b_{j-2} = 0, \quad j \geq 1, \quad (2)$$

characterizes the coefficients b_j together with the initial values $b_0 = 1$ and $b_{-1} = 0$ The impulse response coefficients h_k that are the coefficients of the power-form representation

$$H_{N,K,d}(z) = \sum_{k=0}^{N} h_k z^{-k}, \quad (3)$$

then become [3]

$$h_k = \sum_{j=0}^{N-K} \sum_{p=0}^{k} \sum_{i=0}^{j} \frac{(-1)^{j-i+p} \binom{\frac{N}{2}-d}{i} \binom{\frac{N}{2}+d}{j-i} \binom{j}{p} \binom{N-j}{k-p}}{2^N}, \quad (4)$$

$k = 0, \ldots, N$. Let us write (1) in the traditional Bernstein form by introducing the sequence b'_j defined by

$$b'_j \stackrel{\text{def}}{=} \begin{cases} \dfrac{b_j}{\binom{N}{j}}, & 0 \leq j \leq N, \\ 0, & \text{otherwise.} \end{cases} \quad (5)$$

Then we have

$$H_{N,K,d}(z) = \sum_{j=0}^{N-K} b'_j \binom{N}{j} \left(\frac{1-z^{-1}}{2}\right)^j \left(\frac{1+z^{-1}}{2}\right)^{N-j}. \quad (6)$$

This is the Bernstein form of the transfer function in the traditional sense. For a treatment of the Bernstein forms and related topics see [6, 7]. The next step is to characterize the coefficients b'_j. This can be done by substituting (5) into (2) and dividing both sides by $j\binom{N}{j} \neq 0$. It follows that

$$b'_j + 2d \frac{\binom{N}{j-1}}{j\binom{N}{j}} b'_{j-1} - (j-N-2) \frac{\binom{N}{j-2}}{j\binom{N}{j}} b'_{j-2} = 0, \quad 1 \leq j \leq N. \quad (7)$$

By invoking the definition of the binomial coefficients [8]

$$\binom{r}{k} \stackrel{\text{def}}{=} \begin{cases} \frac{r^{\underline{k}}}{k!} = \frac{r(r-1)\ldots(r-k+1)}{k(k-1)\ldots(1)}, & \text{integer } k \geq 0 \\ 0 & \text{integer } k < 0 \end{cases} \quad (8)$$

and after a few algebraic steps, we can write

$$b'_j = \frac{-2d}{N-j+1} b'_{j-1} - \frac{j-1}{N-j+1} b'_{j-2}, \quad 1 \leq j \leq N, \quad (9)$$

with the initial values $b'_0 = 1$ and $b'_{-1} = 0$. Thus, for given values of N and d we can compute the values of b'_j recursively by (9). Note that the coefficients b'_j form a sequence of polynomials in d with rational coefficients.

The next step is to convert (6) to the power-form (3). A direct expansion of the summand in (6) using the binomial theorem will introduce additional sums and new binomial coefficients. The other alternative is to use a conversion identity well-known in the mathematics literature and in the field of computer-aided geometric design. The identity is based on taking successive differences of the Bernstein coefficients b'_j. We state this identity as a proposition and provide a proof for it.

Proposition 1 Let the Bernstein form of the polynomial

$$p(x) = \sum_{i=0}^{N} p_i x^i$$

be given by

$$p(x) = \sum_{i=0}^{N} c_i \binom{N}{i} x^i (1-x)^{N-i}. \quad (10)$$

Then

$$p_i = \binom{N}{i} \Delta^i c_0, \quad i = 0, \ldots, N, \quad (11)$$

where the operator Δ is the forward difference operator.

The forward difference operator acts on the "subscript" of a sequence and is defined by the general recurrence

$$\Delta^j c_i = \Delta(\Delta^{j-1} c_i), \quad (12)$$

together with

$$\Delta^1 c_i = \Delta c_i = c_{i+1} - c_i, \quad \Delta^0 c_i = c_i. \quad (13)$$

For instance,

$$\Delta c_0 = c_1 - c_0 \quad (14)$$

and

$$\Delta^2 c_0 = c_2 - 2c_1 + c_0. \quad (15)$$

There exists a well-known explicit formula for $\Delta^j c_i$ that we refrain from quoting here. We however reiterate that, in our notation, the operator only acts on the indices, written as subscripts, of the members of a sequence. Hence, the operator has no effect on the terms involving z or constants. To prove the proposition, we need the assistance of the forward shift operator E. The effect of E on the index i of a sequence c_i is specified by

$$E^j c_i = E(E^{j-1} c_i). \quad (16)$$

together with

$$E^1 c_i = E c_i = c_{i+1}, \quad E^0 c_i = c_i. \quad (17)$$

E is related to Δ by

$$\Delta = E - 1, \quad (18)$$

where 1 denotes the identity operator taking a sequence to the same sequence. It can be readily verified that

$$E^j c_i = c_{i+j}. \quad (19)$$

The following is a proof of the proposition.[1]

Proof. We start with the Bernstein form of $p(x)$ and expand the term $(1-x)^{N-i}$ using the binomial theorem to obtain

$$p(x) = \sum_{i=0}^{N} \sum_{j=0}^{N-i} (-1)^{N-i-j} c_i \binom{N}{i} \binom{N-i}{j} x^{N-j}. \quad (20)$$

[1] An anonymous reviewer has mentioned that an alternative proof can be found in [6].

By the symmetry of the binomial coefficients and the trinomial revision property [8], we have

$$\binom{N}{i}\binom{N-i}{j} = \binom{N}{j}\binom{N-j}{N-i-j}. \tag{21}$$

On the other hand

$$c_i = E^i c_0. \tag{22}$$

Thus, we can write

$$p(x) = \sum_{i=0}^{N}\sum_{j=0}^{N-i}(-1)^{N-i-j}\binom{N}{j}\binom{N-j}{N-i-j} x^{N-j} E^i c_0. \tag{23}$$

The ranges for the indices of the two-fold sum above may be interchanged as

$$\sum_{i=0}^{N}\sum_{j=0}^{N-i} = \sum_{j=0}^{N}\sum_{i=0}^{j} \tag{24}$$

with no effect on the value of the sum. Now, we substitute $N-j$ for j in the sum and obtain

$$p(x) = \sum_{j=0}^{N}\sum_{i=0}^{j}(-1)^{j-i}\binom{N}{j}\binom{j}{j-i} x^j E^i c_0. \tag{25}$$

Note that

$$\sum_{i=0}^{j}(-1)^{j-i}\binom{j}{j-i} x^j E^i c_0 = (E-1)^j c_0. \tag{26}$$

It then follows that

$$p(x) = \sum_{j=0}^{N} x^j \binom{N}{j} (E-1)^j c_0 \tag{27}$$

that is the power form representation of $p(x)$. From the uniqueness of the power-form representation of a polynomial and the relation between the operators E and Δ it follows that $p_i = \binom{N}{i}\Delta^i c_0$. ∎

We now present our main result before proceeding to the development of a variable cellular architecture.

Corollary 1 The transfer function $H_{N,K,d}(z)$ is given by

$$H_{N,K,d}(z) = (\frac{1+E}{2} + \frac{1-E}{2}z^{-1})^N b_0'. \tag{28}$$

Proof. Let

$$z^{-1} = 1 - 2x. \tag{29}$$

Then, from (6), we can write

$$H_{N,K,d}(x) = \sum_{j=0}^{N-K} b_j' \binom{N}{j} x^j (1-x)^{N-j}. \tag{30}$$

By the above proposition, we have

$$H_{N,K,d}(x) = \sum_{j=0}^{N}\binom{N}{j}\Delta^j b_0' x^j. \tag{31}$$

By substituting for x in terms of z^{-1} in the above relation and expanding the summand using the binomial theorem, we obtain

$$H_{N,K,d}(z) = \sum_{j=0}^{N}\binom{N}{j}\Delta^j b_0' \sum_{i\le j}\binom{j}{i} 2^{-j}(-z^{-1})^i. \tag{32}$$

By the trinomial revision identity we have

$$\binom{N}{j}\binom{j}{i} = \binom{N}{i}\binom{N-i}{j-i}. \tag{33}$$

We can also modify the sum indices and write

$$H_{N,K,d}(z) = \sum_{i=0}^{N}\binom{N}{i}(\frac{-\Delta z^{-1}}{2})^i \sum_{i\le j} 2^{-(j-i)}\binom{N-i}{j-i}\Delta^{j-i} b_0'. \tag{34}$$

Using $p = j - i$ as an index for the second sum, we obtain

$$H_{N,K,d}(z) = \sum_{i=0}^{N}\binom{N}{i}(\frac{-\Delta z^{-1}}{2})^i \sum_{p=0}^{N-i}\binom{N-i}{p}(\frac{\Delta}{2})^p b_0'. \tag{35}$$

This simplifies to

$$H_{N,K,d}(z) = \sum_{i=0}^{N}\binom{N}{i}(\frac{-\Delta z^{-1}}{2})^i (1+\frac{\Delta}{2})^{N-i} b_0'. \tag{36}$$

By invoking the binomial theorem, we can further write

$$H_{N,K,d}(z) = (1 + \frac{\Delta}{2} - \frac{\Delta}{2}z^{-1})^N b_0'. \tag{37}$$

By using the relation between Δ and E, (28) follows. ∎

It is worthwhile to note that there is an implicit assumption on the coefficients b_j' above. Specifically, it is assumed that

$$b_j' = 0 \quad \text{for} \quad j > N-K. \tag{38}$$

This explains the reason behind the absence of the parameter K from the surface of (28). The absence of K also contributes to the fact that (28) is an intriguingly concise expression. Unlike the sum-based conventional formulas of [3], (28) is a factored operator-based expression. Moreover, it does not involve binomial coefficients. A direct application of such expressions is in developing dedicated architectures for the filters. The following is a treatment where particular interest is given to the variable and adaptive aspects of the structure.

3. DEDICATED CELLULAR STRUCTURE

Let us assume that an $(N+1) \times (N+1)$ grid of computational nodes is available. Also assume that each node is labeled as $P(i,j)$ for $i,j = 0,\ldots,N$. We can construct a structure based on (28) as a signal flow graph over the nodes of the grid. The structure has two parts. Part I forms the coefficient sequence b_j' upon which an operator will act in part II. By running the recurrence (9) from $j=1$ up to $j = N-K$ the sequence

$$\mathbf{B}' = (1, b_1' \ldots, b_{N-K}', 0, \ldots, 0). \tag{39}$$

is obtained. The entries b_j' of $\mathbf{B}'$ for $N-K < j \le N$ are set to zero following (38). In the form of a signal flow graph, the sequence $\mathbf{B}'$ is realized as $N-K+1$ nodes, $P(0,0)\ldots P(0,N-K)$, that are connected regularly by branches having appropriate weights. These nodes are then followed by K zero-valued nodes $P(0,N-K+1)\ldots P(0,N)$. As can be seen from (9), the weights on the branches are functions of N and d. The explicit relation for the jth node is given by

$$P[0,j] = \frac{-1}{N-j+1}\big(2d\,P[0,j-1] + (j-1)P[0,j-2]\big), \quad (40)$$

where $j = 1, \ldots, N-K$. For a fixed N, the value of d can be controlled by a single multiplier coefficient on each branch.

The second part of the structure consists of N layers of nodes. The ith layer of nodes is formed by the action of the operator

$$\frac{1+E}{2} + \frac{1-E}{2}z^{-1} \quad (41)$$

on the $(i-1)$th layer. The action of operator E is realized by moving from the current node $P(i,j)$ to the neighboring $P(i,j+1)$. Let us assume that the $(i-1)$th layer of nodes has already been formed. For $i = 1$ this is nothing but the 0th layer. The relation specifying the interconnection of the nodes on two subsequent layers is obtained by using (41) to write

$$P(i,.) = \left(\frac{1+E}{2} + \frac{1-E}{2}z^{-1}\right)P(i-1,.) \quad (42)$$

where the value of j is intentionally left undetermined. A more precise expression for the nodes in the ith layer is obtained by taking the action of operator E into account. It follows that

$$\begin{aligned} P(i,j) \;=\; & \frac{P(i-1,j)+P(i-1,j+1)}{2} + \\ & \frac{P(i-1,j)-P(i-1,j+1)}{2}z^{-1}. \end{aligned} \quad (43)$$

The output should then be taken from the Nth layer at the node $P(N,0)$. This is in accordance with (28) that requires the computation of the value of the zero-th coefficient after N consecutive action of the operator. Hence, we do not need to construct all the $N+1$ nodes of each layer. Only those nodes that are needed for the formation of the node $P(0,N)$ are required. Consequently, the recurrence (43) is governed by

$$\begin{aligned} i &= 1, \ldots, N \\ j &= 0, \ldots, N-i. \end{aligned} \quad (44)$$

Each node in the second part together with its incident branches can be integrated into a cell. Hence, the second part of the structure forms a regularly connected triangular cellular structure. An illustrative representation of the structure is presented in Fig. 1 for $N = 4$ and $K = 1$. From (40), it can be seen that the structure may be made variable with respect to d by changing the value of a single multiplier coefficient at each node in part I. For a fixed N the value of K can be varied by adding or deleting appropriate nodes from the second part so that the effect of zero-valued nodes of the first part are taken into account.

4. CONCLUSION AND REMARKS

We have shown that an alternative representation for the transfer function of universal maximally flat FIR filters exists. The representation is an operator-based expression that employs the forward shift operator. The expression is derived by means of a technique referred to as finite calculus in [8]. An important feature of the expression is that it is free from the binomial coefficients. It also enables us to develop a dedicated cellular structure for the filters.

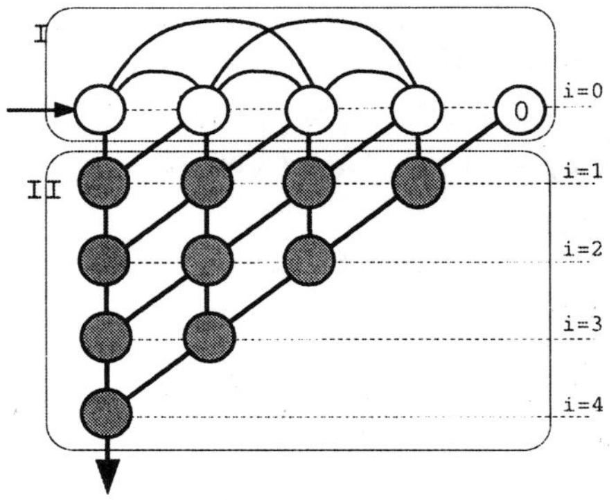

Figure 1: Illustrative representation of the structure for $N = 4$, $K = 1$.

The structure consists of two parts. In the first part, obtained by realizing a two-term recurrence relation, a layer of interdependent nodes is obtained. Each node in this layer is connected to two neighboring nodes by branches with appropriate weights. If the value of the group delay parameter d is a rational number, then the weights of the branches are rational numbers. Moreover, for $d = 0$ every other node is of value zero and no multiplication or addition is needed for its formation. This layer of nodes is then used in a simple N-step process to form a triangular structure. In each step, the nodes are formed by simple additions, subtractions and divisions by 2. The output is taken from the single node in the Nth layer. The value of K can be varied easily by changing the number of zero-valued nodes in the first part of the structure. Equivalently, the same task can be achieved by deleting appropriate nodes from the second part. The value of d can be varied by making the multiplier coefficient that realizes d in the first part a variable multiplier.

5. REFERENCES

[1] O. Herrmann, "On the approximation problem in nonrecursive digital filter design," *IEEE Trans. Circuit Theory,* vol. CT-18, no. 3, pp. 411–413, May 1971.

[2] H. Baher, "FIR digital filters with simultaneous conditions on amplitude and delay," *Electron. Lett.*, vol. 18, pp. 296–297, 1982.

[3] S. Samadi, A. Nishihara and H. Iwakura, "Universal maximally flat lowpass FIR systems," *IEEE Trans. Signal Processing,* vol. 48, pp. 1956–1964, 2000.

[4] S. Samadi and A. Nishihara, "Response of maximally flat lowpass filters to polynomial signals," *In Proc. ISCAS'2001,* Sydney, Australia.

[5] S. Samadi, T. Cooklev, A. Nishihara, and N. Fujii, "Multiplierless structure for maximally flat linear phase FIR digital filters," *Electron. lett.*, Vol. 29, No. 2, pp. 184–185, 1993.

[6] G. G. Lorentz, *Bernstein Polynomials.* University of Toronto Press, 1953

[7] R.T. Farouki and V.T. Rajan, "Algorithms for polynomials in Bernstein form," *Computer Aided Geometric Design,* vol. 5, pp. 1-26, 1988.

[8] R. L. Graham, D. E. Knuth and O. Patashnik, *Concrete Mathematics.* Addison-Wesley, 1989.

DESIGN AND IMPLEMENTATION OF A RECONFIGURABLE FIR FILTER

Kuan-Hung Chen and Tzi-Dar Chiueh

Graduate Institute of Electronics Engineering and Department of Electrical Engineering,
National Taiwan University, Taipei, Taiwan 10617

ABSTRACT

Finite impulse response (FIR) filters are very important blocks in digital communication systems. Many efforts have been made to improve the filter performance, e.g., less hardware and higher speed. In addition, software radio has recently gained much attention due to the need for integrated and reconfigurable communication systems. To this end, reconfigurability has become an important issue for the future filter design. In this paper, we present a digit-reconfigurable FIR filter architecture with the finest granularity. The proposed architecture is implemented in a single-poly quadruple-metal 0.35-μm CMOS technology. Measurement results show that the fabricated chip consumes 16.5 mW of power when operating at 86 MHz under 2.5 V.

1. INTRODUCTION

Finite impulse response (FIR) filters are important blocks in digital communication systems. These filters can be used to perform a wide variety of tasks such as spectral shaping, matched filtering, noise rejection, channel equalization, etc. As such, various architectures and implementation methods have been proposed to improve the performance of these filters in speed and complexity.

However, with explosive proliferation in communication standards, traditional hardwired devices may not be suitable for future communication needs. Software radio [1, 2] has gained much attention from the researchers worldwide due to a unanimous demand for integrated reconfigurable communication systems. Therefore, in the future, not only the speed and the complexity of filters, but also their programmability and reconfigurability should be considered.

It is well known that the canonical signed digit (CSD) representation can be used to reduce the complexity of FIR digital filter implementation [3, 4, 5]. Encoding the filter coefficients using the CSD representation reduces the number of partial products as well as silicon area and power consumption. Hence, it is a useful technique for implementation of FIR filters with fixed coefficients.

This work was supported in part by MediaTek Inc.

While applying the CSD representation to the implementation of programmable FIR filters, it is straightforward to require all filter taps to have the same number of programmable CSDs. However, for most filters, many taps do not require the highest coefficient precision. Valuable hardware resources will be wasted if all taps are implemented with the highest precision. To minimize the computational resources, there have been some works [6, 7] implementing programmable FIR filters by restricting the number of nonzero CSDs in each tap. However, the restriction lowers the coefficient precision and may degrade the performance of the filter. Since there are usually some taps that require fewer CSDs, computational resources are still wasted. Another hardware-efficient implementation of programmable FIR filters with CSD coefficients has been presented in [8]. A 32-tap linear-phase filter, with 2 nonzero CSDs in each tap, is implemented. Additional nonzero CSDs can be allocated to specific filter taps, making it a reconfigurable FIR filter architecture. Nevertheless, some computational resources are unused and the critical path can be quite long in some cases.

Due to the wide range of the filter coefficient precision for different applications, it is not easy to achieve reconfigurability without sacrificing simplicity. As an example, a matched filter that is very important for CDMA-based systems requires only 1-bit coefficient precision but a pulse-shaping filter may require as high a precision as 16 bits. Since the tap complexity in these two cases can be quite different, it is not efficient to construct a tap-based reconfigurable FIR filter.

In this paper, we adopt a fine granularity for filter implementation and develop a digit-reconfigurable FIR filter architecture with extreme flexibility. With this architecture, both the number of taps and the number of nonzero digits in each tap can be arbitrarily assigned given that enough hardware resource is available. The filter implemented with this architecture can be configured as a matched filter, a pulse-shaping filter, or other filters. Furthermore, the proposed FIR architecture also has scalability, modularity, and cascadability, making it amenable to VLSI implementation.

The rest of the paper is organized as follows. In Section 2, the reconfigurable FIR filter architecture is presented. Circuit implementation of the reconfigurable FIR filter chip is then presented in Section 3. In Section 4, physical de-

0-7803-7761-3/03/$17.00 ©2003 IEEE

sign and measurement results of the chip are given and discussed. Finally, Section 5 concludes this paper.

2. RECONFIGURABLE FIR FILTER ARCHITECTURE

At first, note that an N-tap FIR filter can be described as

$$y[n] = \sum_{i=0}^{N-1} h_i \cdot x[n-i]. \qquad (1)$$

If a coefficient, h_i, is expressed in the CSD format as $h_i = \sum_{k=0}^{M_i-1} d_{i,k} \cdot 2^{-p_k}$, we can rewrite (1) as

$$y[n] = \sum_{i=0}^{N-1} \sum_{k=0}^{M_i-1} d_{i,k} \cdot 2^{-p_k} \cdot x[n-i], \qquad (2)$$

where $d_{i,k} \in \{-1, 0, 1\}$, $p_k \in \{0, \ldots, L\}$; $L+1$ is the length of the coefficients; and M_i is the number of nonzero digits in h_i.

2.1. Digit Processing Unit

Most FIR filters that have been proposed are implemented using a tap as the basic building block. A tap is designed to evaluate the term $h_i \cdot x[n-i]$ in (1) and then several taps constitute an FIR filter. After examining (2) carefully, one sees that if a basic building block evaluates the term $d_{i,k} \cdot 2^{-p_k} \cdot x[n-i]$, then the flexibility on the number of taps and the number of nonzero digits in each tap can be achieved.

To meet this requirement, a digit processing unit (DPU), as shown in Figure 1, is designed. Control signals are serially shifted into a serial-in-parallel-out (SIPO) register (REG) array on the top during the initialization. In each DPU, three control signals, **plus**, **zero**, and **shift**, are derived from the corresponding digit in the ith tap coefficient, $d_{i,k} \cdot 2^{-p_k}$. The partial term, $d_{i,k} \cdot 2^{-p_k} \cdot x[n-i]$, is evaluated by the multiplier and the shifter. Another control signal, **config**, controls the multiplexer to select either the buffered or the unbuffered input as the output.

By cascading the DPUs, appropriately configuring the multiplexer in each DPU, and summing up the outputs of the DPUs and the accumulated sum as shown in Figure 2, we can implement an FIR filter with variable number of CSDs in each tap. For the last digit of each tap, the multiplexer in the corresponding DPU selects the buffered input as the output. Even though the architecture depicted in Figure 2 can be implemented directly, usually, it is necessary to insert pipeline registers in the filter to achieve reasonable speed performance.

2.2. Reconfigurable FIR Filter Chip

To illustrate the feasibility of the proposed reconfigurable FIR filter architecture, we designed a reconfigurable 8-DPU FIR chip based on Figure 2. Detailed architecture is shown in Figure 3 and is referred to as a **processing element** (**PE**). In order to reduce the area and the latency required for implementing the addition in the filtering process, eight adders are combined to form a single big adder with nine inputs. With different precision in the output of the DPU (15-bit) and the accumulated sum (24-bit), we introduce a special sign extension generator. It generates the sum of sign extension bits of 8 DPUs' outputs based on the **sign** outputs of all DPUs. Finally, the **addend** outputs of 8 DPUs, the output of the sign extension generator, and the accumulated sum latched in the REG are summed together by an adder. One PE is implemented in the chip so that it is capable of at most 8-digit FIR computation. It is also designed to be cascadable so that the FIR filter with more taps can be implemented.

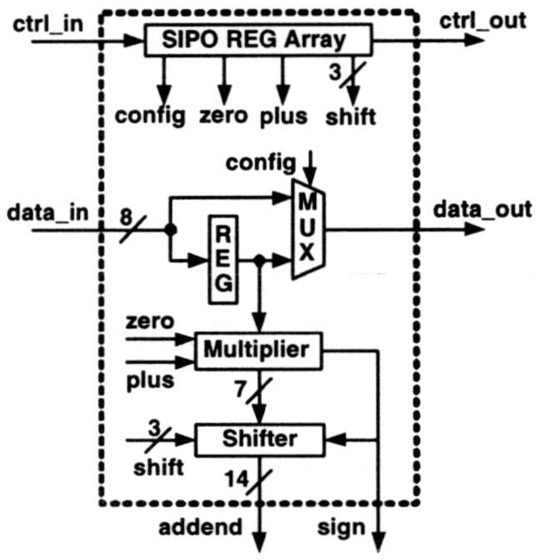

Figure 1: Digit processing unit.

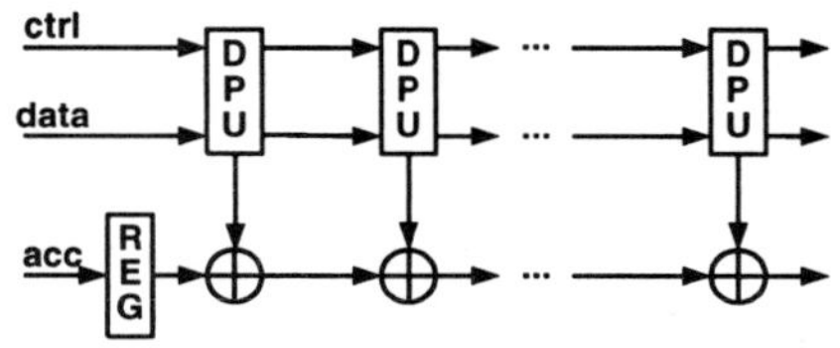

Figure 2: General reconfigurable FIR filter architecture.

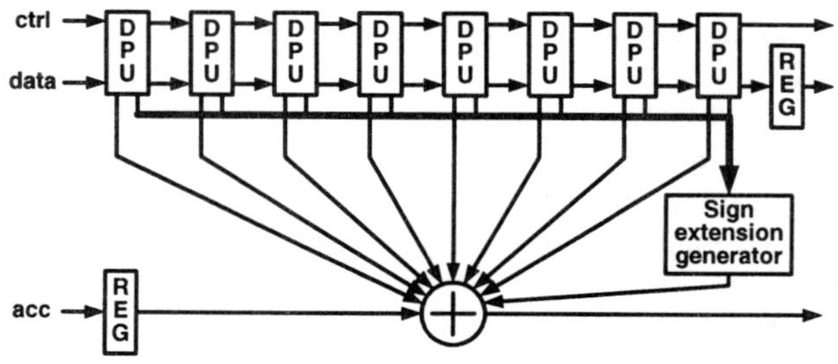

Figure 3: PE architecture.

The reconfigurable FIR chip consists of one PE, one pseudo-random data generator (PRDG), and one test module (see Figure 4). There are two clock signals: **CLK** and **DumpCLK**, where **CLK** controls the operating speed of the filter and **DumpCLK** is used to initialize the chip parameters and to output the results. The parameters in each DPU are pre-calculated based on the filter configuration and then serially shifted into the chip with the **ctrl_ in** signal. Since the chip will be I/O limited, scan-chain is used for both testing and for normal partial sum communication between cascaded chips. The REG in the PE used to store the accumulated sum is replaced by a 24-bit SIPO register array. The **scan_ in** signal is used to store the compensation vector explained below for the first chip or the accumulated sum of the previous chip into the 24-bit SIPO serially while cascading. A pseudo-random data generator is designed to provide the input test patterns in high-speed operation. The data fed into the first DPU can be generated by the pseudo-random data generator or **data_in** signal. Finally, due to the consideration of the output driving capability, a test module accumulates the 24-bit output of the adder in the PE with a 32-bit carry-save adder. The result can be serially scanned out at a slower clock rate to verify the functionality of the chip.

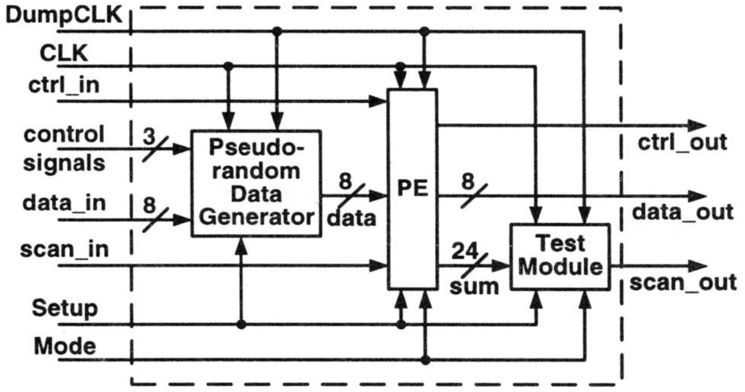

Figure 4: Chip block diagram.

3. CIRCUIT IMPLEMENTATION

3.1. Multiplier and shifter

The multiplier is used to multiply the input data $x[n-i]$ by $d_{i,k}$, which has three possible values: 1, 0, and -1. The operation of each bit can be expressed as

$$output = \overline{zero + (plus \oplus input)}. \quad (3)$$

If $d_{i,k}$ is 0, the **zero** signal will be '1' and force the output to be 0 regardless of the input. Otherwise, the **zero** signal will be '0' and (3) can be rewritten as $output = \overline{plus \oplus input}$. If the CSD coefficient is 1, the **plus** signal will be '1' and the output is the same as the input. If the CSD coefficient is -1, the **plus** signal will be '0' and the output is equivalent to the 1's complement of the input. The one in LSB that needs to be added to form the 2's complement negation is accumulated for all negative digits. This sum forms the compensation vector and will be added by setting the initial value of the accumulator.

A shifter is used to multiply the term $d_{i,k} \cdot x[n-i]$ by 2^{-p_k} where $p_k \in \{0, \ldots, 7\}$. The shifter performs an arithmetic left shift and expands the 7-bit multiplier output (excluding the MSB) into a 14-bit output as the **addend** signal by shifting the input left by $7 - p_k$ bits. Also, zeros are padded at LSB if the CSD coefficient is '1' or '0' and ones are padded if the CSD coefficient is '-1'.

3.2. Sign extension generator

In our architecture, the accumulated sum is 24-bit wide but the term $d_{i,k} \cdot x[n-i] \cdot 2^{-p_k}$ calculated by the DPU is only 15-bit wide (14-bit **addend** signal and 1-bit **sign** signal). While summing them together, it is unwise to extend each of them to the word length of the accumulated sum. It is better to deal with the sign extension bits separately, so that the area and power consumption can be reduced.

A sign extension generator is designed to evaluate the sum of sign extension bits based on eight **sign** signals. By examining the relation between the number of non-negative **sign** signals and the sum of the corresponding sign extension bits, a simple implementation of the sign extension generator is designed. The seven most significant bits can be selected through a MUX by examining if there exists any '1' valued **sign** signal. If the answer is yes, '1111111' will be selected, '0000000' will be selected otherwise. The three least significant bits are equivalent to the three least significant bits of the binary representation of the number of non-negative **sign** signals.

3.3. Adder

The adder is used to sum eight 14-bit **addend** signals from DPUs, one 24-bit **acc** signal, and one 10-bit **sign_ extend** signal from the sign extension generator. The **acc** signal corresponds to the compensation vector or the accumulated sum stored in the REG shown in Figure 3. The **acc** signal is split into two parts where its fourteen LSBs and eight **addend** signals are compressed into four 14-bit signals by five 14-bit carry-save adders in a two-level arrangement. Its ten MSBs are then added with the **sign_ extend** signal and the above four 14-bit signals by a two-level carry-save adder. Finally, an ELM adder [9] modified to reduce the critical path delay is used to compute the final sum.

4. IMPLEMENTATIONS AND MEASUREMENTS

Detail circuit design of the reconfigurable FIR filter chip was described in gate level using a hardware description language. Functional verification of the circuit design was conducted using five filters with different digit configurations. Layout was generated through a standard-cell-based design flow. Functional and timing simulations of the layout were carried out. The final layout is approximately 1.74×1.64 mm^2 in a single-poly quadruple-metal 0.35-μm CMOS technology and contains 27035 transistors. The die photo of the fabricated reconfigurable FIR filter chip is shown in Figure 5. There are 8 DPUs arranged in one row and three other blocks: an adder, a pseudo-random data generator, and a test module.

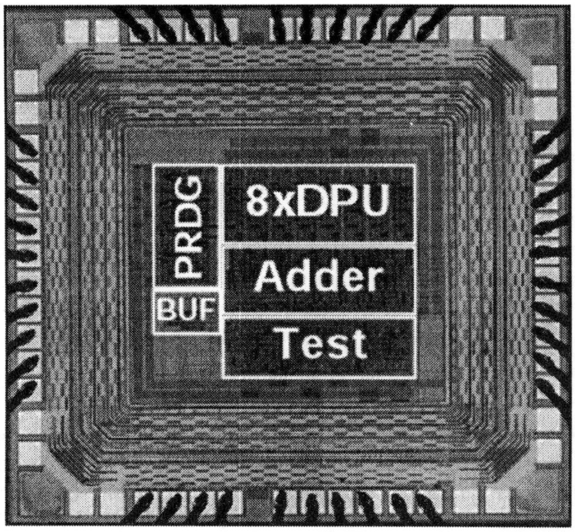

Figure 5: Chip die photo.

The fabricated chip has been tested and its function has been verified. The reconfigurable FIR filter chip can operate correctly at 86 MHz under a supply voltage of 2.5 V and it consumes 16.5 mW. The frequency of the **DumpCLK** is set to one-fourth of the operating frequency. Table 1 summarizes the major features of the reconfigurable FIR filter chip.

Table 1: Summary of the reconfigurable FIR filter chip.

Clock Frequency	86 MHz
Power Supply	2.5 V
Power Dissipation	16.5 mW
Process Technology	0.35-μm 1P4M CMOS
Transistor Count	27K transistors
Die Size	1.74 x 1.64 mm^2
Package	40-pin S/B

5. SUMMARY

In this paper, a digit-reconfigurable FIR filter architecture is proposed. The design concepts of the architecture and the circuit are presented. Testing results show that the fabricated chip draws only 16.5 mW from a 2.5-V power supply while running at 86 MHz.

6. ACKNOWLEDGEMENT

The authors greatly appreciate the Chip Implementation Center (CIC) of the National Science Council, Taiwan, R.O.C., for the fabrication and measurement of the proposed chip.

7. REFERENCES

[1] J. Mitola, "The Software Radio Architecture," *IEEE Communications Magazine,* vol. 33, pp. 26-38, May 1995.

[2] E. Buracchini, "The Software Radio Concept," *IEEE Communications Magazine,* vol. 38, pp. 138-143, Sept. 2000.

[3] R. M. Hewlitt and E. S. Swartzlantler Jr., "Canonical Signed Digit Representation for FIR Digital Filters," in *Proc. of IEEE Workshop on Signal Processing Systems,* 2000, pp.416-426.

[4] M. Tamada and A. Nishihara, "High-Speed FIR Digital Filter with CSD Coefficients Implemented on FPGA," in *Proc. of the ASP-DAC,* 2001, pp. 7-8.

[5] Y. M. Hasan, L. J. Karem, M. Falkinburg, A. Helwig, and M. Ronning, "Canonic Signed Digit Chebyshev FIR Filter Design," *IEEE Signal Processing Letters,* vol. 8, pp. 167-169, June 2001.

[6] T. Zhangwen, Z. Zhanpeng, Z. Jie, and M. Hao, "A High-Speed, Programmable, CSD Coefficient FIR Filter," in *Proc. of 4th International Conference on ASIC,* 2001, pp. 397-400.

[7] K. T. Hong, S. D. Yi, and K. M. Chung, "A High-Speed Programmable FIR Digital Filter Using Switching Arrays," in *Proc. of IEEE Asia Pacific Conference on Circuits and Systems,* 1996, pp. 492-495.

[8] K. Y. Khoo, A. Kwentus, and A. N. Willson Jr., "A Programmable FIR Digital Filter Using CSD Coefficients," *IEEE Journal of Solid-State Circuits,* vol. 31, pp. 869-874, June 1996.

[9] T. P. Kelliher, R. M. Owens, M. J. Irwin, and T. - T. Hwang, "ELM-A Fast Addition Algorithm Discovered by a Program," *IEEE Transactions on Computers,* vol. 41, pp. 1181-1184, Sept. 1992.

DESIGN OF IIR DIGITAL ALLPASS FILTERS USING LEAST *P*TH PHASE ERROR CRITERION

Chien-Cheng Tseng

Department of Computer and Communication Engineering, National Kaohsiung First University of Science and Technology, Kaohsiung, Taiwan

E-mail: tcc@ccms.nkfust.edu.tw

ABSTRACT

In this paper, the problem of designing a digital IIR allpass filter to have an arbitrarily prescribed phase response is considered. The filter coefficients are determined by using least pth phase error criterion. An iterative reweighted least squares method is used to obtain the optimal filter coefficients by solving a set of linear simultaneous equation in each iteration. Design examples of Hilbert transformer and fractional delay filter are demonstrated to illustrate the effectiveness of the new design method.

1. INTRODUCTION

In many applications of signal processing, there is a need for digital allpass filter which possesses unit magnitude response and prescribed phase characteristics. These applications include phase equalization in communication system [1], the designs of notch filter [2], fractional delay filter [3], Hilbert transformer [4], IIR filter with magnitude prescriptions using parallel allpass filters [5], multirate filtering system [6], and the construction of wavelet [7]. An excellent survey of the allpass filter design is described in the papers [8][9]. Generally speaking, the design methods can be classified into three categories, One is maximally flat design [4], another is L_2 approximation [10][11], the other is the minimax or equiripple design [12][13]. The maximally flat design usually has analytic solution, but it is only a pointwise prescription and have limited approximation quality [9]. Two popular methods to achieve the L_2 approximation are eigenfilter method [10] and weighted least squares method [11]. Moreover, minimax design is usually obtained by using linear programming method [12] and generalized exchange algorithm [13].

On the other hand, L_p normed minimization has been proven successfully in numerous applications such as FIR filter design [14][15], adaptive filtering [16], speech processing [17], image vector quantization [18] and sinusoidal frequency estimation [19]. Generally speaking, linear phase FIR filter design usually uses L_∞ norm minimization because it gives the designers a smallest length filter for a given specification, but sinusoidal estimation prefers L_1 norm minimization when signals are contaminated by impulsive noise. Due to the success of L_p norm minimization, it is interesting to design IIR allpass filter based on the least pth power phase error criterion. If we choose $p = 2$, this generalized criterion will reduce to the conventional L_2 approximation. When p approaches to ∞, the design result will be very close to the traditional minimax design.

In the literature, Deczky has synthesized the allpass filter using the minimum p-group delay error criterion [1]. In order to compute the gradient vector, only even power p is considered. In example 1 of [1], the allpass filter has been successfully used to solve the group-delay equalization problem. However, it is desirable to minimize the phase error (rather than group delay error) in many applications such as the design of Hilbert transformer and the design of IIR filter with magnitude prescriptions using parallel allpass filters. Thus, we will consider the synthesis of allpass filter using the least pth phase error criterion in this paper. This paper is organized as follows: The design problem is first described. Then, an iterative reweighted least squares algorithm is proposed to solve this nonlinear problem. Finally, several design examples are presented to show the effectiveness of the proposed method and a conclusion is made.

2. DESIGN METHOD

The transfer function of an Nth order real-coefficient allpass filter is represented by

$$H(z) = \frac{a_N + a_{N-1}z^{-1} + \cdots + a_0 z^{-N}}{a_0 + a_1 z^{-1} + \cdots + a_N z^{-N}} = \frac{z^{-N}A(z^{-1})}{A(z)} \tag{1}$$

where $a_0 = 1$. Because the numerator is the mirror-image polynomial of denominator, the magnitude response $|H(e^{j\omega})|$ is always equal to 1, and the phase response $\theta(\omega) = arg[H(e^{j\omega})]$ is given by

$$\theta(\omega) = -N\omega - 2arg[A(e^{j\omega})] \tag{2}$$

Let $\theta_d(\omega)$ be the desired phase response and $E(\omega)$ be the phase error between $\theta(\omega)$ and $\theta_d(\omega)$, then it can be shown that

$$\begin{aligned} e^{jE(\omega)} &= e^{j(\theta(\omega)-\theta_d(\omega))} \\ &= e^{-jN\omega}\frac{\sum_{n=0}^{N} a_n e^{jn\omega}}{\sum_{n=0}^{N} a_n e^{-jn\omega}} e^{-j\theta_d(\omega)} \\ &= \frac{\sum_{n=0}^{N} a_n e^{jn\omega}}{\sum_{n=0}^{N} a_n e^{-jn\omega}} \frac{e^{-j\frac{N\omega+\theta_d(\omega)}{2}}}{e^{j\frac{N\omega+\theta_d(\omega)}{2}}} \\ &= \frac{\sum_{n=0}^{N} a_n e^{j\phi_n(\omega)}}{\sum_{n=0}^{N} a_n e^{-j\phi_n(\omega)}} \end{aligned} \tag{3}$$

where $\phi_n(\omega) = n\omega - \frac{N\omega+\theta_d(\omega)}{2}$. Define $\Re(\omega) = \cos(\phi_0(\omega)) + \sum_{n=1}^{N} a_n \cos(\phi_n(\omega))$ and $\Im(\omega) = \sin(\phi_0(\omega)) +$

0-7803-7761-3/03/$17.00 ©2003 IEEE

$\sum_{n=1}^{N} a_n \sin(\phi_n(\omega))$, then eq(3) can be rewritten as

$$e^{jE(\omega)} = \frac{\Re(\omega) + j\Im(\omega)}{\Re(\omega) - j\Im(\omega)} \quad (4)$$

Based on this equation, the phase error $E(\omega)$ is given by

$$E(\omega) = 2\tan^{-1}\frac{\Im(\omega)}{\Re(\omega)} \quad (5)$$

In this paper, we will find the filter coefficients a_n to minimize the weighted pth phase error, i.e., the measure

$$\begin{aligned} J &= \int_R W(\omega)|E(\omega)|^p d\omega \\ &= \int_R W(\omega)\left|2\tan^{-1}\frac{\Im(\omega)}{\Re(\omega)}\right|^p d\omega \end{aligned} \quad (6)$$

where R is the frequency band of interest. Evidently, the minimization of the cost function J involves nonlinear programming techniques. However, the series of arctangent function is given by

$$\tan^{-1} x = x - \frac{x^3}{3} + \frac{x^5}{5} - \frac{x^7}{7} + \cdots \quad (7)$$

, so $\tan^{-1} x \approx x$ for small x. Thus, the phase error can be approximated by

$$\begin{aligned} E(\omega) &\approx 2\frac{\Im(\omega)}{\Re(\omega)} \\ &= 2\frac{\sin(\phi_0(\omega)) + \mathbf{a}^t\mathbf{s}(\omega)}{\cos(\phi_0(\omega)) + \mathbf{a}^t\mathbf{c}(\omega)} \end{aligned} \quad (8)$$

where three vectors are defined by

$$\begin{aligned} \mathbf{a} &= [a_1\ a_2 \cdots a_N]^t \\ \mathbf{c}(\omega) &= [\cos(\phi_1(\omega))\ \cos(\phi_2(\omega)) \cdots \cos(\phi_N(\omega))]^t \\ \mathbf{s}(\omega) &= [\sin(\phi_1(\omega))\ \sin(\phi_2(\omega)) \cdots \sin(\phi_N(\omega))]^t \end{aligned}$$

Substituting eq(8) into eq(6), the cost function J can be approximated by

$$J \approx \int_R W(\omega)\left|2\frac{\sin(\phi_0(\omega)) + \mathbf{a}^t\mathbf{s}(\omega)}{\cos(\phi_0(\omega)) + \mathbf{a}^t\mathbf{c}(\omega)}\right|^p d\omega \quad (9)$$

Because of the presence of denominator term $|\cos(\phi_0(\omega)) + \mathbf{a}^t\mathbf{c}(\omega)|^p$, the minimization of J is still a nonlinear problem. To solve this problem, we take the following iterative scheme:

$$J_k = \int_R 2^p W(\omega)\frac{|\sin(\phi_0(\omega)) + \mathbf{a}_{k-1}^t\mathbf{s}(\omega)|^{p-2}}{|\cos(\phi_0(\omega)) + \mathbf{a}_{k-1}^t\mathbf{c}(\omega)|^p} \left|\sin(\phi_0(\omega)) + \mathbf{a}_k^t\mathbf{s}(\omega)\right|^2 d\omega \quad (10)$$

where $\mathbf{a}_k$ is the parameter vector to be determined in the kth iteration. In this paper, the initial polynomial $A_0(z)$ is chosen as $A_0(z) = 1$, i.e., $\mathbf{a}_0 = \mathbf{0}$. At kth iteration, $\mathbf{a}_{k-1}$ is known; hence, cost function J_k can be rewritten as

$$J_k = \int_R W_{k-1}(\omega)\left|\sin(\phi_0(\omega)) + \mathbf{a}_k^t\mathbf{s}(\omega)\right|^2 d\omega \quad (11)$$

where weighting function

$$W_{k-1}(\omega) = 2^p W(\omega)\frac{|\sin(\phi_0(\omega)) + \mathbf{a}_{k-1}^t\mathbf{s}(\omega)|^{p-2}}{|\cos(\phi_0(\omega)) + \mathbf{a}_{k-1}^t\mathbf{c}(\omega)|^p} \quad (12)$$

Clearly, the denominator $|\cos(\phi_0(\omega)) + \mathbf{a}_{k-1}^t\mathbf{c}(\omega)|^p$ and partial numerator $|\sin(\phi_0(\omega)) + \mathbf{a}_{k-1}^t\mathbf{s}(\omega)|^{p-2}$ obtained from the preceding iteration is treated as a part of the weighting function. After some manipulation, J_k can be rewritten as the following quadratic form:

$$J_k = \mathbf{a}_k^t\mathbf{Q}_k\mathbf{a}_k - 2\mathbf{g}_k^t\mathbf{a}_k + d_k \quad (13)$$

where matrix $\mathbf{Q}_k$, vector $\mathbf{g}_k$ and scalar d_k are given by

$$\begin{aligned} \mathbf{Q}_k &= \int_R W_{k-1}(\omega)\mathbf{s}(\omega)\mathbf{s}(\omega)^t d\omega \\ \mathbf{g}_k &= -\int_R W_{k-1}(\omega)\mathbf{s}(\omega)\sin(\phi_0(\omega))d\omega \\ d_k &= \int_R W_{k-1}(\omega)|\sin(\phi_0(\omega))|^2 d\omega \end{aligned} \quad (14)$$

Because J_k is a quadratic function of $\mathbf{a}_k$, the optimal solution is unique and can be obtained by solving simultaneous linear equation:

$$\mathbf{Q}_k\mathbf{a}_k = \mathbf{g}_k \quad (15)$$

Since matrix $\mathbf{Q}_k$ is a positive-definite, real and symmetric matrix, the simultaneous linear equations can be solved by a computationally efficient method, like Cholesky decomposition. Based on the above description, we propose an iterative algorithm for obtaining filter coefficient $\mathbf{a}$ as follows:

Step 1: Given the phase specification $\theta_d(\omega)$, order N of allpass filter, power p, weighting function $W(\omega)$, initial coefficient vector $\mathbf{a}_0 = \mathbf{0}$, and intergal region R. Set $k = 1$.

Step 2: Use the eq(12) to compute the weighting function $W_{k-1}(\omega)$.

Step 3: Use the eq(14) to compute the matrix $\mathbf{Q}_k$, vector $\mathbf{g}_k$ and scalar d_k.

Step 4: Solve the simultaneous equation $\mathbf{Q}_k\hat{\mathbf{a}}_k = \mathbf{g}_k$ to obtain the coefficient $\hat{\mathbf{a}}_k$. Then, new coefficient $\mathbf{a}_k$ is computed by

$$\mathbf{a}_k = \lambda\hat{\mathbf{a}}_k + (1-\lambda)\mathbf{a}_{k-1} \quad (16)$$

Step 5: Terminate the iterative procedure if

$$\frac{|\mathbf{a}_k - \mathbf{a}_{k-1}|}{|\mathbf{a}_k|} \le \epsilon$$

where ϵ is a preset positive number. Otherwise, set $k = k+1$ and go to step 2.

Now, three remarks are made as follows: First, in step 4, λ is a convergence parameter that takes values $0 < \lambda \le 1$. From the discussion in [14], we choose $\lambda = \frac{1}{p-1}$ in this paper. Second, the convergence issue is studied. In the development of design method, the weighting function is generated using the preceding numerator and denominator polynomials. This idea is similar to Levy's algorithm for designing WLS optimal IIR filters [20]. This iterative procedure is also used in system identification context. In 1963, Sanathanan and Koerner used frequency response measurements to identify the unknown system and in 1965, Steiglitz and McBride used time domain measurements [21][22]. Recently, Lu, Pei and Tseng proposed a similar iterative scheme to design IIR filters [23] and some comments are made in [24]. Now, let us study which points the proposed algorithm will converge to. From (11), the mapping relation Ψ between $\mathbf{a}_{k-1}$ and $\mathbf{a}_k$ can be written as

$$\begin{aligned}\mathbf{a}_k &= arg\min_{\mathbf{a}_k} \int_R W_{k-1}(\omega)\left|\sin(\phi_0(\omega)) + \mathbf{a}_k^t\mathbf{s}(\omega)\right|^2 d\omega \\ &= \Psi(\mathbf{a}_{k-1}) \end{aligned} \quad (17)$$

Note that the weighting function $W_{k-1}(\omega)$ depends on $\mathbf{a}_{k-1}$. Let $\mathbf{a}_f$ be the stationary or fixed points of the mapping Ψ, i.e., $\mathbf{a}_f = \Psi(\mathbf{a}_f)$, then the proposed algorithm will converge to $\mathbf{a}_f$. A sufficient condition for the convergence is

$$|\mathbf{a}_{k+1} - \mathbf{a}_k| < |\mathbf{a}_k - \mathbf{a}_{k-1}| \quad (18)$$

,i.e, Ψ is a contraction mapping. Let $\mathbf{a}_o$ denote the optimal solution that minimizes the cost function J in (6). The distance between convergent point $\mathbf{a}_f$ and global minimum point $\mathbf{a}_o$ is not easy to be computed. In the chapter 8 of [25], the system identification examples have been used to show that the global minimum point is quite close to the convergent point of Steiglitz-McBride algorithm. Third, the stability issue is investigated. For a stable allpass filter $H(z)$, its phase response $\theta(\omega)$ has the following three properties: (i) $\theta(0) = 0$ (ii) $\theta(\pi) = -N\pi$. (iii) $\theta(\omega)$ decreases monotonically with frequency ω. Although the phase of the designed allpass filter follows closely a monotonically decreasing function does not imply that the phase itself is monotonic, the designed allpass filter is stable if it has a phase approximation error less than π at $\omega = \pi$ [12]. In our experience, most of the designed allpass filters satisfy this requirement, so the designed allpass filter is stable.

3. DESIGN EXAMPLES

In this section, two design examples performed with MATLAB language are presented to illustrate the effectiveness of the proposed design method. To evaluate the performance, the maximum absolute error e_{max} and root mean squares (RMS) error e_{rms} are defined by

$$\begin{aligned} e_{max} &= \max_{\omega \in R} |E(\omega)| \\ e_{rms} &= \left[\int_R |E(\omega)|^2 d\omega\right]^{\frac{1}{2}} \end{aligned} \quad (19)$$

where phase error $E(\omega) = \theta(\omega) - \theta_d(\omega)$. The smaller errors e_{max} and e_{rms} have, the better designed results we obtain.

Example 1: Hilbert transformer

The Hilbert transformer has been widely used in communication applications to generate single sideband (SSB) signals by splitting the modulating signal into two components which are 90 degrees out of phase. This approach reduces the required bandwidth for transmission of the signal by half. The ideal frequency response of Hilbert transformer is given by

$$H_d(\omega) = \begin{cases} -je^{-j\omega D} & 0 < \omega < \pi \\ je^{j\omega D} & -\pi < \omega < 0 \end{cases} \quad (20)$$

where D is the integer. Thus, the ideal phase response of this specification is

$$\theta_d(\omega) = -D\omega - \frac{\pi}{2} \quad \omega \in R \quad (21)$$

where $R = [\omega_l, \omega_u]$ is the care band. In the following, the design parameters are chosen as $N = 6$, $D = 5$, $R = [0.06\pi, 0.94\pi]$, and $W(\omega) = 1$. When iterative algorithm with $\epsilon = 10^{-4}$ and $p = 3$ is used to design this filter, the algorithm converges after 9 iterations. Fig.1(a) depicts the normalized phase response:

$$PH(\omega) = \frac{arg(H(e^{j\omega})) + D\omega}{\pi} \quad (22)$$

Clearly, $PH(\omega)$ fits the ideal value $\frac{-1}{2}$ very well. The maximum absolute error $e_{max} = 0.0435\pi$ and the RMS error $e_{rms} = 0.0214\pi$. Moreover, the maximum pole radius is 0.8657, so the designed allpass filter is stable. Fig.1(b) also shows the phase error curve of this design result with $p = 3$. It can be found that the errors at band edges 0.06π and 0.94π are large. In fact, we can increase the power p to reduce these band edge errors. Fig.1(c)(d) show the phase error curves for $p = 10$ and $p = 50$. It is clear that peak errors have been reduced and the errors become equiripple for larger power p.

Example 2: Fractional delay filter

In many applications of signal processing, there is a need for a delay which is a fraction of the sampling period. These applications include time adjustment in digital receivers, beam steering of antenna array, speech coding and synthesis, modeling of music instruments, and sampling rate conversion [3]. The desired frequency response of fractional delay filter is given by

$$H_d(\omega) = e^{-j(D+\tau)\omega} \quad (23)$$

where D is an integer and τ is a fractional number. Thus, the ideal phase response of this specification is

$$\theta_d(\omega) = -(D+\tau)\omega \quad \omega \in R \quad (24)$$

where $R = [0, \alpha\pi]$ is the care band. In the following, the design parameters are chosen as $N = D = 6$, $\tau = 0.2$, $\alpha = 0.9$, $W(\omega) = 1$ and $\epsilon = 10^{-4}$. Fig.2 shows the phase error curves for various power p. It is clear that the errors become equiripple when p increases.

4. CONCLUSION

In this paper, the design of allpass filter using least pth phase error criterion has been investigated. An iterative reweighted least squares method is presented to obtain the optimal filter coefficients by solving a set of linear simultaneous equation in each iteration. However, only 1-D allpass filter is considered here. Thus, it is interesting to extend the proposed method to design 2-D allpass filter in future work.

REFERENCES

[1] A.G. Deczky, "Synthesis of recursive digital filters using minimum p-error criterion", *IEEE Trans. Audio Electroacoust.*, vol.20, pp.257-263, Oct. 1972.

[2] S.C. Pei and C.C. Tseng, "IIR multiple notch filter design based on allpass filter", *IEEE Trans. Circuits Syst. II*, vol.44, pp. 133-136, Feb. 1997.

[3] T.I. Laakso, V. Valimaki, M. Karjalainen and U.K. Laine, "Splitting the unit delay: Tools for fractional delay filter design", *IEEE Signal Processing Mag.*, vol. 13, pp. 30-60, Jan. 1996.

[4] S.C. Pei and P.H. Wang, "Maximally flat allpass fractional Hilbert transformers", *IEEE Int. Symp. on Circuits and Systems*, vol.4, pp.701-704, May 2002.

[5] M. Ikehara, H. Tanaka and H. Kuroda, "Design of IIR digital filters using all-pass networks", *IEEE Trans. Circuits Syst. II*, vol.41, pp.231-235, Mar. 1994.

[6] P.P. Vaidyanathan, P.A. Regalia and S.K. Mitra, "Design of doubly complemmentary IIR digital filters using single complex allpass filter, with multirate applications", *IEEE Trans. Circuits Syst.*, vol.34, pp.378-389, Apr. 1987.

[7] X. Zhang, T. Muguruma and T. Yoshikawa, "Design of orthonormal symmetric wavelet filters using real allpass filters", *Signal Processing*, vol.80, pp.1551-1559, 2000.

[8] P.A. Regalia, S.K. Mitra and P.P. Vaidyanathan, "The digital allpass filter: A versatile signal processing building block", *Proc. IEEE*, vol.76, pp.19-37, Jan. 1988.

[9] M. Lang, "Allpass filter design and applications", *IEEE Trans. Signal Processing*, vol.46, pp.2505-2514, Sept. 1998.

[10] T.Q. Nguyen, T.I. Laakso and R.D. Koilpillai, "Eigenfilter approach for the design of allpass filters approximating a given phase response", *IEEE Trans. Signal Processing*, vol.42, pp.2257-2263, Sept. 1994.

[11] S. Sunder, "Weighted least-squares design of recursive allpass filters", *IEEE Trans. Signal Processing*, vol.44, pp.1553-1557, June 1996.

[12] Z. Jing, "A new method for digital allpass filter design", *IEEE Trans. Acoust., Speech, Signal Processing*, vol.35, pp.1557-1564, Nov. 1987.

[13] X. Zhang and H. Iwakura, "Design of IIR digital allpass filters based on eigenvalue problem", *IEEE Trans. Signal Processing*, vol.47, pp.554-559, Feb. 1999.

[14] C.S. Burrus, J.A. Barreto and I.W. Selesnick, "Iterative reweighted least-squares design of FIR filters", *IEEE Trans. Signal Processing*, vol.42, pp.2926-2936, Nov. 1994.

[15] W.S. Lu, "Minimax design of nonlinear-phase FIR filters: A least-pth approach", *IEEE Int. Symp. on Circuits and Systems*, vol.1, pp.409-412, May 2002.

[16] S.C. Pei and C.C. Tseng, "Least mean *p*-power error criterion for adaptive FIR filter", *IEEE Journal on Selected Areas in Communications*, vol.12, pp.1540-1547, Dec. 1994.

[17] E. Denoel and J.P. Solvay, "Linear prediction of speech with a least absolute error criterion", *IEEE Trans. Acoust., Speech, Signal Processing*, vol.33, pp.1397-1403, Dec. 1985.

[18] C. Zhu and Y. Hua, "Image vector quantization with minimax L_∞ distortion", *IEEE Signal Processing Letters*, vol.6, pp.25-27, Feb. 1999.

[19] J. Schroeder, R. Yarlagadda and J.Hershey, "L_p normed minimization with applications to linear predictive modeling for sinusoidal frequency estimation", *Signal Processing*, vol.24, pp.193-216, Aug. 1991.

[20] E.C. Levy, "Complex-curve fitting", *IEEE Trans. Automatic Control*, vol. 4, pp. 37-43, May 1959.

[21] C.K. Sanathanan and J. Koerner, "Transfer function synthesis as a ratio of two complex polynomials", *IEEE Trans. Automatic Control*, vol. 8, pp. 56-58, Jan. 1963.

[22] K. Steiglitz and L.E. McBride, "A technique for the identification of linear systems", *IEEE Trans. Automatic Control*, vol. 10, pp. 461-464, Oct. 1965.

[23] W.S. Lu, S.C. Pei and C.C. Tseng, "A weighted least-squares method for the design of stable 1-D and 2-D IIR digital filters", *IEEE Trans. Signal Processing*, vol. 46, pp. 1-10, Jan. 1998.

[24] P.A. Regalia, "Comments on 'A weighted least-squares method for the design of stable 1-D and 2-D IIR digital filters'," *IEEE Trans. Signal Processing*, vol. 47, pp. 2063-2065, July. 1999.

[25] P.A. Regalia, *Adaptive IIR Filtering in Signal Processing and Control*, Marcel Dekker, 1995.

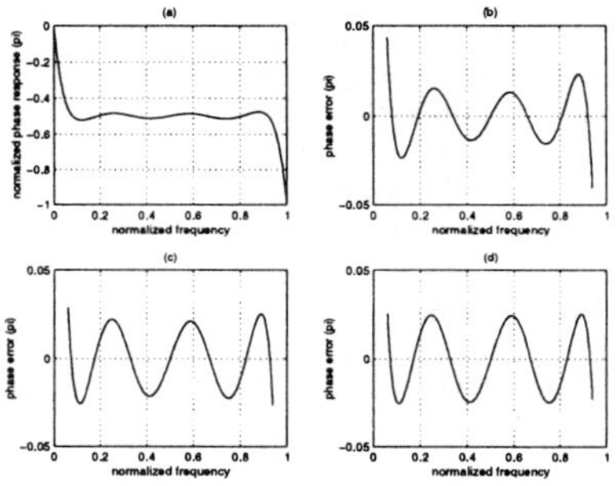

Figure 1. Design results of Hilbert transformer in example 1. (a) Normalized phase response with $p = 3$. (b) Phase error with $p = 3$. (c) Phase error with $p = 10$. (d) Phase error with $p = 50$.

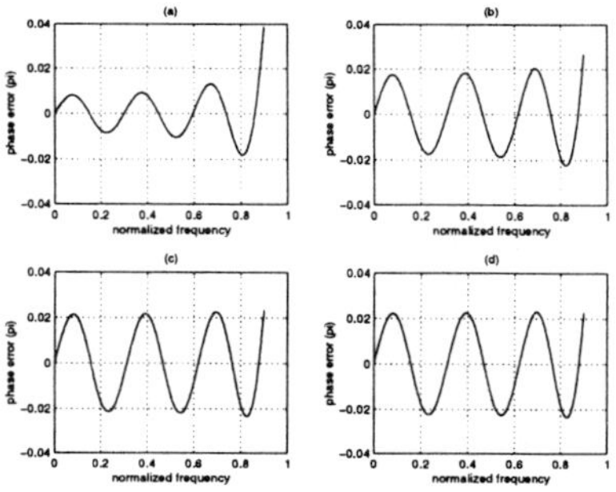

Figure 2. Design results of fractional delay filter in example 2. (b) Phase error with $p = 3$. (b) Phase error with $p = 10$. (c) Phase error with $p = 30$. (d) Phase error with $p = 50$.

DESIGN OF IIR DIGITAL FILTERS IN THE COMPLEX DOMAIN BY TRANSFORMING THE DESIRED RESPONSE

Tatsuya Matsunaga, Masahiro Yoshida, Masaaki Ikehara

Keio University 3-14-1 Hiyoshi Kohoku-ku Yokohama-shi 223 , Japan

ABSTRACT

In this paper, we present a new design method of IIR digital filters with an equiripple magnitude response in the complex domain. This method is based on solving a least square solution iteratively. At each iteration, the desired response is transformed so as to have equiripple error. Hence the algorithm is very simple and does not need a special software. By this method, an equiripple solution is obtained very quickly with less computational complexity. Finally, we show some examples to validate the proposed method.

1. INTRODUCTION

IIR digital filters offer improved selectivity, computational efficiency, and reduced system delay compared to what can be achieved by FIR digital filters of comparable approximation accuracy. However its design is more difficult than FIR digital filters because IIR digital filters have a rational transfer function. A lot of design methods have been proposed to approximate an IIR digital filter in the complex domain. A general method to design an IIR digital filter is to minimize the following L_p norm.

$$E = \left(\frac{1}{2\pi} \int_{-\pi}^{\pi} |H(\omega) - D(\omega)|^p d\omega \right)^{1/p} \qquad (1)$$

where $D(\omega)$ is the desired frequency response, and $H(\omega)$ is the frequency response of the IIR digital filter. In particular, the cases of $p = 2$ and $p = \infty$ are called least squares approximation and complex chebyshev approximation, respectively.

Least-squares methods is easy to design FIR and IIR filters because it only solve a linear equation and a lot of papers have been proposed[3]-[5]. However, the obtained filters are not optimal in the Chebyshev sence. Several methods using nonlinear optimization[1], linear programming [2], semidefinite programming [7] have been suggested to design IIR digital filters in the complex Chebyshev sense. However, the major disadvantages are quite computationally expensive and are to require suitable initial values so that the algorithm converges.

Zhang [6] formulated the complex Chebyshev approximation using the complex Remez algorithm and obtained the equiripple solution by solving the linear equation iteratively. However, this algorithm may not be optimal because the number of local frequency points is limited.

Recently, Lu[7] have presented the design of stable minimax IIR digital filters using semidefinite programming. It is a very interesting paper which can imposes the stability constraint. However even if one could design a stable IIR filter, which is not stable without imposing the stability constraint, the frequency response of the resulting IIR digital filter may be poor. On the other hand, it is known that the stability is guaranteed by changing the group delay[9]. Then our aim in this paper is to design IIR filters in the complex Chebyshev sense without taking account of stability. If a designed IIR filter is unstabel, we should change the group delay.

In this paper, we propose a design method of IIR digital filters in the complex Chebyshev sense. This method is based on solving a least squares solution iteratively. At each iteration, the desired response is transformed so as to have equiripple absolute error. Therefore this algorithm is very simple and easy to implement with less computational complexity. In spite of simplisity, this algorithm converges very quickly without any initial guess and the equiripple solution which is optimal in the Chebyshev sense can be obtained.

The composition of this paper is as follows. Section 2 describes the least squares method, and explains the modified least squares method to design an optimal solution[3]. Section 3 explains the transform of the desired frequency response, which is the focus of this paper and shows the overall design algorithm. Section 4 shows some examples to validate the proposed method by computer simulation. Section 5 is the conclusion.

2. LEAST SQUARE METHOD

The frequency response of an IIR digital filter with the numerator degree N and the denominator degree M is expressed by

$$H(\omega) = \frac{A(\omega)}{B(\omega)} = \frac{\sum_{n=0}^{N} a_n e^{-j\omega n}}{\sum_{m=0}^{M} b_m e^{-j\omega m}}, \quad b_0 = 1, \qquad (2)$$

where a_n, b_m is the set of filter coefficients.

When the desired frequency response of the filter is assumed to be $D(\omega)$ and the weighting function is assumed to be $W_0(\omega)$, the weighted mean square error is defined as follows[3]-[5].

$$E_0 = \frac{1}{L} \sum_{l=0}^{L} W_0(\omega_l) |H(\omega_l) - D(\omega_l)|^2, \qquad (3)$$

where $\omega_l = 2\pi l/L \quad (l = 0, 1, \cdots, L-1)$. L is determined by the specification, in generally L is set to 250.

The least square approximation of the IIR digital filter is to minimize the error function E_0. However, the minimization of E_0 becomes a nonlinear optimization problem because $H(\omega)$ is a rational transfer function. The nonlinear optimization problem includes the initial value problem, and is quite computationally expensive.

0-7803-7761-3/03/$17.00 ©2003 IEEE

To solve this problem, we define the generalized mean square error by only considering the numerator polynomial as

$$E = \frac{1}{L}\sum_{l=0}^{L-1} W_0(\omega_l)\,|A(\omega_l) - D(\omega_l)B(\omega_l)|^2 \quad (4)$$

The error on the discrete frequency point is expressed by

$$E(\omega_l) = A(\omega_l) - D(\omega_l)B(\omega_l) \quad (5)$$

The above equation (5) is written in the matrix form as

$$\mathbf{E} = \mathbf{P}_1\mathbf{a} - \mathbf{D}\mathbf{P}_2\mathbf{b}, \quad (6)$$

where $\mathbf{E} = [E(\omega_0), E(\omega_1), \cdots, E(\omega_{L-1})]^T$, $[\mathbf{P}_1]_{km} = e^{-j\omega_k m}$, $[\mathbf{P}_2]_{kn} = e^{-j\omega_k n}$ (for $k = 0, 1, \cdots, L-1,\ m = 0, 1, \cdots, M,\ n = 0, 1, \cdots, N$). Also $\mathbf{a} = [a_0, a_1, \cdots, a_N]^T$, $\mathbf{b} = [b_0, b_1, \cdots, b_M]^T$, $\mathbf{D} = diag[D(\omega_0), D(\omega_1), \cdots, D(\omega_{L-1})]$.

Therefore, Eq.(4) is rewritten by

$$\begin{aligned} E &= \mathbf{E}^\dagger\mathbf{W}\mathbf{E} \\ &= \mathbf{a}^T\mathbf{P}_1^\dagger\mathbf{W}\mathbf{P}_1\mathbf{a} - 2\mathbf{a}^T Re\{\mathbf{P}_1^\dagger\mathbf{W}\mathbf{D}\mathbf{P}_2\}\mathbf{b} \\ &\quad + \mathbf{b}^T\mathbf{P}_2^\dagger\mathbf{D}^\dagger\mathbf{W}\mathbf{D}\mathbf{P}_2\mathbf{b}, \end{aligned} \quad (7)$$

where superscript † denotes the conjugate transposition and

$$\mathbf{W} = diag[W_0(\omega_0), W_0(\omega_1), \cdots W_0(\omega_{L-1})] \quad (8)$$

The minimization of this cost function is achieved when $\partial E/\partial \mathbf{a} = 0$, namely

$$\frac{\partial E}{\partial \mathbf{a}} = \mathbf{P}_1^\dagger\mathbf{W}\mathbf{P}_1\mathbf{a} - 2Re\{\mathbf{P}_1^\dagger\mathbf{W}\mathbf{D}\mathbf{P}_2\}\mathbf{b} = 0 \quad (9)$$

Therefore, $\mathbf{a}$ is expressed by

$$\mathbf{a} = (\mathbf{P}_1^\dagger\mathbf{W}\mathbf{P}_1)^{-1}Re\{\mathbf{P}_1^\dagger\mathbf{W}\mathbf{D}\mathbf{P}_2\}\mathbf{b} = \mathbf{G}\mathbf{b} \quad (10)$$

Substitute (10) into (7),

$$\begin{aligned} E &= \mathbf{b}^T\mathbf{G}^\dagger\mathbf{P}_1^\dagger\mathbf{W}\mathbf{P}_1\mathbf{G}\mathbf{b} - 2\mathbf{b}^T\mathbf{G}^\dagger\mathbf{P}_1^\dagger\mathbf{W}\mathbf{P}_1\mathbf{G}\mathbf{b} \\ &\quad + \mathbf{b}^T\mathbf{P}_2^\dagger\mathbf{D}^\dagger\mathbf{W}\mathbf{D}\mathbf{P}_2\mathbf{b} \\ &= \mathbf{b}^T\mathbf{K}\mathbf{b} \end{aligned} \quad (11)$$

Here, the error E is written in the quadratic form of $\mathbf{b}$, and the solution which minimizes E can be solved using Rayleigh principle [8]. Therefore, the denominator coefficient $\mathbf{b}$ is decided by solving the eigenvector which corresponds to the minimum eigenvalue.

However, the least square solution of Equation(4) is weighted by $|B(\omega)|^2$ on Eq. (3). Therefore, a solution of Eq.(11) does not always minimize Eq. (3). Then the weighing function of Eq.(4) is expressed by

$$W(\omega) = \frac{W_0(\omega)}{|\hat{B}(\omega)|^2} \quad (12)$$

(where $\hat{B}(\omega)$ is $B(\omega)$ that was obtained in the previous iteration) and solve Equation(11) iteratively by changing $\hat{B}(\omega)$ in order to be close to $B(\omega)$. If $\hat{B}(\omega) = B(\omega)$, the least square solution of Equation(3) is the same as that of Eq.(11). By this iteration scheme, an optimal least square solution can be obtained.

However, we have to solve the inverse matrix in Eq.(10), which is time consuming task. To reduce the computational complexity without solving inverse matrix, we used the Gram Schumidt orthogonalization which is the generalized method of [10]. We show the least square algorithm used in this paper as follows

[**DESIGN ALGORITHM 1**]

1. Decide N, M, and the desired frequency response $D(\omega)$ of the filter
2. Set $\hat{B}(\omega) = 1 (b_0 = 1,\ b_{n(\neq 0)} = 0)$.
3. The filter coefficients are obtained by solving Eq.(11) and (10)
4. If $|\hat{B}(\omega) - B(\omega)|/B(\omega) << 1$, terminate, otherwise go to Step 5
5. Set $\hat{B}(\omega) = B(\omega)$, and return to Step 3

Although we can not show enough pttof of convergence, we have confirmed, through considerable experiences, this algorithm shows good convergence.

3. TRANSFORMING THE DESIRED RESPONSE

In this section, we present a new method to design an IIR digital filter whose absolute magnitude error is equiripple. This method is based on solving the least squares method iteratively while transforming the desired frequency response so that the absolute magnitude error between the desired and the designed frequency response becomes equiripple. The proposed algorithm consists of two iteration algorithms and is very simple.

3.1. Transform of the desired response

First, we design an IIR digital filter using the least squares method explained in the previous section, and then we derive the error $E_m(\omega)$ as follows.

$$E_m(\omega) = H_m(\omega) - D(\omega), \quad (13)$$

where $H_m(\omega)$ is the frequency response obtained by the least squares method at the m-th iteration, $D(\omega)$ is the desired frequency response specified at the beginning and $E_m(\omega)$ is the complex error between the desired and the designed frequency response. Next, we find the local maximum and minimum points of the $|E_m(\omega)|$ and let them be

$$\begin{aligned} &\omega_k^{max} \quad k = 1, \cdots, L_1 \ \text{for local maximum} \\ &\omega_k^{min} \quad k = 1, \cdots, L_2 \ \text{for local minimum} \end{aligned}$$

and $0 < \omega_1^{max} < \omega_1^{min} < \omega_2^{max} < \cdots$. It is clear that the local maximum and minimum points appear alternately. Then, we let the maximum error at the ω_k^{max} be $\delta_m^k = |E(\omega_k^{max})|$. The average of δ_m^k is selected as the maximum error in the next iteration.

$$\delta_m = \frac{\sum_{k=1}^{L_1} \delta_m^k}{L_1} \quad (14)$$

In order to obtain the equiripple error response, the new error response in $\omega_k^{min} \leq \omega \leq \omega_{k+1}^{min}$ is transformed as follows.

$$R_m(\omega) = E_m(\omega) \times \frac{\delta_m}{\delta_m^k} \qquad for \quad \omega_k^{min} \leq \omega \leq \omega_{k+1}^{min} \quad (15)$$

By adding the above error function $R_m(\omega)$ to the original desired function $D(\omega)$, a new desired function with an equiripple magnitude response is obtained.

$$D_m(\omega) = R_m(\omega) + D(\omega) \tag{16}$$

Note that the new desired function is complex. With this new desired response, a least squares solution is solved by algorithm 1. Thus the algorithm is iterated while transforming the desired response until the absolute error between the original desired response and the designed response becomes equiripple.

This procedure is similar to Remez algorithm. The original Remez algorithm interpolates at local points such that the error function has local extrema with alternating sign. On the other hand, this algorithm interpolates at continuous frequency such that a desired response have equiripple magnitude.

3.2. Multiplying the weighting function

When multiplying the weighting functions on the error response of the passband and the stopband, respectively, δ_m in the equations (14) and (15) are transformed as follows. That is, the desired response is transformed such that the new error response has the weighting function. Let the weighting functions on the passband and the stopband be W_p and W_s, respectively, δ_{mP} in the passband and δ_{mS} in the stopband are replaced by

$$\delta_{mP} = \frac{\sum_{\{\delta_m^k \in P\}} \delta_m^k + \frac{W_s}{W_p} \sum_{\{\delta_m^k \in S\}} \delta_m^k}{M} \tag{17}$$

and

$$\delta_{mS} = \delta_{mP} \frac{W_p}{W_s}, \tag{18}$$

where P and S indicate the passband and stopband,respectively. As result, the new error response $R_m(\omega)$ $\omega_k^{min} \leq \omega \leq \omega_{k+1}^{min}$ is rewritten by

$$R_m(\omega) = \begin{cases} E_m(\omega) \times \frac{\delta_{mP}}{\delta_m^k} & \text{in the passband} \\ E_m(\omega) \times \frac{\delta_{mS}}{\delta_m^k} & \text{in the stopband} \end{cases} \tag{19}$$

3.3. Algorithm

Finally, we shows the overall algorithm. The algorithm consists of two iteration parts. Main part is to transform the desired response to get a equiripple solution. Another part is to get a least squares solution by Algorithm 1 in main part.

[**DESIGN ALGORITHM 2**]

1. The filter is designed by the least squares method using Algorithm 1, where the weighting function in the passband and the stopband are W_p and W_s, respectively
2. The error is calculated by $E_m(\omega) = H_m(\omega) - D_0(\omega)$
3. Scaling by $R_m(\omega) = E_m(\omega) \times \frac{\delta_m}{\delta_m^k}$ for $\omega_k^{min} \leq \omega \leq \omega_{k+1}^{min}$
4. If $|\delta_{m-1} - \delta_m|/\delta_m < \alpha << 1$, then the algorithm terminates. Otherwise, go to next step
5. The desired frequency response is transformed by $D_m(\omega) = R_m(\omega) + D_0(\omega)$
6. The filter with the above desired response is designed by the least squares method using Algorithm1, where the weighting function in both passband and stopband is set to 1 in spite of W_p and W_s. Return to 2

Table 1: Comparison with the conventional method

Filter	$(\tau = 9)$	$(\tau = 10)$	Lu[7]
Maximum Passband error in magnitude	0.0101	0.0134	0.0167
Minimum stopband attenuation[dB]	39.86	37.39	35.34
Maximum ripple of group delay	0.056	0.973	0.721
Maximum radius of poles	0.9912	0.9378	0.9440

Step. 1 derives an initial guess for iteration. In this step, we have take account of the weighting function and the weighted mean squares error is minimized to derive an optimal solution in a least squares sense. In step. 6, since the new error response is already weighted, non-weighted means squares error is minimized.

Although we can not prove that this algorithm always converges, we have confirmed that the algorithm converges very quickly and the maximum value of $|E(\omega)|$ monotonically decreases.

4. DESIGN EXAMPLE

In this section, we will show several examples of IIR digital filters designed by the proposed methods.

The specifications of IIR digital filters is $N = 12, M = 6$.

$$D_0(\omega) \;\; = \;\; \begin{cases} e^{-j9\omega} & 0 \leq \omega \leq 0.5\pi \\ 0 & 0.6\pi \leq \omega \leq \pi \end{cases}$$

The weighting function is set to $W(\omega) = 1$ in both passband and stopband, $W(\omega) = 0$ in transition band. This specification is the same as that of [7] for comparison.

In this example, the proposed algorithm converges by ten iterations and can obtain the equiripple solution. Fig.3 shows the frequency response, convergence response and the filter coefficients. The solid line and the dashdotted line show the frequency responses of the proposed and conventional method[7],respectively. This algorithm converges very quickly and the maximum absolute error monotonically decreases. However, as shown in Fig.3(a), a magnitude overshoot appears in the transition band because there is a pair of poles in the transition band near the unit circle. To avoid this overshoot, we changed the group delay from 9 to 10. Tha obtained frequency responses which do not have such a overshoot are shown by dashed line and is still better than [7]. The largest magnitude of the poles is 0.9912 when $\tau = 9$(0.9440 in [7]) and 0.9378 when $\tau = 10$. Table 1 shows the numerical comparison of the proposed method with the conventional method[7].

5. CONCLUSION

In this paper, we proposed a new design method of an IIR digital filter in the complex Chebyshev sense using the least square method by transforming the desired frequency response to obtain equiripple characteristics. With this method, IIR digital filters can be easily designed with a few iterations and without a special optimization algorithm. In spite of a very simple algorithm, its results are much better than the conventional methods.

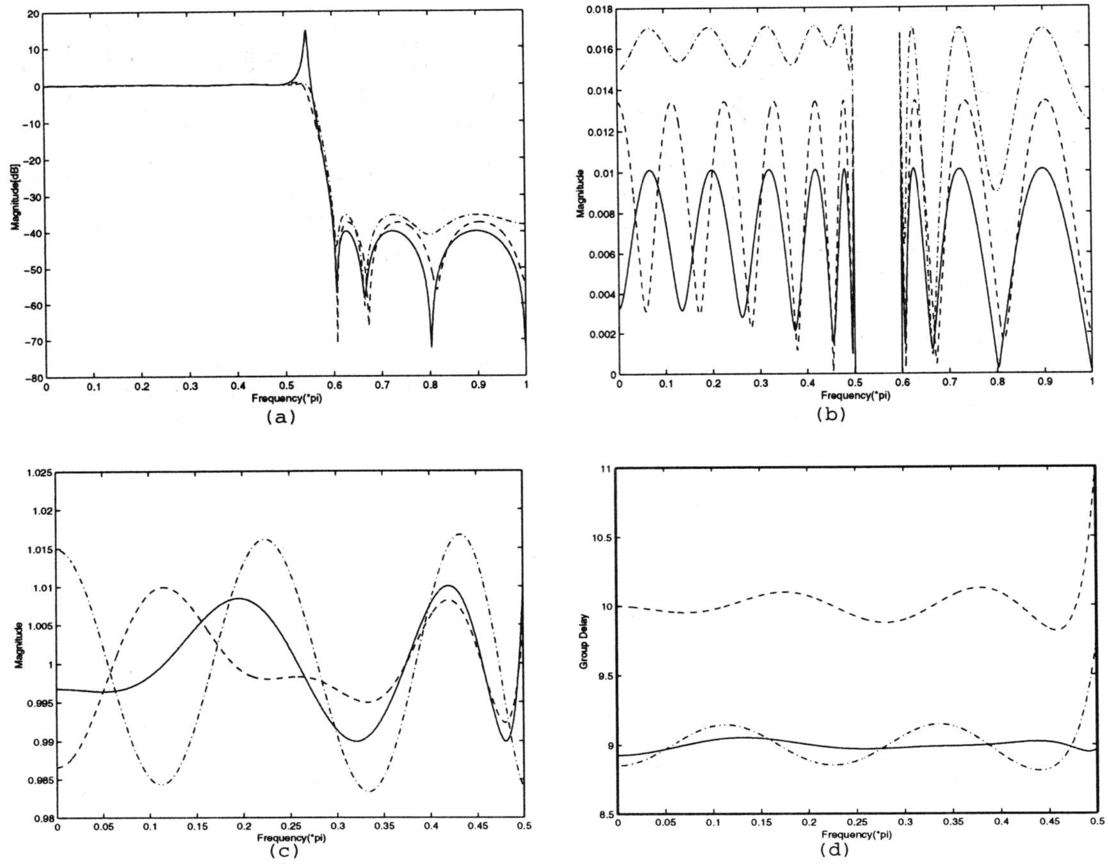

Figure 1: Frequency response of the proposed IIR filter with $\tau = 9$(solid line), $\tau = 10$(dashed line) and [7](dashdotted line) in Example1 (a) Log magnitude response (b)Absolute error response of $E(\omega)$ (c) Magnitude response in passband (d) Group delay in passband

6. REFERENCES

[1] A.G. Deczkey, "Equiripple and minimax (Chebyshev) Approximations for recursive digital filters," *IEEE Trans. Acoust., Speech, Signal Processing.*, Vol.ASSP-22, pp.98-111, 1974.

[2] X. Chen and T. W. Parks: "Design of IIR filters in the complex domain," *IEEE Trans. Acoust., Speech. Signal Processing*, Vol.**ASSP-38**, No.6, p.910-920, 1990.

[3] Y. C. Lim, J.H. Lee, C.K. Chen and R. H. Yang, "A weighted Least Squares algorithm for quasi-equiripple FIR and IIR digital filter design," *IEEE Trans. Acoust., Speech. Signal Processing*, Vol.**ASSP-40**, p.551-558, 1992

[4] M. Lang, "Least-squares design of IIR filters with prescribed magnitude and phase responses and a pole redius constraint ,"*IEEE Trans. Signal Processing*, Vol.48, p.3109-3121, Nov.2000.

[5] W.-S Lu, "Design of stable IIR digital filters with equiripple passbands and peak-constrained least-squares stopbands,"*IEEE Trans. Circuit and Systems-II*, Vol.46, pp.1421-1426, Nov.1998.

[6] X. Zhang, K. Suzuki, and T. Yoshikawa, "Complex Chebyshev Approximation for IIR digital filters Based on Eigenvalue Problem" *IEEE Trans. Circuit and Systems*, Vol.47, No.12 p.371-373, December 2000

[7] W.-S. Lu, "Design of Stable Minimax IIR Digital Filters using Semidefinite Programming ," *IEEE ISCAS2000*, Vol.1, pp.355-358, May 2000.

of allpass filters approximating a given phase response," *IEEE Trans. Signal Processing*, Vol.42, No.9, p.2257-2263, 1994.

[8] P. P. Vaidyanathan and T. Q. Nguyen, "Eigenfilters: A new approach to least squares FIR filter design and applications including nyquist filters," *IEEE Trans. Circuit and Systems*, Vol. CAS-34, No.1, p.11-23, 1987.

[9] R. Vuerinckx, Y. Rolain, J. Schoukens and R.Pintelon, "Design of stable IIR filters in the complex domain by automatic delay selection," *IEEE Trans. Signal Processing*, Vol.44, pp.2339-2344, May 1996.

[10] C.T. Mullis and R.A. Roberts, "The use of second-order information in the approximation of discrete-time linear systems," *IEEE Trans. Acoust., Speech, Signal Processing.*, Vol.24,No.7, pp.226-238, 1976.

A GENERALIZED DIRECT-FORM II TRANSPOSED STRUCTURE FOR IIR FILTER IMPLEMENTATION WITH MINIMAL ROUNDOFF NOISE GAIN

G. Li, Z.X Zhao and J.X Hao

School of EEE
Nanyang Technological University
Singapore 639798

ABSTRACT

In this paper, based on a polynomial operator approach a new structure is derived. This structure is a generalization of the direct-form II transposed structures in the conventional shift-operator and the prevailing delta-operator. The use of polynomial operators provides more degrees of freedom for minimizing the roundoff noise gain of the structure without increasing the computation complexity. Numerical examples are presented to illustrate the design procedure. It is shown that the optimized polynomial operators can yield a much better performance than the shift and delta operators.

1. INTRODUCTION

Finite word length (FWL) effects have been a very important issue in digital filter implementation for more than three decades. The optimal FWL structure design has been considered as one of the most effective methods (see, e.g., [1]-[4]) to minimize the effects of FWL errors on the performance of digital filters. The basic idea behind this approach is that for a given digital filter there exist a number of different structures.[1] Theoretically, they are equivalent since they represent the same system transfer function. However, different structures have different numerical properties and for a given application (measure or criterion) one structure can be better than another. The optimal FWL state-space design is to compute those realizations that minimize the degradation of the filter due to the FWL effects.

It has been noted that the optimal realizations are usually fully parameterized. In practice, it is desired that the filter have a nice performance as well as a very simple structure that possesses many trivial parameters[2], which can be implemented exactly and produce no rounding errors. Noting this fact, a lot of effort has been made to achieve sparse optimal or quasi-optimal realizations (see, e.g., [5]-[6]). It is well known that though having poor numerical properties the direct-forms in the conventional shift operator are the simplest structures. Recently, the direct-forms in delta operator have been studied by researchers (see, e.g., [7]-[11]). An extensive comparative study of different direct forms in delta operator was given in [11], where the transfer function is cascaded into second order sections and each section is implemented with a direct form in delta operator. It was shown there that among all the direct forms, the direct form II transposed (DFIIt) structure has the lowest quantization noise level at output. In [7], the DFIIt structure in δ-operator was investigated for an arbitrary order IIR filter, where the concept of different coupling coefficients at different branch nodes is utilized for better roundoff noise gain suppression.

The use of delta operator, defined as $\delta = \frac{z-1}{T_s}$ with T_s the sampling period, was first promoted by Peterka [12] and Middleton and Goodwin [13] in estimation and control applications. Later on, the numerical properties of the delta operator, where T_s is replaced by a positive factor Δ, were investigated in [4] from a pure algebraic point of view. It was found that one can make the transfer function in delta operator have better numerical properties in the case where the poles of the transfer function are closer to $z = +1$ than $z = 0$. This means that the delta operator based structures have a very good performance for narrow band low-pass filters and may not yield a satisfactory performance for other types of filters. In this paper, based on the concept of polynomial operators a set of special operators is derived, with which a generalized DFIIt structure is obtained. This structure, by a proper choice for the

[1] Here, a structure means a way in which the output of digital filter can be computed with an input given.

[2] By trivial parameters we mean those that are 0 and ± 1. Other parameters are, therefore, referred to non-trivial parameters.

0-7803-7761-3/03/$17.00 ©2003 IEEE

polynomial operators, can be optimized for any given digital filters.

2. A POLYNOMIAL OPERATOR BASED DFIIT STRUCTURE

Consider a single-input-single-output time-invariant linear digital filter $H(z)$ given by

$$H(z) = \frac{b_0 z^p + b_1 z^{p-1} + ... + b_{p-1} z + b_p}{z^p + a_1 z^{p-1} + ... + a_{p-1} z + a_p}. \quad (1)$$

This filter can be implemented with many different structures. For example, it can be implemented with its state space equations:

$$\begin{aligned} x(n+1) &= Ax(n) + Bu(n) \\ y(n) &= Cx(n) + du(n), \end{aligned} \quad (2)$$

where $u(n)$ and $y(n)$ are the scalar input and output of the filter, respectively. $R \triangleq (A, B, C, d)$ with $A \in \mathcal{R}^{p\times p}, B \in \mathcal{R}^{p\times 1}, C \in \mathcal{R}^{1\times p}$ and $d \in \mathcal{R}$ is called a realization of $H(z)$, satisfying

$$H(z) = d + C(zI - A)^{-1}B. \quad (3)$$

2.1. Reparametrization with polynomial operators

Define

$$\rho_k \triangleq \frac{z - \gamma_k}{\Delta_k}, \; k = 1, 2, ..., p, \quad (4)$$

where $\{\gamma_k\}$ and $\{\Delta_k > 0\}$ are two sets of constants to be discussed later.

It can be shown that $H(z)$ can be reparametrized with $\{\alpha_m, \beta_m\}$ in the polynomials $\{\rho_k\}$, called polynomial operators:

$$H(z) = \frac{\beta_0 + \beta_1 \rho_1^{-1} + ... + +\beta_p \prod_{k=1}^{p} \rho_k^{-1}}{1 + \alpha_1 \rho_1^{-1} + ... + \alpha_p \prod_{k=1}^{p} \rho_k^{-1}}. \quad (5)$$

Denoting

$$\begin{aligned} V_a &\triangleq \begin{pmatrix} 1 & \cdots & a_p \end{pmatrix}^T, \; V_b \triangleq \begin{pmatrix} b_0 & \cdots & b_p \end{pmatrix}^T \\ V_\alpha &\triangleq \begin{pmatrix} 1 & \cdots & \alpha_p \end{pmatrix}^T, \; V_\beta \triangleq \begin{pmatrix} \beta_0 & \cdots & \beta_p \end{pmatrix}^T, \end{aligned}$$

one has

$$\left\{ \begin{array}{l} V_a = \mathcal{K} M V_\alpha \\ V_b = \mathcal{K} M V_\beta \end{array} \right. \quad or \quad \left\{ \begin{array}{l} V_\alpha = \mathcal{K}^{-1} M^{-1} V_a \\ V_\beta = \mathcal{K}^{-1} M^{-1} V_b \end{array} \right., \quad (6)$$

where $\mathcal{K} = \prod_{k=1}^{p} \Delta_k$ and $M \in R^{(p+1)\times(p+1)}$ is a lower triangular matrix whose mth column is determined by the coefficients of the polynomial $\prod_{k=m}^{p} \rho_k$ for $m = 1, 2, ..., p$ and $M(p+1, p+1) = 1$.

It can be shown that the output can be computed with the following equations

$$\begin{aligned} y(n) &= \beta_0 u(n) + w_1(n) \\ w_k(n) &= \rho_k^{-1}[\beta_k u(n) - \alpha_k y(n) + w_{k+1}(n)] \\ w_p(n) &= \rho_p^{-1}[\beta_p u(n) - \alpha_p y(n)] \end{aligned} \quad (7)$$

with $w_{p+1}(n) = 0$. Fig. 1 shows the corresponding realization structure to (7), where $w_k(n)$ is the output of the operator ρ_k^{-1}.

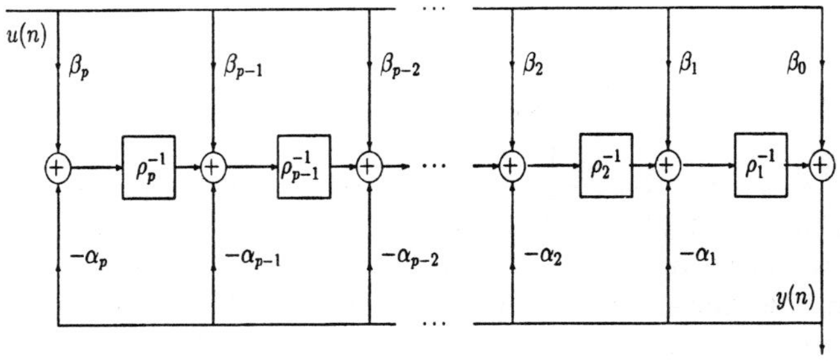

Fig. 1. A generalized DFIIt structure with polynomial operators

Clearly, when $\gamma_k = 0$, $\Delta_k = 1$, $\forall k$, Fig. 1 is the conventional DFIIt, and when $\gamma_k = 1$, $\forall k$, one gets the generalized δDFIIt structure studied in [7]. In this paper, we just consider the cases for which γ_k takes values from the set $\{-1, 0, 1\}$.

2.2. Equivalent state-space realization

One can implement ρ_k^{-1} with the realization depicted in Fig. 2.

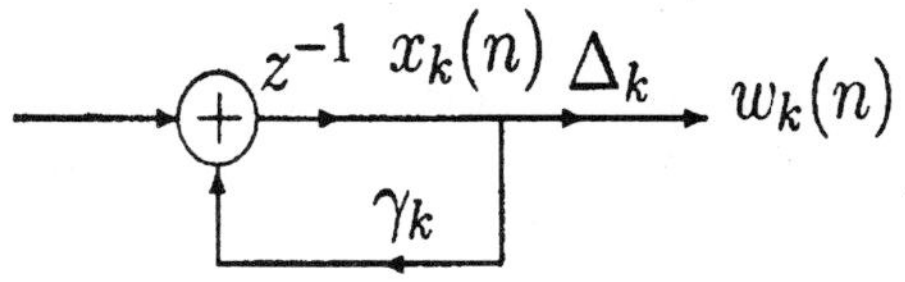

Fig. 2. A realization of operator ρ_k^{-1}

We choose $\{x_k(n)\}$ indicated in Fig. 2 as the state variables and denote $x(n)$ as the state vector. It can be shown that the proposed structure is equivalent to the following state-space realization

$$\begin{aligned} x(n+1) &= A_\rho x(n) + B_\rho u(n) \\ y(n) &= C_\rho x(n) + \beta_0 u(n), \end{aligned} \quad (8)$$

where

$$\begin{aligned} B_\rho &= \bar{\beta} - \beta_0 \bar{\alpha}, \; C_\rho = \begin{pmatrix} \Delta_1 & 0 & \cdots & 0 \end{pmatrix} \\ A_\rho &= D_\gamma + M_\alpha \end{aligned} \quad (9)$$

with $D_\gamma = diag(\gamma_1, \cdots, \gamma_p)$, M_α is the $p \times p$ zero matrix except $M(k,1) = -\Delta_1 \alpha_k, \forall k$ and $M_\alpha(k, k+1) = \Delta_{k+1}$ for $k = 1, \cdots, p-1$, and

$$\bar{\beta} \triangleq \begin{pmatrix} \beta_1 & \cdots & \beta_p \end{pmatrix}^{\mathcal{T}}, \quad \bar{\alpha} \triangleq \begin{pmatrix} \alpha_1 & \cdots & \alpha_p \end{pmatrix}^{\mathcal{T}}.$$

Denote $(\bar{A}_\rho, \bar{B}_\rho, \bar{C}_\rho, \beta_0)$ as the realization given in (9) but corresponding to $\Delta_k = 1, \forall k$. It can be shown that the realization $(A_\rho, B_\rho, C_\rho, \beta_0)$ with any given $\{\Delta_k\}$ can be obtained from $(\bar{A}_\rho, \bar{B}_\rho, \bar{C}_\rho, \beta_0)$ with a diagonal similarity transformation, denoted as T_{sc},

$$A_\rho = T_{sc} \bar{A}_\rho T_{sc}^{-1}, \; B_\rho = T_{sc} \bar{B}_\rho, \; C_\rho = \bar{C}_\rho T_{sc}^{-1}, \quad (10)$$

where

$$T_{sc} = diag(d_1, d_2, \cdots, d_p), \quad d_k = \prod_{m=1}^{k} \Delta_m^{-1}, \; \forall k. \quad (11)$$

3. OPTIMIZED POLYNOMIAL OPERATORS

It is well known (see, e.g., [3], [7]) that when two's complement fixed-point arithmetic is used, summation nodes are allowed to overflow except before multiplier. Under the assumption that the input $u(n)$ and the output $y(n)$ of the initial filter are properly pre scaled, the only signals which may have overflow are the states $\{x_k(n)\}$, which have to be scaled.

There exist different scaling schemes for preventing variables from overflow. The popularly used ones are the l_2- and l_∞- scalings. In what follows, we will concentrate on the l_2-scaling scheme.

The l_2-scaling means that each state variable should have a unit variance when the input is a white noise with a unit variance. This can be achieved if

$$W_c(k,k) = 1, \forall k, \quad (12)$$

where W_c is given by

$$W_c = \sum_{k=0}^{+\infty} A_\rho^k B_\rho B_\rho^{\mathcal{T}} (A_\rho^{\mathcal{T}})^k, \quad (13)$$

called the controllability gramian of $(A_\rho, B_\rho, C_\rho, \beta_0)$.

Denote $\bar{W}_c$, corresponding to $(\bar{A}_\rho, \bar{B}_\rho, \bar{C}_\rho, \beta_0)$. It follows from (10)-(11) that $W_c = T_{sc} \bar{W}_c T_{sc}^{\mathcal{T}}$. Therefore, the l_2-scaling can be achieved if $d_k^2 \bar{W}_c(k,k) = 1, \forall k$, which leads to

$$\begin{aligned} \Delta_1 &= \sqrt{\bar{W}_c(1,1)}, \; \Delta_k = \sqrt{\frac{\bar{W}_c(k,k)}{\bar{W}_c(k-1,k-1)}} \\ k &= 2, 3, ..., p. \end{aligned} \quad (14)$$

Clearly, the proposed structure will be l_2-scaled if the coupling coefficients are given by (14). In the sequel, the structure is assumed to have been well scaled by one of the schemes.

Roundoff noise occurs in those variables computed with multiplications if less-than-double precision fixed-point arithmetic and rounding are utilized. Here, we assume that rounding occurs after multiplication, which means that there is a quantizer after each non-trivial coefficient in Fig. 1 and 2. Since $\{\gamma_k\}$ are trivial parameters, no rounding occurs at all after these parameters, it is easy to understand that all the results for roundoff noise gain obtained in [7] hold here since by forcing $\gamma_k = 0, \pm 1, \forall k$, these parameters will cause no roundoff noise at all. These results are summarized below.

Let W_o be the observability gramian of the realization $(A_\rho, B_\rho, C_\rho, \beta_0)$, which is given by

$$W_o = \sum_{k=0}^{+\infty} (A_\rho^{\mathcal{T}})^k C_\rho^{\mathcal{T}} C_\rho A_\rho^k. \quad (15)$$

Denote G_α, G_β and G_Δ as the roundoff noise gain due to $\{\alpha_k\}$, $\{\beta_k\}$ and $\{\Delta_k\}$, respectively, the total roundoff noise gain is given by

$$\begin{aligned} G &= G_\alpha + G_\beta + G_\Delta \\ &= 3tr(W_o) + 2(1 + \bar{\alpha}^{\mathcal{T}} W_o \bar{\alpha}) - W_o(p,p). \end{aligned} \quad (16)$$

As mentioned before, γ_k takes value from the finite set S_γ:

$$S_\gamma \triangleq \{-1, \; 0, \; 1\}. \quad (17)$$

For a given $\{\gamma_k\}$, one can compute $(\bar{A}_\rho, \bar{B}_\rho, \bar{C}_\rho, \beta_0)$ and hence $(\bar{W}_c, \bar{W}_o)$. Then, the scaling factors $\{\Delta_k\}$ can be computed with either (14) and hence T_{sc} can be determined with (11). It can be shown that $W_o = T_{sc}^{-\mathcal{T}} \bar{W}_o T_{sc}^{-1}$. Therefore, the total roundoff noise gain G can be evaluated with (16). The interesting problem is to find the optimal $\{\gamma_k\}$, which is the solutions to

$$\min_{\gamma_k \in S_\gamma, \; \forall k} G. \quad (18)$$

Though G is a highly nonlinear function of $\{\gamma_k\}$, the problem can be solved easily since the space $\{\gamma_k : \gamma_k \in S_\gamma\}$ is finite.

4. DESIGN EXAMPLES

In this section, we present two design examples to illustrate the performance of the proposed structure. In these examples, l_2-scaling is used.

Example I: This is a fourth order low-pass Butterworth filter but with a normalized $3dB$ frequency $f_c = 0.125$. The corresponding poles are located at $p_{1,2} = 0.5565 \pm j0.5142$, $p_{3,4} = 0.4277 \pm j0.1637$ with $|p_{1,2}| = 0.7577$, $|p_{3,4}| = 0.4576$.

Table I shows the total roundoff noise gain for 4 particular sets of polynomial operators, in which {0, 1, 0, 1} yields the optimal polynomial operators.

Table I

γ_1	γ_2	γ_3	γ_4	G
-1	-1	-1	-1	5.3056×10^3
0	0	0	0	52.4953
0	1	0	1	13.9221
1	1	1	1	27.1187

Example II: The second example is a bandpass Butterworth filter of order eight. The bandpass is $[0.175, 0.375]$. The corresponding poles are located at $p_{1,2} = -0.5925 \pm j0.6018$, $p_{3,4} = 0.3553 \pm j0.7263$, $p_{5,6} = -0.3257 \pm j0.4258$, $p_{7,8} = 0.1006 \pm j0.4633$.

Table II shows the total roundoff noise gain for 4 particular sets of polynomial operators. The optimal operators are given by $\{-1, 0, -1, 0, 0, 0, 0, 0\}$.

Table II

γ_1	γ_2	γ_3	γ_4	γ_5	γ_6	γ_7	γ_8	G
-1	-1	-1	-1	-1	-1	-1	-1	8.3695×10^4
-1	0	-1	0	0	0	0	0	18.3920
0	0	0	0	0	0	0	0	20.0009
1	1	1	1	1	1	1	1	2.5096×10^6

Comment 4.1: In both examples, the delta operator based structure is far away from the one based on the optimal polynomial operators. In fact, the optimal polynomial operators in the first example yields a roundoff noise gain which is just half of that by the delta-operator and the corresponding structure is more efficient in terms of computation. In Example II, the shift-operator based structure is very closed to that based on the optimal polynomial operators, they all yield a much smaller roundoff noise gain than the delta-operator. It has been found that for low pass filters of very narrow bandwidth, the optimal polynomial operators are very close to the delta operator.

5. REFERENCES

[1] S.Y. Hwang, "Minimum uncorrelated unit noise in state-space digital filtering," IEEE Trans. on Acoust., Speech, and Signal Processing, Vol. ASSP-25, No. 4, pp. 273-281, Aug. 1977.

[2] L. Thiele, "On the sensitivity of linear state-space systems," IEEE Trans. on Circuits and Systems, Vol. CAS-33, pp. 502-510, May 1986.

[3] R.A. Roberts and C.T. Mullis, Digital Signal Processing, Reading, MA: Addison Wesley, 1987.

[4] Michel Gevers and Gang Li, Parametrizations in Control, Estimation and Filtering Problems: Accuracy Aspects, Springer Verlag London, Communication and Control Engineering Series, 1993.

[5] G. Amit and U. Shaked, "Small roundoff realization of fixed-point digital filters and controllers," IEEE on Trans. on Acoust., Speech, and Signal Processing, Vol. ASSP-36, no. 6, pp. 880-891, Jun. 1988.

[6] G. Li, M. Gevers, and Y.X. Sun, "Performance analysis of a new structure for digital filter implementation," IEEE on Trans. Circuits and Systems-I, Vol. 47, pp. 474-482, April 2000.

[7] N. Wong and T.S. Ng, "A generalized direct-form delta operator-based IIR filter with minimum noise gain and sensitivity," IEEE Trans. on Circuits and Systems - II, Vol. 48, pp. 425-431, April 2001.

[8] N. Wong and T.S. Ng, "Roundoff noise minimization in a modified direct form delta operator IIR structure," IEEE Trans. on Circuits and Systems - II, Vol. 47, pp. 1533-1536, Dec. 2000.

[9] J. Kauraniemi and T.I. Laakso, "Roundoff noise analysis of modified delta operator direct form structures," in Proc. IEEE Int. Symp. Circuits Syst., 1997.

[10] J. Kauraniemi, T.I. Laakso, I. Harttimo, and S.J. Ovaska, "Roundoff noise minimization in a direct form delta operator structure," in Proc. IEEE Int. Acoust., Speech, and Signal Processing, Atlanta, GA, May, 1996.

[11] J. Kauraniemi, T.I. Laakso, I. Harttimo, and S.J. Ovaska , "Delta operator realizations of direct-form IIR filters," IEEE Trans. Circuits and Systems-II, Vol. 45, pp. 41-45, Jan. 1998.

[12] V. Peterka, "Control of uncertain processes: applied theory and algorithms," Kybernetika, Vol. 22, pp. 1-102, 1986.

[13] R.H. Middleton and G.C. Goodwin, Digital Control and Estimation: A Unified Approach, Prentice Hall, Englewood Cliffs, New Jersey, 1990.

M-CHANNEL LIFTING-BASED DESIGN OF PARAUNITARY AND BIORTHOGONAL FILTER BANKS WITH STRUCTURAL REGULARITY

Ying-Jui Chen
Massachusetts Inst. of Technology
Cambridge, MA 02139, USA
yrchen@mit.edu

Soontorn Oraintara
Univ. of Texas at Arlington
Arlington, TX 76010, USA
oraintar@uta.edu

Kevin Amaratunga
Massachusetts Inst. of Technology
Cambridge, MA 02139, USA
kevina@mit.edu

ABSTRACT

This paper presents a lifting-domain design of filter banks with a given McMillan degree. It is based on the M-channel lifting factorizations of the degree-0 and 1 building blocks $\mathbf{I} - 2\mathbf{u}\mathbf{v}^\dagger$ and $\mathbf{I} - \mathbf{u}\mathbf{v}^\dagger + z^{-1}\mathbf{u}\mathbf{v}^\dagger$, with $\mathbf{v}^\dagger\mathbf{u} = 1$. Paraunitariness further requires $\mathbf{u} = \mathbf{v}$. The proposed lifting factorization has a unity diagonal scaling throughout, and guarantees perfect reconstruction (PR) even when the parameters are quantized. It is shown to be minimal in terms of the minimum number of delays required. Based on the lifting factorization, regularity of the FB can be structurally imposed, and reversible, possibly multiplierless, implementation of the FB can readily be derived. Design examples are given to illustrate the versatility of the proposed approach.

1. INTRODUCTION

The lifting scheme, or the ladder structure [7], has been widely used in both factorization [2] and design [3, 4] of two-channel FBs. It features efficient and in-place computation [2]. Reversible, and possibly integer, implementation can be readily derived from the lifting structure [8, 13, 11]. Fig. 1 shows a series of two lifting steps and their easy-to-compute inverses. In this paper, the M-channel lifting factorization [9] is employed to facilitate FB design.

Consider an M-channel FB. Let $\mathbf{E}(z)$, $M \times M$, be the polyphase matrix. It is said to be paraunitary (PU) if $\widetilde{\mathbf{E}}(z)\mathbf{E}(z) = \mathbf{I}$, where the $\widetilde{\ }$ operation stands for conjugate transpose $\dagger$ and time-reversal ($z \to z^{-1}$). In general, if $\mathbf{E}(z)$ is nonsingular, it is said to be biorthogonal (BO).

1.1. Paraunitary Filter Bank (PUFB) Construction

Any degree-N PUFB $\mathbf{E}(z)$ can be expressed as [1]

$$\mathbf{E}(z) = \mathbf{V}_N(z)\,\mathbf{V}_{N-1}(z)\cdots\mathbf{V}_1(z)\,\mathbf{R}$$

where $\mathbf{V}_m(z) = \mathbf{I} - \mathbf{v}_m\mathbf{v}_m^\dagger + z^{-1}\mathbf{v}_m\mathbf{v}_m^\dagger$ for some $\|\mathbf{v}_m\| = 1$ is the degree-1 PU building block, and $\mathbf{R}$ is unitary: $\mathbf{R}^\dagger\mathbf{R} = \mathbf{I}$. $\mathbf{R}$ can be factored [1] into Householder matrices $\mathbf{H}_m = \mathbf{I} - 2\,\mathbf{p}_m\mathbf{p}_m^\dagger$ with $\|\mathbf{p}_m\| = 1$ for $m = 1, 2, \ldots, M-1$. This structure is complete for any degree-N PUFB, and covers a larger class than the GenLOT [6].

1.2. Biorthogonal Filter Bank (BOFB) Construction

The above PU construction can be extended to the BO context. The following construction spans a large useful class of degree-N BOFBs:

$$\mathbf{E}(z) = \mathbf{W}_N(z)\,\mathbf{W}_{N-1}(z)\cdots\mathbf{W}_1(z)\,\mathbf{E}_0$$

where $\mathbf{W}_m(z) = \mathbf{I} - \mathbf{u}_m\mathbf{v}_m^\dagger + z^{-1}\mathbf{u}_m\mathbf{v}_m^\dagger$ for some $\mathbf{v}_m^\dagger\mathbf{u}_m = 1$ is the degree-1 BO building block, and $\mathbf{E}_0$ is nonsingular. $\mathbf{E}_0$ can be factored into a product of an upper triangular matrix and the generalized Householder matrices $\mathbf{G}_m = \mathbf{I} - 2\mathbf{u}_m\mathbf{v}_m^\dagger$ with $\mathbf{v}_m^\dagger\mathbf{u}_m = 1$ for $m = 1, 2, \ldots, M-1$. The normalization $\det(\mathbf{E}_0) = 1$ can be assumed in constructing $\mathbf{E}_0$.

1.3. Comparison of Dyadic and Lifting Implementations

As shown in [1], the orthogonality of $\mathbf{V}_m(z)$ and $\mathbf{H}_m$ can be guaranteed in their dyadic implementations when $\mathbf{v}_m$ and $\mathbf{p}_m$ are quantized; however, the property of unity scaling is lost. On the other hand, the proposed lifting implementation of $\mathbf{V}_m(z)$ and $\mathbf{H}_m$ ensures PR and unity scaling, but the orthogonality is lost when the lifting multipliers are quantized, since the corresponding $\mathbf{v}_m$ and/or $\mathbf{p}_m$ will no longer be unit-norm. In fact, the operations at the lifting steps can also be non-linear. For $\mathbf{W}_m(z)$ and $\mathbf{G}_m$, PR and unity scaling are always guaranteed when the lifting multipliers are quantized. The same is true for their dyadic implementations, except the scaling can not be preserved.

2. M-CHANNEL LIFTING FACTORIZATION

In the following treatment, the subscript m will be omitted from the building blocks to simplify notation, and $\overline{x} \triangleq -x$ is used.

2.1. Degree-1 Paraunitary Building Block $\mathbf{V}(z)$

In [9], it has been shown that $\mathbf{V}(z)$ can be lifting factorized as

$$\mathbf{V}(z) = \begin{bmatrix} 1 & & & \alpha_1 & & & \\ & \ddots & & \vdots & & & \\ & & 1 & \alpha_{r-1} & & & \\ & & & 1 & & & \\ & & & \alpha_{r+1} & 1 & & \\ & & & \vdots & & \ddots & \\ & & & \alpha_M & & & 1 \end{bmatrix} \begin{bmatrix} 1 & & & & & & \\ & \ddots & & & & & \\ & & 1 & & & & \\ \overline{\beta}_1 & \cdots & \overline{\beta}_{r-1} & z^{-1} & \overline{\beta}_{r+1} & \cdots & \overline{\beta}_M \\ & & & & 1 & & \\ & & & & & \ddots & \\ & & & & & & 1 \end{bmatrix}$$

$$\begin{bmatrix} 1 & & & & & & \\ & \ddots & & & & & \\ & & 1 & & & & \\ \beta_1 & \cdots & \beta_{r-1} & 1 & \beta_{r+1} & \cdots & \beta_M \\ & & & & 1 & & \\ & & & & & \ddots & \\ & & & & & & 1 \end{bmatrix} \begin{bmatrix} 1 & & & \overline{\alpha}_1 & & & \\ & \ddots & & \vdots & & & \\ & & 1 & \overline{\alpha}_{r-1} & & & \\ & & & 1 & & & \\ & & & \overline{\alpha}_{r+1} & 1 & & \\ & & & \vdots & & \ddots & \\ & & & \overline{\alpha}_M & & & 1 \end{bmatrix} \quad (1)$$

The structure is shown in Fig. 2, with $\alpha_i = v_i^*/v_r^*$, $\beta_i = v_i v_r^*$ for some $r \in \{1, 2, \ldots, M\}$ with $v_r \neq 0$. The structure requires

0-7803-7761-3/03/$17.00

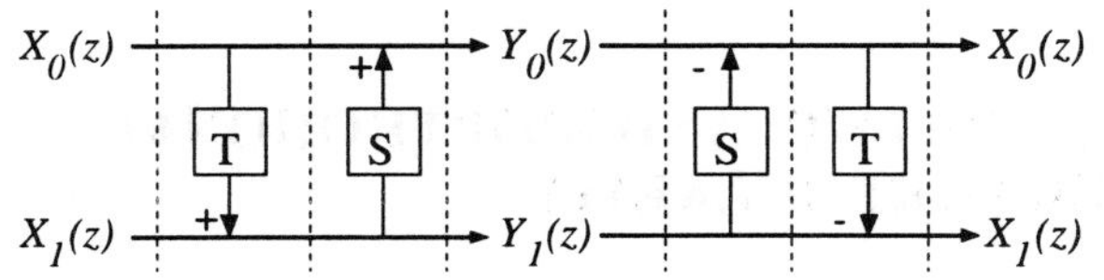

Figure 1: Two lifting steps with lifting multipliers $\mathbf{T}$ and $\mathbf{S}$ which can be any time-invariant, even nonlinear, systems. To invert, just subtract what was added in the reverse order.

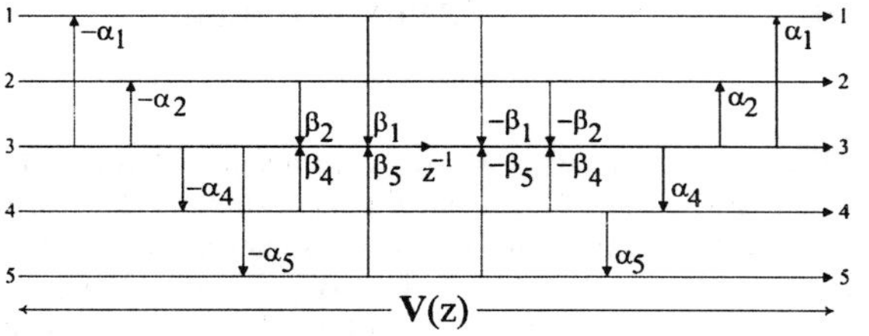

Figure 2: The M-channel lifting factorization of $\mathbf{V}(z)$, drawn for $M = 5$ and $r = 3$. The lifting factorization is minimal in both the McMillan sense and the number of independent lifting multipliers. When $z = 1$, all the lifting steps cancel each other, and the system reduces to the identity matrix, as dictated by $\mathbf{I} - \mathbf{v}\mathbf{v}^\dagger + z^{-1}\mathbf{v}\mathbf{v}^\dagger$. It has a unity scaling throughout, suitable for reversible integer implementation of $\mathbf{V}(z)$.

$4(M-1)$ lifting steps and only 1 delay to implement, and thus is minimal in the McMillan sense. It is also minimal because exactly $2(M-1)$ design variables (or $M-1$ for the real case) are needed. The α_i are the design variables, and the β_i are related to α_i according to

$$\beta_i = \frac{\alpha_i^*}{1 + \sum_{k=1,k\neq r}^{M} |\alpha_k|^2} \tag{2}$$

to ensure paraunitariness. Reversible, possibly integer, implementation of $\mathbf{V}(z)$ is readily available under the lifting structure, as in [14, 13, 12, 11].

2.2. Householder Matrix H

An $M \times M$ Householder matrix is a matrix of the form $\mathbf{H} = \mathbf{I} - 2\mathbf{p}\mathbf{p}^\dagger$, where $\|\mathbf{p}\| = 1$ reflects a vector in $\mathbb{C}^M$ with respect to a plane having $\mathbf{p}$ as normal. $\mathbf{H}$ is a special case of $\mathbf{V}(z)$ when $z = -1$, and its lifting factorization reads

$$\mathbf{H} = \begin{bmatrix} 1 & & \alpha_1 & & \\ & \ddots & \vdots & & \\ & & 1\ \alpha_{r-1} & & \\ & & 1 & & \\ & & \alpha_{r+1}\ 1 & & \\ & & \vdots & \ddots & \\ & & \alpha_M & & 1 \end{bmatrix} \begin{bmatrix} 1 & & 2\overline{\beta}_1 & & \\ & \ddots & \vdots & & \\ & & 1\ 2\overline{\beta}_{r-1} & & \\ & & -1 & & \\ & & 2\overline{\beta}_{r+1}\ 1 & & \\ & & \vdots & \ddots & \\ & & 2\overline{\beta}_M & & 1 \end{bmatrix}^T \begin{bmatrix} 1 & & \overline{\alpha}_1 & & \\ & \ddots & \vdots & & \\ & & 1\ \overline{\alpha}_{r-1} & & \\ & & 1 & & \\ & & \overline{\alpha}_{r+1}\ 1 & & \\ & & \vdots & \ddots & \\ & & \overline{\alpha}_M & & 1 \end{bmatrix} \tag{3}$$

containing $3(M-1)$ lifting steps. The $\alpha_i = p_i^*/p_r^*$ are the design variables, and the β_i are related to α_i as in (2) to ensure paraunitariness.

2.3. Degree-1 Biorthogonal Building Block W(z)

Similarly to $\mathbf{V}(z)$ above, $\mathbf{W}(z)$ can be lifting factorized as

$$\mathbf{W}(z) = \begin{bmatrix} 1 & & & & \\ & \ddots & & & \\ & & 1 & & \\ \alpha_1 & \cdots\ \alpha_{r-1} & 1 & \alpha_{r+1}\ \cdots & \alpha_M \\ & & & 1 & \\ & & & \ddots & \\ & & & & 1 \end{bmatrix} \begin{bmatrix} 1 & & \lambda_1 & & \\ & \ddots & \vdots & & \\ & & 1\ \lambda_{r-1} & & \\ & & 1 & & \\ & & \lambda_{r+1}\ 1 & & \\ & & \vdots & \ddots & \\ & & \lambda_M & & 1 \end{bmatrix}$$

$$\cdot \begin{bmatrix} 1 & & \overline{\lambda}_1 & & \\ & \ddots & \vdots & & \\ & & 1\ \overline{\lambda}_{r-1} & & \\ & & z^{-1} & & \\ & & \overline{\lambda}_{r+1}\ 1 & & \\ & & \vdots & \ddots & \\ & & \overline{\lambda}_M & & 1 \end{bmatrix} \begin{bmatrix} 1 & & & & \\ & \ddots & & & \\ & & 1 & & \\ \overline{\alpha}_1 & \cdots\ \overline{\alpha}_{r-1} & 1 & \overline{\alpha}_{r+1}\ \cdots & \overline{\alpha}_M \\ & & & 1 & \\ & & & \ddots & \\ & & & & 1 \end{bmatrix} \tag{4}$$

It requires $4(M-1)$ lifting steps and only 1 delay to implement, and is minimal in the McMillan sense. Unlike for the PU case, now both α_i and λ_i in (4) are design variables. They are related to $\mathbf{v}$ and $\mathbf{u}$ by $\alpha_i = -v_i^*/v_r^*$ and $\lambda_i = u_i v_r^*$ for some $r \in \{1, 2, \ldots, M\}$ with $v_r \neq 0$. The condition $\mathbf{v}^\dagger\mathbf{u} = 1$ is structurally imposed by the lifting factorization. Again, reversible integer implementation of $\mathbf{W}(z)$ is readily available based on the structure.

2.4. Generalized Householder Matrix G

The biorthogonal matrix $\mathbf{G} = \mathbf{I} - 2\mathbf{u}\mathbf{v}^\dagger$ with $\mathbf{v}^\dagger\mathbf{u} = 1$ is a generalization of the Householder matrix. Its lifting factorization can be found similarly as for $\mathbf{W}(z)$ with $z = 1$, and reads

$$\mathbf{G} = \begin{bmatrix} 1 & & \alpha_1 & & \\ & \ddots & \vdots & & \\ & & 1\ \alpha_{r-1} & & \\ & & 1 & & \\ & & \alpha_{r+1}\ 1 & & \\ & & \vdots & \ddots & \\ & & \alpha_M & & 1 \end{bmatrix}^T \begin{bmatrix} 1 & & 2\overline{\lambda}_1 & & \\ & \ddots & \vdots & & \\ & & 1\ 2\overline{\lambda}_{r-1} & & \\ & & -1 & & \\ & & 2\overline{\lambda}_{r+1}\ 1 & & \\ & & \vdots & \ddots & \\ & & 2\overline{\lambda}_M & & 1 \end{bmatrix} \begin{bmatrix} 1 & & \overline{\alpha}_1 & & \\ & \ddots & \vdots & & \\ & & 1\ \overline{\alpha}_{r-1} & & \\ & & 1 & & \\ & & \overline{\alpha}_{r+1}\ 1 & & \\ & & \vdots & \ddots & \\ & & \overline{\alpha}_M & & 1 \end{bmatrix}^T \tag{5}$$

where $\alpha_i = -v_i^*/v_r^*$ and $\lambda_i = u_i v_r^*$. It requires $3(M-1)$ lifting steps.

3. IMPOSITION OF REGULARITY AND LINEAR PHASE

The Householder/generalized Householder factorization of a unitary/nonsingular matrix [1] and the corresponding lifting structures (3) and (5) allow for structurally imposing at least one degree of regularity. The idea is to enforce the synthesis low-pass filter $F_0(z)$ to have zeros at $z = \exp(j\frac{2\pi k}{M})$, $k = 1, 2, \cdots, M-1$. Then the analysis band-pass and high-pass filters $H_k(z)$, $k = 1, 2, \cdots, M-1$, automatically have zero DC responses. This ensures no DC-leakage in the analysis bank. The following theorems are proved in [10].

Theorem 1 *Given an M-channel PUFB $\mathbf{E}(z) = \prod \mathbf{V}_m(z)\mathbf{R}$, the Householder-based lifting factorization of $\mathbf{R}$ as shown in (6) for $M = 4$. Then, setting $x_1 = \cdots = x_{M-1} = \frac{1}{\sqrt{M+1}}$ structurally imposes one regularity on the FB.*

$$\mathbf{E}(z) = \begin{bmatrix} 1 & & & \\ x_7 & 1 & & \\ x_8 & & 1 & \\ x_9 & & & 1 \end{bmatrix} \begin{bmatrix} z^{-1} & \overline{y}_7 & \overline{y}_8 & \overline{y}_9 \\ & 1 & & \\ & & 1 & \\ & & & 1 \end{bmatrix} \begin{bmatrix} 1 & y_7 & y_8 & y_9 \\ & 1 & & \\ & & 1 & \\ & & & 1 \end{bmatrix} \begin{bmatrix} 1 & & & \\ \overline{x}_7 & 1 & & \\ \overline{x}_8 & & 1 & \\ \overline{x}_9 & & & 1 \end{bmatrix} \left(\begin{bmatrix} 1 & & & \\ x_1 & 1 & & \\ x_2 & & 1 & \\ x_3 & & & 1 \end{bmatrix} \begin{bmatrix} -1 & y_1 & y_2 & y_3 \\ & 1 & & \\ & & 1 & \\ & & & 1 \end{bmatrix} \right.$$

$$\left. \begin{bmatrix} 1 & & & \\ \overline{x}_1 & 1 & & \\ \overline{x}_2 & & 1 & \\ \overline{x}_3 & & & 1 \end{bmatrix} \begin{bmatrix} 1 & & & \\ & 1 & & \\ & x_4 & 1 & \\ & x_5 & & 1 \end{bmatrix} \begin{bmatrix} 1 & & & \\ & -1 & y_4 & y_5 \\ & & 1 & \\ & & & 1 \end{bmatrix} \begin{bmatrix} 1 & & & \\ & 1 & & \\ & \overline{x}_4 & 1 & \\ & \overline{x}_5 & & 1 \end{bmatrix} \begin{bmatrix} 1 & & & \\ & 1 & & \\ & & 1 & \\ & & x_6 & 1 \end{bmatrix} \begin{bmatrix} 1 & & & \\ & 1 & & \\ & & -1 & y_6 \\ & & & 1 \end{bmatrix} \begin{bmatrix} 1 & & & \\ & 1 & & \\ & & 1 & \\ & & \overline{x}_6 & -1 \end{bmatrix} \right)^T . \quad (6)$$

$$\mathbf{E}(z) = \begin{bmatrix} 1 & x_{19} & x_{20} & x_{21} \\ & 1 & & \\ & & 1 & \\ & & & 1 \end{bmatrix} \begin{bmatrix} 1 & & & \\ x_{22} & 1 & & \\ x_{23} & & 1 & \\ x_{24} & & & 1 \end{bmatrix} \begin{bmatrix} z^{-1} & & & \\ \overline{x}_{22} & 1 & & \\ \overline{x}_{23} & & 1 & \\ \overline{x}_{24} & & & 1 \end{bmatrix} \begin{bmatrix} 1 & \overline{x}_{19} & \overline{x}_{20} & \overline{x}_{21} \\ & 1 & & \\ & & 1 & \\ & & & 1 \end{bmatrix} \left(\begin{bmatrix} -1 & x_1 & x_2 & x_3 \\ & 1 & & \\ & & 1 & \\ & & & 1 \end{bmatrix} \begin{bmatrix} 1 & & & \\ x_4 & 1 & & \\ x_5 & & 1 & \\ x_6 & & & 1 \end{bmatrix} \begin{bmatrix} 1 & \overline{x}_1 & \overline{x}_2 & \overline{x}_3 \\ & 1 & & \\ & & 1 & \\ & & & 1 \end{bmatrix} \right.$$

$$\left. \begin{bmatrix} 1 & & & \\ & -1 & x_7 & x_8 \\ & & 1 & \\ & & & 1 \end{bmatrix} \begin{bmatrix} 1 & & & \\ & 1 & & \\ & x_9 & 1 & \\ & x_{10} & & 1 \end{bmatrix} \begin{bmatrix} 1 & & & \\ & 1 & \overline{x}_7 & \overline{x}_8 \\ & & 1 & \\ & & & 1 \end{bmatrix} \begin{bmatrix} 1 & & & \\ & 1 & & \\ & & -1 & x_{11} \\ & & & 1 \end{bmatrix} \begin{bmatrix} 1 & & & \\ & 1 & & \\ & & 1 & \\ & & x_{12} & 1 \end{bmatrix} \begin{bmatrix} 1 & & & \\ & 1 & & \\ & & 1 & \overline{x}_{11} \\ & & & 1 \end{bmatrix} \begin{bmatrix} 1 & x_{13} & x_{14} & x_{15} \\ & 1 & x_{16} & x_{17} \\ & & 1 & x_{18} \\ & & & -1 \end{bmatrix} \right)^{-1} . \quad (7)$$

x_1	1/3	x_4	-43/64	x_7	47/128
x_2	1/3	x_5	-173/256	x_8	21/128
x_3	1/3	x_6	1/64	x_9	-9/128

Table 1: The design variables for the 4-channel PUFB shown in Fig. 3. The resulting FB is structurally 1-regular due to the predetermined x_1, x_2 and x_3. The coding gain is 8.0287 dB.

Theorem 2 *Given an M-channel BOFB $\mathbf{E}(z) = \prod \mathbf{W}_m(z)\mathbf{E}_0$, $\mathbf{E}_0$ can be lifted as shown in (7) for $M = 4$. Then, setting $x_1 = \cdots = x_{M-1} = \frac{1}{\sqrt{M-1}}$ and $x_M = \cdots = x_{2M-1} = \frac{1}{\sqrt{M}}$ structurally imposes one regularity on the analysis bank of the FB.*

The imposition of linear-phase property has been proposed in [6]. However, this remains an open question when the factorization using degree-1 building blocks are used. The following theorem gives a necessary condition for the case of minimal degree-1 filter bank.

Theorem 3 *A filter bank whose analysis polyphase matrix is of the form $\mathbf{E}(z) = \mathbf{V}_1(z)\,\mathbf{R}$ cannot have linear-phase impulse responses except for $M = 2$.*

4. DESIGN EXAMPLE

4.1. Paraunitary Filter Bank

The first example is a real-valued PUFB based on the proposed lifting factorization. A 4-channel polyphase matrix of the form $\mathbf{E}(z) = \mathbf{V}_1(z)\,\mathbf{R}$ is parameterized as given in (6). This is a degree-1 PUFB with length-8 analysis filters. In (6), the x_i are the design variables, and the y_i are chosen so as to satisfy paraunitariness. Having the structure in place, an unconstrained optimization of stopband attenuation and the coding gain is performed over the x_i, with x_1, x_2, and x_3 predetermined as in Theorem 1. The resulting x_i are floating-point numbers. A good binary or dyadic approximation of them can be found using the algorithm in [11]. Table 1 consists of the resulting lifting multipliers x_i. The frequency response is shown in Fig. 3 with coding gain 8.0285dB. As a comparison, the coding gains of 4-pt DCT and 4-channel LOT [5] are 7.5701 and 7.9259dB, respectively. Furthermore, the lifting structure exhibits fast convergence: starting with $x_4 = x_5 = \cdots = x_9 = 0$ and $x_1 = x_2 = x_3 = 1/3$ and an error tolerance of 1e-6, it takes 13 iterations to converge. A good dynamic range performance results since all lifting multipliers are all bounded by 1 in magnitude, suitable for reversible integer implementation of the system.

x_1	1	x_9	-1/64	x_{17}	-1/8
x_2	1	x_{10}	-97/128	x_{18}	-7/128
x_3	1	x_{11}	1	x_{19}	147/64
x_4	1/2	x_{12}	1	x_{20}	5/4
x_5	1/2	x_{13}	0	x_{21}	39/64
x_6	1/2	x_{14}	0	x_{22}	-33/128
x_7	1/16	x_{15}	0	x_{23}	-21/128
x_8	-1/2	x_{16}	0	x_{24}	-5/64

Table 2: The lifting multipliers for the 4-channel BOFB shown in Fig. 4. $x_1, x_2, \cdots, x_6$ are predetermined so that the analysis bank is 1-regular. The coding gain is 8.2429 dB.

4.2. Biorthogonal Filter Bank

We now consider a degree-1, real-valued, BOFB with four channels. The target polyphase matrix takes the form $\mathbf{E}(z) = \mathbf{W}_1(z)\mathbf{E}_0$. The lifting structure of $\mathbf{E}(z)$ as given in (7) is used to design the filter bank, where $\det(\mathbf{E}_0) = 1$ is assumed. It has 24 design parameters $x_1, \cdots, x_{24}$, of which $x_1, \cdots, x_6$ are so chosen according to Theorem 2. Then, an unconstrained optimization is performed over the remaining design variables, the objective being the coding gain and the analysis and synthesis stopband attenuations. Again, the resulting x_i are subject to the multiplierless approximation [11] as above. Table 2 consists of these multiplierless x_i. Fig. 4 shows the frequency response of the FB, with coding gain 8.2429 dB, higher than DCT, LOT and the PUFB in the previous example.

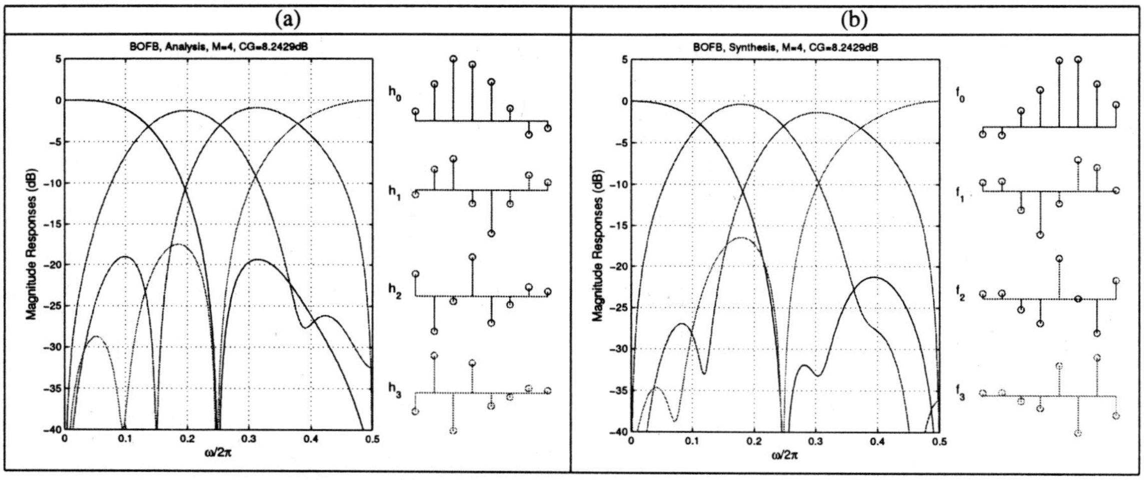

Figure 4: Basis functions and their magnitude responses of the 4-channel BOFB in Table 2: (a) analysis, (b) synthesis. The analysis bank is structurally 1-regular due to the special choice of lifting multipliers $x_1, x_2, \cdots, x_6$. The coding gain is 8.2429 dB.

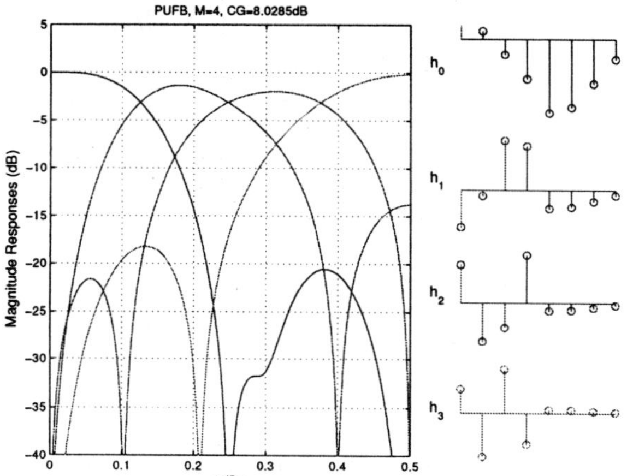

Figure 3: Basis functions and their magnitude responses of the 4-channel PUFB in Table 1. The FB is structurally 1-regular due to the special choice of lifting multipliers x_1, x_2, and x_3. The coding gain is 8.0285 dB.

5. CONCLUSION

A novel filter bank design approach is presented based on the modular, implementationally efficient, and reversible M-channel lifting factorization. Paraunitary and biorthogonal cases are discussed and design examples given. Regularity of the filter bank can be structurally imposed through the lifting multipliers. Furthermore, filter banks are designed directly in the lifting domain, avoiding the need for later factorization. Future research will focus on imposing more properties into the lifting structure.

6. REFERENCES

[1] P. P. Vaidyanathan, *Multirate Systems and Filter Banks*, Prentice Hall, 1992.

[2] I. Daubechies and W. Sweldens, *Factoring Wavelet Transforms into Lifting Steps*, J. Fourier Anal. Appl., vol. 4, no. 3, pp. 247-269, 1998.

[3] W. Sweldens, *The lifting scheme: A custom-design construction of biorthogonal wavelets*, Appl. Comput. Harmon. Anal., vol. 3, no. 2, pp. 186-200, 1996.

[4] G. Strang and T. Nguyen, *Wavelets and Filter Banks*, Wellesley-Cambridge, 1996.

[5] H.S. Malvar, *Signal Processing with Lapped Transforms*, Artech House, Norwood, MA 1992.

[6] R.L. de Queiroz, T.Q. Nguyen and K.R. Rao, *The GenLOT: Generalized linear-phase lapped orthogonal transform*, IEEE Trans. Signal Processing, vol. 44 no. 3 pp. 497-507, 1996.

[7] Fons A. M. L. Bruekers and Ad W. M. van den Enden, *New Networks for Perfect Inversion and Perfect Reconstruction*, IEEE Journal on Selected Areas in Communications, vol. 10, no. 1, Jan. 1992.

[8] R. Calderbank and I. Daubechies and W. Sweldens and B.-L. Yeo, *Wavelet transforms that map integers to integers*, Appl. Comput. Harmon. Anal., vol. 5, no. 3, pp. 332-369, 1998.

[9] Y.-J. Chen and K. Amaratunga, *M-channel Lifting factorization of perfect reconstruction filter banks and reversible M-band wavelet transforms*, submitted to IEEE Trans. Circuits and Systems—II,Oct. 2002.

[10] Y.-J. Chen, S. Oraintara, and K. Amaratunga, *Regular Paraunitary Filter Banks and M-band Orthogonal Wavelets With Structural Vanishing Moments*, in preparation.

[11] Y.-J. Chen, S. Oraintara, T. Tran, K. Amaratunga, and T. Q. Nguyen, *Multiplierless Approximation of Transforms with Adder Constraint*, IEEE Signal Processing Letters, vol. 9, no. 11, Nov. 2002.

[12] S. C. Chan, W. Liu, and K. L. Ho, *Multiplierless perfect reconstruction modulated filter banks with sum-of-powers-of-two coefficients*, IEEE Signal Processing Letters, pp. 163-166, June 2001.

[13] S. Oraintara, Y.-J. Chen, and T. Q. Nguyen, *Integer Fast Fourier Transform*, IEEE Trans. Signal Processing, Mar. 2002.

[14] J. Liang and T. D. Tran, *Fast multiplierless approximations of the DCT with the lifting scheme*, IEEE Trans. on Signal Processing, Dec. 2001.

FILTER STRUCTURES FOR DECIMATION: A COMPARISON

Artur Wróblewski and Josef A. Nossek

Munich University of Technology
Arcisstr. 21, 80333 Munich, Germany
e–mail: Artur.Wroblewski@ei.tum.de

ABSTRACT

In this paper three different digital filter structures for decimation by a factor of 64 are compared. They are based on a commercially available decimator AD7722 by Analog Devices. The AD 7722 is a complete low power $\Sigma\Delta$ analog-to-digital converter with the application in digital audio processing. The applied filter is a two-stage decimator which decreases the sample rate by 32 and 2 respectively. This structure is then compared to a state-of-the-art approach of a FIR decimator with six stages, each of them performing decimation by a factor of 2, and to two different approaches to an almost linear phase IIR decimator based on an very efficient class of lattice wave digital filters with also six down-samplers. The advantages and disadvantages of the selected structures are described.

1. INTRODUCTION

In today's advanced technologies power consumption has become the main limiting factor for many applications. Starting with processors for personal computing through terminals for mobile communications systems, down to applications like bionic ear. In many of those applications digital filtering is a very important issue and represents one of the most power consuming subsystems. Therefore it is important to compare different digital filter architectures to find the one that best fits the requirements for low-power design while maintaining all properties of a good filtering operation. The AD7722 analog-to-digital converter (ADC) [3] employs a $\Sigma\Delta$ conversion technique that converts the analog input into a digital pulse train. Due to high oversampling rate which spreads the quantization noise from 0 to $f_{CLKIN}/2$, the noise energy contained in the band of interest is reduced. The digital filter that follows removes the out of band quantization noise and reduces the data rate from f_{CLKIN} at the input of the filter to $f_{CLKIN}/64$ at the output. The output data rate is a little over twice the signal bandwidth, which guarantees that there is no loss of data in the signal band. The following filter specifications are required:

Input Sampling Frequency: 12.5 MHz
Output Sampling Frequency: 195.3 kHz
Passband(0kHz-90.625kHz) Ripple: 0.005
Stop-band(104.6875kHz-12.395Mhz) Attenuation: 90 dB

In the following different realizations of this digital filter are presented. In Section 2 a two-stage Finite Impulse Response (FIR) structure based on original solution by Analog Devices is given. In Section 3 a more efficient realization based on six half-band FIR filters is described. Section 4 shows two six-stage almost linear phase Infinite Impulse Response (IIR) realizations based on bireciprocal lattice Wave Digital Filters (WDF). The experimental results of Section 5 are followed by conclusions.

2. TWO-STAGE FIR DECIMATOR

The original solution employs two FIR filters in a cascade. The first filter is a 384-tap filter that samples the output of the modulator at f_{CLKIN}. The second filter is a 151-tap half-band filter that samples the output of the first filter at $f_{CLKIN}/32$ and decimates by 2. Half-band filters mark a special case of filters for which

$$\delta_s = \delta_p = \delta$$

$$\omega_s = \pi - \omega_p$$

where δ_p and δ_s are passband and stop-band ripples, ω_p and ω_s are passband and stop-band frequencies, respectively. Then the resulting equiripple optimal solution to the approximation problem has the property that

$$H(e^{j\omega}) = 1 - H(e^{j(\pi-\omega)})$$

which means that the frequency response is symmetric around $\omega = 1/2$ and we get

$$H(e^{j\pi/2}) = 0.5$$

It can be shown, that for such a filter every other impulse response coefficient is exactly 0. Thus, an additional factor of two reduction in computation is obtained [2].

The implementation of the original solution results in a filter with a group delay of 81 sampling periods. The filters are of course linear phase and thus symmetric, which reduces the number of multiplications per sample by a factor of two. Moreover, half of the coefficients of the half-band filter equal 0 which again reduces the number of multiplications. Nevertheless the great disadvantage of this structure is that the first filter runs at a very high sampling rate (f_{CLKIN}). Thus the computational effort per output sample is rather high.

3. SIX-STAGE HALF-BAND FILTER

To improve the computational efficiency of the decimator a cascade of six half-band FIR filters, as in the Fig. 2, can be applied.

The resulting filter is linear phase with a group delay of 81.5 and the passband behaviour is even better than in case of the two-stage filter. There are obvious advantages. The total filter order

0-7803-7761-3/03/$17.00 ©2003 IEEE

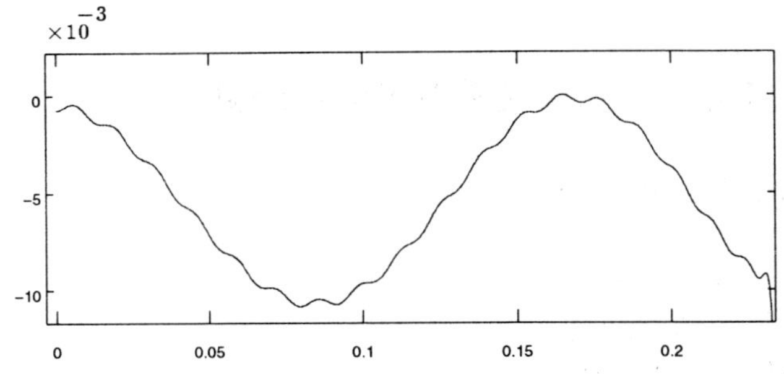

Figure 1: Passband ripple(dB) of the two-stage FIR decimator.

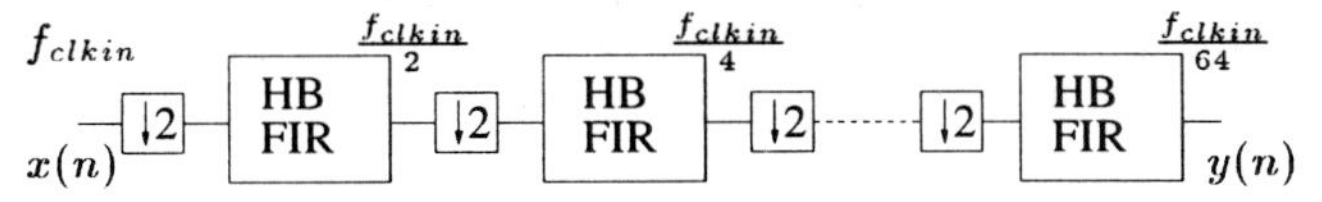

Figure 2: Decimator based on six half-band filters.

could be lowered from 535 to 196.0 All filters in the cascade are half-band, which allows them to run at the lower output sample rate. Thus, even the first filter (order 6) performs computations at $f_{CLKIN}/2$. There's, of course, no impact on the performance of the filter.

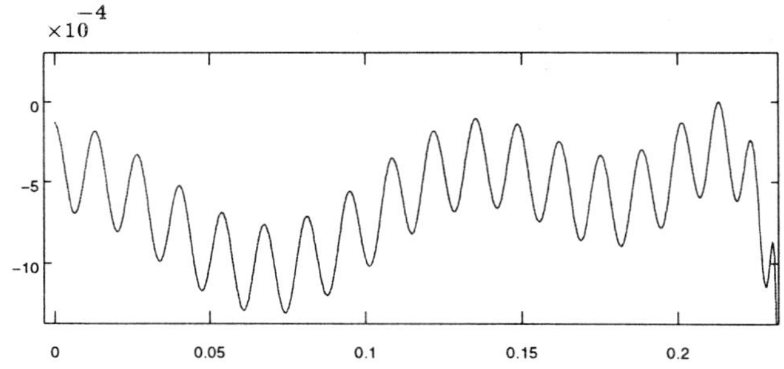

Figure 3: Passband ripple(dB) of the six-stage FIR half-band decimator.

4. SIX-STAGE WAVE DIGITAL FILTER

Wave digital filters (WDFs) are known to have many advantageous properties. They have low coefficient sensitivity, good dynamic range, and especially, good stability properties under quantization effects. Out of all wave digital filters the lattice wave digital filter is the most attractive one. Each WDF has a corresponding filter in a reference domain. The design can therefore be carried out in the analog domain using classical filter approximations. Then a transformation from analog to digital domain can be performed. For lattice WDF explicit formulae are given in [7]. However there exist no closed form solutions for filters satisfying given requirements on both magnitude and phase response.

A lattice WDF is a two-branch structure where each branch realizes an all-pass filter [5]. Out of several ways of realizing them [4] the most attractive one is to use cascaded first-order and second-order sections. They are realized using symmetric two-port adaptors. A bireciprocal (half-band) lattice WDF is a special case of lattice WDF. In this case every other coefficient of the filter becomes 0. This results in a structure shown in Fig. 4. Moreover, when the application is in a decimator by a factor of 2, the filter can run at the output sampling rate [6].

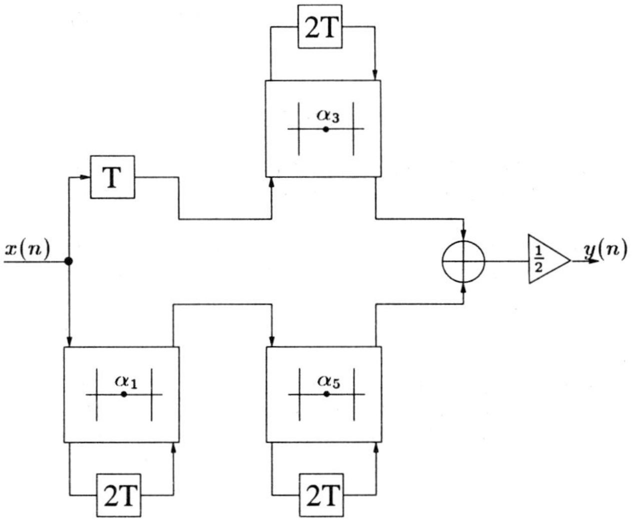

Figure 4: A 7th-order bireciprocal lattice wave digital filter.

The transfer function of a bireciprocal lattice WDF can be written as

$$H(z) = \frac{1}{2}(H_0(z^2) + z^{-1}H_1(z^2))$$

where the transfer function $H_0(z^2)$ corresponds to the lower branch in Fig. 4. The transfer function of the filter and its complementary transfer function are power complementary. For bireciprocal lattice WDFs we therefore have

$$|H(e^{j\omega T})|^2 + |H(e^{j\omega T - \pi})|^2 = 1$$

which means that the passband and stop-band edges are related by $\omega_p T + \omega_s T = \pi$, with T being the sampling time normalized to 2π. The consequence is that the passband ripple will be extremely small for practical requirements on the stop-band attenuation. Thus the bireciprocal WDFs have the efficiency of a FIR half-band filter in terms of reduced computational effort, while preserving the main advantages of IIR filters over FIR, which are sharp transitions for low order filters. Moreover, as will be discussed later, the six-stage WDF solution is more efficient than its FIR counterpart. The plots of magnitude, phase and group delay responses of the not equalized filter are shown in Fig. 5 and 6.

The main drawback of lattice WDFs is the non-linear phase response. However, many methods for obtaining almost linear phase of IIR filters have been presented in the past [9][11][1]. The first method used here is based on all-pass approximation presented in [10].

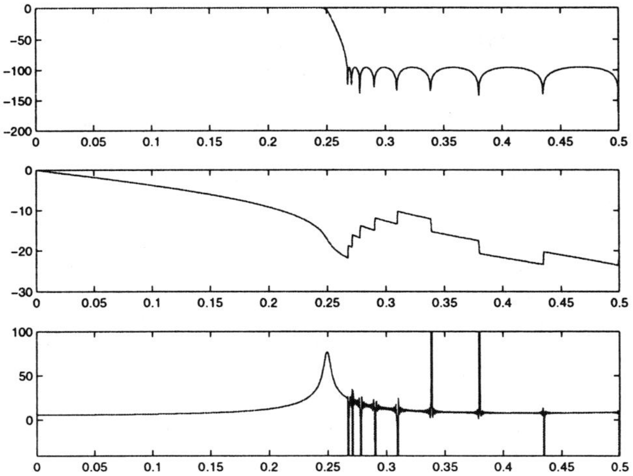

Figure 5: Magnitude(dB), phase(radians) and group delay(sampling intervals) of the six-stage bireciprocal lattice WDF decimator

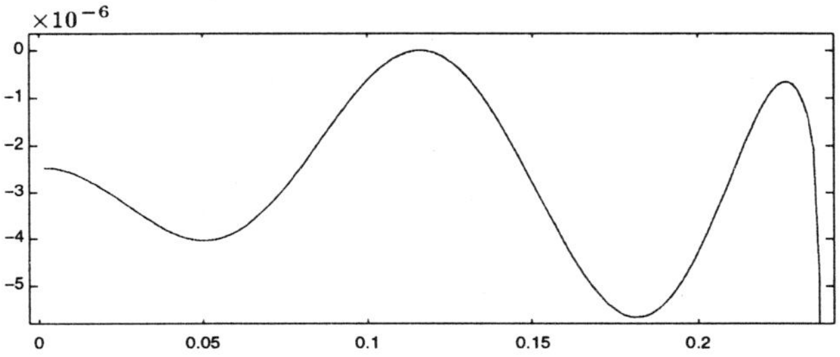

Figure 6: Passband ripple(dB) of the six-stage bireciprocal lattice WDF decimator.

As can be seen in Fig. 7 even with a low order all-pass-equalizer a very good approximation of phase linearity can be obtained. For an order 8 all-pass phase error is less than 0.13 radians, and for order 22 - below $4 \cdot 10^{-3}$ radians. Going as high as for order 40 the equalizer still offers savings of more than 20% accompanied by a meaningless phase error of $4.78 \cdot 10^{-5}$ radians. For the realization of the equalizer we again propose to apply wave digital filters. Also for this purpose they are very efficient and only one multiplier per equalizer order is required. The further advantage of the filter is a low group delay which ranges from 35 with an order 8 equalizer to 55 with order 22 equalizer. Even with an equalizer of order 40 the group delay is only slightly higher than in the case of the FIR.

In this paper we concentrate on bireciprocal lattice wave digital filters since they represent the most efficient, in terms of computational effort, family of IIR filters and are therefore of great interest. It is therefore very important to take a look at the methods dedicated to the design of linear phase bireciprocal lattice WDFs to be able to compare this solution to all-pass equalization. It is possible to obtain a bireciprocal lattice WDF with approximately linear phase by letting one of the branches in Fig. 4 consist of pure

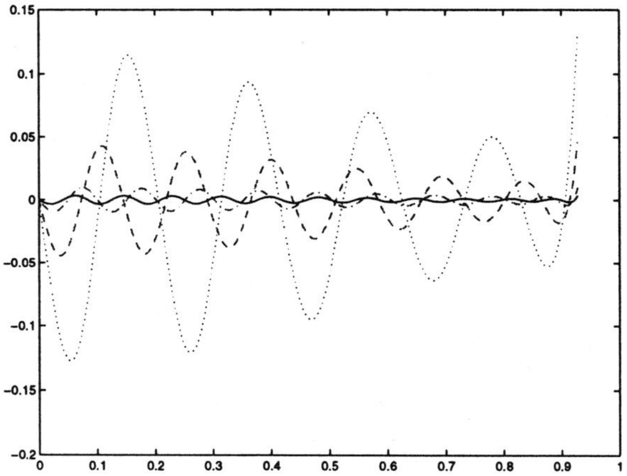

Figure 7: Phase error(radians) in the passband of the equalized bireciprocal lattice WDF decimator for different orders of the equalization all-pass (dotted - order 8, dashed - order 12, dash-dotted - order 18, solid - order 22).

delays [9][11][8]. The other branch is a general all-pass function in z^2, which can be realized using cascaded first and second orders sections (Fig. 8). Even if the order of such designed filter is

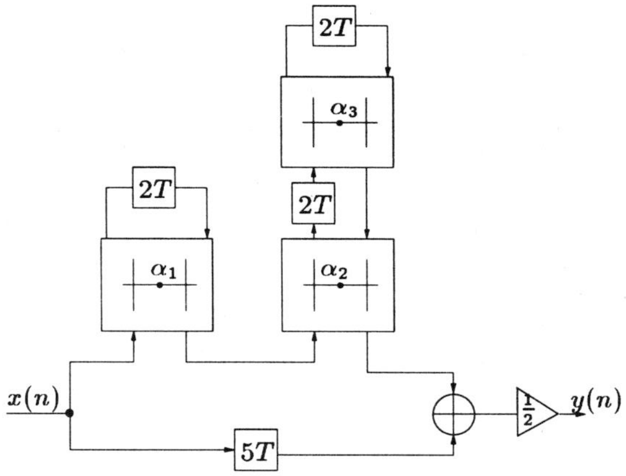

Figure 8: Structure of an 11th order almost linear phase bireciprocal lattice WDF.

much higher than that of the corresponding minimum phase solution, one has to take into account that the number of multipliers in this structure is only $(order + 1)/4$. These filters are thus as efficient, when comparing effort per filter order, as FIR half-band filters and a comparison to FIR solution is straightforward. And of course no additional equalizer is required.

Moreover, it is a well known fact, that wave digital filters have very low coefficient sensitivity. Thus it is possible to represent filter coefficients utilizing only a few bits. This could allow for decreasing the size of applied multipliers or even replacing them by shift and add operations.

5. EXPERIMENTAL RESULTS

The comparison of the computational effort of the described filter architectures is given in Table 1. For each filter, order, number of multiplications per output sample, phase error and group delay are given. Also shown is the reduction in computational complexity compared to FIR half-band approach. Since the FIR half-band solution is the state-of-the-art approach and its phase is linear, it has been taken as a benchmark. It is no surprise that the minimum phase WDF is much more efficient than FIR. But even with an excellent phase linearity given with an order 22 equalizer (phase error of only $4 \cdot 10^{-3}$ radians) the number of multiplications per output sample could be considerably reduced (by 30 %). Even better phase linearity (phase error in the range of 10^{-5} radians) has been achieved with the almost linear phase WDF of Fig. 8 with no loss in computational efficiency. However, only the all-pass equalization gives more degrees of freedom allowing the designer to influence the complexity by choosing how good or bad the approximation of the linear phase will be. Also only one equalizer is needed for all filters in the cascade. In many practical cases the requirements on the phase error will be orders magnitude lower than 10^{-5} and applying an equalizer could lead to an advantageous solution.

Architecture	Filter Order Per Stage	MPS (SAV)	Phase Error	Group Delay
6-Stage HB FIR	6/6/6/10/ 18/150	235 (0%)	0	81.5
6-Stage WDF (no equalizer)	5/5/5/7/ 9/17	141 (40%)	1.97	24
6-Stage WDF (order 8 eq) phase equalizer)	5/5/5/7/ 9/25	149 (36%)	0.129	31
6-Stage WDF (order 14 eq) phase equalizer)	5/5/5/7/ 9/31	155 (34%)	$2 \cdot 10^{-2}$	39
6-Stage WDF (order 22 eq) phase equalizer)	5/5/5/7/ 9/39	163 (30%)	$4 \cdot 10^{-3}$	55
6-Stage WDF (order 40 eq) phase equalizer)	5/5/5/7/ 9/57	181 (23%)	$4 \cdot 10^{-5}$	92
6-Stage almost linear phase WDF	8/8/8/8/ 16/120	158 (33%)	$8 \cdot 10^{-5}$	78

Table 1: Comparison of the three architectures (Phase error in radians, group delay in sampling intervals). MPS - Multiplications Per (Output) Sample, SAV - Savings in percent compared to half-band FIR.

6. CONCLUSION

In this paper a comparison of different decimator architectures has been presented. The results show significant differences in computational effort for their realization. Despite its non-linear phase response the proposed realization based on lattice wave digital filters seems to be advantageous. The computational effort can be varied depending on requirements on phase linearity. For strong linearity contraints almost linear phase WDF makes a better choice, otherwise phase equalization should be preferred.

It should be noted, that this comparison takes into account only the number of multiplications per sample. No attempt has been made here to introduce and compare general methods for low power design like numbers representation, retiming or transistor sizing. Most of these methods are independent of the chosen filter family and can be applied to FIR as well as IIR solutions. The WDF implementation presented here seems to gain on importance from the point of view of low-power design and almost linear phase IIR filters may represent a good alternative to today's standards. Of course the applications are not limited to filters for $\Sigma\Delta$ audio ADC like the one presented here. In fact, it could be applied in any kind of portable devices ranging from MP3-Players to terminals for mobile communication systems, where computational efficiency is extremely important and power consumption the main limiting factor. Especially in mobile communications, where external interferers like multi-path propagation do not allow for perfect symbol synchronization, strict phase linearity may not be a required feature.

7. REFERENCES

[1] M. Abo-Zahhad, M. Yaseen, and T. Henk. Design of Lattice Wave Digital Filters with Prescribed Loss and Phase Specifications. *ECCTD'95 Istanbul, Turkey*, 1:761–764, 1995.

[2] R. E. Crochiere and L. E. Rabiner. Multirate Digital Signal Processing. *Prentice Hall*, 1983.

[3] Analog Devices. AD7722 16-bit, 195kSPS CMOS, Sigma-Delta ADC. *Technical Specification*, 1996.

[4] A. Fettweis. Wave Digital Filters: Theory and Practice. *Proc. IEEE*, 74(2):270–327, February 1986.

[5] A. Fettweis, H. Levin, and A. Sedlmeyer. Wave Digital Lattice Filters. *International Journal on Circuit Theory and Applications*, 2:203–211, June 1974.

[6] A. Fettweis, J. A. Nossek, and K. Meerkötter. Reconstruction of Signals after Filtering and Sampling Rate Reduction. *IEEE Transactions on Acoustics, Speech and Signal Processing*, ASSP-33(4):893–902, August 1985.

[7] L. Gazsi. Explicit Formulas for Lattice Wave Digital Filters. *IEEE Transactions on Circuits and Systems*, CAS-32(1):68–87, January 1985.

[8] H. Johansson and L. Wanhammar. Design of Bireciprocal Linear-Phase Lattice Wave Digital Filters. *Report LiTH-ISY-R-1877*, August 1996.

[9] I. Kunold. Linear Phase Realization of Wave Digital Filters. *IEEE Transactions on Acoustics, Speech and Signal Processing*, pages 1455–1458, 1988.

[10] T. Q. Nguyen, T. I. Laakso, and R. D. Koilpillai. Eigenfilter Approach for the Design of Allpass Filters Approximating a Given Phase Response. *IEEE Transactions on Signal Processing*, 42(9):2257–2263, September 1994.

[11] M. Renfors and T. Saramäki. Recursive Nth-Band Digital Filters Part I: Design and Properties. *IEEE Transactions on Circuits and Systems*, CAS-34(1):24–39, January 1987.

AN OPTIMAL ENTROPY CODING SCHEME FOR EFFICIENT IMPLEMENTATION OF PULSE SHAPING FIR FILTERS IN DIGITAL RECEIVERS

A.P. Vinod, A.B.Premkumar, E.M-K.Lai

School of Computer Engineering, Nanyang Technological University
Nanyang Avenue, Singapore 639798

ABSTRACT

The most computationally intensive part of wide-band receivers is the IF processing block. Digital filtering is the main task in IF processing. Infinite precision filters require complicated digital circuits due to coefficient multiplication. This paper presents an efficient method to implement pulse shaping filters for a dual-mode GSM/W-CDMA receiver. We use an arithmetic scheme, known as pseudo floating-point (PFP) representation to encode the filter coefficients. By employing a span reduction technique, we show that the filters can be coded using an optimal entropy scheme employing PFP which requires only considerably fewer bits than conventional 24-bit and 16-bit fixed-point filters. Simulation results show that the magnitude responses of the filters coded in PFP meet the attenuation requirements of GSM/W-CDMA specifications.

1. INTRODUCTION

A first order estimate of the resources required to implement the wideband receiver of a software radio shows that the IF processing block to be the most computationally intensive part since it operates at the highest sampling rate [1]. Digital filtering is the core function in the IF processing block which is accomplished by FIR filters due to its linear phase characteristics. Hardware implementation of FIR filters generally requires large multipliers if they are realized with continuous-valued coefficients. For a multiplierless filter implementation, the hardware complexity increases with the number of the adders, which is proportional to the filter wordlength [2]. The fixed-point arithmetic implementation of channel filters in digital wideband receivers requires 24-bit wordlength to meet the channel specifications [3, 4]. It has been reported that the 16-bit implementation results in a significant degradation in stop-band attenuation preventing the required spectral mask requirements from being met [3]. The entropy of a coefficient is a measure of the information content that can be coded in minimum number of bits. The objective of efficient entropy coding scheme is to represent the filter coefficients using fewer bits while retaining its frequency response characteristics. Optimal coefficient entropy is a desirable feature in ASIC and FPGA implementation of pulse shaping FIR filters, where look-up tables are employed to store all the possible partial products formed when filter coefficients are convolved with the input signal [5, 6]. The memory required in look-up table implementation is a linear function of coefficient wordlength [6]. Efficient methods to implement fixed-point channel filters using fewer bits that meet the required spectral mask characteristics are hardly discussed in the literature. This paper presents the implementation of pulse shaping filters in digital receivers using an arithmetic scheme known as the Pseudo Floating-Point (PFP) representation. It has been shown that the PFP representation can be employed to implement the quadrature mirror filters (QMF) using the lower bound value of 11 bits in [7]. In this paper, we show that the coefficients of the pulse shaping filters can be coded using a wordlength which is considerably lower than conventional 24-bit and 16-bit implementations discussed in [3, 4]. The resulting filters can be implemented using fewer number of adders and is ideal for the look-up table implementation due to its low coefficient entropy.

This paper is organized as follows. Section II gives a brief review the PFP representation that is suitable for programmable Multiply-and-Accumulate (MAC) operations. We discuss the PFP coding scheme for implementation of pulse shaping filters for a dual-mode W-CDMA/GSM receiver in section III. A span reduction algorithm to obtain the optimal entropy PFP coefficients will also be described. In section IV, we illustrate the implementation of pulse shaping filters for W-CDMA/GSM receiver using several examples. Section V provides our conclusions.

2. THE PSEUDO FLOATING-POINT REPRESENTATION

The general representation of sum-of-powers-of-two (SOPOT) terms for the ith filter coefficient is $h_i = \sum_{j=0}^{B-1} 2^{a_{ij}}$, where B is the number of digits in the power-of-two representation. The expression for h_i can be rewritten as, $h_i = 2^{a_{i0}}.\sum_{j=0}^{B-1} 2^{a_{ij}-a_{i0}} = 2^{a_{i0}}.\left[\sum_{j=0}^{B-1} 2^{c_{ij}}\right]$, where $c_{ij} = a_{ij} - a_{i0}$. The term a_{i0} is known as the *shift* and the upper limit value, $(a_{i(B-1)} - a_{i0})$, is known as the *span*. The bracketed term is known as the normalised value (n value). The shift and the normalised value are analogous to the exponent and mantissa in true floating-point representations. Instead of expressing the coefficients as a 16-bit integer, it can be expressed as a *(shift, n-value)* pair – this is the definition of the pseudo floating point representation. For a given coefficient set, let L and M be the number of bits needed to encode the shift and n-value respectively. Then,

$$L = \max_{0 \le i \le N-1} shift\,(h_i) \quad (1)$$

$$M = \max_{0 \le i \le N-1} span\,(h_i) \quad (2)$$

0-7803-7761-3/03/$17.00

The following example illustrates this concept. Consider the coefficient $h(n)$, whose 16-bit SOPOT representation is given by $h(n) = 2^{-6} + 2^{-8} + 2^{-9} + 2^{-14}$. This can be written as $2^{-6}(2^0 + 2^{-2} + 2^{-3} + 2^{-8})$. In this expression, the term 2^{-6} is the *shift* part (implying 'right shift by 6'), and the bracketed term is the *span* part. The shifts are less complex since they can be hardwired. Therefore, only 3 bits are needed for storing the shift value and correspondingly, $L = 3$. The span value, $M = 8$, is obtained from the bracketed term. Hence the coefficient can be represented in PFP using $L + M = 11$ bits.

In the case of filter implementation in [3, 4], the L and M values of 24-bit fixed-point coefficients are 5 and 23 respectively. Hence, 28 bits are needed by the PFP for general coefficient sets. For the 16-bit coefficients, L and M are 4 and 15 respectively and thus require a total of 19 bits in PFP representation. It would seem that the PFP representation might not be an optimal representation. However, it would be interesting to investigate if the actual coefficient sets would require less than the 28 bits and 19 bits in these cases. It should be noted that the span contributes significantly more to the wordlength requirement than the shift. The shifts are well distributed across coefficients and so is not a parameter that could be optimized further. Therefore, it is beneficial to explore some efficient means of reducing the span without considerable implication on the magnitude response of the filter. In the following section, we show that by employing a span reduction technique, the wordlength requirement of the pulse shaping filters for a dual-mode GSM/W-CDMA receiver can be significantly reduced.

3. PFP CODING SCHEME FOR W-CDMA/GSM CHANNEL FILTERS

We consider the IF architecture for dual-mode GSM/W-CDMA receiver presented in [8] to show the filter implementation using PFP coefficients. The dual-mode architecture of IF processing for GSM/W-CDMA is shown in Fig. 1. The IF block performs frequency conversion for W-CDMA and channel extraction from a wide-band received signal for GSM. The input bandwidth of an IF signal covers one channel of 5 MHz in W-CDMA. In the W-CDMA mode, the filter $H_1(z)$, performs pulse shaping to achieve an attenuation of –40 dB at 5 MHz as in the W-CDMA specification. The output signal at 15.36 MHz, which is four times the W-CDMA chip-rate, is fed to base-band processing. The purpose of IF processing for GSM is to extract a single channel with a bandwidth of 200 kHz from the 5 MHz received signal. Therefore, in the GSM mode, $H_1(z)$ functions as a high decimation filter and $H_2(z)$ performs pulse shaping to attenuate the block signal at 200 kHz as per GSM specifications. Our focus is to realize the PFP implementation of the pulse shaping filters, $H_1(z)$, in W-CDMA mode and $H_2(z)$, in GSM mode.

In our attempt to achieve a minimum wordlength for any coefficient set, we fix the *shift* to the maximum value, l, corresponding to the worst-case coefficient set using (1). The span value is progressively reduced by throwing away the power-of-two terms and checking whether the resulting filter response meets the filter specifications at each stage. We can expect distortion in the frequency response characteristics when such a span reduction technique is employed to all the "offending coefficients". Our observation in employing the span reduction technique is that the pass-band response of the resulting filter does not change. It has also been noted that the effect of span reduction on stop-band attenuation and peak stop-band ripple is minimal in the case of filters having relatively few number of taps (filter length, $N < 40$). The reason for this behaviour can be explained as follows. Let the span value after performing the reduction be $\hat{m}$.

1. In the case of short length filters, the spans are more closely distributed around $\hat{m}$, whereas for long filters spans are sparsely distributed. Hence, the magnitudes of those terms whose span exceed $\hat{m}$, which are thrown away are considerably smaller for short length filters when compared to that of filters having large number of taps. As a result, the sensitivity of PFP coefficients to span reduction is very low. Sensitivity is a measure of the degree of influence on the frequency response of a digital filter when any one of the coefficients is quantized. The sensitivity can be computed by setting each coefficient, in turn, to its nearest power of two, yielding in each case a response $H_q(\omega_i)$.

$$s(n) = \frac{1}{M}\sum_{i=1}^{M}[H_q(\omega_i) - H(\omega_i)]^2 \qquad (3)$$

where $H(\omega_i)$ and $H_q(\omega_i)$ are the frequency responses of the infinite-precision and quantized coefficients respectively at M finite number of frequencies ω_i [9]. The equivalent time-domain expression for sensitivity is given by

$$s(n) = \frac{1}{M}\sum_{n=0}^{N-1}[h_q(n) - h(n)]^2 \qquad (4)$$

where $h_q(n)$ and $h(n)$ represent the impulse responses of the quantized and infinite-precision coefficients respectively. In the case of filters with relatively fewer number of taps, $[h_q(n) - h(n)]$ is considerably small due to the distribution of spans close to $\hat{m}$. Hence, the sensitivity, which is a square function is minimal. This will be illustrated in the examples provided in section IV.

2. The span deviation from $\hat{m}$ is relatively uniform across the coefficient grid in the case of filters with fewer taps when compared to that with larger taps. As a result, applying span reduction to the filter is similar to scaling the entire coefficient. Scaling the coefficient set will not affect the frequency response shape; instead it only changes the filter gain. In the case of PFP representation for short length filters, the deviation in gain is minimal since the span deviation from $\hat{m}$ is minimal. It is known that for fixed-point FIR filter coefficients, the worst-case quantization error usually occurs for the larger valued coefficient. However, in the design examples of raised cosine

filters illustrated in section IV it can be seen that the powers-of-two representation of larger valued coefficient has fewer number of non-zero digits. The span values of these coefficients are less than $\hat{m}$. In the case where the span of largest valued coefficient exceeds $\hat{m}$, the power of two is extremely small. Hence the response deterioration is within the filter stop-band attenuation specification limits when the PFP span reduction technique is employed.

It must be noted that the span reduction technique is limited to filters having fewer number of taps. The raised cosine filters employed as pulse shaping filters in digital receivers are short length filters whose taps is normally less than 40. Hence the proposed span reduction is suitable for this application.

3.1 PFP Filter Implementation Algorithm

The steps for PFP coding of filters using the span reduction approach are presented below.

Step 1: Design the raised cosine filter, $h(n)$, with infinite precision as in the specification for pulse shaping filter of W-CDMA/GSM. Determine the frequency response of the unquantized coefficients, $H_d(\omega) = \sum_{n=0}^{N-1} h(n)e^{-j\omega n}$.

Step 2: Set all the coefficients to their closest sum of power-of-two (SOPOT) coefficients using 16-bits and represent them as a *(shift, span)* pair in PFP. Fix the shift to the maximum value l, corresponding to the worst-case coefficient set. Determine the maximum span value, M. Set iteration $k=0$.

Step 3: Decrease the span to $M-1$ by throwing away the power-of-two terms of offending coefficients and obtain the new set of coefficients, $h_q(n)$. Determine $h_q(n) - h(n)$.

Step 4: Determine the frequency response of the quantized filter whose span is reduced to $M-1$ using:

$$H_q(\omega_i) = H_d(\omega_i) + H_e(\omega_i) = \sum_{n=0}^{N-1} [h(n) + \sum_{n=0}^{N-1} h_q(n) - h(n)]\, e^{-j\omega_i n}$$

where ω_i represents frequency samples in the stop-band.

Step 5: If $|H_{qs}(\omega_i)| \le |H_s(\omega_i)|$, where $H_{qs}(\omega_i)$ represents the stop-band response of the PFP filter and $H_s(\omega_i)$ as in the stop-band specification of the pulse shaping filter, set $k=k+1$ and go to step 3. Otherwise, terminate the program and choose the PFP coefficients, $h_q(n)$, corresponding to the 'k'th iteration.

4. DESIGN EXAMPLE

In this section, we implement the pulse shaping filters, $H_1(z)$, employed in W-CDMA mode and $H_2(z)$, employed in GSM mode. The infinite-precision filter, $h(n)$, is generated by the raised cosine FIR filter design program provided by the MATLAB® "firrcos" function. The roll-off factor is selected as 0.22 for bandwidth efficiency in 3G cellular applications.

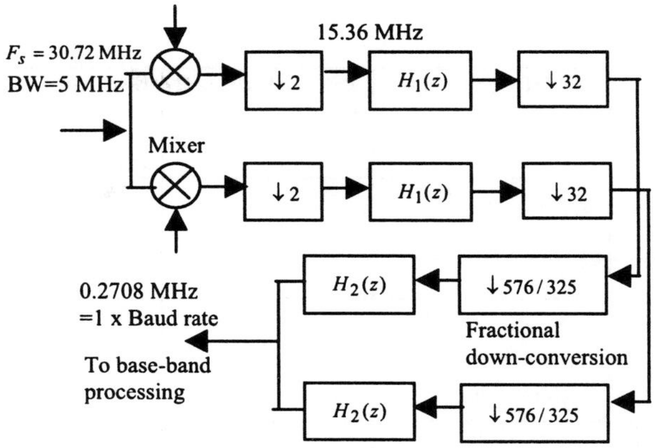

Fig. 1. IF architecture for dual-mode W-CDMA/GSM

Example 1: A raised cosine filter of length 19 is employed as the pulse shaping filter, $H_2(z)$, in GSM mode. The 16-bit SOPOT coefficients and 6-bit PFP coefficients obtained after span reduction are listed in Table 1. The sensitivity values of PFP coefficients are also shown in Table 1. The shift value is fixed at $l = 4$ bits based on the worst-case coefficients, $h(1)$ and $h(3)$. The lower bound of span obtained by employing the proposed algorithm is 2 bits. Hence the coefficient set is represented using 6 bits in PFP. The magnitude responses of the filters are shown in Figure 2. Response of the 8-bit PFP filter shows close resemblance to that of the 16-bit SOPOT filter. Though the attenuation of the PFP filter at 200 kHz (-37.7 dB) is slightly lower than the attenuation of the 16-bit SOPOT filter (-41.7 dB), it is still higher than the GSM specification of –20 dB. Both filters offer identical peak pass-band ripple (PPR) of 0.1 dB and peak stop-band ripple (PSR) of –35.80 dB. These comparisons show that there is practically no difference in the response of filters obtained using the proposed representation and the normal methods. The considerably low sensitivity values of PFP coefficients shown in Table 1 account for this achievement.

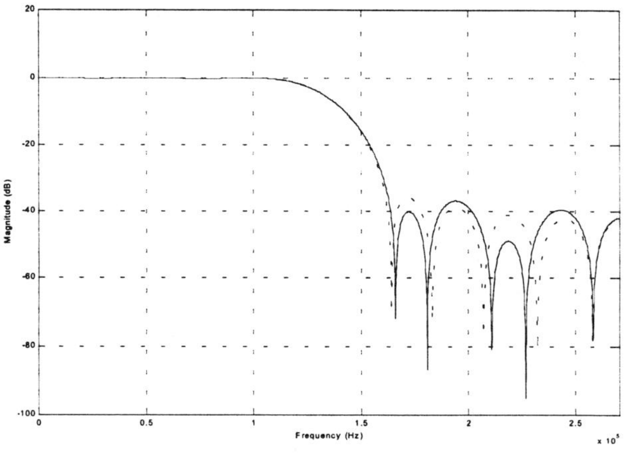

Fig. 2. Filter frequency responses in example 1.
Solid: 6-bit PFP, Dot: 16-bit SOPOT
(Frequency scale indicated is 0 - 270.833 kHz)

$h(n)$	SOPOT Coefficients (16-bit)	PFP Coefficients (6-bit)	Sensitivity of PFP coefficients (M=128)
$h(0)$	$2^{-7}+2^{-8}+2^{-12}+2^{-13}$	$2^{-7}+2^{-8}$	1.02e-009
$h(1)$	2^{-15}	2^{-15}	0
$h(2)$	$-2^{-6}-2^{-7}-2^{-10}-2^{-12}-2^{-13}-2^{-14}-2^{-16}$	$-2^{-6}-2^{-7}$	1.56e-008
$h(3)$	$-2^{-15}-2^{-16}$	$-2^{-15}-2^{-16}$	0
$h(4)$	$2^{-5}+2^{-6}+2^{-11}+2^{-15}+2^{-16}$	$2^{-5}+2^{-6}$	2.27e-009
$h(5)$	2^{-14}	2^{-14}	0
$h(6)$	$-2^{-4}-2^{-5}-2^{-10}-2^{-11}-2^{-12}-2^{-13}-2^{-14}-2^{-15}-2^{-16}$	$-2^{-4}-2^{-5}$	2.97e-008
$h(7)$	$-2^{-14}-2^{-16}$	$-2^{-14}-2^{-16}$	0
$h(8)$	$2^{-2}+2^{-4}+2^{-9}+2^{-12}+2^{-16}$	$2^{-2}+2^{-4}$	3.83e-008
$h(9)$	$2^{-1}+2^{-13}+2^{-16}$	2^{-1}	1.48e-010

Table 1. Coefficients of pulse shaping filter in example 1.

Example 2: For the W-CDMA mode, a raised cosine filter of length 33 is designed. The lower bound PFP obtained is 8-bit. The magnitude responses of the filters are shown in Figure 3. Both the filters meet the desired attenuation of –40 dB as in W-CDMA specifications. The 8-bit PFP filter response shows close resemblance to that of the 16-bit SOPOT filter.

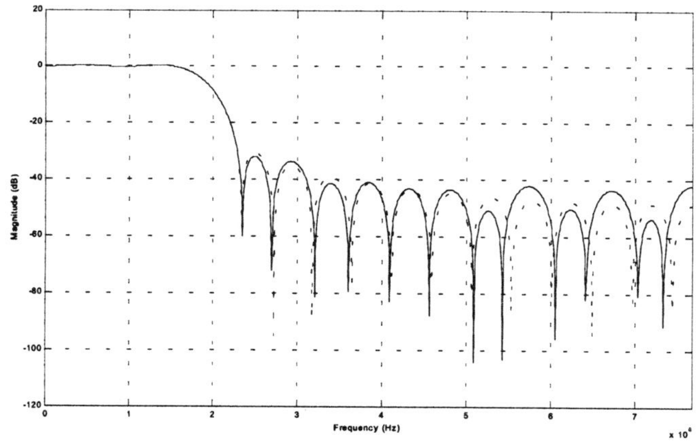

Fig. 3. Filter frequency responses in example 2.
Solid: 8-bit PFP, Dot: 16-bit SOPOT
(Frequency scale indicated is 0 – 7.68 MHz)

5. CONCLUSIONS

We have presented an efficient coefficient coding scheme using pseudo floating-point representation for implementation of pulse shaping filters in digital receivers. The span reduction algorithm reduces the PFP coefficient entropy to considerably lower bits when compared to the fixed-point realizations discussed in the literature. The computational complexity of the algorithm is relatively less since it is applied only to the filter stop-band response samples. Simulation results clearly show that the proposed method results in filters with good frequency response characteristics. The proposed method can be used to implement any FIR filters provided the number of taps is less than 40. It is also worth noting that the proposed PFP filter is hardware efficient due to its minimum wordlength requirement. The PFP representation requires less memory when look-up tables are employed in ASIC/FPGA implementation architecture for pulse shaping FIR filters in 3G mobile communications.

6. REFERENCES

[1] J.Mitola, *Software Radio Architecture.* New York: Wiley, 2000.

[2] Yong Ching Lim, Y.Sun, and Ya Jun Yu, "Design of discrete-coefficient FIR filters on loosely connected parallel machines," *IEEE Transactions on Signal processing,* vol. 50, no. 6, pp. 1409-1416, June 2002.

[3] H.R.Karimi, N.W.Anderson, and P.McAndrew, "Digital signal processing aspects of software definable radios," *IEE Colloquium, Adaptable and Multistandard Mobile Radio terminals,* March 1998.

[4] K.Kalbasi, "Performance tradeoffs in a dual-mode, W-CDMA/EDGE digital IF receiver," Tech. Rep., *Agilent Ees of Electronics Design Automation,* 2001.

[5] W.P. Zhu, M.O. Ahmad and M.N.S.Swamy, "ASIC implementation architecture for pulse shaping FIR filters in 3G mobile communications," *Proc. 2002 IEEE Int. Symp. on Circuits and Systems,* pp. 433-436, vol. 2, May 2002.

[6] Zheng Wu, Cheng Luo, Xin Su, and Xibin Xu, "Digital filter implementation for software radio," *53rd IEEE Vehicular Technology Conference,* pp. 1902-1906, vol. 3, 2001 (Spring).

[7] A.B.Premkumar, C.T.Lau, and A.P.Vinod, "High performance architectures for QMF banks," *NASA 9th Symposium on VLSI Architecture,* November 2000.

[8] M.Jian, W.H.Yung, and B.Songrong, "An efficient IF architecture for dual-mode GSM/W-CDMA receiver of a software radio," *IEEE International Workshop on Mobile Multimedia Communications,* pp. 21-24, November 1999.

[9] Xiaojuan Hu, L.S.DeBrunner, and Victor DeBrunner, "An efficient design for FIR filters with variable precision," *Proc. 2002 IEEE Int. Symp. on Circuits and Systems,* pp. 365-368, vol. 4, May 2002.

EXPONENTIALLY-MODULATED FILTER BANK-BASED TRANSMULTIPLEXER

Juuso Alhava and Markku Renfors

Tampere University of Technology
Institute of Communications Engineering
P.O. Box 553, FIN-33101 Tampere, FINLAND
e-mail: jualhava@cc.jyu.fi, markku.renfors@tut.fi

ABSTRACT

A complex filter bank divides the spectrum of a complex (I/Q) signal into several low-rate subbands (analysis bank) or combines the low-rate signals into a single high-rate signal (synthesis bank). These operations are often cascaded to form analysis-synthesis (AS) or transmultiplexer (TMUX) configurations. Such systems find applications in many areas, especially in communications signal processing. In this paper we develop further the lapped transform theory to cover also the cases of complex, exponentially modulated filter banks. Using the shift-orthogonality of lapped transforms, we show how to design complex critically sampled AS and TMUX systems such that the perfect-reconstruction condition is satisfied.

1. INTRODUCTION

Lapped transforms are non-square transform matrices, that satisfy the shift-orthogonality condition [1]. Commonly they are considered as real-valued and then suitable for subband processing of real signals. Figure 1 shows an analysis-synthesis system based on lapped transforms: The forward transform ^{T} splits the input signal into subbands, the subsignals are filtered in the processing stage, and the inverse transform reconstructs the signal. The forward transform is computed from overlapping data blocks of length N . Alternatively, the system in Figure 1 can be viewed as critically sampled (real) filter bank. The basis functions of the transform (columns of) are the synthesis filter impulse responses for the filter bank.

The Complex Lapped Transform (CLT) we proposed in [2] is a 2 -band transform for complex signals. There we used -band Extended Lapped Transform (ELT) in Cosine-Modulated Filter Bank (CMFB) and Sine-Modulated Filter Bank (SMFB) configurations as basic components of the transform. Fast ELT is an efficient implementation for the computation of the transform [1]. The defined CLT was shown to satisfy the Perfect Reconstruction (PR) condition in analysis-synthesis configuration, and the basis functions of the transform correspond to the subfilters of the critically sampled Exponentially-Modulated Filter Bank (EMFB). Similar subfilter definition was used in [3], where Modulated Complex Lapped Transform (MCLT) was applied in subband processing of real-valued signals. However, in MCLT the passbands of the subfilters cover the spectrum range [0,], and we must add subbands to the negative side [− , 0] to process complex signals.

This research was carried out in the project "Advanced Multicarrier Techniques for Wireless Communications" funded by the Academy of Finland.

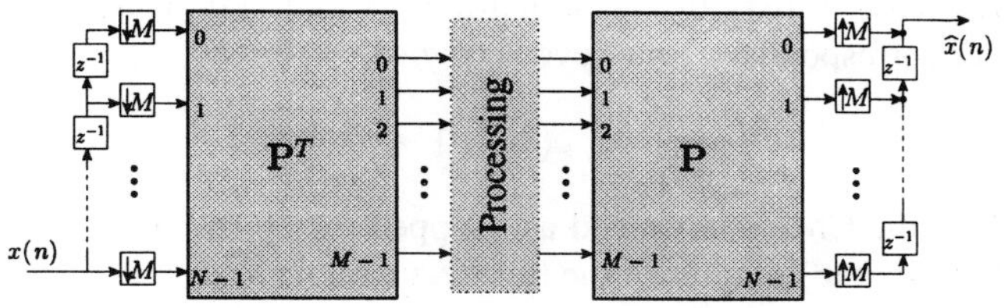

Figure 1: Lapped transform.

In this paper we extend the proof of the PR condition to the complex filter bank based TMUX case. Such a complex TMUX has an important application as multicarrier modulation method for wireless communication systems. Real CMFBs have been considered for baseband communications systems (the Discrete Wavelet Multitone, DWMT, concept is based on this idea), but to extend the idea efficiently to radio communication systems, complex PR filter banks are needed. In Section 2 we review the PR conditions in CMFBs and define sine-modulated filter bank based on ELT subfilter equations. It is shown, that the SMFB satisfies the PR-condition. Section 3 reviews shortly the EMFB equations from [2] that are necessary for this paper. The transmultiplexer structure is given with both filter bank and lapped transforms notations. We use the shift-orthogonality property to show that the system is a perfect reconstruction transmultiplexer.

0-7803-7761-3/03/$17.00 ©2003 IEEE

2. MODULATED FILTER BANKS AS LAPPED TRANSFORMS

2.1. Perfect Reconstruction Condition for Cosine-Modulated Filter Bank

Cosine-modulated filter banks are -channel uniform filter banks. The synthesis subfilters $_k(n)$ are generated by modulating low-pass prototype filter $h\ (n)$ with cosine-sequences:

$$[\ ]_{nk} = \ _k(n) = \sqrt{\frac{2}{}}h\ (n)\cos\ \ n + \frac{+1}{2}\ \ k + \frac{1}{2}\ \frac{}{}\ , \tag{1}$$

where $n = 0, 1, \ldots, N-1$ and $k = 0, 1, \ldots, \ \ -1$. This is the ELT definition for subfilters. The prototype filter is here assumed to satisfy the PR condition [1]. The kth analysis filter corresponds to time-reversed synthesis filter

$$h_k(n) = \ _k(N-1-n). \tag{2}$$

When the CMFB is viewed as a lapped transform, its subfilters are the columns of the transform matrix

$$[\ ]_{nk} = \ _k(n). \tag{3}$$

The CMFB is a PR filter bank, and thus when we partition ^{T} into submatrices

$$^T = \ \ ^T_0 \vdots\ ^T_1 \vdots \cdots \vdots\ ^T_{2\ \ -1} \qquad N$$

they satisfy the shift-orthogonality conditions

$$\sum_{=0}^{2\ \ -1-} \ \ ^T \qquad = \delta(\ )\mathbf{I} \tag{4}$$

$$\sum_{=0}^{2\ \ -1-} \qquad ^T \ \ = \delta(\ )\mathbf{I}, \tag{5}$$

where $\delta(\ ) = 1$ when $= 0$ and $\delta(\ ) = 0$ when $= 0$. $\mathbf{I}$ is the identity matrix. The parameter is the overlapping factor of the CMFB such that the subfilter length $N = 2$.

2.2. Sine-Modulated Filter Bank

The reason for defining sine-modulated FB is to use it as a component for EMFB. The subfilters of SMFB are

$$[\ ]_{nk} = \ _k(n) = \sqrt{\frac{2}{}}h\ (n)\sin\ \ n + \frac{+1}{2}\ \ k + \frac{1}{2}\ \frac{}{}\ , \tag{6}$$

where $n = 0, 1, \ldots, N-1$ and $k = 0, 1, \ldots, \ \ -1$. We claimed in [2], that this subfilter definition leads also to PR-FB (after all, we only change the phases of the CMFB subfilters). Here we verify the shift-orthogonality to obtain a complete proof for the PR property.

The CMFB and SMFB synthesis subfilters can be connected with time-reversing and sign modification [2]

$$_k(n) = (-1)^k \qquad _k(N-1-n). \tag{7}$$

Then the matrix can be written with reversed columns of . We need here matrices and :

$$= \begin{matrix} & & \cdot^{\cdot} \\ & 1 & \\ 1 & & \\ \cdot^{\cdot} & & \end{matrix} \qquad = \begin{matrix} +1 & & & \\ & -1 & & \\ & & +1 & \\ & & & -1 \\ & & & & \ddots \end{matrix} .$$

When is multiplied from the left by $_{N\ N}$, the order of elements in columns of is reversed. Right-multiplication by changes the signs of odd columns. Then we can write

$$= (-1) \qquad . \tag{8}$$

To show the shift-orthogonality condition

$$\sum_{=0}^{2\ \ -1-} \qquad ^T \qquad = \delta(\ )\mathbf{I} \tag{9}$$

we partition ^{T} to submatrices

$$^T = \ \ ^T_0 \vdots \cdots \vdots\ ^T_{2\ \ -2} \vdots\ ^T_{2\ \ -1}$$

$$= (-1) \ \ (\ \ _{2\ \ -1}\ )^T \vdots \cdots \vdots (\ \ _1\ )^T \vdots (\ \ _0\ )^T .$$

Matrices and have dimensions in the above partitioning. Then we substitute the submatrices to (9):

$$\sum_{=0}^{2\ \ -1-} \qquad ^T$$

$$= (-1)^2 \sum_{=0}^{2\ \ -1-} (\ \ _{2\ \ -1-}\ )^T (\ \ _{2\ \ -1-\ \ -}\ )$$

$$= \sum_{=0}^{2\ \ -1-} \ ^T\ ^T_{2\ \ -1-} \underbrace{\ ^T\ }_{\mathbf{I}} \ _{2\ \ -1-\ \ -}$$

$$= \sum_{=0}^{2\ \ -1-} \ ^T\ ^T_{2\ \ -1-} \ \ _{2\ \ -1-\ \ -}$$

The order of summation may be changed through the change of variables: $= 2 \ \ -1-i-$

$$\sum_{=0}^{2\ \ -1-} \qquad ^T \qquad = \ ^T \sum_{i=0}^{2\ \ -1-} \ ^T_i \ \ _i$$

$$= \ ^T(\sum_{i=0}^{2\ \ -1-}(\ ^T_i \ \ _i\ )^T = \ ^T \underbrace{\sum_{i=0}^{2\ \ -1-} \ ^T_i \ \ _i}_{(\ )\mathbf{I}} \ ^T$$

$$= \delta(\ )\ ^T\mathbf{I}^T = \delta(\ )\ ^T = \delta(\ )\mathbf{I}$$

This completes the proof.

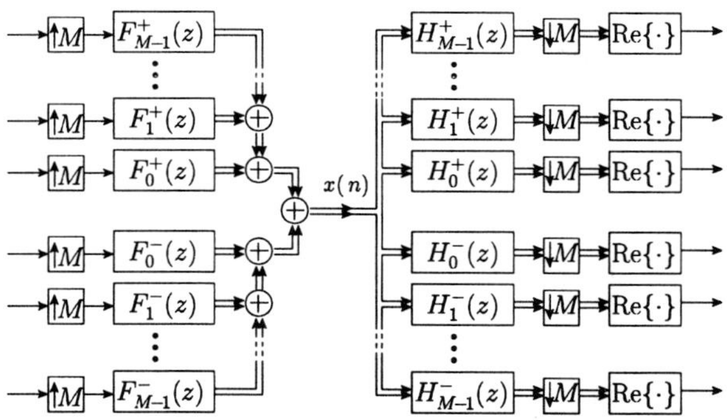

Figure 2: Complex FB-based transmultiplexer.

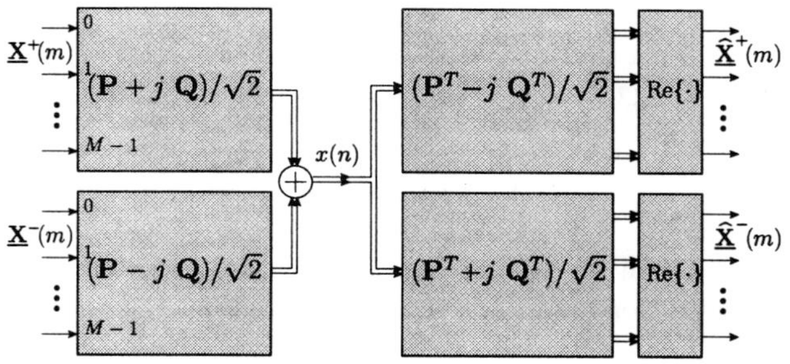

Figure 3: Equivalent representation for EMFB-based transmultiplexer with complex LTs.

3. EXPONENTIALLY-MODULATED FILTER BANK

3.1. Subfilter Definition

The subfilters of the CMFB and SMFB were applied in the definition of EMFB subfilters in [2]. To simplify the PR proofs with LTs, we separated positive and negative side subfilters

$$\ _k(n) = \ _k(n) + \ \cdot \ _k(n) \quad \overline{2} \qquad (10)$$

$$\ \bar{_k}(n) = \ _k(n) - \ \cdot \ _k(n) \quad \overline{2} \qquad (11)$$

The corresponding analysis filters are

$$h_k(n) = \ _k(N-1-n)^* \qquad (12)$$

$$h_k^-(n) = \ \bar{_k}(N-1-n)^* \ . \qquad (13)$$

The subband organization with these subfilters is shown in Figure 4. It was shown in [2], that the signal can be perfectly reconstructed from the real parts of the analysis bank outputs in analysis-synthesis configuration. In a way, the transmultiplexer is its dual system and it should satisfy the PR condition as well. However, when we deal with this PR transmultiplexer case using LTs, it leads to different shift-orthogonality requirements.

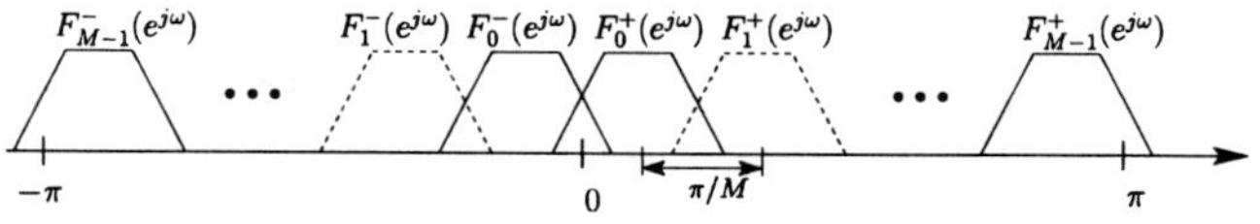

Figure 4: Subband organization with EMFB subfilters.

3.2. Transmultiplexer Case

When the order of filter banks is reversed to synthesis-analysis configuration, we have a complex FB-based transmultiplexer. It has 2 synthesis filters at the transmitter side, while the interpolation factor is half of the number of subfilters. This restricts the input signal vector $\_(\ ) = \frac{\underline{X}(\ )}{\underline{X}^-(\ )}$ to real-valued. If $\_(\ )$ is complex, the analysis bank at the receiver can not separate the transmitted signal perfectly from $x(n)$. In fact, with real subchannel inputs, the system can be regarded as a critically sampled system since the total number of real samples at the subchannel inputs is equal to two times to the number of complex samples at the high-rate synthesized signal.

The transmultiplexer structure is shown in Figure 2 and we would like to have $\_(\ ) = \_(\ )$, when the transmission channel is ideal. We can do the perfect reconstruction proof by using the properties of real lapped transforms and . Figure 3 shows the transmultiplexer when the synthesis and analysis filters are replaced by inverse and forward transform matrices respectively. We can simply describe this system with matrices, when we collect the EMFB subfilters in transform matrix . The inverse transform is thus

$$= \quad + \quad \overline{2} \ \vdots \quad - \quad \overline{2} \ . \qquad (14)$$

and forward transform matrix is obtained by taking the conjugate transpose of the above

$$^H = \frac{\ ^T - \ ^T \quad \overline{2}}{\ ^T + \ ^T \quad \overline{2}} \qquad (15)$$

The shift-orthogonality condition for the system in Figure 3 requires partitioning the matrix to 2 submatrices

$$^H = \ ^H_0 \,|\, ^H_1 \,|\, \cdots$$

$$= \frac{1}{\overline{2}} \begin{array}{c|c|c} ^T_0 + \ ^T_0 & ^T_1 + \ ^T_1 & \cdots \\ \hline ^T_0 - \ ^T_0 & ^T_1 - \ ^T_1 & \cdots \end{array} \ . \qquad (16)$$

The matrix ^{H} has partitioning

$$= \begin{array}{c} \_0 \\ \hline \_1 \\ \hline \vdots \end{array} = \frac{1}{\overline{2}} \begin{array}{cc|cc} \ _0 - & _0 & _0 + & _0 \\ \hline \ _1 - & _1 & _1 + & _1 \\ \hline \vdots & & \vdots & \vdots \end{array} \ . \qquad (17)$$

$$\sum_{=0}^{2\ -1-} \quad {}^{H} \quad = \frac{1}{2}\sum_{=0}^{2\ -1-} \left[\frac{{}^{T} - {}^{T}}{{}^{T} + {}^{T}} \right] \quad + \quad \Big| \quad -$$

$$= \frac{1}{2}\sum_{=0}^{2\ -1-} \left[\frac{{}^{T} + {}^{T}}{{}^{T} - {}^{T}} \,\middle|\, \frac{{}^{T} - {}^{T}}{{}^{T} + {}^{T}} \right]$$

$$= \frac{1}{2} \left[\frac{\sum_{=0}^{2\ -1-} {}^{T} + \sum_{=0}^{2\ -1-} {}^{T}}{\sum_{=0}^{2\ -1-} {}^{T} - \sum_{=0}^{2\ -1-} {}^{T}} \,\middle|\, \frac{\sum_{=0}^{2\ -1-} {}^{T} - \sum_{=0}^{2\ -1-} {}^{T}}{\sum_{=0}^{2\ -1-} {}^{T} + \sum_{=0}^{2\ -1-} {}^{T}} \right]$$

$$= \frac{1}{2} \left[\frac{2\,\delta(\ )\mathbf{I}}{} \,\middle|\, \frac{}{2\,\delta(\ )\mathbf{I}} \right] = \delta(\ )\mathbf{I}.$$

Figure 5: Perfect reconstruction proof of the EMFB-based transmultiplexer using the shift-orthogonality of lapped transforms.

Now the shift-orthogonality formula for the transmultiplexer is

$$\sum_{=0}^{2\ -1-} \quad {}^{H} \quad = \delta(\ )\mathbf{I}_2 \quad {}_2 \ . \qquad (18)$$

This can be proved, when we just substitute (16) and (17) to this summation formula and calculate the real terms. Then we may apply the shift-orthogonality of real LTs and . The details of calculations are shown in Figure 5. Finally, it should be noticed that if we take the imaginary part (instead of real) in equation (18) the result is not a zero matrix.

$$\mathrm{m} \sum_{=0}^{2\ -1-} \quad {}^{H} \quad = \qquad (19)$$

4. CONCLUSIONS

In this paper we completed the perfect reconstruction proof of the sine-modulated filter banks through the use of partitioned matrices. SMFB can be used as a building block for complex FBs together with CMFBs. We can call it as the exponentially-modulated filter bank, because the subfilters are generated by modulating the prototype filter with complex exponential sequences. The EMFB can be used as an efficient transmultiplexer for wireless communications that satisfies the PR condition when the transmission channel is ideal. In future work we will show that oversampling the EMFB analysis filters (taking both real and imaginary parts) gives us the possibility to equalize the transmultiplexer subsignals without cross-filters from the neighboring subchannels. This provides an efficient approach to channel equalization of filter bank based multicarrier systems [5].

5. REFERENCES

[1] H. S. Malvar, *Signal Processing with Lapped Transforms*. Norwood, MA: Artech House, 1992.

[2] J. Alhava and M. Renfors, "Complex lapped transforms and modulated filter banks," in *Proc. Int. TICSP Workshop on Spectral Methods and Multirate Signal Processing*, Tolouse, France, Sept. 7–8, 2002, pp. 87–94.

[3] H. S. Malvar, "A modulated complex lapped transform and its applications to audio processing," in *Proc. IEEE Int. Conf. on Acoustics, Speech, and Signal Processing*, 1999, vol. 3, pp. 1421–1424.

[4] A. Viholainen, T. Hidalgo Stitz, J. Alhava, T. Ihalainen, and M. Renfors, "Complex modulated critically sampled filter banks based on cosine and sine modulation," in *Proc. IEEE Int. Symp. on Circuits and Systems*, Scottsdale, Arizona, USA, May 26–29, 2002, vol. 1, pp. 833–836.

[5] J. Alhava and M. Renfors, "Adaptive sine-modulated/ cosine-modulated filter bank equalizer for transmultiplexers," in *Proc. European Conf. on Circuit Theory and Design*, Espoo, Finland, Aug. 28–31, 2001, pp. III / 337–340.

ALLPASS STRUCTURES FOR MULTIPLIERLESS REALIZATION OF RECURSIVE DIGITAL FILTERS

Mrinmoy Bhattacharya and Tapio Saramäki

Institute of Signal Processing
Tampere University of Technology
P. O. Box 553, Tampere, FIN 33101, Finland
E-mail: {mrinmoy,ts}@cs.tut.fi

ABSTRACT

Under certain conditions an odd-order lowpass or highpass recursive digital filter can be decomposed into a sum of two allpass filters with real coefficients exhibiting a low passband sensitivity. This decomposition has the attractive property that there exist for its implementation structures where both the number of delays and the number of multipliers are equal to the filter order, thereby making the overall implementation very efficient. This paper develops some allpass structures that combine this property for generating multiplierless implementations for odd-order recursive digital filters. Utilizing these structures along with allowing some marginally insignificant deviations in the specifications such as in the passband and stopband tolerances, the total number of nonzero bits for multiplier coefficients, i.e., those of shifts and adds and/or subtracts, becomes quite small, making this approach very attractive. Alternatively, the overall filter can be designed with marginally stricter tolerances than the desired specifications in such a manner that it meets the criteria after quantizing the filter coefficients.

1. INTRODUCTION

In multiplierless implementations of digital filters the minimum number of signed powers of two (MNSPT) or canonic signed digits (CSD) representations of binary digits are extensively used for representing the multiplier coefficients An MNSPT representation of a coefficient value is given by $\sum_i a_i 2^{-t_i}$, where each a_i is either 1 or –1 and t_i is a positive or negative integer. For instance, 1.93359375 can be realized as $2-2^{-4}-2^{-8}$; here, the multiplication is achieved by three bit shifts and two subtracts.

One major approach for multiplierless implementations comprises of that of optimization [5, 6, 11, 15, 16], i.e., searching for the coefficients such that they can be implemented in MNSPT forms and the given criteria are still met. Optimization methods are used to find the optimal transfer functions under the given constraints with the filter design being basically a problem of approximation due to the tolerances in specifications. In general, the methods of optimizations are considered to be quite satisfactory. However, one may not assure or guarantee that the optimal solution will always be found under the given constraints. The solution can be unsatisfactory, for example, in terms of the filter order, the given wordlength of the multipliers, or the specified number of shifts and adds (in the case of the multiplierless implementation), or some combination of them. Under such conditions, some parameters or characteristics of the filter will have to be relaxed to obtain an acceptable design for use in the intended system.

In the case of IIR filters, the structures such as a sum of two allpass filters, including attractive lattice wave digital (LWD) filters, coupled with optimization methods have shown to yield good results for multiplierless implementations [8, 15,16]. These sums of allpass filters are characterized by the attractive property that there exist structures with the number of required multipliers being equal to the filter order, thereby decreasing the number of multipliers compared with conventional realization forms.

Another interesting approach is the one that stems from designing an odd-order elliptic minimal Q-factor analog filter (EMQF) that has some special properties. Using the bilinear transformation these filters can be implemented as a sum of two allpass filters [8] along with an expanded design parameters space as the passband (stopband) tolerances, the edges, and the filter order.

Especially, in the case of FIR filters, another approach is based on combining simple sub-filters that can be implemented using only a few shifts and adds and/or subtracts [1, 13]. Although quite attractive, to make this approach as a viable one, a large database of such filters will have to be generated and some optimization method will have to be evolved in order to combine some of them to meet the desired specifications.

In [2–4], the feasibility of implementing multiplierless recursive digital filters based on low-sensitivity structures

0-7803-7761-3/03/$17.00 ©2003 IEEE

has been demonstrated. It was realized that the development of the allpass structure that combines the advantages of the minimum number of multiplier and delays and the low passband sensitivity of the allpass structure, along with the advantages of those described in [2–4], a gross reduction in the total number of nonzero bits (effectively the number of shifts and adds and/or subtracts required) would be possible. Firstly, the number of multipliers would be less in the case of the sum of two allpass filters and secondly the number of nonzero bits per multiplier would be less due to low values of the modified coefficients. Of course, the transfer function to be realized would have to be of odd order and should fulfill the necessary conditions so as to be decomposable into a sum of the two allpass filters [12, 14]. The next section demonstrates that such an approach is realizable.

2. STRUCTURES FOR THE IMPLEMENTATION

If the conditions given in [12, 14] are fulfilled, then an odd-order transfer function $H(z)$ can be decomposed as

$$H(z) = \frac{1}{2}\left(H_0(z) + H_1(z)\right), \tag{1}$$

where $H_0(z)$ and $H_1(z)$ are allpass transfer functions with their orders differing by one. This necessitates the requirement of first-order and second-order allpass structure that could be cascaded as required to implement the two channels.

The developed first-order and second-order allpass sections that combines the advantages as mentioned in the previous section are depicted in Figures 1 and 2 respectively.

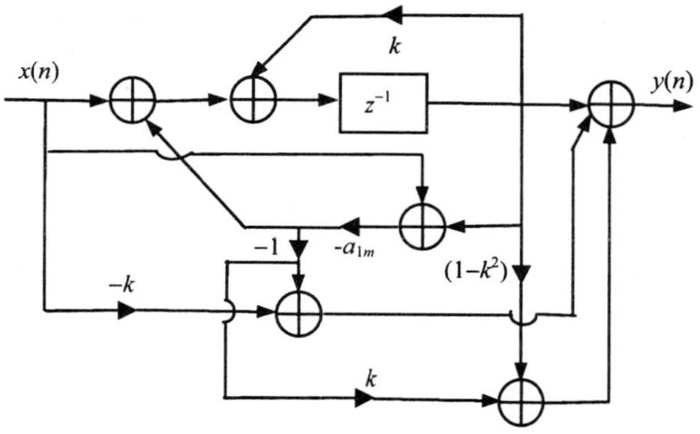

Fig. 1. The first-order allpass section

The first-order section realizes the following allpass transfer function:

$$H_{1st}(z) = \frac{a_1 z + 1}{z + a_1} \tag{2}$$

and the modified coefficient in Figure 1 is given by

$$a_{1m} = k + a_1 \tag{3}$$

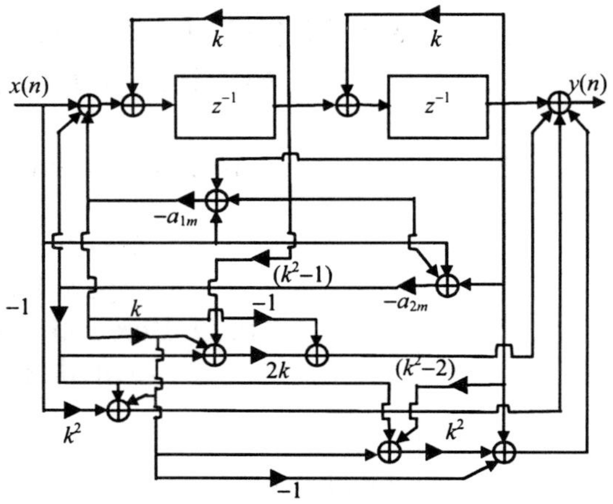

Fig 2. The second-order allpass section.

The second-order section realizes the following second-order allpass transfer function:

$$H_{2nd}(z) = \frac{a_2 z^2 + a_1 z + 1}{z^2 + a_1 z + a_2} \tag{4}$$

and the modified coefficients are given by

$$a_{1m} = 2k + a_1 \quad \text{and} \quad a_{2m} = k^2 + ka_1 + a_2 \tag{5}$$

Here, k is a one-bit multiplier like $\pm 1, \pm 0.5, \ldots$ (whose maximum magnitude could be unity). The value is chosen depending on the pole locations and to achieve the lowest possible magnitude of the modified coefficients [2–4]. If k is zero, then the resulting structures become similar to those considered in [9].

3. RESULTS AND DISCUSSIONS

Quite a few odd-order elliptic filters were realized as a sum of two allpass filters with the aid of both the unmodified [9] and the modified structures as depicted in Figures 1 and 2. Table 1 illustrates the results for the implementation for some of the filters. This table shows the passband and stopband tolerances and the requirement of nonzero bits along with the type of structure used. Figure 3 shows the amplitude responses of Filter 2 of Table 1 in two cases. The dashed and solid line show the responses for the initial design realized with unmodified allpass structures and the realization with revised specifications and using modified allpass structures requiring fourteen nonzero bits only for five multipliers, respectively.

It is mentioned that $k = 1$ in all the example cases in Table 1 leading to simplified modified allpass structures.

Table 1. Requirement of nonzero bits for some filters realized as sum of allpass structures

Filter Characteristics	Details of performance achieved			
	Number of nonzero bits for five multipliers	Tolerances achieved		
		Passband	Stopband	Comments
Filter 1: 5th-order, passband=0.1π, stopband=0.15π, passband ripple=0.6 dB, stopband attenuation=50 dB	(a) 26	0.6 dB	50.0 dB	Unmodified allpass structure is used; design is based on the initial specifications of the filter.
	(b) 21	0.603 dB	49.4 dB	.....".....
	(c) 20	0.605 dB	47.8 dB	.....".....
	(d) 18	0.631 dB	49.8 dB	.....".....
	(e) 16	0.750 dB	48.6 dB	.....".....
	(f) 13	0.52 dB	51.26 dB	Modified allpass structure is used; design is based on the revised specifications of passband ripple = 0.5 dB and stopband attenuation = 51 dB.
Filter 2: 5th-order, passband=0.025π, stopband=0.05π, passband ripple=0.05 dB, stopband attenuation=50 dB	(a) 36	0.05 dB	50.0 dB	Unmodified allpass structure is used; design is based on the initial specifications of the filter.
	(b) 33	0.051 dB	49.99 dB	.....".....
	(c) 30	0.054 dB	49.96 dB	.....".....
	(d) 27	0.055 dB	49.63 dB	.....".....
	(e) 23	0.068 dB	43.7 dB	.....".....
	(f) 14	0.042 dB	51.8 dB	Modified allpass structure is used; design is based on the revised specification of passband ripple = 0.03 dB and stopband attenuation = 51 dB.
Filter 3: 5th-order, passband=0.05π, stopband=0.1π, passband ripple=0.1 dB, stopband attenuation=50 dB	(a) 34	0.1 dB	50.0 dB	Unmodified allpass structure is used; design is based on initial specifications of the filter.
	(b) 31	0.1 dB	49.98 dB	.....".....
	(c) 29	0.1 dB	49.9 dB	.....".....
	(d) 23	0.106 dB	49.95 dB	.....".....
	(e) 18	0.125 dB	46.82 dB	.....".....
	(f) 17	0.021 dB	50.55 dB	Modified allpass structure is used; design is based on the revised specification of passband ripple = 0.02 dB and stopband attenuation = 51 dB.
Filter 4: 5th-order, passband=0.2π, stopband=0.35π, passband ripple=0.3 dB, stopband attenuation=50 dB	(a) 33	0.3 dB	50.01 dB	Unmodified allpass structure is used; design is based on initial specifications of the filter.
	(b) 28	0.3 dB	49.97 dB	.....".....
	(c) 24	0.307 dB	49.7 dB	.....".....
	(d) 22	0.31 dB	49.3 dB	.....".....
	(e) 20	0.32 dB	47.95 dB	.....".....
	(f) 18	0.25 dB	50.75 dB	Modified allpass structure is used; design is based on the revised specifications of passband ripple = 0.25 dB and stopband attenuation = 51 dB.

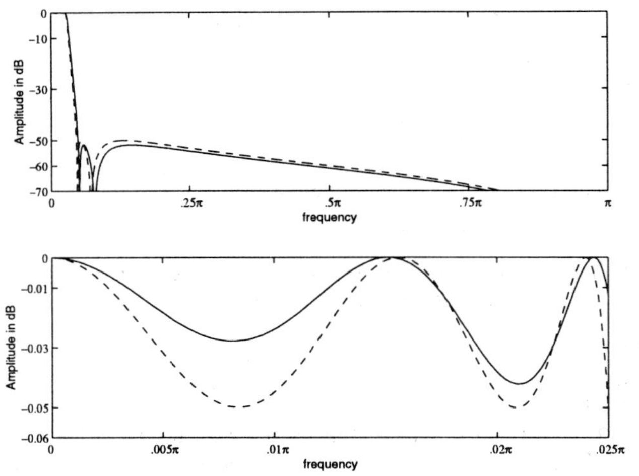

Fig. 3. Amplitude responses for Filter 2 in Table 1 in two cases. The dashed and solid lines represent the initial design realized with the unmodified allpass structure and the realization with revised specifications and modified allpass structures requiring fourteen nonzero bits, respectively.

According to Table 1, the low-sensitivity property of the sum of the proposed allpass filter structures is well evidenced. It is seen from the slow degradation of passband and stopband tolerances with the reduction of number of nonzero bits (that reflects the effective quantization levels).

As stated earlier, the performance table reflects the combined advantages of the proposed structures. These benefits include a reduction in the number of multipliers (N multipliers compared $(3N-1)/2$ multipliers for an odd N-th order elliptic filter) along with a reduction in the total number of nonzero bits required for the multiplierless implementation.

Also, it is observed that the average number of nonzero bits per multiplier is slightly lower than that achieved earlier in [2–4]. Here, the average is approximately three and a half nonzero bits compared with approximately four nonzero bits that has been achieved earlier.

4. CONCLUSIONS

This paper has demonstrated how a concept developed earlier by the authors (using low-sensitivity structures for multiplierless implementations of recursive filters) for utilization in multiplierless implementation of odd-order recursive digital filter can be exploited in developing low-sensitivity allpass filter structures. The resulting structures have been used for implementing odd-order lowpass filters as a parallel connection of two allpass filters. Utilizing the resulting overall structures, in addition to a reduction in the total number of multiplier coefficients, a gross reduction in total number of nonzero bits is achievable, making this approach very attractive. Several examples have indicated that utilizing the approach outlined, multiplierless realizations can be achieved by using around three and half nonzero bits per multiplier on the average, without any increase in the filter order. Future work is devoted to applying optimization techniques to further reducing the number of nonzero bits.

5. ACKNOWLEDGEMENT

This work was supported by the Academy of Finland, project No. 44876 (Finnish Centre of Excellence program (2000-2005)).

6. REFERENCES

[1] J. A. Adams and A. N. Willson, Jr., "Some efficient digital prefilter structures," *IEEE Trans. Circuits Syst.*, vol. CAS-31, pp. 260-266, March 1984.

[2] M. Bhattacharya and J. Astola, "Multiplierless implementation of recursive digital filters based on coefficient translation methods in low sensitivity structures," in *Proc. IEEE Intl. Symp. Circuits Syst., (ISCAS 2001)*, Sydney, Australia, vol.- II, pp. 697-700, 2001.

[3] M. Bhattacharya, T. Saramäki, and J.Astola, "Multiplierless realization of recursive digital filters," in *Proc. 2nd Intl. Symp. Image and Signal Processing and Analysis, (ISPA 2001)*, Pula, Croatia, pp. 469-474, 2001.

[4] M. Bhattacharya and T. Saramäki, "Multiplierless implementation of all-pole digital filters," in *Proc. IEEE Intl. Symp. Circuits Syst., (ISCAS 2002)*, Phoenix, Arizona, USA, vol.- II, pp. 696-699, 2002.

[5] Y. C. Lim and S. R. Parker, "FIR filter design over a discrete powers-of-two coefficient space," *IEEE Trans. Acoust. Speech, Signal Processing*, vol. ASSP-31, pp. 583-591, June 1983.

[6] Y. C. Lim, "Design of discrete-coefficient-value linear phase FIR filters with optimum normalized peak ripple magnitude," *IEEE Trans. Circuits Syst.*, vol. CAS-37, pp. 1480-1486, Dec. 1990.

[7] M. D. Lutovac, D. V. Tosic, and B. L. Evans, *Filter Design for Signal Processing Using Matlab and Mathematica*. Prentice-Hall, New Jersey, 2001.

[8] Lj. D. Milic and M. D. Lutovac, "Design of multiplierless elliptic IIR filters with a small quantization error," *IEEE Trans. Signal Processing,* vol. 47, pp. 469-479, Feb. 1999.

[9] S. K. Mitra and K. Hirano, "Digital all-pass networks," *IEEE Trans. Circuits Syst.*, vol CAS-21, pp. 688-700, Sept. 1974.

[10] R. A. Regalia, Sanjit K. Mitra, and P. P. Vaidyanathan, "The digital all-pass filter: A versatile signal processing building block," *Proc. IEEE,* vol. 76, pp. 19-37, Jan 1988.

[11] H. Samueli, "An improved search algorithm for the design of multiplierless FIR filters with powers-of-two coefficients," *IEEE Trans. Circuits Syst.*, vol. CAS-36, pp. 1044-1047, July 1989.

[12] T. Saramäki, " On the design of digital filters as a sum of two all-pass filters," *IEEE Trans. Circuits Syst.*, vol. CAS-32, pp. 1191-1193, Nov. 1985.

[13] P. P. Vaidyanathan and G. Beitman, " On prefilters for digital filter design," *IEEE Trans. Circuits Syst.*, vol. CAS-32, pp. 494-499, May 1985.

[14] P. P. Vaidyanathan, "Robust Digital Filter Structure," Chapter 7 in *Handbook for Digital Signal Processing,* Mitra S. K. and Kaiser J.F. Eds., John Wiley & Sons, New York, 1993.

[15] J. Yli-Kaakinen and T. Saramäki, "Design of very low-sensitivity and low-noise recursive filters using a cascade of low-order lattice wave digital filters," *IEEE Trans. Circuits Syst. II,* vol.46, pp. 906-914, July 1999.

[16] ....".................., "An algorithm for the design of multiplierless approximately linear-phase lattice-wave digital filters," in *Proc. IEEE Intl. Symp. Circuits Syst., (ISCAS 2000)*, Geneva, Switzerland, vol. 2, pp. 77-80, 2000.

JOINT OPTIMIZATION OF ERROR FEEDBACK AND COORDINATE TRANSFORMATION FOR ROUNDOFF NOISE MINIMIZATION IN STATE-SPACE DIGITAL FILTERS

Takao Hinamoto, Hiroaki Ohnishi and Wu-Sheng Lu†

Graduate School of Engineering, Hiroshima University, Japan
hinamoto@hiroshima-u.ac.jp
†Dept. Elect. Comput. Eng., University of Victoria, Canada
wslu@ece.uvic.ca

ABSTRACT

An iterative approach for joint optimization of a scalar error-feedback matrix and a coordinate transformation matrix is developed to minimize the roundoff noise subject to the l_2-norm dynamic-range scaling constraints. When the iterative algorithm converges and the optimal coordinate transformation matrix is obtained, the diagonal error-feedback matrix is derived to minimize the noise gain in the optimal state-space realization. This diagonal error-feedback matrix enables one to produce more reduction of the noise gain. Finally, a numerical example is given to illustrate the utility of the proposed technique.

1. INTRODUCTION

As is well known, error feedback is an effective technique for the reduction of the roundoff noise due to signal quantizations [1]-[4]. Alternatively, the problem of synthesizing state-space digital filter structures which minimize roundoff noise under l_2-norm scaling constraints has been investigated [5],[6]. Moreover, it has been shown that the output quantization noise can be reduced more effectively by choosing the filter structure and applying the concept of error feedback [7]-[9].

This paper investigates the problem of minimizing roundoff noise under l_2-norm dynamic-range scaling constraints in state-space digital filters by means of joint optimization of error feedback and coordinate transformation. An iterative procedure is proposed for the joint optimization. When the iterative algorithm converges, the optimal coordinate transformation matrix is obtained, which is used to construct the optimal state-space realization. Then, the diagonal error-feedback matrix is derived to minimize the noise gain in the optimal realization. Our simulation results demonstrate the validity of the proposed technique.

2. ERROR FEEDBACK IN STATE-SPACE DIGITAL FILTERS

Consider a stable state-space digital filter $(\boldsymbol{A}, \boldsymbol{b}, \boldsymbol{c}, d)_n$ described by

$$\begin{aligned} \boldsymbol{x}(k+1) &= \boldsymbol{A}\boldsymbol{x}(k) + \boldsymbol{b}u(k) \\ y(k) &= \boldsymbol{c}\boldsymbol{x}(k) + du(k) \end{aligned} \tag{1}$$

where $\boldsymbol{x}(k)$ is an $n \times 1$ state-variable vector, $u(k)$ is a scalar input, $y(k)$ is a scalar output, and $\boldsymbol{A}, \boldsymbol{b}, \boldsymbol{c}$ and d are real constant matrices of appropriate dimensions. The filter (1) is assumed controllable and observable.

Taking into account the quantizations performed before matrix-vector multiplication, one can express an finite-word-length (FWL) implementation of (1) with error feedback as

$$\begin{aligned} \tilde{\boldsymbol{x}}(k+1) &= \boldsymbol{A}\boldsymbol{Q}[\tilde{\boldsymbol{x}}(k)] + \boldsymbol{b}u(k) + \boldsymbol{D}\boldsymbol{e}(k) \\ \tilde{y}(k) &= \boldsymbol{c}\boldsymbol{Q}[\tilde{\boldsymbol{x}}(k)] + du(k) \end{aligned} \tag{2}$$

where $\boldsymbol{e}(k) = \tilde{\boldsymbol{x}}(k) - \boldsymbol{Q}[\tilde{\boldsymbol{x}}(k)]$ and $\boldsymbol{D}$ is referred to as an *error-feedback matrix* of dimension $n \times n$.

All coefficient matrices $\boldsymbol{A}, \boldsymbol{b}, \boldsymbol{c}$, and d are assumed to have an exact fractional B_c bit representation. The FWL state-variable vector $\tilde{\boldsymbol{x}}(k)$ and the output $\tilde{y}(k)$ all have a B bit fractional representation, while the input $u(k)$ is a $(B - B_c)$ bit fraction. The quantizer $\boldsymbol{Q}[\cdot]$ in (2) rounds the B bit fraction $\tilde{\boldsymbol{x}}(k)$ to $(B - B_c)$ bits after completing the multiplications and additions, where the sign bit is not counted. It is assumed that the roundoff error $\boldsymbol{e}(k)$ can be modeled as a zero-mean noise process with covariance $\sigma^2 \boldsymbol{I}_n$. Subtracting (2) from (1) yields

$$\begin{aligned} \Delta\boldsymbol{x}(k+1) &= \boldsymbol{A}\Delta\boldsymbol{x}(k) + (\boldsymbol{A} - \boldsymbol{D})\boldsymbol{e}(k) \\ \Delta y(k) &= \boldsymbol{c}\Delta\boldsymbol{x}(k) + \boldsymbol{c}\boldsymbol{e}(k) \end{aligned} \tag{3}$$

0-7803-7761-3/03/$17.00 ©2003 IEEE

where

$$\Delta\boldsymbol{x}(k) = \boldsymbol{x}(k) - \tilde{\boldsymbol{x}}(k), \qquad \Delta y(k) = y(k) - \tilde{y}(k).$$

Taking the z-transform on both sides of (3) and setting $\Delta\boldsymbol{x}(0) = \boldsymbol{0}$ gives

$$\begin{aligned} \Delta Y(z) &= \boldsymbol{G}_D(z)\boldsymbol{E}(z) \\ \boldsymbol{G}_D(z) &= \boldsymbol{c}(z\boldsymbol{I}_n - \boldsymbol{A})^{-1}(\boldsymbol{A} - \boldsymbol{D}) + \boldsymbol{c} \end{aligned} \tag{4}$$

where $\Delta Y(z)$ and $\boldsymbol{E}(z)$ represent the z-transforms of $\Delta y(k)$ and $\boldsymbol{e}(k)$, respectively.

The normalized noise gain $I(\boldsymbol{D}) = \sigma_{out}^2/\sigma^2$ is then defined by

$$I(\boldsymbol{D}) = \mathrm{tr}[\boldsymbol{W}_D] \tag{5}$$

where

$$\boldsymbol{W}_D = \frac{1}{2\pi j}\oint_{|z|=1} \boldsymbol{G}_D^*(z)\boldsymbol{G}_D(z)\frac{dz}{z}.$$

By utilizing the Cauchy integral theorem, matrix $\boldsymbol{W}_D$ defined in (5) can be expressed in closed form as

$$\boldsymbol{W}_D = (\boldsymbol{A} - \boldsymbol{D})^T\boldsymbol{W}_o(\boldsymbol{A} - \boldsymbol{D}) + \boldsymbol{c}^T\boldsymbol{c} \tag{6}$$

where $\boldsymbol{W}_o$ is the observability Gramian of the filter that can be obtained by solving the Lyapunov equation

$$\boldsymbol{W}_o = \boldsymbol{A}^T\boldsymbol{W}_o\boldsymbol{A} + \boldsymbol{c}^T\boldsymbol{c}. \tag{7}$$

Alternatively, the controllability Gramian $\boldsymbol{K}_c$ can be obtained by solving the Lyapunov equation

$$\boldsymbol{K}_c = \boldsymbol{A}\boldsymbol{K}_c\boldsymbol{A}^T + \boldsymbol{b}\boldsymbol{b}^T. \tag{8}$$

The filter (1) is changed via coordinate transformation $\overline{\boldsymbol{x}}(k) = \boldsymbol{T}^{-1}\boldsymbol{x}(k)$ to a new realization $(\overline{\boldsymbol{A}}, \overline{\boldsymbol{b}}, \overline{\boldsymbol{c}}, d)_n$ with

$$\begin{aligned} &\overline{\boldsymbol{A}} = \boldsymbol{T}^{-1}\boldsymbol{A}\boldsymbol{T}, \quad \overline{\boldsymbol{b}} = \boldsymbol{T}^{-1}\boldsymbol{b}, \quad \overline{\boldsymbol{c}} = \boldsymbol{c}\boldsymbol{T} \\ &\overline{\boldsymbol{W}}_o = \boldsymbol{T}^T\boldsymbol{W}_o\boldsymbol{T}, \quad \overline{\boldsymbol{K}}_c = \boldsymbol{T}^{-1}\boldsymbol{K}_c\boldsymbol{T}^{-T}. \end{aligned} \tag{9}$$

The problem considered here is to jointly optimize a scalar error-feedback matrix $\alpha\boldsymbol{I}_n$ and a coordinate transformation matrix $\boldsymbol{T}$ for roundoff noise minimization under l_2-norm dynamic-range scaling constraints:

$$(\overline{\boldsymbol{K}}_c)_{ii} = (\boldsymbol{T}^{-1}\boldsymbol{K}_c\boldsymbol{T}^{-T})_{ii} = 1, \quad i = 1, 2, \cdots, n. \tag{10}$$

If the unit noise matrix $\boldsymbol{W}_D$, (6), with $\boldsymbol{D} = \alpha\boldsymbol{I}_n$ is denoted by $\boldsymbol{W}_\alpha$, then the noise gain $I(\boldsymbol{D})$ defined in (5) is written as $\mathrm{tr}[\boldsymbol{T}^T\boldsymbol{W}_\alpha\boldsymbol{T}]$ under joint optimization of scalar error-feedback and coordinate transformation.

The proposed joint optimization will be carried out in an iterative manner.

3. AN ITERATIVE PROCEDURE FOR JOINT OPTIMIZATION

In order to minimize $\mathrm{tr}[\boldsymbol{T}^T\boldsymbol{W}_\alpha\boldsymbol{T}]$ (with α fixed) over an $n \times n$ nonsingular matrix $\boldsymbol{T}$ subject to the constraints shown in (10), we define the Lagrange function

$$J(\alpha, \boldsymbol{P}, \lambda) = \mathrm{tr}[\boldsymbol{W}_\alpha\boldsymbol{P}] + \lambda(\mathrm{tr}[\boldsymbol{K}_c\boldsymbol{P}^{-1}] - n) \tag{11}$$

where $\boldsymbol{P} = \boldsymbol{T}\boldsymbol{T}^T$ and λ is a Lagrange multiplier. We compute

$$\begin{aligned} \frac{\partial J(\alpha, \boldsymbol{P}, \lambda)}{\partial \alpha} &= 2(\alpha\,\mathrm{tr}[\boldsymbol{W}_o\boldsymbol{P}] - \mathrm{tr}[\boldsymbol{W}_o\boldsymbol{A}\boldsymbol{P}]) \\ \frac{\partial J(\alpha, \boldsymbol{P}, \lambda)}{\partial \boldsymbol{P}} &= \boldsymbol{W}_\alpha - \lambda\boldsymbol{P}^{-1}\boldsymbol{K}_c\boldsymbol{P}^{-1} \\ \frac{\partial J(\alpha, \boldsymbol{P}, \lambda)}{\partial \lambda} &= \mathrm{tr}[\boldsymbol{K}_c\boldsymbol{P}^{-1}] - n. \end{aligned} \tag{12}$$

Letting $\partial J(\alpha, \boldsymbol{P}, \lambda)/\partial\alpha = 0$ yields

$$\alpha = \frac{\mathrm{tr}[\boldsymbol{W}_o\boldsymbol{A}\boldsymbol{P}]}{\mathrm{tr}[\boldsymbol{W}_o\boldsymbol{P}]}. \tag{13}$$

Letting $\partial J(\alpha, \boldsymbol{P}, \lambda)/\partial\boldsymbol{P} = \boldsymbol{0}$ and $\partial J(\alpha, \boldsymbol{P}, \lambda)/\partial\lambda = 0$,

$$\boldsymbol{P}\boldsymbol{W}_\alpha\boldsymbol{P} = \lambda\boldsymbol{K}_c, \qquad \mathrm{tr}[\boldsymbol{K}_c\boldsymbol{P}^{-1}] = n. \tag{14}$$

It follows from (14) that

$$\begin{aligned} &\boldsymbol{P} = \sqrt{\lambda}\boldsymbol{W}_\alpha^{-\frac{1}{2}}[\boldsymbol{W}_\alpha^{\frac{1}{2}}\boldsymbol{K}_c\boldsymbol{W}_\alpha^{\frac{1}{2}}]^{\frac{1}{2}}\boldsymbol{W}_\alpha^{-\frac{1}{2}} \\ &\frac{1}{\sqrt{\lambda}}\mathrm{tr}[\boldsymbol{K}_c\boldsymbol{W}_\alpha]^{\frac{1}{2}} = \frac{1}{\sqrt{\lambda}}\left(\sum_{i=1}^{n}\theta_i\right) = n \end{aligned} \tag{15}$$

where θ_i^2 for $i = 1, 2, \cdots, n$ are the eigenvalues of $\boldsymbol{K}_c\boldsymbol{W}_\alpha$. This can be used to obtain

$$\boldsymbol{P} = \frac{1}{n}\left(\sum_{i=1}^{n}\theta_i\right)\boldsymbol{W}_\alpha^{-\frac{1}{2}}[\boldsymbol{W}_\alpha^{\frac{1}{2}}\boldsymbol{K}_c\boldsymbol{W}_\alpha^{\frac{1}{2}}]^{\frac{1}{2}}\boldsymbol{W}_\alpha^{-\frac{1}{2}}. \tag{16}$$

Substituting (16) into (11) yields the minimum value of $J(\alpha, \boldsymbol{P}, \lambda)$ for a given scalar α as

$$\min_{\boldsymbol{P}, \lambda} J(\alpha, \boldsymbol{P}, \lambda) = \frac{1}{n}\left(\sum_{i=1}^{n}\theta_i\right)^2. \tag{17}$$

This completes the first round of iteration and this process may continue until both $\boldsymbol{P}$ and α converge. Having obtained an $n \times n$ symmetric positive-definite matrix $\boldsymbol{P}$, an improved value of scalar α can be obtained using (13).

This iterative procedure for minimizing (11) with respect to a scalar parameter α as well as matrix $\boldsymbol{P}$ can be summarized as follows:

1) Set $i = 1$ and

$$P(0) = \text{diag}\{(K_c)_{11}^{-1}, (K_c)_{22}^{-1}, \cdots, (K_c)_{nn}^{-1}\}.$$

2) Compute a scalar $\alpha(i)$ using

$$\alpha(i) = \frac{\text{tr}[W_o A P(i-1)]}{\text{tr}[W_o P(i-1)]}.$$

3) Compute

$$I_{min}(\alpha(i) I_n) = (1 - \alpha(i)^2)\,\text{tr}[W_o P(i-1)].$$

4) Replace W_α by $W_{\alpha(i)}$ computed using

$$W_{\alpha(i)} = (1 + \alpha(i)^2) W_o - \alpha(i)(A^T W_o + W_o A).$$

5) Derive matrix P from (17), and take the resulting P as $P(i)$.

6) Compute $\text{tr}[W_{\alpha(i)} P(i)]$.

7) Update i to $i+1$.

8) Repeat from Step 2) until the change in either $I_{min}[\alpha(i) I_n]$ or $\text{tr}[W_{\alpha(i)} P(i)]$ becomes negligible.

Next, the coordinate transformation matrix T will be constructed so that (10) is satisfied. From (16), the optimal coordinate transformation matrix T that minimizes (11) can be obtained in closed form as

$$T = \frac{1}{\sqrt{n}} \left(\sum_{i=1}^{n} \theta_i \right)^{\frac{1}{2}} W_\alpha^{-\frac{1}{2}} [W_\alpha^{\frac{1}{2}} K_c W_\alpha^{\frac{1}{2}}]^{\frac{1}{4}} U \tag{18}$$

where U is an arbitrary $n \times n$ orthogonal matrix. From (18) it follows that

$$\begin{aligned} \overline{K}_c &= T^{-1} K_c T^{-T} \\ &= n \left(\sum_{i=1}^{n} \theta_i \right)^{-1} U^T [W_\alpha^{\frac{1}{2}} K_c W_\alpha^{\frac{1}{2}}]^{\frac{1}{2}} U. \end{aligned} \tag{19}$$

Let us choose the $n \times n$ orthogonal matrix U such that the matrix $\overline{K}_c$ in (9) satisfies the l_2-norm dynamic-range scaling constraints, (10), on the state-variables. To this end, we perform the eigenvalue-eigenvector decomposition

$$[W_\alpha^{\frac{1}{2}} K_c W_\alpha^{\frac{1}{2}}]^{\frac{1}{2}} = R \Theta R^T \tag{20}$$

where $\Theta = \text{diag}\{\theta_1, \theta_2, \cdots, \theta_n\}$ and $RR^T = I_n$. This yields

$$n \left(\sum_{i=1}^{n} \theta_i \right)^{-1} [W_\alpha^{\frac{1}{2}} K_c W_\alpha^{\frac{1}{2}}]^{\frac{1}{2}} = R \Lambda^{-2} R^T \tag{21}$$

where $\Lambda = \text{diag}\{\lambda_1, \lambda_2, \cdots, \lambda_n\}$ and for $i = 1, 2, \cdots, n$, $\lambda_i = ((\theta_1 + \theta_2 + \cdots + \theta_n)/n\theta_i)^{\frac{1}{2}}$. Now an $n \times n$ orthogonal matrix S such that

$$S \Lambda^{-2} S^T = \begin{bmatrix} 1 & * & \cdots & * \\ * & 1 & \ddots & \vdots \\ \vdots & \ddots & \ddots & * \\ * & \cdots & * & 1 \end{bmatrix} \tag{22}$$

can be obtained by numerical manipulations [6, p.278]. By choosing $U = RS^T$ in (18), the optimal coordinate transformation matrix T both satisfying (10) and minimizing (11) can now be constructed as

$$T = \frac{1}{\sqrt{n}} \left(\sum_{i=1}^{n} \theta_i \right)^{\frac{1}{2}} W_\alpha^{-\frac{1}{2}} [W_\alpha^{\frac{1}{2}} K_c W_\alpha^{\frac{1}{2}}]^{\frac{1}{4}} R S^T. \tag{23}$$

Suppose the iterative algorithm converges after N iterations and the optimal coordinate transformation matrix $T(N)$ has been computed from (20)-(23). Then, the diagonal error-feedback matrix $D = \text{diag}\{\alpha_1, \alpha_2, \cdots, \alpha_n\}$ that minimizes

$$\begin{aligned} I(D) = {} & \text{tr}[T^T(N) W_o T(N)] + \text{tr}[T^T(N) W_o T(N) D^2] \\ & -2\text{tr}[T^T(N) A^T W_o T(N) D] \end{aligned} \tag{24}$$

is given by

$$\alpha_i = \frac{(T^T(N) W_o A T(N))_{ii}}{(T^T(N) W_o T(N))_{ii}}, \quad i = 1, 2, \cdots, n. \tag{25}$$

This diagonal error-feedback matrix D makes it possible to produce more reduction of the noise gain, i.e.,

$$I_{min}(D) < I_{min}[\alpha(N) I_n]. \tag{26}$$

4. A NUMERICAL EXAMPLE

Let a state-space digital filter $(A, b, c, d)_3$ be described in a controllable canonical form as

$$A = \begin{bmatrix} 0 & 1 & 0 \\ 0 & 0 & 1 \\ 0.339377 & -1.152652 & 1.520167 \end{bmatrix}$$

$$b = \begin{bmatrix} 0 & 0 & 0.437881 \end{bmatrix}^T$$

$$c = \begin{bmatrix} 0.212964 & 0.293733 & 0.718718 \end{bmatrix}$$

$$d = 6.59592 \times 10^{-2}$$

which satisfies l_2-norm dynamic-range scaling constraints, and yields $I(0) = \text{tr}[W_o] = 11.133150$.

We apply the iterative optimization procedure in Section 3 to this filter. The convergent profile of first 10 iterations is given in Table I, from which we see that the algorithm converges after six iterations to a scalar $\alpha = 0.647686$ and $I_{min}(\alpha \boldsymbol{I}_n) = \mathrm{tr}[\boldsymbol{T}^T \boldsymbol{W}_\alpha \boldsymbol{T}] = 1.450048$. In this case, the coordinate transformation matrix $\boldsymbol{T}^o$ is given by

$$\boldsymbol{T}^o = \begin{bmatrix} -1.973853 & -0.153371 & -2.328357 \\ -0.063334 & -1.398294 & -1.260527 \\ 1.402772 & -0.676604 & -0.969851 \end{bmatrix}.$$

TABLE I

CONVERGENT PROFILE OF FIRST 10 ITERATIONS

i	$\alpha(i)$	$I_{min}[\alpha(i)\boldsymbol{I}_n]$	$\mathrm{tr}[\boldsymbol{W}_{\alpha(i)}\boldsymbol{P}(i)]$
1	0.764400	4.627965	1.482085
2	0.655454	1.451873	1.450188
3	0.648286	1.450059	1.450049
4	0.647733	1.450048	1.450048
5	0.647689	1.450048	1.450048
6	0.647686	1.450048	1.450048
7	0.647686	1.450048	1.450048
8	0.647686	1.450048	1.450048
9	0.647686	1.450048	1.450048
10	0.647686	1.450048	1.450048

If $\alpha = 0.647686$ is rounded to power-of-two representation with 3 bits after binary point, then the noise gain is founded to be $I(\alpha \boldsymbol{I}_n) = 1.451335$ where $\alpha = 0.625$.

Next, a refined solution which offers further reduced noise gain is deduced by applying an optimal diagonal error-feedback matrix to the optimized realization, i.e., $(\boldsymbol{T}^{o-1}\boldsymbol{A}\boldsymbol{T}^o, \boldsymbol{T}^{o-1}\boldsymbol{b}, \boldsymbol{c}\boldsymbol{T}^o, d)_3$. The optimal diagonal error-feedback matrix obtained using (25) is given by

$$\boldsymbol{D} = \mathrm{diag}\{0.705402,\ 0.510713,\ 0.683277\}$$

which yields $I_{min}(\boldsymbol{D}) = 1.433755$.

The above diagonal error-feedback matrix after 3-bit quantization (power-of-two representation with 3 bits after binary point) gives $I_{min}(\boldsymbol{D}) = 1.438801$, which is less than $I_{min}(\boldsymbol{D}) = 1.450049$ in the optimal scalar error-feedback.

To compare the proposed method with those reported in [6]-[8], we choose $\boldsymbol{D} = \boldsymbol{0}$, i.e., $\boldsymbol{W}_D = \boldsymbol{W}_o$ as in [6] or $\boldsymbol{D} = \boldsymbol{I}_n$ as in [7],[8], and minimize $\mathrm{tr}[\boldsymbol{T}^T \boldsymbol{W}_D \boldsymbol{T}]$ with respect to matrix $\boldsymbol{T}$ under the constraints of (10):

$$\min_{\boldsymbol{T}} \ \mathrm{tr}[\ \boldsymbol{T}^T \boldsymbol{W}_o \boldsymbol{T}\ ] = 2.355360$$

$$\min_{\boldsymbol{T}} \ \mathrm{tr}[\ \boldsymbol{T}^T \boldsymbol{W}_D \boldsymbol{T}\ ] = 1.752546$$

which are considerably larger than our results described above.

5. CONCLUSION

The roundoff noise minimization in state-space digital filters has been considered. The noise minimization problem has been addressed in scenario where a scalar error-feedback matrix and a coordinate transformation matrix are jointly optimized subject to usual l_2-norm dynamic-range scaling constraints. Simulation results have been presented to illustrate and support our theoretical analysis and proposed algorithm.

The extension of the results obtained in this paper to multidimensional case will appear elsewhere.

REFERENCES

[1] W. E. Higgins and D. C. Munson, "Noise reduction strategies for digital filters: Error spectrum shaping versus the optimal linear state-space formulation," *IEEE Trans. Acoust. Speech, Signal Processing*, vol. 30, pp. 963-973, Dec. 1982.

[2] W. E. Higgins and D. C. Munson, "Optimal and suboptimal error-spectrum shaping for cascade-form digital filters," *IEEE Trans. Circuits Syst.*, vol.CAS-31, pp.429-437, May 1984.

[3] T. I. Laakso and I. O. Hartimo, "Noise reduction in recursive digital filters using high-order error feedback," *IEEE Trans. Signal Processing*, vol40, pp.1096-1107, May 1992.

[4] P. P. Vaidyanathan, "On error-spectrum shaping in state-space digital filters," *IEEE Trans. Circuits Syst.*, vol.CAS-32, pp.88-92, Jan. 1985.

[5] C. T. Mullis and R. A. Roberts, "Synthesis of minimum roundoff noise fixed point digital filters," *IEEE Trans. Circuits Syst.*, vol.CAS-23, pp.551-562, Sept. 1976.

[6] S. Y. Hwang, "Minimum uncorrelated unit noise in state-space digital filtering," *IEEE Trans. Acoust. Speech, Signal Processing*, vol.ASSP-25, pp.273-281, Aug. 1977.

[7] D. Williamson, "Roundoff noise minimization and pole-zero sensitivity in fixed-point digital filters using residue feedback," *IEEE Trans. Acoust. Speech, Signal Processing*, vol.ASSP-34, pp.1210-1220, Oct. 1986.

[8] G. Li and M. Gevers, "Roundoff noise minimization using delta-operator realizations," *IEEE Trans. Signal Processing*, vol.41, pp.629-637, Feb. 1993.

[9] T. Hinamoto and S. Kanemori, "Error spectrum shaping in state-space digital filters with l_2-scaling constraints," in Proc. *2000 IEEE Int. Symp. Circuits Syst.*, Geneva, Switzerland, vol.2, pp.329-332, May 2000.

IIR EQUALIZER DESIGN BASED ON THE IMPULSE RESPONSE SYMMETRY CRITERION

Mladen Vucic and Hrvoje Babic
Faculty of Electrical Engineering and Computing
Unska 3, Zagreb, HR10000, Croatia

ABSTRACT

An approach for IIR equalizer design based on the impulse response symmetry is presented. Design is based on numerical optimization. General form of the objective function is given and its special cases are analyzed. It is shown that the general form gives the best results while the special cases can be more appropriate for the implementation. Practical aspects of the implementation are considered and the choice of the optimization starting point is discussed. Method's features are illustrated by an example.

1. INTRODUCTION

There are two approaches for the design of IIR systems with approximately linear phase or constant group delay. First approach starts from defining the requirements on the phase response of the system. Since selectivity is usually needed, requirements are set on the amplitude response as well. An appropriate algorithm is then applied to compute the transfer function. The obtained result is always a compromise because the amplitude and the phase response are mutually dependent.

The second approach for the design of systems with linear phase represents cascading the original system with an appropriate all-pass system. Since the all-pass system does not affect the amplitude response, it is designed to improve phase response. Although such technique increases the overall system complexity, it is rather popular because it allows more freedom in design. Furthermore, it is irreplaceable when the original system can not be arbitrarily modified, as it is in the case of phase correction in data transmission systems. Numerous references describing both approaches are available in tutorial papers, as for example in [1].

For the design of digital all-pass systems several methods have been developed. An early, well-known method was proposed by Deczky [2]. He formed a nonlinear error function and then optimized it using the Fletcher-Powel algorithm. The error function was based on the minimum p-th error criterion. Estola [3] proposed an algorithm for the optimization of the all-pass system in the mean square sense. For searching minima, the multiple exchange Remez algorithm was used. Jing [4] developed a robust all-pass system design method based on linearization of nonlinear constraints in nonlinear programming.

By the end of the last century, several new approaches were used in the all-pass system design. Reddy and Swamy [5] presented a method that uses Hilbert transform relations between amplitude and phase of the minimum phase system. Ikehara, Funaishi and Kuruoda [6] formed a complex error function whose amplitude corresponded to the error between the desired and the designed phase of the all-pass system. The optimum solution was than found using the Remez exchange algorithm. Lang and Laakso [7] used a general weighted least-squares equation error type approximation. Chen and Lee [8] considered the minimax design using weighted least squares approach. Nguyen, Laakso and Koilpillai [9] used the eigenfilter method to obtain all-pass filters that approximate the desired phase in the least squares sense. Recently, Haddad, Yang, Galatsanos and Stark [10] have proposed an approach based on the vector spaced projection method.

Common to all of the mentioned methods is that they start from prescribed phase or group delay response. Setting the requirements on the phase or group delay, i.e. force linearity or constant behavior, is suitable for applications such as Fourier analysis. On the other hand, when the time domain distortion is of the major importance the linear phase or the constant group delay requirement will not give the optimal solution. Namely, the question that arises during the equalizer design is what amount of group delay ripple can be tolerated in the particular frequency band, as well as how large ratio between the amplitude and the group delay cutoff should be set. The dilemma can be solved by moving the requirements into the time domain, as in the case of IIR filters with symmetric impulse response [11].

Recently, the impulse response symmetry criterion has been successfully applied in the design of continuous time all-pass systems [12], [13]. In this paper we propose a method for the design of the IIR all-pass systems, which gives the minimum asymmetry of the impulse response when cascaded to a particular discrete time system. We will present the general form of the error function and discuss some special cases. Then we will describe the implementation and illustrate the method's features by an example.

2. IMPULSE RESPONSE SYMMETRY

A real discrete signal $h_d(n)$ has linear phase if and only if the continuos time signal

$$h_a(t) = \sum_{n=-\infty}^{\infty} h_d(n) \frac{\sin(\pi(t-n))}{\pi(t-n)} \tag{1}$$

is symmetric about the line placed in $t=t_s$ [14]. Expression (1) is in fact convolution and the signal $h_d(t)$ should be considered as a sampled equivalent of the bandlimited continuous time signal $h_a(t)$.

A stable causal system with transfer function containing finite poles can not have ideal linear phase and symmetric impulse response [14] but can approximate it. Thus, we used measure of the impulse response asymmetry defined with

$$e_a = \frac{1}{2} \int_{-\infty}^{\infty} [h_a(t) - h_a(2t_s - t)]^2 dt . \tag{2}$$

Using (1) and assuming that $h_d(n)$ is causal, the symmetry error can be written as

0-7803-7761-3/03/$17.00 ©2003 IEEE

$$e_a = \frac{1}{2}\int_{-\infty}^{\infty}\left(\sum_{n=0}^{\infty} h_d(n)\left(\mathrm{sinc}(t-n)-\mathrm{sinc}(2t_s-t-n)\right)\right)^2 dt \quad (3)$$

where sinc(x)=sin(πx)/(πx). After squaring the integrand and rearranging the expression, we obtain

$$e_a = \frac{1}{2}\sum_{n=0}^{\infty}\sum_{k=0}^{\infty} h_d(n)\, h_d(k) \cdot \int_{-\infty}^{\infty}\left(\mathrm{sinc}(t-n)-\mathrm{sinc}(2t_s-t-n)\right)\left(\mathrm{sinc}(t-k)-\mathrm{sinc}(2t_s-t-k)\right)dt \quad (4)$$

The integral in (4) can be solved analytically resulting with

$$e_a = \sum_{n=0}^{\infty}\sum_{k=0}^{\infty} h_d(n)\, h_d(k)\ \left(\mathrm{sinc}(k-n)-\mathrm{sinc}(k+n-2t_s)\right) \quad (5)$$

Assuming

$$\mathrm{sinc}(k-n) = \begin{cases} 1 \text{ for } k=n \\ 0 \text{ for } k \neq n \end{cases} \quad (6)$$

we can express the impulse response asymmetry as

$$e_a = \sum_{n=0}^{\infty}(h_d(n))^2 - \sum_{n=0}^{\infty}\sum_{k=0}^{\infty} h_d(n)\, h_d(k)\, \mathrm{sinc}(k+n-2t_s)\ . \quad (7)$$

Depending on the symmetry line t_s two particular cases of the equation (7) can be observed. If t_s is an integer the symmetry line passes through the sample of the impulse response $h_d(t)$ while if $t_s \in \{0.5,\ 1.5,\ 2.5,\ ...\}$, the symmetry line lies in the middle between two samples. In both cases the factor $\mathrm{sinc}(k+n-2t_s)$ equals to 0 for $k+n \neq 2t_s$. Since it is always true for $k>2t_s$ or $n>2t_s$, the upper summation indexes in the second addend in (7) can be reduced to $2t_s$. Assuming $k=2t_s-n$ the expression (7) can be simplified

$$e_d = \sum_{n=0}^{\infty}(h_d(n))^2 - \sum_{n=0}^{2t_s} h_d(n)\, h_d(2t_s-n)\ . \quad (8)$$

For $t_s \in \{0.5, 1.5, 2.5, ...\}$ second sum of the equation (8) contains even number of addends and can be written in the form

$$e_d = \sum_{n=0}^{\infty}(h_d(n))^2 - 2\sum_{n=0}^{t_s} h_d(n)\, h_d(2t_s-k)\ , \quad (9)$$

which is known from previously presented approach for the low-pass filter design [11].

It is reasonable to expect that optimizing the general form of the symmetry error (7), which is in fact based on the envelope, will give the best possible symmetry. However, the particular case (9), which deals only with final number of the impulse response samples, may also be interesting because of its simplicity. Subsequently we will analyze the all-pass systems that are obtained by minimization of the symmetry errors, (7) and (9).

3. OBJECTIVE FUNCTION

To perform optimization, the errors (7) and (9) should be expressed by the appropriate transfer function parameters. Transfer function of the cascade of an A-th order all-pass system and the N-th order system with M zeros is given by

$$H(z) = (-1)^A \frac{\prod_{q=1}^{A}(d_q - z^{-1})}{\prod_{q=1}^{A}(1-d_q z^{-1})} \cdot H_0 \frac{\prod_{i=1}^{M}(1-c_i z^{-1})}{\prod_{k=A+1}^{A+N}(1-d_k z^{-1})}\ . \quad (10)$$

Collecting factors of the denominator into a common product, the transfer function can be written as

$$H(z) = (-1)^A H_0 \frac{\prod_{q=1}^{A}(d_q - z^{-1})\prod_{i=1}^{M}(1-c_i z^{-1})}{\prod_{k=1}^{A+N}(1-d_k z^{-1})}\ . \quad (11)$$

If poles of H(z) are simple, the transfer function can be expressed as the sum

$$H(z) = \sum_{q=0}^{M-N} B_q z^{-q} + \sum_{r=1}^{A+N}\frac{A_r}{1-d_r z^{-1}}\ . \quad (12)$$

The coefficients B_q can be obtained by division of numerator and denominator of the transfer function. The coefficients A_r are pole residues given by $A_r=(1-d_r z^{-1})H(z)$ and evaluated in $z=d_r$ [15]

$$A_r = (-1)^A H_0 \frac{\prod_{q=1}^{A}(d_q - \frac{1}{d_r})\prod_{i=1}^{M}(1-\frac{c_i}{d_r})}{\prod_{\substack{k=1\\k\neq r}}^{A+N}(1-\frac{d_k}{d_r})}\ . \quad (13)$$

The impulse response of the system (10) is the inverse z-transform of the expression (12)

$$h_d(n) = \sum_{q=0}^{M-N} B_q \delta(n-q) + \sum_{r=1}^{A+N} A_r d_r^n\ \ ,\ n \geq 0\ . \quad (14)$$

4. PRACTICAL IMPLEMENTATION

The first summations in symmetry errors (7) and (9) represents the total energy of the impulse response. Our goal is to optimize the all-pass transfer function connected to the filter output without modifying the filter. Since the energy of the impulse response depends on the filter transfer function only, it will remain constant during the optimization procedure. However, the total energy of the impulse response (14) can be expressed analytically as

$$e_h = \sum_{q=0}^{M-N} B_q^2 + 2\sum_{q=0}^{M-N}\sum_{r=1}^{N} B_q A_r d_r^q + \sum_{q=1}^{N}\sum_{r=1}^{N}\frac{A_q A_r}{1-d_q d_r}\ . \quad (15)$$

For practical implementation of the symmetry error (7), the impulse response should be truncated. Thus, summation in the second addend will run up to some final number of samples, P, rather than to the infinity. The number of samples, P, should be taken high enough to avoid significant numerical error. On the other hand, too long impulse response will increase the time required for calculation of the error. We were using the number of samples that resulted with the energy of the neglected response tail less than ε. We found that ε as low as 1e-7 is appropriate. However, calculations became time consuming only for systems with high quality factors of poles, i.e. very long impulse response.

The symmetry error (9) can be obtained exactly using (14) and (15).

5. OPTIMIZATION PROCEDURE

Pole and zero positions of the all-pass transfer function that will minimize the impulse response asymmetry (7) of the system (11), can be found solving the problem

$$\min_{t_s, d_1, \ldots, d_A} e_a\left[t_s, d_1, \ldots, d_{A+N}, c_1, \ldots, c_M\right] . \qquad (16)$$

Position of the symmetry line t_s is treated as an independent continuous variable in the optimization procedure because it is impossible to know its optimal position in advance.

To minimize the impulse response symmetry given by (9) the objective function should be modified

$$\min_{d_1, \ldots, d_A} e_d\left[t_s, d_1, \ldots, d_{A+N}, c_1, \ldots, c_M\right] . \qquad (17)$$

Although the symmetry line t_s is a continuous variable in the case (16), here in (17) it takes discrete values only. The simplest way to perform the optimization is by treating t_s as a parameter and running the optimization with t_s taken from some presumed set of values. Despite the necessity of multiple runs of the optimization, the procedure is much faster than the one based on the generalized error (16). It is an expected consequence of the simplicity of the analytically expressed error (9).

For practical implementation the complex poles $d_1, \ldots, d_A$, that are in fact the goal function variables, were separated into their real and imaginary parts; $d_k = d_{\sigma k} + i \cdot d_{\varepsilon k}$. The complex poles of real systems always come in conjugate pairs. Thus, for example, the optimization problem (16) can be formed as

$$\min_{\substack{t_s, d_{\sigma 1}, \ldots, d_{\sigma(A/2)}, \\ d_{\varepsilon 1}, \ldots, d_{\varepsilon(A/2)}}} e_a\left[t_s, d_1, \ldots, d_{A+N}, c_1, \ldots, c_M\right] , \qquad (18)$$

when corrector order A is even, and similar when it is odd.

Searching for the minimum was performed by the Quasi-Newton method with BFGS formula for Hessian matrix update [16].

To perform the optimization an appropriate initial points should be chosen. It was experimentally found that initial corrector poles should be placed inside the region that corresponds to the pass-band of the system. For higher corrector orders some poles should be placed in the transitional band as well. For the initial symmetry line t_s we were using the position of the impulse response maximum of the initial cascade. Sometimes several initial points were needed to find the optimal solution.

6. OPTIMIZATION EXAMPLE

To illustrate features of the proposed method, the equalization of the eighth order IIR Butterworth low-pass filter with cutoff frequency $\omega_{3dB}=0.2\pi$ will be described. The optimization (16) was carried out for equalizers from the first, A=1, up to the sixth order, A=6. The case A=0 corresponds to the original filter.

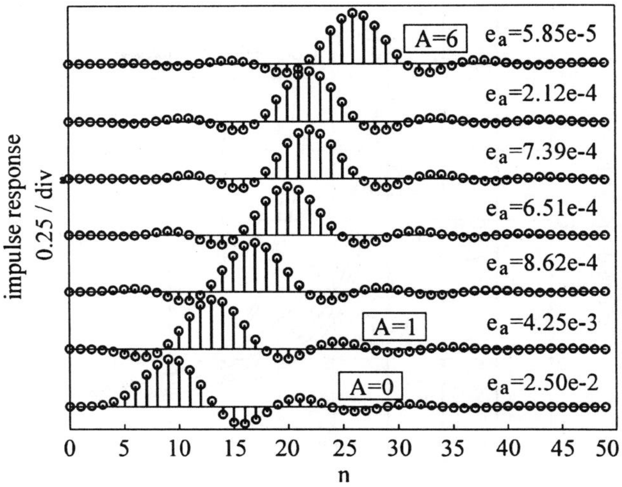

Figure 1. Impulse response of the Butterworth 8-th order filter cascaded with the A-th order optimum corrector.

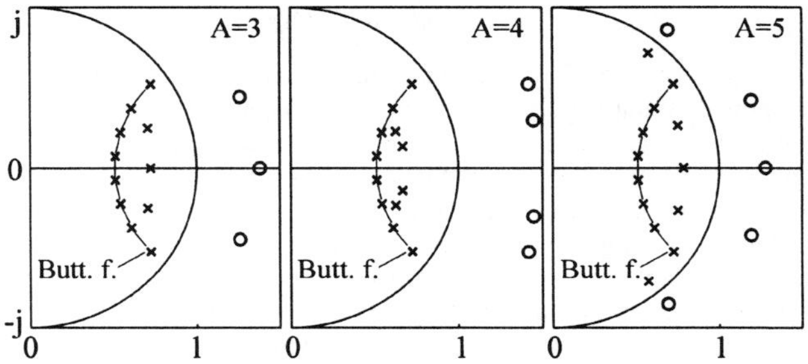

Figure 2. Pole and zero positions of the Butterworth 8-th order filter and the optimum A-th order corrector. (Eight zeros of the filter placed at -1+0j are not shown in the figure.)

Impulse responses of the cascade are shown in Figure 1, together with the symmetry error e_a. By increasing the corrector order up to A=3, the impulse response asymmetry is decreasing and corrector poles are placed inside the pass-band, Figure 2. The case A=4 gives somewhat worse symmetry than the case A=3. Such result is also correct because here we forced four poles to be complex as can be seen in (18). Further increase of the corrector order causes one pair of poles to be placed in the transitional band while symmetry error continues decreasing.

The group delay of the optimum systems is shown in Figure 3. In the pass-band it approximates a constant with ripple which generally decreases by increasing the corrector order. For the fifth and the sixth corrector order the group delay is quasi constant in the pass-band ($\omega_{3dB}=0.2\pi$) and part of transitional band as well, while ripple is somewhat larger. It is the consequence of the previously discussed pole positions.

It is interesting to compare properties of systems with minimum asymmetry of the samples, e_d, (9) and systems with minimum asymmetry of the envelope, e_a, (7). Figure 4 shows the group delay of these systems calculated for A=3 and A=5. The response curves have similar shape. The criterion based on the asymmetry of samples apparently gives somewhat larger group delay ripple, which, depending on application, may also be tolerated.

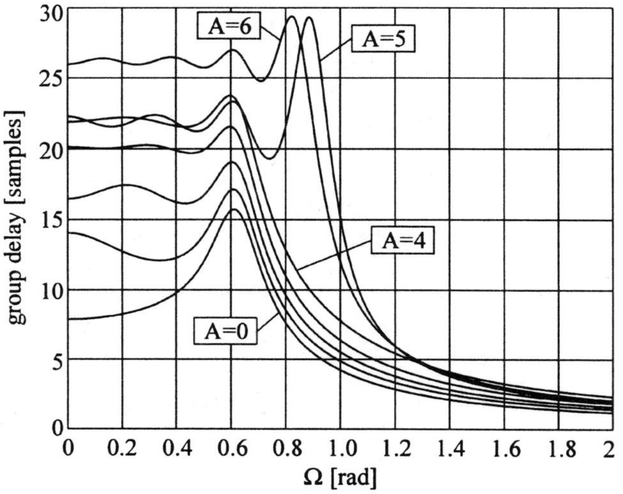

Figure 3. Group delay of the Butterworth 8-th order filter cascaded with the A-th order optimum corrector.

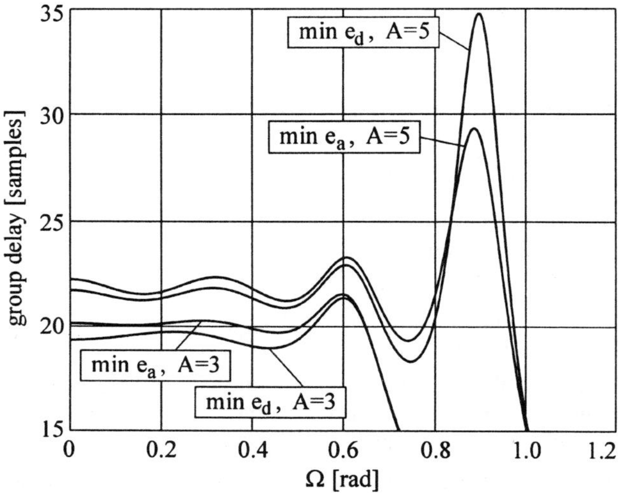

Figure 4. Group delay of systems with minimum asymmetry of the samples, e_d, and systems with minimum asymmetry of the envelope, e_a.

7. CONCLUSION

An approach for IIR equalizer design based on the impulse response symmetry was presented. A general form of the objective function suitable for the numerical optimization was given. Special cases of the function were analyzed in order to increase robustness of the design. It was shown that the general form of the objective function leads to the smallest possible impulse response asymmetry while the special case simplifies design. Described methods give systems that approximate constant group delay with ripple, which generally decreases by increasing corrector order.

8. ACKNOWLEDGEMENT

This study was made at the Department of Electronic Systems and Information Processing of the Faculty of Electrical Engineering and Computing, University of Zagreb, Croatia. It was supported by Ministry of Science and Technology of Croatia, under grant No. 0036029.

9. REFERENCES

[1] Regalia P. A., Mitra S. K., Vaydyanathan P. P., "The Digital All-Pass Filter: A versatile signal processing building Block", *IEEE Proceedings*, Vol. 76, No. 1, January 1988, pages 19-37.

[2] Deczky A. G., "Synthesis of Recursive Digital Filters Using the Minimum p-error Criterion", *IEEE Trans. on Acoustics Speech and Signal Processing*, Vol, AU-20, October 1972, pages 257-263.

[3] Estola K. P., "Design of all-pass filters using a new bandpass transform", *Proc. of 1990 IEEE International Symposium on Circuits and Systems*, San Jose, CA, USA, 5-7 May 1986, pages 1093-1096.

[4] Jing Z., "A new method for digital all-pass filter design", *IEEE Trans. on Acoustics, Speech and Signal Processing*, Vol. ASSP-35, No.11, November 1987, pages 1557-1564.

[5] Reddy G. R., Swamy M. N. S., "Digital all-pass filter design through discrete Hilbert transform", *Proc. of 1990 IEEE International Symposium on Circuits and Systems*, New Orleans, LA, USA, 1-3 May 1990, Vol. 1,pages 646-649.

[6] Ikehara M., Funaishi M., Kuroda H., "Design of all-pass networks using Remez algorithm", *Proc. of 1991 IEEE International Symposium on Circuits and Systems*, Singapore, June 11-14, 1991, Vol. 1, pages 364-367.

[7] Lang M., Laakso T., "Design of allpass filters for phase approximation and equalization using LSEE error criterion", *Proc. of 1992 IEEE Int. Symposium on Circuit and Systems*, San Diego, CA, USA, May 10-13, 1992, pages 2417-2420.

[8] Chen, C. K., Lee J. H., "Design of digital all-pass filters using a weighted least squares approach", *IEEE Trans. on Circuits and Systems II: Analog & Digital Signal Processing*, May 1994, Vol. 41, No. 5, pages 346-351.

[9] Nguyen T. Q., Laakso T. I., Koilpillai R. D., "Eigenfilter Approach for the Design of Allpass Filters Approximating a Given Phase Response", *IEEE Trans. on Signal Processing*, September 1994, Vol. 42, No. 9, pages 2257-2263.

[10] Haddad K. C., Yang Y., Galatsanos N. P., Stark H., "Allpass filter design using projection-based method under group delay constraints", *Proc. of 2001 IEEE International Conference on Acoustics, Speech, and Signal Processing*, Salt Lake City, UT, USA, 7-11 May 2001.

[11] Vucic M., Babic H., "IIR Filters with Maximum Impulse Response Symmetry", *Proc. of 1999 IEEE Symposium on Circuit and Systems*, Orlando, FL, USA, May 30 - June 2, 1999, Vol. 3, pages 319-322.

[12] Carvalho D. B., Filho N. F., Seara R., "Impulse Response Symmetry Error for Designing Phase Equalisers", *IEE Electronics Letters*, Vol. 35, No. 13, June 1999, pages 1052 - 1054.

[13] Vucic M., Babic H., "A Robust Method for Equalizer Design Based on the Impulse Response Symmetry", *Proc. of 2002 IEEE Symposium on Circuit and Systems*, Phoenix, Arizona, USA, May 26-29, 2002, Vol. 3, pages 539-542.

[14] Clements M. A., Pease J. W., "On Causal Linear Phase IIR Digital Filters", *IEEE Trans. on Acoustic, Speech and Signal Processing*, Vol 37, No. 4, April 1989, pages 479-484.

[15] Openheim A. V., Schafer R. W., *Discrete time signal procesing*, Prentice-Hall International, Inc., 1989.

[16] Fletcher R., *Practical Methods of Optimization*, Volume 1, John Wiley & Sons, 1980.

MULTIPLIERLESS IMPLEMENTATION OF BANDPASS AND BANDSTOP RECURSIVE DIGITAL FILTERS USING ALLPASS STRUCTURES

Mrinmoy Bhattacharya and Tapio Saramäki

Institute of Signal Processing
Tampere University of Technology
P. O. Box 553, Tampere, FIN 33101, Finland
E-mail: {mrinmoy,ts}@cs.tut.fi

ABSTRACT

Under certain conditions an odd-order lowpass or highpass recursive digital filter can be decomposed into a sum of two allpass filters with real coefficients. This decomposition has the attractive property that there exist for its implementation structures where both the number of delays and the number of multipliers are equal to the filter order, thereby making the overall implementation very efficient. This paper develops some second- and fourth-order allpass structures that combine this property for generating multiplierless implementations for an odd-order recursive digital filter along with transformations from a prototype lowpass filter to a bandpass or bandstop filter. Utilizing these structures along with allowing some marginally insignificant deviations in the specifications such as in the passband and stopband tolerances, the total number of nonzero bits for multiplier coefficients, i.e., those of shifts and adds and/or subtracts, becomes quite small. This makes the proposed approach very attractive. Alternatively, the prototype lowpass filter can be designed with marginally stricter tolerances than the desired specifications such that it meets the given criteria after quantizing the filter coefficients.

1. INTRODUCTION

Minimum number of signed powers of two (MNSPT) or canonic signed digits (CSD) representations of binary digits are extensively used for representing the multiplier coefficients in multiplierless implementations of digital filters. An MNSPT representation of a coefficient value is given by $\sum_i a_i 2^{-t_i}$, where each a_i is either 1 or −1 and t_i is a positive or negative integer. For instance, 1.93359375 can be realized as $2-2^{-4}-2^{-8}$. In this example case, the multiplication is achieved not by a nine-bit multiplier, but with aid of three bit shifts and two subtracts.

One major approach for multiplierless implementations comprises of that of optimization [6, 7, 14, 18, 19], i.e., searching for the coefficients such that they can be implemented in MNSPT forms and the given criteria are still met. Optimization methods are used to find the optimal transfer functions under the given constraints with the filter design being basically a problem of approximation due to the tolerances in specifications. In general, the methods of optimizations are considered to be quite satisfactory. However, one may not assure or guarantee that the optimal solution will always be found under the given constraints. The solution can be unsatisfactory, for example, in terms of the filter order, the given wordlength of the multipliers, or the specified number of shifts and adds (in the case of multiplierless implementation), or some combination of them. Under such conditions, some parameters or characteristics of the filter will have to be relaxed to obtain an acceptable design and realization specific to the system it is intended to be used.

In the case of IIR filters, the structures such as a sum of allpass filters, including attractive lattice wave digital (LWD) filters, coupled with optimization methods have shown to yield good results for multiplierless implementations [9, 18, 19]. These sums of allpass filters are characterized by the attractive property that there exist structures with the number of required multipliers being equal to the filter order, thereby decreasing the number of multipliers compared with conventional realization forms.

Another interesting approach is the one that stems from designing an odd-order elliptic minimal Q-factor analog filter (EMQF) that has some special properties. Using the bilinear transformation these filters can be implemented as a sum of two allpass filters [8, 9] along with an expanded design parameters space such as the passband (stopband) tolerances, the edges, and the filter order.

Especially, in the case of FIR filters, another approach is based on combining simple sub-filters [1, 2, 16] that can be implemented using only a few shifts and adds and/or subtracts. Although quite attractive, to make this approach as a viable one, a large database of such filters will have to be generated and some optimization method will have to be evolved in order to combine some of them to meet the desired specifications.

In [3], the feasibility of implementing multiplierless recursive digital filters based on low-sensitivity transformation structures has been demonstrated. It has been realized that the development of the allpass structure that combines the advantages of minimum number of multiplier and delays and low passband sensitivity of the allpass structure, along with the advantages of those in [3], a gross reduction in total number of nonzero bits (effectively the number of shifts and adds and/or subtracts required) would be possible. Firstly, the number of multipliers would be less in the case of a sum of allpass filters of the

0-7803-7761-3/03/$17.00 ©2003 IEEE

prototype lowpass filter (and the α-multipliers in the allpass structures to be explained in the next section). Secondly, the number of nonzero bits per multiplier would be less due to low value of the modified coefficients [3]. Of course, the transfer function of the prototype lowpass filter (LPF) to be transformed to a bandpass filter (BPF) or bandstop filter (BSF) would have to be of odd order and should fulfill the necessary condition so as to be decomposable into a sum of two allpass filters [15, 17]. The next section demonstrates that such an approach is realizable.

2. THE STRUCTURES FOR IMPLEMENTATION

For the sake of clarity and appreciation of the advantages, we briefly review some of the aspects from our earlier papers.

One of the design and implementation methods for bandpass filters or bandstop filters comprises that of designing a prototype lowpass filter first followed by an appropriate frequency transformation [5]. In order to gain the advantage of the reduced number of multipliers, and also to avoid delay-free loops due the transformed block, one would prefer using the following substitution [5]:

$$z^{-1} \to -(z^{-2} - \alpha z^{-1})/(1-\alpha z^{-1}) = v_p^{-1} \qquad (1)$$

in the case of BPF's with the passband bandwidth $(\omega_2 - \omega_1)$ being equal to that of the prototype LPF, or

$$z^{-1} \to (z^{-2} - \alpha z^{-1})/(1-\alpha z^{-1}) = v_s^{-1} \qquad (2)$$

in the case of BSF's with $(\omega_2 - \omega_1)$ (i.e., the region including the transition bands and the stopband of the BSF) equaling ($\omega_s / 2$ – the bandwidth of the LPF). In both cases, α is given by

$$\alpha = \cos[(\omega_2 + \omega_1)/2]/\cos[(\omega_2 - \omega_1)/2] = \cos\omega_0, \qquad (3)$$

where $\omega_0, \omega_1, \omega_2$, and ω_s are the center frequency, and the lower and upper passband edges, and the sampling frequency, respectively.

Replacing the unit delay elements in a LPF by the transform block, as given by Eq. (1) or (2), has the inherent advantage of reducing the number of multipliers. For example, consider a second-order section of the prototype LPF with the following transfer function:

$$H_{lp}(z) = \frac{1 + b_1 z^{-1} + z^{-2}}{1 + a_1 z^{-1} + a_2 z^{-2}} \,. \qquad (4)$$

After using the substitution, as given by Eq. (1) or (2), and realizing the resulting BPF or BSF as a conventional cascade of two second-order sections, six multipliers are required. Alternatively, if each unit delay elements in Eq. (4) is replaced in the direct form II structure by using the transformation blocks, as given by Eq. (1) or (2), then we would need only five multipliers, as the transformation blocks can be implemented by one multiplier each. For the LPF with zeros not on unit circle, the transformed implementation requires six multipliers compared to eight multipliers required by the conventional cascade of two second-order sections.

If an Nth-order elliptic LPF with N odd is transformed into a $2N$th-order BPF or BSF implemented as a cascade of second-order sections, then one would require $3N$ multipliers. The requirement would be $(5N-1)/2$ multipliers in the case of substitution of the unit delay by the transformation block as mentioned earlier. It should be mentioned at this stage that one would require only $2N$ multipliers (the least number of multiplier among the three cases) if one decomposes the prototype LPF into a sum of two allpass filters first, followed by the substitution of unit delay by the transformation block. Due to this advantage, it is advantageous to develop second- and fourth-order allpass structures for multiplierless implementations for recursive BPFs or BSFs.

Generalizing the low-sensitivity transformations described in [4] one can express the low-sensitivity transformations for BPFs and BSFs as follows:

$$v_{ps}^{-1} = \frac{k_1(z^{-2} - \alpha z^{-1})}{1 - \alpha(1 - k_1 k) z^{-1} - k_1 k z^{-2}} \qquad (5)$$

where $k_1 = -1$ for BPFs and $k_1 = 1$ for BSFs, and where k is a number that can be represented by one bit or at the most two bits in the MNSPT form. Its absolute value is either equal to or less than unity. For example, k can be equal to ± 1, ± 0.5, etc. For most cases, $k = \pm 1$ will suffice.

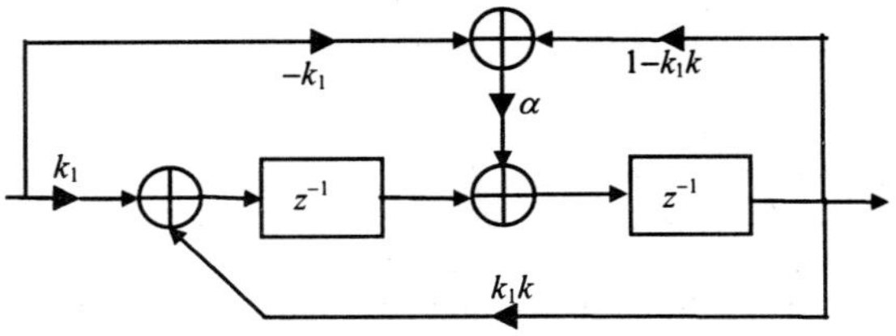

Fig. 1. Transformation block v_{ps}^{-1}.

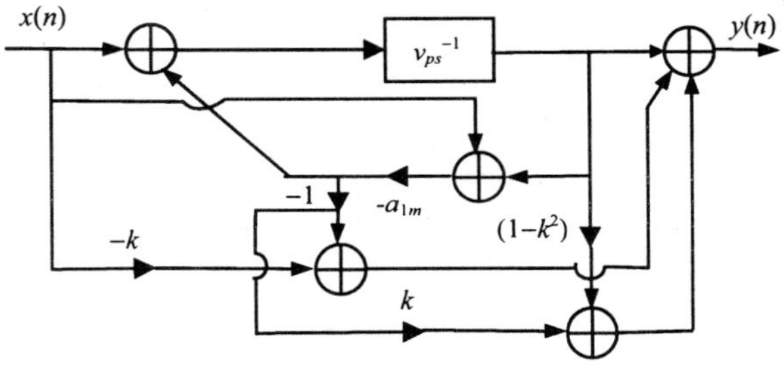

Fig. 2. The second-order allpass structure.

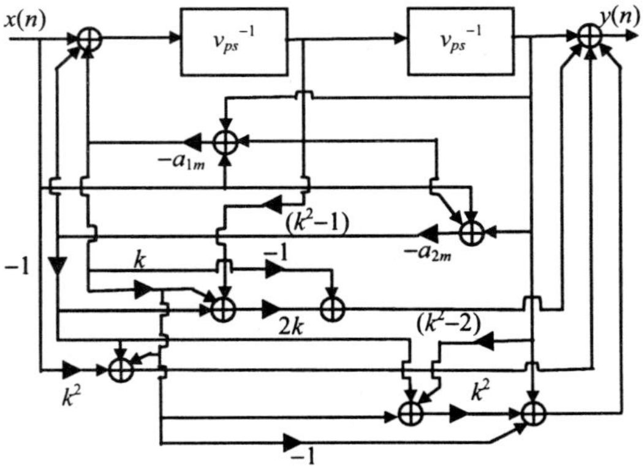

Fig. 3. The fourth-order allpass structure.

Figure 1 shows the generalized transformation block. It can be seen that for $k = 1$ or -1, and $k_1 = 1$ with the transformation given by Eq. (5) is equivalent to Structures *B1* or *B2* for BPFs considered in [4]. Figures 2 and 3 show the modified second- and fourth-order allpass structures, respectively.

The modified coefficients are given by

$$a_{1m} = k + a_1 \qquad (6)$$

and

$$a_{1m} = 2k + a_1 \text{ and } a_{2m} = k^2 + ka_1 + a_2 \qquad (7)$$

for the second-order and fourth-order allpass structures, respectively. The value of k is chosen so as to achieve the lowest possible magnitude of the modified coefficients and depends on pole locations.

3. RESULTS AND DISCUSSIONS

Quite a few BPFs and BSFs were realized based an odd-order prototype LPF. First, the LPF is decomposed into a sum of two allpass filters and then transformed into BPF or BSF by utilizing the appropriate second- or fourth-order allpass structures.

Table 1 illustrates the results for the implementation for one bandpass filter and one bandstop filter. Figure 4 shows the amplitude responses in both the cases of utilizing the unmodified allpass structure with the initial design and the modified allpass structures and with the revised specification with marginally stricter criterion that requires only thirty-seven bits only for ten multipliers.

As will be expected, one can see that the degradation of the passband and stopband tolerances is quite slow with a reduction of nonzero bits that represent the effective quantization level. This is due to the good sensitivity properties of a filter realized as sum of two allpass filters.

The performance table reflects the combined advantages of the reduction of the number of multipliers in the case of sum of allpass filters realization ($2N$ multipliers compared with $(5N-1)/2$ multipliers required when transforming an odd N-th order elliptic filter). Furthermore, a considerable reduction in the total number of nonzero bits required for the multiplierless implementation can be achieved.

Furthermore, it can be observed that the average number of nonzero bits per multiplier is almost of same order as has been observed earlier in [2–4]. Here, the average is less than four nonzero bits.

Table 1. Requirement of nonzero bits for bandpass and bandstop filters utilizing the allpass filter structures.

Filter Characteristics	Details of performance achieved			
	Number of nonzero bits for ten multipliers	Tolerances achieved		
		Passband	Stopband	Comments
Filter 1: Bandpass: 10th-order: passband edges=0.1π, 0.2π, stopband edges=0.08π, 0.24π, passband ripple=0.6 dB, stopband attenuation=50 dB	(a) 66	0.6 dB	50.0 dB	Unmodified allpass structures are used; design is based on initial specifications of the filter.
	(b) 56	0.6 dB	49.99 dB	.....".....
	(c) 46	0.603 dB	49.65 dB	.....".....
	(d) 40	0.606 dB	47.8 dB	.....".....
	(e) 38	0.631 dB	47.75 dB	.....".....
	(f) 37	0.522 dB	51.55 dB	Modified allpass structures are used; design is based on the revised specification of passband ripple = 0.5 dB and stopband attenuation = 51 dB.
Filter 2: Bandstop; 10th-order: passband edges=0.1π, 0.25π, stopband edges=0.12π, 0.21π, passband ripple=0.6 dB, stopband attenuation=50 dB	(a) 77	0.6 dB	50.0 dB	Unmodified allpass structures are used; design is based on initial specifications of the filter.
	(b) 62	0.6 dB	49.99 dB	.....".....
	(c) 45	0.6 dB	49.96 dB	.....".....
	(d) 38	0.602 dB	49.62 dB	.....".....
	(e) 35	0.62 dB	49.2 dB	.....".....
	(f) 33	0.5 dB	50.8 dB	Modified allpass structures are used; design is based on the revised specification of passband ripple = 0.5 dB and stopband attenuation = 51 dB.

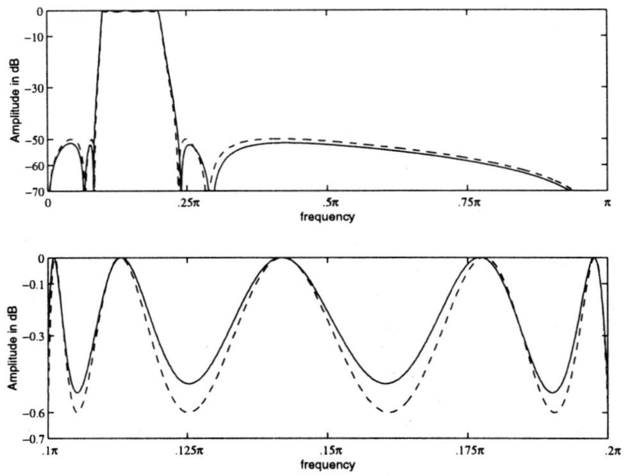

Fig. 4. Amplitude responses for the of BPF in Table 1 in two cases. The dashed and solid lines represent the initial design realized with the unmodified allpass structure and the realization with the revised specification and modified allpass structures requiring thirty-seven nonzero bits, respectively.

Moreover, it can be seen from Table 1 that also by allowing marginal deviations in the bandedges (by reducing the number of nonzero bits for α –multipliers only), an additional reduction in the number of nonzero bits can be achieved. For example, in the case of the BPF, 27 nonzero bits (a reduction of two bits for each of the five α –multipliers) leads to the filter with the passband edges being located at 0.1177π and at 0.2181π.

4. CONCLUSIONS

This paper has demonstrated the usefulness of the proposed second- and fourth-order allpass structures in generating bandpass and bandstop filters as a parallel connection of two allpass filters. For this purpose, a concept developed earlier by the authors (using low-sensitivity structure for multiplierless implementation of recursive filters) for utilization in multiplierless implementation of an even order has been exploited. Utilizing the above-mentioned structures, in addition to a reduction in the total number of multiplier coefficients, a gross reduction in the total number of nonzero bits has been achieved, making the proposed approach very attractive. Numerical examples have indicated that the proposed approach results in multiplierless realizations requiring around three and half nonzero bits per multiplier on the average, without any increase in the filter order. Future work is devoted to applying optimization techniques to further reducing the number of nonzero bits.

5. ACKNOWLEDGEMENT

This work was supported by the the Academy of Finland, project No. 44876 (Finnish Centre of Excellence program (2000-2005)).

6. REFERENCES

[1] John A. Adams and Alan N. Williamson, Jr., "A new approach to FIR digital filters with fewer multipliers and reduced sensitivity," *IEEE Trans. Circuits Syst.*, vol. CAS-30, p. 277-283, May 1983.

[2] ...."............., "Some efficient efficient digital prefilter structures," *IEEE Trans. Circuits Syst.*, vol. CAS-31, pp. 260-266, March 1984.

[3] M. Bhattacharya and T. Saramäki, "Multiplierless Implementation of Bandpass and Bandstop Recursive Digital Filters," *Proc. IEEE Intl. Symp. Circuits Syst., (ISCAS 2002)*, Phoenix, Arizona,USA, vol.-II, pp. 692-695, 2002.

[4] M. Bhattacharya, R. C. Agarwal, and S. C. Dutta Roy ., "Bandpass and bandstop recursive filters with low sensitivity," *IEEE Trans. Acoust., Speech, Signal Processing,*" vol. ASSP-34, pp. 1485-1492, Dec. 1986.

[5] A. G. Constantinides, "Spectral transformations for digital filters," *Proc. IEE,* vol. 117, pp. 1585-1590, Aug. 1970.

[6] Y. C. Lim and S. R. Parker, "FIR filter design over a discrete powers-of-two coefficient space," *IEEE Trans. Acoust. Speech, Signal Processing,* vol. ASSP-31, pp. 583-591, June 1983.

[7] Y. C. Lim, "Design of discrete-coefficient-value linear phase FIR filters with optimum normalized peak ripple magnitude," *IEEE Trans. Circuits Syst.,* vol. CAS-37, pp. 1480-1486, Dec. 1990.

[8] M. D. Lutovac, D. V. Tosic, and B. L. Evans, *Filter Design for Signal Processing Using Matlab and Mathematica.* Prentice-Hall, New Jersey, 2001.

[9] Lj. D. Milic and M. D. Lutovac, "Design of multiplierless elliptic IIR filters with a small quantization error," *IEEE Trans. Signal Processing,* vol. 47, pp. 469-479, Feb. 1999.

[10] S. K. Mitra and K. Hirano, "Digital all-pass networks," *IEEE Trans. Circuits Syst.,* vol CAS-21, pp. 688-700, Sept. 1974.

[11] S. K. Mitra, *Digital Signal Processing, A Computer-Based Approach.* (Second Edition), McGraw-Hill, 2001.

[12] J. G. Proakis and D. G. Manolakis, *Digital Signal Processing: Principles, Algorithms and Applications.* Prentice-Hall, Inc., Upper Saddle River, NJ, 1996.

[13] R. A. Regalia, Sanjit K. Mitra, and P. P. Vaidyanathan, "The digital all-pass filter: A versatile signal processing building block," *Proc. IEEE,* vol. 76, pp. 19-37, Jan 1988.

[14] H. Samueli, "An improved search algorithm for the design of multiplierless FIR filters with powers-of-two coefficients," *IEEE Trans. Circuits Syst.,* vol. CAS-36, pp. 1044-1047, July 1989.

[15] T. Saramäki, " On the design of digital filters as a sum of two all-pass filters," *IEEE Trans. Circuits Syst.,* vol. CAS-32, pp. 1191-1193, Nov. 1985.

[16] P. P. Vaidyanathan and G. Beitman, " On prefilters for digital filter design," *IEEE Trans. Circuits Syst.,* vol. CAS-32, pp. 494-499, May 1985.

[17] P. P. Vaidyanathan, "Robust Digital Filter Structure," Chapter 7 in Chapter 7 in *Handbook for Digital Signal Processing,* Mitra S. K. and Kaiser J.F. Eds., John Wiley & Sons, New York, 1993.

[18] J. Yli-Kaakinen and T. Saramäki, "Design of very low-sensitivity and low-noise recursive filters using a cascade of low-order lattice wave digital filters," *IEEE Trans. Circuits Syst. II,* vol.46, pp. 906-914, July 1999.

[19] ...."................., "An algorithm for the design of multiplierless approximately linear-phase lattice-wave digital filters," in *Proc. IEEE Intl. Symp. Circuits Syst., (ISCAS 2000),* Geneva, Switzerland, vol. 2, pp. 77-80, 2000..

THE DESIGN OF TWO-CHANNEL PREFECT RECONSTRUCTION FIR TRIPLET WAVELET FILTER BANKS USING SEMIDEFINITE PROGRAMMING

K. S. Yeung and S. C. Chan

Department of Electrical and Electronic Engineering,
The University of Hong Kong, Pokfulam Road, Hong Kong.

ABSTRACT

This paper proposes a new method for designing 2-channel PR linear-phase/low-delay FIR triplet wavelet filterbanks with user-defined cutoff frequency and prescribed number of *K*-regularity. The magnitudes of the analysis/synthesis filters pair at $\omega = \pi/2$ can be set to any desired value controlled by a certain number of parameters using the triplet structure. The *K*-regularity conditions are expressed as a set of linear equality constraints in the variables to be optimized. The design method employs the minimax error criteria and solves the design problem as a semidefinite programming (SDP) optimization. By removing the redundant variables, the linear equality constraints are automatically imposed into the design problem. The optimization problem is then formulated as a linear convex objective function subject to a union of affine set, which can be represented by a set of linear matrix inequalities (LMI). Hence they can be solved using existing SDP solver. Design examples are given to demonstrate the effectiveness of the proposed method.

I. INTRODUCTION

Perfect reconstruction (PR) multirate filter banks (FBs) have important applications in signal analysis, signal coding and the design of wavelet bases. One of methods for designing 2-channel PR FBs (wavelets) is to employ the triplet structure [7,8], which can be viewed as three lifting steps. In [8], Ansari et al. proposed two approaches for designing the 2-channel PR linear phase triplet FBs. First of all, the Lagrange halfband filters are employed to obtain the triplet FBs to be maximally flat but without user-defined cutoff frequency. The second approach is to employ the Remez exchange algorithm so that the triplet FBs are equiripple with user-defined frequency but without any *K*-regularity condition. These two approaches are very extreme in the FB characteristics. Recently, Tay [7] showed that the 2-channel PR linear-phase triplet FBs are possible to incorporate the *K*-regularity condition (or equivalently a certain number of zeros of the lowpass analysis filter $H_0(z)$ and the highpass analysis filter $H_1(z)$ at $\omega = \pi$ and $\omega = 0$, respectively) using the Bernstein polynomial. A least squares approach was proposed to find the coefficients of the Bernstein polynomial using an iterative procedure to optimize the objective function, which is a multi-quadratic function. Another method for designing the 2-channel PR FBs is to employ the coiflet structure [1-6], which is the subset of the triplet structure and can be viewed as two lifting steps. In [1], we showed that the structurally 2-channel PR FBs using the coiflet structure can be formulated as a constrained least squares problem as well as a semidefinite programming (SDP) problem. The construction of wavelet basis from these FBs satisfying a certain number of *K*-regularity was also studied. In general, these FBs (wavelets) so obtained can have low delay, a user-defined cutoff frequency, and a prescribed number of *K*-regularity. However, the magnitudes of the lowpass and highpass analysis filters pair $H_m(z)$, $m = 0,1$, at $\omega = \pi/2$ using the coiflet structure are restricted to be 0.5 and 1, respectively. Instead, using the triplet structure, the magnitudes of the analysis filters pair $H_m(z)$ at $\omega = \pi/2$ can be set to any desired value controlled by a certain number of parameters. It is interesting to note that the magnitudes of the analysis filters pair $H_m(z)$ at $\omega = \pi/2$ can be equal by properly choosing the parameters. It is also true for the synthesis filters pair $F_m(z)$ because of the structurally PR property.

In this paper, we propose a new method to design the 2-channel PR linear-phase/low-delay FBs (wavelets) using the triplet structure and incorporating the prescribed number of *K*-regularity. This will enable the wavelet FBs with desired magnitudes at $\omega = \pi/2$ in the transition band of the analysis/synthesis filters pair and different smoothness, delay and time/frequency resolution to be constructed. The design problem can be solved using semidefinite programming (SDP) for the minimax error criteria [10]. SDP has been successfully applied in certain areas such as control, logistic, digital filter design [5], digital signal processing, etc. Thanks to the interior-point method [9], SDP can now be solved efficiently in polynomial time. Design results showed that the SDP design method produces the analysis filters pair with almost equiripple approximation error while satisfying the given *K*-regularity condition. Comparing the result reported in [8], comparable performance of the linear-phase triplet FB was achieved for the same design specifications. Furthermore, it is possible to design a low-delay triplet FB (wavelet) with prescribed number of *K*-regularity. This substantiates the usefulness of proposed SDP method.

The rest of this paper is organized as follows: Section II is devoted to a brief description of the 2-channel PR FIR triplet wavelet FBs to be designed and the problem formulation will be given. Section III presents the *K*-regularity conditions that can be written as a set of linear equality constraints. Section IV describes the proposed SDP design method. This is followed by several design examples in Section V. Finally, conclusion is drawn in Section VI.

II. PROBLEM FORMULATION

The structurally 2-channel PR triplet FBs (wavelets) that we are going to design is shown in Figure 1. It can be seen that the triplet FB is parameterized by three sub-filters $q_i(z)$, three delay parameters N_i, and five constant parameters p_i and C_m with $i = 0,1,2$ and $m = 0,1$. The triplet FB is called PR for arbitrary choice of the sub-filters, which can be chosen as linear-phase/low-delay FIR or all-pass functions to realize FIR and IIR filter banks with very low design and implementation complexities. In [8], two approaches are used for designing the linear phase triplet FBs. One of approaches is to employ the Lagrange halfband filters so that the triplet FBs are maximally flat but without freedom of the cutoff frequency. Another approach is to employ the Remez exchange algorithm so that the triplet FBs are equiripple with the user-defined cutoff frequency but no regularity imposed. These two approaches are very extreme in the FB characteristics. Later in [7], the Bernstein polynomial is used for designing the linear phase triplet FBs with the user-defined cutoff frequency and prescribed *K*-regularity. A least squares approach was proposed to find the coefficients of the Bernstein polynomial with an iterative procedure to optimize the objective function, which is a multi-quadratic function. In this work, the design of the linear-

0-7803-7761-3/03/$17.00 ©2003 IEEE

phase/low-delay triplet FBs (wavelets) are formulated as a SDP problem so that the triplet FBs are equiripple with the user-defined cutoff frequency and prescribed number of *K*-regularity, as we shall see later in Sections III and IV.

First of all, let's consider the structure of the triplet FBs shown in Figure 1. For simplicity, the triplet FBs can be parameterized by only one constant parameter p with proper choices of the constant parameters. They are chosen as $p_0 = -p$, $p_1 = 1/(1+p)$, $p_2 = (p^2 - 1)/2$, $C_0 = (1+p)/2$ and $C_1 = 1/(1+p)$. The z-transform of the analysis filters $H_m(z)$, $m = 0,1$, are given by

$$H_0(z) = \frac{1}{2}(1+p)z^{-(2N_1+1)} + \frac{1}{2}q_1(z^2)\left[z^{-2N_0} - z^{-1}p\cdot q_0(z^2)\right], \quad (1a)$$

$$H_1(z) = \frac{z^{-2N_0} - z^{-1}p\cdot q_0(z^2)}{1+p}z^{-2N_2} - \frac{1-p}{1+p}q_2(z^2)H_0(z)\,. \quad (1b)$$

Using the structurally 2-channel PR property [2], the z-transform of the synthesis filters $F_m(z)$, $m = 0,1$, are then given by

$$F_0(z) = -H_1(-z)\,, \quad (1c)$$

$$F_1(z) = H_0(-z)\,. \quad (1d)$$

For the design of the lowpass analysis filter $H_0(z)$, we first design the sub-filter $q_0(z)$ with a desired frequency response

$$q_0^{(d)}(e^{j2\omega}) = e^{-j\omega(2N_0-1)}, \text{ for } \omega \in [0, \omega_{p0}]\,. \quad (2a)$$

Once $q_0(z)$ is designed, one can design $q_1(z)$ with the following desired frequency response

$$q_1^{(d)}(e^{j2\omega}) = \frac{(1+p)e^{-j\omega(2N_1+1)}}{e^{-j\omega}p\cdot q_0(e^{j2\omega}) - e^{-j\omega 2N_0}}\,, \quad (2b)$$

for $\omega \in [\pi - \omega_{p1}, \pi]$. The lowpass analysis filter $H_0(z)$ can then be obtained. Given $H_0(z)$ and $q_0(z)$, the highpass analysis filter $H_1(z)$ can be obtained by designing $q_2(z)$ with a desired frequency response

$$q_2^{(d)}(e^{j2\omega}) = \frac{e^{-j\omega(2N_0+2N_2)} - e^{-j\omega(2N_2+1)}p\cdot q_0(e^{j2\omega})}{(1-p)H_0(e^{j\omega})}\,, \quad (2c)$$

for $\omega \in [0, \omega_{p2}]$, where ω_{pi} is the passband cutoff frequency of $q_i(z)$. The choices of the desired frequency responses expressed in (2) will give the stopbands of the analysis filters pair to be equiripple. Next, we shall show that the *K*-regularity can be written as a set of linear equality constraints in the coefficients of $q_i(z)$, and the design problem can be solved using a SDP problem.

III. *K*-REGULARITY CONDITIONS

To construct a triplet wavelet FBs, the analysis filters pair $H_m(z)$, $m = 0,1$, should possess at least one zero at $\omega = \pi$ and $\omega = 0$, respectively. Let K_0 and K_1 be the number of zeros to be imposed respectively at $\omega = \pi$ and $\omega = 0$ for $H_m(z)$ with $K_0 \geq K_1 \geq 1$. This is equivalent to

$$\left.\frac{\partial^{k_0}}{\partial\omega^{k_0}}H_0(e^{j\omega})\right|_{\omega=\pi} = \left.\frac{\partial^{k_1}}{\partial\omega^{k_1}}H_1(e^{j\omega})\right|_{\omega=0} = 0\,, \quad (3)$$

for $k_0 = 0,\ldots,K_0 - 1$ and $k_1 = 0,\ldots,K_1 - 1$. In general, the number of zeros imposed for $H_m(z)$ can be observed by halfband filters $H_{qi}(z)$, $i = 0,1,2$, defined as

$$H_{qi}(z) = [z^{-1}q_i(z^2) + z^{-2N_{qi}}]/2\,, \quad (4)$$

where $N_{q0} = N_0$, $N_{q1} = N_1 - N_0 + 1$, $N_{q2} = N_2 + N_0 - N_1$.

Substituting (4) into (1a), one gets

$$H_0(z) = -p\cdot q_1(z^2)H_{q0}(z) + (1+p)z^{-2N_0+1}H_{q1}(z)\,. \quad (5)$$

It can be seen that if the halfband filters $H_{q0}(z)$ and $H_{q1}(z)$ have K_{q0} and K_{q1} zeros at $z = -1$ respectively, $H_0(z)$ also has K_0 zeros with $K_0 = \min\{K_{q0}, K_{q1}\}$. Similarly, substituting (4) into (1c) and using (1b) and (1d), one gets

$$F_0(z) = -\frac{2p\cdot z^{-2N_2}H_{q0}(z) + (1-p)q_2(z^2)H_0(z)}{1+p} + \frac{4(1-p)z^{-2(N_0-1)}H_{q1}(z)H_{q2}(z)}{1+p} - \frac{2(1-p)\left[z^{-2N_1}H_{q2}(z) + z^{-2(2N_0+N_2-N_1-1)}H_{q1}(z)\right]}{1+p} \quad (6)$$

It can be seen that if the halfband filters $H_{qi}(z)$, $i = 0,1,2$, have K_{qi} zeros at $z = -1$ respectively, $F_0(z)$ has K_1 zeros with $K_1 = \min\{K_{q0}, K_{q1}, K_{q2}\}$. Let L_{qi}, $i = 0,1,2$, be respectively the length of $q_i(z)$. For imposing K_{qi} zeros at $z = -1$ in $H_{qi}(z)$, it can be rewritten the halfband filter as follows:

$$H_{qi}(z) = [z^{-2L_{qi}+1}\widetilde{H}_{qi}(z)]/2 \text{ with } L_{qi} > N_{qi}\,, \quad (7)$$

where $\widetilde{H}_{qi}(z) = \sum_{n=0}^{L_{qi}-1} q_{i,n}z^{2(L_{qi}-n-1)} + z^{2(L_{qi}-N_{qi})-1}$. The *K*-regularity condition of imposing K_{qi} zeros at $z = -1$ in $H_{qi}(z)$ is equivalent to imposing the same number of zeros at $z = -1$ in $\widetilde{H}_{qi}(z)$. That is

$$\left.\frac{d^{k_{qi}}}{dz^{k_{qi}}}\widetilde{H}_{qi}(z)\right|_{z=-1} = 0\,. \quad (8)$$

After expanding (8) and slightly manipulating it, a set of linear equality constraints is derived as follows:

$$\sum_{n=0}^{L_{qi}-1-\lceil k_{qi}/2\rceil} q_{i,n}\frac{(2(L_{qi}-n)-2)!}{(2(L_{qi}-n)-2-k_{qi})!} = \frac{(2(L_{qi}-N_{qi})-1)!}{(2(L_{qi}-N_{qi})-1-k_{qi})!}\,, \quad (9a)$$

for $k_{qi} = 0,1,\ldots,K_{qi} - 1$, where the operator $\lceil x \rceil$ denotes the nearest upper integer. (9a) can also be written more compactly in matrix form as

$$\boldsymbol{A}_i \cdot \boldsymbol{q}_i = \boldsymbol{B}_i \quad (9b)$$

where $[\boldsymbol{A}_i]_{k_{qi},n} = \frac{(2(L_{qi}-n)-2)!}{(2(L_{qi}-n)-2-k_{qi})!}$, $[\boldsymbol{q}_i]_n = q_{i,n}$ and $[\boldsymbol{B}_i]_{k_{qi}} = \frac{(2(L_{qi}-N_{qi})-2)!}{(2(L_{qi}-N_{qi})-2-k_{qi})!}$.

IV. SEMIDEFINITE PROGRAMMING (SDP) DESIGN

In this section, we shall show that the design of the sub-filters $q_i(e^{j\omega})$, $i = 0,1,2$, with minimax error criteria and prescribed number of *K*-regularity can be formulated as a semidefinite programming (SDP) problem. First of all, to minimize the maximum ripple of the approximation error is equivalent to the following

$$\min_{\boldsymbol{q}_i}\left|\boldsymbol{q}_i^T\cdot\boldsymbol{e}_i - q_i^{(d)}(e^{j\omega})\right|^2, \text{ for } \omega \in [0, 2\omega_{pi}]\,, \quad (10)$$

where $\boldsymbol{q}_i = [q_{i,0}\ q_{i,1}\ \cdots\ q_{i,L_{qi}-1}]^T$ is the impulse response of $q_i(z)$; $\boldsymbol{e}_i = [1\ e^{-j\omega}\ \cdots\ e^{-j(L_{qi}-1)\omega}]^T$; and $q_i^{(d)}(e^{j\omega})$ is the desired frequency response defined in (2). To solve (10) using the SDP optimization, we densely discretize ω over the band of interest

$\omega \in [0, 2\omega_{pi}]$ into a set of frequency points ω_k, $k = 1, \ldots, K$. The optimization problem (10) is then approximated as:

$$\min \delta_i \text{ subject to}$$
$$\alpha_{R,i}^2(\omega_k) + \alpha_{I,i}^2(\omega_k) \le \delta, \text{ for } k = 1, \ldots, K, \tag{11a}$$

where $\alpha_{R,i}(\omega_k) = \boldsymbol{q}_i^T \cdot \boldsymbol{c}_i - \mathrm{Re}\{q_i^{(d)}(e^{j\omega_k})\}$;

$$\boldsymbol{c}_i = [1 \ \cos(\omega_k) \ \cdots \ \cos((L_{qi}-1)\omega_k)]^T;$$
$$\alpha_{I,i}(\omega_k) = \boldsymbol{q}_i^T \cdot \boldsymbol{s}_i + \mathrm{Im}\{q_i^{(d)}(e^{j\omega_k})\};$$
$$\boldsymbol{s}_i = [0 \ \sin(\omega_k) \ \cdots \ \sin((L_{qi}-1)\omega_k)]^T.$$

Using Schur complement [10], it can be shown that [5] (11a) is equivalent to

$$\min \delta_i \text{ subject to}$$
$$F_k(\boldsymbol{q}_i) \ge 0, \text{ for } k = 1, \ldots, K, \tag{11b}$$

where $F_k(\boldsymbol{q}_i) = \begin{pmatrix} \delta_i & \alpha_{R,i}(\omega_k) & \alpha_{I,i}(\omega_k) \\ \alpha_{R,i}(\omega_k) & 1 & 0 \\ \alpha_{I,i}(\omega_k) & 0 & 1 \end{pmatrix}$. Since $F_k(\boldsymbol{q}_i)$ is affine in $\boldsymbol{q}_i$, it is equivalent to a set of linear matrix inequalities (LMI) [10]. Define the augmented variable $\boldsymbol{x}_i^T = [\delta_i \ \boldsymbol{q}_i^T]$. The optimization problem in (11b) can be cast into the following standard LMI or SDP optimization problem

$$\min \boldsymbol{u}_i^T \cdot \boldsymbol{x}_i \text{ subject to } \boldsymbol{F}(\boldsymbol{x}_i) \ge 0, \tag{11c}$$

where $\boldsymbol{u}_i^T = [1 \ \boldsymbol{O}_{L_{qi}}^T]$; $\boldsymbol{F}(\boldsymbol{x}_i) = diag(F_1(\boldsymbol{x}_i), \ldots, F_K(\boldsymbol{x}_i))$; $\boldsymbol{O}_N$ is a $N \times 1$ zero matrix. Theoretically, it is possible to determine whether a feasible solution exists for the SDP problem such as (11c), and if so, it is possible to determine the global optimal solution, since the problem is convex. In order to simultaneously solve the SDP problem (11c) and the *K*-regularity conditions (9b), the dependent variables defined by the linear equality constraints (9b) can be expressed as a linear combination of independent variables. The number of variables to be optimized is therefore reduced. It not only speeds up the optimization process but also structurally imposes the *K*-regularity conditions. To remove the redundant variables, we rewrite (9b) as follows:

$$\begin{bmatrix} \boldsymbol{A}_{i,L_{qi}-r_i} & \boldsymbol{A}_{i,r_i} \end{bmatrix} \cdot \begin{bmatrix} \boldsymbol{q}_{i,L_{qi}-r_i} \\ \boldsymbol{q}_{i,r_i} \end{bmatrix} = \boldsymbol{B}_i \tag{12}$$

where $\boldsymbol{A}_i = \begin{bmatrix} \boldsymbol{A}_{i,L_{qi}-r_i} & \boldsymbol{A}_{i,r_i} \end{bmatrix}$; $\boldsymbol{q}_i = \begin{bmatrix} \boldsymbol{q}_{i,L_{qi}-r_i} \\ \boldsymbol{q}_{i,r_i} \end{bmatrix}$; and r_i is the number of redundant variables in $q_i(z)$. Using eqn. (12), $\boldsymbol{q}_i$ can be written in terms of $\boldsymbol{q}_{i,L_{qi}-r_i}$ as:

$$\boldsymbol{q}_i = \begin{bmatrix} \boldsymbol{O}_{L_{qi}-r_i} \\ \boldsymbol{A}_{i,r_i}^{-1}\boldsymbol{B}_i \end{bmatrix} + \begin{bmatrix} \boldsymbol{I}_{L_{qi}-r_i} \\ -\boldsymbol{A}_{i,r_i}^{-1}\boldsymbol{A}_{i,L_{qi}-r_i} \end{bmatrix} \boldsymbol{q}_{i,L_{qi}-r_i} \tag{13}$$

where $\boldsymbol{I}_N$ is a $N \times N$ identity matrix. Substituting (13) into (11c) and redefine $\boldsymbol{x}_i^T = [\delta_i \ \boldsymbol{q}_{i,L_{qi}-r_i}^T]$, we still have the objective function and the constraints affine in $\boldsymbol{x}_i$. In other words, it is still a standard SDP problem.

V. DESIGN EXAMPLES

For all examples, the constant parameter p was chosen as $\sqrt{2}-1$. This choice is good enough because the magnitudes of the analysis filters pair $H_m(z)$ at $\omega = \pi/2$ are both equal to $1/\sqrt{2}$. It is able to design the frequency responses of the analysis filters pair as close to symmetry as possible. The frequency ω in the passband was uniformly discretized using $K = 500$ samples. The SDP optimization was then carried out using the MATLAB LMI Toolbox and it took less than 60 iterations to obtain the solution within a minute.

Example 1. Linear Phase Triplet Filter Bank

For comparison purpose, a 2-channel PR linear phase triplet FB with the same specifications as example 2(b) reported in [8] was designed using the proposed SDP method. The specifications of all parameters for designing the sub-filters $q_i(z)$ of the linear phase triplet FB are shown in the second column of Table 1. The frequency responses and the group delays of the analysis filters pair $H_m(z)$ are respectively shown in Figure 2(a) and (b). It should be noted that there is no regularity in the triplet FB. The results and performances of the analysis filters pair are summarized in the second column of Table 2. It can be seen that the passband and stopband deviations of the analysis filters pair using proposed SDP method are comparable to that reported in [8]. When the parameters K_{q0}, K_{q1}, K_{q2} are all changed to one, the analysis filters pair $H_m(z)$ have one zero at $z = -1$ and 1 respectively, i.e. the *K*-regularity condition up to first order moment. Figure 2(c) and (d) show the corresponding analysis scaling and wavelet functions. The methods reported in [7,8] are unable to design the triplet FBs with low-delay and prescribed number of *K*-regularity, which can be solved using the proposed SDP method. This can be shown in the next example.

Example 2. Low Delay Triplet Wavelet Filter Bank

In this example, a 2-channel PR low-delay triplet wavelet FB with prescribed number of *K*-regularity was designed. In order to observe the effect of the low-delay design and the *K*-regularity condition, all the lengths L_{qi} and the cutoff frequencies ω_{pi} of the sub-filters $q_i(z)$ are chosen to be the same as Example 1. The third column of Table 1 shows the specifications of all parameters for designing the $q_i(z)$ of the low-delay triplet wavelet FB. The design results are shown in Figure 3. The results and the performances of the analysis filters pair $H_m(z)$ are summarized in the third column of Table 2. It should be noted that the number of zeros imposed in the analysis filters pair is $K_0 = K_1 = 3$. As expected, the low-delay design and the *K*-regularity condition are traded off the passband and stopband ripple errors.

VI. CONCLUSION

A new method for designing 2-channel PR linear-phase/low-delay FIR triplet wavelet FBs with user-defined cutoff frequency and prescribed number of *K*-regularity is presented. The *K*-regularity conditions are expressed as a set of linear equality constraints and the design problem is formulated as the SDP problem using the minimax error criteria. By removing the redundant variables, the linear equality constraints are automatically imposed into the design problem. Design results showed that the low-delay triplet wavelet FBs with good quality and prescribed number of *K*-regularity can be obtained by the proposed SDP method.

REFERENCES

[1] S. C. Chan, C. K. S. Pun and K. L. Ho, "The design of a class of perfect reconstruction two-channel FIR and wavelets filter banks using constrained least squares method and semidefinite programming," in *Proc. DSP'2002*, vol. 2, pp. 497-500, 2002.

[2] S. M. Phoong, C. W. Lim, P. P. Vaidyanathan and R. Ansari, "A new class of two-channel biothogonal filter banks and wavelet bases," *IEEE Trans. SP*, vol. 43, pp. 649-664, Mar. 1995.

[3] J. S. Mao, S. C. Chan, W. Liu and K. L. Ho, "Design and multiplier-less implementation of a class of two-channel PR FIR filter banks and wavelets with low system delay," *IEEE Trans. SP*, vol. 48, pp. 3379-3394, Dec. 2000.

[4] T. Cooklev and A. Nishihara, "Biorthogonal coiflets," *IEEE Trans. Signal Processing*, vol. 47, no. 9, pp. 2582-2588, Sept. 1999.

[5] W. S. Lu and A. Antoniou, "Design of digital filters and filter banks by optimization: A state of the art review," in *Proc. EUSIPCO'2000*, vol. 1, pp. 78, Sept. 2000.

[6] W. Liu, S. C. Chan and K. L. Ho, "Low-delay perfect reconstruction two-channel FIR/IIR filter banks and wavelet bases with SOPOT coefficients," in *Proc. ICASSP'2000*, vol. 1, pp. 109-112, May 2000.

[7] D. B. H. Tay, "Least squares design of the class of triplet halfband filter banks," in *Proc. IEEE ISCAS'2001*, vol. 2, pp. 481-484, 2001.

[8] R. Ansari, C. W. Kim and M. Dedovic, "Structure and design of two-channel filter banks derived from a triplet of halfband filters," *IEEE Trans. Circuits Systs. II*, vol. 46, no. 12, pp. 1487-1496, Dec. 1999.

[9] Y. Nesterov and A. Nemirovskii, *Interior Point Polynomial Methods in Convex Programming*, SIAM, Philadelphia, 1994.

[10] H. Wolkowicz, R. Saigal and L. Vandenberghe, *Handbook of Semidefinite Programming – Theory, Algorithms, and Applications*, Boston: Kluwer Academic Publishers, 2000.

Parameters	Ex. 1 / Ex. 2(b) in [8]	Ex. 2
ω_{p0}, ω_{p1}, ω_{p2}	0.4π , 0.4π , 0.4π	0.4π , 0.4π , 0.4π
L_{q0}, L_{q1}, L_{q2}	6, 14, 14	6, 14, 14
N_0, N_1, N_2	3, 9, 13	3, 7, 9
K_{q0}, K_{q1}, K_{q2}	0, 0, 0	3, 3, 3

Table 1. Specifications of all parameters for designing the sub-filters $q_i(z)$, $i = 0,1,2$, of the triplet FBs/wavelets in Example 1/Example 2(b) in [8] and Example 2.

	Ex.1 (Ex. 2(b) in [8])	Ex. 2
Length of $H_0(z)$, $H_1(z)$	38, 64	38, 64
Group delay of $H_0(z)$, $H_1(z)$	19, 32	15, 24
Passband deviation of $H_0(z)$	0.03154 (0.0316)	0.04904
Stopband deviation of $H_0(z)$	0.005336 (0.0058)	0.007748
Passband deviation of $H_1(z)$	0.03164 (0.0316)	0.05154
Stopband deviation of $H_1(z)$	0.003915 (0.0043)	0.005749

Table 2. Summarized results of the analysis filters pair $H_m(z)$, $m = 0,1$, of the triplet FBs/wavelets in Example 1/Example 2(b) in [8] and Example 2.

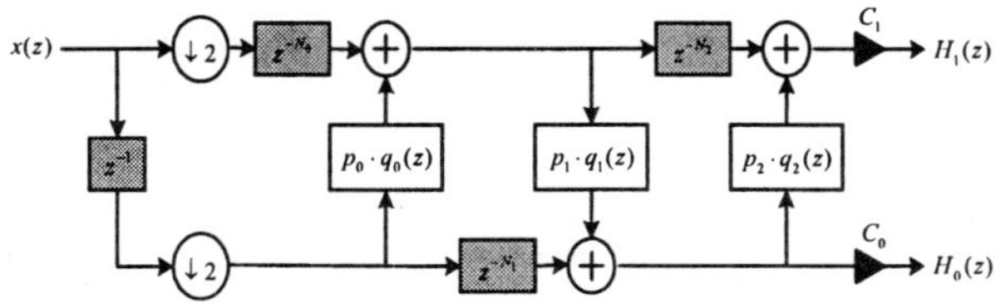

1(a)

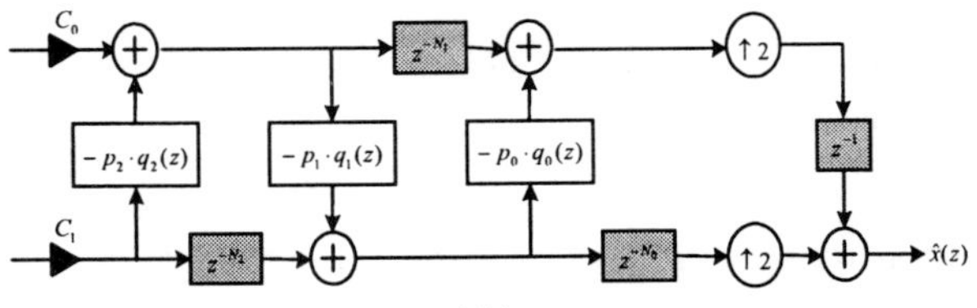

1(b)

Figure 1. Structurally 2-channel PR triplet wavelet FBs: (a) analysis filters pair $H_m(z)$, (b) synthesis filters pair $F_m(z)$, $m = 0,1$.

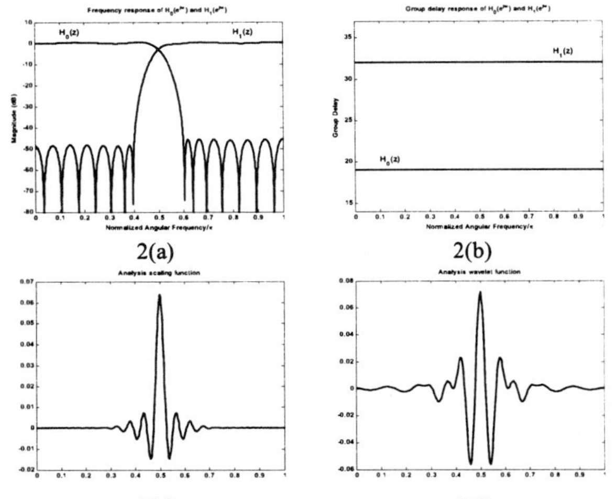

2(a) 2(b)

2(c) 2(d)

Figure 2. (a) Frequency responses and (b) Group delays of the analysis filters pair $H_m(z)$ in Example 1. (c) Analysis scaling function and (d) Analysis wavelet function when imposing one zero respectively at $z = -1$ and 1 of $H_m(z)$ in Example 1.

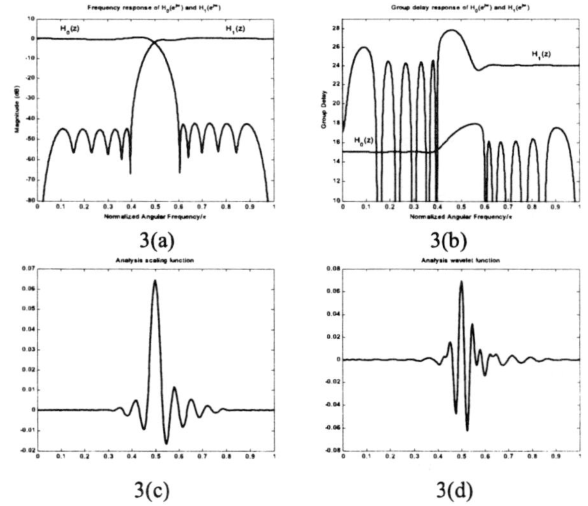

3(a) 3(b)

3(c) 3(d)

Figure 3. Design results of Example 2: (a) Frequency responses and (b) Group delays of the analysis filters pair $H_m(z)$. (c) Analysis scaling function. (d) Analysis wavelet function.

MULTIPLIER-LESS REAL-VALUED FFT-LIKE TRANSFORMATION (ML-RFFT) AND RELATED REAL-VALUED TRANSFORMATIONS

S. C. Chan and K. M. Tsui

Department of Electrical and Electronic Engineering,
The University of Hong Kong, Pokfulam Road, Hong Kong.

ABSTRACT

This paper proposes a new multiplier-less Fast Fourier Transform-like (ML-RFFT) transformation for real-valued sequence. Like the ML-FFT, it parameterizes the twiddle factors in the conventional radix-2^n or split-radix real-valued FFT algorithm as certain rotation-like matrices and approximates the associated parameters inside these matrices by the sum-of-power-of-two (SOPOT) or canonical signed digits (CSD) representations. Because of the symmetry in the algorithm, it only requires about half the number of additions as required by the ML-FFT. Moreover, using the mappings between the DFT and the DCTs and DWTs, new ML-FFT-based transformation called ML-DCTs and ML-DWTs are derived. Design examples of the new transformations are given to demonstrate the proposed approach.

I. INTRODUCTION

The Discrete Fourier Transform (DFT) is an important tool in digital signal processing [2]. A treasure of fast algorithms such as the Cooley-Tukey Fast Fourier Transform (FFT), the prime factor algorithm (PFA) FFT, and the Winogard FFT are available to compute efficiently DFT of different lengths. Further reduction of the arithmetic complexity of FFT is possible when the input sequences are either real-valued or complex conjugate symmetric, due to the symmetry of the Fourier transformed sequences. The basic idea of real-valued FFTs (RFFTs) is to identify the intermediate symmetries that occur in the algorithm when transforming a real sequence and use these symmetries to eliminate redundant operations and the use of the complex storage. The resulting algorithm usually requires approximately half the multiplications and additions, and storage needed by the corresponding algorithm for complex data. Real-valued radix-2, -4, and split-radix FFT algorithms are described in details in [5].

In this paper, the multiplier-less FFT-like (ML-FFT) transformation proposed in [3] is extended to the real-valued case, using the real-valued radix-2 decimation-in-time RFFT (DIT-RFFT) as an example. Generalizations to the real-valued radix-4 and split-radix FFT are straightforward and are omitted here for page limitation. These new transformations will be referred to as ML RFFT-like (ML-RFFT) transformation. Like their ML-FFT counterparts, the ML-RFFTs make use of a special parameterization of the rotation-like matrix and represent its coefficients in the sum-of-power-of-two (SOPOT) or canonical signed digits (CSD) representation [10]. This parameterization allows us to implement both the forward and inverse transforms with the same set of SOPOT coefficients. Since the SOPOT coefficients can be implemented with limited shifts and additions, the ML-FFT and RFFT have an arithmetic complexity of $O(N\log_2 N)$ additions, where N is the size of the transformation. One application of the new ML-RFFT is to derive new multiplier-less approximation of the discrete cosine transforms (DCT's) [1,9] and the DWT's [7], using several classical mappings between the DFT, DCTs, and DWTs. The pre- and/or post- multiplications required in the mapping are again parameterized by the rotation-like matrix and SOPOT coefficients. Design result shows that all the sinusoidal transforms can be approximated well with reasonable arithmetic complexity and they converge to their ideal counterpart if more SOPOT are used. Moreover, the new ML-RFFT and ML-DCTs can be used to approximate the DCTs used in the lapped transforms and the cosine modulated filter banks [4]. The rest of this paper is organized as follows: Section II is devoted to the proposed ML-RFFT algorithm. Sections III and IV describe respectively the mappings and arithmetic complexities of various transforms related to the DFT. The design method and examples demonstrating the effectiveness of the proposed approach are given in Section V. Finally, conclusion is drawn in Section VI.

II. ML-RFFT ALGORITHM

For the sake of presentation, let's consider the decimation-in-time FFT algorithm as follows:

$$X(k) = \sum_{n=0}^{N/2-1} x(2n)\cdot W_{N/2}^{nk} + W_N^k \cdot \sum_{n=0}^{N/2-1} x(2n+1)\cdot W_{N/2}^{nk} \qquad (1)$$

$$X(k+N/2) = \sum_{n=0}^{N/2-1} x(2n)\cdot W_{N/2}^{nk} - W_N^k \cdot \sum_{n=0}^{N/2-1} x(2n+1)\cdot W_{N/2}^{nk}, \; \mathrm{k} = 0,1,\ldots,N/2\text{-}1, \qquad (2)$$

where $\{x(n)\}$ and $\{X(k)\}$ are respectively the input and transformed sequences; N is the transform length and is assumed to be an even number; and $W_N = \exp(-j2\pi/N)$. If $\{x(n)\}$ is real-valued, the DFT coefficients satisfy the following conjugate symmetry:

$X(0)$ and $X(N/2)$ are real;

$$X(k) = X^*(N-k), \qquad 1 \le k \le (N/2)-1. \qquad (3)$$

Using these symmetries, the computational requirements can be reduced by about a factor of 2 as some of the redundant operations at every stage of the FFT can be removed. More precisely, to compute the FFT in place, the real part of the *kth* complex value is placed in the *kth* location in the real array while for the imaginary part, it is stored in the redundant *(N-k) th* location. This approach can be applied at every stage of the FFT by recognizing the fact that the half-length transforms of even- and odd- indexed data samples are again FFT's of real-valued data, therefore the same approach can be recursively applied to shorter FFT's to complete the transform. Figure 1 illustrates an 8-point real-valued radix-2 DIT-FFT. The resulting algorithm requires exactly half of the multiplications and storage, and (N-2) fewer than half of the additions of the algorithm for complex data. Extension to the radix-4 and split-radix FFT and sequences with conjugate symmetric input can be found in [5]. In addition to this direct modification of the FFT algorithms, it is possible to compute the DFTs of two real values sequences by a complex DFT with slightly more additions [2].

Like the ML-FFT in [3], the proposed ML-RFFT algorithm approximates the twiddle factor multiplications W_N^k in (1) and (2) by representing the coefficients of a certain factorization of the rotation using SOPOT coefficients. To start with, let $c = (x + j\cdot y)$ be a complex number. The multiplication of c with $\exp(-j\theta)$, $p = c\cdot\exp(-j\theta)$ can be written as:

0-7803-7761-3/03/$17.00 ©2003 IEEE

$$\begin{bmatrix} p_r \\ p_i \end{bmatrix} = \begin{bmatrix} \cos\theta & \sin\theta \\ -\sin\theta & \cos\theta \end{bmatrix} \cdot \begin{bmatrix} x \\ y \end{bmatrix} = \begin{bmatrix} 1 & 0 \\ 0 & -1 \end{bmatrix} \cdot \boldsymbol{R}_\theta \cdot \begin{bmatrix} x \\ y \end{bmatrix} \quad (4)$$

where $\boldsymbol{R}_\theta = \begin{bmatrix} \cos\theta & \sin\theta \\ \sin\theta & -\cos\theta \end{bmatrix}$ is the rotation-like matrix. However, $\cos\theta$ and $\sin\theta$ in (4) are not suitable for directly conversion to the SOPOT representation since the inverse of $\boldsymbol{R}_\theta$ cannot in general be expressed in terms of SOPOT coefficients. To cope with this, $\mathbf{R}_\theta$ is re-written as follows:

$$\boldsymbol{R}_\theta = \begin{bmatrix} 1 & -\tan(\theta/2) \\ 0 & 1 \end{bmatrix} \cdot \begin{bmatrix} 1 & 0 \\ \sin\theta & 1 \end{bmatrix} \cdot \begin{bmatrix} 1 & \tan(\theta/2) \\ 0 & -1 \end{bmatrix} = \boldsymbol{R}_\theta^{-1} \quad (5)$$

Since the forward and inverse of the matrices involve the same set of coefficients, i.e. $\sin\theta$ and $\tan(\theta/2)$. They can be quantized to the SOPOT coefficients to from

$$\boldsymbol{S}_\theta = \begin{bmatrix} 1 & -\beta_\theta \\ 0 & 1 \end{bmatrix} \begin{bmatrix} 1 & 0 \\ \alpha_\theta & 1 \end{bmatrix} \begin{bmatrix} 1 & \beta_\theta \\ 0 & -1 \end{bmatrix} \approx \boldsymbol{R}_\theta \quad (6)$$

where α_θ and β_θ are respectively the SOPOT approximations to $\sin\theta$ and $\tan(\theta/2)$ having the form $z_\theta = \sum_{k=1}^{t} a_k 2^{b_k}$, where $a_k \in \{-1,1\}$, $b_k \in \{-r,\ldots,-1,0,1,\ldots r\}$; r is the range of the coefficients and t is the number of terms used in each coefficient. As a result, the twiddle factor multiplication can be implemented as simple shift-and-add operations. Unfortunately, direct application of (5) might lead to large dynamic range for $\tan(\theta/2)$ when θ is close to π, so slight adjustment is needed so as to reduce the dynamic range of the coefficients as well as the number of parameters to be optimized (pls. refer to [3] for details). Since all the twiddle factors in the radix-2 DIT FFT are just a subset of those in the final stage and two parameters per rotation are required, there are approximately $2\times(N/8)$ parameters for an N-point ML-RFFT, same as the ML-FFT. Next, we shall introduce the ML -DCTs and -DWTs.

III. ML-DCTS AND ML-DWTS

According to Wang [6], there are four types of DCT's and DWT's. In what follows, we shall discuss the mapping of the DCT's [1,9] and DWT's [7] to real-valued DFT. It should be noted that other maps are also possible as long as the pre- and post- multiplications can be expressed in the form of (4), and the input to the DFT or IDFT are either real-valued or conjugate symmetric.

Mapping for DCT-II

In [9], a simple mapping of DCT-II to DFT is introduced. The algorithm first rewrites the type-II DCT into its odd and even indexed parts as follows:

$$X_c^{II}(k) = \sum_{n=0}^{N-1} x(n)\cos\frac{\pi(2n+1)k}{2N} = \sum_{n=0}^{\lfloor (N-1)/2 \rfloor} x(2n)\cos\frac{\pi(4n+1)k}{2N} + \sum_{n=0}^{\lfloor N/2 \rfloor - 1} x(2n+1)\cos\frac{\pi(4n+3)k}{2N} \quad (7)$$

With the following mapping:

$$\tilde{x}(n) = x(2n),\ n = 0,1,\ldots,\lfloor (N-1)/2 \rfloor$$
$$\tilde{x}(N-n-1) = x(2n+1),\quad n = 0,1,\ldots,\lfloor N/2 \rfloor - 1. \quad (8)$$

(7) can then be converted to a phase modulated DFT:

$$X_c^{II}(k) = \sum_{n=0}^{N-1} \tilde{x}(n)\cos\frac{\pi(4n+1)k}{2N} = \mathrm{Re}\left[W_{4N}^{k} \sum_{n=0}^{N-1} \tilde{x}(n) W_N^{nk}\right] \quad (9)$$

where Re[.] denotes the real part of a complex number. As the sequence $\tilde{x}(n)$ is real-valued, its DFT, $\tilde{X}(k)$, can readily be computed by the N-point real-valued FFT algorithms. The ML-DCT-II can be obtained by replacing the real-valued FFT by the ML-RFFT and the post multiplications W_{4N}^{k} in (9) by the approximation in (6).

Mapping for DCT-III

First of all, consider the equivalent matrix factorization of the aforementioned algorithm as follows:

$$\boldsymbol{C}_N^{II} = \boldsymbol{M}_N \boldsymbol{W}_N \boldsymbol{P}_N, \quad (10)$$

where $\boldsymbol{P}_N$ performs the permutation in (8), $\boldsymbol{W}_N$ is the N-point DFT matrix and $\boldsymbol{M}_N$ performs the post rotation in (9). By taking the transpose of (10), it gives

$$\boldsymbol{C}_N^{III} = \left[\boldsymbol{C}_N^{II}\right]^T = \boldsymbol{P}_N^T \boldsymbol{W}_N^T \boldsymbol{M}_N^T. \quad (11)$$

From which, we obtain the mapping as follows:

$$\tilde{U}(k) = X_c^{III}(2k), \qquad k = 0,1,\ldots,\lfloor (N-1)/2 \rfloor$$
$$\tilde{U}(N-k-1) = X_c^{III}(2k+1), \qquad k = 0,1,\ldots,\lfloor N/2 \rfloor - 1 \quad (12)$$

And the inverse transform of $\tilde{U}(k)$ are given by:

$$\tilde{U}(k) = \sum_{n=0}^{N-1} \tilde{u}(n)\cdot W_N^{-nk}, \quad (13)$$

where $\tilde{u}(n) = \begin{cases} x(0), & n = 0 \\ \frac{1}{2}[x(n) - jx(N-n)]W_{4N}^{-nk}, & n = 1,\ldots,N-1 \end{cases}$.

Since the sequence $\tilde{u}(n)$ is complex conjugate symmetry, the inverse DFT in (13) can also be computed using the RFFT (for complex conjugate symmetry, details omitted here for simplicity).

Mapping for DCT-IV

The type-IV DCT is given by:

$$X_c^{IV}(k) = \sum_{n=0}^{N-1} x(n)\cos[(k+\frac{1}{2})(n+\frac{1}{2})\frac{\pi}{N}]. \quad (14)$$

Following the mapping in [1], the following two sequences *p(n)* and *q(n)* are defined:

$$p(n) = x(2n) \text{ and } q(n) = x(N-1-2n). \quad (15)$$

From which, (14) can be re-written as:

$$X_c^{IV}(k) = \sum_{n=0}^{N/2-1} p(n)\cos\left[(2n+\frac{1}{2})(k+\frac{1}{2})\frac{\pi}{N}\right] + \sum_{n=0}^{N/2-1} q(n)\cos\left[(N-1-2n+\frac{1}{2})(k+\frac{1}{2})\frac{\pi}{N}\right]. \quad (16)$$

Consequently,

$$X_c^{IV}(2k) = \mathrm{Re}\left[W_{2N}^{k} \sum_{n=0}^{N/2-1} [p(n) + jq(n)] W_{8N}^{4n+1} W_{N/2}^{nk}\right] \text{ and} \quad (17a)$$

$$X_c^{IV}(N-1-2k) = -\mathrm{Im}\left[W_{2N}^{k} \sum_{n=0}^{N/2-1} [p(n) + jq(n)] W_{8N}^{4n+1} W_{N/2}^{nk}\right] \quad (17b)$$

A ML-DCT-IV results if the (*N*/2)-point complex DFT is approximated by the ML-FFT. Again the pre- and post-multiplications can be implemented according to (6).

Mapping for DWT-I

The type-I DWT has the same form as the discrete Harley transform (DHT). It can be obtained by subtracting the imaginary part from the real part of the DFT sequence. That is:

$$X_W^I(k) = \mathrm{Re}[X(k)] - \mathrm{Im}[X(k)], \quad k = 0,1,\ldots,N-1 \quad (18)$$

By using the ML-RFFT instead of the FFT, one gets the ML-DHT. Alternatively, another ML-DHT with slightly lower number of additions can be from the fast algorithms in [7,11].

Mapping for DWT-II

Consider the N-point odd DFT associated with the DIT-DFT of a real value sequence $x(n)$ as follows:

$$X_T^o(k) = \sum_{n=0}^{N-1} x(n) \cdot W_{2N}^{(2n+1)k}, \quad k = 0,1,\ldots,N-1. \quad (19)$$

It can be further decomposed into its even and odd parts:

$$X_T^o(k) = W_{2N}^k \sum_{n=0}^{N/2-1} x(2n) W_{N/2}^{nk} + W_{2N}^{3k} \sum_{n=0}^{N/2-1} x(2n+1) W_{N/2}^{nk}. \quad (20)$$

Comparing it with the N-point type-II DWT:

$$X_W^{II}(k) = \sum_{n=0}^{N-1} x(n) cas[\pi(2n+1)k/N], \quad k = 0,1,\ldots,N-1, \quad (21)$$

the following relationship can be easily derived:

$$(1+j)X_T^o(k) = X_W^{II}(k) - jX_W^{II}(N-k). \quad (22)$$

Since the real and imaginary parts of $(1+j)X_T^o(k)$ are related to the first half and the second half of $X_W^{II}(k)$, it is sufficient to calculate (22) for k = 0,1,…,N/2. Therefore, the type-II ML-DWT can be obtained from two (N/2)-point ML-RFFTs with some post-multiplications, which again can be approximated by (6). To save space, we shall only describe the relevant mapping in the following, and it is assumed that similar modified will be applied to the DFT and twiddle factors.

Mapping for DWT-III

Similar to the approach in deriving the type-II DWT, consider the N-point odd DFT associated with the DIF-DFT of a real-value sequence $\{x(n)\}$ as follows:

$$X_F^o(k) = \sum_{n=0}^{N-1} x(n) \cdot W_{2N}^{n(2k+1)}, \quad k = 0,1,\ldots,N-1. \quad (23)$$

This is related to the N-point type-III DWT of $x(n)$:

$$X_W^{III}(k) = \sum_{n=0}^{N-1} x(n) cas[\pi n(2k+1)/N], \quad k = 0,1,\ldots,N-1 \quad (24)$$

by the following formula:

$$(1+j)X_F^o(k) = X_W^{III}(k) + jX_W^{III}(N-k-1). \quad (25)$$

For the case where k is even, (25) becomes:

$$(1+j)X_F^o(2k) = X_W^{III}(2k) + jX_W^{III}(N-2k-1). \quad (26)$$

Therefore, it is sufficient to compute $(1+j)X_F^o(k)$ for k = 0,1,…,N/2-1, as $(1+j)X_F^o(k)$ is associated with the even and odd indexed parts of the type-III DWT. On the other hand, (23) can be written as:

$$X_F^o(2k) = \sum_{n=0}^{N/2-1} [x(n) - jx(n+N/2)] W_{2N}^n W_{N/2}^{nk}. \quad (27)$$

By nesting the multiplication of (1+j) with the twiddle factor W_{2N}^n, the type-III DWT can be obtained by performing a (N/2)-point complex FFT together with twiddle factors of the form $(1+j)W_{2N}^n$.

Mapping for DWT-IV

To find an appropriate mapping for the type-IV DWT, define the following transform:

$$V(k) = \sum_{n=0}^{N-1} x(n) \cdot W_{4N}^{(2n+1)(2k+1)}, \quad k = 0,1,\ldots,N-1. \quad (28)$$

Then the type-IV DWT

$$X_W^{IV}(k) = \sum_{n=0}^{N-1} x(n) cas[\pi(2n+1)(2k+1)/2N], \quad k = 0,1,\ldots,N-1 \quad (29)$$

is related to (28) by the following formula:

$$(1+j)V(k) = X_W^{IV}(k) - jX_W^{IV}(N-k-1). \quad (30)$$

By nesting the multiplication of (1+j) with the twiddle factors in (28), we have:

$$(1+j)V(k) = \left[(1+j)W_{4N}^{2k+1}\right] \cdot \sum_{n=0}^{N-1} x(n) W_{2N}^n \cdot W_N^{nk} \quad (31)$$

Thus, the type-IV DWT can be obtained from a complex-valued FFT with some pre- and post-multiplications.

IV. ARITHMETIC COMPLEXITIES

In general, if S is the average number of SOPOT terms per coefficient, the arithmetic complexity of the radix-2 ML-FFT algorithm, derived from a radix-2 FFT algorithm with an arithmetic complexity of $M_{radix-2}(N) = (N/2)(\log_2 N - 3) + 2$ non–trivial complex multiplications and $A_{radix-2}(N) = N \log_2 N$ complex additions, is approximately $A_{ML-FFT}^{Add}(N) = 3S \cdot [(N/2)(\log_2 N - 3) + 2] + 2N \log_2 N$ real additions. Then the arithmetic complexity of the proposed ML-RFFT is approximately given by:

$$A_{ML-RFFT}^{Add}(N) = 3S \cdot (N/4)(\log_2 N - 3) + 3S + N \log_2 N - N + 2 \quad (32)$$

It follows immediately that we can summarize the arithmetic complexities of the ML-DCTs and ML-DWTs as follows:

$$A_{DCT-II}^{Add}(N) = A_{ML-RFFT}^{Add}(N) + 3S \cdot M_{DCT-II}^{post}(N)$$

$$A_{DCT-III}^{Add}(N) = A_{ML-RFFT}^{Add}(N) + 3S \cdot M_{DCT-III}^{pre}(N)$$

$$A_{DCT-IV}^{Add}(N) = A_{ML-FFT}^{Add}(N/2) + 3S \cdot (M_{DCT-IV}^{pre}(N) + M_{DCT-IV}^{post}(N))$$

$$A_{DWT-I}^{Add}(N) = A_{ML-RFFT}^{Add}(N) + N$$

$$A_{DWT-II}^{Add}(N) = 2A_{ML-RFFT}^{Add}(N/2) + 3S \cdot M_{DWT-II}^{post}(N)$$

$$A_{DWT-III}^{Add}(N) = A_{ML-FFT}^{Add}(N/2) + 3S \cdot M_{DWT-III}^{pre}(N)$$

$$A_{DWT-IV}^{Add}(N) = A_{ML-FFT}^{Add}(N) + 3S \cdot (M_{DWT-IV}^{pre}(N) + M_{DWT-IV}^{post}(N))$$

where $M_D^{pre}(N)$ and $M_D^{post}(N)$ are respectively the total numbers of the non-trivial pre- and post- multiplications, where D represents which transform is performed. Note that since the factor (1+j) involved in the pre- or post- multiplications of ML-DWT-II, -III and -IV is implemented by a rotation factor $e^{j\pi/4}$ in the equation (4). Therefore, the corresponding outputs are assumed to be obtained by a scaling factor $\sqrt{2}$. For simplicity, these multiplications are not counted in the derivation of the above arithmetic complexities.

V. DESIGN METHOD AND EXAMPLES

The SOPOT coefficients in (6) are determined by the random search algorithm in [4]. Its basic idea is to repetitively calculate a candidate SOPOT vector z_c:

$$z_c = \lfloor z + \lambda z_p \rfloor_S, \quad (33)$$

by adding a perturbation vector to the vector containing the real-valued α_θ and β_θ, and then quantizing the elements of the sum to the nearest SOPOT coefficients. Here, $\lfloor \cdot \rfloor_S$ denotes the rounding operator to the nearest SOPOT coefficients. The parameter λ controls the size of the neighborhood to be searched and z_p is a random vector with all its elements bounded by the range ±1. Then, the objective function for each candidate SOPOT vector is evaluated and the one with the best performance for a given arithmetic complexity is recorded. Since this algorithm operates in a random manner, therefore, the longer the searching time, the higher the chance of finding the optimum solution. In this example, the objective function E_{OBJ} used is the Frobenius norm of the error matrix between the candidate transform from its ideal counterpart:

$$E_{OBJ} = \sqrt{\sum_{j=0}^{N-1}\sum_{k=0}^{N-1}\left(f_{j,k} - \hat{f}_{j,k}\right)\cdot\overline{\left(f_{j,k} - \hat{f}_{j,k}\right)}}, \qquad (34)$$

where $\hat{f}_{j,k}$ and $f_{j,k}$ are respectively the (j,k)-th entries of the matrix of the candidate transform and its ideal counterpart. Figure 2 shows the frequency response of various transforms with N=8 obtained by limiting the number of SOPOT terms to 5 for illustration purpose. In some applications, fewer numbers of terms can be used to yield very efficient realization and the number of SOPOT term per coefficients can be made variable without major modification of the design algorithm. We can see that they are almost visually identical to their ideal counterparts. Table 1 shows the corresponding E_{OBJ} and arithmetic complexities of the ML-DCTs and ML-DWTs. For simplicity, the trivial multiplications are excluded.

VI. CONCLUSION

A new ML-FFT for real-valued input, called the ML-RFFT, is presented. Like the ML-FFT, it parameterizes the twiddle factors in the conventional radix-2^n or split-radix real-valued FFT algorithm as certain rotation-like matrices and approximates the associated parameters inside these matrices by the sum-of-power-of-two (SOPOT) or canonical signed digits (CSD) representations. Because of the symmetry in the algorithm, it only requires about half the number of additions as required by the ML-FFT. New ML-FFT-based transformation called ML-DCTs and ML-DWTs are derived, using the mappings between the DFT and the DCTs and DWTs. Design examples of the new transformations are also given.

REFERENCES

[1] H.S. Malvar, "Signal Processing with Lapped Transform," *Artech House*, 1991.

[2] A. V. Oppenheim and R. W. Schafer, "Discrete-Time Signal Processing," *Prentice-Hall,* 1989.

[3] S. C. Chan and P. M. Yiu, "A multiplier-less 1-D and 2-D fast Fourier transform-like transformation using sum-of-power-of-two (SOPOT) coefficients," in *IEEE ISCAS'2002*, Arizona, USA.

[4] S. C. Chan and P. M. Yiu, "Multiplier-less discrete sinusoidal and lapped transforms using sum-of-powers-of-two (SOPOT) coefficients," in *Proc. IEEE-ISCAS, Sydney, Australia* May 6-9, 2001.

[5] H. V. Sorensen, D. L. Jones, M. T. Heideman and C. S. Burrus, "Real-valued fast fourier transform algorithms," *IEEE Trans. ASSP.*, vol. 35, pp. 849-863, June 1987.

[6] Z. Wang, "Fast algorithms for the discrete W transform and for the discrete Fourier transform," *IEEE Trans. ASSP*, vol. 37, pp. 803-816, Aug. 1984.

[7] S. C. Chan, and K.L. Ho, "Fast algorithms for computing the discrete W transform," *IEEE Region 10 Conference on Computer and Communication systems, Hong Kong,* pp. 183-185, Sept. 1990.

[8] S. C. Chan, and K.L. Ho, "Fast algorithms for computing the discrete cosine transform," *IEEE Trans. Circuits Syst. II*, vol. 39, no. 3, March 1992.

[9] M. J. Narasimha and A. M. Peterson, "On the computation of the discrete cosine transform," *IEEE Trans. Commun.*, vol. 26, pp. 934-946, June 1978.

[10] S. C. Chan, W. Liu and K L. Ho, "Perfect Reconstruction Modulated Filter Banks with sum-of-power-of-two coefficients," in *Proc. IEEE-ISCAS, Geneva, Switzerland,* May 28-31, 2000.

[11] S. C. Chan and K. L. Ho, "Split vector-radix fast Fourier transform," *IEEE Trans. SP*, vol. 40, 1992, pp. 2029-2039.

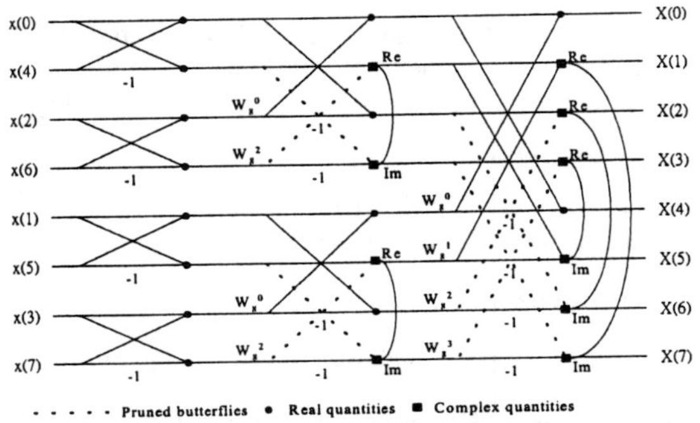

Figure 1: Real-valued symmetry in radix-2 FFT.

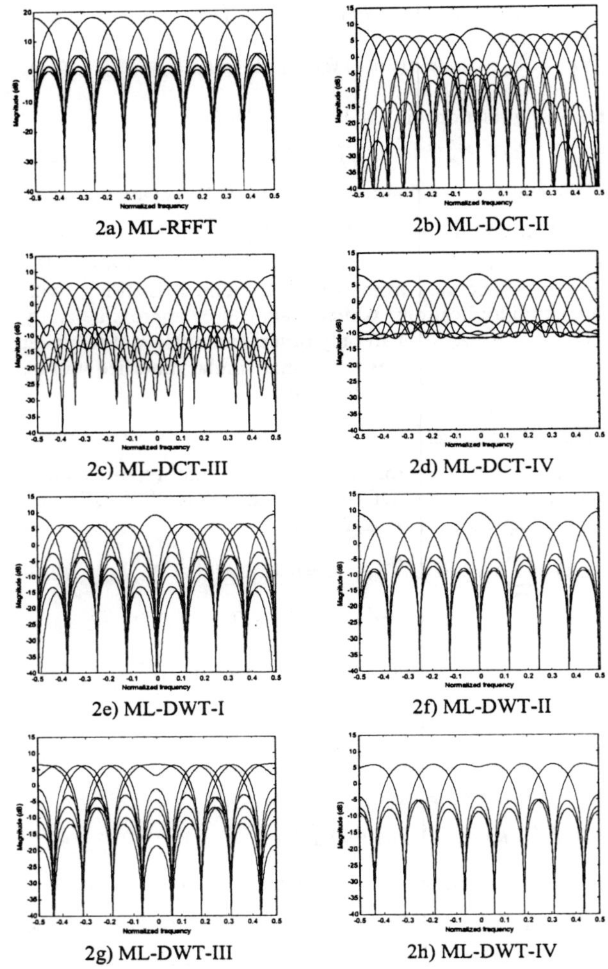

Figure 2: Frequency response of various transforms for N=8 and S=5.

ML Algorithm	E_{OBJ} (in dB)	Number of additions required for ML algorithm FFT or RFFT	Number of non-trivial complex multiplications		Total number of additions (SOPOT representation)
			pre	post	
DCT-II	-60.6885	33	-	7	138
DCT-III	-64.1302	33	7	-	138
DCT-IV	-60.8797	16	4	3	121
DWT-I	-61.7199	33	-	-	41
DWT-II	-60.6393	12	-	6	112
DWT-III	-60.8168	16	3	-	61
DWT-IV	-58.8914	78	6	4	228

Table 1: Frobenius norm and arithmetic complexities of various transforms for N=8 and S=5.

ON UNBIASED PARAMETER ESTIMATION OF AUTOREGRESSIVE SIGNALS OBSERVED IN NOISE

Wei Xing Zheng

School of QMMS, University of Western Sydney
Penrith South DC NSW 1797, Australia

ABSTRACT

In a recent paper, a simple least-squares (LS) based algorithm is introduced for unbiased parameter estimation of autoregressive (AR) signals observed in noise, under the assumption that the ratio between the driving source power and the corrupting noise variance is known. In the present paper, this LS based algorithm is modified with a more computationally efficient algorithmic structure. The mean convergence of the modified algorithm is then investigated. The issue of how the assumption of the known power ratio can be mitigated for practical applications is discussed, which leads to the development of an effective estimation algorithm for noisy AR signals. Theoretical results are validated through computer simulations.

1. INTRODUCTION

Autoregressive (AR) signal models are widely utilized in a broad range of signal processing applications, such as speech analysis, spectral estimation, noise cancellation and digital communications [2]. However, in practical circumstances the signal to be modelled is often observed in noise. Then the AR parameter estimates given by the standard least-squares (LS) method become biased because the observation noise tends to flatten the AR signal spectrum and distort the correlation function [1]. Several unbiased parameter estimation algorithms have been developed for noisy AR signals, such as the modified Yule-Walker (MYW) equations method [2], the maximum likelihood (ML) method [4], the recursive prediction error (RPE) method [3], the γ-least-mean-squares (LMS) and ρ-LMS algorithms [5], and the improved least-squares (ILS) method with direct implementation structure [6] (called the ILSD method for short).

The ILSD method has demonstrated attractive behaviors over the other methods. Recent work on this line is that a simple ILS algorithm is introduced in [7], which is termed the ILSR algorithm as it assumes that the ratio between the driving source power and the corrupting noise variance is known. Compared with the ILSD method, the ILSR algorithm produces better parameter estimates and requires fewer computations. In this paper, the ILSR algorithm will be investigated in several aspects. It is first shown that when the power ratio is given, the corrupting noise variance can be easily estimated even with no calculation of the average LS errors. Accordingly, a more computationally efficient algorithmic structure is established for the ILSR algorithm. The importance of this modification is that it paves the way for investigating the mean convergence of the ILSR algorithm. Note that such an analysis was not made in [7] mainly because the algorithmic structure given there makes it a rather difficult task. Another purpose of this paper is to examine the ILSR algorithm with regard to the key assumption of the known power ratio. As mentioned in [7], this assumption is not very restrictive. A similar assumption is also used for some other methods. For example, the γ-LMS and ρ-LMS algorithms require a priori information of the corrupting noise variance [5]. In spite of this, from the viewpoint of practical applications it is important to investigate whether the assumption of the known power ratio can be mitigated to that of a rough range of this ratio. On this basis, a revised ILSR algorithm is developed for noisy AR signal identification in practical applications.

This work was supported in part by a Research Grant from the Australian Research Council and in part by a Research Grant from the University of Western Sydney, Australia.

2. PROBLEM SETUP AND MOTIVATION

A signal $x(t)$ produced by driving a source signal $v(t)$ through a pth-order AR model can be represented by

$$x(t) = \sum_{i=1}^{p} a_i x(t-i) + v(t) \tag{1}$$

where the driving source $v(t)$ is a white noise with power σ_v^2, the model coefficients $a_1, ..., a_p$ are called the AR parameters, and the model order p is called the AR order.

Introduce the AR parameter vector $\mathbf{a}$ and the signal regression vector $\mathbf{x}_t$ as follows

$$\mathbf{a}^\top = [a_1 \; ... \; a_p] \tag{2}$$

$$\mathbf{x}_t^\top = [x(t-1) \; ... \; x(t-p)]. \tag{3}$$

The AR model (1) can be recast as

$$x(t) = \mathbf{x}_t^\top \mathbf{a} + v(t). \tag{4}$$

0-7803-7761-3/03/$17.00 ©2003 IEEE

Minimization of the mean squared error criterion $J_x(\mathbf{a}) = E[(x(t) - \mathbf{x}_t^\top \mathbf{a})^2]$ with respect to $\mathbf{a}$ gives the optimal AR parameter estimate

$$\mathbf{a}^* = \mathbf{R}_{xx}^{-1}\mathbf{r}_x \tag{5}$$

where $\mathbf{R}_{xx} = E[\mathbf{x}_t\mathbf{x}_t^\top]$ and $\mathbf{r}_x = E[\mathbf{x}_t x(t)]$. Since $E[\mathbf{x}_t v(t)] = \mathbf{0}$, $\mathbf{a}^*$ is an unbiased estimate.

In practice, observations of the AR signal $x(t)$ are usually corrupted by noise. The noisy observation $y(t)$ can be expressed as

$$y(t) = x(t) + w(t) \tag{6}$$

where the corrupting noise $w(t)$ is a white noise with unknown variance σ_w^2. Combining (1) and (6) together, a linear regression model of the noisy AR signal is obtained as follows

$$y(t) = \mathbf{y}_t^\top \mathbf{a} + \varepsilon(t) \tag{7}$$

where

$$\mathbf{y}_t^\top = [y(t-1) \ldots y(t-p)] \tag{8}$$

$$\varepsilon(t) = v(t) + w(t) - \mathbf{w}_t^\top \mathbf{a} \tag{9}$$

$$\mathbf{w}_t^\top = [w(t-1) \ldots w(t-p)]. \tag{10}$$

In order to estimate the AR parameters $a_1, ..., a_p$ from noisy observations $y(t)$, the mean squared error criterion $J_y(\mathbf{a}) = E[(y(t) - \mathbf{y}_t^\top \mathbf{a})^2]$ is minimized, which yields the LS estimate

$$\mathbf{a}_{LS} = \mathbf{R}_{yy}^{-1}\mathbf{r}_y \tag{11}$$

where $\mathbf{R}_{yy} = E[\mathbf{y}_t\mathbf{y}_t^\top]$ and $\mathbf{r}_y = E[\mathbf{y}_t y(t)]$. Since $v(t)$ and $w(t)$ are orthogonal to each other and $E[\mathbf{w}_t\mathbf{w}_t^\top] = \sigma_w^2\mathbf{I}_p$, we have

$$\mathbf{R}_{yy} = E[(\mathbf{x}_t + \mathbf{w}_t)(\mathbf{x}_t + \mathbf{w}_t)^\top] = \mathbf{R}_{xx} + \sigma_w^2\mathbf{I}_p \tag{12}$$

$$\mathbf{r}_y = E[(\mathbf{x}_t + \mathbf{w}_t)(x(t) + w(t))] = \mathbf{r}_x \tag{13}$$

where $\mathbf{I}_p$ is a $p \times p$ identity matrix. Substituting (12) and (13) into (11) leads to

$$\mathbf{a}_{LS} = (\mathbf{R}_{xx} + \sigma_w^2\mathbf{I}_p)^{-1}\mathbf{r}_x. \tag{14}$$

A quick comparison between (14) and (5) reveals that the corrupting noise $w(t)$ causes the LS estimate $\mathbf{a}_{LS}$ to be biased. Certainly, the noise-induced bias can be removed if an estimate of the corrupting noise variance σ_w^2 is available. This is the motive of the ILS based methods.

3. THE ILSR1 ALGORITHM

Let the ratio between the driving source power σ_v^2 and the corrupting noise variance σ_w^2 be defined by

$$\kappa^2 = \frac{\sigma_v^2}{\sigma_w^2}. \tag{15}$$

Under the assumption that this power ratio κ^2 is known, the ILSR algorithm is introduced in [7]. However, implementation of the ILSR algorithm requires computing the average LS errors, which unavoidably causes added computations. To improve the computational efficiency, we take a look at $r_y(0) = [y(t)^2]$, the element on the diagonal of the Toeplitz covariance matrix $\mathbf{R}_{yy}$.

Since

$$E[x(t)\varepsilon(t)] = E[x(t)(v(t) + w(t) - \mathbf{w}_t^\top \mathbf{a})] = \sigma_v^2 \tag{16}$$

$$E[w(t)\varepsilon(t)] = E[w(t)(v(t) + w(t) - \mathbf{w}_t^\top \mathbf{a})] = \sigma_w^2 \tag{17}$$

we get

$$E[y(t)\varepsilon(t)] = E[(x(t) + w(t))\varepsilon(t)] = \sigma_v^2 + \sigma_w^2. \tag{18}$$

It follows from (7) and (18) that

$$r_y(0) = E[y(t)(\mathbf{y}_t^\top \mathbf{a} + \varepsilon(t))] = \mathbf{r}_y^\top \mathbf{a} + \sigma_v^2 + \sigma_w^2. \tag{19}$$

Substituting $\sigma_v^2 = \kappa^2\sigma_w^2$ into (19) and making some rearrangements, the driving source power σ_w^2 is expressible as

$$\sigma_w^2 = \frac{r_y(0) - \mathbf{r}_y^\top \mathbf{a}}{1 + \kappa^2}. \tag{20}$$

Consequently, the key formulae in the ILSR algorithm introduced in [7] can be replaced by

$$\hat{\sigma}_w^2(k) = \frac{\hat{r}_y(0) - \hat{\mathbf{r}}_y^\top \hat{\mathbf{a}}_{ILS}(k-1)}{1 + \kappa^2} \tag{21}$$

$$\hat{\mathbf{a}}_{ILS}(k) = \hat{\mathbf{a}}_{LS} + \hat{\sigma}_w^2(k)\hat{\mathbf{R}}_{yy}^{-1}\hat{\mathbf{a}}_{ILS}(k-1) \tag{22}$$

where

$$\hat{\mathbf{R}}_{yy} = \frac{1}{N}\sum_{t=1}^{N} \mathbf{y}_t\mathbf{y}_t^\top, \qquad \hat{\mathbf{r}}_y = \frac{1}{N}\sum_{t=1}^{N} \mathbf{y}_t y(t) \tag{23}$$

$$\hat{\mathbf{a}}_{LS} = \hat{\mathbf{R}}_{yy}^{-1}\hat{\mathbf{r}}_y, \qquad \hat{r}_y(0) = \hat{\mathbf{R}}_{yy}(1,1). \tag{24}$$

It is important to note that the new ILSR1 algorithm, which is comprised of (21)-(24), is fully based upon the standard LS method in the sense that it makes use of the covariance matrix $\hat{\mathbf{R}}_{yy}$ and the covariance vector $\hat{\mathbf{r}}_y$ only. Hence it can be implemented at a lower computational cost than the previous ILSR algorithm presented in [7].

We now proceed to establish the mean convergence of the new ILSR1 algorithm, namely, $E[\hat{\mathbf{a}}_{ILS}(k)]$ will converge to $\mathbf{a}^*$. First, it is obvious that $E[\hat{\mathbf{R}}_{yy}] = \mathbf{R}_{yy}$ and $E[\hat{\mathbf{r}}_y] = \mathbf{r}_y$, so $E[\hat{\mathbf{a}}_{LS}] = \mathbf{R}_{yy}^{-1}\mathbf{r}_y$ and $E[\hat{r}_y(0)] = r_y(0)$. Suppose that $\hat{\mathbf{r}}_y$ and $\hat{\mathbf{a}}_{ILS}(k-1)$ are independent. It follows from (21) that

$$E[\hat{\sigma}_w^2(k)] = \frac{r_y(0) - \mathbf{r}_y^\top E[\hat{\mathbf{a}}_{ILS}(k-1)]}{1 + \kappa^2}. \tag{25}$$

Noting that in the steady state $E[\hat{\mathbf{a}}_{ILS}(k-1)])\approx E[\hat{\mathbf{a}}_{ILS}(k)]$ and then combining (25) with (20) yields

$$E[\hat{\sigma}_w^2(k)] = \sigma_w^2 + \frac{\mathbf{r}_y^\top(\mathbf{a} - E[\hat{\mathbf{a}}_{ILS}(k)])}{1 + \kappa^2}. \tag{26}$$

Similarly, assuming that $\hat{\mathbf{R}}_{yy}$ and $\hat{\mathbf{a}}_{ILS}(k-1)$ are independent, it is obtained from (22) that

$$E[\hat{\mathbf{a}}_{ILS}(k)] = \mathbf{a}_{LS} + E[\hat{\sigma}_w^2(k)]\mathbf{R}_{yy}^{-1}E[\hat{\mathbf{a}}_{ILS}(k-1)] \tag{27}$$

or equivalently in the steady state,

$$\mathbf{R}_{yy}E[\hat{\mathbf{a}}_{ILS}(k)] = \mathbf{r}_y + E[\hat{\sigma}_w^2(k)]E[\hat{\mathbf{a}}_{ILS}(k)]. \tag{28}$$

Substituting (12), (13) and (26) into (28), we may get

$$\mathbf{R}_{xx}E[\hat{\mathbf{a}}_{ILS}(k)] = \mathbf{r}_x + \frac{\mathbf{r}_x^\top(\mathbf{a} - E[\hat{\mathbf{a}}_{ILS}(k)])}{1 + \kappa^2}E[\hat{\mathbf{a}}_{ILS}(k)]. \tag{29}$$

Using (5) and noticing that $\mathbf{a}^* = \mathbf{a}$, a rearrangement of (29) gives

$$\mathbf{F}(\mathbf{a}^* - E[\hat{\mathbf{a}}_{ILS}(k)]) = \mathbf{0} \tag{30}$$

where

$$\mathbf{F} = \mathbf{I}_p + \frac{\mathbf{R}_{xx}^{-1} E[\hat{\mathbf{a}}_{ILS}(k)]\mathbf{r}_x^\top}{1+\kappa^2}. \tag{31}$$

It is apparent that the $p \times p$ matrix $\mathbf{F}$ is nonsingular. Thus it follows immediately from (30) that $E[\hat{\mathbf{a}}_{ILS}(k)] = \mathbf{a}^*$. It should be mentioned that it is the newly derived formula (21) that greatly facilitates the convergence analysis.

4. THE ILSR2 ALGORITHM

As mentioned in [7], the assumption of the known power ratio is not a very restrictive condition, and may conform to a number of signal processing application cases. However, it is still of great interest to know how the ILSR1 algorithm may behave if an exact knowledge of the power ratio κ^2 is not available. The practical importance of relaxing this known power ratio assumption is self-evident.

Assume that only a partial information of the power ratio κ^2 is accessible, with a rough guess $\hat{\kappa}^2$ obtained for κ^2. Given that $\hat{\kappa}^2 \neq \kappa^2$, we consider two scenarios.

(a) If $\hat{\kappa}^2 < \kappa^2$, then κ^2 is underestimated. Since $\hat{\kappa}^2$ appears in the denominator of (21), the corrupting noise variance σ_w^2 will be likely overestimated. A bigger $\hat{\sigma}_w^2(k)$ will in turn affect the AR parameters to be overestimated via (22). In the meantime, by (15) the signal power σ_v^2 will be underestimated.

(b) Conversely, if $\hat{\kappa}^2 > \kappa^2$, then κ^2 is overestimated. Since $\hat{\kappa}^2$ appears in the denominator of (21), the corrupting noise variance σ_w^2 will be likely underestimated. A smaller $\hat{\sigma}_w^2(k)$ will in turn affect the AR parameters to be underestimated via (22). In the meantime, by (15) the signal power σ_v^2 will be overestimated.

The above analysis of the effect of the power ratio κ^2 on the ILSR1 algorithm is very useful as it motivates to develop another revised ILSR algorithm by mitigating the assumption of the known power ratio κ^2 to that of a rough range of κ^2. Suppose that the rough range of the power ratio κ^2 is determined as follows

$$\kappa_l^2 < \kappa^2 < \kappa_r^2 \tag{32}$$

where κ_l^2 and κ_r^2 are given. Applying the ILSR1 algorithm with κ_l^2 in noisy AR signal identification yields the AR parameter estimate $\hat{\mathbf{a}}_{ILS}^l(k)$, which is an overestimate of $\mathbf{a}$ due to $\kappa_l^2 < \kappa^2$. On the other hand, applying the ILSR1 algorithm with κ_r^2 yields the estimate $\hat{\mathbf{a}}_{ILS}^r(k)$, which is an underestimate of the AR parameter vector $\mathbf{a}$ due to $\kappa_r^2 > \kappa^2$. Clearly, the overestimate $\hat{\mathbf{a}}_{ILS}^l(k)$ may be offset by the underestimate $\hat{\mathbf{a}}_{ILS}^r(k)$, and vice versa. Hence, taking advantage of $\hat{\mathbf{a}}_{ILS}^l(k)$ and $\hat{\mathbf{a}}_{ILS}^r(k)$, we may devise a new ILSR estimate of the AR parameter vector $\mathbf{a}$ in the following

$$\hat{\mathbf{a}}_{ILS}^{av}(k) = \frac{1}{2}(\hat{\mathbf{a}}_{ILS}^l(k) + \hat{\mathbf{a}}_{ILS}^r(k)). \tag{33}$$

The equalized ILSR estimate $\hat{\mathbf{a}}_{ILS}^{av}(k)$ will be better than each individual estimate $\hat{\mathbf{a}}_{ILS}^l(k)$ or $\hat{\mathbf{a}}_{ILS}^r(k)$. This is called the ILSR2 algorithm. Because the assumption of a rough range of the power ratio κ^2 is a much more relaxed assumption than that of an exact knowledge of κ^2, the ILSR2 algorithm will be more practically attractive than its predecessors. Note that the increased computational cost associated with the ILSR2 algorithm is rather marginal when compared with the ILSR1 algorithm, since it essentially needs to perform the standard LS estimation only once.

5. SIMULATION EXAMPLE AND EVALUATION

To validate the preceding theoretical results, several performance measures are first introduced. Denote by $\hat{\mathbf{a}}_m$ an estimator of the AR parameter vector $\mathbf{a}$ in the mth test over a total of M independent tests. The relative error (RE) and the normalized root mean squared error (RMSE) of $\hat{\mathbf{a}}_m$ with respect to the true value $\mathbf{a}$ are defined as follows:

$$\text{RE} = \frac{\|\mathbf{m}(\hat{\mathbf{a}}) - \mathbf{a}\|}{\|\mathbf{a}\|}, \quad \text{RMSE} = \sqrt{\frac{1}{M}\sum_{m=1}^{M}\frac{\|\hat{\mathbf{a}}_m - \mathbf{a}\|^2}{\|\mathbf{a}\|^2}} \tag{34}$$

where $\mathbf{m}(\hat{\mathbf{a}}) = \frac{1}{M}\sum_{m=1}^{M}\hat{\mathbf{a}}_m$. While the RE and RMSE are used to assess estimation accuracy, the MATLAB code `flops` is used to measure the numerical cost involved.

In the simulated example, the underlying AR model is parametrized by

$$\mathbf{a}^\top = [1.352 \quad -1.338 \quad 0.662 \quad -0.24] \tag{35}$$

and the driving source power is $\sigma_v^2 = 1.0$. The corrupting noise $w(t)$ is with variance $\sigma_w^2 = 0.38$. This leads to a signal-to-noise (SNR) ratio of about 10dB. For this noisy AR model, the exact power ratio is given by $\kappa_0^2 = 2.6316$. The model is simulated to generate $M = 200$ independent tests of $N = 2000$ data points each, and then is identified by applying the standard LS method, the MYW method, the ML method, and the ILSD method, the ILSR algorithm [6], the ILSR1 algorithm and the ILSR2 algorithm. Table 1 lists the comparative estimation performances.

The following comments are made based on Table 1.

(a) Using the exact power ratio $\kappa_0^2 = 2.6316$, the ILSR and ILSR1 algorithms have achieved almost the same estimation accuracy. However, due to removal of calculation of the average LS errors, the ILSR1 algorithm has gained a significant saving of about 25% in the computational cost over the ILSR algorithm.

(b) The results obtained by the ILSR1 algorithm have confirmed that the corrupting noise variance σ_w^2 and the AR parameters are overestimated for the case of $\hat{\kappa}^2 = 2.3 < \kappa_0^2 = 2.6316$, and underestimated for the case of $\hat{\kappa}^2 = 3 > \kappa_0^2 = 2.6316$.

(c) Given two rough ranges (2.3, 3) and (1.8, 4.2) for the power ratio κ^2, the ILSR2 algorithm has worked well. Its estimation accuracy is virtually identical to

Table 1. Comparative Estimation Performances
(SNR ≈ 10dB, N = 2000 data points, M = 200 independent tests, NFPT = Number of flops per test)

method	$\hat{\kappa}^2$	a_1	a_2	a_3	a_4	σ_v^2	σ_w^2	RE	RMSE	NFPT
LS		0.8302	−0.5311	−0.0369	0.0416			60.20%	60.26%	56277
		±0.0211	±0.0307	±0.0297	±0.0206					
MYW		0.7661	−0.5943	0.1893	−0.0498			53.00%	661.07%	128113
		±8.3164	±8.8621	±5.2961	±1.6978					
ML		1.0117	−0.9214	0.3760	−0.1465			30.38%	61.39%	7014621
		±0.5848	±0.7289	±0.5077	±0.1991					
ILSD		1.3458	−1.3311	0.6596	−0.2401	1.0217	0.3683	0.46%	13.70%	83230
		±0.1181	±0.1825	±0.1601	±0.0648	±0.2339	±0.0887			
ILSR	2.6315	1.3473	−1.3287	0.6548	−0.2362	0.9993	0.3797	0.64%	6.59%	76302
		±0.0422	±0.0809	±0.0840	±0.0481	±0.0408	±0.0155			
ILSR1	2.6315	1.3464	−1.3271	0.6533	−0.2355	0.9986	0.3794	0.76%	6.61%	57189
		±0.0423	±0.0809	±0.0840	±0.0482	±0.0407	±0.0154			
ILSR1	2.3	1.4103	−1.4401	0.7606	−0.2862	0.9106	0.3959	7.90%	10.28%	57215
		±0.0419	±0.0807	±0.0842	±0.0487	±0.0386	±0.0168			
ILSR1	3	1.2873	−1.2241	0.5565	−0.1906	1.0842	0.3614	8.63%	10.76%	57150
		±0.0418	±0.0794	±0.0820	±0.0467	±0.0424	±0.0141			
ILSR2	(2.3, 3)	1.3488	−1.3321	0.6586	−0.2384	0.9974	0.3786	0.37%	6.50%	57813
		±0.0419	±0.0800	±0.0831	±0.0477	±0.0405	±0.0154			
ILSR2	(1.8, 4.2)	1.3429	−1.3292	0.6600	−0.2424	1.0250	0.3641	0.63%	5.92%	57821
		±0.0386	±0.0725	±0.0751	±0.0435	±0.0394	±0.0147			
true value	2.6315	1.352	−1.338	0.662	−0.24	1.0	0.38			

that of the ILSR1 algorithm with the exact power ratio $\kappa_0^2 = 2.6316$. Moreover, the increase in the numerical cost caused by carrying out twice ILSR estimation in the ILSR2 algorithm (see (33)) is negligible as it is just about 1% increase from that of the ILSR1 algorithm. Like the ILSR and ILSR1 algorithms, the ILSR2 algorithm is more accurate than the ILSD method in terms of smaller variance. Also, the ILSR2 algorithm has attained an around 30% reduction in the computational load in comparison with the ILSD method.

6. CONCLUSIONS

In this paper the ILSR1 algorithm has been introduced for noisy AR signal identification. In contrast to the previous ILSR algorithm, there is no need to compute the average LS errors in implementation of the ILSR1 algorithm, thus the numerical efficiency being significantly upgraded. The mean convergence of the ILSR1 algorithm has been established. Another contribution is that the assumption of the known power ratio has been mitigated to that of a rough range of the ratio. Based on this, the ILSR2 algorithm has been proposed, which has proved to be practically effective. However, we should note that the way of selecting the rough range for κ^2 as shown in (32) may be delicate for the ILSR2 algorithm sometimes, so a practical guideline for the selection of that rough range is needed. This issue is currently under investigation.

7. REFERENCES

[1] S. M. Kay, "Noise compensation for autoregressive spectral estimates," *IEEE Trans. Acoustics, Speech, and Signal Processing*, Vol.28, pp.292-303, 1980.

[2] S. M. Kay, *Modern Spectral Estimation*. Englewood Cliffs, NJ: Prentice-Hall, 1988.

[3] A. Nehorai and P. Stoica, "Adaptive algorithms for constrained ARMA signals in the presence of noise," *IEEE Trans. Acoustics, Speech, and Signal Processing*, Vol.36, pp.1282-1291, 1988.

[4] H. Tong, "Autoregressive model fitting with noisy data by Akaike's information criterion," *IEEE Trans. Information Theory*, Vol.21, pp.476-480, 1975.

[5] W.-R. Wu and P.-C. Chen, "Adaptive AR modeling in white Gaussian noise," *IEEE Trans. Signal Processing*, Vol.45, pp.1184-1192, 1997.

[6] W. X. Zheng, "Autoregressive parameter estimation from noisy data," *IEEE Trans. Circuits and Systems-II: Analog and Digital Signal Processing*, Vol.47, pp.71-75, 2000.

[7] W. X. Zheng, "Unbiased parameter identification for noisy autoregressive signals," in *Proc. 34th IEEE International Symposium on Circuits and Systems*, Sydney, Australia, May 2001, Vol.II, pp.121-124.

PAIRING AND ORDERING TO REDUCE HARDWARE COMPLEXITY IN CASCADE FORM FILTER DESIGN

Hyeong-Ju Kang and In-Cheol Park

Division of EE, Dept. of EECS
Korea Advanced Institute of Science and Technology
373-1, Guseong-dong, Yuseong-gu, Daejeon, Korea

ABSTRACT

This paper presents an algorithm that explores all the combinations of sub-modules in the cascade form filter to reduce hardware complexity under design constraints. Though the cascade form structure has freedom in pairing and ordering of its sub-modules, the hardware complexity is subject to the pairing and ordering if the optimization based on the multiplier block concept is applied. The proposed algorithm selects the pairing and ordering that results in the minimal hardware complexity among all the cases that satisfy the frequency response specification. To cope with the case that the objective filter has many taps and the exploration time is too long, a clustering method is also developed. Experimental results on several filters show that the proposed algorithm reduces the hardware complexity by about 10% on the average, while satisfying the filter specification.

1. INTRODUCTION

Digital filters are frequently used in digital signal processing. In applications demanding high throughput and low power, application specific filters are frequently adopted to meet the constraints of performance and power consumption. In the digital filter implementation, the direct form structure and the cascade form structure are usually used by virtue of simplicity. The direct form structure has less hardware complexity, and the cascade form structure is robust to the quantization and roundoff noise [1][2]. The hardware complexity of the digital filters can be reduced by using the concept of multiplier blocks [3]-[5]. In such approaches, all the coefficient multiplications are decomposed and considered as a whole to construct a hardware block called a multiplier block that implements all the coefficient multiplications. In a multiplier block, the adders used in one multiplication can be shared with other multiplications.

In the cascade form structure, a variety of theoretically equivalent implementations can be obtained by simply pairing the poles and zeros and ordering the sections in different ways. In the previous works, a number of factors that affect the functionality, such as coefficient quantization, roundoff noise, and scaling, are considered to determine the pairing and ordering [1] [2] [6]-[10]. However, they do not consider the hardware complexity because the pairing and ordering does not affect the complexity of filters. Although the assumption is valid for the case that the filters are implemented with multipliers, the pairing and ordering has a significant effect on the hardware complexity if the concept of multiplier block is employed.

In this paper, a pairing and ordering algorithm is suggested for the cascade form structure, which considers the hardware complexity in implementing multiplier blocks. Given a filter equation, the proposed algorithm divides it into second-order polynomials and determines a set of second-order polynomials to be implemented in a multiplier block. When there are too many second-order polynomials, the polynomials are clustered into medium-sized groups, and the optimum ordering and pairing is applied to each group.

This paper is organized as follows. In Section 2, the direct form structure and the cascade form structure are explained, and in Section 3, the concept of a multiplier block is explained. Section 4 describes the proposed algorithm in detail. Experimental results are shown in Section 5, and concluding remarks are made in Section 6.

2. FILTER REALIZATIONS

In IIR filter design, the cascade form structure is preferred to the direct form structure. In fact, the cascade form structure requires more hardware complexity than the direct form structure. The cascade form structure, however, has several advantages that compensate the hardware overhead. One of them is that it is generally much less sensitive to coefficient quantization [1]. Another advantage is that its input, x[n], and output, y[n], have less number of fan-outs, leading to a significant profit in the FIR filter system that has dozens of taps.

In implementing the cascade form structure, we have to solve two problems, pairing and ordering. The pairing deals with how to pair a second-order denominator polynomial and a second-order numerator polynomial to form a second-order section. The ordering decides the order of the second-order sections. In the previous works, many algorithms have been proposed for the problems to improve the roundoff noise [1][6]-[10]. An efficient method is presented in [1] as a rule of thumb. The second-order denominator polynomial whose poles are closer to the unit circle should be paired with the second-order numerator polynomial whose zeros are closer to those poles, and the second-order sections should be ordered such that the second-order section whose poles are closer to the unit circle is located closer (or farther) to the filter-input.

3. MULTIPLIER BLOCK

If the coefficients of a filter are constant, each constant multiplication can be decomposed into addition, subtraction, and

0-7803-7761-3/03/$17.00 ©2003 IEEE

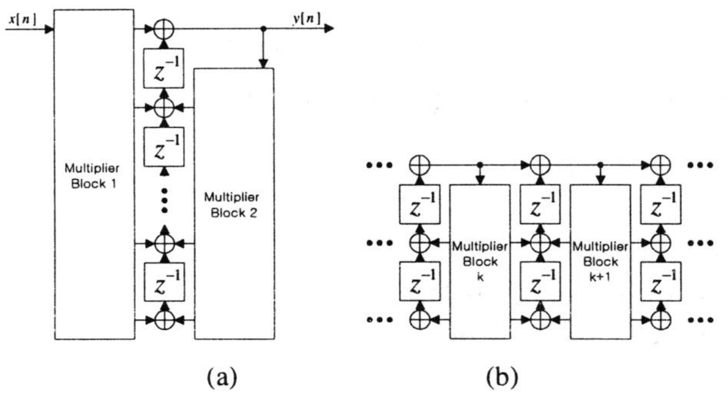

Figure 1. Multiplier blocks (a) in the transposed direct form structure and (b) in the transposed cascade form structure.

shift. The complexity of filters in this case is dominated by the number of additions/subtractions used to implement the coefficient multiplications as the number is proportional to the number of two-input adders and the shifting can be implemented by wire connections. To reduce the complexity, the coefficients can be restricted to powers-of-two or expressed in canonical signed-digit (CSD) representation. However, there is another approach in which coefficient multiplications are considered as a whole. The hardware block called a multiplier block as in Figure 1 is used to implement all coefficient multiplications. Exploiting the concept of a multiplier block enables the additions/subtractions used in one multiplication to be used in other multiplications. Therefore, it can reduce the number of additions/subtractions. Many algorithms have been proposed to make the multiplier block as simple as possible [3-5] In this paper, BHM algorithm [3] is used because it is fast and produces the minimal multiplier block among the algorithms.

4. PAIRING AND ORDERING ALGORITHM

If the concept of multiplier blocks is used, the pairing and ordering affects the hardware complexity of cascade form filters. By changing the pairing and ordering, each multiplier block has a different set of coefficients, and thus it can be implemented with different hardware complexity. In this section, we present the proposed algorithm with considering IIR filters. The algorithm, however, can be easily expanded for FIR filters by eliminating the denominator part. The proposed algorithm assumes that a filter is designed with infinite-precision coefficients and the numerator and the denominator polynomials are factorized into first- or second-order polynomials.

4.1 Clustering

If the target filter is very complex, it will take very long time to search all the possible structures. Therefore, the large problem should be divided into several clusters. There are two issues related with the clustering. The first one is how many polynomials a cluster can have. Since the optimal pairing and ordering is searched in each cluster, increasing the number of polynomials in a cluster expands the search space and leads to better results. The processing time of a cluster, however, increases exponentially according to the cluster size. Experiments show that a 3-cluster (a cluster that has three second-order numerator and denominator polynomials) takes about two hours to explore all the structures, and a 4-cluster takes more than two days. The proposed algorithm, therefore, uses 3 or less sized clusters.

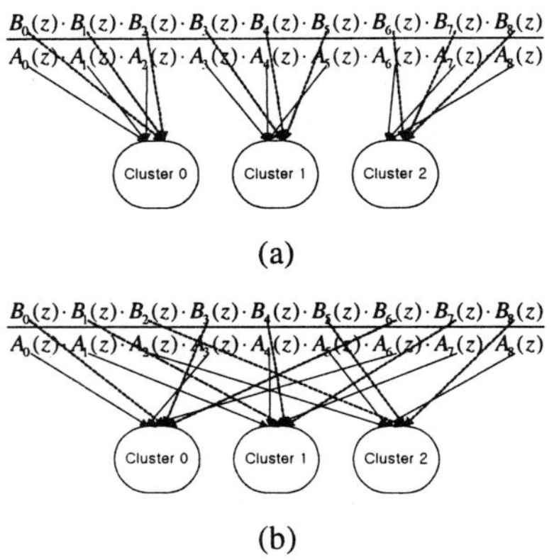

Figure 2. Clustering schemes: (a) Scheme 1 and (b) Scheme 2.

The other issue is how the polynomials are clustered. Since the ordering may degrade the frequency response severely, the clustering method must consider the resulting frequency response. In addition, it is better to put the second-order polynomials that are likely to be merged to a forth-order polynomial in a cluster because a forth-order polynomial may have less hardware complexity than two second-order polynomials. The proposed algorithm has two clustering schemes. The first is that the polynomials are ordered as in the conventional method [1] and then the first three second-order sections are grouped into a cluster, the next three sections are grouped, and so on. This clustering scheme is illustrated in Figure 2(a). This method guarantees that the final frequency response is better than that obtained by the conventional ordering. The other scheme is based on the fact that a polynomial with distant roots suffers less frequency response degradation after quantization. In the conventional ordering, the second-order sections are ordered in order of increasing closeness of the poles to the unit circle or in order of decreasing closeness to the unit circle [1]. The second-order polynomials whose roots are close to each other, therefore, may neighbor with each other. Assigning those polynomials to different clusters can increase the possibility of forth-order polynomials. If N polynomials in the conventional order are to be assigned to M clusters, the first M polynomials are assigned to M clusters one by one, and the next M polynomials are assigned one by one, and so on, as shown in Figure 2(b). As it is not clear which method is better, experimental results will be compared later.

4.2 Scaling and Coefficient Rounding

Many works have treated how to scale each section in filters. The proposed algorithm is based on the scaling method used in [6]. The proposed algorithm uses different rounding schemes according to the order of polynomials. Each coefficient of the second-order polynomials is rounded to the nearest integer. For the forth-order polynomials, the algorithm selects optimal integer values from a small search space. After rounding each coefficient to the nearest integer, it compares the frequency responses of the

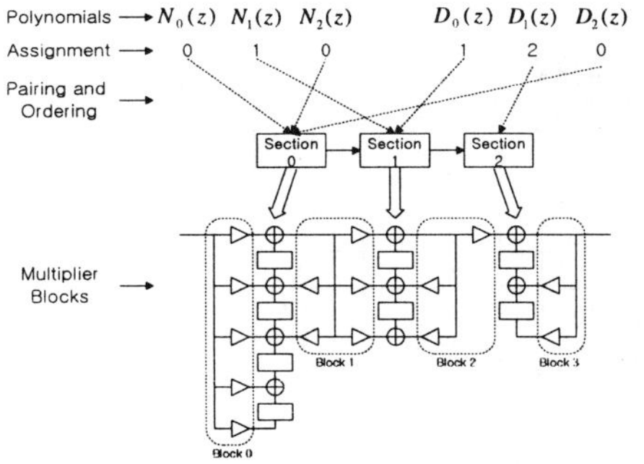

Figure 3. Pairing and ordering.

polynomials obtained by changing the integer values by 0~2. Then it selects the polynomial that has the smallest deviation from the frequency response of the original infinite-precision polynomial. It is not allowed to use more-than-forth-order polynomials for denominators because they degrade the frequency response severely. If higher-order polynomials are used for numerators, each coefficient is rounded to the nearest integer.

4.3 Pairing and Ordering

In a cluster, there are many choices of pairing and ordering. The original pairing problem means only how to pair a denominator polynomial and a numerator polynomial. In this paper, however, the pairing problem also includes how to combine denominator polynomials or numerator polynomials into higher-order polynomials. Combining lower-order polynomials into a higher-order one may reduce the hardware complexity but may deteriorate the frequency response. It is also important to search the best ordering of denominator polynomials and numerator polynomials. The ordering has a significant effect on the hardware complexity and the frequency response.

The cluster located closer to the filter input is searched earlier. Initially, a cluster has as many sections as its second-order denominator polynomials, and each section has its own number. One of the numbers is assigned to each polynomial, as shown in Figure 3. Polynomials are paired and ordered according to those section numbers. The polynomials assigned to the same section are merged to a higher-order polynomial. The proposed algorithm explores all the possible assignments. It can be done in reasonable time since the filter is clustered considering the execution time. It is prohibited to make a more-than-forth-order denominator polynomial because the polynomial degrades the frequency response severely. For each assignment, multiplier blocks are constructed and optimized. This procedure is illustrated in Figure 3, where 3 numerator polynomials and 3 denominator polynomials are paired and ordered to construct a structure. As an example, section number 0, 1, and 0 are assigned to numerator polynomials and section number 1, 2, and 0 are assigned to denominator polynomials. Then the first and third numerator polynomials are inserted into Section 0, the second numerator polynomial into Section 1, the first denominator polynomial into Section 0, and so on. The first and third numerator polynomials are merged into a forth-order polynomial. From this structure, 4 multiplier blocks are constructed.

For each pairing and ordering case, the entire frequency response is estimated and checked whether it satisfies the filter specifications. Since the pairings and orderings of the earlier clusters are already determined, their frequency responses can be obtained. The pairings and orderings of the later clusters, however, are not determined, and thus their frequency responses should be estimated. The proposed algorithm estimates the responses with assuming that the pairings and orderings of those clusters are subject to the conventional method [1].

In the proposed algorithm, the frequency response must satisfy three specifications: passband ripple, stopband ripple, and roundoff noise variance. If not, the pairing and ordering is rejected. The algorithm compares the areas of the structures that satisfy the specifications and selects the pairing and ordering that produces the minimal area.

5. EXPERIMENTAL RESULTS

The proposed algorithm has been applied to a set of IIR filters. The specifications of the filters are shown in Table Ⅰ, where *N* is the number of taps, *Rp* and *Rs* are the ripples in the passband and stopband, and *Wn* is the normalized cut-off frequency. All the IIR filters are elliptic filters that are designed with infinite-precision coefficients by using MATLAB. The proposed algorithm is performed assuming that the coefficients are to be quantized to 11-bit fixed-point values.

The results are summarized in Table Ⅱ. The column of *Cascade* shows the results obtained by using the cascade form structure, and *Direct* is obtained by using the direct form structure. The last two columns are for the proposed algorithm, where *Proposed1* shows the results obtained with the first clustering scheme, and *Proposed2* shows the results with the second clustering scheme. In Table Ⅱ, the resulting ripples in the passband and stopband are denoted as *Rp* and *Rs* , respectively, and the roundoff noise normalized with respect to the rounding noise variance $\sigma^2 = 2^{-2B}/12$ is *Ri*. The resulting area is estimated as a weighted sum of the number of adders and the number of registers. Given a filter specification, the proposed algorithm selects the minimal-area structure under the condition that its passband ripple, stopband ripple, and roundoff noise can be changed by 0.1dB, 2dB, and 6dB, respectively, with respect to those of the initial cascade form structure.

The direct form structure gives the minimum hardware complexity. As there are more coefficients in a section of the direct form structure, the multiplier block synthesis algorithm can make more adders be shared. However, the direct form structure is not usually used in IIR filter design because it is heavily affected by the coefficient quantization and roundoff noise. Coefficient rounding degrades the frequency response of the direct form structure so severely that the response is far from the original filter specification. The empty columns in Table Ⅱ denote that the frequency response is beyond the specification. On the contrary, the proposed algorithm generates filters whose passband and stopband ripples and roundoff noises are similar to those of the floating-point or the cascade form filters. It means that the proposed algorithm generates feasible filters, while reducing the area by 10%. Compared to the direct form structure that fails to meet the frequency specification, the proposed

Table I Filter Specifications

	N	Rp(dB)	Rs(dB)	Wn		N	Rp(dB)	Rs(dB)	Wn
IIR1	6	1.00	60.0	0.05	IIR7	10	0.67	80.0	0.05
IIR2	6	1.00	60.0	0.10	IIR8	10	0.67	80.0	0.10
IIR3	6	1.00	60.0	0.15	IIR9	10	0.67	80.0	0.15
IIR4	8	0.83	60.0	0.05	IIR10	12	0.50	80.0	0.05
IIR5	8	0.83	60.0	0.10	IIR11	12	0.50	80.0	0.10
IIR6	8	0.83	60.0	0.15	IIR12	12	0.50	80.0	0.15

Table II Results for IIR Filters

	Cascade				Direct				Proposed1				Proposed2			
	Rp	Rs	Ri	Area	Rp	Rs	Ri	Area	Rp	Rs	Ri	Area	Rp	Rs	Ri	Area
IIR1	1.75	60.0	37.4	346	3.26	13.6	64.1	259	1.76	60.1	39.1	304	1.76	60.1	39.1	304
IIR2	1.04	59.9	26.2	297	6.20	54.6	56.0	280	1.06	59.8	26.3	269	1.06	59.8	26.3	269
IIR3	1.03	60.0	20.6	318	1.65	59.5	39.6	238	1.06	60.0	26.5	288	1.06	60.0	26.5	288
IIR4	1.61	60.2	42.9	424				381	1.39	60.1	41.1	403	1.42	60.1	42.8	389
IIR5	0.91	60.0	30.1	431				367	0.91	60.0	28.8	389	0.96	59.8	30.3	382
IIR6	0.93	59.8	23.7	389	4.26	57.4	60.2	353	0.95	59.9	29.6	366	0.88	59.6	27.7	356
IIR7	1.19	78.3	43.8	523				353	1.16	80.2	41.8	495	1.09	78.5	37.7	490
IIR8	0.86	79.7	32.0	537				391	0.94	79.9	29.5	494	0.94	78.9	28.9	460
IIR9	0.84	80.0	26.0	488				384	0.79	79.9	31.7	458	0.82	79.0	31.9	413
IIR10	2.93	81.5	47.3	601				520	2.98	81.4	47.2	580	2.21	80.2	45.6	559
IIR11	0.76	78.6	34.3	608				527	0.84	78.9	31.6	573	0.76	79.2	31.6	552
IIR12	0.83	80.0	27.9	608				478	0.72	80.0	32.1	536	0.68	79.9	30.6	526
Avg.	-	-	-	100%	-	-	-	81.7%	-	-	-	92.3%	-	-	-	89.6%

algorithm enables a trade-off between area and frequency response.

6. CONCLUSION

In this paper, we have proposed a new pairing and ordering algorithm to determine cascade form filter structures. In the previous works, the pairing and ordering problem is considered to improve the frequency response. However, the proposed algorithm explores the pairing and ordering to select a structure that has the least hardware complexity and satisfies the frequency-response specification. The target filter is clustered to reduce the execution time, and the proposed algorithm takes into account higher-order polynomials to broaden the design space. The experimental results show that the proposed algorithm reduces the hardware complexity by 10% on the average, while achieving almost the same frequency response. In other words, the proposed algorithm gives a structure that is closer to the optimal structure.

7. ACKNOWLEDGMENT

This work was supported by the Korea Science and Engineering Foundation through the MICROS center, by the Ministry of Science and Technology and the Ministry of Commerce, Industry, and Energy through the project System IC 2010, and by IC Design Education Center (IDEC).

8. REFERENCES

[1] A. V. Oppenheim and R. W. Schafer, *Discrete-Time Signal Processing*, Prentice Hall, 1989.

[2] L. S. DeBrunner, V. DeBrunner, and P. Pinault, "Variable wordlength IIR filter implementations for reduced space designs," in *Proc. IEEE Workshop on Signal Processing Systems*, 2000, pp. 326-335.

[3] A. G. Dempster and M. D. Macleod, "Use of minimum adder multiplier blocks in FIR digital filters," *IEEE Trans. Circuits Syst. II*, vol. 42, pp. 569-577, Sept. 1995.

[4] A. G. Dempster and M. D. Macleod, "IIR digital filter design using minimum adder multiplier blocks," *IEEE Trans. Circuits Syst. II*, vol. 45, pp. 761-763, June 1998.

[5] R. I. Hartley, "Subexpression sharing in filters using canonic signed digit multipliers," *IEEE Trans. Circuits Syst. II*, vol. 43, pp. 677-688, Oct. 1996.

[6] L. B. Jackson, "On the interaction of roundoff noise and dynamic range in digital filters," *Bell Sys. Tech. J.*, vol. 49, no. 2, Feb. 1970.

[7] L. B. Jackson, "Roundoff-Noise Analysis for Fixed-Point Digital Filters Realized in Cascade or Parallel Form," *IEEE Trans. Audio Electroacoust.*, vol. AU-18, pp. 107-122, June 1970.

[8] S. Y. Hwang, "On Optimization of Cascade Fixed-Point Digital Filters," *IEEE Trans. Circuits Syst.*, vol. 23, pp. 163-166, Jan. 1974.

[9] B. Liu and A. Peled, "Heuristic Optimization of the Cascade Realization of Fixed-Point Digital Filters," *IEEE Trans. Acoust., Speech, Signal Processing*, vol. ASSP-23, pp. 464-473, Oct. 1975.

[10] G. Dehner, "On the noise behaviour of a digital filter in cascade structure," in *Proc. Int. Symp. Circuits Syst.*, 1976, pp. 348-351.

Direct Recursive Structures for Computing Radix-r Two-Dimensional DCT

Che-Hong Chen, Bin-Da Liu, and Jar-Ferr Yang
Department of Electrical Engineering
National Cheng Kung University
Tainan, Taiwan 70101, R.O.C.
Tel: +886-6-2762330, Fax: +886-6-2345482
Email: bdliu@spic.ee.ncku.edu.tw

ABSTRACT

In this paper, new recursive structures for computing radix-*r* two-dimensional discrete cosine transform (2-D DCT) are proposed. Based on the same indices of transform bases, the regular pre-add preprocess is established and the recursive structures for 2-D DCT, which can be realized in a second-order infinite-impulse response (IIR) filter, are derived without involving any transposition procedure. For computation of 2-D DCT, the recursive loops of the proposed structures are less than that of one-dimensional DCT recursive structures, which need data transposition to achieve the so-called row-column approach. With advantages of fewer recursive loops and no transposition, the proposed recursive structures achieve more accurate results than the existed methods. The regular and modular properties are suitable for VLSI implementation.

1. INTRODUCTION

The DCT is widely used in areas of digital signal processing and signal compression. It, in particular, is considered as the most effective scheme for transform coding. Hence, the 2-D DCT are adopted by many image/video coding compression standards for various applications.

In past decades, one-dimensional (1-D) recursive transform algorithms were developed for simple VLSI implementation [1]-[8]. Goertzel initially used the periodicity of the finite trigonometric sequence to reduce the computation of discrete Fourier transform [1]-[2]. His recursive structure not only saves the computation but also simplifies the realization complexity. The recursive algorithms in [3]-[8] for computing 1-D DCT and 1-D IDCT are highly regular and modular. A recursive DCT architecture, which requires less chip area and power consumption, was proposed in [9]. For computing 2-D DCT by row-column approaches, the RAM is usually adopted as the transposition memory. However, the large RAM, which has disadvantages of larger power consumption and longer access time, is not applicable to mobile and portable information appliances. In hardware implementation, an even more complicated transposition is required for the parallel 1-D DCT recursive architecture.

In this paper, new 2-D DCT recursive structures with fast and regular recursion are proposed to achieve fewer recursive cycles without using any transposition memory. The 2-D recursive DCT algorithms are developed from the concept that data with the same transform base can be pre-added such that we can reduce the recursive cycles. Based on the use of Chebyshev polynomial, the efficient transform kernels can be obtained. By further folding the inputs of transform kernels, the fewer recursive cycles are achieved.

This work was in part supported by the National Science Council, Republic of China, under grant NSC 91-2218-E-006-006 and NSC 91-2218-E-006-004

2. CONDENSED 1-D TRANSFORM FOR 2-D DCT

Neglecting the scale factor for convenience, the 2-D DCT of $X(k_1,k_2)$ is given by

$$X(k_1,k_2) = \sum_{n_1=0}^{N-1}\sum_{n_2=0}^{N-1} x(n_1,n_2)\cos\frac{2\pi(2n_1+1)k_1}{4N}\cos\frac{2\pi(2n_2+1)k_2}{4N} \quad (1)$$

for $0 \le k_1,k_2 \le N-1$. By using triangular equalities, Equation (1) can be rewritten as

$$X(k_1,k_2) = \frac{1}{2}\sum_{n_1=0}^{N-1}\sum_{n_2=0}^{N-1} x(n_1,n_2)\cdot\Big\{\cos\frac{\pi(n_1k_1+n_2k_2)}{N}\cos\frac{\pi(k_1+k_2)}{2N}$$

$$-\sin\frac{\pi(n_1k_1+n_2k_2)}{N}\sin\frac{\pi(k_1+k_2)}{2N}+\cos\frac{\pi(n_1k_1-n_2k_2)}{N}\cos\frac{\pi(k_1-k_2)}{2N}$$

To develop the 2-D DCT recursive structures, we first assume that $N=r^P$, where r is a prime integer. For convenience, a set $D = \{0, 1, \hat{D}\}$ is defined, where $\hat{D}$ is a subset, whose elements d_i is relatively prime to r and less than N. It is noted that any integer k could be expressed as $d \cdot r^p$ where d is the element of the set D for $0 \le k < N$ and $0 \le p \le P$, where 0 is defined as $0 \cdot r^P$. Finally, to compute $X(k_1,k_2)$, the 2-D DCT can be divided into two cases according to the relationship of k_1, k_2 and N.

Case 1: $k_1 = d_1r^{p_1}$, $k_2 = d_2r^{p_2}$ where $p_1 = 0$ or $p_2 = 0$. (k_1 or k_2 is prime to *N, and* $p = \min(p_1, p_2)$).

In this case, (2) can further be rewritten as an *M*-point transform in the following formula by replacing k_1, k_2 and N as w_1, w_2 and M.

$$X(w_1,w_2) = \frac{1}{2}\sum_{n_1=0}^{M-1}\sum_{n_2=0}^{M-1} x(n_1,n_2)$$

$$\cdot\Big\{(-1)^{s_1}\cos\frac{\pi[(w_1n_1+w_2n_2)\bmod M]}{M}\cos\frac{\pi(w_1+w_2)}{2M}$$

$$-(-1)^{s_1}\sin\frac{\pi[(w_1n_1+w_2n_2)\bmod M]}{M}\sin\frac{\pi(w_1+w_2)}{2M}$$

$$+(-1)^{s_2}\cos\frac{\pi[(w_1n_1-w_2n_2+2M^2)\bmod M]}{M}\cos\frac{\pi(w_1-w_2)}{2M}$$

$$-(-1)^{s_2}\sin\frac{\pi[(w_1n_1-w_2n_2+2M^2)\bmod M]}{M}\sin\frac{\pi(w_1-w_2)}{2M}\Bigg\} \tag{3}$$

where $s_1=\lceil(w_1n_1+w_2n_2)/M\rceil$ and $s_2=\lceil(w_1n_2-w_1n_2+2MM)/M\rceil$. In this paper, $\lceil\cdot\rceil$ and mod denote the rounding and modulo operators, respectively.

By using number theory and index exchange of transform bases, Equation (3) can be further rewritten as a simpler form. A set of U integers $a_1, a_2, \ldots, a_U$ is a complete residue system modulo U if and only if $a_i \neq a_j (\bmod U)$ for $i \neq j$. Let a and U be relatively prime, b be an integer, and $a_1, a_2, \ldots, a_k$ be a complete residue system in modulo U, then, $\{aa_1+b,\ aa_2+b,\ \ldots,\ aa_k+b\}$ is also a complete residue system in modulo U. If w_1 is relatively prime to M ($r_1=0$), let $a=w_1$, $b=w_2n_2$ and $U=M$, then $(w_1n_1+w_2n_2)$ mod M and $(w_1n_1-w_2n_2+2MM)$ mod M make the complete residue system in modulo M, which performs $\{0,1,2,\ldots,M-1\}$ for $n_1=0,1,2,\ldots,M-1$; otherwise, if w_2 is relatively prime to M ($p_2=0$), let $a=w_2$, $b=w_1n_1$ and $M=U$, then $(w_1n_1+w_2n_2)$ mod M and $(w_1n_1-w_2n_2+2MM)$ mod M also make the complete residue system in modulo M, which performs $\{0,1,2,\ldots,M-1\}$ for $n_2=0,1,2,\ldots,M-1$. Thus, the index of the transform bases in (3) can be replaced as

$$m=w_1n_1+w_2n_2 \bmod M \tag{4a}$$

and

$$m=w_1n_1-w_2n_2+2M^2 \bmod M \tag{4b}$$

for $m=0,1,2,\ldots,M-1$. According to the new transform index, the input $x(n_1,n_2)$s with the same transform base are added and the corresponding inputs $x_a(m)$ and $x_s(m)$ are obtained as follows.

$$x_a(m)=\begin{cases} p_1=0: & \sum_{n_2=0}^{M-1}(-1)^{s_1}x(n_1,n_2) \\ p_2=0: & \sum_{n_1=0}^{M-1}(-1)^{s_1}x(n_1,n_2) \end{cases} \tag{5}$$

where n_1 and n_2 satisfy $m=w_1n_1+w_2n_2$ mod M for $n_1=0,1,\ldots,M-1$ and $n_2=0,1,\ldots,M-1$.

$$x_s(m)=\begin{cases} p_1=0: & \sum_{n_2=0}^{M-1}(-1)^{s_2}x(n_1,n_2) \\ p_2=0: & \sum_{n_1=0}^{M-1}(-1)^{s_2}x(n_1,n_2) \end{cases} \tag{6}$$

where n_1 and n_2 satisfy $m=w_1n_1-w_2n_2+2MM$ mod M for $n_1=0,1,\ldots,M-1$ and $n_2=0,1,\ldots,M-1$. By replacing $w_1n_1+w_2n_2$ mod M and $w_1n_1-w_2n_2+2MM$ mod M with m, Equation (3) can be rewritten as the summation of four condensed 1-D DCTs and 1-D DSTs multiplied by multiplication factors $\cos\pi(w_1\pm w_2)/2M$ and $\sin\pi(w_1\pm w_2)/2M$.

$$X(w_1r^p,w_2r^p)=\frac{1}{2}\Bigg\{\sum_{m=0}^{M-1}x_a(m)$$

$$\cdot\left[\cos\frac{m\pi}{M}\cos\frac{\pi(w_1+w_2)}{2M}-\sin\frac{m\pi}{M}\sin\frac{\pi(w_1+w_2)}{2M}\right]$$

$$+\sum_{m=0}^{M-1}x_s(m)\left[\cos\frac{m\pi}{M}\cos\frac{\pi(w_1-w_2)}{2M}-\sin\frac{m\pi}{M}\sin\frac{\pi(w_1-w_2)}{2M}\right]\Bigg\} \tag{7}$$

Case 2: $k_1=d_1r^{p_1}$, $k_2=d_2r^{p_2}$ where $p_1\neq 0$ and $p_2\neq 0$. (k_1, k_2 and N have the common divisor r^p, where $p=\min\{p_1,\ p_2\}$)

In this case, the size of transform in (3) can be further reduced by the scaling factor, r^p. The scaled concept is discussed as follows. If $ca\equiv cb(\bmod m)$ and c is a divisor of m, then $a\equiv b(\bmod m/c)$. When $c=r^p$ and $m=N$, we can replace $(k_1n_1\pm k_2n_2)$ mod N with $(k_1r^{-p}n_1\pm k_2r^{-p}n_2)$ mod Nr^{-p}. Thus, Equation (3) can be rewritten as

$$X(k_1,k_2)=\frac{1}{2}\sum_{n_1=0}^{Nr^{-p}-1}\sum_{n_2=0}^{Nr^{-p}-1}x_p(n_1,n_2)$$

$$\cdot\Bigg\{(-1)^{s_1}\cos\frac{\pi[(k_1r^{-p}n_1+k_2r^{-p}n_2)\bmod Nr^{-p}]}{Nr^{-p}}\cos\frac{\pi(k_1+k_2)}{2N}$$

$$-(-1)^{s_1}\sin\frac{\pi[(k_1r^{-p}n_1+k_2r^{-p}n_2)\bmod Nr^{-p}]}{Nr^{-p}}\sin\frac{\pi(k_1+k_2)}{2N}$$

$$+(-1)^{s_2}\cos\frac{\pi[(k_1r^{-p}n_1-k_2r^{-p}n_2+2N^2r^{-2p})\bmod Nr^{-p}]}{Nr^{-p}}$$

$$\cdot\cos\frac{\pi(k_1-k_2)}{2N}$$

$$-(-1)^{s_2}\sin\frac{\pi[(k_1r^{-p}n_1-k_2r^{-p}n_2+2N^2r^{-2p})\bmod Nr^{-p}]}{Nr^{-p}}$$

$$\cdot\sin\frac{\pi(k_1-k_2)}{2N}\Bigg\} \tag{8}$$

where

$$x_p(n_1,n_2)=\sum_{i_1=0}^{r^p-1}\sum_{i_2=0}^{r^p-1}(-1)^{s_t}x(n_1+i_1Nr^{-p},n_2+i_2Nr^{-p}) \tag{9}$$

and $s_t=i_1k_1r^{-p}+i_2k_2r^{-p}$.

The scaled-down $Nr^{-p}\times Nr^{-p}$ DCT is identical to Equation (3) and can be computed as (7) when $k_1r^{-p}=w_1$, $k_2r^{-p}=w_2$ and $Nr^{-p}=M$. Therefore, according to Equation (9), $x_a(m)$ and $x_s(m)$ are rewritten as follows for Case 2.

$$x_a(m)=\begin{cases} p_1=p: & \sum_{n_2=0}^{M-1}\sum_{i_1=0}^{r^p-1}\sum_{i_2=0}^{r^p-1}(-1)^{s_a}x(n_1+i_1M,n_2+i_2M) \\ p_2=p: & \sum_{n_1=0}^{M-1}\sum_{i_1=0}^{r^p-1}\sum_{i_2=0}^{r^p-1}(-1)^{s_a}x(n_1+i_1M,n_2+i_2M) \end{cases} \tag{10}$$

where n_1 and n_2 satisfy $m=w_1n_1+w_2n_2$ mod M for $n_1=0,1,\ldots,M-1$ and $n_2=0,1,\ldots,M-1$, and $s_a=s_1+s_t=\lceil(w_1r^pn_1+w_2r^pn_2)/M\rceil+i_1w_1+i_2w_2$.

$$x_s(m)=\begin{cases} p_1=p: & \sum_{n_2=0}^{M-1}\sum_{i_1=0}^{r^p-1}\sum_{i_2=0}^{r^p-1}(-1)^{s_s}x(n_1+i_1M,n_2+i_2M) \\ p_2=p: & \sum_{n_1=0}^{M-1}\sum_{i_1=0}^{r^p-1}\sum_{i_2=0}^{r^p-1}(-1)^{s_s}x(n_1+i_1M,n_2+i_2M) \end{cases} \tag{11}$$

where n_1 and n_2 satisfy $m=w_1n_1-w_2n_2+2MM$ mod M for $n_1=0,1,\ldots,M-1$ and $n_2=0,1,\ldots,M-1$, and $s_s=s_2+s_t=\lceil(w_1r^pn_2-w_1r^pn_2+2MM)/M\rceil+i_1w_1+i_2w_2$.

3. RECURSIVE STRUCTURE OF 2-D DCT

From Section II, it is obvious that the corresponding recursive kernels for condensed 1-D DCTs and 1-D DSTs in (7) are needed. Furthermore, by utilizing the input folding technique, the efficient recursive structures are obtained.

For simplicity, Equation (7) is rewritten as follows.

$$X(w_1 r^p, w_2 r^p) = \frac{1}{2}\left\{X_{ac}(w_1,w_2)\cos\frac{\pi(w_1+w_2)}{2M} - X_{as}(w_1,w_2)\sin\frac{\pi(w_1+w_2)}{2M}\right.$$

$$\left. + X_{sc}(w_1,w_2)\cos\frac{\pi(w_1-w_2)}{2M} - X_{ss}(w_1,w_2)\sin\frac{\pi(w_1-w_2)}{2M}\right\} \tag{12}$$

where

$$X_{ac}(w_1,w_2) = \sum_{m=0}^{M-1} x_a(m)\cos\frac{\pi m}{M} \tag{13}$$

$$X_{as}(w_1,w_2) = \sum_{m=0}^{M-1} x_a(m)\sin\frac{\pi m}{M} \tag{14}$$

$$X_{sc}(w_1,w_2) = \sum_{m=0}^{M-1} x_s(m)\cos\frac{\pi m}{M} \tag{15}$$

$$X_{ss}(w_1,w_2) = \sum_{m=0}^{M-1} x_s(m)\sin\frac{\pi m}{M}. \tag{16}$$

If $X(w_1 r^p, w_2 r^p)$ satisfies Case 1, $x_a(m)$ and $x_s(m)$ are defined in (5) and (6) respectively; otherwise, $x_a(m)$ and $x_s(m)$ satisfying Case 2 are defined in (10) and (11). The addition of inputs $x(n_1,n_2)$ for obtaining $x_a(m)$ and $x_s(m)$ is the pre-add procedure of the proposed structure. By folding the input, only half of the summation terms are required to compute the 1-D DCT. If $r \neq 2$, the input-folding formula for (13) and (15) is expressed as

$$X_{xc}(w_1,w_2) = \sum_{m=0}^{(M+1)/2-1} x_f(m)\cos\frac{\pi m}{M} \tag{17}$$

where

$$x_f(m) = \begin{cases} x_x(0), & m=0. \\ x_x(m) - x_x(M-m), & m=1,2,\ldots,(M+1)/2-1. \end{cases} \tag{18}$$

Then, $X_{ac}(w_1,w_2)$ and $X_{sc}(w_1,w_2)$ are obtained when $x_x(m)$ in (18) is replaced by $x_a(m)$ and $x_s(m)$, respectively. The recursive formula for (17) can be expressed as

$$Y_j(w_1,w_2) = \sin\frac{\theta}{2}\left(x_f(j) - x_f(j-1)\right)$$

$$+ 2\cos\theta Y_{j-1}(w_1,w_2) - Y_{j-2}(w_1,w_2) \tag{19}$$

where $\theta = \pi/M$. The recursive kernel for condensed input-folding 1-D DCT, which combines the multiplication factor $\cos\pi(w_1 \pm w_2)/2M$, is shown in Fig. 1. After $(M+1)/2-1$ recursive cycles, the output $X_{xc}(w_1,w_2) = Y_{(M+1)/2-1}(w_1,w_2)$ is obtained.

The input-folding formula for (14) and (16) is expressed as

$$X_{xs}(w_1,w_2) = \sum_{m=0}^{(M+1)/2-1} x_f(m)\sin\frac{\pi m}{M}, \tag{20}$$

where

$$x_f(m) = \begin{cases} 0, & m=0. \\ x_x(m) + x_x(M-m), & m=1,2,\ldots,(M+1)/2-1. \end{cases} \tag{21}$$

Then, $X_{as}(w_1,w_2)$ and $X_{ss}(w_1,w_2)$ are obtained when $x_x(m)$ is replaced by $x_a(m)$ and $x_s(m)$, respectively. The recursive formula for (20) is written as

$$Y_j(w_1,w_2) = \cos\frac{\theta}{2}\left(x_f(j) - x_f(j-1)\right)$$

$$+ 2\cos\theta Y_{j-1}(w_1,w_2) - Y_{j-2}(w_1,w_2) \tag{22}$$

where $\theta = \pi/M$. The recursive kernel for condensed input-folding 1-D DST, which combines the multiplication factor $\sin\pi(w_1 \pm w_2)/2M$, is shown in Fig. 2. After $(M+1)/2-1$ recursive cycles, the output $X_{xs}(w_1,w_2) = Y_{(M+1)/2-1}(w_1,w_2)$ is obtained.

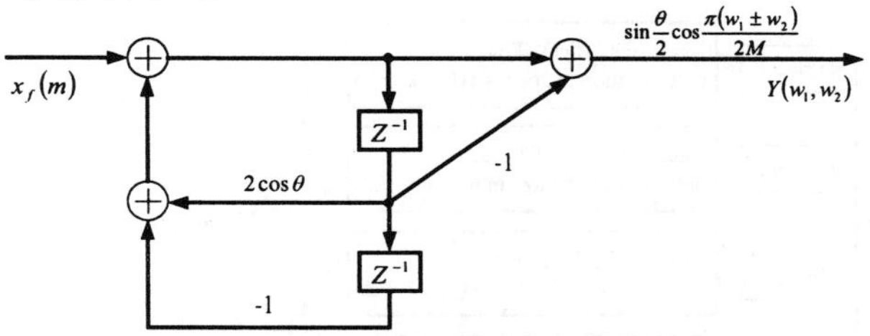

Figure 1. Recursive kernel for condensed input-folding 1-D DCT when $r \neq 2$.

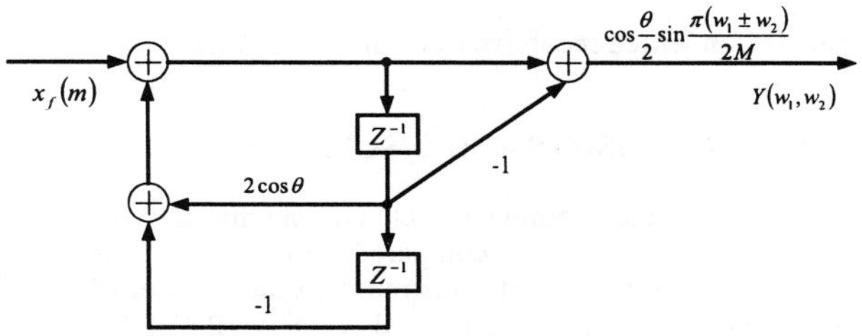

Figure 2. Recursive kernel for condensed input-folding 1-D DST when $r \neq 2$.

If $r = 2$, the input-folding formula for (13)-(16) is expressed as

$$X(w_1,w_2) = \sum_{m=0}^{M/2-1} x_f(m)\cos\frac{\pi m}{M}. \tag{23}$$

$X_{ac}(w_1,w_2)$ and $X_{sc}(w_1,w_2)$ are obtained when

$$x_f(m) = \begin{cases} x_x(0), & m=0, \\ x_x(m) - x_x(M-m), & m=1,2,\ldots,M/2-1, \end{cases}$$

and $x_x(m)$ is replaced by $x_a(m)$ and $x_s(m)$, respectively. $X_{as}(w_1,w_2)$ and $X_{ss}(w_1,w_2)$ are obtained when

$$x_f(m) = \begin{cases} x_x(M/2), & m=0, \\ x_x(M/2-m) + x_x(M/2+m), & m=1,2,\ldots,M/2-1, \end{cases}$$

and $x_x(m)$ is replaced by $x_a(m)$ and $x_s(m)$, respectively. The recursive formula for (23) is written as

$$Y_j(w_1,w_2) = \sin\theta x_f(j) + 2\cos\theta Y_{j-1}(w_1,w_2) - Y_{j-2}(w_1,w_2) \tag{24}$$

where $\theta = \pi/M$. The recursive kernel for condensed input-folding 1-D DCT, which combines the multiplication factor $\cos\pi(w_1 \pm w_2)/2M$, is shown in Fig. 3. Replacing the multiplication factor with $\sin\pi(w_1 \pm w_2)/2M$ in Fig. 3, the recursive kernel for condensed input-folding 1-D DST is also obtained. Finally, the condensed recursive structure for 2-D

DCT is shown in Fig. 4. After $M/2-1$ recursive cycles, the output $X_{xs}(w_1,w_2)=Y_{M/2-1}(w_1,w_2)$ is achieved.

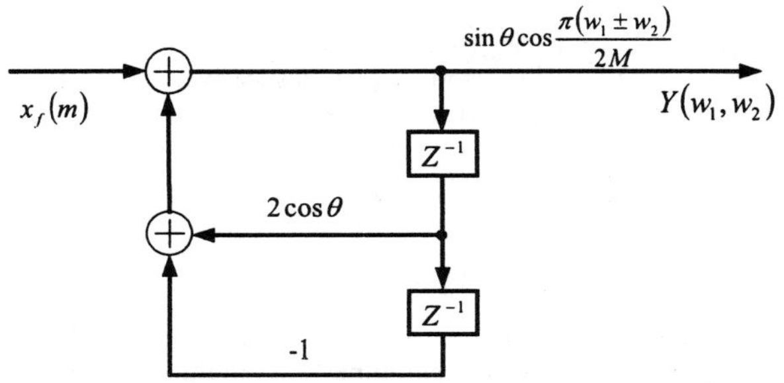

Figure 3. Recursive kernel for condensed input-folding 1-D DCT when $r=2$.

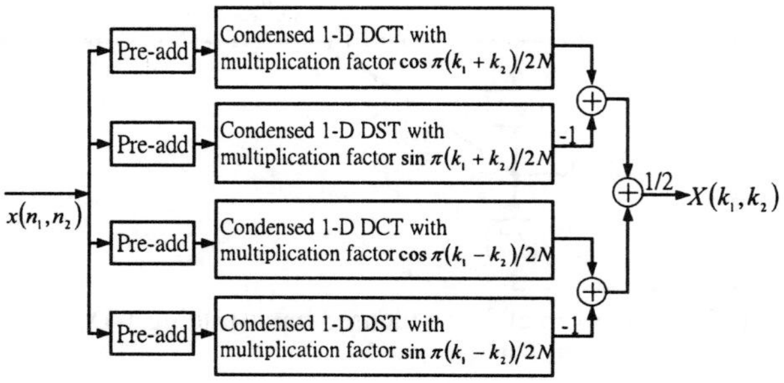

Figure 4. Condensed recursive structure for 2-D DCT.

4. COMPARISONS AND DISCUSSION

In this section, the computational complexity of the proposed recursive structures is compared with those of the existing ones. From Tables I, it is apparent that the number of recursive cycles required in the proposed recursive 2-D DCT are fewer than those required in the existing methods. The significant advantage of the proposed recursive 2-D DCT is that there is no transposition memory requirement in this scheme, and therefore, access time is not required. For the row-column method, the hardware complexity of RAM is $O(N^2)$. As transformation size grows up, the size of RAM will increase dramatically which will be the critical factor for VLSI implementation. On the other hand, due to the fewer recursive cycles, there is a lower round-off error during transformation, producing more accurate results. In addition, the efficient computation without transposition memory requirement is of great importance for portable application because of less power consumption. In this scheme, the desired DCT coefficients can be obtained directly instead of completing the $N\times N$ DCT coefficients. Finally, the proposed recursive derivation for 2-D DCT can also be applied to obtain the recursive structures of the 2-D IDCT, DST, IDST, DFT and IDFT. The regular and modular properties of the proposed recursive architectures are suitable for VLSI implementation.

5. CONCLUSION

In this paper, the condensed recursive structures for 2-D DCT have been proposed. By simple pre-process, the 2-D transformations can be decomposed into four 1-D DCTs and 1-D DSTs. Using Chebyshev polynomials, 1-D DCTs and 1-D DSTs have been derived and realized as condensed recursive kernels. Based on the pre-add procedure of input data with the same transform base, the proposed recursive computation needs fewer recursive loops than other algorithms. The most significant feature of the proposed algorithm is that the coefficients of transformation can be computed directly without transposition memory. In addition, the regular and modular structures are suitable for parallel VLSI implementation.

Table I. Number of recursive cycles and size of transposition memory for *N*x*N* DCT recursive structures.

*N*x*N*	Row-column method with transposition memory			Proposed algorithm
	[4]	[7]	[8]	
8x8	1024	800	256	220
16x16	8192	5952	2048	1756
32x32	65536	45696	16384	14044
64x64	524288	357632	131073	112348
128x128	4194304	2828800	1048576	898780
Size of transposition memory	$O(N^2)$	$O(N^2)$	$O(N^2)$	0

6. REFERENCES

[1] A. V. Oppenheim and R. W. Schafer, *Discrete-Time Signal Processing*. Englewood Cliffs, NJ: Prentice-Hall, 1989.

[2] G. Goertzel, " An algorithm for the evaluation of finite trigonometric series," *Amer. Math. Monthly*, vol. 65, pp. 34-35, Jan. 1958.

[3] L. P. Chau and W. C. Siu, "Recursive algorithm for the discrete cosine transform with general length," *Electron. Lett.*, vol 30, no. 3, pp. 197-198, Feb. 1994.

[4] Z. Wang, G. A. Jullien and W. C. Miller, "Recursive algorithms for the forward and inverse discrete cosine transform with arbitrary length," *IEEE Signal Processing Lett.*, vol. 1, no. 7, pp. 101-102, July 1994.

[5] M. F. Aburdene, J. Zheng, and R. J. Kozick, "Computation of discrete cosine transform using Clenshaw`s recurrence formula," *IEEE Signal Processing Lett.*, vol. 2, no. 8, pp. 155-156, Aug. 1995.

[6] Y. H. Chan, L. P. Chau, and W. C. Siu, "Efficient implementation of discrete cosine transform using recursive filter structure," *IEEE Trans. Circuits Syst. Video Technol.*, vol. 4, no. 6, pp. 550-552, Dec. 1994.

[7] J. F. Yang and C. P. Fan, "Compact recursive structures for discrete cosine transform", *IEEE Trans. Circuits Syst. II*, vol. 47, no.4, pp. 314-321, Apr. 2000.

[8] J. L. Wang, C. B. Wu, B. D. Liu, and J. F. Yang, "Implementation of the discrete cosine transform and its inverse by recursive structures," in *Proc. IEEE Workshop on Signal Processing Systems*, Oct. 1999, pp. 120-130.

[9] S. S. Demirsoy, R. Beck, I. Kale, and A. G. Dempster, "Novel Recursive-DCT Implementations: A Comparative Study", in *Proc. International Workshop on Intelligent Data Acquisition and Advanced Computing Systems: Technology and Applications*, July 2001, pp. 120-123.

A SIMPLIFIED LATTICE FACTORIZATION FOR LINEAR-PHASE PARAUNITARY FILTER BANKS WITH PAIRWISE MIRROR IMAGE FREQUENCY RESPONSES

*Lu Gan and Kai-Kuang Ma**

School of Electrical and Electronic Engineering, Block S2
Nanyang Technological University, Singapore 639798
∗Email: ekkma@ntu.edu.sg

ABSTRACT

In this paper, by simplifying the lattice proposed by Nguyen *et al.*, a new factorization for linear-phase paraunitary filter banks (LPPUFBs) with pairwise mirror image (PMI) frequency responses is developed. The new structure covers the same class of PMI LPPUFBs as the original lattice while substantially reducing the free parameters involved in nonlinear optimization. A design example is presented to demonstrate the effectiveness of the new structure.

1. INTRODUCTION

Linear-phase paraunitary filter banks (LPPUFBs) have been extensively used in the application of image processing. Over the past decade, many works have exploited the design of M-channel LPPUFBs through the lattice factorization [1]-[8]. In [3], Soman *et al.* first developed a complete and minimal factorization for even-channel LPPUFBs. An alternative, but equivalent form of this structure with slightly fewer parameters was presented in [4]. Simplified structures of [3], [4] were recently reported in [5], [6], which result in a considerable reduction on the number of free parameters, while retaining the generality of the factorizations in [3], [4]. This facilitates both the design and the implementation of LPPUFBs.

Even after simplification, the number of free parameters can be still quite large with the increase of channel number and filter length. In this paper, we aim to further reduce the design complexity by adding the pairwise mirror image (PMI) property to LPPUFBs, in which the frequency responses of each pair of filters are symmetric with respect to $\pi/2$. Several works on this topic have been reported by other researchers in the past. A factorizaton for even-channel system was presented in [2] and [3], while odd-channel factorizations can be found in [7] and [8]. However, in these factorizations, the number of free parameters are nearly equal to that of simplified factorizations for general LPPUFBs in [5], [6]. Intuitively, by imposing the PMI property, fewer parameters are required than those of the general LPPUFBs. Therefore, the method introduced in [6] is extended in this paper to simplify the factorization of PMI LPPUFBs. For simplicity, the channel number is assumed to be even and the simplification is based on the structure in [2]. After our simplification, the degree of design freedom is reduced by 50% or so for large M, while the design space is not affected at all. This can significantly simplify the design complexity.

Notations: For a real number x, $\lceil x\rceil$ and $\lfloor x\rfloor$ represent the ceiling and the floor of x, respectively. Vectors and matrices are indicated in bold-faced letters. Subscripts will be provided only if their sizes are not clear from the context. Superscript T stands for transposition. Special matrices used extensively throughout this paper are: the identity matrix $\mathbf{I}$, the reversal matrix $\mathbf{J}$, and the null matrix $\mathbf{0}$. Besides, $\mathbf{W}_{2n}$ and $\hat{\mathbf{W}}_{2n}$ are $2n \times 2n$ butterfly matrices as follows:

$$\mathbf{W}_{2n} = \begin{bmatrix} \mathbf{I}_n & \mathbf{I}_n \\ \mathbf{I}_n & -\mathbf{I}_n \end{bmatrix}, \quad \hat{\mathbf{W}}_{2n} = \begin{bmatrix} \mathbf{I}_n & \mathbf{J}_n \\ \mathbf{I}_n & -\mathbf{J}_n \end{bmatrix}.$$

2. EXISTING STRUCTURE

Consider an M-channel ($M = 2m$) LPPUFB with all filters of the same length $L = KM$ each. Suppose the PMI property is further imposed, i.e., the analysis filters $H_i(z)$ satisfy $H_{M-1-i}(z) = H_i(-z)$ (for $i = 0, \cdots, m-1$) [2]. Let $\mathbf{E}(z)$ be the corresponding polyphase matrix. It was proved in [2] (pp. 322) that $\mathbf{E}(z)$ can be realized as in Fig. 1(a), i.e.,

$$\mathbf{E}(z) = \mathbf{G}_{K-1}(z)\mathbf{G}_{K-2}(z)\cdots\mathbf{G}_1(z)\mathbf{E}_0 \quad (1)$$

where each propagation matrix $\mathbf{G}_k(z)$ (for $k = 1, \cdots, K-1$) and the initial matrix $\mathbf{E}_0$ can be written into

$$\mathbf{G}_k(z) = \frac{1}{2}\begin{bmatrix} \mathbf{U}_k & \mathbf{0} \\ \mathbf{0} & \mathbf{V}_k \end{bmatrix}\begin{bmatrix} \mathbf{I} & \mathbf{I} \\ \mathbf{I} & -\mathbf{I} \end{bmatrix}\begin{bmatrix} \mathbf{I} & \mathbf{0} \\ \mathbf{0} & z^{-1}\mathbf{I} \end{bmatrix}\begin{bmatrix} \mathbf{I} & \mathbf{I} \\ \mathbf{I} & -\mathbf{I} \end{bmatrix} \triangleq \frac{1}{2}\mathbf{\Phi}_k\mathbf{W}\mathbf{\Lambda}(z)\mathbf{W}, \quad (2)$$

$$\mathbf{E}_0 = \frac{1}{\sqrt{2}}\begin{bmatrix} \mathbf{U}_0 & \mathbf{0} \\ \mathbf{0} & \mathbf{V}_0 \end{bmatrix}\begin{bmatrix} \mathbf{I} & \mathbf{J} \\ \mathbf{I} & -\mathbf{J} \end{bmatrix} \triangleq \frac{1}{\sqrt{2}}\mathbf{\Phi}_0\hat{\mathbf{W}}, \quad (3)$$

0-7803-7761-3/03/$17.00 ©2003 IEEE

in which $\mathbf{V}_k$ (for $k = 0, \cdots, K-1$) are $m \times m$ arbitrary orthogonal matrices, and $\mathbf{U}_k$ take the following form

$$\mathbf{U}_k = \begin{cases} \Gamma\mathbf{V}_k\Gamma, & k = 0, \cdots, K-2; \\ \mathbf{J}\mathbf{V}_k\Gamma, & k = K-1, \end{cases} \quad (4)$$

where Γ is a diagonal matrix whose entry is $\Gamma(l,l) = (-1)^l$, for $l = 0, \cdots, m-1$.

In the above lattice structure, each propagation matrix $\mathbf{G}_k(z)$ contains one $m \times m$ arbitrary orthogonal matrix $\mathbf{V}_k$. According to [5], [6], in a general LPPUFB, only one free orthogonal matrix (i.e., either $\mathbf{U}_k$ or $\mathbf{V}_k$) is needed for each propagation building block. By adding the PMI constraint, fewer parameters should be needed, which implies that (1) is a redundant structure.

3. SIMPLIFIED STRUCTURE

In this section, through trivial matrix manipulation, we first arrive at a new representation of (1), where each order-one building block contains one special orthogonal matrix. The cosine-sine (C-S) decomposition [9] is then investigated to parameterize the special orthogonal matrix, which leads to significant parameter reduction.

3.1. A new and equivalent factorization

The approach taken here is just a slight modification as that in [6]. Note that the cascade of any two adjacent stages $\mathbf{G}_k(z)\mathbf{G}_{k-1}(z)$ can be re-formulated into

$$\begin{aligned}
&\mathbf{G}_k(z)\mathbf{G}_{k-1}(z)\\
&=\frac{1}{2}\begin{bmatrix}\mathbf{U}_k & \mathbf{0}\\ \mathbf{0} & \mathbf{V}_k\end{bmatrix}\begin{bmatrix}\mathbf{I}+z^{-1}\mathbf{I} & \mathbf{I}-z^{-1}\mathbf{I}\\ \mathbf{I}-z^{-1}\mathbf{I} & \mathbf{I}+z^{-1}\mathbf{I}\end{bmatrix}\\
&\times\frac{1}{2}\begin{bmatrix}\mathbf{U}_{k-1} & \mathbf{0}\\ \mathbf{0} & \mathbf{V}_{k-1}\end{bmatrix}\begin{bmatrix}\mathbf{I}+z^{-1}\mathbf{I} & \mathbf{I}-z^{-1}\mathbf{I}\\ \mathbf{I}-z^{-1}\mathbf{I} & \mathbf{I}+z^{-1}\mathbf{I}\end{bmatrix}\\
&=\frac{1}{2}\begin{bmatrix}\mathbf{I} & \mathbf{0}\\ \mathbf{0} & \mathbf{V}_k\mathbf{U}_k^T\end{bmatrix}\begin{bmatrix}\mathbf{I}+z^{-1}\mathbf{I} & \mathbf{I}-z^{-1}\mathbf{I}\\ \mathbf{I}-z^{-1}\mathbf{I} & \mathbf{I}+z^{-1}\mathbf{I}\end{bmatrix}\begin{bmatrix}\mathbf{I} & \mathbf{0}\\ \mathbf{0} & \mathbf{U}_k\mathbf{V}_k^T\end{bmatrix}\\
&\times\frac{1}{2}\begin{bmatrix}\mathbf{U}_k\mathbf{U}_{k-1} & \mathbf{0}\\ \mathbf{0} & \mathbf{V}_k\mathbf{V}_{k-1}\end{bmatrix}\begin{bmatrix}\mathbf{I}+z^{-1}\mathbf{I} & \mathbf{I}-z^{-1}\mathbf{I}\\ \mathbf{I}-z^{-1}\mathbf{I} & \mathbf{I}+z^{-1}\mathbf{I}\end{bmatrix}.
\end{aligned}$$

The above manipulation implies that $\mathbf{U}_k$ and $\mathbf{V}_k$ can be moved across the matrices $\mathbf{W}$'s and the delay chain $\Lambda(z)$ to combine with $\mathbf{U}_{k-1}$ and $\mathbf{V}_{k-1}$, respectively. Likewise, the products $\mathbf{U}_k\mathbf{U}_{k-1}$ and $\mathbf{V}_k\mathbf{V}_{k-1}$ can be moved across the lattice again and combined with $\mathbf{U}_{k-2}$ and $\mathbf{V}_{k-2}$, respectively. By iteratively applying this process from $k = K-1$ to $k = 1$, we can get a new and equivalent structure of (1) as follows

$$\mathbf{E}(z) = \mathbf{G}'_{K-1}(z)\mathbf{G}'_{K-2}(z)\cdots\mathbf{G}'_1(z)\mathbf{E}'_0, \quad (5)$$

$$\mathbf{G}'_k(z) = \frac{1}{2}\text{diag}(\mathbf{I},\mathbf{X}_k)\mathbf{W}\Lambda(z)\mathbf{W}\text{diag}(\mathbf{I},\mathbf{X}_k^T), \quad (6)$$

$$\mathbf{E}'_0 = \frac{1}{\sqrt{2}}\text{diag}(\mathbf{U}'_0,\mathbf{V}'_0)\hat{\mathbf{W}}, \quad (7)$$

where the new matrices $\mathbf{X}_k$ (for $k = 1, \cdots, K-1$), $\mathbf{U}'_0$ and $\mathbf{V}'_0$ can be expressed as

$$\mathbf{X}_k = \mathbf{V'}_k\mathbf{U'}_k^T, \quad (8)$$

$$\mathbf{U}'_k = \prod_{i=K-1}^{k}\mathbf{U}_i, \quad \mathbf{V}'_k = \prod_{i=K-1}^{k}\mathbf{V}_i \quad k = 0, \cdots, K-1. \quad (9)$$

The implementation of (5) is shown in Fig. 1(b). It is clear that this new structure has the same implementation cost as the original lattice in Fig. 1(a). However, unlike $\mathbf{V}_k$ in $\mathbf{G}_k(z)$, $\mathbf{X}_k$ in $\mathbf{G}'_k(z)$ is *not* an arbitrary orthogonal matrix. The relation between $\mathbf{U}_k$ and $\mathbf{V}_k$ in (4) put certain constraints on $\mathbf{X}_k$. Note that from (4) and (9), we have

$$\begin{aligned}
\mathbf{U}'_k &= \prod_{i=K-1}^{k}\mathbf{U}_i = \mathbf{J}\mathbf{V}_{K-1}\Gamma\prod_{i=K-2}^{k}\Gamma\mathbf{V}_i\Gamma\\
&= \mathbf{J}\prod_{i=K-1}^{k}\mathbf{V}_i\cdot\Gamma = \mathbf{J}\mathbf{V}'_k\Gamma.
\end{aligned} \quad (10)$$

Then, substituting (10) into (8) yields

$$\mathbf{X}_k = \mathbf{V}'_k\mathbf{U'}_k^T = \mathbf{V}'_k\Gamma\mathbf{V'}_k^T\mathbf{J} = \mathbf{X}'_k\mathbf{J}, \quad (11)$$

in which

$$\mathbf{X}'_k = \mathbf{V'}_k\Gamma\mathbf{V'}_k^T. \quad (12)$$

One can verify that the component $\mathbf{X}'_k$ in $\mathbf{X}_k$ is a special symmetric orthogonal matrix, i.e., $\mathbf{X}'_k = \mathbf{X'}_k^T$. Obviously, $\mathbf{X}_k$ should contain fewer degrees of design freedom than the general orthogonal matrix $\mathbf{V}_k$. Section 3.2 will discuss the parameterization of $\mathbf{X}_k$.

3.2. Matrix parameterization

Let $m_0 = \lceil m/2 \rceil$ and $m_1 = \lfloor m/2 \rfloor$. As Γ has m_0 1's and m_1 -1's, there exists a permutation matrix $\mathbf{P}$ such that $\mathbf{P}\Gamma\mathbf{P}^T = \mathbf{B}$, in which $\mathbf{B} = \text{diag}(\mathbf{I}_{m_0}, -\mathbf{I}_{m_1})$. Define a new orthogonal matrix $\hat{\mathbf{V}}_k = \mathbf{V}'_k\mathbf{P}^T$, i.e., $\mathbf{V}'_k = \hat{\mathbf{V}}_k\mathbf{P}$. From (11), one can derive that $\mathbf{X}_k$ can be written into

$$\mathbf{X}_k = \mathbf{V'}_k\Gamma\mathbf{V'}_k^T\mathbf{J} = \hat{\mathbf{V}}_k\mathbf{P}\Gamma\mathbf{P}^T\hat{\mathbf{V}}_k^T\mathbf{J} = \hat{\mathbf{V}}_k\mathbf{B}\hat{\mathbf{V}}_k^T\mathbf{J}. \quad (13)$$

With this new form of $\mathbf{X}_k$, the parameterization can be obtained by using the C-S decomposition of $\hat{\mathbf{V}}_k$. According to [9], $\hat{\mathbf{V}}_k$ can be completely characterized as

$$\hat{\mathbf{V}}_k = \text{diag}(\mathbf{Y}_{k,0}, \mathbf{Y}_{k,1})\Sigma_k\text{diag}(\mathbf{Z}_{k,0}^T, \mathbf{Z}_{k,1}^T), \quad (14)$$

where $\mathbf{Y}_{k,0}$ and $\mathbf{Z}_{k,0}$ are $m_0 \times m_0$ general orthogonal matrices, while $\mathbf{Y}_{k,1}$ and $\mathbf{Z}_{k,1}$ are $m_1 \times m_1$ general orthogonal

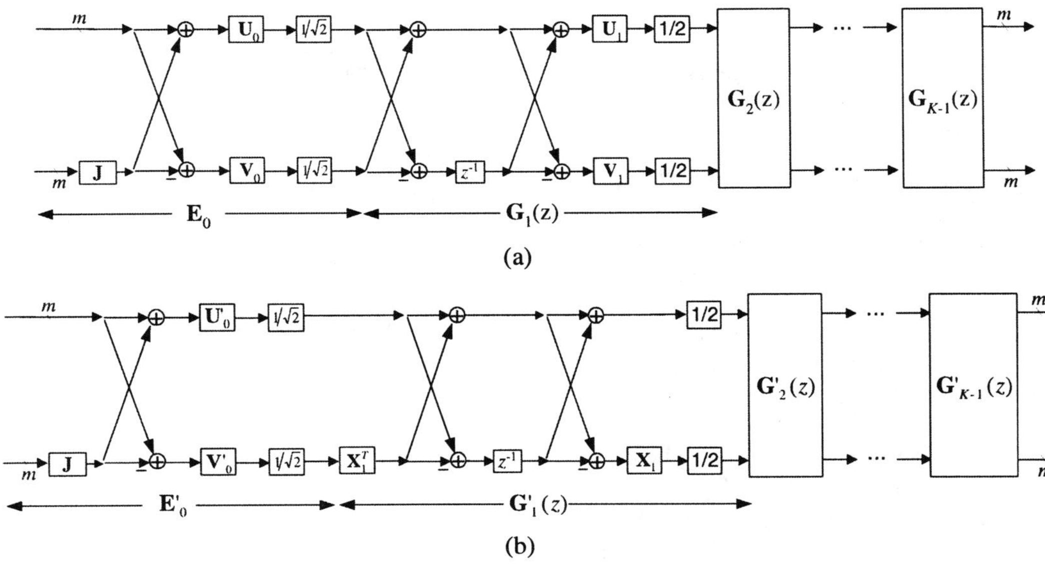

Figure 1: Implementation of two equivalent structure for PMI-LPPUFBs. (a) Structure of (1). (b) Simplified structure in (5).

matrices and Σ_k is a $m \times m$ special orthogonal matrix defined as

$$\Sigma_k = \begin{cases} \begin{bmatrix} \mathbf{C}_k & -\mathbf{S}_k \\ \mathbf{S}_k & \mathbf{C}_k \end{bmatrix}, & m \text{ even;} \\ \begin{bmatrix} 1 & \mathbf{0} & \mathbf{0} \\ \mathbf{0} & \mathbf{C}_k & -\mathbf{S}_k \\ \mathbf{0} & \mathbf{S}_k & \mathbf{C}_k \end{bmatrix}, & m \text{ odd,} \end{cases} \tag{15}$$

in which $\mathbf{C}_k$ and $\mathbf{S}_k$ are $m_1 \times m_1$ diagonal matrices with entries are $\mathbf{C}_k(l,l) = \cos\alpha_{k,l}$ and $\mathbf{S}_k(l,l) = \sin\alpha_{k,l}$ (for $l = 0, \cdots, m_1 - 1$ and $0 \le \alpha_{k,l} \le \pi/2$). Substituting (14) into (13) produces

$$\mathbf{X}_k = \text{diag}(\mathbf{Y}_{k,0}, \mathbf{Y}_{k,1})\Sigma'_k \text{diag}(\mathbf{Y}^T_{k,0}, \mathbf{Y}^T_{k,1})\mathbf{J}, \tag{16}$$

where $\Sigma'_k = \Sigma_k \mathbf{B} \Sigma^T_k$ is given by

$$\Sigma'_k = \begin{cases} \begin{bmatrix} \mathbf{C}'_k & \mathbf{S}'_k \\ \mathbf{S}'_k & -\mathbf{C}'_k \end{bmatrix}, & m \text{ even;} \\ \begin{bmatrix} 1 & \mathbf{0} & \mathbf{0} \\ \mathbf{0} & \mathbf{C}'_k & \mathbf{S}'_k \\ \mathbf{0} & \mathbf{S}'_k & -\mathbf{C}'_k \end{bmatrix}, & m \text{ odd,} \end{cases} \tag{17}$$

in which $\mathbf{C}'_k$ and $\mathbf{S}'_k$ are two diagonal matrices whose entries are $\mathbf{C}'_k(l,l) = \cos\alpha'_{k,l}$ and $\mathbf{S}'_k(l,l) = \sin\alpha'_{k,l}$, respectively, with $\alpha'_{k,l} = 2\alpha_{k,l}$ $(0 \le \alpha'_{k,l} \le \pi)$. Up to this stage, we arrive at the following Theorem:

Theorem 1. *Any $\mathbf{E}(z)$ in (1) can be always written as in (5), where each order-one building block $\mathbf{G}'_k(z)$ and the initial matrix $\mathbf{E}'_0$ are shown in (6) and (7), respectively. In $\mathbf{G}'_k(z)$, each $m \times m$ orthogonal matrix $\mathbf{X}_k$ can be represented as in (16). While in $\mathbf{E}'_0$, $\mathbf{V}'_0$ is an $m \times m$ arbitrary orthogonal matrix, and $\mathbf{U}'_0$ can be expressed as $\mathbf{U}'_0 = \mathbf{J}\mathbf{V}'_0\Gamma$.*

Now, let us compare the degrees of design freedom of (1) and (5). It can be readily seen that $\mathbf{E}_0$ and $\mathbf{E}'_0$ hold the same number of free parameters. The difference lies in $\mathbf{G}_k(z)$ and $\mathbf{G}'_k(z)$. In $\mathbf{G}_k(z)$ of (2), each $\mathbf{V}_k$ requires $\binom{m}{2} = \frac{m(m-1)}{2}$ rotation angles and m sign parameters for a complete parameterization. On the other hand, in $\mathbf{G}'_k(z)$, eq. (16) implies $\mathbf{X}_k$ can be fully characterized by two general orthogonal matrices $\mathbf{Y}_{k,i}$ $(i = 0, 1)$ and one special orthogonal matrix Σ'_k. As $\mathbf{Y}_{k,0}$, $\mathbf{Y}_{k,1}$ and Σ'_k contain $\binom{m_0}{2}$, $\binom{m_1}{2}$ and m_1 rotation angles, respectively, the total number of free rotation angles in $\mathbf{X}_k$ is $n' = \binom{m_0}{2} + \binom{m_1}{2} + m_1$. One can calculate that $n' = \frac{m^2}{4}$ for even m and $n' = \frac{m^2-1}{4}$ for odd m. Note that except for $m = 2$ (where $n' = n = 1$), n' is always less than n. The reduction is nearly 50% for large m. Besides, in (16), if each $\alpha'_{k,l}$ is allowed to take arbitrary values in parameterization, the m sign parameters in $\mathbf{Y}_{k,i}$ $(i = 0, 1)$ can be discarded. Therefore, in our new structure of (5), the design complexity is much less than that in (1).

4. DESIGN EXAMPLE

This section presents a design example of PMI LPPUFB with $M = 8$ and $K = 5$. The chosen criterion is a weighted combination of the coding gain, DC leakage and stopband attenuation as follows

$$C = 0.1C_{\text{coding gain}} + 0.2C_{DC} + 0.7C_{stop}, \tag{18}$$

where the definition of coding gain $C_{\text{coding gain}}$ and DC attenuation C_{DC} are the same as that in [10], while C_{stop} is the sum of stopband energy of $H_i(z)$ (for $i = 0, \cdots, M/2 - 1$)

as follows:

$$C_{stop} = 2 \sum_{i=0}^{M/2-1} \int_{\omega \in \Omega_i} |H_i(e^{j\omega})|^2 d\omega, \qquad (19)$$

where $\Omega_i = [0, \omega_{i,L}] \bigcup [\omega_{i,H}, \pi]$ denotes $H_i(z)$'s stopband, with $\omega_{i,L}$ and $\omega_{i,H}$ defined as $\omega_{i,L} = \max(0, (i-0.6)\pi/M)$ and $\omega_{i,H} = \min(\pi, (i+1.6)\pi/M)$, respectively. Note that due to the PMI property, only the first half filters are included in (19).

The design was carried out for both (1) and (5) through the Matlab function *fminunc*. To have a fair comparison of both structures, the initial values are chosen so that the optimizations start from $\mathbf{E}(z) = z^{-2}\mathbf{WHT}$, in which $\mathbf{WHT}$ denotes the 8×8 Walsh-Hadamard transform. To be more specific, both $\mathbf{E}_0$ and $\mathbf{E}'_0$ are chosen as $\mathbf{E}_0 = \mathbf{E}'_0 = \mathbf{WHT}$. For the order-one propagation matrix, the free matrix $\mathbf{V}_k$ in $\mathbf{G}_k(z)$ (for $k = 1, \cdots, 4$) are set to be $\mathbf{V}_k = \mathbf{J}$. While in $\mathbf{G}'_k(z)$, $\mathbf{X}_k = (-1)^k\mathbf{I}$. These $\mathbf{X}_k$ can be obtained by setting $\mathbf{Y}_{k,0} = \mathbf{I}$, $\mathbf{Y}_{k,1} = \mathbf{J}$ and $\alpha'_{k,l} = (-1)^k \frac{\pi}{2}$.

On a Pentium IV, 1.9 GHz computer, it turns out that the optimization time for (1) and (5) are 23 and 16 seconds, respectively. Fig. 2(a) and Fig. 2(b) show the frequency responses of the optimized PMI LPPUFBs based on (1) and (5), respectively. Table 1 documents the numerical results. One can see that the PMI LPPUFB based on our structure outperforms that of the original lattice in all accounts. Thus, by discarding redundant parameters, our lattice structure can not only speed up the optimization, but also more effectively to avoid being trapped in the local minimum.

Table 1: Comparison of two optimized PMI LPPPUFBs

Filter Banks	Results of (1)	Results of (5)
Coding Gain	9.3587dB	9.3856 dB
DC Attenuation	-29.515 dB	-35.4457dB
Stopband Attenuation	-24.3415dB	-24.4712 dB

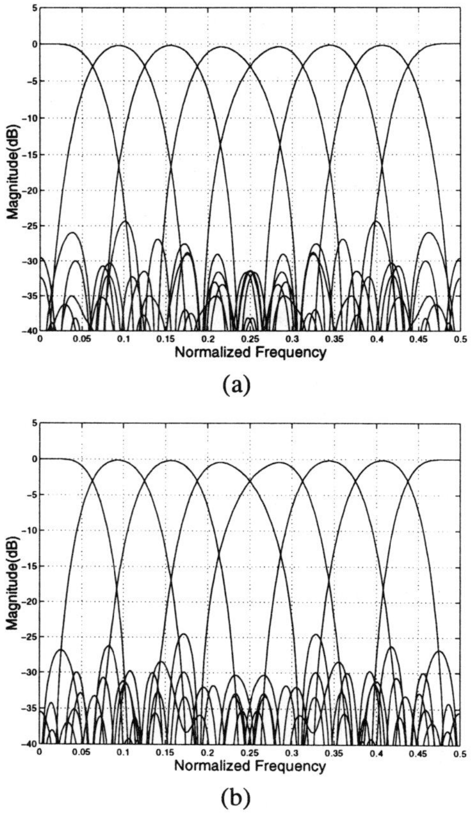

Figure 2: Optimized PMI LPPUFBs with $M = 8$ and $K = 5$. (a) Results of (1). (b) Results of (5).

5. CONCLUSION

This paper proposes a new structure for PMI LPPUFBs with even-channel, which is a simplified version of the lattice in [2]. The new structure spans the same class of PMI LPPUFBs as the original lattice, while the number of free parameters are significantly reduced. Through this way, better results with faster convergence in the optimization can be achieved. A design example is presented to demonstrate the efficiency of the proposed lattice structure.

6. REFERENCES

[1] P.P. Vaidyanathan, *Multirate Systems and Filter Banks*. Englewood Cliffs, NJ: Prentice-Hall, 1993.

[2] G. Strang and T.Q. Nguyen, *Wavelets and Filter Banks*, Wellesley-Cambridge Press, 1997.

[3] A.K. Soman, P.P. Vaidyanathan and T.Q. Nguyen, "Linear-phase paraunitary filter banks: theory, factorizations and designs," *IEEE Trans. Signal Processing*, vol. 41, pp. 3480-3496, Dec. 1993.

[4] R.L. de Querioz, T.Q. Nguyen, and K.R. Rao, "The GenLOT: Generalized linear-phase lapped orthogonal transform," *IEEE Trans. Signal Processing*, vol. 44, no. 3, pp. 497-507, Mar. 1996.

[5] X.Q. Gao, T.Q. Nguyen and G. Srang, "On factorization of M-channel paraunitary filter banks," *IEEE Trans. Signal Processing*, vol. 49, pp. 1433-1446, Jul. 2001.

[6] L. Gan and K.-K. Ma, "A simplified lattice factorization for linear-phase perfect reconstruction filter bank," *IEEE Signal Processing Lett.*, vol. 8, pp. 207-209, Jul. 2001.

[7] K.P. Chan, T.Q. Nguyen, L. Chen, "A new algorithm for linear-phase paraunitary filter banks with pairwise mirror-image frequency responses," *Signal Processing*, vol. 80, pp. 2589-2595, Dec, 2000.

[8] C.W. Kwok, T. Nagai, and T.Q. Nguyen, "Lattice structures of linear phase paraunitary matrices with pairwise mirror-image symmetry in the frequency domain with odd number of rows," *IEEE Trans. Signal Processing*, vol. 47, pp. 3315-3325, Dec. 1999.

[9] G.H. Golub, C.F. Van Loan, *Matrix Computations*, Johns Hopkins University Press, 1983.

[10] T.D. Tran and T.Q. Nguyen, "A progressive transmission image coder using linear phase uniform filter banks as block transforms," *IEEE Trans. on Image Processing*, vol. 8, pp. 1493-1507, Nov. 1999.

A DESIGN FLOW FOR LINEAR-PHASE FIXED-POINT FIR FILTERS: FROM THE NPRM SPECIFICATIONS TO A VHDL CODE

Chia-Yu Yao, Chin-Chih Yeh, Tsuan-Fan Lin, Hsin-Horng Chen, and Chiang-Ju Chien

Department of Electronic Engineering
Huafan University, Taiwan 223, ROC
chyao@huafan.hfu.edu.tw

ABSTRACT

This work combines two distinct research efforts, the coefficient design and the adder-number reduction, of fixed-point FIR filters into an automatic design flow. Given the normalized peak-ripple-magnitude (NPRM) specifications, the canonic-signed-digit (CSD) filter coefficients are calculated by the partial mixed-integer-linear-programming (PMILP) algorithm. Then a signed common subexpression sharing (SCSS) algorithm is used to reduce the number of adders required to implement the FIR filter. Finally a VHDL code that describes the FIR filter hardware with SCSS is generated.

1. INTRODUCTION

There are two major steps to realize a fixed-point FIR filter: the coefficient design and the reduction of adder numbers. Each coefficient of a fixed-point FIR filter is usually represented in canonic-signed-digit (CSD) format whose number of signed-powers-of-two (SPT) terms is minimum. There have been several algorithms dealing with CSD coefficient design for FIR filters. These include the mixed-integer-linear-programming (MILP) algorithm [1, 2], the local search method [3, 4], the trellis search algorithm [5], and the partial MILP (PMILP) method [6].

After the CSD coefficients are obtained, the number of adders to realize the coefficients must be taken in account. There were many works aimed at reducing the number of adders. For examples, the common-subexpression-elimination (CSE) methods [7]–[9] and the graph dependence algorithms [10]–[12].

The n-dimensional reduced-adder-graph (RAGn) method [11] produces the minimum-adder-number structure; however, it does not consider the adder steps [12] (or called the logic depth in [9]). Therefore, the structure generated by the RAGn method may suffer from speed problem. Kang and Park's work [12] cures the speed problem suffered by the RAGn method. However, they did not touch the coefficient design category. In this paper, we combine the coefficient design and the reduction of adder numbers together to complete a fixed-point FIR filter design flow. Given the normalized peak-ripple-magnitude (NPRM) specifications of an FIR filter, we use the PMILP algorithm to find the CSD coefficients. Then we develop a signed common-subexpression-sharing (SCSS) algorithm to reduce the number of adders. Finally an RTL VHDL code with SCSS that realizes the fixed-point FIR filter is created.

This paper is organized as follows. In Section 2, we present our PMILP algorithm shortly for the CSD coefficient design. In Section 3, we demonstrate our SCSS method for the reduction of adder numbers. In Section 4, we show some comparisons of design examples with other methods. Finally Section 5 summarizes our work.

This work was partially supported by the National Science Council of Taiwan, ROC, under Grant NSC 90-2215-E-211-001.

2. THE PMILP ALGORITHM

Based on the observation of the SPT-term distribution of FIR filter coefficients [13], we proposed a PMILP algorithm that minimizes the number of SPT terms for some least significant digits (LSDs) given the NPRM specifications [6]. The problem formulation is as follows. The M-digit CSD coefficients of a length-N FIR filter is

$$h_{\mathrm{spt}}(n) = \sum_{m=1}^{M} (a^{+}_{m,n} - a^{-}_{m,n})2^{-m} \tag{1}$$

where $a^{+}_{m,n}, a^{-}_{m,n} \in \{0,1\}$ and $a^{+}_{m,n} + a^{-}_{m,n} \leq 1$. Ignoring the linear phase factor, the frequency response becomes

$$H_{\mathrm{spt}}(\omega) = \sum_{n=0}^{\lfloor \frac{N-1}{2} \rfloor} c_n \sum_{m=1}^{M} \frac{a^{+}_{m,n} - a^{-}_{m,n}}{2^m} T(n,\omega). \tag{2}$$

For a symmetrical impulse response, $T(n,\omega) = \cos\Big((\frac{N-1}{2} - n)\omega \Big)$ and $c_n = \Big(2 - \delta_{n,\frac{N-1}{2}}\Big)$ ($\delta_{n,\frac{N-1}{2}}$ is the Kronecker delta symbol). For an antisymmetrical impulse response, $T(n,\omega) = \sin\Big((\frac{N-1}{2} - n)\omega \Big)$ and $c_n = 2$.

We define the *fixed variables* as those $a^{\pm}_{m,n}$'s that have been assigned 0 or 1 already. The other variables to be determined are defined as the *unfixed variables*. We use $M_U(n)$ as the collection of m that $a^{\pm}_{m,n}$'s are unfixed variables and $M_F(n)$ as the collection of m that $a^{\pm}_{m,n}$'s are fixed variables. Then the frequency response is divided into two parts:

$$H_o(\omega) = \sum_{n=0}^{\lfloor \frac{N-1}{2} \rfloor} c_n \sum_{m \in M_F(n)} \frac{a^{+}_{m,n} - a^{-}_{m,n}}{2^m} T(n,\omega) \tag{3}$$

and

$$G(a^{\pm}_{m,n},\omega) = \sum_{n=0}^{\lfloor \frac{N-1}{2} \rfloor} c_n \sum_{\substack{m \in \\ M_U(n)}}^{M} \frac{a^{+}_{m,n} - a^{-}_{m,n}}{2^m} T(n,\omega). \tag{4}$$

0-7803-7761-3/03/$17.00 ©2003 IEEE

$H_o(\omega)$ is the coarse frequency response calculated from the fixed variables of $h_{\text{spt}}(n)$'s. $G(a^{\pm}_{m,n}, \omega)$ is the amount contributed by the unfixed variables of $h_{\text{spt}}(n)$'s. In the beginning, $M_F(n) = \{1, 2, \ldots, M-D\}$ and $M_U(n) = \{M-D+1, \ldots, M\}$ for all n. D is chosen not greater than 3. Denote $\omega_{\mathbf{p}}$ as the passband frequency range and $\omega_{\mathbf{s}}$ as the stopband frequency range. Let $b(a^{\pm}_{m,n})$ denote the passband gain. It is defined as

$$b(a^{\pm}_{m,n}) = \sqrt{\max_{\omega \in \omega_{\mathbf{p}}}(H_{\text{spt}}(\omega)) \min_{\omega \in \omega_{\mathbf{p}}}(H_{\text{spt}}(\omega))}. \tag{5}$$

Roughly speaking, the PMILP algorithm is to minimize the number of SPT terms of the last D digits of $h_{\text{spt}}(n)$'s such that the NPRM specifications of the normalized frequency response

$$H_N(\omega) = \frac{H_o(\omega) + G(a^{\pm}_{m,n}, \omega)}{b(a^{\pm}_{m,n})} \tag{6}$$

are satisfied. Because the NPRM specifications are usually given in decibel, we use $1 + \delta_{\text{NPRM}}$ and $1/(1 + \delta_{\text{NPRM}})$ to specify the passband upper and lower limits, respectively, instead of the conventional $1 + \delta_{\text{NPRM}}$ and $1 - \delta_{\text{NPRM}}$ specifications. Since $1/(1 + \delta_{\text{NPRM}}) > 1 - \delta_{\text{NPRM}}$, our description is more rigorous than the conventional one.

Let L_T denote the total number of SPT terms of the CSD coefficients. In our algorithm, we can specify the maximum number of SPT terms per coefficient ($L_{\max}$) such that the throughput rate is under controlled for high-speed application. The PMILP algorithm is given as follows.

1. Find the optimum linear phase FIR filter coefficients $h_o(n)$, $n = 0, 1, \ldots, \lfloor (N-1)/2 \rfloor$, by the Remez exchange algorithm [14].
2. Find a proper scaling factor b_D (which is also the passband gain) such that the quantized CSD coefficients $Q[b_D h_o(n)]$, $n = 0, 1, \ldots, \lfloor (N-1)/2 \rfloor$, produce a satisfactory frequency response and have as small amount SPT terms as possible. In this step, the filter length N can be increased to meet the tradeoff between NPRM requirements and the number of SPT terms.
3. Solve the following MILP problem:

$$\min \sum_{n=0}^{\lfloor \frac{N-1}{2} \rfloor} \sum_{m=M-D+1}^{M} (a^{+}_{m,n} + a^{-}_{m,n}) \tag{7}$$

subject to

$$\begin{aligned}
&G(a^{\pm}_{m,n}, \omega) - \frac{b}{1 + \delta_{\text{NPRM}}} \geq -H_o(\omega), \omega \in \omega_{\mathbf{p}} \\
&G(a^{\pm}_{m,n}, \omega) - b(1 + \delta_{\text{NPRM}}) \leq -H_o(\omega), \omega \in \omega_{\mathbf{p}} \\
&G(a^{\pm}_{m,n}, \omega) + b\delta_{\text{NPRM}} \geq -H_o(\omega), \omega \in \omega_{\mathbf{s}} \\
&G(a^{\pm}_{m,n}, \omega) - b\delta_{\text{NPRM}} \leq -H_o(\omega), \omega \in \omega_{\mathbf{s}} \\
&a^{+}_{m,n} + a^{-}_{m,n} \leq 1, \text{ and } a^{\pm}_{m,n} \in \{0, 1\} \\
&\sum_{m=M-D+1}^{M} (a^{+}_{m,n} + a^{-}_{m,n}) \\
&\qquad \leq L_{\max} - \sum_{m=1}^{M-D} (a^{+}_{m,n} + a^{-}_{m,n}) \\
&b_D - \epsilon_1 \leq b \leq b_D + \epsilon_2.
\end{aligned}$$

The MILP problem in Step 3 is a 0-1 problem and is solved by the branch and bound (B&B) method. Three branches are created at a time. One corresponding to $a^{+}_{i,j} = 0, a^{-}_{i,j} = 0$, another corresponding to $a^{+}_{i,j} = 1, a^{-}_{i,j} = 0$, and the other corresponding to $a^{+}_{i,j} = 0, a^{-}_{i,j} = 1$. Each time when a branch is created, the CSD coefficients have to be reshaped so the linear programming (LP) subproblem will change too. The formulation of the LP sub-problem is

$$\min \sum_{n'=0}^{\lfloor \frac{N-1}{2} \rfloor} \sum_{m' \in M_U(n')} (a^{+}_{m',n'} + a^{-}_{m',n'}) \tag{8}$$

subject to

$$\begin{aligned}
&G(a^{\pm}_{m',n'}, \omega) - \frac{b}{1 + \delta_{\text{NPRM}}} \geq -H_o(\omega), \omega \in \omega_{\mathbf{p}} \\
&G(a^{\pm}_{m',n'}, \omega) - b(1 + \delta_{\text{NPRM}}) \leq -H_o(\omega), \omega \in \omega_{\mathbf{p}} \\
&G(a^{\pm}_{m',n'}, \omega) + b\delta_{\text{NPRM}} \geq -H_o(\omega), \omega \in \omega_{\mathbf{s}} \\
&G(a^{\pm}_{m',n'}, \omega) - b\delta_{\text{NPRM}} \leq -H_o(\omega), \omega \in \omega_{\mathbf{s}} \\
&a^{+}_{m',n'} + a^{-}_{m',n'} \leq 1, \text{ and } a^{\pm}_{m',n'} \geq 0 \\
&\sum_{m' \in M_U(n')} (a^{+}_{m',n'} + a^{-}_{m',n'}) \\
&\leq L_{\max} - \sum_{m' \in M_F(n')} (a^{+}_{m',n'} + a^{-}_{m',n'}), \forall n' \\
&b_D - \epsilon_1 \leq b \leq b_D + \epsilon_2.
\end{aligned}$$

Figure 1 shows the procedure for solving the MILP problem (7). When $a^{\pm}_{i,j}$ are selected to be branched, three LP sub-problems are created. If an LP sub-problem has a solution, it is fathomed. First compute T as is shown in FIgure 1. If the solution is also a solution of the original problem (7) and $T \leq L_T$, then update L_T and save the results. Otherwise, if $T < L_T$, the current fixed variables are pushed into the stack. Then the algorithm solves the next LP subproblem. If the fathoming procedure comes from the left branch, then the algorithm will go to node A next. If the fathoming procedure comes from the middle branch, then the algorithm will go to node B next. Similarly, node C follows the fathoming procedure coming from the right branch.

3. THE PROPOSED SCSS METHOD

The PMILP method produces a set of satisfactory filter coefficients in the CSD format. We next find out the common subexpression between these coefficients. We develop three main steps to do so:

1. Classify all CSD coefficients according to the their minimum adder steps (MAS). An adder step (AS) of a coefficient is the number of adders along the critical path when that coefficient is realized. The MAS of a coefficient is the number of adders along the shortest critical path among all possible realizations of that coefficient. Figure 2 is an example of AS and MAS of a coefficient $10\underline{1}001001$. ($\underline{1}$ means -1.) Let s_{ma} represent the minimum adder step and l represent the number of SPT terms of a coefficient. Apparently, $s_{ma}(l) = \lceil \log_2 l \rceil$. Denote F as the set of all coefficients and let A_i be the set that is composed of the coefficients with i MAS; i.e., $A_i = \{c | c \in F \,\&\, s_{ma}(l(c)) = i\}$. We further define the filter adder step (FAS) as the longest MAS among all CSD coefficients.
2. Decompose the coefficients with larger MAS by smaller MAS coefficients or 1 according to the frequency of appearance of

the smaller MAS coefficients. Let $c_i \in A_i$. What this step is doing is

for j = FAS down to 2 {
$c_j = 2^{p1} c_{j-1} \pm 2^{p2} c_k$, $k \leq j - 1$, or
$c_j = 2^{p1} c_{j-1} \pm 2^{p2}$ }

for some proper $p1$'s and $p2$'s. Searching for all possible decompositions and counting the frequency of appearance for all c_{j-1}'s, we sort c_{j-1}'s according to their frequency of appearance. Thus, we determine the sequence of subexpression sharing.

3. If there are some coefficients that cannot be decomposed in Step 2, then create new auxiliary coefficients with smaller MAS that can decompose the coefficients in the form of Step 2. Then repeat Step 2 and 3 until all coefficients are processed.

The following simple example explains our SCSS method. Let the coefficients be

$$h(0) = h(7) = \underline{1}0\underline{1} = -5, \quad h(2) = h(5) = \underline{1}0\underline{1}0\underline{1}01 = -83$$
$$h(1) = h(6) = 1001 = 9, \quad h(3) = h(4) = 10010101 = 149$$

Taking the absolute value of all $h(k)$'s, then $A_1 = \{5, 9\}$ and $A_2 = \{83, 149\}$. Decompose 149 as a combination of 5 and 9 as

$$149 = 9 \times 2^4 + 5.$$

However, 83 cannot be represented as a combination of 5 and 9. Therefore, we begin to search a new number with MAS=1 that can cooperate with original elements of A_1 to decompose 83. The number we find is 3 such that

$$83 = 5 \times 2^4 + 3$$

Once the common subexpressions are extracted, the filter structure can be translated into a VHDL code straightforwardly. Because the direct form cannot be used in high-throughput scenario without pipelining, we use transpose form FIR filter as our design target for the VHDL code.

4. DESIGN EXAMPLES

We redesign the transposed filters of [8] using our method. The same anonym for each example as in [8] are employed. $S1$ and $S2$ are examples 1 and 2 of [4]. $L1$, $L2$, and $L3$ are examples 1, 2, and 3 of [15]. In our design process, we can choose carry ripple adders (CRA) or carry save adders (CSA) to realize the filters. Table 1 summarizes the results of five filters designed in our algorithm and in the CSE method of [8]. In Table 1, N is the filter length, R_i is the number of adders before any manipulation of subexpression sharing. R_o is the number of adders after the manipulation for both algorithms. It can be seen that our method can produce filters with equal of less number of adders than the CSE method of [8].

After the PMILP and SCSS algorithms, our design program translates the design result into a VHDL code automatically. The wordlength of the input signal is assumed to be sixteen. We then use the SYNOPSYS design compiler to synthesize the filter. Table 2 shows two reports of SYNOPSYS for a short-filter-length ($S1$) and a long-filter-length $L1$ examples in rigorous timing constraint. The data arrival time in nano-seconds means the estimated time for signal travelling through the critical path. The area is measured in equivalents of 2-input NAND gates. It is noticed that under rigorous timing constraint, the structure using CRA's does not have any area advantage over the structure using CSA's.

Table 1: Comparison of our algorithm and the CSE method of [8].

Filter	N	R_i	R_o([8])	R_o(proposed)
$S1$	25	11	6	6
$S2$	60	57	32	28
$L1$	121	145	58	54
$L2$	63	49	23	23
$L3$	36	16	5	5

Table 2: Comparison of structures using CRA's and CSA's in rigorous timing constraint. The filter input is assumed to be 16-bit wide.

Filter	Performance	CRA type	CSA type
$S1$	data arrival time	5.74	2.51
	area	16258.7	17281.6
$L1$	data arrival time	7.84	4.3
	area	125945.1	103331.4

5. SUMMARIES

In this paper, we combine two algorithms for designing a fixed-point linear-phase FIR filters. The first method is the PMILP algorithm that produces the CSD coefficients of an FIR filter. The second method is a SCSS algorithm for adder-number reduction. Thus, a design flow, from the filter's NPRM specifications to the detail filter structure, is complete. We also develop a subprogram that translates the transposed form filter structure to a VHDL code. Using the design examples of [8], we show that our method can produce filters with equal or less number of adders than the CSE method of [8]. Using SYNOPSYS to synthesize the FIR filter VHDL codes produced by our program, we also find that under the stringent timing constraint, the filter structure employing carry ripple adders does not have any area advantage over the filter structure employing carry save adders.

6. REFERENCES

[1] Y. C. Lim and S. R. Parker, "FIR filter design over a discrete powers-of-two coefficient space," *IEEE Trans. Acoust., Speech, Signal Processing*, vol. ASSP-31, pp. 583–591, June 1983.

[2] Y. C. Lim and S. R. Parker, "Design of discrete-coefficient-value linear phase FIR filters with optimum normalized peak ripple magnitude," *IEEE Trans. Circuits Syst.*, vol. 37, pp. 1480–1486, Dec. 1990.

[3] Q. Zhao and Y. Tadokoro, "A simple design of FIR filters with powers-of-two coefficients," *IEEE Trans. Circuits Syst.*, vol. 35, pp. 566–570, May 1988.

[4] H. Samueli, "An improved search algorithm for the design of multiplierless FIR filters with powers-of-two coefficients," *IEEE Trans. Circuits Syst.*, vol. 36, pp. 1044–1047, July 1989.

[5] C.-L. Chen and A. N. Willson, Jr., "A trellis search algorithm for the design of FIR filters with signed-powers-of-two coeffi-

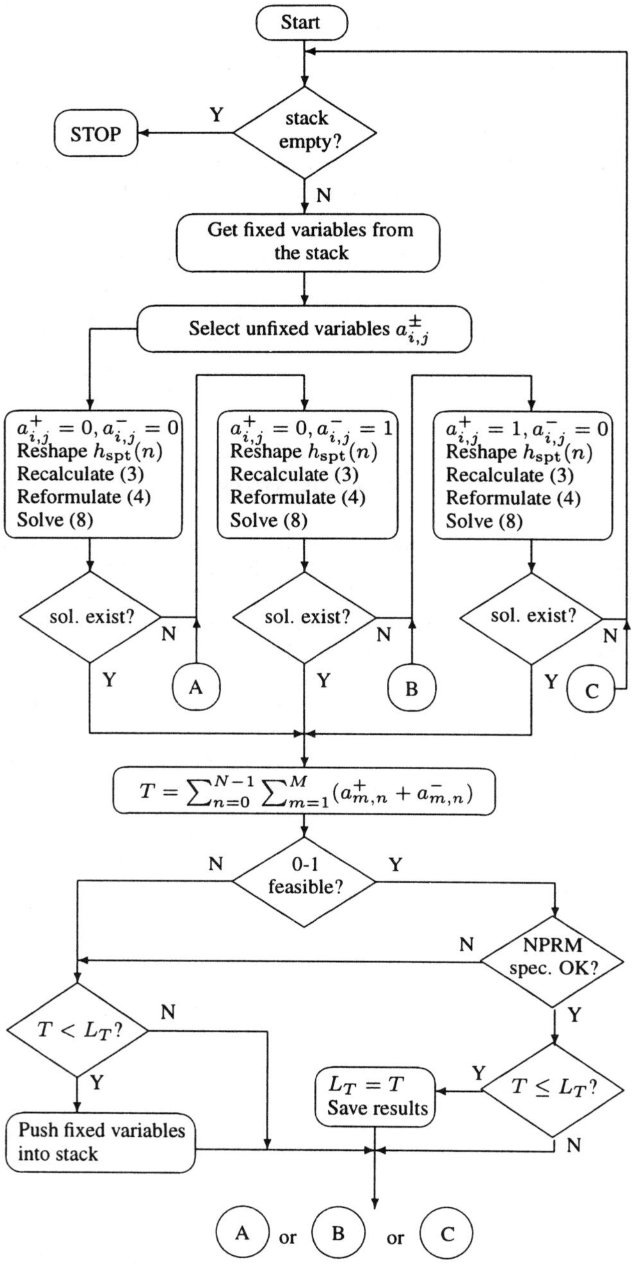

Figure 1: Flowchart of solving (7).

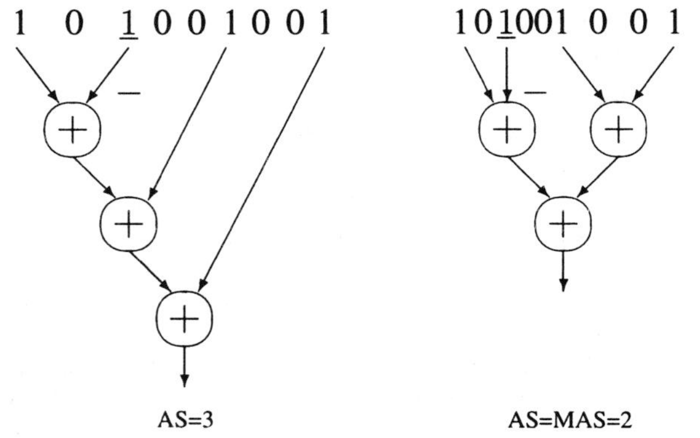

Figure 2: An example of realization a coefficient in AS=3 and AS=MAS=2.

cients," *IEEE Trans. Circuits Syst. II*, vol. 46, pp. 29–39, Jan. 1999.

[6] C.-Y. Yao and C.-J. Chien, "A partial MILP algorithm for the design of linear phase FIR filters with SPT coefficients," to appear in *IEICE Trans. Fundamentals*.

[7] R. I. Hartley, "Subexpression sharing in filters using canonic signed digit multipliers," *IEEE Trans. Circuits Syst. II*, vol. 43, pp. 677–688, Oct. 1996.

[8] R. Paško, P. Schaumont, V. Derudder, S. Vernalde, and D. Ďuračková, "A new algorithm for elimination of common subexpressions," *IEEE Trans. Computer-Aided Design*, vol. 18, pp. 58–68, Jan. 1999.

[9] M. Martínez-Peiró, E. I. Boemo, and L. Wahammar, "Design of high-speed multiplierless filters using a nonrecursive signed common subexpression algorithm," *IEEE Trans. Circuits Syst. II*, vol. 49, pp. 196–203, Mar. 2002.

[10] D. R. Bull and D. H. Horrcks, "Primitive operator digital filter," *Proc. Inst. Elect. Eng. – Circuits Devices Systems*, vol. 138, pp. 401–412, June 1991.

[11] A. Dempster and M. D. Macleod, "Use of minimum adder multiplier blocks in FIR digital filters," *IEEE Trans. Circuits Syst. II*, vol. 42, pp. 569–577, Sept. 1995.

[12] H.-J. Kang and I.-C. Park, "FIR filter synthesis algorithms for minimizing the delay and the number of adders," *IEEE Trans. Circuits Syst. II*, vol. 48, pp. 770–777, Aug. 2001.

[13] C.-Y. Yao, "A study of SPT-term distribution of CSD numbers and its application for designing fixed-point linear phase FIR filters," Proc. ISCAS'01, Sydney, Australia, pp. II-301–II-304, May 2001.

[14] J. H. McClellan, T. W. Parks, and L. R. Rabiner, "A computer program for designing optimum FIR linear phase digital filters," IEEE Trans. Audio Electroacoust., vol. AU-21, pp. 506–526, Dec. 1973.

[15] Y. C. Lim and S. R. Parker, "Discrete coefficient FIR digital filter design based upon an LMS criteria," *IEEE Trans. Circuits Syst.*, vol. CAS-30, pp. 723–739, Oct. 1983.

IMPROVING THE FILTER BANK OF A CLASSIC SPEECH FEATURE EXTRACTION ALGORITHM

Mark D. Skowronski and John G. Harris

Computational Neuro-Engineering Lab
University of Florida, Gainesville, FL, USA
markskow,harris@cnel.ufl.edu

ABSTRACT

The most popular speech feature extractor used in automatic speech recognition (ASR) systems today is the mel frequency cepstral coefficient (mfcc) algorithm. Introduced in 1980, the filter bank-based algorithm eventually replaced linear prediction cepstral coefficients (lpcc) as the premier front end, primarily because of mfcc's superior robustness to additive noise. However, mfcc does not approximate the critical bandwidth of the human auditory system. We propose a novel scheme for decoupling filter bandwidth from other filter bank parameters, and we demonstrate improved noise robustness over three versions of mfcc through HMM-based experiments with the English digits in various noise environments.

1. INTRODUCTION

Davis and Mermelstein (D&M) coined the term 'mel frequency cepstral coefficients' (mfcc) in 1980 when they combined nonuniformly-spaced filters with the discrete cosine transform (DCT) as a front-end algorithm for automatic speech recognition (ASR) [1]. The algorithm can be summarized as follows: a signal passes through a triangular filter bank, spaced on a linear-log frequency axis, and the energy output from each filter is log-compressed and transformed via the DCT to cepstral coefficients. Previous work by Pols [2] showed that the eigenvectors of the log-energy output from his filter bank resembled cosine basis vectors for Dutch vowel data; hence, the DCT provides a quasi-PCA decorrelation of the log-energy, which allows for low-time liftering in the cepstral domain (dimension reduction). The log function provides compression of the dynamic range of filter bank output energies while also making the distribution of output energies more Gaussian. Therefore, all other functionality of mfcc is attributed to characteristics of the bank of triangular filters (see Figure 1). The energy calculation of each filter smooths the speech spectrum, reducing the effects of pitch, while the warped frequency scale provides variable sensitivity to the speech spectrum inspired by the human auditory system.

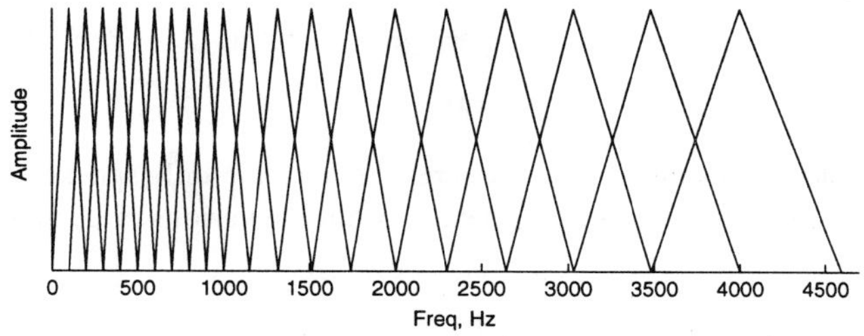

Figure 1: Filter bank of Davis and Mermelstein's mfcc algorithm. The filter bank is comprised of 10 linearly-spaced centers below 1 KHz and 10 log-spaced filters above 1 KHz. The base of each triangular filter is determined by the center frequencies of the neighboring filters, and all filters are unity height.

As seen in Figure 1, the bandwidth of each filter (the principle factor determining spectral smoothing) is arbitrarily set by fixing the base of each triangular filter by the center frequencies of the neighboring filters. Furthermore, popular variations of the mfcc filter bank, in an effort to accommodate data of sampling frequencies greater than 8 KHz, have increased the number of filters present and changed the function for frequency warping *without regard to changes in filter bandwidth that these modifications incur.* For example, Malcolm Slaney's Matlab version of mfcc [3] doubles the number of filters, effectively halving the bandwidth of D&M's filters, and Steve Young's HMM Toolkit (HTK) [4], a principle tool in C/C++ for large vocabular ASR for labs throughout the world, features an mfcc function that allows the user to select frequency range and number of filters for the filter bank (but not bandwidth!). These methods, as well as D&M's original version, are limited by the fact that filter bandwidth is not an independent design parameter; instead, bandwidth is determined by the filter spacing. Bandwidth should at least be related to filter center frequency, as inspired by the critical bands of the human auditory system.

In this paper we introduce a novel scheme for determining filter bandwidth, based on the approximation of critical

0-7803-7761-3/03/$17.00 ©2003 IEEE

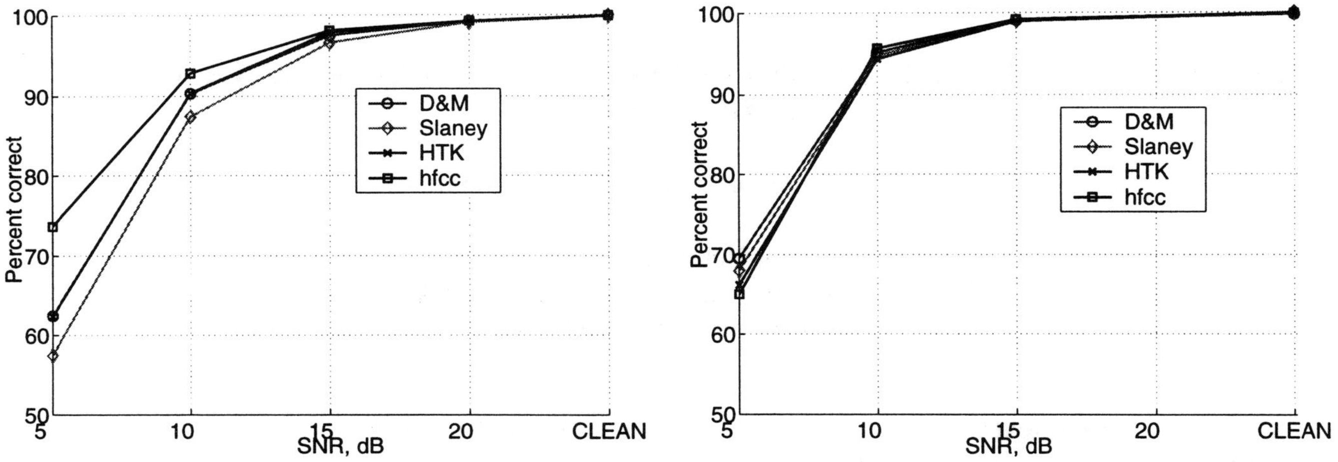

(a) (b)

Figure 2: Absolute recognition results for male-only speakers (50% test/train), averaged over 10 trials, for (a) white noise and (b) pink noise. See text for details about the various mfcc algorithms.

band equivalent rectangular bandwidth (ERB) from Moore and Glasberg [5]. The new scheme, called human factor cepstral coefficients (hfcc), decouples bandwidth from other filter bank design parameters (frequency range, number of filters), allowing for independent design and optimization of bandwidth. We show through ASR experiments that hfcc, using Moore and Glasberg's expression for critical band ERB, produces recognition results at or above those of three popular versions of mfcc (D&M, Slaney, and HTK) in various noise environments. We go on to show that, since bandwidth is an independent design parameter in hfcc, we can improve performance even more by *increasing* filter bandwidth beyond that of Moore and Glasberg's ERB expression.

2. FILTER DESIGN IN HFCC

The key aspect to hfcc is the decoupling of filter bandwidth. The given design parameters are sampling frequency f_s (frequency range $[f_{\min}, f_{\max}]$), number of filters N, and frequency warping function. We use Fant's expression [6] relating mel frequency $\hat{f}$ to linear frequency f:

$$\hat{f} = 2595 \log_{10}(1 + \frac{f}{700}). \quad (1)$$

Let f_{l_i}, f_{c_i}, and f_{h_i} be the low, center, and high frequencies for the ith filter in linear frequency, and let $f_{\min}$ and $f_{\max}$ define the frequency range for the entire filter bank. We require that center frequencies are equally-spaced in mel frequency and that the filters are equilateral in mel frequency. That is,

$$\hat{f}_{c_i} = \frac{1}{2}(\hat{f}_{h_i} + \hat{f}_{l_i}). \quad (2)$$

The steps for filter bank design are summarized as follows (see [7] for complete details):

1. Determine the first and last filter's center frequency. The two equations needed to solve for f_{c_i} come from Equation 2 as well as from the expression of ERB for a triangular function and Moore and Glasberg's ERB expression:

$$\begin{aligned} (700 + f_{c_i})^2 &= (700 + f_{h_i})(700 + f_{l_i}) \\ a f_{c_i}^2 + b f_{c_i} + c &= \frac{1}{2}(f_{h_i} - f_{l_i}) \end{aligned} \quad (3)$$

   where f_{c_i} in Hz and $a = 6.23 \ 10^{-6}, b = 93.39 \ 10^{-3}$, and $c = 28.52$ [5].

2. Find the remaining center frequencies:

$$\hat{f}_{c_i} = \hat{f}_{c_1} + (i-1)\frac{\hat{f}_{\max} - \hat{f}_{\min}}{N-1} \quad (4)$$

3. Find lower and upper frequencies:

$$\begin{aligned} (700 + f_{c_i})^2 &= (700 + f_{l_i} + 2\mathrm{ERB}_i)(700 + f_{l_i}) \\ f_{h_i} &= f_{l_i} + 2\mathrm{ERB}_i \end{aligned} \quad (5)$$

4. Construct filter in frequency domain by connecting straight lines between f_{l_i} and f_{c_i} and between f_{c_i} and f_{h_i}. The triangle has zero height at each end and unity height at f_{c_i}.

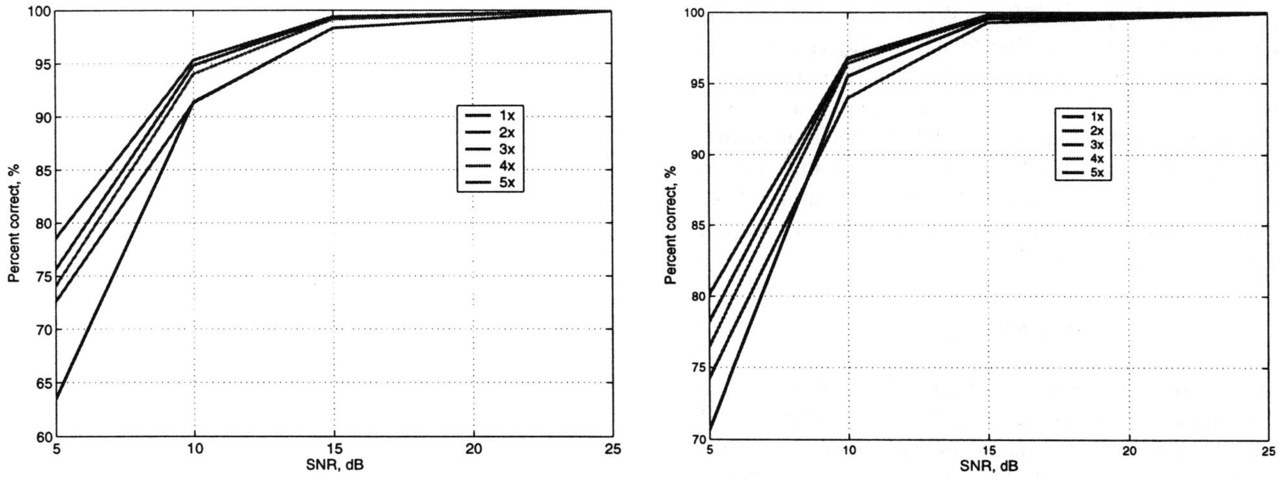

(a) (b)

Figure 3: Absolute recognition results for male-only speakers (50% test/train), averaged over 6 trials for (a) white noise and (b) pink noise. Filter bandwidth (ERB expression) is scaled by a constant (1x, 2x, ...) during filter bank construction.

3. EXPERIMENTS

We evaluate the performance of hfcc and mfcc through ASR experiments. HMM word models for each of the English digits 'zero' through 'nine' are constructed from utterances taken from the TI-46 corpus of isolated digits. Three versions of filter banks for mfcc are included in the tests: 1) D&M's original scheme, 2) Malcolm Slaney's lin-log-spaced Matlab function, and 3) HTK's C++ function [7].

Cepstral mean subtraction is applied to all feature vectors, and Δ coefficients with a lag of ± 4 frames are appended to the original 13 cepstral coefficients (26 coefficients total per window). HMMs using Gaussian mixture models are constructed with Compaq's Probabilistic Model Toolkit for Matlab [8]. Models are trained with noise-free utterances, while noisy test utterances are generated at various signal-to-noise ratios (SNR) by adding noise from the Noisex92 database [9]. For our experiments, we chose white, pink, and babble noise sources.

4. RESULTS

Figure 2(a) shows the absolute results for male-only speakers in white noise, while (b) shows the absolute results in pink noise. Results in babble noise showed no significant difference between all feature extraction algorithms and are not included here. This is expected, since smoothing of the spectrum does not reduce the effects of the non-stationary babble noise. When compared relative to D&M over each trial in white noise, hfcc increases recognition by up to 12 ± 5 percentage points at 5 dB SNR (10 trials of random test/train speakers). Notice that all algorithms perform near-perfectly when no noise is present in the utterance for this vocabulary. Also, hfcc has made no assumptions about the various noise sources–robustness is increased due to the emphasis of frequency ranges in the noisy spectrum. The hfcc algorithm, using Moore and Glasberg's ERB, shows significant robustness to white noise over the other mfcc methods, while the performance of hfcc in pink noise is slightly worse than that of mfcc, particularly D&M's version.

In a second experiment, we scaled Moore and Glasberg's ERB by integer factors between 1.0 and 5.0. Previous experiments with widening schemes showed improved noise robustness [10] for the wider filters,yet with the log compression in the algorithms, analytic analysis is difficult. Figure 3 shows recognition results for the same ASR experimental setup as before using various ERB scale factors in white and pink noise. Marked improvement, up to 15 ± 6 percentage points for white noise and 10 ± 5 percentage points for pink noise above the filter bank of unscaled bandwidth (1x) relative for each trial were achieved with a scale factor of 3 (6 trials).

5. CONCLUSIONS

We have introduced a novel scheme for designing the filter bank in mfcc that decouples filter bandwidth from other filter bank design parameters. By creating filter bandwidth as an independent design parameter, hfcc allows one to increase ASR performance by improving the tradeoff between noise smoothing and resolution of spectral characteristics [11]. As filter bandwidth increases, the more samples are present in the filter to estimate log energy. Therefore, the variance of the estimate decreases. Concurrently, spectral details are blurred by wide filters. Without analytic expressions for the distributions (due to the nonlinear complexity of the feature extraction algorithm), ASR experiments are one of the few useful tools in characterizing this tradeoff. Experiments with simplified vowel models indicate that filter bandwidth affects the trajectory of the feature pdf as SNR decreases, though it currently remains a topic of research to characterize the trajectory behavior as a function of filter bank design parameters.

6. REFERENCES

[1] Steven B. Davis and Paul Mermelstein, "Comparison of parametric representations for monosyllabic word recognition in continuously spoken sentences," *IEEE Trans. Acoust., Speech, Signal Processing*, vol. 28(4), pp. 357–366, 1980.

[2] L. C. W. Pols, *Spectral analysis and identification of Dutch vowels in monosyllabic words*, Ph.D. thesis, Free University, Amsterdam, The Netherlands, 1977.

[3] Malcolm Slaney, *Auditory Toolbox, Version 2, Technical Report No: 1998-010*, Internal Research Corporation, 1998.

[4] S. J. Young et. al., *The HTK Book*, Entropics Cambridge Research Lab, 1995.

[5] B. C. J. Moore and B. R. Glasberg, "Suggested formula for calculating auditory-filter bandwidth and excitation patterns," in *J. Acoust. Soc. America.*, 1983, vol. V74, pp. 750–753.

[6] C. G. M. Fant, "Acoustic description and classification of phonetic units," *Ericsson Technics*, vol. 15, no. 1, 1959, reprinted in *Speech Sound and Features*, MIT Press, Cambridge, 1973.

[7] M. D. Skowronski and J. G. Harris, "Human factor cepstral coefficients," *IEEE Trans. Speech and Audio Processing*, Submitted July 2002.

[8] http://research.compaq.com/downloads.html.

[9] http://spib.rice.edu/spib/select_noise.html.

[10] M. D. Skowronski and J. G. Harris, "Increased mfcc filter bandwidth for noise-robust phoneme recognition," *International Conference on Acoustics, Speech, and Signal Processing*, vol. 1, pp. 801–4, 2002.

[11] M. D. Skowronski and J. G. Harris, "Human factor cepstral coefficients," *Acoustical Society of America First Pan-American/Iberian Meeting on Acoustics*, December 2002.

A New Heuristic Signed-Power of Two Term Allocation Approach for Designing of FIR filters

Tetsuya FUJIE† Rika ITO†† Kenji SUYAMA††† Ryuichi HIRABAYASHI††

† School of Economics and Business Administration, Kobe University of Commerce
††Faculty of Engineering, Science University of Tokyo
†††School of Engineering, Tokyo DENKI University

Abstract

In this paper, we consider design problems of linear phase FIR filter with CSD(or SP2) coefficients. When the total number of non zero SP2 terms is given for the design problem, we have to determine the number of non zero SP2 terms allocated for each filter coefficient respectively, while keeping the total number. However, it is considered this problem NP-hard problem. Hence, Lim et al. [4] developed a heuristic method for this allocation problem. In this paper, we propose a new heuristic method for this problem comparing it with Lim et al.'s heuristic method through numerical experiments.

1 Introduction

In these decades, several methods have been proposed for the design of FIR filters with SP2(CSD) coefficients [1] - [9]. It is well known that the filters whose coefficients are represented as SP2 terms enable implementation of the filters without using multipliers. However, it is difficult to design such filters since it results in an integer programming problem (IP), which is well-known as one of the NP-hard problems [10]. It has been demonstrated that the advantage can be achieved if the coefficient values are allocated with different number of SP2 terms while keeping the given total number of SP2 terms. However this optimization problem is one of the most difficult problems called allocation problems, since we have to obtain the optimal filter coefficients based on that allocated SP2. Hence, it can be effective to use heuristic method for this optimization problem. A well known heuristic method for this allocation problem is given by Lim, Yang, Li and Song [4]. In this paper we propose a new improved heuristic method for this allocation problem. We compare it with Lim et al.'s method through numerical experiments and we show our heuristic method is effective for both relaxation and the original problem.

2 Allocation Problem of m_k

In this paper, we consider a design problem of FIR filter with an odd length and even symmetric linear phase low pass filter for simplicity. The optimization problem is formulated as a problem to approximate the frequency response to the desired frequency response function. In the first, we consider the continuous coefficients case. Then the transfer function of the FIR filter that minimizes an continuous error function e is denoted as

$$H_c(z) = \sum_{k=0}^{N-1} \bar{h}_k z^{-k}. \quad (1)$$

The total number of non-zero SP2 coefficients is given as M, and m_k represents the number of non-zero SP2 terms for the kth filter coefficient. The SP2 coefficients are represented as d_k for each coefficients $\bar{h}_k$. Then the transfer function with d_k is as follows,

$$H(z) = \sum_{k=0}^{N-1} d_k z^{-k}. \quad (2)$$

The value $|d_k|$ of each SP2 coefficient ranges between 2^0 and 2^{-U}. Here U is a natural number. For example, if m_k is given, then d_k in (2) is represented as follows,

$$d_k = \sum_{i=1}^{m_k} b_i^{(k)} 2^{-q_i^{(k)}}. \quad (3)$$

Here $b_i^{(k)} \in \{-1, 1\}$ and $q_i^{(k)} \leq U$ $(1 \leq i \leq m_k,\ 0 \leq k \leq N-1)$.

Lim et al. proposed a heuristic method to determine m_k from $\bar{h}_k$.

[Lim et al.'s method]
Step 1: Let m_k be the number of SP2 terms. $m_k = 0$, $c_k = 0.36 \log_2(|\bar{h}_k|)$ for all of k. $c_0 = c_0 - 0.126$.
Step 2: Let c_i be the largest of c_n for all n.
Step 3: $m_i = m_i + 1$, $c(i) = c(i) - 1$ and $M = M - 1$.
Step 4: If $M = 0$, stop; otherwise, go to Step 2.

Our proposed heuristic method to determine m_k from $\bar{h}_k$ is as follows:

0-7803-7761-3/03/$17.00 ©2003 IEEE

[Our proposed method]
Let $\bar{\boldsymbol{h}} = [\bar{h}_0, \ldots, \bar{h}_{N-1}]^T$ and $\boldsymbol{e}_k$ a unit vector where the k th element is 1 and the rest is 0. Our basic idea is as follows. We aim to obtain an SP2 solution $\boldsymbol{d}$ from $\bar{\boldsymbol{h}}$ so that its error function f doesn't increase so much (see Section 3). For this, we change the value of k th element by λ for each k, and calculate $f(\bar{\boldsymbol{h}} + \lambda \boldsymbol{e}_k)$. Here $\lambda = \lceil \bar{h}_k \rceil - \bar{h}_k$, $\lambda = \bar{h}_k - \lfloor \bar{h}_k \rfloor$ and, $\lceil \bar{h}_k \rceil$ is the least SP2 upper bound and $\lfloor \bar{h}_k \rfloor$ is the largest SP2 lower bound for each continuous coefficient $\bar{h}_k$ in (1). Our selection of k depends on the calculation of $f(\bar{\boldsymbol{h}} + \lambda \boldsymbol{e}_k)$. After selecting k, we add 1 to m_k. The value of $\lceil \bar{h}_k \rceil - \bar{h}_k$ and $\bar{h}_k - \lfloor \bar{h}_k \rfloor$ for selected k approaches $\bar{h}_k$ more and more. If f is a convex function, it is expected to obtain SP2 solutions so as not to increase the value much beside $\bar{\boldsymbol{h}}$.
Step 1: For all of k, if $|\bar{h}_k| < \varepsilon$, $m_k = 0$; otherwise $m_k = 1$.
Step 2: Evaluate the following equations for all of k

$$f_k^{\mathrm{U}} = f(\bar{\boldsymbol{h}} + (\lceil \bar{h}_k \rceil - \bar{h}_k)\boldsymbol{e}_k),$$
$$f_k^{\mathrm{L}} = f(\bar{\boldsymbol{h}} + (\bar{h}_k - \lfloor \bar{h}_k \rfloor)\boldsymbol{e}_k).$$

Step 3: Calculate f_k from f_k^{U}, f_k^{L} (see below) and $f_{k^*} = \max_k\{f_k\}$.
Step 4: Let $m_{k^*} = m_{k^*} + 1$ and $M = M - 1$.
Step 5: If $M = 0$, stop; otherwise, go to Step 2.

In general, the computational cost is not large to calculate $f(\boldsymbol{d})$, so it is considered the computational cost of our algorithm is as much as that of Lim et al. We propose the following approaches as a calculation of f_k (Step 3).
`mode1`: (max-max rule) $f_k = \max\{f_k^{\mathrm{U}}, f_k^{\mathrm{L}}\}$.
`mode2`: (max-min rule) $f_k = \min\{f_k^{\mathrm{U}}, f_k^{\mathrm{L}}\}$.
`mode3`: (average rule) $f_k = f_k^{\mathrm{U}} + f_k^{\mathrm{L}}$.
`mode4`: $f_k = \lceil d_k \rceil - \lfloor d_k \rfloor$.

3 Determination of the Filter Coefficients

When $m_0, \ldots, m_N$ are given, the abstract representation is of the optimization problem to determine the filter coefficients $\boldsymbol{d}$ is as follows. Here $\boldsymbol{d} = [d_0, \ldots, d_N]^T$.

$$\begin{array}{ll} \min & f(\boldsymbol{d}) \\ \text{sub. to} & \text{the number of SP2 terms for } d_k \\ & \quad \text{is at most } m_k \quad (k = 0, \ldots, N). \end{array} \tag{4}$$

For the weighted least square error problem, $f(\boldsymbol{d})$ is

$$f(\boldsymbol{d}) = \boldsymbol{d}^T \boldsymbol{Q} \boldsymbol{d} - 2\boldsymbol{q}^T \boldsymbol{d} + \text{const.}, \tag{5}$$

where

$$\begin{aligned} \boldsymbol{Q} &= \int_0^{\pi} W(\omega)[\boldsymbol{c}(\omega)\boldsymbol{c}^T(\omega)]d\omega, \\ \boldsymbol{q} &= \int_0^{\pi} W(\omega)[\boldsymbol{c}(\omega)H_d(\omega)]d\omega, \\ \boldsymbol{c}(\omega) &= [1, \cos\omega, \ldots, \cos N\omega]^T. \end{aligned} \tag{6}$$

Here, $H_d(\omega)$ is a desired frequency response function.

It is reported that if a suitable frequency response weighting function $W(\omega)$ is used, a quasi-equiripple design can be obtained [3], [12].

4 How to Solve the Problem(4)

4.1 Lu's method

For designing FIR filters with SP2 coefficients, several relaxation problems are proposed. The following relaxation problem of (4) and the polynomial time approximated algorithm are proposed by Lu [11].

$$\begin{array}{ll} \min & f(\boldsymbol{d}) \\ \text{sub. to} & d_k \in \{\lceil \bar{h}_k \rceil, \lfloor \bar{h}_k \rfloor\} \ (k = 0, \ldots, N). \end{array} \tag{7}$$

The detail of this algorithm is skipped here. We executed numerical experiments in [13] to show our proposed method is effective for obtaining both the solution for (4) and the solution for (7).

4.2 B&B technique

The algorithm to solve for (4) by using B & B is represented as follows. Each subproblem is expressed as

$$\begin{array}{ll} \min & f(\boldsymbol{d}) \\ \text{sub. to} & \ell_k \le d_k \le u_k \quad (k = 0, \ldots, N), \\ & \text{the number of SP2 terms for } d_k \\ & \quad \text{is at most } m_k \quad (k = 0, \ldots, N). \end{array} \tag{8}$$

$\ell_k = 0, u_k = 1$ $(k = 0, \ldots, N)$ in the problem (8) corresponds to (4) (original problem). The following relaxation problem whose constraints on SP2 are eliminated is solved.

$$\begin{array}{ll} \min & f(\boldsymbol{d}) \\ \text{sub. to} & \ell_k \le d_k \le u_k \quad (k = 0, \ldots, N). \end{array} \tag{9}$$

Since the relaxation problem(9) is a convex problem with lower and upper bound constraints, it can be solved easily. Let $\overline{\boldsymbol{d}}$ be an optimal solution of (9). Then we calculate $\lceil \overline{d}_k \rceil$ and $\lfloor \overline{d}_k \rfloor$ for $\overline{d}_k$, and operate as follows.
(i) If $\lceil \overline{d}_k \rceil = \lfloor \overline{d}_k \rfloor$ $(k = 0, \ldots, N)$, $\overline{\boldsymbol{d}}$ is an optimal solution of the subproblem. Hence, if the objective function value $f(\overline{\boldsymbol{d}})$ is smaller than the value of the incumbent best solution, we let $f(\overline{\boldsymbol{d}})$ be a new incumbent value.
(ii) If k exists such that $\lceil \overline{d}_k \rceil$ is bigger than $\lfloor \overline{d}_k \rfloor$, we select a k and generate a subproblem in which u_k is changed to $\lfloor \overline{d}_k \rfloor$ and a subproblem in which ℓ_k is changed to $\lceil \overline{d}_k \rceil$.

We continue these operations until we finish searching all of subproblems. We adopted best lower bound first search rule as a rule of selection of unsearched subproblems. There are some selection rules of k in (ii). For example we can consider $\max_k\{\lceil \overline{d}_k \rceil - \lfloor \overline{d}_k \rfloor\}$, $\max_k\{u_k - \ell_k\}$, and their combination. However we adopt the rule $\max_k\{\lceil \overline{d}_k \rceil - \lfloor \overline{d}_k \rfloor\}$ since it is found that this rule is superior to others as a result of preliminary experiments.

5 Numerical Experiments

5.1 Comparison of Problems (4) and (7)

We demonstrate the comparison of solutions for (4) and (7) for the weighted least square error problem. We assumed $M = 2 \times (N+1)$. Moreover, we set $N = 7$, $\omega_p = 0.450$, $\omega_s = 0.550$, $U = 20$ and $W(\omega) = 1$ $(\omega \in [0, \omega_p])$, $W(\omega) = \beta$ $(\omega \in [\omega_s, \pi])$. β is changed in several pattern. Figure 1 shows each optimal value of (4) and (7) for all $(m_0, \ldots, m_N)$ which satisfy $\sum_{k=0}^{N} m_k = M, m_k \geq 1$ $(k = 0, \ldots, N)$. Here the abscissa is the optimal value of (4), and the ordinate is the optimal value of (7). In Figure 1, it is

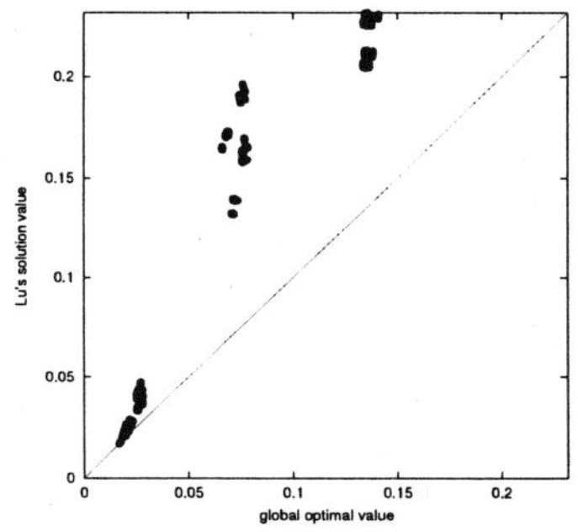

Figure 1: $N = 7, \beta = 10.0$

clear that the optimal value spreads throughout and the optimal value depends on the allocation of m_k. Hence these results show that the allocation of m_k is essential for designing FIR filters.

5.2 Comparison of m_k Allocation Methods (1)

Let `all2` denote an allocation of two non-zero SP2 terms for each d_k, i.e., $(m_0, \ldots, m_N) = (2, \ldots, 2)$. The following figures (Figures 2, 3 and 4) show the plot of optimal solutions of (4) and (7) by using `all2`, Lim et al.'s algorithm (`Lim`) and our proposed algorithm (`mode2`).

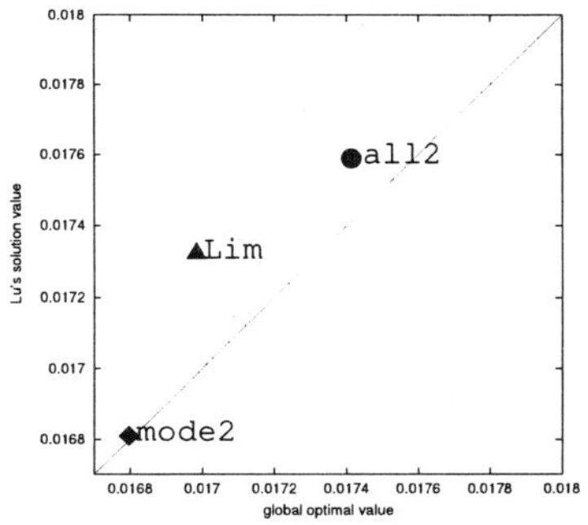

Figure 2: $N = 7, \beta = 10.0$ (extended figure)

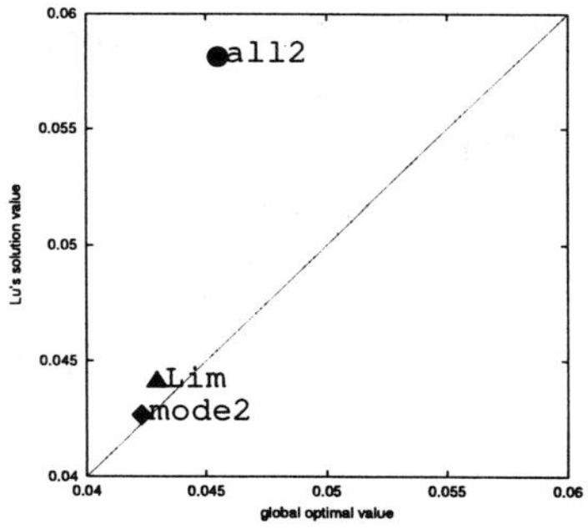

Figure 3: $N = 7, \beta = 100.0$ (extended figure)

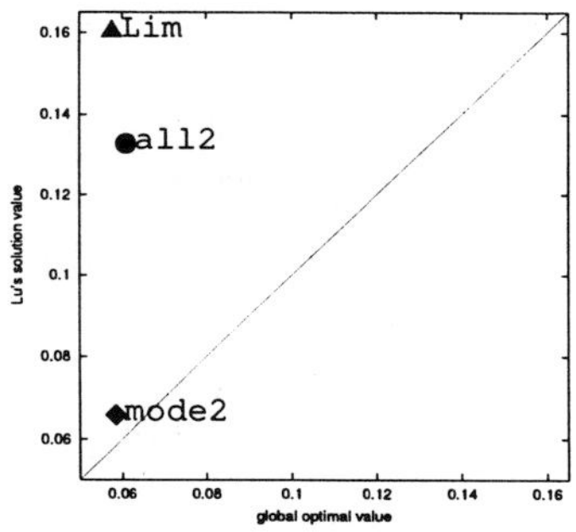

Figure 4: $N = 7, \beta = 500.0$ (extended figure)

From these results, it is demonstrated that the optimal value of our proposed method `mode2` is the smallest of all, and `mode2` is the most qualified solution for (4) and (7). Furthermore it is also successful in obtaining the smallest value even if β is changed while `Lim` not.

5.3 Comparison of m_k Allocation Methods (2)

The following figures (Figures 5 and 6) show the continuous solution of (4) $(\beta = 10.0, N = 5, \ldots, 40)$, and the optimal solution of (4) by `all2`, `Lim` and `mode1`,...,`mode4`. A transition of the optimal value corresponding to each N is shown in these figures. An abscissa is N, and an ordinate is the optimal value. From these figures, it is demonstrated that the optimal value by `mode2` is smaller than that of all others for almost all N $(N = 5 \ldots 40)$. According to this numerical experiments, it is indicated that our proposed allocation method is more successful in improving the performance of the filter than traditional allocation method. And it is also shown that our allocation algorithm `mode2` is the most effective even for (7).

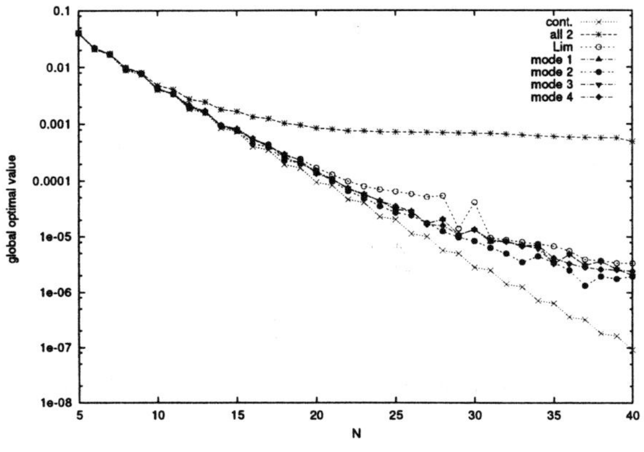

Figure 5: $\beta = 10.0$

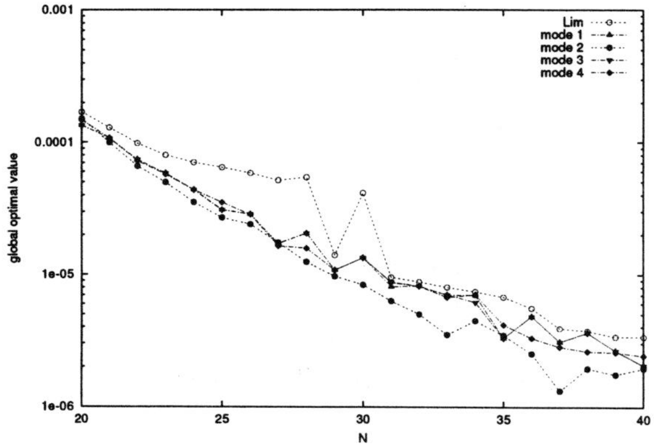

Figure 6: $\beta = 10.0$ (2)

6 Conclusion

In this paper, we proposed a new heuristic SP2 term allocation method for the design of FIR filters with SP2 coefficients. We executed numerical experiments to verify its effectiveness for improving the performance of the filter.

References

[1] D.M. Kodek and K. Steiglitz, "Finite-length wordlength tradeoffs in FIR digital filter design," *IEEE Trans. Acoustics, Speech, Signal Processing,* vol. ASSP-28, pp.739-744, December 1980.

[2] Y.C. Lim and S.R. Parker, "FIR filter design over a discrete power-of-two coefficient space," *IEEE Trans. Acoustics, Speech, Signal Processing,* vol. ASSP-31, pp.583-591, June 1983.

[3] Y.C. Lim and S.R. Parker, "Discrete coefficient FIR digital filter design based upon LMS criteria," *IEEE Trans. Circuits, Syst.,* vol. CAS-30, pp.723-739, Oct. 1983.

[4] Y.C. Lim, R. Yang, D. Li and J. Song, "Signed power-of-two (SPT) term allocation scheme for the design of digital filters," *Proc. ISCAS,* Monterey, CA., June 1998.

[5] Y.C. Lim and Y.J. Yu, "A successive reoptimization approach for the design of discrete coefficient perfect reconstruction lattice filter bank," *Proceedings of ISCAS'2000,* vol. 2, pp.69-72, Geneva, JUne 2000.

[6] S.C. Pei and S.B. Jaw, "Efficient design of 2-D mulitiplierless FIR filters by transformation," *IEEE Transactions on Circuits and Systems,* vol. 34, pp. 436-438, April 1987.

[7] M. Suk and S.K. Mitra, "Computer-aided design of digital filters with finite wordlength," *IEEE Trans. Audio Electroacoust.,* vol. AU-20, pp. 356-366, December 1972.

[8] P.P. Vaidyanathan, "Efficient and multiplierless design of FIR filters with very sharp cutoff via maximally flat building blocks," *IEEE Transaction on Circuits and Systems,* vol. 32, pp. 236-244, March 1985.

[9] J.T. Yli-Kaakinen and T.A. Saramaki, "An Algorithm for the design of multiplierless approximately linear-phase lattice wave digital filters," *Proc. ISCAS'2000,* vol. II, pp. 77-80, Geneva, June 2000.

[10] C.H. Papadimitriou and K. Steiglits, *Combinatorial Optimzation,* Prentice-Hall, 1982.

[11] W.-S. Lu, "Design of FIR Filters with discrete coefficients: A Semidefinite Programming Relaxation Approach", ISCAS-2001.

[12] Y.C. Lim, J. Lee, C.K. Chen and R. Yang, "A weighted least squares algorithm for quasi-equiripple FIR and IIR digital filter design," *IEEE Transactions on Sigmal Processing,* vol. 40, pp.551-558, March. 1992.

[13] R. Ito, T. Fujie, K. Suyama and R. Hirabayashi, "New Design Methods of FIR Filters with Sined Power of Two Coefficients using A New Linear Programming Relaxation with Triangle Inequalties," *Proc. ISCAS'2002,* pp.I-813-816, 2002.

RECONFIGURABLE IMPLEMENTATION OF RECURSIVE DCT KERNELS FOR REDUCED QUANTIZATION NOISE

Süleyman Sırrı Demirsoy, Robert Beck, Andrew G. Dempster and Izzet Kale

Applied DSP and VLSI Research Group, Department of Electronic Systems
University of Westminster, 115 New Cavendish St, London, W1W 6UW, UK
Tel: +44 20 7911 5000 - 3611(ext) e-mail: demirss@cmsa.wmin.ac.uk

ABSTRACT

Time multiplexed implementations of the recursive DCT processors are widely used in many multimedia and compression applications. Recently proposed three Goertzel kernels offer significant improvement (up to 90 %) in the noise performance of the time-multiplexed architecture to allow word-length specifications get reduced. In this paper, a highly optimized reconfigurable DCT architecture is proposed that can perform the function of three different kernels (Type A, B and C) on Virtex FPGA.

1. INTRODUCTION

Recursive DCT implementations are attractive due to their regular structure and reduced computational complexity. The transfer function of the DCT given in (1) can be implemented with a second order IIR filter. The kernel employed in most of the designs in the literature [6], [7] is a resonator of Type B configuration as shown in Figure 1(a). In [2], two alternative forms, Type A and Type C were proposed (Figure 1b and 1c).

$$H(z) = \frac{P_k(1 - z^{-1})}{1 - 2\beta_k z^{-1} + z^{-2}} \qquad (1)$$

$P_k = (4L_k / N)Cos(k\pi / N)$, k: frequency bin index

$L_k = 1/\sqrt{2}$ for k=1 , $L_k = 1$ for all other k

$2\beta_k = 2Cos(k\pi / N)$ N: transform length

The benefits of these structures are investigated in [1] and substantial area gains in a fully parallel implementation were demonstrated. In that design, Type A, Type B and Type C structures were all employed for different frequency bins to have the optimum coefficient magnitude. Figure 2 shows how these magnitudes vary for different frequencies. By using Type A for the first one-third of the filter bins, Type B for the next one-third and Type C for the last one-third of the filter bins, the optimum coefficient values are used.

For a time-multiplexed system, where only one recursive structure is used for all frequency bins by multiplexing it in time, Type B is the natural choice to employ for having the average coefficient magnitude among the three types. However, if a reconfigurable structure that can be transformed into three different types for different frequencies is developed with minimum reconfigurability overhead, a significant improvement on internal quantization noise figures can be achieved.

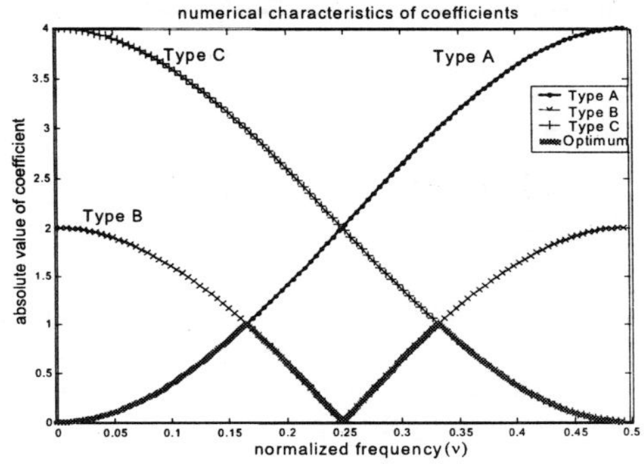

Figure 2 Coefficient magnitude behaviour of Type A, B and C structures vs. frequency. The optimum behaviour is also marked.

FPGAs are attractive platforms for easy and fast implementation and offer high performance when the available resources are used efficiently. The effective use of the Look-Up Tables (LUT) available on-board in Xilinx Virtex FPGAs is investigated in [3].

In this paper, a reconfigurable recursive kernel that is highly optimized for the Xilinx Virtex FPGA series is given and investigated for its noise performance. Section 2 describes the details of the Virtex slice architecture and the optimization steps of the recursive kernels. Section 3 gives the noise performance analysis and Section 4 concludes the paper.

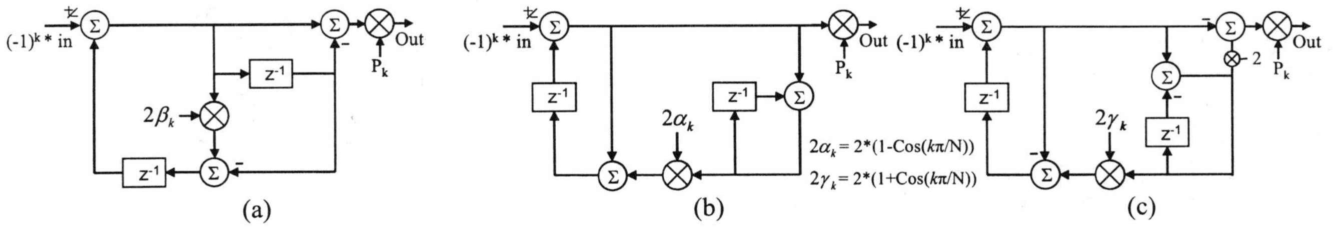

Figure 1 Recursive Goertzel filters, a) Type B, b) Type A, c) Type C

0-7803-7761-3/03/$17.00 ©2003 IEEE

2. OPTIMIZATION

The Virtex FPGA series comprise Configurable Logic Blocks (CLB), which are the main programmable functional blocks. Each CLB contains two slices each of which has two of the structures given in Figure 3a. Fast carry logic is implemented in each half-slice by MUXCY. The Look-up Table (LUT) is a 16-bit SRAM with 4 inputs. It can be programmed to realise any logic function having up to four inputs. Figure 1b and 1c shows two examples for a 1-bit adder subtractor implementation on each half-slice, and a subtractor with a multiplexer choosing one of its two inputs for subtraction from A. The half-slice also has a D-type flip-flop that can function like an output register.

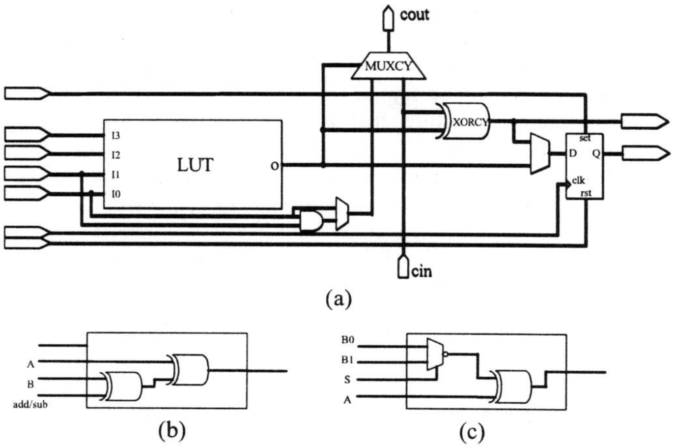

Figure 3 (a) Half of the Virtex Slice diagram (simplified), (b) A combination of XOR gates implemented in the LUT to make an adder/subtractor, (c) a multiplexer choosing the input to subtract from A

Any circuit optimization for a given device with limited resources would be a job of facilitating the available resources as much as possible, bearing in mind that the required operation can be satisfied by a modification on the structure that allows fitting the resources in a better way while maintaining the functionality. For the reconfigurable DCT kernel of this paper given in Figure 4a, there are several multiplexers (MX1 to MX5) which allow us to use the same components with different inputs. The letters A, B and C indicate the Type of the kernel a line is required for. The signs on top of the letters are the functionality that is required in the adder/subtractors (A1 to A4) for that input when the specified kernel is configured. The loop multiplier (M1) takes the coefficient of all kernels. The optimization is performed as follows:

The adder/subtractor A1 and the delay element D2 can fit into a half-slice if the multiplexer MX1 does not exist. Moreover, the adder/subtractor A4 consumes half a slice by itself without making use of the delay element. By replicating D2 as D2b and storing the output of A4 in D2b, it is possible to get rid of MX1. (See Figure 4b)

MX3 requires an extra LUT. However by simply feeding A3 with '0' from MX5, the output for Type A can be achieved via A3 from the same output as Type B and C. This implies we have three different outputs from MX5 (Figure 4c).

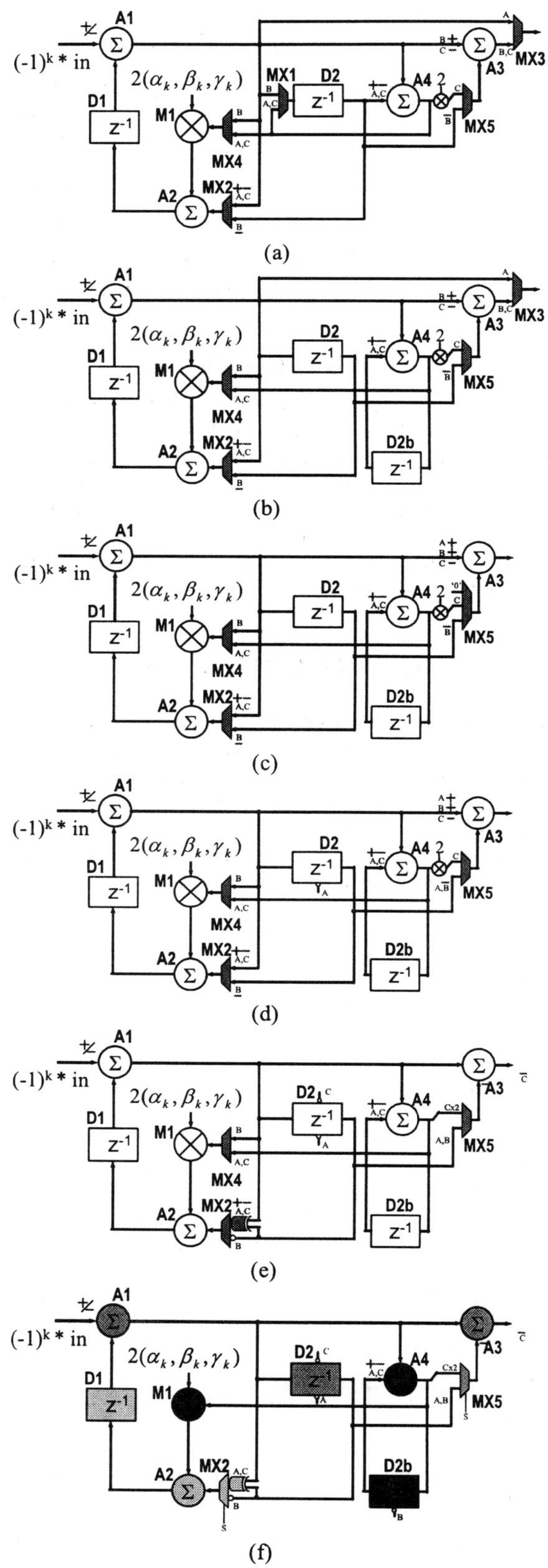

Figure 4 Optimization steps of the reconfigurable kernel

It is possible to generate the '0' signal by clearing D2. This reduces the number of necessary inputs to MX5 down to two. (Figure 4d)

The input to A5 coming out from A1 requires negation for type C and the other input needs to be negated for Type B too. This functionality of negation on both inputs for different kernels is not possible to implement in a half-slice. However by performing negation for Type C at the same input as Type B, it is possible to fit MX5 and A3 to the same half-slice. Hence the output of A3 for the Type C kernel needs to be negated. It is done by using the negated coefficients in the Pk multiplier outside the recursive loop shown in Figure 1.

The multiplication by 2 is performed by hardwiring the output of A4 one bit shifted toward the MSB.

The output of the MX2 needs to be negated for Type B and C. In its current form, it is not possible to fit MX2 with A2 to the same half-slice. It is known that for Type A, D2 generates a '0' signal for MX5. By generating a '1' signal on D2, the output of D2 can be used as a select signal for the operation on the other input to MX2. This modification allows fitting MX2, A2 and D1 to the same half-slice. (See Figure 4e)

MX4 requires an extra LUT. It is possible to get rid of this multiplexer by using the output of A4 for Type B kernel as well. A '0' signal generated by D2b would allow having the A1 signal at the output of A4, after adding a small amout of delay. (Figure 4f)

After all the modifications are performed, the reconfigurable DCT kernel occupies four half-slices where Type B kernel uses three half slices. This overhead is easily compensated by the reduction in the coefficient word-length of multiplier M1. Hence, if a general multiplier is used for M1, the reconfigurable structure needs the same number of half-slice as Type B. For the choice of fixed or Reconfigurable Multiplier Blocks (ReMB) [8], the necessary coefficients are smaller in magnitude and are constructed using less adder stages. An example to this fact is given in [8] for a coefficient word-length of 12-bits. The loop multiplier M1 for Type B kernel is constructed with four basic structures, whereas the loop multiplier for the reconfigurable kernel required only three basic structures.

3. NOISE PERFORMANCE

The quantization noise exists in a circuit due to the fixed word-length of the intermediate and output signals. The performance of the system is affected by the choice of the signal word-lengths. Several restrictions on the maximum quantization noise are defined by the standards to avoid performance degrade below a limit. For example, MPEG-4 standard defines the maximum allowed MSE, mean error and magnitude error that occurs in an 8 by 8 block of pixels in a video data for Two Dimensional Inverse DCT (2D-IDCT) [5].

The coefficient values required by the Type B kernel for the frequencies close to DC and half-Nyquist are higher than the values required by Type A near DC and Type C near half-Nyquist (see Figure 2). An extra bit for the integer part of the coefficient is required if just the Type B coefficients are used. Therefore by using the Type A and C kernels for these frequency bins, we save one bit for the same fractional precision. On top of that, the analysis of the integer parts of the intermediate signals shows that Type B kernel requires more bits for the integer parts of the signals than the Type ABC kernel does. Table 1 shows how many bits are required to represent the integer parts of the signals for a transform length of N=16. The reconfigurable kernel, Type ABC, would spare more bits for the representation of the fractional parts at the output of A1, A2, and M1, hence would lead to better noise performance.

Type	A1	A2	A3	M1	Loop	Pk	out	A4
B	15	15	16	16	2	0	11	-
ABC	13	13	16	13	1	0	11	15

TABLE 1 Number of bits required for the integer parts of the signals at the output of the specified components for N=16. Loop and Pk are the coefficient values for the loop multiplier (M1) and Pk multiplier (see Figure 1)

A set of experiments were performed to demonstrate the enhanced performance. For various word-length specifications, both Type B and Type ABC design were stimulated with a thousand uniformly distributed random numbers between –300 and 300. A reference design with floating point precision was used to find the quantization noise generated by the two designs. Our error measure for these experiments was the Mean Square Error (MSE). Table 2 shows the results for different specifications and transform-lengths (N). The loop coefficient word-length for Type B design (loopb) is always kept one bit more than the loop coefficient of the Type ABC design (loopa) to maintain the same area consumption when a general-purpose multiplier is used. Pk is the coefficient word-length for the Pk multiplier. The frequency bins k=0 and k=4 are neglected during the calculation of the percentage reduction in noise because these frequency bins only have noise contributions from the Pk multiplier which is outside the recursive kernel. The gain for each frequency bin is different. The maximum and minimum gains that occurred for a set are also shown in Table 2. For particular frequency bins, there is 93 % decrease in the noise. The average gain is a reasonable measure of the overall noise performance enhancement. It is observed that, there is up to 67 % decrease in the MSE levels among the given sets.

Word-length specifications	N	Max	Min	Avg
1. wl=18; loopa=17; loopb=18; Pk=16	8	74%	41%	60%
2. wl=20; loopa=18; loopb=19; Pk=16	8	71%	28%	35%
3. wl=20; loopa=18; loopb=19; Pk=16	16	88%	59%	67%
4. wl=22; loopa=21; loopb=22; Pk=18	32	93%	17%	67%

TABLE 2 The percentage decrease in the MSE noise with reconfigurable Type ABC design as opposed to Type B design.

Figure 5 shows the MSE noise levels for the design given in the 1st set of Table 2. As seen from the figure, although the loop coefficient precision for k=3 and k=5 are same for Type B and Type ABC designs, the difference of the magnitude of the intermediate signals leads to an enhancement in the noise performance of Type B kernel outputs in the reconfigurable design.

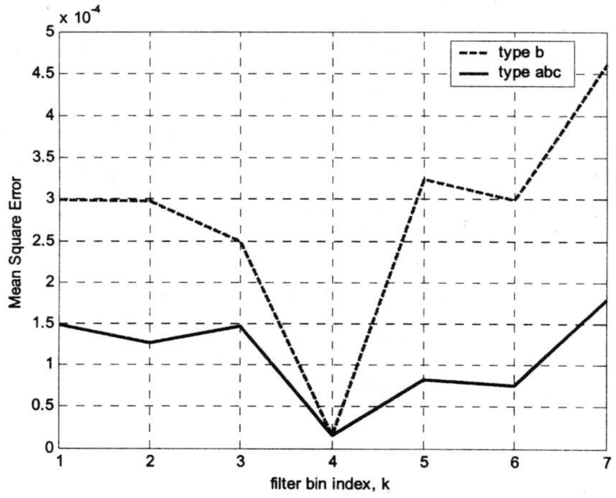

Figure 5 Comparison of MSE noise figures for N=8 and word-length of 18 bit. The coefficient word-lengths in Type B-only and reconfigurable kernel are 18 bit and 17 bit respectively. An average decrease of 67% is achieved.

4. CONCLUSION

A reconfigurable recursive DCT processor, that can compute three kernels (Type A, B and C) has been designed and optimized for Xilinx Virtex FPGA series. It occupies only one more half-slice than Type B kernel does for a single bit, due to high utilization of the slice resources. The loop multiplier coefficient requires one bit less for the same precision. For the same half-slice count, the MSE level is reduced by 67 % on average of all frequency bins in the proposed design when compared to the Type B kernel. The optimized structure includes design ideas that can be applicable to VLSI implementation too. Future work will focus on the pipelined implementation of this reconfigurable structure to increase the operating clock frequency.

5. REFERENCES

[1] Demirsoy S., et al., "Novel recursive DCT implementations: A comparative study", *IEEE Int. Conf. on Intelligent Data Acqusition and Advanced Computing Systems (IDAACS'2001)*, pp.120-123, Ukraine, July 2001

[2] Beck, R. "An Investigation of Finite-Precision Digital Resonators", PhD Thesis, University of Westminster, June 2002

[3] Turner R. H., T. Courtney and R. Woods, "Implementation of fixed DSP functions using the reduced coefficient multiplier", *IEEE Proc. of ICASSP'2001*, vol. 2, pp. 881-884, May 2001, USA

[4] http://www.xilinx.com/xlnx/xil_prodcat_landingpage.jsp?title=Virtex_Series

[5] ISO/IEC, "Information technology-Coding of audio-visual object: Visual ISO/IEC 14496-2 Final Proposed Draft",14496-2, July 1999

[6] Wang J.L. et al, "Implementation of the DCT and its inverse by recursive structures", *IEEE Workshop on Signal Processing Systems,* pp 120-130, Oct 1999

[7] Kozick R.J., and M.F. Aburdene, "Methods for designing efficient parallel –recursive filter for computing discrete transforms", *Telecommunication Systems*, vol. 13, no.1, 2000, pp.69-80

[8] Demirsoy S. S., A. Dempster and I. Kale, "Design Guidelines for Reconfigurable Multiplier Blocks", *submitted to IEEE ISCAS 2003*

DESINGN GUIDELINES FOR RECONFIGURABLE MULTIPLIER BLOCKS

Süleyman Sırrı Demirsoy, Andrew G. Dempster and Izzet Kale

Applied DSP and VLSI Research Group, Department of Electronic Systems
University of Westminster, 115 New Cavendish St, London, W1W 6UW, UK
Tel: +44 20 7911 5000 - 3611(ext) e-mail: demirss@cmsa.wmin.ac.uk

ABSTRACT

The newly proposed reconfigurable multiplier blocks offer significant savings in area over the traditional multiplier blocks for time-multiplexed digital filters or any other system where only a subset of the coefficients that can be produced by the multiplier block is needed in a given time. The basic structure comprises a multiplexer connected to at least one input of an adder/subtractor that can generate several partial products, leading to better area utilization. The multiplier block algorithm complexity of a design increases logarithmically as the number of the multiplexers is increased. Design guidelines for the maximum utilization of the reconfigurable multiplier block structures are also presented.

1. INTRODUCTION

Multiplier blocks are very efficient ways of implementing fixed multipliers for digital filters. The multiplication can be performed by means of shifts and additions of the multiplicand defined by the multiplier, or coefficients in the case of filters. By exploring the common sub-expressions in terms of numbers forming the coefficients, the redundancy in the filter bank implementation can be significantly reduced [1],[2].

Another common sub-expression elimination method employs Canonical Signed Digit (CSD) representation to reduce the number of '1's in the coefficient representation, and seeks common bit patterns among the coefficients [3].

These techniques mainly target the problem of parallel implementation where several multiplications are performed in parallel. For time-multiplexed filtering, although they still can produce better solutions than a general purpose multiplier, there will be redundant parts in the multiplier block that are not active during a certain time slot. This problem is illustrated in Figure 1. Figure 1a depicts a fully parallel FIR filter implementation. Its corresponding time-multiplexed implementation is given in Figure 1b. The multiplier block implementation for all coefficients shown in Figure 1c is time multiplexed to choose a specific partial product. Therefore all partial products except the one that is selected remain unused.

The Reduced Coefficient Multiplier (RCM) technique proposed in [4] offers a very efficient solution to this problem for the XILINX Virtex FPGA implementations where a Look-up Table (LUT) is programmed to have some extra combinational logic including a multiplexer at one input of the adder/subtractor (called a cell) without any extra area cost.

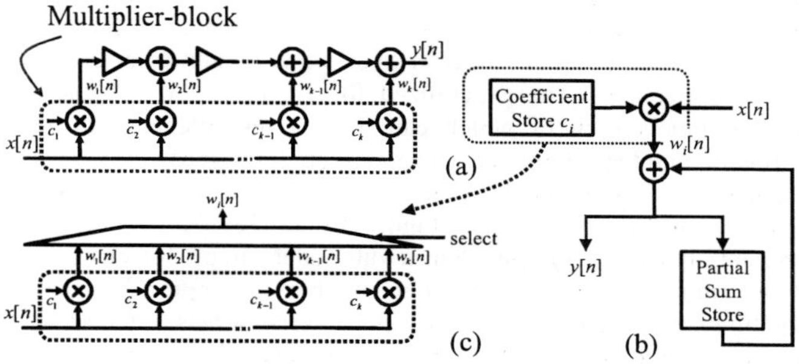

Figure 1 (a) Fully Parallel FIR filter implementation, (b) Time-multiplexed implementation (c) implementation of multiplier in 1b as a multiplier block

Although there are certain limitations in connection with multiplexer, its cost is effectively reduced by distributing the multiplexer stage at the output of the multiplier block in Figure 1c to the individual adder stages inside the multiplications. This idea is demonstrated in Figure 2. The edge values represent multiples of input after shifting is performed. Three LUTs have to be used in Figure 2a where the partial products 5 and 9 are generated separately and multiplexed afterwards. When the multiplexer is placed at one input of the adder, two partial products are generated by the same adder and the resulting circuit costs only one LUT.

The method reported in [4] to group the partial products together to form all the multiplications needed at the output of the last adder stage is based on Hartley's method [3]. Different Signed Digit (SD) representations for the coefficients are investigated and common sub-expressions that can be grouped in a single stage are identified by considering different cell definitions. Fifty eight (58) possibly useful cell implementations using a single LUT for the purpose of RCM design have also been reported in [4].

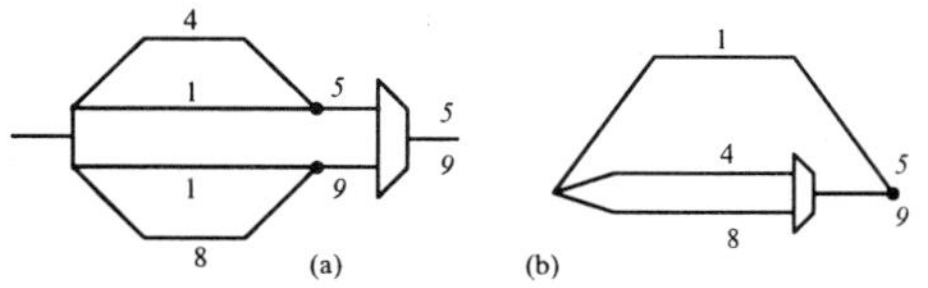

Figure 2 Two alternative implementations of a multiplier block that generates partial products for a time-multiplexed implementation

This paper describes the Reconfigurable Multiplier Blocks (ReMB) as an extension of RCM that target custom VLSI implementation as well as FPGAs. The design guidelines for the ReMB technique are also discussed. Section 2 gives the details of the structure of ReMB. Section 3 investigates several approaches for an algorithm and discusses design guidelines for an optimal result. Section 4 gives an example implementation. Section 5 concludes the paper.

2. RECONFIGURABLE MULTIPLIER BLOCKS

The basic structure of the ReMB is given Figure 3. It consists of a multiplexer connected to at least one input of the adder (in the rest of the text, an adder will refer to an adder or subtractor or an adder/subtractor). The generalized form consists of an adder with n inputs each of which can have a multiplexer with a different number of inputs. The dashed lines mean optional elements. Two particularly useful forms deduced from the generalized form are given in Figure 3b and 3c. The area of a 2x1 multiplexer is smaller than a full-adder circuit for custom VLSI implementation. As the number of inputs of the multiplexer or the number of the multiplexers increases, the area overhead increases as well as the functionality. For example the basic structure given in Figure 3b can produce double the partial results produced by the one given in Figure 3c.

A subset of the basic structures is suitable for Virtex FPGA implementation where the extra area cost of the multiplexer is zero. These forms are three cell definitions proposed in [4] for the RCM technique having the form of Figure 3c with inputs B, C (the inputs to the multiplexer) and A. Their functionality is shown in Table 1 for the different values of the select line.

The inputs to the basic structure are either the input ($x[n]$) to the multiplier block or another partial product generated by a similar basic structure or the shifted forms of them. The construction of the coefficients is similar to the normal multiplier blocks where partial products generated at each node (output of an adder) build up to coefficients. The difference of the ReMB is that the ReMB can produce more than one partial product at the output of the basic structure depending on the select signal for the multiplexer and the inputs to the multiplexer. Figure 4 shows an example design of a single basic structure, capable of producing a combination of four different partial results from signal X.

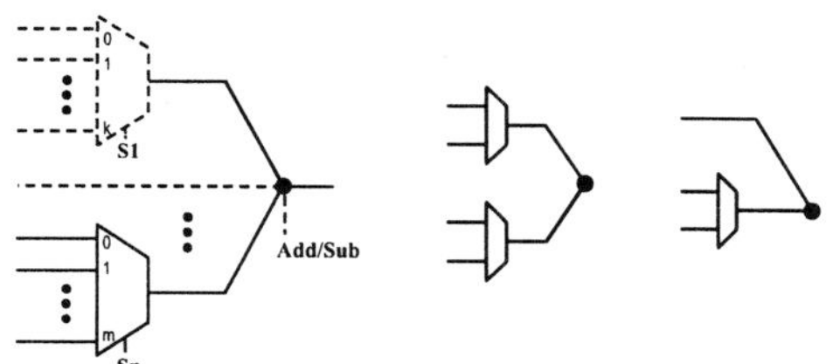

(a) (b) (c)

Figure 3 The basic structure of the ReMB (a) generalized form, S1-Sn being the select lines of the multiplexers, add/sub being the operation select signal of the adder/subtractor., (b) and (c) are mostly employed forms.

Select	Cell 1	Cell 2	Cell 3
0	A+B	A-B	A+B
1	A+C	A-C	A-C

TABLE 1 The operation of cells for different select values

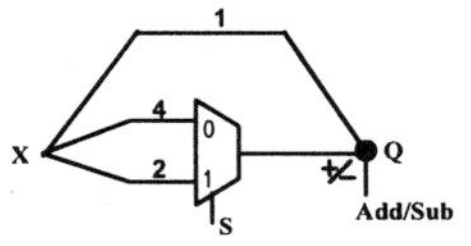

S	Add/sub	Q
0	0	5X
0	1	-3X
1	0	3X
1	1	-X

(a) (b)

Figure 4 (a) A basic structure with all of its inputs connected to X with the specified shift values, (b) Different output (Q) values as the select signals change.

The cascade of two basic structure, one of them taking all of its inputs from the input signal to the ReMB ($x[n]$), can be in several different forms as depicted in Figure 5. These structures are all topologically different and can generate distinct combinations of partial products that are not covered by another cascaded form. If we assume that the adder can perform only addition, the first stages in the cascaded forms would have two different partial products at their outputs. Having two different partial products as their inputs, the second stages of the cascaded forms would generate four different partial products at their outputs for Figure 5a, b, c, e and three different partial products for Figure 5d. Adding one more basic structure to the cascade line would result in producing around twice as many partial products. However, the number of topologically different connections of three basic structures as used in Figure 5 would be a hundred. This figure would continue to exponentially increase if the number of inputs to the multiplexers was increased or a different basic structure was used (i.e. Figure 3b).

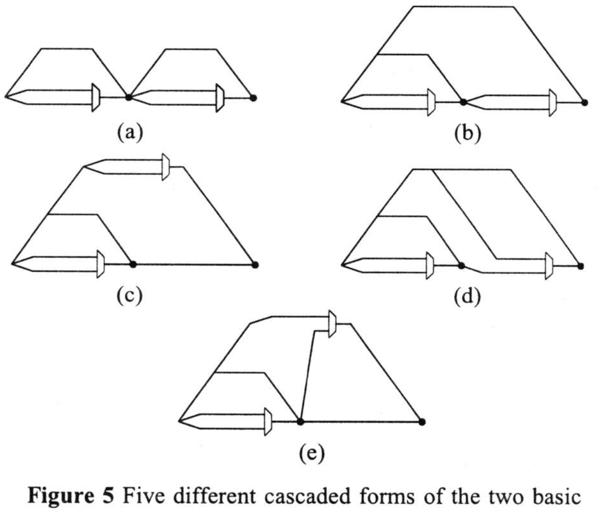

Figure 5 Five different cascaded forms of the two basic structures given in Figure 3c.

The cascaded forms are classified into two groups as regular forms and hybrid forms. Regular forms are the ones with the multiplexers having their input from the same node (Figure 5a, 5b and 5c). Hybrid forms have multiplexers taking their inputs from different nodes (Figure 5d and 5e). For bigger cascaded structures, the number of hybrid forms is much more than the regular forms. Regular forms are easy to handle during an automated design process. However hybrid forms offer a greater degree of flexibility in the design solution space for a given set of coefficients as will be shown in the next section.

3. DESIGN GUIDELINES

The optimal ReMB design for a given coefficient set requires a well defined procedure that can explore the capabilities of a particular basic structure in full. The partial product space gets extremely large as the size of the coefficient and/or coefficient set increases. The main problem is to identify the common partial products or sub-expressions that can be mapped to a single basic structure and use the minimum of such basic structures to build all the coefficient values.

The application of the ReMB to the time-multiplexed filter structures requires the implementation of multiple coefficient values with a single output as shown in Figure 1c. However, it is also possible to apply it to filter architectures with multiple outputs as well. The procedures to build the ReMB designs for these cases have to be different to get an optimal outcome.

The case of Multiple Input Single Output (MISO) applications has been investigated for RCM designs in [4]. The algorithm proposed starts combining sub-expressions into groups that suit one of the 58 cell definition until no sub-expression is left out. It tries to minimize the number of groups of sub-expressions hence the number of cells used. The flexibility of having 58 possible cell combinations (most of them operate on two inputs A,B and generate an arithmetical combination of them like A+B, 2A, 0, A-B) simplifies the grouping of sub-expressions but makes the decision of the minimisation of the number of groups more difficult. For the cells that accept three inputs as shown in Table 1, B and C inputs come from the same node [4]. In other words, the regular structures are taken into account but the hybrid forms are not utilized due to computational complexity.

One of the multiplier block algorithms Reduced Adder Graph (RAG-n), performs an exhaustive search for all possible partial results that can be formed with one adder and then continues to do so until the targets are formed. The number of adders being the cost function, the algorithm gives optimal (fewest adders) results for the sets that are covered by exhaustive search. The second part of the algorithm tries to add the coefficients to the multiplier block using the minimum number of adders individually. This part is not optimal, however gives good results up to 12-bits word-length [2]. RAG-n is still the best available algorithm for the parallel implementation of fixed multipliers.

A similar approach of starting with an exhaustive search of all partial products that can be combined on a structure and then adding the groups of partial products using the minimum number of basic structures would lead to a better design than trying to group the partial products and coefficients for the start. The limitation here is that the size of the set to undertake an exhaustive search increases exponentially as the number of basic structures in the design increases.

Table 2 displays the set sizes of exhaustive search spaces of one (Figure 4a) and two (Figure 5) basic structures for the integer numbers up to a word-length of 10-bits. The exhaustive search for all possible outputs is performed for both positive and negative numbers and recorded with the type of structure and the inputs to the structure. The sets are then filtered for repetitive and unwanted outputs. The basic structures are restricted to the subset that fits into a single half slice in a Virtex FPGA. As observed, the set size dramatically increases for the cascaded structures. For the cascade of three basic structures, the number of different forms becomes 100 as mentioned earlier. It is not feasible to do an exhaustive search for this large number of structure, although if done, it would benefit to achieve a smaller design.

Structure	Set Size	Percent of Total
Figure 4a	471	N/A
Figure 5a	22754	3.8 %
Figure 5b	62003	10.6 %
Figure 5c	126850	21.6 %
Figure 5d	259295	44.4 %
Figure 5e	114335	19.6 %
Total of Figure 5	585237	N/A

TABLE 2 The set sizes of the exhaustive search space of the numbers up to 10 bits word-length for the given structures.

The percentage column on Table 2 indicated the contribution of a particular cascaded form to the total number of sets of two basic structures. The outputs from regular structures add up to 36 % of all possible outputs where only a single hybrid structure contributes 44.4 %. As the number of basic structures increase, the percentage of the hybrid forms will become greater. Therefore, it is essential to incorporate to hybrid forms into the design methodology effectively.

In the context of multiplier blocks, the graph that produces the coefficient was not of importance since the primary goal was to construct the coefficients by sharing intermediate results or partial products [5]. Two different graphs generating the same coefficient were not considered separately. Secondly, the even coefficient values were treated after dividing them by two until an odd number is reached. The outputs from a multiplier block are always odd numbers that are shifted afterwards to generate the even coefficient [2], [5].

Both of the above issues have to be taken into consideration in the design of the ReMB. The structures are of primary importance, since the input value and the shift value on the common input line (the input line without the multiplexer) and the operation type of the basic structure affects the decision of the other inputs to the basic structure. Shifting the output of ReMB at an extra stage is not feasible due to the nature of the ReMB where a single node produces more than one partial product. Shifting only one of the partial products and not the others requires a separate multiplexer to select between the original and shifted forms. Instead, the even numbers can be handled like the odd numbers and generated throughout the ReMB design.

An efficient ReMB design should employ the aforementioned guidelines. The search of grouping of partial products that builds up the coefficients would cover all the different possible ways of constructing a coefficient. After identifying these possible ways for all coefficients, the partial products that have common input values and shift values in their formation should be grouped in a basic structure. As the number of basic structures increase, the possible outputs from the nodes in the ReMB should be taken into account to construct the remaining coefficients. The number of output lines from the ReMB design effects the decision of combining the partial products. Two coefficients that reside at different output nodes of the ReMB design should be constructed in such a way that, the last stage of the cascaded structure should be different for them. However, separating the generation of partial products before the last stage should give better results for some cases and should be considered.

4. ReMB DESIGN EXAMPLES

An example to demonstrate the effectiveness of the proposed ReMB technique is given below. Figure 6 shows the recursive loop of an 8-point Goertzel recursive DCT structure given in [6]. This multiplier has 8 different coefficient values, characterized as:

$$2\beta_k = 2Cos(k\pi / 8) \quad \text{for } k \in [0,7] \quad (1)$$

The coefficient values for k=5, 6 and 7 are the negatives of the coefficients for k=3, 2 and 1 respectively. Therefore the multiplier can generate the multiplications for $k \in [1,3]$ and the results can be negated afterward at the next adder.

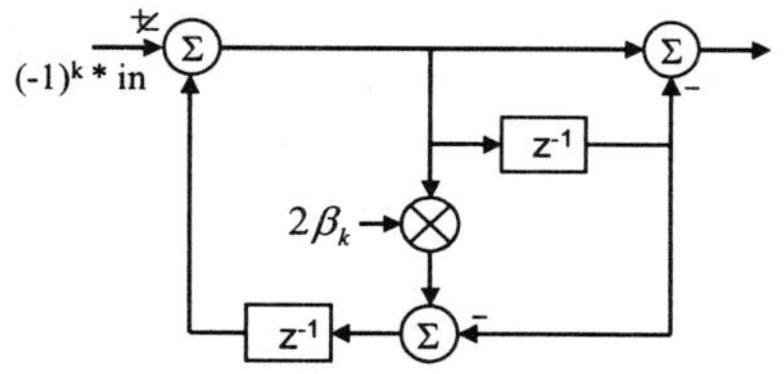

Figure 6 Recursive part of a Goertzel DCT implementation

The ReMB design of this multiplier for a coefficient word-length of 12-bit is shown in Figure 7. The outputs are shifted 9-bits to right to get the result of multiplication with the coefficients. It consists of four basic structures that occupy four half slices for a single bit input in a Virtex FPGA implementation. This design is highly efficient when compared to a classical multiplier block implementation by the RAG-n algorithm [2] that would occupy seven adders and extra multiplexer stages at the output. It also occupies one less half slice in comparison to the RCM design of the same coefficients given in [4]. This saving is due to the exploitation of the hybrid forms of cascades in the ReMB designs as depicted in Figure 7.

Another ReMB design is for the newly proposed reconfigurable recursive DCT kernel given in [6] where the 12-bit coefficients are smaller in magnitude due to the realization of different Goertzel configurations for different frequency bins. As seen from the Figure 8, this ReMB design has three basic structures, again connected in hybrid form cascades. It occupies three half-slices for a single bit input in Virtex FPGA.

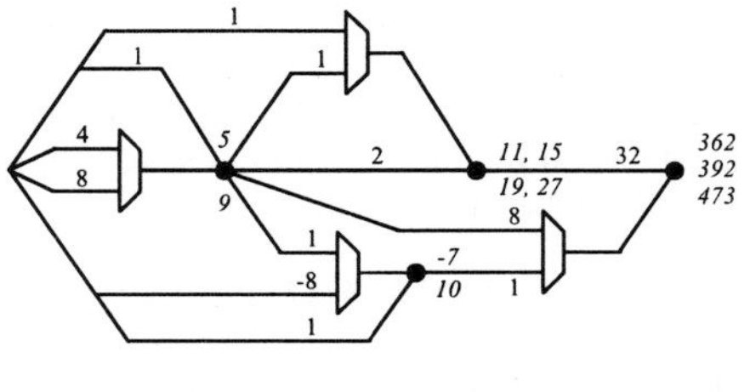

Figure 7 ReMB design of the DCT loop multiplier. The italic numbers are the partial products generated at each node.

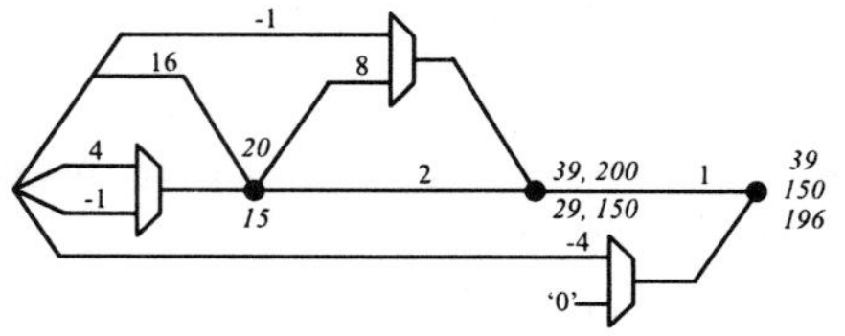

Figure 8 ReMB design for the reconfigurable DCT kernel in [6].

5. CONCLUSION

Reconfigurable Multiplier Blocks (ReMB) offer significant savings in the area of fixed multipliers for time-multiplexed systems for both custom VLSI and FPGA implementations. The cascade forms of the basic structures have to be fully utilized for achieving smallest area. The choice of the basic structure is crucial to reduce the area for FPGA environments. Several factors that are not of importance in the multiplier block design have to be taken into account when designing ReMB, such as even number generation and the structure that generates the partial products. A design methodology employing an exhaustive search as a start would give better results as shown with an example in this paper.

6. ACKNOWLEDGEMENT

The authors would like to thank Richard H. Turner for his generosity in providing us with a copy of his PhD thesis.

7. REFERENCES

[1] Bull D.R. and D.H. Horrocks, "Primitive operator digital filters", *IEE Proceedings-G*, vol. 138, no.3, pp.401-412, June 1991

[2] Dempster A. G. and Macleod M. D., "Use of minimum-adder multiplier blocks in FIR digital filters", *IEEE Trans. CAS-II*, vol. 42, no. 9, pp. 569-577, Nov. 1995

[3] Hartley R., "Optimization of canonic signed digit multipliers for filter design", *IEEE Proc. of Int. Symp. On Circuits and Sytems*, vol. 4, pp. 1992-1995, July 1991

[4] Turner R. H., "Functionally diverse programmable logic implementations of digital signal processing algorithms", PhD Thesis, Queen's University of Belfast, August 2002

[5] Gustafsson O., A. Dempster and L. Wanhammar, "Extended results for minimum-adder constant integer multipliers", *IEEE Proc. of ISCAS'2002*, vol.1, pp. 73-76, May 2002

[6] Demirsoy S. S., et. al, "Reconfigurable implementation of Recursive DCT kernels with reduced quantization noise", *submitted to IEEE ISCAS'2003*

AN EFFICIENT INTERLEAVED TREE-STRUCTURED ALMOST-PERFECT RECONSTRUCTION FILTERBANK

Nan Li and Behrouz Nowrouzian

Department of Electrical and computer Engineering
University of Alberta
Edmonton, Alberta, CANADA T6G-2V4

ABSTRACT

In the conventional DFT filterbanks, the required transition bandwidth of the constituent prototype filter becomes narrower with increasing the number of subbands. This happens at the expense of making the prototype filter length longer, rendering the corresponding hardware implementation impractical beyond a certain number of subbands. This paper presents an alternative approach to the design of almost-perfect reconstruction (almost-PR) filterbanks which circumvents the problem associated with a large number of subchannels. The resulting filterbank has a multi-stage tree-structured configuration, and can be efficiently implemented by using interleaving techniques (to eliminate all identical short-length digital filters). Unlike the conventional DFT filterbanks, the resulting filterbank requires much smaller number of non-zero multiplier coefficients, and when the number of subchannels changes, the design of this tree-structured filterbank needs to be modified only slightly.

1. INTRODUCTION

Multirate filterbanks finds a wide range of applications from data compression to multicarrier modulation and feature detection. Recently, cosine modulated filterbanks have become very popular, mainly due to the fact that they lend themselves to perfect reconstruction (PR). However, the main drawback of these filternaks is their inherent non-linear subchannel phase characteristics. In contrast, Discrete Fourier Transform (DFT) filterbanks [1] preserve linear subchannel phase characteristics while permitting high computational efficiency.

In the conventional DFT filterbanks, the constituent analysis and synthesis filters are all derived by shifting the frequency response of the same prototype filter along the frequency axis (so as to cover the entire frequency band). In DFT filterbanks incorporating a large number of subchannels, the prototype filter is required to possess a very narrow transition band, making the lengths of the constituent analysis and synthesis filters prohibitively long. In addition, in order to achieve PR property, the lengths of the synthesis filters need to be made much larger than those of the analysis filters [2][1].

This paper presents a tree-structured critically-decimated almost-PR filterbank lending itself to a large number of subchannels without any recourse to the problems of the type mentioned above. The signal processing operations in the constituent tree-structured analysis filterbank include, the decomposition of the input signal into the required number of subband signals, the expansion of the frequency spectra of the subband signals through decimation, and the subsequent processing by analysis filters (which are only required to satisfy mild conditions on their transition bandwidths). The signal processing operations in the constituent tree-structured synthesis filterbank, on the other hand, takes place in such a manner as to ensure almost-PR property while using synthesis filters which are the same as the corresponding analysis filters.

2. THE PROPOSED TREE-STRUCTURED ALMOST-PR FILTERBANK

The proposed tree-structured almost-PR filterbank is as shown in Fig. 1, consisting of a L-stage analysis filterbank (Fig. 1a) and a corresponding mirror image L-stage synthesis filterbank (Fig. 1b).

The internal structures of decomposition *Stage 1* and reconstruction *Stage 1'* are as shown in Figs. 1a and 1b, respectively. The structures of the 2^{l-1} intermediate decomposition *Stage l* modules and the 2^{l-1} reconstruction *Stage l'* modules are as shown in Figs. 2a and Fig. 2b, respectively, where $l = 2, 3, \ldots, L-1$, and where H_l represents an analysis filter in *Stage l* modules, and F_l represents the corresponding synthesis filter in *Stage l'* modules. In order to ensure the linear-phase property in each subchannel, all the analysis filters H_l in *Stage l* modules and all the synthesis filters F_l *Stage l'* modules are assumed to be symmetric FIR filters.

The analysis filterbank: The purpose of *Stage 1* decomposition is to shift the frequency band into two bands through

[1] Needless to say, a large number of subchannels also increases the computational complexity of the DFT filterbank.

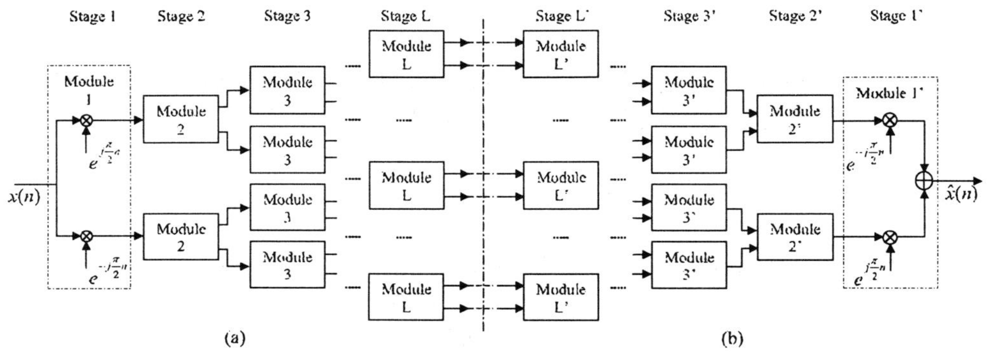

Figure 1: The Tree-Structured Almost-PR Filterbank (a) the Analysis, and (b) the Synthesis Filterbanks

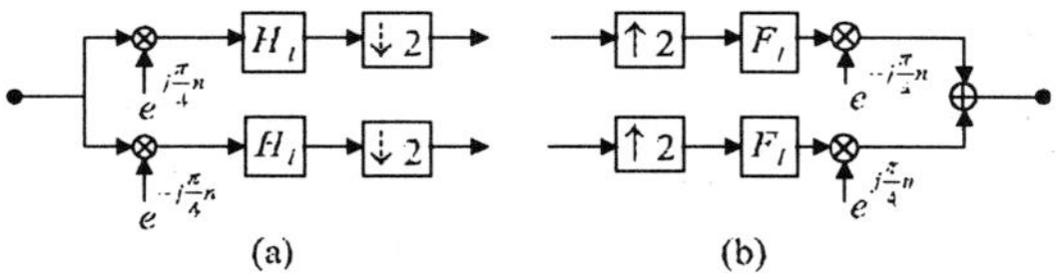

Figure 2: The Intermediate *Stage l* Modules, (a) for Analysis, and (b) for Synthesis Filterbanks

modulation by $e^{\pm j\frac{\pi}{2}n}$, with the resulting positive and negative frequency shifted bands to be processed in *Stage 2* modules. In the intermediate decomposition *Stage l* modules, the input signal to each module which occupies a frequency band of $[-\frac{\pi}{2}, \frac{\pi}{2}]$ is first modulated by $e^{\pm j\frac{\pi}{4}n}$, and then the resulting two modulated signals are made to occupy the frequency band $[-\frac{\pi}{4}, \frac{\pi}{4}]$ by filters H_l. Finally, after a two-fold decimation, the module output signals will occupy the frequency band $[-\frac{\pi}{2}, \frac{\pi}{2}]$. Due to the absence of decimation in *Stage 1*, the whole system operates at a rate twice that of the input sampling rate. An additional 2-fold decimation operation is included in *Stage L* modules as shown in Fig. 3(a) to expand the frequency band of each subchannel signal to $[-\pi, \pi]$ (for critical decimation [3]).

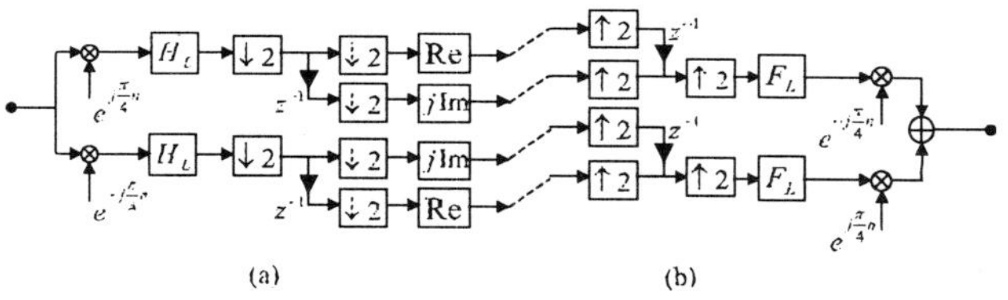

Figure 3: The Internal Structure of (a) *Stage L*, and (b) *Stage L'* Modules

The synthesis filterbank: An additional 2-fold upsampling operation is included in *Stage L'* modules as shown in Fig. 3(b), to shrink the frequency band of the combination of the two subchannel input signals into an output signal occupying the frequency band $[-\frac{\pi}{2}, \frac{\pi}{2}]$. In the intermediate reconstruction *Stage l'* modules, the two input signals to each module are first upsampled by a factor of 2 to occupy the frequency band $[-\frac{\pi}{4}, \frac{\pi}{4}]$, and then the resulting image components are eliminated by the filters F_l. Finally, the resulting two signals are modulated by $e^{\pm j\frac{\pi}{4}n}$, and combined to form the module output signal which occupies the frequency band $[-\frac{\pi}{2}, \frac{\pi}{2}]$. The purpose of *Stage 1'* is to demodulate the two input signals by $e^{\pm j\frac{\pi}{2}n}$, and to combine the resulting two signals to obtain the reconstructed output signal.

3. DESIGN OF ANALYSIS AND SYNTHESIS FILTERS FOR ALMOST-PR PROPERTY

The design of the *Stage l* analysis filters H_l and the *Stage l'* synthesis filters F_l is achieved recursively, beginning with the design of the analysis filter H_L and the synthesis filter F_L. As a result, with proper design of H_L and F_L, the proposed filterbank achieves the almost-PR property.

3.1. Design of H_l and F_l

In order to eliminate the image frequency components in the proposed filterbank, let us first consider a subsystem between *Stage L-1* decomposition module and the corresponding *Stage L-1'* reconstruction module via *Stage L* and *Stage L'* modules as shown in Fig. 4.

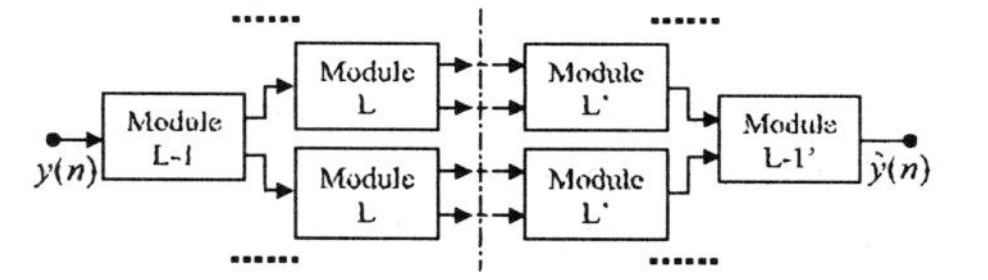

Figure 4: A Subsystem between the Last Two Decomposition Stages and the Corresponding Reconstruction Stages

In order to eliminate the image components in the subsystem in Fig. 4, it is necessary, first to cancel the image components generated by *Stage L* modules, and next to eliminate the image components generated by *Stage L-1* module.

Design of H_L and F_L: Through a detailed analysis of the subsystem in Fig. 4 it can be shown that the image components generated by *Stage L* modules can be cancelled if a) $H_L = F_L$, and b) the stopband edge frequency of H_L, $\omega_{sH,L} < \pi/2$, where $\omega_{sH,l}$ ($\omega_{sF,l}$) represents the stopband edge frequency of H_l (F_l).

Design of H_{L-1} and F_{L-1}: Having determined H_L and F_L as above, it can be shown that the image components generated by *Stage L-1* module can be eliminated if

$$\omega_{pH,L-1} > (\omega_{sH,L} + \pi/4)/2 \quad (1)$$

$$\omega_{sH,L-1} < \pi - (\omega_{sH,L} + \pi/4)/2 \quad (2)$$

where $\omega_{pH,l}$ ($\omega_{pF,l}$) represent the passband edge frequency of H_l (F_l). The conditions in Eqns. 1 and 2 are to make sure, a) that the passband frequency region of H_{L-1} extends over the frequency band of *Stage L* modules, and b) that the image frequency components generated by the 2-fold decimation of H_{L-1} do not fall within the passband frequency region of decimated H_{L-1}, respectively (c.f. Fig. 5).

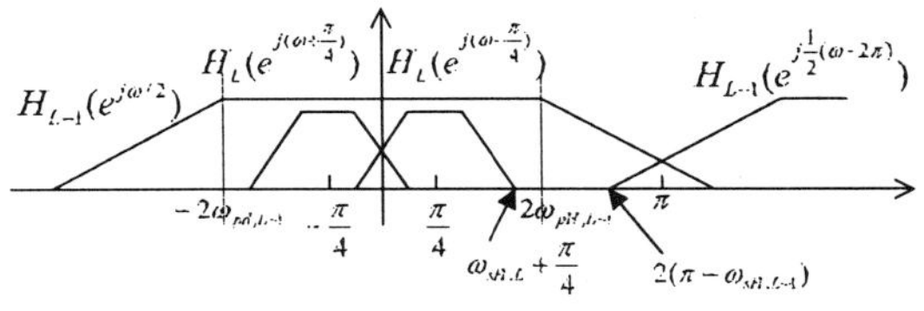

Figure 5: Relationships Between the Frequency Characteristics of H_{L-1} and H_L

Similarly, it can be shown that the image components generated by the 2-fold upsampling in *Stage L-1'* module can be eliminate if

$$\omega_{pF,L-1} \geq (\omega_{sF,L} + \pi/4)/2 \quad (3)$$

$$\omega_{sF,L-1} \leq \pi - (\omega_{sF,L} + \pi/4)/2 \quad (4)$$

The conditions in Eqns. 3 and 4 are to ensure, a) that the passband frequency region of F_{L-1} extends over the frequency band of *Stage L'* modules, and b) that the image frequency components generated by the 2-fold upsampling in *Stage L-1'* are eliminated (c.f. Fig. 6).

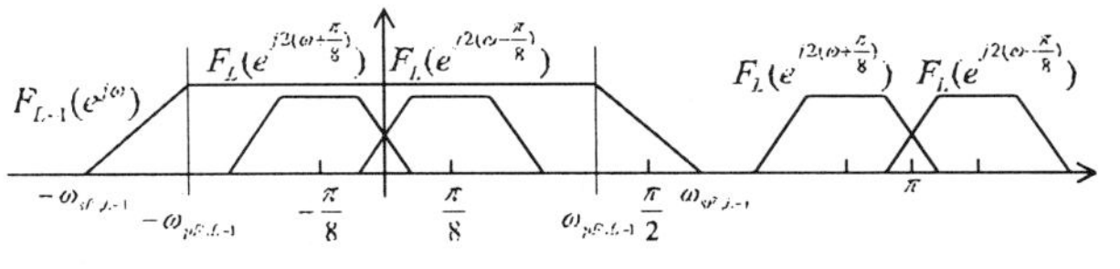

Figure 6: Relationships Between the Frequency Characteristics of F_{L-1} and F_L

Design of intermediate H_l and F_l: In the similar way, it can be verified that the image components generated by *Stage l* can be eliminated if

$$\omega_{pH,l} \geq (\omega_{pH,l+1} + \pi/4)/2, \quad \text{for } 2 \leq l \leq L-2 \quad (5)$$

$$\omega_{sH,l} \leq \pi/2 + (\pi/2 - \omega_{pH,l}) = \pi - \omega_{pHd,l}, \quad \text{for } 2 \leq l \leq L-2 \quad (6)$$

The conditions in Eqns. 5 and 6 are to make sure, a) that the passband frequency region of H_l extends over the passband frequency region of *Stage l+1* modules, and b) that the image frequency components generated by the 2-fold decimation of H_l do not fall within the passband frequency region of decimated H_l, respectively.

In the intermediate *Stage l'* modules, it can be shown that the image components generated by the 2-fold upsampling can be eliminate if

$$\omega_{pF,l} \geq (\omega_{pF,l+1} + \pi/4)/2, \quad 2 \leq l \leq L-2 \quad (7)$$

$$\omega_{sF,l} \leq \pi - (\omega_{sF,l+1} + \pi/4)/2 \quad 2 \leq l \leq L-2 \quad (8)$$

The conditions in Eqns. 7 and 8 are to make sure, a) that the passband frequency region of F_l extends over the frequency band of *Stage l+1'* modules, and b) that the image frequency components generated by the 2-fold upsampling in *Stage l'* are eliminated.

From Eqns. 5, 6, and 7, 8, one can select, a) $F_l = H_l$, and b) H_l as a half-band digital filter for each l.

3.2. Almost-Perfect Reconstruction

Through the above image cancellations/eliminations, the subsystem in Fig. 4 can be characterized by a transfer function

$$T_{L-1}(z) = \acute{Y}(z)/Y(z) = \frac{z^{-4}}{8}\sum_{k=0}^{3} F_{L-1}(zW_8^{a_k}) H_{L-1}(zW_8^{a_k})F_L(z^2W_8^{2k-3})H_L(z^2W_8^{2k-3}) \quad (9)$$

where $a_0 = -2$, $a_1 = a_2 = 0$ and $a_3 = 2$. In accordance with Eqn. 9, if the product $H_L(z)F_L(z)$ ($=H_L^2(z)$) is made a 4th-band digital filter[4], then $T_{L-1}(z)$ will become a half-band digital filter satisfying the relationship [2]

$$T_{L-1}(e^{j(\omega+\frac{\pi}{2})}) + T_{L-1}(e^{j(\omega-\frac{\pi}{2})}) = \frac{z^{-(4+s_{L-1})}}{8}. \quad (10)$$

Consequently, any subsystem from a *Stage l* decomposition module to the corresponding *Stage l'* reconstruction module can again be characterized recursively by the transfer function

$$T_l(z) = [H_l(zW_8^{-1})F_l(zW_8^{-1})T_{l+1}(z^2W_8^{-2}) + H_l(zW_8)F_l(zW_8)T_{l+1}(z^2W_8^2)] \quad \text{for } 2 \leq l \leq L-2 \quad (11)$$

implying that $T_l(z)$ will also represent a half-band digital filter. In this way, the transfer function of the overall tree-structured filterbank can be obtained as

$$T(z) = T_2(zW_4^{-1}) + T_2(zW_4) = \frac{z^{-(4+s)}}{2^L} \quad (12)$$

rendering the filterbank as almost-PR.

4. PRACTICAL IMPLEMENTATION OF THE TREE-STRUCTURED FILTERBANK

For a practical hardware implementation of the proposed tree-structured almost-PR filterbank, one has to resolve two different problems. The first problem relates to modulations by the complex exponentials $e^{\pm j\frac{\pi}{4}n}$ in all the intermediate stage modules, which involve multiplications by the irrational numbers $\pm\frac{\sqrt{2}}{2}$. The second problem, on the other hand, relates to the fact that the number of intermediate stage modules increases exponentially with the numbers of stages, with the sampling rate being halved from one stage to the next (potentially reducing hardware utilization).

The above problems can be resolved, a) by exploiting a polyphase decomposition of the analysis and synthesis digital filters, and b) by the application of interleaving techniques to the tree-structured filterbank.

[2] s_{L-1} here and s in the following depend on analysis and synthesis filter lengths.

4.1. Polyphase decomposition for analysis and synthesis filters

By using the equivalence in Fig. 7, the modulations by the complex exponentials $e^{\pm j\frac{\pi}{2}n}$ at *Stage l* decomposition modules reduce to multiplication by 0, 1 and -1, where

$$H_l(z) = H_{l0}(z^2) + z^{-1}H_{l1}(z^2) \tag{13}$$

and where $H_{l0}(z)$ and $H_{l1}(z)$ are the polyphase components of $H_l(z)$. Since $H_l(z)$ represents a half-band digital filter, only one multiplier coefficient of $H_{l1}(z)$ will be non-zero (precisely 0.5), implying that $H_{l1}(z)$ consists of a single multiplication. Therefore, the new required multiplication by $e^{-j\frac{\pi}{4}}$ can be efficiently absorbed into $H_{l1}(z)$.

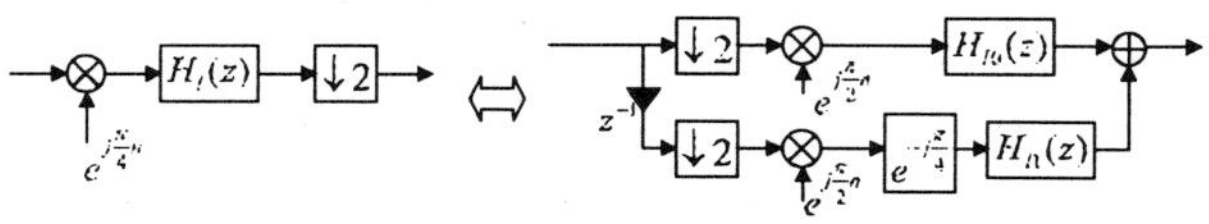

Figure 7: Equivalence for *Stage l* Decomposition Modules

Similarly, one can use the Fig. 8 for *Stage l'* reconstruction modules.

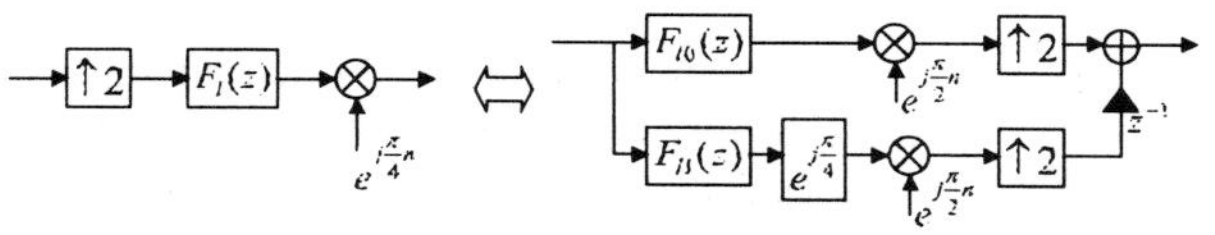

Figure 8: Equivalence for *Stage l'* Reconstruction Modules

4.2. Interleaving the tree-structured filterbank

In the proposed tree-structured filterbank, there are 2^{l-1} identical modules at *Stage l* (or *Stage l'*), which can be efficiently implemented by interleaving all of the $2 \times 2^{l-1}$ constituent identical digital filters H_l (or F_l) by one digital filter $\tilde{H}_l$ (or $\tilde{F}_l$) only. The interleaving digital filter $\tilde{H}_l$ (or $\tilde{F}_l$) is the same as the filters H_l (or F_l) except that each unit-delay component is replaced by $2 \times 2^{l-1}$ unit-delay components. Correspondingly, the outputs of all module in the each stage will be multiplexed into one signal as the input for the following stage. This multiplexing operation can be performed by adding a 2-fold interleaver in each stage. This leads to an interleaved [5] realization of the tree-structured almost-PR filterbank as shown in Fig. 9, where each stage consists of one digital filter operating at a rate twice that of the original input sampling frequency.

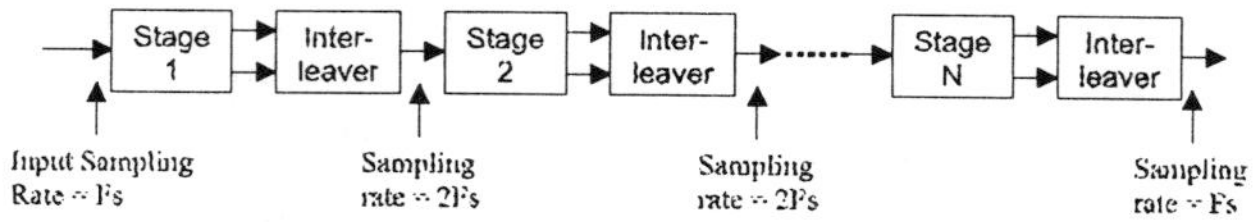

Figure 9: Pipelined Implementation of the Proposed Filterbanks

5. SIMULATION RESULTS

Let us consider a pair of tree-structured filterbanks, one having 3 and the other having 5 decomposition (or reconstruction) stages. By selecting *Stage L* decomposition filter H_L to have a stopband attenuation of 60dB and a (gradual) transition bandwidth of 0.33π, and by selecting the intermediate *Stage l* decomposition filters H_l to have the same stopband attenuations of 60dB, the total distortion between the input signal and the reconstructed output signal is obtained as shown in Figs. 10 and 11.

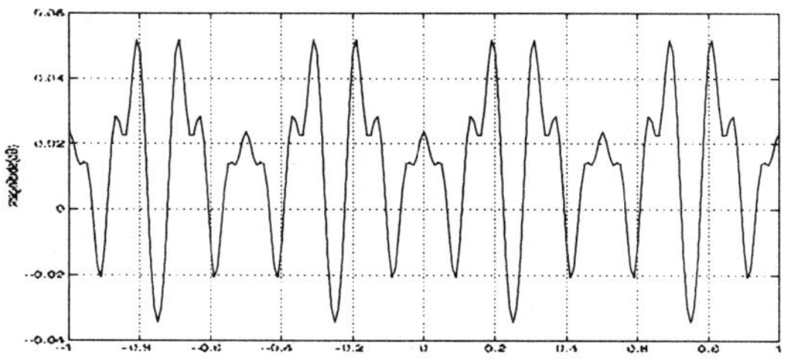

Figure 10: Distortion for the 3-Stage Filterbank

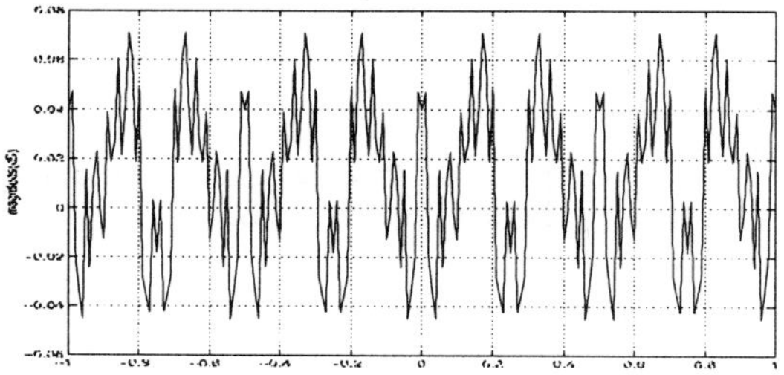

Figure 11: Distortion for the 5-Stage Filterbank

By comparing the distortions in Figs. 10 and 11, it is observed that the distortions are well within tolerable levels of around 0.06 dB, and that the distortion is increased only slightly when the number of stages increases from three to five.

6. ACKNOWLEDGEMENT

This work was supported, in part, by the Natural Sciences and Engineering Research Council of Canada under the Operating Grant #A6715 and a Strategic Grant, by Micronet, and by Nortel Networks.

7. REFERENCES

[1] M. G. Bellanger and J. L. Daguet, "TDM-FDM transmultiplexer: Digital polyphase and FFT," *IEEE Transactions on Communications*, vol. CON-22, pp. 1199–1204, Sept. 1974.

[2] K. Swaminathan and P. Vaidyanathan, "Theory and design of uniform DFT, parallel, quadrature mirror filter banks," *IEEE Trans. Circuits and Systems*, vol. CAS-33, pp. 1170–1190, Dec. 1986.

[3] T. Karp and N. Fliege, "Modified DFT filter banks with perfect reconstruction," *IEEE Transactions on Circuits and Systems-II*, vol. 46, pp. 1404–1414, Nov. 1999.

[4] P. P. Vaidyanathan, *Multirate Systems and Filter Banks*. Prentice Hall, 1993.

[5] RF Engines Limited, "The pipelined frequency transform," *RF Engines Limited White Paper*, Feb. 2002.

THE INTEGER MDCT AND ITS APPLICATION IN THE MPEG LAYER III AUDIO

Soontorn Oraintara and Tharakram Krishnan

University of Texas at Arlington, Electrical Engineering Dept., Arlington, TX
Emails: oraintar@uta.edu, tharak@msp.uta.edu

ABSTRACT

The MPEG audio layer III is an efficient audio coding scheme which gives perceptually high quality audio while achieving very high compression. The modified discrete cosine transform (MDCT) is a filterbank used by the MPEG audio layer III to provide finer spectral resolution by subdividing in frequency the subband outputs from the previous layers. Recently fast and lossless structures for the MDCT used in the layer III have been developed which use integer coefficients for their implementation. In this paper we evaluate the performance of the integer implementation of the MDCT by incorporating in the MPEG layer III codec and compare with that of a standard implementation. The comparison is done based on the decoded waveform, spectrogram and by conducting a subjective survey. The complexity of the lossless structure is also evaluated.
Keywords— Modified Discrete Cosine Transform, Discrete Cosine transform, Discrete Sine transform, Lifting scheme.

1. INTRODUCTION

The MPEG/Audio is a lossy coding technique, which exploits the limitations of the human auditory system to achieve perceptually lossless compression [1]. MPEG/Audio results from more than three years of collaboration within the Moving Pictures Experts Group (MPEG) within the International Standards for Organization (ISO). The standard defines the structure of the coded bitstream and decoding process. This ensures that any decoder will be able to decode an MPEG compliant bitstream. The standard offers flexibility in the design of the encoder and the decoder. Based on tradeoffs between the compressed audio quality and the codec complexity, the MPEG has been standardized to operate in three layers. Out of the three layers, the layer III of the MPEG offers highest audio fidelity at compression rates of 10:1. This is achieved at the cost of complexity of the codec.

The modified discrete cosine transform (MDCT) is a filterbank used by the MPEG layer III to provide finer spectral resolution by subdividing in frequency the subband outputs from the previous layers. The MDCT can be viewed as a lapped transform of which the number of inputs N is twice the number of outputs [2]. The MPEG audio standards use two sizes of MDCT, 6×12 ($N = 12$) and 18×36 ($N = 36$).

Recently, new fast algorithms for the forward and inverse MDCT computation has been proposed [3]. It is based on DCT-II/DST-II fast algorithms and their inverses, which have real (irrational) coefficients. Despite the efficiency of the proposed fast algorithms, the internal operations require real multiplications, which are not preferable in applications that run on batteries such as mobile multimedia communications. In practice, these transforms are often approximated by using fixed-point arithmetic. However, this type of implementation does not preserve the invertibility property of the transforms.

A solution to this problem is presented in [4], where the MDCT is approximated using integer or dyadic coefficients while maintaining its reversibility. Lifting scheme [5] is used to calculate orthogonal matrices. The possibility of lossless implementation of the MDCT with integer coefficients using liftings, is illustrated in [4]. In this paper, an implementation of the lossless MDCT in an MPEG layer III commercial encoder and decoder is presented. Its performance is evaluated and compared with an encoder and decoder using the standard implementation of the MDCT.

1.1. MDCT in MPEG layer III

MPEG/Audio achieves very good compression by exploiting the perceptual limitations of the human auditory system. Figure 1 shows the functional block diagram of the MPEG layer III [6]. The time to frequency transformation is achieved by using a 32-channel polyphase filterbank and the MDCT filterbank. The MDCT performs finer resolution from the 32 subband outputs from the polyphase filterbank. The psychoacoustic model which is a model of the human auditory perception supplies the nonuniform quantization block with information on how the frequencies obtained from the MDCT block are to be quantized and scaled based on their perceptual relevance. The scaled and quantized frequency lines are Huffman coded based on static Huffman tables. The bitstream obtained is formatted as prescribed by the MPEG standard and an error checking code is introduced along with the artist's name, album and other information, which constitute the ancillary data. In the decoder the reverse of the process described above, bitstream unpacking, descaling, inverse quantization, inverse MDCT and the inverse polyphase filterbank are applied. The decoder does not use the psychoacoustic model.

The MPEG layer III specifies two MDCT block lengths: a long block of size 18 and a short block of size 6. Since there is 50% overlap between the successive blocks, the window sizes for the input data are 36 and 12 respectively. The long block length is used for slowly varying signals, whereas the short block is used for transients. There are two other windows, long to short and short to long specified for transition between the slow signals and transients. The psychoacoustic model decides which window type to apply before applying the MDCT on the subband samples. This decision is based on the differences in the FFT spectra of the present samples and the spectra of the previous samples. For perfect reconstruction, the window $h(n)$ must comply with two constraints as follows [2]:

$$h(N-1-n) = h(n), \quad \text{where } n = 0, 1, ..., N-1,$$

$$\text{and } h^2(n) + h^2(n + \frac{N}{2}) = 1, \quad \text{where } n = 0, 1, ..., \frac{N}{2} - 1.$$

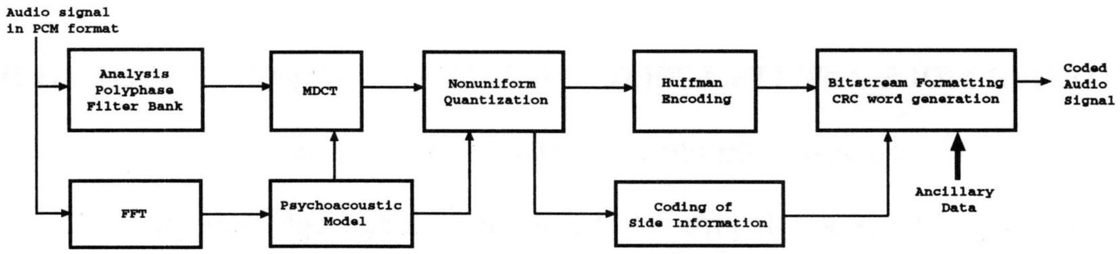

Figure 1: MPEG layer III encoder functional block diagram

A window which matches these constraints is the sine window which is given by:

$$h(n) = \sin[\frac{\pi}{N}(n - \frac{1}{2})], \quad \text{where } n = 0, 1, ..., N-1. \qquad (1)$$

The window is applied twice, once before the forward transform in the encoder and once again after the inverse transform in the decoder.

The MDCT and its inverse (IMDCT) for a windowed input sequence are defined as [7]:

$$Z_k = \sum_{n=0}^{N-1} x_n \cos\left[\frac{\pi(2n+1+\frac{N}{2})(2k+1)}{2N}\right], \qquad (2)$$

$$\hat{x}_n = \frac{2}{N}\sum_{k=0}^{\frac{N}{2}-1} Z_k \cos\left[\frac{\pi(2n+1+\frac{N}{2})(2k+1)}{2N}\right], \qquad (3)$$

where $k = 0, 1, \cdots, N/2-1$ and $n = 0, 1, \cdots, N-1$. The MDCT coefficients posses an even antisymmetry property so that we obtain $N/2$ MDCT coefficients Z_k from N input samples. The input window which is of length N is shifted by $N/2$ samples to obtain the next set of coefficients. Thus the length of the input data vector remains the same after transformation. The data vector is reconstructed by overlapping and adding the IMDCT coefficients $\hat{x}_n$.

Recently fast structures for the MDCT based on DCT/DST of type II have been obtained [8]. The fast structure can be represented by equations (4) and (5), where a_n and b_n are described in (6) and (7), respectively, and $n, k = 0, 1, \cdots, N/4-1$. In the above equations, the factors $\cos\left[\frac{\pi(2n+1)}{2N}\right]$ and $\sin\left[\frac{\pi(2n+1)}{2N}\right]$ represents Given's plane rotations. The factors $\cos\left[\frac{\pi(2n+1)k}{2(N/4)}\right]$ and $\sin\left[\frac{\pi(2n+1)k}{2(N/4)}\right]$ are recognized as the $N/4$-point DCT and DST of type II kernels.

2. FAST INTEGER MDCT

The rotations in the fast implementation can be represented by an orthonormal matrix as:

$$\mathbf{R} = \begin{bmatrix} c & s \\ -s & c \end{bmatrix}, \text{ where } c = \cos\theta, \text{ and } s = \sin\theta$$

Figure 2(a) shows a block diagram for computing $\mathbf{R}$ and its inverse. This matrix $\mathbf{R}$ can be factorised into a product of upper and lower triangular matrices as follows:

$$\mathbf{R} = \begin{bmatrix} 1 & \frac{c-1}{s} \\ 0 & 1 \end{bmatrix}\begin{bmatrix} 1 & 0 \\ s & 1 \end{bmatrix}\begin{bmatrix} 1 & \frac{c-1}{s} \\ 0 & 1 \end{bmatrix},$$

$$\text{where} \quad a = \frac{c-1}{s}, \; b = s, \text{ and } g = \frac{c-1}{s}$$

The coefficients c, s and $-s$ are called *butterfly* coefficients whereas

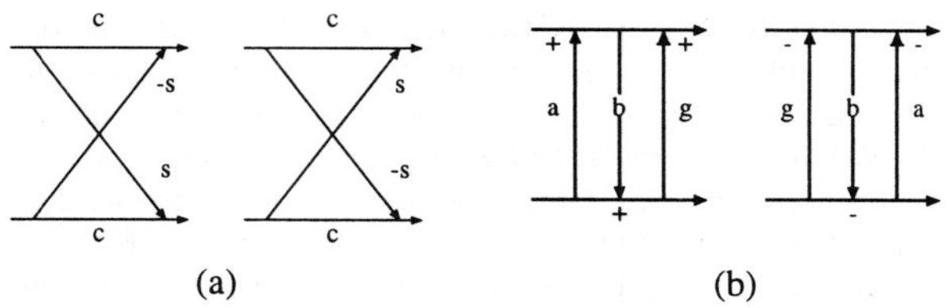

Figure 2: Forward and inverse (a) butterflies and (b) lifting steps.

the coefficients a, b and g are called *lifting* coefficients in this paper. The advantage of lifting is that the quantization introduced in the forward structure can be cancelled out in the inverse structure making the structure invertible. Moreover, nonlinear operations can be used at the liftings without violating its invertibility as long as same operations are used in its inverse. This lifting conversion also reduce the number of multiplications needed to compute the matrix $\mathbf{R}$ from four to three. Using the technique described above, the fast MDCT structure can be made lossless by using the lifting steps as presented in Figure 3 for the 12 point MDCT. The lifting coefficients are presented in Table 1. We note that the butterflies with coefficients 1's and -1's at the frontend are not converted since there are no multiplications involved. The new orthogonal structure for the 9 point DCT-II and the fast, integer structure for the 36 point MDCT are presented in [4].

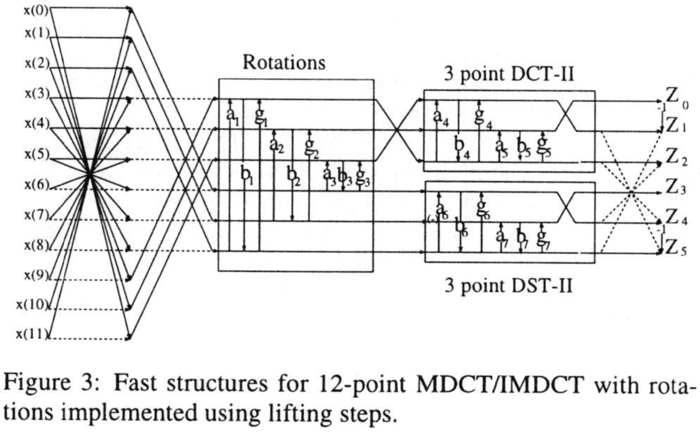

Figure 3: Fast structures for 12-point MDCT/IMDCT with rotations implemented using lifting steps.

3. COMPLEXITY OF THE INTEGER MDCT

In this section, we measure the complexity of the integer MDCT by counting the number of adders used to calculate the multipli-

$$Z_{2k} = (-1)^k \frac{\sqrt{2}}{2} \sum_{n=0}^{\frac{N}{4}-1} \left(a_n \cos\left[\frac{\pi(2n+1)k}{2(N/4)}\right] - b_n \sin\left[\frac{\pi(2n+1)k}{2(N/4)}\right]\right), \tag{4}$$

$$Z_{2k+\frac{N}{2}} = (-1)^{k+\frac{N}{4}} \frac{\sqrt{2}}{2} \sum_{n=0}^{\frac{N}{4}-1} (-1)^{n+1} \left(a_n \cos\left[\frac{\pi(2n+1)k}{2(N/4)}\right] - b_n \sin\left[\frac{\pi(2n+1)k}{2(N/4)}\right]\right), \tag{5}$$

$$a_n = (x'_n - x''_{\frac{N}{2}-n-1}) \cos\frac{\pi(2n+1)}{2N} - (x''_n - x'_{\frac{N}{2}-n-1}) \sin\frac{\pi(2n+1)}{2N}, \tag{6}$$

$$b_n = (x'_n - x''_{\frac{N}{2}-n-1}) \sin\frac{\pi(2n+1)}{2N} + (x''_n - x'_{\frac{N}{2}-n-1}) \cos\frac{\pi(2n+1)}{2N}. \tag{7}$$

a_1	$\frac{\cos\frac{5\pi}{24}-1}{\sin\frac{5\pi}{24}}$	b_1	$\sin\frac{5\pi}{24}$	g_1	$\frac{\cos\frac{5\pi}{24}-1}{\sin\frac{5\pi}{24}}$
a_2	$\frac{\cos\frac{3\pi}{24}-1}{\sin\frac{3\pi}{24}}$	b_2	$\sin\frac{3\pi}{24}$	g_2	$\frac{\cos\frac{3\pi}{24}-1}{\sin\frac{3\pi}{24}}$
a_3	$\frac{\cos\frac{\pi}{24}-1}{\sin\frac{5\pi}{24}}$	b_3	$\sin\frac{\pi}{24}$	g_3	$\frac{\cos\frac{\pi}{24}-1}{\sin\frac{5\pi}{24}}$
a_4, a_6	$\frac{1}{\sqrt{2}}-1$	b_4, b_6	$\frac{1}{\sqrt{2}}$	g_4, g_6	$\frac{1}{\sqrt{2}}-1$
a_5, a_7	$1-\sqrt{\frac{3}{2}}$	b_5, b_7	$\sqrt{\frac{2}{3}}$	g_5, g_7	$1-\sqrt{\frac{3}{2}}$

Table 1: Lifting Coefficients used in the 12 point MDCT.

cation at each lifting step and other parts of each transform. Each coefficient is quantized to a fixed number of bits and implemented by bit-shifting and additions using the canonic signed digit (CSD) representation [9]. An algorithm for calculating the number of adders for a given integer multiplication is presented in [10]. Figures 4(a) and (b) present the number of adders used in the implementation of the 12- and 36-point MDCT, respectively, with the number of bits used in quantization of the lifting coefficients varying from 1 to 8.

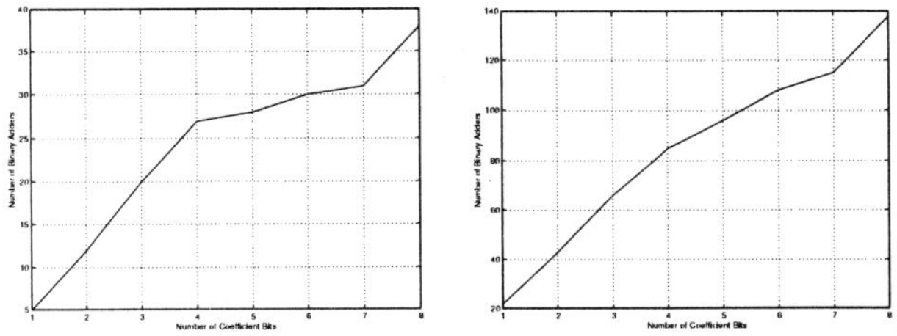

(a) (b)

Figure 4: Number of coefficient bits N_c versus the minimum number of binary adders for (a) $N = 12$ and (b) $N = 36$.

According to Figure 4, it is evident that the total complexity of each transform grows approximately linearly as the quantization resolution is increased.

4. IMPLEMENTATION IN THE MPEG LAYER III

The fast integer MDCT structure presented in the previous section is implemented in a standard MPEG layer III codec. We assess its performance by quantizing the lifting coefficients by increasing the number of bits used until we achieve satisfactory performance, as good as or better than the original floating point transform. The lifting coefficients are quantized to N_c bits. The nodes are quantized at N_l bits, N_l being sufficiently greater than N_c. In the simulation, N_l is set at 25 bits and N_c is varied from 1 to 8 bits. There are different criteria for assessing the performance of the integer MDCT based encoder or the decoder. The integer MDCT based encoder should be capable of producing a bit stream, which can be decoded by any standard decoder. Moreover the decoded bit stream must be perceptually as good as or better than that produced by the original floating point encoder/decoder combination. Similarly, the integer MDCT based decoder should be able to produce sound that is perceptually as good as the standard decoder while decoding a bit stream encoded using the standard encoder. Finally the integer MDCT based encoder bit stream when decoded by an integer MDCT based decoder should produce perceptually lossless audio. By varying the number of bits used to quantize the lifting coefficients N_c, we can find the minimum value of N_c that satisfies each of the criteria listed above.

The mean square error between a given audio signal and a reference one, which is obtained by using the true MDCT in there encoder and decoder, is used to quantitatively assess the performance of the codec. We consider the three scenarios:

(i) integer encoder/integer decoder combination,

(ii) standard encoder/integer decoder and

(iii) the integer encoder/standard decoder,

where the integer encoder/decoder refers to the encoder/decoder with the MDCT block implemented using quantized lifting coefficients and the standard version refers to that uses the floating point transform. We also compare with the conventional fixed-point implementation using the existing fast structures with butterflies and quantized coefficients.

Figures 4(a), (b) and (c) show the differences between the signals decoded using the approximated integer MDCT and the floating-point MDCT for scenarios (i), (ii) and (iii), respectively. From Figures 4(a) and (c), where the integer decoder is used, the error decreases slowly initially and then at $N_c = 17$ it quickly converges to zero. When both encoder and decoder are quantized (Figure 4(a)), the lifting implementation significantly outperforms the butterfly implementation, especially at very low N_c and the lines start to cross at $N_c = 17$, at which point causes no perceptual difference. According to Figure 4(a), at $N_c = 2$ bits, the integer codec yields approximately 25 dB improvement from the butterfly codec quantized at the same value of N_c. When only the decoder is quantized (Figure 4(c)), the lifting yields noticeably better reconstruction that the butterfly when $N_c \leq 17$, which implies standard compliances of the integer transform. From Figure 4(b), when only the encoder is approximated, the error decreases almost linearly as N_c is increased. The integer encoder yields noticeably higher fidelity when compared with the conventional fixed-point version.

It is also confirmed by a subjective test and is concluded that the integer encoder/standard decoder combination requires at least $N_c = 3$ bits to yield perceptually lossless reconstruction whereas it is required only $N_c = 2$ bits for the integer encoder/decoder combination to yield the result.

5. CONCLUSIONS

In this paper an implementation of the MDCT, a filterbank, which provides finer spectral resolution in the MPEG layer III, using a fast structure with integer coefficients is presented. The implementation is far less complex than the conventional implementation in terms of the number of adders used. The implementation was found to be perceptually lossless when the MDCT coefficients were quantized to 2 bits. A comparison with the conventional quantized fast structure using butterfly implementation is also presented. The results show that the proposed structures outperform the conventional method for most practical cases.

6. REFERENCES

[1] D. Pan, "A tutorial on mpeg/audio compression," *IEEE Multimedia*, pp. 60–74, Summer 1995.

[2] H.S.Malvar, *Signal Processing with Lapped Transforms*, Archtech House, 1992.

[3] V. Britanak and K.R. Rao, "An efficient implementation of the forward and inverse MDCT in MPEG audio coding," *IEEE Singal Processing Letters*, vol. 8, no. 2, pp. 48–51, February 2001.

[4] T. Krishnan and S. Oraintara, "A fast and lossless forward and inverse structure for the mdct in mpeg audio coding," *Proc. of the International Symposium on Circuits and Systems*, May 2002.

[5] I. Daubechies and W. Sweldens, "Factoring wavelet transforms into lifting steps," Technical report, Bell Laboratories, Lucent Technologies, 1996.

[6] K. Salomonsen, S. Søgaard, and E.P. Larsen, "Design and implementation of an MPEG/Audio layer III bitstream processor," Thesis report, Institute of Electronic Systems, Aalborg University, 1997.

[7] K.R. Rao and J.J. Hwang, *Techniques and Standards for Digital Image/Video/Audio Coding*, Prentice Hall, 1996.

[8] V. Britanak and K.R. Rao, "New fast algorithms for the unified forward and inverse MDCT/MDST computation," Technical report, Institute of Control Theory and Robotics, Slovak Academy of Sciences, May 2000.

[9] R.I. Hartley, "Subexpression sharing in filters using Canonic Signed Digit multipliers," *IEEE Trans. on Circuits and Systems*, vol. 43, no. 10, pp. 677–688, October 1996.

[10] Y.J. Chen, S. Oraintara, T. D. Tran, K. Amaratunga, and T.Q. Nguyen, "Multiplierless approximations of transforms using lifting scheme and coordinate descent with adder constraint," *IEEE International Conf. on Acoustics, Speech, and Signal Processing,*, May 2002.

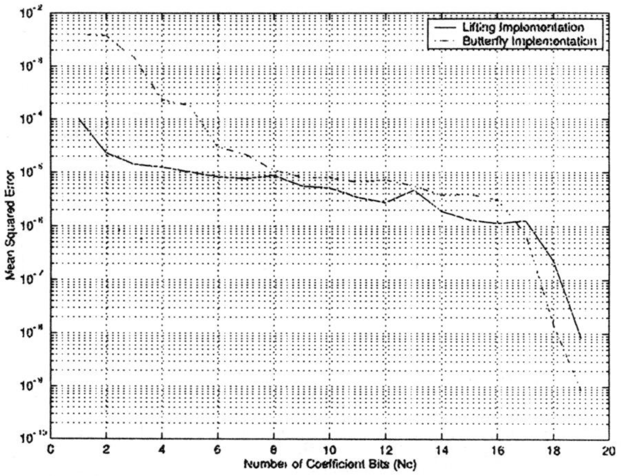

(a)

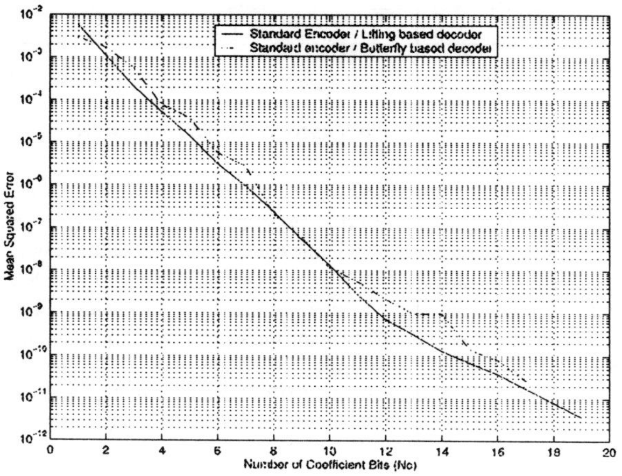

(b)

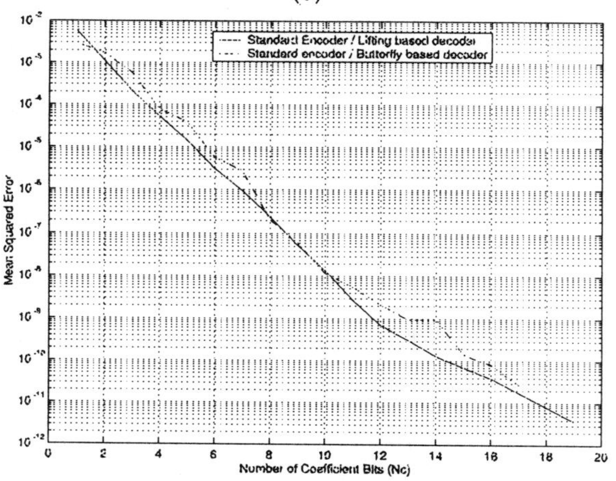

(c)

Figure 5: The mean square error versus N_c, number of coefficient bits for the three scenarios: (a) integer encoder/integer decoder combination, (b) standard encoder/integer decoder and (c) the integer encoder/standard decoder, as N_c varies from 1 to 20 bits.

MINIMAL ARMA LATTICE DIGITAL FILTER REALIZATION

H. K. Kwan

Department of Electrical and Computer Engineering
University of Windsor
401 Sunset Avenue, Windsor, Ontario, Canada N9B 3P4
kwan1@uwindsor.ca

ABSTRACT

In this paper, a minimal and normalized ARMA lattice filter structure for the realization of arbitrary IIR digital filters is presented. The filter structure is minimal in terms of the numbers of coefficients and delays, and exhibits a high degree of structural regularity and modularity. A Levinson-type algorithm is used to compute the reflection coefficients. The magnitude of the reflection coefficients of such a stable filter is bounded by unity and consequently stability can be conveniently checked. Lowpass, highpass, bandpass, and bandstop digital filter realization examples are given.

1. INTRODUCTION

Based on a general set of recursions obtained by a geometric approach, an ARMA lattice-modeling algorithm is formulated by Lee-Friedlander-Morf [1]. Lim-Parker [2] has developed a robust method based on partial fraction expansion for synthesizing AR and MA lattice filters, and their two-channel AR lattice model can be adapted for ARMA lattice filter synthesis. These two algorithms are formulated for models with equal AR-order and MA-order. Equal-order is not required by Benveniste-Chaure [3] in their lattice form algorithm, which is capable of estimating models of arbitrary AR-order and MA-order. Further development is made by Miyanaga-Nagai-Miki [4], who have derived a lattice algorithm that allows arbitrary arrangement of AR-type and MA-type order update recursions, and every estimation error obtained in each cascaded section satisfies orthogonal conditions. Recently, we have developed normalized ARMA lattice modeling algorithms [5-7] that share similar flexibility of the above lattice modeling algorithms and with a regular and modular structure. The algorithms are capable of computing models of any AR-order and MA-order with flexible arrangement of order-update recursions. The algorithms are suitable for applications where the data samples of the input and output processes are available. In [5], a Levinson-type algorithm is proposed for estimating the ARMA lattice model from observed inputs and outputs of an unknown system. When the input is a white process, the algorithms yield a lattice model that is canonic with respect to the number of parameters. A Schur-type algorithm is presented in [6] for computing the ARMA lattice model where the correlations of input and output processes are available. The computation architecture of the algorithm contains a substantial amount of parallelism that can be exploited by using a suitable processing structure.

An approach to design a lattice digital filter for realizing a given digital filter transfer function is to determine the reflection coefficients of the lattice digital filter such that the impulse response of the given transfer function is whitened. The structure of the digital filter for the realization of the given transfer function is then obtained by proper redirection of the signal flow of the whitening filter. This approach has been used by Gray-Markel to design the classical AR lattice filter [8] and by Lim-Parker [2] to design ARMA lattice digital filters. These methods share the advantage that the realized digital filter structure usually exhibits a high degree of regularity and modularity. However, the resultant realization structures of this approach may contain a lot of redundancy as dictated by their modeling algorithms that are designed under various constraints. In [7], we have introduced a minimal and normalized ARMA lattice digital filter structure for digital filter realization in which flexible AR-order and MA-order modeling is allowed. It can be shown that the filter realization structure is minimal and shares the desirable properties of the classical AR lattice filters. Both the Levinson-type algorithm [5] and the Schur-type algorithm [6] can be used to compute the reflection coefficients. Other applications of our ARMA lattice model include speech recognition [9] and speech analysis and synthesis [10]. In this paper, our emphasis is not on ARMA lattice system modeling but on designing minimal and normalized ARMA lattice digital filter from a given digital filter transfer function or impulse response. The modeling and synthesis filters are described in Section 2. Section 3 presents the results on filter realization, and the paper is concluded with a summary in Section 4.

2. ARMA LATTICE DIGITAL FILTER

2.1 Analysis Filter

$$y(n) = \sum_{p=1}^{P} a_p \, y(n-p) + \sum_{q=0}^{Q} b_q \, x(n-q) \qquad (1)$$

In the context of ARMA system modeling, we try to describe an unknown system by an ARMA model of order (P, Q) in the following form:

The x(n) and y(n) are respectively the observed input and output processes of the unknown system. The P and Q are respectively the AR-order and MA-order of the model, a_p (p = 1, 2, ..., P) and b_q (q = 0, 1, ..., Q) are the coefficients of the model. The modeling process performs a whitening filter operation. Given the input and output of the system, we can use the modeling

0-7803-7761-3/03/$17.00 ©2003 IEEE

algorithm to identify the system characteristic in term of the reflection coefficients.

The ARMA lattice structure developed in [5, 7] consists of a set of modular computational blocks and will be used for digital filter realization in this paper. To obtain order update recursions for the ARMA lattice model, two basic modules (Modules A and B), two starting blocks (I_AR, I_MA), and four regular blocks (AR_AR, MA_MA, AR_MA, and MA_AR) are defined. Modules A and B are the basic analysis and synthesis building blocks for the two starting blocks and the four regular blocks. Reflection coefficients inside each lattice stage are computed as model parameters. The error fields upon which the algorithm operates form two groups, one for AR order update, the other for MA order update. Each group consists of 4 error fields, i.e., $[e_x, \varepsilon_y, \tau_x, r_y]^T$ for AR lattices like AR_AR, AR_MA blocks; and $[e_x, \varepsilon_y, r_x, \tau_y]^T$ for MA lattices like MA_AR, MA_MA blocks. The first two fields are the same for both sets; the last two fields will be changed in outputs when the order update needs to shift from AR to MA or from MA to AR. To build an ARMA lattice model, we always start from one of the starting blocks. Other regular order update blocks are then connected according to the rules of error field's compatibility. Basically, An ARMA lattice model of any order can be obtained as long as the inputs and outputs of adjacent blocks are compatible. For an example, starting from I_AR, the next block's inputs must be in the same format of the I_AR's outputs. That is to say, either AR_MA or AR_AR can follow the I_AR block. If the second block is AR_MA, then the third block must be MA_AR or MA_MA. For any (P, Q)th-order ARMA lattice filter, there is more than one way to build it. They could involve different types of blocks and have different block sequences. Among all kinds of possible connections, there is one special arrangement that we call minimal. The minimal ARMA lattice model is to be adopted for building the analysis and synthesis ARMA lattice filters. In the analysis model, an impulse and the impulse response of a filter are used as the inputs to the ARMA lattice analysis filter to compute the reflection coefficients for a specified order and model structure. The reflection coefficients are then used in the ARMA lattice synthesis filter to filter the impulse to obtain the impulse response of the designed filter.

2.2 Synthesis Filter

By reversing the direction of signal flow of the forward-y error in the whitening filter, an ARMA lattice synthesis filter is obtained which synthesizes the process y(n) from x(n) and $\varepsilon_y^{(P,Q)}(n)$ [5, 7]. If the input to $\varepsilon_y^{(P,Q)}(n)$ is zero, then the synthesis filter realizes the transfer function of the model with x(n) as the input and y(n) as the output. The two backward-x errors of order (P, Q) are considered as the dummy outputs of the synthesis filter [5, 7]. All the type-A modules connected to $\varepsilon_y^{(P,Q)}(n)$ in the whitening filter are converted to the corresponding type-B modules in the synthesis filter [5, 7]. Hereafter, each section in the synthesis filter structure will be referred to as a filter-section to distinguish from the order-update section of the corresponding type in the whitening filter structure. In a system-modeling problem, the complexity of the structure is the price to pay for the flexibility of the modeling algorithm. For filter realization purpose, such flexibility is not needed as the order of the transfer function is known beforehand. Therefore, the simplest possible structure should be used, this corresponds to the case where x(n) is white and the arrangement of order-update sections is minimal. For a given transfer function of order (P,Q), the whitening filter structures [5, 7] for the cases P = Q, P>Q and P<Q can be derived and expressed in a general form [7].

3. RESULTS

The ARMA lattice analysis and synthesis filters are used to model and synthesize or realize four eighth-order lowpass, highpass, bandpass, and bandstop elliptic IIR digital filters with specifications as summarized in Table 1. The analog prototype lowpass filters and their filter numbers are selected from the filter handbook [11]. The analog frequency transformation [12] and the bilinear transform (with T=2) are used to obtain the digital prototype filters for realization. The highpass digital filter is obtained by applying lowpass to highpass digital frequency transformation [13] with wpT transformed from 0.5π to 0.8π. Each of the four digital prototype digital filters has a total of 17 coefficients. For each of these four filters, an ARMA lattice digital filter with AR-order and MA-order both equal to 8 and a total of 16 coefficients is used. The lattice coefficients of the four realized ARMA lattice digital filters are listed in Tables 2-5. The magnitude and phase responses of these four realized ARMA lattice digital filters are shown in Figs. 1-4(a)-(b). The magnitude and phase responses differences between the prototype and the realized digital filters are negligibly small.

4. SUMMARY

In this paper, the ARMA lattice model has been used to realize lowpass, highpass, bandpass, and bandstop IIR digital filters examples. For IIR digital filter realization, equal AR-order and MA-order is an obvious choice but arbitrary AR-order and MA-order can be chosen as desired. This ARMA lattice filter requires a minimal number of coefficients and delays; is modular and orthogonal; exhibits low sensitivities, low round-off noise, absence of internal overflow; and with reflection coefficients bounded by unity value for ease of stability check.

5. REFERENCES

[1] Lee D. T. L., Friedlander B., and Morf M., "Recursive ladder algorithms for ARMA modeling". *IEEE Transactions on Automatic Control*, vol. AC-27, pp. 753-764, August 1982.

[2] Lim Y. C. and Parker S. R., "On the synthesis of lattice parameter digital filters". *IEEE Transactions on Circuits and Systems*, vol. CAS-31, pp. 593-601, July 1984.

[3] Benveniste A. and Chaure C., "AR and ARMA identification algorithms of Levinson type: An innovations approach". *IEEE Transactions on Automatic Control*, vol. AC-26, pp. 1243-1260, December 1981.

[4] Miyanaga Y., Nagai N., and Miki N., "ARMA digital lattice filter based on new criterion". *IEEE Transactions on Circuits and Systems*, vol. CAS-34, pp. 617-628, June 1987.

[5] Kwan H. K. and Lui Y. C., "Normalized ARMA Levinson algorithm". *Proceedings of International Symposium on Circuits and Systems*, Oregon, 1989, vol. 1, pp. 234-237.

[6] Kwan H. K. and Lui Y. C., "Normalized ARMA Schur algorithm". *Proceedings of 32nd Midwest Symposium on Circuits and Systems*, Urbana-Champaign, Illinois, August 14-16, 1989, vol. 2, pp. 1103-1106.

[7] Kwan H. K. and Lui Y. C., "Minimal normalized lattice structure for ARMA digital filter realization". *Proceedings of IEEE International Conference on Acoustics, Speech and Signal Processing*, Toronto, Ontario, Canada, May 14-17, 1991, vol. 3, pp. 1629-1632.

[8] Gray A. H. and Markel J. D., "A normalized digital filter structure". *IEEE Transactions on Acoustics, Speech, and Signal processing*, vol. ASSP-23, pp. 268-277, June 1975.

[9] Kwan H. K. and Li Tracy X, "ARMA lattice modeling for isolated word speech recognition". *Proceedings of 43rd Midwest Symposium on Circuits and Systems*, Michigan, August 8-11, 2000, vol. 3, pp. 1186-1190.

[10] Kwan H. K. and Wang M., "Pitch-excited ARMA lattice model for speech analysis and synthesis". *Proceedings of 2001 IEEE Pacific Rim Conference on Communications, Computers and Signal Processing (PACRIM'01)*, Victoria, Canada, August 26-28, 2001, vol. II, pp. 659-662.

[11] Saal R.. *Handbook of Filter Design*. AEG-Telefunken, Berlin, 1979.

[12] Rabiner L. R. and Gold, B. *Theory and Application of Digital Signal Processing*. Prentice-Hall, New Jersey, 1975.

[13] A.G. Constantinides, A. G., "Spectral transformations for digital filters". IEE Proceedings, vol. 117, no. 8, pp. 1585-1590, August 1970.

Table 1 Specifications of the four filter examples.

	LP	HP	BP	BS
Prototype	C0820c66	C0820c66	C0425b32	C0420c31
WpT	0.50000π	0.80000π	-	-
WsT	0.53336π	-	-	-
wp1T	-	-	0.41111π	0.32894π
wp2T	-	-	0.46666π	0.46538π
ws1T	-	-	0.38393π	0.36416π
ws2T	-	-	0.49526π	0.42620π

Table 2 ARMA lattice coefficients of Example 1.

I-AR	K0 = 0.050709
AR-MA	KY2(0,1) = 0.631490
MA-AR	KX2(1,1) = 0.257760
AR-MA	KY2(1,2) = -0.686581
MA-AR	KX2(2,2) = 0.539541
AR-MA	KY2(2,3) = 0.739000
MA-AR	KX2(3,3) = 0.719623
AR-MA	KY2(3,4) = -0.771955
MA-AR	KX2(4,4) = 0.774608
AR-MA	KY2(4,5) = 0.789323
MA-AR	KX2(5,5) = 0.805615
AR-MA	KY2(5,6) = -0.814820
MA-AR	KX2(6,6) = 0.906929
AR-MA	KY2(6,7) = 0.740306
MA-AR	KX2(7,7) = 0.441861
AR-MA	KY2(7,8) = -0.479062
MA-AR	KX2(8,8) = 1.000000

Table 3 ARMA lattice coefficients of Example 2.

I-AR	K0 = 0.006674
AR-MA	KY2(0,1) = -0.933466
MA-AR	KX2(1,1) = -0.057267
AR-MA	KY2(1,2) = -0.946148
MA-AR	KX2(2,2) = 0.137889
AR-MA	KY2(2,3) = -0.948988
MA-AR	KX2(3,3) = -0.301463
AR-MA	KY2(3,4) = -0.947323
MA-AR	KX2(4,4) = 0.659678
AR-MA	KY2(4,5) = -0.692398
MA-AR	KX2(5,5) = 0.640009
AR-MA	KY2(5,6) = 0.377153
MA-AR	KX2(6,6) = 0.995019
AR-MA	KY2(6,7) = 0.523188
MA-AR	KX2(7,7) = 0.775043
AR-MA	KY2(7,8) = -0.034258
MA-AR	KX2(8,8) = 0.999998

Table 4 ARMA lattice coefficients of Example 3.

I-AR	K0 = 0.003175
AR-MA	KY2(0,1) = 0.190245
MA-AR	KX2(1,1) = 0.001782
AR-MA	KY2(1,2) = -0.993249
MA-AR	KX2(2,2) = -0.076431
AR-MA	KY2(2,3) = 0.191317
MA-AR	KX2(3,3) = -0.023615
AR-MA	KY2(3,4) = -0.994576
MA-AR	KX2(4,4) = 0.216580
AR-MA	KY2(4,5) = 0.180568
MA-AR	KX2(5,5) = -0.032895
AR-MA	KY2(5,6) = -0.955551
MA-AR	KX2(6,6) = 0.758046
AR-MA	KY2(6,7) = -0.274188
MA-AR	KX2(7,7) = -0.341287
AR-MA	KY2(7,8) = -0.201827
MA-AR	KX2(8,8) = 1.000000

Table 5 ARMA lattice coefficients of Example 4.

I-AR	K0 = 0.645929
AR-MA	KY2(0,1) = -0.056970
MA-AR	KX2(1,1) = -0.308907
AR-MA	KY2(1,2) = 0.157216
MA-AR	KX2(2,2) = 0.964124
AR-MA	KY2(2,3) = 0.187569
MA-AR	KX2(3,3) = -0.311760
AR-MA	KY2(3,4) = -0.209324
MA-AR	KX2(4,4) = 0.989699
AR-MA	KY2(4,5) = -0.305246
MA-AR	KX2(5,5) = -0.298184
AR-MA	KY2(5,6) = 0.145997
MA-AR	KX2(6,6) = 0.990054
AR-MA	KY2(6,7) = 0.655684
MA-AR	KX2(7,7) = -0.239611
AR-MA	KY2(7,8) = -0.283680
MA-AR	KX2(8,8) = 1.000000

Fig. 1a Magnitude Response of ARMA Lattice Digital Filter (Example 1)

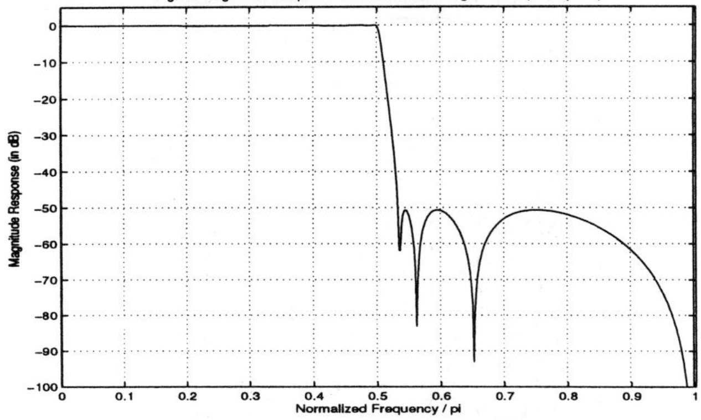

Fig. 1b Phase Response of ARMA Lattice Digital Filter (Example 1)

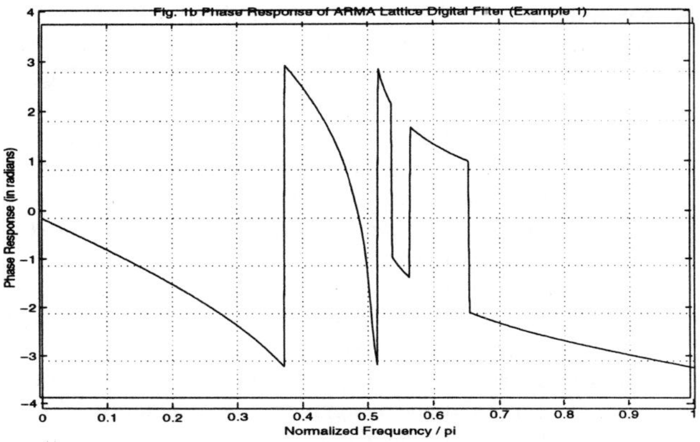

Fig. 2a Magnitude Response of ARMA Lattice Digital Filter (Example 2)

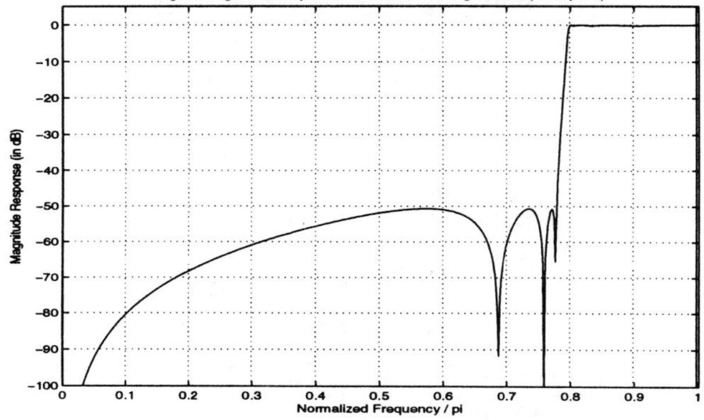

Fig. 2b Phase Response of ARMA Lattice Digital Filter (Example 2)

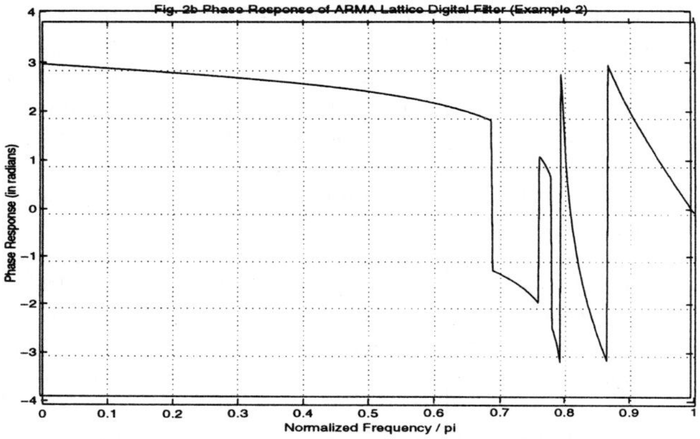

Fig. 3a Magnitude Response of ARMA Lattice Digital Filter (Example 3)

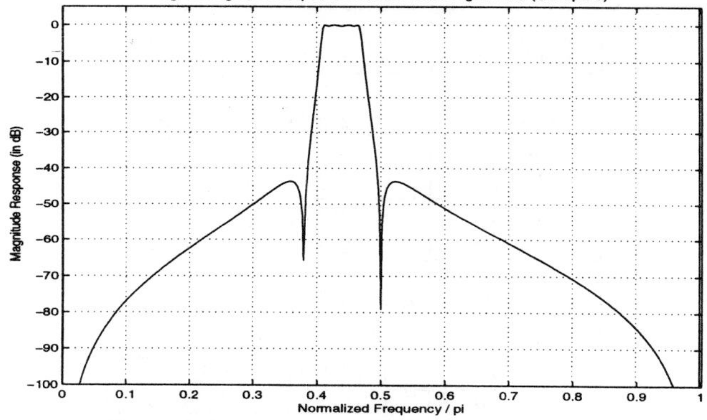

Fig. 3b Phase Response of ARMA Lattice Digital Filter (Example 3)

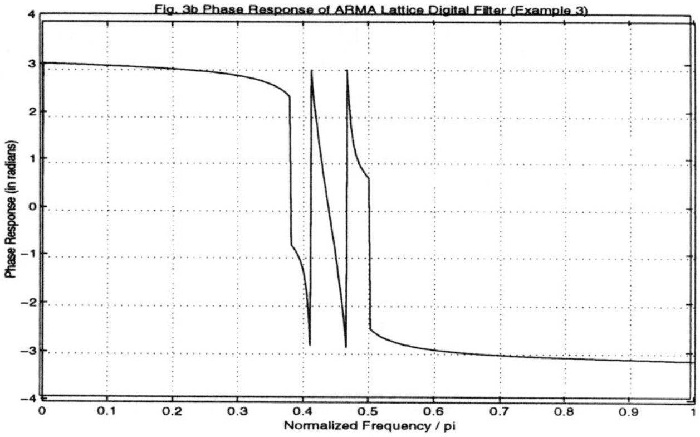

Fig. 4a Magnitude Response of ARMA Lattice Digital Filter (Example 4)

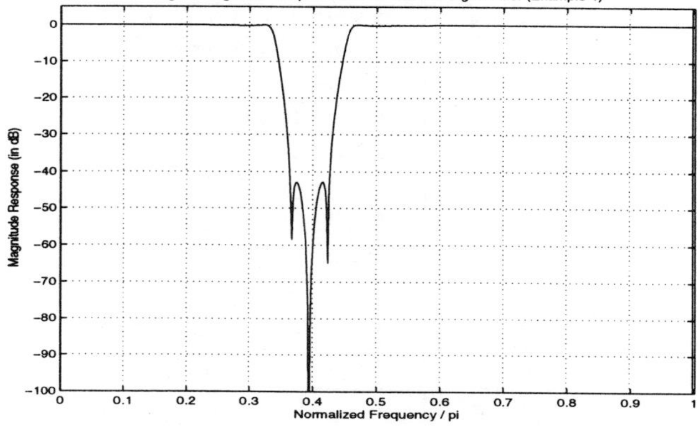

Fig. 4b Phase Response of ARMA Lattice Digital Filter (Example 4)

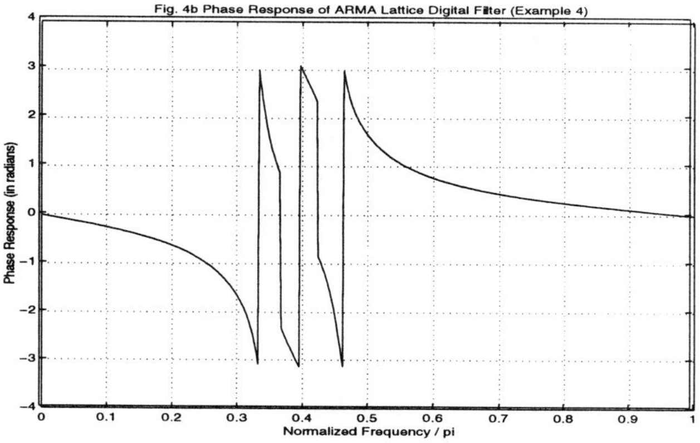

FAST CHARACTERIZATION OF THE NOISE BOUNDS DERIVED FROM COEFFICIENT AND SIGNAL QUANTIZATION

J.A. López, C. Carreras, G. Caffarena and O. Nieto-Taladriz

Departamento de Ingeniería Electrónica, ETSI Telecomunicación
Universidad Politécnica de Madrid
Ciudad Universitaria, s/n, 28040 Madrid, SPAIN
{juanant | carreras | gabriel | nieto} @ die.upm.es

ABSTRACT

This paper presents a new method for computing the absolute noise bounds caused by quantization of coefficients and signals in fixed-point implementation of digital filters. A tool based on refined interval-based computations has been enhanced to calculate with reduced computational cost the implications of both effects in the finite wordlength behavior of the specific realizations. Finally, a particular filter realization is quantized using different wordlengths and the computed bounds are presented for comparison.

1. INTRODUCTION

Since the pioneer work by Jackson [1], quantization or finite wordlength (FWL) effects in IIR digital filters have been extensively studied in the literature during the past decades [2]-[11]. These effects are typically divided in coefficient quantization noise (CQN) [5],[6], roundoff noise (RON) [7],[8], and overflow and underflow limit cycles [9]-[11]. RON and CQN produce small variations around the unquantized result and they can be modeled as small noise sources, independent among them and from the input signal. Therefore, these two effects are labeled as *linear effects*. Limit cycles are autonomous oscillations of variable amplitude and frequency, caused by quantization, and whose values are no longer independent of the signal to be quantized. Therefore, these effects are labeled as *nonlinear effects*.

Nonlinear effects are caused by the use of excessively small wordlengths and they can be avoided by increasing their size above a certain threshold. On the contrary, CQN and RON are unavoidable in the quantization process, even though they can be bounded according to the wordlength of the fractional part. Therefore, the output may vary within some bounds around the ideal response. The aim of this paper is to show how interval simulation [12]-[15] can provide the bounds for CQN and RON in filters implemented with a given number of bits.

The paper is divided as follows: Section 2 presents a brief introduction about the foundations of coefficient quantization and roundoff noise. Section 3 provides the basic rules of interval computation and a brief description of the process followed by the interval simulation tool. Section 4 presents some examples using infinite precision and applying quantization in both, coefficient and intermediate signals. These results are discussed in Section 5. Finally, Section 6 presents the conclusions of this work.

2. COEFFICIENT QUANTIZATION AND ROUNDOFF NOISE

Let an N-order IIR stable filter be described by

$$y(k) = \sum_{i=0}^{M} b_i u(k-i) + \sum_{j=1}^{N} a_j y(k-j)$$

where N and M are respectively the number of poles and zeros of the filter, $u(k)$ and $y(k)$ are the (quantized) input and output of the filter and a_j, b_i are the unquantized coefficients. Both the coefficients and the results of the multiplications must be quantized before FWL implementation of the filter.

CQN refers to the deterministic changes occurring in $h(n)$ or $H(z)$ due to approximation of the real coefficients when using FWL representation. The effect of quantization can be modeled as [1],[4]

$$y(k) = \sum_{i=0}^{M} \hat{b}_i u(k-i) + \sum_{j=1}^{N} \hat{a}_j y(k-j)$$

where $\hat{b}_i = b_i + \Delta b_i$ and $\hat{a}_j = a_j + \Delta a_j$. If the filter does not meet the specifications, we can recalculate the coefficients and/or allocate more bits until the specifications are completely satisfied.

RON refers to the quantization of the products within the realization. Each quantization is modeled as a white noise source, uncorrelated from the input and from the rest of the signals of the circuit. Its effect is best described using the state-space representation of the filter [7],[8]

$$x(k+1) = Ax(k) + bu(k) + e(k)$$
$$y = cx(k) + du(k)$$

where $x(k)$ and $e(k)$ are $N\times 1$ state vectors and A,b,c,d are matrices of appropriate dimensions. Each noise source $e_i(k)$ is assumed to be uniformly distributed on an interval of the size of the quantization step. The exact bounds of the error intervals depend on the representation and quantization strategies.

The previous effects have two important characteristics: (i) They are highly dependent on the structure of the filter network, and (ii) CQN provides a lower bound on RON [2],[6]. Interval simulation provides bounds on the variations of the filter output for implementations using a given number of bits. As it will be shown in the following sections, both characteristics are readily confirmed.

0-7803-7761-3/03/$17.00 ©2003 IEEE

3. REFINED INTERVAL SIMULATION

A given interval I_x is completely specified by two of the following elements: upper bound $\overline{I_x}$, lower bound $\underline{I_x}$, midpoint $(\overline{I_x} + \underline{I_x}) / 2$, and width $(\overline{I_x} - \underline{I_x})$. The basic operation rules between two input intervals, I_1 and I_2, are listed in table I.

The tool developed for interval-based simulations is a modified version of the tool presented in [14]. It takes the specification of the algorithm under test (i.e. the digital filter) and a description of the input data spaces to provide a description of the outputs. Supported types of input spaces [14] include standard numeric traces, interval traces, and interval-based domain specifications, mainly used for range computation. In this study, results are derived mainly from simulating interval traces. Intervals can also be associated to probabilities to obtain estimates of the output probability density functions, although this feature is not used in this study. Besides the output space description, the tool provides detailed statistics of the variables of the algorithm which can also be used in the wordlength selection process.

In the following parts of this section two types of simulation are discussed according to the accuracy required for the results: single and multi-interval simulation.

3.1 Single-Interval Simulation

The basic computation procedure is as follows: First, the algorithm under study is specified by means of a multiple precision language, and the input spaces are described in corresponding data files. In the case of single-interval simulation (i.e. standard range computation), each input space is described by a single interval. The designer has additional control over the input wordlengths by assigning specific fixed-point data types to the inputs in the algorithm specification. Then, a single run of the algorithm is required to obtain estimates of the ranges and precisions of the intermediate and output variables. Noise analysis and wordlength assignment can be performed by repeating this simulation process with refined fixed-point declarations of the intermediate variables of the algorithm.

3.2 Multi-Interval Simulation

While interval analysis provides correct bounds for the output data spaces (i.e. any numeric result is guaranteed to be included in the output interval description), data dependencies in the algorithm may cause that such interval bounds are too conservative in some cases [13]. This situation is comparable qualitatively to scaling using absolute bounds (ℓ_1-scaling [4]).

TABLE I

BASIC OPERATION RULES FOR INTERVAL COMPUTATIONS

OP.	$\overline{I_{RES}}$	$\underline{I_{RES}}$
$I_1 + I_2$	$\overline{I_1} + \overline{I_2}$	$\underline{I_1} + \underline{I_2}$
$I_1 - I_2$	$\overline{I_1} - \underline{I_2}$	$\underline{I_1} - \overline{I_2}$
$I_1 \times I_2$	$max(\overline{I_1 I_2}, \overline{I_1}\underline{I_2}, \underline{I_1}\overline{I_2}, \underline{I_1 I_2})$	$min(\overline{I_1 I_2}, \overline{I_1}\underline{I_2}, \underline{I_1}\overline{I_2}, \underline{I_1 I_2})$
I_1 / I_2	$max \begin{cases} \overline{I_1}/\overline{I_2}, \overline{I_1}/\underline{I_2}, \\ \overline{I_1}/\underline{I_2}, \underline{I_1}/\underline{I_2}, \end{cases} \quad 0 \notin I_2$	$min \begin{cases} \overline{I_1}/\overline{I_2}, \overline{I_1}/\underline{I_2}, \\ \overline{I_1}/\underline{I_2}, \underline{I_1}/\underline{I_2}, \end{cases} \quad 0 \notin I_2$
$I_1 \cap I_2$	$max(\overline{I_1}, \overline{I_2})$	$min(\underline{I_1}, \underline{I_2})$
$I_1 \cup I_2$	$min(\overline{I_1}, \overline{I_2})$	$max(\underline{I_1}, \underline{I_2})$

Multi-interval simulation has been shown to provide more accurate results, while still avoiding other costly methods based on the numeric (exhaustive) simulation of the input data space [14]. Therefore, it can be applied in those cases where single-interval simulation provides oversized results. The calculation process is as follows: (i) Each input data space is described as a set of k_i disjoint (adjacent) intervals covering the input range. (ii) As many single-interval simulations as vectors in the cartesian product of the interval descriptions of all inputs are then performed. (iii) The interval results from these simulations for each particular output variable are finally merged into a single description. Accuracy is improved with respect to standard range computation because the reduced size of the intervals used in each simulation diminishes the impact of the data dependencies causing oversized results.

4. EXAMPLES

Figures 1.a, 1.b and 1.c show the three types of asymptotic stability. They correspond to a second-order Butterworth filter, implemented in transposed direct form II (DFIIt), with cutoff frequencies w_n / w_o = 0.5, 0.30408 and 0.3, respectively. The corresponding coefficients of $H(z)$ are given in table II. The effect of multi-interval simulation is also shown in figure 1.d for $w_n / w_o = 0.5$ and $k_i = 10$. In these cases, the input of the filter is set to $u(k) = (-1, 1)$ at $k = 0$ and 0 otherwise (i) to obtain guaranteed bounds (based on interval arithmetic) for the output signal. (ii) to illustrate that the output range can show a monotonic increasing or decreasing behavior even if the input is set to zero. It can also be observed how multi-interval simulation significantly reduces the oversized bounds.

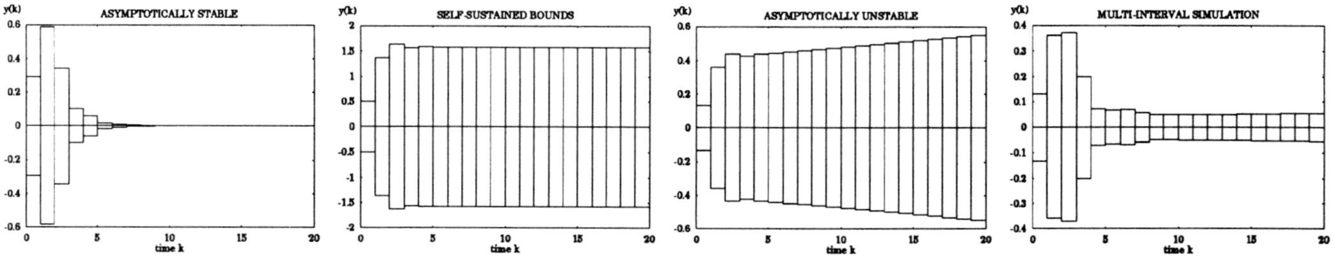

Figure 1. Output responses of the filter using interval simulation: (a) asymptotically stable, (b) self-sustained bounds, (c) asymptotically unstable, (d) use of multi-interval simulation to obtain refined results (with respect to case c).

TABLE II

COEFFICIENTS OF 2ND ORDER BUTTERWORTH FILTER OF FIGURE 1

w_n / w_o	0.5	0.30408	0.3
a1	0	-0.7320	-0.7477
a2	0.1716	0.2679	0.2722
b0=b2	0.2929	0.5	0.1311
b1	0.5858	1	0.2622

TABLE III

COEFFICIENTS AND QUANTIZATION INTERVALS FOR THE FILTER USED IN FIGURES 2 AND 3

COEFS.	*a1*	*a2*	*b0=b2*	*b1*
Ideal case	0	0.1716	0.2929	0.5858
7-bit interval	[0, 0.0078)	[0.1640, 0.1719)	[0.2891, 0.2969)	[0.5781, 0.5859)
Quantized coefficients	0	0.1640	0.2891	0.5781
Error interval	[0,0]	[0.1640, 0.1716]	[0.2891, 0.2929]	[0.5781, 0.5858]

Table III shows the ideal and quantized coefficients of a second-order Butterworth filter with cutoff frequency $w_o / w_n = 0.5$. Quantization has been performed using truncation to 7 fractional bits. The interval of values that contains each unquantized coefficient (i.e. the endpoints are the values represented with 7 fractional bits that contain the coefficients) and the error interval generated when using 7-bit truncation are also given.

Figures 2 and 3 show the effects of quantization on the impulse response, $h(n)$, and the transfer function, $H(z)$, if the filter is implemented in DFIIt with 7 fractional bits. Interval simulation provides a range of values (rectangles in the figures) assigned to each sample. $h(n)$ has been obtained simulating the description of the filter using a delta function as the input signal, and $H(z)$ is the 64-point FFT of $h(n)$, also computed using interval arithmetic. Figures 2.a and 3.a show the response of the filter using truncation to 64 bits, so the quantization effects can be neglected in this case. The maximum variations in $h(n)$ and $H(z)$ caused by quantization of coefficients, quantization of products, and by the combination of both effects are shown in figures 2.b and 3.b, figures 2.c and 3.c, and figures 2.d and 3.d, respectively. Finally, figure 4 illustrates the reduced bounds of $h(n)$ and $H(z)$, with respect to cases 2.d and 3.d, if wordlengths are increased to 10 fractional bits.

In cases 2.b - 2.d, the whole effect of quantization (i.e. maximum deviation with respect to the ideal response) is bounded by 0.0114, 0.0395 and 0.0407, respectively. Similarly, the variations in the filter gain are bounded by 0.0297, 2.4219, and 2.4375. In the 10-fractional-bit case, quantization errors in $h(n)$ are bounded by 0.0052, and the variations in the filter gain are always below 0.3046.

5. DISCUSSION

From the previous analysis, the following conclusions are drawn:

- Linear and nonlinear effects caused by quantization modify the operation of filter realizations. Using interval arithmetic, these effects (roundoff noise, variations in the impulse response, underflow and overflow limit cycles) can be bounded within known limits. It has also been observed that computed intervals bound all linear effects in the frequency domain, and they detect possible oscillations as long as they are not masked by the interval representation in the time domain.

- Interval arithmetic produces oversized results (causes are analogous to ℓ_1-scaling). The multi-interval computation method presented in section 3.2 provides refined results, if required, at the cost of longer computational times.

- Quantization effects are more significant in the RON case than in the CQN case. This fact has been known for a long time [2],[6] since the sensitivity of coefficients provides a lower bound on the noise gain of a properly scaled filter (for ℓ_2-scaled filters, $\|G_i\|_2 \geq \|S_{ij}\|_1$, [2]). In this sense, interval simulation also shows that gain variations from product quantization are almost two orders of magnitude greater than those due to coefficient quantization.

- If the filter is implemented using a very small number of bits, the effect of RON spreads through all the frequencies in the spectrum and the operation of the filter cannot be guaranteed (figures 3.c and 3.d). Obviously, this problem must be solved by increasing the number of bits.

- Computation of RON and CQN requires asymptotically stable responses from the point of view of the interval simulations.

- Finally, interval simulation supplies a fast method for the simulation of the filter response, including coefficient quantization and roundoff effects, because the output bounds can be calculated in a single execution of the algorithm.

6. CONCLUSIONS

This paper has presented a novel approach to study the quantization problems in digital filters based on interval simulation. This technique is widely used in modeling uncertainty and range computation but, up to knowledge of the authors, it has never been used in characterization of the FWL effects in digital filters. The interval computation method presented here has been optimized to provide FWL information in IIR filters and, in general, systems containing feedback loops.

One of the main advantages of this interval-based approach is the reduced computational cost required to obtain FWL information, making it particularly valuable in complex structures. Results providing the noise bounds caused by quantization for both $h(n)$ and $H(z)$ using interval computations have also been presented. These results are in concordance with the existing analytical ones, and they also provide further quantitative details. Consequently, the application of all the aspects presented in this paper provides significant information and a new perspective for the analysis and design of IIR digital filters.

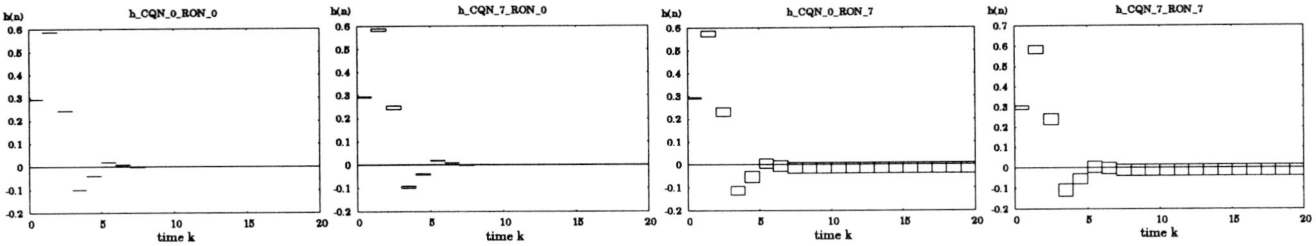

Figure 2. Effects of quantization on $h(n)$: (a) unquantized $h(n)$, (b) worst case CQN: $h(n)$ using the 7-bit coefficient quantization interval, (c) RON: $h(n)$ with products quantized to 7 bits, (d) worst case CQN + RON.

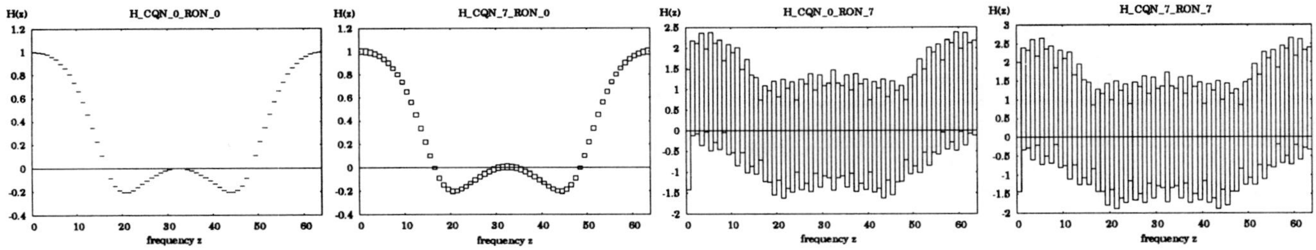

Figure 3. Effects of quantization on $H(z)$: (a) unquantized $H(z)$, (b) worst case CQN: $H(z)$ using the 7-bit coefficient quantization interval, (c) RON: $H(z)$ with products quantized to 7 bits, (d) worst case CQN + RON.

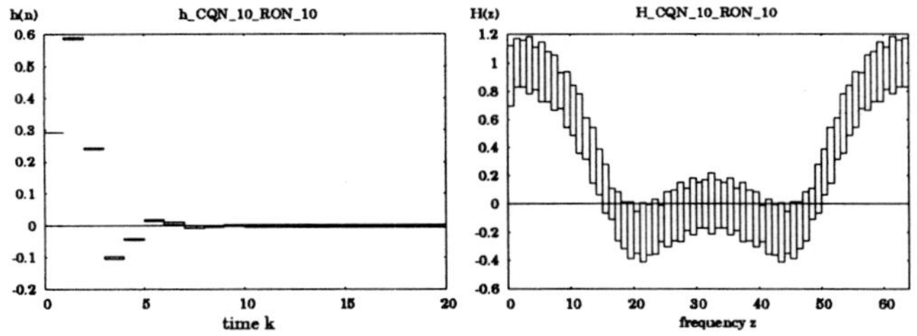

Figure 4. Reduced bounds if coeffcients and signals are quantized with 10 fractional bits: (a) $h(n)$, (b) $H(z)$

ACKNOWLEDGEMENTS

This work has been partially supported by the spanish *CICYT - Subdirección General de Proyectos de Investigación*, under project number TIC2000-1395-C02-01.

7. REFERENCES

[1] L.B. Jackson: "On the interaction of roundoff noise and dynamic range in digital filters", *Bell System Technical Journal,* vol.49, pp.159-184, Feb. 1970.

[2] R.A. Roberts, C.T. Mullis: *Digital Signal Processing* Addison-Wesley, Reading, MA, 1987.

[3] A.V.Oppenheim, R.W.Schafer: *Discrete-Time Signal Processing,* Prentice-Hall, Englewood Cliffs, NJ, 1989.

[4] L.B. Jackson: *Digital Filters and Signal Processing,* Kluwer Academic, Boston, MA, 1989.

[5] J.F. Kaiser: "Digital Filters". Ch. 7 in *Systems Analysis by Digital Computer*, F.F. Kuo, J.F. Kaiser (eds.), Wiley, 1966.

[6] L.B.Jackson: "Roundoff Noise Bounds Derived From Coefficient Sensitivities for Digital Filters". *IEEE Trans. Circuits and Systems,* vol. CAS-23, pp. 481-485, 1976.

[7] C.T. Mullis, R.A. Roberts: "Synthesis of minimum roundoff noise fixed point digital filters", *IEEE Trans. Circuits and Systems,* vol. CAS-23, pp. 551-562, Sept., 1976.

[8] S.Y. Hwang: "Minimum uncorrelated unit noise in state-space digital filtering", *IEEE Trans. Acoustics, Speech, Signal Processing,* vol. ASSP-25, pp.273-281, Aug., 1977.

[9] W.L. Mills, C.T.Mullis, R.A. Roberts: "Digital filter realizations without overflow oscillations". *IEEE Trans. Acoustics, Speech, Signal Processing.,* vol. ASSP-26, pp.334-338, 1978.

[10] K.T.Erikson, A.N.Michel: "Stability analysis of fixed-point digital filters using computer generated lyapunov functions -- Part I: direct form and coupled filters", *IEEE Trans. Circuits and Systems,* vol. CAS-32, pp.113-132, Feb., 1985.

[11] P.H. Bauer, J. Leclerc: "A computer-aided test for the absence of limit cycles in fixed point digital filters". *IEEE Trans. Signal Processing,* vol. SP-39, pp.2400-2410, Nov., 1991.

[12] G. Alefeld, J. Herzberger, *Introduction to Interval Computations,* AP, 1983.

[13] E. Hyvonen: "Evaluation of Cascaded Interval Functions". *Proc. Int. Workshop on Constrain-Based Reasoning, 8th Florida AI Research Symposium.* April, 1995.

[14] C. Carreras, J.A. Lopez, O. Nieto: "Bit-width selection in data-path implementations ", *Proc. 12th IEEE Int. Symp. System Synthesis,* San Jose, pp.114-119, Nov 1999.

[15] C. Carreras, I.D. Walker: "Interval methods for fault-tree analysis in robotics", *IEEE Trans. Reliability,* Vol. 50 nº1 , pp. 3-11, March 2001.

N Stage Non-Separable Two Dimensional Wavelet Transform for Reduction of Rounding Errors

Masahiro IWAHASHI, Munkhbaatar DELGERMAA, Koji UENO and Noriyoshi KAMBAYASHI
Department of Electrical Engineering, Nagaoka University of Technology, Nagaoka, Niigata 940-2188, JAPAN
iwahashi@nagaokaut.ac.jp, http://tech.nagaokaut.ac.jp/

ABSTRACT

This paper proposes an N stage "non-separable" two-dimensional (2D) wavelet transform (WT) for lossless / lossy unified coding of digital images. The proposed method achieves better coding performance especially at high bit rate since it has fewer rounding operations. We convert the conventional "separable" 2D WT into "non-separable" one without changing its frequency characteristics so that the number of the rounding operations is reduced and also the proposed method has compatibility with the conventional one. Effectiveness of the proposed coding system is confirmed experimentally and theoretically.

1. INTRODUCTION

Recently the compression technology is widely used in the field of transmission and storage of digital image data. According to the commonly used JPEG international standard algorithm [1], different hardware or software must be prepared for the lossy coding and the lossless coding respectively. On the other hand, the lossless / lossy "unified coding system" can provide both lossless and lossy coding using the same WT. Over the last few years, several unified coding systems have been proposed based on the integer orthogonal transforms [2,3] or the integer wavelet transforms [4,5]. These integer transforms include rounding operations which truncate a real number into an integer number just after each filtering in the lifting structure [6,7]. These operations, different from the quantization, cause signal to noise ratio (SNR) degradation especially at high bit rate.

To solve this problem, we propose a new N stage "non-separable" two-dimensional (2D) wavelet transform (WT) which achieves better coding performance at high bit rate and also compatibility with the conventional "separable" 2D WT. The conventional WT is converted into "non-separable" without changing its frequency characteristics so that the number of the rounding operations is reduced. Effectiveness of the proposed coding system is confirmed experimentally and theoretically in **4.**.

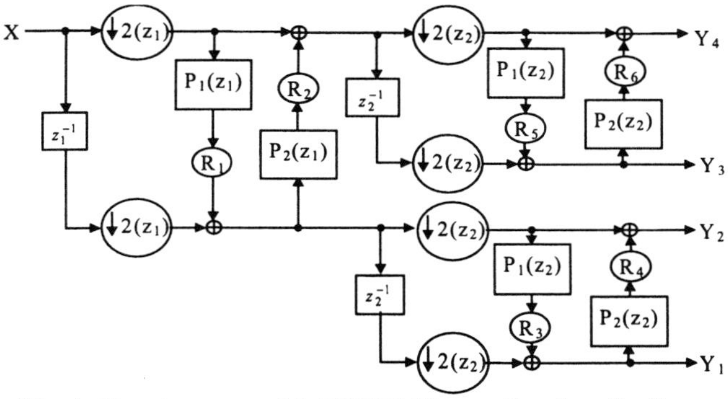

Fig. 1 One stage separable 2D WT (Conventional method).

2. SIGNAL PROCESSING

In this paper, an image signal with brightness $x(k_1,k_2)$ where $k_1 \in [0,N_1)$ and $k_2 \in [0,N_2)$ is expressed by

$$X(\mathbf{z}) = \sum_{k_1=0}^{N_1-1} \sum_{k_2=0}^{N_2-1} x(k_1,k_2) z_1^{-k_1} z_2^{-k_2} \quad (1)$$

$$(\mathbf{z}) = (z_1, z_2)$$

2.1 Separable 2D WT (conventional method) [6]

Figure 1 shows the analysis part of the one stage conventional separable 2D WT. Input signal $X(\mathbf{z})$ is decomposed into four band signals $Y_1(\mathbf{z})$, $Y_2(\mathbf{z})$, $Y_3(\mathbf{z})$ and $Y_4(\mathbf{z})$ with delays z_1, z_2 and FIR filters $P_1(z)$ and $P_2(z)$ [6]. The down-samplers $\downarrow 2(z_1)$ and $\downarrow 2(z_2)$ are defined by

$$\downarrow 2(z_1)[X(\mathbf{z})] = \frac{1}{2}\sum_{p=0}^{1} X\left((-1)^p z_1^{1/2}, z_2\right)$$
$$\downarrow 2(z_2)[X(\mathbf{z})] = \frac{1}{2}\sum_{q=0}^{1} X\left(z_1, (-1)^q z_2^{1/2}\right) \quad (2)$$

The rounding operations denoted by R_i, $i \in [1,2,\cdots,6]$, truncate a real number into an integer number. These make it possible to achieve effective lossless coding by applying entropy coding to each band signals without quantization. The lossy coding is also possible by inserting quantization before the entropy coding [4,5]. The band decomposition procedure in Fig. 1 is cascaded as illustrated in Fig.2 for better coding performance.

2.2 Problem of the conventional method

Figure 3 illustrates the rate-distortion curves of the three stage conventional 2D WT. Without the rounding operations as indicated with broken line, entropy rate at lossless coding (SNR=∞) is not preferable. On the contrary, employing the rounding operations causes SNR degradation at high bit rate as indicated with solid line.

The objective of this paper is to reduce the number of rounding operations under the same frequency characteristics of the band pass filters so that SNR at high bit rate is increased and compatibility with the conventional WT is maintained.

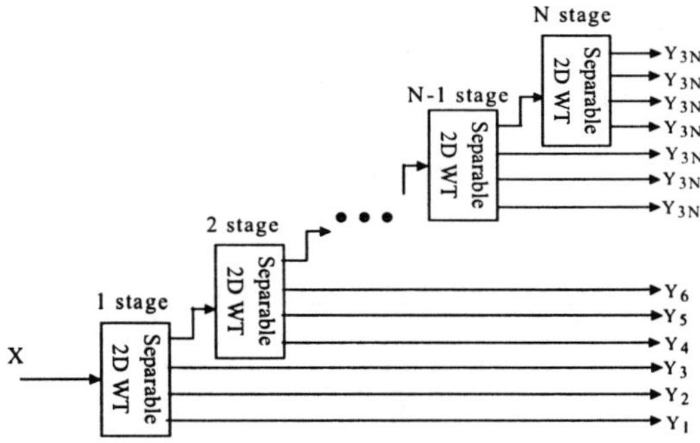

Fig.2 N stage separable 2D WT (conventional method).

0-7803-7761-3/03/$17.00 ©2003 IEEE

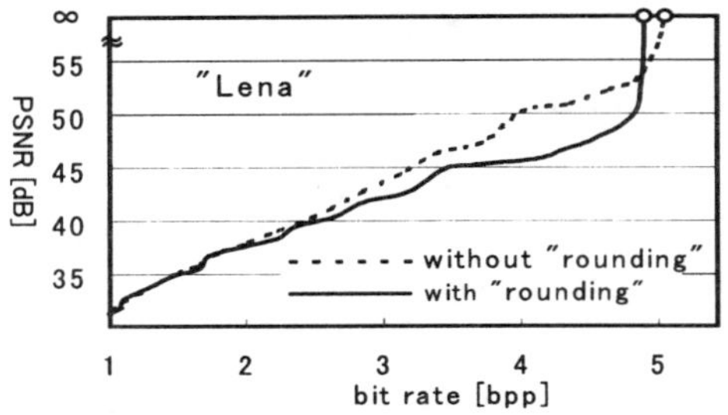

Fig. 3 Rate distortion curves of three stage separable 2D WT.

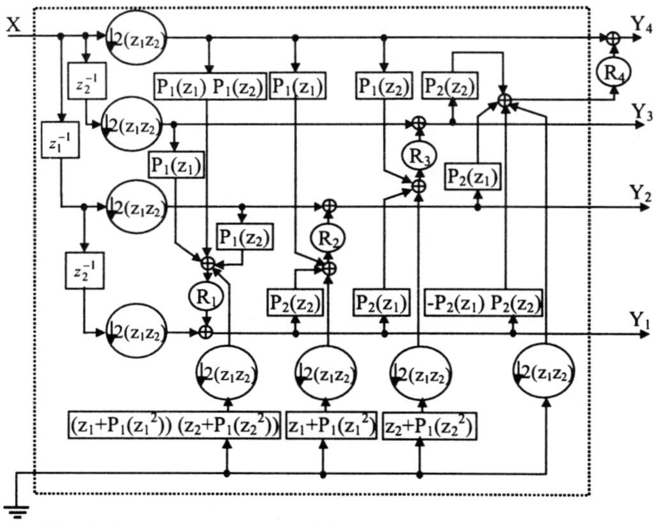

Fig. 4 One stage non-separable 2D WT (proposed method).

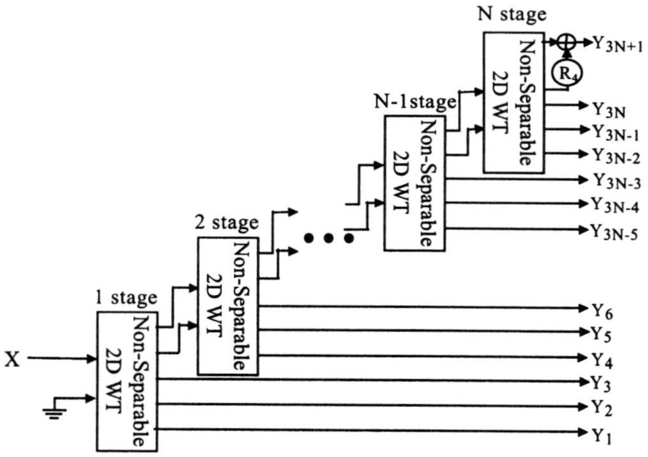

Fig. 5 N stage non-separable 2D WT (proposed method).

Table. 1 Total number of the rounding operations.

Number of stages	1	2	3	4	N
Conventional method	12	24	36	48	12N
Proposed method	**8**	**14**	**20**	**26**	**6N+2**

2.3 Non-separable 2D WT (Proposed method)

The proposed "non separable" 2D WT is illustrated in Fig. 4. It employs FIR filters $P_i(z_1,z_2)$, $i\in[1,2]$, in Fig. 1 and the down-sampler $\downarrow 2(z_1,z_2)$ defined by

$$\downarrow 2(z_1,z_2)[X(\mathbf{z})]=\frac{1}{4}\sum_{p=0}^{1}\sum_{q=0}^{1}X\left((-1)^p z_1^{1/2},(-1)^q z_2^{1/2}\right) \quad (3)$$

Fig. 5 explains how to extend the "one-stage" WT in Fig. 4 to "N-stage" decomposition. Grounding means zero input. The proposed method is driven in such a way that (i) each of the band signals or frequency characteristics is exactly same as that of the conventional method when all the rounding operations are removed and (ii) only one rounding operation is used for each band channel. As a result, the number of the rounding operations of the N stage 2D WT is reduced to 6N+2 from 12N. Table. 1 shows total number of the rounding operations including analysis part and synthesis part. It is confirmed that the number of the rounding operations is reduced.

3. THEORETICAL ANALYSIS

Variance of the errors caused by the rounding operations is theoretically evaluated in order to confirm that reduction of the number of the operations brings about SNR improvement. The following analysis assumes the rounding errors to be (i) uncorrelated each other, (ii) addictive white noises, (iii) with the probability density function of zero centered uniform distribution.

3.1 Signal of the rounding errors

[A] One stage 1D WT

To begin with, the error signal defined as difference between input signal X and decoded signal X' is described as the function of the rounding errors $R_1(z)$, $R_2(z)$, $S_1(z)$ and $S_2(z)$ for one stage 1D WT illustrated in figure 6. In the analysis part, band signals $Y_1(z)$ and $Y_2(z)$ include rounding errors $R_1(z)$ and $R_2(z)$ which propagate through filters $P_1(z)$ and $P_2(z)$. Namely,

$$\mathbf{Y}=\mathbf{T}_2\mathbf{T}_1\mathbf{Z}\cdot X(z)+\mathbf{T}_2\mathbf{R}_1+\mathbf{R}_2 \quad (4)$$

where

$$w=2$$

$$\mathbf{Y}=\begin{bmatrix}Y_1(z)\\Y_2(z)\end{bmatrix},\quad \mathbf{Z}=\begin{bmatrix}1\\z^{-w/2}\end{bmatrix}$$

$$\mathbf{R}_1=\begin{bmatrix}0\\R_1(z)\end{bmatrix},\quad \mathbf{R}_2=\begin{bmatrix}R_2(z)\\0\end{bmatrix} \quad (5)$$

$$\mathbf{T}_1=\begin{bmatrix}1&0\\P_1(z^w)&1\end{bmatrix},\quad \mathbf{T}_2=\begin{bmatrix}1&P_2(z^w)\\0&1\end{bmatrix}$$

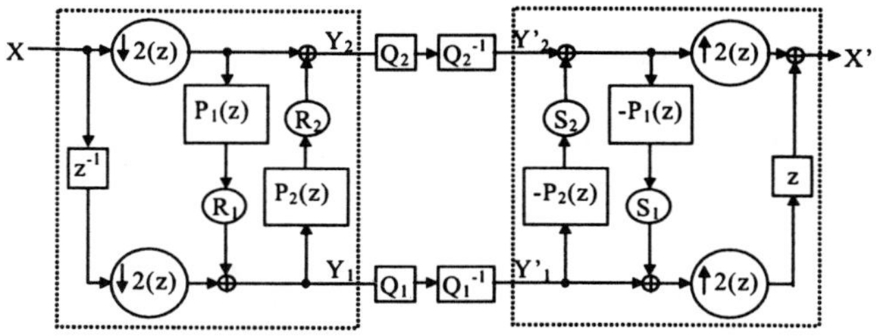

Fig. 6 Equivalent circuit of the 1D WT coding system.

After quantization, $Y'_1(z)$ and $Y'_2(z)$ contain quantization errors $Q_1(z)$ and $Q_2(z)$. Namely,

$$\mathbf{Y}' = \mathbf{Q} + \mathbf{Y} \tag{6}$$

where

$$\mathbf{Y}' = \begin{bmatrix} Y'_1(z) \\ Y'_2(z) \end{bmatrix}, \quad \mathbf{Q} = \begin{bmatrix} Q_1(z) \\ Q_2(z) \end{bmatrix}, \quad \mathbf{Y} = \begin{bmatrix} Y_1(z) \\ Y_2(z) \end{bmatrix}$$

Similarly, rounding errors $S_1(z)$ and $S_2(z)$ in the synthesis part are added to the decoded image X'(z). Therefore,

$$X'(z) = \mathbf{Z}^{-1}\left(\mathbf{T}_1^{-1}\mathbf{S}_1 + \mathbf{T}_1^{-1}\mathbf{T}_2^{-1}\mathbf{S}_2 + \mathbf{T}_1^{-1}\mathbf{T}_2^{-1}\mathbf{Y}'\right) \tag{7}$$

where

$$\mathbf{Z}^{-1} = [1 \quad z], \quad \mathbf{S}_1 = \begin{bmatrix} 0 \\ S_1(z) \end{bmatrix}, \quad \mathbf{S}_2 = \begin{bmatrix} S_2(z) \\ 0 \end{bmatrix} \tag{8}$$

Substituting Eq.(6) and value of Y from Eq.(4) into Eq.(7), X'(z) is represented by X(z) as

$$\begin{aligned} X'(z) = &\mathbf{Z}^{-1}\{ \mathbf{T}_1^{-1}\mathbf{S}_1 + \mathbf{T}_1^{-1}\mathbf{T}_2^{-1}\mathbf{S}_2 \\ &+ \mathbf{T}_1^{-1}\mathbf{T}_2^{-1}(\mathbf{T}_2\mathbf{R}_1 + \mathbf{R}_2 + \mathbf{Q} + \mathbf{T}_2\mathbf{T}_1\mathbf{Z}\cdot X(z)) \} \\ = &\mathbf{Z}^{-1}\{ \mathbf{T}_1^{-1}(\mathbf{S}_1 + \mathbf{R}_1) + \mathbf{T}_1^{-1}\mathbf{T}_2^{-1}(\mathbf{S}_2 + \mathbf{R}_2) \\ &+ \mathbf{T}_1^{-1}\mathbf{T}_2^{-1}\mathbf{Q} + \mathbf{Z}\cdot X(z) \} \end{aligned} \tag{9}$$

As a result, the error signal E(z) is expressed by

$$\begin{aligned} E(z) = &X'(z) - X(z) \\ = &\mathbf{Z}^{-1}\mathbf{T}_1^{-1}(\mathbf{S}_1 + \mathbf{R}_1) + \mathbf{Z}^{-1}\mathbf{T}_1^{-1}\mathbf{T}_2^{-1}(\mathbf{S}_2 + \mathbf{R}_2) \\ &+ \mathbf{Z}^{-1}\mathbf{T}_1^{-1}\mathbf{T}_2^{-1}\mathbf{Q} \end{aligned} \tag{10}$$

The first and the second term of the right side of Eq. (10) are the rounding errors on the decoded signal. The third is the quantization errors. Transmission matrices $\mathbf{T_1}$ and $\mathbf{T_2}$ which are functions of $P_1(z)$ and $P_2(z)$ in Fig.6 indicate how each variance of the rounding errors $R_1(z)$, $R_2(z)$, $S_1(z)$ and $S_2(z)$ are amplified.

[B] N stage non-separable 2D WT (proposed method)

The rounding errors of the 2D WT can be similarly driven with the equivalent circuit of the proposed method in Fig. 7. As a result, rounding error signal $E^{(i-1)}(z)$ becomes

$$\begin{aligned} E^{(i-1)}(\mathbf{z}) = &\mathbf{Z}^{(i)} \sum_{j=1}^{3} \prod_{k=1}^{j} \mathbf{T}_k^{(i)} \left(\mathbf{R}_j^{(i)} + \mathbf{S}_j^{(i)}\right) \\ &+ \mathbf{Z}^{(i)} \prod_{k=1}^{3} \mathbf{T}_k^{(i)} \mathbf{U} \cdot E^{(i)}(\mathbf{z}) \end{aligned} \tag{11}$$

where

$$\begin{aligned} \mathbf{R}_1^{(i)} &= [0 \quad 0 \quad 0 \quad R_1^{(i)}(\mathbf{z})]^T & \mathbf{S}_1^{(i)} &= [0 \quad 0 \quad 0 \quad S_1^{(i)}(\mathbf{z})]^T \\ \mathbf{R}_2^{(i)} &= [0 \quad 0 \quad R_2^{(i)}(\mathbf{z}) \quad 0]^T & \mathbf{S}_2^{(i)} &= [0 \quad 0 \quad S_2^{(i)}(\mathbf{z}) \quad 0]^T \\ \mathbf{R}_3^{(i)} &= [0 \quad R_3^{(i)}(\mathbf{z}) \quad 0 \quad 0]^T & \mathbf{S}_3^{(i)} &= [0 \quad S_3^{(i)}(\mathbf{z}) \quad 0 \quad 0]^T \\ \mathbf{R}_4^{(i)} &= [R_4^{(i)}(\mathbf{z}) \quad 0 \quad 0 \quad 0]^T & \mathbf{S}_4^{(i)} &= [S_4^{(i)}(\mathbf{z}) \quad 0 \quad 0 \quad 0]^T \end{aligned}$$

$$\mathbf{Z}^{(i)} = [1 \quad z_2^{w/2} \quad z_1^{w/2} \quad z_1^{w/2} z_2^{w/2}], \quad \mathbf{U} = [1 \quad 0 \quad 0 \quad 0]^T$$

$$\mathbf{T}_1^{(i)} = \begin{bmatrix} 1 & 0 & 0 & 0 \\ 0 & 1 & 0 & 0 \\ 0 & 0 & 1 & 0 \\ -P_1(z_1^{w_i})P_1(z_2^{w_i}) & -P_1(z_1^{w_i}) & -P_1(z_2^{w_i}) & 1 \end{bmatrix}$$

$$\mathbf{T}_2^{(i)} = \begin{bmatrix} 1 & 0 & 0 & 0 \\ 0 & 1 & 0 & 0 \\ -P_1(z_1^{w_i}) & 0 & 1 & -P_2(z_2^{w_i}) \\ 0 & 0 & 0 & 1 \end{bmatrix} \quad \mathbf{T}_3^{(i)} = \begin{bmatrix} 1 & 0 & 0 & 0 \\ -P_1(z_2^{w_i}) & 1 & 0 & -P_2(z_1^{w_i}) \\ 0 & 0 & 1 & 0 \\ 0 & 0 & 0 & 1 \end{bmatrix}$$

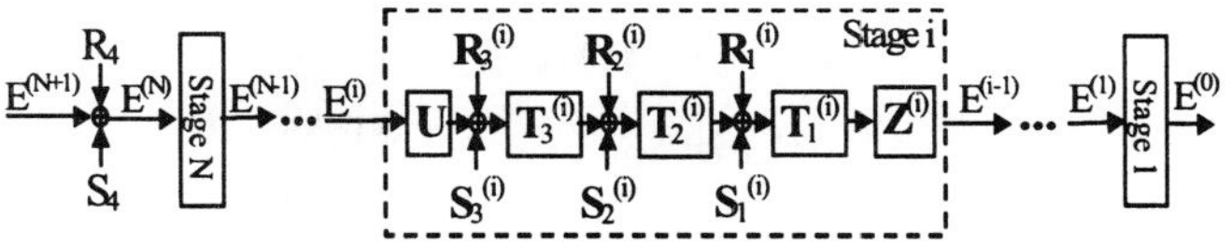

Fig. 7 Equivalent circuit of the 2D WT (proposed method).

The weighting factor w_i for i^{th} stage is given by

$$w_i = 2^i \tag{12}$$

Substituting $E^{(N+1)}(\mathbf{z})=0$ into Eq. (11), theoretical expression of the rounding error signal $E^{(0)}(\mathbf{z})$ is obtained as

$$\begin{aligned} E^{(0)}(\mathbf{z}) = &\mathbf{Z}^{(1)} \sum_{j=1}^{3} \prod_{k=1}^{j} \mathbf{T}_k^{(1)} \left(\mathbf{R}_j^{(1)} + \mathbf{S}_j^{(1)}\right) \\ &+ \sum_{i=2}^{N} \left(\left(\prod_{j=1}^{i-1} \left(\mathbf{Z}^{(j)} \prod_{k=1}^{3} \mathbf{T}_k^{(j)} \mathbf{U} \right) \right) \mathbf{Z}^{(i)} \sum_{j=1}^{3} \prod_{k=1}^{j} \mathbf{T}_k^{(i)} \left(\mathbf{R}_j^{(i)} + \mathbf{S}_j^{(i)}\right) \right) \\ &+ \prod_{i=1}^{N} \left(\mathbf{Z}^{(i)} \prod_{k=1}^{3} \mathbf{T}_k^{(i)} \mathbf{U} \right) \left(R_4(\mathbf{z}) + S_4(\mathbf{z})\right) \end{aligned} \tag{13}$$

Using the equation above, variance of the error is calculated next.

3.2 Variance of the rounding error

[A] One stage 1D WT

As indicated in Eq. (10), the rounding error signal in the decoded image is

$$E(z) = \mathbf{Z}^{-1}\mathbf{T}_1^{-1}(\mathbf{S}_1 + \mathbf{R}_1) + \mathbf{Z}^{-1}\mathbf{T}_1^{-1}\mathbf{T}_2^{-1}(\mathbf{S}_2 + \mathbf{R}_2) \tag{14}$$

Based on the assumptions of the error signals, theoretical expression of variance of the rounding error signal is

$$\sigma_{\text{round}}^2 = \frac{1}{2} \sum_{i=1}^{2} \|G_i\|^2 \left(\sigma_{S_i}^2 + \sigma_{R_i}^2\right) \tag{15}$$

where

$$G_1(z) = z, \quad G_2(z) = 1 - zP_1(z^2), \quad \sigma_{S_i}^2 = \sigma_{R_i}^2 = \frac{1}{12},$$

$$\|G\|^2 = \frac{1}{2\pi} \int_0^{2\pi} \left|G(e^{j\omega})\right|^2 d\omega$$

Therefore variance of the rounding errors can be theoretically estimated with the equations above.

[B] N stage non-separable 2D WT (proposed method)

Similarly, variance of the rounding errors is calculated by

$$\begin{aligned} \sigma_{\text{round}}^2 = &\sum_{i=1}^{N} w_i \sum_{j=1}^{3} \left\|G_j^{(i)}\right\|^2 \left(\sigma_{R_j^{(i)}}^2 + \sigma_{S_j^{(i)}}^2\right) \\ &+ w_N \left\|G_3^{(N)}\right\|^2 \left(\sigma_{R_4^{(i)}}^2 + \sigma_{S_4^{(i)}}^2\right) \end{aligned} \tag{16}$$

where

$$G_j^{(i)}(z) = \begin{cases} G_j^{(i)}(z) & (\text{if} \quad i = 1) \\ G_j^{(i)}(z) \prod_{k=1}^{i-1} G_3^{(k)}(z) & (\text{if} \quad i > 1) \end{cases} \tag{17}$$

$$\begin{aligned} G_1^{(i)}(\mathbf{z}) &= z_1^{w_i/2} z_2^{w_i/2} \\ G_2^{(i)}(\mathbf{z}) &= z_1^{w_i/2}\left(1 - z_2^{w_i/2} P_1(z_2^{w_i})\right) \\ G_3^{(i)}(\mathbf{z}) &= z_2^{w_i/2}\left(1 - z_1^{w_i/2} P_1(z_1^{w_i})\right) \end{aligned}$$

4. EVALUATION RESULTS

4.1 Theoretical Evaluation

Fig. 8 shows difference of SNR in [dB] of the decoded image between the conventional method and the proposed method. In this example, three-stage decomposition with the 5/3 filter [6] is used. Solid line indicates theoretically calculated variance in the manner described in **3.**. As the figure indicates, SNR improvement is confirmed at high bit rate where the quantization step size is relatively small. Accuracy of the theoretical evaluation is also confirmed since experimental results are close to theoretical one. According to the calculation in **3.**, variances of the rounding errors are decreased by 41 to 58 [%] as indicated in Table. 2.

4.2 Experimental Evaluation

Figure 9 shows experimentally calculated rate-distortion curves for "Lena" and "Baboon" (256^2 pixels, gray scale) for three-stage 2D WT with the 5/3-filter. Effectiveness of the proposed method at high bit rate is also confirmed experimentally. However, it is difficult to perceptually recognize this improvement. When a procedure of coding and decoding is repeated, the errors build up and the difference emerges as illustrated in Fig. 10. In this case, the procedure is repeated five times with 5/3, 9/7-M, 13/11, 9/3-S, 13/7-T filters [3].

Table. 2 Variance of the rounding errors.

Number of stages	Existing method σ^2_{round}	Proposed method σ^2_{round}	Improvement ratio [%]
1	0.441	0.262	40.5
2	0.739	0.357	51.7
3	1.008	0.453	55.1
4	1.274	0.551	56.8
5	1.548	0.655	57.7

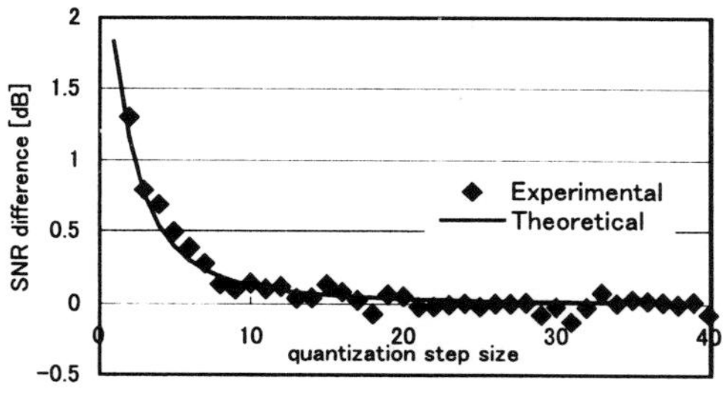

Fig. 8 SNR improvement for three stage 2D WT.

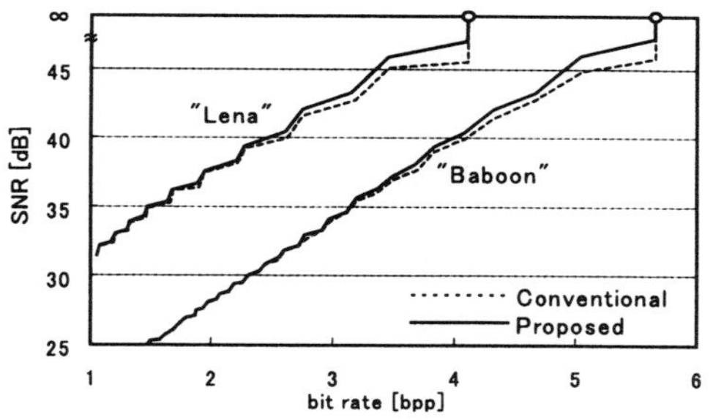

Fig. 9 Rate-distortion curves of the three stage 2D WT.

(a) Conventional method SNR=35.95[dB] (b) Proposed method SNR=36.88[dB]

Fig.10 Examples of the decoded images

5. CONCLUSION

In this paper, we proposed a new N stage "non-separable" two dimensional (2D) wavelet transform (WT) for high quality lossless / lossy unified coding of digital images. We converted the conventional "separable" 2D WT into "non-separable" 2D WT under the same frequency characteristics so that the number of the rounding operations is reduced and compatibility is maintained. We also indicated the way of theoretical evaluation of the rounding errors as well as the quantization errors.

As a result of theoretical evaluation, variance of the rounding errors decreased by 41 to 58 [%] compared to that of the conventional method. Experimental evaluation results confirmed SNR improvement by up to 1.3 [dB].

This research was supported in part by the Grants-in-Aid for Scientific Research in Japan No.14750284.

REFERENCES

[1] JPEG CD10918-1, "Digital compression coding of continuous-tone still images, JPEG-9-R6," Jan.1991.

[2] S. Chokchaitam, M. Iwahashi, P. Zavarsky, N. Kambayashi, "Integrated Lossy and Lossless Image Coding based on Lossless Wavelet Transform and Lossy-Lossless Multi Channel Prediction," IEICE Trans. on Fundamentals, E84-A, vol.5, pp.1326-1338, 2001.

[3] M. D. Adams, F. Kossentini, "Reversible Integer-to-Integer Wavelet Transform for image Compression: Performance Evaluation and Analysis," IEEE Transactions on Image Processing, vol. 9, pp 1010 - 1024, no. 6, June 2000.

[4] K. Komatsu, K. Sezaki, "2D Lossless Discrete Cosine Transform," IEEE International Conference on Image Processings (ICIP), pp.466-469, 2001.

[5] S. Chokchaitam, M. Iwahashi, P. Zavarsky, N. Kambayashi, "A Bit-Rate Adaptive Coding System Based on Lossless DCT," IEICE Trans. on Fundamentals, Vol.E85-A, No.2, pp. 403 -413, Feb. 2002.

[6] D. S. Taubman, M. W. Marcellin, *JPEG 2000 - Image compression fundamentals, standards and practice*, Kluwer Academic Publishers, 2002.

[7] W. Sweldens, "The lifting scheme: a construction of second generation wavelets," Tech. Rep. 1995:6, Industrial Math. Initiative, Dept. of Math., Univ. of South Carolina, 1995.

PROLONGED TRANSPOSED POLYNOMIAL-BASED FILTERS FOR DECIMATION

Djordje Babic, Tapio Saramäki, and Markku Renfors*

Institute of Communications Engineering, Tampere University of Technology
*Institute of Signal Processing, Tampere University of Technology
P. O. Box 553, Tampere, FINLAND
Tel: +358 3 365 3910; fax: +358 3 3653808; E-mail: {babic, ts, mr,}@cs.tut.fi

ABSTRACT

If sample rate conversion (SRC) is performed between arbitrary sample rates, then the SRC factor can be a ratio of two very large integers or even an irrational number. An efficient way to reduce the implementation complexity of a SRC system in those cases is to use polynomial-based interpolation filters. The impulse response of these filters is of a finite duration and piecewise polynomial so that it is expressible in each subinterval of the same length T by means of a polynomial of a low order. Here, T can be equal to, a multiple of, or a fraction of either the input or output sample period. The actual implementation of the polynomial-based filters can be performed directly in the digital domain effectively by using the Farrow structure or its modifications.

This paper introduces for an arbitrary sampling rate reduction a novel implementation form referred to as the prolonged transposed modified Farrow structure. For this structure, T is an integer multiple of the output sampling period. Compared with the modified transposed Farrow structure, it has a narrowed pass-band region with almost the same complexity. In addition, a decimator structure consisting of a cascade of the prolonged transposed Farrow structure and a fixed linear-phase finite-impulse response decimator is introduced in order to reduce the overall computational complexity.

1. INTRODUCTION

The sample rate conversion (SRC) is utilized in many DSP applications where two signals or systems having different sample rates are to be interconnected. The SRC factor is determined by

$$R = F_{out}/F_{in} = T_{in}/T_{out}, \tag{1}$$

where $F_{in} = 1/T_{in}$ (T_{in}) and $F_{out} = 1/T_{out}$ (T_{out}) are the original input sample rate (period) and the sample rate (period) after the conversion, respectively. In general, the SRC factor can be an integer, one divided by an integer, a ratio of two integers, or an irrational number. The SRC can be divided into two general cases. For $R < 1$ ($R > 1$), the sample rate is reduced (increased) and this process is known as decimation (interpolation)

In the most general case, the sampling rate conversion can be regarded as a process of re-sampling an analogue signal according to the hybrid analogue/digital model as shown in Fig. 1 [1]. In this system, the discrete-time input sequence $x(n)$ with the sampling rate equal to $F_{in}=1/T_{in}$ is first converted into an analogue signal $x_s(t)$ using the ideal digital-to-analog converter (DAC), resulting in the following sum of the weighted and shifted impulses:

$$x_s(t) = \sum_{k=-\infty}^{\infty} x(k)\delta_a(t - kT_{in}), \tag{2}$$

This work was carried out in the project entitled "Advanced Transceiver Architectures and Implementations for Wireless Communications" supported by the Academy of Finland. This work was also supported by the Academy of Finland, project No. 44876 (Finnish Centre of Excellence Program (Institute of Signal Processing), 2000–2005), Tampere Graduate School in Information Science and Engineering (TISE), and NOKIA foundation.

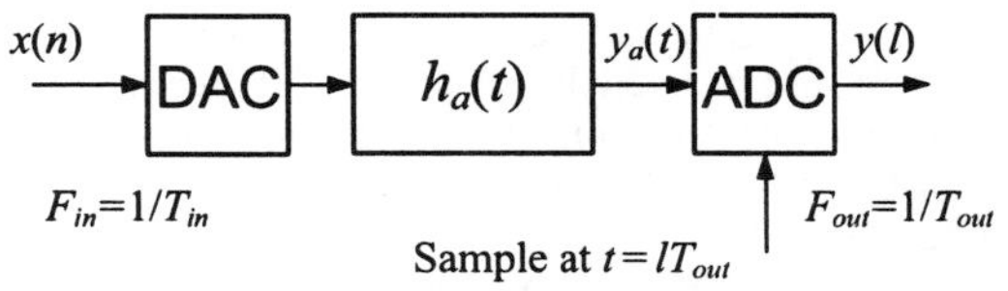

Fig. 1. Hybrid analogue/digital model to be mimicked.

where $\delta_a(t)$ is the analogue Dirac delta function. This sequence is then filtered with the aid of an analogue filter with the impulse response $h_a(t)$ to generate the continuous-time signal as given by

$$y_a(t) = \sum_{k=-\infty}^{\infty} x(k)h_a(t - kT_{in}). \tag{3}$$

Finally, the output $y_a(t)$ is sampled at the time instants $t=lT_{out}$ using the ideal analog-to-digital converter (ADC) producing the following discrete-time output sequence $y(l)$:

$$y(l) = y_a(lT_{out}) = \sum_{k=-\infty}^{\infty} x(k)h_a(lT_{out} - kT_{in}). \tag{4}$$

The remaining problem in the model of Fig. 1 is to determine $h_a(t)$ such that it satisfies the desired frequency-domain requirements and the overall system has an efficient implementation directly in the digital domain. It should be pointed out that in the decimation case, the filter with the impulse response $h_a(t)$ acts as an anti-aliasing filter rejecting the frequency components aliasing onto the new base-band region [0, $F_{out}/2$], while preserving the frequency components in this band. On the other hand, in the interpolation case, the role of this filter is to preserve the original baseband region [0, $F_{in}/2$] and to eliminate the imaging components. Therefore, it acts as an anti-imaging filter.

When the decimation factor $1/R$ or the interpolation factor R is an integer or a ratio of two relatively small prime integers, then the sample rate conversion can be performed conveniently with the aid of fixed digital filters [1]. If these factors are irrational, then fixed digital filters cannot be directly used. Furthermore, if these factors are ratios of two relatively large prime integers, then the required filter orders become very large. In these cases, an efficient overall implementation based on mimicking the system of Fig. 1 can be achieved by generating $h_a(t)$ such that it is of a finite duration and piecewise polynomial so that it is expressible in each subinterval of the same length T by means of a polynomial of a low order [2], [3]. T can be selected to be equal to, a multiple of, or a fraction of either the input or output sample period. The advantage of mimicking the above system lies in the fact that the actual implementation can be efficiently performed by using the Farrow structure [4] or its modifications [2], [3], [5]–[8].

The main advantage of the Farrow structure and its modifications is that they consist of fixed finite-impulse response (FIR) filters and there is only one changeable parameter, the so-called fractional interval μ. The control of μ is easier during the operation

0-7803-7761-3/03/$17.00 ©2003 IEEE

than in the corresponding coefficient memory implementations [3], and the resolution of μ is limited only by the precision of arithmetic and not by the size of the memory. These characteristics of the Farrow structure and its modifications make them very attractive for VLSI and signal processor implementations [3].

This paper introduces a novel implementation form for decimation purposes referred to as the prolonged transposed modified Farrow structure. For this structure, the basic polynomial subinterval is a multiple of the output sample period. In addition, based on the principle presented in [6] for constructing digital interpolators (a cascade of fixed linear-phase finite-impulse response (FIR) interpolator and a modified Farrow structure), a novel decimator structure being a cascade of the prolonged transposed modified Farrow structure and a fixed linear-phase FIR decimator is proposed. Finally, examples are included that illustrate the low computational complexities of the proposed structures compared with other existing structures.

2. POLYNOMIAL-BASED FILTERS

This section forms the framework for the following sections by introducing the basic form for the impulse response of the analogue filter in Fig. 1.

As shown in [2], [3], when deriving the modified Farrow structure for interpolation, it is beneficial to construct $h_a(t)$ as follows:

$$h_a(t) = \sum_{n=0}^{N-1}\sum_{m=0}^{M} c_m(n) f_m(n,T,t), \quad (5)$$

where N is an even integer,

$$f_m(n,T,t) = \begin{cases} \left((2t - nT)/T - 1\right)^m & \text{for } nT \le t < (n+1)T \\ 0 & \text{otherwise} \end{cases} \quad (6)$$

for $n = 0, 1, \ldots, N-1$ and $m = 0, 1, \ldots, M$ are the basis functions, and the $c_m(n)$'s are the adjustable parameters being related as follows:

$$c_m(N-1-n) = \begin{cases} c_m(n) & \text{for } m \text{ even} \\ -c_m(n) & \text{for } m \text{ odd} \end{cases} \quad (7)$$

As shown in Fig. 2, the resulting $h_a(t)$ is characterized by the following properties:

1) $h_a(t)$ is nonzero for $0 \le t < NT$ and zero elsewhere.
2) $h_a(t)$ is a piecewise-polynomial for $0 \le t < NT$ and is of degree M in each subinterval $nT \le t < (n+1)T$ for $n = 0, 1, \ldots, N-1$, where it is expressible as $h_a(t) = \sum_{m=0}^{M} c_m(n) f_m(n,T,t)$.
3) $h_a(t)$ is symmetric around $t = NT/2$, that is, $h_a(NT-t) = h_a(t)$ except for the time instants $t = nT$ for $n = 0, 1, \ldots, N/2-1$ and $n = N/2+1, N/2+2, \ldots, N$.

Property 3 guarantees that the resulting overall system has a linear phase that is a very attractive property in many applications. Furthermore, generating $h_a(t)$ in the above manner ensures that the zero-phase frequency response (the frequency response omitting the linear-phase term) is expressible as (see [2] and [3] for details)

$$H_a(j2\pi f) = \sum_{n=0}^{N/2-1}\sum_{m=0}^{M} c_m(n) G_m(n,T,f), \quad (8)$$

where $G_m(n, T, f)$ is the Fourier transform of

$$g_m(n,T,t) = (-1)^m f_m(n,T,t - NT/2) + f_m(N-1-n,T,t - NT/2). \quad (9)$$

The above form is a direct consequence of the symmetry properties of the $c_m(n)$'s. Since the above approximating function is linear with respect to the unknowns, it enables one to optimize the overall filter to meet the given criteria in a manner similar to that used for synthesizing various types of linear-phase FIR filters. In the above, T, the length of polynomial segments, is not fixed. It can be chosen as $T = \beta T_{in}$ or $T = \beta T_{out}$, where β is unity, an integer, one divided by an integer [7]. The selection depends on whether decimation or interpolation is under consideration and on the structure performing the desired sampling rate conversion. This paper concentrates mainly on the $T = JT_{out}$ case, where J is an integer, and develops a novel decimation structure referred to as the prolonged transposed Farrow structure, as will be discussed in the following section.

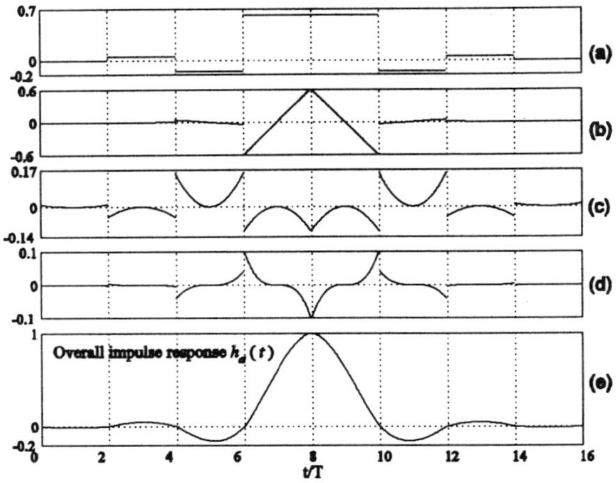

Fig. 2. Construction of the overall impulse response $h_a(t)$ for $N = 16$ and $M = 3$. $h_m(t) = \sum_{n=0}^{N-1} c_m(n) f_m(n,T,t)$ for (a) $m = 0$, (b) $m = 1$, (c) $m = 2$, and (d) $m = 3$. (e) The resulting overall $h_a(t) = \sum_{m=0}^{M} h_a(m,t)$.

3. PROLONGED TRANSPOSED MODIFIED FARROW STRUCTURE FOR DECIMATION

When generating the prolonged transposed modified Farrow structure, the desired $h_a(t)$ is obtained from Eq. (5) by selecting T to be an integer multiple of the output sampling period in Fig. 1, that is, $T = JT_{out}$ with J being an integer, giving

$$h_a(t) = \sum_{n=0}^{N-1}\sum_{m=0}^{M} c_m(n) f_m(n, JT_{out}, t), \quad (10)$$

where

$$f_m(n, JT_{out}, t) = \begin{cases} \left(\dfrac{(2t - nJT_{out})}{JT_{out}} - 1\right)^m & \text{for } nJT_{out} \le t < (n+1)JT_{out} \\ 0 & \text{otherwise.} \end{cases} \quad (11)$$

By substituting the resulting $h_a(t)$ into Eq. (4), the lth output sample is expressible, after some manipulations, as

$$y(l) = \sum_{k=-\infty}^{\infty} x(k)\left[\sum_{n=0}^{N-1}\sum_{m=0}^{M} c_m(n) f_m(n, JT_{out}, kT_{in} - lT_{out})\right] = \sum_{m=0}^{M}\left[\sum_{n=0}^{N-1} c_m(n) \sum_{k=k_{low}(l,n)}^{k_{up}(l,n)} x(k) f_m(n, JT_{out}, kT_{in} - lT_{out})\right], \quad (12)$$

where

$$k_{low}(l,n) = \lceil (l/J - n - 1 + N/2) JT_{out}/T_{in} \rceil \quad (13)$$

and

$$k_{up}(l,n) = \begin{cases} s(l,n) - 1 & \text{if } s(l,n) \text{ is an integer} \\ \lfloor s(l,n) \rfloor & \text{otherwise} \end{cases} \quad (14)$$

with

$$s(l,n) = (l/J - n + N/2) JT_{out}/T_{in}. \quad (15)$$

In the above $\lfloor x \rfloor$ is the largest integer less than or equal to x, whereas $\lceil x \rceil$ is the smallest integer larger than or equal to x. The inner summation in Eq. (12) corresponding to the polynomial interval of length JT_{ou} can be decomposed into J terms, with each corresponding to an interval of length equal to T_{out}, as follows:

$$y_a(l) = \sum_{m=0}^{M} \sum_{n=0}^{N-1} c_m(n) t(m,n,l), \qquad (16)$$

where

$$t(m,n,l) = \sum_{j=0}^{J-1} \sum_{k=k_{low}(l,n,j)}^{k_{up}(l,n,j)} x(k)(2\mu_k(j)-1)^m. \qquad (17)$$

Here, the borders for the innermost summation are given by

$$k_{low}(l,n,j) = \lceil (l-j-1/J-n-1+N/2)JT_{out}/T_{in} \rceil \qquad (18)$$

and

$$k_{up}(l,n,j) = \begin{cases} s(l,n,j)-1 & \text{if } s(l,n,j) \text{ is an iteger} \\ \lfloor s(l,n,j) \rfloor & \text{otherwise,} \end{cases} \qquad (19)$$

where

$$s(l,n,j) = ((l-j)/J - n + N/2)JT_{out}/T_{in}. \qquad (20)$$

After some rigorous manipulations, the fractional intervals $\mu_k(j)$ for $j = 0, 1, \ldots, J-1$ in Eq. (17) can be computed as

$$\mu_k(j) = (kT_{in}/T_{out} + j - \lfloor kT_{in}/T_{out} \rfloor)/J. \qquad (21)$$

Hence, there exist J different fractional intervals for a single input sample. These fractional intervals are computed by using an overflowing accumulator (it takes only the decimal part of the sum as result) as follows:

$$\begin{aligned} \mu_k(0) &= (\mu_{k-1}(0) + T_{in}/JT_{out}) \\ \mu_k(1) &= \mu_{k-1}(1) + 1/J \\ &\vdots \\ \mu_k(J-1) &= \mu_{k-1}(J-1) + (J-1)/J = \mu_{k-1}(J-2) + 1/J \end{aligned} \qquad (22)$$

with the initial value being $\mu_0(0) = 0$. The final implementation form for the prolonged transposed modified Farrow structure is given in Fig. 4. This structure has the same number of fixed coefficients as the original transposed modified Farrow structure described in [5]. Hence, there are $(M+1)N/2$ multipliers working at the lower output sampling rate F_{out}. The main difference is that J extra multipliers are required due to J different fractional intervals $\mu_k(j)$ corresponding to each input sample.

4. DECIMATION STRUCTURE CONSISTING OF A CASCADE OF A PROLONGED TRANSPOSED MODIFIED FARROW STRUCTURE AND A LINEAR-PHASE FIR DECIMATOR

This section introduces a decimator structure being a cascade of a prolonged transposed modified Farrow structure and a fixed linear-phase FIR decimator. This alternative is motivated by the duality between the fixed digital interpolators and decimators and the efficiency of building an interpolator as a cascade of a fixed linear-phase FIR interpolator and a modified Farrow structure as introduced in [6].

The proposed structure is shown in Fig. 3. The first block is the prolonged transposed Farrow structure that generates, based on the discrete-time input sequence $x(n)$, the output sequence, denoted by $z(j)$, such that the output sample rate is an integer multiple of the desired final output sampling rate F_{out}, that is, the output sample rate is LF_{out} with L being an integer. This block can be implemented using the structure of Fig. 3 with the following differences. First, the output sample rate is now LF_{out}, instead of F_{out}. Second, the desired $h_a(t)$ is obtained from Eq. (10) by selecting T to be $T = JT_{out}/L$.

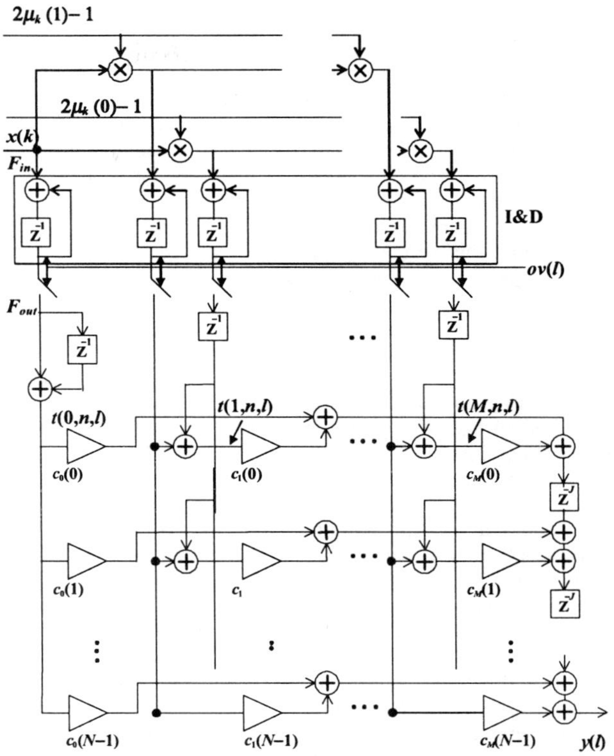

Fig. 4. Prolonged transposed modified Farrow structure for J=2.

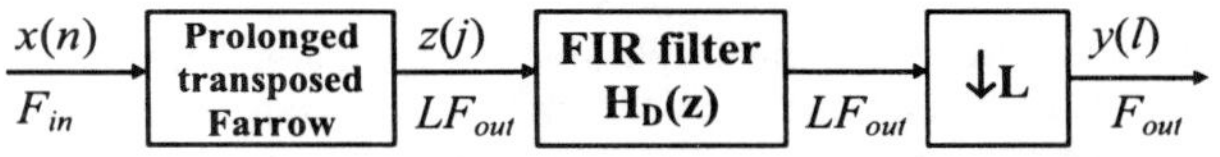

Fig. 3. Prolonged transposed modified Farrow structure in cascade with a linear-phase FIR decimator filter.

In the second block, the sampling rate is reduced by the integer factor L in order to generate the desired output discrete-time sequence $y(l)$ with the sample rate equal to F_{out}. The desired sampling rate reduction can be performed using a single-stage or a multistage linear-phase FIR decimator. The overall decimator transfer function decimating by the integer factor L is given by

$$H_D(z) = \sum_{k=0}^{K_D} h_D(k) z^{-k}, \qquad (23)$$

where the impulse-response of $H_D(z)$ is symmetric, that is,

$$h_D(K_D - k) = h_D(k) \text{ for } k = 0, 1, \ldots, K_D. \qquad (24)$$

The resulting overall system corresponds to that of Fig. 1, where $h_a(t)$ is replaced by

$$g_a(t) = \sum_{l=0}^{K_D} h_D(l) h_a(t - lT_{out}/L). \qquad (25)$$

This $g_a(t)$ is characterized by the following properties:

1) $g_a(t)$ is formed as a shifted and weighted sum of the start-up piecewise polynomial $h_a(t)$ with the length of polynomial segments being equal to T_{out}/L.
2) The resulting $g_a(t)$ is nonzero for $0 \le t < (K_D+NJ)T_{out}/L$ and of order M in each of the subintervals $nT_{out}/L \le t < (n+1)T_{out}/L$ for $n = 0, 1, \ldots, K_D+NJ-1$.

The main advantage of using the structure of Fig. 3, instead of the direct prolonged transposed modified Farrow structure of Fig. 4, is that when jointly optimizing the two building blocks the computational complexity for generating practically the same filtering performance is drastically decreased. This fact has been pointed out in [6] in the dual interpolation case. The computational complexity is reduced due to the following facts. First, the implementation of a fixed linear-phase FIR decimator is not very costly. Second, most importantly, the requirements for implementing the prolonged transposed modified Farrow structure become significantly milder. First of all, its sampling rate becomes L times higher and, due to this fact, its relative pass-band region becomes L times narrower. Second, this structure should only take care of attenuating narrow images that are approximately of width F_{out} and are centered at the frequencies $f = rLF_{out}/2$ for $r = 1, 2, 3...$

5. DESIGN EXAMPLES

This section compares, by means of an example, the three decimation structures proposed in Sections 3, 4, and [5].

It is desired to convert the sample rate between F_{in} and F_{out} so that the decimation factor is equal to $R = 5.5125$. The pass-band and stop-band edges are located at $f_p = 0.4F_{out}$ and $f_s = 0.5F_{out}$, respectively, whereas the minimum stop-band attenuation is 60 dB and the maximum allowable magnitude deviation from unity in the pass-band is 0.01. For simplicity, we concentrate in the sequel on designing filters in such a manner that the pass-band average is scaled to be unity.

When using the minimax optimization procedure proposed in [2], [3], the transposed modified Farrow (TMF) structure proposed in [5] meets the given criteria by $N = 28$ and $M = 4$. For this structure, the overall number of multipliers is $(M+1)N/2+M = 74$. The same requirements are met by the prolonged transposed modified Farrow (PTMF) structure of Section 4 by $N = 14$, $M = 6$, and $J = 2$, requiring $(M+1)N/2+JM = 61$ multipliers. The magnitude responses for these two structures are shown in Fig. 5(a). For the cascade of PTMF structure and a linear-phase FIR decimator considered in Section 4, $L = 2$ minimizes, in the case of simultaneous optimization of the two filter parts, the computational complexity of the overall system. For $L = 2$, the sample rate is first decreased by a non-integer factor 2.7562 by using the PTMF structure with $N = 4$ and $M = 3$. The resulting sample rate is two times the desired one and the desired output sample rate is achieved by using a fixed linear-phase FIR decimator of order $K_D = 49$ for further decreasing the sample rate by a factor of two. The resulting overall structure requires $(M+1)N/2+2M+(K_D+1)/2 = 39$ multipliers. The magnitude responses for the sub-filters as well as for that for the overall filter are shown in Fig. 5(b). The realization requirements for the above structures are shown in Table I in terms of the required number of multiplications per second, the required number of multipliers, and the required number of I&D circuits.

Table I. Requirements for the structures under consideration

	Multiplications/s	No. of multipliers	No. of I&Ds
TMF	$4F_{in}+70F_{out}$	74	5
PTF	$12F_{in}+49F_{out}$	61	13
PTF+FIR	$6F_{in}+41F_{out}$	39	8

TMF stands for the transposed modified Farrow structure, PMTF for the prolonged modified transposed Farrow structure, and PMTF+FIR for a cascade of the prolonged transposed Farrow structure and a fixed FIR decimator filter.

6. CONCLUSIONS

Two computationally efficient structures for arbitrary sample rate reduction have been proposed based on mimicking the hybrid analogue/digital model of Fig. 1. It has been shown that the key idea in generating these structures is to form the impulse response $h_a(t)$ such that it is nonzero for an interval $0 \le t < NT$ with N being an even integer and is expressible in each subinterval of length T by

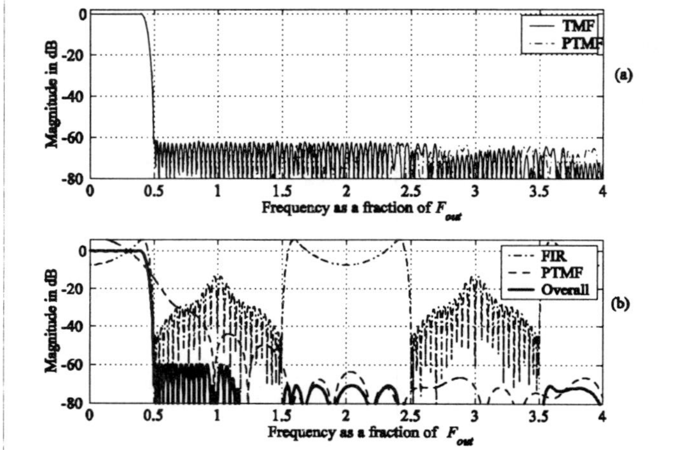

Fig. 5.(a) Magnitude responses of the transposed modified Farrow structure (TMF) and the prolonged transposed modified Farrow structure (PTMF). (b) Magnitude response of the FIR filter, prolonged tranposed modified Farrow structure (PTMF), and overall response of cascade.

means of a polynomial of the given low order. The length of polynomial segments in this paper has been selected to be an integer multiple of the output sampling period. The resulting implementation form, referred to as prolonged transposed modified Farrow structure, has been explained in details. In addition, the novel efficient decimation structure has been introduced that consists of the cascade of a prolonged transposed modified Farrow structure and a fixed linear-phase FIR decimator. It has been shown, by means of an example, that the prolonged versions meet same requirements with a fewer number of multipliers compared to the normal transposed modified Farrow structure. Furthermore, it has been observed that the number of multipliers can be reduced even further using the proposed cascaded structures.

REFERENCES

[1] R. E. Crochiere and L. R. Rabiner, *Multirate Digital Signal Processing*. Englewood Cliffs, NJ: Prentice-Hall, 1983.

[2] J. Vesma and T. Saramäki, "Interpolation filters with arbitrary frequency response for all-digital receivers," in *Proc. 1996 IEEE Int. Symp. Circuits and Systems*, Atlanta, Georgia, May 1996, pp. 568–571.

[3] J. Vesma, *Optimization and Applications of Polynomial-Based Interpolation Filters*. Doctoral Thesis, Tampere University of Technology, Publications 254, 1999.

[4] C. W. Farrow, "A continuously variable digital delay element,"in *Proc. 1988 IEEE Int. Symp. Circuits and Systems*, Espoo, Finland, June 1988, pp. 2641–2645.

[5] D. Babic, J. Vesma, T. Saramäki, and M. Renfors, "Implementation of the transposed Farrow structure," in *Proc. 2002 IEEE Int. Symp. Circuits and Systems*, Scottsdale, Arizona, USA, 2002, vol. 4, pp. 4–8.

[6] T. Saramäki and M. Ritoniemi, "An efficient approach for conversion between arbitrary sampling frequencies," in *Proc. 1996 IEEE Int. Symp. Circuits and Systems*, Atlanta, Georgia, May 1996, pp. 285–288.

[7] D. Babic, T. Saramäki, and M. Renfors, "Conversion between arbitrary sampling frequencies using polynomial-based interpolation filters," in *Proc. Int. Workshop on Spectral Methods and Multirate Signal Processing, SMMSP'02*, Toulouse, France, September 2002, pp. 57–64.

[8] T. Hentschel, and G. Fettweis, "Continuous-time digital filters for sample-rate conversion in reconfigurable radio terminals," in *Proc. the European Wireless 2000*, Dresden, Germany, Sept. 2000, pp. 55–59.

DISCRETE-TIME MODELING OF POLYNOMIAL-BASED INTERPOLATION FILTERS IN RATIONAL SAMPLING RATE CONVERSION

Djordje Babic, Vesa Lehtinen, Markku Renfors

Institute of Communications Engineering, Tampere University of Technology
P.O. Box 553, Tampere, FINLAND
Tel: +358 3 365 3910; fax: +358 3 3653808; E-mail: {babic, mr} @cs.tut.fi

ABSTRACT

If sampling rate conversion (SRC) is performed between arbitrary sampling rates, then the SRC factor can be a ratio of two very large integers or even an irrational number. An efficient way to reduce the implementation complexity of a SRC system in those cases is to use polynomial-based interpolation filters that mimic digitally the hybrid analogue/digital system. In practice, the sampling rate conversion is approximated with a rational factor. In this case, the hybrid analogue/digital model used to represent SRC process may be represented by an equivalent discrete-time model. The discrete-time modeling of the rational SRC has been used earlier for the zeroth order interpolation. This paper extends this idea to arbitrary polynomial-based interpolation. Furthermore, this paper derives the relation between various polynomial-based interpolation filters (Farrow structure and its modifications) and polyphase FIR model filters. This paper observes possible applications of these relations, such as filter design, implementation complexity reduction, and response distortion analysis.

1. INTRODUCTION

Sampling rate conversion (SRC) is utilized in many DSP applications where two signals or systems having different sampling rates are to be interconnected. The SRC factor in general can be an integer, one divided by an integer, a ratio of two integers, or an irrational number. The SRC factor is determined by

$$R = F_{out}/F_{in} = T_{in}/T_{out}, \tag{1}$$

where $F_{in} = 1/T_{in}$ and $F_{out} = 1/T_{out}$ are the original input sampling rate and the sampling rate after the conversion, respectively. The sampling rate conversion can be divided into two general cases. For $R < 1$, the sampling rate is reduced and this process is known as decimation. For $R > 1$, the sampling rate is increased and this process is known as interpolation [1]-[6].

In practical realizations, the SRC factor R is implemented as a ratio of two relatively prime integers, i.e., $R=L/K$. This means that in practice an irrational factor is not achievable, thus it is approximated by a rational number which is determined by the precision of the arithmetic used. In [7], a discrete time model was derived for the zeroth-order polynomial interpolation. In this paper, the discrete-time modeling of the rational SRC is extended to the polynomial-based interpolators of arbitrary order. Further, it is shown that for any modification of the Farrow structure there exists a discrete-time FIR model filter. This paper gives mathematical relations between various polynomial-based filters and the equivalent FIR filters. It is also observed that it is possible to perform transformation in the opposite direction, that is, it is possible to convert a FIR filter into the form of a Farrow structure. Finally, we observe possible applications of the presented relations, such as filter design, response analysis, and reducing the realization complexity for special structures. In [8], it is shown how the presented relations effectively simplify analysis of the frequency response distortion in practical SRC systems.

This work was carried out in the project "Advanced Transceiver Architectures and Implementations for Wireless Communications" supported by the Academy of Finland. This work is also supported by Tampere Graduate School in Information Science and Engineering TISE, and NOKIA foundation.

2. HYBRID ANALOGUE/DIGITAL MODEL FOR SRC

In the most general case, the SRC can be regarded as a process of resampling an analogue signal according to the hybrid analogue/digital model shown in Fig. 1 [1]. In this system, the discrete-time input sequence $x(n)$ with the sampling rate equal to $F_{in}=1/T_{in}$ is first converted into an analog continuous-time signal $x_s(t)$ using an ideal digital-to-analog converter (DAC). The resulting signal is the following sum of the weighted and shifted impulses:

$$x_s(t) = \sum_{k=-\infty}^{\infty} x(k)\delta_a(t-kT_{in}), \tag{2}$$

where $\delta_a(t)$ is the analogue Dirac delta function. This sequence is then filtered using an analogue filter with the impulse response $h_a(t)$ to generate the continuous-time signal given by

$$y_a(t) = \sum_{k=-\infty}^{\infty} x(k)h_a(t-kT_{in}). \tag{3}$$

Finally, the output $y_a(t)$ is sampled at the time instants $t=lT_{out}$ using the ideal analog-to-digital converter (ADC) to produce the following discrete-time output sequence $y(l)$:

$$y(l) = y_a(lT_{out}) = \sum_{k=-\infty}^{\infty} x(k)h_a(lT_{out}-kT_{in}). \tag{4}$$

It should be pointed out that in the decimation case, the filter with the impulse response $h_a(t)$ acts as an anti-aliasing filter rejecting the frequency components aliasing onto the new baseband. On the other hand, in the interpolation case, the role of this filter is to preserve the original baseband region and to eliminate the imaging components. Hence, it acts as an anti-imaging filter.

When the decimation factor $1/R$ or the interpolation factor R is an integer or a ratio of two small relatively prime integers, then the SRC can be performed conveniently with the aid of fixed digital filters [1]. If these factors are irrational, then fixed digital filters cannot be directly used. Furthermore, if R is a ratio of two large relatively prime integers, then the required filter orders become very large. In these cases, an efficient overall implementation based on mimicking the system of Fig. 1 can be achieved by generating $h_a(t)$ to have the following properties [2],[3]. First, $h_a(t)$ is nonzero only in the interval $0 \le t < NT$ with N being an even integer. Second, in each subinterval $nT \le t < (n+1)T$ for $n = 0, 1, \ldots, N-2$, $h_a(t)$ is expressible as a polynomial of a given low order M. Third, $h_a(t)$ is symmetric around $t = NT/2$ to guarantee the phase linearity of the resulting overall system, which is an useful property in many prac-

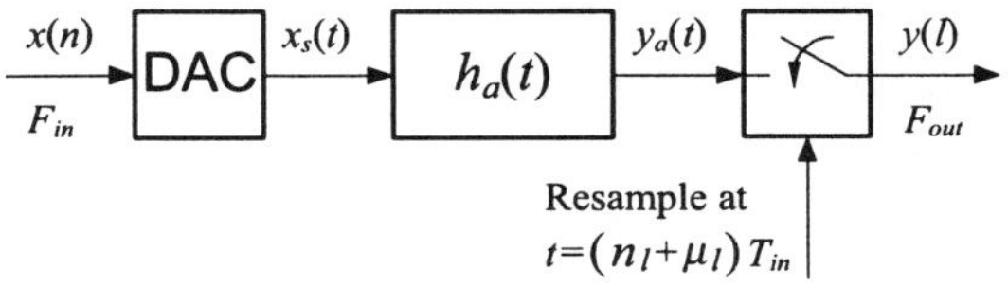

Fig. 1. Hybrid analogue/digital model to be mimicked.

0-7803-7761-3/03/$17.00 ©2003 IEEE

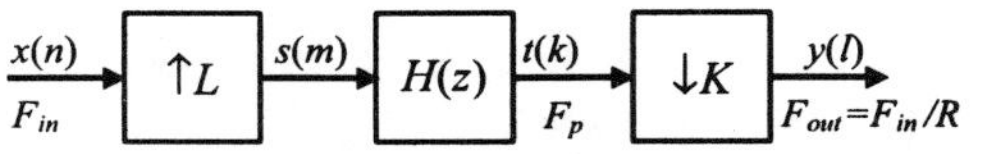

Fig. 2. Discrete-time model for SRC by rational factor.

tical implementations. The length of polynomial segments T can be selected to be equal to the input or output sampling interval, a fraction of the input or output sampling interval, or an integer multiple of the input or output sampling interval. The advantage of mimicking the above system lies in the fact that the actual implementation can be efficiently performed by using the Farrow structure [4] or its modifications [2], [3], [5], and [6]. The main advantage of the Farrow structure lies in the fact that it consists of fixed finite-impulse response (FIR) filters and there is only one changeable parameter, the so-called fractional interval μ. Besides this, the control of μ is easier during operation than in the corresponding coefficient memory implementations, and the resolution of μ is limited only by the precision of arithmetic and not by the size of the memory. These characteristics of the Farrow structure make it a very attractive structure to be implemented using a VLSI circuit or a signal processor [3].

3. DISCRETE-TIME MODEL FOR RATIONAL SRC

This part overviews the discrete-time model for the SRC. The model is valid for the case when the SRC factor is a ratio of two relatively prime integers, i.e., $R=L/K$.

When using the hybrid analogue/digital model directly for system analysis, aliasing may become a problem. Because of that, frequencies far higher than F_{in} and F_{out} must be taken into consideration, leading to a high computational complexity. It may also be difficult to determine how high frequencies should be included in the analysis in order to keep the approximation error, caused by the ignored aliases, at an acceptable level. The aliasing problem can be avoided by using the discrete-time model shown in Fig. 2. This is used earlier in [7] to model the zeroth order interpolation.

Sampling rate conversion by $R=L/K$ can be implemented as a cascade of interpolation by L and decimation by K. As shown in Fig. 2, the interpolation and decimation filters can be combined into a single filter, $H(z)$ [1]. In the general case, the sampling rate is first increased by L in the upsampler block. After that, the signal is filtered with transfer function $H(z)$ and, finally, the sampling rate is reduced by K using a downsampler block. The role of the filter $H(z)$ is to perform both aliasing and imaging attenuation. The key idea for the derivation of the relation between the impulse response of the analogue filter $h_a(t)$, from hybrid model of Fig. 1, and discrete-time filter $H(z)$, in the discrete-time model of Fig. 2, is to sample the continuous-time impulse response $h_a(t)$ at the intermediate rate $F_p=LF_{in}=KF_{out}$. This sampling rate does not exist physically in a polynomial interpolator. In the discrete-time model, the period of the frequency response of the FIR filter $H(e^{j\omega})$ is F_p, that is, all frequency domain information is always contained within a finite band.

4. RELATION BETWEEN POLYNOMIAL-BASED INTERPOLATORS AND FIR FILTERS

This section explains the relation between several modifications of the polynomial based filters and discrete-time model. Further, it presents equivalent discrete-time polyphase FIR structures.

4.1. Modified Farrow structure

This chapter introduces the relation between the modified Farrow structure (which mimics the analogue/digital model given in Fig. 1), and the FIR polyphase filter model.

Consider the system of Fig. 3, where the input sampling period

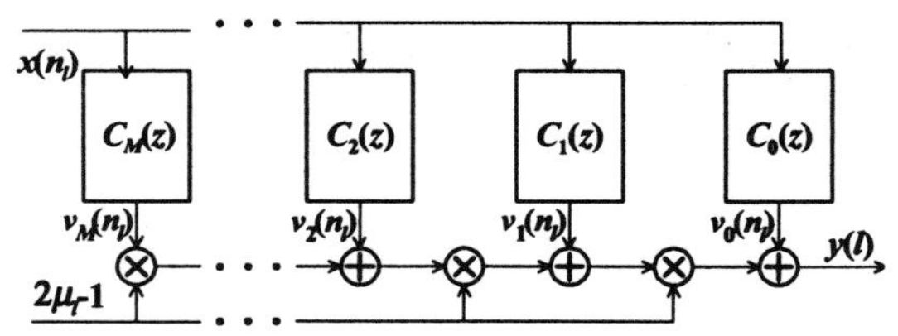

Fig. 3. Modified Farrow structure.

is equal to T_{in}. The structure is known as the modified Farrow structure, and it is represented by [2], [3]:

$$y(l)=\sum_{n=0}^{N-1}\sum_{m=0}^{M}x(n_l-N/2+n)c_m(n)(2\mu_l-1)^m. \tag{5}$$

where

$$n_l=\lfloor l/R+\varepsilon\rfloor,\text{ and }\mu_l=l/R+\varepsilon-\lfloor l/R+\varepsilon\rfloor. \tag{6}$$

Here, N is the number of polynomial segments in $h_a(t)$, M is the order of polynomials, μ_l is the fractional interval, and ε is an offset value in the computation of μ_l. In the case of the rational sampling rate conversion factor $R=L/K$, the fractional interval μ_l has a certain (finite) number of values. It can be shown that in this case, μ_l is a periodic function, having a period of L. It can be concluded that the modified Farrow structure has an equivalent FIR representation. The equivalent FIR filter can be divided into L polyphase branches. Each polyphase branch corresponds to one value of the fractional interval μ_l. The overall transfer function of the equivalent polyphase FIR filter is given by

$$H(z)=\sum_{l=0}^{L-1}z^{-l}G_l(z^L), \tag{7}$$

where

$$G_l(z^L)=\sum_{n=0}^{N-1}g_l(n)z^{nL}. \tag{8}$$

The coefficients of the polyphase branches are computed from the coefficients of the original modified Farrow structure as follows:

$$g_l(n)=\sum_{m=0}^{M}c_m(n)(2\mu_l-1)^m, \tag{9}$$

for $l=0,\ 1,\ \ldots,\ L-1$, and $n=0,\ 1,\ldots,\ N-1$. The relation between two sets of coefficients can be expressed in matrix form as:

$$\begin{bmatrix} g_0(n)\\ g_1(n)\\ \vdots\\ g_{L-1}(n)\end{bmatrix}=\begin{bmatrix}1 & (2\mu_0-1) & \cdots & (2\mu_0-1)^M\\ 1 & (2\mu_1-1) & \cdots & (2\mu_1-1)^M\\ \vdots & \vdots & \cdots & \vdots\\ 1 & (2\mu_{P-1}-1) & \cdots & (2\mu_{P-1}-1)^M\end{bmatrix}\cdot\begin{bmatrix}c_0(n)\\ c_1(n)\\ \vdots\\ c_M(n)\end{bmatrix}, \tag{10}$$

for $n=0,\ 1,\ldots,\ N-1$. Finally,

$$\begin{bmatrix}\mathbf{G}_0\\ \mathbf{G}_1\\ \vdots\\ \mathbf{G}_{L-1}\end{bmatrix}=\begin{bmatrix}1 & (2\mu_0-1) & \cdots & (2\mu_0-1)^M\\ 1 & (2\mu_1-1) & \cdots & (2\mu_1-1)^M\\ \vdots & \vdots & \cdots & \vdots\\ 1 & (2\mu_{L-1}-1) & \cdots & (2\mu_{L-1}-1)^M\end{bmatrix}\cdot\mathbf{C}^{\mathrm{T}}. \tag{11}$$

Here $\mathbf{G}_l$, for $l=0,\ 1,\ldots,\ L-1$, represents the row vector of polyphase branch coefficients. We conclude that in the case of the sampling rate conversion by a rational factor, it is equivalent to implement the required filter either by using the Farrow structure, or the corresponding polyphase FIR filter. The equivalent polyphase FIR representation using commutators is shown in Fig. 4. The commutator switches from one polyphase branch output to another according to cyclic pattern, jumping with a step length of K branches. This corresponds to decimation in the discrete-time model. The polyphase FIR branches work at the input sampling rate F_{in}, and the commutator switches at the output sampling rate F_{out}. The

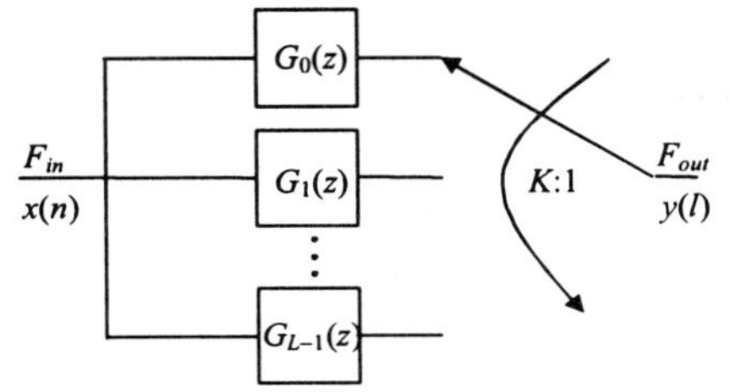

Fig. 4. Polyphase FIR filter equivalent to the modified Farrow structure.

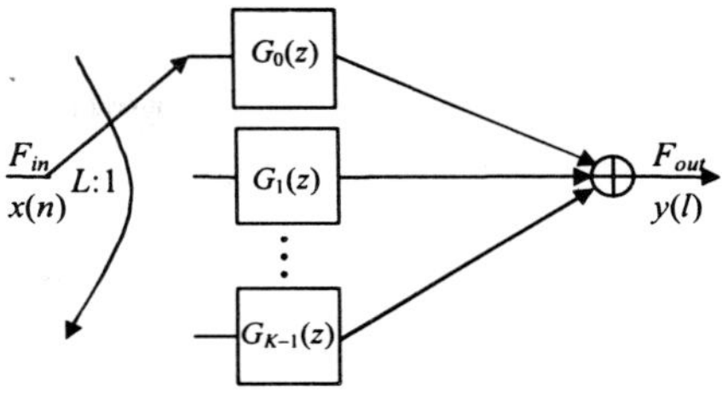

Fig. 5. Polyphase FIR filter equivalent to the transposed modified Farrow structure.

overall structure contains L polyphase branches of length N, where N is the length of the analogue polynomial-based filter (subfilter length in the Farrow structure). Thus, the overall number of multipliers is NL. The FIR model filter can be made symmetric if initial value of fractional interval (offset value) is $\varepsilon=1/(2L)$. In this case, the number of multipliers in FIR model filter is equal to $NL/2$.

4.2. The Transposed Farrow structure

This part introduces the relation between transposed modified Farrow structure and a FIR filter according to the discrete-time model in a special case when $R=K/L$. This case is dual to the previous, explained in Section 4.1.

The transposed Farrow structure is explained in details in [5]. In this case, the fractional interval μ_k depends on the input sample timing and it is given by:

$$\mu_k = kT_{in}/T_{out} + \varepsilon - \lfloor kT_{in}/T_{out} + \varepsilon \rfloor. \tag{12}$$

We conclude that in the case when $R=K/L$ there are K different values for μ_k, and similarly to the previous case, μ_k is a periodic sequence, having a period of K. It can be concluded that the transposed modified Farrow structure can be equally well described by a polyphase FIR filter where each polyphase branch corresponds to one value of the fractional interval μ_k. The set of equations is similar to the one presented in previous Section 4.1. The main difference is in the computation of μ_k and in the number of different values, i.e., polyphase branches. The transposed modified Farrow structure can be equally well described by the system shown in Fig. 5. The commutator switches to a new input jumping each time in cyclic pattern with a step length of L branches. This corresponds to zero insertion due to interpolation by L, and the overall polyphase structure is used for decimation by K. The FIR polyphase branches work at the output sampling rate F_{out}, and commutator switches at input sampling rate F_{in}. The overall structure contains K polyphase branches of length N, thus the number of multipliers is NK. Similarly, as in previous case, the FIR model filter can be made symmetric, by setting the corresponding offset value to be $\varepsilon=1/(2L)$.

4.3. Cascade of fixed linear-phase interpolator and modified Farrow structure

This section considers equivalent FIR representation for the cascade of fixed linear phase interpolator and Farrow structure, used for interpolation [6]. The main result is valid for a dual structure used for decimation, which is formed as cascade of the transposed Farrow structure and fixed linear phase decimator. Using already described principle, the equivalent FIR structure is directly derived. The overall system contains a fixed FIR interpolator followed by a polyphase FIR filter described in Section 4.1 (Section 4.2 for dual decimator case).

4.4. Prolonged polynomial-based filters

This section introduces the relation between prolonged polynomial-based filter and FIR interpolator (FIR decimator in dual transposed prolonged case). The prolonged polynomial-based filters, introduced in [6], can be represented by the following set of equations

$$y(l) = \sum_{n=0}^{N-1} \sum_{m=0}^{M} c_m(n) \sum_{j=0}^{J-1} x(n_l - nJ - j + NJ/2)(2\mu_l(j)-1)^m , \tag{13}$$

where the $\mu_l(j)$'s for $j = 0, 1, \ldots, J-1$ are J fractional intervals depending on the μ_l included in Eq. (6) as follows:

$$\mu_l(j) = (j+\mu_l)/J . \tag{14}$$

It can be observed that the system can be represented by an equivalent FIR polyphase system. In this case, there is one polyphase branch corresponding to each value of $\mu_l(0)$. There are L different values of $\mu_l(0)$, thus there are L polyphase branches. The main difference, compared to the case explained in Section 4.1, is in that the length of polyphase filters is JN. These polyphase filters may be further decomposed, using properties of the prolonged structure. Therefore, each polyphase branch can be seen as a non-decimating polyphase FIR filter of length NJ, having J polyphase branches. Here, N is the number of polynomial segments in the analogue impulse response of the polynomial-based filter $h_a(t)$. Finally, the equivalent FIR filter is given by

$$H(z) = \sum_{l=0}^{L-1} z^{-l} \sum_{j=0}^{J-1} z^{-j} G_l^j(z^{JL}), \tag{15}$$

where polyphase branches are presented by

$$G_l^j(z^L) = \sum_{n=0}^{N-1} g_l^j(n) z^{nJL}, \tag{16}$$

with

$$g_l^j(n) = \sum_{m=0}^{M} c_m(n)(2\mu_l(j)-1)^m . \tag{17}$$

This can be expressed in matrix form as:

$$\begin{bmatrix} g_0^0(n) \\ g_0^1(n) \\ \vdots \\ g_0^{J-1}(n) \\ g_1^0(n) \\ g_1^1(n) \\ \vdots \\ g_1^{J-1}(n) \\ \vdots \\ g_{L-1}^0(n) \\ g_{L-1}^1(n) \\ \vdots \\ g_{L-1}^{J-1}(n) \end{bmatrix} = \begin{bmatrix} 1 & 2\mu_0(0)-1 & \cdots & (2\mu_0(0)-1)^M \\ 1 & 2\mu_0(1)-1 & \cdots & (2\mu_0(1)-1)^M \\ \vdots & \vdots & \ddots & \vdots \\ 1 & 2\mu_0(J-1)-1 & \cdots & (2\mu_0(J-1)-1)^M \\ 1 & 2\mu_1(0)-1 & \cdots & (2\mu_1(0)-1)^M \\ 1 & 2\mu_1(1)-1 & \cdots & (2\mu_1(1)-1)^M \\ \vdots & \vdots & \ddots & \vdots \\ 1 & 2\mu_1(J-1)-1 & \cdots & (2\mu_1(J-1)-1)^M \\ \vdots & \vdots & \ddots & \vdots \\ 1 & 2\mu_{L-1}(0)-1 & \cdots & (2\mu_{L-1}(0)-1)^M \\ 1 & 2\mu_{L-1}(1)-1 & \cdots & (2\mu_{L-1}(1)-1)^M \\ \vdots & \vdots & \ddots & \vdots \\ 1 & 2\mu_{L-1}(J-1)-1 & \cdots & (2\mu_{L-1}(J-1)-1)^M \end{bmatrix} \begin{bmatrix} c_0(n) \\ c_1(n) \\ \vdots \\ c_M(n) \end{bmatrix} \tag{18}$$

for n=0, $J-1$, $2J-1$,..., $NJ-1$. The prolonged polynomial-based filter has an equivalent FIR filter shown in Fig 6. The overall structure contains LJ polyphase branches of length NJ, with N nonzero coefficients. Thus the overall number of multipliers is NLJ. The resulting FIR model filter can be made symmetric using offset

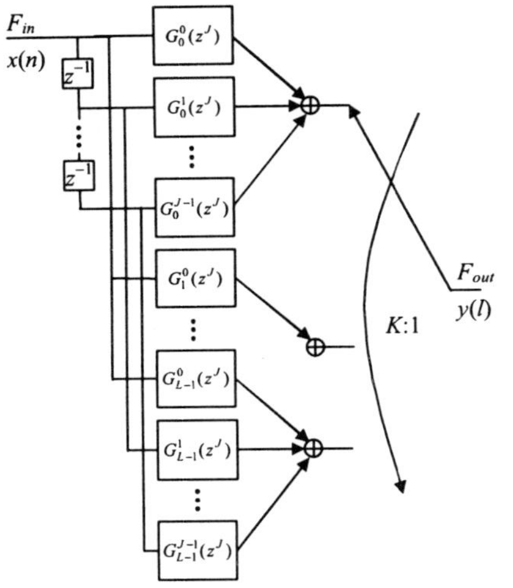

Fig 6. Polyphase FIR equivalent of the prolonged polynomial-based interpolator.

tuning as described in Section 4.1. The commutator at the output switches at the output sampling rate F_{out}, jumping for K branches in a cyclic pattern.

5. CASE STUDIES

In this section, we give a simple illustrative example and we investigate possible applications of the derived relations.

Let us consider an integer interpolation by 4 using the modified Farrow structure. In this case $R=L/K=4$, thus there are $L=4$ polyphase branches (corresponding to L different values of μ_l). The requirements for the interpolation filter are as follow: stopband attenuation at least A_s=60 dB, passband distortion at most A_p=0.001, passband edge $f_p=0.4*F_{in}$, and stopband edge $f_s=0.5*F_{in}$. These requirements are met by using the modified Farrow structure having N=28 and M=4. The equivalent FIR model filter, obtained using Eq. (11), has a length equal to NL=112. We observe that the frequency response (see Fig 7) of the equivalent FIR model filter corresponds to the frequency response of the original modified Farrow structure (polynomial-based filter). The difference is due to aliasing, which is taken into account in the frequency response of the FIR model filter. The FIR model filter frequency response depends on the offset value ε, see Eq. (6). Figure 7(b) shows frequency responses of FIR model filter for two different values of ε. It is clear that the response is better in the case when $\varepsilon=1/(2L)$ (solid line in Fig 7(b)), this corresponds to the symmetric FIR model filter as explained in Section 4.1. We conclude that it is possible to control the aliasing by tuning the offset value ε in computation of fractional interval μ_l, but this is true only for applications where phase offset is not so critical. This aliasing is consequence of resampling in model of Fig. 1, and it is not considered in frequency response calculation of the polynomial-based continuous-time filter. One can observe how derived relation can be used effectively for distortion analysis of the polynomial-based filter response in practical digital implementation.

We can design an FIR filter satisfying the same requirements as in previous case, and then using the inverse of Eq. (11), we can convert it into the modified Farrow structure. The length of designed filter should be, in the most general case, NJL, where J=1 for the regular non-prolonged case (NJK in dual decimation case). However, in order to have unique relation between FIR and corresponding modified Farrow structure, it is required that the polynomial order M in the Farrow structure satisfies $M=L-1$ ($M=K-1$ in dual case). This fact can be used for fast and efficient design of the Farrow structure. This is a topic of our future research.

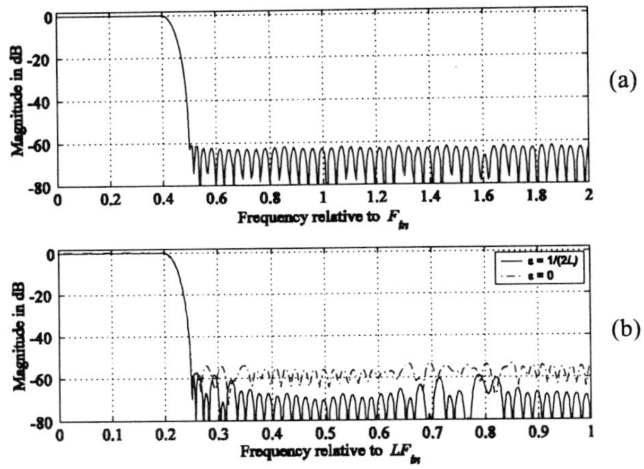

Fig 7. Frequency responses: (a) polynomial-based filter, (b) equivalent FIR filters.

6. CONCLUSIONS

We have developed the relation between various types of polynomial-based filters and the model FIR filter. It has been shown that in the case of SRC by a rational factor, for any polynomial-based filter of arbitrary order there exists an equivalent time-invariant FIR filter. This discrete-time filter can be used in distortion analysis to overcome the aliasing problems associated with the hybrid analog/digital model. The Farrow structure and its modifications can be considered as effective implementation form for linear periodically time varying (LPTV) filters. It has been observed that the derived relation can be used to find equivalent modification of the Farrow structure for the given polyphase FIR filter. Possible applications of these are filter design, implementation complexity reduction, and response distortion analysis.

REFERENCES

[1] R. E. Crochiere and L. R. Rabiner, *Multirate Digital Signal Processing*. Englewood Cliffs, NJ: Prentice-Hall, 1983.

[2] J. Vesma and T. Saramäki, "Interpolation filters with arbitrary frequency response for all-digital receivers," in *Proc. 1996 IEEE Int. Symp. Circuits and Systems*, Atlanta, Georgia, May 1996, pp. 568–571.

[3] J. Vesma, *Optimization and Applications of Polynomial-Based Interpolation Filters*. Doctoral Thesis, Tampere University of Technology, Publications 254, 1999.

[4] C. W. Farrow, "A continuously variable digital delay element," in *Proc. 1988 IEEE Int. Symp. Circuits and Systems*, Espoo, Finland, June 1988, pp. 2641–2645.

[5] D. Babic, J. Vesma, T. Saramäki, M. Renfors, "Implementation of the transposed Farrow structure," in *Proc. 2002 IEEE Int. Symp. Circuits and Systems*, Scotsdale, Arizona, USA, 2002, vol. 4, pp. 4–8.

[6] D. Babic, T. Saramäki and M. Renfors, "Conversion between arbitrary sampling frequencies using polynomial-based interpolation filters," in *Proc. Int. Workshop on Spectral Methods and Multirate Signal Processing, SMMSP'02*, Toulouse, France, September 2002, pp. 57–64.

[7] W. H. Yim, "Distortion analysis for multiplierless sampling rate conversion using linear transfer functions," in *IEEE Signal Processing Letters*, vol. 8, No. 5, May 2001, pp. 143-144.

[8] V. Lehtinen, M. Renfors, "Analysis of rational sample rate conversion using image response combining," this conference.

ANALYSIS OF RATIONAL SAMPLE RATE CONVERSION USING IMAGE RESPONSE COMBINING

Vesa Lehtinen, Markku Renfors

Tampere University of Technology
Institute of Communications Engineering
P.O.Box 553
FIN-33101 Tampere, Finland
vesa.lehtinen@tut.fi, markku.renfors@tut.fi

ABSTRACT

Using discrete-time response modelling, we analyse the complicated frequency domain input-output relationship of sample rate conversion with a rational, non-integer factor and address the ambiguity of the concept of frequency response in fractional multirate systems. The dependence of a system response on the phase of a real signal is shown. We develop a method to translate a magnitude response from a higher sample rate to a lower one to allow better comparison between filter properties and system specifications when rational sample rate conversion is used. The method is based on the idea of combining the responses of all (aliased) images of an input frequency into a scalar quantity. The method is hence referred to as *image response combining* (IRC) and the resulting response function as *image-combined magnitude response.* Two different IRC schemes are proposed. To overcome the sometimes very high computational complexity of frequency domain IRC, we derive a drastically faster time domain IRC algorithm referred to as *sampled autocorrelation of impulse response* (SACIR). The paper concentrates on decimation, but the analysis and proposed methods can be easily applied to interpolation, as well.

1. INTRODUCTION

In multirate systems with integer sample rate conversion (SRC) factors, frequency domain specifications are usually given at the highest sample rate. For example, in a communications receiver, oversampling is used in analog to digital conversion, and the signal is decimated to a lower sample rate for detection. In decimation, some frequencies alias to other frequencies [1]. In decimation with integer factors, each input frequency aliases to one output frequency only. Therefore, the concept of frequency response in terms of input frequencies remains unambiguous; the response can be obtained by translating all filters to the input sample rate by applying the well-known identity shown in Figure 1.1(a) [1]. This is important because system specifications for a receiver are given in terms of input frequencies. In a similar manner, in interpolation with integer factors, each output frequency is an image of one input frequency only, and the frequency response of the system can be unambiguously expressed in terms of output frequencies by translating all filter stages to the output sample rate using the identity shown in Figure 1.1(b).

When sample rate conversion with a noninteger factor is used, the above does not hold, as shown in Section 3. To help the analysis, we explain the use of discrete-time modelling [2][3] of rational SRC in Section 2. In Section 4, we introduce a method for translating filter responses to lower sample rates. Due to the sometimes very high computational complexity of the frequency domain method explained in Section 4, a faster time-domain response translation method is presented in Section 5. The theory is compared to simulation results in Section 6. Finally, conclusions are drawn in Section 7.

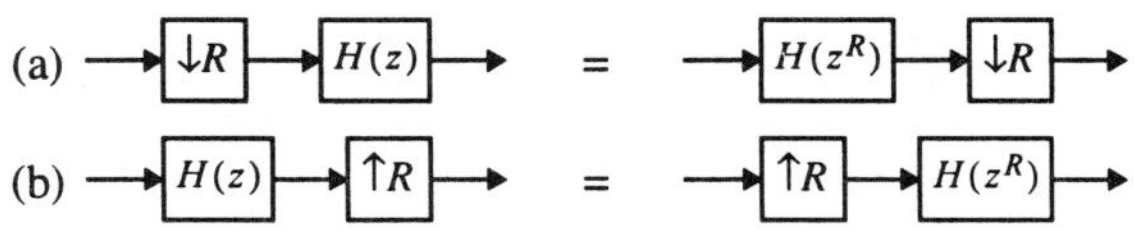

Figure 1.1. Translating a transfer function to a higher sample rate.

2. RESPONSE MODELLING OF POLYNOMIAL INTERPOLATORS

In [2], a discrete-time model was derived for rational SRC with zeroeth-order polynomial interpolators. In [3], the method is extended to arbitrary-order polynomial interpolators implemented with the Farrow structure [4] and its modifications [5],[6]. The discrete-time model is shown in Figure 2.1. The polynomial-based rational sample rate converter with continuous-time transfer function [1] $H_2^{\mathrm{CT}}(s)$ is equivalent to a cascade of an upsampler, a discrete-time model filter $H_2(z)$, and a downsampler. The discrete-time filters $H_1(z)$ and $H_3(z)$ are included in Figure 2.1 to emphasize the fact that in rational SRC, due to the complicated imaging and aliasing phenomena, analysing the response of a single filter may be misleading; ignoring any stage in the design of other stages can easily result in a suboptimal or inadequate design. The overall SRC factor is

$$R = L/K = F_{\mathrm{out}}/F_{\mathrm{in}} \tag{2.1}$$

where L and K are relatively prime integers. The discrete-time impulse response is obtained by sampling the continuous-time impulse response:

$$h_2[n] = h_2^{\mathrm{CT}}(nT_{\mathrm{p}} + \Delta t), \tag{2.2}$$

where $T_{\mathrm{p}} = 1/F_{\mathrm{p}}$ is the pseudo sample interval of $H_2(z)$ and Δt is constant to be selected by system designer; usually $\Delta t = 0$. Notice that the pseudo sample rate $F_{\mathrm{p}} = LF_{\mathrm{in}} = KF_{\mathrm{out}}$ does not exist physically in the system.

The discrete-time modelling method resembles, e.g., that of [7]. However, because of the linear time-invariant (LTI) model filter, the analysis becomes easier to perform and understand.

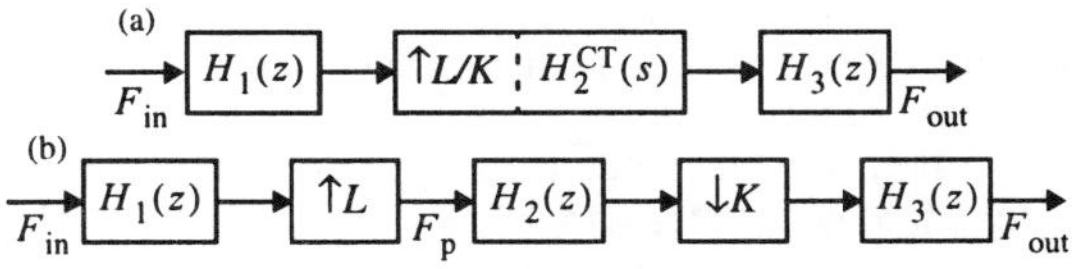

Figure 2.1. (a) A cascade of a rational decimator and two discrete-time filters on both sides of it. (b) The equivalent discrete-time model.

3. INPUT-OUTPUT FREQUENCY RELATIONSHIP IN RATIONAL SRC

The input-output frequency relationship of rational SRC can be studied using bifrequency maps [1]. In Figure 3.1(a), the output frequencies of a rational sample rate converter with SRC factor

0-7803-7761-3/03/$17.00 ©2003 IEEE

$R = L/K = 4/7$ are plotted as functions of input frequencies for complex signals. It can be seen that each input frequency has L images, each of which aliases to a different output frequency on a uniform grid. The input and output sample rates of this example are 14 and 8, respectively, in some (irrelevant) unit.

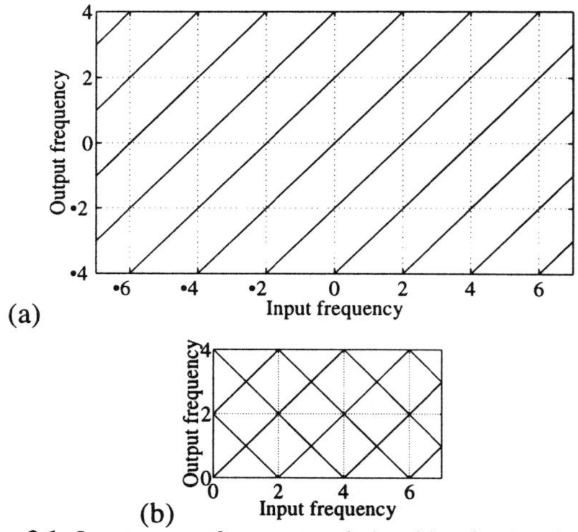

Figure 3.1. Input-output frequency relationship of rational sample rate conversion for (a) complex and (b) real signals.

3.1 Coaliasing

Consider a real sinusoid $x(t) = \sin(2\pi t/T + \phi)$ with period T being sampled at time instants $t = nT/2$. In such a case the amplitude of the sampled signal depends on the phase difference between the sinusoid and the sampling clock. For example, if $\phi = 0$, the signal vanishes because samples are taken exactly at its zero crossings. If $\phi = \pm\pi/2$, samples are taken at the peaks and valleys of the signal, giving the maximum output amplitude. We can see that the phase of the signal affects the response of the sampling process at this particular frequency.

In rational SRC of real signals, the same phase dependence phenomenon occurs at several input and output frequencies. In the bifrequency map shown in Figure 3.1(b) it can be seen that in both input and output frequency domains there is a uniform grid of frequencies at which two images of an input frequency alias to the same output frequency. This is due to the fact that in a real-valued signal each positive frequency has a negative-frequency mirror image and vice versa. The lines tilted left in Figure 3.1(b) represent images of negative frequencies aliased to positive frequencies, and vice versa. If the model filter $H_2(z)$ is real and has a linear phase response, the summation of these two aliased images is either constructively or destructively coherent depending on the phase of the signal and the sign of the zero-phase response of $H_2(z)$ at the image frequencies. As a result, the bifrequency magnitude response at these frequencies becomes dependent on the phase of the signal.

In this phenomenon, hereafter referred to as *coaliasing*, (an image of) a positive frequency aliases to the same frequency with (an image of) its negative-frequency mirror image. At the corner points of Figure 3.1(b), an image Òcoaliases with itselfÓ, making the sine component vanish. Notice that this phenomenon does not occur when the signal is complex-valued (see Figure 3.1(a)). There is an exception, however: if a complex signal has been obtained from a real signal through, e.g., complex mixing (multiplication with a complex exponential), the mirror image frequency pairs remain, now only shifted along the frequency axes, and the coaliasing phenomenon remains (see Figure 3.2).

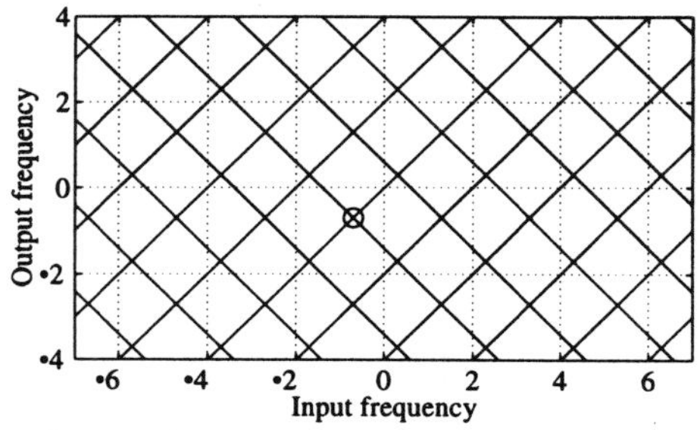

Figure 3.2. Input-output frequency relationship of rational SRC for a complex signal obtained from a real-valued source. The circle shows the zero frequency of the original real signal.

Causing different responses for the sine and cosine signal components, coaliasing also affects the phase of a signal. Figure 3.3 shows the output phase of a coaliasing input-output frequency pair as a function of input phase for a linear-phase $H_2(z)$. The staircase curves correspond to cases when the sine (continuous line) or cosine (dashed) component vanishes; at their discontinuities, the frequency component can vanish entirely. A straight line corresponds to a case that the amplitude ratio of the coaliasing image pair is large.

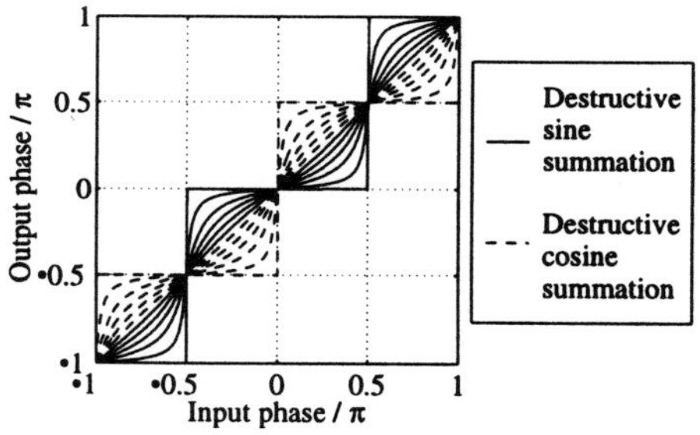

Figure 3.3. Input-output phase relationship of coaliasing image pairs.

Coaliasing effects, i.e., dependence of magnitude and phase response on signal phase, are usually weak within the passband of a properly designed system due to the large magnitude ratio between the coaliasing frequencies Ð one from the passband and one from the stopband of $H_2(z)$. However, if the target signal-to-interference ratio is small, the phenomenon may become significant.

Notice that in SRC with an irrational factor, coaliasing occurs at all input-output frequency combinations. However, when approximating an irrational factor with a time-varying rational factor, the behaviour may be different.

Coaliasing should be taken into account in applications that are sensitive to the amplitude and/or phase of a single sinusoid, such as a pilot tone in a communication system.

4. TRANSLATION OF RESPONSES TO LOWER SAMPLE RATES

Using the identities of Figure 1.1, the overall response of a multistage decimator can be calculated. In the case of rational sample rate conversion, however, this response is obtained at the high pseudo sample rate and cannot be directly compared to system specifications. To enable meaningful evaluation of the system response, the response should be expressed in the same domain with the specifications. For example, specifications of a communications receiver are given in terms of input frequencies. Therefore, we need a means to translate a filter response from a higher sample rate to a lower one. The identities of Figure 1.1 cannot be used in this. Furthermore, the translation problem becomes ambiguous as images of an input frequency alias to multiple output frequencies. Thus, we must define a way to combine image responses to obtain a suitable response measure. Response translation based on this idea is referred to as *image response combining* (IRC). There exist several possible definitions for such a scheme. The choice of an IRC scheme is based on system requirements and properties of the signal to be processed. Notice that IRC is closely related to the concept of *folded spectrum* used in communications literature.

The *image-combined magnitude response* can be divided into three parts. First, the stopband response $H_{SB}(\omega)$ is obtained by combining each stopband frequency with its images. Second, passband response $H_{PB}(\omega)$ is obtained directly from the non-translated passband response without IRC. Third, the images of passband frequencies are combined to obtain the *passband self-interference response* $H_{SI}(\omega)$. For all IRC schemes, the overall passband response is

$$H_{PB}(\omega) = |H_1(e^{j\omega})H_{23}(e^{j\omega/L})|/L, \tag{4.1}$$

where $H_{23}(z) = H_2(z)H_3(z^K)$.

One way to combine image responses is to sum up their absolute values:

$$H_{SB}^{SA}(\omega) = |H_1(e^{j\omega})| \cdot \sum_{i=0}^{L-1} |H_{23}(e^{j\Omega(\omega,i)})|/L, \tag{4.2}$$

$$H_{SI}^{SA}(\omega) = |H_1(e^{j\omega})| \cdot \sum_{i=1}^{L-1} |H_{23}(e^{j\Omega(\omega,i)})|/L, \tag{4.3}$$

where

$$\Omega(\omega, i) = (2\pi\lceil i/2\rceil + (-1)^i\omega)/L \tag{4.4}$$

and

$$\Omega(\omega, i) = (2\pi i + \omega)/L \tag{4.5}$$

are the ith image of angular frequency ω for real and complex signals, respectively. The zeroeth image is the input frequency itself, translated to the pseudo sample rate F_P: $\Omega(\omega, 0) = \omega/L \ \forall\omega$. We refer to the above as the *sum-absolute* (SA) image response combining scheme.

Another IRC scheme can be developed by considering input and output powers. The images of an input frequency alias to different output frequencies. Therefore, the output power produced by a sinusoidal input is the sum of the powers of the input frequency and its aliased images. Thus, we define the *root-sum-square* (RSS) response combining scheme:

$$H_{SB}^{RSS}(\omega) = |H_1(e^{j\omega})| \cdot \sqrt{\sum_{i=0}^{L-1} |H_{23}(e^{j\Omega(\omega,i)})|^2/L}, \tag{4.6}$$

$$H_{SI}^{RSS}(\omega) = |H_1(e^{j\omega})| \cdot \sqrt{\sum_{i=1}^{L-1} |H_{23}(e^{j\Omega(\omega,i)})|^2/L}. \tag{4.7}$$

Scaling with L in (4.1)-(4.3), (4.6), (4.7) is required due to the scaling effect of upsampling by L.

The analysis of IRC schemes is simple for sinusoidal inputs. More complex input spectra introduce another ambiguity: images of multiple input frequencies alias to the same output frequency. If there is coherence between images aliasing to the same output frequency, as described in Section 3.1, the RSS combining scheme fails. However, the RSS scheme can be used successfully for many communication signals because they have been made noise-like through, e.g., source coding and pseudo-random spreading (in spread-spectrum systems). In such cases, it can be expected that summation of aliased images is noncoherent and a single frequency contains a negligible portion of the total signal power.

5. A FAST TIME DOMAIN ALGORITHM FOR RSS IMAGE RESPONSE COMBINING

Even for moderate values of SRC factor $R = L/K$, L and K can be large. In the frequency domain method, the complexity of computing the overall frequency response with a given frequency grid density is proportional to L or even L^2. Frequency-domain IRC of a rather simple decimator chain can take tens of minutes or even hours on a 500 MHz Pentium III processor. This raises a need to increase the speed of computation. Some acceleration can be achieved by computing separately the responses of all filter stages over one (half)period of their (periodic) frequency responses, and using table look-up to reuse those values for frequencies higher than half of their sample rate. This method requires a uniform grid of $1 + kLK$ frequencies from zero up to $F_p/2$, with k being any positive integer. Still, execution times can remain long.

Below, we introduce a drastically faster time domain method for RSS-IRC, based on the aliasing phenomenon. This method is equivalent to the frequency domain RSS-IRC algorithm. Consider the discrete-time impulse response $h_{23}[n]$. The corresponding transfer function is

$$H_{23}(z) = \sum_n h_{23}[n]z^{-n}. \tag{5.1}$$

Let us define another filter as the convolution of the previous filter and its time-domain conjugate mirror image, i.e., the autocorrelation sequence of $h_{23}[n]$:

$$h_{23}^{AC}[n] = h_{23}[n] * h_{23}^*[-n] = \sum_k h_{23}[k]h_{23}^*[k+n], \tag{5.2}$$

where x^* denotes the complex conjugate of x. Equivalently,

$$H_{23}^{AC}(z) = H_{23}(z)H_{23}^*(z) = |H_{23}(z)|^2. \tag{5.3}$$

Thus, $H_{23}^{AC}(z)$ is a noncausal, delay-free, symmetric filter with a non-negative zero-phase frequency response. Now, let us define a third filter with impulse response

$$h_{23}^{SAC}[n] = h_{23}^{AC}[nL]/L. \tag{5.4}$$

Notice that the centre tap of $h_{23}^{AC}[\bullet]$ is always included in $h_{23}^{SAC}[\bullet]$. According to the sampling theorem and (5.3),

$$H_{23}^{SAC}(e^{j\omega}) = \sum_{i=0}^{L-1} |H_{23}(e^{j\Omega(\omega,i)})|^2/L^2. \tag{5.5}$$

Hence, $\sqrt{H_{23}^{SAC}(e^{j\omega})}$ gives the RSS-combined magnitude response of $H_{23}(z)$ in terms of input frequencies. We name this algorithm as *sampled autocorrelation of impulse response*, or SACIR for short.

Note that SACIR does not take into account the coaliasing phenomenon. Phase-dependent frequencies must therefore be treated separately, if required. SACIR also does not distinguish the actual passband response from passband self-interference. The passband self-interference is obtained from

$$H_{SI}^{RSS}(\omega) = |H_1(e^{j\omega})| \cdot \sqrt{H_{23}^{SAC}(e^{j\omega}) - |H_{23}(e^{j\omega/L})|^2/L^2}. \tag{5.6}$$

For the stopband,

$$H_{SB}^{RSS}(\omega) = |H_1(e^{j\omega})| \cdot \sqrt{H_{23}^{SAC}(e^{j\omega})}. \tag{5.7}$$

Thus, SACIR yields the system model illustrated in Figure 5.1 (with up- and downsampling ignored).

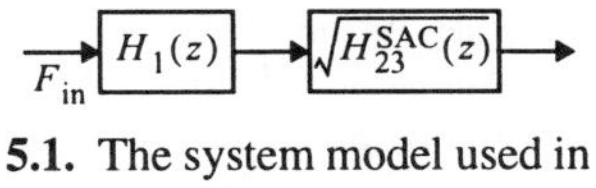

Figure 5.1. The system model used in SACIR.

SACIR can be efficiently implemented with the following optimizations. Consider the system in Figure 2.1(a). It consists of an interpolator performing rational sample rate conversion, and two discrete-time filters, one on both sides of the interpolator. The discrete-time model of the system is shown in Figure 2.1(b). To reduce memory requirements and avoid unnecessary computations, we perform jointly the convolution of the autocorrelations of $h_2[\bullet]$ and $h_3[\bullet]$ and downsampling by L. To allow efficient vectorisation of computations in software such as Matlab or Octave, $h_2^{AC}[\bullet]$ is divided into segments of length L, indexed with variable s. To better reflect properties of programming languages, we use different impulse response indexing here than in (5.2) and (5.4), always starting indices from zero. Below, N_2, N_3, N_{23}^{AC} and N_{23}^{SAC} denote the orders of $h_2[\bullet]$, $h_3[\bullet]$, $h_{23}^{AC}[\bullet]$ and $h_{23}^{SAC}[\bullet]$, respectively. The optimized SACIR algorithm is:

 # Order of convolved autocorrelation of $h_2[\bullet]$ *and* $h_3[\bullet]$ *:*

$N_{23}^{AC} = 2N_2 + 2KN_3$

 # Order of sampled version of convolved autocorrelation of $h_2[\bullet]$ *and* $h_3[\bullet]$ *:*

$$N_{23}^{SAC} = 2\left\lfloor \frac{N_{23}^{AC}}{2L} \right\rfloor$$

 # Index of first sample of $h_{23}^{AC}[\bullet]$ *to be sampled to* $h_{23}^{SAC}[n]$ *:*

$$s_1 = \frac{N_{23}^{AC}\,\text{Ð}\,LN_{23}^{SAC}}{2}$$

 # Number of L-length segments in $h_2^{AC}[\bullet]$ *:*

$$N_S = \left\lceil \frac{1 + 2N_2}{L} \right\rceil$$

 # Initialize:

$h_{23}^{SAC}[n] \leftarrow 0 \quad \forall n$

 # Extend $h_2^{AC}[\bullet]$ *to make its length a multiple of L:*

$h_2^{AC}[n] \leftarrow 0, \quad n = 2N_2 + 1, \ldots, LN_S\,\text{Ð}1$

 # Compute the left half of $h_{23}^{SAC}[\bullet]$ *:*

for $s = 0, \ldots, N_S\,\text{Ð}1$

$$L_3 \leftarrow \max\left\{0, \left\lceil \frac{s_1 + 1\,\text{Ð}L(s+1)}{K} \right\rceil\right\}$$

$$U_3 \leftarrow \min\left\{2N_3, \left\lfloor \frac{s_1 + L(N_{23}^{SAC}/2\,\text{Ð}s)}{K} \right\rfloor\right\}$$

for $i_3 = L_3, \ldots, U_3$

$$i_{23} \leftarrow \left\lceil \frac{Ki_3\,\text{Ð}s_1}{L} \right\rceil + s$$

$$i_2 \leftarrow s_1 + Li_{23}\,\text{Ð}Ki_3$$

$$h_{23}^{SAC}[i_{23}] \leftarrow h_{23}^{SAC}[i_{23}] + h_2^{AC}[i_2] \cdot h_3^{AC}[i_3]/L$$

end

end

for $n = 1, \ldots, N_{23}^{SAC}/2$

$$h_{23}^{SAC}[N_{23}^{SAC}/2 + n] \leftarrow h_{23}^{SAC}[N_{23}^{SAC}/2\,\text{Ð}n]$$

end

After calculating $h_{23}^{SAC}[\bullet]$, the overall image-combined magnitude response is obtained from (4.1), (5.6) and (5.7).

The key to the efficiency of the above algorithm is that $h_{23}^{AC}[\bullet]$ is not computed as such. Its length can be in the order of millions, taking lots of time to compute and memory to store. Instead, we perform the convolution and sampling in a combined manner to obtain $h_{23}^{SAC}[\bullet]$ directly, avoiding the computation of unwanted taps of $h_{23}^{AC}[\bullet]$. Furthermore, the sparsity of $h_{23}^{SAC}[\bullet]$ is exploited by computing only taps that can be non-zero. Also the symmetry of $h_{23}^{SAC}[\bullet]$ is exploited. To gain the full benefit of SACIR, the often occurring sparsity of $h_{23}^{SAC}[\bullet]$ can be exploited when computing its magnitude response, easily resulting in tens of times shorter computation times. Notice that the SACIR algorithm hardly takes a second to run. Most of the execution time is spent on evaluating $H_{23}^{SAC}(e^{j\omega})$.

6. SIMULATIONS

To verify that RSS-IRC can be used for communication signals, a simulation was run. Figure 6.1 shows the power contribution of a sinusoid embedded in a 4-QAM signal as a function of the frequency of the sinusoid. The power contribution is obtained from

$$(E[(y_{\sin} + y_{QAM})^2]\,\text{Ð}E[y_{QAM}^2])/E[x_{\sin}^2] \qquad (6.1)$$

where $x_{\sin}$, $y_{\sin}$ and y_{QAM} are the sinusoidal input, output produced by it and output produced by a QAM signal, respectively. It can be seen that on the average, the simulated curves follow that of RSS-IRC. Within the band of the QAM signal, however, there are variations caused by slight coherence between the sinusoid and QAM signal. Coaliasing is observed as sharp spikes in the response.

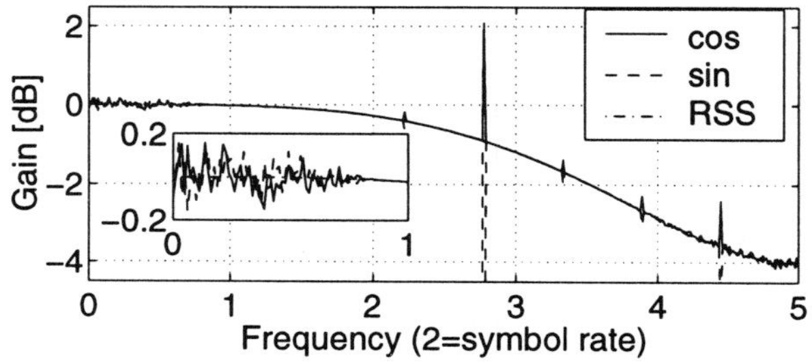

Figure 6.1. Simulation results for RSS-IRC for a 4-QAM signal.

7. CONCLUSIONS

The frequency domain input-output relationship of rational SRC was analysed. We defined a way to translate a magnitude response of a (system of) filter(s) to a lower sample rate to allow meaningful comparisons between filter properties and system specifications.

The very high computational complexity of frequency domain IRC was overcome in the case of RSS-IRC by using an optimized time domain algorithm based on the aliasing phenomenon. In experiments, the computation time of the overall image-combined magnitude response was reduced by 1Ð2 orders of magnitude.

8. REFERENCES

[1] R.E. Crochiere and L.R. Rabiner, Multirate Digital Signal Processing, Prentice Hall, Inc., Englewood Cliffs, New Jersey, USA, 1983.

[2] W. H. Yim, ÒDistortion Analysis for Multiplierless Sampling Rate Conversion Using Linear Transfer Functions,Ó *IEEE Signal Proc. Letters*, Vol. 8, No. 5, May 2001.

[3] D. Babic, V. Lehtinen, and M. Renfors, ÒDiscrete-time modeling of polynomial-based interpolation filters in rational sampling rate conversion,Ó in *Proc. IEEE Int. Symp. Circ. Syst.*, Bangkok, May 2003.

[4] C.W. Farrow, ÒA continuously variable digital delay element,Ó in *Proc. IEEE Int. Symp. Circ. Syst.*, Vol. 3, pp. 2641 -2645, Espoo, Finland, June 1988.

[5] J. Vesma and T. SaramŠki, ÒInterpolation filters with arbitrary frequency response for all-digital receivers,Ó in *Proc. IEEE Int. Symp. Circ. Syst.*, pp. 568-571, Atlanta, GA, USA, May 1996.

[6] D. Babic, J. Vesma, T. SaramŠki, and M. Renfors, ÒImplementation of the transposed farrow structure,Ó in *Proc. IEEE Int. Symp. Circ. Syst.*, Vol. 4, pp. 5 - 8, May 26-29, 2002.

[7] R.L. Reng and H.W. SchŸ§ler, ÒMeasurement of aliasing distortions and quantization noise in multirate systems,Ó in *Proc. IEEE Int. Symp. Circ. Syst.*, pp. 2328 -2331, San Diego, CA, USA, 3-6 May 1992.

Optimally Weighted Local Discriminant Bases

Kamyar Hazaveh, Student Member IEEE and Kaamran Raahemifar, Member IEEE

Electrical and Computer Engineering Dept., Ryerson University
Toronto, Ontario, Canada M5B 2K3

Abstract-Local discriminant bases method is a powerful algorithmic framework for feature extraction and classification applications that is based on supervised training. It is considerably faster compared to more theoretically ideal feature extraction methods such as principal component analysis or projection pursuit. In this paper an optimization block is added to original local discriminant bases algorithm to promote the difference between disjoint signal classes. This is done by optimally weighting the local discriminant basis using steepest decent algorithms. The proposed method is particularly useful when background features in the signal space show strong correlation with regions of interest in the signal, *i.e.* mammograms.
Keywords: Best basis, Local discriminant basis, Local feature extraction, Steepest decent method, Pattern recognition, Time-frequency analysis, Wavelet packet and Mammography.

1. Introduction

In 1994, Coifman and Saito [8], [9], developed the local discriminant bases (LDB) method as Wickerhauser's best-basis algorithm [1], [2] counterpart in feature extraction and signal classification applications. Both best basis and LDB methods use tree-structured collections of basis functions (atoms) called dictionaries. These are redundant sets of basis functions that are localized in time and frequency. Examples of time-frequency dictionaries include wavelet packets and local trigonometric bases. For a complete discussion of these basis function collections and their properties the interested reader is referred to [2], [20].

The best basis method can be tuned to adaptively choose the best set of basis functions so that the entropy of the signal coordinates is minimized. The same basis selection task is done by Karhunen-Loeve transform (KLT) [4] developed in 1965 that is better known as Principal Component Analysis (PCA). The advantage of best basis methods over KLT is twofold. By exploiting the structured nature of time-frequency dictionaries, Coifman and Wickerhauser developed a divide-and-conquer algorithm for best basis search with a computational cost of $O(n \log n)$ that is superior to KLT computational cost *i.e.* $O(n^3)$ [4]. Besides, in contrast to KLT, the best basis method is localized in time and frequency making it suitable for totally non-stationary process analysis. A local KLT has been developed by Coifman and Saito [5] but it suffers from the same high computational cost $O(n^3)$.

Later on, best basis method was used for de-noising [6], [7]. In 1994, Saito [8, Chapter 3] developed an algorithm for simultaneous de-noising and compression of signals. De-noising was the start of a new generation of best basis algorithms for signal classification problems. Compared to other supervised feature extraction solutions such as projection pursuit [23], [24], LDB is modest in the sense that it picks a set of good coordinates from a finite collection rather than a sequence of the absolutely best projections without constraints. The LDB concept has increasingly gained popularity and has been applied to a variety of classification problems including biomedical [7], geophysical [11], sonar [12], radar [10], [22] and military [13], [14], [22] application areas. In 1996, Coifman and Saito discovered a counterexample in which, LDB was unable to distinguish between two classes of synthetic signals. Thereafter several improved versions of the original LDB were developed [21].

The authors have noticed that an additional stage in LDB algorithm for assigning weights to selected basis functions helps to improve the accuracy of LDB method. This feature boosting is especially useful when background data has strong correlation with regions of interest in signal space under study. The method of gradient decent is employed to find the optimal weight for the selected basis functions. The effectiveness of the proposed optimization block is proved by experiment, *i.e.* near 40% misclassification improvement. Yoshida [25] has employed the same technique for improving the performance of matching pursuit method [26] for extraction of microcalcifications from mammograms. The usefulness of our proposed scheme is currently being tested on a collection of mammograms.

This is how this paper is organized. In section 2 the LDB algorithm and its improved version are reviewed. Section 3 is devoted to the development of the new idea of boosting features by using optimally weighted basis functions. Simulation results are given in section 4. Section 5 discusses conclusions and future work.

2. Local Discriminant Bases Algorithm

In this section we review the general problem of feature extraction. Suppose that we have a space $X \subseteq R^n$ of input signals and a space Y of class labels. The goal is to construct a classifier $d: X \rightarrow Y$ that assigns the correct class label to each input signal. The optimal classifier is known to be the so-called Bayes classifier. However, Bayes classifier is impossible to construct due to high dimensionality of the real signals [21]. Examples of high dimensional signals are medical X-ray tomography $(n=512^2)$, seismic signals $(n=4000)$ and a speech segment $(n=1024)$. Faced with the dimensionality and having such difficulty in constructing the Bayes classifier, the extraction of important features becomes essential. As Scott mentions in [3, Chapter 7] multivariate data in R^n are almost never n-dimensional and there often exists lower dimension structures of data. So based on the application whether compression or classification, the problem always has a lower intrinsic dimension. It is important to note that intrinsic dimension is an application-oriented quantity. Coifman and Wickerhauser basis selection scheme followed by a simple thresholding on the amplitude of the coefficients can result in significant signal dimension reduction. Saito's LDB algorithm helps to reduce the dimensionality of the problem for the feature extraction and classification tasks. The feature extractor in LDB is formulated as $d = g o \Theta_m o \Psi$, where Ψ is an orthogonal transformation, Θ_m is a projection operator into m most important coordinates and g is a standard classifier. By a dimension reducer engine such as LDB, *i.e.* $\Theta_m o \Psi$, we select and keep the most important basis vectors according to the classification task and discard the nonessential coordinates. The classifier, g, can be *Linear Discriminant Analysis* (LDA) [15], *Classification and Regression Trees* (CART) [16], *k-nearest neighbour* (k-NN) [17] or *artificial neural networks* (ANN) [18].

LDB method first decomposes available training signals from different classes in a time-frequency dictionary, which is a large collection of bases functions, *i.e.* wavelet packets or local trigonometric basis. Then signal energies are accumulated for each class separately to form a time-frequency energy distribution per class. We assume that there are only two classes of signals. The generalization of the method to more signal classes is straightforward.

Let us assume that ω is a typical basis function in a time-frequency dictionary. The time-frequency distribution energies of class *1* and class *2* along ω are designated by Γ_ω^1 and Γ_ω^2 respectively.

In the original LDB algorithm, the tree-structured time-frequency dictionary is pruned by using a discrimination measure such as

- *l^2-distance:*

$W(\Gamma_\omega^1, \Gamma_\omega^2) = \| \Gamma_\omega^1 - \Gamma_\omega^2 \|^2,$

- *Relative entropy (or Kullback-Leibler divergence):*

$D(\Gamma_\omega^1, \Gamma_\omega^2) = \Gamma_\omega^1 \log(\Gamma_\omega^1 / \Gamma_\omega^2),$

- *Symmetric relative entropy (or J-divergence):*

$J(\Gamma_\omega^1, \Gamma_\omega^2) = D(\Gamma_\omega^1, \Gamma_\omega^2) + D(\Gamma_\omega^2, \Gamma_\omega^1),$

In the first step LDB is the children nodes at the lowest part of the tree in Fig.1. Then the discrimination measures of each two children nodes are compared to their parent's. If the sum of the discrimination measures of the children nodes is higher than their parent's, we keep the children nodes. Otherwise, the parent node is chosen as the LDB. Once a complete basis (LDB) is selected, we further choose $m(<n)$ atoms from the LDB. A typical m would be $n/10$. The simplest way of choosing m atoms from a collection of n atoms is to sort them in the order of decreasing discrimination power and to retain the first m atoms. It's important to note that the functionality of LDB lies in the over-complete nature of

Fig.1. A binary tree structured dictionary

binary tree dictionaries [2]. This redundancy enables the introduction of different search algorithms with different discriminant measures to prune the binary tree in an optimal manner that is tailored to the particular application.

It is possible to construct a simple classification problem that is intractable by original LDB algorithm [21]. Therefore it is sometimes necessary to consider the distribution of expansion coefficients for individual coordinates. The original LDB algorithm measure is based on the differences of mean class energy of projections. It is possible to use a measure based on the differences between probability distribution functions (pdf). For a complete treatment of such problems that lead to the introduction of Type II, Type III and Type IV LDB algorithms the interested reader is referred to [21]. Type II LDB has been applied to real world applications [10], [12]. The initial problem of Type II would be the estimation of pdf's from the available database. Suppose that p_ω^1 and p_ω^2 are the pdf's of *class 1* and *class 2* signals in ω direction respectively. Here is a discrimination measures proposed by Saito [21].

- *Relative entropy (or Kullback-Leibler divergence):*

$D(p_\omega^1, p_\omega^2) = \int p_\omega^1 \log(p_\omega^1 / p_\omega^2)\, dx.$

3. Optimally Weighted Basis Functions

The basis functions selected by LDB capture some of the common features or background components as well. This becomes particularly important when background structures are highly correlated to the features of interest. In this section we introduce a method of boosting desired distinguishing features among different classes by optimally weighting the basis functions.

Let us assume that $\psi_\gamma, 1 \le \gamma \le N$ is the collection of basis functions that are obtained as the output of LDB algorithm. We study the effect of assigning a weight sequence such as $\omega_\gamma, 1 \le \gamma \le N$ to these basis functions. The goal is to promote the discrimination power of $\omega_\gamma \psi_\gamma, 1 \le \gamma \le N$ hence improving the classification accuracy. To this end, we minimize the difference between samples of each signal class with the so-called *teacher signals* [25]. Teacher signals for each class are the average of the signal class windowed at the regions of interest; therefore there are a limited number of them. The error function is given in (1). In this expression, α_γ^k's are the LDB coordinated of signal s^k. The corresponding teacher signal is designated by T^k.

$$E(\omega) = \frac{1}{K}\sum_k\sum_x (S^k(\omega, x) - T^k(x))^2 \quad (1)$$

$$S^k(\omega, x) = \sum_\gamma \alpha_\gamma^k \omega_\gamma \psi_\gamma(x) \quad (2)$$

In (1) and (2) k ranges over the samples used in the optimization phase whereas x ranges over time or pixels in an image processing application and K is the number of samples used for optimization. The error function $E(\omega)$ can be minimized by a gradient decent algorithm to yield an optimal set of weights that maximally separate different signal classes. The partial derivative of $E(\omega)$ in terms of a weight ω_γ is given by (3). Therefore the weight update formula for gradient decent algorithm will be as in (4).

$$\frac{\partial E}{\partial \omega_\gamma} = \frac{2}{K}\sum_k\sum_x (S^k(\omega, x) - T^k(x))\alpha_\gamma^k \psi_\gamma(x) \quad (3)$$

$$\omega_\gamma \rightarrow \omega_\gamma - \eta\frac{\partial E}{\partial \omega_\gamma} \quad (4)$$

Here η is a user-defined learning rate. For the purpose of illustration we adapt example 5.2 form [9]. The problem is to classify three classes of the signals generated by the following process.

$$c(i) = (6 + \varsigma).\chi_{[a,b]}(i) + \varepsilon(i)$$

$$b(i) = (6 + \varsigma).\chi_{[a,b]}(i).\frac{i-a}{b-a} + \varepsilon(i)$$

$$f(i) = (6 + \varsigma).\chi_{[a,b]}(i).\frac{b-i}{b-a} + \varepsilon(i)$$

a is an integer-valued uniform random variable on the interval [16,32]. $b-a$ also obeys an integer-valued uniform distribution on [32,96]. ς and ε are the standard normal variates, and $\chi_{[a,b]}(i)$ is the characteristic function on $[a,b]$. $c(i)$ is called the 'cylinder' class whereas $b(i)$ and $f(i)$ are known as 'bell' and 'funnel' classes. The statistical averages of the each class in the above process over a set of 300 training signals are depicted in Fig.2. The statistical average of each class will be windowed around the regions of interest to be used as teacher signal in the proposed optimization block. The same procedure can be applied to obtain teacher images that locate microcalcifications in mammograms.

4. Optimally Weighted LDB Simulation Results

In the beginning, local discriminant analysis was used to capture LDB for a set of 300 training signals described above. The wavelet packet was generated by a Coiflet 4 mother wavelet. Symmetric relative entropy (J-divergence) was used as discrepancy measure in our study. Fisher's Linear Discriminant Analysis [15] was fixed as the classifier in this study. Two sets of teacher signals were generated by windowing the statistical averages in the intervals [40,55] and [40,60] to study the effect of window size. These two sets of teacher signals are referred to as 15-point and 20-point windows in all the graphs. The number of samples used in optimization phase was fixed to 21 equally distributed between three classes (K=21). Another set of 300 signals was generated as the test bed signals for examining the effectiveness of optimization process. Fig.3 shows the results of using the first 12 most discriminative LDB vectors. The weight update process was performed on 15 and 20 first most discriminative LDB vectors as opposed to 128 to save processor's time. The experiment continued up to 100 iterations. Fig.4 and Fig.5 show similar results for classification based on 10 and 8 most discriminative LDB vectors respectively.

5. Conclusions and Future Work

The original and improved versions of local discriminant bases algorithm are reviewed. An additional feature boosting stage is added to the algorithm. The authors have applied the new technique on classification of synthetic data. Optimal weighting of local discriminant basis improved the misclassification error with reasonable number of iterations. The optimization block can further improve the efficiency and accuracy of the local discriminant basis algorithm specially when the number of selected LDB vectors is small and background data has strong correlation with signal space. The explicit improvement in accuracy proves the process worthwhile as a one-time optimization done right after training phase. Authors are studying the effect and importance of varying different parameters in the experiment such as the learning rate η, number of sample signals used in the optimization phase (K) as well as other optimization algorithms, different wavelet packet dictionaries and discrepancy measures. The usefulness of the feature-boosting module is being studied in classification of mammograms where background structure shows considerable correlation with microcalcification patterns.

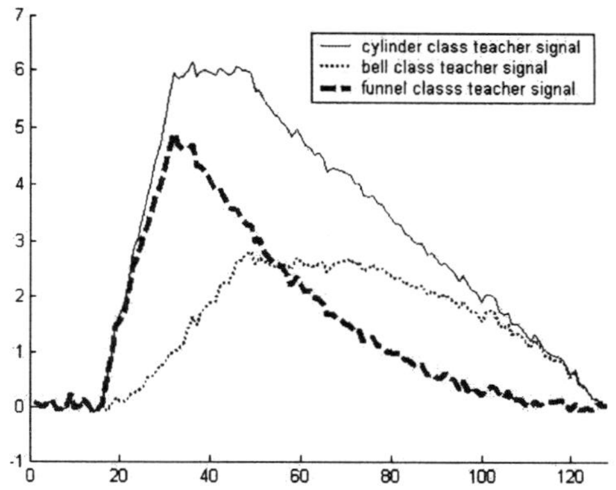

Fig.2. Statistical averages of different signal classes.

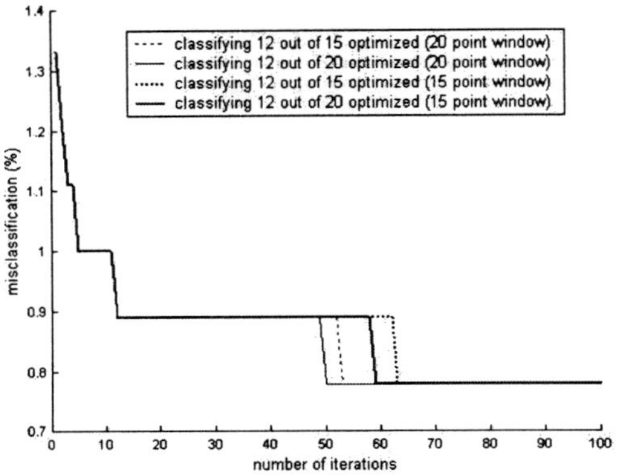

Fig.3. Thick curves show the misclassification rate when teacher signals were windowed at [40,55]. Others are obtained when windowing is done at [40,60].

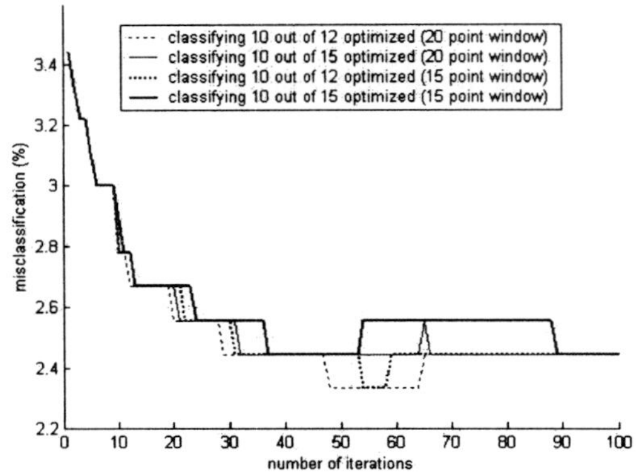

Fig.4. Repeating the experiment with 10 LDB vectors for classification.

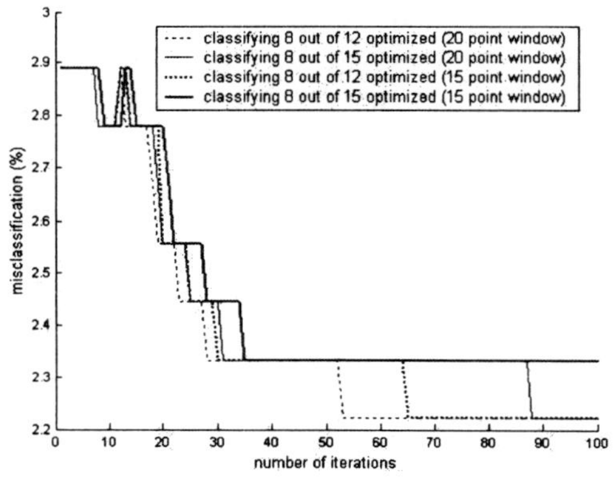

Fig.5. Repeating the experiment with 8 LDB vectors for classification.

6. References

[1] R. R. Coifman and M.V. Wickerhauser, "Entropy-based algorithms for best basis selection," *IEEE Trans. Inform. Theory* 38 (1992), no.2, pp 713-719.

[2] M.V. Wickerhauser, *Adapted Wavelet Analysis from Theory to Software*, A K Peters, Ltd. Wellesley, MA, 1994.

[3] D. W. Scott, *Multivariate Density Estimation: Theory, Practice, and Visualization*, John Wiley & Sons, New York, 1992.

[4] S. Watanabe, "Karhunen-Loeve Expansion and Factor Analysis: Theoretical Remarks and Applications," in *Trans. 4th Prague Conf. Inform. Theory*, Publishing House of Czechoslovak Academy of Sciences, pp 635-660, 1965.

[5] R. R. Coifman, N. Saito, "The Local Karhunen-Loeve Bases," *Time-Frequency and Time-Scale Analysis, Proceedings of the IEEE-SP International Symposium on*, pp 129 –132, 1996.

[6] P. Qua, Z. Lei, Z. Hongcai, D. Guanzhong, "Adaptive wavelet based spatially de-noising," Signal *Processing Proceedings, 1998 Fourth International Conference on*, pp 486-489,1998.

[7] R. R. Coifman and M.V. Wickerhauser, "Adapted waveform "de-noising" for medical signals and images," IEEE Engineering in Medicine and Biology Magazine, v.14, n.5, pp 578 –586, 1995.

[8] N. Saito, *Local Feature Extraction and Its Application Using a Library of Bases*, Ph.D. Thesis, Dept. of Math, Yale University, New Haven, CT 06520 USA, Dec. 1994.

[9] N. Saito and R. R. Coifman, "Local discriminant bases and their applications," J. Mathematical Imaging and Vision 5 (1995), no.4, pp 337-358, Invited paper.

[10] G. Kronquist and H. Storm, "Target Detection with Local Discriminant Bases and Wavelets," *Proc. of SPIE*, v.3710, pp 675-683, 1999.

[11] N. Saito, R. R. Coifman, "Extraction of Geological Information from Acoustic Well-Logging Waveforms Using Time-Frequency Wavelets," *Geophysics*, v.62, n.6, 1997.

[12] L. S. Rogers, C. Johnston, "Land Use Classification of SAR Images Using a Type II Local Discriminant Basis for Preprocessing," *Proc. of IEEE Conference on Acoustics, Speech and Signal Processing*, v.5, pp 2729-2732, 1998.

[13] M. L. Cassabaum, H. A. Schmitt, H. W. Chen, J. G. Riddle, "Application of Local Discriminant Bases Discrimination Algorithm for Theater Missile Defense," *Proc. of SPIE*, v.4119, pp 886-893, 2000.

[14] M. L. Cassabaum, H. A. Schmitt, H. W. Chen, J. G. Riddle, "Fuzzy Classification Algorithm Applied to Signal Discrimination for Navy Theater Wide Missile Defense," *Proc. of SPIE*, v.4120, pp 134-145, 2000.

[15] R. A. Fisher, "The Use of Multiple Measurements in Taxonomic Problems," *Ann. Eugenics*, Vol. 7, pp 179-188, 1936.

[16] L. Breiman, J. H. Friedman, R. A. Olshen, and C. J. Stone, *Classification and Regression Trees*, Chapman and Hall, Inc., New York, 1993.

[17] K. Fukunaga, *Introduction to Statistical Pattern Recognition*, 2nd ed., Academic Press, San Diego, 1990.

[18] B. D. Ripley, "Statistical aspects of neural networks," *Networks and Chaos: Statistical and Probabilistic Aspects*, O. E. Barndorff-Nielson, J. L. Jensen, D. R. Cox, and W. S. Kendall, eds., pp 40-123, Chapman and Hall, Inc., New York, 1993.

[19] P. Frossard, P. Vandegheynst, "Redundancy in non-orthogonal transforms," *Proc. of IEEE International Symposium on Inform. Theory*, n.01CH3725, p 196, 2001.

[20] Mallat, S.G., *A Wavelet Tour of Signal Processing*, San Diego Toronto: Academic Press, 1998.

[21] N. Saito and R. R. Coifman, "Improved Local Discriminant Bases Using Empirical Probability Density Estimation," Proc. *on Statistical Computing*, American Statistic. Assoc., 1996.

[22] C. M. Spooner, "Application of local discriminant bases to HRR-based ATR," *Signals, Systems and Computers, Conference Record of the Thirty-Fifth Asilomar Conference on*, v.2, pp 1067 -1073, 2001.

[23] J. H. Friedman and J. W. Tukey, "A projection pursuit algorithm for exploratory data analysis," *IEEE Trans. Comput.* 23, pp 881-890, 1974.

[24] J. B. Kruskal, "Toward a practical method which helps uncover the structure of a set of multivariate observations by finding the linear transformation which optimizes a new 'index of condensation'", *Statistical Computation* (R. C. Milton and J. A. Nelder, eds.), Academic Press, New York, pp 427-440, 1969.

[25] H. Yashida, "Matching pursuit with optimally weighted wavelet packets for extraction of microcalcifications in mammograms," *Applied Signal Processing*, no.5, pp 127-141, 1998.

[26] S. G. Mallat, Z. Zhang, "Matching pursuits with time-frequency dictionaries," *IEEE Trans. Signal Processing*, vol. 41, pp. 3397-3415, 1993.

REDUCED ORDER RLS POLYNOMIAL PREDISTORTION

Minglu JIN, Sooyoung Kim Shin, Deockgil. Oh, and Jaemoung Kim

Dept. of Broadband Wireless Communication, ETRI

Daejeon, Korea, mljin@etri.re.kr

ABSTRACT

In this paper, a modified recursive least square (RLS) based polynomial predistortion algorithm is proposed by reducing the order of the polynomials used in the RLS algorithm. The basic idea of the reduced order RLS based polynomial predistortion algorithm is to approximate the target function with two simpler functions (with a smaller number of coefficients) instead of a single function. A complex function is decomposed into two simpler functions so that we can compute the RLS algorithm with much less complexity. The proposed reduced order RLS polynomial predistortion show superior performance compared to the conventional RLS polynomial predistortion with the same number of coefficients or with the same computational complexity.

1. INTRODUCTION

Power efficiency and spectrum efficiency are the main concern in modern digital communication systems. The spectral efficiency requires the system to be more linear while the power efficiency requires the system using an amplifier with nonlinear characteristics. Therefore, how to tradeoff between these two conflict factors is a challenge to communication engineers, and the linearization of the power amplifier is the most effective method to overcome this difficulty. Over the years, a number of linearization technologies have been developed and predistortion has been the most common approach developed in new systems today.

Basically, these techniques aim to introduce "inverse" nonlinearities that can compensate the AM/AM and AM/PM distortions generated by the nonlinear amplifier. The predistorter uses a memoryless nonlinear device placed between the shaping filter and the high power amplifier (HPA). This nonlinear device can be designed digitally using a mapping predistorter (a look-up table) or a polynomial function based on Cartesian or polar representation.

The adaptive polynomial predistorter presented in [1], shows a good convergence rate and a high reduction in spectrum spreading. However, the proposed RLS adaptation method is complex and requires a huge number of computations. An evaluation of a predistortion technique should take into account both its performance and computational complexity.

To reduce the complexity of polynomial predistortion, H. Besbes proposed least mean square (LMS) based polynomial predistorter [2], but the convergence rate is very slow compared to the RLS based polynomial predistortion.

The computational complexity of the RLS algorithm is proportional to the number of coefficients to be modified simultaneously and its computational load is about $O(N^2)$ if there are N coefficients to be modified. Because of the property of the square function, a smaller value of N result in high decrease in computational complexity of the RLS algorithm. However, a smaller value of N also results in performance decrease of the RLS based polynomial predistortion.

Although there are a number of Fast RLS algorithms which can result in $O(N)$ computational complexity, this algorithm only work for the data with shifting input.[3]

In this paper, we shall propose a novel predistortion scheme with the reduced complexity RLS adaptation algorithm. We decompose the predistortion polynomial into two sub functions, and at each iteration time we alternatively perform RLS adaptation operation with the coefficients of each sub function. In section 2, the conventional RLS based polynomial predistortion system model including a HPA model is introduced, and then a new reduced order RLS polynomial predistortion is proposed. The performance simulation results of the proposed predistortion scheme are given in section 3 and finally conclusion remarks are addressed in the last section.

2. RLS BASED POLYNOMIAL PREDISTORTION

A general diagram of an RLS based adaptive polynomial predistortion scheme is shown in Figure 1 [1].

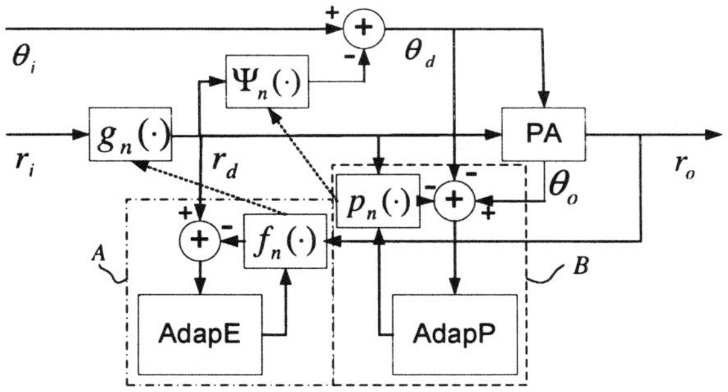

Figure 1. Block diagram of adaptive polynomial predistortion.

The signals in Figure 1 have the relationships as follows:

$$r_d = g_n(r_i) \quad (1)$$

$$\theta_d = \theta_i - \Psi_n(r_i) \quad (2)$$

$$r_o = G(r_d) \quad (3)$$

$$\theta_o = \Phi(r_d) \quad (4)$$

The functions $G(.)$ and $\Phi(.)$ denote the amplitude and phase transfer function of the amplifier, and $g_n(\cdot)$ and $\Psi_n(\cdot)$ denote the amplitude and phase transfer function of the predistortion and used in order to invert the nonlinearity introduced by the amplifier, i.e

$$G(g_n(r_i)) = Kr_i \quad (5)$$

$$\Phi(r_d) - \Psi_n(r_d) = \theta_i \quad (6)$$

The functions $g_n(\cdot)$ and $\Psi_n(\cdot)$ are generally modeled by

$$g_n(r_i) = \sum_{k=1}^{N} a_{g,k,n} r_i^{k-1} \quad (7)$$

$$\Psi_n(r_d) = \sum_{k=0}^{M-1} a_{\Psi,k,n} r_d^{k} \quad (8)$$

where M and N are the order of polynomial and $a_{g,k,n}$ and $a_{\Psi,k,n}$ are coefficients of the polynomial functions. The functions $f_n(\cdot)$ and $p_n(\cdot)$ is used in adaptation algorithm and defined as follows:

$$f_n(r_0) = \sum_{k=1}^{N} a_{f,k,n} r_0^{k-1} \quad (9)$$

$$P_n(r_d) = \sum_{k=0}^{M-1} a_{p,k,n} r_d^{k} \quad (10)$$

It is noted that the functions in (7) and (8) are almost the same as in (9) and (10) with the relationship in (11) and (12)

$$\alpha_{g,k,n} = \alpha_{f,k,n-1}, k = 1,2,\ldots\ldots, N \quad (11)$$

$$\alpha_{\varphi,k,n} = \alpha_{p,k,n-1}, k = 0,1,\ldots. M-1 \quad (12)$$

The RLS adaptation algorithm is used to modify these coefficients by minimizing the errors as follows:

$$e_r(l) = r_d(l) - f_n(r_o(l)) \quad (13)$$

$$e_\theta(l) = \theta_o(l) - \theta_d(l) - P_n(r_d(l)) \quad (14)$$

where l represents the iteration time index.

The RLS algorithm have fast convergence rate, but have very expensive computational load. The basic principle of the RLS based polynomial predistortion is to approximate the inverse transfer functions of the amplifier with polynomial functions based on RLS adaptation algorithm.For Nth order polynomial functions, the adaptation operation have about $O(N^2)$ computational complexity and thus the whole computational complexity for RLS based polynomial predistortion is $O(N^2)$ + $O(M^2)$.

In order to reduce the computational load, we decompose the Nth order polynomial function $f_n(x)$ into two polynomial functions as follows:

$$\begin{aligned} f_n(x) &= a_1x + \cdots + a_{N/2}x^{N/2} + a_{N/2+1}x^{N/2+1} + \cdots + a_Nx^N \\ &= f_n^{(1)}(x) + f_n^{(2)}(x) \end{aligned} \quad (15)$$

where

$$f_n^{(1)}(x) = a_1 x + \cdots + a_{N/2} x^{N/2} \quad (16)$$

$$f_n^{(2)}(x) = a_{N/2+1} x^{N/2+1} + \cdots + a_N x^N \quad (17)$$

By this way, the adaptation block A in Figure 1 modified as shown in Figure 2. The error signal used in adaptation process is the same as in Figure 1, but the error signal is feedback to AdapE1 block or AdapE2 block alternatively. At every iteration time, we process only one adaptation process modify the corresponding coefficients while the other coefficients remain unchanged. Therefore, computational load by the RLS algorithm to modify the coefficients of $f_n(x)$ is reduced to $O(N^2/4)$.

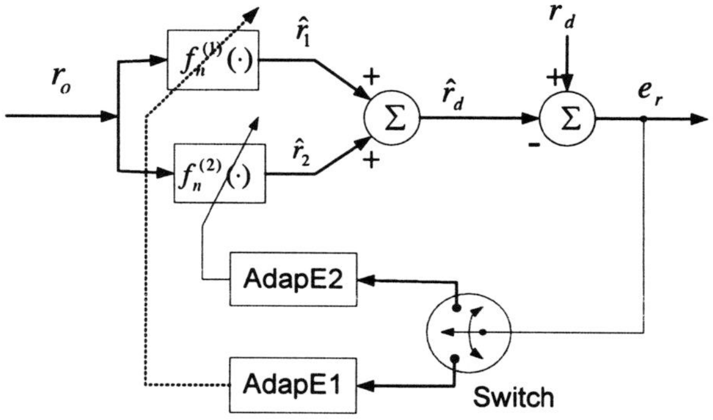

Figure 2. Modification of Block 1 in Figure 1.

The polynomial function $P_n(x)$ can also be decomposed into two sub functions $P_n^{(1)}(x)$ and $P_n^{(2)}(x)$,and the adaptation block B in Figure 1 can be modified in the same manner as block A in Figure 1, and thus whole computational load can be reduced from $O(N^2)$ + $O(M^2)$ to $O((N/2)^2)$ + $O((M/2)^2)$ and we refer to this scheme as the reduced order RLS polynomial predistortion.

The idea of the reduced order RLS polynomial predistortion algorithm is to approximate a function with two simpler functions (with a smaller number of coefficients), or decompose a complex function into two simpler functions so that the RLS operation needs less computation.

In the following section, we investigate the performance of the reduced order RLS polynomial predistortion compared to that of the conventional RLS based polynomial predistortion.

3. COMPUTER SIMULATION

To investigate the performance of the reduced order RLS predistortion algorithm given above, we have performed the computer simulation. To measure the convergence performance, often the normalized mean square error (MSE) was used which is defined as follows

$$MSE = \frac{E[| v_o - v_i |^2]}{E[| v_o |^2]}, \quad (18)$$

where $v_o = r_o e^{j\theta_o}$ represents the output of the amplifier and $v_i = r_i e^{j\theta_i}$ denotes the input to the predistortion, with uniform envelope distribution in the interval [0, 1] and uniform phase distribution in the interval [0, 2π].

In the simulation a normalized Saleh model of the amplifier [4] is used as follows:

$$G(r_d) = \frac{2r_d}{1 + r_d^2} \quad (19)$$

$$\Phi(r_d) = \frac{\pi}{3} \frac{r_d^2}{1 + r_d^2} \quad (20)$$

Because of symmetry property of AM/AM and AM/PM transfer functions of amplifier, we used an odd polynomial function to approximate the inverse AM/AM transfer function of amplifier and used an even polynomial function to approximate the inverse AM/PM transfer function of amplifier

We first performed simulation on the reduced order RLS polynomial predistortion with N=7 and M=6 compared to that of the conventional RLS polynomial predistortion with the same computational load, that is, N=3 and M=2. Figure 3 shows the error performance of the proposed reduced order RLS polynomial predistortion compared with that of the conventional scheme. The converged predistortion polynomial used in Figure 3 are in (21)~(26), where $f_n(x)$ and $P_n(x)$ are for the conventional scheme and $f_n^{(1)}(x)$, $f_n^{(2)}(x)$, $P_n^{(1)}(x)$ and $P_n^{(2)}(x)$ are for the proposed scheme.

$$f_n(x) = 0.31x^3 + 0.45x \quad (21)$$

$$P_n(x) = 0.72x^2 + 0.01 \quad (22)$$

$$f_n^{(1)}(x) = -0.52x^3 + 0.57x \quad (23)$$

$$f_n^{(2)}(x) = -0.68x^7 + 1.41x^5 \quad (24)$$

$$P_n^{(1)}(x) = 0.98x^2 + 0.002 \quad (25)$$

$$P_n^{(2)}(x) = 0.36x^6 - 0.74x^4 \quad (26)$$

The coefficients value in (21) ~ (26) is converged after 1000 iterations in the simulation.

In Figure 3 the symbol "Old MSE" and "New MSE" stands for the MSE error of the conventional method and the proposed method respectively, and the symbol "AE" and "PE" represents the envelope error and phase error between input to the predistorter and output of the amplifier.

It is noted that, there is a little improvements in terms of MSE in envelope but the improvements in phase is evident (about an order). We believe this improvement on approximation performance results from the added higher order terms in predistortion polynomial. In other words, in the reduced order scheme we used the 7th order and the 6th order predistortion polynomial function to approximate the inverse AM/AM and AM/PM transfer functions of amplifier, respectively. While we used the 3rd and the 2nd order predistortion polynomial functions in the conventional scheme. Nevertheless, the computational load of the reduced order RLS predistortion is the same as that of the conventional RLS predistortion.

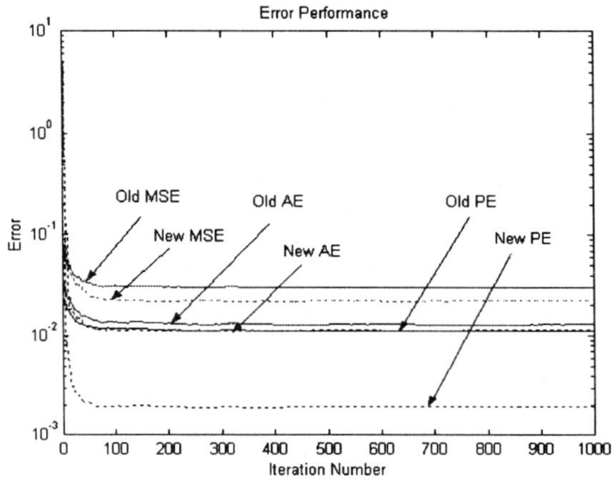

Figure 3. Error Performance.

Now, we compare the MSE performance by increasing the order of the conventional scheme up to the one used in the reduced scheme. We found two predistortion polynomial pairs of $g(x)$ and $\Psi(x)$ with 5th and 4th order (N=5, M=4), and 7th and 6th order (N=7, M=6), respectively. The converged results of the polynomials are in (27)-(30).

$$f_n(x) = 0.39x^5 - 0.09x^3 + 0.53x \quad (27)$$

$$P_n(x) = -0.53x^4 + 0.96x^2 + 0.002 \quad (28)$$

$$f_n(x) = 0.71x^7 - 0.70x^5 + 0.38x^3 + 0.48x \quad (29)$$

$$P_n(x) = 0.71x^6 - 0.85x^4 + 1.03x^2 + 0.0003 \quad (30)$$

Figure 4 compares the performance of the reduced scheme in (23)-(24) with those of the conventional scheme in (21)~(22) and (27)-(30). In Figure 4, "Old (21)~(22)" stands for the conventional method based on the polynomial functions in (21)-(22), "Old (27)~(28)" stands for the conventional method based on the polynomial functions in (27)-(28), and "Old (29)~(30)" stands for the conventional method based on polynomial functions in (29)-(30), and "New (23)~(26)" stands for the proposed method based on polynomial functions in (23)-(26) as in Figure 3.

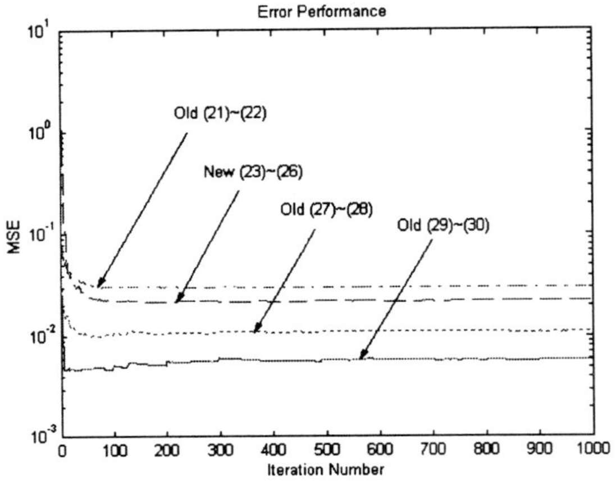

Figure 4. Comparison of Error Performance.

As seen in Figure 4, the proposed method results in worse performance than that of the conventional methods with the increased order of polynomial functions. This is partly because we lost the correlation information between the two decomposed sub-functions, and partly because of the price of the reduced computational complexity.

To investigate the bit error rate (BER) performance, we have performed simulation on the 16-QAM OFDM environment with 512 sub carriers. Figure 5 Shows the BER performances of the reduced order RLS polynomial predistortion and the conventional RLS polynomial predistortion at different output Backoff.

In Figure 5, we used the same legend as in Figure 4 for the conventional scheme and for the proposed scheme with different output Backoff value.

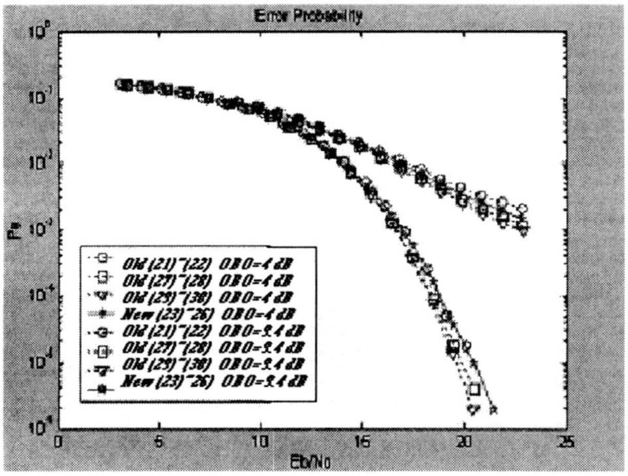

Figure 5. BER Performances at Different output Backoff.

As shown in Figure 5, at both low and high output Backoff value, the performance of the reduced order RLS polynomial predistortion superior to that of the conventional RLS polynomial predistortion with same computational complexity, but worse than that of the conventional RLS polynomial predistortion that based on higher order polynomial function, but the performance difference is not big. Thus, we can get almost the same BER performance by using the proposed method compared to the conventional scheme with much less complexity.

4. CONCLUSIONS

We proposed a new polynomial predistortion scheme with reduced computational load in the RLS adaptation algorithm. By decomposing the polynomial used in the predistortion process, we could reduced the computational burden in the RLS algorithm.

The reduced order RLS polynomial predistortion have superior performance compared to that of the conventional RLS polynomial predistortion under the same number of coefficients which is modified by RLS adaptation algorithm, but with the same computational complexity.

It would be also interested to study on various combination methods of sub-polynomials and their MSE performances.

5. REFERENCES

[1] Mohammad Ghader. "Adaptive linearization of efficient high power amplifiers using polynomial predistortion with global optimization," *Thesis of University of Saskatchewan*, November 1994.

[2] H. Besbes, T. Le-Ngoc and H. Lin, "A fast adaptive polynomial predistorter for power amplifiers," *Global Telecommunications Conference*, 2001. GLOBECOM '01. IEEE , Volume: 1 ,pp659-663, 2001.

[3] Simon Haykin. "Adaptive Filter Theory," Forth Edition, Prentice Hall, 2002.

[4] Saleh, A. A. M., "Frequency-independent and frequency-dependent nonlinear models of TWT amplifiers," *IEEE Trans. On Communications*, Vol. COM-29, November 1981, pp1715-1720

WAVELET-TRANSFORM-BASED STRATEGY FOR GENERATING NEW CHINESE FONTS

Jiu-chao Feng[†], *Chi K. Tse*[#] *and Yuhui Qiu*[‡]

[†]School of Electronic & Information Engineering, South China University of Technology, Guangzhou 510641, China
[#]Department of Electronic & Information Engineering, Hong Kong Polytechnic University, Hong Kong, China
[‡]Department of Computer Science, Southwest China Normal University, Chongqing 400715, China

ABSTRACT

Based on the cubic B-spline curve, new Chinese fonts are generated by wavelet transforms in this paper. The outlines of Chinese fonts are first transformed into B-spline curves. Then, using wavelet transforms, the control points of each curve are decomposed into hierarchies containing the detailed features of the Chinese fonts. Using the synthesis procedure of wavelet transforms, new fonts can be generated by modifying details at selected hierarchies.

1. INTRODUCTION

Being a mathematical tool for hierarchical function decompositions, wavelet transformation allows a function to be described in terms of coarse overall shape, plus features that range from broad to narrow. Regardless of whether the function of interest is an image, curve or surface, wavelets offer an elegant technique for representing the levels of details present [1]. Although wavelets have their roots in approximation theory [2] and signal processing [1], they have been recently applied to solve problems in computer graphics, for example, image editing, image compression, etc.[3]. In this paper we use the cubic B-spline wavelet transform to generate new Chinese fonts. The basic idea is to describe curves by hierarchies and details under a multiresolution analysis (MA). Making use of the synthesis procedure of the B-spline wavelet transforms for some modified and/or combined details, new fonts can be generated.

2. B-SPLINE WAVELETS AND THEIR CONSTRUCTIONS

Consider a function in some approximation space Γ^n. Let us assume that we have the coefficients of the function in terms of some scaling function basis. We can write these coefficients as a column vector $\mathcal{C}^n = [c_0^n\, c_1^n \cdots c_{m^n-1}^n]^T$, where n is the level of the transform (as is customarily used in the wavelet literature) and m defines the size (fineness) of the decomposition [4]. Essentially, these coefficients can be viewed as the control points of a curve in $\mathbf{R}^2$, and for level n, there are m^n control points. A low-resolution version $\mathcal{C}^{n-1}$ of $\mathcal{C}^n$ can be created by using linear filtering and down-sampling on the m^n entries of $\mathcal{C}^n$. This process can be expressed as

$$\mathcal{C}^{n-1} = \mathcal{A}^n \mathcal{C}^n, \tag{1}$$

where $\mathcal{A}^n$ is an $m^{n-1} \times m^n$ matrix of constants. Since $\mathcal{C}^{n-1}$ contains fewer entries than $\mathcal{C}^n$, this filtering process clearly loses some details. For many choices of $\mathcal{A}^n$, it is possible to capture the lost details in another column vector $\mathcal{D}^{n-1}$, which can be computed by

$$\mathcal{D}^{n-1} = \mathcal{B}^n \mathcal{C}^n, \tag{2}$$

where $\mathcal{B}^n$ is an $m^{n-1} \times m^n$ matrix of constants related to $\mathcal{A}^n$. The pair of matrices $\mathcal{A}^n$ and $\mathcal{B}^n$ are called *analysis filters*. If $\mathcal{A}^n$ and $\mathcal{B}^n$ are chosen appropriately, then the original $\mathcal{C}^n$ can be recovered from $\mathcal{C}^{n-1}$ and $\mathcal{D}^{n-1}$ by using the matrices $\mathcal{P}^n$ and $\mathcal{Q}^n$, i.e.,

$$\mathcal{C}^n = \mathcal{P}^n \mathcal{C}^{n-1} + \mathcal{Q}^n \mathcal{D}^{n-1}. \tag{3}$$

Recovering $\mathcal{C}^n$ from $\mathcal{C}^{n-1}$ and $\mathcal{D}^{n-1}$ is called *synthesis* or *reconstruction*. In this context, $\mathcal{P}^n$ and $\mathcal{Q}^n$ are called *synthesis filters*. Note that the procedure for splitting $\mathcal{C}^n$ into a "low-resolution part" $\mathcal{C}^{n-1}$ and a "detailed part" $\mathcal{D}^{n-1}$ can be applied recursively to the low-resolution version $\mathcal{C}^{n-1}$. Thus, the original coefficients can be expressed as a hierarchy of lower resolution versions $\mathcal{C}^0$, $\mathcal{C}^1 \cdots$, $\mathcal{C}^{n-1}$ and details $\mathcal{D}^0$, $\mathcal{D}^1 \cdots$, $\mathcal{D}^{n-1}$. The decomposition and synthesis processes are shown in Fig. 1.

Consider a curve which can be represented by

$$F^n(u) = \Omega^n(u)\mathcal{C}^n, \tag{4}$$

where $u \in [0, 1]$, and $\Omega^n(u)$ is a scaling function row matrix of level n, consisting of $(\omega_0^n(u), \omega_1^n(u), \cdots, \omega_{m^n-1}^n(u))$. In this paper, $\omega_j^n(u)$ is the B-spline basis function and $F^n(u)$ is the spline curve. Let Γ^j be the vector space spanned by $(\omega_0^j(u), \omega_1^j(u), \cdots, \omega_{m^j-1}^j(u))$, where j is an integer $(j \in [0, n])$. It should be noted that $\mathcal{P}^j$ satisfies

$$\Omega^{j-1}(u) = \Omega^j(u)\mathcal{P}^j. \tag{5}$$

This work was supported in part by the Applied Fundamental Research fund of Chongqing Science and Technology Committee, China.

0-7803-7761-3/03/$17.00 ©2003 IEEE

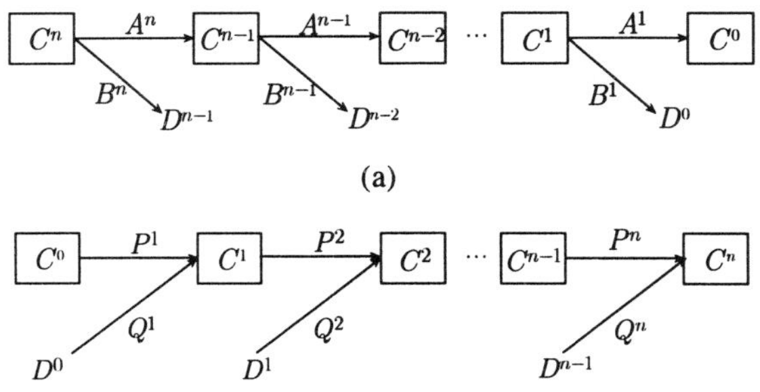

(b)

Figure 1: Wavelet transformation. (a) Decomposition process; (b) synthesis process.

Equation (5) indicates that each scaling function at level $j-1$ must be expressible as a linear combination of the finer scaling functions at level j, and that the subspace Γ^j is nested by $\Gamma^0 \subset \Gamma^1 \cdots \subset \Gamma^n$. Since the wavelet space Σ^{j-1} is the orthogonal complement of Γ^{j-1} in Γ^j, we can write the wavelets $\Psi^{j-1}(u)$ as linear combinations of the scaling functions $\Omega^j(u)$, i.e.,

$$\Psi^{j-1}(u) = \Omega^j(u)\mathcal{Q}^j(u) \tag{6}$$

where $\Psi^{j-1}(u) = (\psi_0^{j-1}(u), \cdots, \psi_{m^{j-1}-1}^{j-1}(u))$. Note that Eqs. (5) and (6) can be expressed, by using block-matrix notation, as

$$[\Omega^{j-1}|\Psi^{j-1}] = \Omega^j[\mathcal{P}^j|\mathcal{Q}^j]. \tag{7}$$

Combining (5), (6) and (7), we have

$$\left[\frac{\mathcal{A}^j}{\mathcal{B}^j}\right] = [\mathcal{P}^j|\mathcal{Q}^j]^{-1}. \tag{8}$$

The MA framework presented above is very general. In practice, one often has the freedom to design an MA which is specifically suited to a particular application. Here, for the purpose of this paper, we summarize the construction of B-spline wavelets as follows:

- Select the endpoint interpolating cubic B-spline basis functions as the scaling functions. For any j ($j \in [0, n]$), construct the $\mathcal{P}^j$ in Γ^j.

- Determine the subspace Ψ^{j-1} in the light of the inner product formula $\langle \omega_k^{j-1} | \psi_l^{j-1} \rangle = 0$ (for all k and l). Determine $\mathcal{Q}^j$ by using

$$[\langle \Omega^{j-1} | \Omega^j \rangle]\mathcal{Q}^j = 0. \tag{9}$$

- Compute the left-hand part of Eq. (8) by using the computed $\mathcal{P}^j$ and $\mathcal{Q}^j$.

3. SYNTHESIS OF CHINESE FONTS

A single Chinese font consists of many closed outlines, each of which can be represented by a B-spline curve. Therefore, a Chinese font can be described hierarchically by decomposition and synthesis of wavelet transforms for the selected B-spline curves.

3.1. Control Points of the B-spline Curve

A B-spline curve is described by a set of control points. To represent a Chinese font by B-spline curves, we need to decompose the control points hierarchically and in an exponential fashion. Here, the number of sections of a B-spline curve required to realize this operation for the jth level (layer) is 2^j. Hence, the number of the control points is also 2^j, due to the closed curve.

In our implementation, to achieve uniformly spaced B-spline, we choose $k = 2^j + d - 1$ ($d = 4$) to produce 2^j equally spaced interior intervals, thus giving $2^j + d$ B-spline basis functions for degree d and level j. Specifically, in this paper, we re-sample the B-spline curve in order to satisfy the requirements for the sections. The exact procedures for extending control points are summarized as follows:

- Given N points, compute a B-spline curve $F^n(u) = (f_1^n(u), f_2^n(u), \cdots, f_N^n(u))$, where $f_i^n(u)$ is the ith section B-spline curve expressed as

$$f_i^n(u) = \sum_{j=1}^{4} c_{i+j-2}^n \mathcal{N}_{j,4}(u), \tag{10}$$

  where $i = 1, 2, \cdots, N$, $u \in [0, 1]$, and $\mathcal{N}_{j,4}(u)$ is the cubic B-spline basis function.

- According to the actual length of the outline, determine the number of levels (layers) for decomposing, which usually requires that $N \leq L$, $L = 2^n + 3$.

- Re-sample $F^n(u)$ with an interval of $\Delta t = \frac{N}{2^n - 1}$, and the number of points is accordingly assigned as $2^n - 1$.

- Determine the new control point c_{i-1}^n by (10).

3.2. Wavelet Transform of Fonts: Examples

Fig. 2 shows the transformed results of four different fonts for the Chinese character "计", namely, 'Song Ti', 'Fang Song Ti', 'Kai Shu' and 'Hei Ti', each of which is decomposed into seven layers. The left most is the original, the second from left is the layer 1 image which has removed $\mathcal{D}^{n-1}$, the third from left is the layer 2 image which has removed $\mathcal{D}^{n-1}$ and $\mathcal{D}^{n-2}$, the fourth from left is the layer 3 image which has removed $\mathcal{D}^{n-1}$, $\mathcal{D}^{n-2}$ and $\mathcal{D}^{n-3}$, etc. As

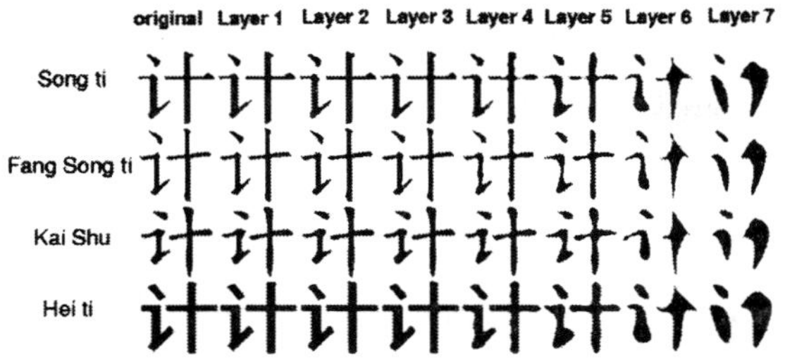

Figure 2: Wavelet transforms when multiple details are removed.

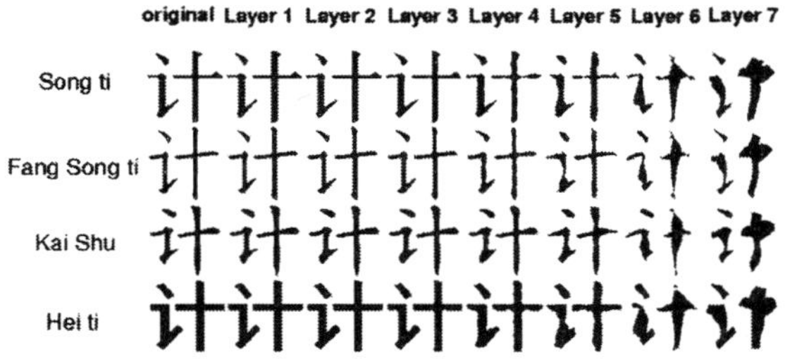

Figure 3: Wavelet transforms when details of one layer are removed.

we can see, progressing from left to right in Fig. 2, more details are lost. Finally, at layer 7, the transforms of the four fonts converge to the same basic image.

Fig. 3 shows the results when only one layer is removed, i.e., omitting $\mathcal{D}^{n-1}$ for layer 1, omitting $\mathcal{D}^{n-2}$ for layer 2, etc. We can see from Fig. 3 that, from layer 1 to layer 7, the appearance of a font is significantly different.

3.3. Algorithm for Generating New Fonts

From the above simulation results, we can see that the B-spline wavelet transform can decompose the outlines of Chinese fonts into different hierarchies and details, which represent some specific features of Chinese characters. In the following, we modify the details on some layers, and then synthesize a new Chinese font.

Two methods can be used to alter the font details. One is to add or reduce details on some selected layers. As shown in Fig. 2, by removing the details of the first four layers, and/or adding new features, we can synthesize a new font. Another method is to combine the details of two or more fonts to produce new fonts. The algorithm for generating new fonts in this paper is summarized as follows:

1. Decompose a Chinese character into different closed areas, from which edges can be extracted and outlines can be tracked.
2. For each outline, repeat steps 3 to 7.

Figure 4: New fonts generated by combining two fonts.

Figure 5: New fonts generated based on multiple fonts.

3. Express the outline as B-spline curves with $2^n + 3$ control points.
4. By using B-spline wavelet transform, decompose the control points $\mathcal{C}^n$ into details $\mathcal{D}^{n-1}$, $\mathcal{D}^{n-2}$, $\cdots$, $\mathcal{D}^1$, and the control point $\mathcal{C}^0$.
5. Let $\breve{\mathcal{D}}^j$ be $s_1\mathcal{D}_1^j + s_2\mathcal{D}_2^j + s_3\mathcal{D}_3^j + s_4\mathcal{D}_4^j$, which stands for the combined details of the newly generated outline (which is generated by combining different fonts with the same topological structure), where $0 \le s_i \le 1$, $\mathcal{D}_i^j$ $(i = 1, 2, 3, 4)$ stands for the details at layer j, and $\breve{\mathcal{D}}^j$ for the new details at layer j.
6. By using the new details $\breve{\mathcal{D}}^j$ together with the original details and the lower layer control point, generate new control points after the synthesis procedure.
7. Re-sample the newly generated curve $\breve{F}^n$, which consists of $\breve{\mathcal{C}}^n$.
8. Fill the connecting area consisting of the new outline, and combine the connecting area in order to obtain new fonts.

4. RESULTS

The above algorithm is used to generate new Chinese fonts. In our simulations, a 200 × 200 Chinese character array,

"计算机", as shown in Figs. 4 and 5, is used to evaluate the generating algorithm.

Fig. 4 shows the synthesis result when two fonts 'Song Ti' and 'Hei Ti' are used as the seed fonts.

- New fonts (A) and (D) are generated by removing the details in layers 1 to 4;
- New fonts (B) and (C) are formed by adding details $\breve{\mathcal{D}}^j$, where the coefficients $s_1 = 0.5$ and $s_2 = 0.5$.

Fig. 5 shows the synthesis results when four fonts 'Song Ti', 'Fang Song Ti', 'Kai Shu' and 'Hei Ti' are used as the seed fonts.

- New font (A) is generated by combining 'Fang Song Ti', 'Kai Shu' and 'Song Ti', with the coefficients $s_1 = 0.2$, $s_2 = 0.4$, $s_3 = 0.4$ and the lowest layer control point $\mathcal{C}^{n-4}$ from 'Fang Song Ti'.
- New font (B) is generated by combining 'Fang Song Ti', 'Kai Shu', 'Song Ti' and 'Hei Ti' with the coefficients $s_1 = 0.1$, $s_2 = 0.3$, $s_3 = 0.3$, $s_4 = 0.3$ and the lowest layer control point $\mathcal{C}^{n-4}$ from'Fang Song Ti'.
- New font (C) is generated by combining 'Kai Shu', 'Fang Song Ti', 'Song Ti' and 'Hei Ti', with the coefficients $s_1 = 0.1$, $s_2 = 0.3$, $s_3 = 0.3$, $s_4 = 0.3$ and the lowest layer control point $\mathcal{C}^{n-4}$ from 'Kai Shu'.
- New font (D) is generated by combining 'Fang Song Ti', 'Kai Shu' and 'Song Ti', with the coefficients $s_1 = 0.2$, $s_2 = 0.4$, $s_3 = 0.4$ and the lowest layer control point $\mathcal{C}^{n-4}$ from 'Kai Shu'.

From these examples, we can see that there exist localized detailed differences between the new fonts and the original fonts, and the basic structure of the new fonts are mainly determined by the lower layer's control points.

5. CONCLUSION

A synthesis method of new Chinese fonts by using wavelet transforms has been described in this paper. The basic idea is to decompose a font into layers of differing details. Then, by deleting, incorporating and altering details of selected layers, new fonts can be generated. In the process of generating new fonts, details from several existing fonts can be used to produce hybrid fonts. Essentially, the process involves transforming the outlines of a Chinese fonts into the B-spline curves, and decomposing the control points of each curve into the hierarchies and the detailed features. Once the decomposition is completed, re-ordering from one or more fonts can be performed to generate new fonts. The method can be extended to other image processing applications for creating new images from existing images.

6. REFERENCES

[1] S. G. Mallat, "A theory for multiresolution signal decomposition: The wavelet representation", *IEEE Trans. Pattern Analysis and Machine Intelligence*, Vol. 11, No. 7, pp. 674–693, 1989.

[2] I. Daubechies, "Orthonormal bases of compactly supported wavelets", *Communication on Pure and Applied Mathematics*, Vol. 41, pp. 909–996, 1988.

[3] S. G. Mallat, *A Wavelet Tour of Signal Processing,* San Diego: Academic Press, 1998.

[4] C. K. Chui, *An Introduction to Wavelets,* San Diego: Academic Press, 1992.

Local Discriminant Basis Algorithm – A Review of Theory and Applications in Signal Processing

K. Hazaveh and K. Raahemifar

Electrical and Computer Engineering Dept., Ryerson University
Toronto, Ontario M5B 2K3, Canada

Abstract

Local Discriminant Basis (LDB) algorithm is a powerful algorithmic framework that was originally developed by Coifman and Saito as a technique for analyzing object classification problems. Prior to the development of LDB, an adapted waveform framework called best basis algorithm had been developed mainly for signal compression problems. The main advantage of LDB over other similar techniques such as Karhunen-Loeve transform (KLT) also known as Principal Component Analysis (PCA) is its lower computational cost of $O(n \log n)$ order. This paper is the outcome of a literature review on theory and applications of LDB in signal processing.

Keywords: Best basis, Local discriminant basis, Local feature extraction, Pattern recognition, Time-frequency analysis, Signal processing, Image processing, Wavelet transform, Wavelet packet, Local trigonometric basis.

1. Introduction

In 1992 Coifman and Wickerhauser [1] introduced the best basis algorithm for signal compression applications. The best basis method can be tuned to adaptively choose the best set of basis functions (atoms) so that the entropy of the signal coordinates is minimized [2]. As Scott mentions in [3, Chapter 7] multivariate data in R^n are almost never n-dimensional and there often exists lower dimensional structures of data. So based on the application whether compression or classification, the problem might have a lower intrinsic dimension. It is important to note that intrinsic dimension is an application-oriented quantity. Coifman and Wickerhauser basis selection dilemma followed by a simple thresholding on the amplitude of the coefficients can result in significant signal dimension reduction. Examples of high dimensional signals are medical X-ray tomography $(n=512^2)$, seismic signals $(n=4000)$ and a speech segment $(n=1024)$.
The same basis selection task is done by Karhunen-Loeve transform (KLT) [4] developed in 1965 which is today's Principal Component Analysis (PCA). The advantage of best basis methods over KLT is twofold. By the time that best basis method was being developed a good collection of redundant sets of basis functions in the form of binary trees [1] had been discovered. Examples of such basis sets include wavelet packets and local trigonometric basis [2]. By exploiting the structured nature of these libraries Coifman and Wickerhauser developed a divide-and-conquer algorithm for best basis search with a computational cost of $O(n \log n)$ that is superior to KLT computational cost *i.e.* $O(n^3)$ [4]. Besides, in contrast to KLT, the best basis method is localized in time and frequency making it suitable for non-stationary process analysis. A local KLT was later developed by Coifman and Saito [5] but it suffers from the same high computational cost $O(n^3)$.

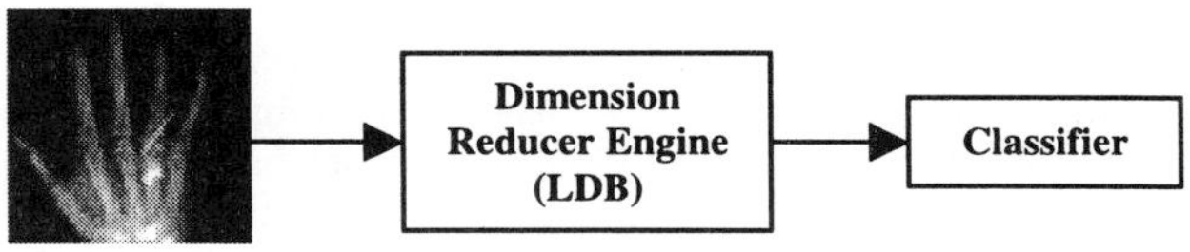

Fig.1. The role of LDB algorithm in a classification application

Later on, best basis method was used for de-noising [6], [7]. In 1994, Saito [8, Chapter 3] developed an algorithm for simultaneous de-noising and compression of signals. De-noising was perhaps the start of a new generation of best basis algorithms for signal classification problems, since it is based on the separation of noise from the actual desired signal. LDB algorithm can be considered as the classification applications counterpart of best basis algorithm of Coifman and Wickerhauser. Since its introduction in 1994 as a Ph.D. thesis by Saito [8], [9], LDB has gained considerable attention from signal and image processing societies. As an efficient classification tool for totally non-stationary signals, LDB has the potential of a feature extraction tool in biomedical [7], geophysical [11], sonar [12], radar [10], [22] and military [13], [14], [22] application areas.
In this paper we review the theory and applications of LDB in signal processing. This paper is organized as follows. Section 2 gives the necessary background to understand the algorithms later introduced in the paper. In section 3 the theory of LDB algorithm is developed. Section 4 discusses some improvements made to the original LDB method. Section 5 is devoted to the applications of LDB in different areas.

2. Background

Fig.1 is a general view of the role of LDB in a signal classification application. The input signal in Fig.1 is of very high dimension (512^2). By a dimension reducer engine such as LDB we select and keep the most important basis vectors according to the classification task and discard the nonessential coordinates. The classifier in Fig.1 can be *Linear Discriminant Analysis* (LDA) [15], *Classification and Regression Trees* (CART) [16], *k-nearest neighbor* (k-NN) [17] or *artificial neural networks* (ANN) [18]. To understand the mechanics of LDB we informally state the following definitions.

- *Over-complete (redundant) basis-* is a collection of vectors in which the number of linearly independent vectors is more than signal space dimension. An interesting recent paper by Frassard and Vandegheynst [19] gives a measure of redundancy for over-complete sets of basis vectors.
- *Dictionary of basis-* is an over-complete (redundant) set of basis functions which are used for signal representation. Examples of such dictionaries are wavelet packets and local

trigonometric bases. For a complete review of these categories of basis functions the interested reader is referred to [2]. Fig.2 (a) shows the time-frequency tiling as the result of Discrete Wavelet Transform (DWT) [20]. For a large variety of applications the tiling offered by DWT is optimal in the sense that it gives the desired time and frequency resolution (in stationary signal analysis). In most real world applications high frequencies have a transient or short duration and they fall into the category of unwanted signals or noise. So DWT can accurately locate this part in time domain (good time resolution for high frequencies) and analyze the low-pass part (the desired signal) accurately in frequency domain.

However, if the signal is representing a non-stationary phenomenon, this type of analysis may not be optimal anymore. Discrete Wavelet Packets (DWP) are redundant basis function sets, *i.e.* dictionaries. Fig.2 (b) is a representative time-frequency tiling obtainable by basis selection from DWP. Local trigonometric dictionaries are also over-complete sets of basis functions. They are the time domain counterparts of DWP. Both local sine transform (LST) and local cosine transform (LCT) are possible.

- *Binary tree-* is a dictionary isomorphic to a disjoint cover of the unit interval on real axis and their adjacent unions [2]. Moreover the space spanned by two basis functions corresponding to adjacent intervals should be equal to the direct sum of the subspaces spanned by the individual functions. Fig.3 helps to understand the above definition. Each node in the tree corresponds to a basis function and the subspace that it generates. The horizontal line in this figure can be the time axis in the case of local trigonometric bases. In the case of DWP it represents the frequency axis. The multiple cover of the horizontal line shows the redundancy. By adaptive selection of basis functions a total of 2^n covers of the horizontal line are possible, where $n=2^{j-1}$ and j is the binary tree depth [2, page 256]. This includes the usual DWT as well.
- *Library of bases-* is a collection of different dictionaries.

3. Local Discriminant Basis Algorithm

The over-complete nature of DWP means that an input signal can be decomposed into a multitude of possible bases. This redundancy enables the introduction of different search algorithms with different discriminant measures to prune the DWP tree in an optimal manner that is tailored to the particular application.

3.1. Description of the method

Suppose that we have a space $X \subseteq R^n$ of input signals and a space Y of class labels. Each input signal $x \in X$ has a unique corresponding label $y \in Y$. The goal is to construct a classifier $d: X \rightarrow Y$ that assigns the correct class label to each input signal. The method could be summarized as:

- Measure the distance between the classes using training data
- Extract the best or more discriminating features
- Reduce the problem dimension
- Construct a standard classifier on the reduced data set

Thus one can write $d = g \circ \Theta_m \circ \Psi$, where Ψ is an orthogonal transformation, Θ_m is a projection operator into m most important coordinates and g is a standard classifier.

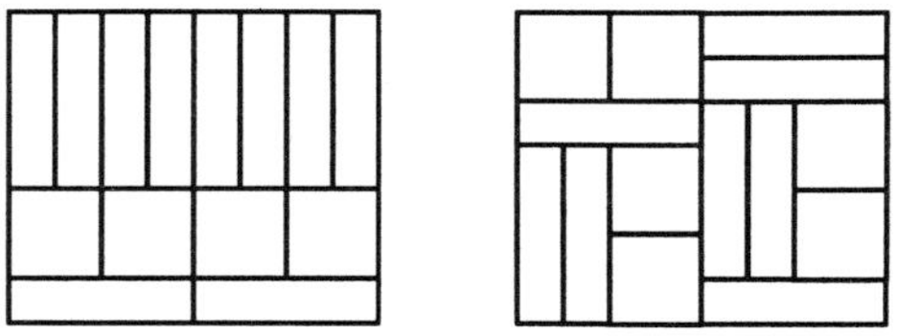

(a) (b)

Fig.2. Time-frequency tiling (a) DWT (b) DWP

Fig.3. A binary tree structure

3.2. Discriminant based binary tree pruning

In order to select such a basis from a binary tree, we need to prune the binary tree: evaluate the nodes (*i.e.* the subspaces and their bases) and remove the nonessential ones for discrimination tasks and only retain the useful ones whose union still spans the input signal space. The original LDB method uses distance measures among time-frequency energy distributions of signal classes for basis evaluation [8], [9]. Let Γ_ω^y be the normalized total energy of class y signals along the direction $\boldsymbol{\omega}$. Then,

$$\Gamma_\omega^y = (\Sigma_{N_y} |\boldsymbol{\omega} . \boldsymbol{x}_i^y|^2)/(\Sigma_{N_y} \| \boldsymbol{x}_i^y \|^2).$$

In the above expression, $\{ \boldsymbol{x}_i^y \}$ is the set of all class y signals (N_y vectors) in the training database. Here is a list of the entropy-based [9] distance measures used in original LDB for computing the discrimination power of $\boldsymbol{\omega}$ between *class 1* and *class 2*.

- *Relative entropy (or Kullback-Leibler divergence):*
  $$D(\Gamma_\omega^1, \Gamma_\omega^2) = \Gamma_\omega^1 \log(\Gamma_\omega^1/\Gamma_\omega^2),$$
- *Symmetric relative entropy (or J-divergence):*
  $$J(\Gamma_\omega^1, \Gamma_\omega^2) = D(\Gamma_\omega^1, \Gamma_\omega^2) + D(\Gamma_\omega^2, \Gamma_\omega^1),$$
- *Hellinger distance:*
  $$H(\Gamma_\omega^1, \Gamma_\omega^2) = (sqrt(\Gamma_\omega^1) - sqrt(\Gamma_\omega^2))^2,$$

where *sqrt* stands for square root. Now assume that $B=\{\boldsymbol{\omega}_1, \boldsymbol{\omega}_2, \ldots, \boldsymbol{\omega}_N\}$ is a basis. We use the summation of the discriminant measures of B individual vectors as the discriminant measure of basis B between *class 1* and *class 2* signals. Another possible approach could be to sum only the discriminant measures that are larger than a certain threshold. LDB algorithm compares the discriminant power of each parent node with that of the union of the two corresponding children nodes (Fig.3). If the parent node carries more discriminant information, we retain it and prune the children nodes and vise versa. Once a complete basis (LDB) is selected, we further choose $m(< n)$ atoms from the LDB. A typical m would be $n/10$. The simplest way of choosing m atoms from a collection of n atoms is to sort them in the order of decreasing discrimination power and to retain the first m atoms.

4. Improved Local Discriminant Basis Algorithm

4.1 Improvements in discriminant measure

It is possible to construct a simple classification problem that is intractable by original LDB algorithm. Suppose *class 1* of input signals consists of a single basis function in a waveform dictionary with amplitude *10* buried in white Gaussian noise with zero mean and unit variance. *Class 2*, is the same as *class 1*, except for the amplitude of waveform that is *–10*. Time-frequency energy distributions for the two classes are identical and LDB is unable to find the right discriminant coordinate. This example taken from [21] suggests that it is sometimes necessary to consider the distribution of expansion coefficients for individual coordinates. Here is a coordinate wise discrimination measure.

Let's call the projection of input signal $\boldsymbol{x}$ onto a unit vector $\boldsymbol{\omega}$ in the dictionary Z_ω. Then Z_ω^y is the projection of class y signals. We are interested in the manner in which Z_ω is distributed for each signal class so that we can quantify the discrimination measure of Z_ω. The original LDB algorithm measure is based on the differences of mean class energy of projection Z_ω. Other possibilities of such a measure are:

- *Type II LDB*: A measure based on the differences among the probability distribution functions (pdf's) of Z_ω,
- *Type III LDB*: A measure based on the differences among the cumulative distribution functions (cdf's) of Z_ω,
- *Type IV LDB*: A measure based on the actual classification performance (*e.g.* a rate of correct classification) using the projection of the training signals.

Type II LDB has already been applied to real world applications [10], [12]. The initial problem of Type II would be the estimation of pdf's from the available database. Suppose that p_ω^1 and p_ω^2 are the pdf's of *class 1* and *class 2* signals in $\boldsymbol{\omega}$ direction respectively. Here are some possible discrimination measures proposed by Saito [21].

- *Relative entropy (or Kullback-Leibler divergence):*

$$D(p_\omega^1, p_\omega^2)= \int p_\omega^1 \log(p_\omega^1/p_\omega^2)\, dx,$$

- *Hellinger distance:*

$$H(p_\omega^1, p_\omega^2)= \int (sqrt(p_\omega^1)-sqrt(p_\omega^2))^2 dx.$$

The feasibility of Type III and Type IV LDB's is still under study.

4.2. Overcoming the lack of translation invariance

The lack of translation invariance is a major problem of the binary trees that are used for LDB divide-and-conquer algorithm. Suppose that we have trained the system to classify the input signals into *class 1* and *class 2* with a synchronized training data. Now if the input to be classified is for instance delayed about *20* samples, a wrong classification occurs, since the DWP, LST and LCT are not translation invariant. For some one-dimensional problems, that the signals are generally well synchronized, we can expand the training data in the following manner. We translate each training signal over $-\tau, -\tau+1, \ldots, \tau-1, \tau$. This way the number of training data signals is multiplied by $2\tau+1$. This is practical if τ is small. For higher dimensional signals this method fails to be practical. The following method has been proposed by Kronquist and Storm [10] for an image processing application.

- Train the algorithm on fixed position (time/space synchronized) data.
- When classifying, use some method to determine the position/localization of the interesting signal.
- When position determined, reposition the signal in the original (training) position and classify in the normal way.
- Finally translate the signal back.

This method tracks both the class and the synchronization of the signal. The method to determine the position of the signal is of course problem dependant. One possible method is to localize a specific feature in the class of signals, for example the point with the maximum amplitude or zero crossing or something that is easily observed in the typical class and then choose the correct translation by aligning the selected feature with the training data feature. If the noise is significant, and the selected LDB basis is highly dependent on small translations, we may combine the two techniques by doing a $-\tau, -\tau+1, \ldots, \tau-1, \tau$ translation of the repositioned signal.

5.Applications

5.1. Target detection in sonar images

Kronquist and Storm [10] have employed LDB algorithm to detect targets in sonar images. The target-like background in sonar image makes it difficult to localize a target (*e.g.* a mine). The combined technique described in section 4.2 was employed to overcome the lack of translation invariance. The use of Type II LDB resulted in better mine detection. They had about *10%* error in terms of missing the target and *0.8%* error in terms of false target detection. The results are based upon a set of *33* training target signals that seems insufficient for evaluation of LDB potential in this area.

5.2. Classification of geophysical acoustic waveforms

Acoustic measurements are used in geological well logging to infer petrophysical properties or the lithology of subsurface formations [11]. For a brief review of the methods and measurement techniques and interpretations the reader is referred to [8, Chapter 6]. The measured signal is one dimensional that is used to infer petrophysical/lithologic properties of the surrounding formations such as porosity, mineralogy, grain contacts, fluid saturation, volume percentage of various rocks such as sandstone, shale, and limestone, etc. Saito [11] studied the problem of inferring lithologic information from the entire waveform as a classification problem. The original LDB algorithm was used to classify rock types by using waveform shape information. Since the time and velocity information of this type of signals are of particular importance, LST was chosen as the time-frequency dictionary. He examined different discriminant measures and also varied the number of LDB coordinates from *5* to *100* to study the effects on the classification accuracy [11]. He observed that the choice of discriminant measure has negligible effect on the final results. However, the best classification (*0.24%* misclassification error) was obtained by using the top *35* LDB vectors chosen by the Hellinger distance criterion.

5.3. Land use classification of Synthetic Aperture Radar (SAR) images

Rogers and Johnson [12] used Type II LDB algorithm to determine land use in SAR images. They implemented the algorithm by first decomposing the input signal to a typical two-dimensional DWP. An object-oriented software in C++ was designed to compute the Type II LDB for two-dimensional samples. They studied the effect of varying the size of training blocks and the amount of overlap between extracted image blocks along with LDB output signal dimension. The result of their investigation is shown by SAR sample images in [12]. Spooner [22] has also employed LDB algorithm for automatic target detection (ATR) using radar-return data. The experiment objects have been military vehicles. The results are tabulated according to the classifier used in general scheme depicted in Fig.1 and the dimension of the analysis, *i.e.* whether one-dimensional or two-dimensional [22]. It seemed that the best classification results were obtained with four level Coifman 05 two-dimensional wavelet packet [20] and the *Binary Hierarchical Classifier with K-vectors* (BHCK) [16].

5.4. Theater missile defense

Cassabaum *et al.* [13], [14] have examined the efficiency of LDB algorithm by using one-dimensional Infra Red (IR) sensor data. The objects consist of: a re-entry vehicle (RV), an associated object (AO), a booster, and burnt solid fuel (debris). The data had time-varying IR signature, which possesses unique spectral patterns. The fluctuations in signature intensity resulted from the angular motion of the objects and their projected area and/or temperature relative to the receiving sensor. Since the four objects underwent different impulsive forces according to their function and different inertial characteristics, the features were well separated among classes. Cassabaum *et al.* [13] used over *30,000* training signals. Test data set was *500* for each of four classes. They classified the signals to two classes: RV and other. DWP dictionaries were constructed by using Daubechies 04 and Coifman 06 mother wavelets. Nearly *100%* classification accuracy was obtained with either DWP. In another experiment the performance of Fuzzy logic classifier together with LDB was studied [14].

6. Conclusion

LDB is a powerful algorithmic framework for the analysis of non-stationary signals. The time-frequency tiling resulting from LDB algorithm is optimal in the sense that it is the most discriminating decomposition for the classification task under study. As a result of exploiting the binary tree structure of basis function dictionaries, LDB computational complexity is considerably lower than existing classification tools such as PCA. A review of the theory and applications of LDB is presented. The evolution of the theory and its improved versions are briefly discussed. Potential application areas are introduced and some examples of the existing practical applications are explained.

7. References

[1] R. R. Coifman and M.V. Wickerhauser, "Entropy-based algorithms for best basis selection," *IEEE Trans. Inform. Theory* 38 (1992), no.2, pp 713-719.

[2] M.V. Wickerhauser, *Adapted Wavelet Analysis from Theory to Software*, A K Peters, Ltd. Wellesley, MA, 1994.

[3] D. W. Scott, *Multivariate Density Estimation: Theory, Practice, and Visualization*, John Wiley & Sons, New York, 1992.

[4] S. Watanabe, "Karhunen-Loeve Expansion and Factor Analysis: Theoretical Remarks and Applications," in *Trans. 4th Prague Conf. Inform. Theory*, Publishing House of Czechoslovak Academy of Sciences, pp 635-660, 1965.

[5] R. R. Coifman, N. Saito, "The Local Karhunen-Loeve Bases," *Time-Frequency and Time-Scale Analysis, Proceedings of the IEEE-SP International Symposium on*, pp 129 –132, 1996.

[6] P. Qua, Z. Lei, Z. Hongcai, D. Guanzhong, "Adaptive wavelet based spatially de-noising," Signal *Processing Proceedings, 1998 Fourth International Conference on*, pp 486-489,1998.

[7] R. R. Coifman and M.V. Wickerhauser, "Adapted waveform "de-noising" for medical signals and images," IEEE Engineering in Medicine and Biology Magazine, v.14, n.5, pp 578 –586, 1995.

[8] N. Saito, *Local Feature Extraction and Its Application Using a Library of Bases*, Ph.D. Thesis, Dept. of Math, Yale University, New Haven, CT 06520 USA, Dec. 1994.

[9] N. Saito and R. R. Coifman, "Local discriminant bases and their applications," J. Mathematical Imaging and Vision 5 (1995), no.4, pp 337-358, Invited paper.

[10] G. Kronquist and H. Storm, "Target Detection with Local Discriminant Bases and Wavelets," *Proc. of SPIE*, v.3710, pp 675-683, 1999.

[11] N. Saito, R. R. Coifman, "Extraction of Geological Information from Acoustic Well-Logging Waveforms Using Time-Frequency Wavelets," *Geophysics*, v.62, n.6, 1997.

[12] L. S. Rogers, C. Johnston, "Land Use Classification of SAR Images Using a Type II Local Discriminant Basis for Preprocessing," *Proc. of IEEE Conference on Acoustics, Speech and Signal Processing*, v.5, pp 2729-2732, 1998.

[13] M. L. Cassabaum, H. A. Schmitt, H. W. Chen, J. G. Riddle, "Application of Local Discriminant Bases Discrimination Algorithm for Theater Missile Defense," *Proc. of SPIE*, v.4119, pp 886-893, 2000.

[14] M. L. Cassabaum, H. A. Schmitt, H. W. Chen, J. G. Riddle, "Fuzzy Classification Algorithm Applied to Signal Discrimination for Navy Theater Wide Missile Defense," *Proc. of SPIE*, v.4120, pp 134-145, 2000.

[15] R. A. Fisher, "The Use of Multiple Measurements in Taxonomic Problems," *Ann. Eugenics*, Vol. 7, pp 179-188, 1936.

[16] L. Breiman, J. H. Friedman, R. A. Olshen, and C. J. Stone, *Classification and Regression Trees*, Chapman and Hall, Inc., New York, 1993.

[17] K. Fukunaga, *Introduction to Statistical Pattern Recognition,* 2nd ed., Academic Press, San Diego, 1990.

[18] B. D. Ripley, "Statistical aspects of neural networks," *Networks and Chaos: Statistical and Probabilistic Aspects*, O. E. Barndorff-Nielson, J. L. Jensen, D. R. Cox, and W. S. Kendall, eds., pp 40-123, Chapman and Hall, Inc., New York, 1993.

[19] P. Frossard, P. Vandegheynst, "Redundancy in non-orthogonal transforms," *Proc. of IEEE International Symposium on Inform. Theory*, n.01CH3725, p 196, 2001.

[20] Mallat, S.G., *A Wavelet Tour of Signal Processing*, San Diego Toronto: Academic Press, 1998.

[21] N. Saito and R. R. Coifman, "Improved Local Discriminant Bases Using Empirical Probability Density Estimation," Proc. *on Statistical Computing*, American Statistic. Assoc., 1996.

[22] C. M. Spooner, "Application of local discriminant bases to HRR-based ATR," *Signals, Systems and Computers, Conference Record of the Thirty-Fifth Asilomar Conference on,* v.2, pp 1067 -1073, 2001.

NEXT CANCELLATION IN xDSL SYSTEMS USING VARIABLE-LENGTH CANCELLERS

R. C. Nongpiur, D. J. Shpak, and A. Antoniou

Dept. of Elec. and Comp. Eng., University of Victoria
Victoria, BC, Canada V8W 3P6
{rnongpiu, dale, aantoniou}@ece.uvic.ca

ABSTRACT

An algorithm that reduces the computational load in near-end crosstalk (NEXT) cancellation in xDSL systems is proposed. The algorithm adjusts the length of each adaptive filter according to the strength of the NEXT and thus results in a minimum usage of filter taps.

The algorithm starts the NEXT cancellation by using adaptive filters with minimum filter lengths. As the adaptation progresses, the algorithm adjusts the filter length of each adaptive filter according to the magnitude of the NEXT. And when all the adaptive filters have fully converged another algorithm is used to readjust the filter length of those adaptive filters that might be too long or too short.

1. INTRODUCTION

Telephone loops were designed and built to carry voice services, and at these frequencies crosstalk is insignificant. However, at higher frequencies crosstalk becomes a major impairment. The worst-case crosstalk scenario used in simulations is often too pessimistic, and having crosstalk cancellers of fixed length for each line [1] is a waste if little or no crosstalk is present on the line. In [2] the complexity is reduced by first detecting the major NEXT sources using a low-complexity cross-correlation algorithm [3] and then deploying fixed-length NEXT cancellers to cancel the detected NEXT. By further allowing the number of filter taps to be variable, taps can be transferred to those cancellers where the crosstalk strength is high from cancellers where the crosstalk strength is low. This can result in improved usage of hardware in terms of the total number of filter taps which also translates into significant savings in computation.

In this paper, an algorithm that adjusts the length of each adaptive filter according to the strength of the NEXT is proposed. By normalizing and then combining many NEXT impulse responses for a particular cable, a normalized NEXT profile is created. Using this profile, a plot of the tap limits for a certain maximum noise threshold is generated. The algorithm starts the NEXT cancellation by using adaptive filters with minimal filter length. The taps are positioned along the indices where the maximum absolute value of the NEXT impulse responses occurs. As the adaptation progresses, the absolute peak value of each adaptive filter is periodically sought. This is used to denormalize the NEXT profile so that the tap limits for each adaptive filter can be found. And when all the adaptive filters have fully converged, another algorithm is used to readjust the length of any adaptive filter with a NEXT impulse response that may have deviated from the crosstalk profile.

The paper is organized as follows. Section 2 derives the conditions for optimizing the number of taps in each adaptive filter. Section 3 presents an algorithm that minimizes the number of filter taps under the constraint that the maximum crosstalk noise be below a given threshold.

2. FILTER-TAP MINIMIZATION

The echo-cancelled received signal on a single twisted pair can be expressed as

$$y(n) = r(n) + v(n) + \sum_i \sum_{k=0}^{\infty} h_i(k) u_i(n-k) \qquad (1)$$

where $y(n)$ is the noisy observation, $r(n)$ is the received message signal, $v(n)$ is additive background noise, $u_i(n)$ is the signal transmitted on pair i and $h_i(n)$ is the crosstalk coupling function between pair i and the considered pair.

For the purpose of crosstalk cancellation, we assume that the echo at the hybrid is already cancelled by the echo canceller before crosstalk detection is performed. In most cases, the impulse response of the channel can be readily estimated [4]; hence the received signal can also be removed leaving behind the crosstalk signals. With the received signal and the transmitted signal echo removed, the remaining signal $y(n)$ can be expressed as

$$y(n) = v(n) + \sum_i \sum_{k=0}^{\infty} h_i(k) u_i(n-k) \qquad (2)$$

Suppose an adaptive filter of length N_i is used to cancel the near-end crosstalk from line i. The instantaneous error $e(n)$ due to all the NEXTs in the bundle, upon convergence, can be written as

$$e(n) = y(n) - \sum_i \sum_{k=0}^{N_i-1} \hat{h}_i(k) u_i(n-k) \qquad (3)$$

where $\hat{h}_i(k)$ is the estimated NEXT impulse response of the ith adaptive filter upon convergence.

Assuming that the transmitted data are statistically independent and wide-sense stationary with zero mean, the mean-square error upon convergence can be expressed as

$$\begin{aligned} E[|e(n)|^2] &= \sum_i \sigma_{u_i}^2 \sum_{k=0}^{N_i-1} |h_i(k) - \hat{h}_i(k)|^2 \\ &\quad + \sum_i \sigma_{u_i}^2 \sum_{k=N_i}^{\infty} |h_i(k)|^2 + \sigma_v^2 \qquad (4) \end{aligned}$$

where $h_i(n)$ is the near-end crosstalk impulse response from the ith line, $\hat{h}_i(n)$ represents the tap weights of the ith adaptive canceller of length N_i, $\sigma_{u_i}^2$ is the variance of the transmitted data and σ_v^2 is the variance of the noise in the system.

0-7803-7761-3/03/$17.00

Assuming that all the lines use the same DSL type, the variance of the transmitted data over the various lines can be considered equal to each other. This simplifies (4) to

$$E[|e(n)|^2] = \sigma_u^2 \sum_i \sum_{k=0}^{N_i-1} |h_i(k) - \hat{h}_i(k)|^2 + \sigma_u^2 \sum_i \sum_{k=N_i}^{\infty} |h_i(k)|^2 + \sigma_v^2 \quad (5)$$

where $\sigma_u = \sigma_{u_1} = \sigma_{u_2} = \cdots = \sigma_{u_i} = \cdots$. Upon convergence, the first term in (5) is much smaller than the background noise σ_v^2 and hence

$$E[|e(n)|^2] = \sigma_u^2 \sum_i \sum_{k=N_i}^{\infty} |h_i(k)|^2 + \sigma_v^2 \quad (6)$$

Suppose that $E[|e(n)|^2]$ is fixed at λ_{max}, which is the maximum tolerable noise, we can minimize $\sum_i N_i$ subject to the constraint

$$\lambda_{max} = \sigma_u^2 \sum_i \sum_{k=N_i}^{\infty} |h_i(k)|^2 + \sigma_v^2 \quad (7)$$

This is an optimization problem of finding the length N_i for each adaptive filter under the constraint given in (7).

Suppose now that the total number of taps of all the NEXT cancellers is fixed, that is, $\sum N_i$ = constant, a length N_i can be found for each NEXT canceller such that $E[|e(n)|^2]$ is at a minimum, that is, we solve the optimization problem

$$\begin{array}{ll} \underset{N_i}{\text{minimize}} & E[|e(n)|^2] \\ \text{subject to:} & \sum N_i = \text{constant} \end{array} \quad (8)$$

Using (6) and neglecting the constant terms σ_u and σ_v, we can state the problem as

$$\begin{array}{ll} \underset{N_i}{\text{minimize}} & \sum_i \sum_{k=N_i}^{\infty} |h_i(k)|^2 \\ \text{subject to:} & \sum N_i = \text{constant} \end{array} \quad (9)$$

or

$$\begin{array}{ll} \underset{N_i}{\text{maximize}} & \sum_i \sum_{k=0}^{N_i-1} |h_i(k)|^2 \\ \text{subject to:} & \sum N_i = \text{constant} \end{array} \quad (10)$$

3. OPTIMIZING THE FILTER LENGTHS

In order to estimate the length N_i for adaptive filter i, we create a NEXT impulse-response profile for the given cable type as

$$C_p(k) = \max \left[\frac{|h_1(k)|}{h_{\text{MAX}1}}, \frac{|h_2(k)|}{h_{\text{MAX}2}}, \ldots, \frac{|h_n(k)|}{h_{\text{MAX}n}} \right] \quad (11)$$

where h_{MAX_i} is the normalizing factor given as

$$h_{\text{MAX}_i} = \max[|h_i(1)|, |h_i(2)|, \ldots, |h_i(\infty)|] \quad (12)$$

and $h_i(n)$ denotes the NEXT impulse response inclusive of the receive and transmit filters. The NEXT impulse responses used for generating $C_p(k)$, can either be obtained through crosstalk impulse-response measurements for the particular cable type, or can be generated through simulations [5].

For a cable of certain specifications, the envelopes of the NEXT impulse responses are similar to one another although of varying magnitudes. The major NEXT between any two twisted pairs occurs in the initial portion of the cable [6] and hence the maximum absolute value of the NEXT impulse responses will have a small delay that cluster around a small value on the time scale. From the NEXT profile $C_p(k)$, the peak value of the NEXT impulse responses can be found to cluster around k_{max} where

$$C_p(k_{max}) = \max_i \; [C_p(i)] \quad (13)$$

Since the peak values of all the crosstalk impulse responses occur around k_{max}, we can start with adaptive filters having filter taps around k_{max}. Assuming that all the peak values occur between $(k_{max} - M/2)$ and $(k_{max} + M/2)$, we can have NEXT cancellers with filter taps positioned between $(k_{max} - M/2)$ and $(k_{max} + M/2)$. Thus the minimum number of filter taps that each of the NEXT cancellers can have is M.

For the ith filter, the maximum absolute value α_i of the M filter taps is defined as

$$\alpha_i = \max_i \; [|h_i(k_{max} - \tfrac{M}{2})|, \ldots, |h_i(k_{max})|, \ldots, |h_i(k_{max} + \tfrac{M}{2})|] \quad (14)$$

The NEXT profile in (11) gives the maximum absolute value at each of the filter taps, with the peak value normalized to 1. As will be shown, we can use the profile to obtain the minimum tap limits for a specified maximum noise tolerance.

An adaptive canceller with taps starting at $n_{start} + 1$ and terminating at $n_{stop} - 1$ will not be able to cancel the NEXT impulse response before $n_{start} + 1$ and after $n_{stop} - 1$. Thus, the maximum possible NEXT power that will not be cancelled is given by

$$\eta_{max} = \sum_{k=0}^{n_{start}} C_p^2(k) + \sum_{k=n_{stop}}^{\infty} C_p^2(k) \quad (15)$$

Accordingly, the maximum power η_l that will not be cancelled before n_{start} is given by

$$\eta_l(n_{start}) = \sum_{k=0}^{n_{start}} C_p^2(k) \quad (16)$$

and the maximum power η_h that will not be cancelled after n_{stop} is

$$\eta_h(n_{stop}) = \sum_{k=n_{stop}}^{\infty} C_p^2(k) \quad (17)$$

Using (16) and (17), we can write (15) as

$$\eta_{max}(n_{start}, n_{stop}) = \eta_l(n_{start}) + \eta_h(n_{stop}) \quad (18)$$

From (16), we can get a plot of η_l versus n_{start}. With η_l being monotonically increasing with n_{start}, an inverse function $g_l(\eta_l)$ is obtained from the plot such that

$$g_l(\eta_l) = n_{start} \quad (19)$$

Similarly, (17) gives the plot of η_h versus n_{stop}, and from the plot an inverse function $g_h(\eta_l)$ is obtained such that

$$g_h(\eta_h) = n_{stop} \quad (20)$$

Using (19) and (20), the filter length is thus obtained as

$$n_{stop} - n_{start} = g_h(\eta_h) - g_l(\eta_l) \quad (21)$$

For a fixed uncancelled power λ that lies before n_{start} and after n_{stop}, finding the tap limits n_{start} and n_{stop} such that (21) is at a minimum is an optimization problem [7] of minimizing (21) under the constraint

$$\eta_l + \eta_h = \lambda \tag{22}$$

Thus the solution of the optimization problem gives the starting and final tap indices, wherein the filter length is at a minimum for the given value of λ.

Let the solution of the optimization problem in (21) and (22) be two functions $f_{start}(\lambda_i)$ and $f_{stop}(\lambda_i)$ such that

$$n_{start,i} = f_{start}(\lambda_i) \tag{23}$$

and

$$n_{stop,i} = f_{stop}(\lambda_i) \tag{24}$$

where $n_{start,i}$ and $n_{stop,i}$ are the optimum starting and ending tap limits for adaptive filter i and λ_i is scaled according to the value of α_i defined in (14) such that

$$\lambda_i = \frac{\lambda}{\alpha_i^2} \tag{25}$$

Rescaling λ is necessary because $f_{start}(\lambda_i)$ and $f_{stop}(\lambda_i)$ are obtained from the crosstalk profile $C_p(k)$ which has been normalized so that its absolute peak value is unity. In order to find the tap limits for an impulse response with a peak absolute value of α_i we need to rescale λ according to (25).

We start the NEXT cancellation with each adaptive filter having M taps positioned between $(k_{max} - M/2)$ and $(k_{max} + M/2)$. The maximum uncancelled NEXT impulse power that can be tolerated is fixed at λ for each line. To find the maximum uncancelled power from all the lines, we define a_i and $\bar{a}_i$ such that

$$a_i = \begin{cases} 1 & \text{if } \alpha_i[\sum_k C_p^2(k) - M] > \lambda \\ 0 & \text{otherwise} \end{cases} \tag{26}$$

$$\bar{a}_i = \text{complement of } a_i \tag{27}$$

Thus $a_i = 1$ implies that the maximum power of the NEXT impulse response at all but the M tap positions is greater than λ. The power at the M tap positions is considered to be cancelled by the adaptive filter. To bring down the maximum uncancelled power to λ the filter length will have to be increased beyond M. This can be done by using (23) and (24) which gives the starting and ending tap limits for each adaptive filter for a given value of λ and α_i.

When $a_i = 0$, the maximum uncancelled NEXT impulse-response power for an adaptive filter with M taps is less than λ. Hence the maximum uncancelled power when $a_i = 0$ is given as $\alpha_i^2[\sum_k C_p^2(k) - M]$.

The total uncancelled NEXT impulse power Γ_{max} from all the lines is given by

$$\Gamma_{max} = \sum_i a_i\lambda + \sum_i \bar{a}_i\alpha_i^2\left[\sum_k C_p^2(k) - M\right] \tag{28}$$

Solving (28) for λ can be complicated since a_i and $\bar{a}_i$ are also dependent on λ. The solution can be simplified if a_i and $\bar{a}_i$ are approximated by b_i and $\bar{b}_i$ such that

$$b_i = \begin{cases} 1 & \text{if } \alpha_i[\sum_k C_p^2(k) - M] > \frac{\Gamma_{max}}{N} \\ 0 & \text{otherwise} \end{cases} \tag{29}$$

$$\bar{b}_i = \text{complement of } b_i \tag{30}$$

Hence,

$$\Gamma_{max} \approx \sum_{i=1}^{N} b_i\lambda + \sum_{i=1}^{N} \bar{b}_i\alpha_i^2\left[\sum_k C_p^2(k) - M\right] \tag{31}$$

Solving for λ, we get

$$\lambda = \frac{\Gamma_{max} - \sum_i \bar{b}_i\alpha_i^2\left[\sum_k C_p^2(k) - M\right]}{\sum_i b_i} \tag{32}$$

where

$$\sum_i b_i \neq 0 \tag{33}$$

Thus by knowing α_i for each adaptive filter, we can use (32) to find λ for a given value of Γ_{max}. Once λ is obtained, we can use (25) to find λ_i for each adaptive filter. Putting λ_i in (23) and (24) gives the starting and final tap indices for each adaptive filter.

3.1. Adjusting the tap weights on convergence

Once all the adaptive cancellers have fully converged, the tap limits can be adjusted further to reduce the mean residual error while keeping the total number of filter taps fixed.

An algorithm that minimizes $E(|e(n)|^2)$ in (6) by adjusting the filter lengths while keeping the total number of filter taps fixed is used. The algorithm scans the sum of the squares of the last n filter taps on all the adaptive filters, and compares the maximum value with the minimum. Parameter n can be varied from 1 to the minimum filter length M. If the ratio of the maximum to the minimum exceeds a certain threshold value, the adaptive canceller having the minimum value is reduced in length by one tap, and the one having the maximum value is increased by one tap. The algorithm assumes that the absolute value of the new filter length will converge to an absolute value which is greater than the previous minimum value, thus maximizing $\sum_i \sum_{k=1}^{N_i} |h_i(k)|^2$, which is the same as solving the maximization problem in (10). The same procedure is repeated again when the adjusted filter tap attains convergence. It stops when the ratio of the maximum to the minimum value is below the prescribed threshold.

4. SIMULATION RESULTS

In the simulations, the crosstalk impulse-response measurements were taken from measurements performed on a University of Ottawa cable [8]. The d.c. component was removed from the impulse response, and was convolved with the transmit and received filters of an HDSL2 system [9]. A total of 50 crosstalk impulse responses were used to generate the crosstalk profile given in (11). The plot is shown in Fig. 1. From the crosstalk profile,

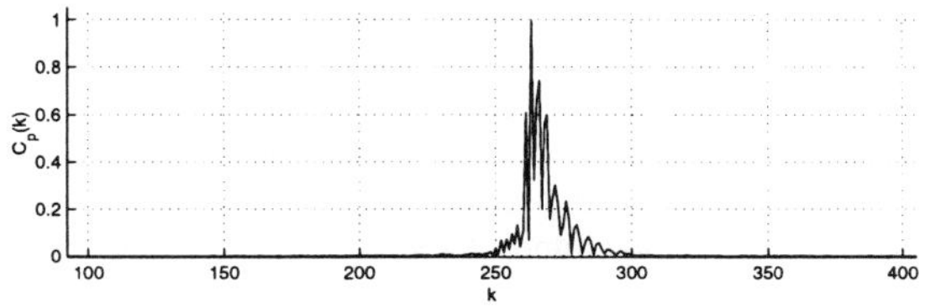

Figure 1: Near-end crosstalk profile generated from 50 NEXT impulse responses.

$C_p(k)$ defined in (11), functions $g_l(\eta_l)$ and $g_h(\eta_h)$ were found using (19) and (20), respectively. Their plots are given in Figs. 2 and 3. Function $g_l(\eta_l)$ gives the starting tap index n_{start} for a given η_l. Function $g_h(\eta_h)$ gives the ending tap index for a given maximum impulse power η_h that can be present after n_{stop}.

Once $g_l(\eta_l)$ and $g_h(\eta_h)$ are known, we can minimize (21) with respect to λ defined in (22). This is an optimization problem which can be solved to give two functions $f_{start}(\lambda_i)$ and $f_{stop}(\lambda_i)$ as given in (23) and (24), respectively. Functions $f_{start}(\lambda_i)$ and $f_{stop}(\lambda_i)$ give the starting and ending tap limits for a given value of λ_i. Their plots are shown in Figs. 4 and 5.

During the NEXT cancellation process, we can use $f_{start}(\lambda_i)$ and $f_{stop}(\lambda_i)$ to find the starting and ending tap limits for each adaptive filter. Eqn. (14) gives α_i for each adaptive filter and (32) gives the value of λ. Using (25), we can obtain λ_i for each adaptive filter.

The NEXT cancellers used in this paper were normalized least mean-square (NLMS) adaptive filters [10] with the initial number of taps M set to 3. Filter length adjustments were done after every 500 iterations of the adaptive filters. After 2000 iterations, the adaptive filters were assumed to have attained convergence. The second algorithm described in section 3.1 that optimizes N_i was then used. The algorithm readjusts the filter lengths while keeping the total length of all the adaptive filters fixed.

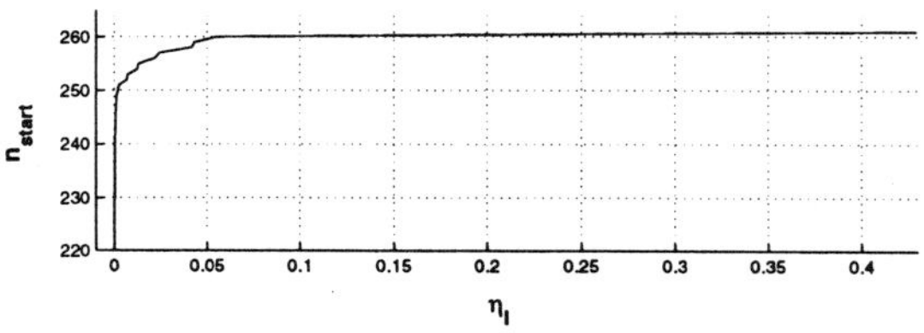

Figure 2: Plot of $g_l(\eta)$ versus η_l.

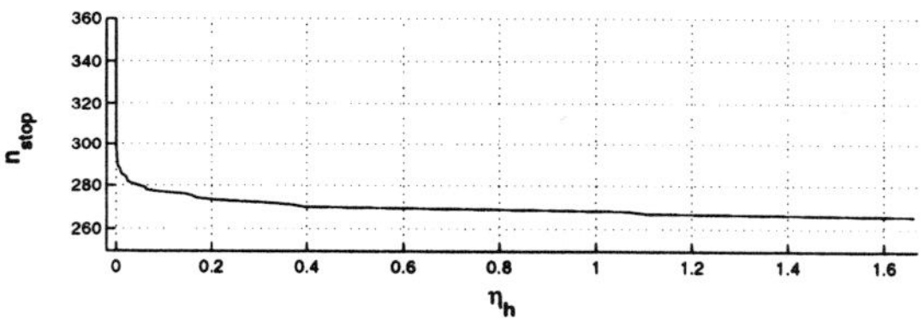

Figure 3: Plot of $g_h(\eta)$ versus η_h.

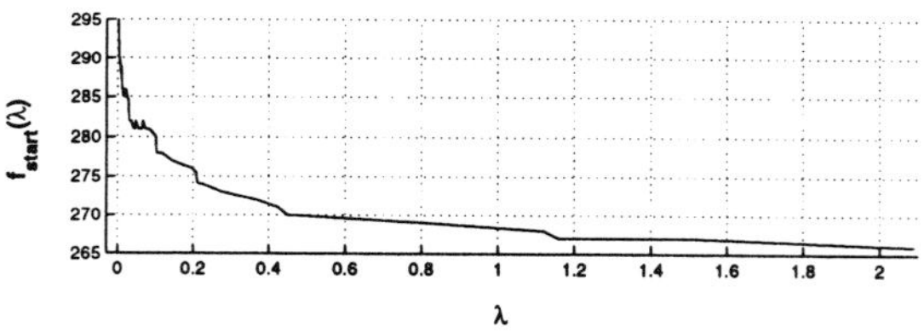

Figure 4: Plot of $f_{start}(\lambda)$ versus λ.

5. CONCLUSIONS

A new method that minimizes the length of each adaptive canceller according to the magnitude of the near-end crosstalk was proposed.

The impulse responses of the NEXT in a bundle have similar envelopes but of varying strength. The crosstalk profile was created by combining a large number of measured values of the NEXT impulse response. Using the profile, two plots that give the starting and ending tap limits were obtained. By taking the maximum absolute value α_i from each adaptive filter, the maximum

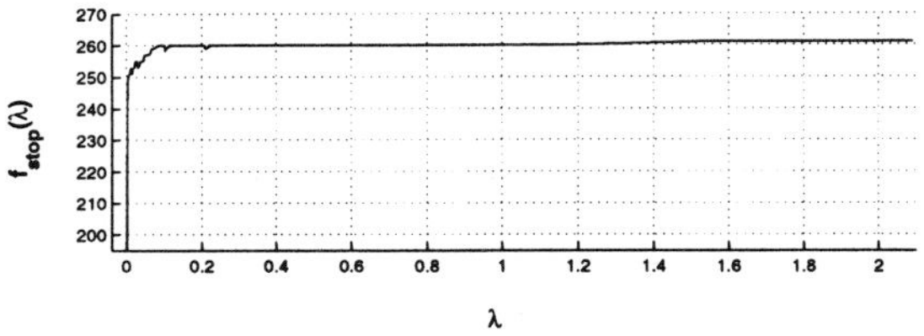

Figure 5: Plot of $f_{stop}(\lambda)$ versus λ.

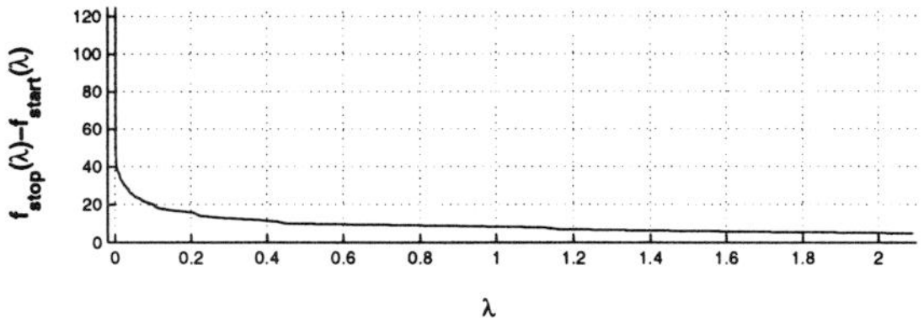

Figure 6: Plot of $f_{stop}(\lambda) - f_{start}(\lambda)$ versus λ.

threshold λ is rescaled to λ_i for each adaptive filter. Knowing λ_i, the functions plotted in Figs. 4 and 5 can be used to get the starting and final tap indices for each adaptive filter.

Once all the adaptive cancellers have attained convergence, another algorithm is then used to periodically adjust the tap limits of the adaptive cancellers while keeping the total length of all the adaptive cancellers constant. This algorithm corrects any major inaccuracies in the prediction of the tap limits that might result from a NEXT impulse response that deviates from the crosstalk profile,[1] or in the event that the magnitude and length of the NEXT impulse response changes over time.

Acknowledgement

The authors are grateful to PMC-Sierra, Micronet, NCE Program, and the Natural Sciences and Engineering Research Council of Canada for supporting this work.

6. REFERENCES

[1] M. L. Honig, K. Steiglitz, B. Gopinath, "Multichannel signal processing for data communication in the presence of crosstalk," *IEEE Transactions on Communications*, vol. 38, no. 4, pp. 551-558, April 1990.

[2] R. C. Nongpiur, D. J. Shpak, A. Antoniou, "NEXT cancellation in xDSL systems," *36th Asilomar Conference on Signals, Systems and Computers, Pacific Grove, CA., Nov. 3-6, 2002.*

[3] T. J. Shan and T. Kailath, "Adaptive algorithms with an automatic gain control feature," *IEEE Transactions on Circuits and Systems*, vol. 35, no. 1, pp. 122-127, Jan. 1988.

[4] C. Zeng, C. Aldana, A. A. Salvekar, and J. M. Cioffi, "Crosstalk identification in xDSL systems," *IEEE Journal on Selected Areas in Communication*, vol. 19, no. 8, pp. 1488-1496, Aug. 2001.

[5] Walter Y. Chen, *Simulation Techniques and Standards Development for Digital Subscriber Lines*, Macmillan Technical Publishing 1998.

[6] J. A. Bingham, *ADSL, VDSL, and Multicarrier Modulation*, Wiley, 1999.

[7] A. Antoniou, *Optimization: Theory and Practice*, University of Victoria, Mar. 2000.

[8] "The University of Ottawa Cable," *http://www.ee.unb.ca/petersen/lib/data/cable_1989/description.html*

[9] "Draft for HDSL2 Standard," *T1E1.4/99-006R5.*

[10] S. Haykin, *Adaptive Filter Theory*, Third Edition, Prentice Hall 1996.

[1] By using a larger number of NEXT impulse responses to create the crosstalk profile, the deviation is reduced.

NORM AND COEFFICIENT CONSTRAINTS FOR ROBUST ADAPTIVE BEAMFORMING

Qiyue Zou, Zhu Liang Yu and Zhiping Lin

Centre for Signal Processing, School of Electrical and Electronic Engineering
Nanyang Technological University, Singapore 639798
Email: eqyzou@ntu.edu.sg

ABSTRACT

Norm-constrained and coefficient-constrained adaptive filters are often used in robust adaptive beamforming to mitigate the target-signal cancellation problem caused by DOA (Direction of Arrival) mismatch. The norm and coefficient constraints are usually selected numerically based on computer simulations or real-time experiments, which is not enough for theoretical analysis that requires explicit mathematical expressions. This paper analytically derives these constraints using the Taylor series expansion of array correlation matrix. The constraints that are expressed in terms of maximal DOA mismatch agree well with the simulation results, and most importantly they are quite useful in practical beamformer design and performance analysis.

1. INTRODUCTION

In the past decades, adaptive array processing has received considerable attention in many areas, e.g. wireless communications and microphone arrays. An adaptive array is able to adjust its beampattern in real time to introduce deep nulls in the directions of arrival (DOA) of strong interferences, while maintaining a chosen frequency response in the look direction along which the target signal propagates.

The beamformer based on a generalized sidelobe canceller (GSC) has been extensively studied in the literature [1,2]. The block diagram of a GSC with K sensors is shown in Fig. 1. It consists of a presteering front end, a fixed beamformer, a blocking matrix and an adaptive canceller. The presteering front end is composed of variable time delays, with which the main lobe of the beamformer can be steered to the desired direction. The fixed beamformer is implemented by the conventional approach, such as the delay-and-sum method, and is used to enhance the signal from the look direction. The blocking matrix aims to totally reject the desired signal, so that the output from the blocking matrix contains only interferences and noises. The adaptive canceller adaptively adjusts its taps so that the interferences and noises are subtracted from the fixed beamformer output in an optimal way. As the blocking matrix output contains only interferences and noises that are assumed to be uncorrelated with the desired signal, the adaptation of the canceller is performed to minimize the output power. The unconstrained Least-Mean-Square (LMS) algorithm is usually employed for tap adaptation.

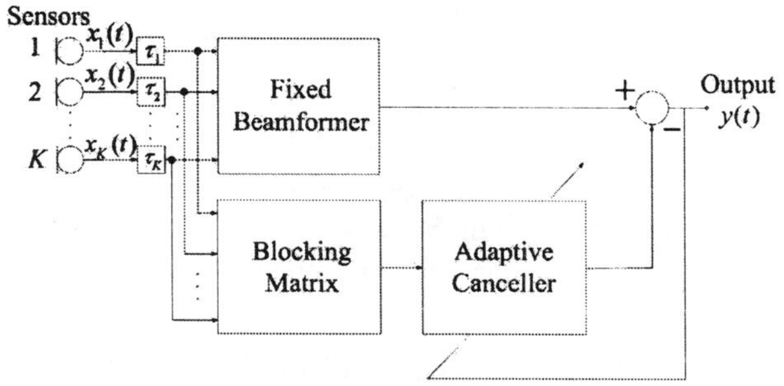

Fig. 1. Block diagram of a generalized sidelobe canceller with K sensors

Unfortunately, in the presence of DOA mismatch, the blocking matrix cannot perfectly block the desired signal, and hence target-signal cancellation will occur at the adaptive canceller. Robust beamforming techniques have been proposed to solve this problem [3–5]. In some algorithms, the norm-constrained adaptive filter (NCAF) is used instead of the traditional LMS filter, and it can effectively reduce the target-signal cancellation caused by the leakage of the blocking matrix [6]. Recently, Hoshuyama proposed a GSC structure with a blocking matrix using coefficient-constrained adaptive filters (CCAF) [7]. The methods using NCAF and CCAF are attractive to real-time applications because they are easy to implement and the performance is good. However, those norm and coefficient constraints for robust beamforming are experimentally determined, and a formal derivation is not available in the literature. Also, only qualitative description on the use of NCAF and CCAF is not convincing for the performance analysis. In this paper, explicit expressions for the constraints are derived in terms of maximal allowed DOA mismatch, which clearly shows the selection of the norm and coefficient constraints is feasible. The results are useful for the design of robust beamformers with NCAF and CCAF.

This paper is organized as follows. A GSC beamformer using NCAF or CCAF is presented in Section 2. The norm

0-7803-7761-3/03/$17.00 ©2003 IEEE

and coefficient constraints are analytically derived in Section 3. Section 4 illustrates the results of computer simulations, and Section 5 gives a brief conclusion.

2. GENERALIZED SIDELOBE CANCELLER WITH NCAF OR CCAF

Consider a uniform linear array with K sensors depicted in Fig. 1. Assume that the look direction of the beamformer is steered to 0°, i.e. the direction perpendicular to the array axis. The array signal vector of size $KJ \times 1$ at instance n is defined as

$$X(n) = [\ x_1(n) \ \ \dots \ \ x_1(n-J+1) \ \ x_2(n) \ \ \dots \ \ x_2(n-J+1) \ \ \dots \ \ x_K(n) \ \ \dots \ \ x_K(n-J+1)\ ]^T$$

where $x_k(m)$ ($k = 1, 2, \dots, K$, $m = n-J+1, n-J+2, \dots, n$) is the digital signal received by the kth sensor at instance m, J is the tap length per sensor, and $(\cdot)^T$ denotes matrix transpose. In [1], the optimization problem of GSC is formulated as

$$\min_{W_M} P = \min_{W_M} (W_F - B_0^T W_M)^T R_{XX} (W_F - B_0^T W_M).$$

Here P is the expected beamformer output power, $W_F \in \mathrm{M}^{KJ\times 1}$ is the weight vector of the fixed beamformer, and W_M is the weight vector of the adaptive canceller. $R_{XX} \in \mathrm{M}^{KJ\times KJ}$ is the array correlation matrix, i.e. $R_{XX} = E\left[X(n)X(n)^T\right]$. B_0 is the blocking matrix of compatible size; when K is even, an example of B_0 is given as

$$B_0 = [\ b_1 \ \ b_2 \ \ \dots \ \ b_j \ \ \dots \ \ b_J\ ]^T, \tag{1}$$

where

$$[b_j]_l = \begin{cases} (-1)^{i-1}, & l = (i-1)J + j, \\ 0, & l \neq (i-1)J + j, \end{cases} \quad i = 1, 2, \dots, K. \tag{2}$$

Here, $[b_j]_l$ represents the lth element of the column vector b_j. We can see that the blocking matrix B_0 acts as a simple spatial filter with a null at 0°. The optimal W_M is solved using the method of Lagrange multipliers, i.e.

$$W_M^{(opt)} = (B_0 R_{XX} B_0^T)^{-1} B_0 R_{XX} W_F. \tag{3}$$

The motivation of NCAF and CCAF is stated here. In the presence of small steering errors, the fixed blocking matrix B_0 can not perfectly reject the target signal, and this leakage will cause target-signal cancellation by the unconstrained LMS algorithm. As the leakage is usually quite small compared with the fixed beamformer output, the adaption of filter coefficients towards target-signal cancellation will cause the filter coefficients to have a much larger norm or amplitude than those in the interference cancellation. Therefore, constraints on the norm or amplitude of the coefficients of LMS algorithm are able to reduce the target-signal cancellation without degrading its interference cancellation ability. The next section derives those constraints for an allowed DOA mismatch level.

3. ANALYTICAL RESULTS

In the following analysis, we assume that only the target-signal source is present, and this assumption allows us to see the close relation between the DOA mismatch error and the optimal taps of the adaptive canceller given by (3). The look direction of the beamformer is assumed to be 0°, but the actual DOA of the target signal may not be exactly 0° and it is denoted by θ_s. It is reasonable to assume that the absolute value of θ_s is small, i.e. $|\theta_s| \leq \theta_{stop}$, where θ_{stop} is a small positive number representing the allowed maximal DOA mismatch.

When there exists only a signal source from θ_s, the correlation matrix of array signals can be written as

$$R_{XX} = \sigma^2 I + R_{SS}(\theta_s) \tag{4}$$

where $\sigma^2 I$ represents the array correlation matrix of additive white sensor noises, and $R_{SS}(\theta_s)$ represents the array correlation matrix of the desired signal. $R_{SS}(\theta_s)$ is a matrix function of θ_s. When θ_s is small, the second order approximation of $R_{SS}(\theta_s)$ using Taylor series expansion in the vicinity of $\theta_s = 0$ is given by

$$R_{SS}(\theta_s) \approx R_{SS}(0) + \left.\frac{\partial R_{SS}(\theta_s)}{\partial \theta_s}\right|_{\theta_s=0} \theta_s + \left.\frac{\partial^2 R_{SS}(\theta_s)}{\partial \theta_s^2}\right|_{\theta_s=0} \frac{\theta_s^2}{2!}. \tag{5}$$

Using (4) and (5), the optimal weight vector of the adaptive canceller in (3) becomes

$$W_M^{(opt)} \approx \left(B_0 \left(\sigma^2 I + R_{SS}(0) + \left.\frac{\partial R_{SS}(\theta_s)}{\partial \theta_s}\right|_{\theta_s=0} \theta_s + \left.\frac{\partial^2 R_{SS}(\theta_s)}{\partial \theta_s^2}\right|_{\theta_s=0} \frac{\theta_s^2}{2!} \right) B_0^T \right)^{-1} B_0 \cdot \left(\sigma^2 I + R_{SS}(0) + \left.\frac{\partial R_{SS}(\theta_s)}{\partial \theta_s}\right|_{\theta_s=0} \theta_s + \left.\frac{\partial^2 R_{SS}(\theta_s)}{\partial \theta_s^2}\right|_{\theta_s=0} \frac{\theta_s^2}{2!} \right) W_F. \tag{6}$$

For further simplification, it is noted that the blocking matrix B_0 is able to perfectly block the signal from 0°, that is

$$B_0 R_{SS}(0) = 0, \text{ and hence } B_0 R_{SS}(0) B_0^T = 0. \tag{7}$$

Moreover, as $B_0 R_{SS}(\theta_s) B_0^T$ is always positive semidefinite for any θ_s, we have for any vector y of compatible length

$$y^T \left(B_0 R_{SS}(\theta_s) B_0^T\right) y \geq 0.$$

From (7) it is obvious that

$$y^T \left(B_0 R_{SS}(0) B_0^T\right) y = 0.$$

Thus, $\theta_s = 0$ attains the minimum of $y^T\left(B_0 R_{SS}(\theta_s) B_0^T\right) y$, which indicates the first order derivative of $y^T\left(B_0 R_{SS}(\theta_s) B_0^T\right) y$ with respect to θ_s is zero at $\theta_s = 0$ due to its continuity, i.e.

$$\frac{\partial}{\partial \theta_s}\left(y^T\left(B_0 R_{SS}(\theta_s) B_0^T\right) y\right)\Big|_{\theta_s=0} = y^T\left(B_0 \left.\frac{\partial R_{SS}(\theta_s)}{\partial \theta_s}\right|_{\theta_s=0} B_0^T\right) y = 0.$$

As y is an arbitrary vector, we must have

$$B_0 \left.\frac{\partial R_{SS}(\theta_s)}{\partial \theta_s}\right|_{\theta_s=0} B_0^T = 0. \tag{8}$$

By removing the zero terms (see (7), (8)), (6) now becomes

$$W_M^{(opt)} \approx \left(\sigma^2 B_0 B_0^T + B_0 \left.\frac{\partial^2 R_{SS}(\theta_s)}{\partial \theta_s^2}\right|_{\theta_s=0} B_0^T \frac{\theta_s^2}{2!}\right)^{-1} \cdot \left(\sigma^2 B_0 W_F + B_0 \left.\frac{\partial R_{SS}(\theta_s)}{\partial \theta_s}\right|_{\theta_s=0} W_F \theta_s + B_0 \left.\frac{\partial^2 R_{SS}(\theta_s)}{\partial \theta_s^2}\right|_{\theta_s=0} W_F \frac{\theta_s^2}{2!}\right). \tag{9}$$

There are two significant quantities in (9): σ^2 and θ_s^2. In most cases, the matrix $\left.\frac{\partial^2 R_{SS}(\theta_s)}{\partial \theta_s^2}\right|_{\theta_s=0}$ is nonzero and the amplitude of its elements is comparable to that of the identity matrix I. When $\theta_s^2 << \sigma^2$, the noise components are dominant in the canceller compared to the blocking matrix leakage, and hence the target-signal cancellation problem is partially alleviated. On the other hand, when $\theta_s^2 >> \sigma^2$, the blocking matrix leakage is dominant and the target-signal cancellation is severe. To reveal the target-signal cancellation problem, we only consider the case that the noise components are negligible compared to the target-signal leakage at the blocking matrix output. In this case, the approximate optimal weight is given by

$$\begin{aligned} W_M^{(opt)} \approx & \left(B_0 \left.\frac{\partial^2 R_{SS}(\theta_s)}{\partial \theta_s^2}\right|_{\theta_s=0} B_0^T\right)^{-1} B_0 \cdot \left.\frac{\partial R_{SS}(\theta_s)}{\partial \theta_s}\right|_{\theta_s=0} W_F \left(\frac{\theta_s}{2}\right)^{-1} + \\ & \left(B_0 \left.\frac{\partial^2 R_{SS}(\theta_s)}{\partial \theta_s^2}\right|_{\theta_s=0} B_0^T\right)^{-1} B_0 \left.\frac{\partial^2 R_{SS}(\theta_s)}{\partial \theta_s^2}\right|_{\theta_s=0} W_F \\ \triangleq & \left(\frac{2}{\theta_s}\right) G + H, \end{aligned}$$

where G and H are defined as

$$G = \left(B_0 \left.\frac{\partial^2 R_{SS}(\theta_s)}{\partial \theta_s^2}\right|_{\theta_s=0} B_0^T\right)^{-1} B_0 \cdot \left.\frac{\partial R_{SS}(\theta_s)}{\partial \theta_s}\right|_{\theta_s=0} W_F, \tag{10}$$

and

$$H = \left(B_0 \left.\frac{\partial^2 R_{SS}(\theta_s)}{\partial \theta_s^2}\right|_{\theta_s=0} B_0^T\right)^{-1} B_0 \cdot \left.\frac{\partial^2 R_{SS}(\theta_s)}{\partial \theta_s^2}\right|_{\theta_s=0} W_F.$$

When the matrix inverse of $B_0 \left.\frac{\partial^2 R_{SS}(\theta_s)}{\partial \theta_s^2}\right|_{\theta_s=0} B_0^T$ doesn't exist, its pseudoinverse is used instead.

When θ_s approaches to zero, the term H in the expression of $W_M^{(opt)}$ can be ignored as the term $(\frac{2}{\theta_s})G$ is dominant. Then, it is clear that for quite small DOA mismatch error θ_s, the norm of the optimal weight vector $W_M^{(opt)}$ is nearly proportional to $\left|\frac{2}{\theta_s}\right|$, and this term becomes quite large as θ_s approaches to zero. In other words, target-signal cancellation becomes severe in the presence of small DOA mismatch only when the norm and amplitude of the adaptive weight vector are too large. Therefore, we can reduce the degree of target-signal cancellation by limiting the norm or amplitude of the optimal weight vector. Such constraints can be mathematically expressed either in an l_2-norm form as

$$\| W_M^{(opt)} \| \leq \left|\frac{2}{\theta_{stop}}\right| \| G \|, \tag{11}$$

or more strictly as

$$-\left|\frac{2}{\theta_{stop}}\right| \left|[G]_l\right| \leq \left[W_M^{(opt)}\right]_l \leq \left|\frac{2}{\theta_{stop}}\right| \left|[G]_l\right|. \tag{12}$$

Here, $[G]_l$ and $[W_M^{(opt)}]_l$ ($l = 1, 2, \ldots$) denote the lth elements of the vector G and $W_M^{(opt)}$ respectively. $\|\cdot\|$ and $|\cdot|$ represent the l_2-norm and the absolute value respectively.

4. SIMULATION RESULTS

In this section, the GSC algorithm with the derived coefficient constraints in (12) is evaluated by computer simulations. The performance of the norm constraint given by (11) is similar and hence omitted here. The conventional GSC algorithm is also implemented for comparison. A four-element uniform linear microphone array is used in this simulation, i.e. $K = 4$. The spacing between adjacent microphones is 4 cm, and the sampling rate is 8 kHz. Bandlimited white Gaussian signals (0.1-3.4 kHz) are used as source signals, and the assumed look direction is 0°. The length of tap line for each sensor is $J = 49$, and the number of delayed samples between the fixed beamformer output and the blocking matrix output is 24. The signal is normalized at 0 dB, and the additive senor noises are at -30 dB. The delay-and-sum algorithm is used for the fixed beamformer, and the blocking matrix B_0 is given by (1) and (2).

To calculate the norm and coefficient constraints numerically, (11) and (12) require an explicit expression of array

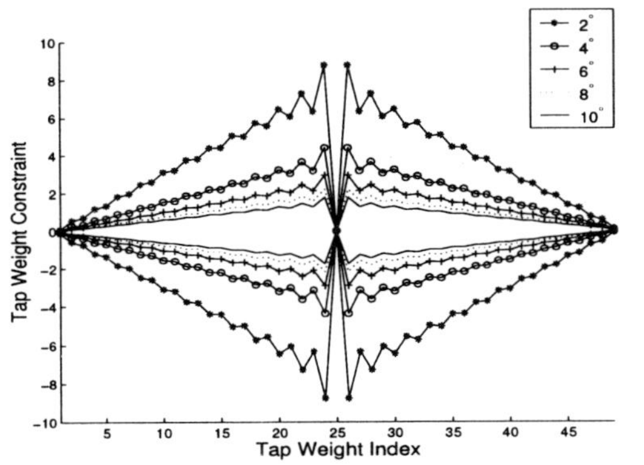

Fig. 2. Theoretically determined coefficient constraints as functions of maximal DOA mismatch

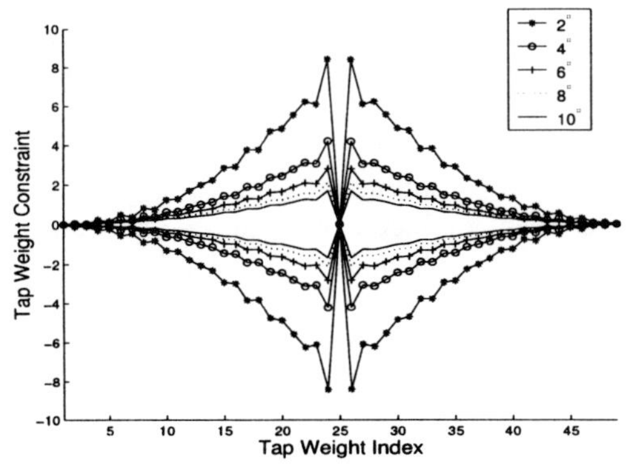

Fig. 3. Experimentally determined coefficient constraints as functions of maximal DOA mismatch

correlation matrix $R_{SS}(\theta_s)$ so that G can be calculated according to (10). Assume the signal source is white Gaussian and normalized, the correlation matrix can be written as

$$R_{SS}(\theta_s) = \begin{bmatrix} R_{11}(\theta_s) & \cdots & R_{1K}(\theta_s) \\ \vdots & \ddots & \vdots \\ R_{K1}(\theta_s) & \cdots & R_{KK}(\theta_s) \end{bmatrix},$$

where $R_{ij}(\theta_s)$ ($i = 1, 2, \ldots, K$, $j = 1, 2, \ldots, K$) is a J-by-J matrix. $R_{ij}(\theta_s)$ is derived as

$$[R_{ij}(\theta_s)]_{pq} = \frac{\sin[\pi(p - q + \frac{(j-i)d\sin(\theta_s)/c}{T_s})]}{\pi(p - q + \frac{(j-i)d\sin(\theta_s)/c}{T_s})},$$

where $[\cdot]_{pq}$ denotes the matrix element in the pth row and qth column. Here d is the array inter-element spacing, c is the sound speed and T_s is the sampling interval.

The coefficient constraints determined by (12), i.e. $\pm\left|\frac{2}{\theta_{stop}}\right| |[G]_l|$, are plotted as a function of tap index (1-49) for various maximal DOA mismatch (2°-10°) in Fig. 2. The constraint for each tap has the maximal amplitude around tap 25 due to the added delays between the fixed beamformer output and the blocking matrix output at the adaptive canceller, i.e. 24. Fig. 3 shows the values of these constraints

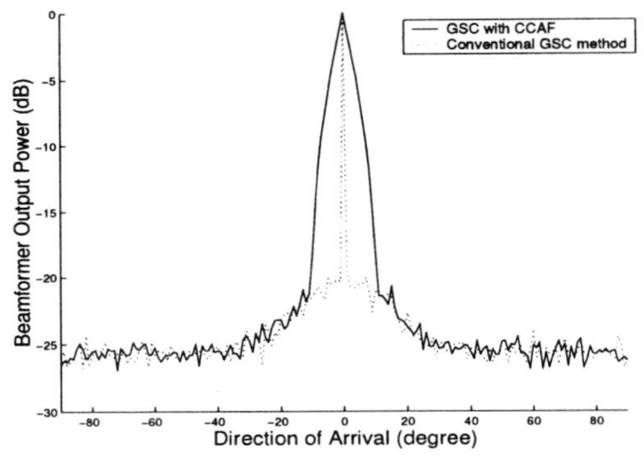

Fig. 4. Spatial response as a function of DOA angle

determined by practical experiments that heuristically set the weight threshold based on the observations on the conventional GSC algorithm. It can be seen that the theoretically determined constraints are quite close to those determined by experimental methods. The beampatterns of the GSC with CCAF for maximal 10° DOA mismatch and the conventional GSC algorithm are plotted in Fig. 4. It is shown that the coefficient constraints are able to improve the robustness of adaptive beamforming to small DOA mismatch. The acceptance angle of the conventional GSC algorithm is quite narrow, which is significantly widened by the use of CCAF.

5. CONCLUSIONS

In this paper, explicit expressions of the norm constraints and coefficient constraints are derived for robust adaptive beamforming, which analytically proves the effectiveness of the norm and coefficient constraints for steering errors. The results are important for beamformer design and performance analysis, and simulations have verified the correctness.

6. REFERENCES

[1] L.J. Griffiths and C.W. Jim, "An alternative approach to linearly constrained adaptive beamforming," *IEEE Trans. Antennas Propagat.*, vol. AP-30, pp. 27–34, Jan. 1982.

[2] K.M. Buckley and L.J. Griffiths, "An adaptive generalized sidelobe canceller with derivative constraints," *IEEE Trans. Antennas Propagat.*, vol. AP-34, pp. 311–319, Mar. 1986.

[3] M.H. Er and A. Cantoni, "A new approach to the design of broad-band element space antenna array processors," *IEEE J. Oceanic Eng.*, vol. OE-10, pp. 231–240, Jul. 1985.

[4] D. Nunn, "Suboptimal frequency-domain adaptive antenna processing algorithm for broadband environments," *Proc. Inst. Elec. Eng. F*, vol. 134, no. 4, pp. 341–351, Jul. 1987.

[5] S. Zhang and I.L. Thng, "Robust presteering derivative constraints for broadband antenna arrays," *IEEE Trans. Signal Processing*, vol. 50, pp. 1–10, Jan. 2002.

[6] H. Cox, R.M. Zeskind, and M.M. Owen, "Robust adaptive beamforming," *IEEE Trans. Acoust., Speech, Signal Processing*, vol. ASSP-35, pp. 1365–1376, Oct. 1987.

[7] O. Hoshuyama, A. Sugiyama, and A. Hirano, "A robust adaptive beamformer for microphone arrays with a blocking matrix using constrained adaptive filters," *IEEE Trans. Signal Processing*, vol. 47, pp. 2677–2684, Oct. 1999.

Integrated Active Noise Control Communication Headsets

Woon S. Gan[1] and Sen M. Kuo[2]

[1] School of Electrical & Electronic Engineering, Nanyang Technological University, Singapore
[2]Department of Electrical Engineering, Northern Illinois University, DeKalb, IL, 60115, USA

ABSTRACT

In this paper, we present the development and evaluation of an integrated active noise control (ANC) communication headsets. This integrated system provides cancellation of noise that corrupted both the transmitted near-end speech and the received far-end speech. It utilizes an adaptive feedback ANC filter to reduce acoustic noise that corrupts the far-end signal inside the ear cups of headsets, and uses an adaptive noise-canceling filter to enhance the near-end speech before sending it to the far-end. This integrated approach minimizes the required signal processing components, and optimizes the functionality of each processing block for combining noise cancellation tasks in the communication headsets. Computer simulations using recorded noise, speech signals, and measured headsets impulse response are carried out to evaluate the performance of this integrated system.

1. INTRODUCTION

As communication devices such as cellular phones, PDA, and wireless communication headsets become more widely used. Designers will need to address the usage of these devices in noisy environments, which corrupts normal voice communications. In addition, existing signal processing techniques such as speech coding, automatic speech recognition, speaker identification, channel transmission, and echo cancellation are usually developed for the communication devices under noise-free assumption. Therefore it is important for the front-end noise cancellation algorithms to remove noise effectively in both the received far-end and transmitted near-end speeches.

Acoustic noise problems become more serious as increased numbers of industrial equipment such as engines, blowers, fans, transformers, and compressors are in use in many outdoor installations, planes, and automobiles. The traditional passive earmuffs are valued for their high attenuation over a broad frequency range; however, they are relatively large, costly, and ineffective at low frequencies. ANC [1-3] systems cancel the unwanted noise based on the principle of superposition. Specifically, an anti-noise of equal amplitude and opposite phase is generated and combined with the primary noise, thus resulting in the cancellation of both noises. The ANC system efficiently attenuates low frequency noise, where passive methods are ineffective, bulky in size, and tend to be very expensive. ANC is developing rapidly because it permits improvement in noise reduction, which results in potential benefits in weight, volume, and cost. A better approach is to use a combination of passive and ANC technique.

Previously, analog controller using feedback configuration commonly referred to as active noise reduction [4] has been employed to cancel noise in the headsets. Since this is a non-adaptive approach, no on-line modeling of the ear-cup transfer function can be carried out in real time. Current research in ANC for communication headsets focuses on using adaptive feedforward technology [2-3]. In practice, however, the feedforward ANC systems for headsets have to handle causality and performance deficiencies caused by non-stationary reference inputs, measurement noise, acoustic feedback [1, 5], and higher cost of using additional reference microphones. More recently, technique has been developed to combine the analog feedback with digital feedforward [4] to achieve better noise canceling performance. However, the limited flexibility in using the analog filter can be a restriction for further improvement, such as the on-line modeling of the secondary path.

In response to these issues, an adaptive feedback ANC communication headset that integrates on-line secondary path modeling using far-end speech with the adaptive noise canceller for enhancing near-end speech transmission is proposed in this paper. Fig. 1 shows the functional blocks required to be integrated in the communication headsets for communicating between two parties. This integrated system handles the following major functions:

- Adaptive feedback ANC to attenuate the near-end noise based on the error microphones inside the ear-cup.
- Off-line and on-line modeling of the secondary path inside the ear-cup using the training signal and far-end speech signal, respectively.
- An additional adaptive filter which functions as an adaptive noise canceller to enhance the near-end speech before transmitting to the far-end.

The main challenge in integrating these components is to minimize the overall system cost, and optimize the usage of existing signals and algorithms within the system.

In this paper, the emphasis is on the algorithm development and verification of the integrated communication headsets. The principle of the main processing blocks is first explained in the next section. This is followed by Section 3, which describes the integration of these processing blocks into integrated ANC-communication headsets. Section 4 presents simulation results that validate the performance of the integrated system. Finally, the concluding section summaries the overall performance of the integrated ANC-communication headsets.

0-7803-7761-3/03/$17.00 ©2003 IEEE

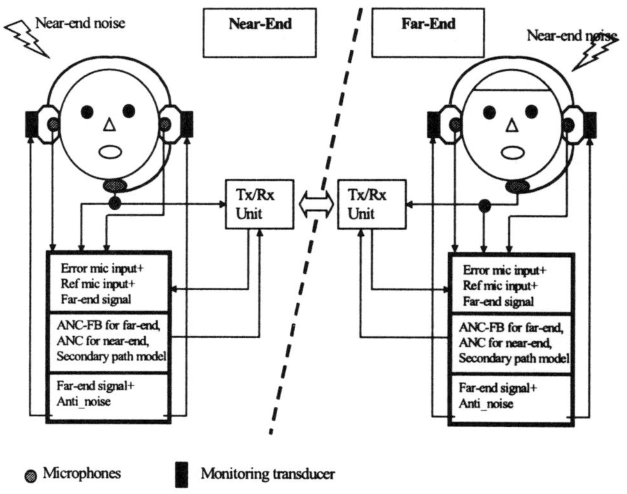

Figure 1. Integrated ANC communication headsets

2. MAIN PROCESSING BLOCKS

2.1 Feedback ANC System

In an ANC application, the primary near-end noise $d(n)$ is not available during the operation of ANC because it was canceled by the secondary noise. Therefore the basic idea of an adaptive feedback ANC is to estimate the primary noise, and use it as a reference signal $x(n)$ for the ANC filter, $W(z)$. An adaptive feedback ANC system is required for applications include spatially incoherent noise generated from turbulence; noise generated from many sources, and propagation paths, and induced resonance where no coherent reference signal is available

The complete adaptive feedback ANC system using the filtered-x least-mean-square (FXLMS) algorithm is illustrated in Fig. 2, where $\hat{S}(z)$ is required to compensate for the secondary path. The reference signal $x(n)$ is synthesized as an estimate of $d(n)$, which is expressed as

$$x(n) \equiv \hat{d}(n) = e(n) + \sum_{m=0}^{M-1} \hat{s}_m y(n-m)\,, \qquad (1)$$

where $\hat{s}_m, m = 0, 1, ..., M-1$ are the coefficients of the Mth order FIR filter $\hat{S}(z)$ used to estimate the secondary path.

The secondary signal $y(n)$ is generated as:

$$y(n) = \sum_{l=0}^{L-1} w_l(n) x(n-l)\,, \qquad (2)$$

where $w_l(n), l = 0, 1, ... L-1$ are the coefficients of $W(z)$ at time n, and L is the order of the FIR filter $W(z)$. The filter coefficients are updated by the FXLMS algorithm expressed as follows:

$$w_l(n+1) = w_l(n) + \mu x'(n-l)e(n), \quad l = 0, 1, ..., L-1 \quad (3)$$

where μ is the step size, and

$$x'(n) \equiv \sum_{m=0}^{M-1} \hat{s}_m x(n-m) \qquad (4)$$

is the filtered reference signal.

The adaptive feedback ANC algorithm summarized in (1) to (4) is very effective and compact since no reference sensor is required at the external of the headsets. However, in some cases, large noise level difference can make the algorithm unstable. An effective solution is to employ the normalized FXLMS algorithm as follows:

$$w_l(n+1) = w_l(n) + \frac{\mu}{\hat{P}_x(n) + c} x'(n-l)e(n), \qquad (5)$$

for $l = 0, 1, ..., L-1$, where the power estimate

$$\hat{P}_x(n) = (1-\alpha)\hat{P}_x(n-1) + \alpha x^2(n) \qquad (6)$$

is based on the first-order recursive filter with $\alpha \approx 0.99$, and c is a small constant to prevent using a large step size.

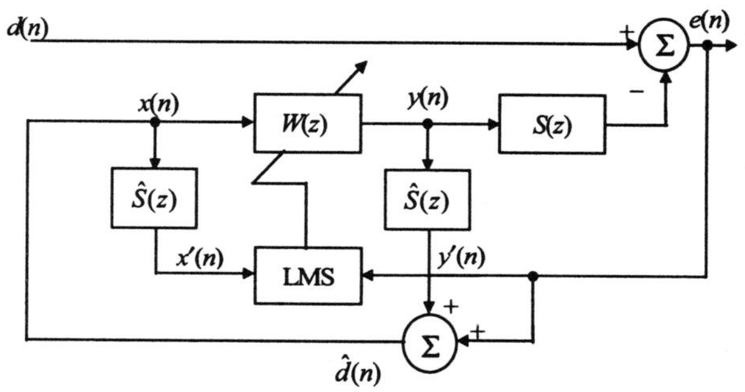

Figure 2. *Block diagram of adaptive feedback ANC system*

2.2 Modeling of Secondary Paths

As mentioned in the previous section, the secondary path $S(z)$ needs to be estimated and used in the updating process of the FXLMS algorithm. Estimation of $S(z)$ is usually performed off-line, followed by on-line modeling [1]. Off-line initialization is carried out initially when a training signal is injected into both the adaptive filter $\hat{S}(z)$, and the actual secondary path $S(z)$. This becomes the traditional adaptive system identification. It is important to select a training signal that is persistently excited so as to model the secondary path over the entire frequency range of interest. A possibility is to use music with wide bandwidth excitation [1] to model the secondary path over a period of time.

On-line modeling of $S(z)$ can be performed after the off-line modeling has converged to the required error level. The received far-end speech replaces the training signal to continuously track the variation of the secondary path transfer function. The far-end speech $a(n)$ must be of persistent excitation, and uncorrelated with the primary noise $d(n)$. However a smaller step-size must be used during on-line modeling compared to off-line modeling to ensure good performance.

2.3 Noise Cancellation for Near-End Speech

In speech communication, there is a need to remove the near-end noise before sending it to the far-end. As shown in Fig. 3, a simple adaptive noise canceling filter $H(z)$ using the LMS algorithm can be used to remove the near-end noise. The microphone used to pick up the desired near-end signal also sensed the undesired near-end noise. $P(z)$ is the primary path from the noise source to the microphones. However, a correlated noise input must be used to train the noise canceling filter $H(z)$ using the LMS algorithm. The adaptive LMS algorithm is given as:

$$h_l(n+1) = h_l(n) + \mu y(n-l)e(n), \quad l = 0,1,...,L-1 \quad (7)$$

where $y(n)$ is correlated with the reference near-end noise.

The above three processing blocks will be combined in the integrated system, and explained in the next section.

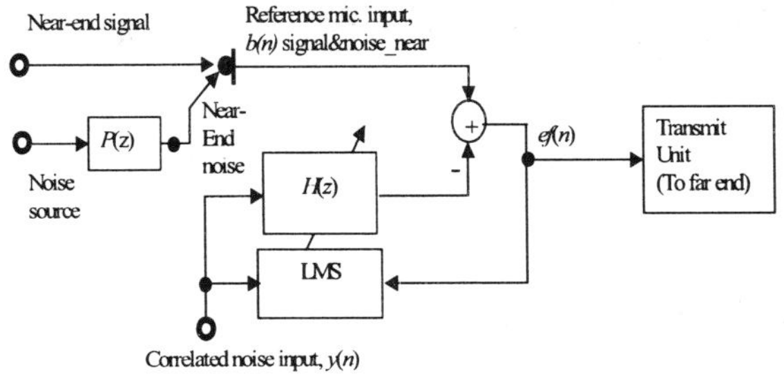

Figure 3. *Adaptive noise cancellation filter for far-end transmission*

3. Integrated ANC-Communication Systems

This section describes the integration of the feedback ANC filter $W(z)$, the secondary path modeling filter $\hat{S}(z)$, and the adaptive noise canceling filter $H(z)$. Three microphones are used in this integrated ANC-communication headsets as shown in Fig. 1. One microphone is placed close to the transducer inside each ear-cups. These microphones are assigned as the error microphones to pick up the noise entering the ear-cup. A communication microphone, which is located external of the ear-cup, is used to pick up the near-end speech, but corrupted by the near-end noise. A variation of the same system was previously described in [5] for entertainment headsets.

In the proposed integrated ANC-communication system, the residual noise picked up by the error microphone is used to synthesize the primary noise $x(n)$ for updating the adaptive filter coefficients using the FXLMS algorithm. The far-end speech component will also become the interference to this integrated system, and a method is developed to combine the communication and the ANC systems.

The detailed integration of adaptive feedback ANC with communication system is shown in Fig. 4. The error sensor output signal $e(n)$ contains both the residual noise and the desired speech signal. The speech interference cancellation filter $\hat{S}(z)$ uses the far-end speech signal $a(n)$ as the reference signal to estimate and then remove the speech components in $e(n)$. The difference error signal, $e'(n)$, consists of the residual noise only, is used to update the adaptive noise control filter $W(z)$. Note that the update of $\hat{S}(z)$ using the LMS algorithm is located at the output of the feedback ANC, and can be used for both off-line and on-line modeling as explained in the previous section.

It has been shown analytically in [5] that the regenerated reference signal $x(n)$ derived from the adaptive filter $W(z)$ and the error signal, is a close estimate of the primary noise $d(n)$. That is, we are able to obtain an accurate reference signal for the noise control filter $W(z)$. The noise cancellation filter $H(z)$ can be neatly integrated into the system by channeling the anti-noise signal $y(n)$ into its input. The signal $y(n)$ is highly correlated to the near-end noise $d(n)$, and therefore it is a good candidate for the input of $H(z)$. This filter provides good cancellation of the near-end noise to produce a cleaner near-end speech for transmission. Furthermore, no additional microphone is required to pick up the reference noise, thus solving the crosstalk interference between microphones.

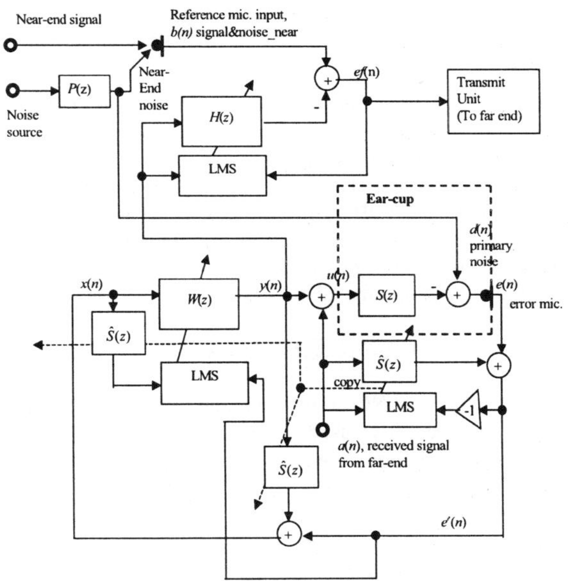

Figure 4. An integrated communication and ANC system

We can summarize several advantages of the integrated ANC-communication system as follows: (1) Good estimation of the true residual noise $e'(n)$ without interfering with the speech signal $a(n)$, (2) large step size can be used in adapting the cancellation filter $W(z)$ since the difference error signal $e'(n)$ used by the FXLMS algorithm is not corrupted by the high volume speech signal, (3) the adaptive feedback ANC technique provides a more accurate noise cancellation since the microphone is placed inside the ear-cup of the headsets, (4) the system uses single microphone per ear cup, thus produces a compact, lower power consumption, and a cheaper solution, (5) the audio signal can be neatly used to drive both on-line and off-line modeling of the secondary path transfer function, and (6) the use of adaptive noise cancellation filter enhances the near-end speech before sending to the far-end. The next section will examine some of the performance results obtained using this integrated system.

4. SIMULATION RESULTS

Computer simulation was conducted to examine the performance of the integrated ANC-communication headsets. The impulse response of the headsets was measured using a dummy head, and recorded speech and noise signals were used in these computer simulations.

Experiment was carried out in order to evaluate the capabilities of the integrated system. An engine noise with prominent harmonics at 61 Hz, 122 Hz and 183 Hz was used as the noise source. The adaptive filters $W(z)$, $H(z)$, and $\hat{S}(z)$ used in this simulation are 128-tap, 64-tap, and 80-tap FIR filters, respectively. Step sizes of 0.01, and 0.05 were used to adapt $W(z)$ and $H(z)$ respectively. Off-line modeling of $S(z)$ was first performed to obtain the secondary path estimation $\hat{S}(z)$ by using a wide spectrum signal [1] with a step size of 0.01, and followed by on-line modeling using the speech signal with a small step size of 0.0005.

Fig. 5 shows the case when the speech signal was corrupted by the engine noise. Note that the engine noise was recorded from the welding power generator running at 3700 rpm. The narrowband harmonics (61 Hz, 122 Hz, and 183 Hz) were canceled by more than 30 dB, while the near-end speech signal was enhanced by more than 25 dB. Fig. 6 shows the original signal corrupted by engine noise, and the enhanced near-end signal after passing through the adaptive filter.

5. CONCLUSION

A novel integrated ANC-communication algorithm was presented in this paper. It is used to reduce the environmental interference without canceling the received far-end signal. At the same time, it enhances the near-end speech signal before sending to the far-end. This system also enables both off-line and on-line modeling to be integrated intelligently within the system. This provides a practical, cheap, and compact solution where low-cost DSP processor can be used to implement it.

We also investigated the performance of the integrated system using different disturbing noises. We found that in general, this system provides a good noise canceling performance for both far-end and near-end speech communication. There are several novel features built into this integrated system, include on-line updating of the secondary path, using a difference error to update the adaptive filter, and reusing the anti-noise signal to train the adaptive cancellation filter. These features provide a robust implementation and faster adaptation to track the changes in the headsets' transfer function and surrounding noise.

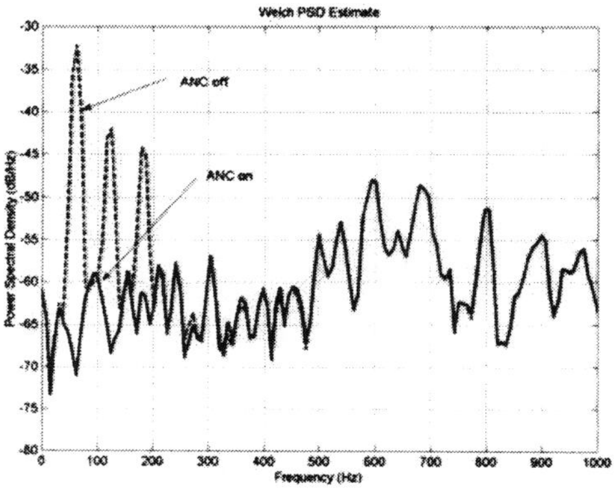

Figure 5. Noise spectral for the error signals without (dotted line) and with (solid line) using ANC filter under an engine disturbance.

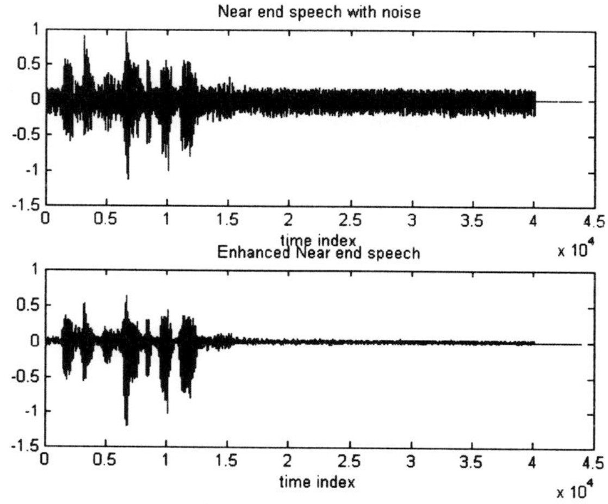

Figure 6. The top and bottom plots show the original signal corrupted by engine noise and enhanced near-end speech respectively.

6. REFERENCES

[1] Sen M. Kuo and Dennis R. Morgan, *Active noise control systems: Algorithms and DSP implementations*, John Wiley & Sons, Inc., New York, 1996.

[2] C. H. Hansen and S. D. Snyder, *Active control of noise and vibration*, E&FN Spon, London, 1997.

[3] B Widrow and E Walach, *Adaptive inverse control*, Prentice Hall, Inc., 1996.

[4] B Rafaely, Active noise reducing headset, OSEE Online Symposium, 2001, pp 1-8.

[5] W.S. Gan and S.M. Kuo, An integrated audio active noise control headsets, *IEEE Transaction on Consumer Electronics*, Vol. 48, No. 2, May 2002, pp 242-247.

Minimum Selection GSC and Adaptive Low-Power Rake Combining Scheme

Suk Won Kim [1], Dong S. Ha[2], and Jeffrey H. Reed[3]

[1]System LSI Division, Device Solution Network
Samsung Electronics Co., Ltd.
San #24 Nongseo-Ri, Giheung-Eup
Yongin-City, Gyeonggi-Do, Korea 449-711

[2]VTVT Laboratory and [3]MPRG
Department of Electrical and Computer Eng.
Virginia Tech
Blacksburg, Virginia 24061 USA

E-mail: sukim4@samsung.co.kr, ha@vt.edu, and reedjh@vt.edu

Abstract

In this paper, we investigate a new generalized selection combining (GSC) technique and an adaptive low-power rake combining scheme to save the power consumption of mobile rake receivers for wideband CDMA systems. The new GSC technique called minimum selection GSC (MS-GSC) selects a minimum number of rake fingers as long as the combined SNR is larger than a given threshold. The proposed rake combining scheme adaptively adjusts the threshold value to maintain the desired BER, in which a GSC dynamically selects rake fingers to meet the given threshold condition. The proposed MS-GSC shows a low standard deviation in bit error statistics and advantages practical implementation. The proposed adaptive scheme works well with existing GSC schemes as well as the proposed MS-GSC. The proposed scheme reduced the power consumption of a mobile rake receiver up to 67.8 % by turning off unselected rake fingers.

1. Introduction

A rake receiver adopts multiple fingers to exploit diversity of multipath signals called diversity combining. In general, a larger number of fingers would improve the SNR (signal to noise ratio) at the cost of higher circuit complexity and hence larger power dissipation. In practice, the number of rake fingers is in the rage of two to five. Since a rake receiver operates at the chipping rate, it is one of the most power consuming blocks in a baseband signal processor for a CDMA (code division multiple access) receiver.

Instead of selecting all the fingers as the maximal ratio combining (MRC), generalized selection combining (GSC) methods choose the best m fingers out of L fingers depending on the SNR or the signal strength [1]-[4]. Note that the MRC is a special case of a GSC where the number of selected fingers m is fixed to L. The number of selected fingers m is decided *a priori* in [1]-[4], while it varies dynamically in [5]-[7]. For the latter approach, selection of fingers whose SNRs are larger than a given threshold is proposed in [5] and [6], and it is called absolute threshold GSC (AT-GSC). Alternatively, selection of a finger whose relative SNR over the maximum SNR among all fingers is larger than a threshold is proposed in [5] and [7]. This method is called normalized threshold GSC (NT-GSC).

GSC methods intend to save hardware and/or power dissipation. If m is fixed and less than L, it reduces the complexity of the rake receiver and hence the power dissipation of the rake receiver circuit. Since m changes dynamically in the range of 1 to L for the AT-GSC and the NT-GSC, the two schemes do not save hardware. In fact, increased hardware complexity is necessary to be able to change m. However, the AT-GSC and the NT-GSC can save power dissipation by turning off unselected fingers. Two major design considerations regarding the AT-GSC and the NT-GSC are:

(i) determination of threshold values, and

(ii) effectiveness of the two methods in terms of power saving and practical implementation.

A threshold value should be set to meet the required QoS (quality of service), and a maximal number of fingers should be turned off as long as it meets the required QoS. The BER (bit error rate) is often used as the metric for the QoS. For example, a BER of 10^{-3} may be necessary for voice communications. This suggests that if the combined SNR is over a certain threshold, then the BER is below a certain level to meet the required QoS.

In this paper, we propose a new GSC method called minimum selection GSC (MS-GSC) and an adaptive low-power rake combining scheme to determine the threshold values for GSCs. Our MS-GSC selects a minimum number of fingers as long as the combined SNR is maintained larger than a given threshold. Our proposed adaptive scheme is applicable to the three GSC methods - the AT-GSC, the NT-GSC, and the proposed MS-GSC. Through simulation, we estimated the effectiveness of the proposed scheme for a mobile rake receiver of a wideband CDMA (WCDMA) system.

The paper is organized as follows. The proposed MS-GSC is presented in Section 2. An adaptive low-power rake combining scheme is proposed in Section 3. The simulation results applied to a mobile rake receiver of a WCDMA system are presented in Section 4. Finally, Section 5 concludes the paper.

2. New GSC Method

The proposed MS-GSC method selects a minimum number of fingers as long as the combined SNR is maintained larger than a given threshold. The proposed MS-GSC denoted as MS-GSC (T_m, L) selects a minimum number of fingers whose combined SNR is larger than a given threshold T_m.. In contrast to the AT-GSC and the NT-GSC, the MS-GSC performs better as the threshold value becomes larger. The SNR of the MS-GSC for a given threshold is ideally independent of the average finger SNRs.

For a practical implementation, the signal level or strength is used instead of a finger SNR, which is difficult to measure. Since the received signal contains the desired signal as well as noise, the signal strength represents signal plus noise (S+N) value. Even if the signal strength is used for the AT-GSC and the NT-GSC, the trend remains the same. However, the trend for the MS-GSC changes as shown in Figure 1. When the SNR of each finger is low/high, the amount of the noise is relatively large/small. Thus, the combined (S+N) shows a low/high SNR.

0-7803-7761-3/03/$17.00 ©2003 IEEE

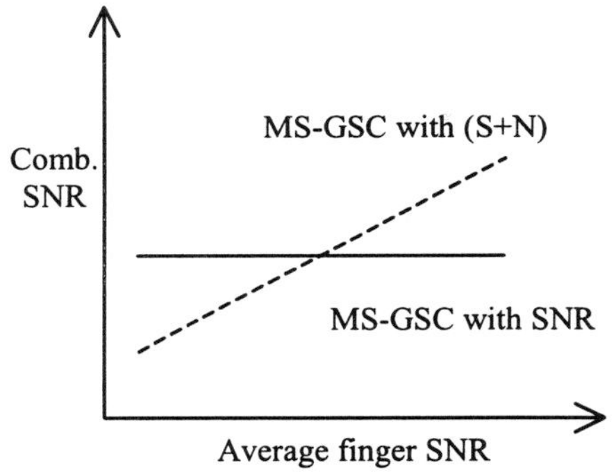

Figure 1. Combined SNR for MS-GSC

The operation of the MS-GSC is as follows. The MS-GSC starts with a given threshold T and k rake fingers. The combined signal strength is periodically measured. If the combined signal strength is less than the threshold, then one more rake finger is activated in the next period. Since the signal with the smaller delay is stronger in general, the finger with the smallest delay is selected. Note that the delay information is provided by a cell searcher. If the combined signal strength is marginally greater than the threshold, then a rake finger with the lowest signal strength is turned off in the next period. Otherwise, the current rake combining with k fingers is maintained.

Comparing with the AT-GSC and the NT-GSC, the proposed MS-GSC has two benefits. Since the combined signal strength in the MS-GSC is maintained constantly, the statistics of erroneous bits have a low standard deviation, which will be presented in Section 4. This may result in less burst errors, which leads to a better error correction for a channel decoder. The second benefit is that the MS-GSC enables power reduction by turning off unselected fingers. The MS-GSC can meet the given threshold condition by turning on another finger with the smallest delay or turning off a finger with the lowest signal strength. In contrast, the AT-GSC and the NT-GSC necessitate activation of each finger momentarily to measure the signal strength, so that it can determine if the signal strength is above the threshold value or not.

3. Adaptive Rake Combining Scheme

We determine a set of N threshold values $\{T(1), T(2), \ldots, T(N)\}$ for each GSC using a system simulation, where each GSC performs better as the index of the threshold value increases. This implies that $T(1) > T(2) > \ldots > T(N)$ for the AT-GSC and the NT-GSC and $T(1) < T(2) < \ldots < T(N)$ for the MS-GSC. A larger N leads to, on average, a larger number of fingers turned off, but it results in more complex hardware and more frequent changes of threshold values. Therefore, a larger N does not guarantee more power saving. The *SNR range* of a threshold value $T(i)$ is the range in which the required BER is met for $T(i)$, but not with $T(i-1)$.

We propose an adaptive scheme to determine threshold values for the three GSC methods, the AT-GSC, the NT-GSC, and the proposed MS-GSC. Threshold values of a GSC method should be adjusted dynamically to turn off a maximal number of fingers, while maintaining the required BER. Since we use BERs as the metric for the QoS, a mechanism is necessary to estimate the current BER. When a Viterbi decoder is employed for the system, the error metric of the survived path at the end of the forward processing is the current BER. For the case of turbo decoders, the number of iterations or the rate of convergence can be used as the metric. The block diagram of the proposed system is presented in Figure 2, in which the control logic adjusts the threshold value dynamically based on the inputs from the channel decoder.

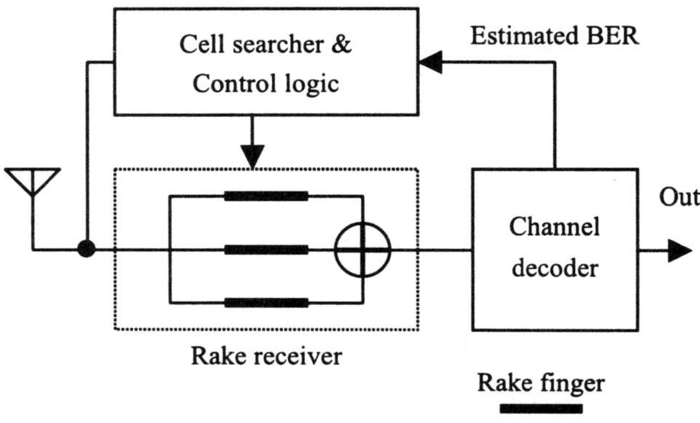

Figure 2. Block Diagram of the Proposed Adaptive Scheme

The proposed adaptive scheme consists of two loops, an outer loop and an inner loop. The outer loop adaptively adjusts the threshold value as described next. The inner loop dynamically changes the finger selection to meet the required condition with the threshold value provided from the outer loop.

Suppose that the current threshold of a GSC method is $T(i)$. The current BER is periodically estimated at the end of each frame. If the estimated BER is higher than the required BER, then the threshold index is increased to $T(i+1)$ in order to lower the BER in the next frame. If the BER is marginally lower than the required BER, then the threshold index is decreased to $T(i-1)$. Otherwise, the current threshold $T(i)$ is maintained.

The power saving of the proposed adaptive scheme is highly dependent on the operating condition and the employed threshold set. The average number of rake fingers activated $\overline{m}$ is obtained as $\overline{m} = \sum_{i=1}^{N} A_i P_i$, where N is the number of threshold values $\{T(i)\}$, A_i is the average number of rake fingers activated for a threshold value $T(i)$, and P_i is the probability of the proposed adaptive scheme operating in the SNR range of the threshold value $T(i)$. Then, the power saving with the proposed adaptive scheme over the MRC rake combiner is obtained as $\frac{L-\overline{m}}{L}$. Thus, if we know each probability P_i, then the power saving with the proposed adaptive scheme could be estimated. The probability P_i depends on the channel condition and the power control strategy. Detailed analytical results are available in 0.

4. Simulation Results

To verify the validity of the adaptive scheme, the proposed scheme with three GSC methods are applied to a mobile rake receiver of a WCDMA system. System models and parameters considered in our simulation are typical for the 3GPP WCDMA system 0 except only one transmit antenna is used at a base station. Eight users' signals with a spreading factor 32 and the common pilot channel (CPICH) signal with a spreading factor 256 are modulated, channelized, combined, scrambled, pulse-shaped, and transmitted through the channel. Twenty percent of the total transmitted power is allocated to the CPICH, and the remaining 80% of the power is allocated equally to each user signal. For the channel profile (such as delay and average power), the ITU (International Telecommunication Union) channel profiles described in [10] are applied. Four or six multipath signals (*M*) are generated in the wireless channel depending on the channel type. Each multipath signal is experienced an independent Rayleigh fading. For the Vehicular channel model, the mobile velocity is assumed to be 50 km/hr, which results in 99.1 Hz of maximum Doppler frequency for a 2.14 GHz carrier frequency. The mobile velocity for the Pedestrian channel is assumed to be 3 km/hr, which results in 5.9 Hz of maximum Doppler frequency. The despread CPICH signal of each multipath signal is utilized to estimate the channel condition, i.e., the amplitude and the phase, and thus is used as a weighting factor of each rake finger to combine the selected rake finger outputs. To reduce the simulation time, channel encoding and decoding is not included. Thus, a hard decision is made at the output of rake combiner and the output is compared with the original data bits to evaluate the BER performance. This BER is fed back to the control logic for the proposed adaptive rake combiners.

4.1 MS-GSC, AT-GSC, and NT-GSC

First, we present the BER performances of the MS-GSC, the AT-GSC, and the NT-GSC. Figure 3 shows the BER performance of the proposed MS-GSC with a threshold set of δ = 0.5, where δ *is defined as the difference between the average numbers of rake fingers activated for two consecutive threshold values*. As shown in the figure, the MS-GSC (T_m, 5) performs better as the threshold T_m becomes large. The AT-GSC and the NT-GSC show almost the same BER performance as that of the MS-GSC with a threshold set of δ = 0.5.

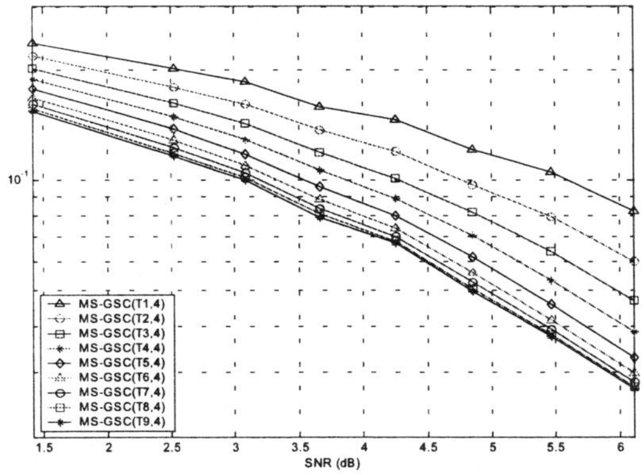

Figure 3. BER Performance with Pedestrian B Channel

Next, we present the BER performance under the ITU Vehicular A channel profile. The BER performance of the AT-GSC with a threshold set of δ = 0.5 is presented in Figure 4. As shown in the figure, the AT-GSC (T_a, 4) performs better as the threshold T_a becomes smaller. We observed that the MS-GSC with a threshold set of δ = 0.5 shows almost the same BER performance as that of the AT-GSC. The NT-GSC performs better than the AT-GSC and the MS-GSC using a threshold set such that the average number of rake fingers activated is same. When the Vehicular B channel profile is applied, each GSC method shows the same trend as for the case of the Vehicular A channel profile. The only difference lies in the combined SNR and the BER performance, since each channel profile has a different power profile.

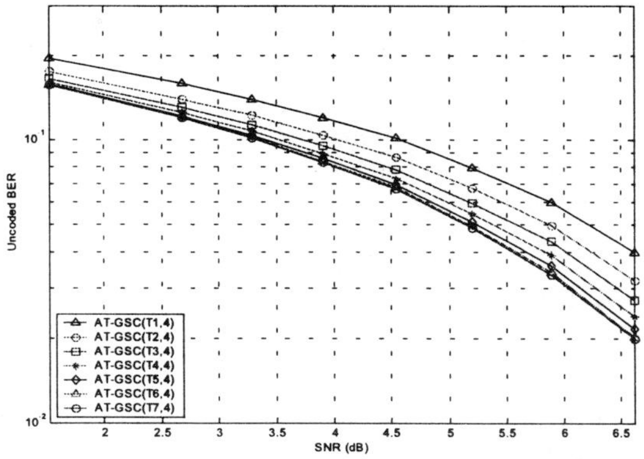

Figure 4. BER Performance with Vehicular A Channel

4.2 Adaptive Rake Combiners

The BER performance as well as the power saving of the proposed adaptive scheme are presented next. To generate the feedback information for the outer loop of the proposed adaptive rake combiners, the current BER performance is estimated on every frame at the rate of 100 Hz. To adjust the number of rake fingers activated for the inner loop of the adaptive rake combiner, the combined signal quality is evaluated in order to check whether it meets the given threshold value provided from the outer loop at the pilot symbol rate of 15 kHz (= 3.84 MHz / 256). The control logic to evaluate the BER and the combined signal quality operates at a much lower frequency compared with the chipping rate (3.84 MHz), at which rake fingers operate. Hence, the power dissipation due to the control logic would be small when implemented in the CMOS technology. (Note that the power dissipation is roughly proportional to the operating frequency in the CMOS.) Thus, the power saving of the adaptive rake combiner can be represented as the average number of rake fingers deactivated.

The performance of adaptive rake combiners under various conditions for the ITU Vehicular A channel profile with a fixed amount of noise is summarized in Table 1. The first column under "Condition" represents operating conditions, in which the desired target BER (TBER) and the average combined SNR with the MRC are presented. A fixed amount of noise is added to the received signal at the receiver, and we obtain an average combined SNR for five fingers with the MRC. The relative noise power to the signal power of the first multipath is –12.0 dB, –11.5 dB, -11.0 dB, and -10.5 dB, which yields 3.90 dB, 4.54 dB, 5.20 dB, and 5.89 dB of the combined SNR, respectively. The second column under "Perform" has three items such that an average BER, a normalized standard deviation of the BER (N-STD), and an average number of rake fingers deactivated (F-saving). The normalized standard deviation is computed as the ratio of standard deviation over the mean of erroneous bits. The column "MRC" represents the performance of the conventional MRC rake combiner. The last three columns (MS-GSC, AT-GSC, and NT-GSC) represent the performance of the adaptive rake combiner employing MS-GSC, AT-GSC, and NT-

GSC, respectively. For brevity, we call them as MS-GSC, AT-GSC, and NT-GSC, respectively, in the following.

As presented in Table 1, all three GSCs achieve the required BER performance for the most cases except the AT-GSC under the target BER of 8 %. However, the deviation is less than 0.25 %, which may be insignificant for a channel decoder. The NT-GSC shows the best performance in terms of power saving. The power saving with the NT-GSC ranges from 44.5 % (= 1.78/4) to 67.8 % (= 2.71/4). The MS-GSC shows the smallest normalized standard deviation, which would result in the least burst errors. Meanwhile, the AT-GSC shows the largest normalized standard deviation. The simulation results under the ITU Pedestrian B channel profile with a fixed amount of noise show the same trend as that under the ITU Vehicular A channel profile.

Table 1. Performance of Adaptive Rake Combiners (Vehicular A)

Condition		Perform	MRC	MS-GSC	AT-GSC	NT-GSC
TBER	SNR					
10 %	3.90 dB	BER	8.48 %	9.67 %	9.93 %	9.59 %
		N-STD	22.19 %	21.86 %	27.25 %	25.21 %
		F-saving	0	1.88	1.91	2.12
	4.54 dB	BER	6.63 %	8.98 %	8.96 %	8.76 %
		N-STD	23.48 %	21.58 %	26.57 %	23.95 %
		F-saving	0	2.62	2.54	2.67
8 %	4.54 dB	BER	6.66 %	7.65 %	8.02 %	7.68 %
		N-STD	23.61 %	22.94 %	29.27 %	27.09 %
		F-saving	0	1.84	1.90	2.13
	5.20 dB	BER	4.89 %	7.05 %	7.05 %	6.87 %
		N-STD	28.27 %	25.93 %	31.87 %	28.95 %
		F-saving	0	2.66	2.58	2.71
6 %	5.20 dB	BER	4.88 %	5.27 %	5.71 %	5.35 %
		N-STD	27.44 %	26.91 %	35.86 %	31.59 %
		F-saving	0	1.39	1.50	1.78
	5.89 dB	BER	3.41 %	4.97 %	5.15 %	4.94 %
		N-STD	30.13 %	28.57 %	36.88 %	31.89 %
		F-saving	0	2.55	2.48	2.68

To verify the ability of the proposed adaptive rake combiner to dynamically adapt to the environment, a variable amount of noise is also considered for the simulation. The result is almost the same as with a fixed amount of noise. Each adaptive rake combiner achieves the required BER performance for most cases. With the exception of a few cases, the maximum deviation is less than 0.5 %. The NT-GSC shows the best performance in terms of power saving, in which the maximum power saving is 65.8 % (= 2.63/4). Like for the case of a fixed amount of noise, the MS-GSC and the AT-GSC show the smallest and the largest normalized standard deviation, respectively. More numerical results are available in [8].

In summary, the simulation results indicate that the proposed adaptive scheme works well with all GSC methods to maintain the required BER performance. The adaptive scheme with the NT-GSC shows good performance in terms of finger saving, and the power saving is as high as 67.8 %. The adaptive scheme with the MS-GSC shows the smallest normalized standard deviation for the all cases, which is somewhat expected.

5. Conclusion

In this paper, we propose a new generalized selection combining (GSC) technique and an adaptive low-power rake combining scheme to save the power consumption of mobile rake receivers for a WCDMA system. The new GSC technique called minimum selection GSC (MS-GSC) selects a minimum number of rake fingers as long as the combined SNR is larger than a given threshold. Since the MS-GSC tries to maintain the combined signal strength, the normalized standard deviation of the BER is the lowest among all the GSCs considered. This results in better error correction when a channel decoder is applied. In addition, the MS-GSC is advantageous in terms of power saving.

The proposed rake combining scheme adaptively adjusts the threshold value to maintain the desired BER, in which each GSC dynamically selects *m* rake fingers among *L* rake fingers to meet the given threshold condition, while turning off unselected fingers to save the power dissipation. The simulation results indicate that the proposed adaptive scheme works well with all GSC methods to maintain the required BER performance. The adaptive scheme with the NT-GSC shows good performance in terms of finger saving. The proposed scheme reduced the power consumption of a mobile rake receiver up to 67.8 %.

References

[1] N. Kong, T. Eng, and L. B. Milstein, "A Selection Combining Scheme for RAKE Receivers," *IEEE International Conference on Universal Personal Communications*, pp. 426-430, November 1995.

[2] M. Z. Win and J. H. Winters, "Virtual Branch Analysis of Symbol Error Probability for Hybrid Selection/Maximal-Ratio Combining in Rayleigh Fading," *IEEE Transactions on Communications*, Vol. 49, No. 11, pp. 1926-1934, November 2001.

[3] M.-S. Alouini and M. K. Simon, "An MGF-based Performance Analysis of Generalized Selection Combining over Rayleigh Fading Channels," *IEEE Transactions on Communications*, Vol. 48, No. 3, pp. 401-415, March 2000.

[4] C. M. Lo and W. H. Lam, "Approximate BER Performance of Generalized Selection Combining in Nakagami-*m* Fading," *IEEE Communications Letters*, Vol. 5, No. 6, pp. 254-256, June 2001.

[5] L. Yue, "Analysis of Generalized Selection Combining Techniques," *IEEE Vehicular Technology Conference*, pp. 1191-1195, May 2000.

[6] A. I. Sulyman and M. Kousa, "Bit Error Rate Performance of a Generalized Diversity Selection Combining Scheme in Nakagami Fading Channels," *IEEE Wireless Communications and Networking Conference*, pp. 1080-1085, September 2000

[7] M. K. Simon and M.-S. Alouini, "Performance Analysis of Generalized Selection Combining with Threshold Test per Branch (T-GSC)," *IEEE Global Telecommunications Conference*, pp. 1176-1181, November 2001.

[8] Suk Won Kim, "Smart Antennas at Handsets for the 3G Wideband CDMA Systems and Adaptive Low-Power Rake Combining Schemes," Ph. D. Dissertation, Department of Electrical and Computer Engineering, Virginia Polytechnic Institute and State University, July 2002.

[9] http://www.3gpp.org/.

[10] ITU ITU-R M.1225, "Guidelines for Evaluations of Radio Transmission Technologies for IMT-2000," 1997.

ADAPTIVE IIR NOTCH FILTER WITH CONTROLLED BANDWIDTH FOR NARROW-BAND INTERFERENCE SUPPRESSION IN DS CDMA SYSTEM

Aloys Mvuma

Graduate School of Engineering,
Hiroshima University,
Higashi-Hiroshima, 739-8527 Japan.
mvuma@ecl.sys.hiroshima-u.ac.jp
hinamoto@ecl.sys.hiroshima-u.ac.jp

Shotaro Nishimura, Takao Hinamoto*

*Department of Electronic and
Control Systems Engineering,
Shimane University,
Matsue-Shi, 690-8504 Japan.
nisimura@riko.shimane-u.ac.jp

ABSTRACT

Suppression of NBI in a DS CDMA system using a fully adaptive IIR notch filter is investigated. An algorithm that requires no gradient signal generating filters for tracking the center frequency of the NBI is proposed. A gradient algorithm for adapting the notch bandwidth coefficient to a value that results in maximum SNR improvement factor and minimum BER is presented. Convergence behaviors of the algorithms are analyzed. Closed-form expressions for SNR improvement factor and BER of the system are derived. Simulations are shown to verify the analyses and compare the performance of the proposed scheme with that of a linear adaptive FIR prediction filter.

1. INTRODUCTION

Adaptive notch filters are favorably applicable in situations where detection and enhancement or suppression of narrow-band signals in broadband signals are required. Such requirements are commonly encountered in many electronic systems. A direct-sequence code-division multiple-access (DS CDMA) overlay system where DS CDMA system shares frequency spectrum with existing narrow-band users is a typical example. Such an overlay alleviates the problem of spectra congestion and is made possible by low power spectral density and inherent immunity to narrow-band interference (NBI) properties of DS CDMA systems [1].

Digital signal processing of DS CDMA signals prior to despreading is normally used to reduce the degradation on the performance of DS CDMA systems caused by NBI. The processing takes advantage of the disparity between bandwidths of the DS CDMA signals and NBI. Adaptive linear finite impulse response (FIR) predictors and interpolators with least-mean-square (LMS) algorithms were proposed in the past. An increase in signal-to-noise ratio (SNR) and a reduction in bit error rate (BER) were realized. Nonlinear methods have also been proposed for improving the performance of conventional predictors and interpolators at the expense of increased complexities [2], [3].

The use of adaptive infinite impulse response (IIR) notch filters for the suppression of NBI in a DSSS system was proposed in [4], [5]. Adaptive IIR notch filters are known for their computational simplicity when compared to their FIR counterparts with same level of performance. We propose a fully adaptive IIR notch filter for suppression of NBI from a DS CDMA system as an extension to results presented in [4], [5]. A simple normalized notch frequency coefficient updating algorithm with reduced computational and hardware complexities is proposed. A gradient algorithm for adapting the notch bandwidth coefficient to a value that results in maximum SNR improvement factor and minimum BER is presented and the convergence characteristics analyzed. Analytical expressions for the SNR improvement factor and BER of the DS CDMA system are then derived. Simulations to support the analytical results and compare the performance of the proposed NBI suppression scheme with that of an adaptive linear prediction filter are presented.

2. SYSTEM MODEL

2.1. Background

A synchronous, biphase spreading, binary phase shift keying (BPSK) data modulation DS CDMA system model with K simultaneous users is used. The transmitted signal $s(t)$ is modeled as

$$s(t) = \sum_{j=1}^{K} \sqrt{2P} b_j(t) c_j(t) \cos(\omega_c t + \phi_j). \tag{1}$$

$b_j(t)$ and $c_j(t)$ are data and pseudo-noise (PN) signals of the jth user with amplitudes ± 1, bit duration T_b, and chip duration T_c, respectively, where $T_c \ll T_b$. ω_c is the carrier frequency. ϕ_j is the phase angle introduced by the BPSK modulator, which is assumed to be zero for simplicity. P is the transmitted signal power.

Received signal $r(t)$ (see Fig. 1) is a sum of the transmitted signal $s(t)$, NBI $i_r(t)$, and channel additive white Gaussian noise (AWGN) $n_r(t)$ with zero mean and two-sided power spectral density $\frac{\eta}{2}$ and is expressed as

$$r(t) = s(t) + i_r(t) + n_r(t). \tag{2}$$

Sampling the correlator output (perfect carrier synchronization is assumed) $u(t)$ at the chip rate, a discrete-time signal

$$u(k) \quad = \quad m(k) + i(k) + n(k) \tag{3}$$

0-7803-7761-3/03/$17.00 ©2003 IEEE

is obtained. $m(k)$ is a zero mean white random process DS CDMA data signal with a binomial density function [3]

$$p\Big(m(k)\Big) = 2^{-K}\sum_{l=0}^{K}\left(\begin{array}{c}K\\l\end{array}\right)\delta\Big(m(k)-(K-2l)S\Big), \quad (4)$$

$-SK, -S(K+2), -S(K+4), ..., S(K-4), S(K-2), SK$ its possible values and variance KS^2, where $S=\sqrt{PT_c}$.

$i(k)$ is a NBI sequence with power σ_i^2, which is modeled as an autoregressive (AR) process of second order [3]

$$i(k) = v(k) - a_1 i(k-1) - a_2 i(k-2) \quad (5)$$

where $v(k)$ is a white Gaussian process of zero mean and variance σ_v^2. $n(k)$ is a white Gaussian random process with zero mean and variance σ_n^2. It can be shown that $\sigma_n^2 = \frac{\eta}{2}T_c$.

2.2. Adaptive IIR Notch Filter

To notch out the NBI, a second-order adaptive IIR notch filter characterized by a transfer function $H(z)$ where

$$H(z) = \frac{1-2\beta z^{-1}+z^{-2}}{1-\beta(1+\alpha)z^{-1}+\alpha z^{-2}} \quad (6)$$

with $-1.0 \leq \beta \leq 1.0$ and $0 \leq \alpha < 1.0$ is used. Coefficients β and α determine the notch frequency ω_n and the 3 dB attenuation notch bandwidth Ω, respectively. A normalized notch frequency coefficient update algorithm formulated as

$$\beta(k+1) = \beta(k) + \mu_\beta \frac{y(k)x(k)}{\Psi(k)} \quad (7)$$

$$\Psi(k) = \rho\Psi(k-1) + (1-\rho)x^2(k) \quad (8)$$

is proposed. μ_β is a step-size constant for controlling the convergence rate, $y(k)$ is the notch filter output signal, and $x(k)$ is an internal state of the structure implementing the notch filter. $\Psi(k)$ is an instantaneous estimate of power in $x(k)$ where $0 \ll \rho < 1$ is the forgetting factor. Transfer function of the circuit from $u(k)$ to $x(k)$ is

$$F(z) = \frac{z^{-1}}{1-\beta(1+\alpha)z^{-1}+\alpha z^{-2}}. \quad (9)$$

$F(z)$ is a bandpass filter with phase response of $-\pi/2$ and maximum magnitude response at the notch frequency.

An algorithm that searches for an optimum value of α, α_{opt}, which results in maximum SNR improvement factor and minimum BER is proposed. It is given by

$$\alpha(k+1) = \alpha(k) - \mu_\alpha y(k)\psi(k). \quad (10)$$

μ_α is a step-size adaptation constant and $\psi(k) = \dfrac{\partial y(k)}{\partial \alpha(k)}$. It can be shown that

$$\begin{aligned}\psi(k) = \; & \beta(k)\Big(1+\alpha(k)\Big)\psi(k-1) - \alpha(k)\psi(k-2)\\ + \; & \beta(k)y(k-1) - y(k-2). \end{aligned} \quad (11)$$

Thus, transfer function of a circuit generating $\psi(k)$ is given by

$$W(z) = \frac{\beta z^{-1} - z^{-2}}{1-\beta(1+\alpha)z^{-1}+\alpha z^{-2}} \quad (12)$$

with $y(k)$ its input.

3. CONVERGENCE ANALYSIS

Signals $y(k)$ and $x(k)$ in Eq. (7) are given by

$$y(k) = m_H(k) + i_H(k) + n_H(k) \quad (13)$$

$$x(k) = m_F(k) + i_F(k) + n_F(k) \quad (14)$$

where $m_H(k)$ and $m_F(k)$, $i_H(k)$ and $i_F(k)$, and $n_H(k)$ and $n_F(k)$ are terms due to $m(k)$, $i(k)$, and $n(k)$, respectively. The normalizing signal $\Psi(k)$ is approximated by a constant Ψ assuming that its values vary slowly where

$$\begin{aligned}\Psi = \; & \frac{1}{2\pi j}\oint \Phi_{ii}(z)F(z)F(z^{-1})z^{-1}dz\\ + \; & \frac{1}{(1-\alpha^2)(1-\beta^2)}(\sigma_n^2 + KS^2). \end{aligned} \quad (15)$$

$\Phi_{ii}(z)$ is z-transform of the autocorrelation function of $i(k)$. Substituting $y(k)$, $x(k)$, and $\Psi(k)$ in Eq. (7), an average step size for each iteration is obtained and is given by

$$E[\Delta\beta(k)] = \frac{\frac{\mu_\beta}{2\pi j}\oint \Phi_{ii}(z)H(z)F(z^{-1})z^{-1}dz}{\Psi} \quad (16)$$

where $\Delta\beta(k) = \beta(k+1) - \beta(k)$. A decrease in $E[\Delta\beta(k)]$ and convergence speed of the algorithm in Eq. (7) with number of active users K is inferred from Eqs. (15)-(16).

An average step size for each iteration of the algorithm in Eq. (10) can be shown to be given by

$$\begin{aligned}E[\Delta\alpha(k)] = \; & -\frac{\mu_\alpha}{2\pi j}\oint \Phi_{ii}(z)V(z)V(z^{-1})W(z^{-1})z^{-1}dz\\ + \; & \mu_\alpha \frac{\sigma_n^2 + KS^2}{(1+\alpha)^2} \end{aligned} \quad (17)$$

where $\Delta\alpha(k) = \alpha(k+1) - \alpha(k)$. The algorithm in Eq. (10) converges when the terms on the right-hand side of Eq. (17) are equal. Analytical convergence behaviors for $\beta(k)$ and $\alpha(k)$ are calculated by simultaneous equations

$$\left.\begin{aligned}\beta(k+1) = \beta(k) + E[\Delta\beta(k)]\\ \alpha(k+1) = \alpha(k) + E[\Delta\alpha(k)]\end{aligned}\right\} \quad (18)$$

with $E[\Delta\beta(k)]$ and $E[\Delta\alpha(k)]$ as given in Eqs. (16) and (17), respectively.

4. SYSTEM PERFORMANCE ANALYSIS

4.1. Mean and Variance of the Decision Variable

In Fig.1, a decision variable v for the 1st user is shown to be obtained by correlating $y(k)$ with a locally generated PN sequence $c_1(k)$ synchronized with that of the transmitter. Thus,

$$v = \sum_{k=1}^{L} c_1(k)y(k). \quad (19)$$

Here, L is the number of PN sequence chips per data bit or processing gain (PG). By superposition v becomes

$$v = Sb_1 h(0)L + Sb_1 \sum_{l=1}^{\infty} h(l)C_L(l)$$

$$+ \quad S\sum_{j=2}^{K} b_j \sum_{l=0}^{\infty} h(l) \sum_{k=1}^{L} c_1(k)c_j(k-l)$$

$$+ \quad \sum_{k=1}^{L} c_1(k) \sum_{l=0}^{\infty} h(l)i(k-l)$$

$$+ \quad \sum_{k=1}^{L} c_1(k) \sum_{l=0}^{\infty} h(l)n(k-l) \qquad (20)$$

where $h(l)$ is an impulse response of $H(z)$. The first term on the right-hand side of Eq. (20) is due to the reference user's data signal. The second term is due to the multiple-access interference (MAI). The third term is due to the NBI, and the fourth term is due to $n(k)$. $C_L(l)$ is the autocorrelation function of the PN sequence $c_1(k)$. The time index k was dropped out from $b_1(k)$ and $b_j(k)$ since $b_j(k)$ is constant over the correlation interval.

Assuming $c_j(k)$ to be statistically independent random processes each with equally probable amplitudes $[+1, -1]$, mean value and variance of v can be shown to be

$$E[v] = \begin{cases} +LS, & b_1 = +1 \\ -LS, & b_1 = -1 \end{cases} \qquad (21)$$

$$var[v] = L\Big(S^2(1+\frac{1}{N})(1-\frac{L}{N})\frac{1-\alpha}{1+\alpha} + \sigma_{i_H}^2 + \frac{2}{1+\alpha}\big(\sigma_n^2 + (K-1)S^2\big)\Big) \qquad (22)$$

where N is the period of the PN sequence. The first term on the right-hand side of Eq. (22) is the notch filter generated noise. The second term $\sigma_{i_H}^2$ is the steady-state notch filter output power due to $i(k)$ which is given by

$$\sigma_{i_H}^2 = \frac{1}{2\pi j}\oint \Phi_{ii}(z)H(z)H(z^{-1})z^{-1}dz. \qquad (23)$$

Remaining part is a sum of terms due to AWGN and MAI.

4.2. SNR Improvement Factor

SNR improvement factor G, defined as the ratio of SNR of the decision variable with and without the interference suppression filter, is given by

$$G = \frac{\sigma_i^2 + \sigma_n^2 + (K-1)S^2}{S^2(1+\frac{1}{N})(1-\frac{L}{N})\frac{1-\alpha}{1+\alpha} + \sigma_{i_H}^2 + \frac{2}{1+\alpha}(\sigma_n^2 + (K-1)S^2)}. \qquad (24)$$

For the case where $1 \ll N$ and $L \ll N$ (long PN sequence), the inverse of G can be expressed in terms of $E[y^2(k)]$ as

$$G^{-1} = \frac{1}{\sigma_i^2 + \sigma_n^2 + (K-1)S^2}\Big(E[y^2(k)] - S^2\Big) \qquad (25)$$

where $E[y^2(k)] = \frac{2}{(1+\alpha)}(\sigma_n^2 + KS^2) + \sigma_{i_H}^2$. By referring to Eq. (25)

$$\frac{dG^{-1}}{d\alpha}\Big|_{\alpha_{opt}} = \frac{dE[y^2(k)]}{d\alpha}\Big|_{\alpha_{opt}} = 0. \qquad (26)$$

It follows from Eq. (26) that the value of α at which the SNR improvement factor G attains its maximum (minimum G^{-1}) α_{opt} is the same as that at which the minimum point of $E[y^2(k)]$ occurs. Hence, the algorithm proposed in Eq. (10) maximizes the performance of the NBI suppression adaptive IIR notch filter by converging to α_{opt}.

4.3. Bit Error Rate

Assuming that the decision variable v is Gaussian distributed, the BER of the DS CDMA system with the proposed NBI suppression adaptive IIR notch filter becomes

$$P_e = Q\Big(\sqrt{\frac{(E[v])^2}{var[v]}}\Big) \qquad (27)$$

where $Q(\epsilon) = \frac{1}{\sqrt{2\pi}}\int_{\epsilon}^{\infty} e^{\frac{-\lambda^2}{2}} d\lambda$ and $E[v]$ and $var[v]$ are as given in Eqs. (21) and (22). It can be shown that minimum BER occurs at α_{opt}.

5. SIMULATION RESULTS

Simulation results are presented to support the analyses. a_1 and a_2 were set to -1.60 and 0.9801, respectively. σ_v^2 was set to obtain ISR $= 10\log\frac{\sigma_i^2}{S^2}$ of 10 dB and σ_n^2 was fixed at 0.1. Initial value of $\Psi(k)$, $\Psi(0)$ was set to 1.0 with μ_β, μ_α, and ρ fixed at 0.0005, 0.000025, and 0.95, respectively. PN m-sequences with $N = 8191$ were used with L set to 255 and $S = 1.0$. Each user was assigned a unique sequence.

Figure 2 shows results for G plotted as a function of α. Simulation results were obtained by averaging over 20 independent computer runs after the adaptation algorithm for the notch frequency had reached its steady-state. G is shown to decrease with K. Dependence of G on α is also shown by the figure. It is noted that there exists a value of α that results in maximum G, which depends on K.

Figure 3 shows convergence behaviors of the algorithm proposed in Eq. (10). Initial values $\beta(0)$ and $\alpha(0)$ were set to 0.75 and 0.90, respectively. Analytical trajectories were obtained from Eq. (18). 50 independent trials were averaged to obtain simulated trajectories. The figure shows good convergence characteristics where mean values of $\alpha(k)$ converge to α_{opt}, the values of α at which the maximum G occur as a comparison between Fig. 2 and Fig. 3 shows.

Results of BER versus energy-per-bit-to-noise spectral density $\frac{E_b}{\eta}$ where $\frac{E_b}{\eta} = \frac{LS^2}{2\sigma_n^2}$ for 10 active users is shown in Fig. 4 with E_b the transmitted signal energy per bit. Analytical results were obtained using Eq. (27). Here 20 independent computer runs each with 50,000 data bits were averaged to obtain simulation results. BER without NBI and BER without NBI suppression filter are included for comparison purpose. Reduction in BER afforded by the NBI suppression scheme is demonstrated by the figure.

In Fig. 5, BER versus K are shown for $\frac{E_b}{\eta} = 10$ dB with results for the FIR prediction filter with 10 taps using the normalized least-mean-square (NLMS) algorithm included for comparison purpose. The curves show how the number of active users permissible for a given threshold BER is increased using the proposed NBI suppression filter. It is also noted that the BER achieved by the proposed adaptive IIR notch filter with 15 additions, 19 multiplications and 1 division is comparable to that of a prediction filter of length 10 with 25 additions, 24 multiplications and 1 division.

6. CONCLUSION

Suppression of NBI in a DS CDMA system using a fully adaptive IIR notch filter has been presented. Compared with the adaptive FIR prediction filter, the proposed scheme has been shown to have less computational complexity for the same BER performance level.

REFERENCES

[1] J. Wang and L. B. Milstein, "CDMA overlay situations for microcellular mobile communications," *IEEE Trans. Commun.*, vol. 43, pp. 603-614, Feb. 1995.

[2] J. D. Laster and J. H. Reed, "Interference rejection in digital wireless communications," *IEEE Signal Processing Magazine,* vol. 14, no. 3, pp. 37-62, 1997.

[3] H. V. Poor and L. A. Rusch, "Narrowband interference suppression in spread spectrum CDMA," *IEEE Personal Commun.*, vol. 1, no. 3, pp. 14-27, 1994.

[4] S. Nishimura, A. Mvuma, and T. Hinamoto, "Performance improvement of DSSS communication systems using an adaptive IIR notch filter," in *2001 Proc. IEEE Int. Symp. Circuits and Syst.,* Australia, May 2001, vol. II, pp. 813-816.

[5] A. Mvuma, S. Nishimura, and T. Hinamoto, "Adaptive optimization of notch bandwidth of an IIR filter used to suppress narrow-band interference," in *2002 Proc. IEEE Int. Symp. Circuits and Syst.,* Scottsdale, Arizona, May 2002, vol. V, pp. 341-344.

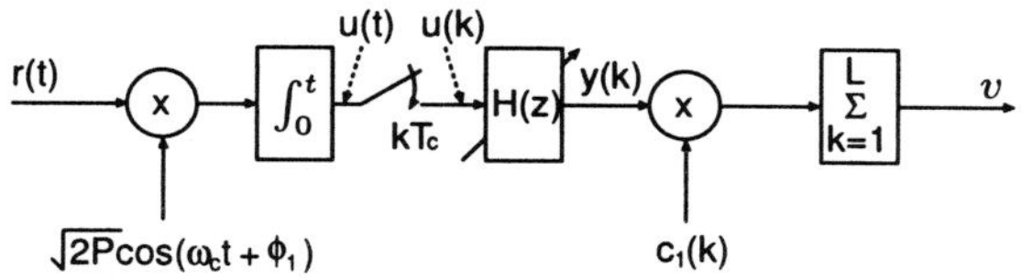

Figure 1: Block diagram of a receiver for the reference user.

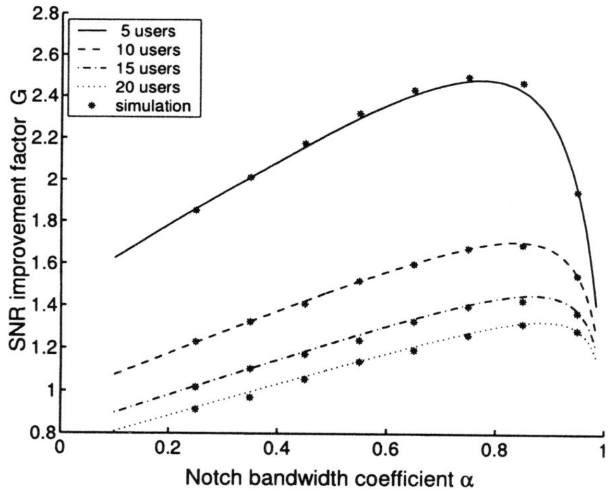

Figure 2: SNR improvement factor as a function of α for ISR = 10 dB.

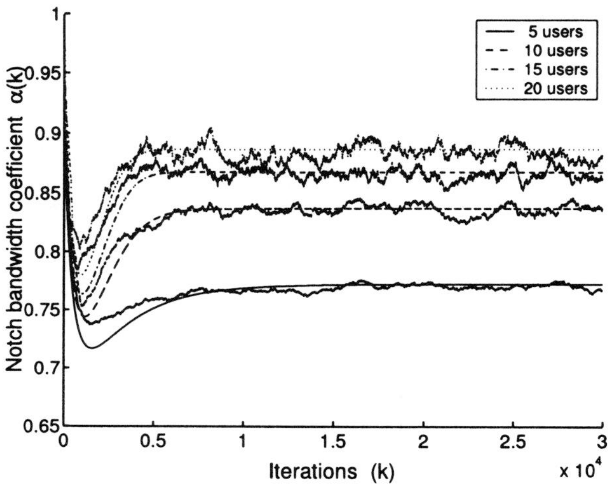

Figure 3: Convergence characteristics of the notch bandwidth coefficient updating algorithm for ISR = 10 dB.

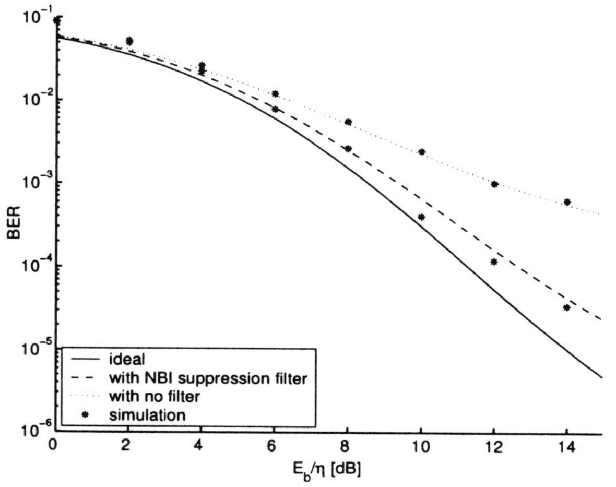

Figure 4: Bit error rate for 10 users and ISR = 10 dB.

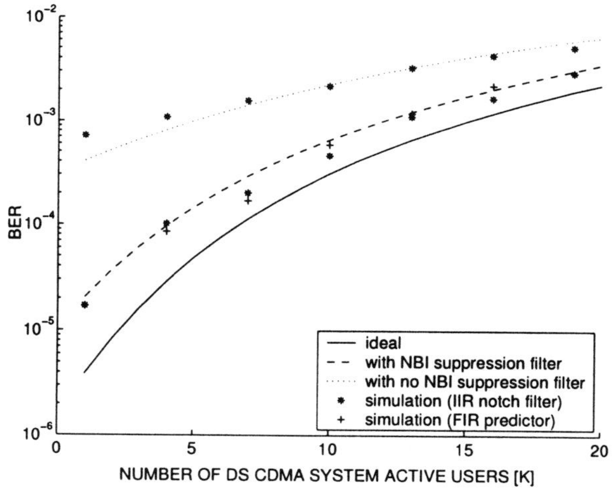

Figure 5: Bit error rate as a function of number of active users K for ISR = 10 dB and $\frac{E_b}{\eta}$ = 10 dB.

A FINITE PRECISION LMS ALGORITHM FOR INCREASED QUANTIZATION ROBUSTNESS

F. Lindström

Konftel Technology AB
Research and Development
Box 268, S-90106, Umeå, Sweden

M. Dahl, I. Claesson

Blekinge Institute of Technology
Department of Signal Processing
S-37225, Ronneby, Sweden

ABSTRACT

The well known Least Mean Square (LMS) algorithm, or variations thereof are frequently used in adaptive systems. When the LMS algorithm is implemented in a finite precision environment it suffers from quantization effects. These effects can severely degrade the performance of the algorithm. This paper proposes a modification of the LMS algorithm that reduces the impact of quantization at virtually no extra computational cost. The paper contains an off-line evaluation of a system identification scheme where the presented algorithm outperforms the classical LMS algorithm yielding a better modelling of the unknown plant. This approach is well suited for adaptive system identification, e.g. beamforming, electrocardiography, and echo cancelling.

1. INTRODUCTION

Adaptive systems can be found in many different signal processing areas, e.g. communications, radar, sonar, navigation systems, seismology, mechanical design and biomedical electronics, [1], [2]. Least Mean Square (LMS) or LMS based algorithms are common in adaptive signal processing systems. When the LMS algorithm is implemented in a finite precision environment, the algorithm suffers from quantization effects. In-depth analysis of the infinite precision LMS algorithm can be found in [1] and [2]. An early treatment of the finite precision effects of the LMS algorithm and the stalling phenomena, i.e. a state where the convergence of the LMS algorithm is very slow or has stopped, can be found in [3]. Analysis of the steady-state behavior of the finite precision LMS algorithm is presented in [4]-[6], where [6] also contains a treatment of the transient performance. Some additional remarks on the convergence rate of the LMS algorithm in a stalling state is given in [7].

Due to quantization effects the performance in a finite precision environment can differ significantly from that of the infinite precision counterpart. The choice of the precision is therefore of outmost importance. In fix-point digital signal processors the precision of internally generated parameters and operations can be increased, e.g. by representing an internal parameter with two words instead of one. A software solution to this problem will however most often lead to an increase in computational load.

This paper proposes a way to increase the roubustness to quantization effects of the LMS algorithm in fix-point systems with a given wordlength by the means of signal processing. The concept of the proposed algorithm is to detect stalling and to use computational resources more efficiently in situations of stalling. The extra processing required to implement the proposed algorithm is insignificant as compared to that of the LMS algorithm.

2. THE FINITE PRECISION LMS ALGORITHM

Generally fix-point systems have a binary number representation using the two's complement format, [8]. In this paper it is assumed that the system at hand is a fix-point two's-complement binary system using q bits to represent numbers in the range [-1,1), and that round-off is used. A detailed description of the binary number system used and finite arithmetics is given in [9]. The representation of an arbitrary infinite precision number, a, in finite precision is denoted a_q, where the subindex 'q' denotes the precision in number of bits. The value of a_q is given by $a_q = Q_q[a]$, where

$$Q_q[a] = (-b_0 + \sum_{i=1}^{q-1} b_i 2^{-i}) \quad , \tag{1}$$

$b_i \epsilon [0,1]$, $i = 0, \cdots, q-1$. The value of the elements b_i are chosen so that they minimize the expression $|a - Q_q[a]|$.

Digital signal processors, e.g. [10], have generally the possibility of representing scalar products temporarily with higher precision and thus an inner product multiplication can be performed without significant quantization loss of the individual scalar products. This will also be valid for the systems in this paper. Under the assumption that the input signals are properly scaled, i.e. that no overflow occurs, the quantized LMS algorithm can be described mathematically[1] as

$$y_q(n) = Q_q[\mathbf{w}_q(n)^T \mathbf{x}_q(n)] \tag{2}$$

$$e_q(n) = d_q(n) - y_q(n) \tag{3}$$

$$\mathbf{w}_q(n+1) = \mathbf{w}_q(n) + Q_q[Q_q[\beta_q e_q(n)]\mathbf{x}_q(n)] \tag{4}$$

where n is the sample index, $d_q(n)$ is the desired signal, $y_q(n)$ is the estimated signal, $e_q(n)$ is the error signal, $\mathbf{w}_q(n) = [w_{q,0}(n), w_{q,1}(n), \cdots, w_{q,N-1}(n)]^T$ is a column vector containing the filter coefficients, $\mathbf{x}_q(n) = [x_q(n), x_q(n-1), \cdots, x_q(n-N+1)]^T$ is an column vector containing the last N samples of the input signal, and β_q is the adaptation step-size.

When the update value for a coefficient in the adaptive filter, $\mathbf{w}_q(n)$, is less than the Least Significant Bit (LSB) used to represent the filter coefficients that coefficient is not updated. This

[1] Observe that a product, p_q of two arbitrary q bits numbers, a_q and b_q, suffers from quantization effects, i.e. $p_q = Q_q[a_q b_q]$, while a sum s_q of the same numbers has no quantization providing that no overflow occurs, i.e. $s_q = a_q + b_q$.

0-7803-7761-3/03/$17.00 ©2003 IEEE

phenomena is called stalling. When stalling occurs it seriously degenerates the performance of the LMS algorithm as compared to the infinite precision algorithm, [3]. From equation (4) stalling for the i:th filter coefficient occurs when

$$|Q_q[Q_q[\beta_q e_q(n)]x_q(n-i)]| < 2^{1-q}. \tag{5}$$

If sufficiently many of the filter coefficients stalls all significant adaptation of the filter ceases. To prevent a certain filter coefficient from stalling two different approaches may be used; the value of the step-size β_q can be limited by a lowest allowed value, or the number of bits, q, used in the quantization $Q_q[\cdot]$ in (5) can be increased, i.e.increaing the number of bits used to represent the coefficients of the adaptive filter. Limiting β_q will imply a limit for the best possible steady-state performance, [1]. However, increasing the number of bits will not imply such a limit. This is the approach taken in this paper.

Further, by the result of [5] and [6] the effect of quantization in non-stalling situations has also been improved by the increased number of bits used to represent the adaptive filter coefficients. In [5] and [6] it was shown that it is the quantization of the adaptive filter coefficients that is dominant in steady-state performance for reasonable values of the step-size β_q, i.e. it is the quantization of the adaptive filter coefficients that is the dominant contributor to the steady-state error signal.

3. THE PROPOSED ALGORITHM

When the LMS algorithm enters a state of stalling all significant adaptation of the filter cease and computational resources used for the update of the adaptive filter are wasted. The main idea of the proposed algorithm is to detect stalling and to use the computational resources more efficiently in these situations. If stalling is detected the updating is done for every other second sample and the resources freed thereby is used to increase the precision. The increase in precision will imply that the update of the filter resumes. The proposed algorithm uses a two state approach, see Fig.1. When no significant stalling is present, i.e.the filter is adapting well, the conventional LMS-algorithm given in equations (2)-(4) is used, which is denoted state A. If a slowdown of adaptation due to stalling is detected, the adaptation of the filter $\mathbf{w}_q(n)$ is freezed and a secondary adaptive filter, $\mathbf{v}_q(n)$, is invoked, which is denoted state B. In state B both filters $\mathbf{w}_q(n)$ and $\mathbf{v}_q(n)$ are used in parallel to model the unknown plant. The output of $\mathbf{v}_q(n)$ is attenuated by a factor 2^{-k}. This causes the optimal setting for the coefficients of $\mathbf{v}_q(n)$ to be gained with 2^k, i.e. the effective precision of the coefficients in $\mathbf{v}_q(n)$ is increased by k bits. The step-size β_q is attenuated with 2^{-k} as well. This unwanted effect on the step-size is avoided by inserting a corresponding gain 2^k, see Fig.1. If any of the coefficients in the secondary filter overflows, the system is switched back to state A. The concept is to adapt the more significant bits of the adaptive filter in state A. If the adaption in A is stopped or slowed down due to stalling the algorithm is switched into state B, where the less significant bits are adapted. If the filter needs to be readapted the algorithm is switched back to state A. The state B processing is defined as

$$y_q(n) = Q_q[\mathbf{w}_q(n)^T\mathbf{x}_q(n) + 2^{-k}\mathbf{v}_q(n)^T\mathbf{x}_q(n)] \tag{6}$$

$$e_q(n) = d_q(n) - y_q(n) \tag{7}$$

$$\mathbf{v}_q(n+1) = \mathbf{v}_q(n) + Q_q[Q_q[2^k\beta_q e_q(n)]\mathbf{x}_q(n)] \text{ if } n \text{ odd} \tag{8}$$

$$\mathbf{w}_q(n+1) = \mathbf{w}_q(n) \tag{9}$$

where k is an positive integer, $\mathbf{v}_q(n) = [v_{q,0}(n),\ v_{q,1}(n),\ \cdots,\ v_{q,N-1}(n)]^T$ is a column vector containing the filter coefficients of the secondary filter. Thus the filter coefficients of $\mathbf{v}_q(n)$ is updated at every other second sample.

To clarify the result of the state B processing an equivalent description of equations (6)-(9) is derived. First define $\mathbf{h}_{q+k}(n) = \mathbf{w}_q(n)+2^{-k}\mathbf{v}_q(n)$, and note that the elements of $\mathbf{h}_{q+k}$ are in $q+k$ bits precision. From equations (8) and (9) it follows that for odd n

$$\begin{aligned}\mathbf{h}_{q+k}(n+1) &= \mathbf{w}_q(n+1) + 2^{-k}\mathbf{v}_q(n+1)\\ &= \mathbf{w}_q(n) + 2^{-k}\mathbf{v}_q(n) + 2^{-k}Q_q[Q_q[2^k\beta_q e_q(n)]\mathbf{x}_q(n)]\\ &= \mathbf{h}_{q+k}(n) + Q_{q+k}[Q_{q+k}[\beta_q e_q(n)]\mathbf{x}_q(n)]\end{aligned} \tag{10}$$

Replacing $\mathbf{w}_q(n) + 2^{-k}\mathbf{v}_q(n)$ with $\mathbf{h}_{q+k}(n)$ in equation (6) and using (10) gives that an equivalent description of the processing of state B is given by

$$y_q(n) = Q_q[\mathbf{h}_{q+k}(n)^T\mathbf{x}_q(n)] \tag{11}$$

$$e_q(n) = d_q(n) - y_q(n) \tag{12}$$

$$\begin{aligned}\mathbf{h}_{q+k}(n+1) &= \mathbf{h}_{q+k}(n)\\ &+Q_{q+k}[Q_{q+k}[\beta_q e_q(n)]\mathbf{x}_q(n)] \text{ if } n \text{ odd}\end{aligned} \tag{13}$$

Comparing equations (2)-(4) with (11)-(13) yields that the state B processing update the adaptive filter only for every other second sample. However, the state B processing will lead to an increase in the precision of the adaptive filter and the update vector of k bits as compared to the LMS algorithm. Thus from the results presented in [5] and [6] the proposed algorithm yields a lower mean square steady-state error than the LMS algorithm.

The parameter k could be set to integer values $0 \leq k \leq q$. When k is increased the precision of the adaptive filter is increased as well. This implies that the quantization effects due to the adaptive filter becomes less dominant, but also that the quantization of $x_q(n)$ and $d_q(n)$ becomes more dominant. Thus increasing k beyond a certain number of bits, will not make any significant impact on the steady-state performance for reasonable values of β_q, [5]. The $q-k$ *least* significant bits of $\mathbf{w}_q(n)$, will correspond to the $q-k$ *most* significant bits of filter $\mathbf{v}_q(n)$. This implies that the value of k determines when a switch from state A to B can be done. Since the k most significant bits of $\mathbf{w}_q(n)$ have no counterpart in filter $\mathbf{v}_q(n)$, they need to be adapted before switching to state B. To summarize k should be large enough to provide an increased precision, but limited to allow a switch from state A to B.

When switching from state B to A the secondary filter $\mathbf{v}_q(n)$ is turned off. The overlapping well adapted bits in filter $\mathbf{v}_q(n)$ should be transferred to their corresponding positions in filter $\mathbf{w}_q(n)$. Hence, equation (4) in the first iteration following a switch from B to A is replaced with

$$\mathbf{w}_q(n+1) = Q_q[\mathbf{w}_q(n) + 2^{-k}\mathbf{v}_q(n)] \tag{14}$$

$$\mathbf{v}_q(n+1) = \mathbf{z} \tag{15}$$

where $\mathbf{z}$ is a vector of length N containing zeros.

When the average value of the expression $|Q_q[Q_q[\beta_q e_q(n)]|$ decreases, the risk for stalling coefficients increases, see equation (5). The concept of the proposed algorithm is that if the algorithm is in state A and if an average, $\overline{u}(n)$ of the expression

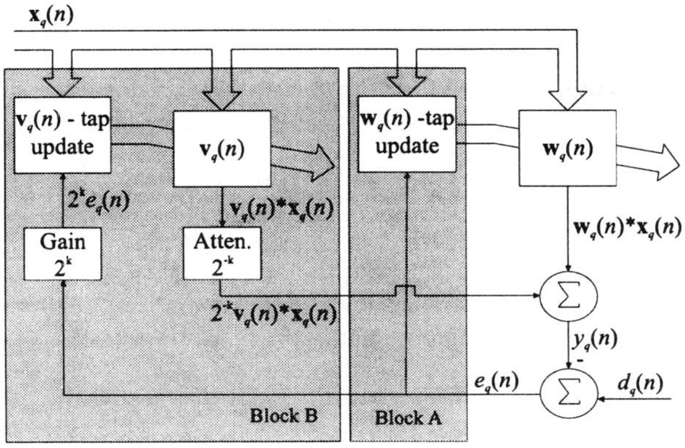

Fig. 1. The proposed algorithm. When in state A the processing of Block B is omitted, and when in state B the processing of Block A is omitted.

$|Q_q[Q_q[\beta_q e_q(n)]]|$ decreases under a certain preset threshold, l_A, the risk of stalling is high and the algorithm is switched to state B. More precisely if the system is in state A, state B is declared if

$$\overline{u}(n) < l_A \tag{16}$$

where $\overline{u}(n)$ is defined as

$$\overline{u}(n) = (1-\gamma)\overline{u}(n-1) + \gamma|Q_q[Q_q[\beta_q e_q(n)]]| \tag{17}$$

where γ is a constant with $0 < \gamma < 1$. The performance of the algorithm is determined by the threshold l_A. The value of k will impose an upper limit for l_A, since the algorithm should not be switched into state B unless the k most significant bits of the elements in $\mathbf{w}_q(n)$ have converged. Further, an optimal setting of l_A is strongly dependent on the input signal $x(n)$, and the statistics of $x(n)$ should be taken into account when determining l_A. Reducing l_A might reduce the convergence time at the expense of an increased risk for stalling. For an increment of l_A the trade-off situation is the opposite.

State B constitutes an update of the filter for every other second sample, thus it is desirable that state B is selected only when the effects of quantization is significant. However, state B implies a higher precision in the adaptive filter and a shift from state B to A should not be done unless the more significant bits in the filter $\mathbf{w}_q(n)$ needs to be readapted. Digital signal processors, in general, automatically detects an overflow in an arithmetic operation, [10]. This implies that the maximal absolute value of the coefficients of the filter $\mathbf{v}_q(n)$ provides a natural way to define the detection of a switch from state B to state A. If

$$\max_{i\epsilon[0,\cdots,N-1]}(|v_{q,i}(n)|) > 1 \tag{18}$$

the system is switched to state A, i.e. if any of the filter coefficients of $\mathbf{v}_q(n)$ overflows. This detector allows the system to stay in the higher precision state B as long as possible.

4. COMPLEXITY

In this section the extra processing required for the proposed algorithm as compared to the LMS algorithm is evaluated.

The processing in state A is the same as for the LMS algorithm with the exception of the processing for the detector given by equations (16)-(17). Thus the extra processing for state A requires one comparison, two multiplications, one addition and one absolute value. Typically this would require about 10-15 digital signal processor instructions.

In state B the processing is given by the equations (6)-(9), and (18). The processing of the LMS algorithm is given by equations (2)-(4). Equations (2) and (7) requires the same amount of processing and equation (9) requires no processing, so these equations can be disregarded in a comparison. Digital signal processors, in general, have the possibility of conducting both an addition and a multiplication in the same instruction when calculating inner products. This implies that the inner products, $\mathbf{w}_q(n)^T\mathbf{x}_q(n)$ and $\mathbf{v}_q(n)^T\mathbf{x}_q(n)$ will require N combined multiplications/additions each. Thus equations (2) and (4) will require N combined multiplications/additions, N additions and $N+1$ multiplications, in all $3N+1$ arithmetical operations. Equation (6) and (8) of state B will require $2N$ combined multiplications/additions, $N/2$ additions and $N/2+2$ multiplications, in all $3N+2$ arithmetical operations. Digital signal processors, in general, set a flag if an arithmetic operation overflows, a flag that can be set in a latch mode, [10]. This implies that the equation (18) can be implemented by checking the overflow flag immediately after the processing of equation (8) has been executed. Thus the extra processing for state B is one arithmetical operation and the check operation. Typically this would require 7-10 digital signal processor instructions.

The proposed algorithm introduces at most 10-15 extra digital signal processor instructions for the state A processing and 7-10 instructions for the state B processing. This should be compared to the instructions of the LMS algorithm which is of order $3N$, where N is the length of the adaptive filter.

5. PERFORMANCE EVALUATION

To demonstrate the performance of the proposed algorithm three system identification schemes using different algorithms were implemented. The three schemes are denoted S1, S2, and S3 and correspond to the classic LMS algorithm (S1), the proposed algorithm, (S2), and the classic LMS algorithm with infinite internal precision, (S3). S1 was implemented as given in equations (2)-(4). S2 was implemented according to the algorithm given in section 3 with, $k = 8$, $\gamma = 0.05$, and $l_A = 0.0073$, these parameters are not optimal, but were considered sufficient for demonstrating the virtues of the proposed algorithm. S3 was also implemented according to equations (2)-(4), but with infinite precision in the representation of internal operations and parameters, i.e. the only quantization in S3 is that of the input signals $x_q(n)$ and $d_q(n)$. S3 is thus used as an reference for the optimal performance possible if computational power was a free resource.

The wordlength q was set to $q = 12$ bits, corresponding to the typical number of effective bits in a 16 bits fix-point processor. The signal $x_q(n)$ was random noise with gaussian distribution. The unknown plant in the system identification scheme consists of a linear finite impulse response filter of length 200, where the values of the filter coefficients were chosen randomly with gaussian distribution. The signal $d_q(n)$ was obtained as the sum of a random gaussian noise signal, i.e. measurement noise, and the plant output. The power ratio between the measurement noise and the plant output was -40dB. The length of the adaptive filter was set to 200 for all three implementations. The implementations were simulated for two different values of the step-size β_q, i.e. $\beta_q = [0.04, 0.02]$. For each β_q the simulation was repeated 50 times in order to obtain the Mean Square Deviation (MSD) of each implementation. The MSD, denoted $\mathcal{D}(n)$ is defined as

$$\mathcal{D}(n) = \mathrm{E}\{\Delta \mathbf{w}^T(n)\Delta \mathbf{w}(n)\} \qquad (19)$$

where $\mathrm{E}\{\cdot\}$ denotes expectation and $\Delta \mathbf{w}(n)$ is the difference between the plant impulse response and the adaptive filter, [1]. In Fig.2 MSD learning curves, i.e. $\mathcal{D}(n)$ as a function of sample index, are shown for the different implementations and the two different values of β_q. From these it can be observed that the proposed algorithm (S2) outperforms the classical LMS algorithm (S1), in the sense that it has a lower MSD.

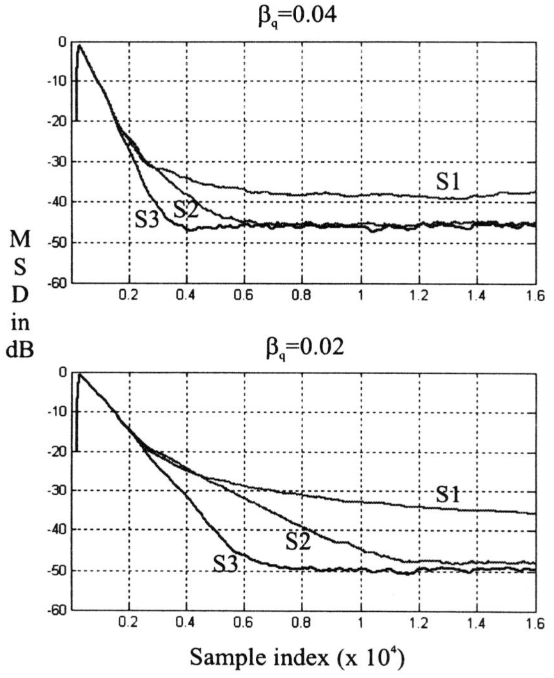

Fig. 2. The MSD for implementation S1, S2, and S3 vs sample index for two different values of the step-size, $\beta_q = 0.04$ and $\beta_q = 0.02$. S1 the classic LMS algorithm, S2 the proposed algorithm, S3 the classic LMS algorithm with infinite precision of internal operations and parameters.

6. CONCLUSIONS

This paper proposes a finite LMS based algorithm. The essence of the algorithm is to avoid stalling effects. The paper explores the fact that when stalling occurs for a finite precision implementation of the classic LMS algorithm the updating process of the adaptive filter coefficients is ineffective. The proposed algorithm detects stalling situations and uses a secondary adaptive filter to increase the precision in these situations. The algorithm reduces the update of the coefficients to every other second sample, and the computational resources that is freed thereby is used for the increased precision. Thus the computational load of the algorithm is essentially the same as that of the LMS algorithm. It was shown analytically that the proposed algorithm corresponds to an increase in the precision of the classical LMS algorithm adaptive filter. Off-line calculations was used to show that the proposed algorithm outperforms the classic LMS algorithm in a lower MSD sense. The proposed algorithm can thus be used to meet specific design requirements with a lower demand on the wordlength of the processor, or a lower demand of computational load, e.g. it can replace a classic LMS algorithm in double precision. This implies that the proposed algorithm can significantly reduce the cost of implementation of adaptive systems.

7. REFERENCES

[1] S. Haykin, *Adaptive filter theory*, 4th ed., Prentice-Hall, NJ, 2002.

[2] B. Widrow, S. D. Stearns, *Adaptive signal Processing*, Prentice-Hall, NJ, 1985.

[3] R. D. Gitlin, J. E. Mazo, M. G. Taylor, "On the design of gradient algorithms for digitally implemented filters", *IEEE Trans. Circuit Theory*, vol. CT-20, 1973, pp.125-136.

[4] A. Weiss, D. Mitra, "Digital adaptive filters: Conditions for convergence, rates of convergence, effects of noise and errors arising from the implementation", *IEEE Trans. Information Theory*, vol. IT-25, 1979, pp. 637-652.

[5] C. Caraiscos, B. Liu, "A roundoff error analysis of the LMS adaptive algorithm", *IEEE Trans. Acoust., Speech, Sig. Proc.*, vol. ASSP-32, no. 1, 1984, pp. 34-41.

[6] S. T. Alexander, "Transient weight misadjustment properties for the finite precision LMS algorithm", *IEEE Trans. Acoust., Speech, Sig. Proc.* vol. ASSP-35, no. 9, 1987, pp. 1250-1258.

[7] N. J. Bershad, J. C. M. Bermudez, "New insights on the transient and steady-state behavoir of the quantized LMS algortihm", *IEEE Trans. on Sig. Proc.*, vol. 44, no. 10, 1996, pp. 2623-2625.

[8] A. V. Oppenheim, R. W. Schafer, *Discrete-time signal processing*, Prentice-Hall, NJ, 1989.

[9] D. E. Knuth, *The art of computer programming: Seminumerical algorithms*, 2nd ed., Addison-Wesley Publishing Co., 1981.

[10] ADSP-2100 Family User's Manual, 3ed., Analog Devices, 1995.

A NEW LMS-BASED FOURIER ANALYZER IN THE PRESENCE OF FREQUENCY MISMATCH

Yegui Xiao† [1], *Rabab Kreidieh Ward*†, *Li Xu*††

† Institute for Computing, Information and Cognitive Systems
University of British Columbia
2356 Main Mall, Vancouver, BC, Canada V6T 1Z4
E-mail: xiao@ece.ubc.ca

†† Akita Prefectural University
Honjo, Akita Prefecture, Japan 015-0055

ABSTRACT

Adaptive Fourier analyzers estimate the coefficients of the sine and cosine terms of a noisy sinusoidal signal assuming the frequencies are known. In real-life applications though, the frequencies may vary from their supposed values. This is referred to as frequency mismatch (FM). In this paper, we analyze the performance of the conventional LMS Fourier analyzer under existence of the FM. The dynamics and steady-state properties of the LMS algorithm are derived in detail. An optimum step size parameter is also derived, which minimizes the influence of the FM in the mean square error (MSE) sense. Based on the insights provided by the analysis, we then introduce a novel LMS-based Fourier analyzer which simultaneously estimates the discrete Fourier coefficients (DFCs) and accommodates the FM. This new LMS-based algorithm has very simple structure, and hence introduces a small increase in computations compared with the conventional LMS algorithm. However, it can compensate, almost completely, for the performance degeneration due to the FM. Simulations are conducted to show the validity of the analytical results and the excellent performance of the new LMS-based algorithm.

1. INTRODUCTION

In many real-life applications in digital communications, power systems, control including active noise /vibration control, biomedical engineering, pitch detection in automated transcription, etc., we are concerned with the analysis of a sinusoidal signal in additive noise [1]-[6]. The frequencies of the sinusoidal signal are arbitrary, and are known or estimated in advance. Furthermore, the signal is nonstationary for most of the time. The discrete Fourier transform (DFT) and its variants may be considered for the analysis of the signal. However, there are two major problems that make them basically awkward; (1) first, the signal frequencies are arbitrary and may not be an integer multiples of the fundamental frequency of the DFT, (2) second, the signal is nonstationary, and it is difficult to find a window of proper length that fits the degree of nonstationarity of the signal.

To overcome the above-mentioned difficulties with the DFT-type algorithms, many adaptive algorithms have been proposed. These algorithms can handle arbitrary frequencies which are not necessarily integer multiples of the fundamental frequency of the DFT, and are capable of operating even when the signal being analyzed is heavily nonstationary or time-varying. Some of these are the Kalman filtering based techniques, the recursive least square (RLS) algorithm, the simplified RLS algorithm (SRLS), the LMS-like algorithms and so on, see e.g. [2]-[6] and references therein. The LMS algorithm [6] requires few computations, but works reasonably well in terms of estimation mean square error (MSE). Moreover, it has been shown analytically that the LMS algorithm presents a uniform convergence, and a nice tracking performance as well [5, 6].

In all the above-mentioned adaptive algorithms, the signal frequencies are provided in advance. However, the frequencies of the signal may show some differences from the given ones. That is, a frequency mismatch (FM), large or small, may exist in real applications. For example, in automated transcription of electronic piano sounds, the frequencies of each note of a piano may be slightly different from the ones specified by the international standard, due to the variance of product quality. If the signal frequencies are estimated by a statistical or an adaptive estimator, they may be biased, and are thus not the same as what the real signal possesses. Then,

[1]The 1st author is on leave from the Hiroshima Prefectural Women's University, Hiroshima, Japan 734-8558.

0-7803-7761-3/03/$17.00 ©2003 IEEE

this poses two issues that need be addressed. First, it is necessary to figure out the damage that might be done by the FM to the performance of the LMS. Second, a new algorithm is required that can remedy the negative influence caused by the FM. This new algorithm will be of tremendous importance in real-life applications, since assuming the existence of the FM is more realistic.

In this paper, we first analyze the influence of the FM on the LMS algorithm when it is used to analyze a single noisy sinusoid for the sake of analysis simplicity. As a result, the dynamics of the LMS in the presence of FM are derived, and its steady-state properties are also evaluated. As a by-product, an optimum step size parameter is also derived, which minimizes the total MSE of the algorithm. We then focus on compensating the performance degeneration of the LMS algorithm. We propose a novel scheme which can simultaneously estimate the DFCs of the signal as well as depress the undesired FM. Simulations are provided to present the excellent performance of our new analyzer even when the FM is not small.

2. PERFORMANCE ANALYSIS OF CONVENTIONAL LMS FOURIER ANALYZER

The discrete sinusoidal signal to be analyzed is given by

$$d(n) = \sum_{i=1}^{q} \{a_{0,i}\cos(\omega_{0,i}n) + b_{0,i}\sin(\omega_{0,i}n)\} + v(n) \quad (1)$$

where q is the known number of frequency components of the sinusoidal signal, $\omega_{0,i}$ is the frequency of the i-th component (assumed known), $v(n)$ is a zero-mean additive noise with variance σ_v^2. The purpose of the LMS Fourier analyzer is to estimate the discrete Fourier coefficients (DFCs) of each component, on a real time basis. Fig. 1 depicts the conventional real-valued LMS algorithm [6] which is derived from the complex-valued LMS algorithm [2, 4]. The recursion of the (real-valued) LMS algorithm is given by

$$\hat{\mathbf{D}}_i(n+1) = \hat{\mathbf{D}}_i(n) + \mu_i e(n)\mathbf{X}_i(n), \; i = 1, 2, \cdots, q \quad (2)$$

where

$$\hat{\mathbf{D}}_i(n) = [\hat{a}_i(n) \;\; \hat{b}_i(n)]^T \quad (3)$$

$$\mathbf{X}_i(n) = [\cos(\omega_i n) \;\; \sin(\omega_i n)]^T \quad (4)$$

$$e(n) = d(n) - y(n) \quad (5)$$

$$y(n) = \sum_{i=1}^{q} \hat{\mathbf{D}}_i(n)^T \mathbf{X}_i(n) \quad (6)$$

Here, μ_i is the step size parameter for the DFCs of the i-th component, which controls the magnitude of the recursion. ω_i is a user-specified frequency for the i-th frequency component, which is supposed to be the same as the real frequency $\omega_{0,i}$. However, as mentioned in section 1, in real applications, there may be a frequency mismatch, FM, between them. That is, the FM

$$\Delta\omega_i = \omega_{0,i} - \omega_i, \quad i = 1, 2, \cdots, q \quad (7)$$

may be not zero. Extensive simulations have revealed that when the FM vector $\boldsymbol{\Delta\Omega}$ $(= [\Delta\omega_1, \Delta\omega_2, \cdots, \Delta\omega_q]^T)$ corresponding to the signal frequency vector $\boldsymbol{\Omega}$ $(= [\omega_1, \omega_2, \cdots, \omega_q]^T)$ is not a $q \times 1$ zero vector, the performance of the LMS may degrade largely. An example is provided in Fig. 2 to show the influence of the FM. It is obviously seen that 5% of FM gives rise to a significant performance degeneration. This implies that the conventional LMS is not robust in the presence of FM, which might limit its usefulness in real-life applications. At this point, we clearly have two issues that need to be addressed. One is the analysis of performance of the LMS algorithm under the existence of the FM. The other is to propose a new algorithm to compensate for the performance deterioration with the LMS due to the FM. Here, we first analyze the LMS for a single sinusoid to see how the FM influences its performance. The i subscript in the above equations is omitted for notational simplicity. From (2), the estimation errors for the DFCs are derived as

$$\delta_a(n) = a_0\cos(\Delta\omega n) + b_0\sin(\Delta\omega n) - \hat{a}(n) \quad (8)$$

$$\delta_b(n) = b_0\cos(\Delta\omega n) - a_0\sin(\Delta\omega n) - \hat{b}(n) \quad (9)$$

Clearly, due to the existence of the FM, the DFC estimates $\hat{a}(n)$ and $\hat{b}(n)$ will try to track the time-varying signals $a_0\cos(\Delta\omega n) + b_0\sin(\Delta\omega n)$ and $b_0\cos(\Delta\omega n) - a_0\sin(\Delta\omega n)$, respectively, rather than the constants a_0 and b_0, in order to achieve minimum MSE $E[e^2(n)]$. This implies that for the LMS, the estimation problem with FM eventually becomes a tracking problem which is generally more difficult. This is thought to attribute to the poorer performance of the LMS.

After complex calculations not shown here, the difference equations for the convergences in the mean and mean square of the errors are derived as

$$E[\delta_a(n+1)] = \left(1 - \frac{1}{2}\mu\right) E[\delta_a(n)] + f_a(n) \quad (10)$$

$$E[\delta_b(n+1)] = \left(1 - \frac{1}{2}\mu\right) E[\delta_b(n)] + f_b(n) \quad (11)$$

$$E[\delta_a^2(n+1)] = \left(1 - \mu + \frac{3}{8}\mu^2\right) E[\delta_a^2(n)] + \frac{1}{8}\mu^2 E[\delta_b^2(n)] + (2-\mu) f_a(n) E[\delta_a(n)] + \frac{1}{2}\mu^2\sigma_v^2 + f_a^2(n) \quad (12)$$

$$E[\delta_b^2(n+1)] = \frac{1}{8}\mu^2 E[\delta_a^2(n)] + \left(1 - \mu + \frac{3}{8}\mu^2\right) E[\delta_b^2(n)] + (2-\mu) f_b(n) E[\delta_b(n)] + \frac{1}{2}\mu^2\sigma_v^2 + f_b^2(n) \quad (13)$$

where

$$f_a(n) = -\xi_a\sin(\Delta\omega n) + \eta_a\cos(\Delta\omega n) \quad (14)$$

$$\xi_a = a_0\sin\Delta\omega + b_0(1 - \cos\Delta\omega) \quad (15)$$

$$\eta_a = b_0\sin\Delta\omega - a_0(1 - \cos\Delta\omega) \quad (16)$$

$$f_b(n) = -\xi_b \sin(\Delta\omega n) - \eta_b \cos(\Delta\omega n) \quad (17)$$
$$\xi_b = b_0 \sin \Delta\omega - a_0(1 - \cos \Delta\omega) = \eta_a \quad (18)$$
$$\eta_b = a_0 \sin \Delta\omega + b_0(1 - \cos \Delta\omega) = \xi_a \quad (19)$$

Now, it is clear form (10) and (11) that: 1) it can be seen that the LMS algorithm is no longer un biased due to the existence of the FM; 2) the con vergence of the LMS algorithm in the mean is still uniform despite the FM; and 3) for a very small FM, one gets $\xi_a = \eta_b \approx a_0\Delta\omega$, and $\eta_a = \xi_b \approx b_0\Delta\omega$, which imply that the amplitudes of $f_a(n)$ and $f_b(n)$ are also very small. Therefore the fluctuations of the tracking errors are approximately proportional to the FM, and will be very small as well. Form (12) and (13), one can see the following: 1) due to the FM, the difference equations for the con vergence in the mean square become correlated with the trac king errors. See the 3rd and 5th terms in the RHS of (12) and (13), which vanish when the FM does not exist; 2) the MSEs at steady state, $E[\delta_a^2(n)]|_{n\to\infty}$ and $E[\delta_b^2(n)]|_{n\to\infty}$, will present fluctuations with a frequency that is twice of the FM; and 3) from the four linear difference equations for the convergence in the mean and in the mean square, a stability bound for the step size parameter μ may be obtained. Because $f_a(n)$ and $f_b(n)$ are time-varying, this stability bound may also be time-varying. It seems that the analytical derivation of this bound is quite difficult, but a numerical solution can be easily obtained based on a grid search.

Because both the tracking errors and the MSEs fluctuate sinusoidally at their steady states, we introduce time averaging ($E_T[\cdot]$) to them. Then, from (10) and (11) one easily get $E_T[\delta_a(n)]|_{n\to\infty} = E_T[\delta_b(n)]|_{n\to\infty} = 0$. After complicated and lengthy calculations based on (12) and (13), one ultimately reaches

$$E_T[\delta_a^2(\infty)] \quad (= E_T[\delta_b^2(\infty)])$$
$$= \frac{\mu\sigma_v^2}{2-\mu} + \frac{2(2-\mu)F_a(\mu,\Delta\omega) + 2(1-\cos\Delta\omega)A_0^2}{\mu(2-\mu)} \quad (20)$$

where

$$F_a(\mu,\Delta\omega) = E_T\left[f_a(n)E[\delta_a(n)]|_{n\to\infty}\right] \quad (21)$$
$$= -\frac{1}{\beta}(\alpha - \cos\Delta\omega)(1-\cos\Delta\omega)A_0^2$$
$$\beta = \alpha^2 - 2\alpha\cos\Delta\omega + 1 \quad (22)$$
$$A_0^2 = a_0^2 + b_0^2 \quad (23)$$

From (20), one sees that: 1) the time-a veraged tracking MSEs for the DFCs have two terms: the first term on the RHS of the above equation stems from the additiv e noise, which is approximately proportional to the noise variance σ_v^2 if the step size parameter $\mu \ll 2$; the remaining term is due to the FM, whic h decreases with the increase of the step size parameter; 2) because of the above relationship of the two terms with respect to the step size parameter, there exists an optim um step size value that makes the tracking MSEs minimum; and 3) the 2nd term of the MSEs is an ev en function of the FM. After some appro ximations, minim izing (20) yields the optimum step size

$$\mu_{opt} = 2\sqrt[3]{\frac{\Delta\omega^2 A_0^2}{\sigma_v^2}} \quad (24)$$

which shows that 1) the larger the FM and/or the signal amplitude, the larger the optim um step size that gives the algorithm more po wer to track the two time-varying signals $f_a(n)$ and $f_b(n)$ derived from the FM; 2) the larger the noise variance, the smaller the step size that allows the algorithm to adapt slowly to depress the portion of MSE due to the additive noise; and 3) the optimum step size is such a one that compromises the above two opposite tendencies and minim izes the MSE totally.

It has been found that the above analytical results show excellent agreement with the simulations.

3. A NEW LMS-BASED FOURIER ANALYZER

The above performance analysis provides us with much insight into the LMS algorithm operating under the existence of the FM. It shows that an optimum step size value can be set for the algorithm in pursuit of the minimum tracking MSEs, if the signal amplitude, the FM and the noise variance are provided in advance. However, this is highly likely to be restrictive for most of the time in real applications. Then, what we need to do now is to explore the possibility of designing a new scheme that can accommodate the negativ e influence or performance degeneration the FM causes.

Here, we propose a new LMS-based Fourier analyzer of steepest descent nature, as shown in Fig.3, which minim izes the mean square error, sim ultaneously estimates the DFCs and reduces the FM. The new algorithm is given by

$$\hat{\mathbf{a}}(n+1) = \hat{\mathbf{a}}(n) + \mathbf{U}_{dfc}e(n)\mathbf{X}_a(n) \quad (25)$$
$$\hat{\mathbf{b}}(n+1) = \hat{\mathbf{b}}(n) + \mathbf{U}_{dfc}e(n)\mathbf{X}_b(n) \quad (26)$$
$$\mathbf{c}(n+1) = \mathbf{c}(n) - \mathbf{U}_c e(n)\mathbf{W}(n) \quad (27)$$

where

$$\hat{\mathbf{a}}(n) = [\hat{a}_1(n),\ \hat{a}_2(n),\ \cdots,\ \hat{a}_q(n)]^T$$
$$\hat{\mathbf{b}}(n) = [\hat{b}_1(n),\ \hat{b}_2(n),\ \cdots,\ \hat{b}_q(n)]^T$$
$$\mathbf{c}(n) = [c_1(n),\ c_2(n),\ \cdots,\ c_q(n)]^T$$
$$\mathbf{X}_a(n) = -\mathbf{C}_{diag}(n)\mathbf{X}_a(n-1) - \mathbf{X}_a(n-2),\ n \geq 2$$
$$\mathbf{X}_a(0) = \mathbf{1},\ \mathbf{X}_a(1) = \cos\mathbf{\Omega}(1)$$
$$\mathbf{X}_b(n) = -\mathbf{C}_{diag}(n)\mathbf{X}_b(n-1) - \mathbf{X}_b(n-2),\ n \geq 2$$
$$\mathbf{X}_b(0) = \mathbf{0},\ \mathbf{X}_b(1) = \sin\mathbf{\Omega}(1)$$
$$\mathbf{X}_a(n) = [x_{a_1}(n),\ x_{a_2}(n),\ \cdots,\ x_{a_q}(n)]^T$$
$$\mathbf{X}_b(n) = [x_{b_1}(n),\ x_{b_2}(n),\ \cdots,\ x_{b_q}(n)]^T$$

$$
\begin{aligned}
\mathbf{W}(n) &= diag\{W_{11}(n),\ W_{22}(n),\ \cdots,\ W_{qq}(n)\} \\
W_{ii}(n) &= \hat{a}_i(n)x_{a_i}(n-1) + \hat{b}_i(n)x_{b_i}(n-1) \\
e(n) &= d(n) - \hat{\mathbf{a}}(n)^T\mathbf{X}_a(n) - \hat{\mathbf{b}}(n)^T\mathbf{X}_b(n) \\
\mathbf{C}_{diag}(n) &= diag\{c_1(n),\ c_2(n),\ \cdots,\ c_q(n)\} \\
& \qquad c_i(n) = -2\cos(\omega_i(n)) \\
\mathbf{U}_{dfc} &= diag\{\mu_1,\ \mu_2,\ \cdots,\ \mu_q\} \\
\mathbf{U}_c &= diag\{\mu_{c_1},\ \mu_{c_2},\ \cdots,\ \mu_{c_q}\}
\end{aligned}
$$

Simulations have revealed that the algorithm is capable of removing the FM excellently, and performing almost as good as the original LMS algorithm without the FM. A comparison among the LMS without the FM and the new algorithm with the FM is shown in Fig. 4, where excellent performance of the new algorithm can be identified.

4. CONCLUSIONS

In this paper, the LMS algorithm has been analyzed in detail under the existence of the FM. A new LMS-based Fourier analyzer has been proposed that simultaneously estimates the DFCs and accommodates the FM excellently. Simulations have been conducted that support both the analysis and the excellence of the new algorithm.

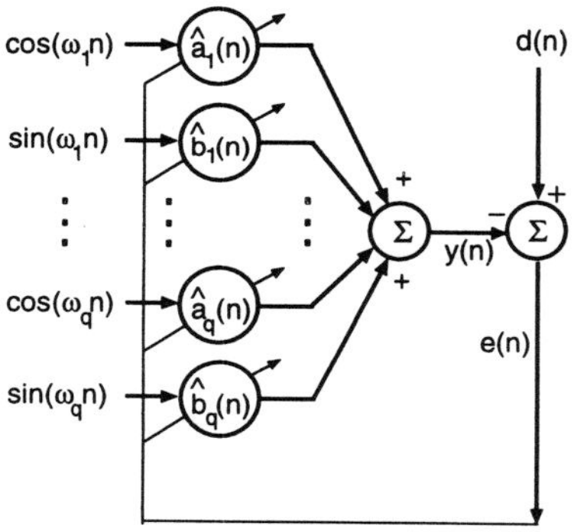

Fig. 1 The conventional LMS.

5. REFERENCES

[1] Int. Conf. Harmonics Power Syst., Inst. Sci. Technol., Univ. Manchester (Power Syst. Eng. Series), Manchester, U.K., 1981.

[2] C. A. Vaz and N. V. Thakor, *"Adaptive Fourier estimation of time-varying evoked potentials,"* IEEE Trans. Biomed. Eng., vol.36, pp.448-455, 1989.

[3] H. C. So, *"Adaptive algorithm for sinusoidal interference cancellation,"* Electron. Lett., vol.33, no.5, pp.356-357, 1997.

[4] T. Umemoto and N. Aoshima, *"The adaptive spectrum analysis for transcription,"* Trans. Soc. Instrum. Contr. Engineers (SICE), vol.28, no.5, pp.619-625, 1992(in Japanese).

[5] Y. Xiao and Y. Tadokoro, *"LMS-based notch filter for the estimation of sinusoidal signals in noise,"* Signal Process., vol.46, no.2, pp.223-231, 1995.

[6] Y. Xiao, Y. Tadokoro, and K. Iwamoto, *"Real-valued LMS Fourier analyzer for sinusoidal signals in noise,"* Signal Process., vol.69, no.2, pp.131-147, 1998.

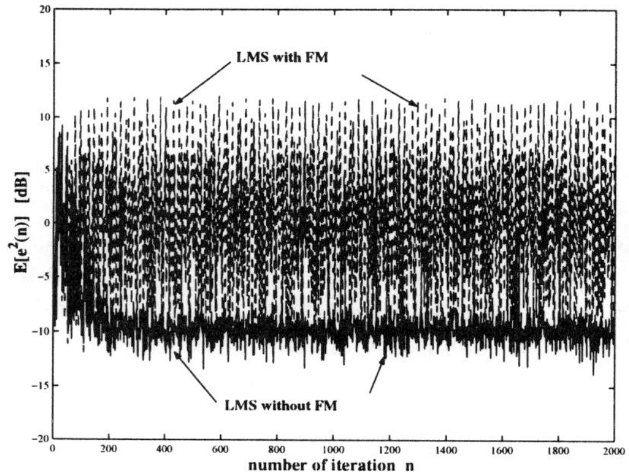

Fig. 2 Comparison between the LMS with and without the FM (true signal frequency: $\mathbf{\Omega}_0 = [0.105\pi,\ 0.190\pi,\ 0.315\pi]^T$, user-specified initial frequency: $\mathbf{\Omega}(1) = [0.100\pi,\ 0.200\pi,\ 0.300\pi]^T$, $a_i = b_i = 1.0$, $\mu_1 = \mu_2 = \mu_3 = 0.025$, $SNR = 10\ [dB]$, 40 runs).

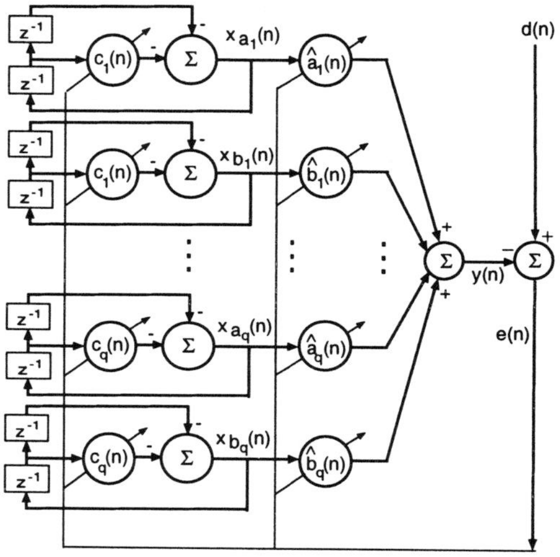

Fig. 3 The new LMS-based Fourier analyzer.

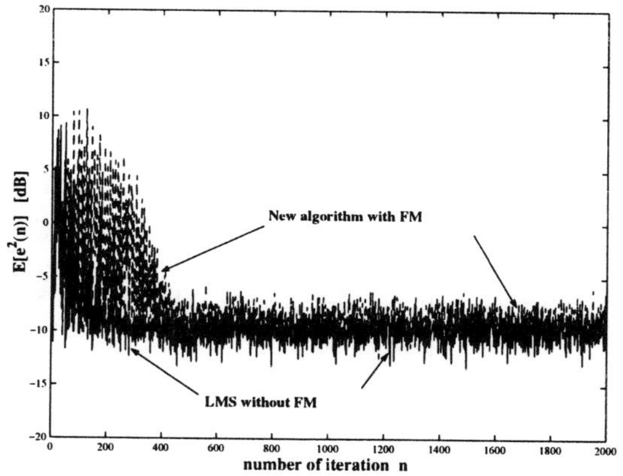

Fig. 4 Comparison between the LMS without FM and the new LMS-based algoritm with FM ($\mu_{c_1} = \mu_{c_2} = \mu_{c_3} = 0.001$, other conditions the same as Fig. 2).

TRANSFORM DOMAIN APPROXIMATE QR-LS ADAPTIVE FILTERING ALGORITHM

Yang Xin-xing and S. C. Chan

Department of Electrical and Electronic Engineering,
The University of Hong Kong, Pokfulam Road, Hong Kong.
Email: {xxyang, scchan}@eee.hku.hk

Abstract: An improved approximate QR-least squares (A-QR-LS) algorithm, called the transform domain A-QR-LS (TA-QR-LS) algorithm, is introduced. The input signal vector is approximately decorrelated by some unitary transformations before applying the A-QR-LS, which is shown to improve considerably the convergence speed of the A-QR-LS algorithm recently proposed by Liu [4]. Further, it is possible to reduce the arithmetic complexities ($O(N)$) of the A-QR-LS, and TA-QR-LS algorithms by using Givens rotations instead of the Householder transformation. Simulation results show that the proposed TA-QR-LS algorithm is a good alternative to the conventional recursive least squares (RLS) algorithm in adaptive filtering applications involving multiple channels, acoustic modeling, and fast parameter variations.

1. INTRODUCTION

Efficient recursive least squares (RLS) algorithms using the QR decomposition (QRD) is well known for their good numerical property [1]-[3], because the condition number of the system is much lower than that of the input correlation matrix. The QRD-based algorithms usually consist of the following two separate parts: 1) recursive updating of the triangular matrix and 2) backsolving of the filter parameters. Although the matrix-updating step can be efficiently performed with $O(N)$ arithmetic operation in single input adaptive filtering situation, the backsolving step still requires $O(N^2)$ operation. Therefore, in applications where the filter parameters have to be computed, the entire algorithm still requires $O(N^2)$ arithmetic operations. Recently, an approximate QR-based LS (A-QR-LS) fast adaptive parameter estimation algorithm with $O(N)$ complexity was presented [4], where a special structured matrix is used and the triangularzation and backsolving step are considered together. In this paper, we propose an improved transform-domain approximate QR-LS (TA-QR-LS) algorithm, where the input signal vector is first approximately decorrelated by some unitary transformations such as the discrete Fourier transform (DFT) or the discrete cosine transform (DCT) before carrying out the normalization in the A-QR-LS algorithm. This considerably improves the convergence speed of the latter algorithm. Further, we show that it is possible to reduce the arithmetic complexities of the A-QR-LS, and TA-QR-LS algorithms by using Givens rotations instead of the Householder transformation, as only one input signal vector is processed as a time. This not only reduces the arithmetic complexity by a factor of two, but also enables us to develop square root-free versions of the algorithms, in the same spirit as the classical square root-free Givens-based QR decomposition LS algorithm [5]. The resultant algorithms are much simpler to implement in either software or hardware. This algorithm can also be used in the QR-LMS algorithm recently proposed in [6].

2. THE TRANSFORM DOMAIN APPROXIMATE QR-LS ALGORITHM

Consider the estimation of the N-dimensional parameter vector $\boldsymbol{\theta}$ for the following linear model

$$d(n) = \boldsymbol{x}_N^T(n)\boldsymbol{\theta} + v(n) \tag{1}$$

where $d(n)$ and $\boldsymbol{x}_N^T(n) = [x_1(n), x_2(n), \ldots, x_N(n)]$ are the desired (observed) signal and input vectors, respectively, and $v(n)$ is an additive white Gaussian noise sequence with zero mean. Let $\boldsymbol{\theta}_n$ be the estimated parameter vector at time n, the estimation error at time instant n is thus given by

$$e(n) = d(n) - \boldsymbol{x}_N^T(n)\boldsymbol{\theta}_n . \tag{2}$$

In least squares parameter estimation, the following time-averaged squared magnitude error $\zeta(n)$ is minimized

$$\xi(n) = \sum_{j=0}^{n} \lambda^{n-j} |e(j)|^2 = \sum_{j=0}^{n} \lambda^{n-j} \left| d(j) - \boldsymbol{x}_N^T(j)\boldsymbol{\theta}_j \right|^2 , \tag{3}$$

where the constant λ is the forgetting factor with a value between 0 and 1. (1) can be written in matrix form as

$$\boldsymbol{e}(n) = \boldsymbol{d}(n) - \boldsymbol{X}_N(n)\boldsymbol{\theta}_n , \tag{4}$$

where
$$\boldsymbol{d}(n) = [d(0), d(1), \ldots, d(n)]^T , \tag{5}$$
$$\boldsymbol{x}_N(n) = [x_1(n), x_2(n), \ldots, x_N(n)]^T$$
and
$$\boldsymbol{X}_N(n) = [\boldsymbol{x}_N(0), \boldsymbol{x}_N(1), \ldots, \boldsymbol{x}_N(n)]^T . \tag{6}$$

$\boldsymbol{x}_N(n)$ and $\boldsymbol{X}_N(n)$ are the received signal vector and the data matrix, respectively. Then, the least squares objective function $\xi(n)$ in (3) becomes

$$\xi(n) = \boldsymbol{e}^H(n)\boldsymbol{W}^2(n)\boldsymbol{e}(n) = \| \boldsymbol{W}(n)\boldsymbol{e}(n) \|^2 , \tag{7}$$

where $\boldsymbol{W}(n)$ is a diagonal weighting matrix given by $\boldsymbol{W}(n) = diag(\sqrt{\lambda^n}, \sqrt{\lambda^{n-1}}, \ldots, \sqrt{\lambda}, 1)$. The QR-LS method solves the least squares problem by performing the QR decomposition (QRD) of the matrix $\boldsymbol{W}(n)\boldsymbol{X}_M(n)$

$$\boldsymbol{Q}(n)\boldsymbol{W}(n)\boldsymbol{e}(n) = \begin{bmatrix} \hat{\boldsymbol{d}}_N(n) - \boldsymbol{\Re}_N(n)\boldsymbol{\theta}_n \\ \boldsymbol{c}_{n+1-N} \end{bmatrix} , \tag{8}$$

where $\boldsymbol{Q}(n)$ is some $(n+1)\times(n+1)$ unitary matrix and $\boldsymbol{\Re}_N(n)$ is an $(N \times N)$ upper triangular matrix, and $\boldsymbol{Q}(n)\boldsymbol{W}(n)\boldsymbol{d}(n) = [\hat{\boldsymbol{d}}_N(n) \quad \boldsymbol{c}_{n+1-N}]^T$. Since $\boldsymbol{Q}(n)$ is an unitary matrix, the square of the Euclidean norm on the left hand side in (8) is equal to $\zeta(n)$. The 2-norm on the right hand side of (8) achieves its minimum value when $\boldsymbol{\theta}_n$ is chosen as

$$\boldsymbol{\Re}_N(n)\boldsymbol{\theta}_n = \hat{\boldsymbol{d}}_N(n) . \tag{9}$$

The QRD can be implemented recursively. Let

$$\boldsymbol{D}_N(n-1) = \boldsymbol{W}(n-1)[\boldsymbol{X}_N(n-1) \quad \boldsymbol{d}_N(n-1)] .$$

be the augmented data matrix. Suppose that we have computed the QRD of $\boldsymbol{W}(n-1)\boldsymbol{X}_N(n-1)$, then we have

$$\boldsymbol{D}_N^*(n-1) = \boldsymbol{Q}(n-1)\boldsymbol{D}_N(n-1) = \begin{bmatrix} \boldsymbol{\Re}_N(n-1) & \hat{\boldsymbol{d}}_N(n-1) \\ \boldsymbol{0} & \boldsymbol{c}_{n-N} \end{bmatrix}$$

Given the new data vector $\boldsymbol{\psi}_n = [\boldsymbol{x}_N^T(n), d(n)]^T$, than we have

$$\boldsymbol{D}_N(n) = \boldsymbol{W}(n)[\boldsymbol{X}_N(n) \quad \boldsymbol{d}(n)] = \begin{bmatrix} \sqrt{\lambda}\boldsymbol{D}_N(n-1) \\ \boldsymbol{\psi}_n^T \end{bmatrix} . \tag{10}$$

Multiply (10) by the following augmented matrix

0-7803-7761-3/03/$17.00 ©2003 IEEE

$$Q'(n) = \begin{bmatrix} Q(n-1) & 0 \\ 0^T & 1 \end{bmatrix},$$

one gets
$$Q'(n)D_N(n) = \begin{bmatrix} \sqrt{\lambda}\Re_N(n-1) & \sqrt{\lambda}\hat{d}_N(n-1) \\ 0 & \sqrt{\lambda}c_{n-N} \\ x_N^T(n) & d(n) \end{bmatrix}. \quad (11)$$

Since $\Re_N(n-1)$ is an upper triangular matrix, the new QRD can be obtained by zeroing out $x_N^T(n)$ by a series of N Givens rotations or Householder reflections. Let $Q^{(i)}(n)$ be the Givens rotation or Householder matrix used to zero out the element $x_i(n)$ at the i-th stage. After performing M stages of Givens rotations or Householder reflections, the first N elements in the last row of (11) becomes zero and we obtain

$$Q^{(N)}(n)\cdots Q^{(1)}(n)Q'(n)D_N(n) = \begin{bmatrix} \Re_N(n) & \hat{d}_N(n) \\ 0 & c_{n-N} \\ 0^T & d^{N+1}(n) \end{bmatrix}$$

Although it is possible to recursively update the triangular matrix with $O(N)$ arithmetic operations, direct back solving of the optimal parameter vector θ^o still requires about $O(N^2)$ arithmetic complexity for multi-channel applications. In [4], Liu proposed an approximate QR-LS algorithm, which combines the updating and the back solving processes together using the Householder transformation and a special structure of the upper triangular matrix. This yields a very efficient algorithm requiring N square roots, $17N$ multiplications and $9N$ additions. Assume that the upper triangular matrix $\Re(n-1)$ is known at time instant $n-1$, then the parameter vector can be computed by back-substitution as follows

$$\hat{\theta}_{n-1}(1) = -[r_{N,N+1}(n-1)]/r_{N,N}(n-1),$$

$$\hat{\theta}_{n-1}(i) = -[r_{i,N+1}(n-1) + \sum_{j=i+1}^{N} r_{i,j}(n-1)\hat{\theta}_{n-1}(j)]/r_{i,j}(n-1), \quad (12)$$

$i = N-1,\cdots,1$, where $\theta_{n-1}(i)$ is the i-th element of the vector θ_{n-1}. Let the term inside the square bracket for $\theta_{n-1}(i)$ in (12) be $s_i(n-1)$. Then

$$s_N(n-1) = -r_{N,N+1}(n-1),$$
$$s_i(n-1) = -[r_{i,N+1}(n-1) + \sum_{j=i+1}^{N} r_{i,j}(n-1)\hat{\theta}_{n-1}(j)], \quad (13)$$
$$i = N-1, N-2,\cdots 1,$$

and (12) can be rewritten as

$$r_{i,j}(n-1)\hat{\theta}_{n-1}(i) = s_i(n-1), \quad i = 1,\dots,N. \quad (14)$$

Given the values of $s_i(n-1)$ and $r_{i,j}(n-1)$, (14) can be viewed as a system of linear equations in θ_n, the parameter estimate of θ at time instant n. This, together with (1), yields the following equation for solving θ_n

$$w\cdot r_{i,i}(n-1)\theta_n(i) = w\cdot s_i(n-1), \quad i=1,\dots,N, \quad (15)$$
$$x_N^T(n)\theta_n = d(n),$$

where w is the square root of the forgetting factor. Note, because of the approximation just mentioned for (14), $r_{i,j}(n-1)$'s in (15) will be different from those of the QRD. To solve (15), let's rewrite it in matrix form as follows:

$$\Phi_n\theta_n = b_n, \quad (16)$$

where

$$\Phi_n = \begin{bmatrix} wD_{n-1} \\ x_n^T \end{bmatrix}, \quad b_n = \begin{bmatrix} wD_{n-1}\hat{\theta}_{n-1} \\ d(n) \end{bmatrix},$$

$$D_{n-1} = diag\{r_{1,1}(n-1),\cdots,r_{N,N}(n-1)\}.$$

(16) can be solved by computing the QRD of Φ_n using say the Householder transformation. However, because the matrix D_{n-1} in Φ_n is a diagonal matrix, the system can be solved using the QRD in order $O(N)$ arithmetic complexity. More precisely, we can construct the appended matrix $D_N(n) = [\Phi_n, b_n]$ as follows [4]

$$D_N(n) = \begin{bmatrix} wr_{1,1}(n-1) & & & & -ws_1(n-1) \\ & wr_{2,2}(n-1) & & 0 & -ws_2(n-1) \\ 0 & & \ddots & & \vdots \\ & & & wr_{N,N}(n-1) & -ws_N(n-1) \\ x_1(n) & x_2(n) & \cdots & x_N(n) & d(n) \end{bmatrix} \quad (17)$$

By zeroing out the elements $x_i(n)$ successively from i=1 to N using the Householder transformation, we obtain an upper triangular matrix, which can be solved using back substitution for the parameter θ_n at the n-th iteration. This yields the algorithm for solving θ_n in [4]. In fact, it can be shown in [7] that this algorithm is actually a variable step size LMS algorithm with individual normalization of the elements in the input signal vector. Thus, its convergence speed is sensitive to the eigenvalue spread of the input autocorrelation matrix. To remedy this problem, we propose to transform the input signal vector by some unitary or orthogonal matrix to approximate decorrelate the input signal vector before applying the A-QR-LS algorithm. This considerably improves the convergence speed as we shall see later in the simulation section. Further, the algorithm converges in the mean if $0 < w < 1$.

3. THE SQUARE ROOT-FREE ALGORITHM

Here a new square root-free Given-based A-QR-LS algorithm will be derived. Let's consider the trangularzation of the following $(N+1)\times(N+1)$ matrix having the same structure as (17):

$$F = \begin{bmatrix} f_{1,1} & & & & f_{1,N+1} \\ & f_{2,2} & & 0 & f_{2,N+1} \\ 0 & & \ddots & & \vdots \\ & & & f_{N,N} & f_{N,N+1} \\ f_{N+1,1} & f_{N+1,2} & \cdots & f_{N+1,N} & f_{N+1,N+1} \end{bmatrix}. \quad (18)$$

Owing to the special structure of matrix F in (6), in applying Givens rotation to matrix F, we note that only two rows, the ith row and the $(N+1)$ th row, are changed during loop i, i.e.

$$f_{i,j}^{(k)} = \begin{cases} f_{i,j} & \text{if } k \le i \\ f_{i,j}^{*} & \text{otherwise} \end{cases}$$

where $f_{i,j}^{*}$ denotes the element of the upper triangular matrix transformed from matrix F.

In order to avoid computing square roots during triangularization process, we use a similar approach as in the square-root free Givens QRD algorithm [5]. More precisely, F is rewritten as follows:

$$F = \begin{bmatrix} \sqrt{\delta_1} & & & & \sqrt{\delta_1}\tilde{f}_{1,N+1} \\ & \sqrt{\delta_2} & & 0 & \sqrt{\delta_2}\tilde{f}_{2,N+1} \\ 0 & & \ddots & & \vdots \\ & & & \sqrt{\delta_N} & \sqrt{\delta_N}\tilde{f}_{N,N+1} \\ \alpha_0\tilde{f}_{N+1,1} & \alpha_0\tilde{f}_{N+1,2} & \cdots & \alpha_0\tilde{f}_{N+1,N} & \alpha_0\tilde{f}_{N+1,N+1} \end{bmatrix}$$

where $\sqrt{\delta_i} = f_{i,i}$ $f_{i,N+1} = \sqrt{\delta_i}\tilde{f}_{i,N+1}$ $i = 1,2,\cdots,N$,

$$f_{N+1,j} = \alpha_0 \tilde{f}_{N+1,j} \quad j = 1,2,\cdots,N+1 .$$

Furthermore, we may express the quantities $f^{(i)}_{N+1,j}$ as

$$f^{(i)}_{N+1,j} = \sqrt{\alpha_{i-1}}\tilde{f}^{(i)}_{N+1,j} \quad j = i+1,\cdots,N+1$$

where the auxiliary parameters α_i, $i = 1,2,\cdots,N$ are

$$\alpha_0 = 1,\ \alpha_i = \alpha_{i-1}\delta_i / \delta_i^* .$$

Consider the ith Givens rotation:

$$\begin{bmatrix} c_i & s_i \\ -s_i & c_i \end{bmatrix} \begin{bmatrix} 0 \cdots 0 & \sqrt{\delta_i} & 0 & \cdots & 0 & \sqrt{\delta_i}\tilde{f}_{i,N+1} \\ 0 \cdots 0 & \sqrt{\alpha_{i-1}}\tilde{f}^{(i)}_{N+1,i} & \sqrt{\alpha_{i-1}}\tilde{f}^{(i)}_{N+1,i+1} & \cdots & \sqrt{\alpha_{i-1}}\tilde{f}^{(i)}_{N+1,N} & \sqrt{\alpha_{i-1}}\tilde{f}^{(i)}_{N+1,N+1} \end{bmatrix}$$
$$= \begin{bmatrix} 0 \cdots 0 & \sqrt{\delta_i^*} & \sqrt{\delta_i^*}\tilde{f}_{i,i+1} & \cdots & \sqrt{\delta_i^*}\tilde{f}_{i,N} & \sqrt{\delta_i^*}\tilde{f}^*_{i,N+1} \\ 0 \cdots 0 & 0 & \sqrt{\alpha_i}\tilde{f}^{(i+1)}_{N+1,i+1} & \cdots & \sqrt{\alpha_i}\tilde{f}^{(i+1)}_{N+1,N} & \sqrt{\alpha_i}\tilde{f}^{(i+1)}_{N+1,N+1} \end{bmatrix}. \tag{19}$$

Since the Givens rotation annihilates the first nonzero element of the second row, it is easy to show that

$$\delta_i^* = \delta_i + \alpha_{i-1}\left[\tilde{f}^{(i)}_{N+1,i}\right]^2, \ c_i = \frac{\sqrt{\delta_i}}{\sqrt{\delta_i^*}}, \ s_i = \frac{\sqrt{\alpha_{i-1}}\tilde{f}^{(i)}_{N+1,i}}{\sqrt{\delta_i^*}} . \tag{20}$$

Combining (19) and (20), we have

$$\sqrt{\alpha_i}\tilde{f}^{(i+1)}_{N+1,j} = \frac{\sqrt{\delta_i}\sqrt{\alpha_{i-1}}\tilde{f}^{(i)}_{N+1,j}}{\sqrt{\delta_i^*}} + \frac{\sqrt{\delta_i}\sqrt{\alpha_{i-1}}\tilde{f}_{i,j}\tilde{f}^{(i)}_{N+1,i}}{\sqrt{\delta_i^*}}, \tag{21}$$

and $\sqrt{\delta_i^*}\tilde{f}^*_{i,j} = \delta_i \tilde{f}_{i,j} / \sqrt{\delta_i^*} + \alpha_{i-1}\tilde{f}^{(i)}_{N+1,j}\tilde{f}^{(i)}_{N+1,i} / \sqrt{\delta_i^*}$. (22)

Since $\tilde{f}_{i,j} = 0, i = 1,\cdots,N,\ j = i+1,\cdots,N$, (21) and (22) can be written after some algebra as

$$\tilde{f}^{(i+1)}_{N+1,j} = \tilde{f}^{(i)}_{N+1,j},\ \tilde{f}^*_{i,j} = \rho_i \tilde{f}^{(i)}_{N+1,j}$$

where $\rho_i = \alpha_{i-1}\tilde{f}^{(i)}_{N+1,i} / \delta_i^*$. Similarly, we have

$$\tilde{f}^{(i+1)}_{N+1,N+1} = \tilde{f}^{(i)}_{N+1,N+1} - \tilde{f}_{i,N+1}\tilde{f}^{(i)}_{N+1,i},\ \tilde{f}^*_{i,N+1} = \sigma_i \tilde{f}_{i,N+1} + \rho_i \tilde{f}^{(i)}_{N+1,N+1}$$

where $\sigma_i = \delta_i / \delta_i^*$. Thus, the triangularization formula for (18) using the square root-free Givens rotation is

$$\delta_i = f^2_{i,i}, \alpha_0 = 1$$
$$\tilde{f}_{N+1,i} = f_{N+1,i}, \tilde{f}_{i,N+1} = f_{i,N+1} / f_{i,i}$$

For $i = 1,2,\cdots,N$ *Loop*

$$\delta_i^* = \delta_i + \alpha_{i-1}\tilde{f}^2_{N+1,i}, \sigma_i = \delta_i / \delta_i^*$$
$$\rho_i = \alpha_{i-1}\tilde{f}_{N+1,i} / \delta_i^*, \quad \alpha_i = \alpha_{i-1}\sigma_i$$
$$\tilde{f}^*_{i,N+1} = \sigma_i \tilde{f}_{i,N+1} + \rho_i \tilde{f}^{(i)}_{N+1,N+1}$$

For $j = i+1, i+2,\cdots,N$ *Loop*

$$\tilde{f}^*_{i,j} = \rho_i \tilde{f}_{N+1,j}$$

End of Loop j (23)

$$\tilde{f}^{(i+1)}_{N+1,N+1} = \tilde{f}^{(i)}_{N+1,N+1} - \tilde{f}_{i,N+1}\tilde{f}_{N+1,i}$$

End of Loop i

Let $$\gamma_i = \sum_{j=i+1}^{N} \tilde{f}_{N+1,j}\theta(j) . \tag{24}$$

Therefore, the backsolving formulas can be expressed by

$$\gamma_N = 0, \theta(N) = -\tilde{f}^*_{N,N+1}$$

For $i = N-1, N-2,\cdots,1$ *Loop*

$$\gamma_i = \gamma_{i+1} + \tilde{f}_{N+1,i+1}\theta(i+1)$$
$$\theta(i) = -\tilde{f}^*_{i,N+1} - \rho_i \gamma_i \tag{25}$$

End of Loop i

Combining (23) and (25), we get the recursive parameter estimation formulas for the proposed square root-free Givens-based transform approximate QR-LS algorithm (TA-QR-LS). It is also summarized in (26). ***Table 1*** compares the arithmetic complexities of various algorithms. $M\_T(N)$ and $A\_T(N)$ are respectively the multiplications and additions required for the discrete cosine transform (DCT). If N is a power of two number, the fast DCT algorithm in [8] has the following complexity,

$$M\_DCT(N) = (N/2)\log_2 N$$
$$A\_DCT(N) = (3/2)N\log_2 N - N + 1 .$$

1. *Initialization* (26)

$$\boldsymbol{\theta} = [0,0\cdots,0]^T$$
$$\boldsymbol{\delta} = [1,1\cdots1]^T,\ \alpha_0 = 1$$

2. *Recurisive Operations*

$$\boldsymbol{f} = -\boldsymbol{\theta}, \boldsymbol{\delta}' = w\boldsymbol{\delta}$$

Add New Equation and DCT Transform

$$\boldsymbol{f}' = DCT(\boldsymbol{x}_d)$$

Transformation

For $i = 1,2,\cdots,N$ *Loop*

$$\alpha' = \alpha_{i-1}f_{N+1,i}, \quad \delta_i = \delta_i' + \alpha' f_{N+1,i}$$
$$\sigma_i = \delta_i' / \delta_i, \quad \rho_i = \alpha' / \delta_i$$
$$\alpha_i = \alpha_{i-1}\sigma_i, \quad \tau = f_{i,N+1}$$
$$f^*_{i,N+1} = \sigma_i f_{i,N+1} + \rho_i f^{(i)}_{N+1,N+1}$$
$$f^{(i+1)}_{N+1,N+1} = f^{(i)}_{N+1,N+1} - \tau f_{N+1,i}$$

End of Loop

$$\gamma_N = 0,\ \tilde{\theta}(N) = -f^*_{N,N+1}$$

For $i = N-1, N-2,\cdots,1$ *Loop*

$$\gamma_i = \gamma_{i+1} + f_{N+1,i+1}\theta(i+1)$$
$$\tilde{\theta}(i) = -f^*_{i,N+1} - \rho_i \gamma_i$$

End of Loop i

Invert DCT Transform

$$\boldsymbol{\theta} = IDCT(\tilde{\boldsymbol{\theta}})$$

3. *Termination Control*

If T.Flag = True, Then Stop

Else go to 2

Table 1

	Proposed Givens-based A-QR-LS algorithm	The A-QR-LS algorithm in [4]	Proposed TA-QR-LS
MULT	$10N$	$17N$	$2\cdot M\_T(N) + 10N$
ADD	$5N$	$9N$	$2\cdot A\_T(N) + 5N$
Square root	*No*	N	*No*

4. EXPERIMENTAL RESULTS

The first test system is the second-order system used in [4]

$$y(k) = a(k)y(k-1) - 0.98y(k-2) + u(k)$$
$$+ 2.0u(k-1) + 0.5u(k-2) + \zeta(k),\ k = 1,2,\cdots,500$$

where $\zeta(k)$ is a zero mean white Gaussian noise with variance σ^2, $u(k)$ is a zero mean white Gaussian process with unit variance. ***Table 2*** and ***Table*** 3 list the maximum estimation errors of each parameter for $a = -1.96$, obtained by the TA-QR-LS and A-QR-LS algorithms, respectively, after the algorithms have converged. The initial model parameters for all the algorithms are initialized to zero. The inverse of the correlation matrix in the RLS algorithm is initialized to constant multiple of the identity matrix. The tracking properties of the new algorithm are evaluated using the same testing system. The parameter $a(k)$ is assumed to be time-varying, and is modeled by a slowly varying sinusoidal: $a(k) = 2\sin(2\pi fk)$, with $f = 0.01Hz$. The estimates of $a(k)$ obtained by the TA-QR-LS, TLMS, A-QR-LS and RLS algorithms are shown in ***Figure 1***. It can be seen that the best performances are obtained by the TA-QR-LS. There is a considerable fluctuation in the parameter estimate obtained by the TLMS algorithm at the beginning of the simulation. The A-QR-LS and the RLS algorithm are unable to track the slowly time-varying parameter, and the performance of the RLS algorithm seems to be the worse. In order to evaluate the performance of the various algorithms to sudden change of system parameter, the value of parameter $a(k)$ is suddenly changed from -1.96 to -1.5 at $k = 780$. The results estimated by TA-QR-LS, A-QR-LS and RLS algorithms are shown in ***Figure 2***, and it was obtained using 100 Monte Carlo simulations. It can be seen that initial convergence of the RLS algorithm is better than the A-QR-LS and TA-QR-LS. However, its response to the sudden change of system parameter is rather poor. The second test system is an acoustic echo cancellation system in which input speech signal is modeled as an 5-th order AR model characterized by poles at: $z_1 = 0.5$, z_2 and $z_3 = 0.85e^{\pm j\pi/3}$, z_4 and $z_5 = 0.7e^{\pm j2\pi/3}$. The acoustic path of the echo is modeled as a linear time invariant system using an exponentially weighted model of order 60. ***Figure 3*** shows the error norms for various algorithms. It can be seen that the initial convergence of the RLS algorithm is inferior to the other algorithms. The performance of the TA-QR-LS algorithm is considerable better than the A-QR-LS algorithm, due to the decorrelating property of the transformation.

5. CONCLUSIONS

A transform-domain approximate QR-LS (TA-QR-LS) algorithm is presented for speeding up the convergence speed of the A-QR-LS algorithm proposed in [4]. Further, efficient square-root free realizations of the A-QR-LS and TA-QR-LS algorithms using the Givens rotation are introduced. Simulation results show that the proposed TA-QR-LS algorithm is a good alternative to the RLS algorithm in adaptive filtering applications involving multiple channels, acoustic modeling, and fast parameter variations.

REFERENCES

[1] J. G. McWhirter, "Recursive least squares minimization using a systolic array," in Proc. SPIE Int. Soc. Opt. Eng., vol. 431, 1983, pp. 105-112.

[2] J. M. Cioffi, "A fast adaptive ROTOR's RLS algorithm," IEEE Trans. ASSP, vol. 38, 1990, pp. 631-653.

[3] D. T. M. Slock and T. Kailath, "Numerically stable fast recursive least squares transversal filters," in Proc. IEEE ICASSP, New York, 1988, pp. 1369-1372.

[4] Z. S. Liu, "QR methods of $O(N)$ complexity in adaptive parameter estimation," IEEE Trans. SP, vol. 43, 1995, pp. 720-729.

[5] W. M. Gentleman and H. T. Kung, "Matrix triangularization by systolic array," in Proc. SPIE Int. Soc. Opt. Eng., vol. 298, 1981, pp. 298-303.

[6] Z. S. Liu, "A QR-based least mean squares algorithm for adaptive parameter estimation," IEEE Trans. CAS-II, vol 45, 1998, pp. 321-329.

[7] S. C. Chan and X. X. Yang, "Improved approximate QR-LS algorithms for adaptive filtering," submitted to IEEE Trans. CAS-II, AUG. 2002.

[8] S. C. Chan and K. L. Ho, "Direct methods for computing the discrete sinusoidal transforms," IEE Proceedings, Part F, vol. 137, 1990, pp. 433-442.

Table 2

parameters	Maximum errors of the estimates		
	σ =0.1	σ =0.2	σ =0.3
-1.96	0.00406	0.00691	0.00459
-0.98	0.00428	0.00621	0.00437
1	0.00670	0.02278	0.02265
2.0	0.02224	0.07924	0.09550
0.5	0.01139	0.04975	0.07336

Table 3

parameters	Maximum errors of the estimates		
	σ =0.1	σ =0.2	σ =0.3
-1.96	0.04999	0.05418	0.05154
-0.98	0.05316	0.05318	0.05094
1	0.01086	0.00205	0.00389
2.0	0.06597	0.05838	0.05309
0.5	0.03369	0.04651	0.05546

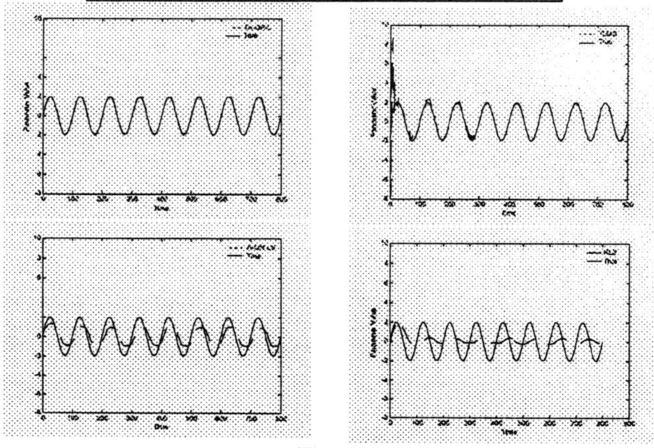

Figure 1

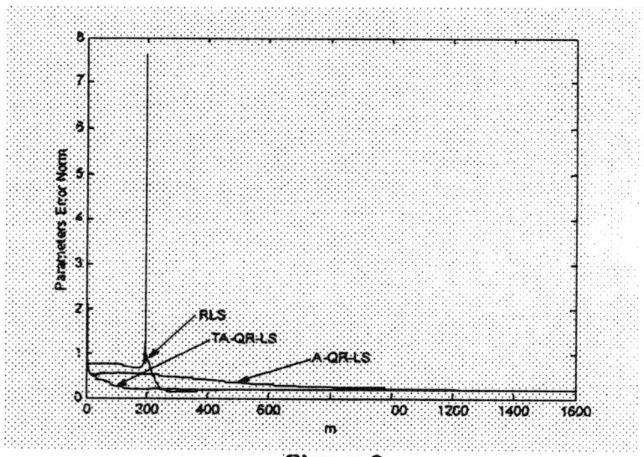

Figure 2

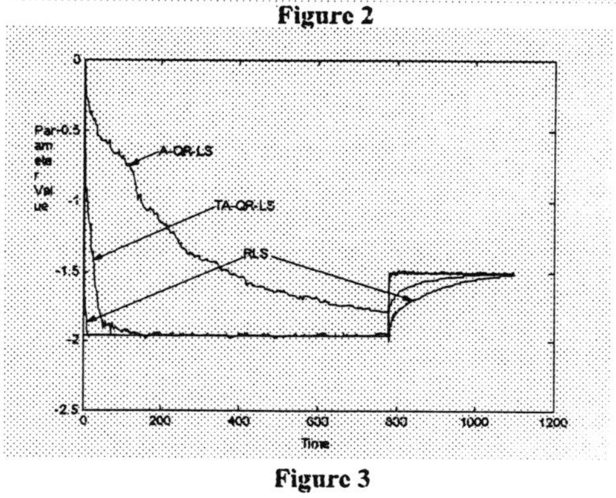

Figure 3

PERFORMANCE ANALYSIS OF ADAPTIVE IIR NOTCH FILTERS BASED ON LEAST MEAN P-POWER ERROR CRITERION

Maha Shadaydeh *Masayuki Kawamata*

Electronic Engineering Department, Graduate School of Engineering
Tohoku University, Sendai 980-8579, Japan.
maha@mk.ecei.tohoku.ac.jp, kawamata@mk.ecei.tohoku.ac.jp

ABSTRACT

In this paper, we present the steady state analysis of adaptive IIR notch filters based on the least mean p-power error criterion. We consider the cases when the sinusoidal signal is contaminated with white Gaussian noise and $p = 3, 4$. We first derive two difference equations for the convergence of the mean and the Mean Square Error (MSE) of the adaptive filter's notch coefficient, and then give the steady state estimation bias and MSE. Stability conditions on the step size value are also derived. Simulation experiments are presented to confirm the validity of the obtained analytical results. It is shown that the notch coefficient steady state bias of the $p-$power algorithm for small step size values is independent of the step size value and is equal for $p = 1, 2, 3$ and 4. However, for larger step size values, the p-power algorithm with $p = 3$ provides the best performance in term of the MSE.

1. INTRODUCTION

Adaptive IIR notch filters have been successfully used for detecting sinusoidal signals in wide-band noise. So far, several adaptive IIR notch filtering algorithms based on least Mean Square Error (MSE) criteria have been proposed [1]- [3]. However, MSE criteria do not always provide the best performance, and accordingly there has been increased interest in developing adaptive algorithms based on L_p normed minimization. So far several L_p norm based adaptive IIR notch filtering algorithms have been proposed, such as the Sign Algorithm (SA) [4], [5] and the p-power algorithm [6]. However, the question to be asked here is: for which value of p does the $p-$power algorithm provide the best performance? The answer to this question, as simulation experiments show, depends on the nature of additive noise. The SA algorithm seems to provide the best performance when the sinusoidal signal is contaminated with impulsive noise [5]. However, for the case when the additive noise is Gaussian, Pei et al. [6] have shown by intensive simulations that the performance of the $p-$power algorithm for $p = 3$ is better than that of the LMS algorithm (i.e. $p = 2$) and the SA (i.e. $p = 1$). The performance analysis for SA and LMS algorithm have been intensively studied in the literature [1]- [3], [7]. However for $p > 2$, no performance analysis has been presented so far.

In this paper, we present the steady state analysis of the p-power algorithm [6] for the cases when $p = 3$ and $p = 4$, and the additive noise is white Gaussian. After a short review of the p-power algorithm, two difference equations for the convergence of the mean and MSE are derived and stability bounds on the step size value are discussed. Closed form expressions for the steady state bias and MSE are then concluded. Simulation experiments that confirm the obtained analytical results are presented with some comparison remarks on the performance of the p-power algorithm for different values of p.

2. THE P-POWER ALGORITHM

In this paper the second-order IIR notch filter with constrained poles and zeros [8] is considered. Its transfer function is given by

$$H(z) = \frac{1 + az^{-1} + z^{-2}}{1 + \rho az^{-1} + \rho^2 z^{-2}} \tag{1}$$

where ρ is the pole radius of the adaptive filter which is restricted to the range $[0, 1)$ to insure stability of the IIR filter. The parameter a in (1) is the filter notch coefficient; its true value is calculated by $a_0 = -2\cos\omega_0$, where ω_0 is the frequency of the input sinusoidal signal

$$x(n) = A\cos(\omega_0 n + \theta) + v(n). \tag{2}$$

The additive noise $v(n)$ in (2) is assumed to be zero mean white Gaussian noise with variance σ_v^2; A and θ are the unknown signal amplitude and phase.

The $p-$power algorithm [6] updates the notch coefficient a such that the mean $p-$power of the notch filter's output signal $e(n)$, that is $\mathrm{E}(|e(n)|^p)$, is minimized. Accordingly, using the steepest descent algorithm, the update

0-7803-7761-3/03/$17.00 ©2003 IEEE

equation of the filter's notch coefficient estimation error $\delta_a(n) = \hat{a}(n) - a_0$ is given by

$$\delta_a(n+1) = \delta_a(n) - \mu_p e^{p-1}(n)\ \mathrm{sign}(e(n))\ s(n) \quad (3)$$

for p odd, and by

$$\delta_a(n+1) = \delta_a(n) - \mu_p\ e^{p-1}(n) s(n) \quad (4)$$

for p even. μ_p is the step size value, and $s(n)$ is the gradient signal calculated by

$$s(n) = \frac{\partial e(n)}{\partial \hat{a}} \approx x(n-1) - \rho e(n-1). \quad (5)$$

3. PERFORMANCE ANALYSIS

At the steady state, the filter's notch coefficient $\hat{a}(n)$ becomes close enough to its true value a_0. Thus, using Taylor series expansion of the notch filter transfer function (1) in the vicinity of a_0, the output and gradient signals can be calculated by [3]

$$\begin{aligned} e(n) &= AB\cos(\omega_0 n + \theta - \phi)\delta_a(n) && (6)\\ &\quad -\rho AB^2\cos(\omega_0 n + \theta - 2\phi)\delta_a^2(n) + v_1(n) \\ s(n) &= A\cos(\omega_0 n + \theta - \omega_0) && (7)\\ &\quad -\rho AB\cos(\omega_0 n + \theta - \omega_0 - \phi)\delta_a(n) \\ &\quad +\rho AB^2\cos(\omega_0 n + \theta - \omega_0 - 2\phi)\delta_a^2(n) + v_2(n) \end{aligned}$$

where $v_1(n)$ and $v_2(n)$ are the additive noise in the filter's output and gradient signal respectively. ϕ and B are defined as

$$B = \frac{1}{(1-\rho)\sqrt{(1+\rho)^2 - 4\rho\cos^2\omega_0}} \quad (8)$$

$$\phi = \begin{cases} \phi_0, & \omega_0 \le \frac{\pi}{2} \\ \phi_0 + \pi, & \omega_0 > \frac{\pi}{2} \end{cases} ;\ \phi_0 = \tan^{-1}\frac{(1+\rho)\sin(\omega_0)}{(1-\rho)\cos(\omega_0)}.$$

In our analysis for $p = 3$, to handle the sign function in the update equation (3), we use similar approach to that used in the analysis of the SA [7] which is based on the assumption that the output signal $e(n)$ is Gaussian distributed with mean value μ_e and variance σ_e^2 that can be calculated directly from (6), and that $e(n)$ and $\delta_a(n)$ are jointly Gaussian distributed. This assumption has been tested in [7] and proved to hold as long as the noise $v(n)$ is white and not necessarily Gaussian.

3.1. Convergence of the Mean

Substituting (6) and (7) in (3)((4)), applying the expectation operator E, and after long mathematical work, we can get the following difference equation for the convergence of the mean:

$$\begin{aligned} \mathrm{E}[\delta_a(n+1)] &= (1 - \mu_p A_{p,1})\,\mathrm{E}[\delta_a(n)] && (9)\\ &\quad -\mu_p B_{p,1}\,\mathrm{E}[\delta_a^2(n)] - \mu_p C_{p,1}, \quad p = 3, 4, \end{aligned}$$

where for $p = 3$,

$$\begin{aligned} A_{3,1} &= \frac{3}{\sqrt{2\pi}}\sigma_{v_1} A^2 B\cos(\omega_0 - \phi) \\ B_{3,1} &= -\frac{3}{\sqrt{2\pi}}\sigma_{v_1}\rho A^2 B^2(\cos(\omega_0 - 2\phi) + \cos(\omega_0)) \\ &\quad + \frac{3}{\sqrt{2\pi}} A^2 B^2 \frac{R_{1,2}}{\sigma_{v_1}} \\ C_{3,1} &= \frac{6}{\sqrt{2\pi}}\sigma_{v_1} R_{1,2}, && (10) \end{aligned}$$

and for $p = 4$,

$$\begin{aligned} A_{4,1} &= 1.5\sigma_{v_1}^2 A^2 B\cos(\omega_0 - \phi) \\ B_{4,1} &= -1.5\sigma_{v_1}^2\rho A^2 B^2(\cos(\omega_0 - 2\phi) + \cos(\omega_0)) \\ &\quad +1.5 A^2 B^2 R_{1,2} \\ C_{4,1} &= 3\sigma_{v_1}^2 R_{1,2}. && (11) \end{aligned}$$

$\sigma_{v_1}^2$, $\sigma_{v_2}^2$, and $R_{1,2}$ are respectively the variance of $v_1(n)$, the variance of $v_2(n)$, and the correlation between $v_1(n)$ and $v_2(n)$, and can be calculated using the theory of residues [3].

In the calculation of (9) and (12) (presented in the following subsection), we need to go through the calculations of many terms of the general form $\mathrm{E}[\delta_a^m(n)\mathrm{sign}(e(n))v_1^l(n)v_2^k(n)]$ for $p = 3$ or of the form $\mathrm{E}[\delta_a^m(n)\ v_1^l(n)\ v_2^k(n)]$ for $p = 3, 4$, where $m = 0, \cdots, 16(p-1)$, $l = 0, \cdots, 2(p-1)$ and $k = 0, 1, 2$. These terms are calculated using the Gaussian factoring theorem, the relations between higher order cumulants of random signals and their moments, and the property that higher order cumulants of Gaussian signals equal zero. The terms of $\delta_a^m(n)$, with $m \ge 3$ are ignored. It is also assumed that the estimation error $\delta_a(n)$ is uncorrelated with the noise signals $v_1(n)$ and $v_2(n)$. For $p = 3$, the joint moments of $\mathrm{sign}(e(n))$ and each of $\delta_a(n)$, $v_1(n)$ and $v_2(n)$ are calculated using the Gaussian probability distribution function of the output signal $e(n)$ [3]. Calculation details are omitted here due to space limitation.

3.2. Convergence of the MSE

Squaring both sides of (3)((4)), using (6) and(7), and then averaging, we can after long calculations, derive the following difference equation for the convergence of the MSE:

$$\begin{aligned} \mathrm{E}[\delta_a^2(n+1)] &= (1 - \mu_p B_{p,2})\,\mathrm{E}[\delta_a^2(n)] && (12)\\ &\quad -\mu_p A_{p,2}\,\mathrm{E}[\delta_a(n)] + \mu_p^2 C_{p,2}, \quad p = 3, 4 \end{aligned}$$

where for $p = 3$,

$$\begin{aligned} A_{3,2} = \frac{12}{\sqrt{2\pi}}\sigma_{v_1} R_{1,2} - \mu_p\big(&12 A^2 B R_{1,2}\sigma_{v_1}^2\cos(\omega_0 - \phi) \\ &-3\rho A^2 B\sigma_{v_1}^4\cos(\phi)\big) \end{aligned}$$

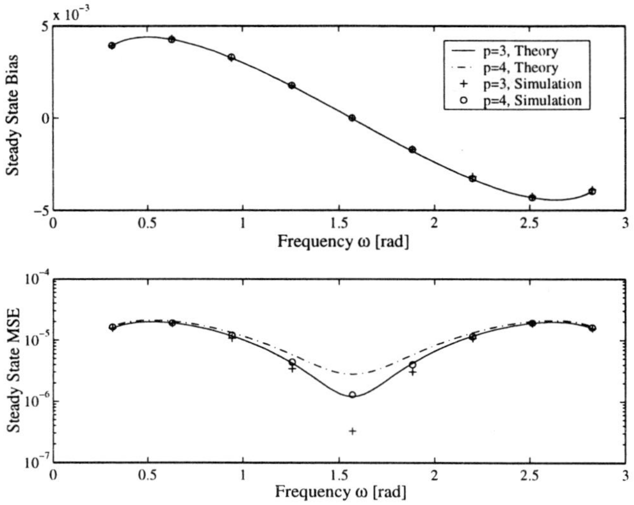

Figure 1: Comparison between theoretical and simulated steady state estimation bias and MSE versus the sinusoidal frequency of the input signal ω_0 ($\mu_p = 0.00005$, $\rho = 0.9$, $A = \sqrt{2}$, SNR=5).

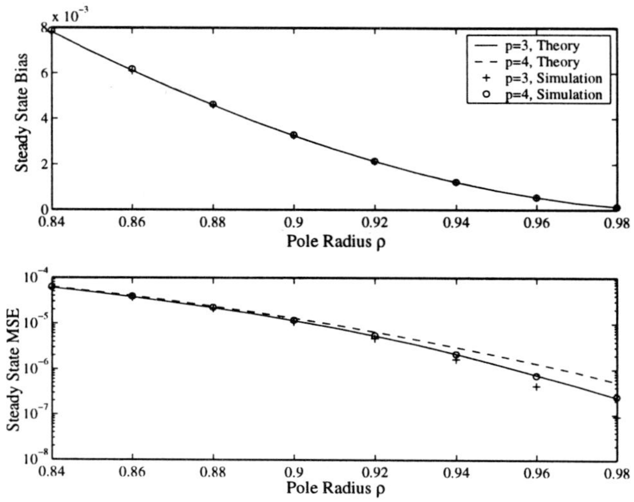

Figure 2: Comparison between theoretical and simulated steady state bias and MSE versus the pole radius ρ ($\mu_p = 0.00005$, $\omega_0 = 0.3\pi$, $A = \sqrt{2}$, SNR=5).

$$
\begin{aligned}
B_{3,2} &= \frac{6}{\sqrt{2\pi}} A^2 B \sigma_{v_1} \cos(\omega_0 - \phi) \\
&\quad -\mu_p \big(1.5 A^4 B^2 \sigma_{v_1}^2 (0.5 + \cos^2(\omega_0 - \phi)) \\
&\quad +3A^2 B^2 \sigma_{v_1}^2 \sigma_{v_2}^2 + 6A^2 B^2 R_{1,2}^2 \\
&\quad +3\rho A^2 B^2 \sigma_{v_1}^4 (\cos(2\phi) + 0.5\rho) \\
&\quad -12 R_{1,2} \rho A^2 B^2 \sigma_{v_1}^2 (\cos(\omega_0 - 2\phi) + \cos(\omega_0))\big) \\
C_{3,2} &= \frac{3}{2} A^2 \sigma_{v_1}^4 + 3\sigma_{v_2}^2 \sigma_{v_1}^4 + 12 R_{1,2}^2 \sigma_{v_1}^2, \qquad (13)
\end{aligned}
$$

and for $p = 4$,

$$
\begin{aligned}
A_{4,2} &= 6\sigma_{v_1}^2 R_{1,2}^2 - \mu_p \big(90 A^2 B \sigma_{v_1}^4 R_{1,2} \cos(\omega_0 - \phi) \\
&\quad - 15\rho \sigma_{v_1}^6 A^2 B \cos(\phi)\big) \\
B_{4,2} &= 3A^2 \sigma_{v_1}^2 B \cos(\omega_0 - \phi) \\
&\quad -\mu_p \big((45/4) A^4 B^2 \sigma_{v_1}^4 (0.5 + \cos^2(\omega_0 - \phi)) \\
&\quad +(15/2) A^2 B^2 \sigma_{v_1}^6 \rho^2 - 15\rho A^2 B \sigma_{v_1}^6 \cos(\phi) \\
&\quad -90\rho A^2 B^2 \sigma_{v_1}^4 R_{1,2} (\cos(\omega_0) + \cos(\omega_0 - 2\phi)) \\
&\quad +(45/2) A^2 B^2 \sigma_{v_2}^2 \sigma_{v_1}^4\big) \\
C_{4,2} &= \frac{15}{2} A^2 \sigma_{v_1}^6 + 15\sigma_{v_2}^2 \sigma_{v_1}^6. \qquad (14)
\end{aligned}
$$

3.3. Stability Bounds

Stability bounds on the step size value can now be easily obtained. In fact if the influence of the the second term in (9) is ignored, the sufficient condition for the convergence of the mean is then given by

$$|1 - \mu_p A_{p,1}| < 1, \qquad p = 3, 4. \qquad (15)$$

Providing that (15) holds, the sufficient condition for the convergence of the MSE can then be deduced from (12) as

$$|1 - \mu_p B_{p,2}| < 1, \qquad p = 3, 4. \qquad (16)$$

3.4. Steady State Estimation Bias and MSE

At the steady state, we have

$$
\begin{aligned}
\mathrm{E}[\delta_a^2(n+1)]_{n\to\infty} &= \mathrm{E}[\delta_a^2(n)]_{n\to\infty} = \mathrm{E}[\delta_a^2(\infty)] \\
\mathrm{E}[\delta_a(n+1)]_{n\to\infty} &= \mathrm{E}[\delta_a(n)]_{n\to\infty} = \mathrm{E}[\delta_a(\infty)]. \qquad (17)
\end{aligned}
$$

Using (17) in (9) and (12), solving the resulting two equations simultaneously, the following closed form expressions for the steady state bias and MSE can be obtained

$$\mathrm{E}[\delta_a(\infty)] = \frac{B_{p,2} C_{p,1} + \mu_p B_{p,1} C_{p,2}}{A_{p,2} B_{p,1} - A_{p,1} B_{p,2}}, p = 3, 4 \qquad (18)$$

$$\mathrm{E}[\delta_a^2(\infty)] = \frac{A_{p,2} C_{p,1} + \mu_p A_{p,1} C_{p,2}}{A_{p,1} B_{p,2} - A_{p,2} B_{p,1}}, p = 3, 4. \qquad (19)$$

3.5. Simulation Results

To confirm the obtained analytical results, we have conducted several experiments. Figure 1 shows comparison between simulated and theoretical steady state bias and MSE versus the sinusoidal signal frequency ω_0 for $p = 3, 4$.

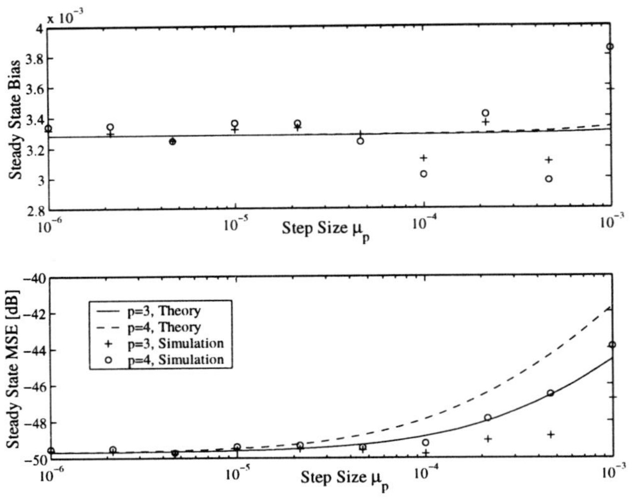

Figure 3: Comparison between theoretical and simulated steady state bias and MSE versus the step size value μ_p ($\rho = 0.9, \omega_0 = 0.3\pi, A = \sqrt{2}$, SNR=5).

It can be observed that the theoretical results match the simulations very well except in the the neighborhood of $\omega_0 = 0.5\pi$. Figure 2 shows comparison between simulated and theoretical steady state bias and MSE versus the pole radius ρ for $p = 3, 4$. As expected, the bias and MSE decrease as the pole radius ρ increases. Figure 3 shows comparison between simulated and theoretical steady state bias and MSE versus the step size value μ_p. This figure indicates that the $p - power$ algorithm results in similar steady state bias for both $p = 3$ and $p = 4$ for sufficiently small step size values. In fact, for small step size values the second term in the numerator of (18) can be neglected and it can be verified that the p-power algorithm has similar expression for the steady state bias for both $p = 3$ and $p = 4$ which is independent of the step size value and is mainly due to the correlation $R_{1,2} = \mathrm{E}[v_1(n)v_2(n)]$. Interestingly, it is similar to the bias expression for the SA (i.e. $p = 1$) [7] and the LMS (i.e. $p = 2$) [3]. However, as it can be observed from Figure 3, the $p-$power algorithm with $p = 3$ performs better than that with $p = 4$ for larger step size values. It has been observed through many experimental results that the convergence speed of the p-power algorithm with $p = 3$ is better than that with $p = 1, 2, 4$. Detailed comparison of the performance of this algorithm for different values of p is out of scope of this paper and will be presented later on.

4. CONCLUSION

This paper presented steady state analysis for constrained adaptive IIR notch filters based on least mean p-power error criterion for the cases when $p = 3, 4$ and the sinusoidal signal is contaminated with white Gaussian noise. Closed form expressions for the steady state estimation bias and MSE have been derived and step size stability bounds have been presented. Simulation results confirm the analytical results. It has been shown that for small step size values, the p-power algorithm produces similar estimation bias for all values of p. However, for larger step size values, the $p-$power algorithm with $p = 3$ performs better than that with $p = 4$.

Acknowledgment

The authors would like to thank Assoc. Prof. Yegui Xiao at University of British Columbia for his close collaboration.

5. REFERENCES

[1] P. Stoica and A. Nehorai, "Performance analysis of an adaptive notch filter with constrained poles and zeros," IEEE Trans. Acoust., Speech, Signal Processing, vol. ASSP-36, no. 4, pp. 911-919, 1988.

[2] M. R. Petraglia, J. J. Shynk and S. K. Mitra, "Stability bounds and steady state coefficient variance for a second-order adaptive IIR notch filter," IEEE Trans. on Signal Processing, vol. 42, no. 7, pp. 1841-1845, 1994.

[3] Y. Xiao, Y. Takeshita and K. Shida, "Steady state analysis of a plain gradient algorithm for a second-order adaptive IIR notch filters with constrained poles and zeros," IEEE Trans. on Circuits, Syst.-II, vol.48, no. 7, pp. 733-740, 2001.

[4] K. Martin and M. T. Sun, "Adaptive filters suitable for real-time spectral analysis," IEEE Trans. on Circuits, Syst., vol. CAS-33, no. 2, pp. 218-229, 1986.

[5] J. Schroeder, R. Yargaladda, and J. Hershely, "L_p normed minimization with applications to linear predictive modeling for sinusoidal frequency estimation," Signal Processing, vol. 24, pp. 193-216 (1991).

[6] S. C. Pei and C. C. Tseng, "Adaptive IIR notch filter based on least mean p-power error criterion," IEEE Trans. on Circuits, Syst., vol. 40, no. 8, pp. 525-529, 1993.

[7] Y. Xiao, A. Ikuta, and R. K. Ward, " Steady state properties of the sign algorithm for the constrained adaptive IIR notch filter," Submitted to IEEE Int. Symp. Circuits and Systems, 2003.

[8] A. Nehorai, "A minimal parameter adaptive notch filter with constrained poles and zeros," IEEE Trans. Acoust., Speech, Signal Processing, vol. ASSP-33, no. 4, pp. 983-996, 1985.

RAMP: AN ADAPTIVE FILTER WITH LINKS TO MATCHING PURSUITS AND ITERATIVE LINEAR EQUATION SOLVERS

John Håkon Husøy

Department of Electrical and Computer Engineering, Stavanger University College,
Post Box 8002, N-4068 Stavanger, Norway. (email: John.H.Husoy@tn.his.no)

ABSTRACT

Using the *matching pursuit* principle we present *Recursive Adaptive Matching Pursuit* (RAMP), a new fast converging adaptive filter having a complexity that is linear in the number of filter coefficients. We show that the algorithm provides for an explicit tradeoff between complexity and performance. The algorithm's relation to a Gauss-Seidel iterative scheme applied to the normal equations is uncovered. This is useful in establishing convergence results. Finally, before presenting an example simulation showing the performance of the algorithm, we point out that the presented algorithm can be viewed as a generalization of the recently introduced *Fast Euclidian Direction Search* (FEDS) algorithm.

1. INTRODUCTION

In this paper a new adaptive algorithm, *Recursive Adaptive Matching Pursuit* (RAMP), is given a concise treatment. While the fundamental idea underpinning the algorithm has been presented in [1], we here present the algorithm in a more general framework that facilitates several contributions which are summarized below:

- We show that RAMP algorithms using either sliding rectangular or exponentially weighted windowed error functions can be easily derived.
- A more efficient implementation of the algorithm is presented. It is shown that RAMP allows for an explicit tradeoff between computational complexity and performance. This flexibility can, if desired, be invoked dynamically while the algorithm is in operation.
- We show that the algorithm is intimately related to a Gauss-Seidel iterative scheme [2] incorporating *dynamic optimal reshuffling of the order of equations* applied to the underlying Wiener-Hopf equations. This this can be used to predict the algorithm's convergence behavior.
- Finally, we point out that RAMP can be viewed as a generalization of the recently introduced *Fast Euclidian Direction Search* (FEDS) algorithm presented in [3, 4, 5] and further elaborated on in a forthcoming book by Bose [6]. Since the RAMP algorithm is always guaranteed to converge faster than FEDS, all convergence results obtained for this algorithm can be viewed as conservative estimates of lower bounds of the convergence performance of RAMP.

We have organized this paper as follows: In the next section we give a short review of the *matching pursuit* (MP) principle in the context of signal representation, which is the most common application of MP in signal processing. Following this we formulate the adaptive filtering problem in such a way as to motivate the use of the MP principle for its solution. Subsequently, we develop an efficient implementation of the MP solution. In Section 4 we show how RAMP relates to the a Gauss-Seidel like iterative scheme applied to the Wiener-Hopf equations, and we also point out that the FEDS algorithm can be viewed as a special case of RAMP. Finally, we present an example simulation result illustrating the behavior of RAMP.

This work was performed while on sabbatical leave with the School of Information Technology and Engineering, University of Ottawa, Ontario, Canada. Support from the Centre for Information and Communication Technology (SIKT) at Stavanger University College and the Norwegian Research Council is gratefully acknowledged. Fruitful discussions with former students O. Mæstad, B.I. Helland and J. Ommundsen are much appreciated.

2. BACKGROUND ON MATCHING PURSUIT

Matching pursuit [7] is a class of greedy algorithms commonly used for finding approximate sparse solution vectors, $\underline{w}$, to

$$\min_{\underline{w}} \|\underline{x} - \mathbf{F}\underline{w}\|. \tag{1}$$

Typically, $\mathbf{F}$ is a matrix of dimension $L \times M$ with $M > L$, whose columns, $\{\underline{f}_j\}_{j=0}^{M-1}$, are designed or selected in such a way that sparse linear combinations, $\sum_{j=0}^{M-1} w_j \underline{f}_j = \mathbf{F}\underline{w}$, i.e linear combinations where most elements of $\underline{w}$ are zero, are good representations/approximations of signal vectors, $\underline{x}$, drawn from some signal class. One way of tackling Eq. 1, referred to as *Basic Matching Pursuit* (BMP) [7], can be stated as shown in Algorithm 1 and explained below.

Algorithm 1 Basic Matching Pursuit (BMP)

1: $w_j = 0, j = 0, \ldots, M-1$
2: $\underline{e} = \underline{x}$
3: **for** $i = 0 : S-1$ **do**
4: $\quad j_i = \arg\max_j \mid \underline{e}^T \underline{f}_j \mid \|\underline{f}_j\|^{-1}$
5: $\quad \Delta w_{j_i} = \underline{e}^T \underline{f}_{j_i} \|\underline{f}_{j_i}\|^{-2}$
6: $\quad w_{j_i} = w_{j_i} + \Delta w_{j_i}$
7: $\quad \underline{e} = \underline{e} - \Delta w_{j_i} \underline{f}_{j_i}$
8: **end for**

What happens here is as follows: We start out with some signal vector, $\underline{x}$, whose sparse representation we want to find. First (line 1) we initialize all weights, $\{w_j\}_{j=0}^{M-1}$, to zero. Next the approximation error, $\underline{e}$, is initialized (line 2). The loop in which the sparse

0-7803-7761-3/03/$17.00 ©2003 IEEE

representation is computed is executed S times ensuring that no more than a maximum of S elements of $\underline{w}$ will be different from zero. The selection of S is guided by the required accuracy of the representation and the degree of sparsity desired. The first step inside the loop (line 4) computes the magnitudes of the projection of $\underline{e}$ onto unit vectors in the direction of each column of $\mathbf{F}$. The index of the column vector onto which the projection of $\underline{e}$ has its largest magnitude is found and the corresponding column vector, $\underline{f}_{j_i}$ is selected for inclusion in the representation. The corresponding coefficient, w_{j_i}, is updated according to line 5 in Algorithm 1. Finally, we update the approximation error, $\underline{e}$. Note that most coefficients, as a consequence of the sparsity constraint, will remain zero after the iterations are done. This does, however, not preclude the possibility that some coefficients may be updated more than once. For more material on the matching pursuit principle and its numerous applications we refer to [7] and the references therein.

In the next section we will show how the MP principle described above can be used to solve the adaptive filtering problem. The major differences from the above in what follows are: 1) The sparsity constraint is no longer relevant. 2) The *time varying* data matrix of the signal to be adaptively filtered takes the place of matrix $\mathbf{F}$. 3) A vector of what we in adaptive filter theory refer to as the *desired signal* takes the place of vector $\underline{x}$ above, and the role of vector $\underline{w}$ is taken by the vector of *time varying* filter coefficients of the adaptive filter. These points will be made clearer in the following section.

3. ALGORITHM DESCRIPTION AND IMPLEMENTATION

The adaptive filtering problem [8] entails the continual adjustment of filter coefficients $\underline{h}(n) = [h_0(n), h_1(n), \ldots, h_{M-1}(n)]^T$ in such a way that when the filter is applied to an input signal, $x(n)$, a best possible estimate of some *desired signal*, $d(n)$, is produced. The signal produced by the filter at time n is given by

$$\sum_{k=0}^{M-1} h_k(n)x(n-k). \tag{2}$$

The vector of signal samples output by $\underline{h}(n)$ applied to the input signal for all time instants from 0 to n can be written as:

$$\begin{bmatrix} x(n) & x(n-1) & \cdots & x(n-M+1) \\ x(n-1) & x(n-2) & \cdots & x(n-M) \\ \vdots & & \cdots & \vdots \\ x(0) & x(-1) & \cdots & x(-M+1) \end{bmatrix} \begin{bmatrix} h_0(n) \\ h_1(n) \\ \vdots \\ h_{M-1}(n) \end{bmatrix}. \tag{3}$$

Denoting the data matrix above by $\mathbf{X}(n)$ and defining the vector of desired signal samples as $\underline{d}(n) = [d(n), d(n-1), \ldots, d(0)]^T$, the adaptive filtering problem, using the mean squared error criterion, can be formulated as: *For each time instant n, find the filter vector $\underline{h}(n)$ that solves*

$$\min_{\underline{h}(n)} \|\underline{d}(n) - \mathbf{X}(n)\underline{h}(n)\|_2. \tag{4}$$

The solution of this problem is the same as the solution to the normal equation[1]:

$$\mathbf{X}^T(n)\mathbf{X}(n)\underline{h}(n) = \mathbf{X}^T(n)\underline{d}(n) \tag{5}$$

[1]We assume $n \geq M - 1$.

whose solution can be obtained directly for each n through:

$$\underline{h}(n) = (\mathbf{X}^T(n)\mathbf{X}(n))^{-1}\mathbf{X}^T(n)\underline{d}(n). \tag{6}$$

This problem can be solved more efficiently by applying the *Recursive Least Squares* (RLS) algorithm [8] with a forgetting factor $\lambda = 1$.

What follows[2] can be described as a "poor man's RLS" in which we allow only a subset of (at most) P filter coefficients to be updated at time n. The key observation here is that Eqs. 1 and 4 are essentially the same if some obvious mappings between the involved quantities are made. Having made such a mapping we can, at each time instant n, make updates to the elements of the filter vector $\underline{h}(n)$ in the same way as the updates of $\underline{w}$ were found in Algorithm 1. There is only one important difference that has to be considered: At each time instant we do exactly the same thing as we do inside the loop of Algorithm 1, however, instead of the weight initialization of the first line of that algorithm, *we keep the weights obtained from running the loop at the previous time instant, $n - 1$.* With these observations in mind the adaptive algorithm, which we have named *Recursive Adaptive Matching Pursuit* (RAMP), shown as Algorithm 2, results. Note that the

Algorithm 2 Recursive Adaptive Matching Pursuit (RAMP)

1: Algorithm initialization
2: **for** $n = 0, 1, \ldots$ **do**
3: $\quad \underline{e}(n) = \underline{d}(n) - \mathbf{X}(n)\underline{h}(n-1)$
4: $\quad$ **for** $i = 0 : P - 1$ **do**
5: $\quad\quad j_i(n) = \arg\max_j \mid \underline{e}^T(n)\underline{x}_j(n) \mid \|\underline{x}_j(n)\|^{-1}$
6: $\quad\quad \Delta h_{j_i(n)} = \underline{e}^T(n)\underline{x}_{j_i(n)}(n)\|\underline{x}_{j_i(n)}(n)\|^{-2}$
7: $\quad\quad h_{j_i(n)} = h_{j_i(n)} + \Delta h_{j_i(n)}$
8: $\quad\quad \underline{e}(n) = \underline{e}(n) - \Delta h_{j_i(n)}\underline{x}_{j_i(n)}(n)$ %Skip for $i = P - 1$.
9: $\quad$ **end for**
10: **end for**

inner loop of Algorithm 2 is the same as the loop of Algorithm 1 with obvious symbol name changes. We will refer to one execution of the inner loop of the RAMP algorithm as an *MP iteration*. Note that the error update of line 8 need not be computed at iteration $P-1$ since a new error vector is computed in line 3 when a new signal sample arrives[3]. Although the conceptual structure of RAMP is well represented in Algorithm 2, the computational complexity of the algorithm and its efficient implementation are not immediately evident. In establishing this, we point out that the key quantity in Algorithm 2 is the term $\underline{e}^T(n)\underline{x}_j(n)$ which has to be computed for all values of $j = 0, 1, \ldots, M - 1$. Making use of the fact that the j-th column of $\mathbf{X}(n)$ can be written as $\underline{x}_j(n) = [x(n-j), x(n-1-j), \ldots x(-j)]^T = \underline{x}(n-j)$, what we have to consider is

$$\underline{e}^T(n)\underline{x}(n-j). \tag{7}$$

We start out by examining the computation of $\underline{e}(n)\underline{x}(n-j)$ for the first MP-iteration. For clarity, we shall introduce the notation $\underline{e}_i(n)$ to mean the error vector $\underline{e}(n)$ at time n *prior* to the entry of

[2]We shall do all the derivations assuming, – for simplicity and economy of space, that $\lambda = 1$. At the end of the section we point out the modifications required when using an exponentially weighted window or a sliding rectangular window.

[3]I will be shown shortly, that in an *efficient* implementation, this step, in the context of an inner product update, will actually be performed. More on that later.

the inner loop of Algorithm 2 for the i'th MP-iteration. Thus, from line 3 of Algorithm 2 we have

$$\underline{e}_0(n) = \underline{d}(n) - \mathbf{X}(n)\underline{h}(n-1), \tag{8}$$

which gives

$$\underline{e}_0^T(n)\underline{x}(n-j) = [\underline{d}(n) - \mathbf{X}(n)\underline{h}(n-1)]\underline{x}(n-j) =$$

$$\{\begin{bmatrix} d(n) \\ \underline{d}(n-1) \end{bmatrix} - \begin{bmatrix} \underline{x}^{(M)^T}(n) \\ \mathbf{X}(n-1) \end{bmatrix} \underline{h}(n-1)\}^T \begin{bmatrix} x(n-j) \\ \underline{x}([n-1]-j) \end{bmatrix},$$

where $\underline{x}^{(M)}(n)$ denotes $[x(n), x(n-1), \ldots, x(n-M+1]^T$, i.e. the vector of input samples having a length given by the number of filter coefficients. In the equation above, we note that the following quantities are involved:

$$d(n) - \underline{x}^{(M)^T}(n)\underline{h}(n-1) = e_a(n) \tag{9}$$

which is the a priori error, and

$$\underline{d}(n-1) - \mathbf{X}(n-1)\underline{h}(n-1) = \underline{e}_P(n-1), \tag{10}$$

which is the error vector at time $n-1$ after the *last* MP-iteration has been performed. From this we get the recursion

$$\underline{e}_0^T(n)\underline{x}(n-j) = e_a(n)x(n-j) + \underline{e}_P^T(n-1)\underline{x}([n-1]-j). \tag{11}$$

For subsequent MP-iterations at time n, the following recursion is easily derived:

$$\underline{e}_{i+1}^T(n)\underline{x}(n-j) = \underline{e}_i^T(n)\underline{x}(n-j) - \Delta h_{j_i(n)}\underline{x}^T(n-j_i(n))\underline{x}(n-j). \tag{12}$$

Note now that line 3 in Algorithm 2 can be substituted with Eqs. 9 and 11, whereas line 8 of the same algorithm can be substituted by Eq. 12. Note that, since the result of Eq. 12 for $i = P-1$ at time $n-1$ is required in Eq. 11 at time n, this recursion is also computed for $i = P-1$. Finally, realizing that the quantities $\underline{x}^T(n)\underline{x}(n-j)$ for $j = 0, 1, \ldots, M-1$ which are needed in Eq. 12, can be found recursively by

$$\underline{x}^T(n)\underline{x}(n-j) = \underline{x}^T([n-1])\underline{x}([n-1]-j) + x(n)x(n-j), \tag{13}$$

we are now in a position to examine the overall computational complexity of the algorithm.

The key computations are given by Eqs. 9, 11, 12 and 13. Focusing on the multiplicative complexity, these equations require M multiplications each, when results available from the previous time instant $n-1$ are fully utilized. Since Eq. 12 is computed P times for each time instant, the total multiplicative complexity of the part of the algorithm involving above mentioned equations is given by $(3+P)M$. The remaining parts of the algorithm are discussed with reference to the algorithm as presented in Algorithm 2: Line 5 involves finding some absolute values and performing $M-1$ comparisons. This can be cheaply implemented. M divisions by $\|\underline{x}_j(n)\|$ for $j = 0, 1, \ldots, M-1$ is also implied. Since we can assume that $\|\underline{x}_j(n)\|$ varies little for different values of j, we need not perform these divisions in determining the index of the filter coefficient to update. The worst possible consequence of this is that, from time to time, we may chose to update another coefficient than the optimal one. We have not observed any problems by not performing the division. Finally, in line 6 a division is performed. Based on the above, we conclude that the multiplicative complexity for each time instant n is given by $(3+P)M$ multiplications and P divisions.

Exponentially weighted and rectangular sliding windows

Adaptive filters with growing, non-weighted memory, as presented above, are rather uncommon. More commonly an exponentially weighted or a sliding rectangular window is applied to the squared error when deriving the adaptive filter algorithm. This is easily accomplished by slightly modifying $\mathbf{X}(n)$ and $\underline{d}(n)$ in the equations presented in the beginning of the present section.

In deriving an exponentially weighted window algorithm, row and element number i of $\mathbf{X}(n)$ and $\underline{d}(n)$, respectively, are multiplied by $\lambda^{\frac{n-i}{2}}$. The recursions corresponding to Eqs. 11 and 13 can now easily be found as

$$\underline{e}_0^T(n)\underline{x}(n-j) = e_a(n)x(n-j) + \lambda\underline{e}_P^T([n-1])\underline{x}([n-1]-j) \tag{14}$$

and

$$\underline{x}^T(n)\underline{x}(n-j) = \lambda\underline{x}^T([n-1])\underline{x}([n-1]-j) + x(n)x(n-j), \tag{15}$$

respectively. The multiplicative complexity is increased to $(5+P)M$. For the sliding rectangular window case it is also easy to show that slightly modified recursions lead to the same increase in multiplicative complexity.

We close this section by noting that the complexity is linear in the number of filter coefficients. We stress that each increment of P beyond $P = 1$ requires only M additional multiplications and one division. Finally, we point out that while in operation, there is nothing preventing P to change dynamically, thus making a tradeoff between complexity and performance possible.

4. RELATION TO ITERATIVE LINEAR EQUATION SOLVERS AND OTHER ADAPTIVE ALGORITHMS

In establishing the relation of the above to iterative linear equation solvers and subsequently the FEDS algorithm mentioned in the introduction, we shall have a brief look at the *Gauss-Seidel iterative equation solver*. More details on Gauss-Seidel iterations can be found in [2]. In solving the linear system $\mathbf{A}\underline{x} = \underline{b}$, with $\mathbf{A}$ assumed to have dimension $M \times M$, the Gauss-Seidel iterative scheme proceeds cyclically through the unknowns, x_j for $j = 0, 1, \ldots M-1$, assuming that when we want to compute (update) unknown x_i, all unknowns from previous computations are available for use. Thus, it can be shown that [2] the update equation for x_i resulting in $x_i^{(new)}$ based on previously computed results, $\{x_j^{(prev)}\}_{j=0}^{M-1}$, is given by

$$x_i^{(new)} = x_i^{(prev)} + \frac{1}{a_{i,i}}\{b_i - \sum_{l=0}^{M-1} a_{i,l}x_l^{(prev)}\}. \tag{16}$$

Applying this to the normal equation (Eq. 5), we have that a Gauss-Seidel update to element i of filter vector[4] $\underline{h}$ is given by

$$h_i^{(new)} = h_i^{(prev)} + \frac{\underline{x}_i^T(n)}{\|\underline{x}_i(n)\|^2}\{\underline{d}(n) - \sum_{j=0}^{M-1} h_j^{(prev)}\underline{x}_j(n)\}. \tag{17}$$

Observing that the expression in curly brackets is $\underline{e}(n)$, we see that Eq. 17 expresses exactly the same thing as lines 6 and 7 of Algorithm 2, i.e. the coefficient update of an MP iteration. From

[4]We suppress the explicit dependency of n for convenience.

this we may conclude that RAMP corresponds to the application of Gauss-Seidel iterations to the normal equation of the problem with the important distinction that *whereas Gauss-Seidel updates the coefficients in a cyclical sequence, we determine which coefficient update contributes maximally to the decrease of the error norm* $\|e(n)\|$. This coefficient is subsequently updated. It is interesting to note that the sequence in which the equations of a linear system is ordered has great impact on how fast a Gauss-Seidel iterative scheme obtains a good solution [2]. The way in which we dynamically select the sequence in which the elements of the filter vector is to be updated, really corresponds to a *dynamic reshuffling of the equation order*. This improves convergence.

In studying the convergence properties of RAMP it would make sense to view the normal equation, Eq. 5, as an approximation to the Wiener-Hopf equation, i.e. we let $\mathbf{X}^T(n)\mathbf{X}(n) \rightarrow \mathbf{R}$ and $\mathbf{X}^T(n)\underline{d}(n) \rightarrow \underline{r}$, where $\mathbf{R}$ is the autocorrelation matrix of the input signal and $\underline{r}$ is the cross correlation vector between the input and desired signals. Since convergence results are readily available for the Gauss-Seidel iterative scheme [2], applying these to the Wiener-Hopf equation should make it possible to establish convergence results that are directly applicable to a RAMP scheme *without the dynamic equation reshuffling scheme.* These results would also be applicable as lower bounds for the performance of the true RAMP algorithm.

Finally, we note that the Fast Euclidian Direction Search (FEDS) algorithm presented in [3, 4, 5] and further elaborated on in a forthcoming book by Bose [6], – although derived in a manner completely different from the derivation given above, essentially corresponds to a RAMP algorithm, but with the exception that the coefficients are updated sequentially in the same fashion as is done in a standard Gauss-Seidel scheme. We also point out that by modifying the objective function to be minimized, – they use a *block exponentially weighted* window, rather than an ordinary exponentially weighted window, their algorithm has a multiplicative complexity of only $4M$. If similar modifications to the objective function for RAMP can give rise to such complexity reductions remains to be investigated. From the above it should be evident that we can view the RAMP algorithm as a generalization of FEDS in the sense that *1) We select the filter coefficient to be updated optimally*, and *2) We are allowed to perform an arbitrary number of coefficient updates at each time instant, whereas the FEDS algorithm updates only one coefficient.*

5. AN EXAMPLE SIMULATION

Due to limited space, we restrict ourselves to presenting a simulation using the popular channel equalization example from Haykin's book [8], see page 285 of that book for a full description of the setup. The channel impulse response is assumed to be

$$w_k = \begin{cases} \frac{1}{2}(1 + \cos(2\pi(k-2)/b)), & k = 1,2,3 \\ 0, & \text{otherwise,} \end{cases} \tag{18}$$

where b determines the eigenvalue spread of the autocorrelation matrix of the signal at the output of the channel, $x(n)$. The input to the channel is a random binary sequence $z(n) = \pm 1$, and the equalizer is an 11 tap FIR filter. The noise added at the equalizer input has a variance of $\sigma_v^2 = 0.001$. Finally, the desired signal, $d(n)$, is given by $z(n-7)$. We set $b = 3.5$ corresponding to the largest eigenvalue spread described in [8]. We apply *sliding window versions*, window size $L = 64$, of the RLS, RAMP (with $P = 1$ and $P = 2$), and FEDS algorithms. Fig. 1 shows the mean squared error as a function of iteration number. Each curve is averaged over 200 independent runs for each algorithm. As expected, RLS performs best. Also, as expected, the RAMP algorithm converges faster that FEDS. We also observe that RAMP with $P = 2$ converges significantly faster that RAMP with $P = 1$. This is rather pleasant given the very low extra computational cost associated with increasing P from 1 to 2.

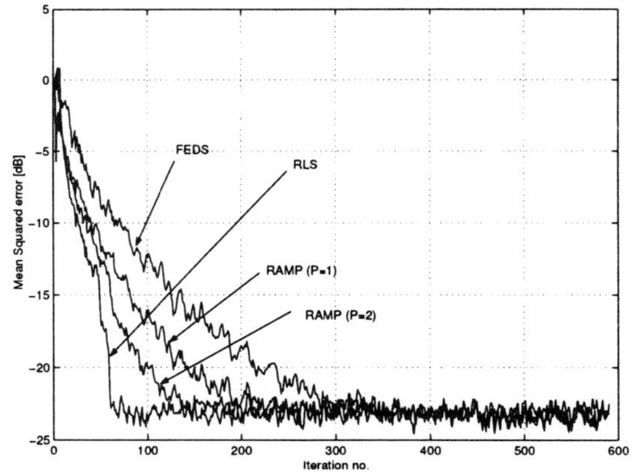

Figure 1: Learning curves for sliding window versions (window size $L = 64$) of RLS, RAMP (with $P = 1$ and $P = 2$), and FEDS.

6. REFERENCES

[1] J. Ommundsen and J. H. Husøy, "An adaptive filter based on matching pursuits," in *Proc. NORSIG*, (Trondheim, Norway), pp. 40–44, Oct. 2001.

[2] Y. Saad, *Iterative Methods for Sparse Linear Systems.* PWS Publishing, 1996.

[3] G. F. Xu, T. Bose, and J. Schroeder, "The Euclidian direction search algorithm for adaptive filtering," in *Proc. ISCAS*, (Orlando, Fl., USA), pp. 146 –149 (Vol. III), May 1999.

[4] G. F. Xu and T. Bose, "Analysis of the Euclidian direction set adaptive algorithm," in *Proc. ICASSP*, (Seattle, Washington, USA), pp. 1689–1692, May 1998.

[5] M.-Q. Chen, T. Bose, and G. F. Xu, "A direction set based algorithm for adaptive filtering," *IEEE Trans. Signal Processing*, vol. 47, pp. 535–539, Feb. 1999.

[6] T. Bose, *Digital Signal and Image Processing.* John Wiley, 2003.

[7] S. F. Cotter, J. Adler, B. D. Rao, and K. Kreutz-Delgado, "Forward sequential algorithms for best basis selection," *IEE Proc. Vis. Image, Signal Process*, vol. 146, pp. 235–244, Oct. 1999.

[8] S. Haykin, *Adaptive Filter Theory.* Upper Saddle River, NJ, USA: Prentice Hall, Fourth ed., 2002.

EXTENDED RLS LATTICE ADAPTIVE FILTERS

Ricardo Merched *

Dept. of Electronics and Computer Engineering
Federal University of Rio de Janeiro, Brazil
E-mail: merched@lps.ufrj.br

ABSTRACT

This paper solves the problem of developing exact fast weighted RLS lattice adaptive Þlters for the class of input signals induced by the so-called general orthonormal Þlter models. The resulting algorithm can be viewed as a counterpart of the *extended fast fixed-order* RLS adaptive Þlters recently derived.

1. INTRODUCTION

For over two decades, there has been a common belief in the literature that fast recursive-least-squares (RLS) adaptive algorithms were only possible when the underlying input data to the adaptive Þlter is originated from a traditional tapped-delay-line structure (see,e.g., [1, 2]).

Recently, however, it has been proved that fast *fixed-order* RLS algorithms are indeed possible even when shift structure is not present (see for example [3], and the references therein). In particular, the work [3] extends such concept to an important class of Þlter structures which are based on the so-called *IIR orthonormal models* [4]. The result was the development of the Þrst fast RLS adaptive Þlters for orthonormal-based implementations in both array (the so-called *extended Chandrasekhar recursions*) and explicit forms (the *extended FTF algorithm*).

This motivates us to pursue in this paper the development of fast *order-recursive*, as opposed to Þxed-order, adaptive algorithms for Þlter structures other than the conventional FIR model. The primary concern of our presentation is to introduce a framework which allows us to obtain both fast Þxed-order and order-recursive RLS algorithms in a uniÞed manner[1]. Hence, the contributions of this paper are organized as follows:

1. First, regularization is incorporated *implicitly* by stating all regularized LS problems as pure LS problems. This is in contrast to [5] where all regularized problems are solved explicitly from the start[2].
2. The derivation that follows takes into account the fact that *data structure will affect the initial conditions of all variables of the lattice recursions.* This is in contrast to lattice Þlters based on FIR models, where initialization can be inferred from the regularization term almost by inspection.
3. We provide a new set of recursions in order to propagate efÞciently the only variable that is intimately affected by data structure. The result is the Þrst exponentially-weighted RLS lattice algorithm based on orthonormal models. In addition, the latter is obtained within the same framework of the extended explicit and array RLS algorithmic forms of [3]. This fact will become apparent from our development.

2. REGULARIZED RLS PROBLEM

Thus given a complex column vector $d_{M,N}$ and data matrix $\mathcal{H}_{M,N}$, the regularized exponentially-weighted least squares problem seeks the column vector $w_M \in \mathbb{C}^M$ that solves

$$\min_{w_M} \left[\lambda^{N+1} w_M^* \Pi_M^{-1} w_M \;+\; \|d_N - \mathcal{H}_{M,N} w_M)\|^2_{\mathcal{W}_N}\right] \qquad (1)$$

where Π_M is a positive deÞnite regularization matrix, and $\mathcal{W}_N = (\lambda^N \oplus \lambda^{N-1} \oplus \cdots \oplus 1)$ is a weighting matrix deÞned in terms of a forgetting factor λ, satisfying $0 \ll \lambda < 1$. The symbol * denotes complex conjugate transposition. The individual entries of the measurement vector d_N are denoted by $\{d(i)\}$, and the individual rows of the data matrix $\mathcal{H}_{M,N}$ by $\{u_{M,i}\}$. We also denote each column of $\mathcal{H}_{M,N}$ by $\{h_{m,N}\}$, for $m = 0, \ldots, M-1$.

Before continuing, we shall redeÞne the regularized problem in (1) equivalently as

$$\min_{w_M} \left\| \begin{bmatrix} 0 \\ d_N \end{bmatrix} - \begin{bmatrix} \mathcal{A}_{M,L} \\ \mathcal{H}_{M,N} \end{bmatrix} w_M \right\|^2_{W_N} \qquad (2)$$

where $W_N = (\lambda^{N+L} \oplus \lambda^{N+L-1} \oplus \cdots \oplus 1)$, and where we have factored Π_M^{-1} as $\Pi_M^{-1} = \mathcal{A}^*_{M,L}\mathcal{W}_L\mathcal{A}_{M,L}$. The regularized LS problem stated as in (1) can be equivalent viewed as the pure LS problem in (2), where it is assumed that the incoming data has started at some point in the past, depending on the number of rows L of $\mathcal{A}_{M,L}$. We thus deÞne the following extended quantities

$$H_{M,N} \triangleq \begin{bmatrix} \mathcal{A}_{M,L} \\ \mathcal{H}_{M,N} \end{bmatrix} = \begin{bmatrix} x_{0,-1} & x_{1,-1} & \cdots & x_{M-1,-1} \\ h_{0,N} & h_{1,N} & \cdots & h_{M-1,N} \end{bmatrix}$$
$$\triangleq \begin{bmatrix} x_{0,N} & x_{1,N} & \cdots & x_{M-1,N} \end{bmatrix} \qquad (3)$$

and $y_N = [0 \;\; d_N]^T$, so that (1) becomes

$$\min_{w_M} \|y_N - H_{M,N} w_M\|^2_{W_N} \qquad (4)$$

R. Merched is an associate professor at the Department of Electronics and Computer Engineering of Federal University of Rio de Janeiro, Brazil. His research is partially funded by CNPq and FAPERJ, Brazil.

[1] A related work was proposed in [5], where the authors develop RLS lattice recursions for the special case of Laguerre models and where the forgetting factor must be chosen as $\lambda = 1$. Our contribution in this paper overcome these drawbacks, and allows us to obtain exponentially-weighted lattice recursions for general orthonormal models under no such restrictions.

[2] The reason for this is two fold. First, it provides us with valuable insight on the initialization step of all lattice recursions. Second, the orthogonality properties of least-squares problems cannot be used directly when working with regularization explicitly. The orthogonality principle represents a powerful tool in our derivation.

0-7803-7761-3/03/$17.00 ©2003 IEEE

The signiÞcance of working with the cost (2), is that it will allow us to make full usage of the orthogonality principle inherent to least squares formulations of the form (4).

3. LATTICE RECURSIONS

We now derive several order-recursive relations. Most of these recursions might be familiar to the reader. Our presentation, however, differs in a fundamental aspect. That is, while in traditional derivations for FIR lattice structures almost all variables are set to zero at time $N = -1$ [except for the minimum costs of forward and backward prediction problems (to be introduced next)], here they must be properly initialized in order to account for the effect of general data structure.

3.1. Joint Process Estimation

Thus consider the data matrix in (3). The projection of y_N onto the range space of $H_{M,N}$, denoted by $\mathcal{R}(H_{M,N})$, is given by

$$\begin{aligned} y_{M,N} &= H_{M,N}(H^*_{M,N}W_N H_{M,N})^{-1}H^*_{M,N}W_N y_N \\ &= H_{M,N}P_{M,N}H^*_{M,N}W_N y_N \end{aligned}$$

Now suppose that one more column is appended to $H_{M,N}$, i.e.,

$$H_{M+1,N} = [\; H_{M,N} \quad x_{M,N} \;] \tag{5}$$

The projection of y_N onto $\mathcal{R}(H_{M+1,N})$ is now

$$y_{M+1,N} = H_{M+1,N}P_{M+1,N}H^*_{M+1,N}W_N y_N \; . \tag{6}$$

We can relate both projections of y_N by noting that

$$P_{M+1,N} = \begin{bmatrix} P_{M,N} & 0 \\ 0 & 0 \end{bmatrix} + \frac{1}{\zeta^b_M(N)}\begin{bmatrix} -w^b_{M,N} \\ 1 \end{bmatrix}\begin{bmatrix} -w^{b*}_{M,N} & 1 \end{bmatrix} \tag{7}$$

where the vector $w^b_{M,N}$ is given by

$$w^b_{M,N} = P_{M,N}H^*_{M,N}W_N x_{M,N} = P_{M,N}c_{M,N} \tag{8}$$

and $\zeta^b_M(N)$ is the corresponding minimum cost of projecting $x_{M,N}$ onto $\mathcal{R}(H_{M,N})$. Let $b_{M,N} = x_{M,N} - H_{M,N}w^b_{M,N}$ denote the resulting (backward) estimation error vector. Substituting (7) into the expression for $y_{M,N}$ and subtracting y_N from both sides of the resulting equation we get

$$\boxed{e_{M+1,N} = e_{M,N} - \kappa_M(N)b_{M,N}} \tag{9}$$

where we deÞned the scalar coefÞcient

$$\kappa_M(N) \triangleq b^*_{M,N}W_N y_N/\zeta^b_M(N) \triangleq \rho_M(N)/\zeta^b_M(N) \; . \tag{10}$$

<u>Initial Conditions</u>

DeÞning the quantities

$$\pi_M(N) \triangleq x^*_{M,N}W_N x_{M,N} \tag{11}$$

$$c_{M,N} \triangleq H^*_{M,N}W_N x_{M,N} \tag{12}$$

we have the following conditions at time -1:

$$\begin{cases} \rho_M(-1) = 0 \\ w^b_{M,-1} = \Pi_M c_M \\ \zeta^b_M(-1) = \pi_M - c^*_M\Pi_M c_M \end{cases} \tag{13}$$

where we have deÞned $\pi_M \triangleq \pi_M(-1)$ and $c_M \triangleq c_{M,-1}$.

The recursion (9) for $e_{M+1,N}$ depends on $b_{M,N}$. We are thus motivated to order-update $b_{M,N}$.

3.2. Backward Estimation Problem

By partitioning $H_{M+1,N} = [\; x_{0,N} \mid \bar{H}_{M,N} \;] = [\; H_{M,N} \mid x_{M,N} \;]$ and using arguments similar to those that led to (9), it is straightforward to verify that

$$\boxed{b_{M+1,N} = \bar{b}_{M,N} - \kappa^b_M(N)f_{M,N}}$$

where the scalar coefÞcient $\kappa^b_M(N)$ is deÞned as

$$\kappa^b_M(N) \triangleq f^*_{M,N}W_N x_{M+1,N}/\zeta^f_M(N) \triangleq \delta_M(N)/\zeta^f_M(N)$$

and $f_{M,N}$ is the residual error from projecting $x_{0,N}$ onto $\mathcal{R}(\bar{H}_{M,N})$. We denote the corresponding minimum cost by $\zeta_M(N)$. Likewise, $\bar{b}_{M,N}$ is the residual error that results from projecting $x_{M+1,N}$ onto $\mathcal{R}(\bar{H}_{M,N})$, whose minimum cost we denote by $\zeta^{\bar{b}}_M(N)$.

<u>Initial Conditions</u>

The initial conditions for $\{\zeta^f_M(N), w^f_{M,N}, w^{\bar{b}}_{M,N}, \delta_M(N)\}$ are obtained by deÞning

$$w^f_{M,N} = \bar{P}_{M,N}\bar{H}^*_{M,N}W_N x_{0,N} \triangleq \bar{P}_{M,N}s_{M,N} \tag{14}$$

$$\zeta^f_M(N) = x_{0,N}W_N f_{M,N} \triangleq \mu(N) - s^*_{M,N}w^f_{M,N} \tag{15}$$

$$w^{\bar{b}}_{M,N} = \bar{P}_{M,N}\bar{H}^*_{M,N}W_N x_{M+1,N} \triangleq \bar{P}_{M,N}\bar{c}_{M,N} \tag{16}$$

$$\zeta^{\bar{b}}_M(N) = \pi_{M+1}(N) - \bar{c}^*_{M,N}\bar{P}_{M,N}\bar{c}_{M,N} \tag{17}$$

so that at time $N = -1$ we have

$$\begin{cases} w^f_{M,-1} = \bar{\Pi}_M s_M \\ w^{\bar{b}}_{M,-1} = \bar{\Pi}_M \bar{c}_M \\ \zeta^f_M(-1) = \mu - s^*_M\bar{\Pi}_M s_M \\ \zeta^{\bar{b}}_M(-1) = \pi_{M+1} - \bar{c}^*_M\bar{\Pi}_M\bar{c}_M \\ \delta_M(-1) = x^*_{0,-1}W_L x_{M+1,-1} - s^*_M\bar{\Pi}_M\bar{c}_M \end{cases} \tag{18}$$

where we have further deÞned $\bar{\Pi}_M \triangleq \bar{P}_{M,-1}$, $\bar{c}_M \triangleq \bar{c}_{M,-1}$, $s_M \triangleq s_{M,-1}$, and $\mu \triangleq \mu(-1)$.

3.3. Forward Estimation Problem

By similar arguments, $f_{M,N}$ can be updated as follows:

$$\boxed{f_{M+1,N} = f_{M,N} - \kappa^f_M(N)\bar{b}_{M,N}}$$

where $\kappa^f_M(N) \triangleq \frac{\bar{b}^*_{M,N}W_N x_{0,N}}{\zeta^{\bar{b}}_M(N)} = \frac{\delta^*_M(N)}{\zeta^{\bar{b}}_M(N)}$.

3.4. Reflection Coefficients Updates

In [5], we have shown that, except for $\bar{b}_{M,N}$, all quantities deÞning the reßection coefÞcients in a lattice network can be recursively updated regardless of the data structure assumed for the underlying data model. The arguments in [5] thus lead to recursions for $\{\delta_M(N), \rho_M(N), \zeta^b_M(N), \zeta^f_M(N), \zeta^{\bar{b}}_M(N), \gamma_M(N), \bar{\gamma}_M(N)\}$ which are listed in Table I.

The reason why it is possible to obtain $\bar{b}_M(N)$ efÞciently for FIR models is that the rows of $\bar{H}_{M,N}$ are identical to the rows of $H_{M,N-1}$, so that $\bar{b}_M(N) = b_M(N-1)$. However, this equality is only true for tapped-delay-line models, and it does not hold for more general data structures. Now, what if two successive *rows* of $H_{M,N}$ are not shifted versions of each other, but are instead related by a more general matrix Φ? Would it still be possible to derive a fast lattice algorithm? Next, we show that fast lattice algorithms are indeed possible for more general orthonormal models where shift structure no longer holds.

4. GENERAL ORTHONORMAL FILTER STRUCTURES

Consider the orthonormal Þlter structure of Figure 1 with transfer function (from $s(N)$ to $\hat{d}(N)$):

$$G(z) = \sum_{i=0}^{M-1} w_i \frac{\sqrt{1-|a_i|^2}}{1-a_i z^{-1}} \prod_{k=0}^{i-1} \frac{z^{-1}-a_k^*}{1-a_k z^{-1}}, \quad |a_k| < 1. \quad (19)$$

The input to the orthonormal Þlter at time N is denoted by $s(N)$, and the coefÞcients that combine the outputs of the successive low-pass sections are denoted by $\{w_i\}$.

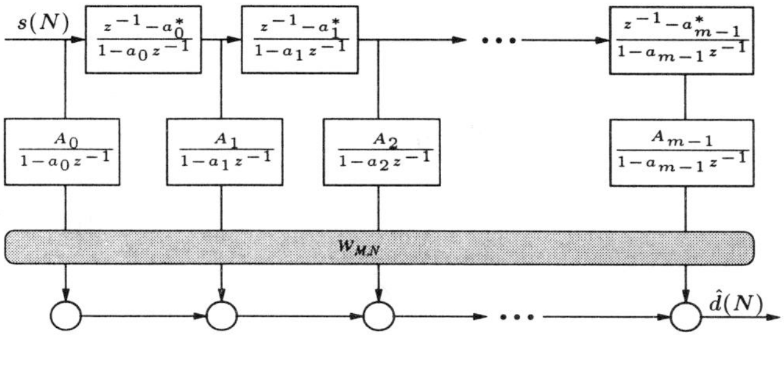

Figure 1: *Transversal orthonormal structure for adaptive filtering.*

Let

$$\begin{aligned} u_{M+1,N} &\triangleq [u(N,0) \ \ u(N,1) \ \ \cdots \ \ u(N,M)] \quad (20) \\ &= [u(N,0) \ \ \bar{u}_{M,N}], \quad (21) \end{aligned}$$

denote the regression vector at time N. Now suppose we set the Þrst pole as $a_0 = 0$. Then, it can be shown that[3].

$$\bar{H}_{M,N} = \begin{bmatrix} 0 \\ H_{M+1,N-1} \end{bmatrix} \Phi_{M+1} \quad (22)$$

where Φ_{M+1} is an $(M+1) \times M$ matrix given by (i.e., for $M = 4$)

$$\Phi_{M+1} = \begin{bmatrix} A_1 & -A_2 a_1^* & A_3(a_2 a_1)^* & -A_4(a_3 a_2 a_1)^* \\ a_1 & A_2 A_1 & -A_3 A_1 a_2^* & A_4 A_1 (a_3 a_2)^* \\ 0 & a_2 & A_3 A_2 & -A_4 A_2 a_3^* \\ 0 & 0 & a_3 & A_4 A_3 \\ 0 & 0 & 0 & a_4 \end{bmatrix}$$

and $A_i = \sqrt{1-|a_i|^2}$. Moreover, the Þlter difference equations further imply that

$$x_{i+1,N} = a_{i+1} \begin{bmatrix} 0 \\ x_{i+1,N-1} \end{bmatrix} + \frac{A_{i+1}}{A_i} \left(\begin{bmatrix} 0 \\ x_{i,N-1} \end{bmatrix} - a_i^* x_{i,N} \right) \quad (23)$$

where $x_{i,N}$, $i = 0, \ldots, M$ are the columns of $H_{M+1,N}$.

[3]The choice $a_0 = 0$ does not bring any degradation to the algorithm performance, and provides valuable insight into the connection between the extended lattice algorithm of this paper and the extended fast Þxed-order algorithms developed in [3].

5. EXPLOITING DATA STRUCTURE

We now return to the main problem, which is to compute the residual vector $\bar{b}_{M,N}$. Thus recall the deÞnition of $\bar{b}_{M,N}$:

$$\begin{aligned} \bar{b}_{M,N} &= x_{M+1,N} - \bar{H}_{M,N} w^{\bar{b}}_{M,N} \\ &= \bar{\mathcal{P}}^{\perp}_{M,N} x_{M+1,N} \quad (24) \end{aligned}$$

where $\bar{\mathcal{P}}^{\perp}_{M,N}$ denotes the projection matrix onto the orthogonal subspace of $\bar{H}_{M,N}$. Using (22) we have
$\bar{\mathcal{P}}^{\perp}_{M,N} =$

$$I - \begin{bmatrix} 0 \\ H_{M+1,N-1} \end{bmatrix} \Phi_{M+1} \bar{P}_{M,N} \Phi^*_{M+1} \begin{bmatrix} 0 & H^*_{M+1,N-1} \end{bmatrix} W_N \quad (25)$$

We are now in position to state a main result in our presentation. Due to lack of space we omit the details of its derivation.

(General rank-one relation) *Consider input regression vectors $u_{M,N}$ arising from the orthonormal filter structure of Figure 1. Then, for $M = 0, 1, \ldots, M_{max}$ it holds that*

$$\Phi_{M+1} \bar{P}_{M,N} \Phi^*_{M+1} =$$

$$P_{M+1,N-1} - \zeta^{\bar{b}}_M(N) P_{M+1,N-1} \phi_{M+1} \phi^*_{M+1} P_{M+1,N-1} \quad (26)$$

where the vector ϕ_{M+1} can be recursively computed by

$$\phi_{M+1} = \begin{bmatrix} -a^*_M \phi_M \\ A_M \end{bmatrix}, \quad \phi_1 = 1 \quad (27)$$

and $\zeta^{\bar{b}}_m(N)$ is the minimum cost that results from the LS problem of projecting $\breve{x}_{M+1,N} = \begin{bmatrix} 0 \\ x_{M,N-1}/A_M \end{bmatrix}$ onto $\mathcal{R}(\bar{H}_{M,N})$, say,

$$\begin{aligned} w^{\bar{b}}_{M,N} &= \bar{P}_{M,N} \bar{H}_{M,N} W_N \breve{x}_{M+1,N} \triangleq \bar{P}_{M,N} \breve{c}_{M,N}, \quad (28) \\ \zeta^{\bar{b}}_M(N) &\triangleq \breve{\pi}_{M+1}(N) - \breve{c}^*_{M,N} \bar{P}_{M,N} \breve{c}_{M,N} \quad (29) \end{aligned}$$

with $\breve{\pi}_{M+1}(N) = \pi_M(N-1)/A^2_M$, and π_M defined in (11).

Referring back to the projection matrix of Eq. (25), and substituting Eqs. (23) and (26) into (24) we obtain, after some algebra,

$$\boxed{\bar{b}_M(N) = a_{M+1} b_{M+1}(N-1) + \kappa^{\bar{b}}_M(N) v_{M+1}(N-1)} \quad (30)$$

where we deÞned a new reßection coefÞcient $\kappa^{\bar{b}}_M(N)$ as

$$\kappa^{\bar{b}}_M(N) \triangleq \zeta^{\bar{b}}_M(N) \chi_{M+1}(N-1),$$

in addition to the following quantities:

$$\begin{aligned} \chi_M(N) &\triangleq a_M \phi^*_M w^b_{M,N} + A_M = \phi^*_{M+1} \begin{bmatrix} -w^b_{M,N} \\ 1 \end{bmatrix} \\ v_M(N) &\triangleq u^*_{M,N} P_{M,N} \phi_M = g^*_{M,N} \phi_M \end{aligned}$$

The quantity $v_M(N)$ is the so-called *rescue variable* which appears in the *fast a posteriori error sequential technique* of [1], and which was further extended to the FTF algorithm for orthonormal models in [3] (except that in the latter it is based on *a priori quantities*). Moreover, a fundamental difference is that here $v_M(N)$ must be deÞned for every M-th order least-squares problem. Due to space limitations, we simply mention that the following recursions can be established for $\{v_M, \chi_M(N), \zeta^{\bar{b}}_M(N)\}$:

$$\begin{aligned} v_{M+1}(N) &= -a^*_M v_M(N) + \kappa^v_M(N) b_M(N) \\ \chi_M(N) &= \chi_M(N-1) + a_M v^*_M(N) \beta_M(N) \\ \zeta^{\bar{b}}_M(N) &= \bar{\gamma}_M(N) \zeta^f_M(N-1) \sigma_M \end{aligned}$$

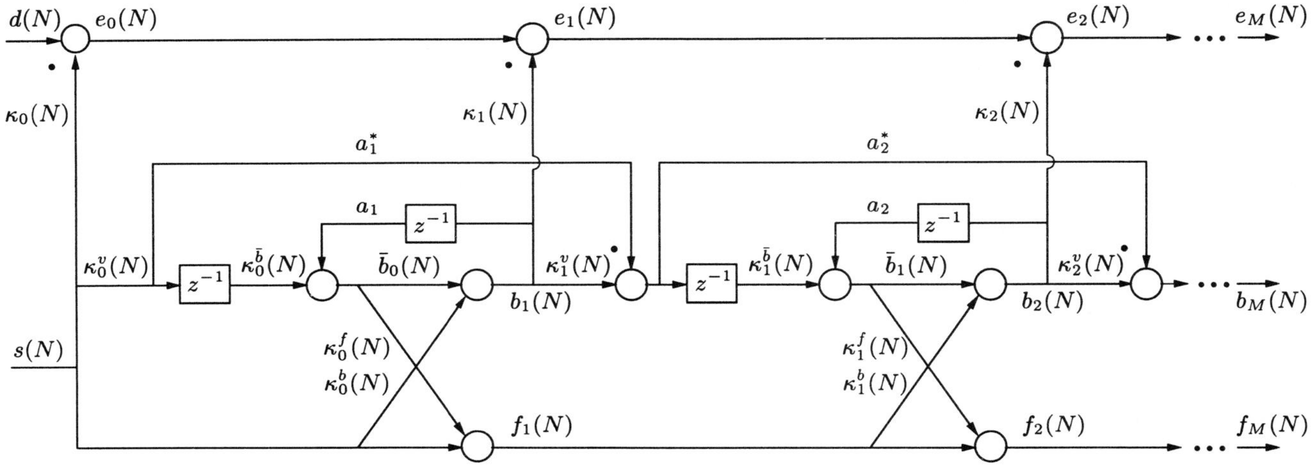

Figure 2: *The RLS lattice network for orthonormal models.*

where $\kappa_M^v(N) \triangleq \dfrac{\chi_M^*(N)}{\zeta_M^b(N)}$ and $\sigma_M \triangleq \dfrac{\lambda\zeta_M^{\breve{b}}(-1)}{\zeta_M^f(-1)}$.

Figure 2 shows the resulting lattice structure for orthonormal models. The initial conditions $\{\Pi_m, \bar{\Pi}_m\}$, $m = 0, ..., M$, needed in the computation of the initial variables deÞned in Sec. 3, as well as variables $\{v_m(-1), \chi_m(-1), \zeta_m^{\breve{b}}(-1)\}$ (deÞned above), can be computed by solving Eq. (26) at time $N = -1, \forall m$. The resulting algorithm is listed in Table I. In comparison with the lattice network for FIR models, the recursions of the new algorithm contributes with two new reßection coefÞcients, in addition to a feedforward and a feedback path connecting two consecutive lattice sections. By setting all poles to zero, it is easy to see that the general Þlter collapses to the usual FIR lattice structure. In addition, several simulations have been carried out to validate the algorithm. Lack of space forbids including these results here.

Initialization

For $m = 0$ to M set:

$\zeta_m^f(-1) = \mu$ (small positive number)
$\delta_m(-1) = \rho_m(-1) = v_m(-1) = b_m(-1) = 0$
$\zeta_m^b(-1) = \pi_m - c_m^*\Pi_m c_m$; $\zeta_m^{\breve{b}}(-1) = \breve{\pi}_{m+1} - \breve{c}_m^*\bar{\Pi}_m\breve{c}_m$
$\sigma_m = \lambda\zeta_m^{\breve{b}}(-1)/\zeta_m^f(-1)$; $\chi_m(-1) = a_m\phi_m^*\Pi_m c_m + A_m$
$\zeta_m^{\bar{b}}(-1) = \zeta_{m+1}^b(-1)$

For $N \geq 0$, repeat:

$\gamma_0(N) = \bar{\gamma}_0(N) = 1$ $\quad$ $f_0(N) = b_0(N) = s(N)$
$v_0(N) = 0$ $\quad$ $e_0(N) = d(N)$

For $m = 0$ to $M - 1$, repeat:

$\zeta_m^{\breve{b}}(N) = \sigma_m\bar{\gamma}_m(N)\zeta_m^f(N-1)$
$\kappa_m^{\bar{b}}(N) \triangleq \zeta_m^{\breve{b}}(N)\chi_{m+1}(N-1)$
$\bar{b}_m(N) = a_{m+1}b_{m+1}(N-1) + \kappa_m^{\bar{b}}(N)v_{m+1}(N-1)$
$\zeta_m^f(N) = \lambda\zeta_m^f(N-1) + |f_m(N)|^2/\bar{\gamma}_m(N)$
$\zeta_m^b(N) = \lambda\zeta_m^b(N-1) + |b_m(N)|^2/\gamma_m(N)$
$\zeta_m^{\bar{b}}(N) = \lambda\zeta_m^{\bar{b}}(N-1) + |\bar{b}_m(N)|^2/\bar{\gamma}_m(N)$
$\chi_m(N) = \chi_m(N-1) + a_m v_m^*(N)\beta_m(N)$
$\delta_m(N) = \lambda\delta_m(N-1) + f_m^*(N)\bar{b}_m(N)/\bar{\gamma}_m(N)$
$\rho_m(N) = \lambda\rho_m(N-1) + e_m^*(N)b_m(N)/\gamma_m(N)$
$\gamma_{m+1}(N) = \gamma_m(N) - |b_m(N)|^2/\zeta_m^b(N)$
$\bar{\gamma}_{m+1}(N) = \bar{\gamma}_m(N) - |\bar{b}_m(N)|^2/\zeta_m^{\bar{b}}(N)$
$\kappa_m^v(N) = \chi_m^*(N)/\zeta_m^b(N)$ $\quad$ $\kappa_m(N) = \rho_m(N)/\zeta_m^b(N)$
$\kappa_m^b(N) = \delta_m(N)/\zeta_m^f(N)$ $\quad$ $\kappa_m^f(N) = \delta_m^*(N)/\zeta_m^{\bar{b}}(N)$
$v_{m+1}(N) = -a_m^*v_m(N) + \kappa_m^v(N)b_m(N)$
$e_{m+1}(N) = e_m(N) - \kappa_m(N)b_m(N)$
$b_{m+1}(N) = \bar{b}_m(N) - \kappa_m^b(N)f_m(N)$
$f_{m+1}(N) = f_m(N) - \kappa_m^f(N)\bar{b}_m(N)$

Table I: The standard Extended RLS adaptive lattice filter.

6. REFERENCES

[1] G. Carayannis, D. Manolakis, and N. Kalouptsidis, ÒA fast sequential algorithm for least squares Þltering and prediction,Ó *IEEE Trans. on Acoust., Speech, Signal Proc.*, vol. ASSP-31, pp. 1394Ð1402, December 1983.

[2] B. Friedlander, ÒLattice Þlters for adaptive processing,Ó *Proc. IEEE*, vol. 70, pp. 829Ð867, Aug. 1982.

[3] R. Merched and A. H. Sayed, ÒExtended fast Þxed order RLS adaptive Þltering,Ó *IEEE Transactions on Signal Processing*, vol. 49, no. 12, pp. 3015Ð3031, Dez. 2001.

[4] B. Ninness and F. Gustafsson, ÒA unifying construction of orthonormal bases for system identiÞcation,Ó *IEEE Transactions on Automat. Control*, vol. 42, pp. 515-521, Apr. 1997.

[5] R. Merched and A. H. Sayed, ÒOrder-recursive RLS Laguerre adaptive Þltering,Ó *IEEE Transactions on Signal Processing*, vol. 48, no. 11, pp. 3000-3010, Nov. 2000.

ON USE OF AVERAGING IN FxLMS ALGORITHM FOR SINGLE-CHANNEL FEEDFORWARD ANC SYSTEMS

Akhtar M. Tahir, Masahide Abe and Masayuki Kawamata

Department of Electronics Engineering, Graduate School of Engineering,
Tohoku University, Aoba Yama-05, Sendai 980-8579, JAPAN
Phone: +81–22–217–7095, Fax: +81–22–263–9169
Email: akhtar@mk.ecei.tohoku.ac.jp

ABSTRACT

In this paper an active noise control (ANC) algorithm is proposed. This algorithm is based on adaptive filtering with averaging (AFA) and uses a similar structure as that of the FxLMS ANC system. The proposed algorithm, which we call FxAFA algorithm, uses averages of both data and correction terms to find the updated values of the tap weights of the ANC controller. The computer simulations are conducted for single-channel feedforward ANC systems. It is shown that the proposed algorithm gives fast convergence as compared with the FxLMS algorithm, for both broadband and narrowband noise signals. The comparison with the FxRLS algorithm shows that the proposed FxAFA algorithm is a better choice for low computational complexity and stable performance.

1. INTRODUCTION

In contrast to passive noise control (PSN), where absorbing materials are used to "absorb" the unwanted noise signal, the active noise control (ANC) [1] is based on the simple principle of destructive interference of propagating acoustic waves. The concept that acoustic wave interference can be controlled to produce zones of quietness was first proposed by P. Lueg in 1936 for an analogue ANC system. However, success with the early analogue controllers was very limited and in the recent years powerful DSP devices have made possible the development of real time ANC systems with wide range of applications including air conditioning ducts, cars, aircrafts, and so on [2].

The most popular adaptation algorithm used for ANC applications (both broadband & narrowband) is the FxLMS algorithm, which is a modified version of the LMS algorithm [3]. The schematic diagram for a single-channel feedforward ANC system using the FxLMS algorithm is shown in Figure 1. The acoustic path between the reference noise source and the error microphone is called a primary path and is denoted by $P(z)$. The reference noise signal is filtered through the primary path $P(z)$ and appears as a primary noise signal at the error microphone. The objective of the adaptive controller $W(z)$ is to generate an appropriate antinoise signal $y(n)$ propagated by the secondary loudspeaker. This antinoise signal combines with the primary noise signal to create a zone of silence in the vicinity of the error microphone. The error microphone measures the residual noise $e(n)$, which is used by $W(z)$ for its adaptation to minimize the sound pressure at error microphone. Here $\hat{S}(z)$ accounts for the model of the secondary path $S(z)$ between the output $y(n)$ of the controller and that of the error microphone $e(n)$. The secondary path $S(z)$ comprises the digital-to-analog (D/A) converter, reconstruction filter, power amplifier, loudspeaker, acoustic path from loudspeaker to error microphone, error microphone, preamplifier, anti-aliasing filter, and analog-to-digital (A/D) converter. The filtering of the reference signal $x(n)$ through the secondary-path model $\hat{S}(z)$ is demanded by the fact that the output $y(n)$ of the adaptive controller $W(z)$ is filtered through the secondary path $S(z)$ [3].

Although the FxLMS algorithm is computationally simple but its convergence speed is slow and signal dependent. This has motivated researchers to look for fast convergence algorithms. If FIR filter structure is used, the convergence rate can be improved by using variable-step-size LMS, Newton algorithm [3], Kalman algorithm [4], or recursive-least-square (RLS) algorithm. The other approach is to condition the reference signal by employing different filter structures such as lattice filter, subband filter, or orthogonal transform. These different approaches have resulted in a number of ANC algorithms (with improved convergence properties); viz., lattice-ANC system [5], frequency-domain-ANC systems (see [6] and references there in), RLS based algorithms called Filtered-x RLS (FxRLS) [1] and Filtered-x Fast-Transversal-Filter (FxFTF) [7], and IIR-filter-based LMS algorithms called Filtered-u Recursive LMS (FuRLMS) [8], and filtered-v algorithms [9].

The potential instability of IIR filters structures and increased computational demands of other ANC algorithms mentioned above still make FxLMS a good choice for ANC applications. The need for a fast convergence yet a computationally simple algorithm for ANC applications is the main motivation for the investigation conducted in this paper.

Here we explore the realization of an ANC algorithm using adaptive filtering with averaging (AFA). The adaptive filtering with averaging stems from the numerical techniques using

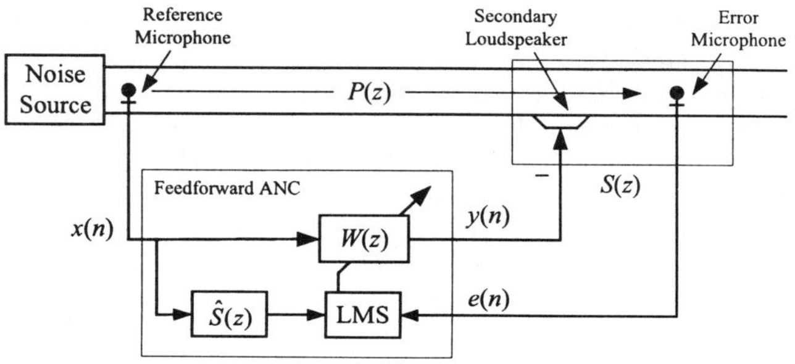

Figure 1. Schematic diagram of a single-channel feedforward ANC system using FxLMS algorithm.

averaging to accelerate the stochastic approximations (see Ref. [10] and references there in). The simulation results show that this averaging based ANC algorithm, which we call filtered-x AFA (FxAFA) algorithm, provides better results for both broadband and narrowband noise signals.

In Section 2, the proposed FxAFA algorithm is explained in connection with the FxLMS algorithm. Moreover the computational complexity issue is discussed. Sections 3 details the computer experiments performed, and in Section 4 concluding remarks are presented.

2. FxAFA ALGORITHM

In Figure 1 the secondary-path model $\hat{S}(z)$ is obtained offline and kept fixed during the online operation of ANC. The secondary signal $y(n)$ is expressed as

$$y(n) = \boldsymbol{w}^{\mathrm{T}}(n)\,\boldsymbol{x}(n) \tag{1}$$

where $\boldsymbol{w}(n)=[w_0(n)\ w_1(n)\ \cdots\ w_{L-1}(n)]^{\mathrm{T}}$ is the tap weight vector and $\boldsymbol{x}(n)=[x(n)\ x(n-1)\ \cdots\ x(n-L+1)]^{\mathrm{T}}$ is the reference signal picked by the reference microphone. This secondary signal $y(n)$ is filtered through the secondary path $S(z)$ and then combines with the primary noise signal to generate the residual noise $e(n)$. These operations can be expressed as

$$y'(n) = s(n) * y(n) \tag{2}$$

$$d(n) = p(n) * x(n) \tag{3}$$

$$e(n) = d(n) - y'(n) \tag{4}$$

where $y'(n)$ is the secondary signal $y(n)$ filtered through $S(z)$, $d(n)$ is the primary noise signal at the error microphone, $*$ is convolution operation, and $p(n)$ and $s(n)$ are impulse responses of the primary path $P(z)$ and secondary path $S(z)$, respectively. It is important to notice that all these operations, (2)–(4), are carried out internally in the system and we have access only the residual noise signal $e(n)$ being picked up by the error microphone.

It is assumed that there is no acoustic feedback from the secondary loudspeaker to the reference microphone. The FxLMS update equation for the coefficients of $W(z)$ is given as

$$\boldsymbol{w}(n+1) = \boldsymbol{w}(n) + \mu\, e(n)\, \boldsymbol{x}'(n) \tag{5}$$

where μ is the step size and $\boldsymbol{x}'(n)$ is the reference signal $\boldsymbol{x}(n)$ filtered through the secondary-path model $\hat{S}(z)$:

$$x'(n) = \hat{\boldsymbol{s}}^{\mathrm{T}}\boldsymbol{x}(n). \tag{6}$$

Here $\hat{\boldsymbol{s}} = [\hat{s}_0\ \hat{s}_1\ \cdots\ \hat{s}_{L-1}]^{\mathrm{T}}$ is impulse response of the secondary-path model $\hat{S}(z)$. In Ref. [10] two averaging based adaptive filtering algorithms are proposed. The first algorithm uses averaging in iterates only and in the second algorithm averaging is incorporated with both iterates and observations. Motivated by the second approach, we incorporate averaging with both the iterates, $\boldsymbol{w}(n)$, and the correction term (the observation vector), $\mu e(n)\,\boldsymbol{x}'(n)$, and propose the following algorithm:

$$\boldsymbol{w}(n+1) = \overline{\boldsymbol{w}(n)} + \overline{\mu\, e(n)\boldsymbol{x}'(n)} \tag{7}$$

where

$$\overline{\boldsymbol{w}(n)} = \frac{1}{n}\sum_{k=1}^{n}\boldsymbol{w}(k) \tag{8}$$

$$\overline{\mu\, e(n)\boldsymbol{x}'(n)} = \frac{1}{n^{\gamma}}\sum_{k=1}^{n}\mu\, e(k)\,\boldsymbol{x}'(k),\quad 1/2 < \gamma < 1. \tag{9}$$

Here computing the running average of the data does not put so much computational burden since averages can be calculated recursively. For example, (8) can be recursively computed as

$$\overline{\boldsymbol{w}(n)} = \frac{1}{n}\Big((n-1)\overline{\boldsymbol{w}(n-1)} + \boldsymbol{w}(n)\Big). \tag{10}$$

Similarly (9) can be computed as

$$\overline{\mu\, e(n)\boldsymbol{x}'(n)} = \frac{1}{n^{\gamma}}\Big((n-1)^{\gamma}\ \overline{\mu\, e(n-1)\boldsymbol{x}'(n-1)} + \mu\, e(n)\,\boldsymbol{x}'(n)\Big). \tag{11}$$

Referring to the feedforward ANC marked in Figure 1, (1), (6), (7), (10), and (11) are combined to give the proposed FxAFA algorithm. We see that the introduction of averaging in the FxLMS update equation results in a multistep algorithm (proposed FxAFA algorithm). Hence an increased computational burden is expected as discussed later in this section. The signal flow diagrams for two algorithms are shown in Figure 2 that depicts that the proposed FxAFA algorithm requires two extra storage bins for previous averaged vectors.

In Section 3 we compare the performance of the proposed FxAFA algorithm with that of FxLMS algorithm and FxRLS algorithm, so for convenience we summarize the FxRLS algorithm here.

$$\boldsymbol{z}(n) = \lambda^{-1}\boldsymbol{\Phi}(n)\ \boldsymbol{x}'(n) \tag{12}$$

$$\boldsymbol{k}(n) = \boldsymbol{z}(n)[1 + \boldsymbol{x}'^{\mathrm{T}}(n)\,\boldsymbol{z}(n)]^{-1} \tag{13}$$

$$\boldsymbol{w}(n+1) = \boldsymbol{w}(n) + \mu\, e(n)\,\boldsymbol{k}(n) \tag{14}$$

$$\boldsymbol{\Phi}(n+1) = \lambda^{-1}\boldsymbol{\Phi}(n) - \boldsymbol{k}(n)\,\boldsymbol{z}^{\mathrm{T}}(n) \tag{15}$$

It is important to note that a scalar gain μ is introduced in the update equation of the FxRLS algorithm [7]. The factor λ in (12) and (15) is a "forgetting factor" typically in the range $0.9<\lambda<1$. Here $\boldsymbol{k}(n)$ is an $L\times 1$ gain vector and $\boldsymbol{\Phi}(n)$ is an $L\times L$ inverse correlation matrix. The inverse correlation matrix $\boldsymbol{\Phi}(n)$ is initialized by $\delta^{-1}\mathbf{I}$, where δ is a small positive constant and $\mathbf{I}$ is an $L\times L$ identity matrix.

In Table 1 the computational complexity of three algorithms (FxLMS, FxAFA and FxRLS) is compared on the basis of multiplications required per iteration. In this analysis both ANC controller W(z) and the secondary-path model $\hat{S}$(z) are assumed to be FIR filters of length L. It is seen that the computational burden of the proposed FxAFA algorithm is between those of the FxLMS algorithm and FxRLS algorithm.

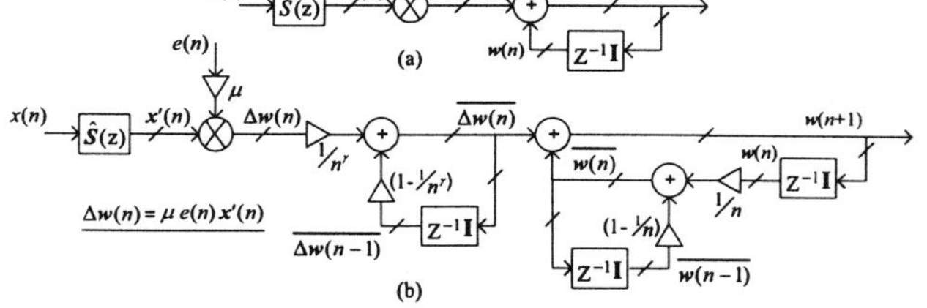

Figure 2. Signal flow diagrams: (a) FxLMS algorithm, (b) FxAFA algorithm.

Table 1. Computational complexity comparison between FxLMS, FxAFA and FxRLS algorithms.

	Number of Multiplications per Iteration		
	FxLMS Algorithm	**FxAFA Algorithm**	**FxRLS Algorithm**
Analytical Expression	$3L+1$	$7L+1$	$3L^2+6L+1$
L=128	385	897	49921
L=256	769	1793	198145

Now we present some comments on the choice of value of γ. For convenience, we rewrite the proposed algorithm in a compact form:

$$w(n+1) = \overline{w(n)} + \alpha(n) \sum_{k=1}^{n} e(k)\, x'(k), \tag{16}$$

where $\alpha(n) = \mu / n^{\gamma}$ is a slowly varying gain parameter. In adaptive algorithms it is desirable to have a large step gain at the startup for fast convergence. As the time increases the gain is desirable to slowly decrease so that the misadjustment is small. The time varying gain $\alpha(n)$ indeed exhibits these properties and $\lim_{n\to\infty} \alpha(n) \to 0$. We observe that $\gamma = 1$ will rapidly decrease the gain but this may reduce update to zero without optimal solution (say w_{opt}) being achieved. Therefore one may wish to choose $\gamma < 1$ [10]. On contrary if $\gamma \to 0$ is selected then $\alpha(n)$ is very slowly decreasing. This is also not desirable for slow convergence. Hence $1/2 < \gamma < 1$ is the recommended range for the values of γ [10].

3. SIMULATIONS

In this section the performance of the proposed FxAFA algorithm is demonstrated using computer simulations. The performance of the proposed algorithm is compared with that of FxLMS algorithm and FxRLS algorithm on the basis of noise reduction R(in dBs) as follows:

$$R(\text{dB}) = -10\log_{10}\left(\frac{\sum e^2(n)}{\sum d^2(n)}\right). \tag{17}$$

The large positive value of R indicates that more noise reduction is achieved at the error microphone. For the primary acoustical path $P(z)$ and the secondary path $S(z)$ the data provided by [1] is used, where both are modeled by IIR filters of order 25. The impulse responses of the primary and secondary paths are shown in Figure 3. The secondary-path model $\hat{S}(z)$ is an FIR filter of order 128, and is identified offline. The adaptive controller $W(z)$ is also an FIR filter of order 128. Since industrial noise often has significant power in the frequency range between 50–250Hz [11], all simulations are carried with the signals having frequency falling in this range. The sampling frequency of 2kHz is used. The parameters for the algorithms are adjusted for fast and stable convergence and are given in Table 2.

3.1 Case 1

Here comparative results of FxLMS, FxAFA and FxRLS are presented for narrowband noise that is a 200Hz sinusoidal signal with additive white Gaussian noise of zero mean and variance 0.05. This signal is used with the feedforward ANC system of Figure 1. Simulation parameters for three algorithms are given in Table 2. The noise reduction curves for performance measure, R, are shown in Figure 4. It is observed that FxRLS algorithm gives fastest convergence speed as compared with those of the FxLMS and FxAFA algorithms, but it diverges towards the higher number of iterations. In Ref. [7] it is suggested that the stability of the FxRLS algorithm can be improved by adding random noise to the input of the FxRLS algorithm. The price paid for the improved stability is the decreased convergence speed [7]. Although the FxLMS algorithm gives stable performance but its convergence speed is slower than that of the proposed FxAFA algorithm.

Table 2. Simulation parameters for FxLMS, FxAFA and FxRLS algorithms.

	FxLMS Algorithm (μ)	**FxAFA Algorithm** (μ,γ)	**FxRLS Algorithm** (μ,λ,δ)
Case 1	1×10^{-4}	1×10^{-2}, 0.6	0.1, 0.999, 0.04
Case 2	1×10^{-5}	1×10^{-3}, 0.6	0.1, 0.999, 5
Case 3	1×10^{-4}	1×10^{-2}, 0.6	0.075, 0.999, 5

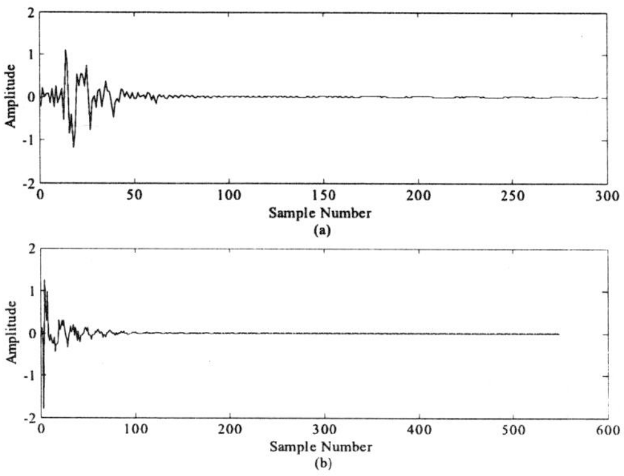

Figure 3. Impulse responses of acoustic paths: (a) Impulse response of primary path $P(z)$, (b) Impulse response of secondary path $S(z)$.

3.2 Case 2

In this case, the reference noise is a broadband signal and is a sinusoid containing five harmonics (of equal power) with the fundamental frequency of 50Hz. The white Gaussian noise of zero mean and variance 0.05 is added for any measurement noise present. The performance, R, of the proposed FxAFA algorithm is compared with that of FxLMS algorithm and FxRLS algorithm in Figure 5. It is observed that initially the FxRLS algorithm converges very fast but then its convergence speed drops dramatically and finally it diverges again showing the problem of numerical instability. The proposed FxAFA algorithm gives stable and fast convergence as in Case 1.

3.3 Case 3

Here we filter a white Gaussian noise of zero mean and unit variance through a bandpass filter with passband 50–250Hz. The simulation parameters are adjusted for fast and stable performance and are given in Table 2. The performance, R, of FxLMS, FxAFA and FxRLS is presented in Figure 6. It is observed that the proposed FxAFA algorithm gives the best performance of all. The FxRLS algorithm not only appears as a slowest convergence algorithm but it diverges also towards the higher number of iterations.

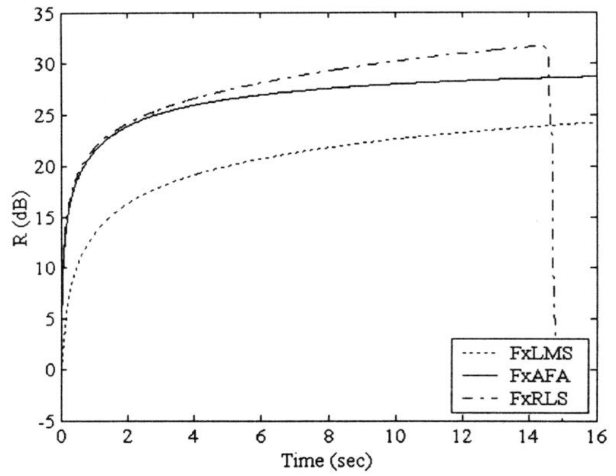

Figure 4. Noise reduction achieved by FxLMS, FxAFA and FxRLS in Case 1.

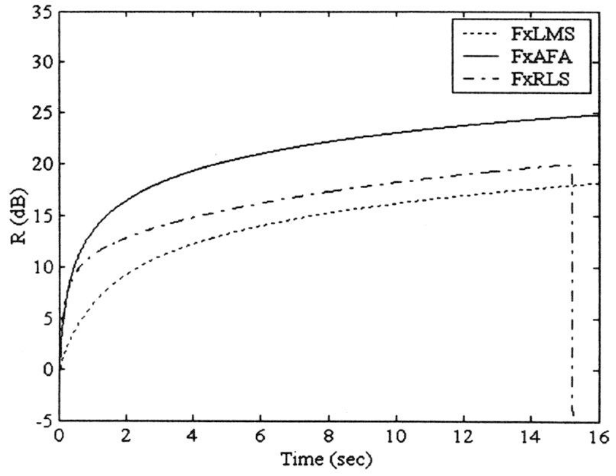

Figure 5. Noise reduction achieved by FxLMS, FxAFA and FxRLS in Case 2.

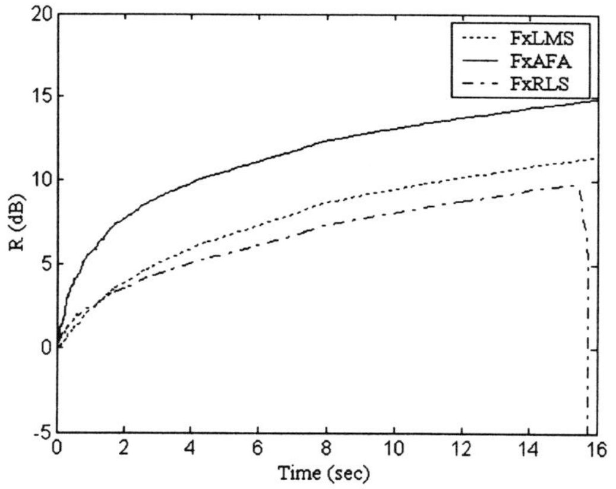

Figure 6. Noise reduction achieved by FxLMS, FxAFA and FxRLS in Case 3.

4. CONCLUSIONS

Here a new ANC algorithm based on adaptive filtering with averaging is presented. Computer simulations are conducted for single-channel feedforward ANC systems using both broadband and narrowband signals. In comparison to the FxLMS algorithm the proposed algorithm achieves faster convergence, but at the expense of slightly increased computational complexity. It has two adjustable parameters γ and μ, and requires more care in selecting their values. It is seen that by proper choice of γ the larger value for μ can be selected, and thus faster convergence can be achieved.

Experiments conducted with the FxRLS algorithm show that it is not possible to fully stabilize the algorithm and to achieve a fast convergence speed simultaneously. This demonstrates the superiority of the proposed FxAFA algorithm over FxRLS algorithm in both computational complexity and numerical stability.

In this paper offline secondary path modeling is used and it is assumed that the secondary path remains fixed all the time. The development of an ANC algorithm with online secondary path modeling, incorporating adaptive filtering with averaging is a task of future work.

5. REFERENCES

[1] S.M. Kuo and D.R. Morgan, *Active Noise Control Systems–Algorithms and DSP Implementation.* New York: Wiley, 1996.

[2] S.M. Kuo and D.R. Morgan, "Active noise control: A tutorial review," *Proc. IEEE*, vol. 87, no. 6, pp. 943–973, June 1999.

[3] B. Widrow and S.D. Stearns, *Adaptive Signal Processing.* Prentice Hall, New Jersey, 1985.

[4] S. Haykin, *Adaptive filter theory*, 3rd edition, Prentice Hall, Englewood Cliffs, New Jersey, 1996.

[5] Y.C. Park, and S.D. Sommerfeldt, "A fast adaptive noise control algorithm based on lattice structure," *Applied Acoustics*, vol. 47, Issue 1, pp. 1-25, 1996.

[6] S.M. Kuo, and M. Tahernezhadi, "Frequency-domain periodic active noise control and equalization," *IEEE Trans. Speech Audio Processing,* vol. 5, no. 4, pp. 348-358, July 1997.

[7] M. Bouchard and S. Quednau, "Multichannel recursive-least-squares algorithms and fast-transversal-filter algorithms for active noise control and sound reproduction systems," *IEEE Trans. Speech, Audio Proc.*, vol. 8, no. 5, pp. 606–618, September 2000.

[8] L.J. Eriksson, M.C. Allie and R.A. Greiner, "The selection and application of an IIR adaptive filter for use in active sound attenuation," *IEEE Trans. Acoust., Speech Signal Processing*, ASSP-35, pp. 433–437, April 1987.

[9] D.H. Crawford, and R.W. Stewart, "Adaptive IIR filtered-v algorithms for active noise control," *J. Acoust. Soc. Am.*, vol. 101, issue 4, pp. 2097-2103, April 1997.

[10] G. Yin, "Adaptive filtering with averaging," in *Adaptive Control, Filtering, and Signal Processing*, K.J. Åström, G.C. Goodwin, and P.R. Kumar, Ed. New York: Springer-Verlog, 1995.

[11] K.K. Shyu and C.Y. Chang, "Modified FIR filter with phase compensation technique to feedforward active noise controller design," *IEEE Trans., Ind. Electronics*, vol. 47, no. 2, pp. 444–453, April 2000.

HARDWARE IMPLEMENTATION OF EVOLUTIONARY DIGITAL FILTERS

Masahide Abe and Masayuki Kawamata

Department of Electronic Engineering,
Graduate School of Engineering, Tohoku University,
Aoba-yama 05, Sendai, 980-8579, Japan

ABSTRACT

This paper designs and implements a hardware-based evolutionary digital filter (EDF). The EDF is an adaptive digital filter which is controlled by adaptive algorithm based on evolutionary computation. The hardware-based EDF consists of two submodules, that is, a filtering and fitness calculation (FFC) module and a reproduction and selection (RS) module. The FFC module has high computational ability to calculate the output and the fitness value since its submodules run in parallel. A synthesis result of the designed chip shows the clock frequency is 20.0MHz and the maximum sampling rate of the EDF is 3.7kHz. Moreover, the hardware-based EDF with 21 submodules of the FFC is 2.2 times faster than the software-based EDF.

1. INTRODUCTION

Several researchers have proposed adaptive algorithms for digital filtering, which are all based on the Darwinian concept of "natural selection." These include the adaptive algorithm based on the genetic algorithm (GA) [1–3], the new learning adaptive algorithm [4], and Darwinian approach to adaptive notch filters [5].

The authors have already proposed evolutionary digital filters (EDFs) [6–8]. The EDF is an adaptive digital filter (ADF) which is controlled by adaptive algorithm based on evolutionary computation. The advantages of the EDF are summarized as follows:

1. The adaptive algorithm of the EDF is a population-based and robust optimization method, especially used to tackle high-dimensional and multi-modal search space problems. It is a non-gradient and multi-point search algorithm. Thus, it is not susceptible to local minimum problems that arise from a multiple-peak surface.
2. The EDF can adopt the various error functions as the fitness function according to application, for example, the p-power norm error function, the maximum error function and so on.
3. The adaptive algorithm of the EDF has a self-stabilizing feature whereby unstable poles have a tendency to migrate back into the stable region. In addition, the EDF can search the poles which are near the unit circle.

Numerical examples in Refs. [6–8] show that the EDF has a higher convergence rate and smaller steady-state value of the square error than the LMS adaptive digital filter (LMS-ADF).

However, the EDF has the following disadvantage: the number of multiplication of the EDF is greater than that of the LMS-ADF, since the EDF consists of many inner digital filters. Thus, we implement the EDF on parallel processors.

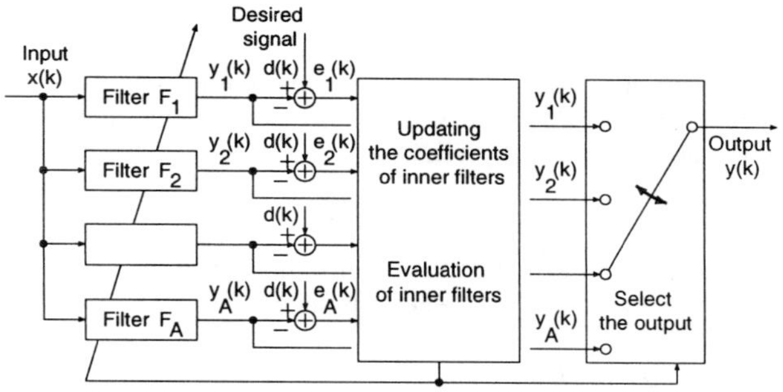

Figure 1: Block diagram of an evolutionary digital filter.

In order to implement the EDF in parallel, we present a hardware implementation of the distributed EDF, which consists of the modified structure and adaptive algorithm.

This paper is organized as follows: Section 2 summarizes the overall structure and the adaptive algorithm of EDFs. Section 3 describes the detailed structure of the proposed hardware-based EDF and its synthesis result. Section 4 gives concluding remarks.

2. EVOLUTIONARY DIGITAL FILTERING

In this section, we summarize the filter structure and the adaptive algorithm of EDFs. Figure 1 shows the block diagram of an EDF. The EDF consists of many linear/time-variant inner digital filters F_i's which correspond to individuals. Inner digital filter coefficients $\boldsymbol{W}$ which correspond to the feature of individuals are controlled by the following adaptive algorithm.

2.1. Adaptive Algorithm of Evolutionary Digital Filters

The adaptive algorithm of the EDF is similar in concept to GA. These concepts are based on the mechanics of natural selection and genetics to emulate the evolutionary behavior of biological systems. However, the adaptive algorithm of the EDF is different from the GA in the genetic operator and the representation of strings.

In the following sections, we use the following notations:

P population of individuals,
N the number of individuals.

The subscripts in the symbols P, N and $\boldsymbol{W}$ are denoted as follows:

0-7803-7761-3/03/$17.00 ©2003 IEEE

Table 1: Numbers of multiplications per iteration in the EDF and the LMS-ADF.

Algorithm (structure)	Number of multiplications for the filtering process	Number of multiplications for the adaptive process
EDF	$A(N+M+1)$	$(A - N_{ap} - \frac{1}{2}N_{sp})(N+M+1)/T_O$
LMS-ADF	$N+M+1$	$3N+M+2$

a the cloning method (the asexual reproduction),
s the mating method (the sexual reproduction),
p parent,
c offspring (child).

In the EDF, the adaptive algorithm updates the inner digital filter coefficients every T_O samples. Thus, the relation between the generation t and the time k is given by

$$k = T_O t,\ T_O t+1,\ \cdots,\ T_O t + (T_O - 1) \tag{1}$$

where k denotes the time in the filtering operation and T_O denotes the period of the evaluation of one generation.

2.1.1. Cloning Method

Each parent in the population P_{ap}, with high fitness value within the population $P(t)$, creates the offspring population P_{ac} using the cloning method. In the cloning method, one parent creates N_{ac} offsprings, and forms a family $P_{af,i}$ which contains itself and its offsprings, where $i = 1, 2, \cdots, N_{ap}$. N_{ap} is the number of parents which use the cloning method. We assume that the proposed cloning method corresponds to transcribing the coefficient vector $\boldsymbol{W}_{ap,i}$ as the parent feature into coefficient vectors as the offspring feature $\boldsymbol{W}_{ac,i,j}$, where $i = 1, 2, \cdots, N_{ap}$, and $j = 1, 2, \cdots, N_{ac}$. Thus, the proposed cloning method updates the inner digital filter coefficients as individual feature according to

$$\boldsymbol{W}_{ac,i,j} = \boldsymbol{W}_{ap,i} + r \cdot \boldsymbol{n}_{i,j} \tag{2}$$

where the scalar r denotes the cloning fluctuation, and $\boldsymbol{n}_{i,j}$ is a Gaussian random variable vector with zero mean and unit variance.

In this algorithm, the cloning method corresponds to the local search. Therefore, this method is provided with the following strategy to select the candidate population for the next generation. In this method, one individual, of which fitness is maximum in each family $P_{af,i}$, is selected. These individuals form the candidate population P_a for the next generation. The population P_a of the best individuals is selected among each family $P_{af,i}$, that is, the coefficient vector of the inner filter with the highest fitness is selected among the $(N_{ac}+1)$ coefficient vectors. These coefficients are scattered on the narrow area. Thus, this operation corresponds to the local search.

2.1.2. Mating Method

If parents with low fitness value in population create the offsprings using the above cloning method, these offsprings may have low fitness value and can not be selected as candidates for the next generation. Therefore, parents in the population P_{sp}, with low fitness value within the population $P(t)$, create the offspring population P_{sc} using the mating method. $N_{sp}/2$ pairs among the N_{sp} parents are randomly selected for mating. In the mating method, each pair of parents creates one offspring, and they form a family $P_{sf,m}$ which contains themselves and their offspring, where $m = 1, 2, ..., N_{sp}/2$. We assume that the proposed mating method corresponds to calculating the middle point $\boldsymbol{W}_{sc,m}$ as the offspring feature of two coefficient vectors $\boldsymbol{W}_{sp,k(m)}$ and $\boldsymbol{W}_{sp,l(m)}$ as parent feature. Thus, this method updates the inner digital filter coefficients as individual feature according to

$$\boldsymbol{W}_{sc,m} = \frac{1}{2}(\boldsymbol{W}_{sp,k(m)} + \boldsymbol{W}_{sp,l(m)}) + s \cdot \boldsymbol{n}_m \tag{3}$$

where $k(m)$ and $l(m)$ are selected in $\{1, 2, ..., N_{sp}\}$ without duplicating, and $m = 1, 2, \cdots, N_{sp}/2$. The scalar s denotes the mating fluctuation, and $\boldsymbol{n}_m$ is a Gaussian random variable vector with zero mean and unit variance.

In this algorithm, the mating method corresponds to the global search and keeps various features of individuals. Therefore, this method is provided with the following strategy to select the candidate population for the next generation. In this method, one parent with higher fitness value in each family $P_{sf,m}$ is selected and the other parent dies out. In order to keep various features of individuals, the offspring in each family $P_{sf,m}$ is always selected regardless of their fitness values.

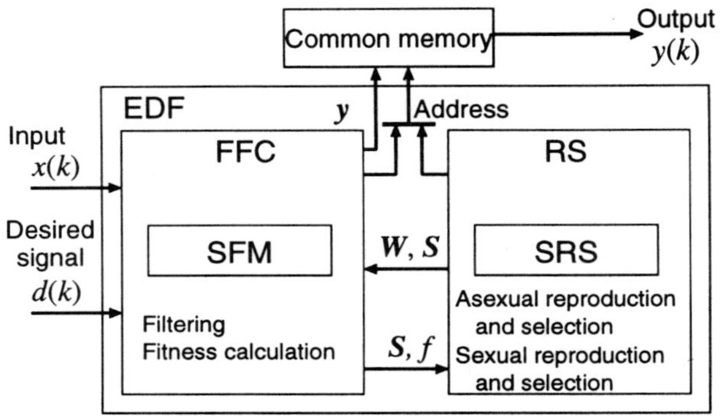

Figure 2: Block diagram of the hardware-based EDF.

2.2. Computational Complexity

The EDF requires $(A - N_{ap} - \frac{1}{2}N_{sp})(N+M+1)/T_O$ multiplications per iteration for the adaptive process, where A is the total number of the evaluated individuals, that is, $A = N_{ap}(N_{ac}+1) + \frac{3}{2}N_{sp}$. In the adaptive algorithm of the EDF, the inner digital filter coefficients are updated every T_O samples. Table 1 shows that the number of multiplications of the EDF is larger than that of the LMS-ADF.

Table 2: Specifications of the hardware-based EDF.

EDF	Fixed-point format	Q14
	Bit width of data	16 bits
	Bit width of instructions	16 bits
	Number of individuals	$N_{ap} + N_{sp} \leq 64$
	Order of filters	$N \leq 3, M \leq 3$
SFM	Number of instructions	45
	Program memory size	256×16 bits
	Data memory size	128×16 bits ×2
RS	Individual memory size	1,024×16 bits
-	Common memory size	11,040×16 bits

3. HARDWARE-BASED EVOLUTIONARY DIGITAL FILTERS

3.1. Hardware Structure of Evolutionary Digital Filters

In order to implement the EDF on parallel processors, we design and implement a hardware-based EDF.

Figure 2 shows the block diagram of the hardware-based EDF. The EDF module consists of two submodules, that is, a filtering and fitness calculation (FFC) module and a reproduction and selection (RS) module. This structure can perform parallel processing efficiently, since these modules work in parallel. Moreover, using the proposed structure, it is easy to design these modules and write HDL code for them.

The output of the EDF is the output of an inner filter for which fitness value is maximum. Therefore, the output of the EDF is selected after all fitness values of inner filters is evaluated. Thus, the EDF module has a common memory to keep output signals of all inner digital filters throughout T_O samples every iteration as shown in Figure 2.

Table 2 shows specifications of the hardware-based EDF. Format of signals and coefficients on the hardware-based EDF is "Q14," that is, 16-bit fixed-point format with an integer part in the high-order 2bits and a fractional part in the low-order 14bits in consideration of the range of the coefficients. The minimum size of the common memory required to keep the output signals is $11,040 \times 16$ bits in the case of $A = 1104$ and $T_O = 10$.

3.1.1. Filtering and Fitness Calculation Module

The FFC module has high computational requirement, since the FFC module performs filtering and fitness calculation of a large number of individuals. Thus, the FFC module has single filtering modules (SFMs) which are submodules and perform filtering and fitness calculation per individual. Figure 3 shows the block diagram of the FFC module. The FFC module proposed here has the high computational ability to calculate the output and the fitness value since the SFMs run in parallel.

3.1.2. Reproduction and Selection Module

Figure 4 shows the block diagram of the reproduction and selection module. This module consists of the following modules: a single reproduction and selection (SRS) module which perform a reproduction and selection opereation every individual, and an SRS control module.

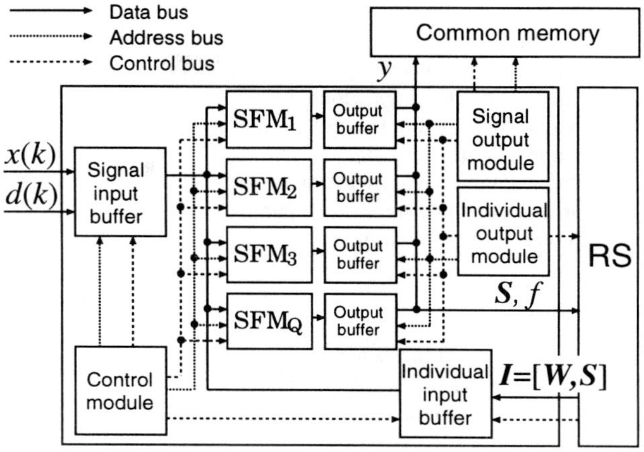

Figure 3: Block diagram of the FFC module.

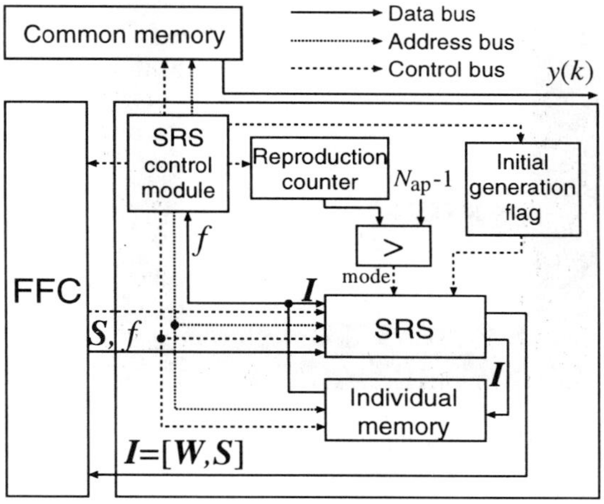

Figure 4: Block diagram of the RS module.

The RS module, first, repeats the following steps in parallel until fitness values of all individuals in population are evaluated.

Step 1. Reproduce an individual according to fitness value in population.

Step 2. Send an information of the individual to the FFC module.

Step 3. Receive a fitness value of the individual from the FFC module.

Second, the individual which has the maximum fitness value in population is selected. Finally, the output for the EDF is selected from the common memory.

3.2. Chip Implementation

A chip of the proposed structure of the EDF is implemented on a silicon area of 4.93×4.93-mm^2 in Rhom 0.35-μm CMOS process. Table 3 shows parameters for the EDF. The implemented chip has only one SFM since the chip is restricted in size. Synplicity and Avanti Apollo are used to synthesize and implement the proposed

Table 3: Parameters for the EDF.

N_{ap}	Number of parents using the cloning method	32
N_{ac}	Number of offsprings which are created by one parent using the cloning method	32
N_{sp}	Number of parents using the mating method	32
N	Order of the regressive average part	3
M	Order of the moving average part	2
T_0	Period of the evaluation	10

Table 4: Synthesis result of the EDF.

Clock frequency	Size
20.0 MHz	63,652 gates

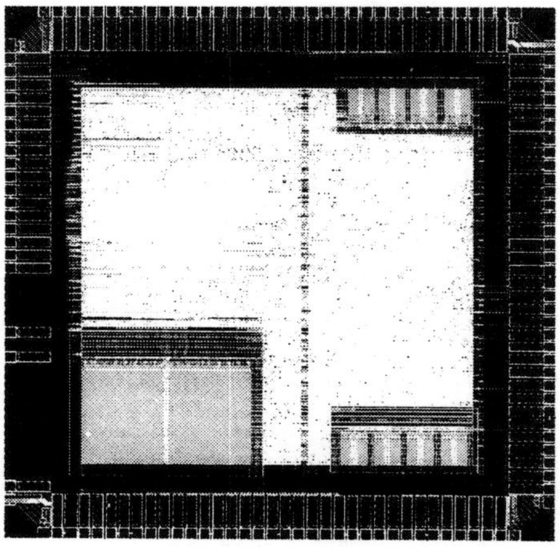

Figure 5: Chip layout of the EDF.

structure of the EDF written in Verilog HDL. The performance of the chip is analyzed by performing a post-layout simulation. Figure 5 shows the layout of the EDF.

Table 4 shows a synthesis result of the designed chip without the memories. Table 5 shows the memory size of the FFC module and the RS module.

Table 6 shows the performances of the FFC module and the RS module. They need 76.8 and 3.6 clocks per individual for processing one sample, respectively. Therefore, the maximum sampling rate of the implemented EDF chip with one SFM is 232.5Hz.

Moreover, Table 6 shows the number of clocks of the FFC module is 21.3 times that of the RS module. Thus, the FFC module needs the 21 SFMs in order that the number of clocks of the FFC module equals that of the RS module when the chip is not restricted in size. In that case, the maximum sampling rate of the EDF is 3.7kHz.

In order to evaluate the performance of the implemented chip, we compare the sampling rate of the hardware-based EDF with that of the software-based EDF. The software-based EDF is written in C and is compiled by gcc on Solaris 8. In that case, the maximum sampling rate of the software-based EDF is 1.7kHz which is executed on the Ultra SPARC III 900MHz. Therefore, the hardware-based EDF with 21 SFMs is 2.2 times faster than the software-based EDF.

Table 5: Memory cell utilization.

Module	Size	Number of cells
FFC	128× 8 bits	8
RS	2,048× 16 bits	1

Table 6: Clocks per individual for processing one sample.

Module	Number of clocks
FFC	76.8
RS	3.6

4. CONCLUDING REMARKS

In this paper, the hardware-based EDF has been designed and implemented. A synthesis result of the designed chip shows the clock frequency is 20.0MHz and the maximum sampling rate of the EDF is 3.7kHz. Moreover, the hardware-based EDF with 21 SFMs of the FFC is 2.2 times faster than the software-based EDF.

ACKNOWLEDGMENT

The authors would like to thank the master student N. Tsushima at Tohoku University for his contributions to the implementation of the proposed hardware-based EDF.

5. REFERENCES

[1] S. D. Stearns, R. A. David, and D. M. Etter, "A survey of IIR adaptive filtering algorithms," Proc. IEEE International Symposium on Circuits & Syst., pp. 709–711, May 1982.

[2] S. J. Flockton and M. S. White, "Pole-zero system identification using genetic algorithms," Proc. the Fifth International Conference on Genetic Algorithm, pp. 531–535, July 1993.

[3] Q. Ma and C. F. N. Cowan, "Genetic algorithms applied to the adaptation of IIR filters," Signal Processing, vol. 48, no. 2, pp. 155–163, Jan. 1996.

[4] S. C. Ng, S. H. Leung, C. Y. Chung, A. Luk, and W. H. Lau, "The genetic search approach — a new learning algorithm for adaptive IIR filtering —," IEEE Signal Processing Mag., vol. 13, no. 6, pp. 38–46, Nov. 1996.

[5] G. D. Cain, A. Yardim, J. Brun, and B. Summers, "Real-time IIR notch filtering using Darwinian adaption," Proc. IEEE International Symposium on Circuits & Syst., pp. 432–435, June 1991.

[6] M. Abe, M. Kawamata, and T. Higuchi, "Convergence behavior of evolutionary digital filters on a multiple-peak surface," Proc. IEEE International Symposium on Circuits & Syst., vol. 2, pp. 185–188, May 1996.

[7] M. Abe and M. Kawamata, "Evolutionary digital filtering for IIR adaptive digital filters based on the cloning and mating reproduction," IEICE Trans. Fundamentals, vol. E81–A, no. 3, pp. 398–406, March 1998.

[8] M. Abe and M. Kawamata, "Comparison of convergence behavior of distributed evolutionary digital filters," Proc. IEEE International Symposium on Circuits & Syst., vol. 2, pp. 729–732, May 2001.

Turbo Coded Multiple Symbol Differential Detection for Correlated Rayleigh Fading Channel

Pisit Vanichchanunt, Chantima Sritiapetch, Suvit Nakpeerayuth, and Lunchakorn Wuttisittikulkij

Department of Electrical Engineering, Chulalongkorn University, Bangkok, Thailand
Tel. (662)218-6908, Fax (662)218-6915, Email: d1pisit, suvit, lunch@ee.eng.chula.ac.th

Abstract- **In this paper, the Multiple Symbol Differential Detection (MSDD) with iterative decoding (turbo decoding) of code rate 1/2 is modified to work under both correlated slow and correlated fast Rayleigh fading channels.**

I. INTRODUCTION

Digital signal transmission over fading channels suffers not only from varying loss but also from phase ambiguity. At the receiver, the fading process is needed to be known or estimated in order to recover carrier and compensate the corrupted signal, and it refers to coherent detection. Alternative approaches without using carrier acquisition are so called non-coherent detection. Conventionally, in the differential phase shift keying (DPSK) system with non-coherent detection, the transmitted signal is differentially encoded (modulated) and then it is differentially detected by comparing the phases between two adjacent symbols without using recovered carrier. This is referred to differentially encoded differentially detected PSK (DDPSK). Coherent detection can still be applied, resulting in differentially encoded coherently detected PSK (DCPSK). Although the channel state information is not needed for differential detection, its performance is degraded comparing to that of coherent detection. To overcome this problem by extending the approach of the conventional differential detection, that is, more than two of consecutive symbols are used and this refers to multiple symbol differential detection (MSDD). Because of using more information, MSDD can bridge the gap of performances between the DDPSK system and the DCPSK system, depending on the number of observed symbols and the number of phases in multiple-phase shift keying (MPSK) systems. Example works of MSDD for AWGN channel can be found in [1,2] and for Rayleigh fading channel can be found in [3].

Research in the error-correction coding area is now taken much attention after the successfulness of the near Shannon limit performance of turbo codes over an AWGN channel [4]. This invites researchers to investigate its application for digital communication over fading channels. Early researches begin with the assumption of the perfect knowledge of uncorrelated fading channels [5] and followed with the assumption of correlated slow fading channels [6]. In the latter case, the channel information is estimated through the channel characteristic model of the fading process. More recently, very interesting works of combining iterative decoding/detection with MSDD are studied for a correlated slow fading channel in [7] with convolutional codes, and for a correlated fast Rayleigh fading channel in [8] with turbo codes. The work in [7] is assumed that the amplitude of fading channel is constant over a block of transmitted symbols and the phase of the channel is constant or changes very slowly. These assumptions are valid only for slow fading channel. The fading channel model in [8] is more general. The amplitude and the phase of fading channel can both be varied according to the Jake's Doppler power spectrum. This is allowed the system in [8] to work well for fast fading channel where the amplitude and the phase of the channel are varied rapidly. Another key that the system in [8] can work wellunder fast correlated fading channel, is to transmit the data bit and the associated parity bit of the same branch of the trellis diagram of constituent encoders into the same QPSK modulation symbol. This allows the decoding/detection system to jointly match between the state transition of a constituent encoder and its associated channel metric function of a received symbol which comprises the information of both data bit and its associated parity bit from one of two constituent encoders (just one of two constituent encoders because of puncturing). This scheme can compensate the effect of the poor performance of fading prediction especially the channel phase estimation under fast fading channel and it may be viewed as the direct assistance for the channel phase estimation by constituent decoders. However, the performance of the system in [8] may be improved for slow fading channel. This is because transmitting each data bit and its associated party bit into the same QPSK symbol is to risk to entirely lose received symbols directly relevant to the information of a data bit or its parity bit, available for the receiver. In slow fading, the channel phase estimation can be performed effectively and the direct assistance for the channel phase estimation by constituent decoders as in [8] may not be needed.

In this paper, we modify a turbo coded QPSK MSDD system for code rate 1/2 in [8] to work better for both correlated slow and correlated fast Rayleigh fading channels. Two key issues of our work are as follows: one is to transmit each data bit and its associated parity bit into different QPSK symbols when the fading is slow. In this scheme, an appropriate metric calculation for decoding is also analyzed. The other scheme is to transmit each data bit and its associated parity bit into the same modulation symbol as work in [8] when the fading is very fast where the effect of phase ambiguity is dominant to the performance of the decoding system.

The remainder of this paper is organized as follows: the modified encoding system is shown in Section II, the channel model is defined in Section III, the modified decoding system is explained and analyzed in Section IV, and the performance results are shown and discussed in Section V. Finally, the conclusions are in Section VI.

II. ENCODING SYSTEM

In this section, we modify the encoding system in [8] as shown in the Figure 1. The encoding system comprises a parallel concatenated convolutional encoder (turbo encoder) punctured for code rate 1/2, a parity bit interleaver (PI) Π, a signal mapper (SM), a channel interleaver (CI) Λ, and a differential encoder (DE). A data bit sequence $a_1, a_2, \ldots, a_{N_b}$ denoted $\underline{a}_1^{N_b}$ where N_b is the data block length, is encoded twice by two identical recursive systematic convolutional encoders RSC1 and RSC2 of the turbo encoder with different orders through the parallel convolutional interleaver (PCI) π. A simile odd-even helical interleaver is used for the PCI to ensure that both RSC encoders can be driven to the zero state with the same tail bit sequence [9]. The parity bit sequence of the RSC2 is reordered to match the order of data bit sequence $\underline{a}_1^{N_b}$ by the parallel convolutional

This work was supported by the Royal Golden Jubilee Ph.D. Program of the Thailand Research Fund.

0-7803-7761-3/03/$17.00 ©2003 IEEE

deinterleaver π^{-1}. A parity bit of the same branch of a data bit will be called *the associated parity bit* with the data. Next, two parity bit sequences of the RSC1 and the RSC2 are punctured at odd and even positions respectively, to be a sequence of p_n's in order to achieve overall code rate of 1/2. The puncturing and the interleaving/deinterleaving of the turbo encoder are performed in such a fashion that for each data bit, there is an associated parity bit from the same trellis branch of one of the RSC1 and the RSC2, to be transmitted. Then the punctured parity bit sequence of p_n's is shuffled to be a sequence of p'_n's by the parity bit interleaver Π so that each data bit and its associated parity bit will not be mapped by the SM into the same QPSK symbol. The scheme of transmitting each data bit and its associated parity bit into the same QPSK symbol can also be performed by using an identity interleaver for the parity bit interleaver. The mapped QPSK symbol sequence is shuffled to be a sequence $\underline{I}_1^N$ by the CI Λ where $N = N_b + L$ and L is the number of tail bits. Then, it is differentially encoded to be a DQPSK symbol sequence $\underline{D}_0^N$ by the DE as follows:

$$D_n = D_{n-1} I_n . \quad (1)$$

The channel interleaver Λ is used to decorrelate the fading effect of the channel whereas the DE is used to provide that MSDD can be performed at the receiver.

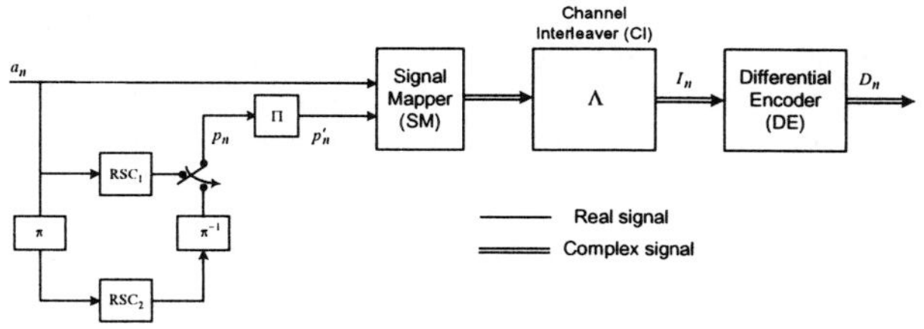

Figure 1: Encoding system

III. CHANNEL MODEL

The discrete-time, complex-baseband equivalent model of the channel is assumed to be a correlated flat Rayleigh fading channel with additive white Gausssian noise (AWGN) as follows

$$R_n = F_n D_n + N_n . \quad (2)$$

The channel fading process $\underline{F}_0^N$ is modeled by a zero-mean complex Gaussian discrete random process satisfied to the autocovariance $\phi_F(m) \equiv E\{F_n^* F_{n+m}\} = J_0(2\pi BTm)$ where $E\{\cdot\}$ is the expectation, $J_0(\cdot)$ is the zero-order Bessel function of the first kind, B is the Doppler spread, and T is the symbol time duration. $\underline{N}_0^N$ is the AWGN process and also modeled by a zero-mean Gaussian discrete random process with the autocorrelation $E\{N_n^* N_{n+m}\} = N_0 \delta(m)$, $n \in [0, N]$, where $\delta(\cdot)$ is the Kronecker delta function. N_0 is the single-sided noise power spectral density.

IV. DECODING SYSTEM

Our modified decoding system comprising a primary metric calculation unit (MCU), and two decoding units is shown in Figure 2. The primary MCU calculates the reduced-complexity channel metric of maximum likelihood sequence estimation (MLSE) for correlated Rayleigh fading channel. In each decoding unit, the secondary MCU receives the primary metric sequence and generates the appropriate metric sequence of received DQPSK symbols called the secondary metric sequence for the constituent decoder (CD) with two different orders. One is matched to the order of the data bit sequence. The other is matched to the order of the parity bit sequence by the parity bit deinterleaver Π^{-1}. Each CD is used to calculate the *a posteriori* probability (APP) of data bits and parity bits for the other CD, and also used to calculate the APP of QPSK symbols for the secondary MCU in the other decoding unit. The information is calculated in such a way that the extrinsic information is only exchanged between two decoding units.

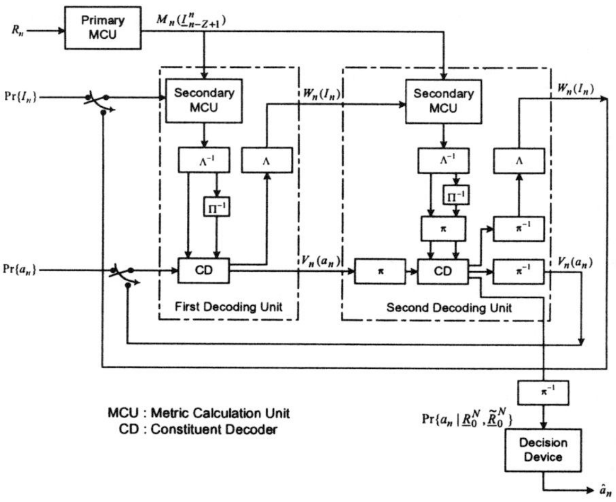

Figure 2: The modified decoding system

A. Review of primary metric calculation unit (Primary MCU)

The primary metric of MSDD is defined as the conditional probability of a received symbol R_n given by the previously transmitted QPSK symbol sequence of length Z or $\underline{I}_{n-Z+1}^n$ and all previously received symbols $\underline{R}_0^{n-1}$ as follows [8]:

$$M_n(\underline{I}_{n-Z+1}^n) \equiv \Pr\{R_n \mid \underline{I}_{n-Z+1}^n, \underline{R}_0^{n-1}\} \quad (3)$$

$$= \frac{1}{\pi\sigma_Z^2} \exp\left\{-\frac{1}{\sigma_Z^2}\left|R_n - \left(\sum_{z=1}^{Z} P_z R_{n-z} \prod_{k=1}^{z-1} I_{n-k}\right) I_n\right|^2\right\} \quad (4)$$

where P_z is the linear prediction coefficient and σ_Z^2 is the variance of the minimum mean-squared prediction error.

B. Review of secondary metric calculation unit (Secondary MCU)

The primary metric is used to calculate the secondary metric $\Gamma_n(I_n)$ for the turbo decoder as follows [8]:

$$\Gamma_n(I_n) \equiv \Pr\{R_n \mid I_n, \underline{R}_0^{n-1}\} = \sum_{\underline{I}_{n-Z+1}^{n-1}} M_n(\underline{I}_{n-Z+1}^n) \Psi_{n-1}(\underline{I}_{n-Z+1}^{n-1}) \quad (5)$$

where

$$\Psi_{n-1}(\underline{I}_{n-Z+1}^{n-1}) = \frac{\sum_{I_{n-Z}} M_{n-1}(\underline{I}_{n-Z}^{n-1}) \Psi_{n-2}(\underline{I}_{n-Z}^{n-2}) \Pr\{I_{n-1}\}}{\sum_{\underline{I}_{n-Z+1}^{n-1}} \sum_{I_{n-Z}} M_{n-1}(\underline{I}_{n-Z}^{n-1}) \Psi_{n-2}(\underline{I}_{n-Z}^{n-2}) \Pr\{I_{n-1}\}} . \quad (6)$$

C. Analysis of Constituent Decoder

There are two constituent decoders and each is in a decoding unit. All constituent decoders are analyzed based on the BCJR algorithm [10]. This algorithm includes the forward recursion and the backward recursion. Before the process of the forward/backward recursions, the branch metric functions of all discrete times must be calculated. The branch metric function at each time comprises the *a priori* probability of the state transition of a constituent encoder and the channel metric functions of received signals of a data bit and a parity bit associated with the state transition. The *a priori* probability of the state transition of a constituent encoder is determined by the extrinsic information of a data bit driving the encoder state transition from the other constituent decoder whereas the channel metric function is determined by the secondary metric. In the scheme that a data bit a_n and its associated parity p_n are not transmitted into the same QPSK symbol, the received signal R_n and the secondary metric $\Gamma_n(I_n)$ must be used with two different orders in decoding process. The first two orders of the sequences are matched to the order of data bit sequence driving the constituent encoder and denoted them $\underline{R}_0^N$ and $\underline{\Gamma}(I_1^N)$ respectively. The second two orders of the sequences are matched to the order of parity bit sequence and denoted them $\underline{\tilde{R}}_0^N$ and $\underline{\tilde{\Gamma}}(\tilde{I}_1^N)$ respectively. So, the channel metric function of a data bit a_n is determined by the secondary metric $\Gamma_n(I_n)$ of the received symbol R_n given by the transmitted QPSK symbol I_n of (a_n, p'_n), whereas the channel metric function of the associated parity bit p_n is determined by the secondary metric $\tilde{\Gamma}_n(\tilde{I}_n)$ of the received symbol $\tilde{R}_n$ given by the transmitted QPSK symbol $\tilde{I}_n$ of (a'_n, p_n) where a'_n is the deinterleaved version of a_n with the parity bit deinterleaver.

In this subsection, the summary of the analysis of constituent decoders is presented. For simplicity of expression, only the first constituent decoder is considered. The second constituent decoder can be analyzed by taking the effect of the PCI into account. The APP of a data bit can be calculated as follows:

$$\Pr\{a_n \mid \underline{R}_0^N, \underline{\tilde{R}}_0^N\} = \sum_{(S_n,S_{n+1}):a_n} \Pr\{S_n, S_{n+1} \mid \underline{R}_0^N, \underline{\tilde{R}}_0^N\} \quad (7)$$

$$= \frac{\sum_{(S_n,S_{n+1}):a_n} \alpha_n(S_n)\,\gamma_n(S_n,S_{n+1})\,\beta_{n+1}(S_{n+1})}{\sum_{a_n}\sum_{(S_n,S_{n+1}):a_n} \alpha_n(S_n)\,\gamma_n(S_n,S_{n+1})\,\beta_{n+1}(S_{n+1})} \quad (8)$$

where $(S_n, S_{n+1}): a_n$ is state transition from S_n to S_{n+1} where a data bit a_n drives the constituent encoder.

The APP of a parity bit can be calculated as follows:

$$\Pr\{p_n \mid \underline{R}_0^N, \underline{\tilde{R}}_0^N\} = \sum_{(S_n,S_{n+1}):p_n} \Pr\{S_n, S_{n+1} \mid \underline{R}_0^N, \underline{\tilde{R}}_0^N\} \quad (9)$$

$$= \frac{\sum_{(S_n,S_{n+1}):p_n} \alpha_n(S_n)\,\gamma_n(S_n,S_{n+1})\,\beta_{n+1}(S_{n+1})}{\sum_{p_n}\sum_{(S_n,S_{n+1}):p_n} \alpha_n(S_n)\,\gamma_n(S_n,S_{n+1})\,\beta_{n+1}(S_{n+1})} \quad (10)$$

where $(S_n, S_{n+1}): p_n$ is state transition from S_n to S_{n+1}, whose parity bit is p_n. $\alpha_n(S_n)$ and $\beta_n(S_n)$ are the forward probability and the backward probability respectively [10].

In the scheme of transmitting each data bit and its associated parity bit into different symbols (denoted scheme 1). The branch metric function is calculated by

$$\gamma_n(S_n,S_{n+1}) = \Pr\{a_n\}\ \Pr\{R_n \mid S_n, S_{n+1}, \underline{R}_0^{n-1}\}\ \Pr\{\tilde{R}_n \mid S_n, S_{n+1}, \underline{\tilde{R}}_0^{n-1}\} \quad (11)$$

where

$$\Pr\{R_n \mid S_n, S_{n+1}, \underline{R}_0^{n-1}\} = \sum_{p'_n} \Pr\{R_n \mid a_n, p'_n, \underline{R}_0^{n-1}\}\Pr\{p'_n\} \quad (12)$$

$$\cong \sum_{p'_n} \Gamma_n(a_n, p'_n)\Pr\{p'_n\}, \quad (13)$$

$$\Pr\{\tilde{R}_n \mid S_n, S_{n+1}, \underline{\tilde{R}}_0^{n-1}\} = \sum_{a'_n} \Pr\{\tilde{R}_n \mid a'_n, p_n, \underline{\tilde{R}}_0^{n-1}\}\Pr\{a'_n\} \quad (14)$$

$$\cong \sum_{a'_n} \tilde{\Gamma}_n(a'_n, p_n)\Pr\{a'_n\}, \quad (15)$$

and a_n and p_n are a data bit and a parity bit of branch (S_n, S_{n+1}), respectively. If p_n is punctured, $\Pr\{\tilde{R}_n \mid S_n, S_{n+1}, \underline{\tilde{R}}_0^{n-1}\}$ will not be used in (11).

When each data bit and its associated parity bit are transmitted into the same symbol (denoted scheme 2), the branch metric function is calculated by [8]

$$\gamma_n(S_n,S_{n+1}) = \Pr\{a_n\}\ \Pr\{R_n \mid S_n, S_{n+1}, \underline{R}_0^{n-1}\} \quad (16)$$

where

$$\Pr\{R_n \mid S_n, S_{n+1}, \underline{R}_0^{n-1}\} = \sum_{p_n} \Pr\{R_n \mid a_n, p_n, \underline{R}_0^{n-1}\}\ \Pr\{p_n \mid S_n, a_n\} \quad (17)$$

$$\cong \sum_{p_n} \Gamma_n(a_n, p_n)\ \Pr\{p_n \mid S_n, a_n\}, \quad (18)$$

and a_n is a data bit of branch (S_n, S_{n+1}). From (18) if the associated parity bit p_n is punctured, $\Pr\{p_n \mid S_n, a_n\}$ will be substituted by 0.5. If p_n is not punctured, $\Pr\{p_n \mid S_n, a_n\}$ will be one for each possible event otherwise it will be zero. Equations (8) can also be used to calculate the APP of a data bit in the scheme 2. The APP of a parity bit is not needed for this scheme.

From (11), if p_n is not punctured, the branch metric function $\gamma_n(S_n, S_{n+1})$ at a time in the scheme 1 contains the *a priori* information of a data bit, and two channel metric functions of two received symbols R_n and $\tilde{R}_n$ whereas from (16), the branch metric function in the scheme 2 contains the *a priori* information of a data bit, and only one channel metric function of a received symbols R_n. Because the scheme 1 uses more number of the channel metric functions of received symbols than the scheme 2, the feasibility that the branch metric function of the scheme 1 has low reliability from no effectively received symbol corrupted by fading, is less comparing to that of the scheme 2. (The probability that two received symbols R_n and $\tilde{R}_n$ which are used at a time in the scheme 1 are both entirely corrupted, is less than that of one received symbol R_n in the scheme 2.) If the associated received symbols are totally corrupted, there shall be no effective channel information directly available for calculating the branch metric function at that time. In this case, the forward and the backward recursions at that time also have low reliabilities unless the previous forward/backward probabilities, and the *a priori* information of a data bit at that time from the other CD are sufficiently high reliable.

Although separately transmitting each data bit and its associated parity bit into different symbols reduces the occurrence that the branch metric function $\gamma_n(S_n, S_{n+1})$ at each time has low reliability from the entire deterioration of the channel information associated with the branch (S_n, S_{n+1}) (measured from $\Pr\{R_n \mid S_n, S_{n+1}, \underline{R}_0^{n-1}\}$ and $\Pr\{\tilde{R}_n \mid S_n, S_{n+1}, \underline{\tilde{R}}_0^{n-1}\}$ in (11)), each channel metric function of a received symbol can just partially be matched between the channel information of only one associated bit (data bit a_n in (13) or parity bit p_n in (15)) and the associated branch of a constituent encoder. The channel information of the other bit of the received symbol is of a non-associated bit (data bit p'_n in (13) or parity bit a'_n in (15)) and it is lost by summation as in (13) and (15) without being matched to the branch of the constituent encoder, so it may be accounted as the uncertainty of phase ambiguity which the branch can not effectively clean. This may lead the performance of the decoding system degraded when the fading is very fast, where the ambiguity of phase is dominant to the degradation of the decoding performance and the prediction of the fading process by using (4) is less effective. In this situation, transmitting both data bit and its associated parity bit into the same modulation symbol may give more attractive performance.

D. The Extrinsic Information

For information exchange in the iteration process of turbo decoding, the *a priori* probabilities of data bits and parity bits must be calculated by their extrinsic information. For the scheme 1, the extrinsic information of a data bit can be calculated by

$$V_n(a_n) = \frac{\Pr\{a_n \mid \underline{R}_0^N, \tilde{\underline{R}}_0^N\}}{\Pr\{a_n\}\sum_{p'_n} \Pr\{p'_n\}\Gamma_n(a_n, p'_n)} . \qquad (19)$$

The extrinsic information of a parity bit can be calculated as follows:

$$V_n(p_n) = \frac{\Pr\{p_n \mid \underline{R}_0^N, \tilde{\underline{R}}_0^N\}}{\sum_{a'_n} \Pr\{a'_n\}\Gamma_n(a'_n, p_n)} . \qquad (20)$$

In the scheme 1 that each data bit and its parity bit are not transmitted into the same QPSK symbol, it may be also assumed that each data bit and its non-associated parity bit transmitted in the same QPSK symbol are statistically independent at the outputs of the CD's. Thus the extrinsic information of a QPSK symbol I_n can be calculated from the product of the extrinsic information of a data bit and a parity bit of the same QPSK symbol I_n as follow:

$$W_n(I_n) = W_n(a_n, p'_n) = V_n(a_n)\, V_n(p'_n) . \qquad (21)$$

For the scheme 2 that each data bit and its associated parity bit are transmitted into the same symbol, the extrinsic information of a data bit and of a symbol I_n is presented in [8].

V. PERFORMANCE RESULTS

The performance of the decoding system is evaluated based on computer simulation. Two RSC's are identical with the feed forward polynomial $1+D^4$ and the feedback polynomial $1+D+D^2+D^3+D^4$. The data block size N_b is 930. The channel interleaver is the odd-even block interleaver of size 41x23 as in [8]. The performances of the decoding system with the two schemes are measured in terms of the bit error rate (BER) of decoded data bits.

In Figure 3, the comparison of the performances of our modified decoding system with the scheme 1 and the decoding system in [8] (scheme 2) are compared. The normalized Doppler frequency is varied as 0.01, 0.125 and 0.20 in 3 (a), 3 (b) and 3 (c) respectively. The number of observed symbols (Z+1) is three or four. The number of iteration is five. For Z = 2, it is shown that the performance of the modified system with the scheme 1 is better than that with the scheme 2 when the fading is slow to medium fast as in 3 (a) and 3 (b) respectively, as the signal to noise ratio is greater than a value. However, when the fading is very fast as in Figure 3 (c), the performance of the system with the scheme 2 is better than that with the scheme 1. At a given Doppler frequency, the performance of the modified system may be improved if the number of observed symbol is increased. For Z = 3 in Figure 3 (c), the performance of the modified system with the scheme 1 is better than that with the scheme 2 for very high E_b/N_0 region. This is because of better fading prediction and it allows the decoding system to take the time diversity advantage of the scheme 1 effectively.

VI. CONCLUSIONS

In this paper, a turbo coded QPSK MSDD system is modified to work under correlated slow and correlated fast Rayleigh fading channels by using two schemes. In the first scheme of the system, each data bit and its associated parity bit are not transmitted into the same QPSK modulation symbols when the fading is slow. This reduces the occurrence that the branch metric function has low reliability from no effectively received symbol for calculating the branch metric at each time. At the decoding system, the metric calculation is analyzed appropriately for utilizing in each constituent decoder. In the other scheme, each data bit and its associated parity bit are transmitted into the same modulation symbol when the fading is very fast, where the phase ambiguity is dominant to the performance of the decoding system. At a given Doppler frequency, the performance of the system with the first scheme may be improved better than that with the second scheme for very fast fading channel if the number of observed symbols is sufficiently large.

REFERENCES

[1] D. Divsalar, and M. K. Simon, "Multiple-symbol Differential Detection of MPSK," *IEEE Trans. Commun.*, vol.38, No. 3, pp. 300-308, March 1990.

[2] F. Edbauer, "Bit Error Rate of Binary and Quaternary DPSK Signals with Multiple Differential Feedback Decision," *IEEE Trans. Commun.*, vol. 40, No. 3, pp. 457-460, March 1992.

[3] D. Makrakis, P. T. Mathiopoulos, and D. P. Bouras "Optimal Decoding of Coded PSK and QAM signals in Correlated Fast Fading Channels and AWGN: A Combine Envelope, Multiple Differential and Coherent Approach," *IEEE Trans. Commun.*, vol. 42, No.1 , pp. 300-308, January 1994.

[4] C. Berrou, A. Glavieux, and P. Thitimajshima, "Near Shannon Limit Error-Correcting Coding and Decoding: Turbo Codes," *in Proceedings of IEEE ICC'93*, Geneva, Switzerland, May 1993, pp. 1064-1070.

[5] J. Hagenauer, E. Offer, and L. Papke, "Iterative Decoding of Binary Block and Convolutional Codes," *IEEE Trans. Inform. Theory*, vol. 42, No. 2, pp. 429-445, March 1996.

[6] E. K. Hall, and S. G. Wilson, "Design and Analysis of Turbo Codes on Rayleigh Fading Channels," *in Proceedings of IEEE GLOBECOM'96*, pp. 16-20, November 1996.

[7] P. Hoeher, and J. Lodge, "Turbo DPSK: Iterative Differential PSK Demodulation and Channel Decoding," *IEEE Trans. Commun.*, vol. 47, No. 6, pp. 837-843, June 1999.

[8] I. D. Marsland, and P. T. Mathiopoulos, "Multiple Differential Detection of Parallel Concatenated Convolutional (Turbo) Codes in Correlated Fast Rayleigh Fading," *IEEE Journal on Selected Areas in Commun.*, vol. 16, No. 2, pp. 265-275, February 1998.

[9] A. S. Barbulescu, and S. S. Pietrobon, "Terminating the Trellis of Turbo-Codes in the Same State," *IEE Electronic Letters*, vol. 31, pp. 22-23, January 1995.

[10] L. R. Bahl, J. Cocke, F. Jelinek, and J. Raviv, "Optimal Decoding of Linear Codes for Minimizing Symbol Error Rate," *IEEE Trans. Inform. Theory*, vol. 20, pp. 284-287, March 1974.

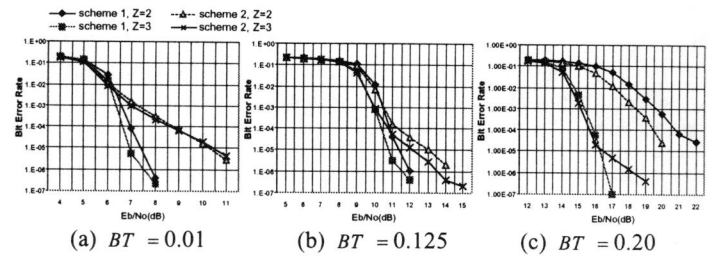

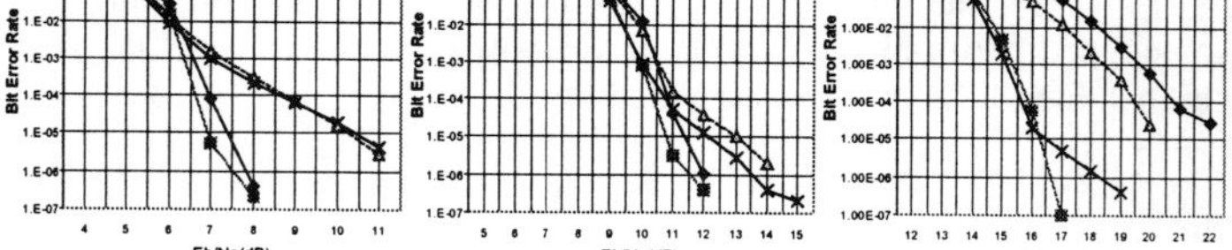

(a) $BT = 0.01$ (b) $BT = 0.125$ (c) $BT = 0.20$

Figure 3: Performance comparison between the decoding system with the scheme 1 and the scheme 2.

A STUDY ON THE STEP SIZE OF CASCADED ADAPTIVE NOTCH FILTER UTILIZING ALLPASS FILTER

Yasutomo Kinugasa† *Yoshio Itoh*†† *Masaki Kobayashi*††† *Yutaka Fukui*†† *James Okello*††††

†Matsue National College of Technology, Matsue, 690-8518 Japan
E-mail:kinugasa@ee.matsue-ct.ac.jp
††Faculty of Engineering, Tottori University, Tottori, 680-8552 Japan
††† Faculty of Engineering, Chubu University, Aichi, 487-8501 Japan
†††† Faculty of Computer Science and Systems Engineering,
Kyushu Institute of Technology, Fukuoka, 820-8502 Japan

ABSTRACT

A cascaded second order notch filter implemented using an allpass filter, for elimination of multiple sinusoids, have been proposed. Various adaptive algorithms for the adaptive notch filters are used. As the one of them, an adaptive algorithm which has a reduced bias in comparison to the gradient based algorithm has been proposed. It has the problem that the convergence speed deteriorates when the number of the narrow band signal increases. Therefore small step size is required. However, the range of the step size for the adaptive notch filters to be stable has not been derived yet. The purpose of this paper is the derivation of it by using the principle of the contraction mapping. Finally, computer simulation results are presented to confirm the convergence characteristics.

1. INTRODUCTION

Adaptive IIR notch filter has been widely used in various applications such as detection of sinusoids in noise or canceling a periodic interference from signal measurements [1]-[6]. As the one of the kind of the adaptive notch filter, there is the cascaded notch filter implemented using allpass filter [7]-[10]. When updated using a simple algorithm that employs the output of each notch filter, the algorithm converges but with a bias. An alternative algorithm, which uses the overall output of the notch filter, has also been proposed [11]. In this method, bias of frequency estimation is reduced, but the convergence speed slightly decreases. In both cases, the convergence speed considerably decreases when the bandwidth of the notch filter become narrow. Moreover, it is known that the convergence characteristics deteriorates because of the small step size as the number of the narrow band signals increases. In such case, the range of the step size which guarantee the convergence is not cleared. Therefore, it is necessary to be cleared.

The purpose of this paper is the derivation of the convergence condition which makes the adaptive algorithm stable. By using the proposed method, it is not necessary to find the optimum step size like a conventional method. Moreover, though the conventional method has the possibility of the divergence, when the narrow band signal changes on the way, the proposed method is always stable and settled.

This paper is organized as follows. In section 2, we explain the conventional method for the cascaded adaptive notch filter and point out the problems. Next, using the principle of the contraction mapping, we derived the converge condition of the adaptive notch filter in section 3. In section 4, we confirm the convergence characteristics through the computer simulations. Section 5 concludes this paper.

2. PREVIEW OF CONVENTIONAL METHOD

The transfer function of a notch filter implemented using cascaded second order notch filters, each of which has been realized by second order allpass filter is given by

$$\begin{aligned} H(z) &= \prod_{k=1}^{N} H_k(z) \\ H_k(z) &= (1 + U_k(z))/2 \end{aligned} \tag{1}$$

where $H_k(z)$ is a second order notch filter and N is number of cascaded second order notch filters. $U_k(z)$ can be realized using the direct form allpass filter, which is shown in Fig.(2), and expressed as

$$U_k(z) = \frac{\rho^2 + a_k z^{-1} + z^{-2}}{1 + a_k z^{-1} + \rho^2 z^{-2}} \tag{2}$$

0-7803-7761-3/03/$17.00

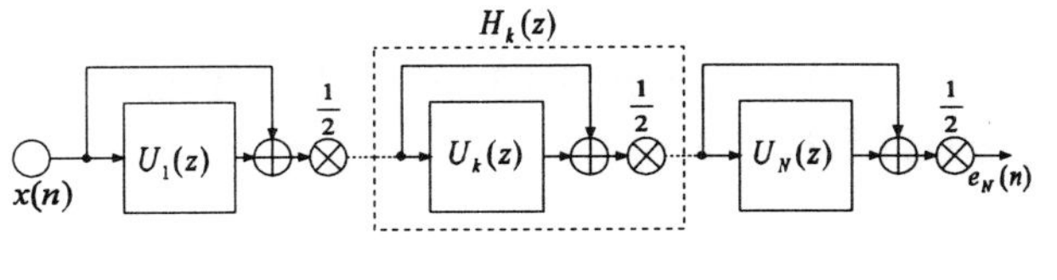

Fig. 1. Structure of an adaptive notch filter.

is shown in Fig.(2). From this figure, the transfer function of the kth cascade stage can be given by

$$H_k(z) = \frac{1}{2}\frac{(1+\rho^2)+2a_kz^{-1}+(1+\rho^2)z^{-2}}{1+a_kz^{-1}+\rho^2z^{-2}} \quad (3)$$

If we define the radian frequency ω along the unit circle on the z-plane such that $z = e^{j\omega}$, then the radian frequency ω_k, which corresponds to the zero of $H_k(\omega)$ will be given by $a_k(n) = -(1+\rho^2)\cos(\omega_k)$. This radian frequency ω_k is referred to as the notch frequency of the kth stage of the cascaded notch filter. The value of ρ determines the bandwidth of notch filter [3]. Conventionally, ρ is set to the close to one.

The simplified adaptive algorithm used to update this kind of notch filter is given by

$$\begin{aligned}
\Phi(\mathbf{a}(n)) &= \mathbf{a}(n+1) = \mathbf{a}(\mathbf{n}) - \mu\mathbf{f}(\mathbf{a}(n)) && (4)\\
\mathbf{a}(n) &= [a_1(n),\cdots,a_N(n)]^T \\
&: \textit{Tap vector} && (5)\\
\mathbf{u}(n) &= [u_1(n),\cdots,u_N(n)]^T \\
&: \textit{Tap input vector} && (6)\\
\mathbf{f}(\mathbf{a}) &= E\left[e_N(n)\mathbf{u}(n)\right] && (7)
\end{aligned}$$

where μ is the step size of adaptation, $e_N(n)$ is the output of the cascaded second order notch filter, T denotes transponse, and $u_k(n)$ is given in Fig. (2). This algorithm can be considered as minimizing the mean square of $e_N(n)$ based on the forward coefficient $a_k(n)$. In order to minimize the bias, the value of ρ must be kept close but less than one. However, as explained earlier, which this kind of setting, the convergence speed reduces significantly due to the small ranges of step size that ensures the convergence. Therefore, the resulting range of the step size for stabilizing the algorithm is desired.

3. PROPOSED METHOD

In this section, we derive the range of the step size for the adaptive algorithm to be stable. From the principle of the contraction mapping, the sufficient condition as the algorithm converges is given by

$$\|\Phi'(\mathbf{a})\|_2 < 1 \quad (8)$$

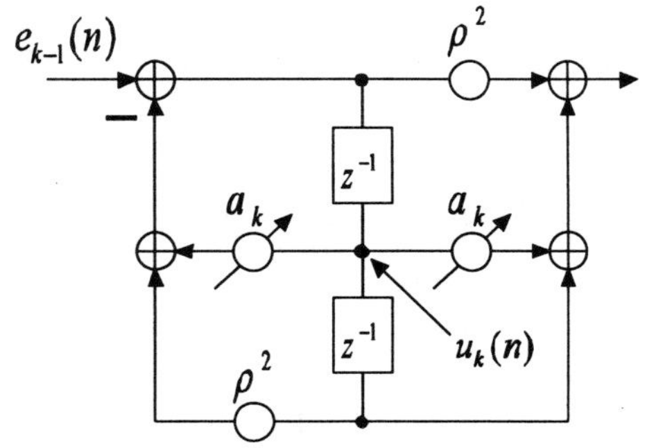

Fig. 2. Structure of a seconf order allpass filter.

where, $\|\cdot\|_2$ represents the matrix norm. $\Phi'(\mathbf{a})$ is expanded as follows:

$$\Phi'(\mathbf{a}) = \mathbf{I} - \mu\mathbf{f}'(\mathbf{a}) \quad (9)$$

Where,

$$\begin{aligned}
&\mathbf{f}'(\mathbf{a}) = \mathbf{v}\left\{E\left[e_N(n)\mathbf{u}^T(n)\right]\right\}\\
&\mathbf{v} = [\partial/\partial a_1,\cdots,\partial/\partial a_N]^T \\
&\mathbf{I} : \textit{Identity matrix}
\end{aligned} \quad (10)$$

Assuming $\rho \simeq 1$ in the previous expression, we can obtain

$$\begin{aligned}
&\partial\left\{e_N(n)u_i(n)\right\}/\partial a_j = u_j(n)u_i(n)/2,\\
&i,j = 1,2,\cdots,N
\end{aligned} \quad (11)$$

and Eq. (10) becomes

$$\mathbf{f}'(\mathbf{a}) = \frac{1}{2}\mu\mathbf{R}, \quad (12)$$

$$\mathbf{R} = E[\mathbf{u}(n)\mathbf{u}^T(n)]. \quad (13)$$

Substituting Eqs.(12) and (13) for Eq. (9), we have

$$\Phi'(\mathbf{a}) = \mathbf{I} - \mu\mathbf{R}/2. \quad (14)$$

From Eqs.(8) and (14), let eigenvalue of the auto-correlation matrix $\mathbf{R}$ denote λ_i $(i = 1,\cdots,N)$. Then

$$\|\Phi'(\mathbf{a})\|_2 = \max_i |1-\mu\lambda_i/2| < 1. \quad (15)$$

Therefore, we can obtain the region of convergence

$$0 < \mu < \frac{4}{\lambda_{max}} \quad (16)$$

If the step size μ satisfy the above expression for all tap coefficients in the stable allpass system, it follows that the adaptive algorithm will converge. Where, λ_{max} denotes maximum value of an eigenvalue λ_i. Hence, it is used for the adaptive algorithm to be stable as much as possible. Generally, it is difficult to obtain the maximum value of the eigenvalue accurately. Regarding λ_{max}, it is cleared that

$$Trace(\mathbf{R}) = \sum_{i=1}^{N} \{E[u_i^2(n)]\} > \lambda_{\max}. \quad (17)$$

From Eqs.(16) and (17), we can obtain

$$0 < \mu < \frac{4}{Trace(\mathbf{R})}. \quad (18)$$

If the step size μ satisfy the equation, the adaptive algorithm converges.

4. SIMULATIONS

In this section, we verify the performance characteristics by the computer simulation. The input signal $x(n)$ is the sum of narrow band signal $s(n)$ and white signal $w(n)$. In the first simulation, the narrow band signal $s(n)$ was

$$s(n) = \cos(0.3\pi n)+\cos(0.5\pi n)++\cos(0.7\pi n)+\cos(0.9\pi n) \quad (19)$$

The above sinusoids were mixed with a white signal $w(n)$ such that the signal to noise ratio (SNR) was 40dB. SNR is expressed as

$$SNR = 10\log_{10}\frac{E[s^2(n)]}{E[w^2(n)]}(\text{dB}) \quad (20)$$

We used a performance function defined using the function EA. EA is given by

$$EA = 10\log_{10}\frac{E[s^2(n)]}{E[s_h^2(n)]}(\text{dB}) \quad (21)$$

where, $s_h(n)$ is the component of the narrow band signal $s(n)$ within the output $e_N(n)$.

As the the step size $\mu(n)$ of the proposed method is used and given by

$$\mu(n) = \frac{4\alpha}{Trace(\mathbf{R})} \quad (22)$$

where, α is used for $\mu(n)$ to satisfy the region of the convergence in Eq.(18). In this simulation, $\alpha = 0.90$.

On the other hand, μ_c is used as the one of the conventional method. The value of ρ and μ_c were set to 0.99 and 6.0×10^{-6}. Figure 3 shows the result obtained. From this result, we can say that proposed method have the good convergence speed than the conventional method. Figure 4 shows the amplitude response of the proposed notch filter.

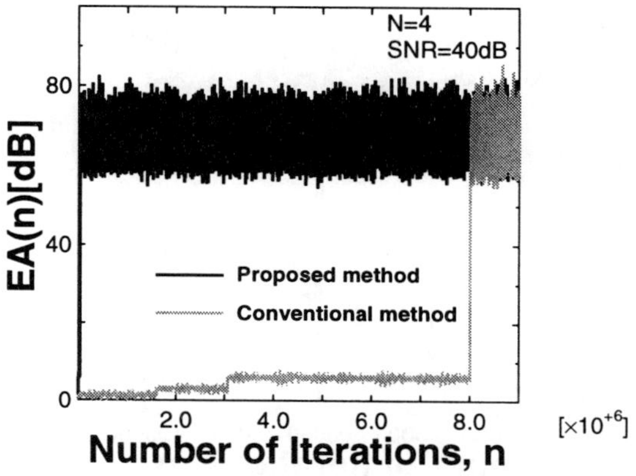

Fig. 3. Convergence performance of estimation accuracy in first case condition.

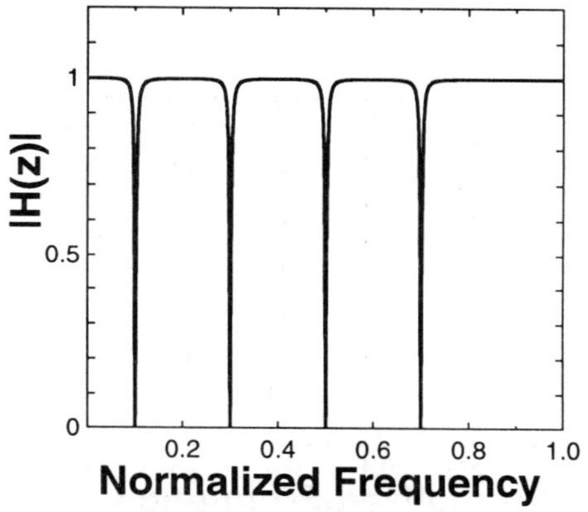

Fig. 4. Amplitude Response in first case condition.

Next, consider the case in which the narrow band signal $s(n)$ changes on the way. In this simulation, when the iteration number lager than 60000, it changes from Eq.(21) to as follows.

$$s(n) = \cos(0.2\pi n)+\cos(0.4\pi n)++\cos(0.6\pi n)+\cos(0.8\pi n) \quad (23)$$

The other parameters are used same one. Figure 5 shows the results obtained. From these results, our proposed method has good convergence characteristics, whereas the conventional one does not converge. Figure 6 shows the amplitude response of the proposed notch filter. From these results, we can see that the adaptive algorithm stably operates, as long as the condition of Eq.(8) is satisfied, and it has been settled even if the conventional method does not converge. Moreover, it has been shown that when tracking multiple sinusoids,

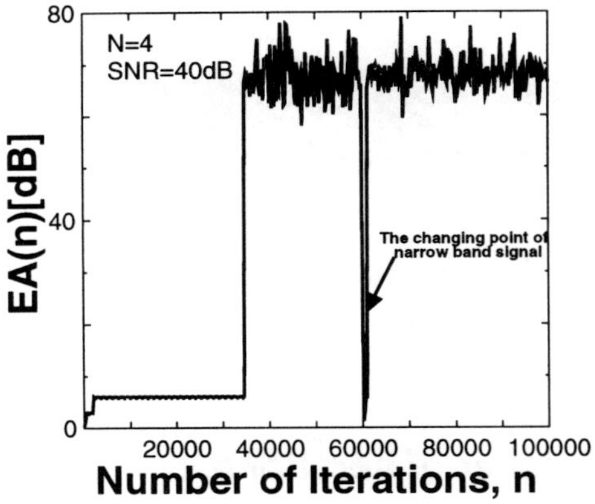

Fig. 5. Convergence performance of estimation accuracy in second case condition.

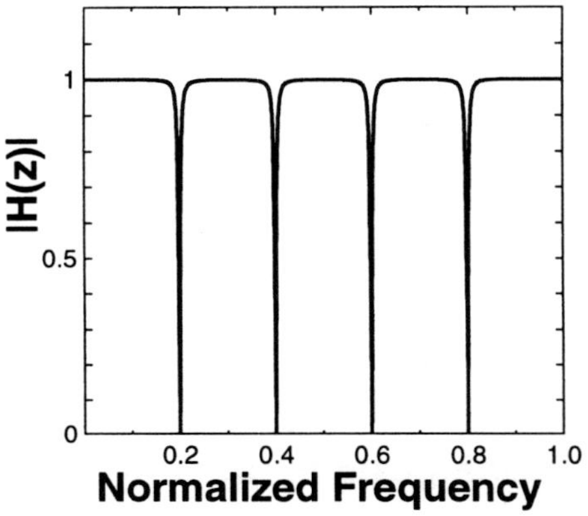

Fig. 6. Amplitude Response after the narrow band signal changed.

the proposed method exhibits the superior convergence speed in comparison to the conventional one. Consequently, the validity of our method is verifiable.

5. SUMMARY

In this paper, we have derived the convergence condition of adaptive notch filter theoretically and evaluated the convergence speed of our proposed and conventional method from the principle of the contraction mapping. In the conventional method, it is necessary to find the optimum step size. While, it has been shown that the adaptive algorithm is stable as long as the convergence condition is satisfied. Moreover, the convergence speed is very fast than the conventional method. Considerations on the convergence condition of the colored input signal is considered as a future work.

6. REFERENCES

[1] R.Carney, " Design of a digital notch filter with tracking requirements," IEEE Trans. Space Electron. Telem., vol. SET-9, pp. 109-114, Dec.1963.

[2] J. C. Huhta and J. G. Webster," 60-Hz interference in electrocardiograph," IEEE Trans. Biomed. Eng., vol. BME-20, pp. 91-101, Mar. 1973.

[3] A. Nehorai, " A minimal parameter adaptive notch filter with constrained poles and zeros," IEEE Trans. Acoust., Speech & Signal Process., vol.ASSP-33, no.4, pp.983-996, Aug. 1985.

[4] J. M. Travassos-Romano and M. Bellanger, " Fast least squares adaptive notch filtering," IEEE Trans. Acoust., Speech & Signal Process., vol.36. no.9, pp.1536-1540, Sept. 1988.

[5] S. Nishimura, " An improved adaptive notch filter for detection of multiple sinusoids," IEICE Trans. Fundamentals, vol.E77-A, no.6, pp.950-955, June 1994.

[6] P. A. Regalia, Adaptive IIR Filtering in Signal Processing and Control. Marcel Dekker Inc., 1995.

[7] T. Kwan and K. Martin, " Adaptive detection and enhancement of multiple sinusoids using a cascade IIR filter," IEEE Trans. Circuits & Syst., vol.36, no.7, pp.937-947, July 1989.

[8] S.-C. Pei and C.-C. Tseng, " IIR multiple notch filter based on allpass filter," IEEE Trans. Circuits & Syst., Pt.U, vol. 44, no.2, pp. 133-136, Feb. 1997.

[9] M. Kobayashi, T. Akagawa, Y. Itoh, " A Study on an Algorithm and a Convergence Performance for Adaptive Notch Filter Utilizing an Allpass Filter," IEICE Vol. J82-A, No. 3, pp. 325-332, March, 1999.

[10] M. Kobayashi, I. Komatuzaki, Y. Itoh, " A Study on an Algorithm and a Convergence Performance for Adaptive Notch Filter Using Tandem Connection of Second-Order System," IEICE Vol. J83-A, No. 5, pp. 594-598, March, 1999.

[11] J. Okello, S. Arita, Y. Itoh, Y. Fukui, and M. Kobayashi, " An Adaptive Algorithm for Cascaded Notch Filter with Reduced Bias," IEICE Trans. Fundamentals, vol. E84-A, no.2, pp.589-596, Feb. 2001.

Polyphase IIR Filter Banks for Subband Adaptive Echo Cancellation Applications

Artur Krukowski and Izzet Kale

Applied DSP and VLSI Research Group, University of Westminster
115 New Cavendish Street,
London W1W 6UW, United Kingdom

ABSTRACT

Polyphase IIR structures are known to be very attractive for very high performance filters that can be designed using very few coefficients. This combined with their reduced sensitivity to coefficient quantization in comparison to standard FIR and IIR structures makes them very applicable to very fast filtering especially when implemented in fixed point arithmetic. In this paper we suggest replacing standard FIR filter banks used in Subband adaptive polyphase FFT echo cancellation applications with such a structure. We demonstrate here that such an alternative approach results in a much more computationally efficient implementation combined with more accurate channel detection and improvement in the adaptation speed.

1. INTRODUCTION

Adaptive signal processing applications such as adaptive equalization or adaptive wideband active noise and echo cancellation involve filters with hundreds of taps required for accurate representation of the channel impulse response. The computational burden associated with such long adaptive filters and their implementation complexity is very high. In addition adaptive filters with many taps may also suffer from long convergence time, especially when the reference signal has a large dynamic range. It is well known that subband adaptive techniques are well suited for high-order adaptive FIR filters, with a reduction in the number of calculations by approximately the number of subbands, whereby both the number of filter coefficients and the weight update rate can be decimated in each subband. Additionally faster convergence is possible as the spectral dynamic range can be greatly reduced in each subband [1] and [2]. A number of subband techniques have been developed in the past that uses a set of bandpass filters, block transforms [3] or hybrids [4], which introduce path delays dependent on the complexity of subband filters employed. The architecture proposed in [1] avoids signal path delay while retaining the computational efficiency and convergence speed of the subband processing. The architecture of the delay-less subband acoustic echo cancellation employing LMS in each subband is shown in Figure 1 [1].

In this structure *x(k)* is interpreted as a far-end signal, which contains some echoes after passing through the channel. The signal *d(k)* is interpreted as the signal received by the microphone and e(k) is the error (de-echoed return signal as defined in [1]). Both *x(k)* and *e(k)* are decomposed into 16 bands and decimated down by 16. The LMS algorithm calculates 16 individual filters for each of the subbands, which are then re-composed back into one in the frequency domain to obtain the coefficients of the high-order Adaptive FIR Filter. The update of the weights, like in the original design in [1], is computed every 128 samples (at 128 times lower rate than the input rate)..

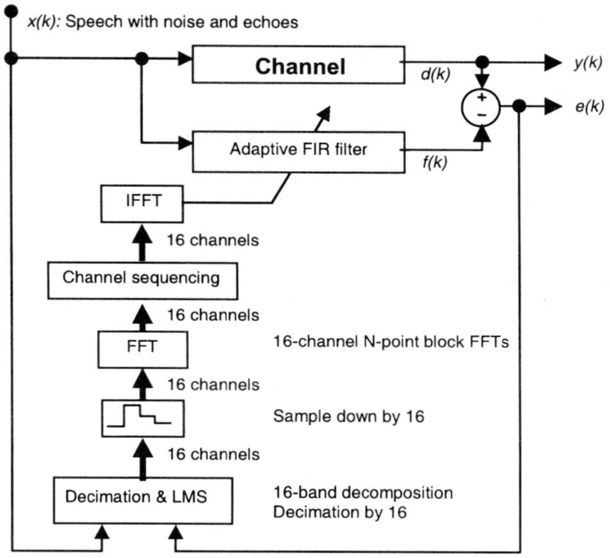

Figure 1. Subband adaptive echo canceller structure.

The computational requirements can be separated into four sections: subband filtering, LMS, composition of the wideband filter and signal convolution by the wideband filter. Naylor and Constantinides first proposed to replace the FIR filter bank with polyphase IIR structures for subband echo cancellation [5]. The work reported in this paper presents further improvement of the computational efficiency of the subband filtering stage of the polyphase FFT subband echo cancellation system. Consider the 16-band polyphase FFT adaptive system with the Adaptive FIR Filter having 1024 taps, in which each subband filter is based on the 128-coefficient prototype FIR filter. The number of Multiply/Add operations (MAC) required for the structure with the FIR filter bank was estimated to be 1088 per input sample. In order to lower the number of calculations per input sample Mullis suggested updating the weights of the high-order Adaptive FIR filter every 128 samples [1]. He argued that the output of the adaptive filter could not change faster than the length of its impulse response. Applying polyphase IIR filter structures to perform the subband filtering makes it possible to reduce the number of calculations, improving the convergence, accuracy of adaptation and allows efficient implementation.

0-7803-7761-3/03/$17.00

2. POLYPHASE IIR FILTER BANK

The two-path polyphase IIR structures as given in [6] and [7] can be modified in such a way to perform simultaneously both the lowpass and highpass filtering operation (Figure 2a). The output of the adder returns the lowpass filtered signal and the output of the subtractor returns the highpass filtered signal. It should be noted that both the lowpass and highpass filtering actions are complementary, i.e. they result in perfect reconstruction giving zero reconstruction error. Adding a sample rate decreaser by two gives rise to achieving a two-band subband filter. Shifting the sample rate decreaser to the input can further modify this basic building block [8]. This modification results in half the number of calculations per input sample and half the storage requirements.

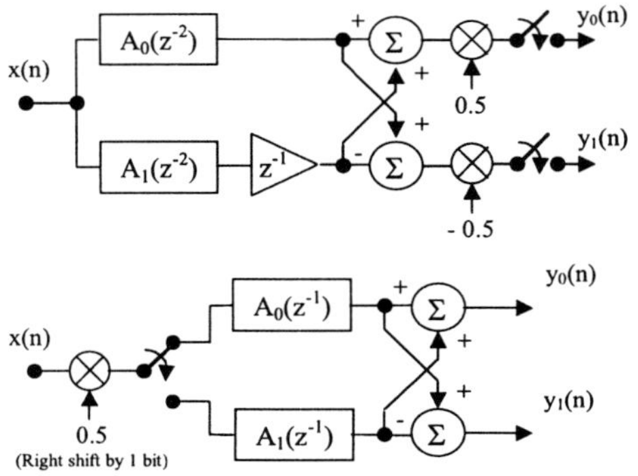

(a)

(b)

Figure 2. Two-band polyphase IIR filter bank with sample rate decrease at the output, (a), at the input, (b).

$$H(z) = \frac{1}{2}\sum_{n=0}^{1} A_n\left(z^{-2}\right) z^{-n} \tag{1a}$$

$$A_n\left(z^{-2}\right) = \prod_{k=1}^{K_n} A_{n,k}\left(z^{-2}\right) = \prod_{k=1}^{K_n} \frac{\alpha_{n,k} + z^{-2}}{1+\alpha_{n,k} z^{-2}} \tag{1b}$$

The polyphase IIR structure incorporates 2nd-order allpass sub-filters, $A_i(z)$, as given by (1b), and has a possible structure as shown in Figure 3.

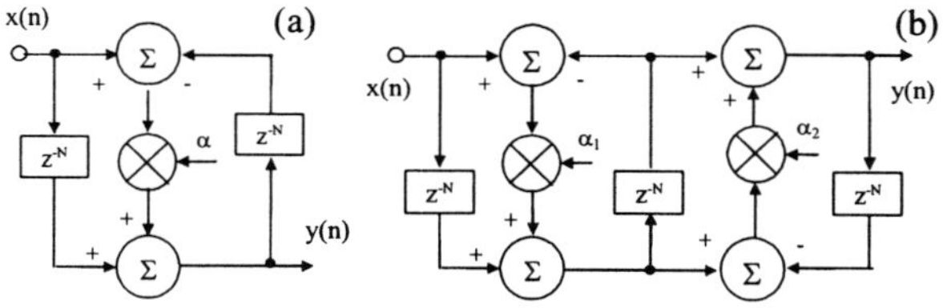

Figure 3. Allpass filter structures: the second-order filter, (a), and the fourth-order filter, (b).

Because of the small number of calculations required per filter order and very high performance, such a structure is very attractive for filtering requiring high speed of operation and high levels of integration.

The 16-channel subband filtering and 16-times decimation was achieved by incorporating the polyphase IIR filtering block from Figure 2 in the structure shown in Figure 4 where LPF stands for the LowPass Filter (LPF) and the number indicates the number of coefficients. This four-stage structure splits the signal frequency response into equal size bands followed by decimation by two. Each of the resulting signals undergoes a similar operation at the next stage.

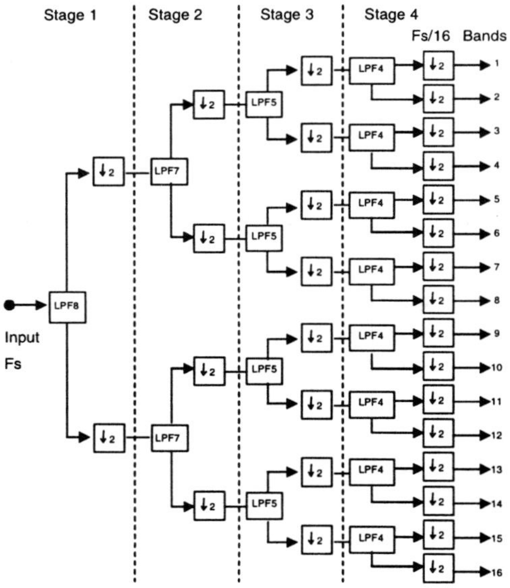

Figure 4. Polyphase IIR four-stage 16-bank filter structure combined with 16-times sample rate reduction.

All filters were designed for the same 70dB of attenuation for achieving an appropriate separation from the neighboring bands. The transition bandwidths were different, as they had to cater for the decrease of the sampling frequency at which they were operating. The required Transition Band width, TB, and resulting filter coefficients for each stage of the filter bank are given in Table 1.

Stage	TB	Filter coefficients	
		Top branch	Bottom branch
1	Fs/2	0.06420239969799 0.42467376298912 0.74596983516706 0.91928513358894	0.22691819065413 0.60517810579077 0.84726220095035 0.97462589167035
2	Fs/4	0.05896569792564 0.40122185929635 0.73335764717435 0.95211781083584	0.21086585025121 0.58276390614624 0.85256695362631
3	Fs/8	0.07294398933668 0.48088373734321 0.89653509769761	0.25672675991015 0.69620335291973
4	Fs/16	0.06883322519472 0.50599695800030	0.25221956029507 0.81339466660703

Table 1. Parameters of the polyphase subband structure.

Each output from the filter bank is applied to each of the 16 LMS blocks (Figure 5), each operating on a small fraction of overall input frequency response, thus achieving fast operation speed.

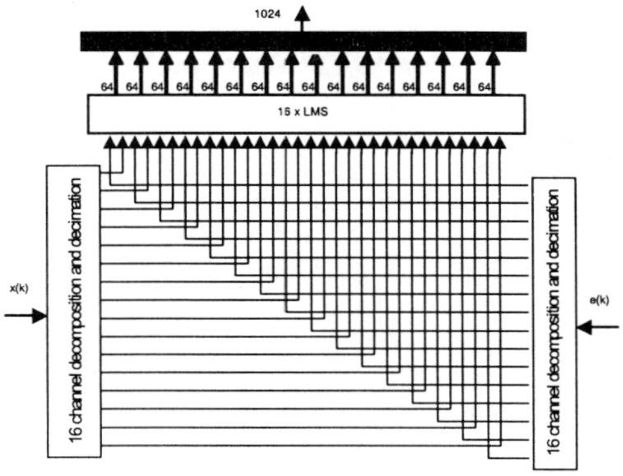

Figure 5. "Decomposition and LMS" block contents.

The error of channel approximation is split into 16 frequency bands in the same way as the input signal was. This way the phase non-linearity of the polyphase IIR filter bank does not cause errors as both the input signal and the error are subject to the same group delays. Each of the LMS block returns a 64-tap FIR filter in order to decrease the error of channel approximation to an acceptable level in its frequency band. The output of each LMS block is then applied to an N point FFT, where $N=64$.

Note that the output bands of the filter bank were not in an increasing order. Additionally the sample rate decrease of the output of the highpass filter causes a flip of the frequency response. Therefore channel re-ordering and re-flipping is necessary before doing the IFFT operation (Figure 6).

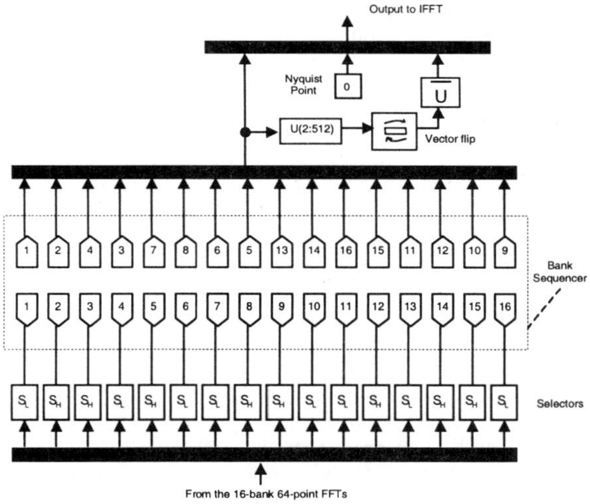

Figure 6. Channel re-ordering and the reconstruction of the adaptive filter frequency response.

The bank re-ordering block is responsible for arranging the subbands in an increasing order of frequencies. The SL and SH (2) operators are the frequency selectors returning the lower half or the upper-half of the FFT output respectively. This was necessary as some outputs of the 16-channel subband decomposition had their frequency responses flipped.

$$\begin{cases} S_L = x(1,\ldots,N/2) \\ S_H = x(N/2+1,\ldots,N) \end{cases} \qquad (2)$$

The bank re-ordering creates the positive part of the frequency response of the approximated channel filter. Flipping and conjugating the positive part of the frequency response inherently calculated the negative part of the frequency response. This means that the output of the IFFT returns the real FIR filter, which accurately approximates both the magnitude response of the channel and its bulk delay.

3. FIR VS POLYPHASE ADAPTATION

The performance of channel identification when using the polyphase IIR filter bank was compared to Morgan's FIR approach [1] both in theory and in simulation . Analysis included the comparison of frequency responses from both subband decomposition methods in terms of passband and stopband ripples and channel overlap. Morgan suggested using the 'fir1' Matlab routine for designing the prototype FIR filter for the subband filtering, which resulted in a filter as shown in Figure 7 [1]. It has -6dB at the crossover point to the next band while -3dB was required for zero reconstruction error at this point. It had 50dB of attenuation at the center of the next band and 70dB, required for the aliasing noise floor to be below -116dB, at the second band.

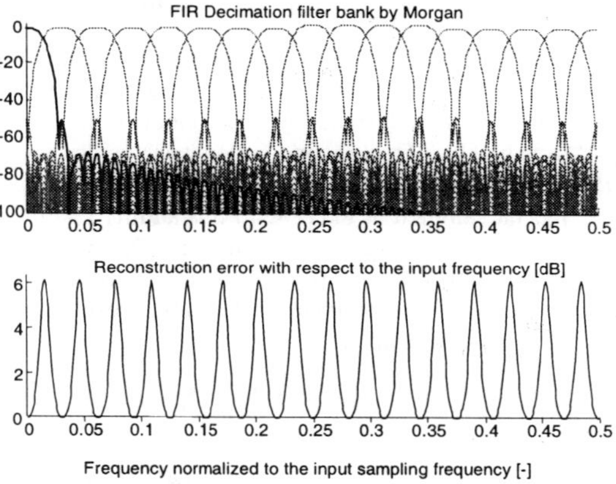

Figure 7. Performance of a 16-bank FIR subband filter.

Our polyphase structure achieves the -3dB at the Fs/2 point by definition. It reached 45dB attenuation at the center of the next band and 70dB before the end of the next band, giving a good separation from all the bands except the two adjacent ones. The polyphase filter bank achieved an overall reconstruction error below 10-13dB, which is close to the arithmetic noise level of the simulation platform (Figure 8).

One of the main advantages of the polyphase approach is in its low number of multiplications required per input sample. The FIR approach as suggested by Morgan requires 32-tap filters and 17 bands giving an overall MAC requirement of 1088 for subband filtering before the 16 times sample rate decrease. In comparison, the use of the multi-stage multi-rate polyphase IIR structure allowed a decrease in the number of MACs (excluding the trivial subtractions) per input sample to 24 if implemented as in Figure 2a, and only 14 when using the structure in Figure 2b. Half of the calculations for the structure in Figure 2b are done at the odd sample intervals and the rest of them at the even sample intervals.

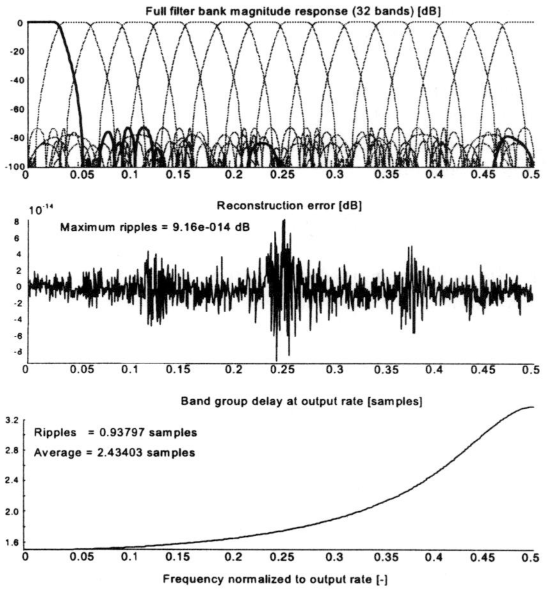

Figure 8. Performance of a 16-bank polyphase structure.

Practical tests were carried out to compare the channel approximation when using both the FIR and the Polyphase based approaches. The first one was designed in accordance to the one reported by Morgan [1]. The second one was using the Polyphase IIR filter bank as described in this paper. The input test signal was speech sampled at 8kHz with 5% additive white noise. The channel was a 50th-order least-squared FIR bandpass filter positioned from 0.0625 to 0.375 on the normalized frequency axes with 50dB of stopband attenuation using an additional bulk delay of 400 samples.

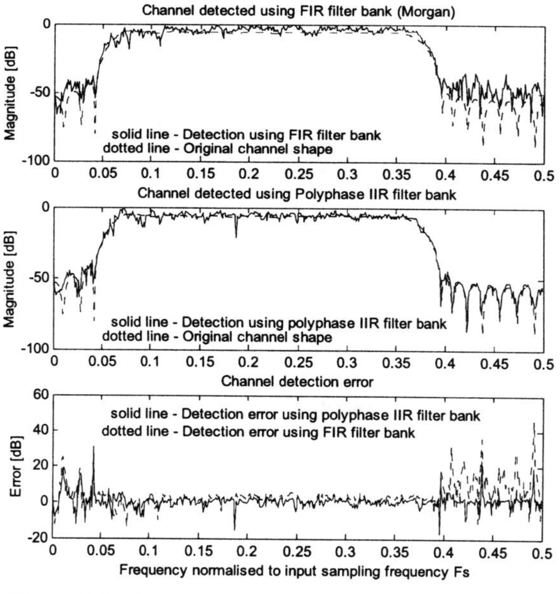

Figure 9. Performance comparison of channel adaptation for FIR and polyphase IIR filter banks.

The performance comparison between the polyphase approach and the standard one employing the FIR filter bank is shown in Figure 9. Both approaches accurately detected the 400-sample bulk delay proving the viability of the method for echo cancellation. Their adaptation speed was similar. The estimated least squared error of channel approximation for the polyphase approach was 6.2dB in the passband and 9.4dB in full range. This shows an improvement in comparison to the FIR approach giving 7.5dB error in the passband and 13.5dB in full range. The superior results for the polyphase IIR filter bank can be attributed to its steeper transition bands, perfect reconstruction (zero error), good channel separation and very flat passband response within each band. For an input signal rate of 8kHz the response time to the changes in the channel is 0.064 seconds. The adaptation time for the given channel and input signal was measured to be below 0.2 seconds. The channel approximation error fell below 10% in approximately 0.5 seconds. The times were estimated assuming that all calculations were completed within one sample period.

4. SUMMARY

In this paper we have presented the application of polyphase IIR filters for subband filtering of the polyphase FFT adaptive echo cancellation architecture. We showed comparative results to the standard FIR version as was reported in [1]. Our alternative multi-stage multi-rate polyphase IIR approach for the design of the subband filter-bank gives an almost ten-fold decrease in the number of MACs required, which can be easily translated into an increased number of bands for higher fidelity for the same computational cost as that of the FIR or a low-power subbands adaptive echo canceller. Additionally the polyphase IIR filter structure used here is are not very sensitive to coefficient quantization [7], which makes a fast fixed-point implementation of the echo cancellation algorithm an attractive option. Applying the polyphase IIR filters in the filter-bank demonstrated more accurate channel detection than was possible with the FIR version suggested by Morgan [1], with much reduced computational complexity.

5. REFERENCES

[1] Morgan D and C. Thi, "A delayless subband adaptive filter architecture", *IEEE Trans. on Signal Proc.*, Vol. 43, No. 8, 1995.

[2] Chen J.D., H. Bes, J. Vandewalle et al, "A zero-delay FFT-based subband acoustic echo canceller for teleconferencing and hands-free telephone systems", *IEEE TCAS II*, vol. 43, no. 10, pp. 713(7), 1996.

[3] Sondhi, M. M. and W. Kellerman, "Adaptive echo cancellation for speech signals", Advances in Speech Signal Processing, S. Furui and M. M. Sondhi, Eds, NY: M. Dekker, Ch. 1, 1992.

[4] Gerald J., N. L. Esteves and M. M. Silva, "A new IIR echo canceller structure", *IEEE TCAS II*, vol. 42, no. 12, pp. 818-821, 1995.

[5] Naylor P. A., O. Tanrýkulu and A. G. Constantinides, "Subband adaptive filtering for acoustic echo control using allpass polyphase IIR filterbanks", *IEEE Trans. on S&P*, vol. 6, no. 2, 1998.

[6] Valenzuela, R. A. and A. G. Constantinides, "Digital signal processing schemes for efficient interpolation and decimation", *IEE Proc.*, Pt. G., vol. 130, no. 6, pp. 225-235, 1983.

[7] Vaidyanathan P. P.,"Multirate digital filters, filter banks, polyphase networks, and applications: A tutorial", *IEEE Proc.*, vol. 78, no. 1, 1990.

[8] Kale, I., R. C. S. Morling and A. Krukowski, "A high-fidelity decimator chip for the measurement of Sigma-Delta modulator performance, *IEEE Trans. on I&M*, vol. 44, no. 5, 1995.

A NEW ALGORITHM FOR HOWLING DETECTION

Jianqiang Wei, Limin Du, Zhe Chen, Fuliang Yin

Institute of Acoustics
Chinese Academy of Sciences
Beijing 100080, China

ABSTRACT

In this paper, we propose a new algorithm for the detection of howling signal employing LMS adaptive notch filter and digital phase locked loop. The main advantages of the proposed algorithm are that it has very low computational complexity and has a short detection delay. Computer simulations are given to demonstrate the performance of this method.

1. INTRODUCTION

In many acoustic equipments such as hand-free telephone and videoconference, the channel, i.e. acoustic couple path, forms a closed loop. The basic structure of acoustic couple path in acoustic equipments is shown in Figure 1, which includes acoustic equipment, loudspeaker and microphone. High volume level output at the loudspeaker can cause the loop to get into an oscillatory mode, usually referred to as howling. It not only disturbs normal communication, but also damages power amplifier for overload. This is undesirable and this problem can be combated by reducing the total loop gain with the insertion of an attenuation unit if howling is detected. Hence, it is crucial to detect howling signal in these cases.

Traditional algorithms for the detection of howling signal are usually designed using adaptive notch filter. A recent paper on howling control by S. M. Kou and J. Chen [4] generalizes the conventional methods. Their approach is based on the LMS algorithm and IIR adaptive notch filter. In this paper, we propose a novel algorithm for the detection of howling signal based on LMS adaptive notch filter and digital phase locked loop. The proposed algorithm has not only very low computational complexity but also a short detection delay. Intensive computer simulations illustrate that it really works well.

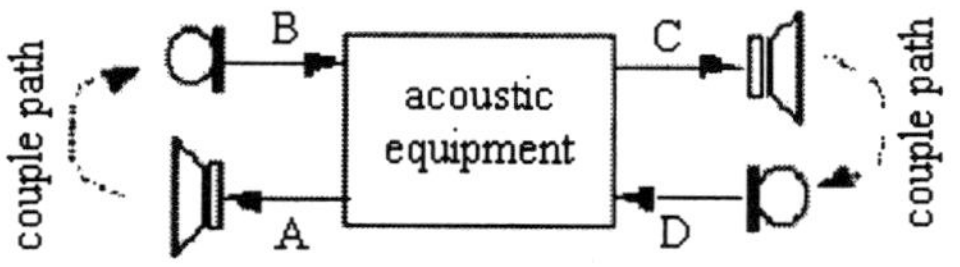

Figure 1. The basic structure of acoustic couple path in acoustic equipment.

2. HOWLING

Howling phenomenon can be explained as signal feedback in-phase at a particular frequency that is further amplified until it can be perceived by the listeners. It is evident that, under the howling condition, both near-end signals and received signals consist of a strong (unwanted) sinusoid waveform at the oscillation frequency, along with noise/speech. That is to say, when being howling, the frequency of signals in location A, B, C and D shown in Figure 1 are same, but their phase are different. Moreover, the phase difference changes slowly and the change rate is about second-level.

3. HOWLING DETECTION BASED ON LMS ADAPTIVE NOTCH FILTER

Howling detection can be considered as the detection of narrowband (sinusoid) noise from the broadband speech signal and can be achieved by using an adaptive notch filter [5]. Several forms of the adaptive notch filter can be found in previous papers [4][5]. In this paper, we consider a LMS adaptive notch filter with only two adaptive parameters. The structure of howling detection based on LMS adaptive notch filter is shown in Figure 2, where the howling detector is composed of LMS adaptive filter and 90^0 phase shifter.

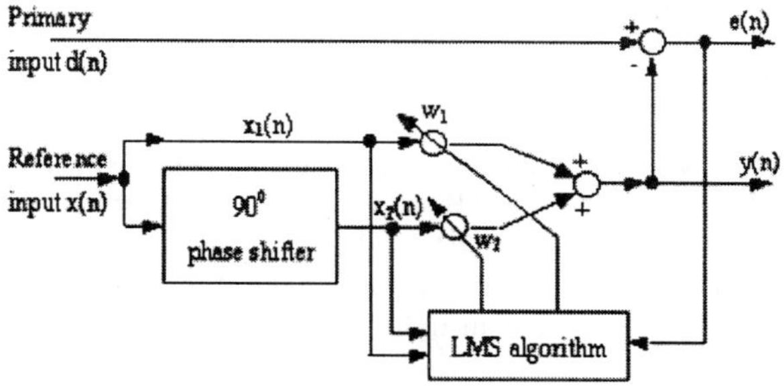

Figure 2. Howling detection based on LMS adaptive notch filter.

One input of the adaptive filter is $x_1(n)$, and after undergoing 90^0 phase shift giving another input $x_2(n)$. The output of the adaptive filter is given by

$$y(n) = w_1(n)x_1(n) + w_2(n)x_2(n) \quad (1)$$

Consequently, the error signal $e(n)$ is given by

$$e(n) = d(n) - y(n) \quad (2)$$

The value of $w_1(n)$ and $w_2(n)$ are chosen using the LMS algorithm, so that the mean squared error $E[e^2(n)]$ is

0-7803-7761-3/03/$17.00 ©2003 IEEE

minimized [1][2]. Thus, the update equation for weights $w_1(n)$ and $w_2(n)$ can be written as

$$w_1(n+1) = w_1(n) + 2\mu e(n)x_1(n) \quad (3)$$

$$w_2(n+1) = w_2(n) + 2\mu e(n)x_2(n) \quad (4)$$

where μ is the step size.

Based on howling signal property mentioned above, we know that, under the howling condition, the frequency of signal in location A, B, C and D shown in Figure 1 are same, however, their phase are different. Consequently, let $A\cos(\omega_0 t + \varphi_a)$ be the signal in location A, then the signal in location B is $B\cos(\omega_0 t + \varphi_b)$. The reference input $x(n)$ of Figure 2 is assumed to be the signal in location A shown in Figure 1, i.e.

$$x(n) = A\cos(\omega_0 t + \varphi_a) \quad (5)$$

Then a Hilbert transform (90^0 phase shifter) is introduced to give the signal $x_2(n)$ which has 90^0 phase difference with $x_1(n)$, i.e.

$$x_1(n) = A\cos(\omega_0 t + \varphi_a) \quad (6)$$

$$x_2(n) = A\sin(\omega_0 t + \varphi_a) \quad (7)$$

The primary input $d(n)$ of Figure 2 is assumed to be the signal in location B shown in Figure 1, i.e.

$$d(n) = B\cos(\omega_0 t + \varphi_b) \quad (8)$$

According to the triangle equations, $B\cos(\omega_0 t + \varphi_b)$ which is included in $d(n)$ can be entirely cancelled under the condition that $w_1(n) = \frac{B}{A}\cos(\varphi_b - \varphi_a)$ and $w_2(n) = -\frac{B}{A}\sin(\varphi_b - \varphi_a)$.

In conclusion, when being howling, the howling signal will come about in both location A and B shown in Figure 1. After a period of adapting, the output $e(n)$ of LMS adaptive notch filter depicted in Figure 2 will become very little, that is to say, the energy of $e(n)$ is far less than that of $d(n)$. In contrast, when no howling presents, the energy of $e(n)$ is almost equal to that of $d(n)$. Hence, the howling signal can be effectively detected as long as a proper threshold is set.

4. HOWLING DETECTION BASED ON LMS ADAPTIVE NOTCH FILTER AND DIGITAL PHASE LOCKED LOOP

In Figure 2, a 90^0 phase shifter can be realized by using a FIR filter which is designed by a few methods [6]. In these cases, the excellent phase shifting effect can only be achieved under the condition that the order of corresponding filter designed is high enough, but its computation cost is too much to bear. In order to reduce computation complexity, digital phase locked loop is introduced to realize the 90^0 phase shifter in this work.

An analog phase locked loop includes three parts [3]: phase detector (PD), loop low pass filter (LPF) and voltage controlled oscillator (VCO). The principle of analog phase locked loop is that: the reference input signal and the output signal of voltage controlled oscillator (VCO) is compared in phase detector (PD), in turn, the phase detector (PD) feeds back an error voltage signal in proportion to the phase difference. Then loop low pass filter (LPF) selects the voltage element which changes slowly, uses it to control voltage controlled oscillator (VCO), and makes its output frequency is as same as the reference input signal's. The math model of an analog phase locked loop is shown in Figure 3.

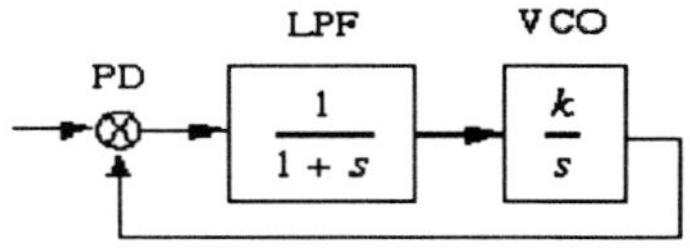

Figure 3. The math model of analog phase locked loop.

It is easy to realize a digital phase locked loop based on this math model. The details of digital phase locked loop are as follows:

Phase detector (PD) can be realized by a simple multiplier, i.e.

$$\mathrm{pd}(n) = \mathrm{ref}(n)\cos[2\pi\varphi(n)] \quad (9)$$

where $\mathrm{ref}(n)$ is the reference input signal at time n, $\mathrm{pd}(n)$ is the input signal of loop low pass filter at time n and $\varphi(n)$ is the phase of VCO output signal at time n.

Loop low pass filter (LPF) can be realized by a digital low pass filter which is designed properly, i.e.

$$\mathrm{lpf}(n) = (1-\alpha)\mathrm{lpf}(n-1) + \alpha\,\mathrm{pd}(n) \quad (10)$$

where $\mathrm{lpf}(n)$ is the output signal of loop low pass filter at time n and α is the parameter that can decide the frequency property of loop low pass filter ($0 < \alpha < 1$).

It is little difficult to realize the integral section of voltage controlled oscillator (VCO). As depicted in Figure 3, this section has the property of low pass filtering. Consequently, it can be realized using a low pass filter, i.e.

$$\varphi(n+1) = \varphi(n) + \frac{f_{VCO}(n) + k\,\mathrm{lpf}(n)}{f_s} \quad (11)$$

$$f_{VCO}(n+1) = (1-\beta)f_{VCO}(n) + \beta\,\mathrm{lpf}(n) \quad (12)$$

where $f_{VCO}(n)$ is the frequency of VCO at time n, k is the parameter that can decide the change rate of VCO, f_s is the sampling frequency and β is the parameter that can decide the frequency property of integral section of VCO ($0 < \beta < 1$).

As shown in equation (9), when $\varphi(n) = 1/4$ (i.e. the phase difference between the reference input signal and the output signal of voltage controlled oscillator is 90^0), the input signal $\mathrm{pd}(n)$ of loop low pass filter at time n is minimized, that is to say, digital phase locked loop is locked. Hence, digital phase locked loop can be used to realize the 90^0 phase shifter.

The signal flow graph of digital phase locked loop is depicted in Figure 4.

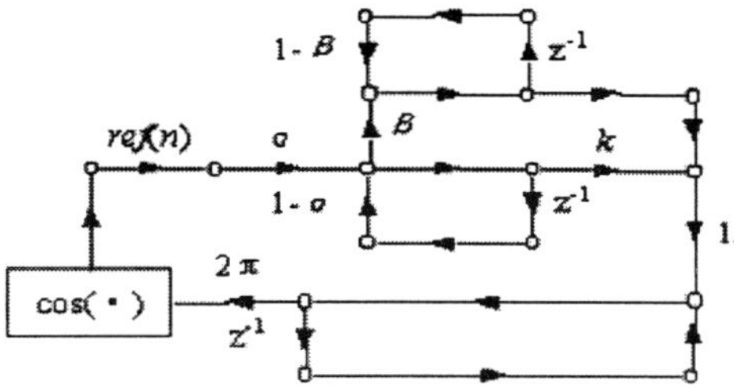

Figure 4. The signal flow graph of digital phase locked loop.

5. SIMULATION RESULTS

To verify the effectiveness of the proposed algorithm, we acquire some howling signals by computer sound card, where sample rate f_s = 8 kHz and A/D conversion is 16 bits. For the LMS adaptive notch filter, we choose initial weights $w_1(0) = 0.0$, $w_2(0) = 0.0$ and step size $\mu = 0.25$. While for digital phase locked loop, we use $\alpha = 2^{-11}$ and $\beta = k = 2^{-9}$. After about 20 samples, the adaptive notch filter shown in Figure 2 converges. Detect results are depicted in Figure 5, where the transversal axis represents sampling number of input data and the vertical axis represents amplitude of input data. The rectangular direct line indicates detect results (the higher level shows the existence of howling signal). Figure 5(b) is the first 200 samples zoomed from that of Figure 5(a). The float-point computation cost of the proposed algorithm is about 0.4 MIPS. The detect delay is about 0.2 s and it is acceptable to most of practical applications.

6. CONCLUSIONS

In acoustic equipments and communication systems, it gradually becomes one of the basic performance demands to detect the existence of howling signal quickly and exactly with very low computation cost. To detect howling signals, a new algorithm based on LMS adaptive notch filter and digital phase locked loop is presented. Computer simulations illustrate that this method is very effective in detecting howling signal and has the advantage of low computational complexity.

7. REFERENCES

[1] Haykin S. *Adaptive filter theory*. Prentice-Hall, 1993.

[2] Widrow B. and Stearns S. D. *Adaptive signal processing*. Prentice-Hall, 1985.

[3] Hardy J. K., *et al. High frequency circuit design*. Reston Publishing Co., 1979.

[4] Kou S. M. and Chen J. "New adaptive IIR notch filter and its application to howling control in speakerphone system". *Electronics Letters*, vol. 28, no. 8, pp. 764-766, April 1992.

[5] Widrow B., *et al.* "Adaptive noise canceling: principles and applications". *Proc. IEEE*, pp. 1692-1316, December 1975.

[6] Oppenheim A. V., Schafer R. W. and Buck J. R. *Discrete-time signal processing (2ed.)*. Prentice-Hall, 1999.

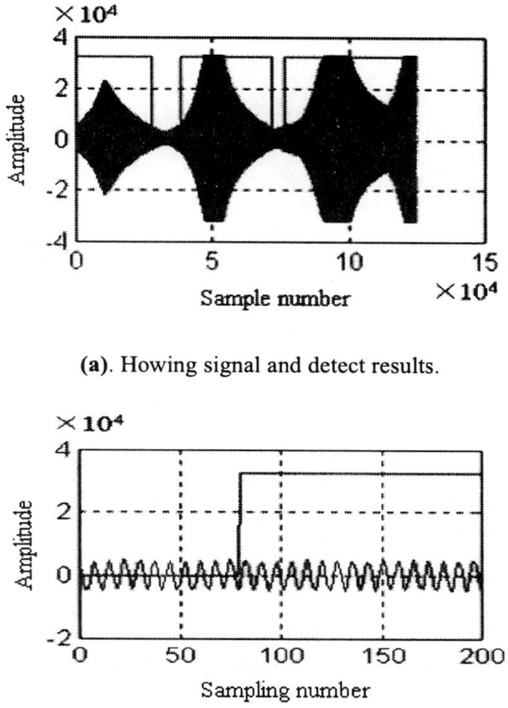

(a). Howing signal and detect results.

(b). The first 200 samples of Figure 5(a).

Figure 5. Detect results of the proposed algorithm.

Integrated Near-End Acoustic Echo and Noise Reduction Systems

S. M. Kuo[1], D. W. Sun[1], and W. S. Gan[2]

[1]Dept. of Electrical Engineering
Northern Illinois University
DeKalb, IL 60115

[2]School of Elec. & Electronics Engineering
Nanyang Technological University
Singapore (639798)

Abstract – **This paper presents the development of integrated near-end acoustic echo and noise reduction algorithm. The modified frequency-sampling filter (FSF) provides an effective filterbank for splitting signal into equally spaced frequency channels. The center clipper and attenuator are employed at each frequency bin for attenuating near-end acoustic echo and noise. The adaptive clipping threshold and dynamic attenuator value are updated based on the optimized voice-activity detector (VAD). Computer simulations using speech and measured room impulse response show that the near-end acoustic echo and noise are suppressed successfully.**

I. INTRODUCTION

The wide spread use of cellular phones has significantly increased the use of communication devices in high noise environments. Intense background noise, however, often corrupts speech, degrading the performance of communication systems. Existing digital signal processing techniques such as speech coding, automatic speech recognition, speaker identification, channel transmission, and echo cancellation are usually developed under noise-free assumption. These techniques could be employed in noisy environments if a front-end noise reduction algorithm can sufficiently reduce an additive noise. The noise reduction is becoming increasingly important with the development of hands-free and voice-activated cellular phones.

Traditional acoustic echo cancellation deals with the acoustic echo generated by the far-end talker, which is broadcasted in a room using speakerphones or hands-free phones. In general, a room or a vehicle compartment is a reverberated chamber. When the near-end talker inside the room talks, far-end listeners located at a remote location will receive not only the desired direct path energy, but also many delayed replicas with varying amplitudes, the so-called reverberation. This undesired effect makes the near-end speech sound hollow. Depending on the microphone location, the energy of these near-end acoustic echoes may be large enough to degrade the intelligibility of speech. Near-end acoustic echo reduction is critical for teleconferencing, multimedia, public addressing, mobiles telephony, audio system correction, recorded signal quality improvement, speech recognition, and many other voice communications applications. In this paper, we focus on acoustic echoes produced by the near-end talker. This is different from the traditional acoustic echo cancellation, which focuses on the acoustic echo produced by the far-end talker through the loudspeaker in the room.

Signal processing techniques [1] can be applied to reduce acoustic noise and reverberation picked up by the microphone. Acoustic noise reduction and de-reverberation can be classified as speech enhancement problems. However, previous research works in this field treats these two problems separately. To solve these problems effectively and provide a quality that is sufficient for telecommunications, combined reduction of these disturbances is required.

The de-reverberation method developed by Allen [2] was based on voice production model. In this method, the reverberated speech was first analyzed to estimate the linear predictive coding parameters of clean speech, and then the de-reverberated speech was reconstructed. In reference [3], the adaptive algorithms in conjunction with other techniques were used in improving the intelligibility of corrupted speech in the reverberated environment. Based on the similarity of smoothed spectral magnitudes of measured room impulse responses, Cole and Moody proposed the spectral-subtraction-type method for the enhancement of reverberated speech [4]. The idea of using spectral subtraction is adapted and improved by using center clipper in this paper for reducing reverberation.

There are four speech enhancement techniques; each has its own set of assumptions, advantages, and limitations [5]. The first class of technique is based on the short-term spectrum. These techniques suppress noise by subtracting the noise spectrum estimated during non-speech activity. The second class of technique is based on speech modeling using iterative methods. These systems estimate speech parameters based on autoregressive or autoregressive-moving-average models, followed by re-synthesis of the noise-free speech using the non-causal Wiener filtering. The third class of system is based on adaptive noise canceling (ANC) using a dual-channel system with the least-mean-square algorithm. The last area of speech enhancement is based on the periodicity of voiced speech. These methods

0-7803-7761-3/03/$17.00 ©2003 IEEE

employ fundamental frequency tracking using either single-channel ANC or adaptive comb filtering of the harmonic noise. In this research, we modified spectral subtraction algorithm [6] by estimating the short-term spectral magnitude of the noise and noisy speech spectra through the FSF instead of using fast Fourier transform.

The objective of this paper is to develop and integrate digital signal processing techniques for reducing acoustic noise and near-end acoustic echoes. The algorithm consists of four elements: a frequency-sampling filter, a voice activity detector, a dynamic attenuator, and an adaptive center clipper. Attenuator reduces the noise and center clipper removes the reverberated echoes. These two operations are performed at every frequency channel formed by the frequency-sampling filter [7].

II. FREQUENCY-SAMPLING FILTER

The FSF is based on sampling a desired amplitude spectrum and obtaining filter coefficients. In this approach, the desired frequency response $H(\omega)$ is first uniformly sampled at L equally spaced points $\omega_k = 2\pi k / L$, $k = 0, 1, \ldots, L-1$. The frequency sampling technique is particularly useful when several bandpass functions are desired simultaneously. Another unique attraction of the frequency sampling method is that it allows recursive implementation of filters, leading to computationally efficient algorithms.

The transfer function of FSF can be expressed as [1]

$$H(z) = \frac{2}{L}\left(1 - r^L z^{-L}\right) \sum_{k<L/2} H_k \frac{1 - r\cos(2\pi k / L) z^{-1}}{1 - 2r\cos(2\pi k / L) z^{-1} + r^2 z^{-2}}, \qquad (1)$$

where r is the pole radius that is slightly less than one, and H_k is the gain at frequency channel k. As illustrated in Fig. 1, the FSF can be realized by cascading a comb filter with a bank of second-order resonators.

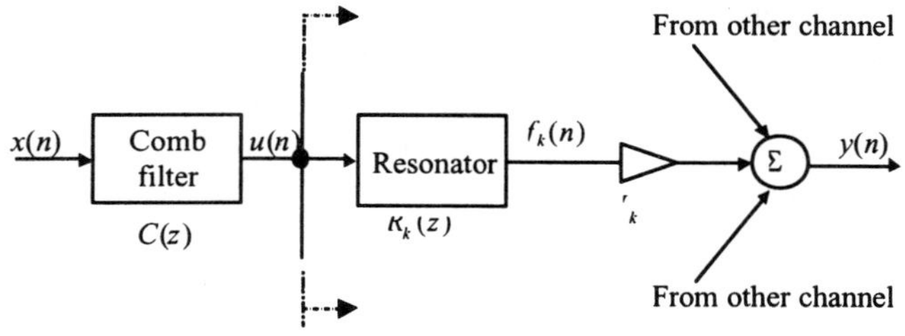

Fig. 1 Block diagram of FSF structure, channel k.

In the FSF structure shown in Fig. 1, each resonator effectively acts as a narrowband filter and passes only the frequencies centered at and close to the resonant frequency. By choosing the coefficient H_k according to given specifications, a time-domain filter with an arbitrary magnitude response can be obtained.

III. VOICE ACTIVITY DETECTOR

Near-end acoustic echo and noise suppression algorithms based on the FSF have a unique way to distinguish the speech and non-speech by evaluating the power in some specific channels. It is well known that voiced speech has considerable energy at the first formant (250-800 Hz) and a majority of noises do not have strong resonant frequencies in this region. Based on this principle, an optimized VAD using an adaptive noise floor threshold scheme is developed in this paper.

The decision whether a given signal is considered as speech or non-speech is made based on the power in the first formant region. A dynamic noise floor estimate $\hat{P}_f(n)$ used for estimating the background noise level is updated using a very long window (4 seconds) during the speech periods, but a median window (32 ms) is used for the non-speech periods. An adaptive threshold $T(n)$ for the speech/non-speech decision is based on the time-varying noise floor estimate. A decision that the speech signal is present is made if the power estimate $\hat{P}_s(n)$ exceeds the adaptive threshold $T(n)$.

IV. ACOUSTIC NOISE ATTENUATION

The FSF structures discussed in Section II can be used to replace FFT in the speech enhancement technique based on spectral subtraction. In this case the input signal is split into K channels by the FSF instead of fast Fourier transform. Speech enhancement algorithm is applied to determine the dynamic attenuator value $H_k(n)$ based on the power of noise and noisy speech, and these processed channel signals are processed by the center clipper for removing reverberation, and then recombined to form the overall output signal.

The block diagram of the FSF-based speech enhancement system is illustrated in Fig. 2. The noise and noisy signal levels are individually estimated in each band based on VAD output. The noise reduction (NR) algorithm is identical at each channel and is summarized as follows:

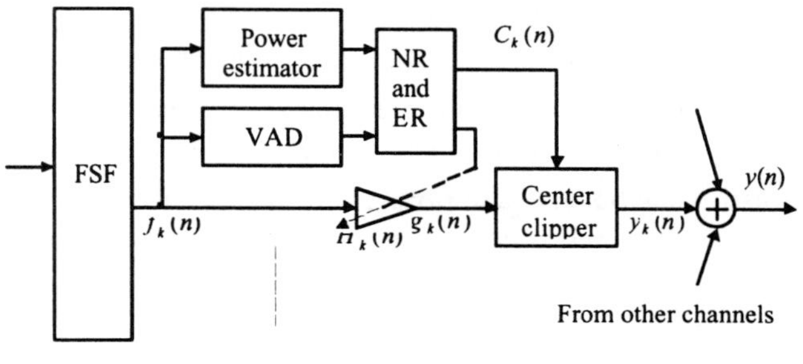

Fig. 2 Integrated acoustic noise and near-end echoes reduction system using the FSF, channel k.

1. Estimate the power of noisy speech signal $\hat{P}_{f,k}(n)$ for speech segments at each channel as follows:

$$\hat{P}_{f,k}(n) = (1-\alpha)\hat{P}_{f,k}(n-1) + \alpha f_k^2(n), \quad (2)$$

where $0<\alpha<1$.

2. Estimate the power of noise for noise segments at each channel as follows:

$$\hat{P}_{V,k}(n) = (1-\beta)\hat{P}_{V,k}(n-1) + \alpha f_k^2(n), \quad (3)$$

where $0<\beta<1$.

3. Compute dynamic attenuator value $H_k(n)$ for each channel k using the principle of spectral subtraction algorithm expressed as follows:

$$H_k(n) = 1 - \frac{\hat{P}_{V,k}(n)}{\hat{P}_{f,k}(n)}. \quad (4)$$

4. Attenuate noise at each channel as follows:

$$g_k(n) = \hat{f}_k(n)H_k(n). \quad (5)$$

This channel output is then used by the center clipper for further reducing near-end acoustic echoes.

V. NEAR-END ACOUSTIC ECHO REDUCTION

For near-end acoustic echo reduction, the center clipper after dynamic attenuator shown in Fig. 2 is an effective technique to suppress undesired echo components. In general, there are two different kinds of center clippers. Center clipper A chops the signal components of the near-zero region (central region) and reduces the signal amplitude of non-central region with the desired amplitude. Center clipper B leaves the non-central region unchanged. The most popular clipper B can be expressed as

$$y_k(n) = \begin{cases} 0, & |g_k(n)| \le C_k(n) \\ g_k(n), & |g_k(n)| > C_k(n) \end{cases}, \quad (6)$$

where $C_k(n)$ is the adaptive clipping level. This center clipper completely eliminates signals below the clipping level, but leaves instantaneous signal values greater than the clipping level unaffected. Thus large signals go through unchanged but small signals are eliminated. Since small signals are consistent with reverberated elements, the clipper achieves the function of reducing near-end acoustic echoes. The threshold parameter, $C_k(n)$, determines how "choppy" the speech will sound with respect to the reverberation level. A large value of $C_k(n)$ suppresses the entire near-end acoustic echo but also deteriorates the quality of the speech.

In the proposed echo reduction (ER) algorithm illustrated in Fig. 2, a center clipper for reducing near-end acoustic echo follows the time-varying attenuator $H_k(n)$ for attenuating acoustic noise. The adaptive clipping threshold is defined by

$$C_k(n) = \beta(n) \times \hat{P}_{f,k}(n), \quad (7)$$

where $\beta(n)$ is a scaling factor, and $\hat{P}_{f,k}(n)$ is power estimation at each channel of the FSF, which is implemented by an effective moving-average technique.

Because reverberate signal components are located at the tail portions of voice samples, $C_k(n)$ can be further optimized by updating $\beta(n)$ according to VAD result. The parameter $\beta(n)$ takes larger value when a silence sample is detected and takes smaller value when a voice sample is presented.

VI. SIMULATION RESULTS

For simulation purpose, a room impulse response was measured in a conference room using the maximum-length sequence technique. The dean speech is then convoluted with the measured room impulse response to simulate the signal picks up by a microphone in that conference room. In this way, we have the original speech to evaluate the success of near-end acoustic echo reduction. The reverberated signal consists of two major components: the desired speech directly from the near-end talker and the undesired near-end echoes reflected from the walls and floor. As shown in Fig. 3, the shaded line (yellow) represents the reverberated speech; and the dark line (blue) represents the original speech.

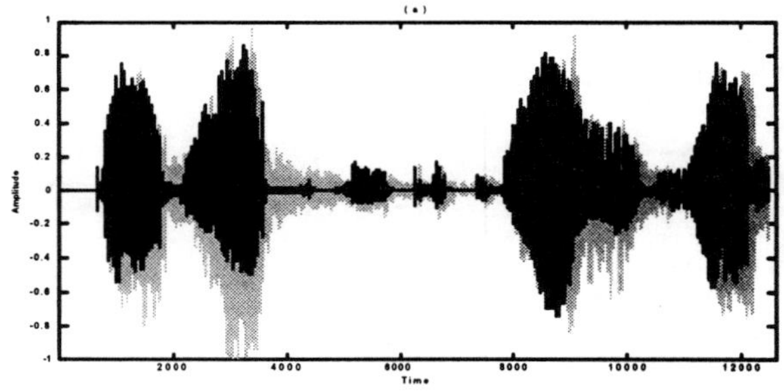

Fig. 3 Original speech (blue dark line) and the reverberated speech (yellow line).

The technique given in Section V was used to reduce the near-end acoustic echoes embedded in the desired speech. The FSF with $L = 256$ and $r = 0.995$ was used for simulations. The result is shown in Fig. 4, where the reverberated speech is shown as shaded line (yellow) and the dark line (blue) shows the near-end acoustic echoes were reduced. Comparing the de-reverberated signal (blue line in Fig. 4) with the original signal (blue line in Fig. 3), we showed that the near-end acoustic echo was significantly reduced, and the nonlinear distortion caused by the center clipper was minimized.

The performance of near-end acoustic echo reduction can be further evaluated in frequency domain. Figure 5 shows the comparison of the spectrograms of the original speech (a), the reverberated speech (b), and de-reverberated speech (c). Figure 5(b) shows the near-end acoustic echoes blurred the gaps between the original speeches shown in Fig. 5(a). Figure

5(c) showed the center clipper effectively cleans up the blurred regions. Comparing the playback of original, reverberated, and processed signals shows that reverberated speech is suppressed significantly without noticeable distortion.

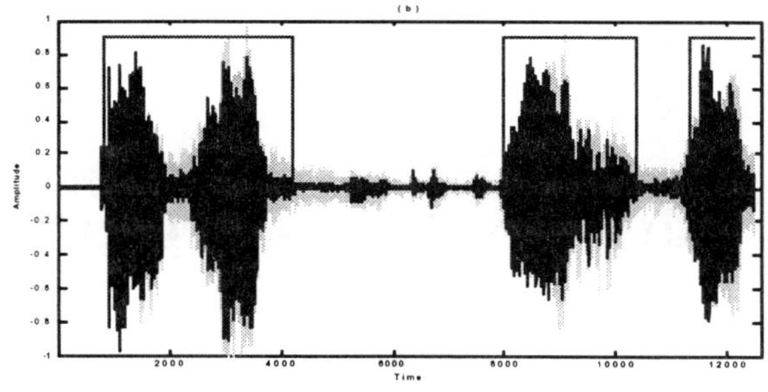

Fig. 4 De-reverberated speech (blue line) and reverberated speech (yellow line).

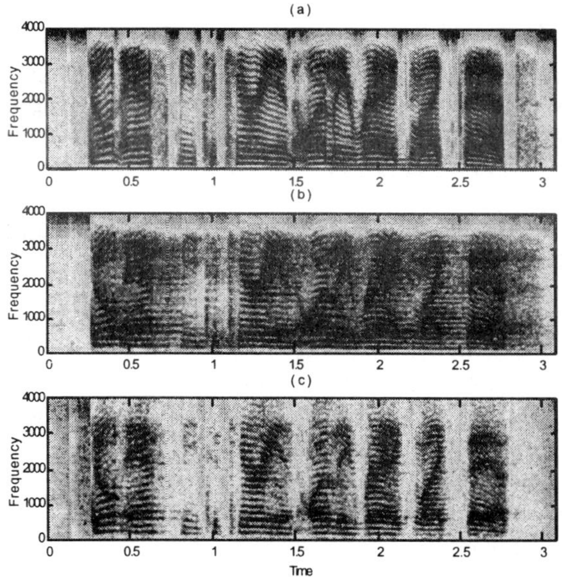

Fig. 5 Spectrograms of the (a) original, (b) reverberated, and (c) de-reverberated speech using center clipper

For evaluating the performance of acoustic noise reduction introduced in Section IV, the noisy speech was recorded in a machinery room. As shown in the top graph of Fig. 6, the desired speech is embedded in broadband noise. The acoustic noise reduction algorithm was used to reduce the undesired noise. The enhanced speech is shown at the bottom plot of Fig. 6. Comparing the top and bottom waveforms, especially during the non-speech periods, we show that the algorithm able to reduce broadband noise significantly. Subjective evaluation by listening the original noisy speech and the enhanced speech shows the distortion of speech quality is very low.

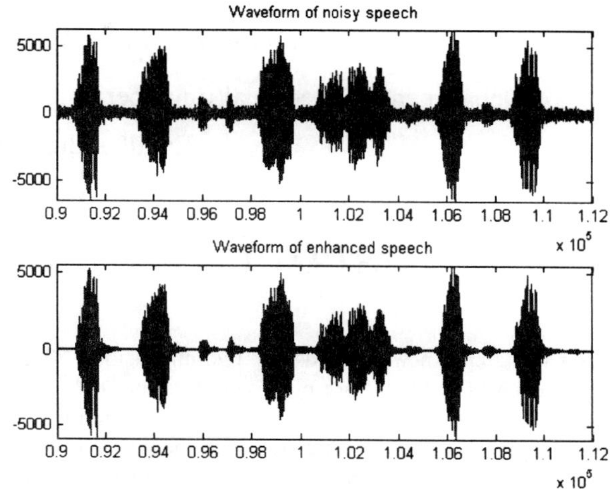

Fig. 6 Performance of acoustic noise reduction. The original noisy speech is shown on the top graph, and the enhanced signal is displayed at the bottom graph.

REFERENCES:

[1] S. M. Kuo and B. H. Lee, *Real-Time Digital Signal Processing*, John Wiley, New York, 2001.

[2] J. B. Allen, "Synthesis of Pure Speech from a Reverberant Signal," US Patent No. 311,731, Jan. 15, 1974.

[3] P. W. Shields, Douglas R. Campbell, "Intelligibility Improvements Obtained by an Enhancement Method Applied to Speech Corrupted by Noise and Reverberation," *Speech Communication,* vol. 25, pp. 165-175, 1998.

[4] D. Cole, M. Moody, "Position-Independent Enhancement of Reverberant Speech," *J. Audio Eng. Soc.*, vol. 45, pp. 142-146, March 1997.

[5] J. R. Deller, JR., J. G. Proakis, and J. H. L. Hanson, *Discrete-Time Processing of Speech Signals*, Macmillan, Englewood Cliffs, New Jersey, 1993.

[6] S. F. Boll, "Suppression of acoustic noise in speech using spectral subtraction," *IEEE Trans. Acoustics, Speech, and Signal Processing*, vol. ASSP-27, pp. 113-120, April 1979.

[7] C. P. Liu, *Digital Signal Processing Technique for Audio Applications*, Master Thesis, Northern Illinois University, August 1998.

[8] D. W. Sun, *Speech De-reverberation by Digital Signal Processing Techniques*, Master Thesis, Northern Illinois University, August 200

SIMPLIFIED STRUCTURES FOR TWO-DIMENSIONAL ADAPTIVE NOTCH FILTERS

Soo-Chang Pei, Chang-Long Wu, and Jian-Jiun Ding

Department of Electrical Engineering, National Taiwan University, Taipei, Taiwan, R.O.C.

Email address: pei@cc.ee.ntu.edu.tw

ABSTRACT

In this paper, three new methods for two-dimensional (2-D) adaptive notch filter (ANF) are proposed. They not only contain similar and simpler structures but also can be derived by simpler adaptive algorithms compared with the previous ones. They are helpful for 2-D signal processing. Simulation results show that they can eliminate 2-D sinusoidal signals successfully.

1. INTRODUCTION

The Adaptive notch filter (ANF) is used to remove narrowband or sinusoidal interference embedded in broadband signal. During these years, many structures and methods for one-dimensional adaptive notch filters have been proposed [1][2]. For two-dimensional case, however, there haven't been many articles talking about it. Although [3]-[5] proposed several techniques for designing 2-D notch filters, and, a method was proposed in [6] for 2-D adaptive notch filter recently, their structures were not simple enough.

Here, we propose two types of structures for 2-D adaptive notch filters whose complexities and orders were lower than the ones in the former workings and have satisfying performances when tracking 2-D sinusoidal signals. First type is based on the outer product expansion [5][6] but with a few changes so as to cancel the redundant part and simplify the structure of the 2-D notch filter (NF). The other type simply was derived from the frequency-shift method, which moves a notch at zero frequency toward the desired frequencies, respectively. The NFs coming from these two methods have lower orders, and we can find new 2-D ANFs based on the proposed NFs and Burg algorithm [6][7].

This paper is organized as follows. First we express our new 2-D notch filter structures. Then, based on it, the adaptive algorithm is proposed. Finally, simulations are given to evaluate their performances.

2. TWO-DIMENSIONAL NOTCH FILTERS

The frequency response of an ideal real-coefficient 2-D NF is

$$H_n(e^{j\omega_1}, e^{j\omega_2}) = \begin{cases} 0, & (\omega_1,\omega_2) = (\omega_1',\omega_2'). \\ 0, & (\omega_1,\omega_2) = (-\omega_1',-\omega_2'). \\ 1, & otherwise. \end{cases} \quad (1)$$

where (ω_1',ω_2') is the notch frequency, and for real 2-D NF, there is another notch at $(-\omega_1',-\omega_2')$.

The main method to construct the two-dimensional notch filter (2-D NF) $H_d(z_1,z_2)$ is subtracting the two-dimensional diagonal bandpass filter (BPF) $H_c(z_1,z_2)$ from (1):

$$H_n(z_1,z_2) = 1 - H_b(z_1,z_2) \quad (2)$$

In [5], the outer product expansion method for finding the 2-D BPF is given by

$$H_b(z_1,z_2) = \frac{1}{2}H_{c1}(z_1)H_{c2}(z_2) - \frac{1}{2}H_{s1}(z_1)H_{s2}(z_2) \quad (3)$$

where $H_{ci}(z_i)$ and $H_{si}(z_i)$ are 1-D filters:

$$H_{ci}(e^{j\omega_1}) = \begin{cases} 1, & \omega_i = \omega_i' or - \omega_i'. \\ 0, & otherwise. \end{cases} \quad (4)$$

and

$$H_{si}(e^{j\omega_2}) = \begin{cases} -j, & \omega_i = \omega_i'. \\ j, & \omega_i = -\omega_i'. \\ 0, & otherwise. \end{cases} \quad (5)$$

Furthermore, if we replace $H_{si}(z_i)$ by

$$H_{si}(z_i) = H_{ci}(z_i)H_{ai}(z_i) \quad (6)$$

where

$$H_{ai}(e^{j\omega_i}) = \begin{cases} -j, & \omega_i = \omega_i'. \\ j, & \omega_i = -\omega_i'. \\ anyvalue, & otherwise. \end{cases} \quad (7)$$

The 2-D NFs obtained from Eqs. (2), (3) and (6) are:

$$H_n(z_1,z_2) = 1 - \frac{1}{2}H_{c1}(z_1)\cdot H_{c2}(z_2)\cdot(1 - H_{a1}(z_1)\cdot H_{a2}(z_1)) \quad (8)$$

0-7803-7761-3/03/$17.00 ©2003 IEEE

where $H_{ci}(z_i)$'s are 1-D 2nd-order BPFs and $H_{ai}(z_i)$'s are 1-D 1st-order allpass filters (APFs).

2.1 Two-dimensional NF from simplified band-pass filters

The following is the derivation of our type-1 NF. First consider the 1-D BPF (based on APF) in Eq.(8) to be

$$H_{ci}(z_i)=\frac{1}{2}\left(1-\frac{r_i^2-(1+r_i^2)\cos(\omega_i')z_i^{-1}+z_i^{-2}}{1-(1+r_i^2)\cos(\omega_i')z_i^{-1}+r_i^2z_i^{-2}}\right) \qquad (9)$$

$$=\frac{1-r_i^2}{2}\frac{1-z_i^{-2}}{1-(1+r_i^2)\cos(\omega_i')z_i^{-1}+r_i^2z_i^{-2}}$$

where $0<r_i<1$ is the pole radius. Then assume the corresponding $H_{si}(z_i)$'s in Eq.(3) to be

$$H_{si}(z_i)=\frac{1-r_i^2}{2}\frac{X_i(z_i)}{1-(1+r_i^2)\cos(\omega_i')z_i^{-1}+r_i^2z_i^{-2}} \qquad (10)$$

From previous discussion, we know if

$$H_{ai}(e^{j\omega_i})=\left[\frac{X_i(z_i)}{1-z_i^{-2}}\right]_{z_i=e^{j\omega_i}}=-j \qquad (11)$$

is satisfied, the $H_{si}(z_i)$'s in Eq.(10) will satisfy the desired frequency response in Eq.(5). After simple manipulations, we get

$$X_i(z_i)=2z_i^{-1}\sin(\omega_i') \qquad (12)$$

Therefore

$$H_{si}(z_i)=\frac{1-r_i^2}{2}\frac{2\sin(\omega_i')z_i^{-1}}{1-(1+r_i^2)\cos(\omega_i')z_i^{-1}+r_i^2z_i^{-2}} \qquad (13)$$

It is observed that Eq.(9) and (13) have the same denominators, and we can construct a new 2-D NF from Eq.(3), (10) and (13). Here we call this type of 2-D NFs *NF1*, and its block diagram is presented in Fig. 1 with a simpler structure than the conventional ones [4][5][6].

2.2 Two-dimensional NFs by frequency-shift method

Here we introduce a simpler method to find the 2-D NF. First we consider a 2-D BPF denoted by $H_{b,0}(z_1, z_2)$ with only a spike at origin, that is, $(\omega'_1, \omega'_2) = (0, 0)$. Next, we shift the spike of $H_{b,0}(z_1, z_2)$ from (0, 0) to (ω'_1, ω'_2) or $(-\omega'_1, -\omega'_2)$ respectively, which means generating the 2-D BPF from:

$$H_b(z_1,z_2)=\left[H_{b,0}(z_1,z_2)\Big|_{\substack{z_1=z_1\cdot e^{j\omega_1'}\\ z_2=z_2\cdot e^{j\omega_2'}}}+H_{b,0}(z_1,z_2)\Big|_{\substack{z_1=z_1\cdot e^{-j\omega_1'}\\ z_2=z_2\cdot e^{-j\omega_2'}}}\right] \qquad (14)$$

Based on this method, we can found two 2-D BPFs $H_{b,0}(z_1, z_2)$ by applying different $H_{b,0}(z_1, z_2)$ and construct the 2-D NFs by Eq.(2).

First, for

$$H_{b,0}(k)=\left(1-\frac{1-z_1^{-1}}{1-r_1z_1^{-1}}\right)\left(1-\frac{1-z_2^{-1}}{1-r_2z_{22}^{-1}}\right) \qquad (15)$$

we get the 2-D BPF

$$H_b(z_1,z_2)=\frac{2(1-r_1)(1-r_2)z_1^{-1}z_2^{-1}(\cos(\omega_1'+\omega_2')-r_1\cos\omega_2'z_1^{-1}-r_2\cos\omega_1'z_2^{-1}+r_1r_2z_1^{-1}z_2^{-1})}{(1-2r_1\cos\omega_1'z_1^{-1}+r_1^2z_1^{-2})(1-2r_2\cos\omega_2'z_2^{-1}+r_2^2z_2^{-2})} \qquad (16)$$

Form Eq.(2), we get a new 2-D NF (*NF2*). Next, for

$$H_{b,0}(z_1,z_2)=\frac{1}{2}\left(1-\frac{r_1-z_1^{-1}}{1-r_1z_1^{-1}}\right)\cdot\frac{1}{2}\left(1-\frac{1-z_2^{-1}}{1-r_2z_2^{-1}}\right) \qquad (17)$$

we get another one

$$H_b(z_1,z_2)=\frac{(1-r_1)(1-r_2)}{2}\cdot\frac{G(z_1,z_2)}{(1-2r_1\cos\omega_1'z_1^{-1}+r_1^2z_1^{-2})(1-2r_2\cos\omega_2'z_2^{-1}+r_2^2z_2^{-2})} \qquad (18)$$

where

$$\begin{aligned}G(z_1,z_2)&=1+(1-r_1)\cos\omega_1'z_1^{-1}+(1-r_2)\cos\omega_2'z_2^{-1}\\&+[(1+r_1r_2)\cos(\omega_1'+\omega_2')-(r_1+r_2)\cos(\omega_1'-\omega_2')]z_1^{-1}z_2^{-1}-r_1z_1^{-2}-r_2z_2^{-2}\\&+r_1(r_2-1)\cos\omega_2'z_1^{-2}z_2^{-1}+r_2(r_1-1)\cos\omega_1'z_2^{-2}z_1^{-1}+r_1r_2z_1^{-2}z_2^{-2}\end{aligned} \qquad (19)$$

Similarly, we can get another NF (*NF3*) from Eq.(2).

These two NFs have the same order as NF1 and similar denominators. Their block diagram is depicted in Fig. 2 with $F(z_1,z_2) = 4\cdot z_1^{-1}z_2^{-1}(\cos(\omega_1+\omega_2)-r_1\cos\omega_2 z_1^{-1}-r_2\cos\omega_1 z_2^{-1}+r_1r_2z_1^{-1}z_2^{-1})$ for NF2 and $F(z_1,z_2) = G(z_1,z_2)$ for NF3.

3. TWO-DIMENSIONAL ADAPTIVE NOTCH FILTERS

Since the structures of the NFs proposed in the previous section are much similar to the one proposed in [6], we apply the technique used in the 2-D ANF in [6] here.

First, to guarantee the stability, we implement the denominators of the proposed NFs in lattice form shown in Fig. 3 where the coefficients ($i = 1,2$)

$$a_{0i}(k)=-\cos(\hat{\omega}_i(k)),\quad a_{1i}(k)=r_i^2 \qquad (20)$$

for NF1 and

$$a_{0i}(k)=-\cos(\hat{\omega}_i(k))\cdot\frac{2\cdot a_{1i}(k)}{1+a_{1i}^2(k)},\quad a_{1i}(k)=r_i^2 \qquad (21)$$

for NF2 and NF3.

Then the Burg algorithm [7] can be used here as well. Because $a_{11}(k)$ and $a_{12}(k)$ are related to the bandwidths merely, here we only have to use the algorithm to adapt $a_{01}(k)$ and $a_{02}(k)$ with little changes in their update equations. The new algorithm is shown as follows:

Initial condition:

$$((1+r_i^2)\cdot|D_i(0)|>2r_i\cdot|C_i(0)|)\qquad i=1,2$$

Main loop:

$$D_1(k+1) = \lambda_1 D_1(k) + 2\varepsilon_1 x_1(m-1,n)^2$$

$$C_1(k+1) = \lambda_1 C_1(k) + \varepsilon_1 x_1(m-1,n)[x_1(m,n) + x_1(m-2,n)]$$

$$a_{01}(k+1) = -\frac{C_1(k+1)}{D_1(k+1)}\frac{2r_1}{1+r_1^2}$$

$$D_2(k+1) = \lambda_2 D_2(k) + 2\varepsilon_2 x_2(m,n-1)^2$$

$$C_2(k+1) = \lambda_2 C_2(k) + \varepsilon_2 x_2(m,n-1)[x_2(m,n) + x_2(m,n-2)]$$

$$a_{02}(k+1) = -\frac{C_2(k+1)}{D_2(k+1)}\frac{2r_2}{1+r_2^2}$$

$$a_{0i}(k+1) = \begin{cases} 1 & , a_{0i}(k+1) > 1 \\ a_{0i}(k+1) & , |a_{0i}(k+1)| \le 1 \\ -1 & , a_{0i}(k+1) < -1 \end{cases} \qquad i = 1,2$$

$$\hat{\omega}_i(k+1) = \cos^{-1}(-a_{0i}(k+1)) \qquad i = 1,2$$

$$0 < \lambda_i < 1 \text{ (forgetting factor)}, \quad \varepsilon_i = 1 - \lambda_i \quad i = 1,2$$

By the above algorithm, we can construct *ANF1*, *ANF2* and *ANF3*, which represent the adaptive notch filters based on NF1, NF2 and NF3, respectively.

4. SIMULATION

Two examples are given in this chapter. First we use the proposed ANF1, ANF2 and ANF3 to eliminate 2-D sinusoidal signals embedded in additive white Gaussian noise (AWGN) and compare the results with the one's given in [6] (*ANF0*). Next, we display an application of the 2-D ANFs by recovering a contaminated image from 2-D sinusoidal interference.

Example1: Estimation of input frequencies

Assume that the input is given by

$$u(m,n) = 20 \cdot \sin(\omega_1' \cdot m + \omega_2' \cdot n) + v(m,n)$$

where $(0,0) \le (m,n) \le (150,150)$ and $v(m,n)$ is AWGN with zero mean and unit variance. In this example, we set the input frequency (normalized by 2π) is (0.20, 0.15). We use the proposed ANF1, ANF2 and ANF3 to eliminate the input signal with the settings pole radiuses: $r_1 = r_2 = 0.95$, the forgetting factors: $\lambda_1 = \lambda_2 = 0.9999$, initial conditions: $\hat{\omega}_1(0) = \hat{\omega}_2(0) = \pi/2$, $D_1(0) = D_2(0) = 0.01$, $C_1(0) = C_2(0) = 0.005$. Then, we compare their adaptive results with ANF0.

The differences between the results are so slight in visual sight that we only present the results of ANF1 in the following. Fig. 4 is the convergence curves of the coefficients where the upper and lower lines represent $\hat{a}_{01}(k)$ and $\hat{a}_{02}(k)$, respectively. The coefficients finally converged to $a_{01} = -0.3086$ and $a_{02} = -0.5870$, and resulting estimated frequencies are $\hat{\omega}_1' = 0.40013534\ \pi$ and $\hat{\omega}_2' = 0.30029418\pi$. Fig. 5(a) is the output where we see the signal along n time-axis decays faster than m time-axis because the red line in Fig. 4 converges faster than the blue one. Fig. 5(b) is the magnitude response. The numerical comparison between these NFs is shown in Table 1 where BIAS denotes the estimation error and STD denotes the standard deviation. We see the performances are quite close.

Example 2: Application to image processing

Assume that the input is given by

$$u(m,n) = 50\sin(0.4\pi m + 0.3\pi n) + v(m,n)$$

where $(0,0) \le (m,n) \le (255,255)$ and $v(m, n)$ is a pixel of original image, as Fig. 6(a). The image degraded by sinusoidal noise with $\omega_1' = 0.4\pi$ and $\omega_2' = 0.3\pi$ is shown in Fig. 6(b). Now let the input to the 2-D adaptive notch filter be $u(m,n) - \bar{u}$ where

$$\bar{u} = \frac{1}{256^2}\sum_{m=0}^{255}\sum_{n=0}^{255} u(m,n).$$

If we assume the output is $y(m,n)$, then $y(m,n) + \bar{u}$ shall be the recovered image.

We set $\lambda_1 = \lambda_2 = 0.9999$, $r_1 = r_2 = 0.95$, and all initial conditions are the same as those in Example 1. Using the proposed ANFs, we can get the recovered image in Fig. 6(c) where the sinusoidal noise is gradually eliminated. The coefficients finally converged to $a_{01} = -0.3101$ and $a_{02} = -0.5870$, and resulting estimated frequencies are $\hat{\omega}_1' = 0.3996521\ \pi$ and $\hat{\omega}_2' = 0.30030439\pi$. From this example, we see the performance is still good even though the original image is not white.

5. CONCLUSION

In this paper, we proposed three new structures for two-dimensional adaptive notch filter. The three structures were mainly composed of one all-zero and two 1-D all-pole filters cascaded together. They were simpler than pervious structures. When implementing the two all-pole filter with lattice form, we can apply an algorithm similar to the one proposed in [6]. In addition, the estimated notch frequencies were found from the two 1-D filters separately. Simulation result demonstrated the validity of the proposed 2-D adaptive notch filter.

6. REFERENCES

[1] N. I. Cho, C. H. Choi and S. U. Lee, "Adaptive line enhancement by using an IIR lattice Notch filter", IEEE Trans. ASSP, vol.37, no.4, pp.585-589, Aprl. 1989.

[2] P. A. Regalia, "An improved lattice-based adaptive IIR Notch filter", IEEE Trans. SP, vol.39, no.9, pp.2124-2128, Sept. 1991.

[3] S. C. Pei and C. C. Tseng, "Two dimensional IIR digital Notch filter design", IEEE Trans. Circuits Syst. II, vol. 41, no.3, pp.227-231, Mar. 1997.

[4] S. C. Pei and C. C. Tseng, "Corrections to two dimensional IIR digital Notch filter design", IEEE Trans. Circuit Syst. II, vol.41, no.9, pp.630, Sept. 1994.

[5] S. C. Pei, W. S. Lu and C. C. Tseng, "Analytical two-dimensional IIR Notch filter design using outer product expansion", IEEE Trans. Circuits Syst. II, vol.44, no.9, pp.765-768, Sept. 1997.

[6] T. Hinamoto, N. Ikeda, S. Nishimura and A. Doi, "Design of two dimensional adaptive digital Notch filters", Proceedings of ICSP2000, pp.538-542.

[7] N. I. Cho, C. H. Choi and S. U. Lee, "Adaptive line enhancement by using an IIR lattice Notch filter", IEEE Trans. ASSP, vol.37, np.4, pp.585-589, April 1989.

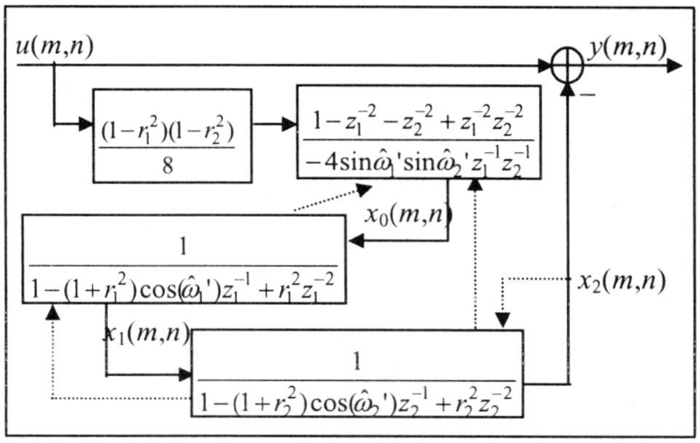

Fig. 1. Block diagram of NF1

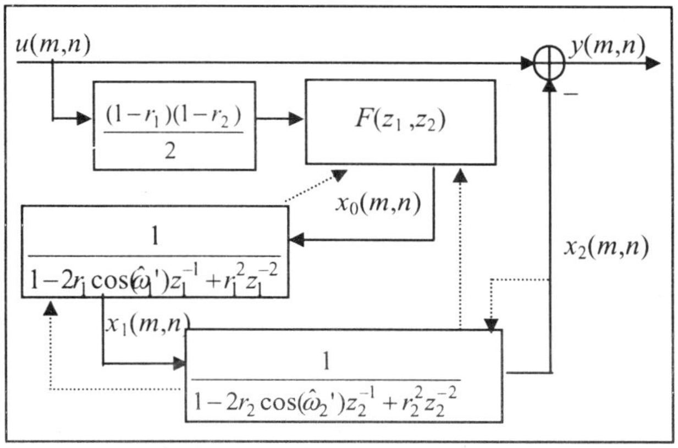

Fig. 2. Block diagram of NF2 or NF3

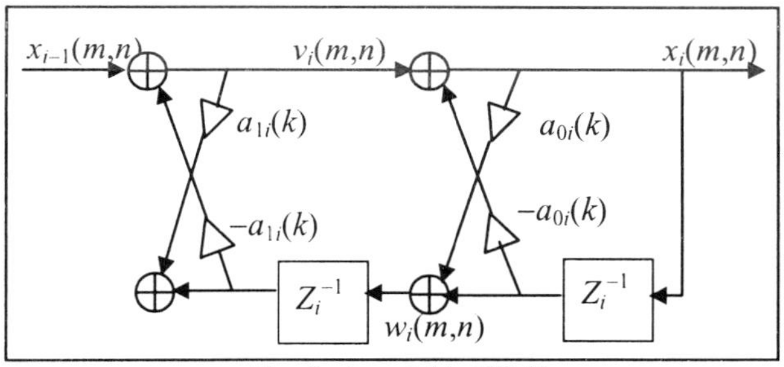

Fig. 3. A variable IIR filter

	BIAS1	BIAS2	STD1	STD2
ANF0	-1.5808e-004	-1.8367e-005	4.6970e-006	2.3783e-006
ANF1	-2.0804e-005	1.4015e-004	3.2487e-006	2.1935e-006
ANF2	-1.2543e-004	-1.2087e-005	2.7791e-006	2.2496e-006
AMF3	-8.9263e-005	-1.1628e-005	3.0736e-006	1.8559e-006

Table 1. Performance Comparison between the ANFs

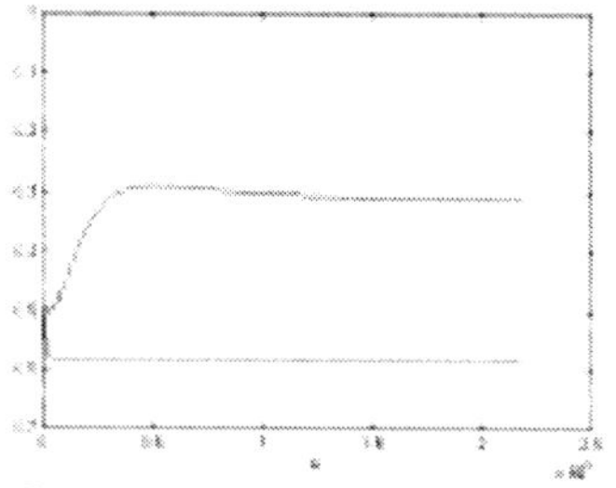

Fig. 4. Convergence of the coefficients of NF1

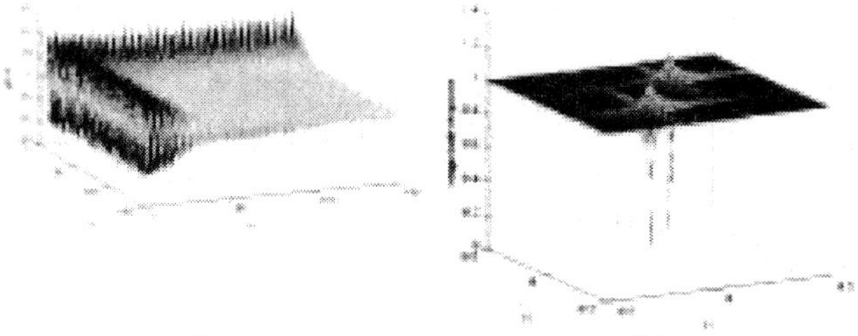

(a) (b)

Fig. 5. (a)The output of NF1 (b) The magnitude response of NF1

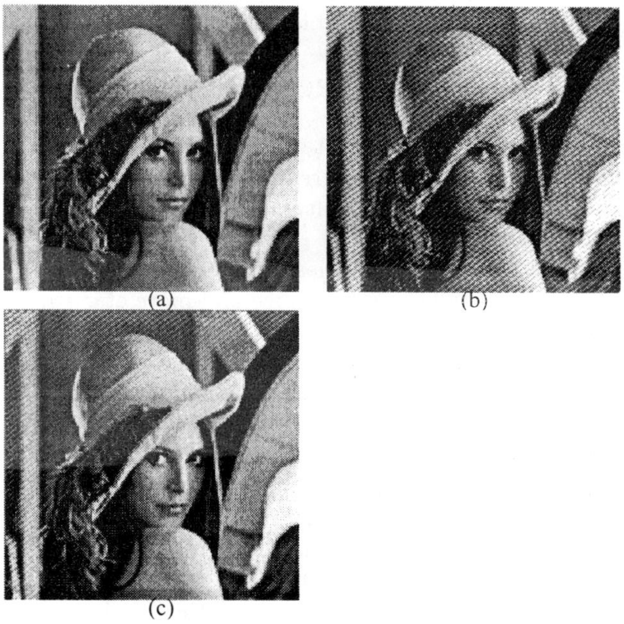

Fig. 6. Application of the 2-D ANF (a) the original image, (b) the degraded image (c) the recovered image.

INPUT BALANCED STATE-SPACE REALIZATION BASED ADAPTIVE RECURSIVE FILTERS

J. Zhou and G. Li

School of EEE
Nanyang Technological University
Singapore 639798

ABSTRACT

It is well known that adaptive infinite impulse response (IIR) filters can offer high performance in many applications and that the conventional direct-form parameterization yields a simple implementation but suffers from potential instability problems and slow convergence. In this paper, based on an input balanced state-space realization a new adaptive IIR filter is derived. Besides other nice properties, this realization is characterized with a parametrization under which the stability of IIR filters is ensured all the time and therefore, like the lattice parametrization based adaptive IIR filters, no stability monitoring is needed. Simulations show that the proposed adaptive IIR filters over-perform those parameterized with the direct-form realization and the normalized lattice parametrization.

1. INTRODUCTION

The use of infinite impulse response (IIR) filters in adaptive filtering has drawn great attention among researchers during the last twenty years or so (see, e.g., [1]-[7]). This is due to the fact that IIR filters have superior system modeling abilities and hence the potential of achieving higher performance and more efficient implementation compared to their FIR counterparts. The main difficulties that prevent the adaptive IIR filters from being used in real applications are the slow convergence and more seriously, the instability during the adaptation process.

The stability issue is strongly related to the parametrization which is used to characterize the IIR filters. It is well known that a time invariant digital filter can be characterized in many different parametrizations and that these parametrizations are theoretically equivalent since they represent the one and the same system. The important point is, however, that different parametrizations have different properties and for a given application, one parametrization can be better than another. Traditionally, the adaptive IIR filters are implemented with the so-called direct-forms which require high computation costs for stability monitoring and have slow convergence and poor numerical properties (see, [4],[5]). The most successful structures in adaptive IIR filter design are those with lattice based parametrizations. See, e.g., [1] - [5]. One of the nice properties of the lattice based parametrizations is that the stability can be easily monitored and hence the corresponding implementation is very simple. It was shown in [3] that the normalized lattice structure is also stable in time-varying environment.

A linear system can be parametrized with its state-space realizations. It seems that there are a few of work reported on design of adaptive recursive filters using state-space realizations [6]. How to take advantage of the degrees of freedom provided by the infinite set of realizations is still an open problem. One of the main problems in designing adaptive recursive state-space filters is that it is usually very difficult to control the stability if the parameters for adaptation are taken directly from the realization since the relationship between stability and those parameters is very complicated. The main objective in this paper is to develop an adaptive filter structure implemented with an input balanced state-space realization. This realization is characterized with a new parametrization under which the stability is automatically guaranteed.

2. INPUT BALANCED STATE-SPACE REALIZATIONS

A nth order IIR filter can be implemented with the following state-space equations:

$$\begin{aligned} x(m+1) &= Ax(m) + Bu(m) \\ y(m) &= Cx(m) + b_0 u(m) \end{aligned} \qquad (1)$$

where $u(m)$ and $y(m)$ are the input and output of the filter, respectively, and $A \in \mathcal{R}^{n\times n}, B \in \mathcal{R}^{n\times 1}, C \in \mathcal{R}^{1\times n}$. $R \triangleq (A, B, C, b_0)$ is called a state-space realiza-

0-7803-7761-3/03/$17.00 ©2003 IEEE

tion of $H(z)$, satisfying $H(z) = b_0 + C(zI - A)^{-1}B$. The realizations are not unique. In fact, $R_c \triangleq (A_c, B_c, C_c, b_0)$ given below is one of them

$$A_c = \begin{pmatrix} -a_1 & -a_2 & \cdots & -a_{n-1} & -a_n \\ 1 & 0 & \cdots & 0 & 0 \\ & & \vdots & & \\ 0 & 0 & \cdots & 1 & 0 \end{pmatrix}$$
$$B_c = (1 \ 0 \ \cdots \ 0)^{\mathcal{T}}, \ C_c = (c_1 \ \ldots \ c_n) \quad (2)$$

with $\mathcal{T}$ denoting the transpose operator. All the realizations can be characterized with

$$A = T^{-1}A_cT, \quad B = T^{-1}B_c, \quad C = C_cT$$

for any non-singular T. R_c is usually refereed as a direct form or controllable realization of $H(z)$.

Let $\theta_d \triangleq \begin{pmatrix} \bar{a}^{\mathcal{T}} & C_c & b_0 \end{pmatrix}^{\mathcal{T}}$, where $\bar{a}$ is the vector contains $\{a_k\}$. Many direct-forms are closely related to this parametrization.

The controllability gramian of realization R is defined as

$$W_c \triangleq \sum_{k=0}^{\infty} A^kBB^{\mathcal{T}}(A^k)^{\mathcal{T}}, \quad (3)$$

which satisfies

$$W_c = AW_cA^{\mathcal{T}} + BB^{\mathcal{T}}. \quad (4)$$

A realization is said input balanced if its controllability gramian is equal to the identity matrix.

2.1. Normalized lattice structure

In [3], it is shown that the normalized lattice filter is equivalent to

$$y(m) = h^{\mathcal{T}} \begin{pmatrix} x(m+1) \\ w(m) \end{pmatrix}$$
$$\begin{pmatrix} x(m+1) \\ w(m) \end{pmatrix} = Q \begin{pmatrix} x(m) \\ u(m) \end{pmatrix}, \quad (5)$$

where $w(m)$ is an intermediate (auxiliary) signal, $x(m)$ is the state vector, and $h \in \mathcal{R}^{(n+1)\times 1}, Q \in \mathcal{R}^{(n+1)\times(n+1)}$, where Q is an orthogonal matrix and has the following structure

$$Q = Q_1Q_2 \cdots Q_k \cdots Q_{n-1}Q_n \quad (6)$$

with Q_k the identity matrix except

$$Q_k(k : k+1, k : k+1) = \begin{pmatrix} -sin\phi_k & cos\phi_k \\ cos\phi_k & sin\phi_k \end{pmatrix}, \ \forall k. \quad (7)$$

It is easy to see that the corresponding state-space realization to (5), denote as $R_l \triangleq (A_l, B_, C_l, b_0)$, is

$$A_l = Q(1:n, 1:n), \ B_l = Q(1:n, n+1)$$
$$C_l = h^{\mathcal{T}}(1:n)A_l + h(n+1)Q(n+1, 1:n). \quad (8)$$

Since Q is orthogonal, it can be shown that the pair (A_l, B_l) satisfies

$$I = A_lA_l^{\mathcal{T}} + B_lB_l^{\mathcal{T}}, \quad (9)$$

which means that R_l is an input balanced realization.

Denote $\theta_l \triangleq \begin{pmatrix} \phi_1 & \cdots & \phi_n & h^{\mathcal{T}} \end{pmatrix}^{\mathcal{T}}$. Clearly, the normalized lattice structure is implemented with an input balanced realization parametrized with θ_l. Since the filter is stable if and only if $|cos\phi_k| > 0, \ \forall k$, the stability can be checked very easily in this parametrization.

2.2. An input balanced state-space realization in a new parametrization

Now, consider the realization $(A_{in}, B_{in}, C_{in}, b_0)$ given below, which was used for digital filter implementation in [8]:

$$A_{in} = (I + \Phi_{in})(I - \Phi_{in})^{-1}, \ B_{in} = \frac{\sqrt{2}}{2}(I + A_{in})K_{in} \quad (10)$$

with C_{in} having no special structure, where Φ_{in} is a zero matrix except

$$\Phi_{in}(n,n) = -\alpha_1, \ \Phi_{in}(k, k+1) = \alpha_{n+1-k}$$
$$\Phi_{in}(k+1, k) = -\alpha_{n+1-k}, \ \forall k$$
$$K_{in} = \begin{pmatrix} 0 & \cdots & 0 & \sqrt{2\alpha_1} \end{pmatrix}^{\mathcal{T}} \quad (11)$$

with all α_k real.

This is an input-balanced realization since based on $\Phi_{in} + \Phi_{in}^{\mathcal{T}} + K_{in}K_{in}^{\mathcal{T}} = \mathbf{0}$, it can be shown that $I = A_{in}A_{in}^{\mathcal{T}} + B_{in}B_{in}^{\mathcal{T}}$. Clearly, this input balanced realization and hence $H(z)$ can be parameterized with

$$\theta_\alpha \triangleq \begin{pmatrix} \alpha_1 & \cdots & \alpha_n & C_{in} & b_0 \end{pmatrix}^{\mathcal{T}} \quad (12)$$

Denote

$$S_{\theta_\alpha} \triangleq \{\theta_\alpha : \alpha_k > 0, \forall k\}. \quad (13)$$

The parameterization θ_α has such a property that the set of all stable transfer functions $H(z)$ (of order n) is one-to-one mapped into the subset S_{θ_α} of the parameter space. See [8]. This implies that S_{θ_α} defined in (13) is complete and compact. Since this subset is characterized with n linear constraints, which are very easy to implement or check. This is one of the important advantages of this parameterization in designing IIR adaptive filters. In [7], this parametrization was proposed for adaptive IIR filtering.

Define

$$\alpha_k = e^{\beta_k},\ \forall k \tag{14}$$

and hence a new parametrization can be obtained

$$\theta_\beta \triangleq (\ \beta_1 \ \cdots \ \beta_n \ \ C_{in} \ \ b_0\ )^{\mathcal{T}}. \tag{15}$$

Clearly, the stable region S_α in parametrization θ_α is one-to-one mapped into the entire parameter space of this new parametrization θ_β. Therefore, any transfer function parametrized with (finite) θ_β is stable without constraint at all! In the next section, based on this new parametrization, an adaptive algorithm will be developed.

3. AN ADAPTIVE IIR ALGORITHM

As mentioned before, a transfer function $H(z)$ can be characterized with many different parameterizations. Let θ be the parameter vector of a parameterization. We use $H(\theta, z)$ to denote the fact that the transfer function is parameterized with this parameterization. Denote $S_\theta \triangleq \{\theta :\ H(\theta, z)\ stable\}$. The error signal is defined as the difference between the reference signal $r(m)$ and $y(m)$, the output of $H(\theta, z)$ excited by $u(m)$:

$$e(m) \triangleq r(m) - y(m) = r(m) - H(\theta, z)u(m). \tag{16}$$

The identification problem is to find out a parameter vector $\theta^0 \in S_\theta$ such that a defined cost function (error surface) is minimized.

There exist many stochastic adaptive schemes in which the parameters can be updated with a given numerical algorithm. Our objective here is to study the behavior of different parametrizations. Therefore, in what follows we will concentrate on the output error scheme in which the recursive Gauss-Newton prediction error (RPE) method is used to update the parameter vector. The corresponding adaptive algorithm for parametrization θ is of the following form

$$\theta(m+1) = \theta(m) + e(m)P(m)\psi(m), \tag{17}$$

where $\psi(m)$ is the gradient vector (regressor) of the error signal: $\psi(m) = -de(m)/d\theta$ evaluated at $\theta = \theta(m)$ and $P(m)$ is a matrix updated with

$$\begin{aligned} P(m) &= \lambda^{-1}(m)[P(m-1) \\ &\quad - \frac{P(m-1)\psi(m)\psi^{\mathcal{T}}(m)P(m-1)}{\lambda(m) + \psi^{\mathcal{T}}(m)P(m-1)\psi(m)}] \\ \lambda(m+1) &= \lambda_\infty - [\lambda_\infty - \lambda(m)]\lambda_0 \end{aligned} \tag{18}$$

with λ_∞ and λ_0 two constants. It is desired that $\theta(\infty) \in S_\theta$.

One can compute $\psi(m) = -de(m)/d\theta = dy(m)/d\theta$ for $\theta = \theta_d,\ \theta_l,\ \theta_\alpha,\ \theta_\beta$ to obtain the corresponding adaptive algorithms. Due to the limited space, only $dy(m)/d\theta_\beta$ is given.

Noting $y(m) = [d + C_{in}(zI - A_{in})^{-1}B_{in}]u(m)$ and $\frac{\partial(zI-A_{in})^{-1}}{\partial p} = (zI - A_{in})^{-1}\frac{\partial A_{in}}{\partial p}(zI - A_{in})^{-1}$, where p is a variable, it follows from (14) that

$$\begin{aligned} \frac{\partial y(m)}{\partial \beta_k} &= \alpha_k C_{in}(zI - A_{in})^{-1}[\frac{\partial A_{in}}{\partial \alpha_k}x(m) \\ &\quad + \frac{\partial B_{in}}{\partial \alpha_k}u(m)],\ \forall k \\ \frac{\partial y(m)}{\partial C_{in}} &= x^{\mathcal{T}}(m),\quad \frac{\partial y(m)}{\partial d} = u(m). \end{aligned} \tag{19}$$

It turns out from (10) that

$$\begin{aligned} \frac{\partial B_{in}}{\partial \alpha_k} &= \frac{\sqrt{2}}{2}[\frac{\partial A_{in}}{\partial \alpha_k}K_{in} + (I + A_{in})\frac{\partial K_{in}}{\partial \alpha_k}] \\ \frac{\partial A_{in}}{\partial \alpha_k} &= (I + A_{in})\frac{\partial \Phi_{in}}{\partial \alpha_k}(I - \Phi_{in})^{-1}, \forall k. \end{aligned} \tag{20}$$

Denote e_k the k-th elementary (column) vector, whose elements are all zero except the k-th one which is 1. All the derivatives can be further specified noting the following

$$\begin{aligned} \frac{\partial \Phi_{in}}{\partial \alpha_k} &= \begin{cases} -e_n e_n^{\mathcal{T}} & k = 1 \\ e_{n+1-k}e_{n+2-k}^{\mathcal{T}} - e_{n+2-k}e_{n+1-k}^{\mathcal{T}} & k > 1 \end{cases} \\ \frac{\partial K_{in}}{\partial \alpha_k} &= \begin{cases} \frac{1}{\sqrt{2\alpha_1}}e_n & k = 1 \\ 0 & k > 1 \end{cases} \end{aligned} \tag{21}$$

With (A, B, C, b_0) replaced by $(A_{in}(m), B_{in}(m), C_{in}(m), b_0(m))$, (1) is an adaptive filter implemented with the input-balanced realization, where $C_{in}(m)$ and $b_0(m))$ can be updated with θ_β directly, while $A_{in}(m)$ and $B_{in}(m)$ are computed with (10) via $\Phi_{in}(m)$ and $K_{in}(m)$, both of which are updated with $\{\alpha_k\}$ computed with (14).

As to θ_β, it can be updated using (17) with $P(m)$ given by (18). The gradient vector $\psi(m) = \frac{dy(m)}{d\theta_{in}}$ are computed with (19) - (21) at $\theta_\beta(m)$.

Due to the limited space, the performance analysis is omitted here.

4. NUMERICAL EXAMPLE AND SIMULATIONS

In this section, we present one numerical example and the simulation results. The system to be identified is a third order Butterworth low-pass filter obtained with

MATLAB command *butter*(3, 0.10), corresponding to a bandwidth of 0.05. The denominator coefficients are $a_1^0 = -2.3741$, $a_2^0 = 1.9294$, $a_3^0 = -0.5321$. The output of the system, denoted as $y_0(m)$, corresponds to such an input signal $u(m)$ that the output has an unit variance. The measurement noise $v(m)$ is a white noise sequence such that the reference signal $r(m) = y_0(m) + v(m)$ has a signal-to-noise ratio of $40dB$. Five different initial points for parametrization θ_β are used, which are generated randomly. Their equivalents are also computed for parametrizations θ_d, θ_l and θ_α. The initial conditions for (18) are exactly the same in all the simulations.

For the direct parametrization θ_d, with the five different sets of initial conditions only the first three sets lead to convergence (to the true system), while the algorithm becomes unstable for the other two sets of initial conditions! The evolution of parameters (for the parameters $a_1(m)$, $a_2(m)$ and $a_3(m)$ only due to the limited space) is depicted in Fig. 1 for the first set of initial conditions. For the normalized lattice parametrization θ_l, though no instability occurs this parametrization shows surprisingly a very slow convergence behavior. In fact, with the five sets of initial conditions there is none leading to convergence within the 4,000 iterations. For the parametrizations θ_α and θ_β, like θ_l they exhibit very good stability performance, as expected, and yield much better convergence than θ_l. With the five sets of initial conditions, the algorithm with θ_α converges within 4,000 iterations for the first 3 sets, while the algorithm with the newly proposed θ_β can converge to the true system within 4,000 iterations for all the five sets. In Fig. 2 and 3, the parameter evolution, corresponding to the first set of initial conditions, is shown for θ_α and θ_β, respectively. For comparison, in these two figures the equivalent denominator parameters $a_1(m)$, $a_2(m)$ and $a_3(m)$ are computed from $\theta_\alpha(m)$ and $\theta_\beta(m)$, respectively, and then plotted.

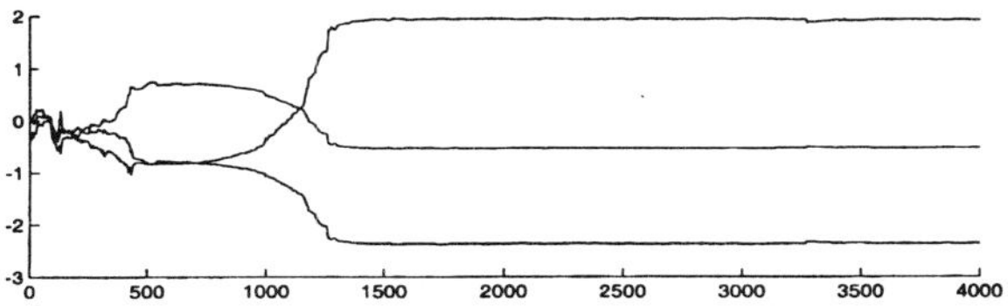

Fig. 1. Evolution of parameters for a_1, a_2 and a_3 using θ_d.

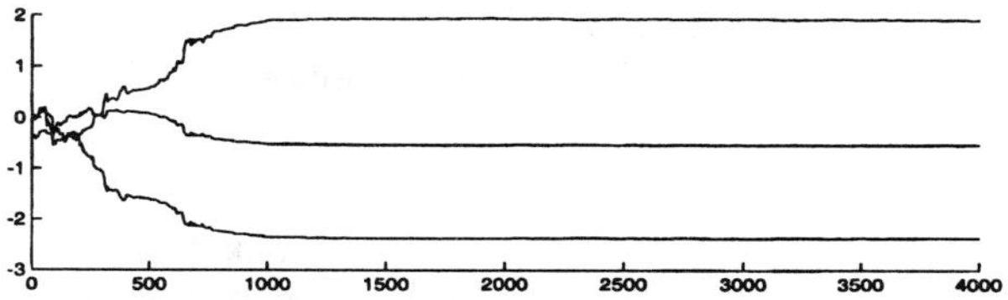

Fig. 2. Evolution of parameters for the equivalent a_1, a_2 and a_3 using θ_α.

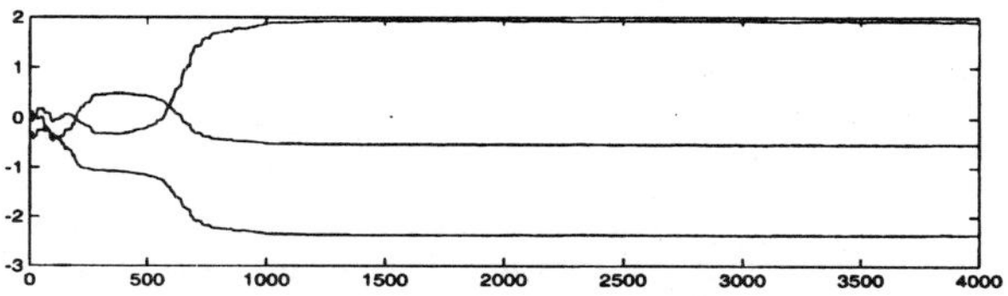

Fig. 3. Evolution of parameters for the equivalent a_1, a_2 and a_3 using θ_β.

5. REFERENCES

[1] A. Gray, "Passive cascaded lattice digital filters," IEEE Trans. on Circuits and Systems, vol. CAS-27, pp. 337-344, may, 1980.

[2] D. Parikh, N. Ahmed, and S.D. Stearns, "An adaptive lattice algorithm for recursive filters," IEEE Trans. on Acoust., Speech, Signal Processing, vol. ASSP-28, pp. 110-112, 1980.

[3] P.A. Regalia, "Stable and efficient lattice algorithms for adaptive IIR filtering," IEEE Trans. on Signal Processing, vol. 40, no. 2, pp. 375-388, Feb. 1992.

[4] H. Fan, "A structural view of asymptotic convergence speed of adaptive IIR filtering algorithms: Part I - infinite precision implementation," IEEE Trans. on Signal Processing, vol. 41, no. 4, pp. 1493-1517, Apr., 1993.

[5] K.X. Miao, H. Fan and M. Doroslovacki, "Cascade lattice IIR adaptive filters," IEEE Trans. on Signal Processing, vol. 42, no. 4, pp. 721-741, Apr., 1994.

[6] D. A. Johns, W. M. Snelgrove and A. S. Sedra, "Adaptive recursive state-space filters using a gradient-based algorithm," IEEE on Trans. Circuits and Systems, vol. 37, no. 6, pp. 673-684, Jun., 1990.

[7] G. Li and L.J. Qiu, "A new parametrization for designing stable adaptive IIR filters," Proc. IEEE Conf. Acoust., Speech, Signal Processing, vol. II, pp. 1333-1336, 2002.

[8] G. Li, M. Gevers and Y.X Sun, "Performance analysis of a new structure for digital filter implementation," IEEE Trans. on Circuits and Systems I, vol. 47, no. 4, pp. 474-482, Apr. 2000.

ROBUST RECURSIVE BI-ITERATION SINGULAR VALUE DECOMPOSITION (SVD) FOR SUBSPACE TRACKING AND ADAPTIVE FILTERING

Y. Wen, S. C. Chan, and *K. L. Ho*

Department of Electrical and Electronic Engineering
The University of Hong Kong, Pokfulam Road, Hong Kong
email: ywen@eee.hku.hk, scchan@eee.hku.hk, klho@eee.hku.hk

ABSTRACT

The recursive bi-iteration singular value decomposition (Bi-SVD), proposed by Strobach [1], is en efficient and well-structured algorithm for performing subspace tracking. Unfortunately, its performance under impulse noise environment degrades substantially. In this paper, a new robust-statistics-based bi-iteration SVD algorithm (robust Bi-SVD) is proposed. Simulation results show that the proposed algorithm offers significantly improved robustness against impulse noise than the conventional Bi-SVD algorithm with slight increase in arithmetic complexity. For nominal Gaussian noise, the two algorithms have similar performance.

1. INTRODUCTION

Subspace decomposition is a very valuable tool in signal processing, digital communications, and array signal processing. The singular value decomposition, SVD, is a known to be a good method for such computation because of its exact numerical properties. However, direct computation of the SVD is an iterative process and its complexity is very high in real-time applications. In many practical applications, only a sub-set of the singular vectors and singular values of the SVD have to be computed from a recursively updated correlation matrix. These properties can be utilized to obtain efficient algorithms for tracking the sub-set of singular vectors required. A number of recursive eigen-decomposition algorithms have also been proposed for recursive subspace tracking [1-4]. In particular, the Bi-Iteration Singular Value Decomposition (Bi-SVD) proposed by Stobach [1], which is based on an extension of Bauer's classical bi-iteration [16] to sequential SVD updating, has a comparable performance to the well-known reference algorithm of Karasal [4], but the Bi-SVD has a regular structure and much lower complexity. Unfortunately, when it is applied to subspace tracking applications where impulsive noise sporadically intrudes the system, say in a communication channel, its performance degrade substantially. The deeper reason for this fragility is that the conventional autocorrelation matrix estimate: $R_{\hat{x}\hat{x}}(n)=\sum_{i=1}^{n}\lambda^{n-i}\hat{x}\hat{x}^T$, where λ is the forgetting factor and $\hat{x}$ is the input signal vector, is not a robust estimate of the underlying autocorrelation $R_{xx}=E[xx^T]$, if x is corrupted by the impulse noise to form $\hat{x}$ [11-13]. This problem has been studied in robust statistics [14] and the minimum volume ellipsoid (MVE) or other robust estimators should be used. However, the computational complexity is usually prohibitive for real-time applications. In recursive subspace estimation, a rough prior knowledge of the subspace estimate is available from previous iterations. Therefore, it is easier to detect whether the incoming signal vector is potentially corrupted by impulse noise or not. This idea happens to coincide with the M-estimators or Maximum likelihood like estimators of the correlation matrix. It is therefore not surprising that the Bi-SVD algorithm and other RLS-based subspace tracking algorithms [11-13] are sensitive to impulsive or non-Gaussian noise, as we shall see later from the simulation results.

In this paper, a new robust Bi-SVD recursive subspace tracking algorithm using a robust statistics approach similar to [5,14] is proposed. In particular, a method for detecting an impulse-corrupted data vector is developed based on the statistics of the orthogonal innovation vector, which is the complement vector of the orthogonal projection of data vector on "old " singular subspace. Data samples, which are potentially corrupted by impulse noise, are then replaced by appropriate estimate from previously "cleaned" data samples. The layout of the paper is as follows: Section 2 is a brief description of the Bi-SVD subspace tracking problem. The proposed robust Bi-SVD subspace tracking algorithm is introduced in Section 3. Simulation results and comparison with the conventional method are presented in Section 4. And finally, conclusions are drawn in Section 5.

2. BI-ITERATION SVD ALGORITHM

The Bi-SVD subspace tracking algorithm [1] computes the r dominant singular values and corresponding right singular vectors from the following growing data matrix $X(t)$:

$$X(t)=\begin{bmatrix}(1-\alpha)^{1/2}x^T(t)\\ \alpha^{1/2}X(t-1)\end{bmatrix} \tag{1}$$

where $X(t-1)$ is the old data matrix at time t-1, $x(t)=[x(t),x(t-1),...,x(t-N-1)]^T$ is the newly incoming data vector of dimension N at time t, $x(t)$ $(t=1,2,3,...)$ is the received data samples, and $0\le\alpha\le1$ is a non-negative exponential forgetting factor. The classical bi-iteration SVD algorithm applied to the $L\times N$ real data matrix X is shown in the following:

Classical Bi-Iteration SVD Algorithm

Initialize $Q_A(0)=\begin{bmatrix}I_r\\0\end{bmatrix}$

FOR $\mu=1,2,....$ **DO (Until Converge)**

$B(\mu)=XQ_A(\mu-1)$,

$B(\mu)=Q_B(\mu)R_B(\mu)$, ($L\times r$ QR -Factorization)

$A(\mu)=X^TQ_B(\mu)$,

$A(\mu)=Q_A(\mu)R_A(\mu)$, ($N\times r$ QR -Factorization).

It has been proved that the orthonormal columns of the $L\times r$ matrix $Q_B(\mu)$ will converge to the r dominant left singular vectors, and the orthonormal columns of the $N\times r$ matrix $Q_A(\mu)$ will converge to the r dominant right singular vectors of the SVD of the data matrix X. At the same time, both triangular matrices $R_A(\mu)$ and $R_B(\mu)$ will converge to the $r\times r$ diagonal matrix of dominant singular values of X. Having replaced the index μ with the discrete time index t and made a *consistent approximation* of,

$$\hat{X}(t)=Q_B(t)R_B(t)Q_A^T(t-1) \tag{2}$$

Strobach [1] extends this classical bi-iteration SVD to a sequential square-root type bi-iteration SVD subspace-updating algorithm as follows:

0-7803-7761-3/03/$17.00 ©2003 IEEE

Bi-Iteration SVD Subspace Tracking

Initialize $Q_A(t-1)=\begin{bmatrix} I_r \\ 0 \end{bmatrix}$; $R_B(t-1)=\theta_A(t-1)=I_r$;

$R_A(t-1)=[0]$; r is rank.

FOR $t=1,2,....$ **DO**

Input $x(t)$

$$h(t)=Q_A^T(t-1)x(t),\; x_\perp(t)=x(t)-Q_A(t-1)h(t),$$

$$e_x(t)=x_\perp^T(t)x_\perp(t),\; \bar{x}_\perp(t)=e_x^{-1/2}(t)x_\perp(t),$$

$$H(t)=R_B(t-1)\theta_A(t-1),$$

$$\begin{bmatrix} R_B(t) \\ --- \\ 0\cdots 0 \end{bmatrix}=G_B(t)\begin{bmatrix} (1-\alpha)^{1/2}h^T(t) \\ ------ \\ \alpha^{1/2}H(t) \end{bmatrix},$$

$$h_R^T(t)R_B(t)=h^T(t) \xrightarrow{\text{back substitution}} h_R^T(t),$$

$$H_R(t)R_B(t)=H(t) \xrightarrow{\text{back substitution}} H_R(t),$$

$$\begin{bmatrix} R_A(t) \\ --- \\ 0\cdots 0 \end{bmatrix}=G_A(t)\begin{bmatrix} \alpha R_A(t-1)H_R(t)+(1-\alpha)h(t)h_R^T(t) \\ ------------------ \\ (1-\alpha)e_x^{1/2}(t)h_R^T(t) \end{bmatrix}$$

$$G_A^T(t)=\begin{bmatrix} \theta_A(t) & \begin{matrix} * \\ \vdots \\ * \end{matrix} \\ ------- & \\ f^T(t) & * \end{bmatrix} \xrightarrow{\text{extract}} \theta_A(t); f(t),$$

$$Q_A(t)=Q_A(t-1)\theta_A(t)+\bar{x}_\perp(t)f^T(t),$$

END

TABLE 1.

The compressed data vector $h(t)$ of dimension r is actually the *projection* of the input data vector $x(t)$ onto the right singular subspace basis $Q_A^T(t-1)$; $x_\perp(t)$ is the complement of this orthogonal *projection*; $e_x(t)$ the squared norm of the complement vector $x_\perp(t)$; and $\bar{x}_\perp(t)$ is the normalized complement vector which acts as the *innovation* for subspace updating. The $r\times r$ cosine matrix $\theta_A(t)=Q_A^T(t-1)Q_A(t)$ describes the *distance* between consecutive subspaces, while $G_B(t)$ and $G_A(t)$ represent a sequence of orthonormal Givens plane rotations, which are determined in such a way that the rotated old triangular matrix on its right hand side is transformed into a strictly upper-right triangular matrix on its left hand side. $f(t)$ is the *compressed innovation* vector defined as $f(t)=Q_A^T(t)\bar{x}_\perp(t)$. $H(t)$, $h_R^T(t)$, $H_R(t)$ are vectors and matrixes defined respectively as $H(t)=R_B(t-1)\theta_A(t-1)$, $h_R^T(t)=h^T(t)R_B^{-1}(t)$, and $H_R(t)=H(t)R_B^{-1}(t)$. Like the PAST algorithm, the overall computation can also be viewed as a mean of minimizing $e_x(t)$, the squared norm of the complement vector $x_\perp(t)$. As mentioned earlier, the performance of this algorithm is very sensitive to impulse noise, because the *squared norm* of $x_\perp(t)$ is minimized, which is not a robust estimator. In fact, if $x(t)$ is corrupted by impulse noise, $h(t)$, $x_\perp(t)$, $R_B(t)$, $h_R^T(t)$, $H_R(t)$, $R_A(t)$, $G_A(t)$, $\theta_A(t)$, and $f(t)$ will all be affected in turn. The new subspace estimate will hence be perturbed to a point which is far away from the true subspace, even though the impulse noise power is merely $20\,dB$ to $25\,dB$ higher than nominal Gaussian background noise. More importantly, the corrupted matrices, $R_B(t)$, will be used to compute the new $R_B(t)$'s, which takes the Bi-SVD algorithm many iterations to recover, especially when α is close to one.

3. ROBUST BI-SVD SUBSPACE TRACKING

We now consider the proposed robust statistics-based Bi-SVD algorithm. First of all, we note that when $x(t)$ is corrupted by one or more impulses, $e_x(t)$, which is the squared norm of complement vector $x_\perp(t)$, in the Bi-SVD algorithm mentioned above will become very large. Instead of using the squared norm, an M-estimator $E[e_x(t)]=E[q(e_x(t))\cdot e_x(t)]$ can be used, where $q(e_x(t))$ is a weighting factor which de-emphasis error norm with exceptional large value. For simplicity, we consider the modified Huber function, where $q(e_x(t))$ is equal to 0 and 1 respectively when $e_x(t)$ is greater than or smaller than a threshold *T*, which is to be estimated continuously. Though the exact distribution of the $e_x(t)$ is unknown, it is assumed for simplicity to be Gaussian distributed but corrupted by additive impulse noise to simplify the detection of the impulses (note also that $e_x(t)$ is always positive). It then follows that the probability for $|\Delta e_\mu(t)|=|e_x(t)-\hat{\mu}(t)|$ to be greater than a given threshold $T(t)$ is

$$p_T=P_r\{|\Delta e_\mu(t)|>T(t)\}=erfc(T(t)/\hat{\sigma}(t)), \tag{3}$$

where $erfc(r)=(2/\sqrt{\pi})\int_r^\infty e^{-x^2}dx$ is the complementary error function. $\hat{\mu}(t)$ and $\hat{\sigma}(t)$ are the estimated mean and standard deviation of the squared norm $e_x(t)$ of the "impulse free" $x_\perp(t)$. Using different threshold parameters $T(t)$, we can detect whether the incoming vector is potentially corrupted by impulse noise with different degrees of confidence. In this work, p_T is chosen to be 0.05 so that we have 95% confidence in saying that the current data vector $x(t)$ is corrupted by impulse noise. The selection of this threshold is a tradeoff between impulse suppression and signal distortion. The larger the threshold, the smaller will be the signal distortion, however, at the expense of less immunity to impulse noise. In practice, a threshold value corresponding to a confidence of 93% to 97% works well. For 95% confidence, the corresponding threshold parameter $T(t)$ is determined to be $T(t)=1.96\cdot\hat{\sigma}(t)$. A commonly used estimate for $\hat{\sigma}^2(t)$, and $\hat{\mu}(t)$ are $\hat{\sigma}^2(t)=\lambda_\sigma\hat{\sigma}^2(t-1)+(1-\lambda_\sigma)(\Delta e_\mu(t))^2$, and $\hat{\mu}(t)=\lambda_\mu\hat{\mu}(t-1)+(1-\lambda_\mu)e_x(t)$, respectively, where λ_μ and λ_σ are some forgetting factors. It is, however, not robust to impulse noise. In fact, a single impulse with large amplitude can substantially increase the value of $\hat{\sigma}(t)$ and $\hat{\mu}(t)$, and hence the values of $T(t)$. Following [6], we employ the following robust estimates for $\hat{\sigma}^2(t)$ and $\hat{\mu}(t)$

$$\hat{\sigma}^2(t)=\lambda_\sigma\hat{\sigma}^2(t-1)+1.483\left(1+\frac{5}{N_w-1}\right)(1-\lambda_\sigma)\text{med}\left(A((\Delta e_\mu(t))^2)\right)$$

$$\text{and } \hat{\mu}(t)=\lambda_\mu\hat{\mu}(t-1)+(1-\lambda_\mu)\text{med}\left(A(e_x(t))\right), \tag{4}$$

where $A(x(t))=\{x(t),\cdots,x(t-N_w+1)\}$, N_w is the length of the estimation window, and med(.) is the median operation. λ_μ and λ_σ are the forgetting factors. It can be seen from the above discussion that the arithmetic complexity of the proposed robust Bi-SVD algorithm is very close to that of the conventional Bi-SVD algorithm. For very large values of N_w, the complexity for performing the medium filter can be significantly reduced by computing the pseudo median [17], instead of the median. The pseudo median is an approximation to the median with much lower complexity and can be efficiently implemented in a pipeline structure. In practice, N_w is limited to 5 to 11, therefore the increase in complexity is quite acceptable. Our robust Bi-SVD algorithm updates $T(t)=1.96\cdot\hat{\sigma}(t)$, $\hat{\sigma}^2(t)$ and $\hat{\mu}(t)$ at each iteration. If $|\Delta e_\mu(t)|>T(t)$, the incoming received data sample $x(t)$ in the new date vector $x(t)=[x(t),x(t-1),...,x(t-N-1)]^T$ is treated as a impulse-corrupted data sample. To prevent it from entering the subspace updating iteration, $x(t)$ has to be replaced by certain estimate such as the linear predictor from previous "impulse free" samples. For simplicity, the robust mean estimate of the received data sample $\hat{\mu}_s(t)$ is used:

$$\hat{\mu}_s(t)=\beta_s\hat{\mu}_s(t-1)+(1-\beta_s)med(A(x(t))), \quad (5)$$

where β_s is its forgetting factor. This scheme is shown to be very effective in blocking the impulse noise from entering the Bi-SVD algorithm. Finally, we obtain the robust Bi-SVD subspace tracking algorithm in TABLE 2.

4. SIMULATION RESULTS

The performance of the proposed robust Bi-SVD algorithm is evaluated under the ε -contamination noise model, and it is compared to the conventional Bi-SVD algorithm. The probability density function (pdf) of the ε -contamination noise model is

$$f=(1-\varepsilon)N(0,\sigma^2)+\varepsilon N(0,\sigma_I^{\,2}), \quad (6)$$

where $0\le\varepsilon\le1$, $N(0,\sigma^2)$ is the pdf of the Gaussian background noise, $N(0,\sigma_I^{\,2})$ is the pdf of the impulse noise. In the simulation, the impulse noise occurrence probability ε was set to 0.1, with $10\cdot\log_{10}\frac{\sigma_I^{\,2}}{\sigma^2}=25dB$. The two algorithms are applied to track the signal subspace of the received data sequence, which is the superposition of two temporary monochromatic sources and noise. After each subspace update, adaptive subspace filtering is implemented to reconstruct the two source signals from the corrupting noise. The adaptive subspace filtering performed is:

$$\hat{s}(t)=Q_A(t)Q_A^T(t)x(t), \quad (7)$$

and the actual sample $\hat{s}(t)$ of the reconstructed signal sequence is extracted via the bottom pinning of the vector $s_\Sigma(t)$, which is the sum of successively shifted vector $\hat{s}(t)$ as:

$$s_\Sigma(t)=Ds_\Sigma(t-1)+\frac{1}{N}\hat{s}(t),$$

$$D=\begin{bmatrix}0\cdots0 & 0\\ & 0\\ I_{N-1} & \vdots\\ & 0\end{bmatrix},\quad \hat{s}(t)=[0\cdots0,1]s_\Sigma(t). \quad (8)$$

Two temporary monochromatic source signals are sinusoidal sequences with normalized frequencies of $\omega_1=10^o$, and $\omega_2=12^o$, respectively. Each source signal is $-4.88\,dB$ below the Gaussian background noise. The parameters for the two algorithms are: order of data vector: $N=181$, forgetting factor: $\alpha=0.993$, rank of signal subspace: $r=4$. For the robust Bi-SVD, the length of the median filter N_w is set to 9, and the forgetting factors λ_μ, λ_σ, and β_s are all set equal to 0.97. The initial value $\hat{\mu}_s(0)$ is chosen to be zero. Both $\hat{\sigma}^2(0)$ and $\hat{\mu}(0)$ are set to 10, a relatively large number to their normal value, to initialize system adaptation. Fig.1 shows the test data: Fig.1 (a) is the first sinusoidal sequence with normalized frequency $\omega_1=10^o$. Fig.1 (b) is the second sinusoidal sequence with normalized frequency $\omega_2=12^o$. Fig.1 (c) is the transmitted signal constituting of the sum of these two sinusoidal sequences. Fig.1 (d) is the channel noise, where impulse noise intrudes the channel during the time interval from the 1000th sample to the 1500th sample. Fig.1 (e) shows the received samples, which is the sum of the transmitted signal and channel noise. Fig.2 depicts the signal reconstruction process using Bi-SVD based adaptive subspace filtering. From Fig.2 (b), we can see that the reconstructed samples are severely distorted by the impulse noise in the interval between the 1000th sample and the 1500th sample. Moreover, the reconstruction error during this interval has a shape resembling the impulsive noise, which suggests that they are malign errors. In addition, even though the impulsive noise passes away after the 1500th sample, the reconstruction error couldn't converge to the normal level until the 2000th sample. Fig.3 shows the mean maximum principal angle (a measurement of "distance") between the Bi-SVD subspace estimate and the true subspace derived from the batch eigendecomposition of noise-free data covariance matrix. This mean maximum principal angle was obtained by averaging over 100 independent Monte Carlo trials. It can be observed in Fig.3 that the Bi-SVD subspace estimate is significantly affected by the impulse noise (very large maximum principle angle) and deviates far away from its true value. It is unable to re-converge to true signal subspace (small maximum principle angle) until 2000th sample. In contrast, Fig.4 shows that reconstruction error of the proposed robust Bi-SVD is much smaller than that of the Bi-SVD algorithm, and it is similar to the transmitted signal, despite the presence of the impulse noise. Besides, the reconstruction error re-converged to the normal level almost right after the impulse noise was turned off. Fig.5 substantiates the robustness of the proposed robust Bi-SVD algorithm, where it can be seen that the subspace estimate remains largely unaffected and is always close to the true subspace in the presence of the impulses.

5. CONCLUSION

A new robust-statistics-based Bi-SVD subspace tracking algorithm in impulse noise environment is presented. Simulation results using adaptive subspace-filtering show that the proposed algorithm offers improved robustness than the conventional Bi-SVD algorithm under contaminated Gaussian noise environment. The computational complexity of the robust Bi-SVD algorithm is slightly higher than that of the conventional Bi-SVD algorithm

6. REFERENCES

[1] P. Strobach, "Bi-Iteration SVD Subspace Tracking Algorithms", *IEEE Trans SP*, vol.45, no.5, pp.1222-1240, May 1997.

[2] P. Strobach, "Low-Rank Adaptive Filters", *IEEE Trans SP*, vol.44, no.12, pp.2932-2947, December 1996.

[3] P. Strobach, "Bi-Iteration Recursive Instrumental Variable Subspace Tracking and Adaptive Filtering", *IEEE Trans SP*, vol.46, no.10, pp.2708-2725, October 1998.

[4] I. Karasalo, "Estimating The Covariance Matrix By Signal Subspace Averaging", *IEEE Trans Acoustl Speech, Signal Processing*, ASSP-34, pp.8-12, Feb. 1986.

[5] Y. X. Zou, S. C. Chan, and T. S. Ng, "A Robust M-Estimate Adaptive Filter for Impulsive Noise Suppression", *Proceedings of ICASSP'99*, Vol.4, Phoenix, Arizona, USA, March 1999.

[6] Y. X. Zou, S. C. Chan and T. S. Ng, "A Recursive M-estimate Algorithm for Robust Adaptive Filtering in Impulse Noise," *IEEE Sig. Proc. Letter*, Nov. 2000, pp. 324 –326.

[7] Y. X. Zou, S. C. Chan and T. S. Ng, "Fast Least Mean *M*-estimate Algorithms for Robust Adaptive filtering in impulse noise," *IEEE Circuits and Systems II*, Dec. 2000, pp. 1564-1569.

[8] Y. X. Zou, S. C. Chan and T. S. Ng, "Robust M-Estimate Adaptive Filtering," *IEE Proc. Signal Processing*, vol.148, no.4, pp. 289-294, Aug. 2001.

[9] Y. X. Zou, S. C. Chan, and T. S. Ng, "A Robust Statistics Based Adaptive Lattice-Ladder Filter In Impulsive Noise, " in *Proc. IEEE ISCAS'2000*, Geneva.

[10] Y. X. Zou, S. C. Chan and T. S. Ng, "Fast Least Mean M-Estimate Algorithms for Robust Adaptive Filtering in Impulse Noise," in *Proc. EUSIPCO'2000*, Tampere, Finland.

[11] Y.Wen, S.C. Chan, and K.L. Ho, "Robust Subspace Tracking In Impulsive Noise" in *Proc. ICC'2001*, Helsinki, Finland.

[12] Y.Wen, S.C. Chan, and K.L. Ho, "Robust Subspace Tracking Base Blind Channel Identification In Impulsive Noise Environment" in *Proc. EUSIPCO2002*, Toulouse, France.

[13] Y.Wen, S.C. Chan, and K.L. Ho, "A Robust Subspace Tracking Algorithm For Subspace-Based Blind Multiuser Detection In Impulsive Noise" in *Proc. DSP2002*, Santorini, Greece.

[14] P. J. Huber, "Robust statistics", John Wiley, New York, 1981.

[15] G. H. Golub, C. F. Van Loan, "Matrix Computations", The Johns Hopkins Uni. Press, Baltimore and London, 1996.

[16] F. L. Bauer, "Das Verfahren der Treppeniteration und verwandte Verfahren zur L¨osung algebraischer Eigenwertprobleme" Z. Angew. Math. Phys., vol. 8, pp. 214–235, 1957.

[17] W. K. Pratt, Digital Image Processing, 2nd Edition. John Wiley & Sons, Inc. 1991.

Bi-Iteration SVD Subspace Tracking

Initialize $Q_A(t-1)=\begin{bmatrix} I_r \\ 0 \end{bmatrix}$; $R_B(t-1)=\theta_A(t-1)=I_r$;

$R_A(t-1)=[0]$; $\hat{\sigma}^2(0)$, $\hat{\mu}(0)$, and $\hat{\mu}_s(0)$; r is rank.

FOR $t=1,2,....$ **DO**

Input $x(t)$

$$h(t)=Q_A^T(t-1)x(t),\ x_\perp(t)=x(t)-Q_A(t-1)h(t),$$

$$e_x(t)=x_\perp^T(t)x_\perp(t),$$

IF $|\Delta e_\mu(t)|>T(t)$ **THEN**

$$x(t)=\hat{\mu}_s(t-1),\ h(t)=Q_A^T(t-1)x(t),$$

$$x_\perp(t)=x(t)-Q_A(t-1)h(t),\ e_x(t)=x_\perp^T(t)x_\perp(t),$$

$$\bar{x}_\perp(t)=e_x^{-1/2}(t)x_\perp(t),\ H(t)=R_B(t-1)\theta_A(t-1),$$

$$\begin{bmatrix} R_B(t) \\ --- \\ 0\cdots0 \end{bmatrix}=G_B(t)\begin{bmatrix} (1-\alpha)^{1/2}h^T(t) \\ ------ \\ \alpha^{1/2}H(t) \end{bmatrix},$$

$$h_R^T(t)R_B(t)=h^T(t)\ \xrightarrow{\text{back substitution}}\ h_R^T(t),$$

$$H_R(t)R_B(t)=H(t)\ \xrightarrow{\text{back substitution}}\ H_R(t),$$

$$\begin{bmatrix} R_A(t) \\ --- \\ 0\cdots0 \end{bmatrix}=G_A(t)\begin{bmatrix} \alpha R_A(t-1)H_R(t)+(1-\alpha)h(t)h_R^T(t) \\ ------------------ \\ (1-\alpha)e_x^{1/2}(t)h_R^T(t) \end{bmatrix}$$

$$G_A^T(t)=\begin{bmatrix} \theta_A(t) & \begin{matrix} * \\ \vdots \\ * \end{matrix} \\ ------- & \\ f^T(t) & * \end{bmatrix}\xrightarrow{\text{extract}}\theta_A(t);f(t),$$

$$Q_A(t)=Q_A(t-1)\theta_A(t)+\bar{x}_\perp(t)f^T(t),$$

update $\hat{\sigma}^2(i)$, $\hat{\mu}(i)$ and $\hat{\mu}_s(0)$ using(4)(5), $T(i)=1.96\cdot\hat{\sigma}(i)$,

END

TABLE 2.

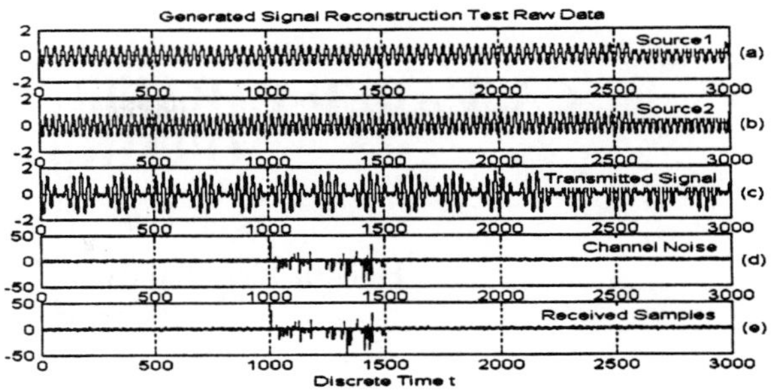

Fig. 1

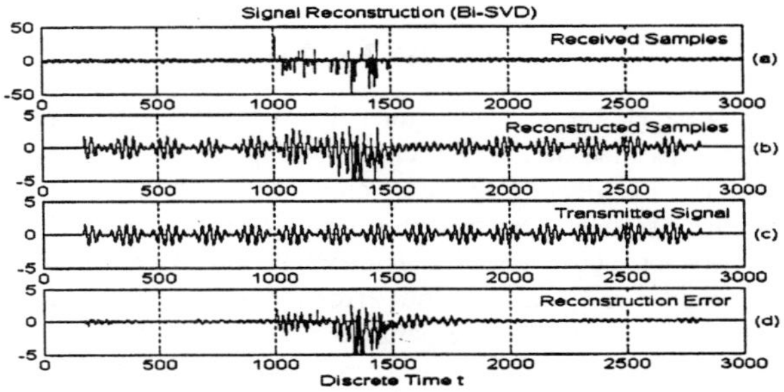

Fig. 2

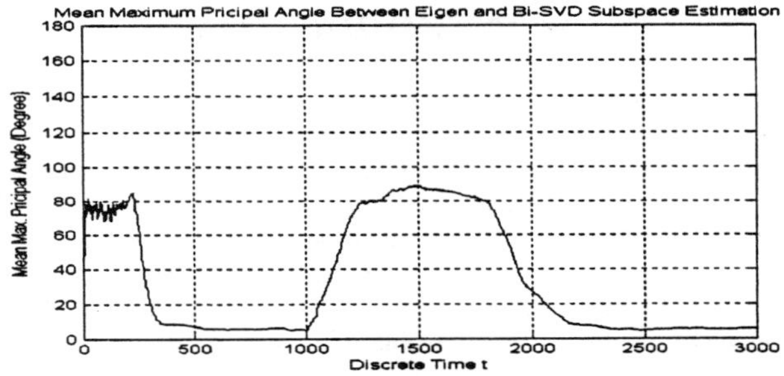

Fig. 3

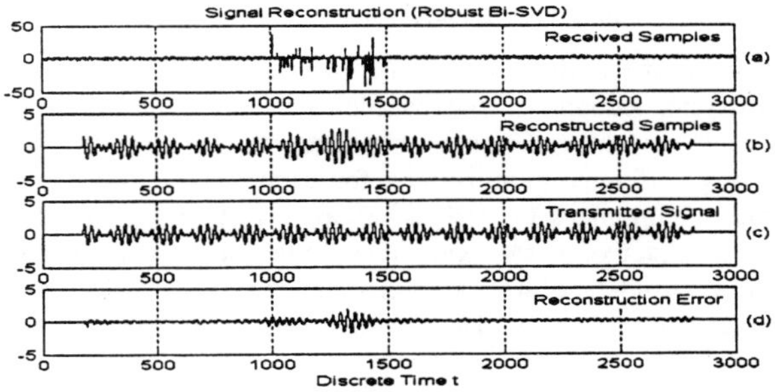

Fig. 4

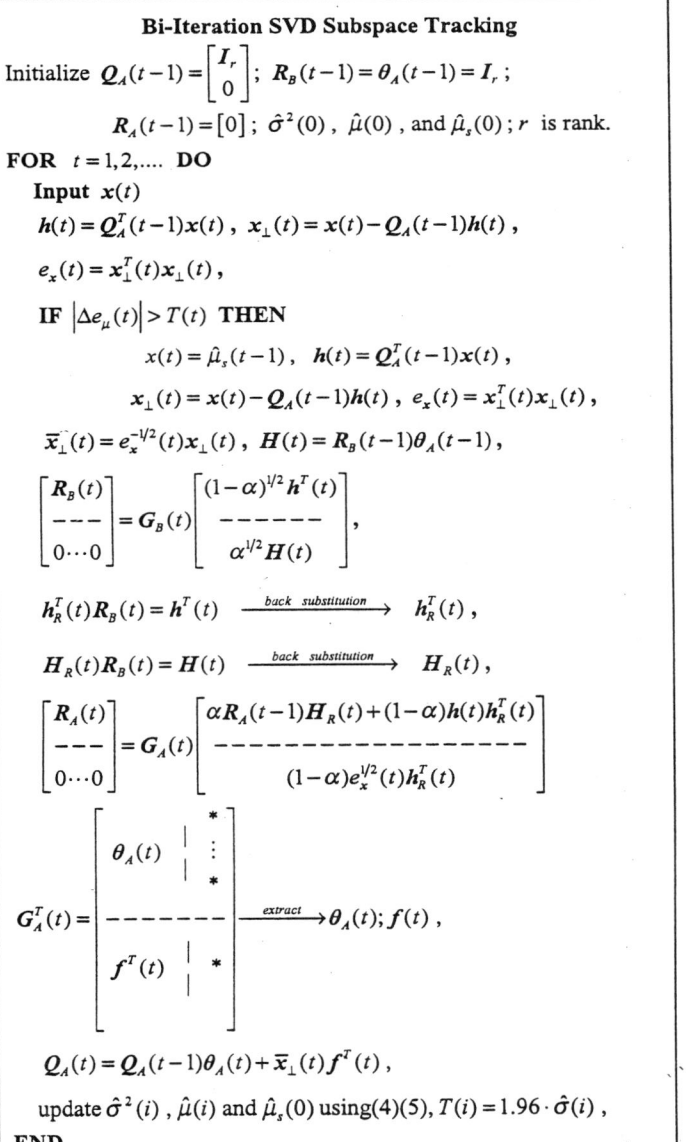

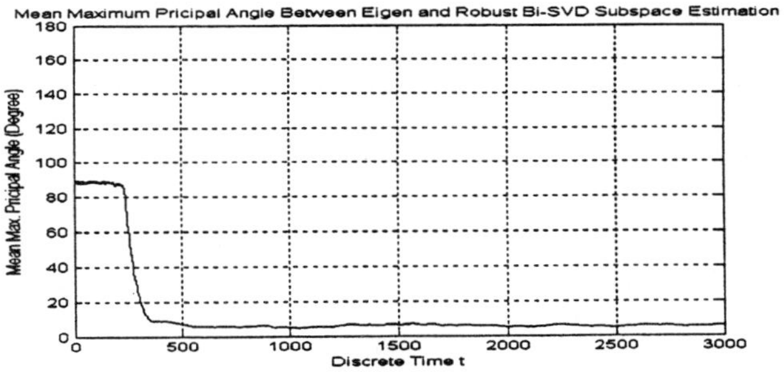

Fig. 5

CASCADED-PARALLEL ADAPTIVE NOTCH FILTER BASED ON ORTHOGONAL DECOMPOSITION

Y. Liu[1], *P. S. R. Diniz*[2], *and T. I. Laakso*[1]
[1]Helsinki University of Technology, Signal Processing Laboratory
P.O.Box 3000, FIN-02150, Finland
[2]COPPE / Universidade Federal do Rio de Janeiro
Caixa Postal 68504, RJ, Brazil, 21945-970

ABSTRACT

In this paper, rational orthogonal basis functions are introduced to realize cascaded-parallel structure of adaptive notch filters (ANF) in the case of identifying and separating multiple sinusoids from wideband signals. The structural orthogonality provides good trade-off between convergence speed and computational complexity. A simplified mean square output error (MSOE) updating algorithm is derived for the cascade-parallel realization. Simulation results confirm that the proposed structure converges faster compared with the cascade realizations with the same computational complexity.

1. INTRODUCTION

In broadband communication systems, the signal is often corrupted by narrowband or sinusoidal sources across the spectrum. Notch filters are widely used to cancel such interference. When using ANF to suppress multiple sinusoids, we often consider a cascade or parallel connections of second order notch sections [1]. Two considerations arise when designing multiple notch filters. The convergence point of each notch filter section should match the frequency of one of the sinusoids. If the output of each notch filter section contains one sinusoid less, the filtered output from each notch filter section should remain close to white for the next stage frequency estimate to be unbiased.

In case of parallel realization for the ANF, an apparent risk is that different sections of the parallel structure might follow identical evolutions since the input presented to each section is identical. Using different initializations for each section reduces but by no means eliminates the risk. The cascade structure would then seem more attractive, to the extent that a given section converges to one of the sinusoids and the remaining sections have no choice but to seek other sinusoids [2,3]. However, if the bandwidth of each cascaded section is wide, varying the notch frequency parameter varies the suppression to all the sinusoidal components, such that the local minima do not necessarily coincide with the sinusoidal frequencies, introducing bias into the local minima of the cost function. In principle, by sufficiently narrowing the bandwidth of each notch section in the cascade reduce the bias. However, the narrow bandwidth in notch adaptation algorithms causes slow convergence, because the gradient is "small" if the notch frequency is far from one of the sinusoidal frequencies [4].

To increase the convergence speed, we can constrain the cascaded notch sections to model different sets of poles and zeros, and the ambiguity of the notching stable points can be avoided. Utilizing the orthogonality of the notch sections can realize such constraints and find good tradeoff between computational complexity and convergence speed.

The paper is organized as follows. Section 2 defines the system model for ANF with multiple notches. In Section 3, the orthonormal rational basis function is introduced. The ANF constructed on the orthonormal basis leads to a cascaded-parallel filter structure that is presented in Section 4. A simplified MSOE updating algorithm is consequently derived in Section 5. In Section 6, simulation examples are provided. Finally section 7 concludes the paper.

2. SYSTEM MODEL

2.1 Received signal

Many communication scenarios consist of sinusoidal components of unknown frequencies immersed in white noise. Sometimes the broadband signal is the desired one while the sinusoids are interferences, *e.g.*, broadband very high-speed digital subscriber line (VDSL) signal can be corrupted by narrowband amateur radio interferences [5]. ANF is designed for such signal scenarios. Assume that the received signal is of the form

$$u(n)+\xi(n)=\sum_{k=1}^{M} p_k \sin(\omega_k n+\phi_k)+\xi(n) \tag{1}$$

where p_k is the amplitude of the kth sinusoid (M total) and ϕ_k is its phase. $\xi(n)$ is a white noise process which is assumed to be independent of the sinusoid terms.

2.2 Lattice notch filter

In order to model the sinusoidal frequencies ω_k and consequently eliminate the corresponding sinusoids, the following constrained form of the notch filter has been considered corresponding to a lattice realization. The above notch filter consists of cascades of M second-order filters having zeros on the unit circle, resulting in exactly zero gain at each notch frequency.

$$\hat{H}(z)=\prod_{k=1}^{M}\frac{1+\sin\theta_{1k}}{2}\,\frac{1+2\sin\theta_{1k}z^{-1}+z^{-2}}{1+\sin\theta_{1k}(1+\sin\theta_{2k})z^{-1}+\sin\theta_{2k}z^{-2}} \tag{2}$$

Note that the second-order lattice notch filter is built around the second-order all-pass lattice filter [6], whose transfer function is

$$V(z)=\frac{\sin\theta_2+\sin\theta_1(1+\sin\theta_2)z^{-1}+z^{-2}}{1+\sin\theta_1(1+\sin\theta_2)z^{-1}+\sin\theta_2 z^{-2}}X(z) \tag{3}$$

Since $|V(e^{j\omega})| = 1$ for all ω, we may write $V(e^{j\omega}) = e^{j\phi(\omega)}$, where $\phi(\omega)$ is the phase response versus frequency. Since $V(z)$ is causal all-pass

0-7803-7761-3/03/$17.00 ©2003 IEEE

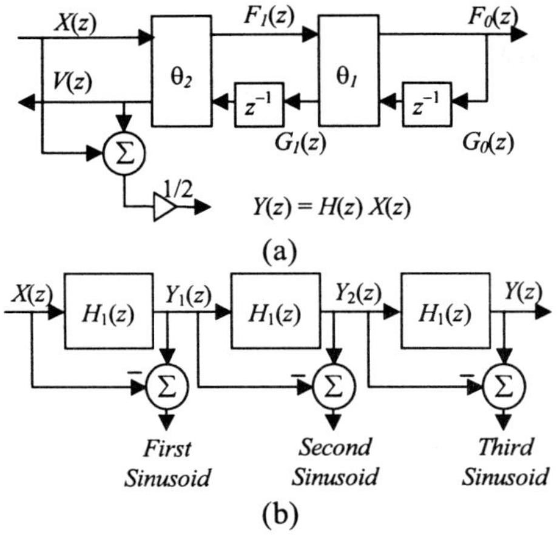

Figure 1. (a) Second-order lattice notch filter realization; (b) Cascade of second-order notch sections

function for all $|\sin\theta_1|<1$, the phase function $\phi(\omega)$ is a monotonically increasing function of ω, which satisfies the boundary constraints $\phi(0)=0$ and $\phi(\pi)=2\pi$. Letting $z=e^{-j\omega_0}$, comparing the input and the output of the all-pass filter $V(z)$ at frequency ω_0, if ω_0 satisfies the condition

$$\omega_0=\theta_1+\frac{\pi}{2} \qquad \text{for}\,|\theta_1|<\frac{\pi}{2} \tag{4}$$

then the amplitude remains the same, a single notch at ω_0 is obtained by adding the input and output,

$$\hat{H}(z)=\frac{1}{2}[1+V(z)] \tag{5}$$

The filter is depicted in Figure 1(a). Note that the notch frequency ω_0 depends on $\sin\theta_1$ by a simple formula in (4) and it is independent of θ_2 [8]. By cascading this second order notch filter, the filter structure in Figure 1(b) can model multiple notches.

3. ORTHONORMAL BASIS

In a multiple notch filter, the convergence point of each second-order notch section should match the frequency of each sinusoid, as the rest part of the spectrum at the output of each notch section remains unaltered. Considering the parallel realization, if we force each notch section to converge to a different frequency, our target is met. The construction of rational orthonormal basis functions can realize such constraints [7]. Since the second order lattice notch filter is derived from the all-pass filter, the orthogonality of the all-pass filter can be set to design multiple notch filters with constrained notches. Given two rational functions $\mathcal{F}_k(z)$ and $\mathcal{F}_l(z)$, we define that they are orthogonal if their inner product is zero. i.e.,

$$\langle F_k(z),F_l(z)\rangle=\frac{1}{2\pi j}\oint F_k(z)F_l(z^{-1})\frac{dz}{z}=0 \tag{6}$$

A generic stable linear IIR filter $H(z)$ can be expressed with the state-space representation

$$\begin{bmatrix}\mathbf{x}(n+1)\\ y(n)\end{bmatrix}=\begin{bmatrix}\mathbf{A} & \mathbf{b}\\ \mathbf{c} & d\end{bmatrix}\begin{bmatrix}\mathbf{x}(n)\\ u(n)\end{bmatrix} \tag{7}$$

$H(z)$ is orthogonal and all-pass if and only if the matrix $(\mathbf{A},\mathbf{b},\mathbf{c},d)$ is an orthonormal matrix. The orthonormality condition is satisfied if

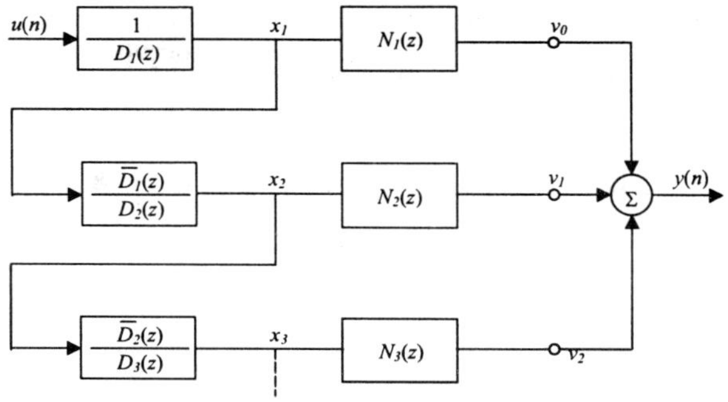

Figure 2 Second-order based orthogonal adaptive IIR filter.

$\mathbf{APA}^T+\mathbf{bb}^T=\mathbf{P}$ has solution $\mathbf{P}=\mathbf{I}$. If we express $H(z)=\Sigma_{k=0}^{\infty}v_k\mathcal{F}_k(z)$, the basis function of $\mathcal{F}_k(z)$ can be chosen as

$$\mathcal{F}_k(z)=\begin{cases}\mathbf{e}_k^T(z\mathbf{I}-\mathbf{A})^{-1}\mathbf{b} & k=0,\dots M-1\\ z^{M-k}V(z) & k\geq M\end{cases} \tag{8}$$

where $\mathbf{e}_k=[0,\dots,1,\dots0]^T$ is a unit vector with a one at k^{th} position and $V(z)=z^{-M}A(z^{-1})/A(z)$ is an all-pass filter. An important characteristic of the orthogonality of all-pass filter can be used for our purpose. If $H_1(z)$ and $H_2(z)$ are two orthogonal all-pass filters, the serial connection $H_1(z)H_2(z)$ of the two filters remains orthogonal to both $H_1(z)$ and $H_2(z)$ and it is still all-pass [7].

We consider here a basis function

$$F_k(z)=\frac{N_k(z)}{D_k(z)}\prod_{i=1}^{k-1}\frac{\check{D}_i(z)}{D_i(z)} \tag{9}$$

where $D_k(z)$ is a second-order monic polynomial, $\check{D}_i(z)=z^2D_i(z)$ and $N_k(z)$ are chosen to maintain the orthogonality of the basis functions $\mathcal{F}_k(z)$, depending on the coefficients of $D_k(z)$, i.e., $\|\mathcal{F}_k(z)\|=1$, and $<\mathcal{F}_k(z),\ \mathcal{F}_l(z)>=\delta_{k,l}$ [1]. In this particular case, $V(z)=\Pi_{i=1}^{k}\check{D}(z^{-1})/D(z)$ is the cascaded all-pass section. Consider the filter structure in Figure 2, if each branch is forced to be a notch transfer function, the realized basis $F_k(z)$ is approximately orthogonal. There are two reasons. First, notch filter will notch out some energy of the received signal, $\|\mathcal{F}_k(z)\|<1$. Second, the orthogonality of the all-pass part of the filter has forced the notch frequency to be different for each branch. The advantage of such a structure is that each branch sees an all-passed but unbiased received signal.

4. ORTHOGONAL ANF

4.1 ANF structure

The second order lattice notch filter is built on two basic blocks, which is called *Schur* recursion [5]. According to Figure 1(a), the input and output relationship can be expressed in z domain as

$$\begin{bmatrix}F_{k-1}(z)\\ G_k(z)\end{bmatrix}=\begin{bmatrix}\cos\theta_k & -\sin\theta_k\\ \sin\theta_k & \cos\theta_k\end{bmatrix}\begin{bmatrix}F_k(z)\\ z^{-1}G_{k-1}(z)\end{bmatrix} \tag{10}$$

By adapting $\{\theta\}$ in such a way that $|\sin\theta|\neq1$, we are ensured of the stability of the lattice filter output. Applying *Schur* recursion, it's straightforward to derive in z domain the z-transform of the received signal seen at node $F_0(z)$ as

$$F_0(z)=\frac{\cos\theta_2\cos\theta_1}{D(z)}X(z) \tag{11}$$

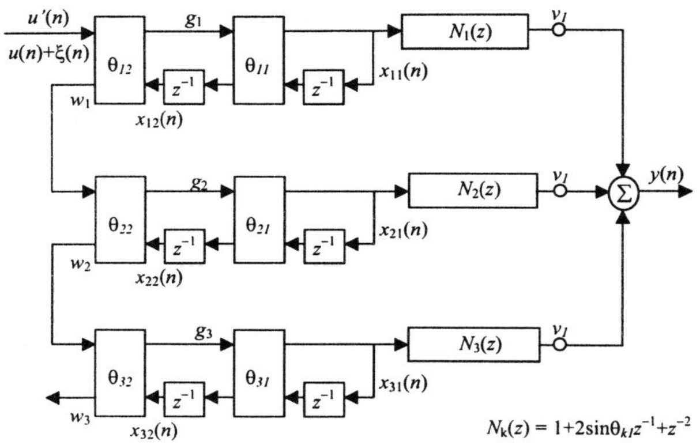

Figure 3. Proposed cascade-parallel structure of orthonormal lattice ANF

Where $D(z) = 1+2\sin\theta_1(1+\sin\theta_2)z^{-1}+\sin\theta_2 z^{-2}$. It is the denominator of the second order notch transfer function. As mentioned earlier, to model multiple notches, rather than cascading the notch filter output, we can cascade the all-pass filter, so that, the second notch filter sees the input signal is an unbiased but all-passed version of the input signal. The new proposed structure is depicted in Figure 3. If we define $N_1(z) = 1+2\sin\theta_{1k}+z^{-2}$ and $v = (1+\sin\theta_1)/2\cos\theta_1\cos\theta_1$, the output of each branch becomes a notch filter transfer function.

4.2 Partial gradient algorithm

For the second lattice notch filter, the MSOE $E[\hat{y}^2(k)]$ has a single minimum point which depends on the notch frequency θ_1

$$\theta_1(k+1) = \theta_1(k) - \mu\frac{\partial \hat{y}(k)}{\partial \theta_1} \qquad (12)$$

The gradient of the notch transfer function is complicated to implement. Straightforward derivation yields [5]

$$\frac{\partial \hat{y}(k)}{\partial \theta_1} = \frac{1+\sin\theta_2}{\cos\theta_2}\Psi(z)x_1(k) \qquad (13)$$

where $\Psi(z)$ is a function of $1/D(z)$. We can simplify the algorithm by using directly the internal state $x_1(k)$ of the lattice filter in Figure 1(a). We summarize the algorithm in Table 1.

Input:	$u(n) = u'(n)+\zeta(n)$	
At time instant n	$\theta_1(n)$: Notch frequency parameter $x_1(n)$ and $x_2(n)$: filter states	
Filter loops	$\begin{bmatrix} g_1 \\ w \end{bmatrix} = \begin{bmatrix} \cos\theta_k & -\sin\theta_k \\ \sin\theta_k & \cos\theta_k \end{bmatrix}\begin{bmatrix} u(n) \\ x_2(n) \end{bmatrix}$	(1)
	$\hat{y}(n) = \frac{1}{2}[u(n)+w]$	(2)
	$\theta_1(n+1) = \theta_1(n) - \mu\hat{y}(n)x_1(n)$	(3)
	$\begin{bmatrix} x_1(n+1) \\ x_2(n+1) \end{bmatrix} = \begin{bmatrix} \cos\theta_1(n+1) & -\sin\theta_1(n+1) \\ \sin\theta_1(n+1) & \cos\theta_1(n+1) \end{bmatrix}\begin{bmatrix} g_1 \\ x_1(n) \end{bmatrix}$ end	(4)

Table 1 Simplified adaptation algorithm for Second order lattice notch filter

Input:	$u'(n) = u(n)+\zeta(n)$	
At time instant n	θ_{k1}: Notch frequency parameter $x_{kl}(n)$ are filter states	
Filter loops	$u'(n) = w_0$ for $k = 1\ldots M$; $\begin{bmatrix} g_k \\ w_k \end{bmatrix} = \begin{bmatrix} \cos\theta_{k2} & -\sin\theta_{k2} \\ \sin\theta_{k2} & \cos\theta_{k2} \end{bmatrix}\begin{bmatrix} w_{k-1} \\ x_{k2}(n) \end{bmatrix}$ end	(1)
	$y(n) = \sum_{k=1}^{M} v_k N_k(z)x_{k1}(n)$ where $z^{-1}x(n) = x(n-1)$	(2)
	$\theta_{k1}(n+1) = \theta_{k1}(n) - \mu\hat{y}(n)x_{k1}(n)$	(3)
	for $k = 1\ldots M$; $\begin{bmatrix} x_{k1}(n+1) \\ x_{k2}(n+1) \end{bmatrix} = \begin{bmatrix} \cos\theta_{k1}(n+1) & -\sin\theta_{k1}(n+1) \\ \sin\theta_{k1}(n+1) & \cos\theta_{k1}(n+1) \end{bmatrix}\begin{bmatrix} g_k \\ x_{k1}(n) \end{bmatrix}$ end	(4)

Table 2 Simplified adaptation algorithm for Second order lattice notch filter

Using the same simplified gradient algorithm, we can derive the corresponding algorithm for the proposed filter structure. Note that this algorithm has the same computational complexity as the cascaded structure. Define the estimated parameters are $\theta_{kl}(n)$ where n is the time index; k is notch number index (k=1,2;...,M) and l is the Lattice parameter index when l=1, θ represent the notch frequency parameter and when l=2, θ is the notch bandwidth parameter. The detailed algorithm is summarized in Table 2.

5. SIMULATIONS

The input signal consists of 3 sinusoids with noise and their frequencies are 0.1π, 0.25π and 0.4π. We compare the convergence of the two algorithms. The filter bandwidth parameter $\sin\theta_2$ affects the convergence speed. So, it is a common practice to apply the algorithm with a small value for $\sin\theta_2$ at the beginning of the algorithm's operation and increase its value later. A simple way to do this is to let $\sin\theta_2$ grow exponentially from a desired value $r(1)$ to $r(\infty)$ according to $\sin\theta_{2,n+1} = r_0\sin\theta_{2,n} + (1-r_0)r_\infty$. In this simulation, $\sin\theta_2$ is set varying from 0.7 to 0.99. The fixed updating stepsize is set to be $\alpha = 0.0025$ to guarantee the stability of the estimate.

In the first experiment, the sinusoid power is 20 dB higher than the noise power. The frequency estimation of the cascaded structure is showed in Figure 4(a) while the corresponding estimate of the proposed algorithm is shown in Figure 4(b). As can be seen, the variances of the estimate are in the same order while the proposed algorithm converges at 750 iterations and the cascaded structure needs more than 2 times of the time to converge.

The second experiment is done with the sinusoid power only 5 dB higher than the noise power. In this case, the cascaded algorithm's frequency estimates do not converge to corresponding frequency due to the altered noise gain after previous cascading section, The proposed algorithm still converges to the desired frequency due to the reason that their input at each branch are unbiased. They are depicted in Figure 5 (a) and (b) respectively.

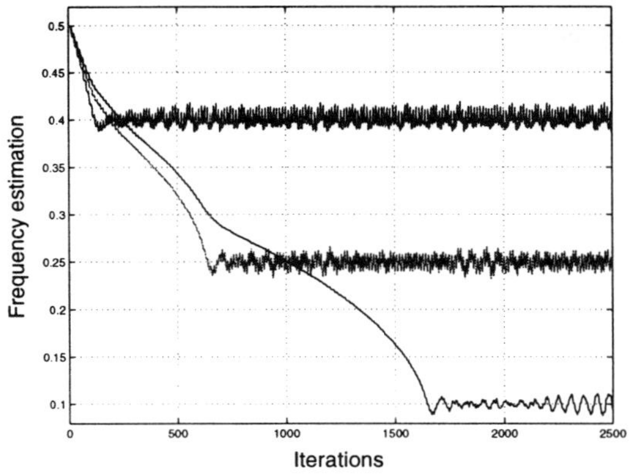

(a)

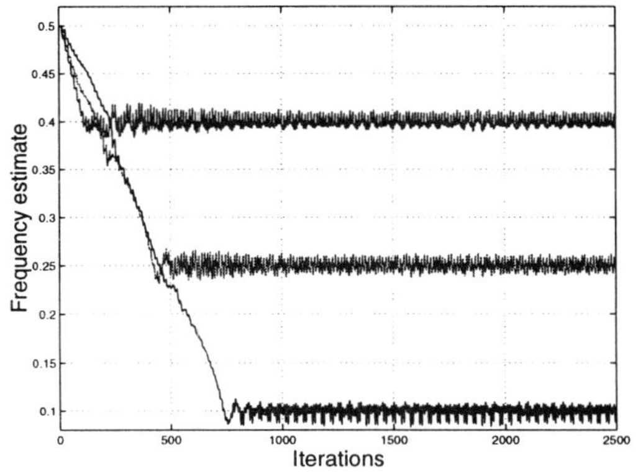

(b)

Figure 4. High SNR case: (a) frequency estimates of cascaded filter structure, (b) Frequency estimate of the proposed parallel-cascaded structure

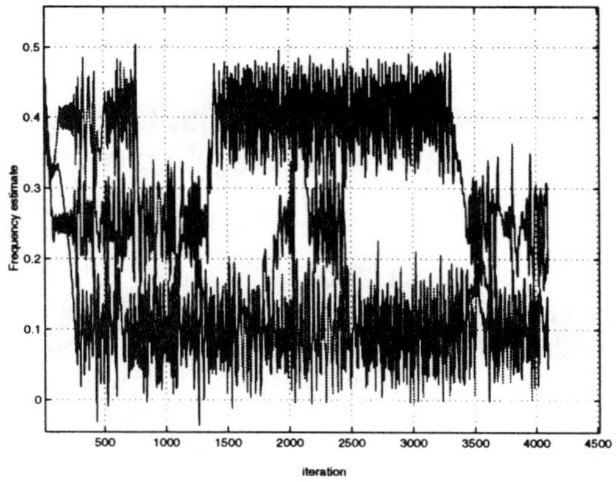

(a)

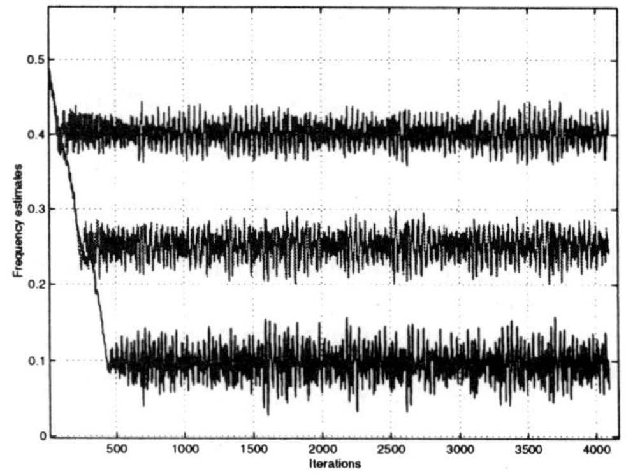

(b)

Figure 5. Low SNR case (a) frequency estimates of cascaded filter structure, (b) Frequency estimate of the proposed parallel-cascaded structure

6. CONCLUSIONS

In this paper, a cascade-parallel multiple adaptive lattice notch filter structure is proposed based on orthogonality of the rational basis function. The advantage of such a structure is that each branch sees an all-pass filtered and unbiased received signal. The notch frequencies are forced to match different sinusoidal frequencies. A simplified gradient algorithm is derived for the proposed algorithm. Simulation shows that the proposed filter structure has better frequency estimation compared with the traditional cascade lattice adaptive notch filters while their complexity remains comparable.

7. REFERENCES

[1] Regalia P. A. *Adaptive IIR Filtering in Signal Processing and Control*, Marcel Dekker, New York, 1994.

[2] Kwan T. and Martin K. "Adaptive detection and enhancement of multiple sinusoids using a cascade of IIR filter" *IEEE Trans. CAS.* Vol. 36 no. 7, pp. 937-947, July 1989

[3] Nishimura S. "An improved adaptive notch filter for detection of multiple sinusoids" *IEICE Trans. Fundamentals* Vol.E77-A pp.950-955, June 1994.

[4] Okello J. and et.al An adaptive notch filter for eliminating multiple sinusoids with reduced bias, *in Proceedings of IEEE International Symposium on Circuits and Systems* (ISCAS '2000), pp. 551 –554, May 2000.

[5] Liu Y. *Adaptive radio frequency interference cancellation in very high speed digital subscriber line systems*, Licentiate Thesis, Helsinki University of Technology, Espoo, Finland, December, 2001.

[6] Cho N.I. and Lee S. U., "On the adaptive lattice notch filter for the detection of sinusoids" *IEEE Transactions on Circuits and Systems II: Analog and Digital Signal Processing*, Vol. 40, No. 7, pp. 405 –416, July 1993.

[7] Cousseau J, Diniz P. S. R, Sentoni G, and Agamennoni O. "On orthogonal realization for adaptive IIR filters," *Int. J. Circ. Theor. Appl.*, Vol.28, pp. 481-500, April. 2000.

[8] Cousseau J. and, Diniz P. S. R. "A Consistent Steiglitz-McBride Algorithm", *in Proceedings of IEEE International Symposium on Circuits and Systems* (ISCAS '93), pp. 52 –55, May 1993.

ON PRE-WHITENED SIGN ALGORITHMS

S. Ben Jebara and H. Besbes

Sup'Com, MASC Department, Cité Technologique des Communications, Ariana, TUNISIA.
E-mail: sofia.benjebara@supcom.rnu.tn, hichem.besbes@supcom.rnu.tn

ABSTRACT

This paper adresses the problem of improving convergence rate of sign algorithms. For such purpose, the idea of decorrelating signals used to pilot the adaptive algorithm is investigated. More precisely, two major modifications are carried : both input and error signals are filtered and the adaptation process is carried each two iterations. The novel algorithm is called "Pre-whitened Sign Algorithm". To justify the idea, we develop an analytical formulation and we consider the particular case of the Dual Sign Algorithm. We prove that the proposed algorithm has low complexity and provides good convergence rate, with acceptable steady state performances. Simulations results are presented to support the theoretical analysis.

1. INTRODUCTION

The signed variants of stochastic gradient adaptive algorithms are proposed in order to improve many aspects: complexity, normalization, robustness, ... These variants include the Sign Algorithm (SA) [1], the Signed Regressor Algorithm (SRA) [2], the Sign-Sign Algorithm (SSA) [3], the Dual Sign Algorithm (DSA) [4]. The main idea is to use the polarity of the error and/or the input to update the filter coefficients.

The main drawback of signed algorithms is the slow convergence rate, especially when colored signals are used as inputs. We think that, as it was applied with basic LMS version [5, 6, 7], we can improve the convergence speed by decorrelating signals used to pilot the adaptive filter.

At our knowledge, there is no pre-whitened version of sign algorithms. We propose to developp a novel family of signed algorithm which is called "Pre-whitened Sign Algorithms".

The main idea of the proposed algorithms is to use both input and error signals in the adaptation process and to carry the adaptation process each two iterations. The input is pre-whitened using an appropriate adaptive predictor and the error is filtered using the same pre-whitener.

The paper is organized as follows : in the next section, we will present the problem, the notations and the motivation of our work. In section 3, we present the main idea, we justify the pre-whitening approach and we illustrate in the case of the Dual Sign Algorithm. In section 4, we present an analytical analysis on the Pre-whitened Dual Sign Algorithm (PDSA), we present transient and steady states performances and compare the proposed algorithm to DSA. Finally, in section 5, some concluding points and perspectives are drawn.

2. PRELIMINARIES AND NOTATIONS

2.1. Sign algorithms

Before presenting the proposed idea, it is necessary to give a brief outline of the adaptation process. The classical block-diagram of

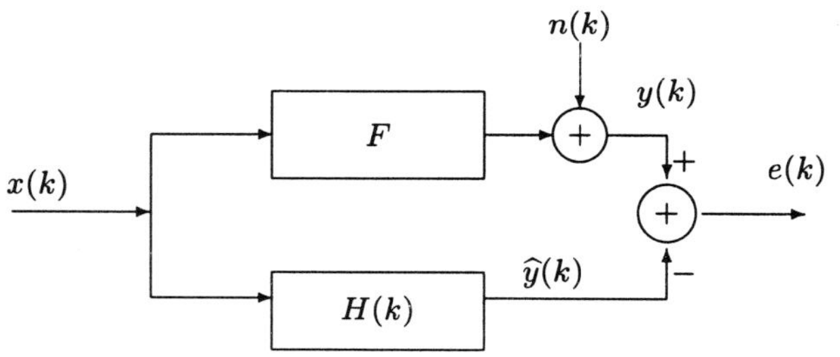

Figure 1: Identification system.

adaptive identification system is depicted in Figure 1. The input/output equation of the system is given by:

$$y(k) = F^T X(k) + n(k), \quad (1)$$

where $F = [f_0, \cdots, f_{L-1}]^T$ is the system impulse response of length L, $(.)^T$ denotes the transpose operator, $X(k) = [x(k), \cdots, x(k-L+1)]^T$ is the input observation vector and $n(k)$ is an additive white Gaussian noise.

The adaptive filter $H(k)$ is governed by the following equations:

$$\begin{array}{l} e(k) = y(k) - H(k)^T X(k) \\ H(k) = H(k) + \mu f(X(k), ..., X(k-m))\, r(e(k), ..., e(k-l)), \end{array} \quad (2)$$

where $e(k)$ is the error signal, μ is a positive step size, $f(.)$ and $r(.)$ are two functions characterizing the used algorithm.

The main sign algorithms are described by the following equations :

- Sign Data Algorithm (SDA):

$$\begin{array}{l} f\left(X(k), , X(k-m)\right) = \text{sign}(X(k)), \\ r\left(e(k), , e(k-l)\right) = e(k). \end{array}$$

- Sign Algorithm (SA):

$$\begin{array}{l} f\left(X(k), , X(k-m)\right) = X(k) \\ r\left(e(k), , e(k-l)\right) = \text{sign}\left(e(k)\right). \end{array}$$

- Sign-Sign Algorithm (SSA):

$$\begin{array}{l} f\left(X(k), , X(k-m)\right) = \text{sign}\left(X(k)\right) \\ r\left(e(k), , e(k-l)\right) = \text{sign}\left(e(k)\right). \end{array}$$

- Dual Sign Algorithm (DSA):

$$\begin{array}{l} f\left(X(k), , X(k-m)\right) = X(k) \\ r\left(e(k)\right) = \begin{cases} \text{sign}\left\{e(k)\right\} & |e(k)| \leq \tau \\ \gamma\text{sign}\left\{e(k)\right\} & |e(k)| > \tau. \end{cases} \end{array}$$

0-7803-7761-3/03/$17.00

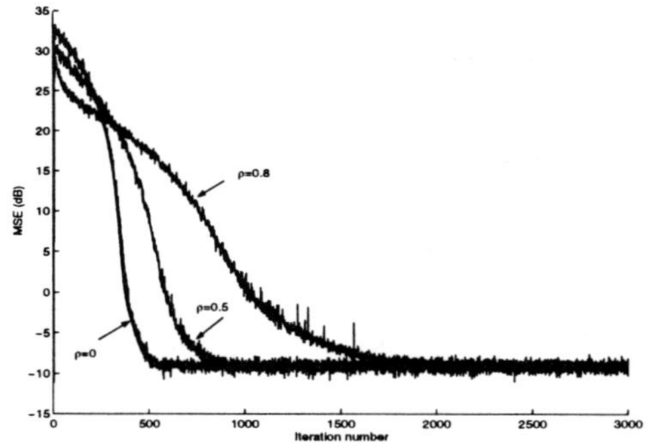

Figure 2: DSA performances with white and correlated inputs.

2.2. Performances criteria

The behavior of the algorithm is usually analyzed through the evolution of the deviation vector $V(k)$ defined as follows:

$$V(k) = H(k) - F. \tag{3}$$

The performances are deduced from the values of the Mean Square Deviation (MSD) and the Mean Square Error (MSE), defined as follows:

$$\begin{aligned} MSD(k) &\triangleq E\left\{V(k)^T V(k)\right\} \\ MSE(k) &\triangleq E\left\{e(k)^2\right\}. \end{aligned} \tag{4}$$

By defining:

$$\mathcal{V}(k) \triangleq E\left\{V(k)V(k)^T\right\}, \tag{5}$$

and using the independence assumption between the observation vector $X(k)$ and the deviation vector $V(k)$, the MSD and MSE can be approximated by:

$$\begin{aligned} MSD(k) &= trace(\mathcal{V}(k)) \\ MSE(k) &\approx trace(R_x \mathcal{V}(k)) + P_n. \end{aligned} \tag{6}$$

were R_x is the auto-correlation matrix of the input vector $X(k)$.

2.3. Motivations

To point out the effect of input correlation on the rate of convergence, we illustrate the performances of the Dual Sign Algorithm. We present in Figure 2 the evolution of $MSE(k)$ versus iteration number for different values of input correlation. In this simulation context, we consider the following case: the input is a first order autoregressive process $AR(1)$:

$$x(k) = \rho x(k-1) + g(k), \tag{7}$$

where $g(k)$ is a Gaussian white noise. We have chosen three values of ρ, namely, $\rho = 0$, $\rho = 0.5$ and $\rho = 0.8$. The system parameters are $L = 10$, $F = [1, 0, 10, -6, -1, 4, 0.1, 5, -2, -0.1]^T$, $P_n = 0.1$ and $P_x = 10$. The DSA parameters are $\tau = 2$, $\gamma = 2^4$ and $\mu = 10^{-3}$. The simulation results are averaged over 200 independent runs using 3000 samples.

Figure 2 shows that, for the same steady state, the convergence time increases when the input correlation increases. Best convergence rate is obtained for white input. The signal pre-whitening approach can then be applied in order to enhance the convergence rate of the classical DSA algorithm.

3. PRE-WHITENED SIGN ALGORITHMS

3.1. Main idea

In this section, we try to develop a pre-whitened version of an existing algorithm given by:

$$H(k+1) = H(k) + \mu f\left(X(k)\right) r(e(k)). \tag{8}$$

Without loss of generality, let consider the input signal modeled by an AR(1). By filtering the error signal as follows:

$$e^f(k) = e(k) - \rho e(k-1), \tag{9}$$

we can show easily that $e^f(k)$ satisfies the following relationship:

$$\begin{aligned} e^f(k) = &\left[F - H(k)\right]^T G(k) + n^f(k) \\ &-\rho\left[H(k) - H(k-1)\right]^T X(k-1) \end{aligned} \tag{10}$$

where $G(k) = X(k) - \rho X(k-1)$ is the observation vector of the white signal $g(k)$.The term $[F - H(k)]^T G(k) + n^f(k)$ corresponds to the residual error when a white input is used and when the additive noise is the filtered version of $n(k)$:

$$n^f(k) = n(k) - \rho n(k-1). \tag{11}$$

It is interesting to note that since $n(k)$ is a white noise, $n^f(k)$ and $n^f(k-2)$ are decorrelated.

The term $-\rho\left[H(k) - H(k-1)\right]^T X(k-1)$ of $e^f(k)$ corresponds to an augmented error due to the adaptation procedure.

If we assume that the adaptive filter is constant during iterations k and $k-1$, and it is equal to $H(k-1)$, the filtered error will be equal to:

$$e^f(k) = -V(k)^T G(k) + n^f(k). \tag{12}$$

At this point, we can note that the filtered error is equivalent to the classical error obtained when the filter and the algorithm are excited by the white input $G(k)$ and when the additive noise is $n^f(k)$. The filtered input $G(k)$ leads to better convergence rate. However, during steady state, the error is amplified because the filtered noise power is greater than the non filtered additive noise: $E({n^f}^2) = (1+\rho^2)E(n^2) > E(n^2)$.

As a result, we can conclude that by updating the adaptive filter, one each two iterations, using a function of the whitened input signal and a function of the filtered error, we will watch the same behavior as exciting the filter by a white signal with an amplified additive noise.

We remark that this idea was carried with the LMS algorithm and was shown to give good performances [7].

3.2. The Pre-whitened Dual Sign Algorithm (PDSA)

In this work, we will carry out our analysis based on the DSA. The adaptation process of Pre-whitened Dual Sign Algorithm (PDSA) is defined as follows:

$$\left\{\begin{array}{ll} H(2k+1) & = H(2k) \\ H(2k+2) & = H(2k) + \mu_f r^f(2k) X^f(2k), \end{array}\right. \tag{13}$$

where

$$r^f(2k) = \left\{\begin{array}{ll} sign\left\{e^f(2k)\right\} & |e^f(2k)| \leq \tau_f \\ \gamma_f sign\left\{e^f(2k)\right\} & |e^f(2k)| > \tau_f, \end{array}\right. \tag{14}$$

where γ_f and τ_f are the new parameters characterizing the $PDSA$.

The filtered signals are defined by:

$$\begin{cases} x^f(k) & = x(k) - p(k)x(k-1) \\ e^f(k) & = e(k) - p(k)e(k-1). \end{cases} \quad (15)$$

The coefficient $p(k)$ estimates the first order correlation of $x(k)$. Since the input statistics are unknown, we use an adaptive algorithm, for example, the NLMS algorithm to estimate the unknown value of ρ:

$$p(k) = p(k-1) + \mu_P \frac{x^f(x-1)x(k-1)}{x(k-1)^2 + \epsilon_p} \quad (16)$$

where μ_P is the predictor step size and ϵ_p is a regularization parameter.

We note that we limit our algorithm to a one tap predictor for two reasons. The first one is to reduce the computational requirements imposed by the additional pre-whitening filter and the second reason is to reduce the noise enhancement due to the filtering process.

4. PDSA PERFORMANCES EVALUATION

4.1. Analytical evaluation

The behavior of the deviation vector is described by the following relationship:

$$V(2k+2) = V(2k) + \mu_f r^f(2k) X^f(2k). \quad (17)$$

The deviation matrix follows the recurrent relationship:

$$\begin{aligned} \mathcal{V}(2k+2) &= \mathcal{V}(2k) + \mu_f^2 E\left\{r^f(2k)^2 X^f(2k) X^f(2k)^T\right\} \\ &+ \mu_f E\left\{r^f(2k)\left[V(2k)X^f(2k)^T + X^f(2k)V(2k)^T\right]\right\}. \end{aligned} \quad (18)$$

The analysis of the proposed algorithm is inspired from the work developed by Mathews [8] on the dual algorithm, it uses the following assumptions:

- the filtered signal is Gaussian,
- the filtered signal $x^f(k)$ is white,
- the deviation vector $V(k)$ is independent of the input signal $X(k)$,
- the step size is too small.

Using the above mentionned assumptions, and applying the Price theorem [9], we get:

$$\begin{aligned} & E\left\{r^f(2k)^2 X^f(2k) X^f(2k)^T\right\} = \\ & \sqrt{\tfrac{2}{\pi}}\left(\gamma_f^2 - 1\right) R_{x^f} \mathcal{V}(2k) R_{x^f} \frac{\tau_f}{E\left\{e^f(2k)^2\right\}^{\frac{3}{2}}} e^{-\frac{\tau_f^2}{2E\left\{e^f(2k)^2\right\}}} \\ & + \left[\gamma_f^2 - \left(\gamma_f^2 - 1\right) erf\left(\frac{\tau_f}{\sqrt{E\left\{e^f(2k)^2\right\}}}\right)\right] R_{x^f}, \end{aligned} \quad (19)$$

$$\begin{aligned} & E\left\{r^f(2k)\left[V(2k)X^f(2k)^T + X^f(2k)V(2k)^T\right]\right\} = \\ & -\sqrt{\frac{2}{\pi}} \frac{\left\{R_{x^f}\mathcal{V}(2k) + \mathcal{V}(2k)R_{x^f}\right\}}{\sqrt{E\left\{e^f(2k)^2\right\}}} \left\{1 + (\gamma_f - 1)\, e^{-\frac{\tau_f^2}{E\left\{e^f(2k)^2\right\}}}\right\} \end{aligned} \quad (20)$$

where $erf(.)$ is the error function defined by:

$$erf(s) = \sqrt{\frac{2}{\pi}} \int_0^s e^{\frac{-t^2}{2}} dt.$$

The power of the filtered error

$$E\left\{e^f(2k)^2\right\} \triangleq E\left\{\left[-V(2k)^T X^f(2k) + n^f(2k)\right]^2\right\}$$

is expressed by :

$$E\left\{e^f(2k)^2\right\} = \left(1 - \rho^2\right) P_x trace(\mathcal{V}(2k)) + \left(1 + \rho^2\right) P_n. \quad (21)$$

Using the three previous expressions, we obtain:

$$\begin{aligned} \mathcal{V}(2k+2) &= \mathcal{V}(2k) + \\ & \mu_f^2 \left[\sqrt{\tfrac{2}{\pi}}\left(\gamma_f^2 - 1\right)\left(1 - \rho^2\right)^2 P_x^2 \frac{\tau_f}{E\left\{e^f(2k)^2\right\}^{\frac{3}{2}}} e^{-\frac{\tau^2}{2E\left\{e^f(2k)^2\right\}}} \mathcal{V}(2k) \right. \\ & \left. + \left(\gamma_f^2 - \left(\gamma_f^2 - 1\right) erf\left(\frac{\tau_f}{\sqrt{E\left\{e^f(2k)^2\right\}}}\right)\right)\left(1 - \rho^2\right) P_x I_L\right] \\ & -\mu_f \sqrt{\frac{2}{\pi}} \frac{2\left(1 - \rho^2\right) P_x}{\sqrt{E\left\{e^f(2k)^2\right\}}} \left(1 + (\gamma_f - 1)\, e^{-\frac{\tau_f^2}{E\left\{e^f(2k)^2\right\}}}\right) \mathcal{V}(2k), \end{aligned} \quad (22)$$

where I_L is the $L \times L$ identity matrix.

The behavior of the algorithm is described by resolving iteratively equations (21) and (22).

We note that by replacing filtered parameters, in equations (19) to (22) by the original ones, we get the performances of the classical Dual Sign Algorithm [8].

4.2. Validity of the theoretical expression

Figure 3 compares the theoretical and the simulation curves of $MSE(k)$ for two values of step size $\mu_{fx} = 0.002$ and $\mu_{fx} = 0.003$. The PDSA parameters are $\gamma_f = 2^5$ and $\tau_f = 0.5$. The simulation context is the same as previous. This figure shows that simulation curves closely match the corresponding theoretical ones, validating the proposed analysis. In the next subsections, we use theoretical curves to analyze algorithm behavior.

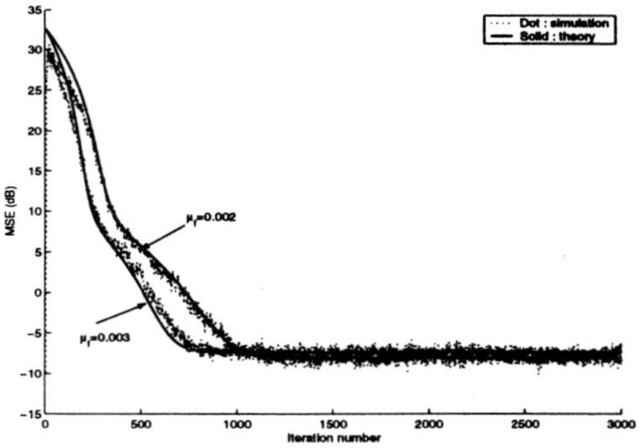

Figure 3: Comparison between theory and simulation results for PDSA algorihm.

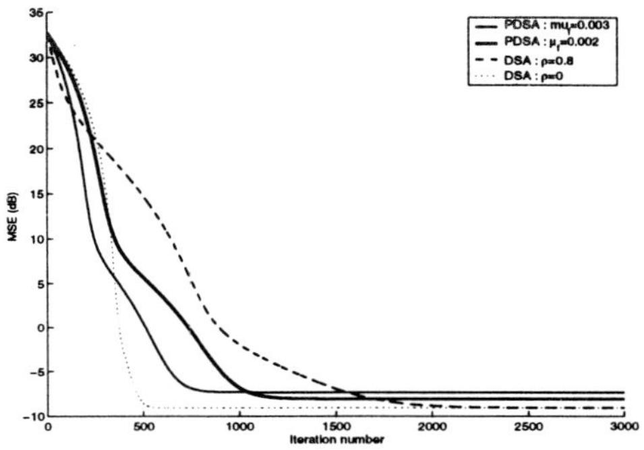

Figure 4: Convergence analysis of PDSA algorihm.

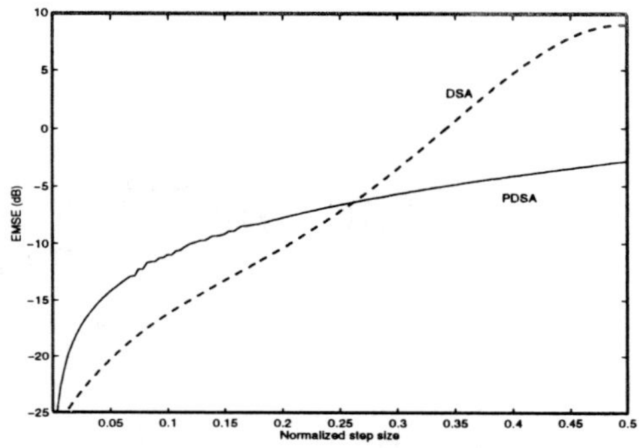

Figure 5: Evolution of the $EMSE$.

4.3. Transient analysis

Figure 4 shows the evolution of the theoretical curves of $MSE(k)$ using DSA and PDSA, we show that effectively, the transient state is improved when we use the PDSA. When $\mu_f = 0.002$, the convergence time is twice the one obtained using the DSA algorithm when the input is white ($\rho = 0$). In fact, this was expected since the PDSA is updated one time each two iterations. We notice therefore, that steady state becomes worse, if the step size is greater ($\mu_f = 0.003$). The adaptation step size should result from a good tradeoff between good convergence rate and acceptable steady state.

It is important to note that the same kind of results are obtained with the other pre-whitened sign algorithms (for example Pre-whitened Sign Algorithm). Due to lack of space, results are not presented in this paper.

4.4. Steady state behavior

Figure 5 shows the evolution of the Excess Mean Square Error ($EMSE \triangleq MSE - P_n$) versus the normalized step size ($\nu = \mu L P_x$ for DSA and ($\nu_f = \mu_f L P_{x^f}$ for PDSA), using the same simulation conditions as previous, with $\rho = 0.8$. This figure shows that for both algorithms, steady state degradates when the step size increases. For small step size, the DSA outperforms, it is due to additive noise amplification in the term $e^f(k)$ ($P_{n^f} > P_n$). We therefore remind that PDSA convergence rate is better. One advantage of $PDSA$ is the increasing slope, it is slower than that of DSA. This means that, if we increase the step size, the $PDSA$ diverges slower than DSA, this is explained by the fact that power of the pre-whitened signal is smaller than the original signal.

5. CONCLUSION

In this paper, we have applied the pre-whitening concept for sign algorithms. We derived an algorithm using both input and error pre-whitening in the adaptation process, which is carried each two iterations. We emphasis our analysis on Pre-whitened Dual Sign Algorithm. The Transient behavior and steady state are then derived. We justify that the proposed concept has low complexity and provides good convergence rate, with acceptable steady state performances. As perspectives, we think about using this algorithm in acoustic echo cancellation, in order to improve algorithm robusteness during double talk. In fact, classical sign algorithms fail during crossing from single talk to double talk.

6. REFERENCES

[1] S. H. Cho and V. J. Mathews, "Tracking analysis of the sign algorithm in non stationary environments," *IEEE Trans. Acoust., Speech, Signal Process.*, vol. 38, no. 12, pp. 2046-2057, Dec. 1990.

[2] E. Eweda, "Analysis and design of a signed regressor LMS algorithm for stationary and non stationary adaptive filtering with correlated Gaussian data," *IEEE Trans. Circuits Syst.*, vol. 37, no. 11, pp. 1367-1374, Nov. 1990.

[3] E. Eweda, "Convergence analysis of an adaptive filter equipped with the sign-sign algorithm," *IEEE Trans. Automat. Contr.*, vol. 40, no. 10, pp. 1807-1811, Oct. 1995.

[4] C. P. Kwong, "Dual sign algorithm for adaptive filtering," *IEEE Trans. Comm.*, vol. 34, pp. 1272-1275, Dec. 1986.

[5] B. Widrow and S. D. Stearns, "*Adaptive Signal Processing,*" Englewood Cliffs, NJ:Prentice-Hall, 1985.

[6] M. Mboup, M. Bonnet and N. Bershad, "LMS coupled adaptive prediction and system identification: a statisical model and transient mean analysis," *IEEE Trans. on Signal Processing*, vol. 42, no 10, pp. 2607-2615, Oct. 1994.

[7] H. Besbes, S. Ben Jebara "The pre-whitened NLMS: a promising algorithm for acoustic echo cancellation," *Proc. IEEE International Conference on Electronics, Circuits and Systems*, Dubrovnik, Croatia, 2002.

[8] V. J. Mathews, "Performance analysis of adaptive filters equipped with the dual sign algorithm," *IEEE Trans. on signal processing*, vol. 39, no. 1, pp. 85-91, Jan. 1991.

[9] R. Price, "A useful theorem for nonlinear devices having Gaussian inputs," *IEEE Trans. on Information theory* , vol. IT-4, pp. 69-72, June 1958.

OPTIMUM MIMO TRANSMIT-RECEIVER DESIGN IN PRESENCE OF INTERFERENCE

S. U. Pillai

Department of Electrical Engineering
Polytechnic University
Brooklyn, New York 11201
Email: pillai@hora.poly.edu

H. S. Oh

Samsung Electronics Co., Ltd.
Suwon P.O.Box 105
Suwon, Korea 442-742
Email: hs910.oh@samsung.com

ABSTRACT

In presence of channels that respond to transmitter diversity to maximize the output SINR, it becomes necessary to jointly optimize the transmit signal vector and the receiver bank of filters in presence of signal dependent interference and noise. This multichannel matched transmitter-receiver structure is superior to its scalar counterpart, since several aspects of the signal dependent interference spectrum are brought to light in this case. The cross-interference spectral matrices infact contain much more information compared to the single channel case, and despite the highly nonlinear nature of the problem, it is possible to derive a closed form solution for the optimum matched receiver filter bank. In addition, an iterative algorithm that appears to converge for all SINR is proposed to determine the optimum transmit signal vector.

1. INTRODUCTION

The problem of jointly optimizing the transmitter and receiver so as to maximize the output signal-to-interference plus noise ratio (SINR) is an important one in many communication scenes where signal dependent interference or multipath is a leading source of interference. To excite various hidden modes of the channel and signal dependent interference for better channel detection in presence of signal dependent interference and noise, it is desirable to illuminate the channel with both horizontally and vertically polarized signals, and in that context the design of an optimal transmit signal vector is quite meaningful. In radar, for example, such multi-mode transmit signal vectors induce in addition to channel returns various aspects of signal dependent interference returns – horizontally and vertically polarized signal dependent interference returns – that possess their own respective power spectra as well as cross-spectra. At the receiver an array of sensors is used to collect the data that is passed through a bank of filters to obtain a single output. In the classical detection problem, at a specified instant, this output is used to decide the presence or absence of a channel, and the problem is to design the optimal transmit-receiver pair so as to maximize the output signal-to-interference plus noise ratio (SINR) at the decision instant.

This research was partially supported by the Office of Naval Research under contract N00014-02-1-0083.

This multi-mode pulse shaping is a highly nonlinear problem that depends on the multichannel channel response, the multichannel interference spectral and cross spectral matrices and the noise spectra. It is shown here that explicit analytic solution to this nonlinear problem involves a Fredholm matrix integral operator of the first kind whose kernel is formed from a "whitened form" of the impulse response matrix. Discussions on practical scenarios are presented to demonstrate the effectiveness of the multi-mode approach in unfriendly terrains.

2. PROBLEM FORMULATION

Consider an $m \times 1$ real transmit signal vector[1]

$$\begin{aligned} \mathbf{f}(t) &= [f_1(t), f_2(t), \cdots, f_m(t)]' \\ &\leftrightarrow \mathbf{F}(\omega) = [F_1(\omega), F_2(\omega), \cdots, F_m(\omega)]'. \end{aligned} \quad (1)$$

of finite duration t_0 and constant energy E that simultaneously illuminates both a stationary (nonmoving) channel and surrounding signal dependent interference. The returns are collected at m distinct sensors at the receiver in a noisy environment. Let

$$\mathbf{q}(t) = \begin{bmatrix} q_{11}(t) & q_{12}(t) & \cdots & q_{1m}(t) \\ q_{21}(t) & q_{22}(t) & \cdots & q_{2m}(t) \\ \vdots & \vdots & \ddots & \vdots \\ q_{m1}(t) & q_{m2}(t) & \cdots & q_{mm}(t) \end{bmatrix} \leftrightarrow \mathbf{Q}(\omega),$$

where

$$\mathbf{Q}(\omega) = \begin{bmatrix} Q_{11}(\omega) & Q_{12}(\omega) & \cdots & Q_{1m}(\omega) \\ Q_{21}(\omega) & Q_{22}(\omega) & \cdots & Q_{2m}(\omega) \\ \vdots & \vdots & \ddots & \vdots \\ Q_{m1}(\omega) & Q_{m2}(\omega) & \cdots & Q_{mm}(\omega) \end{bmatrix} \quad (2)$$

represent the impulse response matrix of the channel which is assumed to be square integrable, real, causal and deterministic, while the signal dependent interference returns are

[1] In general, lower case and upper case bold letters are used to represent vectors and matrices respectively. Thus a, $\mathbf{a}$, $\mathbf{A}$ represent scalar, vector and matrix respectively, while $\mathbf{A}'$, $\overline{\mathbf{A}}$, $\mathbf{A}^*$ represent the transpose, complex conjugate and complex conjugate transpose of $\mathbf{A}$.

0-7803-7761-3/03/\$17.00 ©2003 IEEE

assumed to be stationary stochastic processes. Let $\mathbf{w}_i(t)$ represen t the signal dependent interference response to the input signal $f_i(t)$, so that with $\mathbf{n}(t)$ and $\mathbf{r}(t)$ represen ting the noise and received signal vectors respectively, we get

$$\begin{aligned} \mathbf{r}(t) &= \mathbf{q}(t) * \mathbf{f}(t) + \sum_{i=1}^{m} \mathbf{w}_i(t) * f_i(t) + \mathbf{n}(t) \\ &\triangleq \mathbf{s}(t) + \mathbf{w}_f(t) + \mathbf{n}(t). \end{aligned} \quad (3)$$

Let

$$\mathbf{h}(t) = \begin{bmatrix} h_1(t) \\ h_2(t) \\ \vdots \\ h_m(t) \end{bmatrix} \leftrightarrow \mathbf{H}(\omega) = \begin{bmatrix} H_1(\omega) \\ H_2(\omega) \\ \vdots \\ H_m(\omega) \end{bmatrix} \quad (4)$$

represen t the causal receiver impulse response vector to be determined, and $y(t)$ the final receiver output signal. Then

$$y(t) = \mathbf{h}'(t) * \mathbf{r}(t) \triangleq y_s(t) + y_{i+n}(t),$$

where $y_s(t)$ represen ts the output channel signal component, and $y_{i+n}(t)$ the total in terference component.

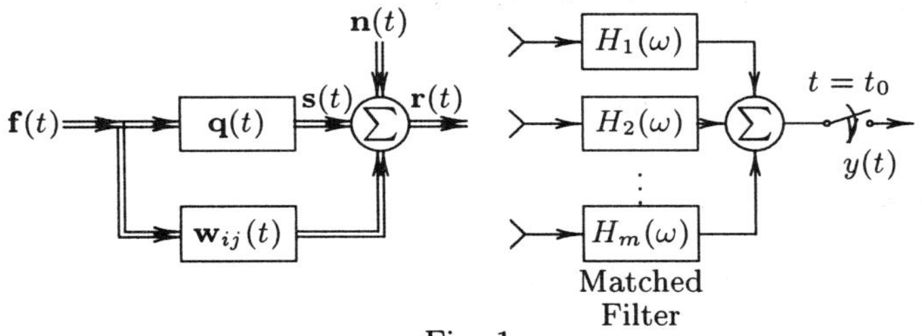

Fig. 1

The output signal component in (3) is given b y

$$y_s(t) = \int_0^t \mathbf{h}'(t-\tau)\mathbf{s}(\tau)\,d\tau = \frac{1}{2\pi}\int_{-\infty}^{+\infty} \mathbf{H}'(\omega)\mathbf{S}(\omega)e^{j\omega t_0}\,d\omega \quad (5)$$

where

$$\mathbf{S}(\omega) = \mathbf{Q}(\omega)\mathbf{F}(\omega). \quad (6)$$

Let $\mathbf{G}_{ij}(\omega)$ represen t the $m \times m$ cross signal dependent interference spectral density matrix betw een the signal dependent in terference responses $\mathbf{w}_{c,i}(t)$ and $\mathbf{w}_{c,j}(t)$. Th us with

$$\mathbf{G}_{ij}(\omega) \leftrightarrow \mathbf{R}_{ij}(\tau) = E[\mathbf{w}_{c,i}(t+\tau)\mathbf{w}^*_{c,j}(t)], \; i,j = 1 \to m, \quad (7)$$

and from (3), we get the total signal dependent interference pow er spectral density to be

$$\mathbf{G}_F(\omega) = \sum_{i=1}^{m}\sum_{j=1}^{m} \mathbf{G}_{ij}(\omega)F_i(\omega)F_j^*(\omega) > 0. \quad (8)$$

Notice that although $\mathbf{G}_F(\omega)$ is $m \times m$, all m^2 signal dependent interference cross-spectral matrices in (7) participate in its formulation. With $\mathbf{G}_n(\omega)$ represen ting the noise spectral density, we get the receiver output interference plus noise po wer to be

$$E[y^2_{i+n}(t)] = \frac{1}{2\pi}\int_{-\infty}^{+\infty} \mathbf{H}'(\omega)[\mathbf{G}_F(\omega) + \mathbf{G}_n(\omega)]\overline{\mathbf{H}}(\omega)\,d\omega. \quad (9)$$

Our objective is to determine an admissible transmit-receive pair $\mathbf{f}(t)$, $\mathbf{h}(t)$ that maximizes the output signal to interference plus noise ratio

$$SINR = \rho(t_0) = \frac{y_s^2(t_0)}{E[y^2_{i+n}(t)]} \quad (10)$$

at some prescribed detection instant t_0, subject to an energy constraint on $\mathbf{f}(t)$.

More precisely , w e seek to achieve the iterated maxim um

$$\gamma_0 = \underset{\mathbf{f}}{\text{Max}}\;\underset{\mathbf{h}}{\text{Max}}\;\rho(t_0), \quad (11)$$

and the first step is to maximize $\rho(t_0)$ ov er $\mathbf{h}(t)$ with $\mathbf{f}(t)$ held fixed. To make further progress, let

$$\mathbf{G}_0(\omega) = \mathbf{G}_F(\omega) + \mathbf{G}_n(\omega) \quad (12)$$

represent the total interference and noise spectrum. There exists a matrix $\mathbf{L}(s)$, real for real s, unique up to sign and analytic together with its inverse in $Re\, s > 0$, such that

$$\mathbf{G}_0(\omega) = \mathbf{G}_F(\omega) + \mathbf{G}_n(\omega) = \mathbf{L}(j\omega)\mathbf{L}^*(j\omega). \quad (13)$$

$\mathbf{L}(s)$ is known as the left-inv erse Wiener factor for $\mathbf{G}_0(s)$. Using (13), we obtain

$$\rho(t_0) = \frac{\left|\frac{1}{2\pi}\int_{-\infty}^{+\infty} \mathbf{H}'(\omega)\mathbf{L}(j\omega)\mathbf{L}^{-1}(j\omega)\mathbf{S}(\omega)e^{j\omega t_0}\,d\omega\right|^2}{\frac{1}{2\pi}\int_{-\infty}^{+\infty} \mathbf{H}'(\omega)\mathbf{L}(j\omega)\mathbf{L}^*(j\omega)\overline{\mathbf{H}}(\omega)\,d\omega}, \quad (14)$$

a form which suggests an obvious use of Sch warz's inequality. However, the direct application of this inequality leads to $\overline{\mathbf{H}}(\omega) = \mathbf{L}^{*-1}(j\omega)[\mathbf{L}^{-1}(j\omega)\mathbf{S}(\omega)e^{j\omega t_0}]$, which does not correspond to a causal receiver, and hence it must be postponed.

T o obtain the best causal receiver, we need to ov ercome this difficulty. T ow ards this let

$$\mathbf{v}(t) \leftrightarrow \mathbf{L}'(j\omega)\mathbf{H}(\omega), \quad \mathbf{g}(t) \leftrightarrow \mathbf{L}^{-1}(j\omega)\mathbf{S}(\omega). \quad (15)$$

Clearly, since $\mathbf{L}(s)$ and $\mathbf{L}^{-1}(s)$ are analytic in $Re\, s > 0$ and bounded in the closure $Re\, s \geq 0$, both $\mathbf{v}(t)$ and $\mathbf{g}(t)$ are causal and if

$$\mathbf{g}(t_0 - t)u(t) \leftrightarrow \mathbf{K}(\omega), \quad (16)$$

represen t the transform of the 'causal pulse' $\mathbf{g}(t_0 - t)u(t)$. Then Parsev al's theorem applied to (14) yields

$$\Delta = \int_0^{t_0} \mathbf{v}'(t)\mathbf{g}(t_0 - t)\,dt = \frac{1}{2\pi}\int_{-\infty}^{+\infty} \mathbf{H}'\mathbf{L}(j\omega)\overline{\mathbf{K}}(\omega)\,d\omega. \quad (17)$$

Hence by Sch w arz' inequality, we obtain

$$\rho(t_0) \leq \frac{1}{2\pi}\int_{-\infty}^{+\infty} \mathbf{K}^*(\omega)\mathbf{K}(\omega)\,d\omega \quad (18)$$

with equality iff

$$\mathbf{L}'(j\omega)\mathbf{H}(\omega) = \mu\mathbf{K}(\omega), \text{ or } \mathbf{H}(\omega) = \mu\mathbf{L}'^{-1}(j\omega)\mathbf{K}(\omega), \quad (19)$$

μ any real nonzero normalization constant. The impulse response $\mathbf{h}(t)$ of this matched filter with transfer function $\mathbf{H}(\omega)$ given in (19) is real and causal. In addition,

$$\underset{\mathbf{f}}{\text{Max}}\ \underset{\mathbf{h}}{\text{Max}}\ \rho(t_0) = \underset{\mathbf{f}}{\text{Max}} \int_0^{t_0} \mathbf{g}'(t)\mathbf{g}(t)\,dt, \quad (20)$$

where

$$\mathbf{g}(t) \leftrightarrow \mathbf{L}^{-1}(j\omega)\mathbf{Q}(\omega)\mathbf{F}(\omega). \quad (21)$$

Maximization over $\mathbf{f}(t)$ is achieved by choosing an $\mathbf{F}(\omega)$ in (20) whose associated $\mathbf{g}(t)$ possesses the largest possible energy over the interval $0 \leq t \leq t_0$. Unfortunately, when the signal dependent interference is significant, the Wiener-Hopf equation in (13) imposes a nonlinear dependence of $\mathbf{f}(t)$ on $\mathbf{L}(j\omega)$ as well as on $\mathbf{g}(t)$, thus severely complicating the problem.

However, in the absence of signal dependent interference it is possible to obtain a solution to the problem presented here. In section 3, we first discuss this particular case and use that solution as a guideline for the general case.

3. THE OPTIMUM TRANSMIT PULSE

In the absence of signal dependent interference and in presence of white noise, we have $\mathbf{G}_{ij}(\omega) \equiv 0$ and $\mathbf{G}_n(\omega) \equiv \sigma_n^2\mathbf{I}_m$. Thus $\mathbf{L}(j\omega) \equiv \sigma_n\mathbf{I}_m$ and it follows from (21) that

$$\mathbf{g}(t) \leftrightarrow \frac{1}{\sigma_n}\mathbf{Q}(\omega)\mathbf{F}(\omega). \quad (22)$$

In addition, for an ideal stationary point channel, $\mathbf{q}(t) = c\delta(t)\mathbf{I}_m$, c a real nonzero constant, so that $\mathbf{Q}(\omega) \equiv c\mathbf{I}_m$, $\mathbf{g}(t) = c\mathbf{f}(t)/\sigma_n$ and the desired maximum in (11) is $\gamma_0 = c^2E/\sigma_n^2$, where E is the energy of the transmit pulse $\mathbf{f}(t)$. In this limiting case the shape of $\mathbf{f}(t)$ is theoretically irrelevant and may be chosen chirp-like to combine the need for increased range and enhanced resolution that is made possible by the compressive properties of the matched receiver with impulse response

$$\mathbf{h}(t) = \frac{\mu c}{\sigma_n}\mathbf{f}(t_0 - t). \quad (23)$$

However, the freedom in selecting the input transmit signal is restricted as soon as we move away from this ideal scene. Before we discuss the general case, another interesting special case that affords exact solution is of interest. This situation also corresponds to a signal dependent interference free scene, but with nonwhite noise spectral density and a nonimpulsive channel response.

Thus, in the absence of signal dependent interference, $\mathbf{G}_{ij}(\omega) \equiv 0$ and $\mathbf{L}(s) = \mathbf{L}_n(s)$, where $\mathbf{L}_n(s)$ is the known Wiener-Hopf factor of $\mathbf{G}_n(\omega) = \mathbf{L}_n(j\omega)\mathbf{L}_n^*(j\omega)$. From (21), $\mathbf{g}(t) \leftrightarrow \mathbf{L}_n^{-1}(j\omega)\mathbf{Q}(\omega)\mathbf{F}(\omega)$. Let $\mathbf{\Phi}(t) \leftrightarrow \mathbf{L}_n^{-1}(j\omega)\mathbf{Q}(\omega)$ and rewrite $\mathbf{g}(t)$ as $\mathbf{g}(t) = \int_0^t \mathbf{\Phi}(t-\tau)\mathbf{f}(\tau)\,d\tau$. Introduce the kernel

$$\mathbf{\Omega}(\tau_1, \tau_2) = \int_0^{t_0} \mathbf{\Phi}'(t-\tau_1)\mathbf{\Phi}(t-\tau_2)\,dt, \quad (0 \leq \tau_1, \tau_2 \leq t_0), \quad (24)$$

and the associated linear operator T defined by the mapping

$$\mathbf{f} \rightarrow \mathbf{Tf} = \int_0^{t_0} \mathbf{\Omega}(\tau_1, \tau_2)\mathbf{f}(\tau_2)\,d\tau_2, \quad (0 \leq \tau_1 \leq t_0). \quad (25)$$

Then[2]

$$\underset{\mathbf{h}}{\text{Max}}\ \rho(t_0) = \int_0^{t_0} \mathbf{g}'(t)\mathbf{g}(t)\,dt = (\mathbf{f}, \mathbf{Tf}) \quad (26)$$

is computed as the inner-product of $\mathbf{f}(t)$ and $\mathbf{Tf}(t)$.

Since $\mathbf{\Omega}(\tau_1, \tau_2) = \mathbf{\Omega}(\tau_2, \tau_1)$, $\mathbf{T}$ is symmetric, completely continuous and nonnegative-definite. As such, the integral equation [1]

$$\int_0^{t_0} \mathbf{\Omega}(\tau_1, \tau_2)\mathbf{\Psi}_r(\tau_2)\,d\tau_2 = \lambda_r\mathbf{\Psi}_r(\tau_1), \quad (0 \leq \tau_1 \leq t_0), \quad (27)$$

admits a denumerable set of eigenfunction solutions $\mathbf{\Psi}_r(\tau_1)$ of $\mathbf{T}$, orthonormal and kernel mean-square complete over $0 \leq \tau_1 \leq t_0$. The eigenvalues λ_r are positive and when arranged as a monotonically nonincreasing sequence

$$\lambda_1 \geq \lambda_2 \geq, \cdots, \cdots, \quad (28)$$

have zero as sole possible limit point. In view of (26),

$$\underset{\mathbf{f}}{\text{Max}}\ \underset{\mathbf{h}}{\text{Max}}\ \rho(t_0) = \underset{\mathbf{f}}{\text{Max}}\ (\mathbf{f}, \mathbf{Tf}) = \lambda_1 E. \quad (29)$$

is attained by choosing

$$\mathbf{f}(t) = \sqrt{E}\mathbf{\Psi}_1(t), \quad (0 \leq t \leq t_0). \quad (30)$$

From (16) and (19), the receiver transfer function matched to this optimum transmit signal is given by

$$\mathbf{H}(\omega) = \mu\mathbf{L}_n'^{-1}(j\omega)\mathbf{K}(\omega) \quad (31)$$

in which $\mathbf{g}(t) \leftrightarrow \sqrt{E}\mathbf{L}_n'^{-1}(j\omega)\mathbf{Q}(\omega)\mathbf{\Gamma}_1(\omega)$ and $\mathbf{\Psi}_1(t) \leftrightarrow \mathbf{\Gamma}_1(\omega)$ represents the eigenvector associated with the largest eigenvalue in (27). In the absence of signal dependent interference, equations (29)–(31) are exact and represent generalizations to [2]–[4].

In the most general situation, signal dependent interference is not negligible and the equation in (13) for the Wiener-Hopf factor $\mathbf{L}(j\omega)$ permits no simplification. Nonetheless, it is possible to construct a successful iterative procedure for the determination of an optimal pair $\mathbf{f}(t)$, $\mathbf{h}(t)$ that extends the integral equation approach developed in the signal dependent interference free case and seems to converge robustly over a wide range of input SINR. Given the channel impulse response $\mathbf{q}(t)$, or its transform $\mathbf{Q}(\omega)$, and the noise and signal dependent interference spectral densities

---

[2] For any two functions $\alpha(t)$ and $\beta(t)$ square-integrable over $0 \leq t \leq t_0$,

$$(\alpha, \beta) \stackrel{\Delta}{=} \int_0^{t_0} \alpha^*(t)\beta(t)\,dt.$$

Also, $\| \alpha \| \stackrel{\Delta}{=} (\alpha, \alpha)^{1/2}$.

$\mathbf{G}_n(\omega)$ and $\mathbf{G}_{ij}(\omega)$, let t_0 and E represent the duration and energy of the desired optimum pulse $\mathbf{f}(t)$.

1. To begin with, for $k=0$ let $\mathbf{f}_0(t)$ represent any real causal signal of duration t_0 and energy E.

2. At stage k, we assume $\mathbf{f}_k(t)$ is known. Let $\mathbf{f}_k(t) \leftrightarrow \mathbf{F}_k(\omega)$ and obtain the Wiener-Hopf solution of the equation

$$\mathbf{L}_k(j\omega)\mathbf{L}_k^*(j\omega) = \mathbf{G}_n(\omega) + \sum_{i=1}^{m}\sum_{j=1}^{m} \mathbf{G}_{ij}(\omega) F_i^{(k)}(\omega) F_j^{(k)*}(\omega), \tag{32}$$

where $F_i^{(k)}(\omega)$ represents the ith component of $\mathbf{F}_k(\omega)$.

3. Let

$$\Phi_k(t) \leftrightarrow \mathbf{L}_k^{-1}(j\omega)\mathbf{Q}(\omega) \tag{33}$$

and compute

$$\mathbf{\Omega}_k(\tau_1,\tau_2) = \int_0^{t_0} \Phi_k'(t-\tau_1)\Phi_k(t-\tau_2)\,dt, \quad (0 \le \tau_1,\tau_2 \le t_0); \tag{34}$$

4. Find the largest eigenvalue $\lambda_1^{(k)}$ and the corresponding eigenfunction $\Psi_1^{(k)}(t)$ of the integral equation

$$\int_0^{t_0} \mathbf{\Omega}_k(\tau_1,\tau_2)\Psi_1^{(k)}(\tau_2)\,d\tau_2 = \lambda_1^{(k)}\Psi_1^{(k)}(\tau_1), \quad (0 \le \tau_1 \le t_0). \tag{35}$$

5. Ideally for some k we should have

$$\mathbf{f}_k(t) = \sqrt{E}\Psi_1^{(k)}(t).$$

If such is not the case, define the error function

$$\mathbf{e}_k(t) = \mathbf{f}_k(t) - \sqrt{E}\Psi_1^{(k)}(t).$$

The energy of the above error signal is given by

$$\epsilon_k^2 = \| \mathbf{e}_k(t) \|^2 = E + E - 2\sqrt{E}\underbrace{(\mathbf{f}_k, \Psi_1^{(k)})}_{c_1^{(k)}} = 2\sqrt{E}(\sqrt{E} - c_1^{(k)}). \tag{36}$$

The positive quantity ϵ_k is a meaningful measure of the error at stage k, and can be used to weigh the desired solution $\Psi_1^{(k)}(t)$ and provide the correction factor for the next stage. Thus we define the new update rule

$$\mathbf{f}_{k+1}(t) = \mathbf{f}_k(t) + \epsilon_k \Psi_1^{(k)}(t), \tag{37}$$

and equal energy constraint at all stages gives the final update rule to be

$$\mathbf{f}_{k+1}(t) = \frac{\mathbf{f}_k(t) + \epsilon_k \Psi_1^{(k)}(t)}{\sqrt{\left(1 + \frac{\epsilon_k}{\sqrt{E}}\right)^2 - \left(\frac{\epsilon_k}{\sqrt{E}}\right)^3}}; \tag{38}$$

Notice that the update rule in (38) only requires the knowledge of $c_1^{(k)}$ in (36) in addition to $\Psi_1^{(k)}(t)$.

6. Let $\mathbf{f}_{k+1}(t) \leftrightarrow \mathbf{F}_{k+1}(\omega)$, and at the next stage the above steps (2–5) are repeated until ϵ_k becomes acceptably small. Then

$$\mathbf{f}(t) = \underset{k\to\infty}{\text{limit}}\ \mathbf{f}_k(t). \tag{39}$$

The optimum receiver impulse response $\mathbf{h}(t)$ is computed accordingly from (19) or (31), and

$$\underset{\mathbf{f}}{\text{Max}}\ \underset{\mathbf{h}}{\text{Max}}\ \rho(t_0) = E\lambda_1, \tag{40}$$

where $\lambda_1 = \text{limit}\ \lambda_1^{(k)}$ as $k \to \infty$.

4. CONCLUSIONS

In the absence of signal dependent interference and in the presence of white noise, for ideal impulse channels the shape of the transmit waveform is irrelevant, and it may be chosen to be chirp-like for enhanced range resolution. However, when signal-dependent signal dependent interference is present and it is comparable to channel noise, in any channel scene the chirp is almost invariably suboptimal and its use often entails a drastic reduction in output SINR.

Unlike the classical radar case, the choice of transmit pulse shape can be critically important for the detection of extended channels in presence of additive channel noise and signal-dependent signal dependent interference. Our analysis shows an optimal transmit-receiver pair that realizes maximum output SINR exists even for the multichannel case and is uniquely determined up to a real constant by a specification of channel impulse response matrix, nontrivial signal dependent interference and noise spectral densities and transmit signal energy. When multiple transmitters and receivers are used to capture otherwise missing information such as polarization, it is possible to excite certain modes both for the channel and signal dependent interference that are unavailable in the case of a single sensor. This additional information can be put to use in designing the optimal transmit-receiver vector pair so as to further improve the obtainable SINR.

5. REFERENCES

[1] F. G. Tricomi, *Integral Equations*, Interscience Publishers, Inc., New York, 1957.

[2] J. H. H. Chalk, "The Optimum Pulse Shape for Pulse Communication," *Proc. Inst. Elec. Eng. London*, vol. 87, pp.88-92, 1950.

[3] J. R. Guerci and P. Grieve, "Optimum Matched Illumination-Reception Radars," U. S. Patent # 5,121,125, June 1992, and U. S. Patent# 5,175,552, Dec. 1992.

[4] S. U. Pillai, H. S. Oh, D. C. Youla and J. R. Guerci, "Optimum Transmit-Receiver Design in the Presence of Signal-Dependent Interference and Channel Noise," *IEEE Transactions on Information Theory*, vol.46, no.2, pp.577-584, March 2000.

ESTIMATION OF TRANSMISSION LINE PARAMETERS BY ADAPTIVE INVERSE SCATTERING

Akihiro YONEMOTO, Takashi HISAKADO and Kohshi OKUMURA

Department of Electrical Engineering, Kyoto University, Japan
{yone,hisakado,kohshi}@kuee.kyoto-u.ac.jp

ABSTRACT

An adaptive inverse scattering for estimation of transmission line parameters is presented. In order to reduce estimation errors caused by noise in measured data, we analyze error propagation of inverse scattering with layer-peeling algorithm and propose the iteration of the layer-peeling algorithm with adaptively reconstructing input signal waveforms. The error reduction is achieved by trying to vanish error propagation to 0 based on the error analysis. Numerical simulations show that the proposed method can improve errors.

1. INTRODUCTION

Layer-peeling algorithm for inverse scattering problems is a very efficient method to identify parameters of layered wave propagation medium such as transmission lines, earth, vocal tract [1, 2]. However, although the algorithm works very well under a noise free assumption, if initial data which the algorithm starts with are contaminated by noise, the algorithm may cause severely large errors especially in latter stages because of its sequential nature.

In order to enable the layer-peeling algorithm under noisy environment, some methods used for stochastic signal processing such as Levinson algorithm, Burg algorithm, maximum likelihood method have been applied [3, 4]. Also, a detailed error analysis has been reported in terms of inequalities of norms [5]. In these papers, input signals are considered to be given and fixed.

Unlike the given and fixed input signals, real input signals contain some sorts of noise. For such input signals, we give a method to refine the estimation by rearranging input signal waveforms. We first analyze error propagation of the layer-peeling algorithm analytically and show its dependency on the input waveform. Then, based on the dependence, we propose an iterative layer-peeling algorithm which adaptively reconstructs the input waveforms to reduce the error propagation and refine the estimation. Numerical simulations show that the proposed method can refine the estimation.

2. INVERSE SCATTERING FOR LOSSLESS TRANSMISSION LINES

2.1. Discrete non-uniform lossless transmission lines

Let us consider a discrete transmission line, as shown in Fig. 1, which is composed of N uniform lossless transmission line segments. The discretization is performed in such a way that the travel time for propagating waves from one end of each segment to the other end of the segment is unity, so that z denotes the travel time for the right-propagating wave originated at the left end at $t = 0$. The parameter $Z_n (n = 0, 1, \dots, N-1)$ is the characteristic impedance of each segment. The line is driven by a current source $J(t)$, and for simplicity, the source impedance is assumed to be 0 without the loss of generality within the scope of following discussions.

The right-propagating wave variable $w_R(n,t)$ and left-propagating wave variable $w_L(n,t)$ are defined by

$$w_R(n,t) = \frac{1}{2}\left[\frac{v(n,t)}{\sqrt{Z_n}} + \sqrt{Z_n}\, i(n,t)\right] \quad (1)$$

$$w_L(n,t) = \frac{1}{2}\left[\frac{v(n,t)}{\sqrt{Z_n}} - \sqrt{Z_n}\, i(n,t)\right] \quad (2)$$

where $v(n,t)$ and $i(n,t)$ are respectively the voltage and the current on the line at "position" $z = n+0$ and at time t. Since each line segment is uniform and thereby it can simply be considered as a time delay element of unit time delay, the recurrence relation for the propagating waves becomes, for

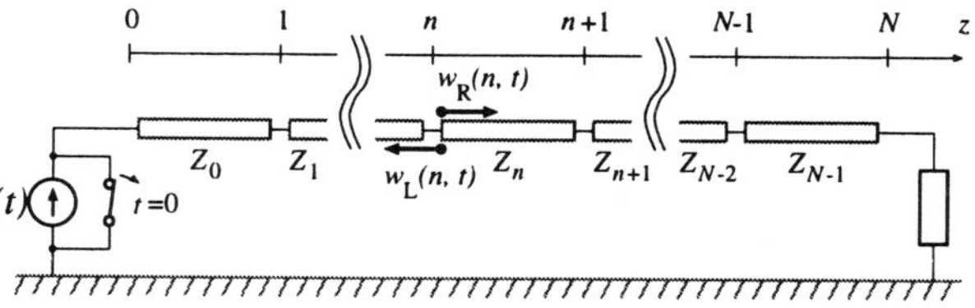

Figure 1: A discrete non-uniform transmission line

0-7803-7761-3/03/$17.00 ©2003 IEEE

$n = 1, 2, \ldots, N-1$,

$$\begin{bmatrix} w_R(n,t) \\ w_L(n,t) \end{bmatrix} = \begin{bmatrix} a_n & b_n \\ b_n & a_n \end{bmatrix} \begin{bmatrix} \Delta & 0 \\ 0 & \Delta^{-1} \end{bmatrix} \begin{bmatrix} w_R(n-1,t) \\ w_L(n-1,t) \end{bmatrix} \tag{3}$$

$$a_n = \frac{Z_{n-1} + Z_n}{2\sqrt{Z_{n-1} Z_n}}, \quad b_n = \frac{Z_{n-1} - Z_n}{2\sqrt{Z_{n-1} Z_n}} \tag{4}$$

where Δ denotes a unit time delay operator defined by

$$\Delta f(t) = f(t-1). \tag{5}$$

Defining the reflection coefficient r_n at $z = n$ by

$$r_n = \frac{Z_n - Z_{n-1}}{Z_n + Z_{n-1}}, \quad (n = 1, 2, \ldots, N-1) \tag{6}$$

equation (3) can be rewritten as follows.

$$\begin{bmatrix} w_R(n,t) \\ w_L(n,t) \end{bmatrix} = \Theta_n \begin{bmatrix} \Delta & 0 \\ 0 & \Delta^{-1} \end{bmatrix} \begin{bmatrix} w_R(n-1,t) \\ w_L(n-1,t) \end{bmatrix} \tag{7}$$

$$\Theta_n = \frac{1}{\sqrt{1 - r_n^2}} \begin{bmatrix} 1 & -r_n \\ -r_n & 1 \end{bmatrix} \tag{8}$$

2.2. Inverse scattering with layer-peeling algorithm

Assuming the initial distribution of the line to be 0, the inverse scattering method with layer-peeling algorithm recursively estimates the reflection coefficients r_n and so the characteristic impedances Z_n according to (7). Specifically, given the value of the first characteristic impedance Z_0 and a response $v(0,t)$, $i(0,t)$ obtained for a certain current waveform $J(t)$ applied to the line for $t \geq 0$, the algorithm proceeds in the following steps.

S0. Calculate $w_R(0,t)$ and $w_L(0,t)$ using (1), (2).

S1. $n := 0$.

S2. Estimate the reflection coefficient r_{n+1} by

$$\hat{r}_{n+1} = \frac{w_L(n, t+2)}{w_R(n,t)}, \quad n < t < n+2. \tag{9}$$

S3. Estimate the characteristic impedance Z_{n+1} by

$$Z_{n+1} = \frac{1 + \hat{r}_{n+1}}{1 - \hat{r}_{n+1}} Z_n. \tag{10}$$

S4. Compute $w_R(n+1,t)$, $w_L(n+1,t)$ by (7) using the estimated value $\hat{r}_n$ instead of r_n.

S6. $n := n+1$. If $n = N-1$ then end, otherwise go to S2.

For example, for $n = 1$, the reflection coefficient r_1 is estimated in the algorithm by

$$\hat{r}_1 = \frac{w_L(0, t+2)}{w_R(0,t)}, \quad 0 < t < 2. \tag{11}$$

This can be explained as follows. With the zero initial distribution, we have by causality

$$w_L(n,t) = 0, \quad 0 < t < n+2 \tag{12}$$

and thus equation (7) for $n = 1$, $1 < t < 3$ becomes

$$\begin{bmatrix} w_R(1,t) \\ 0 \end{bmatrix} = \Theta_n \begin{bmatrix} w_R(0, t-1) \\ w_L(0, t+1) \end{bmatrix}. \tag{13}$$

The equation in the second row leads to the relation

$$r_1 = \frac{w_L(0, t+1)}{w_R(0, t-1)}, \quad 1 < t < 3. \tag{14}$$

which is essentially the same with (11).

The algorithm recursively estimates the reflection coefficient r_{n+1} in this manner by **S2** and "peels off" the n-th line segment by **S4**.

3. ERROR ANALYSIS

If no error is incurred in the initial values $v(0,t)$ and $i(0,t)$, the algorithm will work very well. However, if just a small amount of error is involved in the initial values, it may cause severely large errors in the estimated reflection coefficients $\hat{r}_n$.

Suppose a transmission line whose characteristic impedance is given by

$$Z_n = n + 1 \quad (n = 0, 1, \ldots, N-1), \tag{15}$$

and that one of the initial values, $i(0,1)$, has relative error of 1% by letting

$$i(0,1) \longleftarrow 1.01 \times i(0,1). \tag{16}$$

With this small error involved, Fig. 2 shows the estimated reflection coefficients $\hat{r}_n$ with the ramp input waveform $J(t) =$

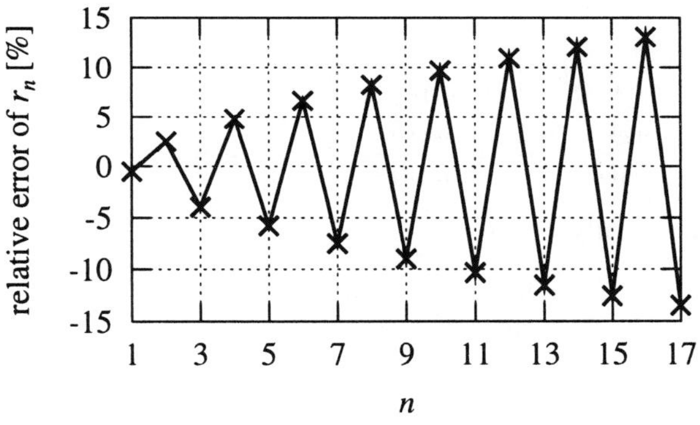

Figure 2: The relative error of estimated reflection coefficients with the ramp input waveform $J(t) = t$.

t, which function has wide generality and applicability. It can be seen that the precision of the reflection coefficients are not acceptable.

In Fig. 2, we can see that the estimation errors $\hat{r}_n - r_n$ appear like a geometric series. Hence, we develop an expression for $\mathrm{d}r_{n+1}/\mathrm{d}r_n$ assuming that there is no error in $w_R(n,t)$ and $w_L(n,t)$ other than caused by $\mathrm{d}r_n$. By differentiating (9) with respect to r_n, we have for $n < t < n+2$

$$\frac{\mathrm{d}r_{n+1}}{\mathrm{d}r_n} = \frac{\mathrm{d}}{\mathrm{d}r_n}\frac{w_L(n,t+2)}{w_R(n,t)}$$

$$= \frac{1}{w_R^2(n,t)}\begin{bmatrix} w_R(n,t) \\ w_L(n,t+2)\end{bmatrix}^T \begin{bmatrix} 0 & \Delta^{-2} \\ -1 & 0\end{bmatrix} \frac{\mathrm{d}}{\mathrm{d}r_n}\begin{bmatrix} w_R(n,t) \\ w_L(n,t)\end{bmatrix} \quad (17)$$

where T denotes transpose. Differentiating (7) and then reusing (7) to eliminate the term $[w_R(n-1,t), w_L(n-1,t)]^T$ gives

$$\frac{\mathrm{d}}{\mathrm{d}r_n}\begin{bmatrix} w_R(n,t) \\ w_L(n,t)\end{bmatrix} = -\frac{1}{1-r_n^2}\begin{bmatrix} 0 & 1 \\ 1 & 0\end{bmatrix}\begin{bmatrix} w_R(n,t) \\ w_L(n,t)\end{bmatrix}. \quad (18)$$

Substituting this to (17) gives

$$\frac{\mathrm{d}r_{n+1}}{\mathrm{d}r_n} = -\frac{1}{(1-r_n^2)w_R^2(n,t)}\begin{bmatrix} w_R(n,t) \\ w_L(n,t+2)\end{bmatrix}^T \begin{bmatrix} w_R(n,t+2) \\ -w_L(n,t)\end{bmatrix}. \quad (19)$$

By the causality (12), for $n < t < n+2$, this leads to

$$\frac{\mathrm{d}r_{n+1}}{\mathrm{d}r_n} = -\frac{1}{1-r_n^2}\frac{w_R(n,t+2)}{w_R(n,t)}. \quad (20)$$

Furthermore, the causal expression of (7) is given by

$$\begin{bmatrix} w_R(n,t) \\ w_L(n-1,t)\end{bmatrix} = \begin{bmatrix} 1 & 0 \\ 0 & \Delta\end{bmatrix}\begin{bmatrix} t_n & -r_n \\ r_n & t_n\end{bmatrix}\begin{bmatrix} \Delta & 0 \\ 0 & 1\end{bmatrix}\begin{bmatrix} w_R(n-1,t) \\ w_L(n,t)\end{bmatrix}, \quad (21)$$

$$t_n = \sqrt{1-r_n^2}. \quad (22)$$

From this relation and the causality (12), we can derive for $n < t < n+2$

$$\begin{bmatrix} w_R(n,t) \\ w_R(n,t+2)\end{bmatrix} = \begin{bmatrix} t_n & 0 \\ -r_n r_{n+1} t_n & t_n\end{bmatrix}\begin{bmatrix} w_R(n-1,t-1) \\ w_R(n-1,t+1)\end{bmatrix}, \quad (23)$$

$$\therefore \frac{w_R(n,t+2)}{w_R(n,t)} = \frac{w_R(n-1,t+1)}{w_R(n-1,t-1)} - r_n r_{n+1}. \quad (24)$$

Therefore, finally we have

$$\frac{\mathrm{d}r_{n+1}}{\mathrm{d}r_n} = -\frac{u_n}{1-r_n^2}, \quad (25)$$

$$u_n = u_{n-1} - r_n r_{n+1}, \quad (26)$$

$$u_0 = \frac{w_R(0,t+2)}{w_R(0,t)}, \quad 0 < t < 2. \quad (27)$$

This set of equations suggests that the error propagation depends on $w_R(0,t)$, so the input waveform $J(t)$.

4. ADAPTIVE INVERSE SCATTERING

From (25), it is expected that if we can make $\mathrm{d}r_{n+1}/\mathrm{d}r_n = 0$ by rearranging the input waveform $J(t)$, the error involved in the estimation of r_{n+1} will be reduced. The value of the differentiation can be controlled by setting u_0 to the appropriate value in accordance with (26), if we know the value of the reflection coefficients required in doing so in advance. The value of u_0 can also be specified by appropriately arranging the input waveform $J(t)$.

Thus we propose the following algorithm which adaptively reconstructs $J(t)$ using the reflection coefficients that are already estimated in the preceding steps, and iterates the layer-peeling algorithm to reduce the estimation errors.

S'0. With the layer-peeling algorithm, calculate $\hat{r}_1$ and $\hat{r}_2$, and let $\hat{r}'_1 = \hat{r}_1$.

S'1. $n := 1$.

S'2. Compute the targeted value of u_0 to achieve $u_n = 0$ based on (26) by

$$u_0 = \left(\sum_{k=1}^{n-1} \hat{r}'_k \hat{r}'_{k+1}\right) + \hat{r}'_n \hat{r}_{n+1}. \quad (28)$$

S'3. Reconstruct $J(t; u_0)$ so as to attain the specified value of u_0, and measure the data $v(0,t), i(0,t)$.

S'4. Peel off the line segments from 0 to $n-1$ with $\hat{r}'_1, \ldots, \hat{r}'_n$ by (7) to get $w_R(n,t), w_L(n,t)$.

S'5. Refine $\hat{r}_{n+1}$ by

$$\hat{r}'_{n+1} = \frac{w_L(n,t+2)}{w_R(n,t)}, \quad n < t < n+2. \quad (29)$$

Also, if $n < N-2$, compute $\hat{r}_{n+2}$ by the layer-peeling algorithm starting with $w_R(n,t), w_L(n,t)$.

S'6. $n := n+1$. If $n = N-1$ then end, otherwise go to **S'2**.

The prime means refinements, though $\hat{r}_1$ is not refined actually.

5. SIMULATION RESULTS

We tried the above proposed algorithm to improve the estimation errors in the case of the ramp input waveform. For $J(t; u_0)$, we used

$$J_r(t; u_0) = \begin{cases} t & (t < 2) \\ (u_0 - \hat{r}_1)(t-2) & (t \geq 2)\end{cases} \quad (30)$$

as a ramp-like input waveform. With this input waveform, given a value of u_0, we can obtain a waveform $w_R(0,t)$ which satisfies (27).

Figure 4 shows the relative errors of reflection coefficients estimated by the proposed method under the same conditions as in the previous section. For the ramp-like input waveform, the estimation is improved for $n \leq 4$ and becomes worse for $n > 5$.

The proposed method yields non-acceptable estimations especially after several first reflection coefficients. We consider the main reason is our assumption in the error analysis that there is no error in $w_R(n,t), w_L(n,t)$ other than caused by the estimation error of r_n. Indeed, the initial values of the algorithms, or the measured values, $v(0,t), i(0,t)$ generally contain errors, and hence $w_R(0,t), w_L(0,t)$ also contain errors.

6. CONCLUSION

We studied the errors involved in the inverse scattering with the layer-peeling algorithm. Taking notice of the fact that the error propagation depends on the input waveform, we proposed a method to adaptively reconstruct input waveforms in order to reduce the errors. By the proposed method, estimations of several first reflection coefficients for the ramp input waveform were improved. However, further elaboration on the error analysis and more numerical simulations for various kinds of input waveforms are needed to make the estimation much better.

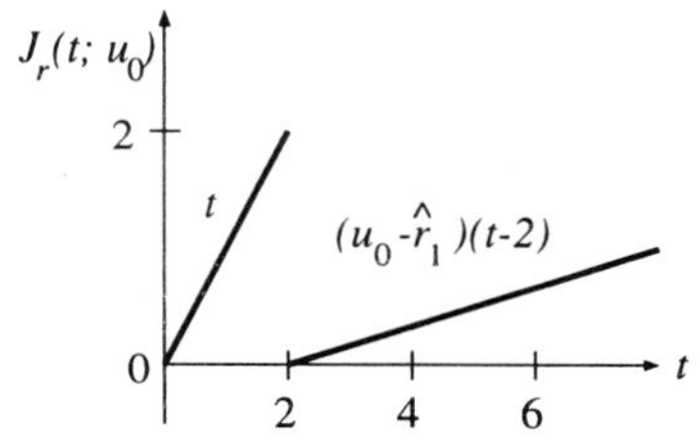

Figure 3: The ramp-like input waveform $J_r(t; u_0)$.

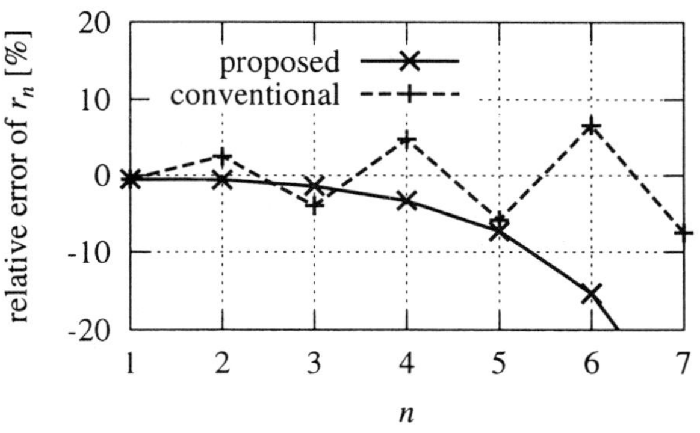

Figure 4: The relative error of estimated reflection coefficients by the proposed method with $J_r(t; u_0)$.

7. REFERENCES

[1] A. M. Bruckstein and T. Kailath, "Inverse scattering for discrete transmission-line models," *SIAM Review*, vol. 29, no. 3, pp. 359–389, 1987.

[2] A. E. Yagle and B. C. Levy, "The Schur algorithm and its applications," *Acta Applicandae Mathematicae*, no. 3, pp. 255–284, 1985.

[3] J. M. Mendel and F. Habibi-Ashrafi, "A survey of approaches to solving inverse problems for lossless layered media systems," *IEEE Trans. Geosci. Remote Sensing*, vol. 18, no. 4, pp. 320–330, 1980.

[4] F. Habibi-Ashrafi and J. M. Mendel, "Estimation of parameters in lossless layered media systems," *IEEE Trans. Automat. Control*, vol. 27, no. 1, pp. 31–49, 1982.

[5] A. M. Bruckstein, I. Koltracht, and T. Kailath, "Inverse scattering with noisy data," *SIAM J. Sci. Stat. Comput.*, vol. 7, no. 4, pp. 1331–1349, 1986.

A LEAST-SQUARES BASED ALGORITHM FOR FIR FILTERING WITH NOISY DATA

Wei Xing Zheng

School of QMMS, University of Western Sydney
Penrith South DC NSW 1797, Australia

ABSTRACT

This paper is concerned with finite impulse response (FIR) filtering with noisy input and output measurements. A new least-squares (LS) based algorithm is proposed to estimate the FIR filter coefficients. It is shown that the noise-induced bias can be removed once the variances of the input noise and output noise are obtained. A simple procedure is presented for estimating these variances by taking advantage of the FIR filter structure. The proposed LS based algorithm is easy to implement. Numerical results that illustrate the attractive properties of the new FIR filtering algorithm are presented.

1. INTRODUCTION

Finite impulse response (FIR) identification is aimed at estimating the impulse response of an unknown system from measurements of the system input and output. This problem is of fundamental significance in digital communications and signal processing [4], [5]. In many applications, not only the system output is corrupted by measurement noise, but also the measured input signal may often be corrupted by additive noise due to such as sampling error, quantization error and wide-band channel noise. In such FIR filtering with noisy input and output measurements, the standard least-squares (LS) method is known to yield biased estimates of the filter coefficients [6]. Application of the total least-squares (TLS) method in noisy FIR filtering has thus received much attention. Among the existing TLS based methods for adaptive FIR filtering, there are the recursive TLS (RTLS) method [1], the constrained anti-Hebbian algorithm [3] and the total least mean squares (TLMS) algorithm [2]. In particular, it is shown in [1] that the TLS criterion for noisy FIR filtering leads to a generalized eigenvalue decomposition problem. Using a priori knowledge of the variances ratio of the input and output noises, the TLS method produces unbiased FIR coefficient estimates.

In this paper, a new LS based algorithm is proposed to estimate the FIR filter coefficients from noisy input and output measurements. It is demonstrated that the noise-induced bias in the standard LS estimate can be removed once the variances of the input noise and output noise are obtained in a certain manner. A simple procedure is presented for estimating these variances by taking advantage of the FIR filter structure. The proposed LS based algorithm is easy to implement, and is less computationally demanding than the TLS method presented in [1]. More importantly, it can achieve estimation unbiasedness without a priori knowledge of the variances ratio of the input and output noises. Note that though noisy FIR filtering may be viewed as a special case of infinite impulse response (IIR) filtering with noisy data as studied in [7], the particular form of the FIR filter can be exploited to achieve simpler algorithmic structure and better performance as will be seen later in this paper.

2. NOISY FIR FILTERING PROBLEM

Consider the system configuration for noisy FIR filtering depicted in Fig.1. The unknown FIR system is described by

$$d(t) = H(z)s(t) \tag{1}$$

where $s(t)$ is the input signal, $d(t)$ is the filter output response, and $H(z)$ is the finite impulse response (FIR) filter defined by

$$H(z)=\sum_{i=0}^{M-1} h_i z^{-1} = h_0 + h_1 z^{-1} + \ldots + h_{M-1} z^{-(M-1)}. \tag{2}$$

Here the filter order M is assumed to be known.

The noiseless input $s(t)$ and the noiseless output $d(t)$ are usually not available for filtering. Instead, their noisy measurements $u(t)$ and $y(t)$ are obtained respectively as

$$u(t) = s(t) + n_i(t), \qquad y(t) = d(t) + n_o(t) \tag{3}$$

where the input noise $n_i(t)$ is zero-mean white noise with unknown variance $\sigma^2_{n_i}$ and the output noise $n_o(t)$ is zero-mean white noise with unknown variance $\sigma^2_{n_o}$. Assume that the measurement noises $n_i(t)$ and $n_o(t)$ are mutually uncorrelated, and independent of the input signal $s(t)$.

The purpose of noisy FIR filtering is to estimate the filter coefficient vector

$$\mathbf{h} = [h_0 \; h_1 \; \ldots \; h_{M-1}]^\top \tag{4}$$

from noisy measurements $u(t)$ and $y(t)$ under the assumptions stated above.

This work was supported in part by a Research Grant from the Australian Research Council and in part by a Research Grant from the University of Western Sydney, Australia.

0-7803-7761-3/03/$17.00 ©2003 IEEE

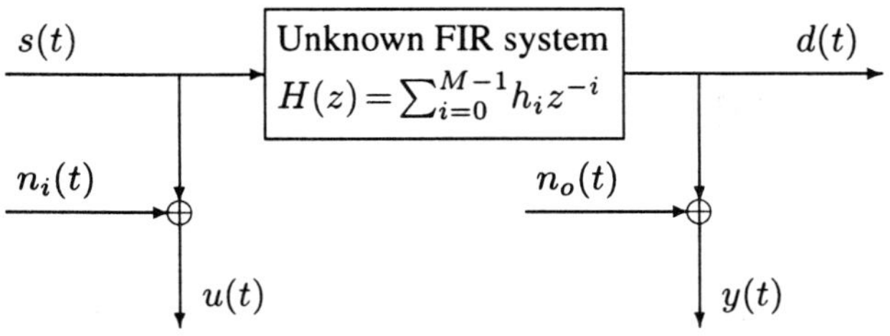

Figure 1: System configuration for noisy FIR filtering

The noisy FIR system described by (1)-(3) can be expressed in linear regression form as

$$y(t) = \mathbf{u}_t^\top \mathbf{h} + e(t) \tag{5}$$

where the noisy input signal vector $\mathbf{u}_t$ is given by

$$\mathbf{u}_t = [u(t)\ u(t-1)\ ...\ u(t-M+1)]^\top \tag{6}$$

and the equation error $e(t)$ is given by

$$e(t) = n_o(t) - \mathbf{n}_{i,t}^\top \mathbf{h} \tag{7}$$

$$\mathbf{n}_{i,t} = [n_i(t)\ n_i(t-1)\ ...\ n_i(t-M+1)]^\top. \tag{8}$$

The standard LS estimate of the filter coefficient vector $\mathbf{h}$ is obtained by minimizing the mean squared error criterion

$$J(\mathbf{h}) = E[(y(t) - \mathbf{u}_t^\top \mathbf{h})^2] \tag{9}$$

with respect to $\mathbf{h}$, which leads to

$$\mathbf{h}_{LS} = \mathbf{R}_{uu}^{-1}\mathbf{R}_{uy} \tag{10}$$

where $\mathbf{R}_{uu} = E[\mathbf{u}_t\mathbf{u}_t^\top]$ and $\mathbf{R}_{uy} = E[\mathbf{u}_t y(t)]$.

Introduce the noiseless input signal vector

$$\mathbf{s}_t = [s(t)\ s(t-1)\ ...\ s(t-M+1)]^\top. \tag{11}$$

It is easy to see that $\mathbf{u}_t = \mathbf{s}_t + \mathbf{n}_{i,t}$. Owing to the independence between $s(t)$, $n_i(t)$ and $n_o(t)$ and the whiteness of $n_i(t)$, it follows from (7) that

$$E[\mathbf{u}_t e(t)] = E[(\mathbf{s}_t + \mathbf{n}_{i,t})(n_o(t) - \mathbf{n}_{i,t}^\top \mathbf{h})] = -\sigma_{n_i}^2 \mathbf{h}. \tag{12}$$

Using (5) and (12), $\mathbf{R}_{uy}$ is found to be

$$\mathbf{R}_{uy} = E[\mathbf{u}_t y(t)] = E[\mathbf{u}_t(\mathbf{u}_t^\top \mathbf{h} + e(t))] = \mathbf{R}_{uu}\mathbf{h} - \sigma_{n_i}^2 \mathbf{h}. \tag{13}$$

Combining (13) with (10) gives

$$\mathbf{h}_{LS} = \mathbf{h} - \sigma_{n_i}^2 \mathbf{R}_{uu}^{-1}\mathbf{h}. \tag{14}$$

The above equation shows that the LS estimate $\mathbf{h}_{LS}$ is biased in noisy FIR filtering. The resulting bias term is seen to be

$$\mathbf{h}_{bias} = -\sigma_{n_i}^2 \mathbf{R}_{uu}^{-1}\mathbf{h}. \tag{15}$$

It is important to note that the magnitude of $\mathbf{h}_{bias}$ is proportional to the input noise variance $\sigma_{n_i}^2$. In particular, $\mathbf{h}_{bias} = \mathbf{0}$ whenever $\sigma_{n_i}^2 = 0$ (i.e. no presence of the input noise). Meanwhile, the influence of the output noise $n_o(t)$ on the bias is nil.

3. ALGORITHM DEVELOPMENT AND ANALYSIS

Assume that the input noise variance $\sigma_{n_i}^2$ can be estimated in a certain manner as $\hat{\sigma}_{n_i}^2(k)$. Then the bias term $\mathbf{h}_{bias}$ will become computable. Using (14), the LS estimate $\mathbf{h}_{LS}$ of the filter coefficient vector $\mathbf{h}$ can be compensated for the bias so as to achieve unbiased parameter estimation:

$$\hat{\mathbf{h}}_{CLS}(k+1) = \mathbf{h}_{LS} + \hat{\sigma}_{n_i}^2(k)\mathbf{R}_{uu}^{-1}\hat{\mathbf{h}}_{CLS}(k). \tag{16}$$

This method is called the compensated least-squares (CLS) algorithm. There is no doubt that feasible implementation of the bias compensation scheme (16) depends upon the availability of an estimate of $\sigma_{n_i}^2$.

For this purpose, we introduce the covariances

$$r_{yy}(j) = E[y(t)y(t-j)] \tag{17a}$$

$$r_{uy}(j) = E[u(t)y(t-j)] \tag{17b}$$

$$r_{uu}(j) = E[u(t)u(t-j)]. \tag{17c}$$

Because of the independence between $s(t)$, $n_i(t)$ and $n_o(t)$ and the whiteness of $n_o(t)$, it is obtained from (7) and (3) that

$$\begin{aligned} & E[e(t)y(t-1)] \\ &= E[(n_o(t) - \mathbf{n}_{i,t}^\top \mathbf{h})(d(t-1) + n_o(t-1))] \\ &= 0. \end{aligned} \tag{18}$$

With (5) and (18), we can get

$$r_{yy}(1) = E[y(t)y(t-1)] = E[(\mathbf{u}_t^\top \mathbf{h} + e(t))y(t-1)] = \mathbf{R}_{uy,1}^\top \mathbf{h} \tag{19}$$

where

$$\begin{aligned} \mathbf{R}_{uy,1} &= E[\mathbf{u}_t y(t-1)] \\ &= [r_{uy}(1)\ r_{uy}(0)\ r_{uy}(-1)\ ...\ r_{uy}(-M+2)]^\top. \end{aligned} \tag{20}$$

Premultiplying (14) with $\mathbf{R}_{uy,1}^\top$ yields

$$\mathbf{R}_{uy,1}^\top \mathbf{h}_{LS} = \mathbf{R}_{uy,1}^\top \mathbf{h} - \sigma_{n_i}^2 \mathbf{R}_{uy,1}^\top \mathbf{R}_{uu}^{-1}\mathbf{h}. \tag{21}$$

Substituting (19) into (21) and making some rearrangements leads to

$$\sigma_{n_i}^2 = \frac{r_{yy}(1) - \mathbf{R}_{uy,1}^\top \mathbf{h}_{LS}}{\mathbf{R}_{uy,1}^\top \mathbf{R}_{uu}^{-1}\mathbf{h}}. \tag{22}$$

This newly derived equation can be used to estimate the input noise variance $\sigma_{n_i}^2$ in conjunction with (16).

Next we consider estimating the output noise variance $\sigma_{n_o}^2$. Again, due to the independence between $s(t)$, $n_i(t)$ and $n_o(t)$ and the whiteness of $n_o(t)$, it follows from (7) and (3) that

$$E[e(t)y(t)] = E[(n_o(t) - \mathbf{n}_{i,t}^\top \mathbf{h})(d(t) + n_o(t))] = \sigma_{n_o}^2. \tag{23}$$

Applying (5) and (23), we have

$$r_{yy}(0) = E[y(t)^2] = E[(\mathbf{u}_t^\top \mathbf{h} + e(t))y(t)] = \mathbf{R}_{uy}^\top \mathbf{h} + \sigma_{n_o}^2. \tag{24}$$

From (24) it is obtained that

$$\sigma_{n_o}^2 = r_{yy}(0) - \mathbf{R}_{uy}^\top \mathbf{h} \tag{25}$$

which can be utilized to estimate the output noise variance $\sigma_{n_o}^2$.

In order to save the numerical cost in computing the covariance vector $\mathbf{R}_{uy,1}$, we take a look at $\mathbf{R}_{uy}$. Since

$$\mathbf{R}_{uy} = [r_{uy}(0)\ r_{uy}(-1)\ r_{uy}(-2)\ \dots\ r_{uy}(-M+1)]^{\top} \quad (26)$$

$\mathbf{R}_{uy,1}$ is expressible as

$$\mathbf{R}_{uy,1} = [r_{uy}(1)\ \ \mathbf{R}_{uy}(0:M-1)]^{\top}. \quad (27)$$

where the Matlab notations are used for $\mathbf{R}_{uy}(0 : M-1)$. This shows that $\mathbf{R}_{uy,1}$ can be formed largely by using the elements of $\mathbf{R}_{uy}$ except for extra computation of one covariance $r_{uy}(1)$. Furthermore, we notice that the covariance matrix $\mathbf{R}_{uu}$ is a Toeplitz matrix spanned by the autocovariances $\{r_{uu}(0), r_{uu}(1), ..., r_{uu}(M-1)\}$. So the proposed algorithm needs to compute the following covariances:

$$\begin{aligned} &r_{uu}(j),\ j = 0, 1, ..., M-1 \\ &r_{uy}(j),\ j = -M+1, -M+2, ..., -1, 0, 1 \\ &r_{yy}(j),\ j = 0, 1 \end{aligned} \quad (28)$$

which amounts to a total of $2M + 3$ distinct covariances.

The above derivation may be summarized as the following CLS algorithm which can estimate the filter coefficient vector $\mathbf{h}$ and the noise variances $\sigma^2_{n_i}$ and $\sigma^2_{n_o}$ simultaneously.

The CLS Algorithm

Step 0. Initialization:

(i) Compute the estimates of the $2M + 3$ covariances listed in (28) by using noisy measurements $\{u(t), y(t), t = 1, ..., N\}$. Then utilize them to form the covariance matrix and vector estimates $\hat{\mathbf{R}}_{uu}$, $\hat{\mathbf{R}}_{uy}$ and $\hat{\mathbf{R}}_{uy,1}$.

(ii) Evaluate the standard LS estimate

$$\hat{\mathbf{h}}_{LS} = \hat{\mathbf{R}}^{-1}_{uu}\hat{\mathbf{R}}_{uy}. \quad (29)$$

(iii) Calculate the following variables

$$\hat{f} = \hat{r}_{yy}(1) - \hat{\mathbf{R}}^{\top}_{uy,1}\hat{\mathbf{h}}_{LS} \quad (30)$$

$$\hat{\mathbf{g}} = \hat{\mathbf{R}}^{-1}_{uu}\hat{\mathbf{R}}_{uy,1}. \quad (31)$$

(iv) Set $k = 0$ and $\hat{\mathbf{h}}_{CLS}(0) = \hat{\mathbf{h}}_{LS}$.

Step 1. Compute the estimates of the input noise and output noise variances

$$\hat{\sigma}^2_{n_i}(k) = \hat{f}/(\hat{\mathbf{g}}^{\top}\hat{\mathbf{h}}_{CLS}(k)) \quad (32)$$

$$\hat{\sigma}^2_{n_o}(k) = \hat{r}_{yy}(0) - \hat{\mathbf{R}}^{\top}_{uy}\hat{\mathbf{h}}_{CLS}(k). \quad (33)$$

Step 2. Calculate the estimate of the filter coefficient vector via bias compensation

$$\hat{\mathbf{h}}_{CLS}(k+1) = \hat{\mathbf{h}}_{LS} + \hat{\sigma}^2_{n_i}(k)\hat{\mathbf{R}}^{-1}_{uu}\hat{\mathbf{h}}_{CLS}(k). \quad (34)$$

Step 3. Terminate the iteration procedure according to the preselected convergence criterion. Otherwise, set $k = k + 1$ and go to step 1.

The rationale behind the convergence of the proposed CLS algorithm is briefly mentioned. Since $\mathbf{R}_{uu} = \mathbf{R}_{ss} + \sigma^2_{n_i}\mathbf{I}_M$, where $\mathbf{R}_{ss} = E[\mathbf{s}_t\mathbf{s}_t^{\top}]$, it is quite reasonable to expect that the matrix $\hat{\sigma}^2_{n_i}(k)\hat{\mathbf{R}}^{-1}_{uu}$ will have its spectral radius less than unity for large N as long as $\hat{\sigma}^2_{n_i}(k)$ obtained via (32) is a reasonably good estimate of $\sigma^2_{n_i}$. Then $\hat{\sigma}^2_{n_i}(k)\hat{\mathbf{R}}^{-1}_{uu}$ will be a contraction mapping. The CLS algorithm may be expected to be contractive and cannot diverge. Given the fact that the bias compensation scheme (34) is entirely based on asymptotic analysis, the CLS algorithm is expected to not only converge but possess desired unbiasedness as well.

The implementation of the proposed FIR filtering algorithm is very straightforward. As stated before, the CLS algorithm makes use of $2M + 3$ distinct covariances, which is in contrast with $2M$ covariances required by the standard LS method. Moreover, estimating the input noise variance $\sigma^2_{n_i}$ via (32) involves simple linear operations. Hence, the computational load of the CLS algorithm can be kept small.

The CLS algorithm outperforms the TLS method presented in [1] in two aspects. First, it requires no a priori knowledge of the variances ratio of the input and output noises, thus being more practically useful than the TLS method. Second, as a linear regression based algorithm, it requires fewer computations than the TLS method since the latter conducts numerically costly generalized eigenvalue decomposition. In the meantime, the use of the bias compensation scheme ensures that its estimation accuracy can be highly comparable to that of the TLS method.

4. PERFORMANCE ASSESSMENT

Numerical results are now presented to illustrate the attractive properties of the proposed CLS algorithm. The unknown system has the finite impulse response

$$\mathbf{h} = [-0.3\ \ -0.9\ \ 0.8\ \ -0.7\ \ 0.6]^{\top}. \quad (35)$$

The noiseless input signal $s(t)$ is a zero-mean random signal with unit variance. The variances of the white input and output noises are $\sigma^2_{n_i} = 0.4$ and $\sigma^2_{n_o} = 1.0$, respectively. So the signal-to-noise (SNR) ratios at the system input and output are given by

$$\text{SNR}_i = 10\log_{10}\frac{\sigma^2_s}{\sigma^2_{n_i}} \approx 3.98\text{dB} \quad (36a)$$

$$\text{SNR}_o = 10\log_{10}\frac{\sigma^2_d}{\sigma^2_{n_o}} \approx 3.78\text{dB}. \quad (36b)$$

The estimation accuracy is measured in terms of the mean square error (MSE) [3]:

$$\text{MSE} = 10\log_{10}\frac{\|\hat{\mathbf{h}}_m - \mathbf{h}\|^2}{M}\ \text{(dB)} \quad (37)$$

where $\hat{\mathbf{h}}_m$ stands for an estimate of the filter coefficient vector $\mathbf{h}$ in the mth Monte-Carlo run. The computational complexity is measured approximately in terms of the MATLAB code `flops` (count of floating point operations). The identification results obtained by the standard LS method, the

Table 1. Idenfication Results
(SNR$_i \approx$3.98dB, SNR$_o$ $\approx$3.78dB, 200 Monte-Carlo runs, MSE = mean square error, NFPR = number of flops per run)

method	N	h_0	h_1	h_2	h_3	h_4	σ_v^2	σ_w^2	MSE	NFPR
LS	200	-0.2169	-0.6426	0.5704	-0.5012	0.4239			-14.60dB	8497
		±0.0700	±0.0810	±0.0857	±0.0778	±0.0755				
LS	500	-0.2165	-0.6378	0.5736	-0.5039	0.4266			-14.70dB	20496
		±0.0511	±0.0487	±0.0484	±0.0468	±0.0470				
LS	1000	-0.2126	-0.6460	0.5689	-0.4979	0.4276			-14.72dB	40496
		±0.0343	±0.0333	±0.0341	±0.0308	±0.0335				
TLS	200	-0.3023	-0.9057	0.7994	-0.7052	0.5979	0.3891	0.9728	-20.92dB	32553
		±0.1137	±0.1139	±0.1193	±0.1207	±0.1165	±0.0482	±0.1205		
TLS	500	-0.3015	-0.8915	0.8006	-0.7048	0.5966	0.3915	0.9789	-25.33dB	44452
		±0.0795	±0.0674	±0.0691	±0.0667	±0.0700	±0.0292	±0.0731		
TLS	1000	-0.2983	-0.9062	0.7976	-0.6987	0.5998	0.3988	0.9971	-28.29dB	64825
		±0.0524	±0.0503	±0.0496	±0.0486	±0.0477	±0.0210	±0.0526		
CLS	200	-0.3104	-0.9147	0.8020	-0.7068	0.6029	0.3847	0.9621	-19.98dB	10340
		±0.1365	±0.1579	±0.1288	±0.1345	±0.1399	±0.1334	±0.2423		
CLS	500	-0.3027	-0.8910	0.7989	-0.7028	0.5961	0.3853	0.9850	-23.92dB	24070
		±0.0875	±0.0870	±0.0807	±0.0781	±0.0782	±0.0799	±0.1474		
CLS	1000	-0.2994	-0.9081	0.7983	-0.6995	0.6009	0.3981	0.9935	-26.93dB	47025
		±0.0584	±0.0701	±0.0578	±0.0584	±0.0582	±0.0576	±0.1085		
true value		-0.3	-0.9	0.8	-0.7	0.6	0.4	1.0		

TLS method [1] and the CLS algorithm are summarized in Table 1 for three different sample sizes N, based on ensemble averaging of 200 Monte-Carlo runs.

The proposed CLS algorithm exhibits the superior performance. The filter coefficient estimates produced by the CLS algorithm are unbiased, with the estimation accuracy being significantly improved in cases of larger sample sizes N. Moreover, the convergence speed of the CLS algorithm is pretty fast as the averaged number of iterations that are needed to get the results of Table 1 is about four iterations per Monte-Carlo run. The implementation of the TLS method requires the known variances ratio of the input and output noises $\beta = \sigma_{n_o}^2/\sigma_{n_i}^2 = 2.5$. With this extra information, the TLS method performs slightly more accurately as its MSE is just about -1dB below that of the CLS algorithm. However, it is found that the TLS method is much computationally intensive due to the need for making generalized eigenvalue decomposition. For example, compared with the standard LS method, the TLS method increases the computational load by roughly 283% for $N = 200$ and by roughly 60% for $N = 1000$, whereas the CLS algorithm causes an increase in the numerical cost only by roughly 21% for $N = 200$ and by roughly 15% for $N = 1000$.

5. CONCLUDING REMARKS

The CLS algorithm has been established for FIR filtering with noisy data. It can produce unbiased estimates of the filter coefficients. The numerical results have shown that the proposed algorithm is much more cost effective in terms of estimation accuracy and computational load than the TLS method. Due to its algorithmic structure, the CLS algorithm has the great potential to be applicable in adaptive FIR filtering. Investigation on this line is currently under way.

6. REFERENCES

[1] C. E. Davila, "An efficient recursive total least squares alogrithm for FIR adaptive filtering," *IEEE Trans. Signal Processing*, Vol.42, No.2, pp.268-280, 1994.

[2] D. Z. Feng, Z. Bao and L. C. Jiao, "Total least mean squares alogrithm," *IEEE Trans. Signal Processing*, Vol.46, No.8, pp.2122-2130, 1998.

[3] K. Gao, M. O. Ahmad and M. N. S. Swamy, "A constrained anti-Hebbian learning alogrithm for total least-squares estimation with applications to adaptive FIR and IIR filtering," *IEEE Trans. Circuits and Systems-II*, Vol.41, No.11, pp.718-729, 1994.

[4] S. M. Kay, *Fundamentals of Statistical Signal Processing: Estimation Theory*. Englewood Cliffs, NJ: Prentice-Hall, 1993.

[5] J. G. Proakis, *Digital Communications*, 3rd ed. New York: McGraw Hill, 1995.

[6] P. A. Regalia, "An unbiased equation error identifier and reduced-order approximations," *IEEE Trans. Signal Processing*, Vol.42, No.6, pp.1397-1412, 1994.

[7] W. X. Zheng, "Transfer function estimation from noisy input and output data," *Int. J. Adaptive Control and Signal Processing*, Vol.12, No.4, pp.365-380, 1998.

A SPACE-TIME DECORRELATING RAKE RECEIVER FOR DS-CDMA COMMUNICATIONS OVER FAST FADING CHANNELS

C. Y. Fung, S. C. Chan and K. W. Tse

Department of Electrical and Electronic Engineering
The University of Hong Kong, Pokfulam Road, Hong Kong.

ABSTRACT

This paper proposes a new space-time decorrelating RAKE (ST-DRAKE) receiver for CDMA systems in fast fading channel. It is based on the ST-DRAKE receiver previously proposed in [2], which is a generalization of the DRAKE receiver originally proposed by Liu *et al.* [1] for blind CDMA reception over frequency-selective fading channels. The performances of these traditional DRAKE CDMA receivers, which are based on slowly time varying channel, are significantly affected by fast fading channels. The proposed receiver employs a Kalman filter to estimate the impulse response of the fast fading channel so that the transmitted symbols can be estimated after coherent combining. Computer simulations show that the proposed receiver has comparable but slightly inferior performance to the space-time MMSE receiver.

1. INTRODUCTION

Multiple access interference (MAI) and multipath fading are two major impairments that affect the capacity of direct sequence code division multiple access (DS-CDMA) systems. MAI results from the sharing of the same frequency band by multiple users communicating with each other. Whereas, multipath fading is caused by the movement of the mobile stations and the scattering of radio signals by the surroundings. Conventional matched filtering technique suffers from the near-far problem and user's signature waveform mismatch of fading channels, resulting in poor performance. Multiuser detection (MUD) [6] and multiple antennas [9] have been shown to be effective means for suppressing the MAI and the adverse effect of multipath fading.

In [3,4], Honig *et al.* have developed a blind multiuser detector based on the minimum output energy (MOE) criteria. It requires only the knowledge of the desired user's signature waveform, timing, and channel response in order to demodulate the user's signal. Although the channel characteristics can be estimated periodically using training sequence or constantly using pilot signal, it reduces the data transmission rate and may not be affordable in high-rate wireless transmission. Blind or semi-blind techniques for estimating the fast fading channels are thus highly desirable.

In [1], a decorrelating RAKE (DRAKE) receiver for blind MUD in frequency selective channel was proposed. It has a complexity and performance comparable to the adaptive MOE receiver without the knowledge of the desired user's channel response. The receiver structure is similar to the conventional RAKE receiver [5], where the outputs from a set of adaptive MOE receivers are combined to give the signal estimate. Further complexity reduction of the DRAKE receiver was considered in [6] by employing the reduced rank multistage Wiener (MSW) filter [7]. A space-time DRAKE receiver with improved performance was also considered in [2] by using multiple antennas.

The DRAKE receiver is capable of dealing with frequency-selective and slowly time-varying radio channel. However, when the channel is undergoing fast fading, its performance can degrade significantly because the channel response is estimated implicitly by an eigendecomposition [1], which is unable to follow the rapidly changing channel characteristics. In this paper, a new DRAKE receiver incorporating a channel-tracking algorithm for fast fading channel is proposed. The basic idea is to estimate, through some kind of initial training, a state-state model which models the dynamic of the fading channel [8]. The channel estimation problem can then be formulated as a state-estimation problem and the Kalman filter can be used to estimate the state, i.e. the channel response, of the time-varying channel. This new DRAKE receiver is further generalized to include multiple receive antennas for better bit-error-rate (BER) performance. The rest of the paper is organized as follows. First of all, the system model will be described in section 2. The DRAKE receiver and its space-time generalization are briefly reviewed in section 3. The incorporation of the Kalman filter-based channel estimator for fast fading channels is then introduced in Section 4. Simulation results are presented in section 5.

2. SYSTEM MODEL

Consider an asynchronous DS-CDMA system with K users. The transmitted baseband signal of the kth user is given by

$$x_k(t) = \sum_{m=-\infty}^{\infty} b_k(m) p_k(t - mT_s) \qquad (1)$$

where T_s is the symbol period, $b_k(m) \in \{-1,+1\}$ is the mth data symbol, and $c_k(t)$ is the spreading waveform given by

$$p_k(t) = \sum_{i=1}^{L_c} c_k(i) \psi(t - iT_c) \qquad (2)$$

where $c_k(i) = \{\pm 1/\sqrt{L_c}\}$ is the normalized spreading code assigned to the kth user, T_c is the chip duration, $L_c = T_s / T_c$ is the process gain, and $\psi(t)$ is the chip waveform. For simplicity, $\psi(t)$ is assumed to be a rectangular pulse on the interval $[0, T_c]$.

0-7803-7761-3/03/$17.00 ©2003 IEEE

We consider a receiver with a uniform linear array consisting of M receive antennas. The spacing between each antenna elements is $\lambda/2$, where λ is the wavelength of the carrier. The received signal $y^{(m)}(t)$ at the mth antenna is given by

$$y^{(m)}(t)=\sum_{k=1}^{K}\sum_{i=-\infty}^{\infty}h_k^{(m)}(t-iT_c-\tau_k)x_k(iT_c)+u^{(m)}(t) \quad (3)$$

where τ_k is the delay of the kth user's signal, $h_k^{(m)}$ is the channel response from the kth user to the mth receive antenna. For a multipath channel with delay spread $L_m \geq 1$, the channel impulse response for the kth user and the mth antenna is given by

$$h_k^{(m)}(t)=\sum_{l=1}^{L_m}\alpha_{k,l}^{(m)}(t)\delta(t-\tau_{k,l}^{(m)})\exp[-j\pi(m-1)\sin\theta_{k,l}^{(m)}], \quad (4)$$

where $\alpha_{k,l}^{(m)}(t)$, $\tau_{k,l}^{(m)}$ and $\theta_{k,l}^{(m)}$ are the channel coefficient, delay and direction of arrival (DOA), respectively, of the lth path of the kth user's signal received at the mth antenna. The DOA $\theta_{k,l}^{(m)}$ for the kth user is assumed to be constant and uniformly distributed in the interval $(-\pi/2,\pi/2)$. The channel impulse response $h_{k,l}^{(m)}(t)$ for each path is assumed to be independent of each other.

The received signal at the mth antenna is sampled at the chip rate over a symbol duration of $N=L_c+L_m-1$. The N-vector of chip-sampled signal samples is written as

$$\begin{aligned}\mathbf{y}^{(m)}(n)&=[y^{(m)}(nL_c)\ldots y^{(m)}(nL_c+N-1)]^T\\&=\sum_{k=1}^{K}[\mathbf{C}_k^{+}(n)\mathbf{h}_k^{(m)}(n)b_k(n)\\&\quad+\mathbf{C}_k^{-}(n)\mathbf{h}_k^{(m)}(n-1)b_k(n-1)]+\mathbf{u}^{(m)}(n),\end{aligned} \quad (5)$$

where $\mathbf{u}^{(m)}(n)$ is a complex white Gaussian noise with covariance matrix $\sigma_u^2\mathbf{I}$. The N-vectors $\mathbf{C}_k^{+}$ and $\mathbf{C}_k^{-}$ are associated with the kth interfering user due to asynchronous transmission. Without loss of generality, we assume that the desired user has a value of k equal to 1 and is synchronized to the receiver. The received signals $\mathbf{y}^{(m)}(n)$ from all antennas can be rewritten as

$$\begin{aligned}\mathbf{y}(n)&=[\mathbf{y}^{(1)}(n)\ldots\mathbf{y}^{(M)}(n)]^T\\&=\mathbf{C}\mathbf{h}_1(n)b_1(n)+\mathbf{i}(n)+\mathbf{u}(n),\end{aligned} \quad (6)$$

where $\mathbf{C}=\mathbf{I}_{MN\times ML_m}\cdot\mathbf{C}_1$, (7)

$$\mathbf{C}_1=[\mathbf{c}_{1,1}\ldots\mathbf{c}_{1,l}\ldots\mathbf{c}_{1,L_m}]=\begin{bmatrix}c_1(1) & & \mathbf{0}\\ \vdots & \ddots & c_1(1)\\ c_1(L_c) & & \vdots\\ \mathbf{0} & \ddots & c_1(L_c)\end{bmatrix}, \quad (8)$$

and $\mathbf{h}_1(n)=[\mathbf{h}_1^{(1)}(n)\ldots\mathbf{h}_1^{(M)}(n)]^T$, (9)

$$\mathbf{h}_1^{(m)}(n)=[h_{1,1}^{(m)}(n)\ldots h_{1,L_m}^{(m)}(n)]^T. \quad (10)$$

$\mathbf{I}_{m,n}$ and $\mathbf{i}(n)$ denote, respectively, the $m\times n$ identity matrix and the multiple access interference. $\mathbf{u}(n)$ is the noise vector.

3. SPACE-TIME DRAKE RECEIVER

The structure of the ST-DRAKE receiver is shown in Figure 1. For the rest of this paper, the desired user's subscript is dropped for simplicity. As shown in Figure 1, the received signal at the mth antenna is first filtered by weight vectors $\{\mathbf{w}_{m,l}\}_{l=1}^{L_m}$ such that the desired signal along each code vector $\mathbf{c}_l$ is extracted while suppressing the MAI and noise. It can be readily shown that this constrained optimization criterion, as specified in the MOE receiver, can be formulated as follows [10]

$$\mathbf{w}_{m,l}=\arg\min_{\mathbf{w}_{m,l}}\mathbf{w}_{m,l}^H\mathbf{R}_{\mathbf{y}^{(m)}\mathbf{y}^{(m)}}\mathbf{w}_{m,l} \quad \text{s.t.} \quad \mathbf{C}\mathbf{w}_{m,l}=\mathbf{1}_l, \quad (11)$$

where $\mathbf{R}_{\mathbf{y}^{(m)}\mathbf{y}^{(m)}}$ is the covariance matrix of $\mathbf{y}^{(m)}$ and the weight vector $\mathbf{w}_{m,l}$ is given by

$$\mathbf{C}\mathbf{w}_{m,l}=[0\cdots1\cdots0]^T=\mathbf{1}_l, \quad l=1,\ldots L_m. \quad (12)$$

The optimal solution to the problem in (11) is given by

$$\widetilde{\mathbf{w}}_{m,l}=\mathbf{R}_{\mathbf{y}^{(m)}\mathbf{y}^{(m)}}^{-1}\mathbf{C}(\mathbf{C}^H\mathbf{R}_{\mathbf{y}^{(m)}\mathbf{y}^{(m)}}^{-1}\mathbf{C})^{-1}\mathbf{1}_l. \quad (13)$$

Each weight vector in the lth arm accounts for signal received from different delayed paths. The output of the lth arm of the adaptive filter is given by

$$x_{m,l}(n)=\mathbf{w}_{m,l}^H\mathbf{y}^{(m)}(n), \quad m=1,\ldots,M. \quad (14)$$

From (4) and (11), (14) can be rewritten to emphasize the desired user's signal as

$$\begin{aligned}x_{m,l}(n)&=h_l^{(m)}(n)b(n)+\mathbf{w}_{m,l}^H\mathbf{u}^{(m)}(n)\\&=h_l^{(m)}(l)b(n)+\varepsilon_{m,l}(n),\end{aligned} \quad (15)$$

where $\varepsilon_{m,l}(n)$ is the effective MAI and noise after filtering. With a total of $L_m\times M$ paths, all signals scattered through different multipaths are extracted and stacked together such that

$$\begin{aligned}\mathbf{x}(n)&=[\mathbf{x}^{(1)}(n)\cdots\mathbf{x}^{(M)}(n)]^T\\&=[\mathbf{h}^{(1)}(n)\cdots\mathbf{h}^{(m)}(n)]^T\cdot b(n)+[\boldsymbol{\varepsilon}^{(1)}(n)\cdots\boldsymbol{\varepsilon}^{(1)}(n)]^T\\&=\mathbf{h}(n)\cdot b(n)+\boldsymbol{\varepsilon}(n)\end{aligned} \quad (16)$$

where $\mathbf{x}^{(m)}(n)=[x_{m,1}(n)\cdots x_{m,L_m}(n)]^T$, $\mathbf{h}^{(m)}(n)$ is given by (10), and $\boldsymbol{\varepsilon}^{(m)}(n)=[\varepsilon_{m,1}(n)\cdots\varepsilon_{m,L_m}(n)]^T$.

The filter output vectors $\{\mathbf{x}^{(m)}(n)\}_{m=1}^{M}$ are coherently combined to obtain the estimate of transmitted symbols. The optimum combining vector $\mathbf{w}_{coh}(n)$ is given by

$$\mathbf{w}_{coh}(n)=\mathbf{R}_{\mathbf{xx}}^{-1}\mathbf{h}(n). \quad (17)$$

Since MAI and noise are decorrelated and suppressed in constrained adaptive filter, the signal-to-interference-and noise ratio (SINR) of the filter outputs is high. For slowly time-varying channel the optimum combining vector can be approximated by the principal eigenvector of $\mathbf{R}_{\mathbf{xx}}^{-1}$ [1]. The combined output $z\{n\}$ is given by

$$z(n)=\mathbf{w}_{coh}^H\mathbf{x}(n). \quad (18)$$

By making a decision on $z\{n\}$, the estimate of the transmitted symbol is then obtained. For BPSK systems, this estimate is

$$\hat{b}(n) = \text{sgn}[\text{Re}\{z(n)\}]. \tag{19}$$

For MMSE receiver, the optimal weighting vector is given by [11]

$$\mathbf{w}_{MMSE} = \arg\min_{\mathbf{w}}\{E\|\mathbf{w}^H y(n) - b\ (n)\|^2\}. \tag{20}$$

In section 5, the performance of the proposed ST-DRAKE receiver is compared with the MMSE receiver, where the channel is assumed known.

4. CHANNEL TRACKING USING KALMAN FILTER

In section 3, the filtered outputs from the MOE detectors are coherently combined. However, if the channel is fast fading, the channel impulse response may vary too rapidly that the eigendecomposition approach becomes ineffective. As a result, the performance of the ST-DRAKE receiver will be degraded. In this work, Kalman filtering is employed to estimate the channel impulse response for such a fast fading channel.

In [12], it is suggested that a fast fading channel can be represented by an autoregressive (AR) model. More precisely, the time-varying component of each channel multipath is modeled as:

$$h_l^{(m)}(n) = \alpha_1 h_l^{(m)}(n-1) + \cdots + \alpha_R h_l^{(m)}(n-R) + v_l^{(m)}(n), \tag{21}$$

where $\{\alpha_i\}_{i=1}^R$ are the AR parameters which are specified by the Doppler spread [13]. The process noise for the AR process is assumed to be a zero-mean independent and identical distributed complex Gaussian random process with known covariance σ_v^2. For simplicity, we assume that a first-order AR process is precise enough to model the fading channel. We also assume the order of the AR process R is known to the receiver.

The AR model of the fast fading channel in (21) admits the following state-space representation:

$$\mathbf{h}^{(m)}(n) = \mathbf{A}\mathbf{h}^{(m)}(n-1) + \mathbf{v}^{(m)}(n) \tag{22}$$

where $$\mathbf{A} = diag\{\alpha_1 \cdots \alpha_R\}, \tag{23}$$

and $\mathbf{v}^{(m)}(n) = [v_1^{(m)}(n) \cdots v_{L_m}^{(m)}(n)]^T$ is the process noise vector. Note that the channel is assumed as a wide-sense stationary uncorrelated scattering channel, such that the coefficients of each multipath fading channel is uncorrelated with each other.

Similar to (22), the channel vector for the user $\mathbf{h}(n)$ can be represented by

$$\mathbf{h}(n) = \mathbf{\Gamma}\mathbf{h}(n-1) + \mathbf{v}(n) \tag{24}$$

where $$\mathbf{\Gamma} = \mathbf{I}_M \otimes \mathbf{A}, \tag{25}$$

and $\mathbf{v}(n) = [\mathbf{v}^1(n) \cdots \mathbf{v}^M(n)]^T$. Once the state space model is obtained, the channel can be formulated as a state-estimation problem and it can be solved using the Kalman filter as shown in Table 1 below. The measurement noise consists of the white Gaussian noise as well as the residue of the MOE detector, which depends on the received signals of other users. In practice, the number of users in the system is time varying. Hence, the residue of the MOE detector and the covariance of the measurement noise are also time varying. In this paper, we employ the estimation algorithm proposed in [14] to estimate the covariance of the measurement noise.

Initial conditions:
$\hat{\mathbf{h}}(0) = \mathbf{0}$
$\mathbf{P}(0) = \delta^{-1}\mathbf{I}$, $\delta = \text{small positive constant}$
For each $n = 1,2,\ldots$, compute:
$e(n) = y(n) - \mathbf{C}(n)\hat{\mathbf{h}}(n \mid n-1)$
$\mathbf{K}(n) = \mathbf{P}(n \mid n-1)\mathbf{C}^H(n)[\mathbf{C}(n)\mathbf{P}(n \mid n-1)\mathbf{C}^H(n) + \sigma_v^2\mathbf{I}]^{-1}$
$\mathbf{P}(n \mid n) = \lambda^{-1}[\mathbf{I} - \mathbf{K}(n)\mathbf{C}(n)]\mathbf{P}(n \mid n-1)$
$\mathbf{P}(n+1 \mid n) = \mathbf{\Gamma}\mathbf{P}(n \mid n)A^H + \sigma_u^2\mathbf{I}$
$\hat{\mathbf{h}}(n \mid n) = \hat{\mathbf{h}}(n \mid n-1) + \mathbf{K}(n)e(n)$
$\hat{\mathbf{h}}(n+1 \mid n) = \mathbf{\Gamma}\hat{\mathbf{h}}(n \mid n)$

Table 1.

5. SIMULATION RESULTS

We consider an asynchronous CDMA system with BPSK modulation. The processing gain of the system is 31 and the Gold code is used. The number of multipaths L_m to all users is 4 and they are generated with delays chosen from $\{0, T_c, 2T_c, 3T_c\}$. It is also assumed that the AOA of each interfering users are randomly distributed over $[0, \pi]$. We employ the adaptive MSW filter as proposed in [15] to implement the MOE detector. The rank of the MSW filter is chosen as 8.

The covariance matrix of the received signal is obtained recursively by

$$\mathbf{R}_{yy}(n+1) = \lambda\mathbf{R}_{yy}(n) + (1-\lambda)\mathbf{y}(n)\mathbf{y}^H(n) \tag{26}$$

where λ is a forgetting factor chosen as 0.995. To obtain the steady state performance of the proposed receiver, the first 50 symbols of the transmitted data are not counted in the calculation of the receiver's performance.

The fading rate $f_d T$ of the system is 0.0083, which corresponds to users traveling at 100 km/hr with a carrier frequency at 900 MHz and a signaling rate of 10^4 Hz. The parameter α_1 of the first-order AR process is given by $J_0(2\pi f_d T)$, where $J_0(\cdot)$ is the zeroth-order Bessel function of the first kind.

In Figure 3, the performance of the ST-DRAKE receiver with Kalman filter is compared with the space-time MMSE receiver. The number of users, including the desired user, is 10. Results are shown for both single-antenna and multiple-antenna cases. As shown in Figure 3, the ST-DRAKE has better BER performance than the DRAKE receiver when antenna array is used. Although the channel is highly time varying and unknown to the receiver, the proposed ST-DRAKE receiver in general has a comparable but slightly inferior performance to the MMSE receiver. We also compare the result to that of the DRAKE receiver suggested in [6]. Simulation shows that both the proposed receiver and the MMSE receiver offer significantly better performance that the DRAKE receiver.

Figure 4 shows the performance of the ST-DRAKE receiver versus the number of users in the system. The bit-energy-to-noise rate E_b / N_o is chosen as 10 dB. Both the performances of the ST-DRAKE and MMSE receiver degrade when the number of users in the system increases. It is also shown that the proposed receiver has better performance than the DRAKE receiver in [6].

6. CONCLUSIONS

A new space-time decorrelating RAKE receiver for CDMA systems for fast fading channel is presented. It is based on the ST-DRAKE receiver in [2]. The proposed receiver employs a Kalman filter to estimate the channel impulse response of the fast fading channel so that the transmitted symbols can be estimated by coherently combining. Computer simulation shows that the performance of the proposed receiver is only slightly inferior to that of the space-time MMSE receiver.

REFERENCES

[1] H. Liu and K. Li, "A decorrelating RAKE receiver for CDMA communications over frequency-selective fading channels," *IEEE Trans. Commun.*, vol. 47, pp. 1036-1045, Jul. 1999.

[2] C. Y. Fung, S. C. Chan and K. W. Tse, "Space-time decorrelating RAKE receiver in multipath CDMA channel," in *Proc. EUSIPCO*, vol. 3, pp. 511-514, Sep. 2002.

[3] S. Verdu, *Multiuser Detection.* New York: Cambridge University Press, 1998.

[4] M. L. Honig, U. Madhow and S. Verdu, "Blind adaptive multiuser detection," *IEEE Trans. Inform. Theory*, vol. 41, pp. 944-960, Jul. 1995.

[5] J. G. Proakis, *Digital Communications*, New York: McGraw Hill, 2001.

[6] O. Ozdemir and M. Torlak, " A reduced rank decorrelating RAKE receiver for CDMA communications over frequency-selective channels," *Proc. 11th Signal Processing Workshop on Statistical Signal Processing*, pp. 98-101, 2001.

[7] J. S. Goldstein, I. S. Reed and L. L. Scharf, "A multistage representation of Wiener filter based on orthogonal projections," *IEEE Trans. Inform. Theory*, vol. 44, pp. 2943-2959, Nov. 1998.

[8] S. Haykin, *Adaptive Filter Theory.* New Jersey: Prentice Hall, 2002.

[9] A. J. Paulraj and C. B. Papadias, "Space-time processing for wireless communications," *IEEE Signal Processing Mag.*, vol. 13, pp. 49-83, Nov. 1997.

[10] J. B. Schodorf and D. B. Williams, "A constrained optimization approach to multiuser detection," *IEEE Trans. Signal Processing*, vol. 45, pp. 258-262, Jan. 1996.

[11] U. Madhow and M. L. Honig, "MMSE interference suppression for direct-sequence spread-spectrum CDMA," *IEEE Trans. Commun.*, vol. 42, pp. 3178-3188, Dec. 1994.

[12] M. K. Tsatsanis, G. B. Giannakis and G. Zhou, "Estimation and equalization of fading channels with random coefficients," in *Proc. ICASSP*, vol. 2, pp. 1093-1096, May 1996.

[13] W. C. Jakes, *Microwave Mobile Communications.* New Jersey: IEEE Press, 1993.

[14] A. Moghaddamjoo and R. L. Kirlin, "Robust adaptive Kalman filtering with unknown inputs," *IEEE Trans. Acoustic, Speech and Sig. Proc.*, vol. 37, pp. 1166-1175, Aug. 1989.

[15] M. L. Honig and J. S. Goldstein, "Adaptive reduced-rank residual correlation algorithms, for DS-CDMA interference suppression," in *Proc. 32nd Asilomar Conf. Signals, Systems and Computing*, vol. 2, pp. 1106-1110, Nov. 1998.

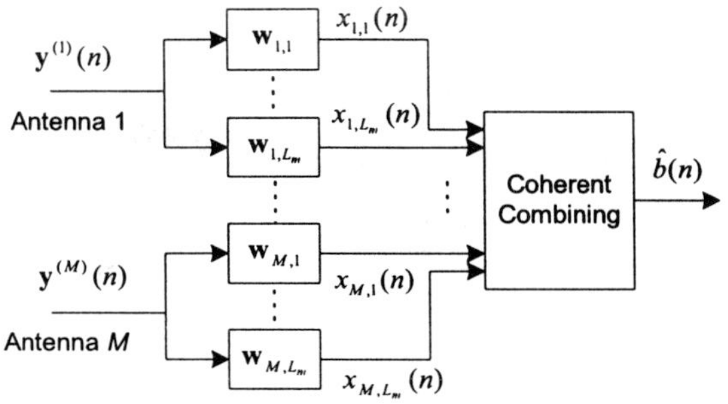

Figure 1. Space-time decorrelating RAKE receiver.

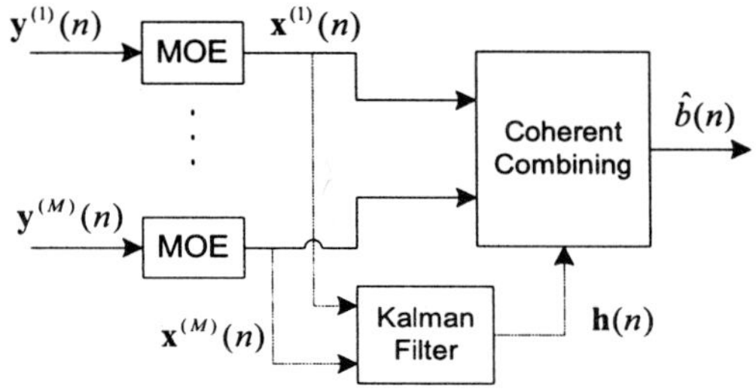

Figure 2. Space-time decorrelating RAKE receiver with Kalman filter

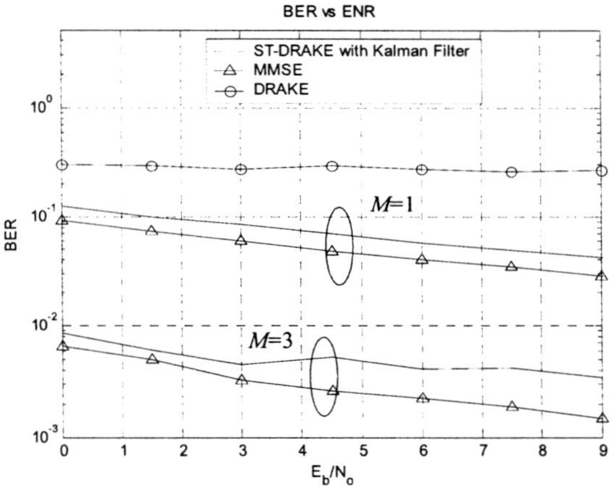

Figure 3. Bit-error rate versus energy-to-noise ratio.

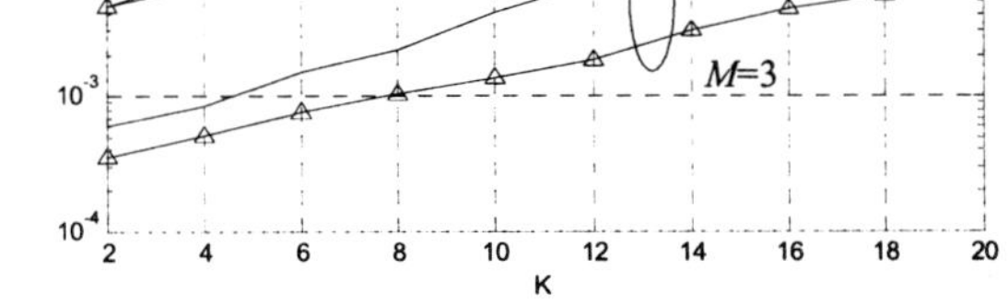

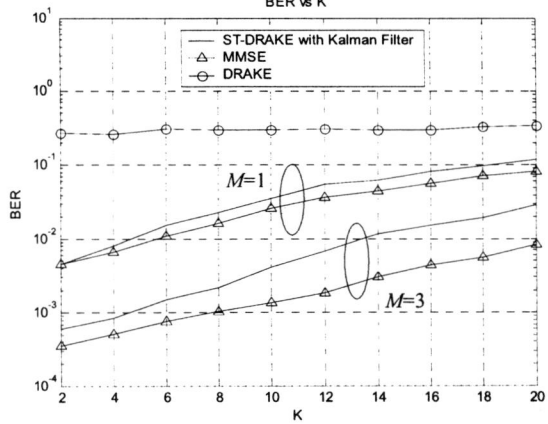

Figure 4. Bit-error rate versus number of users.

REALIZATION OF THE NLMS BASED TRANSVERSAL ADAPTIVE FILTER USING BLOCK FLOATING POINT ARITHMETIC

Abhijit Mitra and Mrityunjoy Chakraborty

Electronics and Electrical
Communication Engineering Department,
I.I.T. Kharagpur, India.
email:(abhijit, mrityun)@ece.iitkgp.ernet.in

ABSTRACT

This paper presents a novel scheme to implement the normalized LMS algorithm in block floating point (BFP) format which permits processing of data over a wide dynamic range at a processor cost marginally higher than that of a fixed point processor. Appropriate BFP formats for both the data and the filter coefficients have been adopted and adjustments made in filtering as well as weight updatation processes so as to sustain the adopted formats and to prevent overflow in these two operations jointly. This is achieved by restricting the step size control parameter available in the NLMS algorithm to lie within an upper bound, which is less than the upper bound for convergence only slightly and thus has marginal effect on convergence speed.

1. INTRODUCTION

The block floating point (BFP) format, proposed first by Wilkinson [1] as an alternative to the fixed point (FxP) and the floating point (FP) systems, has been used successfully in recent years in a wide class of signal processing applications ([2]-[7]). Under this scheme, the incoming data ia partitioned in non-overlapping blocks and based on the relative magnitudes of data samples in each block, a common exponent is assigned. This permits a FP like representation of the data, with FxP like computation within every block, thereby enabling the user to handle data over a wide dynamic range at hardware and power requirement comparable to that of FxP based systems. In the context of digital filters, the BFP method has been successfully used for efficient implementation of fixed coefficient filters ([2], [3], [5]-[7]), but, to the best of our knowledge, no effort has so far been made to extend this treatment to adaptive filters which present more complex structures including error feedback. A BFP treatment to adaptive filters faces certain difficulties, not encountered in the fixed coefficient case, namely, (a) unlike a fixed coefficient filter, the filter coefficients in an adaptive filter can not be represented in the simpler fixed point form, as the coefficients in effect evolve from the data by a time update relation; (b) the two principal operations in an adaptive filter, namely, filtering and weight updating, are mutually coupled, thus requiring an appropriate arrangement for joint prevention of overflow.

In this paper, we present a novel scheme for BFP-based realization of an important class of transversal adaptive filter, namely, the normalized LMS (NLMS) algorithm [9]. The presented method can be viewed as an extension of the scheme [8] we have recently come up with for the case of the LMS algorithm [9]. The proposed approach adopts appropriate BFP formats for the data and the filter coefficients separately and makes necessary adjustments in both the filtering as well as weight update equations so as to sustain the adopted format and also to prevent overlow jointly in both these operations. For this, we restrict the step size control parameter available in the NLMS algorithm to lie within an upper bound, which is, however, only slightly smaller than its universal upper bound and thus has negligible effect on convergence speed.

2. THE BFP ARITHMETIC

The BFP representation can be considered as a special case of FP format, where every block of N incoming data has a joint scaling factor corresponding to the largest (magnitude) data sample in the block. In other words, given a block $[x_1, ..., x_N]$, we represent it as

$$[x_1, ..., x_N] = [\overline{x}_1, ..., \overline{x}_N].2^{\gamma} \quad (1)$$

where $\overline{x}_l (= x_l.2^{-\gamma})$ represents the mantissa for $l = 1, 2, ..., N$ and the block exponent γ is defined as

$$\gamma = \lfloor log_2 Max \rfloor + 1 + S \quad (2)$$

0-7803-7761-3/03/$17.00

where $Max = max(|x_1|, ..., |x_N|)$, '$\lfloor . \rfloor$' is the so-called floor function, meaning rounding down to the closest integer and the integer S is a scaling factor which is needed to prevent overflow during filtering operation. For the presence of S, the range of each mantissa is given as $|\overline{x}_l| \in [0, 2^{-S})$. The scaling factor S can be calculated from the inner product computation representing filtering operation. An inner product is calculated in BFP format as

$$\begin{aligned} y(n) &= <\mathbf{w}, \mathbf{x}(n)> = \mathbf{w}^T\mathbf{x}(n) \\ &= [w_0\overline{x}(n) + ... + w_{L-1}\overline{x}(n-L+1)].2^{\gamma} \\ &= \overline{y}(n).2^{\gamma} \end{aligned} \quad (3)$$

where $\mathbf{w}$ is an Lth order fixed point filter coefficient vector and $\mathbf{x}(n)$ is the data vector at the nth index, represented in BFP format. For no overflow in $y(n)$, we need $|\overline{y}(n)| \leq 1$ at every time index, which can be satisfied by selecting [3]

$$S \geq S_{min} = \lceil log_2(\sum_{k=0}^{L-1} |w_k|) \rceil \quad (4)$$

where '$\lceil . \rceil$' is the so-called ceiling function, meaning rounding up to the closest integer.

3. THE PROPOSED IMPLEMENTATION

Consider a Lth order NLMS based adaptive filter that takes an input sequence $x(n)$ and updates the weights [9] as

$$\mathbf{w}(n+1) = \mathbf{w}(n) + \mu(n)\mathbf{x}(n)e(n) \quad (5)$$

where $\mathbf{w}(n) = [w_0(n)w_1(n)...w_{L-1}(n)]^T$ is the tap weight vector at the nth index, $\mathbf{x}(n) = [x(n)x(n-1)...x(n-L+1)]^T$, $e(n) = d(n) - \mathbf{w}^T(n)\mathbf{x}(n)$ is the error signal with $d(n)$ being the desired response available during the initial training period and $\mu(n)$ denoting the so-called time varying step-size paramemter. In the most commonly used form [9] of the NLMS algorithm, $\mu(n)$ is taken as : $\mu(n) = \tilde{\mu}/(\sigma_n^2 + \theta)$, where $\sigma_n^2 = \mathbf{x}^T(n)\mathbf{x}(n)$, $\tilde{\mu}$: a step size control parameter, used to control the speed of convergence and chosen according to : $0 < \tilde{\mu} < 2$ for convergence and θ : an appropriate positive number introduced to avoid divide-by-zero like situations which arise when σ_n^2 becomes very small. It may be noted that the original NLMS algorithm is a special case of the above corresponding to $\tilde{\mu} = 1$ and $\theta = 0$.

The proposed scheme consists of two simultaneous BFP representations, one for the filter coefficient vector $\mathbf{w}(n)$ and the other for the given data, namely, $x(n)$ and $d(n)$. These are as follows :

(a)**BFP representation of the filter coefficient vector**:

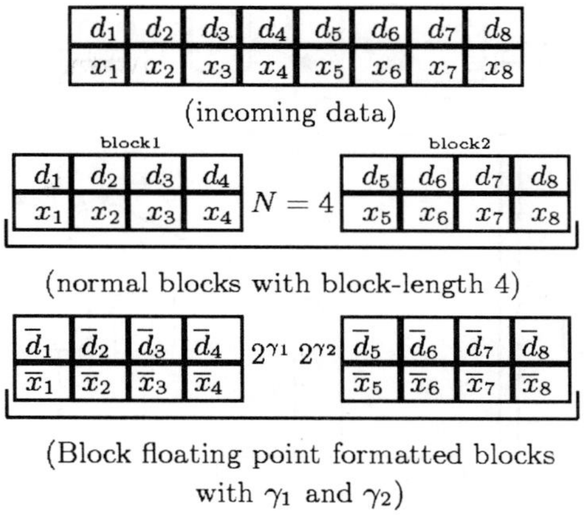

Figure 1: Block floating point joint scaling mechanism.

In this, at each index of time, we have a scaled representation of the filter coefficient vector as

$$\mathbf{w}(n) = \overline{\mathbf{w}}(n).2^{\psi_n} \quad (6)$$

where ψ_n is a time-varying block exponent that needs to be updated at each index n and is chosen to ensure that each $|\overline{w}_k(n)| < \frac{1}{2}$ for $k = 0, 1, ..., L-1$. If a data vector $\mathbf{x}(n)$ is given in the aforesaid BFP format as $\mathbf{x}(n) = \overline{\mathbf{x}}(n).2^{\gamma}$, where $\gamma = ex + S$, $ex = \lfloor log_2 M \rfloor + 1$, $M = max(|x(n-k)| \mid k = 0, 1, ..., L-1)$ and S is an appropriate scaling factor, then, the filter output $y(n)$ can be expressed as

$$y(n) = \overline{y}(n).2^{\gamma + \psi_n} \quad (7)$$

with $\overline{y}(n) = \overline{\mathbf{w}}^T(n)\overline{\mathbf{x}}(n)$ denoting the output mantissa. To prevent overflow in $\overline{y}(n)$, it is required that $|\overline{y}(n)| < 1$. However, in the proposed scheme, we restrict $\overline{y}(n)$ to lie between $+\frac{1}{2}$ and $-\frac{1}{2}$, i.e., $|\overline{y}(n)| < \frac{1}{2}$ for reasons explained later. Since $|\overline{y}(n)| \leq \sum_{k=0}^{L-1} |\overline{w}_k(n)||\overline{x}(n-k)|$, $0 \leq |\overline{x}(n-k)| < 2^{-S}$ and $|\overline{w}_k(n)| < \frac{1}{2}$, this implies a lower limit of S as

$$S_{min} = \lceil log_2 L \rceil. \quad (8)$$

(b) **BFP representation of the given data**:
The input data $x(n)$ and the desired response sequence $d(n)$ are partitioned jointly into non-overlapping blocks of N samples each as shown in Fig. 1, with the ith block ($i \in \mathsf{Z}$) consisting of $x(n)$, $d(n)$ for $n \in \mathsf{Z}_i = \{iN, iN+1, ..., iN+N-1\}$. Further, both $x(n)$ and $d(n)$ are jointly scaled so as to have a common BFP representation within each block. This means that, for $n \in \mathsf{Z}_i$, $x(n)$ and $d(n)$ are expressed as

$$x(n) = \overline{x}(n).2^{\gamma_i}, \quad d(n) = \overline{d}(n).2^{\gamma_i} \quad (9)$$

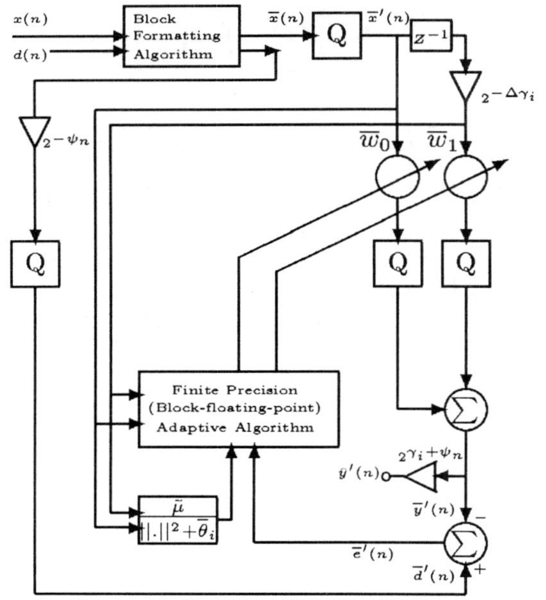

Figure 2: Block floating point implementation of a 2-tap NLMS based adaptive filter.

where γ_i is the common block exponent for the ith block and is given as

$$\gamma_i = ex_i + S_i \tag{10}$$

where

$$ex_i = \lfloor log_2 M_i \rfloor + 1 \tag{11}$$

and

$$M_i = max\{|x(n)|, |d(n)| \mid n \in Z_i\}. \tag{12}$$

The scaling factor S_i is assigned as per the following algorithm:

Algorithm: *Assign* $S_{min} = \lceil log_2 L \rceil$ *as the scaling factor to the first block and for any (i-1)-th block, assume* $S_{i-1} \geq S_{min}$.
Then, if $ex_i \geq ex_{i-1}$, *choose* $S_i = S_{min}$ *(i.e.,* $\gamma_i = ex_i + S_{min}$*)*
else (i.e., $ex_i < ex_{i-1}$*)*
choose $S_i = (ex_{i-1} - ex_i + S_{min})$, *s.t.* $\gamma_i = ex_{i-1} + S_{min}$.

Note that when $ex_i \geq ex_{i-1}$, we can either have $ex_i + S_{min} \geq \gamma_{i-1}$ (Case A) implying $\gamma_i \geq \gamma_{i-1}$, or, $ex_i + S_{min} < \gamma_{i-1}$ (Case B) meaning $\gamma_i < \gamma_{i-1}$. However, for $ex_i < ex_{i-1}$ (Case C), we always have $\gamma_i \leq \gamma_{i-1}$.
Additionally, we rescale the elements $\overline{x}(iN-L+1), \cdots, \overline{x}(iN-1)$ by dividing by $2^{\Delta\gamma_i}$, where

$$\Delta\gamma_i = \gamma_i - \gamma_{i-1}. \tag{13}$$

Equivalently, for the elements $x(iN-L+1), \cdots, x(iN-1)$, we change S_{i-1} to an effective scaling factor of $(S_{i-1} + \Delta\gamma_i)$. This permits a BFP representation of the data vector $\mathbf{x}(n)$ with common exponent γ_i during block-to-block transition phase too, i.e., when part of $\mathbf{x}(n)$ comes from the $(i-1)$-th block and part from the i-th block. In practice, such rescaling is effected by passing each of the delayed terms $\overline{x}(n-j)$, $j = 1, ..., L-1$, through a rescaling unit that applies $\Delta\gamma_i$ number of right or left shifts on $\overline{x}(n-j)$ depending on whether $\Delta\gamma_i$ is positive or negative respectively. For a 2-tap NLMS based adaptive filter that has only one update factor, this is illustrated in Fig. 2. This is, however, done only at the beginning of each block, i.e., at indices $n = iN$, $i \in Z$. Also, note that though for the case (A), $\Delta\gamma_i \geq 0$, for (B) and (C), however, $\Delta\gamma_i < 0$, meaning that in these cases, the aforesaid mantissas from the (i-1)-th block are actually scaled up by $2^{-\Delta\gamma_i}$. It is, however, not difficult to see that the effective scaling factor $(S_{i-1}+\Delta\gamma_i)$ for the elements $x(iN-L+1), \cdots, x(iN-1)$ still remains lower bounded by S_{min}, thus ensuring no overflow during filtering operation.

The output error $e(n)$ is then evaluated as $e(n) = \overline{e}(n).2^{\gamma_i+\psi_n}$ where the mantissa $\overline{e}(n)$ is given as

$$\overline{e}(n) = \overline{d}(n).2^{-\psi_n} - \overline{y}(n). \tag{14}$$

Clearly, computation of $\overline{e}(n)$ involves an additional step of right-shift operation on $\overline{d}(n)$ — an operation that comes up frequently in FP arithmetic. However, since in an adaptive filter, filter coefficients are derived from data and thus can not be represented in the FxP format when data is given in a scaled form, such a step seems to be unavoidable. It is easy to check that $|\overline{e}(n)| < 1$, since,

$$\begin{aligned} |\overline{e}(n)| &\leq |\overline{d}(n)|.2^{-\psi_n} + |\overline{y}(n)| \\ &< 2^{-(S_i+\psi_n)} + \frac{1}{2} \leq \frac{2^{-\psi_n}}{L} + \frac{1}{2} \end{aligned} \tag{15}$$

as $2^{-S_i} \leq \frac{1}{L}$. Except for the case: $\psi_n = 0$ and $L = 1$, the R.H.S.$\leq$ 1. For the above description of $e(n)$, $\mathbf{x}(n)$, $d(n)$ and $\mathbf{w}(n)$, the weight update equation (5) can now be written as

$$\mathbf{w}(n+1) = \overline{\mathbf{v}}(n).2^{\psi_n} \tag{16}$$

where

$$\overline{\mathbf{v}}(n) = \overline{\mathbf{w}}(n) + \frac{\overline{\mu}}{\overline{\sigma}_n^2 + \overline{\theta}_i}\overline{\mathbf{x}}(n)\overline{e}(n), \tag{17}$$

with $\overline{\sigma}_n^2 = \overline{\mathbf{x}}^T(n)\overline{\mathbf{x}}(n)$ and $\overline{\theta}_i = \theta 2^{-2\gamma_i}$, i.e., the mantissa of θ corresponding to the i-th block.

To satisfy $|\overline{w}_k(n+1)| < \frac{1}{2}$ for $k = 0, 1, \cdots, L-1$, we first limit each $|\overline{v}_k(n)| < 1$, $k = 0, 1, \cdots, L-1$ with $\overline{v}_k(n)$ denoting the k-th component of $\overline{\mathbf{v}}(n)$. Then, if each $\overline{v}_k(n)$ happens to be lying within $\pm\frac{1}{2}$, we make the assignments:

$$\overline{\mathbf{w}}(n+1) = \overline{\mathbf{v}}(n), \quad \psi_{n+1} = \psi_n. \tag{18}$$

Otherwise, we scale down $\overline{\mathbf{v}}(n)$ by 2, in which case,

$$\overline{\mathbf{w}}(n+1) = \frac{1}{2}\overline{\mathbf{v}}(n), \quad \psi_{n+1} = \psi_n + 1. \tag{19}$$

Since each $|\overline{w}_k(n)| < \frac{1}{2}$ for $k = 0, 1, \cdots, L-1$, it is sufficient to have

$$\frac{\tilde{\mu}}{\overline{\sigma}_n^2 + \overline{\theta}_i}|\overline{x}(n-k)||\overline{e}(n)| < \frac{1}{2} \tag{20}$$

in order to satisfy the relation $|\overline{v}_k(n)| < 1$ for $k = 0, 1, \cdots, L-1$. Noting that $|\overline{x}(n-k)|/(\overline{\sigma}_n^2 + \overline{\theta}_i) < 1$ and taking the upper limit of $|\overline{e}(n)|$ from (15) as $(\frac{2^{-\psi_n}}{L} + \frac{1}{2})$, it implies

$$\tilde{\mu} \leq \frac{L}{2^{-\psi_n+1} + L}. \tag{21}$$

Since $\psi_n \geq 0$ for all n, this then results in the following global upper bound : $\tilde{\mu} \leq \frac{L}{L+2}$.

It is also required that no overflow takes place during computation of $(\overline{\sigma}_n^2 + \overline{\theta}_i)$ in (17). Noting that $|\overline{x}(n-k)| < 2^{-S_{min}} \leq \frac{1}{L}$, we have, $\overline{\sigma}_n^2 + \overline{\theta}_i < L2^{-2S_{min}} + \overline{\theta}_i \leq \frac{1}{L} + \overline{\theta}_i$. Thus, for no overflow, it is sufficient to have $\overline{\theta}_i \leq \frac{L-1}{L}$ (≈ 1 for large L). In the proposed scheme, even though θ is fixed, $\overline{\theta}_i$ is to be updated once for each block by appropriate right/left shift operations on its previous value. Since both right and left shifts may be in operation, a practical approach will be to set its initial value by placing a binary 1 at the central bit location of the corresponding register, with all other bits set to binary 0. In that case, the initial value of γ may be taken to be the average block exponent, say γ_{av}, which needs to be estimated beforehand heuristically from some *a priori* knowledge of input statistics and S_{min}. Finally, we evaluate $\overline{\sigma}_n^2$ as

$$\overline{\sigma}_{n+1}^2 = \overline{x}^2(n+1) + \overline{\sigma}_n^2 - \overline{x}^2(n-L+1), \tag{22}$$

with a slight modification for the case when $\overline{x}(n+1)$ belongs to one block while all other entries of $\overline{\mathbf{x}}(n+1)$ to the previous block. In such case, the modified recursion is given by : $\overline{\sigma}_{n+1}^2 = \overline{x}^2(n+1) + 2^{-2\Delta\gamma_i}\overline{\sigma}_n^2 - \overline{x}^2(n-L+1)$, where $\overline{x}(n-L+1)$ denotes the rescaled mantissa for $x(n-L+1)$.

4. DISCUSSIONS AND CONCLUSIONS

This paper has presented a novel scheme for implementing the NLMS algorithm in a BFP format. The main computation involved, as given by equations (14), (17) and (22) is based on fixed point arithmetic only and thus can be realized using hardware that is simple, cheap and fast. However, since the underlying representation model is a BFP system, we can also simultaneously enjoy floating point like wide dynamic range by means of a block exponent. The price paid for this has been a reduced range for the step-size control parameter governed by a new upper limit, which is, however, not overly restrictive when compared against the conventional bound for convergence. In other words, it has marginal effect on convergence speed, as is also borne out by several simulation studies of the proposed method (not included here).

5. REFERENCES

[1] J. H. Wilkinson, *Rounding Errors in Algebraic Processes.* Englewood Cliffs, NJ: Prentice-Hall, 1963.

[2] K. R. Ralev and P. H. Bauer, "Realization of Block Floating Point Digital Filters and Application to Block Implementations," *IEEE Trans. Signal Processing*, vol. 47, no. 4, pp. 1076-1086, April 1999.

[3] K. Kalliojärvi and J. Astola, "Roundoff Errors in Block-Floating-Point Systems," *IEEE Trans. Signal Processing*, vol. 44, no. 4, pp. 783-790, April 1996.

[4] A. Erickson and B. Fagin, "Calculating FHT in Hardware", *IEEE Trans. Signal Processing*, vol. 40, pp. 1341-1353, June, 1992.

[5] S. Sridharan and D. Williamson, "Implementation of High Order Direct Form Digital Filter Structures", *IEEE Trans. Circuits Syst.*, vol. CAS-33, pp. 818-822, Aug., 1986.

[6] F. J. Taylor, "Block Floating Point Distributed Filters," *IEEE Trans. Circuits Syst.*, vol. CAS-31, pp. 300-304, Mar. 1984.

[7] A. V. Oppenheim, "Realization of digital filters using block floating point arithmetic," *IEEE Trans. Audio Electroaccoust.*, vol. AE-18, no. 2, pp. 130-136, June 1970.

[8] M. Chakraborty, A. Mitra and H. Sakai, "A Block Floating Point Realization of the LMS Algorithm," communicated to *IEEE Trans. Signal Processing.*

[9] S. Haykin, *Adaptive Filter Theory.* Englewood Cliffs, NJ: Prentice-Hall, 1986.

CONVERGENCE ANALYSIS OF A CORDIC-BASED GRADIENT ADAPTIVE LATTICE FILTER

Shin'ichi Shiraishi, Miki Haseyama, and Hideo Kitajima

School of Engineering, Hokkaido University
N-13 W-8 Kita-ku, Sapporo 060-8628, Japan

ABSTRACT

This paper presents a theoretical analysis of a CORDIC-based gradient adaptive lattice filter. First, we provide a convergence model and reveal convergence properties of filter coefficients. Second, we derive a steady-state model based on a Markov chain. By using the steady-state model, the relation between the step size parameter and the variance of the estimation error is clarified. The results of the analysis facilitate an efficient hardware design of the filter.

1. INTRODUCTION

With the advent of the mobile phone and wireless LAN, wireless communications have become immediately popular. The increasing bandwidth requirement in the wireless communication systems makes adaptive filters and adaptive beamforming, which can be used for channel equalization and interference canceling, attractive. The adaptive filters are implemented either as transversal or lattice filters. The lattice filter structure can be a good candidate for the hardware implementation because of its low sensitivity to the quantization error of filter coefficients.

In the hardware implementation of the lattice filter, low-cost and high-precision architectures are required. As one of such architectures, we have proposed a cost-effective architecture for a gradient adaptive lattice filter [1]. According to [1] we can obtain a simple filter architecture based on the CORDIC (COordinate Rotation DIgital Computer) [2] and GAL (Gradient Adaptive Lattice) algorithms. In our previous attempt [1], convergence properties of the filter are considered by some computer simulations, whereas no theoretical analysis is mentioned.

The convergence analysis is an important task because it helps to set an appropriate step size parameter for desirable properties: high convergence-speed and small variance of the estimation error. Therefore, we try to inspect the convergence properties of the CORDIC gradient adaptive lattice (CORDIC-GAL, hereafter) algorithm. First, we provide a convergence model to analyze the behavior of filter coefficients in the non-steady state. By using the convergence model, the relation between the convergence speed of the coefficients and the step size parameter can be clarified. Second, we derive a steady-state model based on a Markov chain. The obtained steady-state model shows the relation between the step size parameter and the variance of the estimation error of the coefficients.

2. PREVIOUS ALGORITHMS

2.1. The CORDIC Algorithm

The CORDIC algorithm [2] performs rotation operations of a vector in the circular, linear, and hyperbolic coordinate systems via iterative shift-and-add elementary steps. For example, the iteration equations at ith iteration in the hyperbolic coordinate system are[1]

$$\begin{cases} x_{i+1} = \left\{1 + (2|\delta_i| - 1)2^{-2i-4}\right\} x_i - \delta_i 2^{-i-1} y_i \\ y_{i+1} = -\delta_i 2^{-i-1} x_i + \left\{1 + (2|\delta_i| - 1)2^{-2i-4}\right\} y_i \\ z_{i+1} = z_i - 2\delta_i \psi_i, \quad i = 0, 1, \ldots, N-1 \end{cases} \tag{1}$$

where $\delta_i \in \{-1, 0, 1\}$ is the rotation direction parameter and determined to reduce the remaining angle $|z_{i+1}|$ at each iteration. The basis angles $\psi_i (i = 0, 1, \ldots, N-1)$ are precomputed constants and defined as

$$\psi_i = \tanh^{-1} 2^{-i-2} .$$

After N iterations according to Eq. (1), we have

$$\begin{bmatrix} x_N \\ y_N \end{bmatrix} = K \begin{bmatrix} \cosh\theta & -\sinh\theta \\ -\sinh\theta & \cosh\theta \end{bmatrix} \begin{bmatrix} x_0 \\ y_0 \end{bmatrix} \tag{2}$$

where $K = \prod_{i=0}^{N-1} \left(1 - 2^{-2i-4}\right)$ is the scaling factor. From Eqs. (1) and (2), it can be seen that the CORDIC algorithm decomposes a given rotation angle θ into a weighted sum of the basis angles as follows:

$$\theta = \sum_{i=0}^{N-1} 2\delta_i \psi_i \ . \tag{3}$$

2.2. A CORDIC-Based Gradient Adaptive Lattice Filter

As mentioned in Section 1, the CORDIC gradient adaptive lattice algorithm [1] produces a simple CORDIC-based architecture. The target of the CORDIC-GAL algorithm is a variance normalized lattice filter specified by the following recursive equations:

$$\begin{bmatrix} e_s^{nf}(k) \\ e_s^{nb}(k) \end{bmatrix} = \frac{1}{\sqrt{1 - (\tau_s(k))^2}} \begin{bmatrix} 1 & -\tau_s(k) \\ -\tau_s(k) & 1 \end{bmatrix} \begin{bmatrix} e_{s-1}^{nf}(k) \\ e_{s-1}^{nb}(k-1) \end{bmatrix} \tag{4}$$

where $e_s^{nf}(k)$ and $e_s^{nb}(k)$ are the normalized forward and backward prediction errors at time k, respectively. The variable $\tau_s(k)$ indicates the reflection coefficient in the sth elementary section. Let

[1]These equations are dedicated to an efficient implementation of the lattice filter and slightly different from those in the original algorithm [2].

0-7803-7761-3/03/$17.00 ©2003 IEEE

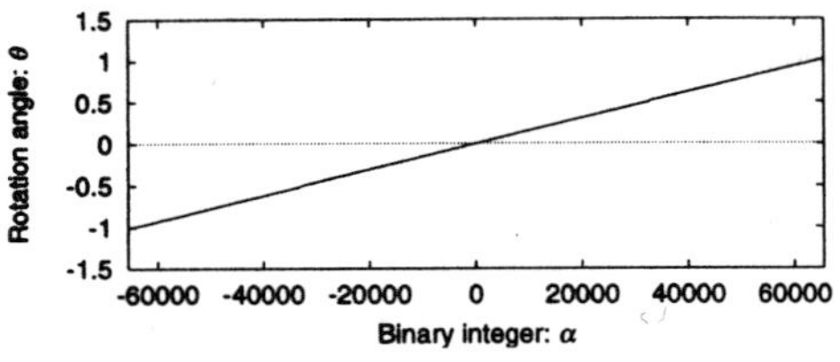

Figure 1: The relation between α and θ, $N = 16$.

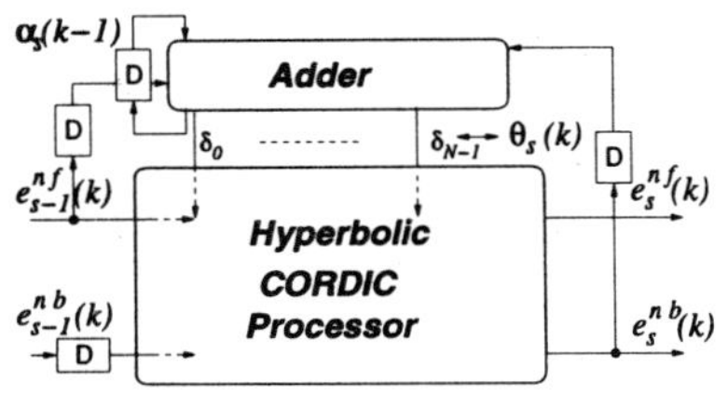

Figure 2: The structure of the CORDIC lattice section [1].

$\theta_s(k) = \tanh^{-1} \tau_s(k)(|\tau_s(k)| < 1)$, and Eq. (4) can be alter into a hyperbolic rotation of $\left[e^{nf}_{s-1}(k) \quad e^{nb}_{s-1}(k-1)\right]^T$ by the angle $\theta_s(k)$. That is to say, the order-update operation (Eq. (4)) can be executed by using a hyperbolic CORDIC processor.

For the sake of data adaptive operation, we must time-update the rotation angle $\theta_s(k)$. The angle decomposition nature of the CORDIC algorithm depicted in Eq. (3) motivates us to time-update the set of δ_i $(i = 0, 1, \ldots, N-1)$ rather than $\theta_s(k)$. To acquire such a direct time-update of the set of δ_i $(i = 0, 1, \ldots, N-1)$, we may use the following time-update formula:

$$\alpha_s(k+1) = \alpha_s(k) + \mu \cdot \mathrm{sgn}\left[e^{nb}_s(k) \cdot e^{nf}_{s-1}(k)\right] \cdot$$
$$\alpha = \alpha_{\mathrm{sgn}}\alpha_0\alpha_1 \ldots \alpha_{N-1}, \ \alpha_i \in \{0, 1\}\,. \quad (5)$$

In these equations, $\alpha_s(k)$ is a $N+1$-bit signed binary representation of $\theta_s(k)$ as shown in Fig. 1 and generates δ_i $(i = 0, 1, \ldots, N-1)$ as follows:

$$\delta_i = \left(1 - 2\alpha_{\mathrm{sgn}}\right)\alpha_i, \ i = 0, 1, \ldots, N-1\,. \quad (6)$$

The constant μ is a positive integer and called the step size parameter. The recursive equation (5) is composed of integer additions, so that it can be easily implemented with an adder.

According to the above procedure, we can obtain a CORDIC-implemented lattice section which consists of simple components: a CORDIC processor for Eq. (4) and an adder for Eq. (5). Due to its simplicity, the lattice section (Fig. 2) is certainly attractive from a hardware point of view. However, to implement the filter in hardware, we need to know how to set the step size parameter adequately. It is obvious that results of a theoretical convergence analysis help us to obtain such an appropriate step size parameter. Therefore, we try to clarify convergence properties of the filter in the following sections.

3. CONVERGENCE ANALYSIS OF THE CORDIC GRADIENT ADAPTIVE LATTICE FILTER

In this section, convergence properties of the CORDIC-GAL are analyzed. First, we explain some approximations and assumptions required for our analysis. Then, we derive a convergence model to analyze the behavior of the filter coefficient $(\alpha_s(k))$ until they converge to its steady-state value. Finally, we provide a steady-state model to consider fluctuations of the coefficient in the steady state.

3.1. Preliminaries

Since the filter coefficient $(\alpha_s(k))$ is a binary integer, it is difficult to obtain a convergence model from Eq. (5). Fortunately, Fig. 1 shows that $\alpha_s(k)$ can be approximated as follows:

$$\vartheta = c_N \cdot \alpha$$

where ϑ is a quantized representation of $\theta_s(k)$ and c_N is a constant of proportion. By using this approximation, we substitute $\vartheta_s(k)$ for $\alpha_s(k)$, and then we may have the following time-update formula:

$$\vartheta_s(k+1) = \vartheta_s(k) + \mu_\vartheta \cdot \mathrm{sgn}\left[e^{nb}_s(k) \cdot e^{nf}_{s-1}(k)\right] \quad (7)$$

where $\mu_\vartheta = c_N \cdot \mu$. This time-update formula allows us to derive a formulation of the convergence model, therefore, we will use it in the following sections.

As in [3], we assume that the mean-square value of $e^{nf}_{s-1}(k)$ is equal to that of $e^{nb}_{s-1}(k-1)$. In addition, we assume that $e^{nf}_{s-1}(k)$ and $e^{nb}_{s-1}(k-1)$ are jointly Gaussian with zero mean. According to these two assumptions, the instantaneous reflection coefficient $\frac{e^{nf}_{s-1}(k)\cdot e^{nb}_{s-1}(k-1)}{(e^{nf}_{s-1}(k))^2} = \frac{e^{nb}_{s-1}(k-1)}{e^{nf}_{s-1}(k)}$ is a random variable with the following Cauchy distribution:

$$F(u) = P\left(\frac{e^{nb}_{s-1}(k-1)}{e^{nf}_{s-1}(k)} < u\right)$$
$$= \frac{1}{2} + \frac{1}{\pi}\tan^{-1}\left((u - \tanh\phi_s(k))\cosh\phi_s(k)\right) \quad (8)$$

where $\phi_s(k) = \tanh^{-1} \frac{E\left\{e^{nf}_{s-1}(k)\cdot e^{nb}_{s-1}(k-1)\right\}}{E\left\{\left(e^{nf}_{s-1}(k)\right)^2\right\}}$.

3.2. A Convergence Model

In order to obtain a convergence model, we need to take statistical expectations of both sides of Eq. (7). To take the statistical expectation requires the probabilities $P(e^{nb}_s(k) \cdot e^{nf}_{s-1}(k) > 0)$ and $P(e^{nb}_s(k) \cdot e^{nf}_{s-1}(k) < 0)$. On the other hand, the product $e^{nb}_s(k) \cdot e^{nf}_{s-1}(k)$ can be expressed in another form:

$$e^{nb}_s(k) \cdot e^{nf}_{s-1}(k) = \cosh\vartheta_s(k)\left(e^{nf}_{s-1}(k)\right)^2\left\{\frac{e^{nb}_{s-1}(k-1)}{e^{nf}_{s-1}(k)} - \tanh\vartheta_s(k)\right\}. \quad (9)$$

Therefore, from Eqs. (8) and (9), the required probabilities are described as

$$\begin{cases} P(e^{nb}_s(k) \cdot e^{nf}_{s-1}(k) > 0) & = P(e^{nf}_{s-1}(k)/e^{nb}_s(k) > \tanh\vartheta_s(k)) \\ & = 1 - F(\tanh(\vartheta_s(k))) \\ P(e^{nb}_s(k) \cdot e^{nf}_{s-1}(k) < 0) & = P(e^{nf}_{s-1}(k)/e^{nb}_s(k) < \tanh\vartheta_s(k)) \\ & = F(\tanh(\vartheta_s(k)))\,. \end{cases} \quad (10)$$

Taking the expectation of both sides of Eq. (7) by using these probabilities, we have

$$E\{\vartheta_s(k+1)\} = E\{\vartheta_s(k)\} + \frac{2\mu_\vartheta}{\pi} \cdot E\left\{\tan^{-1}\left((\tanh\phi_s(k) - \tanh\vartheta_s(k))\cosh\phi_s(k)\right)\right\} .$$

Exchanging the expectation operator and functions yields the following convergence model:

$$E\{\vartheta_s(k+1)\} = E\{\vartheta_s(k)\} + \frac{2\mu_\vartheta}{\pi} \cdot \tan^{-1}\left((\tanh\phi_s(k) - \tanh E\{\vartheta_s(k)\})\cosh\phi_s(k)\right) . \quad (11)$$

This convergence model provides the trajectory of $E\{\vartheta_s(k)\}$. Moreover, applying some rough approximations to Eq. (11), we get

$$E\{\vartheta_s(k+1)\} - \phi_s = \left\{1 - \frac{2\mu_\vartheta}{\pi}\cosh\phi_s\right\} \cdot \{E\{\vartheta_s(k)\} - \phi_s\} \quad (12)$$

where $\phi_s = \lim_{k\to\infty}\phi_s(k)$. From the above equation, we obtain the following time-constant:

$$\lambda_s = \frac{\pi}{2\mu_\vartheta \cosh\phi_s} . \quad (13)$$

Thus, the time-constant λ_s is inversely proportional to the step size parameter μ_ϑ and $\cosh\phi_s$. In other words, the convergence speed is in proportion to μ_ϑ and $\cosh\phi_s$.

Experimental Results

In order to verify the proposed convergence model, we provide some experimental results by computer simulations. In the experiments 20,000 samples of the data are used. The number of iterations in the CORDIC algorithm (N) is set at 16. Figures 3(a)-(c) show the comparisons of our proposed convergence models and simulated trajectories obtained by averaging 100 trials. Figures 4(a) and (b) present the coefficient learning curves.

Figures 3 (a)-(c) show that our convergence models can predict the experimental results closely. Moreover, from Figs. 4 (a) and (b), it can be seen that the time-constant λ_s indicates precise convergence speed.

3.3. A Steady-State Model

In Section 3.2, we derived a convergence model and revealed the convergence behavior of $\vartheta_s(k)$. On the other hand, another model is required to analyze the behavior of $\vartheta_s(k)$ in the steady state. The coefficient $\vartheta_s(k)$, or $\alpha_s(k)$, is time-updated by Eq. (5) and takes a distinct value. Therefore, $\vartheta_s(k)$ is a Markov chain and we can use the Markov chain as the steady-state model.

To inspect the behavior of $\vartheta_s(k)$ in the steady state, we should consider properties of the Markov chain. For convenience, we will assume $\mu = 1$ in the following discussion. From Eq. (10), the state transition probability of the Markov chain is

$$P(\vartheta_j|\vartheta_i) = \begin{cases} 1 - F_\infty(\tanh\vartheta_i) & j = i+1 \text{ or } i = j = 2^N - 1 \\ F_\infty(\tanh\vartheta_i) & j = i-1 \text{ or } i = j = -2^N + 1 \\ 0 & \text{otherwise} \end{cases}$$
$$-2^N + 1 \le i, j \le 2^N - 1 \quad (14)$$

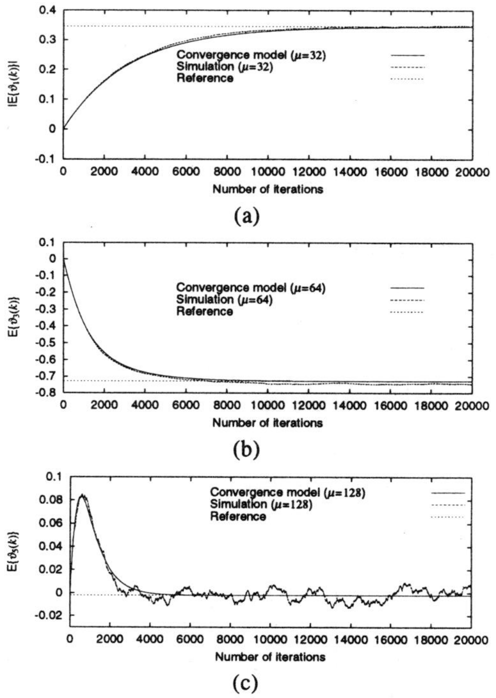

Figure 3: The comparisons of convergence models and simulated trajectories. (a) $E\{\vartheta_1(k)\}$, $\mu = 32$. (b) $E\{\vartheta_3(k)\}$, $\mu = 64$. (c) $E\{\vartheta_5(k)\}$, $\mu = 128$.

where ϑ_i denotes a rotation angle corresponding to $\alpha_s(k) = i$. Function F_∞ is a Cauchy distribution in the steady state and described as follows:

$$F_\infty(\tanh\vartheta_i) = \lim_{k\to\infty} F(\tanh\vartheta_i) = \frac{1}{2} + \frac{1}{\pi}\tan^{-1}\left((\tanh\vartheta_i - \tanh\phi_s)\cosh\phi_s\right) . \quad (15)$$

Using Eqs. (14) and (15), we have the following inequalities:

$$\begin{cases} P(\vartheta_{i+1}|\vartheta_i) > P(\vartheta_i|\vartheta_{i+1}) & \phi_s > \vartheta_{i+1} > \vartheta_i \\ P(\vartheta_{i+1}|\vartheta_i) < P(\vartheta_i|\vartheta_{i+1}) & \phi_s < \vartheta_i < \vartheta_{i+1} . \end{cases} \quad (16)$$

In addition, the Markov chain based on Eq. (14) has a limiting state probability distribution. In the limiting state probability distribution, the following equation is held:

$$P(\vartheta_{i+1}|\vartheta_i)\pi_i = P(\vartheta_i|\vartheta_{i+1})\pi_{i+1} \quad (17)$$

where π_i is the limiting state probability corresponding to ϑ_i. From Eqs. (16) and (17), we get

$$\begin{cases} \pi_{i+1} > \pi_i & \phi_s > \vartheta_{i+1} > \vartheta_i \\ \pi_{i+1} < \pi_i & \phi_s < \vartheta_i < \vartheta_{i+1} . \end{cases}$$

This implies that the nearest state to the optimal rotation angle ϕ_s has the maximum state probability.

In order to analyze fluctuations of the coefficient in the steady state, we need the variance of $\vartheta_s(k)$. Unfortunately, the quantization interval of $\vartheta_s(k)$ is not constant as shown in Eq. (3), so that it is difficult to get a formulation of the variance directly from Eq. (7).

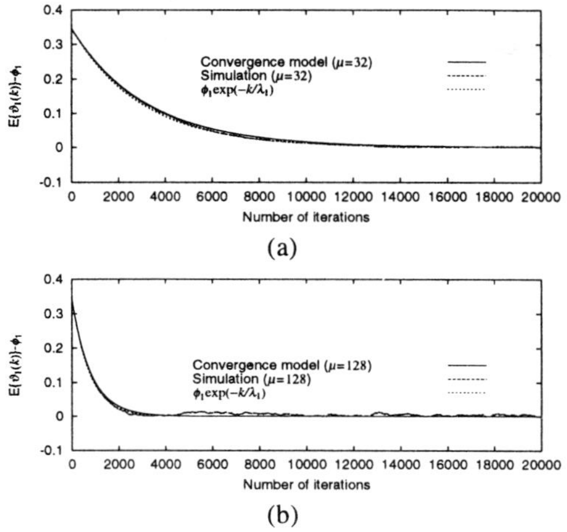

Figure 4: The learning curves of $\vartheta_s(k)$. (a) $\vartheta_1(k)$, $\mu = 32$. (b) $\vartheta_1(k)$, $\mu = 128$.

However, by utilizing the limiting state probability distribution, the variance of the estimation error of the coefficient is expressed as

$$\lim_{k\to\infty} \mathrm{Var}\{\vartheta_s(k)\} = \sum_{i=-2^N+1}^{2^N-1} \pi_i \left(\vartheta_i - \phi_s\right)^2 . \tag{18}$$

According to the above equation, we can compute theoretical variances of $\vartheta_s(k)$.

The steady-state model described in this section is inspired by the previous attempt [4]. In the previous method [4] a Markov chain is used as a convergence model. On the other hand, in our method, we use the Markov chain as a steady-state model.

Experimental Results

In this section, we provides some experimental results in order to verify the proposed steady-state model. Figures 5(a)-(c) show the comparisons of filter coefficient distributions based on our proposed steady-state models and computer simulations. Figure 6 presents the relation between the theoretical variance of the estimation error and the step size parameter.

From Figs. 5 (a)-(c), we note that the steady-state models provide good estimation of the simulation results. In addition, from Fig. 6 we can assume that the variance is proportional to the step size parameter.

4. CONCLUSION

In this paper, a theoretical analysis of the CORDIC gradient adaptive lattice filter [1] is presented. In the analysis, a convergence model and a steady-state model are derived. The convergence model shows that the convergence speed is proportional to the step size parameter. In addition, the steady-state model indicates that the error variance is also proportional to the step size. These results of the analysis help to determine an adequate step size in a hardware implementation of the filter.

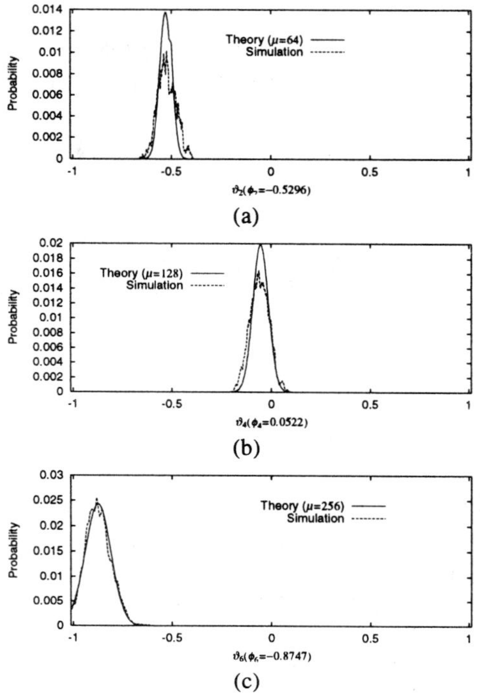

Figure 5: The comparisons of convergence models and simulation results. (a) $E\{\vartheta_2(k)\}$, $\mu = 64$. (b) $E\{\vartheta_4(k)\}$, $\mu = 128$. (c) $E\{\vartheta_6(k)\}$, $\mu = 256$.

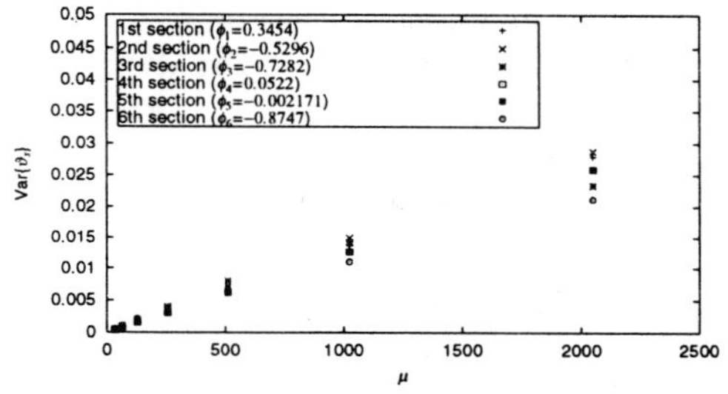

Figure 6: The relation between μ and $\mathrm{Var}\{\vartheta_s(k)\}$.

5. REFERENCES

[1] Shin'ichi Shiraishi, Miki Haseyama, and Hideo Kitajima, "A cost-effective and high-precision architecture for CORDIC-based adaptive lattice filters," in *Proc. ISCAS*, May 2002, vol. V, pp. 297–300.

[2] J. S. Walther, "A unified algorithm for elementary functions," in *Proc. AFIPS Spring Joint Computer Conf.*, May 1971, pp. 379–385.

[3] Youji Iiguni, Hideaki Sakai, and Hidekatsu Tokumaru, "Convergence properties of simplified gradient adaptive lattice algorithms," *IEEE Trans. Acoust., Speech, Signal Processing*, vol. ASSP-33, no. 6, pp. 1427–1434, 12 1985.

[4] Yu Hen Hu, "On the convergence of the CORDIC adaptive lattice filtering (CALF) algorithm," *IEEE Trans. Signal Processing*, vol. 46, no. 7, pp. 1861–1871, Jul. 1998.

ON THE IMPROVEMENT OF BLIND MOE DETECTOR VIA *A POSTERIORI* ADAPTATION AND ADAPTIVE STEP-SIZE

Samphan Pampichai[1] *and Phaophak Sirisuk*[1,2]

[1]Department of Electronic Engineering
[2]Thai-Paht Satellite Research Centre
Mahanakorn University of Technology, Bangkok 10530, THAILAND
Phone: +66 2988 3655, Fax: +66 2988 4040
Email: samphan,phaophak@mut.ac.th

ABSTRACT

This paper presents a blind multiuser detector for Direct Sequence-Code Division Multiple Access (DS-CDMA) systems. The adaptation of the proposed detector is based upon the adaptive step-size Minimum Output Energy (MOE) algorithm, which can achieve improvements over the original MOE detector. In this work, we develop *a posteriori* adaptation for the adaptive step-size MOE detector to further improve performance of the technique. Simulations show that the proposed detector yields significant improvements over the existing techniques especially in dynamic environments where the number of interferers are time-varying.

Keywords: Blind Multiuser Detection, *A Posteriori* Adaptation, Adaptive Step-Size, DS-CDMA, MOE

1. INTRODUCTION

Direct sequence Code Division Multiple Access (DS-CDMA) is promising technology for wireless environments with multiple simultaneous transmission. For the last few years, it is well accepted that Multiuser Detection (MUD) technique [1] can efficiently suppress Multiple Access Interference (MAI) and substantially increase the capacity of a CDMA system. During the last few years, *blind* adaptive MUD has received considerable interest for its application in DS-CDMA system. The main motivation of the blind algorithm is to make the requirements of a training sequence superfluous. By "*blind*" it is meant that only the signature waveform and timing of the desired user is available at a receiver. The problem is then to correctly detect the desired transmitted symbol, which experiences MAI caused by interfering users using this information only.

One of the most representative methods for blind MUD is a Minimum Output Energy (MOE) [2]. This algorithm is in fact the Least Mean Square (LMS) algorithm that minimises the mean output energy of the detector subject to a linear constraint on its coefficient. The problem is analogous to that of a Generalised Sidelobe Cancellor (GSC) widely found in adaptive filtering literatures, see e.g. [3]. As similar to the LMS algorithm, the performance the MOE algorithm heavily depends upon a choice of step-size. Recently, an adaptive step-size algorithm that involves cross coupling two LMS algorithms has been proposed for blind MOE detection [4]. In the sequel, we will refer to this algorithm as the Adaptive Step-size MOE (AS-MOE) algorithm. It was shown that the technique yields an improvement over the conventional MOE in dynamic environments where the number of users is time varying.

In many adaptive filtering applications that rely upon the LMS algorithm, an *instantaneous* error might be too noisy to provide a reliable *mean* square error estimate required for computing an optimal solution. This results in slow convergence of the adaptive algorithm. A means to avoid this problem is to involve some sort of regularisation through the use of prior knowledge by data reusing. The technique is referred to as *a posteriori* adaptation or data-reusing technique [5], [6]. The so-called *a posteriori* error estimates provide us with some information after computation in each recursion and can significantly improve convergence speed of the adaptive algorithm. The methods have been applied to several areas in adaptive filtering. Recently, *a posteriori* adaptation has been applied to the MOE detector [7]. It was shown that the performance of the proposed technique outperforms that of the original MOE.

In this paper, we adopt *a posteriori* adaptation for the AS-MOE detector. It is anticipated that the combination of *a posteriori* adaptation and adaptive step-size will further improve both transient and steady-state performance of the standard AS-MOE. The remaining of this paper is organised as follows. In Section 2, a mathematical framework is developed for a synchronous DS-CDMA system. In section 3, the MOE algorithm and its adaptive step-size variation are presented. In Section 4, *a posteriori* adaptation is introduced and the novel algorithms are proposed accordingly. Simulation results are then shown in Section 5. Finally, conclusions are drawn in Section 6.

0-7803-7761-3/03/$17.00 ©2003 IEEE

2. SIGNAL MODEL

Throughout the paper, we invoke the signal model that is identical to that of [1], [2], and [4]. Consider a synchronous baseband DS-CDMA system with K users. It is easily verified that a vector of chip-matched filter output samples within a symbol interval T is

$$\mathbf{r}(n) = \sum_{k=1}^{K} A_k b_k(n)\mathbf{s}_k + \mathbf{n}(n) \tag{1}$$

, where $\mathbf{s}_k$ is the normalised signature waveform vector of the kth user, i.e. $\|\mathbf{s}_k\| = 1$; A_k is the received amplitude of the kth user; $b_k(n)$ is the data bit of the kth user at time n and $\mathbf{n}(n)$ is a noise vector. Note that the received signal vector is of length N, where N is a processing gain. It is assumed that the $b_k(n)$, $k = 1,\ldots,K$ are mutually independent and uncorrelated with the noise vector $\mathbf{n}(n)$. To estimate the desired user, a linear detector is applied at the matched-filter output. Therefore, the output of the detector can be written as an inner product of the received signal vector $\mathbf{r}(n)$ and a detector coefficient vector $\mathbf{w}$. That is

$$z = \mathbf{w}^T\mathbf{r} = \sum_{k=1}^{K} A_k b_k \mathbf{w}^T\mathbf{s}_k + \mathbf{w}^T\mathbf{n}\,. \tag{2}$$

Without loss of generality, it is also assumed that user 1 is the desired user such that the decision on $b_1(n)$ is

$$\hat{b}_1(n) = \operatorname{sgn}\left(\mathbf{w}^T\mathbf{r}(n)\right). \tag{3}$$

3. THE AS-MOE DETECTOR

3.1 The Original MOE Detector

In [2], a linear blind MUD equivalent to a Minimum Mean Square Error (MMSE) detector was presented. The detector is known as a Minimum Output Energy (MOE) detector. The term MOE stems from the fact that the receiver attempts to minimise the mean output energy $E\{z^2\} = E\{(\mathbf{w}^T\mathbf{r})\}$, where z, $\mathbf{w}$ and $\mathbf{r}$ denotes the detector output, detector vector and received signal vector, respectively. To avoid a trivial solution $\mathbf{w} = \mathbf{0}$, an additional constraint must be imposed. By virtue of canonical representation [1], [2], a linear detector can be decomposed into two orthogonal components, i.e. $\mathbf{w} = \mathbf{s}_1 + \mathbf{x}_1$, where $\mathbf{s}_1$ and $\mathbf{x}_1$ is the signature waveform of the desired user and its orthogonal component, respectively. By minimising the MOE cost function at time n, $E\{z^2(n)\}$, subject to the orthogonal constraint described by $\mathbf{s}_1{}^T\mathbf{x}_1(n) = 0$, we arrive at the following adaptation:

$$\mathbf{x}_1(n+1) = \mathbf{x}_1(n) - \mu\, z(n)[\mathbf{r}(n) - z_{\text{MF}}(n)\mathbf{s}_1] \tag{4}$$

, where $z_{\text{MF}}(n) = \mathbf{s}_1{}^T\mathbf{r}(n)$ and $z(n) = \mathbf{w}^T(n)\mathbf{r}(n)$ denote the matched filter and detector output, respectively.

3.2 The AS-MOE Detector

Since the MOE is essentially the LMS algorithm, its performance will be heavily dependent on the choice of step-size. Recently, an adaptive step-size technique has been applied to the MOE algorithm [4]. It involves cross coupling two algorithms. In particular, while the standard MOE algorithm adapts a time-varying tap coefficient $\mathbf{w}(n)$ by minimising output energy of the linear detector, the other LMS algorithm will adaptively compute an optimal step-size for the MOE algorithm. The step-size adaptation $\mu(n)$ is obtained by the following estimation of step-size

$$\mu(n+1) = \left[\mu(n) - \alpha z(n)\mathbf{r}^T(n)\mathbf{Y}(n)\right]_{\mu-}^{\mu+} \tag{5}$$

, where $\alpha > 0$ denotes learning rate (step-size) and $\mathbf{Y}(n)$ denotes the derivative $\partial\mathbf{w}(n)/\partial\mu\,|_{\mu=\mu(n)}$. Furthermore, the bracket with μ^- and μ^+ denotes truncation of step-size. The update of $\mathbf{Y}(n)$ with constraint $\mathbf{w}^T\mathbf{s}_1 = 1$ yields.

$$\begin{aligned}\mathbf{Y}(n+1) = &\left[\mathbf{I} - \mu(n)\mathbf{r}(n)\mathbf{r}^T(n)\right]\mathbf{Y}(n) + \\ &\mu(n)\mathbf{r}^T(n)\mathbf{Y}(n)\left(\mathbf{r}^T(n)\mathbf{s}_1\right)\mathbf{s}_1 - \\ &z(n)\left(\mathbf{r}(n) - \left(\mathbf{r}^T(n)\mathbf{s}_1\right)\mathbf{s}_1\right)\end{aligned} \tag{6}$$

Finally, by replacing the fixed step-size μ in (4) with its adaptive counterpart given by (5), we arrive at the following update equations for the AS-MOE detector.

$$\mathbf{x}_1(n+1) = \mathbf{x}_1(n) - \mu(n)z(n)\left[\mathbf{r}(n) - z_{\text{MF}}(n)\mathbf{s}_1\right] \tag{7}$$

, where $z_{\text{MF}}(n)$ and $z(n)$ are the matched filter and detector outputs, respectively.

4. THE PROPOSED ALGORITHM

In *a posteriori* adaptive filters, the procedure of calculating an instantaneous error, output and weight update may be repeated several times, keeping the same input vector and training signal. Direct application of this concept to the MOE leads us the following adaptation [7]:

$$\begin{aligned}\mathbf{x}_{1,i+1}(n) &= \mathbf{x}_{1,i}(n) - \mu z_i(n)\left[\mathbf{r}(n) - z_{\text{MF}}(n)\mathbf{s}_1\right] \\ \mathbf{w}_{i+1}(n) &= \mathbf{s}_1 + \mathbf{x}_{1,i+1}(n)\end{aligned} \tag{8}$$

, where $\mathbf{x}_{1,1}(n) = \mathbf{x}_1(n)$ and $\mathbf{x}_1(n+1) = \mathbf{w}_{1,L+1}(n)$ (c.f. (4)). In addition, $z_i(n) = \mathbf{w}_i^T(n)\mathbf{r}(n)$ and $z_{\text{MF}} = \mathbf{s}_1{}^T\mathbf{r}(n)$. The subscript $i = 1,\ldots,L$ indicates an index for data reuse, where L is the number of iterations, and is intentionally used to distinguished itself from the time index n. Note that an instantaneous error in the standard LMS algorithm translates to the detector output $z_i(n)$ in the MOE (constrained LMS) algorithm.

Here the concept of *a posteriori* adaptation is extended to the AS-MOE. Clearly, we can replicate the coefficient update given by (8) for the novel algorithm. Nevertheless, the technique cannot be applied to the adaptation of the step-size since it relies on *a priori* knowledge of input statistics [4]. A natural approach is to *freeze* the step size μ until the next time index (n+1). Therefore, we the following update equations can be adapted.

$$\mathbf{x}_{1,i+1}(n) = \mathbf{x}(n) - \mu(n)z_i(n)\left[\mathbf{r}(n) - z_{MF}(n)\mathbf{s}_1\right]$$

$$\mu(n+1) = \left[\mu(n) - \alpha z(n)\mathbf{r}^T(n)\mathbf{Y}(n)\right]_{\mu-}^{\mu+}$$

$$\begin{aligned}\mathbf{Y}(n+1) = &\left[\mathbf{I} - \mu(n)\mathbf{r}(n)\mathbf{r}^T(n)\right]\mathbf{Y}(n) + \\ &\mu(n)\mathbf{r}^T(n)\mathbf{Y}(n)\left(\mathbf{r}^T(n)\mathbf{s}_1\right)\mathbf{s}_1 - \\ &z(n)\left(\mathbf{r}(n) - \left(\mathbf{r}^T(n)\mathbf{s}_1\right)\mathbf{s}_1\right)\end{aligned} \quad (9)$$

, where $z_i(n) = \mathbf{w}_i^T(n)\mathbf{r}(n)$. An algorithm can repeatedly update their detector coefficient vectors for a certain number of times L. It has been shown that, by adjusting the coefficients of an adaptive filter in *a posteriori* manner, a convergence speed of the adaptive filter can be improved [5]. This effect will be seen later in simulations.

5. SIMULATIONS

Simulations have been conducted to evaluate the performance of all algorithms investigated in this paper. These include the original MOE [2], AS-MOE [4] and our proposed AS-MOE using *a posteriori* adaptation. To access the performance of each algorithm, the average Signal-to-Interference Ratio (SIR) defined by [2]

$$\mathrm{SIR}_{\mathrm{av}}(n) = \frac{\sum_{r=1}^{M}\left[\mathbf{w}_r^T(n)\mathbf{s}_1\right]^2}{\sum_{r=1}^{M}\left[\mathbf{w}_r^T(n)\left(\mathbf{r}_r(n) - b_{1,r}(n)\mathbf{s}_1\right)\right]^2} \quad (10)$$

, where $b_{1,r}(n)$ is the transmitted symbol of the first user, is employed. A subscript r denotes the index of a particular run. In addition, M is the number of total runs. A synchronous DS-CDMA system with processing gain N = 31 was considered. It was assumed that user one is the desired user, i.e. k = 1 with unit power. The background noise was zero mean AWGN with SNR = 20dB.

Case 1- MAI Suppression:

First, the performance of the proposed algorithm is investigated in comparison with the existing algorithms under a static environment. Simulation parameters were set in accordance with the system described in [8]. In particular, the user of interest has an SNR of 20dB. There are seven multiple access interferers - six with an SNR of 20dB, and one with an SNR of 40dB - were in the channel. A signature waveform mismatch between transmitter and receiver was also set by means of ten multipath rays as described in [2], the filter was intitialised such that $\mathbf{w}(0) = \mathbf{s}_1$ for all algorithms. An appropriate step size μ of 5×10^{-5} were chosen for the original MOE to ensure a stable convergence and good steady state SIR performance. For the AS-MOE and the proposed algorithm, the initial step-size $\mu(0) = \mu = 5\times10^{-5}$. Moreover, the upper and lower step size limits μ^+ and μ^- were 10^{-3} and 0, respectively.

Fig.1 illustrates the averaged SIR over 100 Monte-Carlo runs. Clearly, *a posteriori* adaptation with AS-MOE achieves an improvement on the convergence speed over the standard AS-MOE. More specifically, while the standard AS-MOE (L=1) reaches its steady state at n = 350, the novel algorithm reach their steady state at n = 200, and n = 150 for L=3 and 7 respectively. Note that the original MOE is inferior to all algorithms regardless the number of iterations L.

Case 2- Effect of new users with strong MAI:

This experiment aims to compare tracking capability of each algorithm in a dynamic environment where the number of users is time varying. Simulation parameters were identical to those of case 1 except that the system started with five users at n = 0. All users including the desired user k = 1 had SNRs or 20dB. At time n = 400 and n = 800 other five users with SNR of 45dB and 30dB were present in addition to the existing users. These represent the system with strong MAI. At time n = 1,200 four of the 20 dB SNR users and one of the 30 dB SNR user were absent from the system. Note that data reusing factor L = 3 was used throughout the simulations.

Fig. 2(a) illustrates averaged SIR over 100 Monte-Carlo runs for each algorithm. At time n = 400 and n = 1,200 the standard AS-MOE is able to track the channel variation but with noticeable degradation in SIR level while the fixed μ MOE algorithm did not reach its steady state. It is the proposed algorithm that can successfully track the abrupt changes with the highest SIR upon convergence. Fig.2 (b) shows the average of step size of the AS-MOE and the novel algorithm. For the sudden changes in the strong MAI, the trajectories of step size of the novel algorithm decays to optimal value faster than that of the AS-MOE algorithm. In summary, the simulations confirm our expectation in the improvement over the existing algorithms in terms of both transient and steady-state performance.

6. CONCLUSIONS

In this paper, the problem of blind multiuser detection has been addressed. The blind adaptive MOE detector and its adaptive step-size variant namely, AS-MOE algorithm, were studied. Subsequently, *a posteriori* adaptation for adaptive filters has been introduced and adopted for the AS-MOE detector. Simulations have shown that the combination of adaptive step-size and *a posteriori* adaptation can achieve the improvement in the detector performance in terms of tracking capabilities and steady-state SIR even in the presence of strong MAI.

7. REFERENCES

[1] S. Verdu, *Multiuser Detection*, Cambridge University Press, 1998.

[2] M. Honig, U. Madhow and S. Verdu, "Blind Adaptive Multiuser Detection," *IEEE Trans. Inform. Theory*, vol.41 pp. 944-960, July 1995.

[3] S. Haykin, *Adaptive Filter Theory,* 3rd ed., Prentice Hall, 1996.

[4] V. Krishnamurthy, G. Yin and S. Singh, "Adaptive Step-Size Algorithm for Blind Interference Suppression in DS-CDMA", *IEEE Trans. Signal Processing,* vol. 49, pp. 190-201, Jan. 2001.

[5] D. P. Mandic and J. A. Chambers, *Recurrent Neural Networks for Prediction: Learning Algorithms, Architectures and Stability*, John Wiley & Sons, 2001.

[6] S. C. Douglas and M. Rupp, "A posteriori updates for adaptive filters," in *Proc. 31st Asilomar Conf. Signals, Systs., Comput.*, Vol. 2, pp.1641-1645, 1997.

[7] S. Pampichai and P. Sirisuk, "On the use of *a posteriori* adaptation for blind multiuser detection," in *Proc. IEEE ISCIT'2002*, pp. 186-189, Pattaya, Thailand, 2002.

[8] V. Krishnamurthy, "Averaged stochastic gradient algorithms for adaptive blind multiuser detection in DS/CDMA systems," *IEEE Trans. Commun.*, Vol.48, No.1, pp.125-134, Jan 2000.

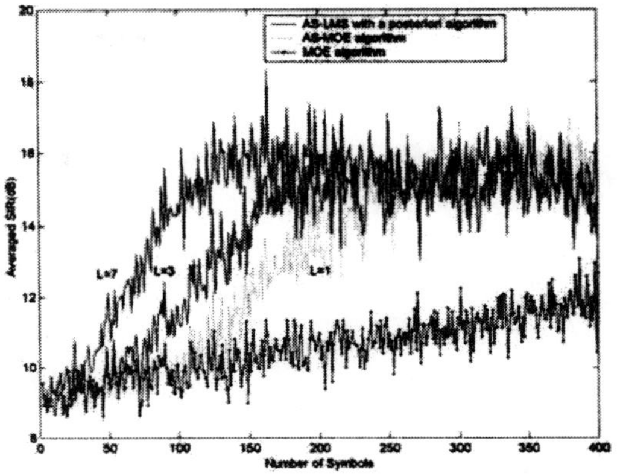

Figure 1 Averaged SIR trajectories of the original MOE, AS-MOE and the proposed AS-MOE using *a posteriori* adaptation with $L = 3$ and 7 in a static environment.

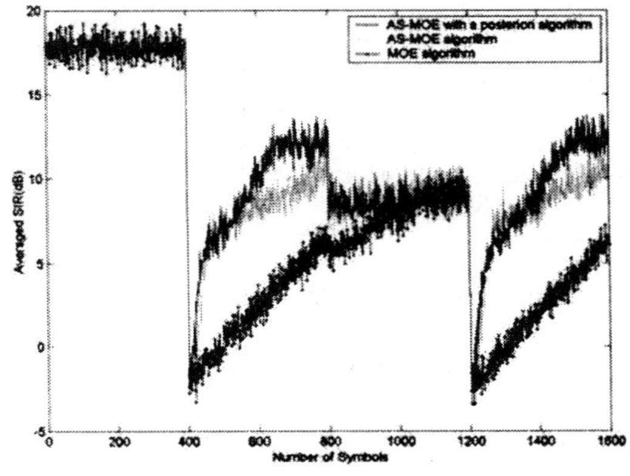

(a)

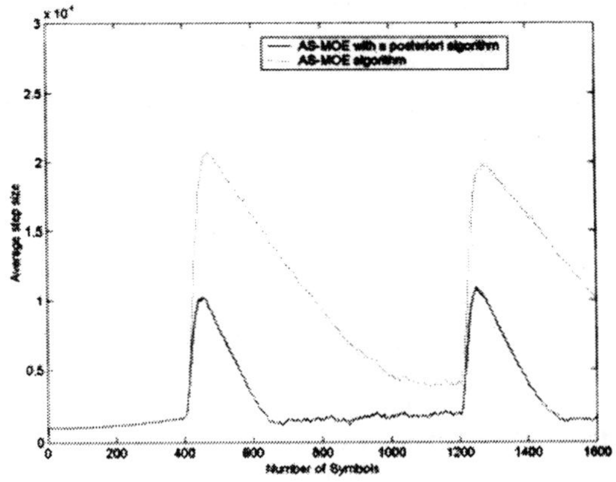

(b)

Figure 2 (a) Averaged SIR trajectories and (b) Averaged step size with sudden arrival of users for the original MOE, AS-MOE and the proposed AS-MOE using *a posteriori* adaptation ($L = 3$) detector under the strong MAI

A HIGH-SPEED BLIND DFE EQUALIZER USING AN ERROR FEEDBACK FILTER FOR QAM MODEMS

Jung Hoo Lee, Weon Heum Park, Ju Hyung Hong, Myung H. Sunwoo, Kyung Ho Kim***

*School of Electrical and Computer Engineering. Ajou University
San 5, Wonchun-Dong, Paldal-Ku, Suwon, 442-749 KOREA
** Telecommunications R&D Center, SAMSUNG Electronics
Email : sunwoo@ajou.ac.kr

ABSTRACT

This paper proposes a DFE (Decision Feedback Equalizer) equalizer with an error feedback filter using the Multi-Modulus Algorithm (MMA). The proposed equalizer has been designed for 64/256 QAM constellations. The existing MMA equalizer uses two transversal filters or feedforward and feedback filters, while the proposed equalizer uses feedforward, feedback and error feedback filters to improve the channel adaptive performance and to reduce the number of taps. The architecture has been modeled by VHDL and logic synthesis has been performed using the 0.25 μm Faraday CMOS standard cell library. The total number of the gates is about 190,000 gates. The proposed equalizer operates at 15 MHz and provides a symbol rate up to 64 Mbps which is higher than the DOCSIS recommendation.

Keywords : Equalizer, MMA, LMS, DFE, error feedback

1. INTRODUCTION

The fading channel and ISI (InterSymbol Interference) caused by multipath in bandlimtted time dispersive channels distorts the transmitted signal, causing bit errors at the receiver in high-speed digital transmission. The equalizer improves the communication quality without increasing the signal power or widening the bandwidth because it compensates the signal amplitude and the delay characteristics of the received signal [1][2]. The general operating modes of equalizers include training and tracking [3]. In the training mode, the transmitter sends a known, fixed-length training sequence to the receiver and the equalizer initializes tap weights for the known training sequence. When the weights are converged, the equalizer switches to the tracking mode, the equalizer assumes the decision logic output as the correct transmitted data for tracking of residual channel variation [1][3].

However, several standards require channel adaptation without the training sequence and pilot channel, such as HDTV, LMDS (Local Multipoint Distribution Service), DSL(Digital Subscriber Line), the downstream of the MCNS (Multimedia Cable Network System) standard of CATV network [3][4], etc. Thus, blind adaptive algorithms are needed to exploit known statistics of the transmitted data, that is, the probability distribution and to adapt to the equalizer weights [1]. One of blind algorithms is MMA (Multi-Modulus Algorithm) [5], which uses the distribution of the transmitted and received signals without a training sequence. The MMA algorithm is more flexible than both RCA (Reduced Constellation Algorithm) [6] and CMA (Constant Modulus Algorithm) [6] and it can easily adapt to systems which use non-square and very dense signal constellations [4][7].

In the proposed equalizer, MMA is used when the modem is just starting acquisition or when the channel is degraded, and the LMS (Least Mean Square) algorithm is used after the equalizer converges the steady state. This algorithm is particularly well suited to the applications which use two-dimensional transmission schemes, such as CAP (Carrierless AM/PM) and QAM. The value of MMA error function is too large to converge the channel in high-order QAM, so we use the generalized MMA (GMMA), which is well suited to very dense signal constellations. [4][7].

In general, the existing MMA equalizer [4] uses two transversal filters for In-phase and Quadrature-phase channels to reduce residual ISI. To reduce the number of taps and the acquisition latency, we have already proposed the DFE equalizer [2]. In this paper, we propose the equalizer uses feedforward, feedback, and error feedback filters to improve the channel adaptive performance.

Error feedback is a general method that can be used to reduce errors in quantization and is widely used in applications, such as predictive speech coding, predictive image coding and sigma-delta analog-to-digital conversion [8]. The proposed equalizer using error feedback reduces the correlation of residual error, MSE(Mean Square Error) and the initial error propagation.

This paper proposes the blind equalizer that uses the DFE architecture using an error feedback filter and the MMA algorithm to reduce the acquisition latency and the hardware complexity. The proposed equalizer can support a symbol rate up to 64 Mbps in 256-QAM demodulation. We have simulated the proposed equalizer using VHDL models and performed logic synthesis using the 0.25 μm Faraday CMOS standard cell library. The implemented equalizer contains about 190,000 gates and the chip operates at 15 MHz. The chip can be used for wire and wireless communications, such as HDTV, LMDS, cable modem, DSL, etc.

The remainder of this paper is organized as follows. Section 2 introduces the algorithm of the proposed equalizer and describes the existing equalizer architectures, Section 3 shows the architecture of the proposed equalizer. Section 4 explains the simulation and the implementation results. Finally, Section 5 contains concluding remarks.

* This work was supported in part by the SystemIC2010 program, in part by the National Research Laboratory (NRL) program and in part by the IDEC.

0-7803-7761-3/03/$17.00 ©2003 IEEE

2. EQUALIZATION ALGORITHMS AND ARCHITECTURES

This section describes the algorithms of the proposed equalizer and the existing equalizer architectures.

2.1 Adaptive Algorithms for Equalization

We explain two adaptive algorithms, that is, MMA [2][4] and LMS [1] algorithms.

2.1.1 The LMS Algorithm

The LMS algorithm has been proposed to simplify the Wiener-Hopf solution [1]. The Wiener-Hopf solution is difficult for real-time processing because it requires a massive amount of computations.

The LMS algorithm [1] is described as follows. Assume that x_n is the received data vector for time $t = n$ and w_n is the tap weight vector, $w_n = \{w_{n,1}, w_{n,2}, \ldots\ldots, w_{n,M}\}$ where M is the number of taps. The output of the equalizer, y_n, is given by

$$y_n = \sum_{i=1}^{M} x_{n-i} \cdot w_{n-i}^{T}. \qquad (1)$$

The error function, ε_n^{LMS}, for the LMS algorithm can be expressed as

$$\varepsilon_n^{LMS} = d_n - y_n \qquad (2)$$

where d_n is the decision data vector at time $t = n$ and ε_n^{LMS} is negative when y_n is greater than d_n and positive when y_n is smaller than d_n. The weight updating function is given by equation (3)

$$w_{n+1} = w_n - \mu^{LMS} \cdot \varepsilon_n \cdot x_n^{*} \qquad (3)$$

where μ^{LMS} is called the stepsize of the equalizer and x_n^{*} is the complex conjugate form of the received data vector at time $t = n$. The stepsize decides the acquisition speed. The large stepsize leads to fast acquisition but gives a large residual error, while the small stepsize leads to slow acquisition but gives a small residual error. The proposed equalizer uses $\mu^{LMS} = 0.000065035$ as an optimized stepsize value for the LMS algorithm after extensive simulations.

2.1.2 The MMA (Multi-Modulus Algorithm)

The MMA algorithm is proposed in [4] and is considered as one of blind algorithms that use the probability distribution of both the transmitted and received signals without a training sequence. MMA is used when a modem is just starting the acquisition or when the channel is degraded, and then LMS is used after the equalizer goes to the steady state. This algorithm is particularly well suited to the applications that use two-dimensional transmission schemes, such as CAP (Carrierless AM/PM) and QAM [4]. The equation of the MMA error function [4] is given by

$$\varepsilon_n^{MMA} = \left(y_{n,r}^2 - R_M^2\right) \cdot y_{n,r} + \left(y_{n,i}^2 - R_M^2\right) \cdot y_{n,i} \qquad (4)$$

where $y_{n,r}$ and $y_{n,i}$ are the real and the imaginary part of the equalizer output, R_M is a constant in equations (4) and (5) for acquisition in unknown channel and is expressed as

$$R_M = \frac{E\left[|a_n|^{4L}\right]}{E\left[|a_n|^{2L}\right]} \qquad (5)$$

where L is equal to 2 for MMA and a_n is the transmitted signal. R_M is the function of statistics of the symbols in the signal constellation, which takes a fixed constant value when the modulation method is determined. We use $R_M = 6.08$ for 64 QAM, $R_{M1} = 6.08$, $R_{M2} = 10.25$ and $R_{M3} = 14.17$ for 256 QAM.

The weight updating function of MMA is expressed by (6)

$$W_{n+1} = W_n - [\mu^{MMA} \cdot \varepsilon_{n,r}^{MMA} + j\mu^{MMA} \cdot \varepsilon_{n,i}^{MMA}] \cdot x_n^{*} \qquad (6)$$

where μ^{MMA} is the stepsizes, $\varepsilon_{n,r}^{MMA}$ and $\varepsilon_{n,i}^{MMA}$ are the real and the imaginary parts of the error function. The proposed equalizer uses $\mu_1^{MMA} = 0.0000976562$, $\mu_2^{MMA} = 0.001953125$ as the optimized stepsize value of MMA.

2.1.3 The GMMA (Generalized MMA)

The MMA has difficulties with very dense constellations. The value of the MMA error function is too large to converge channel in high-order QAM. The GMMA (Generalized MMA) algorithm is proposed for high-order QAM that has dense signal constellation and generally indicates QAM above 128-constellation [4]. GMMA is much more flexible than both RCA and CMA [6], and can easily be adapted to the systems that use non-square and very dense signal constellation.

The basic concept of GMMA is to divide the constellation into small subsets, and each subset has its own MMA cost function. Figure 1 describes the principle of GMMA in 256-constellation.

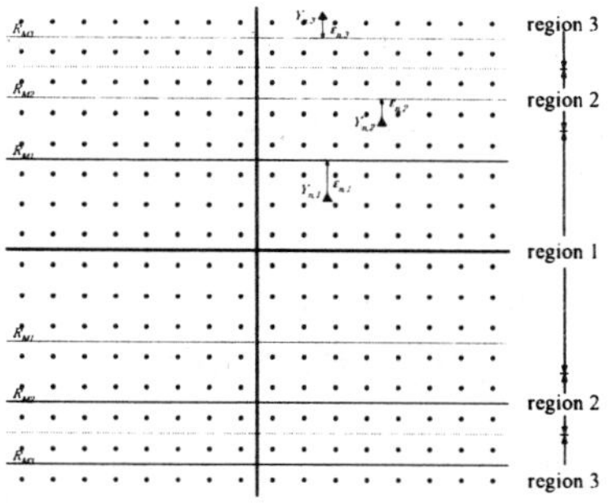

Figure 1. GMMA sample subsets and moduli for 256-point constellation

In Figure 1, the 256-point constellation is divided into 3 regions, each region has its own R_M (R_{M1}, R_{M2}, R_{M3}) value. GMMA selects the value of R_M according to the equalizer output (y_n). The selection method of R_M is described by

$$\begin{cases} If & y_n \le 8 & R_M = R_{M1} \\ Else\ if & 8 < y_n \le 12 & R_M = R_{M2} \\ Else\ if & y_n > 12 & R_M = R_{M3} \end{cases} \qquad (7)$$

If the equalizer output is in region 1, region 2 or region 3, then the value of the GMMA error function, R_M in equation (4), is replaced by one value of R_{M1}, R_{M2} and R_{M3}, respectively.

2.2 Existing Equalizer Architectures

Figure 2 shows the existing MMA equalizer architecture in [4]. The existing architecture consists of two transversal filters, such as an In-phase filter (real) and a Quadrature-phase (imaginary) filter. The two filters operate independently.

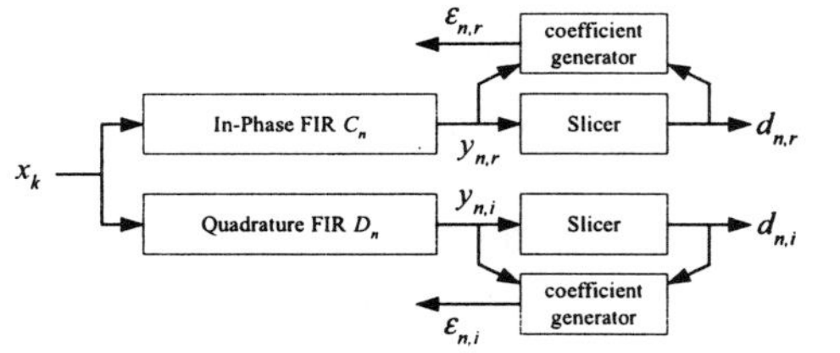

Figure 2. The existing MMA equalizer

Figure 3 shows our previous DFE equalizer using the MMA algorithm [2] which consists of feedforward and feedback filters. This architecture improves the BER performance and reduces residual ISI efficiently compared with the transversal equalizer [4]. However, this architecture still has the residual error and propagates the initial error by an inaccurate decision value.

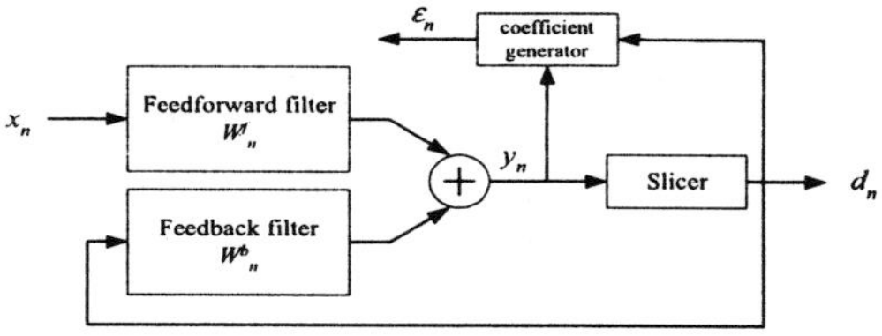

Figure 3. The existing MMA equalizer (DFE)

The proposed equalizer employs the error feedback filter to our previous DFE equalizer [2], which reduces the residual error and the initial propagation error even further.

3. The PROPOSED EQUALIZER ARCHITECTURE

The proposed equalizer consists of feedforward, feedback, and error feedback filters. The output signal, y_n is determined by the equation (8)

$$y_n = \sum_{i=0}^{M-1} x_{n-i} \cdot W_{n-i}^{f^T} + \sum_{j=0}^{N-1} x_{n-j} \cdot W_{n-j}^{b^T} + \sum_{k=0}^{L-1} \varepsilon_{n-k} \cdot W_{n-k}^{e^T} \qquad (8)$$

where $W_n^{f^T}$, $W_n^{b^T}$, $W_n^{e^T}$ are the weight transposed vectors for the feedforward filter, feedback filter and error feedback filter, respectively.

Figure 4. shows the error function generator. It generates the error functionss for MMA and LMS algorithms. Its output, ε_n, is fed into each filter.

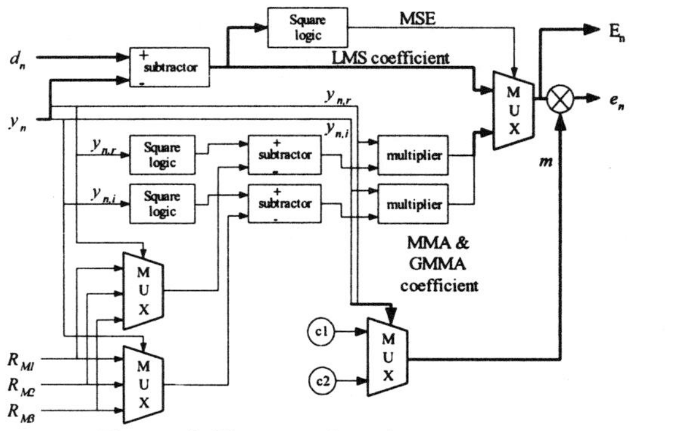

Figure 4. The error function generator

We propose the switching method between two algorithms as follows. MMA initializes the equalizer tap weight without a training sequence, a pilot channel and any other information, and then the LMS algorithm is used after the equalizer goes to the steady state.

The selection operation between two algorithms is expressed by equation (9). The adaptive algorithm is selected by the proper value of MSE. MUX selects the MMA coefficients when MSE is higher than 1.5 (threshold2), selects the LMS algorithm coefficients when MSE is lower than 0.5 (threshold1) and holds the previous coefficients when MSE is between two thresholds. Thresholds are important parameters for the convergence of the equalizer. We decide threshold values by extensive simulations.

$$\begin{cases} \textit{If} \quad MSE \le 0.5 & \text{select LMS} \\ \textit{Else if} \quad 0.5 < MSE \le 1.5 & \text{Last selection algorithm} \\ \textit{Else if} \quad MSE > 1.5 & \text{select MMA(64 - QAM) or GMMA (256 - QAM)} \end{cases} \qquad (9)$$

Figure 5 shows the overall architecture of the proposed equalizer, which consists of three filters, i.e., (the 5-tap feedforward filter and the 4-tap feedback filter, the 2-tap error feedback filter), the slicer (decision logic) and the coefficient generation block.

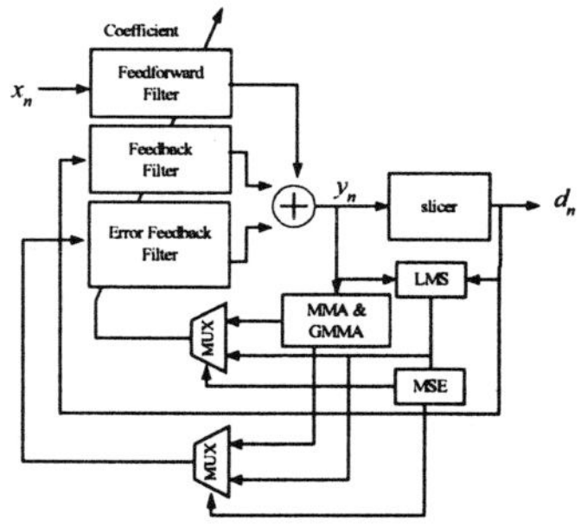

Figure 5. The overall block diagram of the proposed equalizer

The slicer output, d_n, is fed into the feedback filter to reduce the acquisition time and to improve the channel adaptive performance. In addition, the error function generator output, E_n, is fed into the error feedback filter to reduce the residual error, MSE and initial error propagation.

The proposed architecture has better channel adaptive performance and lower BER than the existing equalizers [2][4].

4. THE SIMULATION AND IMPLEMENTATION RESULT

This section explains simulation and implementation results. We have simulated the proposed equalizer and estimated performance. We have performed logic synthesis with the 0.25 μm Faraday CMOS standard cell library.

We first have simulated the proposed equalizer using the SPW™ CAD tool to estimate the performance. The simulations of the proposed equalizer and the existing equalizers in [2][4] with the AWGN and multipath channels [9] have been performed. We verified the proposed DFE equalizer using the floating-point model and the fixed-point model using the SPW™ CAD tool. All the bit precisions in the proposed equalizer are decided by the SPW™ fixed-point simulation results.

Figure 6 shows the BER performance of the proposed DFE equalizer using error feedback and the existing equalizers [2][4] in 64-QAM. As shown in Figure 6, the transversal equalizer [4] cannot converge channel when SNR is below 13dB. The BER performance of the proposed equalizer improves compared with the existing equalizer [2][4]. The performance improvement of the proposed equalizer is measured about 0.5 dB at 10^{-5} BER compared with the transversal equalizer [4] and 0.3dB at 10^{-5} BER compared with the DFE equalizer [2].

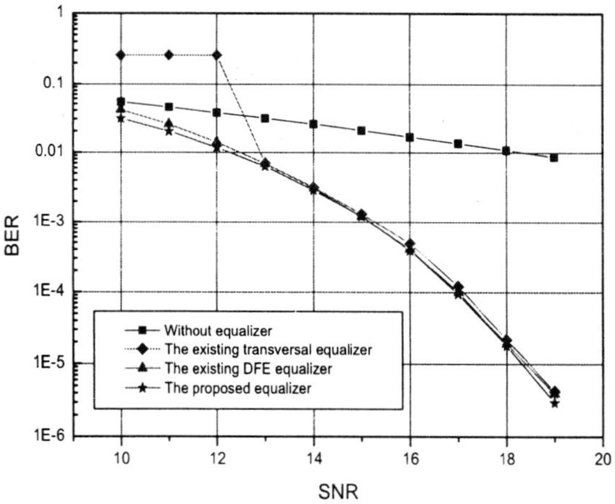

Figure 6. BER performance comparisons in the AWGN and multipath channels.

The proposed equalizer has been modeled by VHDL and logic synthesis has been performed using the 0.25μm Faraday CMOS standard cell library. The total gate count is about 190,000 and operates at 15MHz. Figure 7 shows the synthesis result of the proposed equalizer.

The transversal architecture consists of a 48 tap transversal filter [4] and the DFE architecture [2] consists of a 5 tap feedforward filter and a 6 tap feedback filter. The proposed architecture consists of a 5-tap feedforward filter, a 4-tap feedback filter and a 2-tap error feedback filter. It reduces the number of taps compared with the transversal architecture [4] and has the same number of taps with our previous DFE architecture [2]. However, the proposed architecture improves the BER performance and has the fast acquisition time.

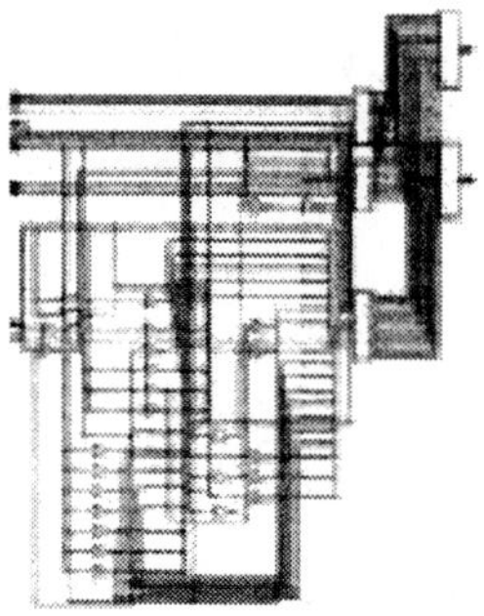

Figure 7. The synthesis result of the proposed equalizer

5. CONCLUSIONS

This paper proposed the high speed blind DFE architecture with an error feedback filter. The proposed equalizer combines the MMA algorithm and the DFE architecture with the error feedback filter to reduce the acquisition time and the residual error. The proposed equalizer shows the improved BER performance and the fast acquisition time compared with the the existing equalizers. We used the UMC 0.25 μm Faraday CMOS standard cell library. The total gate count is about 190,000 and operates at 15 MHz. The proposed equalizer can be used for various QAM modems, such as HDTV, LMDS, cable modem, DSL, etc.

REFERENCES

[1] Haykin, *Adaptive Filter Theory - third edition,* Prentice-Hall, 1996.

[2] D. K. Shin, S. J. Hwang, and M. H. Sunwoo, "Design of a DFE equalizer ASIC chip using the MMA algorithm," in *Proc. IEEE Workshop on Signal Processing Systems (SiPS)*, Oct. 2000, pp. 200-209.

[3] Cheng-I Hwang, David W. Lin, "Joint low-complexity blind equalization, carrier recovery and timing recovery with application to cable modem transmission," *IEICE Trans. Commun.*, vol. E82-B, no. 1, pp. 120 - 128, Jan. 1999.

[4] L. M. Garth, "A dynamic convergence analysis of blind equalization algorithms," *IEEE Trans. Commun.*, vol. 49, no. 4, pp. 455-466, Apr. 2001.

[5] J. Yang, J.J. Werner, and G. A. Dumont, "The multimodulus blind equalization and its generalized algorithms," *IEEE Journal Selected Area in Commun.* vol. 20, no. 5, pp. 997-1014, June 2002.

[6] J. Yang, J.J. Werner, and Jr., G. A. Dumont, "The multimodulus blind equalization algorithm," in *Proc. Int. Conf. on Digital Signal Processing*, Santorini, Greece, July 2-4. 1997. pp. 127-130.

[7] A. P. des Rosiers, and P. H. Siegel, "Effect of varying source kurtosis on the multimodulus algorithm," in *Proc. IEEE Int. Conf. on Commun.*, vol. 2, 1999, pp. 1300 -1304.

[8] T. I. Laakso, and I. O. Hartimo. "Noise reduction in recursive digital filters using high-order error feedback," *IEEE Trans. Sig. Proc.* vol. 40, pp. 1096-1107, May. 1992.

[9] John R. Treichler, Michael G. Larimore, and Jeffrey C. Harp, " Practical blind demodulators for high-order QAM signals," in *Proceedings of the IEEE,* vol. 86, no. 10, pp. 1907-1926, Oct. 1998.

Stability, Controllability and Observability of 2-D Continuous-Discrete Systems*

Yang Xiao

Institute of Information Science, Northern Jiaotong University
Beijing 100044, China E-mail: yxiao@center.njtu.edu.cn

Abstract: Based on the model of 1-D complex systems, the criteria of stability, controllability and observability of 2-D continuous-discrete are derived. Controllability and observability of 2-D continuous-discrete systems are then defined in terms of s domain.

KeyWords: 2-D continuous-discrete systems, 2-D controllability, 2-D observality

1. Introduction

In this paper we consider 2-D continuous-discrete systems [1-7]. This study comes from both practical and theoretical reasons. From a practical viewpoint 2-D continuous-discrete systems have many fields of application. In fact 2-D control problems have been the subject of several investigations in recent years, including [8-21], but only Ref. [1-7] latter dealt with continuous-discrete case and, however, the theory of 2-D continuous-discrete systems deserves to be deeply studied. These systems play a fundamental role in 1D case where multi input-output digital controllers.

2-D continuous-discrete systems arise also in signal processing whenever the output of a 2-D filter is sampled in one direction on the discrete plane. Besides all the control and/or modelling problems where 2-D continuous-discrete systems arise the most compelling reason justifying the study of the subject is the theoretical issue. With this motivation we will show a theory to treat 2-D continuous-discrete systems independent of the particular field of application and taking into account both 1-D results and 2-D invariant results.

2. Description of 2-D continuous-discrete system model

Some practical systems need to be expressed by the model of 2-D continuous-discrete systems, such as some hybrid systems[1]. Similar to other systems, there are also controlabity and reachability [2-4]. Consider following the Fornasini-Marchesini model of 2-D Continuous-Discrete Systems[4]

$$\frac{\partial \mathbf{x}(t,n+1)}{\partial t} = \mathbf{A}_0\mathbf{x}(t,n) + \mathbf{A}_1\frac{\partial \mathbf{x}(t,n)}{\partial t} + \mathbf{A}_2\mathbf{x}(t,n+1) + \mathbf{B}\mathbf{u}(t,n) \tag{1a}$$

$$\mathbf{y}(t,n) = \mathbf{C}\mathbf{x}(t,n) \tag{1b}$$

where $\mathbf{x}(t,n) \in R^N$ is state vector of the system, $\mathbf{u}(t,n) \in R^M$ is the input vector of the system, $\mathbf{y}(t,n) \in R^K$ is output input vector of the system, $\mathbf{C} \in R^{K\times N}$.

Take Laplace transformation to the first variable t of Eq. (1), we get

$$s\mathbf{X}(s,n+1) - \mathbf{x}(0,n+1) = \mathbf{A}_0\mathbf{X}(s,n) + \mathbf{A}_1[s\mathbf{X}(s,n) - \mathbf{x}(0,n)] + \mathbf{A}_2\mathbf{X}(s,n+1) + \mathbf{B}\mathbf{U}(s,n) \tag{2a}$$

$$\mathbf{Y}(s,n) = \mathbf{C}\mathbf{X}(s,n) \tag{2b}$$

Arrange Eq. (2a), we get

$$\mathbf{X}(s,n+1) = (s\mathbf{I} - \mathbf{A}_2)^{-1}(s\mathbf{A}_1 + \mathbf{A}_0)\mathbf{X}(s,n) + (s\mathbf{I} - \mathbf{A}_2)^{-1}[\mathbf{x}(0,n+1) - \mathbf{A}_1\mathbf{x}(0,n)] + (s\mathbf{I} - \mathbf{A}_2)^{-1}\mathbf{B}\mathbf{U}(s,n) \tag{3}$$

From Eq. (1a), we can also get

$$\mathbf{x}(0,n+1) = -\mathbf{A}_2^{-1}\mathbf{A}_0\mathbf{x}(0,n) - \mathbf{A}_2^{-1}\mathbf{B}\mathbf{u}(0,n) \tag{4}$$

Substitute Eq.(4) into Eq.(3), we get a complex system

$$\mathbf{X}(s,n+1) = (s\mathbf{I} - \mathbf{A}_2)^{-1}(s\mathbf{A}_1 + \mathbf{A}_0)\mathbf{X}(s,n) - (s\mathbf{I} - \mathbf{A}_2)^{-1}[\mathbf{A}_2^{-1}\mathbf{A}_0 + \mathbf{A}_1]\mathbf{x}(0,n) - (s\mathbf{I} - \mathbf{A}_2)^{-1}\mathbf{A}_2^{-1}\mathbf{B}\mathbf{u}(0,n) + (s\mathbf{I} - \mathbf{A}_2)^{-1}\mathbf{B}\mathbf{U}(s,n) \tag{5a}$$

$$\mathbf{Y}(s,n) = \mathbf{C}\mathbf{X}(s,n) \tag{5b}$$

* This work is supported by NSFC of China, under grant 69971002.

From Eq. (5), we can get the solution of the complex system (5) as following

$$\begin{aligned}\mathbf{X}(s,n) &= (s\mathbf{I}-\mathbf{A}_2)^{-n}(s\mathbf{A}_1+\mathbf{A}_0)^n\mathbf{X}(s,0)\\ &-\sum_{k=0}^{n-1}(s\mathbf{I}-\mathbf{A}_2)^{-(n-1-k)}(s\mathbf{A}_1+\mathbf{A}_0)^{n-1-k}(s\mathbf{I}-\mathbf{A}_2)^{-1}\\ &\cdot[\mathbf{A}_2^{-1}\mathbf{A}_0+\mathbf{A}_1]\mathbf{x}(0,k)\\ &-\sum_{k=0}^{n-1}(s\mathbf{I}-\mathbf{A}_2)^{-(n-1-k)}(s\mathbf{A}_1+\mathbf{A}_0)^{n-1-k}(s\mathbf{I}-\mathbf{A}_2)^{-1}\\ &\cdot\mathbf{A}_2^{-1}\mathbf{B}\mathbf{u}(0,k)\\ &+\sum_{k=0}^{n-1}(s\mathbf{I}-\mathbf{A}_2)^{-(n-1-k)}(s\mathbf{A}_1+\mathbf{A}_0)^{n-1-k}(s\mathbf{I}-\mathbf{A}_2)^{-1}\\ &\cdot\mathbf{B}\mathbf{U}(s,k)\end{aligned} \tag{6}$$

To solve above equation is still difficult, we use it as a tool to analyse the stability and controllability of the 2-D invariant continuous-discrete systems.

3. Stability of 2-D continuous-discrete systems

Definition 1 The 2-D continuous-discrete system (1) is asymptotically stable if assuming $\mathbf{u}(t,n)\equiv 0$, then $\lim\limits_{\substack{t\to\infty\\ n\to\infty}}\mathbf{x}(t,n)=0$.

Theorem 1: Given initial condition $\mathbf{x}(t,0)$ and $\mathbf{x}(0,n)$, the 2-D continuous-discrete system (1) is asymptotically stable, if and only if

$$\begin{aligned}&\det[z\mathbf{I}-(s\mathbf{I}-\mathbf{A}_2)^{-1}(s\mathbf{A}_1+\mathbf{A}_0)]\neq 0\\ &\operatorname{Re}s\geq 0, |z|\geq 1\end{aligned} \tag{7}$$

Proof: Clearly when asymptotic stability of system (1) occurs it implies stability of system (5) and of all others descriptions associated to system (1), also the vice versa holds true.
We suppose complex system (5) asymptotically stable, for fixed complex variable s, from the condition (8), $(s\mathbf{I}-\mathbf{A}_2)^{-n}(s\mathbf{A}_1+\mathbf{A}_0)^n$ is of roundedness,

$$\left\|(s\mathbf{I}-\mathbf{A}_2)^{-n}(s\mathbf{A}_1+\mathbf{A}_0)^n\right\|<1 \tag{8}$$

and

$$\lim_{n\to\infty}\left\|(s\mathbf{I}-\mathbf{A}_2)^{-n}(s\mathbf{A}_1+\mathbf{A}_0)^n\right\|=0 \tag{9}$$

which means

$$\begin{aligned}\|\mathbf{X}(s,n)\| &= \left\|(s\mathbf{I}-\mathbf{A}_2)^{-n}(s\mathbf{A}_1+\mathbf{A}_0)^n\mathbf{X}(s,0)\right\|\\ &<\|\mathbf{X}(s,0)\|\end{aligned} \tag{10}$$

If we fix $\varepsilon>0$ there's $\delta>0$ such that $\|\mathbf{X}(s,n)\|<\varepsilon$ with $n>\delta$.

Due to Eq. (5b), $\lim\limits_{n\to\infty}\|\mathbf{Y}(s,n)\|=0$.

This argument still holds true for any invariant description of (1), in view of polynomial stability condition for 2-D systems expressed in terms of zeros on s-z space [14], it is easy to prove the following part of the theorem. Q.E.D.

4. Controllability of 2-D continuous-discrete systems

In 2-D invariant continuous-discrete systems, a local controllability assumption is defined in terms of the possibility of driving to zero initial local states, for example $\mathbf{x}(0,0)$, within a finite distance in the future.

Definition 2 System (1) is locally controllable if any of its complex descriptions (5) is locally controllable, i.e. $\forall\mathbf{x}(t_0,n_0)\in R^N$, there exists $\mathbf{u}(t,n)$ such that system(1) in finite time can reach given state $\mathbf{x}(t_1,n_1)$.

Next theorem ensures the controllability for complex system (5).

Theorem 2: Given initial condition $\mathbf{X}(s,n_0)$, the complex system (5) is controllable, if and only if the following matrix

$$\mathbf{M}[\mathbf{A}(s),\mathbf{B}(s)]=[\mathbf{B}(s),\mathbf{A}(s)\mathbf{B}(s),\dots,\mathbf{A}(s)^{N-1}\mathbf{B}(s)] \tag{11}$$

with the rank of N for $\operatorname{Re}s\geq 0$, where

$$\mathbf{A}(s)=(s\mathbf{I}-\mathbf{A}_2)^{-1}(s\mathbf{A}_1+\mathbf{A}_0) \tag{12}$$

and

$$\mathbf{B}(s)=(s\mathbf{I}-\mathbf{A}_2)^{-1}\mathbf{B} \tag{13}$$

Proof: The easiest way to understand controllability is to consider the time-invariant complex discrete-time systems, i.e., a system

$$\mathbf{X}(s,n+1) = \mathbf{A}(s)\mathbf{X}(s,n) + \mathbf{B}(s)\mathbf{U}(s,n)$$

n=0, 1, 2, ...

$$\mathbf{X}(s,0) = 0$$

with the following state solution due to Eq. (6),

$$\mathbf{X}(s,n) = \sum_{k=0}^{n-1} \mathbf{A}(s)^{-(n-1-k)}\mathbf{B}(s)\mathbf{U}(s,k)$$

where $\mathbf{A}(s)$ and $\mathbf{B}(s)$ are defined by Eq. (12) and (13).

The controllability problem is to make $\mathbf{X}(s,n)$ equal to any pre-assigned vector in R^N by choosing appropriate controls, $\mathbf{U}(s,0)$, $\mathbf{U}(s,1)$, ..., $\mathbf{U}(s,n-1)$. $\mathbf{X}(s,n)$ is simply a linear combination of vectors,

$$\{\mathbf{B}(s), \mathbf{A}(s)\mathbf{B}(s),..., \mathbf{A}(s)^{n-1}\mathbf{B}(s)\}.$$

In order for $\mathbf{X}(s,n)$ to be equal to any vector in R^N, the vectors above must span the whole space R^N, hence they must satisfy the condition

$$\text{Rank}[\mathbf{B}(s), \mathbf{A}(s)\mathbf{B}(s),..., \mathbf{A}^{n-1}(s)\mathbf{B}(s)] = n, \ \text{Re}\, s \geq 0.$$

For vector control, this requires $n \geq N$, otherwise the number of columns in the matrix above is less than N.

In case the control $\mathbf{U}(s,k)$, is a M-vector, M>1, the right hand side of the formula for $\mathbf{X}(s,N)$ is still a linear combination of columns of the matrix

$$[\mathbf{B}(s), \mathbf{A}(s)\mathbf{B}(s),..., \mathbf{A}^{n-1}(s)\mathbf{B}(s)]$$

where the coefficients of the linear combination are now individual components of vectors $\mathbf{U}(s,0)$, $\mathbf{U}(s,1)$, ..., $\mathbf{U}(s,M)$. Since M>1, it may no longer be necessary to have n≥N. By the Caley-Hamilton Theorem, $\mathbf{A}^n(s)$, and thus $\mathbf{A}^{n+j}(s)$ for j≥0, can be expressed as linear combination of $\{\mathbf{I}, \mathbf{A}(s),..., \mathbf{A}^{n-1}(s)\}$. It follows that the number of linearly independent columns in matrix (11) will not increase with n if n≥N. Therefore

$$\text{Rank}[\mathbf{B}(s), \mathbf{A}(s)\mathbf{B}(s),..., \mathbf{A}(s)^{N-1}\mathbf{B}(s)] = N,$$

for $\text{Re}\, s \geq 0$.

This is a necessary and sufficient condition for controllability of system (5). Q.E.D

Now, we can prove the following theorem resorting to Theorem 2.

Theorem 3 Given initial condition $\mathbf{x}(0,n)$, the 2-D continuous-discrete system system (1) is controllable, if and only if the matrix (11) with the rank of N for $\text{Re}\, s \geq 0$.

Proof: Since $\mathbf{X}(s,n)$, the solutions of the complex system (5) have following relation with $\mathbf{x}(t,n)$, the solutions of the 2-D continuous-discrete system system (1),

$$\lim_{s\to 0}\left[s\mathbf{X}(s,n) - \mathbf{x}(0^-,n)\right] = \lim_{s\to 0}\int_{0^-}^{\infty}\frac{\partial \mathbf{x}(t,n)}{\partial t}e^{-st}dt$$

$$= \lim_{s\to 0}\int_{0^-}^{\infty}\frac{\partial \mathbf{x}(t,n)}{\partial t}e^{-0}dt = \mathbf{x}(\infty,n) - \mathbf{x}(0^-,n)$$

thus

$$\lim_{s\to 0} s\mathbf{X}(s,n) = \lim_{t\to\infty}\mathbf{x}(t,n).$$

Similarly, the input of complex system (5) and the input of the 2-D continuous-discrete system (1) have the relation

$$\lim_{s\to 0} s\mathbf{U}(s,n) = \lim_{t\to\infty}\mathbf{u}(t,n).$$

The above properties hold when poles of $s\mathbf{X}(s,n)$ and $s\mathbf{U}(s,n)$ are in the left-half plane (except for a simple pole at the origin), which can be ensured by the matrix (11) with the rank N. The properties also establish the relation of system (5) and systems (1). Thus, the controllability of 2-D continuous-discrete system (1) can be guaranteed by that of complex system (5) . Q.E.D.

Definition 3 System (1) is locally observable if any of its complex descriptions (5) is locally controllable, i.e. $\forall \mathbf{x}(t_0, n_0) \in R^N$, there exists $\mathbf{y}(t,n)$ such that system(1) in finite time can be observed by given state $\mathbf{x}(t_1, n_1)$.

Theorem 4 Given initial condition $\mathbf{x}(t,0) \neq \mathbf{0}$ and $\mathbf{x}(0,n) = \mathbf{0}$, system (1) is obervable, if and only if

$$\mathbf{M}[\mathbf{A}(\mathrm{s}),\mathbf{C}] = \begin{bmatrix} \mathbf{C} \\ \mathbf{CA}(\mathrm{s}) \\ \vdots \\ \mathbf{CA}^{N-1}(\mathrm{s}) \end{bmatrix} \tag{14}$$

with the rank of N for $\text{Re}\, s \geq 0$, where where $\mathbf{A}(s)$ and $\mathbf{C}$ are defined by Eq. (12) and (1b) respectively.

Proof: If the system kas a solution for $\mathbf{x}(t,0)$ (which is so by hypothesis) then this solution must unique if and only if the matrix $\mathbf{M}[\mathbf{A}(\mathrm{s}),\mathbf{C}]$ has rank *N*.

If we define the deviation

$$\mathbf{V}(s,n)=\mathbf{Y}(s,n)-\mathbf{C}\mathbf{A}^{n-1}(\mathrm{s})\mathbf{X}(s,0),$$

then the equation amounts to $\mathbf{V}(s,n)=0,\ 1\le n\le N$.

If the equations are not consistent we could still define a 'least-squares' solution to them by minimizing any positive-definite quadratic form in these deviations with respect to $\mathbf{x}(t,0)$. In particular, we could minimize

$$\sum_{n=0}^{N-1}\mathbf{V}^T(s,n)\mathbf{V}(s,n)=0.$$

This minimization gives

$$\mathbf{X}(s,0)=(\mathbf{M}^T\mathbf{M})^{-1}\sum_{n=0}^{N-1}(\mathbf{A}^T(s))^{n-1}\mathbf{C}^T\mathbf{Y}(s,n) \quad (15)$$

where $\mathbf{M}=\mathbf{M}[\mathbf{A}(\mathrm{s}),\mathbf{C}]$.

If the following equation

$$\mathbf{Y}(s,n)=\mathbf{C}(s\mathbf{I}-\mathbf{A}_2)^{-n}(s\mathbf{A}_1+\mathbf{A}_0)^n\mathbf{X}(s,0)$$

indeed has a solution and this is unique then Eq. (15) must equal this solution, the actual value of $\mathbf{X}(s,0)$, which is corresponding to $\mathbf{x}(t,0)$. The criterion for uniqueness of the least-squares solution is that $\mathbf{M}^T\mathbf{M}$ should be nonsingular, which is the condition of the theorem. Q.E.D

4. Conclusions

Based on the model of 1-D complex systems, the criteria of stability, controllability and observability of 2-D continuous-discrete are derived. Controllability and observability of 2-D continuous-discrete systems are then defined in terms of s domain.

References

[1] D. Frank, 2D Analysis of Hybrid of Hybrid Systems. in *Advances in Control*, Springer, London, 1999, pp.291-299.

[2] T. Kaczorek, "Stabilization of Singular 2–D Continuous-Discrete Systems by State-Feedbacks Controllers". IEEE Trans. Automatic Control, AC-41, No.7, pp.1007-1009, 1996.

[3] T. Kaczorek, "Stabilization of Singular 2–D Continuous-Discrete Systems by Output-Feedbacks Controllers". SAMS, Vol.28, pp.21-30, 1997.

[4] T. Kaczorek. "Positive 2D Continuous-Discrete Linear Systems". in *Advances in Control*, Springer, London, 1999, pp.309-316.

[5] Y. Xiao, "Robust Hurwitz-Schur Stability Conditions of Polytopes of 2-D Polynomials". Proceedings of 40th IEEE Conference on Decision and Control, Orlando, Florida, USA, 2001, pp. 3643-3648.

[6] Y. Xiao. "Hurwitz-Schur Stability of Interval of Bivariate Polynomials". Proceedings of IEEE International Symposium on Circuits and Systems (ISCAS2001), 2001. I: 829-832.

[7] Y. Xiao. "Stability Test for 2-D Continuous-Discrete Systems". Proceedings of 40th IEEE Conference on Decision and Control, Orlando, Florida, USA, 2001, pp. 3649-3654.

[8] Xiao Y, Du X, Unbehauen R, "Improved 2-D Stability Margin Test for 2-D Discrete Systems". Journal of Systems Science and Systems Engineering, Vol.7, No.2, 1998, pp.228-233.

[9] Xiao Y, Unbehauen R and Du X, "Sufficient Conditions of Robust Schur Stability for 2-D Polynomials". Journal of Systems Science and Systems Engineering,Vol.8, No.3, 1999, pp.368-374.

[10] Y. Xiao, R. Unbehauen, X. Du.,"Schur Stability of Polytopes of Bivariate Polynomials". Proceedings of 6th IEEE International Conference on Electronics Circuits and Systems(ICECS'99), pp.1269-1272.

[11] Y. Xiao, R. Unbehauen, X. Du. "Robust Hurwitz Stability of Conditions of Polytopes of Bivariate Polynomials". Proceedings of 38th IEEE Conference on Decision and Control, 1999, pp.5034-5035.

[12] Y. Xiao, "Schur Stability of Interval of Bivariate Polynomials". Proceedings of IEEE International Symposium on Circuits and Systems (ISCAS2000), 2000, pp.I-527-530.

[13] DU Xiyu, XIAO Yang et al. *Multidimensional Digital Filters*. National Defense Industry Publishing House, Beijing, 1995. pp.18-21.

[14] XIAO Yang and R. Unbehauen, "New Stability Test Algorithm for Two-Dimensional Digital Filters". IEEE Trans. Circuits and Systems I, 1998, 45(7): 739-740.

[15] D. L. Davis. "A Correct Proof of Huang's Theorem on Stability". IEEE Trans. on Acoustics, Speech, and Signal Processing, Vol. ASSP-24, No.10, pp425-426, Oct., 1976.

[16] E. Fornasini and G. Marchesini, "Stability Analysis of 2D Systems," *IEEE Trans. Circuit and Systems*, vol. 27, 1980, pp. 1210–1217.

[17] M. Bisiacco, "On the Algebraic Theory of 2D State Observers," *Systems and Control Letters*, vol. 5, 1985, pp. 347–353.

[18] M. Bisiacco, "State and Output Feedback Stabilizability of 2D Systems," *IEEE Trans. Circuit and Systems*, vol. 32, 1985, pp. 1246–1254.

[19] M. Bisiacco and E. Fornasini, "Optimal Control of Two Dimensional Systems," *SIAM J. Contr. and Optim.*, vol. 28, 1990, pp. 582–601.

[20] E. Fornasini and G. Marchesini, "Doubly Indexed Dynamical Systems: State Space Models and Structural Properties," *Math. Systems Theory*, vol. 12, 1978, pp. 59–72.

ROUNDOFF NOISE MINIMIZATION IN TWO-DIMENSIONAL STATE-SPACE DIGITAL FILTERS USING ERROR FEEDBACK

Takao Hinamoto, Keisuke Higashi and Wu-Sheng Lu†

Graduate School of Engineering, Hiroshima University, Japan
hinamoto@hiroshima-u.ac.jp

†Dept. Elect. Comput. Eng., University of Victoria, Canada
wslu@ece.uvic.ca

ABSTRACT

This paper considers the problem of minimizing roundoff noise in two-dimensional (2-D) state-space digital filters subject to L_2-norm dynamic-range scaling constraints. The minimization will be achieved by using error feedback. Several techniques for the determination of the optimal full-scale, block-diagonal, diagonal, and scalar error-feedback matrices for a given 2-D state-space digital filter are proposed. A numerical example is presented to illustrate the utility of the proposed techniques.

I. INTRODUCTION

One of the primary finite-word-length (FWL) register effects in fixed-point digital filters is the roundoff noise caused by the rounding of products/summations within the realization. One can reduce the roundoff noise at the filter output using error feedback, which is achieved by extracting the quantization error after multiplications and additions, and then feeding the error signal back to a certain point through a simple circuit. Several techniques for error feedback have been presented in the past for 1-D digital filters [1]-[5], and more recently for 2-D digital filters [6]-[9].

This paper proposes several new algorithms for the reduction of roundoff noise in 2-D state-space digital filters. Several closed-form formulas for evaluating the optimal full-scale, block-diagonal, diagonal, and scalar error-feedback matrices for a given 2-D state-space digital filter are derived. A numerical example is presented to illustrate the algorithms proposed and to demonstrate their performance.

II. 2-D STATE-SPACE DIGITAL FILTERS WITH ERROR FEEDBACK

Consider the Roesser local state-space (LSS) model $(\boldsymbol{A}, \boldsymbol{b}, \boldsymbol{c}, d)_{m,n}$ which is stable, separately locally controllable and separately locally observable:

$$\begin{aligned} \boldsymbol{x}_{11}(i,j) &= \boldsymbol{A}\boldsymbol{x}(i,j) + \boldsymbol{b}u(i,j) \\ y(i,j) &= \boldsymbol{c}\boldsymbol{x}(i,j) + du(i,j) \end{aligned} \tag{1}$$

where

$$\boldsymbol{x}_{11}(i,j) = \begin{bmatrix} \boldsymbol{x}^h(i+1,j) \\ \boldsymbol{x}^v(i,j+1) \end{bmatrix}, \quad \boldsymbol{x}(i,j) = \begin{bmatrix} \boldsymbol{x}^h(i,j) \\ \boldsymbol{x}^v(i,j) \end{bmatrix}$$

$$\boldsymbol{A} = \begin{bmatrix} \boldsymbol{A}_1 & \boldsymbol{A}_2 \\ \boldsymbol{A}_3 & \boldsymbol{A}_4 \end{bmatrix}, \quad \boldsymbol{b} = \begin{bmatrix} \boldsymbol{b}_1 \\ \boldsymbol{b}_2 \end{bmatrix}, \quad \boldsymbol{c} = \begin{bmatrix} \boldsymbol{c}_1 & \boldsymbol{c}_2 \end{bmatrix}.$$

Here, $\boldsymbol{x}^h(i,j)$ is an $m \times 1$ horizontal state vector, $\boldsymbol{x}^v(i,j)$ is an $n \times 1$ vertical state vector, $u(i,j)$ is a scalar input, $y(i,j)$ is a scalar output, and $\boldsymbol{A}_1$, $\boldsymbol{A}_2$, $\boldsymbol{A}_3$, $\boldsymbol{A}_4$, $\boldsymbol{b}_1$, $\boldsymbol{b}_2$, $\boldsymbol{c}_1$, $\boldsymbol{c}_2$, and d are real constant matrices of appropriate dimensions.

Carrying out the quantization before matrix-vector multiplication, an FWL implementation of (1) can be expressed as

$$\begin{aligned} \tilde{\boldsymbol{x}}_{11}(i,j) &= \boldsymbol{A}\boldsymbol{Q}[\tilde{\boldsymbol{x}}(i,j)] + \boldsymbol{b}u(i,j) \\ \tilde{y}(i,j) &= \boldsymbol{c}\boldsymbol{Q}[\tilde{\boldsymbol{x}}(i,j)] + du(i,j) \end{aligned} \tag{2}$$

where each component of $\boldsymbol{A}, \boldsymbol{b}, \boldsymbol{c}$, and d assumes an exact fractional B_c bit representation. The FWL local state vector $\tilde{\boldsymbol{x}}(i,j)$ and the output $\tilde{y}(i,j)$ all have a B bit fractional representation, while the input $u(i,j)$ is a $(B - B_c)$ bit fraction.

The quantizer $\boldsymbol{Q}[\cdot]$ in (2) rounds the B bit fraction $\tilde{\boldsymbol{x}}(i,j)$ to $(B - B_c)$ bits after multiplications and additions, where the sign bit is not counted. The quantization error

$$\boldsymbol{e}(i,j) = \tilde{\boldsymbol{x}}(i,j) - \boldsymbol{Q}[\tilde{\boldsymbol{x}}(i,j)] \tag{3}$$

coincides with the residue left in the lower part of $\tilde{\boldsymbol{x}}(i,j)$. The roundoff error $\boldsymbol{e}(i,j)$ is modeled as a zero-mean noise process of covariance $\sigma^2 \boldsymbol{I}_{m+n}$ with

$$\sigma^2 = \frac{1}{12} 2^{-2(B-B_c)}.$$

0-7803-7761-3/03/$17.00 ©2003 IEEE

To reduce the filter's roundoff noise, the quantization error $\boldsymbol{e}(i,j)$ is fed back to each input of delay operators through an $(m+n)\times(m+n)$ constant matrix $\boldsymbol{D}$ in the FWL filter (2). The 2-D filter with error feedback can be characterized by

$$\begin{aligned}\tilde{\boldsymbol{x}}_{11}(i,j) &= \boldsymbol{A}\boldsymbol{Q}[\tilde{\boldsymbol{x}}(i,j)] + \boldsymbol{b}u(i,j) + \boldsymbol{D}\boldsymbol{e}(i,j)\\ \tilde{y}(i,j) &= \boldsymbol{c}\boldsymbol{Q}[\tilde{\boldsymbol{x}}(i,j)] + du(i,j)\end{aligned} \tag{4}$$

where $\boldsymbol{D}$ is referred to as an *error-feedback matrix*.

Subtracting (4) from (1) yields

$$\begin{aligned}\Delta\boldsymbol{x}_{11}(i,j) &= \boldsymbol{A}\Delta\boldsymbol{x}(i,j) + (\boldsymbol{A}-\boldsymbol{D})\boldsymbol{e}(i,j)\\ \Delta y(i,j) &= \boldsymbol{c}\Delta\boldsymbol{x}(i,j) + \boldsymbol{c}\boldsymbol{e}(i,j)\end{aligned} \tag{5}$$

where

$$\begin{aligned}\Delta\boldsymbol{x}(i,j) &= \boldsymbol{x}(i,j) - \tilde{\boldsymbol{x}}(i,j)\\ \Delta\boldsymbol{x}_{11}(i,j) &= \boldsymbol{x}_{11}(i,j) - \tilde{\boldsymbol{x}}_{11}(i,j)\\ \Delta y(i,j) &= y(i,j) - \tilde{y}(i,j).\end{aligned}$$

Let $\boldsymbol{G}_D(z_1,z_2)$ be the 2-D transfer function from the quantization error, $\boldsymbol{e}(i,j)$, to the filter output, $\Delta y(i,j)$. Then, we obtain

$$\boldsymbol{G}_D(z_1,z_2) = \boldsymbol{c}(\boldsymbol{Z}-\boldsymbol{A})^{-1}(\boldsymbol{A}-\boldsymbol{D}) + \boldsymbol{c} \tag{6}$$

where $\boldsymbol{Z} = z_1\boldsymbol{I}_m \oplus z_2\boldsymbol{I}_n$. The noise variance gain $I(\boldsymbol{D}) = \sigma^2_{out}/\sigma^2$ is then defined by

$$I(\boldsymbol{D}) = tr[\boldsymbol{W}_D] \tag{7}$$

where σ^2_{out} denotes noise variance at the output, and

$$\boldsymbol{W}_D = \frac{1}{(2\pi j)^2}\oint_{\Gamma_1}\oint_{\Gamma_2}\boldsymbol{G}^*_D(z_1,z_2)\boldsymbol{G}_D(z_1,z_2)\frac{dz_1dz_2}{z_1z_2}$$

with $\Gamma_i = \{z_i : |z_i| = 1\}$ for $i = 1,2$. By applying the 2-D Cauchy integral theorem, we obtain

$$\boldsymbol{W}_D = (\boldsymbol{A}-\boldsymbol{D})^T\boldsymbol{W}_o(\boldsymbol{A}-\boldsymbol{D}) + \boldsymbol{c}^T\boldsymbol{c} \tag{8}$$

where $\boldsymbol{W}_o$ is called the *local observability Gramian* of the 2-D filter, and is defined by

$$\begin{aligned}\boldsymbol{W}_o &= \frac{1}{(2\pi j)^2}\oint_{\Gamma_1}\oint_{\Gamma_2}(\boldsymbol{Z}^*-\boldsymbol{A}^T)^{-1}\boldsymbol{c}^T\boldsymbol{c}(\boldsymbol{Z}-\boldsymbol{A})^{-1}\\ &\quad\cdot\frac{dz_1dz_2}{z_1z_2} = \sum_{i=0}^{\infty}\sum_{j=0}^{\infty}\boldsymbol{g}(i,j)^T\boldsymbol{g}(i,j).\end{aligned} \tag{9}$$

If there is no error feedback in the 2-D filter, then the noise variance gain $I(\boldsymbol{D})$ with $\boldsymbol{D} = \boldsymbol{0}$ becomes

$$\begin{aligned}I(\boldsymbol{0}) &= \mathrm{tr}[\boldsymbol{A}^T\boldsymbol{W}_o\boldsymbol{A} + \boldsymbol{c}^T\boldsymbol{c}]\\ &= \mathrm{tr}[\boldsymbol{W}_o].\end{aligned} \tag{10}$$

The l_2-norm dynamic-range scaling constraints on the local state vector involves the *local controllability Gramian* of the 2-D filter, which is defined by

$$\begin{aligned}\boldsymbol{K}_c &= \frac{1}{(2\pi j)^2}\oint_{\Gamma_1}\oint_{\Gamma_2}(\boldsymbol{Z}-\boldsymbol{A}^T)^{-1}\boldsymbol{b}\boldsymbol{b}^T(\boldsymbol{Z}^*-\boldsymbol{A}^T)^{-1}\\ &\quad\cdot\frac{dz_1dz_2}{z_1z_2} = \sum_{i=0}^{\infty}\sum_{j=0}^{\infty}\boldsymbol{f}(i,j)\boldsymbol{f}(i,j)^T.\end{aligned} \tag{11}$$

The problem considered is to design the error-feedback matrix $\boldsymbol{D}$ that minimizes (7), where matrix $\boldsymbol{W}_D$ is specified by (8), subject to that all the diagonal elements of $\boldsymbol{K}_c$ equal unity.

III. DETERMINATION OF OPTIMAL ERROR FEEDBACK MATRICES

In this section, we derive closed-form formulas for the determination of the optimal full-scale, block-diagonal, diagonal, and scalar error-feedback matrix $\boldsymbol{D}$ to minimize $I(\boldsymbol{D}) = \mathrm{tr}[\boldsymbol{W}_D]$ for a given 2-D state-space digital filter.

Case 1: $\boldsymbol{D}$ is a general matrix

Substituting (8) into (7), we obtain

$$\begin{aligned}I(\boldsymbol{D}) &= \mathrm{tr}[\boldsymbol{c}^T\boldsymbol{c} + (\boldsymbol{A}-\boldsymbol{D})^T\boldsymbol{W}_o(\boldsymbol{A}-\boldsymbol{D})]\\ &= \mathrm{tr}[\boldsymbol{W}_o] + \mathrm{tr}[\boldsymbol{D}^T\boldsymbol{W}_o\boldsymbol{D}] - 2\,\mathrm{tr}[\boldsymbol{D}^T\boldsymbol{W}_o\boldsymbol{A}].\end{aligned} \tag{12}$$

Differentiating (12) with respect to the error-feedback matrix $\boldsymbol{D}$ yields

$$\frac{\partial I(\boldsymbol{D})}{\partial \boldsymbol{D}} = 2\boldsymbol{W}_o(\boldsymbol{D}-\boldsymbol{A}). \tag{13}$$

By choosing the error-feedback matrix as $\boldsymbol{D} = \boldsymbol{A}$, the noise gain $I(\boldsymbol{D})$ in (12) achieves its minimum value

$$\begin{aligned}I_{min}(\boldsymbol{D}) &= \mathrm{tr}[\boldsymbol{W}_o] - \mathrm{tr}[\boldsymbol{A}^T\boldsymbol{W}_o\boldsymbol{A}]\\ &= \mathrm{tr}[\boldsymbol{c}^T\boldsymbol{c}].\end{aligned} \tag{14}$$

Case 2: $\boldsymbol{D}$ is a block-diagonal matrix

In this case, matrix $\boldsymbol{D}$ assumes the form

$$\boldsymbol{D} = \boldsymbol{D}_1 \oplus \boldsymbol{D}_4 \tag{15}$$

where $\boldsymbol{D}_1$ and $\boldsymbol{D}_1$ are $m\times m$ and $n\times n$ matrices, respectively, which leads (12) to

$$\begin{aligned}I(\boldsymbol{D}) = {} & \mathrm{tr}[\boldsymbol{W}_o] + \mathrm{tr}[\boldsymbol{D}_1^T\boldsymbol{W}_{o1}\boldsymbol{D}_1] + \mathrm{tr}[\boldsymbol{D}_4^T\boldsymbol{W}_{o4}\boldsymbol{D}_4]\\ & -2\,\mathrm{tr}[\boldsymbol{D}_1^T(\boldsymbol{W}_{o1}\boldsymbol{A}_1 + \boldsymbol{W}_{o2}\boldsymbol{A}_3)]\\ & -2\,\mathrm{tr}[\boldsymbol{D}_4^T(\boldsymbol{W}_{o3}\boldsymbol{A}_2 + \boldsymbol{W}_{o4}\boldsymbol{A}_4)]\end{aligned} \tag{16}$$

where

$$W_o = \begin{bmatrix} W_{o1} & W_{o2} \\ W_{o3} & W_{o4} \end{bmatrix}.$$

Letting $\partial I(D)/\partial D_i = 0$ for $i = 1, 4$ yields

$$\begin{aligned} D_1 &= A_1 + W_{o1}^{-1} W_{o2} A_3 \\ D_4 &= A_4 + W_{o4}^{-1} W_{o3} A_2. \end{aligned} \tag{17}$$

By substituting (17) into (16), we obtain the minimum value of the noise variance gain $I(D)$ as

$$\begin{aligned} I_{min}(D) &= \mathrm{tr}[W_o] - tr[D_1^T (W_{o1} A_1 + W_{o2} A_3)] \\ &\quad - tr[D_4^T (W_{o3} A_2 + W_{o4} A_4)]. \end{aligned} \tag{18}$$

Case 3: D is a diagonal matrix

In this case, matrix D assumes the form

$$\begin{aligned} D_1 &= \mathrm{diag}\{\alpha_1, \alpha_2, \cdots, \alpha_m\} \\ D_4 &= \mathrm{diag}\{\beta_1, \beta_2, \cdots, \beta_n\} \end{aligned} \tag{19}$$

which leads (16) to

$$\begin{aligned} I(D) = {} & \mathrm{tr}[W_o] + \mathrm{tr}[D_1^2 W_{o1}] + \mathrm{tr}[D_4^2 W_{o4}] \\ & -2\,\mathrm{tr}[D_1(W_{o1} A_1 + W_{o2} A_3)] \\ & -2\,\mathrm{tr}[D_4(W_{o3} A_2 + W_{o4} A_4)]. \end{aligned} \tag{20}$$

This implies that if α_i's and β_i's satisfy

$$\begin{aligned} &\alpha_i \left(\alpha_i - 2\frac{(W_{o1} A_1 + W_{o2} A_3)_{ii}}{(W_{o1})_{ii}} \right) < 0, \ i = 1, 2, \cdots, m \\ &\beta_i \left(\beta_i - 2\frac{(W_{o3} A_2 + W_{o4} A_4)_{ii}}{(W_{o4})_{ii}} \right) < 0, \ i = 1, 2, \cdots, n \end{aligned} \tag{21}$$

then $I(D) = \mathrm{tr}[W_D] < \mathrm{tr}[W_o]$. Letting $\partial I(D)/\partial \alpha_i = 0$ for $i = 1, 2, \cdots, m$ and letting $\partial I(D)/\partial \beta_i = 0$ for $i = 1, 2, \cdots, n$ gives

$$\begin{aligned} \alpha_i &= \frac{(W_{o1} A_1 + W_{o2} A_3)_{ii}}{(W_{o1})_{ii}}, \ i = 1, 2, \cdots, m \\ \beta_i &= \frac{(W_{o3} A_2 + W_{o4} A_4)_{ii}}{(W_{o4})_{ii}}, \ i = 1, 2, \cdots, n \end{aligned} \tag{22}$$

at which $I(D)$ achieves its minimum as

$$\begin{aligned} I_{min}(D) &= \mathrm{tr}[W_o] - \sum_{i=1}^{m} \frac{(W_{o1} A_1 + W_{o2} A_3)_{ii}^2}{(W_{o1})_{ii}} \\ &\quad - \sum_{i=1}^{n} \frac{(W_{o3} A_2 + W_{o4} A_4)_{ii}^2}{(W_{o4})_{ii}} \end{aligned} \tag{23}$$

where $(A)_{ii}$ denotes the ith diagonal element of a square matrix A.

Case 4: D_1 and D_4 are scalar matrices αI_m and βI_n

If $D_1 = \alpha I_m$ and $D_4 = \beta I_n$ with scalars α and β, then (20) becomes

$$\begin{aligned} I(D) = {} & \mathrm{tr}[W_o] + \mathrm{tr}[W_{o1}]\alpha^2 + \mathrm{tr}[W_{o4}]\beta^2 \\ & -2\mathrm{tr}[W_{o1} A_1 + W_{o2} A_3]\alpha \\ & -2\mathrm{tr}[W_{o3} A_2 + W_{o4} A_4]\beta. \end{aligned} \tag{24}$$

Hence, if α and β satisfy

$$\begin{aligned} &\alpha \left(\alpha - 2\frac{\mathrm{tr}[W_{o1} A_1 + W_{o2} A_3]}{\mathrm{tr}[W_{o1}]} \right) < 0 \\ &\beta \left(\beta - 2\frac{\mathrm{tr}[W_{o3} A_2 + W_{o4} A_4]}{\mathrm{tr}[W_{o4}]} \right) < 0 \end{aligned} \tag{25}$$

then $I(D) = \mathrm{tr}[W_D] < \mathrm{tr}[W_o]$. Moreover, from $\partial I(D)/\partial \alpha = 0$ and $\partial I(D)/\partial \beta = 0$, it follows that

$$\begin{aligned} \alpha &= \frac{\mathrm{tr}[W_{o1} A_1 + W_{o2} A_3]}{\mathrm{tr}[W_{o1}]} \\ \beta &= \frac{\mathrm{tr}[W_{o3} A_2 + W_{o4} A_4]}{\mathrm{tr}[W_{o4}]} \end{aligned} \tag{26}$$

which lead (24) to

$$\begin{aligned} I_{min}(D) &= \mathrm{tr}[W_o] - \frac{(\mathrm{tr}[W_{o1} A_1 + W_{o2} A_3])^2}{\mathrm{tr}[W_{o1}]} \\ &\quad - \frac{(\mathrm{tr}[W_{o3} A_2 + W_{o4} A_4])^2}{\mathrm{tr}[W_{o4}]}. \end{aligned} \tag{27}$$

IV. A NUMERICAL EXAMPLE

Let a 2-D state-space digital filter $(A, b, c, d)_{3,3}$ with $d = 0.0$ be described by

$$A_1 = \begin{bmatrix} 0.621553 & 0.014666 & -0.476979 \\ -0.081625 & 0.621548 & -0.181986 \\ 0.181983 & 0.476990 & 0.663600 \end{bmatrix}$$

$$A_2 = \begin{bmatrix} 0.059369 & -0.004829 & -0.024002 \\ -0.646852 & 0.061969 & 0.227715 \\ -0.229635 & 0.021958 & 0.076674 \end{bmatrix}$$

$$A_3 = \begin{bmatrix} 0.000378 & 0.000763 & 0.001503 \\ -0.000463 & -0.001501 & 0.000812 \\ -0.000021 & -0.000219 & 0.000908 \end{bmatrix}$$

$$A_4 = \begin{bmatrix} 0.620418 & 0.016504 & -0.479313 \\ -0.083124 & 0.620420 & -0.181961 \\ 0.181967 & 0.479315 & 0.661692 \end{bmatrix}$$

$$b_1 = \begin{bmatrix} -0.007708 & 0.081835 & 0.028969 \end{bmatrix}^T$$

$$b_2 = \begin{bmatrix} -0.079883 & 0.846271 & 0.294745 \end{bmatrix}^T$$

$$c_1 = \begin{bmatrix} -0.766526 & 0.072050 & 0.267706 \end{bmatrix}$$

$$c_2 = \begin{bmatrix} -0.074064 & 0.007031 & 0.026238 \end{bmatrix}.$$

which is stable, separately locally controllable and separately locally observable. This corresponds to the *optimal realization* with minimum roundoff noise $I(\mathbf{0}) = 4.927082$, subject to the l_2-norm dynamic-range scaling constraints.

In case $\boldsymbol{D}$ is allowed to be a general matrix, then (13) suggests that we should choose $\boldsymbol{D} = \boldsymbol{A}$ which yields $I_{min}(\boldsymbol{D}) = 0.670643$. Suppose the elements of matrix $\boldsymbol{D}$ are rounded to power-of-two quantization with 3 bits after binary point (integer quantization), then the noise gain is given by

$$I(\boldsymbol{D}) = 0.700468 \qquad (1.726719).$$

If $\boldsymbol{D}$ is constrained to be a block-diagonal matrix, then the optimal $\boldsymbol{D} = \boldsymbol{D}_1 \oplus \boldsymbol{D}_4$ is calculated using (17), which gives

$$\boldsymbol{D}_1 = \begin{bmatrix} 0.621550 & 0.014658 & -0.476971 \\ -0.081619 & 0.621565 & -0.181992 \\ 0.181976 & 0.476968 & 0.663613 \end{bmatrix}$$

$$\boldsymbol{D}_4 = \begin{bmatrix} 0.618351 & 0.016703 & -0.478595 \\ -0.082737 & 0.620382 & -0.182106 \\ 0.181250 & 0.479384 & 0.661937 \end{bmatrix}$$

$$I_{min}(\boldsymbol{D}_1 \oplus \boldsymbol{D}_4) = 1.331653.$$

After 3-bit quantization (integer quantization), this block-diagonal matrix $\boldsymbol{D} = \boldsymbol{D}_1 \oplus \boldsymbol{D}_4$ gives

$$I(\boldsymbol{D}_1 \oplus \boldsymbol{D}_4) = 1.351427 \qquad (2.247670).$$

If $\boldsymbol{D}$ is constrained to be a diagonal error-feedback matrix, then it can be calculated using (22) as

$$\boldsymbol{D}_1 = \text{diag}\{0.523405,\ 0.934410,\ 0.859620\}$$

$$\boldsymbol{D}_4 = \text{diag}\{0.521174,\ 0.934021,\ 0.858810\}$$

which yields $I_{min}(\boldsymbol{D}) = 1.833208$. After 3-bit quantization (integer quantization), this diagonal matrix $\boldsymbol{D} = \boldsymbol{D}_1 \oplus \boldsymbol{D}_4$ yields

$$I(\boldsymbol{D}_1 \oplus \boldsymbol{D}_4) = 1.840195 \qquad (2.247670).$$

If a scalar error-feedback matrix is calculated using (26), then we obtain

$$\alpha = 0.772479, \qquad \beta = 0.771335$$

which yields $I_{min}(\boldsymbol{D}) = 1.991329$. After 3-bit quantization (integer quantization), this scalar matrix $\boldsymbol{D}$ results in

$$I(\boldsymbol{D}) = 1.993695 \qquad (2.247670).$$

From these results, it is observed that the utilization of an optimal error feedback matrix leads to considerable reduction in roundoff noise, even when a scalar $\boldsymbol{D} = \alpha \boldsymbol{I}_m \oplus \beta \boldsymbol{I}_n$ with quantized α and β.

V. CONCLUSION

In this paper, the problem of minimizing roundoff noise in 2-D state-space digital filters has been investigated by means of error feedback. General, block-diagonal, diagonal, and scalar error-feedback matrices for minimizing the noise variance gain in a given 2-D state-space digital filter have been derived. Simulation results have been presented to illustrate the validity of our theoretical analysis and proposed algorithms.

1. REFERENCES

[1] W. E. Higgins and D. C. Munson, "Noise reduction strategies for digital filters: Error spectrum shaping versus the optimal linear state-space formulation," *IEEE Trans. Acoust. Speech, Signal Processing*, vol. 30, pp. 963-973, Dec. 1982.

[2] W. E. Higgins and D. C. Munson, "Optimal and suboptimal error-spectrum shaping for cascade-form digital filters," *IEEE Trans. Circuits Syst.*, vol. 31, pp. 429-437, May 1984.

[3] T. I. Laakso and I. O. Hartimo, "Noise reduction in recursive digital filters using high-order error feedback," *IEEE Trans. Signal Processing*, vol. 40, pp. 1096-1107, May 1992.

[4] P. P. Vaidyanathan, "On error-spectrum shaping in state-space digital filters," *IEEE Trans. Circuits Syst.*, vol. 32, pp. 88-92, Jan. 1985.

[5] D. Williamson, "Roundoff noise minimization and pole-zero sensitivity in fixed-point digital filters using residue feedback," *IEEE Trans. Acoust., Speech, Signal Processing*, vol. 34, pp. 1210-1220, Oct. 1986.

[6] T. Hinamoto,S. Karino and N. Kuroda,"Error spectrum shaping in 2-D digital filters," *Proc. IEEE Int. Symp. Circuits Syst.* (ISCAS'95), vol. 1, pp. 348-351, May 1995.

[7] P. Agathoklis and C. Xiao, "Low roundoff noise structures for 2-D filters," *Proc. IEEE Int. Symp. Circuits Syst.* (ISCAS'96), vol. 2, pp. 352-355, May 1996.

[8] T. Hinamoto,S. Karino and N. Kuroda,"2-D state-space digital filters with error spectrum shaping," *Proc. IEEE Int. Symp. Circuits Syst.* (ISCAS'96), vol. 2, pp. 766-769, May 1996.

[9] T. Hinamoto, S. Karino, N. Kuroda and T. Kuma, "Error spectrum shaping in two-dimensional recursive digital filters," *IEEE Trans. Circuits Syst.*, vol. 46, pp. 1203-1215, Oct. 1999.

A 4D Frequency-Planar IIR Filter and its Application to Light Field Processing

by Don Dansereau and Len Bruton

Dept. of Electrical and Computer Engineering,
University of Calgary, Alberta, Canada

ABSTRACT

A 4D frequency-planar filter is described, having a steady-state frequency response with a pass band that approximates that of a 4D plane. It is shown that a light field containing Lambertian surfaces with no occlusions may be selectively filtered for depth using such a filter. Examples are given showing the effectiveness of the filter in scenes with and without occlusions. A technique for effectively reducing the length of the transient response of the filter is also proposed and the effectiveness of this technique demonstrated.

1. INTRODUCTION

Image-based rendering has gained attention as a fast alternative to geometric model-based rendering. Light field rendering and its variants [1][2] seek to model a 4D subset of the more general 7D plenoptic function [3] associated with a scene. In this way, the set of light rays permeating a scene are represented, rather than the geometry of the objects within the scene.

The 7D plenoptic function describes the light rays in a scene as a function of position, orientation, spectral content, and time. This can be simplified to a 4D function by considering only the color of each ray as a function of its position and orientation in a static scene, and by constraining each ray to have the same value at every point along its direction of propagation. This disallows scenes in which the medium attenuates light as it propagates, and it fails to completely model the behavior of rays as they are occluded. These limitations are not an issue for scenes in clear air, and for which the camera is not allowed to move behind occluding objects.

The 4D light field typically parameterizes light rays using the two-plane parameterization (2PP), as depicted in Fig. 1. Each ray is described by its point of intersection with two reference planes: the s,t plane given by z=0, and the u,v plane which is parallel to the s,t plane at some positive depth $z=d$. A full light field may consist of multiple sets of such planes, though this paper will deal only with a single set of reference planes. Note also that each sample of a light field can be taken as a grayscale intensity, though extension to utilize color samples given as red, green and blue components is a simple matter of keeping one light field for each color channel, and repeating each operation accordingly.

This paper deals exclusively with 4D light fields, and so all signals will be assumed to be 4-dimensional. The continuous-domain light field will be denoted as $L_{cont}(s,t,u,v)$, and the discrete-domain version as $L(n_s,n_t,n_u,n_v)$, where $\boldsymbol{n}$ is the discrete domain index of the signal. The continuous Fourier transform of the light field will be denoted as $L_{freq}(\Omega_s,\Omega_t,\Omega_u,\Omega_v)$.

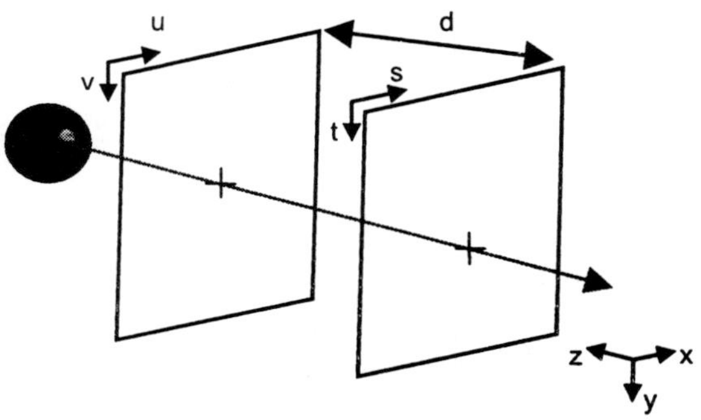

Figure 1. Two-plane parameterization of light rays.

This paper is organized as follows: Section 2 shows that a Lambertian scene can be filtered for depth using a 4D frequency-planar filter. Section 3 shows how to construct such a filter, including a technique for building a zero-phase version. Section 4 gives experimental results, and Section 5 discusses conclusions and indicates future work.

2. SPECTRAL CHARACTERISTICS OF LAMBERTIAN SCENES

2.1 The Omni-Directional Point Light Source

Fig. 2 depicts a 2D slice along s and u of a subset of the rays emanating from an omni-directional point source of light. It is clear from this figure that for any given point on the s,t plane there is only one point on the u,v plane for which a ray will intersect the light source. The result is that an s,u slice of the corresponding continuous-domain light field $L_{cont}(s,t,u,v)$ takes the form of a line, as depicted in Fig. 2b. The equation of this line is given by

$$(d/P_z - 1)\cdot s + u = P_x d/P_z \ , \qquad (1)$$

where the point $\boldsymbol{P}$ gives the 3D position of the point light source, and d is the separation of the reference planes. The behavior in the t *and* v dimensions is similar, and can be expressed as

$$(d/P_z - 1)\cdot t + v = P_y d/P_z \ . \qquad (2)$$

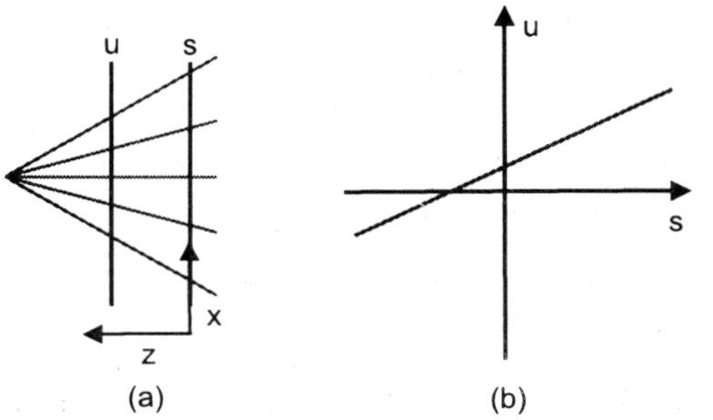

Figure 2. a) 2D slice of a point source of light shown with the two reference planes; b) 2D slice of the corresponding light field $L_{cont}(s,t,u,v)$.

In 4D space, (1) and (2) are the equations of two hyperplanes with normals in the directions $\boldsymbol{D_1}=[d/P_z\text{-}1,\ 0,1,0]$ and $\boldsymbol{D_2}=[0,d/P_z\text{-}1,0,1]$, respectively. The set of points which satisfies both (1) and (2) belongs to a plane defined by the intersection of these two hyperplanes. Only points belonging to this plane of intersection correspond to rays emanating from the point source – all other points in the light field will have a value of zero. Because the point source is omni-directional, every point in the plane will have the same value – equal to the color of the point light source. Thus, an omni-directional point light source is a plane of constant value in the light field $L_{cont}(s,t,u,v)$, where the plane is the solution of (1) and (2).

The normals of the hyperplanes (1) and (2) depend only on the distance P_z of the light source from the s,t plane, implying that a set of omni-directional point sources at a single depth z exists as a set of parallel planes in the light field $L_{cont}(s,t,u,v)$, each with constant value. The region of support (ROS) of the Fourier transform of such parallel planes can be shown to be a single frequency-domain plane through the origin, given as the intersection of the two frequency-domain hyperplanes

$$\Omega_s + (1 - d/P_z)\cdot\Omega_u = 0\text{, and} \tag{3}$$

$$\Omega_t + (1 - d/P_z)\cdot\Omega_v = 0\,. \tag{4}$$

Thus, the frequency-domain region of support (ROS) of a collection of omni-directional point sources at some depth $z=P_z$ is a single plane through the origin, where the plane is the solution of (3) and (4).

2.2 Lambertian Surfaces

A Lambertian surface is one which presents the same luminance independent of viewing angle [4], and thus behaves similarly to an omni-directional point light source. As a consequence, a Lambertian surface which exists at some constant depth $z=P_z$ will exist as a set of parallel planes in a light field, each of constant value, the frequency-domain ROS of which is a plane through the origin with an orientation which depends only on the z depth of the surface in the scene. This result is similar to that shown in Section 2.1 for a collection of omni-directional point light sources at a constant depth, and was also shown mathematically in [5].

Based on the above, we propose to design a filter to extract portions of a Lambertian scene at some depth by extracting the appropriately oriented plane in the frequency domain. Note that the frequency-planes at other orientations, corresponding to surfaces at other depths, all pass through the frequency-domain origin and thus intersect the planar passband of the filter at the origin. As a result, the low-frequency components of undesired signals (at stop band depths) will be transmitted. Note also that the filter functions with input scenes that do have occlusions and specular reflections. However, the desired occluded points in the input will assume values which depend on the value of the occluding surface, the desired specular reflections may vanish, and undesired specular reflections may interfere with desired surfaces.

3. 4D FREQUENCY-PLANAR FILTER

3.1 4D Frequency-Hyperplanar Filter

In order to construct a planar passband in 4D, two hyperplanar passbands, with appropriate orientations, are designed to intersect using the basic approach suggested in [6] for the 3D case. The design of the hyperplanar filters is a simple matter of extending the 3D frequency-planar filter presented in [6] by including an extra spatial variable. The input-output difference equation of the resulting filter is

$$y(\boldsymbol{n}) = \frac{1}{b_{0000}}\left[\begin{array}{l}\sum\limits_{i=0}^{1}\sum\limits_{j=0}^{1}\sum\limits_{k=0}^{1}\sum\limits_{l=0}^{1} L(n_s - i, n_t - j, n_u - k, n_v - l) - \\ \qquad \sum\limits_{\substack{i=0\\ i+j+k+l\neq 0}}^{1}\sum\limits_{j=0}^{1}\sum\limits_{k=0}^{1}\sum\limits_{l=0}^{1} b_{ijkl}\cdot y(n_s - i, n_t - j, n_u - k, n_v - l)\end{array}\right], \tag{5}$$

where the b_{ijkl} coefficients are found by iterating through the 16 sign configurations of $b_{ijkl} = 1 + (\pm N_v \pm N_u \pm N_t \pm N_s)/B$, with $\boldsymbol{N}$ as the normal of the passband hyperplane, and with B as the -3dB bandwidth of the filter. The continuous-domain Laplace transform transfer function of the proposed filter is given as

$$T(s_s, s_t, s_u, s_v) = \frac{1}{1 + (N_s s_s + N_t s_t + N_u s_u + N_v s_v)/B}\,, \tag{6}$$

which approximates a hyperplanar passband with its normal given by $\boldsymbol{N}$. This filter is practical-BIBO stable, but only for values of $\boldsymbol{N}$ which are positive along all 4 dimensions. In order to filter outside the first hexadecimant, the direction of iteration is reversed for all dimensions along which the normal is initially negative, then the sign of the normal is changed to be positive along those dimensions [6].

3.2 4D Frequency-Planar Filter

In order to construct the frequency-planar filter, two of the above frequency-hyperplanar filters are arranged in a cascaded configuration. The normal of each filter is selected to satisfy (3) and (4), respectively, resulting in the unit normals

$$\boldsymbol{N_1} = \frac{[1,0,1-d/P_z,0]}{\sqrt{1+(1-d/P_z)^2}}\text{, and } \boldsymbol{N_2} = \frac{[0,1,0,1-d/P_z]}{\sqrt{1+(1-d/P_z)^2}}\,. \tag{7}$$

The overall transfer function is given by

$$T(s_s, s_t, s_u, s_v) = T_1(s_s, s_t, s_u, s_v) \cdot T_2(s_s, s_t, s_u, s_v) \ , \tag{8}$$

where T_1 and T_2 are the transfer functions of the filters with normals N_1 and N_2, respectively. The magnitude frequency response of (8) is unity only where both filters have a magnitude frequency response of unity – that is, only where equations (3) and (4) are both satisfied. Thus, a filter that approximates a frequency-planar passband in 4D has been designed. The bandwidth of this filter is adjusted with B, and the orientation of the plane is adjusted to extract objects at a given depth in the light field by selecting N_1 and N_2 using (7).

3.3 Zero-phase filtering

The filter described above yields good depth discrimination but with a long non-ideal transient response in the output light field, the effects of which are darkening of the light field at the edges, and smearing of the images in the direction of iteration. This is especially significant for light fields which have low sample rates in s and t – a typical light field might have only tens of samples in these dimensions – and so a way of reducing the effects of the transient response is desirable.

The proposed technique is to employ zero-phase filtering: that is, to re-filter in a second pass the output signal with the direction of iteration reversed along each dimension. This is equivalent to flipping the normal of the frequency-hyperplanar passband of each frequency-hyperplanar filter in the second pass. This has no effect on the magnitude frequency response of the second-pass filter, but enforces a zero-phase frequency response, leading to a much shorter transient response at the output of the second-pass filter. By keeping extra output data as they are smeared off the edge of the light field by the first-pass filter, then utilizing them with the second-pass filter, the darkening of the edges of the light field can also be reduced significantly. Furthermore, because a second-pass filter is being applied, the overall magnitude frequency response is squared. The disadvantages of this zero-phase filtering technique are increased processing time and increased memory requirements.

4. RESULTS

The frequency-planar filter designed in Section 3.2 was applied with a bandwidth of B=0.05 to two light fields with geometric parameters as summarized in Table 1. The first light field was constructed using a raytracer and does not contain occlusions, while the second was measured using a gantry system and does contain occlusions. Each scene was filtered twice: once to extract foreground objects, and once to extract background objects. The filtering operation took about 10 minutes on a Pentium IV running at 1.3 GHz. The zero-phase technique described in Section 3.3 was also applied to the second light field, increasing the memory requirements by a factor of about 1.7, and the processing time to about ½-hour.

Color channels	3	s,t size (cm)	45
s,t samples	32	u,v size (cm)	36
u,v samples	256	separation d (cm)	57

Table 1. Light field parameters.

The results for the first light field can be seen in Fig. 3. Parts (a) through (c) are slices through u and v for values of s and t near the center of the light field. It is clear from the figure that the filter successfully extracted the two triangles, which are parallel with the s,t plane at depths of 60 and 50 cm. The transient response of the filter can be seen for the foreground and background triangles, respectively, in (d). These images are projections of the plane in which a single point of the triangle exists onto the u and v dimensions. The startup transient lasts about 7 samples in each dimension for this bandwidth.

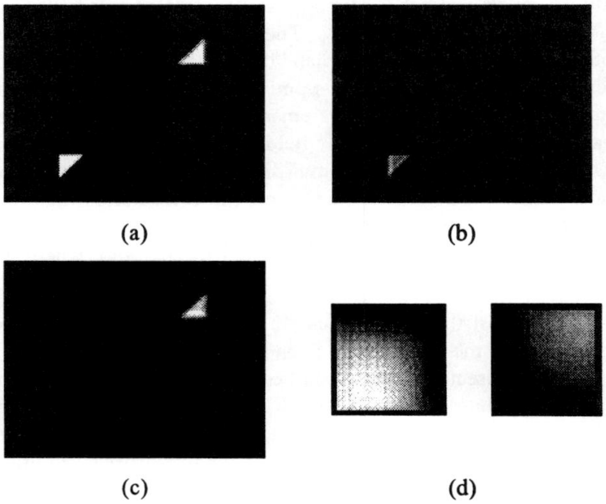

Figure 3. Light field of raytraced scene with no occlusions. a) Unfiltered; b) Filtered for objects at 60 cm; c) Filtered for objects at 50 cm; d) Transient response as projected onto u,v.

The second light field, shown in Fig. 4, was acquired using a camera gantry. The background is a poster of a supernova imaged by the Dominion Radio Astrophysical Observatory in British Columbia, Canada. The foreground is a beer coaster – a wooden dowel can be seen holding the coaster in place. The poster and coaster are at depths of 66 and 48 cm, respectively. The filter was reasonably successful at extracting both objects, though the effects of the occlusion are clear near the center of the supernova in b), and the startup transient significantly darkens the edges of the images.

The effect of applying the zero-phase filtering technique described in Section 3.3 to the second light field is shown in Fig. 5. The startup transient is significantly reduced, as is most clearly seen near the top-right of both images. The selectivity has also noticeably increased, as the undesired portions of the images appear more blurred. The low-frequency components of the undesired signals remain, as anticipated.

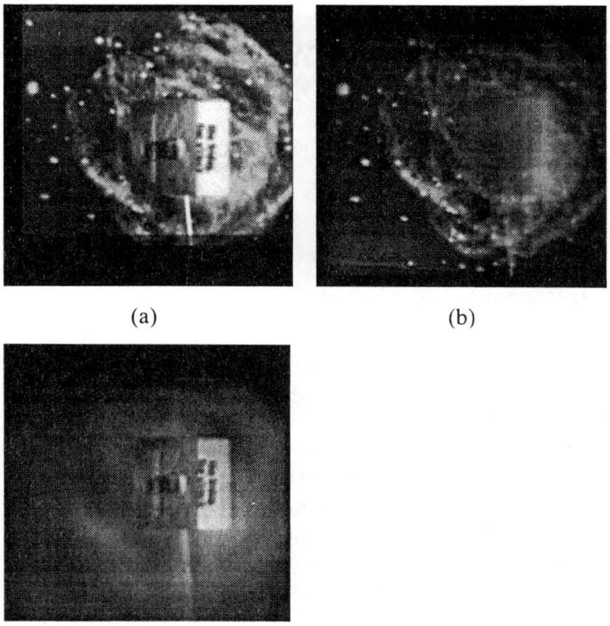

(a) (b)

(c)

Figure 4. Measured scene with occlusions. a) Unfiltered; b) Filtered for objects at 66 cm; c) Filtered for objects at 48 cm.

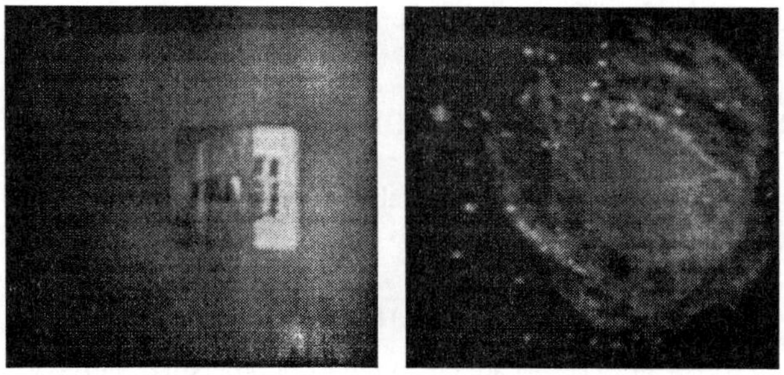

(a) (b)

Figure 5. Measured light field filtered using zero-phase filter. a) Filtered for objects at 48 cm; b) Filtered for objects at 66 cm.

5. CONCLUSIONS AND FUTURE WORK

It has been shown that an omni-directional point source of light exists as a plane of constant intensity in a light field. As a result, a Lambertian surface at a constant depth in a scene has a frequency-domain ROS which is a plane through the origin. The construction of a 4D 1st-order frequency-planar recursive filter to extract this planar ROS was accomplished by intersecting two appropriately designed hyperplanar filters. The hyperplanar filters were constructed by extension of the method used to design a 3D frequency-planar filter. Finally, it was shown that zero-phase filtering can be accomplished by applying a second-pass filter with reversed directions of iteration.

The 4D planar filter was tested on two light fields – one created using raytracer output, and one measured using a gantry system. It was shown that the filter could successfully isolate elements at different depths. It was also shown that the proposed zero-phase technique significantly reduces the effects of the long transient response of the planar filter, while increasing selectivity, though at the cost of increased memory requirements and processing time.

Image-based techniques seem destined to become popular in processing and analysis applications such as object recognition [7], scene modeling [8], and robot navigation [9]. The possibility of applying simple techniques, such as the 1st-order filter presented in this paper, to accomplish complex tasks is exciting.

For future work, it should be possible to optimize the filter described in this paper significantly, increasing speed through parallelization, for example. A more thorough exploration of the impact of occlusions and specular reflection might also be interesting. Other possibilities include the exploration of non-linear techniques for processing light fields, possibly exploiting the correspondence between points in space and planes in light fields.

6. REFERENCES

[1] Levoy, M., and Hanrahan, P. "Light Field Rendering". *Proceedings SIGGRAPH '96*, 1996, pages 31-42.

[2] Gortler, S.J., Grzeszczuk, R., Szeliski, R., and Cohen, M. "The Lumigraph". *Proceedings SIGGRAPH '96*, 1996, pages 43-54.

[3] Adelson, E.H., and Bergen, J.R. "The Plenoptic Function and the Elements of Early Vision." *Computation Models of Visual Processing"*, 1991, pages 3-20.

[4] Hearn,Donald and M. Pauline Baker. *Computer Graphics, C version, 2nd Edition*, Prentice Hall, Inc, 1997.

[5] Chan, S.C., and Shum, H.Y. "A Spectral Analysis for Light Field Rendering". *Proceedings 2000 International Conference on Image Processing*, v 2, 2000, pages 25-28.

[6] Bruton, L.T., and Bartley, N.R. "Three-Dimensional Image Processing Using the Concept of Network Resonance". *IEEE Transactions on Circuits and Systems*, v CAS-32, n 7, Jul 1985, pages 664-672.

[7] Heigl, B., Denzler, J., and Niemann, H. "On the Application of Light Field Reconstruction for Statistical Object Recognition." *European Signal Processing Conference (EUSIPCO)*, 1998, pages 1101-1105.

[8] Vogelgsang, C., Heigl, B, Greiner, G., and Niemann, H. "Automatic Image-Based Scene Model Acquisition and Visualization". *Workshop Vision, Modeling and Visualization, Saarbrücken*, Germany, 2000, pages 189-198.

[9] Heigl, Denzler, and Niermann. "Combining Computer Graphics and Computer Vision for Probabilistic Visual Robot Navigation". *Enhanced and Synthetic Vision 2000*, volume 4023 of *Proceedings of SPIE*, April 2000, pages 226-235.

REALIZATION OF HIGH ACCURACY 2-D VARIABLE IIR DIGITAL FILTERS BASED ON THE REDUCED-DIMENSIONAL DECOMPOSITION FORM

Hyuk-Jae Jang and Masayuki Kawamata

Graduate School of Engineering, Tohoku University
Aoba-yama 05, Sendai 980-8579, JAPAN
jang@mk.ecei.tohoku.ac.jp, kawamata@ecei.tohoku.ac.jp

ABSTRACT

This paper proposes a realization method of high accuracy 2-D variable IIR digital filters with real coefficients. The proposed method uses the reduced-dimensional decomposition form as a prototype structure of 2-D variable digital filters to suppress the degradation of tuning accuracy due to the linear approximation of variable coefficients. Furthermore, the 1-D subfilters of the prototype filter are realized as cascade connections of low sensitivity second-order variable sections to suppress the degradation of tuning accuracy due to finite coefficient wordlength. Numerical examples show that the proposed 2-D variable digital filter has high frequency tuning accuracy under finite coefficient wordlength and the linear approximation.

1. INTRODUCTION

In many applications of digital signal processing, such as telecommunications, digital audio equipment and adaptive systems, it is sometimes required to change the frequency characteristic of a digital filter during its operation. One of the filters that satisfy this requirement is a variable digital filter (VDF). The VDF is a frequency selective filter that can control frequency characteristics using variable parameters included in the filter coefficients. A number of design methods of the VDFs have been published recently and most of them are classified and compared by G. Stoyanov and M. Kawamata [1].

In the design of variable IIR digital filters, the frequency transformation [2] is usually applied to a prototype filter in order to obtain frequency variable characteristic as

$$H_{var}(z) = H(z)\big|_{z^{-1}=T(z)}. \qquad (1)$$

After applying the frequency transformation, the transformed prototype transfer function is synthesized as a new VDF circuit with the filter coefficients that are complicated functions of a variable parameter. This fact means that the realized VDF requires a large amount of calculations for coefficient update, so that it is very difficult to perform high speed signal processing. The well known methods of reducing the amount of calculations are the limitation of coefficient wordlength and linear approximation of variable coefficients. However, these methods have a serious problem that they cause the degradation of frequency tuning accuracy.

In order to overcome this problem, in the 1-D case, the structures with low sensitivity characteristic such as the parallel allpass filter and wave digital filter have been used as prototype filters of the variable IIR digital filters [3, 4]. Since low sensitivity ensures high tuning accuracy, the above VDFs have good tuning accuracy under finite coefficient wordlength and the linear approximation of variable coefficients.

In the 2-D case, the authors have proposed 2-D VDFs based on the 2-D parallel complex allpass structure [5, 6]. The parallel complex allpass 2-D VDF possesses several advantages such as high frequency tuning accuracy, simple stability monitoring and a small amount of calculations for coefficient update. In spite of those good characteristics, however, the parallel complex allpass 2-D VDF has a disadvantage that it requires a complicated filter structure to perform complex signal processing because it has complex filter coefficients.

The aim of this paper is to realize high accuracy 2-D variable IIR digital filters with real coefficients. In this paper, restricting ourselves to the 2-D separable denominator case, we will present a realization method of high accuracy 2-D VDFs based on the reduced-dimensional decomposition form [7].

2. REDUCED-DIMENSIONAL DECOMPOSITION FORM

Consider a stable and causal separable denominator 2-D digital filter

$$H(z_1, z_2) = \frac{N(z_1, z_2)}{D_1(z_1)D_2(z_2)} \qquad (2)$$

where $D_1(z_1)$ and $D_2(z_2)$ are polynomials of orders M_1 and M_2 in z_1 and z_2, respectively. In this paper, we restrict ourselves to a class of 2-D separable denominator digital filters with even orders M_1 and M_2. 2-D digital filters with separable denominators can be decomposed into the reduced-dimensional decomposition form [7]. The realization procedure of the reduced-dimensional decomposition form is as follows: The numerator polynomial $N(z_1, z_2)$ in (2) can be expressed in the matrix form as

$$N(z_1, z_2) = \mathbf{Z}_1\mathbf{N}\mathbf{Z}_2 \qquad (3)$$

where

$$\mathbf{Z}_1 = [1 \; z_1^{-1} \; z_1^{-2} \; \cdots \; z_1^{-M_1}] \qquad (4)$$

$$\mathbf{N} = \begin{bmatrix} b_{00} & b_{01} & \cdots & b_{0M_2} \\ b_{10} & b_{11} & \cdots & b_{1M_2} \\ \vdots & \vdots & & \vdots \\ b_{M_10} & b_{M_11} & \cdots & b_{M_1M_2} \end{bmatrix} \qquad (5)$$

$$\mathbf{Z}_2 = [1 \; z_2^{-1} \; z_2^{-2} \; \cdots \; z_2^{-M_2}]^t. \qquad (6)$$

The matrix $\mathbf{N}$ can be decomposed into the product form of two matrices $\mathbf{N}_f$ and $\mathbf{N}_g$ by using the singular value decomposition

0-7803-7761-3/03/$17.00 ©2003 IEEE

theory as

$$\mathbf{N} = \sum_{i=1}^{r} \mathbf{u}_i s_i \mathbf{v}_i = \sum_{i=1}^{r} \mathbf{u}_i s_i^{\frac{1}{2}} \cdot s_i^{\frac{1}{2}} \mathbf{v}_i = \mathbf{N}_f \cdot \mathbf{N}_g \tag{7}$$

where $s_1 \geq s_2 \geq \cdots \geq s_r > 0$ are singular values of the matrix $\mathbf{N}$, r is the rank of the matrix $\mathbf{N}$, $\mathbf{u}_i$ and $\mathbf{v}_i$ are normalized eigenvectors of the matrices $\mathbf{N}\mathbf{N}^t$ and $\mathbf{N}^t\mathbf{N}$ respectively.

Substituting (7) for (2) and (3), we obtain

$$H(z_1, z_2) = \frac{\mathbf{Z}_1 \mathbf{N}_f}{D_1(z_1)} \cdot \frac{\mathbf{N}_g \mathbf{Z}_2}{D_2(z_2)} = \mathbf{F}(z_1) \cdot \mathbf{G}(z_2) \tag{8}$$

where $\mathbf{F}(z_1)$ is an r-input/1-output 1-D filter, and $\mathbf{G}(z_2)$ is a 1-input/r-output 1-D filter

$$\mathbf{F}(z_1) = [f_1(z_1) \; f_2(z_1) \; \cdots \; f_r(z_1)] \tag{9}$$

$$\mathbf{G}(z_2) = [g_1(z_2) \; g_2(z_2) \; \cdots \; g_r(z_2)]^t. \tag{10}$$

In (9) and (10), $f_1(z_1), \cdots, f_r(z_1)$ and $g_1(z_2), \cdots, g_r(z_2)$ are 1-D subfilters of $\mathbf{F}(z_1)$ and $\mathbf{G}(z_2)$, respectively. Consequently, the separable denominator 2-D filter $H(z_1, z_2)$ can be represented as

$$H(z_1, z_2) = \mathbf{F}(z_1) \cdot \mathbf{G}(z_2) = \sum_{i=1}^{r} f_i(z_1) g_i(z_2). \tag{11}$$

3. LOW SENSITIVITY SECOND-ORDER SECTIONS

It is well known that a second-order section, i.e., a biquad section

$$H_b(z) = \frac{b_0 + b_1 z^{-1} + b_2 z^{-2}}{1 + a_1 z^{-1} + a_2 z^{-2}} \tag{12}$$

is a strong candidate for a basic section of high order digital filters with a cascade or a parallel form. A primary drawback of the second-order section in (12) is that the density of possible pole locations is very low at low and high frequency regions. As a result of this, the frequency characteristic of the second-order section that has poles in the vicinity of low frequency region are degraded under finite coefficient wordlength.

In order to overcome this problem, a new second-order section with reduced coefficient sensitivity in the case of low pole frequency has been proposed by D. Schlichthärle [8]. A stable and casual transfer function of this section is defined as

$$H_b(z) = \frac{d_0 - (2d_0 + d_1 + d_2)z^{-1} + (d_0 + d_1)z^{-2}}{1 - (2 - sv - tv)z^{-1} + (1 - sv)z^{-2}} \tag{13}$$

where the numerator coefficients are given by

$$d_0 = b_0, \; d_1 = b_2 - b_0, \; d_2 = -b_0 - b_1 - b_2. \tag{14}$$

By an appropriate choice of the denominator coefficients s, t and v, we can obtain a variety of transfer functions with reduced sensitivity in the case of low pole frequency. In the following, transfer functions of several second-order sections using the above approach [8, 9] are introduced.

- *Kingsbury's section*

$$H_b(z) = \frac{d_0 - (2d_0 + d_1 + d_2)z^{-1} + (d_0 + d_1)z^{-2}}{1 - (2 - c_1 c_2 - c_1^2)z^{-1} + (1 - c_1 c_2)z^{-2}} \tag{15}$$

$$c_1 = \sqrt{1 + a_1 + a_2}, \; c_2 = (1 - a_2)/c_1 \tag{16}$$

$$s = c_2, \; t = c_1, \; v = c_1 \tag{17}$$

- *Avenhaus' section*

$$H_b(z) = \frac{d_0 - (2d_0 + d_1 + d_2)z^{-1} + (d_0 + d_1)z^{-2}}{1 - (2 - c_1)z^{-1} + (1 - c_1 + c_1 c_2)z^{-2}} \tag{18}$$

$$c_1 = a_1 + 2, \; c_2 = (1 + a_1 + a_2)/(2 + a_1) \tag{19}$$

$$s = 1 - c_2, \; t = c_2, \; v = c_1. \tag{20}$$

Each second-order section has its own distribution pattern of possible pole locations.

4. PROPOSED 2-D VDF

4.1. Second-Order Variable Section

By applying the frequency transformation [2] to the second-order section in (13), we can obtain the low sensitivity second-order variable section as

$$H_{v,b}(z) = H_b(z)|_{z^{-1}=T(z)} = \frac{\hat{d}_0 + \hat{B}_1 z^{-1} + \hat{B}_2 z^{-2}}{1 + \hat{A}_1 z^{-1} + \hat{A}_2 z^{-2}} \tag{21}$$

where $T(z)$ represents a frequency transformation. We know that there are various kinds of frequency transformation. For example, in the case of the lowpass to lowpass transformation, $T(z)$ defined as

$$T(z) = \frac{z^{-1} - \alpha}{1 - \alpha z^{-1}}, \; \alpha = \frac{\sin[(\omega_p - \omega_d)/2]}{\sin[(\omega_p + \omega_d)/2]}, \quad -1 < \alpha < 1 \tag{22}$$

where ω_p and ω_d are cutoff frequencies of a prototype filter and a desired filter, respectively. In (21), the variable coefficients are given by

$$\hat{d}_0 = \frac{d_0 - B_1\alpha + B_2\alpha^2}{1 - A_1\alpha + A_2\alpha^2} \tag{23}$$

$$\hat{B}_1 = \frac{B_1 - 2(d_0 + B_2)\alpha + B_1\alpha^2}{1 - A_1\alpha + A_2\alpha^2} \tag{24}$$

$$\hat{B}_2 = \frac{B_2 - B_1\alpha + d_0\alpha^2}{1 - A_1\alpha + A_2\alpha^2} \tag{25}$$

$$\hat{A}_1 = \frac{A_1 - 2(1 + A_2)\alpha + A_1\alpha^2}{1 - A_1\alpha + A_2\alpha^2} \tag{26}$$

$$\hat{A}_2 = \frac{A_2 - A_1\alpha + \alpha^2}{1 - A_1\alpha + A_2\alpha^2} \tag{27}$$

and

$$A_1 = -(2 - sv - tv), \; A_2 = 1 - sv \tag{28}$$

$$B_1 = -(2d_0 + d_1 + d_2), \; B_2 = d_0 + d_1. \tag{29}$$

As shown in (23)–(29), the coefficients of the realized second-order variable section are complicated functions of the variable parameter α. Therefore, they require a large amount of calculations for coefficient update. To overcome this problem, we take the Taylor series expansion of the filter coefficients with respect to the variable parameter α. If we assume α to be very small ($|\alpha| \ll 1$) and truncate the expansion after a linear term, we can obtain new variable coefficients of which amount of calculations is reduced as

$$\hat{d}_0 \approx d_0 + \alpha(d_0 A_1 - B_1) = d_0 + \alpha K_0 \tag{30}$$

$$\hat{B}_1 \approx B_1 + \alpha\{B_1 A_1 - 2(d_0 + B_2)\} = B_1 + \alpha K_1 \tag{31}$$

$$\hat{B}_2 \approx B_2 + \alpha(B_2 A_1 - B_1) = B_2 + \alpha K_2 \tag{32}$$

$$\hat{A}_1 \approx A_1 + \alpha\{A_1^2 - 2(1 + A_2)\} = A_1 + \alpha L_1 \tag{33}$$

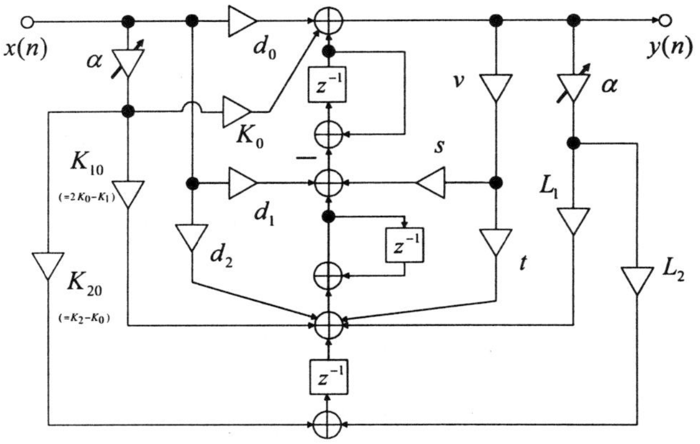

Figure 1: Proposed second-order variable section.

$$\hat{A}_2 \approx A_2 + \alpha(A_2 A_1 - A_1) = A_2 + \alpha L_2. \tag{34}$$

From the above realization procedure, the transfer function of the second-order variable section is given by

$$H_{v,b}(z) = \frac{(d_0 + \alpha K_0) + (B_1 + \alpha K_1)z^{-1} + (B_2 + \alpha K_2)z^{-2}}{1 + (A_1 + \alpha L_1)z^{-1} + (A_2 + \alpha L_2)z^{-2}}. \tag{35}$$

The block diagram of the proposed second-order variable section is shown in Fig. 1. From Fig. 1, we can see that the proposed second-order variable section requires two variable multipliers, 11 fixed multipliers, 6 adders and 3 delay elements.

4.2. Stability Condition

Since the variable coefficients are approximated, the range of the variable parameter for the second-order variable section to be stable is restricted. For this reason, we need to determine a stability region of the variable parameter. The necessary and sufficient condition for the second-order variable section to be stable is given by

$$|A_2 + \alpha L_2| < 1 \tag{36}$$
$$A_2 + \alpha L_2 > -(A_1 + \alpha L_1) - 1 \tag{37}$$
$$A_2 + \alpha L_2 > (A_1 + \alpha L_1) - 1. \tag{38}$$

From the above conditions, the range of α are determined as follows: From (36),

$$\begin{cases} \frac{-1-A_2}{L_2} < \alpha < \frac{1-A_2}{L_2} & \text{for } L_2 > 0 \\ \frac{1-A_2}{L_2} < \alpha < \frac{-1-A_2}{L_2} & \text{for } L_2 < 0 \\ \alpha \text{ is unconditional} & \text{for } L_2 = 0. \end{cases} \tag{39}$$

From (37),

$$\begin{cases} \alpha > \frac{-A_1-A_2-1}{L_1+L_2} & \text{for } L_1 + L_2 > 0 \\ \alpha < \frac{-A_1-A_2-1}{L_1+L_2} & \text{for } L_1 + L_2 < 0 \\ \alpha \text{ is unconditional} & \text{for } L_1 + L_2 = 0. \end{cases} \tag{40}$$

From (38),

$$\begin{cases} \alpha < \frac{-A_1+A_2+1}{L_1-L_2} & \text{for } L_1 - L_2 > 0 \\ \alpha > \frac{-A_1+A_2+1}{L_1-L_2} & \text{for } L_1 - L_2 < 0 \\ \alpha \text{ is unconditional} & \text{for } L_1 - L_2 = 0. \end{cases} \tag{41}$$

The second-order variable section is stable when α satisfies conditions (39), (40) and (41) at the same time.

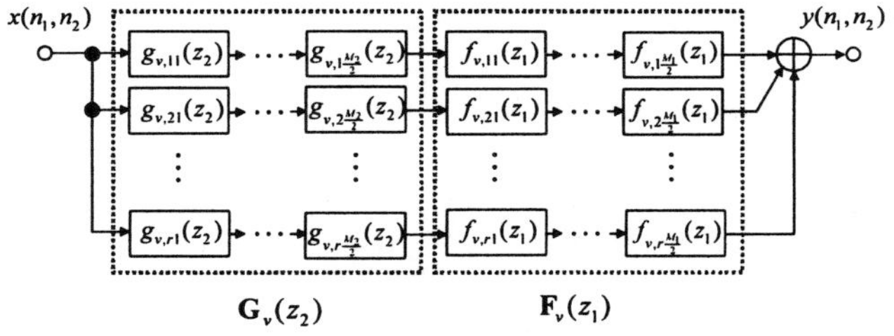

Figure 2: Proposed 2-D variable digital filter structure.

4.3. Realization of Proposed 2-D VDF

This section describes the realization procedure of the proposed 2-D VDF. In our approach, we use the reduced-dimensional decomposition form as a prototype filter of 2-D VDFs. Then, the 1-D subfilters of the prototype filter are realized as cascade connections of the proposed second-order variable sections to obtain high frequency tuning accuracy. In the following, the realization procedure is described in detail.

Consider 1-D filters $\mathbf{F}(z_1)$ and $\mathbf{G}(z_2)$

$$\mathbf{F}(z_1) = [f_1(z_1) \; f_2(z_1) \; \cdots \; f_k(z_1) \; \cdots \; f_r(z_1)] \tag{42}$$
$$\mathbf{G}(z_2) = [g_1(z_2) \; g_2(z_2) \; \cdots \; g_k(z_2) \; \cdots \; g_r(z_2)]^t \tag{43}$$

of the reduced-dimensional decomposition form. First, the 1-D subfilters in (42) and (43) are realized as cascade connections of second-order sections as

$$f_k(z_1) = \prod_{l=1}^{M_1/2} f_{kl}(z_1), \quad g_k(z_2) = \prod_{l=1}^{M_2/2} g_{kl}(z_2). \tag{44}$$

Then, we apply the 2-D frequency transformation [10] to the second-order sections in (44) as

$$f_k(z_1)|_{z_1^{-1}=T_1(z_1)} = f_{v,k}(z_1) = \prod_{l=1}^{M_1/2} f_{v,kl}(z_1) \tag{45}$$

$$g_k(z_2)|_{z_2^{-1}=T_2(z_2)} = g_{v,k}(z_2) = \prod_{l=1}^{M_2/2} g_{v,kl}(z_2). \tag{46}$$

In the case of the lowpass to lowpass frequency transformation, $T_1(z_1)$ and $T_2(z_2)$ is defined as

$$T_1(z_1) = \frac{z_1^{-1} - \alpha}{1 - \alpha z_1^{-1}}, \quad T_2(z_2) = \frac{z_2^{-1} - \beta}{1 - \beta z_2^{-1}}. \tag{47}$$

Next, according to the realization procedure described in Sect. 4.1, all the second-order variable sections $f_{v,kl}(z_1)$ and $g_{v,kl}(z_2)$ with the direct form are re-synthesized as the proposed second-order variable section.

From the above realization method, we can obtain the 1-D VDFs $\mathbf{F}_v(z_1)$ and $\mathbf{G}_v(z_2)$ as

$$\mathbf{F}_v(z_1) = [f_{v,1}(z_1) \; f_{v,2}(z_1) \; \cdots \; f_{v,r}(z_1)] \tag{48}$$
$$\mathbf{G}_v(z_2) = [g_{v,1}(z_2) \; g_{v,2}(z_2) \; \cdots \; g_{v,r}(z_2)]^t. \tag{49}$$

Finally, the 2-D VDF based on the reduced-dimensional decomposition form is realized as

$$H_v(z_1, z_2) = \mathbf{F}_v(z_1) \cdot \mathbf{G}_v(z_2). \tag{50}$$

The block diagram of the proposed 2-D VDF structure is shown in Fig. 2. From Fig. 2, we can see that the proposed 2-D VDF consists of two 1-D VDFs $\mathbf{F}_v(z_1)$, an r-input/1-output 1-D VDF, and $\mathbf{G}_v(z_2)$, a 1-input/r-output 1-D VDF. Furthermore, the 1-D VDFs are realized as the parallel connection of cascaded second-order variable sections.

5. NUMERICAL EXAMPLES

In order to verify the effectiveness of the proposed method, we design lowpass 2-D VDFs through the proposed realization method and compare them with 2-D VDFs with the direct form. The transfer function of the prototype separable denominator filter used in numerical examples is as follows: From (2)–(6),

$$\mathbf{N} = 10^{-2} \cdot \begin{bmatrix} -0.021 & 0.365 & 0.336 & 0.126 & -0.401 \\ 0.365 & 0.207 & 0.375 & 0.152 & -0.235 \\ 0.336 & 0.375 & 0.520 & 0.214 & -0.419 \\ 0.126 & 0.152 & 0.214 & 0.087 & -0.178 \\ -0.401 & -0.235 & -0.419 & -0.178 & 0.267 \end{bmatrix} \quad (51)$$

$$D(z_1) = 1 - 1.5706z_1^{-1} + 0.8544z_1^{-2} - 0.1236z_1^{-3} - 0.0281z_1^{-4} \quad (52)$$

$$D(z_2) = 1 - 1.5706z_2^{-1} + 0.8544z_2^{-2} - 0.1236z_2^{-3} - 0.0281z_2^{-4}. \quad (53)$$

From (52) and (53), the poles of the prototype 2-D filter are given by

$$z_1 = z_2 = 0.65335, -0.11515, 0.51622 \pm 0.32789i. \quad (54)$$

In order to suppress the degradation of frequency characteristics due to finite coefficient wordlength, we need to select the best second-order section providing the highest pole density at the frequency regions where poles are located. For this reason, since the Kingsbury's section has the highest pole density on the real axis of the z-plane, we select it as the second-order section with real poles

$$z_1 = z_2 = 0.65335, -0.11515. \quad (55)$$

Similarly, we also select the Avenhaus' section as the second-order section with complex poles

$$z_1 = z_2 = 0.51622 \pm 0.32789i. \quad (56)$$

In numerical examples, filter coefficients have been quantized as short as 8 bits in order to show numerical results more clearly. Figure 3 shows the frequency tuning errors of $H_{df}(\omega_1, \omega_2)$ and $H_{pf}(\omega_1, \omega_2)$ to $H_{id}(\omega_1, \omega_2)$, which is the 2-D VDF with no coefficient approximation and infinite coefficient wordlength, when both variable parameters α and β are 0.1. Figure 4 shows the frequency tuning errors of $H_{df}(\omega_1, \omega_2)$ and $H_{pf}(\omega_1, \omega_2)$ to $H_{id}(\omega_1, \omega_2)$ when both variable parameters α and β are -0.1. From Figs. 3 and 4, we can clearly observe that the frequency tuning errors of the proposed 2-D VDF are by far less than that of the direct form 2-D VDF. In other words, the tuning accuracy of the proposed 2-D VDF is higher than that of the direct form 2-D VDF.

6. CONCLUSIONS

In this paper, we have proposed the high accuracy 2-D variable IIR digital filters with real coefficients based on the reduced-dimensional decomposition form. In order to improve frequency tuning accuracy, the proposed 2-D VDF has been realized as the cascade connections of new second-order variable sections with low sensitivity characteristic. Numerical examples have shown that the proposed 2-D VDFs have higher frequency tuning accuracy than the 2-D VDFs with the direct form.

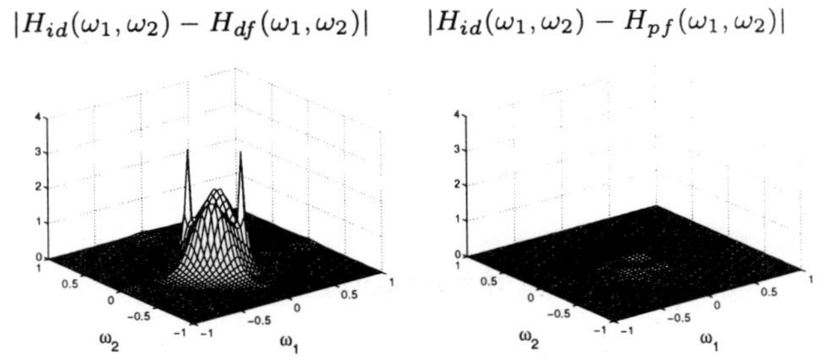

Figure 3: Frequency tuning errors ($\alpha = \beta = 0.1$).

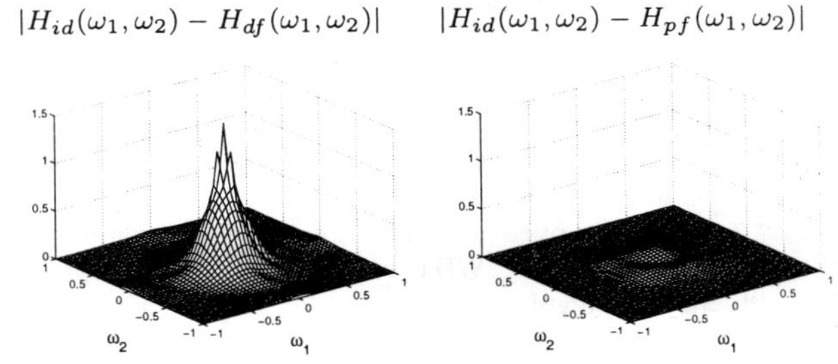

Figure 4: Frequency tuning errors ($\alpha = \beta = -0.1$).

7. REFERENCES

[1] G. Stoyanov and M. Kawamata, "Variable digital filters," *Journal of Signal Processing*, vol. 1, no. 4, pp. 275–290, Jul. 1997.

[2] A. G. Constantinides, "Spectral transformations for digital filters," *Proceedings of the IEE*, vol. 117, pp. 1585–1590, Aug. 1970.

[3] S. K. Mitra, Y. Neuvo and H. Roivainen, "Design of recursive digital filters with variable characteristics," *International Journal of Circuit Theory and Applications*, vol. 18, pp. 107–119, 1990.

[4] W. Watanabe, M. Ito, N. Murakoshi and A. Nishihara, "A synthesis of variable wave-digital filters", *IEICE Trans. Fundamentals*, vol. E77-A, no. 1, pp. 263-271, 1994.

[5] H. Jang and M. Kawamata, "Realization of high accuracy 2-D variable IIR digital filters," *IEICE Trans. Fundamentals*, vol. E85-A, no. 10, pp. 2293-2301, Oct. 2002.

[6] H. Jang and M. Kawamata, "Realization of 2-D variable IIR digital filter structures with a small amount of calculations for coefficient update," *International Midwest Symposium on Circuits and Systems*, vol. 3, pp. 684-687, Aug. 2002.

[7] T. Lin, M. Kawamata and T. Higuchi, "Decomposition of 2-D separable denominator systems: existence, uniqueness and applications," *IEEE Trans. Circuits Syst.*, vol. CAS-34, no. 3, pp. 292-296, Mar. 1987.

[8] D. Schlichthärle, "Digital filters," *Springer*, Berlin, 2000.

[9] N. G. Kingsbury and P. Rayner, "Second-order recursive digital filters element for poles near the unit circle and the real Z axis," *Electronic Letters*, vol. 8, no. 6, pp. 155-156, 1972.

[10] N. A. Pendergrass, S. K. Mitra and I. Jury, "Spectral transformations for two-dimensional digital filters," *IEEE Trans. Circuits Syst.*, vol. CAS-23, no. 1, pp. 26-35, Jan. 1976.

A NEW RECURSIVE FORMULATION FOR 2-D WHT

Ayman Elnaggar

Department of Information Engineering,
Sultan Qaboos University, P.O. Box 33,
Muscat, Oman 123

Mokhtar Aboelaze

Department of Computer Science,
York University, Toronto, Ontario,
Canada M3J 1P3

ABSTRACT

This paper presents a new recursive formulation for computing the Walsh-Hadamard Transform (WHT) that allows the generation of higher order (longer size) 2-D WHT architectures from four lower order (shorter sizes) WHT architectures. Our methodology is based on manipulating tensor product forms so that they can be mapped directly into modular parallel architectures. The resulting WHT circuits have very simple modular structure and regular topology.

1. INTRODUCTION

The Walsh-Hadamard Transform (WHT) has been used in many DSP, image, and video processing applications such as filter generating systems [9], block orthogonal transforms (BOTs) [5], and block wavelet transforms [2]. Other applications in communications are in CDMA [1] and spread spectrum [6].

This paper proposes an efficient and cost-effective methodology for mapping WHT onto VLSI structures. The main objective of this paper is to derive a design methodology and recursive formulation for computing the multidimensional (*m*-d) WHT which is useful for the true modularization and parallelization of the resulting computation.

The main result reported in this paper shows that a large two-dimensional 2-D WHT computation on an $n \times n$ input image can be decomposed recursively into three stages as shown in Fig. 1 for the case $n = 4$. The second stage is constructed recursively from four parallel (data-independent) blocks each realizing a smaller-size WHT. The pre-additions and the post-permutations stages serve as "glue" circuits that combine the 2^2 lower order WHT blocks to construct the higher order WHT architecture.
Observe that, we have drawn our networks such that data flows from right to left. We chose this convention to show the direct correspondence between the derived algorithms and the proposed VLSI networks.
Although, as far as we know from the literature, the recursive 1-D WHT algorithm is widely presented in the literature [8], [11], neither the proposed 2-D WHT algorithm nor the modular forms were previously derived.

Our work is based on a non-trivial generalization of the 1-D WHT using tensor product. When coupled with permutation matrices, tensor products provide a unifying framework for describing a wide range of fast recursive algorithms for various transform [3], [4], [10].

Some of the tensor product properties that will be used throughout this paper are [3], [10]:

$$AB \otimes CD = (A \otimes B)(C \otimes D) \tag{1}$$

$$(A \otimes B) \otimes C = A \otimes (B \otimes C) \tag{2}$$

If $n = n_1 n_2$, then

$$A_{n_1} \otimes B_{n_2} = P_{n,n_1}(I_{n_2} \otimes A_{n_1})\, P_{n,n_2}(I_{n_1} \otimes B_{n_2}) \tag{3}$$

If $n = n_1 n_2 n_3$, then

$$I_{n_1} \otimes A_{n_2} \otimes I_{n_3} = P_{n,n_1 n_2}(I_{n_1 n_3} \otimes A_{n_2}) P_{n,n_3} \tag{4}$$

If $2n = n_1 n_2$, then

$$P_{n,2} = P_{n,n_1} P_{n,n_2} \tag{5}$$

Where $\otimes$ denotes the tensor product, I_n is the identity matrix of size *n*, and $P_{n,s}$ is an $n \times n$ binary matrix specifying an *n*/*s* shuffle (or *s*-stride) permutation.

This paper is organized as follows. In Section 2 we modify the original 1-D WHT. In Section 3 we then propose the 2-D WHT recursive algorithm. Finally, we conclude our results.

2. THE MODIFIED FORMULATIONS OF THE 1-D WHT

In this section, we modify the original 1-D WHT to the iterative form that allows a hardware saving without affecting the processing speed.
The original 1-D WHT transform matrix is defined as [8], [11]

$$W_n = \begin{pmatrix} W_{n/2} & W_{n/2} \\ W_{n/2} & -W_{n/2} \end{pmatrix}, \quad W_2 = \begin{pmatrix} 1 & 1 \\ 1 & -1 \end{pmatrix} \tag{6}$$

where W_2 is the 2-point WHT.

0-7803-7761-3/03/$17.00 ©2003 IEEE

2.1 The 1-D WHT Iterative Formulation

Let $k = \log_2 n$, we can write equation (6) in the iterative tensor-product form

$$W_n = W_2 \otimes W_{n/2} = W_2 \otimes W_2 \otimes \cdots \otimes \cdots W_2$$
$$= \prod_{i=0}^{k-1} (I_{2^i} \otimes W_2 \otimes I_{2^{k-i-1}}) \quad (7)$$

which using property (4), can be modified to

$$W_n = \prod_{i=0}^{k-1} P_{n,2^{i+1}} (I_{2^{k-1}} \otimes W_2) P_{n,2^{k-i-1}} \quad (8)$$

As an example, we can express W_8 as

$$W_8 = \lfloor P_{8,2} (I_4 \otimes W_2) P_{8,4} \rfloor . \left[P_{8,4} (I_4 \otimes W_2) P_{8,2} \right] \left[P_{8,8} (I_4 \otimes W_2) P_{8,1} \right] \quad (9)$$

The realization of W_8 is shown in Fig. 2 (a).
Applying property (5) to equation (9), now the adjacent permutations $P_{8,2}$ $P_{8,8}$ (from the first and the second stages) will be replaced by the single permutation $P_{8,2}$ and the adjacent permutations $P_{8,4}$ $P_{8,4}$ (from the second and the third stages) will be replaced by the single permutation $P_{8,2}$ as shown in Fig. 2 (b).
Similarly, equation (8) can be simplified to

$$W_n = \prod_{i=0}^{k-1} P_{n,2} (I_{2^{k-1}} \otimes W_2) \quad (10)$$

Thus, W_n can be computed by the cascaded product of *k* similar stages (independent of *i*) of double matrix products instead of the triple matrix products in equation (8). Alternatively, we can realize (10) by a single block of $P_{n,2} (I_{2^{k-1}} \otimes W_2)$ and take the output after k iterations as shown in Fig. 3 for the case $n = 8$.

2.2 The 1-D WHT Recursive Formulation

Applying property (1), equation (7) can be modified to

$$W_n = W_2 \otimes W_{n/2} = I_2 W_2 \otimes W_{n/2} I_{n/2}$$
$$= (I_2 \otimes W_{n/2})(W_2 \otimes I_{n/2}) \quad (11)$$
$$= (I_2 \otimes W_{n/2}) Q_n$$

where

$$Q_n = (W_2 \otimes I_{n/2}) \quad (12)$$

Equation (11) represents the two-stage recursive tensor product formulation of the 1-D WHT in which the first stage is the pre-additions (Q_n), followed by the second stage of the core computation $(I_2 \otimes W_{n/2})$ that consists of a parallel stage of two identical smaller WHT computations each of size $n/2$ as shown in Fig.4.

3. THE PROPOSED FORMULATION OF THE 2-D WHT

This section will develop two recursive methodologies for realizing the 2-D WHT computations from smaller WHT computations. First, we present a direct method for realizing the 2-D WHT computations using the conventional row-column decomposition of the 1-D WHT in a tensor product form. The second methodology provides a truly 2-D recursive structures that employ one stage of smaller 2-D WHTs to realize the large 2-D WHT.

3.1 Conventional Row-Column Decomposition

Since the WHT matrix is separable [7], the 2-D WHT for an input image of dimension $n_1 \times n_2$ can be computed by a stage of n_2 parallel 1-D WHT computations on n_1 points each, followed by another stage of n_1 parallel 1-D WHT computations on n_2 points each. This can be represented by the matrix-vector form

$$X = W_{n_1,n_2} x , \quad (13)$$

where W_{n_1,n_2} is the 2-D WHT transform matrix for an $n_1 \times n_2$ image, X and x are the output and input column-scanned vectors, respectively.
For separable transforms, the matrix W_{n_1,n_2} can be represented by the tensor product form [7]

$$W_{n_1,n_2} = W_{n_1} \otimes W_{n_2} \quad (14)$$

where W_{n_1} and W_{n_2} are the row and column 1-D WHT operators, as defined by equation (7), on *x*, respectively.

By substituting (14) in (13), we have

$$X = (W_{n_1} \otimes W_{n_2}) x . \quad (15)$$

Which using equation (7) can be expressed as

$$X = W_{n_1 \times n_2} x \quad (16)$$

Therefore, the 2-D WHT on an $n_1 \times n_2$ input is equivalent to a 1-D WHT on a 1-D input vector of size $n_1 \times n_2$ that can be implemented using either the modified 1-D iterative algorithm given by (10) or the modified 1-D recursive algorithm given by (11).

3.2 The Truly Recursive Formulation of the 2-D WHT

Now we will derive a truly 2-D recursive formulation of the WHT by further manipulation of equation (15). Substituting (11) in (15), the 2-D WHT transform matrix can be written as

$$W_{n_1,n_2} = [(I_2 \otimes W_{n_1/2}) Q_{n_1}] \otimes [(I_2 \otimes W_{n_2/2}) Q_{n_2}] \quad (17)$$

Applying property (1), we can write W_{n_1,n_2} as

$$W_{n_1,n_2} = [(I_2 \otimes W_{n_1/2}) \otimes (I_2 \otimes W_{n_2/2})] [Q_{n_1} \otimes Q_{n_2}] = [C_{n_1,n_2} Q_{n_1,n_2}] \quad (18)$$

where

$$C_{n_1,n_2} = [(I_2 \otimes W_{n_1/2}) \otimes (I_2 \otimes W_{n_2/2})], \quad Q_{n_1,n_2} = [Q_{n_1} \otimes Q_{n_2}] \quad (19)$$

Now, from property (2), we can write C_{n_1,n_2} in the form

$$C_{n_1,n_2} = (I_2 \otimes W_{n_1/2} \otimes I_2) \otimes W_{n_2/2} \quad (20)$$

Applying property (4), we can write (20) in the form

$$\begin{aligned} C_{n_1,n_2} &= [P_{2n_1,n_1}(I_4 \otimes W_{n_1/2})\, P_{2n_1,2}] \otimes [W_{n_2/2}] \\ &= [P_{2n_1,n_1}(I_4 \otimes W_{n_1/2})\, P_{2n_1,2}] \otimes [I_{n_2/2} W_{n_2/2} I_{n_2/2}] \\ &= [P_{2n_1,n_1} \otimes I_{n_2/2}][(I_4 \otimes W_{n_1/2}) \otimes W_{n_2/2}][P_{2n_1,2} \otimes I_{n_2/2}] \\ &= [P_{2n_1,n_1} \otimes I_{n_2/2}][(I_4 \otimes W_{n_1/2,\,n_2/2}][P_{2n_1,2} \otimes I_{n_2/2}] \end{aligned} \quad (21)$$

where

$$W_{n_1/2\,,\,n_2/2} = W_{n_1/2} \otimes W_{n_2/2} \quad (22)$$

Finally, substituting (21) in (18), we have

$$W_{n_1,n_2} = [\widetilde{R}_{n_1,n_2}(I_4 \otimes W_{n_1/2\,,\,n_2/2})\, \widetilde{Q}_{n_1,n_2}] \quad (23)$$

where

$$\widetilde{Q}_{n_1,n_2} = (P_{2n_1,2} \otimes I_{n_2/2})\, Q_{n_1,n_2}, \quad \widetilde{R}_{n_1,n_2} = (P_{2n_1,n_1} \otimes I_{n_2/2}). \quad (24)$$

Equation (23) represents the truly recursive 2-D WHT in which $\widetilde{Q}_{n_1,n_2}$ and $\widetilde{R}_{n_1,n_2}$ are the pre- and post-processing glue structures, respectively, that combine 2^2 identical lower-order 2-D WHT modules each of size $n_1/2 \times n_2/2$ in parallel, to construct the higher order 2-D WHT of size $n_1 \times n_2$ as shown previously in Fig. 1.

4. CONCLUSIONS

In this paper, we showed that a large two-Dimensional 2-D WHT computation on an $n \times n$ input image can be decomposed recursively into three stages. The middle stage is constructed recursively from four parallel (data-independent) blocks each realizing a smaller-size WHT. The pre-additions and the post-permutations stages serve as "glue" circuits that combine the $\mathbf{2^2}$ lower order WHT blocks to construct the higher order WHT architecture. The resulting networks have a very simple modular structure and highly regular topology.

5. REFERENCES

[1] F. Adachi, K. Ohno, A. Higashi, T. Dohi, and Y. Okumura, "Coherent Multicode DS-CDMA Mobile Radio Access", *IEICE Trans. on Communications*, Vol. E79-B, 1996, pp. 1316-1325.

[2] A. E. Cetin, O. N. Gerek, and S. Ulukus, "Block Wavelet Transforms for Image Coding", *IEEE Trans. on Circuits and systems for Video Technology*, Vol. 3, 1993, pp. 433-435.

[3] A. Elnaggar and M. Aboelaze, "An Efficient Architecture for Multi-Dimensional Convolution", *IEEE Trans. on Circuits and Systems II*, Vol. 47, No. 12, 2000, pp. 1520-1523.

[4] A. Elnaggar and M. Aboelaze, "A Modified Shuffle Free Architecture for Linear Convolution, *IEEE Trans. on Circuits and Systems II*, Vol. 48, No. 9, 2001, pp. 862-866.

[5] Makur, "BOT's Based on Non-uniform Filter Banks", *IEEE Trans. on Signal Processing*, Vol. 44, 1996, pp. 1971-1981.

[6] R. L. Peterson, R. E. Ziemer, and D. E. Borth, *Introduction to Spread Spectrum Communications*, Prentice Hall, 1995.

[7] W. K. Pratt, *Digital Image Processing*, John Wiley & Sons, Inc., 1991.

[8] S. Rahardja and B.J. Falkowski, "Family of unified complex Hadamard transforms", *IEEE Trans. On Circuits and Systems II*, Vol. 46, 1999, pp. 1094-1100.

[9] S. Samadi, A. Nishihara and H. Iwakura, "Filter Generating Systems", *IEEE Trans. on Circuits and Systems II*, Vol. 47, No. 8, 2000, pp. 214-221.

[10] R. Tolimieri, M. An, C. Lu, *Algorithms for Discrete Fourier Transform and Convolution*, Springer-Verlag, New York 1989.

[11] Yarlagadda, RKE and Hershey, JE, Hadamard Matrix Analysis and Synthesis With Applications to Communications and Signal/Image Processing, Kluwer Academic Publishers, Boston 1997.

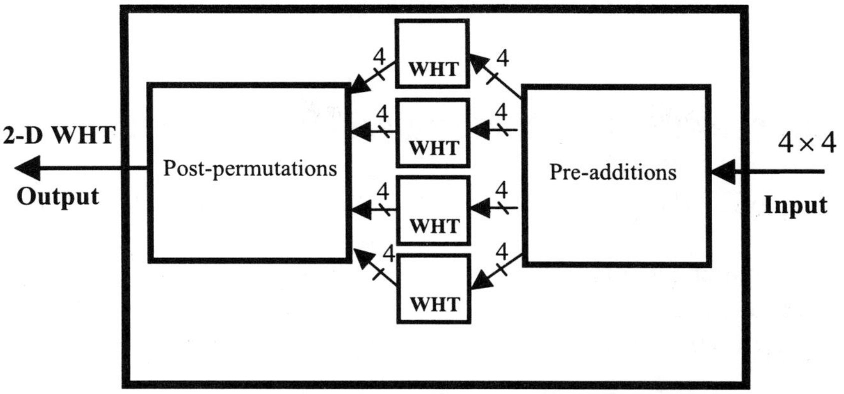

Figure 1 The proposed 2-d WHT recursive realization for a 4×4 input image

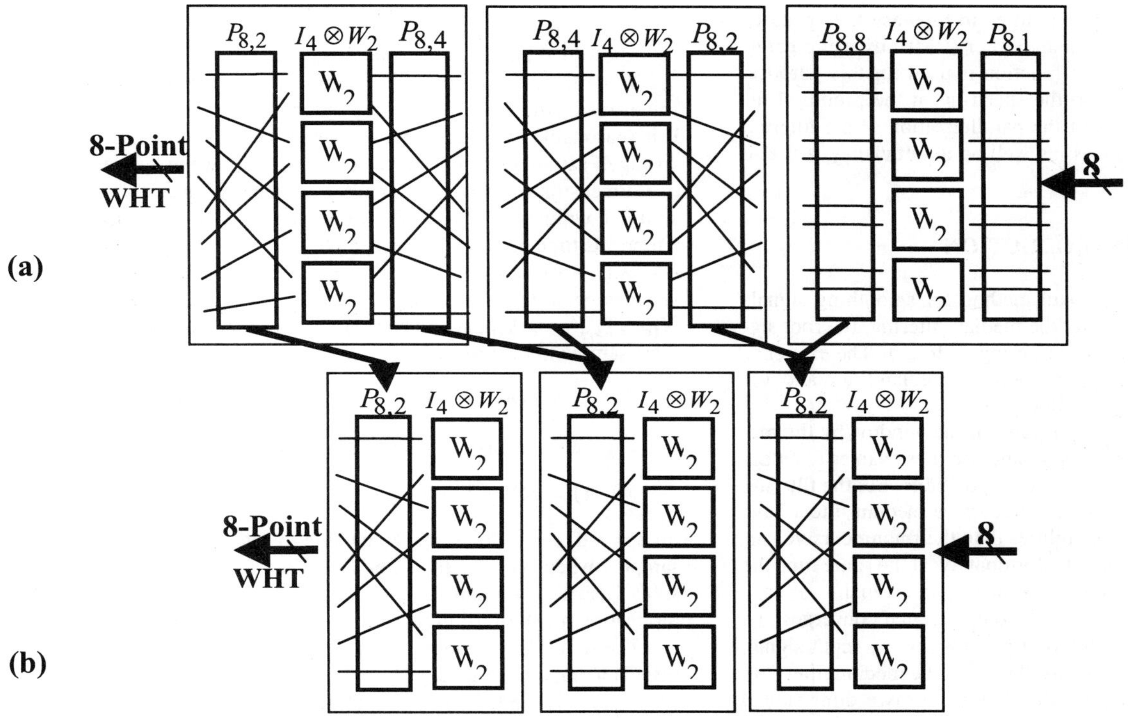

Figure 2 The realization of W_8 : **(a)** The original 1-d WHT iterative algorithm. **(b)** The modified 1-d WHT algorithm.

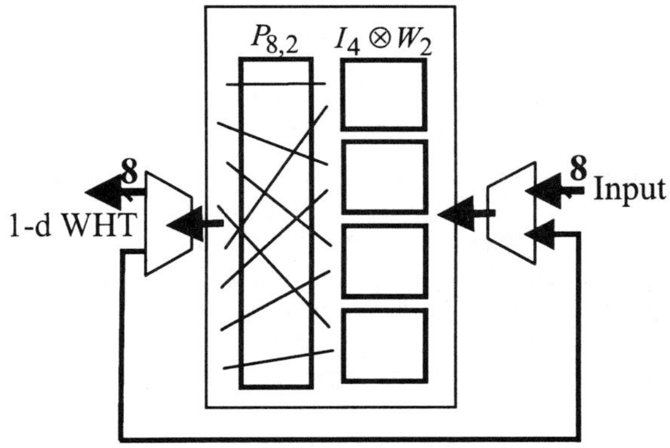

Figure 3 The reduced hardware realization of the modified 1-d WHT algorithm

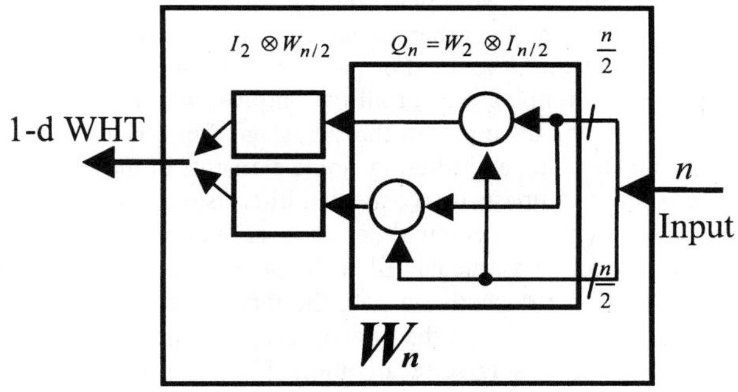

Figure 4 The realization of the recursive 1-d WHT

PARALLEL ITERATIONS FOR RECURSIVE MEDIAN FILTER

Adrian Burian, Jarmo Takala

Institute of Digital and Computer Systems
Tampere University of Technology
PO Box 553, FIN-33101, Tampere, Finland

Marina Dana Ţopa

Bases of Electronics Department
Technical University of Cluj-Napoca
Baritiu 26-28, Cluj-Napoca, Romania

ABSTRACT

In this paper we study the possibilities to increase the speed of convergence of the recursive median filter (RMF) by using parallel iterations. The median cost function is used for this purpose. The convergence behavior and the speed-up of the proposed filters are studied. The effects of the parallelization of the filtering process over the performances, as well as its efficiency, are also considered.

1. INTRODUCTION

Median filters provide a powerful method for smoothing signals and images [1, 5]. By repeating the median filtering, the root signal, which is invariant to further filtering, is found. The existence of root signals is a fundamental property of median filter, and it is used in characterizing these nonlinear filters.

The RMFs replace the center point of the window by the median value of all the points inside it, and use these values to calculate the medians in subsequent window positions. So, the filtering operation is performed 'in-place'. Recursive median filters have stronger noise attenuation capabilities than their nonrecursive version, and a faster convergence. Unfortunately, at the same time the recursive operation destroys more of the original signal.

A filter is called idempotent if its output signal converges to a root in only one pass of the filtering process for any input signal. For one-dimensional (1-D) signals, the recursive median filters are idempotent, but this is not generally the case for two-dimensional (2-D) signals [8].

There are two common ways of filtering a binary sequence. In a parallel iteration all the window's outputs are computed simultaneously. In a serial iteration we filter one element of the input binary sequence at a time; i.e. we replace the current value with the filtered one, while leaving the other values unchanged. Then, in either case, we repeat the process. For parallel iterations, we repeat the simultaneous computations of all the outputs, with the previous output result being the input for the next stage. For serial iteration, we keep on choosing and filtering single elements of the binary sequence. Clearly, a nonrecursive median filter uses a parallel iteration, while a recursive median filter uses a serial iteration. This is because for the recursive median filter the pixel's value of the filtered binary sequence depends on both the input sequence and the filtered sequence itself. All the already proposed implementations of the RMF are using serial iterations [3, 6]. In this paper we study the possibilities to obtain a parallelization of the recursive median filtering operation. In this way, more than one single output of the RMF are computed simultaneously. While it is straightforward that for the non-recursive median filter a parallel iteration gives the same result as a serial iteration, this is obviously not the case for recursive filtering structures.

A filter is understood as an operation whose window slides over the signal and at every time instant t operates on signal values inside the filter window. For simplicity, the number of signal values inside the filter window is assumed to be odd $N = 2k + 1$. It is reasonable to assume that the signal is of finite length, consisting of samples from $X(0)$ to $X(L-1)$. So, when applied to a one-dimensional (1-D) signal, a filter of length N is calculated by sliding a window of length N over the input signal one sample at a time. At time instant t let the samples $X(t-k)$, $X(t-k+1)$, $\cdots$, $X(t+k)$ of the 1-D signal be in the filter window. Let these samples be sorted into increasing (or decreasing) order ($X_{(t-k)}$, $X_{(t-k+1)}$, $\cdots$, $X_{(t+k)}$). Then the sample median is the centermost sample in the ordered set:

$$\text{MED}[X(t-k), \cdots, X(t+k)] = X_{(t)}. \quad (1)$$

The standard median filter is defined by:

$$Y(t) = \text{MED}[X(t-k), \cdots, X(t), \cdots, X(t+k)], \quad (2)$$

for $t = \{0, 1, \cdots, L-1\}$. To be able to filter the outmost input samples when parts of the filter's window fall outside the input signal, usually the input signal is appended to the required size by replicating the outmost input sample as many times as needed.

The standard median filter is scale invariant and nonlinear with respect to superposition [1]. A very important property of median filters, namely the threshold decomposition and stacking property, was observed in [4]. By using this property, the input signal can be decomposed into several binary signals that are easier to analyze. The median filter can then be applied to the binary signals and the binary outputs can be added together to produce the final result. By using the threshold decomposition technique, the analysis of these filters is reduced to studying their effects on binary signals.

The standard RMF is defined as follows:

$$Y(t) = \text{MED}[Y(t-k), \cdots, Y(t-1), X(t), \cdots, X(t+k)], \quad (3)$$

where $Y(t-k), \cdots, Y(t-1)$ are the already computed outputs. RMFs, though useful in practice are in general more difficult to handle theoretically. With the same amount of operations, recursive filters can usually provide better smoothing capability than nonrecursive filters, at the expense of increased distortion.

2. MEDIAN COST FUNCTION

The median cost function was firstly introduced in [7]. Its definition was based on a combination of two terms. The first one measures the smoothness between the median filter output and its neighbor points within the operation window. The second term measures the discrepancy between the filter output and its original signal. We used it in [2] to study the smoothing capabilities of the RMF. In this paper, the median cost function will be used to establish the range of the used parameters, in order to evaluate the performances of the parallel iterations.

For introducing the median cost function all the $\{0,1\}$ binary values are transferred into the binary domain $\{-1,1\}$. Let us use the following notations: $\mathbf{O}=2\cdot\mathbf{Y}-1$, $\mathbf{Z}=2\cdot\mathbf{X}-1$. Because of the scale invariance property, finding the median sequence $\mathbf{Y}$ of the original binary sequence $\mathbf{X}$, is equivalent to finding the median sequence $\mathbf{O}$ of the transformed binary sequence $\mathbf{Z}$.

The median cost function is defined by:

$$M(t) = -\sum_{i=-k,i\neq 0}^{k} Z(t+i)\cdot O(t) - Z(t)\cdot O(t). \quad (4)$$

Let us denote $S(t) = \sum_{i=-k}^{k} Z(t+i)$. For the output of the median filter, we have that if $S(t) > 0$ then $O(t) = 1$, otherwise $O(t) = -1$. Equation (4) can be rewritten as: $M(t) = S(t)O(t)$.

Theorem 2.1 *Recursive median filtering process never increases the median cost function of the input signal.*

Proof: Because the process is sequential and at any time only one output is changed, the changes of the median cost function from one instant time step to another are given by the changes at the output of each filtering window. This is because if the output of the filtering window remains unchanged, it means that also the $S(t)$ quantity is the same, so their product is the same too. We can distinguish only the following two situations, that implies an instant change in the cost function:

$$\text{Case 1: } Z(t) = -1; S(t) > 0 \Rightarrow O(t) = 1. \quad (5)$$

In this case, the change of the value $O(t)$ that replaces $Z(t)$ implies an increase of all the sums that contain this value by 2. Let us consider first the effect on the window that is centered in $O(t)$:

$$\Delta M_1 = -O(t)\cdot[S(t)+2] + Z(t)\cdot S(t) = \quad (6)$$

$$-1\cdot[S(t)+2] - 1\cdot S(t) = -2[S(t)+1] < 0. \quad (7)$$

The effect of the other windows that contain the value $O(t)$ is given by the increase with 2 of all the corresponding $S(t+i)$ values for $i \in \{-k,\cdots,k\} - \{0\}$. In conclusion, we have

$$\Delta M_2 = -\sum_{i=-k}^{-1} O(t+i)\Delta S(i) - \sum_{i=1}^{k} Z(t+i)\Delta S(i) = \quad (8)$$

$$-2\sum_{i=1}^{k}[Z(t+i) + O(t-i)] = \quad (9)$$

$$-2[S(t) - Z(t)] = -2[S(t)+1] < 0. \quad (10)$$

The total value of the change of the cost function is

$$\Delta M = \Delta M_1 + \Delta M_2 = -4[S(n)+1] < 0. \quad (11)$$

$$\text{Case 2: } Z(t) = 1; S(t) < 0 \Rightarrow O(t) = -1. \quad (12)$$

In a similar manner it can be shown that

$$\Delta M = 4[S(t) - 1] < 0. \quad (13)$$

The value of ΔM is equal to zero when the corresponding output remains the same, and when is changed it is equal to

$$\begin{cases} +4[S(t)-1] & \text{if } S(t) < 0; Z(t) = 1 \\ -4[S(t)+1] & \text{if } S(t) > 0; Z(t) = -1. \end{cases} \quad (14)$$

In conclusion, the values of ΔM are less than or equal to zero, and this proves our theorem.

The maximum value of changes is given by

$$|\Delta M_{\text{MAX}}| = -4(2k - 1 + 1) = 4(N-1). \quad (15)$$

3. PARALLEL ITERATIONS

We start this section with a few remarks about the impact of the parallelization over the recursivity of the filtering. We note that we cannot compute all the outputs in parallel; otherwise we end up with a nonrecursive median filter. So, we can implement the parallelism only to some degree. The first solution is to force the parallel iteration to be recursive. This can be done by advancing over the signal in the same way as the RMF - step by step. We called this solution the multiple windows approach. The second solution is to advance by jumping over all the already computed outputs; we called this solution the parallel iterations approach. The difference consists in the fact that while in the first case the recursivity of the process is preserved, in the second is perturbed at the moment of the jump, when the process becomes temporarily nonrecursive.

In order to evaluate the performances of these two parallel approaches, similar parameters as the ones used for parallel algorithms will be defined. If we denote by U the number of parallel processes (each process computes one filter's window output) the changes of the median cost function for the RMF are given by

$$\Delta M = -\sum_{i=1}^{U} \Delta M_i. \quad (16)$$

The maximum value of changes in this case is

$$|\Delta M_{\text{MAX}}| = 4U(N-1). \quad (17)$$

We also note that for parallel iterations, the recursivity of the filtering process is stopped at the moment of jump, when the maxim value of changes can attain its maximum possible value:

$$|\Delta M_{\text{MAX}}| = 4UN. \quad (18)$$

In order to evaluate the speed of the convergence of the proposed filter structures, we define the speed-up parameter:

$$S_U = \frac{N_O}{N_U}, \quad (19)$$

where N_U, N_O represents the total number of iterations needed for filtering to the root signal when using parallel iterations and serial iterations, respectively. We note from this definition, that S_U parameter is in direct relation with the changes of the median cost function. So, in the case of the multiple windows we have that $1 \leq S_U < U$ since both processes are recursive, while for parallel iterations $1 \leq S_U < [1 + 2/(N-1)]U$. The upper bound is bigger in the later case because of the disturbance induced into recursivity of the filtering process by the jump points.

The efficiency parameter is defined by the ratio:

$$E = \frac{S_U}{U} \tag{20}$$

This parameter is the equivalent of an efficiency indicator. The better usage is made of the parallelism, the bigger E is. We note that $0 < E < 1$ for multiple windows, and $0 < E < 1 + 2/(N-1)$ for parallel iterations.

4. EXPERIMENTAL RESULTS

For the experiments we have used the well-known Lena image corrupted by impulsive noise. The used values for the impulses were $\{0, 255\}$ (salt-and-pepper noise). The used image contains 256×256 pixels with 8 bits resolution per pixel. A filtering window of size $N = 3 \times 3$ was used and the threshold decomposition technique was applied. The raster scan was used - the recursive filter advanced from left to right on every row, proceeding on columns from top to bottom.

The obtained results for 25% of noise corruption for multiple windows with $2 \div 40$ component windows are plotted in Figure 1. The plots represent -from left to right - the S_U and E parameters, and the filtering performance in terms of the mean square error (MSE). For obtaining these plots the distance between successive window centers has been $D_c = 1, 2, 3, 4, 5$ pixels, represented by continuous, dashed, dash-dot, dotted, and spotted lines, respectively. The column parallelism was used. We note that in the case of multiple windows the usage of parallelism in the same direction with the filtering is useless because the 1-D RMF is idempotent. The only possibility to obtain an efficient usage of the parallelism in this case is to separate the input signal into segments and to choose the D_c value such that to filter each segment with only one component window. We consider this case to be trivial.

The biggest values of the speed-up were obtained for overlapped windows (the distance in pixels between the centers of two successive component windows is $D_c = 1$). We note that in this case the effect of the recursivity is increased, so also the smoothing capability increases. As a result the filtering performances are degraded (the single line at 194 in the MSE plot). Visually, the increase of the MSE in this situation means a smoother filtered image, comparatively with the one obtained by the classical RMF. In all the other cases the filtering performances are exactly the same with the performances of the RMF (the line around 188.2 in the MSE plot). Also visually there is no difference between the filtered images (multiple and classical single window). The only difference is the increase of the speed of the convergence. The speed-up parameter is decreasing with the distance D_c between two successive component windows, and is increasing with the number of used windows U if this number is small ($U < 10$). For bigger values, the efficiency becomes too small to justify the use of a bigger number of component windows.

The performance results for parallel iterations are given in Figure 2 plotted with continuous line for column parallelism, and with dash-dot line for row parallelism. The number of parallel process has been varied between 2 and 256. The other initial conditions and the order of the plots are the same as for Figure 1. We note that in this case we can use a parallelism in the same direction with the filtering, because the jumps cancel the idempotent property of the 1-D RMF. Better convergence performances are obtained when using a parallelism in the complementary direction to the filtering.

The way the filtering performance of parallel iterations is affected by the disturbances of recursivity is illustrated in Figure 3. The two plots of this figure give the obtained MSE values when the noise corruption is varied between 0 and 90% at different scales of representation. The MSE values for RMF and a multiple window with 10 component windows and $D_c = 2$ are represented with continuous line. The filtering performances for these two cases are the same, as we expected. With dash-dot line the obtained MSE values when using 128 parallel column processes are represented. The stronger smoothing capabilities of RMF can be observed from both plots: for small percent of noise corruption the performances are slightly worst because the original signal is also affected, while for high percentage the better noise attenuation gives the increase of performances.

5. CONCLUSIONS

Median cost function can be an useful tool to analyze and understand some of the properties of the median-type filtering operation. The convergence properties of RMFs were studied using the median cost function concept. By using parallel iterations a more abrupt slope of the median cost function was obtained. The results of the simulations for multiple windows case illustrate the improving of the speed of the convergence for the same filtering performances. The efficiency of the parallelism has acceptable values only for a small number of component windows (< 10). The usage of parallel iterations allows the usage of a bigger number of parallel filtering windows, but it affects the recursivity nature of the RMF. Good performances are obtained if the parallelism is not implemented in the direction of filtering, and if the degree of parallelism is much smaller than the size of the input (e.g., 256 parallel processes for 256×256 images).

6. REFERENCES

[1] J. Astola, P. Kuosmanen, *Fundamentals of Nonlinear Digital Filtering*, CRC Press, New York, 1997.

[2] A. Burian, P. Kuosmanen, "Tuning the Smoothness of the Recursive Median Filter", *IEEE Trans. Signal Processing*, vol. 50, no. 7, pp.1631-1639, July 2002.

[3] C. Chakrabarti, "VLSI Architectures for Recursive Median Filters", *IEEE Int. Conf. Acoust., Speech, Signal Processing*, vol. 1, pp. 413-416, 1993.

[4] J.P. Fitch, E.J. Coyle, N.L. Gallagher Jr., "Median Filtering by Threshold Decomposition", *IEEE Trans. Acoust., Speech, Signal Processing*, vol. 32, no. 6, pp. 1183-1188, Dec. 1984.

[5] —, "Root Properties and Convergence Rates of Median Filters", *IEEE Trans. Acoust., Speech, Signal Processing*, vol. 33, no. 1, pp. 230-239, Feb. 1985.

[6] S.J. Ko, Y.H. Lee, A.T. Fam, "Efficient Implementation of One-Dimensional Recursive Median Filters", *IEEE Trans. Circuits and Systems*, vol. 37, no. 11, pp. 1447-1450, Nov. 1990.

[7] G. Qiu, "Functional Optimization Properties of Median Filtering", *IEEE Signal Processing Letters*, vol.1, no.4, pp. 64-66, April 1994.

[8] B. Zeng, "Convergence Properties of Median and Weighted Median Filters", *IEEE Trans. Signal Processing*, vol. 42, no. 12, pp. 3515-3519, Dec. 1994.

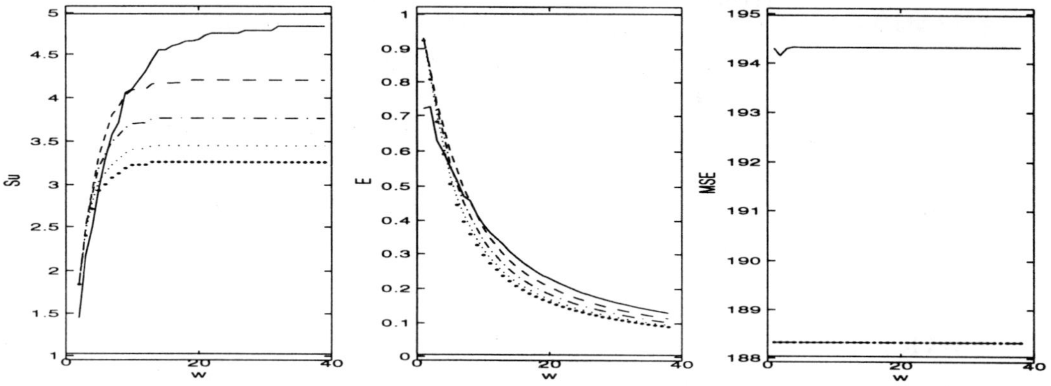

Figure 1: Performance of multiple windows.

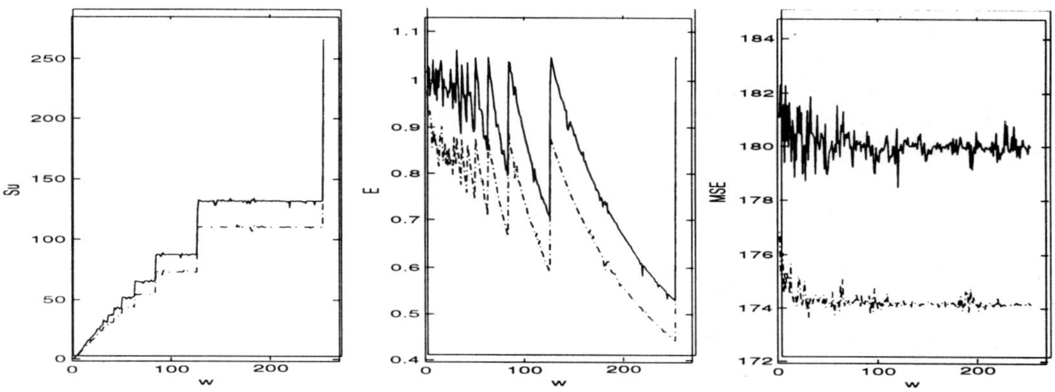

Figure 2: Performance of parallel iterations.

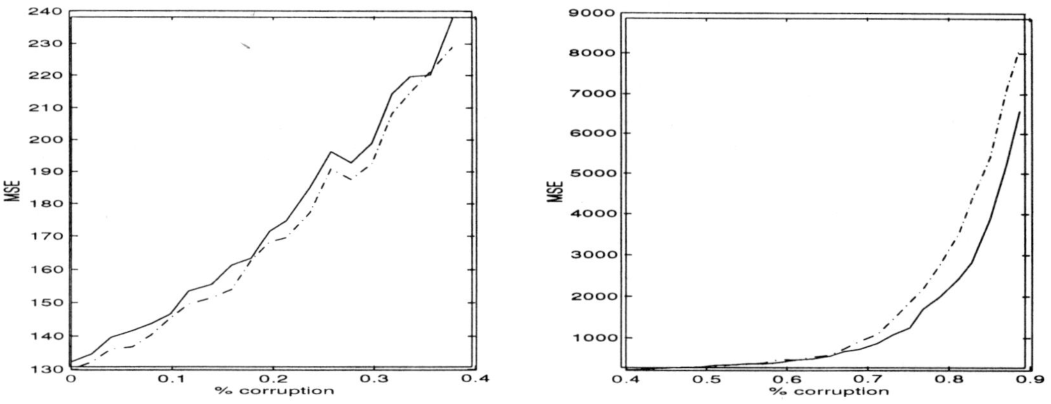

Figure 3: Comparison of MSE values.

AN IMPLEMENTATION OF NUMERICAL INVERSION OF LAPLACE TRANSFORMS ON FPGA

Akihiro YONEMOTO, Takashi HISAKADO and Kohshi OKUMURA

Department of Electrical Engineering, Kyoto University, Japan
{yone, hisakado, kohshi}@kuee.kyoto-u.ac.jp

ABSTRACT

This paper presents an implementation of numerical inversion of Laplace transforms on FPGA. To make real-time transient analysis possible using Laplace transforms, it is desirable to implement hardware of numerical inversion of Laplace transforms. However, the implementation of the FFT-based numerical inversion method with fixed-point numbers results in severely large errors due to the exponential function which appears in the inversion formula. In order to overcome this difficulty, we propose the use of block floating-point FFT in the implementation. As a result, the errors are remarkably reduced.

1. INTRODUCTION

Many implementations of FFT (Fast Fourier Transform) have been devised for hardware from LSI to recent DSP and FPGA. This is because FFT has many applications and some of them require real-time data processing such as spectrum analysis, frequency-domain filtering, convolution computation and so on. To deal with transient signals, however, Laplace transforms are more appropriate since the discrete Fourier transform is theoretically a method to analyze periodic signals. In order to realize a real-time system with Laplace transforms, we propose implementation of their numerical inversion on FPGA.

Many algorithms for the numerical inversion have been proposed thus far [1]. We have focused on the FFT-based method [2–6] and have proposed effective error reduction method [8] since only with the FFT-based method is the corresponding numerical Laplace transform observed [7]. Hence we also adopt this method in the implementation.

One of the difficulties in implementing the FFT-based inversion is that the simple implementation of this method with fixed-point numbers results in severely large errors due to the exponential function appearing in the Laplace inversion. Although the implementation with floating-point numbers evidently solves this problem, it requires an excessive number of logic gates on FPGA. To overcome this difficulty, we propose the use of block floating-point implementation of the FFT block [9].

2. NUMERICAL INVERSION OF LAPLACE TRANSFORMS

Durbin [2] and some other authors [3–6] have proposed the following formula as a numerical inversion method of a Laplace transform $F(s)$: for $0 \le t \le 2T$,

$$f_N(t) = \frac{e^{at}}{2T} \sum_{k=-N+1}^{N-1} F\left(a + \frac{j\pi k}{T}\right) e^{j\pi kt/T}. \tag{1}$$

This formula is obtained by applying the trapezoidal rule to the well-known Bromwich integral for the inversion of Laplace transforms and truncating the resulting infinite sum. For discrete values of $t = 2nT/N\,(n = 0, \cdots, N-1)$, the formula leads to

$$f_N\left(\frac{2nT}{N}\right) = \frac{e^{2nTa/N}}{T} \times \left\{ \mathrm{Re}\left[\sum_{k=0}^{N-1} F\left(a + \frac{j\pi k}{T}\right) e^{j2\pi kn/N}\right] - \frac{F(a)}{2} \right\}. \tag{2}$$

This formula can be computed fast with the FFT algorithm.

Since the truncation error becomes so large in the latter half of the inversion $f_N(t)$ for some kinds of Laplace transforms, only the former half ($n = 0, \cdots, N/2 - 1$) is acceptable.

3. NUMERICAL INVERSION OF LAPLACE TRANSFORMS WITH FIXED POINT NUMBERS

In this section, we present an implementation of the numerical inversion of Laplace transforms (2) with 16-bit fixed-point numbers. The fixed-point numbers are of 2's complement notation, and the decimal point is placed at right after the first digit as follows

$$d_0.d_1d_2\cdots d_{15} \tag{3}$$

where d_0 is a sign bit. This notation can represent numbers from -1 to slightly less than 1.

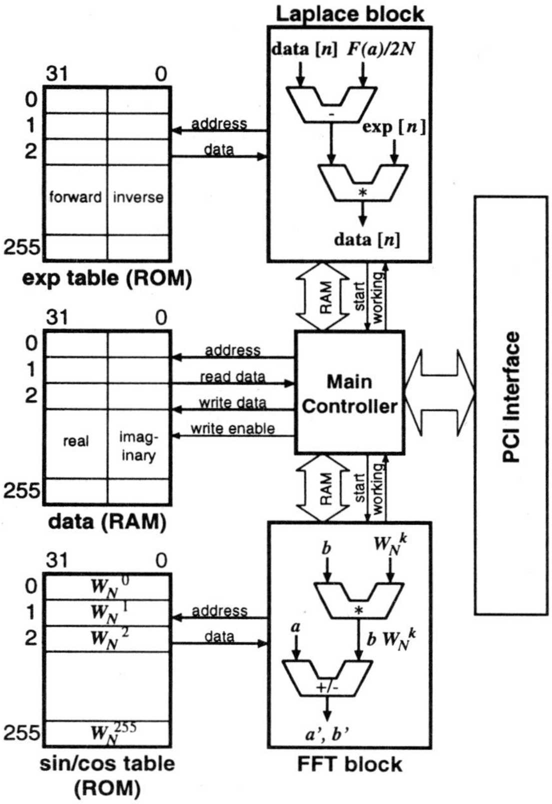

Figure 1: The block diagram of designed Laplace transformation processor.

3.1. Block diagram

Figure 1 illustrates a block diagram of the Laplace transformation processor we designed. This circuit is downloaded to an FPGA device implemented on a PCI board, so that the control and data transfer is carried out via PCI interface of a standard PC/AT compatible computer.

Description of the diagram follows below. Before the inversion, the data to be inverted are stored in data memory. Then, FFT block computes the inverse discrete Fourier transform of the data using sin/cos table memory, i.e. it computes the summation in (2), and write back the result into data. After that, Laplace block does remained operations in (2), that is, the subtractions of $F(a)/2$ and multiplications of $\exp(2nTa/N)$. In the multiplications, the multiplicands are computed in advance, and stored in exp table memory. The result is finally written back to data.

3.2. FFT block

We employ the radix-2 decimation-in-time algorithm for the FFT block. The FFT block contains an arithmetic unit which computes

$$a' = \frac{1}{2}(a + bW), \quad b' = \frac{1}{2}(a - bW) \tag{4}$$

where W is an integral power of $W_N = e^{-j2\pi/N}$ and the division by two is implemented by a 1-bit right shift operation. Since every variable is represented by the fixed-point number (3), the addition and subtraction of a and bW may overflow. In order to prevent these overflows, the result of $a \pm bW$ is divided by two as in (4).

Thus, for an input sequence $X(k)$ ($k = 0, \cdots, N-1$), the FFT block outputs

$$x(n) = \frac{1}{N}\sum_{k=0}^{N-1} X(k) W_N^{-nk} \tag{5}$$

where $n = 0, \cdots, N-1$.

3.3. Laplace block

The Laplace block computes following two operations:

1. Subtract $F(a)/2N$ from each term
2. Multiply each term by $e^{2(n-N/2)Ta/N}$

These operations are executed only for former half terms ($n = 0, \cdots, N/2-1$). We discard the latter half terms since they contain large errors as stated earlier.

In the operation 1, we subtract $F(a)/2N$, not $F(a)/2$, from each term because the output sequence of the FFT block is divided by N as (5). In the operation 2, the multiplicand is scaled down so that it can be represented by (3).

Thus, this block outputs the scaled numerical inversion

$$K f_N\left(\frac{2nT}{N}\right), \quad K = \frac{1}{Ne^{Ta}} \tag{6}$$

for $n = 0, \cdots, N/2-1$.

3.4. Example and Problem

Figure 2 illustrates the process of the numerical inversion of $F(s) = 3/s$. To see the accuracy of the intermediate and the final data, the right vertical axis shows 16-bit integers that are values of the fixed-point variables if they are viewed as 16-bit integers.

The parameters are $N = 256$, $T = 1$ and $a = 3.5$, where $aT = 3 \sim 5$ is the standard range for the FFT-based method. The scaling factor K is then

$$K = 0.000118 \tag{7}$$

and the expected result is

$$Kf(t) = 3K = 0.000354 = 11.6/32768. \tag{8}$$

The bottom figure shows that we seemingly have the expected result, however, the values are quite small if viewed as 16-bit integers. Indeed, they are just 4-bit numbers.

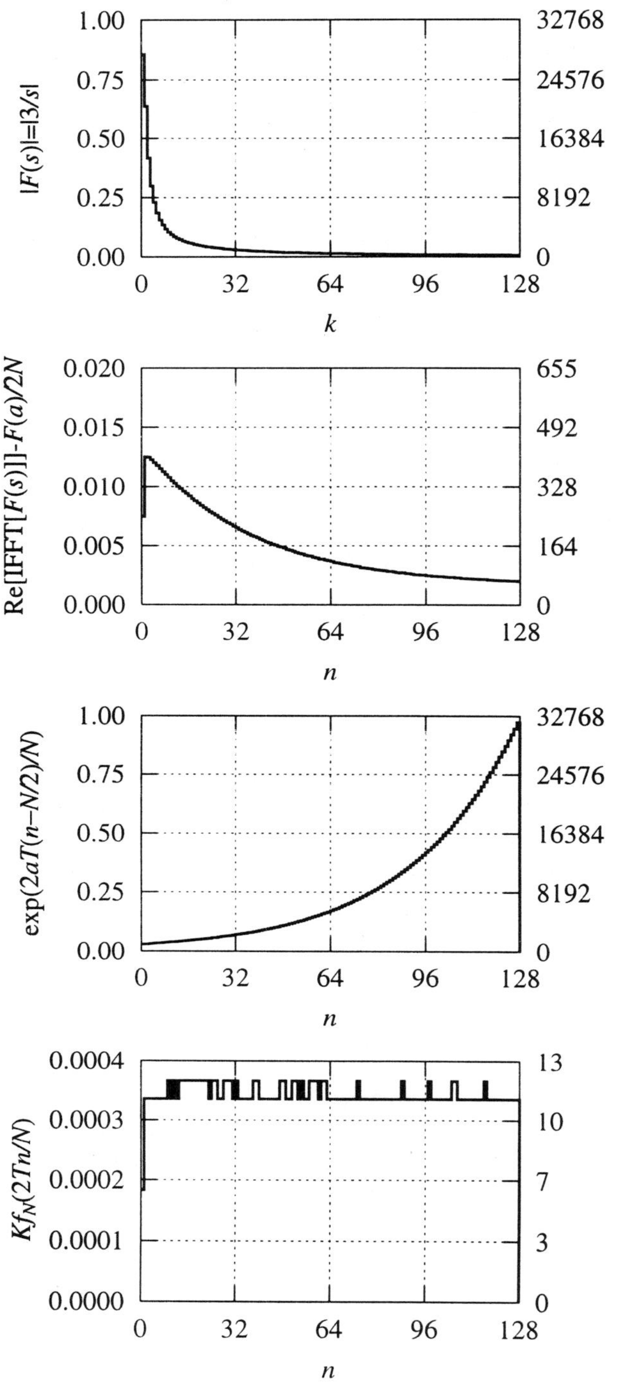

Figure 2: Numerical inversion of the step function $F(s) = 3/s$. ($N = 256$, $T = 1$, $a = 3.5$, $K = 0.000118$).

4. INTRODUCING BLOCK FLOATING-POINT NUMBERS

From the second figure of Fig. 2, we can see that the values are already fairly small after the operation 1 in section 3.3. This is because (4) always divide the result of $a \pm bW$ by two to avoid overflows. However, if a, b are small enough to ensure that $a \pm bW$ will be within the range of (3), the division is not needed.

Moreover, the small values in the second figure are then multiplied by the exponential function shown in the third figure of Fig. 2. Since the multiplication is indispensable and the dynamic range of the exponential function is very wide, the redundant divisions are critical to the accuracy of the result in the bottom figure.

In order to eliminate the redundant divisions, we introduce block floating-point numbers to the FFT block, and revise the FFT block as follows.

1. $m := 1, \quad c := 2, \quad d := 1$
2. Do butterfly operations of m-th stage using
$$a' = \frac{1}{c}(a + bW_N{}^k), \quad b' = \frac{1}{c}(a - bW_N{}^k).$$
3. If, in step 2, all a', b' satisfy
$$|a'| < \frac{1}{2} \quad \text{and} \quad |b'| < \frac{1}{2}, \tag{9}$$
then set $c := 1$, otherwise set $c := 2$, $d := d + 1$.
4. $m := m + 1$
5. If $m \leq \log_2 N$ then go to 2, otherwise end.

The variable d signifies the place of the decimal point.

Accodingly, the output and the scaling factor K are revised as

$$x(n) = \frac{1}{2^d} \sum_{k=0}^{N-1} X(k) W_N^{-nk}, \tag{10}$$

$$K = \frac{1}{2^d e^{Ta}}. \tag{11}$$

Comparing these with (5), (6), the improvement achieved by our modification is up-scaling by the factor $N/2^d$.

5. EXAMPLE

Figure 3 presents the improved result of numerical inversion of $F(s) = 3/s$. For this case, d was 3, and thereby the improvement was the factor $N/2^d = 32$.

The scaling factor K and the expected result are

$$K = 0.00377, \tag{12}$$

$$Kf(t) = 3K = 0.0113 = 371/32768. \tag{13}$$

The bottom figure shows that we have the expected result. The values are improved from 4-bit numbers to 9-bit numbers if viewed as 16-bit integers.

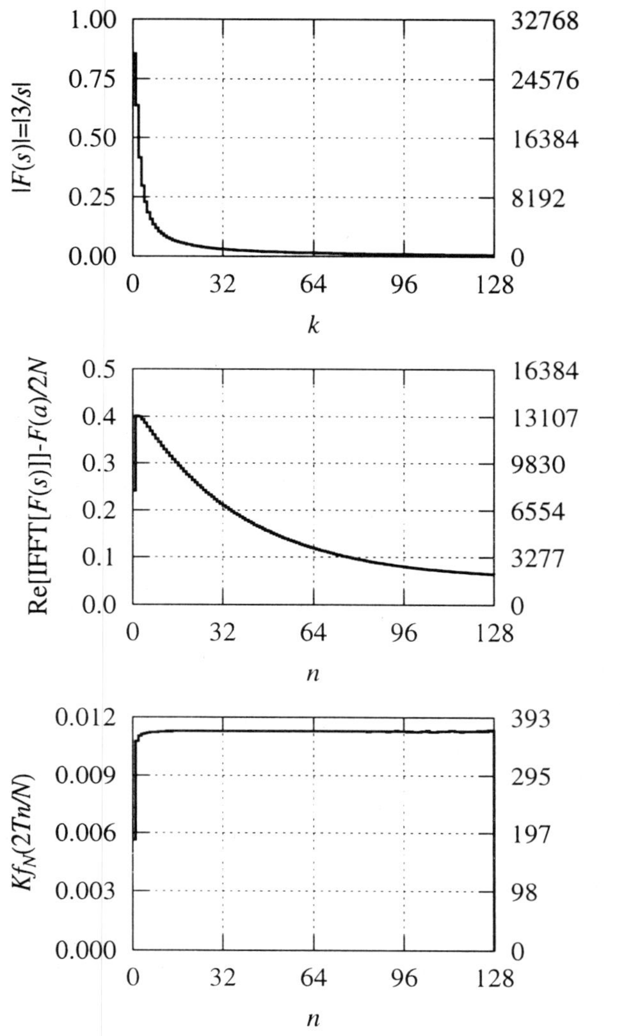

Figure 3: Numerical inversion of the step function $F(s) = 3/s$. ($N = 256$, $T = 1$, $a = 3.5$, $d = 3$, $K = 0.00377$).

6. DESIGN RESULT

This processor is implemented on a 100,000-gate FPGA device EPF10K100ARC240-1 manufactured by Altera, and the hardware description is written in VHDL. The result of the placing and routing is shown in Tab.1.

At the clock frequency 33 MHz, which is a standard clock frequency of the PCI interface, it takes 173 μs to invert one Laplace transform after the data to be inverted are stored in the data memory.

Table 1: Design result

Clock Frequency	36.5MHz
LE usage	38% (38,400 gates)
EAB usage	100% (24Kb)

7. CONCLUSION

We presented an implementation of numerical inversion of Laplace transforms on FPGA. We showed that the straightforward implementation of the FFT-based method with fixed-point numbers causes severely large errors due to the wide dynamic range of the exponential function. To solve this problem, we proposed the use of block floating-point numbers and removed redundant divisions involved in the FFT block with fixed-point numbers. This modification has successfully reduced the errors.

8. REFERENCES

[1] B. Davies and B. Martin, "Numerical inversion of the Laplace transform: a survey and comparison of methods," *J. Comp. Phys.*, vol. 33, pp. 1–32, 1979.

[2] F. Durbin, "Numerical inversion of Laplace transforms: an efficient improvement to Dubner and Abate's method," *Computer J.*, vol. 17, no. 4, pp. 371–376, 1974.

[3] Ichikawa S. and Kishima A., "Applications of Fourier series technique to inverse Laplace transform II," *Mem. Fac. Eng. Kyoto Univ.*, vol. 35, pp. 393–400, 1973, (Japanese).

[4] F. Veillon, "Numerical inversion of Laplace transform," *Commun. ACM*, vol. 17, no. 10, pp. 587–591, 1974.

[5] Kenny S. Crump, "Numerical inversion of Laplace transforms using a Fourier series approximation," *J. ACM*, vol. 23, no. 1, pp. 89–96, 1976.

[6] J. T. Hsu and J. S. Dranoff, "Numerical inversion of certain Laplace transforms by the direct application of fast Fourier transform (FFT) algorithm," *Comput. & Chem. Eng.*, vol. 11, no. 2, pp. 101–110, 1987.

[7] Kohshi Okumura, Akira Kishima, and Setsuo Tokoro, "A method for computing electrical transients of transmission lines by numerical Laplace transform," *IECE Trans. A*, vol. J68-A, no. 2, pp. 107–114, 1985, (Japanese. English translation available).

[8] A. Yonemoto, T. Hisakado, and K. Okumura, "An improvement of convergence of FFT-based numerical inversion of Laplace transforms," *IEEE Proc. ISCAS 2002*, vol. V, pp. 769–772, 2002.

[9] Alan V. Oppenheim and Clifford J. Weinstein, "Effects of finite register length in digital filtering and the fast Fourier transform," *Proc. IEEE*, vol. 60, pp. 957–975, 1972.

A NEW MEMORY REFERENCE REDUCTION METHOD FOR FFT IMPLEMENTATION ON DSP

Yiyan Tang[1], Lie Qian[1], Yuke Wang[1] and Yvon Savaria[2]

1. Dept. of Computer Science
University of Texas at Dallas
Richardson, TX 75083, USA
{yiyan, lqian, yuke}@utdallas.edu

2. Electrical Engineering Dept.
École Polytechnique de Montréal,
P.O.Box 6079, Station, Centre-ville,
Montréal, QC, Canada, H3C 3A7
savaria@vlsi.polymtl.ca

ABSTRACT

Memory reference in digital signal processors (DSP) is among the most costly operations due to its long latency and substantial power consumption. In this paper, we present a new method to minimize memory references due to twiddle factors for implementing any existing fast Fourier transform (FFT) algorithms on DSP processors. The new method takes advantage of previously proposed twiddle factor reduction method (TFRM) and twiddle-factor-based butterfly grouping method (TFBBGM). It can compute two butterflies in one stage of any FFT diagram by loading only one twiddle-factor. Further memory reference reduction is done by computing butterflies with the same twiddle factor at the same time in different stages of the FFT diagram. We have applied the new method to implement radix-2 DIT FFT algorithm on TI TMS320C64x DSP. While using only 50% memory space for storing twiddle factors compared to the conventional DIT FFT implementation, the new method achieves an average reduction in the number of memory references by 79% for accessing the twiddle factors, and 17.5% reduction in the number of clock cycles.

1. INTRODUCTION

The discrete Fourier transform (DFT) plays a vital role in design, analysis, and implementation of digital signal processing algorithms and systems [1]. DFT of a discrete signal $x(n)$ can be directly computed by:

$$X(k)=\sum_{n=0}^{N-1} x(n)W_N^{nk} \qquad k=0,1,\dots,N-1 \qquad (1)$$

where $W_N = e^{-j2\pi/N}$, which is known as the phase or twiddle factor, $j^2=-1$, $x(n)$ and $X(k)$ are sequences of complex numbers.

The fast Fourier transform (FFT) algorithms are a class of efficient algorithms for DFT, which have been studied for a long time. An efficient FFT algorithm [1] was first discovered by Guess and rediscovered by Cooley and Tukey [2] in 1960s. Since then, research in FFT has proliferated. Significant advances include higher radix algorithms [3], mixed-radix [4], prime-factor [5], Winograd (WFTA) [6], split-radix Fourier transform algorithms [7][8], recursive FFT algorithm [9], and the combination of decimation-in-time (DIT) and decimation-in-frequency (DIF) FFT algorithms [10].

The structures of the FFT algorithms are all organized in a similar way as defined in [10], which are based on the principle of decomposing the computation of the DFT into a sequence of smaller size DFTs. For example, to complete the diagram of an *N*-pt radix-2 DIF FFT, the output of the *N*-pt DFT is recursively decomposed until it reaches 2-pt DFTs.

FFT has been implemented on application specific integrated circuits (ASIC) as FFT processors [11] for high speed or low power hardware designs. However, hardware designs of FFT processor are often tailored to fit specific application and lack flexibility. FFT has also been implemented in software on general-purpose processors as part of simulation or data processing systems [12]. Software implementations on general processors are typically much slower than hardware implementations based on comparable technologies. Digital signal processors (DSPs) are a special class of processors that have been optimized for various signal-processing applications including the FFT. Software implementations of FFT algorithms on DSPs are popular because they offer an excellent tradeoff between cost, performance, flexibility and implementation complexity.

However, implementing the FFT effectively on DSPs is not trivial. It has been recognized that the memory references in DSP are expensive due to their long latencies and high power consumption. For example, in the TI TMS320C64x DSP [15], the load operation requires five pipeline execution phases to complete, which takes four delay slots in execution time. The FFT implementation on DSP involves many memory references to access inputs and twiddle factors.

To minimize the memory references to twiddle factors in FFT, two reduction methods called twiddle factor reduction (TFRM) and twiddle-factor-based butterfly grouping (TFBBGM) have been proposed in [13] and [14] respectively. TFRM method reduces the number of necessary twiddle factors from *N*/2 to *N*/4 by the following properties, where $W_N^{nk}=W_N^{(nk)\bmod N}=W_N^m$, $nk \bmod N = m,\ m=0,\dots,N-1$.

$$W_N^m=\begin{cases} W_N^m & m\in(0,N/4) \\ W_N^{N/4}\cdot W_N^{m-N/4}=-j\cdot W_N^{m-N/4} & m\in(N/4,N/2) \\ W_N^{N/2}\cdot W_N^{m-N/2}=-W_N^{m-N/2} & m\in(N/2,3N/4) \\ W_N^{3N/4}\cdot W_N^{m-3N/4}=j\cdot W_N^{m-3N/4} & m\in(3N/4,N) \end{cases} \qquad (2)$$

The above property of twiddle factors can be applied to any FFT algorithm and reduce the number of twiddle factors needed to store in memory.

One consequence of applying the property in Equation (2) is that any FFT algorithm implemented with TFRM can compute two butterflies in one stage of the FFT algorithm diagram by loading only one twiddle factor. The detail of this step is discussed in the next section.

The other memory reduction method, TFBBGM [14], computes many butterflies that share the same twiddle factor in different

0-7803-7761-3/03/$17.00 ©2003 IEEE

stages together. By this means, each twiddle factor is loaded only once to complete the computation for the whole FFT diagram. Details are discussed in the next section as well.

The new memory reference reduction method proposed in this paper takes advantage from both previous methods and reduces the memory references even further. Experiment to apply the new method to radix-2 DIT FFT algorithm implementation shows that much fewer memory references and memory spaces for the twiddle factors are required than any existing FFT algorithms, and the number of clock cycles required to compute the result could be reduced compared to the conventional radix-2 DIT FFT.

In the following, Section 2 describes how to apply the new method to implement radix-2 DIF FFT and DIT FFT algorithms. Experiments results are shown in Section 3. Finally, conclusions are given in Section 4.

2. IMPLEMENT FFT ALGORITHMS WITH REDUCED TWIDDLE FACTORS

As pointed out earlier, the proposed new method works for the implementation of all FFT algorithms on DSP processors. Due to space limitations, we only demonstrate the use of the new method on the two most popular FFT algorithms: DIF FFT algorithms and DIT FFT algorithms.

2.1 Application of the New Method to Implement the DIF FFT Algorithm

The conventional DIF FFT algorithm computes an N-pt radix-2 FFT in $\log_2 N$ stages according to the classic FFT diagrams. The new memory reduction method is applied to the DIF FFT in two major steps.

Step 1: the new method can compute two butterflies together in one stage of the DIF FFT diagram by loading one twiddle factor.

On the stage s, there are $N/2^s$ different twiddle factors, which can be represented by:

$$W_N^{m\times 2^{s-1}} \quad \text{where} \quad m\in[0, N/2^s-1] \qquad (3)$$

Any butterfly in the stage s is composed with the input $x[n]$ and $x[n+N/2^s]$, which is shown in Figure 1 with the corresponding twiddle factor.

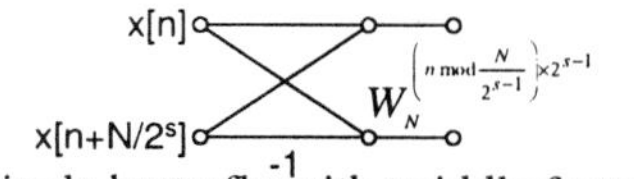

Figure 1 Single butterfly with twiddle factor at stage s

For example, in the second stage of a 8-pt DIF FFT diagram as shown in Figure 2, the butterfly with the input $x[1]$ and $x[1+8/2^2]=x[3]$ uses the twiddle factor $W_8^{\left(1 \bmod \frac{8}{2^{2-1}}\right)\times 2^{2-1}} = W_8^2$.

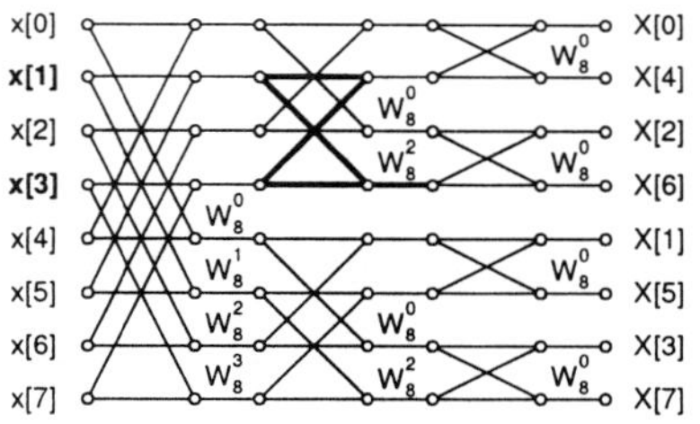

Figure 2 The conventional 8-pt DIF FFT diagram

Theorem: Considering the two butterflies of stage s shown in Figure 3.a, both butterflies can be computed together by loading only one twiddle factor $W^m, m=(n \bmod \frac{N}{2^{s-1}})\times 2^{s-1}$, as shown in Figure 3.b.

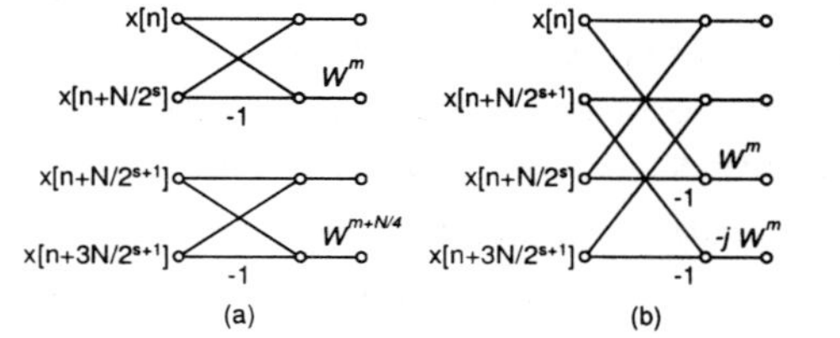

Figure 3 Computing two butterflies together in one stage

Proof: Based on Figure 1, $x[n+N/2^{s+1}]$ in stage s pairs with $x[n+3N/2^{s+1}]$ to form one butterfly, and the twiddle factor used in the butterfly is

$$W_N^{\left(\left(n+\frac{N}{2^{s+1}}\right)\bmod\frac{N}{2^{s-1}}\right)\times 2^{s-1}} = W_N^{\left[\left(n \bmod\frac{N}{2^{s-1}}+\frac{N}{2^{s+1}}\bmod\frac{N}{2^{s-1}}\right)\bmod\frac{N}{2^{s-1}}\right]\times 2^{s-1}} \qquad (4)$$

Since $\left(n \bmod \frac{N}{2^{s-1}}\right) < \frac{N}{2^{s-1}}$ and $\frac{N}{2^{s+1}} < \frac{N}{2^{s-1}}$, we have

$$W_N^{\left[\left(n \bmod\frac{N}{2^{s-1}}+\frac{N}{2^{s+1}}\bmod\frac{N}{2^{s-1}}\right)\bmod\frac{N}{2^{s-1}}\right]\times 2^{s-1}} = W_N^{\left(n \bmod\frac{N}{2^{s-1}}+\frac{N}{2^{s+1}}\right)\bmod\frac{N}{2^{s-1}}\times 2^{s-1}} \qquad (5)$$

if $\left(n \bmod \frac{N}{2^{s-1}} + \frac{N}{2^{s+1}}\right) \le \frac{N}{2^{s-1}}$, the result of the above equation is

$$W_N^{\left(n \bmod\frac{N}{2^{s-1}}\right)\times 2^{s-1}+\frac{N}{4}} = W_N^{\left(n \bmod\frac{N}{2^{s-1}}\right)\times 2^{s-1}} W_N^{\frac{N}{4}} = -j\cdot W_N^{\left(n \bmod\frac{N}{2^{s-1}}\right)\times 2^{s-1}} \qquad (6)$$

else if $\left(n \bmod \frac{N}{2^{s-1}} + \frac{N}{2^{s+1}}\right) > \frac{N}{2^{s-1}}$, the result becomes

$$W_N^{\left(n \bmod\frac{N}{2^{s-1}}\right)\times 2^{s-1}+\frac{N}{4}-N} = W_N^{\left(n \bmod\frac{N}{2^{s-1}}\right)\times 2^{s-1}} W_N^{\frac{N}{4}} W_N^{-N} = -j\cdot W_N^{\left(n \bmod\frac{N}{2^{s-1}}\right)\times 2^{s-1}} \qquad (7)$$

Therefore we have $W_N^{\left(\left(n+\frac{N}{2^{s+1}}\right)\bmod\frac{N}{2^{s-1}}\right)\times 2^{s-1}} = -j\cdot W_N^{\left(n \bmod\frac{N}{2^{s-1}}\right)\times 2^{s-1}}$, which implies that

$$\mathrm{Re}\left[W_N^{\left(n \bmod\frac{N}{2^{s-1}}\right)\times 2^{s-1}}\right] = -\mathrm{Im}\left[W_N^{\left(\left(n+\frac{N}{2^{s+1}}\right)\bmod\frac{N}{2^{s-1}}\right)\times 2^{s-1}}\right] \qquad (8)$$

and

$$\mathrm{Im}\left[W_N^{\left(n \bmod\frac{N}{2^{s-1}}\right)\times 2^{s-1}}\right] = \mathrm{Re}\left[W_N^{\left(\left(n+\frac{N}{2^{s+1}}\right)\bmod\frac{N}{2^{s-1}}\right)\times 2^{s-1}}\right] \qquad (9)$$

The twiddle factors $W_N^{\left(n \bmod\frac{N}{2^{s-1}}\right)\times 2^{s-1}}$ and $W_N^{\left(\left(n+\frac{N}{2^{s+1}}\right)\bmod\frac{N}{2^{s-1}}\right)\times 2^{s-1}} = -j\cdot W_N^{\left(n \bmod\frac{N}{2^{s-1}}\right)\times 2^{s-1}}$ are complex numbers with separated real and imaginary parts, which are stored separately in memory. Therefore by loading $\mathrm{Re}\left[W_N^{\left(n \bmod\frac{N}{2^{s-1}}\right)\times 2^{s-1}}\right]$ and $\mathrm{Im}\left[W_N^{\left(n \bmod\frac{N}{2^{s-1}}\right)\times 2^{s-1}}\right]$ we can compute both butterflies.

Step 2: the new method can compute more butterflies in different stages together by loading only one twiddle factor.

In the first stage of the DIF FFT diagram, there are *N/2* different twiddle factors expressed as W_N^m where $m \in [0, N/2^s - 1]$. Among them, those with twiddle factors as odd number W_N^m where $m = 1,3,5,\ldots,N/2-1$ will not occur anymore in any later stage, due to the fact that at any stage *s*, there are $N/2^s$ different twiddle factors numbered by $W_N^{m\times 2^{s-1}}$ which are even number index. Therefore we load W_N^m where $m = 1,3,5,\ldots,N/2-1$ at the first stage of the computation and compute *N/4* butterflies.

Secondly, we compute the butterflies with twiddle factors W_N^m where $m = 2,6,10,\ldots,(N/4-1)\times 2$. Those butterflies contain *N/8* butterflies in the first stage and *N/4* butterflies in the second stage in the DIF FFT diagram.

At the *s*th step, the new method computes *N/4* butterflies in the stage *s*, *N/8* butterflies in the stage *s*-1, …, and $N/2^{s+1}$ butterflies in the first stage of the DIF FFT diagram by loading only $N/2^{s+1}$ twiddle factors, i.e. W_N^m where $m = 2^{s-1}, 3\times 2^{s-1}, 5\times 2^{s-1}, \ldots, (N/2^s - 1)\times 2^{s-1}$.

After applying the new method to implement the 16-pt DIF FFT, we can redraw the FFT diagram in Figure 4, in which we can notice that the butterflies with the twiddle factor $W_N^{N/4} = -j$ and $W_N^0 = 1$ can be computed without multiplication.

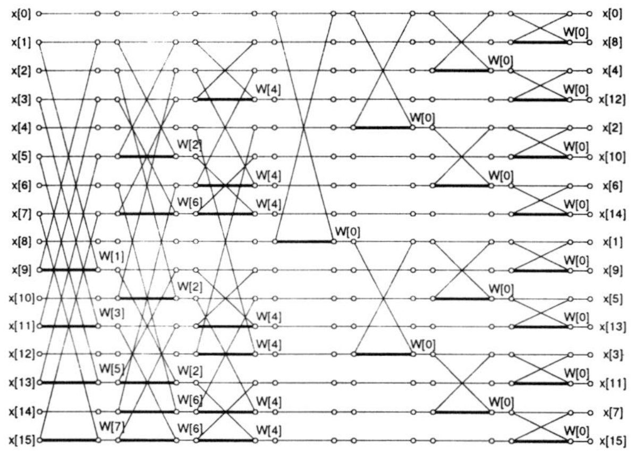

Figure 4 Redraw the DIF FFT diagram after the new method is applied

2.2 Application of the New Method to DIT FFT

The above new method can be applied to implement any existing FFT algorithms. We further show its application on implementing DIT FFT algorithm as an example. The conventional radix-2 DIT FFT diagram is shown in Figure 5.

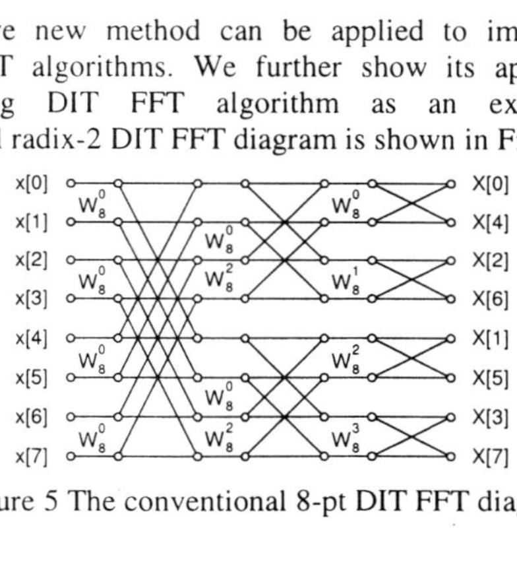

Figure 5 The conventional 8-pt DIT FFT diagram

By applying the new method to DIT FFT diagram, both butterflies in Figure 6.a can be computed together by loading only one twiddle factor $W^m, m = (n \bmod 2^s)\times 2^{s-1}$, shown in Figure 6.b.

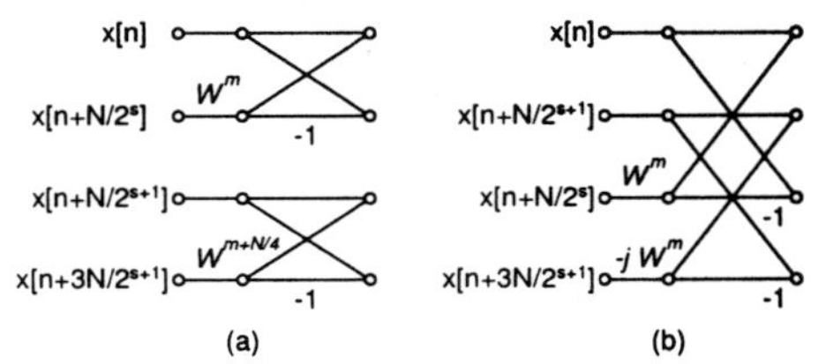

Figure 6 Computing two butterflies together in one stage

Similar to Figure 4 a 16-pt DIT FFT diagram can be redrawn in Figure 7 after applying the new memory reference reduction method.

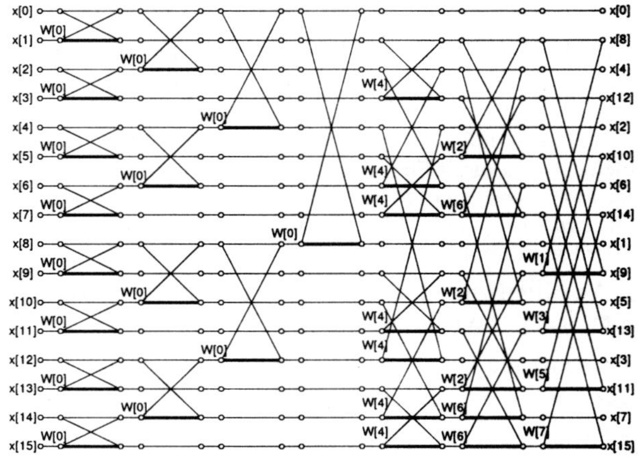

Figure 7 Redraw the DIT FFT diagram after the new method is applied

3. PERFORMANCE EVALUATION OF THE NEW METHOD

The number of memory references to twiddle factors in conventional *N*-pt radix-2 FFT algorithm is $\log_2 N \times N/2$. Applying TFRM alone reduces the memory reference to twiddle factors in each stage by half, and computes the last two stages without multiplication, so the number of memory references for twiddle factor is $(\log_2 N - 2)\times N/4$. On the other hand by applying TFBBGM alone, the twiddle factors are loaded only once and the butterflies that multiply by 1 are computed separately. Therefore, the number of memory references is reduced to $N/2-1$.

After applying the new method, only *N/4* twiddle factors are loaded once to compute all the butterflies. The number of memory references to twiddle factors for a 1024-pt FFT is listed in Table 1,where the new method use 95% fewer memory references for twiddle factor than the conventional one, 87% less than TFRM alone, and 50% less than TFBBM alone.

Table 1 Memory references of twiddle factors for 1024-pt FFT

	Conventional	TFRM	TFBBGM	New Method
Mem. Ref.	5120	2048	511	255

We have applied the new method to implement the radix-2 DIT FFT. The chosen platform is the TI TMS320C64x DSP core, which is a fixed-point DSP core with enhanced VLIW (Very Long

Instruction Word) architecture. The C64x core has eight functional units that can work in parallel to complete a maximum of eight operations; two register files with each 32 32-bit registers, 32-bit internal communication bandwidth and maximum 600MHz clock frequency.

Two pieces of C codes are compiled with the different compilation options in the Code Composer Studio (CCS) v1.2 for C6000. The results of executions of the codes are summarized in Table 2. The classic code is the DIT FFT code from TI's DSP library. The proposed code is written based on the result of applying the new method to conventional DIT FFT algorithm implementation. The codes are run in the CCS as sub-functions and the number of clock cycles needed to run the DIT FFT sub-function is precisely measured with different number of input points using the break point function in CCS.

The experimental results show that the radix-2 DIT FFT algorithm implementation with the new memory reference reduction method can produce 88.2% fewer memory references, and consume 21.5% fewer clock cycles. This is associated with 50% saving in the memory space for twiddle factors.

Table 2 Summary of the experimental results

# pt	Classic (cycles)	Proposed (cycles)	Clock cycle reduction	# of memory access for twiddle factor	Memory access reduction
Optimization option -o3 file					
16	1423	1216	14.5%	3	62.5%
32	3133	2883	8.0%	7	85.4%
64	6859	6646	3%	15	88.2%
Optimization option –o2 function					
16	1423	1216	14.5%	3	62.5%
32	3133	2883	8.0%	7	85.4%
64	6859	6646	3%	15	88.2%
Optimization option –o1 local					
16	2462	1901	22.7%	3	62.5%
32	5872	4768	18.8%	7	85.4%
64	13682	11511	15.9%	15	88.2%
Optimization option –o0 register					
16	2724	2258	17.1%	3	62.5%
32	6470	5608	13.3%	7	85.4%
64	15016	13422	10.6%	15	88.2%
No Optimization					
16	5241	4112	21.5%	3	62.5%
32	12339	10220	17.2%	7	85.4%
64	28445	24444	14%	15	88.2%

4. CONCLUSION

In this paper, we have presented a new memory reference reduction method for the twiddle factors to implement existing FFT algorithms. The new method takes advantage from two previous memory reference reduction methods called TFRM and TFBBGM. The new method can compute two butterflies in one stage of any FFT diagram by loading only one twiddle-factor. Further memory reference reduction is done by computing butterflies with the same twiddle factor at the same time in different stages of the FFT diagram. Experimental results on the TI TMS320C6414 DSP showed that the radix-2 DIT FFT algorithm implementation with the new memory reference reduction method requires much fewer memory references to twiddle factors compared to the conventional radix-2 DIF FFT implementation. Among the benefits from the combination of the reduction methods, power efficient implementations on hardware platforms of FFT processors can be expected.

REFERENCE

[1] C.S. Burrus and T.W. Parks, "DFT/FFT and Convolution Algorithms and Implementation," NY John Wiley & Sons, 1985.

[2] J.W. Cooley and J.W. Tukey, "An algorithm for the machine calculation of complex Fourier series," *Math. Compu.*, vol. 19, pp. 297-301, 1965.

[3] G.D. Bergland, "A Radix-Eight Fast-Fourier Transform Subroutine for Real-Valued Series," *IEEE Trans. Electroacoust.*, vol. 17, no. 2, pp. 138-144, June 1969.

[4] R.C. Singleton, "An Algorithm for Computing the Mixed Radix Fast Fourier Transform," *IEEE Trans. Audio Electroacoust.*, vol. 1, no. 2, pp. 93-103, June 1969.

[5] D.Pl. Kolba and T.W. Parks, "A Prime Factor FFT Algorithm Using High-Speed Convolution," *IEEE Trans. Acoustic, Speech, Signal Processing*, vol. 25, no. 4, pp. 281-294, Aug. 1977.

[6] S. Winograd, "On Computing the Discrete Fourier Transform," *Math. Comput.*, vol. 32, no. 141, pp. 175-199, Jan. 1978.

[7] P. Duhamel, and H. Hollmann, "Split Radix FFT Algorithm," *Electronics Letters*, vol. 20, pp. 14-16, Jan. 5, 1984.

[8] D. Takahashi, "An Extended Split-Radix FFT Algorithm," *IEEE Signal Processing Letters*, vol. 8, no. 5, pp. 145-147, May 2001.

[9] A. R. Varkonyi-Koczy, "A Recursive Fast Fourier Transform Algorithm," *IEEE Trans. Circuits and Systems, II*, vol. 42, pp. 614-616, Sep. 1995.

[10] A. Saidi, "Decimation-in-Time-Frequency FFT Algorithm," *Proc. ICAPSS*, pp. III:453-456, April 1994.

[11] B.M. Baas, "A low-power, high-performance, 1024-point FFT processor," *IEEE J. Solid-State Circuits*, vol. 34, issue 3, pp. 380 -387, March 1999.

[12] Mathwork Inc., Matlab function reference – FFT, http://www.mathworks.com/access/helpdesk/help/techdoc/ref/fft.shtml?BB=1

[13] Y. Jiang, Y. Tang, and Y. Wang, "Twiddle-factor-based FFT algorithm with reduced memory access," Proc. IDPDS, pp. 653-660, 2002.

[14] Y. Tang, L. Qian, Y. Wang, and Y. Jiang, "Twiddle Factor Based Memory Reduction Method for FFT Implementation on DSP," to be published.

[15] Texas Instrument, "TMS320C64x DSP Library Programmers'Reference (Rev. A)," *SPRU565A*, 2002.

REAL-TIME ACQUISITION AND TRACKING FOR GPS RECEIVERS

Abdulqadir Alaqeeli, Janusz Starzyk, Frank van Graas

School of Electrical Engineering and Computer Science
Ohio University
Athens, OH 45701

ABSTRACT

Current GPS receivers spend much time in base-band processing performing acquisition and tracking. This is due to the large number of required operations in the software-based signal processing. This paper presents a novel signal acquisition and tracking method that reduces the number of operations, simplifies hardware implementation and decreases the acquisition time. The implementation of this method in an FPGA provides very fast processing of incoming GPS samples that satisfies real-time requirements.

1. INTRODUCTION

The Global Positioning System (GPS) has been widely used in civilian and military positioning, velocity, and timing applications. GPS transmits its standard positioning service (SPS) on the L1 carrier frequency of 1575.42 MHz. Each GPS satellite (or transmitter) has a unique spreading gold code (C/A code) that is orthogonal to all the other satellites' codes. GPS receivers, on the other hand, must search for these C/A codes to know which satellites are available to the user. For each code, a receiver must perform a 2-D search for carrier frequency offset and code shift, or in other words acquire the C/A code. Then it should track (or lock in) the signal. Acquisition is the most time consuming operation in the GPS receiver [1-2]. A fast search algorithm was presented by Van Nee in [1]. This method searches all possible code shifts in one step using the well-known FFT-based correlator. It performs the correlation using frequency domain multiplication [1].

When a SPS GPS signal is received, it goes through many steps of filtering and amplification. Then the signal is digitized using at least a 5MHz sampling rate which obeys Nyquist's law for the C/A code, which has a bandwidth of 2.046 MHz [3]. This leaves the GPS signal as 5,000 samples of C/A code for each 1-ms period of the code, which could be downconverted to an intermediate frequency of 1.25 MHz [4]. From this point, the GPS receiver starts its digital signal processing (DSP) algorithms to find the C/A codes. Different search algorithms were presented in [1][3-5]. Assuming that the receiver knows the C/A code and the exact intermediate frequency, it needs to do a carrier wipe-off before performing the FFT-based correlation function with the locally-generated C/A code.

The FFT-based correlator is chosen for the implementation of the acquisition process as part of the block processing method [6]. This way, the gained advancements in block processing could be used. Building power-of-two-based FFTs is achieved by using a uniform butterfly structure. However, 5,000-point FFTs are required to implement the C/A code correlator. The size of this FFT is not a power-of-two, so it will be very difficult to implement since it requires a mixed-radix algorithm that also includes non–power-of-two small FFTs [7]. The use of smaller FFTs is needed to map the acquisition process in the FPGA. A novel method for solving this problem was first presented by Starzyk and Zhen in [8]. This method is called acquisition of C/A code using averaging correlators, and it uses FFTs with a similar size as the C/A code (1,023 instead of 5,000-point FFTs).

2. AVERAGING CORRELATOR

The averaging method averages the incoming 5,000 samples to become 1,023 averaged samples. If the starting point of the averaging operation is chosen in the right place, the 1,023 new samples may represent the original chips of the C/A code. In other words, one can say that a full C/A code is presented by 5,000 samples and that its chips are represented by either 4 or 5 samples each. Since the C/A code is circular, therefore, the right averaging starting point is one of 5 successive samples. This will generate 5 of the 1,023 averaged-samples code. One of the generated sequences is considered a good approximation of the original C/A code (or the best recovery) [8]. This best recovered averaged sequence contains the strongest peak among the other four and estimates the code phase in chips (1/1023 ms). The resultant five correlation functions are a good approximation of the 5,000-point correlation function and thus can be used with a triangle fitting to refine the code phase.

This method did not reduce the peak to the second peak value[8]. As a result, detection probability is not affected by replacing the 5,000-point FFT-based method with the averaging method. The acquisition time is reduced by using this method, because calculating five 1,023-point FFTs and IFFTs requires less time (in software and in hardware) than the 5,000-point FFTs and IFFTs. However, implementing 1,023-point FFT (or IFFT) is not an easy task since it is not a power of two. A zero-padding-based solution was introduced by Zhen in [9]. The signal energy loss on average was acceptable. However, the computation time or the necessary hardware resources are increased to a certain level that makes it useless, especially when real-time acquisition is required. One possible solution to the problem of size mismatching of the code and the available Xilinx's 1024-point FFT core is presented in the next section. The new approach will show a fast acquisition method that can be implemented much easier than the 5,000-point FFT-based method.

0-7803-7761-3/03/$17.00 ©2003 IEEE

3. MODIFIED C/A CODE

The size incompatibility of C/A code and the available FFT core can be solved by changing the down-sampling rate from 1,023 to 1,024. So, the 5,000 samples will be down sampled (or averaged) to 1,024 points. A similar procedure will be done to the local code. Therefore, the local code will be up-sampled to 5,000 and then down sampled to 1,024 points. Therefore, the averaging correlator here will use 1,024 averaged samples and 1,024-point modified C/A code. The resultant code has changed from a binary code to a multilevel code which contains values as ±1.0, ±0.8, ±0.6, ±0.5, ±0.4, ±0.2, and 0.0. However, this modified C/A code is still considered a unique code related to the selected C/A code. It cannot be generated from a different C/A code.

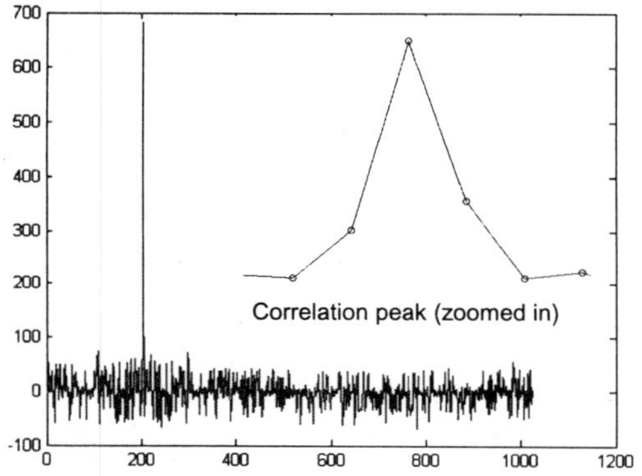

Figure 1. Winner autocorrelation function of modified C/A code

As stated above, the averaging correlator needs to be applied five times with five successive starting samples. When the five correlation functions for this averaged C/A code are generated, the correlation function that contains the strongest peak keeps the important information necessary for acquisition. Fig-1 shows the winner autocorrelation function of the modified C/A code. The peak is not based on equilateral triangle, as what it should be in a normal C/A code, but it is good enough to roughly estimate the code phase for acquisition. One may notice that each point in this correlation function is one out of 1,024 modified chips. Each modified chip is a 1/1024 ms while the original chip width is a 1/1023 ms. To estimate the code phase, one needs to consider this change in chip width and should correct the value of the code phase. For example, if the peak is at location 120 out of 1,024, then the estimated code phase equals to (120/1024)*1023 original chips, or in other words, the code phase is (120/1024)*1ms.

The averaged-code correlator needs five times 2*1024*10 plus 5,120 multiplications and five times 2*1024*10 additions. Therefore, the total required number of operations equals to 107,520 multiplications and 102,400 additions. Also, the averaging computation requires about 25,000 additions and 5,120 divisions. The division operations will not be counted because they can be avoided. Therefore, the total number of operations is 107,520 multiplications and 127,400 additions. This does not reduce the computation time much compared to the 5,000-point FFT-based correlator, but the implementation of 1,024-point FFT is much simpler than the 5,000-point FFT. This will lead to significant simplification in the hardware implementation.

In order to use this method, the effects on signal-to-noise ratio and the other characteristics should be studied. The next section presents the characteristics of the modified-code averaging correlation method in terms of signal-to-noise ratio (SNR) loss, code phase accuracy and carrier phase accuracy.

4. CHARACTERISTICS OF MODIFIED CODE AVERAGING CORRELATION

In the previous section, the modified code averaging correlation method was described. However, in order to use it for the acquisition process and the tracking loops, it is necessary to check its effect on signal power and on calculation accuracy of code and carrier phases. For this reason, a 200ms of GPS data is used to test the acquisition using the averaged-code averaging method. The peak-to-peak ratio was chosen as a measure of signal strength. Matlab simulation showed that the average loss in peak-to-peak ratio is about 12.7%, as shown in Fig-2. This loss is about 0.5dB, and is acceptable in most of the cases, except in the case of weak signal acquisition.

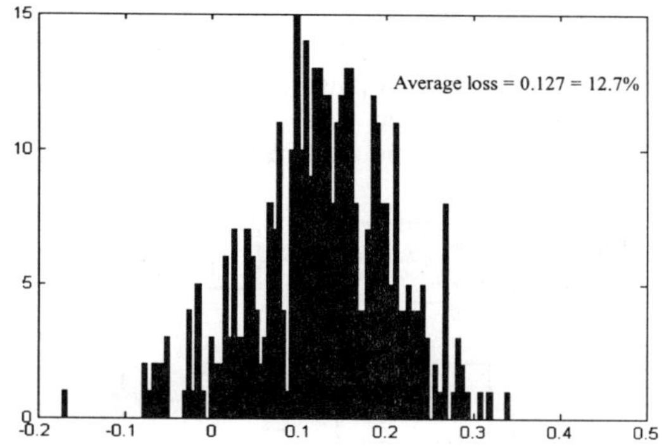

Figure 2. peak-to-peak loss using modified C/A code for 400ms

For the acquisition process, code phase estimation accuracy needs to be within ± ½ chip. The averaging method was able to produce the right code phase with accuracy of 100ns (± 0.1 of a chip) as seen in Fig-3a. Therefore, the averaging method is acceptable for acquisition. However, tracking requires that the block-processing produce refined code phase with high accuracy. For this reason, the triangle fitting approach was applied to the peaks of the averaging method. This approach was able to refine the code phase estimation and the average error in code phase estimation using triangle fitting was 42.7ns. A modified triangle fitting with acquisition history feedback was developed and showed an improvement over triangle fitting (with average error of 10ns) in the code phase calculation accuracy. However, none of these results has sufficient accuracy to replace the real-time tracking. Moreover, the carrier phase estimation based on the block processing using averaging method was tested. The accumulated carrier phase error was 0.03 radians when computed from Fig-3b. The accumulated carrier phase error was considered acceptable for C/A code tracking. However, the limiting factor is the code phase error, which was found not acceptable. Therefore, the modified-code averaging correlator method was selected for the implementation of the acquisition process only.

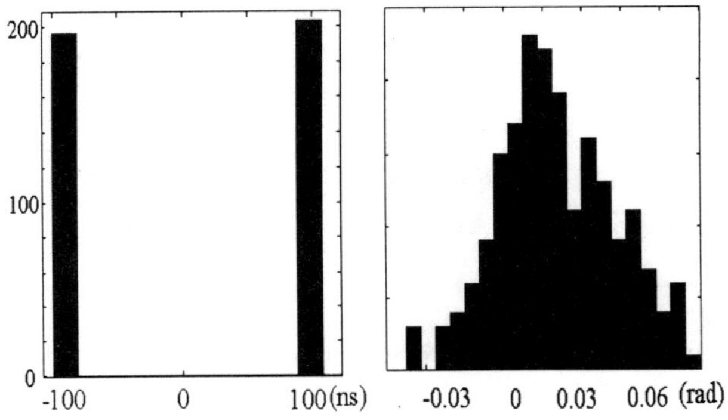

Figure 3. (a) code phase error (b) carrier phase error

5. IMPLEMENTATION OF BASE-BAND PROCESS OF GPS RECEIVER

In the previous section, we showed that the computational effort of the averaging correlator with the modified-code approach is small compared to regular 5,000-point FFT-based correlators. However, the implementation of 5 1024-point FFTs is much simpler than designing a single 5000-point FFT. Therefore, the averaging correlator is our choice for the implementation of the acquisition process.

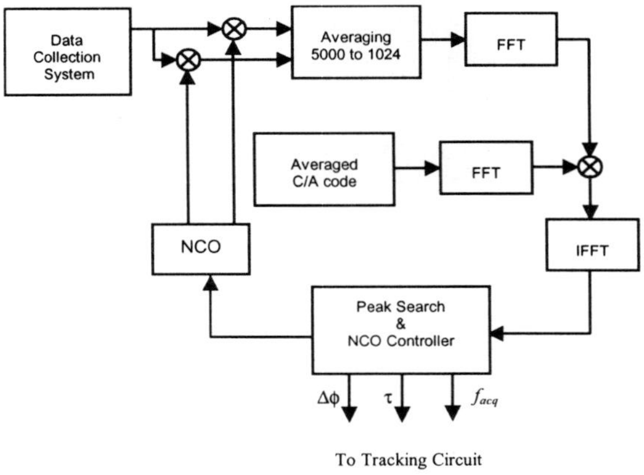

Figure 4. Acquisition using averaging-correlator

A block diagram of the averaging-correlator-based acquisition is shown in Fig-4. First, the samples are multiplied by in-phase and quad-phase components of the carrier signal. The I and Q channels are each averaged to 1,024 points. Then they are converted to frequency domain using 1,024-point complex FFT and multiplied by the conjugate of the FFT of the averaged-local-code. A 1,024-point complex IFFT is used then to return back to the time domain. A peak searcher inspects the 1,024 outputs of the IFFT and stores the location of the peak and its value. This process is repeated four times more, each with a different starting point, as described in the previous section. After checking all five loops, the peak searcher compares the peak value to a threshold to decide if the peak (or a GPS code) is detected. If the searcher does not detect a peak, then a new search cell with different frequency bin is inspected using the above- described process. This is repeated until a true peak is found or until all frequency search bins are completed. When the GPS signal is acquired, the frequency is selected and the search is conducted only in the code-phase dimension.

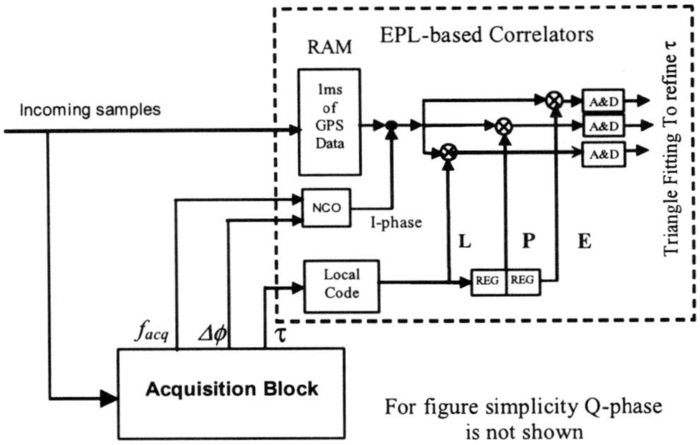

Figure 5. Block diagram of the implemented base-band

Every 1ms the acquisition process will produce estimate of the code phase, carrier phase, frequency, and peak value. These values are accurate enough to guide three 5000-point serial correlators for the early-prompt-late (EPL) implementation of delay-locked-loop (DLL) [2-3]. Fig-5 shows the EPL-based correlators and how they communicate in the whole system. They provide the three correlation points around the true peak and then used with triangle fitting technique to refine code and carrier phases. This method enhances the code phase accuracy to few nano-seconds [6]. One of the reasons behind using the serial correlators is that they occupy small silicon area or fit easily in an FPGA. These correlators do not require closed loops, like the ones in the standard tracking loops, because the acquisition component provides all the necessary information to keep the serial correlators in-lock with the signal.

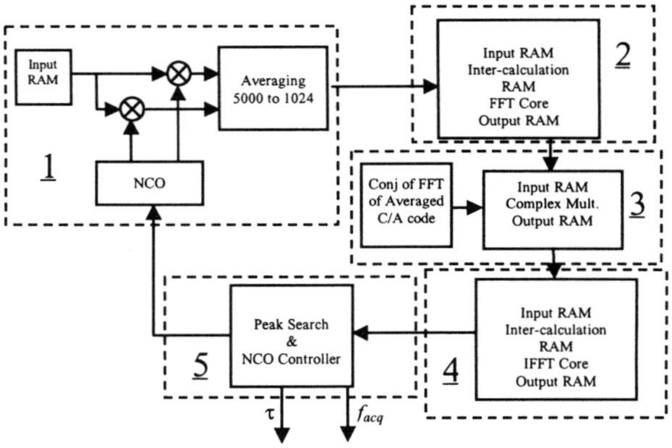

Figure 6. Partitions of Acquisition Process

The FPGA platform we used has a Virtex FPGA that provides 800k logic gates. This FPGA cannot implement the whole base-band system. Therefore, the system was partitioned into smaller parts as shown in Fig-6. Each block was mapped to the FPGA and was tested separately. These blocks were also tested in sequence with real GPS data and acquisition and tracking were achieved. Fig-7 shows the hardware simulation for the correlation functions of the averaging-based acquisition. Five peaks are shown and they are identical to Matlab simulation using fixed-point operations.

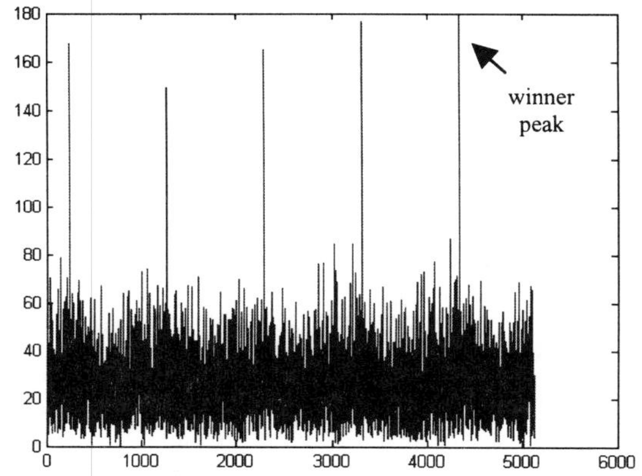

Figure 7. Averaging Correlation Hardware Results

The required FPGA resources (or implementation area) for each part of the design are shown in Table-1. In order to estimate the required FPGA resources for the whole system, we cannot simply add of the numbers of the logic slices used and the used block RAMs of all the blocks. This is because each block contains data collection, intermediate calculation handling, and result storage that can be reduced for the complete system implementation to avoid redundancy. Therefore, estimation of the required FPGA resources of the whole system which implements acquisition and tracking is about 8,000 programmable logic slices and 114 Block RAMs. With the availability of such resources in one FPGA such as the VirtexE1600, the implementation of such system is possible. The whole design may take about 50 % of the available slices and 80 % of the available Block RAMs. This leaves more room for routing that can directly affect the maximum clock speed. Routing in FPGA is one of the main problems since it is not predictable. However, based on the already implemented blocks and the above estimations, a system clock of 45MHz may be used with no timing problems. The whole system requires about 45,000 clock cycles. If the clock speed is equal to or more than 45MHz, the averaging correlation of 1-ms GPS data is computed in 1-ms or less. Assuming that the carrier frequency is known, the acquisition is conducted in the code phase dimension and therefore processing blocks of 1-ms GPS data can be achieved in real time.

Table 1. Implementation Cost (Virtex Resources)

	Acquisition					Serial Correlators (zooming)
Part.#	1	2	3	4	5	
Configurable Logic Slices (max. 9408)	704	2288	649	2288	862	697
Block RAMs (max. 28)	25	24	18	24	8	17

6. Summary

This paper presented the current challenges for designing a real-time base-band block processor for a GPS receiver using FPGA Technology. The two main processes, acquisition and tracking, were described. Speeding up the GPS block processing was the goal of this paper. A fast and simple-to-implement acquisition algorithm based on averaging-correlation method was presented. This gave acceptable accuracy to guide serial correlators to zoom in for better accuracy similar to block processing technique. The algorithm was implemented using acquisition-driven-correlators architecture for block processing and was mapped to an FPGA. The implementation architecture satisfies real-time processing and tracking accuracy that makes it a useful core in a GPS receiver for real-time applications.

7. REFERENCES

[1] Van Nee, D. and Coenen, A., "New fast GPS code-acquisition technique using FFT," *Electronics Letters*, vol. 27, no. 2 pp. 158-160, January 17, 1991.

[2] M. Braasch and A. J. Van Dierendonck, "GPS Receiver Architectures and Measurements," *Proc. of The IEEE,* vol. 87, no. 1, pp. 48-64, January 1999.

[3] Akos, D., "A Software Radio Approach to Global Navigation Satellite System Receiver Design," *Ph.D. Dissertation*, Ohio University, August 1997.

[4] Tsui, J., *Fundamentals of Global Positioning System Receivers: A Software Approach*, John Wiley & Sons, Inc., 2000.

[5] D. Lin and J. Tsui, "Comparison of Acquisition Methods for Software GPS Receiver," *ION GPS-2000*, pp. 2385-2390, Salt Lake City, UT, September 2000.

[6] G. Feng, and F. van Graas, "GPS Receiver Block Processing", *Proc. ION GPS-99*, pp. 307-316, Nashville, September 1999.

[7] Sanjeev, G., "Feasibility Study for the Implementation of Global Positioning System Block Processing Techniques In Field Programmable Gate Arrays," *MS. Thesis*, Ohio University, June 2000.

[8] J. Starzyk and Z. Zhu, "Averaging Correlation for C/A code Acquisition and tracking in Frequency Domain," *MWSCS,* Fairborn, Ohio August 2001.

[9] Zhu Z., " Averaging Correlation for Weak GPS Signal Processing," *MS. Thesis*, Ohio University, April 2002.

A Novel Sampling Process and Pulse Generator for a Low Distortion Digital Pulse-Width Modulator for Digital Class D Amplifiers

Bah-Hwee Gwee, Joseph S. Chang, Victor Adrian and Haryanto Amir
School of Electrical and Electronic Engineering
Nanyang Technological University, Singapore 639798

ABSTRACT

We propose a novel Direct Interpolation (DI) sampling process and novel hybrid pulse generator for a Pulse Width Modulator (PWM) for a digital Class D amplifier. The DI sampling process features the lowest harmonic distortion compared to other algorithmic sampling processes. The computation for the DI process is simple, leading to small IC area and very low power dissipation (7μW@1.1V). We analytically derive the double Fourier expression for the DI process and show that the THD is very low (<0.03%). The novel hybrid pulse generator embodies a counter, noise shaper, and novel 1-bit frequency doubler. The complete PWM features a simple circuit implementation (small IC area), micropower low voltage operation (~60μW@1.1V), low sampling rate (48kHz) and low harmonic distortion (~0.03%).

1. INTRODUCTION

Digital Class D amplifiers (amps) are increasingly prevalent as power amps in audio applications, in particular audio portable devices where the critical parameters including low voltage (1.1V – 1.4V) and low power (<1mA) operation and small Integrated Circuit (IC) area. An example of such an application includes the digital hearing instrument (hearing aid) whose total quiescent current budget is approximately 1,000μA@1.1V, and because of this tight current budget, most of the power should be allocated to complex signal processing (such as noise reduction [1]) as opposed to signal conditioning. The Class D amp is particularly advantageous in this application because when properly designed [2], it features high power efficiency (of the order of 90%) over a large modulation index range (signal swing). The digital Class D amp is also advantageous when interfaced to a digital processor (for example, in a digital hearing instrument) because the need for a digital-to-analog (D/A) converter is eliminated, hence the immediate power savings and reduced hardware.

In general, the Class D amp, as depicted in Fig. 1, comprises a Pulse Width Modulator (PWM) and an output stage. The output stage drives a load consisting of a low-pass filter and an output transducer. We have earlier described a methodology [2] to optimize the design of the output stage of the Class D amp for maximum power efficiency (and that results in a small IC area). In general, the PWM output can be generated by a direct approach or an indirect approach. The direct approach usually involves two steps as depicted in Fig. 1. The first step is a sampling process to determine the digital value of the PWM pulse width that corresponds to the sampled value of the input signal. The second step is to generate the corresponding PWM pulses based on the digital value provided by the sampling process. The indirect approach [3] usually involves a modulation process including oversampling by interpolation and delta-sigma modulation. The required computation for the indirect method is rather intensive and does not render well for portable power sensitive applications.

In this paper, we will restrict our work to the direct PWM approach because of its simpler computation, low power dissipation and small IC area. We will first provide a brief overview of the sampling process and we will describe our proposed direct interpolation (DI) sampling process (an

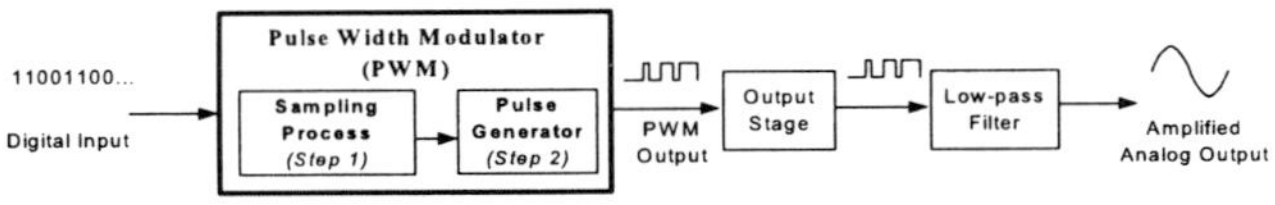

Fig. 1 A digital Class D amplifier

algorithmic process) that features simple computation (comparable to other sampling processes in the direct approach and much simpler than the indirect approach) and has very low distortion (~0.03% and superior to other direct PWM approaches, and comparable (if not superior) to indirect PWM approaches). We will further analytically derive the double Fourier series expression for our DI sampling process to show that the theoretical THD is very low. In this paper, we will also describe a novel hybrid pulse generator embodying a noise shaper, a novel frequency doubler and a medium wordlength (10-bit) counter. Of particular interest, our proposed frequency doubler effectively increases the resolution of the preceding counter by one bit and we achieve this with virtually no overhead. The overall implementation yields a pulse generator with an equivalent 16-bit precision.

We verify our proposed DI sampling process and novel hybrid pulse generator by computer simulations and on the basis of experimental measurements using a Complex Programmable Logic Device (CPLD) and a prototype IC embodying a Class D output stage. Our experimental results show that the measured THD is ~0.03% and agrees well with our theoretical derivations. Based on computer simulations using parameters from a commercial 0.3μm CMOS fabrication process, our Class D amplifier embodying the proposed DI sampling process cum pulse generator is micropower (~60μW) and operates at low voltage (1.1V), rendering it suitable for a digital hearing instrument.

2. PROPOSED DI SAMPLING PROCESS

2.1 Review of Digital PWM Sampling Processes

In this section, we will briefly review the reported sampling processes. This review provides an insight into the differences between the various PWM sampling processes and our proposed DI sampling process, and how we reduce the harmonic distortion in our DI sampling process to a very low level (<0.03%).

At the outset, it is worth noting that an ideal PWM [4] of the Class D amp outputs a PWM signal with zero Total Harmonic Distortion (THD). Reported PWM sampling processes include the Natural Sampling (NS), Uniform Sampling (US), Linear Interpolation (LI) Sampling [5] and more recently, our Delta-

0-7803-7761-3/03/$17.00

Compensation (δC) Sampling [6]. The NS and US sampling processes are rather academic as the former features a near ideal process but requires an inordinately high sampling rate (e.g. for a 16 bit resolution at 48kHz sampling ⇒ 48kHz x $2^{16} \approx$ 3.15GHz is required) while the latter process suffers from intolerably high THD (typically 2%) at low sampling rates (e.g. 48kHz). The LI and δC processes, being algorithmic sampling processes, are more practical (same low sampling rate as US) and the LI process exhibited a slightly lower THD but at the cost of more complex computation and hardware [6]. In [6], we have summarized and derived the double Fourier expressions depicting the harmonic distortion coefficients of the NS, US and δC processes. We can compare these coefficients with our proposed DI sampling process that is derived in this paper.

2.2 Proposed Direct Interpolation Sampling Process

At the outset, we state that the objective of our DI process as in other algorithmic-based sampling processes (including [5,6,7]) is to obtain a pulse width that is as close to the Natural sampling process as possible, but at substantially reduced sampling rate (as the US) and with simple computation.

We depict in Fig. 2 how the sampled points are obtained for the different sampling processes. The magnitude of sampled points S_1 and S_2 as well as the sampling period, T, are known quantities. We propose that the DI sampling algorithm interpolates the modulating signal (arc S_1S_2) by connecting the two sampled points with a straight line BE. The resultant intersection of the straight line with the carrier signal (S_{DI}) is a close estimation of the Natural sampling (S_{NS}) process. Based on S_{DI}, we can derive the pulse width, t_p, in terms of the sampling period, T, and we will now describe this.

By noting that the length of CH and CJ are t_p and $T\text{-}t_p$ respectively, we use a pair of similar triangles, i.e. triangles CHB and CJE to obtain

$$\frac{t_p}{T-t_p} = \frac{CH}{CJ} = \frac{CB}{CE} \qquad \text{- (1)}$$

Now, since CB is in triangle CBA and CE is in triangle CED and both triangles are another pair of similar triangles, we obtain

$$\frac{t_p}{T-t_p} = \frac{BA}{DE} \qquad \text{- (2)}$$

$$\Rightarrow \quad t_p(DE) = (T - t_p)(BA) \qquad \text{- (3)}$$

$$\Rightarrow \quad t_p = T\frac{BA}{DE+BA} \qquad \text{- (4)}$$

where BA is the magnitude between the minimum normalization point (i.e. 0) and S_1, and DE is the magnitude between the maximum normalization point (i.e. 1) and S_2, that is the magnitude of DE is 1- S_2.

We finally obtain the pulse width t_p of the DI process:

$$t_p(DI) = T\frac{S_1}{1+S_1-S_2} \qquad 0 \le V_c \le 1 \qquad \text{- (5)}$$

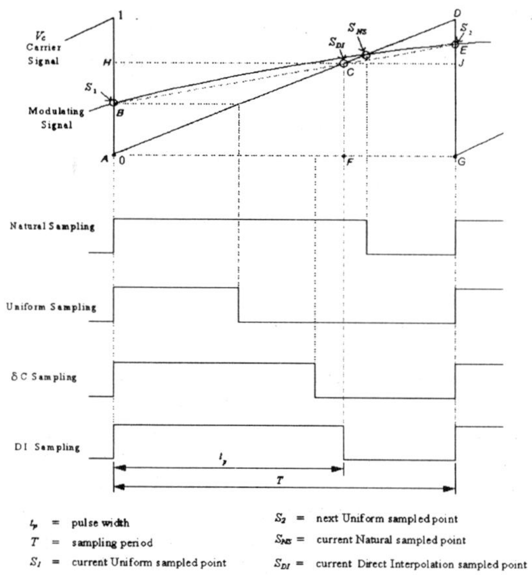

Fig. 2 The Natural, Uniform, δC and DI sampling processes.

It is instructive at this juncture to note the simplicity of the proposed DI sampling process. It is also instructive to note that the same pulse width will be obtained for the case where the modulating input signal has a negative gradient.

We depict in Fig. 3 our design of the block diagram hardware to realize the abovementioned simple arithmetic operation of the DI process. The result of $S_1/(1+(S_1-S_2))$ will be available in the quotient output of the divider. We remark that the actual divider circuit required can be greatly simplified compared to a normal divider but due to the brevity of this paper, we will not discuss its details here.

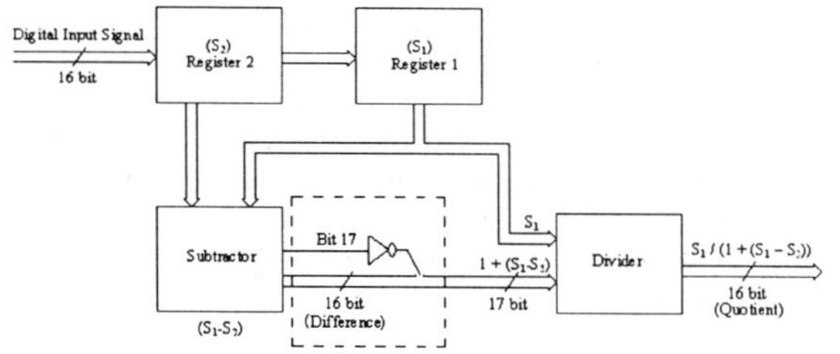

Fig. 3 Hardware implementation for the DI sampling process

2.3 Spectral Analysis of the Proposed DI Sampling Process

In view of the simplicity of the DI sampling process, it is instructive to analytically determine the extent of its harmonic distortion. By means of the double Fourier series [4] analysis, we derive the resultant double Fourier series expression for the DI single-sided trailing edge PWM (of a cosine input modulating waveform) sampling process as (refer to [6] for the definition of symbols)

$$\begin{aligned} F_{DI}(t) = K_{00} &- \sum_{n=1}^{\infty} \frac{\omega_c I_{0n}}{2\omega_s n\pi^2} \sin(n\omega_s t - 2n\frac{\omega_s}{\omega_c}\pi k) \\ &+ \sum_{m=1}^{\infty}\left[\frac{\sin m\omega_c t}{m\pi} - \frac{I_{m0}}{2m\pi^2}\sin(m\omega_c t - 2m\pi k)\right] \\ &- \sum_{m=1}^{\infty}\sum_{n=\pm 1}^{\pm\infty} \frac{I_{mn}}{2(m+\frac{n\omega_s}{\omega_c})\pi^2}\sin(m\omega_c t + n\omega_s t - 2m\pi k) \end{aligned} \qquad \text{- (6)}$$

where

$$I_{0n} = \int_0^{2\pi} \exp\left\{ -jn\Phi \left[\frac{2Q\pi\cos\Phi - 2BQ\sin\frac{\pi}{p}\sin(\Phi + \frac{\pi}{p})}{2\pi + 2Q\sin\frac{\pi}{p}\sin(\Phi + \frac{\pi}{p})} \right] \right\} d\Phi$$

$$I_{m0} = \int_0^{2\pi} \exp\left\{ -jm\left(\frac{2Q\pi\cos\Phi - 2BQ\sin\frac{\pi}{p}\sin(\Phi + \frac{\pi}{p})}{2\pi + 2Q\sin\frac{\pi}{p}\sin(\Phi + \frac{\pi}{p})} \right) \right\} d\Phi$$

$$I_{mn} = \int_0^{2\pi} \exp\left\{ -j\left[n\Phi + \frac{(m + \frac{n}{p})(2Q\pi\cos\Phi - 2BQ\sin\frac{\pi}{p}\sin(\Phi + \frac{\pi}{p}))}{2\pi + 2Q\sin\frac{\pi}{p}\sin(\Phi + \frac{\pi}{p})} \right] \right\} d\Phi$$

We interpret the double Fourier series expression in Eq. (6) as follows. The first term is the DC component. The third term corresponds to the carrier and its associated harmonics. The fourth term represents the input modulating signal and its harmonics inter-modulated with the carrier and its harmonics. These third and fourth terms are of little consequence in practice because the low-pass filter at the load of the Class D amp effectively eliminates these terms. The second term is the term of interest as it comprises the modulating signal and its harmonics, and is the source of the harmonic distortion.

To evaluate the extent of this harmonic distortion, we note that the integral term in the second term of Eq. (6) is a complex term comprising real and imaginary components. As this term appears to be mathematically intractable, we can employ the numerical integration method to determine the magnitude of the individual signal harmonics, and eventually determine the THD. We can show that the THD in second term is very small and to appreciate the extent of the harmonic distortion of the DI process, we will later compare the THD of the proposed DI sampling process with other sampling processes in Sec. 4.

3. PROPOSED PWM PULSE GENERATOR

Pulse Generators for the PWM based on the direct approach are often realized by using the fast-clock-counter [8] or tapped-delay-line methods [9]. The fast-clock-counter usually requires its clock freq to be 2^n x sampling freq (f_s) where n is the number of bits. The clock freq will subsequently be very high if n is large or if f_s is high, resulting in undesirable high power dissipation. For example, if n=16 bits and f_s=48kHz are specified, the resultant inordinately high clock freq of 3.15GHz would be required.

The reported methods to circumvent this high clock freq problem include a combination of the fast-clock-counter and tapped-delay-line techniques, that is allocating the n-bit as an p-bit counter and q-bit delay line, where $n=p+q$. For example, we [6] had previously realized a 12-bit pulse generator as a 9-bit counter and 3-bit delay line. It is worth noting that the delay-line suffers from supply voltage variations and transistor W/L ratio variations from the fabrication process, and hence delay variations. As these delay variations may introduce some harmonic non-linearities, we avoid the delay line in our present design. We instead propose a hybrid pulse generator embodying a 16-bit to 11-bit Noise Shaper, a proposed 1-bit frequency doubler and a 10-bit fast clock counter.

The 16-bit to 11-bit noise shaper is depicted in Fig. 4. The 16-bit digital pulse width determined by the DI sampling process will be added with the 5-bit quantization error obtained from the previous output of the noise shaper. The 16-bit digital value from this addition is then quantized into an 11-bit output signal by removing the least significant 5 bits. This 11-bit quantized output is also input to a converter that inserts 5 zero bits to obtain a 16-bit representation. This 16-bit data is then subtracted from the 16-bit output directly obtained from the adder to obtain the 5-bit quantization error. The error is then fedback to the adder to correct the next input. This noise shaping technique [10] allows digital signal to be represented with fewer bits, thereby enabling us to reduce the clock freq to a lower rate for lower power.

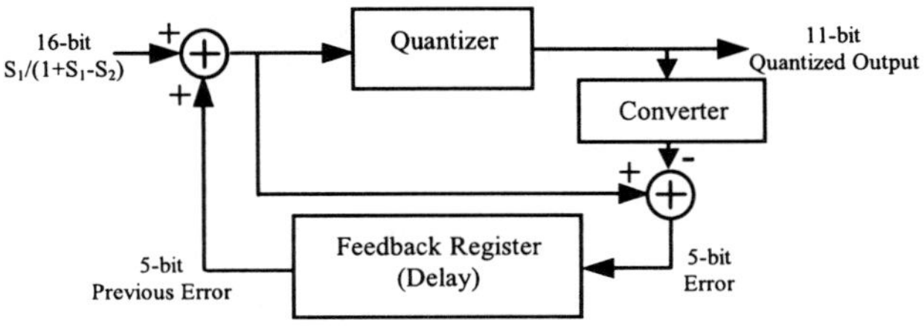

Fig. 4 16-bit to 11-bit Noise Shaper

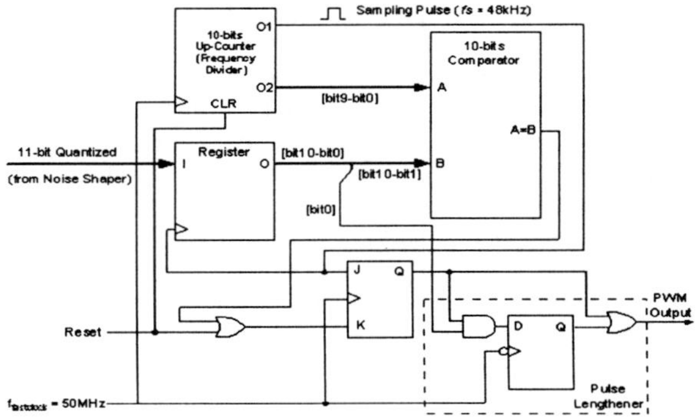

Fig. 5 Block diagram of the 10-bit counter, comparator, and the 1-bit frequency doubler

Despite reducing the number of bits from 16 to 11 bits (@ f_s=48kHz), we still require a relatively high clock freq of ~100MHz. In our pulse generator design, we propose a novel frequency doubler to achieve an 11-bit resolution with only 50MHz; a frequency easily accommodated at 1.1V operation for a typical 0.3μm CMOS process. The frequency doubler is depicted in Fig. 5. The first 10 bits of the 11-bit data obtained from the noise shaper is compared with the 10-bit output generated from the up-counter to provide the timing for the initial pulse width. Depending on the least significant bit of the 11-bit data, we lengthen the width of the PWM pulse using the Pulse Lengthener by half clock period of the 50MHz clock.

It is worthwhile to note that the counter in our pulse generator design also serves as a frequency divider. This novelty of sharing effectively reduces the power dissipation of our entire hybrid pulse generator by ~50%. We will later show that our pulse generator dissipates very low power (~53μW).

4. SIMULATION AND EXPERIMENTAL RESULTS

To verify our PWM embodying our proposed DI sampling process and proposed pulse generator, we simulate our PWM using MATLAB and construct it using a CPLD (the output stage of the Class D is realized in a prototype IC shown in Fig.

6). For comparison, we repeat the same simulations and hardware for the Natural, Uniform and the δC sampling processes. We use a 1kHz 16-bit digital audio data sampled at 48kHz for simulating and measuring the THD performance for all the sampling processes with the exception of the Natural sampling process where we use a 512kHz sampling freq. We summarize in Fig. 7 the THD of the different digital Class D amp realizations.

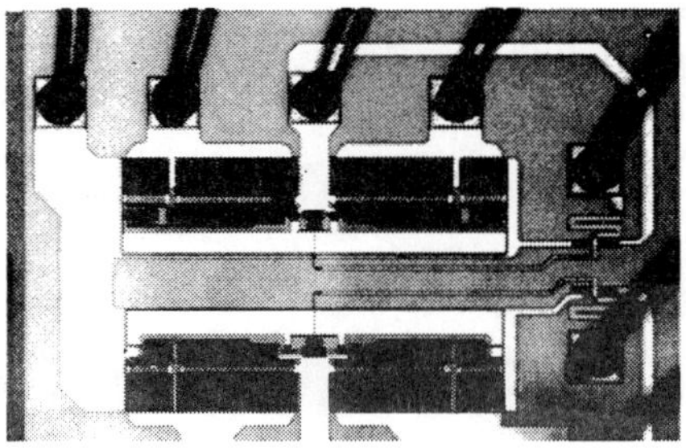

Fig. 6 Microphotograph of the Class D output stage

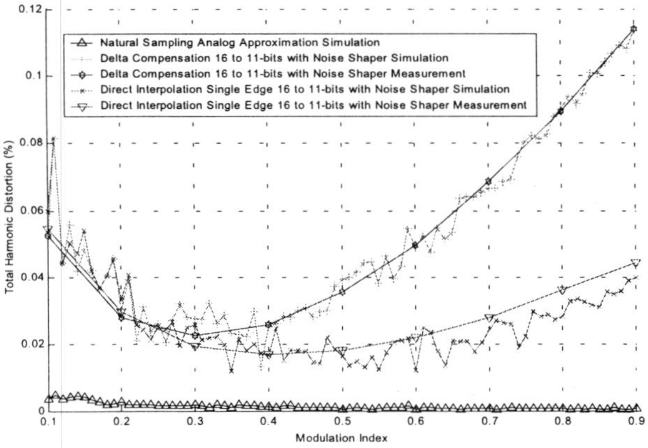

Fig. 7 A comparison of THD for various realizations (note that the THD of the US process is from 0.4%-2.8% and is beyond the scales shown)

We remark from Fig. 7 that the our proposed 16-bit DI sampling with our proposed 16-bit hybrid PWM pulse generator features a much lower average THD (~0.03%) than US and δC sampling; LI for ε=0.35 and δC sampling have similar THD [6]. For completeness, we remark that the THD of PWMs based on the indirect approach is typically 0.08% and is higher than our design here.

We will now present in Table 1 the power dissipation simulations based on NANOSIM of the different sampling processes-cum-hybrid pulse generator in a 16-bit digital PWM. At 1.1V operation and at modulation index = 0.5, the 16-bit hybrid pulse generator dissipates 53μW; as the US process effectively has no sampling process, the power dissipated is solely due to the pulse generator. The 53μW power dissipated by the pulse generator is distributed as follows: 50MHz clock freq divider dissipates 44μW and the remaining dissipates 9μW. The DI sampling process dissipates very low power - 7μW only.

Table 1 Power dissipation simulations of different 16-bit PWM sampling processes-cum-pulse generator @V_{DD}= 1.1V

16-bit Digital PWM	Power Dissipation (μW)
Uniform Sampling (US)	53
δC Sampling (δC)	55
Direct Inter. Sampling (DI)	60

In summary, the total power dissipation of the 16-bit DI sampling process-cum-hybrid pulse generator is a low 60μW. We remark that because the pulse generator dominates the power dissipation, the power dissipation of the PWM embodying the proposed DI and hybrid pulse generator is comparable to other PWMs (that embodies either the δC or US sampling processes)-cum-hybrid pulse generator. However, its THD is significantly better.

In summary, the PWM embodying our proposed DI sampling process and proposed pulse generator is suitable for low voltage, micropower Class D amps, rendering it suitable for a practical digital hearing instrument.

5. CONCLUSIONS

We have described a low voltage micropower digital PWM embodying the proposed DI sampling process and the proposed hybrid pulse generator that based on a noise shaper, a frequency doubler and a counter. We have shown that the DI sampling process features the lowest harmonic distortion compared to other algorithmic sampling processes while requiring simple computation leading to small IC area and very low power dissipation. We have analytically derived the double Fourier expression for the DI process. We have verified our design via computer simulations and experiments.

6. REFERENCES

[1] B.L. Sim, Y.C. Tong, J.S. Chang and C.T. Tan, "A Parametric Formulation of the Generalized Spectral Subtraction Method," *IEEE Trans. Speech & Audio Proc.*, v6, n4, pp. 328-337, 1998.

[2] J.S. Chang, M.T. Tan, Z.H. Cheng and Y.C. Tong, "Analysis and Design of Power Efficient Class D Amplifier Output Stages," *IEEE Trans. Cir. & Sys. I*, v47, n6, pp. 897-902, 2000.

[3] J.L. Melanson, "Delta Sigma PWM DAC to Reduce Switching," *AudioLogic, U.S. Patent*, 5,815,102, (incorporated in Cirrus Logic CS44210), 1998.

[4] H.S. Black, *Modulation Theory*, Van Nos., pp. 263-281, 1953.

[5] P.H. Mellor, S.P. Leigh and B.M.G. Cheetham, "Improved Sampling Process for a Digital, Pulse-width Modulated Class D Power Amplifier," *IEE Col. on DAS Proc.*, pp. 3/1-3/5, 1991.

[6] B.H. Gwee, J.S. Chang and H. Li, "A Micropower Low-Distortion Digital Pulse Width Modulator for a Digital Class D Amplifier," *IEEE Trans. Cir. & Sys. II*, v 49, n5, pp. 1-13, 2002.

[7] V.V. Mananov, "Single-polarity Pulse-width Modulation of the Third Kind," *Telecomm. Rad. Eng.*, v22, n2, pp. 67-71, 1967.

[8] G.Y. Wei and M. Horowitz, "A Low Power Switching Power Supply for Self-clocked Systems," *Int'l Symp. Low Power Elect. & Design*, pp. 313-318, 1996.

[9] A. Dancy and A.P. Chandrakasan, "Ultra Low Power Control Circuits for PWM Converters," *IEEE Power Elect. Spec. Conf.*, pp. 21-27, 1997.

[10] J.M. Goldberg and M.B Sandler, "Noise Shaping and Pulse-width Modulation for an All-digital Audio Power Amplifier," *J. Audio Eng. Soc.*, v39, n6, pp. 449-460, 1991.

A COMPARISON OF ALGORITHMS FOR SOUND LOCALIZATION

Pedro Julián[*], *Andreas G. Andreou*, *Larry Riddle*, *Shihab Shamma*[†], *Gert Cauwenberghs*

Johns Hopkins University
Electrical and Computer Engineering Dept.
3400 North Charles St., Baltimore, MD 21218, USA

ABSTRACT

In this paper, we compare the performance of four algorithms for sound localization: one-bit correlation, one-bit correlation derivative, and two methods inspired from biology, namely, gradient flow and sterausis. We employ real-data recorded from four microphones to compare the localization performance.

1. INTRODUCTION

We present a comparison of four algorithms for sound localization using four microphones and data experimentally recorded in a natural environment. We consider the situation in which the sensors are passive; in our case, a pair of microphones to sense the signal and estimate the source position. Two of the algorithms employ the classical approach of cross-correlation [1]; the other two are bio-inspired: spatial-temporal gradients techniques [2], [3] and the stereausis network architecture proposed in [4]. The comparison study presented in this paper is aimed at a micropower sound localizer in CMOS technology. The companion paper [5], discusses the implementation and testing of a micropower binary cross-correlation architecture.

2. SETUP AND DATA COLLECTION

The localization setup under consideration consists of four microphones, as shown in Fig. 1, with an effective distance $L = 15.87cm$. We are assuming that the sound source is far away from the microphones ($L << L_s$), and is also limited in frequency from $20Hz$ to $300Hz$. To compare the algorithms in a natural environment a series of experiments were performed in an open field with a speaker set at a distance of approximately 18 m from the microphones. For each angular location of the speaker, 30 seconds of data were emitted and recorded simultaneously by the four microphones. This experiment was repeated for nineteen angles between 0° and 180° in steps of 10°. As all algorithms were designed to produce an estimation after one second, for every angle we obtained a set of 30 readings of time delay.

*P. Julián is with CONICET (Consejo Nacional de Investigaciones Cientificas y Técnicas), Av. Rivadavia 1917, 1033 Cap. Fed., Argentina; and on leave from the Departamento de Ingenieria Eléctrica y Computadoras, Universidad Nacional del Sur, Av. Alem 1253, 8000 Bahia Blanca, Argentina. E-mail: pjulian@ieee.org

†S. Shamma is with the Department of Electrical Engineering, University of Maryland at College Park

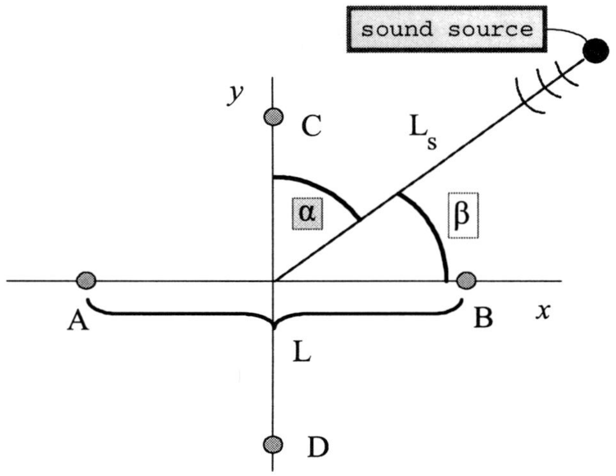

Figure 1: Microphones setup to measure the bearing angle.

3. DESCRIPTION OF THE DIFFERENT METHODS

In this section, we describe four algorithms employed to estimate the bearing angle. Two of them are algorithms previously reported in the literature, based on correlation [1] and spatial gradients techniques [2]. The other two are new approaches: the first one is based on a modification of the correlation approach, and the second is based on a neuromorphic approach [4].

3.1. Correlation Approach

This is a standard approach that has been extensively reported in the literature [1], [6]. Let us consider one pair of microphones and assume that the signals $x_A(t)$, $x_B(t)$ entering that pair of microphones are described by:

$$\begin{aligned} x_A(t) &= s(t) + n_A(t) \\ x_B(t) &= s(t-D) + n_B(t) \end{aligned} \tag{1}$$

where $s(\cdot)$ is the signal emitted by the source, $n_A(\cdot)$ and $n_B(\cdot)$ are uncorrelated noise signals and D is the time delay between microphones. Under the assumption that the source is far away, the wave arriving at the microphones can be considered as plane, and then the following relation

0-7803-7761-3/03/$17.00 ©2003 IEEE

holds:

$$D = L/c\cos(\beta) = D_{\max}\cos(\beta), \quad (2)$$

where $c = 345m/s$ is the speed of sound in air at ambient temperature and $D_{\max} = 460\mu s$ is the maximum delay. Considering that n_A and n_B are uncorrelated, the correlation between signals x_A and x_B is given by:

$$R_{x_A x_B}(\tau) = \int_{-\infty}^{\infty} s(t)\, s(t - D + \tau)\, dt$$

This function will exhibit a maximum at $\tau = D$. Therefore, one way to estimate the time delay is to generate the correlation function numerically and calculate the time where the maximum is achieved. In practice, the signal is sampled at a certain frequency $f_s = 1/T_s$ and the correlation is approximated using a discrete time sum:

$$\tilde{R}_{x_A x_B}(iT_s) = \sum_{k=0}^{K} x_A(kT_s)\, x_B((k-i)T_s), \quad (3)$$

where K is such that $K \cdot T_s$ is the time window under consideration. Operation (3) can be implemented in a digital fashion after quantization of the signals. Using experimental data, we found that a one bit quantization leads to accurate estimations, as we will show later. From a hardware prospective, coding the signal with just one bit produces a dramatic reduction in the density and complexity of the design. The associated structure is composed of a number of stages

$$y(i) = \sum_{k=0}^{K} x_A(k)\, x_B(k-i), \quad (4)$$

where i is an index to the stage number. As was explained in [5], a sampling frequency of $200KHz$ permits to estimate angles in the range $\{\alpha \in [0, 50] \cup [+130, +180]\}$, with an accuracy of one degree. This choice of sampling frequency implies that every discrete time delay is $T_s = 5\mu s$. It also implies that the maximum possible delay –corresponding to an angle $\beta = 90°$– is $D_{\max} = 460\mu s$, so that it is necessary to implement 92 stages. Accordingly, index i in (4) ranges from 0 to 91. From a hardware viewpoint, the digital implementation of (4) requires shift registers to generate the delayed versions of x_B, a counter implementing the correlation operation and finally one block to determine where the maximum has occurred.

A drawback to this approach is that once the signal is quantized with one bit, the information corresponding to the time delay between signals is encoded in the changes of state from zero to one, and viceversa. No information is contained in those parts of the signal where there are no state changes. However, every stage (4) is counting all the time at the frequency clock, regardless of the input values. As the frequency of the clock is much higher than the frequency of the signal, this architecture will dissipate power at a much higher rate than what is actually necessary. This observation motivated the approach presented in the following sub-section. An additional disadvantage of this approach is the need to calculate the occurrence of the maximum of (4), which would require the implementation of additional circuitry (a winner-takes-all circuit or an equivalent digital circuit).

3.2. Correlation Derivative Approach

As we said, the maximum of the correlation occurs when the delay produced by the shift register chain coincides with the relative delay between signals. Mathematically, detecting the maximum of the correlation function is equivalent to detecting the zero-crossing of its derivative when the second derivative is negative. This methodology has several advantages that we will describe now. If we consider (4) and calculate the discrete difference between adjacent elements, we get for every stage

$$\Delta y(i) := y(i) - y(i-1) = \sum_{k=0}^{l} x_A(k)\left[x_B(k-i) - x_B(k-(i-1))\right] \quad (5)$$

Careful observation of (5) reveals an UP/DOWN counter, which counts up when $x_A(k) = 1$, $x_B(k-i) = 1$ and $x_B(k-(i-1)) = 0$, and counts down when $x_A(k) = 1$, $x_B(k-i) = 0$ and $x_B(k-(i-1)) = 1$. In this case, the count is only updated when one of the two signals changes its state, and it is idle the rest of the time. This mode of operation reduces the circuit activity and therefore the power consumption, and also reduces the size of the counters. In addition, to obtain the value of the delay it is just necessary to read the position of the stage where the zero-crossing occurred, eliminating the need for searching the maximum of the outputs.

3.3. The Stereausis Approach

This approach is inspired in the stereausis network described in [4] and uses two cochlea channels to pre-process the microphones input signals. Indeed, the sound from the left and right microphones are fed to two cochlea channels. All outputs are quantized to 1 bit and the outputs of every stage of one channel are digitally correlated with the outputs of the other channel. In this way, a spatial arrangement of elements results, which can be associated to an image C, whose (i,j) element $C_{i,j} \geq 0$ represents the correlation between the output of the i-th element of the left cochlea and the output of the j-th element of the right cochlea (see Fig. 2). When the left and right signals are equal, the resulting image C will have a high density of nonzero elements along the main diagonal. However, if there is a delay in one of the signals, the image C will show a shift of the main diagonal towards one of the sides. The network that we used in the simulations consists of a 32–stages cochlea with cut-off frequencies between $252Hz$ and $618Hz$. Notice that as a delay of τ seconds is equivalent to a phase shift of $\phi(f) = 2\pi f\tau$, the higher the frequency the more noticeable the unbalance of the image with respect to the main diagonal (see [4]). The indication of time delay is calculated by measuring the unbalance of the image C with respect to the main diagonal. This is done by computing the difference between the sum of upper diagonal elements and lower diagonal elements, i.e.,

$$\Upsilon = \sum_{i<j} C_{i,j} - \sum_{i>j} C_{i,j}.$$

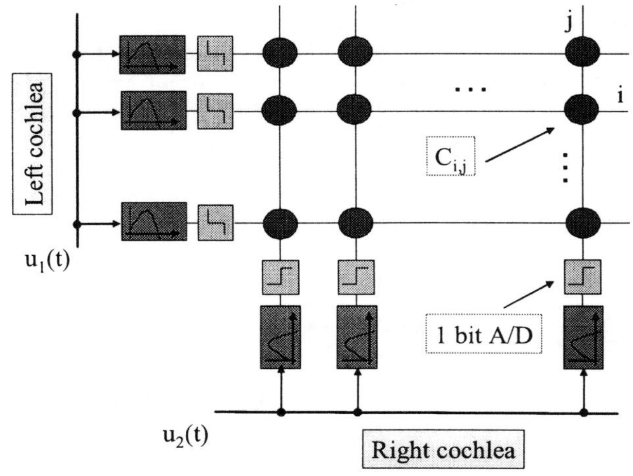

Figure 2: Architecture for estimation based on the stereausis approach.

3.4. Spatial Gradients Approach

In this approach, the signals recorded by the microphones are interpreted as samples of a sound field $s(\cdot)$ and the bearing angle is estimated using first order derivatives ([2]). This algorithm, in contrast to the previous cases, takes full advantage of the four microphones for the time delay estimation. For the present situation let us consider the position of the microphones with respect to the center of the array. We will assume that for any given location $\mathbf{r}$ in the plane, where $\mathbf{r} \in \mathbf{R}^2$, the magnitude $\tau(\mathbf{r})$ represents the time delay between the wavefront of the sound wave at $\mathbf{r}$ and the wavefront of the sound wave at the center of the array. Using this definition and a Taylor series, we can express the field $s(t+\tau(\mathbf{r}))$, $s : \mathbf{R}^1 \mapsto \mathbf{R}^1$ in a neighborhood of the origin as

$$s(t+\tau(\mathbf{r})) = s(t) + \tau(\mathbf{r}) \frac{d}{dt} s(t) + \frac{1}{2}\tau(\mathbf{r})^2 \frac{d^2}{dt} s(t) + O\left(\tau(\mathbf{r})^3\right)$$

To first order, and after geometric considerations, it can be easily seen that

$$x_A(t) \approx s(t) + \tau_2 \dot{s}, \quad x_D(t) \approx s(t) + \tau_1 \dot{s}$$
$$x_B(t) \approx s(t) - \tau_2 \dot{s}, \quad x_C(t) \approx s(t) - \tau_1 \dot{s}$$

where $\tau_1 = \frac{1}{2}\frac{L}{c}\cos(\alpha)$, $\tau_2 = \frac{1}{2}\frac{L}{c}\cos(\beta)$ are the delays with respect to the coordinate axes. Then, a simple manipulation of the variables leads to

$$\begin{array}{c} s(t) = \frac{1}{4}\left(x_A(t) + x_B(t) + x_C(t) + x_D(t)\right) \\ \tau_1 \dot{s} = \frac{1}{2}\left(x_D(t) - x_C(t)\right), \tau_2 \dot{s} = \frac{1}{2}\left(x_A(t) - x_B(t)\right) \end{array} \quad (6)$$

If we sample the signals with a sampling time $T_s = 1/f_s$, and assume that $ds(t)/dt$ at $t = kT_s$ can be adequately measured by filtering $s(kT_s)$, then (6) is a standard least squares problem and τ_1, τ_2 can be obtained independently after collecting $N+1$ samples.[1] This approach heavily relies on the accuracy of the signals measurement, especially to calculate the derivative with precision. Due to this, in this case the amplitude cannot be quantized. In practice, the original signal was used with the original sampling rate of 2048 samples per second, and the derivative was calculated using finite differences.

[1] Similar results can be obtained using adaptive algorithms.

Table 1: Accuracy of the algorithms (STD) in degrees

	Corr.	Ster.	Spatial Gr.
STD	1.18	1.47	0.87

4. COMPARISON OF RESULTS

Based on the collected data, we used the mean to define the transfer curve time delay–angle, and the standard deviation to quantify the precision. As was shown in [5], the time delay variation corresponding to a change of one degree at an angle β^* is

$$\Delta D|_{1^o} = D_{\max} \sin(\beta^*)\, \pi/180. \quad (7)$$

In the present case, it is useful to quantify the error in degrees. This requires a conversion of the measurement from seconds to degrees. Accordingly, if the reading of a certain time delay has a standard deviation of σ_T, then the standard deviation in degrees is given by

$$\sigma_D = \frac{\sigma_T}{D_{\max} \sin(\beta^*)\, \pi/180}$$

The correlation and correlation derivative approach give indistinguishable results, therefore, in what follows we are only referring to the latter approach. Figure 3 shows the mean value of the output corresponding to the three algorithms in the range $\{\alpha \in [0,50] \cup [+130,+180]\}$; Fig. 4 shows the standard deviation corresponding to the range $\{\alpha \in [0,90]\}$[2]. From this figure, it can be seen that whenever the pair of microphones $A-B$ is used, the precision of the estimation deteriorates in the range $\{\alpha \in [50,130]\}$. However, the other pair of microphones can be used in this range to obtain the same accuracy. Accordingly, Table 1 summarizes the average standard deviation of the three algorithms in the range $\{\alpha \in [0,50]\}$.

5. CONCLUSIONS

We have compared four different algorithms for sound localization. Two of the algorithms were previously reported in the literature, and the other two were developed specifically for this application. The spatial gradients method shows the best accuracy results, but from an implementation viewpoint it requires a sampled data analog architecture able to solve adaptively an LMS problem. The stereausis based approach shows acceptable results but requires a two dimensional array of correlators in addition to the two

[2] This plot is symmetric with respect to $\alpha = 90^o$, so for the sake of clarity only the portion between 0° and 90° is shown.

cochlea filter channels. Finally, the correlator derivative approach shows an accuracy very close to the spatial gradients approach, but it offers a very convenient architecture, evidenced not only by its simplicity but also by the associated low power consumption due to the low temporal activity. Experimental results of an integrated circuit that implements this approach can be found in the companion paper [5].

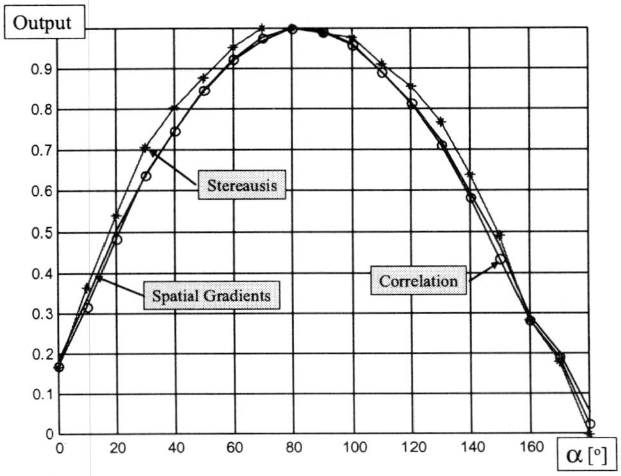

Figure 3: Mean value of the output for the three algorithms in the range of interest given by $\{\alpha \in [0, 180]\}$.

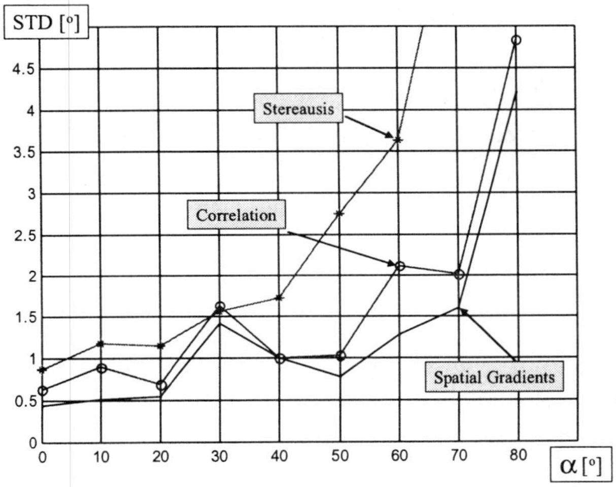

Figure 4: Standard deviation in degrees for the three algorithms in the range of interest given by $\{\alpha \in [0, 90]\}$.

6. REFERENCES

[1] G. C. Carter, "Time delay estimation for passive sonar signal processing," *IEEE Trans. Acoustics, Speech, Signal Processing*, vol. ASSP-29, pp. 463–470, June 1981.

[2] G. Cauwenberghs, M. Stanacevic, and G. Zweig, "Blind broadband source localization and separation in miniature sensor arrays," in *Proc. of the IEEE Int. Symp. on Circuits and Syst. (ISCAS)*, vol. III, pp. 193–196, 2001.

[3] M. Stanacevic and G. Cauwenberghs, "Mixed-signal gradient flow bearing estimation," in *Proc. of the IEEE Int. Symp. on Circuits and Syst. (ISCAS)*, 2003.

[4] S. Shamma, N. Shen, and P. Gopalaswamy, "Stereausis: Binaural processing without neural delays," *J. Acoust. Soc. Am.*, vol. 86, pp. 989–1006, 1989.

[5] P. Julian, A. G. Andreou, P. Mandolesi, and D. Goldberg, "A low-power CMOS integrated circuit for bearing estimation," in *Proc. of the IEEE Int. Symp. on Circuits and Syst. (ISCAS)*, 2003.

[6] G. C. Carter, "Coherence and time delay estimation," *Proc. IEEE*, vol. 75, pp. 236–255, February 1987.

Moduli Selection in RNS for Efficient VLSI Implementation

Wei Wang*, M.N.S. Swamy**, Fellow, IEEE, and M.O. Ahmad**, Fellow, IEEE

* Department of Electrical & Computer Engineering
The University of Western Ontario
London, Ontario, Canada N6G 1H1

** Department of Electrical & Computer Engineering
Concordia University
Montreal, Quebec, Canada H3G 1M8

ABSTRACT

In this paper, we carry out a study on an important issue concerning the use of residue numbers in the design of digital systems, namely, the moduli selection. Based on a new formulation of the Chinese remainder theorem and the efficient residue-to-binary (R/B) converters designed therefrom, we propose a guideline for the selection of the *low-cost moduli sets* for different dynamic ranges. These sets consist of the *low-cost moduli* of the form 2^n, (2^n-1) or (2^n+1), which can offer simplified modulo adders and multipliers. It is shown that for medium dynamic ranges (less than 22 bits), the three-moduli set $\{2^n, 2^n+1, 2^n-1\}$ is the most efficient one in terms of the design of a complete RNS system. On the other hand, for large dynamic ranges (equal to or larger than 22 bits), the general-moduli set in the form $\{2^{n_1}, 2^{n_1}+1, 2^{n_1}-1, 2^{n_2}\pm 1, ..., 2^{n_i}\pm 1\}$, with a length greater than three, is the most efficient one. These results provide the possibility of a wide range of applications of the residue number system in the design of DSP, telecommunication and cryptography systems.

1. INTRODUCTION

VLSI digital systems are used in numerous applications in DSP, telecommunication and cryptography. The objective of VLSI digital design is to continuously reduce the cost and improve the performance of the VLSI digital systems in terms of complexity, speed and power [1], [2]. One method of designing high-speed, low-power VLSI digital systems is by using the residue number system (RNS). The RNS has no carry chain and offers high-speed operations [3]. The high-speed gained by the RNS parallelism can then be traded-off for low power consumption. In a low-power ASIC or FPGA implementation of VLSI digital systems, the RNS-based structures provide a promising future.

The challenges of the RNS system design lie in the choice of the moduli set and in the conversion of the residue numbers to binary numbers (R/B). Both these challenges are associated with the Chinese remainder theorem (CRT) algorithm. The direct usage of the CRT requires a large-valued modulo operation that is not efficient for VLSI implementation. We have modified the formulation of the Chinese remainder theorem, to reduce the size of modulo operations [4], [5]. Based on the new formulation, we have proposed efficient residue-to-binary conversion algorithms and designed several efficient R/B converters [4]-[6]. The new formulation of the CRT and the R/B conversion study can also be used in resolving the important issue of moduli selection. To represent the binary numbers in a certain range, there are many different choices of the moduli sets for the RNS. Since the speed and the hardware complexity of a system employing residue numbers is related to the choice of the moduli set, such a study is of crucial importance [2].

In this paper, we propose an efficient scheme for moduli selection to determine the *low-cost moduli sets* for different dynamic ranges. We show that for a medium dynamic range (less than 22 bits), the three-moduli set $\{2^n, 2^n+1, 2^n-1\}$ is the most efficient one, whereas for large dynamic ranges (equal to or larger than 22 bits), the general-moduli set in the form of $\{2^{n_1}, 2^{n_1}+1, 2^{n_1}-1, 2^{n_2}\pm 1, ..., 2^{n_i}\pm 1\}$, with a length greater than three, is the most efficient one. These sets consist of the low-cost moduli only in the form of 2^n, (2^n-1) and (2^n+1), which can offer simplified modulo adders and multipliers. Also, these sets can be used to obtain very efficient R/B converters. Thus, the VLSI implementation of a complete RNS system is improved. The proposed moduli selection scheme provides the possibility of a wide range of applications of RNS in the design of DSP, telecommunication and cryptography systems.

2. BACKGROUND MATERIAL

A residue number system is defined in terms of a relatively prime moduli set $\{P_1, P_2, ..., P_m\}$, that is, $\mathrm{GCD}(P_i, P_j) = 1$ for $i \neq j$. A binary number X can be represented as $U = (u_1, u_2, ..., u_m)$, where $u_i = |U|_{P_i}$, $0 \le u_i < P_i$. Such a representation is unique for any integer $U \in [0, M-1]$, where $M = P_1 P_2 ... P_m$ is the dynamic range of the moduli set $\{P_1, P_2, ..., P_m\}$.

The RNS is a carry-free system for addition, subtraction, and multiplication operations. Hence, a large dynamic range binary system can be partitioned into several small-wordlength channels in parallel. Thus, the RNS can result in a parallel and high-speed operation. In order to convert numbers from binary to residue form and vice-versa, a binary-to-residue (B/R) converter is required at the front-end of the system and a R/B converter at the

0-7803-7761-3/03/$17.00 ©2003 IEEE

back-end. The B/R converter consists of several modulo adders, whereas the R/B converter involves a lot of modulo operations. Thus, the R/B converter is a crucial part of the RNS system. To perform the R/B conversion, that is, to convert the residue number $(x_1, x_2, ..., x_m)$ into the binary number X, the CRT is generally used [1]-[3].

Chinese Remainder Theorem: Given the moduli set $\{P_1, P_2, ..., P_m\}$ and the dynamic range $M = P_1 P_2 ... P_m$, the residue number $(x_1, x_2, ..., x_m)$ is converted into the binary number X by

$$X = \left| \sum_{i=1}^{m} N_i \left| N_i^{-1} \right|_{P_i} x_i \right|_M \quad (1)$$

where $n > 1$, $N_i = \frac{M}{P_i}$, and $\left|N_i^{-1}\right|_{P_i}$ is the multiplicative inverse of $\left|N_i\right|_{P_i}$ defined by $\left| \left|N_i^{-1}\right|_{P_i} N_i \right|_{P_i} = 1$.

The CRT requires a modulo-M (large valued) operation and it is not efficient for the implementation. Thus, we apply the following modified format of the CRT to reduce the modulo operation from modulo $M = P_1 P_2 ... P_m$ to modulo $P_2 ... P_n$ [4], [5].

Theorem 1: Given the moduli set $\{P_1, P_2, ..., P_m\}$, the residue number $(x_1, x_2, ..., x_m)$ is converted into the binary number X by

$$X = x_1 + P_1 \left| \sum_{i=1}^{m} w_i x_i' \right|_{P_2 ... P_m} \quad (2)$$

where $m > 1$, $w_1 = \frac{N_1 \left|N_1^{-1}\right|_{P_1} - 1}{P_1}$, $w_i = \frac{N_i}{P_1}$, for $i = 2,3,...,m$,

$x_1' = x_1$, and $x_i' = \left|N_i^{-1} x_i\right|_{P_i}$, for $i = 2,3,...,m$.

Since P_1 is generally chosen as 2^n, the last step in (2), that is, the multiplication by P_1 and the binary addition of u_1, can be reduced to a simple concatenation. Based on the new formulation, we have proposed efficient residue-to-binary conversion algorithms and designed several efficient R/B converter techniques [4]-[6]. The new formulation of the CRT and the R/B conversion study can be used to resolve the important issue of the moduli selection in such designs. To represent the binary numbers in a certain range, there are many different choices of moduli sets for the RNS. The importance of the moduli selection is due to the fact that the dynamic range, the speed, as well as the VLSI implementation of RNS systems depend on the form as well as the number of the moduli chosen. In the literature, there are two kinds of moduli sets. A set of any given moduli is called a general-moduli set, as it is efficient for RNS systems with a large dynamic range. The three-moduli sets, $M_1 = \{2^n, 2^n+1, 2^n-1\}$, $M_2 = \{2n, 2n+1, 2n-1\}$, $M_3 = \{2^n, 2^n - 1, 2^{n-1} - 1\}$, and $M_4 = \{2^{2n}+1, 2^n+1, 2^n-1\}$, are special cases of the general-moduli sets, and these are widely used for RNS systems with a medium dynamic range [3], [8]. However, in the literature, no detailed study has been carried out in developing a method for the moduli selection. There exists no definition for the large and medium dynamic ranges.

3. MODULI SELECTION SCHEME

The moduli set chosen for an RNS system affects its complexity. In general, it is beneficial to make the largest modulus as small as possible, since it is the magnitude of the largest modulus that dictates the speed of the arithmetic operations. However, the speed and cost depend not only on the wordlengths of the residues, but also on the moduli chosen. For example, modulus 16 might be better than a smaller modulus, say, 13. The reason is that modulus 2^n simplifies the arithmetic operations and is referred to as the *low-cost modulus*. The modulo adders and multipliers for the modulus 2^n are just the binary adders and multipliers with truncation. Further, the moduli of the forms ($2^n - 1$) and ($2^n + 1$) are desirable and are also referred to as the *low-cost moduli* [3], [7], [8]. The addition and multiplication operations based on these moduli can be performed using the simplified modulo arithmetic units. The best modulo adder and multiplier for the modulus ($2^n - 1$) have been proposed in [7], whereas those for the modulus ($2^n + 1$) have been proposed in [8]. Based on these discussions, we are justified in choosing the general-moduli set to include only the low-cost moduli in the form of 2^n, ($2^n - 1$) or ($2^n + 1$). Thus, we propose the moduli format for the general moduli set as $M_G = \{2^{n_1}, 2^{n_1}+1, 2^{n_1}-1, 2^{n_2} \pm 1, ..., 2^{n_t} \pm 1\}$, where $n_1 > n_i$, and n_1 as well as n_i's ($i = 2,3,..,t$) need to be chosen such that all these moduli are co-prime numbers.

By using the proposed moduli format, we can design efficient RNS subsystems, B/R converters and R/B converters. In the proposed moduli format, each modulus is a low-cost modulus that can provide simplified modulo adder and multiplier. Since RNS subsystems and B/R converters consist of only modulo adders and multipliers, their designs get simplified. Based on the new formulation of the CRT (Theorem 1) and the general-moduli R/B converter study of [5], we already know that the general-moduli sets with the modulus 2^{n_1} can provide an efficient R/B converter. Hence, the proposed moduli format is efficient for the design of the R/B converter for the general-moduli set. In [4], we have carried out a comprehensive study on R/B converters for the three-moduli sets $M_1 = \{2^n, 2^n+1, 2^n-1\}$, $M_2 = \{2n, 2n+1, 2n-1\}$, $M_3 = \{2^n, 2^n-1, 2^{n-1}-1\}$, and $M_4 = \{2^{2n}+1, 2^n+1, 2^n-1\}$. Through this study, it has been found that amongst these three-moduli sets, the set M_1 provides the fastest R/B converter with the smallest area. It is noted that this moduli set M_1 is a particular case of the proposed general-moduli set M_G.

For a given dynamic range, we now give a scheme to obtain the most efficient set of the form $M_G = \{2^{n_1}, 2^{n_1}+1, 2^{n_1}-1, 2^{n_2}\pm 1, ..., 2^{n_t}\pm 1\}$. First, we choose a number of sets with different lengths (that is, the three-moduli set, the four-moduli set, etc.). Then, for each of these moduli sets, we determine the n_i's ($i = 1,..,t$) such that the resulting dynamic range is the smallest one, that is, at least equal to the given dynamic range. Thus, we can determine n_i's ($i = 2,3,..,t$) for the various moduli sets of different lengths. We now compare these sets from the point of view of efficiency of their R/B converters to determine the most efficient set. We implement and compare the R/B converters for various lengths of M_G. For example, given a 20-bit dynamic range, we have the following results:

Three-moduli set:	{128, 129, 127}
Four-moduli set:	{64, 65, 63, 17}
Five-moduli set:	{32, 33, 31, 17, 5}
Six-moduli set :	{32, 33, 31, 17, 7 ,5}

Since the six-moduli set is the five-moduli set with an additional modulus 7, its dynamic range is larger than that of the five-moduli set. Hence, the six-moduli set is not as efficient as the five-moduli set for a 20-bit dynamic range. At the same time, we cannot choose $n_1 = 4$ for the six-moduli set, since we will not be able to obtain six moduli of the form M_G. The sets with lengths greater than six are even less efficient compared to the six-moduli set. Thus, we consider only the three-moduli, four-moduli and five-moduli sets for a 20-bit dynamic range. In the next section, we will implement the R/B converters for these three sets. From the implementation results, it will be seen that for a 20-bit dynamic range, the R/B converter for the moduli set {128, 129, 127} is the most efficient one in terms of complexity and speed.

4. MEDIUM AND LARGE DYNAMIC RANGES

It has been seen in Section 2 that the general-moduli set is efficient for RNS systems with a large dynamic range, and the three-moduli sets are widely used for RNS systems with a medium dynamic range. However, no definition of medium or large dynamic range was given. By using the moduli selection method proposed in Section 3, we now carry out a study to identify medium and large dynamic ranges. It has been stated in [9] that the upper bound of the dynamic range for using the three-moduli sets is around 24 bits. Thus, we choose the 20-bit, 22-bit, 24-bit, 28-bit and 32-bit dynamic ranges for our study. We obtain the most efficient set for each of these ranges using the proposed moduli selection method. The dynamic range for which the three-moduli sets are efficient will be considered as belonging to the medium dynamic range. A range larger than this medium range will be considered as a large dynamic range.

For each of the five dynamic ranges mentioned above, we choose a number of sets with different lengths. The n_i's, ($i = 2,...,t$), are then determined for the various moduli sets of different lengths. These sets are given in Table 1. Next, we implement and compare the R/B converters for the various moduli sets for each of the five dynamic ranges. The R/B converter design for the three moduli set $\{2^n, 2^n+1, 2^n-1\}$ is the one proposed in [10], whereas that for the general-moduli set is the one proposed in [5].

For the example considered, the proposed converter is implemented using 0.5-micrometer CMOSIS VLSI technology. The design flow is as follows. First, we construct a new library to include all the building blocks. Then, the architectures of the converters are coded in the VHDL language using this library. Next, the codes are executed at the register transfer level (RTL) to verify the correctness of the designs. The logic synthesis is carried out to optimize the designs, and the gate-level simulation performed. Finally, the placement and routing are carried out automatically to generate the layout. A performance evaluation in terms of the area and delay is carried out at the layout level. The software packages used include Cadence V4.4.1 (Release 9504), Synopsys V3.4b, CMC Generic Environment for Cadence V1.7, CMC CMOS5 Design Kit V2.2 for Cadence and CMC CMOS5 Design Kit V1.0 for Synopsys.

The implementation results in terms of the area and delay performance of all these R/B converters are obtained using Cadence V4.4.1 (Release 9504) tools and summarized in Table 2. The area refers to the core area of the layout without the input/output pads, whereas the delay is the timing of the critical path in nanoseconds.

It is seen from Table 2 that for the 20-bit dynamic range, the R/B converter for the three-moduli set $\{2^n, 2^n+1, 2^n-1\}$ is the fastest and requires the smallest area amongst the converters for the various moduli sets. As a consequence, this converter has the lowest power consumption ($POW_{avg} \approx fCV^2$). Similarly, for the 22-bit, 24-bit, 28-bit, and 32-bit dynamic ranges, the R/B converters for the moduli sets M_G with lengths greater than three are more efficient in terms of area, speed and power consumption compared to the three-moduli set. For each of these five dynamic ranges, the most efficient set in terms of the area, speed and power consumption is obtained and highlighted in Table 1.

Based on these results, we conclude that a dynamic range less than 22 bits belongs to the medium dynamic range, whereas a dynamic range greater than or equal to 22 bits belongs to the large dynamic range.

TABLE 1
SPECIFIC SETS FOR 20-BIT, 22-BIT, 24-BIT, 28-BIT AND 32-BIT RANGES

Moduli sets	20-bit range	22-bit range	24-bit range	28-bit range	32-bit range
Three-moduli	***{128, 129,127}***	{256, 257,255}	{512, 513,511}	{1024, 1025,1023}	{2048, 2049,2047}
Four-moduli	{64,65,63,17}	***{64,65,63,17}***	***{128,129,127,17}***	{256,257,255,17}	{512,513,511,65}
Five-moduli	{32,33,31,17,5}	-	{64,65,63,31,17}	***{128,129,127,31,5}***	***{128,129,127,65,31}***
Six-moduli	-	-	{32,33,31,17,7,5}	-	{128,129,127,31,17,7}

TABLE 2
IMPLEMENTATION RESULTS OF R/B CONVERTERS FOR THESE SETS

	20-bit		22-bit		24-bit		28-bit		32-bit	
Moduli sets M_i	Area (μm^2)	Time (ns)	Area (μm^2)	Time (ns)	Area (μm^2)	Time (ns)	Area (μm^2)	Time (ns)	Area (μm^2)	Time (ns)
Three-moduli	***280×398***	***17.65***	368×578	19.65	415×762	20.75	553×897	22.98	713×987	26.54
Four-moduli	312×413	18.16	***356×549***	***19.34***	***403×704***	***19.98***	583×889	22.48	641×945	26.21
Five-moduli	342×434	18.65	-	-	424×749	20.67	***483×789***	***21.87***	***511×869***	***25.20***
Six-moduli	-	-	-	-	445×791	20.91	-	-	597×948	25.20

5. CONCLUSION

In this paper, we have focused on one important issue of an RNS design: the moduli selection. Based on a new formulation of the CRT and the efficient R/B converters designed using this formulation, we have proposed a guideline for the selection of the low-cost moduli sets for different dynamic ranges. These sets consist of only the low-cost moduli in the form of 2^n, (2^n-1) and (2^n+1), which can offer simplified modulo adders and multipliers. For a medium dynamic range (less than 22 bits), the three-moduli set $\{2^n, 2^n+1, 2^n-1\}$ is the most efficient one in terms of the design of the B/R converters, RNS sub-systems and R/B converter. On the other hand, for large dynamic ranges (equal to or larger than 22 bits), the general-moduli set in the form of $\{2^{n_1}, 2^{n_1}+1, 2^{n_1}-1, 2^{n_2}\pm 1, \ldots, 2^{n_i}\pm 1\}$, with a length greater than three, is the most efficient one. These results provide the possibility of a wide range of applications of RNS in the design of DSP, telecommunication and cryptography systems.

ACKNOWLEDGMENT

This work was supported by NSERC (Canada), and FCAR (Quebec) grants.

REFERENCES

[1]. M. A. Soderstrand, W. K. Jenkins, G. A. Jullien, and F. J. Taylor, *Residue Number System Arithmetic: Modern Applications in Digital Signal Processing.* New York: IEEE Press, 1986.

[2]. B. Parhami, *Computer Arithmetic: Algorithms and Hardware Designs*, Oxford: Oxford University Press, 2000.

[3]. W. K. Jenkins and B. J. Leon, "The use of residue number systems in the design of finite impulse response digital filter," *IEEE Transactions on Circuits and Systems*, vol. CAS-24, pp. 191-201, April 1977.

[4]. Wei Wang, M.N.S. Swamy, M. O. Ahmad and Yuke Wang, "A study of residue-to-binary converters for three-moduli sets," to appear in *IEEE Transactions on Circuits and Systems I.*

[5]. Wei Wang, M.N.S. Swamy, and M.O. Ahmad, "An efficient residue-to-binary converter", in *Proceedings of the IEEE Midwest Circuits and Systems Conference*, Lansing, MI, USA, pp.904-907, Aug. 2000.

[6]. Wei Wang, M.N.S. Swamy, M.O. Ahmad, and Yuke Wang, "A high-speed residue-to-binary converter for three-moduli RNS and a scheme of its VLSI implementation," *IEEE Trans. on Circuits and Systems II,* vol. 47, pp. 1576-1581, Dec. 2000.

[7]. A.A. Hiasat, "High-speed and reduced-area modular adder structures for RNS," *IEEE Trans. on Computers*, vol. 51, pp. 84-89, Jan. 2002.

[8]. Y. Ma, "A simplified architecture for modulo (2^n+1) multiplication", *IEEE Transactions on Computers*, vol. 47, no. 3, pp. 333-337, Mar. 1998.

[9]. A. Garcia, U. Meyer-Base, A. Lloris, and F. J. Taylor, "RNS implementation of FIR filters based on distributed arithmetic using field-programmable logic," in *Proceedings of the IEEE International Symposium on Circuits and Systems*, pp. 486-489, May/June 1999.

[10]. Y. Wang, X. Song and M. Aboulhamid, "Adder based residue to binary number converters for $(2^n, 2^n-1, 2^n+1)$," *IEEE Transactions on Signal Processing*, vol. 50, pp. 1772-1779, July 2002.

NONUNIFORM SAMPLING DRIVER DESIGN FOR OPTIMAL ADC UTILIZATION

F. Papenfuß [1], *Y. Artyukh* [2], *E. Boole* [2], *D. Timmermann* [1]

[1] Inst. of Applied Microelectronics and Computer Science, University of Rostock, Germany,
EMail: frank.papenfuss@etechnik.uni-rostock.de, dirk.timmermann@etechnik.uni-rostock.de
[2] Institute of Electronics and Computer Science (IECS), Riga, Latvia,
EMail: artyukh@edi.lv, buls@edi.lv

ABSTRACT

Deliberate nonuniform sampling promises increased equivalent sampling rates with reduced overall hardware costs of the DSP system. The equivalent sampling rate is the sampling rate that a uniform sampling device would require in order to achieve the same processing bandwidth. Equivalent bandwidths of realizable systems may well extend into the GHz range while the mean sampling rate stays in the MHz range. Current prototype systems (IECS) have an equivalent bandwidth of 1.6GHz at a mean sampling rate of 80MHz, achieving 40 times the bandwidth of a classic DSP system that would operate uniformly at 80MHz (cf. [1]). Throughout the literature on nonuniform sampling (e. g. [2] and [3]) different sampling schemes have been investigated. This paper focuses on nonuniform sampling schemes optimized for fast and efficient hardware implementations. To our knowledge this is the first proposal of an efficient nonuniform sampling driver (SD) design in the open literature.

1. INTRODUCTION

Nonuniform sampling circumvents traditional sampling limitations requiring a sampling rate of at least twice the input signal bandwidth. A special unit, the SD, generates the sampling pulse train used to digitize the analog signal. To realize a SD in digital circuits obviously a synchronous design is desirable keeping the design process simple. According to sampling theory a straightforward implementation of a SD produces sampling instances deliberately jittered around a fixed system clock. A pseudo random number generator (PRNG) generates numbers passed to a digitally controllable delay line (DCDL) delaying pulses produced by a central controlling unit. Though each digital circuit driving an ADC performs, strictly speaking, periodic sampling with jitter (due to phase noise) a simple SD realization depicted in Fig. 1 does it deliberately. One can consider the time axis being separated into time slots having system clock duration T_{clk}. Inside each slot a sampling instance t_k is produced. For the SD design to be successful it must realize a sampling instance with equal probability anywhere in the k-th time slot in order to achieve a constant probability to produce sampling points anywhere at the time axis. Failure to do so will result in an undesired spectrum of the sampled signal containing spurious frequencies (cf. [4]). Therefore, sampling algorithm, architecture and SD hardware implementation have to be carefully aligned to obtain maximum benefit from nonuniform sampling.

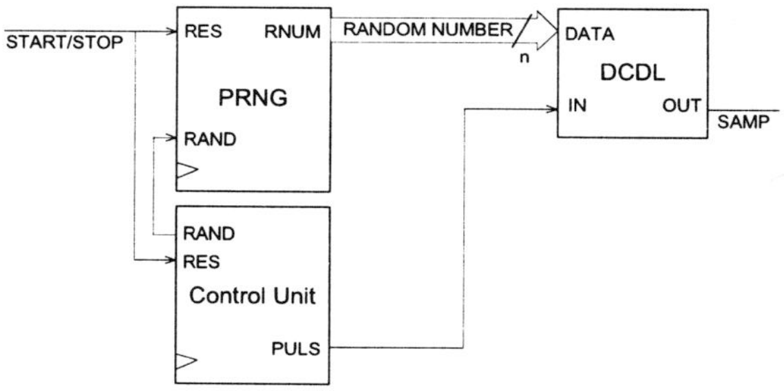

Fig. 1: General synchronous SD building block.

This is due to the convolution of sampling process spectrum and signal spectrum. The process is illustrated in Fig. 3 showing the spectrum of a signal with one component. The figure is added to stress the importance of matching the probability density function (PDF) of a sampling instance to the SD system clock period.

Real circuits will not produce sampling points with infinite accuracy but will realize time increments of so called time quantum size T_Q. This renders the sampling instance PDF discrete (see Fig. 2). The equivalent sampling rate is given by the inverse time quantum. The limited amount of time increments in a matched time slot is expressed by the system clock period to time quantum ratio M

$$M = \frac{T_{clk}}{T_Q}. \tag{1}$$

This is a key parameter of a sampling driver since it represents the factor by which the processing bandwidth of the digital system is increased. It is convenient to keep

0-7803-7761-3/03/$17.00 ©2003 IEEE

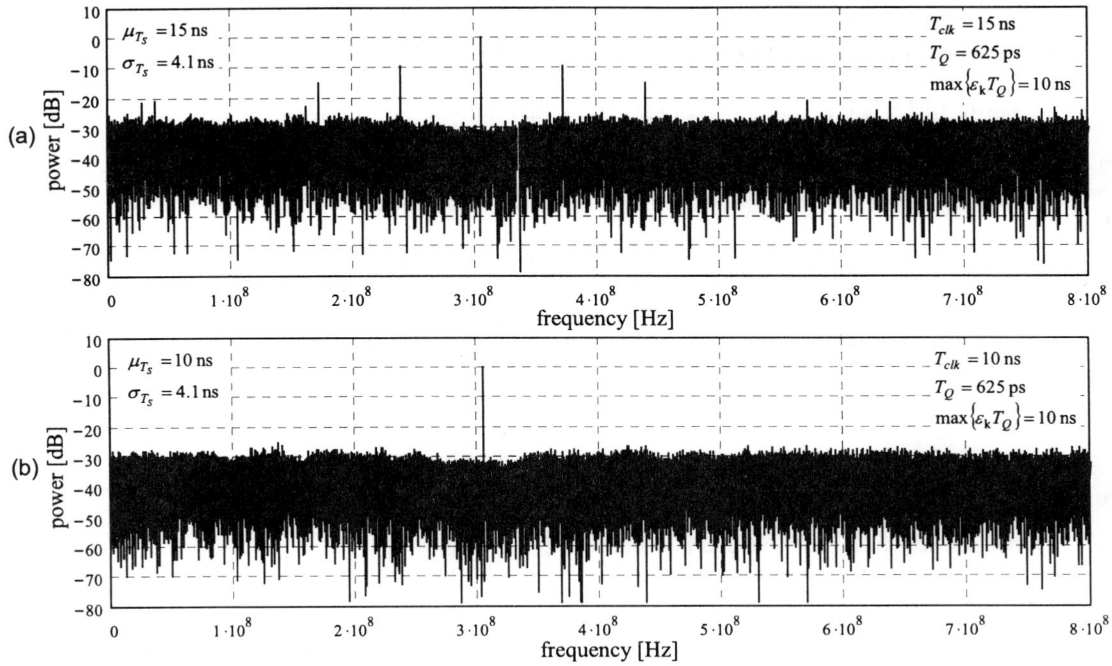

Fig. 3: Power density spectra (via DFT) of a signal containing exactly one frequency at 305MHz. The PDF of a sampling instance is (a) not matched and (b) matched to SD system clock period.

M at a power of two to fully utilize the bits of the data vector entering the DCDL. The process of sampling instant generation is well known as periodic sampling with jitter (cf. [1]) and can be described by

$$t_k = kT_{clk} + \varepsilon_k T_Q \qquad k, \varepsilon_k \in \mathbf{N} \qquad 0 \le \varepsilon_k < M \, , \tag{2}$$

where ε_k is a pseudo random number produced by the PRNG at the k-th time slot. Unfortunately equation (2)

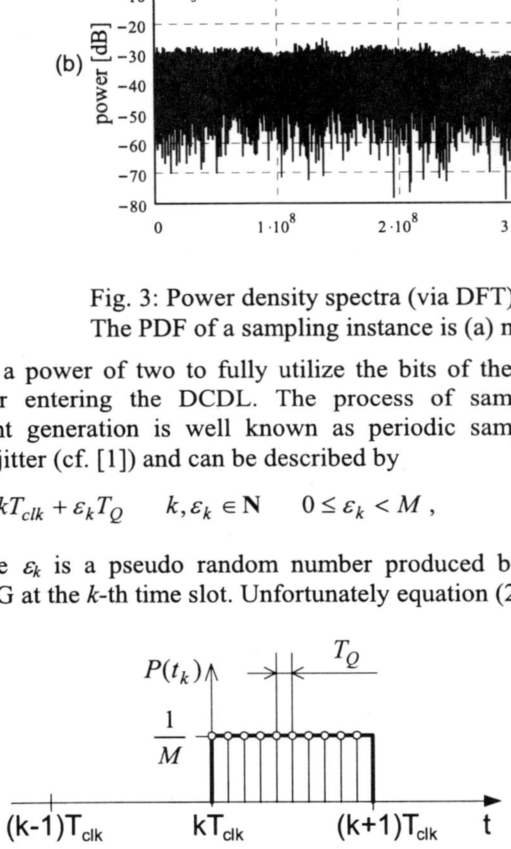

Fig. 2: PDF of sampling instances at k-th time slot.

has a bad property. Two successive samples may be separated by only the time quantum T_Q. It therefore seems to be desirable to define a setup of a random experiment that will serve to assess the quality of generated sampling sequences. Let T_s be the time between consecutive samples, the intersample time

$$T_s = t_k - t_{k-1} \, . \tag{3}$$

Thus, the intersample time is a derived random variable. For convenience we define the Laplacian random experiment E_0

$$\begin{aligned}
&\Omega_0 = \{\omega_0^{(0)}, \omega_1^{(0)}, \ldots, \omega_n^{(0)}, \ldots, \omega_{M^2-1}^{(0)}\} \\
&\Omega_0 = \{(0,0),(0,1),\ldots(i_n, j_n),\ldots,(M-1, M-2),(M-1, M-1)\} \\
&i_n, j_n, n \in \mathbf{N}; \quad 0 \le i_n, j_n, n < M \\
&\omega_n^{(0)} \equiv (i_n, j_n) \equiv \quad (\varepsilon_{k-1} = i_n \quad and \quad \varepsilon_k = j_n) \\
&P(\omega_n^{(0)}) = \frac{1}{M^2}
\end{aligned} \tag{4}$$

where (i_n, j_n) denotes the event that ε_{k-1} takes on value i_n and ε_k takes on value j_n. It is easy to see that there are M^2 such events. Assuming that both ε_{k-1} and ε_k have uniform distribution and are statistically independent, it immediately follows that the events (i_n, j_n) have equal probability $1/M^2$. Observing that, given (2) and (3) T_s will never become larger than $2M$ one can define a different random experiment E_1 with a set Ω_1 of $2M$ elementary events

$$\begin{aligned}
&\Omega_1 = \{\omega_0^{(1)}, \omega_1^{(1)}, \ldots, \omega_l^{(1)}, \ldots, \omega_{2M-1}^{(1)}\} \\
&\Omega_1 = \{0, T_Q, 2T_Q, \ldots, lT_Q, \ldots, (2M-1)T_Q\} \qquad l \in \mathbf{N}, 0 \le l < 2M \, , \\
&\omega_l^{(1)} \equiv T_s = lT_Q
\end{aligned} \tag{5}$$

where the l-th event in Ω_1 denotes the event that T_s takes on value lT_Q. Unlike the events in Ω_0 the events in Ω_1 do not occur with equal probability. However, these probabilities can be obtained from events in E_0 by

$$\begin{aligned}
&P(\omega_l^{(1)}) = \sum_{\omega_n^{(0)} \in \omega_l^{(1)}} P(\omega_n^{(0)}) = \frac{1}{M^2} \sum_{\omega_n^{(0)} \in \omega_l^{(1)}} 1 \, . \\
&n, l \in \mathbf{N}; \quad 0 \le n < M^2; \quad 0 \le l < 2M
\end{aligned} \tag{6}$$

Using (1), (2) and (3) we can say when an event in Ω_0 is said to be a favorable event in terms of an event in Ω_1

$$\omega_n^{(0)} \in \omega_l^{(1)} \quad if \quad l = j_n + M - i_n \,. \tag{7}$$

Applying (6) and (7) the probabilities for all events in Ω_1 and hence the discrete PDF of T_s can be estimated. It is sketched in Fig. 4.

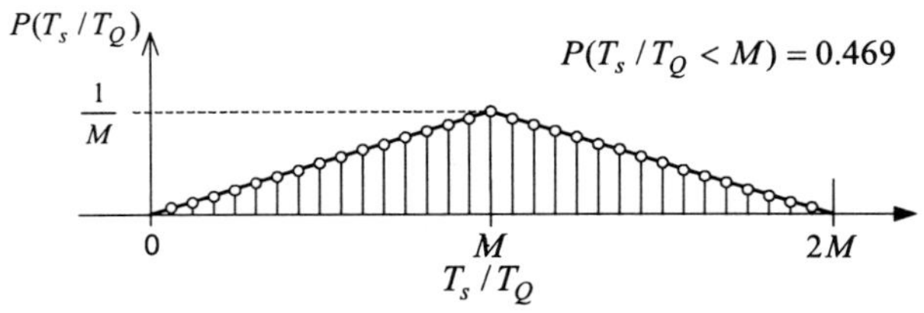

Fig. 4: PDF of intersample time.

In an optimal SD design for full ADC utilization the system clock period T_{clk} is usually matched to the minimum conversion time of the attached ADC

$$T_{clk} = \min\{T_{ENCODE}\} \,. \tag{8}$$

This is justified by the design decision to operate the sampling driver also in a uniform mode (ε_k constant) in which case the ADC should be fully utilized. Hence the intersample time constraint

$$\begin{aligned} t_k - t_{k-1} &\geq T_{clk} \\ T_s &\geq MT_Q \end{aligned} \tag{9}$$

must always be met. Given (6) and (7) we can calculate the probability that (9) is not met in case of this straightforward design. It is about 47% and we conclude that such a straightforward design is not usable as a sampling driver.

2. PHASE SHIFTING

To avoid too short intersample times we propose a different sampling scheme that introduces phase shifts at times when consecutive samples occur too close for the ADC to handle. The modified sampling scheme can be described recursively as described in (10). Only the control unit of the design shown in Fig. 1 needs to be changed. A phase shift of the sampling pulse means deferring it one SD system clock period (i. e. 360°).

$$t_k = t_{k-1} - \varepsilon_{k-1} T_Q + T_{clk} + \begin{Bmatrix} 0 & if \quad \varepsilon_{k-1} < \dfrac{M}{2} \\ T_{clk} & otherwise \end{Bmatrix} + \varepsilon_k T_Q \,. \tag{10}$$

$$k, \varepsilon_k \in \mathbf{N}; \quad 0 \leq \varepsilon_k < M$$

In the modified design the control unit of the sampling driver constantly checks the random numbers that have been and are produced by the PRNG and introduces phase shifts described by (10). Deliberate phase shifting fundamentally changes the sampling scheme. Periodic sampling with deliberate jitter becomes additive random sampling. A single sampling instance has still an evenly distributed PDF but is now stretched over two SD system clock periods because of the introduced phase shift.

A well-known property of the additive random sampling scheme is that it produces a constant valued sampling point density function (SPDF) after a transient phase. This property (based on the central limit theorem) is extensively treated in [1]. The PDF of the derived random variable T_s looks different than in the previous Section 1. Using (3) and (10) one can write

$$T_s = \left(M - \varepsilon_{k-1} + \begin{Bmatrix} 0 & if \quad \varepsilon_{k-1} < \dfrac{M}{2} \\ M & otherwise \end{Bmatrix} + \varepsilon_k \right) T_Q \,. \tag{11}$$

$$k, \varepsilon_k \in \mathbf{N}; \quad 0 \leq \varepsilon_k < M$$

We use the same method as in Section 1 to determine the probabilities $P(T_s / T_Q = l)$ a priori. The result is depicted in Fig. 5. The probability to produce intersample times less than MT_Q is around 11% and thus non-zero.

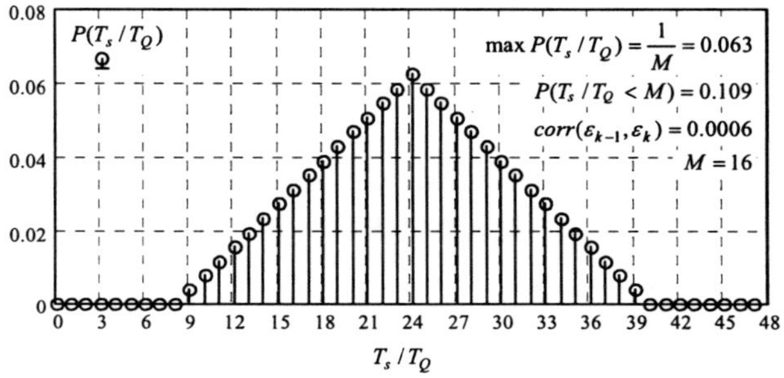

Fig. 5: PDF of intersample time with phase shift.

The property of such a sampling driver is certainly better but would still be too demanding for an attached ADC being operated at its limits as described above. A solution will be presented in the next Section.

3. RANDOM NUMBER CORRELATION

When generating pseudo random numbers maximum length linear feedback shift registers (LFSR) are commonly used (see [5] and [6]). Using a slice of bits from a longer LFSR one can write for consecutive random numbers ε_{k-1} and ε_k

$$\varepsilon_k = (2\varepsilon_{k-1} + \tau_k) \bmod 2^n \qquad k, \varepsilon_k \in \mathbf{N} \quad \tau_k \in \{0,1\}, \tag{12}$$

where τ_k is a binary random number assumed to be evenly distributed and n is the dimension of the vector passed as random number to the DCDL. It is important to note that, given (12), the probabilities for events in Ω_0 are no longer evenly distributed. Through computer simulation the distribution of T_s was determined given that two successive random numbers are correlated as defined in

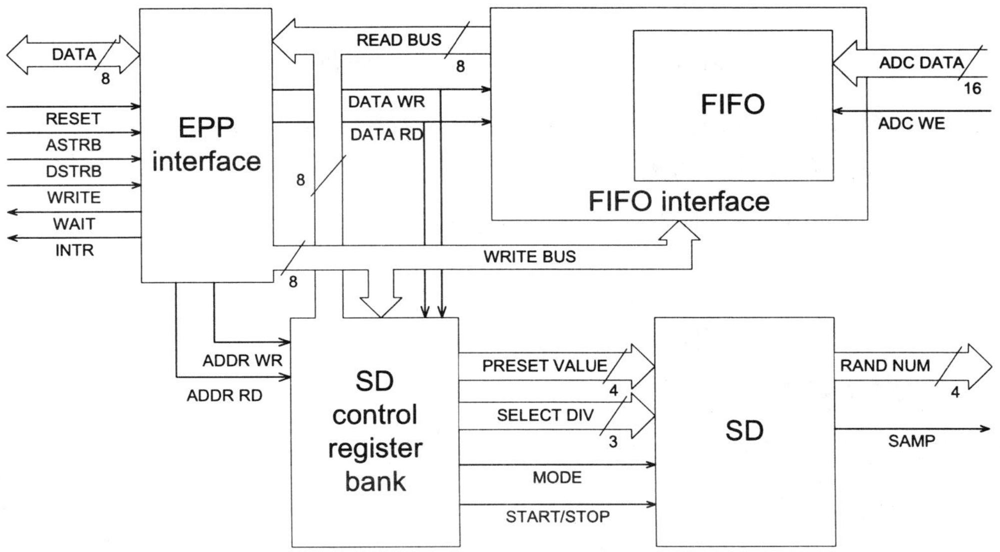

Fig. 6: Overall sampling driver architecture (main units).

(12). The simulation results clearly reveal that the PDF of the intersample time now satisfies the constraints (Fig. 7).

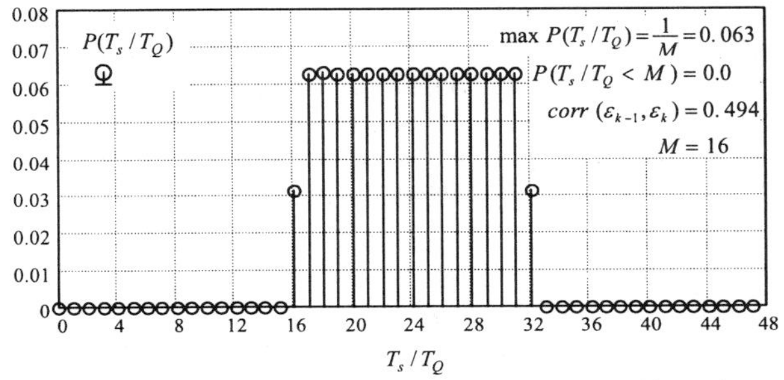

Fig. 7: PDF of intersample time with phase shift and correlated random numbers.

4. SD INTEGRATION

Obviously the SD cannot stand alone. It has to be integrated into a larger architecture. In Fig. 6 an integration of a SD design into a sample recording architecture is presented. The extended parallel port (EPP) interface is used to program the SD. Functions such as starting and stopping as well as uniform and nonuniform sampling are realized. Sampled data is buffered in the FIFO and read via the EPP. The design was tested and implemented in a FPGA using VHDL.

5. CONCLUSIONS

In this paper we have derived an efficient sampling algorithm for deliberate additive random sampling. The algorithm is well suited to fully utilize the minimum conversion time of an ADC. Cost reduction is achieved because cheaper ADCs can be used instead of expensive ones while processing the same or even higher bandwidth than the comparable traditional system. Alternatively it is possible that a system utilizing the suggested design is used to process GHz signals fully digital with higher bandwidth and/or resolution than possible today in a traditional design using a cutting edge ADC.

It was shown that introducing deliberate correlation into the random number generation process is beneficial. It will create exactly the sampling pulse train that best utilizes the ADC. In our case the introduced correlation coefficient of consecutive random numbers is 0.5.

6. REFERENCES

[1] Y. Artyukh, A. Ribakov, V. Vedin, "Evaluation of Pseudorandom Sampling Processes". In *Proceedings of the 1997 Workshop on Sampling Theory and Applications,* pp. 361-363, SampTA '97.

[2] I. Bilinskis and A. Mikelsons, *Randomized Signal Processing*, Prentice Hall International, 1992, ISBN 0137510748.

[3] J. J. Wojtiuk, *Randomised Sampling for Radio Design*, PhD Thesis, University of South Autralia, 2000

[4] D. M. Bland and A. Tarczynski, "The effect of sampling jitter in a digitized signal". In *Proceedings of the 1997 IEEE International Symposium on Circuits and Systems,* Vol. 4, pp. 2685-2688, ISCAS '97.

[5] M. Abamovici, M. A. Breuer and A. D. Friedman, *Digital Systems Testing and Testable Design*, IEEE Press, 1990, ISBN 0780310624.

[6] P. Alfke, "Efficient Shift Registers, LFSR Counters, and Long Pseudo-Random Sequence Generators", *XILINX XAPP 052,* July 7, 1996 (Version 1.1)

DESIGN OF A HIGH SPEED REVERSE CONVERTER FOR A NEW 4-MODULI SET RESIDUE NUMBER SYSTEM

Bin Cao, Thambipillai Srikanthan and Chip-Hong Chang

Centre for High Performance Embedded Systems
Nanyang Technological University
N4-B3b-06, Nanyang Avenue, Singapore 639798

ABSTRACT

This paper presents an elegant residue-to-binary algorithm for a new 4-moduli set $\{2^n - 1, 2^n, 2^n + 1, 2^{2n} + 1\}$ Residue Number System. Our reverse conversion algorithm takes advantage of the special number properties of the proposed moduli set. The recently introduced New CRT theorem has been exploited to simplify the costly and time consuming modular corrections. The resulting architecture is notably simple and can be realized in hardware with only bit reorientation operation and a Multi-Operand Modular Adder. In terms of area-time complexity, the performance of the new reverse converter has surpassed the previously reported reverse converters for several celebrated four-moduli sets.

1. INTRODUCTION

The intense drive for low power-delay product has been at the forefront of drastic and escalated development for the leading-edge VLSI and giga-scale-integration (GSI) circuits. Residue Number System (RNS), with the inherent carry-free operations, parallelism and fault-tolerance properties, has emerged as a prerogative candidate for low power, high speed and specialized high precision digital signal processing [1][2]. By decomposing large binary numbers into smaller residues, addition and subtraction in RNS arithmetic have no inter-digit carries or borrows, and multiplication can be performed without the need to generate partial products.

As the peripheral interfaces of digital systems are accustomed to the weighted number system, the overhead incurred in the conversions into and out of the residue number system has been the major barrier that prohibits the pervasion of RNS-based arithmetics. It is well acknowledged that the forward conversion from binary number to residues is conceivably simple and some efficient architectures have been devised for residue generator based on special moduli of the *2^n-type* (i.e., expressible in the form 2^n or $2^n \pm 1$) [3]. The performance bottleneck lies in the reverse conversion from residues to binary number, which involves computational intensive modular multiplications and additions that annihilate the performance gained in the RNS.

During the past several decades, the reverse conversion algorithms are based primarily on the Chinese Remainder Theorem (CRT) [4][5] or Mixed-Radix Conversion (MRC) [6][7]. The use of CRT entails a large modular adder, whereas MRC is a sequential process that often requires a number of look-up tables. Both of them are not efficient for practical systems with large dynamic ranges.

In order to reduce the area-time complexity in the implementation of the reverse converters, special moduli sets have been used extensively. The triple moduli set $\{2^n - 1, 2^n, 2^n + 1\}$ is by far the most pervasive RNS due to its inherent number theoretic properties in CRT algorithm. Several worthful architectures of reverse converters based on CRT have been proposed. Andraos *et al.* [8] derived the closed form expressions of the moduli inverses, and used them to reduce the conversion complexity. Carry Save Adders (CSA) with End Around Carry (EAC) were introduced by Piestrak [9] to allow a very efficient implementation of the reverse converters based on CRT. Recently, Y. Wang [10][11] devised the revolutionary New CRT theorems to capitalize the merits of the classical CRT and MRC algorithms. If the New CRT I and II are applied to the special triple moduli set, the implementation of the reverse converter can be conceivably simplified, and the conversion speed surpasses the converters designed with the celebrated traditional CRT[11].

In this paper, we leverage on the New CRT to explore new moduli set that has increased parallelism and extended dynamic range. The proposed new four-moduli set, $\{2^n - 1, 2^n, 2^n + 1, 2^{2n} + 1\}$ for any integer n, retains the simplistic forward conversion and efficient residue arithmetic. Its ensuing residue-to-binary converter benefits from the same efficiency betterment as that of the triple moduli set, but its dynamic range has improved from $3n - 1$ bits to $5n - 1$ bits. We will compare our reverse converters with other four-moduli set reverse converters. One of the newest reverse converters based on the four-moduli set was proposed by Bhardwaj *et al.* [12] which used the moduli set $\{2^n - 1, 2^n, 2^n + 1, 2^{n+1} + 1\}$ for odd n. The dynamic range is $4n + 1$ bits and the converter is constructed with linear ROM and modular adders. An improved memoryless reverse converter for four-moduli set of $\{2^n - 1, 2^n, 2^n + 1, 2^{n+1} - 1\}$ with an overall latency of $14n + 8$ full adder delays was recently proposed by Vinod and Premkumar [13]. This four-moduli set is valid only for even values of n, which unnecessarily increases the interconnects and adder complexity as a consequence of the increased ratio of the bus widths of RNS to weighted binary number for several commonly encountered binary IO interfaces. Performance analysis of these converters will be presented in this paper.

2. BACKGROUND

In a Residue Number System (RNS), an integer X can be represented by an n-tuple of residues, $(x_1, x_2, \ldots, x_n)$ defined over a set of relatively prime moduli $\{P_1, P_2, \ldots, P_n\}$, where $gcd(P_i, P_j) = 1$ for $1 \leq i, j \leq n$, and $i \neq j$. The term forward and reverse conversions are commonly used to describe the conversions of

0-7803-7761-3/03/$17.00 ©2003 IEEE

the weighted binary representation X to and fro its residue set (x_1, x_2, …, x_n), respectively. The bottleneck operation in a RNS is the reverse conversion, which can be calculated using the Chinese Remainder Theorem (CRT) [1] as follows:

$$X = \left| \sum_{i=1}^{n} \left| \frac{x_i}{\hat{M}_i} \right|_{P_i} \cdot \hat{M}_i \right|_M \tag{1}$$

where $M = \prod_{i=1}^{n} P_i$, $\hat{M}_i = M / P_i$.

The weighted binary number X can be also calculated by the New CRT-I [10]:

$$X = x_1 + P_1 \left| \begin{matrix} k_1(x_2 - x_1) + k_2 P_2 (x_3 - x_2) + \ldots + \\ k_{n-1} P_2 P_3 \ldots P_{n-1} (x_n - x_{n-1}) \end{matrix} \right|_{P_2 P_3 \ldots P_{n-1} P_n} \tag{2}$$

where

$$\begin{aligned} &|k_1 P_1|_{P_2 P_3 \ldots P_n} = 1 \\ &|k_2 P_1 P_2|_{P_3 P_4 \ldots P_n} = 1 \\ &\ldots \\ &|k_{n-1} P_1 P_2 \ldots P_{n-1}|_{P_n} = 1 \end{aligned} \tag{3}$$

The product of all moduli, M is called the dynamic range. For any integer X within the dynamic range, i.e., $0 \le X < M$, its residue representation is canonical.

3. REVERSE CONVERSION ALGORITHM

The following proposition is needed for our new theorem. The proof has been omitted due to the page constraint. It can be established based on the principle of divisibility that for natural numbers a, b and c, if $a|b$ and $a|(b + c)$, then $a|c$.

Proposition 1: The moduli from the four-moduli set $\{2^n - 1, 2^n, 2^n + 1, 2^{2n} + 1\}$ are pairwise relatively prime for any natural number n.

With the pairwise relatively prime four-moduli set $\{2^n - 1, 2^n, 2^n + 1, 2^{2n} + 1\}$, the reverse conversion algorithm from the residues (x_1, x_2, x_3, x_4) to the weighted binary number X can be derived from the New CRT-I [10].

Let $P_1 = 2^n$, $P_2 = 2^n + 1$, $P_3 = 2^{2n} + 1$ and $P_4 = 2^n - 1$. By Eqn. (3), we have

$$\left| k_1 \cdot 2^n \right|_{(2^n+1)(2^{2n}+1)(2^n-1)} = 1 \tag{4}$$

$$\left| k_2 \cdot 2^n (2^n + 1) \right|_{(2^n-1)(2^{2n}+1)} = 1 \tag{5}$$

$$\left| k_3 \cdot 2^n (2^{2n} + 1)(2^n + 1) \right|_{2^n-1} = 1 \tag{6}$$

The three multiplicative inverses, k_1, k_2 and k_3 are given as follows:

$$k_1 = 2^{3n} \tag{7a}$$

$$k_2 = 2^{3n-2} + 2^{2n-1} - 2^{n-2} \tag{7b}$$

$$k_3 = 2^{n-2} \tag{7c}$$

Proof of (7a):

$$\left| k_1 \cdot 2^n \right|_{(2^n+1)(2^{2n}+1)(2^n-1)} = \left| 2^{3n} \cdot 2^n \right|_{2^{4n}-1} = \left| 2^{4n} \right|_{2^{4n}-1} = 1 \quad \square$$

Proof of (7b):

$$\begin{aligned} &\left| k_2 \cdot 2^n (2^n + 1) \right|_{(2^n-1)(2^{2n}+1)} \\ &= \left| \begin{matrix} -2^{2n-2}(2^{3n} - 2^{2n} + 2^n - 1) + 2^{5n-2} + 2^{4n-1} \\ -2^{3n-2} + 2^{4n-2} + 2^{3n-1} - 2^{2n-2} \end{matrix} \right|_{2^{3n}-2^{2n}+2^n-1} \quad \square \\ &= \left| 2^{3n} - 2^{2n} + 2^n \right|_{2^{3n}-2^{2n}+2^n-1} = 1 \end{aligned}$$

Proof of (7c):

$$\begin{aligned} &\left| k_3 \cdot 2^n (2^{2n} + 1)(2^n + 1) \right|_{2^n-1} = \left| 2^{n-2} \cdot 2^n (2^{2n} + 1)(2^n + 1) \right|_{2^n-1} \\ &= \left| 2^{n-2} \cdot 1 \cdot 2 \cdot 2 \right|_{2^n-1} = \left| 2^n \right|_{2^n-1} = 1 \quad \square \end{aligned}$$

Theorem 1: In a RNS defined by the four moduli set $\{2^n - 1, 2^n, 2^n + 1, 2^{2n} + 1\}$, the weighted binary number X can be calculated from the residues (x_1, x_2, x_3, x_4) by:

$$X = x_2 + 2^n \cdot \left| A \cdot x_1 + B \cdot x_2 + C \cdot x_3 + D \cdot x_4 \right|_{2^{4n}-1} \tag{8}$$

where

$$A = 2^{n-2}(2^{3n} + 2^{2n} + 2^n + 1) \tag{9a}$$

$$B = -2^{3n} \tag{9b}$$

$$C = 2^{3n} - (2^n + 1)(2^{3n-2} + 2^{2n-1} - 2^{n-2}) \tag{9c}$$

$$D = (2^n + 1)(2^{3n-2} + 2^{2n-1} - 2^{n-2}) - 2^{n-2}(2^n + 1)(2^{2n} + 1) \tag{9d}$$

Proof: Let $P_1 = 2^n$, $P_2 = 2^n + 1$, $P_3 = 2^{2n} + 1$ and $P_4 = 2^n - 1$, then the corresponding residues are x_2, x_3, x_4 and x_1. Substituting P_1 to P_4 and their residues, and the values of k_1 to k_3 from Eqn. (7a) to (7c) into Eqn. (2), we have

$$X = x_2 + 2^n \left| \begin{matrix} 2^{3n}(x_3 - x_2) + (2^{3n-2} + 2^{2n-1} - 2^{n-2})(2^n + 1)(x_4 - x_3) \\ + 2^{n-2}(2^n + 1)(2^{2n} + 1)(x_1 - x_4) \end{matrix} \right|_{2^{4n}-1}$$

The result follows directly by expanding the terms and simplifying the coefficients of x_1, x_2, x_3 and x_4. □

4. HARDWARE REALIZATION

Theorem 1 forms the basis of the implementation of a VLSI efficient reverse converter. To simplify the modular arithmetics, the closed form expressions of Eqn. (8) and (9) can be reduced to a form that can be used to substantially lower the hardware complexity.

Let the residues x_1, x_2, x_3 and x_4 be represented by binary strings of different lengths:

$$\begin{aligned} x_1 &= (X_{1,n-1} X_{1,n-2} \ldots X_{1,1} X_{1,0})_2 \\ x_2 &= (X_{2,n-1} X_{2,n-2} \ldots X_{2,1} X_{2,0})_2 \\ x_3 &= (X_{3,n} X_{3,n-1} \ldots X_{3,1} X_{3,0})_2 \\ x_4 &= (X_{4,2n} X_{4,2n-1} \ldots X_{4,1} X_{4,0})_2 \end{aligned}$$

Furthermore, let $Y = \left| \sum_{i=1}^{4} \beta_i \right|_{2^{4n}-1}$

where $\beta_1 = |A \cdot x_1|_{2^{4n}-1}$, $\beta_2 = |B \cdot x_2|_{2^{4n}-1}$, $\beta_3 = |C \cdot x_3|_{2^{4n}-1}$ and $\beta_4 = |D \cdot x_4|_{2^{4n}-1}$.

According to Theorem 1, X can be calculated by $X = x_2 + 2^n \cdot Y$. The following property from [1] can be used to evaluate Y.

Property 1: Multiplying an integer, x by 2^r modulo $(2^p - 1)$ can be accomplished by expressing x as a p bits binary representation and then shifting it circularly by r bits to the left.

As an example, let $x = 6$, $p = 5$ and $r = 3$, then we have

$$\left|x \cdot 2^r\right|_{2^p-1} = \left|6 \cdot 2^3\right|_{2^5-1} = CLS(00110,3) = (10001)_2 = 17$$

where the function $CLS(x, r)$ is used to denote a circular shift of the binary number x by r bits to the left. In the sequent, we apply Property 1 recursively to replace β_i by the arrangement of bits from the residues with interleaving strings of constants '0' and '1'.

Evaluating β_1

$$\beta_1 = \left|2^{n-2}\left(2^{3n} + 2^{2n} + 2^n + 1\right) x_1\right|_{2^{4n}-1}$$
$$= \left|CLS\{(CLS(x_1,3n) + CLS(x_1,2n) + CLS(x_1,n) + x_1), n-2\}\right|_{2^{4n}-1} \quad (10)$$
$$= \underbrace{X_{1,1}X_{1,0}}_{2}\underbrace{X_{1,n-1}\ldots X_{1,0}}_{n}\underbrace{X_{1,n-1}\ldots X_{1,0}}_{n}\underbrace{X_{1,n-1}\ldots X_{1,0}}_{n}\underbrace{X_{1,n-1}\ldots X_{1,2}}_{n-2}$$

Evaluating β_2

$$\beta_2 = \left|-2^{3n} \cdot x_2\right|_{2^{4n}-1} = \underbrace{\overline{X}_{2,n-1}\ldots\overline{X}_{2,0}}_{n}\underbrace{11\ldots11}_{3n} \quad (11)$$

Evaluating β_3

$$\beta_3 = \left|C \cdot x_3\right|_{2^{4n}-1} = \left|\left\{\left(2^{4n-2} + 2^{2n-2}\right) + \left(2^{3n-2} + 2^{n-2}\right)\right\} x_3\right|_{2^{4n}-1}$$
$$= \left|\beta_{3,1} + \beta_{3,1}\right|_{2^{4n}-1}$$

where

$$\beta_{3,1} = \underbrace{\overline{X}_{3,1}\overline{X}_{3,0}}_{2}\underbrace{1\ldots1}_{n-1}\underbrace{\overline{X}_{3,n}\ldots\overline{X}_{3,0}}_{n+1}\underbrace{1\ldots1}_{n-1}\underbrace{\overline{X}_{3,n}\ldots\overline{X}_{3,2}}_{n-1} \quad (12)$$

$$\beta_{3,2} = 0\underbrace{X_{3,n}\ldots X_{3,0}}_{n+1}\underbrace{0\ldots0}_{n-1}\underbrace{X_{3,n}\ldots X_{3,0}}_{n+1}\underbrace{0\ldots0}_{n-2} \quad (13)$$

Evaluating β_4

$$\beta_4 == \left|\left(2^{3n-1} - 2^{n-1}\right) \cdot x_4\right|_{2^{4n}-1} = \left|\beta_{4,1} + \beta_{4,2}\right|_{2^{4n}-1}$$

where

$$\beta_{4,1} = \underbrace{X_{4,n}X_{4,n-1}\ldots X_{4,0}}_{n+1}\underbrace{0\ldots0}_{2n-1}\underbrace{X_{4,2n}X_{4,2n-1}\ldots X_{4,n+1}}_{n} \quad (14)$$

$$\beta_{4,2} = \underbrace{1\ldots1}_{n}\underbrace{\overline{X}_{4,2n}\ldots\overline{X}_{4,0}}_{2n+1}\underbrace{1\ldots1}_{n-1} \quad (15)$$

Now Y can be expressed as the sum of the binary strings given by Eqn. (10) to (15) as follows:

$$Y = \left|\beta_1 + \beta_2 + \beta_{3,1} + \beta_{3,2} + \beta_{4,1} + \beta_{4,2}\right|_{2^{4n}-1} \quad (16)$$

From Eqn. (8) and (16), it can be seen that the calculation of X is magically simple and elegant. It involves only modulo $2^{4n} - 1$ adder, which can be realized by equally efficient architecture as that used in the triple moduli set of $\{2^n - 1, 2^n, 2^n + 1\}$. The complicated arithmetics in the algorithm of [12][13] have been eliminated with the aid of Property 1. With larger dynamic range and higher parallelism, more promising avenues for optimization at the Carry Save Adder (CSA) tree than that of the triple moduli set are envisaged.

Fig. 1 shows the architecture of our reverse converter for the four-moduli set $\{2^n - 1, 2^n, 2^n + 1, 2^{2n} + 1\}$. The bits orientation block generates β_1, β_2, $\beta_{3,1}$, $\beta_{3,2}$, $\beta_{4,1}$, $\beta_{4,2}$ by simply manipulating the routings of the bits from the input residue numbers of x_1, x_2, x_3 and x_4. The summation can be done by one $(6, 2^{4n} - 1)$ MOMA [9], which consists of a $4n$-bit 3-level CSA with EAC, and a $4n$-bit 1's complement adder. It should be noted that as Y is weighted by 2^n, addition of x_2 in Eqn. (8) incurs no additional hardware and computation cost as it can be directly wired to the right of Y.

5. PERFORMANCE EVALUATION

In this section, we will estimate the hardware costs and evaluate the delay of the reverse converter. For ease of reference to relevant reverse converters, the standard practice of measuring complexity in terms of the number and delay of fundamental logic units like full adder (FA), standard logic gates, etc., is adopted for the reverse converters with generic n. The wire loads are normally neglected.

In the bits orientation block of Fig. 1, there are n inverters used for β_2, $n + 1$ for $\beta_{3,1}$ and $2n + 1$ for $\beta_{4,2}$. The total number of inverters is $4n + 2$. The delay of this block is equal to t_{INV}, which is just the delay of an inverter.

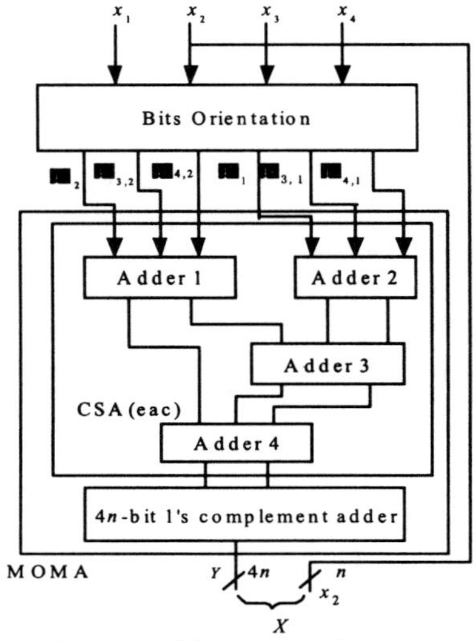

Fig. 1 *Realization of the proposed reverse converter*

It is also observed that there are strings of consecutive '1's and '0's embedded in the binary expressions of β_2, $\beta_{3,1}$, $\beta_{3,2}$, $\beta_{4,1}$ and $\beta_{4,2}$. Without annihilating the basic simplistic architecture of Fig. 1, a first cut simplification of the Carry-Save-Adders can be accomplished by manipulating those embedded constant strings. The fundamental idea is: a FA with a constant input '1' can be reduced to a pair of two-input XNOR and OR gates, a FA with a constant input '0' can be reduced to a pair of two-input XOR and AND gates, and a FA with an input '0' and an input '1' can be reduced to an inverter.

After this simplification, the hardware cost of our converter for the proposed four-moduli set is equivalent to $7n + 6$ FAs, $2n - 1$ pairs of 2-input XOR/AND gates, $4n$ pairs of 2-input XNOR/OR gates, $6n - 1$ inverters, and one $4n$-bit carry propagate adder (CPA). It should be noted that further reduction in hardware complexity is possible if similarly weighted bits are swapped across β_i to optimize the use of constant inputs. The total delay of the proposed reverse converter is $t_{INV} + 3t_{FA} + 2t_{CPA(4n)}$, if the CPA of [9] is employed, and we call this low complexity reverse converter the cost-effective version (CE). The delay can be lowered to $t_{INV} + 3t_{FA} + t_{CPA(2n)} + t_{MUX} + 2t_{NAND}$ at the expense of increasing hardware cost when the CPA circuit of [14] is adopted, and we call this faster converter design the high speed version (HS).

In [12], a residue-to-binary converter for the four-moduli set $\{2^n - 1, 2^n, 2^n + 1, 2^{n+1} + 1\}$ is proposed. Excluding the delay of multiplexer, ROM lookup tables and XOR gates, the delay of this converter is approximately $(10n + 13)t_{FA}$. A different four-moduli set $\{2^n - 1, 2^n, 2^n + 1, 2^{n+1} - 1\}$ reverse converter has been proposed in [13]. The total delay is $(14n + 8)t_{FA}$, and the design consists of 13 adders/subtractors with bit width ranging from $n + 1$ to $5n - 1$ bits. Table 1 summarizes the comparisons of the reverse converters for these four-moduli sets. Only the number of adders and their operand width are considered for the converter of [13] and ours. This is because for these two converters, the total hardware costs are contributed predominantly by the adders. The same simplification method of FAs in our converter is adopted for the evaluation of the area complexity of [13] so that any discrepancy is attributed to the differences in algorithm and design rather than the inconsistency in analysis or biased use of computing primitives. If we assume that the modular adder/subtractor in [13] uses the modular adder proposed in [3], then the total area of this converter is estimated to be approximately $O(37n)$. Due to heterogynous operand width and varying types of adders and subtractors used in [13], it is conjectured that the scalability and modularity of [13] are inferior to ours. The reverse converter in [12] is implemented with ROM and modular adders, the area complexity of $O(n^2)$ has been reported as the equivalent area occupied by the full adders from the fabricated circuits, but the area complexity of ours is $O(18n)$ for HS version and $O(14n)$ for CE version.

Table 1: Comparisons of Reverse Converters of 4-Moduli Sets

Converters		Area	Time
[12]		$O(n^2)$	$t_{MUX} + t_{LUT} + t_{XOR} + (10n+13)t_{FA}$
[13]		$O(37n)$	$(14n+8)t_{FA}$
Ours	HS	$O(18n)$	$t_{MUX} + t_{INV} + 2t_{NAND} + (2n+3)t_{FA}$
	CE	$O(14n)$	$t_{INV} + (8n+3)t_{FA}$

6. CONCLUSION

In the past decade, the triple moduli set $\{2^n -1, 2^n, 2^n + 1\}$ has been the prerogative choice of RNS due to its relative efficiency in the reverse converter design. The revolutionary New CRT I have rekindled the interest of exploring new moduli sets that have never been thought of before. In this paper we have discovered a new four-moduli set $\{2^n -1, 2^n, 2^n + 1, 2^{2n} + 1\}$ with a larger dynamic range. We show that this new moduli set possesses a number of interesting characteristics that make its reverse conversion algorithm under the New CRT I amenable to efficient VLSI implementation. The new reverse converter completely eliminates the need for modulo addition and allows further optimization opportunity in a simple MOMA realization. With an area complexity of $O(18n)$ and a delay of approximately $2n + 3$ FA's for the HS version, and with an area complexity of $O(14n)$ and a delay of approximately $8n + 3$ FA's for the CE version, they are more efficient than the reverse converters for the four-moduli sets $\{2^n - 1, 2^n, 2^n + 1, 2^{n+1} + 1\}$ and $\{2^n - 1, 2^n, 2^n + 1, 2^{n+1} - 1\}$, both in hardware area and computation delay.

7. REFERENCES

[1] N. S. Szabo and R. I. Tanaka, *Residue Arithmetic and its Applications to Computer Technology*. New York: McGraw Hill, 1967.

[2] M. A. Soderstrand, W. K. Jenkins, G. A. Jullien and F. J. Taylor, *Residue Number System Arithmetic: Modern Applications in Digital Signal Processing*. New York: IEEE Press, 1986.

[3] S. J. Piestrak, "Design of residue generators and multioperand modular adders using carry-save adders". *IEEE Trans. Comput.*, vol. 423, no. 1, pp. 68-77, 1994.

[4] K. M. Elleithy and M. A. Bayoumi, "Fast and flexible architectures for RNS arthmetic decoding". *IEEE Trans. Circuits Syst.*, vol. 39, no. 4, pp. 226-235, 1992.

[5] F. Barsi and M. C. Pinotti, "A fully parallel algorithm for residue to binary conversion". *Information Proc. Lett.*, vol. 50, pp. 1-8, 1994.

[6] C. H. Huang, "A fully parallel mixed radix conversion algorithm for residue number applications". *IEEE Trans. Comput.*, vol. 32, no. 4, pp. 398-402, 1983.

[7] H. M. Yassine and W. R Moore, "Improved mixed-radix conversion for residue number system architectures". *IEE Proc.-G*, vol. 1338, no. 1, pp. 120-124, 1991.

[8] S. Andraos and H. Ahmad, "A new efficient memoryless residue to binary converter". *IEEE Trans. Circuits Syst.*, vol. 35, no. 11, pp. 1441-1444, 1988.

[9] S. J. Piestrak, "A high speed realization of residue to binary number system converter". *IEEE Trans. Circuits Syst. -II* , vol. 42, no. 10, pp. 661-663, 1995.

[10] Y. Wang, "New Chinese Remainder Theorems". *Proc. 32th Asilomar Conf. Signals, Syst., Comput.*, vol. 1, pp. 165-171, 1998.

[11] Y. Wang, X. Song, M. Aboulhamid and H. Shen, "Adder based residue to binary number converters for $(2^n - 1, 2^n, 2^n + 1)$". *IEEE Trans. Signal Processing*, vol. 50, no. 7, pp. 1772-1779, 2002.

[12] M. Bhardwaj, T. Srikanthan and C. T. Clarke, "A reverse converter for the 4-moduli superset $\{2^n - 1, 2^n, 2^n + 1, 2^{n+1} + 1\}$". *Proc. of 14th IEEE Symposium on Computer Arithmetic*, Adelaide, Australia, pp. 168-175, Apr., 1999.

[13] A. P. Vinod and A. B. Premkumar, "A memoryless reverse converter for the 4-moduli superset $\{2^n - 1, 2^n, 2^n + 1, 2^{n+1} - 1\}$". *Journal of Circuits, Systems, and Computers*, vol. 10, no. 1&2, pp. 85-99, 2000.

[14] M. Bhardwaj, A. B. Premkumar and T. Srikanthan, "Breaking the $2n$-bit carry propagation barrier in residue to binary conversion for the $\{2^n - 1, 2^n, 2^n + 1\}$ moduli set". *IEEE Trans. Circuits Syst. -I*, vol. 45, no. 9, pp. 998-1002, 1998.

CONFLICT-FREE PARALLEL MEMORY ACCESS SCHEME FOR FFT PROCESSORS

Jarmo H. Takala, Tuomas S. Järvinen, and Harri T. Sorokin

Tampere University of Technology, P.O.B. 553, FIN-33101 Tampere, Finland

ABSTRACT

In this paper, a parallel access scheme for constant geometry FFT algorithms is proposed, which allows conflict-free access of operands distributed over parallel memory modules. The scheme is a linear transformation and the address generation is performed with the aid of bit-wise XOR operations. Different FFT lengths can be supported with the aid of a simple address rotation unit. The scheme is general supporting several radices in FFT computations and different numbers of parallel memory modules. The scheme allows parallel butterfly computations independent of the FFT length.

1. INTRODUCTION

Due to the symmetric structure of Cooley-Tukey fast Fourier transform (FFT) algorithm, it lends itself to VLSI implementations and several realizations with varying level of parallelism have been proposed over the years. In parallel implementations, memory bandwidth may limit the performance of the system; in radix-S FFT algorithm, the butterfly operation consumes S operands and produces S results, thus S memory accesses should be performed in parallel. Often the memory bandwidth is increased by partitioning the memory into S independent memory modules, which can be accessed simultaneously. Such a memory architecture is referred to as an interleaved memory system and it is illustrated in Fig. 1. If the FFT architecture contains d radix-S butterfly units, dS parallel memory modules are needed.

Often the operands to be accessed simultaneously lie in the same memory module thus the parallel access cannot be performed. In FFT algorithms, such conflicts are due to the data reordering between the computational columns as seen in Fig. 2 where signal flow graphs of 16-point constant geometry algorithms for radix-2 and radix-4 FFT are shown. Therefore, the principal problem in interleaved memory systems is to find a method to distribute data over the memory modules in such a way that the conflicts are avoided.

The conflict-free access can arranged with the aid of double-buffering but this is extremely expensive when long FFTs are to be computed. In order to minimize the memory consumption, the results of computation should be written into the same memory locations where the operands were read. In [1], it was found that in radix-2 FFT the operands can be distributed over two memory modules based on the parity of the operand index. This observation was exploited in [2] where an address generator for 2-memory radix-2 FFT systems were proposed. In [3], the address generation was simplified at the expense of additional registers in the interconnection network. The registers are used to delay the conflicting write accesses. In [4], the conflicts were avoided by allocating an additional memory module for delaying certain write accesses. The previous solutions covered only radix-2 FFTs and assume that a single radix-2 butterfly is computed at a time.

A general solution for radix-S FFTs was given in [5] but again it was assumed that the computations are performed a single butterfly at a time. Memory addressing in FFT systems containing parallel butterfly computations is considered in [6] where the access conflicts are avoided by reordering the operands in an interconnection network. The reordering is performed in time and space, which requires additional operand registers.

The memory distribution in FFT computation has also been considered in supercomputing area. In [7], an access scheme based on linear transformation is proposed but the described address generation is complex requiring binary matrix multiplication. A far more simple scheme was proposed in [8]. However, access conflicts cannot be completely avoided in these schemes.

In this paper, a parallel access scheme is proposed, which allows conflict-free access of operands distributed over the parallel memories. The scheme supports several radices in FFT computations and it allows parallel butterfly computations independent of the FFT length.

2. PRELIMINARIES

FFT algorithms can be scheduled into a form where the interconnections between the processing columns of the signal flow graph are stride permutations. In general, interconnections in radix-2^s FFT algorithm are related to stride-by-2^s permutation. Such permutations can be described with the aid of matrix transpose; stride-by-S permutation of an N-element vector can be performed by dividing the vector into S-element sub vectors, organizing them into $S \times (N/S)$ matrix form, transposing the obtained matrix, and rearranging the result back to the vector representation [9]. Another interpretation is to use indexing functions as used in the following formal definition.

Definition 1 (Stride Permutation) *Let us assume a vector* $X = (x_0, x_1, \ldots, x_{N-1})$. *Stride-by-$S$ permutation reorders* X *as* $Y = (x_{f_{N,S}(0)}, x_{f_{N,S}(1)}, \ldots, x_{f_{N,S}(N-1)})^T$ *where the index function* $f_{N,S}(i)$ *is given as*

$$f_{N,S}(i) = (iS \bmod N) + \lfloor iS/N \rfloor \mid N \text{ rem } S = 0,\ i = 0, 1, \ldots, N-1 \quad (1)$$

where mod *is the modulus operator,* $\lfloor\cdot\rfloor$ *is the floor function, and* rem *is remainder.*

The stride permutation of an array X can also be expressed in matrix form as $Y = P_{N,S}X$ where $P_{N,S}$ is stride-by-S permutation matrix of order N.

In this paper, we limit ourselves to practical cases where array lengths and strides are powers of two, $N = 2^n, S = 2^s$. A property of stride permutations in such cases is given in the following.

0-7803-7761-3/03/$17.00 ©2003 IEEE

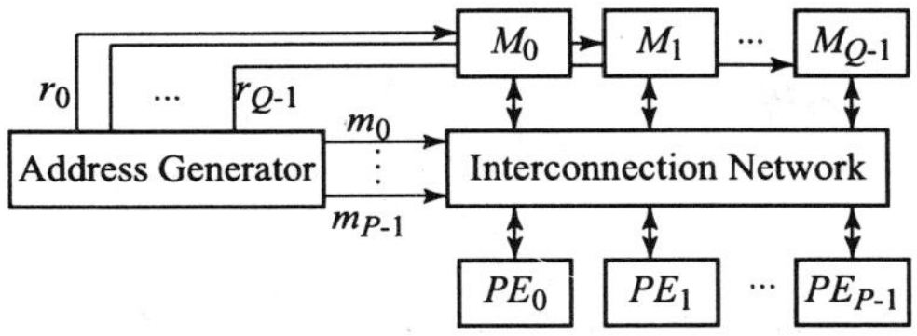

Figure 1: Interleaved memory system. PE_i: Processing element. M_i: Memory module. m_i: Module address. r_i: Row address.

Theorem 1 (Factorization of stride permutations) *Let $ab \leq N$, then*

$$P_{N,ab} = P_{N,a}P_{N,b} = P_{N,b}P_{N,a} \quad (2)$$

The proof for the previous theorem can be found, e.g., from [9].

Let us assume that an N-element array is to be distributed over Q independent memory modules, $N = 2^n, Q = 2^q$. In such a case, an access scheme performs two mappings; the index address of an element, $a = (a_{n-1}, a_{n-2}, \ldots, a_0)^T$, is mapped onto a module address, $m = (m_{q-1}, \ldots, m_0)^T$, and a row address, $r = (r_{n-q-1}, \ldots, r_0)^T$. It should be noted that in this representation the least significant bit of a is in the bottom of the vector. If the access scheme is a linear transformation, the address arithmetic is based on modulo-2 arithmetic, which implies that the arithmetic is realized with bit-wise XOR operations. The address mappings in the linear transformation can be expressed with binary transformation matrices as

$$r = Ka\,;\; m = Ta \quad (3)$$

The matrices K and T are the row and module transformation matrix, respectively. Often K is defined as

$$K = \left(I_{n-q}0_{(n-q),q}\right) \quad (4)$$

where I_k denotes the identity matrix of order k and $0_{i,k}$ denotes an $i \times k$ matrix of zeros. Then the row address is obtained simply by extracting the $(n-q)$ most significant bits of the address a.

3. CONFLICT-FREE ACCESS SCHEME FOR STRIDE PERMUTATION

Let us first investigate the read and write accesses in radix-2 FFT illustrated in Fig. 2(a). In order to minimize memory consumption, the results should be stored into the same memory locations where the operands were obtained. In the first iteration, operands are read in linear order, i.e., according to $P_{16,1}$. The results need also to be written in linear order, although they should be permuted according to stride-by-2, $P_{16,2}$. This implies that, in the second iteration, the operands should be read and written according to $P_{16,2}$. In the third iteration, the accesses are according to $P_{16,4}$, which is due to Theorem 1. Eventually we find that $\log_2 N$ different strides are needed, i.e., all the strides of power-of-two from 1 to $N/2$.

In [10], it was shown that an access scheme supporting several strides cannot be designed for matched memory systems, if the access should be conflict-free regardless of the array length and initial address. A matched memory system refers to an organization where the number of memory modules is the same as the number of operands accessed in a cycle by the processing elements. We may, however, relax the requirements. We make the following assumptions: a) the array length is constant and power-of-two, $N = 2^n$, b) the array is stored in 2^n-word boundaries, c) the number of memory modules is a power-of-two, $Q = 2^q$, and d) the strides in stride permutation access are powers-of-two, $S = 2^s$. Assumption a) implies that constraints on the initial address need to be set resulting in the assumption b). Such an constraint has already been used in several commercial DSP processors for performing circular addressing. Assumption c) is actually practical in digital systems. Assumption d) implies that the address mapping should produce a q-bit memory module address and a $(n-q)$-bit row address. All these assumptions can be considered practical.

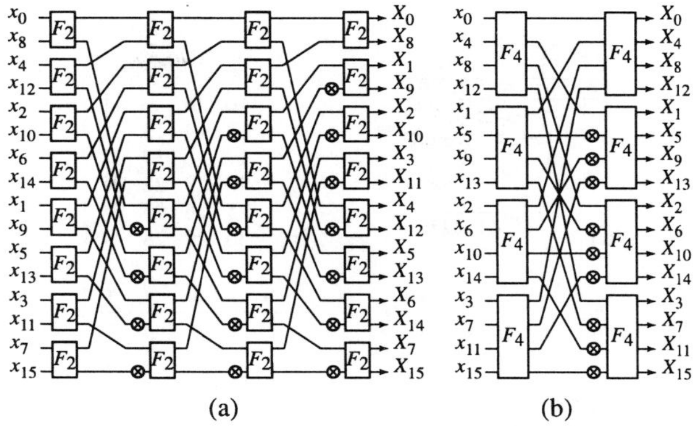

Figure 2: Signal flow graphs of FFT algorithms: (a) radix-2 and (b) radix-4 algorithm. F_k: k-point FFT.

3.1. Access Scheme

We propose an access scheme, which can be used under the previous assumptions. The row and module address mappings are defined as follows

$$\begin{aligned} r_i &= a_{i+q}, i = 0, 1, \ldots, n-q-1; \\ m_i &= \bigoplus_{k=0}^{l_{n,q}(i)} a_{(jq+i) \bmod n},\; i = 0, 1, \ldots, q-1\,; \\ l_{n,q}(i) &= \lfloor (n+q-\gcd(q, n \bmod q) - i - 1)/q \rfloor \end{aligned} \quad (5)$$

where $\oplus$ is bit-wise XOR operation and $\gcd(\cdot)$ is the greatest common denominator.

The module transformation is dependent on the array length N and the number of memory modules Q, thus we introduce a new notation for the module transformation matrix: $T_{N,Q}$. The module transformation matrix $T_{32,4}$ is the following

$$T_{32,4} = \begin{pmatrix} 0 & 1 & 0 & 1 & 1 \\ 1 & 0 & 1 & 0 & 1 \end{pmatrix}. \quad (6)$$

The contents of the memory modules stored according to $T_{32,4}$ is illustrated in Fig. 3(b) and the module address can be computed with the circuit shown in Fig. 3(a). All the possible parallel accesses in this case are listed in the following:

$$\begin{aligned} P_{32,1} &: ([0,1,2,3],[4,5,6,7],\ldots,[28,29,30,31]) \\ P_{32,2} &: ([0,2,4,6],[8,10,12,14],\ldots,[25,27,29,31]) \\ P_{32,4} &: ([0,4,8,12],[16,20,24,28],\ldots,[19,23,27,31]) \\ P_{32,8} &: ([0,8,16,24],[1,9,17,25],\ldots,[7,15,23,31]) \\ P_{32,16} &: ([0,16,1,17],[2,18,3,19],\ldots,[14,30,15,31]) \end{aligned}$$

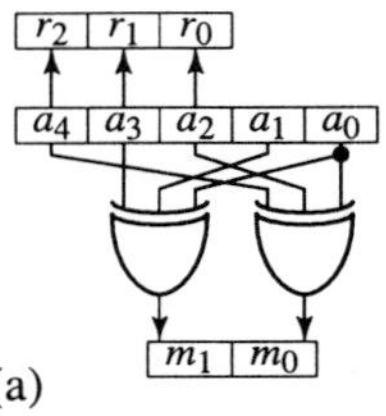

(a)

(b)

m	0	1	2	3
	0	3	2	1
	7	4	5	6
	10	9	8	11
	13	14	15	12
	19	16	17	18
	20	23	22	21
	25	26	27	24
	30	29	28	31

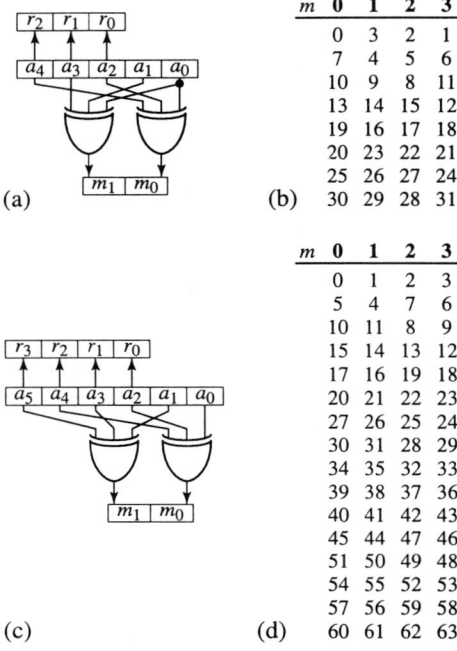

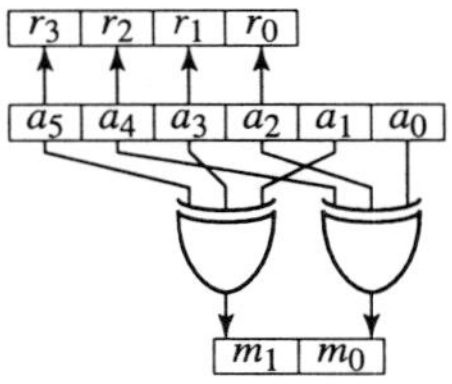

(c)

(d)

m	0	1	2	3
	0	1	2	3
	5	4	7	6
	10	11	8	9
	15	14	13	12
	17	16	19	18
	20	21	22	23
	27	26	25	24
	30	31	28	29
	34	35	32	33
	39	38	37	36
	40	41	42	43
	45	44	47	46
	51	50	49	48
	54	55	52	53
	57	56	59	58
	60	61	62	63

Figure 3: Access scheme for 4-module system: (a) module address generation and (b) contents of memory modules for transform matrix in (6) and (c) module address generation and (d) contents of memory for matrix in (7).

It can be seen that all the possible stride permutation accesses are conflict-free. The transformation matrix $T_{64,4}$ would be

$$T_{64,4} = \begin{pmatrix} 1 & 0 & 1 & 0 & 1 & 0 \\ 0 & 1 & 0 & 1 & 0 & 1 \end{pmatrix} \qquad (7)$$

and the module address generation is shown in Fig. 3(c) and the storage is depicted in Fig. 3(d). Once again, all the possible power-of-two stride accesses are conflict-free.

3.2. Validation

The presented access scheme has been validated with the aid of computer simulations by generating several storage organizations and verifying that each access is conflict-free. For a given array length $N = 2^n$, the number of memory modules Q was varied to cover all the possible numbers of powers-of-two, i.e., $Q = 2^0, 2^1, \ldots, 2^{n-1}$. For each parameter pair (N, Q), all the stride permutation accesses were performed with strides covering all the possible powers-of-two: $S = 2^0, 2^1, \ldots, 2^{n-1}$ and each parallel access was verified to be conflict-free. The power-of-two array lengths were iterated from 2^1 to 2^{20}. The extensive simulation showed that the presented access scheme provides conflict-free parallel stride permutation access in practical cases, i.e., array lengths up to 2^{20}, for all the possible power-of-two strides on matched memory systems where the number of memory modules is a power-of-two.

$$T_{32,16} = \begin{pmatrix} 0 & 1 & 1 & 0 & 0 \\ 0 & 0 & 1 & 1 & 0 \\ 0 & 0 & 0 & 1 & 1 \\ 1 & 0 & 0 & 0 & 1 \end{pmatrix}$$

$$T_{64,16} = \begin{pmatrix} 0 & 0 & 1 & 0 & 1 & 0 \\ 0 & 0 & 0 & 1 & 0 & 1 \\ 1 & 0 & 0 & 0 & 1 & 0 \\ 0 & 1 & 0 & 0 & 0 & 1 \end{pmatrix}$$

$$T_{128,16} = \begin{pmatrix} 0 & 0 & 0 & 1 & 0 & 0 & 1 \\ 1 & 0 & 0 & 0 & 1 & 0 & 0 \\ 0 & 1 & 0 & 0 & 1 & 1 & 0 \\ 0 & 0 & 1 & 0 & 0 & 1 & 1 \end{pmatrix}$$

$$T_{256,16} = \begin{pmatrix} 1 & 0 & 0 & 0 & 1 & 0 & 0 & 0 \\ 0 & 1 & 0 & 0 & 0 & 1 & 0 & 0 \\ 0 & 0 & 1 & 0 & 0 & 0 & 1 & 0 \\ 0 & 0 & 0 & 1 & 0 & 0 & 0 & 1 \end{pmatrix}$$

Figure 4: Module transformation matrices in 16-memory module system.

3.3. Address Generation

Before going into implementations, we may investigate the structure of $T_{N,Q}$ when N is varied. In practical systems, the number of memory modules , Q, is constant; Q is only a design time parameter. As an example transformation matrices for 64-module systems are illustrated in Fig. 4 and few observations can be made from the structure of these matrices. The off-diagonal ones affect at most the $q - 1$ least significant bits of the address a as defined in (5). In addition, the structure of off-diagonals depends on the relation between n and q but in practical systems q is constant, thus the structure depends only on the array length. However, there are only q different structures; the off-diagonal structure has periodic behavior when the array length is increasing. In Fig. 4, one complete period is shown and $T_{512,16}$ would have the same off-diagonal structure as $T_{32,12}$. The structure of off-diagonals implies that several array lengths can be supported if a predetermined control word configures additional hardware to perform the functionality of the off-diagonals. Such a configuration is actually simple by noting that the form of off-diagonals in different array lengths indicates rotation of the least significant bits in a. The number of bits rotated is dependent on the relation between n and q.

According to the previous observations, the computation of the module address m can be interpreted as follows. First, the address a is divided into q-bit fields, F^i, starting from the least significant bit of a. If $e = n \bmod q > 0$, the e most significant bits of a exceeding the q-bit block border are extracted as a bit vector L. Next, a q-bit field X is formed by extracting the $(q - \gcd(q, e))$ least significant bits of the address a and placing zeros to the most significant bits. The bit vector X is rotated $g = (n - q \bmod q)$ bits to the left to obtain a bit vector O. Finally the module address m is obtained by performing bit-wise XOR operation between the vectors F_i, L, and O. A principal block diagram of the module address generation according to the previous interpretation is illustrated in Fig. 5(a). This block diagram contains a rotation unit shown in Fig. 5(b), which computes the vector O.

The main advantage of the presented scheme can be seen from the block diagram in Fig. 5. In the address generation, each individual XOR is performed on at most $\lfloor n/q \rfloor + 2$ bit lines while

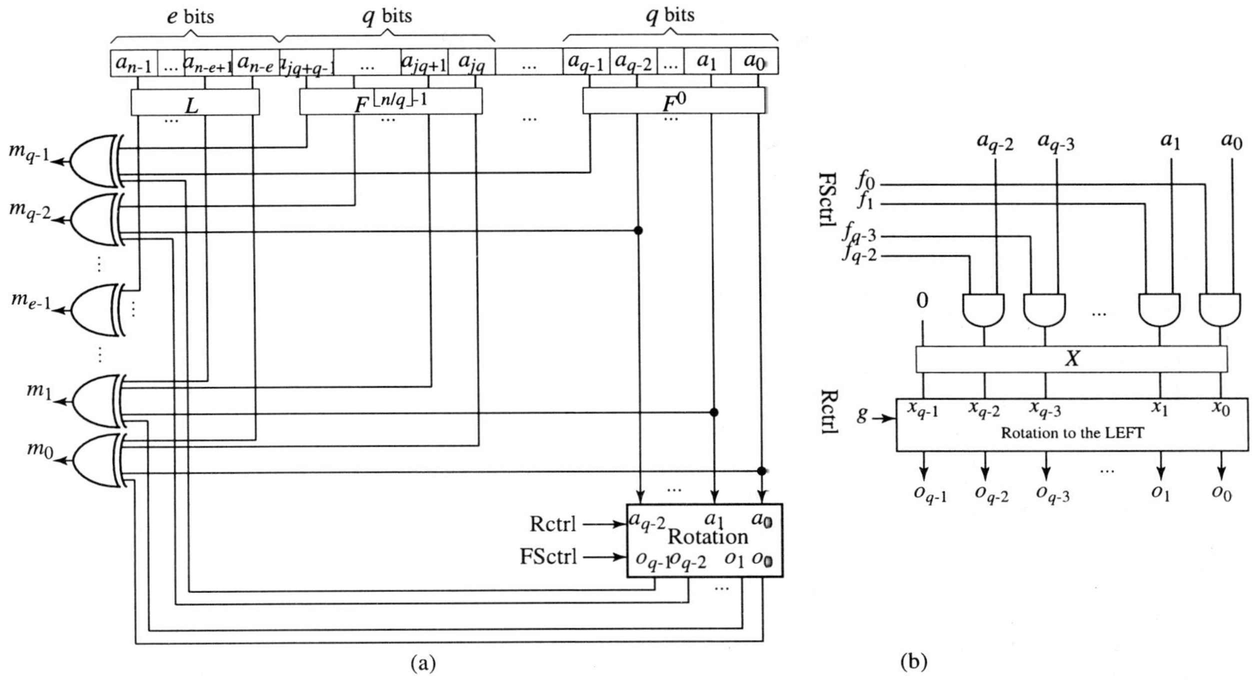

Figure 5: Principal block diagram of (a) module address generation and (b) rotation unit. Rctrl: Rotation control. FSctrl: Field selection control.

in other schemes, e.g., in [7], some XORs require all the n address bits, which complicates implementations when several array lengths need to be supported. The support for different array lengths requires only a single predetermined control word, which defines the bit selection and rotation. This control word needs to be modified only when the array length is changed.

4. CONCLUSIONS

In this paper, a conflict-free stride permutation access scheme for constant geometry FFT computations was presented. It was assumed that 2^n data elements are distributed over 2^q independent memory modules. The performed simulations showed that all the possible power-of-two stride permutation accesses are conflict-free. The module address generation is simple requiring only bit-wise XOR operations. It was shown that several array lengths can be supported by including a q-bit rotation into the module address generator. In this case, all the additional operations are performed on the $q-1$ least significant bits of the index address.

5. REFERENCES

[1] M. C. Pease, "Organization of large scale Fourier processors," *J. Assoc. Comput. Mach.*, vol. 16, no. 3, pp. 474–482, July 1969.

[2] D. Cohen, "Simplified control of FFT hardware," *IEEE Trans. Acoust., Speech, Signal Processing*, vol. 24, no. 6, pp. 255–579, Dec. 1976.

[3] Y. Ma, "An effective memory addressing scheme for FFT processors," *IEEE Trans. Signal Processing*, vol. 47, no. 3, pp. 907–911, Mar. 1999.

[4] C.-H. Chang, C.-L. Wang, and Y.-T. Chang, "A novel memory-based FFT processor for DMT/OFDM applications," in *Proc. IEEE ISCAS*, Orlando, FL, U.S.A., May 30 –June 2 1999, vol. 4, pp. 1921–1924.

[5] L. G. Johnson, "Conflict free memory addressing for dedicated FFT hardware," vol. 39, no. 5, pp. 312–316, May 1992.

[6] J. A. Hidalgo, J. Lopéz, F. Argüello, and E. L. Zapata, "Area-efficient architecture for fast Fourier transform," *IEEE Trans. Circuits Syst. II*, vol. 46, no. 2, pp. 187–193, Feb. 1999.

[7] A. Norton and E. Melton, "A class of boolean linear transformations for conflict-free power-of-two stride access," in *Proc. Int. Conf. Parallel Processing*, St. Charles, IL, U.S.A., Aug. 17–21 1987, pp. 247–254.

[8] D. T. Harper III, "Block, multistride vector, and FFT accesses in parallel memory systems," *IEEE Trans. Parallel and Distrib. Syst.*, vol. 2, no. 1, pp. 43–51, Jan. 1991.

[9] J. Granata, M. Conner, and R. Tolimieri, "Recursive fast algorithms and the role of the tensor product," *IEEE Trans. Signal Processing*, vol. 40, no. 12, pp. 2921–2930, Dec. 1992.

[10] D. T Harper III, "Increased memory performance during vector accesses through the use of linear address transformations," *IEEE Trans. Comput.*, vol. 41, no. 2, pp. 227–230, Feb. 1992.

Parallel *Sub-Convolution* Filter Bank Architectures

Andrew A. Gray
Jet Propulsion Laboratory, California Institute of Technology
4800 Oak Grove Drive Pasadena, CA 91101

***Abstract*-This paper provides an overview of the design and properties of parallel discrete-time filter bank architectures based on the concept of frequency-domain *sub-convolution* developed by the author. It will be demonstrated that this lossless filter bank method is an excellent choice for implementing many signal processing functions in real-time high rate systems. These filter bank architectures incorporate vector processing, the discrete Fourier transform-inverse discrete Fourier transform (DFT-IDFT) overlap-and-save convolution method [1], and the sub-convolution method. The parallel processing architectures presented here facilitate processing for very high rate sampled systems (multi-giga-samples per second) with lower rate complementary metal oxide semiconductor (CMOS) hardware with relatively low complexity (low transistor count). Complexity comparisons will demonstrate that the sub-convolution filter bank results in less complex concurrent implementations than parallel time-domain convolution and conventional frequency-domain convolution methods for many processing rate reductions and filter orders. In addition, the sub-convolution filter bank may be used to provide intermediate computation gain, with computation requirements lying in between those of these two conventional methods. As such, the parallel sub-convolution filter bank may also be useful in low-power hardware realizations.**

I. Introduction

Figure 1 illustrates a parallel DFT-IDFT filtering architecture for frequency domain filtering using the overlap and save method. The DFT-IDFT length is L+1 (L is odd), and M, the downsample rate, is the number of samples the input window "slides". The architecture in Figure 1 has 50% input vector overlap, that is the downsample rate, is equal to half the input vector length, $M = (L+1)/2$.

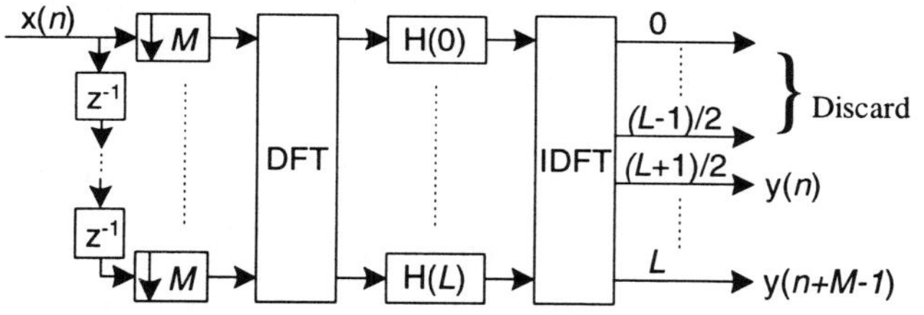

Figure 1. Overlap-and-Save FIR Filter

The research described in this publication was carried out as part of a task funded by the *Technology and Applications Program* (TAP) at the Jet Propulsion Laboratory, California Institute of Technology, under contract with the National Aeronautics and Space Administrations

With such an architecture a M+1 coefficient filter may be implemented in the frequency domain. The filter, $h(n)$, is zero-padded to length L+1 and then transformed to the discrete frequency domain via the DFT, to obtain the frequency domain coefficients, $H(k)$=DFT{$h(n)$} [2]. It is well known that a FIR filter with an order M or less can be used with this architecture. Similar derivations for FIR filters have been developed in [2,3,4] for implementation in software and hardware. The limitation to all of these methods is that the DFT-IDFT, or fast Fourier transform-inverse fasts Fourier transform (FFT-IFFT) lengths are increased to increase the order of the FIR filter to be implemented. To implement very large filters with such methods in VLSI application specific integrated circuits (ASICs) or field programmable gate arrays (FPGAs) becomes very complex. However, the total number of computations to perform the convolution decreases and the processing rate may decrease as well. Alternative approaches include subband convolution, using analysis and synthesis filter banks; subband convolution may be used to implement arbitrarily long FIR filters [3]. However, due to the complexity of the hardware implementations for these methods more simple approaches are desirable. The methods presented here target low-complexity VLSI architectures in which parallel processing or processing rate reduction is required.

II. Parallel Sub-convolution Filter Banks

Consider the simple convolution sum of equation (1). The convolution may be broken into numerous sub-convolutions, each time shifted input convolved with a *sub-filter*, as indicated.

$$y(n) = \sum_{k=0}^{N-1} x(n-k)h(k) = \sum_{k=0}^{j_1} x(n-k)h(k) + \sum_{k=j_1+1}^{j_2} x(n-k)h(k) + \ldots + \sum_{k=j_{R-1}+1}^{N-1} x(n-k)h(k) \quad (1)$$

First, observe that each sample vector input to the DFT of Figure 1, and therefore the frequency domain vector, is a time delay of M samples from the next sample vector input. From (1), it is obvious that each of the sums are themselves a convolution with a block of the filter or *sub-filter*, the *sub-convolutions* are performed with sub-filters, the sum of their outputs is equal to the convolution of the input, $x(n)$, with the filter $h(n)$. Each of these sub-convolutions may be implemented in the frequency domain using the technique illustrated in Figure 1, then the results summed to yield the convolution output. To break a convolution up into R equal length sub-convolutions, each (L+1) in length, using this method would require R DFTs, R IDFTs, and R filter banks. Assuming 50% overlap, the DFT-IDFT pairs would each be in (L+1) length, however simplifications requiring only one DFT-IDFT pair are possible with one additional constraint. The constraint is derived by simply realizing that each input vector to the DFT of Figure 1 is a shift in time of M samples, therefore each frequency domain vector is separated in time from the previous or next vector by M sample periods. From (1), if $j_i + M = j_{i+1} \ \forall\ i$, that is the time delay between each sub-filter is equal to the time delay between time-consecutive input vectors, then the convolution of (1) may be calculated in the frequency domain by simply delaying the frequency domain vectors and multiplying by the appropriate frequency domain sub-filter. These sub-filters are generated as follows.

$$H_k(i) = DFT\{h_k(n)\} \quad i = 0,\dots,L, \ k = 1,\dots,R \\ n = 0,\dots,L \tag{2}$$

and $h_k(n)$ is the kth zero padded sub-filter given by:

$$\begin{aligned} h_k(n) &= h(n+(k-1)M) \quad n = 0,\dots,\frac{L-1}{2}, \quad k = 1,\dots,R \\ &= 0 \quad n = \frac{L+1}{2},\dots,L, \quad k = 1,\dots,R \end{aligned} \tag{3}$$

Using simple properties of linearity only one DFT-IDFT pair of this length is required as all of the frequency domain sub-convolutions may be calculated then summed in the frequency domain then transformed back into the time domain. The resulting architecture is illustrated in Figure 2. This system performs convolution at a rate of $1/M$ that of the sample rate. It is clear that the length of the DFT-IDFT pairs may be chosen with rate reduction as the principal design criterion independent of FIR filter length. This simple architecture then allows relatively short DFT-IDFT lengths to be used to reduce the processing rate of high order FIR filtering or correlation operations, yielding overall simple designs. Rate reductions of between 8 and 16 are extremely useful as this is approximately the difference between the clock rates of the fastest high rate digital-to-analog converters and those of commercial CMOS processors [5].

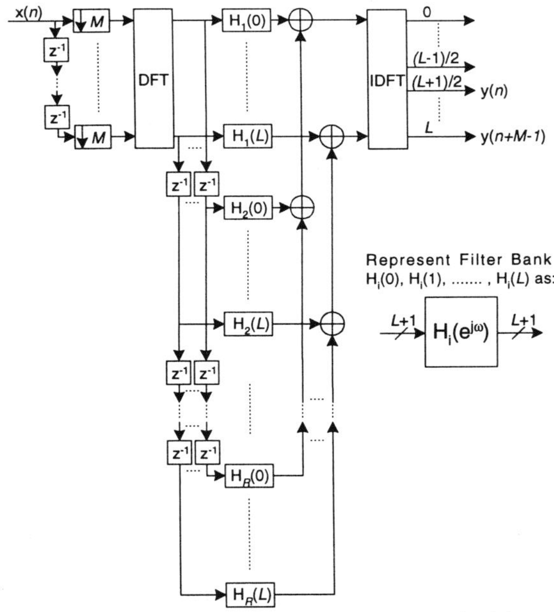

Figure 2. Parallel Sub-convolution Filter Bank Architecture

III. Complexity Comparisons

Here a comparison is made between the number of complex multiplies required for a parallel hardware implementation of the frequency domain sub-convolution method, frequency domain overlap-and-save method, and time-domain parallel method [5] for various processing rate reductions M. This comparison is the primary metric for determining the appropriate architecture for concurrent VLSI hardware implementation where the goal is to maximize operations per second while minimizing hardware required for a given maximum clock rate. However, a comparison between the computational efficiency, measured as the number of multiples per filtered output, of the sub-convolution method, conventional convolution, and frequency domain overlap-and-save convolution is also useful. This comparison is valuable when the goal is to maximize filter output rate where a full-parallel architecture is not possible and hardware reuse is employed. The extreme example of hardware reuse is an implementation in a microprocessor (software implementation) where perhaps one hardware multiplier is used for all multiplications required in a filtering architecture. This metric is also useful in determining the appropriate architecture to minimize power consumed in a CMOS hardware implementation.

III.A Complexity of Parallel Hardware Implementation

The number of complex multiplies is used as the basis for complexity comparison. Further it is assumed the input $x(n)$, and filter coefficients, $h(n)$, are complex. When L+1 is the FFT-IFFT length and $L+1 = 2^{\nu}$, the number of multipliers required by the FFT is given by $\mu(L+1) = \frac{(L+1)(\nu-1)}{2}$ [1]. Note that there are numerous types of FFT-IFFT algorithms that could be used for comparison. The radix two algorithms are chosen for convenience. The amount of overlap is determined to minimize multiplier count for a given processing rate reduction. The maximum number of coefficients (constrained that they be a power of 2) that can be implemented in the architecture of Figure 1 is $N = ((L+1) - M)$. Without this constraint the total number of coefficients is one more; $N = ((L+1) - M + 1)$. The FFT-IFFT lengths as a function of rate reduction M and filter length N is then $(L+1) = N + M$. The multipliers required for the frequency domain filtering of Figure 1 is given by:

$$\begin{gathered} 2\mu(L+1) + (L+1) = (L+1)(\nu-1) + (L+1) = \\ (N+M)(\log_2(N+M)-1) + (N+M) = \\ (N+M)(\log_2(N+M). \end{gathered} \tag{4}$$

For the parallel sub-convolution filter bank method the FFT-IFFT lengths are set to be twice the rate reduction (yielding the most computationally efficient 50% overlap [4]), that is $(L+1)=2M$. It is also assumed that each filter length increment is an integer multiple of $(L+1)/2$. The number of frequency domain multiples for each sub-filter of length M is $2M$ and there are

$R=N/M$ filter banks. The total number of multipliers required for the filtered vector output of the PSFB in Figure 2 is then

$$2\mu(2M)+(2M)R = 2M(\log_2(2M)-1)+2N \quad (5)$$

The multiplies required by parallel time-domain convolution is given by [5]:

$$MN \quad (6)$$

For a parallel (concurrent) VLSI implementation, what values of M and N the PSFB require the least hardware? First, determine for which values of M and N (5) is less than (6); that is $2M(\log_2(2M)-1)+2N < MN$. This simplifies to $M^{2M/(M-2)} < 2^N$, and finally to

$$\left(2/\left(1-\frac{2}{M}\right)\right)\log_2(M) < N. \quad (7)$$

The inequality of (7) holds for processing rate reduction of 1/4th or less (M great than or equal to 4) for all filters with greater than 8 coefficients ($N > 8$). Figures 3.a-c illustrates the multiplier savings for several values of M and N, where TC indicates parallel time-domain convolution and parallel sub-convolution filter banks are represented by PSFB.

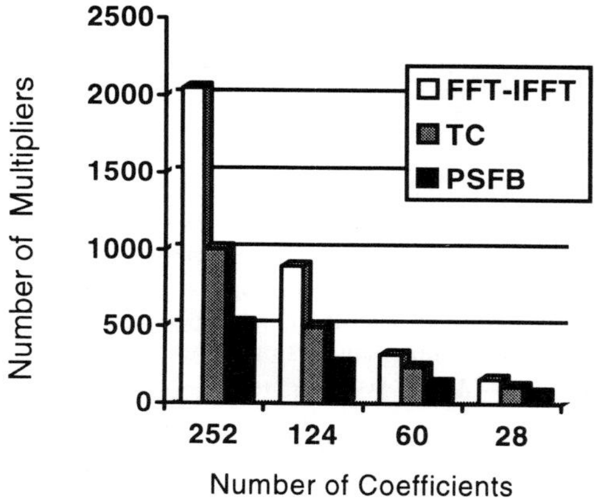

Figure 3.a. 1/4th Rate Processing Architectures (M=4)

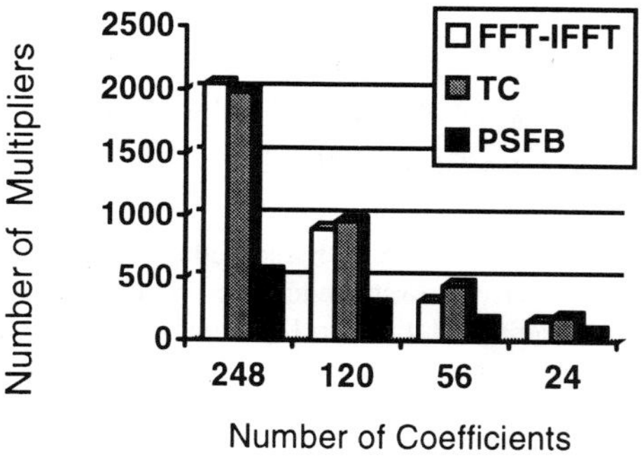

Figure 3.b. 1/8th Rate Processing Architectures (M=8)

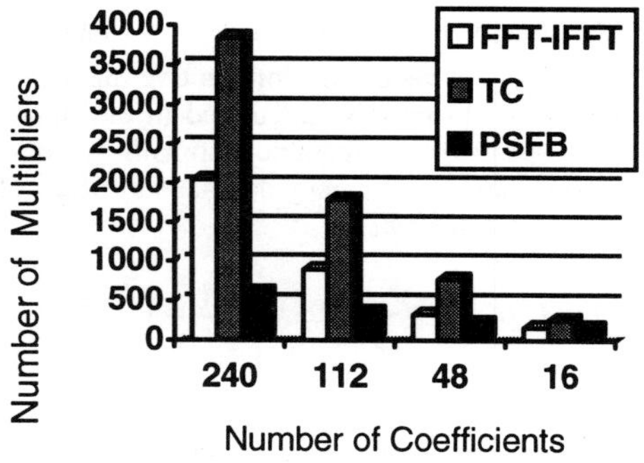

Figure 3.c. 1/16th Rate Processing Architectures (M=16)

Now determine for what values of M and N (5) is less than (4); that is:

$$2M(\log_2(2M)-1)+2N < (M+N)(\log_2(N+M)) \quad (8)$$

Letting $N=M+K$ where K is some integer and substituting in (8) yields

$$2M\log_2(2M)-2M+2(M+K) < (2M+K)\log_2(2M+K)$$
$$2M\log_2(2M)+2K < (2M+K)\log_2(2M+K)$$
$$2M\log_2(2M)+2K < 2M\log_2(2M+K)+K(\log_2(2M+K)).$$

(9)

Now it is clear that for any value of M and all $K \geq 0$

$$2M\log_2(2M) < 2M\log_2(2M+K). \quad (10)$$

Given this constraint what values of M and K yield

$$2K < K\log_2(2M+K)? \quad (11)$$

This inequality is true when $M \geq 2$ and $K \geq 1$, that is $N>M$. Summarizing, the inequality of (8) holds for processing rate reduction of 1/2 or less, M great than or equal to 2, and for all filters with the number of coefficients N greater than M, $N=M+K>3$. Figures 3.a-c illustrates the multiplier savings for several values of N and M, where FFT-IFFT indicates parallel frequency domain convolution and parallel sub-convolution filter banks are represented by PSFB.

III.B Computational Complexity of Filtering Implementations

Generally the overlap-and-save method of convolution using 50% overlap is the most computationally efficient, measured as number of multiplies per filtered output, method of the methods considered here [6]. In an implementation employing hardware reuse, such as a software implementation, the FFT-IFFT length can be determined from the filter length to minimize the number of multiplies per filtered output. The PSFB may be used to provide intermediate computation gain, between that of the overlap-and-save method and conventional convolution. The number of complex multiplies per filtered output in the PSFB is

$$\frac{2M(\log_2(2M)-1)+2N}{M} \tag{12}$$

For conventional convolution obviously the number of multiplies per filtered output is N. The PSFB design method provides intermediate computation gain when the following inequality holds:

$$\begin{aligned} &\frac{2M(\log_2(2M)-1)+2N}{M} < N \\ &2(\log_2(2M)-1) < N\left(1-\frac{2}{M}\right) \\ &2\log_2(M) < N\left(1-\frac{2}{M}\right) \\ &\frac{2M}{M-2}\log_2(M) < N. \end{aligned} \tag{13}$$

Figure 4 is a plot of the number of multiplies required per filter output for traditional serial convolution, and four different parallel sub-convolution filter banks (PSFB). The four FFT-IFFT lengths are $L+1$=8, 16, 32, and 64 (Figure 2). These four architectures are labeled as PSFB(8), PSFB(16), PSFB(32), and PSFB(64) respectively.

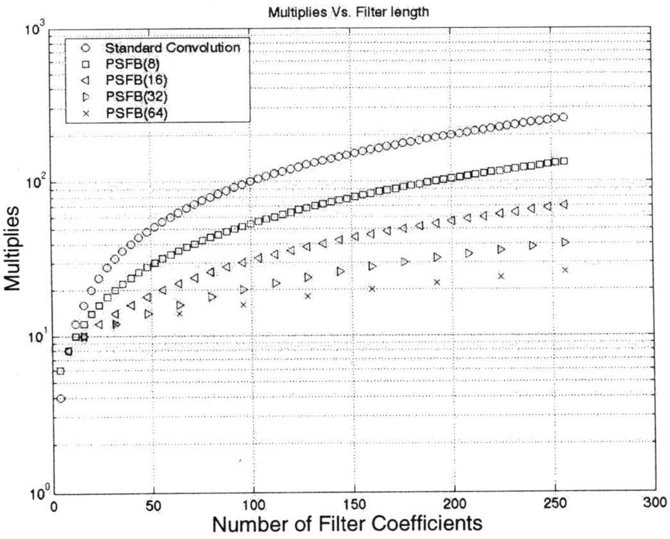

Figure 4. Computation Comparison

It is evident in Figure 4 that a reduction in the number of multiplies per filtered output sample is a possible result for parallel sub-convolution filter banks. From this high-level analysis, the sub-convolution filter bank may be a useful design technique for low power CMOS applications.

V. Conclusion

A design method has been presented for realizing (lossless) convolution/FIR filtering using parallel sub-convolution filter bank architectures. These architectures are derived from methods of separating a convolution into sub-convolutions and deriving sub-filters for these from the filter to be realized. These methods are then combined with overlap-and-save convolution techniques to create the parallel (frequency-domain) sub-convolution filter bank architecture. Using these architectures, the FFT-IFFT length is determined by the desired rate reduction and is independent of the FIR filter length to be implemented. A complexity and computation comparison between the parallel sub-convolution method and conventional methods was made. Significant complexity reduction over time-domain convolution and conventional overlap-and-save frequency domain convolution was demonstrated for certain classes of parallel filters. The design method may be used to provide intermediate computation gain, with computation requirements between those of time-domain convolution and the traditional all-frequency domain convolution. Finally, an example application of the parallel sub-convolution filter bank design method in the design of a high rate fast LMS equalizer is given in [7].

VI. References

[1] A.V. Oppenheim, R. W. Schafer, *Discrete-Time Signal Processing*, Premtice-Hall, Englewood Cliffs, N.J. 1989.

[2] P. P. Vaidyanathan, *Multirate Systems and Filter Banks*, Printice Hall, Englewood Cliffs, New Jersey 1993

[3] R. Sadr, P. P. Vaidyanathan, D. Raphaeli, S. Hinedi, "Multirate Digital Modem Using Multirate Digital Filter Banks", JPL Publication 94-20, August 1994.

[4] J. Shynk, *"Frequency-Domain and Multirate Adaptive Filtering,"* IEEE Signal Processing Magazine, Jan. 1992.

[5] Andrew A. Gray, "*Very Large Scale Integration Architectures for Nyquist-Rate Digital Communication Receivers*", PhD Dissertation, University of Southern California, Los Angeles, CA, May 2000

[6] Simon Haykin, *Adaptive Filter Theory*, Prentice Hall, Upper Saddle River, New Jersey, 1996

[7] A. Gray, S. Hoy, P. Ghuman, "*Parallel VLSI Architectures for Multi-Gbps Satellite Communicaitons*" IEEE GlobeComm, November, 2001

Dual Clock Rate Block Data Parallel Architecture

An-Te Deng and Winser E. Alexander

Abstract— **This paper presents a multiprocessor system with a dual clock rate. We used the block data overlap-save algorithm [1] and the Block Data Parallel Architecture (BDPA) to implement the two dimensional (2D) FIR filter. We were able to significantly improve the performance of the Block Data Parallel Architecture (BDPA) multiprocessor system by using a dual rate clock as compared to the performance using a single rate clock. We designed a 2D-FIR filter system with a four processor module array to demonstrate the improvement in performance. It had a throughput performance of 7.975 samples per processor clock cycle and the processor utilization was 78.53%.**

Keywords— **FIR, BDPA, Dual clock rate.**

I. Introduction

Sung, et al. [2] introduced a digital filtering algorithm using a parallel block processing method in early 1990s. A ring type network, one dimensional (1D) processor array was investigated. We have modified this method by using algorithm partitioning, block data input/output distribution, hierarchical data flow control (HDFC), and the asynchronously task processing mechanism to form the Block Data Flow Paradigm (BDFP).

The speed with which data flows into the processor array is very important. The key to the BDPA is to provide processors with enough data to keep them busy at all times. This is the basic requirement for a BDPA to generate a linear speedup (Amdahl's law) characteristic, high utilization, and near supercomputer performance. Several different applications have been implemented on the BDPA with linear speedup [3], [4]. Unfortunately, with hardware implementation for these example BDPA, the linear speedup was limited because eventually, the data flow bottleneck was reached in each case. We present the use of a dual clock rate in this paper to reduce the impact of this data flow bottleneck problem in the BDPA. We obtained results that were almost forty times faster using this approach than we obtained with a single clock rate system.

The paper starts with this introduction describing the data flow bottleneck limitation of the BDPA and our proposed solution. It then describes the concept of the BDFP and the system architecture of the BDPA. An example and detail description of the Dual Clock Rate mechanism is given in section three. Section four gives the analysis of the performance for two BDPA applications and the improvement that the dual clock rate has made. The last section concludes this paper and gives potential future work.

An-Te is now with the High Performance Digital Signal Processing Laboratory, NC State University, Raleigh, U.S.A. E-mail: adeng@nc.rr.com.

Winser is Professor of ECE at NC State University,Raleigh, NC, U. S. A., senior member of IEEE, E-mail: winser@ncsu.edu.

II. Block Data Flow Paradigm

The Block Data Flow Paradigm (BDFP) involves partitioning data into large granularity blocks, scheduling and processing the data within different processors which implement partitioned algorithms. The Block Data Parallel Architecture (BDPA) is an example of the implementation of the BDFP [3]. It reduces the communication overhead problem between memory and processors. It uses a asynchronous data flow protocol so that the control overhead is reduced. The data communication time is overlapped with the computing time. The BDPA adopts the 1D linear array pipeline characteristic so that the system utilization is very high and easy to scale. The number of input/output pins per-processor will not increase as the number of processors increases. The power dissipation is lower than for the 2D array system due to the asynchronous data flow protocol, yet the performance is similar [5] to that of synchronous systems. The BDPA has the following special characteristics:

- Massive data is partitioned geometrically into relatively small blocks that fit the size of each identical processor. A different block of data is assigned to each processor.
- The architecture of each processor is optimized for high throughput for the algorithm to be implemented. We emphasize the use of pipeline computations rather than the use of small grain parallel computations.
- The processors are arranged in a unidirectional linear array pipeline.
- The hierarchical data flow control concept is implemented to enable individual processors to operate concurrently and asynchronously. The processing of a block begins as soon as it is available. No global control or complicated scheduling is needed to indicate the explicit start and stop times for block processing.
- Each processor computes the output derived from its assigned input block and transmits this output to the output device as soon as it is ready.
- Partial or intermediate results, which must be exchanged between processors, pass through a point-to-point simple (single stage) interconnection network.

A. Block Data Parallel Architecture (BDPA)

The theme of the BDPA is to process the data in large granularity blocks. We developed the BDPA to implement the BDFP for a variety of one-dimensional and two-dimensional DSP applications. The architecture is composed of three main parts, the Input Module (IM), the Processor Module Array (PMA), and the Output Module (OM) in order to manage the block flow smoothly through the processors.

In figure 1, the IM receives the data string coming from the input device. The input device could be an audio/video

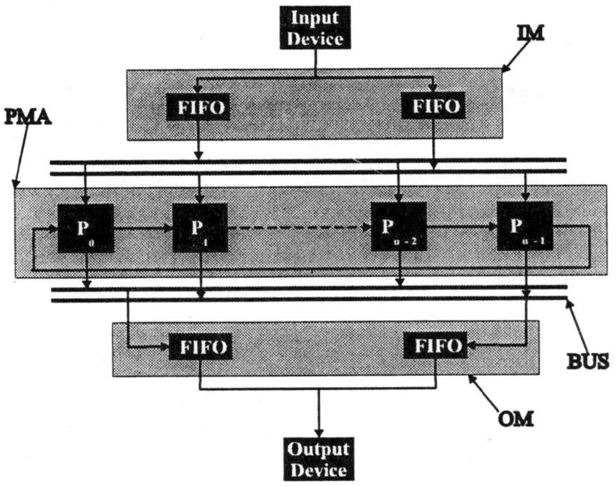

Fig. 1. Block Data Parallel Architecture

recorder, a sound/graphic file installed on a disk, or a host system data file. The IM acts as an I/O buffer which splits and formats the incoming data string into data blocks. It has a distributor at the front end so that the data blocks can be distributed into the corresponding FIFO buffer. The IM transfers the data blocks from the FIFO into each of the different processor modules without interference. A data flow regulator is responsible for the handshaking protocols between the IM and both the input device and the PMA.

The PMA contains an appropriate number of processors to meet the data processing time requirements. The IM sequentially sends the designated blocks of data to the corresponding PMA processor. Each processor has a separate time-multiplexed input channel and output channel. The processors are divided into two processor groups, namely, the odd processor group and the even processor group. Each processor group is directly connected to one of the input FIFO buffers and one of the output FIFO buffers. The purpose of dividing the processors into two groups is to ensure that the two IM FIFO buffers alternatively write and read blocks of data. This provides double buffering for the input device.

Each of the individual processors is designed by our algorithm mapping methodology [3], [4] and performs the required mathematical operations (small programs) on its assigned block of data and updates the intermediate values as well as the output values until the whole block of data processing has been processed. The partitioned algorithm and the partitioned data determine the time required for processing for each block of data. FIFO buffers are directly connected between the adjacent processors for transfer of intermediate data as required for a particular application. The transferring of the intermediate values is unidirectional.

The OM has two FIFO buffers to receive the output blocks of data as they are available from the processors in the PMA. The even group of processors uses one FIFO buffer and the odd group uses the other FIFO. The output blocks of data coming from the different groups of processors are multiplexed into a synchronized output data stream. The OM may contain a post-processing sub-module as needed for a particular application. This post-processing sub-module may also contain different functional modules that adapt to different applications (i.e., wavelet coding or voice identification).

III. Dual Clock Rate Multi-processor system

Our goal in developing the BDPA was to achieve linear speedup and high throughput performance. In order to reach this goal, it is necessary to provide the processor modules with data as necessary for them to stay busy at all times. However, most of the time, a multiprocessor system bottleneck occurs either due to the speed of the data bus which is not fast enough or instruction scheduling (i.e., branch prediction) causes interruption of the data flow. We categorize this kind of bottleneck as a data flow bottleneck. This data flow bottleneck will limit the multiprocessor operation and prevent the system from increasing throughput. In other words, no matter how many processors are added to the multiprocessor system, there is no increase in throughput.

Another category of bottleneck is the processing bottleneck. Assume that the data flow bottleneck problem has been solved and the processors are provided with as much data as they need. In this case, the processors are busy all the time, and the throughput cannot be improved by providing more data. This is called the processing bottleneck. Normally, it is easier to solve the processing bottleneck problem in most multiprocessor systems by adding more processors. Multiprocessor systems with inter-processor data transfers are an exception and do not provide increased system throughput by adding more processors.

It is more difficult to solve the data flow bottleneck problem than to solve the processing bottleneck problem. The use of two FIFOs in the IM and the OM was utilized to reduce the data flow bottleneck problem. The use of a dual rate clock multiprocessor can further reduce this problem. A data flow bottleneck example for a 2D-FIR filter is shown in figure 2. The simulator output signal waveforms in row

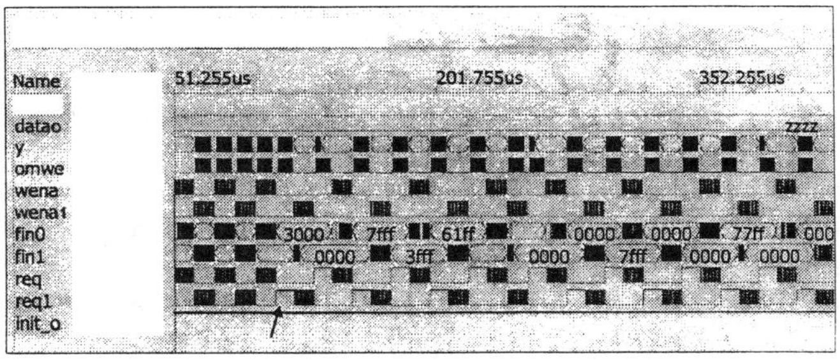

Fig. 2. 2D-FIR Filter Data Flow Bottleneck

8 and row 9 in figure 2 are the PMA request signals. The normal operation is to provide a processor its requesting data block (shaded blocks). Now, every request signal is suspended for a block period after the first five processors obtain their data blocks. This means a data flow bottleneck exist between the data source and the IM FIFO.

Several possible solutions to this type of data flow bottleneck were considered. The traditional way to solve this problem is to enlarge the IM FIFO. This solution simply delays the occurrence of the data flow bottleneck. Eventually, the IM FIFO will still be empty. The second solution is to change the FIFO memory into a dual port RAM, since dual port RAM can be read from and written to at the same time. However, the read/write controller for the dual port RAM must not read and write at the same address at the same time.

A third solution is to use the dual clock rate multiprocessor system. The sequence for using a dual clock rate system is to use a very high clock rate to acquire the data block, process the data at the slower clock rate required in the processor, and output the processed data block from the processor FIFO to the OM at the original high clock rate. Since system clock rates are increasing at a very fast pace, it is simple and easy to add a clock divider in our design to implement the dual clock rate multiprocessor system to obtain high throughput. This method is better than adding complicated control circuitry or a large memory to the system. The linear speedup characteristic of the whole system can be improved dramatically by using a dual clock rate and a smaller number of multiprocessor system.

The BDPA processor architecture and its HDFC architecture are well designed to accommodate globally asynchronous, locally synchronous processing, and it is totally modularized. After the processor gets its own data block, it is isolated from the outside environment. The only interface of a processor to the outside environment is the on board FIFO. Therefore, it is feasible to use the system clock, which is faster, to input/output data blocks into/from the on board processor FIFO while using another slower rate clock to drive the data through the processor.

Since the system data flow speed is faster than the data processing speed, the data flow bottleneck problem is resolved. Details of the potential improvement is found in the performance analysis section of this paper. This performance improvement also demonstrates that the data flow bottleneck problem has a significant impact on the system throughput. The concept of using dual clock rates to drive the BDPA is similar to increasing the data block buffer without actually adding any memory or a complicated controller to the system circuitry. Hence, we call this technique **phantom memory**. The ratio of data flow speed to data processing speed can be adjusted. Therefore by varying this ratio and adjusting the number of processors in the system, we can obtain the targeted performance and processor utilization for the BDPA. The dual clock rate multiprocessor system concept can be extended to a multirate clock system on chip (SoC).

IV. Performance Analysis

A. *1D-FIR Filter*

We validated the performance and the timing information for the hardware implementations discussed in this paper by simulation using the Verilog Hardware Description Language (HDL). Table I shows the results for two versions of the 1D-FIR system using a single clock rate. The

	12 proc/4 prim	4 proc/64 prim
Throughput	0.3673 s/c	0.4093 s/c
Utilization	88.92%	39.26%

TABLE I
1D-FIR Filter System Performance Table

second column in the table represents the performance of a system with 12 processors. Each processor has a four multiply/add units and generates one sample output every 16-clock cycles. The utilization is high because the processors are working most of the times. All the multiply/add units in this system are the same.

The third column represents a version with 4 processors and each processor has 64 multiply/add units. The processor is supposed to generate one sample for each clock cycle if the IM provides sufficient data blocks. There are three different multiply/add units in this version. The second version has better performance, but lower processor utilization. This means the data flow speed can not match the data processing speed and the processors in the PMA are idle most of the time.

The system performance of the second column version in table I is shown in figure 3. This 1D-FIR filter system

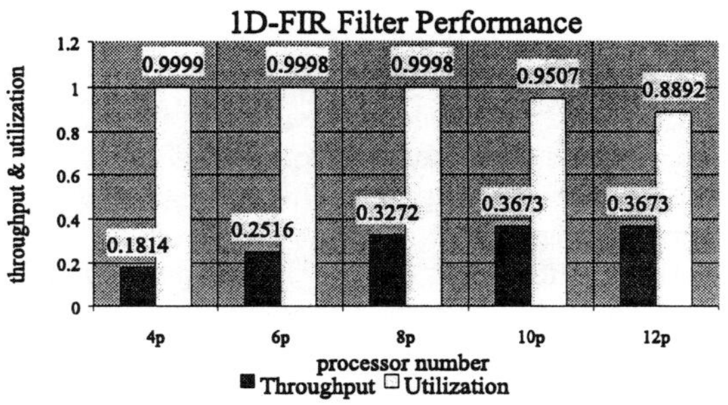

Fig. 3. 1D-FIR Filter Performance

also uses a mono-clock rate system. If the system clock frequency is 100 Mhz, the system throughput can reach more than 36 million samples per second. The small scale linear speedup characteristic is observed for 4-processors to 10-processors. After 10-processor, the system throughput will not be able to increase by adding more processors; the utilization of the processor is going down. This means there exists a data flow bottleneck at the input/output area at this point. The data flow bottleneck problem can be improved by using a dual clock rate multiprocessor system. This performance improvement was verified by the 2D-FIR filter system simulation.

B. *2D-FIR Filter*

A 128 x 128 pixels image was block formatted in the image pre-processing module. A 6 x 6 filter block and a

32 x 32 block size were used. The same sized sub-images were sent through different 2D-FIR filtering processor. The timing information from the result of a simulator gives us the throughput:

$$throughput = \frac{total\ image\ pixels}{\frac{system\ clock\ cycles}{clock\ ratio}}. \tag{1}$$

Total image pixels is the total pixels that have been processed. System clock cycles is the total number of system clock cycles that was used and the clock ratio is the ratio for the system clock rate and the processor clock rate. The utilization is:

$$Utilization = 1 - \frac{total\ idle\ time}{total\ image\ pixel\ processing\ time}, \tag{2}$$

where the total idle time is the accumulated all processor idle time, and the total image pixel processing time is the time elapsed for all processors during the total image pixel processing.

Figure 4 shows a 2D-FIR filter system performance chart. The simulation systems are dual clock rate mul-

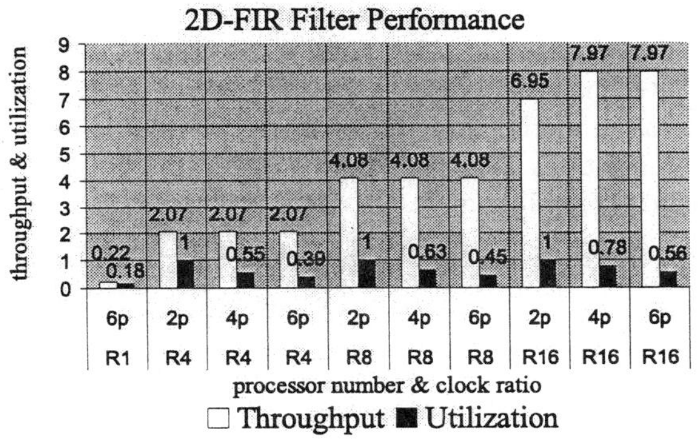

Fig. 4. 2D-FIR Filter Performance

tiprocessor systems. There are four different clock ratio simulations (R1, R4, R8, and R16), which use different numbers of processors. R1 stands for a clock ratio equal to one, R4 stands for a clock ratio equal to four, and so on. The two digit linear speedup characteristic can be observed through increasing the clock ratio. For example, the R4 simulation has system throughput almost ten times as fast as the R1 simulation and the system throughput of the R8 simulation is double that for the R4 simulation. The highest system throughput happens at a clock ratio equal to 16 with more than four processors. If the processor clock rate is 100 Mhz, then the system throughput obtained here is 800 million pixels per second, which is far more than sufficient to process 1024 x 1024 pixel images and 60 frames per second. The 1024 x 1024 x 60 pixel/second processing standard for our 2D-FIR filter system example needs a computing speed more than 6.6 GFLOPs. A TMS320C6713 DSP processor running at 225 MHz has a 1350 MFLOPS computing speed. It is obvious that more than four such DSP processors are needed to achieve this processing speed. Besides, the system designed with TMS320C6713 processors can not guarantee with 800 million pixel/second system data throughput due to its data flow bottleneck.

Notice that the two simulations when the clock ratios are four and eight, and the number of processors in the system is four, the utilization of the processors decreases compared to the two processors system's utilization. This means the system clock ratio, which is to drive data blocks into the processors as fast as possible and keep them busy at all times, is not large enough. Another interesting simulation is when the system has two processors and the clock ratio equals 16. The processors are fully utilized yet the system throughput has not reached its linear speedup limit, which is 8 pixels per processor clock cycle. Obviously the clock ratio is big enough driving sufficient data blocks into the processors. Only, the processors can't handle this data speed and obtain the maximum throughput. However, after adding another two processors in the PMA, the data flow bottleneck appears again and the throughput reaches its limit. In order to obtain the highest throughput performance, the PMA architecture should have block size, clock ratio, and the number of processor as large as possible.

V. Conclusion

In order to solve the processing and data flow bottleneck problem, we developed the dual clock rate multiprocessor system. We verified by simulations that the circuit we designed for our system has high throughput performance and high processor utilization. We obtained very high throughput with few processors for our dual clock rate multiprocessor system.

Since the clock ratio is variable and can be tailored to meet different application specifications, a further investigation is necessary to detail the optimization of resources (i.e., block size, clock ratio, and processor number) to accommodate the different applications' requirements. Further investigation in a multiple clock rate SoC in order to coordinate different system's processing speed and to obtain the best system throughput is another challenging topic in the future.

References

[1] Alan V. Oppenheim, Ronald W. Schafer, and John R. Buck, Discrete–Time Signal Processing, Prentice Hall, Upper Saddle River, New Jersey 07458, 1999.

[2] Wonyong Sung, Sanjit K. Mitra, and Branko Jeren, "*Multiprocessor Implementation of Digital Filtering Algorithms Using a Parallel Block Processing Method*," IEEE Transactions on Parallel and Distributed Systems, vol. 3, no. 1, pp. 110–120, Jan. 1992.

[3] Winser E. Alexander, Douglas Reeves, and Clay Gloster Jr., "*Parallel Image Processing with the block data parallel architecture*," Proceedings of the IEEE, vol. 84, no. 7, pp. 947–968, July 1996.

[4] An-Te Deng and Winser E. Alexander, "*Configurable/Scalable Block Data Flow Paradigm(CSBDFP)*," Proceedings of World Multiconference on Systemics, Cybernetics and Informatics, vol. 6, no. 1, pp. 57–62, July 2001.

[5] Dan W. Hammerstrom and Daniel P. Lulich, "*Image Processing Using One-Dimensional Processor Arrays*," Proceedings of the IEEE, vol. 84, no. 7, pp. 1005–1017, July 1996.

NEW EFFICIENT RESIDUE-TO-BINARY CONVERTERS FOR 4-MODULI SET $\{2^n - 1, 2^n, 2^n + 1, 2^{n+1} - 1\}$

Bin Cao, Chip-Hong Chang and Thambipillai Srikanthan

Centre for High Performance Embedded Systems
Nanyang Technological University
N4-B3b-06, Nanyang Avenue, Singapore 639798

ABSTRACT

This paper describes a new algorithm of residue-to-binary conversion for the 4-moduli set $\{2^n - 1, 2^n, 2^n + 1, 2^{n+1} - 1\}$ RNS which is valid for even n. The number theoretic properties of this moduli set are exploited to realize an efficient reverse converter. By using the most efficient residue-to-binary conversion algorithm for the triple moduli set $\{2^n - 1, 2^n, 2^n + 1\}$, the problem can be reduced to a simpler residue-to-binary converter for two moduli set RNS amenable to fast mixed-radix conversion. Two different versions of architecture based on this new formulation are proposed. Comparing to the fastest reverse converter reported for the same moduli set, our proposed converters surpass it in both the area and time complexities.

1. INTRODUCTION

The inherent properties of carry-free operations, parallelism and fault-tolerance have made residue number system (RNS) a promising candidate for high speed arithmetic and specialized high-precision digital signal processing applications [1][2]. By decomposing large binary numbers into smaller residues, addition and subtraction in RNS arithmetic have no inter-digit carries or borrows, and multiplication can be performed without the need to generate partial products. For elegantly balanced moduli set, the advantages offered by very large scale integration (VLSI) technology can be commendably achieved in the implementation of RNS-based architectures.

As the peripheral interfaces of most digital systems are still based on the weighted number system, the overhead incurred in the conversions into and out of the residue number system has been the major critique limiting the diffusion of RNS-based processors. It is well established that the forward conversion from the binary number to its residues can be implemented quite efficiently, but the reverse conversion from the residues to the binary number is more difficult [3]. For RNS with large dynamic range, a significant proportion of the resources and computation time of the total RNS are committed to the reverse conversion. RNS processor becomes profitable in vector-like processing where a large number of intermediate arithmetic operations are carried out in the RNS domain before the final results are obtained and converted to binary numbers. During the past several decades, the residue-to-binary conversion algorithms are based primarily on the Chinese Remainder Theorem (CRT) [3][4] or Mixed-Radix Conversion (MRC) [5]. The use of CRT involves a large modular addition, whereas MRC is a sequential process that often requires a number of look-up tables. Recently, Y. Wang proposed New CRT theorems, which have inherited the merits of the classical CRT and MRC algorithms [6], but their converters for general moduli set are still relatively complex.

Special moduli sets have been used extensively to reduce the hardware complexity in the implementation of the residue-to-binary converters. Furthermore, it is easier to design and implement RNS processors based on special moduli set because there exists a clear relationship between the moduli set and the dynamic range. Among the special moduli sets, the triple moduli set $\{2^n - 1, 2^n, 2^n + 1\}$ has gained unprecedented popularity by virtue of its inherent number theoretic properties in CRT algorithm. However, this special moduli set is insufficient to support signal processing applications that demands high performance and greater precision, which implies higher parallelism and larger dynamic range in RNS. The necessity for increased parallelism and larger dynamic range is the main motivation to expand the cardinality of the conventional special moduli sets.

Vinod and Premkumar proposed a new moduli set $\{2^n - 1, 2^n, 2^n + 1, 2^{n+1} - 1\}$ for even values of n, and the corresponding residue-to-binary converter [7]. This paper presents a new approach to the residue-to-binary conversion for the same moduli set. The four moduli set is decomposed into the triple moduli set $\{2^n - 1, 2^n, 2^n + 1\}$ and a single modulus $2^{n+1} - 1$. The binary output composed by the most efficient reverse converters for the triple moduli set is combined with the remaining residue based on the MRC algorithm. Two versions of architecture are suggested in this paper. They are not only faster, but

0-7803-7761-3/03/$17.00 ©2003 IEEE

also more hardware efficient than the fastest converter [7] reported in the literature for pragmatic dynamic range of less than 208 or 160 bits.

2. BACKGROUND

In a Residue Number System (RNS), an integer X can be represented by an n-tuple of residues, $(x_1, x_2, \ldots, x_n)$ defined over a set of relatively prime moduli $\{P_1, P_2, \ldots, P_n\}$, where $gcd(P_i, P_j) = 1$ for $1 \le i, j \le n$, and $i \ne j$. The residue x_i is the least positive residues of X modulo P_i. For any integer X within the dynamic range, i.e, $0 \le X < M$, where M is the product of all moduli, the residue representation of X is unique.

To convert a residue-represented number $(x_1, x_2, \ldots, x_n)$ into its binary number X, Chinese Remainder Theorem (CRT) has been used traditionally. The CRT requires modulo addition and multiplication, which is not amenable to efficient hardware realization. Alternatively, Mixed-Radix Conversion (MRC) can be used. In MRC, the integer X can be calculated by

$$X = a_n \prod_{i=1}^{n-1} P_i + \cdots + a_3 P_2 P_1 + a_2 P_1 + a_1 \quad (1)$$

where a_is are called the mixed-radix coefficients, and can be obtained from the residues by the following equations:

$$\begin{aligned} a_1 &= x_1 \\ a_2 &= \left| (x_2 - a_1) \left| \frac{1}{P_1} \right|_{P_2} \right|_{P_2} \\ a_3 &= \left| \left((x_3 - a_1) \left| \frac{1}{P_1} \right|_{P_3} - a_2 \right) \left| \frac{1}{P_2} \right|_{P_3} \right|_{P_3} \\ &\cdots \\ a_n &= \left| \left(\left((x_n - a_1) \left| \frac{1}{P_1} \right|_{P_n} - a_2 \right) \left| \frac{1}{P_2} \right|_{P_n} - \cdots - a_{n-1} \right) \left| \frac{1}{P_{n-1}} \right|_{P_n} \right|_{P_n} \end{aligned} \quad (2)$$

For a simple 2-moduli set $\{P_1, P_2\}$, the integer X can be converted from its residue representation (x_1, x_2) by

$$X = a_1 + a_2 P_1 = x_1 + P_1 \cdot \left| (x_2 - x_1) \left| 1/P_1 \right|_{P_2} \right|_{P_2} \quad (3)$$

where $|1/P_1|_{P_2}$ is the multiplicative inverse of P_1 modulo P_2.

3. NEW CONVERSION ALGORITHM

Consider the 4-moduli set $\{2^n - 1, 2^n, 2^n + 1, 2^{n+1} - 1\}$, where n is any even natural number, the residue representation of the integer X is (x_1, x_2, x_3, x_4), and the binary representations of the residues are as follows: $x_1 = (X_{1,n-1}, X_{1,n-2}, \ldots, X_{1,0})_2$, $x_2 = (X_{2,n-1}, X_{2,n-2}, \ldots, X_{2,0})_2$, $x_3 = (X_{3,n}, X_{3,n-1}, \ldots, X_{3,0})_2$, and $x_4 = (X_{4,n}, X_{4,n-1}, \ldots, X_{4,0})_2$. We propose a new algorithm for the residue-to-binary conversion by first generating the integer $X^{(1)} = (x_1, x_2, x_3)$ corresponding to the moduli set $\{2^n-1, 2^n, 2^n+1\}$ using the efficient triple moduli set reverse converter, then using MRC method to calculate the final integer $X = (X^{(1)}, x_4)$ corresponding to the two moduli set $\{2^n(2^{2n}-1), 2^{n+1}-1\}$.

The following property and propositions are needed for the derivation of our new algorithm. The proofs have been omitted due to the page constraint.

Property 1: Multiplying an integer, x by 2^r modulo (2^p-1) can be accomplished by expressing x in a p bits binary representation and then shifting it circularly by r bits to the left.

Property 1 is originally proposed in [1]. As an example, let $x = 6$, $p = 5$ and $r = 3$, then we have

$$\left| x \cdot 2^r \right|_{2^p-1} = \left| 6 \cdot 2^3 \right|_{2^5-1} = CLS(00110, 3) = (10001)_2 = 17,$$

where the function $CLS(x, r)$ is used to denote a circular shift of the binary number x by r bits to the left.

Proposition 1: For even nature number n, if k_0 is the multiplicative inverse of $(5 \cdot 2^{n-2} - 1)$ modulo $(2^{n+1} - 1)$, i.e.,

$$\left| k_0 \left(5 \cdot 2^{n-2} - 1 \right) \right|_{2^{n+1}-1} = 1 \quad (4)$$

then, $k_0 = \frac{1}{3}\left(2^{n+2} - 10\right)$.

Proposition 2: For even number of n, the multiplicative inverse of 3 modulo $(2^{n+1} - 1)$ is $(2^{n+2} - 1)/3$, i.e.,

$$\left| \frac{1}{3} \right|_{2^{n+1}-1} = \frac{1}{3}\left(2^{n+2} - 1\right) = \sum_{i=0}^{\frac{n}{2}} 2^{2i} \quad (5)$$

Based on the above property and propositions, a new algorithm can be developed for the reverse conversion. First consider moduli set $\{2^n - 1, 2^n, 2^n + 1\}$ and $X^{(1)} = (x_1, x_2, x_3)$. A triple moduli set converter is needed to obtain $X^{(1)}$ from its residues. Using the method of [6],

$$X^{(1)} = x_2 + 2^n Y \quad (6)$$

where

$$Y = \left| (A_1 + A_2 + C_{2n}) + 2^n (B_1 + B_2) \right|_{2^{2n}-1} \quad (7)$$

The calculations of A_1, A_2, B_1, B_2 and C_{2n} can be found in [6], and Y is a $2n$-bit number. Let

$$Y = Y_1 + 2^n Y_2 = \left(Y_{2n-1} Y_{2n-2} \ldots Y_1 Y_0 \right)_2 \quad (8)$$

where Y_1 is the n lower order bits of Y and Y_2 is the n higher order bits of Y. Thus, Y can be calculated by $2n$ full adders (FAs) and one $2n$-bit 1's complement adder.

Next, consider the moduli set $\{2^n(2^{2n} - 1), 2^{n+1} - 1\}$ and $X = (X^{(1)}, x_4)$. Using the MRC algorithm of (3), X can be calculated by

$$X = X^{(1)} + 2^n \left(2^{2n} - 1 \right) \left| k_0 \left(x_4 - X^{(1)} \right) \right|_{2^{n+1}-1} \quad (9)$$

where $k_0 = \left| \frac{1}{2^n \left(2^{2n} - 1 \right)} \right|_{2^{n+1}-1}$

So we have

$$\left| k_0 \cdot 2^n \left(2^{2n} - 1 \right) \right|_{2^{n+1}-1} = 1 \quad (10)$$

It is difficult to calculate k_0 directly from (10). However, the left side of (10) can be simplified as follows:

$$\left|k_0 2^n \left(2^{2n}-1\right)\right|_{2^{n+1}-1} = \left|k_0 2^n \left|\left(2^{n-1}\left(2^{n+1}-1\right)+2^{n-1}-1\right)\right|_{2^{n+1}-1}\right|_{2^{n+1}-1}$$

$$= \left|k_0 \left(5 \cdot 2^{n-2}-1\right)\right|_{2^{n+1}-1}$$

Therefore, (10) can be re-expressed as:

$$\left|k_0 \left(5 \cdot 2^{n-2}-1\right)\right|_{2^{n+1}-1} = 1$$

According to Proposition 1, $k_0 = 1/3\times(2^{n+2}-10)$. Substituting it into (9), we have

$$X = X^{(1)} + 2^n\left(2^{2n}-1\right)\left|\frac{1}{3}\left(2^{n+2}-10\right)\left(x_4 - X^{(1)}\right)\right|_{2^{n+1}-1}$$

$$= X^{(1)} + 2^n\left(2^{2n}-1\right)Z \tag{11}$$

where Z is a $(n+1)$-bit number, and

$$Z = \left|\frac{1}{3}\left(2^{n+2}-10\right)\left(x_4 - X^{(1)}\right)\right|_{2^{n+1}-1}$$

From (6), we have

$$Z = \left|\frac{1}{3}\left(2^{n+2}-10\right)\left(x_4 - x_2 - 2^n Y\right)\right|_{2^{n+1}-1}$$

$$= \left|\left|\frac{1}{3}\right|_{2^{n+1}-1} \left|(H+I+J)\right|_{2^{n+1}-1}\right|_{2^{n+1}-1} \tag{12}$$

where

$$H = 2^{n+2}x_4 - 10x_4$$
$$I = -2^{n+2}x_2 + 10\cdot x_2 \tag{13}$$
$$J = 2^{n+3}Y$$

Let

$$Q = \left|H+I+J\right|_{2^{n+1}-1} \tag{14}$$

We can use Property 1 to simplify the calculation of Q.

$$|H|_{2^{n+1}-1} = \left|2^{n+2}x_4 - 10x_4\right|_{2^{n+1}-1} = \underbrace{\overline{X}_{4,n-3}\ldots\overline{X}_{4,0}\overline{X}_{4,n}\overline{X}_{4,n-1}\overline{X}_{4,n-2}}_{n+1} \tag{15a}$$

$$|I|_{2^{n+1}-1} = \left|-2^{n+2}x_2 + 10\cdot x_2\right|_{2^{n+1}-1} = \underbrace{X_{2,n-3}\ldots X_{2,0}0X_{2,n-1}X_{2,n-2}}_{n+1} \tag{15b}$$

$$|J|_{2^{n+1}-1} = \left|2^{n+3}Y\right|_{2^{n+1}-1} = \left|J_1 + J_2\right|_{2^{n+1}-1}$$

where

$$J_1 = CLS(Y_1,2) = \underbrace{Y_{n-2}\ldots Y_0 0 Y_{n-1}}_{n+1} \tag{15c}$$

$$J_2 = CLS(Y_2,1) = \underbrace{Y_{2n-1}\ldots Y_n 0}_{n+1} \tag{15d}$$

Now, Q can be easily calculated by equations from (14) to (15), using only one $(4, 2^{n+1}-1)$ Multi-Operand Modulo Adder (MOMA).

By applying *Proposition 2* and (14) to (12), Z can be evaluated as:

$$Z = \left|\sum_{i=0}^{n/2} 2^{2i}\cdot Q\right|_{2^{n+1}-1} = \left|S\cdot Q\right|_{2^{n+1}-1} \tag{16}$$

where

$$S = \left|2^0 + 2^2 + 2^4 + \ldots + 2^n\right|_{2^{n+1}-1} \tag{17}$$

So we have

$$Z = \left|\left(2^0 + 2^2 + 2^4 + \ldots + 2^n\right)Q\right|_{2^{n+1}-1} = \left|Q^{(0)} + Q^{(2)} + \ldots + Q^{(n)}\right|_{2^{n+1}-1}$$
$$= \left|CLS(Q,0) + CLS(Q,2) + \ldots + CLS(Q,n)\right|_{2^{n+1}-1} \tag{18}$$

In (18), $Q^{(0)}$, $Q^{(2)}$,…, $Q^{(n)}$ are all of $(n+1)$ bits. As they are circularly left-shifted versions of Q, Z can be easily implemented by one $(n/2 + 1, 2^{n+1} - 1)$ MOMA [8].

Finally, X can be calculated by (12) as follows:

$$X = x_2 + 2^n\left(2^{2n}Z + Y - Z\right) = x_2 + 2^n\left(Z^{(1)} - Z\right) = x_2 + 2^n T \tag{19}$$

where

$$Z^{(1)} = 2^{2n}Z + 2^n Y_2 + Y_1 = \underbrace{Z_n\ldots Z_0}_{n+1}\underbrace{Y_{2n-1}Y_{2n-2}\ldots Y_0}_{2n} \tag{20}$$

One $(3n+1)$-bit subtractor is needed to calculate T, and the resulting $(3n+1)$-bit T is concatenated to the n-bit x_2 to obtain the final value of X.

4. HARDWARE ARCHITECTURES

Fig. 1 shows the hardware implementation ported directly from the algorithm given in Section 3. The first stage is an efficient three-moduli set residue-to-binary converter [6]. The second stage is the calculation of Q, which involves only a $(4, 2^{n+1} - 1)$ MOMA, as shown in Fig. 2. The third stage is the calculation of Z, which is realized with a single $(n/2 + 1, 2^{n+1} - 1)$ MOMA, as shown in Fig. 3. The final stage is a $(3n+1)$-bit subtractor. Each stage in Fig. 1 contains one 1's complement adder and the four stages are sequential, therefore the total delay of the converter is the sum of the delay of each individual stage.

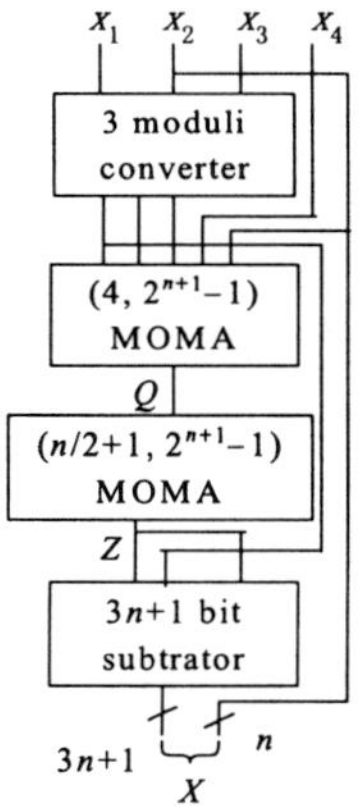

Fig. 1 *Hardware scheme of new converter*

The new residue-to-binary converter for 4-moduli set $\{2^n - 1, 2^n, 2^n + 1, 2^{n+1} - 1\}$ can have two different implementations by using two variants of CPA proposed in [8] and [9]. The high speed and cost effective versions of the architecture are abbreviated as HS and CE, respectively.

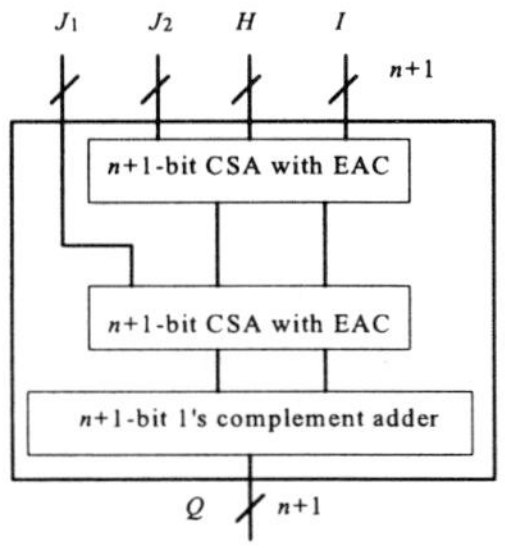

Fig. 2 *Calculation of Q*

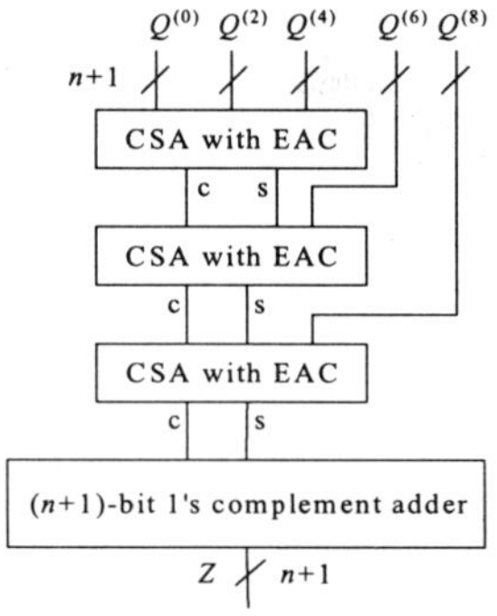

Fig. 3 *Calculation of Z*

5. PERFORMANCE ANALYSIS

The area and time complexities of the proposed reverse converter architectures are analyzed. The hardware complexities for CE and HS versions in terms of primitive arithmetic cells are tabulated in Table 1. Different bit-width CPAs are used in each implementation, which is indicated in the sub-columns of column 1 and 3 of Table 1. Table 2 shows the delays of the two implementations where l is the number of levels of the CSA tree in Fig. 3.

Table 1: Hardware complexity analysis

CE			HS		
Inverter		$3n+2$	Inverter		$3n+2$
HA		1	HA		1
FA		$n^2/2+7n/2+1$	FA		$n^2/2+7n/2+1$
1-bit MUX		2	1-bit MUX		$4n+4$
2-input AND		0	2-input AND		3
2-input OR		0	2-input OR		6
CPA	$n+1$	2	CPA	$n/2+1$	4
	$2n$	1		$n/2$	4
	$3n+1$	1		n	4
				$3n+1$	1

Table 2: Delay of the proposed converters

Type	Delay
CE	$t_{INV} + t_{MUX} + (3+l)t_{FA} + 2t_{CPA(2n)} + 4t_{CPA(n+1)} + t_{CPA(3n+1)}$
HS	$t_{INV} + 4t_{MUX} + 6t_{NAND} + (3+l)t_{FA} + t_{CPA(n)} + 2t_{CPA(n/2+1)} + t_{CPA(3n+1)}$

In [7], Vinod and Premkumar proposed a residue-to-binary converter for this 4-moduli set, and it is reported as the best converter by the author for this moduli set. If we assume that the modular adder/subtractor in [7] uses the modular adder proposed in [3], then the total area of this converter is estimated to be approximately $O(37n)$. Table 3 shows the comparisons of our proposed reverse converters with theirs. It is evident that both versions of our new converter are faster than the converter in [7]. Moreover, both implementations of our converter are more cost effective than theirs within the word length of most practical DSP applications. Our reverse converters are more hardware efficient when the dynamic range, DR $\leq$ 208 bits for CE, and when DR $\leq$ 160 for HS.

It should be noted that the architecture of Fig. 1 is pipelinable. By pipelining, a very high throughput is achievable as the delay is dominated by only the slowest last stage once the pipeline is filled.

Table 3: Comparisons of area-time complexities

Converters	Area	Delay
[7]	$(37n + 14)$ FAs	$(14n + 8)t_{FA}$
CE	$(n^2/2 + 11n + 3)$ FAs	$(11n + l + 8)t_{FA}$
HS	$(n^2/2 + 17n + 6)$ FAs	$(5n + l + 5)t_{FA}$

6. CONCLUSION

We have presented a new residue-to-binary converters for the 4-moduli set $\{2^n - 1, 2^n, 2^n + 1, 2^{n+1} - 1\}$, where n is an even number. The conversion algorithm is based on the MRC of two moduli where one modulus is generated by the most efficient reverse conversion algorithm from the first three moduli of the 4-moduli set. A cost-effective and a high-speed architectures are proposed and their area-time complexities are analyzed in details. The analysis shows that the two proposed new converters are faster and have lower complexity than the fastest converter [7] for the same moduli set.

7. REFERENCES

[1] N. S. Szabo and R. I. Tanaka, *Residue Arithmetic and its Applications to Computer Technology*. New York: McGraw Hill, 1967.

[2] M. A. Soderstrand, W. K. Jenkins, G. A. Jullien and F. J. Taylor, *Residue Number System Arithmetic: Modern Applications in Digital Signal Processing*. New York: IEEE Press, 1986.

[3] S. J. Piestrak, "Design of residue generators and multioperand modular adders using carry-save adders". *IEEE Trans. Comput.*, vol. 423, no. 1, pp. 68-77, Jan. 1994.

[4] K. M. Elleithy and M. A. Bayoumi, "Fast and flexible architectures for RNS arithmetic decoding". *IEEE Trans. Circuits Syst.II*, vol. 39, no. 4, pp. 226-235, Apr. 1992.

[5] H. M. Yassine and W. R Moore, "Improved mixed-radix conversion for residue number system architectures". *IEE Proc.-G*, vol. 138, no. 1, pp. 120-124, Feb. 1991.

[6] Y. Wang, X. Song, M. Aboulhamid and H. Shen, "Adder based residue to binary number converters for $(2^n - 1, 2^n, 2^n + 1)$". *IEEE Trans. Signal Processing*, vol. 50, no. 7, pp. 1772-1779, Jul. 2002.

[7] A. P. Vinod and A. B. Premkumar, "A memoryless reverse converter for the 4-moduli superset $\{2^n - 1, 2^n, 2^n + 1, 2^{n+1} - 1\}$". *Journal of Circuits, Systems, and Computers*, vol. 10, no. 1&2 (2000), pp. 85-99, 2000.

[8] S. J. Piestrak, "A high-speed realization of a residue to binary number converter". *IEEE Trans. Circuits Syst. II* , vol. 42, no. 10, pp. 661-663, Oct. 1995.

[9] M. Bhardwaj, A. B. Premkumar and T. Srikanthan, "Breaking the $2n$-bit carry propagation barrier in residue to binary conversion for the $\{2^n - 1, 2^n, 2^n + 1\}$ moduli set". *IEEE Trans. Circuits Syst. I*, vol. 45, no. 9, pp. 998-1002, Sept. 1998

A Parallel/Pipelined Algorithm for the Computation of MDCT and IMDCT

N. Rama Murthy
Centre for Artificial Intelligence and Robotics
Bangalore, India – 560 001

M. N. S. Swamy
Concordia University
Montreal, Canada, H3G 1M8

ABSTRACT

In this paper, we propose a parallel/pipelined algorithm for the high-speed computation of the transforms used in wide band audio coding.

I. Introduction

Advanced Audio Compression (AAC) defined in the MPEG standards [1, 2] and Audio Compression-3 (AC-3) defined in ATSC standard [3] are two of the widely used algorithms for the compression of wide band audio signals.

Above two standards use oddly stacked analysis / synthesis filter banks based on time-domain aliasing cancellation (TDAC) technique [4] and employ the modified discrete cosine transform (MDCT) and the inverse modified discrete cosine transform (IMDCT) for the subband/ transform coding. But for an insignificant difference in their definition of these transforms, both these standards employ the MDCT for the transformation of the time-domain signal samples into frequency-domain coefficients (in the audio encoder) and, the IMDCT for the transformation of frequency-domain coefficients back into the time-domain signal samples (in the audio decoder).

We now proceed to give the definition of the MDCT and the IMDCT as in AAC.

Let $\bar{x}(n)$, n = 0, 1, 2,..., (N-1) represent time domain signal samples in a signal block. These signal samples are windowed by an N sample real window $w(n)$ to obtain N windowed time domain signal samples $x(n)$, n = 0, 1, 2,..., (N-1). The MDCT of $x(n)$ is defined as

$$X(k) = \sum_{n=0}^{N-1} x(n) \cos\left[\frac{\pi\ (2n+1+N/2)\ (2k+1)}{2N}\right]$$

where k = 0, 1, 2, ..., (N/2) – 1. (1)

The IMDCT is defined as

$$\hat{x}(n) = \frac{2}{N} \sum_{k=0}^{\frac{N}{2}-1} X(k) \cos\left[\frac{\pi\ (2n+1+N/2)\ (2k+1)}{2N}\right]$$

where n = 0, 1,...., (N-1). (2)

$\hat{x}(n)$ in Equation (2) emphasizes the fact that the recovered data sequence by the inverse transform does not correspond to the original data sequence and is time-domain aliased.

Efficient methods for computing the MDCT and the IMDCT have been proposed in [5-7]. The focus in the development of the algorithms therein has been to achieve efficient implementation on a standard DSP chip as in the case of [5, 6] or minimizing the total number of multiplications as in the case of [7]. However, for very high-speed computation in real-time, a parallel/pipelined architecture is desirable. Hence, we propose a parallel/ pipelined algorithm that would lead to a common high-speed architecture for the computation of both the MDCT and the IMDCT

II. Computation of the MDCT through FFT

Due to contribution from researchers interested in VLSI implementation of computational algorithms, efficient hardware architectures in the form of systolic arrays and constant geometry structures exist for FFT. Hardware solutions for FFT are now available commercially, in both ASIC form and as a soft core on FPGA. Hence, logically, it is a good proposition to develop a parallel/pipelined algorithm that would lead to a parallel/pipelined hardware architecture for the MDCT/IMDCT based on the FFT. Towards this, we observe that Equation (1) can be written as

$$X(k) = \text{Re}\,[Z(k)] \quad (3)$$

where the notation Re[ ] denotes the real part of the complex quantity [ ], and

0-7803-7761-3/03/$17.00 ©2003 IEEE

$$Z(k)=\sum_{n=0}^{N-1} x(n)\exp\left(\frac{-j\pi\ (2n+1+N/2)(2k+1)}{2N}\right)$$

for $k=0,1,2,\ldots\ldots,(N/2)-1.$ (4)

Partitioning the N-point sequence $x(n)$ into two $(N/2)$-point sequences $x_1(n)$ and $x_2(n)$, such that

$x_1(n) = x(n)$ and
$x_2(n) = x(n+N/2) \quad 0 \le n \le (N/2)-1$ (5)

Equation (4) can now be written as

$Z(k)=\phi_1(k)+\phi_2(k)$, where

$$\phi_1(k)=\sum_{n=0}^{(N/2)-1} x_1(n)\exp\left(\frac{-j\pi(2n+1+\frac{N}{2})(2k+1)}{2N}\right),$$

$$\phi_2(k)=\sum_{n=0}^{(N/2)-1} x_2(n)\exp\left(\frac{-j\pi(2n+1+\frac{3N}{2})(2k+1)}{2N}\right)$$

and $k = 0, 1, 2,\ldots, (N/2) - 1$ (6)

$\phi_1(k)$ and $\phi_2(k)$ can now be written as

$\phi_1(k)=\psi_1(k)\,X_1(k),$
$\phi_2(k)=\psi_2(k)\,X_2(k)$, where

$\psi_1(k)=\exp(-j\pi\ (2k+1)/4)$
$\psi_2(k)=\exp(-3j\pi\ (2k+1)/4)$

$$X_i(k)=\sum_{n=0}^{(N/2)-1} x_i(n)\exp\left(\frac{-j\pi\ (2n+1)(2k+1)}{2N}\right)$$

for $i = 1,\ 2$ and $0 \le k \le (N/2)-1$ (7)

$X_1(k)$ and $X_2(k)$ in equation (7) can be written as

$X_1(k)=\lambda(k)\,U_1(k),$
$X_2(k)=\lambda(k)\,U_2(k)$

where $\lambda(k)=\exp(-j\pi\ (2k+1))/N$,

$$U_i(k)=\sum_{n=0}^{(N/2)-1} u_i(n)\exp(-2j\pi\ n\ k\ /\ N)\ \text{for}$$

$0 \le k \le (N/2)-1$ and

$u_i(n)=x_i(n)\exp(-j\pi\ n/N),\quad i=1,2$ (8)

It may now be recognized that $U_i(k)$, i = 1, 2 are the $(N/2)$-point DFTs of the two sequences $u_i(n)$, i = 1, 2. Thus, the computation of the N-point MDCT can be done through the following steps.

Step 1: Windowing of the N-point real sequence $\bar{x}(n)$ to obtain the N-point real sequence $x(n)$ and formation of the two $(N/2)$-point real sequences $x_1(n)$ and $x_2(n)$ as in Equation (5).

Step 2: time-domain modulation of sequences $x_1(n)$ and $x_2(n)$ by $\exp(-j\pi\ n/N)$ to obtain two $N/2$ point complex sequences $u_i(n)$, $i=1,2$ as in Equation (8).

Step 3: Computation of the two $(N/2)$-point DFTs of the complex sequences $u_1(n)$ and $u_2(n)$ to obtain the frequency-domain transformed sequences $U_1(k)$ and $U_2(k)$ respectively.

Step 4: Frequency-domain modulation of the two $(N/2)$-point sequences $U_1(k)$ and $U_2(k)$ with $\lambda(k)$ to obtain the $(N/2)$-point complex sequences $X_1(k)$ and $X_2(k)$ respectively.

Step 5: Modulating $X_1(k)$ with $\psi_1(k)$ and $X_2(k)$ with $\psi_2(k)$ to obtain the $(N/2)$-point sequences $\varphi_1(k)$ and $\phi_2(k)$ respectively.

Step 6: Combining $\varphi_1(k)$ and $\phi_2(k)$ to get $Z(k)$ as per Equation (6) and thus obtain the desired N-point real sequence $X(k)$ as per Equation (3).

In the context of MDCT computation for TDAC filter banks used in AAC and AC-3, operations in the above 6-step procedure can be combined/sequenced to obtain a parallel/pipelined algorithm as shown in Section III.

III. Development of Proposed Algorithm

AAC and AC-3 algorithms implement a critically sampled, oddly stacked TDAC filter bank and the necessary transforms are computed on successive data blocks each of size N with adjacent data blocks overlapping by fifty percent as shown in Figure 1. Using superscript *b, b*+1, *b*+2 …to reference successive data blocks, subscripts 1 and 2 to indicate first-half and second-half of signals in a signal block, a 'bar' to indicate an un-windowed signal in the case of time domain signals, it is seen that

$$\bar{x}_1^{b+1}(n) = \bar{x}_2^{b}(n) \quad \text{and} \quad \bar{x}_2^{b+1}(n) = \bar{x}_1^{b+2}(n)$$
$$b = 1, 2, \ldots \quad \text{and} \quad n = 0, 1, \ldots, (N/2-1) \tag{9}$$

We now propose the following 4–step algorithm that envisages a sequence of computations be carried out on a frame by frame basis with each frame being processed in one computation cycle. Referring to Figure 1, where the time-index is in ascending order from right to the left, successive frames and computation cycles are shown as *b*-1, *b*, *b*+1,…. A computation cycle extends over *N*/2 sample period duration (same as frame duration) and involves following steps to be carried out.

Step 1: Windowing and modulation

Let $w(n)$ be an *N*-point real symmetric window. Define $w_1(n) = w(n)$ and $w_2(n) = w(n+N/2)$, n = 0, 1, 2,…., (*N*/2) – 1. Compute the (*N*/2)-point complex sequences $u_1(n)$ and $u_2(n)$, where for $0 \le n \le (N/2)-1$

$$u_1(n) = \bar{x}_2^{b}(n) * w_1(n) * \exp(-j\,\pi\, n/N) \quad \text{and}$$
$$u_2(n) = \bar{x}_2^{b}(n) * w_2(n) * \exp(-j\,\pi\, n/N)$$

Step 2: Computation of the (*N*/2)-point DFTs

Compute the (*N*/2)-point DFTs $U_1(k)$ and $U_2(k)$ of the complex sequences $u_1(n)$ and $u_2(n)$ respectively.

Step 3: Computation of $X_2^b(k)$ and $X_1^{b+1}(k)$

It may be recognized that

$$X_2^b(k) = \lambda(k)\,U_2(k) \quad \text{and} \quad X_1^{b+1}(k) = \lambda(k) U_1(k)$$

Store $X_1^{b+1}(k)$ for usage during next computation cycle. Retrieve the stored $X_1^b(k)$ (stored while performing this step during the previous computation cycle (*b*-1).

Step 4: Computation of *N*/2)-point sequences *A*(*k*) and *B*(*k*)

A(k) and *B*(k) are defined by

$A(k) = \text{Re}[\phi_1^b(k)] = \psi_1(k) X_1^b(k)$ and

$B(k) = \text{Re}\ [\phi_2^b(k)] = \psi_2(k) X_2^b(k)$.

Add, on a sample by sample basis, *A*(*k*)+*B*(*k*) to get $X^b(k)$ the transform coefficient block corresponding to the time-domain signal block *b*.

The proposed algorithm is parallel in that it computes $X_2^b(k)$ and $X_1^{b+1}(k)$ in parallel. It is pipelined as it sequences computation of $X_1^{b+1}(k)$ one computation cycle ahead.

The total number of real multiplications involved per computation cycle is 8*N* plus the number of real multiplications required for computing the two (*N*/2)-point DFTs, which depend on the choice of the FFT algorithm.

For the parallel computation of the IMDCT, in Equation (2), compute $\hat{x}_1(n)$ and $\hat{x}_2(n)$ in parallel where, for n = 0, 1,…., (*N*/2 –1), $\hat{x}_1(n) = \hat{x}(n)$ and $\hat{x}_2(n) = \hat{x}(n+N/2)$. From Equation (2), expressions for $\hat{x}_1(n)$ and $\hat{x}_2(n)$ (ignoring the scaling factor of 2/N which can be absorbed into computations at a later stage) are

$$\hat{x}_1(n) = \sum_{k=0}^{N/2-1} X(k)\cos\left(\frac{\pi\,(2n+1+N/2)(2k+1)}{2N}\right)$$

$$\hat{x}_2(n) = \sum_{k=0}^{N/2-1} X(k)\cos\left(\frac{\pi\,(2n+1+3N/2)(2k+1)}{2N}\right)$$

where $n = 0, 1, \ldots, (N/2)-1$.

Now, $\hat{x}_1(n) = \text{Re}[\bar{\phi}_1(n)]$ and $\hat{x}_2(n) = \text{Re}[\bar{\phi}_2(n)]$ with

$$\bar{\phi}_1(n) = \sum_{n=0}^{N/2-1} X(k)\exp\left(\frac{-j\pi\,(2n+1+\frac{N}{2})(2k+1)}{2N}\right),$$

$$\bar{\phi}_2(n) = \sum_{n=0}^{N/2-1} X(k)\exp\left(\frac{-j\pi(2n+1+\frac{3N}{2})(2k+1)}{2N}\right)$$

for n = 0, 1, 2,… (*N*/2) – 1.

The process of computation of $\bar{\phi}_1(n)$ and $\bar{\phi}_2(n)$ from the (*N*/2)-point real sequence *X*(*k*) is similar to that for $\phi_1(k)$ and $\phi_2(k)$ from $x_1(n)$ and $x_2(n)$ in the case of the MDCT. Hence, we omit further details on computation.

In the context of TDAC filter bank, inverse transforming of the consecutive input transform coefficient blocks are done on a frame by frame basis as in the case of MDCT with input block $X^b(k)$ being processed during computation cycle b. The two $N/2$ point aliased sequences $\hat{x}_1^b(n)$ and $\hat{x}_2^b(n)$ are obtained. $\hat{x}_2^b(n)$ is stored for use in the next computation cycle and $\hat{x}_1^b(n)$ is over lapped and added with $\hat{x}_2^{b-1}(n)$ (computed and stored during the previous computation cycle) to get the alias-cancelled time domain sequence $x^b(n)$.

The total number of real multiplications involved per computation cycle is $5N$ plus the number of real multiplications for the two $N/2$ point DFTs.

The proposed algorithm can easily be adopted for the computation of the MDCT/IMDCT for both the *long* and *short* transforms defined in AAC and AC-3 standards.

IV. Conclusions

A novel parallel/pipelined algorithm that would lead to a scalable parallel/pipelined hardware architecture for the computation of the MDCT and the IMDCT required for AAC has been presented. The development of the algorithms for the MDCT and the IMDCT for AC-3 as well as the *short* and *long* transforms in the two wide band audio coding standards follow the approach delineated in this paper.

V. References

[1] " Information Technology – Coding of moving pictures and associated audio for digital storage media at up to about 1.5 Mb/s Part 3: Audio ", ISO / IEC JTC1 / SC29 / WG11 MPEG, International Standard IS 11172- 3, (MPEG-1), 1992

[2] " Information Technology – Generic coding of moving pictures and associated audio - Part 3: Audio ", ISO / IEC / JTC1 / SC29 / WG11 MPEG, International Standard IS 13818 – 3 (MPEG-2), 1994.

[3] " Digital audio compression (AC-3) standard ", Document of Advanced Television Systems Committee (ATSC), Audio Specialist Group T3/S7, Revision A, 20 August 2001.

[4] J. P. Princen, A. W. Johnson, and A. B. Bradley, "Subband / Transform Coding Using Filter Bank Designs Based on Time - Domain Aliasing Cancellation ", in Proc. IEEE Int. Conf. Acoustics, Speech, and Signal Processing '87, Dallas, TX, Apr. 1987, pp 2161-2164.

[5] P. Duhamel, Y. Mahieux, and J.P. Petit, " A Fast Algorithm for the Implementation of Filter Banks Based on Time - Domain Aliasing Cancellation ", in Proc. IEEE International Conf. Acoustics, Speech, and Signal Processing '91, Toronto, Canada, May 1991, pp. 2209-2212.

[6] R. Gluth, "Regular FFT - Related Transform Kernels for DCT / DST – Based Polyphase Filter Banks", *ibid.*, pp. 2205-2208.

[7] V. Britanak and K.R. Rao, "An Efficient Implementation of the Forward and Inverse MDCT in MPEG Audio Coding ", IEEE Signal Processing Letters, Vol. 8, No 2, February 2001, pp 48-51.

Acknowledgements

The first author would like to thank the Director, CAIR, for granting permission to submit this research work carried out at Concordia University, Canada, as part of his post doctoral research work. The second author would like to thank Natural Sciences and Engineering Research Council of Canada, and the Faculty of Engineering and Computer Science for financial support for this work.

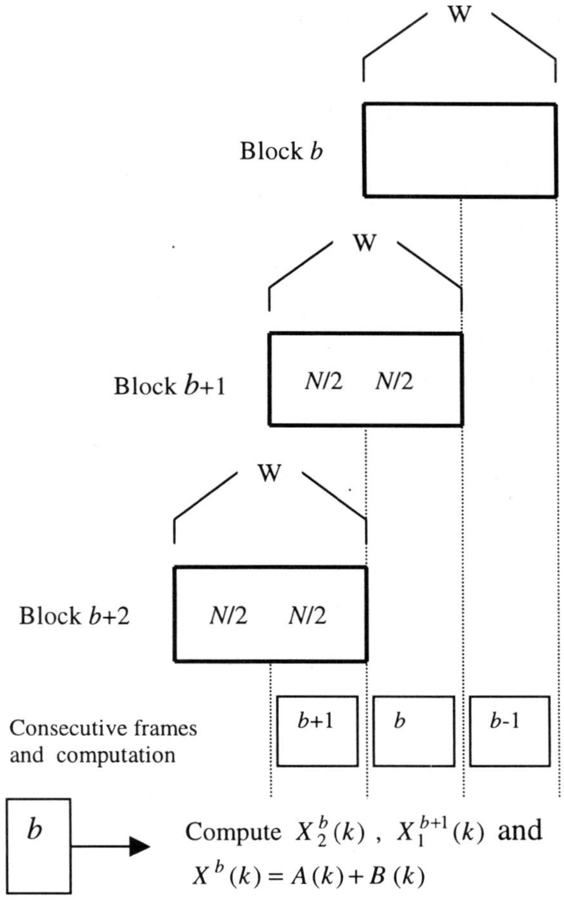

Fig 1: Position of windows/data blocks

ARCHITECTURE FOR CORDIC ALGORITHM REALIZATION WITHOUT ROM LOOKUP TABLES

Chuen-Yau Chen and Wen-Chih Liu

Department of Electronic Engineering, I-Shou University
Kaohsiung, 84008, Taiwan, Republic of China
Email: cychen@ieee.org

ABSTRACT

This paper proposed architecture for implementing the CORDIC algorithm. By decomposing an arbitrary rotation angle into a sequence of coarse angles (each of whose tangent value is a power of 2) and a fine angle which is small enough, this architecture can be implemented by performing a sequence of shift-and-add operations in the radix-2 system without any ROM lookup table or real multiplication requirement. It is suitable to be designed in pipelined architecture for performing the high-speed operations.

1. INTRODUCTION

Sine and cosine are the trigonometric functions which play the important roles in many fields of applications, especially in the control systems and communication systems. For examples, it might be treated as a carrier to modulate the signal for transmitting on a specific channel.

With regard to the generation or computation of the sine and cosine functions, there have been many schemes proposed [1]. Most of these schemes are developed on the COordinate Rotation Digital Computer (CORDIC) algorithm [3], [4]. CORDIC algorithm computes the values of sine and cosine based on the property of rotating a given angle from an initial point around the origin on the X-Y plane. Then, the X-coordinate and Y-coordinate of the resulting point are the sine and cosine of the given angle, respectively. Schemes based on CORDIC algorithm can be classified into two categories. The first one is called the mixed-hybrid CORDIC method, and the second one is called the partitioned-hybrid CORDIC method [6]. In the mixed-hybrid CORDIC methods, the real multiplications are required to accomplish the ArcTangent Radix (ATR) part subrotations, which degrades the operating speed of the whole stage. In additional, from the hardware viewpoints, the multipliers induce a large hardware overhead. In the partition-hybrid CORDIC methods, the multiplications of the ATR part are replaced by a ROM lookup table which contains the pre-computed values of all the possible summations of the ATR terms as mentioned in the mixed-hybrid CORDIC method. Although the quart-wave symmetry properties of the sine/cosine functions have been adopted to compress the size of ROM [2], the size of the ROM lookup table grows exponentially with the increasing of the precision of the angle.

The work in this paper is motivated by the development of the partitioned-hybrid CORDIC methods and concerned with the reduction of the size of the ROM lookup table even removing the ROM lookup table completely. In this proposed architecture, all the multiplications become the radix-2 multiplications which can be easily realized by the shift-and-add operations in the radix-2 system.

2. SINE/COSINE COMPUTATIONS

In the X-Y plane, a point $(X, Y) = (r\cos\theta, r\sin\theta)$ can be viewed as by rotating an angle of θ counterclockwise along the circle of a radius of r centered at the origin from the point $(X_0, Y_0) = (r\cos 0, r\sin 0) = (r, 0)$. This operation can be expressed as

$$\begin{aligned}\begin{bmatrix} X \\ Y \end{bmatrix} &= \begin{bmatrix} \cos\theta & -\sin\theta \\ \sin\theta & \cos\theta \end{bmatrix}\begin{bmatrix} X_0 \\ Y_0 \end{bmatrix} \\ &= \cos\theta\begin{bmatrix} 1 & -\tan\theta \\ \tan\theta & 1 \end{bmatrix}\begin{bmatrix} X_0 \\ Y_0 \end{bmatrix}. \end{aligned} \quad (1)$$

In (1), if the arbitrary angle θ is expressed as the linear combination of the angle θ_k where $\tan\theta_k$ is known in priori, the resulting point $(X, Y) = (r\cos\theta, r\sin\theta)$ can be obtained by decomposing the rotation of θ into a sequence of subrotations of θ_k, and $(\cos\theta, \sin\theta)$ will be obtained in turn. That is,

$$\theta = \sum_{k=0}^{N} \sigma_k \theta_k \quad (2)$$

and

$$\begin{bmatrix} X \\ Y \end{bmatrix} = K(\prod_{k=0}^{N} \begin{bmatrix} 1 & -\tan\sigma_k\theta_k \\ \tan\sigma_k\theta_k & 1 \end{bmatrix}) \begin{bmatrix} X_0 \\ Y_0 \end{bmatrix} \quad (3)$$

where $K = \cos\sigma_0\theta_0 \cdots \cos\sigma_N\theta_N$ is a constant, N is the number of subrotations, and $\sigma_k \in \{-1, 0, 1\}$ determines the direction of each subrotation [3].

2.1. The CORDIC Algorithm

From the digital computation viewpoint, CORDIC algorithm takes advantage of the property that decomposes a rotation into a sequence of subrotations to develop an approach that reaches the target angle θ by applying successive subrotations [3], [4]. In CORDIC algorithm, a positive angle θ is represented in an N-bit binary number as

$$\theta = \sum_{k=1}^{N} b_k\theta_k \quad (4)$$

where b_k is the bit in the radix-2 numbering system with the weight of 2^{-k} and $\theta_k = 2^{-k}$ represents the positional power-of-two weight. By applying the angle recoding process, θ can be rewritten as [5]

$$\theta = \sum_{k=1}^{N} b_k\theta_k = \phi_0 + \sum_{k=2}^{N+1} r_k\theta_k \quad (5)$$

where ϕ_0 is constant and $r_k \in \{-1, 1\}$.

2.2. The Coarse-Fine Rotation Method

In this method, an arbitrary angle is first mapped from the full range $[0, 2\pi]$ to $\theta \in [0, \frac{\pi}{4}]$ according to the quarter-wave symmetry property of the sine/cosine functions. Then, the angle θ is partitioned into two terms expressed as [10]

$$\theta = \theta_M + \theta_L \quad (6)$$

where θ_M is the coarse subangle that can be expressed as

$$\theta_M = \sum_{k=1}^{\frac{N}{3}} b_k 2^k \quad (7)$$

and θ_L is the fine subangle that can be expressed as

$$\theta_L = \sum_{k=\frac{N}{3}+1}^{N} b_k 2^k. \quad (8)$$

Accordingly, the rotation in (1) can be partitioned into two cascaded stages. The first stage is

$$\begin{aligned} X_M &= X_0 - Y_0 \tan\sigma_M\theta_M \\ Y_M &= Y_0 + X_0 \tan\sigma_M\theta_M, \end{aligned} \quad (9)$$

and the second stage is

$$\begin{aligned} X &= X_M - Y_M \tan\sigma_L\theta_L \\ Y &= Y_M + X_M \tan\sigma_L\theta_L. \end{aligned} \quad (10)$$

Clearly, the second stage computation can be easily realized by a sequence of shift-and-add operations in radix-2 numbering system because θ_L is small enough so that the value of $\tan\theta_L$ can be approximated by θ_L that is a power of 2 [6]. However, the first stage computations involve multiplying $\tan\theta_M$ on the datapath, which becomes the bottleneck of this method. These multiplication operations are usually replaced by a ROM lookup table.

2.3. The Proposed Modified Coarse-Fine Rotation Method

In order to reduce the size of the ROM lookup table in the coarse-fine rotation method, we treat ϕ_0 in (5) as the initial angle and decompose the coarse angle θ_M as follows:

$$\theta_M = \sum_{k=2}^{\frac{N}{3}} \theta_{Mk} = \sum_{k=2}^{\frac{N}{3}} (\theta_{Hk} + \theta_{Lk}) \quad (11)$$

where θ_{Mk} is defined as

$$\theta_{Mk} = 2^{-k}, \quad (12)$$

θ_{Hk} is the angle that should satisify

$$\tan\theta_{Hk} = 2^{-k} \quad (13)$$

and

$$\theta_{Lk} = \theta_{Mk} - \theta_{Hk} = \theta_{Mk} - \arctan 2^{-k} \quad (14)$$

is treated as the error correction term. Substitute (11) into (6), and the angle θ can be rewritten as

$$\begin{aligned} \theta &= \sum_{k=2}^{\frac{N}{3}} \theta_{Hk} + \theta_{Lk} + \theta_L \\ &= \sum_{k=2}^{\frac{N}{3}} \theta_{Hk} + \theta_{\Sigma L} \end{aligned} \quad (15)$$

where $\theta_{\Sigma L}$ is defined as

$$\theta_{\Sigma L} = \sum_{k=2}^{\frac{N}{3}} \theta_{Lk} + \theta_L \quad (16)$$

Table 1: The size of the ROM lookup table versus the precision.

Number of bits N	Number of address bits	Address of the first word	Address of the final word	Number of words
16	4	0000	1100	13
17	4	0000	1100	13
18	5	00000	11001	26
19	5	00000	11001	26
20	5	00000	11001	26
21	6	000000	110010	51
22	6	000000	110010	51
23	6	000000	110010	51
24	7	0000000	1100100	101
25	7	0000000	1100100	101
26	7	0000000	1100100	101

for convenience. With regard to θ_{Hk}, it is inherently a power of 2. As to $\theta_{\Sigma L}$, it may be either small enough or with a carry in the $\frac{N}{3}$-th stage. If it is small enough, of course it can be realized by a sequence of shift-and-add operations; otherwise, the $\theta_{H\frac{N}{3}}$ should be rotated again to realize the carry, and the remain of $\theta_{\Sigma L}$ can be realized by a sequence of shift-and-add operations. As a whole, the rotation of θ_M in (6) can be realized by a sequence of shift-and-add operations in the radix-2 system. Therefore, neither the ROM lookup table nor the real multiplication operation is required.

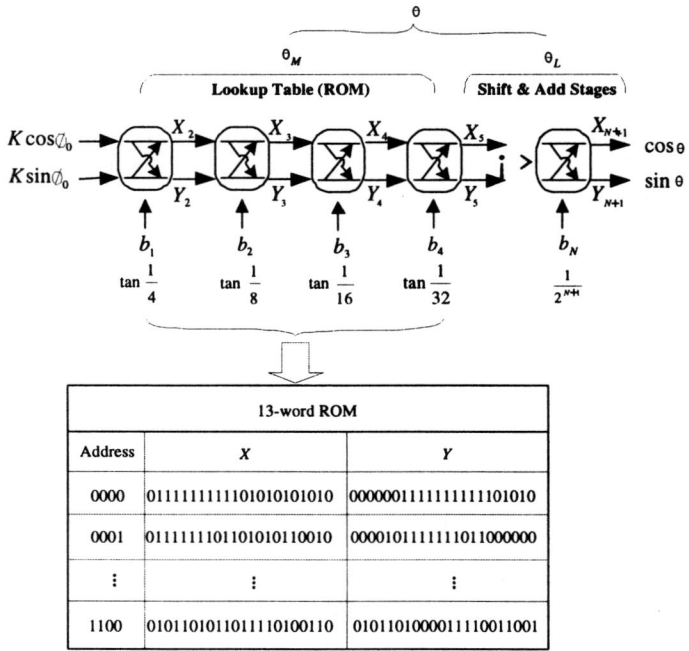

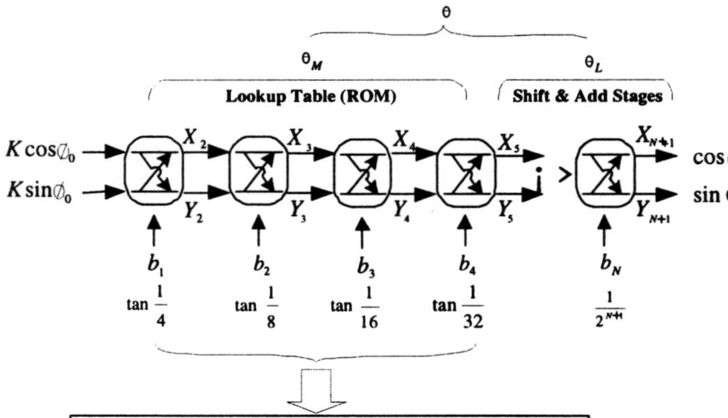

13-word ROM		
Address	*X*	*Y*
0000	0111111111101010101010	0000001111111111101010
0001	0111111101101010110010	0000101111111011000000
⋮	⋮	⋮
1100	0101101011011110100110	0101101000011110011001

Figure 1: The ROM-based architecture for the coarse-fine rotation method [5].

3. ARCHITECTURE

3.1. The ROM-Based Architecture for the Coarse-Fine Rotation Method

In the coarse-fine rotation method, the bottleneck is that the requirement of multipliers to generate the product terms $X_0 \tan\theta_M$ and $Y_0 \tan\theta_M$ where the value of $\tan\theta_M$ is itself needed to be computed. Figure 1 shows a ROM-Based architecture [5]. In this architecture, ϕ_0 in (5) is treated as the initial angle, and the information for the rotation of the coarse angle θ_M is built in a 13-word ROM, each word contains one value of $\tan\theta_M$ for 13 possible θ_M's. Obviously, as shown in Table 1, for the precision improvement, the number of bits used should increase, the size of ROM will grows exponentially which becomes a large overhead from the hardware implementation viewpoint. Figure 2 is a graphic representation showing this trend.

3.2. The ROM-less Architecture for the Modified Coarse-Fine Rotation Method

In the modified coarse-fine rotation method, since we decomposed θ into the sum of $\sum_{k=2}^{5}\theta_{Hk}$ and $\theta_{\Sigma L}$ where $\tan\theta_{Hk}$ is a power of 2 and $\theta_{\Sigma L}$ is small enough such that $\tan\theta_{\Sigma L}$ can be approximated by $\theta_{\Sigma L}$, we can realize the whole rotation by an all radix-2 shift-and-add datapath architecture as shown in Fig. 3. In this fig-

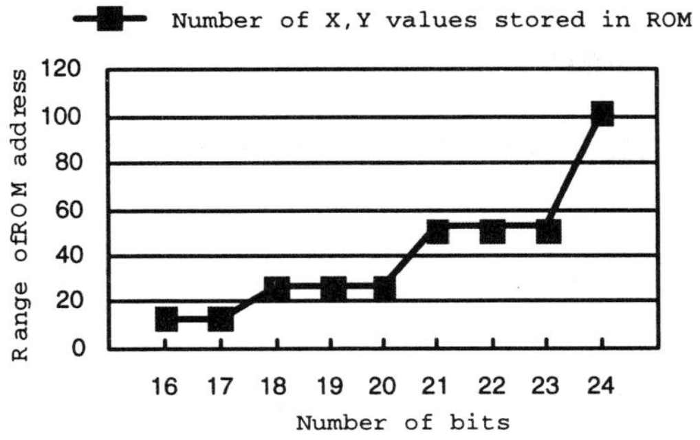

Figure 2: The size of ROM lookup table increases with the precision in the ROM-based architecture.

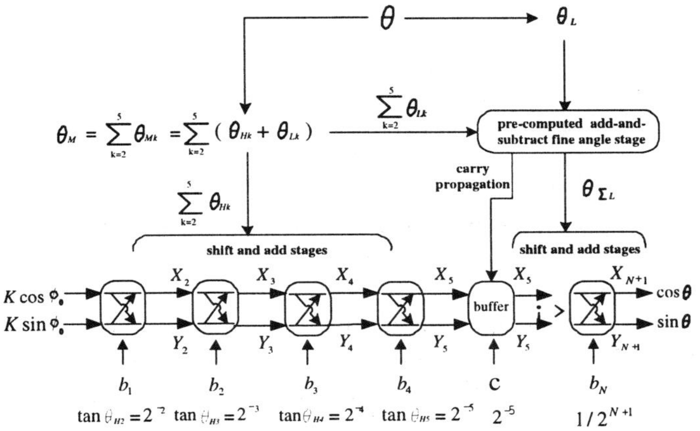

Figure 3: The proposed ROM-less all radix-2 stages architecture for the modified coarse-fine rotation method.

ure, the multiplication of $\tan\theta_{Mk}$ that was implemented by a 13-word ROM lookup table as shown in Fig. 1 is replaced by $\frac{N}{3}$ stages of shift-and-add operations. From the hardware realization viewpoint, since the whole data path are operated with shift-and-add operations without any real multiplication or ROM lookup table, it is real suitable to realize the whole system in pipelined architecture. Therefore, the operating speed will be faster than the ROM-based one, and the hardware overhead will be also smaller.

4. CONCLUSION

We have proposed architecture to implement the CORDIC algorithm in a more efficient manner in this paper. By decomposing the rotation angle into a sequence of sub-angles that are either small enough or with a power-of-2 tangent amplitude, the whole rotation can be accomplished by a sequence of shift-and-add operations in the radix-2 system instead of constructing the ROM lookup table or performing the real multiplication operations on the datapath. The hardware realization can be arranged in pipelined architecture to speed up the operation. The precision can be increased to a satisfactory level without a large hardware overhead. In the future, by suitably configuring this design with other functional cores, such as D/A, low-pass filter, we can implement a direct digital frequency synthesizer.

5. REFERENCES

[1] J. Vankka, "Methods of mapping from phase to sine amplitude in direct digital synthesis", *IEEE Trans. Ultrasonics, Ferroelectrics, and Frequency Control*, vol. 44, no. 2, pp. 526–534, Mar. 1997.

[2] H. T. Nicholas, H. Samueli, and B. Kim, "The optimization of direct digital frequency synthesizer performance in the presence of finite word length effects", in *Proc. 42nd Annu. Freq. Contr. Symp.*, 1988, pp. 357–363.

[3] J. Volder, "The CORDIC trigonometric computing technique", *IEEE Trans. Comput.*, vol. 8, pp. 330–334, 1959.

[4] J. Walther, "A unified algorithm for elementary functions", in *Proc. Spring Joint Computer Conf.*, 1971, pp. 379–385.

[5] A. Madisetti, A. Y. Kwentus, and A. N. Willson, Jr., "A 100-MHz ,16-b, direct digital frequency synthesizer with a 100-dBc spurious-free dynamic range", *IEEE J. Solid-State Circuits*, vol. 34, no. 8, pp. 1034–1043, Aug. 1999.

[6] S. Wang, V. Piuri, and E. E. Swartzlander, Jr., "Hybrid CORDIC algorithms", *IEEE Trans. Comput.*, vol. 46, no. 11, pp. 1202–1207, Nov. 1997.

[7] J. A. Lee and T. Lang, "Constant-factor redundant CORDIC for angle calculation and rotation", *IEEE Trans. Comput.*, vol. 41, pp. 1016–1025, Aug. 1992.

[8] M. D. Ercegovac and T. Lang, "Redundant and on-line CORDIC: application to matrix triangularization and SVD", *IEEE Trans. Comput.*, vol. 39, pp. 725–740, June 1990.

[9] N. Takagi, T. Asada, and S. Yajima, "Redundant CORDIC methods with a constant scale factor for sine and cosine computation", *IEEE Trans. Comput.*, vol. 40, pp. 989–995, Sept. 1991.

[10] D. Fu and A. N. Willson, Jr., "A high-speed processor for digital sine/cosine generation and angle rotation", in *Proc. IEEE 32 Asilomar Conf. Signals, System, and Computers*, Nov. 1988, pp. 177–181.

RELATIONSHIP BETWEEN HAAR WAVELET AND REED-MULLER SPECTRA

Bogdan J. Falkowski
Nanyang Technological University
School of Electrical and Electronic Engineering
Block S1, Nanyang Avenue, Singapore 639798

Abstract: *The mutual relations between Haar Wavelet and Reed-Muller spectral domains are presented in the form of matrix decomposition and as layered vertical and horizontal Kronecker matrices. The new relations apply to an arbitrary dimension of the transform matrices and allow to perform direct conversions between Reed-Muller and Haar Wavelet spectra.*

INTRODUCTION

During the last decade there has been increasing interest in applications of different transforms used in digital signal processing in spectral analysis, synthesis and testing of logic functions [1-5, 7-10]. The recent interest in application of various transforms in Computer-Aided Design tools is also caused by development of efficient methods of their calculation and storage [2, 8, 10]. While majority of research has been done using the Reed-Muller [1, 3, 8, 10] and the Walsh transform [1, 4, 5, 10], dependent on a case some transforms are better suited than others. Frequently it is advantageous to apply more than one transform in a given task based on the local properties of a data function [2, 4, 7]. For example if the logic function has many zeros in its truth vector it is better to apply local transform that has non-zero entries in its own transformation matrix that almost overlap with non-zero elements in the truth vector. In the latter case it is also of interest to investigate mutual relations between various local discrete transforms such as, for example, the Haar Wavelet and the Reed-Muller transform. Both the Haar Wavelet transform (unnormalized version of the transform where only signs are entered into the transform matrix) and the Reed-Muller transform have been used in many applications of logic design [1-5, 7, 8, 10]. Therefore it is not only interesting theoretically but also practical to state their mutual relations. It should be noticed that the Reed-Muller transform is carried over Galois Field 2 - GF(2), while the Haar Wavelet transform is carried over integers. The case of one way conversion from the Walsh to the Reed-Muller spectrum was discussed in [9]. In this work, mutual relations between the Reed-Muller and the Haar Wavelet transforms are presented for arbitrary transform size in the form of matrix decomposition and as layered vertical and horizontal Kronecker product based matrices for an arbitrary transform matrix order size. Due to the properties of the Reed-Muller transform some extra operations based on GF(2) algebra are necessary.

REED-MULLER AND HAAR TRANSFORMS

The matrix RM_N of order $N = 2^n$ for *Reed-Muller Transform* in Hadamard ordering is defined as [1, 3, 8, 10]:

$$RM_N = \begin{bmatrix} RM_{\frac{N}{2}} & 0 \\ RM_{\frac{N}{2}} & RM_{\frac{N}{2}} \end{bmatrix}, RM_1 = 1, \; N = 2, 3, \ldots \text{ Also}$$

$RM_N = RM_2 \otimes RM_{\frac{N}{2}}$ for $N = 2, 3, \ldots$

The *unnormalized Haar Transform matrix* H_N of order $N = 2^n$ in Hadamard ordering can be defined recursively as [4, 10] :

$$H_N = \begin{bmatrix} H_{\frac{N}{2}} \otimes \begin{bmatrix} 1 & 1 \end{bmatrix} \\ I_{\frac{N}{2}} \otimes \begin{bmatrix} 1 & -1 \end{bmatrix} \end{bmatrix} \text{ and } H_1 = 1$$

where $I_{\frac{N}{2}}$ is an identity matrix of order $N/2$. In the above equations, the symbol '$\otimes$' denotes the Kronecker direct product [11].

For an n- variable Boolean function $F(x_1, x_2, \ldots, x_n)$, the Haar and the Reed-Muller spectrum (a column vector of dimension 2^n x 1) is given by $\overrightarrow{H_N} = H_N \vec{F}$ and $\overrightarrow{RM_N} = RM_N \oplus \vec{F}$ where $\overrightarrow{H_N}$ is the Haar spectrum and $\overrightarrow{RM_N}$ is the Reed-Muller spectrum, accordingly. It should be noticed that calculation of the Haar transform is performed in standard algebra and of the Reed-Muller spectrum is performed over GF(2) represented by symbol '$\oplus$'.

Example 1. Let $f(\vec{x_3}) = \bar{x}_3\bar{x}_2 \vee x_3\bar{x}_2 \vee x_3\bar{x}_1$ with $\vec{F} = [1,1,0,0,1,1,1,0]^T$. The Reed-Muller and the Haar spectra for this Boolean function are as follows.

$$\overrightarrow{RM_8} = RM_8 \oplus \vec{F} = \begin{bmatrix} 1&0&0&0&0&0&0&0 \\ 1&1&0&0&0&0&0&0 \\ 1&0&1&0&0&0&0&0 \\ 1&1&1&1&0&0&0&0 \\ 1&0&0&0&1&0&0&0 \\ 1&1&0&0&1&1&0&0 \\ 1&0&1&0&1&0&1&0 \\ 1&1&1&1&1&1&1&1 \end{bmatrix} \oplus \begin{bmatrix} 1 \\ 1 \\ 0 \\ 0 \\ 1 \\ 1 \\ 1 \\ 0 \end{bmatrix} = \begin{bmatrix} 1 \\ 0 \\ 1 \\ 0 \\ 0 \\ 0 \\ 1 \\ 1 \end{bmatrix}$$

$$\overrightarrow{H_8} = H_8 \vec{F} = \begin{bmatrix} 1&1&1&1&1&1&1&1 \\ 1&1&1&1&-1&-1&-1&-1 \\ 1&1&-1&-1&0&0&0&0 \\ 0&0&0&0&1&1&-1&-1 \\ 1&-1&0&0&0&0&0&0 \\ 0&0&1&-1&0&0&0&0 \\ 0&0&0&0&1&-1&0&0 \\ 0&0&0&0&0&0&1&-1 \end{bmatrix} \begin{bmatrix} 1 \\ 1 \\ 0 \\ 0 \\ 1 \\ 1 \\ 1 \\ 0 \end{bmatrix} = \begin{bmatrix} 5 \\ -1 \\ 2 \\ 1 \\ 0 \\ 0 \\ 0 \\ 1 \end{bmatrix}$$

0-7803-7761-3/03/$17.00 ©2003 IEEE

The above spectra can be used to represent Boolean functions in terms of polynomial expansions as well as to investigate their different properties such as linearity, symmetry, classification, testing, etc [1-5, 7, 8, 10]. Calculation of both spectra and their storage can be efficiently done using reduced representations for both original Boolean functions and spectra themselves with the help of various types of decision diagrams [8, 10].

RELATIONS BETWEEN HAAR AND REED-MULLER SPECTRA AND FUNCTIONS

The following general relations are valid between the Reed-Muller and non normalized Haar Wavelet spectra: $\overrightarrow{H_N} = H_N\left[RM_N \oplus \overrightarrow{RM_N}\right]$ and $\overrightarrow{RM_N} = \left(RM_N^{-1} H_N^{-1} \overrightarrow{H_N}\right)$ where H_N^{-1} is inverse of the Haar transform matrix and RM_N^{-1} is inverse of the Reed-Muller transform in standard and not GF(2) algebra. It should be noticed that the inverse of Reed-Muller transform matrix is the same as its forward matrix in GF(2), the symbol '$\oplus$' represents operations in GF(2) and bracket ( ) means taking modulo-2 of the absolute values of the result.

Example 2. Let us convert the Reed-Muller and the Haar spectra directly for the Boolean function from Example 1.

$$\overrightarrow{H_8} = H_8\left[RM_8 \oplus \overrightarrow{RM_8}\right] = \begin{bmatrix} 1 & 1 & 1 & 1 & 1 & 1 & 1 & 1 \\ 1 & 1 & 1 & 1 & -1 & -1 & -1 & -1 \\ 1 & 1 & -1 & -1 & 0 & 0 & 0 & 0 \\ 0 & 0 & 0 & 0 & 1 & 1 & -1 & -1 \\ 1 & -1 & 0 & 0 & 0 & 0 & 0 & 0 \\ 0 & 0 & 1 & -1 & 0 & 0 & 0 & 0 \\ 0 & 0 & 0 & 0 & 1 & -1 & 0 & 0 \\ 0 & 0 & 0 & 0 & 0 & 0 & 1 & -1 \end{bmatrix} \times$$

$$\left[\begin{bmatrix} 1 & 0 & 0 & 0 & 0 & 0 & 0 & 0 \\ 1 & 1 & 0 & 0 & 0 & 0 & 0 & 0 \\ 1 & 0 & 1 & 0 & 0 & 0 & 0 & 0 \\ 1 & 1 & 1 & 1 & 0 & 0 & 0 & 0 \\ 1 & 0 & 0 & 0 & 1 & 0 & 0 & 0 \\ 1 & 1 & 0 & 0 & 1 & 1 & 0 & 0 \\ 1 & 0 & 1 & 0 & 1 & 0 & 1 & 0 \\ 1 & 1 & 1 & 1 & 1 & 1 & 1 & 1 \end{bmatrix} \oplus \begin{bmatrix} 1 \\ 0 \\ 1 \\ 0 \\ 0 \\ 0 \\ 1 \\ 1 \end{bmatrix}\right] = \begin{bmatrix} 5 \\ -1 \\ 2 \\ 1 \\ 0 \\ 0 \\ 0 \\ 1 \end{bmatrix}$$

$$\overrightarrow{RM_8} = \left(RM_8^{-1} H_8^{-1} \overrightarrow{H_8}\right) = \begin{bmatrix} 1 & 0 & 0 & 0 & 0 & 0 & 0 & 0 \\ -1 & 1 & 0 & 0 & 0 & 0 & 0 & 0 \\ -1 & 0 & 1 & 0 & 0 & 0 & 0 & 0 \\ 1 & -1 & -1 & 1 & 0 & 0 & 0 & 0 \\ -1 & 0 & 0 & 0 & 1 & 0 & 0 & 0 \\ 1 & -1 & 0 & 0 & -1 & 1 & 0 & 0 \\ 1 & 0 & -1 & 0 & -1 & 0 & 1 & 0 \\ -1 & 1 & 1 & -1 & 1 & -1 & -1 & 1 \end{bmatrix} \times$$

$$\tfrac{1}{8} \times \begin{bmatrix} 1 & 1 & 2 & 0 & 4 & 0 & 0 & 0 \\ 1 & 1 & 2 & 0 & -4 & 0 & 0 & 0 \\ 1 & 1 & -2 & 0 & 0 & 4 & 0 & 0 \\ 1 & 1 & -2 & 0 & 0 & -4 & 0 & 0 \\ 1 & -1 & 0 & 2 & 0 & 0 & 4 & 0 \\ 1 & -1 & 0 & 2 & 0 & 0 & -4 & 0 \\ 1 & -1 & 0 & -2 & 0 & 0 & 0 & 4 \\ 1 & -1 & 0 & -2 & 0 & 0 & 0 & -4 \end{bmatrix} \times \begin{bmatrix} 5 \\ -1 \\ 2 \\ 1 \\ 0 \\ 0 \\ 0 \\ 1 \end{bmatrix} =$$

$$\left(\tfrac{1}{8} \times \begin{bmatrix} 1 & 1 & 2 & 0 & 4 & 0 & 0 & 0 \\ 0 & 0 & 0 & 0 & -8 & 0 & 0 & 0 \\ 0 & 0 & -4 & 0 & -4 & 4 & 0 & 0 \\ 0 & 0 & 0 & 0 & 8 & -8 & 0 & 0 \\ 0 & -2 & -2 & 2 & -4 & 0 & 4 & 0 \\ 0 & 0 & 0 & 0 & 8 & 0 & -8 & 0 \\ 0 & 0 & 4 & -4 & 4 & -4 & -4 & 4 \\ 0 & 0 & 0 & 0 & -8 & 8 & 8 & -8 \end{bmatrix} \times \begin{bmatrix} 5 \\ -1 \\ 2 \\ 1 \\ 0 \\ 0 \\ 0 \\ 1 \end{bmatrix}\right) = \left(\begin{bmatrix} 1 \\ 0 \\ -1 \\ 0 \\ 0 \\ 0 \\ 1 \\ -1 \end{bmatrix}\right)$$

$$= \begin{bmatrix} 1 \\ 0 \\ 1 \\ 0 \\ 0 \\ 0 \\ 1 \\ 1 \end{bmatrix}$$

Due to the fact that both the Reed-Muller and the Haar Wavelet transform matrices are defined through right Kronecker products, their spectra can be calculated through fast transforms for an arbitrary N using Good's theorem [1, 3, 4, 10, 11].

Let us now introduce the mutual relations between the Haar and the Reed-Muller transforms for a general case of an arbitrary N using recursive definition for the transformation matrices. It should be noticed that presented in this article relations apply to conversions not only between the Haar Wavelet and the Reed-Muller spectra but also between the Haar Wavelet and the Reed-Muller functions. Therefore in the following developments the symbols for functions instead of spectra will be used. However, the Reed-Muller and the Haar Wavelet functions can be freely replaced with the Reed-Muller $\overrightarrow{RM_N}$ and Haar $\overrightarrow{H_N}$ spectra when needed. It is trivial to modify the presented equations for a case of normalized Haar Wavelet functions [2, 4, 10] by adding normalizing factors. From the definition presented earlier for unnormalized Haar transform matrix it is obvious that the first two rows of H_N are global basis functions $H_0(x)$ and $H_1(x)$, respectively. All subsequent rows are constituted by local basis functions $H_l^{(k)}(x)$ in an ascending order of l and k. $l=1, 2,..$ is known as a degree of the Haar function describing the number of zero crossings, $k = 1,..,2^l$ is an order of the Haar function describing the position of the subset l within a function [2, 4, 10]. In the Reed-Muller transformation matrix RM_N, all but the last row are local basis functions and RM_i denotes an i-th Reed-Muller function. The symbol RM_i^* denotes the Reed-Muller spectrum obtained from the calculation using inverse Reed-Muller transformation matrix RM_N^{-1} in standard and not GF(2) algebra. The conversions for higher n ($N = 2^n$) are shown for expressing non normalized Haar functions by Reed-Muller functions and vice verse in the form of layered Kronecker product as follows.

$$\begin{bmatrix} H_0 \\ H_1 \\ H_2^{(1)} \\ H_2^{(2)} \\ H_3^{(1)} \\ H_3^{(2)} \\ H_3^{(3)} \\ H_3^{(4)} \\ \vdots \\ \vdots \\ \vdots \end{bmatrix} = \begin{bmatrix} \begin{bmatrix} 2 & 1 \\ 0 & -1 \end{bmatrix} & \otimes & \left(\overset{n-1}{\otimes}\begin{bmatrix} 2 & 1 \end{bmatrix}\right) & & \\ \hdashline \begin{bmatrix} 1 & 0 \\ 1 & 1 \end{bmatrix} & \otimes & \begin{bmatrix} 0 & -1 \end{bmatrix} & \otimes & \left(\overset{n-2}{\otimes}\begin{bmatrix} 2 & 1 \end{bmatrix}\right) \\ \hdashline \left(\overset{2}{\otimes}\begin{bmatrix} 1 & 0 \\ 1 & 1 \end{bmatrix}\right) & \otimes & \begin{bmatrix} 0 & -1 \end{bmatrix} & \otimes & \left(\overset{n-3}{\otimes}\begin{bmatrix} 2 & 1 \end{bmatrix}\right) \\ \hdashline \left(\overset{3}{\otimes}\begin{bmatrix} 1 & 0 \\ 1 & 1 \end{bmatrix}\right) & \otimes & \begin{bmatrix} 0 & -1 \end{bmatrix} & \otimes & \left(\overset{n-4}{\otimes}\begin{bmatrix} 2 & 1 \end{bmatrix}\right) \\ \vdots & \vdots & \vdots & \vdots & \vdots \\ \vdots & \vdots & \vdots & \vdots & \vdots \\ \vdots & \vdots & \vdots & \vdots & \vdots \end{bmatrix} \begin{bmatrix} RM_0^* \\ RM_1^* \\ RM_2^* \\ RM_{12}^* \\ RM_3^* \\ RM_{13}^* \\ RM_{23}^* \\ RM_{123}^* \\ \vdots \\ \vdots \\ \vdots \end{bmatrix}$$

$$\begin{bmatrix} RM_0 \\ RM_1 \\ RM_2 \\ RM_{12} \\ RM_3 \\ RM_{13} \\ RM_{23} \\ RM_{123} \\ \vdots \\ \vdots \\ \vdots \end{bmatrix} = \frac{1}{8} \times \begin{bmatrix} \begin{bmatrix} 1 & 1 \\ 0 & -2 \end{bmatrix} & \otimes & \left(\overset{n-1}{\otimes}\begin{bmatrix} 1 \\ 0 \end{bmatrix}\right) & \vdots & & \\ \begin{bmatrix} 1 & 0 \\ -1 & 1 \end{bmatrix} & \otimes & \left(2\begin{bmatrix} 1 \\ -2 \end{bmatrix}\right) & \otimes & \left(\overset{n-2}{\otimes}\begin{bmatrix} 1 \\ 0 \end{bmatrix}\right) & \vdots \\ \left(\overset{2}{\otimes}\begin{bmatrix} 1 & 0 \\ -1 & 1 \end{bmatrix}\right) & \otimes & \left(4\begin{bmatrix} 1 \\ -2 \end{bmatrix}\right) & \otimes & \left(\overset{n-3}{\otimes}\begin{bmatrix} 1 \\ 0 \end{bmatrix}\right) & \vdots \\ \left(\overset{3}{\otimes}\begin{bmatrix} 1 & 0 \\ -1 & 1 \end{bmatrix}\right) & \otimes & \left(8\begin{bmatrix} 1 \\ -2 \end{bmatrix}\right) & \otimes & \left(\overset{n-4}{\otimes}\begin{bmatrix} 1 \\ 0 \end{bmatrix}\right) & \vdots \\ \vdots & & \vdots & & \vdots & \\ \vdots & & \vdots & & \vdots & \\ \vdots & & \vdots & & \vdots & \end{bmatrix} \times \begin{bmatrix} H_0 \\ H_1 \\ H_2^{(1)} \\ H_2^{(2)} \\ H_3^{(1)} \\ H_3^{(2)} \\ H_3^{(3)} \\ H_3^{(4)} \\ \vdots \\ \vdots \\ \vdots \end{bmatrix}$$

In the above equations, the symbols "$\overset{n}{\otimes}$"and "$\otimes$" represent the Kronecker direct product of n and two matrices respectively. It should be noticed that the second equation needs taking modulo-2 of the absolute values of the result to obtain the Reed-Muller spectrum. The vertical dashed lines denote the layered vertical Kronecker matrices, and the horizontal dashed lines denote the layered horizontal Kronecker matrices, accordingly. A layered horizontal Kronecker matrix is defined as the horizontal sum of Kronecker matrices [11]. In a similar manner, the notion of a layered vertical Kronecker matrix is introduced that can be defined as the vertical sum of Kronecker matrices. By the used convention when the Kronecker direct product of i matrices happens for the above equations for $i \leq 0$ then the term $\overset{i}{\otimes}$ disappears from them.

Let us apply the above formulae for n=3 in Example 3. The results are exactly the same as those presented in Example 2.

It should be noticed that the calculation of the Haar spectral coefficients from RM_i^* spectral coefficients can be also done directly from RM_i spectral coefficients as shown in Example 2. As mentioned earlier, the same matrices can be used for conversions between the Haar Wavelet and the Reed-Muller functions and their corresponding spectra.

Example 3:

For n=3, the above relations become:

$$\begin{bmatrix} H_0 \\ H_1 \\ H_2^{(1)} \\ H_2^{(2)} \\ H_3^{(1)} \\ H_3^{(2)} \\ H_3^{(3)} \\ H_3^{(4)} \end{bmatrix} = \begin{bmatrix} \begin{bmatrix} 2 & 1 \\ 0 & -1 \end{bmatrix} & \otimes & \begin{bmatrix} 2 & 1 \end{bmatrix} & \otimes & \begin{bmatrix} 2 & 1 \end{bmatrix} \\ \hdashline \begin{bmatrix} 1 & 0 \\ 1 & 1 \end{bmatrix} & \otimes & \begin{bmatrix} 0 & -1 \end{bmatrix} & \otimes & \begin{bmatrix} 2 & 1 \end{bmatrix} \\ \hdashline \left(\overset{2}{\otimes}\begin{bmatrix} 1 & 0 \\ 1 & 1 \end{bmatrix}\right) & \otimes & \begin{bmatrix} 0 & -1 \end{bmatrix} & & \end{bmatrix} \times \begin{bmatrix} RM_0^* \\ RM_1^* \\ RM_2^* \\ RM_{12}^* \\ RM_3^* \\ RM_{13}^* \\ RM_{23}^* \\ RM_{123}^* \end{bmatrix} =$$

$$= \begin{bmatrix} 8 & 4 & 4 & 2 & 4 & 2 & 2 & 1 \\ 0 & 0 & 0 & 0 & -4 & -2 & -2 & -1 \\ 0 & 0 & -2 & -1 & 0 & 0 & 0 & 0 \\ 0 & 0 & -2 & -1 & 0 & 0 & -2 & -1 \\ 0 & -1 & 0 & 0 & 0 & 0 & 0 & 0 \\ 0 & -1 & 0 & -1 & 0 & 0 & 0 & 0 \\ 0 & -1 & 0 & 0 & 0 & -1 & 0 & 0 \\ 0 & -1 & 0 & -1 & 0 & -1 & 0 & -1 \end{bmatrix} \times \begin{bmatrix} RM_0^* \\ RM_1^* \\ RM_2^* \\ RM_{12}^* \\ RM_3^* \\ RM_{13}^* \\ RM_{23}^* \\ RM_{123}^* \end{bmatrix} =$$

$$= \begin{bmatrix} 8 & 4 & 4 & 2 & 4 & 2 & 2 & 1 \\ 0 & 0 & 0 & 0 & -4 & -2 & -2 & -1 \\ 0 & 0 & -2 & -1 & 0 & 0 & 0 & 0 \\ 0 & 0 & -2 & -1 & 0 & 0 & -2 & -1 \\ 0 & -1 & 0 & 0 & 0 & 0 & 0 & 0 \\ 0 & -1 & 0 & -1 & 0 & 0 & 0 & 0 \\ 0 & -1 & 0 & 0 & 0 & -1 & 0 & 0 \\ 0 & -1 & 0 & -1 & 0 & -1 & 0 & -1 \end{bmatrix} \times \begin{bmatrix} 1 \\ 0 \\ -1 \\ 0 \\ 0 \\ 0 \\ 1 \\ -1 \end{bmatrix} = \begin{bmatrix} 5 \\ -1 \\ 2 \\ 1 \\ 0 \\ 0 \\ 0 \\ 1 \end{bmatrix}$$

$$\begin{bmatrix} RM_0 \\ RM_1 \\ RM_2 \\ RM_{12} \\ RM_3 \\ RM_{13} \\ RM_{23} \\ RM_{123} \end{bmatrix} = \frac{1}{8} \begin{bmatrix} \begin{bmatrix} 1 & 1 \\ 0 & -2 \end{bmatrix} & \otimes & \begin{bmatrix} 1 \\ 0 \end{bmatrix} & \otimes & \begin{bmatrix} 1 \\ 0 \end{bmatrix} \vdots \\ \begin{bmatrix} 1 & 0 \\ -1 & 1 \end{bmatrix} & \otimes & \left(2 \begin{bmatrix} 1 \\ -2 \end{bmatrix} \right) & \otimes & \begin{bmatrix} 1 \\ 0 \end{bmatrix} \vdots \\ \left(\overset{2}{\otimes} \begin{bmatrix} 1 & 0 \\ -1 & 1 \end{bmatrix} \right) & \otimes & \left(4 \begin{bmatrix} 1 \\ -2 \end{bmatrix} \right) & & \end{bmatrix} \times$$

$$\begin{bmatrix} H_0 \\ H_1 \\ H_2^{(1)} \\ H_2^{(2)} \\ H_3^{(1)} \\ H_3^{(2)} \\ H_3^{(3)} \\ H_3^{(4)} \end{bmatrix} =$$

$$= \left(\frac{1}{8} \begin{bmatrix} 1 & 1 & 2 & 0 & 4 & 0 & 0 & 0 \\ 0 & 0 & 0 & 0 & -8 & 0 & 0 & 0 \\ 0 & 0 & -4 & 0 & -4 & 4 & 0 & 0 \\ 0 & 0 & 0 & 0 & 8 & -8 & 0 & 0 \\ 0 & -2 & -2 & 2 & -4 & 0 & 4 & 0 \\ 0 & 0 & 0 & 0 & 8 & 0 & -8 & 0 \\ 0 & 0 & 4 & -4 & 4 & -4 & -4 & 4 \\ 0 & 0 & 0 & 0 & -8 & 8 & 8 & -8 \end{bmatrix} \times \begin{bmatrix} H_0 \\ H_1 \\ H_2^{(1)} \\ H_2^{(2)} \\ H_3^{(1)} \\ H_3^{(2)} \\ H_3^{(3)} \\ H_3^{(4)} \end{bmatrix} \right)$$

$$= \left(\frac{1}{8} \begin{bmatrix} 1 & 1 & 2 & 0 & 4 & 0 & 0 & 0 \\ 0 & 0 & 0 & 0 & -8 & 0 & 0 & 0 \\ 0 & 0 & -4 & 0 & -4 & 4 & 0 & 0 \\ 0 & 0 & 0 & 0 & 8 & -8 & 0 & 0 \\ 0 & -2 & -2 & 2 & -4 & 0 & 4 & 0 \\ 0 & 0 & 0 & 0 & 8 & 0 & -8 & 0 \\ 0 & 0 & 4 & -4 & 4 & -4 & -4 & 4 \\ 0 & 0 & 0 & 0 & -8 & 8 & 8 & -8 \end{bmatrix} \times \begin{bmatrix} 5 \\ -1 \\ 2 \\ 1 \\ 0 \\ 0 \\ 0 \\ 1 \end{bmatrix} \right) = \begin{bmatrix} 1 \\ 0 \\ 1 \\ 0 \\ 0 \\ 0 \\ 1 \\ 1 \end{bmatrix}$$

CONCLUSION

In this article mutual conversions between the Haar Wavelet and the Reed-Muller transforms for an arbitrary transform size have been investigated. The Reed-Muller transform can represent an arbitrary logic function by using EXOR and AND gates only. The recent interest in applications of such expressions in logic synthesis is caused by their excellent properties for testability and the fact that many practical functions have a big content of strongly nonunate functions (e.g. parity, addition and multiplication) that are best realized by EXOR and AND expressions. Also many multilevel circuits based on EXOR elements are more advantageous when area, speed and testability are of main concern. Many other applications and achievements of the Reed-Muller transform in error correcting codes are also well known [6]. The tutorial on recent applications of the Haar Wavelet transform in logic design is available in [2].

The mutual relations between both transforms are introduced here through recursive equations in the form of layered vertical and horizontal Kronecker products. The presented relations allow to transfer known results of spectral logic design in the Reed-Muller domain [1, 3, 8, 10] to the Haar Wavelet domain and vice verse [1, 2, 4, 5, 7, 10] and compare efficiency of both approaches in different applications for large Boolean functions. Finally it should be also noticed that presented derivations based on layered matrices can be efficiently implemented in the form of operations on spectral decision diagrams for both the Reed-Muller and the Haar Wavelet transforms.

Acknowledgments: Agency for Science, Technology and Research Singapore grant no. 0121060053 supported this study.

References

1. S. Agaian, J. Astola, K. Egiazarian, *Binary Polynomial Transforms and Nonlinear Digital Filters*, New York, Marcel Dekker, 1995.
2. B.J. Falkowski, "Haar transform: calculation and applications in logic design", *Proc. 2nd Int. Workshop on Transforms and Filter Banks*, Brandenburg, Germany, pp. 101-120, March 1999.
3. D. Green, *Modern Logic Design*, Wokingham, England, Addison-Wesley, 1986.
4. M.G. Karpovsky, *Finite Orthogonal Series in the Design of Digital Devices*, New York, Wiley, 1976.
5. M. G. Karpovsky, R. S. Stankovic, J. T. Astola, "Spectral techniques for design and testing of computer hardware", *Proc. 1st Int. Workshop on Spectral Techniques and Logic Design for Future Digital Systems*, Tampere, Finland, pp. 9-42, June 2000.
6. A. Poli, L. Huguet, *Error Corecting Codes: Theory and Applications*, Hertfordshire, Masson and Prentice Hall Int. (UK), 1992
7. G. Ruiz, J.A. Michell, and A. Buron, Switch-level fault detection and diagnosis environment for MOS digital circuits using spectral techniques, *IEE Proc. Computers, Digital Techniques*, **139**, (4), pp. 293-307, 1992.
8. T. Sasao, and M. Fujita (eds.), *Representations of Discrete Functions*, Boston, Kluwer Academic, 1996.
9. R.S. Stankovic, "A note on relations between Reed-Muller and Walsh transforms", *IEEE Trans. on EMC*, **24**, pp. 68-70, 1982
10. R. S. Stankovic, M. Stankovic, D. Jankovic, *Spectral Transforms in Switching Theory: Definitions and Calculations*, Belgrade, Nauka, 1998.
11. L.P. Yaroslavsky, *An Introduction to Digital Picture Processing*, Berlin, Springer-Verlag, 1985

IMPROVEMENTS ON LAYOUT OF GARMENT PATTERNS FOR EFFICIENT FABRIC CONSUMPTION

Sophon Vorasitchai and Suthep Madarasmi

Department of Computer Engineering, Faculty of Engineering
King Monkut's University of Technology Thonburi
Pracha-Uthit Road, Bangkok 10140, Thailand
Tel. +66-2-470-9085, Fax.: +66-2-872-5050

ABSTRACT

This paper deals with the problem of optimal layout for the placement of garment patterns. The proposed genetic algorithm is adapted for use to minimize fabric waste in garment production industry. Each garment piece is modeled as a polygonal object with a default orientation, but may be rotated 180 degrees. The layout of all the pieces is realized in yielding some successive partial solutions until the order book is empty. The efficiency is measured as the length of fixed-width fabric used. Results provided compare our genetic algorithm to human experts. The results from our experiments show that our genetic algorithm can improve the efficiency of almost all production quality markers, shirts, trousers and other garments.

1. INTRODUCTION

The reduction of production cost is one of the major issues in manufacturing industries such as the garment industry, the sheet metal industry, the paper industry, and the leather industry. High material utilization is of particular interest to industries with mass-production, since small improvements of the layout can result in large savings of raw material, considerably reducing production cost.

The placement of garment pieces or markers to use the minimal amount of fabric is an important concern for the garment production industry, since fabric cost can be relatively high. Each piece of garment part such as collar, pocket, or arm comes in different shapes and sizes, some symmetrical and some asymmetrical, presenting a challenge to finding a good placement solution, even for an expert. Much effort has been devoted to automate this process [8-12] by using artificial intelligence and optimization techniques.

In our related earlier work, Sirivarothakul and Madarasmi [13] uses a Genetic Algorithm [1-5] to find the optimal layout of garment patterns on fabric to minimize fabric waste. This paper presents the improvements made to the previous Genetic Algorithm in [13]. The garment pieces are 2-dimensional patterns placed on a fixed width fabric to create strips of variable length on the fabric. All the pieces in the order book are placed on the fabric while allowing 2 placement orientations: 0 degrees and 180 degrees. The efficiency is measured by the length of fabric used once all the pieces in the order book have been placed. The results from their experiments on various pattern designs indicate that their genetic algorithm can effectively be used to obtain highly efficient solutions.

In this proposed research work, we examine a different approach to solving the same problem earlier done in [13] by modifying some of the GA parameters to obtain a better solution. In [13] during crossover, when a new individual was created by choosing good chromosomes or strips from the available population, we noted that only a few new good strips could be created. After obtaining a few good strips, the pieces in the order book were used up and, thus, the new individual started to have pieces that were already available in the other good strips among the available population. Since we do not allow for an individual solution to have more cut pieces than the order book, the good strips available in the current population soon became inefficient strips and useless. In this research work we propose to use a new crossover scheme that will avoid this problem by keeping the good partial solutions or chromosomes. Once good solutions are found, they will be kept aside and the order book will be reduced by the pieces already in the partial solution. The GA will be restarted using the pieces in the revised order book to obtain a new partial solution. By keeping partial solutions, we hope that the system will be able to generate better solutions throughout its lifetime, not only during the initial few chromosomes.

2. PROBLEM REPRESENTATION

In our experiment, each garment piece is modeled as a polygonal object. Objects that have smooth boundaries such as quadratic splines, must be approximated by a polygonal surface such as convex hull [7] surrounding the object prior to starting, since we model each garment piece as a polygonal object. In each placement, the marker has a default initial orientation, but may be rotated 180 degrees. A solution or individual is a structure with the following format:

$$S = [(F1, O1), (F2, O2), (F3, O3), \ldots, (Fn, On), L]$$

where S is the Solution from the order book, F represents each garment piece, O is the piece's orientation: 0 for 0 degrees or 1 for 180 degrees, and L is the length (cost) of the solution.

Figure 1 shows a particular individual where each chromosome is denoted by a strip of a particular length. Items are randomly selected from the order book and placed on the fabric. This model of chromosome is similar to that in [8]. Each piece is placed starting at upper-left edge of the fabric. If there is no space to place the piece, we move downwards until there is space or we run out of fabric width. If we reach the bottom of the fabric width without finding space, then we attempt to place the piece towards the right. When placing a new piece we must check that the space is available. We do this by a simple 2-D graphics algorithm of

0-7803-7761-3/03/$17.00 ©2003 IEEE

checking that none of the vertices of our polygon is inside another, previously placed polygon [6,7]. Thus, we do assume that all pieces are made of convex polygons. For non-convex shapes, we convert them to 2 or more pieces of convex shapes before we begin our procedure (see [7] for converting a non-convex polygon to 2 or more convex polygons).

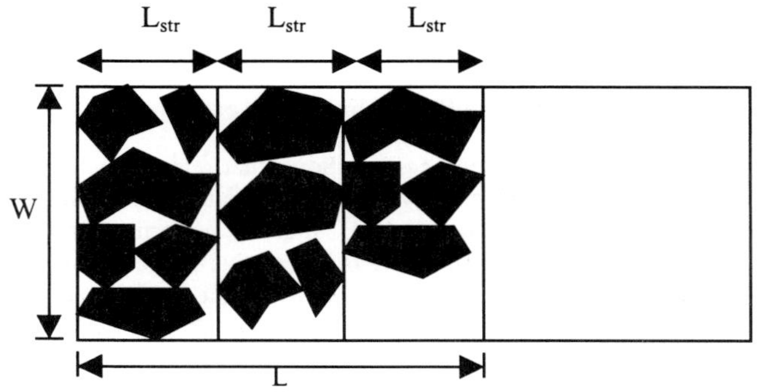

Figure 1. Placement on fabric of fixed width

- *Chromosomes and An Individual Solution* The chromosomes are composed of two parts: the object part and the group part. The object part has one gene for each object (garment piece). The group part contains all of the garment pieces in any order. Note that since the number of groups is not constant, the chromosomes have a varying length. Crossover and mutation work with the group part directly, using the object part indirectly.
- *Population* Genetic algorithms rely on a population of individuals, each individual consisting of a group of chromosomes. Each member or individual of the population represents a possible solution in the search space. The individuals in the new generation come from two sources: some are the offspring of recombination or crossover while others are the product of mutation of the individuals in the current population.
- *Selection* Selection is based on the fitness value of each individual. Individuals with high fitness are more likely to be selected to generate offspring. However, weaker individuals will be eliminated from the population. Selection is achieved by limiting the population size to N and ensuring that only the individuals with the top N fitness levels survive.
- *Crossover* After selection, crossover may proceed. Crossover is an operator applied to chromosome (strips) in the current generation in order to produce new chromosomes or offspring. In our GA model, a child is obtained by collecting good chromosomes or efficient strips in the current population. We do not choose the best chromosome, but bias our selection based on the fitness level of each chromosome or strip.
- *Mutation* In the ordering GA, the mutation operator modifies the order of the genes on the chromosome (strip). We select the genes and reinsert them in random order or permute at random within the solution.

3. OUR GENETIC ALGORITHM

The genetic algorithm used in this paper is outlined in Table 1.

3.1 The Initial Population

Information used in this process includes the width of fabric, the shape of the pieces, the number of those pieces that need to be placed. The initial population is created by randomly generating N individuals. Each individual is created by randomly picking items from the Order Book and placing them on the fixed width fabric.

3.2 The Fitness Values

Each strip shown in figure 1 represents a chromosome. The fitness of each chromosome is defined as percent area efficiency of each chromosome. For each individual, the fitness of the individual is the length of the fabric used. Only the fittest N individuals are maintained in the population.

3.3 Reproduction: Creating a New Individual

A new individual may be created by either crossover or mutation. In the crossover step, we start with a new individual with no chromosomes, corresponding to a fabric with no marker pieces. We then use a linearly biased random number generator [8] to pick a chromosome or strip section from our population. The effect is that the probability of selecting a solution is linearly proportional to its relative position in the population; the better the solution, the higher the probability of its being selected. After selecting a chromosome to become part of a new individual, the pieces in that chromosome will decrease the pieces required in our order book. Thus, we must compute the efficiency of each strip or chromosome in the remaining population again. For example, it is possible that our new individual already has all the collar pieces needed and any population containing a chromosome with a collar will now be rendered unfit and undesirable. Once the efficiency is recomputed, we can again sample for a fit chromosome to continue to create our new individual.

After selecting each chromosome, we must check for overlap in the order book and re-compute the efficiency of our chromosome pool in the population. Thus, it is clear that as chromosomes are selected, the available chromosome pool becomes less and less efficient. We have found this to be a weakness in our previous work [13]. We noted that the new individual solution was only good for about 4-5 chromosomes, after which 'holes' were created in our chromosome pool. Thus, the GA in [13] converged rather quickly since no new individual with improved fitness was being created. We propose here to modify this section by finding partial solutions (say the 4-5 chromosomes with high efficiency) and then restarting the entire GA. After the partial good solution is found, we will remove those pieces in our solution from the order book and use the new, revised order book to run the GA again. It is expected that the smaller order book will yield a few highly efficient chromosomes again which will again be kept aside and the GA restarted. By restarting we expect to improve the solution since we do not get stuck at merely generating solutions that are not any better in the new individual.

3.4 Reporting a Solution

Usually, the algorithm is terminated by specifying a maximum number of generations. Some additional termination criteria may be added, such as maximum number of generations for which the best value remains the same. Once the system has converged, we pick the individual with the best fitness and report its configuration as the solution.

Table 1. The Genetic Algorithm used in this paper

```
1. Set and Initialize the population
2. Compute the efficiency of each strip.
3. Crossover, Mutation, and Selection:
Repeat
   If (r = random (0, 1)) < (p = crossover
probability)
       Create a New Individual by Crossover:
       Repeat
       •  Sample to obtain a chromosome from
          the solution that has high
          efficiency
       •  Re-compute the efficiency of all
          strips in the population
       Until No Efficient Chromosome
       Available
       Fill Remaining Solution Randomly by
       Items remaining in Order Book
  Else
       Create New Individual by Mutation:
       Fill Solution Randomly by Order Book
  Insert the solution into the population
Until(Population has converged) or (No
improvement in Length of Best Solution) or
(NumIterations > MaxIterations)
4. Keep the good strips (efficiency over a
   certain threshold) as partial solution.
   Use items not in the strips as a new,
   revised Order Book. Run the GA again by
   Restarting at step 1 using the new Order
   Book.
```

4. EXPERIMENTAL RESULTS

A case study of our genetic algorithm is considered. One typical set of genetic parameters used in testing is population size n = 15, crossover probability = 0.9 and the number of strips to select to crossover at one time = 3.

Two simulations are reported, one without biased area (case 1) and another with biased area (case 2), with the two resulting shown in figure 2 (a) and (b), respectively

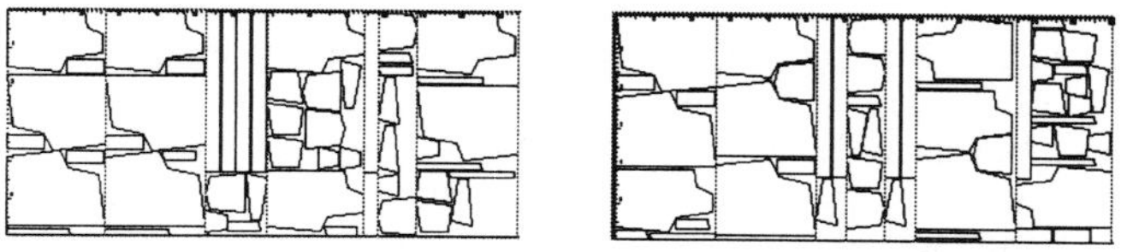

(a) Case 1- efficiency 77.573% *(b) Case 2 - efficiency 80.365%*

Figure 2. Results of the GA (a) without biased area and (b) with biased area

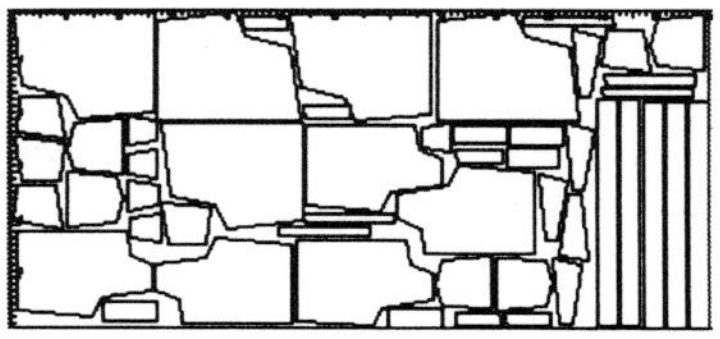

Shorts, 50 pieces, efficiency 80.676%

Figure 3. Shorts result done by human expert

The results between case 1 and case 2 shows that our GA with biased area obtains a better result which happens to be very close to the human expert shown in figure 3. After we performed an analysis using several experiments, we observed that the result of the GA with biased area usually improves the efficiency compared to that in [13]. Thus, we used this to test all of our experiments.

We first experimented with 20 marker pieces for an actual shirt, used to produce 10 shirts. Data used for this comparison is obtained from [13]. Figure 4 and 5 shows our algorithm results vs. Human Expert with 77.0% vs. 79.3% efficiency, respectively.

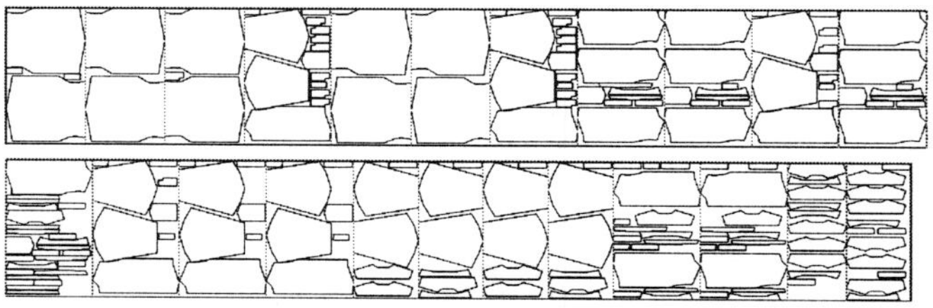

Shirts, 200 pieces, efficiency 77.0%

Figure 4. Actual shirt parts experimental result

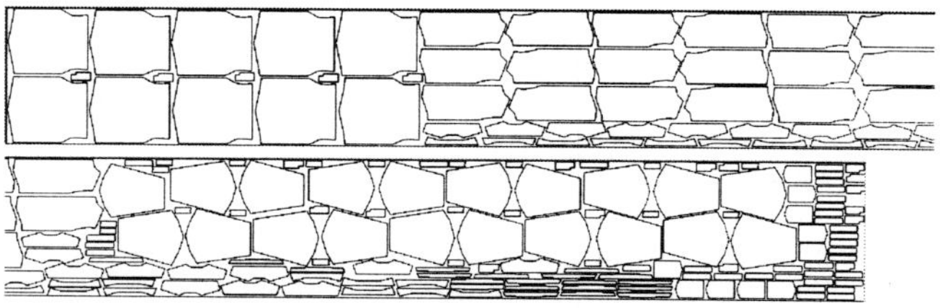

Shirts, 200 pieces, efficiency 79.3%

Figure 5. Actual shirt result done by human expert

We also experimented with 10 marker pieces for trousers, used to produce 10 trousers. Figure 6 and 7 compares GA vs. Human Expert with efficiency 80.7 % vs. 82.7%, respectively.

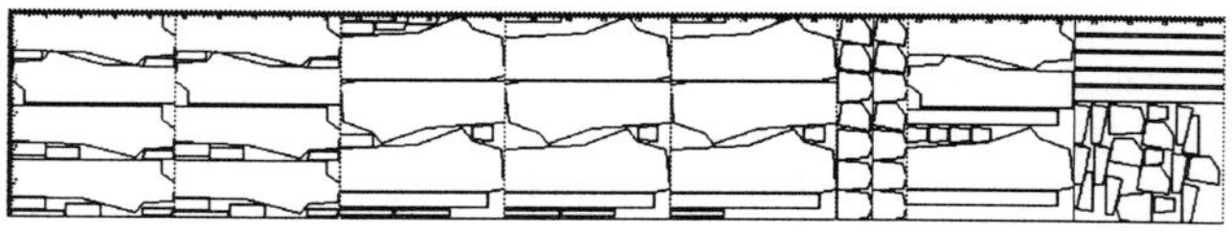

Trousers, 100 pieces, efficiency 80.7%

Figure 6. Trousers parts experimental result

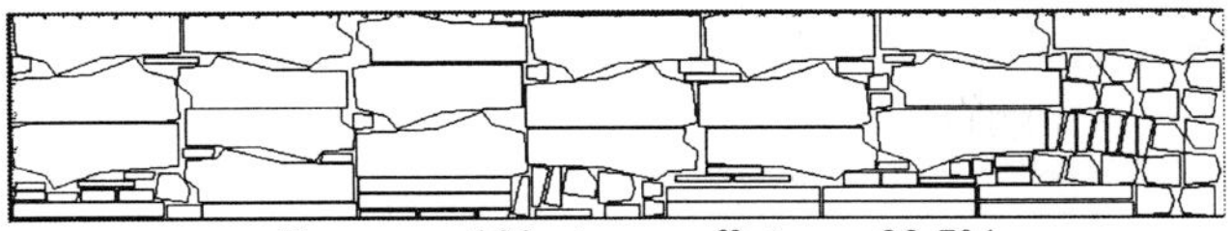

Trousers, 100 pieces, efficiency 82.7%

Figure 7. Trousers result done by human expert

5. DISCUSSION AND CONCLUSION

Finding good parameter settings that work for a particular problem is not a trivial task. There are many parameters that need to be determined for our algorithm, the two most important being the population size and the number of strips to crossover. If the population is too small, the genetic algorithm will converge too quickly to a local optimal point and may not find the best solution. On the other hand, too many members in the population result in long waiting times for significant improvement. The fitter member will have a greater chance of reproducing. The members with lower fitness are replaced by the offspring. Thus in successive generations, the members on average are fitter as solutions to the problem.

In this study, before deciding these values we performed an analysis using several experiments. We concluded that a population of 15 and 3 strips to crossover was most efficient.

Figure 8 shows the results obtained in running our algorithm with 10 different shapes in 5 sets. We present the performances of the algorithm when it processes the partial solutions compared with the algorithm presented in [13], where the partial solutions and biased area were not implemented.

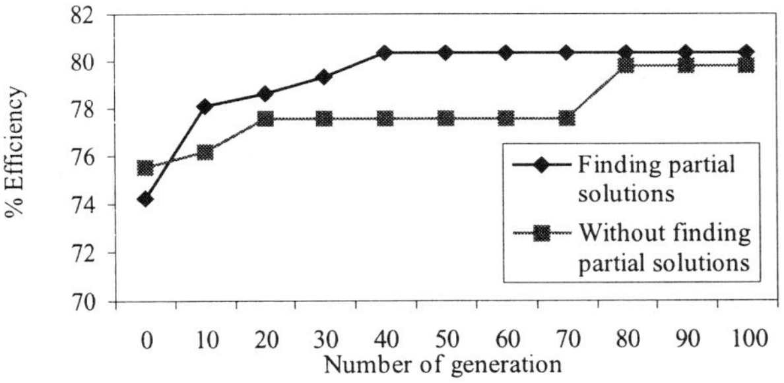

Figure 8. Comparison with or without finding partial solutions

From the graph, we observed that after finding partial solutions using the good strips or chromosomes and then restarting the GA to find good solutions for the remaining garment pieces in the order book that have not been used, we usually get higher efficiency.

Finally, as expected, our experiment results indicate that our GA can improve the efficiency of almost all production quality markers, although human experts generally obtain a 1%-3% higher efficiency. In the future we plan to further fine tune our system and attempt to improve the efficiency to achieve human expert result.

6. REFERENCES

[1] T.M. Michell, *Machine Learning*, New York, McGraw-Hill, pp, 249-250, 1997.

[2] E. Falkenauer, 1998, *Genetic Algorithms and Grouping Problems*, Chichester, John Wiley & Sons, pp. 25-58, 1998

[3] Z. Michalewiez, *Genetic Algorithms / Data Structures = Evolution Programs*, Berlin, Springer-Verag, 1992.

[4] D. E. Goldberg, *Genetic Algorithms in Search, Optimization, and Machine Learning*, Reading MA, Addison-Wesley, pp. 1-23, 1989.

[5] A. M. S. Zalzala and P. J. Fleming, *Genetic Algorithms in Engineering systems*, London, The Institution of Electrical Engineers, pp. 1-19, 1997.

[6] S. Harrington, *Computer Graphics*, New York, McGraw-Hill, pp. 3-5, 70-76, 1987.

[7] D. Hearn and M.P. Baker, 1994, *Computer Graphics C Version*, NJ, Prentice-Hall, pp. 235-236, 1994.

[8] R. P. Pargas and R. Jain "A Parallel Stochastic Optimization Algorithm for Solving 2D Bin Packing Problems" *IEEE Proc. of the 9th Int. Conf. on Artificial Intelligence for Applications*, pp. 18-25, 1993.

[9] G. Roussel and S. Maouche "Improvements About Automatic Lay-Planning For Irregular Shapes on Plain Fabric" *IEEE Proc. of Int. Conf. on Systems Man and Cybernetics, System Engineering in the Service of Humans*, Vol. 3., pp. 90-97, 1993.

[10] C. Bounsaythip, S. Maouche and M. Neus "Evolutionary Search Techniques Application in Automated Lay-Planning Optimization Problem" *IEEE International Conference on Intelligent Systems for the 21st Century*, Vol. 5, pp. 4497-4502, 1995

[11] H.S. Ismail and K.B. Hon, "The Nesting of two dimensional Shapes Using Genetic Algorithms", *Proceedings of the Institution of Mechanical Engineers Part B, Journal of Engineering Manufacture*, v.209 (B2), 1995, pp. 115-24.

[12] C. Bounsaythip and S. Maouche "Irregular Shape Nesting And Placing With Evolutionary Approach" *IEEE International Conference on Systems Man and Cybernetics, Computational Cybernetics and Simulation*, Vol. 4, pp. 3425-3430, 1997.

[13] S. Madarasmi and P. Sirivarothakul, "Layout of Garment Patterns for Efficient Fabric Consumption", *International Technical Conference On Circuits/Systems, Computers and Communication*, Vol. 2, 2002

PROPERTIES OF FASTEST LIA TRANSFORM MATRICES AND THEIR SPECTRA

Bogdan J. Falkowski and Cicilia C. Lozano

School of Electrical and Electronic Engineering
Block S1, Nanyang Technological University
50 Nanyang Avenue, Singapore 639798

ABSTRACT

Two fastest Linearly Independent Arithmetic (LIA) transforms, which have most efficient computational complexity among arithmetic transforms, have been identified recently. In this paper, their various properties are presented. Some experimental results for standard benchmark functions and comparison with multi-polarity Arithmetic transform are also discussed.

1. INTRODUCTION

The concept of Linearly Independent Arithmetic (LIA) transforms based on the inverse matrices from Linearly Independent (LI) logic transformations [3, 5] and forward matrices calculated in standard arithmetic has recently been introduced [8]. Various classes of LIA transforms and their basic properties have been discussed [8]. In the recent article, two fastest and most efficient basis matrices for LIA transforms in terms of the number of arithmetic operations to be performed were identified together with their application for Boolean functions verification [2]. Before these transforms have been introduced, Arithmetic transform [1, 4, 6, 9, 10] was known as the most computationally efficient transform in standard algebra. The Arithmetic transform itself belongs to one type of LIA transform which has been widely used in many applications, such as testing [1, 6] and probabilistic analysis of logical circuits [4] as well as the underlying transform to develop new types of word decision diagrams [9, 10]. Comparisons among various LIA transforms show that for many classes of logical functions some LIA transforms outperform Arithmetic transform when applied to fault detection [7].

In this paper, the detailed properties of matrices describing two fastest LIA transforms are presented. Some experimental results for standard benchmark functions showing properties of the two fastest LIA spectra and standard Arithmetic spectra are also included.

2. BASIC DEFINITIONS FOR LIA TRANSFORMS

The following definitions are valid for both generalized multi-polarity Arithmetic transform and all LIA transforms.

Definition 1: An LI transformation matrix T_n is defined as a 2^n x 2^n matrix with its columns (basis functions) corresponding to truth vectors of some n-variable binary functions that are linearly independent when they are bit-by-bit XOR-ed.

Definition 2: A spectral coefficient vector, $\vec{C} = [c_0, c_1, ..., c_{2^n-1}]^T$, is a column vector obtained by equation $\vec{C} = T_n \vec{F}$ where $\vec{F}$ is the truth vector of the binary function $f(x)$. $\vec{F}$ can be obtained back from $\vec{C}$ by equation $\vec{F} = T_n^{-1} \vec{C}$ where T_n^{-1} is the inverse of the corresponding LI transformation matrix T_n and all the calculations are performed in standard algebra.

Definition 3: An n-variable binary function can be expressed as an arithmetic polynomial expansion $f(x) = c_0 g_0 + c_1 g_1 + c_2 g_2 + ... + c_{2^n-1} g_{2^n-1}$ where g_i represents a particular n-variable binary function whose truth vector is given by i-th column of the corresponding LI transformation matrix and c_i represents i-th spectral coefficient.

It should be noticed that only when basis functions in above definition have some special properties, i.e. they are all only AND logic functions of various variables then it is possible to introduce the concept of multi-polarity generalized transform and it is the case for standard Arithmetic transform. Also in order to calculate spectral coefficients the inverse Arithmetic transform can be obtained directly from the forward transform based on basis functions but for generalized multi-polarity Arithmetic transform it is easy to define it in regular structure as well [10].

Definition 4: The generalized multi-polarity Arithmetic transform matrix AR of order $N=2^n$ in polarity ω is given by:

$$AR_N^{\omega} = \bigotimes_{j=1}^{n} Arr_{\omega_j} = Arr_{\omega_n} \otimes Arr_{\omega_{n-1}} \otimes \cdots Arr_{\omega_2} \otimes Arr_{\omega_1}$$

where ⊗ denotes Kronecker product [10],

$$Arr_0 = \begin{bmatrix} 1 & 0 \\ -1 & 1 \end{bmatrix} \text{ and } Arr_1 = \begin{bmatrix} 0 & 1 \\ 1 & -1 \end{bmatrix}.$$

The polarity number, $\omega = \omega_n \omega_{n-1 ..} \omega_2 \omega_1$ is a binary string computed by taking the n-bit straight binary code formed by writing a 1 or a 0 for each literal according to whether this literal is used in negative ($\bar{x}$) or positive (x) form in all the terms corresponding to all spectral coefficients.

Definition 5 [2]: The fastest LIA transformation matrix can be constructed recursively from smaller transformation matrices by using either equation (1) or equation (2).

$$T_n^{(1)} = \begin{bmatrix} T_{n-1}^{(1)} & O_{n-1} \\ Y_{n-1} & T_{n-1}^{(1)} \end{bmatrix} \quad (1) \qquad T_n^{(2)} = \begin{bmatrix} T_{n-1}^{(2)} & Z_{n-1} \\ O_{n-1} & T_{n-1}^{(2)} \end{bmatrix} \quad (2)$$

where T_{n-1} is a transformation matrix of size 2^n*-1 x* 2^n*-1*, O_{n-1} is zero matrix of size 2^n*-1 x* 2^n*-1*, Y_{n-1} is a matrix of size 2^n*-1 x* 2^n*-1* with all its elements equal to zero except one at the bottom left corner that is equal to –1 and Z_{n-1} is a 2^n*-1 x* 2^n*-1* matrix with all its elements equal to zero except one element at the top right corner that is equal to –1.

There are two types of coding that can be used for binary functions before their LIA spectra are calculated. In R-coding, true minterms are represented by 1, false minterms by 0 and don't care minterms by 0.5. In S-coding, they will be represented by –1, 1, and 0, respectively.

0-7803-7761-3/03/$17.00 ©2003 IEEE

3. PROPERTIES OF FASTEST LIA TRANSFORM MATRICES

Let us define a general matrix form that applies to both fastest LIA transforms:

$$T_n = \begin{bmatrix} T_{0,0} & T_{0,1} & \cdots & T_{0,2^n-1} \\ T_{1,0} & \ddots & & T_{1,2^n-1} \\ \vdots & & \ddots & \vdots \\ T_{2^n-1,0} & \cdots & \cdots & T_{2^n-1,2^n-1} \end{bmatrix}.$$

In following properties $T_n^{(1)}$ and $T_n^{(2)}$ correspond to fast transform matrices defined by equations (1) and (2), respectively.

Property 1: Total number of nonzero elements in T_n is equal to $2^{n+1}-1$, which consists of 2^n '1's and 2^n-1 '-1's.

Property 2: The row i of $T_n^{(1)}$ will have one '1' element and d '-1' elements, where d is the number of zeros behind the last one in the binary representation of $(i+1)$.

If D is defined as $\{1, 2, \ldots, d\}$ then the '-1's inside the row i are located at the columns: $i-(2^1-1)$, $i-(2^2-1)$, …, $i-(2^d-1)$.

Property 3: The column i of $T_n^{(1)}$ will have one '1' element and h '-1' elements, where h is the number of zeros behind the last one in the binary representation of i.

If H is defined as $\{0, 1, 2, \ldots, h-1\}$ then the '-1's inside the column i are located at the rows: $i+2^0$, $i+2^0+2^1$, $i+2^0+2^1+2^2$, …, $i+2^0+2^1+2^2+\ldots+2^{h-1}$.

Example 1:

$$T_3^{(1)} = \begin{bmatrix} 1 & 0 & 0 & 0 & 0 & 0 & 0 & 0 \\ -1 & 1 & 0 & 0 & 0 & 0 & 0 & 0 \\ 0 & 0 & 1 & 0 & 0 & 0 & 0 & 0 \\ -1 & 0 & -1 & 1 & 0 & 0 & 0 & 0 \\ 0 & 0 & 0 & 0 & 1 & 0 & 0 & 0 \\ 0 & 0 & 0 & 0 & -1 & 1 & 0 & 0 \\ 0 & 0 & 0 & 0 & 0 & 0 & 1 & 0 \\ -1 & 0 & 0 & 0 & -1 & 0 & -1 & 1 \end{bmatrix}$$

Number of '-1's in the row 3 is equal to 2, which is equal to the number of zeros behind the last one in 100 (d=2). According to Property 2, $D = (1, 2)$ and the '-1's inside the row 3 are located at the columns: 3-(2-1) = 2, and 3-(4-1) = 0.

Number of '-1's in the column 4 is equal to 2, which is equal to the number of zeros behind the last one in 100 (h=2). According to Property 3, $H = (0, 1)$ and the '-1's in the column 4 are located at the rows: 4+1 = 5 and 4+1+2 = 7

Property 4: $T_n^{(1)} = [T_n^{(2)}]^T$, where the superscript T indicates matrix transpose operation.

Due to Property 4, the positions of nonzero elements in rows and columns for the matrix $T_n^{(2)}$ can be described in a similar manner to Properties 2 and 3.

Property 5: The sum S of spectral coefficients for both fastest LIA transforms can be calculated using $\vec{F}_1$, a subset of $\vec{F}$. For an n-variable binary function f, $\vec{F}_1$ has $\frac{3}{4}x2^n$ elements for n>1, and 1 element for n=1

Proof:

$$S = \sum_{i=0}^{2^n-1} c_i = \sum_{i=0}^{2^n-1}\sum_{j=0}^{2^n-1} T_{n_{i,j}} f_j = \sum_{j=0}^{2^n-1}\sum_{i=0}^{2^n-1} T_{n_{i,j}} f_j = \sum_{j=0}^{2^n-1} f_j \left(\sum_{i=0}^{2^n-1} T_{n_{i,j}} \right)$$

From Property 3, sum of all the elements in a column of T_n is zero if the column has only one '-1' element. This condition is true for the columns whose last two bits of their n binary representation is either 10 for $T_n^{(1)}$ or 01 for $T_n^{(2)}$. Since the last two bits of the columns fulfilling the above condition are fixed, there are 2^{n-2} columns that satisfy it.

Let A contains the numbers corresponding to the columns for which sums of their elements are zero. Then for all $j \in A$, $f_j \sum_{i=0}^{2^n-1} T_{n_{i,j}}$ is zero irrespective to the value of f_j. Thus, the number of $\vec{F}$ elements that contribute to S is $2^n-2^{n-2} = \frac{3}{4}x2^n$ for n>1, and 1 for n=1.

Property 6: Only a subset $\vec{F}_2$ of $\vec{F}$ is necessary to obtain $S(k)$ that is the sum of the upper 2^{n-k} spectral coefficients of $T_n^{(1)}$ ($0 \le k \le n$). $\vec{F}_2$ has $\left[\frac{3}{4}x2^{n-k}\right]+k$ elements for $k<n-1$, n elements for $k=n-1$, and $n+1$ elements for $k = n$.

Proof:

$$S(k) = \sum_{i=2^{n-1}+2^{n-2}+\ldots+2^{n-k}}^{2^n-1} c_i = \sum_{i=2^{n-1}+2^{n-2}+\ldots+2^{n-k}}^{2^n-1} \sum_{j=0}^{2^n-1} T_{n_{i,j}} f_j$$

$$= \sum_{j=0}^{2^n-1} \sum_{i=2^{n-1}+2^{n-2}+\ldots+2^{n-k}}^{2^n-1} T_{n_{i,j}} f_j = \sum_{j=0}^{2^n-1} f_j \left(\sum_{i=2^{n-1}+2^{n-2}+\ldots+2^{n-k}}^{2^n-1} T_{n_{i,j}} \right)$$

The bottom half of the matrix $T_n^{(1)}$ is $\left[Y_{n-1} T_{n-1}^{(1)} \right]$ and for the left side of the bottom half of this matrix the part of the following equation $\left(\sum_{i=2^{n-1}+2^{n-2}+\ldots+2^{n-k}}^{2^n-1} T_{n_{i,j}} \right)$ is equal to –1 for the first column and equal to 0 for the remaining columns. For the right side of the bottom half of $T_n^{(1)}$, the number of columns for which $\left(\sum_{i=2^{n-1}+2^{n-2}+\ldots+2^{n-k}}^{2^n-1} T_{n_{i,j}} \right)$ is not equal to zero corresponds to the number of the elements in $\vec{F}_1$ of $T_{n-i}^{(1)}$ plus k-1. Hence by Property 5, the number of elements in $\vec{F}_2$ is equal to $\left[\frac{3}{4}x2^{n-k}\right]+k$ for $k<n$, n for $k=n-1$, and $n+1$ for $k=n$.

Property 7: Only a subset $\vec{F}_3$ of $\vec{F}$ is necessary to obtain the sum of the upper 2^{n-k} spectral coefficients of $T_n^{(2)}$ ($0 \le k \le n$). $\vec{F}_3$ has $\left[\frac{3}{4}x2^{n-k}\right]$ elements for $k<n-1$ and 1 element for $k=n-1$ and $k=n$.

Proof: Similar method used to prove Property 6 can be used to obtain Property 7.

For the following two properties, it is assumed that the binary function $f(x)$ for which the spectrum is calculated is completely specified and that the R-coding is used for the encoding of its truth vector values.

Property 8: Sum of all absolute values for the fastest LIA spectra is always less than or equal to $2^{n+1}-1$.

Proof: $$S=\sum_{i=0}^{2^n-1}|c_i|=\sum_{i=0}^{2^n-1}\left|\sum_{j=0}^{2^n-1}T_{n_{i,j}}f_j\right|\le\sum_{i=0}^{2^n-1}\sum_{j=0}^{2^n-1}\left|T_{n_{i,j}}f_j\right|\le\sum_{i=0}^{2^n-1}\sum_{j=0}^{2^n-1}\left|T_{n_{i,j}}\right|$$

By Property 1, the upper bound is $2^{n+1}-1$.

Property 9: Sum of the squares of the spectral coefficients for the fastest LIA transforms of an n-variable binary function is always less than or equal to

$$\left(\sum_{k=1}^{2}2^{n-k}\right)+\left(\sum_{k=3}^{n}2^{n-k}x(k-1)^2\right)+n^2.$$

Proof:

$$S=\sum_{i=0}^{2^n-1}c_i^2=\sum_{i=0}^{2^n-1}\left(\sum_{j=0}^{2^n-1}T_{n_{i,j}}f_j\right)^2\le\sum_{i=0}^{2^n-1}\left(\max\left|\sum_{j=0}^{2^n-1}T_{n_{i,j}}\right|\right)^2$$

The maximum absolute value of sums of all the elements in the row will be either number of '-1's or the number of '1's in this row. The latter case applies to the rows with either zero or one '-1' element, while the former case applies to the remaining rows. Based on Property 2, the number of the rows having (k-1) '-1' elements is 2^{n-k}. This is valid for the rows having up to n-1 '-1' elements and expressed by the first two entries in the equation. In addition, there will be a row having n '-1' elements, which is represented by the last part of the equation.

4. EXPERIMENTAL RESULTS

The calculation of the coefficients of two fastest LIA transforms and of multi-polarity generalized Arithmetic transform has been coded using C language and run on Pentium III 500 MHz computer with 128 MB RAM.

Tables 1 and 2 show the results for several MCNC benchmark functions. Table 1 presents the total number of nonzero coefficients of the fastest LIA transforms based on equation (1) and equation (2) and Arithmetic transform for both R-coding and S-coding. Table 2 lists values of the worst absolute coefficients for all the transforms when only the R-coding is used.

Comparing the numbers given in the columns of Table 1, it can be seen that two fastest LIA transforms give significantly smaller number of nonzero coefficients than Arithmetic transform for some benchmark functions. It should be also noticed that in order to obtain the best polarity result for Arithmetic transform all 2^n spectra have to be calculated, also based on properties of multi-polarity Arithmetic transform the results using the S-coding and the R-coding differ only by one coefficient for the same benchmark function. On the other hand the two fastest LIA transforms do not have polarity so the calculation needs to be performed only once for each of them and their results differ much more for the S-coding and the R-coding. The numbers of nonzero spectral coefficients in columns 1-4 in Table 1 that are equal or smaller than the corresponding numbers for the best multi-polarity Arithmetic transform are written in italic.

Besides giving smaller number of nonzero coefficients for some input files and much faster calculation, the two fastest LIA transforms generally also give smaller worst absolute coefficient values when compared to Arithmetic transform. In fact the worst absolute coefficient values of two fastest LIA transforms are equal or smaller than n, the number of variables in the benchmark function. This can be easily noticed from Table 2 where for almost all benchmarks, the largest number in each entry of the first and the second column are equal or smaller than the ones in the third or the fourth column.

5. CONCLUSION

Various properties of two fastest LIA transform matrices have been presented here together with some experimental results. The two fastest LIA transforms discussed in this paper have the least computational complexity when compared to other known arithmetic transforms. From experimental results, it can also be seen that two fastest LIA transforms give less number of nonzero coefficients than the best polarity Arithmetic transform for some cases and its computation is much faster that the one necessary to calculate all Arithmetic spectral coefficients. Due to the importance of Arithmetic transform for functional verification of circuits under the error modeling of their spectra, the presented transform properties and advantages of spectra for fastest LIA transforms can be very useful in these applications. All of the results presented in our article can also be applied to the development of novel classes of word level decision diagrams.

6. REFERENCES

[1] T.H. Chen, *Fault diagnosis and fault tolerance: systematic approach to special topics*. Berlin: Springer-Verlag, 1992.

[2] B.J. Falkowski and S. Rahardja, "Boolean verification with fastest LIA transform", *Proc. 35th IEEE International Symposium on Circuits and Systems*, Scottsdale, Arizona, vol. 5, pp. 321-324, May 2002.

[3] B.J. Falkowski and S. Rahardja, "Classification and properties of fast linearly independent logic transformations". *IEEE Trans. on Circuits and Systems II: Analog and Digital Signal Processing,* Vol. 44, No. 8, pp.646-655, Aug. 1997.

[4] S.L. Hurst, *Custom VLSI microelectronics*. Hertfordshire: Prentice Hall, 1992.

[5] M.A. Perkowski, "A fundamental theorem for EXOR circuits," *Proc. IFIP WG 10.5 Workshop on Applications of the Reed-Muller expansion in Circuit Design*, U. Kebschull, E. Schubert, W. Rosenstiel, Eds., Hamburg, Germany, pp.52-60, Sept. 1993.

[6] K. Radecka and Z. Zilic, "Using arithmetic transform for verification of datapath circuits via error modeling", *Proc. 18th IEEE VLSI Test Symposium*, Montreal, Canada, pp. 271-277, Apr. 2000.

[7] S. Rahardja and B.J. Falkowski, "Application of linearly independent arithmetic transform in testing of digital circuits". *IEE Electronics Letters*, Vol. 35, No. 5, pp. 363-364, Mar. 1999.

[8] S. Rahardja and B.J. Falkowski,, "Fast linearly independent arithmetic expansions", *IEEE Trans. on Computers*, Vol. 48, No. 9, pp. 991-999, Sep. 1999.

[9] R.S. Stankovic, "Some remarks about spectral transform interpretation of MTBDDs and EVBDDs", *Proc. 1st IEEE/ ACM Asia and South Pacific Design Automation Conference*, Makuhari, Chiba, Japan, pp. 385-390, Aug. 1995.

[10] R.S. Stankovic, M. Stankovic and D. Jankovic, *Spectral transforms in switching theory: definitions and calculations*. Belgrade: IP Nauka, 1998.

Table 1. Number of nonzero spectral terms of fastest LIA and Arithmetic transforms for some benchmark functions

Input files	$(T_n^{(1)})$		$(T_n^{(2)})$		Arithmetic transform	
	Nonzero terms (R)	Nonzero terms (S)	Nonzero terms (R)	Nonzero terms (S)	Best polarity nonzero terms (R)	Best polarity nonzero terms (S)
xor5	*21*	*32*	*21*	*32*	31	32
squar5	32	31	31	32	23	24
rd53	*28*	*32*	32	*32*	31	32
bw	26	30	*22*	30	22	23
con1	104	102	94	103	18	18
5xp1	128	128	128	128	62	62
z5xp1	128	128	128	128	62	62
rd73	*122*	*128*	128	*128*	127	128
inc	92	108	89	107	49	50
rd84	*249*	*256*	256	*256*	255	256
misex1	105	198	104	198	20	21
ex5	240	240	223	223	113	113
9sym	*344*	398	*344*	398	352	352
z9sym	*344*	398	*344*	398	352	352
clip	494	502	493	502	255	255
apex4	495	495	493	493	445	445
sao2	1023	1023	511	1023	127	127
ex1010	*130*	*798*	*125*	*797*	1022	1021

Table 2. Largest absolute spectral coefficients of fastest LIA and Arithmetic transforms for some benchmark functions

Input files	$(T_n^{(1)})$	$(T_n^{(2)})$	Arithmetic transform	
	Largest (absolute) coefficients of output 1, output2, ...	Largest (absolute) coefficients of output 1, output2, ...	Largest (absolute) coefficients of output 1, output2, ...(polarity zero)	Largest (absolute) coefficients of output 1, output2, ...(best polarity)
xor5	2	3	16	16
squar5	2,3,2,1,1,1,1,1	2,2,2,2,2,2,1	1,2,2,4,2,2,1,1	1,2,2,2,4,2,2,1
rd53	1,2,2	2,3,2	4,16,2	4,16,2
bw	2,3,2,2,1,4,2,4,2, 4,1,2,4,1,1,2,4,2,1, 3,3,1,3,2,4,2,2,1	1,1,1,1,2,1,2,1, 2,1,1,1,1,2,1,1,1,2,1, 1,2,1,1,2,1,2,2,1	3,1,3,4,2,2,2,1, 3,1,2,2,1,3,2,2,1,3,3, 3,4,1,4,3,1,2,2,1	3,1,3,4,2,2,2,2,3, 1,2,2,1,3,3,2,1,3,3, 4,4,1,5,5,1,2,2,1
con1	2,5	4,2	1,2	1,3
5xp1	2,1,2,3,4,4,4,2,4,1	1,1,5,4,6,5,3,2,2,5	2,2,4,4,6,2,2,2,1,1	2,3,6,6,6,4,2,2,1,1
z5xp1	3,4,3,4,3,2,1,1,1,7	3,4,3,4,3,4,4,6,1,1	1,2,2,4,4,6,2,2,2,1	1,2,3,6,6,6,4,2,2,1
rd73	2,3,2	4,4,4	8,64,20	8,64,20
inc	4,3,3,2,3,3,2,3,3	4,3,4,4,3,2.5,1.5,4,3	1,2,4,4,1,2,1.5,1.5,1	2,2,4,5,1.5,2,1.5,2,1
rd84	4,4,1,4	4,4,1,4	8,128,1,34	8,128,1,34
misex1	3,4,4,5,5,4,4	3,4,3,3,5,5,4	1,2,3,2,3,3,2	1,3,3,2,3,4,2

Fast Linearly Independent Ternary Arithmetic Transforms

Bogdan J. Falkowski and Cheng Fu
School of Electrical and Electronic Engineering
Block S1, Nanyang Technological University
50 Nanyang Avenue, Singapore 639798

ABSTRACT

In this paper, family of fast Linearly Independent Ternary Arithmetic (LITA) transforms, which possesses fast forward and inverse butterfly diagrams have been identified. This family is recursively defined and has consistent formulas relating forward and inverse transform matrices. Computational costs of the calculation for presented transforms are also discussed.

1. INTRODUCTION

Two closely related transforms are frequently used in logic design and testing. They are the Reed-Muller transform based on Galois Field (2) and the arithmetic transform performed in standard arithmetic [1, 4, 8]. The concept of Linearly Independent Arithmetic (LIA) transforms based on the inverse matrices from Linearly Independent (LI) logic transformations [5] and forward matrices calculated in standard arithmetic has recently been introduced [7]. Various classes of LIA transforms and their basic properties have been discussed [7]. While the majority of applications of these transforms are still for binary case due to increased interest in multiple-valued circuit design and testing [3, 4], these transforms can also be modified for multiple-valued switching functions and the simplest case of such generalization is ternary case where these transforms are based on GF (3). LIA transforms are also used as the underlying transforms to develop new types of word decision diagrams [6]. Recently Arithmetic Transform Ternary Decision Diagrams with the sign alternating factor have been developed for generation of arithmetic expressions and compared with other known Ternary Decision Diagrams [9]. In this article, a new approach to the development of the arithmetic transform for ternary functions is proposed. The fast Linearly Independent Ternary Arithmetic (LITA) transforms are introduced here that can be created efficiently in the form of matrices for n=3 and such matrices can be extended further to the transform matrices with higher dimensions. Relations and properties between different matrix definitions are also discussed.

2. BASIC DEFINITIONS OF LITA TRANSFORM

Definition 1: Let M_n be a $N \times N$ $(N = 3^n)$ matrix with rows corresponding to minterms and columns corresponding to some switching ternary functions of n variables. If the sets of rows are linearly independent over GF (3), then M_n has only one inverse and is said to be *ternary linearly independent.* If M_n is ternary linearly independent in GF(3), then M_n is a non-singular square matrix with respect to standard arithmetic with integer operations and has a unique inverse M_n^{-1}.

Definition 2: The LITA expansion for any n-variable ternary function $f(\vec{x_n})$ is

$$f(\vec{x_n}) = \sum_{j=0}^{3^n-1} A_j g_j , \tag{1}$$

where g_j is any set of n-variable ternary switching functions such that the matrix $M_n = \left[\vec{g_0}, \vec{g_1}, \ldots \vec{g_{3^n-1}}\right]$, $\vec{g_j}$ represents the truth vector of the ternary functions, $0 \le j \le 3^n - 1$, A_j is the respective coefficient for the particular transform matrix M_n with arithmetic inverse M_n^{-1}, and the symbol $\sum$ is the standard arithmetic addition.

Definition 3: The LITA transform based on Definition 1 and 2 can be described by the following general formulae performed in standard arithmetic algebra:

$$\vec{F} = M_n \vec{A} \text{ and} \tag{2}$$

$$\vec{A} = M_n^{-1} \vec{F}, \tag{3}$$

where $\vec{F} = \left[F_0, F_1, \ldots, F_{3^n-1}\right]^T$ is a column vector defining the truth vector of a switching function $f(\vec{x_n})$ in a natural ternary ordering, M_n is a *ternary linearly independent* matrix of order $N = 3^n$ and $\vec{A} = \left[A_0, A_1, \ldots A_{3^n-1}\right]^T$ is the spectral coefficient column vector for the particular LITA transform matrix M_n with arithmetic inverse M_n^{-1}.

Definition 4: Let M_n be a $3^n \times 3^n$ square matrix as specified in Definition 1. Then M_n can be recursively defined by

$$M_n = \begin{bmatrix} {}^kM_{n-1}^{(1)} & O_{n-1} & {}^kM_{n-1}^{(2)} \\ O_{n-1} & {}^kM_{n-1}^{(5)} & O_{n-1} \\ {}^kM_{n-1}^{(3)} & O_{n-1} & {}^kM_{n-1}^{(4)} \end{bmatrix}, \tag{4}$$

where each submatrix $M_{n-1}^{(j)}$, $j = \{1,2,3,4,5\}$, has a dimension of $3^{n-1} \times 3^{n-1}$, and O_{n-1} is a $3^{n-1} \times 3^{n-1}$ matrix with all it's elements 0, $k \in \{1,2\}$ and denotes the value of each element in M_n when n=1.

Definition 5: The Rotation operator R on the square recursive matrix M_n is recursively defined as 4^{n-r} clockwise rotation involving 4^{n-r+1} submatrices each of order 2^{r-1} for $r = n, n-1, \ldots, 2, 1$.

Definition 6: The inverse Rotation operator R^{-1} on the square recursive matrix M_n is recursively defined as 4^{n-r}

0-7803-7761-3/03/$17.00 ©2003 IEEE

counterclockwise rotation involving 4^{n-r+1} submatrices each of order 2^{r-1} for $r = n, n-1, \ldots, 2, 1$.

Property 1: Let M_n be a $3^n \times 3^n$ nonsingular square matrix with inverse M_n^{-1}. Defining the integer power of the rotation operator as $R^2(M_n) = R[R(M_n)]$, the following properties may be derived:

$$R^r(M_n) = R^{r-4}(M_n) \text{ for any } r \in \{0,1,2,3\}, \tag{5}$$

$$R^{-1}[R(M_n)] = R[R^{-1}(M_n)] = M_n, \tag{6}$$

$$[R(M_n)]^{-1} = R^{-1}(M_n^{-1}). \tag{7}$$

Definition 7: Let M_n be a $3^n \times 3^n$ nonsingular square matrix, which is partitioned into appropriate $3^{n-1} \times 3^{n-1}$ dimensional submatrices, as shown in Equation 4. If $M_{n-1}^{(j^*)} = O_{n-1}$, $j^* = \{1,2,3,4\}$, then the operator α_A on M_n is defined as selecting the submatrix $M_{n-1}^{(5-j^*)}$, which is opposite to $M_{n-1}^{(j^*)}$, and changing it to $-M_{n-1}^{(5-j^*)}$.

The following notation for submatrices will also be used: I_{n-1} is a $3^{n-1} \times 3^{n-1}$ identity matrix, $J_{n-1} (= R(I_{n-1}))$ is a $3^{n-1} \times 3^{n-1}$ reverse-identity matrix and X_{n-1} is either I_{n-1} or J_{n-1}.

3. NEW FAST LITA TRANSFORMS

A new class of fast LITA transforms is introduced here. The transform matrices are constructed by the basic recursive submatrices O_{n-1}, M_{n-1}, and X_{n-1}, where at most one O_{n-1} and one X_{n-1} could appear in the recursive definition (4) in locations $M_{n-1}^{(j)}$, $j = \{1,2,3,4,5\}$. Table 1 shows all basic LITA transforms for this class with low computation cost. The solid and dotted lines correspond to addition and subtraction, the narrow and broad lines represent the value 1 and 2, respectively. They are categorized as Class A and separated to two sub-classes: A1 and A2.

Let M_n be a nonsingular $3^n \times 3^n$ square matrix with only $M_{n-1}^{(5)}$ equal to X_{n-1}, and $M_{n-1}^{(j)} = O_{n-1}$ for j=2 or 3. Then there are eight LITA transform matrices belonging to Class A1 which are defined by

$$M_n = \begin{bmatrix} {}^1M_{n-1} & O_{n-1} & O_{n-1} \\ O_{n-1} & {}^1X_{n-1} & O_{n-1} \\ {}^kM_{n-1} & O_{n-1} & {}^1M_{n-1} \end{bmatrix} \text{ or} \tag{8}$$

$$M_n = \begin{bmatrix} {}^1M_{n-1} & O_{n-1} & {}^kM_{n-1} \\ O_{n-1} & {}^1X_{n-1} & O_{n-1} \\ O_{n-1} & O_{n-1} & {}^1M_{n-1} \end{bmatrix}. \tag{9}$$

Their inverse transform matrices are defined by

$$M_n = \begin{bmatrix} {}^1M_{n-1} & O_{n-1} & O_{n-1} \\ O_{n-1} & {}^1X_{n-1} & O_{n-1} \\ -{}^kM_{n-1} & O_{n-1} & {}^1M_{n-1} \end{bmatrix} \text{ or} \tag{10}$$

$$M_n = \begin{bmatrix} {}^1M_{n-1} & O_{n-1} & -{}^kM_{n-1} \\ O_{n-1} & {}^1X_{n-1} & O_{n-1} \\ O_{n-1} & O_{n-1} & {}^1M_{n-1} \end{bmatrix}. \tag{11}$$

Property 2: Let $M_{A1} = M_n$ be any $3^n \times 3^n$ square matrix in Class A1. Then, the inverse of M_{A1} is given by

$$[M_{A1}]^{-1} = \alpha_A[M_{A1}]. \tag{12}$$

Example 1: Let us consider factorization of one transform from (8) for which $X_{n-1} = I_{n-1}$ and k=2, and its inverse matrix is obtained from (10), where $X_{n-1} = I_{n-1}$ and k=2, as

$$\begin{bmatrix} {}^1M_{n-1} & O_{n-1} & O_{n-1} \\ O_{n-1} & {}^1I_{n-1} & O_{n-1} \\ {}^2M_{n-1} & O_{n-1} & {}^1M_{n-1} \end{bmatrix} \underset{M_n \leftrightarrow M_n^{-1}}{\overset{\alpha_A}{\Leftrightarrow}} \begin{bmatrix} {}^1M_{n-1} & O_{n-1} & O_{n-1} \\ O_{n-1} & {}^1I_{n-1} & O_{n-1} \\ -{}^2M_{n-1} & O_{n-1} & {}^1M_{n-1} \end{bmatrix}.$$

For n=3, $M_n = C(n) \cdot B(n) \cdot A(n)$,

$$A(1) = \begin{bmatrix} 1 & 0 & 0 \\ 0 & 1 & 0 \\ 2 & 0 & 1 \end{bmatrix}, \quad A(n) = \begin{bmatrix} A(n-1) & O_{n-1} & O_{n-1} \\ O_{n-1} & I_{n-1} & O_{n-1} \\ O_{n-1} & O_{n-1} & A(n-1) \end{bmatrix},$$

$$B(2) = \begin{bmatrix} 1 & 0 & 0 & 0 & 0 & 0 & 0 & 0 & 0 \\ 0 & 1 & 0 & 0 & 0 & 0 & 0 & 0 & 0 \\ 0 & 0 & 1 & 0 & 0 & 0 & 0 & 0 & 0 \\ 0 & 0 & 0 & 1 & 0 & 0 & 0 & 0 & 0 \\ 0 & 0 & 0 & 0 & 1 & 0 & 0 & 0 & 0 \\ 0 & 0 & 0 & 0 & 0 & 1 & 0 & 0 & 0 \\ 1 & 0 & 0 & 0 & 0 & 0 & 1 & 0 & 0 \\ 0 & 1 & 0 & 0 & 0 & 0 & 0 & 1 & 0 \\ 0 & 0 & 1 & 0 & 0 & 0 & 0 & 0 & 1 \end{bmatrix},$$

$$B(n) = \begin{bmatrix} B(n-1) & O_{n-1} & O_{n-1} \\ O_{n-1} & I_{n-1} & O_{n-1} \\ O_{n-1} & O_{n-1} & B(n-1) \end{bmatrix}, \quad C(n) = \begin{bmatrix} I_{n-1} & O_{n-1} & O_{n-1} \\ O_{n-1} & I_{n-1} & O_{n-1} \\ I_{n-1} & O_{n-1} & I_{n-1} \end{bmatrix}.$$

Based on the above matrix decomposition, the forward and inverse butterfly diagrams for n=3 are shown in Fig. 1 and Fig. 2.

There are sixteen transform matrices that belong to Class A2. The location of the submatrix O_{n-1} changes from j=2 or 3 in Class A1 to j=1 or 4 in Class A2. The eight forward transform matrices are defined by

$$M_n = \begin{bmatrix} O_{n-1} & O_{n-1} & {}^1M_{n-1} \\ O_{n-1} & {}^1X_{n-1} & O_{n-1} \\ {}^1M_{n-1} & O_{n-1} & {}^kM_{n-1} \end{bmatrix} \text{ or} \tag{13}$$

$$M_n = \begin{bmatrix} {}^kM_{n-1} & O_{n-1} & {}^1M_{n-1} \\ O_{n-1} & {}^1X_{n-1} & O_{n-1} \\ {}^1M_{n-1} & O_{n-1} & O_{n-1} \end{bmatrix}, \tag{14}$$

and the other eight inverse transform matrices are defined by

$$M_n = \begin{bmatrix} -{}^{k}M_{n-1} & O_{n-1} & {}^{1}M_{n-1} \\ O_{n-1} & {}^{1}X_{n-1} & O_{n-1} \\ {}^{1}M_{n-1} & O_{n-1} & O_{n-1} \end{bmatrix} \text{ or} \quad (15)$$

$$M_n = \begin{bmatrix} O_{n-1} & O_{n-1} & {}^{1}M_{n-1} \\ O_{n-1} & {}^{1}X_{n-1} & O_{n-1} \\ {}^{1}M_{n-1} & O_{n-1} & -{}^{k}M_{n-1} \end{bmatrix}. \quad (16)$$

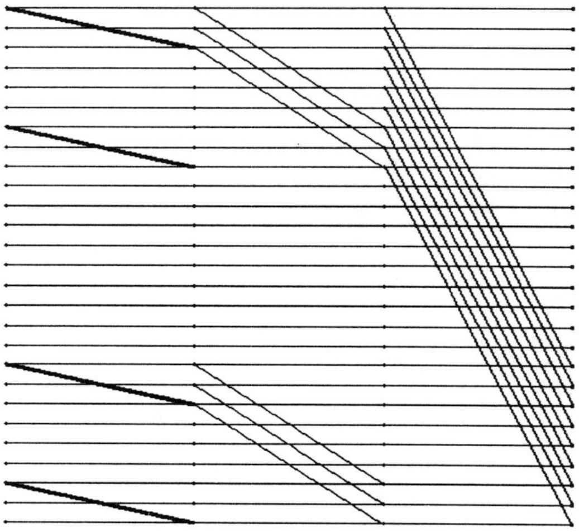

Figure 1. Forward butterfly diagram.

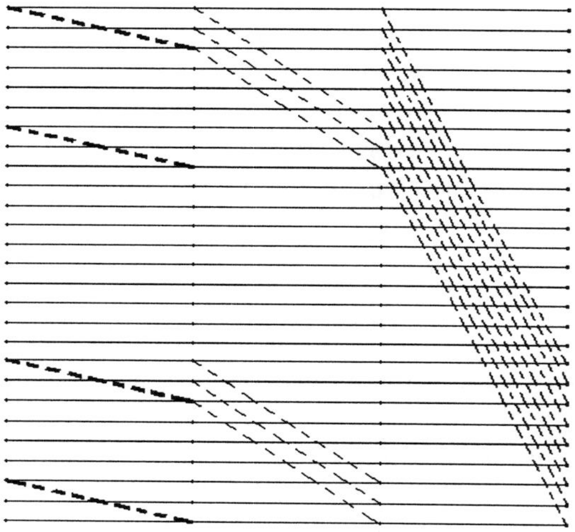

Figure 2. Inverse butterfly diagram.

Property 3: Let $M_{A2} = M_n$ be a $3^n \times 3^n$ square matrix from Class A2. Then, the inverse of M_{A2} is given by

$$[M_{A2}]^{-1} = \alpha_A R^2 [M_{A2}]. \quad (17)$$

Example 2: Let M_n be a $3^n \times 3^n$ square matrix according to (13) with $X_{n-1} = J_{n-1}$ and k=1. The inverse transform matrix is obtained by (15) with $X_{n-1} = J_{n-1}$ and k=1. According to Property 3,

$$\begin{bmatrix} O_{n-1} & O_{n-1} & {}^{1}M_{n-1} \\ O_{n-1} & {}^{1}J_{n-1} & O_{n-1} \\ {}^{1}M_{n-1} & O_{n-1} & {}^{1}M_{n-1} \end{bmatrix} \underset{M_n \leftrightarrow M_n^{-1}}{\overset{\alpha_A R^2}{\Leftrightarrow}} \begin{bmatrix} -{}^{1}M_{n-1} & O_{n-1} & {}^{1}M_{n-1} \\ O_{n-1} & {}^{1}J_{n-1} & O_{n-1} \\ {}^{1}M_{n-1} & O_{n-1} & O_{n-1} \end{bmatrix}.$$

The forward and inverse butterfly diagrams for n=2 are shown in Fig. 3 and Fig. 4, respectively.

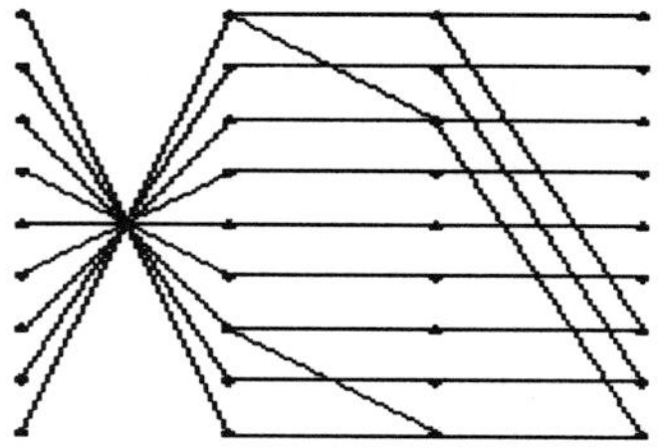

Figure 3. Forward butterfly diagram.

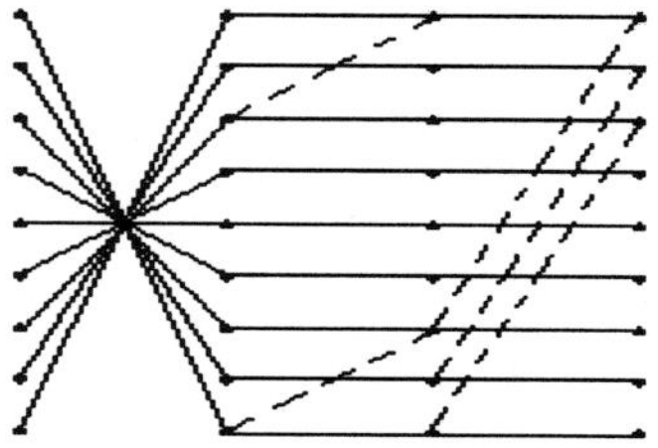

Figure 4. Inverse butterfly diagram.

For each transform matrix in Class A2, its butterfly diagram has a vertical-flipping part J_n and the other parts are the same as another unique transform from Class A1.

The introduced class has very low computational complexity. As shown in Fig. 1 and Fig. 2, the fast transforms have very regular patterns for higher n. The cost can be recursively derived by

$$S_n = 2S_{n-1} + 3^{n-1}. \quad (18)$$

From (18),

$$S_n = 2^{n-1} + 2 \cdot 3^{n-1} - 3, \; n \geq 2$$

$$S_1 = 1.$$

Class A2 has the same computation complexity as Class A1 even though each transform matrix of it has an additional vertical-flipping part. The price of hardware overhead for Class A2 consists only of the circuitry for permutation of input data.

4. CONCLUSION

The concept of Linearly Independent Ternary Arithmetic transforms is considered for the first time in this paper. Various properties of a new class for LITA transforms have been described. In order to make the calculation of these ternary expansions efficient, only such expansions that have fast transforms are considered here. In a similar manner to the binary case [6, 7], all the presented results in our article can also be applied to the development of novel classes of word level decision diagrams for LITA. The unified approach to butterfly decomposition presented here has also allowed the identification of fast class for LITA transforms. Since presented techniques may have applications to other types of transforms [1, 2, 4, 8], the different butterfly structures introduced here should be considered as more important than simply as fast algorithms to calculate new LITA transforms and expansions.

Acknowledgments: Agency for Science, Technology and Research, grant no. 0121060053 supported this study.

REFERENCES

[1] S. Agaian, J. Astola, K. Egiazarian, *Binary Polynomial Transforms and Nonlinear Digital Filters*. New York: Marcel Dekker, 1995.

[2] A, Drygajlo, "Butterfly orthogonal structure for fast transforms, filter banks and wavelets", *Proc. IEEE Int. Conf. Acoustic, Speech, Signal Processing*, San Francisco, CA, vol. 5, pp. 81-84, March 1992.

[3] S. Hurst, "Multiple-valued logic – its status and its future", *IEEE Trans. on Computers*, Vol. 31, pp. 260-264, 1982.

[4] G. A. Kukharev, V. P. Shmerko, E.N. Zaitseva, *Multiple-Valued Data Processing Algorithms and Systolic Processors*. Minsk: Science and Engineering, 1990.

[5] M.A. Perkowski, "A fundamental theorem for EXOR circuits", *Proc. IFIP WG 10.5 Workshop on Applications of the Reed-Muller expansion in Circuit Design*, U. Kebschull, E. Schubert, W. Rosenstiel, Eds. Hamburg, Germany, pp.52-60, Sept. 1993.

[6] M.A. Perkowski, B.J. Falkowski, M. Chrzanowska-Jeske, and R. Drechsler, "Efficient algorithms for creation of Linearly-Independent Decision Diagrams and their mapping to regular layouts", *VLSI Design*, Vol. 14, No. 1, pp. 35-52, February 2002.

[7] S. Rahardja and B.J. Falkowski, "Fast linearly independent arithmetic expansions", *IEEE Trans. on Computers,* Vol. 48, No. 9, pp. 991-999, Sept 1999.

[8] R.S. Stankovic, M. Stankovic, and D. Jankovic, *Spectral Transforms in Switching Theory, Definitions and Calculations*. Belgrade: IP Nauka, 1998.

[9] R.S. Stankovic, "Arithmetic transform TDDs for generation of arithmetic expressions", *1st International Workshop on Spectral Techniques and Logic Design for Future Digital Systems,* Tampere, Finland, June 2000, published as TICSP Series #10, pp.79-94, December 2000.

Table 1. List of Transformation Matrices and Butterfly Diagrams of Class A

	Forward butterfly	Forward $M_n \Leftrightarrow M_n^{-1}$ Inverse	Inverse butterfly		Forward butterfly	Forward $M_n \Leftrightarrow M_n^{-1}$ Inverse	Inverse butterfly
A1	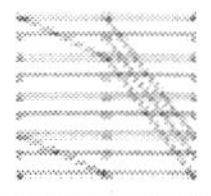	$\begin{bmatrix} {}^{1}M_{n-1} & & \\ & {}^{1}I_{n-1} & \\ {}^{1}M_{n-1} & & {}^{1}M_{n-1} \end{bmatrix} \overset{\Rightarrow}{\underset{\Leftarrow}{\alpha_A}} \begin{bmatrix} {}^{1}M_{n-1} & & \\ & {}^{1}I_{n-1} & \\ -{}^{1}M_{n-1} & & {}^{1}M_{n-1} \end{bmatrix}$	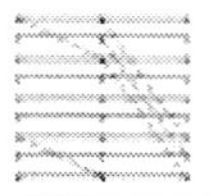		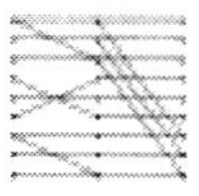	$\begin{bmatrix} {}^{1}M_{n-1} & & \\ & {}^{1}J_{n-1} & \\ {}^{1}M_{n-1} & & {}^{1}M_{n-1} \end{bmatrix} \overset{\Rightarrow}{\underset{\Leftarrow}{\alpha_A}} \begin{bmatrix} {}^{1}M_{n-1} & & \\ & {}^{1}J_{n-1} & \\ -{}^{1}M_{n-1} & & {}^{1}M_{n-1} \end{bmatrix}$	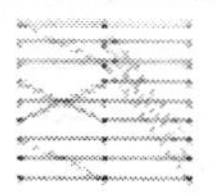
	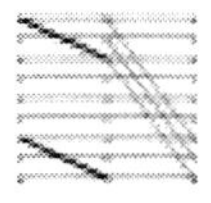	$\begin{bmatrix} {}^{1}M_{n-1} & & \\ & {}^{1}I_{n-1} & \\ {}^{2}M_{n-1} & & {}^{1}M_{n-1} \end{bmatrix} \overset{\Rightarrow}{\underset{\Leftarrow}{\alpha_A}} \begin{bmatrix} {}^{1}M_{n-1} & & \\ & {}^{1}I_{n-1} & \\ -{}^{2}M_{n-1} & & {}^{1}M_{n-1} \end{bmatrix}$	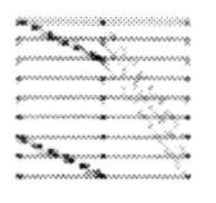		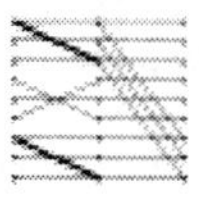	$\begin{bmatrix} {}^{1}M_{n-1} & & \\ & {}^{1}J_{n-1} & \\ {}^{2}M_{n-1} & & {}^{1}M_{n-1} \end{bmatrix} \overset{\Rightarrow}{\underset{\Leftarrow}{\alpha_A}} \begin{bmatrix} {}^{1}M_{n-1} & & \\ & {}^{1}J_{n-1} & \\ -{}^{2}M_{n-1} & & {}^{1}M_{n-1} \end{bmatrix}$	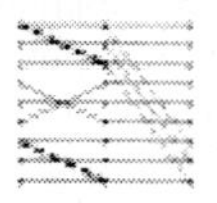
	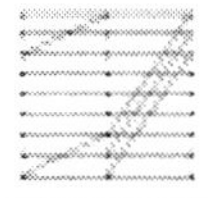	$\begin{bmatrix} {}^{1}M_{n-1} & & {}^{1}M_{n-1} \\ & {}^{1}I_{n-1} & \\ & & {}^{1}M_{n-1} \end{bmatrix} \overset{\Rightarrow}{\underset{\Leftarrow}{\alpha_A}} \begin{bmatrix} {}^{1}M_{n-1} & & -{}^{1}M_{n-1} \\ & {}^{1}I_{n-1} & \\ & & {}^{1}M_{n-1} \end{bmatrix}$	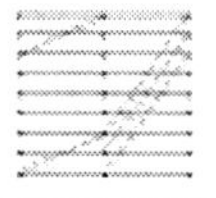		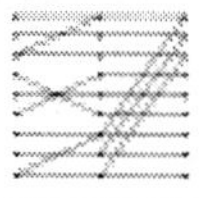	$\begin{bmatrix} {}^{1}M_{n-1} & & {}^{1}M_{n-1} \\ & {}^{1}J_{n-1} & \\ & & {}^{1}M_{n-1} \end{bmatrix} \overset{\Rightarrow}{\underset{\Leftarrow}{\alpha_A}} \begin{bmatrix} {}^{1}M_{n-1} & & -{}^{1}M_{n-1} \\ & {}^{1}J_{n-1} & \\ & & {}^{1}M_{n-1} \end{bmatrix}$	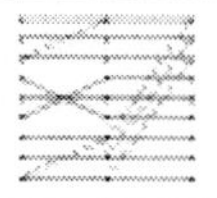
	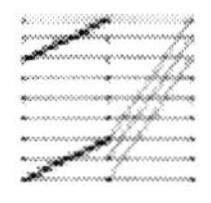	$\begin{bmatrix} {}^{1}M_{n-1} & & {}^{2}M_{n-1} \\ & {}^{1}I_{n-1} & \\ & & {}^{1}M_{n-1} \end{bmatrix} \overset{\Rightarrow}{\underset{\Leftarrow}{\alpha_A}} \begin{bmatrix} {}^{1}M_{n-1} & & -{}^{2}M_{n-1} \\ & {}^{1}I_{n-1} & \\ & & {}^{1}M_{n-1} \end{bmatrix}$	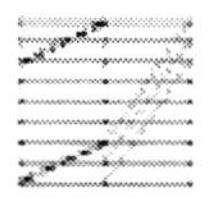		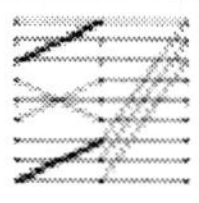	$\begin{bmatrix} {}^{1}M_{n-1} & & {}^{2}M_{n-1} \\ & {}^{1}J_{n-1} & \\ & & {}^{1}M_{n-1} \end{bmatrix} \overset{\Rightarrow}{\underset{\Leftarrow}{\alpha_A}} \begin{bmatrix} {}^{1}M_{n-1} & & -{}^{2}M_{n-1} \\ & {}^{1}J_{n-1} & \\ & & {}^{1}M_{n-1} \end{bmatrix}$	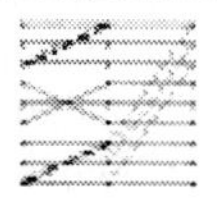
A2	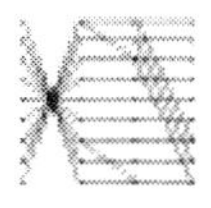	$\begin{bmatrix} & & {}^{1}M_{n-1} \\ & {}^{1}J_{n-1} & \\ {}^{1}M_{n-1} & & {}^{1}M_{n-1} \end{bmatrix} \overset{\Rightarrow}{\underset{\Leftarrow}{\alpha_A R^2}} \begin{bmatrix} -{}^{1}M_{n-1} & & {}^{1}M_{n-1} \\ & {}^{1}J_{n-1} & \\ {}^{1}M_{n-1} & & \end{bmatrix}$	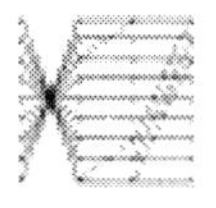		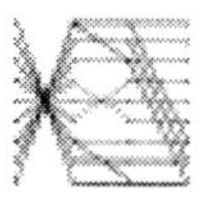	$\begin{bmatrix} & & {}^{1}M_{n-1} \\ & {}^{1}I_{n-1} & \\ {}^{1}M_{n-1} & & {}^{1}M_{n-1} \end{bmatrix} \overset{\Rightarrow}{\underset{\Leftarrow}{\alpha_A R^2}} \begin{bmatrix} -{}^{1}M_{n-1} & & {}^{1}M_{n-1} \\ & {}^{1}I_{n-1} & \\ {}^{1}M_{n-1} & & \end{bmatrix}$	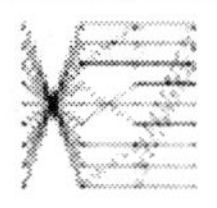
	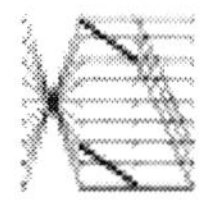	$\begin{bmatrix} & & {}^{1}M_{n-1} \\ & {}^{1}J_{n-1} & \\ {}^{1}M_{n-1} & & {}^{2}M_{n-1} \end{bmatrix} \overset{\Rightarrow}{\underset{\Leftarrow}{\alpha_A R^2}} \begin{bmatrix} -{}^{2}M_{n-1} & & {}^{1}M_{n-1} \\ & {}^{1}J_{n-1} & \\ {}^{1}M_{n-1} & & \end{bmatrix}$	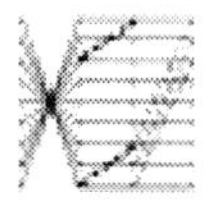		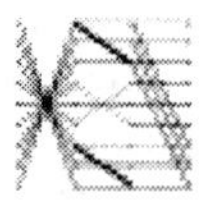	$\begin{bmatrix} & & {}^{1}M_{n-1} \\ & {}^{1}I_{n-1} & \\ {}^{1}M_{n-1} & & {}^{2}M_{n-1} \end{bmatrix} \overset{\Rightarrow}{\underset{\Leftarrow}{\alpha_A R^2}} \begin{bmatrix} -{}^{2}M_{n-1} & & {}^{1}M_{n-1} \\ & {}^{1}I_{n-1} & \\ {}^{1}M_{n-1} & & \end{bmatrix}$	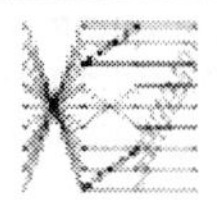
	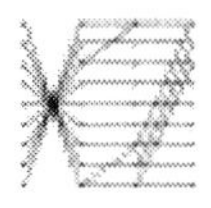	$\begin{bmatrix} {}^{1}M_{n-1} & & {}^{1}M_{n-1} \\ & {}^{1}J_{n-1} & \\ {}^{1}M_{n-1} & & \end{bmatrix} \overset{\Rightarrow}{\underset{\Leftarrow}{\alpha_A R^2}} \begin{bmatrix} & & {}^{1}M_{n-1} \\ & {}^{1}J_{n-1} & \\ {}^{1}M_{n-1} & & -{}^{1}M_{n-1} \end{bmatrix}$	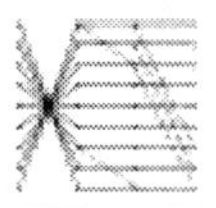		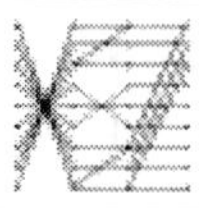	$\begin{bmatrix} {}^{1}M_{n-1} & & {}^{1}M_{n-1} \\ & {}^{1}I_{n-1} & \\ {}^{1}M_{n-1} & & \end{bmatrix} \overset{\Rightarrow}{\underset{\Leftarrow}{\alpha_A R^2}} \begin{bmatrix} & & {}^{1}M_{n-1} \\ & {}^{1}I_{n-1} & \\ {}^{1}M_{n-1} & & -{}^{1}M_{n-1} \end{bmatrix}$	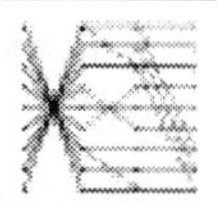
	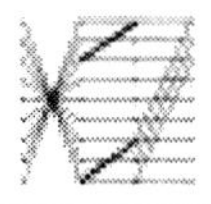	$\begin{bmatrix} {}^{2}M_{n-1} & & {}^{1}M_{n-1} \\ & {}^{1}J_{n-1} & \\ {}^{1}M_{n-1} & & \end{bmatrix} \overset{\Rightarrow}{\underset{\Leftarrow}{\alpha_A R^2}} \begin{bmatrix} & & {}^{1}M_{n-1} \\ & {}^{1}J_{n-1} & \\ {}^{1}M_{n-1} & & -{}^{2}M_{n-1} \end{bmatrix}$	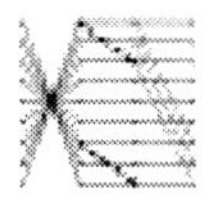		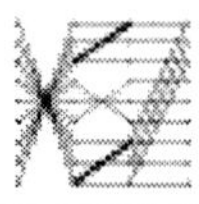	$\begin{bmatrix} {}^{2}M_{n-1} & & {}^{1}M_{n-1} \\ & {}^{1}I_{n-1} & \\ {}^{1}M_{n-1} & & \end{bmatrix} \overset{\Rightarrow}{\underset{\Leftarrow}{\alpha_A R^2}} \begin{bmatrix} & & {}^{1}M_{n-1} \\ & {}^{1}I_{n-1} & \\ {}^{1}M_{n-1} & & -{}^{2}M_{n-1} \end{bmatrix}$	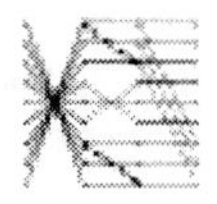

Ternary Arithmetic Polynomial Expansions Based on New Transforms

Bogdan J. Falkowski and Cheng Fu
School of Electrical and Electronic Engineering
Block S1, Nanyang Technological University
50 Nanyang Avenue, Singapore 639798

ABSTRACT

New classes of Linearly Independent Ternary Arithmetic (LITA) transforms being the bases of ternary arithmetic polynomial expansions are introduced here. Recursive equations defining the LITA transforms and the corresponding butterfly diagrams are shown. Computational costs to calculate LITA transforms and applications of corresponding polynomial expansions in logic design are also discussed.

1. INTRODUCTION

The concept of Linearly Independent Logic [6] generalizes all possible expansion of binary logic circuits that are realized in GF(2) algebra. Processing the computation in normal arithmetic has lead to an analogous concept of Linearly Independent Arithmetic (LIA) expansions [7]. Similarly, the frequently used transform in logic design, Arithmetic transform [1, 5, 9] falls into the broad definition of LIA logic. It has been shown [1, 5, 9] that in many applications where logic functions need to be analyzed, it is useful to transform a Boolean function into the equivalent Arithmetic expansion. In particular, such Arithmetic expansions are used in probabilistic verification of a pair of functions, to describe stochastic behavior of a logical circuit under random values of input variables and in testing [4, 8]. In this paper we extend previous binary results to the ternary case where these transforms are based on GF(3). New ternary polynomial expansions based on Linearly Independent Ternary Arithmetic (LITA) transforms are introduced here. They are based on the bases from two new classes of LITA transforms that are defined in a recursive way and their computational costs are also discussed. Similarly to the binary case these new expansions can have applications to describe stochastic behavior of ternary functions and in testing where ternary logic has already been used for sequential design testing [3].

2. DEFINITIONS OF TERNARY ARITHMETIC POLYNOMIAL EXPANSIONS

Definition 1: Let M_n be a $N \times N$ $(N = 3^n)$ matrix with rows corresponding to minterms and columns corresponding to some switching ternary functions of *n* variables. If the sets of rows are linearly independent with respect to *Ternary Galois Field*, then M_n has only one inverse in GF(3) and is said to be *ternary linearly independent.* If M_n is ternary linearly independent in GF(3), then M_n is a non-singular square matrix with respect to standard arithmetic and has a unique inverse M_n^{-1}.

Definition 2: The LITA polynomial expansion for any *n*-variable ternary function $f(\overrightarrow{x_n})$ is

$$f(\overrightarrow{x_n}) = \sum_{j=0}^{3^n-1} A_j g_j, \tag{1}$$

where g_j is any set of *n*-variable ternary functions such that the matrix $M_n = [\overrightarrow{g_0}, \overrightarrow{g_1}, \ldots \overrightarrow{g_{3^n-1}}]$, $\overrightarrow{g_j}$ represents the truth vector of the ternary functions, $0 \le j \le 3^n - 1$, A_j is the respective coefficient for the particular transform matrix M_n with arithmetic inverse M_n^{-1}, and the symbol $\sum$ is the standard arithmetic addition.

Definition 3: The LITA transform based on Definition 1 and 2 can be described by the following general formulae performed in standard arithmetic algebra:

$$\vec{F} = M_n \vec{A}, \tag{2}$$

and

$$\vec{A} = M_n^{-1} \vec{F}, \tag{3}$$

where $\vec{F} = [F_0, F_1, \ldots, F_{3^n-1}]^T$ is a column vector defining the truth vector of a ternary function $f(\overrightarrow{x_n})$ in a natural ternary ordering, M_n is a *linearly independent ternary* matrix of order $N = 3^n$ defined by any *n*-variable switching functions and $\vec{A} = [A_0, A_1, \ldots A_{3^n-1}]^T$ is the coefficient column vector for the particular transform matrix M_n with arithmetic inverse M_n^{-1}.

Definition 4: Let M_n be a $3^n \times 3^n$ square matrix as specified in Definition 1. Then M_n can be recursively defined by

$$M_n = \begin{bmatrix} M_{n-1}^{(1)} & M_{n-1}^{(2)} & M_{n-1}^{(3)} \\ M_{n-1}^{(4)} & M_{n-1}^{(5)} & M_{n-1}^{(6)} \\ M_{n-1}^{(7)} & M_{n-1}^{(8)} & M_{n-1}^{(9)} \end{bmatrix}, \tag{4}$$

where each submatrix $M_{n-1}^{(j)}$, $j = \{1,2,3,\ldots,8,9\}$ has a dimension of $3^{n-1} \times 3^{n-1}$.

Definition 5: Let M_n be a $3^n \times 3^n$ nonsingular square matrix that is partitioned into appropriate $3^{n-1} \times 3^{n-1}$ dimensional submatrices. The μ_{EH} operator on M_n is defined as grouping the recursive equations in the submatrices vertically and interchanging them in the submatrices horizontally.

$$\mu_{EH}[M_n] = \begin{bmatrix} M_{n-1}^{(3)} & M_{n-1}^{(2)} & M_{n-1}^{(1)} \\ M_{n-1}^{(6)} & M_{n-1}^{(5)} & M_{n-1}^{(4)} \\ M_{n-1}^{(9)} & M_{n-1}^{(8)} & M_{n-1}^{(7)} \end{bmatrix} \tag{5}$$

0-7803-7761-3/03/$17.00 ©2003 IEEE

Definition 6: Let M_n be a $3^n \times 3^n$ nonsingular square matrix that is partitioned into appropriate $3^{n-1} \times 3^{n-1}$ dimensional submatrices, as shown in Equation 4.

The μ_{EV} operator on M_n is defined as grouping the recursive equations in the submatrices horizontally and interchanging them in the submatrices vertically

$$\mu_{EV}[M_n] = \begin{bmatrix} M_{n-1}^{(7)} & M_{n-1}^{(8)} & M_{n-1}^{(9)} \\ M_{n-1}^{(4)} & M_{n-1}^{(5)} & M_{n-1}^{(6)} \\ M_{n-1}^{(1)} & M_{n-1}^{(2)} & M_{n-1}^{(3)} \end{bmatrix} \quad (6)$$

3. NEW CLASSES OF LITA TRANSFORMS

In this section, we introduced two new classes of fast ternary transforms that are denoted as Class S and P. The basic transformation matrices are constructed by the basic recursive submatrices O_{n-1}, X_{n-1} and M_{n-1}, where O_{n-1} is a $3^{n-1} \times 3^{n-1}$ matrix with all its elements 0, I_{n-1} is a $3^{n-1} \times 3^{n-1}$ identity matrix, J_{n-1} is a $3^{n-1} \times 3^{n-1}$ reverse-identity matrix, i.e., it has elements in the reverse-diagonal position equal to 1 while all other elements are 0 and X_{n-1} is either I_{n-1} or J_{n-1}.

Let M_n be a $3^n \times 3^n$ square matrix as specified in Definition 4 having one submatrix M_{n-1} in one corner $M_{n-1}^{(1)}$, three submatrices I_{n-1} are at the place of $M_{n-1}^{(6)}$, $M_{n-1}^{(8)}$ and $M_{n-1}^{(9)}$ and the submatrix I_{n-1} which is opposite to M_{n-1} is multiplied by -1. The remaining five submatrices of M_n are O_{n-1}. Then the first basic LITA transform is defined by

$$M_n = \begin{bmatrix} M_{n-1} & O_{n-1} & O_{n-1} \\ O_{n-1} & O_{n-1} & I_{n-1} \\ O_{n-1} & I_{n-1} & -I_{n-1} \end{bmatrix}, \quad (7)$$

For n=2,

$$M_2 = \begin{bmatrix} 1 & 0 & 0 & 0 & 0 & 0 & 0 & 0 & 0 \\ 0 & 0 & 1 & 0 & 0 & 0 & 0 & 0 & 0 \\ 0 & 1 & -1 & 0 & 0 & 0 & 0 & 0 & 0 \\ 0 & 0 & 0 & 0 & 0 & 0 & 1 & 0 & 0 \\ 0 & 0 & 0 & 0 & 0 & 0 & 0 & 1 & 0 \\ 0 & 0 & 0 & 0 & 0 & 0 & 0 & 0 & 1 \\ 0 & 0 & 0 & 1 & 0 & 0 & -1 & 0 & 0 \\ 0 & 0 & 0 & 0 & 1 & 0 & 0 & -1 & 0 \\ 0 & 0 & 0 & 0 & 0 & 1 & 0 & 0 & -1 \end{bmatrix}$$

Its forward butterfly diagram is shown in Table 1. In all the butterfly diagrams presented in this article, the solid and dotted lines represent addition and subtraction while the narrow and broad lines represent the value 1 and 2, respectively. It is obvious that due to the fact that many of the connections in that butterfly diagram are direct connections not requiring any calculation the whole butterfly diagram can be executed in one time period which speeds up its computation and the final butterfly diagram can be represented as shown below,

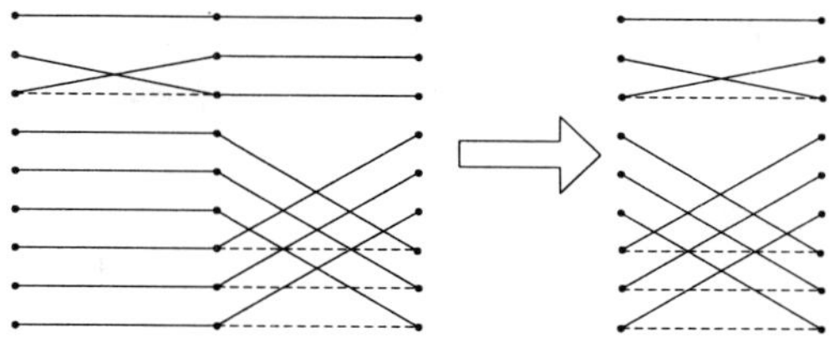

Due to the recursive definition, similar factorization and appearance of butterfly diagrams applies for higher dimensions. For example for any n, the butterfly diagram of n-1 variables will stay on the top and other two parts that have the same size of n-1 will be placed below and there is the crossing between them.

The inverse matrix in standard arithmetic algebra for (7) is shown by

$$M_n = \begin{bmatrix} M_{n-1} & O_{n-1} & O_{n-1} \\ O_{n-1} & I_{n-1} & I_{n-1} \\ O_{n-1} & I_{n-1} & O_{n-1} \end{bmatrix}. \quad (8)$$

Other forward transform matrix of this class can be derived from (7) by multiplying the submatrix $-I_{n-1}$ by 2, which is opposite to M_{n-1}. Then

$$M_n = \begin{bmatrix} M_{n-1} & O_{n-1} & O_{n-1} \\ O_{n-1} & O_{n-1} & I_{n-1} \\ O_{n-1} & I_{n-1} & -2I_{n-1} \end{bmatrix}. \quad (9)$$

Two more forward transform matrices can be derived from (7) and (9) by flipping the matrices along the reverse-diagonal, as shown in Table 1.

From (7), by interchanging the central submatrix $M_{n-1}^{(5)}$ and the submatrix $M_{n-1}^{(9)}$ that is opposite to M_{n-1}, the fifth basic forward transform matrix is defined by

$$M_n = \begin{bmatrix} M_{n-1} & O_{n-1} & O_{n-1} \\ O_{n-1} & -I_{n-1} & I_{n-1} \\ O_{n-1} & I_{n-1} & O_{n-1} \end{bmatrix}. \quad (10)$$

Similarly, other three basic forward transform matrices can be derived from (10) by using the following operations together or separately: multiplying the central submatrix by 2 or flipping along the reverse-diagonal.

These eight basic forward transform matrices are categorized as Class S. Their inverse matrices can be derived from (8) in a similar manner that the forward matrices are obtained.

Property 1: Let $M_S = M_n$ be one of the square matrices in Class S. Then, the inverse of M_S can be obtained by selecting the central submatrix and the submatrix that is opposite to M_{n-1}, after that they are multiplied by -1 and interchanged with each other.

The entire basic forward and inverse transform matrices of Class S and their fast butterfly diagrams are shown in Table 1.

Horizontal-flipping of the transform matrix based on (7) will bring it into Class P. The new transform matrix is defined by

$$M_n = \begin{bmatrix} O_{n-1} & O_{n-1} & M_{n-1} \\ J_{n-1} & O_{n-1} & O_{n-1} \\ -J_{n-1} & J_{n-1} & O_{n-1} \end{bmatrix}, \qquad (11)$$

and its inverse transform matrix can be derived from the transform matrix of (8) by vertical-flipping as follows

$$M_n = \begin{bmatrix} O_{n-1} & J_{n-1} & O_{n-1} \\ O_{n-1} & J_{n-1} & J_{n-1} \\ M_{n-1} & O_{n-1} & O_{n-1} \end{bmatrix}. \qquad (12)$$

Table 2 shows the list of 8 LITA transforms in Class P. The following properties of Class P may be derived.

Property 2: Let $M_P = M_n$ be one of the square matrices in Class P. Then, the inverse of M_P is given by selecting the central submatrix and the submatrix that is opposite to M_n, then multiplying both of them by -1 and interchanging them. After that the resulting matrix is flipped along the diagonal.

Property 3: Let M_S and M_P represent any square matrix in Class S and Class P, respectively. Then,

$$\mu_{EH}[M_S] \in M_P \qquad (13)$$

and

$$\mu_{EV}[M_S] \in M_P. \qquad (14)$$

These two classes of LITA transform matrices have very similar basic constructions. As shown in the previous section, they all have a same simple way to extend to higher n. The total number S of arithmetic additions/subtractions required to compute the transform of an n-variable ternary function is

$$S_n = S_{n-1} + 3^{n-1}. \qquad (15)$$

From (15), since $S_1 = 1$, then for arbitrary n

$$S_n = (3^n - 1)/2 \qquad (16)$$

Let us show application of one of the presented transforms as ternary polynomial arithmetic expansion for some ternary switching function.

Example 1:

Let the ternary switching function be: $\vec{F} = [0 \;\; 2 \;\; 1 \;\; 2 \;\; 2 \;\; 2 \;\; 1 \;\; 1 \;\; 1]^T$, and the LITA transform matrix being the bases of ternary expansion be:

$$M_n = \begin{bmatrix} M_{n-1} & O_{n-1} & O_{n-1} \\ O_{n-1} & 2I_{n-1} & I_{n-1} \\ O_{n-1} & I_{n-1} & O_{n-1} \end{bmatrix}.$$

For n=2,

$$\vec{A} = M_2^{-1}\vec{F} = \begin{bmatrix} 1 & 0 & 0 & 0 & 0 & 0 & 0 & 0 & 0 \\ 0 & 0 & 1 & 0 & 0 & 0 & 0 & 0 & 0 \\ 0 & 1 & -2 & 0 & 0 & 0 & 0 & 0 & 0 \\ 0 & 0 & 0 & 0 & 0 & 0 & 1 & 0 & 0 \\ 0 & 0 & 0 & 0 & 0 & 0 & 0 & 1 & 0 \\ 0 & 0 & 0 & 0 & 0 & 0 & 0 & 0 & 1 \\ 0 & 0 & 0 & 1 & 0 & 0 & -2 & 0 & 0 \\ 0 & 0 & 0 & 0 & 1 & 0 & 0 & -2 & 0 \\ 0 & 0 & 0 & 0 & 0 & 1 & 0 & 0 & -2 \end{bmatrix} \begin{bmatrix} 0 \\ 2 \\ 1 \\ 2 \\ 2 \\ 2 \\ 1 \\ 1 \\ 1 \end{bmatrix} = \begin{bmatrix} 0 \\ 1 \\ 0 \\ 1 \\ 1 \\ 1 \\ 0 \\ 0 \\ 0 \end{bmatrix}.$$

It should be noticed that the ternary arithmetic expansion for this particular function has only four spectral coefficients not equal to zero and the value of all of them is 1. From Definition 2,

$f = A_1g_1 + A_3g_3 + A_4g_4 + A_5g_5 = g_1 + g_3 + g_4 + g_5$, where $g_1 = 2 \cdot x^{[1]} \cdot y^{[0]} + x^{[2]} \cdot y^{[0]}$, $g_3 = 2 \cdot x^{[0]} \cdot y^{[1]} + x^{[0]} \cdot y^{[2]}$, etc.

In the above equations symbols x and y represent ternary variables and their superscripts refer to their corresponding ternary values.

4. CONCLUSION

In this article novel arithmetic expansions for ternary functions have been introduced. They are based on new classes of LITA transforms. Recursive matrix equations defining the LITA transforms and their corresponding butterfly diagrams are also shown. Similarly to known polynomial expansions based on binary and multiple-valued logic [5, 9], the new expansions can have applications in spectral representations of ternary logic functions, and calculation of their stochastic behavior [2, 8]. They can also be the bases of new ternary word decision diagrams in a manner similar to the ones developed in [10].

5. REFERENCES

[1] S. Agaian, J. Astola, K. Egiazarian, *Binary Polynomial Transforms and Nonlinear Digital Filters.* New York: Marcel Dekker, 1995.

[2] D. Bochmann and Ch. Posthoff, *Binare dynamishe Systeme.* Berlin: Springer-Verlag, 1981. (in German)

[3] J.A. Brzozowski and C-J.H. Seger, *Asynchronous Circuits.* New York: Springer-Verlag, 1995.

[4] T.H. Chen, *Fault Diagnosis and Fault Tolerance: Systematic Approach to Special Topics.* Berlin: Springer-Verlag, 1992.

[5] G.A. Kukharev, V.P. Shmerko, E.N. Zaitseva, *Multiple-Valued Data Processing Algorithms and Systolic Processors.* Minsk: Science and Engineering, 1990.

[6] M.A. Perkowski, "A fundamental theorem for EXOR circuits", *Proc. IFIP WG 10.5 Workshop on Applications of the Reed-Muller expansion in Circuit Design*, U. Kebschull, E. Schubert, W. Rosenstiel, Eds. Hamburg, Germany, pp.52-60, Sept. 1993.

[7] S. Rahardja and B.J. Falkowski, "Fast linearly independent arithmetic expansions", *IEEE Trans. on Computers,* Vol. 48, No. 9, pp. 991-999, Sept 1999.

[8] W.G. Schneeweiss, *Boolean Function with Engineering Applications and Computer Programs.* Berlin: Springer-Verlag, 1989.

[9] R.S. Stankovic, M. Stankovic, and D. Jankovic, *Spectral Transforms in Switching Theory, Definitions and Calculations.* Belgrade: IP Nauka, 1998.

[10] R.S. Stankovic, "Arithmetic transform TDDs for generation of arithmetic expressions", *1st International Workshop on Spectral Techniques and Logic Design for Future Digital Systems,* Tampere, Finland, June 2000, published as TICSP Series #10, pp.79-94, December 2000.

Table 1. Transform Matrices and Butterfly Diagrams for Class S

$M_n \Leftrightarrow M_n^{-1}$			$M_n \Leftrightarrow M_n^{-1}$		
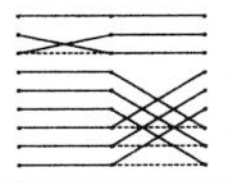	$\begin{bmatrix} M_{n-1} & O_{n-1} & O_{n-1} \\ O_{n-1} & O_{n-1} & I_{n-1} \\ O_{n-1} & I_{n-1} & -I_{n-1} \end{bmatrix} \Leftrightarrow \begin{bmatrix} M_{n-1} & O_{n-1} & O_{n-1} \\ O_{n-1} & I_{n-1} & I_{n-1} \\ O_{n-1} & I_{n-1} & O_{n-1} \end{bmatrix}$	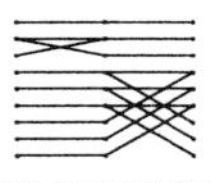	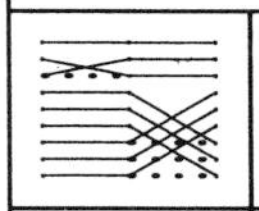	$\begin{bmatrix} M_{n-1} & O_{n-1} & O_{n-1} \\ O_{n-1} & O_{n-1} & I_{n-1} \\ O_{n-1} & I_{n-1} & -2I_{n-1} \end{bmatrix} \Leftrightarrow \begin{bmatrix} M_{n-1} & O_{n-1} & O_{n-1} \\ O_{n-1} & 2I_{n-1} & I_{n-1} \\ O_{n-1} & I_{n-1} & O_{n-1} \end{bmatrix}$	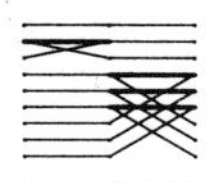
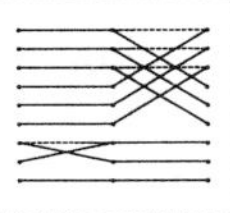	$\begin{bmatrix} -I_{n-1} & I_{n-1} & O_{n-1} \\ I_{n-1} & O_{n-1} & O_{n-1} \\ O_{n-1} & O_{n-1} & M_{n-1} \end{bmatrix} \Leftrightarrow \begin{bmatrix} O_{n-1} & I_{n-1} & O_{n-1} \\ I_{n-1} & I_{n-1} & O_{n-1} \\ O_{n-1} & O_{n-1} & M_{n-1} \end{bmatrix}$	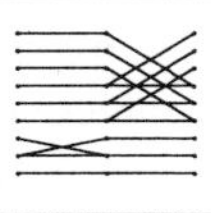	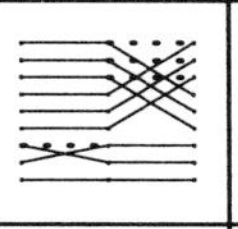	$\begin{bmatrix} -2I_{n-1} & I_{n-1} & O_{n-1} \\ I_{n-1} & O_{n-1} & O_{n-1} \\ O_{n-1} & O_{n-1} & M_{n-1} \end{bmatrix} \Leftrightarrow \begin{bmatrix} O_{n-1} & I_{n-1} & O_{n-1} \\ I_{n-1} & 2I_{n-1} & O_{n-1} \\ O_{n-1} & O_{n-1} & M_{n-1} \end{bmatrix}$	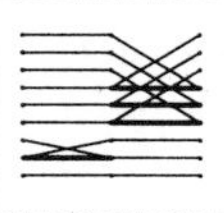
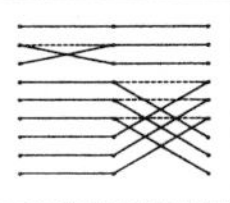	$\begin{bmatrix} M_{n-1} & O_{n-1} & O_{n-1} \\ O_{n-1} & -I_{n-1} & I_{n-1} \\ O_{n-1} & I_{n-1} & O_{n-1} \end{bmatrix} \Leftrightarrow \begin{bmatrix} M_{n-1} & O_{n-1} & O_{n-1} \\ O_{n-1} & O_{n-1} & I_{n-1} \\ O_{n-1} & I_{n-1} & I_{n-1} \end{bmatrix}$	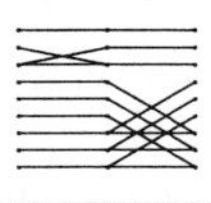	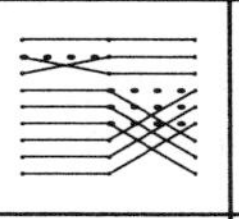	$\begin{bmatrix} M_{n-1} & O_{n-1} & O_{n-1} \\ O_{n-1} & -2I_{n-1} & I_{n-1} \\ O_{n-1} & I_{n-1} & O_{n-1} \end{bmatrix} \Leftrightarrow \begin{bmatrix} M_{n-1} & O_{n-1} & O_{n-1} \\ O_{n-1} & O_{n-1} & I_{n-1} \\ O_{n-1} & I_{n-1} & 2I_{n-1} \end{bmatrix}$	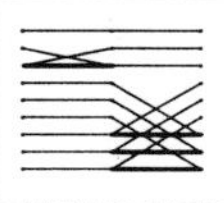
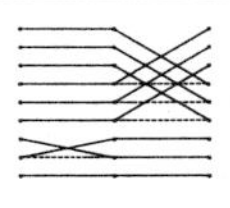	$\begin{bmatrix} O_{n-1} & I_{n-1} & O_{n-1} \\ I_{n-1} & -I_{n-1} & O_{n-1} \\ O_{n-1} & O_{n-1} & M_{n-1} \end{bmatrix} \Leftrightarrow \begin{bmatrix} I_{n-1} & I_{n-1} & O_{n-1} \\ I_{n-1} & O_{n-1} & O_{n-1} \\ O_{n-1} & O_{n-1} & M_{n-1} \end{bmatrix}$	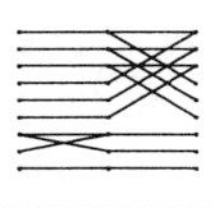	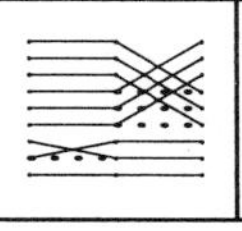	$\begin{bmatrix} O_{n-1} & I_{n-1} & O_{n-1} \\ I_{n-1} & -2I_{n-1} & O_{n-1} \\ O_{n-1} & O_{n-1} & M_{n-1} \end{bmatrix} \Leftrightarrow \begin{bmatrix} 2I_{n-1} & I_{n-1} & O_{n-1} \\ I_{n-1} & O_{n-1} & O_{n-1} \\ O_{n-1} & O_{n-1} & M_{n-1} \end{bmatrix}$	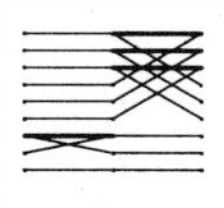

Table 2. Transform Matrices and Butterfly diagrams for Class P

$M_n \Leftrightarrow M_n^{-1}$			$M_n \Leftrightarrow M_n^{-1}$		
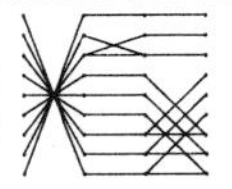	$\begin{bmatrix} O_{n-1} & O_{n-1} & M_{n-1} \\ J_{n-1} & O_{n-1} & O_{n-1} \\ -J_{n-1} & J_{n-1} & O_{n-1} \end{bmatrix} \Leftrightarrow \begin{bmatrix} O_{n-1} & J_{n-1} & O_{n-1} \\ O_{n-1} & J_{n-1} & J_{n-1} \\ M_{n-1} & O_{n-1} & O_{n-1} \end{bmatrix}$	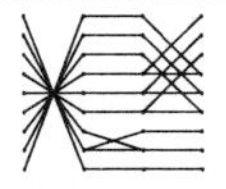	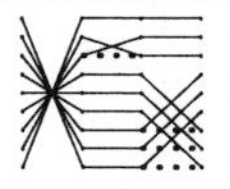	$\begin{bmatrix} O_{n-1} & O_{n-1} & M_{n-1} \\ J_{n-1} & O_{n-1} & O_{n-1} \\ -2J_{n-1} & J_{n-1} & O_{n-1} \end{bmatrix} \Leftrightarrow \begin{bmatrix} O_{n-1} & J_{n-1} & O_{n-1} \\ O_{n-1} & 2J_{n-1} & J_{n-1} \\ M_{n-1} & O_{n-1} & O_{n-1} \end{bmatrix}$	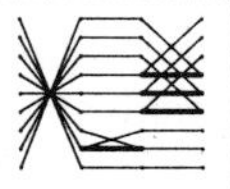
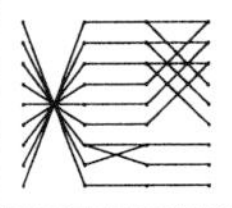	$\begin{bmatrix} O_{n-1} & J_{n-1} & -J_{n-1} \\ O_{n-1} & O_{n-1} & J_{n-1} \\ M_{n-1} & O_{n-1} & O_{n-1} \end{bmatrix} \Leftrightarrow \begin{bmatrix} O_{n-1} & O_{n-1} & M_{n-1} \\ J_{n-1} & J_{n-1} & O_{n-1} \\ O_{n-1} & J_{n-1} & O_{n-1} \end{bmatrix}$	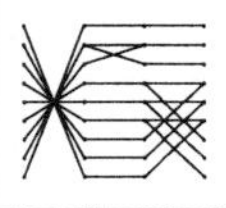	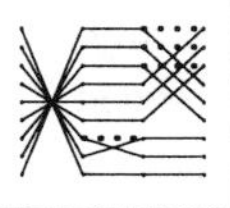	$\begin{bmatrix} O_{n-1} & J_{n-1} & -2J_{n-1} \\ O_{n-1} & O_{n-1} & J_{n-1} \\ M_{n-1} & O_{n-1} & O_{n-1} \end{bmatrix} \Leftrightarrow \begin{bmatrix} O_{n-1} & O_{n-1} & M_{n-1} \\ J_{n-1} & 2J_{n-1} & O_{n-1} \\ O_{n-1} & J_{n-1} & O_{n-1} \end{bmatrix}$	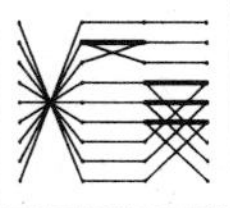
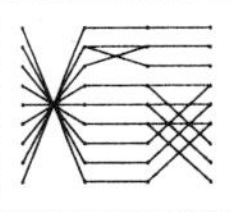	$\begin{bmatrix} O_{n-1} & O_{n-1} & M_{n-1} \\ J_{n-1} & -J_{n-1} & O_{n-1} \\ O_{n-1} & J_{n-1} & O_{n-1} \end{bmatrix} \Leftrightarrow \begin{bmatrix} O_{n-1} & J_{n-1} & J_{n-1} \\ O_{n-1} & O_{n-1} & J_{n-1} \\ M_{n-1} & O_{n-1} & O_{n-1} \end{bmatrix}$	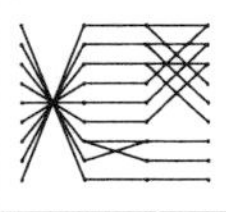	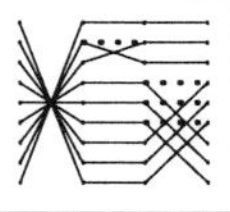	$\begin{bmatrix} O_{n-1} & O_{n-1} & M_{n-1} \\ J_{n-1} & -2J_{n-1} & O_{n-1} \\ O_{n-1} & J_{n-1} & O_{n-1} \end{bmatrix} \Leftrightarrow \begin{bmatrix} O_{n-1} & J_{n-1} & 2J_{n-1} \\ O_{n-1} & O_{n-1} & J_{n-1} \\ M_{n-1} & O_{n-1} & O_{n-1} \end{bmatrix}$	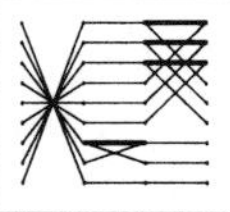
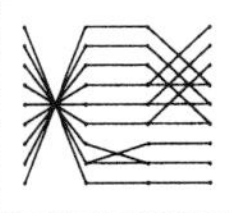	$\begin{bmatrix} O_{n-1} & J_{n-1} & O_{n-1} \\ O_{n-1} & -J_{n-1} & J_{n-1} \\ M_{n-1} & O_{n-1} & O_{n-1} \end{bmatrix} \Leftrightarrow \begin{bmatrix} O_{n-1} & O_{n-1} & M_{n-1} \\ J_{n-1} & O_{n-1} & O_{n-1} \\ J_{n-1} & J_{n-1} & O_{n-1} \end{bmatrix}$	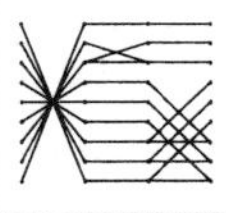	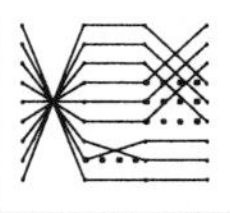	$\begin{bmatrix} O_{n-1} & J_{n-1} & O_{n-1} \\ O_{n-1} & -2J_{n-1} & J_{n-1} \\ M_{n-1} & O_{n-1} & O_{n-1} \end{bmatrix} \Leftrightarrow \begin{bmatrix} O_{n-1} & O_{n-1} & M_{n-1} \\ J_{n-1} & O_{n-1} & O_{n-1} \\ 2J_{n-1} & J_{n-1} & O_{n-1} \end{bmatrix}$	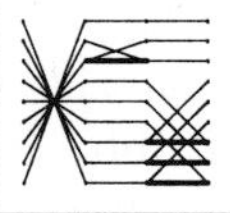

STABILITY ROBUSTNESS OF INTERCONNECTED DISCRETE TIME SYSTEMS WITH SYNCHRONIZATION ERRORS

Peter H. BAUER *and Cédric* LORAND

Dept. of Electrical Engineering
University of Notre Dame
Notre Dame, IN 46556,USA
pbauer@mars.ee.nd.edu,clorand@nd.edu

Kamal PREMARATNE

Dept. of Electrical and Computer Engineering
University of Miami
Coral Gables, Fl. 33124
kamal@miami.edu

Abstract

This paper addresses the effects of synchronization errors in two interconnected discrete time systems represented in state space with small clock frequency mismatches. An event based discrete time index is used to capture the dynamics of the arising system. In a second step, stability of the overall system is analyzed and compared to stability of both, the perfectly synchronized system (with identical clock signals) and the synchronized system with a non-zero phase difference between the two clock signals.

1. INTRODUCTION

Practically all design methods for discrete time systems (in digital signal processing and control) assume the entire system to change state at certain time instances that are typically defined by the rising or falling edge of one common clock signal[1, 2]. However, there is a growing number of applications where this assumption is routinely violated. A typical example are systems that are connected through communication networks (e.g. sensor networks, congestion control systems), and high speed discrete-time systems where propagation of clock signals cannot be neglected anymore. In these systems synchronization at the clock speed of the discrete time system is either expensive (requiring significant overhead) or impossible to achieve, especially if the clock rates are high.

There has been a substantial amount of work addressing the case of two/multi - rate feedback systems, see for example [6, 7]. Most available work concentrated on designing and analyzing systems with components that run at significantly different clock rates. On the other hand, there has recently been strong interest in distributed, networked control systems, where sensors, actuators and controllers are at different locations [5].In most articles in the literature, synchronization of all subsystems was assumed for simplicity, even though this is often hard or impossible to achieve. Previous work on a state space approach to synchronization is reported in [8]. Our work improves on these results by providing explicite stability conditions for the asynchronous case.

In this paper, we will therefore focus on the case of two interconnected discrete time systems that are running at non-matching clock frequencies. In section 2, both systems are modeled and represented in state space. In the developed model, states evolve on an event based time scale. Section 3 addresses the problem of stability robustness with respect to synchronization errors. A necessary and sufficient stability condition is introduced and simple to check sufficient conditions are also presented. Finally, section 4 provides an example with concluding remarks presented in section 5.

2. SYNCHRONIZATION ERROR MODEL

In this section we present a model for discrete time systems in which not all states are switched at the same time. For reasons of simplicity and brevity, we focus on the case of two subsystems. Each subsystem is partially characterized by its sampling period (T_1 and T_2 for subsystem 1 and respectively 2), which is assumed to remain constant. We will consider three cases:

(a) $T_1 > T_2$. Without loss of generality, we assume system 2 has a higher clock rate than system 1. This system will be referred to as type (a).

(b) $T_1 = T_2$ with clock signals out of phase, i.e. transitions in both systems are not coinciding. This system will be referred to as type (b).

(c) $T_1 = T_2$ with clock signals in phase. This system will be referred to as type (c).

The discrete time index n describes a switching event that either occurs in subsystem 1, or subsystem 2 or in both. The time index is incremented after a single switching event has occurred in either one of the two systems. If two coinciding switching events occur, the time-index is increased by 2. This definition of time is illustrated in Figure 1. Throughout this paper, we will make use of the following notation:
$\underline{x}(n) \in R^m$: State vector of the composite system
$\underline{x}_1(n) \in R^{m_1}$: State vector of subsystem 1
$\underline{x}_2(n) \in R^{m_2}$: State vector of subsystem 2
$A \in R^{m \times m}$: System matrix of the fully synchronized com-

0-7803-7761-3/03/$17.00 ©2003 IEEE

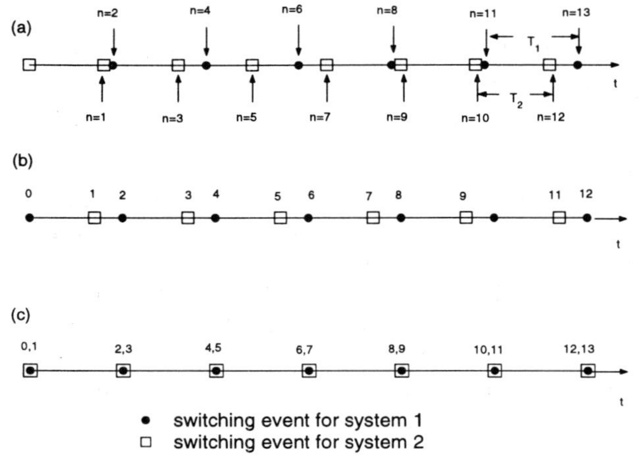

Figure 1: *Time index and switching events for the three cases: (a) $T_1 > T_2$, (b)$T_1 = T_2$ with phase error in clock signal, (c) $T_1 = T_2$. without phase error.*

posite system, $m_1 + m_2 = m$
$A_{11} \in R^{m_1 \times m_1}$: System submatrix for subsystem 1 (fully synchronized case)
$A_{22} \in R^{m_2 \times m_2}$: System submatrix for subsystem 2 (fully synchronized case)
$A_{21} \in R^{m_2 \times m_1}, A_{12} \in R^{m_1 \times m_2}$: Interconnection submatrices
$u_1(n) \in R$: System input of subsystem 1
$u_2(n) \in R$: System input of subsystem 2
T_1, T_2: Sampling/switching periods of subsystems 1 and 2 respectively.

Finally, we make the following assumption:

$$\frac{T_1}{T_2} = \frac{q+1}{q} = 1 + \frac{1}{q} \qquad , q \geq 2$$

For systems with almost identical clock frequencies, this is the most simple and analytically tractable model, resulting in large integer values of q. Note that $\frac{1}{q}$ measures the frequency error. The switching patterns repeat periodically every $qT_1 = (q+1)T_2$ time periods which is illustrated by Figure 1a for $q = 4$. Let us first derive the state space transition matrix for two consecutive state transitions (zero input case), i.e. one transition in subsystem 1 and a consecutive transition in subsystem 2 (see Figure 2).The corresponding state transition (in subsystem 1) from n to $n+1$ is then described by equations (1,2):

$$\underline{x}_1(n+1) = A_{11}\underline{x}_1(n) + A_{12}\underline{x}_2(n) \qquad (1)$$

$$\underline{x}_2(n+1) = \underline{x}_2(n) \qquad (2)$$

Similarly, we have for the transition (in subsystem 2) from time instance $n+1$ to $n+2$:

$$\underline{x}_1(n+2) = \underline{x}_1(n+1) \qquad (3)$$

$$\underline{x}_2(n+2) = A_{21}\underline{x}_1(n+1) + A_{22}\underline{x}_2(n) \qquad (4)$$

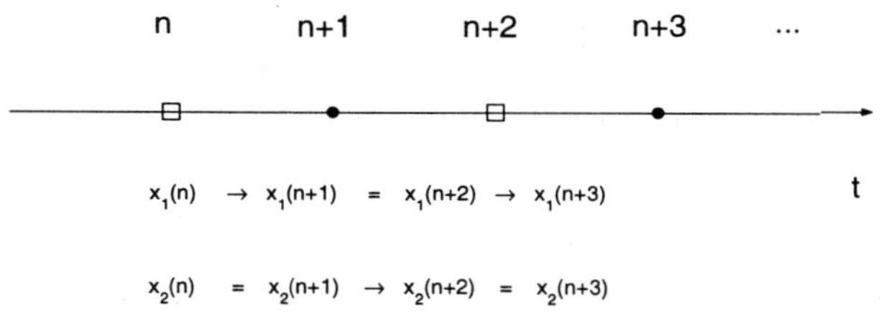

Figure 2: *Two consecutive switching events*

Rewriting equations (1) -(4) in state space form we obtain:

$$\underline{x}(n+2) = \begin{bmatrix} \underline{x}_1(n+2) \\ \underline{x}_2(n+2) \end{bmatrix} = \left(\begin{array}{c|c} A_{11} & A_{12} \\ \hline A_{21}A_{11} & A_{21}A_{12} + A_{22} \end{array} \right) \begin{bmatrix} \underline{x}_1(n) \\ \underline{x}_2(n) \end{bmatrix} \qquad (5)$$

Equation (5) describes two consecutive state transitions of the composite state vector (one for each state partition), as shown in Figure 1a and 1b. In case subsystem 2 switches first, a similar expression as in (5) is obtained. The eigenvalues of both representations are identical. From Figure 1 it is obvious, that two other types of transitions may also occur. The first one is shown in Figure 1c and corresponds to full synchronization (case c). Both sub-systems undergo a transition at the same time and this case therefore corresponds to the classical state space form:

$$\underline{x}(n+2) = \begin{bmatrix} \underline{x}_1(n+2) \\ \underline{x}_2(n+2) \end{bmatrix} = \left(\begin{array}{c|c} A_{11} & A_{12} \\ \hline A_{21} & A_{22} \end{array} \right) \begin{bmatrix} \underline{x}_1(n) \\ \underline{x}_2(n) \end{bmatrix} \qquad (6)$$

One could have chosen to move the time-index by one instead of two time periods, but the introduced notation is fully consistent with the fact that two transitions occur (one in each subsystem). This is important in comparative studies with case (a). It should be noted that the A matrices for the two cases given by (5) and (6) differ only in one term: The state $x_1(n+1)$ in (4) becomes $x_1(n)$ in (6). Therefore the two systems differ by a delay in the propagation from state $x_1(n)$ to state $x_2(n)$. Let us now turn to the third type of possible transitions. Such a transition occurs between switching times $n = 9$ and $n = 12$ in Figure 1(a). This transition is comprised of three individual switching events (two events in subsystem 2, one in subsystem 1) and therefore advances the time index by three. For the first switching event in subsystem 2 we can write:

$$\underline{x}_1(n+1) = \underline{x}_1(n) \qquad (7)$$

$$\underline{x}_2(n+1) = A_{21}\underline{x}_1(n) + A_{22}\underline{x}_2(n) \qquad (8)$$

For the second switching event in subsystem 1 we have:

$$\underline{x}_1(n+2) = A_{11}\underline{x}_1(n) + A_{12}\underline{x}_2(n+1) \qquad (9)$$

$$\underline{x}_2(n+2) \quad = \quad \underline{x}_2(n+1) \tag{10}$$

Similarly, we write for the 3rd switching event in subsystem 2

$$\underline{x}_1(n+3) \quad = \quad \underline{x}_1(n+2) \tag{11}$$

$$\underline{x}_2(n+3) \quad = \quad A_{21}\underline{x}_1(n+2) + A_{22}\underline{x}_2(n+2) \tag{12}$$

Now define the following matrices:

$$\tilde{A} = \left(\begin{array}{c|c} \tilde{A_{11}} & \tilde{A_{12}} \\ \hline \tilde{A_{21}} & \tilde{A_{22}} \end{array}\right) \tag{13}$$

$$\tilde{A_{11}} = A_{11} + A_{12}A_{21} \tag{14}$$

$$\tilde{A_{12}} = A_{12}A_{22} \tag{15}$$

$$\tilde{A_{21}} = A_{21}A_{11} + A_{21}A_{12}A_{21} + A_{22}A_{21} \tag{16}$$

$$\tilde{A_{22}} = A_{21}A_{12}A_{22} + A_{22}^2 \tag{17}$$

$$\hat{A} = \left(\begin{array}{c|c} A_{11} & A_{12} \\ \hline A_{21}A_{11} & A_{21}A_{12} + A_{22} \end{array}\right) \tag{18}$$

$$A = \left(\begin{array}{c|c} A_{11} & A_{12} \\ \hline A_{21} & A_{22} \end{array}\right) \tag{19}$$

with the definitions above (13)-(17) and equations (7) - (12), we obtain the following state space formulation:

$$\left[\begin{array}{c} \underline{x}_1(n+3) \\ \underline{x}_2(n+3) \end{array}\right] = \left(\begin{array}{c|c} \tilde{A_{11}} & \tilde{A_{12}} \\ \hline \tilde{A_{21}} & \tilde{A_{22}} \end{array}\right) \left[\begin{array}{c} \underline{x}_1(n) \\ \underline{x}_2(n) \end{array}\right] \tag{20}$$

It is now obvious, that the dynamical system properties of type (b) and (c) systems are defined by the matrices $\hat{A}$ and A respectively. For a type (a) system, the zero input state response is given by the following sequence of matrices:

$$\begin{array}{l} \underline{x}(M(2q+1)) = B^M \underline{x}(0), \quad B \in S \ with \\ S = \{P \in R^{m \times m} \mid P = \hat{A}^{\mu}\tilde{A}\hat{A}^{\nu}, \mu + \nu = q - 1\} \end{array} \tag{21}$$

3. ASYMPTOTIC STABILITY

Since type (b) and (c) systems have a linear time-invariant description, the stability problem is trivial and easily solved through evaluating the spectral radius [1] of the associated system matrix $\tilde{A}$ or A. In case (a), i.e. type a systems, a periodically time-variant system description arises. It is well known that asymptotic stability of this type of system is determined by the transition matrix over a full period of the system. Also since the system is linear, asymptotic stability and global asymptotic stability (of the origin) are equivalent [4].

[1] The spectral radius of a square matrix refers to the absolute value of its greatest magnitude eigenvalue.

Theorem 1:

The type (a) system with $T_1/T_2 = 1 + 1/q$ is globally asymptotically stable, iff:

$$\rho(B) < 1, \ \forall B \in S \tag{22}$$

Proof:

Note that for linear systems, global asymptotic stability and asymptotic stability of the origin are equivalent. The proof of Theorem 1 follows directly from (22) and the fact that a linear periodic system is stable, iff its transition matrix over an entire system period is stable, i.e. has a spectral radius smaller than unity [4]. All matrices in S have the same eigenvalues [3], and hence it suffices to check any matrix $B \in S$.

Corollary 2:

A sufficient condition for global asymptotic stability of the type (a) system is given by the norm condition:

$$\| \tilde{A} \| < \frac{1}{\| \hat{A}^{q-1} \|} \tag{23}$$

$\| . \|$ being any vector induced matrix norm.

Proof:

From (22) and the fact that any vector induced matrix norm is larger than or equal to the spectral radius, we have:

$$\rho(B) \leq \| B \| \leq \| \hat{A}^{\mu+\nu}\tilde{A} \| \leq \| \hat{A}^{q-1} \| . \| \tilde{A} \| < 1 \tag{24}$$

If the right most inequality in (24) is satisfied, then by Theorem 1, the system is globally asymptotically stable. The right most inequality in (24) can be rewritten as:

$$\| \tilde{A} \| \leq \frac{1}{\| \hat{A}^{q-1} \|} \tag{25}$$

which provides the desired result.

Corollary 3:

If the type (b) system is globally asymptotically stable, then the type (a) system can always be made asymptotically stable by choosing q sufficiently large.

Proof:

If the type (b) system is g.a.s., then the spectral radius of $\hat{A}$ must be strictly smaller than 1. This means, that the norm value $\| \hat{A}^{\nu} \|$ can be made arbitrarily small if ν is chosen sufficiently large. Now using Corollary 2 rewrite (23) as:

$$\| \hat{A}^{q-1} \| < \frac{1}{\| \tilde{A} \|} \tag{26}$$

Therefore given an $\tilde{A}, \exists\ q$ sufficiently large such that (26) is satisfied. Hence condition (23) is also satisfied.

Comments:

(1) Since a stable matrix A (with spectral radius smaller than 1) does not imply stability of the system matrices, there is no stability robustness of the fully synchronized system with respect to even the slightest differences in clock frequency between the two systems.

(2) If the system matrix $\hat{A}$ (type (b)) is stable and if the

difference in clock frequency between the two systems is sufficiently small, then by Corollary 3 stability of the type (b) system also guarantees stability of the type (a) system. Therefore, the type (b) system shows some stability robustness with respect to synchronization errors.

(3) It is straight forward to extend these results to the general case of any rational value of T_1/T_2. Stability of a type (a) system can still be determined by the spectral radius of a product of matrices involving $\tilde{A}$ and $\hat{A}$ only.

(4) The case of non-rational ratios T_1/T_2 is more difficult to describe. However, since the rational numbers are dense, a sufficiently accurate rational approximation of the non-rational switching period ratio can always be found. Note that there are no simple mathematical expressions relating spectral radii of A, $\hat{A}$ and $\tilde{A}$.

4. EXAMPLE

In this example we consider a closed-loop system with the transfer functions: $F(z) = \frac{5}{4z-3}$ and $G(z) = \frac{-2}{4z+3}$, in the forward and feedback path, respectively. Therefore the state transition matrix for the fully synchronized closed-loop system is given by:

$$A = \begin{pmatrix} 3/4 & 5/4 \\ -1/2 & -3/4 \end{pmatrix}$$

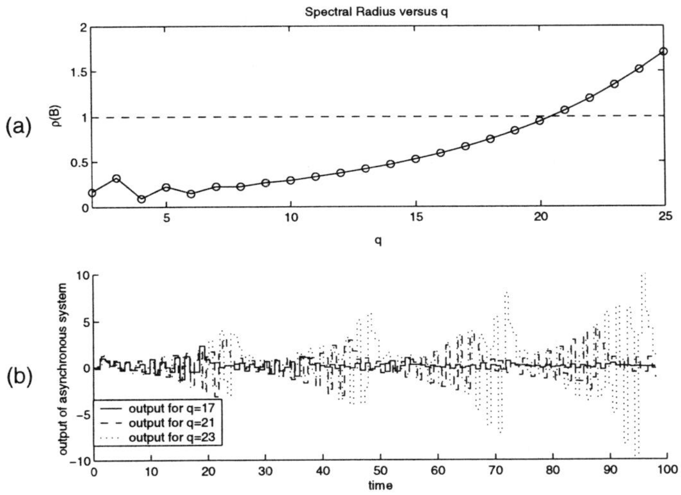

Figure 3: *(a) $\rho(B)$ as a function of q; (b) solid line: output for $q = 17$ (stable), dashed line: output for $q = 21$ (unstable); dotted line: output for $q = 23$ (unstable)*

the spectral radius of which is $\rho(A) = 1/4$. Consequently the output of the synchronously working feedback system is always stable. Using equation (18), $\rho(\hat{A}) = 1.125 > 1$, i. e. type (b) system is unstable. This is in agreement with the spectral radius of B being larger than 1 for $q > 20$, which is shown in Figure 3a.

Figure 3b shows the system impulse response for q =17, 21 and 23. As it is expected from the Figure 3a, the impulse response converges towards 0 for $q = 17$, and diverges for $q = 21$ and 23.

5. CONCLUSION

A state space synchronization error model was derived for linear time-invariant systems. Based on this model, the stability robustness of discrete time systems with respect to synchronization errors was investigated. It was shown that the fully synchronized system has zero stability robustness with respect to only the smallest perturbations in switching times. A modified system (characterized by both systems having the same switching period but not the same switching times) is shown to exhibit this type of stability robustness. If the switching periods of the two interconnected systems are sufficiently close and this modified system is stable, then stability in the presence of synchronization errors can be guaranteed. Future work will address the fact that in reality, sampling periods T_1 and T_2 are time-variant and uncertain.

6. REFERENCES

[1] O. Katsuhiko, "Discrete-Time Control Systems", Prentice Hall, 1995.

[2] A. Antoniou, "Digital Filters: Analysis, Design, and Applications", McGraw-Hill, New York, 1993

[3] M. Marcus and H. Ming, "A survey of Matrix Theory and Matrix Inequalities", Allyn and Bacon, INC., Boston, 1964.

[4] P. J. Antsaklis and A. N. Michel "Linear Systems", McGraw-Hill, New York, 1997.

[5] J. Elson and D. Estrin, "Time synchronization for wireless sensor networks", *IPDPS'01*, 2001.

[6] P. Voulgaris, "Control of asynchronous sampled data systems", *IEEE Transactions on Automatic Control* , Vol. 39, pp. 1451 -1455, Issue 7, July 1994.

[7] V.S. Ritchey and G.F. Franklin, " A stability criterion for asynchronous multirate linear system", *IEEE Transactions on Automatic Control* , Vol. 34 ,pp.529 -535, Issue 5 , May 1989

[8] A. F. Kleptsyn, V.S. Kozyakin, M. A. Krasnosel'skii, ams N. A.Kuznetsov, "Effect of small synchronization errors on stability of complex systems II", *Avtomatika i Telemekhanika.* Translated in: *Automation and Remote Control*,pp 309-314, 1984.

MICROWAVE AMPLIFIER DESIGN FOR MOBILE COMMUNICATION VIA IMMITTANCE DATA MODELLING

A. Kılınç, H. Pınarbaşı, B.S. Yarman, A. Aksen

Isik University, Istanbul, Turkey

ABSTRACT

In this paper, a practical broadband microwave amplifier design algorithm based on immittance data modelling is presented. In the course of design, first, the optimum input and output terminations for the active device are produced employing the real frequency technique. Then, these terminations are modelled utilizing the new immittance-modeling tool to synthesize the front-end and back-end matching networks. An example is included to exhibit the implementation of the proposed design algorithm to construct a single stage wideband microwave amplifier over a wide frequency band. It is expected that the proposed design algorithm will find applications in the design of microwave amplifiers put on MMIC for mobile communication.

1. INTRODUCTION

One of the fundamental problems in the design and development of communication systems is to match a given device to the system via coupling circuits so as to achieve optimum performance over the broadest possible frequency band. This problem inherently involves the design of an equalizer network to match the given complex impedances, and usually referred as *impedance matching* or *equalization*.

Recently introduced immittance data modeling tool can be employed successfully to design broadband matching networks for microwave amplifiers [1]. When a broadband microwave amplifier is designed, optimum termination immittances for the active device can be generated point by point employing the Carlin's Real Frequency Line Segment Technique [2-5]. Then, the data for the terminations are modelled by means of the immittance modeling tool. Eventually, Positive Real (PR) immittances are synthesized to yield the front-end and the back-end matching networks which completes the design.

In this presentation, first the immittance based modeling tool is summarized. In section III, Generalized Real Frequency Technique (GRFT) is outlined. The complete design algorithm is given in Section IV. Finally, utilization of the design algorithm is exhibited with an example.

2. THE IMMITTANCE BASED DATA MODELLING TOOL [1]

Any positive real rational immittance function $F(s)$ can be written in terms of its minimum and the Foster parts;

$$F(s) = F_m(s) + F_f(s) \tag{1}$$

where $s=\sigma+j\omega$ is the complex domain variable, $F_m(s)$ is the minimum part which is free of $j\omega$ poles, and $F_f(s)$ is the Foster part which includes only $j\omega$ poles. On the real frequency axis $j\omega$, one has

$$F(j\omega)=R(\omega)+jX(\omega) \tag{2a}$$

$$F_m(j\omega)=R_m(\omega)+jX_m(\omega) \tag{2b}$$

$$F_f(j\omega)=jX_f(\omega) \tag{2c}$$

It is clear that

$$R(\omega)=R_m(\omega) \tag{3a}$$

$$X(\omega)=X_m(\omega)+X_f(\omega) \tag{3b}$$

Since $F_m(s)$ is a positive real minimum, which contains no poles on the $j\omega$ axis, its imaginary part $X_m(\omega)$ is related to the real part $R_m(\omega)$ by the Hilbert transformation relation;

$$X_m(\omega) = H\{R(\omega)\} \tag{4}$$

where $\boldsymbol{H}\{.\}$ designates the Hilbert Transformation operation.

In the immittance based modelling technique, the crux of the matter is to decompose the given data into its minimum part and Foster part. Hence, the modelling process is carried out within two major steps: model for the minimum part and the Foster part.

To model the minimum part, it is sufficient to match an analytic form $R(\omega^2)$ for the real part data. Then the complete minimum function $F(s)$ can easily be generated from $R(-s^2)$ by means of Gewertz procedure [4].

The real part forms are classified based on the selection of the transmission zeros of the matching networks. Let $R(\omega^2) = \dfrac{N(\omega^2)}{D(\omega^2)}$, in this case regarding the zeros of $N(\omega^2)$, the real part forms are described as follows:

For modelling Form-A: $N(\omega) = \omega^{2k}$

For modelling Form-B: $N(\omega) = \omega^{2k} \cdot \prod_{p=1}^{m} (\omega^2 - \omega_p^2)^2$

For modelling Form-C:

$$N(\omega) = \omega^{2k} \cdot \prod_{p=1}^{m} (\omega^2 - \omega_p^2)^2 \cdot \prod_{t=1}^{m_t} (\sigma_t^2 + \omega^2) \cdot \prod_{r=1}^{m_r} \left\{\sigma_r^4 + 2\sigma_r^2(\omega^2 + \beta_r^2) + (\omega^2 - \beta_r^2)^2\right\}$$

These choices will be picked in accordance with the given data for $R(\omega)$. In order to extract the Foster part from the original measured data, one has to generate $X_m(\omega)$ using the Hilbert Transformation relation [3]. Eventually, realisable analytical forms for the minimum immittance function and the Foster function are obtained by means of an appropriate curve fitting or interpolation algorithms and they are synthesized to yield the desired model under consideration.

0-7803-7761-3/03/$17.00 ©2003 IEEE

3. GENERALIZED LINE SEGMENT TECHNIQUE FOR MATCHING A COMPLEX LOAD TO A RESISTIVE GENERATOR

Consider the single matching circuit arrangement shown in Figure 1.

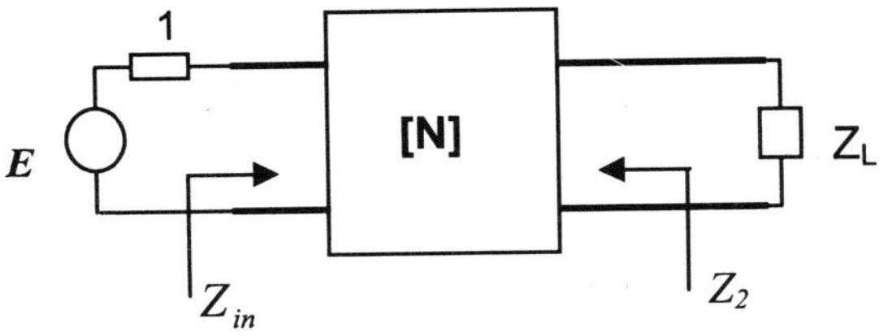

Figure 1: Single matching problem

Let the load impedance Z_L and the equalizer back impedance Z_2 be written in terms of their real and imaginary parts on the real frequency axis as

$$Z_L(j\omega)=R_L(j\omega)+jX_L(j\omega)\ ,$$
$$Z_2(j\omega)=R_2(j\omega)+jX_2(j\omega)\ , \qquad (5)$$

Basic idea is the use of a piecewise linear approximation to represent the unknown real part $R_2(\omega)$ as a number of straight-line segments as shown in Figure 2.

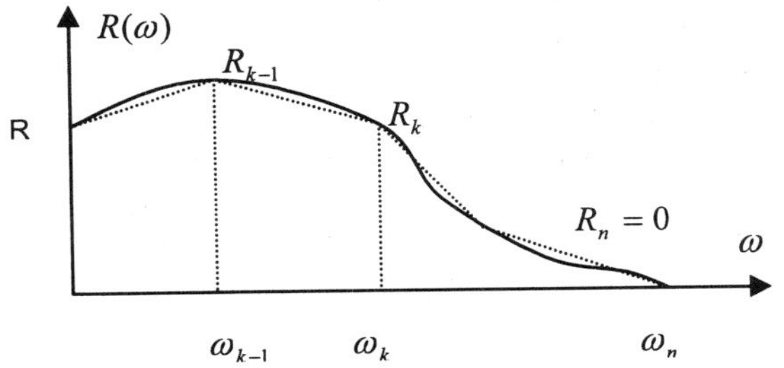

Figure 2: Line segment approximation of the real part

$$R_2(\omega)=R_0+\sum_{i=1}^{n}a_i(\omega)R_i \quad ,X_2(\omega)=\sum_{i=1}^{n}b_i(\omega)R_i \qquad (6)$$

The coefficients $a_i(\omega)$ in (6) can be expressed directly in terms of sampling frequencies (ω_i, $i=1,2,3,...n$) as follows:

$$a_i(\omega)=\begin{cases}1 & \omega\geq\omega_i\,,\\ \dfrac{\omega-\omega_{i-1}}{\omega_i-\omega_{i-1}} & \omega_{i-1}\leq\omega\leq\omega_i\,,\\ 0 & \omega\leq\omega_{i-1}\,,\end{cases}$$

$b_i(\omega)$ in (6) can be expressed using Hilbert transform techniques as

$$b_i(\omega)=\frac{1}{\pi(\omega_i-\omega_{i-1})}\int_{\omega_{i-1}}^{\omega_i}\ln\left|\frac{y+\omega}{y-\omega}\right|dy$$

In the Generalized Real Frequency Technique (GRFT), $Z_2(j\omega)$ can be determined as

$$Z_2(j\omega)=R_2(\omega)+j\lfloor H\{R_2(\omega)\}+X_{2f}\rfloor \qquad (7)$$

where, X_{2f} designates the Foster part of the equalizer impedance. It is also noted that X_{2f} is among the unknowns of the problem.

The Transducer Power Gain (TPG) of the system can be written in terms of the reflection coefficient at port 2 as

$$T(\omega)=1-|\rho_2|^2\ ;\quad \rho_2=\frac{Z_2-Z_L^{\,*}}{Z_2+Z_L} \qquad (8)$$

(8) can directly be expressed in terms of the real and imaginary parts of the load impedance Z_L and the back-end impedance Z_2 of the equalizer [2].

$$T(\omega)=\frac{4R_2(\omega)R_L(\omega)}{(R_2(\omega)+R_L(\omega))^2+(X_2(\omega)+X_L(\omega))^2} \qquad (9)$$

If the real frequency load data is given as $Z_L(j\omega)=R_L(\omega)+jX_L(\omega)$, then the matching problem becomes essentially to that of finding $Z_2(j\omega)$ point by point such that $T(\omega)$ is maximized over the band of operation.

Once $Z_2(j\omega)=R_2(\omega)+jX_2(\omega)$ is determined point by point employing the Generalized Real Frequency Technique, it is modelled as a positive real function by means of the *"immittance based data modelling tool[1]"*.

In the following section, we will introduce the new microwave amplifier design technique via the immittance modelling tool.

3.1 Extension of Immittance Based Data Modelling Tool to the Design of Amplifiers

Let us consider the single stage amplifier configuration shown in Figure 3 where the active two-port device is denoted by [A]. The lossless two-ports N_1 and N_2 designate the front-end and back-end matching networks respectively. A single stage microwave amplifier can conceptually be constructed within two steps by using the Real Frequency Technique.

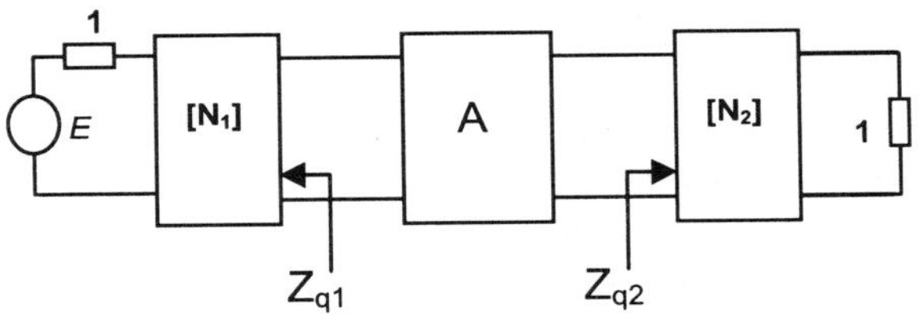

Figure 3: Single stage amplifier equalized at both input and output

In the first step, the optimum immittance data Z_{q1} for the front-end matching network is generated point by point over the band of operation. In this step, we presume that the output port of the active is closed with unit termination (i.e. 50 ohms). Hence, the input impedance of the transistor is given by

$$Z_{in}=\frac{1+S_{11}}{1-S_{11}} \qquad (10)$$

and it is considered as the termination of the input matching network. In this case, we face a single matching problem. Thus, employing the GRFT, optimum impedance data Z_{q1} for the front-end matching network is generated. The gain of the system is given by

$$T_1(\omega)=P_1(\omega)\frac{4R_{q1}R_{L1}}{(R_{q1}+R_{L1})^2+(X_{q1}+X_{L1})^2} \qquad (11)$$

P1 in (11), can be regarded as a weight function and it is

$$P_1(\omega)=\left\{\frac{|S_{21}|^2}{1-|S_{11}|^2}\right\} \qquad (12)$$

The driving point input impedance of the front-end equalizer is

$$Z_{q1}(\omega) = R_{q1}(\omega) + jX_{q1}(\omega) \tag{13}$$

The load impedance $Z_{L1}(j\omega)=R_{L1}(\omega)+jX_{L1}(\omega)$ is set to Z_{in} which is specified by (10).

In this step, the negative slope of the gain is compensated by optimizing T_1 to a flat gain level T_{01}.

In the second step of the conceptual design, the back-end matching network will be generated as set of data. In this case, the gain $T_2(\omega)$, which is subject to optimization, is expressed in terms of the driving point impedance Z_{q2} of the output- matching network N_2.

$$T_2(\omega) = \frac{T_1(\omega)}{(1-|S_2|^2)} \cdot \frac{4R_{q2}R_{L2}}{(R_{q2}+R_{L2})^2 + (X_{q2}+X_{L2})^2} \tag{14}$$

In (14), the term S_2 is the reflection coefficient of the active device seen at the output when the front-end matching network is present. Hence,

$$S_2 = S_{22} + \frac{S_{12}S_{21}S_{q1}}{1 - S_{11}S_{q1}} \tag{15}$$

In (15), S_{q1} is the input reflection coefficient of the front-end equalizer and it is given by

$$S_{q1} = \frac{Z_{q1}+1}{Z_{q1}-1} \tag{16}$$

Furthermore, the output impedance of active device when its input is loaded by the front matching circuit is given by

$$Z_{L2} = \frac{1+S_2}{1-S_2} = R_{L2} + jX_{L2} \tag{17}$$

Finally, optimization of $T_2(\omega)$ to a flat gain level T_{02} yields the Thevenin's impedance Z_{q2} as set of points.

In the first step, it would be wise to select T_{01} as the minimum value of $P_1(\omega)$ over the operation band.

Similarly in the second step, one can choose T_{02} as the minimum value of the term P_2 :

$$P_2 = \frac{|S_{21}|^2}{(1-|S_{11}|^2)(1-|S_{22}|^2)} \tag{18}$$

which is available maximum gain for the given transistor over the specified frequencies.

In the course of the optimization $Z_{q1}(j\omega)=R_{q1}(\omega)+jX_{q1}(\omega)$ and $Z_{q2}(j\omega)=R_{q2}(\omega)+jX_{q2}(\omega)$ are computed point by point as described in the Generalized Real Frequency Technique (GRFT). To improve the optimization, the imaginary parts X_{qi} can be computed as $X_{qi}=\boldsymbol{H}\{R_{qi}\}+X_{qf}$, $i=1,2$ where X_{qf} designates the Foster parts of driving point impedance Z_{qi}.

The above-mentioned process is summarized in the following algorithm.

4. THE AMPLIFIER DESIGN ALGORITHM

Inputs: S_{11}, S_{12}, S_{21}, S_{22}: The Scattering parameters of the active element over the prescribed frequency band.

Part I: Design of front-end equalizer

Step I: Construct the weight function $P_1(\omega)$ in (12).

Step II: Construct the gain function $T_1(\omega)$ in (11). Here, the terms R_L and X_L refer to the real and imaginary parts of the input impedance of the transistor when its output is loaded by 1-ohm resistor, which is given by the equation.

Step III: Optimize the gain function $T_1(\omega)$ to obtain a flat gain level of $min(P_1(\omega))$ and determine the break points for Z_{q1} as described in the GRTF at section 3. (Optimisation parameters are: R_{q1})

Step IV: calculate S_2 and Z_{L2} by using (15) and (17).

Part II: Design of back-end equalizer

Step I: Construct the function $P_2(\omega)$ in (18).

Step II: Construct the gain function $T_2(\omega)$ in (14).

Step III: Optimize the gain function $T_2(\omega)$ to obtain a flat gain level of $min(P_2)$. And determine Z_{q2} point by point to optimize T_2 employing GRFT. (Optimisation parameters are: R_{q2})

Part III. Optimize the final gain.

Step I: Repeat Part I by using new values of R_L and X_L, the real and imaginary parts of the input impedance of the transistor when its output is loaded by Z_{q2}, which is calculated at the end of Part II.

Step II: Repeat Part II.

Step III: Optimize the gain function $T_2(\omega)$ to obtain a flat gain level of $min(P_2)$. and determine Z_{q1} and Z_{q2} point by point to optimize T_2 employing GRFT. (Optimisation parameters are: R_{q1} , R_{q2})

Step IV: Having obtained the data for Z_{q1}, generate the analytic form for it using immittance based modelling tool and synthesize it.

Step V: Having obtained the data for Z_{q2} generate the analytic form for it using immittance based modelling tool and synthesize it.

Now, let us introduce an example to design a single stage amplifier.

5. EXAMPLE

In this example, we wish to design a microwave amplifier employing the immittance-based data-modelling tool. For this purpose commercially available transistor AM012MXQF from AMCOM Communications Inc. was selected and its biasing conditions are $V_{CE}=5V$, $I_C=150mA$, $Z_0=50$ and Bandwidth = 4 GHz. (1GHz-5GHz)

Table 1: Typical Scattering Parameters for AM012MXQF

f	S_{11}	S_{21}	S_{12}	S_{22}
1.0	0.008-0.912i	-3.930+6.294i	0.044+0.035i	0.002-0.301i
1.2	-0.186-0.864i	-2.753+6.198i	0.050+0.031i	-0.041-0.306i
1.4	-0.325-0.805i	-1.808+5.919i	0.056+0.029i	-0.078-0.301i
1.6	-0.439-0.736i	-1.051+5.586i	0.061+0.026i	-0.111-0.292i
1.8	-0.529-0.662i	-0.448+5.220i	0.064+0.022i	-0.138-0.276i
2.0	-0.597-0.589i	0.038+4.859i	0.067+0.018i	-0.160-0.262i
2.2	-0.651-0.519i	0.429+4.506i	0.069+0.015i	-0.181-0.246i
2.4	-0.696-0.453i	0.753+4.173i	0.070+0.011i	-0.199-0.229i
2.6	-0.730-0.389i	1.017+3.854i	0.071+0.008i	-0.214-0.211i
2.8	-0.757-0.328i	1.231+3.552i	0.071+0.004i	-0.228-0.194i
3.0	-0.777-0.269i	1.408+3.264i	0.071+0.001i	-0.241-0.176i
3.2	-0.790-0.212i	1.553+2.992i	0.071-0.002i	-0.252-0.157i
3.4	-0.802-0.158i	1.668+2.737i	0.071-0.005i	-0.262-0.139i
3.6	-0.808-0.104i	1.763+2.497i	0.070-0.008i	-0.278-0.122i
3.8	-0.812-0.052i	1.841+2.272i	0.068-0.010i	-0.281-0.107i
4.0	-0.813-0.001i	1.901+2.060i	0.068-0.013i	-0.286-0.094i
4.2	-0.828+0.057i	1.923+1.834i	0.065-0.014i	-0.295-0.080i

f	S_{11}	S_{21}	S_{12}	S_{22}
4.4	-0.824+0.102i	1.960+1.655i	0.064-0.016i	-0.298-0.069i
4.6	-0.815+0.149i	1.991+1.485i	0.063-0.018i	-0.300-0.057i
4.8	-0.804+0.195i	2.014+1.322i	0.062-0.020i	-0.301-0.046i
5.0	-0.789+0.242i	2.033+1.165i	0.060-0.021i	-0.300-0.035i

In this design, there was no need to employ foster part for Z_{q1}, Z_{q2} .

Part I: The Transducer power gain T_1 is compensated to a flat gain level T_{01} =12.4dB. Hence, as the result of optimization R_{q1} is computed

R_{q1}= 0.1*[1.87 1.78 1.78 1.76 1.74 1.72 1.70 1.68 1.65 1.62 1.57 1.57 1.49 1.56 1.36 1.56 1.39 1.28 1.15 1.17 0.934];

Part II: In this part the back-end matching network is constructed when the front-end is present. Similarly, supplying the initial guess values for the resistive excursions R_{q2}, T_2 is optimised to a flat gain level T_{02}=12.8dB. As the result of optimisation R_{q2} is found as:

R_{q2}= 0.1*[9.95 11.1 8.87 8.76 8.52 8.36 8.22 8.09 8.01 7.95 7.89 7.85 7.79 7.71 7.67 7.7 7.51 7.58 7.27 7.13 10.4];

Part III: At all steps T_1 and T_2 are optimised to a flat gain level T_{02} =12.8dB. At the result of optimisations R_{q1} , R_{q2} and their imaginary parts are found as:

R_{q1}= 0.1*[0.763 0.915 1.39 1.32 1.44 1.53 1.59 1.65 1.7 1.73 1.79 1.82 1.83 1.86 1.86 1.85 1.8 1.8 1.35 2.02 1.31];

R_{q2}= 0.1*[10.6 25.7 9.69 9.04 7.68 7.06 6.71 6.47 6.29 6.15 6.02 6.02 5.94 6.04 5.86 6.11 6.31 5.48 6.58 3.62 7.78];

X_{q1}= 0.01*[7.52 8.69 7.58 5.7 5.3 4.36 3.43 2.53 1.64 0.711 -0.169 -1.16 -2.25 -3.4 -4.66 -6.01 -7.25 -9.32 -8.19 -10 -13.1];

X_{q2}= 0.1*[8.39 -0.598 -8.69 -5.48 -5 -4.38 -3.99 -3.73 -3.54 -3.4 -3.26 -3.16 -3.09 -3.1 -3.05 -3 -3.4 -3.25 -3.76 -2.57 -1.79];

By using the immittance based data modelling tool, the minimum reactance functions can be calculated analytically and this leads to the synthesis of the equalizer circuits. For both front-end and back-end matching networks, modelling form A is selected for $R(\omega^2)$

The program code was run and at the end the minimum reactance functions for the input and the output equalizers were found to be

$$Z_{front} = \frac{1.151s^3 + 1.705s^2 + 1.311s}{10.96s^4 + 16.25s^3 + 19.58s^2 + 10.5s + 1.998} + 0.013s$$

$$Z_{back} = \frac{14.96s^2 + 12.20s + 11.86}{25.51s^3 + 20.81s^2 + 28.84s + 7.018} + 0.44s$$

The final amplifier configuration and overall performance curve are given in Figure 5 and Figure 6 respectively.

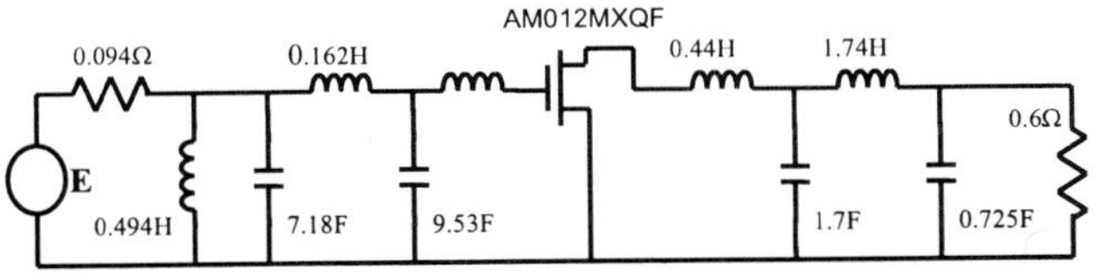

Figure 5: Designed amplifier configuration

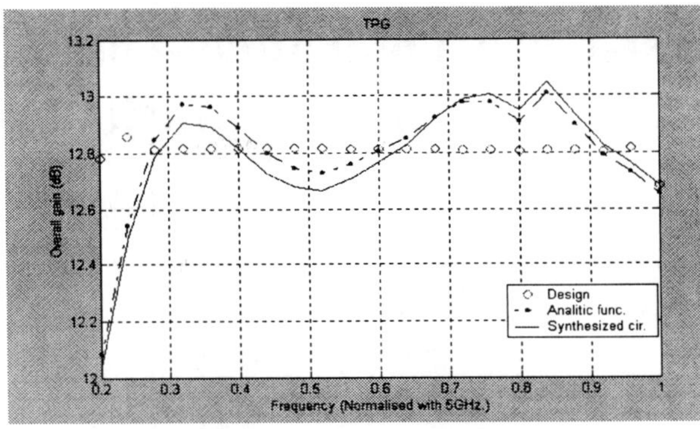

Figure 6: Overall TPG performance of the amplifier

6. CONCLUSION

In this paper, the immittance data-modelling tool is applied to design single stage microwave amplifiers. Optimum immittance terminations for the active device are generated employing the Generalized Real Frequency Technique. An algorithm is presented to ease the understanding of the design process introduced in this paper. Implementation of the algorithm has been exhibited by means of an example. It can readily be seen that the single stage microwave amplifier design algorithm involves only simple linear arithmetic computations during optimization routine, while processing numerically defined load impedances of any complexity. The gain function is quadratic in the unknowns, and hence the problem reduces to that of a quadratic optimisation. The design algorithm presented here, can easily be extended to construct microwave amplifiers with mixed, lumped and distributed elements, employing realizable, two variable, driving point network functions [6].

7. REFERENCES

[1] Yarman, B.S, Aksen, A., Kılınç, A. "An Immittance Based Tool for Modelling Passive One-Port Devices by Means of Darlington Equivalents" *A.E.Ü 55 No.6*, December 2001, pp.443-451

[2] Carlin, H.J. "A New Approach to Gain-Bandwidth Problems" *IEEE Trans. Cas,* Vol.23, April 1977, pp.170-175

[3] Yarman, B.S., Carlin, H.J. "A Simplified Real Frequency Technique Applied to Broadband Multistage Microwave Amplifiers" *IEEE Trans.on MTT 30 1983, pp.15-28*

[4] Aksen A. "Design of lossless two ports with mixed, lumped and distributed elements for Broadband Matching", PhD.Dissert., Lehrsthunl fuer Nachrich- tentechnic, Ruhr-Universitaet Bochum, 1994.

[5] Yarman, B.S. "Broadband Networks", Wiley Encyclopedia of Electrical and Electronics Engineering, Vol. II, pp.589-604, 999.

[6] Aksen, A., Yarman, B.S. "A Real Frequency Approach to Describe Lossless Two-Ports Formed With Mixed Lumped and Distributed Elements" *A.E.Ü 55 No.6,* December 2001, pp.389-396

Worst-Case Tolerance Analysis of Non-Linear Systems Using Evolutionary Algorithms

B.De Vivo[#], G.Spagnuolo[#], M.Vitelli[*]

[#] Dipartimento di Ingegneria dell'Informazione ed Ingegneria Elettrica
Università di Salerno
I-84084, Fisciano (SA) – ITALY
Phone: +39 089 964258, Fax: +39 089 964218
e-mail: bdevivo@unisa.it, spanish@ieee.org

[*] Dipartimento di Ingegneria dell'Informazione
Seconda Università di Napoli
Real Casa dell'Annunziata, Aversa (CE), Italy
Phone: +39 081 5010240
e-mail: vitelli@unina.it

Abstract - In this paper an evolutionary algorithm-based approach to the worst-case analysis of non-linear systems is presented. The evolutionary algorithm is used to minimize the underestimation error which affects classical Monte Carlo-based methods. The method seems to be useful especially whenever large parameters variations need to be taken into account and performance functions that are non-linear with respect to parameters are considered. An application example involving a lumped non-linear circuit modeling a cable termination equipped with a stress control tube is used to introduce the method.

I. Introduction

In recent years Worst Case Analysis (WCA) has become a crucial task in system design because of the increasing demand of high-performances and high reliability devices. Modern WCA imposes the adoption of new CAD tools featuring efficient handling capabilities of the large variations and uncertainties which affect the system parameters. As for the uncertainties, large ones come from several sources. Firstly, we have parasitic parameters. Their uncertainty is intrinsically large as it depends on system manufacturing technologies and on operating factors. In fact, the "small" factory tolerance guaranteed by vendors for virgin components and devices is a limited part only of the total uncertainty affecting the parameters in real operating conditions. A ±5% factory tolerance can easily degenerate into a ±30% worst case tolerance, due to aging, temperature stress, field stress, radiation, distorted operation, and other factors. On the other side, a worst case analysis is also indispensable in order to investigate the system reliability with respect to parameters' variations. This may be the case, for example, of a circuit whose ordinary operation mode provide for large load variations.

Performing an effective and reliable WCA in presence of large parameter variations means to be able to provide a neither too optimistic nor too pessimistic foreseeing. A too optimistic True-Worst-Case (TWC) foreseeing in systems design can compromise the compliance with regulations and/or performance constraints. On the other side, a too pessimistic foreseeing can lead to badly sized components and/or to products less competitive for the market.

In WCA problems one firstly defines an *evaluation function* (e.f.) of interest, $f:(x,p_1,p_2,\ldots,p_n)\in\Re^{n+1}\rightarrow\Re$, which depends on the variable x (e.g. time, frequency) and on the system parameters $p_1, \ldots, p_n$, each one spanning an interval of variation, $p_i\in P_i$, $i=1,..,n$, $P_i=[P_{i,min},P_{i,max}]$. The goal of WCA is to evaluate the range the e.f. spans because of the variations of the parameters. At least, three factors make WCA a demanding task. Firstly, the *type* of e.f. is always a strongly conditioning factor. In general, the e.f. is not available in explicit form with respect to the system parameters p_k. Even if the e.f. is available in explicit form it is often an involved non linear function of the circuit parameters, so that it can be evaluated in a numerical way only.

Secondly, a further source of difficulty is the width of the variation interval of the parameters: the wider the range the higher the probability the e.f. is non-monotonic with respect to the parameters and the lower the chances to get the TWC using Monte Carlo techniques.

Thirdly, in many cases parameters' values are uniformly distributed within their respective assigned intervals. This is the case, for example, of parasitics, of parameters deriving from experimental measurements and/or from data interpolation, or, more in general, whenever no prior information about a statistical distribution were available. Furthermore, in many applications no correlation among the values assumed by different system parameters can be considered, either because it is not available, or because the WCA, oriented to the most reliable design, is needed.

A consolidated approach for solving WCA problems is the stochastic Monte-Carlo one, which provides an underestimation of the TWC. The main limit of such method consists in the big amount of trials required to get a "good solution", namely close to the TWC, in presence of large parameters' variations.

In this paper it is shown that the calculation of an underestimation of the TWC in WCA problems can be improved by means of Evolutionary Algorithms (EA). The solution is built up as a combination of parts of the solutions

0-7803-7761-3/03/$17.00 ©2003 IEEE

obtained selecting the values of parameters within their own uncertainty ranges in a stochastic way. This approach, which can be viewed as a variant of Monte-Carlo (MC) method, is best suited for WCA with the e.f. in implicit form, which is mostly used in the analysis not only of non linear circuits but also of linear ones.

A non-linear circuit modeling a high voltage cable termination has been considered to illustrate the approach.

II. THE EVOLUTIONARY ALGORITHM

Let it be considered a system whose analytical model includes M varying parameters, each one characterized by a given range of variation. In the sequel we suppose that the values assumed by the parameters are uniformly distributed and there is no correlation among them. In classical MC-like WCA, the analysis of the system is carried out by taking a certain number of different combinations of values of parameters, named *trials*, thus calculating the value of the e.f.. The worst case is thus obtained by taking the minimum and maximum values assumed by the e.f. along the trials run. Nevertheless, often the e.f. is non linear, or worse non monotonic, within the parameters' ranges of variation. Moreover, the e.f. does not always assume an explicit form; in general, it is an implicit function of parameters, so that its non-monotonicity and non-linearity is complicated to be foreseen. Consequently, it becomes very difficult to catch by MC analysis the combination of parameters corresponding to maxima and minima of the e.f. and it is not easy to foresee which number of MC trials allows detecting a quite good estimation of the TWC. The only way of approaching the problem seems to be that one that increases step-by-step the number of trials until the discrepancy between the results of two successive analyses falls below a fixed threshold. Nevertheless, if the set of parameters to be tested is not chosen purely randomly, as in the MC analysis, but trying to select the combinations which likely give the solutions closest to the TWC, the analysis should be more fruitful. To this aim, an index expressing the contribution given to the WC by the solution obtained taking each combination of parameters needs to be defined. Such an index can be useful to drive a global optimization method in pursuing the parameters' sets giving the upper and the lower values of the performance function.

Evolutionary Algorithms (EA's) can be of great help in performing this task. They are based on the principles of biological evolution [1]. An *evolutionary pressure* is driven mainly by the genetic operators of cross-over and mutation acting on a population of individuals, each one representing a possible solution for the given problem. EA's ensure a good balance between the need of an exhaustive visit of the space of the solutions and the will of tracking "good" solutions toward the search of the best one.

II.a A practical non-linear TWC problem

To the aim of illustrating the application of EA's to WCA, we refer to a practical problem. In fig.1a a Stress Control Tube (SCT) in a high voltage cable termination is depicted. SCT's are tubes of suitable stress grading materials characterised by a rather high permittivity and a non linear resistivity. Such materials are used in order to reduce the electric field enhancement in high voltage cables-terminations, to achieve a *resistive-capacitive* field control. Stress control materials are obtained by loading the polymeric matrix with suitable fillers such as carbon-black. SCT's can be modelled with the simple transmission line RC network [2] shown in fig.1b. The cable termination is divided in N elements of length Δz, characterised by a transversal capacitance C_t, a longitudinal capacitance C_l and a longitudinal conductance G. Such parameters assume the following expressions:

$$C_t = \frac{2\pi\varepsilon_0\varepsilon_r\Delta z}{\ln\frac{r_2}{r_1}};C_l = \frac{\pi\varepsilon_0\varepsilon_{rt}(V)((r_2+\delta)^2-r_2^2)}{\Delta z};G = \frac{\pi((r_2+\delta)^2-r_2^2)}{\rho(V)\Delta z}$$

It has been assumed r_1=0.5cm and r_2=1.5cm. The capacitance C_t is associated to the elementary flux tube inside the cable primary insulation; ε_r is the relative permittivity of the cable main insulation while r_1 and r_2 represent its inner and outer radii. Capacitance C_l and conductance G are associated to the SCT; in particular δ is its thickness and ρ(V) and ε_{rt}(V) are respectively the nonlinear resistivity and the nonlinear relative permittivity of the SCT. The nodes (1,2,...,N) in fig.1.b represent the sections where the potential has to be evaluated. Although this model suffers from some theoretical drawbacks, it can be used for the approximate evaluation of the distribution of the electrical field in a termination. The design of a reliable cable termination takes advantage of an accurate evaluation of the electrical field distribution inside the component performed by computer simulation. Nevertheless, due to the production process, the exact knowledge of the materials characteristics is practically impossible. This brings about variation of the potential and field distribution along the cable with respect to the "nominal" one obtained giving "nominal" values to the SCT parameters. These latter include dimensional parameters (diameter, length and so on) and electrical parameters (permittivity, resistivity).

II.b EA details

The varying parameters considered in the WCA are the insulation relative permittivity ε_r spanning the range [2,7] and the SCT thickness δ in the range [0.1,0.3]cm.

Let it be given the circuit of fig.1b consisting in N cells, with N node potentials. We want to determine the upper and lower bounds for such potentials embedding all the possible solutions admitted by the variations of, say M, parameters.

The EA starts with a randomly generated population of "individuals", with each individual is identified by a vector of M real values, each one chosen within the ranges of variation initially settled for the varying parameters. The real-valued chromosome is built by collecting M real numbers $[p_1,p_2,...,p_M]\in\Re^M$, each one representing one of the possible values assumed by the corresponding varying parameter in the related variation interval, $[p_1,p_2,...,p_M]\subset[P_1,P_2,...,P_M]$. The adopted real-coded representation of the genes is dictated by the nature of the problem and by the need of overcoming some problems connected to the use of a binary representation of the individuals (low precision and long computation time required in multidimensional, high precision numerical problems). Each set of parameters $[p_1,p_2,...,p_M]$ corresponds to a set of N node potentials. To each individual the *performance index*, or *fitness*, has been assigned: it is based on the values the potential assumes in the N nodes by taking the related set of parameters. In particular, the fitness is calculated by considering the number of nodes where the potential is better than the best solution obtained up to the current generation. To clarify this concept, let us refer to the evaluation of the upper bound of the worst case; the same operation can be repeated for the lower bound, thus getting the whole inner solution.

At the beginning, the current upper bound, collecting the potentials of the N nodes, is fixed at the one obtained with the nominal values of the parameters. The fitness assigned to each individual is an integer number in the range [0,N]: it is the number of nodes where the individual overcomes the current upper bound. At this point, the current upper bound is updated by taking, for each node, the maximum value among the ones given by the individuals of the current population and the nominal solution. Afterwards, the population is evolved (selection+crossover+mutation) and the new population is obtained. Thus, the fitness of each individual is updated looking at the number of nodes where the individual overcomes the current upper bound and the upper bound itself is newly updated and the evolution continues. In brief, in place of blindly taking sets of parameters within the tolerance ranges as in MC WCA, the new trial sets of parameters are generated using the chromosomes' fitness to drive the reproduction of the existing individuals: the higher the fitness the higher the surviving probability of the genetic wealth of the individual. This makes the proposed EA-based approach a kind of intelligent MC analysis. The peculiarity of this approach consists in the fact that the current upper bound does not correspond to a given set of parameters: it is a collage of the best parts of all the solutions explored during the evolution. This increases notably the selection capability of the EA.

The number of cells chosen to model the SCT is a factor that greatly influences the convergence of the algorithm. A too coarse discretization (N small) leads to a poor diversification of the individuals and to a flattening of the population with a weak evolutionary pressure. On the other side, a high N involves a long simulation time for each individual. Consequently, the number of cells can be treated as a tuning parameter of the EA. In the final paper a tradeoff solution will be suggested and the results obtained using a different definition of the fitness function will be presented.

Standard EA operators (selection, crossover and mutation) have been used: some detail will be provided in the final version of the paper.

The algorithm has been implemented in MATLAB®. In the next section numerical results are presented and discussed.

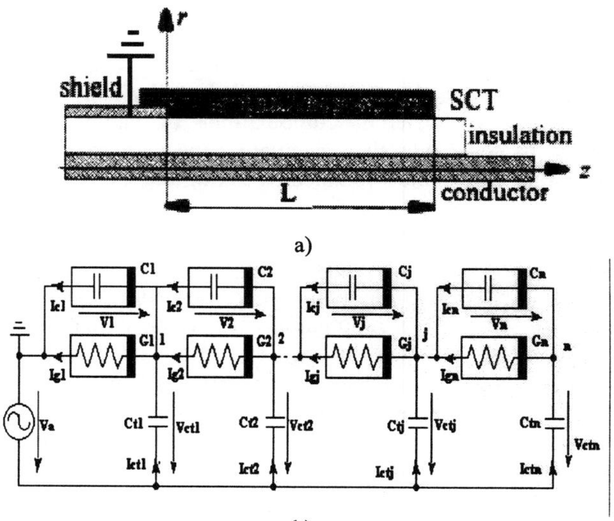

b)

Figure 1. a) Schematic setup of a cable termination based on a heat shrinkable SCT; b) lumped parameter network adopted for the simulation.

III. EXAMPLE

In fig.2 the non linear functions used to model the resistivity and permittivity of the SCT are shown. Such curves have been obtained by interpolating experimental data.

Fig.3 shows intermediate results obtained by MC and EA approaches. Fig.3a shows the improvement of the upper and lower bounds obtained by MC WCA. Each marker indicates a node of the equivalent circuit model; it has been used a number of nodes equal to N=30. Fig.3b shows the enlargement of the range obtained at each node after 20 generations of a population of 25 individuals. In this figure the nominal solution has been also plotted, to put in evidence the large variation the potential sustains due to the variations regarding the relative permittivity of the insulation and the SCT thickness. A comparison between the final results obtained by 500 MC trials and 20 EA generations of a 25 individuals population. The meaningful comparison has been performed for the same number of evaluation of the e.f, namely the N nodes potential: the computation time required by the evaluation of the e.f. is in fact quite high with respect to the one the EA needs to perform selection, mutation and crossover. Thus, the comparison we made is for the same computation time. Fig.4 shows that the EA approach offers better performances, since it allows obtaining a better evaluation of the TWC, namely an interval of variation of the potential that in every node is larger than the one achieved by the MC analysis. Such discrepancy is more sensible for the lower bound, as can be deduced from fig.4. In the final paper a further investigation of the parameters that much more affects the potential distribution will be provided.

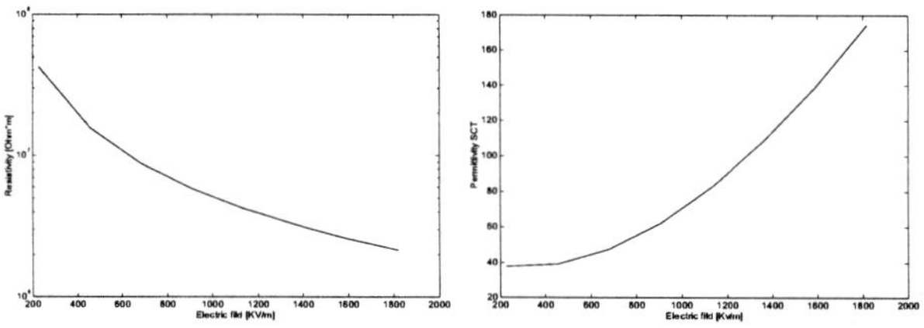

Figure 2. Non linear characteristics adopted for resistivity and permittivity.

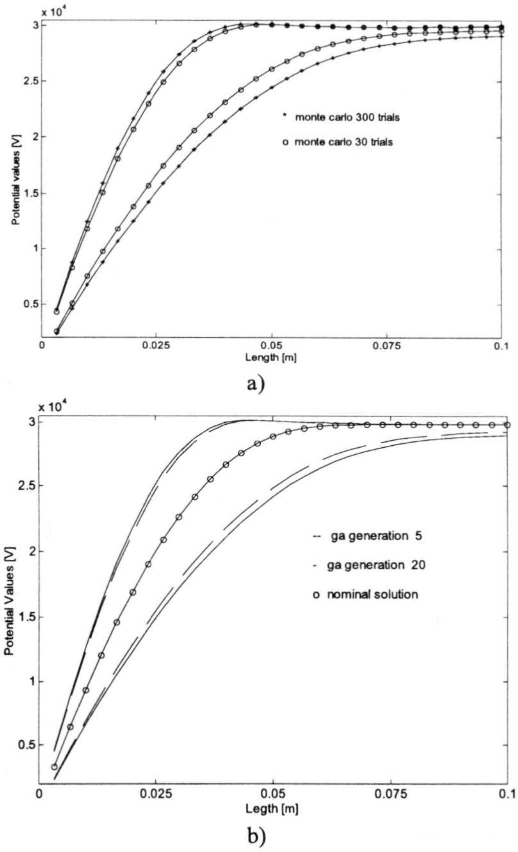

Figure 3. Solution improvement during: a) the EA run, b) the MC simulation.

IV. Conclusions

In this paper a new approach to the worst-case tolerance analysis of non-linear circuits is presented, which is best suited for studying systems represented by implicit models/equations and whose parameters spans wide ranges. An underestimation of the true worst case is calculated by means of an evolutionary algorithm. It seems to be an interesting alternative to the Monte Carlo methods because it is possible to achieve, with less evaluations of the performance function of interest, the same accuracy ensured by the Monte Carlo approach. In a forthcoming paper the effect of the dispersion of such data on the parameters involved in the interpolating functions and consequently on the potential distribution will be also investigated.

V. References.

[1] Michalewickz, Z.: Genetic Algorithms + Data Structures = Evolution Programs, Artificial Intelligence, Springer-Verlag, 1992.

[2] Lupò, G. et al.: “Electric field calculation in HV cable terminations employing heat-shrinkable composites with non linear characteristics“, Proc. of the 4th Int. Conf. on Properties and Applications of Dielectric Materials, 1994, Vol.1, pp. 278 -281.

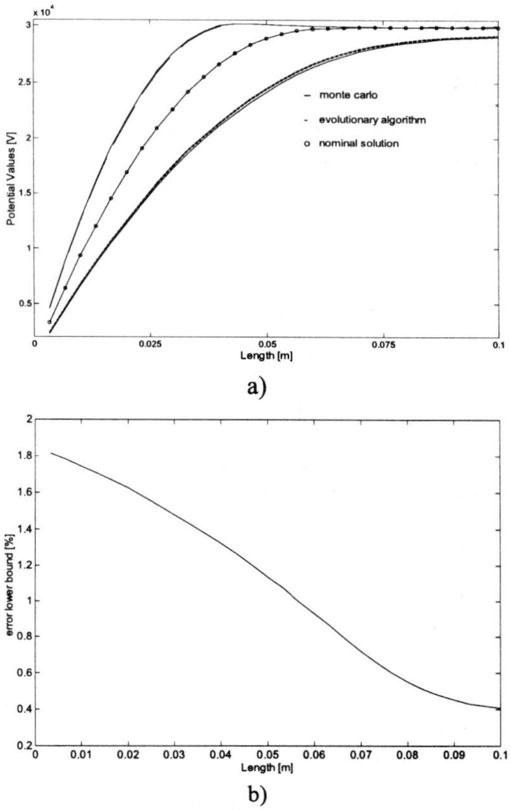

Figure 4. a) Final results by EA and MC; b) lower bound discrepancy between EA and MC results.

Computation of Projections and Eigen Distribution in Half-Planes and Disks

Mohammed A. Hasan† and Ali. A. Hasan‡
†Department of Electrical & Computer Engineering, University of Minnesota Duluth
‡College of Electronic Engineering, Bani Waleed, Libya

Abstract

This paper provides ecient solutions to the problem of computing projections onto invariant subspaces of a nonsingular matrix corresponding to eigenvalues in specified region in the complex plane such as the inside or the outside of a circle, half planes or horizontal and vertical strips. Additionally, new form of matrix decompositions called "sector factorization" is introduced in which the polar and sign decompositions are special cases. The main results are based on properties of matrices whose squares are the identity matrix. Several power-like methods for computing the proposed factorizations, particularly the sign and polar decompositions are developed.

1. Introduction

Zero distribution of a polynomial is a problem of importance in several engineering applications. The relative stability of matrices and polynomials is a problem is of interest in control theory and filter design in signal/image processing. There are several tests in the literature to determine the number of zeros in some specified regions without actually solving for zeros of real or complex polynomial. In most traditional methods some tables are formed and some determinants are computed. Then the distribution of zeros are related to the number of sign changes in these determinants. There are many cases where these tests fail especially when some of these determinants are zero. In these situations a procedure called the -method is normally used. This method is generally inconvenient to use and its success for a generic problem is not guaranteed especially for systems of large dimensions. In this paper, we develop numerical techniques for testing stability inside circles, rectangle and sectors. The matrix sector function provides a natural setting for studying this problem. These techniques have the advantage of alleviating the zero array elements which are associated with Routh criterion and Jury test.

In the literature, projections into certain subspaces are associated with the matrix sign function [1], [2]. The concept of the matrix sign function has also been used for block diagonalization of complex matrices [3]. The matrix sign function was utilized in [4] for eigenvalue separation in dierent regions in the complex plane. A comprehensive treatment of the matrix sign function and its applications can be found in [1]-[3]. The theory and algorithms for the matrix sign function are presented in [5]-[10]. Numerically ecient methods for computing the polar decomposition of nonsingular complex matrices can be found in [11]-[15].

2. Projections and Eigen Distribution in Half-Planes Using Involutions

In this section we describe two results that demonstrate the basic role of involutions in block diagonalization of matrices in general.

Proposition 1. *Let S and $A \in \mathcal{C}^{p\times p}$ be nonsingular matrices such that $S^2 = I_p$, and $N = SA = AS$, then for any $\in \mathcal{C}$, the following holds*

$$H_1 = \begin{bmatrix} I\ _p & A \\ A & I\ _p \end{bmatrix} = \begin{bmatrix} I_p & I_p \\ S & -S \end{bmatrix} \begin{bmatrix} I\ _p + N & 0 \\ 0 & I\ _p - N \end{bmatrix} \begin{bmatrix} I_p & I_p \\ S & -S \end{bmatrix} \quad (1a)$$

Similarly,

$$H_2 = \begin{bmatrix} A & I\ _p \\ I_p & A \end{bmatrix} = \begin{bmatrix} I_p & I_p \\ S & -S \end{bmatrix} \begin{bmatrix} A\ + S & 0 \\ 0 & A\ - S \end{bmatrix} \begin{bmatrix} I_p & I_p \\ S & -S \end{bmatrix}^{-1} \quad (1b)$$

Proposition 2. *Let $A \in \mathcal{C}^{p\times p}$ be nonsingular matrix, then there exists a unique matrix $S \in \mathcal{C}^{p\times p}$ satisfying $S^2 = I_p$, $SA = AS$ such that $|\ _i(A + S)| > |\ _i(A - S)|$, for $i = 1\ 2 \cdots p$.*

From these propositions, it follows that for any method which is based on eigenvalue dominance such as the power method or the QR algorithm, one expects that such algorithms will converge if there is a matrix S for which $|\ _i(A\ + S)| > |\ _i(A\ - S)|$ if applied to H_1 or the matrix S for which $|\ _i(I\ _p + AS)| > |\ _i(I\ _p - AS)|$ if applied to H_2. Clearly, when > 0, the resulting matrix S will be the matrix sign function of A. If is any complex number, then the computed matrix satisfies $S^2 = I_p$, $SA = AS$ and all eigenvalues of $N = AS$ are in one side of the half plane determined by the straight line passing through the origin and perpendicular to the straight line passing through and $-$. Particularly, the number of positive eigenvalues of A in the right half plane will be equal to the algebraic multiplicity of the eigenvalue one of S. This can be shown to be $\frac{p+trace(S)}{2}$.

Hence, one can obtain the matrix S from the QR factorization of large powers of H_1 or H_2. Specifically, let Y be either of the matrices H_1 or H_2, and form the QR factorization $Y^k = Q_k R_k$, where Q_k is a unitary matrix and R_k is upper triangular matrix. As k tends to ∞, the matrix Q_k can be shown to converge to a matrix Q such that $Q^* Y Q$ is upper triangular. Moreover, if Q is partitioned so that $Q = \begin{bmatrix} Q_{11} & Q_{12} \\ Q_{21} & Q_{22} \end{bmatrix}$, where Q_{ij} are of size p, then the matrix S can be approximated as $S \approx Q_{11}^{-1} Q_{21} = -Q_{12}^{-1} Q_{22}$. The limiting matrices Q_{11} and Q_{12} can be shown to be invertible. Once the matrix S is computed, then the matrix A can be factored as $A = SN = NS$, where $N = AS = SA$. The desired projections are then obtained as $\frac{I_p \pm S}{2}$. Bases of the

0-7803-7761-3/03/$17.00 ©2003 IEEE

orthogonal subspaces can be obtained from the QR decomposition of these projections, i.e., if $\frac{I_p+S}{2} = QR$, then the first $\frac{p+trace(S)}{2}$ columns and the last $\frac{p-trace(S)}{2}$ columns of Q span the positive and negative invariant subspaces of A, respectively.

It is interesting to note that several useful variations of Newton's method for computing invariant subspaces of a matrix in dierent half-planes can be obtained. For example, if A has no eigenvalues on the real line, then the iteration

$$X_0 = A \quad X_{k+1} = \frac{1}{2}\{X_k - X_k^{-1}\} \tag{2a}$$

converges to a matrix S satisfying $S^2 = I_p$, $SA = AS$, and all eigenvalues of SA are in the upper half plane. This observation can be extended to find invariant subspaces of dierent regions of the complex plane. For example, the iteration

$$X_0 = A \quad \neq 0 \quad X_{k+1} = \frac{1}{2}\{X_k + \ {}^2 X_k^{-1}\} \tag{2b}$$

converges to a matrix S satisfying $S^2 = \ {}^2 I_p$, $SA = AS$, and all eigenvalues of SA are all in one side of the half-plane defined by the line passing through the origin and . Note that if A has no eigenvalues on the boundary of this half plane, then the number of positive and negative eigenvalues of S represent the number of eigenvalues of A in the upper and lower half planes, respectively. Propositions 3 and 4 can be utilized to derive algorithms for computing S as follows. Assume that is any complex number and for each positive integer k set

$$\begin{bmatrix} I_p & A \\ A & I_p \end{bmatrix}^k = \begin{bmatrix} H_1(k) & H_2(k) \\ H_2(k) & H_1(k) \end{bmatrix} \tag{3}$$

It follows from (3) that as $k \to \infty$, we have the following limits:

$$H_1(k)^{-1}H_2(k) \to S$$

$$H_1(k)^{-1}H_1(k+1) \to I_p + N$$

where $N = AS$. Stable and ecient implementations of these limits will be developed.

3. Higher Order Methods for Subspace Computation in and Outside a Circle

The general eigen-problem is one of the fundamental problems in matrix computations. A number of methods have been developed in the last few decades for computing the eigenvalues and eigenvectors of a dense matrix, notably, the QR method [16]-[18], Cuppen's divide-and-conquer method for Hermitian matrices [19], and Jacobi method [20, Chapter 8]. In recent years, there has been intensive interest in developing algorithms that are iterative in nature and rely on a few operations as their building blocks at each iteration. Those operations have already been successfully implemented on many parallel architectures, and portable parallel library is now available for their ecient computation [21].

We will use $P_{|\lambda|<|b|}$, and $P_{|\lambda|>|b|}$ to denote the orthogonal projections onto the regions described by $|z| < |b|$ and $|z| > |b|$ in the complex plane, respectively. An elegant method for computing those invariant subspaces is the matrix sign function method, a simple scheme of which is the following iteration [3]:

$$A_{j+1} = \frac{A_j + A_j^{-1}}{2} \quad j = 0\ 1 \quad A_0 = A \tag{4}$$

In this paper, new implementation of (4) will be developed for ill-conditioned matrices and other matrices which have eigenvalues near the imaginary axis. Specifically, for each integer $r \geq 2$, an rth order convergent algorithm will be derived. To establish this goal, let A be a nonsingular matrix and consider the sequence of matrices defined by

$$S_k = (b^k I_p - A^k)(b^k I_p + A^k)^{-1} \tag{5}$$

then the eigenvalues of S_k given by $\{\frac{b^k-\lambda_j^k}{b^k+\lambda_j^k}\}_{j=1}^p$ converge to 1 or -1 as $k \to \infty$. Thus S_k is bounded for all suciently large k. It can be shown that the sequence S_k converges to a matrix S satisfying $S^2 = I_p$, and $SA = AS$, where $S = P_{|\lambda|<|b|} - P_{|\lambda|>|b|}$. Moreover, S and A have the the same invariant subspaces inside and outside a circle of radius $|b|$ and centered at the origin. If (5) is computed directly using power matrices, over- and under-flow will occur. Fast implementation of computing the limit of the sequence $\{S_k\}_{k=1}^{\infty}$ which also avoids the problem of over- and under-flow will be given next.

$$\begin{aligned} S_0 &= (bI_p - A)(bI_p + A)^{-1} \\ S_{k+1} &= \{(I_p + S_k)^r - (I_p - S_k)^r\}\{(I_p + S_k)^r + (I_p - S_k)^r\}^{-1} \\ k &= 0\ 1\ \cdots\ K \end{aligned} \tag{6}$$

It can be shown that S_k satisfies the following elegant error formula

$$\begin{aligned} (S_{k+1} + S)^{-1}(S_{k+1} - S) &= \{(S_k + S)^{-1}(S_k - S)\}^r \\ &= \{(S_0 + S)^{-1}(S_0 - S)\}^{r^k} \end{aligned} \tag{7}$$

From several numerical experiments, it was observed that for $r = 2$, a suitable $K = 5$, while $K = 3$ if $r = 3$. Once the desired convergence is obtained, the subspace projections are computed as $P_{|\lambda|<|b|} = \frac{I_p+S_{K+1}}{2}$ and $P_{|\lambda|>|b|} = \frac{I_p-S_{K+1}}{2}$. Note that the iteration in (6) is dependent on the threshold b through the initial guess S_0.

It is shown in [22] that matrix inversion in (6) can be avoided by transforming these algorithms to other forms that depend only on the computation of matrix-matrix multiplication and QR factorization.

4. Sector Factorization

Let $A \in \mathcal{C}^{p\times p}$ be a nonsingular complex matrix, where $\mathcal{C}$ is the field of complex numbers, then there exists a matrix $W_n(A)$ satisfying $W_n^n(A) = I_p$ such that $A = A_n W_n(A)$, $AW_n(A) = W_n(A)A$ and all eigenvalues of A_n are in one sector of the complex plane. Here I_p denotes the identity matrix of size p. Examples of such decompositions and their applications are the well known polar and the sector factorization which will be introduced and investigated in this section. One of the useful applications of these decompositions is the parallel computation of the eigendecomposition and singular value decomposition of very large matrices. The main additional advantage of the arising methodology is that EVD and SVD can be obtained using parallel algorithms that depend only on matrix multiplications and the QR factorization [22].

It will be shown that if A is a nonsingular matrix which does not have eigenvalues on the boundary of the sectors $\frac{(2k-1)\pi}{n} < \ < \frac{(2k+1)\pi}{n}$, for $k = 0\ 2\ \cdots\ n-1$, then A can be factored as

$$A = A_n W_n(A) \tag{8}$$

where all eigenvalues of A_n are in one sector and W_n is an nth root of the identity matrix such that $AW_n = W_n A$. We will next present several identities which can be utilized to solve for A_n and W_n.

$$S_n(\begin{bmatrix} 0 & I_p & 0 & \cdots & 0 \\ 0 & 0 & I_p & \cdots & 0 \\ \cdots & \cdots & \cdots & \cdots & \cdots \\ 0 & 0 & \cdots & 0 & I_p \\ A^n & 0 & 0 & \cdots & 0 \end{bmatrix}) = \begin{bmatrix} 0 & A_n^{-1} & 0 & \cdots & 0 \\ 0 & 0 & A_n^{-1} & 0 & \cdots \\ \cdots & \cdots & \cdots & \cdots & \cdots \\ 0 & 0 & \cdots & 0 & A_n^{-1} \\ A_n^{n-1} & 0 & 0 & \cdots & 0 \end{bmatrix} \quad (9)$$

where $AA_n = A_nA$, and $A_n^n = A^n$. Note also that $A^n - A_n^n = {}^{n}_{i=1}(A - w^{i-1}A_n) = 0$ for some w, an nth root of 1. We also has developed the following link between W_n and the sector function

$$S_n(\begin{bmatrix} 0 & A & 0 & \cdots & 0 \\ 0 & 0 & A & \cdots & 0 \\ \cdots & \cdots & \cdots & \cdots & \cdots \\ 0 & 0 & \cdots & 0 & A \\ A & 0 & 0 & \cdots & 0 \end{bmatrix}) = \begin{bmatrix} 0 & W_n^{-1} & 0 & \cdots & 0 \\ 0 & 0 & W_n^{-1} & 0 & \cdots \\ \cdots & \cdots & \cdots & \cdots & \cdots \\ 0 & 0 & \cdots & 0 & W_n^{-1} \\ W_n & 0 & 0 & \cdots & 0 \end{bmatrix} \quad (10)$$

where $W_nA = AW_n$, $W_n^n = I_p$, $W_n^{n-1} = S_n(A)$, and all eigenvalues of AW_n lie in the sector $\frac{-\pi}{n} < \quad < \frac{\pi}{n}$. The importance of this identity is that any method for computing W_n will automatically translates into a method of computing sector factorization of A and the principal nth root of A.

It is interesting to note that the block Toeplitz matrix of the left hand side of (10) has the following factorization in terms of block Vandermonde and block diagonal matrices:

$$\begin{bmatrix} 0 & A & 0 & \cdots & 0 \\ 0 & 0 & A & \cdots & 0 \\ \cdots & \cdots & \cdots & \cdots & \cdots \\ 0 & 0 & \cdots & 0 & A \\ A & 0 & 0 & \cdots & 0 \end{bmatrix} = \begin{bmatrix} I_p & I_p & \cdots & I_p \\ S_n & wS_n & \cdots & w^{n-1}S_n \\ \cdots & \cdots & \cdots & \cdots \\ S_n^{n-1} & w^{n-1}S_n^{n-1} & \cdots & wS_n^{n-1} \end{bmatrix}$$

$$\times \begin{bmatrix} AS_n & 0 & \cdots & 0 \\ 0 & wAS_n & \cdots & 0 \\ \cdots & \cdots & \cdots & \cdots \\ 0 & 0 & \cdots & w^{n-1}AS_n \end{bmatrix} \begin{bmatrix} I_p & I_p & \cdots & I_p \\ S_n & wS_n & \cdots & w^{n-1}S_n \\ \cdots & \cdots & \cdots & \cdots \\ S_n^{n-1} & w^{n-1}S_n^{n-1} & \cdots & wS_n^{n-1} \end{bmatrix}^{-1} \quad (11)$$

Consequently, a power method can be developed to compute the matrix sector function. Once the matrix $S_n(A)$ is obtained, then projections onto each sector can be computed. The columns of the orthogonal factor of the QR decomposition of these projections form bases for corresponding eigenspaces.

Next we consider two familiar types of decompositions.

4.1 Computation of the Sign Decomposition

This is a special case of the sector factorization when $n = 2$. The concept of the matrix sign function can be used to uniquely decompose the matrix A as $A = A_2S_2(A)$, where A_2 is a matrix with eigenvalues in the right half plane and S_2 is involution. Iterations for computing A_2 will be developed using the identity

$$S_2(\begin{bmatrix} 0 & I_p \\ A^2 & 0 \end{bmatrix}) = \begin{bmatrix} 0 & A_2^{-1} \\ A_2 & 0 \end{bmatrix} \quad (12)$$

from which $S_2(A) = AA_2^{-1}$.

4.2 Computation of the Polar Decomposition

The polar decomposition is a useful tool in many applications. For example, in situations where it is desirable to minimize

$$||A - BQ||_F$$

over unitary matrices, where A and B are known matrices and $||X||_F$ denotes the Frobenius norm of X. It can be shown that the solution matrix Q is the polar factor of B^*A. The techniques of this research plan will be employed to compute the polar decomposition of any matrix. To establish this objective, let A be an $p \times p$ complex matrix, then A can uniquely be decomposed as $A = UH$, where U is unitary and H is positive semidefinite. We will use the following identity

$$S_2(\begin{bmatrix} 0 & A \\ A^* & 0 \end{bmatrix}) = \begin{bmatrix} 0 & U \\ U^* & 0 \end{bmatrix} \quad (13)$$

to develop linearly convergent method for U. This can be seen from the relations $AU^* = UA^*$, and $A^*U = U^*A$ which indicate that both are similar to H. Since H is positive definite, all its eigenvalues are in the right half plane. Analogous relations describing the relation between the matrix sign function and the polar decomposition are as follow:

$$S_2(\begin{bmatrix} 0 & I_p \\ AA^* & 0 \end{bmatrix}) = \begin{bmatrix} 0 & UH^{-1}U^* \\ UHU^* & 0 \end{bmatrix} \quad (14)$$

and

$$S_2(\begin{bmatrix} 0 & I_p \\ A^*A & 0 \end{bmatrix}) = \begin{bmatrix} 0 & H^{-1} \\ H & 0 \end{bmatrix} \quad (15)$$

These two relations will be utilized to compute the factors U and H of the polar decomposition of A.

Since the matrix $\begin{bmatrix} 0 & A \\ A^* & 0 \end{bmatrix}$ is Hermitian for any matrix A, we can apply the matrix inverse-free algorithm developed in [22] for computing the polar decomposition of any matrix A be it rectangular or singular.

Next, we present a power like method for block decomposition of a matrix. This could be useful in the derivation of fast algorithms for computing the polar factors of a matrix.

Theorem 5. *Let A be nonsingular matrix of size p such that $A = UH$, where $U^*U = I_p$, and H is positive definite, then for any complex number the following identity holds*

$$\begin{bmatrix} I_p & A^* \\ A & I_p \end{bmatrix} = \begin{bmatrix} I_p & I_p \\ U & -U \end{bmatrix} \begin{bmatrix} I_p + H & 0 \\ 0 & I_p - H \end{bmatrix} \begin{bmatrix} I_p & I_p \\ U & -U \end{bmatrix}^{-1} \quad (16)$$

Since $(I_p + H) > (I_p - H)$ for any positive value , the block QR algorithm applied to the matrix $\begin{bmatrix} I_p & A^* \\ A & I_p \end{bmatrix}$ will converge to $\begin{bmatrix} I_p + H & 0 \\ 0 & I_p - H \end{bmatrix}$, which reveals the factor H of the polar decomposition of A. Using (16) along with the identity $\begin{bmatrix} I_p & I_p \\ U & -U \end{bmatrix}^{-1} = \frac{1}{2}\begin{bmatrix} I_p & U^* \\ I_p & -U^* \end{bmatrix}$, we obtain

$$\begin{bmatrix} I_p & A \\ A^* & I_p \end{bmatrix}^k = \begin{bmatrix} I_p & I_p \\ U & -U \end{bmatrix} \begin{bmatrix} (I_p + H)^k & 0 \\ 0 & (I_p - H)^k \end{bmatrix} \begin{bmatrix} \frac{1}{2}I_p & \frac{1}{2}U^* \\ \frac{1}{2}I_p & \frac{-1}{2}U^* \end{bmatrix}$$

$$= \begin{bmatrix} \frac{1}{2}\{(I_p + H)^k + (I_p - H)^k\} & \frac{1}{2}\{(I_p + H)^k - (I_p - H)^k\}U^* \\ \frac{1}{2}U\{(I_p + H)^k - (I_p - H)^k\} & \frac{1}{2}U\{(I_p + H)^k + (I_p - H)^k\}U^* \end{bmatrix} \quad (17)$$

Now assume that is a positive number and let

$$\begin{bmatrix} I_p & A \\ A^* & I_p \end{bmatrix}^k = \begin{bmatrix} H_1(k) & H_2(k) \\ H_2^*(k) & H_3(k) \end{bmatrix} \quad (18)$$

It follows from (17) that as $k \to \infty$, we have the following limits:

$$\begin{aligned} & H_1(k)^{-1}H_2(k) \to U^* \quad H_2^*(k)H_1(k)^{-1} \to U \\ & H_3(k)H_2^*(k)^{-1} \to U \\ & H_1(k)^{-1}H_1(k+1) \to I_p + H \\ & H_3(k)^{-1}H_3(k+1) \to I_p + H \end{aligned} \quad (19)$$

If (19) is implemented directly, over- or under-flow will occur when k is large. It turns out that alleviating this overflow will lead to ecient algorithms for the computation of U and H. It is our goal in this research plan to develop ecient implementation of the above theorem and the limits in (19).

4.3 Example

In this example, we compute the polar decomposition of a 6×6 matrix using the limits in (19). Let A be a matrix given by

$$A = \begin{bmatrix} 1\ 1045 & -1\ 0008 & 0\ 1803 & -0\ 5446 & -0\ 2729 & 0\ 3308 \\ 0\ 1870 & -1\ 0230 & 1\ 4922 & -0\ 6000 & 0\ 5425 & -0\ 1397 \\ 1\ 0858 & 0\ 0945 & 0\ 5829 & -1\ 6362 & 0\ 9970 & 1\ 6424 \\ -0\ 5982 & 0\ 2179 & 0\ 9747 & 0\ 8756 & -0\ 2455 & -1\ 4850 \\ 0\ 4898 & -0\ 7644 & 0\ 3195 & -0\ 6168 & -1\ 1470 & -0\ 9670 \\ 0\ 3975 & -0\ 1166 & -1\ 5252 & -1\ 0984 & -0\ 4976 & -0\ 9235 \end{bmatrix}$$

This matrix is generated using the Matlab function *rand*. The eigenvalues of the matrix A are given by the set $\{1\ 4027 + 1\ 9351i\ 1\ 4027 - 1\ 9351i\ 0\ 5902\ -1\ 8665\ -1\ 0299 + 0\ 0493i\ -1\ 0299 - 0\ 0493i\}$. The computation of the decomposition $A = UH$ where U is unitary and H is positive definite, is carried out using Matlab. The computed matrix H is

$$H = \begin{bmatrix} 1\ 2784 & -0\ 5241 & 0\ 0027 & -0\ 9928 & -0\ 0967 & 0\ 5345 \\ -0\ 5241 & 1\ 3885 & -0\ 4528 & 0\ 4281 & 0\ 2702 & 0\ 1962 \\ 0\ 0027 & -0\ 4528 & 2\ 3583 & 0\ 1549 & 0\ 4292 & 0\ 0496 \\ -0\ 9928 & 0\ 4281 & 0\ 1549 & 2\ 0685 & -0\ 1959 & -0\ 4210 \\ -0\ 0967 & 0\ 2702 & 0\ 4292 & -0\ 1959 & 1\ 3781 & 0\ 8846 \\ 0\ 5345 & 0\ 1962 & 0\ 0496 & -0\ 4210 & 0\ 8846 & 2\ 3523 \end{bmatrix}$$

The eigenvalues of H are given by the set $\{3\ 4826\ 0\ 3818\ 0\ 6854\ 2\ 6507\ 1\ 0951\ 2\ 5285\}$. The matrix U is orthogonal satisfying $||U^* U - I_6||_2 = 1\ 2043(10^{-4})$ and with eigenvalues given by $\{1\ 0000\ 0\ 5687 + 0\ 8226i\ 0\ 5687 - 0\ 8226i\ -0\ 8837 + 0\ 4680i\ -0\ 8837 - 0\ 4680i\ -1\ 0000\}$. It can be easily checked that all these eigenvalues are on the unit circle as expected.

5. Conclusion

The main objective of this paper is to develop higher order numerical methods of computing projections into invariant subspaces of a non-singular matrix. These projections can be used to determine the number of zeros or eigenvalues in a given region of the complex plane without actually computing any zero or eigenvalue. Power-like iterations for computing the matrix factorizations such as the matrix sign function, the polar decomposition and the sector factorization of complex matrices in general are developed. It should be noted that the matrix S defined in Sections 2-4 can also be used for block diagonalization of a matrix A. For example if $\frac{I_p+S}{2} = QR$ is the a QR factorization, then Q^*AQ is block diagonal. This issue in addition to rigorous numerical stability of the algorithms stated in this paper will be dealt with in a forthcoming paper. Simulations and numerical evaluation of some of the algorithms will also be established.

References

[1] C. S. Kenney and A. J. Laub, "The Matrix Sign Function," *IEEE Trans. Automatic Control,* Vol. 40, No. 8, pp. 1330-1348, August 1995.

[2] J. D. Roberts, "Linear Model Reduction and Solution of the Algebraic Riccati Equation by Use of The Sign Function," Internat. J. of Control, 32, pp. 677-687, 1980. (Reprint of Technical Report No. TR-13, CUED/B-Control, Cambridge University, Engineering Department, 1971).

[3] E. D. Denman and A. N. Beavers, "The Matrix Sign Function and Computation of Systems," *Appl. Math. Comput.*, Vol. 2, pp. 63-94, 1976.

[4] E. U. Stickel, "Separating Eigenvalues Using the Matrix Sign Function," *Linear Algebra Appl.*, 148, pp. 75-88, 1991.

[5] C. S. Kenney, A. J. Laub, and P. M. Papadopoulos, "A Newton-Squaring Algorithm for Computing the Negative Invariant Subspace of a Matrix," *IEEE Trans. Automatic Control,* Vol. 38, No. 8, pp. 1284-1289, August 1993.

[6] C. Kenney, and A. J. Laub, "Rational Iterative Methods for the Matrix Sign Function", *SIAM J. Matrix Anal. Appl.*, Vol 12., pp. 237-291, April 1991.

[7] R. Byers, "Numerical Stability and Instability in Matrix Sign Function Based Algorithms," In C. I. Byrnes and A. Lindquist, editors, Computational and Combinatorial Methods in Systems Theory, pp. 185-200. Elsevier (North-Holland), New York, 1986.

[8] R. Byers, C. He, and V. Mehrmann, "The Matrix Sign Function Method and the Computation of Invariant Subspaces," SIAM J. Matrix Anal. Appl., 18 pp. 615-632, 1997.

[9] P. Pandey, C. Kenney, and A. B. Laub, "A Parallel Algorithm for the Matrix Sign Function," International Journal of High Speed Computing, Vol. 2, No. 2, (1990) 181-191.

[10] C. Kenney, and A. J. Laub, "On Scaling Newton's Method for Polar Decomposition and the Matrix Sign Function, "SIAM J. Matrix Anal. Appl., 13, pp. 688-706, 1992.

[11] N J. Higham and P. Papadimitriou, "A parallel Algorithm for Computing the Polar Decomposition," Parallel Computing, 20, pp. 1161-1173, 1994.

[12] W. Gander, "Algorithms for Polar Decomposition", *SIAM J. Sci. Stat. Comput.*, Vol 11., No. 6, pp. 1102-1115, Nov. 1990.

[13] N. J. Higham and R. S. Schreiber, "Fast Polar Decomposition of an Arbitrary Matrix", *SIAM J. Sci. Stat. Comput.*, Vol 11., No. 4, pp. 648-655, July 1990.

[14] N. J. Higham, "Computing the Polar Decomposition with Applications", *SIAM J. Sci. Stat. Comput.*, Vol 7., No. 4, pp. 1160-1174, Oct. 1986.

[15] C. Kenney, and A. J. Laub, "Polar Decomposition and Matrix Sign Function Condition Estimates", *SIAM J. Sci. Stat. Comput.. Appl.*, Vol 12., No. 3, pp. 488-504, May 1991.

[16] J. Stoer and Bulirsch, Introduction to Numerical Analysis, Springer-Verlag, New York 1980.

[17] P. Henrici, Applied and Computational Complex Analysis, Vol. 1, John Wiley & Sons, New York, 1974.

[18] A. S. Householder, The Numerical Treatment of a Single Nonlinear Equation, McGraw-Hill, New York, 1970.

[19] J. Cuppen, A divide and Conquor Method for the Symmetric Tridiagonal Eigen-problem, Numer. Math., 36 (1981), pp. 177-195.

[20] G. H. Golub and C. G. Van Loan, Matrix Computations, 2nd ed., the John Hopkins University Press, Baltimore, 1989.

[21] ScaLAPACK, available online from http://www.netlib.org/scalapack/index.html

[22] M. A. Hasan and A. A. Hasan, "Fast Approximated Subspace Algorithms," Proceedings of the SSAP-2000, August 14-16, Pocono, Pennsylvania.

Fitting Considerations of Polynomial Device Models

Timo Rahkonen, Antti Heiskanen

Electronics Laboratory
Dept. of Electrical and Information Engineering, and Infotech Oulu
University of Oulu, PO Box 4500, 90014 Oulu, FINLAND
timo.rahkonen@ee.oulu.fi

ABSTRACT

In this paper, fitting and usefulness of polynomial device models are discussed. It is shown that a polynomial model can provide an accurate AC fit of device nonlinearities and hence give accurate results in distortion simulations up to -35 dBc IM3 levels. Moreover, polynomial modeling is applicable in Volterra analysis, which allows the visualization of the fine structure of the distortion. The effects of the fitting range are also discussed.

1. INTRODUCTION

For accurate distortion simulations, we need a device simulation model that has accurate higher order derivatives for its I-V and Q-V characteristics [1]. However, many of the earlier simulation models were written to model only the average current or charge, and even the first derivatives could have errors of tens of percents. Even with the latest simulation models, it is still a fact that the model functions that have been chosen to present the I-V and Q-V curves fix also the derivative functions, and the model may simply not have enough degrees of freedom to model some specific but real form of nonlinearity like non-constant Early voltage, or charge dependency of multiple voltages.

Polynomial models have not been very popular in device simulation models, mainly because their response grows towards infinity outside the fitting range. However, they have some very useful features, including:

- Fitting of a polynomial to measured data is a single-step operation requiring no optimization.
- The degree of modelling is easily extended, if needed.
- The normal function-based device models try to cover the entire I-V plane with limited number of control parameters. A locally fitted polynomial can provide more accurate higher-degree curvatures, that are needed for distortion simulations.
- Volterra analysis is based on the use of polynomial models, because the spectral regrowth is easiest to calculate using them. This study aims in the use of Volterra analysis, because it is the only analysis method known to authors that can give a per-component information of the different contributions of distortion.

2. POLYNOMIAL DEVICE MODELS

The idea of polynomial modeling is simply that each I-V and Q-V nonlinearity of the device is modeled as a polynomial function of the AC controlling voltages and junction temperature. For example, a 3rd-degree polynomial model of the I_D-V_{GS}-V_{DS} nonlinearity can be written as

$$\begin{aligned} i_D = {} & K_{100}v_{GS} + K_{200}\cdot v_{GS}^2 + K_{300}\cdot v_{GS}^3 \\ & + K_{010}v_{DS} + K_{020}\cdot v_{DS}^2 + K_{030}\cdot v_{DS}^3 \\ & + K_{110}\cdot v_{GS}\cdot v_{DS} + K_{210}\cdot v_{GS}^2\cdot v_{DS} + K_{120}\cdot v_{GS}\cdot v_{DS}^2 + \\ & K_{001}\cdot t_J + K_{101}\cdot t_J\cdot v_{GS} + K_{011}\cdot t_J\cdot v_{DS} \end{aligned} \tag{1}$$

where $v_{GS}= V_{GS}-V_{GSQ}$ and $v_{DS}=V_{DS}-V_{DSQ}$ are the AC voltages around the chosen quiescent point Q, K's are coefficients and the indexes in the coefficients K_{ijk} corresponds to the powers of the controlling voltages v_{GS}, v_{DS} and the junction temperature t_J, respectively [2]. Similarly, a 3rd-degree model of the gate charge is written as

$$\begin{aligned} Q_{GS} = {} & K_{10}\cdot v_{GS} + K_{20}\cdot v_{GS}^2 + K_{30}\cdot v_{GS}^3 \\ & + K_{01}\cdot t_J + K_{11}\cdot t_J\cdot v_{GS} \end{aligned} \tag{2}$$

Note that this equation describes charge, and the device current is a time derivative of this. In practice, charge can not be measured, but a polynomial model is built by fitting a polynomial to the measured capacitance values and then integrating the polynomial with respect to the controlling voltage.

3. ELECTRO-THERMAL FITTING

The fitting starts from a data file, where the currents and capacitances (or charges, if available) are stored as a function of bias point and junction temperature. At least two measurement temperatures and 10-20 bias points are needed. In this study, over 3000 bias points were available to study the behavior at different quiescent points and signal levels.

The fitting is based on a normal LMSE (least mean square error) polynomial fit,

$$c = (M^T \cdot M)^{-1} \cdot (M^T \cdot y) \tag{3}$$

where c is the vector of the solved coefficients, y is the vector containing the measured values from points i=1,2,...K, and M is the (typically non-square) matrix containing all the polynomial model functions evaluated at points i=1,2,...K. For example, matrices

$$M = \begin{bmatrix} v_1 & v_1^2 & v_1^3 & t_1 & v_1t_1 \\ v_2 & v_2^2 & v_2^3 & t_2 & v_2t_2 \\ : & : & : & : & : \\ v_K & v_K^2 & v_K^3 & t_K & v_Kt_K \end{bmatrix} \quad y = \begin{bmatrix} q_1 \\ q_2 \\ : \\ q_K \end{bmatrix} \quad c = \begin{bmatrix} K_{10} \\ K_{20} \\ K_{30} \\ K_{01} \\ K_{11} \end{bmatrix} \tag{4}$$

can be used to fit the polynomial shown in Eq. (2) to measured charge data $q_1, q_2, ..., q_K$. To make the fitting procedure flexible, the fitting Matlab program is written so that the chosen fitting function is given as vector of powers, so that the model in Eq. (2) is defined as an array

pw = [1, 2, 3, 0, 1;
0, 0, 0, 1, 1]

where the top row corresponds to the powers of the 1st variable (v_{GS}) and bottom row the powers of the 2nd variable (t_J). The ith column in M is built by filling it with a model function of form v_{GS}^pw(1,i).*t_J^pw(2,i), calculated for each selected pair of V_{GS} and t_J.

Normalization of the input variables can improve the condition number of M by several decades. This reduces the sensitivity to the numerical noise and allows also the use of a narrower fitting range.

4. SELECTION OF THE FITTING RANGE

The polynomial model is used as an AC model, and there is no need to cover the entire bias space. Instead, it is beneficial to select the measurement points only from the expected signal trajectories. The selection of the fitting range is a compromise in many ways, however, as:

- There must be at least as many data points as there are unknown coefficients.
- The fitting range should match the expected maximum signal swing in the simulated device.
- The data points in a multi-dimensional case must be chosen so that the model functions do not correlate strongly, as this results in an ill-conditioned matrix M and non-solvable group of equations.
- Last, the fitting range must not be too large, as the fit gets worse, if the polynomial is forced to model nonlinearities outside the actual operating range.

The obvious solution is to select the data points along the dynamic load line with maximum signal level, as shown in Fig. 1. However, the points can not be chosen from a single line alone, as this causes strong correlation between the v_{GS} and v_{DS} variables and makes model functions like v_{GS}^2 and $v_{GS}*v_{DS}$ linearly dependent. Thus, the fitting range must be chosen as a wider area along the expected load line.

Moreover, both the v_{DS} and v_{GS} ranges must be large enough to cover all voltages that will be visited during the simulation. For example, if the V_{GS}-I_D characteristic shown in Fig. 2 were fitted over the range of 3.5 to 5 V only, the fitted polynomial would most probably give a good fit over this range but would also suggest a quite large, but totally non-physical current at a gate voltage of 2.5 V. This would then give wrong results in deep class-AB or B where the gate voltage sweeps deep below the threshold voltage.

Based on the above, the points for the fitting are chosen as a logical AND operation of the following three rules, where v_{GS} and v_{DS} are the changes from the quiescent point, v_{GSpeak} and v_{DSpeak} are the gate and drain amplitudes, K is the slope of the load line (i.e. voltage gain) and W is the width of the fitting area (here K ~ -8).

$$\begin{aligned} |v_{GS}| &< v_{GSpeak} \\ |v_{DS}| &< v_{DSpeak} \\ |v_{DS} - K \cdot v_{GS}| &< W \end{aligned} \tag{5}$$

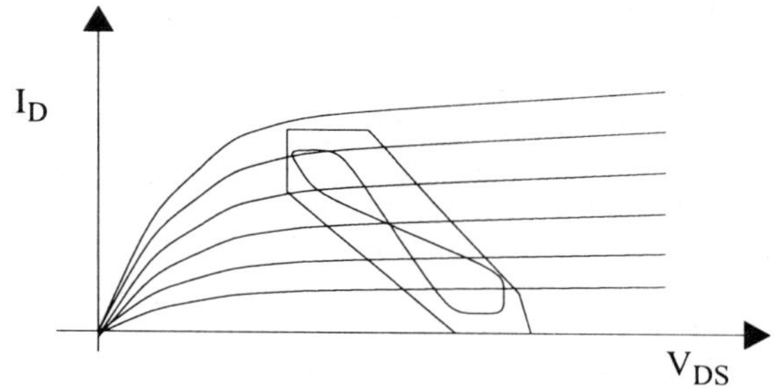

Figure 1. Dynamic load line and fitting range over ID-VDS coordinates.

Fig. 2 shows the I_D-V_{GS} curve of the modeled power transistor, fitted at V_{GSQ}=3.7V and V_{DSQ}=28V, calculated here along the load line. The figure illustrates a couple of typical features of polynomial models. First, the LMSE fit does not necessarily pass through any of the measured points. Second, the function is seen to diverge outside the fitting range below V_{GS} of 2.4 V. Third, the modeling of the cut-off region calls for high nonlinearity, and this results in slight oscillations in the response: a small bump at V_{GS}=2.7V, and rounding of the steep corner at V_{GS}=3.7V. However, the overall matching accuracy looks quite good.

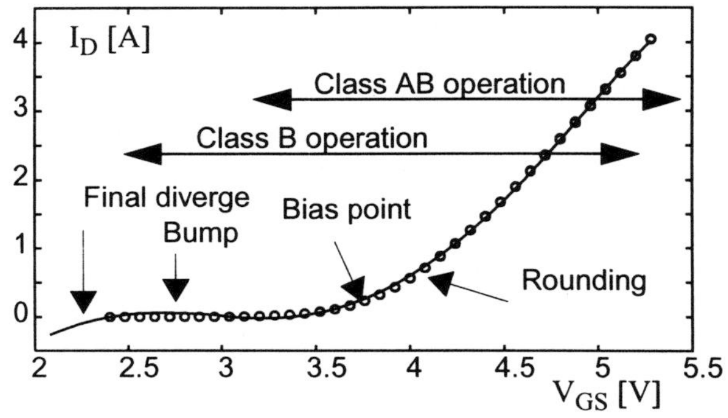

Figure 2. I_D-V_{GS} curve fitted along the load line using a 30-term, 5th-degree polynomial model for I_D-V_{GS}-V_{DS}-T nonlinearity. V_{GSQ}=3.7V, I_{DQ}=0.17A.

5. EXAMPLE

As an example, a polynomial model of a 30-W LDMOS MRF21030 power transistor is fitted. Instead of measured data, the current and capacitance values were simulated using Motorola's MET model [3] over a V_{DS} range of 0,0.5,...35V and V_{GS} range of 2.4, 2.48,... 5.5V, all at two temperatures. Fifth order electro-thermal polynomial models were fitted to the simulated I-V and C-V curves, so that the I-V_{GS}-V_{DS}-T polynomial has 30 terms and coefficients, and the C-V-T nonlinearities (C_{GS}, C_{GD}, C_{DS}) are all modeled by 10 coefficients. For reference, the Motorola MET model has ca. 20 coefficients affecting the shape of the I-V curve and 5-7 coefficients for each capacitance. Thus, the number of control parameters in the polynomial models is slightly larger, but the irrelevant terms can very easily be dropped out from the characterization procedure.

The fitting procedure was verified so that the transistor was biased to different operating points and the amplitude range of the fitting area was gradually increased, so that the maximum amplitude corresponds to a 5 to 6W average output power. Fig. 3 shows the peak and rms errors in class B biasing over the entire fitting range for different ranges of input amplitude at an average voltage gain K of -4. The rms error of the current over largest fitting range is seen to be ca. 15 mA and peak errors close to 50 mA. However, on the 4 A maximum output current shown in Fig. 2 this does not look very bad. The fit of the capacitances was much more accurate, as the peak errors remained below 1% of their quiescent value. The bias point and the largest signal swing corresponds to a class-B operation, where polynomial models are normally thought to be completely insufficient. For a class-AB biasing with I_{DQ}=0.75A and the same signal swings, the peak error of the I-V curve remains below 8% and the rms error below 2% of I_{DQ}.

It is interesting to study how the linearity of the device depends on the signal amplitude. In most papers about Volterra modeling (e.g. [4]), analytically solved higher order derivatives are used as the coefficients of the I-V and Q-V polynomials. In such a case, the coefficients are independent of the signal amplitude, and for increased amplitude, more terms are needed in the series expansion. With the curve fitting approach presented here, the degree of the model is fixed, but selection of the fitting area affects the values of the polynomial coefficients, as the lower-degree terms try to achieve the best possible match to the data. Surprisingly, the simulated distortion remains quite accurate, due to good overall matching of the model.

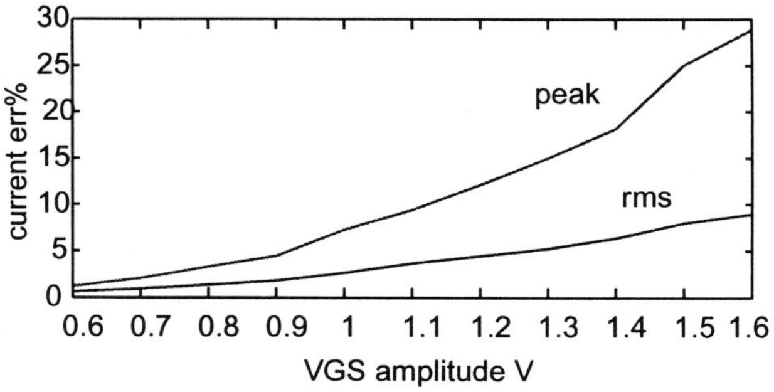

Figure 3. Peak and rms errors of the I-VGS-VDS-T model in class B over the entire fitting area, compared to I_{DQ}=0.17A.

Fig. 3 is drawn so that a new set of coefficients is calculated for each amplitude, and Fig. 4 shows some of the fitted coefficients of the I-V nonlinearity as functions of the increased input amplitude. For example, the linear gain K_{100} (corresponding to term $v_{GS}{}^1 v_{DS}{}^0 t_J{}^0$) increases from 1.0 to 1.07 S when the signal amplitude is increased. Note also that the values of the higher-degree coefficients decrease with increased signal level, as the small-signal coefficients would cause excessive curvature and poor match at high signal levels. Traditionally, large-signal effects are compensated by adding more fixed-coefficient high-degree terms, but here we try to make a best possible match with a fifth-degree model.

6. POWER SWEEP SIMULATIONS

As a verification of the model we will see how a 2-tone power sweep of the polynomial model matches with that of the original MET model. When the device is biased in class AB (V_{GS}=4.1, I_{DQ}=0.78A) the amplitudes of the fundamental, IM3 and IM5 match within 1 dB up to the input power of 12 dBm, corresponding to an output power of ca. 2 W per tone and -35 dBc IM3 level. Also the phases of the tones simulated by the MET and the polynomial model match within 10 degrees. Using bias-dependent polynomial models did not help to increase the power range any further from the 12 dBm input level, but this accuracy is very good, and the fitted polynomial model can be used in numerical Volterra analysis to find out the dominant causes of distortion, as described in [5] and [6]. In this case, the dominant cause of nonlinearity is the 3rd-degree I-V nonlinearity K_{300}.

When the device is biased to class B (V_{GSQ}=3.7V, I_{DQ}=0.17A), the device is already heavily nonlinear. The 5th order model should be sufficient, however, because a comparison between the MET model I_D-V_{GS} equation

$$i_D = \beta \cdot \left(V_{ST} \cdot \log\left(\exp\left(\frac{V_{GS} - V_{TO}}{V_{ST}}\right) + 1\right)\right)^r \quad (6)$$

and a 1-dimensional 5th-degree polynomial resulted in a sub-1dB error in IM3 and sub-3dB error in IM5 over the entire input range in class B operation. However, when the complete amplifier was simulated with the MET model and a 3-D polynomial model, the amplitude errors of the fundamental, IM3 and IM5 tones were 2, 9 and 15 dB, correspondingly. There are two reasons for this discrepancy: First, the selection of the fitting range for each power level was not automated but seems to be slightly misplaced, and second, also the 5th order harmonic balance simulation begins to break down due to higher order distortion.

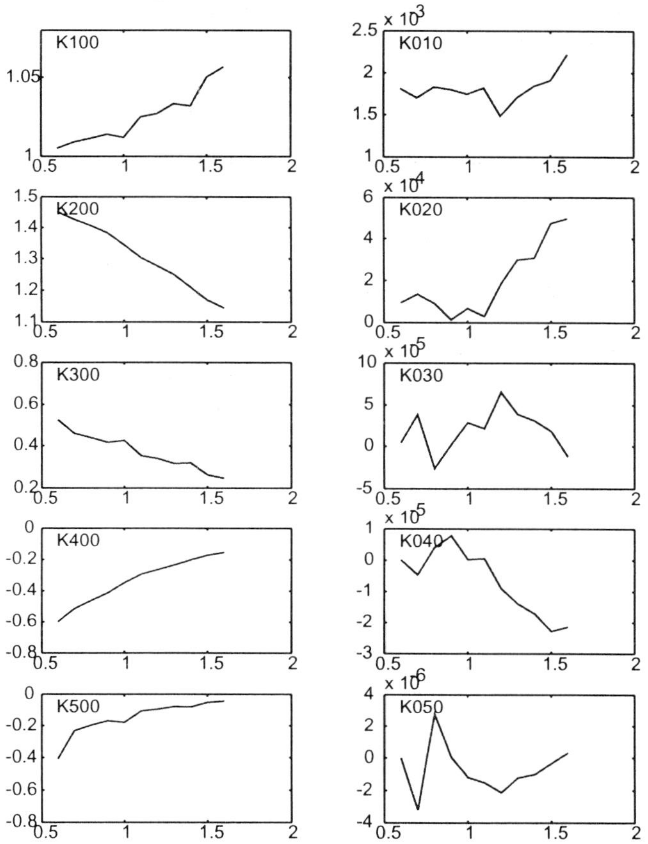

Figure 4. Fitted coefficients of the I-V nonlinearity. The x-axis is the amplitude at the gate and y-axis is the best-fit value of the coefficient K_{ijl} of term $v_{gs}^{i} v_{ds}^{j} t^{l}$.

According to the Volterra analysis the dominant cause of IM3 in class B operation at the maximum power level is the 3rd-degree I-V nonlinearity K_{300}, but due to the strong clipping, also the 2nd harmonic generated by K_{400} couples back to input and gets down-converted to IM3 in the 2nd-degree nonlinearity K_{200}. Another interesting phenomenon is that the sign of the extracted K_{300} changes between V_{GS}=3.7 and 4.1V, and is practically zero at V_{GS}=3.85V. This does not reduce IM3 dramatically, however, as now the down-converted 2nd harmonics dominate IM3.

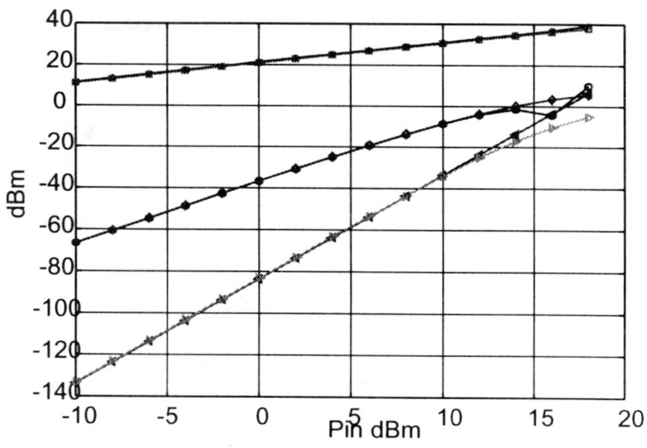

Figure 5. Comparison of the MET model and 5th-degree polynomial model in class AB. The model is valid up to -35 dBc IM level.

7. SUMMARY

In this paper, a polynomial curve-fitting technique was applied to modelling of an RF power transistor, resulting in a quick modeling approach that does not require numerical optimization. Further, it makes Volterra analysis possible, as it necessarily requires polynomial device models. The trade-offs related to the selection of the fitting range were discussed, and it was seen that the polynomial coefficients are not necessarily fixed to values of the higher order derivatives but may depend on the chosen fitting range. This is because the curvature of the low-order terms is needed to compensate the effects of the excluded high-degree nonlinearities.

The polynomial model was compared to the Motorola's large-signal MET model. When the device is reasonably linear, biased in class AB, the results match very well. A simplified nonlinear analysis suggests good matching even when biased in highly nonlinear operation in class B or C, but this has not yet been verified with simulations of the complete amplifier.

8. REFERENCES:

[1] S. Maas: Nonlinear Microwave Circuits. Artech House, 1988

[2] J. Vuolevi, T. Rahkonen: Distortion in RF Power Amplifiers. Artech House 2003.

[3] Motorola's Electro Thermal (MET) LDMOS Model, available at http://e-www.motorola.com/collateral/MET_LDMOS_MODEL_DOCUMENT_0502.pdf

[4] P. Wambacq, W. Sansen: Distortion Analysis of Analog Integrated Circuits. Kluwer Academic Publishers, 1998.

[5] A. Heiskanen, T. Rahkonen: 5th order multi-tone Volterra Simulator with Component-wise Output. IEEE ISCAS02, Phoenix,AZ, May 26-29,2002, Vol. 3, pp. 591-594.

[6] A. Heiskanen, J. Aikio, T. Rahkonen, A 5th order Volterra study of a 30W LDMOS power amplifier. Accepted to IEEE ISCAS 2003.

Modeling Skin Effect With Reduced Decoupled R-L Circuits

S. Mei and Y. I. Ismail

Electrical and Computer Engineering
Northwestern University
2145 Sheridan Road
Evanston, Illinois 60208, USA

ABSTRACT

On-chip conductors such as clock and power distribution networks require accurately modeling skin effect. Furthermore, to incorporate skin effect in the existing generic simulation tools such as SPICE, simple frequency independent lumped-element circuit models are needed. A rule-based R-L circuit model is proposed in this paper that predicts the skin effect in the entire frequency range accurately. This circuit model only contains a few parallel branches of resistors and inductors. A maximum error in impedance values of less than 3% is achieved using only two or three constant element branches.

1. INTRODUCTION

Multiple metal layers are used for interconnect in high performance VLSI circuits, with thicker layers on the top. To maintain low clock skew and low electromigration, the clock and power distribution networks in the interconnect layers are made wide. With clock frequencies in the gigahertz range, it is necessary to use accurate skin effect models to capture the behavior of the clock and power distribution networks, especially the parts in the top layers.

Skin effect is a well-known physical phenomenon. Several models [1]-[3] aim at finding the values of inductance and resistance as functions of frequency. Although these models are accurate, they are difficult to use with most available simulators [4]. Several other models aim at finding frequency independent lumped-element circuits to replace the original frequency dependent elements. Among these models are the volume filament model [5], ladder model [6], and the compact circuit models [7]-[8]. All these models can be directly used in generic simulators such as SPICE. However, the volume filament model and the ladder model have large number of elements and are expensive in terms of computational time. The compact circuit model in [7] needs an iterative procedure to find the best circuit elements. Although the model in [8] gives accurate results for several important interconnect structures, its accuracy for a general interconnect wire is not demonstrated.

The method described here starts with the volume filament model and proceeds to produce a reduced realizable and decoupled R-L equivalent circuit. Since the self/mutual inductance matrix is positive semi-definite [11], all the eigenvalues are non-negative. It is very easy to group the branches of new resistances and inductances and to find the equivalent resistance and inductance of each group. Furthermore, only a few branches of the equivalent resistance and inductance are much smaller than the rest. That makes it feasible to accurately approximate the original wire by a few resistor and inductor branches in parallel.

The rest of the paper is organized as follows. The approach to obtain decoupled R-L branches via similarity transformation is explained in section 2. Section 3 explains the rule of reduction and compares the frequency dependent resistance and inductance calculated from the reduced R-L circuits and those from the volume filament model. Conclusions are given in section 4.

2. SKIN EFFECT MODEL

At DC, the current in a conductor is evenly distributed over the cross section. Skin effect occurs when alternating current flows through a conductor. The alternating current induces a time varying magnetic field, which in turn induces electrical field and causes an uneven distribution of current over the cross section of the conductor. The electrical current tends to crowd toward the surface of the conductor, leading to an increase in the resistance and a decrease in the internal inductance.

If the cross section of the conductor is divided into much smaller sections, then the current distribution in each filament can be regarded as uniform. The frequency dependent resistance and inductance can be obtained by solving the currents in the inductively coupled R-L branches, which is the main idea of the volume filament model.

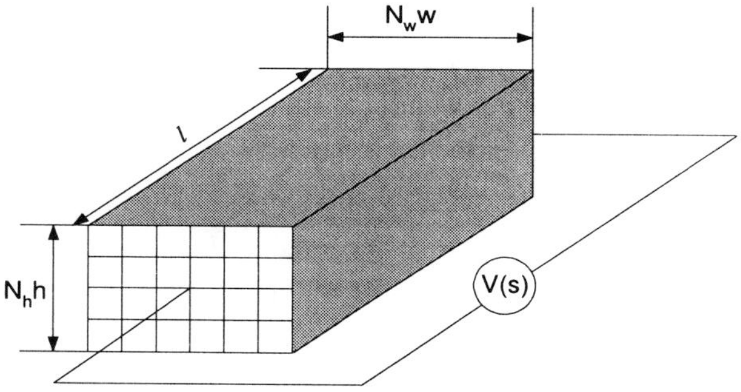

Fig. 1. A rectangular conductor driven by voltage $V(s)$

Fig. 1 shows a rectangular conducting wire driven by any voltage $V(s)$. To apply the volume filament model, the cross section is divided into $N_w \times N_h$ identical filaments with width w, height h, and length l for each filament. Assume N_w and N_h are large enough, so that the current density in each filament is essentially uniform. Denoting σ the conductivity of the wire, the resistance of each filament is given by

$$r = \frac{l}{\sigma w h} \tag{1}$$

The self-inductance for each filament is [9]

0-7803-7761-3/03/$17.00

$$L_{self} = 0.2l[\ln(\frac{2l}{w+h}) + 0.5 + 0.2235\frac{w+h}{l}]\mu H \quad (2)$$

Denoting d_{ij} the distance between the center axes of the ith and the jth filaments ($i \neq j$), the mutual inductance between the two filaments is [10]

$$M_{ij} = 0.2l[\ln(\frac{l}{d_{ij}} + \sqrt{1+\frac{l^2}{d_{ij}^2}}) - \sqrt{1+\frac{d_{ij}^2}{l^2}} + \frac{d_{ij}}{l}]\mu H \quad (3)$$

With these parameters, the conductor in Fig. 1 is equivalent to the circuit in Fig. 2 where $N = N_w N_h$.

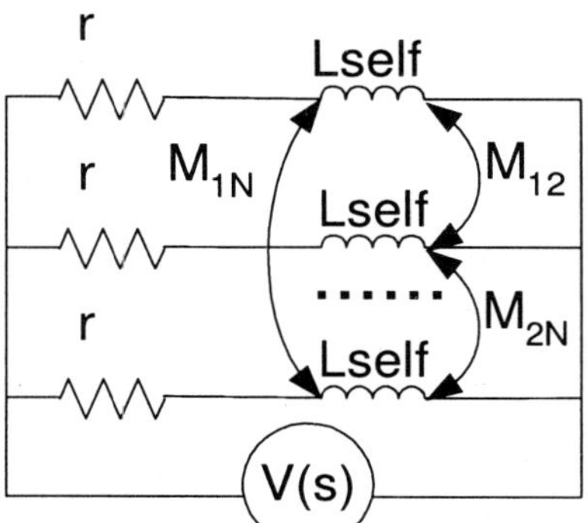

Fig. 2. Coupled R-L circuit

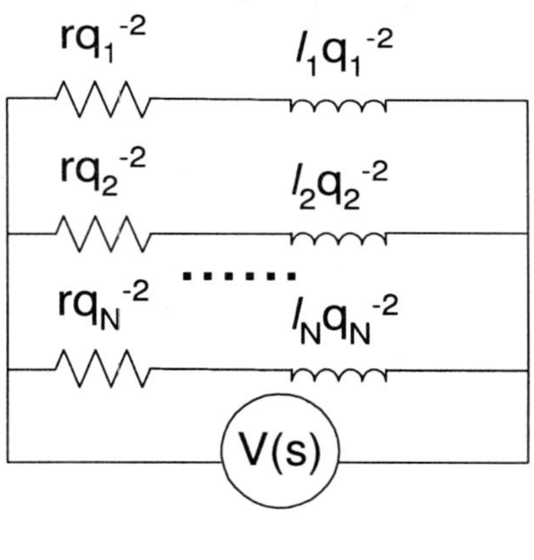

Fig. 3. Equivalent R-L circuit

To guarantee an almost even distribution of current in each filament, the dimension of the filament w and h are selected to be smaller than the skin depth at the highest frequency of interest. This criterion means large number of filaments, *e.g.*, 20, needed to replace wide conductors with significant skin effect. Non-uniform division of the conductor cross sectional area with fine division near the surface and coarse division away from the surface can reduce the total number of divisions. However, direct application of this volume filament model to simulate large interconnect circuits with significant skin effect is still formidable.

A more efficient way of using the volume filament model is to reduce the original coupled R-L circuit to a circuit of a few, *e.g.*, 3, *decoupled* R-L branches in parallel. The rest of this section and part of next section are dedicated to the explanation of a formal reduction technique to achieve this goal.

The voltage drop on any R-L branch in Fig. 2 is $V(s)$. According to Ohm's law, the voltage drop $V(s)$ equals the current in each R-L branch times the impedance of each branch, which in matrix form is given by

$$V(s)\begin{bmatrix}1\\1\\\vdots\\1\end{bmatrix} = r\begin{bmatrix}i_1\\i_2\\\vdots\\i_N\end{bmatrix} + s\begin{bmatrix}L_{self} & M_{12} & \cdots & M_{1N}\\ M_{21} & L_{self} & \cdots & M_{2N}\\ \vdots & \vdots & \vdots & \vdots\\ M_{N1} & M_{N2} & \cdots & L_{self}\end{bmatrix}\begin{bmatrix}i_1\\i_2\\\vdots\\i_N\end{bmatrix} \quad (4)$$

where i_1, i_2, $\cdots$, and i_N are the currents in each R-L branch. For simplicity, four symbols $\boldsymbol{V}$, $\boldsymbol{I}$, $\boldsymbol{L}$, and $\boldsymbol{R}$ are introduced to refer the voltage vector, current vector, and self/mutual inductance matrix in (4), and the product of the resistance r and an $N \times N$ identity matrix $\boldsymbol{I}_N$, respectively. In terms of $\boldsymbol{V}$, $\boldsymbol{I}$, $\boldsymbol{L}$, and $\boldsymbol{R}$, (4) can be rewritten as

$$\boldsymbol{V} = \boldsymbol{RI} + s\boldsymbol{LI} \quad (5)$$

The symmetric matrix $\boldsymbol{L}$ is positive semi-definite, meaning that all the eigenvalues of $\boldsymbol{L}$ are non-negative real numbers. Because $\boldsymbol{L}$ is real and symmetrical, there always exist normal and orthogonal matrices to diagonalize it. Denote $\boldsymbol{Q}$ any normal orthogonal matrix and $\boldsymbol{L}_{Diag}$ the diagonal matrix, namely

$$\boldsymbol{L}_{Diag} = \boldsymbol{Q}^T\boldsymbol{LQ} = \begin{bmatrix}l_1 & 0 & \cdots & 0\\ 0 & l_2 & \cdots & 0\\ \vdots & \vdots & \vdots & \vdots\\ 0 & 0 & \cdots & l_N\end{bmatrix} \quad (6)$$

where l_1, l_2, $\cdots$, and l_N are non-negative real eigenvalues. Multiplying the matrix $\boldsymbol{Q}$ to the left of both sides of (5) and applying (6) and the normality and orthogonality property ($\boldsymbol{Q}^T\boldsymbol{Q} = \boldsymbol{QQ}^T = \boldsymbol{I}_N$) of the matrix $\boldsymbol{Q}$ yields

$$\boldsymbol{V}' = \boldsymbol{RI}' + s\boldsymbol{L}_{Diag}\boldsymbol{I}' \quad (7)$$

where $\boldsymbol{V}' = \boldsymbol{Q}^T\boldsymbol{V}$ and $\boldsymbol{I}' = \boldsymbol{Q}^T\boldsymbol{I}$.

The original current vector $\boldsymbol{I}$ equals $\boldsymbol{QI}'$ or $\boldsymbol{Q}(\boldsymbol{R}+s\boldsymbol{L}_{Diag})^{-1}\boldsymbol{V}'$. The latter can be rewritten as $\boldsymbol{Q}(\boldsymbol{R}+s\boldsymbol{L}_{Diag})^{-1}\boldsymbol{Q}^T[1 \ 1 \ \cdots \ 1]^T V(s)$. Given the voltage $V(s)$, the total current in all R-L branches should be computed to obtain the total impedance of the R-L circuit in Fig. 2. Denote the total current as I_t that is

$$\begin{aligned} I_t &= [1 \ 1 \ \cdots \ 1]\boldsymbol{I} \\ &= [1 \ 1 \ \cdots \ 1]\boldsymbol{Q}(\boldsymbol{R}+s\boldsymbol{L}_{Diag})^{-1}\boldsymbol{Q}^T[1 \ 1 \ \cdots \ 1]^T V(s) \end{aligned} \quad (8)$$

The product $[1 \ 1 \ \cdots \ 1]\boldsymbol{Q}$ and its transpose $\boldsymbol{Q}^T[1 \ 1 \ \cdots \ 1]^T$ are a row and a column vectors respectively. If the ith element of $[1 \ 1 \ \cdots \ 1]\boldsymbol{Q}$ or $\boldsymbol{Q}^T[1 \ 1 \ \cdots \ 1]^T$ is represented by the symbol q_i, and the ij^{th} element of $(\boldsymbol{R}+s\boldsymbol{L}_{Diag})^{-1}$ is represented by the symbol z_{ij}, then I_t can be re-expressed as

$$I_t = V(s)\sum_{i,j=1}^{N} q_i z_{ij} q_j \quad (9)$$

Since matrix $(\boldsymbol{R} + s\boldsymbol{L}_{Diag})^{-1}$ is diagonal, among z_{ij}'s only z_{ii}'s are nonzero and equal to $1/(r + sl_i)$. So (9) further reduces to

$$I_t = V(s)\sum_{i=1}^{N} \frac{q_i^2}{r + sl_i} \quad (10)$$

Equation (10) implies that the circuit in Fig. 2 can be replaced by N decoupled R-L branches with resistance and inductance equal to r/q_i^2 and l_i/q_i^2 respectively, as shown in Fig. 3.

3. THE RULES OF REDUCTION AND SIMULATED DATA

When skin effect is negligible, to guarantee accurate simulation result, the interconnect wires are divided into smaller sections along its length using π or T sections. When skin effect is prominent, this approach is still needed. The challenge in the latter case is that the resistance and inductance of each section are frequency dependent and their formulae or their equivalent reduced R-L circuits are difficult to find. Section 2 has explained how to find *decoupled* R-L circuits for the frequency dependent resistance and inductance. This section will present the rules to reduce the decoupled R-L circuit and demonstrate the accuracy of the reduced R-L circuits.

Consider three conductor sections all 1μm thick and 20μm long but 2μm, 5μm, and 10μm wide, respectively. Assume that the highest frequency of interest is 30GHz and the conductivity takes the value 3.5×10^7 S/m. According to the skin depth formula $\delta = \sqrt{1/(\pi\mu\sigma f)}$, the skin depth at 30GHz is 0.49μm. To make the volume filament model accurate, the cross section of each conductor is divided into filaments of 0.25μm × 0.25μm.

The new resistance r/q_i^2 and new inductance l_i/q_i^2 ($i = 1,\cdots,N$) are calculated for each filament. These branches are grouped such that the branches in any group are within one order of magnitude from each other. The elements in each group can be combined to get a branch with the new resistance equal to the total parallel resistance and the new inductance equal to the total parallel inductance. The logic behind this reduction is that the resistance of the branch matches the low frequency impedance of the group while the inductance of the branch matches the impedance of the group at high frequency. In the intermediate frequency range, the impedance of the branch more or less matches the impedance of the group. To further reduce the size of the R-L circuit, the conductance of each R-L branch at 30GHz is calculated and those R-L branches whose conductance contribute a little to the overall conductance, *e.g.*, less than 1%, are removed from the R-L circuit.

Table 1 shows the resistance and inductance in the reduced R-L circuits. It should be pointed out that the rules above result in only two significant R-L branches in the frequency of interest for the 2μm wire and three significant R-L branches for the 5 and 10μm wires. Table 2 lists the maximum error in the resistance and inductance in the frequency of interest (up to 30GHz) obtained from the volume filament model and the reduced R-L model. As shown in Table 2, two branches give sufficient accuracy for the 2μm wire up to the frequency of interest (30GHz). However, three branches are required for the 5 and 10μm wires for an error of less than 2.2%.

Table 1. Resistance and inductance values in the reduced R-L circuits

Conductors of 1μm thick and 20 μm long				
# of R-L branches	Element	Conductor Width (μm)		
		2	5	10
2	Resistance (Ω)	$r_1 = 108.12$ $r_2 = 0.287$	$r_1 = 17.92$ $r_2 = 0.115$	$r_1 = 3.875$ $r_2 = 0.058$
	Inductance (pH)	$l_1 = 187$ $l_2 = 12.4$	$l_1 = 124.7$ $l_2 = 9.7$	$l_1 = 60$ $l_2 = 7.3$
3	Resistance (Ω)	Not needed	$r_1 = 375.76$ $r_2 = 18.82$ $r_3 = 0.115$	$r_1 = 60.41$ $r_2 = 4.141$ $r_3 = 0.058$
	Inductance (pH)		$l_1 = 506$ $l_2 = 165$ $l_3 = 9.7$	$l_1 = 187$ $l_2 = 88.5$ $l_3 = 7.3$

Table 2. Accuracy of the reduced R-L circuits as compared to the volume filament model

Conductors of 1μm thick and 20 μm long			
Conductor Width (μm)	# of R-L branches	Error boundary	
		Resistance	Inductance
2	2	0.0%	0.0%
5	2	4.3%	7.5%
	3	1.9%	1.9%
10	2	6.6%	20.7%
	3	2.2%	0.1%

The simulation reveals that the larger the cross section dimension of a wire, the more number of R-L branches needed to capture skin effect. This can be explained intuitively by recalling the skin depth formula $\delta = \sqrt{1/(\pi\mu\sigma f)}$. Apparently, resistances of wider and thicker conductors start to deviate from their DC values at lower frequency and increase with frequency. However, resistances of reduced R-L circuits tend to saturate at high frequency, meaning that the R-L branches that accurately capture skin effect impedance at low frequency are not accurate at high frequency. More R-L branches are needed to improve the accuracy at high frequency. This trend means more R-L sections needed for conductors of bigger cross section dimensions.

Fig. 4 to Fig. 6 compare the impedance calculated from the volume filament model and that from the reduced R-L circuits for the 2μm, 5μm, and 10μm conductors, respectively. As can be seen, the error boundaries, or the maximum errors in Table 2 only exist in a small portion of the whole frequency range. Hence, replacing the original conductors with the reduced R-L circuits will cause much less error in propagation delay, power consumption, and so on.

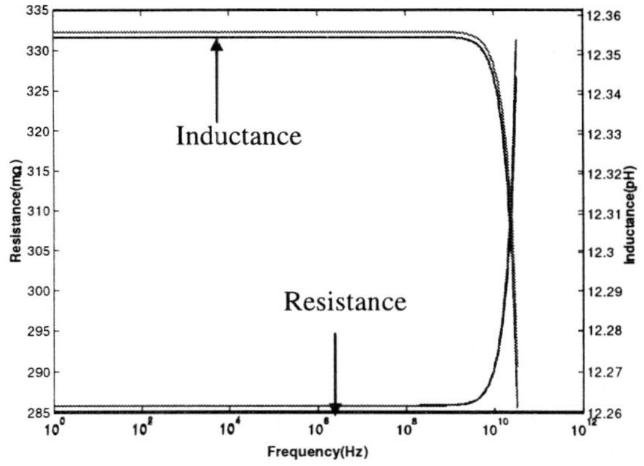

Fig. 4. Resistance and inductance calculated from the volume filament model and the two branches R-L circuit for the 2μm wide conductor.

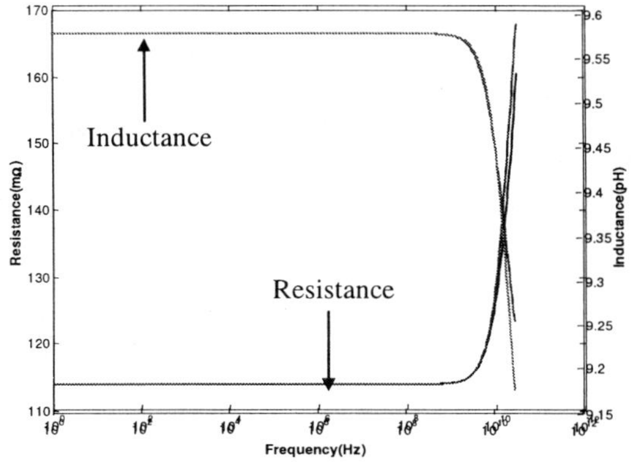

Fig. 5. Resistance and inductance calculated from the volume filament model and the two branches R-L circuit for the 5μm conductor.

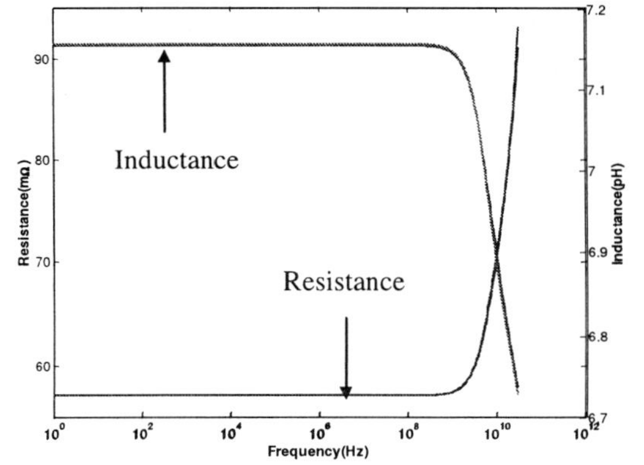

Fig. 6 Resistance and inductance calculated from the volume filament model and the three branches R-L circuit for the 10μm conductor.

4. CONCLUSIONS

A reduced R-L circuit model is developed which accurately captures skin effect. Given a conductor's dimensions, the resistance and inductance of the reduced R-L circuit can be calculated based on the rule explained in this paper. The resulting circuit model only contains a few parallel branches of resistors and inductors. A maximum error in impedance values of less than 3% is achieved using only two or three constant element branches.

Chip level interconnect circuits consist of large number of wire segments with prominent skin effect. Direct application of volume filament model will render state space dimension much larger, e.g., 10 times, than application of the reduced R-L circuits. The major overhead of obtaining reduced R-L circuit is the computation time in orthogonalizing small matrices, e.g, 20×20 matrices. Consequently, computational time saved due to the reduction in matrices dimensions is much more than the total overhead of obtaining reduced R-L circuits, which makes the reduced R-L circuit model very efficient for chip level interconnect circuits simulation.

5. REFERENCES

[1] K. M. Coperich and A. E. Ruehli, "Enhanced skin effect for partial-element equivalent-circuit (PEEC) models," *IEEE Transactions on Microwave Theory and Techniques*, vol. 48, pages 1435-1442, 2000.

[2] M. Xu and L. He, "An efficient model for frequency-dependent on-chip inductance," *Proceedings of the 2001 conference on Great Lakes symposium on VLSI*, pages 115-120, 2001.

[3] M. J. Tsuk and A. J. Kong, "A hybrid method for the calculation of the resistance and inductance of transmission lines with arbitrary cross sections," *IEEE Transactions on Microwave Theory and Techniques*, vol. 39, pages 1338-1347, 1991.

[4] L. T. Pillage and R. A. Rohrer, "Asymptotic waveform evaluation for timing analysis," *IEEE Trans. Computer-Aided Design*, vol. 9, pages 352-366, 1990.

[5] P. Silvester, "Modal network theory of skin effect in flat conductors," *Proceedings of the IEEE*, vol. 54, pages 1147-1151, 1966.

[6] H. A. Wheeler "Formulas for the skin effect". *Proceeding of the institute of radio engineers*, vol. 30, pages 412-424, 1942.

[7] S. Kim and D. P. Neikirk, "Compact equivalent circuit model for the skin effect," *1996 IEEE-MTT-S International Microwave Symposium*, San Francisco, June 1996.

[8] B. Krauter and S. Mehrotra, "Layout based frequency dependent inductance and resistance extraction for on-chip interconnect timing analysis," *Proceedings of the 35th annual conference on design automation*, pages 303-308, 1998.

[9] B. E. Keiser, Principles of Electromagnetic Compatibility, Artech House Inc., Dedaham, Massachusetts, page 102, 1979.

[10] E. B. Rosa "The self and mutual inductance of linear conductors". *Bulletin of the National Bureau of Standards*, vol. 4, pages 301-344, 1908.

[11] M. Kamon, N. Marques, L. M. Silveira, and J. White, "Generating reduced order models via PEEC for capturing skin and proximity effects," in *Electrical Performance of Electronic Packaging* West Point, NY, pages 259-262, 1998.

Efficient Interconnect Modeling by Finite Difference Quadrature Methods

Qinwei Xu and Pinaki Mazumder

Abstract

This paper proposes a numerical approximation technique, called the Finite Difference Quadrature (FDQ) Method, to efficiently model interconnects. A discrete modeling approach, FDQ adapts grid points along the interconnects to compute the finite difference between adjacent grid points by using global approximation frameworks. Similarly to Gaussian Quadrature methods to compute the numerical integrals, FDQ methods use global quadrature method to compute the numerical finite differences. As a global approximation technique, FDQ needs much sparser grid points than the Finite Difference (FD) methods to achieve comparable accuracy. Low order FDQ-based equivalent-circuit models have been incorporated into HSPICE, showing comparable accuracy yet higher efficiency and wider applicability.

I. Introduction

Fast operation and large integration scale have made the interconnect effect an important issue in high-speed systems. In the lump-element RLC equivalent-circuit modeling which is mathematically based on the finite difference (FD) methods, the TL's are segmented into small sections, each of which is represented by lumped elements. As the FD methods have low order accuracy, the electrical length of each section has to be a considerably small fraction (1/12-1/20) of the minimum wave length of the signal, therefore, the equivalent models consist a excessive number of lumped elements. Furthermore, in spite of the numerical flexibility of FD methods, the problem with FD modeling is that, although we are only interested in a small number of grid points, especially the end points, the grid points have to be dense enough to accurately represent the derivatives along the transmission lines (TL's), which leads to large sparse matrices. The drawback of low order finite methods can be removed by using the high order finite methods or pseudospectral methods [1]. The mathematical fundamental of finite difference schemes is the Taylor series expansion. The scheme of low order finite method is determined by low order Taylor series, while the scheme of high order finite method is determined by high order Taylor series. In general, the high order schemes have a high order of truncation error. Thus, to achieve the same order of accuracy, the mesh size used by the high order schemes can be much less than that used by low order schemes. As a result, the high order schemes can obtain accurate numerical solutions using very few mesh points. Chebyshev polynomial representation has been used to model transmission line in literature [2], which serves as an example of high order FD methods.

This work was partially supported by MURI grant and also by an ONR grant under the Dual-Use Program.

On the other hand, numerical integration (quadrature) is more stable and reliable than differentiation. An integral approximation framework includes the global grid points over the entire domain, while a differential one includes only the local grid points. Integral approaches like Gaussian Quadrature generally give more accurate solutions [3]. In this paper, the Finite Difference Quadrature (FDQ) method is proposed to model TL's. The idea of the FDQ method is to quickly compute the finite difference of two neighboring grid points by estimating a weighted linear sum of *derivatives* at a small set of points belonging to the domain. The weighted linear sum is like the numerical integral in Gaussian Quadrature method, yet it is to compute the finite differences rather than the integrals. The FDQ method is developed in the following steps. Starting from TL's equation in the frequency domain, FDQ discretizes the ordinary differential equations (ODE's) as algebraic equations which give the discrete model of the TL's. Due to the globality of FDQ's approximation framework, high accuracy can be obtained using a moderate number of grid points, which reduces the modeling complexity. The transmission line is then modeled as an equivalent circuit to be incorporated into SPICE simulators. Numerical examples are presented and the results FDQ methods have been compared with the results obtained using HSPICE.

II. Finite Difference Quadrature Methods

A. Motivation

The finite difference (FD) methods have been fully developed in the literature and are widely used to numerically solve differential equations. Consider a smooth function $u(x)$ defined on the domain $[0,1]$, which is divided by grid points $\{x_i, i = 1..n\}$, then the central FD framework is given by

$$u(x_{i+1}) - u(x_{i-1}) = \Delta x \frac{d}{dx} u(x_i) \tag{1}$$

where $\Delta x = x_{i+1} - x_{i-1}$ is the distance between two adjacent grid points. This framework is a local approximation in that the finite difference is only represented by the immediately neighboring grid points. Despite its wide popularity and uses, it requires very dense grid points and therefore takes computationally prohibitive time to solve large problems. The general approximation framework of Gaussion Quadrature is shown as

$$u(1) - u(0) = \int_a^b \frac{du}{dx} dx = \sum_{i=1}^{n} c_i \frac{d}{dx} u(x_i), \tag{2}$$

where c_i's are the coefficients which are determined by the orthogonal polynomials in the particular Gaussian rules. Compared to the FD framework in Eqn. 1, the Gaussian Quadrature in Eqn. 2 is a global approximation in that the difference between $u(1)$ and $u(0)$ is represented by all the grid points over the entire domain.

0-7803-7761-3/03/$17.00 ©2003 IEEE

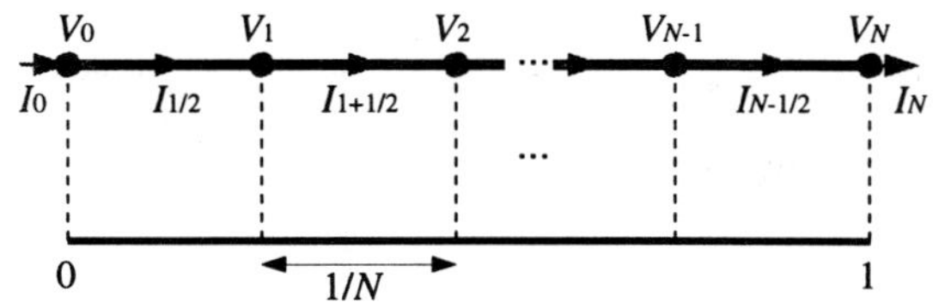

Fig. 1. FDQ frame work.

The proposed method is to integrate the finite difference and quadrature methods, the general framework of which is shown by

$$u(x_{i+1}) - u(x_i) = \sum_{k=1}^{n} c_{ik} \frac{d}{dx} u(x_k). \tag{3}$$

The right-hand-side representation is apparently a global quadrature approximation, which is the same as that in Eqn. 2; however, it is to compute the finite difference at the left-hand-side. The *FDQ coefficients* c_k's are determined by using testing function approach, similarly to the Galerkin's method.

B. FDQ modeling of TL's

For simplicity, the FDQ modeling is developed on a uniform singel transmission line. A direct numerical technique, FDQ methods do not need to decouple the Multi-conductor Transmission Lines (MTL's), therefore its application to non-uniform MTL's is straightforward extended. Assume that a TL stretches from 0 to 1 along the x axis of a Cartesian coordinate system, where the length of the line has been normalized to one unit. With $V(x, s)$ and $I(x, s)$ being respectively the Laplace-domain voltage and current vectors at point x, the normalized Telegrapher's equations in s-domain can be written as:

$$\frac{d}{dx} V(x, s) = -(sL + R) I(x, s) \tag{4}$$

$$\frac{d}{dx} I(x, s) = -(sC + G) V(x, s) \tag{5}$$

where $R(x)$, $L(x)$, $G(x)$, and $C(x)$ are the normalized PUL parameters, representing resistance, inductance, conductance and capacitance, respectively.

As shown in Fig. 1, the TL is equally discretized to small sections. There are two sets of grid points: one set are at integer-spatial positions $x_i = i/N$, $i = 0..N$, the other are at half-spatial positions $x_{i+1/2} = (i + 1/2)/N$, $i = 0..(N-1)$. The grid points are numbered in such a way that the voltages are evaluated at integer-spatial positions as $V_i = V(x_i, s)$, $i = 0..N$ and the currents are evaluated at half-spatial positions as $I_{i+1/2} = I(x_{i+1/2}, s)$, $i = 0..(N-1)$. In addition, the currents $I_0 = I(0, s)$ and $I_N = I(1, s)$ are input and output currents at the port, respectively. For clarity of the intermediate use in the context, we define the intermediate voltages $V_{i+1/2} = V(x_{i+1/2}, s)$, $i = 0..(N-1)$ and the intermediate currents $I_i = I(x_i, s)$, $i = 1..(N-1)$.

For the finite difference of voltage, the FDQ approximation framework is

$$V_{i+1} - V_i = \sum_{k=0}^{N-1} a_{ik} \frac{d}{dx} V_{k+1/2}, \quad i = 0..N-1 \tag{6}$$

For the finite difference of current, the FDQ approximation framework is

$$I_{i+1/2} - I_{i-1/2} = \sum_{k=0}^{N} b_{ik} \frac{d}{dx} I_k, \quad i = 1..N-1 \tag{7}$$

for internal grid points and

$$I_{1/2} - I_0 = \sum_{k=0}^{N} b_{0k} \frac{d}{dx} I_k \tag{8}$$

$$I_N - I_{N-1/2} = \sum_{k=0}^{N} b_{Nk} \frac{d}{dx} I_k \tag{9}$$

for the left and right boundary, respectively.

Substituting Eqns. 4 and 5 into Eqns. 6-9, we obtain:

$$V_{i+1} - V_i = \sum_{k=0}^{N-1} a_{ik} (sL + R) I_{k+1/2}, \quad i = 0..N-1 \tag{10}$$

$$I_{i+1/2} - I_{i-1/2} = \sum_{k=0}^{N} b_{ik} (sC + G) V_k, \quad i = 1..N-1 \tag{11}$$

for the internal grid points and

$$I_{1/2} - I_0 = \sum_{k=0}^{N} b_{0k} (sC + G) V_k \tag{12}$$

$$I_N - I_{N-1/2} = \sum_{k=0}^{N} b_{Nk} (sC + G) V_k \tag{13}$$

for the boundary points.

Eqns. 10-13 are the FDQ approximation framework of the single TL in s-domain, the coefficients of which have yet to be determined.

C. Determination of FDQ coefficients

The key procedure to this technique is to determine the weighting coefficients $a_{ij}, i, j = 0..N-1$ and $b_{ij}, i, j = 0..N$. Following the concept of Weighting Residual Method, a testing-function approach similar to the Galerkin's method is employed. Consider the following function set:

$$1, x, x^2, \ldots, x^n, \ldots \tag{14}$$

defined in the domain $[0, 1]$. In order for Eqn. 6 to be exact with respect to the function set in Eqn. 14, every item in Eqn. 14 serves as a test function to fit Eqn. 6. The coefficient $a_{i0}, a_{i1}, \ldots, a_{i(N-1)}$ are uniquely determined in this way. Repeatedly applying the same process to $i = 0..(N-1)$, an $N \times N$ matrix $\mathbf{A}$ of coefficients is obtained. Similarly, selecting the first $(N+2)$ functions in Eqn. 14 as testing functions, an $(N+1) \times (N+1)$ matrix $\mathbf{B}$ of coefficients is obtained after the same processing.

Once the positions of the grid points are fixed, the above testing-function approach gives constant FDQ coefficient matrices, no matter in what applications the TL equations appear. With the pre-calculated coefficient matrices, Eqns. 10-13 are rewritten as

$$\begin{bmatrix} \mathbf{B}Y(s) & \mathbf{P}^T \\ \mathbf{P} & \mathbf{A}Z(s) \end{bmatrix} \begin{bmatrix} \mathbf{V} \\ \mathbf{I} \end{bmatrix} = \begin{bmatrix} \mathbf{b} \\ \mathbf{0} \end{bmatrix} \begin{bmatrix} I_0 \\ I_N \end{bmatrix} \tag{15}$$

where

$$Z(s) = sL + R, \quad Y(s) = sC + G$$
$$\mathbf{V} = [V_0 \, V_1 \, \ldots \, V_N]^T, \quad \mathbf{I} = \left[I_{1/2} \, I_{1+1/2} \, \ldots \, I_{N-1/2}\right]^T$$

$\mathbf{P}$ is an $N \times (N+1)$ matrix, $\mathbf{P}^T$ is the transpose matrix of $\mathbf{P}$, and $\mathbf{b}$ is an $(N+1) \times 2$ connecting matrix of the external exiting current sources:

$$\mathbf{P} = \begin{bmatrix} -1 & 1 & & \\ & \ddots & \ddots & \\ & & -1 & 1 \end{bmatrix}, \quad \mathbf{b} = \begin{bmatrix} 1 & 0 \\ \vdots & \vdots \\ 0 & -1 \end{bmatrix}$$

Note that I_0 and I_N are the external exciting current sources.

D. Equivalent circuit models

A model practically offers more flexibility if it can be integrated with simulators like SPICE. In order for the FDQ-based models to be incorporated into SPICE simulators, the equivalent circuit models are derived.

In Eqn. 15, if we define

$$\mathbf{V}^e = \mathbf{A} \cdot Z(s) \cdot \mathbf{I}, \quad \mathbf{I}^e = \mathbf{B} \cdot Y(s) \cdot \mathbf{V} \tag{16}$$

where

$$\mathbf{V}^e = \left[V^e_{1/2} \, \ldots \, V^e_{i+1/2} \, \ldots \, V^e_{N-1/2}\right]^T =$$
$$\left[V^e(x_{1/2}) \, \ldots \, V^e(x_{i+1/2}) \, \ldots \, V^e(x_{N-1/2})\right]^T,$$
$$\mathbf{I}^e = [I^e_0 \, \ldots \, I^e_i \, \ldots \, I^e_N]^T =$$
$$[I^e(x_0) \, \ldots \, I^e(x_i) \, \ldots \, I^e(x_N)]^T, \tag{17}$$

then we obtain:

$$V_{i+1} - V_i = V^e_{i+1/2}, \; I_{i+1/2} - I_{i-1/2} = I^e_i \tag{18}$$

Eqn. 18 is represented by the equivalent circuit shown as in Fig. 2. The equivalent circuit model of nonuniform MTL's can be similarly derived.

III. Numerical results

According to [4], the maximum frequency of interest is evaluated as

$$f_{max} = 0.35/t_r \tag{19}$$

where t_r is the rise time of the input waveform. The maximum frequency determines the minimum wavelength within the spectral range of interest.

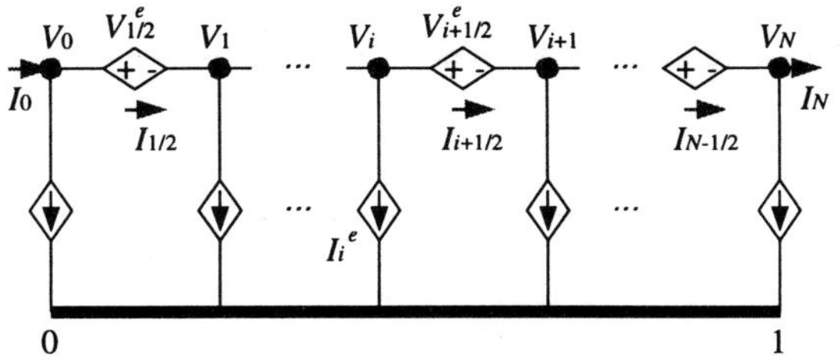

Fig. 2. Equivalent circuit of transmission line.

Referring to the literature on Chebyshev expansion [2], the heuristic for the resolution of FDQ modeling is taking two grid points per wavelength, e.g., $N = 2$ in the equivalent circuit Fig. 2 per wavelength. The following numerical experiment will show the accuracy of this evaluation. It is well established that for the finite difference method, a resolution of more than dozen points per wavelength is needed for required accuracy [3]. Therefore, the resolutions of two points per wavelength is a significant improvement, which approaches the Nyquist limit of two sample points per wavelength. The intrinsic reason for the improvement is the global approximation. The more global an approximation framework is, the closer its resolution approaches the Nyquist limit to achieve the required accuracy.

The key to improving the efficiency of discrete modeling is to reduce the grid points per wavelength while maintaining required accuracy. FDQ carries out the goal of resolution reduction without loss of accuracy by employing the global approximation framework. On the other hand, the approximation globality is achieved at the cost of computational complexity. For example, both FDQ and FD modelings can result in the Modified Nodal Analysis (MNA) equations as Eqn. 15, but the difference is that $\mathbf{A}$ and $\mathbf{B}$ are dense matrices for (global) FDQ modeling, while they are diagonal matrices for (local) FD modeling. Therefore, at the same approximation order, FDQ modeling is more time-consuming than FD modeling, since the matrix equations resulted from the former is denser than those from the latter. However, high order FDQ modelings are generally unnecessary because the required accuracy can be achieved by considerably low order framework. In practical application, the FDQ order is suggested to be no more than six. Under this criterion, the computational costs of FDQ methods are much lower than those of FD methods, since the FD order should be $5-6$ times higher than FDQ order. For the electrically long TL's whose approximation orders need to be larger according to the aforementioned heuristic, the line can be segmented to such smaller sections that each of the sections can be modeled using the low order FDQ.

The example discussed below is about a microstrip TL to test the validation of the FDQ resolution heuristic. The length of this TL is 2 cm, and the PUL parameters are $l = 360$ nH/m, $c = 100$ pF/m, $r = 1$ kΩ/m, and $g = 1$ mS/m. The applied voltage source has an internal resistance of 100 Ω, and the load is a capacitance of 0.5 pF. In the frequency-domain, the FDQ methods with different $N = 1$, 2, and 3 as in Eqn. 15 give frequency responses, which agree with the accurate value from DC to up to 10 GHz (Fig. 3).

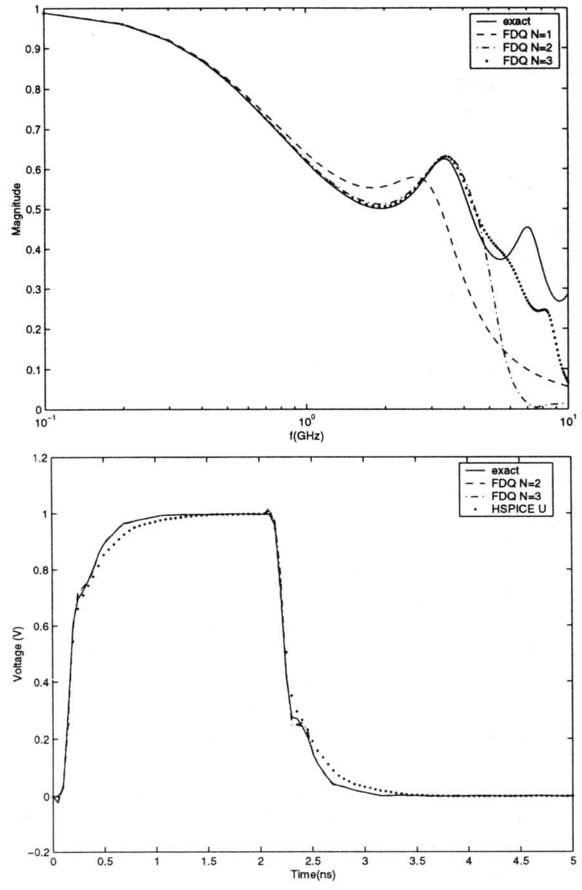

Fig. 3. FDQ modeling accuracy.

The applied input is a pulse with rise/fall time of 50 ps and width of 2 ns. The maximum frequency is determined as $f_{max} = 7$ GHz by using Eqn. 19. The propagation velocity along the line is $v = 1/\sqrt{lc} = 5/3 \times 10^8$ m/s. Therefore, the minimum wavelength of interest is $\lambda_{min} = v/f_{max} = 2$ cm. By the heuristic, discretizing the TL into two sections can give satisfied accuracy. Applying Eqn. 15, the transient responses are shown in Fig. 3 by using FDQ with $N = 2$ and $N = 3$, respectively. As comparison, the results by using HSPICE U model are given. We assume that the FD method with a solution of twelve points per wavelength give the exact value. Fig. 3 shows that both the results by FDQs with $N = 2$ and $N = 3$ match perfectly with the exact result, compared to the results by HSPICE which do not agree so well.

FDQ methods handle uniform and nonuniform TL's in the same way. Next example is a coupled nonuniform interconnect in Fig. 4. The length of the line is 0.5 cm. The input is a step with rise time of 50 ps. Using FDQ model with $N = 2$, the transient result is shown in Fig. 4. As comparison, the result by HSPICE is shown together. As HSPICE does not have a model for nonuniform TL, the inconnect is equally segmented to five sections, each of which is regarded as uniform line, and the U model is applied. The memory used by FDQ modeling is 214 KBytes, compared to 377 KBytes by HSPICE; while

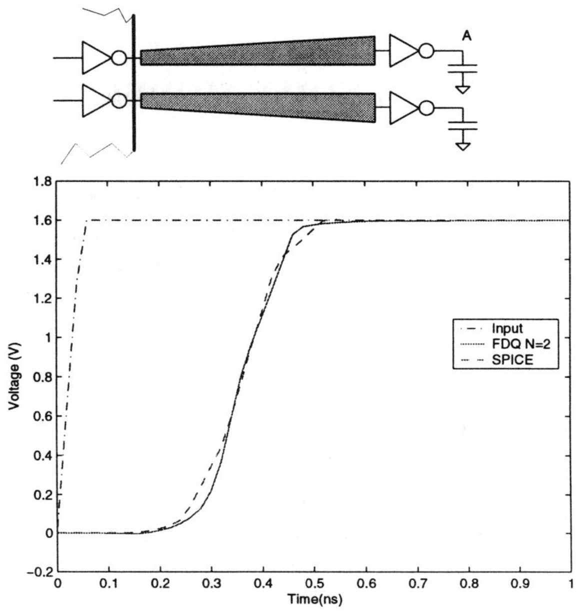

Fig. 4. Transient responses.

the CPU time used by FDQ is 0.97 s, compared to 1.7 s by HSPICE.

IV. Conclusions

An efficient numerical approximation technique, Finite Difference Quadrature (FDQ) Method, is proposed to model transmission lines. A discrete modeling approach, FDQ adapts grid points along the transmission lines, and uses global quadrature method to construct approximation frames to compute the numerical finite differences. As a global approximation technique, FDQ needs much sparser grid points than the Finite Difference (FD) methods to achieve required accuracy. Equivalent circuit models are derived, which can be incorporated into HSPICE simulator. A heuristic of FDQ resolution of two points per wavelength is given to determine the FDQ order in the application of FDQ modeling. Numerical experiments show that FDQ-based modeling is an effective way to model uniform and nonuniform transmission lines.

References

[1] B. Fornberg, *A practical guide to pseudospectral methods*. Cambridge university press, 1996.

[2] M. Celik and A. C. Cangellaris, "Simulation of multiconductor transmission lines using Krylov subspace order-reduction techniques," *IEEE Trans. Computer-Aided Design*, vol. 16, no. 5, pp. 485–496, 1997.

[3] M. N. O. Sadiku, *Numerical techniques in electromagnetics*. Boca Raton, FL: CRC, 2001.

[4] Avant!, *Star-HSPICE Manual*. Avant! Corporation, 2000, 46871 Bayside Parkway, Fremont, CA 94538.

[5] R. Gupta, S.-Y. Kim, and L. T. Pillegi, "Domain characterization of design," *IEEE Trans. Computer-Aided Design*, vol. 16, no. 2, pp. 184–193, 1996.

1/f NOISE MODELING USING DISCRETE-TIME SELF-SIMILAR SYSTEMS[1]

R. Narasimha, S. B. Rachaiah, R. M. Rao, and P. R. Mukund

Department of Electrical Engineering
Rochester Institute of technology
79 Lomb Memorial Drive, Rochester, NY 14623 USA

ABSTRACT

Flicker noise, popularly known as *1/f* noise is a commonly observed phenomenon in semiconductor devices. To incorporate *1/f* noise in circuit simulations, models are required to synthesize such noise in discrete-time. This paper proposes a model based on the fact that *1/f* processes belong to the class of statistically self-similar random processes. The model generates *1/f* noise in the time domain with a simple white noise input and is parameterized by a quantity whose value can be adjusted to reflect the desired 1/f parameter, that is, the slope of the 1/f spectrum. It thus differs from most of the earlier modeling approaches which were confined to the spectral domain. To verify fit between the model and actual *1/f* noise measurements, experiments were conducted with a PIN photodiode at various bias conditions and sampling frequencies. The noise synthesized by the model was found to match the measurements quite well. These models can be easily incorporated with circuit simulation tools to generate *1/f* noise.

1. INTRODUCTION

1/f noise is a universally observed phenomenon and the systems exhibiting this noise range from electronic devices, music, and nerve membranes to economic data[1]. There is considerable amount of documented work in the area of noise modeling in electronic devices. Nemirovsky *et al*[2, 3] model the power spectral density (PSD) of the drain current *1/f* noise in saturation as well as subthreshold conditions of CMOS transistors and compare their results with SPICE models. An on-wafer *1/f* noise measurement and characterization system was proposed in [4]. A unified flicker noise model that incorporates both the number fluctuation and the correlated surface mobility fluctuation mechanism is presented in [5]. A setup that allows observation of *1/f* noise spectra in MOSFETs under switched bias condition over a wide range of frequencies is studied in[6]. Tian et al[7] have investigated a non-stationary time domain model for *1/f* noise. The current bias-dependence of low-frequency noise spectra in single growth planar separate absorption, grading, charge and multiplication avalanche photodiodes was studied in [8]. Williams et al investigated photo-induced excess low frequency noise in HgCdTe photodiodes and concluded that the point at which *1/f* noise is equal to the shot noise, referred to as the *1/f* knee, increases with photocurrent. For image sensors operated in the charge storage mode, noise at low frequencies dominates the total noise so that a quantitative description of the *1/f* noise is required. A new empirical model and a method to calculate the noise in pin diodes were suggested in [9]. Noise measurement methodology in pin diode pixels in imagers at different operating conditions was proposed in [10].

The previous approaches deal with spectral models that do not lead to actual synthesis or they are time-domain models based on physical processes and are not amenable to practical implementation. Instead, we turn to recent studies in digital signal processing literature of statistical self-similar random processes. Self-similar processes have gained a lot of attention in the recent years due to their wide range of application areas. However, *1/f* noise[11, 12] that occurs in devices is a special class of statistical self-similar noise. In this paper we investigate low frequency modeling techniques in photodiodes using discrete-time systems developed from self-similarity considerations in the time domain (TD). The advantage of this model is that it only requires the slope of the measured PSD to estimate the time-domain noise. The versatility of the model has been tested for a range of frequencies and bias conditions. The estimated TD noise is also verified with the actual measured noise current in the time-domain.

The paper is organized as follows. Section 2 gives a brief view of the discrete-time statistically self-similar model and discusses the *1/f* behavior for a range of parameter values. Measurement setup is explained in section 3. Simulation results are provided in section 4 and conclusion and future work are drawn in section 5.

2. DISCRETE-TIME SELF-SIMILARITY

2.1 Model Description

A continuous-time process $y(t)$ is statistically self-similar in the wide-sense if it satisfies

$$\begin{aligned} &E[y(t)] = a^{-H}E[y(at)] \\ &R_{yy}[t,s] = E[y(t)y(s)] = a^{-2H}R_{yy}[at,as],\ t,s \in \mathbf{R},\ a>0 \end{aligned} \quad (1)$$

where E denotes expectation

Thus the definition of self-similarity is connected with time scaling, an operation that is well defined in continuous time. Because of the ambiguity of defining scaling in discrete-time, approaches to studying self-similarity in this domain tend to rely on sample properties of continuous-time self-similar signals and avoid reference to time scaling. However, a continuous dilation,

[1] This work was supported in parts by grants from the Gleason foundation, the Laboratory for Autonomous Co-operative Microsystems at RIT, Photon Vision Systems and Microelectronics Design Center-NYSTAR. We also thank Agilent Technologies for lending measurement equipment.

0-7803-7761-3/03/$17.00 ©2003 IEEE

scaling operator has recently been developed in discrete-time leading to a tangible definition of discrete-time self-similarity.

Let

$$X(\omega) = G[x(n)] \triangleq \sum_n x(n)e^{-j\omega n} \quad (2)$$

where $\omega = [-\pi, \pi]$. Let $\Omega = f(\omega)$ be a warping function that maps $[-\pi, \pi]$ to $[-\infty, \infty]$. The bilinear transform (BLT) for which $\Omega = f(\omega) \equiv 2\tan(\omega/2)$ is used in this paper. Then, the discrete time-scaling operation by a factor of $a > 0$ is defined by an operator $S_a[]$ as [13], [14], [18],

$$x_a(n) = S_a[x(n)] \triangleq aG^{-1}\{X[\Lambda_a(\omega)]\} \quad (3)$$

where G^{-1} is the inverse of the discrete-time Fourier transform in (2), $S_a[]$ denotes scaling by a and $\Lambda_a(\omega) = f^{-1}[af(\omega)]$. Note that a can take any value over the continuum $[0,\infty)$. A process $x(n)$ is now defined as self-similar in the wide-sense if

$$\begin{aligned} E[S_a[x(n)]] &= a^{-H}E[x(n)] \\ S_{a,a}[R_{XX}(n,n)] &= a^{-2H}R_{xx}(n,n'),\ a > 0 \end{aligned} \quad (4)$$

where $R_{xx}(n,n')$ is the autocorrelation function of $x(n)$ and $S_{a,a}$ is a scaling of both dimensions of a two-dimensional random process by a factor a, effected by repeated application of the scaling in (3) to both dimensions. It has been shown that if the power spectrum of a process $x(n)$ is

$$P_X(\omega) = \frac{|f(\omega)|^r}{|f'(\omega)|} \quad (5)$$

then the process satisfies (4), and is self-similar with parameter $H = -(r+1)/2$ [13]. For the BLT this power spectrum can be factored in the z-domain as

$$P(z) = L(z)L(z^{-1}) \quad (6)$$

where

$$L(z) = H_1(z)H_2(z) \quad (7)$$

with

$$H_1(z) = (1-z^{-1})^{r/2}, \quad H_2(z) = 2^{r/2-1}(1+z^{-1})^{1-r/2} \quad (8)$$

Therefore, a discrete-time self-similar process $x(n)$ with parameter H can be generated by applying a white noise input $w(n)$ to a filter whose transfer function is $L(z)$. Discrete-time self-similar processes generated as white-noise driven outputs of the $L(z)$ filter for $-2 < r < 4$ are stationary as they are outputs of a stable linear time-invariant system with a stationary input. Analogous to the continuous-time case [15] we define increments of a discrete-time self-similar process $x(n)$ as

$$y(n) \triangleq x(n) - x(n-1) \quad (9)$$

and call it a second-order self-similar process. It can be obtained by passing white noise through a system whose transfer function is

$$L_s(z) \equiv (1-z^{-1})L(z) \quad (10)$$

where subscript s stands for second-order. The system with transfer function $L_s(z)$ is stable for $|r| < 4$.

2.2 *1/f* behavior of the model

The complex power spectrum of the model output in (7) is

$$R_x(z) = \sigma_w^2 R_h(z) \quad (11)$$

where $R_h(z)=P(z)$ and σ_w^2 is the variance of the white noise input. For $z = e^{j\omega}$ and $d=-r/2$,

$$R_h(z)|_{z=\exp(j\omega)} = \frac{2^{-2(d+1)}}{(2\sin(\omega/2))^{2d}(2\cos(\omega/2))^{-2(d-1)}},\ -\pi < \omega \le \pi \quad (12)$$

As the frequency $\omega \to 0$

$$R_h(e^{j\omega}) \sim \frac{1}{(\omega)^{2d}}, \quad (13)$$

and thus the output conforms to *1/f* characteristics. The autocorrelation of the white noise driven output can be obtained by the inverse Fourier transform of the power spectrum,

$$\begin{aligned} r_{h_1}(l) &= \frac{1}{2\pi}\int_{-\infty}^{\infty} R_{h_1}(e^{j\omega})e^{-j\omega l}d\omega \\ &= \frac{1}{2\pi}\int_0^{\pi}(\cos\omega l)(2\sin\omega/2)^{-2d}d\omega \end{aligned} \quad (14)$$

$$r_{h_2}(l) = \frac{1}{2\pi}\int_0^{\pi}(\cos\omega l)(2\cos\omega/2)^{2(d+1)}d\omega \quad (15)$$

using identities in [16], the above become,

$$r_{h_1}(l) = \frac{(-1)^l\Gamma(1-2d)}{\Gamma(1+l-d)\Gamma(1-l-d)} \qquad l = 0,1,\ldots \quad (16)$$

$$r_{h_2}(l) = \frac{\Gamma(2d+3)}{\Gamma(2+l+d)\Gamma(2-l+d)} \quad (17)$$

where $\Gamma()$ is the gamma function. For $C=2^{-2(d+1)}$,

$$r_h(l) = C\left(r_{h_1}(l) * r_{h_2}(l)\right). \quad (18)$$

At low frequencies, at $\omega \approx 0$, the covariance function for large lags can be written as

$$\rho(l) \sim C_\rho l^{2d-1},\ \text{as } l \to \infty \quad (19)$$

where $\rho(l) = r_h(l)/r_h(0)$ for some $C_\rho \ne 0$ thus indicating *1/f* behavior of the output. For the second-order self-similar process $y(n)$, its power spectrum has the form

$$P(\omega) \propto |\omega|^{r+2} \quad (20)$$

3. MEASUREMENT SETUP

Packaged commercial discrete devices were used for measurements. A Silicon pin photodiode with plastic single in-line package, under reverse bias conditions was the device under test (DUT). The measurement setup is similar to as shown in [4, 8]. The setup assembly comprises of the Agilent[17] 4156C source monitor unit (SMU), a test fixture 16442A that holds the DUT, a low noise current amplifier and the spectrum analyzer. Kelvin triaxial cables were used to connect the test fixture to the SMU. Stanford research system's SR570 low noise trans-resistance amplifier was used for amplifying the diode output current. It also biases the DUT besides amplifying the output current. Agilent's 35670A spectrum analyzer was used for low frequency noise measurements. For dark current measurements, the test fixture can also be used to block unwanted illumination. The measurement setup is shown in Figure 1. The SMU provides ground connection to the anode of the DUT and was used to obtain the TD noise current. The PSD of the output voltage of the amplifier is related to the current PSD as follows,

$$S_{vout} = \frac{S_I}{S^2} \tag{21}$$

where S_{vout} is the PSD of the spectrum analyzer output voltage, S_I is the PSD of current of the DUT and S is the sensitivity of the amplifier.

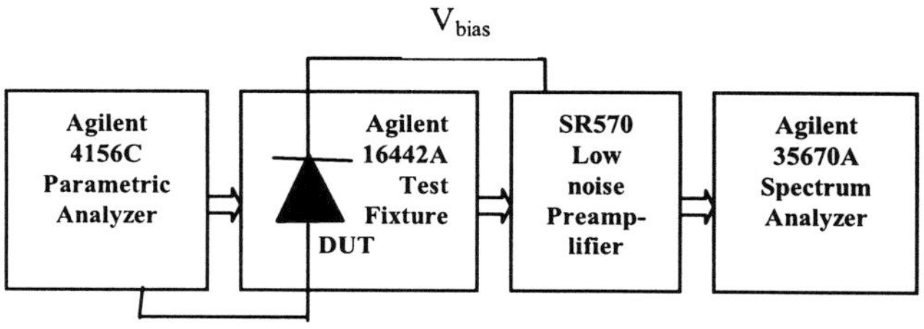

Figure 1. Measurement setup

4. SIMULATION RESULTS

Dark current measurements were done at different reverse bias voltages and at various sampling frequencies. The voltages were varied in the range (1-5)V in steps of 1V and frequencies in the range 1Hz-100KHz. It was observed that with the increase in the reverse bias voltage the noise magnitude increased as shown in [9]. The system noise was neglected since its magnitude was very low compared to the device noise. The top plot in Figure 2 shows the measured slope of the noise PSD. For a self-similar signal $x(n)$, the estimate of the spectral density is proportional to $|\omega|^{1-2H}$ for frequencies close to the origin. Therefore, the slope of the logarithm of the periodogram versus the logarithm of the frequency is $1 - 2H$, where H is the Hurst parameter. Using the relation $H=-(r+1)/2$, TD noise current was simulated at the corresponding fractional parameter r and the estimated variance by applying white noise to the filter in (7). The variance of the noise was estimated through a least-square fit to the autocorrelation of the noise using an inverse Fourier transform operation on the PSD. The periodogram of the synthesized TD noise is depicted in the bottom plot of Figure 2. The slopes of both these plots match very closely. The synthesized TD noise current is shown in Figure 3(b). Figure 3(a) depicts the measured TD noise current using the SMU. Similar procedure was applied to different reverse bias voltages and various sampling frequencies. Figure 4 shows the slope comparison of the measured PSD (top plot) and simulated PSD (bottom plot) for a reverse bias of 1V and sampling frequency of 10Hz. The plots of the measured and the synthesized noise are provided in Figure 5(a) & (b) respectively. The amplitudes of the synthesized and the measured noise match closely in both these cases. The results for different conditions are summarized in Table 1.The Hurst parameter values estimated from the measured noise spectrum and from the synthesized PSD show a close match. The deviation of the Hurst parameter values computed from the measured TD noise can be attributed to using two separate measuring instruments for TD and PSD measurements, although the deviation is within 10% range.

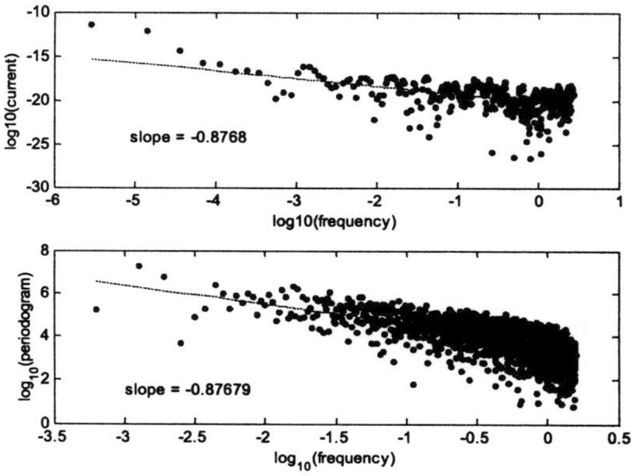

Figure 2. Slope comparison of the measured and synthesized PSD for 4V, 1Hz sampling.

5. CONCLUSION AND FUTURE WORK

The empirical studies reported here have shown that white noise driven DTSS models can synthesize TD noise currents that closely match those of the measured TD noise for varying conditions. *1/f* behavior of the model at low frequencies is presented. The difference between this modeling technique and previous approaches of synthesizing TD noise is that this model is derived from self-similar considerations and it requires only the slope information of the noise spectrum and is not confined just to the spectral domain. A potential application of the research presented in this paper is to develop an equivalent *1/f* noise source that can be incorporated in noise analysis of devices.

	4 Volts, 1 Hz	5 Volts, 1 Hz	1 Volts, 10 Hz
Measured noise spectrum using SMU	0.9384	0.9402	0.9518
Synthesized PSD from TD noise	0.9384	0.9291	0.9599
Measured TD noise	0.8591	0.9079	0.8720

Table 1. Hurst parameter comparison for various conditions

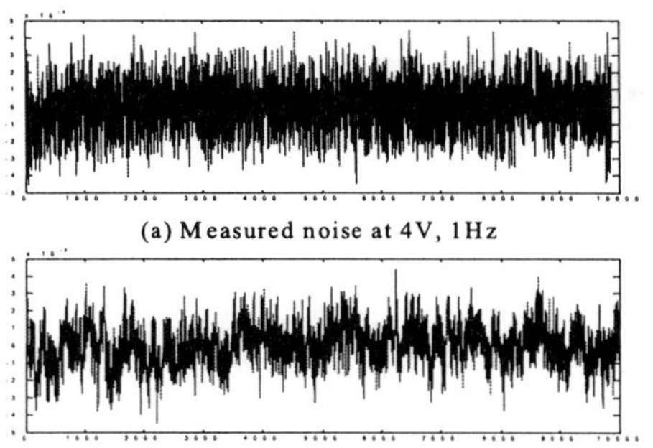

(a) Measured noise at 4V, 1Hz

(b) Synthesized noise at 4V, 1Hz

Figure 3. (a) Measured noise current (b) Synthesized noise current

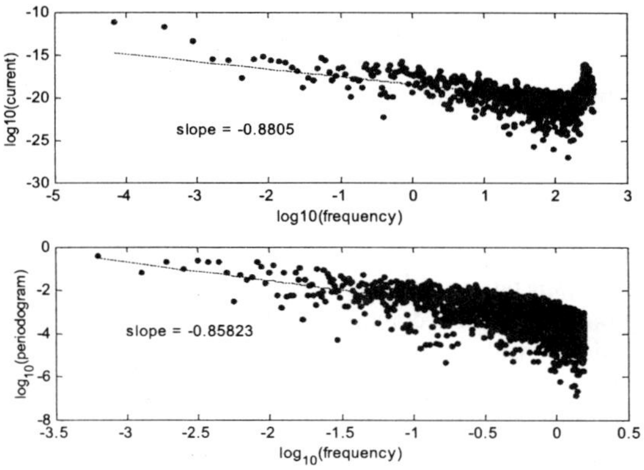

Figure 4. Slope comparison of the measured and synthesized PSD for 1V, 10Hz sampling.

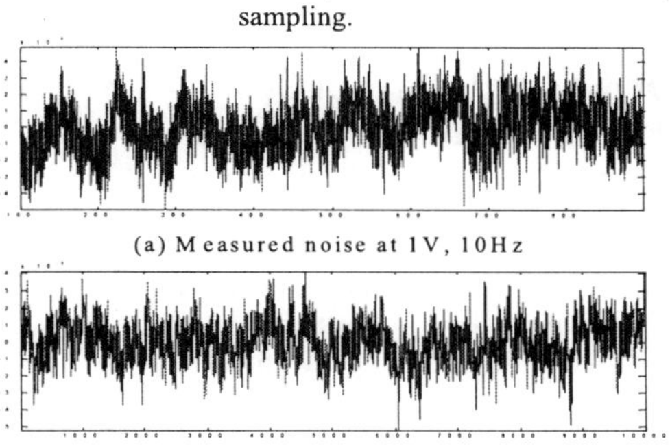

(a) Measured noise at 1V, 10Hz

(b) Synthesized noise at 1V, 10Hz

Figure 5. (a) Measured noise current (b) Synthesized noise current

6. REFERENCES

[1] M. S. Keshner, "1/f Noise," *Proceedings of the IEEE*, vol. 70, pp. 212-218, 1982.

[2] Y. Nemirovsky, I. Brouk, and C. G. Jakobson, "1/f noise in CMOS transistors for analog applications," *IEEE Transactions on Electron Devices*, vol. 48, pp. 921-927, May 2001.

[3] C.Jakobson, I.Bloom, and Y.Nemirovsky, "1/F noise in CMOS transistors for analog applications from subthreshold to saturation," *Solid-State Electronics*, vol. 42, pp. 1807-1817, 1998.

[4] A. Blaum, O. Pilloud, G. Scalea, J. Victory, and F. Sischka, "A new robust on-wafer 1/f noise measurement and characterization system," *Proceedings of IEEE 2001 International Conference On Microelectronic Test Structures*, vol. 14, pp. 125-130, March 2001.

[5] Kwok.K.Hung, Ping.K.Ko, C. Hu, and Yiu.C.Cheng, "A unified model for the flicker noise in metal-oxide-semiconductor field-effect transistors," *IEEE Transactions on Electron Devices*, vol. 37, pp. 654-665, March 1990.

[6] A. P. vanderWel, E. A. M. Klumperink, S. L. J. Gierkink, R. F. Wassenaar, and H. Wallinga, "MOSFET 1/f Noise Measurement Under Switched Bias Condition," *IEEE Electron Device Letters*, vol. 21, pp. 43-46, Jan 2000.

[7] H. Tian and A. E. Gamal, "Analysis of 1/f noise in CMOS APS," *proceedings of SPIE Electronic Imaging*, vol. 3965, Jan 2000.

[8] S. An and M. J. Deen, "Low frequency noise in single growth planar separate absorption, grading, charge and multiplication avalanche photodiodes," *IEEE Transactions on Electron Devices*, vol. 47, pp. 537-543, March 2000.

[9] F.Blecher, K.Seibel, and M.Bohm, "Photo and dark current noise in a Si:H PIN diodes at forward and reverse bias," *Materials Research Society (MRS) spring meeting*, San Francisco April 1998.

[10] F.Blecher, B.Schneider, J.Sterzel, and M.Bohm, "Noise of a Si:H PIN diode pixels in imagers at different operating conditions," *Materials Research Society (MRS) spring meeting*, San Francisco April 1999.

[11] J. A. Barnes and D. W. Allan, "A statistical model of flicker noise," *Proceedings of the IEEE*, vol. PROC-54, pp. 176-178, 1966.

[12] B. B. Mandelbrot, "Some noises with 1/f spectrum, a bridge between direct current and white noise," *IEEE Transactions on Information Theory*, vol. IT-13, pp. 289-298, 1967.

[13] W. Zhao and R. M. Rao, "Continuous-dilation discrete-time self-similar signals and linear scale-invariant systems," presented at ICASSP, 1998.

[14] R.Narasimha and R.M.Rao, "Modeling Variable-Bit Rate video traffic using Linear Scale Invariant Systems," presented at Wavelet Applications, SPIE 02, Orlando, FL, 2002.

[15] J. Beran, *Statistics for Long-Memory Processes*. New York, NY: Chapman & Hall, 1994.

[16] I. S. Gradshteyn and I. M. Ryzhik, "Table of Integrals, Series, and Products,",, 1994.

[17] http://www.agilent.com/.

[18] S.Lee, W.Zao, R.Narasimha and R.M.Rao, "Discrete-Time Models for Statistically Self-Similar Signals," *IEEE Trans. on Signal Processing*, 2003 to appear.

HIGH LEVEL ACCURACY LOSS ESTIMATES FOR A CLASS OF ANALOG/DIGITAL SYSTEMS

C.Alippi, M.Stellini

DEI, Politecnico di Milano
P.za L. Da Vinci 32
20133 Milano, Italy

ABSTRACT

The paper perfects and adapts the methodology developed in [1] to estimate the performance degradation of an application being implemented in a mixed analog/digital technology. Finite precision representations, deviations from the reference computation and fluctuations of physical parameters due to the production process for passive components are suitably abstracted by implementation-related perturbations and associated to the performance loss index. Such index can be used to validate a candidate solution and identify the most sensitive modules composing the circuit/system. The effectiveness of the high level approach is shown on a non-trivial embedded system comprising an analog filter, an ADC and a digital FFT module.

1. INTRODUCTION

A measure of the "performance degradation" of an application at the early steps of the design cycle, namely an index characterizing the discrepancy between the performance of the implemented computation and the reference one, can be of invaluable help in the high level synthesis of complex systems. Effective and reliable application-level performance estimates can be exploited by lower level synthesis layers, e.g., by validating an architectural choice before the final implementation takes place, dimensioning the data word length, deciding whether a floating point or a fixed point operation is needed, even solving the issue of analog vs. digital implementation. Such general performance degradation estimates cannot be extracted by existing CAD tools which, in the best case, can only measure the impact of a finite precision representation of the computational flow on accuracy for a given input and do not provide any confidence index for mixed digital/analog systems.

The paper addresses the performance degradation issue at the very high level by following the guidelines suggested in [1], detailing and applying the theory to mixed analog/digital circuits. The methodological high-level framework for accuracy estimation is based on the intuitive concepts of perturbations (abstractions of physical uncertainties affecting the computation) and perturbed computation. Here perturbations model fluctuations in the production process of passive parameters (e.g., resistors and capacitors) in analog circuits even if the analysis can be extended also to cope with different error-affected entities. On the digital side, a perturbation abstracts the equivalent errors arising within a sub-module and propagating up to its output (the actual error depends on the chosen architecture and quantisation techniques).

The novelty of the suggested performance loss estimation approach resides in weakening the common small perturbation hypothesis of sensitivity analysis (which is true neither for low cost analog implementations, nor for digital circuits where low power and silicon area constraints limit the number of bits to represent the interim variables of the computation) as well as in addressing a large class of systems (those whose computation can be described by a mathematical Lebesgue-measurable function).

In addition, the method provides an effective quantitative measure for the performance loss both in the case the circuit is given and we wish to validate it, or we wish to consider the performance loss in correspondence with an ensemble of circuits. By considering an ensemble of analog circuits we address the fluctuation in parameters issue, fluctuations associated with the production process of passive components. In a digital circuit the ensemble implies considering at the same time the performance loss of all architectures whose computation represents error-affected realizations of the reference computation.

The structure of the paper is as follows. Section II briefly introduces the perturbation model and formalizes the analysis to analog/digital circuits. Section III provides the algorithms based on Randomised Algorithms for estimating the accuracy loss. Experiments are finally given in section IV.

2. MODELLING THE ACCURACY LOSS

Denote by $y = f(x)$, $y \in Y \subset \Re^k$, $x \in X \subseteq \Re^d$ the mathematical description of the reference computation associated with the application to be implemented in mixed analog/digital technology. The designer, on the basis of a priori knowledge, available macrocells and other constraints, partitions the computation in sub-systems and decides whether implementing the i-th sub-system described by function $f_i(x)$ in digital or analog technology. At this abstraction level each analog subsystem can be described with function $f_i(x,V)$, where the parameter vector V accounts for all passive independent components (e.g., resistors and capacitors). On the digital side, partitioning generates atomic sub-modules each of which will be considered as a black-box characterised by $f_i(x)$ (for instance, a multiplier or a butterfly module of a FFT).

The subsequent implementation of analog components and digital sub-systems will introduce errors affecting V and the output of each digital black-box, respectively.

At the very high level we can abstract such sources of error by means of a generic p-dimensional perturbation vector $\Delta \in D \subseteq \Re^p$, having a component for each independent perturbation affecting $f(x)$ (an additive perturbation for each passive component and each digital module output). The

0-7803-7761-3/03/$17.00 ©2003 IEEE

characterisation of the input space X and the perturbation space D is given in terms of the probability density functions pdf_X and pdf_D, respectively. Denote by $f_\Delta(x)$ the perturbed computation. In order to characterise the accuracy of the approximated computation w.r.t. the reference one, we have to compute the discrepancy between $y = f(x)$ and $y_\Delta = f_\Delta(x)$ by means of a suitable –but general- loss function $u(x,\Delta) = u(f(x), f_\Delta(x))$ that here we assume to be Lebesgue measurable. Basically, all functions involved in signal/image processing are measurable according to Lebesgue. The proposed analysis for evaluating the performance loss $u(x,\Delta)$ can be divided into two different cases reflecting the application needs.

The first verification problem addresses the case in which the reference computation is affected by a fixed perturbation $\overline{\Delta}$. The perturbation can be of any nature, even inducing a structural modification of the reference computation (which becomes a different function $g(x)$).

The second verification problem addresses the case in which the perturbation Δ is not fixed but can assume values defined over D according to the pdf_D. The two verification problems model different implementation aspects that, for analog/digital systems, can be summarised as in table 1.

Type	I[st] Verification Problem	IInd Verification Problem
Analog	• Production process (a given circuit: V is fixed); • A given failure; • Components aging effects (at time t)	• Production process (mass production); • Components aging effects (with bounded deviation over time);
Digital	• A given quantisation technique for interim variables; • Failure of a module (e.g., stuck-at faults); • Approximated realisation (e.g., LUT);	• Finite precision representation (we do not specify the quantisation technique); • Failure of a module (e.g., several faults with bounded effect);

Table I: The two verification problems

In the first verification case we have that inputs are the unique degree of freedom being the perturbation fixed. The performance loss function is $u(x) = u(f(x), f_\Delta(x,\overline{\Delta}))$ and our goal is to evaluate the "goodness" of the perturbed system $\forall x \in X$. The second verification case is more complex since we also have to deal with all perturbations Δ floating in D and we are looking for the performance loss $u(x) = u(f(x), f_\Delta(x,\Delta)), \forall x \in X$ and $\forall \Delta \in D$. If we consider mass production each analog sub-system can be seen as the outcome of a stochastic process, being each resistance/capacitance modellable as an independent random variable (e.g., ruled by a gaussian pdf for resistors/capacitors).

Similarly, a digital sub-system can be described as a black box with an additive perturbation at its output belonging to dominion D. Whatever the final architecture, the designer has to grant that the effective error at the sub-system output is within D.

3. MEASURING THE ACCURACY LOSS

3.1 I[st] Verification Problem

The evaluation of the accuracy loss associated with the verification problem requires knowledge of $u(x,\Delta)$ which, for a generic Lebesgue measurable function $y = f(x)$ and a generic $y_\Delta = f_\Delta(x)$ is not available in a closed form. We can characterise the performance loss by following a different approach, which requires testing whether

$$u(x,\overline{\Delta}) \le \gamma \text{ holds or not, } \forall x \in X \tag{1}$$

in correspondence with all positive values γ s. Denote by $\overline{\gamma}$ the minimum value of γ for which $u(x,\overline{\Delta}) \le \overline{\gamma}$ holds $\forall x \in X$. $\overline{\gamma}$ identifies the maximum performance loss induced by $\overline{\Delta}$ on the computation and, hence, it provides an index of performance-accuracy. Despite the fact that identification of the exact $\overline{\gamma}$ might be extremely difficult for a generic function, its value could be too conservative for subsequent analyses. This aspect has been pointed out by other authors in problems related to the identification of the robustness margin index for robust control [2-4]. In fact, the maximum error is excited by particular perturbations that arise, in general, with a very low, almost null, probability. The risk of a deterministic worst-case analysis is to be too conservative ($\overline{\gamma}$ is overdimensioned w.r.t. the application needs). A dual probabilistic problem can therefore be formulated for (1), which requires that inequality $u(x,\Delta) \le \overline{\gamma}$ be attained $\forall x \in X$ with probability one. This framework allows measuring the performance loss associated with the perturbed computation. In fact, it states that the perturbed computation induces a performance loss smaller than $\overline{\gamma}$ with probability one: $\overline{\gamma}$ becomes the performance loss index.

3.2 IInd Verification Problem

Different figures of merit can be envisaged to describe the computational loss associated with perturbations. We feel that a natural characterisation covering a large spectrum of applications can be obtained by first averaging w.r.t. the perturbation space and then considering the maximum error amplified by the inputs:

$$\tilde{\gamma} = \max_{x \in X} E_D[u(x,\Delta)]. \tag{2}$$

The designer could consider different max loss functions such as $\tilde{\gamma} = \max_{x \in X} Var_D[u(x,\Delta)]$ (the attention is on the maximum error of the energy of the loss function w.r.t the perturbation).

The main problem of (2) is related to the computational complexity needed to compute $\tilde{\gamma}$. We can solve the verification problem by resorting to probability [2]: given accuracy ε, we require the estimate of the maximum over M points $\hat{\gamma} = \max_{x_i, i=1,M} E_D[u(x_i,\Delta)]$ to grant that

$$\Pr(\Pr(E_D[u(x,\Delta)] \geq \hat{\gamma}) \leq \varepsilon) = 1. \quad (3)$$

$\overline{\gamma}$ and $\tilde{\gamma}$ can be obtained in a closed form only in toy examples supported by strong hypotheses to make the mathematics amenable. Estimates for $\overline{\gamma}$ and $\tilde{\gamma}$ can be derived by relaxing *ALL* hypotheses assumed in the literature and facing the problem with the recent theories based on Randomised Algorithms [1-4]. Denote by $p_\gamma = \Pr\{u(x,\Delta) \leq \gamma\}$ the probability that the performance loss is satisfied at a given accuracy loss level γ. Extract from *X* a set of *N* independent and identically distributed samples x_i according to pdf_X and generate

$$I(x_i,\Delta) = \begin{cases} 1 & \text{if } u(x_i,\Delta) \leq \gamma \\ 0 & \text{if } u(x_i,\Delta) > \gamma \end{cases} \quad (4)$$

γ is a given –but arbitrary- positive value. The true unknown probability p_γ can be estimated as $\hat{p}_N = \frac{1}{N}\sum_{i=1}^{N} I(x_i,\Delta)$. It is intuitive that the adherence of $\hat{p}_N$ to p_γ depends on the required accuracy level ε (e.g., 0.03) so that $\left|\hat{p}_N - p_\gamma\right| \leq \varepsilon$ and, indirectly, on the number of samples *N* we draw. By introducing a confidence value δ (e.g., 0.01) we can develop the two algorithms solving the verification problems. Details can be found in [1]. The final algorithm for determining the estimate of $\overline{\gamma}$ is given in figure 1.

CHARACTERISE *X*;
SELECT ε, δ;
EXTRACT $N \geq \frac{\ln\frac{2}{\delta}}{2\varepsilon^2}$ POINTS FROM *X* ACCORDING TO pdf_X;
GENERATE THE FUNCTION $\hat{p}_N = \hat{p}_N(\gamma), \forall \gamma \geq 0$;
SELECT THE MINIMUM VALUE $\hat{\gamma}$ SO THAT $\hat{p}_N(\hat{\gamma}) = 1, \forall \gamma \geq \hat{\gamma}$;
$\hat{\gamma}$ IS THE ESTIMATE OF $\overline{\gamma}$.

Fig. 1. The procedure for estimating the performance loss $\overline{\gamma}$

The obtained $\hat{\gamma}$ is an index of the computation accuracy for the embedded system, once affected by the perturbation. Conversely, the final algorithm for determining the estimate of $\tilde{\gamma}$ is given in figure 2. If device D_1 introduces a performance loss $\hat{\gamma}_1$ and D_2 a loss $\hat{\gamma}_2$, D_1 is more accurate in the computation than D_2 if and only if $\hat{\gamma}_1 < \hat{\gamma}_2$. By measuring the performance index $\hat{\gamma}$ of an application at the early stages of the circuit design cycle we can validate/select a solution. Of course by repeating the analysis with different perturbations strength (i.e., by acting on the variance of the pdf_D associated with the most sensitive sub-modules, e.g., changing some components of *D*) we test a different architectural solution. The designer can therefore identify, by trial and error or automatically with a synthesis engine, the most adequate circuit for the given application.

To grant an error below $\hat{\gamma}$ at the system output the designer must grant that each real perturbation associated with the sub-systems implementation belongs to the selected *D*.

CHARACTERISE *X* AND *D*;
IDENTIFY ε, δ;
SELECT $M \geq \frac{\ln\delta}{\ln(1-\varepsilon)}$, $N \geq \frac{\ln\frac{2M}{\delta}}{2\varepsilon^2}$;
EXTRACT M POINTS x_i FROM *X* ACCORDING TO pdf_X;
EXTRACT *N* POINTS Δ_k FROM *D* ACCORDING TO pdf_D;
COMPUTE THE M VALUES $\hat{\gamma}_i = \frac{1}{N}\sum_{k=1}^{N} u(x_i,\Delta_k)$;
SELECT THE ESTIMATE FOR $\tilde{\gamma}$ AS $\hat{\gamma} = \max\{\hat{\gamma}_1,...,\hat{\gamma}_M\}$;

Fig. 2. The procedure for estimating $\tilde{\gamma}$

4. EXPERIMENTAL RESULTS

The envisaged system receives an analog signal, filters and digitizes it and, finally, extracts its frequency components for subsequent usage. This processing core is typical in several signal processing systems and has a large industrial impact, for instance, in embedded systems developed for the automotive and the speech processing fields. The application is composed of a low-pass analog filter followed by an AD converter that includes a sampler and a 12-bit quantizer. The digital output sequence is then processed in blocks of 16 samples by a digital FFT module. In the following experiments data come from a photodiode, which investigates the temperature of the melted material associated with a laser cut of steel/stainless steel.

The low pass filter is implemented as an active fourth-order Butterworth filter designed to attenuate the frequency components above fcut = 4 kHz. Errors due to the production process of analog parameters (8 resistors, 4 capacitors) and their aging effects inevitably affect all passive components inducing a strong impact on the filter response and, consequently, on the subsequent computation. The FFT architecture is composed of 4 stages, each containing 8 butterflies. The designer has decomposed the FFT by considering each butterfly as an independent subsystem (the error contribution originated during the multiply-and-accumulate operations within each butterfly is abstracted by the designer with a behavioural perturbation variable injected at the output of the summing node). Therefore, the digital system has been modelled with 32 independent perturbations with a uniform pdf modelling truncation of fractional values. Conversely, perturbations affecting the passive elements of the analog filter are modelled with gaussian distributions centred on the nominal value of the parameters, hence resembling the nature of the production process.

Note that another error contribution exists in the computation, i.e. that coming from the Analog-to-Digital conversion phase. Anyway, such error is constant since its resolution is fixed to 12 bits. We selected the performance loss function $u(\cdot)$ based on the power spectrum of each coefficient Px_i as

$$u(\Delta) = \max_k \left[\frac{\left|Px_i(k) - \tilde{P}x_i(k,\Delta)\right|}{\max_i\left[Px_i(k)\right]} \right], \; k = 1,\ldots,15$$

where $\widetilde{P}x_i(k,\Delta)$ is the perturbation-affected power spectrum for the i-th signal block, $\max_i[Px_i(k)]$ is the maximum coefficient computed in the *k-th* subband (considering the entire input signal of 16-samples blocks). Here, the DC coefficient of the FFT has not been considered. The loss function addresses the maximum relative variation of the magnitude of the frequency component.

The fixed perturbation case

We consider a case in which the designer wants to test the effect of truncation-like operators applied to the output of each stage inside the FFT module; truncation occurs when the intermediate results of each stage are stored in finite-length memory registers. Three possible architectures will be considered for memory size: 16 bits words (1 bit for the sign; 5 bits for the integer part of each value; 10 bits for the fractional part), 14 bits (8 bits for the fractional part) and 12 bits words (6 bits available for the fractional part). Obviously the effect of truncation is superimposed to the noise coming from the ADC which, in turn, receives its input from a filter where analog components have parameters different from nominal values (the actual values have been extracted from gaussian distributions associated with 1% tolerance for resistors and 2% for capacitors, hence modelling a given "real" circuit). To verify the impact of all these sources of error with varying input, the designer simulates the circuit behavior with a set of N=2950 input blocks ($\varepsilon = 0.03$, $\delta = 0.01$); the estimates $\hat{p}_N$ are represented in Figure 3.

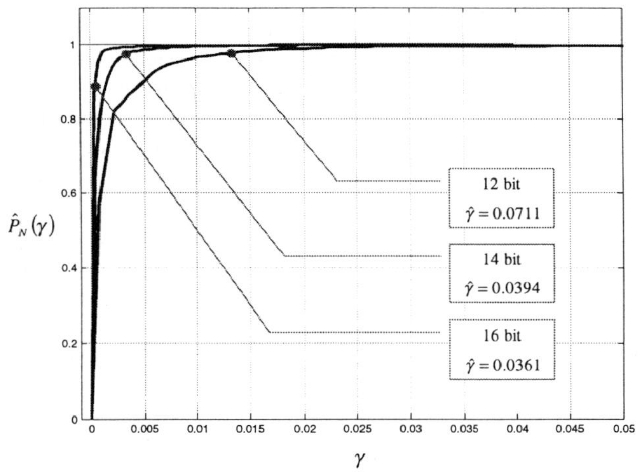

Fig. 3. Performance loss cumulative function

As we can expect, by lowering the number of bits we increase the performance loss estimated with $\hat{\gamma}$: the designer, by providing a tolerated performance loss for the whole system, e.g., 0.08, can establish the dimension of the digital sub-systems registers (12 bits).

The non-fixed perturbation case

The perturbation space is now explored by simulating different possible realizations of the analog circuit (i.e. by considering different extractions for resistors and capacitors from their respective distributions defined over *D*). Moreover, the perturbation that affects the values stored in FFT registers is now randomly generated from uniform distributions in the $[-\alpha,+\alpha]$ interval. In particular, the designer wishes to test the performance loss of the system associated with a mass production with tolerance 1% or 5% for resistors and 2% or 5% for capacitors; perturbation intervals for the butterflies are $\alpha = 2^{-8}$ (equivalent to a 8 fractional digits representation) and $\alpha = 2^{-10}$ (10 fractional bits). Figure 4 shows the estimates of $\hat{p}_N$ for *N*=M=1060.

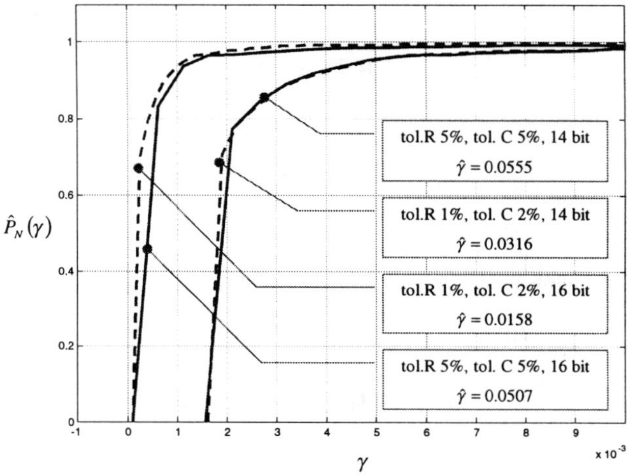

Fig. 4. Performance loss cumulative function

The designer, based on the information given in figure 4, identifies the best analog/digital system on the basis of the tolerated performance loss for the whole system (e.g., tol. 5% for resistors and capacitors and 14 bits for the butterfly outputs.

5. SUMMARY

The paper presents a methodology for estimating "in the large" the effect of perturbations affecting a computational flow. By suitably modeling the perturbations, their injection points and pdfs, a performance loss index can be obtained for a class of mixed analog/digital systems. The analysis can be carried out at the application level and allows validating candidate architectural solutions before the effective implementation or implemented solutions.

6. REFERENCES

[1] C.Alippi "A Probably Approximately Correct Framework to estimate Performance Degradation in Embedded Systems", IEEE-*TCAD*, Vol. 21, No. 7, July 2002

[2] R.Tempo, F.Dabbene, "Probabilistic Robustness Analysis and Design of Uncertain Systems", *Progress in Systems and Control Theory*, Vol. 25, 1999

[3] M.Vidyasagar, "Statistical learning Theory and Randmized algorithms for Control", IEEE-*Control systems*, Dec. 1998

[4] M.Vidyasagar, *A Theory of Learning and Generalisation with Applications to Neural Networks and Control Systems*, Springer-Verlag, Berlin, 1996

Closed Form Metrics to Accurately Model the Response in General Arbitrarily-Coupled *RC* Trees

Dinesh Pamunuwa* and Shauki Elassaad**

**Laboratory of Electronics and Computer Systems, Dept. of Microelectronics and Information Technology Royal Institute of Technology, Stockholm, Sweden. e-mail: dinesh@imit.kth.se*

***Cadence Berkeley Laboratories, Cadence Design Systems, Berkeley, CA. e-mail: shauki@cadence.com*

ABSTRACT

Closed form expressions are presented for the first and second moment of the impulse response for arbitrarily-coupled *RC* trees with multiple drivers, and used to generate accurate second order estimations of the transfer function from any driver to the receiver. The superposition of the waveforms for all switching events allows precise delay and noise calculations for systems of coupled interconnects with different aggressor arrival times, with a minimum of computational complexity.

1. INTRODUCTION

As gate transition times decrease and wiring density increases, particular attention has to be paid to noise modeling and its impact on performance and functionality. Delay models that account for the coupling noise induced by a number of switching neighboring aggressors on a victim signal are necessary. Finding the response of such systems requires the analysis of general arbitrarily-coupled trees. However the computational overhead associated with numeric techniques such as Spice is usually unacceptable for large designs. This has led to a proliferation of research in the past twenty years in the area of delay and noise modeling, with the aim of developing simplified metrics that still give acceptable accuracy. In the rest of this document, the term *simple tree* is used to refer to a tree that has capacitances only to ground, and *coupled tree* to refer to a tree that consists of simple trees coupled to each other through series capacitors.

There is a large body of literature that deals with delay modeling in simple trees. One of the most important and widely used metrics, the first moment of the impulse response, was proposed back in 1948 as an upper bound for the delay in valve circuits[1], and is known as the Elmore delay. Subsequently the authors of [2] developed tighter bounds and metrics that gave an indication of when the Elmore model was poor. Its attraction is that it is very simple, and yet exhibits good *fidelity*, giving results as good as more expensive models when used as a metric in interconnect optimization algorithms. A model based on the first and second moment of the impulse response, and the sum of the open circuit time constants was proposed in [3], which gives a stable approximation to the second order transfer function for simple trees. Since then, generic moment based techniques have been developed which are applicable to any linear circuit, and allow the calculation of an arbitrary number of poles [4]. In today's systems, delay and noise calculations are essential at an early stage in the planning process. However the complexity is such that generic moment matching and model order reduction techniques which require the formulation of nodal matrices and costly matrix manipulations are too expensive. Hence a lot of simplified models have been proposed. The models of [3] represent the minimum complexity for second order approximations of simple trees. In [5] the authors explicitly match the first three moments of the impulse response to a second order model in a methodology that guarantees stability. In [6] a heuristic delay model based on the first two moments was proposed.

In analyzing coupled trees, most research has concentrated on certain simplified configurations of interest. In [7] the authors present two pole delay models for a single π section, and extend it to accommodate multiple segmented aggressors in [8], but the allowed topology is still limited. In [9] the authors use circuit transformations to simplify a general tree to a 2-π model when analytic formulae can be used, but intermediate steps require the calculation of admittances at each branch point and the estimation of equivalent capacitances which increase run time and impact on the accuracy respectively. In [10] a technique is presented to generate the poles of a system with n storage elements, which has long been used in analog design to estimate the bandwidth of amplifiers. The complexity of the computation is proportional to n^{α} where α is the order of the pole. There are works which use this technique to estimate the two lowest frequency poles and use them to model the response for all switching events on the tree [11][12]. This can result in unacceptable accuracy, as the poles which determine the response for different switching events can be very far apart on the frequency axis. If used to generate higher frequency poles, this estimation technique will become prohibitively expensive.

In this paper we propose metrics based on the first and second moment of the impulse response to generate second order transfer functions from each driver to the receiver in arbitrarily-coupled trees, which allow both delay and noise estimations. Our contributions are that the moments are matched to the characteristic time constants in a novel way which minimizes computational complexity while allowing good accuracy, and that the *moments themselves* are calculated from completely accurate closed form expressions which are only slightly more complex than the Elmore delay, while retaining all its elegance. In this paper we are only concerned with the generation of the transfer function, which is the most important aspect of the modeling.

2. MODELING

The modeling requires the matching of easily calculable metrics of the circuit to the system transfer function. This involves

0-7803-7761-3/03/$17.00 ©2003 IEEE

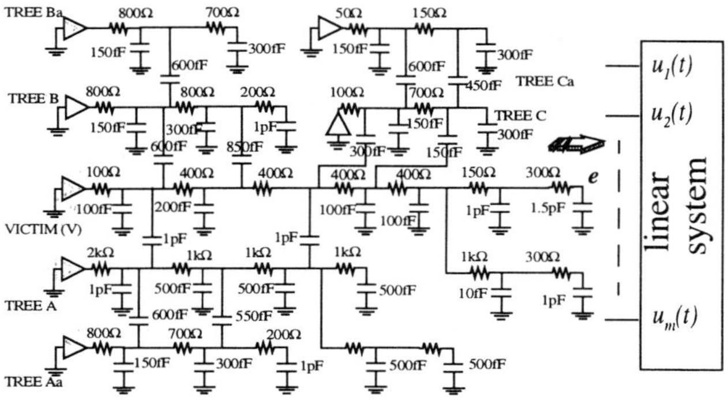

Figure 1. Example coupled *RC* tree

ascertaining the characteristics of the transfer function, calculating the metrics and then matching them to the terms in the transfer function.

A coupled *RC* tree is characterized by a resistive path from the receiver node *e* to the forcing (victim) driver, and series capacitive elements to other (aggressor) drivers. Hence when the victim driver switches the output will always change rails, while it will start and end at the same rail for an aggressor switching. Therefore the transfer functions characterizing the response to the victim switching and any of the aggressors switching are different. The former will have a zero on the negative part of the real axis:

$$H_v(s) = \frac{1 + s\tau_{z,v}}{(1 + s\tau_1)(1 + s\tau_2)} \tag{1}$$

while the latter will have a zero at the origin.

$$H_{a_i}(s) = \frac{s\tau_{z,a_i}}{(1 + s\tau_1)(1 + s\tau_2)} \tag{2}$$

2.1 Calculation of moments

In the following section, expressions are presented for the first and second moment of the impulse response for general coupled trees, which form the core of our models. The derivation is based on Kirchoff's laws and integration by parts, and is omitted due to lack of space. Shown in Fig. 1 is an example of a coupled tree which can be referred to in the following descriptions. First our notation is described below.

CS^p_k = capacitance to ground at node k in pth tree

CC^{pq}_{kj} = capacitance between node k on pth tree and node j on qth tree (first sub(super)script refers to reference tree)

R^p_{ke} = resistance shared on paths between source to nodes e and k on tree p

Υ^n_k = nth moment of the impulse response at the kth node

It should be noted that superscripts always refer to simple trees while subscripts always refer to nodes, except in the definition for moments, where the superscript refers to the order of the moment. Additionally, rail voltages are normalized to 0 and 1, and the expressions always derived for a positive step without loss of generality.

The first moment of the impulse response at the receiver node *e* for the victim driver switching is given by:

$$\Upsilon^1_{e,v} = \int_0^\infty t h^v_e(t)dt \tag{3}$$

Now the impulse response is the first time derivative of the step response, for which an expression can be formulated by summing up all capacitor currents. This can then be integrated by parts to yield (4), where a_1, a_2.. are the aggressors.

$$\Upsilon^1_{e,v} = \sum_{k \in victim} R^v_{ke}[CS^v_k + CC^{va_1}_k + CC^{va_2}_k + \ldots] = \tau^v_{D_e} \text{ say} \tag{4}$$

The second moment of the impulse response at *e* is given by:

$$\Upsilon^2_{e,v} = \int_0^\infty t^2 h^v_e(t)dt \tag{5}$$

Following the procedure described above in two stages, this can be shown to be equivalent to (6).

$$\Upsilon^2_{e,v} = 2\sum_{k \in victim} R^v_{ke}\left\{CS^v_k\tau^v_{D_k} + CC^{va_1}_k\left[\tau^v_{D_k} + \sum_{K \in a_1} R^{a_1}_{Kj}CC^{a_1v}_K\right] + CC^{va_2}_k\left[\tau^v_{D_k} + \sum_{K \in a_2} R^{a_2}_{Kj}CC^{a_2v}_K\right] + \ldots\right\} = 2(\tau^v_{G_e})^2 \text{ say} \tag{6}$$

From an approach identical to the former case, the first moment of the impulse response at node *e* on the victim tree for aggressor a_i switching can be shown to be:

$$\Upsilon^1_{e,a_i} = -\sum_{k \in victim} R^v_{ke}CC^{va_i}_k = -\tau^{a_i}_{D_e} \text{ say} \tag{7}$$

and the second moment:

$$\Upsilon^2_{e,a_i} = -2\sum_{k \in victim} R^v_{ke}\left\{(CS^v_k + CC^{va_1}_k + CC^{va_2}_k + \ldots)\tau^{a_i}_{D_k} + CC^{va_i}_k\left[\sum_{K \in a_i} R^{a_i}_{Kj}(CS^{a_i}_K + CC^{a_iv}_K + CC^{a_ib_1}_K + \ldots)\right]\right\} = -2\left(\tau^{a_i}_{G_e}\right)^2 \text{ say} \tag{8}$$

The expressions in (4), (6), (7) and (8) form the basis of our proposed models. An examination of these reveal their similarity to the Elmore delay, and all the accompanying characteristics that make the estimation algorithms very efficient.

2.2 Matching moments to the characteristic time constants in the circuit

The moments can be matched to the characteristic time constants in the circuit by using the identity that the nth moment of the impulse response is $(-1)^n$ times the nth derivative of the transfer function evaluated at *s*=0. This identity used on (1), (4) and (6) results in:

$$\tau^v_{D_e} = \tau_1 + \tau_2 - \tau_{z,v} \tag{9}$$

$$(\tau^v_{G_e})^2 = (\tau_1 + \tau_2 - \tau_{z,v})(\tau_1 + \tau_2) - \tau_1\tau_2 \tag{10}$$

Now additional information is necessary to solve for the three unknowns in (9) and (10). If the reciprocal pole sum is designated as τ_{sum}, these two equations can be combined to form the following quadratic, which yields two time constants.

$$\tau^2 - \tau_{sum}\tau + \tau^{v}_{D_e}\tau_{sum} - (\tau^{v}_{G_e})^2 = 0 \qquad (11)$$

The mathematical basis of our reduced order model is that geometric attributes (area and first moment of the area) and the values at t=0 and $t \to \infty$ of the estimated and actual waveform are equated. Since the initial and final values are already considered in the formulation of the transfer function, additional information is necessary to solve for the two poles and zeros associated with a switching event. At this point, it is helpful to look at the physical interpretation of the first and second moments of the impulse response. The first moment always considers resistances of the switching line, and either all capacitances connected to the switching line (in the case of the victim driver switching) or capacitances connecting it to a particular line (for the switching of an aggressor driver). The second moment propagates outwards another level, and considers the resistances and capacitances of immediately adjacent lines as well. This intuition is valuable in generating a solution with minimum computational complexity; namely, equation (11) can be used to generate the pole time constants for *all* switching events, by using the appropriate reciprocal pole sum.

For the victim switching, the metric that gives the best solution is the sum of the open circuit time constants with reference to the victim driver, which we shall call τ_p^*. This is simply the summation of the products of all capacitances connected to the victim line with the driving point resistance to each of those capacitors. This is a good approximation for the sum of the pole time constants[10], giving:

$$\tau_p^* = \tau_1 + \tau_2 \qquad (12)$$

To solve for the poles and zeros associated with an aggressor switching, the above identity is used on (2), (7) and (8) to give:

$$\tau^{a_i}_{D_e} = \tau_{z,\,a_i} \qquad (13)$$

$$\left(\tau^{a_i}_{G_e}\right)^2 = \tau_{z,\,a_i}(\tau_1 + \tau_2) \qquad (14)$$

Now the zero time constant is available immediately, and dividing (14) by (13) results in the reciprocal pole sum:

$$\left(\tau^{a_i}_{G_e}\right)^2 / \tau^{a_i}_{D_e} = \tau_1 + \tau_2 \qquad (15)$$

The pole time constants can be obtained by substituting (15) as τ_{sum} in (11). Since minimum information is used for both cases, some slight modification is necessary to guarantee stability, as explained below.

The conditions for potential instability can be identified by analyzing the quadratic which yields the time constants. The first limiting condition is that the magnitude of the square root should be greater than the reciprocal pole sum, which yields:

$$\tau_{sum} > (\tau^{v}_{G_e})^2 / \tau^{v}_{D_e} \qquad (A)$$

The second is that the sign under the radical is negative, which, after some simplification, results in the following:

Table 1. Different values for τ_{sum} for an aggressor switching

Condition	τ_{sum}
no violation	$\left(\tau^{a_i}_{G_e}\right)^2 / \tau^{a_i}_{D_e}$
$\left(\tau^{a_i}_{G_e}\right)^2 / \tau^{a_i}_{D_e} < \left(\tau^{v}_{G_e}\right)^2 / \tau^{v}_{D_e}$	$0.99\left(\tau^{v}_{G_e}\right)^2 / \tau^{v}_{D_e} + 0.02\left[\tau^{v}_{D_e} - \sqrt{\left(\tau^{v}_{D_e}\right)^2 - \left(\tau^{v}_{G_e}\right)^2}\right]$

$$\tau_{sum} < 2\left[\tau^{v}_{D_e} - \sqrt{(\tau^{v}_{D_e})^2 - (\tau^{v}_{G_e})^2}\right] \text{ or}$$
$$\tau_{sum} > 2\left[\tau^{v}_{D_e} + \sqrt{(\tau^{v}_{D_e})^2 - (\tau^{v}_{G_e})^2}\right] \qquad (B)$$

An inspection of the relevant expressions shows that the only possible violation in the case of the victim driver switching is (B). i.e. very occasionally, using τ_p^* can result in complex poles. The physical interpretation of such an occurrence is that the sum of the open circuit time constants underestimates the reciprocal pole sum, which has been unusually escalated by an aggressor or aggressors with exceptionally high parasitics. Because both exponential waveforms are either additive or subtractive unlike when an aggressor switches (where one is additive and the other is subtractive), the higher frequency pole does not have a significant impact. In fact, this form of instability is usually an indication of a very low frequency pole which makes the prediction of the waveform straightforward. The simplest remedy therefore is to consider a single pole response, with the pole time constant being given by $\tau^{v}_{D_e}$, which results in good accuracy as shown in the results section.

For the case of an aggressor driver switching, the only possible violation is (A). This is in fact the more common form of instability encountered and occurs when the dominant poles for the victim and the particular aggressor are very far apart on the frequency axis. Physically, this translates to a situation where the receiver node is charged extremely rapidly by a very strong aggressor (i.e. through a relatively very small time constant), and decays with a very long tail, dictated by the much larger time constant of the victim. Such behavior is common for far end coupling, The instability in the solution predicted by (11) occurs because the reciprocal pole sum given by (15) accurately reflects the high frequency nature of the poles in the aggressor's charging path, but $\tau^{v}_{D_e}$ and $\left(\tau^{v}_{G_e}\right)^2$ reflect the much lower frequency content of the victim's dominant poles, and the gap is too much to bridge. The solution without generating extra information about the circuit, is to accept the next best approximation. That is to say, if τ_{sum} is so small that it violates inequality (A), the simplest and most logical remedy is to increase τ_{sum} so that real roots are generated. Since the equality will generate coincident poles which is not acceptable, the exact value should be chosen so that it is slightly greater than the equality, which can be achieved with a percentage factor, such as 1%. This yields accurate results, because the intention is to generate the *best two pole single zero*

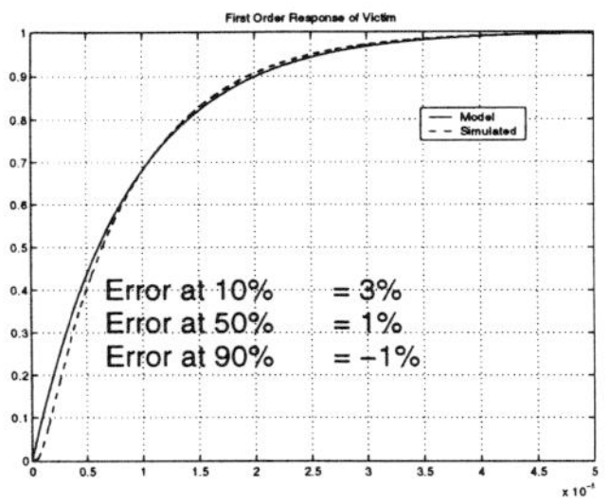

Fig.5.a Victim Switching

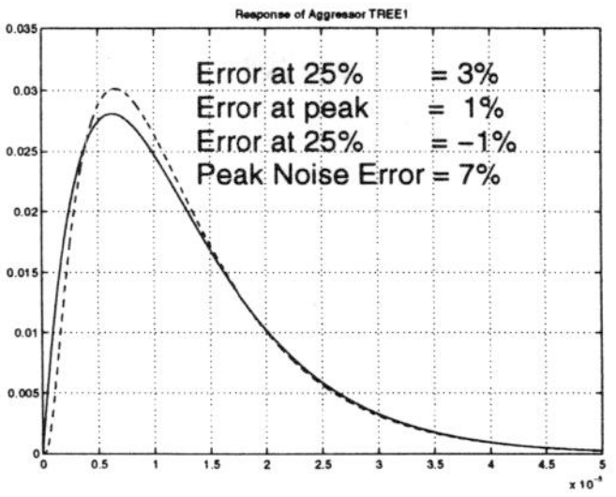

Fig.5.b Tree B Switching

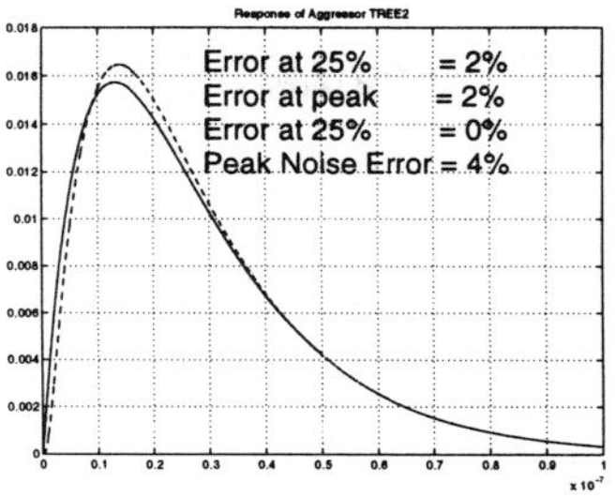

Fig.5.c Tree A Switching

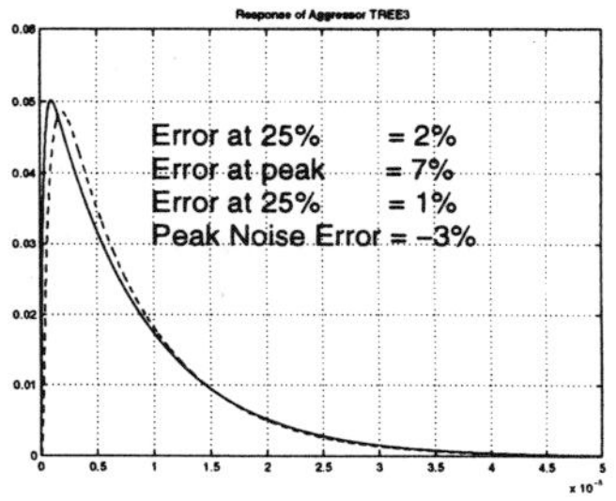

Fig.5.d Tree C Switching

Figure 2. Response to different switching events

model; in other words the poles and zero need not equate to actual poles and zeros of the system, and indeed should differ for a second order approximation. Using the factor of 1% beyond the threshold which yields coincident poles ensures that both the high and low frequency behavior is matched. Following this approach, the values that τ_{sum} should take in the different cases are summarized in Table 1.

3. RESULTS

The proposed metrics were tested on several different test beds which cover a wide range of topologies and all the corner cases where conditions (A) and (B) are violated, by comparing the step response for different switching events with Spectre simulations. Due to space restrictions, only the results pertaining to the tree shown in Fig. 1 consisting of the victim, three primary aggressors, and three secondary aggressors are presented. Shown in Fig. 2 are the waveforms (step response) at the receiver node *e* for each driver switching. It can be seen that the model prediction is very close to the Spectre simulation. In the case of the victim switching, inequality (A) is violated, and the single pole response shows good accuracy. Since the actual and predicted delay at a single threshold can agree very well, and still result in significant deviations along the full waveform, we tested the accuracy at three points along the waveform. For the victim switching, the thresholds are 10%, 50% and 90%, while for the aggressors they are 25%, 50% and 75% of the peak amplitude. The error at different thresholds is given as a fraction of the pulse width from v_1 to v_3 for the aggressors.

4. CONCLUSIONS

Closed form expressions for the first two moments of the impulse response for general coupled *RC* trees were presented, and used to generate second order approximations to the transfer function for any switching event with guaranteed stability. The summation of all waveforms results in the complete response at the node of interest. The new models we propose represent the minimum complexity for second order estimations of coupled trees when generality is not compromised, and in fact subsume a lot of models that address simplified structures. For testing purposes, these expressions were used to derive the time domain waveform for the step response. The accuracy with which the delay at a given threshold is predicted was found to be more than 90%, even for complex circuits such as shown in Fig. 1. The peak noise was predicted with similar accuracy on circuits representative of nano meter interconnect structures. The simplicity and accuracy of the models combined with their generality should make them useful in delay and noise estimations in complex systems, early in the design flow.

5. REFERENCES

[1] W. C. Elmore, "The transient response of linear damped circuits," *J. Appl. Physics*, vol. 19, pp-55-63, Jan. 1948.

[2] J. Rubinstein, P. Penfield, and M. Horowitz, "Signal delay in RC tree networks," *IEEE Trans. Computer-Aided Design*, vol CAD-2, no. 3, pp. 202-211, Jul. 1983.

[3] M. A. Horowitz, "Timing models for MOS circuits," Ph.D. dissertation, Stanford Electronics Laboratories, Stanford University, Stanford, CA, Jan. 1984.

[4] L. T. Pillage and R. A. Rohrer, "Asymptotic waveform evaluation for timing analysis," *IEEE Trans. Computer-Aided Design of ICs and Sys.*, vol. 9, pp. 352-366, Apr. 1990.

[5] B. Tutuianu, F. Dartu and L. T. Pillage, "An explicit RC-circuit delay approximation based on the first three moments of the impulse response," in *Proc. DAC*, 1996, pp. 611-616.

[6] C. J. Alpert, A. Devgan, and C. V. Kashyap, "RC Delay metrics for performance optimization," *IEEE Trans. Computer-Aided Design of ICs and Sys.*, vol. 20, no. 5, pp. 571-582, May 2001.

[7] A. B. Kahng, S. Muddu, and D. Vidhani, "Noise and delay uncertainty studies for coupled RC interconnects," in *Proc. ASIC/SOC*, 1999, pp. 3-8.

[8] A. B. Kahng, S. Muddu, N. Pol, and D. Vidhani, "Noise model for multiple segmented coupled RC interconnects," in *Proc. ISQED*, 2001, pp. 145-150.

[9] M. Takahashi, M. Hashimoto, and H. Onodera, "Crosstalk noise estimation for generic RC trees", in *Proc. ICCD*, 2001, pp. 110-116.

[10]B. L. Cochrun and A. Grabel, "On the determination of the transfer function of electronic circuits," *IEEE Trans. Circuit Theory*, vol. CT-20, pp.16-20, Jan. 1973.

[11]X. Tong and M. Marek-Sadowska, "Efficient delay calculation in presence of crosstalk," in *Proc. ISQED*, 2000, pp. 491-497.

[12]L. H. Chen and M. Marek-Sadowska, "Efficient closed-form cross-talk delay metrics," in *Proc. ISQED*, 2002, pp. 431-436.

PODEA: POwer Delivery Efficient Analysis with Realizable Model Reduction*

Rong Jiang, Tsung-Hao Chen
University of Wisconsin, Madison, 53705, WI

Charlie Chung-Ping Chen
National Taiwan University, Taipei 106, Taiwan

ABSTRACT

The huge number of independent sources of power delivery system prevents the use of traditional model reduction algorithms due to the port domination nature. This paper presents an innovative RC model reduction method, PODEA, to analyze RC linear circuits with many dynamic independent sources. Based on multi-port Norton theorem and model order reduction techniques, we develop and apply current source transformation algorithm to transform attached current sources from one node to its neighboring nodes. Since there is no source attached, general RC reduction algorithms can be applied to eliminate the node. Experimental results demonstrate the efficiency and accuracy of the proposed PODEA algorithm. With linear running time, for case with over 50,000 nodes, our reduction method only takes about 0.6 seconds while maintain with 1% of error and 88% reduction ratio.

1. INTRODUCTION

Due to the higher power dissipation, lower supply voltage and faster operation frequency of VLSI chips nowadays, power grid analysis has become one of the designer's high-priority concerns. Power delivery noises, which include IR-drop, Ldi/dt-drop and resonance, may degrade the system performance or even affect its functionality. Hence extensive analyses of power delivery system are required to ensure that they meet the targeted performance and reliability goals [1] [2] [3].

Size explosion is one of the major obstacles to analyze power delivery system efficiently. Typically the power delivery network has 1 million to 100 million nodes and lots of distributed voltage and current sources. The tremendous amount of power delivery elements prevents the general-purpose circuit simulators such as SPICE to fulfill the demanding task in a timely manner.

Model order reduction has been extensively studied during the last decade to overcome this obstacle, see [4] [5] [6] [7] [8] [9] [10]. The major difficulty of applying model order reduction algorithms to power delivery circuits is the size explosion of port number since the performance and reduction ratio of most reduction algorithms, such as PRIMA [6], are strongly dependent on the number of ports. Each current or voltage source in power delivery system has to be treated as a port in PRIMA and hence it could easily exceed more than millions of ports which makes it computationally prohibitive. [9] and [10] propose Gauss elimination type algorithms to eliminate the insignificant nodes from its time constant. Albeit their great efficiency and accuracy, [9] and [10] can not handle circuits with independent sources such as power delivery circuits.

In this paper, we propose an efficient RC model reduction method, PODEA, with emphasis on power delivery system analysis. We develop CTRAN algorithm, which integrates multi-port Norton theorem with finite time Piece-Wise-Linear (PWL) modeling of independent sources, to transform current source on one node to its neighboring nodes. Once the attached current source is transformed, RC reduction algorithms such as [7] [8] [9] [10] can be applied to eliminate the node. The CTRAN algorithm is easy to calculate and has controllable accuracy. It cooperates well with Gaussian elimination type reduction algorithms which reduce most circuits in linear time.

In the following section, we will present our PODEA reduction method and explain CTRAN algorithm in details. Then we will discuss the computation complexity of PODEA and illustrate it with several examples and report the experimental results. Last we will make some conclusion remarks.

2. POWER DELIVERY ANALYSIS MICROMODELING ALGORITHM

In this section we will present our PODEA algorithm for reducing RC circuits with many internal current sources.

Given an RC circuit with many internal current sources, our reduction method consists of three steps. First, we partition the circuit into two parts: part one includes nodes that connect no current sources, and part two includes nodes with attached current sources. The second step is to reduce nodes in part one using traditional RC model reduction algorithms, such as [7] [8] [9] [10]. In the third step, two phases are applied to each node in part two: first we use CTRAN algorithm to transform the attached current source to its neighboring nodes, then Gaussian elimination type algorithms such as TICER can be applied to reduce the node. Table 2-1 presents the outline of PODEA. Figure 2-1 shows the typical reduction procedure in step three.

Algorithm: PODEA, POwer Delivery Efficient Analysis
Step 1: Partition the given circuit into two parts: part one includes nodes which connect no current sources; part two includes nodes with attached current sources.
Step 2: Reduce part one using traditional realizable RC model order reduction algorithms.
Step 3: For each node in part two, the following two phases are applied to eliminate the node.
- **3-1:** Apply CTRAN to transform its attached current source to the neighboring nodes.
- **3-2:** Apply Gaussian elimination type model order reduction algorithms to eliminate the node.

Table 2-1. Outline of PODEA algorithm

* This work is partially supported by NSF Grant CCR-0093309 and 0204468, Intel Corporation and Faraday Technology Corporation.

0-7803-7761-3/03/$17.00 ©2003 IEEE

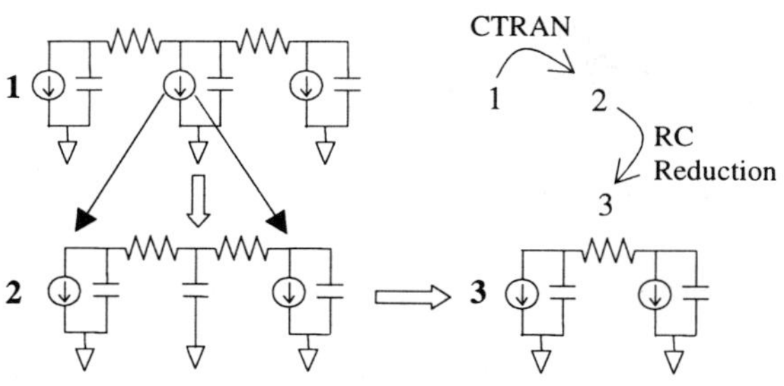

Figure 2-1. Typical reduction procedure in step three.

In the following subsections, we will focus on deriving CTRAN algorithm and explaining how CTRAN can be used to transform current sources. First, we will use an m-terminal star network to model the basic structure in power delivery system and derive the Norton equivalent current sources on its terminals. After that we introduce the finite time PWL waveform of internal current sources. Based on this background, we will derive and present CTRAN algorithm in details. In the last subsection, computation complexity of PODEA is discussed.

2.1 M-terminal Norton Equivalent Circuit and PWL Current Sources

A basic m-terminal star network with central node N can be used to model the basic structure of RC power delivery system, see Figure 2-2 (a). A central current source I_N is attached on central node N. The ith branch consisting of a conductance g_i and a capacitance c_i joins the terminal to the central node N.

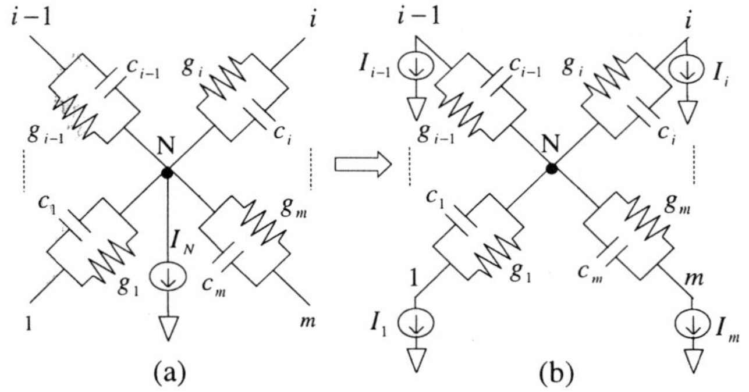

Figure 2-2. An m-terminal star network and its Norton equivalent circuit.

We can obtain the Norton equivalent current sources on each terminal by turning on the central current source and grounding all the terminals. Then the central current source can be replaced by equivalent terminal current sources according to substitution theorem, see Figure 2-2 (b). The equivalent current source on the ith terminal is given by:

$$I_i = \frac{g_i + sc_i}{G + sC} I_N \qquad (2\text{-}1)$$

where $G=\sum_{i=1}^{m} g_i$, $C=\sum_{i=1}^{m} c_i$ are total conductance and capacitance of central node N respectively. Equation (2-1) gives us a way to calculate equivalent current sources. However we also need a method to model the internal current sources. This modeling method should have controllable accuracy and by using this model, Equation (2-1) is easy to calculate.

Finite time PWL sources can fulfill the task of modeling most types of waveforms with controllable accuracy. Given two PWL sources with different set of time delays, it's easy to add them together. Assume I_N is a finite time PWL current source represented by delayed steps and ramps as shown in Figure 2-3. For the ith segment, α_i^N, γ_i^N and τ_i^N stand for the initial value, slope and time delay respectively.

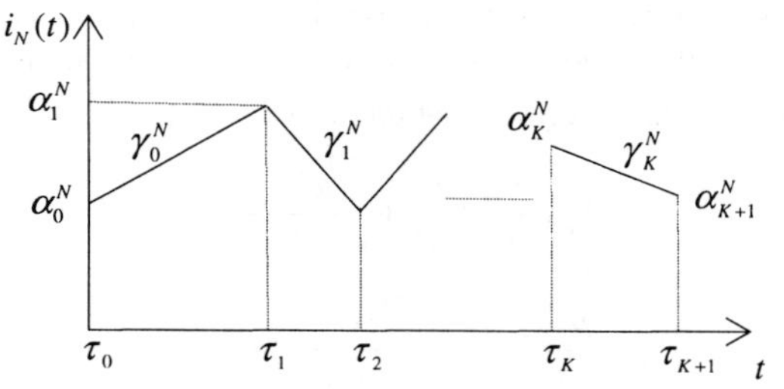

Figure 2-3. Piece wise linear waveform of $i_N(t)$.

$i_N(t)$ can be represented in the time domain as:

$$i_N(t) = \sum_{j=0}^{K}\{[\alpha_j^N + \gamma_j^N(t-\tau_j)]E(t-\tau_j) - [\alpha_{j+1}^N + \gamma_j^N(t-\tau_{j+1})]E(t-\tau_{j+1})\} \qquad (2\text{-}2)$$

Laplace transformation in s-domain is given by:

$$I_N = \sum_{j=0}^{K}[(\frac{\alpha_j^N}{s}+\frac{\gamma_j^N}{s^2})e^{-\tau_j s} - (\frac{\alpha_{j+1}^N}{s}+\frac{\gamma_j^N}{s^2})e^{-\tau_{j+1} s}] \qquad (2\text{-}3)$$

With Equation (2-1) and representation of I_N in Equation (2-3), we can develop our CTRAN algorithm to calculate Norton equivalent current sources.

2.2 CTRAN: An Efficient Way for Transforming the Internal Current Sources

In this subsection, we introduce CTRAN algorithm which transforms current source on one node to its neighboring nodes.

Substitute Equation (2-3) to Equation (2-1), we get the equivalent current source on the ith terminal in s-domain:

$$I_i = (\frac{g_i + sc_i}{G + sC})\sum_{j=0}^{K}[(\frac{\alpha_j^N}{s}+\frac{\gamma_j^N}{s^2})e^{-\tau_j s} - (\frac{\alpha_{j+1}^N}{s}+\frac{\gamma_j^N}{s^2})e^{-\tau_{j+1} s}] \qquad (3\text{-}1)$$

First consider the simplest scenario in which the central node is a resistive node (node with only resistors connected to it). In this case, Equation (3-1) equals to:

$$I_i = \sum_{j=0}^{K}[(\frac{\alpha_j^i}{s}+\frac{\gamma_j^i}{s^2})e^{-\tau_j s} - (\frac{\alpha_{j+1}^i}{s}+\frac{\gamma_j^i}{s^2})e^{-\tau_{j+1} s}] \qquad (3\text{-}2)$$

where

$$\alpha_j^i = \frac{g_i}{G}\alpha_j^N,\ \gamma_j^i = \frac{g_i}{G}\gamma_j^N \tag{3-3}$$

Equation (3-3) can be translated into a procedure for calculating the equivalent current sources on neighboring nodes. The equivalent current source on the ith neighboring node can be obtained by multiplying the initial values of central current source by g_i / G. In the case that the central node is a capacitive node, the equivalent current source on the ith neighboring node equals to multiple each initial value of central current by c_i / C.

Next we consider the general case. To get the equivalent current source on the ith terminal, $i_i(t)$, we need to calculate its values at different time delays. Equation (3.1) can be shown as follows:

$$I_i = \sum_{j=0}^{K}[(\frac{\mu_{j,1}^i}{s} + \frac{v_j^i}{s^2} + \frac{\omega_{j,1}^i}{s+G/C})e^{-\tau_j s}$$

$$-(\frac{\mu_{j,2}^i}{s} + \frac{v_j^i}{s^2} + \frac{\omega_{j,2}^i}{s+G/C})e^{-\tau_{j+1} s} \tag{3-4}$$

where

$$v_j^i = \frac{g_i}{G}\gamma_j^N$$

$$\mu_{j,1}^i = \frac{G(g_i\alpha_j^N + c_i\gamma_j^N) - Cg_i\gamma_j^N}{G^2}$$

$$\mu_{j,2}^i = \frac{G(g_i\alpha_{j+1}^N + c_i\gamma_j^N) - Cg_i\gamma_j^N}{G^2}$$

$$\omega_{j,1}^i = \frac{G^2c_i\alpha_j^N + C^2g_i\gamma_j^N - GC(g_i\alpha_j^N + c_i\gamma_j^N)}{G^2C}$$

$$\omega_{j,2}^i = \frac{G^2c_i\alpha_{j+1}^N + C^2g_i\gamma_j^N - GC(g_i\alpha_{j+1}^N + c_i\gamma_j^N)}{G^2C} \tag{3-5}$$

Inverse Laplace transformation of Equation (3-4) is

$$i_i(t) = \sum_{j=0}^{K}\{[\mu_{j,1}^i + v_j^i(t-\tau_j) + \omega_{j,1}^i e^{-\frac{G}{C}(t-\tau_j)}]E(t-\tau_j)$$

$$-[\mu_{j,2}^i + v_j^i(t-\tau_{j+1}) + \omega_{j,2}^i e^{-\frac{G}{C}(t-\tau_{j+1})}]E(t-\tau_{j+1})\} \tag{3-6}$$

Also we notice the following equations hold for parameters in Equations (3-4) and (3-6):

$$\mu_{j,1}^i + v_j^i\Delta\tau_{j+1,j} = \mu_{j,2}^i \tag{3-7}$$

$$\omega_{j+1,1}^i - \omega_{j,2}^i = \kappa^i\Delta\gamma_{j+1} \tag{3-8}$$

$$\mu_{j+1,1}^i + \omega_{j+1,1}^i = \mu_{j,2}^i + \omega_{j,2}^i \tag{3-9}$$

where $\kappa^i = (Cg_i - Gc_i)/G^2$, $\Delta\tau_{m,n} = \tau_m - \tau_n$, $\Delta\gamma_{j+1}^N = \gamma_{j+1}^N - \gamma_j^N$.

In order to calculate $i_i(\tau_j)$ using Equation (3-6), we need to proof $i_i(t)$ is continuous at τ_j because step function $E(t-\tau_j)$ has no definition at τ_j. This can be done by showing that $i_i(\tau_j^+) = i_i(\tau_j^-)$. For example, at τ_1:

$$i_i(\tau_1^-) = \mu_{0,1}^i + v_0^i\Delta\tau_{1,0} + \omega_{0,1}^i e^{-\frac{G}{C}\Delta\tau_{1,0}} = \mu_{0,2}^i + \omega_{0,1}^i e^{-\frac{G}{C}\Delta\tau_{1,0}}$$

$$i_i(\tau_1^+) = \mu_{1,1}^i + \omega_{1,1}^i - \omega_{0,2}^i + \omega_{0,1}^i e^{-\frac{G}{C}\Delta\tau_{1,0}} = \mu_{0,2}^i + \omega_{0,1}^i e^{-\frac{G}{C}\Delta\tau_{1,0}}$$

It follows $i_i(\tau_1^+) = i_i(\tau_1^-)$. $i_i(t)$ can be proven to be continuous in time range from τ_0^+ to τ_{K+1}^- and hence we can calculate α_j^i that is the exact value of $i_i(t)$ at τ_j by calculating $i_i(\tau_j^-)$ except at τ_0. After indulging in some algebra, we get the following formula for calculating $i_i(t)$ at different time delays.

For $t = \tau_0$, $\alpha_0^i = \frac{c_i\alpha_0^N}{C}$ (3-10)

For $t = \tau_j$, $j = 1, \cdots, K+1$,

$$\alpha_j^i = \mu_{j-1,2}^i + \omega_{0,1}^i e^{-\frac{G}{C}\Delta\tau_{j,0}} + \kappa^i\sum_{k=1}^{j-1}\Delta\gamma_k^N e^{-\frac{G}{C}\Delta\tau_{j,k}} \tag{3-11}$$

The CTRAN algorithm is summarized in Table 2-2.

Algorithm: CTRAN: Current Transformation Algorithm
Input: Central current source $I_N = \{\alpha_i^N, \tau_i\}_{i=0}^{K+1}$
G, total conductance; C, total capacitance
Conductor and capacitor on branches, $\{g_i, c_i\}_{i=1}^M$
M, the number of branches
Output: $I_i = \{\alpha_{0,\cdots,K+1}^i, \tau_{0,\cdots,K+1}\}_{i=1}^M$
Begin
$\sigma = -G/C$
for j = 0 : K
$\gamma_j^N = (\alpha_{j+1}^N - \alpha_j^N)/(\tau_{j+1}^N - \tau_j^N)$
end
for i = 1 : M
$\kappa = (Cg_i - Gc_i)/G^2$
$\omega = \frac{G^2c_i\alpha_0^N + C^2g_i\gamma_0^N - GC(g_i\alpha_0^N + c_i\gamma_0^N)}{G^2C}$
$\alpha_0^i = c_i\alpha_0^N / C$
for m = 1 : K+1
$\mu = \frac{G(g_i\alpha_m^N + c_i\gamma_{m-1}^N) - Cg_i\gamma_{m-1}^N}{G^2}$
$\alpha_m^i = \mu + \omega e^{\sigma(\tau_m - \tau_0)} + \kappa\sum_{n=1}^{m-1}(\gamma_n^N - \gamma_{n-1}^N)e^{\sigma(\tau_m - \tau_n)}$
end
end
End

Table 2-2. CTRAN: current transformation algorithm

In practice, we can simplify calculation by eliminating some exponential terms in Equation (3-11) according to the value of G/C and the time interval between different time delays. For example, if $G >> C$, we can expect the exponential terms in

Equation (3-11) decay very fast. In this case, the following formula can be used to approximate Equation (3-11):

$$\alpha_j^i \approx \frac{g_i \alpha_j^N + c_i \gamma_{j-1}^N}{G} \qquad (3\text{-}12)$$

Hence accuracy is controllable in the process of current source transformation.

2.3 Computation Complexity of PODEA

Our PODEA reduction method has linear running time and controllable accuracy. The running time of CTRAN is proportional to the number of neighboring nodes and the running time of Gaussian elimination type RC reduction is $O(n)$ as reported in [10]. Hence the total running time $O(\alpha n)$ can be achieved, where n is the number of eliminated nodes, and α is the average number of their neighboring nodes.

3. EXPERIMENTAL RESULTS

In this section, we illustrate our PODEA power delivery analysis algorithm with couple of examples. The test circuits are randomly generated RC tree. The value range of R is from 0.1 Ω to 1 Ω and C is from $1fF$ to $1pF$. The proposed algorithm is implemented in C language and run on a PIII 900MHz machine with 256MB memory.

Table 3-1 lists the simulation results of different circuits. For the 51,205-node circuit, the reduction time is about 0.6 secs. The simulation speedup before and after reduction is at least 59X.

Circuit Size		Reduction Time (secs)	Percent Error	Run Time B/A Reduction (secs)		Reduction Ratio
Nodes	Sources					
50,002	30,310	0.561	1.0%	2878	48 (59X)	87.6%
51,205	32,124	0.570	1.1%	3064	49 (62X)	87.6%
65,371	32,534	0.691	N/A	N/A	72	87.4%
72,938	54,765	0.822	N/A	N/A	93	88.7%
81,125	61,205	1.092	N/A	N/A	117	87.9%
120,513	91,125	1.632	N/A	N/A	218	88.6%

Table 3-1. Some experimental results.

Figure 3-1 (a) plots the SPICE simulation of the 51,205-node circuit before and after reduction; (b) shows the error curve. It shows that the error is less than 35 mV.

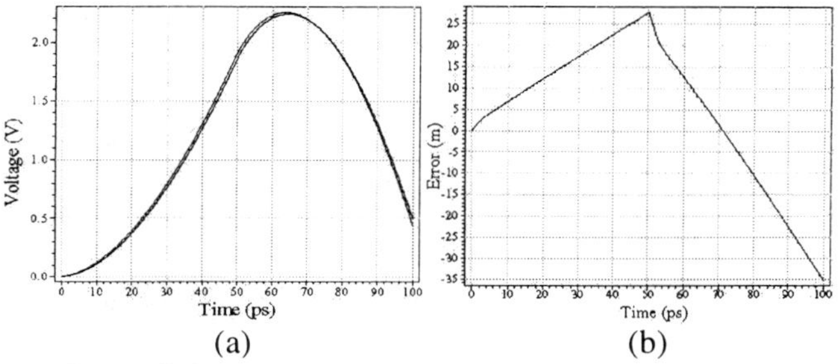

Figure 3-1. (a) Waveforms of the 51205-node circuit B/A reduction; (b) Error curve in terms of mV.

Figure 3-2 shows the running time and approximate memory consumption of different circuits. We can see from Figure 3-2, the running time and memory consumption is roughly linear proportional to the node number.

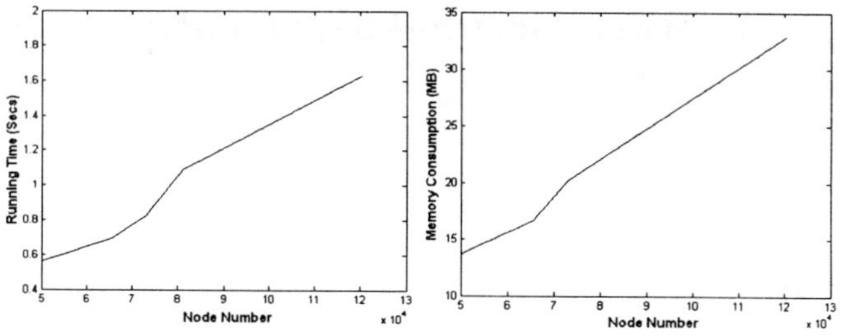

Figure 3-2. Running time and memory consumption versus the number of nodes.

4. SUMMARY

The contributions of this paper are as follows. First, we establish a realizable RC reduction method with independent current sources, PODEA, for power delivery system analysis. It has linear running time and controllable accuracy. Second we develop CTRAN algorithm to transform current sources thus traditional RC model reduction algorithms can be applied to reduce the circuits. The analysis results will help the designers to identify potential power supply noise problems and optimize various design configurations.

5. REFERENCES

[1] M. Zhao, R. V. Panda, S. S. Sapatnekar, T. Edwards, R. Chaudhry and D. Blaauw. Hierarchical analysis of power distribution networks. *DAC*, 2000.

[2] J. M. Wang and T. V. Nguyen. Extended Krylov method for reduced order analysis of linear Circuit with multiple Sources. *DAC*, 2000.

[3] Y. Cao, Y.-M. Lee, T.-H. Chen and C. C.-P. Chen. HiPRIME: hierarchical and passivity reserved interconnect macromodeling engine for RLKC power delivery. *DAC*, June 2002.

[4] L. T. Pillage and R. A. Rohrer. Asymptotic waveform evaluation for timing analysis. *IEEE Trans. Computer Aided Design*, 1990.

[5] A. Odabasioglu, M. Celik and L. T. Pillage. PRIMA: Passive reduced order interconnect macromodeling algorithm. *ICCAD*, pages 58–65, 1997.

[6] P. Rabiei and M. Pedram. Model order reduction of large circuits using balanced truncation. *DAC*, February 1999.

[7] A. Devgan and P. R. O'Brien. Realizable reduction for RC interconnect circuits. *Proc. ICCAD*, 1999.

[8] P. Ganesh and C. C.-P. Chen. RC-in RC-out model order reduction accurate up to second order moments. *ICCAD*, 1999.

[9] A. J. Genderen and N. P. Van Der Meijs. Space user's manual, Space tutorial, Space 3D capacitance extraction user's manual. Delft University of Technology, Dept. of EE, Delft, Netherlands, 1995.

[10] B. N. Sheehan. TICER: Realizable reduction of extracted RC circuits. *Proc. ICCAD*, 1999.

5th Order Electro-Thermal Multi-tone Volterra Simulator with Component-level Output

Antti Heiskanen, Timo Rahkonen

Electronics Laboratory
Dept. of Electrical Engineering and Infotech Oulu
University of Oulu, PO Box 4500, 90014 Oulu, FINLAND

Abstract

This paper describes a method of implementing thermal power feedback (TPF) into a 5th order nonlinear distortion analysis with Volterra-series. The main advantages of the Volterra analysis is that different thermal and electro-thermal contributions of distortion can be separated, giving extensive information about dominant distortion mechanisms. As an example, electro-thermal distortion in a simple common emitter amplifier is studied.

1. Introduction

Operating temperature has a great effect to most electrical parameters in semiconductor devices. The temperature-dependent parameters of the device depend on the external temperature, bias condition, signal-induced power dissipation inside the device, and of thermal couplings with the surroundings [1]. The self-heating process of a semiconductor device can be modeled with a thermal network heated by the dissipated power of the device [2,3,4]. Self-heating affects the electrical parameters of the device and causes distortion to the output signals [5,6].

When studying electrical and electro-thermal distortion mechanisms of an RF power amplifier, a Volterra-series analysis has proven to be effective tool [7,8], as it gives extensive information about the dominant distortion mechanisms.

Here, a method of implementing the self-heating effects into a numerical Volterra analysis is described. The self-heating procedure known also as thermal power feedback (TPF) is implemented into a 5th order numerical Volterra simulator [9] and a simple simulation example is presented.

Section 2 shows the TPF calculation process with numerical Volterra analysis. Section 3 shows a method of how to implement the TPF into a numerical 5th order Volterra simulator [9] in such a way that the rich information of distortion mechanisms is maintained even when using numerical Volterra analysis. Section 4 shows a simple simulation example where the thermally induced distortion causes dominant and strongly bandwidth dependent IM3 distortion.

2. TPF in numerical Volterra analysis

Both numerical and symbolic Volterra-series analysis are well covered e.g. in [10-14]. In these two methods the nonlinear elements can be resistive or reactive and are modeled as polynomial nonlinearities around the DC operating point.

The symbolic Volterra analysis is an effective method to study distortion mechanisms for small circuits with 1 or 2-tone excitation and 1-2 dimensional 3rd order polynomial nonlinearities. When analyzing larger circuits with higher order multi-dimensional nonlinearities or multi-tone multi-input excitations, the symbolic Volterra analysis becomes inefficient. At least a 5th order Volterra analysis is needed to model amplitude-dependent phase rotation of IM3, and at least a 3-tone excitation is needed for studying different predistortion systems, for example. Thus the 3rd order symbolic Volterra analysis is often insufficient for power amplifier distortion studies.

The complexity of symbolic Volterra analysis is avoided by using so called direct nonlinear current method [13]. By using direct method the circuit can have multitone-multi-input excitation and higher order multi-dimensional nonlinearities. Here, a component-wice distortion presentation described in [9] is used to obtain the extensive information of the dominant distortion mechanisms.

Time domain current equations for 1, 2 and 3 dimensional elements are presented in [13]. When TPF is considered, also the junction temperature is considered as a controlling voltage [7] and is fed to one controlling port. For example, the 3rd degree electro-thermal time domain Volterra model for a 3-dimensional transconductance is shown in (1).

$$\begin{aligned} i(v_g, v_d, T) = {} & K_{100} \cdot v_g + K_{200} \cdot v_g^2 + K_{300} \cdot v_g^3 \\ & + K_{010} \cdot v_d + K_{020} \cdot v_d^2 + K_{030} \cdot v_d^3 \\ & + K_{110} \cdot v_g v_d + K_{210} \cdot v_g^2 v_d + K_{120} \cdot v_g v_d^2 \\ & + K_{001} \cdot T + K_{101} \cdot T \cdot v_g + K_{011} \cdot T \cdot v_d \end{aligned} \quad (1)$$

The indexes in the coefficients stand for the powers of 1st, 2nd and 3rd controlling inputs, respectively. Thus the 1st row presents the linear gm element and its 2nd and 3rd degree nonlinearities. The 2nd row presents the same effects for the drain (conductance) and the 3rd and 4th rows presents the effects of pure temperature-dependency and mixing products of temperatures and voltages.

The polynomials include time domain multiplications which correspond to frequency domain convolutions. For sinusoidal signals this is easy to do by just replacing time domain multiplications with frequency domain convolutions of voltage line spectrums. In Volterra analysis the nonlinear currents must be always calculated from lower order voltages and a probing method is used to get all contributions generating nth order currents [10]. For example the $K200 \cdot vg^2$ nonlinearity causes 3rd order distortion by mixing 1st and 2nd order controlling voltages. This is seen from gm's frequency domain current ($K_{200} \cdot ((Vg1 + Vg2 + Vg3) \otimes (Vg1 + Vg2 + Vg3))$). To find out higher order contributions for more complex nonlinearities the probing method is used throughout in this work.

Now let's suppose we have an arbitrary circuit (fig.1.) with several different types of nonlinear elements and sinusoidal excitation. Without loosing generality, we perform a Volterra analysis and show how the TPF can be implemented for a 2-dimensional gm element.

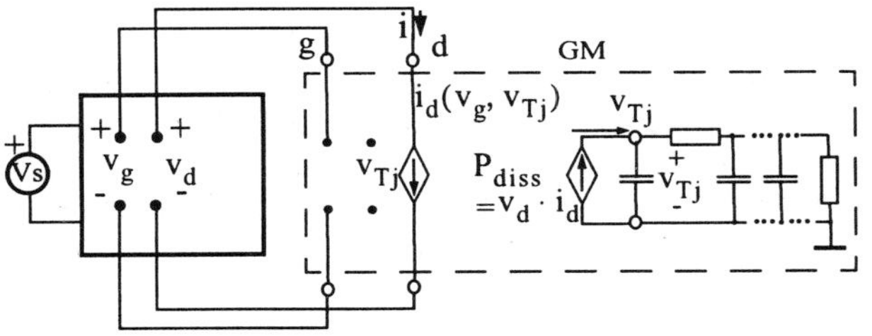

Figure 1. Circuit with TPF.

The Volterra procedure starts by evaluating all 1st order nodal voltages. Then, by using the nonlinear currents (In) and the admittance matrix, all nth order nodal voltages are calculated in increasing order (2). The nonlinear currents are always generated from lower order voltages.

$$[V_n(\omega_{q1} + \omega_{q2} + .. + \omega_{qn})] = [inv(Y(\omega_{q1} + \omega_{q2} + .. + \omega_{qn}))] \cdot [I_n] \quad (2)$$

When the first order nodal voltages are calculated, the 2nd order temperatures in each gm element's thermal network are calculated by convolving 1st order voltage and current spectrums and feeding the resulting power flow as a current to the thermal impedance. For the 2-D gm element the 2nd order temperature is (3).

$$Tj_2 = (V_{d1} \otimes (K_{100} V_{g1})) \cdot Z_{TH} \quad (3)$$

The Volterra procedure continues by evaluating all different contributions that generate 2nd order nonlinear currents. The difference compared to normal Volterra procedure is that now there is also a 2nd order controlling temperature. For the 2nd order case it is easy just to look at the polynomial equations and realize that the second order currents can be generated only by squaring or mixing the 1st order controlling voltages and by linear dependency of the 2nd order temperatures. For the 2-D gm element the 2nd order current is then (4)

$$I_{d2} = K_{200}(V_{g1} \otimes V_{g1}) + K_{001} Tj_2 \quad (4)$$

After calculating all 2nd order currents the 2nd order nodal voltages can be calculated using (2). The calculation of the 2nd order nodal voltages is a plain AC analysis for circuit with 2nd order current excitations. Graphically presented, this means that all original input voltages are shorted and input currents are left open and that each nonlinear element is modeled with their linear part parallel with 2nd order nonlinear current sources.

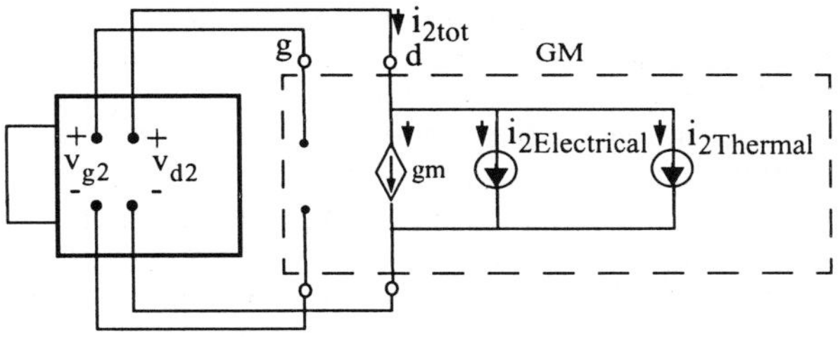

Figure 2. Circuit to calculate 2nd order voltages caused by the 2-D gm element.

Fig. 2 shows the circuit and (5) the corresponding equation which is used to calculate the 2nd order electro-thermal node voltages caused by the 2-D gm element.

$$[V_{2gm}] = [inv(Y)] \cdot [K_{200} V_{g1} \otimes V_{g1} + K_{001} Tj_2] \quad (5)$$

Now the third order temperatures can be calculated from 1st an 2nd order voltage and current spectrums of the gm elements. In power calculations the total voltage and current spectrums must be used. Thus the generated 2nd order voltages in the controlling ports of each gm element induce 2nd order current through the linear gm element which must be taken account. Therefore, the 3rd order temperature for a 2-D gm element is calculated using (6).

$$Tj_3 = (V_{d1} \otimes (I_{d2} + K_{100} V_{g2}) + V_{d2} \otimes I_{d1}) \cdot Z_{TH} \quad (6)$$

Next the 3rd order currents for each nonlinear element can be calculated. The 3rd order current for the 2-D gm element can be written as (7).

$$I_{d3} = K_{300}(V_{g1} \otimes V_{g1} \otimes V_{g1}) + 2K_{200}(V_{g1} \otimes V_{g2}) + K_{001} Tj_3 + K_{101}(V_{g1} \otimes Tj_2) \quad (7)$$

After calculating each 3rd order nonlinear current the 3rd order node voltages can be calculated with (2).

The electro-thermal Volterra analysis can be analyzed forward up to nth order just like described above. The difference to normal Volterra analysis is the nth order temperature calculation before each nth order current calculation. It was seen that the temperature is always calculated from lower order voltages (and currents which are functions of controlling voltages). The TPF implementation to Volterra analysis suits well for numerical Volterra analysis where the admittance matrix (MNA) includes also the thermal networks.

3. A 5th order electro-thermal Volterra simulator

The electro-thermal Volterra-procedure described above is implemented into a 5th order Volterra simulator [9]. By using a method described in [9] the insight to the distortion mechanisms is maintained even when using fully numerical Volterra analysis. The idea is to plot the result vector as a sum of all contributions with symbols telling the cause of each distortion component. The Volterra simulator is coded to label all different distortion mechanisms caused by polynomial nonlinear currents of the circuit. For example, the 3rd order distortion components caused by the 2-D gm-element in (7) can be separated to purely electrical, different mixing products of voltage and temperature spectrums and to pure temperature controlled nonlinearities. The mixing products like $K_{200} \cdot ((Vg1 + Vg2) \otimes (Vg1 + Vg2))$ can be separated further into sub-contributions of mixing envelope and second harmonics with fundamentals, for example. The separation of mixing nonlinearities is done by masking the voltage spectrum before the convolution.

Because of AC analysis, the nth order voltage vectors for each nonlinear current excitation and frequency can be calculated and plotted separately, giving extensive detailed information of different distortion contributions.

The flowchart of the analysis program is shown in Fig.3. First the circuit description (consisting of MNA matrix, list of nonlinear elements, inputs and sweeps) is read and linear node voltages V_1 are evaluated. After that we jump to loop that generates all temperatures, nonlinear currents and voltages up to 5th order: All (n+1)th order temperatures for each electro-thermal gm element are calculated and stored. Then the nonlinear currents I_{n+1} for each nonlinear element are calculated and stored component-wise, using the linear and lower order distortion voltages and temperatures appearing in the controlling nodes. Inside the component loop the different distortion mechanisms are recognized - if needed - by calculating the convolution separately for different portions of the spectrums, selected by frequency masks. When all components and frequency masks are looped, the (n+1)th order voltage spectrums are calculated and stored. This V_{n+1} is used to calculate (n+2)th order components and so on. The loop is continued until V_5 is calculated. After calculating all 5th order voltages we can optionally sweep some wanted parameters (input signals, components, polynomial coefficients etc.) and do the 5th order Volterra procedure again or plot the results. The results for each sweep are stored to five 3-D tables or cubes shown in Fig.3, where the axes are nodes, frequencies and components and different harmonic zones.

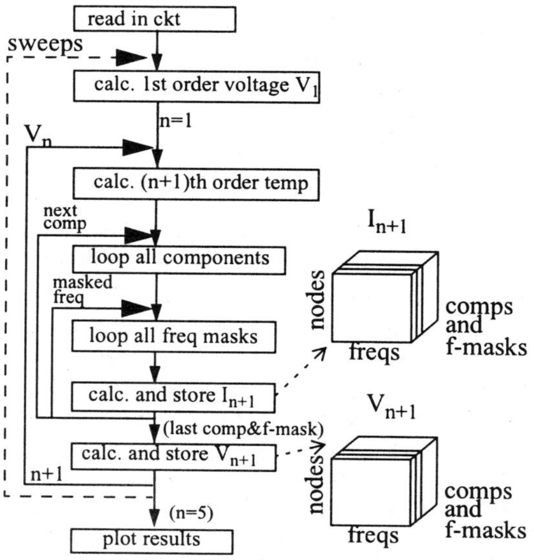

Figure 3. Flowchart of the program

4. Simulation example

A simple CE amplifier with a 5th-order 2-dimensional (Vin,T) gm element is simulated to show how the electro-thermal distortion can be studied using this numerical Volterra simulator. Fig.4 shows the simulated CE amplifier with dominating nonlinearity coefficients.

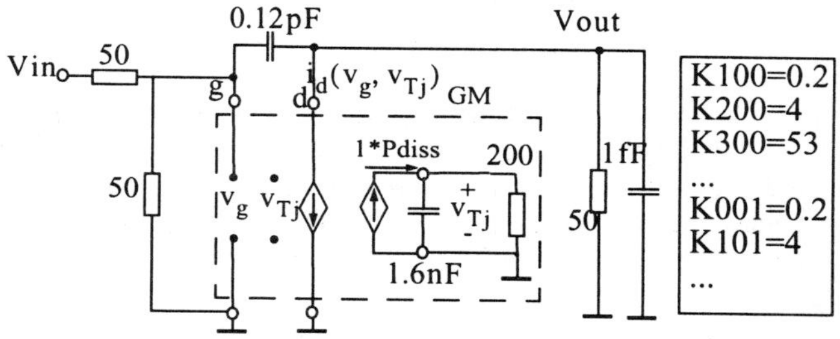

Figure 4. Simulated electro-thermal model for a common emitter amplifier

The gm element is modeled as exponential and is biased to 5 mA. Temperature dependent variables are arbitrarily chosen for illustration purposes -in practice all coefficients are characterized by measurements. A 0.12pF Miller capacitance is between base and collector so that also the mixing effect of K_{200} is seen. Thermal bandwidth is 0.5 MHz, and the thermally introduced distortion is studied.

Two small sinusoidal voltages (25mV/tone) with frequencies 950 MHz and 950.1 MHz are applied to Vin. The higher tone frequency is swept to 954 MHz with 100 kHz steps and the behavior of lower IM3 (2f1-f2) in the output is studied.

Fig. 5 shows the amplitude and phase of the lower IM3

when the tone spacing is swept. It is clear that the changes are stronger for small tone differences.

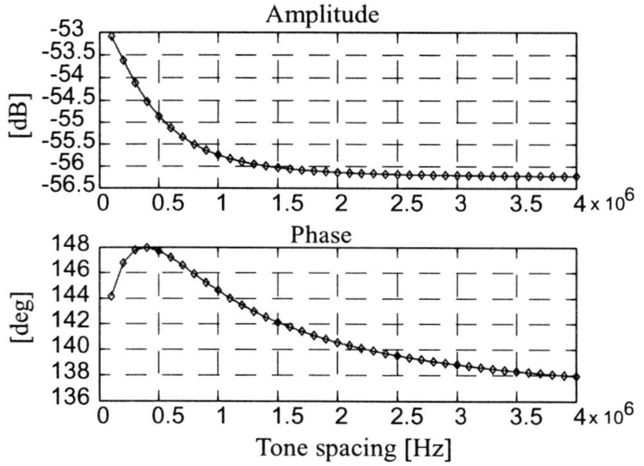

Figure 5. IM3 amplitude and phase.

Fig. 6 shows the vector sum of IM3 (result) for the same sweep. The pure 3rd degree nonlinearity (K_{300}) of the gm element is the dominant IM3 distortion source, but when the tone spacing is swept only the junction temperature has strong change, because of the 0.5 MHz bandwidth of the thermal impedance. Second degree nonlinearity (K_{200}) produces both f2-f1 and 2nd harmonics in the output, but only the 2nd harmonics can couple through the Miller capacitor to the input and mix again to IM3 - hence K_{200}(2f1)(-f2) is visible, but K_{200}(f1)(f1-f2) is very small. 3rd order junction temperature is also almost zero due to the thermal low-pass response. Finally, the thermal envelope tone mixes with the electrical fundamental tones in K_{101}, and produces an IM3 term that depends strongly on the signal bandwidth. Hence, also the total IM3 (marked as result) rotates with signal bandwidth and is very difficult to cancel using predistortion techniques.

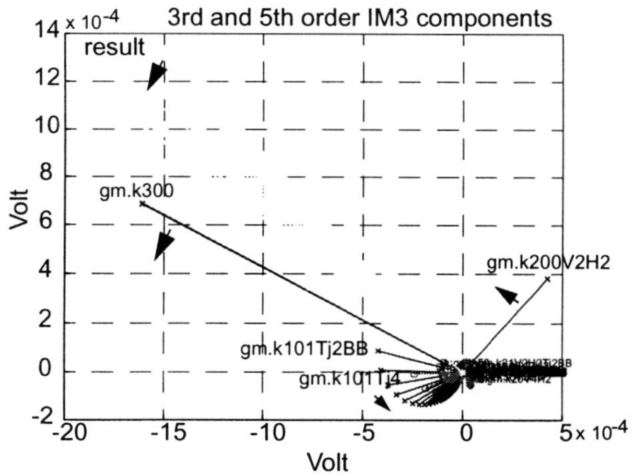

Figure 6. IM3 contributions at tone spacing sweep.

5. Summary

This paper describes a method of implementing thermal power feedback (TPF) into 5th order nonlinear distortion analysis with Volterra-series. Electrical and electro-thermal distortion is calculated using nonlinear current method, where different contributions of distortion can be nicely separated by masking the voltage spectrums before calculating the distortion currents. This gives detailed information about dominant distortion mechanisms, as illustrated by the example studying electro-thermal distortion in a CE RF amplifier.

References:

[1] Hefner, A.R., Blackburn, D.L., Thermal component models for electro thermal network simulation, IEEE Transactions on Components, Packaging, and Manufacturing Technology, Part A, vol. 17 no. 3, Sept. 1994 pp. 413 -424

[2] Hefner, A.R., Blackburn, D.L., Simulating the dynamic electro-thermal behaviour of power electronic circuits and systems, IEEE Trans. on Power Electronics, vol. 8 no. 4, Oct. 1993 pp. 376 -385

[3] Rizzoli, V., Lipparini, A., Esposti, V.D., Mastri, F., Cecchetti, C., Simultaneous thermal and electrical analysis of nonlinear microwave circuits, IEEE Trans. on Microwave Theory and Techniques, vol. 40 no. 7, July 1992 pp. 1446 -1455

[4] APLAC-electro thermal simulation manuals

[5] Anholt, R.E., Swirhun, S.E., Experimental investigation of the temperature dependence of GaAs FET equivalent circuits, IEEE Trans. on Electron Devices, vol. 39 no. 9, Sept. 1992 pp. 2029 -2036

[6] Schurack, E., Rupp, W., Latzel, T., Gottwald, A., Analysis and measurement of nonlinear effects in power amplifiers caused by thermal power feedback, Proceedings., 1992 IEEE International Symposium on Circuits and Systems, 1992. vol. 2, 1992 pp.758 -761

[7] Vuolevi, J., Rahkonen, T., Extracting a polynomial ac FET model with thermal couplings from S-parameter measurements, The 2001 IEEE Int'l Symp. on Circuits and Systems, 2001, vol. 2, pp.461 -464

[8] Fong, L., High-frequency analysis of linearity improvement technique of common-emitter transconductance stage using a low-frequency-trap network, IEEE j. Solid-State Circuits, vol. 35 no 8, Aug. 2000, pp. 1249 -1252

[9] Heiskanen, A.; Rahkonen, T., A 5th Order Multi-tone Volterra Simulator with Component-level Output, Circuits and Systems, 2002. ISCAS 2002. IEEE International Symposium on, Volume: 3, 2002 Page(s): 591 -594.

[10] J. W. Graham, and L. Ehrman, Nonlinear System Modeling and Analysis With Applications to Communications Receivers, Signatron, Inc. 1973.

[11] M. Schetzen, The Volterra and Wiener Theories of Nonlinear Systems, John Wiley & Sons 1980.

[12] S. Maas, Nonlinear Microwave Circuits, Artech House, Norwood, MA. 1988.

[13] P. Wambacq, W. Sansen, Distortion Analysis of Analog Integrated Circuits, Kluwer Academic Publishers, 1998.

[14] D.D. Weiner, and J. E. Spina, Sinusoidal Analysis and Modeling of Weakly Nonlinear Circuits, Van Nostrand, New York, 1980.

A 5th Order Volterra Study of a 30W LDMOS Power Amplifier

Antti Heiskanen, Janne Aikio, Timo Rahkonen

Electronics Laboratory , Dept. of Electrical Engineering and Infotech Oulu,
University of Oulu, FINLAND

ABSTRACT

A 30 W LDMOS is modeled using a 5th order polynomial model. The polynomial model is compared to the large-signal MET model using harmonic balance, and as the results agreed very well, the polynomial model was imported to a numerical Volterra simulator to find out the dominant cause of distortion for a class A biased amplifier. The characterization technique is briefly discussed.

1. INTRODUCTION

The increasing demand for linear mobile communication systems has challenged the scientists and engineers for making better device models and nonlinear simulation techniques. To study distortion mechanisms and possible cancellations the device model must have accurate high order derivatives that give accurate distortion results. Also the simulation method must give insight into distortion mechanisms.

Commonly used nonlinear transient analysis or steady-state harmonic balance displays only the **total** distortion at each node. Dominant distortion contributions and cancelling mechanisms can be derived by using Volterra-series analysis that is well documented in [1-3]. Volterra analysis employs ac models that are fitted in a given operating point. Hence the higher order derivatives of I-V and Q-V curves are not fixed properties of the chosen large-signal equations as they are in conventional models, but can be freely chosen to match the measured data. Hence by using Volterra analysis it is possible to have both accurate predictions of high order derivatives and insight to the distortion mechanisms.

In this work the electrical distortion mechanisms of a class A to moderate class AB mode operated Motorola's 30W LDMOS amplifier [4] is analyzed using 5th order Volterra analysis.

This work shows with harmonic balance simulations that a 5th order polynomial model gives accurate results up to reasonable high power levels. Then by using 5th order Volterra analysis the dominant amplitude and bandwidth dependent distortion mechanisms of IM3 of the amplifier can be found out.

Section 2 discusses how the polynomial model is fitted to the MET model and the amplifier model for the 5th order Volterra analysis is presented. Section 3 discusses the verification of the polynomial and Volterra model against to the MET model. Good accuracy within the fitting range is achieved and discussion of the fitting method is presented. Section 4 presents the dominant amplitude and bandwidth dependent distortion mechanisms of the analyzed amplifier.

2. EXTRACTION OF THE 5TH ORDER POLYNOMIAL AND VOLTERRA MODEL

Before being able to do Volterra simulations the polynomial model of the intrinsic FET is needed. To extract and verify the 5th order polynomial model, a test bench described in [4] is constructed to APLAC simulator environment. The de-embedding method described in [5] was successfully used for extracting the 3rd order polynomials of MRF21030 LDMOS transistor [6]. The same procedure is now used to fit 5th order polynomials to the simulated AC data. Briefly, the fitting procedure includes isothermal measurements of the packaged transistor's S-parameters at different biasing conditions, de-embedding the intrinsic transistors small-signal elements in each bias point at different temperatures, and finally fitting the polynomials to the extracted small-signal elements [6].

After collecting necessary amount of intrinsic FETs ac-values of all nonlinear elements the polynomials and the correct fitting range must by chosen. The optimum output loading resistance principle [7] can be used to estimate the maximum class A drain and gate voltage swings. Here to get more linearity the transistor is backed off so the signal swings are smaller. The circuit studied here is biased to Vgs = 4.1 V and Vds = 28 V, giving 0.780 A DC current. With a trans conductance value at this bias point the needed gate voltage swing for maximum class A drain current is about 0.38 V. This drain current is driven to impedance seen at the drain giving about 4 V signal swing to the drain (@2.14 GHz). The polynomials are fitted over a rectangular region of size (vgs,vds) = (4.1+/- 0.5, 28 +/- 4 V).

The fitted functions for all nonlinear capacitances are 1-dimensional 5th order polynomials. The gm element is fitted to 2-dimensional 5th order polynomial shown in (1).

$$
\begin{aligned}
i_D = {} & g_m v_g + K_{20} v_g^2 + K_{30} v_g^3 + K_{40} v_g^4 + K_{50} v_g^5 \\
& + g_o v_d + K_{02} v_d^2 + K_{03} v_d^3 + K_{04} v_d^4 + K_{05} v_d^5 \\
& + K_{11} v_g v_d + K_{21} v_g^2 v_d + K_{12} v_g v_d^2 + K_{31} v_g^3 v_d + K_{22} v_g^2 v_d^2 \\
& + K_{13} v_g v_d^3 + K_{41} v_g^4 v_d + K_{32} v_g^3 v_d^2 + K_{23} v_g^2 v_d^3 + K_{14} v_g v_d^4
\end{aligned}
\quad (1)
$$

After successful fit procedure the Volterra model is constructed. Figure 1 shows the Volterra test bench that models the entire amplifier. The input and output matching networks and the package of the transistor are transformed to frequency dependent Y-parameters extracted from S-parameter measurements. The shown nonlinear elements are modeled using the coefficients of the fitted polynomials.

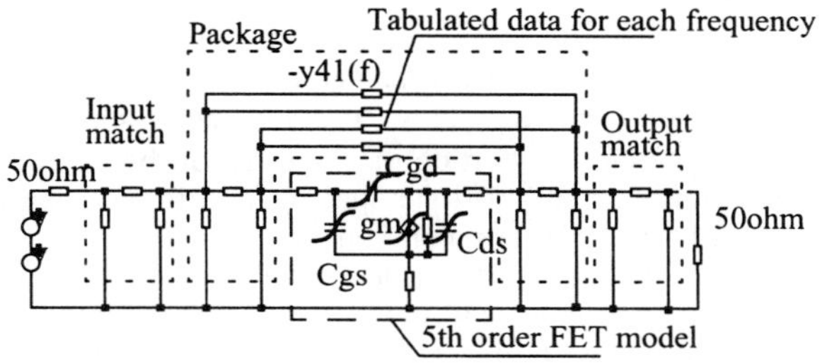

Figure 1. Volterra model for MRF21030 test bench.

The next section compares the fitted polynomial and Volterra model to the simulated Motorola's MRF21030 MET-model. Improvements of the fitting procedure will also be discussed.

3. VOLTERRA AND POLYNOMIAL MODEL VERIFICATION

The polynomial and Volterra model verification is done by comparing the outputs using a 2-tone power sweep within the fitting area using harmonic balance simulations.

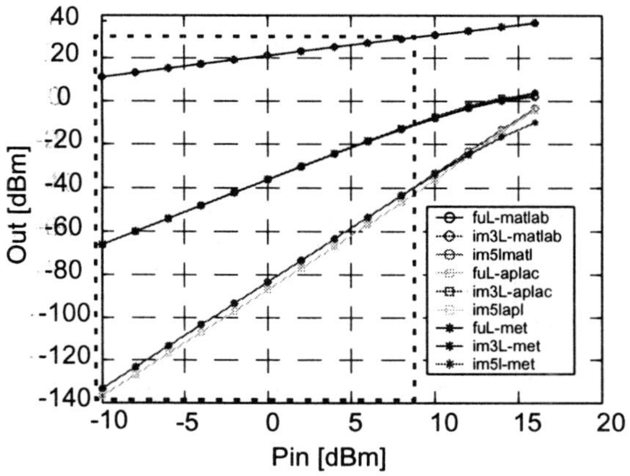

Figure 2. Comparison of three power sweeps.

Figure 2 shows the level of fundamental (f_1), lower IM3 and IM5 of Motorola's MET model, fitted polynomial model and the polynomial model calculated in a Volterra simulator. It can be seen that within the fitting area (shown dotted) the accuracy compared to Motorola's MET model is good. Figure 3 shows the absolute amplitude error and figure 4 the absolute phase error of the Volterra model compared to Motorola's MET model.

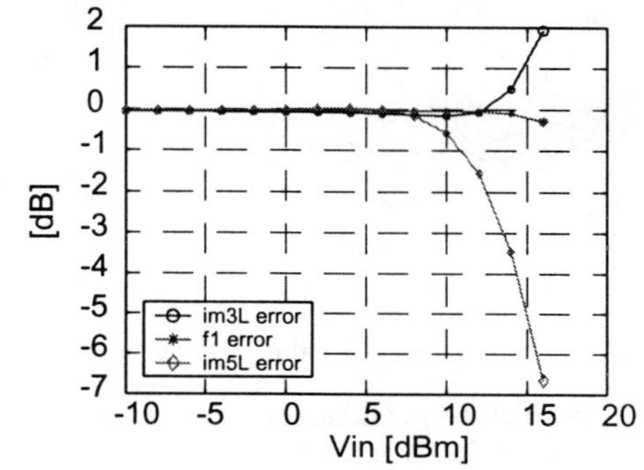

Figure 3. Amplitude difference between the MET and the Volterra model.

The phase difference in Fig. 4 is mostly due to the fact that we are using a quasi-static model for the intrinsic FET while the MET model has some non-quasi-static effects.

It can be seen from both figures that as the signal swings grow beyond the fitting range the results expand and the accuracy is lost. For smaller signal swings the accuracy remains which is expected as we are using smaller signals than assumed during the fitting of the model.

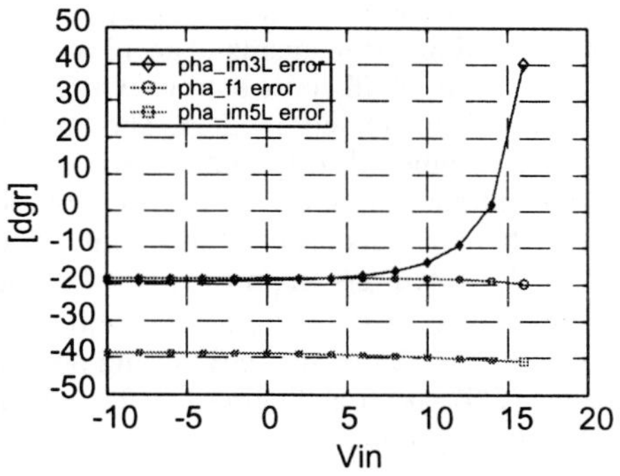

Figure 4. Phase difference between the MET and the Volterra model.

If the fitting is done for larger signal swings some problems will rise up. Figure 5a shows the extracted gm element values at different bias points and figure 5b the measured signal swings at (vgs,vds) coordinates for the simulated power sweep. From figure 5a and 5b we can see that the rectangular fitting range includes fitting points from (vgs,vds) space that the actual signals will not use and this affects to the coefficients. From the gm surface it can also be seen that by increasing the fitting area the old, smaller area must be fitted with fewer points, giving rise to coefficient errors for small signals. From figure 5b we see that the chosen fitting points should track to the real signal swings of the intrinsic transistor.

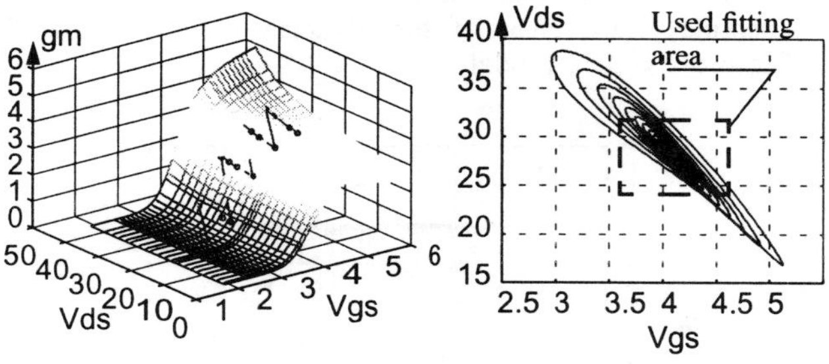

a) b)

Figure 5. a) extracted gm values b) (vgs,vds) swings

Still for the bias point and signal levels used here the rectangularly fitted model gives accurate results and we can say that the Volterra model of the MRF21030 amplifier is verified to be accurate and the Volterra simulation results can now be analyzed.

4. DISTORTION MECHANISMS OF THE TEST CIRCUIT IN CLASS A-AB MODE

After verifying the Volterra model at the chosen bias point and signal levels we can study the 5th order distortion mechanisms of the amplifier. Normal Volterra analysis gives the IM3 result in symbolic form. For circuit of this size the symbolic method is impractical. Here the same result is constructed by a method described in [8], still providing a lot of insight into distortion mechanisms. The idea is to plot the result vector as a sum of all contributions with labels naming the cause of each distortion component. The Volterra simulator [8] is coded to label all different distortion mechanisms caused by polynomial nonlinear currents of the circuit. For example, the 3th degree distortion component caused by gm-element (1) and labeled $gm.K_{30}$ means that the cause of the distortion component is the 3rd-degree nonlinearity of the gm element controlled by its first controlling port with its 1st order voltages (f_1,f_1,-f_2). Label $gm.K_{20}V2H2$ tells us that the 3rd order distortion component is caused by mixing the 1st (-f_2) and 2nd ($2f_1$) order controlling voltages of the 1st controlling port in the second-degree nonlinearity of the gm element.

Figure 6 shows the lower IM3 (IM3L) in the output when the input power level is 8 dBm/tone (being slightly in class AB operation). Only the strongest contributions of the IM3L are plotted. The result vector (IM3L) is a sum of all 3rd and 5th order distortion components caused by different distortion mechanisms.

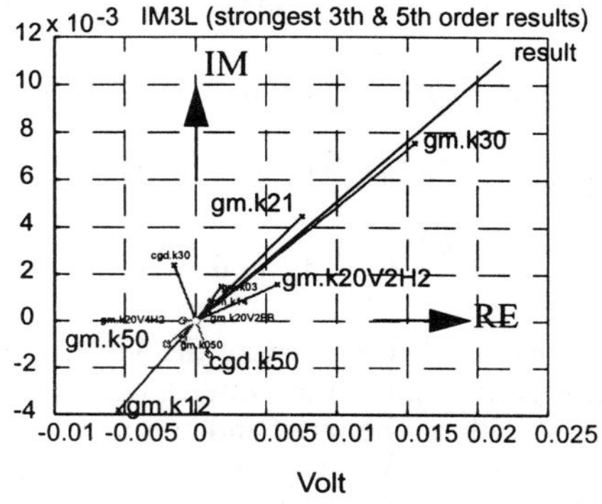

Figure 6. Lower IM3 (8 dBm/tone at input).

It can be clearly seen that the 3rd order product from pure 3rd degree nonlinearity of the gm element ($gm.K_{30}$) is the dominant distortion component at this input level. Some canceling distortion mechanisms can also be seen (for example $gm.K_{12}$ and $gm.K_{21}$). Now to study amplitude and bandwidth dependent distortion mechanisms of the amplifier, the input power level and tone spacing is differentiated to see how different distortion components change.

When the input power is increased, all the 3rd order products change proportional to the 3rd power and the 5th order products to the 5th power, but their phases remain the same. Only the sum of 3rd and 5th order products can change in phase because of different rate of change of the summated products. When the tone spacing is changed, the controlling voltages of the nonlinear elements are changing because of frequency dependency of node impedances. Thus the impedance variations are causing frequency dependent distortion effects.

Now looking back at figure 6 we can predict that small increase in the input power will cause 5:1 (dB) amplitude growth to all 5th order products and 3:1 (dB) growth to all 3rd order products. Only the result (IM3L) will have also amplitude dependent phase shift. Figure 7 shows the lower IM3 of the amplifier when the input power level is 6, 8 and 10 dBm per tone.

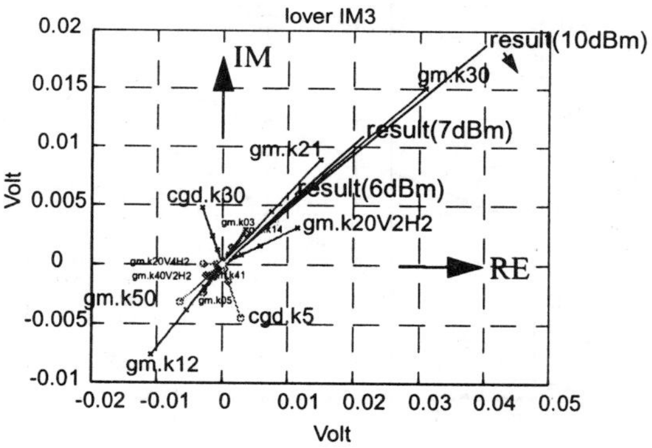

Figure 7. Lower IM3 at input power of 6,7 and 10 dBm.

From figure 7 it can be seen that the gm element seems to be the dominant amplitude dependent distortion source of IM3L. The 3rd degree nonlinearity of the gm element ($gm.K_{30}$) is causing the biggest distortion contribution. The phase difference for the 4 dBm input power difference is 6 degrees. From figure 7 we can see that the pure 5th degree nonlinearity of the gm element ($gm.K_{50}$) starts to cancel the strong effect of the 3rd degree nonlinearity of the gm element ($gm.K_{30}$). Because the cancellation of ($gm.K_{30}$) by ($gm.K_{50}$), the ($cgd.K_{50}$) starts to have an affect to the amplitude dependent phase rotation of IM3L. The strong effect of Cgd can also be explained by its big controlling voltage and high amplification to the output.

Next the phase dependent IM3 effects are studied by sweeping the tone difference from 0.5 MHz to 7 MHz while the input power level is kept at 10 dBm.

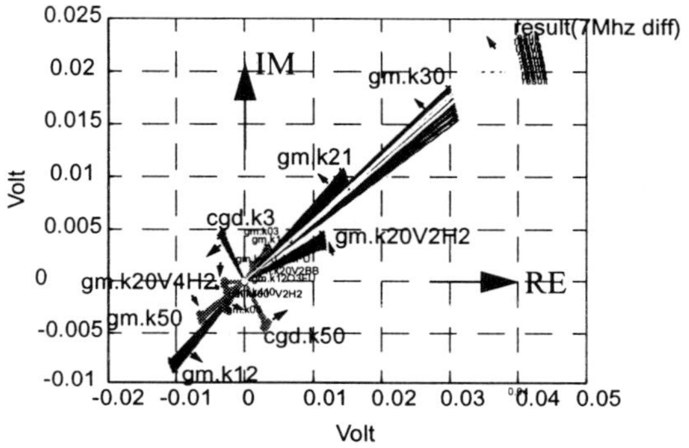

Figure 8. Lower IM3 with different tone spacings.

Figure 8 shows the results of lower IM3 when the tone spacing is varied. The arrows show the rotation direction of each distortion component. All dominant distortion mechanisms are rotating to the same direction but still the overall lower IM3 resultant rotates only 5 degrees. This is a result of cancelling nonlinearities.

There is no clear dominant bandwidth dependent distortion mechanism for the studied amplifier at this bias point. The amplitude growth is very small and both bandwidth dependent amplitude growth and phase rotation are almost linear. This means that the impedances at the dominant controlling nodes are not changing much.

From circuit point of view we can say that there is not much that can be done to the dominant distortion mechanisms of the amplifier because they are controlled by the fundamental signals. The nonlinearities that are controlled by out-of-band voltages can be minimized by manipulating the out-of-band impedances. Thus the amplifier studied here can be linearized by forcing the gate-impedance at 2nd and 4th harmonics (3f1-f2 - 3f2-f1) to zero and by trying to keep the gate and drain impedances maximally flat over the signal bandwidth. Even better would be to force the gate-impedance at 2nd and 4th harmonics to rotate 180 degrees forcing the ($gm.K_{20}V2H2$) to cancel the ($gm.K_{30}$).

Because all the vectors in fig. 8 are stored, it is possible to calculate and see what is the effect if the gate impedance at 2nd and 4th harmonics would be zero while everything else is remaining the same. This gives -1.5 dB improvement to IM3L. By forcing a 180 degree phase shift to the $gm.K_{20}V2H2$, $gm.K_{40}V2H2$ and $gm.K_{20}V4H2$, and by calculating the IM3L, a -3 dB improvement is achieved.

Both studies leads to reverse engineering. The goal is to force the circuit to maintain all old features and try to have the needed phase shift or nullation of 2nd and 4th harmonics of gate impedance.

5. SUMMARY

This work showed by simulations that a 5th order polynomials can be used to model the electrical behavior of the 30W LDMOS transistor. A 5th order Volterra model of the Motorola's MRF21030 power amplifier was presented. Model comparison showed good accuracy and dominant distortion mechanisms of the class A and moderate class AB mode operated 30W LDMOS transistor model were given. Discussion of amplitude and bandwidth dependency of the IM3 and the used fitting method were presented.

6. REFERENCES:

[1] S. Maas: Nonlinear Microwave Circuits. Artech House, Norwood, MA. 1988.

[2] P. Wambacq, W. Sansen: Distortion Analysis of Analog Integrated Circuits. Kluwer Academic Publishers, 1998.

[3] M. Schetzen. The Volterra and Wiener Theories of Nonlinear Systems. John Wiley & Sons 1980.

[4] Motorola's MRF21030 Electro-Thermal (MET) LDMOS model. MRF21030 test bench described at http://search.motorola.com/semiconductors/query.html?qt=MRF21030+datacheet

[5] J.V. Butler et al: 16-term error model and calibration procedure for on-wafer network analysis measurements. IEEE Trans. on MTT, 39(12) 1991. pp.2211-2217.

[6] Vuolevi, J.; Aikio, J.; Rahkonen, T., Extraction of a polynomial LDMOS model for distortion simulations using small-signal S-parameter measurements, Microwave Symposium Digest, 2002 IEEE MTT-S International, Volume: 3, 2002, Page(s): 2157 -2160 vol.3

[7] S. C. Cripps, RF Power Amplifiers for Wireless Communications. Norwood, MA: Artech House, 1999.

[8] Heiskanen, A.; Rahkonen, T., A 5th Order Multi-tone Volterra Simulator with Component-level Output, Circuits and Systems, 2002. ISCAS 2002. IEEE International Symposium on, Volume: 3, 2002 Page(s): 591 - 594.

A SIMULINK-BASED APPROACH FOR FAST AND PRECISE SIMULATION OF SWITCHED-CAPACITOR, SWITCHED-CURRENT AND CONTINUOUS-TIME ΣΔ MODULATORS

Javier Moreno-Reina, José M. de la Rosa, Fernando Medeiro, Rafael Romay, Rocío del Río, Belén Pérez-Verdú and Angel Rodríguez-Vázquez

Instituto de Microelectrónica de Sevilla, IMSE-CNM (CSIC)
Edif. CNM-CICA, Avda. Reina Mercedes s/n, 41012 Sevilla, SPAIN
Phone: +34 95056666, FAX: +34 95056686, E-mail: jrosa@imse.cnm.es

ABSTRACT

This paper describes how to extend the capabilities of SIMULINK for the time-domain simulation of ΣΔ modulators implemented by using switched-capacitor, switched-current and continuous-time circuits, considering the most important error mechanisms. The behavioural models of these circuits are incorporated into the SIMULINK environment by using C-language *S*-function blocks, which leads to a drastic saving in the simulation time as compared to previous approaches based on MATLAB functions. The outcome is a complete SIMULINK block library that allows interactive, fast and accurate simulation of an arbitrary ΣΔ topology (*).

1. INTRODUCTION

Simulation is a critical part of both the top-down synthesis and the bottom-up verification of Integrated Circuits (ICs). Thus, the iterative use of simulators helps designers to explore the design space and to optimize critical trade-offs at different hierarchical levels [1]. In the case of ΣΔ Modulators (ΣΔMs), as a consequence of their sampled-data nature, simulation has to be done in the time-domain. However, transistor-level simulations with SPICE-like simulators yield to excessively long CPU times – typically several days, or even weeks. The reason is that several thousands clock cycles – with small numerical integration steps and complex models– are needed to obtain a realistic evaluation [2].

To overcome this problem, different alternatives for the simulation of ΣΔMs have been proposed, which at the price of losing accuracy in their models, reduce the simulation time [2][3][4]. One of the best accuracy-speed trade-off is achieved by using the so-called *behavioural simulation* technique [1]. In this approach the modulator is broken up into a set of subcircuits, often called building blocks, which are described by explicit equations that relate the outputs in terms of the inputs and the internal state variables. Thus, the accuracy of the simulation depends on how precisely those equations describe the real behaviour of each block.

In case of Discrete-Time (DT) ΣΔMs implemented with either *switched-capacitor* (SC) [3] or *switched-current* (SI) circuits [5], the value of signals is important only at specific time points. Therefore, each building block is defined by a set of finite difference equations which describe its functionality, and the simulation process consists of computing the node voltages and branch currents of the circuit consecutively in each clock phase. The outcome is a drastic saving in CPU time – only a few seconds to evaluate an output spectrum. Recently, behavioural simulation has been applied also to Continuous-Time (CT) ΣΔMs [6]. In this case, model equations are computed analytically instead of numerically, leading to CPU times comparable with the DT case.

(*)This work has been supported by the EU ESPRIT IST Project 2001-34283/TAMES-2 and the Spanish CICYT Project TIC2001-0929/ADAVERE.

In spite of their good trade-off between precision and CPU-time, previously reported behavioural simulators present several drawbacks. On the one hand, there is a limited number of ΣΔM topologies that can be simulated, normally using only one circuit technique. On the other hand, except for [4], the user interface consists of an input netlist with a dedicated syntax, while postprocessing is performed by using commercial tools like MATLAB [7].

The above-mentioned problems can be overcome by implementing the behavioural models in the SIMULINK environment [8]. The benefits are a friendly Graphical User Interface (GUI), high flexibility for the extension of the block library and huge signal processing capabilities. Recently, a set of SIMULINK block models has been proposed for the behavioural simulation of SC ΣΔMs [9]. However, it has two major constraints:

- The block library is limited to SC circuits, using simple models which do not include some important limitations like mismatch and the non-linearities associated to the open-loop opamp DC gain and capacitors. In addition, as models are implemented in the *Z*-domain, the circuit behaviour during different clock phases is not considered, thus leading to an imprecise modelling of some errors like the incomplete settling.
- Block models are realized by using MATLAB functions. This causes the MATLAB interpreter to be called at each time step, slowing down the simulation time drastically [8]. This problem is aggravated as the model complexity increases, yielding to excessive CPU times as compared to C-written simulators. This is true even using the SIMULINK accelerator [8].

This paper presents an interactive and flexible approach for a fast time-domain behavioural simulation of LowPass (LP) and BandPass (BP) ΣΔMs implemented by using not only SC, but also SI and CT circuits. In order to speed up the simulation, ΣΔ-blocks are incorporated in SIMULINK†1 as C-coded *S*-functions [10]. As a result the CPU-time for one 65536-point simulation of a DT/CT ΣΔM is typically less than 5 seconds†2, meaning only a few times slower than C-written simulators, but up to 2 orders of magnitude faster than using MATLAB functions as in [9].

2. DESCRIPTION OF THE ΣΔM-BLOCK LIBRARY

The proposed SIMULINK ΣΔM-block library includes different sublibraries which are classified according to the modulator hierarchy level and the circuit technique. As an illustration, Fig.1 shows some of these sublibraries showing: SI memory cells, SC/CT integrators and resonators (used in BP-ΣΔMs), quantizers,

†1. SIMULINK 5 and MATLAB 6.5 (release 13) were used.

†2. All simulations shown in this paper were done using an Intel Pentium 4 CPU@1.7GHz @256MB RAM PC.

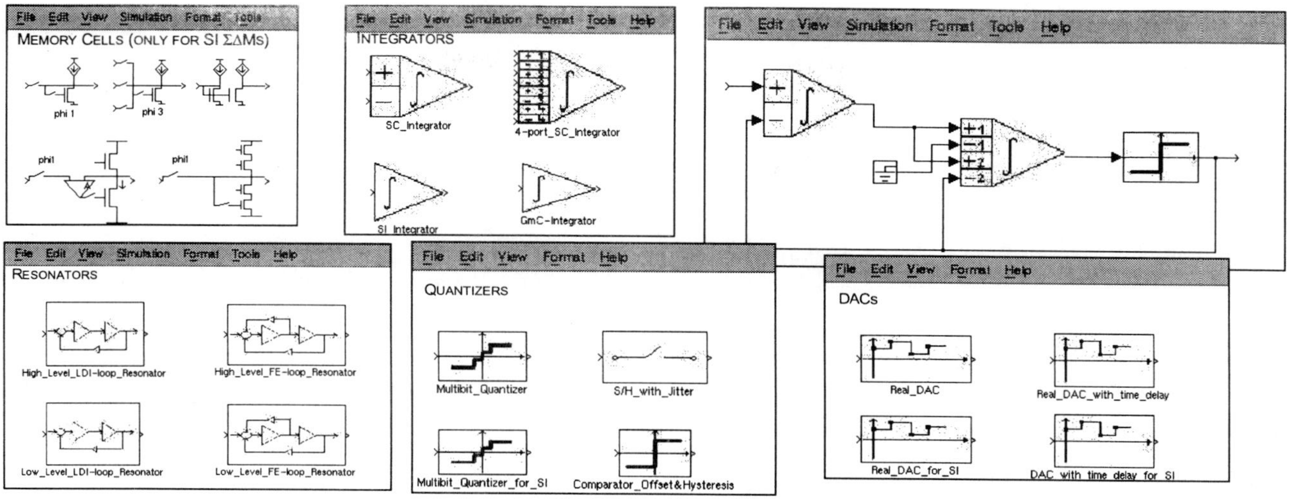

Figure 1. Illustrating some blocks of the proposed SIMULINK ΣΔM-block library.

and both 1-bit and multi-bit (*mb*) Digital-to-Analog Converters (DACs). There is also a sublibrary including the most usual architectures of both LP- and BP-ΣΔMs using SC, SI and CT circuits. For each building block, the ΣΔM-block library provides models with a different abstraction level. The purpose is twofold. First, high level models are suited for system level simulations and initial transmission of specifications. Second, low level accurate models, which takes into account main circuit parasitics, are suited for fine-tuning the specs transmission and circuit validation.

The main circuit non-idealities included in the integrators (and resonators) are:

- *SC circuits*: finite open-loop opamp DC gain, incomplete settling error, mismatch capacitor ratio error, thermal noise; and main non-linear effects, namely: non-linear sampling switch-on resistance, non-linear open-loop opamp DC gain, slew rate and non-linear capacitors.
- *CT circuits*: finite DC gain, integration time constant error, slew rate, finite non-linear transconductance and thermal noise.
- *SI circuits*: linear and non-linear gain error, finite output-input conductance ratio error, charge injection error, incomplete settling error, mismatch error and thermal noise.

Detailed descriptions of these errors as well as their behavioural models – beyond the scope of this paper – can be found in [3], [11] and [5] for SC, CT and SI circuits, respectively.

In addition to integrators and resonators, quantizer and DAC errors have to be considered, specially in SC/SI cascade *mb* architectures and CT single-loop topologies. For this purpose, the following circuit parasitics have been included:

- *quantizers*: offset, both deterministic and random hysteresis, and in *mb* realizations, gain error and integral non-linearity.
- *single-bit and multi-bit DACs*: offset, gain error and integral non-linearity. In case of CT ΣΔMs, a time delay is also included in order to simulate the effect of *excess loop delay* [11].

The behavioural models of the above-mentioned errors have been coded in C language, and incorporated into the SIMULINK environment through the so-called *S*-functions [10]. These are special purpose C source files which allow us to add C algorithms to SIMULINK models. The outcome is a notable saving of simulation time as compared to use MATLAB functions or M-files to code the models, even when the accelerator utility is used [8].

In order to create an *S*-function associated to building blocks like those shown in Fig.1, the following steps have to be followed:

- *Create a C-coded S-function containing the behavioural model*. For this purpose, SIMULINK provides different *S*-function templates which can accommodate the C-coded model of both DT and CT systems. These templates are composed of several routines that perform different tasks required at each simulation stage [10]. Among the others, these tasks include: variable initialization, computing output variables, updating state variables, etc. Thus, programmers' work simply consists in placing the C-coded behavioural model in the different parts of the template file. For illustration purposes, Fig. 2(a) shows the behavioural modeling and some significant sections of the *S*-function file associated to an SC integrator with non-linear opamp DC gain – not included in [9]. It includes model parameters, clock phase diagram, model code, etc.
- *Compiling the C MEX-file S-function*. This is done by using the *mex* utility provided by MATLAB [10]. The resulting object files are dynamically linked into SIMULINK when needed.
- *Incorporating the model into the SIMULINK environment*. This is done by using the *S-function block* of the SIMULINK libraries [8]. Fig.2(b) illustrates this process for the SC integrator of Fig.2(a). A block diagram containing the *S*-function block is created including the input/output pins. The dialogue box is used to specify the name of the underlying *S*-function – in this case *intfescavnl*. In addition, model parameters are also included in this box. In order to facilitate the use of building blocks, the user can insert the values of model parameters from a dialogue box associated to the block.

3. SIMULATION EXAMPLES

An arbitrary modulator architecture can be defined by connecting the building blocks available in the ΣΔM-block library. This can be done by using the SIMULINK Library browser as usual. Alternatively, the ΣΔM-block library can be browsed by using a dedi-

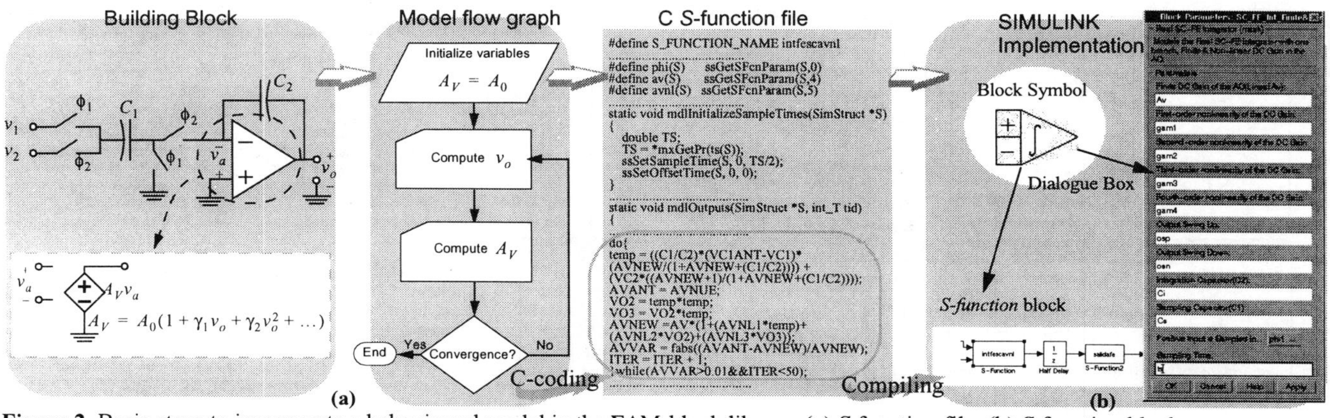

Figure 2. Basic steps to incorporate a behavioural model in the ΣΔM-block library: (a) *S*-function file. (b) *S*-function block.

cated GUI that allows the user to navigate in an easy way through all the steps of the simulation and post-processing of results.

As an illustration of the capabilities of the ΣΔM-block library, this section shows the impact of some circuit parasitics on the performance of the following modulator architectures:

- A CT (Gm-C) 2nd-order LP-ΣΔM (CT 2nd-LPΣΔM).
- A SI 4th-order BP-ΣΔM (SI 4th-BPΣΔM).
- A SC 2-1-1 cascade *mb* (3b) ΣΔM (SC 2-1^2*mb*).

Fig.3 shows the block diagram of these architectures in the SIMULINK environment, including building blocks from the ΣΔM-block library.

3.1 CT 2nd-LPΣΔM example

In the example shown in Fig.3(a), an ideal 1-bit quantizer was used while the other blocks include the following parameters:

- *Gm-C integrators*: finite DC gain, time-constant error, unity gain frequency, slew rate, temperature and output-swing.
- *DAC*: reference voltage and time delay.

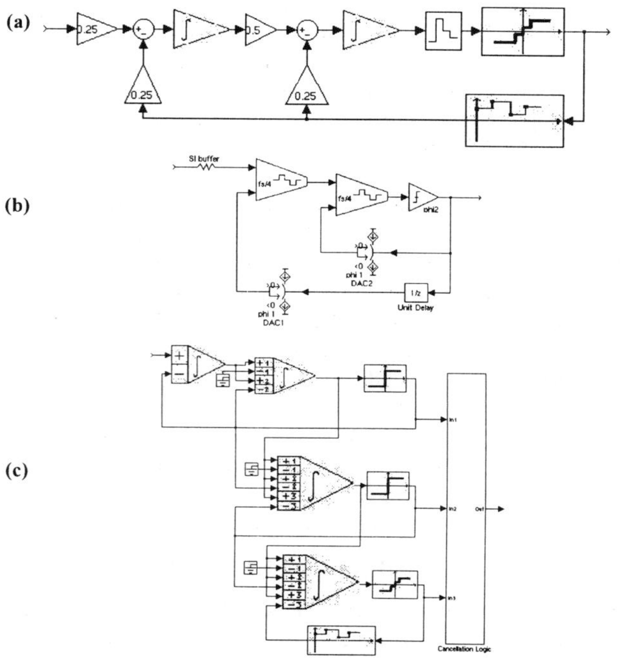

Figure 3. Block diagram of the ΣΔM-library examples. (a) CT 2nd-LPΣΔM. (b) SI 4th-BPΣΔM. (c) SC 2-1^2*mb*.

In high-speed applications, the performance of the modulator can be severely degraded by finite bandwidth and slew rate. These errors cause an increase of both the in-band noise power and the harmonic distortion. This is illustrated in the output spectrum of Fig.4(a), obtained by performing an Hanning-windowed 65536-point FFT to the output bit stream of Fig.3(a), with a half-scale@10-kHz input tone, when clocked at 20 MHz. This simulation takes 3 seconds when the accelerator is used.

In addition to integrator dynamics, one of the most important limiting factors arising in CT-ΣΔMs is the time delay between the quantizer clock edge and DAC response. This delay, referred to as *excess loop delay*, modifies the noise-shaping transfer functions, and may eventually make CT-ΣΔMs unstable [11]. Mathematically speaking, a complex analysis would be required to obtain the stability condition that relates the loop delay, τ_d, with the clock

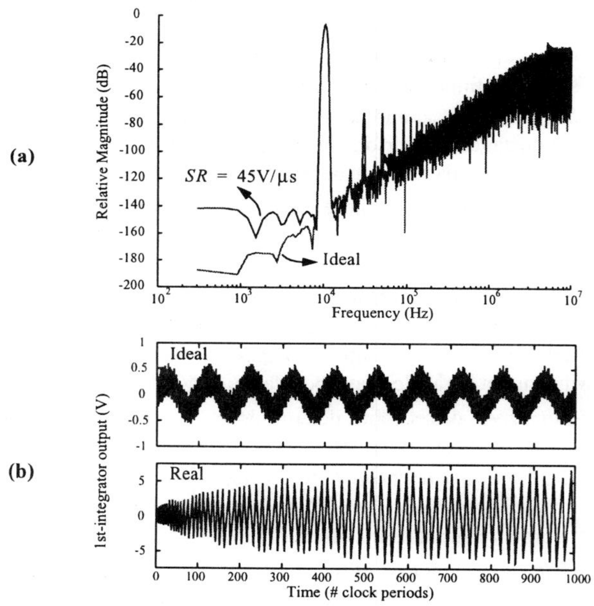

Figure 4. Performance degradation of a CT 2nd-LPΣΔM. (a) Harmonic distortion caused by slew rate. (b) Effect of excess loop delay on the transient response of the first integrator.

period, T_S. Instead of that, simulation-based analyses are normally used. As an illustration, Fig.4(b) shows the first integrator output waveform for different values of the DAC time delay, showing unstable behaviour for $\tau_d \cong 3T_S/2$.

3.2 SI 4th-BPΣΔM example

The SI 4th-BPΣΔM shown in Fig.3(b) has been obtained by applying a LP-to-BP transformation ($z^{-1} \to -z^{-2}$) to a 2nd-LPΣΔM. As a consequence of this transformation, the zeroes of the noise transfer function shift from DC to a quarter of the sampling frequency, f_S. In addition, the integrators in the original LP-ΣΔM become resonators. In this example, resonators are based on lossless direct integrators. Note that a front-end block, named SI buffer, is used to model the voltage-to-current conversion.

One of the most important degrading factors in SI BP-ΣΔMs is the signal-dependent transconductance of memory cells, g_m, which force all errors to be non-linear. As a consequence, in addition to increase the in-band noise power, SI errors cause InterModulation Distortion (IMD). As an illustration, Fig.5 shows the impact of the non-linear settling on the performance of the modulator in Fig.3(b). In this case, the gate-source capacitance of memory transistors, C_{gs}, is varied showing three effects: increase of the in-band noise, third-order IMD and a shift of the quantization noise-filtering notch frequency, δf_n. These output spectra are obtained by running two 65536-point simulations of Fig.3(b), each one taking 4 seconds.

3.3 SC 2-1^2*mb* example

In the case of SC ΣΔMs, the behavioural models of building blocks have been translated from ASIDES, a C-coded time-domain behavioural simulator for SC ΣΔMs [3]. As a consequence of this translation, simulation results obtained with both SIMULINK and ASIDES are practically identical. Compared to ASIDES, the proposed SIMULINK ΣΔ-block library offers a friendly user interface and a great flexibility to simulate an arbitrary SC filter topology, specially but not only, dedicated to ΣΔMs. However, there is a minor CPU-time penalty due to the SIMULINK interface. For instance, a 65536-point simulation of the modulator in Fig.3(c) including the most complex models for building blocks takes 2 seconds using ASIDES and 5 seconds using the ΣΔM-block library. This CPU-time increases up to 415 seconds if M-file building blocks are used – which means about 2 orders of magnitude slower than the approach in this paper.

As an illustration, Fig.6 shows the performance degradation of the 2-1^2*mb* modulator in Fig.3(c) caused by two error mechanisms:

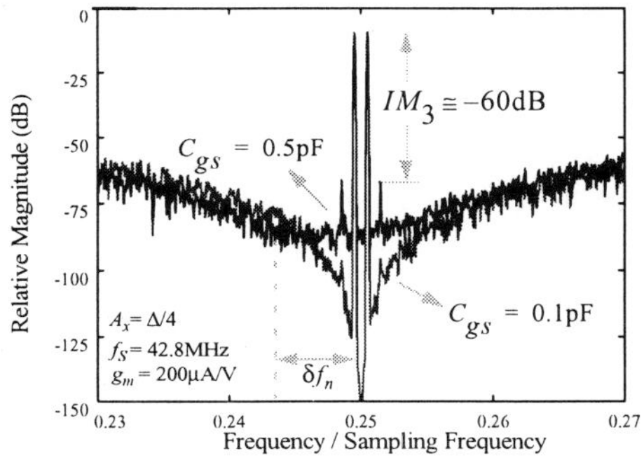

Figure 5. Effect of non-linear settling on SI BP-ΣΔMs.

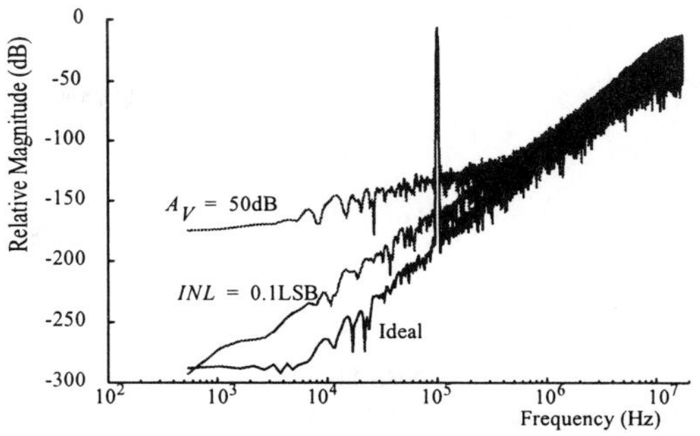

Figure 6. Degradation of an SC 2-1^{2}mb ΣΔM with INL and A_V.

the Integral Non-Linearity (INL) of the 3-bit DAC and the finite DC gain of the opamps, A_V. Both the isolated and the combined effects of these two error mechanisms on the output spectrum are shown in Fig. 6 when the modulator is clocked at f_S = 35.2MHz for INL = 0.1 LSB and A_V = 50dB. In this case, as the INL error is shaped by the filtering performed by previous stages, the main degradation is caused by A_V, basically increasing the quantization noise power in the signal band.

CONCLUSIONS

A complete SIMULINK block library intended for fast and interactive simulation of SC, SI and CT ΣΔMs has been described. The behavioural models of building blocks, including main circuit parasitics, have been incorporated as SIMULINK C-coded *S*-functions. The combination of high accuracy, short CPU-time and interoperability of different circuit models, make the block-library into a valuable instrument to optimize the design of ΣΔ analog-to-digital converters using MATLAB.

REFERENCES

[1] G.G. E. Gielen and R.A. Rutenbar: "Computer-Aided Design of Analog and Mixed-Signal Integrated Circuits", *Proceedings of the IEEE*, Vol. 88, pp. 1825-1852, December 2000.

[2] V. F. Dias, V. Liberali and F. Maloberti: "Design Tools for Oversampling Data Converters: Needs and Solutions", *Microelectronics Journal*, Vol. 23, pp. 641-650, 1992.

[3] F. Medeiro, B. Pérez-Verdú, and A. Rodríguez-Vázquez: *Top-Down Design of High-Performance Modulators*, Kluwer Academic Publishers, 1999.

[4] K. Francken, P. Vancorenland and G. Gielen: "DAISY: A Simulation-Based High-Level Synthesis Tool for ΔΣ Modulators", *Proc. of* 2000 *IEEE International Conference on Computer Aided Design (ICCAD)*, pp. 188-192.

[5] J. M. de la Rosa, B. Pérez-Verdú and A. Rodríguez-Vázquez: *Systematic Design of CMOS Switched-Current Bandpass Sigma-Delta Modulators for Digital Communications Chips*, Kluwer Academic Publisher, 2002.

[6] K. Francken, M. Vogels, E. Martens and G. Gielen: "DAISY-CT: A High-Level Simulation Tool for Continuous-Time ΔΣ Modulators", *Proc. of* 2002 *Design, Automation and Test in Europe Conference (DATE)*, pp. 1110.

[7] The MathWorks Inc.: "Using MATLAB Version 6.5", 2002.

[8] The MathWorks Inc.: "Using Simulink Version 5", 2002.

[9] S. Brigati, F. Francesconi, P. Malcovati, D. Tonieto, A. Baschirotto and F. Maloberti: "Modeling Sigma-Delta Modulator Non-idealities in SIMULINK", *Proc. of* 1999 *IEEE International Symposium on Circuits and Systems (ISCAS)*, pp. II.384-387.

[10] The MathWorks Inc.: "Writing S-Functions Version 5", 2002.

[11] J. A. Cherry and W. M. Snelgrove: *Continuous-Time Delta-Sigma Modulators for High-Speed A/D Conversion: Theory, Practice and Fundamental Performance Limits*, Kluwer Academic Publishers, 2000.

Discrete-Time Modeling and Simulation of Vehicle Audio Systems

F. Piazza [1], S. Bartoloni [1], R. Toppi [2], M. Navarri [2], M. Pontillo [2], F. Bettarelli [3], A. Lattanzi [3]

[1]Dip. di Elettronica e Automatica, University of Ancona, Ancona, Italy
[2]FAITAL Spa, S. Donato Milanese, Milano, Italy
[3]Leaff Engineering S.r.l., Ancona, Italy

ABSTRACT

This paper presents a new integrated set of tools for obtaining both objective and subjective evaluations of car audio systems using a single session of recording with a mannequin in the cockpit. The system should be an important aid for the design of more effective car audio systems, especially loudspeaker subsystems.

A fully automated stimulation, acquisition and identification procedure together with a fully integrated database, simulation and listening procedure has been developed.

The simulator is able to compare in real time different audio systems, change seat and select a head pan and tilt within a limited number of positions. Visual cues make the listening experience more effective: a real car cockpit is shown during the listening with the listener virtually seated on each of the possible positions.

1. INTRODUCTION

In the last years the number of applications of electronic systems in the automotive field is constantly growing. Among the electronic devices the audio subsystem constitutes nowadays the main entertainment system for the passengers and is becoming an ordinary part of each vehicle. However the complex nature of the car audio environment makes the audio systems difficult to design and set up. In particular the interactions between the cockpit and the loudspeaker subsystem are very hard to model and estimate, thus a trial-and-test approach is usually employed. Due to the lack of standards for measuring the perceived audio quality in a car, the only widely accepted assessment procedure is a listening test in the cockpit. The problem of assessing and comparing audio systems is hence of fundamental importance.

To alleviate this problem, we have designed and realized a software tool that is able to model, store, reproduce (with a headphone) and compare car audio subsystems. The software firstly models the whole car audio subsystem, from the source (e.g. an audio CD reader) to the ears of a passenger located in one of the car seats. The model is assumed to be a discrete-time, linear, time-invariant MIMO (Multiple Input Multiple Output) system, realized by means of a matrix of FIR filters. To identify the model parameters, MLS pseudo-random sequences are sent to the car audio system and the resulting stereo signal in the cockpit is recorded by a couple of microphones inside a Bruel and Kjaer head and torso simulator. The obtained stereo signal is sampled by a high quality multichannel audio system (a Echo Layla system connected to a Pentium III PC) and used to estimate the model parameters, which can be stored for late retrieval. Also the microphone-headset chain is modeled with the same method. A fully automated procedure has been developed to make measures exactly repeateble, to avoid errors due to the user and to speed up the acquisition phase.

The system then allows to reproduce in real time through a headphone the listening experience in the car selecting the audio subsystem, the seat, the head pan (with fixed angle increments) and the head tilt (2 positions). The system works on a genuine Pentium III processor, whose multimedia extension is heavily used, and implements a user-friendly graphical interface employing also a real time visual feedback of the cockpit view for each position. The microphone-headset chain is equalized using a constrained optimal approach. The tool is ready to include software modules for the computation of several features of the audio subsystems, which will be presented to the listener.

2. MODELING THE AUDIO SYSTEM

Car interior is a complex acoustic environment that can be very difficult to model [1]. The source signals are transformed into sounds by the loudspeakers, which have their own frequency responses and can exhibit non-linear behaviors, and then propagate in the cabin. Depending on the absorbing or reflecting interior materials, the position of loudspeakers and the shape of the car cabin itself, the reflected sounds attenuate or amplify the direct sound from the loudspeakers, causing distortions in the amplitude and phase responses. A listener in the car also contributes to the final acoustic result, by modifying the sound field with the presence of the body and by adding the effects of auralization to the listening experience. It is known, in fact, that the head and ear shape deeply modify the perceived sounds [2].

Therefore an exact modeling of the vehicle cabin cannot be reasonably obtained. A possible solution to this problem is the so-called *black-box* approach [3]: the effect of the acoustic environment between a source (e.g. a loudspeaker) and a sensor (e.g. a microphone) is modeled only by its input/output behavior. Under the hypotheses of causality, linearity and time-independence, the input/output behavior can be described by a linear stationary transfer function (in an appropriate transform domain). Although the acoustic phenomena are intrinsically continuous in time, they can more effectively described in the discrete-time domain. This means that, being the signals of interest band-limited, we will sample them at an appropriate sampling rate in order to avoid aliasing.

This approach, based on DSP (Digital Signal Processing) techniques, allows to model each of the acoustical effects in the cabin with an appropriate static discrete-time (digital) filter [4]. In fact, the frequency response of each loudspeaker can be described by a discrete-time transfer function, considering a sampling frequency of 44.1 or 48 KHz and neglecting its non-linear behavior. The effects of the cabin can also be modeled by a linear discrete-time transfer function, once the position of the listener is set. The effect of the listener presence is usually modeled again by linear discrete-time transfer functions known as HRTF (Head Related Transfer Functions) [2]. Therefore the overall relationship between an input source signal (a signal

0-7803-7761-3/03/$17.00 ©2003 IEEE

driving a loudspeaker) and an output sensor signal (the signal at one of the listener ear) can reasonably be modeled by a discrete-time filter, which corresponds to the cascade of all the previous transfer functions. Being all the components passive, FIR (Finite Impulse Response) filters can be effectively used; it is well known in fact that they can approximate any stable IIR (Infinite Impulse Response) filter, providing a sufficient number of taps [4].

2.1 Parameters Identification

Given a car interior and a discrete-time MIMO model with suitable dimensions, it is necessary to estimate the parameters of the model which allow for a best fit of the phenomena, at least under some hypotheses and optimization criteria. The parameters of the model are the taps of each FIR filter in the matrix $\mathbf{H}[n]$, therefore MxNxL coefficients when the length of each filter is chosen equal to L. To estimate these parameters, a usual stimulus-response can be followed, which, by minimizing the Mean-Square-Error between the model output and the true output, brings to a linear multidimensional identification problem. It is interesting to note that, due to the linearity and time-invariance of the system, this problem can be seen as a multiple one-dimensional identification procedure. Many techniques [2] can be employed to identify the model parameters, both full band or subband-based.

In the preliminary version of the RT tool (later explained) we used a well-known technique based on the MLS (Maximum Length Sequences) [7]. MLS are pseudo-random binary sequences that behave as stationary white noise but are fully deterministic. They are easy to generate using a multistage binary shift register. A fast implementation of the cross-correlation algorithm allows an easy and robust estimation of the filters coefficients.

The experimental setup we used for practical measurements consists in a Bruel & Kjaer Head & Torso Simulator 4128, with ear simulators and microphones, a Yamaha 01V digital mixer, Crown Macrotech 1200 power amplifiers, a Layla 24 bit multichannel audio acquisition system connected to a Pentium III PC running the software under Microsoft Windows 2000.

2.2 Undesired Effects

Some undesired effects can degrade the quality of the final result: in fact, as we stated, the subjective tests will be performed by listening with a headphone to the output of the identified model, giving suitable input signals. Although the Signal-to-Noise ratio can be kept very high during acquisition by recording in a properly chosen acoustic environment, the effects of the various devices (microphones, mixer, sampler, headphone, etc.) cannot be ignored. These undesired effects should be modeled and removed. Again with the hypotheses of causality, linearity and time-independence, they can be modeled by FIR filters whose coefficients can be estimated as described in the previous section. Under the hypothesis of linearity and using the symbols previously defined:

$$x[k] = h_u[n] * s[k] \qquad X[z] = H_u[z]\, S[z] \quad (7)$$

The estimated transfer function of the undesired effects $H_u[z]$ in the Z transform domain, however, must be inverted to eliminate them.

$$S[z] = G[z]\, X[z] \qquad G[z] = H_u^{-1}[z] = \frac{1}{H_u[z]} \quad (8)$$

The problem of obtaining a stable and robust inverse of a linear discrete-time filter can be quite difficult, especially when the system is non-minimum-phase. However it is always possible to approximate the true inverse by a Finite-Impulse-Response (FIR) filter. To identify the coefficients of these inverse FIR filters, in the preliminary version of the SQ tool two different methods have been implemented: the first is based on the frequency approach reported in [5], while the second implements a constrained LMS adaptive algorithm, as reported in [3] and [4].

It is interesting to note that, once the acquisition chain is definitively set, the only device that could easily change is the headset for listening to the results. Therefore in the following the inverse transfer function used to reduce the undesired effects is often called *headphone driver*.

3. OVERVIEW OF THE SYSTEM

In a typical measurement and identification session, the mannequin sits in a car seat, with a given head pan and tilt angles, during the output and recording of each MLS (few seconds per each speaker). The measurements are repeated for each position and seat in the car. Figure 2 reports some shots taken during an identification session in a FIAT Punto ELX car.

The acquisition process has been fully automated thanks to *Recording Tool*. This software is actually a more friendly interface to Cool Edit Pro (by Syntrillium) which lets anyone make a recording without knowing quite anything about CEP. Two are the steps required to make a recording. The first one is to choose the type of acquisition one wants to perform: car identification, headphone identification or calibration of the acquiring chain. In this phase some information about the session can be tagged (car make, number of positions acquired, headphone model, date, etc...) and stored together with recordings. A graphical interface, shown in Figure 3, helps the user to better set the features. Next step is to properly play MLS and record listening experience by the dummy head. RT sets all CEP parameters such as audio device channels to be used, names and locations of wave files recorded, sessions to be loaded, etc... The user is also guided by the movement of mouse pointer which follows the complete necessary path to deliver a standard recording. At the end of the identification phase, the software module creates a main directory containing a set of sb folders with all the data acquired and all the important information about the car audio system.

The *Sound Quality Tool* is able to read and properly process all this information building an "ambient" file containing the coefficients of the FIR filters that characterize the car examined. A graphical interface developed to help the user during the retriving process. Several different normalization procedures can also be applied to the filter impulse responses. The software is also able to identify the coefficients of the filters that model the acquisition\reconstruction chain with a similar experimental setup using a headphone over the mannequin head instead of using the car loudspeakers. Using the FIR matrix model described in the previous section, a real-time PC based car audio simulator has been implemented. The simulator employs an efficient frequency domain approach to convolution and is able to deal easily in real-time, on a Pentium III processor, with filter matrices whose elements have 1024 taps each.

The software allows to load several different "virtual" car audio systems, choose an equalization filter for the headphone and select an audio file to play in real-time. During playing it's possible to change seat and select a head pan and tilt (the number of positions is limited to eight). It is also possible to select in real-time one of the loaded audio system, passing from a car to another or from a model to another, in order to make better and more accurate comparisons. Figure 4 presents a screenshot from the simulator that shows the "real-time" folder from which all this selection can be made.

To make the listening experience more effective, a visual feedback has been added to the audio simulator, i.e. it is possible to be "virtually seated" in the car seeing the interior from the proper point of view. Figures 5 shows some screenshots from the simulator visual feedback. Note the capability of moving around in the car cockpit using the navigator bar, and the effect of "virtually" moving the head on the right. Although it employs visual processing, this feature seems to greatly increase the listener feeling.

4. CONCLUSION

A new set of tools for both objective and subjective evaluations of car audio systems has been proposed. A fully automated procedure for data recording by a mannequin positioned inside a car cockpit, has been developed. A real-time simulator lets the user compare several different audio systems in the same time. Visual cues make the listening experience more realistic while normalization features let comparisons between different systems more accurate. Future versions of the software will include algorithms for objective measurements and presentation of complex graphical results together with some kind of useful quality index.

5. REFERENCES

[1] W.N. House, "Aspects of the Vehicle Listening Environment", *Proc. of 87th AES*, New York, Oct. 1989

[2] M. Kleiner, B.I. Dalenbäck, P. Svensson, "Auralization – an Overview", *Journal of the Audio Engineering Society*, vol. 41(11), pp. 861-875 (1993).

[3] S. Haykin, "Adaptive Filter Theory", *Prentice Hall Information and System Sciences Series.*

[4] S.T. Neely, J.B. Allen, "Invertibility of a room impulse response", *J.A.S.A.*, vol. 66, pp. 165-169 (1979).

[5] A.V Oppenheim, R.W. Schafer, "Discrete-Time Signal Processing", *Prentice-Hall*, Inc., Englewood Cliffs, N.J., USA 1996

[6] J. Borish, J.B. Angell, "An Efficient Algorithm for Measuring the Impulse Response Using Pseudorandom Noise", *J.A.E.S.*, Vol. 31, n. 7, 1983 July/August, pp. 478-487.

[7] P. Odya et Al., "Determination of Influence of Visual Cues on Perception of Spatial Sound", *Proc. Of 110th AES*, Amsterdam, May 2001.

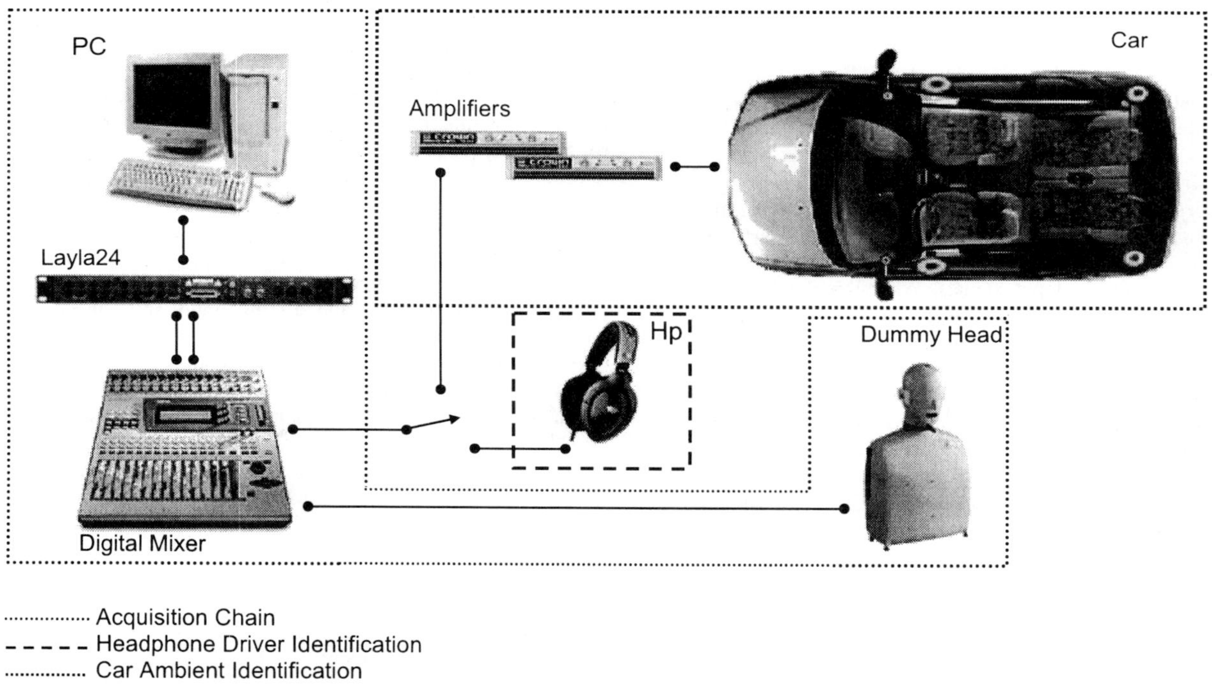

Fig. 1: Acquisition set up

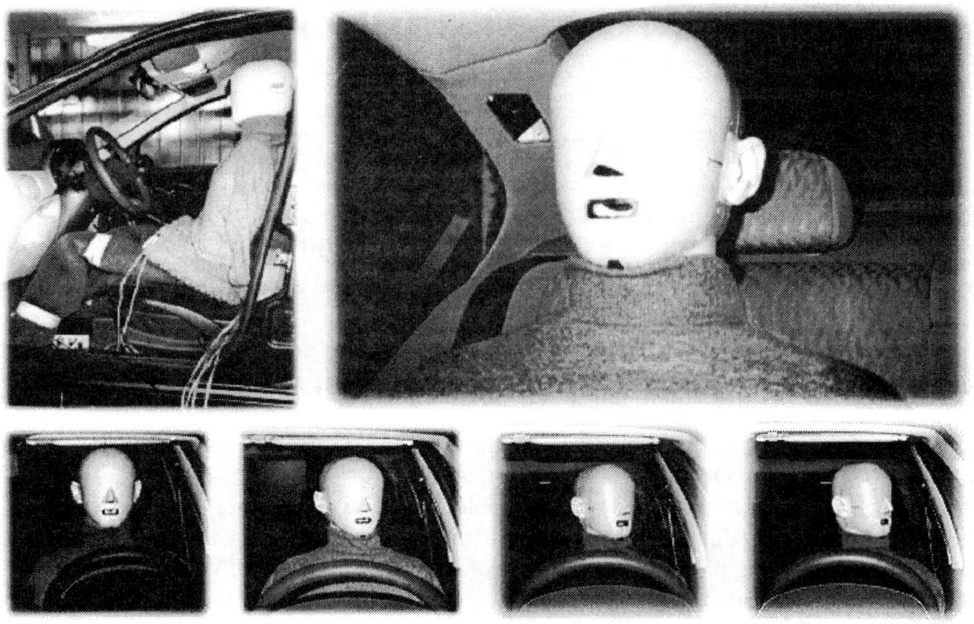

Fig. 2: Shots of the dummy head inside a car cockpit. In cw sense, from top left:

- Dummy in driver position.
- Rear right.
- Driver, 45° left pan.
- Driver, 30° left pan.
- Driver, 15° left pan.
- Driver, 0° left pan.

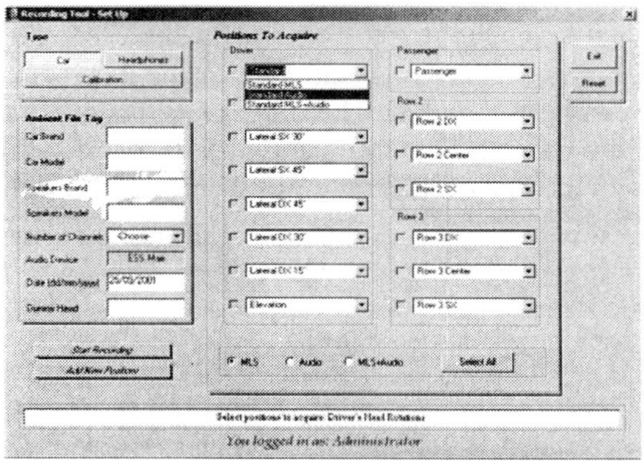

Fig. 3: *Recording Tool - "Set Up"* screen shot.

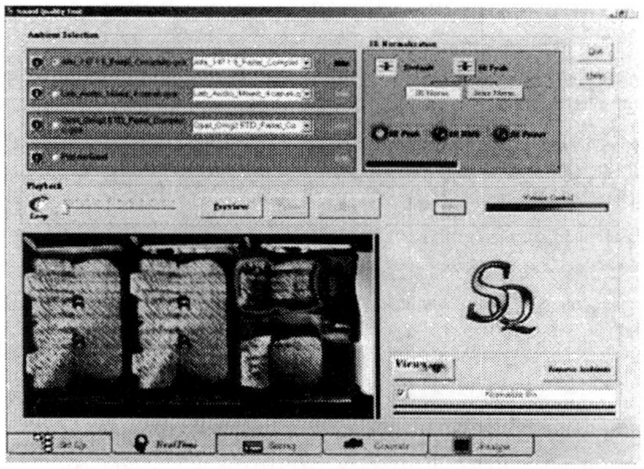

Fig. 4: *SQ Tool – "Real Time"* screen shot.

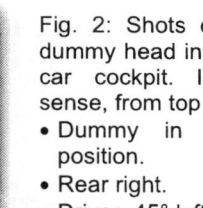

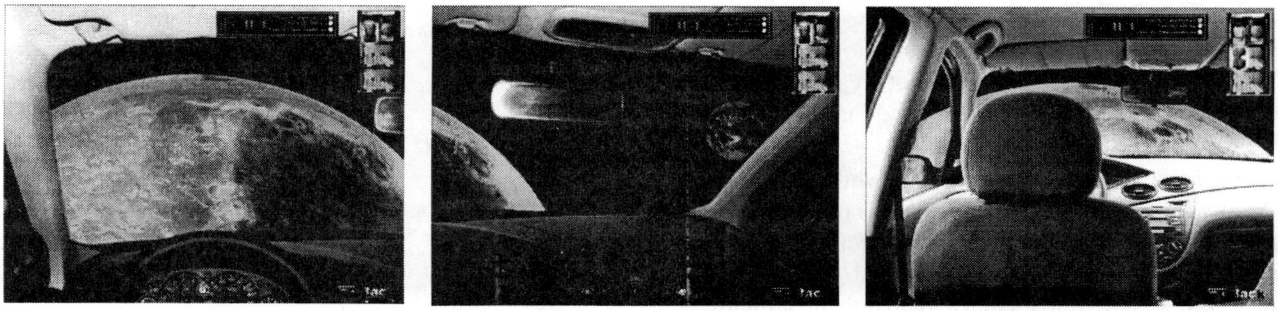

Fig. 5: *SQ Tool - Visual Cues*. Driver position; driver position 45° right pan; rear left passenger position.

AN EFFICIENT REDUCED-ORDER INTERCONNECT MACROMODEL FOR TIME-DOMAIN SIMULATION

Timo Palenius and Janne Roos

Helsinki University of Technology, Department of Electrical and Communications Engineering,
Circuit Theory Laboratory, P.O.Box 3000, FIN-02015 HUT, Finland.
Tel: +358-9-4515036, Fax: +358-9-4514818, E-mail: timppa@aplac.hut.fi

ABSTRACT

As signal speeds grow and feature sizes shrink in digital VLSI circuits, there is an increasing need to correctly model the interconnects between transistors. Since the size of the resulting RLC-interconnect network can be huge, model-reduction algorithms have been developed for replacing the RLC networks with reduced-order frequency-domain models. In this paper, we present an efficient method for interfacing these frequency-domain representations with the time-domain simulation of the original nonlinear circuit.

1. INTRODUCTION

A typical integrated-circuit model consists of nonlinear transistors and large, linear RLC-circuit blocks describing the interconnects. A number of algorithms have been proposed to reduce the order of these linear blocks. In this paper, we use the Passive Reduced-order Interconnect Macromodeling Algorithm (PRIMA) [1] to obtain reduced-order macromodels for the interconnects.

There are several methods of linking these frequency-domain models into the transient analysis of the nonlinear circuit. In order to be general enough, the methods should be able to handle a varying time step, and an estimate for the Local Truncation Error (LTE) should be available. The latter is needed in the automatic control of the step size [2]. Furthermore, since the goal of order reduction is to speed up the simulation, the methods should not create additional nodes; each node increases the time needed for the repeated LU factorization of the system matrix.

Section 2 provides some background, while Sections 3 and 4 briefly review two published reduced-order interconnect macromodels. In Section 5, we propose a new method that meets the requirements stated above. To compare the three methods, we have implemented them into the in-house development version of APLAC circuit simulator [3]. Simulation examples that demonstrate the advantages of the new method and conclusions are given in Sections 6 and 7.

2. STATE-VARIABLE FORMULATION

In PRIMA, the RLC circuit to be reduced is treated as an N-port. Thus, all the reduced-order Y-parameters have the same set of poles. PRIMA also preserves passivity, which is necessary to guarantee the overall stability of the system. The reduced-order admittances of the N-port can be written as [1]:

$$\tilde{Y}_{ij}(s) = k_{ij\infty} + \sum_{m=1}^{q} \frac{k_{ijm}}{s - p_m}, \tag{1}$$

where p_m and k_{ijm} are the mth pole and the corresponding residue of $\tilde{Y}_{ij}(s)$, respectively, and $k_{ij\infty}$ is the direct coupling between ports i and j. The order of reduction is q. If some of the poles in Eq. (1) are complex, they appear in complex-conjugate pairs, $a_m \pm \mathrm{j}b_m$, and their corresponding residues are also complex conjugates: $c_{ijm} \pm \mathrm{j}d_{ijm}$.

The port current produced by one (generic) pole-residue pair can be written as:

$$I(s) = \tilde{Y}(s)U(s) = \frac{k}{s-p}U(s) \mathrel{\hat{=}} kX(s), \tag{2}$$

where state variable $X(s)$ has been defined. Solving for $sX(s)$ yields: $sX(s) = pX(s) + U(s)$. The complete state-space model for a linear, time-invariant system with q_r real poles and q_c complex-conjugate pole pairs can be written in the time domain as [4]:

$$\begin{cases} \dfrac{\mathrm{d}x_m}{\mathrm{d}t} = p_m x_m + u(t), & m = 1, 2, \ldots, q_\mathrm{r}, \\ \dfrac{\mathrm{d}\tilde{\mathbf{x}}_m}{\mathrm{d}t} = \tilde{\mathbf{A}}_m \tilde{\mathbf{x}}_m + \tilde{\mathbf{b}}u(t), & m = 1, 2, \ldots, q_\mathrm{c}, \\ i(t) = k_\infty u(t) + \sum_{m=1}^{q_\mathrm{r}} k_m x_m + \sum_{m=1}^{q_\mathrm{c}} \tilde{\mathbf{c}}_m \tilde{\mathbf{x}}_m, & \end{cases} \tag{3}$$

where

$$\tilde{\mathbf{A}}_m = \begin{bmatrix} a_m & -b_m \\ b_m & a_m \end{bmatrix}, \tilde{\mathbf{b}} = \begin{bmatrix} -\sqrt{2} \\ 0 \end{bmatrix},$$
$$\tilde{\mathbf{c}}_m = \begin{bmatrix} -\sqrt{2}c_m & \sqrt{2}d_m \end{bmatrix}.$$

Altogether, there are Nq state variables for an N-port, when the q poles for each element of the $\tilde{Y}$ matrix are the same.

0-7803-7761-3/03/$17.00 ©2003 IEEE

3. DIFFERENTIAL-EQUATION MACROMODEL

The state-space model of Eq. (3) can be realized with a simple equivalent circuit containing only capacitors, resistors, and time-independent Voltage-Controlled Current Sources (VCCS's) [5]. Fig. 1(a) presents an example of a Differential-Equation Macromodel (DEM) with $N = q = 1$.

Since all the components in the equivalent circuit are time-invariant, DEM can be easily realized as a user-defined netlist-level macromodel. The only dynamic components are the capacitors, so the LTE will be calculated, automatically, by the circuit simulator. The drawback of this method is that it creates Nq additional nodes that can slow down the simulation considerably.

4. RECURSIVE CONVOLUTION

Recursive Convolution (RC) is a method that updates the state variables numerically in the time domain. The equivalent circuit of this method consists of VCCS's and independent current sources at the port nodes, so no additional nodes are created. For a varying time step, RC cannot be realized on the netlist level.

If the solution at the previous time point t_k is known, the exact solution of Eq. (3) at time t_{k+1} can be written as:

$$\mathbf{x}(t_{k+1}) = \mathrm{e}^{\mathbf{A}\Delta t_k}\mathbf{x}(t_k) + \int_0^{\Delta t_k} \mathrm{e}^{\mathbf{A}\tau}\mathbf{b}u(t_{k+1}-\tau)\mathrm{d}\tau. \quad (4)$$

In [4], the voltage waveform $u(t_{k+1}-\tau)$ is assumed piecewise linear. Then the integral in Eq. (4) can be evaluated and a state-updating equation is obtained. However, if nonlinear drivers and loads are used, the voltage waveforms will, obviously, not be piece-wise linear. If the time step is small enough, RC will give a good approximation of the true waveform, but nevertheless, an error is produced. The major drawback of RC is the lack of an estimate for the LTE.

5. TIME-DOMAIN DEM

Here, we will derive the state-updating equations for a new method, which has an equivalent circuit similar to that of RC, but with much simpler equations. We call the new method Time-Domain DEM (TDDEM).

The general form of a numerical integration algorithm for solving initial-value problems can be written as [2]

$$x^{k+1} = \sum_{r=0}^{\rho} a_r x^{k-r} + \Delta t_k \sum_{r=-1}^{\rho} b_r \frac{\mathrm{d}x^{k-r}}{\mathrm{d}t}, \quad (5)$$

where the notation $x(t_{k+1}) = x^{k+1}, x(t_k) = x^k, x(t_{k-1}) = x^{k-1}, \ldots$ has been adopted. The three methods generally used in circuit simulation, Backward Euler (BE), Trapezoidal rule (TR), and Gear-Shichman (GE) [6] methods, can be described with only four non-zero coefficients, a_0, a_1, b_{-1}, b_0. The coefficients for the three methods are collected in Table 1.

Table 1: Coefficients for the integration methods.

Coeff.	BE	TR	GE
a_0	1	1	$\frac{(\Delta t_k + \Delta t_{k-1})^2}{\Delta t_{k-1}(2\Delta t_k + \Delta t_{k-1})}$
a_1	0	0	$-\frac{\Delta t_k^2}{\Delta t_{k-1}(2\Delta t_k + \Delta t_{k-1})}$
b_{-1}	1	$\frac{1}{2}$	$\frac{\Delta t_k + \Delta t_{k-1}}{2\Delta t_k + \Delta t_{k-1}}$
b_0	0	$\frac{1}{2}$	0

The present-time derivative can be solved from Eq. (5):

$$\frac{\mathrm{d}x^{k+1}}{\mathrm{d}t} = \frac{x^{k+1} - a_0 x^k - a_1 x^{k-1} - b_0 \Delta t_k \frac{\mathrm{d}x^k}{\mathrm{d}t}}{b_{-1}\Delta t_k}. \quad (6)$$

State-updating equations are obtained when Eq. (6) is applied to the left-hand side of Eq. (3):

$$x^{k+1} = \frac{\left(a_0 h + \frac{b_0}{b_{-1}} p\right) x^k + a_1 h x^{k-1} + u^{k+1} + \frac{b_0}{b_{-1}} u^k}{h - p} \quad (7)$$

for the real pole and

$$\begin{cases} \tilde{x}_1^{k+1} = \dfrac{c_1 \tilde{x}_1^k - c_2 \tilde{x}_2^k + a_1 h \tilde{x}_1^{k-1} - \frac{a_1 h b}{h-a}\tilde{x}_2^{k-1} + c_3}{h - a + \frac{b^2}{h-a}}, \\ \tilde{x}_2^{k+1} = \dfrac{c_2 \tilde{x}_1^k + c_1 \tilde{x}_2^k + \frac{a_1 h b}{h-a}\tilde{x}_1^{k-1} + a_1 h \tilde{x}_2^{k-1} + c_4}{h - a + \frac{b^2}{h-a}} \end{cases} \quad (8)$$

for the complex-pole pair. Auxiliary coefficients

$$h = \frac{1}{b_{-1}\Delta t_k},$$

$$c_1 = a_0 h + \frac{b_0}{b_{-1}}\left(a - \frac{b^2}{h-a}\right),$$

$$c_2 = b\left(\frac{b_0}{b_{-1}} + \frac{\frac{b_0}{b_{-1}}a + a_0 h}{h-a}\right),$$

$$c_3 = -\sqrt{2}\left(u^{k+1} + \frac{b_0}{b_{-1}}u^k\right),$$

$$c_4 = \frac{b}{h-a}c_3$$

have been used in Eqs. (7) and (8). Inserting the coefficients from Table 1 yields the final state-updating equations for each integration method.

Eqs. (7) and (8) can also be derived from the equivalent circuit produced by DEM. First, the capacitors are replaced with their companion models, which is equivalent to the way how the circuit simulator treats the capacitors in the transient analysis. Next, the port currents are solved as a function of the present and old port voltages, and the resulting independent and controlled current sources are placed at the port nodes. The process of replacing linear circuit blocks with their Norton equivalents can be thought of as an application of hierarchical analysis.

Consider, as an example, the DEM for a one real pole one-port shown in Fig. 1 (a). When the capacitor is replaced with its BE integration method companion model, a resistor and a current source is inserted into the circuit, as shown in Fig. 1 (b). Solving for the port current yields

$$i^{k+1} = kx^{k+1} = k\frac{\frac{1}{\Delta t_k}x^k + u^{k+1}}{\frac{1}{\Delta t_k} - p}, \tag{9}$$

which is Eq. (7) for BE. The final equivalent circuit for TD-DEM is shown in Fig. 1 (c).

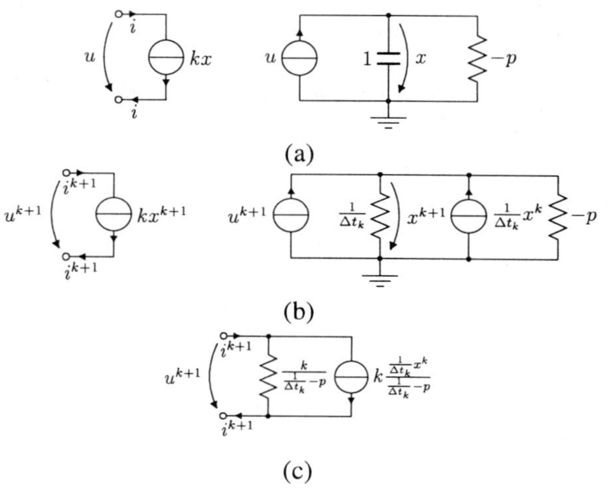

Figure 1: (a) DEM and (b) its companion model. (c) TD-DEM.

In usual transient analysis, the LTE is calculated for the charge of each voltage-controlled charge source (dynamic VCCS). In the case of DEM, all the dynamic VCCS's are 1 F linear capacitors. Since $q = Cu$, the charge is numerically equivalent to the voltage of the capacitor, which, in turn, is numerically equivalent to the state variable x. Therefore, the LTE can be calculated for the state variables using the well-known formulas routinely used in transient analysis. LTE formulas for different integration methods can be found, e.g., in [2], [6].

6. SIMULATION RESULTS

6.1. RL series connection

Consider the simple circuit of Fig. 2. The admittance of the RL series connection is of the form $Y(s) = k/(s - p)$, with $k = 1 \cdot 10^9$ and $p = -5 \cdot 10^9$. The exact solution for the current $i(t)$ can be written as:

$$i_{\text{ex}}(t) = \frac{k}{\omega^2 + p^2}\left[\omega e^{pt} - \omega\cos(\omega t) - p\sin(\omega t)\right]. \tag{10}$$

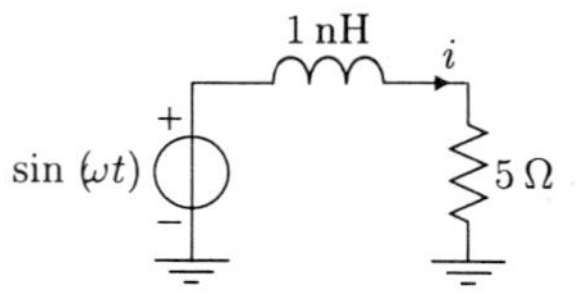

Figure 2: RL series connection.

We simulated the current with RC and TDDEM with $\omega = 2\pi \cdot 10^9$ using a fixed time step of 20 ps. By default, APLAC uses the TR integration method, except in the first two time points, where BE is used. The exact solution i_{ex} and the error $|i_{\text{ex}} - i_{\text{sim}}|$ for both methods are shown in Fig. 3. It can be seen that, at first, the error is large with TD-DEM because BE is used in the first two time points. However, after 0.3 ns the error is in fact a little smaller with TD-DEM, although RC claims to be the most accurate method.

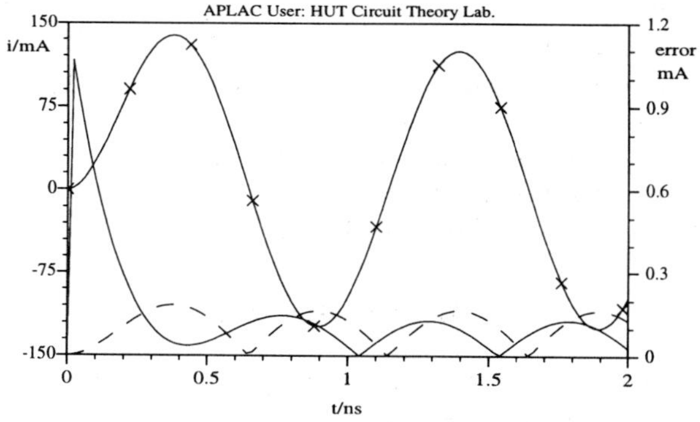

Figure 3: Exact solution (—×—), and error for RC (– – –) and TDDEM (——).

6.2. Circuit with seven transmission lines

We simulated a large interconnect circuit containing seven transmission lines and three inverters as nonlinear driver and loads [7]. After replacing the transmission lines with fifty RLC-sections, we reduced the linear interconnect block containing 728 nodes with PRIMA. The resulting reduced-order model was a three-port with 30 poles.

We simulated the circuit with DEM, RC, and TDDEM. The simulation results were compared to the transient simulation of the original circuit. The excitation was a voltage pulse with 1 ns rise and fall times and the simulation time step was allowed to change between 4 ps–20 ps. The input and output voltages are shown in Fig. 4. The voltage curves of the three methods are identical in the scale of the plot, so only one of them is shown in the figure. It can be seen that the waveforms obtained with the reduced-order models agree very well with the simulation of the original circuit. Therefore, we can conclude that the 30-pole model is sufficient to describe the interconnect circuit in this example.

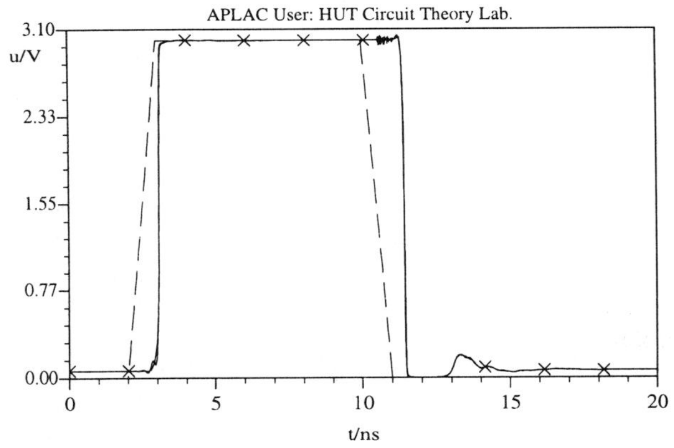

Figure 4: Input voltage (– – –), and APLAC (——) and reduced-order model (—×—) output voltages.

The total number of nodes and VCCS's, as well as the CPU time and memory consumption of the simulations, are presented in Table 2. TDDEM, which is numerically equivalent to DEM, and RC have much shorter simulation CPU times than DEM. The reason is that instead of a 99 × 99-matrix, only a 9 × 9-matrix has to be LU-factorized. RC is a little slower than TDDEM because of the complexity of the state-updating equations. Note that the simulator takes longer time steps in the simulation with RC, as shown in Fig. 5, because of the lack of LTE criterion. Had we not restricted the maximum time step to 20 ps, large errors might have occurred.

Table 2: CPU time comparison.

method	nodes	VCCS's	CPU/s	mem/kB
APLAC	725 + 9	1869	87.3	6388
DEM	90 + 9	654	21.0	3872
RC	0 + 9	75	14.3	3428
TDDEM	0 + 9	75	12.6	3444

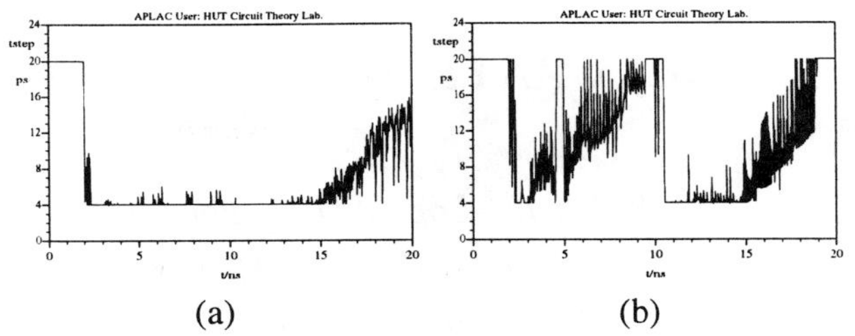

(a) (b)

Figure 5: Simulation time step: (a) TDDEM and (b) RC.

7. CONCLUSIONS

An efficient method, TDDEM, for simulating reduced-order interconnect macromodels in the time domain was presented. The state-updating equations were derived for the commonly used numerical integration schemes. The method proposed can be used with a varying time step, and an estimate for the local truncation error can be easily calculated. The efficiency and accuracy of the method were verified with simulation examples.

8. REFERENCES

[1] A. Odabasioglu, M. Celik, and L. T. Pileggi, "Prima: Passive reduced-order interconnect macromodeling algorithm," *IEEE Trans. Computer-Aided Design*, vol. 17, no. 8, pp. 645–654, Aug. 1998.

[2] L. Chua and P. Lin, *Computed-Aided Analysis of Electronic Circuits: Algorithms and Computational Techniques*, Prentice-Hall, 1975.

[3] Aplac Solutions Corporation, Finland, *APLAC 7.80 Reference Manual and 7.80 User's Manual*, 2002, http://www.aplac.com.

[4] Y. Liu, L. T. Pileggi, and A. J. Strojwas, "ftd: Frequency to time domain conversion for reduced-order interconnect simulation," *IEEE Trans. Circuits Syst. I*, vol. 48, no. 4, pp. 500–506, Apr. 2001.

[5] T. Palenius, J. Roos, and S. Aaltonen, "Development and comparison of reduced-order interconnect macromodels for time-domain simulation," in *ICECS'02*, Dubrovnik, 2002, vol. 2, pp. 757–760.

[6] H. Shichman, "Integration system of a nonlinear network-analysis program," *IEEE Trans. Circuit Theory*, vol. 17, no. 3, pp. 378–386, Aug. 1970.

[7] R. Achar, P. Gunupudi, M. Nakhla, and E. Chiprout, "Passive interconnect reduction algorithm for distributed/measured networks," *IEEE Trans. Circuits Syst. II*, vol. 47, no. 4, pp. 287–301, Apr. 2000.

ACCURATE VHDL-BASED SIMULATION OF ΣΔ MODULATORS

R. Castro-López[1], F. V. Fernández[1,2] F. Medeiro[1,2] and A. Rodríguez-Vázquez[1,2]
[1]IMSE-CNM, Edif. CICA, Avda. Reina Mercedes s/n, E-41012 Sevilla, SPAIN
Phone: +34 955056666, FAX: +34 955056686, e-mail: castro@imse.cnm.es
[2]Escuela Superior de Ingenieros, Isla de la Cartuja s/n, E-41092 Sevilla, SPAIN
Phone: +34 954487378

ABSTRACT

Computational cost of transient simulation of ΣΔ modulators (ΣΔMs) at the electrical level is prohibitively high. Behavioral simulation techniques arise as a promising solution to this problem. This paper demonstrates that both, hardware description languages (HDLs) and commercial HDL simulators, constitute a valuable alternative to traditional special-purpose ΣΔ behavioral simulators. In this sense, a library of HDL building blocks, modeling a complete set of circuit non-idealities which influence the performance of ΣΔMs, is presented. With these blocks, ΣΔM architectures can be described in two different ways, which are analyzed in detail. Experimental results are provided through several simulations of a fourth-order 2-1-1 cascade multi-bit ΣΔM.

1. INTRODUCTION

Among the architectures performing signal conversion between the analog and digital worlds, ΣΔMs have become very popular thanks to their ability to solve problems exhibited by other architectures. For instance, ΣΔMs do not need high-accuracy analog antialiasing filtering and have relatively large insensitivity to circuit imperfections or noisy environments [1],[2].

During the design process of ΣΔMs (actually of any integrated circuit), accurate emulation of the circuit's behavior, by using appropriate simulation techniques, becomes an essential step previous to circuit fabrication. The most accurate one amid available simulation techniques is, undoubtedly, electrical simulation. Unfortunately, although theoretically possible, it may become impractical for circuits like ΣΔMs. For instance, days or weeks of CPU time can be required to estimate the signal-to-noise ratio of a typical ΣΔM architecture [3],[4].

Macromodel descriptions of building blocks, i.e., opamps, can only simplify the problem slightly. This is because the resulting system of differential equations must still be solved numerically. Thus, these techniques are not fast enough for simulation of ΣΔMs. To separate the analog and digital parts of the circuit and simulate them dependently is the basis for mixed-signal (multilevel) simulation. But as long as numerical resolution of differential equations is used for the analog part, the extraction of ΣΔM performances will still be too costly in terms of CPU time.

Event-driven behavioral simulation is a smart solution to further reduce computational cost. This kind of simulation technique involves a circuit partitioning into basic behavioral blocks with fully independent functionality. This means that an instantaneous block output cannot be related to itself, or, in other words, either there is no global feedback loop, or in case such loop exists, there is a delay that avoids the instantaneous dependence. A behavioral simulation of the complete circuit requires, then, a behavioral model for each building block of the circuit. This model takes the form of explicit expressions relating the output variable with the input and internal state variables.

Different approaches to behavioral simulation of ΣΔM have been reported. Special-purpose approaches have been described in [4]-[10]. These are, as their name suggests, simulation techniques specifically devoted to behavioral description and simulation of ΣΔ converters. Among them, the main differences are in the number of topologies/basic blocks included, the accuracy of the models used, the postprocessing capabilities and the friendliness of the user interface. ASIDES [9] is a representative example of this group. In this tool, both, behavioral models of the building blocks composing the ΣΔMs and the simulation engine, are written in C language. This enhances the simulation speed, making ASIDES very appropriate, not only for ΣΔM validation, but also for ΣΔM synthesis (i.e., in simulation-based optimization approaches). A drawback is that ASIDES, like the rest of special-purpose ΣΔ behavioral simulators, is restricted to the simulation of this class of systems.

Simulation approaches based on HDLs (i.e., *VHDL* [11] or *Verilog* [12] and their analog extensions *VHDL-AMS* [13] and *Verilog-AMS*) can overcome the limitation of special-purpose approaches since HDL simulators are found in many commercial design environments. In this way, ΣΔMs modeled with *VHDL* or Verilog can be simulated together with other *VHDL*-modeled blocks and even continuous-time descriptions of blocks, modeled using *VHDL-AMS* or *Verilog-AMS*. Furthermore, different description levels (extracted layout, transistor, macromodel or behavioral levels) can be combined into a single simulation. This is clearly advantageous when compared with special-purpose simulators.

In this paper, the *VHDL* language is used to model high-performance ΣΔMs. Special care has been put to capture all major non-idealities affecting the nominal behavior of these circuits. The paper is organized as follows. An overview of the non-idealities considered and modeled with *VHDL* is provided in Section 2. Section 3 describes two different techniques adopted to describe ΣΔM topologies in detail. Section 4 shows experimental results of a ΣΔM simulation. Concluding remarks are given in Section 5.

2. HDL MODELING OF ΣΔM BUILDING BLOCKS

A basic block representation of a ΣΔM is shown in Figure 1. The non-idealities degrading its behavior can be implemented separately by including imperfections in the performance of each of the three

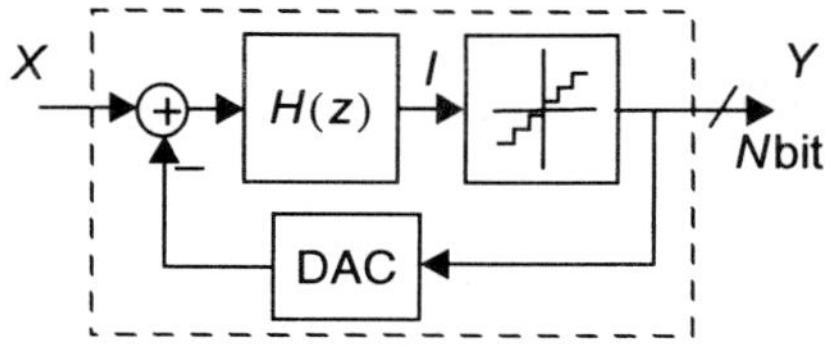

Figure 1. Basic structure of a ΣΔ modulator.

building blocks: the discrete-time filter, $H(z)$, the quantizer and the DA converter (DAC). The ideal behavior is as follows: the output y is subtracted from the input signal, x, which has been sampled at a rate larger than the Nyquist frequency. The result, after passing through $H(z)$, is the input to the quantizer. The goal is to reduce the quantization power spectral density in the low-frequency range. The simplest way to do it is to implement $H(z)$ as a SC integrator, thus performing a shaping on the quantization error.

2.1. SC Integrators

The integrator is the fundamental block because its non-idealities largely affect the performance of ΣΔMs. Figure 2 shows a typical SC integrator used in ΣΔMs.

The set of non-idealities considered and modeled with *VHDL* are: thermal noise, finite and non-linear dc gain, output range, opamp slew-rate and dynamics, switch resistance and capacitor non-linearity and mismatching [2]. These non-ideal effects are included in the complete integrator model, illustrated with the flow graph in Figure 3. During the integration phase, an iterative procedure is started to calculate the integrator output voltage, including the effects of the finite and non-linear opamp, the non-linear capacitors, transient response and the output range limitation [2],[14]. While most integrator models limit the opamp dynamics to the integration phase, introducing these effects also in the sampling phase may become important, especially for high-speed applications. During the sampling phase, the final value of the voltages stored in the input capacitors are calculated considering the values of these and the ON resistance of the switches. The input equivalent thermal noise of the integrator is calculated and added to the sampled voltage. Settling errors are then evaluated.

There is a problem regarding the implementation of the settling error model with *VHDL*. For ΣΔMs of order larger than 1, the calculation of the opamp input and output voltages requires to know the voltage stored in the sampling capacitors of the following integrator during the previous phase. Since each block has to be modeled independently (e.g., its description cannot be related to variables of other blocks), one solution is to accommodate such voltages in the so-called dynamic external ports of the block model with *VHDL*. As these ports can only have only fixed values during a simulation, they have to be transformed into input ports, thus increasing the number of input ports of the integrators. Consequently, block interconnection for complex ΣΔ architectures may become a rather complex task.

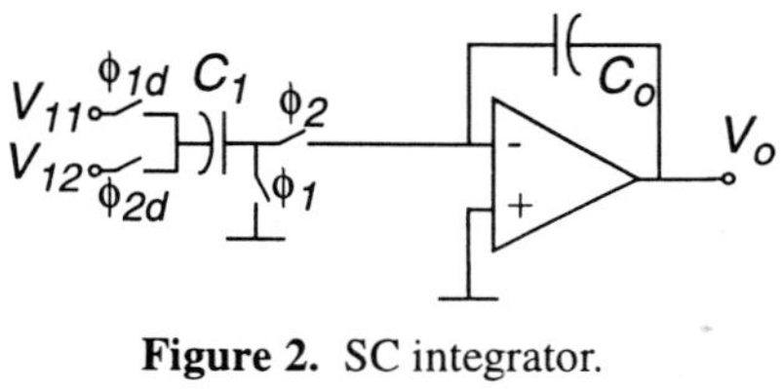

Figure 2. SC integrator.

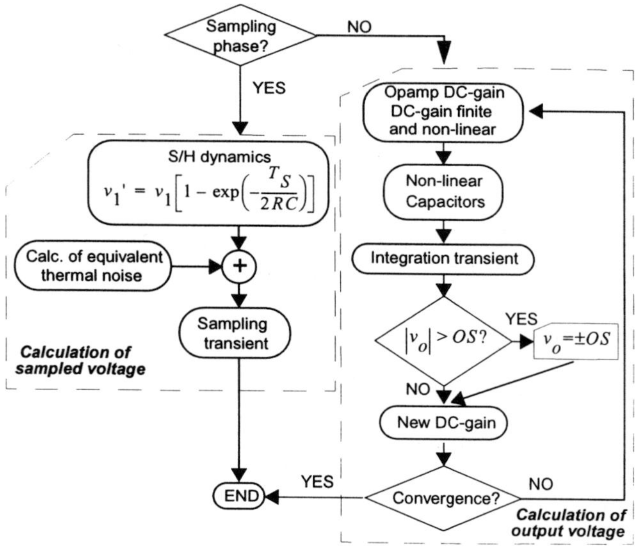

Figure 3. Complete integrator model.

2.2. Quantizers and DA converters

Single-bit architectures incorporate a simple comparator to perform the internal quantization. A simple DA converter is then used in the feedback loop, which does not introduce any non-linearity error. Some comparators also exhibit a hysteresis of random nature. This is illustrated in Figure 4(a), where the transfer curve of such a comparator is shown. The corresponding behavioral model of the comparator with hysteresis is described with the flow diagram in Figure 4(b). It presents three well-differentiated operation zones: in the first of them, in response to an input, the output value (V_{max} or V_{min}) is a function of the sign of $|v_i - V_{off}|$, provided that $|v_i - V_{off}| > h/2$ is fulfilled, where V_{off} and h represent the offset and hysteresis of the comparator, respectively. Otherwise, the output is randomly determined in the dynamic hysteresis model and simply does not change in the deterministic model. For illustration's sake, the *VHDL* implementation of the comparator model is shown in Figure 5.

For multi-bit DA converters, characterized by an offset *off*, a gain error γ, and an integral non-linearity (INL), the model contains first an ideal DA converter, as shown in Figure 6. The result goes through a non-linear block with a third-order non-linearity and a gain error block. Finally, the offset is added [2]. An analogous scheme is used for multi-bit quantizers.

3. HDL MODELING OF ΣΔM ARCHITECTURES

ΣΔM architectures can be described by interconnecting the building blocks above. This can be accomplished by following two different procedures in *VHDL*. The first one consists in defining the ΣΔM architecture through *VHDL* instances [11].

The second procedure, more user-friendly, involves the use of schematic capture tools. With these resources, it is possible to build ΣΔMs just by placing and routing pre-defined symbols. Figure 7 shows the schematic of a 2-1-1 cascade multi-bit ΣΔM. The same architecture built with Mentor Graphics's HDL Designer® sche-

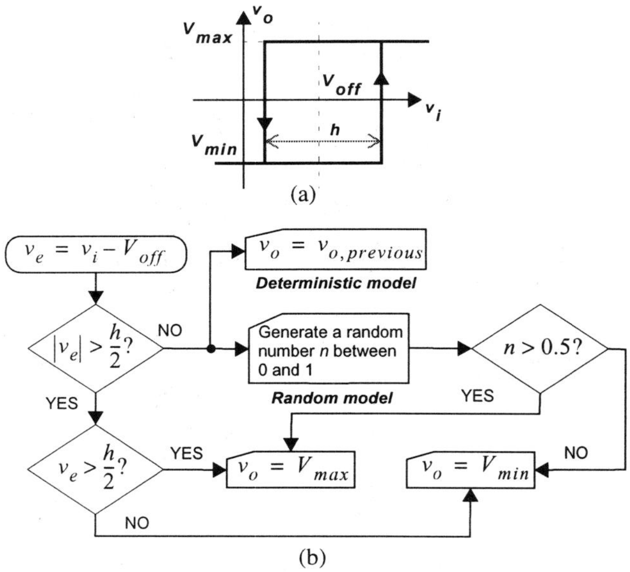

Figure 4. (a) Transfer curve of a comparator with hysteresis and (b) its model flow graph.

```
LIBRARY IEEE;
USE IEEE.STD_LOGIC_1164.ALL;
USE IEEE.MATH_REAL.ALL;
USE IEEE.STD_LOGIC_ARITH.ALL;
ENTITY comparator IS
GENERIC (Vmax: REAL:= 1.5;Vmin: REAL:= -1.5;
         Htype: INTEGER:=0);
PORT (Reset: IN STD_LOGIC;
      Clock: IN STD_LOGIC;
      Xin: IN REAL;
      Xout: OUT REAL:= Vmax;
      voff,h: IN REAL);
END comparator;
ARCHITECTURE function OF comparator IS
BEGIN
 PROCESS
 VARIABLE n: REAL;
 VARIABLE vo: REAL:= Vmin;
 VARIABLE voprev: REAL:= Vmin;
 VARIABLE seed1: INTEGER:= 1;
 VARIABLE seed2: INTEGER:= 2;
 BEGIN
 WAIT UNTIL Xin'EVENT;
 IF Reset='1' THEN
    vo:=Vmax;
    voprev:=Vmax;
    Xout<=Vmax;
 ELSE
     IF clock='0' THEN
        IF Xin >= (h+voff) THEN
           vo:= Vmax;
        ELSE IF Xin <= (voff-h) THEN
           vo:= Vmin;

           ELSE IF Htype=0 THEN
              vo:=voprev;

              ELSE UNIFORM (seed1,seed2,n);
                 IF n>0.5 THEN
                    vo:= Vmax;
                 ELSE vo:= Vmin;
                 END IF;
              END IF;
           END IF;
        END IF;
        voprev:=vo;
        Xout <= vo;
     END IF;
 END IF;
END PROCESS;
END function;
```

comparator reset

output as a function of the input

deterministic hysteresis

random hysteresis

Figure 5. *VHDL* model of comparator with hysteresis.

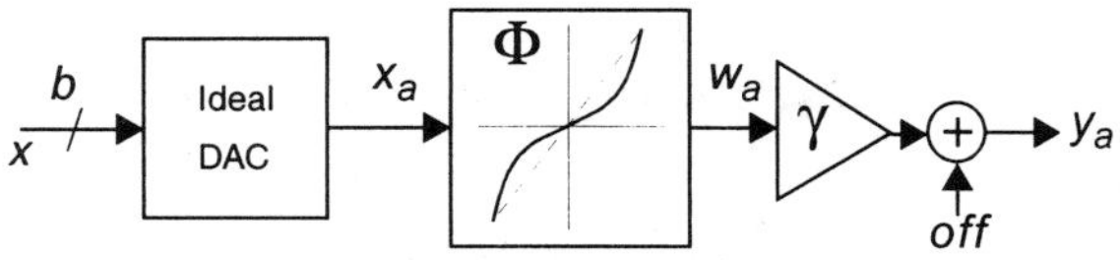

Figure 6. Behavioral model of DA converters.

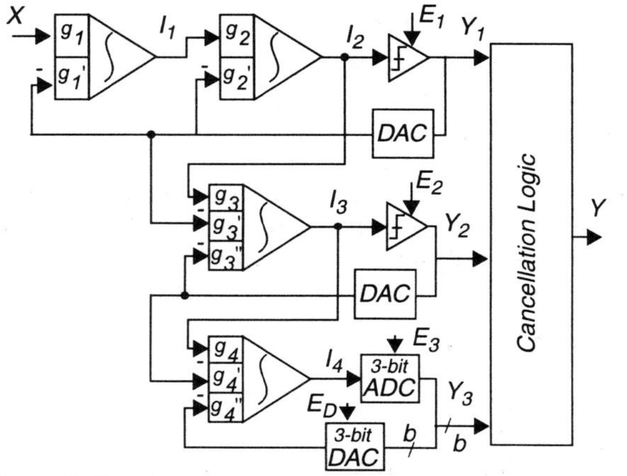

Figure 7. Fourth-order 2-1-1 cascade multi-bit ΣΔ modulator.

matic capture tool is depicted in Figure 8.

In both alternatives, the user must know the number of parameters and inputs of every building block, including the extra ports corresponding to voltages in the sampling capacitors, as explained in Section 2.1. To avoid this extra burden, an optional facility has been developed. It allows the user to introduce just the *VHDL* code or create the schematic of the modulator architecture (with no additional input ports). Then, circuit connectivity is automatically traced and a correct *VHDL* code, with all extra information, is generated.

For subsequent data processing, a number of capabilities have also been implemented as *VHDL* functions. This enable full exploitation of the behavioral simulation results and allows to obtain the output spectrum, the SNDR as a function of the input level or frequency, the in-band error power and the effective resolution. Other *VHDL* functions have been created to perform several types of analysis, such as Monte Carlo or parametric analysis.

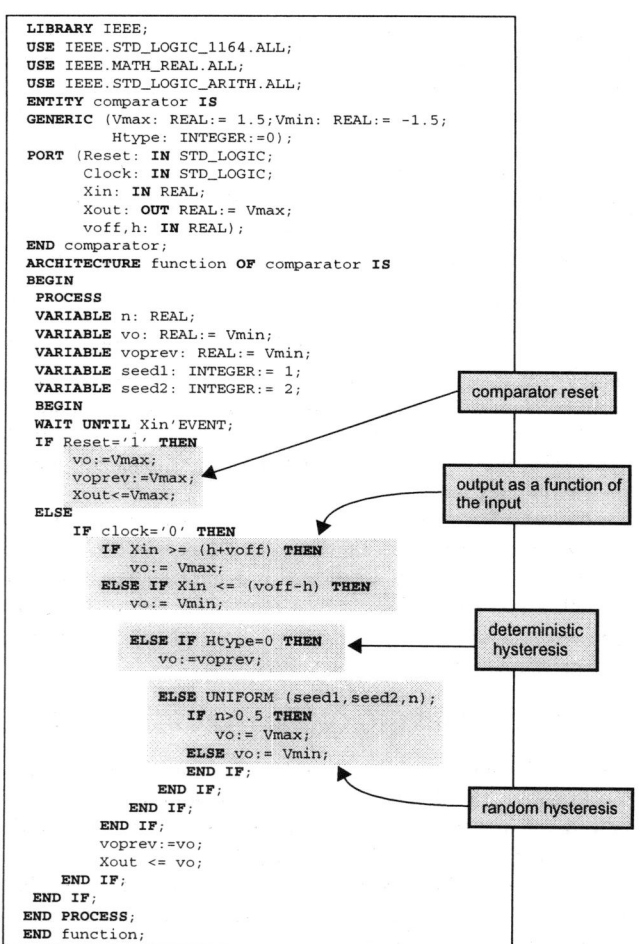

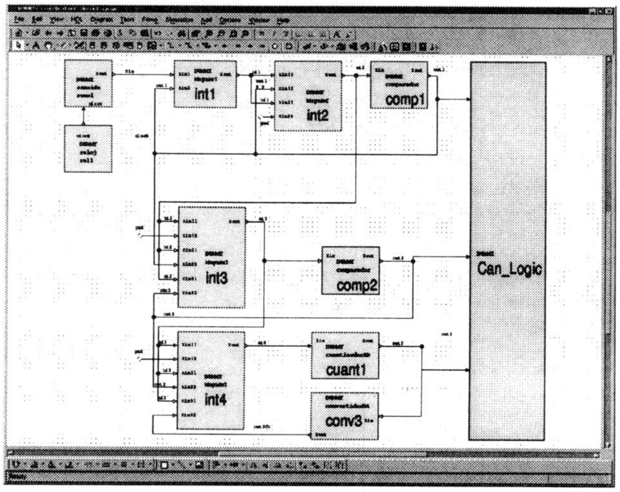

Figure 8. Snapshot of a 2-1-1 cascade multi-bit ΣΔ modulator drawn with HDL Designer®.

4. EXPERIMENTAL RESULTS

HDL simulation of ΣΔM behavior presented here, will be illustrated via behavioral simulations of the 2-1-1 cascade multibit ΣΔM of Figure 7. The result of a *VHDL*-based behavioral simulation is shown in Figure 9. There, the influence of the non-linearity of the DA converter in the in-band error power is analyzed by performing a parametric simulation. Compilation of the *VHDL* ΣΔM architecture took 7.5s. of CPU time and simulation[1] took 203.5s, using Mentor Graphics' Advance-MS®. The same simulation required 25s. in ASIDES [2] to provide identical results (note that block models were exactly the same in both cases). Thus, ASIDES is more suitable for processes where repetitive simulation is required, i.e., in simulation-based iterative optimization tasks. However, the capability to simulate *VHDL*-modeled ΣΔMs together with other *VHDL*- or *VHDL-AMS*-modeled subsystems suggests this as a more flexible methodology for system-level verification.

It is very interesting to evaluate which is the accuracy of the behavioural simulation and how it compares with a conventional electrical simulator. The power spectral density (PSD) plots in Figure 10 were obtained from three different sources. First, it was computed, from 65536 samples, by simulating the *VHDL*-modelled ΣΔM in Figure 7 with Mentor Graphics' Advance-MS®. Simulation time was 3.6s. Second, the PSD was also obtained with HSPICE. The simulation took 5 days of CPU time to get only 8192 samples. It can be observed that HSPICE computes a lower error power because thermal noise cannot be included in the transient simulation. Finally, the PSD was measured directly from a chip prototype (a different signal frequency has been chosen for better visualization)

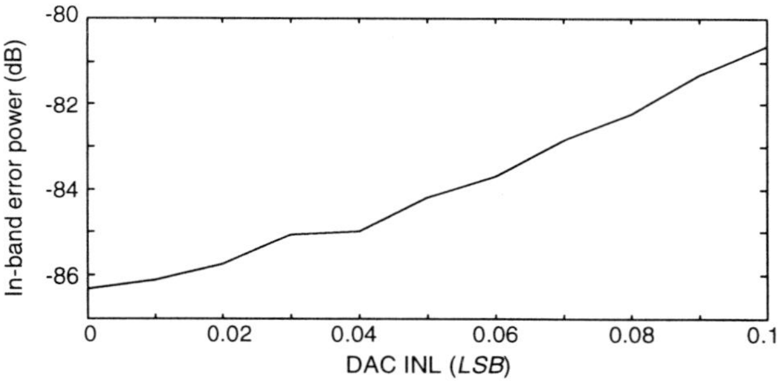

Figure 9. Influence of the DA converter non-linearity on the in-band error power.

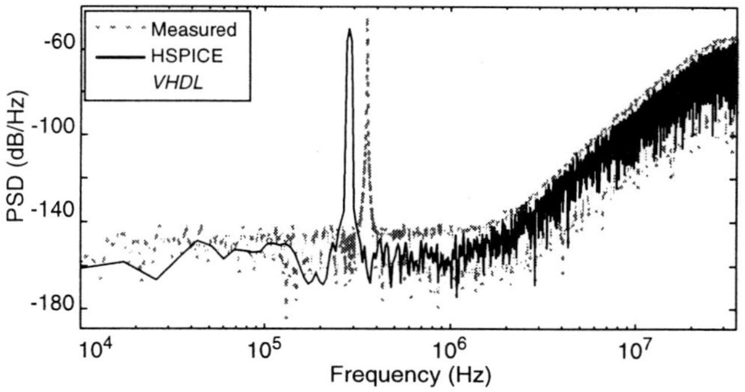

Figure 10. Simulated and measured PSD.

[1] All simulations were performed on a SunFire 3800 @ 750 MHz.

5. CONCLUSIONS

HDLs together with commercial behavioral simulators are efficient resources to perform accurate verification of ΣΔMs. They also allow to combine ΣΔMs with *VHDL* descriptions of digital blocks or *VHDL-AMS* descriptions of other analog blocks (even a *VHDL-AMS* description of any block of the ΣΔM can be used). The capability to simulate ΣΔMs within more complex systems and the integration into commercial design environments makes HDLs valuable alternatives to special-purpose ΣΔ behavioral simulators.

6. REFERENCES

[1] J. Candy and G.C. Temes, "Oversampling Methods for A/D and DA Conversion," in *Oversampling ΣΔ converters*, pp. 1-25, IEEE Press, 1992.

[2] F. Medeiro, B. Pérez-Verdú and A. Rodríguez-Vázquez, *Top-Down Design of High-Performance Sigma-Delta Modulators*. Kluwer Academic Publishers, Boston, 1999.

[3] V. F. Dias, V. Liberali and F. Maloberti: "Design Tools for Oversampling Data Converters: Needs and Solutions", *Microelectronics Journal*, Vol. 23, pp. 641-650, 1992.

[4] C. H. Wolff and L. Carley: "Simulation of Δ-Σ Modulators Using Behavioral Models", *Proc. IEEE Int. Symp. Circuits and Systems*, pp. 376-379, 1990.

[5] V. Liberali, V.F. Dias, M. Ciapponi and F. Maloberti, "TOSCA: a Simulator for Switched-capacitor Noise-shaping A/D Converters," *IEEE Trans. Computer-Aided Design*, Vol. 12, No. 9, pp. 1376-1386, Sept. 1993.

[6] S. R. Norsworthy, I. G. Post and H. S. Fetterman: "A 14-bit 80-kHz Sigma-Delta A/D Converter: Modeling, Design and Performance Evaluation", *IEEE J. Solid-State Circuits*, Vol. 24, pp. 256-266, April 1989.

[7] A. Opal: "Sampled Data Simulation of Linear and Nonlinear Circuits", *IEEE Trans. on Computer-Aided Design*, Vol. 15, pp. 295-306, March 1996.

[8] T. Ritoniemi, T. Karema, H. Tenhunen and M. Lindell: "Fully-differential CMOS Sigma-Delta Modulator for High-Performance Analog-to-Digital Conversion with 5 V Operating Voltage", *Proc. IEEE Int. Symp. Circuits and Systems*, pp. 2321-2326, 1988.

[9] F. Medeiro, B. Pérez-Verdú, A. Rodríguez-Vázquez and J.L. Huertas "A Vertically Integrated Tool for Automated Design of ΣΔ Modulators" *IEEE J.Solid-State Circuits*, pp. 762-777, July 1995.

[10] K. Francken, P. Vancorenland and G. Gielen, "DAISY: a Simulation-based High-level Synthesis Tool for ΔΣ Modulators," *Proc. IEEE/ACM Int. Conf. Computer-Aided Design*, pp. 188-192, 2000.

[11] *IEEE VHDL Language Reference Manual*, IEEE Std 1076-2002.

[12] *IEEE Standard Verilog hardware description language*, IEEE Std 1364-2001.

[13] *IEEE VHDL 1076.1 Language Reference Manual*, IEEE Std 1076.1-1999.

[14] R del Río, F. Medeiro, B. Pérez-Verdú and A Rodríguez-Vázquez "Reliable Analysis of Settling Errors in SC Integrators: Application to ΣΔ Modulators," *IEE Electronics Letters*, pp. 503-504, Vol. 36 Iss. 6, March 2000.

AN ACCURATE BEHAVIORAL MODEL OF PHASE DETECTORS FOR CLOCK RECOVERY CIRCUITS

M. Balsi, F. Centurelli, G. Scotti, P. Tommasino, A. Trifiletti

Dipartimento di Ingegneria Elettronica
Università "La Sapienza" di Roma
Via Eudossiana 18, I-00184 Roma, ITALY

ABSTRACT

We propose in this paper a behavioral modeling technique of analog phase detectors for clock recovery applications, that allows fast but accurate simulations of PLL and clock recovery systems. The model is extracted from transistor-level simulations of the phase detector circuit, thus it can be used for behavioral simulations of the PLL in the design verification phase of a systematic design methodology. The technique is applied to a DRML phase detector for 2.5 Gb/s applications in Si BJT technology (27 GHz f_T), providing a rms error below 5 %.

1. INTRODUCTION

The phase-locked loop (PLL) is a key building block for analog and digital communication and control applications, where it is used for frequency synthesis, clock (timing) recovery, phase synchronization, signal modulation and demodulation. In many of these systems, PLL jitter behavior is a critical performance that is important to predict to achieve a successful system design. PLL simulation at the transistor level requires very long simulation times and a huge amount of memory, since it involves the simulation of large circuits with very small time steps, determined by the VCO period, on a time scale imposed by the PLL time constants, that are orders of magnitude greater (three to four orders of magnitude typically [1]). These simulation problems and the use of systematic design methodologies for design reusability led to a strong interest in the development of behavioral models for the PLL and its building blocks. A design methodology with a top-down design and a bottom-up verification phase is typically used [2], and this requires the development of behavioral models at different levels: both functional models for architecture exploration and system design, and accurate models corresponding to actual circuit implementation, for fast system simulation and performance verification.

Recently, several papers have been presented dealing with PLL behavioral simulation, addressing both model structure [3]-[5] and numerical simulation issues [6]. Whereas some of these papers describe accurate VCO models (e.g. [7]-[8]), very simple models for the phase detector are usually presented, that are not adequate for the bottom-up verification phase: multiplier phase detectors are described in the phase domain [9] or using polynomial VCVS [10], whereas phase-frequency detectors are modeled as state machines [8], sometimes including finite transit time and slew rate effects [11]-[12]. Perrott in [1] also considered the case of digital phase detectors for clock recovery circuits (Hogge detector and band-bang PD), presenting a model where the (ideal) output waveform of the phase detector is sampled to get a discrete-time PLL model. However, in order to uncover potential design problems, the behavioral models used for the PLL blocks should reflect the real circuit behavior as much as possible, including realistic nonlinear and second order effects. This is still more important when clock recovery behavioral simulations are considered, where nonlinear effects like pattern dependent jitter have to be taken into account.

In this paper we present a behavioral modeling technique of analog phase detectors for clock recovery applications from NRZ data, illustrating it in the case of the DRML phase detector [13]. The model is extracted from transistor-level simulations of the phase detector circuit, thus allowing accurate behavioral simulations of the PLL in the bottom-up design verification phase. Such a simulation provides the output time-domain waveforms as in SPICE simulations, but with a much shorter simulation time since the number of circuit's internal nodes is drastically decreased, without a significant loss of information.

2. THE DRML PHASE DETECTOR

We have chosen to check the effectiveness of our modeling approach on the DRML (differentiator-rectifier-mixer-low pass filter) phase detector [13], which is composed of 12 devices in a stacked configuration and suffers from strong data dependency as the operating frequency increases.

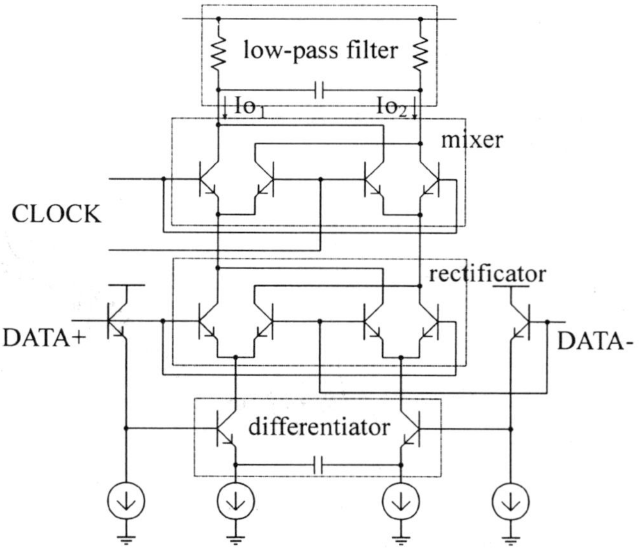

Figure 1. The DRML phase detector.

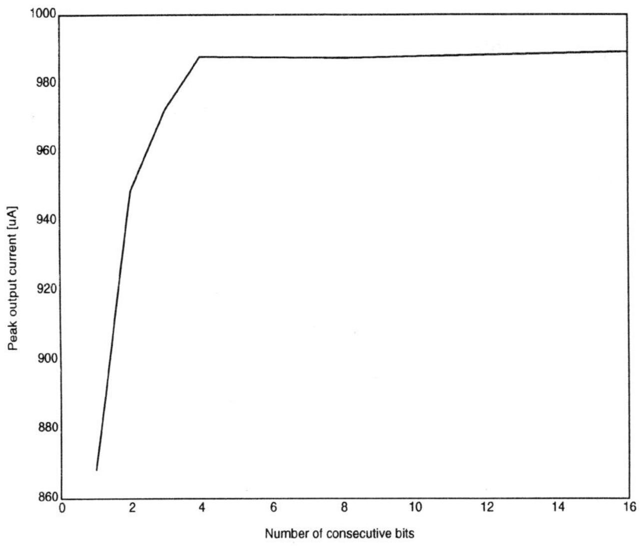

Figure 2. Amplitude of the peak of the output current as a function of the distance between data edges.

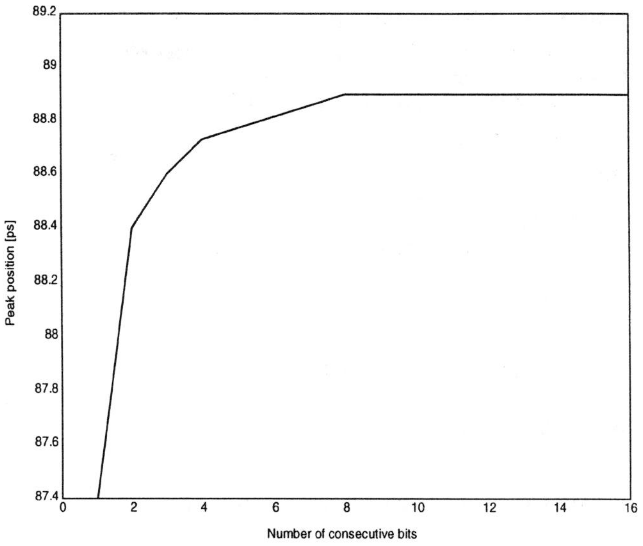

Figure 3. Position of the peak of the output current as a function of the distance between data edges.

Figure 1 depicts the DRML circuit schematic, highlighting its different functional blocks: from bottom to top, a capacitively-coupled current amplifier (C3A) is used to differentiate the NRZ input data stream, providing pulses in correspondence with the data edges. The following Gilbert multiplier provides full wave rectification of such pulses, by exploiting the intrinsic transit time of the differentiator to multiply current pulses by input signals of the same polarity. The upper Gilbert cell multiplies the rectified pulses by the clock signal, providing as output differential current pulses Io_1-Io_2 proportional to the phase difference between clock and data, that are integrated by the low pass filter (a simple RC as in Figure 1, or a more complex structure if needed). The DRML is then composed by a differentiate-and-rectify preprocessor [14] and a sampling phase detector [15]. Its phase characteristic (dc value of the differential output current versus input phase difference) replicates the input clock waveform: a sinusoidal phase characteristic is obtained only if the input clock is sinusoidal and its amplitude does not bring into saturation the differential pairs of the upper Gilbert cell.

When the DRML is operated at high frequency, the time the differentiator needs to recover its steady-state condition is longer than the bit time: when the next data transition arrives, current is still flowing in the capacitor, and different voltage drops are present across the base-emitter junctions of the transistors, determining different values for the nonlinear junction capacitances. The sampling pulse generated by the differentiator is therefore affected by its initial conditions, hence providing different amplitudes and delays in the output current pulses: this memory effect can lead to pattern dependent jitter in a NRZ clock and data recovery system, where the distance between consecutive data edges depends on the transmitted data pattern, and it is important that an accurate behavioral model describes it.

Simulations of a DRML in a 27 GHz-f_T silicon bipolar technology, used with a 2.5 Gb/s data signal, have shown that the output signal characteristics are dependent only on the distance of the current data edge from the previous one (further history can be neglected with small errors), and Figures 2 and 3 show respectively the dependence on such distance of the amplitude and the position of the peak of the output current pulse; the distance between data edges is measured as the number of consecutive identical bits preceding the present data edge. A saturation effect after about 3 ns is present, due to the circuit time constants.

3. THE DRML MODEL

We propose a behavioral model of the phase detector that provides a full reconstruction of the differential output current waveform, hence allowing fast but accurate time-domain simulations where second-order effects such as the memory effect are taken into account. The model is extracted from transistor-level simulations of the phase detector circuit, thus it can be easily used in the design verification phase; the model provides the phase detector response before the low-pass filter, allowing filter design and optimization to be carried out with fast simulations.

The output waveform of the phase detector is reconstructed by simply adjoining the responses of the circuit to single data edges: such responses depend on the time distance of the current data edge from the previous one, T_{del}, thus taking into account the memory effect, and are not impulse responses, since they include the tail of the response to the previous data edge. This approach has been preferred since it allows a simpler implementation of the model and faster simulation times with respect to the use of a model based on the convolution of a time-variant impulse response with a train of pulses corresponding to the data edges. The response of the phase detector to the single data edge has been modeled as the product of the clock waveform, time-shifted of a given amount, times a sampling pulse P(t). The latter is given by a gaussian pulse, used to describe the rising edge of the pulse, and a decreasing exponential tail: the use of two different functions suitably joined allows a better fitting of both the sampling pulse and of its tail, where the clock feedthrough is evident. For a sinusoidal clock waveform whose frequency is f_B, the output of the phase detector is given by

$$Y = P\sin(2\pi f_B t + X_1), \quad (1)$$

where

$$P = X_2 W_1 + Ae^{-t/B} W_2 \quad (2)$$

is the sampling pulse, composed of the gaussian pulse

$$W_1 = \exp\left[-\frac{(t - X_3)^2}{K}\right] \quad (3)$$

and an exponential tail. The function

$$W_2 = \prod_{n=a_1}^{a_2} [1 - W_1(t - n\tau)] \quad (4)$$

is used to obtain a smooth transition between the two parts of the sampling pulse. As is evident from the above equations, the model is described by some parameters that are circuit dependent (the width of the gaussian pulse K and the parameters of the exponential tail), and some others (X_j) that depend on the input phase difference and describe the memory effect. The latter are the amplitude (X_2) and the time-shift (X_3) of the gaussian pulse, that depend on the distance T_{del} between the actual data edge and the previous one, and the time-shift of the clock waveform (X_1), that depends also on the input phase difference $\Delta\phi$.

The model is extracted from transistor-level simulations of the phase detector circuit, under conditions that correspond to selected points $P_i = (N_i, \Delta\phi_i)$ of a bidimensional domain $N x \Delta\phi$, where $\Delta\phi \in [0, 360°)$ is the input phase difference and N is the number of consecutive identical bits preceding the data edge of interest. For each point P_i, the inputs to the phase detector are a sinusoidal clock signal with phase $\Delta\phi_i$ and a square wave with period $N_i T_B/2$. The parameters of the model are obtained by a fitting of the output waveform and then a fitting of the parameters X_j in the bidimensional domain.

4. MODEL IMPLEMENTATION

The phase detector model has been implemented in Simulink® software tool [16], and is composed by two main blocks, as shown in Figure 4. The block PD_CORE models the effective phase detector operation, and implements the model in equations (1)-(4); it receives as input the clock waveform, the phase difference $\Delta\phi$ (i.e. the phase of the clock signal at the instant of the data edge), the time of the data edge T_o and the time distance from the previous data edge T_{del}. The block PD_IN provides such information to the block PD_CORE, and its structure depends on how the clock and data signal are defined in the PLL behavioral model. In case a time-domain waveform representation is used, the block PD_IN computes the phase difference $\Delta\phi$ as the four-quadrant arcsine of the clock value in the instant of the data edge, and the time distance T_{del} as the difference between the time value at the instant of the present data edge and that at the previous data edge, that has to be kept in a memory register.

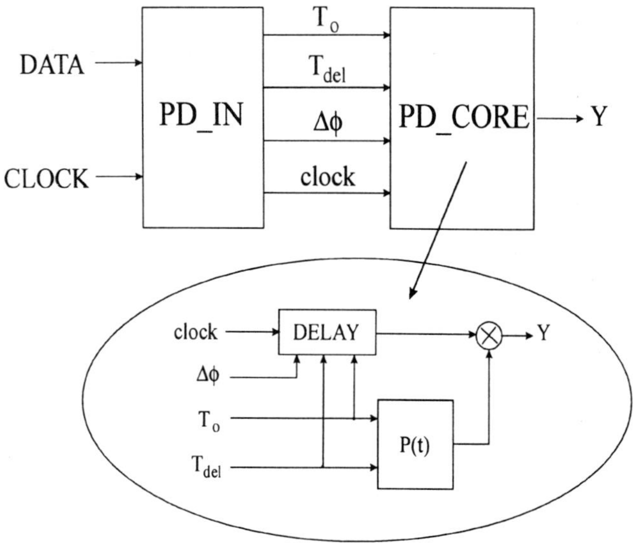

Figure 4. Model implementation block scheme.

5. MODEL VALIDATION

Spice simulations of a 2.5 Gb/s DRML phase detector in silicon bipolar Maxim GST-2 technology (f_T = 27 GHz) have been used to extract a behavioral model as described in the previous Sections. Figure 3 shows that the memory effect saturates for time distances T_{del} above 3 ns (N = 8 at 2.5 Gb/s), therefore transistor-level simulations have been performed for N = 1, 2, 3, 4, 8, 16, and for phase differences between 0° and 360° with a 15° step. From the fitting procedures, the parameters in Table I have been obtained; moreover the X_j parameters are given by the following equations:

$$X_1 = 1.802e^{-2}\Delta\phi + 4.95e^{-2}\,\mathrm{arctg}\left(-\frac{N - 3.055}{0.101}\right) - 0.54, \quad (5)$$

$$X_2 = 4.92e^{-4} + 1.85e^{-5}\,\mathrm{arctg}\left(\frac{N - 2.1}{35e^{-2}}\right), \quad (6)$$

$$X_3 = 122e^{-12} + 3.8e^{-12}\,\mathrm{arctg}\left(\frac{N - 3}{1e^{-11}}\right), \quad (7)$$

where obviously

$$N = T_{del} / 400e^{-12} \quad (8)$$

is the variable used to evaluate the memory effect.

TABLE I: MODEL PARAMETERS.

Parameter	Value
A	6e-5
B	600e-12
K	3e-21
τ	20e-12
a_1	-1
a_2	6

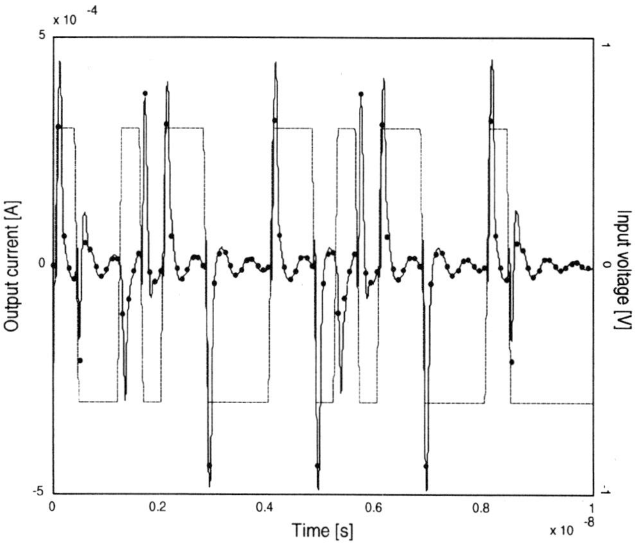

Figure 5. Comparison of the output current given by Spice simulation (solid line) and by the proposed model (dots).

The model has been validated by comparing the output waveforms obtained from transistor-level Spice simulations and from Simulink® simulations at 2.5 Gb/s, for a PRBS input data stream. Figure 5 shows the output waveforms obtained from Spice and Simulink® simulations at 2.5 Gb/s, highlighting the good matching both on the current peaks and on the tail with the ringing due to the clock feedthrough: the rms error normalized to the mean value of the output current is below 5 %.

6. CONCLUSIONS

We have presented a behavioral modeling technique of analog phase detectors for clock recovery applications, that allows to accurately reproduce the output waveform of the phase detector before the loop filter: high frequency components that could affect jitter behavior are then preserved, and filter optimization is possible by performing fast system simulations. This is very important for PLL-based clock recovery circuits, that usually require very long simulation times and where the jitter behavior is of utmost importance. The model is extracted by transistor-level simulation of the phase detector circuit, and can be easily implemented in system-level simulators like Simulink® or in an AHDL environment. The modeling technique has been applied to a 2.5 Gb/s DRML phase detector, showing a rms error below 5%.

7. REFERENCES

[1] M. H. Perrott, "Fast and accurate behavioral simulation of fractional-N frequency synthesizers and other PLL/DLL circuits", Proc. DAC, 2002, pp. 498-503.

[2] B. De Smedt, G. Gielen, "Models for systematic design and verification of frequency synthesizers", IEEE Trans. Circuits and Systems Part II, vol. 46, pp. 1301-1308, Oct. 1999.

[3] P. Acco, M. P. Kennedy, C. Mira, B. Morley, B. Frigyik, "Behavioral modeling of charge pump phase locked loops", Proc. ISCAS, 1999, vol. 1, pp. 375-378.

[4] L. Wu, H. Jin, W. C. Black Jr., "Nonlinear behavioral modeling and simulation of phase-locked and delay-locked systems", Proc. CICC, 2000, pp. 447-450.

[5] D. Andreu, D. Stepan, S. Josse, "Phase loop locked behavioral modeling: from SPICE to VHDL-AMS", Proc. MixDes, 2001, pp. 401-406.

[6] A. Demir, E. Liu, A. L. Sangiovanni-Vincentelli, I. Vassiliou, "Behavioral simulation techniques for phase/delay-locked systems", Proc. CICC, 1994, pp. 453-456.

[7] E. Liu, A. L. Sangiovanni-Vincentelli, "Behavioral representations for VCO and detectors in phase-lock systems", Proc. CICC, 1992, paper 12.3.

[8] B. De Smedt, G. Gielen, "Nonlinear behavioral modeling and phase noise evaluation in phase locked loops", Proc. CICC, 1998, pp. 53-56.

[9] B. A. A. Antao, F. M. El-Turky, R. H. Leonowich, "Mixed-mode simulation of phase-locked loops", Proc. CICC, 1993, paper 8.4.

[10] M. Sitkowski, "The macro modeling of phase locked loops for the Spice simulator", IEEE Circuits and Devices Magazine, vol. 7, n. 2, pp. 11-15, Mar. 1991.

[11] A. Phanse, R. Shirani, R. Rasmussen, R. Mendel, J. S. Yuan, "Behavioral modeling of a phase locked loop", Proc. Southcon, 1996, pp. 400-404.

[12] M. Hinz, I. Könenkamp, E.-H. Horneber, "Behavioral modeling and simulation of phase-locked loops for RF front ends", Proc. MWSCAS, 2000, pp. 194-197.

[13] B. Razavi, "A 2.5 Gb/s 15 mW clock recovery circuit", IEEE J. Solid-State Circuits, vol. 31, pp. 472-480, Apr. 1996.

[14] Z. Wang, "MultiGbits/s data regeneration and clock recovery IC design", Ann. Télécommun., vol. 48, n. 3-4, pp. 132-147, 1993.

[15] J. A. Crawford, "Frequency synthesizer design handbook", Artech House, 1994.

[16] Simulink® for model-based and system-level design, Reference Guide, The Mathworks Inc., 1999.

SYMBOLIC ANALYSIS: A FORMULATION APPROACH BY MANIPULATING DATA STRUCTURES

Tlelo-Cuautle E.[1,2], *Sánchez-López C.*[1]*, *Sandoval-Ibarra F.*[3]

[1]INAOE: Electronics Department, Integrated Circuit Design Group. México
[2]Instituto Tecnológico de Puebla: Electrical and Electronics Department. México
[3]CINVESTAV: Guadalajara Unit, Electronics Design Group. México
e.tlelo@ieee.org, sandoval@ieee.org

ABSTRACT

A formulation approach for symbolic analysis of analog circuits modeled by nullors, is introduced. The proposed technique is based on manipulating data-structures: including only admittances, nullators and norators, instead of the nodal-admittance-matrix manipulation. By using this technique, CPU-time minimization as well as fast manipulation of complex symbolic expressions are tasks highly improved. That way, the reduced system of equations is formulated by fill-in the reduced admittance-matrix using the nullor interconnection-properties.

1. INTRODUCTION

Since symbolic analysis helps the designer to get more insight about the behavior of a circuit [1]-[10], it is quite useful in computing to have relationships that can be used as design-equations suitable for synthesis and optimization procedures. In order to apply a pure-nodal-analysis (PNA) method [3], the complexity in manipulating big matrices, e.g. using the modified-nodal-analysis (MNA) method has been improved by transforming all non-NA-compatible elements into NA-compatibles ones using nullors, i.e. by avoiding the computation of stamps. Most important in using nullors is that the system of equations is reduced in one order for each nullor [3, 4, 9]. However, such a reduction process has been done directly on manipulating the formulated big-admittance-matrix by adding and deleting columns and rows until the reduced matrix is obtained [9]. That is the reason why to improve the reduction process complexity in formulating the reduced set of equations, a formulation approach focused on manipulating two kind of data structures, according to the interconnection relationship of nullator-norator pairs, is presented.

To describe the proposed formulation approach, in section 2 the traditional method used to formulate the reduced set of equations of nullor circuits, is presented. The formulation approach by manipulating data-structures is introduced is section 3. Section 4 presents a resume of the proposed technique. Finally, the conclusions and future work are given in section 5.

*Sánchez-López holds a scholarship from CONACYT-MEXICO

2. TRADITIONAL FORMULATION APPROACH

Analog modeling approaches at different abstraction levels are quite useful and convenient to avoid time-waste if an approximated expression representing the dominant behavior of a circuit is required. That way, the ideal behavior of many analog circuits can be modeled using nullors [11]-[14]. The nullor is composed of two elements (see Fig. 1a): the nullator, where v_1=i_1=0, and the norator where v_2=i_2=undefined. The nullor-based model of an operational transconductance amplifier (OTA) is shown in Fig. 1b [3, 13, 14].

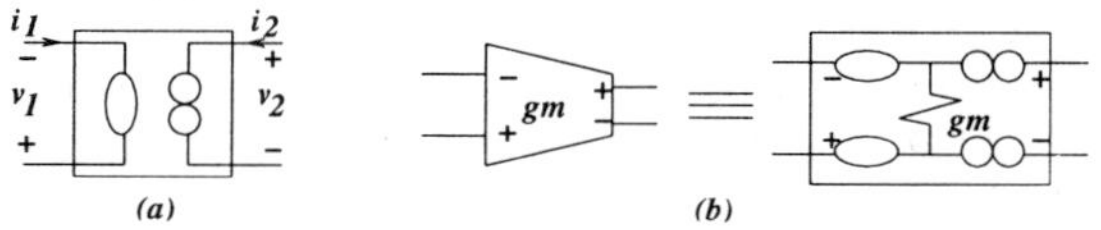

Fig. 1. (a) Nullor and (b) Nullor-based OTA

Let us consider the current-mode low-pass filter shown in Fig. 2 [14]. By applying a transformation process to this OTA-based filter using nullors, the resulting nullor circuit is shown in Fig. 3.

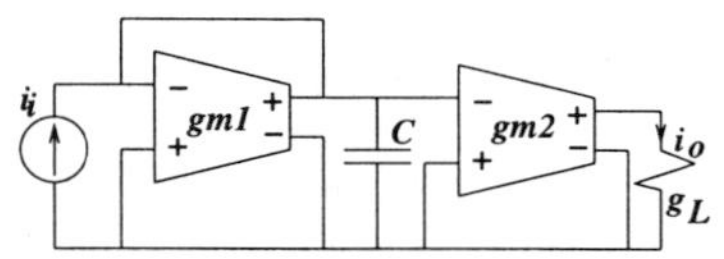

Fig. 2. OTA-based current-mode low-pass filter

For symbolic analysis purposes, the circuit has six nodes. Thus, the computation of the symbolic transfer function (TF) i_o/i_i is done as follows [3, 9]:

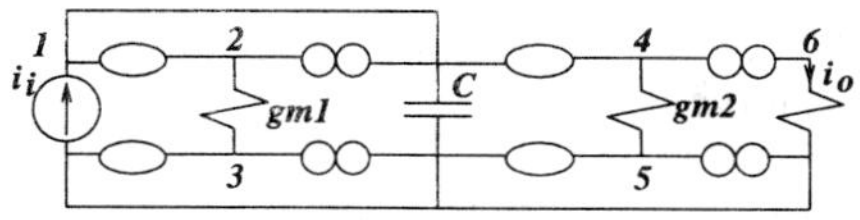

Fig. 3. Nullor equivalent of Fig. 2

(a) Reduce to the small-signal equivalent.
(b) Compute equation i=$Y_{NA}v$.
(c) Compute the reduced set of equations **i**=Y_{PNA}**v**.
(d) Compute the symbolic TF.

The small-signal equivalent is already shown in Fig. 3, from which by applying the PNA method, the formulation of the system of equations is given by:

$$\begin{bmatrix} i_i \\ 0 \\ 0 \\ 0 \\ 0 \\ 0 \end{bmatrix} = \begin{bmatrix} sC & 0 & 0 & 0 & 0 & 0 \\ 0 & g1 & -g1 & 0 & 0 & 0 \\ 0 & -g1 & g1 & 0 & 0 & 0 \\ 0 & 0 & 0 & g2 & -g2 & 0 \\ 0 & 0 & 0 & -g2 & g2 & 0 \\ 0 & 0 & 0 & 0 & 0 & go \end{bmatrix} \begin{bmatrix} v_1 \\ v_2 \\ v_3 \\ v_4 \\ v_5 \\ v_6 \end{bmatrix} \tag{1}$$

The nullator-norator properties are summarized in 4 basic rules related to the manipulation of the matrix Y_{NA}

(1) Delete cols related to a grounded nullator Fig. 4.
(2) Delete rows related to a grounded norator Fig. 4.
(3) Add cols related to a floating nullator Fig. 5.
(4) Add rows related to a floating norator Fig. 6.

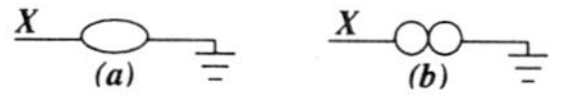

Fig. 4. Grounded (a) nullator and (b) norator

Fig. 5. Floating nullators

By applying this rules to equation (1), the reduced formulation is given by

$$\begin{bmatrix} i_i \\ 0 \end{bmatrix} = \begin{bmatrix} sC + g1 & 0 \\ g2 & go \end{bmatrix} \begin{bmatrix} v_{1,2,4} \\ v_6 \end{bmatrix} \tag{2}$$

Since i_o=$g_o v_6$, the computation of v_6 is:

$$v_6 = \frac{g_2 i_i}{(sC + g_1)g_o} \tag{3}$$

and finally, the fully-symbolic TF becomes:

$$\frac{i_o}{i_i} = \frac{g_2}{sC + g_1} \tag{4}$$

The matrix in equation (1) has an order 6×6, which is reduced to equation (2) having an order 2×2, because there are four-nullors from which cols 1,2,4 were added and the result put in col 1, in order to delete cols 2,3,4,5.

Fig. 6. Floating norators

In the same manner, rows 1,2 and 4,6 were added which results were put in rows 1 and 4 respectively, to delete rows 2,3,5,6. As one sees, most computational time is waste since the reduction process begin with 36 matrix-elements from which 32 are deleted. Additionally, the interconnection of nullators and norators need several computational effort to manipulate cols and rows.

3. PROPOSED FORMULATION APPROACH

To improve the traditional formulation approach, this work is devoted to improve both the CPU-time and memory during the formulation process, avoiding the computation of equation (1) by simply manipulating data-structures considering the interconnection relationships of nullators and norators to fill-in the reduced matrix directly.

3.1. Data-structure of admittance-elements

Let us consider, one more time, the circuit shown in Fig. 3. Instead of computing equation (1), i.e. the matrix of order 6×6 (n^2), one needs to compute a data-structure of only $n\frac{n+1}{2}$ elements, with n as the order of the matrix. The resulting data-structure of the nullor circuit is shown in Table 1. The label *nodes-relationship* referes to the upper triangular part of a matrix. For each pair of nodes (i,j), the column admittance is filled by considering two basic rules:

(a) The resulting element $Y_{(i,j)}$ is positive $\forall(i{=}j)$.
(b) Else $Y_{(i,j)}$ is negative $\forall(i{\neq}j)$.

As one sees, some computational effort is saved compared to the traditional formulation approach. Most important is that the resulting data-structure can be optimized by deletting all zero-elements.

3.2. Data-structure of nullator-norator relationships

Fig. 3 is redrawn in Fig. 7, where each nullator-norator pair has been labeled. Then, the resulting data-structure including nullator-norator interconnection relationships is shown in Table 2.

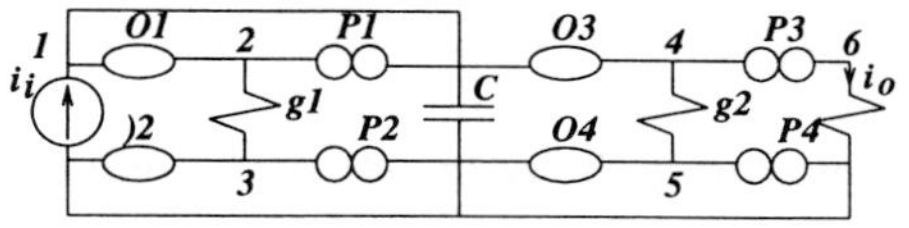

Fig. 7. Nullor circuit labeling nullator-norator pairs.

Table 1. Data-structure of the elements in Fig. 3

nodes-relationship	admittance
1,1	sC
1,2	0
1,3	0
1,4	0
1,5	0
1,6	0
2,2	g1
2,3	-g1
2,4	0
2,5	0
2,6	0
3,3	g1
3,4	0
3,5	0
3,6	0
4,4	g2
4,5	-g2
4,6	0
5,5	g2
5,6	0
6,6	g0

Since the nullor circuit has 6-nodes and 4-nullors, the resulting reduced matrix Y_{PNA} [3], will be of order **M**, which is given by (6-4)×(6-4)=2×2, as it was resulted in (2) using the traditional formulation method.

Table 2. Data-structure of nullator-norator elements

nullator	nodes	norator	nodes
O1	1,2	P1	1,2
O2	0,3	P2	0,3
O3	1,4	P3	4,6
O4	0,5	P4	0,5

According to the nullator and norator properties the following conclusions have been obtained:

1. From rule shown in Fig. 5, the connection of nullators O1,O3 indicates an addition of the elements at nodes 1,2,4. As Fig. 4 shown, the connection of nullators O2,O4 indicates avoiding the inclusion of the elements at nodes 3,5.

2. From rule shown in Fig. 6, the connection of norators P1,P3 indicates an addition of the elements at nodes 1,2 and 4,6, respectively. As Fig. 4 shown, the connection of norators P2,P4 indicates avoiding the inclusion of the elements at nodes 3,5.

3.3. Computing the reduced set of equations

Since nullators become associated with columns and norators with rows, the columns and rows of the reduced matrix formulation are computed as follows:

1. Columns (nullators manipulation): There are 6 node variables, however nodes 1,2,4 are virtually connected to the same node-voltage, nodes 3,5 are virtually connected to the reference node, and node 6 avoids nullator connection. In this manner, the two node variables are associated to the voltage variables $v_{1,2,4}$ and v_6.

2. Rows (norators manipulation): There are 6 rows associated to 6-KCL equations, however the current through nodes 1,2 and 4,6 becomes the same, while nodes 3,5 are virtually connected to the reference node. In this manner the two KCL equations are associated to apply KCL to the pairs of nodes (1,2) and (4,6).

The fill-in of the reduced matrix of order 2×2 is done by combining the resulting variables associated to all columns and rows. This process is done as follows:

(a) The rows and columns are associated by nodes [(1,2),(4,6)] and [(1,2,4),(6)], respectively.

(b) From the relationships given in (a), the resulting combinations associated to the elements of the reduced matrix Y_{PNA}, are given as:

- Y_{11}=Adding the elements at the pair of nodes (1,1), (1,2), (1,4), (2,1), (2,2), (2,4).
- Y_{12}=Adding the elements at the pair of nodes (1,6), (2,6).
- Y_{21}=Adding the elements at the pair of nodes (4,1), (4,2), (4,4), (6,1), (6,2), (6,4).
- Y_{22}=Adding the elements at the pair of nodes (4,6), (6,6).

(c) Table 1 is used to search the corresponding elements associated to the pair of nodes computed in (b), but considering that the combination (i,j) equals to the combination (j,i),

- Y_{11}=sC+0+0+g1+0.
- Y_{12}=0+0.
- Y_{21}=0+0+g2+0+0+0.
- Y_{22}=0+go.

(d) The resulting reduced set of equations is given by equation (5), and by using i_o=$g_o v_6$, the resulting TF is shown by equation (6).

$$\begin{bmatrix} i_i \\ 0 \end{bmatrix} = \begin{bmatrix} sC + g1 & 0 \\ g2 & go \end{bmatrix} \begin{bmatrix} v_{1,2,4} \\ v_6 \end{bmatrix} \quad (5)$$

$$\frac{i_o}{i_i} = \frac{g_2}{sC + g_1} \quad (6)$$

4. PROPOSED TECHNIQUE

As one sees, by comparing both methods (the traditional formulation approach versus the proposed formulation technique) it is clear that manipulating data-structures results in a suitable method to improve both CPU-time and memory. That way, the proposed technique can be summarized as follows:

1. Transform the analog circuit into a suitable nullor circuit to apply it the PNA method [3].
2. Reduce the nullor circuit to its small-signal equivalent.
3. Compute the data-structure of the admittance-elements.
4. Compute the data-structure of the nullators and norators interconnection relationships.
5. Compute the combinations of pair of nodes associated to the elements of the reduced matrix, which has an order equal to the number of nodes minus the number of nullator-norator pairs.
6. Search the resulting elements related to every pair of nodes and add them for every element of the reduced matrix.
7. Compute the vectors **i** and **v** as in the traditional formulation approach, in order to obtain the reduced set of equations, i.e. $\mathbf{i}=Y_{PNA}\mathbf{v}$, by fill-in the elements of the matrix Y_{PNA}.
8. Solve for the desired transfer function.

5. CONCLUSION

A formulation approach suitable for symbolic analysis of analog circuits modeled by nullors, has been introduced. The proposed technique is based on manipulating data-structures instead of manipulating a big-nodal-admittance-matrix as traditional approaches. It has been demonstrated that the proposed technique improves both CPU-time and memory by avoiding the manipulation of complex symbolic expressions.

The reduced set of equations has been formulated by fill-in the reduced admittance-matrix using the nullor interconnection-properties, by generating all combinations of pair of nodes associated to the voltage-variables and KCL equations. The proposed formulation approach can be applied to analog circuits at different abstraction levels, i.e. at the macromodel and transistor circuit level.

Acknowledgment

This work has been partially supported by: CoSNET/México under project 397.02-p and CONACYT/México under project 40321-Y.

6. REFERENCES

[1] Leon O. Chua, Pen-Min Lin, *Computer-aided analysis of electronic circuits*, Prentice Hall 1975.

[2] Sánchez-Sinencio Edgar, *Special section on symbolic circuit analysis techniques*, IEEE TCAS-II Analog and digital signal processing, vol 45., no. 10, October 1998.

[3] Tlelo-Cuautle E., Cid-Monjaráz J., *Computing symbolic transfer functions from SPICE files using nullors*, WSEAS PRESS, Advances in systems theory, mathematical methods and applications, ISBN: 9608052610, pp. 74-79, May 2002.

[4] George S. Moschytz, Linear integrated networks: Fundamentals, Van Nostrand Reinhold Company, 1974.

[5] Vlach J. Singhal K., Computer Methods for Circuit Analysis and Design, New York: Van Nostrand, 1983.

[6] A. F. Schwarz, Computer-Aided Design of Microelectronic Circuits and Systems, Academic Press, 1987.

[7] Georges Gielen and Willy Sansen, Symbolic analysis for automated design of analog integrated circuits, Kluwer Academic Publishers, 1991.

[8] Pen-Min Lin, Symbolic network analysis, Elsevier Science Publishers, 1991.

[9] Henrik Floberg, Symbolic analysis in analog integrated circuit design, Kluwer Academic Publishers, 1997.

[10] F. V. Fernández, A. Rodriguez-Vázquez, J. L. Huertas and G. Gielen, Eds., Symbolic Analysis Techniques. Applications to Analog Design Automation. Piscataway, NJ: IEEE Press, 1998.

[11] H. Schmid, Approximating the universal active element, IEEE TCAS-I, v. 47, no. 11, pp. 1160-1169, Nov. 2000.

[12] Alison Payne and Chris Tomazou, Analog Ampli£ers: Classi£cation and Generalization, IEEE TCAS-I: F. Theory Applications, vol. 43, no. 1, 43-50, 1996.

[13] R. Cabeza and A. Carlosena, On the use of symbolic analyzers in circuit synthesis, Analog IC and Signal Processing, 25, 67-75, 2000.

[14] Tlelo-Cuautle E., Sarmiento-Reyes A., Transforming OTA-C filters from voltage- to current-mode, IEEE ICCDCS2002, pp. C022-1-4, ISBN: 0-7803-7381-2, Aruba, April 2002.

Accurate Compact Model Extraction for On-Chip Coplanar Waveguides

Taeik Kim, Xiaoyong Li, and David J. Allstot[1]

Dept. of Electrical Engineering
Campus Box 352500
University of Washington
Seattle, WA 98195-2500

Abstract

Employing a novel *ABCD* matrix partitioning technique, the proposed method extracts an accurate frequency-dependent distributed circuit model for the broadband characteristics of coplanar waveguide transmission lines using either measured or simulated scattering parameters. A neural network with a training algorithm is then applied to predict CPW performance for a range of geometry sizes. The circuit model validated using EM (*MOMENTUM)* and circuit (*HSPICE)* simulations demonstrates good accuracy, high computational efficiency, and fast model development.

1. INTRODUCTION

Compared to GaAs MESFET technology for wireless communication systems, the RF capabilities of CMOS technology are becoming increasingly attractive because of low cost, high integration levels, process maturity, etc. However, owing to lossy silicon substrates, the performance of integrated RF blocks in CMOS suffers from the relatively poor characteristics of on-chip passive components. Furthermore, as the operating frequencies of RF CMOS components increase to >10GHz frequencies, interconnects begin to show transmission line characteristics [1]. Thus, it is increasingly important in the design of passive components for RF circuits to better understand CMOS transmission line characteristics and to develop accurate circuit models.

Circuit models describing the signal propagation behavior of transmission lines in CMOS have been reported [2]. While some of the approaches are compatible with commercially available circuit simulators, such as *HSPICE*, they are not generally directly applicable in optimization algorithms used to maximize the performance of RF circuits. In this paper, we present a novel and accurate circuit modeling methodology to characterize the properties of on-chip transmission lines, particularly coplanar waveguides (CPW). The solution involves first extracting an accurate circuit model as a function of frequency for a few geometries and then predicting the model parameters versus frequency for other geometry sizes. A novel *ABCD* matrix partitioning technique is applied to extract a frequency-dependent distributed circuit model from the available *S*-parameters. The effects of arbitrary geometry sizes are then characterized and predicted using neural networks and a training algorithm. Comparisons between *MOMENTUM* and *HSPICE* show excellent agreement.

II. FREQUENCY-DEPENDENT CHARACTERISTICS OF TRANSMISSION LINES

A straightforward method to partition a transmission line based on the Telegrapher equation is described in [3]. Specifically, knowing the complex characteristic impedance, *Z,* and the complex propagation constant, γ,

$$Z = \sqrt{\frac{(R + j\omega L)}{(G + j\omega C)}}\,,\; \gamma = \sqrt{(R + j\omega L)(G + j\omega C)} \qquad (1)$$

the series impedance components $R(\omega)$ and $L(\omega)$, and shunt admittance counterparts $G(\omega)$ and $C(\omega)$, per unit length of transmission line are easily derived as

$$R = \mathrm{Re}\{\gamma Z\}\,,\; L = \mathrm{Im}\{\gamma Z\}/\omega \qquad (2)$$

$$G = \mathrm{Re}\{\gamma / Z\}\,,\; C = \mathrm{Im}\{\gamma / Z\}/\omega \qquad (3)$$

Although appealingly simple, the method described above possesses several drawbacks. First, the conventional *RLGC* structure derived from the Telegrapher equation is topologically asymmetric and therefore exhibits asymmetric characteristics at its two ports [4]. Of course, it is easy to convert the structure into a symmetric T-model by replacing $R_1(\omega)=R_2(\omega)=0.5R(\omega)$ and $L_1(\omega)=L_2(\omega)=0.5L(\omega)$. However, transmission lines implemented in a fine-line CMOS process are not necessarily symmetric due to process variations. Moreover, an infinite number of *RLGC* transmission line segments are needed, in theory, to achieve highly accurate results. It is usually found from numerous simulations that N = 10 to 20 segments provides sufficient accuracy. Higher values of N provide greater accuracy at the expense of increased circuit complexity and longer simulation time. While the calculations required for model extraction are straightforward, partitioning based on the Telegrapher equation is inefficient in terms of both accuracy and simulation speed.

III. DISTRIBUTED CIRCUIT MODELING

The *parasitic-aware* methodology for synthesizing RF integrated circuits with embedded passive components is shown in Fig. 1. Based on experience, upwards of one million iterations around the optimization loop requiring several days of computer time are required in a typical RF

[1] Supported by National Science Foundation grants CCR-0086032 and CCR-0120255, Semiconductor Research Corporation grants 2000-HJ-771 and 2001-HJ-926, and grants from National Semiconductor Corp. and Texas Instruments, Inc.

0-7803-7761-3/03/$17.00 ©2003 IEEE

circuit design and optimization. Hence, there is a need for a compact circuit model of coplanar waveguides that can be efficiently evaluated by the standard circuit simulator (*HSPICE*, *SPECTRE*, etc.) used in the optimization loop. The choice of partitioning methods is critical in determining the model complexity.

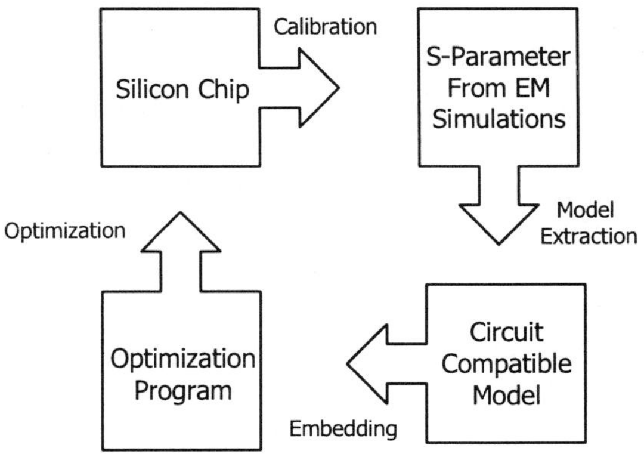

Fig. 1. The parasitic-aware design and optimization flow for RF integrated circuit synthesis.

The compact distributed model of a CPW in CMOS and the proposed *ABCD* matrix modeling methodology are shown in Fig. 2. It is known that if two identical transmission line segments are cascaded, the overall *ABCD* matrix is expressed in terms of individual *ABCD* matrices as

$$[ABCD]_T = [ABCD]_1^2 = [ABCD]_2^2 \tag{4}$$

where, $[ABCD]_T$, $[ABCD]_1$ and $[ABCD]_2$ are the *ABCD* matrices of the total cell, unit cell 1, and unit cell 2, respectively. Note that parameter extraction using an *ABCD* matrix expressed in terms of total voltages and currents is compatible with general-purpose circuit simulators, and that the parameters extracted from the *ABCD* matrix for a T-model cell are asymmetric. Thus, partitioning and parameter extraction using the *ABCD* matrix method is more accurate than conventional methods using the Telegrapher equation.

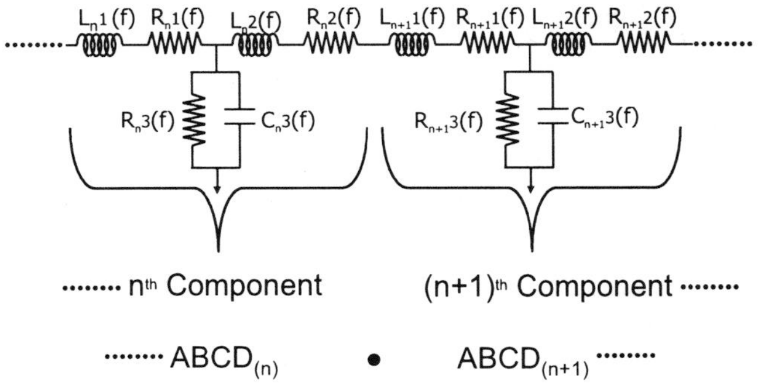

Fig. 2. Compact distributed model of a CPW the proposed *ABCD* matrix-based extraction methodology.

By iterating $n = log_2(N)$ times, the CPW is partitioned into N identical segments. Thus, the partitioning process based on *ABCD* matrix methods is inherently computationally efficient. The *ABCD* matrix of one unit cell, denoted [*abcd*] is easily converted into a corresponding T-model unit cell. The T-model lumped element values needed for an *HSPICE* netlist are extracted using an approximating polynomial function. Specifically, curve-fitting as a function of frequency using a third-order polynomial is applied. A total of 24 coefficients are used to generate the *HSPICE* netlist: six lumped component values for each segment and four coefficients for the third-order polynomial function.

Several CPW configurations in a typical CMOS process technology with a low loss substrate and multi-metal layers are considered. To generate *S*-parameters of the various CPW structures, *MOMENTUM* [5], a tool for simulating planar microwave structures using the Method of Moments, is used. The complete new extraction process is illustrated in Fig. 3.

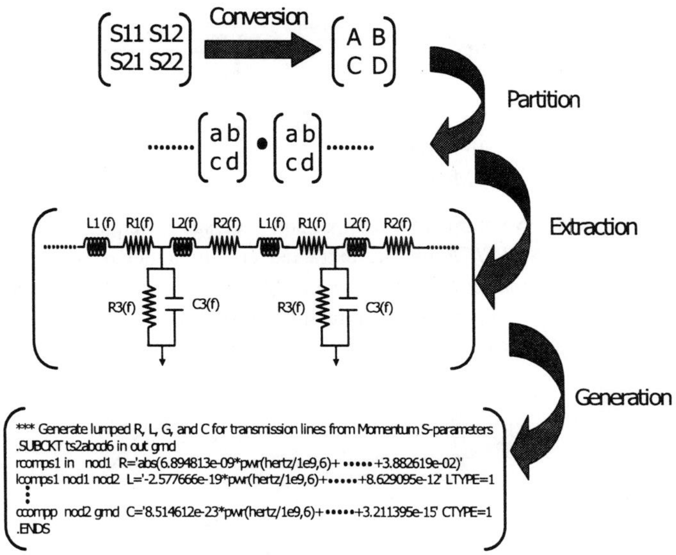

Fig. 3. The *ABCD* matrix partitioning method for CPW circuit model extraction.

In Figure 4, *S*-parameters obtained from *MOMENTUM* simulations are compared with *HSPICE* results of the compact equivalent circuit model generated using the *ABCD* matrix extraction method. Despite the fact that the CPW is modeled using only eight segments in this paper, excellent results are achieved. Detailed error performance for the example over the frequency range from 1GHz to 30GHz is summarized in Table I, where *ws*, *sp*, *wg*, and *L*, represent the width of a signal line, the space between the signal line and the ground line, the width of a ground line, and the length of a CPW, respectively. While N is chosen only large enough to adequately model the distributed effects of the CPW structures when the Telegrapher equation based model is applied, the *ABCD* matrix partitioning method uses a smaller N to achieve similar accuracy as shown in Fig. 5. Hence, the new *ABCD* matrix partitioning method reduces the complexity of circuit models and the required simulation time compared to previous approaches.

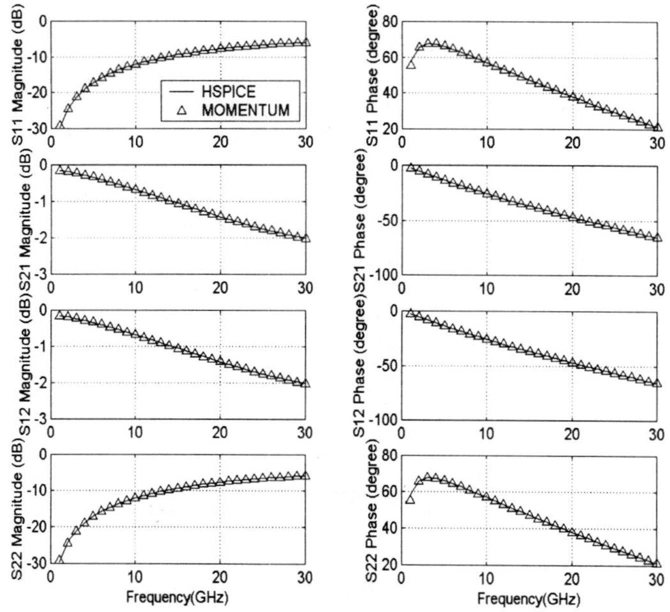

(a) Magnitude (b) Phase

Fig. 4. *S*-parameter comparison versus frequency between *HPSICE* for the extracted circuit model using the *ABCD* matrix methodology and *MOMENTUM* with *ws* = 10μm, *sp* = 10μm, *wg* = 30μm, and *L* = 1000μm.

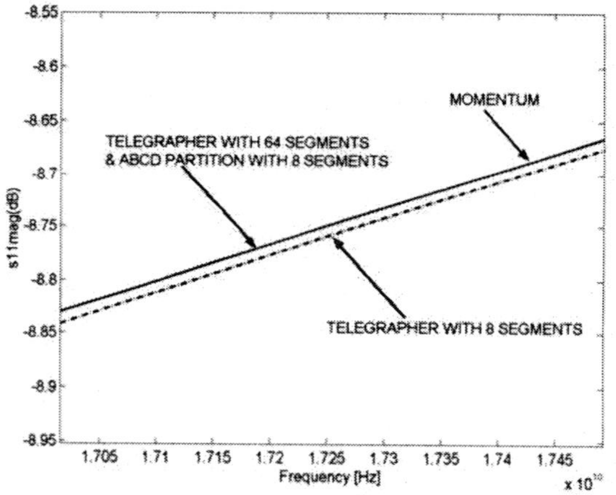

Fig. 5. The number of segment effects between the proposed *ABCD* partitioning technique and the conventional method using telegrapher equation.

TABLE I

ERROR PERFORMANCE OF *ABCD* MATRIX EXTRACTION

(*ws*=10μm, *sp*=10μm, *wg*=30μm, and *L*=1000μm)

S-parameter	$\lvert S_{11} \rvert$ (dB)	$\angle S_{11}$ (°)	$\lvert S_{12} \rvert$ (dB)	$\angle S_{12}$ (°)
Mean Error	0.0265	0.1582	0.0062	0.0087
Max Error	0.8227	3.9263	0.0154	0.0227

IV. NEURAL NETWORK MODELS

As mentioned previously, the circuit model of an arbitrary CPW is a function of both frequency and geometry sizes. The effects of geometry sizes must be taken into account *on the fly* in order to completely model a CPW for parasitic-aware optimization purposes. Based on observations from extensive *MOMENTUM* simulations, the *S*-parameter values depend strongly on frequency, *ws*, *sp*, and *L*, but rather weakly on *wg*, probably because *wg* is at least 3 times larger than *ws* in our examples. Thus, without loss of generality, it is assumed that the *S*-parameters are constant with respect to *wg*.

The discussion focuses on the question of how to efficiently characterize the circuit model, that is, the polynomial curve-fitting coefficients, in terms of *ws*, *sp* and *L*. Since neural networks are known to provide powerful interpolation capabilities [6], especially for highly nonlinear problems, it is appropriate to consider neural network modeling methods for arbitrary CPW structures [7]. Due to the nonlinear nature of the CPW modeling, a multilayer network is used consisting of an input layer, an output layer and several hidden layers. The numbers of nodes in input layers (*ws*, *sp* and *L*) and output layers (coefficients of the third-order polynomial curve fitting) are three and four, respectively. The geometry sizes for the database are chosen so that relationships between inputs and outputs are adequately represented in the training data. A back error propagation algorithm [8] is then executed to train the neural network so that the outputs of neural network approach the targets of the training database. The training process is stopped when the mean square error of the difference between network outputs and targets is below a preset value.

The neural network model represents a generalization of the nonlinear mapping implicit in a training database. To save considerable computation during execution of the training algorithm and subsequent simulations of the neural network, frequency information is temporarily hidden. Inputs to the neural network include *ws*, *sp* and *L*, while outputs are the *coefficients*. A set of verification data taken from *MOMENTUM* simulations is used to determine the modeling errors. More important, the verification data is used to avoid the neural network over-training problem [8].

The complete process of neural network modeling is shown in Fig. 6. In this paper, 125 sets of training data and 120 sets of verification data are used to implement the neural network model. The data covers the range of parameter values shown in Table II. The feedforward neural network model combined with the *ABCD* matrix extraction technique is capable of handling four parameters including frequency, *ws*, *sp* and *L*. Due to the random nature of the training algorithm, the modeling process was repeated to investigate its robustness. The methodology is robust in that different runs of the training algorithm give roughly the same error performance. Figure 7 shows excellent agreement between *MOMENTUM* and verification data sets, and the error performance from verification sets in Table III shows good accuracy.

Since the neural network model training is done only once and is unnecessary during circuit model generation phase, it presents no burden to model generation at all. The CPU time to generate the distributed circuit model of the CPW using the method described above is less than 1ms on a *SUN ULTRA-10* workstation. When compared to time-consuming full-wave EM simulations, the technique presented in this paper can achieve much better computational efficiency.

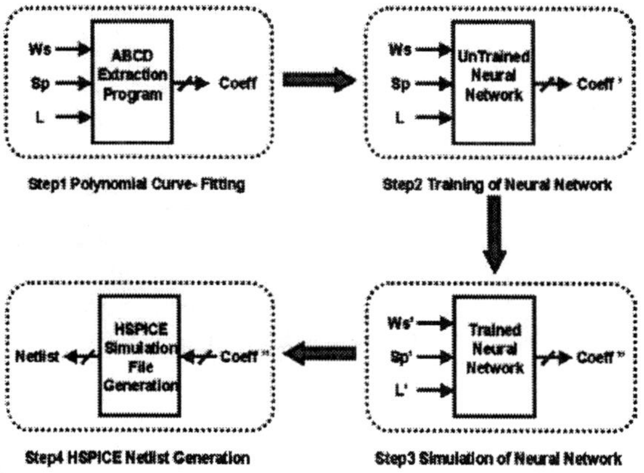

Fig. 6. Processing of the neural network CPW model.

TABLE II

PARAMETER RANGE FOR THE NEURAL NETWORK MODELS

Parameter	Min	Max
Frequency	1GHz	30GHz
ws	10μm	40μm
sp	10μm	40μm
L	200μm	1000μm

TABLE III

ERROR PERFORMANCE OF THE NEURAL NETWORK MODEL

S-parameter	$\lvert S_{11}\rvert$ (dB)	$\angle S_{11}$ (°)	$\lvert S_{12}\rvert$ (dB)	$\angle S_{12}$ (°)
Mean Error	0.0454	0.2155	0.0033	0.0291
S-parameter	$\lvert S_{21}\rvert$ (dB)	$\angle S_{21}$ (°)	$\lvert S_{22}\rvert$ (dB)	$\angle S_{22}$ (°)
Mean Error	0.0033	0.0291	0.0459	0.2184

V. CONCLUSIONS

In this paper, a novel fast and accurate extraction methodology for frequency-dependent circuit models of on-chip CPWs on a heavily doped CMOS process is described. The extraction methodology using an *ABCD* matrix partitioning technique significantly decreases parasitic-aware design and optimization time. Both *MOMENTUM* and *HSPICE* simulation results confirm that the new *ABCD* matrix partitioning methodology, together with a feed forward neural network model, accurately models CPW characteristics to tens of gigahertz frequencies. The model is readily integrated into optimization programs for parasitic-aware RF IC design using transmission line structures.

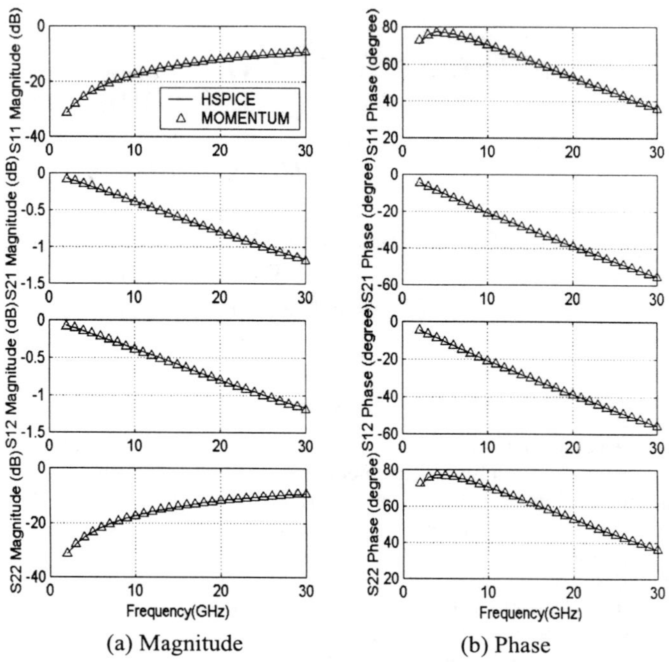

Fig. 7. *S*-parameters using the feedforward neural network model parameters and *MOMENTUM* simulations with *ws* = 24μm, *sp* = 18μm, *wg* = 60μm, and *L* = 800μm.

REFERENCES

[1] T. H. Lee and S. S. Wong, "CMOS RF integrated circuits at 5GHz and beyond," *Proceedings of the IEEE*, vol. 88, pp. 1560-1571, Oct. 2000.

[2] C. Veyres and V. F. Hanna, "Extension of the application of conformal mapping techniques to coplanar lines with finite dimensions," *Int. J. Electronics*, vol. 48, pp. 47–56, 1980.

[3] W.R. Eisenstadt and Y. Eo, "S-parameter-based IC interconnect transmission line characterization," *IEEE Trans. on Components, Hybrids, and Manufacturing Technology*, vol. 15, pp. 483–490, April 1992.

[4] D.M. Pozar, *Microwave Engineering*, Second Edition, New York: Wiley, 1998.

[5] Agilent EEsof Momentum, Electromagnetic Design and Simulation, Advanced Design System 1.5, Agilent Technologies, 395 Page Mill Road, Palo Alto, CA, December 2000.

[6] G. Van der Plas, J. Vandenbussche, G. Cielen, and W. Sansen, "EsteMate: A tool for automated power and area estimation in analog top-down design and synthesis," in *IEEE CICC Dig. Tech. Papers*, May. 1997, pp. 139-142

[8] P.M. Watson and K. C. Gupta "Design and optimization of CPW circuits using EM-ANN models for CPW components," *IEEE Trans. on Microwave Theory and Techniques*, vol. 46, pp. 632-640, May 1998.

[9] S. Haykin, *Neural Networks, a Comprehensive Foundation.* Second Edition, New Jersey: Prentice Hall, 1999.

INTERCONNECT MODELING AND SENSITIVITY ANALYSIS USING ADJOINT NETWORKS REDUCTION TECHNIQUE

Herng-Jer Lee, Chia-Chi Chu, and Wu-Shiung Feng

Chang Gung University, Department of Electrical Engineering, Kwei-San, Tao-Yuan 333, Taiwan
Phone:886-3-3283016-5717, Fax:886-3-3288026
Email:d8821001@stmail.cgu.edu.tw, {ccchu,wsfeng}@mail.cgu.edu.tw

ABSTRACT

An efficient model-order reduction technique for general RLC networks is proposed. The method is extended from the previous projection-base moment matching method with considering both the circuit network and its corresponding adjoint network. By exploring symmetric properties of the MNA formulation, the proposed method needs only one half of the system moment information compared with the previous ones. Passivity of the reduced-order model is still preserved. The relationship between the adjoint network and the sensitivity analysis will also be discussed. Experimental results will demonstrate the accuracy and the efficiency of the proposal method.

1. INTRODUCTION

With considering the issues of the signal integrity in high-speed VLSI designs, interconnects are often modeled as lumped RLC networks. Since the computational cost for simulating such large circuits is indeed tremendously huge, model-order reduction techniques have been proposed recently to reduce the computational complexity [1, 2, 4, 8, 10]. The moment matching techniques based on Padé approximation and Krylov subspace projections take advantage of efficiency and numerical stability [4, 5, 6, 9, 12, 13]. In general, these methods can be divided into two categories: one-sided projection methods [9, 12, 13] and two-sided projection methods [5, 6]. The one-sided projection methods use the congruence transformation to generate passive reduced-order models while the two-sided ones can not be guaranteed.

In this paper, we introduce an efficient technique to further reduce the computational cost of the one-sided projection methods. Recently, a technique was proposed to reduce the computational cost of solving Lyaponov equations by using the concepts of the adjoint network [12]. However, the relationship between a circuit network and its corresponding adjoint network have not been developed completely. By exploring symmetric properties of the MNA formulation, we will show that the transfer functions and system moments of the adjoint network can be directly calculated from those of the original RLC network. The cost about constructing the congruence transformation matrix can be simplified up to 50% of the previous methods. In addition, it will be shown that this can be directly applied to the sensitivity analysis of the original circuits, and to generate the congruence transformation matrices for the sensitivity analysis of the reduced-order system [7, 11].

This work was supported by the National Science Council, Republic of China, under Grants NSC 87-2215-E-182-004, NSC 88-2215-E-182-003, NSC 90-2215-E-182-001 and NSC 91-2218-E-182-005.

The paper is organized as follows. Section 2 reviews the background of the circuit equations and passive reduced-order models techniques. In Section 3, the relationship between a general RLC circuit system and its corresponding adjoint network will be investigated. The connection of the adjoint network with Krylov subspaces and sensitivity analysis of the reduced-order system will also be explored. Simulation results will demonstrated in Section 4. Finally, conclusions are made in Section 5.

2. BACKGROUND

To analyze an RLC linear network, the modified nodal analysis (MNA) can be used as follows [3, 9, 12]:

$$\begin{aligned} M\frac{dx(t)}{dt} &= -Nx(t) + Bu(t) \\ y(t) &= D^T x(t), \end{aligned} \qquad (1)$$

where $M, N \in \mathcal{R}^{n\times n}$, $x, B \in \mathcal{R}^{n\times m}$, $D \in \mathcal{R}^{n\times p}$, and $y \in \mathcal{R}^{p\times m}$. Matrices M and N containing capacitances, inductances, conductances and resistances are positive definite. The state matrix $x(t)$ contains node voltages and branch currents of inductors, and $u(t)$ and $y(t)$ represent inputs and outputs. The adjoint equation associated with the system in (1) is of the form

$$M\frac{dx_a(t)}{dt} = -N^T x_a(t) + Du(t), \qquad (2)$$

which is the modified node equation of the adjoint network [3, 11] (or the dual system [12]). If the m-port transfer functions are concerned, then $p = m$ and $D = B$. The transfer functions of the state variables and of the outputs are $X(s) = (N + sM)^{-1}B$ and $Y(s) = B^T X(s)$. Conversely, those of the corresponding adjoint network is given as $X_a(s) = (N^T + sM)^{-1}B$.

By expanding $Y(s)$ about a frequency $s_0 \in \mathcal{C}$, we have $Y(s) = \sum_{i=-\infty}^{\infty} Y^{(i)}(s_0)(s - s_0)^i = \sum_{i=-\infty}^{\infty} B^T X^{(i)}(s_0)(s - s_0)^i$, where $X^{(i)}(s_0) = (-(N + s_0M)^{-1}M)^i(N + s_0M)^{-1}B$ is the ith-order system moment of $X(s)$ about s_0 and $Y^{(i)}(s_0)$ is the corresponding output moment. Similarly, the ith-order system moment of $X_a(s)$ about s_0, $X_a^{(i)}(s_0) = (-(N^T+s_0M)^{-1}M)^i(N^T + s_0M)^{-1}B$, can be obtained.

One simple way to generate a reduced-order network of (1) is using the one-sided projection method for moment matching [9, 12, 13]. First, a congruence transformation matrix V_q can be generated by the Krylov subspace methods. Let $A = -(N + s_0M)^{-1}M$ and $R = (N + s_0M)^{-1}B$. The kth-order block Krylov subspace generated by A and R is defined as $\mathcal{K}(A, R, k) =$

0-7803-7761-3/03/$17.00 ©2003 IEEE

$colsp[R, AR, A^2R, \cdots, A^{k-1}] = colsp(V_q)$,, where $q = km$. The Krylov subspace $\mathcal{K}(A, R, k)$ is indeed equal to the subspace spanned by system moments $X^{(i)}(s_0)$ for $i = 0, 1, \cdots, k-1$. Matrix V_q can be iteratively generated by the block Arnoldi algorithm and thus be an orthonormal matrix. Next, by applying V_q, n-dimensional state space can be projected onto a q-dimensional space, where $q \ll n$: $x(t) = V_q \hat{x}_q(t)$. Then the reduced-order model can be calculated as

$$\hat{M} = V_q^T M V_q,\ \hat{N} = V_q^T N V_q \text{ and } \hat{B} = V_q^T B. \quad (3)$$

The transfer function of the reduced network is $\hat{Y}(s) = \hat{B}^T(\hat{N} + s\hat{M})^{-1}\hat{B}$. The corresponding ith-order output moment about s_0 is $\hat{Y}^{(i)}(s_0) = \hat{B}^T(-(\hat{N} + s_0\hat{M})^{-1}\hat{M})^i(\hat{N} + s_0\hat{M})^{-1}\hat{B}$. It can be shown that $\hat{Y}^{(i)}(s_0) = Y^{(i)}(s_0)$ for $0 \le i \le k-1$ and the reduced-order model is passive [9, 13].

3. EFFICIENT COMPUTATIONS OF THE REDUCED-ORDER MODELS

3.1. Frequency Response of Adjoint Networks

Suppose that nv and ni are the dimension of the node voltages and the branch currents in $x(t)$. Let each port be connected with a current source so that $B^T = [B_v^T\ 0]$, where $B_v \in \mathcal{R}^{nv \times m}$. Let the signature matrix S be defined as $S = diag(I_{nv}, -I_{ni})$. The symmetric properties of the MNA matrices are as follows [12]:

$$S^{-1} = S,\ SMS = M,\ SNS = N^T \text{ and } SB = B. \quad (4)$$

If port impedance parameters are concerned, each port is connected with a voltage source and thus $B^T = [0\ B_i^T]$, where $B_i \in \mathcal{R}^{ni \times m}$. To preserve the properties in (4), then $\bar{S} = diag(-I_{nv}, I_{ni})$ will be used. If port transmission parameters are concerned, $B^T = [B_v^T\ B_i^T]$, the properties in (4) can still be preserved using superposition principles.

The following lemma represents the relationship between the transfer functions of the original system $X(s)$ and those of the adjoint network $X_a(s)$.

Lemma 1: Substituting (4) into $X_a(s)$, the state variables of the adjoint network and those of the original system have the following relationship:

$$X_a(s) = [S(N + sM)S]^{-1}B = SX(s). \quad (5)$$

Thus $X_a(s)$ can also be calculated from $X(s)$ directly.

3.2. Reduced-Order Models Based on Projection

From Sec. 2, it has been known that the one-sided projection method can preserve output moments. If matrix U is chosen as the congruence transformation matrix such that

$$\{X^{(i)}(s_0), X_a^{(j)}(s_0)\} \in colsp(U)$$
$$\text{for } 0 \le i \le k-1 \text{ and } 0 \le j \le l-1, \quad (6)$$

then $\hat{Y}^{(i)}(s_0) = Y^{(i)}(s_0)$ for $0 \le i \le k+l-1$ [12]. The reduced-order transfer function satisfies $\hat{Y}(s) = Y(s) + \mathcal{O}(s - s_0)^{k+l}$. In particular, if matrix U is built only from $X^{(i)}(s_0)$ with no component from $X_a^{(j)}(s_0)$, then $\hat{Y}^{(i)}(s_0) = Y^{(i)}(s_0)$ for $0 \le i \le k-1$, as in Sec. 2. Although (6) can overcome the numerical instability problem when generating the basis matrix U if order $k + l$ is extremely high, $X_a^{(j)}(s_0)$ and $X^{(i)}(s_0)$ still need to be calculated individually for general RLC networks. The computational cost of generating U can not be reduced.

The following theorem further ensures that $Y^{(i)}(s_0)$ can be matched up to $(2k-1)$st-order by using another congruence transformation matrix $U = [V_q\ SV_q]$.

Theorem 1: Suppose that $X^{(i)}(s_0) \in colsp(V_q)$ for $0 \le i \le k-1$ is a set of moments of $X(s)$ about s_0. V_q is the orthonormal matrix generated iteratively by the block Arnoldi algorithm. Let $U = [V_q\ SV_q]$ be the congruence transformation matrix for model-order reductions in (3), then

$$\hat{Y}^{(i)}(s_0) = Y^{(i)}(s_0), \text{ for } 0 \le i \le 2k-1, \quad (7)$$

Proof: We want to show that $X_a^{(i)}(s_0) \in colsp(SV_q)$ for $0 \le i \le k-1$. Since $X^{(i)}(s_0) \in colsp(V_q)$ for $0 \le i \le k-1$, it is equivalent to show that $X_a^{(i)}(s_0) = SX^{(i)}(s_0)$ for $0 \le i \le k-1$.

It will be shown by induction on i. For $i = 0$ and using (5) in Lemma 1, then $X_a^{(0)}(s_0) = SX^{(0)}(s_0)$. Next, we assume that it is true for $1 \le i \le k-2$ and prove it is also true while $i = k-1$, such that

$$\begin{aligned} X_a^{(i)}(s_0) &= -(N^T + s_0M)^{-1}MX_a^{(i-1)}(s_0) \\ &= -[S(N + s_0M)S]^{-1}SMSSX^{(i-1)}(s_0) \\ &= S[-(N + s_0M)^{-1}MX^{(i-1)}(s_0)] \\ &= SX^{(i)}(s_0) \end{aligned}$$

Therefore, $\{X^{(i)}(s_0), X_a^{(j)}(s_0)\} \in colsp[V_q\ SV_q] = colsp(U)$ for $0 \le i, j \le k-1$. $\hat{Y}^{(i)}(s_0) = Y^{(i)}(s_0)$, for $0 \le i \le 2k-1$. □

Remarks:

1. In general, linear independence of the columns in the block Krylov sequence $R, AR, \cdots, A^{k-1}R, \cdots$, will be lost only gradually [4]. Thus the numerical instablility may occur during the orthonormalization process as k is high. If we perform only half of orthonormalization iterations, which may be more stable than the previous works [4, 9].
2. In spite of the matrix U in Theorem 1 merely partial orthonormal, the first $2k$ output moments of $\hat{Y}(s)$ are still matched to those of $Y(s)$. In addition, even V_q need not be orthonormal and need not be obtained only from the block Arnoldi algorithm. Attempting to further reduce the computational cost for generating U, the matrix V_q generated by SyMPVL algorithm [6] can also be applied, as long as it can span the same subspace $\mathcal{K}(A, R, k)$.

Another way to generate the reduced-order model is using the two-sided projection method. It is based on two Krylov subspaces $\mathcal{K}(A, R, k)$ and $\mathcal{K}(A^T, B, k)$. Let $\mathcal{K}(A, R, k) = colsp(V_q)$ and $\mathcal{K}(A^T, B, k) = colsp(W_q)$. Two bases V_q and W_q are biorthogonal using the block Lanzos algorithm iteratively. If the original system is projected onto $\mathcal{K}(A, R, k)$ and $\mathcal{K}(A^T, B, k)$, that is, the reduced-order model is

$$\hat{Y}(s) = B^T V_q (W_q^T V_q + (s - s_0)W_q^T A V_q)^{-1} W_q^T R.$$

Although $\hat{Y}^{(i)}(s_0) = Y^{(i)}(s_0)$ for $0 \le i \le 2k-1$, $\hat{Y}(s)$ is not guaranteed to be positive real for general RLC networks. By exploring symmetric properties of MNA, it can be found that W_q can be calculated directly from V_q: $W_q = S(N + s_0M)V_q$, which is another expression of the results of SyMPVL algorithm [4, 6].

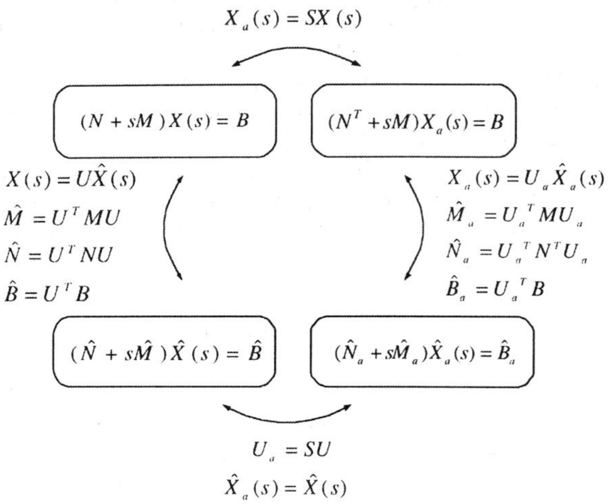

Figure 1: Relationships between the original system, the adjoint networks and their corresponding reduced-order models.

3.3. Sensitivity Analysis

We can also apply $X_a(s) = SX(s)$ to perform the sensitivity analysis. If the sensitivity of the output $Y(s)$ with respect to one circuit parameter λ is concerned, we have [11]

$$\frac{\partial Y(s)}{\partial \lambda} = -X_a^T(s)\frac{\partial(N+sM)}{\partial \lambda}X(s). \quad (8)$$

Substituting the symmetrical property $X_a(s) = SX(s)$ into (8), we get

$$\frac{\partial Y(s)}{\partial \lambda} = -X^T(s)S\frac{\partial(N+sM)}{\partial \lambda}X(s). \quad (9)$$

Thus the computational cost of sensitivity analysis can be reduced about 50% by only solving $X(s)$.

Although we can perform the sensitivity analysis of the original network using (9), it is advisable to perform the sensitivity analysis by applying the model-order reduction techniques. In the previous works [7], the congruent transformation matrices U and U_a such that $\{X^{(i)}(s_0)\} \in colsp(U)$ and $\{X_a^{(i)}(s_0)\} \in colsp(U_a)$ for $0 \le i \le k-1$ are constructed individually. However, with the aid of Theorem 1, it is not hard to see that $U_a = SU$. In addition, the computational cost of generating U can be further reduced by using method in Sec. 3.2. The proposed sensitivity analysis includes the following steps:

1. Calculate the congruence transformation matrix $U = [V_{q/2}\ SV_{q/2}]$ if q is even.
2. Generate the reduced order systems $\{\hat{M}, \hat{N}, \hat{B}\}$ through the congruence transformation (3).
3. Solve $(\hat{N} + s\hat{M})\hat{X}(s) = \hat{B}$ for each frequency.
4. Map $\hat{X}$ back to the original and adjoint state spaces $X(s)$ and $SX(s)$.

Fig. 1 shows the relationship between the original system, the adjoint networks and their corresponding reduced-order models.

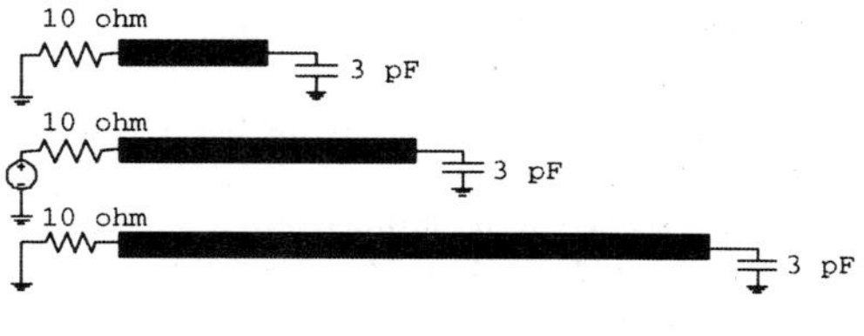

Figure 2: A coupled tree-line circuit.

4. EXPERIMENTAL RESULTS

We provide an example, a coupled three-line circuit in Fig. 2, to show the efficiency of the proposed method. The line parameters are resistance: 3.5 Ω/cm, capacitance: 5.16 μF/cm, inductance: 3.47 nH/cm, coupling capacitance: 6 μF/cm and mutual inductance: 3.47 nH/cm. The short victim net, the aggressor net and the long victim net are divided into 50, 100 and 150 sections, respectively. The dimension of the MNA matrices is 600×600 and the number of ports is 4. Suppose that the block Arnoldi algorithm is chosen to generate the orthonormal basis for the corresponding Krylov subspace during the whole experiment. We set shift frequency $s_0 = 1GHz$ and iteration number k=10. So q=40. The frequency responses of the original model and the reduced-order model generated by the block Arnoldi algorithm with the congruence transformation matrices $U = V_q$, $U = V_{2q}$, and $U = [V_q\ \ SV_q]$ are illustrated in Fig. 3. The time to generate the reduced-order models are with $U = V_q$: 1.50s, $U = V_{2q}$: 3.86s, and the proposed $U = [V_q\ \ SV_q]$: 2.02s by using Matlab 6.1 with Pentium II 450MHz CPU and 128MB DRAM.

In addition, sensitivity analysis results are also compared. We choose λ to be the effective driver impedance at the near end of the aggressor net and total 101 frequency points ranged from 0 to 15GHz to be simulated. The results are generated by the traditional adjoint method, the adjoint method with the 24th-order reduced-order models, and the proposed method are compared in Fig. 4. The simulation time of these models are 555.36s, 24.55s, and 14.15s, respectively. Therefore, it can be observed that the proposed method shows pretty good approximate results and takes less time.

5. CONCLUSIONS

An efficient model-order reduction technique for general RLC networks has been proposed in this paper. Extending the traditional projection method with considering both the original system and the adjoint network, the proposed method only needs to use one half of the original moment information by exploring symmetric properties of the MNA formulation. In addition, moment matching and passivity are preserved. Sensitivity analysis also can be efficiently calculated. Experimental results have demonstrated the accuracy and the efficiency of the proposal method.

6. REFERENCES

[1] Z. Bai, "Krylov subspace techniques for reduced-order modeling of large-scale dynamical systems", *Applied Numerical Mathematics*, Vol. 43, Issues 1-2, pp. 9-44, 2002.

[2] M. Celik, L. T. Pileggi, and A. Odabasioglu, *IC Interconnect Analysis*, Kluwer Academic Publisher, 2002.

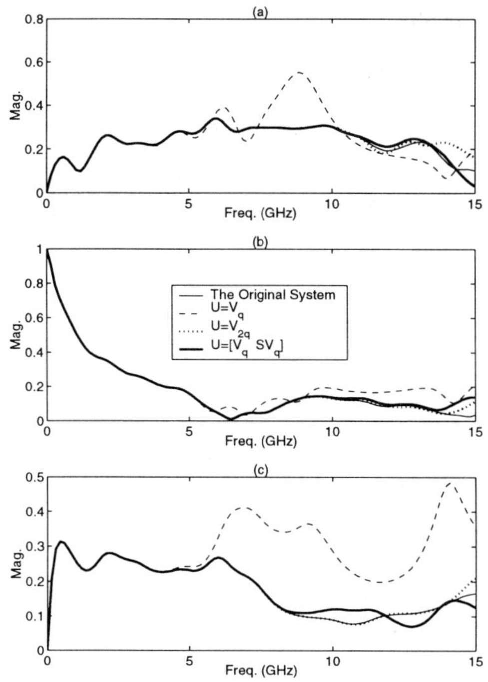

Figure 3: Transfer functions of the far end of the aggressor net and the victim nets: (a) the short victim net; (b) the aggressor net; (c) the long victim net.

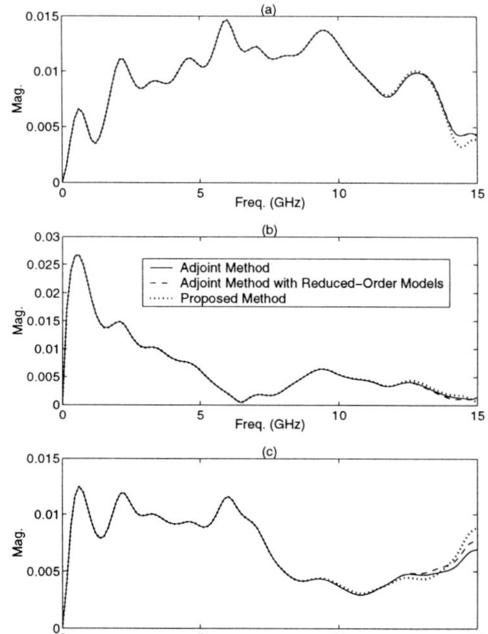

Figure 4: Sensitivity analysis of the far end of the aggressor net and the victim nets: (a) the short victim net; (b) the aggressor net; (c) the long victim net.

[3] L. O. Chua, C. A. Desoer, and E. S. Kuh, *Linear and Nonlinear Circuits*, McGraw-Hill, Inc., 1987.

[4] R. W. Freund, "Krylov-subspace methods for reduced-order modeling in circuit simulation", *Journal of Computational and Applied Mathematics*, vol. 123, pp. 395–421, 2000.

[5] P. Feldmann and R. W. Freund, "Efficient linear circuit analysis by Padé approximation via the Lanczos process", *IEEE Trans. on Computer-Aided Design of Integrated Circuits and Systems*, vol. 14, no. 5, pp. 639-649, 1995.

[6] P. Feldmann and R. W. Freund, "The SyMPVL algorithm and its applications to interconnect simulation", *Proc. 1997 International Conference on Simulation of Semiconductor Processes and Devices*, pp. 1113-116, 1997.

[7] R. Khazaka, P. K. Gunupudi, and M. S. Nakhla, "Efficient sensitivity analysis of transmission-line networks using model-reduction techniques", *IEEE Trans. on Microwave Theory and Techniques*, vol. 48, no. 12, pp. 2345-2351, 2000.

[8] S. Y. Kim, N. Gopal, and L. T. Pillage, "Time-domain macromodels for VLSI interconnect analysis", *IEEE Trans. on Computer-Aided Design of Integrated Circuits and Systems*, vol. 13, no. 10, pp. 1257-1270, 1994.

[9] A. Odabasioglu, M. Celik, and L. T. Pileggi, "PRIMA: passive reduced-order interconnect macromodeling algorithm", *IEEE Trans. on Computer-Aided Design of Integrated Circuits and Systems*, vol. 17, no. 8, pp. 645-653, 1998.

[10] L. T. Pillage and R. A. Rohrer, "Asymptotic waveform evaluation for timing analysis", *IEEE Trans. on Computer-Aided Design of Integrated Circuits and Systems*, vol. 9, no. 4, pp. 352-366, 1990.

[11] J. Vlach and K. Singhal, *Computer Methods for Circuit Analysis and Design*, 2nd ed., Van Nostrand Reinhold, 1993.

[12] J. M. Wang, C. C. Chu, Q. Yu, and E. S. Kuh, "On projection based algorithms for model order reduction of interconnects", *IEEE Trans. on Circuits and Systems-I: Fundamental Theory and Applications*, vol. 49, no. 11, pp. 1563-1585, 2002.

[13] Q. Yu, J. M. Wang, and E. S. Kuh, "Passive multipoint moment matching model order reduction algorithm on multiport destributed interconnect networks", *IEEE Trans. on Circuits and Systems-I: Fundamental Theory and Applications*, vol. 46, no. 1, pp. 140-160, 1999.

MIXED-MODE ESD PROTECTION CIRCUIT SIMULATION-DESIGN METHODOLOGY

H. Feng, R. Zhan, Q. Wu, G. Chen, X. Guan, H. Xie and A. Z. Wang

Department of Electrical and Computer Engineering, Illinois Institute of Technology
3301 S. Dearborn St., Chicago, IL 60616, USA, Phone: (312) 567-6912, Email: awang@ece.iit.edu

ABSTRACT

Trial-and-error approach still dominates in on-chip electrostatic discharge (ESD) protection circuit design. We present a new predictive mixed-mode ESD protection simulation-design methodology, which involves multiple-level electro-thermal-process-device-circuit-layout coupling in ESD protection simulation that solves complex electro-thermal equations self-consistently at process, device and circuit levels, in a coupled fashion, to investigate ESD protection circuit behaviors without any pre-assumption. Practical design example in commercial 0.35μm CMOS is presented.

1. INTRODUCTION

It is reported that up to 35% of all IC field failures are associated with ESD damages [1][2]. Commonly experienced ESD damages are catastrophic failures that lead to immediate malfunction of IC chips caused by either high current induced thermal breakdown in silicon and/or metal interconnects or high voltage caused dielectric rupture. On-chip ESD protection circuits are required on all IC chips. Generally, an ESD protection structure acts as a two-terminal device connected between an I/O pad and a power supply. Typical on-chip ESD protection devices are diode, grounded gate NMOS/PMOS (ggNMOS/ggPMOS), silicon-controlled rectifier (SCR), etc. Figure 1 illustrates a typical snapback I-V characteristic of an ESD protection structure. Under ESD stresses, the terminal voltage increases gradually until reaching its trigger threshold (V_{t1}, I_{t1}) at a trigger time (t_1), then, snaps back into a holding point (V_h, I_h) and creates a low-impedance (R_{on}) shunting path to discharge ESD transients safely and to clamp the pad voltage to a sufficiently low level. ESD protection level is determined by its second breakdown threshold (V_{t2}, I_{t2}). The goal of ESD protection circuit design is to precisely define these critical operational parameters. With continuous scaling down in VLSI IC technologies, ESD protection design becomes a major IC design challenge. Unfortunately, trial-and-error approaches still dominate in current ESD protection design practices that are extremely time-consuming and costly. A rational simulation-based ESD design methodology is therefore highly desirable. Since ESD phenomena involve electro-thermal coupling at process, device, and circuit levels, and are geometry-dependent and device physics-related, a mixed-mode simulation approach must be used in ESD protection circuit design. This concept becomes critical to VDSM ULSI IC technologies where coupling effects dominate. This paper presents a new mixed-mode (electro-thermal-process-device-circuit-layout) ESD simulation-design methodology aiming to design prediction.

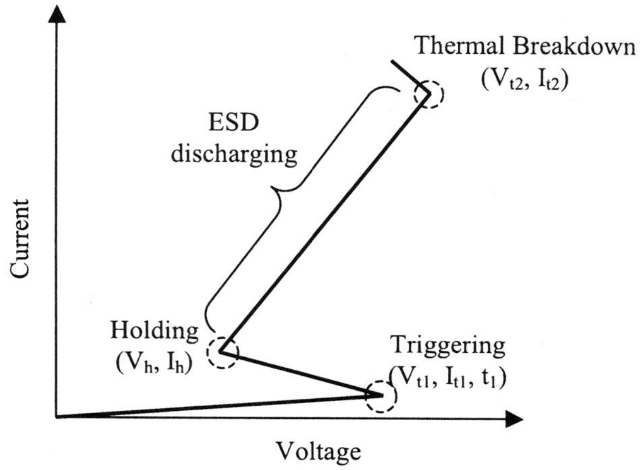

Figure 1. A Typical snapback I-V characteristic for an ESD protection structure.

2. MIXED-MODE ESD SIMULATION-DESIGN METHODOLOGY

Traditional trial-and-error ESD design approaches are experience based with only partial simulation at either device level or circuit level, which requires intensive testing, debugging and design iterations to complete the design. We propose a new mixed-mode ESD protection simulation-design methodology that involves electro-thermal-process-device-circuit-layout coupling and aims to provide design prediction. Figure 2 is a flow chart of the new ESD simulation-design method consisting of following steps:

The very first step of ESD protection circuit design is to set the design specifications as illustrated in Figure 1, i.e., (V_{t1}, I_{t1}, t_1), (V_h, I_h), R_{on} & (V_{t2}, I_{t2}). V_{t1} is typically set to be around 110% of power supply to ensure enough safety margins in order to avoid ESD protection structure being miss-triggered by non-ESD signals. Trigger time t_1 should be significantly shorter than the rising time of the ESD pulse to ensure prompt turn-on of the ESD protection structure under ESD stressing. V_h should be low enough to ensure sufficient voltage clamping in order to avoid dielectric breakdown. I_h plays a role in eliminate post-ESD latch-up. V_{t2} should be greater than V_{t1} in multi-finger ESD protection design to ensure uniform triggering [3][4]. I_{t2} determines the ESD protection level.

After defining the ESD specifications, the next step is to select

0-7803-7761-3/03/$17.00 ©2003 IEEE

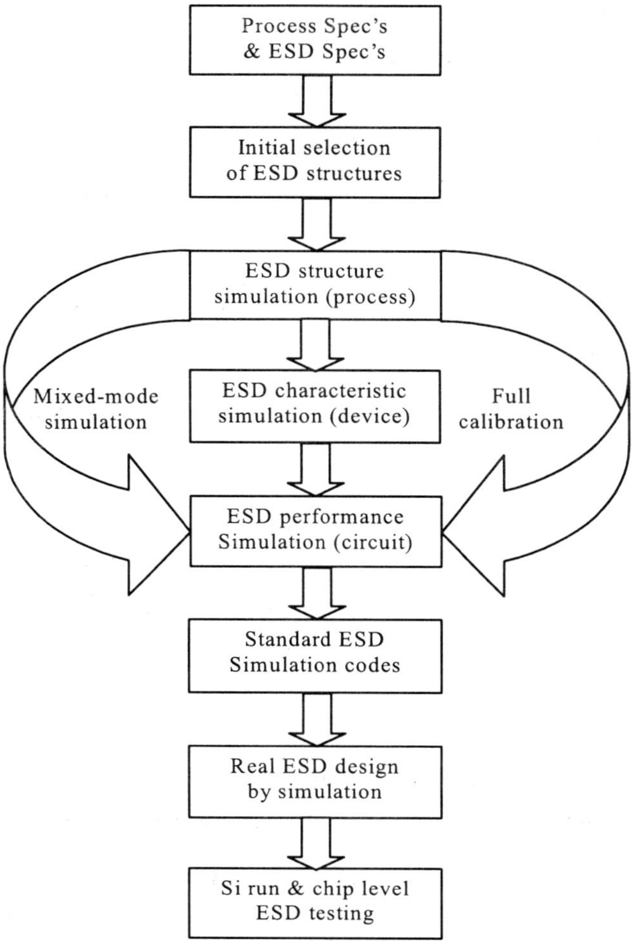

Figure 2. A flow chart for the new mixed-mode ESD simulation-design methodology.

initial ESD protection structures according to core circuit specifications and process features. There are many possible ESD protection structures, i.e., diode, ggNMOS/ggPMOS, SCR, etc. However, not all of them may be applicable in a specific design. One has to carefully choose the most possible structures based upon specific process technologies and circuit features.

Next, the selected possible ESD protection structures are created by process simulation using TCAD tool, TSUPREM4 [5], which solves process model equations at each node of the device mesh and produces detailed impurity doping profiles. It is imperative for one to start the design from process simulation because many advanced process steps have negative impacts on ESD robustness. These types of considerations can be simulated in design phase.

The novelty of this new ESD simulation approach lies in its non-assumption and mixed-mode features that include both static and transient simulation, where a proper ESD model circuit, e.g., a HBM model [6], instead of any artificial waveforms as reported [7], is used to produce real-world ESD pulses to stress ESD protection device under test (DUT) in simulations, as shown in Figure 3. This approach can automatically locate heating sources

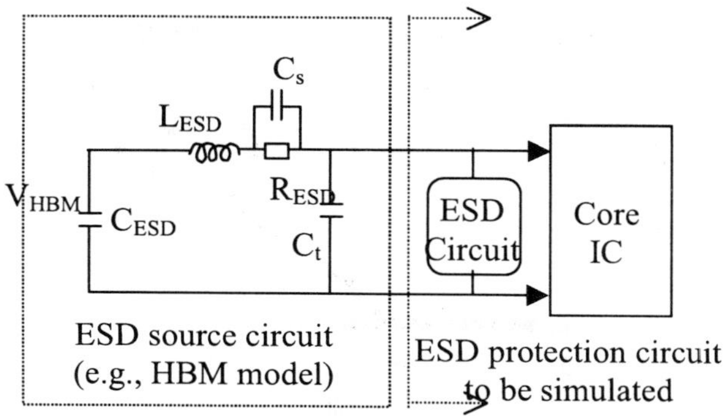

Figure 3 Sample mixed-mode ESD simulation schematic.

in numerical simulation, without assuming its existence as in other work [7]. Hence, there is no assumption throughout this new mixed-mode ESD simulation procedure. The key feature of the new mixed-mode ESD simulation-design methodology is that the complex electro-thermal equations are solved self-consistently at process, device and circuit levels, in a coupled fashion, in simulating ESD protection circuit behaviors. These include process simulation as discussed previously, various device physics equations, i.e., thermal equation, Poisson's equation, continuity equation, as well as Kirchhoff circuit equations. Lattice temperature distribution in ESD devices as well as the distributions of potential and carrier concentrations will be obtained at the end of the simulation. TCAD tool, Medici [8] is used. ESD simulation includes both steady-static and transient simulation. The ESD failure criteria include thermal breakdown where maximum lattice temperature, T_{max}, is higher than material melting temperature, $T_{melting}$, and dielectric rupture where gate voltage, V_g, exceeds its breakdown, BV_g. Steady-static simulation can be done by performing DC sweeping and provide a rough, time-independent characterization of ESD protection behaviors, which provides insightful trend behaviors. Transient ESD simulation uses an ESD equivalent model sub-circuit to generate real ESD pulse to stress an ESD protection sub-circuit as shown in Figure 3. Both single ESD protection structure and sub-circuit can be simulated in transient ESD simulation. Figures 4 & 5 show example I-V curves for a ggNMOS ESD protection structure, implemented in a commercial 0.35μm CMOS, under 2000V HBM ESD stresses for passed and failed cases, respectively.

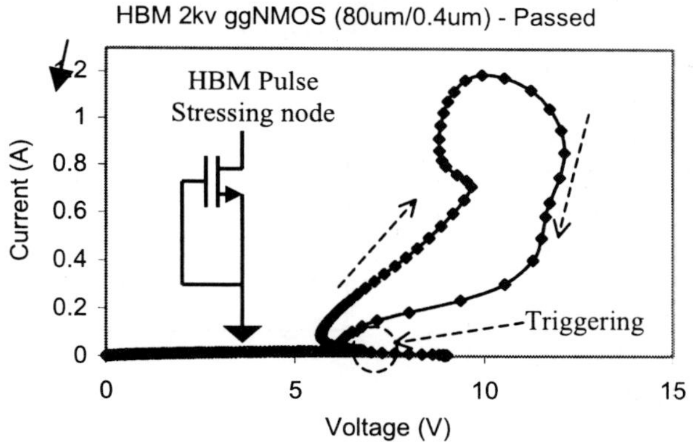

Figure 4 I-V curve for ggNMOS (80μ) – passed 2kV HBM test.

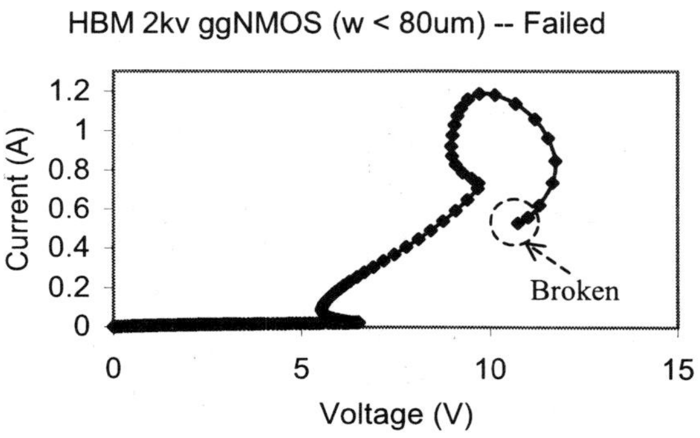

Figure 5 I-V curve for ggNMOS (<80μ) – failed 2kV HBM test.

Calibration is one of the most complicated yet critical steps in mixed-mode ESD simulation, which includes both process calibration and device characteristic calibration. Process calibrations focus on the parameters related to IC processes, e.g., doping profiles and voltage threshold values. Device characteristic calibrations are made by comparing steady-static and transient simulation results with related testing data performed by curve tracer, TLP (transmission line pulsing) and ESD zapping tests. The main calibration covers mobility model parameters, material parameters and thermal coefficients. Impact ionization calibration is a critical step.

Physical layout design follows the mixed-mode ESD simulation. A variety of test patterns for both ESD protection structures and core circuits will be generated in layout for testing and calibration purpose. Generally, these ESD test structures should be as simple as possible in order to avoid any confusion.

Three types of ESD measurements are involved with a universal test set-up illustrated in Figure 6 that includes quasi-DC (Tektronix 370 curve tracer), TLP (TLP 4002 tester), and ESD zapping (IMCS 10000 model) testing. Curve tracers are normally used to collect DC parameters (e.g., V_{t1}, I_{t1} & V_h) that will be matched with steady-static ESD simulation results. The transient parameters (V_{t1}, I_{t1}, t_1, V_h, I_h, V_{t2} & I_{t2}) are collected by conducting non-destructive TLP tests that provides insights into both thermal failure threshold and instantaneous I-V characteristics. These transient testing data will be calibrated with transient ESD simulation results. Finally, ESD zapping is conducted to evaluate full-chip ESD failure threshold levels.

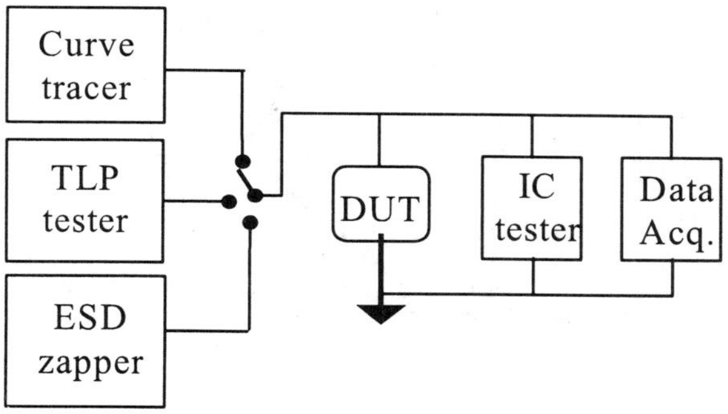

Figure 6 A universal ESD testing diagram.

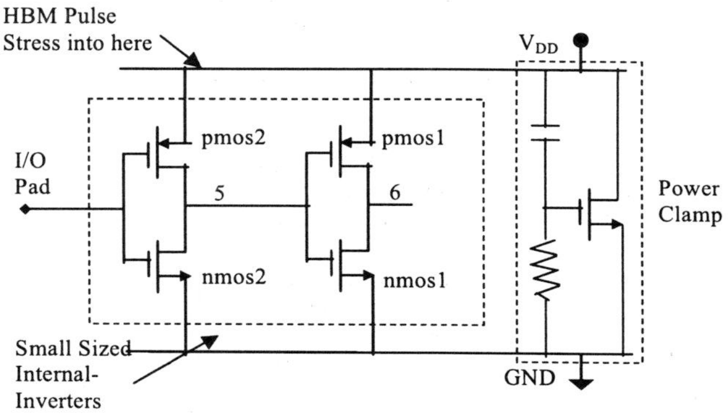

(a) Original 2-stage internal inverter plus the ESD power clamp.

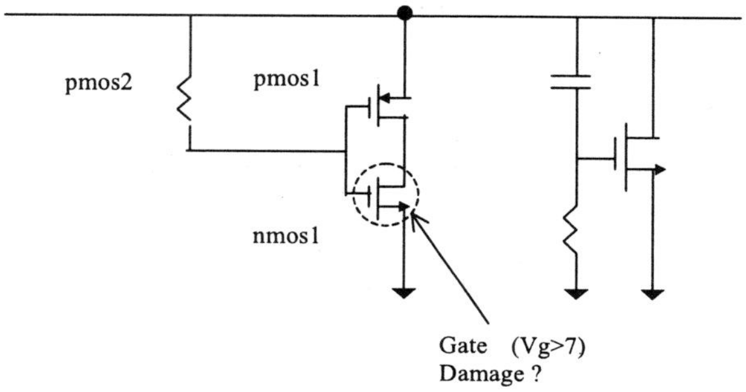

(b) Equivalent circuit with *low* logic input.

Figure 7 Schematics for a full-chip ESD protection circuit.

3. DESIGN EXAMPLE

As a design example, a full-chip ESD protection solution was developed for a commercial 0.35μm 3.3V CMOS. The starting point was an original gcNMOS ESD protection structure with a size of W/L=1120μm/0.35μm for 1000V HBM ESD protection, designed by using the trial-and-error method. The design goal was to shrink the gcNMOS size, to predict ESD performance and to evaluate ESD performance at full-chip level. Firstly, mixed-mode ESD simulation was performed to verify the gcNMOS behaviors and its size was successfully reduced to W/L=80μm/0.35μm that, however, achieves a 2000V HBM ESD protection. Next, mixed-mode ESD simulation was conducted to investigate a special circuit operation case, where a small internal two-stage inverter logic circuit block, with a PMOS of W/L=2μm/0.35μm and an NMOS of W/L=1.5μm/0.35μm, is placed nearby the gcNMOS power clamp, as shown in Figure 7a. When the input pin sees a logic Low signal, its equivalent circuit is given by Figure 7b, since NMOS2 is in off state while PMOS2 acts as an active resistor. The concern is that when an ESD pulse stresses the power rail, will the gate voltage of NMOS1, V(5), exceeds the gate breakdown voltage that was given as 8V for typical and 7V as minimum for the 0.35μm CMOS used, hence, resulting gate dielectric failure in the core circuit? Ideally, the gate voltage of NMOS1, V(5), should be clamped at below 7V in order to avoid any gate ESD damage. The V(5) is probed in mixed-mode ESD simulation. Figure 8 shows the I-V characteristics for both the gcNMOS power clamp, $V_{esd\text{-}drain}$, and V(5) of the NMOS1, denoted as $V_{inv\text{-}gate}$. It is observed that $V_{inv\text{-}}$

$_{gate}$ of the NMOS1 is indeed lower than the gate breakdown of 8V. However, in a very brief period of less than 1ns, the $V_{inv\text{-}gate}$ briefly exceeds minimum gate breakdown of 7V by a fraction of one volt. It would be perfect if one could ensure a $V_{inv\text{-}gate}$ lower than 7V, which is not the case in this design. However, fortunately, data from lifetime analysis suggests that this brief overshoot in V(5) does not cause failure in application. On the other hand, the maximum lattice temperature curves obtained for all MOSFETs involved, as shown in Figure 9, clearly show that significant temperature increase only occurs in the gcNMOS power clamp as it should be, while all other MOSFETs almost remain in room temperature operation during the ESD stressing period. This is because the blunt of ESD transient current discharges into the gcNMOS clamp as designed. Therefore, mixed-mode ESD simulation confirms successful full-chip ESD protection in this design. HBM zapping test shows that the individually optimized gcNMOS clamp of W/L=80μm/0.35μm achieves 2500V HBM ESD protection stressing level.

4. CONCLUSIONS

This paper discusses a new mixed-mode ESD protection simulation-design methodology that allows IC designers to predict ESD protection circuit performance in early design phase. The mixed-mode design approach involves multiple-level electro-thermal-process-device-circuit-layout coupling in ESD protection simulation by solving complex electro-thermal equations self-consistently at process, device and circuit levels, in a coupled way to investigate ESD protection circuit behaviors without any pre-assumption. Practical ESD protection design example, implemented in commercial 0.35μm CMOS technologies, is presented to demonstrate the usefulness of this mixed-mode ESD simulation methodology. This new design approach, used properly, provides IC designers with an efficient way to conduct predictive ESD protection design that delivers adequate ESD protection while suppresses ESD-induced parasitic effects, which is critical to VDSM ULSI IC designs, particularly in mixed-signal and RF IC applications.

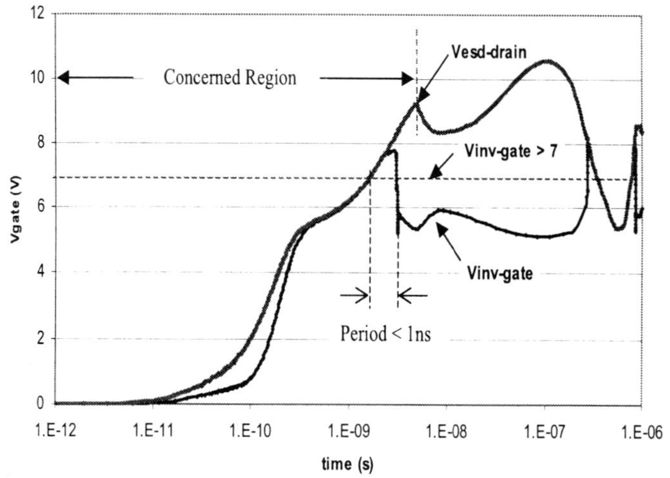

Figure 8 V~t curves from full-chip transient ESD simulation where $V_{esd\text{-}drain}$ is power rail voltage and $V_{inv\text{-}gate}$ is at V(5) node.

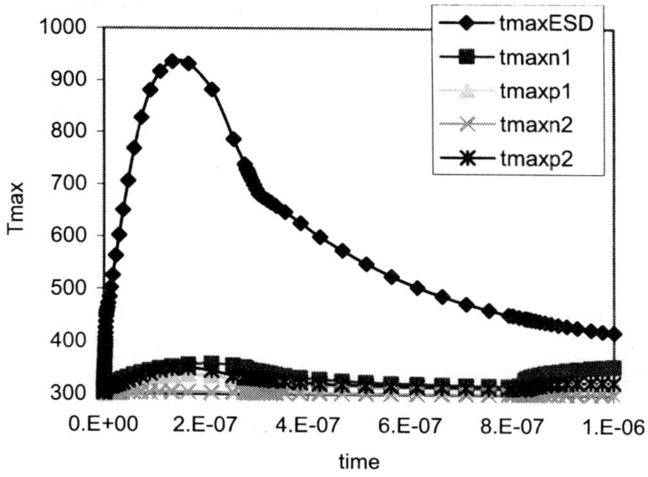

Figure 9 T_{max} ~ t curves for each transistors show that the ESD power clamp is the main ESD discharging device as designed.

5. REFERENCES

[1] Merri R. and Issaq E., "ESD design methodology", *Proc. EOS/ESD Symposium*, pp.233-237, 1993.

[2] Green T., "A review of EOS/ESD field failures in military equipment", *Proc. EOS/ESD Symposium*, pp.7-14, 1988.

[3] Wang A., Tsay C. and Deane P., "A study of NMOS behaviors under ESD stress: simulation and characterization", *Microelectronics Reliability*, 38, Elsevier, pp.1183-1186, 1998.

[4] Wang A., "A new design for complete on-chip ESD protection", *Proc. IEEE Custom Integrated Circuits Conference*, pp. 87-90, May 2000.

[5] "TSUPREM4 2D Semiconductor Process Simulation User's Manual", *Avant!.Corp.*

[6] *MIL-STD-883C, 3015.7*, notice 8, "Military Standard for Test Methods and Procedures for Microelectronics: ESD Sensitivity Classification", 1989.

[7] Diaz C. and Kang S., *Modeling of Electrical Overstress in Integrated Circuits*, Kluwer Academic Publishers, 1994.

[8] "MEDICI 2D Semiconductor Device Simulation User's Manual", *Avant! Corp.*

ACKNOWLEDGMENT

The authors wish to thank Avant! Corp. for CAD software donation and AKM for wafer fabrication. This material is based upon work supported partially by the National Science Foundation under Grant No. ECS-0132869.

A TECHNIQUE FOR REDUCTION OF UNCERTAIN FIR FILTERS

A. G. Lim, V. Sreeram* and E. Zeheb†*

*Department of Electrical and Electronic Engineering
University of Western Australia, 35 Stirling Highway, Crawley,
WA 6009, AUSTRALIA, email:sreeram@ee.uwa.edu.au

†Department of Electrical Engineering,
Technion-Israel Institute of Technology,
Haifa 32000, ISRAEL, email:zeheb@ee.technion.ac.il

ABSTRACT

In this paper, an algorithm for designing an IIR filter from a given uncertain FIR filter is proposed. An approach to properly assign the uncertainties in the FIR polynomial transfer function, to two parts of its factorization, is cleverly used. The resulting approximated IIR filter is inherently stable. The method proposed is in the time domain, based on the LSI approximation. An example of reducing a 16th order interval FIR filter to a 4th order numerator order over a 2nd order denominator order IIR filter is presented to illustrate the design method.

1. INTRODUCTION

FIR (Finite Impulse Response) filters are frequently used in all kinds of digital systems. However, in some cases where the order of the required FIR digital filter is too large, these filters cannot be used because of the large delays introduced by them. In these cases, a reduced order IIR (Infinite Impulse Response) approximated digital filters provide a more satisfactory solution, provided it retains stability. Thus, there are numerous results and methods to carry out FIR order reduction by IIR approximation [1]-[8]. Some of these methods render inherently stable IIR filters, while others require a second stage of stabilizing the IIR filter in case the basic algorithm does not render a stable one.

The significance of the results in this paper is that we consider the problem of FIR order reduction by IIR approximation, where it cannot be assumed that we have a complete knowledge about the model of the original FIR filter. Due to inherent intervals where some physical parameters can take on different values, or due to measurement inaccuracies, round off errors and etc, or due to environmental changes, the coefficients of the high order FIR filters are only known to take on values in given intervals, but the exact fixed values are not known.

The method derived in this paper is an adaptation to interval coefficients, of the method proposed in [8] for fixed coefficients FIR approximation in the time domain. It is based on LSI (Least Squares Inverse) approximation [9], which enables the approximation of the inverse of a polynomial by another polynomial. There are two important advantages that are inherent to this method:

1. The approximated reduced order IIR filter is guaranteed to be stable. No stabilization procedure is needed.
2. The numerator of the reduced order IIR filter is part of the original FIR high order filter.

A major obstacle in trying to adapt the fixed coefficients method in [8] to an interval coefficients method is the assignment of the uncertainties of the FIR polynomial to its factorisation into two parts. This obstacle is overcome using the results in [10], and described in Section II. The uncertain FIR order reduction by IIR approximation is described in Section III. Although the approximation is carried out in the time domain, and therefore frequency domain comparisons between the FIR and the approximated IIR behaviour are not too promising, we provide in Section IV an illustrative example where such frequency response comparisons indicate very good results. Conclusions are given in Section V.

2. ASSIGNING UNCERTAINTIES TO TWO FACTORS OF A POLYNOMIAL

Suppose a nominal polynomial is given by

$$H_o(z) = z^n + \sum_{i=0}^{n-1} a_i z^i = \prod_{i=1}^{n} (z - z_i) \qquad (1)$$

and suppose the coefficients a_i can take on values in the intervals $[a_i \pm \Delta a_i]$, $i = 0, \ldots, n\text{-}1$. Denote the family of obtained polynomials (the "uncertain polynomial") by

$$H(z) = z^n + \sum_{i=0}^{n-1} \alpha_i z^i \qquad (2a)$$

where, if Δa_i is assumed positive,

$$a_i - \Delta a_i \le \alpha_i \le a_i + \Delta a_i \qquad (2b)$$

Denote the extreme displacement of each zero of a polynomial in $H(z)$ by $\pm\Delta z_i$. In other words, if a_i becomes $a_i + \Delta a_i$ ($i = 0, \ldots, n$-1), then the zero z_i ($i = 1, \ldots, n$) of $H_o(z)$ move to $z_i + \Delta z_i$, i.e.,

$$a_i \to a_i + \Delta a_i \Rightarrow z_i \to z_i + \Delta z_i \qquad (3)$$

Then, a first order linear approximation yields [10]

0-7803-7761-3/03/$17.00 ©2003 IEEE

$$\Delta z_k = -\frac{\sum_{i=0}^{n-1}(\Delta a_i) z_k^i}{H'(z_k)} \qquad k = 1,\dots,n \tag{4}$$

where $H'(z_k)$ denotes the derivative of $H(z)$ with respect to z, evaluated at $z = z_k$. The expression in (4) is well defined, assuming the zeros of the nominal polynomial (1) are distinct. In matrix notation, let $\overline{\Delta z}$ and $\overline{\Delta a}$ be the vectors

$$\overline{\Delta z} = \left(\Delta z_1,\dots,\Delta z_n\right)^T \tag{5}$$

$$\overline{\Delta a} = \left(\Delta a_0,\dots,\Delta a_{n-1}\right)^T \tag{6}$$

and let Q be the an $[n \text{ x } n]$ matrix with entries

$$q_{ij} = \frac{-z_i^{j-1}}{H'(z_i)} \tag{7}$$

Then [10]

$$\overline{\Delta z} = (Q)\overline{\Delta a} \tag{8}$$

Now let $H_o(z)$ be factored into

$$H_o(z) = H_{o1}(z)H_{o2}(z) \tag{9}$$

where

$$H_{o1}(z) = \prod_{i=1}^{r}(z - z_i) = \sum_{i=0}^{r} b_i z^i \tag{10}$$

$$H_{o2}(z) = \prod_{i=r+1}^{n}(z - z_i) = \sum_{i=0}^{n-r} c_i z^i \tag{11}$$

and z_i (i = 1, …, r) in (10) are the zeros of $H_o(z)$ which are furthest from the unit circle[1], and z_i $(i = r+1,\dots,n)$ in (12) are the remaining (n-r) zeros of $H_o(z)$.

The assignment of uncertainties to $H_{o1}(z)$ and $H_{o2}(z)$ to be in correspondence with the uncertainties in (2a), (2b) is done using the displacements of the particular zeros belonging to $H_{o1}(z)$ and $H_{o2}(z)$, and calculated by (4) or (8). Explicitly, let

$$\tilde{H}_1(z) = \prod_{i=1}^{r}(z - z_i - \Delta z_i) = \sum_{i=0}^{r} \tilde{b}_i z^i \tag{12}$$

and

$$\tilde{H}_2(z) = \prod_{i=r+1}^{n}(z - z_i - \Delta z_i) = \sum_{i=0}^{n-r} \tilde{c}_i z^i \tag{13}$$

Then the corresponding families of polynomials (the "uncertain polynomials") are:

$$H_1(z) = \sum_{i=0}^{r} \beta_i z^i \tag{14}$$

where

$$2b_i - \tilde{b}_i \le \beta_i \le \tilde{b}_i \qquad \text{if} \quad \tilde{b}_i > b_i \tag{15a}$$

or

$$\tilde{b}_i \le \beta_i \le 2b_i - \tilde{b}_i \qquad \text{if} \quad b_i > \tilde{b}_i \tag{15b}$$

and

$$H_2(z) = \sum_{i=0}^{n-r} \gamma_i z^i \tag{16}$$

where

$$2c_i - \tilde{c}_i \le \gamma_i \le \tilde{c}_i \qquad \text{if} \quad \tilde{c}_i > c_i \tag{17a}$$

or

$$\tilde{c}_i \le \gamma_i \le 2c_i - \tilde{c}_i \qquad \text{if} \quad c_i > \tilde{c}_i \tag{17b}$$

The product of the polynomial families $H_1(z)$ and $H_2(z)$ corresponds to the polynomial family $H(z)$ in (2).

3. UNCERTAIN FIR ORDER REDUCTION BY IIR APPROXIMATION

Suppose the transfer function of the high order FIR digital filter of degree n is the uncertain family of polynomials $H(z)$ defined in (2a) and (2b), where we replaced the delay operator z^{-1} by the variable z. Suppose also that our goal is to approximate the FIR filter by an IIR digital filter with numerator of degree "l" and denominator of degree m. Then we choose

$$r = n - l \tag{18}$$

and form the two families of polynomials $H_1(z)$ and $H_2(z)$ as in (15) and (17). Let the Least Squares Inverse polynomial of $H_1(z)$ be of the desired degree "m", and denoted by

$$\overline{H}_1(z) = \sum_{i=0}^{m} d_i z^i \tag{19}$$

Here the Least Squares Inverse means that the product approximates the zero order unit polynomial 1 by the least squares criterion. Then, following [8], the optimal coefficients d_i (i = 0,…, m) are computed by solving the set of equations

$$\begin{bmatrix} r_0 & r_1 & \cdots & r_m \\ r_1 & r_0 & \cdots & r_{m-1} \\ \vdots & & \ddots & \vdots \\ r_m & r_{m-1} & & r_0 \end{bmatrix} \begin{bmatrix} d_0 \\ d_1 \\ \vdots \\ d_m \end{bmatrix} = \begin{bmatrix} \beta_0 \\ 0 \\ \vdots \\ 0 \end{bmatrix} \tag{20}$$

where

$$r_k = \sum_{i=0}^{r} \beta_i \beta_{i-k}\,; \qquad \beta_{i-k} \triangleq 0 \quad \text{if} \quad k > i \tag{21}$$

[1] This criterion for choosing the r zeros is needed for the derivation in the next section [8]. For the derivation in the present section, an arbitrary choice of zeros is possible.

However, differing from [8], the entries in the matrix in (20) and β_0 are interval coefficients. Thus the solution of (20) should be carried out using the following rules of interval arithmetics [11]:

$$[\underline{x},\overline{x}]+[\underline{y},\overline{y}]=[\underline{x}+\underline{y},\overline{x}+\overline{y}] \quad (22a)$$

$$[\underline{x},\overline{x}]-[\underline{y},\overline{y}]=[\underline{x}-\overline{y},\overline{x}-\underline{y}] \quad (22b)$$

$$[\underline{x},\overline{x}].[\underline{y},\overline{y}]=\left[\min\left(\underline{x}\underline{y},\underline{x}\overline{y},\overline{x}\underline{y},\overline{x}\overline{y}\right),\max\left(\underline{x}\underline{y},\underline{x}\overline{y},\overline{x}\underline{y},\overline{x}\overline{y}\right)\right] \quad (22c)$$

$$[\underline{x},\overline{x}]/[\underline{y},\overline{y}]=[\underline{x},\overline{x}].[1/\overline{y},1/\underline{y}] \qquad \text{provided } \underline{y}\overline{y}>0 \quad (22d)$$

and the resulting d_i ($i = 0,\dots, m$) will be interval coefficients.

A crucial point of this method is that the zeros of each of the polynomials in the family (19) are necessarily in the open unit disk. This property has been proved for the fixed coefficients case in [8], but is retained for the interval coefficients case, as can be readily verified. Choosing an "average" polynomial from the family $\overline{H}_1(z)$ in (19), e.g.

$$\overline{H}_{a1}(z)=\frac{1}{2}\sum_{i=0}^{m}\left(\underline{d}+\overline{d}\right)z^i \quad (23)$$

and an "average" polynomial from the family $H_2(z)$ in (16), e.g. as in (11)

$$H_{a2}(z)=H_{o2}(z)=\sum_{i=0}^{n-r}c_i z^i \quad (24)$$

we arrive at an IIR reduced order transfer function:

$$\hat{H}(z)=\frac{H_{a2}(z)}{H_{a1}(z)} \quad (25)$$

The IIR filter with transfer function $\hat{H}(z)$ in (25) should approximate the uncertain FIR filter with family of transfer functions $H(z)$ in (2), up to a constant multiplier whose purpose is to adjust the energy of the IIR filter. The optimal value of this constant multiplier is found by minimizing the error between $\hat{H}(z)$ and $H(z)$.

4. SIMULATION EXAMPLE

In this section, a 16th order FIR low pass filter given by

$$H(z)=z^{16}+a_0z^{15}+a_1z^{14}+\dots+a_{15}$$

is used as the "uncertain" filter and the coefficients a_i are given in the following table:

a_0	1.643	a_8	156.138
a_1	-5.791	a_9	-43.263
a_2	0	a_{10}	-23.867
a_3	21.843	a_{11}	21.843
a_4	-23.867	a_{12}	0
a_5	-43.263	a_{13}	-5.791
a_6	156.138	a_{14}	1.643
a_7	320.688	a_{15}	1

Table 1: Coefficients of the FIR polynomial.

The magnitude response of this FIR filter (fixed coefficients) is given in Fig. 1. The filter is a low-pass filter with a cut-off frequency of 0.6π rad/s. Fig. 2 shows the magnitude response of the filter with uncertainties, i.e. some of the filter's coefficients are interval coefficients $a_i \pm \Delta a_i$ and some remain fixed.

$a_i \rightarrow a_i \pm \Delta a_i$			
a_0	[1.623, 1.663]	a_8	156.138
a_1	[-5.89, -5.69]	a_9	-43.263
a_2	0	a_{10}	-23.867
a_3	21.843	a_{11}	21.843
a_4	-23.867	a_{12}	0
a_5	-43.263	a_{13}	[-5.89, -5.69]
a_6	156.138	a_{14}	[1.623, 1.663]
a_7	320.688	a_{15}	[0.99, 1.01]

Table 2: Fixed and interval coefficients of the FIR filter.

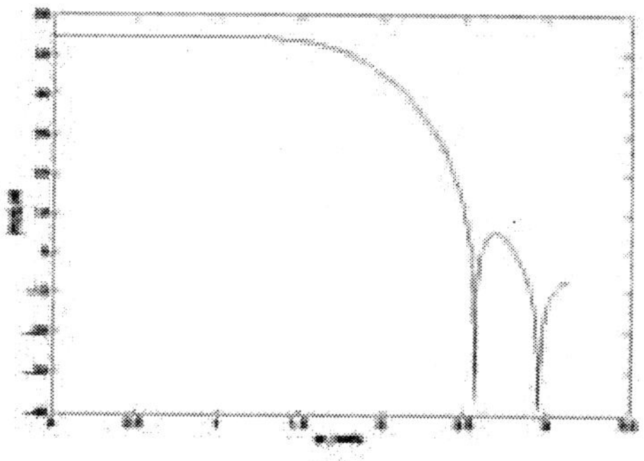

Fig.1: Nominal response of FIR filter (fixed coefficients).

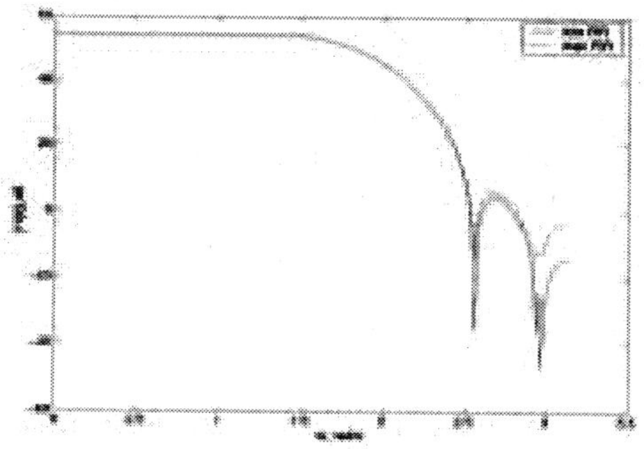

Fig.2: Magnitude response of FIR filter with uncertainties (interval coefficients).
Solid line: Minimum magnitude response.
Dashed line: Maximum magnitude response.

This FIR filter has 17 coefficients (N = 16). In this example, $H_{a2}(z)$ has order, l = 4 and $H_{a1}(z)$ is of order, m = 2. Fig. 3 shows the minimum and maximum magnitude responses of the family of reduced order IIR polynomials while Fig. 4 gives the response of the "average" polynomial from the family of the IIR polynomials. In Fig. 4, the response of the "average" IIR polynomial shows excellent approximation to the original 16th order FIR filter.

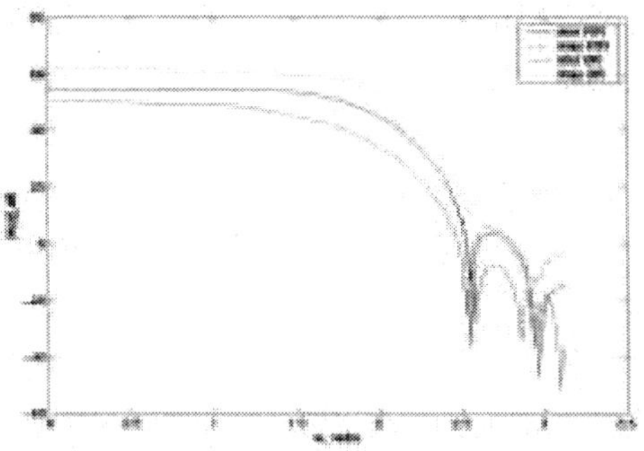

Fig.3: Minimum and maximum magnitude response of the family of IIR polynomials.

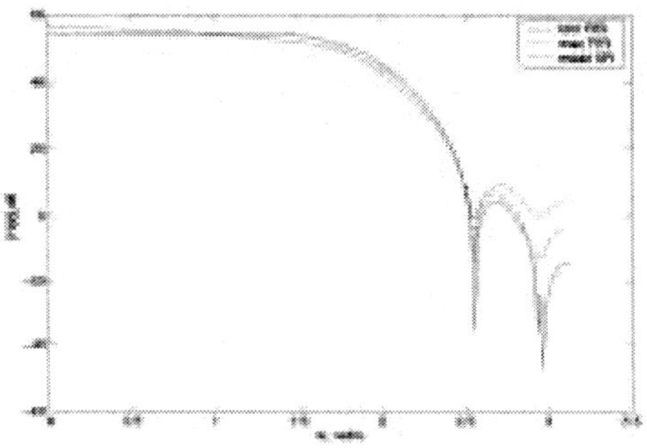

Fig.4: Magnitude response of the "average" IIR filter.

Some general remarks for this example are as follows:

- The family of reduced order IIR polynomials is scaled by a factor of 0.015 to match the response of the FIR polynomials.
- The nominal FIR polynomial has 4 zeros on the unit circle, therefore the algorithm uses the remaining 12 zeros that are not on the unit circle for approximation by poles.

5. CONCLUSION

An efficient method has been proposed to reduce the order of interval polynomial transfer functions of FIR filters by interval or fixed coefficients lower order IIR filters. Although the approximation is in the time domain, very good frequency domain results are obtained in the presented illustrative example. The method has two inherent features that are advantageous; Firstly, the approximated reduced order IIR filter is guaranteed to be stable. Secondly, the numerator of the reduced order IIR filter is part of the original FIR high order filter, except that the uncertainties are adjusted. It should be noted, however, that although the described method generically yields a stable filter, there may be numerical instability when l is too small, which depends on wordlength limitations and the computer capabilities.

Another point which should be noted is that because of the interval arithmetics involved in calculating the assigned uncertainties and the inverse polynomial, the uncertainties set of the resulted IIR filter tends to be larger than the original FIR set of uncertainties. However, this does not necessarily affect the "average" resulted IIR filter.

The further the zeros of the original nominal FIR filter are from the unit circle, both outside and inside the unit disk, the better the approximated IIR results. Further research has to be carried out for cases where too many of the zeros of the original nominal FIR filter are on or close to the unit circle.

6. ACKNOWLEDGMENT

The third author wishes to acknowledge support by the Fund for the Promotion of Research at the Technion-Israel Institute of Technology. He also wishes to acknowledge support by the Distinguished Visitors Fund of the University of Western Australia (UWA) and the Department of Electrical and Electronic Engineering at UWA, where this work has been carried out while the third author has been visiting there.

7. REFERENCES

[1] C. Charalambous and A. Antoniou, "Equalization of recursive digital filters," *Proc. IEE Pt. G*, vol. 127, no. 4, pp. 219-225, 1980.

[2] V. Sreeram and P. Agathoklis, "Design of linear-phase IIR filters via impulse-response Gramians," *IEEE Trans. Signal Processing*, vol. 40, pp. 398-394, Feb. 1992.

[3] M. F. Fahmy, Y. M. Yassin, G. Abdel-Raheem and N. El-Gayed, "Design of linear phase IIR filters from FIR specifications," *IEEE Trans. Signal Processing*, vol. 42, pp. 437-440, Feb. 1994.

[4] S. Holford and P. Agathoklis, "The use of model reduction techniques for designing IIR filters with linear phase in the passband," *IEEE Trans. Signal Processing*, vol. 44, pp. 2396-2404, Oct. 1996.

[5] W.-S. Lu, S.-C. Pei and C.-C. Tseng, "A weighted least-squares method for the design of stable 1-D and 2-D IIR digital filters," *IEEE Trans. Signal Processing*, vol. 46, pp. 1-10, Jan. 1998.

[6] L. Li, L. Xie, W.-Y. Yan and Y. C. Soh, "Design of low-order linear-phase IIR filters via orthogonal projection," *IEEE Trans. Signal Processing*, vol. 47, pp. 448-457, Feb. 1999.

[7] C.-S. Xiao, J. C. Oliver and P. Agathoklis, "Design of linear phase IIR filters via weighted least-squares approximation," *Proc. ICASSP*, vol. 6, pp. 3817-3820, 2001.

[8] Betser and E. Zeheb, "Reduced order IIR approximation to FIR digital filters," *IEEE Trans. Signal Processing*, vol. 39, no. 11, pp. 2540-2544, Nov. 1991.

[9] E. A. Robinson, *Statistical Communication and Detection*, London, England: Griffin, 1967.

[10] G. Martinelli, "On the matrix analysis of network sensitivities," *Proc. IEEE*, vol. 54, pp. 72, 1966.

[11] M. Mansour, "Robust stability in systems described by rational functions," *Academic Press Inc.*, vol. 51, pp. 79-128, 1992.

Concurrent Logic and Interconnect Delay Estimation of MOS Circuits by Mixed Algebraic and Boolean Symbolic Analysis*

Sambuddha Bhattacharya and C-J. Richard Shi
Department of Electrical Engineering, University of Washington
Seattle, WA 98195-2500, USA
{sbb,cjshi}@ee.washington.edu

ABSTRACT: *Accurate estimation of delay in logic-stages and interconnects is of utmost importance in digital VLSI design. Conventional delay estimation techniques are numeric in terms of design parameters for both logic-stages and interconnect trees driven by them. In this paper, we present a symbolic method of computing delay in logic stages followed by interconnect trees. For each stage, our method provides a single analytic delay expression that is symbolic in terms of all input logic assignments as well as transistor and interconnect parameters. The method has been implemented and validated on modern digital VLSI technologies.*

1. Introduction

Efficient and accurate delay estimation for logic stages is fundamental to many VLSI automations like timing analysis and transistor sizing. For MOS circuits, delay computation is traditionally performed on the channel connected regions (CCR) that consist of conducting transistors connected to each other through their drains and sources. Varying input patterns at the gates of the transistors result in changing CCRs causing different signal delays in the logic stage.

Recently, delay estimation by symbolically representing input logic patterns has gained attention. [4] presents a BDD based method of estimating the signal delay in logic stages by replacing the transistors with their equivalent on-resistances and capacitances. A similar scheme of delay estimation based on the more general MTBDDs is presented in [3].

However, these methods suffer several deficiencies. First, these methods symbolically enumerate only the different input logic patterns at the gates of transistors. Changes in the transistors' sizes cannot be handled in these schemes without recreating the BDD/MTBDD structures as their delay calculation is inherently numeric. Second, as these methods are based on RC tree methods, loops of transistors cannot be handled. Third, severe inaccuracies result due to the Elmore model [2] and empirical handling of input rise/fall times. Finally, these methods cannot model the delay due to the interconnect parasitic networks driven by logic gates, which is becoming more important in deep submicron digital MOS design.

This paper introduces a novel approach that symbolically represents the delay of a logic-stage not only for all possible input logic patterns but also for all possible transistor parameters (sizes). We use a technique called multi-terminal determinant decision diagram (*MTDDD*), introduced recently for symbolic circuit analysis [5]. We extend MTDDD for efficient handling of Boolean conditions along with regular algebraic equations.

We use a delay estimate based on higher order circuit moments [6] which renders greater accuracy. Our method extends directly from the modified nodal analysis (MNA) circuit equations and therefore can correctly handle any circuit topology. We use a pre-computed lookup table for the equivalent resistors of conducting transistors and incorporate the slope of the input signal at the gate of the transistor into the table.

We further extend our method to the integrated estimation of signal delay in a logic-stage followed by the interconnect tree driven by the stage. Thus, the interconnect parameters are also symbolic variables along with transistor parameters and input logic patterns.

The paper is organized as follows. The formulation of delay estimation in terms of Boolean-ized moment equations is introduced in Section 2. Section 3 illustrates the computation of circuit moments using MTDDDs. Experimental results are presented in Section 4.

2. Formulation with Boolean-ized Moments

2.1 Background

A MOS transistor is modeled by a voltage controlled on-resistor between the drain and the source and three grounded capacitors at the drain, source and gate nodes. Interconnects are modeled as RC trees. The modified nodal analysis (MNA) based formulation of circuit equations [8] can be written as

$$\mathbf{C}\dot{\mathbf{x}} = \mathbf{G}\mathbf{x} + \mathbf{b}\mathbf{u} \qquad \mathbf{x}\,|_{t=0} = \mathbf{x}_0 \tag{1}$$

where $\mathbf{x}(t) \in \Re^n$ is a vector composed of node voltages and necessary branch currents, $\mathbf{G}$ is the modified conductance matrix, $\mathbf{C}$ is the capacitance and inductance matrix, and $\mathbf{u}$ is due to the system's input. The transfer function of such a linear network at any node can be expressed in terms of circuit moments as

$$H(s) = m_0 + m_1 s + m_2 s^2 + \ldots + m_n s^n + \ldots \tag{2}$$

The circuit moments then can be derived from (1) in a recursive form [6] where $\mathbf{m}_i$ is a vector composed of the ith moments.

$$\mathbf{m}_0 = \mathbf{x}_h(0) \qquad \mathbf{G}\mathbf{m}_{k+1} = \mathbf{C}\mathbf{m}_k \tag{3}$$

The transfer function (2) is then matched to a lower order function, such as the one in (4), using Pade approximation [6].

$$\hat{H}(s) = \hat{m}_0 + \hat{m}_1 s + \hat{m}_2 s^2 \tag{4}$$

The propagation delay for the transfer function (4) can be estimated in terms of the 1st and 2nd moments. The delay metric we use in this work is adopted from [1] and is given as

* Supported by DARPA under Grant No. 66001-01-1-8920.

0-7803-7761-3/03/$17.00 ©2003 IEEE

$$t_{pd} = \ln(2)\frac{m_1^2}{\sqrt{m_2}} \quad (5)$$

2.2 Moment Equations and MTDDDs

Consider the pull-up circuit of an OAI21 and its equivalent RC network of Fig.1. Transistor M_i with Boolean X at its gate in the original circuit is replaced by a resistor R_{Mi} and a Boolean switch $\overline{X}$ (as the transistor is a PMOS). Node 2 in the equivalent circuit has capacitance contribution due to the transistors M_2 and M_3 of the original circuit. Similarly, the capacitance at node 3 is due to the load capacitance C_L and the contribution of the transistors M_1 and M_3.

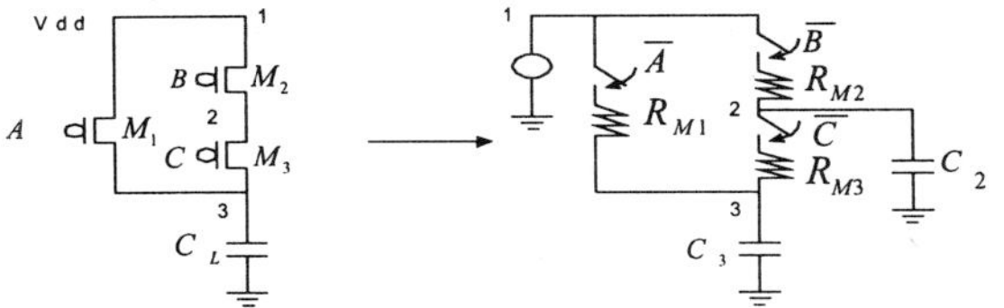

Figure 1: Pullup section of OAI21 and equivalent circuit.

The equation set (6) represents the recursive moment equations (3) for the equivalent RC network of Fig.1 when all the switches are closed. The kth moment at the ith node is represented as $m_k(i)$, whereas the kth moment for current I_P is represented as $m_k(I_P)$. I_4 represents the current in the voltage source, and any other I_P represents the current in R_{MP}. The 3 equations on top represent KCL-type moment equations. The last 4 rows represent the branch equations.

$$\begin{bmatrix} 0 & 0 & 0 & 1 & 1 & 0 & 1 \\ 0 & 0 & 0 & 0 & -1 & 1 & 0 \\ 0 & 0 & 0 & -1 & 0 & -1 & 0 \\ -1 & 0 & 1 & R_{M1} & 0 & 0 & 0 \\ -1 & 1 & 0 & 0 & R_{M2} & 0 & 0 \\ 0 & -1 & 1 & 0 & 0 & R_{M3} & 0 \\ 1 & 0 & 0 & 0 & 0 & 0 & 0 \end{bmatrix} \begin{bmatrix} m_k(1) \\ m_k(2) \\ m_k(3) \\ m_k(I_1) \\ m_k(I_2) \\ m_k(I_3) \\ m_k(I_4) \end{bmatrix} = \begin{bmatrix} 0 \\ C_2 m_{k-1}(2) \\ C_3 m_{k-1}(3) \\ 0 \\ 0 \\ 0 \\ 0 \end{bmatrix} \quad (6)$$

Solving the equation set (6) is the same as repeatedly solving a set of linear equations in the following matrix form:

$$\mathbf{Tx} = \mathbf{b} \quad (7)$$

Similar to symbolic circuit analysis, our procedure for symbolic computation of moments is based on Cramer's rule for solving sets of linear equations. Then, the ith element of **x** is obtained as

$$x_i = \sum_j \mathbf{b}_j \det(\mathbf{T}_{ij}) \Big/ \det(\mathbf{T}) \quad (8)$$

where $det(\mathbf{T})$ is the determinant of the matrix **T**, and $det(\mathbf{T}_{ij})$ is the determinant of the matrix **T** after removing row i and column j, or the cofactor of the matrix **T** with respect to the element at (i,j). Therefore, the key task is the representation of the determinants and the cofactors of a semi-symbolic matrix.

For this purpose, we utilize an efficient technique called *Multi-Terminal Determinant Decision Diagram (MTDDD)*, introduced recently in the context of symbolic circuit analysis [5]. An MTDDD is an ordered, rooted, directed acyclic graph. As illustrated in Fig.2, it consists of some symbolic vertices and a set of terminal vertices, which can be the *0-terminal* vertex, the *1-terminal* vertex and some numeric terminal vertices with non-zero values. A symbolic vertex V, is characterized by a label *(V.label)*, and two edges, namely *1-edge* (solid line) and *0-edge* (dotted line) pointing, respectively, to its *1-child (V.1-child)* and *0-child (V.0-child)*. Thus, a vertex V represents a *semi-symbolic expression V.expr* defined recursively as follows:

If (V is 0-terminal), then V.expr=0
if (V is 1-terminal), then V.expr=1
if (V is numeric terminal), then V.expr=V.value
if (V is a symbolic vertex), then
V.expr=V.label(V.1-child).expr + (V.0-child).expr*

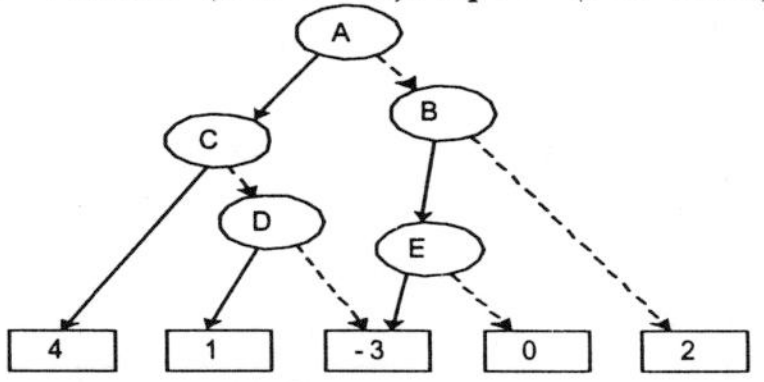

Figure 2: An MTDDD representing 4AC + AD -3A -3BE + 2.

To see how an MTDDD can be used to represent the determinant and cofactors of a circuit matrix, we consider the Laplace expansion of a matrix determinant with respect to a particular element, $t_{i,j}$, at row i and column j. Then, the matrix determinant $det(\mathbf{T})$ can be represented as follows:

$$\det(\mathbf{T}) = (-1)^{(i+j)} t_{i,j} \det(\mathbf{T}_{ij}) + \det(\mathbf{T}_{\overline{ij}}) \quad (9)$$

where, $det(\mathbf{T}_{ij})$ is the cofactor of the matrix **T** with respect to the element $t_{i,j}$ and $\det(\mathbf{T}_{\overline{ij}})$ is the remainder of the matrix **T** with respect to the element $t_{i,j}$, which is defined as the determinant of the matrix **T** after setting $t_{i,j}$ to 0. Clearly, if we can represent this expansion by a vertex with label $(-1)^{(i+j)} t_{i,j}$, the cofactor as the 1-child and the remainder as the 0-child, and then recursively expand the cofactor and the remainder, we can obtain an MTDDD. During this process, all the subgraphs can be shared.

2.3 Boolean-ized Moment Equations

Equation (6) is valid for one input logic pattern. For an ***n***-input logic cell, the 2^n different input pattern sets result in 2^n sets of moment equations like (6). Naturally, creating MTDDDs for each such moment equation set is inefficient. This motivates the need for a single set of moment equations valid for all input patterns. This is accomplished by incorporating the Boolean input variables into the moment equations.

For the circuit of Fig.1, we recognize that the current in a branch is zero when the switch in the branch is open. That leads to the Boolean-ized branch I-V equation shown in Fig.3.

Figure 3. Boolean-ized MOSFET branch equation.

An isolated node results in a zero column in the ***G*** matrix of (6) rendering it singular. A node is isolated if there are no sensitizing paths from the voltage source to that node. In other words, the presence of currents in branches connected to a node and isolation of that node are mutually exclusive events. The current I_2 in branch R_{M2} of Fig.1 exists if $\overline{B}$ is "1". Similarly, the condition for existence of current I_3 is

$$F_{I3} = \overline{A}\,\overline{C} + \overline{B}\,\overline{C} \qquad (10)$$

The condition that node 2 is isolated is then given by

$$F_2 = !(\overline{A}\,\overline{C} + \overline{B}) \qquad (11)$$

If node 2 is isolated, it does not affect any other node or branch that forms the CCR. So replacing the element *(2,2)* in the matrix by F_2 in (11) prevents the singularity in the matrix, and we forcibly set node 2 to 0v if it is isolated. Also, all the elements in the matrix ***G*** corresponding to the currents are replaced by respective Boolean functions. The complete set of changes to be incorporated into the ***G*** matrix of equation (6) is shown in Table 1.

Table 1. Complete set of changes in matrix G of (6).

Loc	Old	New	Loc	Old	New
(1,4)	1	$\overline{A}$	(3.6)	-1	$-(\overline{A}\,\overline{C} + \overline{B}\,\overline{C})$
(1,5)	1	$\overline{B}$	(4,1)	-1	$-\overline{A}$
(2,2)	0	$!(\overline{A}\,\overline{C} + \overline{B})$	(4,3)	1	$\overline{A}$
(2,5)	-1	$-\overline{B}$	(5,1)	-1	$-\overline{B}$
(2,6)	1	$(\overline{A}\,\overline{C} + \overline{B}\,\overline{C})$	(5,2)	1	$\overline{B}$
(3,3)	0	$!(\overline{A} + \overline{B}\,\overline{C})$	(6,2)	-1	$-\overline{C}$
(3,4)	-1	$-\overline{A}$	(6,3)	1	$\overline{C}$

3. MTDDDs for Boolean-ized Moments

The Boolean-ized moment equations thus obtained are now converted to MTDDD structures according to the algorithm presented in Fig. 4. The MTDDD operations *cofactor*, *union, multiply* and *getvertex* described in [5][7], are modified to incorporate the simplification due to the presence of Boolean and numeric elements. The algebraic and Boolean symbols occupy places closer to the root while the numeric elements are pushed down to the leaf terminals. The procedure *getvertex(top,D_1,D_2)* generates an MTDDD vertex for the element *top* with the sub-graphs rooted at D_1 and D_2 as its 1-child and 0-child respectively. The procedure *cofactor(G-{j,k})* returns an MTDDD vertex representing the cofactor of the matrix with respect to the element at *{j,k}*. The *multiply(top,P)* operation returns an MTDDD vertex corresponding to the multiplication of the MTDDD *P* with the numeric element *top*. The MTDDD obtained is similar to the MTDDD *P* except that the terminal vertices are the product of the terminal values of *P* with the value of *top*.

```
Create_mtddd
    List_boolean_conditions
    for (i = 1 to MAX_MOMENT_ORDER)
        moment[i,j] = NULL
        for (j: capacitive node)
            for (k: capacitive node)
                P = cofactor(G - {j,k})
                if (C(k) is SYMBOLIC)
                    Q = getvertex(C(k),P,0-terminal)
                else
                    Q = multiply(C(k),P)
        moment[i,j]=union(moment[i,j],Q)
```

Figure 4: Algorithm for setting up MTDDD for moments.

Procedure *List_boolean_conditions* in *Create_mtddd* lists the Boolean functions in the modified ***G*** matrix. This is different from identification of sensitization conditions for cell level networks as loops of sensitized paths can be obtained for such transistor level networks. An efficient tree-link based implementation for identifying the Boolean conditions is presented in Fig. 5. First, the Boolean conditions along a spanning tree of the switched-resistor network are enumerated. Sensitization through the link branches are considered next and the corresponding Boolean function at each node and branch of the switched-resistor network are updated. For each switch-resistor link branch, an update in the sensitization function at either of its two incident nodes **i,j** triggers further enumeration. In case of an update in the function at node **i**, the branches in the tree connected to the node **j** are updated with the new function at node **i**. The Boolean functions are compactly stored as MTDDD trees.

```
List_boolean_conditions
    create_spanning_tree
    create_list_of_links
    List_boolean_along_tree
    for (; ;)
        foreach branch in list_of_links
            L = branch . get_node1_function()
            R = branch . get_node2_function()
            update_tree (node1, R, branch)
            update_tree (node2, L, branch)
        if (no_update_in_one_pass)
            break

update_tree (node N, function F, branch B )
    if (loop_formed_by_node N)
        return
    G = update_branch_function ( B , F)
    foreach ( tree branch X at node N)
        nextnode = X. get_2nd_node()
        if ( ! X . updated() )
            update_tree ( nextnode, G, X)
    return
```

Figure 5: Algorithm for enumerating Boolean sensitization.

The progression of the Boolean enumeration algorithm for the example circuit graph of Fig. 6 is presented in Table 2. In Fig. 6, the solid and the dashed lines represent the tree and link branches, respectively. Each branch has a Boolean variable associated with it. Node 1 is the start node and node 3 the sink. First, a traversal through the spanning tree from the source to the sink sets up the Boolean conditions at every node (Table 2, column 2). The remaining columns represent the additional Boolean functions at the nodes due to each link branch.

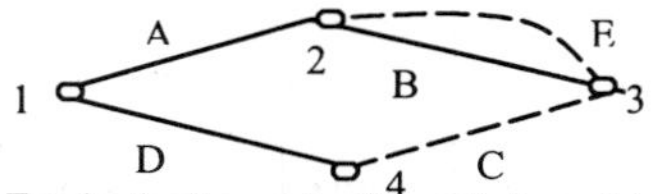

Figure 6: Equivalent graph of a switch-resistor circuit.

Table 2. Progression of Boolean enumeration algorithm.

Node #	Tree Setup	Due to C	Due to E	Due to C
2	A	DCB	DCE	-
3	AB	DC	AE	-
4	D	ABC	-	AEC

Once the MTDDDs for moments are set up, the evaluation of moments involves simple traversals through the

MTDDD tree. The sharing of sub-graphs in the MTDDD trees enables efficient calculation of moments, as sub-graphs evaluated once need not be traversed again during the entire moment calculation process. The generation of expression for the moment at any node in the circuit requires a traversal through the MTDDD tree.

Figure 7 shows a nand2 cell, its equivalent circuit and a part of the corresponding MTDDD for the first moment at node 1.

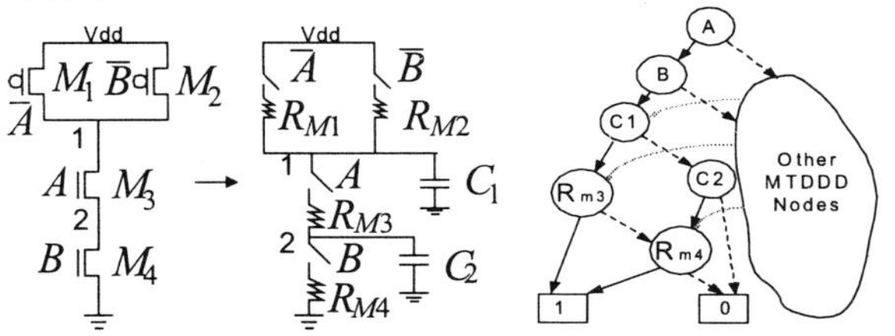

Figure 7: A nand2 gate, its equivalent circuit and MTDDD.

4. Experimental Results

The proposed symbolic procedure has been implemented in the program SAMBA (Symbolic Analysis Mixing Boolean and Algebra). SAMBA generates a single completely symbolic expression for delay under any input pattern. For the nand2 cell of Fig. 7, the 1st moment at the output node under any input logic pattern is given as

$$m_1(1) = n / d$$

$$d = (1-A)BR_{M2}R_{M3} + A(1-B)R_{M1}R_{M4} + ABR_{M1}R_{M2} + (1-A)(1-B)(R_{M1}R_{M3}R_{M4} + R_{M2}R_{M3}R_{M4})$$

$$n = (1-A)BR_{M1}R_{M2}R_{M3}C_1 + (1-A)(1-B)R_{M1}R_{M2}R_{M3}R_{M4}C_1 + A(1-B)(R_{M1}R_{M2}R_{M4}C_1 + R_{M1}R_{M2}R_{M4}C_2) + AB(R_{M1}R_{M2}R_{M4}C_1 + R_{M1}R_{M2}R_{M4}C_2 + R_{M1}R_{M2}R_{M3}C_1)$$

Semi-symbolic expressions with most transistor parameters as numeric and only a few as symbolic can also be obtained.

The computation of moments involves library lookup for the equivalent resistance of transistors. Our library generation scheme [9] uses a novel technique that incorporates effects of circuit topology, input signal slope and events at the input in addition to transistor sizes and loads for the resistance computation and results in excellent accuracy.

We apply our method of delay computation on a standard cell library in TSMC 0.18 micron technology. Experiments are conducted with different p-transistor and n-transistor sizes, loads, input patterns and input signal slopes. The maximum error is less than 5% when the estimated delays are compared to HSPICE. We present the results for a 6-input AOI321 cell and an RC interconnect tree driven by it in Fig. 8. The estimated delay in SAMBA is compared with HSPICE under parameter variation. The results show excellent match with HSPICE values.

5. Conclusion

A symbolic approach for the accurate estimation of delay in a logic stage followed by interconnect tree is presented. Our approach leverages MTDDDs to enable the efficient and compact symbolic representation of delay. Under this scheme, the equivalent resistor and capacitor of transistors, resistors and capacitors of interconnect tree and input logic assignments are all symbolic. For each stage, we provide a single delay expression that holds for all possible input assignments as well as all possible sizes of the transistors. This technique has potential application in behavioral simulation, timing analysis, variational analysis and concurrent transistor and interconnect sizing.

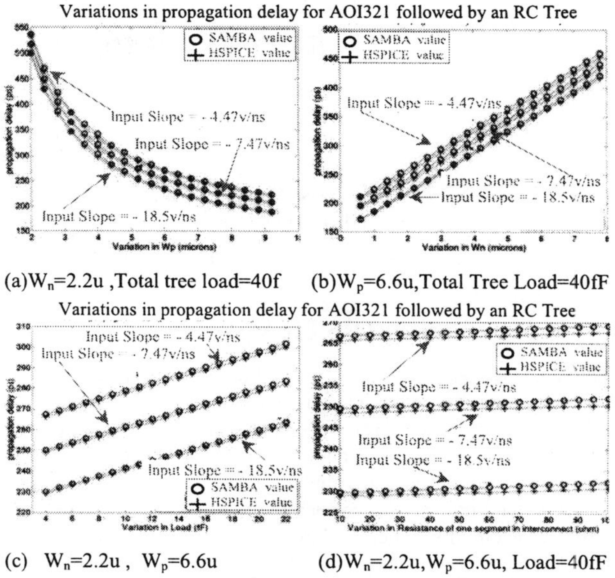

Figure 8: Rise delay variation in SAMBA and HSPICE in AOI321 cell followed by an RC Tree. The sensitized CCR consists of 3 PMOS in series and 3 NMOS. (a) Wp variation (b) Wn variation, (c) Load variation at interconnect sink (d) interconnect Resistance variation. Test results shown for 3 different input signal slopes.

6. References

[1]. C. J. Alpert, A. Devgan and C. V. Kashyap, "RC delay metrics for performance optimization", *IEEE Trans. Computer-Aided Design*, vol. 20, pp. 571-582, May, 2001.

[2]. W. C. Elmore, "The transient response of damped linear networks", *J. Appl. Phys*., vol. 19, pp. 55-63, Jan. 1948.

[3]. C. B. McDonald and R. E. Bryant, "Computing logic-stage delays using circuit simulation and symbolic elmore analysis", *Proc. 38th IEEE/ACM Design Automation Conf., Las Vegas, NV*, June 2001, pp. 283-288.

[4]. M. P. Desai and Y. T. Yen, "A systematic technique for verifying critical path delays in a 300MHz alpha CPU design using circuit simulation", *Proc. 33rd IEEE/ACM Design Automation Conf., Las Vegas, NV, June 1993, pp. 125-130.*

[5]. T. Pi and C-J. R. Shi, "Multi-terminal determinant decision diagrams: A new approach to semi-symbolic analysis of analog integrated circuits", *Proc. 37th IEEE/ACM Design Automation Conf., Los Angeles, CA*, June 2000, pp. 19-22.

[6]. L. T. Pillage and R. A. Rohrer, "Asymptotic waveform evaluation for timing analysis", *IEEE Trans. Computer-Aided Design*, vol. 9, pp. 352-366, Apr. 1990

[7]. C-J. R. Shi and X-D. Tan, "Canonical symbolic analysis of large analog circuits with determinant decision diagrams", *IEEE Trans. Computer-Aided Design*, vol. 19, pp. 1-18, Jan. 2000.

[8]. J. Vlach and K. Singhal, *Computer methods for circuit analysis & design*, New York: Van Nostrand Reinhold, 1983.

[9]. S. Bhattacharya and C-J. R. Shi, "A Table Lookup Method for Effective Resistance Estimation for Digital VLSI Delay Calculation", Department of Electrical Engineering, University of Washington, Technical Report TR-12, Oct. 2002.

AN EFFICIENT SYLVESTER EQUATION SOLVER FOR TIME DOMAIN CIRCUIT SIMULATION BY WAVELET COLLOCATION METHOD

Xuan Zeng[1], Sheng Huang[1], Yangfeng Su[2] and Dian Zhou[3]

ASIC & System State Key Lab, Microelectronics Dept., Fudan University, Shanghai 200433, P. R. China[1]
Mathematics Dept., Fudan University, Shanghai 200433, P. R. China[2]
ECE Dept., University of Texas at Dallas, Richardson, TX75073, USA[3]

ABSTRACT

The Fast Wavelet Collocation Method (FWCM) is a recently proposed circuit simulation approach, which is very promising in achieving uniform error distribution and handling with singularities in high-speed circuits, compared with the conventional time domain simulation methods. In this paper, we aim to extend the wavelet collocation method to the simulation of large-scale circuits, where a large dimension Sylvester equation needs to be solved. The performance of the wavelet simulator is significantly dominated by the time and memory consumption of the Sylvester equation solver. In order to improve the simulation efficiency, we propose a direct method to solve the Sylvester equation by Schur decomposition. Numerical experiments have demonstrated that the proposed simulator can achieve higher computation speed and higher simulation accuracy as well as more robust convergence than SPICE.

1. INTRODUCTION

With the process technology of VLSI stepping into the very deep sub-micron range, the circuit scale has becomes prohibitively large and the circuit speed has gone into GHz range. The fast changing of signals in high-speed circuits implies severe singularities [1,5]. Efficiently and accurately handling the singularities and the problem of non-uniform error distribution have challenged today's circuit simulation techniques, like time domain methods [2] and frequency domain methods [3], etc. As pointed in [1,5], to catch up with the fast changing waveforms, time marching simulators like SPICE need to control the time step small enough to avoid error accumulation and ensure the stability of the numerical integration [2], thus slowing down the simulation. The frequency domain moment-matching approaches, like AWE [3] and PVL [4] have achieved spectacular success in solving many practical VLSI interconnect problems. Unfortunately, the errors in frequency domain may get amplified during the inverse transform from the frequency domain back to the time domain. This has been a key drawback for the frequency domain methods [1].

Recently, the Fast Wavelet Collocation Method [1] (FWCM) has been proposed for high-speed circuit simulation. It can effectively treat the time domain singularities [1,5] and achieve uniform error distribution on the whole approximation region in time domain. Moreover, it has an $O(h^3)$ convergence rate, where h is the step length. However, when applying the wavelet method for large-scale circuit simulation, we encounter with solving a Sylvester equation, which becomes the bottleneck of the simulation speed of FWCM. So far, how to efficiently solve the large-scale Sylvester equation and make the FWCM algorithm fast enough for practical circuit simulation has not been explored in the published literatures.

In this paper, we develop a direct method based on Schur decomposition to solve a matrix equation rather than solving a vector equation. This method works efficiently in the time domain to deal with large-scale circuits and in the meantime obtain high simulation accuracy.

The rest of the paper is organized as follows. In Section II, we derive the Sylvester equation in FWCM. In section III, we propose the new algorithm to solve the Sylvester equation and analyze the algorithm complexity. Numerical experiments are presented in Section IV to demonstrate the promising features of the proposed algorithm. Conclusions are drawn in Section V.

2. SYLVESTER EQUATION IN FWCM

Without loss of generality, a linear dynamic circuit can be characterized by the following state equations:

$$\frac{dx(t)}{dt} = Ax(t) + Bu(t) \tag{1}$$

$$y(t) = Cx(t) + Du(t) \tag{2}$$

$$x(0) = x_0 \tag{3}$$

where $x(t) = [x_1(t), x_2(t) \cdots x_N(t)]^T$ is the unknown N dimensional state vector and $u(t) = [u_1(t), u_2(t) \cdots u_K(t)]^T$ is the K dimensional excitation vector. $y(t) = [y_1(t), y_2(t) \cdots y_P(t)]^T$ is the unknown P dimensional output vector. $x(0)$ is the initial condition of the circuit. A, B, C and D are the coefficient matrixes, whose dimensions are $N \times N$, $N \times K$, $P \times N$ and $P \times K$ respectively.

Equations (1) and (2) are solved by FWCM [1] as follows. At first, we map the equations (1) and (2) defined in time interval $[0,T]$ into the equations (4) and (5) defined in wavelet interval $[0,L]$ by $t = T \times l / L$.

$$\frac{d\hat{x}(l)}{dl} = \hat{A}\hat{x}(l) + \hat{B}\hat{u}(l) \tag{4}$$

$$\hat{y}(l) = C\hat{x}(l) + D\hat{u}(l) \tag{5}$$

where $\hat{x}(l) = x(T \times l / L)$, $\hat{u}(l) = u(T \times l / L)$, $\hat{y}(l) = y(T \times l / L)$, $\hat{A} = A \times T / L$, $\hat{B} = B \times T / L$.

0-7803-7761-3/03/$17.00

Suppose $H^2(I)$ is the Sobolev space, which basically contains functions with square integrable second derivatives. Given an integer $J \geq 0$ and a fixed interval $I=[0,L]$, we have the subspace $V_{bJ} \subset H^2(I)$.

$$V_{bJ} = \{\psi_1(l), \psi_2(l), \cdots, \psi_M(l)\} \tag{6}$$

where $\psi_1(l), \psi_2(l), \cdots, \psi_M(l)$ are all of the basis functions including boundary functions, scaling functions and wavelet functions defined in subspace V_{bJ} and $M = 2^J \times L + 3$.

Expanding each elements of $\hat{x}(l)$ with the wavelet basis functions in subspace V_{bJ}, we have equation (7)

$$\hat{x}(l) = H\Psi(l) \tag{7}$$

where $\hat{x}(l) = \begin{bmatrix} \hat{x}_1(l) \\ \hat{x}_2(l) \\ \vdots \\ \hat{x}_N(l) \end{bmatrix}$ and $H = \begin{bmatrix} h_1^1 & h_2^1 & \cdots & h_M^1 \\ h_1^2 & h_2^2 & \cdots & h_M^2 \\ \vdots & \vdots & \ddots & \vdots \\ h_1^N & h_2^N & \cdots & h_M^N \end{bmatrix}$ is the $M \times N$ unknown coefficients matrix. Also $\hat{x}_i(l) = x_i(T \times l / L)$ ($i = 1,2 \ldots N$) and $\Psi(l) = [\psi_1(l), \psi_2(l), \cdots, \psi_M(l)]^T$. If matrix H is solved, the solutions of each state variable are also obtained.

To solve H, substituting $\hat{x}(l)$ in equation (4) with equation (7), we get equation (8).

$$H\frac{d\Psi(l)}{dl} = \hat{A}H\Psi(l) + \hat{B}\hat{u}(l) \tag{8}$$

Discretizing equation (8) with a number of M collocation points [1] in subspace V_{bJ}, we derive equation (9).

$$H\frac{d[\Psi(l_1) \quad \Psi(l_2) \quad \cdots \quad \Psi(l_M)]}{dl} = \hat{A}H[\Psi(l_1) \quad \cdots \quad \Psi(l_M)] + \hat{B}[\hat{u}(l_1) \quad \cdots \quad \hat{u}(l_M)] \tag{9}$$

where l_i ($i = 1,2 \ldots M$) are the collocation points in subspace V_{bJ}.

Denote $\Phi = [\Psi(l_1) \quad \Psi(l_2) \quad \cdots \quad \Psi(l_M)]$ and $U = [u(l_1) \quad u(l_2) \quad \cdots \quad u(l_M)]$. Note that Φ is an invertible $M \times M$ square matrix with M wavelet base function values at M collocation points. Further denote $\hat{Z} = \frac{d\Phi}{dl}\Phi^{-1}$ and $\overline{B} = \hat{B}U\Phi^{-1}$, we finally derive the Sylvester equation [6] in (10) from equation (9).

$$H\hat{Z} - \hat{A}H = \overline{B} \tag{10}$$

3. SYLVESTER EQUATION SOLVER

The speed of the circuit simulator FWCM is determined by the efficiency of the Sylvester equation solver. When the circuit scale is very large, the dimension of matrix $\hat{A}$ will also become very large. An accurate and faster solver for the large-scale matrix equation (10) is crucial for the practical application of the wavelet collocation method to large-scale circuit simulation.

A general direct method for solving equation (10) is to transform it into a vector equation [7] in (11).

$$(-I_N \otimes \hat{A} + \hat{Z}^T \otimes I_M)vec(H) = vec(\overline{B}) \tag{11}$$

where the problem unknowns $vec(H)$ is a vector with length $M \times N$. Then the vector equation (11) is solved by direct LU decomposition. As will be demonstrated in Section IV, this method is very time consuming and not applicable to large-scale circuit simulation.

3.1 The Schur Decomposition Based Algorithm

In the following we present another direct method [6], which is more efficient and is based on the Schur reduction of a matrix to triangular form by orthogonal similarity transformations. Equation (10) is solved as follows. The matrices $\hat{A}$ and $\hat{Z}$ are reduced to lower real Schur forms $\overline{A}$ and $\overline{Z}$ by orthogonal similarity transformation of matrices P and Q respectively.

$$\overline{A} = P^T\hat{A}P = \begin{bmatrix} \overline{A}_{11} & & 0 \\ \vdots & \ddots & \\ \overline{A}_{\overline{N}1} & \cdots & \overline{A}_{\overline{NN}} \end{bmatrix}, \quad \overline{A}_{ii} \in R^{1\times1} \ or \ R^{2\times2},$$

$$\overline{Z} = Q^T\hat{Z}Q = \begin{bmatrix} \overline{Z}_{11} & \cdots & \overline{Z}_{1\overline{M}} \\ & \ddots & \vdots \\ 0 & & \overline{Z}_{\overline{MM}} \end{bmatrix}, \quad \overline{Z}_{ii} \in R^{1\times1} \ or \ R^{2\times2}$$

Matrices $\hat{A}$ and $\hat{Z}$ can be expressed by

$$\hat{A} = P \cdot \overline{A} \cdot P^{-1} \tag{12}$$

$$\hat{Z} = Q \cdot \overline{Z} \cdot Q^{-1} \tag{13}$$

where $P^T = P^{-1}$ and $Q^T = Q^{-1}$

Equation (14) is obtained after substituting equation (10) with equations (12) and (13).

$$HQ\overline{Z}Q^{-1} - P\overline{A}P^{-1}H = \overline{B} \tag{14}$$

By left multiplying P^{-1} and right multiplying Q to equation (14), we obtain equation (15).

$$P^{-1}HQ\overline{Z} - \overline{A}P^{-1}HQ = P^{-1}\overline{B}Q \tag{15}$$

Let $\tilde{B} = P^{-1}\overline{B}Q = \begin{bmatrix} \overline{B}_{11} & & \overline{B}_{1\overline{M}} \\ \vdots & \ddots & \\ \overline{B}_{\overline{N}1} & \cdots & \overline{B}_{\overline{NM}} \end{bmatrix}$ and $\hat{H} = P^{-1}HQ = \begin{bmatrix} \hat{H}_{11} & & \hat{H}_{1\overline{M}} \\ \vdots & \ddots & \\ \hat{H}_{\overline{N}1} & \cdots & \hat{H}_{\overline{NM}} \end{bmatrix}$, equation (15) is equivalent to equation (16).

$$\hat{H}\overline{Z}-\overline{A}\hat{H}=\widetilde{B} \quad (16)$$

If the partitions are conformal, then

$$\hat{H}_{ij}\overline{Z}_{jj}-\overline{A}_{ii}\hat{H}_{ij}=\widetilde{B}_{ij}+\sum_{k=1}^{i-1}\overline{A}_{ik}\hat{H}_{kj}-\sum_{k=1}^{j-1}\hat{H}_{ik}\overline{Z}_{kj}$$

$$(i=1,2,...\overline{N}; j=1,2,...\overline{M})$$

These equations can be solved successively for $\hat{H}_{11},\hat{H}_{21},...\hat{H}_{N1},\hat{H}_{12},\hat{H}_{22},...$. After $\hat{H}$ is solved, we can get H easily by equation (17).

$$H=P\hat{H}Q^{-1} \quad (17)$$

In the following, we show how to solve $\hat{h}_{ij}$, supposing $\hat{h}_{lm}$ ($l<i$ and $m<j$) have been solved.

1) if $\overline{a}_{i(i+1)}=0$ and $\overline{z}_{(j+1)j}=0$, $\hat{h}_{ij}$ is solved by

$$\hat{h}_{ij}\overline{z}_{jj}-\overline{a}_{ii}\hat{h}_{ij}=\widetilde{b}_{ij}+\sum_{k=1}^{i-1}\overline{a}_{ik}\hat{h}_{kj}-\sum_{k=1}^{j-1}\hat{h}_{ik}\overline{z}_{kj}$$

2) if $\overline{a}_{i(i+1)}=0$ and $\overline{z}_{(j+1)j}\neq 0$, $\hat{h}_{ij}$ and $\hat{h}_{i(j+1)}$ must be solved together by:

$$\hat{h}_{ij}\overline{z}_{jj}+\hat{h}_{i(j+1)}\overline{z}_{(j+1)j}-\overline{a}_{ii}\hat{h}_{ij}=\widetilde{b}_{ij}+\sum_{k=1}^{i-1}\overline{a}_{ik}\hat{h}_{kj}-\sum_{k=1}^{j-1}\hat{h}_{ik}\overline{z}_{kj}$$

$$\hat{h}_{ij}\overline{z}_{j(j+1)}+\hat{h}_{i(j+1)}\overline{z}_{(j+1)(j+1)}-\overline{a}_{ii}\hat{h}_{i(j+1)}$$
$$=\widetilde{b}_{i(j+1)}+\sum_{k=1}^{i-1}\overline{a}_{ik}\hat{h}_{k(j+1)}-\sum_{k=1}^{j-1}\hat{h}_{ik}\overline{z}_{k(j+1)}$$

3) if $\overline{a}_{i(i+1)}\neq 0$ and $\overline{z}_{(j+1)j}=0$, $\hat{h}_{ij}$ and $\hat{h}_{(i+1)j}$ must be solved together by

$$\hat{h}_{ij}\overline{z}_{jj}-\overline{a}_{ii}\hat{h}_{ij}-\overline{a}_{i(i+1)}\hat{h}_{(i+1)j}$$
$$=\widetilde{b}_{ij}+\sum_{k=1}^{i-1}\overline{a}_{ik}\hat{h}_{kj}-\sum_{k=1}^{j-1}\hat{h}_{ik}\overline{z}_{kj}$$

$$\hat{h}_{(i+1)j}\overline{z}_{jj}-\overline{a}_{(i+1)i}\hat{h}_{ij}-\overline{a}_{(i+1)(i+1)}\hat{h}_{(i+1)j}$$
$$=\widetilde{b}_{(i+1)j}-\sum_{k=1}^{j-1}\hat{h}_{(i+1)k}\overline{z}_{kj}+\sum_{k=1}^{i-1}\overline{a}_{(i+1)k}\hat{h}_{kj}$$

4) if $\overline{a}_{i(i+1)}\neq 0$ and $\overline{z}_{(j+1)j}\neq 0$, $\hat{h}_{ij}$, $\hat{h}_{(i+1)j}$, $\hat{h}_{i(j+1)}$ and $\hat{h}_{(i+1)(j+1)}$ must be solved together by:

$$\hat{h}_{ij}\overline{z}_{jj}+\hat{h}_{i(j+1)}\overline{z}_{(j+1)j}-\overline{a}_{ii}\hat{h}_{ij}-\overline{a}_{i(i+1)}\hat{h}_{(i+1)j}$$
$$=\widetilde{b}_{ij}+\sum_{k=1}^{i-1}\overline{a}_{ik}\hat{h}_{kj}-\sum_{k=1}^{j-1}\hat{h}_{ik}\overline{z}_{kj}$$

$$\hat{h}_{(i+1)j}\overline{z}_{jj}+\hat{h}_{(i+1)(j+1)}\overline{z}_{(j+1)j}-\overline{a}_{(i+1)i}\hat{h}_{ij}-\overline{a}_{(i+1)(i+1)}\hat{h}_{(i+1)j}$$
$$=\widetilde{b}_{(i+1)j}+\sum_{k=1}^{i-1}\overline{a}_{(i+1)k}\hat{h}_{kj}-\sum_{k=1}^{j-1}\hat{h}_{(i+1)k}\overline{z}_{kj}$$

$$\hat{h}_{ij}\overline{z}_{j(j+1)}+\hat{h}_{i(j+1)}\overline{z}_{(j+1)(j+1)}-\overline{a}_{ii}\hat{h}_{i(j+1)}-\overline{a}_{i(i+1)}\hat{h}_{(i+1)(j+1)}$$
$$=\widetilde{b}_{i(j+1)}+\sum_{k=1}^{i-1}\overline{a}_{ik}\hat{h}_{k(j+1)}-\sum_{k=1}^{j-1}\hat{h}_{ik}\overline{z}_{k(j+1)}$$

$$\hat{h}_{(i+1)j}\overline{z}_{j(j+1)}+\hat{h}_{(i+1)(j+1)}\overline{z}_{(j+1)(j+1)}-\overline{a}_{(i+1)i}\hat{h}_{i(j+1)}-\overline{a}_{(i+1)(i+1)}\hat{h}_{(i+1)(j+1)}$$
$$=\widetilde{b}_{(i+1)(j+1)}+\sum_{k=1}^{i-1}\overline{a}_{(i+1)k}\hat{h}_{k(j+1)}-\sum_{k=1}^{j-1}\hat{h}_{(i+1)k}\overline{z}_{k(j+1)}$$

w

here $\hat{h}_{ij},\overline{z}_{ij},\overline{a}_{ij}$ and $\widetilde{b}_{ij}$ are the elements of matrix $\hat{H},\overline{Z},\overline{A}$ and $\widetilde{B}$ respectively on the i th row and the j th column.

3.2 Complexity Analysis

There are number of $M\times N$ unknowns in the vector equation (11), where M is the number of collocation points and N is the number of state variables. The time complexity of solving vector equation (11) by direct LU decomposition is $O(M^3\times N^3)$ and the space complexity is $O(M^2\times N^2)$. For the large-scale circuit, the complexity of this method is unbearable.

Considering the matrix equation solver by Schur decomposition, the computation cost is dominated by the Schur decomposition of matrices $\hat{A}$ and Z, where the complexities are $O(\frac{14}{3}N^3)$ and $O(\frac{14}{3}M^3)$ respectively. In large-scale circuit simulations, the circuit size will be much larger than the collocation number, i.e. $N>>M$. In this circumstance, the dimension of Φ or Z is much smaller than $\hat{A}$. Therefore the calculation of Φ^{-1} and the Schur decomposition of Z will only take much less time than the Schur decomposition of $\hat{A}$. As a result, the time complexity is dominated by $O(N^3)$ and the space complexity is dominated by $O(N^2)$ in large circuit simulation.

Another advantage of the proposed method is that, when applying the adaptive scheme of wavelet method, the Schur decomposition of $\hat{A}$ only needs to be performed for one time. So the adaptive scheme [5] can also be used to accelerate the simulation efficiently.

4. NUMERICAL EXPERIMENTS

In this section, a distributed interconnect network which models ten parallel lines (eight buses with power and ground) are tested to demonstrate the effectiveness of our method. Each line is modeled by a number of 32 RLC segments. There are coupling capacitance and mutual inductance between any two parallel segments in any different lines. The original system has 2152 elements and 585 state variables. We test this circuit with different RLC segments in order to compare the efficiency of each method for different circuit size. The circuit is excited at the driving end of one bus line by a step input. The signal received at the output end of this bus line is tested in a simulation interval of one nanosecond. The program runs on a PIII 933 PC with 512M memories.

Listed in Table 1 are the comparison results of three different simulation methods, i.e., direct LU decomposition of vector equation (Method I in Table 1), Schur decomposition of matrix equation (Method II in Table 1) and SPICE. The simulation error denotes the relative mean square error compared with the ideal solution by Matlab step response function for LTI systems. The symbol *** stands for the situation where the simulation cannot finish because of running out of memory.

Table 1. Simulation results of wavelet method by direct LU decomposition of vector equation (Method I), wavelet method by Schur decomposition of matrix equation (Method II) and SPICE.

Circuit Size	State Variables	Methods	Simulation Time (s)	Simulation Error
332	81	Method I	344.7	4.358936e-4
		Method II	6.9	2.549603e-6
		SPICE	7.1	1.250654e-6
592	153	Method I	***	***
		Method II	12.8	1.003606e-6
		SPICE	33.5	1.301111e-6
852	225	Method I	***	***
		Method II	17.2	1.161043e-6
		SPICE	42.8	1.844975e-6
1112	297	Method I	***	***
		Method II	23.2	1.649042e-6
		SPICE	62.9	2.219224e-6
1372	369	Method I	***	***
		Method II	34.8	9.299126e-7
		SPICE	182.4	1.906376e-6
1632	441	Method I	***	***
		Method II	51.3	5.322239e-7
		SPICE	115.0	1.985522e-6
1892	513	Method I	***	***
		Method II	69.5	5.003210e-7
		SPICE	176.6	2.286639e-6
2152	585	Method I	***	***
		Method II	79.5	2.605608e-6
		SPICE	No convergence	***

From Table 1, it is obvious that the efficiency of the Sylvester equation solver directly affects the simulation speed of the FWCM simulator. Method I based on the direct LU decomposition of the vector equation is really time and memory consuming and fails to deal with large-scale circuits. The matrix equation solver by Schur decomposition works successfully for large-scale circuits and the simulation speed and accuracy are much higher than those of SPICE.

In this example, the interconnect circuit presents strong singularities in time domain because of the existence of a large number of inductances and the lower resistance under the copper technology. When dealing with singularities, SPICE needs to control the time step very small to avoid error accumulation. If the smallest time step is still unable to meet the local truncation error requirement, SPICE comes with the convergence problem. In the 2152 circuit size example, SPICE fails to converge while wavelet approach copes with the singularity easily by adaptive scheme [5]. This example demonstrates that the convergence property of the wavelet method with fast Sylvester solver is also much better than SPICE.

5. CONCLUSIONS

In this paper, we propose a fast Sylvester equation solver by Schur decomposition of the matrix equation rather than directly solving the large-scale vector equation. The wavelet collocation method employing the proposed Sylvester equation solver can simulate large-scale circuits very fast. Compared with SPICE, our simulator presents higher simulation speed, higher simulation accuracy as well as more robustness of convergence when handling with the singularities in high-speed circuits.

Our current work is limited to the simulation of linear circuits. However, we'll continue to investigate if this proposed method can be employed to nonlinear circuits simulation, since the nonlinear terms can be linearized during simulation.

We are now working at iterative methods for solving the Sylvester equation, which is expected to be faster than the direct method proposed in this paper for high-speed large scare circuit simulation.

6. ACKNOWLEDGEMENTS

This research is supported by NSFC research project 60176017, NSFC key project 90207002, NSFC oversea's young scientist joint research project 69928402, the doctoral program foundation of Ministry of Education of China 2000024628, Shanghai Science and Technology committee project 01JC14014 and Shanghai AM R&D fund 0107, Science & Technology key project of Ministry of Education of China 02095, National 863 plan projects 2002AA1Z1340 and 2002AA1Z1460.

7. REFERENCES

[1] Dian Zhou and Wei Cai, "A Fast wavelet collocation method for high speed circuit simulation", *IEEE Trans.CAS-I*, 46,8, 920-930, 1999.

[2] L. O. Chua and P. M. Lin, *Computer Aided Analysis of Electronic Circuits: Algorithms, and Computational Techniques*, Englewood Cliffs, NJ: Prentice Hall, 1975.

[3] L. T. Pillage and R. A. Rohrer, "Asymptotic waveform evaluation for timing analysis," *IEEE Trans. Computer-Aided Design*, vol. 9, pp. 352-366, Apr. 1990.

[4] Peter Feldmann and Roland W. Freund, "Efficient linear analysis by Pade approximation via Lanzos process," *IEEE Trans. CAD*, vol. 14, pp. 639-649, May. 1995.

[5] Dian Zhou, Wei Cai and Wu Zhang, "An adaptive wavelet method for nonlinear circuit simulation", *IEEE trans. CAS-I*, 46,8, 931-938, 1999.

[6] R.H.Bartels and G.W.Stewart "Solution of the Matrix Equation AX+XB=C", *Comm. Assoc. Comput*, pp.820-826, March., 1972.

[7] James W. Demmel, *Applied Numerical Linear Algebra*, SIAM, 1997.

Table Look-up Based Compact Modeling for On-chip Interconnect Timing and Noise Analysis

Haitian Hu, ECE Department, University of Minnesota, Minneapolis, MN 55455
David T. Blaauw, EECS Department, University of Michigan, Ann Arbor, MI 48104
Vladimir Zolotov, Kaushik Gala, Min Zhao, Rajendran Panda, Motorola, Inc., Austin, TX 78729
Sachin S. Sapatnekar, ECE Department, University of Minnesota, Minneapolis, MN 55455

Abstract
A compact model for RLC interconnect lines, in the form of a two-path hybrid ladder, is proposed for on-chip interconnect timing and noise analysis. The model parameters are synthesized through constrained nonlinear optimization to directly match the circuit response characteristics over a range of transition times and loads, both at the driving point and at the receiver end. The effect of capacitances on the return current distribution is explicitly considered in our work in obtaining the accurate responses for industrial circuits, and is found to have a significant effect. The parameters for this model are embedded in a table that is characterized once for a design and then used for the analysis of various structured interconnects. Compared with a prior compact modeling approach, our model is demonstrated to accurately predict responses such as the interconnect delay, gate delay, transition times at near and far ends of switching lines as well as the overshoot at the far ends of switching lines.

1. Introduction

On-chip inductance is a growing issue for high-performance circuits, and technology trends indicate that these effects will grow even more prominent in the future.

A commonly used inductance model is the PEEC model [1], which represents a complex multiconductor topology without predetermined current return paths. However, it results in a dense partial inductance matrix that makes simulation computationally expensive. Although the computational cost can be greatly decreased by sparsification techniques [2-6], the PEEC model is still computationally expensive for simulating large industrial circuits. Loop inductance is an alternative way to represent on-chip inductance system [7].

In this paper, we propose a computationally efficient compact model for fast and accurate on-chip interconnect timing and noise analysis, which is valid over a range of typical transition times. The technique utilizes a table look-up of model parameters which are characterized for different layout parameters, such as signal wire width, length, spacing to the nearest power/ground wires, shielding etc.. Parameters for layouts that do not directly correspond to a table entry are interpolated. We demonstrate the viability of our approach on a clock net built to industrial specifications.

When on-chip inductance is not important, a standard model for wire segments is the RC-π model that incorporates the loop resistance, which is dominated by the resistance of the wire segment. The loop inductance, calculated as the sum of the partial self and mutual inductance along a wire and its current return paths, can be introduced into this π model by connecting it in series with the loop resistance. Signals with different transition times τ have different frequency spectrum and will experience different loop electrical characteristics. The frequency dependency of the loop resistance and loop inductance arises primarily due to the proximity effect. As demonstrated in [8], the change in the loop resistance and inductance can be very large over a wide range of frequencies. Compared to its low-frequency value, the loop inductance decreases by about 50% at high frequencies, while the loop resistance increases monotonically as the frequency increases.

An RL ladder circuit, as shown in Figure 1(a), was proposed in [9] to approximate the frequency-dependent resistance and inductance due to the proximity and skin effects. This model was further developed in [8], as shown in Figure 1(b), to synthesize a layout-based hybrid ladder circuit. A shunt circuit of R_2 and L_2 in parallel with L_0 compensates for the additional reduction of the loop inductance at high frequencies. The model is synthesized in frequency domain by fitting input impedance. Specifically, the low-frequency inductance and resistance, the high-frequency inductance and the cross-over frequency (where $R=2\pi f_c L$) are calculated using an RL-only technique. The model parameters R_0, L_0, R_1, and L_1 are then calculated by forcing the low-frequency inductance and resistance, high-frequency inductance and the crossover frequency of the model to match the calculated values above. Next, R_2 and L_2 are obtained from the resistance and inductance of parallel plate return conductors.

This procedure has two limitations: first, in order to compute the loop inductance, it ignores the effect of capacitance on the return current distribution, thereby causing errors in the estimation of the frequency-dependent resistance and inductance. Second, it models the input impedance of the interconnect at the driving point, but not the transfer characteristics.

Our work overcomes both of these limitations and presents an extension of this hybrid ladder model to a two-path hybrid ladder model, using a different characterization technique. Specifically, the parameters are determined in time domain for a wide range of transition times and loads, through a constrained nonlinear optimization to match the response characteristics of the compact model, including the interconnect delay, the gate delay, the transition times at both the near and far ends of switching lines and the overshoot at the far ends of switching lines, to the exact response of the three-dimensional circuits under a comprehensive PEEC model, which includes the mutual inductance between wires, the power grid decoupling capacitance and pad placements. Response characteristics are more directly related to the interconnect timing and noise analysis, accordingly time domain synthesis procedure gives higher accuracy. In addition, in time domain characterization capacitance effects on the estimation of current return paths could be considered. A comparison between the responses from the hybrid ladder model and the accurate response shows that hybrid ladder model could overestimate delay and overshoot by 100%. The experiments also demonstrate that the runtime of constructing circuit models through table lookup is only seconds, while formula-based extraction needs minutes or more than an hour. This table lookup based modeling is easy to be used in design process.

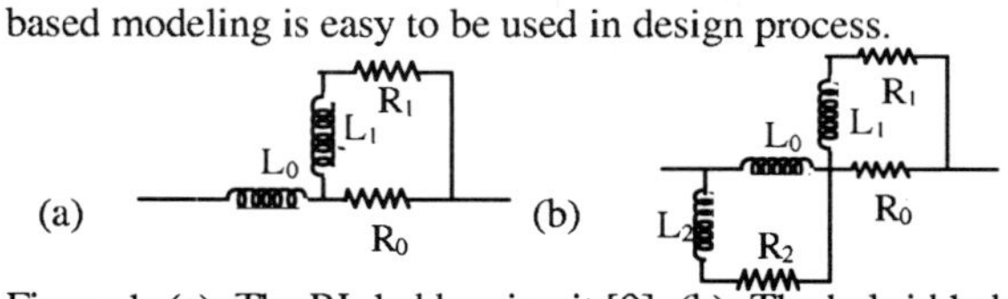

Figure 1: (a): The RL ladder circuit [9]. (b): The hybrid ladder model [8].

0-7803-7761-3/03/$17.00 ©2003 IEEE

2. Two-path hybrid ladder model

A signal making a logic transition has a spectrum of frequencies. Current components corresponding to different frequencies choose different paths through the power grid and must be modeled correctly. We propose a compact model, a two-path hybrid ladder model as shown in Figure 2 (a), that is constructed to separate the paths through which the currents for different frequency components would flow. The path with R_0, L_0, R_3 and L_3, corresponds to the current flow for low-frequency components, and that with R_1, L_1, R_2 and L_2, represents the current flow for high-frequency components. R_2 and L_2 form a branch to compensate for the change in the loop inductance at high frequencies, while R_3 and L_3 compensate for the change of the loop resistance at low frequencies.

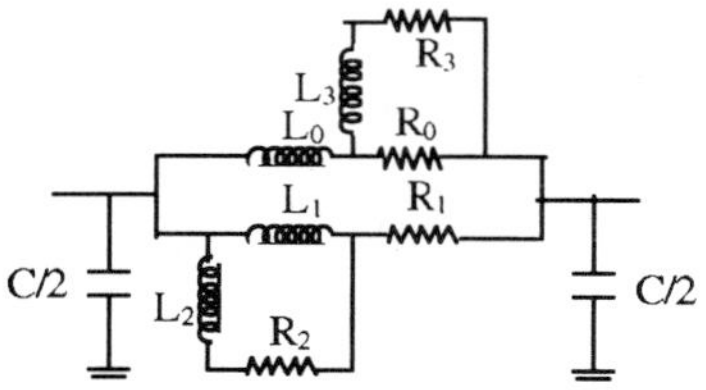

Figure 2: Two-path hybrid ladder model.

The intuition behind the approach may be explained as follows. The structure shown in Figure 1 (a) is known to be adequate for lower frequencies, and forms the upper path in Figure 2 (a). The structure in Figure 1 (b), attempts to perform high-frequency compensation through R_2 and L_2, but R_0 and L_0 are required to be involved in both the high-frequency and low-frequency behavior. We remove this constraint by creating the high-frequency path in parallel with the low-frequency path. In doing so, we use a larger number of parameters, which enables a better fit of the accurate response. At extremely high and low frequencies, both our model and hybrid ladder model behave similarly. At intermediate frequencies our approach provides a greater flexibility for a better fit.

At extremely high frequencies the two-path model is simplified to three parallel inductances L_0, L_1, and L_2, while at extremely low frequencies it is simplified to three parallel resistances R_0, R_1, and R_3, respectively. Note that the parameters R_2 and L_3 do not appear in either the high- or low-frequency reductions, so that they may be tuned to capture the circuit response at intermediate frequencies. The increased number of tunable parameters over hybrid ladder model is expected lead to a higher accuracy for a wide range of layout topologies and a range of frequencies.

3. Outline of the approach

Our approach is based on fitting parameters of our two-path hybrid ladder model by nonlinear optimization to match a set of desired characteristics of transient response from the compact model to that of the response from a comprehensive PEEC model, over the ranges of interest for the driver/receiver sizes and transition times. By performing this optimization for a variety of circuit topologies, representative structures are characterized and stored in a table. Subsequently, a compact model for a given structure may be interpreted by looking up entries in the table.

We study our approach applying to analyze the top level clock tree and signal lines routed on four metal layers, M6, M7, M8 and M9 in a nine-layer structure. The power/ground wires are distributed densely in the four layers. In order to accurately estimate the current return paths and inductance effects, a comprehensive PEEC model, described in [4], is used to determine the responses at the near and far ends of the switching wires. In addition to the interconnect net under consideration, the circuit model includes supply lines, drivers and receivers, vias, pads and intrinsic/explicit decoupling capacitances. The circuit in synthesis procedure, whose top view is shown in Figure 3, includes a switching wire with different wire lengths, widths and spacings to the nearest supply line, corresponding to different sets of model parameters in table. The layout width a chosen so that along the layout length, there are one row of pads on each side of the switching line.

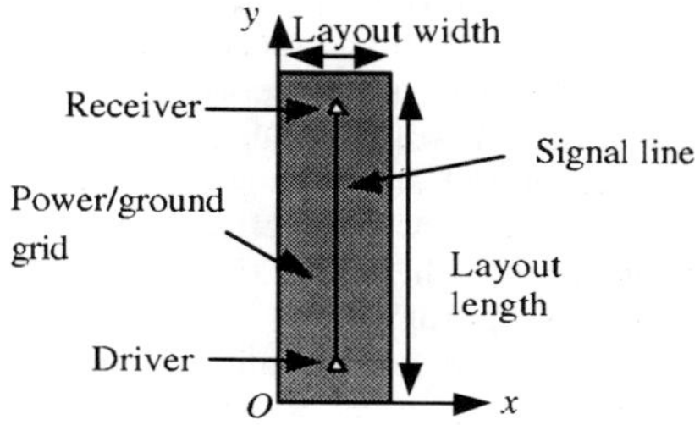

Figure 3: Top view of the layout of a four metal layer structure.

Accurate transient response of this model is computed by our precise inductance analysis tool that uses precorrected-FFT technique [10] to accurately compute inductance effects, together with PRIMA algorithm building reduced order model of the circuit which is then simulated by SPICE.

It is desirable for the compact model to be independent of the gate parameters, so that it can be utilized under any value of the loads and input transition time. The response of a switching line is impacted by the driver size, the receiver size and the transition time at the near end. However, for a given line, only any two of these three parameters are independent. Here, we choose the receiver sizes and the transition times at the near end of the switching line as the independent parameters. For a given receiver size and a given transition time at the near end, the driver size can be calculated to produce the given transition time at the near end for the given receiver size. Let the ranges of interest for the receiver sizes and transition times consist, respectively, of the sets of discrete points:

Receiver sizes: $W = \{w_1, w_2, \ldots, w_m\}$

Transition times: $S = \{s_1, s_2, \ldots, s_n\}$

Thus, the compact model is required to be accurate over all mn combinations of the above parameters.

The compact model is constructed to capture the transient response characteristics of the line. Specifically, we match five response characteristics: the gate delay (from the input to the output of driver), interconnect delay (from the output of driver to the input of receiver), transition time at the near and far ends of the switching line and the peak overshoot at the far end for a transition. For a given interconnect topology with $p=mn$ distinct combinations of receiver sizes and transient times, there will therefore be $5p$ responses which should be matched. However in our work, we choose $p=4$, which corresponds to four combinations of heaviest/lightest receiver sizes and highest/lowest transition times, using constrained nonlinear optimization. The experimental results show that matching the response characteristics for these four circuits will automatically match for the mn circuits. The formulation of the non-linear optimization problems for the set of four objective circuits with *20* response characteristics:

minimize Relative error of *20* response characteristics

subject to all model parameters ≥ 0

We transform the multi-objective optimization problem to a single objective function through a weighted minmax objective. Specifically, the error is calculated as a weighted maximum of the percentage errors in all *20* response characteristics. In practice, we choose the weights to emphasize low errors in delay and overshoot. Thus, in summary, each model synthesis procedure involves a first step of simulating the objective circuits for accurate responses, followed by a second step of fitting the parameter values to the compact model through constrained nonlinear optimization. These model parameters are then put in a table.

Since a real layout consists of a large number of parameters, such as the number of switching lines, the metal layers the switching lines are on, the width, length, spacing of these switching lines, the spacing between the switching lines to the nearest power/ground grid lines or shields, width and spacing of power/ground grid lines, the width of shields, the spacing between the shields and the nearest grid lines and the pad positions. Obviously it is impossible to build simple practical model and manageable table size to capture the effect of all these parameters, therefore we have selected through the experimentation only the most significant. The model is thus simplified by the following:

1. For now, we focus our attention to building a model for one switching line, such as a signal line or clock net. This restriction still yields useful solutions to important problems (for example, to a critical signal line, or a clock network structure). The line may be parameterized by factors such as its width, its length, its distance to the nearest supply line, whether it is shielded or not, etc.
2. For a reasonable design, even significant changes in the structure of the power grid do not noticeably influence the response characteristics of the logic. It has been demonstrated in [11] that a small deviation from the regular power/ground topology will not cause a significant change in the response characteristics. We utilize this fact to work with a representative power grid, under the assurance that to the first order, our model will remain reasonable even if the actual grid is perturbed from the assumptions under which our characterizations are performed. A high level structure for the regular power grid and regular pad locations is provided for characterization, including parameters such as the pitches and widths of power lines and pad spacings. Once the spacing from a signal line or shield to the nearest power/ground lines is given, the power/ground environment for that signal line is well determined. Our approach requires a single characterization step for each design using a model for the power grid.

With this simplified approach, each entry of the table corresponds to a set of values of the following parameters:

1. Metal layer on which the signal line lies.
2. Width of the switching lines.
3. Length of switching lines.
4. Distance to the nearest supply grid line.
5. Shielded and unshielded cases.

4. Experimental results

A multidimensional table was constructed for a 0.18μm technology, with 0.36 μm minimum line width and spacing. The transition times were measured from 10%-to-90% of V_{dd}. The ranges of the receiver sizes and transition times were set to be:

$$W = \{15, 30, 90, 150, 210, 270, 330, 390\}\ \mu m, \text{ and}$$
$$S = \{1000, 800, 600, 400, 200, 100, 80, 60\}\ ps,$$

respectively. Each switching line was modeled by a cascade of two-path hybrid ladder models. The characterization was done on four circuits with receiver sizes of *15/390* μm and transition times of *1000/60* ps. The model was tested on all 64 combinations of *W* and *S* above, and the accuracy results are summarized in Table 1. In addition to the high accuracy, two-path ladder model also gives a high speed. The simulation time for each circuit in Section 4.1 in PEEC model is 16 mins, while it is reduced to seconds with the two-path ladder model.

4.1. Accuracy of the responses for signal lines

A set of experiments is carried out to test the accuracy of the responses of signal lines. The layout structure of this set of experiments is shown in Figure 3. Experiments are performed on various lengths of signal lines: 300 μm, 600 μm, 900 μm, 1200 μm, 1500 μm and 1800 μm, corresponding to six circuits S_{300}, S_{600}, S_{900}, S_{1200}, S_{1500} and S_{1800} respectively. Each circuit is modeled by a cascade of three two-path hybrid ladder model segments.

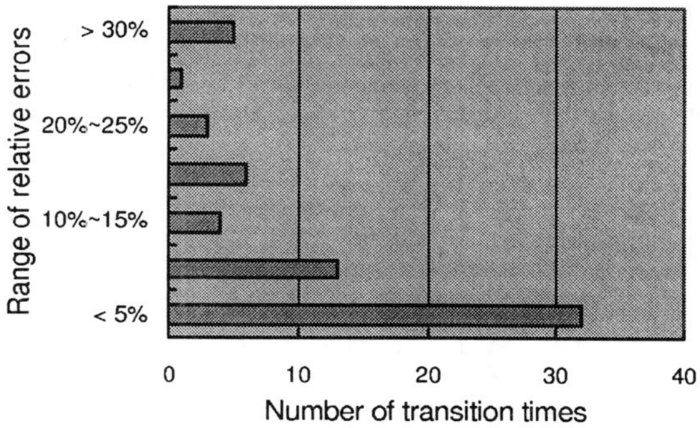

Figure 4: A histogram showing the distribution of errors in the far end transition time for the 64 combinations of *W* and *S* for circuit S_{900}. For example, the lowest bar corresponds to the fact that 32 of the 64 combinations showed errors of < 5%.

For all six circuits, all the five critical response characteristics are matched well. For example, although the maximum error for interconnect delay can reach 11% for the six circuits, the average value is between 2% and 5%. Even the worst-case errors are acceptable in some cases for the accuracy requirements in current timing analysis tools. For the circuit S_{900}, a histogram of the distribution of the transition times at the far end for the 64 possible combinations of *W* and *S* are shown in Figure 4, and are seen to be acceptable.

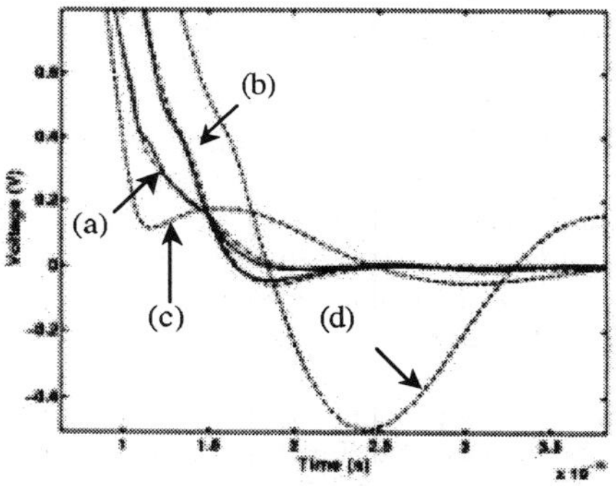

Figure 5: Comparison of the responses from the two-path ladder model, the hybrid ladder model and the PEEC model. (a) near end response under the accurate model and our two-path model (almost identical). (b) far end response under the accurate model and our two-path model (almost identical). (c) near end and (d) far end response for the hybrid ladder model.

A comparison between the accuracy of the two-path ladder model and the hybrid ladder model [8] was also carried out. Figure 5 includes the responses from both of these models, as well as the accurate responses using the full inductance matrix, for the circuit S_{900} under an 80ps input transition time to the driver and 150μm receiver size. The responses from the two-path model are almost indistinguishable from the accurate response, while the responses from hybrid ladder model are seen to overestimate the inductance effects. The interconnect delay error could reach 100%. The error in overshoot is even larger.

4.2. Accuracy of the responses for a clock net

Experiments are carried out to test the accuracy of two clock nets that have multiple switching line segments with non-uniform width, non-uniform spacing to the nearest power/ground grid lines and are on different metal layers. Circuit CLK_H, as shown in Figure 6, is a clock H-tree with one source and sixteen sinks. There are 31 line segments distributed on three metal layers M6, M7 and M8. The responses are measured at the output of the source and the input of the sink at node C, while the sizes of the other sinks are identical and are set to be 100 µm or 900 µm. The length of the path from the source to sink at C is 3900 µm with five line segments that are modeled by fifteen model segments on this path. Circuit CLK_{HBF} is an optimized version of circuit CLK_H with buffers inserted at all nodes marked *D* and *E*.

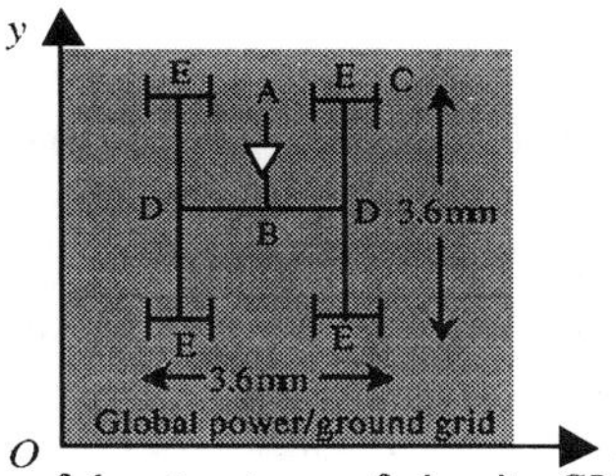

Figure 6: Top views of the structures of circuits CLK_H and CLK_{HBF}. (A: driver input, B: driver output, C: receiver input, D and E: buffer position in circuit CLK_{HBF}.)

The maximum and mean errors of timing characteristics for circuit CLK_H with 900 µm receiver sizes at all sinks except the one at C are 15% and 4%, while those for circuit CLK_{HBF} with the same sink sizes are 11% and 2% respectively. The overshoots in both the circuits are smaller than 50 mV. It is reasonable that the errors for circuit CLK_{HBF} are smaller than those for circuit CL_{KH} because the buffers are intended to reduce the inductance effects and this also has the side effect of making the modeling easier and more accurate.

5. Conclusion

A two-path ladder model for compact modeling of on-chip interconnect timing and noise analysis is proposed in this paper to accurately approximate the proximity effect in high speed circuits. The paths for both the high and low frequency currents are explicitly included in the model. The synthesis procedure uses constrained nonlinear optimization to match the response characteristics of the model to those under an accurate simulation, with comprehensive PEEC model and the effects of capacitances on the current return path estimation. A comparison with the hybrid ladder modeling shows that the proposed modeling results in more accurate responses. Extensive experiments on single signal lines and clock nets demonstrate that the proposed table look-up based compact modeling is a highly accurate and fast approach for on-chip interconnect timing and noise analysis in large circuits. In future work, we expect to extend this work to multiple-line buses.

References:

[1] A. E. Ruehli, "Inductance Calculations in a Complex Integrated Circuit Environment," *IBM Journal of Research and Development*, pp. 470-481, vol. 16, No. 5, September 1972.

[2] B. Krauter and L. T. Pileggi, "Generating Sparse Inductance Matrices with Guaranteed Stability," *Proc. of the IEEE/ACM ICCAD*, pp. 45-52, November 1995.

[3] K. Gala, D. Blaauw, J. Wang, M. Zhao and V. Zolotov, "Inductance 101: Analysis and Design," *Proc. of the ACM/IEEE DAC*, pp. 329-334, June 2001.

[4] A. Devgan, H. Ji and W. Dai, "How to Efficiently Capture On-Chip Inductance Effects: Introducing a New Circuit Element K," *Proc. of the IEEE/ACM ICCAD*, pp. 150-155, November 2000.

[5] K. L. Shepard and Z. Tan, "Return-Limited Inductances: A Practical Approach to On-Chip Inductance Extraction," *Proc. of the IEEE CICC*, pp. 453-456, May 1999.

[6] H. Hu and S. Sapatnekar, 'Efficient PEEC-Based Inductance Extraction using Circuit-Aware Techniques," *Proc. of the IEEE ICCD*, pp. 434-439, September 2002.

[7] M. Kamon, M. J. Tsuk and J. White, "FastHenry: A Multipole-Accelerated 3-D Inductance Extraction Program," *Proc. of the ACM/IEEE DAC*, pp. 678-683, June 1993.

[8] B. Krauter and S. Mehrotra, "Layout Based Frequency Dependent Inductance and Resistance Extraction for On-chip Interconnect Timing Analysis," *Proc. of the ACM/IEEE DAC*, pp. 303-308, June 1998.

[9] H. A. Wheeler, "Formulas for the Skin-Effect," *Proc. of the Institute of Radio Engineers*, pp. 412-424, vol. 30, September 1942.

[10] H. Hu, D. Blaauw, V. Zolotov, K. Gala, M. Zhao, R. Panda and S. Sapatnekar, "A Precorrected-FFT Method for Simulating On-chip Inductance," accepted by *the IEEE/ACM ICCAD*, November 2002.

[11] K. Gala, V. Zolotov, R. Panda, B. Young, J. Wang and D. Blaauw, "On-chip Inductance Modeling and Analysis," *Proc. Of the ACM/IEEE DAC*, pp.63-68, June 2000.

Table 1: Mean and maximum relative errors for all the response characteristics in a set of test circuits.

Circuit	Relative error of Interconnect delay		Relative error of Gate delay		Relative error of Transition time at the far end		Relative error of Transition time at the near end		Relative error of >50mV overshoot at the far end	
	mean	max	mean	max	mean	max	mean	max	mean	max
S_{300}	2.5%	9.3%	1.2%	3.4%	2.3%	8.1%	2.0%	5.0%	15%	30%
S_{600}	3.5%	10%	0.7%	1.3%	2.6%	7.2%	2.5%	7.0%	10%	21%
S_{900}	2.3%	10%	0.9%	1.3%	3.3%	6.9%	3.5%	6.9%	12%	25%
S_{1200}	3.4%	11%	1.6%	2.6%	4.0%	8.5%	2.3%	7.2%	11%	35%
S_{1500}	4.0%	9.6%	1.1%	3.1%	3.7%	7.9%	3.6%	6.3%	10%	20%
S_{1800}	4.2%	8.6%	1.5%	3.6%	5.0%	10%	3.4%	8.0%	11%	30%
CLK_H	4.1%	5.6%	0.7%	2.6%	2.1%	5.7%	4.0%	15%	-	-
CLK_{HBF}	1.2%	4.0%	1.2%	4.2%	2.5%	11%	3.3%	5.8%	-	-

"-" implies that there was no overshoot for this case

Process Variation Dimension Reduction Based on SVD

Zhuo Li, Xiang Lu and Weiping Shi

Department of Electrical Engineering
Texas A&M University
College Station, Texas 77843-3128
{zhuoli, xiang, wshi}@ee.tamu.edu

ABSTRACT

We propose an algorithm based on singular value decomposition (SVD) to reduce the number of process variation variables. With few process variation variables, fault simulation and timing analysis under process variation can be performed efficiently. Our algorithm reduces the number of process variation variables while preserving the delay function with respect to process variation. Compared with the principal component analysis (PCA) method, our algorithm requires less computation time and guarantees the reduced process variation variables are independent. Experimental results on ISCAS85 circuits show that the algorithm works well.

1 INTROCUTION

In modern multi-layer VLSI circuits, the dimension of the process variation is large. For example, in a *k*-metal technology, there are *2k* variables for metal width and thickness, *k* variables for inter-layer dielectric (ILD) thickness, a gate length variable, and some other variables. With such a large number of process variation variables, timing analysis under process variation is very time consuming, and may not find the worst-case process corner [1][2].

In this paper, we propose the novel concept of variation dimension reduction. The idea is originated from the following observation. For different nets, the underlying process variation variables are the same. Therefore, the delays of the nets are correlated. We model the delay from an input pin of a cell to an input pin of a downstream cell, also known as the buffer-to-buffer delay, as a linear function of process variation variables:

$$d = d_{nominal}(1 + a_1x_1 + a_2x_2 + \dots + a_nx_n),$$

where $x_1, x_2, \dots, x_n$ are process variation variables, $a_1, a_2, \dots, a_n$ are their corresponding coefficients, and $d_{nominal}$ is the nominal delay. For different nets, the coefficients are different, but the process variation variables are the same. The assumption that the delay is linear with respect to each process variation variable is supported by SPICE simulation of real circuits within reasonable ranges of the process variation [3].

There are many forms of process variation [4][5]. In this paper, we only consider systematic process variation and assume the process variation variables among different nets are fully correlated. We assume the following variation ranges: metal thickness ±20%, metal width ±10%, ILD thickness ±40%, and gate length ±10%. To obtain the delay function with respect to process variation, we first extract interconnect parasitic from the layout and compute the nominal buffer-to-buffer delay for all nets. Then for each process variation variable, we modify the layout or the technology file by varying a specified amount (for example, +10% for metal1 thickness), and perform RC extraction and delay evaluation for the modified layout. The coefficient for this process variation variable is obtained by interpolating the two delays.

Recently, the principal component analysis (PCA) method was used to reduce the dimension of process variation [7]. It was shown to be effective for analyzing analog circuits under process variation. However, the PCA method is inefficient for digital circuits, because the PCA method constructs and solves a correlation matrix of size $mn \times mn$, where n is the number of process variation variables and m is the number of nets or devices. For digital circuits, m is easily in hundreds or thousands.

In this paper, we propose a new method based on singular value decomposition (SVD) [6]. Our algorithm detects the linear dependence and reduces the number of process variation variables, while minimizing the average error of the delay. Note that although the PCA method also uses SVD, their matrix is size $mn \times mn$ while ours is size $m \times n$. To make the reduced process variables linearly/statistically independent, which is desirable for fault simulation and timing analysis, we also propose a sub-matrix SVD method.

In Section 2, the SVD method and sub-matrix SVD method for dimension reduction are given. In Section 3, we show the experimental results. It shows that if we simply pick the most important process variation variables and discard the

0-7803-7761-3/03/$17.00 ©2003 IEEE

rest, then the average error of net delay is twice as much as the average error produced by our algorithm.

2 ALGORITHM

2.1 Singular Value Decomposition (SVD)

The SVD method is based on the following theorems [6].

Theorem 1 If A is an $m \times n$ matrix, then there exist orthogonal $m \times m$ matrix $U=[u_1,\ldots,u_m]$ and $n \times n$ matrix $V=[v_1,\ldots,v_n]$, such that $U^T \cdot A \cdot V = diag(\sigma_1,\ldots,\sigma_p)$, which is an $m \times n$ matrix, $p=\min\{m,n\}$, and $\sigma_1 \geq \sigma_2 \geq \ldots \geq \sigma_p \geq 0$.

The σ_i's are the singular values of A, and vectors u_i and v_i are the ith left and right singular vector, respectively. The process to find U and V is called SVD. It can be done by the Golub-Reinsch algorithm of time complexity $O(m^2n+mn^2+n^3)$ [6] (pp. 254). To compute $U(:,1:n) =[u_1,\ldots,u_n]$, which is an $m \times n$ $(m>n)$ matrix, the time complexity is $O(mn^2+n^3)$ [6].

Theorem 2 Let the SVD of $m \times n$ matrix A be defined in Theorem 1. If $r <$ rank(A) and let $A^* = \sum_{i=1}^{r} \sigma_i u_i v_i^T$, then

$$\min_{rank(B)=r} \|A-B\|_2 = \|A-A^*\|_2 = \sigma_{r+1}.$$

In other words, A^* is the best 2-norm approximation of A among all rank r matrices.

2.2 Application to Dimension Reduction

We are given the delay of m nets represented as Ax, where A is an $m \times n$ matrix, n is the number of process variation variables, $x=[x_1,\ldots,x_n]$ is the vector of process variation variables, and x_i's are independent of each other.

From Theorem 2, $A^*=U(:,1:r) \cdot S \cdot V(:,1:r)^T$, where $S=diag(\sigma_1,\ldots,\sigma_r)$, $U(:,1:r)=[u_1,\ldots,u_r]$, and $V(:,1:r) =[v_1,\ldots,v_r]$. Therefore, $Ax \approx A^*x = U(:,1:r) \cdot S \cdot V(:,1:r)^T x = Bz$, where $B=U(:,1:r) \cdot S$ is an $m \times r$ matrix and $z=V(:,1:r)^T x$ is an $r \times 1$ vector. Since the dimension of B is much less than the dimension of A, we use Bz to approximate Ax.

If we just want to reduce the number of process variation variables, then Bz is the resulting delay function, and z is the vector of new process variation variables. However, we often have an additional requirement from circuit simulation that the new process variation variables $z_1, \ldots, z_r$ must be independent. For example, if $z_1=x_1-x_2+x_3$ and $z_2=x_1+x_2-x_3$, then z_1 and z_2 are not independent although the number of variables is reduced from 3 to 2.

2.3 Sub-Matrix SVD (SMSVD)

We propose a sub-matrix SVD method (SMSVD) to ensure the independence between z_i's. The basic idea is to partition columns of A into r sub-matrices, and then use SVD on each sub-matrix. The following is the algorithm:

1) Partition A into sub-matrices $A_1, A_2, \ldots, A_r$, where $A=[A_1, A_2,\ldots, A_r]$. A_i is an $m \times q_i$ matrix, $i=1,\ldots, r$.

2) For $i=1,\ldots, r$, compute the approximation matrix A_i^* of A_i through SVD in which the approximation rank is 1. Define SVD of A_i as $U_i^T \cdot A_i \cdot V_i = diag(\sigma_{i1},\ldots,\sigma_{ip})$, then $A_i^*=U_i(:,1) \cdot \sigma_{i1} \cdot V_i(:,1)^T$ and $\|A_i-A_i^*\|_2=\sigma_{i2}$.

3) Construct $B=[U_1(:,1)\sigma_{11}, U_2(:,1)\sigma_{21}, \ldots, U_r(:,1)\sigma_{r1}]$,

$$z = Cx = \begin{bmatrix} z_1 \\ z_2 \\ \vdots \\ z_r \end{bmatrix} = \begin{bmatrix} V_1(:,1)^T & 0 & \ldots & 0 \\ 0 & V_2(:,1)^T & \ldots & 0 \\ \vdots & \vdots & \ddots & \vdots \\ 0 & 0 & \ldots & V_r(:,1)^T \end{bmatrix} \begin{bmatrix} \mathbf{x}_1 \\ \mathbf{x}_2 \\ \vdots \\ \mathbf{x}_r \end{bmatrix},$$

where B is an $m \times r$ matrix, C is an $r \times n$ matrix, $\mathbf{x}_i$ is a $q_i \times 1$ vector, $i=1,\ldots,r$, and $B \cdot C=[A_1^*, A_2^*, \ldots, A_r^*]$.

It is easy to see that z_i's are independent since x_j's are independent for $j=1,\ldots,n$, and $\mathbf{x}_i$'s are independent for $i=1,\ldots,r$.

The error of approximating Ax by Bz can be evaluated through $\|A-B \cdot C\|_2$. We can prove the following upper bound on the error.

Theorem 3 Given net delay matrix Ax, we can use SMSVD to compute B, C and z such that

$$\|A-B \cdot C\|_2 \leq \sqrt{r} \max_i \sigma_{i2},$$

and z_i's are independent. The time complexity is

$$O(r \cdot (m \cdot \max_i(q_i)^2 + \max_i(q_i)^3)).$$

Proof After SMSVD, we have $A_i^*=U_i(:,1) \cdot \sigma_{i1} \cdot V_i(:,1)^T = U_i \cdot diag(\sigma_{i1},0,\ldots,0) \cdot V_i^T$, where U_i is an orthogonal $m \times m$ matrix, V_i is an orthogonal $q_i \times q_i$ matrix and $diag(\sigma_{i1},0,\ldots,0)$ is an $m \times q_i$ diagonal matrix, $i=1,\ldots,r$. Thus,

$B \cdot C=[A_1^*, A_2^*,\ldots, A_r^*]$

$=[U_1 \cdot diag(\sigma_{11},0,\ldots,0) \cdot V_1^T,\ldots,U_r \cdot diag(\sigma_{r1},0,\ldots,0) \cdot V_r^T]$,

and $A-B \cdot C$

$=[U_1 \cdot diag(0,\sigma_{12},\ldots,\sigma_{1p}) \cdot V_1^T,\ldots,U_r \cdot diag(0,\sigma_{r2},\ldots,\sigma_{rp}) \cdot V_r^T]$.

Let

$$D = \begin{bmatrix} diag(0,\sigma_{12},\ldots,\sigma_{1p}) & 0 & \cdots & 0 \\ 0 & diag(0,\sigma_{22},\ldots,\sigma_{2p}) & \cdots & 0 \\ \vdots & \vdots & \ddots & \vdots \\ 0 & 0 & \cdots & diag(0,\sigma_{r2},\ldots,\sigma_{rp}) \end{bmatrix},$$

then $A-B\cdot C=[U_1,\ldots,U_r]\cdot D\cdot \begin{bmatrix} V_1^T & 0 & \cdots & 0 \\ 0 & V_2^T & \cdots & 0 \\ \vdots & \vdots & \ddots & \vdots \\ 0 & 0 & \cdots & V_r^T \end{bmatrix}$. The 2-norm of matrix $\begin{bmatrix} V_1^T & 0 & \cdots & 0 \\ 0 & V_2^T & \cdots & 0 \\ \vdots & \vdots & \ddots & \vdots \\ 0 & 0 & \cdots & V_r^T \end{bmatrix}$ is 1, and 2-norm of matrix D is $\max_i \sigma_{i2}$. Furthermore, $[U_1,\ldots,U_r]=$

$U_1\cdot\underbrace{[I_m,\ldots,I_m]}_{r}\cdot\begin{bmatrix} I_m & 0 & \cdots & 0 \\ 0 & U_1^T U_2 & \cdots & 0 \\ \vdots & \vdots & \ddots & \vdots \\ 0 & 0 & \cdots & U_1^T U_r \end{bmatrix}$, where I_m is an $m\times m$ identity matrix, U_1 and $\begin{bmatrix} I_m & 0 & \cdots & 0 \\ 0 & U_1^T U_2 & \cdots & 0 \\ \vdots & \vdots & \ddots & \vdots \\ 0 & 0 & \cdots & U_1^T U_r \end{bmatrix}$ are orthogonal matrices. Since the 2-norm of $\underbrace{[I_m,\ldots,I_m]}_{r}$ is $\sqrt{r}$, and 2-norm of an orthogonal matrix is 1, then the 2-norm of $[U_1,\ldots,U_r]$ is $\sqrt{r}$.

Therefore, $\|A - B\cdot C\|_2 \le \sqrt{r}\max_i \sigma_{i2}$.

The time complexity is given by [6]. ♦

A good partition method that minimizes the value of $\max_i \sigma_{i2}$ can reduce the error of SMSVD. We will give a partition scheme and prove an upper bound for $\max_i \sigma_{i2}$.

Theorem 4 Let $a_1,\ldots,a_n$ be singular values of each column of matrix A and assume without loss of generality $a_1\ge a_2\ge\ldots\ge a_n$. There exists a partition such that

$$\max_i \sigma_{i2} \le ((a_r^2+a_{r+1}^2+\ldots+a_n^2)/2)^{1/2},$$

where r is the reduced dimension.

Proof Assume the singular values of any $l\times k$ matrix M are $\sigma_1,\ldots,\sigma_k$, where $\sigma_1\ge\ldots\ge\sigma_k\ge 0$, the singular value of column i of matrix M is b_i, $i=1,\ldots,k$. Then it is well known that $b_1^2+b_2^2+\ldots+b_k^2=\sigma_1^2+\sigma_2^2+\ldots+\sigma_k^2\ge 2\sigma_2^2$. Thus,

$$\sigma_2\le((b_1^2+b_2^2+\ldots+b_k^2)/2)^{1/2}.$$

To reduce the dimension of A to r, let A_i contain a single column whose singular value is a_i for $i=1,\ldots,r-1$, and A_r contain all the other columns. Then

$$\max\{0,\ldots,0,\sigma_{r2}\}=\sigma_{r2}\le((a_r^2+a_{r+1}^2+\ldots+a_n^2)/2)^{1/2}. \quad ♦$$

The above proof also implies that to further reduce the error of SMSVD, we need to find a column combination that minimizes $\max_i(a_{i1}^2+a_{i2}^2+\ldots+a_{ik}^2)$. This problem is NP-complete for $r=2$ and strongly NP-complete for $r\ge 3$ [8]. We use the following heuristic:

- Compute the singular value for each column of A.
- Select r columns A_1, A_2, …, A_r that contains the r maximum singular values. Put A_1, A_2, …, A_r into r different groups and delete these columns from A.
- Repeat the above process until A is empty.

3 EXPERIMENTAL RESULTS

We first performed physical layout for ISCAS85 circuits using TSMC 0.25um 5 metal technology. After the preliminary processing, we have 12 process variation variables. Gate length is not included in this experiment since for 0.25um technology, gate length variation dominates all other process variation. Commercial parasitic extraction tools are used to extract parasitic and to compute all buffer-to-buffer net delays in each circuit (only for rising input). The delays of the nets are represented as linear functions of process variation variables.

Using the SMSVD algorithm, we reduce the number of process variation variables from 12 to 3. The 3 new variables are independent of each other. To demonstrate the efficiency of SMSVD, we compare the errors of the delays with the method MAX3, which selects three process variation variables that have the greatest impact on the delays. (In this experiment, the three process variation variables are ILD1, medal2 thickness and medal2 width.) Table 1 compares the error of delay for all nets in the ISCAS85 circuits. The process variation ranges are ±10%. The approximation error is defined as follows. For each net, let the accurate delay function be

$$d=d_{nominal}(1+a_1x_1+a_2x_2+\ldots+a_nx_n).$$

Let the delay of our SMSVD approximation be $d_1=d_{nominal}(1+b_1z_1+b_2z_2+b_3z_3)$. Then the error of SMSVD for the net is $(d-d_1)/d$. Similarly, let the delay of MAX3 be $d_2=d_{nominal}(1+a_1x_1+a_2x_2+a_3x_3)$. Then the error of MAX3 for the net is $(d-d_2)/d$. For each circuit, the CPU time for SMSVD algorithm is 0.1sec to 0.7sec. From the table, we can see that SMSVD reduces the range, mean and standard deviation of relative errors of net delays.

Table 1 Approximation error for ISCAS85 Circuits

Circuit (# of nets)	SMSVD			MAX3		
	Range	Mean	Std Dev	Range	Mean	Std Dev
C432 (335)	−2.9% +1.9%	0.15%	0.28%	−3.4% +3.1%	0.30%	0.52%
C499 (852)	−3.9% +6.0%	0.28%	0.55%	−9.6% +3.1%	0.45%	0.81%
C880 (699)	−9.4% +8.1%	0.26%	0.69%	−12% +6.3%	0.51%	0.90%
C1355 (1074)	−6.9% +6.4%	0.31%	0.74%	−9.6% +5.0%	0.52%	0.94%
C1908 (822)	−3.0% +6.0%	0.20%	0.42%	−4.0% +4.4%	0.42%	0.51%
C2670 (1478)	−5.1% +2.7%	0.18%	0.41%	−13% +3.2%	0.48%	0.82%
C3540 (2044)	−4.0% +5.1%	0.19%	0.39%	−13% +6.3%	0.50%	0.73%
C5315 (3218)	−5.0% +5.0%	0.19%	0.38%	−13% +4.4%	0.50%	0.71%
C6288 (4630)	−6.8% +5.3%	0.18%	0.38%	−9.6% +6.8%	0.41%	0.46%
C7552 (3709)	−5.5% +6.9%	0.39%	0.85%	−6.7% +7.7%	0.63%	0.90%

4 CONCLUSIONS

An algorithm to reduce the dimension of process variation is presented and the error bound is proved. The experiment shows the algorithm successfully reduces the number of process variation variables while preserving the independence among new process variation variables and the range of the delay function. The new method is applied on ISCAS85 circuits and show good performance in both range and accuracy.

The main advantage of our method over the PCA method is that the PCA method performs SVD for the correlation matrix of size $mn \times mn$, while our algorithm performs SVD for the delay matrix of size $m \times n$.

Acknowledgement

This research is supported in part by SRC grant 2000-TJ-844, NSF grants CCR-0098329, CCR-0113668, EIA-0223785, ATP grant 000512-0266-2001, and a fellowship from Applied Materials.

5 REFERENCES

[1] Y. Liu, L. Pileggi and A. J. Strojwas, Model order reduction of RC (L) interconnect including variational analysis, *DAC* 1999, 201-206.

[2] Y. Liu, S. Nassif, L. Pileggi and A. J. Strojwas, Impact of interconnect variations on the clock skew of a gigahertz microprocessor, *DAC* 2000, 168-171.

[3] A. Gattiker, S. Nassif, R. Dinakar and C. Long, Timing and yield estimation from static timing analysis, *ISQED* 2001, 437-442.

[4] B. Stine, et al., Analysis and decomposition of spatial variation in integrated circuit process and devices, *IEEE Trans. Semiconductor Manufacturing*, Vol. 10, No. 1, Feb 1997, 24-41.

[5] S. R. Nassif, Modeling and analysis of manufacturing variations, *CICC* 2001.

[6] G. H. Golub and C. F. Van Loan, *Matrix Computations*, Third Edition, Johns Hopkins University Press, 1996.

[7] C. Guardiani, S. Saxena, P. McNamara, P. Schumaker, D. Coder, An asymptotically constant, linearly bounded methodology for the statistical simulation of analog circuits including component mismatch effects, *DAC* 2000, 15-18.

[8] M. R. Garey and D. S. Johnson, *Computer and Intractability: A Guide to the Theory of NP-completeness*, Freeman, 1979.

Analog IP Design Flow for SoC Applications

Marwa Hamour, Resve Saleh, Shahriar Mirabbasi, André Ivanov
University of British Columbia
Vancouver, BC
Canada

Abstract

The analog/mixed-signal (AMS) portion of the IC design process continues to be a major bottleneck, slowing the progress towards fully integrated system-on-chip (SoC) designs. A clear definition of reusable analog IP and an analog IP authoring flow has not emerged as yet, although many efforts are underway in industry and academia to establish these notions. In this work, practical definitions of analog IP and an associated design process is proposed. A methodology is developed for analog IP hardening. The VCO of a phase locked loop (PLL) is chosen to illustrate the process due to the increasing importance of PLLs in SoC designs.

1. Introduction

Technology scaling following Moore's law has led to designs with multimillion transistors paving the way to the development of SoCs. The complexity of the resulting designs has an impact on design productivity and time-to-volume. This has put pressure on designers and CAD engineers to develop better and more efficient approaches to the design process.

One approach to increasing design productivity is the development of re-usable cores [1]. SoCs are characterized by the heavy reuse of intellectual property (IP). This move has resulted in changes in the design philosophy adopted by integrated circuit (IC) engineers. The goal is to produce *reusable* IP cores to minimize "re-inventing the wheel" in the design process. However, the extra cost of reusability must not exceed the added value. Therefore, return-on-investment (ROI) of a reusable core must be assessed before attempts at its design are initiated.

Great strides are being made in the development of re-usable design methodologies for digital circuits. The development of CAD tools that automate parts of the design process (repetitive tasks) with libraries containing pre-designed re-usable cores has helped to accelerate the development cycle of IP. Unfortunately, this is not true for all parts of SoC design. In particular, analog/mixed signal (AMS) productivity is still lagging significantly behind digital productivity, and the integration of both on a single substrate is not a trivial matter.

Recent reports [2] suggest that 20-25% of all designs today have some analog/mixed-signal (AMS) components integrated with the digital components. This value is expected to grow to over 60% by 2006. With a world-wide shortage in analog designers, this growth must be accelerated with the introduction of reusable analog IP. The ROI of re-usable analog IP can be high since there are limited number of blocks that are used in many designs. However, the specifications in each application may be slightly different and usually involves a redesign of the entire block. Of course, defining the notion of reusable analog IP and establishing an authoring flow is still the subject on ongoing activity in industry and academia.

To illustrate the types of blocks that exist in AMS designs, a high-level diagram of a generic baseband processor for the Bluetooth standard [3] is shown in Figure 1. In this case, the PLL, A/D and D/A blocks are standard in the baseband processor. As a result, these blocks are suitable for the development of reusable analog IP cores. Given that mixed-signal designs such as Bluetooth, 802.11b, etc. are target SoC platforms, there will be an increasing demand for these types of AMS IP blocks. While a viable business model for analog IP has not emerged, a number of companies have been pursuing the development of analog IP authoring flows [2,4].

According to the Reuse Methodology Manual [1] and VSIA [5], IP cores can be classified into soft, firm and hard IP. The difference is in the degree of flexibility of the IP; soft IP is in the form of RTL code while hard IP is in GDS-II format. Firm IP is somewhere in between; it is usually presented as a netlist with a set of additional views and information pertaining to physical design. Digital designs are commonly defined as RTL code or soft IP while analog/mixed signal designs come in the form of hard IP that has been designed and optimized for a specific application and technology.

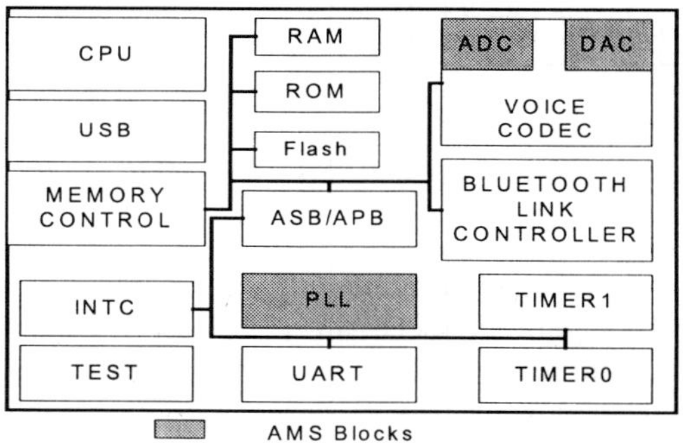

Figure 1. Bluetooth System-on-Chip

In the analog domain, the notion of reusable IP has been the subject of heated debate. At present, analog IP blocks are available in the form of hard IP. That is, the final layout of the block has been defined in GDS-II format. But hard IP is targeted to specific applications, and intended for use in the specific technology associated with it. Therefore, it is not reusable in a more general sense.

To be truly portable, analog IP must be available in at least the firm IP category. This would also facilitate migration from one process to another [6]. Technology migration of analog design is almost as time-consuming as custom designing for the new application. Even higher levels of abstraction, leading to automatic synthesis, is desirable. However, this level is not in widespread use as correct modeling of all effects is complicated and involves modeling of a large number of variables for every technology used. Of course, analog hardware description languages (AHDL) have been

0-7803-7761-3/03/$17.00 ©2003 IEEE

developed to provide a modeling capability at the system level, enabling the co-simulation of analog and digital designs at the behavioral level.

2. Analog and mixed signal IP

As described above, the development of analog and mixed signal design strategy that is able to keep up with the pace of the digital domain tools is greatly needed. In [7], the challenges and possible solution to mixed signal design on SoC are presented. In [8,9], a review of different analog synthesis tools employing different techniques is presented. These synthesis tools target specific applications. Optimization based analog synthesis tools include [10]. In [11] an optimization method that simulates mismatch is presented. The tool provides the experienced designer with the necessary expertise to optimize for worst-case fabrication mismatches. In [12], a complete analog synthesis flow is presented. It does however lack the reusable factor that is deemed an essential part of today's IC development cycle.

A clear definition of analog IP is the starting point for analog reuse. When comparing digital and analog CAD tools, it is unrealistic to expect an exact replica of the digital tools for their analog counterpart. However, there are parts of the reuse methodology that can be applied to analog. Specifically, digital IP has an RTL representation and a corresponding testbench. An analog IP block requires a schematic representation along with a simulation testbench to generate the performance measures. It also requires that certain physical layout constraints, such as symmetry, proximity and common-centroid, be associated with the schematic. Additional layout information for transistor folding and design of guard-rings is needed. This information, along with X-Y location of the cells, should be linked to the schematic. A behavioral view is needed for mixed digital-analog simulation purposes. Analytical models for major performance characteristics are also valuable for initial configuration selection.

In this work, a design approach using *firm* reusable analog IP is presented. In firm IP, the designer is presented with an architecture and netlist for the design, together with a detailed information about the behavior, the layout and simple analytical models that capture that performance characteristics of the block.

In terms of views, firm analog IP components are:

- Schematic View (transistor level description)
- Behavioral View (Verilog-A or VHDL-AMS)
- Analytical View (computational models)
- Physical (layout) View (tied to schematic view)
- Test Benches (simulation scripts)

The next step is to define a flow to *harden* a design block from such an analog IP. The concept of IP hardening can be generically defined as the process of working from an IP block and a set of specifications to the production of the GDS-II layout. Figure 2 shows a typical flow: 1) Architecture/topology is selected using the behavioral and analytical models associated with the IP blocks; the selected blocks must satisfy the specifications supplied by the user, 2) once a topology is selected; the parameterized IP is optimized through sizing using the testbenches to measure performance; the circuit is optimized over process corners with typical input/output loading effects included, 4) the circuit layout is performed; this is followed by parasitic extraction and re-simulation; adjustments to the layout are performed until the specifications are satisfied. For a physical synthesis flow to be effective, it must have the knowledge of the typical layout structure for this type of block.

Most tools take a hierarchical top-down approach to the design process. Thus, there are typically two sets of specifications to be satisfied, one at the system level (upper hierarchy) and then at the sub-block level (low hierarchy). Once the performance targets of the sub-block are specified, a bottom-up approach can be used to construct a working system level design. This top-down/bottom-up approach is useful with analog circuits when clear distinctions can be established in each hierarchical stage.

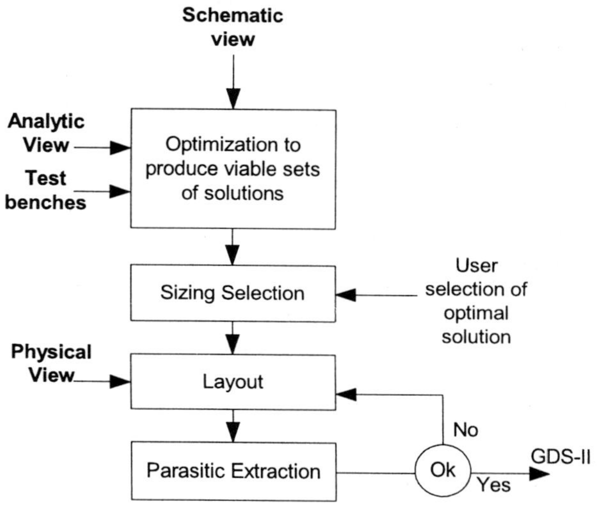

Figure 2 Analog IP hardening flow

3. PLL design

PLLs are a main component in many SoC designs, as seen in the Bluetooth SoC (Figure 1). For Bluetooth, the target frequency range of operation for the baseband PLL is 80 – 300MHz, while the RF front end operates at 2.4GHz. Traditionally, the development process for PLLs involves custom design. The PLL is one of the most challenging blocks to design, perhaps requiring 3 to 6 months worth of trial-and-error iterations by the designer. Nonlinear characteristics of PLLs complicate attempts at their modeling, simulation and design. When attempting to design PLLs as reusable cores, a variety of designs will need to be considered depending on the target application. There do exist tools that are specifically design suites for PLLs [2,4], in which the designer supplies the specifications of the PLL to be designed but has no control over which topology is to be implemented. The final product is hard IP in GDS-II format. In [13] behavioral modeling of phase locked and delay locked systems are presented, and these models were used as a means of automating part of the design process in an effort to cut down on design time. In this work, it is believed that a combination of all these concepts is needed for the development of reusable analog IP.

The main performance metrics of a PLL can be specified in terms of acquisition time, capture range, lock range and phase noise/jitter.

Each of these values is dependent on individual sub-blocks and the resulting system level interaction. Test benches are needed to generate these specifications from simulation runs.

The main contributor of phase noise/jitter within the PLL is the voltage-controlled oscillator (VCO). Simulation of the VCO and design for correct phase noise performance can become one of the most time consuming tasks when designing PLLs. During the early design phase of the PLL, numerical simulations (mostly using Spice) are carried out to decide whether or not phase noise is within acceptable limits. The amount of time spent on simulations is dependent on the expertise of the designer involved.

Taking a hierarchical constraint driven top-down and optimization driven bottom-up approach, each sub-block is designed and then taken to the system level for co-simulation. Because of the generic nature of the components of the PLL excluding the VCO, standard cells were used in the design for the phase frequency detector (PFD) and charge pump (CP). The filter was designed based on the approach proposed by [14].

3.1 Case Study VCO

A VCO with good noise performance and linearity is essential for correct PLL performance. The first required deliverable is a behavioral model of the structure in AHDL. Figure 3 shows a high level Verilog-A code for a generic VCO [15].

```
module vco(in,out);
 input in; output out; voltage in,out;voltage vin;
analog
 begin
   vin = V(in) - vin_offset;
   if (vin > lockrange)    vin = lockrange;
   else
     if (vin < -lockrange) vin = -lockrange;
     w = `M_TWO_PI * (f0 + kf*vin);
   V(out) <+ vout_offset + vout_mag*sin(w*$realtime +
phi0);
 end
endmodule
```

Figure 3. Behavioral view of VCO

The analytical view of the VCO is now to be developed. For accurate simulation of performance values, accurate models for CMOS devices are needed. Advanced models such as BSIM (level 13 and 39) incorporate many parameters for accurate modeling of CMOS within different operating regions. However, as the operating frequency increases, there are behavioral characteristics of CMOS that are not sufficiently modeled [16]. In this work, it has been assumed that the operating frequency range is within limits that make BSIM3 an appropriate model.

The analytic view is a more detailed representation of the circuit. It can be used for specifically targeted simulations and to direct the optimizer towards the solution space. In [13] different techniques for modeling of the delay of each stage of the VCO, and thus the operating frequency, are presented. The method presented in [18] calculates this delay by making only the assumption of sinusoidal input and output. This method was used in the calculation of the delay of each stage. The classical ring oscillator frequency equation is given by $f=1/(2N\tau_{stage})$ where N is the number of stages, and τ_{stage} is the delay of each stage. For an accurate representation of the characteristics of the VCO, parasitics resulting from the layout should be considered in the analysis.

The resulting frequency calculations as compared with Spice simulated results are shown in Figure 4 for varying control voltage when implementing a 3-stage VCO. The percentage error for the model is within 15%.

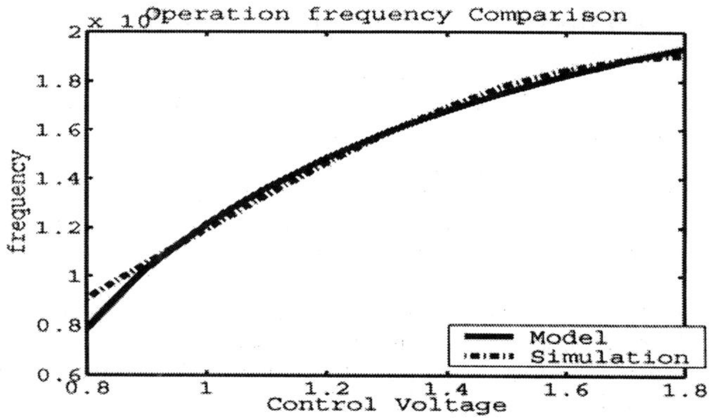

Figure 4. Analytical view of VCO

Using the behavioral and analytical view, the designer or CAD tool can make an educated decision about the choice of VCO to be implemented. In our case, the VCO shown in Figure 5 [19] was used.

The next step is the design of the test benches and optimization. The user must select the parameters that will have the higher weight in the optimization process. Jitter is a critical parameter when determining VCO/PLL's performance. It can be measured in simulation, but obtaining values for optimal performance involves long simulation hours. If however, the tradeoff curves for the different characteristics of the core are available, the designer can make educated decisions about the design. In effect, the expertise of the analog designer is being propagated along with the IP.

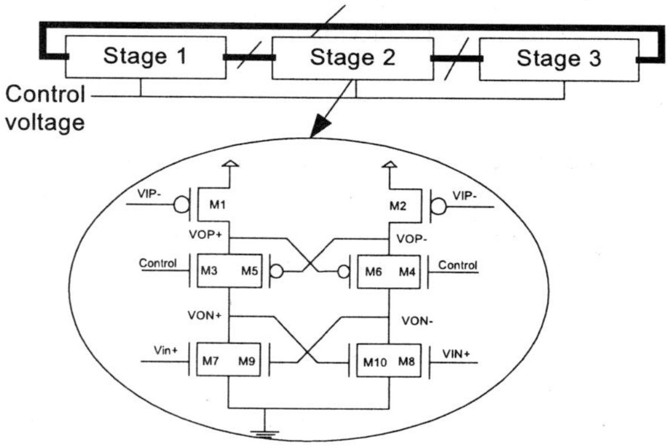

Figure 5. Schematic view of VCO

For this specific core, optimization with respect to cycle-to-cycle jitter, power, center frequency and tuning range was performed. Figure 6 shows the tradeoff curves for the highest scoring optimization run using a commercial program [20].

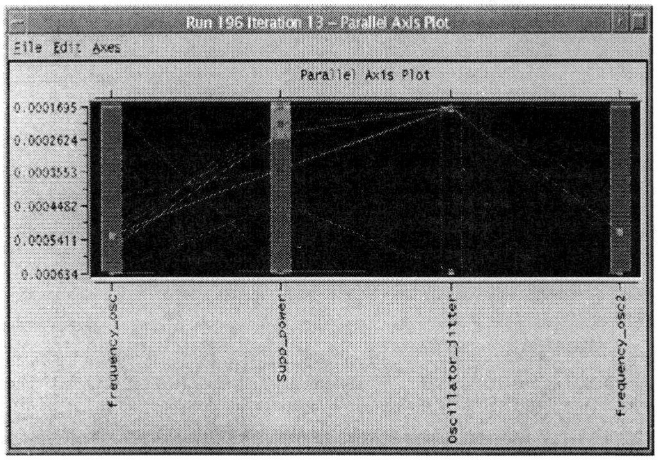

Figure 6. Solution space

The next stage in the IP hardening process is generation of the physical layout using the specified layout view, Figure 7. The layout view comprises the initial floor plan information needed for physical layout. That is comprised of transistor layout coordinates with respect to a defined origin, as well as matching criteria that need to be taken into consideration.

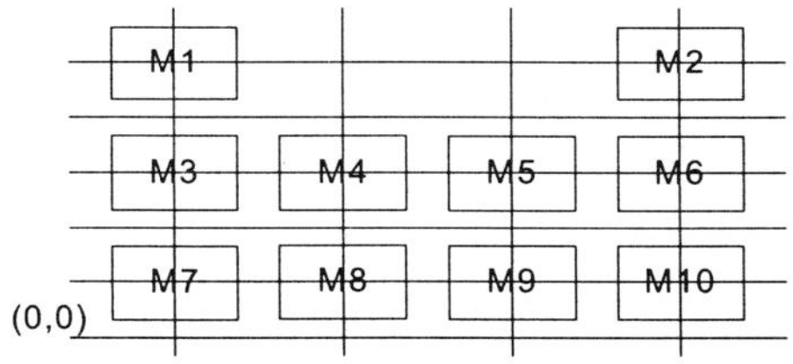

Figure 7. Layout view

In the design of the overall PLL interaction of the sub-blocks has been incorporated into the design flow. This case study presented the design and optimization of one such sub-block, i.e., a VCO. The importance of consistency in circuit macromodeling, as well as a method for automatic consistency checking and optimization of parameterized macromodels was shown in [21]; the same approach was used here. If not properly accounted for, inter-block interaction could cause deterioration in the circuit performance. In the case of a VCO, the effects of the preceding and succeeding sub-blocks should be considered in the design of the testbench. For example, the capacitance at the VCO input node might limit its sensitivity (speed of response). It could also change the effective values for the loop filter, which would affect the overall performance and specifications of the PLL.

Summary

This paper proposes a practical definition of *reusable* analog IP and its associated design process. A methodology is developed for analog IP hardening emphasizing design reuse. The design of the VCO of a PLL in a Bluetooth system-on-a-chip platform was described as a case study.

Acknowledgements

The authors would like to acknowledge NSERC, PMC-Sierra, Vector12, ASI and ADA for their contribution to this work.

References

[1] M. Keating and P. Bricaud, *Reuse Methodology Manual for System-on-a-Chi Designs.* Norwell, MA: Kluwer Academic, 1998.

[2] www.neolinear.com

[3] P. Rashinkar, P. Paterson, L. Singh, *"System-on-a-chip Verification,"* Kluwer Academic Publishers 2001

[4] www.barcelonadesign.com

[5] www.vsi.org

[6] P. Chatterjee, "*Legacy Data: A Structured Methodology for Device Migration in DSM technology*", Kluwer Academic Publishers, 2002.

[7] K. Kundert, H. Chang, D. Jeffries, G. Lamant, E. Malavasi and F. Sending, "*Design of Mixed-Signal System-on-a-Chip*". IEEE trans Computer-Aided Design, vol. 12,December 2000

[8] B. Antao, "*Techniques for synthesis of analog integrated circuits*," IEEE Design and test of computers 1992

[9] G. Gielen, R. Rutenbar, *"Computer-Adied Design of Analog and Mixed-Signal Integrated Circuits"*, Proc of IEEE, vol. 88, No.12, December 2000, pp. 1825-1852

[10] E. Ochotta, R. Rutenbar and L. Carley, "*ASTRX/OBLX: Tools for rapid synthesis of high performance analog circuits*," IEEE trans. Computer Aided Design, vol.15, March 1996

[11] K. Antreich, J. Eckmueller, H. Graeb, M. Pronath, F. Schenkel, R. Schwencker, S. Zizala, *"WiCkeD: analog circuit synthesis incorporating mismatch,"* Custom Integrated Circuits Conference, 2000.pp.511-514

[12] G. Van der Plas, G. Debyser, F Leyn, K Lampaert, J. Vandenbussche, G.G.E. Gielen, W. Sansen, P. Veselinovic and D. Leenarts,"*AMGIE-A Synthesis Environment for CMOS Analog Integrated Circuits*," D. IEEE trans. Computer-Aided Design of Integrated Circuits and Systems, 2001 pp. 1037 –1058

[13] A. Demir, E. Liu, A. Sangiovanni-Vincentelli and I. Vassiliou, *"Behavioral Simulation Techniques for Phase/Delay-Locked Systems,"* IEEE CICC 1994 pp. 453-456

[14] S. Mirabbasi and K. Martin, "*Design of loop filter in phase-locked loops*," Electronic letters vol.35 October 1999 pp.1801-1802

[15] www.eda.org/verilog-ams/models/vco.va

[16] B. Razavi, "*CMOS Technology Characterization for Analog and RF Design*," IEEE ISSCC 1999 pp. 268-276

[17] M. Alioto, G. Di Cataldo and G. Palumbo, "*CML Ring Oscillators: Oscillation Frequency*," IEEE ISCAS 2001 pp. 112-115

[18] S. Docking and M. Sachdev, "*A method to determine the frequency of oscillation of a ring oscillator based VCO*," Micronet Annual Workshop 2002.

[19] I. Hwang and S. Kang, "*A Self regulating VCO with Supply Sensitivity of <0.15%-Delay1/%-Supply*," IEEE ISSCC 2002, pp. 104-422

[20] www.analogsynthesis.com

[21] Y. Ju, B. Rao, R. Saleh, *"Consistency Checking and Optimization of Macromodels,"* IEEE Trans. on CAD of IC and Syst., Vol. 10, No. 8, Aug. 1991, pp. 957-967

SINGLE VARIABLE SYMMETRY CONDITIONS IN BOOLEAN FUNCTIONS THROUGH REED-MULLER TRANSFORM

Sudha Kannurao
School of Engineering/Electronics
Temasek Polytechnic, 21 Tampines Avenue 1,
Singapore 529757

Bogdan J. Falkowski
School of Electrical and Electronic Engineering,
Nanyang Technological University,
Block S1, Nanyang Avenue, Singapore 639798

ABSTRACT

A new method to detect single variable symmetry and complement single variable symmetry in Boolean functions through the Reed-Muller transform is presented. The Reed-Muller spectral coefficients are used to identify all the four types of single variable symmetries. To reduce the time and to increase the efficiency in identifying the symmetries, necessary and spectral conditions are highlighted.

1. INTRODUCTION

Detection of Boolean symmetries has been an important issue as its knowledge leads to a more proficient realization of the functions that are important in modern VLSI systems. Such functions include adders, multipliers and parity checkers. The problem of symmetries is also significant in many other applications, for example in Boolean matching, cell library design, testing, verification, identification of disjunctive decomposition, and computing arithmetic and related functions using threshold circuits [1].

There are twelve basic types of symmetry based on a relation between four cofactors of the Boolean function in respect to two variables. Four of them are two variable symmetries and eight other are single variable symmetries (SVS). Both classical (binary decision diagrams) [5] and spectral (Hadamard and Reed-Muller transform) based methods [2, 3] have been used to identify Boolean symmetries. Originally spectral approach was considered inefficient due to the lack of effective methods to calculate and store spectra in reduced forms. These drawbacks have been overcome recently [7] and the latest spectral methods operate on spectral decision diagrams or cubes rather than on spectral vectors making these methods more suitable to big benchmark functions. In this article, for the first time, conditions for finding SVS and complement SVS in Boolean functions through the Reed-Muller spectrum have been presented. The results show that only a part of the Reed-Muller spectrum is needed to identify the symmetries.

2. REED-MULLER TRANSFORM

Definition 1 [4, 7]: The *n-variable Reed-Muller expansion* takes the form:

$$F(x_n, x_{n-1}, \dots, x_1) = \sum_{i=1}^{2^n-1} c_i x_n^{e_{i,n}} x_{n-1}^{e_{i,n-1}} \dots x_1^{e_{i,1}} .$$

In the above definition, the symbol $\sum$ means summation over the Galois Field (2) (addition modulo-2), and the $e_{i,j}$ are either 0 or 1 so that the literals $x_k^0 = 1$ and $x_k^1 = x_k$. The n-variable Reed-Muller expression has 2^n possible product terms (*piterms* represented by symbol π) selected from n variables. The subscript i in the piterm π_i represents the decimal equivalent of the straight binary code (SBC) formed from the join of symbols $e_{i,n}$, $e_{i,n-1}$, …,$e_{i,1}$, where the Most Significant Bit of the SBC is $e_{i,n}$ and the Least Significant Bit of the SBC is $e_{i,1}$.

The *n-variable Reed-Muller expansion* can be also represented directly through piterms by: $F(x_n, x_{n-1}, \dots, x_1) = \sum_{i=1}^{2^n-1} c_i \pi_i$. The column vector $\vec{C}$ of the size 2^n represents all the Reed-Muller spectral coefficients c_i.

Definition 2 [4, 7]: The Reed-Muller transform matrix T_n of order n can be obtained through recursive formula and is defined over Galois Field (2). The basic matrix T_1 is $\begin{bmatrix} 1 & 0 \\ 1 & 1 \end{bmatrix}$ and T_n is given by $\begin{bmatrix} T_{n-1} & 0 \\ T_{n-1} & T_{n-1} \end{bmatrix}$.

Definition 3: The coefficient c_0 is called the DC coefficient and the *order* of any Reed-Muller Coefficient (RMC) is decided by the number of the variables associated with the RMC in its expansion. The RMC can also be represented as c_I, where I is called the index and is $I = i, i \neq 0$ for the first order coefficients, and is equal to $ij, i \neq 0, j \neq 0, i \neq j$ for the second order coefficients, and is equal to $ijk, i \neq 0, j \neq 0, k \neq 0, i \neq j, i \neq k, j \neq k$ for the third order coefficients and so on.

0-7803-7761-3/03/$17.00 ©2003 IEEE

3. DECOMPOSITION OF BOOLEAN FUNCTIONS

Considering any Boolean function $f(x_n,...x_i,...x_j,...x_1)$, the decomposition of the function around any two arbitrary variables x_i and x_j is given by $\{\overline{x_i x_j}\, f_0(x_n,...0,...0,...x_1)$ + $\overline{x_i} x_j\; f_1(x_n,...0,...1,...x_1)$ $+ x_i \overline{x_j}\; f_2(x_n,...1,...0,...x_1)$ $+ x_i x_j f_3(x_n,...1,...1,...x_1)\}$. The Reed-Muller spectrum of the sub-function $f_0(x_n,...0,...0,...x_1)$ is represented as $\overrightarrow{C_0}$, and the spectrum of the sub-function $f_1(x_n,...0,...1,...x_1)$ is represented as $\overrightarrow{C_1}$, $\overrightarrow{C_2}$ represents the spectrum of the function $f_2(x_n,...1,...0,...x_1)$ and $\overrightarrow{C_3}$ represents the spectrum of $f_3(x_n, ...1,...1,...x_1)$. Also let $\overrightarrow{C^0}$ represent the ordered part of the RMC vector of the function $f(x)$ with the c_I $(Y \notin I)$ where Y is the set of variables around which the symmetry is described; $\overrightarrow{C^1}$ is the ordered part of the RMC with c_I where $(i \notin I, i \in Y)$ and i corresponds to the Least Significant Bit in the binary representation of Y and so on.

The symmetry conditions of the Boolean functions relate to the equality between pairs of the sub-functions. Hence the consideration of relations between the spectrum of $f(x)$ with the spectra of the sub-functions becomes essential for identifying the symmetries using the spectral methods. The definition given below relates the Reed-Muller spectrum of $f(x)$ with that of the sub-functions.

Definition 4: From the known relations between the Reed-Muller and Walsh spectra [6] and from the earlier work by Tokmen on Walsh spectral decomposition [8], the *spectral Reed-Muller decomposition* for a function $f(x_n,...x_i,...x_j,...x_i)$ based on RMC is given by:

$$\left[\overrightarrow{C_0}\,\overrightarrow{C_1}...\overrightarrow{C_\beta}\right] = \left[\overrightarrow{C^0}\,\overrightarrow{C^1}...\overrightarrow{C^\beta}\right]\left\{T^{n-m}\right\}^T \qquad (1)$$

($\beta = 2^{n-m} -1$). In this paper we are analyzing the single variable symmetry in x_i over x_j of the function. Hence $n\text{-}m = 2$ and $\beta = 3$. Rewriting equation (1), we have,

$$\left[\overrightarrow{C_0}\,\overrightarrow{C_1}\,\overrightarrow{C_2}\,\overrightarrow{C_3}\right] = \left[\overrightarrow{C^0}\,\overrightarrow{C^1}\,\overrightarrow{C^2}\,\overrightarrow{C^3}\right]\begin{bmatrix} 1 & 1 & 1 & 1 \\ 0 & 1 & 0 & 1 \\ 0 & 0 & 1 & 1 \\ 0 & 0 & 0 & 1 \end{bmatrix} \qquad (2)$$

Here $\overrightarrow{C_0}$ is the ordered part of the RMC vector $\vec{C}$ of the function $f(x)$ with c_I $(i \notin I, j)$, $\overrightarrow{C^1}$ is with the c_I $(i \in I, j \notin I)$, $\overrightarrow{C^2}$ the ordered part of the RMC where c_I $(i \notin I, j \in I)$ and finally $\overrightarrow{C^3}$ involves the c_I $(i, j \in I)$. In the above equation, all the operations are done over the Galois Field (2).

Definition 5: The *positive function* u_i for any Boolean function $f(x)$ corresponds to the area covered by the variable x_i of $f(x)$ and a_i denotes the number of true minterms of $f(x)$ in this area.

Definition 6: The *negative function* v_i for any Boolean function $f(x)$ corresponds to the area covered by the variable $\overline{x_i}$ of $f(x)$ and b_i denotes the number of true minterms of $f(x)$ in this area.

4. SYMMETRY IN BOOLEAN FUNCTION

In this paper we introduce Reed-Muller spectral conditions for existence of four types of single variable symmetry and complement single variable symmetry in any given function. Only half of the spectral coefficients are needed to identify two types of SVS and for the other two SVS symmetry only one fourth of the spectral coefficients are needed. Using the same spectral coefficients, the complement SVS can also be identified.

4.1 Single Variable Symmetry

Definition 7: A function $f(x_n,...x_i,...x_j,...x_1)$ of n variables possesses single variable symmetry in x_i over x_j, represented as $S(x_i \mid x_j)$, if $f(x_n,...0,...1,...x_1) = f(x_n,...1,...1,...x_1)$.

Definition 8: A function $f(x_n,...x_i,...x_j,...x_1)$ of n variables possesses a single variable symmetry in x_i over $\overline{x_j}$, represented as $S(x_i \mid \overline{x_j})$ if $f(x_n,...0,...0,...x_1) = f(x_n,...1,...0,...x_1)$.

Definition 9: A function $f(x_n,...x_i,...x_j,...x_1)$ of n variables possesses single variable symmetry in x_j over x_i, represented as $S(x_j \mid x_i)$, if $f(x_n,...1,...0,...x_1) = f(x_n,...1,...1,...x_1)$.

Definition 10: A function $f(x_n,...x_i,...x_j,...x_1)$ of n variables possesses single variable symmetry in x_j over $\overline{x_i}$ represented as $S(x_j \mid \overline{x_i})$, if $f(x_n,...0,...0,...x_1) = f(x_n,...0,...1,...x_1)$.

For any function to exhibit $S(x_j \mid x_i)$ and $S(x_i \mid x_j)$ in x_i over x_j, the necessary but not the sufficient condition is that the number of minterms in the positive functions u_i and u_j that are a_i and a_j must be even numbers, respectively. For the function to exhibit $S(x_i \mid \overline{x_j})$ and $S(x_j \mid \overline{x_i})$, b_j and b_i must be even numbers, respectively.

Theorem 1: A function $f(x_n,...x_i,...x_j,...x_1)$ of n variables possesses single variable symmetry in x_i over x_j if $\overrightarrow{C^2} \oplus \overrightarrow{C^3} = [0\,0\,0...0]^T$.

Proof: If a function exhibits single variable symmetry in x_i over x_j then, $f_1(x_n,...0,...1,...x_1) = f_3(x_n,...1,...1,...x_1)$. Hence $\overrightarrow{C_1} = \overrightarrow{C_3}$ where $\overrightarrow{C_1}$ and $\overrightarrow{C_3}$ are the Reed-Muller spectral vectors for the functions f_1 and f_3 respectively.

From equation (2), $\overrightarrow{C_1} = \overrightarrow{C^0} \oplus \overrightarrow{C^1}$ (3)

$$\overrightarrow{C_3} = \overrightarrow{C^0} \oplus \overrightarrow{C^1} \oplus \overrightarrow{C^2} \oplus \overrightarrow{C^3} \quad (4)$$

From (3) and (4), $\overrightarrow{C_1} \oplus \overrightarrow{C_3} = \overrightarrow{C^2} \oplus \overrightarrow{C^3}$ which must be equal to $[0\ 0\ 0\ ...0]^T$. QED

Theorem 2: A function $f(x_n,...x_i,...x_j,...x_1)$ of n variables possesses single variable symmetry in x_i over $\overline{x_j}$, if $\overrightarrow{C^2} = [0\ 0\ 0...0]^T$.

Proof: If a function exhibits single variable symmetry in x_i over $\overline{x_j}$ then, $f_0(x_n,...0,...0,...x_1) = f_2(x_n,...1,...0,...x_1)$. Hence $\overrightarrow{C_0} = \overrightarrow{C_2}$ where $\overrightarrow{C_0}$ and $\overrightarrow{C_2}$ are the Reed-Muller spectral vectors for the functions f_0 and f_2 respectively. From equation (2), $\overrightarrow{C_0} = \overrightarrow{C^0}$ and $\overrightarrow{C_2} = \overrightarrow{C^0} \oplus \overrightarrow{C^2}$ which implies that $\overrightarrow{C^2} = [0\ 0\ 0...0]^T$. QED

Theorem 3: If a function $f(x_n,...x_i,...x_j,...x_1)$ of n variables possesses single variable symmetry in x_j over x_i, then $\overrightarrow{C^1} \oplus \overrightarrow{C^3} = [0\ 0\ 0...0]^T$.

Proof: If a function exhibits single variable symmetry in x_j over x_i then, $f_2(x_n,...1,...0,...x_1) = f_3(x_n,...1,...1,...x_1)$. Hence $\overrightarrow{C_2} = \overrightarrow{C_3}$ where $\overrightarrow{C_2}$ and $\overrightarrow{C_3}$ are the Reed-Muller spectral vectors for the functions f_2 and f_3 respectively.

From equation (2), $\overrightarrow{C_2} = \overrightarrow{C^0} \oplus \overrightarrow{C^2}$ (5)

$$\overrightarrow{C_3} = \overrightarrow{C^0} \oplus \overrightarrow{C^1} \oplus \overrightarrow{C^2} \oplus \overrightarrow{C^3} \quad (6)$$

From (5) and (6), $\overrightarrow{C_2} \oplus \overrightarrow{C_3} = \overrightarrow{C^1} \oplus \overrightarrow{C^3}$ which must be equal to $[0\ 0\ 0\ ...0]^T$. QED

Theorem 4: For a function $f(x_n,...x_i,...x_j,...x_1)$ of n variables to possess single variable symmetry in x_j over $\overline{x_i}$, $\overrightarrow{C^1} = [0\ 0\ 0...0]^T$ should be satisfied.

Proof: If a function exhibits single variable symmetry in x_j over $\overline{x_i}$ then, $f_0(x_n,...0,...0,...x_1) = f_1(x_n,...0,...1,...x_1)$. Hence $\overrightarrow{C_0} = \overrightarrow{C_1}$ where $\overrightarrow{C_0}$ and $\overrightarrow{C_1}$ are the Reed-Muller spectral vectors for the functions f_0 and f_1 respectively. From equation (2), $\overrightarrow{C_0} = \overrightarrow{C^0}$ and $\overrightarrow{C_1} = \overrightarrow{C^0} \oplus \overrightarrow{C^1}$. Hence $\overrightarrow{C_0} \oplus \overrightarrow{C_1} = \overrightarrow{C^0} \oplus \overrightarrow{C^0} \oplus \overrightarrow{C^1} = \overrightarrow{C^1}$ which must be equal to $[0\ 0\ 0\ ...0]^T$. QED

4.2 Complement Single Variable Symmetry

Definition 11: A function $f(x_n,...x_i,...x_j,...x_1)$ of n variables possesses complement single variable symmetry in x_i over x_j, represented as $CS(x_i \mid x_j)$, if $f(x_n,...0,...1,...x_1) = \overline{f(x_n,...1,...1,...x_1)}$.

Definition 12: A function $f(x_n,...x_i,...x_j,...x_1)$ of n variables possesses complement single variable symmetry in x_i over $\overline{x_j}$, represented as $CS(x_i \mid \overline{x_j})$ if $f(x_n,...0,...0,...x_1) = \overline{f(x_n,...1,...0,...x_1)}$.

Definition 13: A function $f(x_n,...x_i,...x_j,...x_1)$ of n variables possesses complement single variable symmetry in x_j over x_i, represented as $CS(x_j \mid x_i)$, if $f(x_n,...1,...0,...x_1) = \overline{f(x_n,...1,...1,...x_1)}$.

Definition 14: A function $f(x_n,...x_i,...x_j,...x_1)$ of n variables possesses complement single variable symmetry in x_j over $\overline{x_i}$ represented as $CS(x_j \mid \overline{x_i})$, if $f(x_n,...0,...0,...x_1) = \overline{f(x_n,...0,...1,...x_1)}$.

The necessary but not the sufficient condition for any function to exhibit $CS(x_j \mid x_i)$ and $CS(x_i \mid x_j)$ in x_i over x_j, is that both a_i and a_j must be equal to 2^{n-2}, respectively. Similarly for the function to exhibit $CS(x_j \mid \overline{x_i})$ and $CS(x_i \mid \overline{x_j})$, then both b_i and b_j must be equal to 2^{n-2}, respectively.

Theorem 5: A function $f(x_n,...x_i,...x_j,...x_1)$ of n variables possesses complement single variable symmetry in x_i over x_j if $\overrightarrow{C^2} \oplus \overrightarrow{C^3} = [1\ 0\ 0...0]^T$.

Proof: If a function exhibits complement single variable symmetry in x_i over x_j then, $f_1(x_n,...0,...1,...x_1) = \overline{f_3(x_n,...1,...1,...x_1)}$. Hence $\overrightarrow{C_1} \oplus \overrightarrow{C_3} = [1\ 0\ 0...0]^T$ where $\overrightarrow{C_1}$ and $\overrightarrow{C_3}$ are the Reed-Muller spectral vectors for the functions f_1 and f_3 respectively. From the proof for Theorem 1, $\overrightarrow{C_1} \oplus \overrightarrow{C_3} = \overrightarrow{C^2} \oplus \overrightarrow{C^3}$ which is equal to $[1\ 0\ 0...0]^T$ for the symmetry conditions. QED

Theorem 6: A function $f(x_n,...x_i,...x_j,...x_1)$ of n variables possesses complement single variable symmetry in x_i over $\overline{x_j}$, if $\overrightarrow{C^2} = [1\ 0\ 0...0]^T$.

Proof: For a function to have complement single variable symmetry in x_i over $\overline{x_j}$, $f_0(x_n,...0,...0,...x_1) =$

$\overline{f_2(x_n,...1,...0,...x_1)}$. Hence $\overrightarrow{C_0} \oplus \overrightarrow{C_2} = [1\,0\,0...0]^T$ where $\overrightarrow{C_0}$ and $\overrightarrow{C_2}$ are the Reed-Muller spectral vectors for the functions f_0 and f_2 respectively. From equation (2), and from the proof for Theorem 2, the above-mentioned condition is satisfied. QED

Theorem 7: If a function $f(x_n,...x_i,...x_j,...x_1)$ of n variables possesses complement single variable symmetry in x_j over x_i, then $\overrightarrow{C^1} \oplus \overrightarrow{C^3} = [1\,0\,0...0]^T$.

Proof: For a function with $CS(x_j \mid x_i)$, $f_2(x_n,...1,...0,...x_1) = \overline{f_3(x_n,...1,...1,...x_1)}$ or $\overrightarrow{C_2} \oplus \overrightarrow{C_3} = [1\,0\,0...0]^T$. From equation (2), and from the proof for Theorem 3, the above-mentioned condition is satisfied. QED

Theorem 8: If a function $f(x_n,...x_i,...x_j,...x_1)$ of n variables possesses complement single variable symmetry in x_j over $\overline{x_i}$, then $\overrightarrow{C^1} = [1\,0\,0...0]^T$.

Proof: If a function exhibits $CS(x_j \mid \overline{x_i})$ then $f_0(x_n,...0,...0,...x_1) = \overline{f_1(x_n,...0,...1,...x_1)}$. Hence $\overrightarrow{C_0} \oplus \overrightarrow{C_1} = [1\,0\,0...0]^T$. From equation (2) and from the proof for Theorem 4, the condition for the existence of the complement single variable symmetry in x_j over $\overline{x_i}$ can be proved. QED

5. DETECTION OF SYMMETRY

SVS and complement SVS can be identified using the Reed-Muller spectral coefficients based on the results from Theorems 1 to 8. Table 1 summarizes the necessary and spectral conditions for the existence of the eight single variable symmetries. Identification of single variable symmetries in any pair of variables for a function of n variables is done in two steps to reduce the time of detection and to improve the efficiency. In the first step, the necessary but not satisfactory condition for the symmetry to exist in any pair of variables is tested. An asymmetry filter is used to identify the existence of symmetry in a pair of variables. In the second step, the symmetry test is conducted on only those pair of variables that satisfy the symmetry condition. In the process of identification of symmetry, the necessary condition can be checked first following the first spectral condition and then the test has to be performed on further spectral conditions. At any stage if the function does not satisfy the condition, then the function does not possess the symmetry for the chosen x_i and x_j.

Example 1: Consider a four variable Boolean function $f(x_4, x_3, x_2, x_1) = \sum m\,(0, 2, 3, 5, 9, 12, 14, 15)$. a_i, a_j, b_i and b_j for i, j = 1 to 4 are even and hence satisfy the necessary symmetry condition. Considering any two variables say x_3 and x_4, $\overrightarrow{C^1} = [1\,0\,0\,0]^T$, $\overrightarrow{C^2} = [1\,0\,0\,0]^T$ and $\overrightarrow{C^3} = [0\,0\,0\,0]^T$. As $\overrightarrow{C^1} \oplus \overrightarrow{C^3} = [1\,0\,0\,0]^T$ from Theorem 7, the function possesses CS ($x_4 \mid x_3$) and also as $\overrightarrow{C^2} = [1\,0\,0\,0]^T$, from Theorem 6, the function possesses complement single variable symmetry in x_3 over $\overline{x_4}$. Similar test could be done for other variables as well.

Type of symmetry	Necessary condition	Type of symmetry	Necessary condition
$S\,(x_i \mid x_j)$	$a_j = Even\ number$	$CS\,(x_i \mid x_j)$	$a_j = 2^{n-2}$
$S\,(x_i \mid \overline{x_j})$	$b_j = Even\ number$	$CS\,(x_i \mid \overline{x_j})$	$b_j = 2^{n-2}$
$S\,(x_j \mid x_i)$	$a_i = Even\ number$	$CS\,(x_j \mid x_i)$	$a_i = 2^{n-2}$
$S\,(x_j \mid \overline{x_i})$	$b_i = Even\ number$	$CS\,(x_j \mid \overline{x_i})$	$b_i = 2^{n-2}$

Table 1: Necessary conditions for symmetry

6. SUMMARY

In this paper, the detection of SVS and complement SVS using the Reed-Muller spectral coefficients have been developed. From the introduced theorems it is obvious that either one half or one fourth of the spectral coefficients are needed to identify any symmetry. As the spectral coefficients could be stored in the form of reduced spectral representations and found using similar reduced Boolean representations in the form of disjoint cubes or decision diagrams [7] our method may also have applications in the fields outside those mentioned in the paper for example to discover symmetries in large databases.

7. REFERENCES

[1] G. DeMicheli, *Synthesis and Optimization of Digital Circuits*, New York: McGraw Hill, 1994.

[2] C.R. Edwards and S.L. Hurst, "A Digital Synthesis Procedure under Function Symmetries and Mapping Methods", *IEEE Trans. Computers*, Vol. 27, No. 11, pp. 985-997, Nov. 1978.

[3] B.J. Falkowski, S. Kannurao, "Identification of Boolean Symmetries in Spectral Domain of Reed-Muller Transform", *Electronics Letters,* Vol. 35, No. 16, pp. 1315-1316, 1999.

[4] D.H. Green, *Modern Logic Design*, Wokingham, MA: Addison-Wesley, 1996.

[5] D. Moller, J. Mohnke and M. Weber, "Detection of Symmetry of Boolean Functions Represented by ROBDDS", *Proc. Int. Conf. on Computer-Aided Design,* pp. 680-684, Nov. 1993.

[6] R.S. Stankovic, "A Note on the Relation between Reed-Muller Expansions and Walsh Transform", *IEEE Trans. Electromagnetic Compatibility*, Vol. 24, No. 1, pp. 68-70, 1982.

[7] R.S. Stankovic, M. Stankovic and D. Jankovic, *Spectral Transforms in Switching Theory: Definitions and Calculations*, Belgrade: IP Nauka, 1998.

[8] V.H. Tokmen, "Disjoint Decomposability of Multi-Valued Functions by Spectral Means", *Proc. IEEE 10th Int. Symp. on Multi-Valued Logic*, pp. 88-93, 1980.

A BOOLEAN EXTRACTION TECHNIQUE FOR MULTIPLE-LEVEL LOGIC OPTIMIZATION

Oh-Hyeong Kwon

Div. of Computer and Multimedia Engineering
Uiduk University
San 50 Yugeom, Gangdong
Gyeongju, Gyeongbook 780-713
Republic of Korea
ohkwon@uiduk.ac.kr

ABSTRACT

Extraction is the most important step in global minimization. Its approach is to identify and extract subexpressions, which are multiple-cubes or single-cubes, common to two or more expressions which can be used to reduce the total number of literals in a Boolean network. Extraction is described as either algebraic or Boolean, according to the trade-off between run-time and optimization. Boolean extraction is capable of providing better results, but difficulty in finding common Boolean divisors arises. In this paper, we present a new method for Boolean extraction to remove the difficulty. The key idea is to identify and extract two-cube Boolean subexpression pairs from each expression in a Boolean network. Experimental results show the improvements in the literal counts over the extraction in SIS for some benchmark circuits.

1. INTRODUCTION

A logic expression can be expressed in various logic forms, which differ in literal counts. In widely-used MOS circuits, the number of transistors to implement a Boolean expression is directly proportional to literal counts in its logic form. Thus, a logic optimization is simply to derive a logic form with the fewest literals. For multiple-level logic design, two basic methodologies have evolved local optimizations and global optimizations. Local optimizations are to optimize a local Boolean expression without affecting the structure of a Boolean network. Global optimizations are to factor Boolean expressions into optimal multiple-level forms with little consideration of the form of an original Boolean network. As a method for a global optimization, extraction is to obtain an equivalent representation of given multiple-output Boolean expressions that are optimal with respect to the literal counts. Extraction is the process of identifying and creating some intermediate Boolean expressions and variables, and re-expressing the original Boolean expressions in terms of original as well as intermediate variables. General approach for extraction is to identify and extract multiple-cube divisors from Boolean expressions in a Boolean network. It terminates when there are no more multiple-cube divisors of any pair of Boolean expressions. Then, single-cube divisors are extracted until no further single-cube divisors exist.

Brayton and McMullen proposed a solution to detect multiple-cube divisors using kernels [1]. Later, Brayton, Rudell, Sangiovanni-Vincentelli, and Wang proposed an algorithm for generating subsets of kernels [2]. Their algorithm is based on the notion of level-0 kernels and kernel intersections. These algebraic extractions in [1, 2], called kernel-based extraction, speed up extracting process, but some optimal results sometimes are not possible. Rajski and Vasudevamurthy introduced an idea of extracting common subexpressions [3]. They considered only algebraic single-cube divisors for multiple-output logic circuits. Singh and Diwan introduced how to extract common subexpressions using a permutation of variables [4]. The extraction depends on the given permutation. Algebraic extraction techniques do not exploit all of the Boolean properties, i.e., idempotence (i.e., $aa = a$) and complement (i.e., $aa' = 0$). Hsu and Shen invented an extended algebraic division, called a coalgebraic division, introducing the Boolean properties [5]. They applied the technique to only resubstitution. Accordingly, Boolean extraction can produce extracted Boolean expressions with fewer literals, compared with algebraic extraction. But difficulty in finding common Boolean divisors arises in

0-7803-7761-3/03/$17.00 ©2003 IEEE

Boolean extraction.

In this paper, we propose a heuristic Boolean extraction, which is a modification of the technique in [6] to multiple-output logic circuits. The key idea is to identify and extract two-cube Boolean subexpression pairs from each original expression in a Boolean network. Because the size of set of all Boolean subexpressions is very large, we restrict our attention to two-cube Boolean subexpressions. After the extraction of two-cube Boolean subexpression pairs, common single-cubes are extracted until no further common single-cube exists.

The rest of this paper is organized as follows. In Section 2, we give some basic definitions and review the kernel-based extraction in [2]. In Section 3, we present our new Boolean extraction. In Section 4, we present our experimental results. Finally in Section 5, we present some conclusions.

2. PRELIMINARIES

2.1. Definitions

In this section, we provide some formal definitions which are helpful in describing extraction. These definitions are consistent with the definitions introduced in [1, 2].

A *variable* or a *Boolean variable* is a symbol representing a single coordinate of Boolean space. A *literal* is a variable or its complement. A *cube* is a set of literals such that if a literal a is present, then a literal a' is not present. An *expression* or *sum-of-products form* is a set of cubes. An expression is *irredundant* if no cube in the expression properly contains another. A *Boolean expression* F is an irredundant expression. A *subexpression* G of a given expression F is another expression, where $G \subseteq F$. The *support* of an expression F, written as $sup(F)$, is the set of variables used in the expression, i.e., $sup(F) = \{x | \exists \text{ cube } C \in F \text{ such that } x \in C \text{ or } x' \in C \}$. An expression is said to be *cube-free* if there is no literal common to all cubes. A *kernel* of an expression is a cube-free quotient of the expression divided by a cube which is called *co-kernel* of the expression. A *Boolean network* is a directed acyclic graph (DAG). Each node i of the graph corresponds to a variable y_i and a representation F_{y_i} of a Boolean expression. The product FG of two expressions F and G is a set $\{C_i \cup D_j | C_i \in F \text{ and } D_j \in G\}$. When F and G have disjoint support, FG is an *algebraic product*. Otherwise, FG is a *Boolean product*. The quotient F/G of an expression F by another expression G is the largest set Q of cubes such that $F = QG + R$ where Q is a quotient and R is a remainder. If QG is restricted to an algebraic product, F/G is an *algebraic quotient*, otherwise F/G is a *Boolean quotient*. If $F/G = Q \neq \phi$, and Q can be obtained using algebraic division, then G is an *algebraic divisor* of F, otherwise G is a *Boolean divisor* of F.

2.2. Kernel-based Extraction

The relations between kernel sets and common multiple-cube expressions are stated precisely by Brayton and McMullen's theorem [1]. From their theorem, we acquire two consequences. First, when two expressions have either no kernel intersections or only single-cube kernel intersections, they cannot have any multiple-cube common subexpression. Second, multiple-cube kernel intersections are common multiple-cube divisors of Boolean expressions corresponding to kernels.

Therefore, we need two steps to extract common subexpressions for multiple-output logic circuits. The computation of the kernel set of expressions in a Boolean network is the first step toward the extraction of multiple-cube expression. The candidate common subexpressions to be extracted are then chosen among kernel intersections. Since there can be an exponential number of kernels in the number of Boolean variables, these operations take an impractical amount of time for many circuits. Then, these operations are restricted to only level-0 kernels in MIS [2]. This problem is particularly evident in the case of symmetric Boolean expressions.

Example 1: Consider the following expressions: $F_0 = ace+bce+de+g$ and $F_1 = ad+bd+cde+ge$. The kernel set $K(F_0)$ of F_0 is $K(F_0) = \{(a+b), (ac+bc+d), (ace+bce+de+g)\}$. The kernel set $K(F_1)$ of F_1 is $K(F_1) = \{(a+b+ce), (cd+g), (ad+bd+cde+ge)\}$. Then, multiple-cube common subexpressions can be extracted from F_0 and F_1. There is only one kernel intersection, namely between $(a+b) \in K(F_0)$ and $(a+b+ce) \in K(F_1)$. The intersection is $a+b$, which can be extracted to yield: $F_0 = wce + de + g$, $F_1 = wd + cde + ge$, and $F_w = a + b$.

3. BOOLEAN EXTRACTION

Our Boolean extraction uses a Boolean division with two-cube Boolean subexpression pairs. If a Boolean expression is a constant or a literal, the trivial extraction is returned. Otherwise, we search two-cube Boolean subexpression pairs for Boolean extraction. In this section, we shall introduce how to find two-cube Boolean subexpression pairs and demonstrate that these pairs provide a framework of Boolean extraction through an example.

3.1. Two-cube Boolean Subexpression Pairs

Two-cube Boolean subexpression pairs are calculated from two algebraic divisor/quotient pairs. The divisor is a cube which divides arbitrary two cubes of a given Boolean expression. C and Q denote sets of algebraic divisors and two-cube quotients, respectively. We express each pair in parentheses. Suppose (c_i, q_i) and (c_j, q_j), where $c_i \in C$, $c_j \in C$, $q_i \in Q$, $q_j \in Q$, and $i \neq j$, are algebraic divisor/quotient pairs. If $c_i \in q_j$, $c_j \in q_i$, and $q_i q_j$ is a subexpression of a given Boolean expression, then (q_i, q_j) is called a two-cube Boolean subexpression pair.

Example 2: Consider three Boolean expressions $F_0 = wx'y + wxz + yz$, $F_1 = wx'y + xz + yz$, and $F_2 = wxy + xz + yz$. Then two-cube algebraic divisor/quotient pairs for F_0, F_1, and F_2 are $\{(w, x'y + xz), (y, wx' + z), (z, wx + y)\}$, $\{(y, wx' + z), (z, x + y)\}$, and $\{(x, wy + z), (y, wx + z), (z, x + y)\}$, respectively. Consider two divisor/quotient pairs, $(y, wx' + z)$ and $(z, wx + y)$, of F_0. The quotient, $wx' + z$, of the first pair contains the divisor, z, of the second pair and the quotient, $wx + y$, of the second pair contains the divisor, y, of the first pair. And $(wx' + z)(wx + y)$ is a subexpression of F_0. Thus, we have a two-cube Boolean subexpression pair, $(wx' + z)(wx + y)$. Similarly, two-cube Boolean subexpression pair, $(wx' + z)(x + y)$, is obtained from two divisor/quotient pairs, $(y, wx' + z)$ and $(z, x + y)$, of F_1.

3.2. Weight Extraction

We can rate each two-cube Boolean pair by its weight which is the number of saved literals if we were to extract that pair. The formula, which is a modification of MIS [2], for computing the weight is as follows:

$$weight(k) = \sum_{i=0}^{1} \{(NF(k_i) - 1)(L(k_i) - 1) - 1\}$$

where $k = k_0 k_1$ is a two-cube Boolean subexpression pair, $NF(k_i)$ is the number of expressions containing two-cube subexpression k_i, and $L(k_i)$ is the number of literals of k_i.

3.3. Algorithm

With the concept of two-cube Boolean subexpression pair, we describe our extraction algorithm as follows:

Algorithm : Boolean extraction.
Input : Set F of Boolean expressions.
Output : Set F of Boolean expressions.
Method :

```
begin
    for each Boolean expression in F do
    begin
        generate algebraic divisor/quotient pairs S;
        find two-cube Boolean pairs P from S;
    end
    for i = 1 to |P| do
    begin
        x = argmax_{k∈P}{weight(k)};
        if (weight(k) ≤ 0) then break;
        /* x is the multiplication of x0 and x1. */
        for each expression f ∈ F for which
                                x0 may be a divisor do
        begin
            substitute f with x0;
        end
        F = F ∪ {x0};
        for each expression f ∈ F for which
                                x1 may be a divisor do
        begin
            substitute f with x1;
        end
        F = F ∪ {x1};
    end
    extract single-cube divisors for F;
end
```

The algorithm takes as an input Boolean expressions F. It generates all divisor/quotient pairs from arbitrary two-cubes of F. Then it generates two-cube Boolean pairs. It picks a two-cube Boolean pair with the maximum weight and substitutes its each two-cube subexpression into all expressions. Finally, the algorithm extracts common single-cubes.

Example 3: Consider again the three Boolean expressions and the two-cube Boolean pairs of Example 2. The two-cube Boolean pairs are $(wx' + z)(wx + y)$ and $(wx' + z)(x + y)$ of F_0 and F_1, respectively. Now, we select a two-cube Boolean pair of F_1, $(wx' + z)(x + y)$, with maximum weight. Since $wx' + z$ is a divisor of F_0 and F_1, $wx' + z$ is substituted into F_0 and F_1. The subexpression, $x + y$, of F_1 is also substituted into F_1 and F_2. Hence, the algorithm generates new Boolean expressions, $F_0 = t(wx + y)$, $F_1 = ts$, $F_2 = wxy + zs$, $F_t = wx' + z$, and $F_s = x + y$. The new Boolean expressions contain 16 literals. However, by algebraic extraction, we have expressions with 17 literals as fol-

Table 1: Experimental results

Circuit	#input	#output	SIS1.2	Proposed method
Example1	4	3	17	16
rd53	5	3	68	73
con1	7	2	19	19
z4ml	7	4	67	61
cmb	16	4	70	59
C17	5	2	9	9
rd73	7	3	160	169
decod	5	16	52	52

lows. $F_0 = wxz + t$, $F_1 = xz + t$, $F_2 = x(wy + z) + yz$, and $F_t = (wx' + z)y$.

4. RESULTS

In this section, we present some results comparing our Boolean extraction to kernel-based extraction in SIS1.2 or MIS. Since we are only interested in the number of literals after extraction, the number of literals reported for SIS1.2 was obtained by issuing the command "gkx -b" and then "gcx -b" on all benchmark circuits. Note that the command "gkx" and "gcx" are to extract common multiple-cube divisors and single-cube divisors, respectively. The experimental results are summarized in Table 1. The column titled Circuit gives the names of the benchmark circuits. The columns #input and #output show the number of primary inputs and outputs, respectively. The column titled SIS1.2 gives the number of literals. The final column gives the number of literals reported from our method. Table 1 shows that SIS1.2 provides fewer literal counts on symmetric expressions such as rd53 and rd73. However, our method provides fewer or equal literal counts on the other benchmark circuits.

5. CONCLUSIONS

Generation of good common subexpressions for extraction is very important for optimizing the global structure of a Boolean network, but it is a computationally hard problem. Algebraic extraction alone does not exploit all of the Boolean properties of Boolean expressions. To improve the results, we also perform Boolean extraction by two-cube Boolean subexpression pairs. Considering only two-cube Boolean subexpression pairs, we can restrict search space. And since we exploit Boolean properties for two-cube pairs, we can do Boolean extraction and get a better Boolean network than algebraic one. From the experimental results, we conclude that we can acquire Boolean networks with significantly fewer literals if our method is used as a supplement for Boolean extraction by SIS.

6. REFERENCES

[1] R. K. Brayton and C. McMullen, "The decomposition and factorization of Boolean expressions," *in Proceedings of ISCAS*, pp. 49-54, 1982.

[2] R. K. Brayton, R. Rudell, A. Sangiovanni-Vincentelli, and A. R. Wang, "MIS: A multiple-level logic optimization system," *IEEE Trans. CAD*, vol. 6, no. 6, pp. 1062-1081, Nov. 1987.

[3] J. Rajski and J. Vasudevamurthy, "The testability-preserving concurrent decomposition and factorization of Boolean expressions," *IEEE Trans. CAD*, vol. 11, no. 6, pp. 778-793, June 1992.

[4] V. K. Singh and A. A. Diwan, "A heuristic for decomposition in multilevel logic optimization," *IEEE Trans. VLSI*, vol. 1, no. 4, pp. 441-445, Dec. 1993.

[5] W.-J. Hsu and W.-Z. Shen, "Coalgebraic division for multilevel logic synthesis," *in Proceedings of DAC*, pp. 438-442, 1992.

[6] O.-H. Kwon, S. J. Hong, and J. Kim, "A Boolean factorization using an extended Boolean matrix," *IEICE Trans. Inf. & Sys.*, vol. E81-D, no. 12, pp. 1466-1472, Dec. 1998.

OPTIMAL USE OF 2-PHASE TRANSPARENT LATCHES IN BUFFERED MAZE ROUTING

S. Hassoun
Tufts University, Medford, MA

Abstract

Clocking frequencies continue to increase due to the demand for higher performance. Together with the larger die sizes, multiple clock cycles are now required to cross a chip. A routing tool must thus insert registers as well as buffers while minimizing the path latency. This paper addresses optimal buffered path construction across multiple clock cycles using 2-phase transparent latches. We demonstrate the benefits of routing using latches over registers, and we present a polynomial routing algorithm. Our results confirm the correctness of our algorithm.

1. INTRODUCTION

Future SoC designs give rise to new problems in routing and buffer insertion. A particular concern is that multiple clock cycles are now required to cross a chip. This is due to the increase in die sizes and the continued demand for high performance and thus higher clocking frequencies. Another concern is the need to route through complicated terrains due to circuit blockages (e.g. IP blocks, memories – typically used and reused to decrease time to market) and wire blockages (e.g. data paths).

Routing and buffering long routes that require multiple clock cycles while using registers was recently addressed [4]. The authors introduce an algorithm that minimizes latency (or number of registers) along a route. The resulting register-to-register delays are less than the clock period. The proposed algorithm is polynomial and builds upon the Fast Path framework proposed in[6].

We address in this paper the problem of buffered routing while using transparent (i.e. level-sensitive) latches. Such latches are typical in high-performance designs, and they can improve on the performance when compared to using registers[3]. We demonstrate the benefits of using latches in routing, and we present a polynomial algorithm to do such optimal routing. To find the optimal buffered-latched path between a sink t and a source s, we explore all routing and latch insertion points within a given routing area while considering both physical and wire obstacles. Our models and algorithms builds on those developed in[6, 4]. We restrict our discussion here to 2-phase clock schedule; the work, however, can be extended to handle any clock clock schedule.

The remainder of the paper is partitioned as follows. Section 2 presents our routing and clocking models. Section 3 demonstrates the use of 2-phase level-sensitive latches in routing and its benefit over registers. Section 4 defines our routing problem and presents our algorithm. We present experiments in section 5 and conclude in section 6.

2. BACKGROUND

2.1. Clock Schedule Model

We use the SMO clocking formulation[5] to model our clocks. A 2-phase clock schedule, Φ, is an ordered collection of 2 periodic signals, ϕ_1 and ϕ_2, having a common period π. Because phases are periodic, a *local time zone* of width π is associated with each phase. Each phase ϕ_i is characterized by two parameters e_i and ω_i. Parameter e_i represents the absolute time when ϕ_i begins (relative to an arbitrary global time reference). Parameter ω_i is the length of time that ϕ_i is active (latch is open). We assume that the design intention and thus the clock schedule specify that a signal departing from a latch k must be captured by the next latching edge (which occurs after the latching edge of k) of the following latch l.

Because we perform our routing starting from the sink and moving towards the source, it is useful to translate a time measurement a from the local time zone of ϕ_i into the *previous* local time zone of ϕ_j. To do so, we subtract from a a phase shift operator $E'_{i,j}$ defined as:

$$E'_{i,j} = \begin{cases} e_j - e_i - \pi & \text{if } i < j \\ e_j - e_i & \text{otherwise.} \end{cases}$$

$E'_{i,j}$ is π less than the original $E_{i,j}$ phase shift operator in[5] which translates a time measurement a from the local time zone of ϕ_i into the *next* local time zone of ϕ_j.

A 2-phase non-overlapping clocking scheme is demonstrated in Figure 1. If the clock period π is 10 time units, $\omega_1 = 7$, $\omega_2 = 3$, $E'_{1,2} = -7$, $E'_{2,1} = -3$, then an arrival of 8 in ϕ_2's time zone translates to an arrival of 11 relative to the *previous* occurrence of ϕ_1's time zone.

The earliest arrival time at the output of a latch k clocked by ϕ_i is at the opening edge: $\pi - w_i + latch\_propagation$. The latest arrival at the input of a latch clocked by ϕ_i is at the closing edge: $\pi - setup\_time$. Both times are in reference to ϕ_i's time zone.

2.2. Routing Model

To model physical and wiring blockages, one may construct a grid (maze) graph $G(V, E)$ over the potential routing

0-7803-7761-3/03/$17.00 ©2003 IEEE

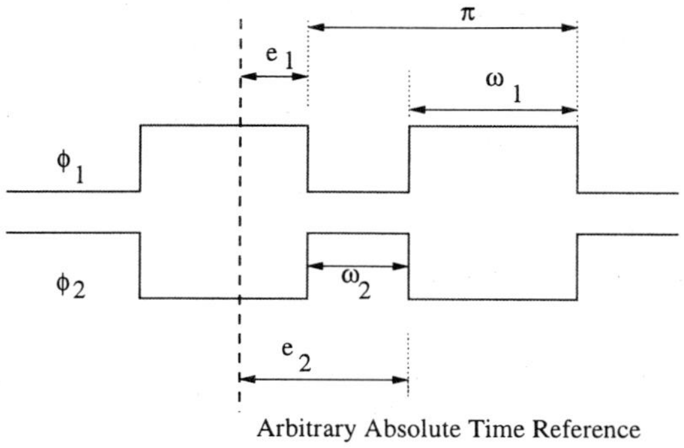

Figure 1. An example 2-phase clock schedule.

area, whereby each node corresponds to a potential insertion point for a buffer or latch element, and each edge corresponds to part of a potential route. Edges in the grid graph which overlap wiring blockages are deleted, and nodes that overlap physical obstacles are labeled blocked. More precisely, we define a label function $p : V \leftarrow \{0,1\}$ where $p(v) = 0$ if $v \in V$ overlaps a physical obstacle and $p(v) = 1$ otherwise.

For each edge $(u,v) \in E$, let $R(u,v)$ and $C(u,v)$ denote the capacitance and resistance of a wire connecting u to v. Let $R(g)$, K(g), and $C(g)$ respectively denote the resistance, propagation delay, and input capacitance of a given buffer or latch element g. We use the resistance-capacitance π-model to represent the wires b, a switch-level model to represent the gates, and the Elmore model to compute path delays.

A *path* from node s to t in the grid graph G is a sequence of nodes $(s = v_1, v_2, \ldots, v_k = t)$. An *optimized path* from s to t is a path plus an additional labeling m of nodes in the path. We assume that the target node has a latch: $m(t) = l$, and a phase assignment, $phase(m(t)) = k$, where $k \in \{\phi_i, \phi_2\}$. We also assume that the signal arrive at the target by the closing edge of ϕ_i. We assume that the source node will also have a latch; however, we do not assign it a phase. We assume that the required time at the source to be the opening edge of the phase clocking the source latch. Let B be a buffer library consisting of non-inverting buffers. Each node $v \in V - s,t$ may be assigned a buffer from B or a latch, or not have a gate (corresponding to $m(v) = 0$). When assigned a latch, v is also assigned a phase. The assumptions regarding the phase assignments and required arrival times at the sink are simply to simplify the presentation in this short paper.

3. ROUTING WITH LATCHES V.S. REGISTERS

To understand the benefits of routing with latches v.s. registeres, consider Figure 2. Assume that the specified clock period is 10 ns, and that the delay through each grid edge is 1 ns. The optimal routed path with registers from source to sink in Figure 2(a) requires 3 registers. Thus, there is a 3 clock period latency, or a 30 ns time latency. The circuit blockage requires placing a register on each side of the blockage. Slack thus exists in the routing segments on both sides of circuit blockage.

Assume that transparent symmetric latches are used in Figure 2(b). In this case, the time latency is from the rising edge

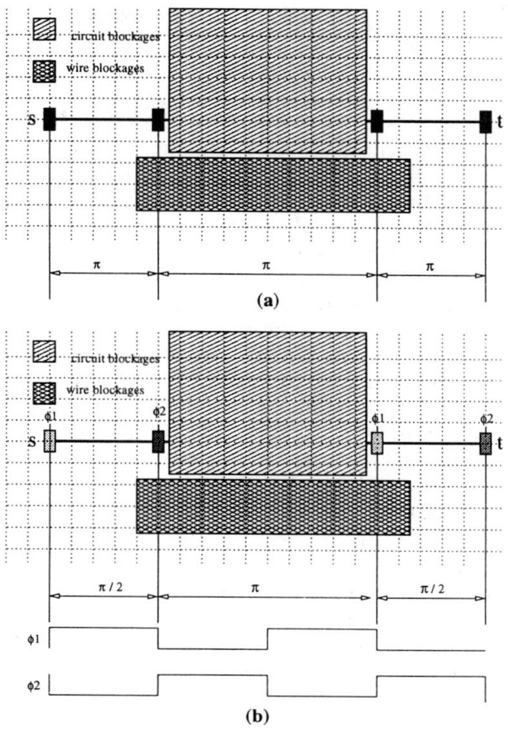

Figure 2. An example of routing within a single-clock domain

of ϕ_1 to the second falling edge of the ϕ_2, for a total time latency of 20 ns. Like in (a), latches are forced to surround the circuit blockage to accommodate the long delay. However, there is no slack along the routing segments on both sides of the circuit blockage. Latches thus offer greater time flexibility which could result in a smaller latency.

4. PROBLEM & SOLUTION

Our general routing problem can be formally defined as follows:

Problem: Given a routing graph $G(V,E)$, a set $I = B \cup \{l\}$, and two nodes $s,t \in V$, find a feasible buffered-latched path from s to t such that the time latency from s to t is minimized.

The time latency here refers to the time from the opening edge of the latch at the source to the closing edge of the latch at the sink.

Several key issues were considered when developing the routing algorithm. First, latches must be inserted periodically whenever the timing constraints dictate that a latch is needed. Furthermore, latches are clocked by different phases and hence latch ordering must be respected. Our example circuit in Figure 2 required alternating ϕ_1 and ϕ_2.

Second, latches cannot be inserted too far apart, otherwise, the routed signal will not be latched properly and timing violations occur. Furthermore, since signals propagate across latches only during latch transparency periods, and we are propagating from sink towards source, we compute *arrival time* at each grid point – in contrast to *delays* that were used in[4, 6].

Finally, since latches have a convenient *local time zone* notion, and a phase shift operator that translates from one local time zone to another, we will use the shift operator to update the arrival time at the latch input upon placing a latch at a grid point. To facilitate comparing partial solution, the expansion of candidate solutions from sink to source will proceed in waves of partial solutions wherein each wave corresponds to a different number of latches.

The Latched-Buffered Path (LBP) algorithm is presented in Figure 3. Like the algorithms in[4, 6], we wish to explore all buffering and latching solutions, and we need to prune solutions as we route the signal from the sink back to the source. The core data structures are two priority queues, Q and Q^*, that sort candidates based on the latest arrival time. We also use two arrays A_1 and A_2 to mark whether a solution with $m(v) = l$ has been generated for node v to prevent multiple candidates with same-phase latches from getting inserted at v.

The algorithm begins by initializing Q to the set containing a single sink candidate, while Q^* is empty. The current phase is set to be the one that must precede the phase of the latch at the sink. Candidates are then iteratively deleted from the Q and expanded either to add an edge (Step 5) or a buffer (Step 7) or a latch (Step 8). Buffers and latches are only inserted when $p(v) = 1$: when there is no overlap with a blocking physical obstacle (Step 6).

When candidate solution expands via edges or buffers, the arrival time is updated by subtracting the necessary RC delay. The new solution is then only added to the *current* queue Q if no timing violation will occur. However, when a new latch is inserted, the arrival time is set to be the minimum of $\pi - setup(l)$, the closing edge of the latch (i.e. a signal cannot possibly arrive at the output of the latch before then), or by subtracting the necessary RC delay. Moreover, to account for transforming into a previous time domain, the arrival time is updated by subtracting the phase shift operator $E'_{current\_phase,previous\_phase}$. This candidate solution is then added to the *next* queue, Q^*, if no timing violation will incur.

If the source is reached, it is pushed onto the Q in Step 5, and when it is eventually popped from the queue, it is returned as the optimum solution (Step 4). With each addition to either queue, candidates for the current node are checked for inferiority and then pruned accordingly. The complexity of the aglorithm is polynomial, as the one presented in [6].

5. EXPERIMENTAL RESULTS

We implemented our algorithm in C and ran the experiments on a Sun Solaris Enterprise 250. We use estimated parameters for a 0.07μ technology as reported by Cong and Pan [2]. We use a single buffer size of 100 times minimum gate width and triple wide wires, and assume delay characteristics for the latches to be identical to that of the buffer. As in [1], we use a 25 by 25 mm chip and place the source and sink 40 mm apart. These choices guarantee a significant number of clock cycles will be required to traverse from s to t. We used a grid separation of $0.5mm$, which corresponds to a 100×100 grid.

Our results are presented in Table 1. For each clock period, we used a 2-phase symmetric schedule. We report the time latency, the number of latches, the number of inserted buffers, and the run time. As π decreases, more latches are needed but less buffers are used as latches are used as buffers to reduce RC delays. The run time decreases as π decreases as more aggressive pruning occurs with shorter clock periods.

We then compare our results with routing using registers and without using any at the bottom part of the table. We see that when π is 250, the latches are able to achieve a smaller clock latency. The run times are smaller in the register case because one can prune more aggressively[4].

Table 1. Buffered Latched Routing statistics as a function of π and grid separation of 0.5 mm and a grid of 100×100.

Routing with latches				
π	Latency	# Latches	# Buffers	Run Time
1000	3000	4	12	100.36
400	2800	12	2	82.3
250	2875	21	0	59.03
Routing with registers				
π	Latency	# Latches	# Buffers	Run Time
1000.00	3000	2	13	49.93
400.00	2800	6	7	36.08
250	3000	11	0	27.20
Routing without registers or latches				
∞	2738.57	-	16	48.65

6. CONCLUSION

Automated buffered routing is a necessity in modern VLSI design. Any CAD tools currently performing buffer insertion will eventually have to deal with synchronizer insertion. This paper described the important problem of routing signals and synchronizing them using transparent latches. The two contributions of this paper are the polynomial algorithm for latched buffered routing and demonstrating that latched buffered routing can achieve better performance than registered buffered routing.

7. REFERENCES

[1] J. Cong. "Timing Closure Based on Physical Hierarchy". In *Proceedings of the International Symposium on Physical Design*, pages 170–174, 2002.

[2] J. Cong and Z. Pan. "Interconnect Performance Estimation Models for Design Planning". *IEEE Transactions on Computer-Aided Design*, 20(6):739–752, June 2001.

[3] C. Ebeling and B. Lockyear. "On the Performance of Level-Clocked Circuits". In *Advanced Research in VLSI*, pages 242–356, 1995.

[4] S. Hassoun, C. Alpert, and M. Thiagarajan. "Optimal Buffered Routing Path Constructions for Single and Multiple Clock Domain Systems". In *Proc. of the IEEE International Conference on Computer-Aided Design (ICCAD)* 2002.

[5] K. Sakallah, T. Mudge, and O. Olukotun. "Analysis and Design of Latch-Controlled Synchronous Circuit". In *Proc. 27th ACM-IEEE Design Automation Conf.*, pages 111–7, 1990.

[6] H. Zhou, D. F. Wong, I.-M. Liu, and A. Aziz. "Simultaneous Routing and Buffer Insertion with Restrictions on Buffer Locations". *IEEE Transactions on Computer-Aided Design*, 19(7):819–824, July 2000.

Latched Buffered Routing Algorithm $(G, B, s, t, m', l, \Phi)$

Input: $G(V,E) \equiv$ Routing grid graph
$B \equiv$ Buffer library
$s \equiv$ source node
$t \equiv$ sink node
$m' \equiv$ initial labeling with $m'(t) = l$
$l \equiv$ latch for clocking signal
$\Phi \equiv$ given clock schedule: $\pi, E'_{1,2}, E'_{2,1}, \omega_1, \omega_2$

Vars: $Q \equiv$ priority queue of candidates
$Q^* \equiv$ queue holding next candidate wave
$\alpha = (c,a,m,v) \equiv$ Candidate at v
$A_i \equiv$ Marking of nodes with inserted $l(\phi_i), \forall_i \in P_\Phi$

Output: $m \equiv$ Labeling of complete s-t path

1. $Q \leftarrow \{(C(r), \pi, predecessor(phase(m'(t))), m', t)\}$.
   $Q^* = \emptyset,\ A_i(v) = 0, \forall v \in V$ and $\forall i \in P_\Phi$
   $current\_phase = predecessor(phase(m'(t)))$
2. **while** ($Q \neq \emptyset$) **or** ($Q^* \neq \emptyset$) **do**
   if ($Q = \emptyset$) **then**
   $Q = Q^*,\ Q^* = \emptyset$.
   $current\_phase = predecessor(current\_phase)$
3. $(c,a,m,u) \leftarrow extract\_min(Q)$
4. **if** $u = s$ **and** $a - K(l) - R(l)\cdot c \geq \pi - \omega_{current\_phase}$ **then**
   return labeling m.
5. **for each** $(u,v) \in E$ **do**
   $c' \leftarrow c + C(u,v)$
   $a' \leftarrow a - R(u,v)\cdot(c + C(u,v))/2$
   if $a' \geq \pi - \omega_{current\_phase} + K(l) + min(R(B \cup l))\cdot c'$ **then**
   push (c', a', m, v) onto Q and prune
6. **if** $p(u) = 1$ **and** $m(u) = 0$ **then**
7. **for each** $b \in B$ **do**
   $c' \leftarrow C(b),\ m(u) = b$
   $a' \leftarrow a - R(b)\cdot c$ - K(b)
   if $a' \geq \pi - \omega_{current\_phase} + K(b) + R(b)\cdot c'$ **then**
   push (c', a', m, u) onto Q and prune
8. **if** $A_{current\_phase}(u) = 0$ **and** $a - K(l) - R(l)\cdot c \geq \pi - \omega_{current\_phase}$ **then**
   $m(u) = l,\ A_{current\_phase}(u) = 1$
   $a' = min(\pi - setup(l), a - K(l) - R(l)\cdot c) - E'_{current\_phase,pred(current\_phase)}$
   push $(C(l), a', m, u)$ onto Q^* and prune

Figure 3. The Latched-Buffered Path (LBP) Algorithm.

CONCURRENT OPTIMIZATION OF PROCESS DEPENDENT VARIATIONS IN DIFFERENT CIRCUIT PERFORMANCE MEASURES

Ayhan A. Mutlu, Norman G. Gunther, and Mahmud Rahman

Electron Devices Laboratory
Department of Electrical Engineering, Santa Clara University
500 El Camino Real, Santa Clara, CA 95053-0569, USA

ABSTRACT

A method for multi-objective circuit variability optimization in the presence of process variations is presented. Critical process parameter variations are identified by determining their correlations to the circuit performance measures of interest. Then, the distributions of these critical process parameters are used to identify the critical designable parameters for variability optimization. Membership functions and fuzzy set intersection operators are used to transform multiple design objectives into a single objective function suitable for optimization. Afterwards, the objective function for variability is minimized. Finally, the mean circuit performance measures are fine tuned for given target specifications.

1. INTRODUCTION

Recent technological advances in lithography leading to rapid shrinking of minimum feature sizes in ICs have not been accompanied by suitable scaling of geometric tolerances. As a consequence, circuit performance is becoming more sensitive to uncontrollable statistical process variations.

Traditionally, Computer Aided Design (CAD) tools have been used to determine nominal design parameters in ICs, such that the nominal circuit response meets the desired performance specifications. Even though one set of parameter values that satisfy performance criteria can be found, due to variations in the manufacturing process, the actual circuit response always shows variations around the nominal set of values. Sometimes the variation is so large that the circuit performance will not meet the target specifications. Consequently, these variations may affect the manufacturing yield.

As the demand for high quality ICs continuously increases, selecting optimal values of circuit elements and parameters to minimize adverse effects of statistical process variations becomes an important engineering design problem. The objective of Design for Quality and Manufacturability is to minimize the performance variability and to keep the mean value of the performance close to the defined target.

Until now, a number of methods [1,2] have been developed to reduce performance variability on discrete and integrated circuits. Due to simplistic approximations of circuit performance sensitivities to element tolerances, most of these methods have had somewhat limited success. Therefore, the work presented here does not attempt to describe the performance dependencies to process variations with simple analytical expressions. Instead, Monte Carlo (MC) estimations are used to describe the variability behavior of the circuit. The MC method is applicable to any type of circuit without requiring simplifying assumptions of any form in the probability distributions, the process parameter variations, and the size of the parameter space. However, MC analysis requires a large number of circuit simulations to have a valuable estimation. We used Latin Hypercube Sampling (LHS) approach [3] for sample generation to overcome this caveat of MC analysis. The number of circuit simulations can considerably be reduced with the LHS technique, since it ensures that each parameter has all portions of its distribution represented on the given sample size.

The method presented here, finds the optimal values of the CMOS circuit designable parameters to minimize the spreads in the critical delay and power dissipation due to the variations in the process. Process variations are expressed by the variations in five variables significant in transistor behavior; channel length, channel width, flat-band voltage, gate oxide thickness, and sheet resistance both for NMOS and PMOS. Since these parameters are determined at different steps of the manufacturing process, they can be assumed to be statistically independent [4,5].

In the following sections, the LHS method is explained. It is followed by the identification of critical process disturbances. Then, the gradient expressions of the variability and the mean with respect to the designable parameters are derived. After that the individual objectives are transferred into appropriate fuzzy membership functions, which then combined with fuzzy intersection operations to obtain a single deterministic objective. Finally, the overall optimization framework is described and applied to a typical delay circuit.

2. METHODOLOGY

2.1 Latin Hypercube Sampling (LHS)

Random samples are easy to generate. The drawback of primitive random sampling is that it requires a large number of samples to estimate the quantities with sufficiently small errors. For example, in a Gaussian distribution, the values near to the mean are more likely to be selected in the sample due to their high probability of occurrence. In other words, the values away from the mean would not be well represented in a small number of samples. The LHS strategy alleviates this problem by partitioning the sampling region into disjoint sub-regions for the range of each random variable. It is computationally inexpensive to generate and can cope with many input variables. To generate

0-7803-7761-3/03/$17.00 ©2003 IEEE

an M point sample from n variables $x_1, x_2, \ldots, x_n$, the range of each variable is partitioned into M non-overlapping intervals of equal marginal probability 1/M. One value from each interval is randomly selected with respect to the probability density of the interval. The M values obtained for x_1 are paired in a random manner with the M values for x_2. These M pairs are randomly combined with M values of x_3. This process continues until a set of M point n-tuples is formed. Five point LHS for two random variables, one with uniform distribution, and one with Gaussian distribution, is shown in Fig. 1.

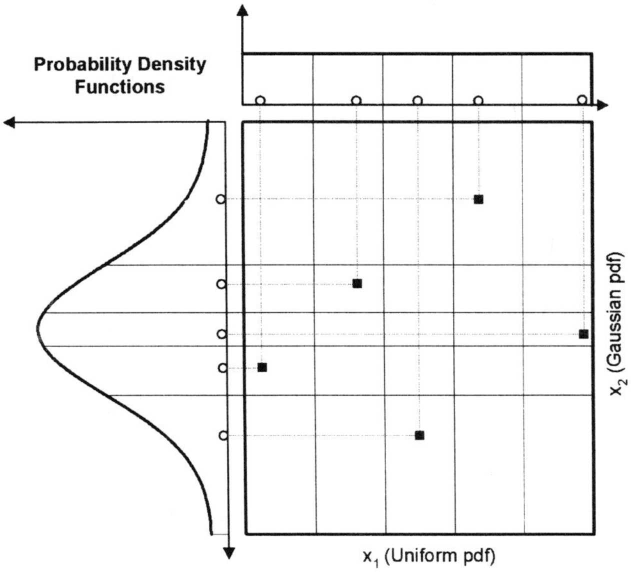

Figure 1. Five point Latin Hypercube Sampling (LHS) for a uniform and a Gaussian random variable

2.2 Critical Process Variations

The solution to IC optimization problems in the presence of process variations requires interaction and parameter correlation analyses. This may form a major bottleneck for the analysis large circuits due to the required number circuit simulations. In order to solve real life statistical IC design problems screening the process variation effects is necessary. This screening enables us to determine the relative effects of these variations on the circuit performance measures of interest. Estimating the correlation coefficients between the disturbances and the performances is one way of identifying the critical variations. The correlation coefficient ρ, between the performance y, and the disturbance ξ_j at nominal values of designable parameters x^0 can be written as,

$$\rho_{y,\xi_j} = \frac{Cov(y,\xi_j)}{\sqrt{Var(y)\times Var(\xi_j)}} \tag{1}$$

The covariance and variances can be found with the following MC estimators,

$$Cov(y,\xi_j) = \frac{1}{N-1}\sum_{i=1}^{N}\left(y(x^0,\xi^i)-\overline{y}(x)\right)\left(\xi_j^i-\overline{\xi}_j\right) \tag{2}$$

$$Var(y) = \frac{1}{N}\sum_{i=1}^{N}\left(y(x^0,\xi^i)-\overline{y}(x^0)\right)^2 \tag{3}$$

and

$$Var(\xi_j) = \frac{1}{N}\sum_{i=1}^{N}\left(\xi_j^i-\overline{\xi}_j\right)^2 \tag{4}$$

In the above equations, N is the total number of LHS samples, $\xi^i = (\xi_1^i, \xi_2^i, \ldots, \xi_m^i)$ is the ith sample of m disturbances, identified to be critical, $y(x^0,\xi^i)$ is the circuit response to the ith sample of disturbances, $\overline{y}(x^0)$ and $\overline{\xi}_j$ are means of y and ξ_j respectively.

2.3 Gradient Analysis

The output sensitivity analysis for small changes of designable parameters is performed using gradient analysis techniques. The gradient analysis estimates the relative effects of the designable parameters on the circuit performance measures, and enables us to sort these parameters by their relative effects to the performance variability and the target mean. The designable parameters with significant output effects are then used for variability minimization and mean performance tuning.

The variability of a performance measure can be expressed by its variance, given in equation 3. Finite-difference method can be used to find the gradient of the variability with respect to designable parameters. In this method, each designable parameter of the circuit is perturbed one at a time. Differencing of the corresponding measurement values and then dividing by the difference interval form each component of the gradient estimate. The MC estimator for the gradient of variability with respect to a designable parameter x_j can be written as,

$$\frac{\partial Var(y)}{\partial x_j} \cong \frac{1}{2\Delta x_j}\frac{1}{N}\sum_{i=1}^{N}\mathrm{K}_y\left(x^0+\Delta x_j,\xi^i\right)-\mathrm{K}_y\left(x^0-\Delta x_j,\xi^i\right) \tag{5}$$

where

$$\mathrm{K}_y\left(x^0,\xi^i\right) = \left[y(x^0,\xi^i)-\overline{y}(x^0)\right]^2$$

Similarly, the MC estimator for the gradient of mean performance with respect to a designable parameter x_j is,

$$\frac{\partial \overline{y}}{\partial x_j} \cong \frac{1}{2\Delta x_j}\frac{1}{N}\sum_{i=1}^{N} y(x^0+\Delta x_j,\xi^i)-y(x^0-\Delta x_j,\xi^i) \tag{6}$$

The variability minimization is more difficult than tuning of the mean performance, because the designable parameters have larger effects on the mean than the variance. Therefore, in the optimization flow, first the variability is minimized with the choice of appropriate designable parameters based on equation 5, then the mean value of the performances are tuned by choosing the suitable designable parameters based on equation 6. Detailed description of this two-step procedure is given in section 3.

2.4 Objective Function

A critical step in multi-objective optimization is to define the function for optimization. In this work, fuzzy set theory is applied to the construction of the single objective function. The major motivation for the application of fuzzy set methodologies lies in the fuzziness involved in the specifications in the circuit design, where the design specifications are usually described in terms of linguistic variables, such as high speed, low power, etc. Each circuit performance y_i is transformed into the corresponding optimization objective by using membership function, which characterizes the degree of membership of y_i in the fuzzy set. The selection of the membership function depends on the optimization objective. Figure 2 shows an example of sigmoidal membership function used in this work to transform standard deviation of the average power to the degree of membership. Since this is a minimization problem, the membership function is getting smaller as we reduce the variability Similar function is used for the input output delay variability.

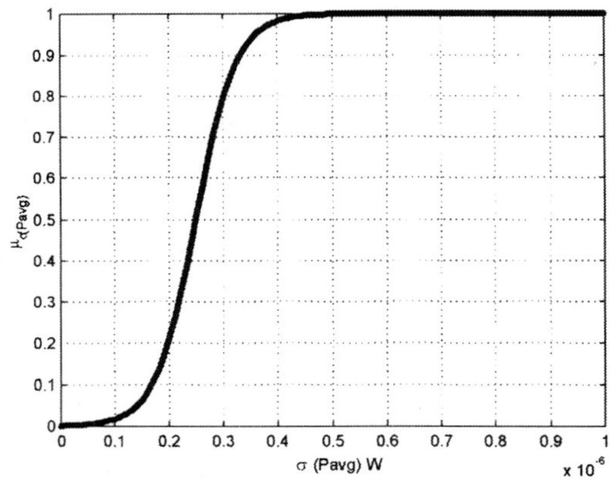

Figure 2. Sigmoidal membership function used in minimization of variability of average power

Once the membership function of each fuzzy set is formed, they are combined using a fuzzy set intersection operation. With the intersection operation we can measure the degree of simultaneous satisfaction of defined design objectives. Figure 3 illustrates the fuzzy set intersection operation used to combine the design objectives into the final objective function.

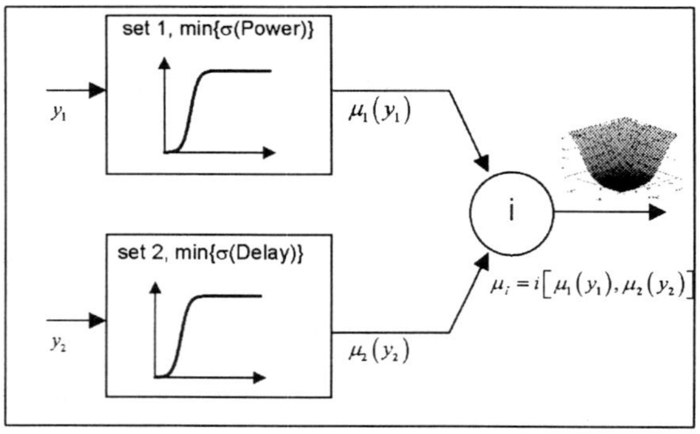

Figure 3. Fuzzy set intersection to obtain final objective

Here, a weighted sum intersection operation is used to move the optimization toward the objective of greater interest. The general form of this intersection operation is,

$$\mu_I(y) = \left(\sum_{i=1}^{n} w_i \mu(y_i)^p \right)^{1/p} \tag{7}$$

where,

$$0 \le w_i \le 1 \text{ and } \sum_{i=1}^{n} w_i = 1$$

2.5 Flowchart

The flowchart of the overall optimization flow is given in figure 4. As seen from the figure the flow has three main steps,

1. Identification of critical process disturbances.
2. Analysis to determine the critical parameters of the design for optimization process.
3. Optimization of power and delay variations

After variability is minimized step 3 is repeated again for the performance tuning by using different designable parameters.

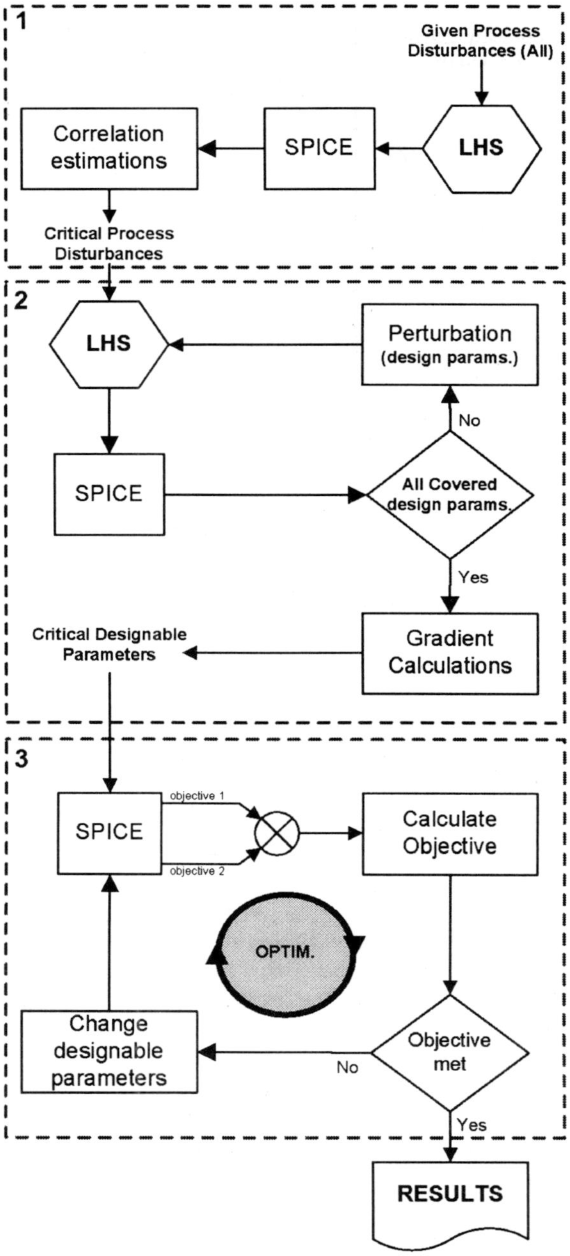

Figure 4. The main steps of the optimization flow

3. SIMULATION RESULTS

The proposed method was applied to a typical delay circuit, shown in Fig. 5.

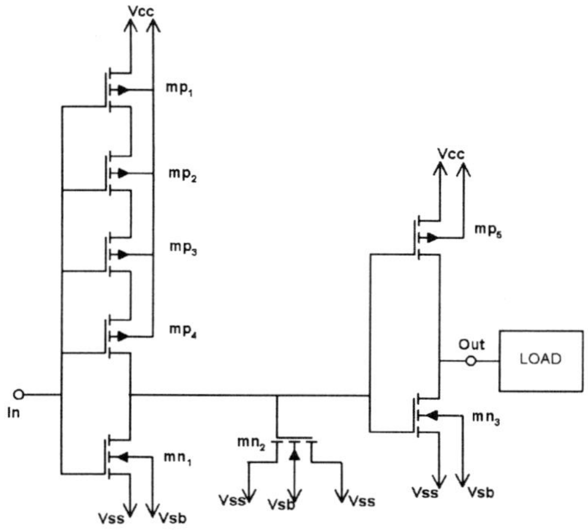

Figure 5. The delay circuit

The delay between out and in nodes, and the average power dissipation are chosen to be the performance measure of interests. Berkeley Predictive Technology Model (BPTM) 0.13 μm device model parameters are used in the simulations. The disturbances used to describe the process variations are listed in Table 1.

Parameter	Mean	Sigma	Description
xl	0 (μm)	0.005 (μm)	Length variation
xwn	0 (μm)	0.03 (μm)	NMOS width variation
xwp	0 (μm)	0.03 (μm)	PMOS width variation
delVton	0 (V)	0.01 (V)	NMOS Vto variation
delVtop	0 (V)	0.01 (V)	PMOS Vto variation
tox	25 (A)	2.25 (A)	Oxide thickness
Rshn	7 (Ohm/Sq.)	0.3 (Ohm/Sq.)	NMOS sheet resistance
Rshp	9 (Ohm/Sq.)	0.5 (Ohm/Sq.)	PMOS sheet resistance

Table 1. Process variations in the delay circuit

To identify the critical ones, the correlation coefficients between variations and output measures are determined. These coefficients are shown in Table 2.

Disturbance	Corr(Pavg)	Corr(Tdly)
xl	0.82479829	-0.1436952
xwn	-0.5294481	0.61806269
xwp	-0.0297072	0.03478332
tox	-0.4296449	-0.4320805
delvton	0.07309311	-0.4436035
delvtop	-0.0383101	0.31254418
rshn	-0.1812213	0.11320898
rshp	-0.1806922	0.09282089

Table 2. The noise correlations to average power dissipation and output fall delay

As seen from Table 2, both sheet resistances and PMOS critical width variations have relatively low correlations to the power and delay. Therefore, these disturbances were not considered in the rest of the flow.

The next step is to find the critical designable parameters for the optimization of power and delay variability. In this analysis, the designable parameters are the widths of the transistors. All designable parameters perturbed by 10% around their nominal values. The MC estimated gradients of these perturbations are given in Table 3.

	Delay (tfall)		Power (Pavg)	
Design Par.	Gradient(Mean)	Gradient(Var)	Gradient(Mean)	Gradient(Var)
dwp1	1.699E-09	-4.450E-22	-1.500E-06	-1.340E-13
dwp2	1.630E-09	-8.410E-22	-2.200E-06	-1.660E-13
dwp3	1.570E-09	-1.188E-21	-2.200E-06	-1.100E-13
dwp4	1.524E-09	-1.466E-21	-1.200E-06	-1.070E-13
dwp5	8.230E-10	-5.535E-22	2.400E-06	1.740E-13
dwn1	3.810E-09	-3.665E-21	-3.000E-06	-2.675E-13
dwn2	3.810E-09	-3.665E-21	-3.000E-06	-2.675E-13
dwn3	3.603E-09	-5.075E-21	5.750E-06	1.120E-12

Table 3. Gradients of means and variances of delay and power with respect to the designable parameters

It is clear from Table 3 that NMOS transistor widths are critical for both power and delay variability. For further minimization of power variability, wp5 and wp2 are also critical. After minimizing the delay and power variability, wp1 and wp2 are good candidates for tuning of the delay mean, because these widths have largest mean delay gradients following NMOS transistor widths.

As a result of simultaneous minimizations of delay and average power variability, the standard deviations of delay and average power dissipation have been reduced from 24.23 ps and 0.636 μW to 12.48 ps and 0.238 μW, respectively. After that the mean of the delay has been reduced from 0.219 ns to 0.175 ns. Power mean was kept close to its original value 10.7 μW, since tuning of the power mean were adversely affecting the variability.

4. SUMMARY

A three stage method for multi objective variability optimization and mean performance tuning was presented. The effects of process variations were minimized by changing the proper designable parameters of the circuit. Once the variability was minimized, the mean performance values were tuned for target specifications. The method has been demonstrated on a simple delay circuit. The approach is general enough, it can deal with any number of disturbances and designable parameters by sorting them according to their relative effects to the performance measures of interests.

5. REFERENCES

[1] A. Ilmoka, and R. Spence, "A Sensitivity Based Approach to Tolerance Assignment", *IEE Proc.* vol. 129, pt G, no. 4, pp. 139-149, August, 1982.

[2] N. Shigyo, "An Analysis of Process Fluctuation Induced Propagation Delay Variation using Analytical Model", Solid-State Electronics, vol. 44, pp. 2183-2191, 2000.

[3] M. Keramat, and R. Kielbasa, "Worst Case Efficiency Latin Hypercube Sampling Monte Carlo (LHSMC) Yield Estimator of Electrical Circuits", *Proc. of IEEE Int.l Symposium on Circuits and Systems*, pp. 1660-1664, Hong Kong, June 1997.

[4] P. Cox, P. Yang, S.S. Mahant-Shetti, and P. Chatterjee, "Statistical modeling for efficient parametric yield estimation of MOS VLSI circuit," *IEEE Trans. on Electron Devices*, Vol. ED-32, no. 2, pp. 471-478, Feburary 1985.

[5] J. Zhang, M. Styblinski, *Yield and Variability Optimization of Integrated Circuits*, Kluwer Academic Publishers, Boston, Publishers. 1995.

TIME DOMAIN RESPONSE AND SENSITIVITY OF PERIODICALLY SWITCHED NONLINEAR CIRCUITS

Quan Li and Fei Yuan

Department of Electrical and Computer Engineering
Ryerson University Toronto, Ontario, Canada M5B 2K3

ABSTRACT

This paper presents a new and efficient method for time domain response and sensitivity analysis of periodically switched nonlinear (PSN) circuits. By representing the nonlinear characteristics using Volterra functional series and interpolating Fourier series, the response and sensitivity of PSN circuits are obtained. The method correctly handles the inconsistent initial conditions that may occur at switching instants by extending the previously proposed two-step algorithm to switched nonlinear circuits. The algorithms have been implemented in a computer program. Example PSN circuits are analyzed and the results are compared with those from PSPICE simulation and the Brute-Force method. An excellent agreement has been obtained.

1. INTRODUCTION

Although general purpose simulation tools, such as PSPICE and SPECTRE that use linear multi-step predictor-corrector (LMS-PC) algorithms as their main simulation engines, have the abilities to handle both mildly and harsh nonlinearities, they can not provide the desired accuracy and efficiency when simulating some special classes of circuits, such as PSN circuits. The main reasons are as follows : (i) Inability to handle inconsistent initial conditions that arise when switches are modeled as an ideal device [1]. (ii) Poor simulation accuracy due to the use of low-order extrapolation schemes in numerical integration. (iii) High computational cost due to the need for Newton-Raphson iterations in every step of integration. In addition to reponse analysis, several methods were proposed to compute sensitivity of nonlinear circuits. The piece-wise linear approach [2] suffers from poor accuracy as the accuracy directly depends upon the number of piecewise-linear sections. The adjoint network approach [3] requires the linearization of nonlinear circuits and construction of corresponding adjoint network at every time point, resulting in poor efficiency. Efficient and unified computer methods for time-domain response and sensitivity analysis of PSN circuits is not available.

In this paper, we present a new, efficient and accurate method for time domain response and sensitivity analysis of PSN circuits. The two main contributions of this paper are : (i) a new extended two-step algorithm is proposed for computing the consistent initial conditions of PSN circuits correctly and efficiently. The algorithm is assessed using the charge conservation before and after switching instants. (ii) a new, efficient and unified method for time-domain response and sensitivity analysis of PSN circuits is proposed. The method is computationally efficient, especially for relatively long time analysis. The sensitivity can be obtained simultaneously with response with little additional computation.

2. RESPONSE OF PSN CIRCUITS

Usually the nonlinearities encountered in telecommunication systems are mildly and can be sufficiently represented by the truncated Taylor series expansion at their DC operating point [4]. The time domain response of the PSN circuits can be solved as follows: within a clock phase, the circuits are nonlinear time-invariant circuits and can be solved using the sampled data simulation for nonlinear circuits given in [5]. At each switching instants, the inconsistent initial conditions are obtained using the extended two-step algorithm for PSN circuits, which will be given in this section. The nonlinear circuits with input $w(t) = e^{j\omega_o t}$ can be formulated using modified nodal analysis(MNA):

$$\mathbf{G}\mathbf{v}(t) + \mathbf{C}\frac{d\mathbf{v}(t)}{dt} = \mathbf{d}w(t) + f[\mathbf{v}(t)], \tag{1}$$

where $\mathbf{v}(t)$ is the network variable vector, and $\mathbf{d}$ is a constant vector specifying the nodes to which the input is connected. $\mathbf{G}$ and $\mathbf{C}$ are the conductance and capacitance matrices, respectively. Their entries are made of the coefficients of the linear elements and the first-order term of the Taylor series expansion of the nonlinear characteristics. The higher order terms of the nonlinear characteristics are embedded in the nonlinear function $f[\mathbf{v}(t)]$. Using Volterra series [7], we get

$$\mathbf{G}\mathbf{v}_m(t) + \mathbf{C}\frac{d\mathbf{v}_m(t)}{dt} = \mathbf{d}_m f_m(t), \quad m = 1, 2, \cdots \tag{2}$$

Equation (2) represents a set of linear circuits that are called the Volterra circuits of m-th order. $\mathbf{v}_m(t)$ and $f_m(t) = f_m[\mathbf{v}_1(t), \mathbf{v}_2(t), \cdots, \mathbf{v}_{m-1}(t)]$ are the response and input of the m-th order Volterra circuit, respectively. $\mathbf{d}_m$ is a constant vector specifying the nodes to which the input $f_m(t)$ is connected. Note that $f_1(t) = w(t)$ and $\mathbf{d}_1 = \mathbf{d}$. The Volterra circuits can be solved efficiently using the sampled data simulation for linear circuits given in [6].

$$\mathbf{v}_1(nT + T) = \mathbf{M}(T)\mathbf{v}_1(nT) + \mathbf{P}(T)e^{j\omega_o nT},$$

$$\mathbf{v}_m(nT + T) = \mathbf{M}(T)\mathbf{v}_m(nT) + \mathcal{R}e\left\{ \frac{a_{m,0}}{2}\mathbf{P}_0(T) \right.$$

$$+\frac{a_{m,N}}{2}\mathbf{P}_N(T)e^{jN\omega_s nT}+\sum_{k=1}^{N-1}a_{m,k}\mathbf{P}_k(T)e^{jk\omega_s nT}\Bigg\}$$

$$+\mathcal{I}m\Bigg\{\sum_{k=1}^{N-1}b_{m,k}\mathbf{P}_k(T)e^{jk\omega_s nT}\Bigg\},\quad m=2,3,\cdots \tag{3}$$

where

$$\mathbf{M}(T)=\mathcal{L}^{-1}[\mathbf{A}^{-1}\mathbf{C}]_{t=T},$$
$$\mathbf{P}(T)=\mathcal{L}^{-1}\Big[\mathbf{A}^{-1}\frac{\mathbf{d}}{s-j\omega_o}\Big]_{t=T},$$
$$\mathbf{P}_k(T)=\mathcal{L}^{-1}\Big[\mathbf{A}^{-1}\frac{\mathbf{d}_m}{s-jk\omega_s}\Big]_{t=T},\quad k=0,1,\cdots,N, \tag{4}$$

$\omega_s=\frac{2\pi}{T_w}$ and $\mathcal{L}^{-1}[.]$ is the inverse Laplace transform operator. $\mathbf{A}=\mathbf{G}+s\mathbf{C}$. The coefficients $a_{m,k}$ and $b_{m,k}$ are determined from Fourier series interpolation of the sampled date $f_m(nT)$, $n=0,1,\cdots,(2N-1)$ in a preselected simulation window T_w. The response of the nonlinear circuit is given by

$$\mathbf{v}(nT)=\sum_{m=1}^{\infty}\mathbf{v}_m(nT). \tag{5}$$

In this section, we propose an extended two-step algorithm to calculate the consistent initial conditions of PSN circuits in case of discontinuities at switching instants. The two-step algorithm [1], which was proposed for switched linear circuits, can be applied separately to each Volterra circuit. The consistent initial conditions of PSN circuit can be obtained by summing up the consistent initial conditions of Volterra circuits.

Suppose the switching action occurs at $t=nT$ and the initial conditions before switching are given by $\mathbf{v}_1(nT^-)$, $\mathbf{v}_2(nT^-)$, $\cdots$, $\mathbf{v}_m(nT^-)$, $\cdots$. The forward step calculates the response of Volterra circuits at $t=nT^-+T$,

$$\mathbf{v}_1(nT^-+T)=\mathbf{M}(T)\mathbf{v}_1(nT^-)+\mathbf{P}(T)e^{j\omega_o nT},$$
$$\mathbf{v}_m(nT^-+T)=\mathbf{M}(T)\mathbf{v}_m(nT^-)+\mathcal{R}e\Bigg\{\frac{a_{m,0}}{2}\mathbf{P}_0(T)+$$
$$\frac{a_{m,N}}{2}\mathbf{P}_N(T)e^{jN\omega_s nT}+\sum_{k=1}^{N-1}a_{m,k}\mathbf{P}_k(T)e^{jk\omega_s nT}\Bigg\}$$
$$+\mathcal{I}m\Bigg\{\sum_{k=1}^{N-1}b_{m,k}\mathbf{P}_k(T)e^{jk\omega_s nT}\Bigg\},m=2,3,\cdots$$

Taking a backward step of identical step size from $t=nT^++T$ to $t=nT^+$, we can get the initial conditions after switching at $t=nT^+$ for each Volterra circuit,

$$\mathbf{v}_1(nT^+)=\mathbf{M}_B(T)\mathbf{v}_1(nT^++T)+\mathbf{P}_B(T)e^{-j\omega_o(nT+T)},$$
$$\mathbf{v}_m(nT^+)=\mathbf{M}_B(T)\mathbf{v}_m(nT^++T)+\mathcal{R}e\Bigg\{\frac{a_{m,0}}{2}\mathbf{P}_{B0}(T)$$
$$+\frac{a_{m,N}}{2}\mathbf{P}_{BN}(T)e^{-jN\omega_s nT}+\sum_{k=1}^{N-1}a_{m,k}\mathbf{P}_{Bk}(T)e^{-jk\omega_s nT}\Bigg\}$$
$$+\mathcal{I}m\Bigg\{\sum_{k=1}^{N-1}b_{m,k}\mathbf{P}_{Bk}(T)e^{-jk\omega_s nT}\Bigg\},m=2,3,\cdots$$

where

$$\mathbf{M}_B(T)=\mathbf{M}(-T),$$
$$\mathbf{P}_B(T)=\mathbf{P}(-T),$$
$$\mathbf{P}_{Bk}(T)=\mathbf{P}_k(-T),\quad k=0,1,\cdots,N. \tag{6}$$

Finally we can obtain the initial conditions of the PSN circuit, immediately after switching ($t=nT^+$),

$$\mathbf{v}(nT^+)=\sum_{m=1}^{\infty}\mathbf{v}_m(nT^+). \tag{7}$$

3. SENSITIVITY OF PSN CIRCUITS

The time domain sensitivity of a nonlinear circuit is obtained by differentiating its response in (5) with respect to an element x.

$$\mathbf{z}(nT)=\sum_{m=1}^{\infty}\frac{\partial\mathbf{v}_m(nT)}{\partial x}=\sum_{m=1}^{\infty}\mathbf{z}_m(nT). \tag{8}$$

This indicates that the sensitivity of a nonlinear circuit is the sum of the sensitivities of all Volterra circuits. The sensitivity of each Volterra circuit is obtained as follows.

$$\mathbf{z}_1(nT+T)=\mathbf{M}_s(T)\mathbf{v}_1(nT)+\mathbf{M}(T)\mathbf{z}_1(nT)+\mathbf{P}_s(T)e^{j\omega_o nT},$$
$$\mathbf{z}_m(nT+T)=\mathbf{M}_s(T)\mathbf{v}_m(nT)+\mathbf{M}(T)\mathbf{z}_m(nT)$$
$$+\frac{1}{2}\mathcal{R}e\Bigg\{[a_{m,0}\mathbf{P}_{0s}(T)+\frac{\partial a_{m,0}}{\partial x}\mathbf{P}_0(T)]$$
$$+[a_{m,N}\mathbf{P}_{Ns}(T)+\frac{\partial a_{m,N}}{\partial x}\mathbf{P}_N(T)]e^{jN\omega_s nT}\Bigg\}$$
$$+\mathcal{R}e\Bigg\{\sum_{k=1}^{N-1}[a_{m,k}\mathbf{P}_{ks}(T)+\frac{\partial a_{m,k}}{\partial x}\mathbf{P}_k(T)]e^{jk\omega_s nT}\Bigg\}$$
$$+\mathcal{I}m\Bigg\{\sum_{k=1}^{N-1}[b_{m,k}\mathbf{P}_{ks}(T)+\frac{\partial b_{m,k}}{\partial x}\mathbf{P}_k(T)]e^{jk\omega_s nT}\Bigg\},m=2,3,\cdots$$

where

$$\mathbf{M}_s(T)=\mathcal{L}^{-1}\Big[\frac{\partial\mathbf{A}^{-1}}{\partial x}\mathbf{C}+\mathbf{A}^{-1}\frac{\partial\mathbf{C}}{\partial x}\Big]_{t=T},$$
$$\mathbf{P}_s(T)=\mathcal{L}^{-1}\Big[\frac{\partial\mathbf{A}^{-1}}{\partial x}\frac{\mathbf{d}}{s-j\omega_o}\Big]_{t=T},$$
$$\mathbf{P}_{ks}(T)=\mathcal{L}^{-1}\Big[\frac{\partial\mathbf{A}^{-1}}{\partial x}\frac{\mathbf{d}_m}{s-jk\omega_s}\Big]_{t=T}.\quad k=0,1,\cdots,N. \tag{9}$$

At the switching instants, the extended two-step algorithm is applied to solve the inconsistent initial conditions for sensitivity analysis of PSN circuits. Assume the switching action happens at $t=nT$. The sensitivities of Volterra circuits at $t=nT^+$ are obtained,

$$
\begin{aligned}
\mathbf{z}_1(nT^+) &= \mathbf{M}_{Bs}(T)\mathbf{v}_1(nT^+ + T) + \mathbf{M}_B(T)\mathbf{z}_1(nT^+ + T) \\
&\quad + \mathbf{P}_{Bs}(T)e^{-j\omega_o nT}, \\
\mathbf{z}_m(nT^+) &= \mathbf{M}_{Bs}(T)\mathbf{v}_m(nT^+ + T) + \mathbf{M}_B(T)\mathbf{z}_m(nT^+ + T) \\
&\quad + \frac{1}{2}\mathcal{R}e\Big\{ [a_{m,0}\mathbf{P}_{B0s}(T) + \frac{\partial a_{m,0}}{\partial x}\mathbf{P}_{B0}(T)] \\
&\quad + [a_{m,N}\mathbf{P}_{BNs}(T) + \frac{\partial a_{m,N}}{\partial x}\mathbf{P}_{BN}(T)]e^{-jN\omega_s nT} \Big\} \\
&\quad + \mathcal{R}e\Big\{ \sum_{k=1}^{N-1} [a_{m,k}\mathbf{P}_{Bks}(T) + \frac{\partial a_{m,k}}{\partial x}\mathbf{P}_{Bk}(T)]e^{-jk\omega_s nT} \Big\} \\
&\quad + \mathcal{I}m\Big\{ \sum_{k=1}^{N-1} [b_{m,k}\mathbf{P}_{Bks}(T) + \frac{\partial b_{m,k}}{\partial x}\mathbf{P}_{Bk}(T)]e^{-jk\omega_s nT} \Big\}, \\
&\qquad m = 2, 3, \cdots
\end{aligned}
$$

where

$$
\begin{aligned}
\mathbf{M}_{Bs}(T) &= \mathbf{M}_s(-T), \\
\mathbf{P}_{Bs}(T) &= \mathbf{P}_s(-T), \\
\mathbf{P}_{Bks}(T) &= \mathbf{P}_{ks}(-T), \quad k = 0, 1, \cdots, N.
\end{aligned} \tag{10}
$$

The sensitivity of PSN circuits immediately after switching is obtained by summing up the sensitivities of all Volterra circuits,

$$
\mathbf{z}(nT^+) = \sum_{m=1}^{\infty} \mathbf{z}_m(nT^+). \tag{11}
$$

4. NUMERICAL EXAMPLES

The algorithms are implemented in a computer program RSPSN (Response and Sensitivity of Periodically Switched Nonlinear circuits). The example circuit is shown in Fig. 1. The clock frequency is 5Hz. Zero initial conditions are assumed for the two capacitors at the onset of simulation. The nonlinear conductor is modeled as $i(t) = v(t) + 0.5v^2(t) + 0.25v^3(t)$. This circuit is simulated using both RSPSN and PSPICE. The response is shown in Fig. 2. It is seen that both of the results match each other well. It should be emphasized that, at the switching instants, inconsistent initial conditions are handled correctly by RSPSN. This is verified by the charge conservation of the circuit before and after switching. The normalized difference between the results from RSPSN and PSPICE is shown in Fig. 3. The maximum normalized difference is below 0.5%.

The sensitivity of the response with respect to G_1 and C_1 are calculated using RSPSN and verified using the Brute-Force (BF) method, which estimates the sensitivity of the response $v(t)$ with respect to a parameter x using $\frac{\partial v(t)}{\partial x} \approx \frac{\Delta v(t)}{\Delta x}$. The results are shown in Fig. 4 and Fig. 5. It is seen that the results from RSPSN and BF match each other very well. The differences between the results from RSPSN and BF are plotted in Fig. 6 and Fig. 7, for 100 clock cycles. The maximum normalized difference is less than 0.3%. The results indicate that the algorithms used in

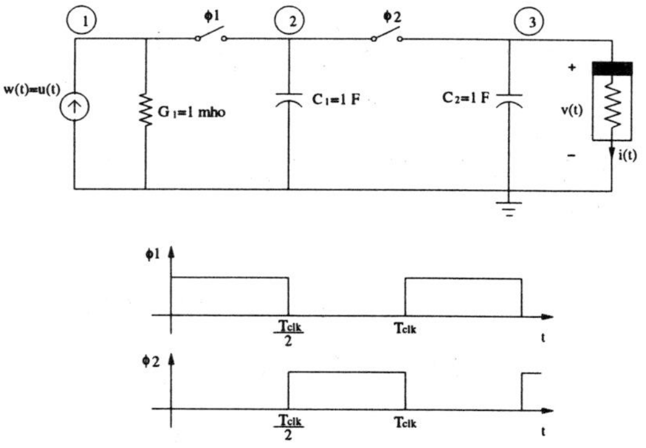

Figure 1: Example circuit

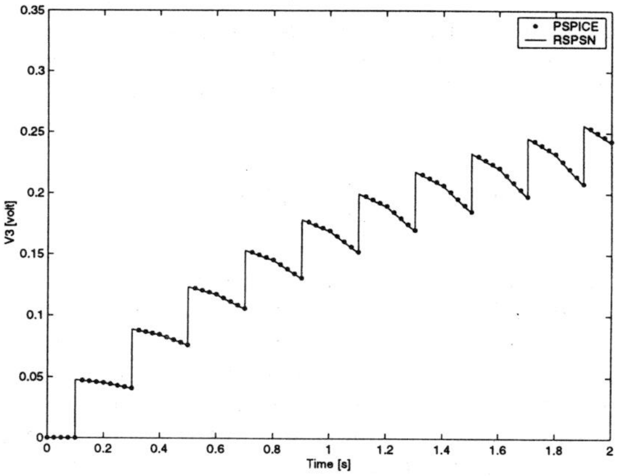

Figure 2: Response

RSPSN are stable and have the ability to correctly handle the inconsistent initial conditions at the switching instants for both response and sensitivity analysis.

5. CONCLUSIONS

In this paper we have presented a new, efficient, and accurate time domain response and sensitivity analysis method for PSN circuits. By representing the mildly nonlinear characteristics using Volterra series expansion and use of interpolating Fourier series, the time-domain response and sensitivity of PSN circuits are obtained by solving a set of Volterra circuits. Computationally intensive Newton-Raphson iterations are avoided. To handle the inconsistent initial conditions encountered in analysis of PSN circuits, a new extended two-step algorithm has been proposed. The accuracy and efficiency of the method have been verified by comparing the simulation results on example circuits with those obtained from PSPICE and the Brute-force method.

6. REFERENCES

[1] A. Opal and J. Vlach, "Consistent initial conditions of linear switched networks," *IEEE Trans. on Circuits and Systems*, Vol. 37, No. 3, pp. 364-372, March 1990.

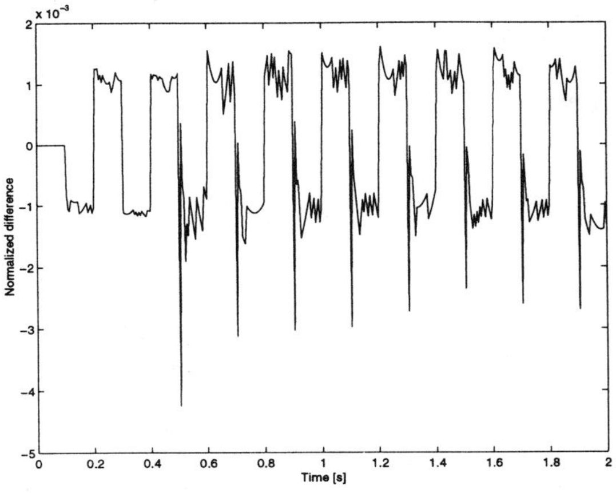

Figure 3: Normalized difference of response between RSPSN and PSPICE

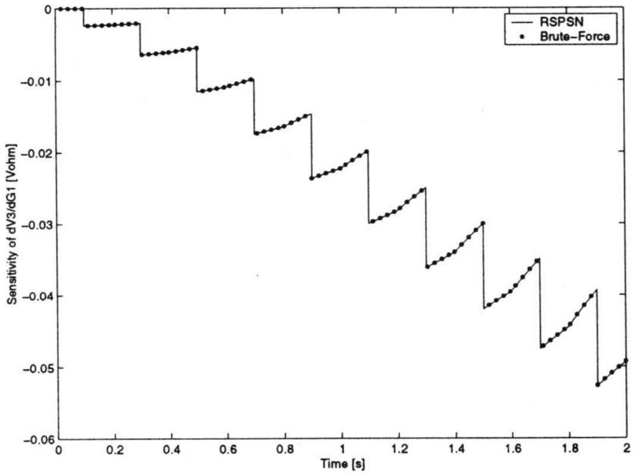

Figure 4: Sensitivity of $v_3(t)$ with respect to G_1

[2] Y. Elcherif and P. M. Lin, "Transient analysis and sensitivity computation in piecewise-linear circuits," *IEEE Trans. on Circuits and Syst.*, vol. 38, No. 12, pp. 1525-1533, December 1991.

[3] S. W. Director and R. A. Rohrer, "The generalized adjoint network and network sensitivity," *IEEE Trans. on Circuit Theory*, vol. 16, pp. 318-323, August 1969.

[4] P. Wampacq and W. Sansen, *Distortion Analysis of Analog Integrated Circuits.* Boston, MA : Kluwer, 1998.

[5] F. Yuan and A. Opal, "An efficient transient analysis algorithm for mildly nonlinear circuits," *IEEE Trans. on Computer-Aided Design of Integrated Circuits and Systems*, Vol. 21, No. 6, pp. 662-673, June 2002.

[6] A. Opal, "Sampled data simulation of linear and nonlinear circuits," *IEEE Trans. on Computer-Aided Design of Integrated Circuits and Systems*, Vol.15, No. 3, pp. 295-307, March 1996.

[7] M. Schetzen, *The Volterra and Wiener theory of nonlinear systems*, New York : John Wiley and Sons, 1981.

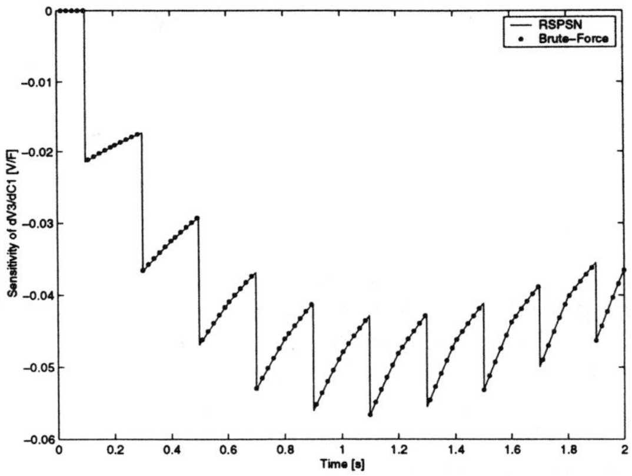

Figure 5: Sensitivity of $v_3(t)$ with respect to C_1

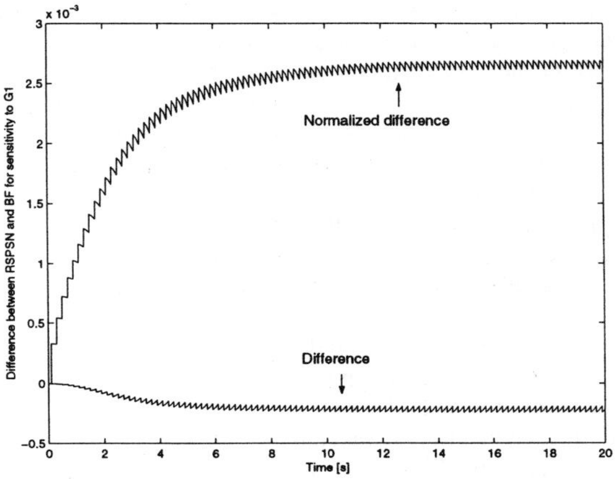

Figure 6: Difference between RSPSN and BF for sensitivity of $v_3(t)$ with respect to G_1

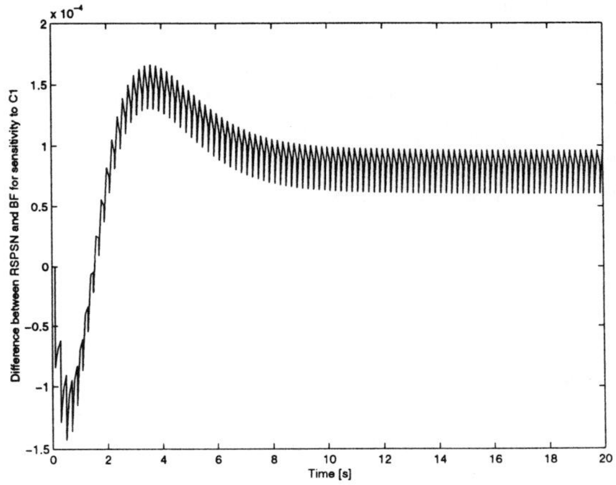

Figure 7: Difference between RSPSN and BF for sensitivity of $v_3(t)$ with respect to C_1

CAD SYSTEM FOR DESIGN AND SIMULATION OF DATA CONVERTERS

P. Estrada[1], F. Maloberti[2]

1. Texas A&M University, College Station, Texas, USA.
2. University of Pavia, Pavia, Italy and University of Texas at Dallas, USA.

Abstract

This work presents a simulation environment for the design of data converters using behavioral modeling in the MATLAB- SIMULINK platform. This environment utilizes a graphic user interface as a CAD tool to design and simulate different data converter architectures. Post-processing analysis tools are included for static and dynamic performance calculation.

Introduction

A data converter system contains many different parts, both digital and analog. For high resolution and for high conversion speed the complexity of systems becomes very high. In order to handle the global design it is therefore necessary to use a top-down approach. Starting from the specifications the architecture and the basic cells are defined and designed. The top-down approach must be followed by a bottom-up verification. If the required specifications are not met, there is an adjustment of individual block specifications, partitioning and architecture. The use of behavioral modeling significantly contributes to the design process: an accurate description of data converter behavior permits us to reduce the computational efforts translating into shorter simulation time.

This paper describes a behavioral design environment addressing and supporting the following design steps:

- Selection of the architecture.
- Verification that the architecture is suitable.
- Estimation of the limits coming from each block.

Simulation Environment Considerations

Behavioral modeling is a problem that can be approached differently depending on the type of system that we want to simulate. In the case of mixed-signal systems, there are several languages or tools available that allow us to construct behavioral models. For instance, MATLAB has the SIMULINK toolbox in order to describe continuous and discrete systems in a graphical environment. The language has also provided the ability to develop in an easy way Graphic User Interfaces (GUI) to interact with the user in a graphical way. By using GUIs, the simulation environment becomes very intuitive and easy to use, as opposed to having to learn a set of commands and functions used in the software.

Data converter simulation requires aside of the model itself, suitable input signal generators as well as post-processing tools for calculating the performance of the converter. The typical test signal used for characterizing data converters is a sinusoid or a ramp, which are easy to implement in any computer simulation environment. Also in many cases white noise sources are required. In fact, to properly model data converters, the thermal noise produced by resistors, switches and operational amplifiers has to be considered [1]. Since a data converter is a sampled data system, the aliasing effect of the thermal noise has to be taken into account.

Once we have a model of the data converter and the proper input signals, a simulation of the circuit in the time domain provides a sequence of sampled data values. Then, a suitable processing of the raw output data permits us to evaluate the performance of the data converter. Typically, depending on the data converter architecture, we are interested in the linearity parameters (integral non-linearity or *INL* and differential non-linearity or *DNL*) or in the resolution parameters (effective number of bits *or* N_{eff}, signal-to-noise ratio or *SNR* and the signal-to-noise and distortion ratio or *SND*R).

Basic Blocks

The most common data converter architectures used are the successive approximation algorithm, the flash approach, the sigma delta technique and all the algorithms that can be achieved with pipeline implementations. The basic blocks that are found in these architectures can also be found in the architectures for other conversion algorithms.

The successive approximation converter commonly uses the charge redistribution architecture. This architecture has two basic blocks, a capacitive array controlled by switches and a comparator. The flash converter utilizes a passive resistive network to generate reference voltages and comparators. The sigma delta uses switched capacitor integrators, a comparator and complex digital circuitry. DAC and pipeline architectures use operational amplifiers (to achieve amplification by a given factor or to subtract voltages) and, again comparators. Therefore we will discuss the operational amplifier, and, in particular, its use in switched capacitor integrators, the comparator and the resistive array used in flash converters.

Operational Amplifier

Operational amplifiers are key components in data converter circuits. Quite often the performance of the operational amplifiers bounds the performance of a complete data converter. It is therefore necessary to include an accurate model of the operational amplifier, considering all of the non-idealities. Linear parameters such as finite gain and bandwidth are considered, as well as the non-linear effects, such as (hard) saturation and slew-rate.

Since data converters are sampled-data systems, there are two possible approaches for modeling operational amplifiers. The first approach is based on traditional models of the operational amplifiers. The models consist of a set of equations and differential equations, which describe the behavior of the circuit. In the simulation, the transient behavior of the circuit is considered for each clock cycle. The simulation obviously allows us to obtain a good accuracy, but at the expense of a long simulation time. The second approach is based on models of the complete sub-circuit (for example an integrator or a buffer), which includes the operational amplifier. The model doesn't perform the time simulation each clock cycle but uses given equations that account for the global effect of

0-7803-7761-3/03/$17.00 ©2003 IEEE

the operational amplifier non-idealities at the end of each clock cycle. Therefore, the use of relatively simple behavioral equations permits us to estimate the error produced at the output of the sub-circuit without going into the details of the transient behavior within the clock cycle. This approach is of course less accurate than the previous one, but much faster [2].

As an example, we can consider a switched capacitor (SC) integrator with transfer function

$$H(z)=\frac{z^{-1}}{1-z^{-1}}. \qquad (1)$$

Analog circuit implementations of the integrator deviate from this ideal behavior due to several non-ideal effects. One of the major causes of performance degradation in the SC integrators is the incomplete transfer of charge. This non-ideal effect is a consequence of the operational amplifier non-idealities, namely finite gain and bandwidth, slew rate and saturation voltages. Fig. 1 shows the model of the integrator including all the non-idealities, which will be considered in detail in the more complex circuits discussed in the next paragraphs. The MATLAB function that implements the Slew Rate, actually comprises Eqns. (4), (5), (6), (7) and (8).

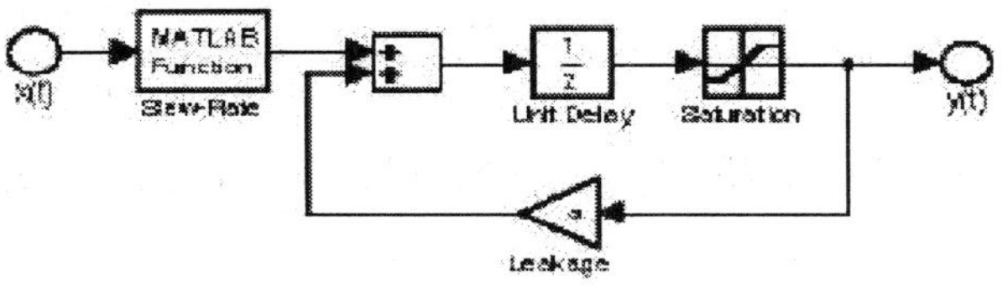

Fig. 1 Simulink model of an SC integrator

The op-amp of the integrator described by Eqn. (1) is ideal. However, the gain of any op-amp is finite and this causes a first limit. The effect is an integrator "leakage": only a fraction of the previous output of the integrator is added to each new input sample. The transfer function of the integrator with leakage becomes

$$H(z)=\frac{z^{-1}}{1-\alpha z^{-1}}. \qquad (2)$$

The dc gain H_0 becomes therefore:

$$H_0=\frac{1}{1-\alpha}. \qquad (3)$$

The effect of the finite bandwidth and the slew-rate are related to each other and may be interpreted as a non-linear gain. The evolution of the output node during the nth integration period is governed by

$$v_0(t)=v_0(nT-T)+\alpha V_s\left(1-e^{-\frac{t}{\tau}}\right) \quad nT-\frac{T}{2}<t<nT, \qquad (4)$$

where $V_S = V_{in}(nT - T/2)$, is the integrator leakage and is the time constant of the integrator (GBW is the unity gain frequency of the operational amplifier when loaded by C_f). The slope of this curve reaches its maximum value when $t=0$, resulting in

$$\left.\frac{d}{dt}v_0(t)\right|_{max}=\alpha\frac{V_s}{\tau}. \qquad (5)$$

We must consider now two separate cases:

1. The value specified by Eqn. (5) is lower than the operational amplifier slew-rate, *SR*. In this case there is not slew-rate limitation and the evolution of v_0 conforms Eqn. (4).
2. The value specified by Eqn. (5) is larger than *SR*. In this case, the operational amplifier is in slewing and, therefore, the first part of the temporal evolution of v_0 $(t < t_0)$ is linear with slope *SR*. The following equations hold (assuming $t_0 < T$

$$t<t_d,\quad v_0(t)=v_0(nT-T)+SRt, \qquad (6)$$

$$t>t_d.\quad v_0(t)=v_0(t_0)+(\alpha V_s-SRt_0)\left(1-e^{-\frac{t-t_0}{\tau}}\right) \qquad (7)$$

Imposing the condition for the continuity of the derivatives of Eqn. (6) and Eqn. (7) in t_0, we obtain

$$t_0=\frac{\alpha V_s}{SR}-\tau \qquad (8)$$

If $t_0 > T$ only Eqn. (6) holds. The MATLAB function in Fig.1 implements the above equations to calculate the value reached by v_0 (t) at time T, which will be different from V_0 due to the gain, bandwidth and slew-rate limitations of the operational amplifier.

Comparator

A widely used configuration of comparator comprises the cascade of a preamplifier and a latch (with hysteresis). The input signal can vary significantly; then, the amplifier can operate in the linear or the overdrive region of operation. Therefore, as we have for the integrator discussed above, the behavioral model should be able to determine the region of operation and to apply the proper behavioral model. The final stage of the comparator is a latch with hysteresis; the hysteresis is included to account for the metastability error due to signals whose magnitude is very close to the comparator reference. Additionally, fixed offset voltage and random offset voltage must be added to the input to model the offset and any time dependent spur signals.

Simulation Environment

The simulation environment uses the MATLAB- SIMULINK platform but it can be easily translated in other simulation frameworks. The environment is flexible enough to cover a wide range of specifications and applications. It covers popular architectures like flash converter, pipeline and sigma-delta are defined in hierarchical SIMULINK descriptions. The system also incorporates a library of behavioral models that represent the basic blocks used for building data converter architectures. The designer can build with these blocks a custom architecture and try new ideas. Each block has a set of parameters that define its specifications so the designer is able to see the effect of each parameter on the whole system performance.

These parameters represent non-idealities that have been included in the models. Dialog masks permit a quick specification of parameters. The behavioral models of each block have been validated with SPICE simulations.

An example of basic block included in the library is the switched capacitor integrator used in sigma-delta modulator. The model accounts for the finite gain, bandwidth and slew-rate of the op-amp, linearity of the capacitors used, and permits to simulate the effect of white noise sources.

Implied in the design activity is the analysis of the data converter performance. For this analysis several functions were developed to calculate static and dynamic performance.
Among them, the tool permits to estimate the INL and the DNL; it is also possible to evaluate the SNR, the SFDR and other dynamic parameters. Moreover, the environment includes a number of options for plotting diagrams, running parametric analysis, where a certain performance can be studied for particular parameter variations.

Beyond the above-described features, the simulation environment provides an additional benefit: the environment includes a specific section for educational purposes. The section provides specific examples and design guidelines for illustrating basic concepts of data converter operations. The educational aid has available the same analysis tools used for design. Several exercises proposed allow the user to study the features and limitations of the architectures included. The designer can also use the predefined architectures as a starting point and then modify it to his particular needs. The education aid section is illustrated in Fig. 2, 3, and 4. Fig. 2 shows the menu that permits the student to study basic concepts like the effect of ideal and non-ideal sampling and the spectrum of the quantization noise for different input frequency-sampling ratios.

Fig. 3 shows the simulation control menu. It includes a Help button and buttons for setting the simulation parameters. Fig. 4 shows the graphic interface for setting parameters as numerical inputs or cursor-like.

Finally the simulation environment has a section where the designer can choose among popular data conversion algorithms. Set the specifications requirements according to his needs and then study the requirements for each individual block that he must be able to satisfy in order to achieve the desired performance. This section is showed in Fig. 5 and 6. Fig. 5 shows the menu where the user can select the desired conversion algorithm. By selecting any of these options, the user is presented with an input menu where specifications of the particular conversion algorithm can be entered. Fig. 6 shows an example where the user has selected the pipeline architecture and entered in the specifications that 3 stages where desired. The software synthesizes a block diagram of the converter with the desired specifications.

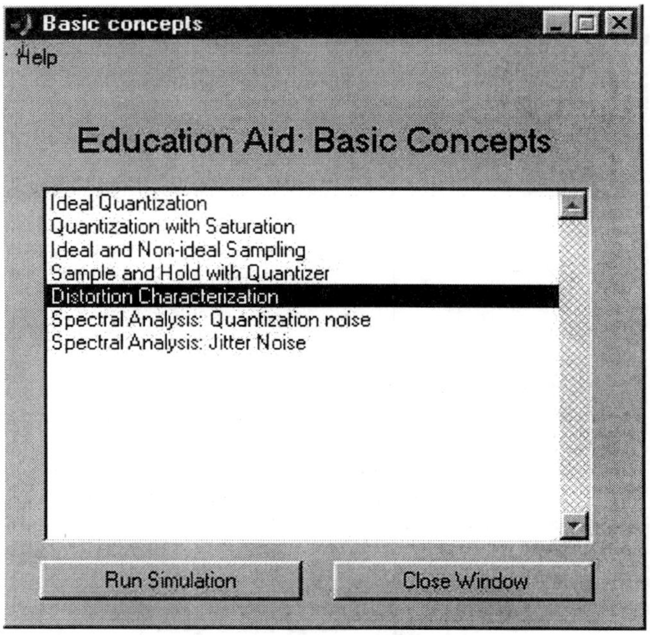

Fig 2. Education aid GUI

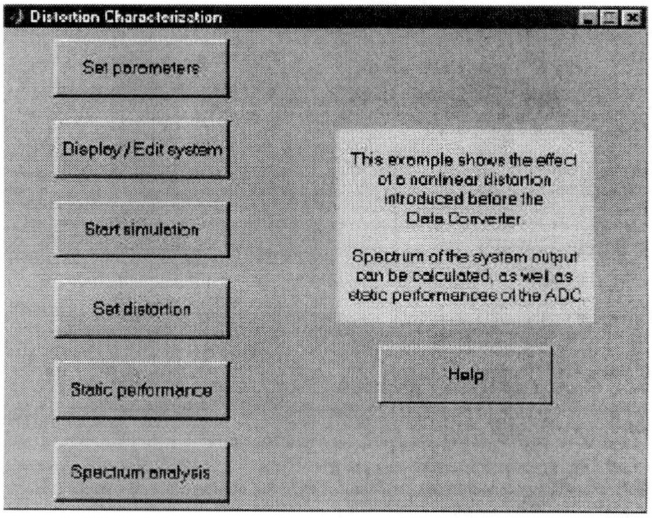

Fig 3. Main menu GUI for an education aid example

We assume some specification requirements for the data converter. An ideal architecture achieves better performance than a real implementation. Therefore, the specification requirements should leave some margin to account for the non-idealities. The designer should use this "budget" properly. The non-critical parameters will utilize a little of the budget, so that the major part of it can be devoted to the real limitations. For example, in a medium resolution high-speed converter, the gain is not very important, the offset won't be a problem (if the *dc* response is not relevant) but the gain-bandwidth and, even more the slew-rate, are very critical. The results of the behavioral simulations quantify the relevance of the above mentioned limitations and permit a proper "budget" assignment for an optimum design trade-off.

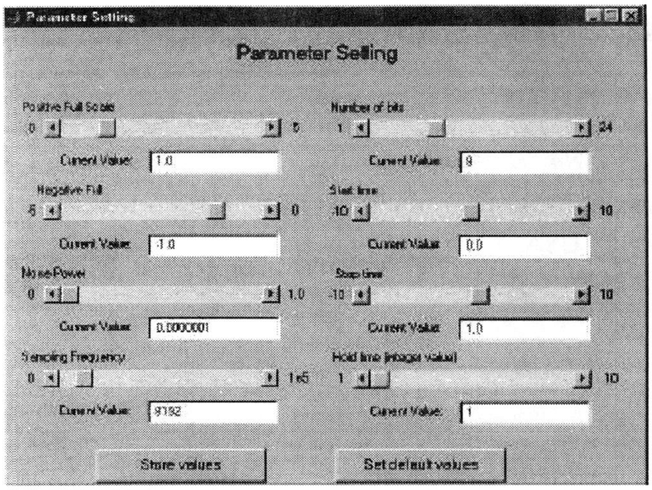

Fig 4. GUI for parameter setting

Fig 5. GUI for design tools

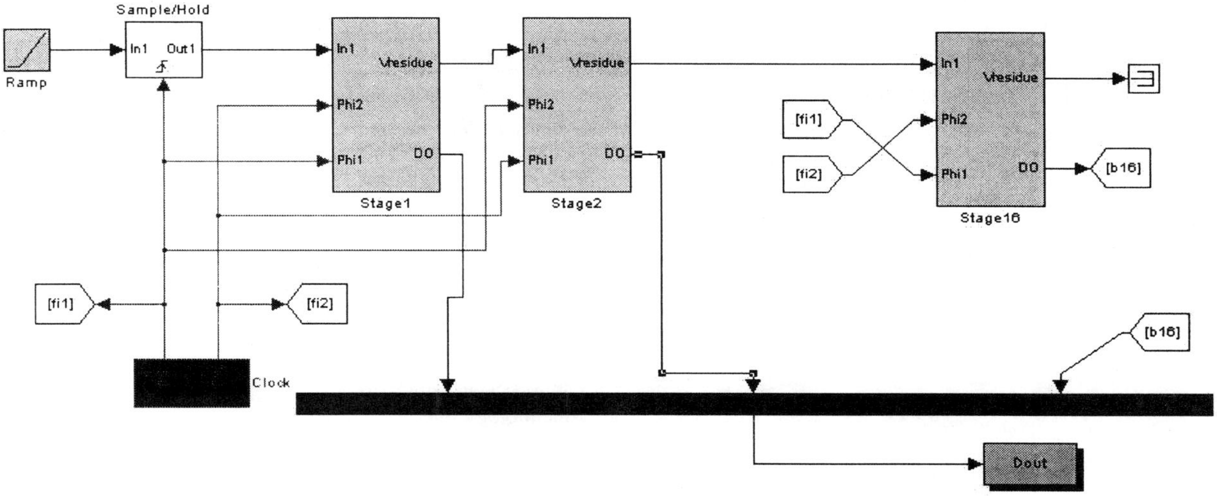

Fig 6. Synthesized 3 stage pipeline behavioral model

Conclusions

In summary, the developed user-friendly environment allows the user to design a data converter and define its basic block specifications at system level. By using behavioral models, it significantly reduces the simulation time yielding a close match in results to low level models and, for educational purposes, with behavioral description it helps in better understanding the data converter architectures and the limitations associated to the real blocks used.

References

[1] *F. Maloberti, P. Estrada, P. Malcovati, A. Valero*, "Behavioral Modeling and Simulations of Data Converters" Journal of the International Measurement Confederation IMEKO, *June. 2001.*

[2] F. Medeiro, B. Perez-Verdu, A. Rodriguez-Vazquez, J. L. Huertas, "Modeling OpAmp-Induced Harmonic Distortion for Switched-Capacitor Modulator Design", *Proceedings of ISCAS '94*, vol. 5, pp. 445-448, London, UK, 1994.

[3] S. Brigati, F. Francesconi, P. Malcovati, D. Tonietto, A. Baschirotto and F. Maloberti", Modeling Sigma-Delta Modulator Non-Idealities in SIMULINK", *Proceedings of IEEE International Symposium on Circuits and Systems (ISCAS '99)*, 2, Orlando, USA, pp. 384-387, 1999.

[4] *Gielen, G. E. – Franca, J. E.* "CAD tools for data converter design: an overview" IEEE Transactions on Circuits and Systems II: Analog and Digital Signal Processing, page(s): 77 - 89 Feb. 1996 Volume: 43 Issue: 2

Automatic Analog Layout Retargeting for New Processes and Device Sizes

Nuttorn Jangkrajarng, Sambuddha Bhattacharya, Roy Hartono, and C.-J. Richard Shi

Department of Electrical Engineering, University of Washington
Seattle, WA 98195 USA
{njangkra,sbb,rhartono,cjshi}@ee.washington.edu

ABSTRACT – *This paper presents an automatic analog layout resizing tool that can generate a new layout incorporating the target technology process and the target transistor sizes. The tool automatically preserves the analog layout integrity by extracting layout symmetry and matching, and then solving the constrained layout generation problem using a combined linear programming and graph-theoretic approach. The tool has been applied successfully to integrate specified transistor sizes and to migrate layouts for various analog designs from TSMC 0.25um CMOS to TSMC 0.18um CMOS process with comparable performances to re-design.*

1. Introduction

The scaling of feature size in VLSI circuits, both digital and analog, has been one of the strongest driving forces toward the rapid development of electronics technology. For digital circuits, the shrinkage of transistor sizes (for example from 0.25um to 0.18um to 0.13um) is the main reason microprocessors rapidly increase in speed. In the analog or mixed signal layouts, the circuit performances also benefit from smaller minimum feature size.

When there is a change in technology process, digital designers can utilize benefits from the new technology without much effort by using existing high-level VHDL or Verilog designs, scalable cell libraries, and readily available automatic place & route tools to generate a new circuit layout with better performances, or by layout retargeting tools. In contrary, analog designers do not have the comparable ability, which means they have to go through a full time-consuming cycle of redesigning, testing and drawing layouts. Therefore, an automated tool for re-layout of analog circuits will be essential in significantly reducing the design time for mixed-signal circuit technology migration.

In this paper, we present, for the first time, a computer-aided design tool that can automatically resize an existing analog layout for some modestly new processes. The tool is developed based on the existing algorithms [2-5] related to layout compaction. Its interface is shown in figure 1. The automatic analog layout resizer reads an original layout and its technology file, a new target technology file, and new transistor sizes. Then, it automatically generates a new layout that satisfies all the design rules while preserving all the analog layout integrities such as matching and symmetry constraints. The methodology we propose here is based on "recycling" already fine-tuned analog layouts. As high performance analog circuits are layout-sensitive and require device/wiring alignment, matching and symmetry, the method of recycling the layouts will be able to conserve the above requirements. Moreover, it will preserve all the unique aspects intended by engineers on any particular layout.

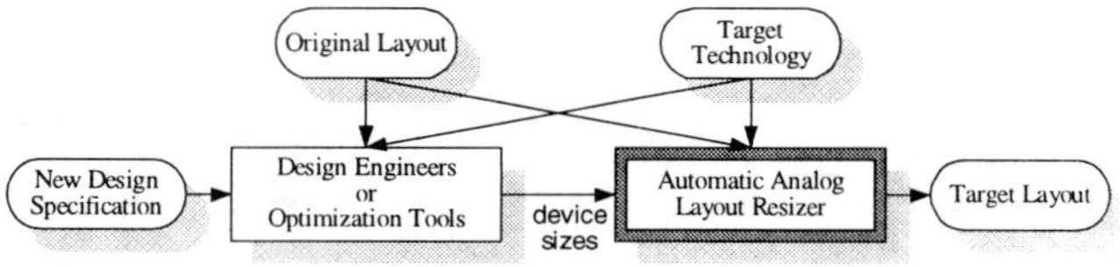

Figure 1: ***Interface of the automatic analog layout retargeting tool.***

This paper is structured as follows. Section 2 presents the proposed analog layout retargeting flow and the implementation of important sections. Section 3 describes results using this new retargeting tool. Section 4 concludes the paper.

2. Automatic Layout Resizing Procedure

In order to automatically retarget an analog layout to a new technology process, three main considerations - namely new technology restriction, new device sizes, and layout structure preservation - have to be taken into account. The original design is used as a starting point for our program, with the purpose of maintaining the layout property. Our approach consists of layout representation and extraction [1,2], symmetry detection [3], technology conversion, circuit components resizing, and layout compaction [2,5]. The flow is shown in Figure 2, which important sections are described as follow.

2.1 Layout Representation and Symmetry Detection

First, the layout is represented by the corner stitching data structure [1], which recognizes each rectangle and its neighbors for every layer. The transistors and nets are then extracted from the layout.

The symmetry axes can be detected automatically between transistor pairs, based on the algorithm from [3]. However, this method may create many unnecessary symmetry axes. To overcome this problem, we introduce a user-specified option, which instructs the detection function to check and compare only

* This research was supported by NSF-ITR and DARPA NeoCAD grant.

between specified transistors or blocks. The symmetry axes then become one of the main criteria that the retargeting tool has to maintain.

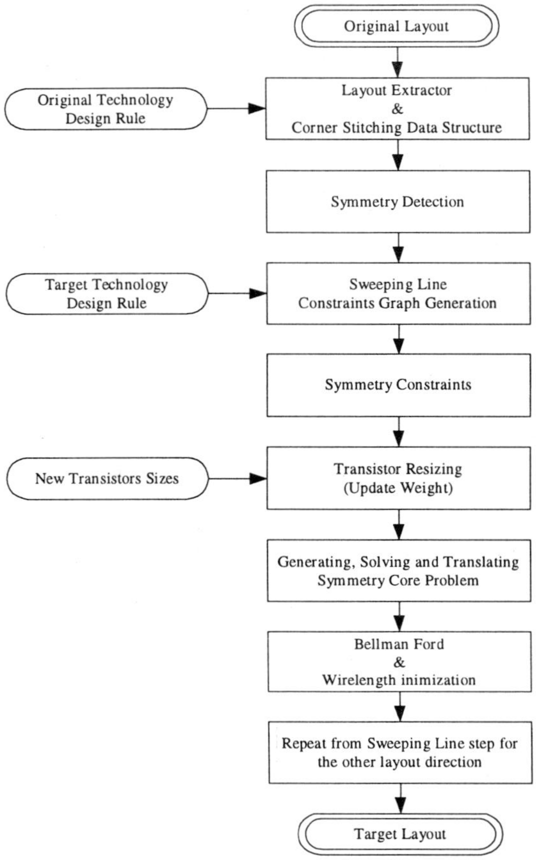

Figure 2: Proposed analog layout resizing flow.

2.2 Technology Migration Constraints

In order to facilitate the layout technology process migration and device resizing, a constraint graph that consists of nodes (representing the rectangles edges) and directed-weighted arcs (representing the constraints between edges) has to be created. One way of obtaining the graph is by using the sweeping line method [2]. First, the design rules for the target technology have to be acquired. Here, we categorize the design rules into three groups – minimum width, spacing, and extension. The sweeping line will start from the most left edge of the layout. While the line is traversing to the right, all required constraints from the current edge to the visible edges on its left will be added. The sweeping line algorithm also reduces redundant constraints, thus speeding up the solution solving. The example of constraints generated is shown in Figure 3.

As the sweeping line algorithm preserves the layout structure, our resizing tool requires two conditions: the target technology that covers all layers employed in the original layout, and the design that can be retargeted to the new process. Therefore, we shall call such process a *modestly* new process.

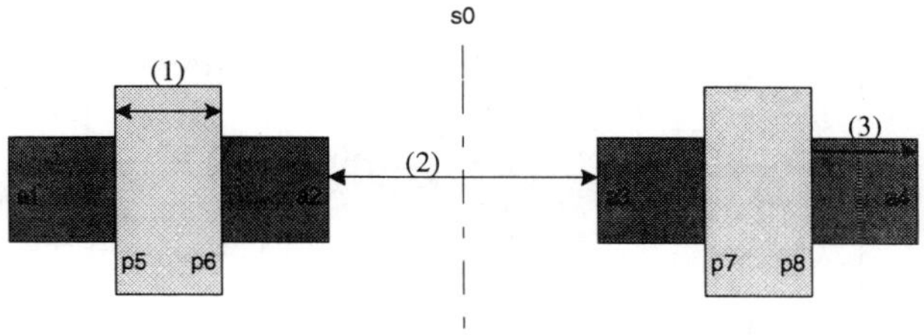

Figure 3:** **A layout of two transistors of only active layer (stripes) and poly layer (gray). Shown in numbers are examples of design rules: (1) minimum-width (2) spacing and (3) extension. The letters denote edges of active (a) rectangles and poly (p) rectangles, where (s) is a symmetry axis. Here are the constraints generated from this figure in horizontal direction. (Example design rules are taken from TSMC 0.25um)

a2-a1≥3	***a4-a3≥3***	***p6-p5≥2***	***p8-p7≥3***
p5-a1≥3	***a2-p6≥3***	***p7-a3≥3***	***a4-p8≥3***
a3-a2≥3	***s0-p6=p7-s0***	***p6-p5=p8-p7***	

2.3 Transistor Resizing

Typically, it is necessary to resize transistors to accomplish the same or better performance when a design is targeted on a new process. The target transistor sizes can be found either by manual simulations or, preferably, by some automatic transistor sizing tools.

Transistor resizing is accomplished by adding to the constraint graph the fixed-width constraints for each transistor to reflect new widths (added to active rectangles) and lengths (added to poly rectangles). This needs to be done on both horizontal and vertical direction. For symmetric transistor pairs, all the pairs have to be resized with exactly the same dimensions.

While performing transistor resizing, there is one difficulty regarding the number of active to metal-one contacts. Since the transistors sizes can be either tightened or widened, when decreasing transistors width, the reduced active area may not be able to fit all existing contacts. Thus a contact removal scheme has to be executed. After the addition of size-constraints, all active to metal-one contacts that reside by the transistors are removed. We, then, need to add two constraints between the metal-one rectangle edges and active rectangle edges, as shown in Figure 4, in order to preserve the connectivity between the two areas. After the constraint graph problem is solved, rows of contacts will be placed back in the right location.

2.4 Constrained Layout Generation

The new layout can be achieved by solving the constrained linear programming problem, which can be mathematically described as

$min\ (x_{r,o} - x_{l,o})$	*subject to layout constraint*
$(a)\ x_i - x_j \geq w$	*min-width, spacing, extension*
$(b)\ x_i - x_j = w$	*fixed-width*
$(c)\ x_i - x_j = 0$	*connectivity*
$(d)\ x_i - sym = sym - x_j$	*symmetry*

where variables $x_{r,0}$, $x_{l,0}$, and x_i or x_j are the most-right edge of the layout, the most-left edge of the layout, and any rectangles edges respectively.

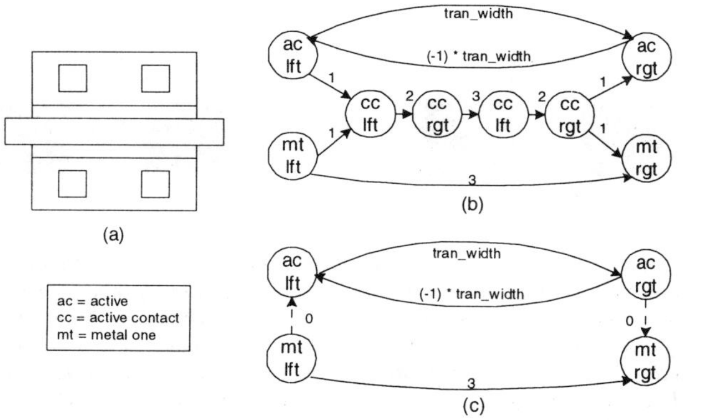

Figure 4: Contact removal in transistor resizing. (a) Transistor layout. (b) Original constraint graphs for only lower drain/source side with active contacts. (c) After removing contacts, two constraints are added for connectivity (metal->active and active->metal). Note: Constraints weights are taken from TSMC 0.25um process and in unit of lambda.

If we ignore the symmetry constraints, the above linear programming problem can be converted to the shortest path problem of the constraint graph represented by nodes as the layout rectangle variables and directed-weighted arcs as the design rule constraints.

From Section 2.2, constraint (a) $x_1 - x_2 \geq w$ can be expressed with a directed arc of weight w from node x_1 to x_2. Constraint (b) and (c) can be divided into two constraints of $x_1 - x_2 \geq w$ and $x_2 - x_1 \geq -w$, which also can be specified in the graph. Without the symmetry constraints, the minimum distance from one side of the layout to the other can be solved quickly with a graph-based shortest path algorithm (i.e. Bellman-Ford [6]).

However, in the presence of symmetry constraints, the linear programming is still necessary. Okuda *et al.* [4] has established an algorithm to solve this problem more efficiently by using advantages from both fast graph-based and linear programming technique. Instead of employing linear programming on a large problem, a smaller equivalent problem consisting of only the layout boundary variables and variables associated with symmetry axes are generated, and then solved with linear programming. The solution will give us the exact location of all variables in the equivalent problem. After that, we can replace each symmetry constraints with two fixed-width constraints and solve the compaction problem with the graph-based shortest path algorithm. Examples are

$$x_1 - sym = sym - x_2 \quad \Rightarrow \quad x_1 - sym = b \;\; and \;\; sym - x_2 = b$$

$$x_3 - x_4 = x_5 - x_6 \quad \Rightarrow \quad x_3 - x_4 = c \;\; and \;\; x_5 - x_6 = c$$

One weakness of the basic shortest path algorithm is that it tries to find the minimum distance from every variable to the starting (most-left) variable, which creates excessive leftward extension in some rectangles. It consequently results in bad performances due to surplus parasitic resistance and capacitance. Therefore, after solving the problem, minimization of individual rectangles as described in [2] or [5] has to be performed to secure good performance.

3. Examples

This section presents the results of applying our resizing tool on a single-ended folded-cascode and a two-stage operational amplifier. Both circuits are initially designed using the TSMC 0.25um CMOS process, and laid-out manually using Cadence's Virtuoso with multi-finger transistor structures. The target technology is the TSMC 0.18um CMOS process.

The CIF format files and the Cadence format technology file are imported from Virtuoso to our program. Once resizing is finished, the target CIF layout is exported back to Virtuoso, and a design-rule checking is performed. Finally, the netlists from both layouts are simulated by Hspice to compare their functionalities and performances.

3.1 Folded Cascode Operational Amplifier

Figure 5 shows the schematic of a folded cascode operational amplifier with 14 transistors (43 transistors if each finger is counted as a separate transistor). The original layout in the TSMC 0.25um process is shown in Figure 6, where the three symmetry axes A, B and C represented are {M1}:{M2}, {M3}:{M13} and {M4,M6,M8,M10}:{M5,M7,M9,M11}.

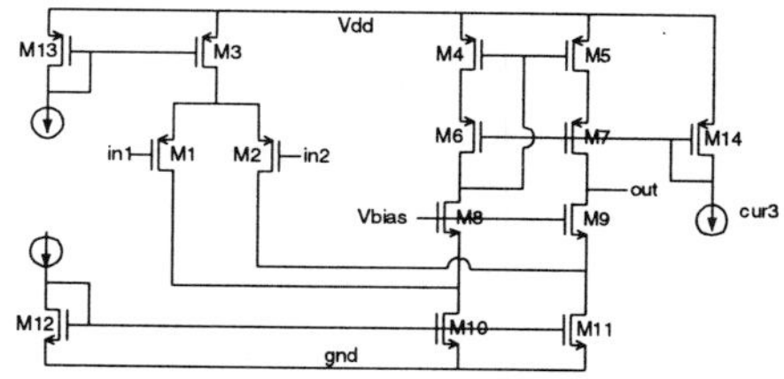

Figure 5: Schematic of a single-output folded-cascode opamp.

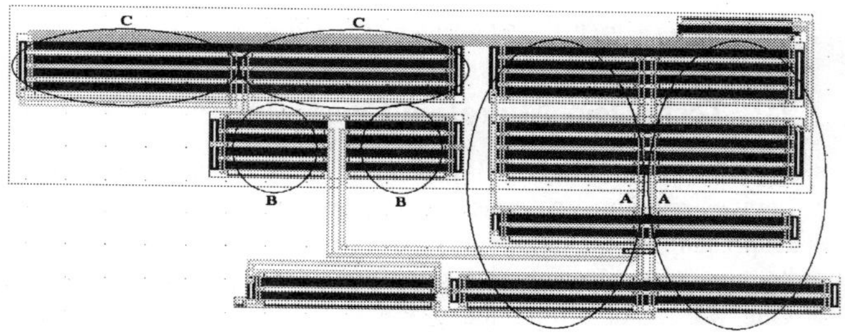

Figure 6: Original layout of the folded cascode operational amplifier in TSMC 0.25um. A, B and C are transistors symmetry blocks.

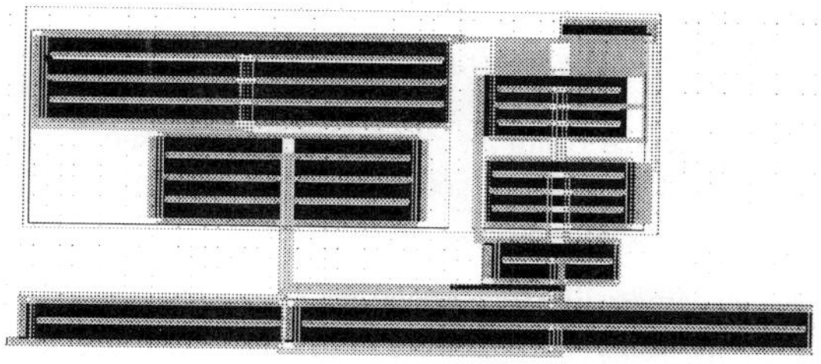

Figure 7: Resized layout of folded cascode opamp in TSMC 0.18um.

Figure 7 shows the resized layout in the TSMC 0.18um process. The transistor sizes are selected such that general operational amplifier specifications are met. The statistics on the performances and silicon area of the resized layout are summarized in Table 1, where "resize" and "no-resize" represent, respectively, the results with resized (the modified width and length from design engineers) transistors and no-resized (the same width and length as in the original layout) transistors.

Table 1: Performances comparison of folded-cascode opamp.

	0.25um	*0.18um no-resize*	*0.18um resize*
Vdd	*2.5 V*	*1.8 V*	*1.8 V*
Load Cap.	*1.0 pF*	*0.7 pF*	*0.7 pF*
Gain	*60.9 dB*	*61.9 dB*	*60.6 dB*
Bandwidth	*51.7 MHz*	*71.7 MHz*	*63.5 MHz*
Phase Margin	*63 deg*	*42 deg*	*71 deg*
Gain Margin	*12.5 dB*	*12.4 dB*	*10.5 dB*
Power	*1.48 mW*	*1.07 mW*	*0.88mW*
Area	*4,813 um²*	*3,000 um²*	*2,083 um²*

3.2 Two-Stage Operational Amplifier

The schematic of a two-stage operational amplifier, shown in Figure 8, comprises of 1 MOS capacitor and 8 transistors (48 as each finger counted). There is one symmetry axis between two pairs of transistors {M1,M4}:{M2:M5}. The original layout (on TSMC 0.25um) is illustrated in Figure 9. The resizing is performed on width and length of all transistors, including the MOS capacitor. The target layout (in TSMC 0.18um) is depicted in Figure 10. The statistics on the performances and silicon area of the resized layout are summarized in Table 2.

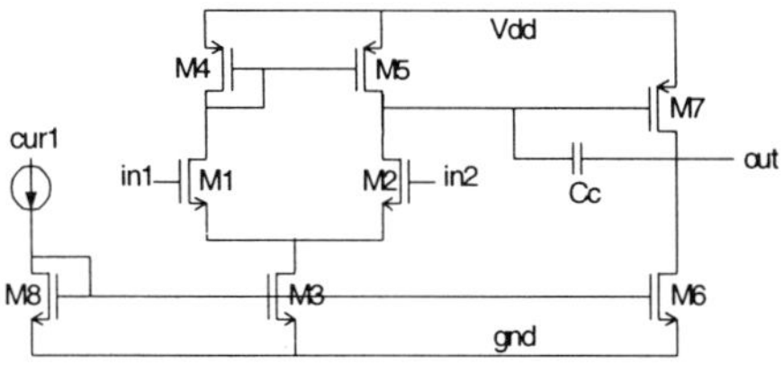

Figure 8: Schematic of a two-stage opamp.

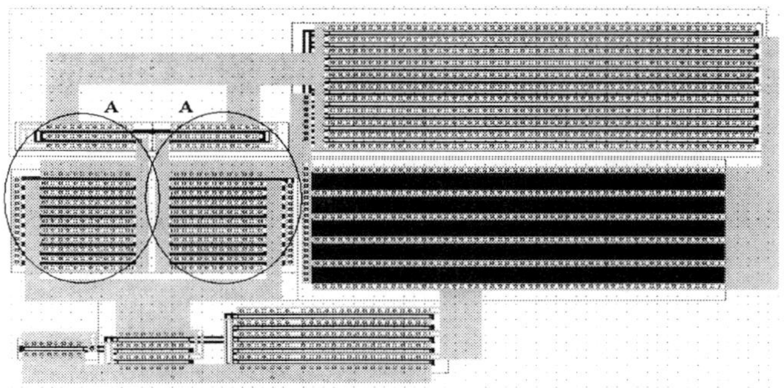

Figure 9: Original layout of the two-stage opamp in TSMC 0.25um.

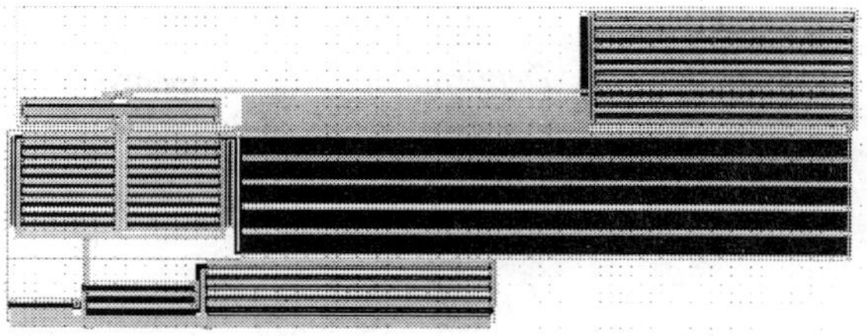

Figure 10: Resized layout of the two-stage opamp in TSMC 0.18um.

Table 2: Performances comparison of two-stage opamp.

	0.25um	*0.18um no-resize*	*0.18um resize*
Vdd	*2.5 V*	*1.8 V*	*1.8 V*
Load Cap.	*1.0pF*	*0.7pF*	*0.7pF*
Gain	*57.7 dB*	*39.6 dB*	*64.4 dB*
Bandwidth	*135 MHz*	*181 MHz*	*104 MHz*
Phase Margin	*50 deg*	*56 deg*	*56 deg*
Gain Margin	*9.6 dB*	*12.5 dB*	*9.2 dB*
Power	*4.82 mW*	*3.56 mW*	*3.46 mW*
Area	*3,648 um²*	*2,664 um²*	*2,745 um²*

The runtime for the folded cascade opamp is 39.2 seconds and the two stage opamp is 37.6 seconds on a 440MHz SUN ultrasparc10 workstation.

4. Conclusion

An automatic tool for re-targeting existing analog layouts to new technology processes and new device sizes is presented. Layout recycling with symmetry detection and layout conservation scheme is applied in order to preserve the properties of analog circuit layout. Additionally, the tool has the ability to consider new transistor sizes to achieve better performances as part of the re-targeting process.

5. References

[1] J. K. Ousterhout, "Corner stitching: A Data-Structuring Technique for VLSI Layout Tools", *IEEE Transactions on Computer Aided-Design of Integrated Circuits and Systems*, pp.87-100, January 1984.

[2] S. L. Lin and J. Allen, "Minplex – A Compactor that Minimizes the Bounding Rectangle and Individual Rectangles in a Layout", *Proceedings of Design Automation Conference*, pp.123-130, 1986.

[3] Y. Bourai and C. J. R. Shi, "Symmetry Detection for Automatic Analog Layout Recycling", *Proceedings of Asian and South Pacific Design Automation Conference*, pp.5-8, 1999.

[4] R. Okuda, T. Sato, H. Onedera and K. Tamaru, "An Efficient Algorithm for Layout Compaction Problem with Symmetry Constraints", *Proceedings of International Conf. on Computer Aided Design*, pp.148-151, Nov. 1989.

[5] G. Lakhani and R. Varadarajan, "A Wire-Length Minimization Algorithm for Circuit Layout Compaction", *Proceedings of International Symposium on Circuits and Systems*, pp.276-279, 1987.

[6] T. H. Cormen, C. E. Leiserson and R. L, Rivest, *Introduction to Algorithms*, MIT Press, 1990.

Acknowledgement: The authors would like to thank Kiyong Choi and Jinho Park, SOC lab, Dept. of Electrical Engineering, University of Washington, for valuable discussions on circuit examples.

Evaluating a Bounded Slice-line Grid Assignment in O(*nlogn*) Time

Song Chen [1], Xianlong Hong[1], Sheqin Dong[1], Yuchun Ma [1], Yici Cai[1], Chung-Kuan Cheng[2], Jun Gu[3]
[1]*Department of Computer Science and Technology, Tsinghua University, Beijing, China*
[2]*Department of Computer Science and Engineering, University of California, San Diego USA*
[3]*Department of Computer Science, Science & Technology University of HongKong*

Abstraction

Bounded Slice-line Grid (BSG) is an elegant representation of block placement/floorplan. All block placement algorithms based on Bounded Slice-line Grid make use of simulated annealing or Solution Space Smoothing where the generation and evaluation of a large number of BSG assignments is required. Therefore, a fast algorithm is needed to evaluate the BSG assignments. We present a very simple and efficient O(*nlogn*) algorithm to evaluate the BSG assignments. Implementation of our algorithm is significantly faster than the original O(p×q) graph-based algorithm, where p×q is BSG size. The graph-based algorithm and our algorithm are embedded in a space smoothing search procedure. The experimental results demonstrated the efficiency of our algorithm.[λ]

1. Introduction

A dramatic increase in the complexity of integrated circuits has taken place because of rapid advances in integrated circuit technology. Therefore, hierarchical design and IP reuse become very important. The placement/floorplan has received much more attention in the latest decade. Many representations of the placement/floorplan have been proposed., such as Sequence Pair[2], Corner Block List[1], O-tree[4], Bounded Slice-line Grid[3], Transitive Closure Graph[5] and B*-tree[6].

The Bounded Slice-line Grid representation is introduced by Nakatake et al. The BSG of size n×n with assignments of blocks to BSG-rooms is used to represent the placement of *n* blocks. This approach has $n!\cdot C(n^2, n)$ combinations, in which the optimal placement is included. It is very intuitionistic. Unfortunately, It takes $O(n^2)$ time to evaluate a BSG assignment.

Compared with other placement/floorplan representations, BSG has an advantage of intuitiveness. Placement constraints such as pre-placed constraint, boundary constraint, abutment constraint, alignment constraint, rectilinear shape constraint, and even rectilinear chip constraint, are easily implemented based on BSG representation. But the evaluation of BSG will cost much more time than that of SP, CBL, and O-tree. Consequently, the people concentrate rarely on the BSG. This paper proposed an O(*nlogn*) algorithm to evaluate a BSG assignment. Implementation of our algorithm is significantly faster than the original graph-based O(p×q) , where p×q is the BSG size. The original algorithm and our algorithm are embedded in space smoothing search[10] procedure. The experimental results demonstrated the efficiency of our algorithm.

The rest of this paper is organized as follows: Section 2 is a brief review of Bounded Slice-line Grid representation. In section 3, some useful conclusion on Bounded Slice-line Grid is presented. The implementation details of the algorithms are shown in section 4, and section 5 provided some experimental results. Section 6 gives the conclusion.

2. Bounded Slice-line Grid

Bounded Slice-line Grid is introduced for the packing of the rectangle blocks. BSG is a topology defined on a plane, and dissects regularly the planar into rectangle rooms denoted as BSG-room by horizontal and vertical segments called BSG-seg. The BSG is infinite, but we restrict it into special region for practical use. The term BSG means BSG with certain size without special explanations in following. The fig.1 (b) gives an image of the BSG of size 6×6. In the procedure of packing, the blocks are assigned to BSG-rooms. Each BSG-room is assigned one block at most and one-dimension horizontal and vertical compaction follows.

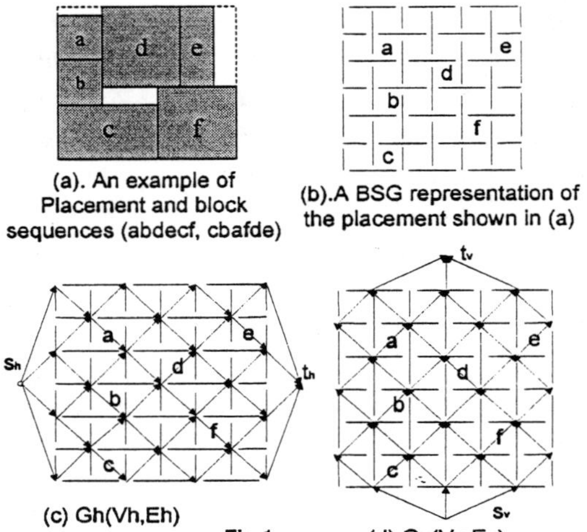

(a). An example of Placement and block sequences (abdecf, cbafde)

(b).A BSG representation of the placement shown in (a)

(c) Gh(Vh,Eh)

(d) Gv(Vv,Ev)

Fig.1

In a placement, block ***a*** is said to be right to block ***b*** if the left side of ***a*** is *right to* the right side of ***b***. Similarly, *left to*, *above*, *below* relations between blocks are defined.

Each BSG-room is bounded by a pair of half-shifted horizontal BSG-segs and a pair of half-shifted vertical BSG-segs. For every adjacent pair of vertical BSG-segs, the relation of right-to (or the inverse, left-to) is defined naturally. These horizontal relations can be represented by a directed acyclic graph $G_h(V_h, E_h)$, called BSG-horizontal Constraint Graph, where a source vertex and a sink vertex are added to represent the far left and far right of the packing. The edges crossing the rooms assigned blocks are weighted by the width of the blocks, and the edges crossing the empty rooms are weighted zero. The definition will be explicit by the examples shown in fig.1 (c). The vertical relations can be represented similarly by a directed acyclic graph $G_v(V_v, E_v)$, called BSG-vertical Constraint Graph, and an example is shown in fig.1.(d). The final packing can be computed by a graph-based O(p×q) longest path algorithm in [9].

3 Traversal on Bounded Slice-line Grid

In this section we will introduce some definitions, analyze the BSG representation and get some useful conclusions, from which we derived an approach to compute the final packing in O(*nlogn*) *time*, where n is the number of the blocks.

Given a BSG whose size is $n \times n$, the left-bottom T-junction combined by two BSG-segs has two types, '├' and '┴'.

Definition 3.1 H-BSG and V-BSG The BSG is defined as **V-BSG** if the left-bottom of the BSG is a '├' T-junction. The BSG is defined as **H-BSG** if the left-bottom of the BSG is a '┴' T-junction.

[λ] This work is supported by the National Natural Science Foundation of China 60121120706 and National Natural Science Foundation of USA CCR-0096383, the National Foundation Research(973) Program of China G1998030403, the National Natural Science Foundation of China 60076016 and 863 Hi-Tech Research & Development Program of China 2002AA1Z1460

0-7803-7761-3/03/$17.00 ©2003 IEEE

Without loss of generality, BSG means V-BSG in the following sections without special declaration.

Definition 3.2 BSG-coordinate: R(r, c) *R is a BSG-room assigned a pair of number (r, c), where r called* ***Row-coordinate*** *is the row number of the BSG-room, and c called* ***Column-coordinate*** *is the column number of the BSG-room.*

As illustrated in fig.2, *R (2, 3)* means that BSG-room R is located in the second row and third column along *x*-coordinate and *y*-coordinate orientation respectively.

Definition 3.3 Horizontal Relation Diagonal (HRD) and **Vertical Relation Diagonal (VRD)**: Have the BSG-rooms with identical sum of **Column-coordinate** and **Row-coordinate** connected, we can get a segment with diagonal orientation. If the connected BSG-rooms have horizontal relations each other, the segment is called a **Horizontal Relation Diagonal.** Otherwise, the segment is called a **Vertical Relation Diagonal.** An example is provided in fig.2.

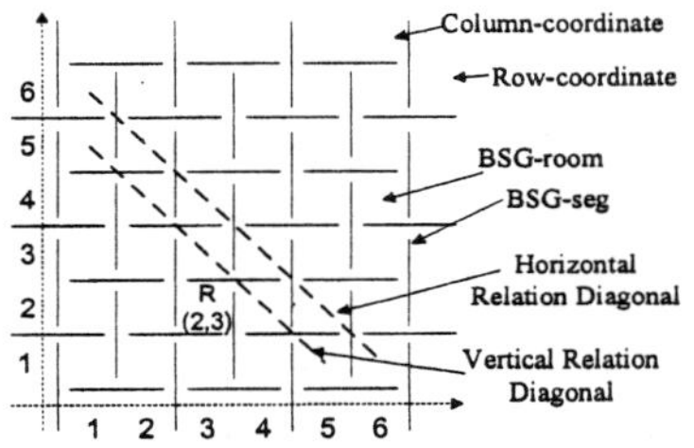

Fig. 2 Diagonals

The properties about two types of diagonals can be easily concluded.

Property 3.1 *For the BSG-rooms R_1 and R_2 in a Horizontal Relation Diagonal, if R_1.c is less than R_2.c, or R_1.r is larger than R_2.r, R_1 is left to r_2, vice versa.*

Property 3.2 *For the BSG-rooms R_1 and R_2 in a Vertical Relation Diagonal, if R_1.r is less than R_2.r, or R_1.c is larger than R_2.c, R_1 is below R_2, vice versa.*

Property 3.3 *For the Horizontal Relation Diagonal D_1^h and the Vertical Relation Diagonal D_2^v, if the sum of BSG-room in D_1^h is less than that of D_2^v, all the BSG-rooms in D_1^h will be left to or below the BSG-rooms from D_2^v.*

Property 3.4 *The Horizontal Relation Diagonal and Vertical Relation Diagonal will present alternately with the increment of the summation of the Row-coordinate and Column-coordinate.*

Definition 3.4 Minus Diagonal Traversal (MDT) *Given a Bounded Slice-line Grid with size of p×q, we define Minus Diagonal traversal on BSG as follow.*

1. Set the BSG-coordinate of the left-bottom BSG-room of BSG to (1,1), and the other BSG-coordinates will increase along *x*-axes and y-axes orientation.
2. For two BSG-rooms R_1 and R_2 in different relational diagonals, if (R_1.r + R_1.c) < (R_2.r+R_2.c) R_1 will be visited before R_2. Otherwise, R_2 will be visited before R_1.
3. For two BSG-rooms R_1 and R_2 in a HRD, if R_1.r > R_2.r (or R_1.c< R_2.c) R_1 will be visited before R_1. Otherwise, R_2 will be visited before R_1.
4. For two BSG-rooms R_1 and R_2 in a VRD, if R_1.r < R_2.r (or R_1.c > R_2.c) R_1 will be visited before R_1. Otherwise, R_2 will be visited before R_1.

An example of MDT is shown in Fig.3 (a). From the above properties, we can easily conclude the following lemmas.

Lemma 3.1 *A BSG-room must be visited after all the BSG-rooms left to or below this BSG-room in a MDT.*

Lemma 3.2 *Given a BSG with size of p×q and the assignment of blocks to BSG-rooms, we can get a block sequence by MDT, denoted as $S^{135°}$, which observing the following property.*

*A block **a** is left to or below another block **b** in the placement, i.e., the block **b** is right to or above the block **b** if **a** is before **b** in the sequence $S^{135°}$.*

Definition 3.5 Plus Diagonal Traversal (PDT) *Given a Bounded Slice-line Grid with size of p×q, we define Plus Diagonal Traversal on BSG as follow.*

1. Set the BSG-coordinate of the upper-left BSG-room of BSG to (1, 1), and the other BSG-coordinates will increase along *x*-axes and reverse y-axes orientation.
2. For two BSG-rooms R_1 and R_2 in different relational diagonals, if (R_1.r + R_1.c) < (R_2.r+R_2.c) R_1 will be visited before R_2. Otherwise, R_2 will be visited before R_1.
3. For two BSG-rooms R_1 and R_2 in a HRD, if R_1.r > R_2.r (or R_1.c< R_2.c) R_1 will be visited before R_1. Otherwise, R_2 will be visited before R_1.
4. For two BSG-rooms R_1 and R_2 in a VRD, if R_1.r < R_2.r (or R_1.c > R_2.c) R_1 will be visited before R_1. Otherwise, R_2 will be visited before R_1.

Lemma 3.3 *A BSG-room must be visited after all the BSG-rooms left to or above this BSG-room in PDT.*

Lemma 3.4 *Given a BSG with size of p×q and the assignment of blocks to BSG-rooms, we can get a block sequence by a PDT, denoted as $S^{45°}$, which observes the following property.*

*A block **a** is left to or above another block **b** in the placement, i.e., the block b is right to or below the block **a** if **a** is before **b** in the sequence $S^{45°}$.*

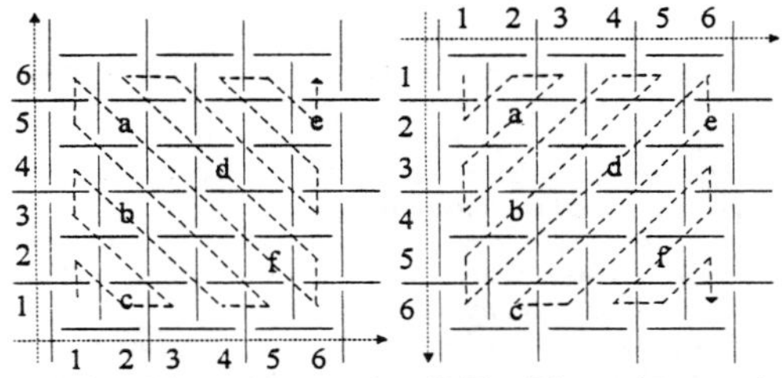

(a) Minus Diagonal Traversal (b) Plus Diagonal Traversal

Fig.3 BSG traversals

Theorem 3.1 *A Bounded Slice-line Grid representation can be uniquely transformed into two block sequence $S^{45°}$ and $S^{135°}$ in O(p×q) time, where the p×q is the BSG size.*

Theorem 3.2 For a block ***a***, any other block b is uniquely one of the following four cases.

1. L^S (a) = {***b*** | ***b*** is before ***a*** in both $S^{45°}$ and $S^{135°}$},
2. R^S (a) = {***b*** | ***b*** is after ***a*** in both $S^{45°}$ and $S^{135°}$}
3. A^S (a) = {***b*** | ***b*** is before ***a*** in $S^{45°}$ and after ***a*** in $S^{135°}$}
4. B^S (a) = {***b*** | ***b*** is after ***a*** in $S^{45°}$ and before ***a*** in $S^{135°}$}

According to lemma 3.2 and lemma 3.4, ***b*** is left to ***a*** in the packing if ***b*** ∈ L^S (a). The conclusion holds replacing the pair words ("L^S (a)" and "left to") with any of ("R^S (a)" and "right to"), ("A^S (a)" and "above") and ("B^S (a)" and "below"). From the above discussion, we can conclude the following theorem.

Theorem 3.3 The sequence $S^{45°}$ and Sequence $S^{135°}$ determined the topological relations among all the blocks, and block placement can be obtained from the two block sequences.

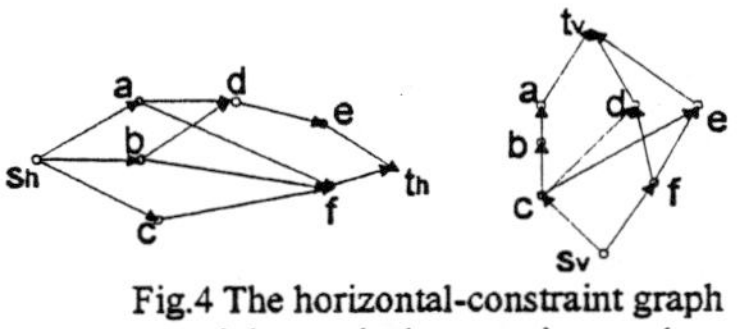

Fig.4 The horizontal-constraint graph and the vertical-constraint graph

Based on L^S, a directed and vertex-weighted graph, called horizontal-constraint graph G_h(V, E) can be constructed as follows.

V: source ***s***, sink ***t***, and ***n*** vertices labeled with block names

E: ***(s,b)*** and ***(b,t)*** for each block ***b***, and ***(b, b')*** if and only if ***b*** ∈ M^L(b')

Vertex-Weight: zero for ***s*** and ***t***, width of block ***b*** for other vertices.

Based on M^B, the vertical-constraint $G_v(E, V)$ can be constructed similarly. The coordinates of blocks can be computed by applying the longest path algorithm on the both constraint graph. An example is shown in fig.4.

4. Fast Evaluation of the BSG assignment

In this section, we described an O(***n***log***n***) time algorithm to evaluate the BSG assignments.

4.1 Fast Block Sequences Computing Algorithm

Only the computation of the sequence $S^{135°}$ is discussed in detail. The other sequence $S^{45°}$ can be computed analogously.

When BSG is traversed, we are only interested in the BSG-rooms assigned blocks. The following approach is addressed to perform MDT on the BSG to compute $S^{135°}$. According to the properties 1, 2 and 3, we can compute the sequence $S^{135°}$ by sort ascending on the summation of the Row-coordinate and Column-coordinate of the BSG-rooms assigned block to avoid traversing all the BSG-room which leads to a large time complexity. The main idea of the algorithm is shown in the following steps.

Step 1.Sort the blocks by sort ascending on the sum of the Row-coordinate and Column-coordinate of the corresponding rooms (Property 3.3).If the sums are equal, the following steps are available.

Step 2.If the BSG-rooms are in a Horizontal Relation Diagonal, sort ascending on the Column-coordinate of blocks (or sort descending on the Row-coordinate) is applied. (Property 3.1)

Step 3.If the BSG-rooms are in a Horizontal Relation Diagonal, sort descending on the Column-coordinate of blocks (or sort ascending on the Row-coordinate) is applied. (Property 3.2)

Step 4.Rotate the BSG planar by 90°. The BSG coordinate of BSG-rooms corresponding blocks are recomputed.

Step 5.Sort the blocks by sort ascending on the sums of the Row-coordinate and Column-coordinate of the corresponding rooms. If the sums are equal, the step 2 and step 3 are available.

From step 1 to step 3, the sort results in the sequence $S^{135°}$, and the sequence $S^{45°}$ is computed by the step 4 and step 5. The implementation details of the algorithm are shown in following.

Assume the blocks are 1…n, and *S_135 and S_45* are two arrays used to record the sequence $S^{135°}$ and $S^{45°}$ respectively. For the block ***b***, ***b***.*Sum* is used to record the sum of the Row-coordinate and the Column-coordinate of the BSG-room assigned block ***b***. The two-dimension array *M_R*[***b***][2], ***b***= 1…n is used to record the Row-coordinate and Column-coordinate of the BSG-room assigned block ***b***, *M_R*[***b***][0] is for the Row-coordinate and *M_R*[***b***][1] for the Column-coordinate. T_M_R is used for computing the block sequence $S^{45°}$. The BSG size is p×q.

```
Algorithm S_Computing (M_R, p, q)
    For i ← 1 to n
        Do S_45[i] ← i;
           S_135[i] ← i;
    QuickSort (S_135, 1, n, M_R, 0);
    For i ← 1 to n
        Do T_M_R[0][i] ← M_R[1][i];
           T_M_R[1][i] ← p – M_R[0][i] +1;
    QuickSort (S_45, 1, n, T_M_R, 1);
```

An improved quick-sort [8] algorithm is used. The function ***QuickSort*** computes $S^{45°}$ and $S^{135°}$ from special BSG-assignment.

```
QuickSort(Seq, low, up, M_R, flag)
    if  low<up
        iSplit=BiSection(Seq, low, up, M_R, flag);
        QuickSort(Seq, low, iSplit, M_R, flag);
        QuickSort(Seq, iSplit+1, up, M_R, flag);
    end if;
BiSection(Seq, low, up, M_R, flag)
    i ← low-1;  j ← up+1;
    while i<j
        i ← i+1;
        while Seq[i].Sum <= Seq[low].Sum
            if(Seq[i].Sum < Seq[low].Sum)  i ← i+1;
        else
            if (Seq[i].Sum+flag) mod 2==0
                if M_R[Seq[i]][0]<M_R[Seq[low]][0]  i ← i+1;
                else  goto L1;
            else
                if (M_R[Seq[i]][1]<M_R[Seq[low]][1])  i ← i+1;
                else  goto L1;
    L1: j ← j – 1;
        while(Seq[j].Sum >= Seq[low].Sum)
          if(Seq[j].Sum > Seq[low].Sum)  j ← j – 1;
          else
            if (Seq[j].Sum + flag) mod 2==0
                if M_R[Seq[j]][0]>M_R[Seq[low]][0]  j ← j – 1;
                else  goto: L2
            else
                if (M_R[Seq[j]][1]>M_R[Seq[low]][1])  j ← j – 1;
                else  goto: L2
    L2: if  i<j   Seq[i] ←→ Seq[j];
        else  return j;
```

As shown in fig.3.(b), blocks *a, b, c, d, e, f* are sorted as $S^{135°}$: *c(1,2), b(3,2), a(5,2), f(2,5), d(4,4), e(5,6) and* $S^{45°}$:*a(5,2), b(3,4), d(4,4), e(5,6), c(1,2), f(2,5)*.

Theorem 4.1 The block sequences $S^{45°}$ and $S^{135°}$ can be computed in O(***nlogn***) time by the above algorithm from the BSG assignment.

4.2 Packing in O(*nlogn*) time

To make the paper self-contained, we introduced an o(***n***loglog***n***) algorithm to compute packing from the block sequences $S^{135°}$ and $S^{45°}$. The final packing will be obtained in O(***nlogn***) time.

Definition 4.1 Given block set ***B***, each element $b_i \in B$ is assigned a weight $\omega(b_i)$ with two candidates, denoted as $w(b_i)$ and $h(b_i)$ which represent the width and the height of the block b_i, respectively. A block sequence is a permutation of some elements in ***B***.

We assume $\forall\, b_i \in B$, $\omega(b_i) > 0$ and $\omega(b_i) \in \{w(b_i), h(b_i)\}$.

Definition 4.2 Given two block sequences *R* and *S*, a sequence *C* is a common subsequence of *R* and *S* if *C* is a subsequence of both *R* and *S*. For example, <2, 3> is the common subsequence of <2, 5, 3> and <1, 2, 3, 5>.

Definition 4.3 The length of a common subsequence $C = \langle b_1, b_2, \ldots, b_n \rangle$ is $\sum_{i=1}^{n} \omega(b_i)$.

The longest common subsequence for two sequences *R* and *S* is the common subsequences of *R* and *S* with maximum length, denoted as *long(R, S)*.

Suppose there is a block *b* in the sequence *R* and *S*. Let $(R, S) = (R_1 b R_2, S_1 b S_2)$. Then $(R^r, S) = (R_2^r b R_1^r, S_1 b S_2)$. We can see a path from S_h to *b* in the horizontal constraint graph introduced in the previous section corresponds to a common subsequence of (R_1, S_1). Similarly, a path from S_v to *b* in the vertical constraint graph corresponds to a common subsequence of (R_2^r, S_1). Note that the coordinates of a block are the coordinates of the lower-left corner of the block. Thus, if $\omega(b_i)=w(b_i)$, $long(R_1, S_1)$ is the x-coordinate of block b_i, and *long(R, S)* is the width of the block packing. Similarly, if $\omega(b_i)=h(b_i)$, $long(R_2^r, S_1)$ is the *y*-coordinate of block b_i, and $long(R^r, S)$ is the height of the block placement.

Only the details of computing *x*-coordinates for the blocks are discussed, *y*-coordinates can be achieved by dealing with the sequences (R^R, S) similarly. *Priority Queue* implemented in [11] is used to manage the array in the algorithm. By using the Priority Queue, the two sequences can be translated into a packing in O(***nloglogn***) time. Each element in the Priority Queue is associated with two keys, (*index*, *length*), and both keys must be kept in non-decreasing order. We use *index* as the primary key, and discard the element whose *length* is not increasing because of its making no contribution to subsequent Long Common Sequence computation. In some pass of iteration, the discarding may take Ω (n) time. However, the total time cost for discarding elements can be **amortized.** [7]

In order to describe the algorithm, we introduce the involved data structures firstly. Assume the blocks are 1 … n, and the input block sequences are *(R, S)*. Both *R* and *S* are a permutation of {1, …, n}. Block array C[b], ***b*** = 1 … n is used to record the *x* or *y* coordinate of block ***b*** depending on the weight ω (*b*) equals to w(*b*) or h(*b*) respectively. The array P[*b*], ***b*** = 1 … n indicate the indices of ***b*** in *R* and *S*, i.e. P[*b*]. *x* = *i* and P[*b*].*y* = *j* if *b* = R[*i*] = *S*[*j*]. Let the complete binary tree be H = {1, …, 2^h + n} where h = $\lceil log(n+1) \rceil$, the bucket list be BL. Each bucket has an index and stores value. Let BL[*index*] denote its value. BL[*index*] records the length of candidates of the longest common subsequence. The algorithm is shown below.

Algorithm Packing(R, S)

```
Initialize index Array P;
Initialize H, insert the initial index 0;
Initialize BL with BL[0] = 0;
FOR i = 1 TO n DO
  b = R[i];
  p = P[b]:y;
  insert p to H and BL;
  C[p] = BL[predecessor(p)];
  BL[p] = C[p] + ω (b);
discard the successors of p from H and BL whose value<= BL[p];
RETURN BL[indexmax];
```

Theorem 4.2[7] Algorithm **Packing** computes *long*(*R*, *S*) and records the position of each block in O(***nloglogn***) time for two sequence (*R*, *S*).

Theorem 4.3 The evaluation of the BSG representations can be performed in time O(***nlogn***).

The procedure ***S_computing*** can be completed in *O(nlogn)* time. And the final packing can be transformed from the block sequences in O(***nloglogn***) time. Consequently, the evaluation of BSG can be achieved in O(***nlogn***) time.

5 Experimental results

Experiments are designed to compare our algorithm with original graph-based algorithm in two aspects. First, we compare the runtimes in obtaining block placement. Second, we compare the runtimes in evaluating a single BSG assignment. Both experiments are completed on Sun-Fire v880.

The experiments are carried out for the MCNC benchmarks and some test cases with 98, 147, 198, 294 blocks generated from ami33 or ami49. Table 1 lists the average running time for the graph-based algorithm and our algorithm. The graph-based algorithms construct the constraint graphs in O(n^2) time, and uses an algorithm similar to that in [9] to find the longest path length to all vertices from source vertex in O($n + m$) time where n is the number of blocks(vertices), m is the number of edges.

From Table 1, we can see that our algorithm is significantly faster than the graph-based algorithms. For example, we achieved at least 4X speedup over the graph-based algorithm when the number of blocks is 147.

The evaluation of a BSG assignment is a weighted sum of the area of the packing and the total wire length based on the half perimeter estimate of bounding box for each net. The cost function C is defined as:

$$Cost = A + \alpha \cdot W + \lambda \cdot Rs$$

Where A is the area, W is the wiring cost and Rs is the factor related the ratio of the Weight and Height of the chip. The weight parameter α and λ are used in the cost function for balancing the factors A, W and Rs.

From the Table 2, we can find out that our algorithm is faster than the graph-based algorithm. For example, we achieved 8X speedup over the original graph-based algorithm when the number of blocks is 294.

6 Conclusion and Acknowledgements

Bounded Slice-line Grid is an elegant topological representation of placement/floorplan. In this paper we have proposed a new approach to transform a BSG assignment to a block placement in O(***nlogn***) time. Experimental results show the efficiency of our approach. Bounded Slice-line Grid has many good qualifications in processing the placement/floorplan with constraints, such as rectilinear placement, etc. Our approach can bring Bounded Slice-line Grid into many more applications.

Table 1 Evaluation of a final assignment

circuit	Blocks	p×q	Original (s)	Ours(s)
Ami33	33	33×33	179	37
Ami49	49	49×49	599	92
98	98	50×50	1109	279
147	147	80×80	2876	648
198	198	100×100	6309	1106
294	294	150×150	27882	3643

Table 2 Time to obtain a single BSG placement

Circuit	Blocks	p×q	Original (s)	Ours(s)
Ami33	33	33×33	2.68e-4	5.50e-5
Ami49	49	49×49	5.91e-4	8.80e-5
98	98	50×50	6.14e-4	1.78e-4
147	147	80×80	1.58e-3	3.08e-4
198	198	100×100	2.48e-3	4.31e-4
294	294	150×150	5.56e-3	6.78e-4

7. References

[1] X. Hong, G. Huang, S. Dong, Y. Cai, C.-K, Cheng, J. Gu, "Corner Block List: An Effective and efficient topological representation of non-slicing floorplan and its applications," Proc. ICCAD, pp.8-12, Nov.2000.

[2] H.Murata, K.Fujiyoshi, S.Nakatake, and Y.Kajitani, "Rectangle-packing based module placement," Proc. ICCAD, pp.472-479, Nov.1995.

[3] S.Nakatake, et al. "Module Packing Based on the BSG-Structure and IC layout Applications", IEEE Trans. on Computer-Aided Design of Integrated Circuits and Systems, Vol.17, No.6, June 1998, pp.261-267.

[4] P.Guo, C.K.Cheng, T.Yoshimura, "An O-tree Representation of Non-Slicing Floorplan and Its Applications", Proc. 36th DAC, 1999, pp.268-273.

[5] J-M.Lin, Y-W.Chang, "TCG:A Transitive Closure Graph-Based Representation for Non-Slicing Floorplans," Proc. 38th DAC, 2001.

[6] Y. Chang, G.Wu, S.Wu, "B*-tree: A new representation for Non-Slicing Floorplans", Proc.37th DAC, 2000, pp.458-463.

[7] X.Tang, M.Wong, "FAST-SP: A Fast Algorithm for Block Placement based on Sequence Pair", Proc. ASP-DAC 2001.

[8] Thomas H. Cormen et al, Introduction to Algorithms. MIT Press.

[9] J.A. McHugh, "Algorithmic Graph Theory", Prentice Hall (1990).

[10] S. Q. Dong, X. L. Hong, S. Zhou, J. Gu, "Efficient VLSI Module Placement with Solution Space Smoothing", Volume II, Inter. Conf. On Communictaions Circuits and Systems and West Sino Expositions Proceedings 1396-1400..

[11] D.B. Johnson. "A priority queue in which initialization and queue operations take O(loglogD) time", Mathematical Systems Theory 15, pp. 295-309, 1982.

Noise Constraint Driven Placement for Mixed Signal Designs

William H. Kao and Wen K. Chu
Cadence Design Systems

Abstract

In this paper we discuss how the problem of substrate-coupled switching noise (*dI/dt* and *dV/dt* noise) in mixed signal designs can be solved by using the substrate analysis capabilities in the SeismIC tool to drive the placement of macro cells using the Virtuoso Custom Placer (VCP) for Mixed Signal designs. An objective function consisting of area, wire length and other constraints such as substrate noise is minimized by VCP's annealing engine and the Constraint Manager.

I. INTRODUCTION

With the increase in levels of integration in current and future chip design there is a larger number of SOCs where digital, analog and RF functions are all integrated on a single chip substrate. Furthermore in the deep submicron regime signal integrity issues are increasingly becoming more critical. One of the significant signal integrity problems in mixed signal designs today is the handling of noise coupled through the common substrate and power supplies which is caused by signal switching inside the digital section and affecting sensitive analog circuitry [1]. The injected switching noise could perturb the substrate potential near the analog circuitry. The body effect modulates the transistor threshold voltage and the device might not perform as designed. Researchers have extended substrate-coupling analysis to floorplanning and placement domains [2]. The coupling effects were considered as a cost function in several simulated annealing based cell placement [3] and power distribution synthesis [4] systems. In this paper, we propose a solution to constrain noise coupling through the substrate based on an analysis-optimization loop between the substrate analysis capabilities of SeismIC [5] and the placement capabilities of Virtuoso Custom Placer [6].

II. SUBSTRATE NOISE NETWORK MODEL AND ANALYSIS

Substrate noise is defined as the voltage deviation in the bulk node of a device caused by currents propagating through a substrate. There are three mechanisms that inject noise into the substrate. Power buses couple noise into the substrate through ohmic contacts, wells couple capacitively through reverse biased bulk/well junctions, and transistors couple capacitively through drain/source diffusions.

Techniques used to mitigate substrate noise include use of a different process, a more costly low-inductance package, separate supplies to isolate sensitive blocks from noisy blocks, guard rings, a Kelvin reference in which the substrate or well contacts are separate from the circuit power or ground connections, or a change in the floorplan. Noise analysis is effective and needed to verify if a design will incur problems, to identify major sources of substrate noise, and to perform trade-off analysis to figure out the optimum strategy for minimizing noise.

A complete analysis of substrate noise requires both extraction of a substrate as well as simulation. Extraction is the process by which an electrical equivalent model of the substrate, including resistance, capacitance is determined. To accurately extract a substrate, the complex geometries of wells, contacts, well taps, diffusions, trenches, etc need to be extracted. Once extraction is completed, simulation can be performed on a circuit including the three dimensional extracted RC substrate network. Simulation also requires the knowledge of the source and location of noise injectors that are causing the noise. A Spice simulation of all devices and substrate parasitics would take an inordinate amount of time. SeismIC performs a faster noise simulation using equivalent noise sources and can be used to analyze chips with one million or more devices, something not practical with Spice.

The substrate noise model is depicted in Figure 1. Current sources (a in Fig. 1) are used to model charge injected into substrate via device drain/source diffusions. The contacts and guard rings are modeled by r_c , r_g and packaging inductors (b in Fig. 1). So the packaging induced noise at the substrate ties can be calculated by the simulation engine. Ideal voltmeters V_n (c in Fig. 1) are used to detect voltages at relevant substrate nodes and are used to calculate the substrate noise related penalty terms in the form min $(0, V_n - V_{target})$.

The substrate itself can generally be represented as a network grid (d in Fig.1) composed of impedances Z_s.

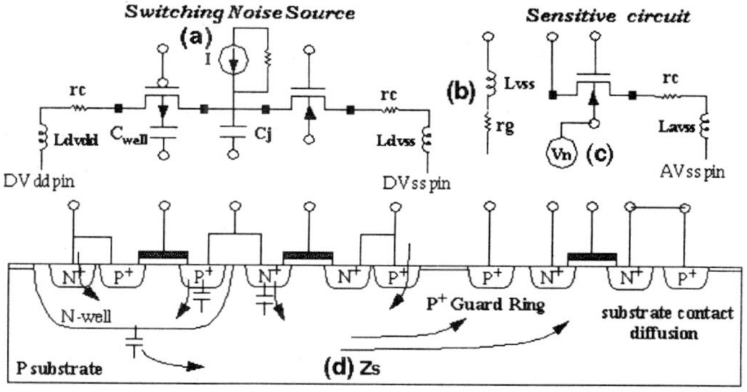

Fig 1. Substrate Noise Network Model

0-7803-7761-3/03/$17.00 ©2003 IEEE

The substrate network is represented as an admittance matrix Y_s and the noise sources represent the current vector I. Then the equation

$$Y_sV=I \quad (1)$$

can be solved through traditional circuit simulation techniques using LU factorization, forward and backward substitution and sparse matrix techniques.

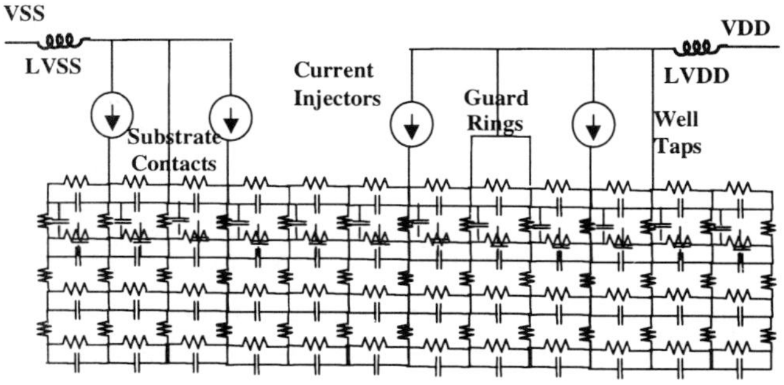

Fig 2. Substrate RC Mesh Network

To see how the substrate RC network (Figure 2) driven by active current sources is analyzed, consider that the voltage response at a bulk node of interest, V_b is desired. The voltage response can be written as follows:

$$V_b(s) = \sum_k z_k(s)i_k(s) \quad (2)$$

where $i_k(s)$ are the current sources at various locations on the substrate and $z_k(s)$ are their corresponding impedances to the substrate node of interest. The current source values, $i_k(s)$ can be determined from a simulation of the original circuit (without parasitics) by observing the currents flowing in the power/ground nodes and the device bulk terminals. This can be accomplished either with a transistor-level circuit simulator or a gate-level event driven simulator in conjunction with pre-characterized cell libraries [7,8]. The currents can be either time domain waveforms or a composition of spectral values at every frequency ($s = j\omega$) of interest. The impedances, $z_k(s)$ can be obtained by inverting the admittance matrix Y_s formed by the RC substrate network and package inductances. The frequency domain response of V_b can be obtained by solving (2) at every frequency of interest. Applying the inverse Laplace transform to this response results in the corresponding time domain waveform.

One advantage of using (2) to calculate the noise response of a bulk node of interest is that each individual noise contributor can be calculated independently. Hence, from (2), the noise contribution at the bulk node of interest from injector 1 is $z_1(s).i_1(s)$. Similarly, $z_2(s).i_2(s)$ is the contribution from injector 2, $z_3(s).i_3(s)$ is from injector 3 and so on. Thus, the most significant noise contributors can be identified and appropriate measures can be taken to minimize their impact.

By employing the method described above to the adjoint circuit equations instead of the direct circuit equations (1), one can also solve the substrate surface noise distribution for a given number of noise sources. Using noise macromodels, one can effectively limit the number of noise sources that appear in (2) making the computational time required for transient noise simulation more acceptable. Macromodels are characterized current injection patterns of a noise contributor cell.

The SeismIC tool provides a complete extraction and analysis solution, which analyzes substrate noise in large mixed-signal designs. It permits design planning trade-off analysis at an early stage and debugs noise coupling problem at a later verification stage.

SeismIC supports the use of macromodels in high-level designs and generates noise distribution contour plots for floorplanning guidance. The digital cells in the design have currents, and power supplies data characterized. A macromodel can be viewed as a buffer which injects current through coupling capacitors and contacts to the substrate.

III. VIRTUOSO CUSTOM PLACER (VCP)

Automatic placement of custom analog, mixed signal and high performance digital circuits is a challenging task. It is also impossible to rely solely on manual custom layout design to remain performance competitive [9]. The Virtuoso Custom Placer (VCP) was developed by combining interactive and automated placement to enhance productivity without sacrificing custom performance.

VCP supports 3 design styles: area based placement for analog layout and blocks of varying sizes and aspect ratios, standard cell row based placement and CMOS N and P transistor rows. VCP's major features include constraint driven placement that ensures the satisfaction of design specifications, connectivity-driven placement which ultimately provides shortest wiring in the finished design, and reentrant placement capability which allows flexibility in switching between interactive and automated modes.

VCP's placement process consists of 4 basic steps:

1. Data Preparation where the netlist and constraints are read in from the database and constraint manager.
2. Global Placement. It starts by using iterative quadratic/mincut [10,11] methods to recursively bipartition components into different regions. The goal is to find an overall violation-free placement and spread all components into the defined placeable area.
3. Detail Placement. Using annealing based methods it refines the initial placement to obtain a better placement based on the costs defined by the user.
4. Congestion Estimation and Spacing Adjustment: To make the placement routable and reduce the overall wire length, there is a need to identify where the congested spots are and make room for routing.

IV. SUBSTRATE NOISE CONSTRAINED PLACEMENT

Many successful placement methodologies for general rectilinear instances have used unconstrained optimization techniques to achieve an optimal placement. Independent placement variables such as area, wire length and other cost functions are used to evaluate the placement quality. Placement objectives such as smaller area and shorter overall wire length are rewarded and noise constraint violations are penalized.

To tackle the problem at hand we propose to solve the following problem:

Minimize Total Area and Wirelength constrained by

$$[\, F_{min}] < [\, F(P)] < [F_{max}] \qquad (3)$$

where P is the placement instance and F(P) represents the noise constraint which is affected by the placement instance P. The constrained multi-objective optimization problem above is converted to the unconstrained single objective problem:

Mimimize:

w_1 Area + w_2 Wire Length + w_3 Constraint Violations,
where w_1, w_2 and w_3 are weighting functions.

The goal is to drive all 3 terms to a minimum. This is accomplished by treating the above parameters of area, wire length and other constraints such as timing, noise, placement overlaps and placement symmetry as variables, have the Constraint Manager pre-process to optimize and to resolve constraint conflicts, and then have VCP optimize the values with its annealing engine. (see upper right hand corner of Figure 3).

V. INTEGRATED SEISMIC - VCP FLOW

In previous sections we covered separately the substrate coupling modelling and analysis capabilities of SeismIC and the placement and optimization strategy for VCP. In this section we will describe the working flow (see Figure 3) between these two tools to come up with a solution for handling substrate coupling induced noise in Mixed Signal designs. Constraint driven technology for Analog and Mixed-Signal IC layout has been developed [12] and has been well integrated in VCP [13]. SeismIC can analyze the impact of instance placement, power supply distribution and other changes in substrate noise distribution across the chip surface.

The proposed SeismIC-VCP flow starts with the user setting up the constraints through the Constraint Manager which pass the timing, power and noise constraints to VCP.

The VCP auto-placement engine initiates a multi-pass placement approach based on quadratic placement and simulated annealing algorithms and comes up with an initial placement (location, shape orientation) and then invokes the SeismIC engine to calculate substrate noise distribution.

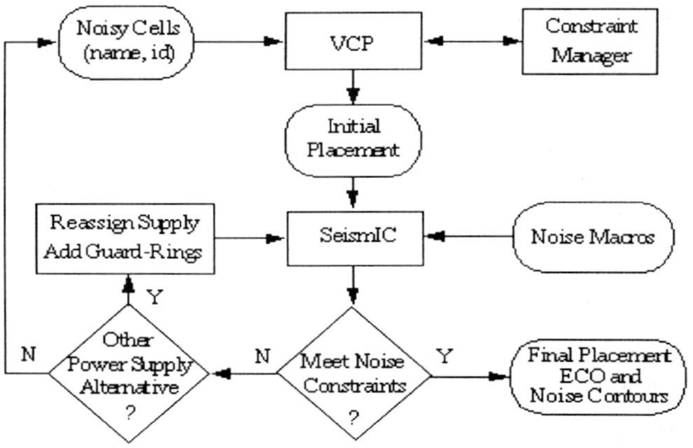

Fig 3. SeismIC-VCP Flow

SeismIC generates the substrate network and hooks it up with packaging network, wells, substrate contacts, noise macros, guard rings and sensitive cells. SeismIC analyzes the complete network and calculates a noise distribution. A noise contour plot is generated and can be super-imposed on the design layout for adjustment guidance [3]. SeismIC compares the calculated noise levels against the specified noise constraints on sensitive cells. If noise levels exceeds the constraints it then tries various ways to quiet substrate noise, which include reassigning power source to other supplies, grouping package pins to reduce *dI/dt*, adding guard-rings around sensitive cells and so on. The SeismIC noise calculation method identifies the significant noise contributors that affect sensitive analog components coming from each particular power supply or particular macro. The tool then uses this information to determine effective strategies to minimize coupling from each noise source. It automatically tries to add guard rings, remove guard rings, and redistribute power pads to see which changes have the most impact on reducing noise coupling. The noise level and noise reduction strategies are then fed to VCP for constraint optimization.

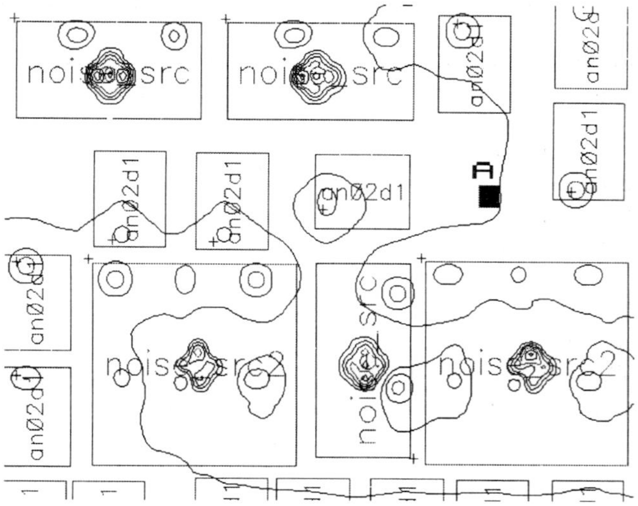

Fig 4. Initial placement (Vn at Cell A =22mV)

Figure 4 shows a study example and its first placement iteration with noise contours superimposed on top of the layout. The peak noise at sensitive cell A is calculated to be at 22mV, far exceeding the specified constraint of 8mV.

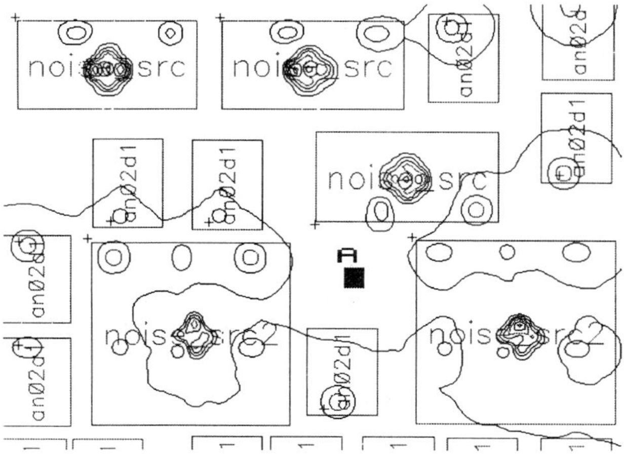

Fig 5. Placement after few Iterations (Vn = 19mV)

Figure 5 shows the results after a few placement iterations, getting the peak noise reduced down to 19mV. Notice that the location of cell A has changed and blocks nearby have been moved. Since the constraint is not yet satisfied, a guard ring is added for cell A and connected to a quiet ground supply (Vss2) bringing the final noise level down to an acceptable 6.2mV (see Figure 6). Runtimes for the 20 blocks example were 5.3 minutes for SeismIC extraction and analysis, and about 2.5 minutes for a VCP placement for reaching each local optimization for a set of regular constraints. The total runtime including data passing was about 8 minutes for one iteration.

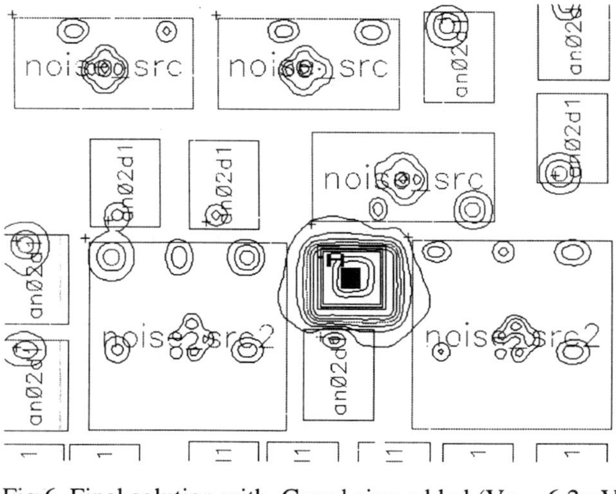

Fig 6. Final solution with Guard-ring added (Vn = 6.2mV).

VI. CONCLUSION AND FUTURE WORK

In this paper, we have presented a constraint driven placement flow which addresses the problem of substrate coupled switching noise in Mixed Signal designs. It is accomplished by utilizing the substrate noise modelling and analysis capabilities in SeismIC and coupling it to the Virtuoso Custom Placer in an analysis and optimization loop. Further work would be to extend this flow to include other signal integrity issues such as crosstalk, EM, IR and thermal effects [14].

REFERENCES

[1] G.H. Warren and C. Jungo, "Noise, crosstalk and distortion in Mixed Analog/digital Integrated Circuit," IEEE Custom Integrated Circuits Conference, pp. 12.1.1-12.1.4, May 1988.

[2] S. Zhao, K. Roy and C. K. Koh, "Decoupling Capacitance Allocation and Its Application to Power-Supply Noise-Aware Floorplanning," IEEE Trans. Computer-Aided Design, vol. 21, no. 1, Jan. 2002, pp. 81-92.

[3] S. Mitra, R. A. Rutenbar, L. R. Carley and D. J. Allstot, "Substrate-Aware Mixed-Signal Macro-Cell Placement in WRIGHT", Proc. IEEE Customer Integrated Circuits Conf., May 1994, pp. 24.2.1-24.2.4.

[4] B. R. Stanisic, N. K. Verghese, R. A. Rutenbar, L. R. Carley and D. J. Allstot, "Addressing Substrate Coupling in Mixed-Mode IC's: Simulation and Power Distribution Synthesis", IEEE J. Solid State Circuits, vol. 29, no. 3, Mar. 1994, pp. 226-238.

[5] Saila Ponnapalli, N. Verghese, W. K. Chu and G. Coram, " Preventing a Noisequake: Substrate analysis identifies Potential problems in Mixed Signal and RF designs,", IEEE Circuits and Devices, Vol. 17, No. 6, Nov. 2001, pp. 19-28.

[6] Doug Curtis, George Gadelkarim, Wai Ho and Joe Wang, "Virtuoso Custom Placer: Interactive and Automated Custom Placement," Proc. Cadence Techn. Conference 2001, pp 24-29.

[7] N. Verghese, D. Allstot, and S. Masui, "Rapid Simulation of Substrate Coupling Effects in Mixed Mode ICs", Proc. of IEEE Custom Integrated Circuits Conference, May 1993, pp. 18.3.1-18.3.4.

[8] E. Charbon, P. Miliozzi, L. P. Carloni, A. Ferrari and A. Sangiovanni-Vincentelli, "Modeling Digital Substrate Noise Injection in Mixed-Signal IC's", *IEEE Transactions on Computer-Aided Design of Integrated Circuits and Systems*, vol. 18, no. 3, March 1999, pp. 301-310.

[9] Enrico Malavasi, Dan Guilin, Kathy Jones and William Kao, "Layout Acceleration for IC Physical Design," Proc. DATE-99, Munich 1999, pp. 7-11.

[10] J. Kleinhans, G. Sigl, and F. Johannes, and K. Antreich, "GORDIAN: VLSI Placement by Quadratic Programming and Slicing Optimization." IEEE Trans. On CAD, vol. 10 no.3, March 1991, pp. 356-365.

[11] C.M. Fiduccia and R. M. Mattheyses, " A linear-time heuristics for improving network partitions," Proc. 19th Design Automation Conference, 1982, pp. 175-181.

[12] E. Malavasi and W. Kao, "Current Issues in a Constraint-Driven Mixed Signal Physical Design Flow", Proceedings of 1997 Intern. Symposium on Circuits and Systems, pp. 133-136.

[13] K. Lampaert, G. Gielen and W. M. Sansen, "A Performance-Driven Placement Tool for Analog Integrated Circuits", IEEE J. Solid State Circuits, vol. 30, no. 7, July 1995, pp. 773-780.

[14] W. Kao and W. K. Chu, "ATLAS: An Integrated Thermal Layout and Simulation System for ICs, Proceedings of 1994 European Design and Test Conference, pp. 43-48.

ALGORITHMS FOR ANALOG VLSI 2D STACK GENERATION AND BLOCK MERGING †

Rui Liu[1, 2], *Sheqin Dong*[2], *Xianlong Hong*[2], *Di Long*[2], *Jun Gu*[3]
[1]Institute of Software, Chinese Academy of Sciences, Beijing, 100080, P.R.China
Email: liurui@isdn.iscas.ac.cn
[2]Department of Computer Science and Technology, Tsinghua University, Beijing, 100084, P.R.China
Email: {hxl-dcs}{dongsq}@tsinghua.edu.cn
[3]Department of Computer Science, Science & Technology University of Hong Kong

ABSTRACT

In analog VLSI design, 2-axial symmetry stack and block merging are critical for mismatch minimization and parasitic control. In this paper, algorithms for analog VLSI 2-axial symmetry stack and block merging are described. We get several theory results by studying symmetric Eulerian graph and symmetric Eulerian trail. Based on those, an *O(n)* algorithm for dummy transistor insertion, symmetric Eulerian trail construction and 2-axial symmetry stack construction are developed. The generated stacks are 2-axial symmetric and common-centroid. Block merging algorithm is described, which is essentially independent to topological representation. Formula for calculating maximum block merging distance is given. Experimental results show effectiveness of our algorithms.

1. INTRODUCTION

Analog blocks typically constitute only a small fraction of components on mixed-signal ICs and emerging systems-on-a-chip (SoC) designs. But due to the increasing levels of integration available in silicon technology and the growing requirement for digital systems to communicate with the continuous-valued external world, there is a growing need for quality of analog integrated circuits. Although many efforts have been done in this field, the analog layout is still a hard and time consuming task which has a considerable impact on circuit performance. Asymmetries and device mismatch can easily upset the critical precision of component and together with the parasitics associated with the interconnections they can introduce intolerable performance degradation. Device merging, i.e. placing devices such that diffusion geometry is shared between electrically connected devices, is a very important technology for analog VLSI layout to dramatically reduce the parasitics as well as area occupation. Merging a series transistor is called *stacking*.

Many researches have been done to explore the device merging optimizations during the analog layout in the past years. KOAN/ANAGRAM II [1] keeps the macro cell style, and merges the common area when the nodes are exactly placed together during the simulated annealing. This kind of approach not only puts heave burden on the simulated annealing, but also can not achieve the interdigitated and common-centroid structure [2], which is often used by manual layout. The stack generator in [3] generates optimum stacks that satisfy performance constraints, using a path partitioning algorithm. However, because it attempts to enumerate all optimal stacks, runtime can be extremely sensitive to the size of the problem. Symmetry and matching constraints can greatly prune the search, but the basic algorithm has exponential time complexity. [4] gives an approach using simulated annealing to randomly generate stack. An algorithm with *O(n)* was present in [5], which can generate optimal stack without symmetry constraints, or approximation solution under symmetry constraints.

However, two problems exist in the previous tools. First, they can only generate 1-D stack, i.e., they can not generate the 2-axial symmetry stack. 1-D stacks are sometimes slim and long, that is not desirable during placement. Furthermore, [6][7] have shown 2-axial symmetry minimum the mismatch as well as cancel out press, process and thermal gradients in every direction. So, 2-axial stack is more desirable. Second, during the stacking, only transistors with similar channel width can be generated in one stack. The different stacks can not be merged during placement.

In this paper, we firstly present an *O(n)* algorithm to generate 2-axial symmetry stacks. Several stacks with different ratio are generated for one group of transistor. These stacks can be chosen during the placement. Then, algorithm for block merging is given, which can be used to merge stacks with different channel width transistors. The block merging algorithm is implemented on Sequence Pair [8], but it is essentially independent to any topology representation. This paper is organized as follows. The 2-axial symmetry stack generation algorithms are described in section 2. Section 3 presents the block merging algorithms. Finally, section 4 gives the experimental results and concludes the paper.

2. STACK GENERATION ALGORITHMS

2.1 Basic stacking strategy

Circuit schematic should be modeled in a format appropriating for a graph algorithm to solve the layout problem effectively. Our strategy is similar to that introduced in [3][6]:

† This work is supported by Fund of Tsinghua University Fundamental Research (Jc2001025973) and 973 National Key Project (No.G1998030403).

0-7803-7761-3/03/$17.00 ©2003 IEEE

1. Divide the circuit into *partitions* with respect to device type and bias node (body node in MOS transistors).
2. Perform device *folding*: split large transistors into smaller parallel transistors.
3. Perform further partitioning to reduce the variation on the module widths in a partition, only the transistor with same with is in the same partition. Only the partition that is fully symmetric and has not self-symmetry edge can generate common-centroid 2-axial symmetry stacks. If there are self symmetry devices, further partition is applied to isolate the self symmetry devices from other devices.
4. *Generate* stacks that implement each partition. The 2-axisle stack generation algorithms (Section 2.2, 2.3) operate on each circuit partition separately. Several stacks of different ratio are generated for each partition. Choosing one of the stacks of different ratio for same partition is added to move set of the Simulated anneal during placement.

2.2 Finding symmetric Eulerian trail

For the partition $G=(V, E)$ that is fully symmetry, the first step to generate the 2-axial symmetry stacks is to construct the symmetry Eulerian graph by adding dummy edges. At most one self symmetric dummy edge can be added. Based on this symmetry Eulerian graph, symmetric Eulerian trail can be build. Now, we give some definition.

Definition 1: *self symmetric vertex* is vertex on the symmetric axle whose symmetric vertex is itself. (Figure 1(a))

Definition 2: *self symmetric edge* is an edge to which the two vertices associated are symmetric to each other. And this edge crosses the symmetric axle and whose symmetric edge is itself. (Figure 1 (a))

Definition 3: *Symmetric Eulerian trail* $(v_1, v_2 \ldots v_n)$ is a trail symmetrically covering the graph, i.e., in the trail v_i and v_{n+1-i} are symmetric to each other.

From definition 3, we can easily get the following lemma that is useful during the Eulerian trail constructing.

Lemma 1: Given the *Symmetric Eulerian trail* $(v_1, v_2 \ldots v_n)$, if n is an odd number, there are n-1 edges in the trail and there is no self symmetric edge. If n is an even number, $(v_{n/2+1}, v_{n/2+2})$ is a self symmetric edge.

Now we show how to construct symmetric Eulerian graph based on that a symmetric Eulerian trail can be build. We first find all the vertices with odd degree. The *degree* of a vertex is the number of edges incident to it. For each odd degree vertex, there

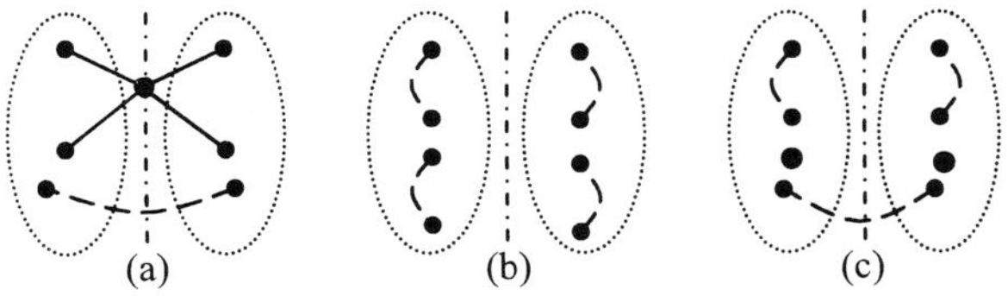

Figure 1. (a) Self symmetric vertex and self symmetric edge. (b) All the odd degree vertices are connected pair by pair. (c) Two odd degree vertices leaved not connected.

must be a symmetric vertex that is also odd degree. The following lemma 2 shows self symmetric vertex must be even degree. Therefore, we only need find out all the odd degree vertices on one side of the symmetric axle. All and only their symmetric vertices are also odd degree on the other side of symmetric axle.

Lemma 2: Given a symmetry graph $G=(V, E)$, and there is no *self symmetric edge*, if there are some self symmetric vertices, their degree must be even.

Proof: The proof is straightforward. Since there is no self symmetric edge, each edge incident to the self symmetric vertex must have a symmetry one incident to the same vertex. So, the degree of self symmetry vertices must be even □

If the number of odd degree vertices on each side of the symmetry axle is even, there are two method can be applied.

Method-1: After connecting these odd degree vertices by adding dummy edges pair by pair on each side, there will no odd degree vertex left in the partition (Figure 1(b)). Because all vertices are even degree and each edge on one side of the symmetric axle has a symmetric edge on the other side, a symmetry Eulerian trail can be constructed symmetrically and synchronously on both sides of the axle. And there are even edges in the trail.

Method-2: Another method to connect these odd degree vertices is as follows. We connect these vertices pair by pair just as previous except the last two pairs. For the last two pair of odd degree vertices, we connect an odd degree vertex with its symmetry one by a self symmetric dummy edge and leave the other odd degree symmetry pair not connected (Figure 1(c)). Let the self symmetry edge be $(v_{n/2+1}, v_{n/2+2})$ in the symmetric Eulerian trail. Then, the two vertices $v_{n/2+1}$and $v_{n/2+2}$ are used as start vertex on each side of the axle to construct symmetric Eulerian trail. There are odd edges in the trail.

Because previous two methods can generate trails with different number of edges, we generate both trails to be used by future stack generation.

On the other hand, if the number of odd degree vertices on each side of symmetric axle is odd, there are also two methods can be applied.

Method-3: After connected them pair by pair as previous, there will be one odd degree vertex leaved unconnected on each side of the symmetric axle, which is symmetry to each other (Figure 2 (a)). The following theorem guaranteed that there must exists a self symmetric vertex. Therefore, we use the self symmetric

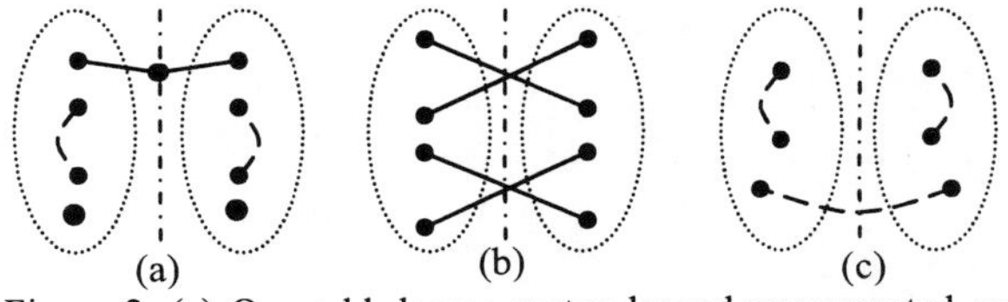

Figure 2. (a) One odd degree vertex leaved unconnected on each side. (b) Edges that crossing symmetry axis. (c) Connect the last odd degree vertex with its symmetric vertex by a self symmetric dummy edge.

vertex as start vertex to construct symmetric Eulerian trail symmetrically and synchronously on both side of symmetrical axle. The trail has even edges.

Theorem 1: Given a symmetry graph $G = (V, E)$, if there is no *self symmetric edge* and the number of odd degree vertices on each side of the symmetric axis is odd, there exist at least one *self symmetric vertex.*

Proof: Suppose for the sake of contradiction that there is a not *self symmetric vertex*, i.e., each vertex has a symmetry one on the other side of symmetry axis.

Since there is neither self symmetric edge nor self symmetric vertex, we can separate the graph into two sub graphs by removing the edges that crossing symmetry axis (Figure 2 (b)). While removing one pair of crossing edges, two vertices' degrees change their parity, i.e., change from odd (even) degree to even (odd) degree. Therefore, the parity of odd degree vertex number on each side of symmetric axle keeps unchanged. Then, in each sub graph, the number of odd degree vertices is still odd, contradicting the corollary of Eulerian theorem that the number of odd degree vertices is even. □

Method-4: After connected them pair by pair as previous, we connect the last odd degree vertex with its symmetric vertex by a self symmetric dummy edge (Figure 2 (c)). Let the self symmetry edge be $(v_{n/2+1}, v_{n/2+2})$ in the symmetric Eulerian trail. Then, the two vertices $v_{n/2+1}$ and $v_{n/2+2}$ are used as start vertex on each side of the axle to construct symmetric Eulerian trail. There are odd edges in the trail.

Following procedure give an explicit description of the previous processes.

Procedure BuildSymmetricEulerianTrail
Begin
  Find all odd degree vertices one side of axle;
  If the number of odd degree vertices is even
    // symmetric Eulerian Trail-1 has even edges
    Use **method-1** to construct symmetric Eulerian Trail-1;
    // symmetric Eulerian Trail-2 has odd edges
    Use **method-2** to construct symmetric Eulerian Trail-2;
  Else if the number of odd degree vertices is odd
    // symmetric Eulerian Trail-1 has even edges
    Use **method-3** to construct symmetric Eulerian Trail-1;
    // symmetric Eulerian Trail-2 has odd edges
    Use **method-4** to construct symmetric Eulerian Trail-2;
End;

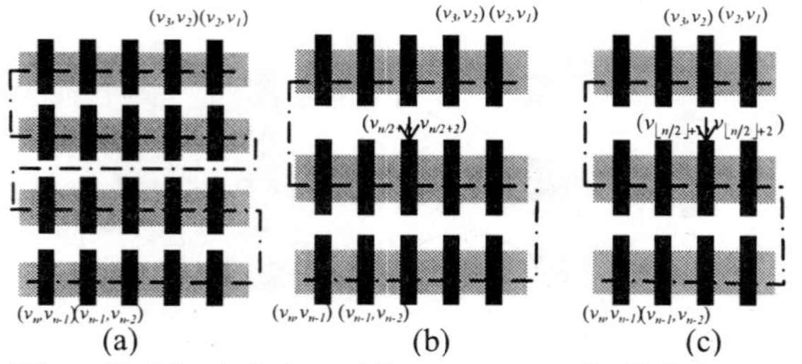

Figure 3. The dash-dotted line represents the Eulerian trail. (a) Even row stack. (b) Odd row odd column stack. (c) Odd row and even column stack.

2.3 2D Stack Generation

We construct several common centroid stacks with different ratio for each partition. These stacks can be selected by simulated annealing placement algorithm. The upper and lower bounds of stack ratio can be specified by user.

In the following procedure, one row stack is constructed first and then two row stack and so on, until the ratio limits is reached. The trail (trail-1 or trail-2) that can be divided exactly by row number is used to construct stack.

Procedure ConstructStack
Begin
  row=1;
  length-1=edge number of trail-1;
  length-2=edge number of trail-2;
  while stack ratio do not reach the limits
    if row is even and length-1 can be divided exactly by row
      Construct stack as Figure 3(a) using Trail-1;
    else if row is odd
      if length-2 can be divided exactly by row
        Construct stack as Figure 3(b) using Trail-2;
      if length-1 can be divided exactly by row
        Construct stack as Figure 3(c) using Trail-1;
    row++;
End.

From the procedure and the definition of symmetric Eulerian trail we can see the centroid of the symmetric devices is at the center of the stack. The only possible self symmetric edge $(v_{n/2+1}, v_{n/2+2})$ appears at the center of the middle row (Figure 3(b)).

There are still some other tricks we have used in our algorithms. Edges including dummy edges incident to the same two vertices are mutually interchangeable. We do not decide which one is chosen during constructing the trail. Decision is postponed to the stack generation. During stack generation, which edge is chosen, dummy edge or real device edge, is decided dynamically based on its position in the stack. If current position is the end of a stack row, a dummy one is preferred [9].

3. BLOCK MERGING ALGORITHMS

Block merging is to further explore the possibility of geometry share between different stacks during placement. It reduces not only the area but also the parasitics.

The block merging has following two main steps which occur during the packing.

1. For each module all the merging candidate modules are found.

2. The maximum merging distance is calculated, according to which the module position is adjusted.

In step 1, because each module a can only be merged to left and bottom, only the module left or below module a is merging candidate for module a. For module a and arbitrary module b,

if $X(a)<X(b)+Width(b)$ and $X(b)<X(a)+Width(a)$, b is a's bottom merging candidate.

if $Y(a)<Y(b)+Height(b)$ and $Y(b)<Y(a)+Height(a)$, b is a's left merging candidate.

In step two, the maximum merging distance is the length that module can horizontally (vertically) merge to left (bottom). Each module can be represented by two rectangles (Figure 4). One is merging area *m*, and the other is solid area *s* that can not be merged. w_1 and w_4 is widths of solid areas. w_2 and w_3 is widths of merging areas. d_1 is distance between merging area. d_2 is distance between solid area. d_3 and d_4 are distances between merging area and solid area. During the module-2 horizontally merged to left, if the two areas will never touch, the distance between them is set to be infinite, for example, d_2 in Figure 4(b) is infinite. Figure 4 is under the situation of horizontally merging to left. The vertically merging is similar.

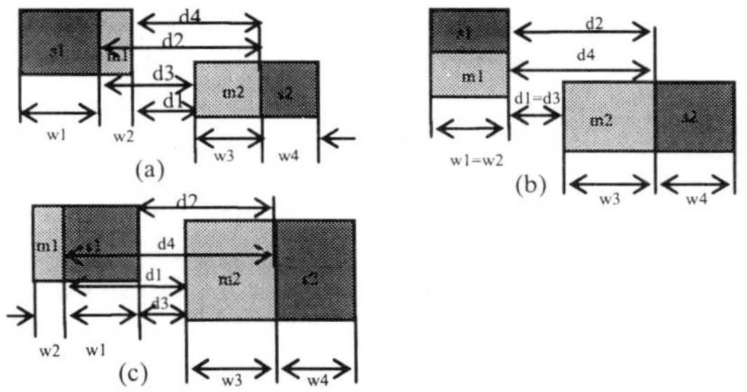

Figure 4. The light rectangle in each module represents the merging area. The dark rectangles represent solid areas.

Theorem 2: The maximum merging distance is

$D_m = min\ (min(w_2,\ w_3)+d_1,\ d_2,\ d_3,\ d_4).$

The proof is omitted for the sake of page limits. Our algorithms are implemented based on sequence pair. However, they are essentially independent to any topological representations.

4. RESULTS AND CONCLUSION

Figure 5 shows a multiplier circuit which has bee used as a benchmark in several previous researches [1][5]. The number of stacks generated is 4. All these stacks are 2-axial symmetric and common-centroid.

The test cases for block merging algorithm are based on MCNC benchmark, and the merging areas are designated randomly. Figure 6 and Table I give the results of test case based on ami33 and ami49. The merged areas have been circled in Figure 6.

In this paper, we proposed the algorithms to generate 2-axial symmetry stack in linear time. Several stacks with different ratio are generated for one group of transistor. These stacks are chosen during the placement. Then, algorithm for block merging is given, which can be used to merge stacks with different channel width transistors during placement. These algorithms have been proved to efficient by the experiment results.

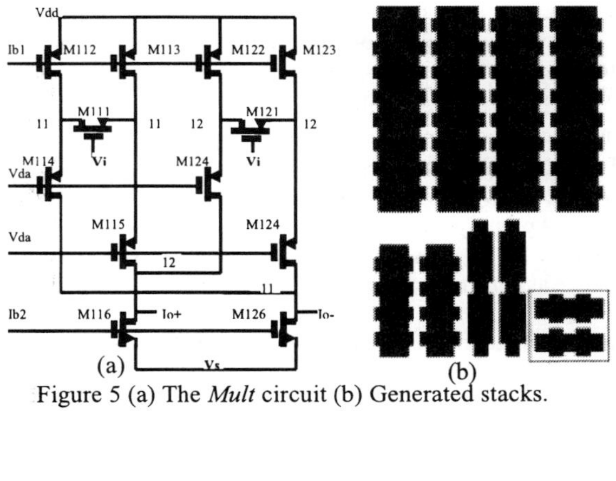

Figure 5 (a) The *Mult* circuit (b) Generated stacks.

TABLE I Experimental Results of Test Cases

Test case	Without merging			With merging		
	Area	Usage	Time(sec)	Area	Usage	Time(sec)
Ami33	1,245,972	92.82%	44.23	1,157,870	99.88%	80.17
Ami49	37,239,020	95.18%	82.12	36,862,560	96.16%	206.56

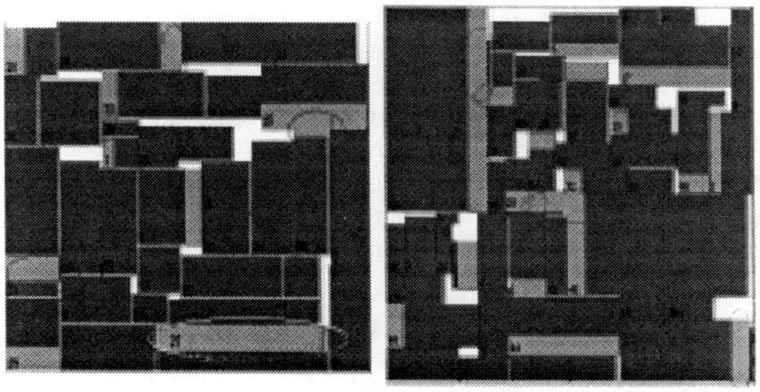
Figure 6. Experimental results of block merging.

5. REFERENCES

[1] J. M. Cohn, D. J. Garrod, R. A. Rutenbar and L. R. Carley, "KOAN/ANAGRAM II: New Tools for Device-Level Analog Placement and Routing", IEEE Journal of Solid State Circuits, vol. 26, n. 3, pp. 330-342, March 1991.

[2] U. Gatti, F. Maloberti, and V. Liberali, "Full stacked layout of analogue cells," in *Proc. IEEE Int. Symp. On Circuits and Systems*, May 1989, pp. 1123-1126

[3] E. Malavasi and D. Pandini, "Optimum CMOS stack generation with analog constraints," *IEEE Trans. Computer-Aided Design*, vol. 14, pp. 107–122, Jan. 1995.

[4] Arsintescu, G.B. S.A. Spanoche, "Global Generation of MOS Transistor Stacks", *IEEE EuroDAC*, sep. 1996

[5] A. Basaran and R. A. Rutenbar, "An *O(n)* algorithm for transistor stacking with performance constraints", in *Proc. Of IEEE/ACM DAC*, pp. 221-226, June 1996.

[6] J. Bastos, M. Steyaert, B. Graindourze, W. Sansen, "Matching of MOS Transistors with Different Layout Styles," *Proc. IEEE Int. Conference on Microelectronic Test Structures*, pp. 17-18, March 1996.

[7] J. Bastos, M. Steyaert, B. Graindourze, W. Sansen, "Influence of Die Bonding on MOS Transistor Matching," *Proc. IEEE Int. Conference on Microelectronic Test Structures*, pp. 17-31, March 1996.

[8] H. Murata, K. Fujiyoshi, S. Nakatake, and Y.Kajitani, "VLSI module placement based on rectangle-packing by the sequence pair," *IEEE Trans. Computer-Aided Design*, vol. 15, pp. 1518-1524, Dec. 1996

[9] M. Ismail and T. S. Fiez, Eds., Analog VLSI: Signal and Information Processing. New York: McGraw-Hill, 1994

COMBINING CLUSTERING AND PARTITIONING IN QUADRATIC PLACEMENT

Yongqiang Lu, Xianlong Hong, Wenting Hou, Weimin Wu, Yici Cai
Dept. of Computer Science and Technology
Tsinghua Univ. Beijing China

ABSTRACT

Because of the computation complexity of large circuits, the quadratic placement (Q-Place) cannot solve the placement problem fast enough without any preprocessing. In this paper, a method of combining the MFFC clustering and hMETIS partitioning based quadratic placement algorithm is proposed. Experimental results show it can gain good results but consume long running time. In order to cut down the running time, an improved MFFC clustering method (IMFFC) based Q-place algorithm is proposed in this paper. Comparing with the combining clustering and partitioning based method, it is much fast but with a little increase in total wire length.*

1. INTRODUCTION

Quadratic Placement (Q-Place) is a classic and efficient placement method based on quadratic programming optimization, adopting the quadratic wire length as the objective function. It first converts the placement problem into a quadratic programming model, and then uses proper means to solve the model. Q-place has been broadly used in nowadays placement tools for standard cell layout design. However, as the scale of ICs increases rapidly, the Q-Place can not solve such large-scale layout problems fast enough without any preprocessing. Hierarchical methodology is a good approach for reducing the size of problem and speeding up the running. Some clustering or partitioning techniques can be used as a preprocessor in hierarchical method. There are a lot of clustering methods, such as density-based clustering algorithm [4], random-walk based clustering [7], greedy clustering [2], and MFFC clustering [1]; and partitioning methods, such as the generalization of the FM-algorithm with look-ahead scheme [6], the spectrum-based partitioning method [5], and some two-way or multilevel partitioning methods. For the purpose of reducing the size of problem without lowering the solution quality, two issues must be taken into account in consideration of clustering or partitioning methods. First, the inner nets in clusters or partitions must be as many as possible. Second, the interconnected nets among clusters or partitions must be as few as possible. As a successful application of clustering scheme in Q-Place, the greedy cluster-based Q-Place algorithm has been proposed in [2]. It gains excellent results but still has two shortcomings. Firstly, the inner nets have not been grouped as many as possible because it ignores the signal flow and logic dependency of the circuits. Secondly, the method has not considered reducing interconnected nets among clusters. In this paper we select the model of the combining MFFC clustering and hMETIS partitioning method not only to group cells in a cluster with connection as close as possible, but also to reduce the interconnected nets among clusters. The MFFC clustering can create little clusters with high quality, and the hMETIS can partition the MFFC-clustered circuits with less interconnected nets (cutsize) among the partitions. The hMETIS is an available partitioning package; it can also take the standard deviation into account. However, the hMETIS leads to a long running time except for the low cutsize. For the efficiency of the model, an improved MFFC clustering algorithm is developed to achieve a fast run with nearly the same total wire length. Experimental results show that the algorithm can cut down the running time of placement remarkably especially with large-scale circuits.

We call the placement used in our experiment the hierarchical placement process. Hierarchical placement process can reduce the placement time and improve the placement quality. Particularly in very-large-scale standard cell placement, it acts as the main placement means instead of the flatten Q-Place or other flatten ways. In this paper, it includes these sub processes: Q-Place, introduced in Section 2; combining clustering and partitioning, introduced in Section 3. At the end of Section 3, the flowchart of our hierarchical placement will be concluded. Section 4 gives the model of the improved MFFC clustering method. The experimental results are presented in Section 5. Finally, conclusions and future research are included in Section 6.

2. Q-Place

The quadratic placement (Q-Place) is a determinable method based on quadratic programming optimization. Its objective function is the quadratic wire length. The Q-Place begins with modeling the problem into a quadratic problem. Then it solves the problem by certain means and obtains the global placement solution.

In the Q-Place used in this paper, the problem is modeled into a linear constraint quadratic problem (***LQP***) [4], and it is solved to obtain a placement solution in each partition level. The partition level is a state of each level of the partition tree, partitioning a parent region into two son regions in each partition step. The number of regions in i^{th} partition level is 2^i.

First of all, a circuit is presented as a hypergraph $G(V, E)$, $V = V_1 \cup P$, $V_1 = \{v_1, v_2, \ldots, v_n\}$, representing all movable cells in the circuit; $P = \{p_1, p_2, \ldots, p_q\}$, representing all fixed cells in the circuit, such as primary outputs and primary inputs. The hyperedge set $E = \{e_1, e_2, \ldots, e_m\}$, representing the connection of the circuit. Every edge is assigned a weight $w(e)$. The coordinate of cell$_i$ is (x_i, y_i).

The objective function of the global placement is the weighted sum of the squared lengths of the nets [4]; form in matrix is:

* This work is supported by the National Natural Science Foundation of China 60121120706 and National Natural Science Foundation of USA CCR-0096383, the National Foundation Research (973) Program of China G1998030403, the National Natural Science Foundation of China 60076016 and 863 Hi-Tech Research & Development Program of China 2002AA1Z1460.

0-7803-7761-3/03/$17.00 ©2003 IEEE

$$\Phi(x, y) = x^T Cx + d_x^T x + y^T Cy + d_y^T y \quad (1)$$

The x and y are the vectors of the cells' coordinates, and they are just to be solved. The $n \times n$ connection matrix $C=(c_{ij})$ for G represents the weight information of edges. For each edge $e_k = (v_i, v_j)$, $c_{ij} = c_{ji} = -w(e_k)$; the diagonal entry c_{ii} is equal to or larger than the sum: $\sum_{j=1, j\neq i}^{n} c_{ij}$, all other entries of C are zeros. The vector d_x and d_y is generated by the connection between the cells and pads. In order to distribute the cells evenly on the chip, distribution constraints are added:

$$\sum_{i\in Sr} x_i / |S_r| = u_r, \quad \sum_{i\in Sr} y_i / |S_r| = v_r \quad (2)$$

S_r is the cell set in the region r; (u_r, v_r) is the center of the region r; and the $|S_r|$ is the number of the cells in S_r.

So, the ***LQP*** obtained is:

$$Min\ \{\Phi(x, y) = 1/2x^T Cx + d_x^T x + 1/2y^T Cy + d_y^T y \mid A^{(m)}x = u^{(m)};\ A^{(m)}y = v^{(m)}\} \quad (3)$$

In m^{th} placement-partition level, matrix A is a $2^m \times n$ matrix. In matrix $A = (a_{ij})_{2^m \times n}$, the a_{ij} equal to $1/|S_j|$ if $cell_i$ belongs to $region_j$ or a_{ij} is zero. And the number of total none-zero entries is equal to n.

The Lagrange Multipliers Method is employed to solve the ***LQP*** [4]. The corresponding Lagrange Function is:

$$min\{\Phi(x) = 1/2x^T Cx + b_x^T x - \lambda\ (A^{(m)}x - u^{(m)})\} \quad (4)$$

The λ is the Lagrange multiplier. Making the gradient equal to zero the linear system will be obtained:

$$\begin{bmatrix} C & -A^{(m)} \\ -A^{(m)T} & 0 \end{bmatrix} \begin{pmatrix} x \\ \lambda \end{pmatrix} = -\begin{pmatrix} b_x \\ \mu^{(m)} \end{pmatrix} \quad (5)$$

To solve this equation the cells optimal location in x-direction can be obtained, and the y-direction optimal solution can be obtained in the same way. The ***LQP*** is solved repeatedly until reaching the desired placement-partition level. After accomplishing solving the ***LQP*** at the desired placement-partition level, the global optimal placement of the cells is acquired.

To solve the problem, the Equation (5) needs to be solved, and the time complexity of solving the equation would amount to at least $O(nlogn)$. So, to decrease the scale of the quadratic programming problem is very important to solve the large-scale standard cell placement.

3. COMBINING CLUSTERING AND PARTITIONING

In order to decrease the problem scale in very large-scale standard cell placement, the method of combining the MFFC clustering and hMETIS partitioning is introduced to simplify the circuits in this paper. The Q-Place runs on the clustered circuits to give the global placement of cells.

3.1 The MFFC Clustering

The MFFC clustering technique is composed of the MFFC decomposition technique and the MFFC splitting technique.

3.1.1 MFFC Decomposition Technique

The MFFC decomposition technique was first proposed for combinational circuits in [8], and was frequently used in later research. For clarity, we define:

- ***output(v)*** as the set of nodes which are the fanouts of node v, that is, the node w in ***output(v)*** must be the terminal of one of the output nets of the node v.
- C_v as the subgraph of the logic gates (excluding primary inputs (PIs)) consisting of v and its predecessors such that any path connecting a node in C_v and v lies entirely in C_v.
- FFC_v as the fanout-free cone (FFC) at node v, it is a cone of v such that for any node w (not v) in FFC_v, ***output(w)*** is included by FFC_v.
- $MFFC_v$ as the FFC of v such that for any non-PI node w, if ***ouput(w)*** is included by $MFFC_v$, then w is in $MFFC_v$. *The node* v is called as the root of the $MFFC_v$.

In general, the gates in a single $MFFC_v$ can be considered closely related because the MFFC clustering technique guarantees it.

The MFFC decomposition technique is to cluster the cells into MFFC sets from the corresponding POs, and POs are the cells with which decompositions begin, initially the primary outputs. The decomposition technique executes as Fig2 illustrates.

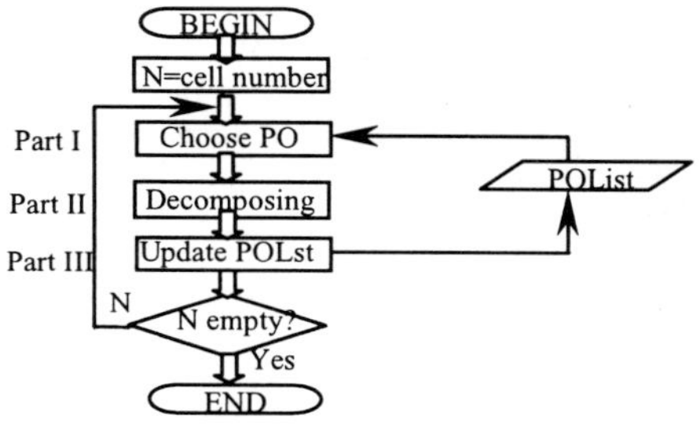

Fig1 MFFC decomposition flow

At the beginning of the flow, let N be the set of the cells of the circuit; let POList be the list of primary outputs in arbitrary sequence initially. In the part I, the PO v selected is to be the cell that decomposition begins with, and is put into the corresponding $MFFC_v$ first. In the part II, trace cells from each input net of v, put the cell w traced into the $MFFC_v$ as soon as ***output(w)*** is included in the $MFFC_v$, and let $N = N - MFFC_v$. In the part III, the new trace-from cells are updated into the POList; they are those that do not need the conditions of clustering into current MFFC.

3.1.2 MFFC Splitting Technique

After MFFC decomposition, we use an MFFC splitting process to split the large MFFC sets into smaller ones so that the area of the clusters could not be too large.

We employ a splitting tree (ST) to complete the MFFC splitting process. The nodes in the ST stand for MFFCs; one node is one MFFC. If node r is $MFFC_v$, sons of r are those decomposed from one of the cells in the ***input(v)*** (the fanins of v). Let the root of the ST be the MFFC to be divided, and keep creating the sons of the nodes in the ST recursively until the area of the MFFC to be split is proper. For example, the splitting tree as the Fig2 shows is to be split as: {u1}, {u2}, {u3, u4}, {u7}, {u5, u6, u8}.

The decomposition technique used in constructing ST is same to the decomposition technique described above. Every time splitting a big MFFC, a one-cell MFFC will appear.

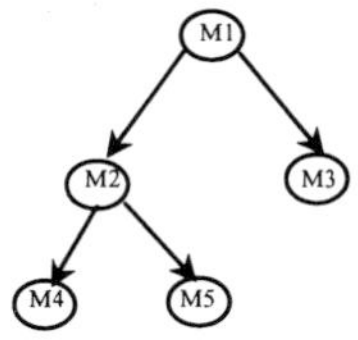

M1: {u1,u2,u3,u4,u5,u6,u7,u8};
M2: {u2, u3, u4, u7};
M3: {u5, u6, u8};
M4: {u3, u4};
M5: {u7}

Fig2 A splitting tree and its splitting result

3.2 Partitioning

The clustered circuits have many little clusters, and the number of the clusters is also too large. So we choose the hMETIS package to serve as the partitioning method, partitioning the MFFC clusters into proper number of partitions with even areas. The details about hMETIS are in the manual which can be downloaded from WWW at ***http://www.cs.umn.edu/~meti***s.

The proper partition number is determined by experiment according to the corresponding circuits, constrained by proper problem scale for Q-Place. By default the parameter is set to one tenth of the total cell number.

The hypergraph of the circuit which will be partitioned by hMETIS is constructed by making the MFFC clusters be the nodes with weights of area and making the nets be the hyperedges with weights of cluster net weights. According to the weight of the nodes, we can control the evenness of the partition obtained.

3.3 Combining the two

In large-scale standard cell placement, a preprocessor should be added before the Q-Place, in which the large circuits are clustered or partitioned into small circuits. So the Q-Place works on the clustered or partitioned circuits. Two ways are often used to preprocess circuits: clustering and partitioning. However, clustering can get high-quality clusters but is not easy to reduce the interconnect nets among clusters; and partitioning can reduce interconnect nets but is not good at reducing the inner nets of partitions. So we think to combine them two together. The preprocessor used in this paper, the combining of MFFC clustering and hMETIS can reduce the inner connection in clusters and interconnected nets among partitions remarkably. From the experimental results in Section 5, the good total wire length proves it. So, the entire hierarchical placement process can be described as Fig3 illustrates:

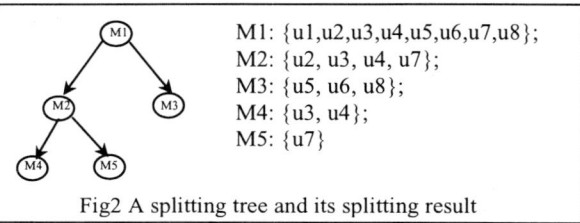

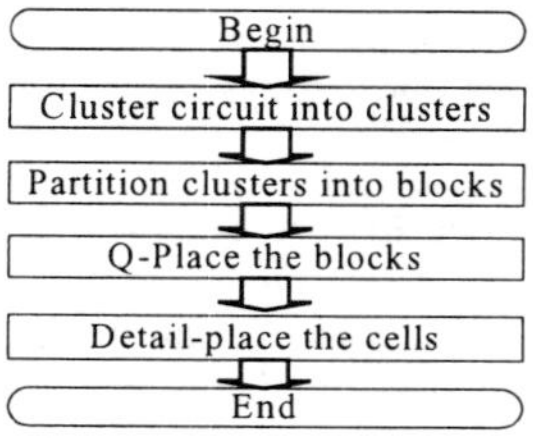

Fig3 Hierarchical placement process

The part of Clustering executes the MFFC clustering, collecting the cells with closest connections. The partitioning part regroups the MFFC clusters into partitions. Then the Q-Place processes the clusters to obtain the global optimal solution.

4. IMPROVED MFFC CLUSTERING

Due to the not too short running time of combining method, we introduce the improved MFFC clustering to preprocess circuits. The MFFC clustering method can create high quality clusters rapidly, but the number of clusters may be too large to do the global placement fast. If the MFFC clustering method can create a proper number of clusters, it can make the placement of large scale run faster. We find that the process of the MFFC decomposition may be too strict because only when the cell ***v*** needs the condition that ***output(v)*** is included in the current MFFC set, can it be include.. In this algorithm, we include ***v*** into the MFFC when the ***output(v)*** is mainly included in it, and we find out that it can help create proper clusters with even area. We call it merging algorithm.

First, let us see these properties of MFFC clustering:

- The output nets of a cluster must begin with the root cell of the cluster. This is obvious because the decomposition technique.
- The input nets of a cluster must begin with the cells that are generated as new POs.

The proof in detail is omitted due to the length restriction.

Knowing that, the connection information of the MFFCs conveniently can be generated when cells are clustered. According to the connection information recorded in the sets, a merging process is appended after the primary MFFC clustering. The merging algorithm used is described as following.

```
Algorithm Merging
while (MergeControl > 0) {
    for each MFFC set
        MergingProperSet();
    end for
    MergeControl = MergeControl – controlStep;
end while
End Algorithm
Process MergingProperSet
for each set connecting with the current
        select the set with maximum connection degree to merge;
        update the connection information;
end for
End Process
```

The MergeControl and controlStep used in the merging algorithm are experiential parameters. The MergeControl is chosen as the average number of the cells in clusters and the controlStep is chosen as 1 in our development.

The connection degree of the set ***i*** and ***j*** is counted as

$$D_{ij} = \sigma C_{in} / C_{out} \qquad (6)$$

C_{in} is the interconnection of the two sets; C_{out} is the outer connection of the two sets; σ is the area control factor, experience determined, and taken as ***1/S***, ***S*** is the total area of the two sets. Merging the proper set is to select the set with maximum connection degree calculated with ***(6)***, and updating connection is to delete the merged connections.

5. EXPERIMENTAL RESULTS

We have implemented the combining and the improved MFFC clustering method on Sun workstation V880 in C language. We obtained the package of hMETIS V1.5.3 on WWW at ***http://www.cs.umn.edu/~metis***. We used 6 LEF/DEF benchmarks obtained from real circuits in industrial fields. All experimental results are obtained after the global placement completes and before the detail placement begins, and the number of clusters is the same to what the greedy cluster-based Q-Place algorithm used. Table 1 shows the characteristics of these benchmarks.

Table 1 The characteristics of the circuits

Case	#cell	#net
U05614	32112	36452
U08421	48168	54741
CPU	48876	49204
U11228	64224	72966
U14035	80280	91127
U28070	160560	181932

First of all, let us see the experiment with only partitioning. Because of the too much cpu time consumed, only the data of case CPU is presented: wire length: 5445162, cpu time: 1265sec. Almost the same short wire length as Table 2 but much more time consumed. Also, the flatten Q-Place would consume even more time on those cases. That is why we use the model of hierarchical placement process with combining of clustering and partitioning.

Then, let us see the comparison between our combining Q-Place and the greedy cluster-based Q-Place [2] as Table 2 illustrates. The comparisons can be presented in three aspects: total wire length, cpu time (running time) and standard deviation.

- ***Total wire length.*** The improvement in total wire length can be seen in the Table 2. We can see the method of combining MFFC clustering and hMETIS (the column of *comb* in the table) can decrease the total wire length 28% in average. On the circuit of CPU, it can even decrease the wire length 39%.
- ***Running time.*** From the Table 2, we see that the running time our method used may be a little long. The running time is mostly taken in running partitioning (about 1/4).
- ***Standard deviation.*** Our cluster area is very even and most clusters have a nearly average area; while that of greedy cluster-based method ranges much. For example, with circuits of CPU, our results range from 10 to 10^2 in order, while the greedy method ranges from 10 to 10^3.

Table 2 The comparison between the combining and [2]

Case	total wire length			cpu time (sec)	
	comb	greedy	- %	comb	Gree
U0561	7612377	1034440	26	459.7	317.2
U0842	1290552	1642805	21	700	491
CPU	5377955	8811934	39	909.5	492.2
U1122	1698727	2284372	26	926.8	650.4
U1403	2137201	2958767	28	1027.	738.5
U2807	4512022	6231557	28	1333.	999.4

At last Table 3 gives the comparison between the IMFFC and the combining:

- ***Total wire length.*** It is because hMETIS is random-number driven in initial partition phase why there appear some decrease items in the column of the wire length increase (+%), such as U08421 and U14035. This may appen when the poor results are obtained in hMETIS run. The IMFFC can only increase 4% wire length in average.
- ***Running time.*** The IMFFC can cut down the running time remarkably. In large circuits, it can even cut down about 50% running time, such as U14035 and U28070.
- ***Area evenness.*** The area is also even, but not as good as the method of combining. For some circuits, its result is not even, such as U08421.

Table 3 The comparison between IMFFC and combining

Case	total wire length			cpu time (sec)	
	IMFFC	comb	+ %	IMFFC	comb
U05614	8267863	7612377	9%	304.4	459.7
U08421	12508793	12905521	-3%	514.5	700
CPU	5626875	5377955	4%	656.4	909.5
U11228	18621614	16987275	9%	371.3	926.8
U14035	20975662	21372019	-2%	527.1	1027.
U28070	48429853	45120227	7%	567.1	1333.

6. CONCLUSIONS AND FUTURE WORK

The method of combining clustering and partitioning can help improve the quality of placement, and it is worth studying further more. In addition, congestion and other kinds of placement such as big macros will be taken into account in future work.

7. REFERENCES

[1] Jason Cong and Dongmin Xu. "Exploiting Signal Flow and Logic Dependency in Standard Cell Placement". Proceedings of the conference on Asia Pacific design automation conference, Mukuhari, Japan, 1995, pages 399-404.

[2] Hong Yu, Xianlong Hong, Changge Qiao, Yici Cai, "CASH: A Novel Quadratic Placement Algorithm for Very Large Standard Cell Layout Design Based on Clustering", Proceedings of the 5th International Conference on Solid-State and Integrated Circuit Technology, 1998, pages 496-501.

[3] Cong. J. and M. Smith, "A Bottom-up Clustering Algorithm with Applications to Circuit Partitioning in VLSI Designs". Proc. ACM/IEEE Trans. on VLSI Systems, June 1994, Vol. 2, pages 137-148.

[4] Wenting Hou, Xianlong Hong, Weimin Wu and Yici Cai , "FaSa: A Fast and Stable Quadratic Placement Algorithm". 2002 International Conference on Communications Circuits and Systems and West Sino Expositions Proceedings, Chengdu, China, Vol 2, pages 1391-1395.

[5] Chan, P., M. Schlag, and J. Zien, "Spectral K-Way Ratio-Cut Partitioning and Clustering", *Proc. 30th ACM/IEEE Design Automation Conf.*, June 1993.

[6] Sanchis, L., "Multiple-Way Network Partitioning", *IEEE Trans. on Computers*, 1989, Vol. 38, pages 62-81.

[7] Hagen, L. And A. B. Kahng, "A New Approach to Effective Circuit Clustering", Int'l Conf. On Computer-Aided Design, 1992, pages 422-427.

[8] Cong, J. and Y. Ding, "On Area/Depth Trade-off in LUT-Based FPGA Technology Mapping", Proc. 30th ACM/IEEE Design Automation Conf., June 1993, pages 213-218

Floorplanning with Performance-based Clustering[1]

Malgorzata Chrzanowska-Jeske, Benyi Wang and Garrison Greenwood

Department of Electrical and Computer Engineering
Portland State University
Portland, Oregon, 97229, USA
Email: jeske@ece.pdx.edu

ABSTRACT

There are many different reasons why it is desirable to keep a set of modules (called a cluster) together during floorplanning. A cluster might be strongly connected, or functionality and testability of a design could be improved. We address the problem of performance constraints in a non-slicing floorplan represented by a sequence pair, and we use an evolutionary algorithm to generate a hard module placement preserving cluster constraints. The main contribution is to define cluster constraints as sequence pair constraints, and therefore reduce significantly the feasible solution search space. We use Lagrangian Relaxation formulation to generate an optimal soft module floorplan. Experimental results on modified MCNC benchmarks show the efficiency of our approach.

1. INTRODUCTION

Continuous advancement in VLSI technology and design techniques has increased circuit size, but has unfortunately increased design complexity as well. Floorplanning, the critical first stage of VLSI physical design, is becoming more important due to hierarchical design styles and the increased use of intellectual property modules. Furthermore, increase in the number of available metalization layers, to five or six in deep-submicron technologies, has caused channel routing to be largely replaced by over-the-cell (area) routing, thereby transforming floorplanning/block placement problems into block packing problems. Rectangle packing is the simplest formulation of the floorplanning problem. The modules are classified as hard or soft; hard modules have fixed dimensions whereas soft modules have a fixed area and a range of acceptable aspect ratios.

This paper describes a performance-constrained floorplanning strategy that allows designers to define a priori clusters as sets of modules that need to be placed together. Clusters can be defined for a number of reasons, including: module connectivity, functionality, testability or power requirements. We developed our approach for a non-slicing floorplan as it offers possibly better solutions compared to slicing floorplans. A number of non-slicing floorplan representations have been proposed recently; we have chosen the sequence pair representation [5] because it shows many advantages over the other listed approaches and a number of fast algorithms for floorplan generation from sequence pair [4], [6], [12] have been developed.

Our approach uses relations between a floorplan s topology and relative positioning of modules in a sequence pair. Other authors have observed relations that do exist between sequence pair ordering and module placement for L-shaped and T-shaped modules [1][3]. In [12], constraints for module linear alignments and for performance-based module grouping have been developed. In [13] conditions for feasibility of a sequence pair for rectangle packing with pre-placed rectangles were proposed. The constraints derived in this paper, however, are more general since we consider arbitrary contiguous rectilinear clusters.

The paper is organized as follows. Section 2 defines the performance-constrained floorplanning problem and provides background information on the sequence-pair representation. Section 3 gives theoretical background for the cluster-constrained approach. Section 4 provides details on ELF-SP floorplanning algorithm and cluster-constrained modifications. Experimental results are presented in Section 5 and Section 6 presents conclusions.

2. BACKGROUND

In our formulation the layout of a cluster is not fixed. We can preserve the structural adjacency/topology of cluster modules or their topology can be optimized when floorplanning is performed. The topology of a cluster is defined by the horizontal and vertical graphs generated from a sequence pair representing an optimal placement of the cluster modules only. The designer determines how strictly adjacency of modules must be preserved.

Cluster-Constrainted Floorplanning Problem, CCFP

Given a set S_n of n rectangular modules with known areas, A_i, aspect ratio ranges $[r_i, R_i]$, a set S_k of intermodule nets, and p disjoint subsets C_l (clusters) of m_j modules in S_n, for j=1,2,...p. Find the minimum area packing that includes all n modules subject to the following constraints:

1. *The packing is feasible i.e., the aspect ratio of the whole packing is within a given range [r, R].*
2. *Each module satisfies its area and aspect ratio constraints*
3. *Cluster constraints for all clusters are satisfied.*
4. *All other objective constraints (e.g., timing or routability) are met*

Sequence Pair

The main idea behind a *sequence pairs* representation is that two permutations (Γ^{+}, Γ^{-}) of a set of modules are sufficient to describe a packing [5]. Given sizes of all modules, and a

[1] This work was supported in part by the NSF grants MIP-9629419 and CCR-9988402

0-7803-7761-3/03/$17.00 ©2003 IEEE

placement algorithm, each permutation imposes a specific optimized placement for each module. Given a sequence pair, one can construct an oblique lattice structure with the lines labeled in the same order as they appear in Γ^+ and Γ^-. Each module is placed at the lattice point that is the intersection of the two lines with the same label as shown in Fig. 1.

Two weighted, acyclic digraphs are constructed in which the modules at the lattice points form the vertex set. A weight is associated with each vertex in the G_h (G_v) graph that indicates the width (height) of the respective module. Source and sink vertices are added to each graph. Fig..2 shows this construction. Generated horizontal and vertical graphs [5] represent topological relations defined by a given sequence pair. These graphs do not have to be planar.

We use the basic ideas of the algorithm from [11] to obtain a floorplan from the sequence pair. This algorithm evaluates the longest common sequence in (Γ^+, Γ^-) within $O(n^2)$ time without having to generate horizontal or vertical graphs. However, we need to generate these two graphs to perform Lagrangian Relaxation to introduce soft modules. This is only done once at the end of the hard module placement. We have developed a fast algorithm for generating these graphs so the additional computational effort is quite small.

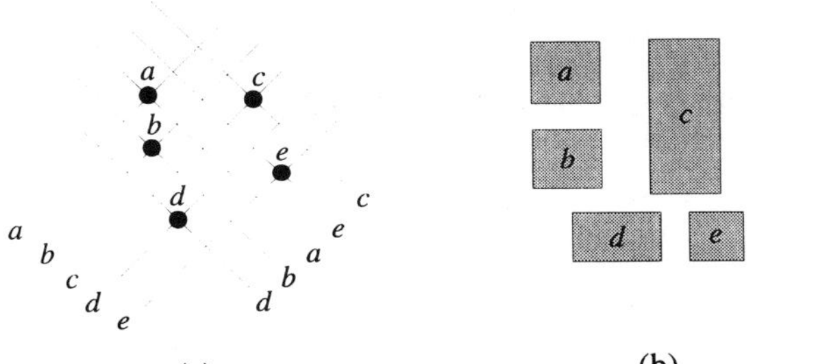

Figure 1: (a) oblique grid for $\Gamma^+ = \{a\ b\ c\ d\ e\}$ and $\Gamma^- = \{d\ b\ a\ e\ c\}$, and (b) resultant placement

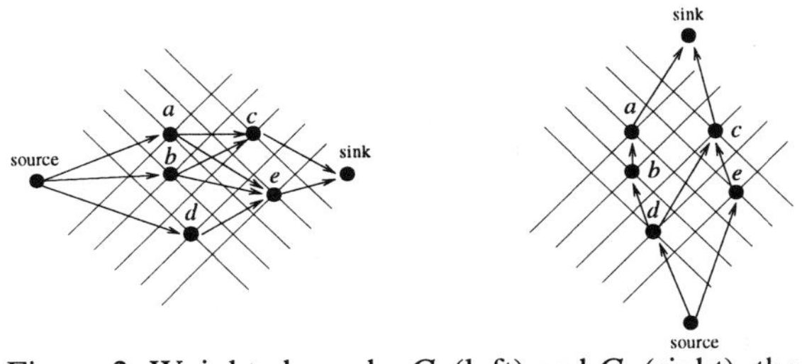

Figure 2: Weighted graphs G_h (left) and G_v (right) the sequence pair given in Figure 1.

3. CLUSTER-CONSTRAINED FLOORPLANNING

In this work, we investigate relations between a sequence pair, its floorplan, and the set of modules forming a cluster. We use these relations to limit the size of the solution space, which allows us to develop an efficient search algorithm. For brevity we do not include proofs.

3.1 Preliminaries

Let Ψ be a set of n rectangular modules and we define a cluster C as a subset of Ψ. The objective is to pack Ψ such that all modules in C are adjacent to each other —i.e., the cluster is contiguous - and if required the specified topology of the cluster is maintained.

Definition 1 (cluster sequence)
Let C be a cluster. The cluster sequence is the minimum length subsequence in a sequence-pair that contains all modules $m \in C$ in a given order. For any cluster C, there is a positive cluster sequence S_C^+ contained in Γ^+ and another negative cluster sequence S_C^- contained in Γ^-.

Definition 2 (cluster-gap module)
Any module located within a cluster sequence that does not belong to a cluster $r \notin C$ is called a cluster gap module.

The order of cluster modules in a positive cluster sequence, S_C^+, and in a negative cluster sequence, S_C^-, defines the topology of a cluster. The layout of an isolated cluster is defined by its topology and also by the sizes of its modules.

If in a given sequence pair both cluster sequences, S_C^+ and S_C^-, contain only cluster modules, then cluster topology will be preserved in all floorplans generated from that sequence pair and that cluster can be substituted with a macro-cell. Once we allow cluster sequences to contain gap modules, only some floorplans will preserve cluster topology depending on module sizes.

In our non-deterministic approach we are interested in rejecting solutions that are not better than those that we choose to further consider. We would like to recognize those solutions as early as possible to improve the performance of our algorithm. Our goal is to define rejection conditions based on the sequence-pair representation. Therefore, when describing a cluster we consider two sets of constraints: topological and geometrical. Topological constraints can be defined based on sequence-pair representation, but geometrical constraints require knowledge of module sizes, and therefore cannot be checked on sequence pair representation.

In this paper we focus on a *flexible cluster* where the cluster layout is not fixed, and is constructed to specifically satisfy performance constraints such as delay or power dissipation. The layout of a flexible cluster is not known a priori even when the cluster sequence and all cluster module dimensions are known. It can change depending on imposed geometrical constraints and the sequence pair of a specific floorplan. We will assume a convex cluster and we establish conditions to limit solution space by eliminating sequence pairs that are infeasible in those conditions.

Theorem 1.
A necessary condition to maintain the topology of cluster C after a perturbation of the sequence pair $\{\Gamma^+, \Gamma^-\}$ is that the order of the cluster modules in the positive and the negative sequences remains unchanged.

We define a *feasible floorplan* as one that meets all usual floorplanning constraints and in addition meets the cluster constraints of cluster C.

Definition 3. A cluster gap module present in S_C^+ or in S_C^- but not both is called a *single cluster-gap module.*

Definition 4. A cluster gap module that is present in S_C^+ and in S_C^-, is called a double cluster-gap module.

Definition 5. A sequence pair is *feasible* in respect to a set of convex cluster constraints if all cluster gap modules are single gap modules (otherwise it is infeasible).

Vertical and horizontal constraint graphs of all sequence pairs with the same ordering of cluster modules and only single cluster-gap modules contain isomorphic sub-graphs, horizontal and vertical, defined on only cluster modules.

Theorem 2.
The topology of a cluster can always be preserved in at least one floorplan generated from a given sequence pair if the orders of cluster modules in both sequences remain unchanged and if the sequence pair contains only single cluster-gap modules.

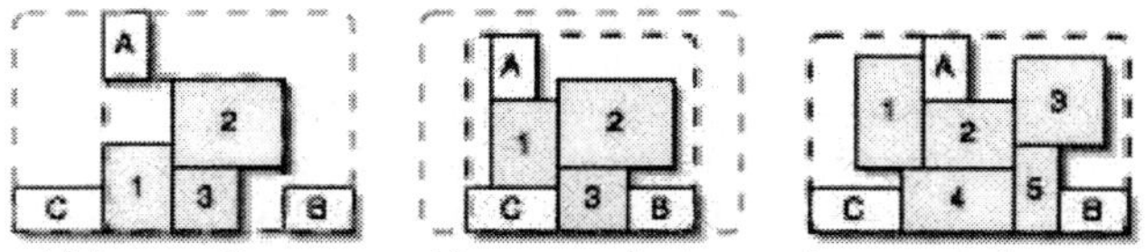

a) Macro-cell (CA123B, C132AB) b) convex cluster (A1C23B, C1A3B2) c) concave cluster (1A2C435B, C412A5B3)

Figure 3. Cluster dead-spaces for convex gap modules

If cluster sequences contain only cluster modules then the area of the cluster layout can be the smallest possible, but no other modules can enter the dead space that exists between the cluster layout and the smallest rectangle containing that layout as shown in Fig 3a. Modules 1, 2 and 3 are cluster modules in Fig.3a and Fig.3b. Modules 1, 2, 3, 4 and 5 are cluster modules in Fig.3c. If some cluster gap modules are allowed in between cluster modules, then these modules will be allowed to enter cluster dead-space subject to size constraints. In the best case, single cluster-gap modules can fill cluster dead-space for a convex cluster as shown in Fig 3b, subject to size constraints. Double cluster-gap modules can fill dead-space for a concave module, subject to size constraints, as shown in Fig3.c.

For convex clusters, we can state that by rejecting infeasible (definition 3) sequence pairs we do not lose any solutions better than those generated with feasible sequence pairs. If, due to small alignments between modules (an unlikely event), a convex cluster changes to concave, we will still do much better than if we had treated a cluster as a macro-cell.

We use the following set of conditions to preserve the topology of convex clusters (if required) during floorplan optimizations.

Necessary conditions:
The order of cluster modules in both cluster sequences have to remain the same.

Rejection conditions:
To limit the search space without reducing the quality of final solutions only single cluster-gap modules are allowed.

If we assume that the topology of a cluster has to remain fixed, we fix the order of cluster modules in both sequences. If we allow the topology of a cluster to change we only consider the rejection conditions.

It is relatively easy to assure that necessary conditions are met during perturbation of sequence pairs. However, checking sequence pairs to see if a feasible floorplan is also an optimal one is complicated and impractical. Consequently, we have developed geometrical constraints to assure that clusters are preserved.

Geometrical constraints are defined in a straightforward manner. The dimensions of the smallest enclosing rectangle containing a cluster layout are calculated (by enclosing, we mean all cluster modules are contained inside the rectangle.) Depending on design criteria a rectangle that is 0%, 5%, 10%, or 20% larger than the smallest enclosing rectangle is chosen. We have generalized this approach to include multiple clusters.

4. ELF-SP FLOORPLANNER

Although Simulated Annealing (SA) is the most prevalent stochastic search method used in floorplanning we have decided to use an evolutionary algorithm (EA). We believe EAs [2] are better suited for dealing with multi-objective optimization such as Pareto and pre-optimal approaches.

Our EA begins with an initial population of $k=20$ randomly generated parents where a parent encodes a sequence pair. This population size provides a reasonable tradeoff between computation time and exploration of the solution space. Indeed, it works well for test problems with up to 200 modules. Three types of mutation operators were used.

Swap operator: exchange the position of two modules in only Γ^+, or only Γ^-, or both Γ^+ and Γ^-.

Inversion operator: two positions in a sequence pair are randomly chosen and the order of modules between these points is reversed.

Insert operator: a randomly chosen module is inserted into another randomly chosen position in the sequence pair.

In the unconstrained approach soft modules are made hard by fixing their aspect ratio. We use a $O(n^2)$ algorithm to get a packing [9]. Lagrangian Relaxation [8] then optimizes the final best evolutionary algorithm solution while respecting all aspect ratio ranges.

Cluster-Constrained Algorithm:

1. Generate an initial random population of 20 feasible sequence pairs.
2. For each parent, generate an offspring using one of the reproduction operators.
3. Check that only single cluster-gap modules are present (rejection conditions). If rejection conditions are not satisfied, assign a high value to the cost function (low fitness).
4. Generate a floorplan using the LCS (longest common subsequence) algorithm [16]
5. Check if the offspring floorplan is feasible by checking the geometrical constraints.
7. If a floorplan is feasible, the cost is equal to its area. Else, the cost function is set to a high value (low fitness).
8. Select the lowest-cost 20 individuals from the population of 20 parents and 20 children.
9. If the stopping criteria are not satisfied, go to Step 2.

5. EXPERIMENTAL RESULTS

All tests were conducted on: 333 MHz UltraSparc-Iii. We averaged all results over 100 runs. The small standard deviations obtained clearly indicate that our EA approach is robust. There are no results for cluster-constraint floorplans we can compare too, therefore we first compare our general results to demonstrate the performance of our tool. Table 1 compares our results for soft module floorplans, generated with our ELF-SP floorplanner, with Young et al.[7]. Our method produces considerably better results (in terms of area) in significantly shorter CPU time. Impressive time reduction was achieved by performing LR only once on the final hard-module placement while it is done at each iteration of Simulated Annealing [7]. The superiority of our approach is demonstrated in the significant reduction of deadspace. When we run Lagrangian Relaxation we intentionally allow more flexibility to improve packing. Therefore, in soft module floorplans some cluster modules may be moved from their original position. This flexibility could be easily removed if stricter geometrical constraints are necessary. We defined the cost function as $cost = Area/10^{scale} + \beta Wirelen$, where $scale = \log(area/wirelen)$. Wire length is calculated using the half perimeter method, and terminals are assumed to be in

File	md	Young s [7]		ELF-SP (Ours)			
		Ds (%)	Time (sec)	Deadspace(%)		s-wire (mm)	Time (sec)
				min/mean	SD		
apte	9	0.54	53.0	1.40/1.45	0.25	403.9	1.04
xerox	10	0.4	71.6	0.00/0.46	0.55	478.5	1.32
Hp	11	1.4	107.3	0.00/0.54	0.81	125.7	1.36
ami33	33	4.3	774.6	0.00/0.47	0.63	46.2	8.09
ami49	49	7.7	2354.	0.00/0.98	0.84	675.4	12.16

Table 1: Modules with aspect ratio [0.5, 2.0]

the center of each block, and the best wire length results for soft-module floorplans (s-wire) are given in Table 1.

To demonstrate cluster-constrained floorplanning we show two experiments on ami33 and ami49. For the first experiment we selected four clusters for ami33, and three clusters for ami49, shown in table 2. Soft module floorplans for ami49 are shown in Fig. 4. In the second experiment (Table 3) we compare floorplans generated by our cluster-constrained floorplanner

	mods	Cluster sequence pair
1 cluster	5	[25 16 23 0 29, 16 23 25 29 0]
2 clusters	5	[25 16 23 0 29, 16 23 25 29 0]
	3	[3 6 31, 31 3 6]

Table 2 Clusters for figure 4

with floorplans where clusters are treated as macro-cells. Please notice that the layouts of all floorplans generated by the cluster-constrained floorplanner have smaller dead space area and comparable CPU time. It is very difficult to compare wire length, because there is no uniform standard on where pins should be located, therefore results for wire length are omitted.

6. CONCLUSIONS

We have defined necessary conditions to preserve a cluster s topology under a sequence pair perturbation and developed rejection conditions for convex clusters that allow us to efficiently limit solution space without reducing the quality of floorplans. Our algorithm preserves continuous clusters as demonstrated by the experiments. Our results reinforce the notion that sequence pair is a very powerful representation and topology checks can be done directly on these sequences without requiring floorplan generation.

The algorithm is designed as an area and total wire length minimization algorithm but can be easily adapted to handle fixed-frame floorplanning and total wire length timing restrictions.

References

1. M. Z. Kang, W. W-M. Dai, Arbitrary rectilinear Block Packing.
2. D. Fogel, Evolutionary Computation: Toward a New Philosophy of Machine Intelligence , *IEEE Press, 2nd edition*, 2000
3. J. Xu, P-N. Guo, C-K. Cheng, Sequence-pair Approach for Rectilinear Module Placement, IEEE Trans. on CAD, v. 18, no. 4, pp. 484-493, 1999.
4. C. Lin, D. M. W. Leenaerts, A New Faster Sequence Pair Algorithm, Proc. of ISCAS. pp. 407-I410, 2000.
5. H. Murata, K. Fujiyoshi, S. Nakatake, and Y. Kajitani, Rectangle packing based module placement , *Proc. of ICCAD* , pp. 472-479, 1995.
6. X. Tang and D. Wong, FAST-SP: A Fast Algorithm for Block Placement Based on Sequence Pair, *Proc ASP-DAC*, pp. 521-526, 2001.
7. F. Young, C. Chu, W. Luk, and Y. Wong, Handling soft modules in general non-slicing floorplan using Lagrangian relaxation , *IEEE Trans.on Computer-Aided Design* 20(5):687-692, 2001.
8. W. S. Yuen and F.Y. Young, Slicing Floorplan with Clustering Constraints, *Proc. of Asia South Pacific Design Automation Conference*, pp.503-508, January 2001,
9. S. Nakatake, K. Fujiyoshi, H. Murata and Y. Kajtani, Module Placement on BSG-structure and IC layout application, *Proc. ICCAD*, pp. 484-491, 1996.
10. B. Wang, M. Chrzanowska-Jeske, G. Greenwood, ELF-SP — Evolutionary Algorithm for Non-Slicing Floorplans with Soft Modules, *Proc. of ICECS-02*, pp. 681-684, 2002.
11. X. Tang, R. Tian, D.F. Wong,, Fast evaluation of sequence pair in block placement by longest common subsequence computation , *Proc. of DATE-00*, pp.106-111, 2000.
12. X. Tang, D.F. Wong,, Floorplanning with Alignment and Performance Constraints, DAC-02, pp. 848-853, 2002.
13. H. Murata, K. Fujiyoshi, M. Kaneko, VLSI/PCB Placement with Obstacles Based on Sequence Pair, *TCAD* vol. 17, no. 1, pp.60-68, 1998.

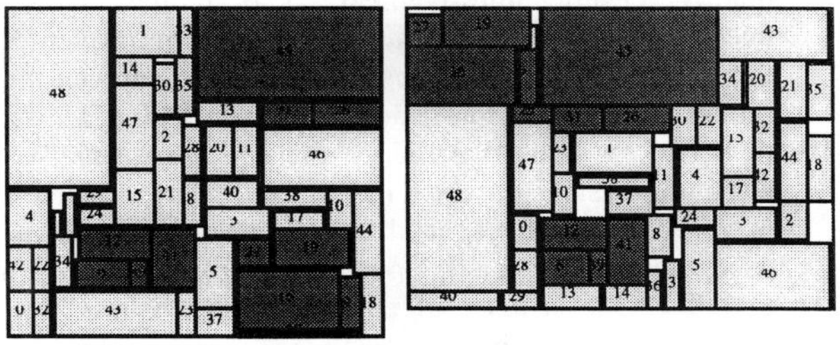

Figure 4 Soft module floorplan of ami49 with multi-cluster constraints

File	cls	mods	Cluster sequence pair
Ami33	4	5	[25 16 23 0 29, 16 23 25 29 0]
		3	[15 32 13, 15 13 32]
		3	[3 6 31, 31 3 6]
		3	[17 22 20, 17 22 20]
Ami49	3	3	[45 31 26, 31 26 45]
		4	[12 6 39 41, 6 39 12 41]
		5	[27 19 16 9 25, 25 16 27 9 19]

Table 3: Clusters for efficiency test

File	Toler (%)	1 Cluster		1 Macrocell		2 Clusters		2 Macrocells	
		Deadsp(%)	CPU(s)	Deadsp(%)	CPU(s)	Deadsp(%)	CPU(s)	Deadsp(%)	CPU(s)
Ami33	0	1.41	9.17	2.61	5.20	1.16	9.19	3.71	4.96
	10	1.22	8.92	2.54	5.44	1.32	9.06	3.76	4.98
	20	1.13	9.20	2.56	5.48	1.04	9.34	3.76	4.97
Ami49	0	2.50	13.80	4.44	11.67	2.87	13.89	6.24	11.25
	10	2.34	13.82	4.50	11.84	2.62	14.34	6.16	11.41
	20	2.24	13.85	4.45	11.89	2.65	14.37	6.15	11.34

Table 4: Results of efficiency

THE STRUCTURE DETERMINATION FOR THE TIME-OPTIMAL SYSTEM DESIGN ALGORITHM

A. Zemliak

Puebla Autonomous University
E-mail: azemliak@fcfm.buap.mx

ABSTRACT

An additional acceleration effect of the design process has been discovered on the basis of the generalized system design methodology by means of various design strategies analysis with the different initial points. Some principal characteristics of this effect were studied to select the optimal position of the initial point of the design process. The special Lyapunov function of the design process was proposed to analyze some conditions to obtain the optimal position of the switching points between the different design strategies. These optimal switching points serve as the basis for the structure determination of the time-optimal design algorithm.

1. INTRODUCTION

The problem of the computer time reduction of a large system design is one of the essential problems of the total quality design improvement. The reduction of the necessary time for the circuit analysis and improvement of the optimization algorithms are two main sources for the reduction of the total design time. There are some powerful methods that reduce the necessary time for the circuit analysis. Because a matrix of the large-scale circuit is very sparse, the special sparse matrix techniques are used successfully for this purpose [1]-[3]. Other approach to reduce the amount of computational required for the linear and nonlinear equations is based on the decomposition techniques. The well-known ideas for partitioning of a circuit matrix into bordered-block diagonal form were described in original works using the branches tearing [4] or nodes tearing [5] and jointly with direct solution algorithms gives the solution of the problem. The extension of the direct solution methods can be obtained by hierarchical decomposition and macro model representation [6]. The optimization technique is developed both for the unconstrained and for the constrained optimization and can be improved in future. Meanwhile there is another way to reduce the total computer design time. This approach consists of the reformulation of the total design problem and generalization of it to obtain a set of different design strategies inside the same optimization procedure [7][8]. This approach serves for the time-optimal design algorithm definition and can be proposed as a theoretical basis for the time-optimal design algorithm searching. The significant computer time gain occurs when the optimal or quasi-optimal design trajectory can be found. An additional acceleration effect of the design process has been discovered by the analysis of various design strategies with the different initial points [9]. This effect provides an additional computer time gain of the design process and serves as the basis for the time-optimal design algorithm construction. On the other hand all of these advantages are realized only when the time-optimal algorithm is already found. One of the main steps of the time-optimal algorithm construction is the optimal switching point definition between the different design strategies. This problem is discussed in the present work.

2. PROBLEM FORMULATION

The design process for any analog system design has been generalized on the basis of the control theory approach [7][8]. In this case the design process is defined by means of the optimization procedure (1) and by the analysis of the physical system (2):

$$X^{s+1} = X^{s} + t_{s} \cdot H^{s} \quad (1)$$

$$\left(1 - u_{j}\right) g_{j}(X) = 0, \quad j = 1,2,\dots,M \quad (2)$$

where M is the number of dependent parameters and H is the vector of directional movement. In this case the vector H depends not only on the optimization procedure and objective function structure but from the vector of special control functions $U = \left(u_1, u_2, \dots, u_m\right)$ that control the design process, where $u_j \in \Omega$; $\Omega = \{0;1\}$. In this case a new generalized objective function is needed to define as $F(X) = C(X) + \psi(X)$ with a special additional penalty function $\psi(X) = \frac{1}{\varepsilon}\sum_{j=1}^{M} u_j \cdot g_j^2(X)$. All control variables u_j are the functions of the current point of the design process. The total number of the different design strategies, which are produced inside the same optimization procedure, is practically infinite. The problem of the optimal design strategy search is formulated as the typical problem for the functional minimization of the control theory. The searching of the optimal behavior of all the control functions is the main stage of the time-optimal algorithm construction. The discovery of the additional acceleration effect of the design process [9] allows to define the essential features of the optimal algorithm. The optimal algorithm consists of one or several trajectory jumps from quasi-modified

0-7803-7761-3/03/$17.00 ©2003 IEEE

traditional strategy to quasi-traditional strategy in terminology of the papers [7][8]. We can obtain a large computer time gain if the acceleration effect is already realized. For instance the time gain for the three-transistor cells amplifier is more than 400 times as shown in [9]. On the other hand the construction of the time-optimal algorithm on the basis of the acceleration effect turns on unknown switching point positions for the control functions u_j. The analysis of the essential characteristics of the acceleration effect, the optimal initial point selection and the optimal switching point definition are discussed below.

3. INITIAL POINT OPTIMAL SELECTION

3.1 Two-dimensional problem

The problem of the initial point selection for the design process is one of the main problems of the time-optimal algorithm construction. The analysis of the design process and acceleration effect for the simplest electronic circuit of the Fig. 1 was provided in [9]. The element r_1 has a non-linear dependency in general case. The vector X of the state variables has two components $X=(x_1,x_2)$, where x_1 is the independent parameter ($x_1 \equiv r_2$) and $x_2 \equiv V_1$. The objective function is defined by the formula $C(X) = (x_2 - k_V)^2$, where k_V has a fixed value. The optimization procedure in accordance with the new design methodology is defined by the equations:

$$x_i^{s+1} = x_i^s + t_s \cdot f_i(X,U), \quad i = 1,2 \qquad (3)$$

where, for the gradient method $f_1(X,U) = -\frac{\delta}{\delta x_1}F(X,U)$,

$$f_2(X,U) = -u_1 \frac{\delta}{\delta x_2}F(X,U) + \frac{(1-u_1)}{t_s}\left[-x_2^s + \eta_2(X)\right].$$

The vector of the control variables U consists on one coordinate u_1 only. The equation (2) is transformed now to the next form:

$$(1-u_1)g_1(X) = 0 \qquad (4)$$

where $F(X,U)$ is the generalized objective function,

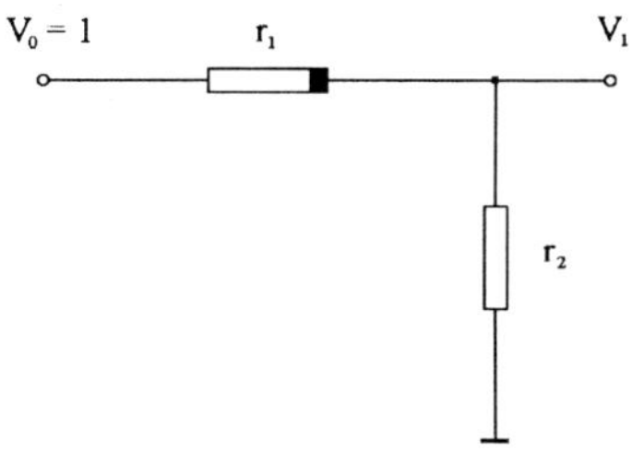

Figure 1. Topology of a simplest electronic circuit

$\eta_2(X)$ is the implicit function $(x_2^{s+1} = \eta_2(X))$ and it gives the value of the parameter x_2 from the equation (4). The vector of the control variables U consists on one coordinate u_1 only. As shown in paper [9] we need to select the initial point of the design process with the negative coordinate x_2. In this case the acceleration process is realized. We need to analyze the characteristics of the acceleration effect to decide what value of the coordinate x_2 is better. The family of the design curves for the circuit on Fig. 1, which corresponds to the modified traditional design strategy (u=1) and the negative initial value of the second coordinate (x_2<0) of the vector X is shown in Fig. 2 for the 2-D phase space. These curves have different start points but the same final point F. The start points were selected on the circle arc and have the different initial coordinates. The special curve S-F, which is marked by thick line, is the separating curve. This curve separates the trajectories that are the candidates for the acceleration effect achievement (all curves that lie under the curve S-F), and the trajectories that cannot produce the acceleration effect (curves that lie over the curve S-F). It is clear that the projections of the final point F to all curves of the first group define the switching point of the optimal trajectory, which produces the acceleration effect. All curves of the first group (1-7) approach to the finish point F from the left side, and on the contrary, all curves of the second group (9-16) approach to the finish point from the right side. The comparison of the relative computer time for all curves of the Fig. 2 is shown in Fig. 3 as the function of the curve number n.

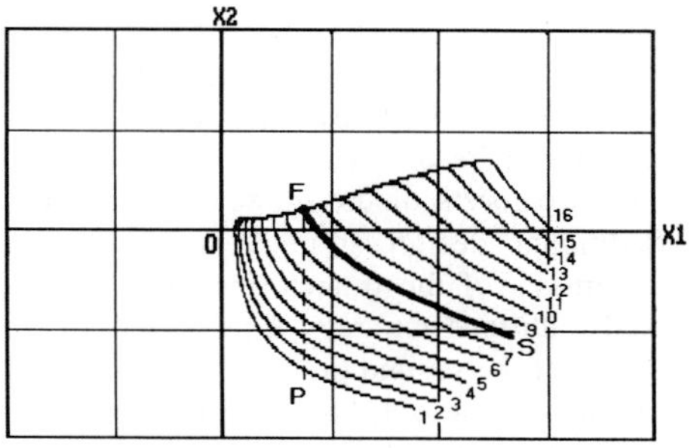

Figure 2. Trajectories of the modified traditional strategy for the different start points with the negative coordinate x_2.

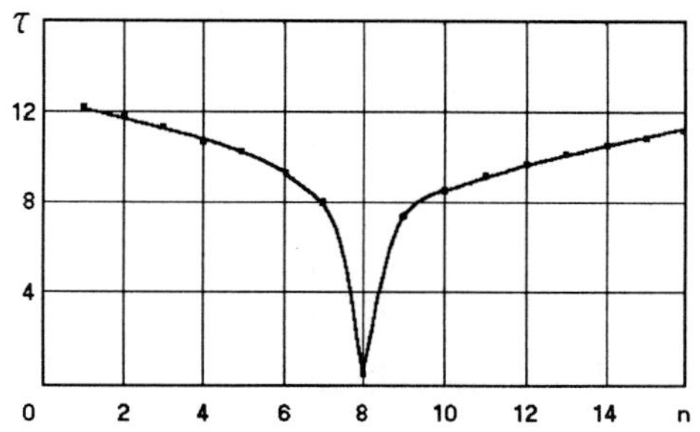

Figure 3. Relative computer time τ as the function of the curve number n.

The separating curve S-F (number 8) has the minimal computer time among all of the trajectories. At the same time this curve cannot be used as the basis for the time-optimal trajectory construction because the projection of the point F to this curve is the same point F, but the movement slows down near this point along the all design trajectories. Only the curves that lie under the curve S-F serve as the first part of the time-optimal trajectory with the following jump to the point F. The relative computer time τ of the optimal trajectories with acceleration effect (on the basis of the curves 1-7, Fig. 2) is shown in Fig. 4 as the function of the curve number n. The curves 9-16 can be optimized too but in this case the time reduction about 10-15% only takes place. Fig. 4 shows that the total computer time increases when the start point approaches to the curve S-F, and on the contrary, the more acceleration can be obtained if the start point lies far from the curve S-F (from curve 7 to curve 1). So, the start point selection with at least one negative initial coordinate of the vector X and the value of this coordinate that gives the start point position under the separating line are the sufficient conditions for the acceleration effect appearance.

3.2 N-dimensional problem

All these conclusions are correct for the N-dimensional problem too. We need to analyze the different projections of the N-dimensional curve in this case. The N-dimensional problem solution gets complicated by a large number of the different admissible trajectories and a large number of the different trajectory projections. In this case we need to choose the most perspective trajectories and analyze them. It can be done by some approximate methods [10]. In this paper it was done by the careful analysis of all possibilities. The total set of the various trajectories can be divided in two different subsets. The first subset consists of the trajectories that are similar to the traditional design strategy trajectory. The second subset consists of the trajectories that are similar to the modified traditional strategy trajectories. In this case the trajectories of the second group serve as the candidates for the first part of the optimal trajectory and the first group trajectories serve as the candidates to the jump produce. Two of these main steps together with the following different trajectory adjustment make up the essence of the optimal algorithm construction. By the experience [8]-[9] we can decide that not all of the feasible projections are important to the acceleration effect obtained. First of all the admittance-voltage two-dimensional projections are more important. Variables that are included in the objective function formula have a greater importance among all of these projections. By this preliminary selection we can reduce the number of the more perspective candidates for the time-optimal algorithm elaboration. This problem final solution will be based on the optimal algorithm intrinsic structure. However, the results obtained here serve as the next step on the way of this problem solution. Now it is clear that the time-optimal algorithm must include the special conditions for the start point selection to the acceleration effect reach. The problem of the switching point position determination is discussed in the next section.

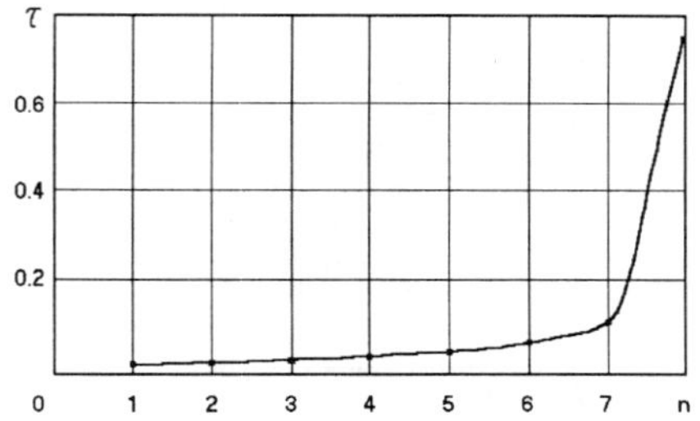

Figure 4. Relative computer time τ of the optimal trajectories with acceleration effect as the function of the curve number n.

4. SWITCHING POINT DEFINITION

On the basis of the analysis in previous section we can conclude that the time-optimal algorithm has one or some switching points where the switching realize from like modified traditional strategy to like traditional strategy with an additional adjusting. At least one negative component of the start value of the vector X is needed for the acceleration effect obtained. The main problem of the time-optimal algorithm construction is the unknown sequence of the switching points during the design process. We need to define a special criterion that permits realizing the optimal or quasi-optimal algorithm by means of the optimal switching points searching. In this paper we propose to use a Lyapunov function of the design process for the optimal algorithm structure revelation, in particular for the optimal switching points searching. There is a freedom of the Lyapunov function choice because of a non-unique form of this function. Let define the Lyapunov function of the design process as:

$$V(Y,U)=\sum_i\left(\frac{\partial F(Y,U)}{\partial x_i}\right)^2 \qquad (5)$$

where $F(Y,U)$ is the generalized objective function of the optimization procedure. This form holds all of the necessary characteristics of the standard Lyapunov function definition. It is supposed that the vector Y is defined as the difference between two vectors X and A, where A is the stationary point of the design process (the final point). First of all the function (5) can be used for the stability analysis of the design process. In this context this function is used for the analysis of the design trajectories behavior with the different switching points. We can define now the system design process as a transition process that provides the stationary point during some time. The problem of the time-optimal design algorithm construction is the problem of the transition process searching with the minimal transition time. There is a well-known idea in control theory [11] to minimize the transition process time by means of the special choice of the right hand part of the principal system of equations, in our case these are the functions $f_i(X,U)$. In this conception it is necessary to change the functions $f_i(X,U)$ by means of the control vector U selection to obtain the maximum speed of the Lyapunov function decreasing (maximum of *-dV/dt*) at each point of the process. Unfortunately the direct using of this idea does not serve well for the time-optimal design algorithm construction. It occurs because the design strategy changing produces not only continuous design trajectories (when we change from the strategy u=0 to the strategy u=1 for the circuit in Fig. 1 for instance) but non-continuous trajectories too (the changing from u=1 to u=0). Non-continues trajectories had never appeared in the control theory for the objects that are described by differential equations, but this is the ordinary case for the design process on the basis of the described design theory. In this case we need to correct the idea to maximize *-dV/dt* at each point of the design process. We define another principle: it is necessary to obtain the maximum speed of the Lyapunov function decreasing for that trajectory part which lies after the switching point. In this case the trajectories with the different switching points are compared to obtain the maximum value of *-dV/dt*. The behavior of the function *dV/dt* for three switching points 1, 2 and 3 that correspond to three consecutive integration steps before (a), in (b) and after (c) the optimal point is

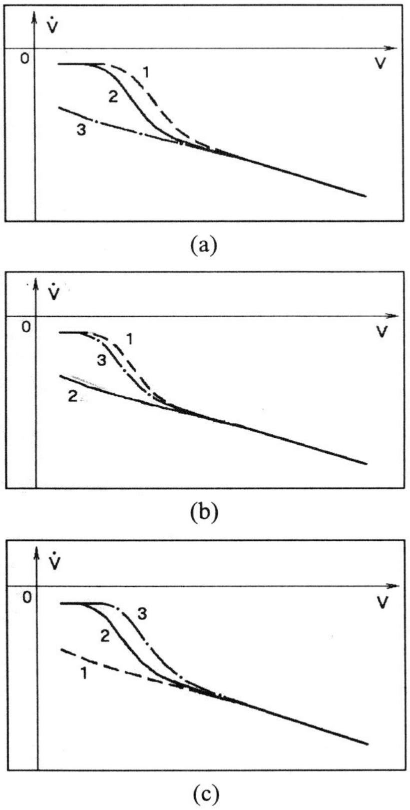

Figure 5. Time derivative of Lyapunov function behavior for three switching points 1,2,3 consecutive integration steps before (a), in (b) and after (c) the optimal point.

shown in Fig. 5 (the one transistor amplifier was designed). The optimal switching point corresponds to the curves 3 of Fig. 5a, or 2 of Fig. 5b, or 1 of Fig. 5c. It is clear that this point corresponds to the maximum negative value of function $\dot{V}$ and at the same time corresponds to the minimum value of the total design steps. In this case the optimal position of the switching point is found. This point serves as the basis to the optimal algorithm construction.

5. CONCLUSIONS

The problem of the time-optimal algorithm construction can be solved as the functional optimization problem of the control theory. The additional acceleration effect of the system design process serves as the basis for the time-optimal algorithm searching. The initial point selection permits to obtain the acceleration effect with a great security. The optimal position of the necessary switching points can be obtained on the basis of the Lyapunov function analysis. The minimization of the time derivative of this function serves as the principal criterion for the optimal switching point definition. Thus the combination of the acceleration effect and the optimal switching point determination serve as the principal ideas for the time-optimal algorithm construction.

6. REFERENCES

[1] Bunch J.R., and Rose D.J., Eds., *Sparse Matrix Computations*, Acad. Press, N.Y., 1976.

[2] Osterby O., and Zlatev Z., *Direct Methods for Sparse Matrices*, Springer-Verlag, N.Y., 1983.

[3] George A., "On Block Elimination for Sparse Linear Systems", *SIAM J. Numer. Anal.* vol. 11, no.3, 1984, pp. 585-603.

[4] Wu F.F., "Solution of Large-Scale Networks by Tearing", *IEEE Trans. Circuits Syst.*, vol. CAS-23, no. 12, 1976, pp. 706-713.

[5] Sangiovanni-Vincentelli A., Chen L.K., and Chua L.O., "An Efficient Cluster Algorithm for Tearing Large-Scale Networks", *IEEE Trans. Circuits Syst.*, vol. CAS-24, no. 12, 1977, pp. 709-717.

[6] Rabat N., Ruehli A.E., Mahoney G.W., and Coleman J.J., "A Survey of Macromodeling", *IEEE Int. Symp. Circuits Systems*, 1985, pp. 139-143.

[7] Zemliak A., "System Design Problem Formulation by Control Theory", *Proc. IEEE Int. Sym. on Circuits and Systems – ISCAS2001*, Sydney, Australia, 2001,Vol. 5, pp. 5-8.

[8] Zemliak A.M., "Analog System Design Problem Formulation by Optimum Control Theory", *IEICE Trans. on Fundam.*, vol. E84-A, No. 8, 2001, pp. 2029-2041.

[9] Zemliak A., "On Start Point Selection for the Time-Optimal System Design Algorithm", *IEEE Int. Sym. on Circuits and Systems – ISCAS2002*, Scottsdale, USA, May, 2002, Vol. IV, pp 465-468.

[10] Pytlak R., *Numerical Methods for Optimal Control Problems with State Constraints*, Springer-Verlag, Berlin, 1999.

[11] Rouche N., Habets P., and Laloy M., *Stability Theory by Liapunov's Direct Method*, Springer-Verlag, N.Y, 1977.

An Efficient Approach for Error Diagnosis in HDL Design

Che-Hua Shi and Jing-Yang Jou

Dept. of Electronics Engineering
National Chiao Tung University
Hsinchu, Taiwan, ROC

ABSTRACT

The growing of the modern design complexity leads the design error diagnosis to be a challenge for designers. In this paper, we propose an efficient approach for design error diagnosis automatically for designs in HDL. This approach can handle multiple errors occurring in a HDL design simultaneously with only one test case by analyzing the simulation outputs of the incorrect implementation. Furthermore, this approach reduces the error space by eliminating those statements that have no or lower possibility to become the error sources with retaining at least one error source in it. Hence, the effort spent on the debugging process can be greatly reduced.

1. INTRODUCTION

Functional mismatches between an implementation and its specification happen frequently during the design stage. Due to the increasing complexity in modern VLSI design, designers may examine thousands of lines in a HDL design for error diagnosis while mismatches occur. As a result, finding out errors in the erroneous implementation becomes a complicated and time-consuming task. Therefore, an efficient automatic error diagnosis technique is needed to speed up the debugging process of the entire design flow.

Many approaches were proposed for the design error diagnosis before. They can be broadly categorized into two classifications: symbolic approaches [1, 2] and simulation-based approaches [3-8]. The symbolic approaches, based on binary decision diagram (BDD), can be easily extended for multiple errors and they are more accurate without error models needed. However, they may encounter the memory explosion problem for large circuits. For the simulation-based approaches, on the other hand, the memory explosion problem vanishes and the run-time for large circuits is decreasing significantly as compared with the symbolic ones. The simulation-based approaches usually exploit a quantity of simulation vectors, which can differentiate the implementation and the design specification, to reduce the error space for identifying locations of possible errors based on the particular error models. The error models simplify the diagnosis problem but may have lower quality diagnosis solutions in multiple-error design cases.

Lots of the previous work focus on the error diagnosis in the gate or lower levels of circuits [1-6]. Some of them target at the combinational circuits [1-4], and some target at the sequential ones [5, 6]. It is obvious that the error diagnosis algorithms for sequential circuits are more applicable to most of the modern VLSI designs. Many of previous researches can just handle single-error designs well; however, an erroneous implementation usually contains many errors. Thus, these approaches are incapable of dealing with practical designs with multiple errors.

The error diagnosis approaches at RT level were proposed in [7-9]. In [7], Boppana et al. proposed a hierarchical error diagnosis approach for RTL components. This approach is based on an error model that is called the Xlists model. The Xlists is used to enhance the stuck-at-fault model. While the stuck-at-1 and stuck-at-0 of a certain node in the RTL design cause different results at the erroneous primary output, this node is an error candidate of stuck-at-X and is included into the set of the Xlists. The Xlists can capture the effects of all possible errors within RTL circuits. Nevertheless, diagnosis with this approach may be too complex for sequential circuits or multiple-error designs.

In [8], Khalil et al. proposed an algorithm containing four hypotheses to diagnose design errors using the HDL information. For systematic analysis, the algorithm classified all possible situations into four hypotheses that are defined from looseness to strictness. The first two hypotheses assume that there is only one erroneous statement in the design. The first and the third hypothesis assume that the executed statements of correct test cases are impossible to be the error sources. By using these two assumptions, error space can be reduced. Since a designer does not know which hypothesis can reflect the actual condition of his erroneous design, the algorithm may obtain an error space without any error in it while the first three hypotheses are applied. It also assumes there are many test cases that can trigger design errors. Realistically, it is not easy to generate lots of test cases that can trigger the same design errors while designers do not know where the errors are located. In fact, designers always diagnose design when an erroneous test case is simulated.

In [9], Jiang proposed an error diagnosis technique for RTL designs in HDL by means of intersecting the executed statements and the relation statements. It eliminated statements that cannot be the error sources to reduce the error space. However, it assumed that all correct values of the registers are given in the test case. Based on this assumption, the error effects cannot be propagated through the registers. In other words, when an error is triggered at a certain clock cycle, its error effect can be observed either at the primary outputs or at the registers of the clock cycle. Therefore, this approach can find errors at the error-occurring clock cycle and acquires a smaller set of error candidates. Nevertheless, this approach is inappropriate while only the correct values of the primary outputs are given in most test cases.

Due to the disadvantages of previous approaches and some limitation of practical diagnosis processes mentioned above, we propose an efficient approach for error diagnosis in HDL. Our approach can handle multiple errors in sequential circuits with only one test case. Furthermore, this approach is error model free.

0-7803-7761-3/03/$17.00 ©2003 IEEE

Since designers often start to re-simulate the erroneous test case after correcting the first error they find, the main goal of our approach is to minimize the error space with containing at least one error source by some heuristics.

This paper is organized as follows. In Section 2, we will introduce some preliminaries and the overview of our work. Our reduction rules will be detailed in Section 3. Section 4 shows the experimental results. Finally, Section 5 concludes this paper.

2. PRELIMINARY

Definition 1: **Correct Primary Output** (CPO) is the primary output whose simulation value is the same as the expected one; otherwise it is called an **Erroneous Primary Output** (EPO).

Definition 2: **Error-Observed Clock Cycle** (EOC) is defined as the clock cycle of simulation in which the values of POs are different from the expected ones.

In most cases, we only have the expected values of POs. Observing the simulation values of POs and comparing them with the expected ones can determine whether the implementation is an erroneous design. When an error causes a PO to be an EPO, the error effect is propagated to the EPO via a path. Each variable on the path has incorrect value and is called an error propagator. Furthermore, if the design is a sequential circuit, the error may be activated before the EOC and its effect can be propagated through registers.

Error space is the set of error candidates. All statements in the error space are potential error sources and may cause design be incorrect. Since the error space is used to identify errors occurred in a design, it has to contain at least one design error. Otherwise, it would just be a meaningless set.

Definition 3: The statements that are executed at a specific clock cycle of the simulation are defined as the **executed statements** of the clock cycle.

Definition 4: Statements in the fan-in cone of one variable are defined as the **relation statements** of the variable.

It is clear that if one statement has not been executed while mismatch occurs, it is impossible to be the error source of the EPO. If one statement is not in the relation statements of the EPO, it is not an error source, either. In [9], Jiang proposed one approach to find error space by back-tracing from the EPO in the executed statements. Error space obtained with this approach is the intersection of executed statements and relation statements.

2.1 The Overview of Our Approach

This section presents the overview of our approach and describes how to construct the error space in sequential circuits without knowing the correct values of internal registers. Our approach is based on the back-tracing technique that was proposed in [9]. The first back-tracing origin is the EPO. Sometimes there are more than one EPO at the EOC, and we will choose one of them to construct the error space.

After selecting one EPO, we back-trace the executed statements from the EPO at the EOC. Since all back-traced statements will be included into the error space, fewer back-tracing operations result in smaller error space. Our approach is to apply some rules to terminate back-tracing procedures earlier and therefore reduces the error space size and shortens the processing time. During the diagnosis process, the back-tracing procedures will stop at primary inputs or constants. If the back-traced variable is a register, we will stop back-tracing at it as well at the current clock cycle. Because the value of this register is set at previous clock cycles, it is marked for further back-tracing in the next iteration, where it is the origin of back-tracing procedures at the previous clock cycle. These back-tracing procedures will be repeated until no marked register remained at the current clock cycle. Then, the error space is obtained. This flow is shown in Figure 1.

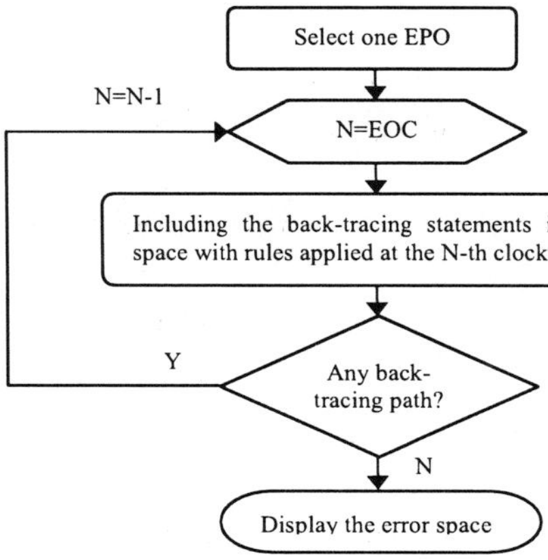

Figure 1. The flow of our approach

3. ERROR SPACE REDUCTION RULES

Since a statement often has more than one right-hand side variable, the error space obtained by back-tracing from the EPO expands exponentially. Thus, to terminate the back-tracing operations earlier can greatly reduce the error space. In this section, we will propose three rules to terminate back-tracing operations at some variables to minimize the size of error space.

Definition 5: **BTS (V)** denotes the set of statements that are back-traced in the executed statements from variable V.

Definition 6: **Reversible statement** is a statement whose right-hand side variable (RV) value can be computed by knowing the value of its left-hand side variable (LV).

Definition 7: **Reversible path** is a path that only consists of reversible statements.

Example 1: "B=~A; C=~B+1; D=C;". These three statements are all reversible statements. The path from variable A to variable D is a reversible path.

Lemma 1

Given a reversible path that ends with a PO. If the PO value and all statements on the reversible path are correct, the simulation value of any variable on the reversible path is correct.

Theorem 1

Given a reversible path that ends with a PO. If the PO value is correct and the simulation value of the beginning variable on the reversible path is incorrect, then there exists at least one erroneous statement on this reversible path.

Rule I

When we back-trace to a variable that is on a reversible path ending with a correct primary output, according to Theorem 1, if the value of the variable is incorrect, there exists at least one erroneous statement on this reversible path. Otherwise, this variable is not an error propagator and the back-tracing operation from this variable is unnecessary.

Therefore, we can terminate the back-tracing operation from this variable by putting all statements on this reversible path into the error space to retain at least one erroneous statement in it.

HDL code:
R1=A+B;
...
PO1=R1*C;
// PO1 is the EPO at simulation
PO2=~R1;
// PO2 is the CPO at simulation

(a) A partial Verilog code

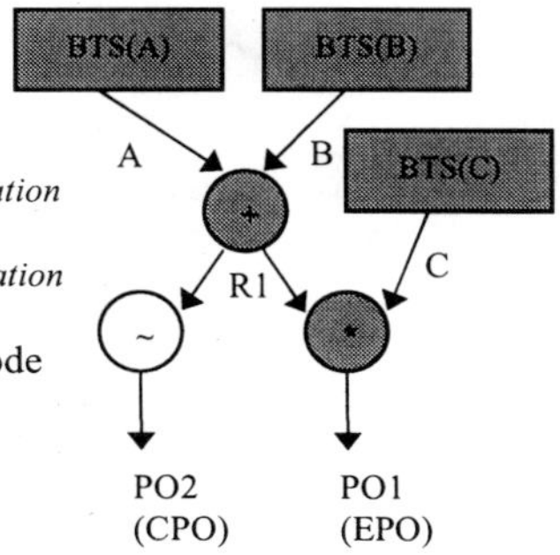

(b) The corresponding CDFG of (a)

Original error space:
PO1=R1*C;
BTS(C)
R1=A+B;
BTS(A);
BTS(B);

Reduced error space:
PO1=R1*C;
BTS(C);
PO2=~R1;

(c) Original and reduced error space

Figure 2. An example of applying Rule I

We use an example shown in Figure 2 to demonstrate Rule I. Figure 2 (a) is a partial Verilog code of a design and its corresponding CDFG (control data flow graph) is shown in Figure 2 (b). When a mismatch happens at PO1, the back-tracing operation from PO1 is conducted. During the process of back-tracing operations, the highlighted parts of Figure 2 (b) are included into the error space. This is because the highlighted parts are the intersection of the relation statements and the executed statements. This original error space is shown in the upper part of Figure 2 (c). However, when Rule I is applied at R1 (because there is a reversible path from R1 to a CPO, PO2), we can stop back-tracing from R1 by means of including the reversible path which contains a single statement"PO2=~R1" to our error space. This is because that R1 is on a reversible path to a CPO. If R1 is incorrect, there must exist errors in this reversible path. The reduced error space is shown in the lower part of Figure 2 (c). We replace the statement "R1=A+B", BTS(A), and BTS(B) with the reversible statement "PO2=~R1". These statements are shown in bold in Figure 2 (c), and we find that the reduced error space is indeed smaller than the original one.

Lemma 2

Assume a statement whose simulation value of LV is incorrect and the statement is correct, there exists a set of values for the RVs that can produce the correct value of LV.

Lemma 3

Given a statement whose erroneous simulation value of LV and correct value of LV are known, if this statement and the simulation value of one target RV are both correct, we can find a set of values for other RVs to produce the correct value of LV while fixing the value of the target one.

Theorem 2

Given a statement whose erroneous simulation value of LV and correct value of LV are known, if there does not exist any set of values for other RVs to produce the correct value of LV while fixing the value of the target RV, the statement is incorrect or the simulation value of the target RV is incorrect.

Rule II

During the back-tracing procedures, when the correct value of LV is known and there does not exist any values of other RVs to produce the correct value of LV while assuming the value of the target RV is correct, according to Theorem 2, the statement is erroneous or the target RV must be incorrect.

Therefore, we can include this statement into the error space and only back-trace at the target RV to guarantee that at least one erroneous statement is included into the error space.

HDL code:
PO=A*B;

The simulation values:
(PO,A,B)=(100,25,4)

case1: If the correct values of PO is 50
➔ stop back-tracing from A
case2: If the correct values of PO is 99
➔ stop back-tracing from A or B

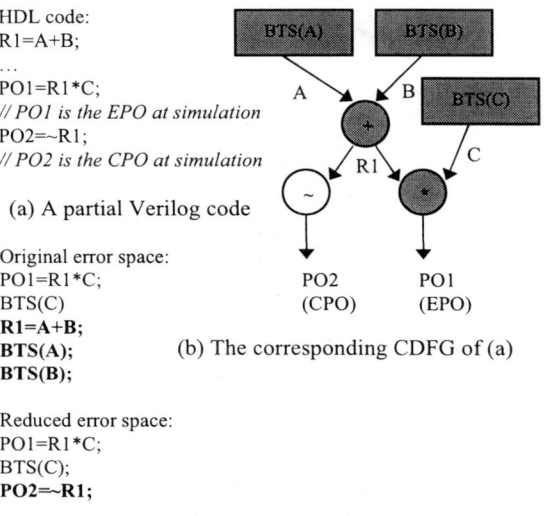

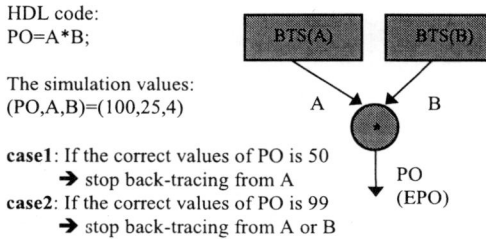

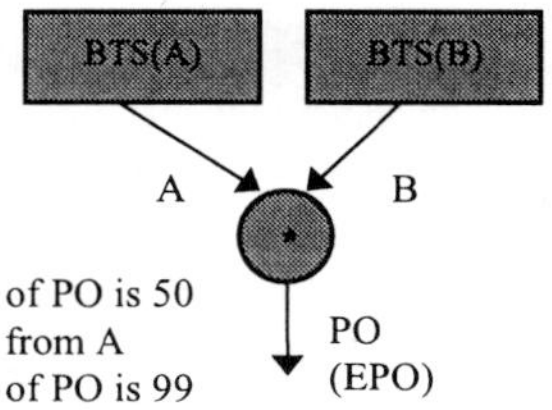

Figure 3. An example of applying Rule II

Figure 3 shows an example of applying Rule II. In case1, the correct value of PO is 50. If we assume the simulation value of B is correct, which is 4, the value of A should be 12.5 to make PO correct. But the value of A cannot be a floating point. Therefore, the assumption that B is correct is contradictory. Thus, by applying Rule II, we include only the statement "PO=A*B" into the error space and back-trace from variable B only. In case2, the correct value of PO is 99. Both variable A and B can be used to apply Rule II. We can only choose either one of them to back-trace from.

It is clear that applying Rule II requires a statement with multiple RVs. Sometimes the first statement back-traced from the EPO has only one RV. Because we only have the expected value of the EPO, it seems that Rule II cannot be used in this kind of situation. However, a statement with only one RV is also a reversible statement. We can get the expected value of its RV by reverse calculation from the value of the LV. The expected value got by reverse calculation may be erroneous only when the reversible statement itself is erroneous. This problem can be solved by including the reversible statement into the error space. Rule II often terminates the back-tracing operations near EPO, so it can reduce the size of error space significantly.

When the expected value of LV is not given, Rule II cannot be applied. At this time, we proposed Rule III that exploits the controlling value idea to terminate the back-tracing procedure. The controlling value of a gate is the value that, if presents at any of the inputs, determines the output value. When the simulation value of one RV is a controlling value of this statement, the other RVs cannot affect the value of the LV. Thus, we can stop back-tracing from these RVs.

Rule III

When the simulation value of one RV is a controlling value of the statement, we can stop back-tracing from the other RVs and still retain at least one erroneous statement in the error space.

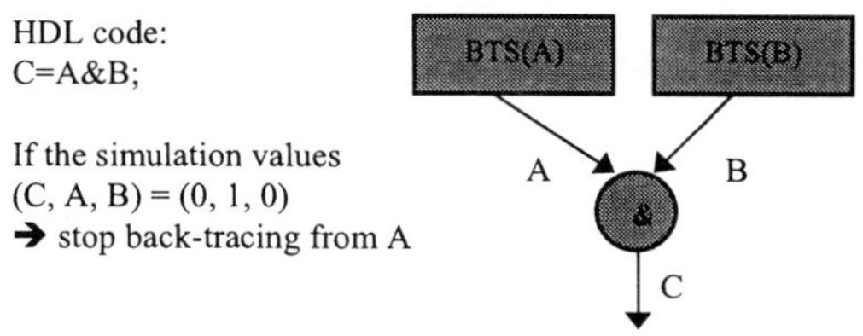

Figure 4. An example of applying Rule III

Figure 4 shows an example of applying Rule III. In this case, variable A cannot change the value of variable C because the value of B is a controlling value of the statement. We can stop back-tracing from A and still guarantee there is at least one error in the error space.

4. EXPERIMENTAL RESULTS

The experiments are conducted over four real sequential designs written in Verilog HDL. The design Matrix is a design for 2x2 matrix multiplication. The design FSM is a finite state machine that is used to control traffic lights. The design DCT is a circuit for discrete cosine transform. The design PCPU is a simple 32-bit pipelined CPU.

Design / Error Insertion	Lines	Error Space Avg./Min./Max.	Ratio
Matrix/Single	80	11.1 / 7 / 15	13.88%
Matrix/Multiple		10.7 / 6 / 13	13.38%
FSM/Single	113	16.6 / 8 / 24	14.69%
FSM/Multiple		15.6 / 10 / 19	13.81%
DCT/Single	713	14.8/ 8 / 34	2.08%
DCT/Multiple		13.1/ 8 / 31	1.84%
PCPU/Single	952	49.1 / 10 / 86	5.16%
PCPU/Multiple		38.4 / 10 / 74	4.03%

Table 1. Experimental results

In our experiments, we inject errors into a design by modifying correct statements. As we insert errors into a correct design, the design becomes an erroneous implementation. We insert one error for single-error and 5 errors for multiple-error experiments. Then, we obtain an erroneous test case for it by simulation. Using the information of this test case, the error space of this erroneous implementation is obtained by our approach. Table 1 summarizes the experimental results. The column "Lines" gives the number of lines in the HDL design. The average, minimum, and maximum lines of error space are reported in column "Error Space". In column "Ratio", we compute the ratio of average lines in the error space to the total lines of the corresponding design.

The experimental results demonstrate that the error candidates are indeed reduced as compared with the total lines of the design by our approach. Thus, the designers can easily locate the error source with a small set of error candidates and therefore, accelerate the diagnosis stage of design flow. As can be seen from this table, the average error space of each multiple-error design is slightly as smaller compared with its corresponding single-error design. It is because the PO mismatch of a multiple-error design often occurs earlier than that of a single error design, and the back-tracing operations within fewer clock cycles may produce smaller error space. These experimental results show that our approach can handle multiple-error designs well.

5. CONCLUSIONS

In this paper, we propose an efficient approach for design error diagnosis in HDL. Our approach is based on the proposed three rules to shrink the error space with retaining at least one error source in it. The approach is not only efficient for single-error designs, but also for multiple-error designs with requiring only one erroneous test case. By conducting some experiments on real designs, it shows that the error space is indeed reduced.

6. REFERENCES

[1] S. Y. Kuo, "Locating Logic Design Errors via Test Generation and Don't-Care Propagation", European Design Automation Conference, Sept. 1992.

[2] P. Y. Chung, Y. M. Wang, and I. N. Hajj, "Logic Design Error Diagnosis and Correction", IEEE Transaction on VLSI System, vol.2, no.3, Sept. 1994.

[3] S. Y. Huang, K. T. Cheng, K. C. Chen, and D. I. Cheng, "Error Tracer: A Fault Simulation-Based Approach to Design Error Diagnosis", International Test Conference, Nov. 1997.

[4] A. Gupta and P. Ashar, "Fast Error Diagnosis for Combinational Verification", the Thirteenth International Conference on VLSI Design, 2000.

[5] M. Fujita, "Methods for Automatic Design Error Correction in Sequential Circuits", European Conference on Design Automation, 1993.

[6] S. Y. Huang, K. T. Cheng, K. C. Chen, and J. Y. Lu, "Fault-Simulation Based Design Error Diagnosis for Sequential Circuits", Design Automation Conference, Jun. 1998.

[7] V. Boppana, I. Ghosh, R. Mukherjee, J. Jain, and M. Fujita, "Hierarchical Error Diagnosis Targeting RTL Circuits", International Conference on VLSI Design, 2000.

[8] M. Khalil, Y. L. Traon, and C. Robach, "Towards an Automatic Diagnosis for High-level Design Validation", International Test Conference, Oct. 1998.

[9] T. Y. Jiang, C. N. Liu, and J. Y. Jou, "Effective Error Diagnosis for RTL Designs in HDLs", the Eleventh Asian Test Symposium, 2002.

Yield Optimization with Correlated Design Parameters and Non-Symmetrical Marginal Distributions

K. Ponnambalam, A. Seifi** and J. Vlach****

*Department of Systems Design Engineering, University of Waterloo, Waterloo, Canada, N2L 3G1
** Department of Industrial Engineering, Amirkabir University of Technology, Tehran 15914, Iran
***Department of Electrical and Computer Engineering, University of Waterloo, Waterloo, Canada, N2L 3G1

ABSTRACT

This paper extends the recently developed hybrid method to find the optimal designs of systems with correlated non-gaussian random parameters. A double-bounded density function is used to approximate marginal distribution and a Frank copula is used to define dependence (a more general concept than correlation) among the random parameters. We use a Piecewise Ellipsoidal method to approximate the constraint region by a set of quadratic functions. The yield is estimated by a joint cumulative density function over a portion of the tolerance body contained in the feasible region. Yield maximization is done for positive and negative correlations and non-symmetrical marginal distributions, and tested on an example using Monte-Carlo simulation.

1. INTRODUCTION

The method combines a constraint region approximation with yield optimization. A Piecewise Ellipsoidal Approximation (PEA) is first found for the constraint region. It is a set of quadratic functions that can work with non-convex situations and cases where no analytical functions are available. Yield maximization is based on this approximation and avoids evaluations of the original functions. Components may have correlated non-symmetrical distributions in addition to Gaussian correlated parameters, a case that has been well studied.

The tolerance body associated with any joint probability density function (PDF) is approximated by a rectangular hypercube. The objective is to maximize yield by shifting and re-scaling the tolerance body [1]. The associated optimization can determine, given the joint PDF of the components, the lower and upper bounds of the random parameters which is the optimal range. If the size of the optimal range is greater than or equal to the given tolerance range, the yield will be 100 %. Otherwise the optimal range will include the higher yield portion of the given tolerance range, which is the best a designer can do

Earlier in [2,3,5], we have introduced the use of a double-bounded probability density function (DB-PDF) that can approximate almost any bounded probability density function (assuming independence) Here, for the first time, we introduce the extension of previous methods to the case of correlated (dependent) design parameters which is needed in the design of integrated circuits..

2. METHODOLOGY

The set of design specifications can be written as [4]:

$$h(x,\omega) \leq 0 , \qquad (1)$$

where h is the vector of constraints, x is the vector of design parameters, and ω is an independent continuous variable (such as time, frequency, temperature, etc). The set of feasible design points in (1) is called the constraint region, R_h, that may not be explicitly known. The aim of the deterministic design is to find a point in R_h which meets all the specifications. The aim of yield optimization is to estimate the optimal range described by the n-dimensional hypercube R_Y, explained later. The PEA method approximates R_h by quadratic ellipsoidal functions; see [2, 7] for details. Once the ellipsoidal approximations have been found, checking specifications reduces to the evaluation of these functions. Since inscribing an n-dimensional cube R_Y in a polytope is easier than fitting it in a region defined by quadratic functions, The AFOSM method (discussed later) is used to construct this polyhedral approximation.

In summary, yield maximization involves the following steps:

- Quadratic approximation of the constraint region using PEA.
- Polyhedral approximation of the quadratic region using AFOSM and adding extra bounds on the design variables if necessary.
- Modeling the joint cumulative distribution using a Frank Copula [6], explained later, and the marginal distributions by DBPDFs.
- Yield calculation over the optimal box defined by R_Y
- Yield optimization by maximizing the volume of R_Y.

3. POLYHEDRAL APPROXIMATION

A polyhedral approximation of the constraint region uses an Advanced First-Order Second Moment (AFOSM) method [1,8]. This method can be applied either to the approximated

0-7803-7761-3/03/$17.00 ©2003 IEEE

region by the PEA or to the original constraints. It takes first-order approximation of each $h_i(x)$ at x^*,

$$h_i(x) \approx h_i(x^*) + g_i(x^*)^T (x - x^*) \qquad (2)$$

where $g_i^* = g_i(x^*) = \dfrac{\partial h_i(x^*)}{\partial x}$ is the gradient vector of $h_i(x)$. The point x^* is on the surface of $h_i(x)=0$ and has the minimal distance from the nominal value of parameters. This is the original AFOSM method which is difficult to solve when $h(x)$ is a non-smooth function. In the hybrid method using PEA's quadratic function, the problem of polyhedral approximation was made considerably easier. Let the resulting polytope be called P.

4. MODELING ARBITRARY MARGINAL DISTRIBUTIONS

For each component we consider a double-bounded probability density function (DB-PDF) of the following form as the marginal distribution,

$$f(z) = abz^{a-1}(1-z^a)^{b-1} \qquad (3)$$

$$z = \frac{x - x^{\min}}{t}, \quad x^{\min} \le x \le x^{\min} + t \qquad (4)$$

where t is the range of variability [3]. Depending on the choice of parameters a and b, the DB-PDF can take various shapes. It can approximate uniform, triangular, tail or almost any unimodal distribution. It can also reproduce the results of a truncated Gaussian or Beta distributions. The function is applicable to any situation where a truncated distribution can be used as an approximation. A significant advantage of the DB-PDF is that its CDF is available in closed form, which will be used in the calculation of yield function.

$$F(z) = 1 - (1 - z^a)^b \qquad (5)$$

5. DEFINING DEPENDENCE STRUCTURE WITH COPULAS

For simplicity, consider only two random variables x and y. Given the joint distribution $H(x,y)$ and the marginal distributions $F(x)$ and $G(y)$, a copula can be defined as:

$$C(u,v) = H(F^{-1}(u), G^{-1}(v)), \qquad (6)$$

which is a bivariate cumulative distribution function (CDF) with Uniform (0,1) marginal distributions [6]. On the other hand, given a copula $C(u,v)$, the function

$$H(x,y) = C(F(x), G(y)), \qquad (7)$$

is a multivariate CDF with arbitrary marginal distributions of $F(x)$ and $G(y)$.

To emphasize that a copula does not depend on the marginal distributions, its arguments are written as (u,v) or $(F(x), G(y))$. There are a large number of choices for copulas. We choose the Frank copula, which has a rather simple form and a single parameter to express the dependence. It can provide full range of positive and negative correlation (explained later) as given by:

$$C(u,v) = -\frac{1}{\alpha}\ln\left[1 + \frac{(e^{-\alpha u}-1)(e^{-\alpha v}-1)}{e^{-\alpha}-1}\right]. \qquad (8)$$

Approximating the marginal distributions by DBPDFs and having assumed the Frank Copula and the dependence parameter α, we can generate correlated (dependent) non-gaussian random parameters to be used both in design and simulation. The simulation however is used only for the verification of the yield.

5.1 Measuring Dependence

In general, the ordinary correlation coefficient (Pearson's product-moment correlation) is a measure of linear dependence (association) between (X,Y) and is not invariant under monotonic transformation of the marginal distributions of X and Y. Kendall's τ and the grade correlation ρ_g are the best-known rank correlation coefficients that measure the correlation between rankings, rather than between the actual values of X and Y, and hence, they are invariant under certain transformations.

Spearman's ρ_S is used for the sample rank correlation and grade correlation ρ_g is used for the population:

$$\rho_g = 12 \iint [C(u,v) - uv] \cdot du \cdot dv \qquad (9)$$

In a manufacturing system, either ρ_g can be pre-specified or ρ_S can be measured. There exists a unique function that maps this correlation to the dependence parameter α.

6. YIELD OPTIMIZATION

We will use x^{min} as an optimization variable while the range t will remain constant. Let

$$R_T(x^{\min}, t) = \{x \in R^n \mid x^{\min} \le x \le x^{\min} + t\} \qquad (10)$$

be the tolerance body. If R_T is inscribed in the constraint region, then we have the worst-case design. If not, then the optimization attempts to overlay the maximum yield part of R_T with the constraint region. This can be done by searching for the maximum yield rectangular n-dimensional cube

$$R_Y(x^{li}, x^{ui}) = \{x \in R^n \mid x^{li} \le x \le x^{ui}\} \qquad (11)$$

so that it is contained in P, *i.e.*, $R_Y \subseteq R_T$. The li and ui, respectively, indicate the lower and upper bounds of the contained box and are also optimization variables. In two-dimensional space, the yield is given by the copula as:

$$Yield(x^{\min}, x^{li}, x^{ui}) = C(u^{ui}, v^{ui}) + C(u^{li}, v^{ui}) - C(u^{li}, v^{ui}) - C(u^{ui}, v^{li}) \qquad (12)$$

where $u = F(x) = 1 - \left[1 - \left(\frac{x - x_{\min}}{t_x}\right)^a\right]^b$ and

$$v = G(y) = 1 - \left[1 - \left(\frac{y - y_{\min}}{t_y}\right)^a\right]^b.$$

The variable x may take the values li or ui and similarly for the variable y. The yield is maximized, subject to the containment condition described above.

7. EXAMPLE

The example, taken from [9], has two design variables and describes a non-convex constraint region by three nonlinear equations. The constraints are:

$$h_1(x_1, x_2) = \frac{\exp(-x_1 + 1)}{[(x_2 - 1)^2 + 1]} - 2$$

$$h_2(x_1, x_2) = \exp(x_1 - 2x_2 + 1) - 2$$

$$h_3(x_1, x_2) = x_1^2 + x_2^2 - 3$$

The final design and the simulated random values are shown in Figures 1 to 3. The Figures shows the following: (i) the original constraint region, (ii) the PEA approximation, (iii) the resulting designs defined by R_Y the inside rectangle, (iv) the tolerance box defined by R_T, and lastly, (v) the Monte-Carlo simulation of the design parameters. Figure 1 corresponds to the case of high positive dependence, Figure 2 to the case of high negative dependence, and Figure 3 shows the case of nearly independent variables. Let us stress the difference of our simulation results with the usual gaussian case where the simulation would have formed an ellipsoid [1]. Here, the marginal distributions are of two kinds: one is a non-symmetric triangular and the second is a truncated gaussian-like. Table 1 presents in the first three rows the results for the three cases of positive, negative dependence and the nearly independent corresponding to Figures 1 to 3. Yield can be increased by redesigning with a smaller value of t in Equation (10), hence with a lower tolerance, as shown in Figure 4 whose results are presented in the fourth row of Table 1. It is interesting to note the corresponding change in the nominal design in rows 1 and 4.

Table 1. Results corresponding to Figures 1 to 4.

Spearman Correlation	Yield from Monte-Carlo	Nominal Design	Tolerance %
0.94	68%	[1.25 1.02]	[94 104]
-0.99	61%	[0 .93 1.22]	[109 95]
-0.02	53%	[1.41 0.96]	[88 107]
0.94	96%	[0.85 0.99]	[79 72]

8. SUMMARY

This paper presents an extension of our previous work [1,5] to the case of dependent random variables with non-symmetrical distributions. A hybrid method is used that combines two methods of PEA and AFOSM for yield maximization, especially in situations where there is a possibility for correlation in the random design parameters. Frank Copula specified the dependence structure. Marginal distributions were modeled by DBPDFs, a versatile function with a closed form for its CDF. The design that results considering correlation might be different from the design considering independence, although a general conclusion cannot be drawn due to problem uniqueness.

9. REFERENCES

[1] A. Seifi, K. Ponnambalam and J. Vlach: "A unified approach to statistical design centering of integrated circuits with correlated parameters", IEEE Transactions on Circuits and Systems I, Vol. 46, No. 1, 1999, pp. 190-196.

[2] Ponnambalam, K., J. Wojciechowski, J. Vlach, 2001. A Hybrid Method for Optimal Design with General Distributions, Proc. of European Conference on Circuit Theory and Design, ECCTD01, Volume 3, 385-388, Helsinki, Finland.

[3] P. Kumaraswamy, "A generalized probability density function for double-bounded random processes", Journal of Hydrology, Vol. 46, 1980, pp.79-88.

[4] J.W. Bandler and S.A. Chen, "Circuit optimization: The state of the art, " IEEE Trans. Microwave Th. Tech., vol. MIT-36, 1988, pp.424-442.

[5] Ponnambalam, K., A. Seifi, and J. Vlach, 2001. Probabilistic design of systems with general distributions of parameters, *Intl' J. of Circuit Theory and Appl.*, 29(6), 527-536.

[6] Nelsen, R. B. (1999), "An introduction to copulas," Lecture Notes in Statistics, Springer-Verlag, New York.

[7] J.M. Wojciechowski and J. Vlach, "Ellipsoidal method for design centering and yield estimation," IEEE Trans. Computer-Aided Design, Vol. 12, 1993, pp. 1570-1579.

[8] A. Seifi, K. Ponnambalam and J. Vlach: Probabilistic design of integrated circuits with correlated input parameters, IEEE Transactions on Computer-Aided Design of Integrated Circuits and Systems, Vol. 18, No. 8, 1999, pp. 1214-1219.

[9] H. Schajer-Jacobsen and K. Madsen, "Algorithms for worst-case tolerance optimization", IEEE Transactions on Computer-Aided Design of Integrated Circuits and Systems, Vol. 26, 1979, pp 775-783.

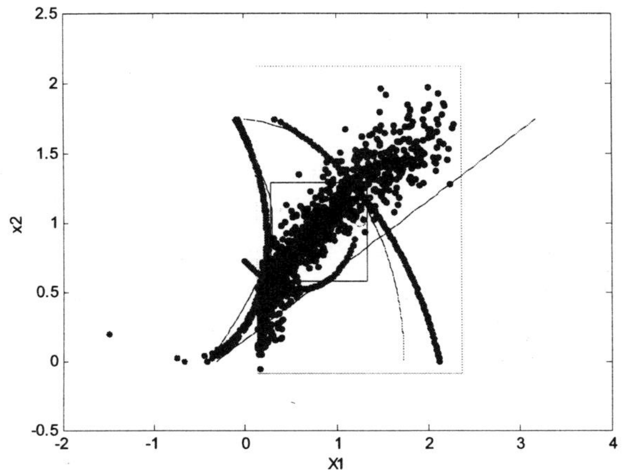

Figure 1. Design for positive dependence

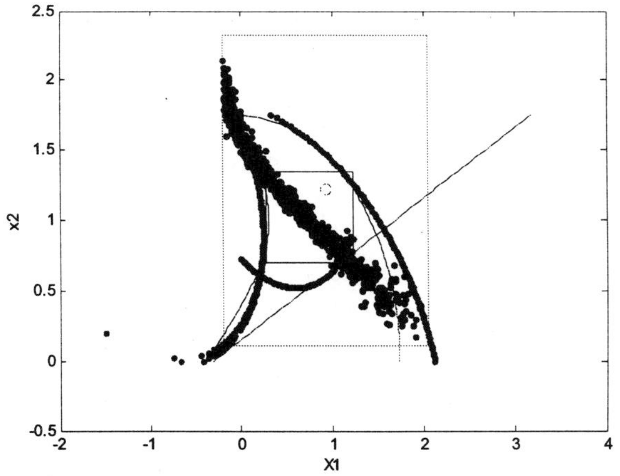

Figure 2. Design for negative dependence

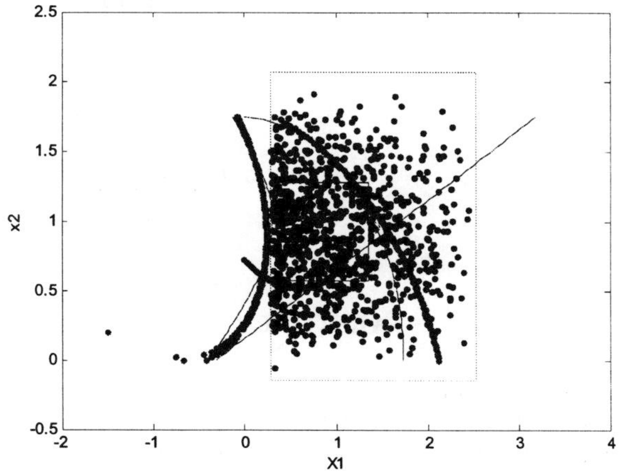

Figure 3. Design for nearly independent variables

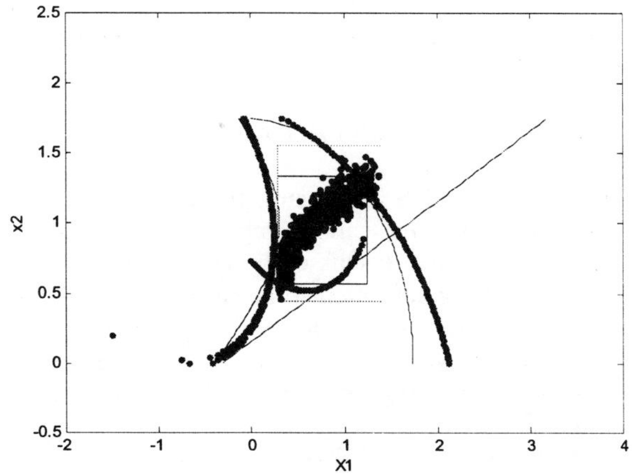

Figure 4. Case 1 (positive dependence) with a reduced range of variability t in Equation (10)

GENERATION AND PROPERTIES OF FASTEST TRANSFORM MATRICES OVER GF(2)

Bogdan J. Falkowski and Cicilia C. Lozano

School of Electrical and Electronic Engineering
Block S1, Nanyang Technological University
50 Nanyang Avenue, Singapore 639798

ABSTRACT

Linearly Independent (LI) transforms in Galois Field (2) algebra that have fastest transform calculation have been investigated recently. It was found that there are some LI transforms having smaller computational cost than the Reed-Muller transform, which was previously known as the most efficient transform over GF(2). This paper discusses various properties of these fastest LI transforms as well as some experimental results for them using standard benchmark functions and their comparison with generalized Reed-Muller transform.

1. INTRODUCTION

Logic synthesis and testing using the Reed-Muller based spectral expansions that can be implemented by AND and EXOR gates has been discussed by many authors [1, 3, 6, 7]. They are of great interest because the Reed-Muller based circuits have many desirable properties, such as simple testability vectors. They can also be implemented easily using Field Programmable Gate Arrays and other programmable logic devices. For many important functions having many EXOR gates such as adders and multipliers, the Reed-Muller implementations are much more efficient than inclusive circuits.

There exist many methods that can be used to obtain the Reed-Muller expansion for a given Boolean function. The most efficient ones operate on reduced representations of Boolean functions, such as set of cubes and decision diagrams [6, 7]. The Reed-Muller transform itself is only one type of big family of Linearly Independent (LI) transforms presented in [2]. Further research showed that there are fastest LI transforms having the smallest computational cost in terms of additions that is lower than that of the Reed-Muller transform [5]. In this paper, properties of these fastest LI transforms are investigated and some experimental results for them and generalized Reed-Muller transform for standard benchmark functions are also discussed.

2. BASIC DEFINITIONS AND PROPERTIES

The following definitions are valid for generalized Reed-Muller transform and all LI transforms.

Definition 1: An LI transformation matrix M_n is a 2^n x 2^n matrix with its columns corresponding to truth vectors of some n-variable binary functions that are linearly independent when they are bit-by-bit XOR-ed.

Definition 2: A spectral coefficient vector $\vec{A} = [A_0, A_1, \dots, A_{2^n-1}]^T$ can be obtained by equation $\vec{A} = M_n^{-1}\vec{F}$. $\vec{F}$ can be obtained back from $\vec{A}$ by equation $\vec{F} = M_n\vec{A}$, where M_n^{-1} is the inverse of corresponding LI transform matrix M_n in GF(2) and $\vec{F}$ is the truth vector of the binary function *f(x)*.

Definition 3: An n-variable Boolean function can be expressed as a polynomial expansion over GF(2) $f(x) = A_0g_0 \oplus A_1g_1 \oplus A_2g_2 \oplus \dots \oplus A_{2^n-1}g_{2^n-1}$ where g_i represents an n-variable switching function whose truth vector is given in column i of corresponding transform matrix M_n and A_i represents i-th spectral coefficient.

Definition 4: The generalized Reed-Muller transform matrix RM of order $N=2^n$ in polarity ω for an n-variable Boolean function is given by:

$$RM_N^{\omega} = \bigotimes_{j=1}^{n} rm_{\omega_j} = rm_{\omega_n} \otimes rm_{\omega_{n-1}} \otimes \dots \otimes rm_{\omega_2} \otimes rm_{\omega_1}$$

where '⊗' denotes Kronecker product operation [3, 6, 7], $rm_0 = \begin{bmatrix} 1 & 0 \\ 1 & 1 \end{bmatrix}$ and $rm_1 = \begin{bmatrix} 0 & 1 \\ 1 & 1 \end{bmatrix}$.

The polarity number ω is the decimal equivalent of $\omega_n\omega_{n-1}\dots\omega_2\omega_1$ where ω_j ($1 \le j \le n$) is either 0 or 1 according to whether the j-th variable appears in positive or negative form throughout the polynomial expression.

Definition 5 [5]: The fastest transform matrices M_n and M_n^* can be constructed recursively from smaller transform matrices by using equation (1) and equation (2), respectively.

$$M_n = \begin{bmatrix} M_{n-1} O_{n-1} \\ Y_{n-1} M_{n-1} \end{bmatrix} \quad (1)$$

$$M_n^* = \begin{bmatrix} M_{n-1}^* Y_{n-1} \\ O_{n-1} M_{n-1}^* \end{bmatrix} \quad (2)$$

0-7803-7761-3/03/$17.00

In the above equations, M_{n-1} and M^{*}_{n-1} are transform matrices of size 2^n-1 x 2^n-1, O_{n-1} is zero matrix of size 2^n-1 x 2^n-1 and Y_{n-1} is a 2^n-1 x 2^n-1 matrix with all its elements equal to zero but one element at the corner (bottom left corner for equation 1, top right corner for equation 2) that is equal to 1.

Definition 6: Due to the recursive definitions, the fastest transform matrices of order $N=2^n$ can be obtained by operations over GF(2) of n factorized transform matrices.

Example 1:

Let us generate matrices M_3 and M_3^{*} from factorized transform matrices. For n=3, $M_3 = A_3\ B_3\ C_3$ where:

$$A_1 = \begin{bmatrix} 1 & 0 \\ 1 & 1 \end{bmatrix},\quad A_n = I_{n-1} \otimes A_1,$$

C_n is a modified identity matrix of size 2^n x 2^n with bottom left corner element replaced by 1,

and $B_n = I_1 \otimes C_{n-1}$ where I_n denotes identity matrix of size 2^n x 2^n and the symbol '$\otimes$' is Kronecker product.

Thus:

$$A_3 = \begin{bmatrix} 1&0&0&0&0&0&0&0 \\ 1&1&0&0&0&0&0&0 \\ 0&0&1&0&0&0&0&0 \\ 0&0&1&1&0&0&0&0 \\ 0&0&0&0&1&0&0&0 \\ 0&0&0&0&1&1&0&0 \\ 0&0&0&0&0&0&1&0 \\ 0&0&0&0&0&0&1&1 \end{bmatrix} \quad B_3 = \begin{bmatrix} 1&0&0&0&0&0&0&0 \\ 0&1&0&0&0&0&0&0 \\ 0&0&1&0&0&0&0&0 \\ 1&0&0&1&0&0&0&0 \\ 0&0&0&0&1&0&0&0 \\ 0&0&0&0&0&1&0&0 \\ 0&0&0&0&0&0&1&0 \\ 0&0&0&0&1&0&0&1 \end{bmatrix}$$

$$C_3 = \begin{bmatrix} 1&0&0&0&0&0&0&0 \\ 0&1&0&0&0&0&0&0 \\ 0&0&1&0&0&0&0&0 \\ 0&0&0&1&0&0&0&0 \\ 0&0&0&0&1&0&0&0 \\ 0&0&0&0&0&1&0&0 \\ 0&0&0&0&0&0&1&0 \\ 1&0&0&0&0&0&0&1 \end{bmatrix} \quad M_3 = \begin{bmatrix} 1&0&0&0&0&0&0&0 \\ 1&1&0&0&0&0&0&0 \\ 0&0&1&0&0&0&0&0 \\ 1&0&1&1&0&0&0&0 \\ 0&0&0&0&1&0&0&0 \\ 0&0&0&0&1&1&0&0 \\ 0&0&0&0&0&0&1&0 \\ 1&0&0&0&1&0&1&1 \end{bmatrix}$$

Similarly, M_3^{*} can be obtained as $M_3^{*} = A_3^{*}\ B_3^{*}\ C_3^{*}$ where:

$$A_1^{*} = \begin{bmatrix} 1 & 1 \\ 0 & 1 \end{bmatrix},\quad A_n^{*} = I_{n-1} \otimes A_1^{*},$$

C_n^{*} is a modified identity matrix of size 2^n x 2^n with top right corner element replaced by 1,

and $B_n^{*} = I_1 \otimes C_{n-1}^{*}$.

It follows that:

$$A_3^{*} = \begin{bmatrix} 1&1&0&0&0&0&0&0 \\ 0&1&0&0&0&0&0&0 \\ 0&0&1&1&0&0&0&0 \\ 0&0&0&1&0&0&0&0 \\ 0&0&0&0&1&1&0&0 \\ 0&0&0&0&0&1&0&0 \\ 0&0&0&0&0&0&1&1 \\ 0&0&0&0&0&0&0&1 \end{bmatrix} \quad B_3^{*} = \begin{bmatrix} 1&0&0&1&0&0&0&0 \\ 0&1&0&0&0&0&0&0 \\ 0&0&1&0&0&0&0&0 \\ 0&0&0&1&0&0&0&0 \\ 0&0&0&0&1&0&0&1 \\ 0&0&0&0&0&1&0&0 \\ 0&0&0&0&0&0&1&0 \\ 0&0&0&0&0&0&0&1 \end{bmatrix}$$

$$C_3^{*} = \begin{bmatrix} 1&0&0&0&0&0&0&1 \\ 0&1&0&0&0&0&0&0 \\ 0&0&1&0&0&0&0&0 \\ 0&0&0&1&0&0&0&0 \\ 0&0&0&0&1&0&0&0 \\ 0&0&0&0&0&1&0&0 \\ 0&0&0&0&0&0&1&0 \\ 0&0&0&0&0&0&0&1 \end{bmatrix} \quad M_3^{*} = \begin{bmatrix} 1&1&0&1&0&0&0&1 \\ 0&1&0&0&0&0&0&0 \\ 0&0&1&1&0&0&0&0 \\ 0&0&0&1&0&0&0&0 \\ 0&0&0&0&1&1&0&1 \\ 0&0&0&0&0&1&0&0 \\ 0&0&0&0&0&0&1&1 \\ 0&0&0&0&0&0&0&1 \end{bmatrix}$$

Let us show in Figures 1 to 3 the butterfly diagrams of size 2^3 for the calculation of Fast Reed-Muller transform for polarity 000 as well as for M_3 and M_3^{*}. All the operations on the butterfly diagrams are performed over GF (2). It can be easily noticed that the shape of Fast Reed-Muller transform for polarity 000 corresponds to the shape of fast M_3 transform and that Fast Reed-Muller transform requires more calculation than the fastest transforms.

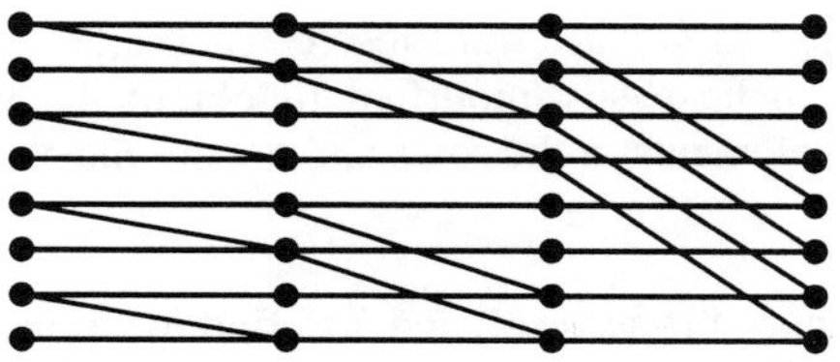

Figure 1. Butterfly diagram for RM_8^0.

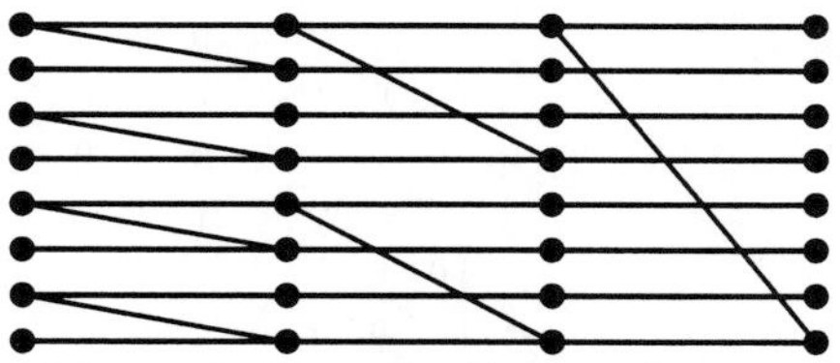

Figure 2. Butterfly diagram for M_3.

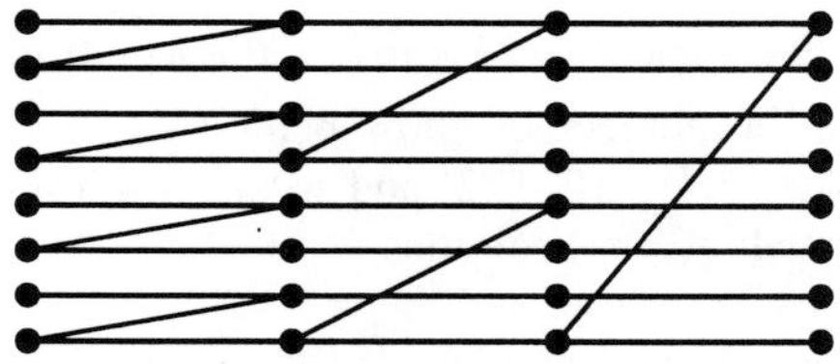

Figure 3. Butterfly diagram for M_3^{*}.

Property 1 [5]: Both M_n and M_n^* are self-inversed. That is $M_n^{-1} = M_n$ and ${M_n^*}^{-1} = M_n^*$.

Property 2: Total number of '1' elements in both M_n and M_n^* is equal to 2^{n+1}-1.

In order to represent the following properties, let M_n and M_n^* are in the form

$$\begin{bmatrix} M_{0,0} & M_{0,1} & \cdots & M_{0,2^n-1} \\ M_{1,0} & \ddots & & M_{1,2^n-1} \\ \vdots & & \ddots & \vdots \\ M_{2^n-1,0} & \cdots & \cdots & M_{2^n-1,2^n-1} \end{bmatrix}$$

Property 3: Row i of M_n^* has 1+d '1' elements, where d is the number of zeros behind the last one in the binary representation of i.

If D is defined as {0, 1, 2,..., d-1} then the '1's inside the row i are located at the columns i, $i+2^0$, $i+2^0+2^1$, $i+2^0+2^1+2^2$, ..., $i+2^0+2^1+2^2+\ldots+2^{d-1}$.

Property 4: Column i of M_n^* has 1+h '1' elements, where h is the number of zeros behind the last one in the binary representation of $(i+1)$.

If H is defined as {0, 1, 2,..., h} then the '1's inside the column i are located at the rows $i-(2^0-1)$, $i-(2^1-1)$, $i-(2^2-1)$, ..., $i-(2^h-1)$.

Since M_n and M_n^* are transpose of each other, Properties 3 and 4 can be easily modified to obtain the positions of nonzero elements in the rows and the columns for M_n.

Example 2:

Let us apply Properties 2 and 3 to find locations of nonzero elements in M_3^*.

$$M_3^* = \begin{bmatrix} 1 & 1 & 0 & 1 & 0 & 0 & 0 & 1 \\ 0 & 1 & 0 & 0 & 0 & 0 & 0 & 0 \\ 0 & 0 & 1 & 1 & 0 & 0 & 0 & 0 \\ 0 & 0 & 0 & 1 & 0 & 0 & 0 & 0 \\ 0 & 0 & 0 & 0 & 1 & 1 & 0 & 1 \\ 0 & 0 & 0 & 0 & 0 & 1 & 0 & 0 \\ 0 & 0 & 0 & 0 & 0 & 0 & 1 & 1 \\ 0 & 0 & 0 & 0 & 0 & 0 & 0 & 1 \end{bmatrix}$$

For M_3^* the number of '1's inside the row 0 is equal to 4, which is the number of zeros behind the last one in 000 (d=3) plus 1. D = (0, 1, 2) and so the '1's inside the row 0 are located at the columns 0, 0+1=1, 0+1+2 =3 and 0+1+2+4 =7.

Number of '1' elements inside the column 3 is equal to 3, which is the number of zeros behind the last one in 100 (h=2) plus 1. H = (0, 1, 2) and the '1's in the column 3 are located at the rows 3 – (1– 1) = 3, 3 – (2 – 1) = 2, and 3 – (4 – 1) = 0.

3. EXPERIMENTAL RESULTS

Table 1. Number of nonzero spectral coefficients for Reed-Muller transform for some benchmark functions.

Input files	Number of nonzero coefficients for best polarity	Number of nonzero coefficients for polarity 00…0	Number of nonzero coefficients for polarity 11…1
xor5	5	5	6
squar5	23	20	30
rd53	20	23	21
bw	22	32	22
con1	17	19	24
5xp1	61	61	72
z5xp1	61	61	72
rd73	63	63	64
inc	49	91	57
rd84	107	107	256
misex1	20	60	20
ex5	113	256	151
9sym	173	210	210
z9sym	173	210	210
clip	206	217	314
apex4	445	445	511
sao2	100	1022	242
ex1010	1010	1023	1021

Table 2. Number of nonzero spectral coefficients for two fastest transforms for some benchmark functions.

Input files	Nonzero coefficients (M_n)	Nonzero coefficients (M_n^*)
xor5	19	20
squar5	30	31
rd53	28	30
bw	26	22
con1	96	91
5xp1	128	128
z5xp1	128	128
rd73	118	121
inc	88	85
rd84	240	246
misex1	99	99
ex5	240	223
9sym	302	302
z9sym	302	302
clip	487	486
apex4	490	488
sao2	874	511
ex1010	895	897

The calculation of spectral coefficients for two fastest transforms and generalized Reed-Muller transform was programmed using C language and run on Pentium III 500 MHz computer with 128 MB RAM.

The total number of nonzero Reed-Muller spectral coefficients in polarity 00…0, polarity 11…1 and polarity having smallest number of nonzero coefficients for MCNC benchmark functions are listed in Table 1 while the total numbers of nonzero spectral coefficients for the fastest transforms using the same benchmarks are given in Table 2. From the results in both tables, it can be seen that the fastest transforms give significantly smaller number of nonzero spectral coefficients for some benchmarks.

Table 3. Execution time to calculate spectral coefficients for polarity zero and all 2^n polarities of Reed-Muller and two fastest transforms for some benchmarks.

Input files	Execution time for polarity zero RM	Execution time for all polarities RM	Execution time for M_n	Execution time for M_n^*
xor5	0	0.06	0	0
squar5	0	0.28	0	0
rd53	0	0.11	0	0
bw	0.06	0.99	0.05	0
con1	0	0.61	0	0
5xp1	0	3.02	0	0
z5xp1	0.06	2.91	0.06	0.05
rd73	0.06	0.99	0	0
inc	0.05	2.64	0	0
rd84	0.06	3.90	0	0.06
misex1	0.05	6.59	0.06	0
ex5	0.38	57.78	0.44	0.38
9sym	0.05	3.95	0.06	0.06
z9sym	0	3.74	0	0
clip	0.11	17.31	0.11	0.11
apex4	0.27	65.20	0.17	0.22
sao2	0.22	54.81	0.06	0.11
ex1010	0.33	134.79	0.28	0.22

Execution time (in seconds) for the spectral coefficients calculation of both fastest transforms, polarity 00…0 and all 2^n polarities of Reed-Muller spectral coefficients using Fast Reed-Muller transform are given in Table 3. The entries in Table 3 indicate that the calculation time for fastest transforms spectral coefficients are generally either the same or shorter than the time required for calculation of the spectral coefficients in polarity 00…0 using the Fast Reed-Muller transform. This is because butterfly diagrams corresponding to the fastest transforms have lesser number of nonzero elements than the butterfly diagram for the Fast Reed-Muller transform (see Figures 1-3) and hence they require smaller number of additions over GF(2) what reduces the computation time. It should be noticed that to be able to obtain the best polarity for the generalized Reed-Muller transform, all 2^n spectra have to be calculated and that increases the computation time.

4. CONCLUSION

The suitability of an LI transform in a given application depends not only on the choice of its basis function but also on the existence of efficient ways of its calculation as well as the complexity of its final polynomial expansion. It should be also noticed that polynomial expansions based on two fastest transforms could be easily implemented in terms of reversible logic gates [4]. In this article experimental results have shown that the two fastest transforms are advantageous over well known generalized Reed-Muller transform in terms of the number of nonzero spectral coefficients as well as the time of their computation. Various properties of two fastest transforms over GF(2) have also been presented here. The pair of transforms discussed in this paper has the least computational complexity among all LI transforms. This is also reflected in the smaller time required to calculate the coefficients. It should be noticed that the same concept of fast LI transforms could be easily applied to logical functions with multiple-valued input and all the presented results are valid for such instances as well.

5. REFERENCES

[1] P. Davio, J.P. Deschamps and A. Thayse, *Discrete and switching functions*. New York: George and McGraw-Hill, 1978.

[2] B.J. Falkowski and S. Rahardja, "Classification and properties of fast linearly independent logic transformations", *IEEE Trans. on Circuits and Systems II-Analog and Digital Signal Processing*, Vol. 44, No. 8, pp. 646-655, Aug. 1997.

[3] D. Green, *Modern logic design*. Wokingham: Addison-Wesley, 1986.

[4] P. Kerntopf, "A comparison of logical efficiency of reversible and conventional gates", *Proc. 9th IEEE International Workshop on Logic Synthesis*, Dana Point, CA, pp. 261-269, May 2000.

[5] S. Rahardja and B.J. Falkowski, "Polynomial expansions over GF(2) based on fastest transformation", *Proc. 35th IEEE International Symposium on Circuits and Systems*, Scottsdale, Arizona, Vol. 3, pp. 377-380, May 2002.

[6] T. Sasao and M. Fujita, *Representations of discrete functions*. Boston: Kluwer, 1996.

[7] R.S. Stankovic, M. Stankovic and D. Jankovic, *Spectral transforms in switching theory: definitions and calculations*. Belgrade: IP Nauka, 1998.

SOC DESIGN INTEGRATION BY USING AUTOMATIC INTERCONNECTION RECTIFICATION

Chun-Yao Wang, Shing-Wu Tung and Jing-Yang Jou

Department of Electronics Engineering
National Chiao Tung University
Hsinchu, Taiwan, R.O.C.

ABSTRACT

This paper presents an automatic interconnection rectification (AIR) technique to correct the misplaced interconnection occurred in the integration of a SoC design automatically. The experimental results show that the AIR can correct the misplaced interconnection and therefore accelerates the integration verification of a SoC design.

1. INTRODUCTION

In the SoC era, system level integration and platform-based design [1] are evolving as a new paradigm in system designs, hence, design reuse and reusable building blocks (cores) trading are becoming popular. However, present design methodologies are not enough to deal with cores which come from different design groups and are mixed and matched to create a new system design. In particular, design verification is one of the most difficult task.

The focus of core-based design verification should be on how the cores communicate with each other [2]. However, prior to the interface verification, the interconnection between the cores in a SoC have to be verified first. This is because the SoC integrator has to connect a large number of ports in a SoC design. The likelihood of interconnection misplacements between the cores is high. Thus, the interconnection verification can be conducted as the first step to the interface verification between the cores in a SoC design.

By creating the testbenches at a high level, a connectivity-based design fault model, port order fault (POF), is proposed in [3]. This fault model is similar to the Type H design error "incorrectly placed wire" in the logic level [4] [5]. The POF-based automatic verification pattern generation (AVPG) are also developed in [6] [7]. The AVPG are effective in generating the verification pattern set for detecting the misplacements of interconnection in a SoC design. However, the diagnosis and correction issues on the misplaced interconnection are even more important for SoC verification. Thus, to accelerate the SoC integration process, this paper presents an automatic interconnection rectification (AIR), which not only detects the erroneous interconnection among the cores, but also diagnoses and corrects them automatically.

Traditional diagnosis and correction algorithms in the logic level can be divided into two categories with respect to the underlying techniques: those based on symbolic techniques [8] $\sim$ [10] and those based on simulation techniques [5] [11] $\sim$ [12]. The approaches based on symbolic techniques can return valid correction and handle circuits with multiple errors well, however, they are not applicable to circuits that have no efficient Ordered Binary Decision Diagram (OBDD) [13] representation. Thus, to verify the interconnection among the IP cores with all description levels (soft, firm, and hard cores) embedded into a system, the AIR algorithm has to deal with IP cores that are described in different levels, for example, logic level, register transfer (RT) level, or even behavioral level. Consequently, the symbolic approach is inadequate to this application and the simulation based AIR algorithm is presented.

2. PRELIMINARY

Definition 1: The type I POF is at least an output misplaced with an input. The type II POF is at least two inputs misplaced. The type III POF is at least two outputs misplaced [3].

It has been proven that the type II POFs **dominate** the other two types of POFs [6]. Thus, the AIR focuses on the **type II POFs**.

Definition 2:A **port sequence** is an input port numbers permutation. The **f**ault **f**ree **p**ort **s**equence (**FFPS**) is a port sequence that none of the input ports is misplaced. For an N-input core, the $N!$ permutations represent the $N!$ port sequences. Except the FFPS, the remaining ($N!$-1) port sequences are called **f**aulty **p**ort **s**equences (**FPSs**).

Example 1: The schematic representation of the FFPS 1234 and the FPS 1423 of BLK2 are shown in Fig. 1(a) and Fig. 1(b), respectively.

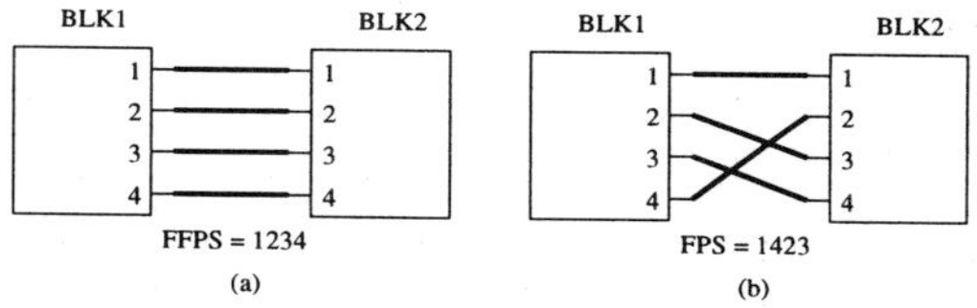

Figure 1: The schematic representation of the FPS

The PUPs representation is a metric used in the AIR to indicate the remaining **p**ossible **u**ncorrected **p**orts (**PUPs**) currently in the integrated design. We use Example 2 to demonstrate the PUPs representation.

Example 2: Given an 8-input core, the inputs are numbered from 1 to 8. These port numbers are all possible uncorrected ports and are placed in a pair of parentheses. If these ports are faulty indeed, they must be misplaced with the other ports in the same group only. The group with only one port number represents the port is correct. For example, the (12345678) represents that the port 1 $\sim$ port 8 are all possible uncorrected ports and they could be misplaced with each other, the |PUPs| = 8. The (1)(234)(567)(8) represents that the port 1 and port 8 are correct ports, the port 2 $\sim$ port 4 and the

0-7803-7761-3/03/$17.00

port 5 ∼ port 7 are the possible uncorrected ports and they could be misplaced with the other ports in the same group only, the |PUPs| = 6. If the PUPs are induced to (1)(2)(3)(4)(5)(6)(7)(8), the |PUPs| = 0, and the interconnection is correct.

Definition 3 defines the cross-group operation, which is used to calculate the updated PUPs.

Definition 3: Given two PUPs representation P1 and P2 with the same number of ports. The cross-group P3 of P1 and P2 is denoted as P1 △ P2 and it satisfies with the following condition: if any two port numbers x and y are both placed in the same group of P1 and P2, they will be placed in the same group of P3; otherwise, they are placed in the different groups of P3.

Example 3: Given two PUPs, P1 = (12)(34) and P2 = (14)(23), the cross-group P3 of P1 and P2 = P1 △ P2 = (1)(2)(3)(4).

We also exploit Example 3 to explain the physical meaning of the cross-group operation. P1 represents that if the port 2 and port 3 in a FPS are faulty, they **must not** be misplaced with each other. On the contrary, P2 represent that if the port 2 and port 3 in the same FPS are faulty, they **must** be misplaced with each other. Since the port 2 and port 3 in this FPS cannot be "misplaced" with each other and "not misplaced" with each other simultaneously, the port 2 and port 3 have not to be faulty, and they have to be placed in single groups in the PUPs, respectively. The proposed cross-group operation accomplishes this object indeed.

The environment and mechanism of POF verification, which exploits the IEEE P1500 SECT [15], can be found in [6]. Thus, we omit describing them here.

3. THE AIR ALGORITHM

The input to the AIR is the simulation model of an IP core. The four stages of AIR are pattern generation, fault detection, fault diagnosis and correction, and PUPs calculation as shown in Fig. 2. The pattern generation stage generates valid patterns Si to differentiate the outputs of fault free interconnection and faulty interconnection. The fault detection stage applies Si into the integrated design to examine whether the interconnection are misplaced or not. If the fault effect appears after the fault detection stage, the misplaced interconnection are identified and rectified in the fault diagnosis and correction stage. Otherwise, the PUPs calculation stage is performed. After the fault diagnosis and correction stage, the same PUPs calculation stage is performed as well. The PUPs calculation stage can figure out the possible faulty ports remained in the integrated design. These remaining possible faulty ports will be verified in the subsequent iterations by additional verification patterns. Since the fault diagnosis and correction stage usually cannot correct all misplaced ports in one iteration, the rectified interconnection has to be verified (detection, diagnosis, and correction) by Si again until the fault effect disappears after the fault detection stage. These iterative procedures are presented with bold lines in the AIR flow as shown in Fig. 2.

3.1. Pattern Generation

Definition 4: The set consists of all patterns with m 1s and (N-m) 0s is denoted as Θ_m^N, where $m \in [0, 1, 2, \cdots, N-1, N]$. The size of Θ_m^N is the number of patterns in Θ_m^N and is denoted as $|\Theta_m^N|$.

Example 4: For a 4-input core, Θ_0^4={0000}, $|\Theta_0^4|$ =1. Θ_1^4={1000, 0100, 0010, 0001}, $|\Theta_1^4|$ =4.

Theorem 1: *Θ_m^N can activate all (N!-1) POFs where m ∈ [1, 2, ⋯, N-2, N-1].* [6]

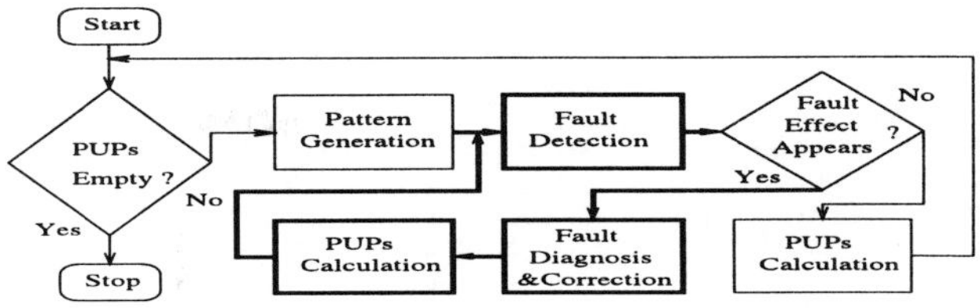

Figure 2: The AIR flow

According to Theorem 1, we arbitrarily apply one Θ_m^N to the inputs of core for $m \in [1, 2, \cdots, N\text{-}2, N\text{-}1]$. Since $|\Theta_m^N|$ is smaller when m is closer to the end points of interval [1, 2, ⋯, N-1], **we select m from 1 up to $\lfloor N/2 \rfloor$ or from N-1 down to $\lfloor N/2 \rfloor$.**

We use an example to demonstrate the pattern generation stage. This example will be used to demonstrate the other stages in the AIR as well later. Given an 8-input combinational core, the initial PUPs are (12345678). Assume the simulation results of Θ_1^8 are shown in Fig. 3 and are represented in symbolic output representation. The patterns with the same output are grouped into one set. Θ_1^8 patterns can be grouped into two sets S1 and S2. Either S1 or S2 can be selected as the verification patterns. Here we select the smaller set S1 as the verification patterns [6].

initial PUPs = (12345678)

Pattern	S1	S2
1 0 0 0 0 0 0 0 -> A0	10000000	
0 1 0 0 0 0 0 0 -> B0		01000000
0 0 1 0 0 0 0 0 -> B0		00100000
0 0 0 1 0 0 0 0 -> B0		00010000
0 0 0 0 1 0 0 0 -> B0		00001000
0 0 0 0 0 1 0 0 -> B0		00000100
0 0 0 0 0 0 1 0 -> B0		00000010
0 0 0 0 0 0 0 1 -> B0		00000001

Figure 3: The simulation outputs of Θ_1^8

The system integrators do not know how the cores are connected exactly in the actual integrated design. However, to demonstrate the fault detection, diagnosis, and correction procedures on the interconnection verification, we assume the FPS λ of this example is given and is 83762451.

3.2. Fault Detection

We apply S1 {10000000} into the design with the FPS λ 83762451, and find that the corresponding output of {10000000} is B0 as shown in Fig. 4(a). Since the fault free output is A0, the fault effect appears and λ is detected by S1.

3.3. Fault Diagnosis and Correction

Definition 5: Given a set of patterns S with the same length, we count the number of digits 1 in the same bit position to form a vector with the same length. This vector is called the characteristic vector (CV) of S and is denoted as CV_S.

We apply S1 into the integrated design, and realize that the real applied input is not {10000000} by observing the unexpected output B0. Since there are seven patterns that produce B0 output, we do not know exactly what the actual applied pattern is. Nevertheless, **we know the actual applied pattern which produces A0 output instead**. Therefore, we simulate Θ_1^8 patterns with the FPS 83762451 and observe the outputs until the output is A0. These results are shown in Fig. 4(a). From Fig. 4(a), we find that when

Pattern	Outputs	
	Fault Free	FPS 83762451
10000000	A0	B0
01000000	B0	B0
00100000	B0	B0
00010000	B0	B0
00001000	B0	B0
00000100	B0	B0
00000010	B0	B0
00000001	B0	A0

(a)

S1' 00000001 —FPS→ S1 10000000

CV_S1'= 00000001 CV_S1= 10000000

13762458 ← 83762451
2_switch(1, 8)

PUPs = (12345678) △ (1234567)(8)
= (1234567)(8)

(b)

S1	Outputs	
	Fault Free	FPS 13762458
10000000	A0	A0

PUPs = (1234567)(8) △ (1)(2345678)
= (1)(234567)(8)

(c)

Figure 4: Rectification processes of Θ_1^8

the last pattern {00000001} is applied, the output becomes A0. This result implies that the FPS λ turns the pattern {00000001} to {10000000}. We put the pattern {00000001} into S1' and assume S1 is {10000000}, and we calculate CV_S1'=00000001 and CV_S1=10000000. Then, the misplaced ports can be identified by comparing CV_S1' and CV_S1.

Definition 6: The exchange of the ith port with the jth port is denoted as 2_switch(i, j).

Comparing CV_S1' and CV_S1, we observe that the 1st digit and the 8th digit of CV_S1' and CV_S1 are different. We know that if the FPS is the FFPS, CV_S and CV_S' will be identical. **The FFPS will keep CV being intact for all patterns. Therefore, switching the ports with different CV values to preserve CV can move the FPS to the FFPS.** Thus, we apply 2_switch(1, 8) on CV_S1 to let CV_S1 be the same as CV_S1' as shown in Fig. 4(b). Now, the corrected FPS becomes 13762458.

We exploit this property of the FFPS to rectify the misplaced ports throughout the AIR algorithm.

The following lemmas and theorems state the convergency of fault correction procedure.

Theorem 2: *The correct ports in a FPS λ will not be rectified to faulty ones in the fault diagnosis and correction stage.*

Definition 7: Given a set of n-bit patterns S', when a 2-switch is applied on S' and CV_S' is invariant, the 2-switch is called a CV invariant fault of S' (CVIF(S')). Otherwise, it is called a CV variant fault of S' (CVVF(S')).

Theorem 3: *Given a set of n-bit patterns S' and a FPS λ. There exists a finite sequence of 2_switches, $2\_switch_1 \sim 2\_switch_l$, to convert the FFPS into λ, and this sequence of 2_switches must be in one of the following three categories:*

(I): $2\_switch_1 \sim 2\_switch_l$ are all CVIFs(S')

(II): $2\_switch_1 \sim 2\_switch_i$ are CVIFs(S') and $2\_switch_{i+1} \sim 2\_switch_l$ are CVVFs(S') where $1 \leq i \leq l-1$

(III): $2\_switch_1 \sim 2\_switch_l$ are all CVVFs(S')

In this example, the 2_switch(1,8) is a CVVF(S1'). Therefore, according to Theorem 3, the corrected FPS is a sequence of CVIFs(S1'). **Theorem 3 guarantees that CVIF(S1') are the only possible faults which we have to deal with in the succeeding iterations.**

3.4. PUPs Calculation

As mentioned above, CVIF(S1') are the remaining possible faults only and they are occurred among the port 1 ~ 7. Thus, the PUPs representation is (1234567)(8). This PUPs has to be cross-grouped with the initial PUPs (12345678) to obtain the updated PUPs. Thus, the updated PUPs = (12345678) △ (1234567)(8) = (1234567)(8). This result is also shown in Fig. 4(b).

Thereafter, we apply S1 into the integrated design again with the corrected FPS 13762458 to examine whether the corrected FPS 13762458 can be further corrected by S1. In Fig. 4(c), we find that the outputs of S1 with the FPS 13762458 are the same as the expected outputs A0, thus, the FPS 13762458 cannot be further corrected by S1. **Please note that this reexamining process also provide the information to reduce the PUPs representation.** The corrected FPS 13762458 does not change the outputs A0, therefore, the actual input pattern is also {10000000} (S1) and the corresponding CV_S1 is unchanged. According to Theorem 3, the remaining possible faults are CVIFs(S1) and they can be expressed as (1)(2345678). Thus, the updated PUPs become (1)(234567)(8), which are obtained from (1234567)(8) △ (1)(2345678). At this time, we move to the next iteration to generate further patterns to detect and rectify the remaining faulty ports among port 2 ~ 7.

3.5. Summary

The FPS is corrected from 83762451 to 13762458, and the PUPs representation is reduced from (12345678) to (1)(234567)(8). These results illustrate that the PUPs representation presents the corrected FPS appropriately. The succeeding iterations in the AIR follow the same flow to rectify the misplaced interconnection. However, due to the page limit, we skip these repeated demonstration of the AIR algorithm here.

The success of the AIR depends on the pattern generation stage strongly. Since the pattern generation stage will search all Θ_m^N, for m=1, 2, $\cdots$, N-1 if necessary, it is a complete algorithm [6] [7]. This complete pattern generation algorithm leads the AIR algorithm to be complete as well.

3.6. The Sequential AIR

The development of the sequential AIR is based on the same assumption as the combinational AIR, i.e., the CUV is pre-verified and fault free. The fault occurs only at the interconnection between the cores. For the testability concern, most sequential cores are designed with scan chains. Thus, here we assume that the sequential cores in the experiments are scan-testable. These sequential cores can be set in arbitrary state and therefore they can be seen as combinational ones. Consequently, the AIR algorithm used in the combinational cores is applicable to the sequential ones. The only difference is that the sequential cores have to be set to a state by sequential AIR before evaluating outputs.

4. EXPERIMENTAL RESULTS

The heuristic AIR, which adds the iteration counter to bound the processing time, has been integrated into the SIS [14] environment. Experiments are conducted over a set of ISCAS-85, 89, and ITC-99 benchmarks. These benchmarks are in BLIF format which is a netlist level design description. However, we only use the simulation information to conduct the experiments and therefore, arbitrary level of design description can be used for conducting POF

bench	parameters		blind connection		
	|PI|	lits.	a/b	|PUPs|	time(s)
$c17$	5	12	4/4	0	0.1
$c880$	60	703	58/58	0	105
$c1355$	41	1032	40/40	0	152
$c1908$	33	1497	32/32	0	149
$c432$	36	372	36/36	0	41
$c499$	41	616	39/39	0	57
$c3540$	50	2934	49/49	0	636
$c5315$	178	4369	178/178	0	10270
$c2670$	233	2043	231/231	0	9373
$c7552$	207	6098	207/207	0	23972
$c6288$	32	4800	32/32	0	425

Table 1: Experimental results of the heuristic AIR on ISCAS-85

verification. The simulation information of the BLIF benchmarks imitate the simulation model of IP cores. The functionalities of these benchmarks include ALU (c5315), multiplier (c6288), processors (b14, b15), and some ASIC designs, thus, the experiments can represent the realistic SoC design appropriately to some degree.

Table 1 summaries the experimental results of the heuristic AIR on ISCAS-85. The |PI| represents the number of inputs. The number of literals (lits.) indicates the scale of a benchmark. The a/b presents "number of corrected ports/number of faulty ports". These faulty ports in the experiments are caused by the blind connection. The blind connection represents the worst case of the SoC integration. To imitate the actual interconnection faults in the integration, the FPS is generated as follows. For each port i, i from 1 to N, we assign a random number ($\in [1 \sim N]$) to it. If the number has been assigned to port j, where $1 \leq j < i$, we generate another one to the port i until it is not repeated. This process is similar to the real interconnection process with blindness. Since the FPSs in the experiments are generated randomly, the generated FPSs quantify the inject out of order permutations.

The iteration bound in the experiment was set to 100. The AIR algorithm will be terminated automatically if the iteration counter is over the bound or the PUPs representation becomes empty. At the end of AIR, the number of corrected ports, |PUPs|, and CPU time are returned. The number of corrected ports is obtained by comparing the final FPS with the FFPS. The |PUPs| is obtained from the final PUPs representation. The CPU time is measured in second on an Ultra Sparc II workstation.

Note that since we greatly concern about how many faulty ports are injected and corrected rather than the number of verification patterns [6] [7] in the experiments, we do not report the number of the verification patterns in the experimental results.

According to Table 1 and 2, the faulty ports of each benchmark **are all corrected**, the |PUPs| of each benchmark is 0 as well, and the processing time of each benchmark is acceptable. These results demonstrate that the heuristic AIR is able to correct the misplaced ports within reasonable efforts.

5. CONCLUSIONS

In the SoC era, the embedded cores are mixed and integrated to create a system chip. System designers integrate those cores manually and have the possibility of incorrect integration due to the misplaced I/O ports. Furthermore, without the knowledge of the internal structures of the embedded cores, system designers have difficult time to locate the position of having erroneous interconnection. The AIR technique provides an efficient solution to integrate the cores with correct interconnection automatically. Therefore this algorithm can reduce the time on design verification in core-based design methodology.

bench	parameters			blind connection		
	|PI|	lits.	FFs	a/b	|PUPs|	time(s)
$s1196$	14	1009	18	13/13	0	37.2
$s1238$	14	1041	18	12/12	0	37.6
$s1488$	8	1387	6	7/7	0	17.1
$s1494$	8	1393	6	8/8	0	18.4
$s15850$	14	13659	597	14/14	0	608
$s27$	4	18	3	3/3	0	0.1
$s5378$	35	4212	164	34/34	0	680
$s641$	35	539	19	35/35	0	136
$s713$	19	591	19	17/17	0	127
$s820$	18	757	5	17/17	0	137
$s832$	18	767	5	16/16	0	147.6
$s9234$	36	7971	211	34/34	0	1961
$s444$	3	352	21	3/3	0	3.4
$s510$	19	424	6	19/19	0	38.5
$s344$	9	269	15	8/8	0	4.9
$s349$	9	273	15	9/9	0	5.5
$s382$	3	306	21	3/3	0	9.2
$s386$	7	347	6	7/7	0	3.8
$s400$	3	320	21	3/3	0	4.2
$s13207$	31	11165	669	29/29	0	2338
$s1423$	17	1164	74	17/17	0	67.4
$s6669$	83	5343	239	82/82	0	6985
$s4863$	49	4092	104	49/49	0	1391
$s1269$	18	1047	37	17/17	0	66.7
$s1512$	29	1264	57	28/28	0	218.2
$s3271$	26	2697	116	25/25	0	403
$s3330$	40	2816	132	40/40	0	820
$s3384$	43	2755	183	41/41	0	1327
$b10$	11	331	17	11/11	0	8.4
$b11$	7	1078	31	7/7	0	52.6
$b12$	5	1887	121	5/5	0	19.3
$b13$	10	507	53	9/9	0	15.8
$b14$	32	11849	245	31/31	0	2634
$b15$	37	15856	449	37/37	0	5062

Table 2: Experimental results of the heuristic AIR on sequential benchmarks

6. REFERENCES

[1] H. Chang, et al., "Surviving the SoC revolution - a guide to platform-based design," Kluwer Academic Publishers, 1999.

[2] J. A. Rowson, et al., "Interface-based design," *in Proc. Design Automation Conference*, pp.178-183, Jun. 1997.

[3] J.-Y. Jou, et al., "A logic fault model for library coherence checking," *Journal of Information Science and Engineering*, pp.567-586, Sep. 1998.

[4] M. S. Abadir, J. Ferguson, and T. E. Kirkland "Logic design verification via test generation," *IEEE Transactions on Computer-Aided Design*, vol.7, no.1, pp.138-148, Jan. 1988.

[5] A. Veneris, and I. N. Hajj, "Design error diagnosis and correction via test vector simulation," *IEEE Transactions on Computer-Aided Design*, vol.18, no.12, pp.1803-1816, Dec. 1999.

[6] J.-Y. Jou, et al., "On automatic-verification pattern generation for SoC with port-order fault model," *IEEE Transactions on Computer-Aided Design*, pp.466-479, vol.21, no.4, Apr. 2002.

[7] J.-Y. Jou, et al., "An automorphic approach to verification pattern generation for SoC design verification using port-order fault model," *IEEE Transactions on Computer-Aided Design*, vol.21, no.10, Oct. 2002.

[8] S.-Y. Huang, K.-C. Chen, and K.-T. Cheng "Incremental logic rectification," *in Proc. VLSI Test Symposium*, pp.143-149, 1997.

[9] M. Fujita, Y. Tamiya, Y. kukimoto, and K.-C. Chen, "Application of Boolean unification to combinational synthesis," *in Proc. IEEE/ACM Int. Conf. Computer-Aided Design*, pp.510-513. 1991.

[10] P.-Y. Chung, Y.-M. Wang, and I. N. Hajj, "Logic design error diagnosis and correction," *IEEE Transaction on VLSI Syst.*, vol.2, pp.320-332. Sep. 1994.

[11] S.-Y. Huang, K.-C. Chen, and K.-T. Cheng "ErrorTracer: design error diagnosis based on fault simulation techniques," *IEEE Transactions on Computer-Aided Design*, vol.18, no.9, pp.1341-1352, Sep. 1999.

[12] I. Pomeranz and S. M. Reddy "On error correction in macro-based circuits," *IEEE Transactions on Computer-Aided Design*, vol.16, no.10, pp.1088-1100, Oct. 1997.

[13] R. E. Bryant "Graph-based algorithms for Boolean function manipulation," *IEEE Transactions on Computer*, vol.C-35, no.8, pp.677-691, 1986.

[14] E. M. Sentovich, et al., "Sequential circuit design using synthesis and optimization," *in Proc. IEEE International Conference on Computer Design*, pp.328-333, Oct. 1992.

[15] E. J. Marinissen, et al., "Towards a standard for embedded core test:An example," *in Proc. IEEE International Test Conference*, pp.616-627, Sep. 1999.

SYNTHESIZING CHECKERS FOR ON-LINE VERIFICATION OF SYSTEM-ON-CHIP DESIGNS

Rolf Drechsler

Institute of Computer Science
University of Bremen
28359 Bremen, Germany
email: drechsle@informatik.uni-bremen.de

ABSTRACT

In modern System-on-Chip (SoC) designs verification becomes the major bottleneck. Since by using state-of-the-art techniques complete designs cannot be fully formally verified, it becomes more and more important to check the correct behaviour during operation. This becomes even more significant in systems that are changed during life-time, like re-configurable systems.

In this paper we present a hardware extension that allows to efficiently synthesize checkers and properties that have been used in the verification process. This allows for an on-line verification of SoC designs. For the verification hardware a regular layout is discussed that can easily be synthesized and has a very low area overhead. The on-line check has (nearly) no effect on the delay of the considered chip.

1. INTRODUCTION

Modern circuits contain up to several million transistors. In the meantime it has been observed that verification becomes the major bottleneck, i.e. up to 80% of the overall design costs are due to verification. This is one of the reasons why recently several methods have been proposed as alternatives to classical simulation, since it cannot guarantee sufficient coverage of the design. E.g. in [2] it has been reported that for the verification of the Pentium IV more than 200 billion cycles have been simulated, but this only corresponds to 2 CPU minutes, if the chip is run with 1 GHz.

As alternatives, formal verification or symbolic simulation have been proposed and in the meantime these have been successfully applied in many projects [6]. But so far, all approaches are based on software solutions and cannot be applied after the chip is fabricated. On-line verification approaches have not been considered. But there is a need for these techniques in at least two problem domains:

- If the properties specified by the designer or verification engineer cannot be proven by the verification tool. This might result from too complex properties or from the difficulty of the circuit considered (e.g. for multipliers).
- If the circuit is re-configured during operation [14,4]. The new programmed hardware has to be checked for correct functional behaviour.

In this paper, we present an approach to synthesize hardware that can check properties that have been applied during the verification process. Often these properties are also available directly from the specification [13]. The approach of adding extra hardware has been very successfully applied in the testing domain for many years (see e.g. [15,9]), while hardware verification is mainly applied on the software level. With new emerging technologies this has to be extended. The synthesized circuits are called *verification hardware* in this paper. Verification hardware allows to check the correct behaviour also later, after the chip production. These techniques become very important in SoC designs that allow parts to be re-configured during operation.

After explaining the underlying principle of our approach, a regular hardware layout is discussed that allows to map properties defined in the verification process directly on the circuit with very low hardware overhead. The extra delay resulting from the verification hardware is small, i.e. only one extra fanout per checked signal.

2. CHECKER STRUCTURES

Even though the formalism to describe properties in verification languages varies a lot (see e.g. [10]), the underlying mechanisms are very similar. In the following we use the notation from the property checker used at Infineon Technologies AG (see e.g. [7,8] for more details).

0-7803-7761-3/03/$17.00 ©2003 IEEE

Notice that it is straightforward to generalize the results to also work for other verification languages.

A property consists of an *assume part* and a *proof part*. If all assumptions hold, the property specified in the proof part must hold.

Example 1: We want to prove a property `test`. The property says that whenever signal `x` becomes `1`, two clock cycles later signal `y` has value `0`. More formally:

```
theorem test is
assume:
  at t: x = 1 ;

prove:
  at t+2: y = 0 ;
end theorem ;
```

In general, each property is of the form that whenever some signals have given values (eventually over several time frames), other – or the same – signals assume specified values. Notice that property languages also allow to argue over time intervals, e.g. a requirement can be that a signal assumes a value within a given time interval, while the concrete time point is not given. It is obvious that each of the properties can easily be transferred to hardware realization based on shift registers and some additional logic (see below).

Remark: This is exactly the method by which efficient property checkers formulate the problem by translating the property to a Boolean network and running Boolean provers, like e.g. SAT and BDD [11]. (For more details on SAT-based property checking see [3].) In contrast to shift registers the solvers "un-roll" the circuit for the maximal number of time frames specified in the property. For our approach shift registers are more adequate, since for simulation only the value at an earlier time frame has to be stored, while the surrounding logic can be ignored.

While property checkers "un-roll" the circuit, we now show by a motivating example a hardware realization:

Example 2: Consider the property from Example 1. Whenever signal `x` is `1` it has to be checked that `y` is `0` two time frames later. This can be easily done by storing the value of `x` for two clock cycles and then computing the result by performing an AND operation of the `x`-signal with the negated value of `y` (`y` has to be negated, since it should have the value 0). The corresponding circuit is shown in Figure 1. If the output of the AND gate is 0, the property is violated.

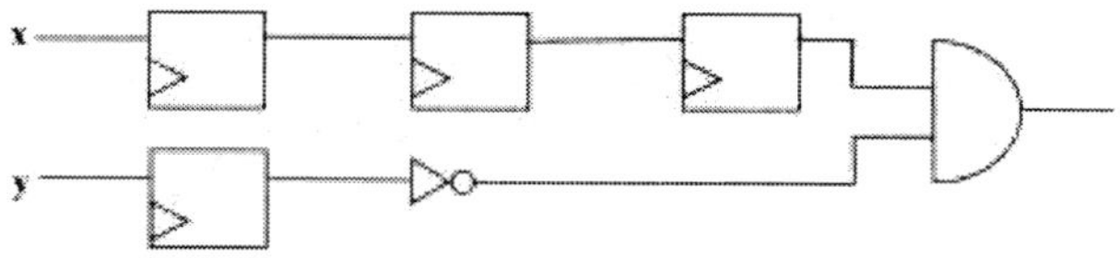

Figure 1. Shift register and logic for Example 1

In this way, it becomes very easy to generate monitors in hardware from given properties. The hardware can be used to check the correct circuit behaviour during operation, i.e. it is applicable for on-line test. This finds application in several scenarios, e.g.:

- "Hard" to proven properties: Some properties turn out to be too difficult to be formally verified. This depends on the property itself and the circuit under consideration. If the design e.g. contains large multipliers, formal verification cannot prove the property. In these cases, simulation is not sufficient (see [2] and the discussion in Section 1).

- In re-configurable computing hardware components are exchanged during normal operation. An "external" verification method based on software cannot be applied in this case. Here verification hardware is a promising alternative. (Similar concepts have been studied in the area of testing for a longer time [15], while verification is so far mainly "software-oriented".)

It is straightforward to also include more complex operations, like reasoning over several time intervals, by simply adding the corresponding signals, i.e. OR-ing the signals.

2.1 Algorithmic Solution Using String Matching

Summing up the observations above, the problem to be solved is a specific type of string matching, i.e. during circuit operation it has to be determined whether a pattern occurred that is defined over a number of signals for the assume and proof part. While exact string matching is a very well studied problem (see [5] for a list of more than 30 algorithms), in our case some further constraints or properties have to be considered:

- Since the realization has to be included in the circuit, we need an efficient hardware description.
- We can make use of parallelism, while "classical" string matching is used in software.
- The signals only assume binary values.

- The layout of the circuit should be regular and scalable.

Due to these reasons, the technique described in the following differs to the software approaches presented earlier, but has some similarities to the *Shift-Or-Algorithm* [1].

We first describe the main idea following Example 2. The signals that are used in the property to be checked are fed into chains of shift registers. The outputs of the flipflops of the chain give the signal values in the corresponding time frame. Then the property can easily be mapped by AND- and OR-gates resulting in the required behaviour.

If several properties are defined over the same set of variables, of course these signals only have to be stored once allowing for an efficient reuse methodology.

3. REGULAR LAYOUT

In this section we show a block diagram, how to realize the concept above using a regular layout. The overall flow is given in Figure 2. Here, the method described in Section 2 is generalized to save hardware (see below).

The core block mainly consists of *shift registers*, that allow to "remember" the signal value of previous time frames. The length of this chains is determined by the maximal time interval a property uses. The block *logic* implements the checks according to the properties specified.

Remark: In a typical application, the maximal time interval of properties is less than 20. Thus, the hardware required is moderate.

The number of scan chains – this determines the height of the block – is given by the number of signals that are used in the properties to be checked. If this number becomes too large, a bus system can be used including the corresponding control logic (blocks *bus* and *control* in Figure 2). For the designs of these blocks standard methods can be applied. By this approach, the number of shift registers can be reduced to the maximal number of signals used in one single property. The control block also has to select the correct property that has to be checked for the signals fed in the shift registers.

The decision, which of the solutions to chose (i.e. with or without a bus and control block), has to be made dependent on the design and the application. In cases where high quality has to be assured the "bus-less" solution should be preferred, since in that case all properties are checked in parallel, while the bus concept is more efficient if the number of properties becomes very large.

The logic block is directly derived from the properties along the lines described in the previous section. To further optimise the hardware, some of the scan chains can be cut, if the corresponding signal is not needed.

Example 3: Consider again Figure 1, where the scan chain of signal y has only length 1, while the one of x has length 3.

Finally, we briefly comment on the extra delay resulting from the proposed approach. For each signal that occurs in a property, the corresponding wire has to be made available. But these signals are directly available in the circuit to be verified and for this only one extra fanout is needed. An extra delay caused by this is usually negligible.

4. RELATED WORK

Independent of this work, in [12] a technique has been proposed to synthesize checkers, but these are declared on a very high level of abstraction and are included in the synthesis process. This makes them difficult to use in re-configurable systems. Furthermore, the layout of the verification hardware is not considered and by this a reuse of hardware components becomes very difficult.

5. CONCLUSIONS

A new approach has been presented for on-line verification based on synthesizing checkers in hardware. Properties originally specified for (formal) verification on a software level can be directly mapped.

A regular layout has been described that allows the implementation with small hardware overhead. The size of the hardware grows linear with the maximal time interval of the longest property.

It is focus of current work to apply the techniques described in this paper to systems containing re-configurable components.

6. REFERENCES

[1] R.A. Baeza-Yates, G.H. Gonnet, A new approach to text searching, Communications of the ACM. 35(10):74-82, 1992

[2] B. Bentley. Validating the Intel Pentium 4 microprocessor. In Design Automation Conf., pp. 244-248, 2001.

[3] A. Biere, A. Cimatti, E. Clarke, M. Fujita, Y. Zhu, Symbolic Model Checking using SAT procedures instead of BDDs, In Design Automation Conference, pp. 317-320, 1999

[4] K. Compton, S. Hauck, Reconfigureable Computing: A Survey of Systems and Software, ACM Computing Surveys, Vol. 34, No. 2, pp. 171-210, 2002

[5] C. Charras, T. Lecroq, Handbook of Exact String-Matching Algorithms, http://www-igm.univ-mlv.fr/~lecroq/string/string.pdf, 2002

[6] R. Drechsler, S. Höreth, Gatecomp: Equivalence Checking of Digital Circuits in an Industrial Environment, International Workshop on Boolean Problems, pp. 195-200, 2002

[7] P. Johannsen, R. Drechsler, Formal Verification on the RTL – Computing One-To-One Design Abstractions by Signal Width Reduction, In 11th IFIP International Conference on Very Large Scale Integration, pp. 127-132, 2001

[8] P. Johannsen, R. Drechsler, Utilizing High-Level Information for Formal Hardware Verification, Advanced Computer Systems, J. Soldek, J. Pejaz (Ed.), Kluwer Academic Publishers, pp. 419-431, 2002

[9] M. Gericota, G. Alves, M. Silva, J. Ferreira, DRAFT : An On-Line Fault Detection Method for Dynamic and Partially Reconfigurable FPGAs, IEEE International On-Line Testing Workshop, 2001

[10] R. Goering, Assertion's Babel tower?, http://www.eedesign.com/columns/tool_talk/OEG20010828S0054, 2001

[11] A. Kuehlmann, M. Ganai and V. Paruthi, Circuit-Based Boolean Reasoning, In Design Automation Conference, pp. 232-237, 2001

[12] M. Oliveira, A. Hu High-Level Specification and Automatic Generation of IP Interface Monitors, In Design Automation Conference, pp. 129-134, 2002

[13] K. Shimuzu, D. Dill, A. Hu, Monitor-based formal specification of PCI, International Conference on Formal Methods in Computer-Aided Design (FMCAD), 2000

[14] R. Tessier, W. Burleson, Reconfigurable Computing and Digital Signal Processing: A Survey, Journal of VLSI Signal Processing, Kluwer, 28, pp. 7-27, May/June 2001

[15] T.W. Williams, K.P. Parker. Design for testability - a survey. IEEE Trans. on Comp., 31(1):2-15, 1982

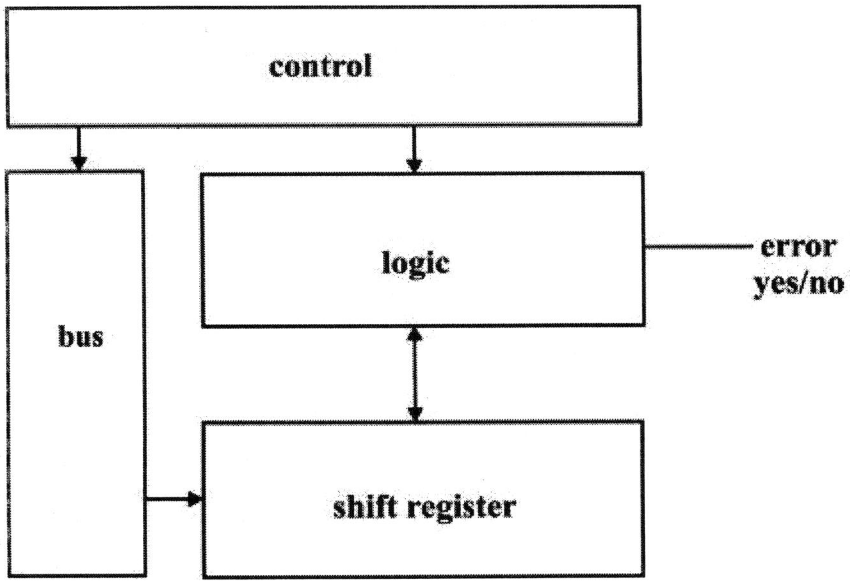

Figure 2. Regular layout for verification hardware

LOAD CELL RESPONSE CORRECTION USING ANALOG ADAPTIVE TECHNIQUES

M. Jafaripanah, B. M. Al-Hashimi and N. M. White

ESD Group, Department of Electronics and Computer Science
University of Southampton, SO17 1BJ, UK
{mj01r,bmah,nmw}@ecs.soton.ac.uk

ABSTRACT

Load cell response correction can be used to speed up the process of measurement. This paper investigates the application of analog adaptive techniques in load cell response correction. The load cell is a sensor with an oscillatory output in which the measurand contributes to response parameters. Thus, a compensation filter needs to track variation in measurand whereas a simple, fixed filter is only valid at one load value. To facilitate this investigation, computer models for the load cell and the adaptive compensation filter have been developed and implemented in PSpice. Simulation results are presented demonstrating the effectiveness of the proposed compensation technique.

1. INTRODUCTION

Load cells are used in a variety of industrial weighing applications. Since information processing and control systems cannot function correctly if they receive inaccurate input data, compensation of the imperfections of the sensors is one of the most important aspects of sensor research. Influence of unwanted signals, non ideal frequency response, parameter drift, non-linearity, and cross-sensitivity are the five major defects in primary sensors [2]. In the new generation of sensors, called intelligent or smart sensors, the influence of these imperfections has been dramatically reduced by using signal processing techniques.

Some sensors such as load cells have an oscillatory output, which need time to settle down. For dynamic measurement, it is important to make a decision on the measurand as fast as possible. Dynamic measurement refers to the ascertainment of the final value of a sensor signal while its output is still in oscillation. It is used to speed up the process of measurement. One example of processing that can be done on the sensor output signal is filtering to achieve response correction. Several methods have been reported addressing this problem. Software techniques for sensor compensation are reviewed in [1]. Digital adaptive techniques have been used in [7] for load cell response correction. An artificial neural network has been proposed for dynamic measurement which needs a learning phase [8]. Other methods such as employing kalman filter [5] and estimation with recursive least square (RLS) procedure [6] have also been applied for dynamic weighing systems. Almost all the above reported methods are based on digital signal processing techniques which need analog-to-digital convertors and powerful signal processors. Although digital techniques have been used efficiently, the aim of this paper is to investigate the possibility of using analog adaptive techniques for load cell response correction. The potential benefits of analog adaptive techniques compared to digital methods include higher signal processing speeds, lower power dissipations, and smaller integrated circuit areas. It should be noted that most applications of analog adaptive techniques have focused on communications and digital magnetic storage [3] and there has been little or no work on application of analog adaptive techniques to intelligent sensors which is the main focus of this paper.

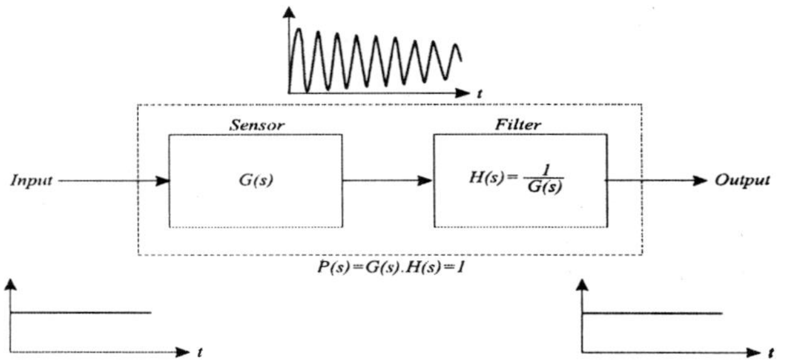

Figure 1: General principle of load cell response correction

2. LOAD CELL RESPONSE CORRECTION

The primary sensor is considered as a system with transfer function $G(s)$. The general principle for eliminating the transient time is shown in Fig.1. A filter having the reciprocal characteristic of the sensor is cascaded with it. Therefore, the transfer function of the whole system is *"unity"* which means that any changes in the input transfer to the output without any distortion. The response of a load cell can change for different measurands. For example, the characteristic of a load cell changes when a load is applied to it because the mass of the load contributes to the inertial parameters of the system. Therefore the transfer function of the filter should change accordingly. In other words, a fixed filter can be used only for one specific load value.

The general equation for the dynamic response of the load cell is given by [7]:

$$(m + m_0) \cdot \frac{d^2y(t)}{dt^2} + c \cdot \frac{dy(t)}{dt} + k \cdot y(t) = F(t) \qquad (1)$$

0-7803-7761-3/03/$17.00 ©2003 IEEE

Where m is the mass being weighed, m_0 is the effective mass of the sensor, c is the damping factor, k is the spring constant, and $F(t)$ is the force function. The Laplace transfer function of this sensor is

$$G(s) = \frac{Y(s)}{F(s)} = \frac{\frac{1}{m+m_0}}{s^2 + \frac{c}{m+m_0}s + \frac{k}{m+m_0}} = \frac{g}{s^2 + \frac{\omega_0}{Q}s + \omega_0^2} \quad (2)$$

This shows that m affects all inertial parameters of the sensor such as gain factor , g , quality factor , Q , and natural frequency , ω_0.
Eq. 2 yields a pair of complex conjugate poles $a \pm jb$ where

$$a = -\frac{c}{2(m+m_0)} \quad (3)$$

and

$$b = \sqrt{\frac{k}{(m+m_0)} - \frac{c^2}{4(m+m_0)^2}} \quad (4)$$

Thus the zeros of the adaptive filter, which are the poles of the sensor can be obtained.

In general, assume $\boldsymbol{w}$ is defined as a vector that contains all of the parameters of adaptive filter i.e.

$$\boldsymbol{w} = [w_1\ w_2\ w_3\ \ldots]^T \quad (5)$$

The elements of $\boldsymbol{w}$ can be calculated for different values of the measurand. To emphasise that $\boldsymbol{w}$ depends on m, it can be written as $\boldsymbol{w}(m)$. m is unknown in the first instance when a new measurement begins. Therefore the parameters of the adaptive filter can not be set to appropriate values in order that the filter behaves as an inverse system. Hence, an adaptive rule is required to modify the parameters of the adaptive filter according to the value of measurand. This rule is a crucial element but there is not a straightforward solution for it. Usually, in classic adaptive techniques, an adaptive algorithm, such as LMS, updates $\boldsymbol{w}$ to minimize a cost function. However, Eq.(2) shows that, for a load cell, the suitable filter has a pair of conjugate zeros, $z_{1,2} = a \pm jb$, which, a and b can be considered as the parameters of adaptive filter and the relationship between them and load can be modelled as in Eqs.(3) and (4). The real-time measurement operation is shown in Fig.2. In this block

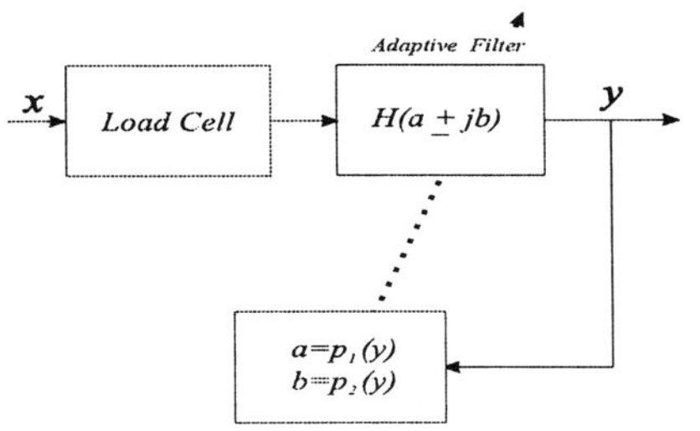

Figure 2: Block diagram showing the adaptive load cell response correction method

diagram m has been substituted with y, the output of the whole system, which is proportional to m. Initially the zeros of the filter are set to arbitrary values. Then the the output y is calculated. This new value of y is used to calculate the zeros of the filter once again. Repeating these steps results in a rapid approach to obtain the steady state value of y.

So far the zeros of a 2nd-order compensation filter have been examined. In order that the analog filter can be realised, it is necessary to add at least two poles to the filter. The values of these poles can be determined practically. For simulation purposes, these poles are selected by trail and error so that the output of the filter quickly reaches its steady-state value with minimum oscillation. The transfer function of the compensation filter is

$$H(s) = \frac{(m+m_0)}{10^{-5}} \cdot \frac{s^2 + \frac{c}{m+m_0}s + \frac{k}{m+m_0}}{s^2 + 600s + 10^5} \quad (6)$$

The transfer functions of the load cell (Eq. 2) and its compensation filter (Eq. 6) are biquadratic functions. There now exists a wealth of theoretical and experimental information on the design of fixed or non-adaptive analog biquads [4]. The problem is how to make a biquad adaptive and it is necessary to have only one filter component to track changes in m without any influence on the other parameters such as damping factor, c, and the spring coefficient, k.

3. LOAD CELL MODEL

Amongst the various biquad structures, the state-variable lowpass filter [4], shown in Fig.3, can be used to model the behaviour of the load cell.

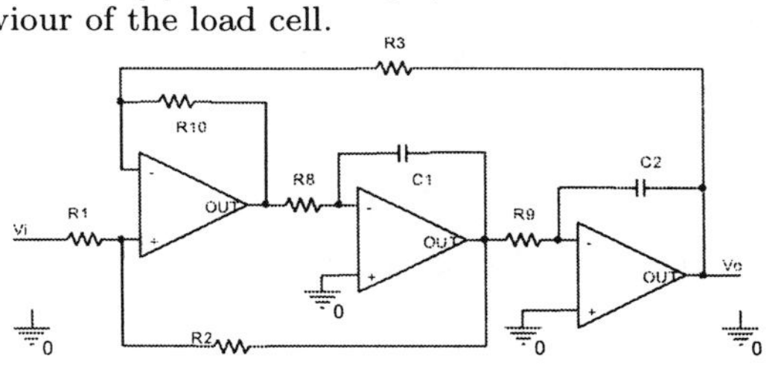

Figure 3: State-variable lowpass filter

The state-variable filter transfer function is:

$$T_{sv}(s) = K1\frac{\frac{1}{R_8R_9C_1C_2}}{s^2 + K1(\frac{R_1}{R_2R_8C_1})s + \frac{R_{10}}{R_3R_8R_9C_1C_2}} \quad (7)$$

where

$$K1 = \frac{R_2(R_3+R_{10})}{R_3(R_1+R_2)} \quad (8)$$

Comparing this transfer function with the load cell transfer function, Eq.(2), shows that R_8 can model $(m+m_0)$. R_8 has to be split into a fixed resistor equal to m_0 and a resistor proportional to m. Since m is the mass being weighed and in the model it is equivalent to stimulating voltage (V_i), the resistor has to be a voltage-controlled device whose resistance can be varied with V_i. With analog behavioural modeling facility in PSpice, it is possible to simulate such a resistor. This is achieved by using the G component(a voltage-controlled current source) and "TABLE" which allows the user to enter different resistors for different voltages. Using this voltage-controlled resistor in the lowpass filter (Fig.3) produces an analog biquadratic filter which can model the behaviour of the load cell. The complete model is depicted in Fig.4. From experimental data for a

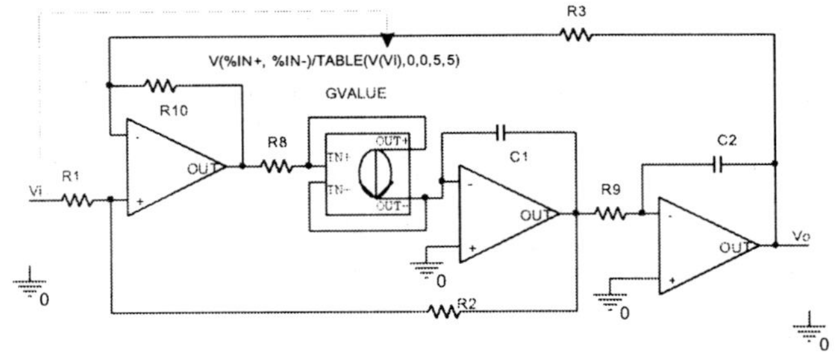

Figure 4: Load cell model

particular load cell [8] the damping factor c, spring constant k, and the effective mass of the load cell m_0, are 3.5, 2700 Pa, and 0.5 kg, respectively. These numbers are used to determine the values for resistors and capacitors in Fig.4.

For step excitation, the input voltage of the model is a step function whose amplitude is proportional to m. The simulation results for two different values of m are shown in Figs.5 and 6 which indicate that changing the input, similar to the practical case, varies all inertial parameters of the output waveform such as the steady state value, resonant frequency and damping factor.

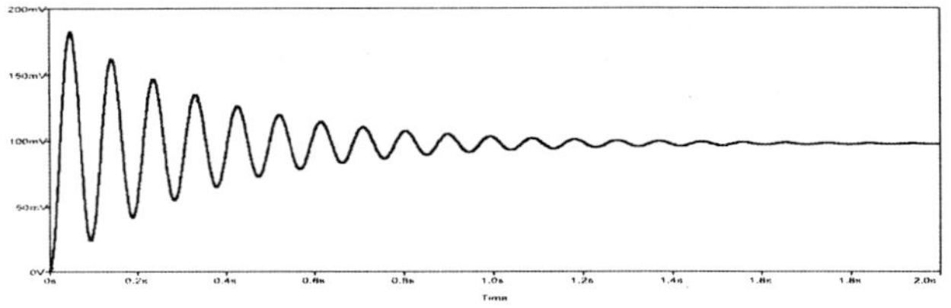

Figure 5: Output of the load cell model for $m = 0.1kg$

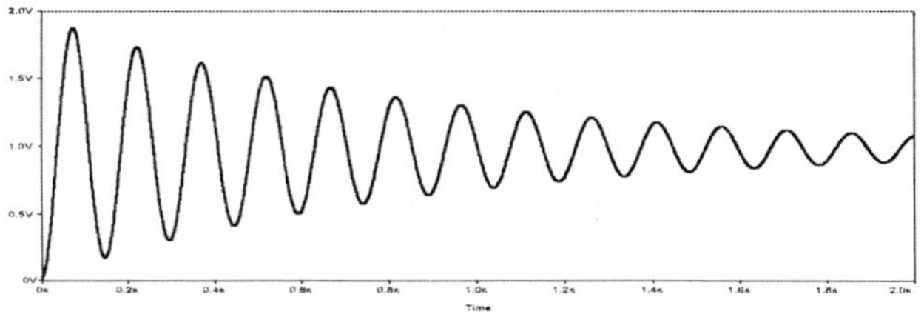

Figure 6: Output of the load cell model for $m = 1kg$

4. ADAPTIVE COMPENSATION FILTER MODEL

Since the transfer function of the compensation filter (Eq. 6) is a biquadratic function, different scaled outputs in the state variable filter, shown in Fig.3, need to be added to form a complete biquad. To make this biquad adaptive, as described in the block diagram of Fig.2, the filter's zeros have to be changed by the output of the biquad. Similar to the sensor model approach, it is possible to use a voltage-controlled resistor in the compensation filter. The filter output voltage is used to control this resistance. The complete adaptive biquad is shown in Fig.7. The transfer

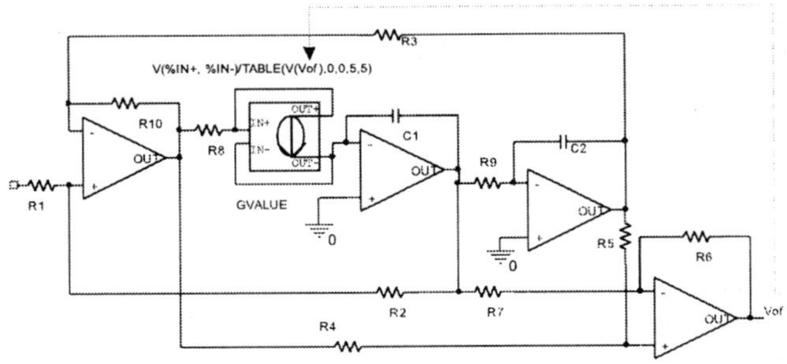

Figure 7: Adaptive compensation filter model

function of this filter is

$$H(s) = K\frac{s^2 + \frac{R_6(R_4+R_5)}{R_5R_8C_1(R_6+R_7)}s + \frac{R_4}{R_5R_8R_9C_1C_2}}{s^2 + K1(\frac{R_1}{R_2R_8C_1})s + \frac{R_{10}}{R_3R_8R_9C_1C_2}} \tag{9}$$

where

$$K = K1\frac{R_5(R_6+R_7)}{R_7(R_4+R_5)} \tag{10}$$

and $K1$ was previously defined in Eq.(8). Similar to the sensor model, R_8 consists of a fixed resistor and a voltage-controlled resistor whose resistance is controlled by the filter's output voltage. In other words R_8 models $(m + m_0)$.

The adaptation sequence can be described as follow. Before stimulating the load cell, filter output voltage is zero and the initial transfer function of the filter will be

$$H_0(s) = g \cdot \frac{(s - a_0 - jb_0)(s - a_0 + jb_0)}{(s - d - je)(s - d + je)} \tag{11}$$

Where g is the gain factor of the filter, a and b are the real and imaginary parts of filter's zeros respectively, d and e are real and imaginary parts of filter's poles respectively, and the subscript ($_0$) denotes the initial values. The zeros of the filter need to cancel the poles of the sensor i. e. a and b are the same as eqs.(3) and (4). Since a and b depend on m (the output of the filter) which is unknown at first, they cannot be fixed values. The initial values for a and b are:

$a_0 = -\frac{c}{2(0+m_0)}$ and $b_0 = \sqrt{\frac{k}{(0+m_0)} - \frac{c^2}{4(0+m_0)^2}}$

When the input is applied to the filter with initial transfer function of $H_0(s)$, it produces an output, say m_1. Since the zeros of the filter change with the output voltage, the new values for a and b will be a_1 and b_1 and then the transfer function of the filter changes to

$$H_1(s) = g \cdot \frac{(s - a_1 - jb_1)(s - a_1 + jb_1)}{(s - d - je)(s - d + je)} \tag{12}$$

With this new transfer function, the filter produces a new output that changes the filter's zeros again and this procedure continues until a and b converge to their final values.

It should be noted that the poles of this compensation filter (Eq. 9) vary as m varies, which is not the case with the filter model (Eq. 6). However, this does not represent a problem because the load cell does not have any zeros and for pole-zero cancellation only the zeros of the filter are important. In addition, the variation of filter's poles can be tolerated as long as the filter remains stable and does not create significant oscillation in the output. To examine the stability of the adaptive compensation filter, its transfer function is considered as follow

$$H(s) = g \cdot \frac{s^2 + \frac{c}{m+m_0} \cdot s + \frac{k}{m+m_0}}{s^2 + \frac{600}{m+m_0} \cdot s + \frac{10^5}{m+m_0}} \qquad (13)$$

The poles of the filter are : $p_{1,2} = d \pm je$ where $d = -\frac{600}{2(m+m_0)}$ and $e = \sqrt{\frac{10^5}{(m+m_0)} - \frac{600^2}{4(m+m_0)^2}}$

As m varies from its initial value, $m = 0kg$, to its final value, for example $m = 1kg$, the real and imaginary parts of the filter's poles, d and e will change. The root locus of the filter's poles in s-plane is shown in Fig.8. This figure

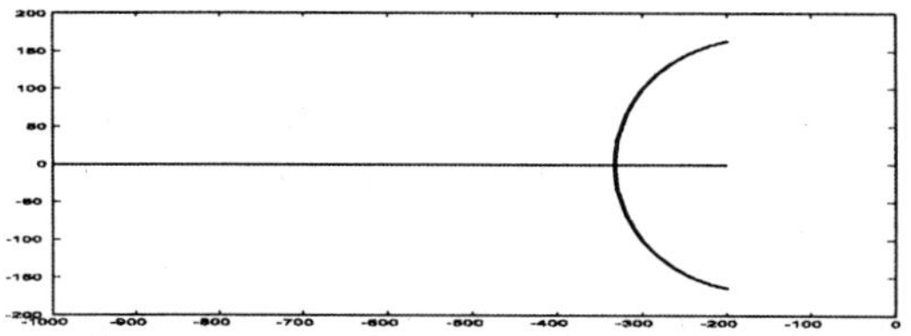

Figure 8: Filter poles root-locus for m from 0 to $1kg$

shows that the poles of the filter remain in the left hand side of $j\omega$ axes for all values of m from 0 to 1kg. Moreover, when $m \longrightarrow \infty$, the real part of poles are still negative and hence the filter remains stable for all values of m.

5. SIMULATION RESULTS

In this section the load cell and the adaptive compensation filter models will be used to examine how analog adaptive techniques can be used for load cell response correction. Fig.9 shows the load cell output and the compensation filter output for $m = 0.1kg$. To illustrate the capability in tracking changing in m, Fig.10 shows the results when $m = 1kg$. Clearly the simulation results show that analog adaptive biquad filter, shown in Fig.7, can be applied for response correction of the second order sensors.

To indicate the necessity for using an adaptive filter, a fixed filter is used for compensation. The filter is adjusted for $m = 0.1kg$. If the sensor is stimulated with $m = 0.1kg$, the result is the same as Fig.9, but for $m = 1kg$ stimulation, the input and output of the filter is depicted in Fig.11, which shows that the fixed filter is unable to perform response correction when m varies.

6. CONCLUDING REMARKS

This paper has addressed response correction of the load cell sensor using analog adaptive techniques. It has been shown that the state-variable biquadratic filter provides accurate and flexible sensor and adaptive compensation filter models. The load cell model in addition to tracking the variation in the mass being weighed, allows the user to vary the other parameters including damping factor and spring coefficient. The effectiveness of the models has been validated by simulation. Further work is aimed at practical implementation of the analog adaptive filter for load cell response correction.

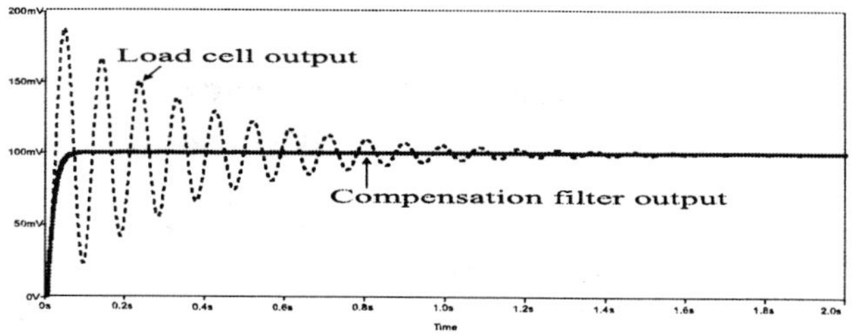

Figure 9: Result of adaptive compensation for $m = 0.1kg$

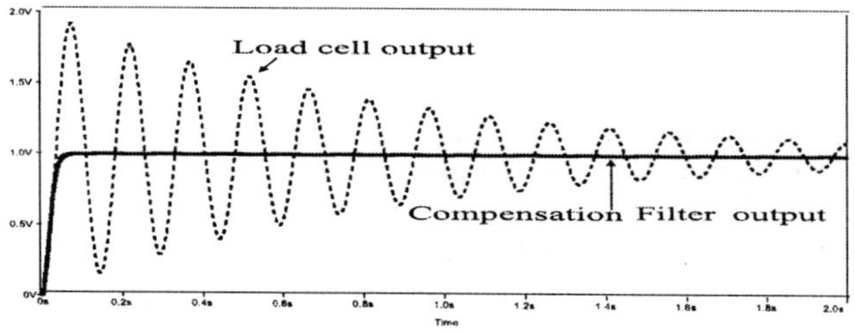

Figure 10: Result of adaptive compensation for $m = 1kg$

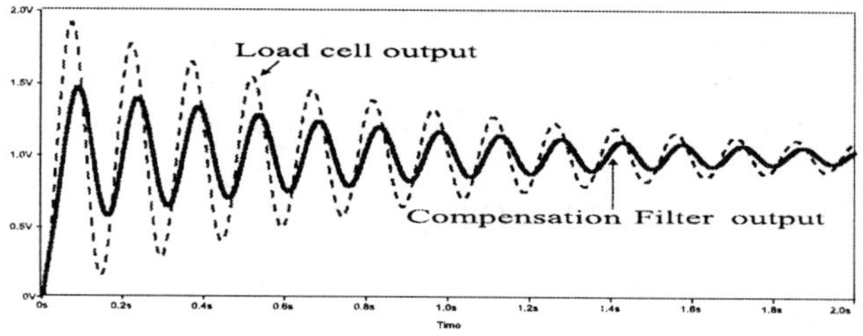

Figure 11: Fixed filter for load cell response correction when m varies from 0.1 to 1kg

7. REFERENCES

[1] J E Brignell. Software techniques for sensor compensation. *Sensors and Actuators A*, 25-27:29–35, 1991.

[2] J E Brignell and N M White. *Intelligent Sensor Systems.* Instiitute of Physics Publishing Ltd, 1994.

[3] A. Carusone and D. A. Johns. Analogue adaptive filters: Past and present. *IEE Proceedings on Circuits, Devices, and Systems*, 47(1):82–90, February 2000.

[4] W.-K. Chen. *Passive and active filters.* John Wiley and sons, Inc., 1986.

[5] M. Halimic and W. Balachandran. Kalman filter for dynamic weighing system. pages 787–791, IEEE Internatiol Symposium on industrial electronics, July 1995.

[6] W-Q. Shu. Dynamic weighing under nonzero initial condition. *IEEE Transaction on Instrumentation and measurement*, 42(4):806–811, August 1993.

[7] W J Shi N M White and J E Brignell. Adaptive filters in load cell response correction. *Sensors and Actuators A*, A 37-38:280–285, 1993.

[8] S M T Alhoseyni Almodarresi Yasin and N M White. The application of artificial neural network to intelligent weighing systems. *IEE proceedings- Science, Measurement and Technology*, 146:265–269, November 1999.

Bias-Adaptive Cross-Coupled CMOS MAGFET Pair for Bipolar Magnetic Field Detection

Z.Q. Li [a], *X.W. Sun* [a,*], *W. Fan* [b], *G.J. Qi* [b]

[a] School of EEE, Nanyang Technological University, Nanyang Avenue, Singapore 639798
[b] Singapore Institute of Manufacturing Technology, 71 Nanyang Drive, Singapore 638075
*Email: exwsun@ntu.edu.sg

ABSTRACT

In the conventional cross-coupled CMOS magnetic field-effect transistor (MAGFET) pair, at least one MAGFET is self-biased. The output swing at the self-biasd end is then inevitably limited to the threshold voltage of the self-biased MAGFET. This problem emerges for bipolar magnetic field sensing. In this paper, we propose a bias-adaptive voltage level shifter to remove the direct self-bias connection between the gate and the drain and to adjust the operating point at the output node, so as to achieve symmetric yet maximum positive and negative output swings. The improvements have been verified by HSPICE simulation.

1. INTRODUCTION

Due to the good linearity and fully-compatibility with standard CMOS process, the MAGFET is often combined together with other integrated electronic circuits to detect the strength of an applied magnetic field [1][2][3] or to realize some magnetic field-related functions [4][5][6].

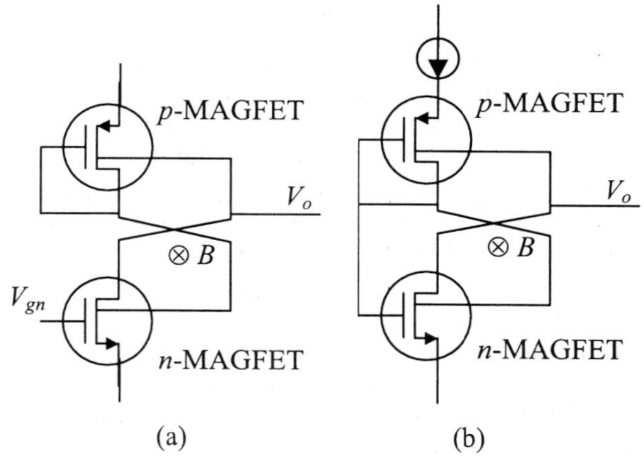

Figure 1. Cross-coupled CMOS MAGFET pair. (a) circuit schematic with *p*-MAGFET self-biased, and (b) circuit schematic with *p*- and *n*-MAGFETs both self-biased

When an *n*-channel MAGFET and a *p*-channel MAGFET are combined in such a way that the drains of the two MAGFETs are cross-connected, a cross-coupled CMOS MAGFET pair (CCMAGFET) is formed [7][8], the basic circuit schematics of which, with *p*-channel MAGFET self-biased and with *p*- and *n*-channel MAGFETs both self-biased, are drawn in Fig. 1.

As described in [7], the CCMAGFET performs the following four operations simultaneously: magnetic field detction, working point control, differential to single-ended conversion, and amplification. Indeed, a high sensitivity, for example, more than 4 V/T from our experience, can be achieved without difficulty, which is more than 100 times as large as that achieved from the so-called *B*-to-*V* converter [9] which consists of a single *n*-MAGFET followed by much more complicated readout circuitry.

The CCMAGFET combines the MAGFET together with the amplifier to achieve high sensitivity with simple electronic circuitry. However, it is obvious from [7] that the negative output swing is much less than the positive one. This is because the n-channel MAGFET was self-biased. With the configuration of Fig. 1(a), the positive output swing will be limited. If the CCMAGFET is configured as in Fig. 1(b), the symmetric positive and negative swings can be achieved, but both are reduced, and finally the total range of the output swing is sacrificed.

In order for the CCMAGFET to achieve as large as possible output swing, we need to set the bias condition such that the absolute drain to source saturation voltages of the *n*- and *p*-channel devices, $|V_{dsat}|=|V_{gs}-V_{th}|$, are as small as possible. This is also preferable for power saving, and not harmful to the absolute magnetic sensitivity, which has been verified by [7] and our experiments. Usually, the *n*- and *p*-MAGFETs need to be matched in size [7][8] for the two devices to work in equal effective voltage at the same time. In such situation, the quiescent operating point at the output node, V_{oq}, for the case of *p*-MAGFET self-biased, will unfortunately be set to $V_{oq}=V_{dd}-|V_{thp}|-(V_{gsn}-V_{thn})$, and the positive swing of v_o is then equal to $(V_{dd}-V_{oq})-(|V_{gsp}|-|V_{thp}|)=|V_{thp}|$. Therefore, for the CCMAGFET with the basic circuit configuration as shown in Fig. 1, the output swing at the self-biased MAGFET side is always limited to its threshold voltage. If the *n*- and *p*-MAGFETs both are self-biased as in Fig. 1(b) to achieve the symmetric positive and negative output swings, the two

0-7803-7761-3/03/$17.00 ©2003 IEEE

swings will be set to $|V_{thp}|$ and V_{thn}, respectively, and thus the magnetic field detection range of the CCMAGFET will be very much limited. In this paper, we present a biasing circuit to detach the direct self-bias connection and to realize maximum yet symmetric positive and negative output swings.

2. PRINCIPLES

In order to extend the output swing of the CCMAGFET while keeping symmetric positive and negative swings, the traditional and classical self-bias connection must be broken up. We here introduce a voltage level shifter, which is inserted in between the gate of the *p*-MAGFET and the node of V_{oq} as shown in Fig. 2, to bias the *p*-MAGFET.

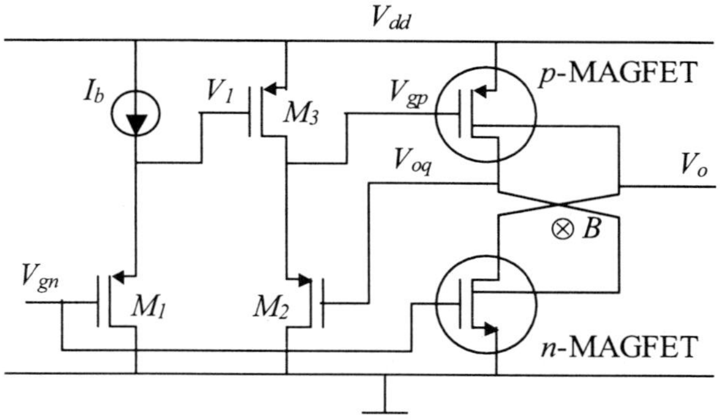

Figure 2. Circuit schematic of ross-coupled CMOS MAGFET pair with bias-adaptive voltage level shifter

Assume firstly the *n*- and *p*-MAGFETs are matched with each other, which means $\mu_p C_{ox}(W/L)_{p-MAGFET} = \mu_n C_{ox}(W/L)_{n-MAGFET}$ with W of the width and L of the length of the respective MAGFETs and all the other parameters of the conventional meanings; secondly the two *p*-MOSFETs, M_2 and M_3, have the same aspect ratios, i.e., $(W/L)_2 = (W/L)_3$; and thirdly the threshold voltages of the n- and p-channel transistors have equal absolute value, i.e., $V_{thn} = |V_{thp}|$. Because the two MAGFETs in Fig. 2 are in series, we have

$$V_{gp} = V_{dd} - V_{gn} \tag{1}$$

The $I-V$ characteristics of the three transistors, M_1, M_2 and M_3, can be expressed as

$$I_1 = -\frac{\mu_p C_{ox}}{2}\left(\frac{W}{L}\right)_1 \left(V_1 - V_{gn} - |V_{thp}|\right)^2 \tag{2}$$

$$I_2 = -\frac{\mu_p C_{ox}}{2}\left(\frac{W}{L}\right)_2 \left(V_{gp} - V_{oq} - |V_{thp}|\right)^2 \tag{3}$$

$$I_3 = -\frac{\mu_p C_{ox}}{2}\left(\frac{W}{L}\right)_3 \left(V_{dd} - V_1 - |V_{thp}|\right)^2 \tag{4}$$

Based on the above assumptions and the facts that $I_1 = -I_b$ and $I_2 = I_3$, we get by solving Eqns. (1) ~ (4)

$$V_{oq} = |V_{thp}| + \sqrt{\frac{I_b}{(\mu_p C_{ox}/2)(W/L)_1}} \tag{5}$$

Thus, we can achieve the most desirable quiescent operating point at the output node which is independent of the voltage bias V_{gn} but of the current bias I_b. If setting $V_{oq} = V_{dd}/2$, the headroom of the negative output swing of the CCMAGFET is now

$$V_{hrn} = V_{oq} - V_{dsatn} = \frac{V_{dd}}{2} - \left(V_{gn} - V_{thn}\right) \tag{6}$$

and the headroom of the positive swing is

$$V_{hrp} = V_{dd} - V_{oq} - |V_{dsatp}| = \frac{V_{dd}}{2} - \left(V_{gn} - V_{thn}\right) \tag{7}$$

The voltage difference between nodes V_{gp} and V_{oq} is

$$V_{gp} - V_{oq} = \frac{V_{dd}}{2} - V_{gn} \tag{8}$$

which is bias (V_{gn})-adaptive. This means that, when V_{gn} changes, V_{gp} also changes according to Eqn. (1), while $V_{gp} - V_{oq}$ tracks the variation of V_{gn}, so as to keep V_{oq} fixed at the most desirable value.

Therefore, we have detached the direct connection between the gate and the drain of the *p*-MAGFET and realized symmetric negative and positive output swings with maximum headroom.

In order for the whole circuit to function well, all the transistors need to work in saturation region, which requires V_{gn} to meet

$$V_{gn} > V_{thn} \tag{9}$$

and

$$V_{dd} - V_1 > |V_{thp}| \tag{10}$$

Inserting Eqn. (2) and $I_1 = -I_b$ into Eqn. (10) gives

$$V_{gn} < V_{dd} - 2|V_{thp}| - \sqrt{\frac{I_b}{(\mu_p C_{ox}/2)(W/L)_1}} \tag{11}$$

By combining Inequalities (9) and (11) we achieve the working condition of V_{gn} as

$$V_{thn} < V_{gn} < V_{dd} - 2|V_{thp}| - \sqrt{\frac{I_b}{(\mu_p C_{ox}/2)(W/L)_1}} \tag{12}$$

under which the CCMAGFET with voltage level shifter can function properly. When setting $V_{oq} = V_{dd}/2$, Inequality (12) reduces to

$$V_{thn} < V_{gn} < V_{dd}/2 - |V_{thp}| \tag{13}$$

3. HSPICE SIMULATION RESULTS AND DISCUSSION

It has been shown that the HSPICE is a reliable tool for the computer simulation of the MAGFET circuits by using the proper macro models of the MAGFETs [9]. We have carried out the HSPICE simulation on the circuit shown in Fig. 2 using the SPICE macro model for the MAGFET [9][10]. The device model used in the simulation was provided by a commercial foundry of their standard 0.6 μm 5 V CMOS process. In addition, the relative sensitivities of the *n*- and *p*-MAGFETs and the absolute sensitivity of the CCMAGFET for the simulation were experimentally determined. In applying the measured sensitivities to the simulation, the current feedthrough between the two drains of the individual MAGFETs has been taken into account due to the fact that, in the application of the CCMAGFET, the voltage difference between the two drains of each MAGFET are significant.

3.1 Operating point simulation

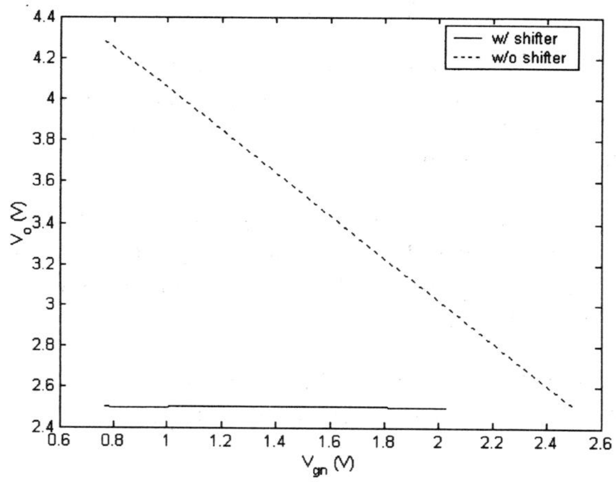

Figure 3. Dependency of operating point of V_o on bias V_{gn} for CCMAGFET with (solid line) or without (dotted line) voltage level shifter

Fig. 3 shows the bias V_{gn}-dependency of the quiescent output voltage V_{oq} of the CCMAGFET with (solid line) and without (dotted line) voltage level shifter. When $V_{gn} < V_{thn}$, the *n*-MAGFET enters subthreshold region; when $V_{gn} > 2$ V, the MOSFET M_3 is cut off. Therefore, Fig. 3 only shows the operating range of V_{gn} from 0.8 V to 2 V for the case of having voltage level shifter. It can be seen from Fig. 3 that, by employing the level shifter, the quiescent output voltage of the CCMAGFET becomes stable (almost fixed at the preset 2.5 V in the simulation) within a relatively large input range of V_{gn} from 0.8 V to 2 V. This complies well with Eqns. (5) and (12).

3.2 DC analysis

The DC analysis result is depicted in Fig. 4. In this figure, the output voltage V_o is plotted as a function of the magnetic flux density *B* applied perpendicularly to the MAGFET for two biasing conditions of V_{gn} = 1 V and 2 V, respectively. The solid line shows the result with level shifter as in Fig. 2, while the dotted line shows that without level shifter.

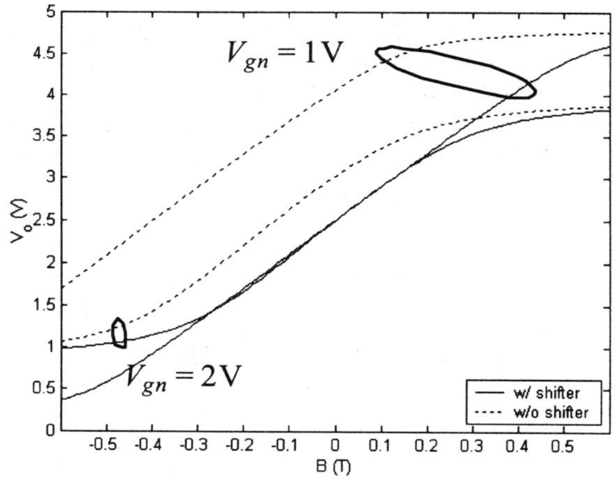

Figure 4. DC analysis at V_{gn} = 1 V and 2 V, respectively, for CCMAGFET with (solid line) or without (dotted line) voltage level shifter

It is evident from Fig. 4 that, when V_{gn} increases, the total output swing of the CCMAGFET decreases accordingly as explained in the introduction. Under the same bias, the CCMAGFET with or without voltage level shifter has the same magnetic sensitivity (the same slope). Also it is shown in Fig. 4 that, without level shifter, the positive output swing of the CCMAGFET is much limited and the operating point shifting of the output is significant. By inserting the level shifter, the performance of the CCMAGFET is greatly improved. The quiescent operating point is compensated and the output swing becomes symmetric. This improvement would not impair the magnetic sensitivity at all.

3.3 Transient analysis

The transient performance of the CCMAGFET with and without level shifter was also simulated. Fig. 5 shows the simulation results. In this figure, the biasing voltage V_{gn} is set to 1 V, and the magnetic flux density is set to be bipolar and varying with time with a triangular waveform. The solid line shows the result with level shifter, while the dotted line shows that without level shifter. For better comparison, the offset of the CCMAGFET without voltage level shifter has been removed. Fig. 5 clearly demonstrates that the output of the CCMAGFET without level shifter is distorted in the positive part due to the limited positive output swing as shown in Fig. 4. With the level shifter, the output follows the varying magnetic flux density well in triangular shape.

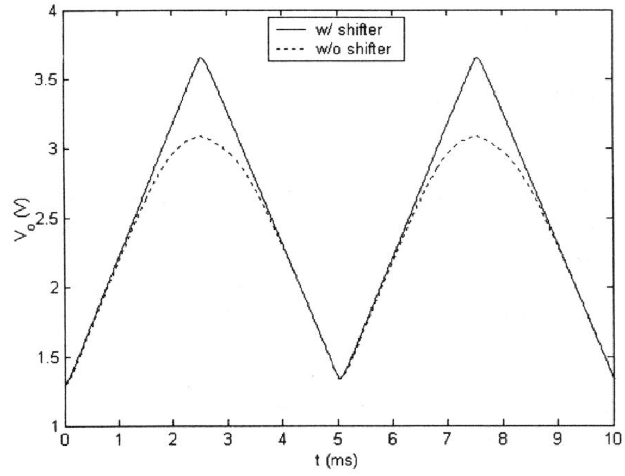

Figure 5. Transient analysis for CCMAGFET with (solid line) or without (dotted line) voltage level shifter

4. CONCLUSION

We have presented a bias-adaptive voltage level shifter for the cross-coupled CMOS MAGFET pair. The classical self-bias configuration for active load has been improved for achieving better output swing performance. The functioning of the CCMAGFET with the level shifter has been verified by HSPICE simulation. The simulation results show significant improvements on shifting and stablizing the operating point and achieving the maximum and symmetric output swing for bipolar magnetic field detection by employing this voltage level shifter. Reduction of power consumption can be achieved simultaneously with extension of output swing. This new circuitry may also be useful in pure electronic circuits for bias-adaptive applications.

5. ACKNOWLEDGEMENT

The authors would like to thank the financial support of this project from the Academic Research Fund RG 16/99 of Nanyang Technological University.

6. REFERENCES

[1] D. Misra and T. R. Viswanathan, "Circuit design for a CMOS magnetic field sensor," in *Proc. IEEE Int. Symp. Circuits Syst.*, 1986, pp. 1183-5, New York, NY, USA.

[2] C. Rubio, S. Bota, J. G. Macias and J. Samitier, "Monolithic integrated magnetic sensor in a digital CMOS technology using a switched current interface system," in *Proc. IEEE 17th Instrum. Meas. Technol. Conf.*, 2000, pp. 69-73, Piscataway, NJ, USA.

[3] D. Misra, "A novel magnetic field sensor array," *IEEE J. Solid-State Circuits*, vol. 25, pp. 623-25, 1990.

[4] A. Nathan, I. A. McKay, I. M. Filanovsky and H. P. Baltes, "Design of a CMOS oscillator with magnetic-field frequency modulation," *IEEE J. Solid-State Circuits*, vol. SC-22, pp. 230-2, 1987.

[5] Michael P. Flynn, Michael Buckley and John Ryan, "A CMOS Magnetic-Field-Controlled Duty-Cycle Oscillator Sensor," in *Proc. IEEE Custom Integrated Circuits Conf.*, 1992, pp. 23.4/1-4, New York, NY, USA.

[6] Carlos A. dos Reis Filho and Hugo Jimenez Grados, "Analog CMOS Tesla-Volt Multiplier Circuit," in *Proc. IEEE Third Int. Caracas Conf. Devices, Circuits Syst.*, 2000, pp. C82/1-4, Piscataway, NJ, USA.

[7] Radivoje S. Popovic and Heinrich P. Baltes, "A CMOS magnetic field sensor," *IEEE J. Solid-State Circuits*, vol. SC-18, pp. 426-8, 1983.

[8] Z. Q. Li, X. W. Sun and S. C. Tan, "Towards Achieving Maximum Magnetic Sensitivity with the Cross-Coupled CMOS MAGFET Pair", *IEE Electronics Letters*, vol. 38, pp. 709-710, 2002.

[9] Shen-Iuan Liu, Jian-Fan Wei and Guo-Ming Sung, "SPICE Macro Model for MAGFET and its Applications," *IEEE Trans. Circuits Syst. II*, vol. 46, pp. 370-5, 1999.

[10] Dirk Killat, Johannes v. Kluge, Frank Umbach, Werner Langheinrich and Richard Schmitz, "Measurement and modelling of sensitivity and noise of MOS magnetic field-effect transistor," *Sensors and Actuators A*, vol. 61, pp. 346-51, 1997.

PHOTO-THERMOELECTRIC POWER GENERATION FOR AUTONOMOUS MICROSYSTEMS

S. Baglio, S. Castorina, L. Fortuna, N. Savalli
Dipartimento di Ingegneria Elettrica Elettronica e dei Sistemi
University of Catania
Viale Andrea Doria 6, 95125 Catania, ITALY

ABSTRACT

This work deals with the feasibility study, the modeling and the design of an integrated photo-thermoelectric micro generator for autonomous microsystems. In particular each single part of such an autonomous system, composed by moving elements, and energy supply system has been just developed by the authors, taking into account a standard CMOS technology and suitable micro-machining compatible processes. Micro lenses have been used to achieve high efficiency for the thermoelectric conversion. The discussed work expands our photo-thermo-mechanical actuation strategy and allows to further enhance the device efficiency.

1. INTRODUCTION

The development of a novel fully integrated and autonomous microsystem, focusing on the concrete limitation represented by the energy supply system is reported in this work. Modeling and design considerations of a system based on both micro-lenses, which focus the light onto the anchoring point of a bilayer moving-element and the hot junction of a thermoelectric generator, will be addressed.

While, in general, sensors and electronics, require low powers, the actuators introduce the main restrictions in the autonomy of the microsystem. The problem of the high power, required by the actuators, has been treated in a previous work [1], where an innovative photo-thermo-mechanical actuation system has been presented. Extending the concepts reported in that work, a micro-system where both the energy for the actuation, and that for the electronics, are provided from the external environment through an external light source have been conceived.
Since sensors and electronics are the low power sections of a microsystem, they could be supplied by micro batteries, however the energy stored in a battery scales down with its volume, and very small batteries could not provide enough energy and/or not guarantee enough autonomy to the microsystem. On the other hand, they could have size and weight bigger than those of the microsystem and rechargeable batteries should be used if autonomous microsystems are taken into account.
Thin Film micro-batteries, realized with microelectronic techniques have been developed, as on-chip power source [2]. However, their fabrication requires non-standard materials and processes and, furthermore, due to their small size, they will need an external power supply source for recharging. Some other problem arises with the use of wires to connect the microsystem to an external macroscopic power supply due to the introduction of interconnections, especially for arrays of microsystems, of noise (due to the stray capacitances) and cross talk between power and signal lines [3]; furthermore, in the case of micro-robots, interconnections wires seriously affect their mobility.
From the above considerations, it arises that an external power source connected by wires will reduce the flexibility of applications for a MEMS autonomous devices, while internal energy cells will require regular recharging. Therefore it is desirable to have a regenerating power source embedded into the system.

Several approaches for supplying energy to a (micro) system from an external power source has been presented in literature [4-6].
In a previous work, the feasibility, the modeling and the design of a photo-thermo-mechanical actuation system [7] that directly exploits the thermal power generation of light and provides the high energy required by thermal actuators have been shown. The system presented can achieve high efficiency values by using micro-machined lenses that locally improve the energy density and, at the same time, allow the use of a low power light source, thus avoiding an excessive heating of the whole system and preventing the on-board microelectronics from damages. Moreover, using lenses the problem of focusing over a small area such that represented by the hot junction of the thermopile can be overcome. In fact, as already pointed out for the photo-thermo-mechanical approach, the whole device is lighted but at a low energy while this increases under the lenses to a level suitable for the thermocouples supplying.
The conceived strategy, couples the photo-thermal conversion system to an integrated thermoelectric generator [8, 9] for the on-chip generation of electric power, as schematically shown in Fig. 1. The photo-thermo-mechanical actuation system can be realized in CMOS-like technology process, that is fully compatible with on-board electronics. In particular a standard CMOS 0.8 μm (AMS) process has been adopted to design and realize the thermal bilayer actuators and the thermopiles.

Since light have been used as external energy source, the comparison with micro-photovoltaic cells as electric power supply system [6], will be considered. Even if, as the miniaturized micro-batteries, they are based on non-standard fabrication process and materials, such suitable figure of merit allow for comparing these two strategies strictly dependent on energies supplied by a light source.

The key point of this work is that the on-chip electrical power generation through thermoelectric generators, i.e. thermopiles, can be added at zero cost to a the photo-thermo mechanical actuation system, because the elements of the thermoelectric generator, the thermocouples, can be realized with any couple of

0-7803-7761-3/03/$17.00 ©2003 IEEE

conducting, but different, materials available in any integrated technology with minimum layout design efforts.

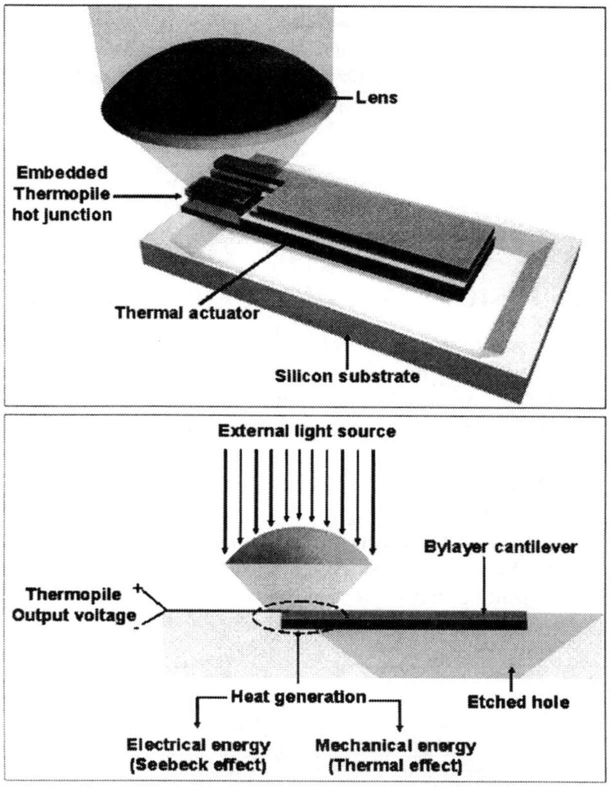

Fig.1: 3D model and Cross section of the combined photo-thermo-mechanical and photo-thermo-electrical energy supply systems.

2. ANALYSIS OF THE PHOTO-THERMO-ELECTRIC GENERATOR

The thermoelectric generator has been modeled as a series connection of an ideal voltage source V_O and its internal resistance R_O (as any real voltage source). The voltage V_O, depending on the temperature difference ΔT that exists between the hot and the cold junction of the thermopile, as a function of the relative Seebeck coefficient α [8, 9], for a thermopile made by N elements is:

$$V_O = N\alpha\Delta T \tag{1}$$

The power P delivered to the load R_L assumes its maximum value P_{MAX} when the load resistance equals the source internal resistance R_O, therefore:

$$P_{MAX} = V_O^{\,2}/4R_O \tag{2}$$

by substituting eq. (1) in eq. (2), the relationship between the maximum power and the temperature difference across the thermoelectric generator is:

$$P_{MAX} = (N\alpha\Delta T)^2/4R_O \tag{3}$$

The internal resistance of the thermoelectric generator is a function of the electrical resistivity ρ_1 and ρ_2 of the thermo elements' materials and their length and cross section. By assuming, without loosing of generality, that both the thermo elements have the same length $\boldsymbol{l}$ and cross section $\boldsymbol{A}$, results:

$$R_O = N(\rho_1 + \rho_2)l/A = N\rho_T\, l/A \tag{4}$$

By substituting eq.(4) in eq.(3), the maximum power can be expressed in terms of the electrical resistivity and the geometry of the thermo elements, resulting in the useful design equation:

$$P_{MAX} = \frac{NA\alpha^2\Delta T^2}{4l\rho_T} \tag{5}$$

In eq.(5), N, A and l are design parameters, while α and ρ_T are process parameters. The temperature difference is a function of the incident power, therefore a thermal analysis should be performed to determine this relationship.

By considering a single thermocouple it can be observed that, from the thermal point of view, the two thermo elements experience the same temperature difference ΔT and there exists a total heat rate Q that flow through them, therefore they are thermally connected in parallel.

The mass and lateral surface of the thermo elements are very small compared to those of the substrate, furthermore for its entire length, except for the two junctions, the thermocouple can be thought as immersed in a thermal insulating medium (or a medium which thermal conductivity is very small compared to that of the substrate); therefore it can be assumed that there are no heat losses due to thermal conduction and convection along the thermocouple. Losses due to thermal radiation can also be neglected due to the fact that the maximum temperature should be maintained below 400-500°C to avoid damages to the materials and the structures. Therefore, the heat rate through the hot and the cold junctions respectively, will be dissipated by conduction through the substrate.

By the above assumptions the two thermo elements can be modeled as a single, equivalent thermo element, subjected to a heat rate Q and with one of its ends (the cold junction of the thermocouple) at a fixed temperature T_0. By the above stated hypothesis the substrate has been reasonably assumed to be a heat reservoir with respect to the thermocouples. Note that T_0 is not the ambient temperature because a certain heating of the substrate cannot be neglected. The average temperature reached by the substrate after a transient must be considered. It has been assumed constant and small enough with respect to the hot junction temperature T_1, if the operating time of the light source is not so long to cause the excessive heating of the whole chip.

Since thermal convection and radiation can be neglected, as stated above, the heat rate Q is transmitted through the equivalent thermo element by thermal conduction, therefore it is related to the temperature gradient that exists across the thermo element through the Fourier's law [10]:

$$Q = -kA(dT/dx) \tag{6}$$

where k is the equivalent thermal conductivity.

Eq. (6) can be written in a more useful form in terms of the heat flux $q = Q/A$:

$$q = -k(dT/dx) \tag{7}$$

Eq. (7) can be solved to determine the temperature of the hot junction T_1 or, equivalently, the temperature difference across the thermo elements ΔT. The heat flux q id directly related to the energy density of the light source and, in the case of pulsed light source (that is, in the most practical case), it corresponds to the root mean square value of the pulsed heat flux.
By integrating eq. (7) along the equivalent thermo element, where the temperature varies from T_1 (hot junction) to T_0 (cold junction), and by solving it for the temperature difference ΔT, results:

$$\Delta T = q\,l/k \tag{8}$$

As it can be easily shown [10], since the thermocouple is composed by the parallel connection of two elements, the equivalent thermal conductivity k is the sum of the thermal conductivity of each element: $k = k_1+k_2$.
By substituting eq. (8) in eq. (5), the relationship between the maximum power deliverable by the thermoelectric generator P_{MAX}, and the energy density of the light source q is given by:

$$P_{MAX} = \frac{NAl\alpha^2 q^2}{4(\rho_1+\rho_2)(k1+k2)^2} \tag{9}$$

This first design-equation shows that the energy conversion method has an intrinsic high efficiency, because the electrical power generated depends on the square power of the incoming energy density.
Further improvements in terms of energy efficiency have been achieved by introducing micro lenses. As shown in [7], the lens collects the energy coming from the light source over a surface area larger than the flat surface of the chip, and concentrates that energy over a smaller spot, thus introducing a gain in terms of energy density, given by the ratio between the convex surface area of the lens and the area of the spot where the light is focused, that is:

$$q = q_s\, A_L / A_H \tag{10}$$

where q_s is the energy density of the light source (or its root mean square value), A_L is the area of the convex surface of the lens and A_H is the area of the spot where light is focused.
In the previous analysis the lens has been considered as an ideal optical element, without losses in the transfer of energy. The introduction of the gain of the lens $G = A_L/A_H$ in eq. (9), gives:

$$P_{MAX} = \frac{NAl\alpha^2 G^2 {q_s}^2}{4(\rho_1+\rho_2)(k_1+k_2)^2} \tag{11}$$

It can be highlighted as the advantage deriving from the introduction of the lens is the improvement of the maximum power, proportionally to the square power of the lens' gain. The use of the micro-lenses will improve the efficiency of the whole energy supply system, and will allow the on-chip generation of a higher electrical power.
Moreover it can be highlighted as this approach produces thermo-electric power without a significant increase in the whole device temperature. In fact, the temperature difference between the thermocouple junctions is always ensured by the use of lenses without the need of large values of energy over the device surface.
Eq. (11) can be rewritten in terms of power per unit of volume $p = P_{MAX}/V$, resulting in a useful equation to compare the power supply system described here to other systems presented in literature. Since the volume of the thermoelectric generator can be expressed as: $V = NAl$, results:

$$p = \frac{\alpha^2 G^2 {q_s}^2}{4(\rho_1+\rho_2)(k_1+k_2)^2} \tag{12}$$

3. DESIGN OF A DEVICE PROTOTYPE

As previously stated, the adopted technology has allowed to design different thermoelectric devices by using several combinations of the electrical conducting materials, as the two poly-silicon layers, the two metal layers and the p+/doped silicon layer.
The poly 1/metal 1 and p+ diffusion/metal 1 combinations has been adopted for the design of two thermopiles prototypes. In Fig.2 a detail of the layout (in particular one of the ends of the realized thermopiles) is shown. Their behaviors have been characterized in terms of the heating power by mean of an integrated heater and a reference PTAT thermometer.

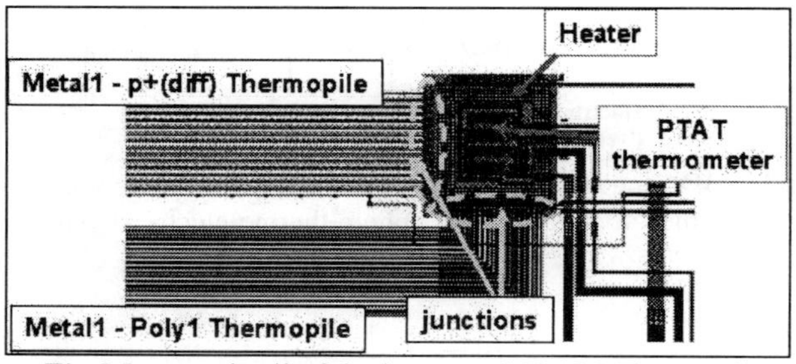

Fig.2 Layout detail of the integrated CMOS Thermopiles.

Both the thermopiles have the symmetrical end to a distance of about 2000 μm. The sensitivity of the proposed poly 1/metal 1 thermopile as a function of the incoming thermal power is reported in Fig. 3. This kind of thermocouple will be considered in the following.

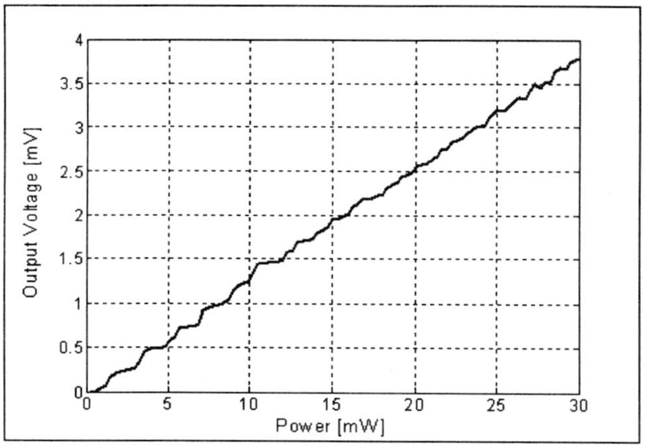

Fig.3 Characterization of the poly 1/metal 1 thermopile.

A value of about *100 μV/K* for the relative Seebeck coefficient is reported in literature [8, 9] for metal 1/poly 1 thermopiles. The electrical resistivity of the first poly-silicon layer results $9 \cdot 10^{-6}\ \Omega m$, while those of the first metal layer can be neglected [11].

It is interesting to compare the proposed power supply system with a different approaches presented in literature, referring to alternative strategies where the light is again used as external energy source. For this reason, even if some other approaches to feed power to autonomous systems as vibrational piezoelectric systems or even wireless RF power transfer systems have been reported in literature, an array of miniaturized photovoltaic cells will be considered as term of comparison.

In [12] an array of 100 photovoltaic cells that can deliver a maximum voltage of *101 V* and a maximum current of *88 μA*, i.e., a maximum power of *8888 μW* when the incident power is *1 W*, are presented. The size of a single cell is reported to be *550 μm × 300 μm*.
If this photovoltaic cells would be used to supply all the required power in a thermal actuator as the one presented in [7] that requires *8,4 mW* per actuation cycle, it results that such a system would require 98 photovoltaic cells for each actuator. Moreover other cells will be required for the electronics. This would result in a large requirement in terms of surface occupation with respect to the proposed approach of combined direct photo-thermo-mechanical and photo-thermo-electrical conversion.
Besides photo-voltaic cells require dedicated processes while here only standard CMOS processes are addressed.
Therefore the combined photo-thermo-mechanical and photo-thermoelectric power supply system presents several advantages in terms of costs and efficiency, with respect to other systems in applications where microelectronics and "strong" microactuators are integrated in the same chip.

4. CONCLUSIONS

An integrated power supply system for low power microelectronics on MEMS has been presented here. This work is an extension of a previous work where a photo-thermo-mechanical actuation system has been presented.
The system presented here is based on a thermo-electric generator realized in the same substrate of the microelectronics. The technology of reference is a standard CMOS. The thermal energy is provided to the thermoelectric generator through light. The combined use of micro-machined lenses allows high-energy efficiency values. Therefore, the resulting device is a photo-thermoelectric integrated generator. The performances of a concrete device prototype have been estimated and compared with those of photovoltaic cells arrays system. The combined photo-thermo-mechanical/photo-thermoelectric energy supply system has several advantages over photovoltaic cells, in terms of cost and energy efficiency for application fields where microelectronics and microactuators are integrated in the same substrate and fully autonomous microsystems are required.

5. REFERENCES

[1] S. Baglio, S. Castorina, L. Fortuna, N. Savalli, *Development of Autonomous, Mobile Micro-Electro-Mechanical Devices*, ISCAS 2002, IEEE Internat. Symp. on Circuits and Systems, Scottsdale, Arizona, USA, May 26-29, 2002.

[2] W.C.West et al., *Lithium micro-battery development at the Jet Propulsion Laboratory*, IEEE Aerospace and Electronics Syst. Mag., vol. 16 no. 8, pp. 31-33, 2001.

[3] P.B. Koeneman et al., *Feasibility of micro power supplies for MEMS*, IEEE J. Microelectromech. Syst., vol. 6, no. 4, pp. 355-362, 1997.

[4] T. Shibata et al., *Microwave energy supply system for in-pipe micromachine*, in: Proc. of The Int. Symp. on Micromechatronics and Human Science, MHS 98, pp. 237-242, Nagoya, November 25-28, 1998.

[5] N. Mayashi et al, *Micro moving robotics*, Proc. of The Int. Symp. on Micromechatronics and Human Science, MHS 98, pp. 41-50, Nagoya, November 25-28, 1998.

[6] J.B. Lee et al., A miniaturized high voltage solar cell array as an electrostatic MEMS power supply, IEEE J. Microelectromech. Syst, vol.4, no.3, 1995, pp. 102-108

[7] S. Baglio, S. Castorina, L. Fortuna, N. Savalli, *Modeling and Design of Novel Photo-Thermo-Mechanical Microactuators*, Sens. and Act. A: Phys., vol. 101, no. 1-2, pp. 185-193, 2002.

[8] H. Baltes et al., *Micromachined thermally based CMOS microsensors*, Proceedings of the IEEE, vol. 86, no. 8, pp.1660-1978,1998.

[9] M. Strasser et al., Miniaturized thermoelectric generators based on polySi and polySi-Ge surface micromachining, Sens. and Act. A: Phys., vol. 97-98, 2002, pp. 535-542

[10] F. P. Incropera et al., Fundamentals of heat and mass transfer, 5th Ed., Wiley, NY, USA.

[11] AMS 0.8 μm CMOS CXQ design parameters.

[12] T. Sakakibara et al., *Multi-source power supply system using micro-photovoltaic devices combined with microwave antenna*, Sens. and Act. A: Phys., vol. 95, no. 2-3, pp. 208-211, 2002.

DEVELOPMENT OF A CMOS LOW-NOISE ANALOG FRONT-END ASIC FOR X-RAY IMAGING APPLICATIONS

Em. Zervakis[1], D. Loukas[2], N. Haralabidis[3], A. Pavlidis[3]

[1] Microelectronic Circuit Design Group, National Technical University of Athens, Athens, GREECE

[2] Institute of Nuclear Physics, National Center of Scientific Research "Demokritos", Athens, GREECE

[3] Athena Semiconductors S.A., Athens, GREECE

ABSTRACT

High density compound materials (CdTe, GaAs, HgI_2) are currently evaluated as direct photon conversion sensors for applications in X-ray imaging. A new front-end stage has been designed with the aim to be implemented as readout chain in a luggage inspection system based on a linear array of CdTe sensors. Both types of electron or hole collecting CdTe sensors will be evaluated. Each detecting element is characterized by leakage current smaller than $5nA/mm^2$ at working point and capacitance in the range of 0.5pF. A low noise front-end chain has been developed based on the voltage mode architecture. The operation of the front-end stage is optimized for a 2pF capacitance, system input capacitance, and for energy range of 20-220keV. A prototype ASIC has been fabricated in a commercial 0.6um CMOS process.

1. INTRODUCTION

The readout architectures for radiation sensors can be divided in two major categories: the voltage mode architecture and the current mode architecture. In the current domain approach, the first stage after the detector is a transimpedance amplifier which conveys the current and then converts it to voltage. When high counting rates are required the use of current mode architecture is preferable [1-2]. In the voltage mode architecture, the signal is processed in the voltage domain and the first stage is a charge amplifier followed by a shaper. This is a widely adopted solution for multichannel VLSI readout electronics in many different fields such as High Energy Physics, medicine and material science. In X-ray imaging applications in particular, where noise performance is of primary concern, the use of voltage mode architecture is the best solution.

Fig. 1 shows the block diagram of a single channel of the front-end electronics. The first stage after the detector is a low noise charge amplifier. In order to provide the ASIC with the ability to process signals of both polarities (electrons or holes collecting sensors), a polarity select circuit is added after the charge amplifier. This makes the read out system suitable for both polarities. The next stage is a semi-Gaussian shaper which optimizes the noise performance of the system. A voltage amplifier is connected after the shaper to further amplify the signal. The final stage of the analog part of the read out system is a voltage discriminator which produces digital pulses fed to the counters. In order to set the discriminator thresholds, a 4-bit DAC per channel has been also implemented within the ASIC.

2. CdTe SENSORS FOR X-RAY LUGGAGE INSPECTION SYSTEM

Current X-ray luggage inspection systems are working with a combination of photodiode arrays followed by scintillator-photodiode arrays. The first layer of photodiodes records the low energy photons emitted by the X-ray generator. For energies over 20keV the detection efficiency of Si drops dramatically. In order to record photons with higher energies, scintillators serve as intermediate devices for X-ray to light conversion. A system based on a direct photon conversion array of sensors would present advantages in terms of linearity, uniformity and resolution. CdTe and CdZnTe sensors are regarded as promising devices for hard X-ray and γ-ray recording [3-4]. The high atomic number of the materials gives high detection efficiency relative to Si. A 2mm CdTe sensor keeps detection efficiency over 50% all the way up to 150 keV. The large band gap energy (1.44 for CdTe, 1.6 for CdZnTe) allows operation of the sensors at room temperatures. Due to low mobility and short life time of holes, in combination with losses due to trapping centers such as dislocations and defects, electron collection mode is preferable in CdTe detector systems. The short propagation length of holes in combination with charge accumulation phenomena, known as polarization effect, in the bulk of the material limits the use of CdTe sensors in high quality spectroscopic application. CdTe sensors are suitable for counting mode applications as luggage inspection systems.

Linear arrays of CdTe sensors are under evaluation as detecting elements for a luggage inspection system. A 20mm x 2mm x 2mm CdTe rod is lithographically patterned on one side in 20 pads measuring 0.9mm x 1.8mm and separated with a gap of 0.1mm. The pads and the entire backside are plated with Pt in order to form ohmic contacts. This way, 20 individual detecting elements with a common backside are formed. Thirty arrays of this type will be placed on a line in order to form a 60 cm long linear system for luggage inspection. Each sensor pad will be routed to the input of an electronics readout chain as depicted in Fig.1. Typical nominal working voltage of the sensors is 100 volts. A good uniformity of leakage currents with values around 5 nA is measured for each pad at 22 °C at 100 volts. The capacitance of each pad is bellow 1pF. With a careful interconnection design, to avoid parasitic effects, a total capacitance smaller than 2 pf can be reached at the input of each readout channel.

0-7803-7761-3/03/$17.00 ©2003 IEEE

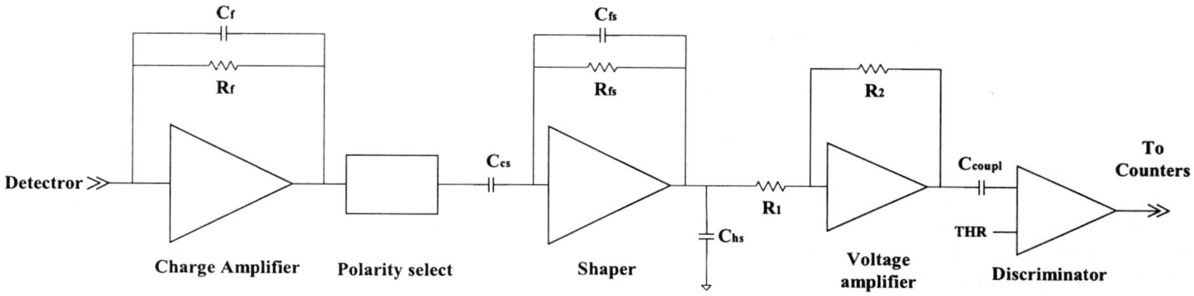

Figure 1. Block diagram of the analog part of one channel.

3. CIRCUIT DESCRIPTION

3.1 Charge amplifier

Fig. 2 shows the complete schematic of the charge amplifier. The single-ended folded cascode configuration is used due to the high gain, broad bandwidth and good stability that exhibits. The noise level of the preamplifier depends mainly on its input stage so the input device is the most important in the whole design [5-7]. A PMOS transistor has been chosen as input device because it gives lower flicker noise compared to an NMOS. In order to optimize the noise behavior, the input PMOS transistor is matched with the detector capacitance [5].

The feedback consists of a capacitor C_f in parallel with a resistance R_f, which is implemented by a PMOS transistor operating in the linear region. The series of current pulses released by the detector are integrated on the feedback capacitance of the charge amplifier producing at its output voltage steps of approximately Q/C_f, where Q is the total charge resulting from integrating the pulse. The feedback capacitor is discharged by the feedback resistance and the decay time is externally controlled through the bias voltage V_{preamp}. The preamplifier is biased at the center of its operating range in order to accept input charge pulses of both polarities.

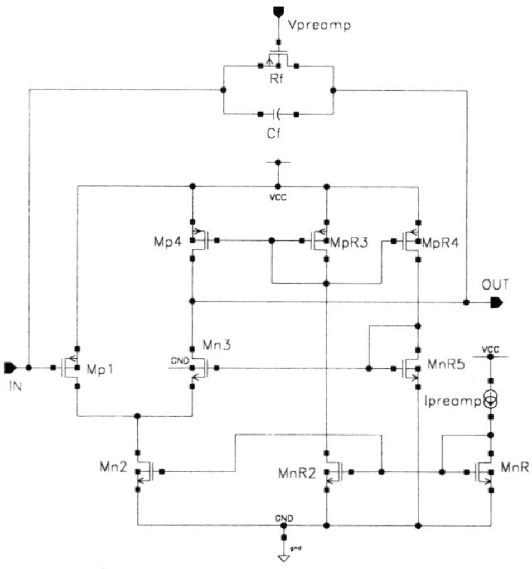

Figure 2. Complete schematic of the charge amplifier.

3.2 Polarity select circuit

The second stage of the front-end chain is a polarity select circuit which is shown in Fig. 3. It consists of the polarity amplifier, which is a typical two-stage amplifier with high gain and broad bandwidth, four switches and two identical resistors. All the switches are implemented with PMOS devices and are controlled externally by the same signal. When the control signal PS is high the switches SW1 and SW2 turn off while the SW3 and SW4 turn on. In that case the gain of the polarity select circuit is A_{pol}=+1 and practically operates like a buffer. When the control signal PS is low the gain of the stage is A_{pol}=-1, as $R_{p1}=R_{p2}$, and the polarity circuit inverts its input signal. In this way at the output of the polarity select circuit the voltage steps have the same polarity regardless the polarity of the input pulses, which can be either positive or negative. So the same front-end chain can be used to process signals produced either from electron collecting sensors or from hole collecting sensors.

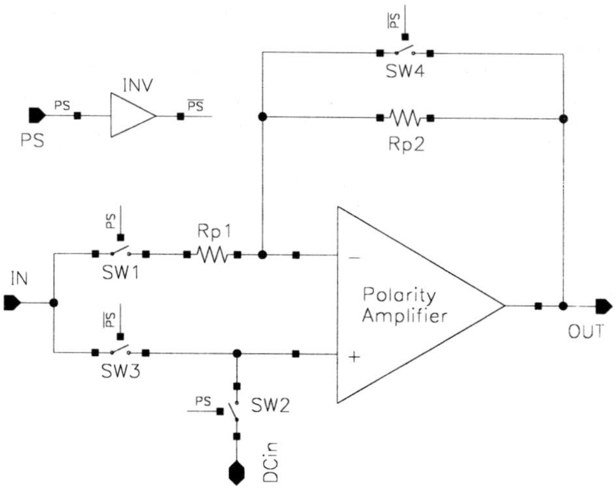

Figure 3. Block diagram of the polarity select circuit.

3.3 Shaper

The stage following the polarity select circuit is the shaper which is a filter that optimizes the noise behavior of the system. As in the charge amplifier circuit, the single-ended folded cascode configuration is used with a PMOS as an input device. It is shown that when increasing the order of the filter only small improvement in noise performance is achieved [5]. Keeping that in mind and in order to have simpler circuit, the order of the shaper is set to 1. In the frequency domain, the shaper acts as a band-pass filter limiting the noise produced by the preamplifier and thus increasing the signal to noise ratio (S/N). In time domain, the voltage steps produced by the preamplifier are converted into semi-Gaussian output pulses. The shaping time is determined by the values of the capacitors, C_{cs}, C_{fs} and C_{hs}, and the feedback resistance R_{fs}, which is implemented by a PMOS transistor. The optimum shaping time is calculated in order to optimize the noise of the system taking also into account the noise of the preamplifier. It is possible to adjust the shaping time through the gate voltage of the PMOS transistor connected in the feedback.

3.4 Voltage Amplifier

In many cases, apart from the gain achieved by the preamplifier and the shaper, a further amplification is needed so a voltage amplifier is also added. The amplifier is a two stage compensated Miller opamp with high open loop gain and good driving capability. The closed loop gain of the preamplifier is set to A_{amp}=-2.

3.5 Discriminator

The final stage of the analog part of the front-end is the discriminator. Fig. 4 shows the complete schematic of the discriminator. The output of the voltage amplifier is AC coupled to the input of the discriminator. This way, the DC level in the input of the discriminator is externally controlled. The discriminator circuit consists of three stages. The first stage is a simple source follower which significantly reduces the load of the voltage amplifier. The second stage, which is the core of the discriminator, is a high gain differential amplifier. Hysteresis is achieved by adding two cross-coupled PMOS devices. Finally, two CMOS inverters are connected to improve the shape of the output pulse. The threshold of the discriminator is fine adjusted through a 4-bit DAC cell which is implemented within the ASIC.

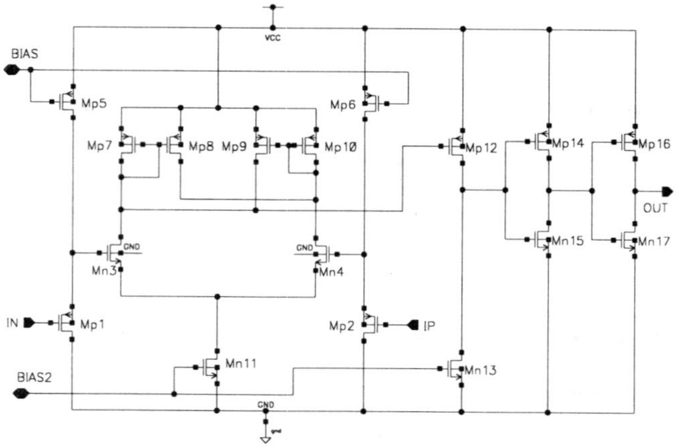

Figure 4. Complete schematic of the discriminator.

4. SIMULATION RESULTS

The front-end electronics have been designed in a 0.6um CMOS process with 3.3V power supply. Fig. 5 depicts the response at the output of the voltage amplifier for three different values of input charges, namely 1fC, 4fC and 8fC. The total gain of the chain is 130mV/fC while the peaking time is 35ns. The front-end stage recovers to the baseline within 400ns. The linearity of the chain is excellent up to 8fC, which corresponds to 220keV. The voltage pulses produced at the output of the discriminator in the case of 4fC input charge and event rate of 2Mevents/s is shown in Fig. 6. These pulses are subsequently fed to the digital part of the system for counting.

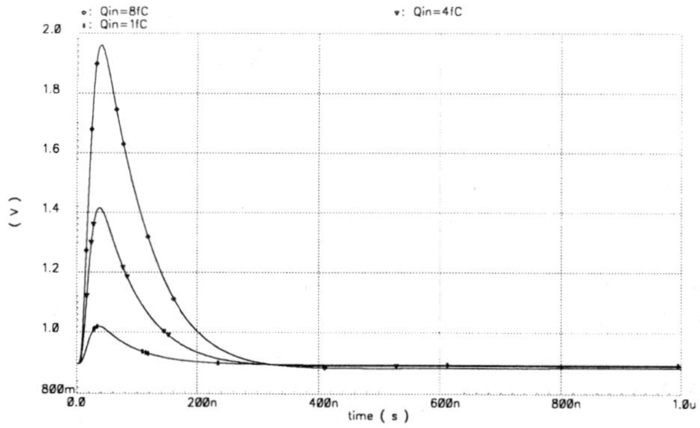

Figure 5. Transient response at the output of the voltage amplifier (input charge 1, 4, 8fC).

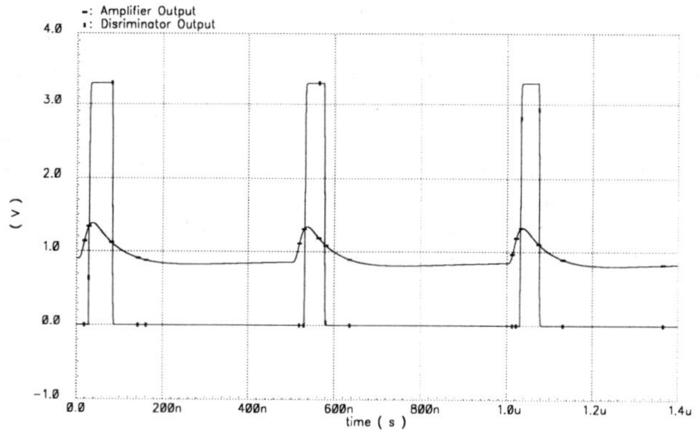

Figure 6. Voltage pulses produced at the output of the discriminator (input charge 4fC, event rate 2Mevents/s).

5. EXPERIMENTAL RESULTS

Fig. 7 shows the microphotograph of the fabricated ASIC. On the left of the chip is the analog part of each channel while on the right the mixed analog-digital part. Special layout techniques, including different power supply rails, have been used in order to isolate the two parts of the ASIC. The prototype chip holds 12 channels with 100um pitch and small variations in the building blocks from channel to channel, which share the same on chip biasing. The total area of the die is 3000x4000um^2.

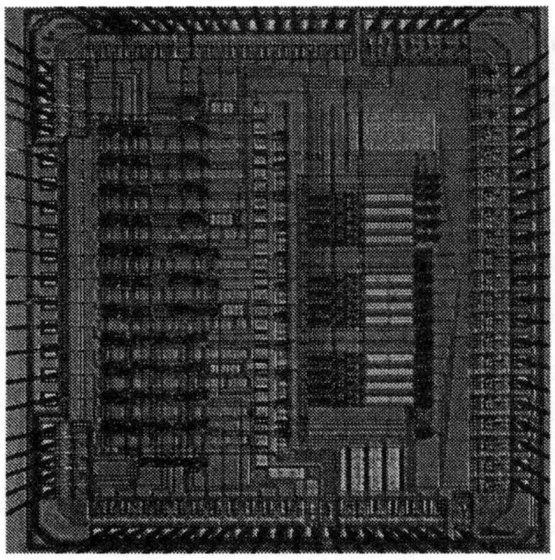

Figure 7. Microphotograph of the fabricated chip.

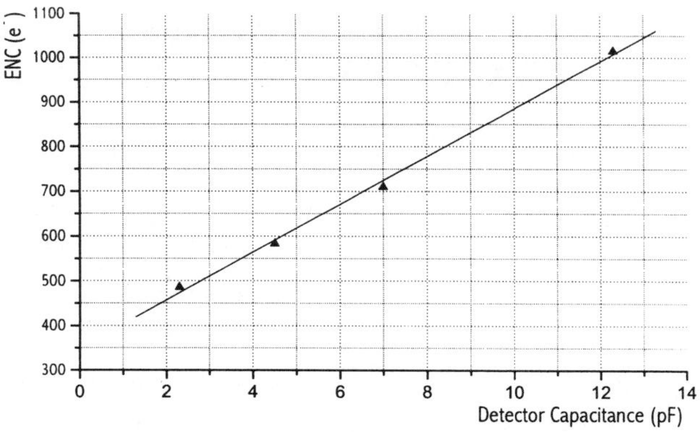

Figure 9. ENC vs detector capacitance.

The ASIC has been tested by injecting charge pulses at its input and simulating the capacitive effects of the detector with capacitors placed between the input of the preamplifier and ground. By placing decoupling capacitors on the testing board the applied voltage steps in its input are becoming charge pulses. Fig. 8 depicts the measured transient response at the output of the amplifier for an input charge of 2fCb. The overall gain of the chain is reduced due to the buffering scheme used on the testing board, which provides a gain of 0.9. Furthermore a slight increase in the peaking time is also noticed due to the same reason. The analog output of the front-end electronics exhibits a peaking time of 45ns. The measured gain is 110mV/fCb and remains almost constant for typical values of input charges, 1-8fCb. Due to the AC coupling the system is insensitive to DC level shift of the analog output.

The ENC has been measured by changing the value of the detector simulating capacitors. The r.m.s. value of the output voltage fluctuations was measured with an oscilloscope and the ENC was calculated by dividing this noise level by the overall gain of the chain. Fig. 9 shows the Equivalent Noise Charge (ENC) vs the detector capacitance. The calculated noise relationship is $350+55e^{-}/pF$ resulting in $460e^{-}$ rms for the 2pF detector capacitance.

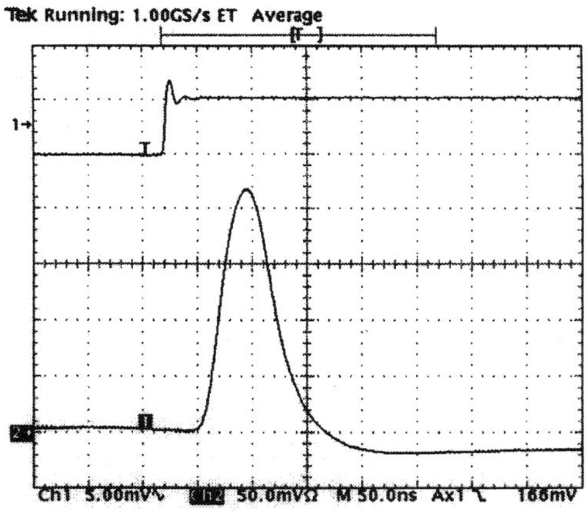

Figure 8. Measured transient response at the output of the voltage amplifier (input charge 2fC).

6. SUMMARY

A new front-end stage has been designed for use in a luggage inspection system based on a linear array of CdTe sensors. The read-out chain, which is based on the voltage mode architecture, consists of a charge amplifier, a polarity select circuit, a semi-Gaussian shaper, a voltage amplifier and a discriminator. A multi-channel prototype, including also DACs for fine adjustment of the discriminator threshold, has been fabricated in a 0.6um CMOS process. The measured gain of the chain is 110mV/fC while the peaking time 45ns. The measured ENC for the case of 2pF detector capacitance is $460e^{-}$. Finally the power consumption per channel is 8mW.

7. REFERENCES

[1] D. Moraes et al.. "CARIOCA - 0.25um CMOS Fast Binary Front-End for Sensor Interface using a Novel Current-Mode Feedback Technique". *Proceedings. of the IEEE ISCAS* 2001, pp. I360-I363.

[2] N. Haralabidis and D. Loukas. "A versatile CMOS Low-Noise Analog Front-End stage for Solid State Detector Interfaces". *Proceedings of IEEE ISCAS* 2001, pp I364-367.

[3] Tadayuki Takahshi and Shin Watanabe, "Recent Progress in CdtE and CdZnTe Detectors". IEEE Transactions on Nuclear Science, vol 48, Aug. 2001, pp. 950-958.

[4] Y.Eisen "Current state of the art industrial and research applications using room temperature CdTe and CdZnTe solid state detectors". Nuclear Instruments and Methods in Physics Research, A380,1996, pp. 431-439.

[5] W. M. C. Sansen and Z. Y. Chang. "Limits of Low Noise Performance of Detector Readout Front Ends in CMOS Technology". IEEE Transactions on Circuits and Systems, Vol. 37, No. 11, Nov. 1990, pp 1375-1383.

[6] G. De Geronimo, P. O'Connor, V. Radeka, B. Yu. "Front-end electronics for imaging detectors". Nuclear Instruments and Methods in Physics Research, A471,2001, pp. 192-199.

[7] P. F. Manfredi, M. Manghisoni. "Front-end electronics for pixel sensors". Nuclear Instruments and Methods in Physics Research, A465, 2001, pp. 140-147.

FLUXGATE MAGNETIC SENSORS WITH A READOUT STRATEGY BASED ON RESIDENCE TIMES MEASUREMENTS

B. Andò[(a], S. Baglio[(a], A.R. Bulsara[(b], L. Gammaitoni[(c]

a) Dipartimento Elettrico Elettronico e Sistemistico
University of Catania, V.le A. Doria 6, 95125, Catania, Italy

b) Space and Naval Warfare Systems Center
Code 2363, 49590 Lassing Road, San Diego, CA 92152-6147, USA

c) Dipartimento di Fisica
University of Perugia, Perugia, Italy

ABSTRACT

We deal with analytical and experimental results on a highly sensitive fluxgate magnetometers wherein we exploit an efficient mode of operation that takes advantage of the inherent nonlinear (bistable) character of the core dynamics itself. Our approach, based on the passage time statistics of transitions to the stable steady states of the hysteresis loop, offers the possibility of using bias signals that have lower amplitude and frequency than those used in conventional fluxgate operation. The experiments are carried out on a laboratory prototype fluxgate, fabricated as a miniaturized device; this device is expected to lead to a MEMs implementation of a single fluxgate as well as a coupled ensemble of fluxgates.

1. INTRODUCTION

Fluxgate magnetometers [1] are currently used as practical and convenient sensors for room temperature magnetic field measurements. Typically, one uses two detection coils wounded on two ferromagnetic cores and connected in a differential arrangement, as shown in Fig.1a.

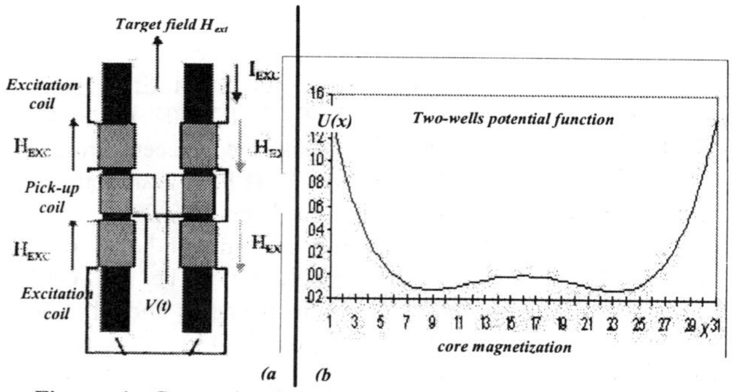

Figure 1. Conventional arrangement for fluxgates (a) and the bistable potential energy function ***U(x)***.

Conventional fluxgate magnetometers show a bistable dynamics that is exploited in order to operate the device as a magnetic field detector. Standard operation is performed by applying a known time-periodic bias signal, having a very large-amplitude, to drive the fluxgate dynamics between the stable states.

Typically single valued transfer characteristic of the magnetic core are considered; however, at a closer view, dynamic effects in the domain wall infrastructure render the transfer characteristic hysteretic [2] with a potential energy function $U(x)$ that is bistable (Fig.1b) where x represents the averaged magnetization.

By using two periodic counter-phased excitation currents, the magnetic cores are periodically driven into saturation in opposing directions. Hence, the output voltages $V_i(t)$ (i=1,2), measured across each detection coil, will oscillate back and forth between the two saturation states at the forcing frequency. In the differential arrangement, the output "voltage signal" is denoted by $V(t)=V_1(t)-V_2(t)$. In the absence of the target signal, the underlying potential energy function is symmetric and the output voltage signal is zero. The presence of a magnetic field target signal H_{ext} (dc or low frequency) skews the hysteresis loop and the potential energy function; now, the voltage signal $V(t)$ is non-vanishing, and its magnitude depends on the asymmetrizing target signal.

Conventional fluxgates present limitations in practical applicability due to the Barkhausen noise produced by large-amplitude and frequency signals [3] and to the power level required to generate large bias signals.

Here we exploit hysteretic behavior in a highly sensitive magnetic field detection strategy. We present an alternate approach where we experimentally measure the residence times in each of the two stable states of the bistable potential energy function that underpins the core dynamics. With a time-periodic deterministic bias signal (H_{exc}) having amplitude just enough sufficient to cause switching between the steady states and in absence of noise, the hysteresis loop, or the underlying potential energy function, is symmetric and therefore the two times will always remain constant and equal to each other. The presence of a target signal (H_{ext}) leads to a skewing of the loop with a direct effect on the residence times: they are no longer the same. In the presence of noise, the residence times must be replaced by their ensemble averages. The difference of the mean residence times can then lead to a quantification of the target signal.

The above method of operation is applicable to conventional fluxgates with the advantages of mitigating the effects of noise, in fact lower amplitude bias signals are required, of improving sensitivity and of using just one excitation coil and one detection coil without any need of differential structures.

0-7803-7761-3/03/$17.00 ©2003 IEEE

2. THE RESIDENCE TIMES APPROACH

The model that most closely describes the dynamical response of the averaged magnetization $x(t)$ of a ferromagnetic core [4] is:

$$\zeta \frac{dx}{dt} = -x + \tanh\left[\frac{x + H_{ext}}{K}\right] \equiv -\frac{\partial U}{\partial x}(x,t) \qquad (1)$$

being ζ a system time constant and K a dimensionless control parameter. $H_{ext}(t)$ is the external magnetic field. Eq. (1) can be also expressed in terms of the gradient of a potential energy function:

$$U(x,t) = \frac{x^2}{2} - \frac{1}{c} \ln \cosh[c\{x + H_{ext}\}] \qquad (2)$$

where $c=K^{-1}$. The potential energy function (2) is bistable for $c>1$ as in the case shown in Fig.1b.

Here we propose a new measurement technique based on the system residence times in its steady states. For a two-state system, the residence time in one of the stable steady states, is defined as the time elapsed between the first crossing of that threshold and the first crossing of the other threshold. In the presence of a noise background, the residence times in the stable states have random components. The residence time statistics in a bistable system were proposed initially [5] as a quantifier for the "Stochastic Resonance (SR)" phenomenon [6,7] which involves *sub*threshold driving signals and noise-assisted switching between the stable steady states (the potential minima).

We start by noting that absent any background noise and with a *supra*threshold bias signal amplitude, one obtains the same residence times in each stable state. With a small d.c. target signal, the potential is skewed. Hence one obtains unequal residence times in the two states. In the presence of weak (compared with the bias signal) noise, one obtains a spread in the residence times which can now be described statistically by using the Residence Times Distributions (RTD). For the case in which the bias signal is *supra*threshold, RTDs for the right and left potential wells will be almost symmetric with a mean value corresponding to the mean residence time. Absent the target dc signal, the distributions coincide. The presence of the external target signal renders the potential asymmetric with a concomitant difference in the mean residence times which, to first order, should be expected to be proportional to the target signal itself. Hence, the difference between the mean residence times in the two states of the system $\langle \Delta T \rangle$ provides an observable that is used as a quantifier for detecting the presence of the target signal.

This procedure has some advantages compared to the standard ones: it can be implemented experimentally without complicated feedback electronics, moreover it is very low power demanding. In fact the residence-times based technique works without the knowledge of the computationally demanding power spectral density of the system output: in most cases a simple averaging procedure on the system output works just fine. Moreover, as in the case of "noise activated sensors", the difference in residence times is quantifiable even in the *absence* of the bias signal, with only noise *driving* the bistable sensor [8].

Let us draw some consideration with reference to Fig. 2. T_+ is the residence time in the right well of the potential $U(x)$ defined as the time window between the first system commutation (time t_1) and the successive one (time t_2); T_- represents the time interval between the second commutation and the successive one (time t_3).

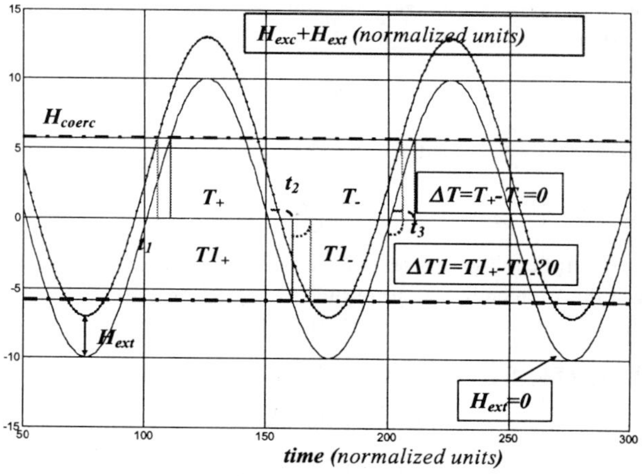

Figure 2. Resident Time readout scheme

As described previously the value of $\Delta T=(T_+-T_-)$ is directly dependent on the external field H_{ext}. In Fig.2 the effect of H_{ext} is shown with reference to $\Delta T1=(T1_+-T1_-)$.

One of the key issues attains sensitivity of the fluxgate magnetometer; in absence of noise this is defined as:

$$S = \frac{\partial}{\partial H_{ext}} \Delta T \qquad (3)$$

The slope of the signal $(H_{ext}+H_{exc})$ at the intersection with the threshold value H_{coerc}, the coercive field of the magnetic core, is therefore considered. In particular a reduction of the bias signal slope at the crossing point corresponds to an increase in the sensor sensitivity. However, as results also from Fig.2, in general sensitivity is a function of the amplitude of the target signal. In order to avoid this drawback a triangular signal has been considered here having slope α and period τ.

$$H_{exc}(t) = \begin{cases} \alpha t & \text{if} \quad -\frac{\tau}{4} < t < \frac{\tau}{4} + N\tau \\ -\alpha\left(t - \frac{\tau}{2}\right) & \text{if} \quad N\tau + \frac{\tau}{4} < t < \frac{3\tau}{4} + N\tau \end{cases} \quad N = 1,2,\ldots \qquad (4)$$

The slope a can be expressed in a convenient way as a function of the excitation field parameters:

$$\alpha = \frac{d}{dt} H_{exc}(t) = \frac{4H^M_{exc}}{\tau} \qquad (5)$$

where H^M_{exc} is the maximum value of the field produced by the excitation coils that depends on the current I_{exc} and on some geometrical parameters. Thus sensor sensitivity is fixed and doesn't change with the H_{ext} signal magnitude. For this signal it is quite immediate to obtain the following conditions

$$t_1 : H_{ext} + H_{exc}(t_1) = H_{coerc} \qquad t_2 : H_{ext} + H_{exc}(t_2) = -H_{coerc}$$

$$t_3 : t_1 + \tau$$

By using eq.(4) these conditions can be rewritten such to obtain

$$T_+ = t_2 - t_1 = \frac{2H_{ext}}{\alpha} + \frac{\tau}{2} \qquad T_- = t_3 - t_2 = \frac{\tau}{2} - \frac{2H_{ext}}{\alpha}$$

leading to the following expression for the time interval ΔT $\Delta T = {4H_{ext}}/{\alpha}$ that can also be expressed as a function of the forcing current parameters and the physical parameters of the FluxGate.

$$\Delta T = \frac{H_{ext}}{H^{M}_{exc}}\tau = \frac{H_{ext}}{H^{M}_{exc}}\frac{2\pi}{\omega} \qquad (6)$$

Therefore sensor sensitivity results:

$$S = \frac{\tau}{H^{M}_{exc}} \qquad (7)$$

As it can be observed sensitivity of the device will increase as both the frequency and amplitude of $\boldsymbol{H_{exc}}$ decrease. Moreover, $\boldsymbol{S}$ is independent on the external target field.

3. THE FLUXGATE PROTOTYPE IN PCB TECHNOLOGY.

In this section an experimental prototype of a Fluxgate sensor, operating on the principle of the residence time, is presented. The design procedure proposed here starts by fixing some geometrical and physical parameters while the device sensitivity represents the critical specification. The amplitude of the current in the excitation coil is determined from the knowledge of the coercive field of the magnetic core, finally the biasing signal period is obtained on the basis of the required sensitivity.

The FluxGate sensor has been realized in PCB technology that we already successfully used for other magnetic sensors [9] and is made up of three layers as shown in Fig.3. The sensor is d=45mm long and W=20mm wide. The outer layers carry the coils while the inner one is a high permeability hysteretic core. Thickness of the whole device is equal to 1mm being 0.2mm the coils structure thickness and 0.6mm the magnetic core thickness.

This structure adopts two excitation coils surrounding one detection coil in order to guarantee the geometric symmetry. An *Amorphous Metal* (also known as metallic glass alloys) has been chosen for the magnetic core due to its suitable hysteretic characteristic. This material has a non-crystalline structure and possesses unique physical and magnetic properties that combine strength and hardness with flexibility and toughness. In particular an as cast Magnetic Alloy 2714A (Cobalt-based), by Metglass® Solutions, has been adopted. Its characteristic is very sharp, as shown in the inset of Fig.4, and can be suitably approximated with a two state bistable hysteresis. Other important parameters of the adopted core are its coercive field, H_{coerc}=0.8A/m, the saturation magnetic flux density, B_{sat}=0.25T and the D.C. permeability μ>80000.

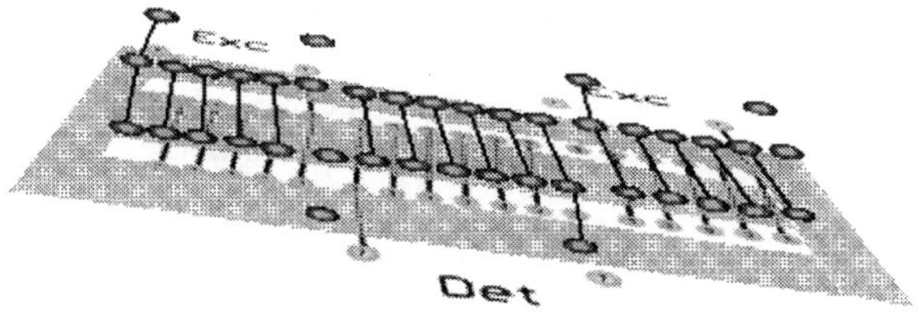

Figure 3. The FluxGate sensor is built up of three layers. The middle layer is a high permeability hysteretic core.

A Simulink model has been developed and its schematic is reported in Fig.4. A very simple readout strategy has been implemented by using a counter whose output gives alternatively information on the amplitude of residence times $\boldsymbol{T_+}$ and $\boldsymbol{T_-}$. The following actual geometric dimensions and electric parameters of the designed sensor have been set:

l=0.01 m; N_{exc}=10; H_{coerc}=0.01 A/m.

Moreover, by fixing the maximum amplitude of external magnetic field to be detected at 100μT and the desired device sensitivity at S=10000 s/T=1/100 s/μT the following excitation current amplitude and frequency have been computed by using the design flow described in section 4: $\boldsymbol{I_{exc}}$=80mA, ω=6.2 rad/s.

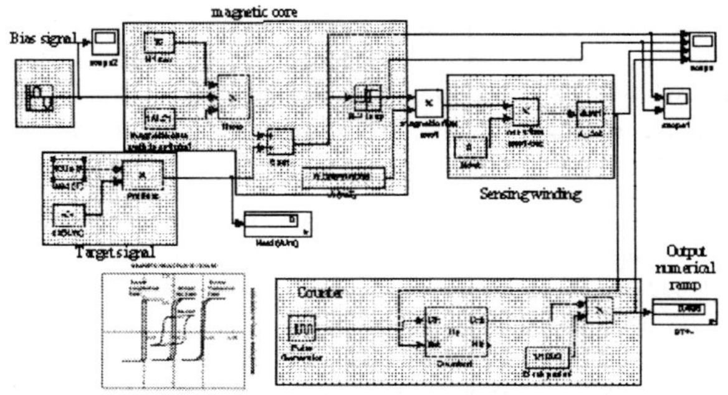

Figure 4. A schematic Simulink model of the residence time based FluxGate.

Several simulations have been performed and some results are shown in Fig.5. From top to bottom: the excitation magnetic field; the excitation magnetic induction and the counter output in the case $\boldsymbol{B_{ext}}$=0μT; the excitation magnetic induction and the counter output in the case $\boldsymbol{B_{ext}}$=50μT. As it can be observed in the case of $\boldsymbol{H_{ext}=0}$ the time intervals T_+ and T_- are ideally equal while they become different in the other case ($\boldsymbol{H_{ext}\neq0}$). It should be observed as the difference between two consecutive counter output gives a direct estimation of the target magnetic field. The pulsed inducted voltage on the detection coil acts as a reset signal for the counter.

In Fig.6 the experimental setup is shown. It can be also observed the sensor prototype and the schematic of the conditioning circuit together with a closer view of the typical output signal.

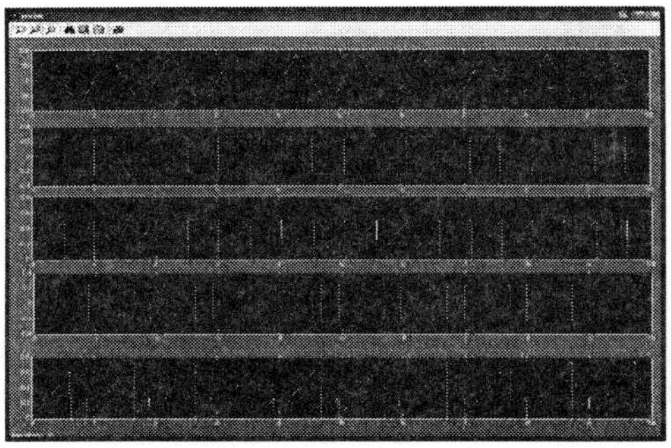

Figure 5. Simulation results. From top to bottom: H_{exc}, excitation magnetic induction and counter output for $\mathbf{B_{ext}}$=0μT, excitation magnetic induction and counter output for $\mathbf{B_{ext}}$=50μT.

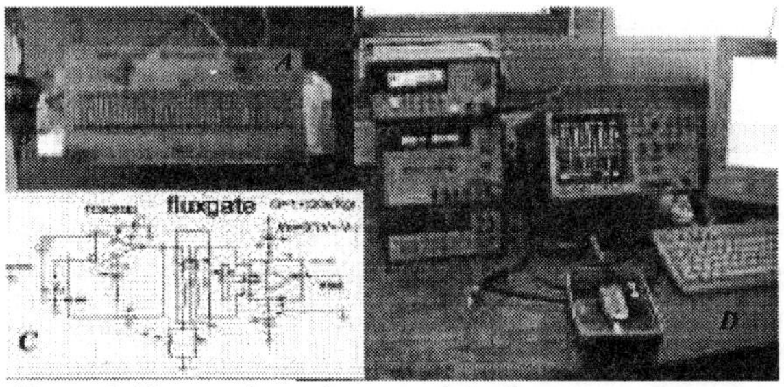

Figure 6. The experimental setup and the fluxgate sensor prototype realized.

4. EXPERIMENTAL RESULTS

Many experiments have been performed; a typical situation is shown in Fig.7 where the acquired excitation current and the corresponding output voltage are reported. Moreover the positions of the "equivalent" threshold values (the coercive field in the case of sharp hysteresis loop) of the hysteresis loop are derived with reference to the excitation current.

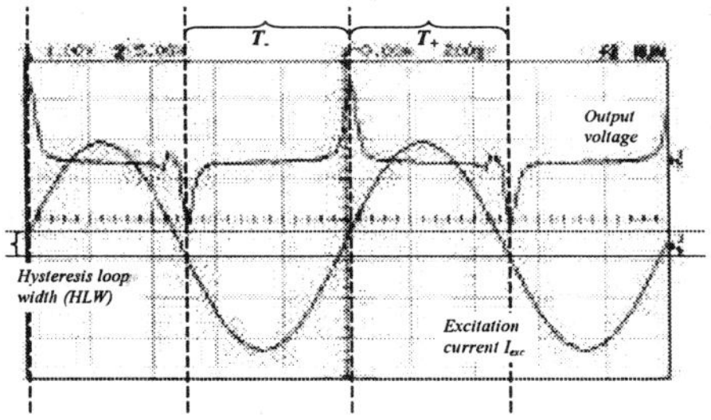

Figure 7. Biasing (sine) and output (spikes) signal for the experimental fluxgate.

The presence of an external field to be measured produces a reciprocal displacement between excitation current signal and these thresholds; such a displacement reflects in a change in the residence times difference T_+-T_-.

In our experiments the external magnetic field has been produced by an external coil placed in a fixed position close to the fluxgate to be characterized as shown in Fig.6. From the experimental measures it resulted a good repeatability of the sensor and a sensitivity of approximately 40 ns/μT.

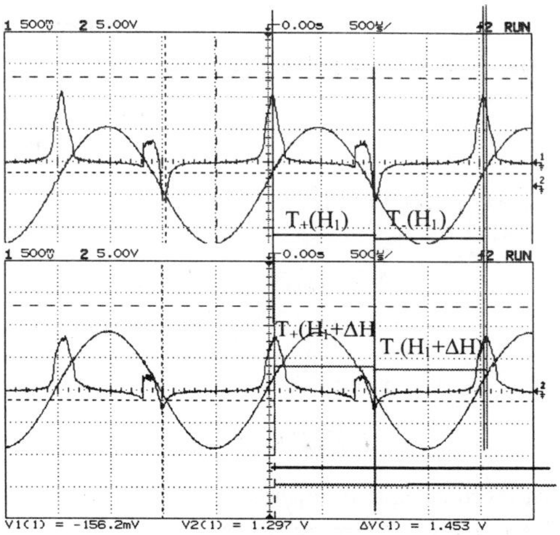

Figure 8. Experimental signals for a 0.96 A/m.

5. CONCLUSIONS

A novel measurement strategy for highly sensitive fluxgate magnetic sensors has been presented in this paper. We exploit here the nonlinear bistable features of the hysteretic magnetic core in order to estimate the differences in resident time in the two stable states of the potential energy function. The proposed approach results greatly efficient under several points of view. In fact it requires a simpler sensor architecture, having only one detection winding instead of two, and a simpler signal conditioning electronics. Moreover the signals involved need has lower amplitude thus reducing the power consumption and therefore broadening the possible applications field.

A prototype device has been designed, modeled and simulated. Moreover a large set of experiments have been performed in order to validate the predicted results. Work is still in progress toward realizing integrated fluxgate sensors with MEMS technologies. Miniaturization is expected to lead significant improvements, especially in terms of noise immunity, and to open to several application fields. Moreover theoretical efforts are also being paid in order to assert the optimal working condition to furtherly enhance sensitivity of the fluxgate sensor

6. REFERENCES

[1] P. Ripka, Review of magnetic fluxgate sensors, Sensors and Actuators A, 33, 129, 1992.

[2] G. Bertotti, Hysteresis in Magnetism, Academic Press, San Diego, 1998.

[3] G. Durin; in Proceedings of the 14th. International Conference on Noise in Physical Systems and 1/f Fluctuations,, World Scientific, Singapore 1997.

[4] A. Bulsara, C. Seberino, L. Gammaitoni, M. Karlsson, B. Lundquist, J. Robinson; Phys. Rev. E, submitted (2002).

[5] L. Gammaitoni, F. Marchesoni, E. Menichella-Saetta, S. Santucci; Phys. Rev. Lett., 62, 349, 1989.

[6] A. R. Bulsara, L. Gammaitoni, Physics Today, March p. 39, 1996.

[7] B. Andò, S. Baglio, S. Graziani, N. Pitrone, "Optimal improvement in bistable measurement device performance, via stochastic resonance", *Int. Journal of Elec.*, July 1999

[8] L. Gammaitoni, A. Bulsara; Phys. Rev. Lett., 88, 230601 (2002).

[9] S. Baglio et al., Micro Inductive Based Biosensor Arrays: Strategies and circuits for the detection of inductance changes due to magnetic nano-particles effect", Eurosensors 2002, Prague, Sept. 2002

A 96 × 64 INTELLIGENT DIGITAL PIXEL ARRAY WITH EXTENDED BINARY STOCHASTIC ARITHMETIC

Tarik Hammadou (†), Magnus Nilson (†), Amine Bermak (‡) & Philip Ogunbona (†)

(†)MOTOROLA LABS, Locked Bag 5028, Botany NSW 1455, Australia
(‡)Hong Kong University of Science and Technology, EEE Department

ABSTRACT

A chip architecture that integrates an optical sensor and a pixel level processing element based on binary stochastic arithmetic is proposed. The optical sensor is formed by an array of fully connected pixels, and each pixel contains a sensing element and a Pulse Frequency Modulator (PFM) converting the incident light to bit streams of identical pulses. The processing element is based on binary stochastic arithmetic to perform signal processing operations on the focal plane VLSI circuit. A 96×64 CMOS image sensor is fabricated using $0.5\mu m$ CMOS technology and achieves $29 \times 29\mu m$ pixel size at 15% fill factor.

1. INTRODUCTION

Pixel level processing and pixel level ADC are at the centre of Smart Integration Solution system research. The introduction of ADC at the pixel level will permit all of the post pixel signal processing to be performed digitally, thereby, significantly reducing the amount of analog circuitry required.

The addition of pixel level memory allows imaging data to be stored locally and accessed in a manner similar to standard DRAM memory. Therefore, the trend towards higher transistor densities with each successive semiconductor technology generation makes it highly probable that image sensors using pixel level ADC will be widespread in the near future. However, the operation voltage decreases accompanied with such a deep sub-micron process, and it may directly affect the signal quality and thus deteriorate a signal-to-noise-ratio (SNR). In addition, the area of a photodiode decreases as the process rule becomes fine, and consequently the signal capacity in the photodiode decreases, which also causes to degrade the SNR.

Thus, the big challenge today is to develop robust smart image sensor architecture that can deal with low operation voltage and reduced photodiode area.

Pulse coded arithmetic and stochastic neural arithmetic are biologically inspired networks that have been extensively studied and have been used for characters recognition and pattern recognition tasks [1]. However, most of the work on stochastic arithmetic is limited to software implementation and simulation and very little work was reported on the hardware implementation.

In this paper we present a robust smart image sensor architecture with fully connected pixel level stochastic arithmetic processor. The sensor is capable of image capture as well as image computation. In Section 2 of this paper we will present the pixel architecture, we will demonstrate that Pulse Frequency Modulation (PFM) is well compatible with digital logic circuits and robust against noise. It is essentially digital circuits, thus deep sub-micron technology is applicable to it and the operation voltage hardly affects the SNR. In Section 3, we present the concept of using stochastic arithmetic on the focal plane VLSI image sensor circuit, simulation results of the MATLAB model of the sensor performing image capture and contrast enhancement is presented. In the next Section we describe the image sensor implementation and the characteristic of the testing prototype. Finally in Section 5 we present the conclusion.

2. PIXEL LEVEL CONVERSION

The main advantage of our pixel level conversion method is that the circuit is essentially digital well suitable for deep-micron technology. Another advantage, the output is pulse train which means the operation voltage hardly affects the SNR. As illustrated in Figure. 1, the pixel architecture consists of a sensing element(n^+psub photodiode PD) a reset circuit, and a clocked comparator.

2.1. Pixel Functionality

The photodiode acts as a variable current source controlled by the input light intensity and is charged through the reset transistor M1. The gate of M1 is switched by the feedback of the inverted comparator output. The analog value of the light intensity is then consequently converted into a pulse train coded signal. Figure. 2 demonstrates SPICE simulation of the functional operation of the pixel. The volt-

The authors would like to thank Farid Boussaid for his contribution to the design of the sensor

0-7803-7761-3/03/$17.00 ©2003 IEEE

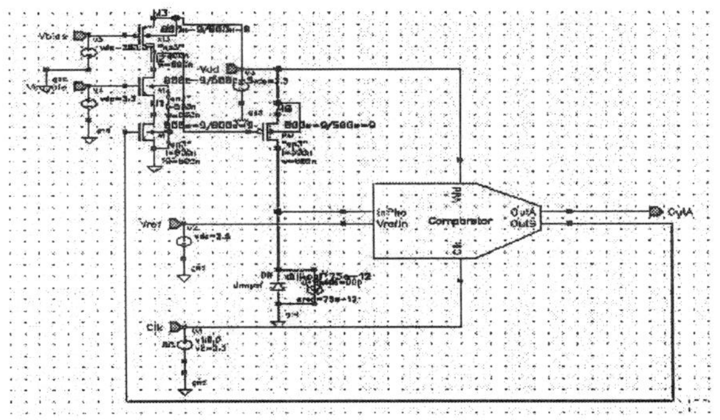

Figure 1: Image capture pixel architecture.

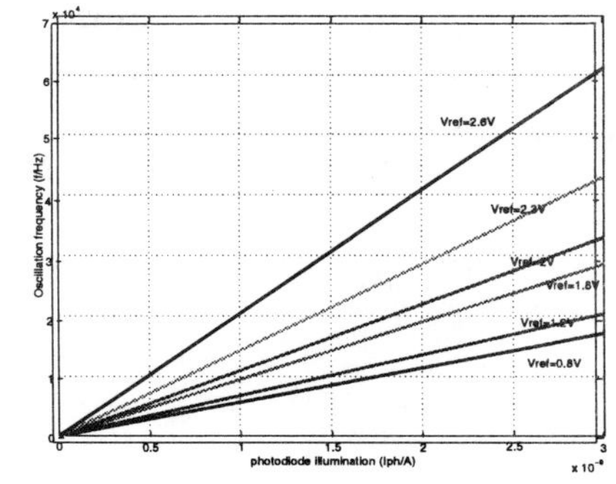

Figure 3: Photodiode linear characteristic.

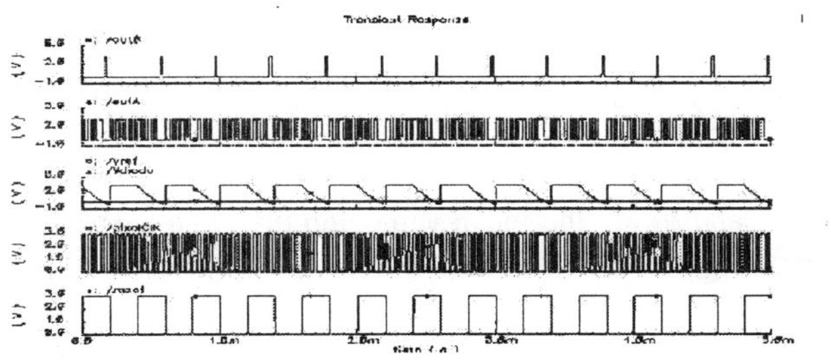

Figure 2: Pixel functionality SPICE simulation.

age V_D across the node of the photodiode is compared to a programmable clocked comparator reference voltage V_{ref}, as light illuminates PD, V_D gradually drops from the bias voltage of the reset transistor M1 by the photocurrent and finally reaches V_{ref}, this operation will switch the output of the comparator and the feedback circuit will reset the photodiode node to V_{dd}. In this case, if I_{ph} and C_D represent the photocurrent and the photodiode internall capacitor, respectively, we can express the oscillating frequency f as:

$$f \approx \frac{1}{T_D} = \frac{I_{ph}}{(V_r - V_{ref}) \times C_D} \tag{1}$$

where T_D represents the integration time, it describes the time needed for the photodiode to discharge from V_r (the bias voltage of the reset transistor) to V_{ref}. From Eq. (1) it can be concluded that the oscillation frequency f, or the firing rate, increases if the input light intensity becomes large and the size of the photodiode becomes small. This therefore makes this type of oscillating pixels suitable for low voltage operations and deep sub-micron technology. A large well capacity is no longer necessary, this is mainly because of the reuse of the well capacity during integration due to these oscillations.

In order to achieve a programmable high dynamic range and adaptation to different lighting conditions, we developed a compact programmable counter and windowing readout strategy to achieve high frame rate imaging capability. It is possible to control the counter to adapt the dynamic range to different lighting condition or to extend the dynamic range to values over 100 db. This can be easily demonstrated by the following equations:

$$N_{pulse} = \frac{T_{count}}{T_D} = \frac{T_{count} \times I_{ph}}{(V_r - V_{ref}) \times C_D} \tag{2}$$

then the counting period can be expressed as follow:

$$T_{count} = N_{pulse} \times \frac{(V_r - V_{ref})}{I_{phmax}} \times C_D \tag{3}$$

in our case $N_{pulse} = 2^n$. By controlling T_{count} we can optimize the dynamic range for different lighting conditions as well as the conversion speed. The minimum detection limit can be determined by the dark current as well as the leak current of the reset transistor, the minimum frequency can be given by:

$$f_{min} = \frac{I_{leakage}}{(V_r - V_{ref}) \times C_D} \tag{4}$$

If we assume that the leakage current of the photodiode is few fA, the minimum frequency can be estimated to $10Hz$. Thus, by exploring Eq. 4, we can conclude that the dynamic range can be expanded by adaptively changing the voltage. However, in order to take full advantage of the pixel architecture, a sensor callibrating faze is very important to generate the external signals needed to achieve the expected dynamic range. Figure. 3 illustrates the linearity of the photodiode.

2.2. Clocked comparator

The heart of the bit stream conversion at the pixel level is a dynamic latch type comparator. This type of circuits are

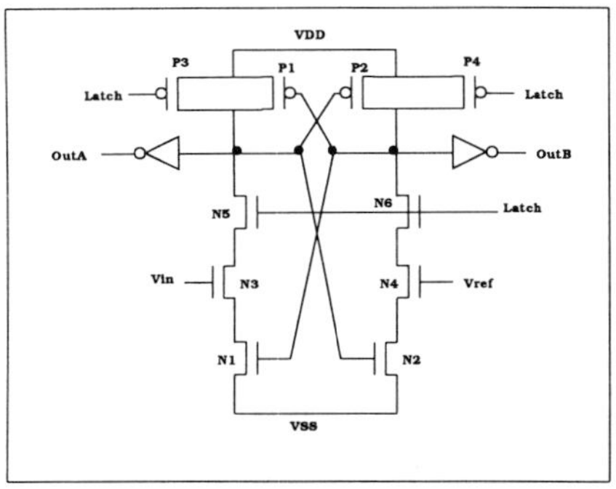

Figure 4: Clocked comparator.

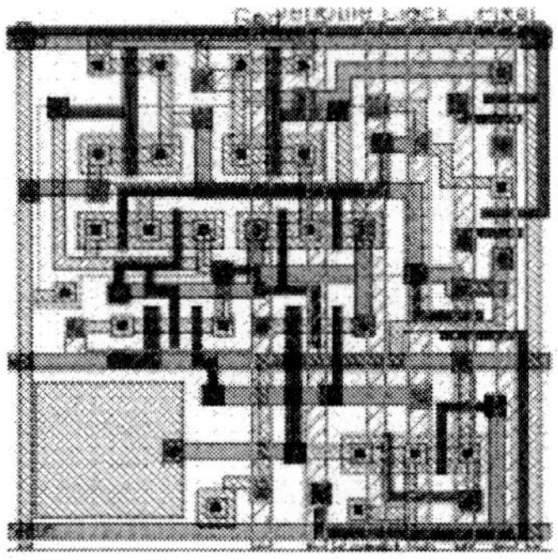

Figure 5: Pixel layout.

commonly used in DRAM's and SRAM's as a sense amplifier. When it's applied as a comparator, a dramatic reduction in the power dissipation can be expected. Figure. 4 shows the schematic of the comparator used. When the clock pulse is low, the supply current is cut off and the outputs are connected to V_{dd} through switches $P3$ and $P4$, the input transistors $N3$ and $N4$ are in triode mode. When a signal is applied to these two MOS gates differentially and the clock pulse goes high, the output of the stronger side $N3$ and $N4$ is pulled down more strongly than the other side. This will cause the latch to flip to one of the two stable states.

3. PIXEL LEVEL STOCHASTIC ARITHMETIC

The main objective of this work is the development of smart imaging devices with programmable processing capability in this section we describe the principle behind pixel level stochastic arithmetic.

Given a set of analog (pixel input) inputs, they are stochastically converted using the frequency modulator, then processed, and finally recovered from the stochastic pulse stream as a digital value. The main advantage of the stochastic processing system is the possibility of doing pseudo-analog functions working with the value of the pulse stream, but with a digital implementation. It is well known that a probability cannot be exactly measured but only estimated as the relative frequency of "high" levels in a long enough sequence. As a consequence, the stochastic computing introduces errors in the form of variance when we attempt to estimate the number from the sequence.
If $Z = \sum_i \frac{X_i}{n}$ the relative frequency of $1s$ in n pulse sequence $X_1, X_2....X_n$. Expectation value and variance of Z are given respectively by

$$E[Z] = \sum_{1}^{n} \frac{\mu_i}{n} \tag{5}$$

$$Var[Z] = E[Z - E[Z]]^2 = \frac{1}{n^2} \sum_i \sigma^2 \tag{6}$$

Where μ_i and σ_i are the mean value and the typical deviation respectively for each X_i.

From figure. 2 the pixel functionality can be described as detection of presence or absence of a signal on each clock cycle. This result on streams of Bernoulli probability $P_i = p(x_i = 1)$ on each clock cycle. As expressed in Eq. 7, P_i is determined by the relative magnitude of the expected signal to the V_{ref} of the comparator, and by the variance of the zero mean gaussian noise on the signal.

$$P_i = \frac{1}{\sqrt{2\pi\sigma_N^2}} \int_{V_{ref}}^{\infty} e^{\frac{(\nu - V_m)^2}{2\sigma_N^2}} d\nu \tag{7}$$

The integration imager communications involve only serial bit streams of identical pulses. The main advantage of stochastic computing is the similarity between Boolean algebra and statistical algebra. When a stochastic signal $Sgn(t)$ is multiplied by weighting coefficient $0 \leq w \leq 1$, the multiplication can be achieved by using a simple AND gate. Addition of two stochastic signals $Sgn1(t)$ and $Sgn2(t)$ can be obtained by combining them in an OR gate [2]. Each pixel involve an AND gate to perform the weighted multiplication. None of these operations affects the SNR of the resulting signal [2]. It is possible by using a suitable distributed architecture at the pixel level and with minimum hardware to perform a weighted sum which is the standard task in image processing, such as low-pass filtering, removal isolated nonzero pixels, or edge detection.

In order to prove the concept of stochastic arithmetic we built a model of the sensor in Matlab (Simulink) to emulate the behavior of the intelligent CMOS digital pixel. The main objective of the model is to generate hyper-spectral test images from synthetic scenes with the introduction of

Figure 6: Simulation of the MATLAB model of the sensor. SICNN using pulse stochastic arithmetic

the physics-based model of the optical and CMOS solid-state elements of the imager. The stochastic arithmetic operations were based on the SICNN contrast enhancement algorithm we introduced in [3]. Equation. 8 illustrates the mathematical equation of SICNN.

$$X_{ij} = \frac{I_{ij}}{a_{ij} + \sum_{k,l \in N_r(C)} w_{ij} I_{kl}} \tag{8}$$

Where x_{ij} represents the SICNN output pixel, I_{ij} is the input pixel, $a_{ij} (= 1)$ is a constant; the term $\sum_{k,l \in N_r(C)} w_{ij} I_{kl}$ represents a convolution operation or the weighted sum. For simplicity reasons the convolution mask used is,

$$W = \begin{bmatrix} 1 & 1 & 1 \\ 1 & 0 & 1 \\ 1 & 1 & 1 \end{bmatrix}$$

Figure. 6 shows the simulation results of the sensor model.

4. VLSI PROTOTYPE

In this design a 96×64 pixel array with counter length of 8bits has been implemented. The chosen configuration is sufficient for a feasibility demonstration. The increase of the precision of the frequency measurement by adding more counters (One counter for each 3×3 pixels) stages in the design is straightforward for technologies $< 0.5\mu m$. Figure. 7 shows the prototype imager fabricated in $0.5\mu m$, 1 poly/4 metals CMOS standard technology. The pixel layout is shown in Figure. 5, it occupies an area of $29 \times 29\mu m^2$ with a fill factor of 15%. The average power consumption per pixel at a frequency of $150khz$ is $78\mu W$. The test chip will allow us to test the expected dynamic range and also evaluate simple image processing operations based on stochastic arithmetic.

Figure 7: The CMOS imager prototype.

5. SUMMARY AND FUTURE WORK

In this paper the authors introduced a robust pixel architecture suitable for deep submicron technology demonstrating an adaptive wide dynamic range exceeding 100 db without affecting the SNR. We demonstrated that it is possible to take advantage of the concept of stochastic arithmetic to implement complex image processing algorithms. To the best of our knowledge this paper is the first in demonstrating a pixel level stochastic arithmetic. As mentioned earlier the main advantage of the stochastic processing system is the possibility of doing pseudo-analog functions using the values of the pulse stream, but with digital implementation.

More issues related to deep sub micron technology need to be addressed in the future. We predict that digital pixel architecture (DPS) technology will flourish with the availability of photodectors characterization data from foundaries, and the development of new sensors introducing pixel level stochastic arithmetic, this situation will lead to the establishment of an advanced technology taking over the dominance of CCDs.

6. REFERENCES

[1] B.D. Brown & H.C. Card "Stochastic neural computation. II. Soft competitive learning", *IEEE Trans. on Computers,* vol 50, No. 9, September 2001, pp 906-920.

[2] J.F. Keane & L.E. Atlas "Impulses and stochastic arithmetic for signal processing", *IEEE International Conference on Acoustics, Speech, and Signal Processing,* Salt Lake City, vol. 2, pp. 1257 -1260, 2001

[3] T. Hammadou & A. Bouzerdoum "Novel Image Enhancement Technique Using Shunting Inhibitory Cellular Neural Networks", *IEEE Trans. on Consumer Electronics,* vol 47, No. 4, November 2001, pp 934-940.

A CMOS IMAGER WITH PIXEL PREDICTION FOR IMAGE COMPRESSION

Daniel León, Sina Balkır, Khalid Sayood, and Michael W. Hoffman

Department of Electrical Engineering, 209N WSEC,
University of Nebraska-Lincoln, Lincoln, NE 68588-0511, USA
E-mail: daniel@torpedo.unl.edu

ABSTRACT

A predictive approach to on-chip focal plane compression is presented. This approach, unlike the transform coding techniques, is based on pixel prediction. In this technique, the value of each pixel is predicted by considering the values of its closest neighbor pixels. An analog circuit has been designed to accomplish a simple prediction function based on the values of two neighbor pixels. To demonstrate the approach, a prototype chip containing an imager and the prediction circuits has been designed, implemented, and tested.

1. INTRODUCTION

Improvements in CMOS imager technology have made possible the creation of pixels with performance challenging that of CCD sensors. A key advantage of CMOS imagers is the integration of processing circuitry in the focal plane. Processing implementations that exploit this integration include the artificial retina, motion detection, feature detection, color segmentation, and sensor output compression.

As the size of the sensing array increases, the pixel readout process demands more bandwidth. Compression of the sensor output is a very desirable feature in bandwidth constrained applications such as wireless devices or satellite imaging.

Efforts in this direction have followed different paths. Most of the attention has been devoted to transform coding compression. On-sensor discrete cosine transform (DCT) has been reported in [1]. In that work the analog signal from the sensor is processed by an analog 2-D DCT processor which is based on a switched capacitor coefficient multiplier array. The processor is followed by an analog-to-digital converter.

Conditional replenishment was used in [2], [3] to compress the video output of a pixel array. In this technique, a pixel value is compared with its last output value. If their difference exceeds a threshold, the new pixel value and its address are output. In their experiments, the authors found that more than 40% of the pixels were active at 33 frames per second. Conditional replenishment can be viewed as a temporal prediction process. Significant performance improvements can be obtained if we also use spatial or pixel prediction [4].

Pixel prediction is usually done using neighboring or past pixels in the current frame that are already available to the decoder. The prediction error is then compressed by means of an entropy encoder. To our best knowledge, this compression scheme has not been implemented on an imager focal plane. In this paper we concentrate on developing an analog focal plane pixel predictor. This will be combined in later work with temporal predictors to compress video.

There are several reasons for the selection of spatial prediction over transform coding. On a theoretical level it can be shown that for popular image models both transform coding and spatial prediction asymptotically provide the same compression gain [5]. On a more practical level spatial prediction fits in well with conditional replenishment to provide compression for video sources. Finally, we believe that the use of analog pixels in the focal plane provides both compression and complexity advantage over first digitizing the pixels and then applying compression.

A 23x20 imager has been designed using a standard CMOS 1.5 μm technology to test this idea.

2. PIXEL PREDICTOR

Pixel prediction is done by estimating the value of a pixel based on the values and relations between its neighbors. Figure 1 shows the pixel being encoded x and its neighbors a and b. Here we assume that the array is being read in a

0-7803-7761-3/03/$17.00 ©2003 IEEE

raster scan fashion, thus, the pixels values of a and b are already available at the decoder. We take a simplified version of the gradient-adjusted prediction (GAP) presented in [6], so that the prediction value $\hat{x}$ is given by:

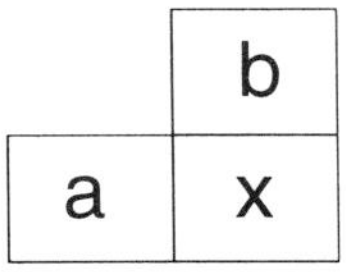

Figure 1: Neighbor pixels used in the prediction.

$$\hat{x} = \frac{a+b}{2} \tag{1}$$

Then, the prediction error will be given by:

$$e = x - \hat{x} \tag{2}$$

To take care of the image boundaries, the pixels in the first row and first column are not predicted. This does not have a significant impact on the overall compression gain. If the predicted value is close to x, the error e will have small values grouped around zero. Compression is obtained when this error is entropy coded. The coded error is then transmitted or stored.

The entropy coder requires as an input a quantized representation of the prediction error e. The quantization process introduces unavoidable distortion. At this point two quantization schemes can be devised and are shown in Figures 2 and 3. In the first one, pixel prediction is performed with analog pixel values and the resulting prediction error is quantized, whereas in the second one, pixel prediction is computed with quantized pixel values. The quantization noise is modeled by q_1 and q_2.

A general model for the predictor P is a weighted sum of the previous N pixel values [7].

$$P(x) = \sum_{i=1}^{N} c_i x[n-i]. \tag{3}$$

Where c_i are the predictor coefficientes. Thus, the prediction error e is given by (see Figure 2)

$$e[n] = x[n] - \sum_{i=1}^{N} c_i x[n-i]. \tag{4}$$

From Figure 3 we have

$$\tilde{e}[n] = x[n] - \sum_{i=1}^{N} c_i \bar{x}[n-i]. \tag{5}$$

but

$$\bar{x}[n] = x[n] + q_2[n]. \tag{6}$$

therefore,

$$\tilde{e}[n] = x[n] - \sum_{i=1}^{N} c_i (x[n-i] + q_2[n-i]). \tag{7}$$

$$\tilde{e}[n] = e[n] - \sum_{i=1}^{N} c_i q_2[n-i]. \tag{8}$$

Taking expected values and assuming $e[n]$ and $\tilde{e}[n]$ are zero mean

$$\sigma_{\tilde{e}}^2 = \sigma_e^2 + \sigma_{q2}^2 \sum_{i=1}^{N} {c_i}^2 \tag{9}$$

From Figure 2

$$\sigma_{\bar{e}}^2 = \sigma_e^2 + \sigma_{q_1}^2 + 2E\{e \cdot q_1\} \tag{10}$$

From Figure 3 and using (9)

$$\sigma_{\hat{e}}^2 = \sigma_e^2 + \sigma_{q_2}^2 + \sigma_{q_2}^2 \sum_{i=1}^{N} {c_i}^2 + 2E\{\tilde{e} \cdot q_2\} \tag{11}$$

Comparing (10) and (11) we note that impact of the quantizer is less severe when it is applied to the prediction error. This fact together with the smaller area required by analog processing circuits serve as the main motivation to implement the pixel prediction in the analog domain.

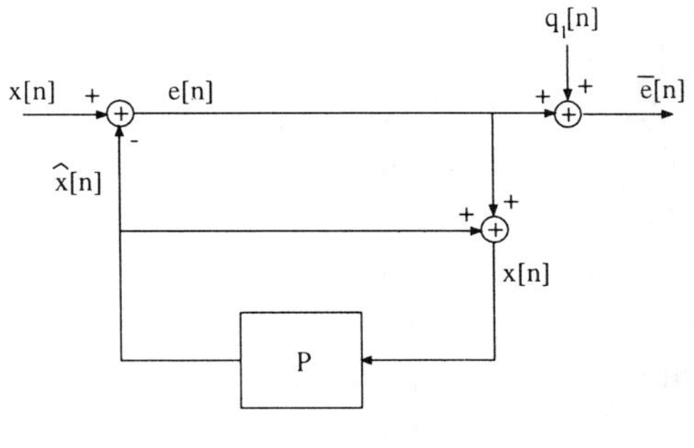

Figure 2: Prediction scheme with quantizer at the error output.

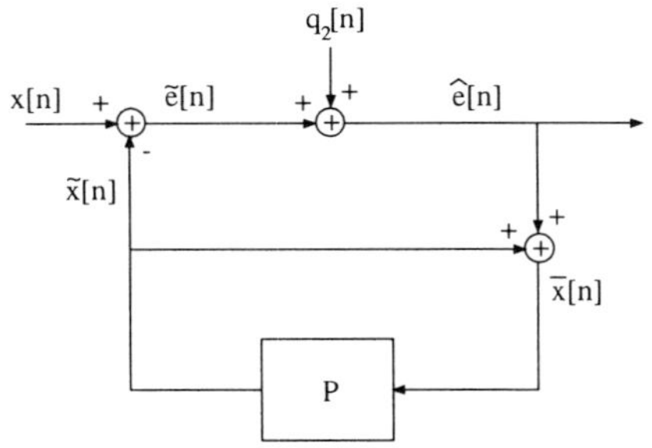

Figure 3: Prediction scheme with quantizer at the pixel output.

3. DESIGN OVERVIEW

A pixel prediction circuit has been implemented using the AMI CMOS 1.5 μm process. The circuit includes a 23x20 pixel array. The pixel design is based on the active-pixel sensor (APS) technology. Each pixel consists of a n-well photodiode, a reset transistor, a source follower transistor, and a row-access transistor with a total area of 39.2x50.5 μm^2 and 20% fill factor.

Each row is sequentially addressed by a 24-bit shift register, the columns are enabled by an external circuit. Since the pixel prediction requires the pixel value in the row above the pixel being predicted (see Figure 1), each previous row is stored by a 20-cell analog memory. Each cell in the analog memory consists of a poly1/poly2 capacitor with their respective write and read transistors (optimized to minimize charge injection). Figure 4 shows a block diagram of the complete circuit.

The prediction circuit computes the prediction error based on its inputs a, x (from the pixel array) and b (from the analog memory). To provide the inputs a and x, two consecutive columns are simultaneously activated during the array read out. Their roles are switched accordingly by a multiplexer in order to present the correct pixel value to the predictor circuit. A timing diagram of the read-out operation is shown in Figure 5.

A schematic diagram of the prediction circuit is shown in Figure 6. It consists of two low-gain differential amplifiers outputting currents proportional to $x - a$ and $x - b$ wich are then added to obtain $I_d = g_m(2x - a - b) + I_{bias}$. Where g_m is the transconductance of the differential amplifiers. The final output is the voltage V_{out}. Changes in V_{out} are proportional to I_d.The transistors in the differential amplifiers are matched by a careful layout design.

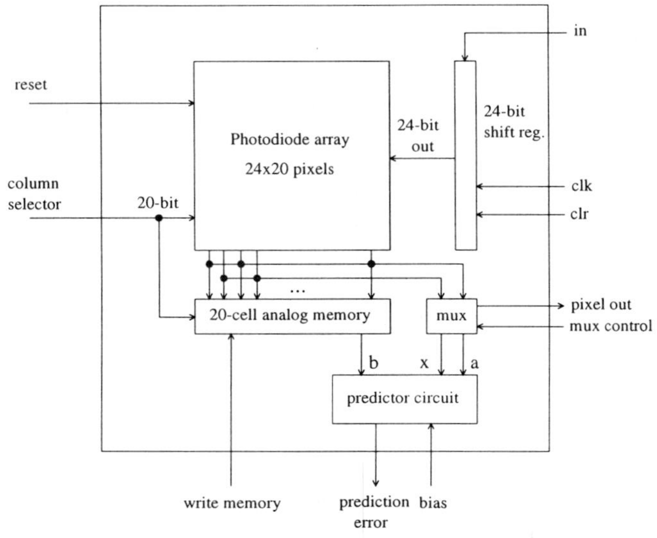

Figure 4: Block diagram of the chip.

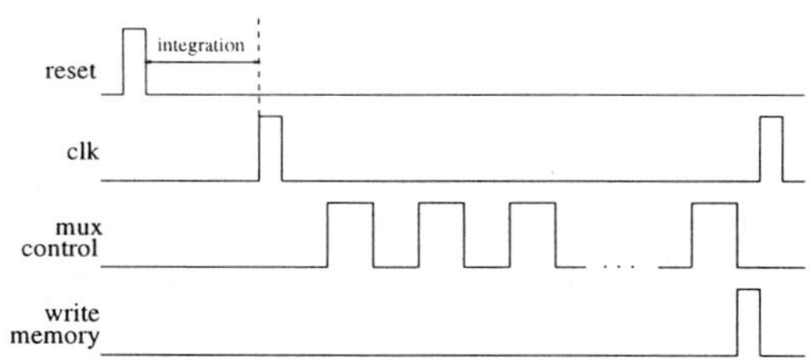

Figure 5: Timing diagram.

Figure 7 shows a sample image and its corresponding error image acquiried from the actual chip. As expected the edges on the image are extracted by this type of prediction. The dynamic range is 48dB. The layout of the completed design is shown in Figure 8.

4. SUMMARY AND FUTURE WORK

A predictive approach to on-chip focal plane compression has been presented. The approach uses analog pixel values directly. A test chip containing a 23x20 imager incorporating the prediction circuits have been designed and fabricated. Functional tests show the validity of the design. In future designs, correlated-double-sampling (CDS) circuits will be included to reduce transistor variation effects. The entropy coder digital circuit will also be implemented.

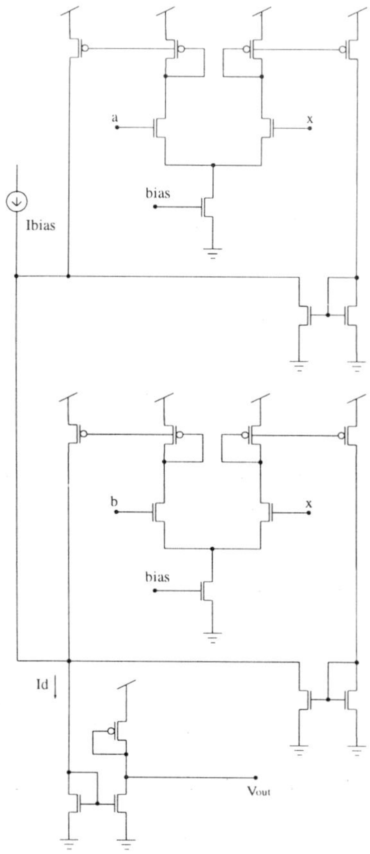

Figure 6: Schematic of the analog predictor circuit.

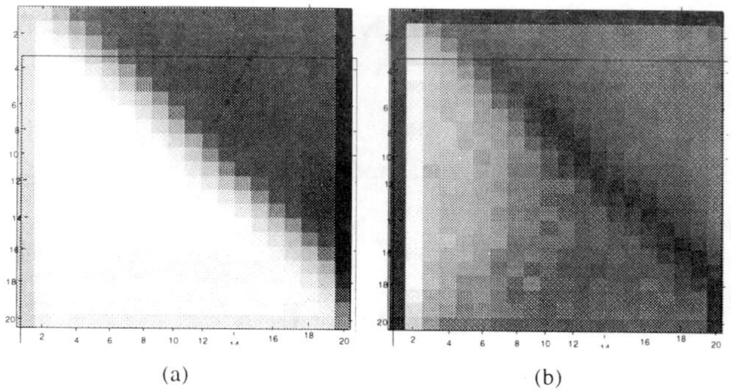

Figure 7: Images acquired from the chip. (a) Pixels without processing. (b) Error image

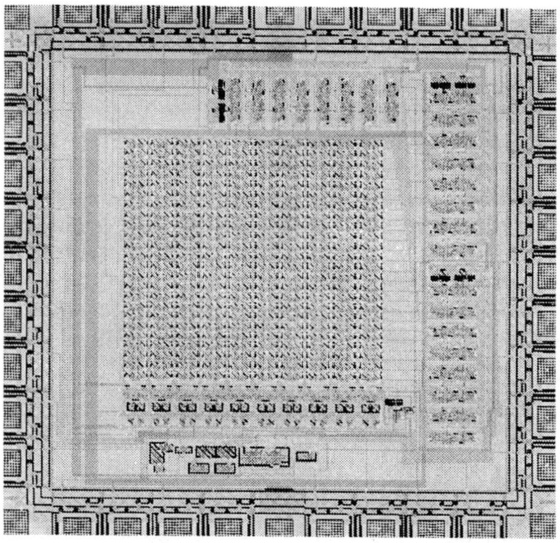

Figure 8: Layout of the complete circuit.

Acknowledgments: This work has been supported by the Catalyst Foundation grant "Focal Plane Video Compression".

5. REFERENCES

[1] S. Kawahito et al, "A CMOS Image Sensor with Analog Two-Dimensional DCT-Based Compression Circuits for One-Chip Cameras," *IEEE J. Solid State Circuits*, vol. 32, no. 12, pp. 2030-2041, Dec. 1997.

[2] K. Aizawa et al, "Computational Image Sensor for On Sensor Compression," *IEEE Trans. On Electron Devices*, vol. 44, no. 10, pp. 1724-1730, Oct. 1997.

[3] T. Hamamoto, K. Aizawa, and M. Hatori, "Focal Plane Compression and Enhancement Sensors," In *Proceedings of the 1997 IEEE Int. Symposium on Circuits and Systems*, pp. 1912-1915, Hong Kong, 1997.

[4] N.D. Memon and K. Sayood, "Lossless Compression of Video Sequences", *IEEE Transactions on Communications*, vol. 44, pp.1340-1345, Oct. 1996.

[5] N. S. Jayant and P. Noll, *Digital Coding of Waveforms. Principles and Applications to Speech and Video*, Prentice Hall, New Jersey, 1984.

[6] X. Wu and N. Memon, "Context-Based, Adaptive, Lossless Image Coding," *IEEE Trans. On Communications*, vol. 45, no. 4, pp. 437-444, April 1997.

[7] K. Sayood, "Analysis of Differential DPCM," *Asilomar IEEE Conf. on Circ., Syst. and Computers*, vol. 1, pp. 218-222, 1985.

ON CHIP GAUSSIAN PROCESSING FOR HIGH RESOLUTION CMOS IMAGE SENSORS

Sanjayan Vinayagamoorthy and Richard Hornsey+*

*Department of Electrical and Computer Engineering
University of Waterloo
Waterloo, Ontario, Canada

+Department of Computer Science
York University
Toronto, Ontario, Canada

ABSTRACT

Spatial image processing chips, known as silicon retinas, are based on the architecture of vertebrate retina and can be mathematically represented as the Laplacian of Gaussian (LOG) and Difference of Gaussian (DOG). In this paper, attention has been paid on implementing a retina function through the LOG model.

Previous implementations have used a hexagonal resistive mesh within the pixel array, which can lead to low resolutions and difficulty of readout. Here, a rectangular resistive array is designed separate from the pixel array. Placing the resistive mesh at the bottom of the array allows for image sensors with high resolution. New circuits designed to accomplish this separation are reported here. Coupled with an array of photo diode pixels, it may be used for image smoothing and edge detection. The image kernel is 5x5 pixels and is implemented in analog CMOS circuits using a standard 0.35-micron process. The IC was fabricated and the hardware implementation was validated through physical testing.

1. INTRODUCTION

Traditional mechanisms for video processing consist of (i) a video camera and (ii) a high speed computers system (or a DSP module) that performs most (if not all) of the processing off-chip. The purpose of this research is to integrate certain aspects of off-chip processing with the image sensor system. The current implementation of most video systems involves a number of different subsystems that rely primarily on digital signal processing. By shifting a few of these simple processes into the video microchip, the complexity of subsequent off-chip processes can be reduced.

The retina is the first element in the visual processing system of vertebrates. It serves 3 purposes: (i) optical sensing, (ii) edge extraction and (iii) motion detection. Analogous systems that perform this same spatial filtering in silicon have been dubbed "silicon retinas"[1][2]. Of particular importance for this research, is the smoothing action performed by spatial filtering and a key feature is the variability in the coarseness and sharpness in the filtering. The resolution of the filtering determines the smallest possible object that may be resolved. The ideal equation used to model the retina-smoothing filter is:

$$\nabla^2 G(r) = \frac{-1}{\pi\sigma^4}\left(1-\frac{r^2}{2\sigma^2}\right)\mathbf{exp}\left(\frac{-r^2}{2\sigma^2}\right)$$

where G(r) is the Gaussian function, ∇^2 is the Laplacian, r is the distance away the origin/node and σ is the variance.

It has been shown in previous work [3] that resistive mesh behaves in manner similar to the Gaussian function. The filtered image is obtained by converting each photo-pixel voltage to a current via a transconductance amplifier (or an equivalent) and injecting this current to the appropriate points along the network. After a known settling time, the voltage from the node is read back out. This collection of node voltages denotes the filtered image. The basis of the filtering characteristic of a resistive mesh lies in the fact that resistors directly connected to a node will be influenced more by the injected current at that node than by resistors that are indirectly connected to it. As one moves further away from the node the effect of the current at the starting node will decrease. When the resistance of the network is relatively low, the effect of the current at one node has a larger effective range compared with a network of high resistance. A network with a low resistance has a wider "kernel" effect at each node than a high resistance network that has a smaller "kernel" effect. This is illustrated in Figure 1.

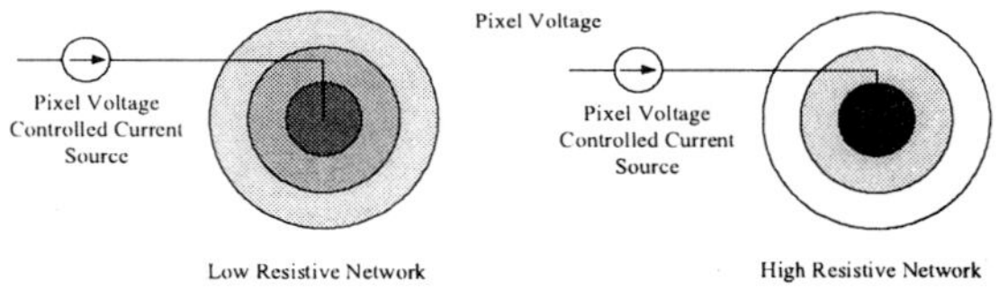

Figure 1 Abstract Filtering Comparison between Two Resistive Networks

In previous designs [4], a hexagonal resistive network is used for smoothing since a small number of transistors are required and because of its inherent symmetry [5]. The biggest disadvantage in this scheme is the loss of resolution. It is also very hard to amalgamate a hexagonal resistive array with a rectangular pixel array. Employing an innovative off-array rectangular mesh separate from the pixels solves these problems. First, by separating the resistive mesh from the pixel array, higher resolutions can be supported. Second, having a square mesh allows for easy implementation in practical image sensors. To ensure a symmetrical smoothing pattern the rectangular array has to be modified. Making the diagonal resistors larger artificially creates this symmetry. The modified square resistive mesh is shown in Figure 2. By using this modified square resistive mesh instead of the traditional hexagonal one, one gains better spatial sampling as well as an easier readout control structure. The better spatial sampling comes at the cost of increased transistor count and thus power. Furthermore, the smoothing function is even

0-7803-7761-3/03/$17.00 ©2003 IEEE

further approximation of the LOG algorithm. This can be alleviated using a negative resistors connected between every other pixel as noted in [1].

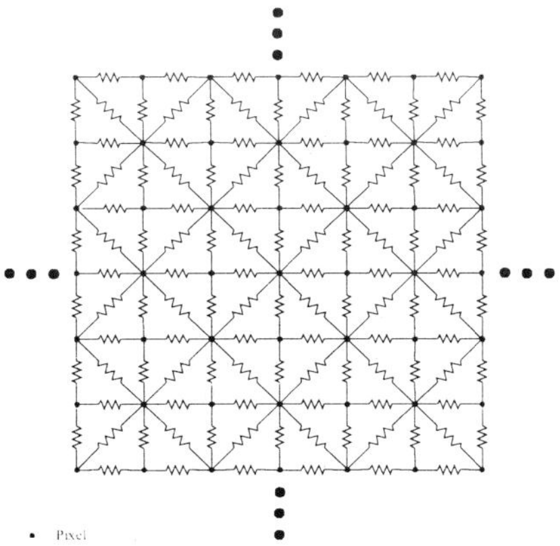

Figure 2 Square Resistive Mesh (Note that the diagonal resistors are $\sqrt{2}$ times larger

Separating the resistive mesh from the pixel array creates a new problem. A method of retaining the pixel voltage and passing to rows further down had to be devised and thus new circuits had to be developed to overcome them.

2. SYSTEM OVERVIEW

In Figure 3 the system overview used to implement the LOG algorithm through a resistive mesh, is shown.

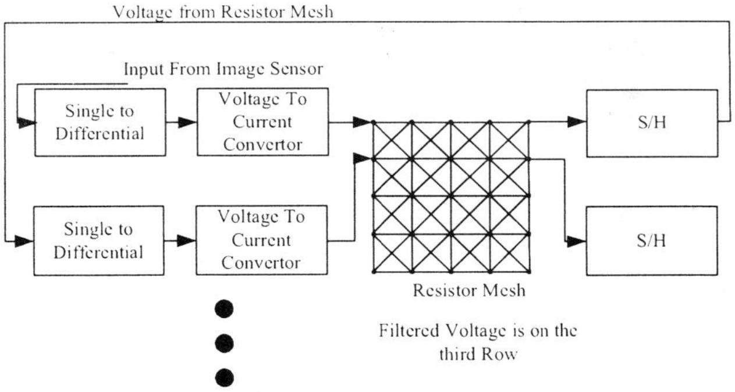

Figure 3 System Overview

Initially the voltage captured from the image sensor is converted to a differential signal. Then a current proportional to this voltage is injected into the square resistive mesh. The voltage is then sampled from each node of this square resistive mesh. This represents 1 cycle. The sampled pixel voltage from the resistive mesh is then passed to the next stage i.e. which begins with a single to differential signal converter. After each cycle is complete, the 3rd row is sampled (this represents the filtered voltage) and is passed to further stages. Note that after the voltage is sampled initially from the image sensor, future circuits must retain this voltage for at least 5 cycles.

This system was simulated in MATLAB and VHDL-AMS and the resulting image is shown in Figure 4.

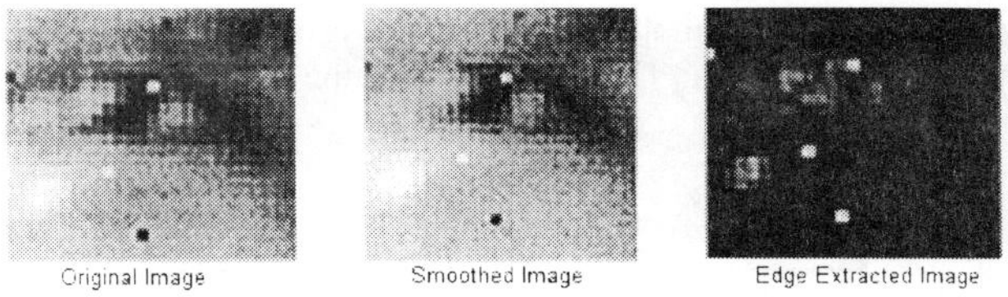

Figure 4 Examples of LOG through Resistive Mesh (Image with Noise)

3. PHYSICAL IMPLEMENTATION

Several circuits were required to implement this algorithm on an IC. In this design, particular attention is paid to the S/H subsystem since this is the most critical. Other systems such as the Voltage to Current Converter are not as critical.

A physical implementation of this system is heavily dependent on ensuring the sampled voltages, which are used in generating the final filtered voltages, are maintained accurately for the required number of cycles. Degradation in a single stage would be amplified once this sampled voltage reached the last row. This would cause poor results in the overall performance of the filter. Thus the most critical component of this system is the S/H circuit. A basic S/H system would not suffice since gain errors and clock feed-through would be hard to overcome. Furthermore a complex system, whereby more than one operation amplifier is necessary would take too much space and power. To this end a modified implementation presented in [6][7] is used as shown Figure 5.

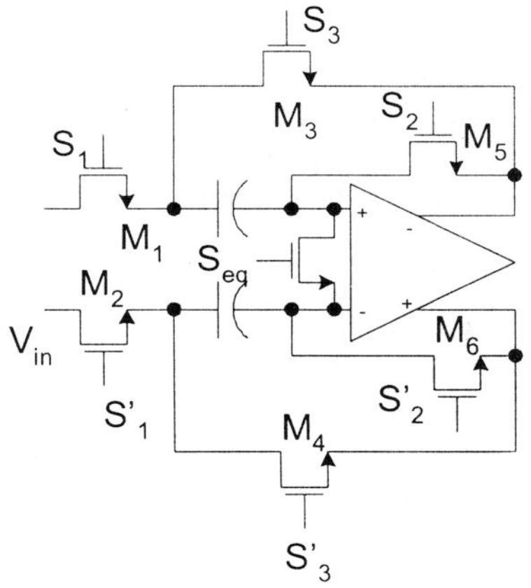

Figure 5 Sample and Hold

In any S/H system there is bound to be a drop in the hold voltage after the initial sampling. This error is even more pronounced when the sampling capacitor is not large. To minimize this drop, the sampling frequency has to be increased and this forces the amplifier to have a high bandwidth as well. Another requirement is that the operational amplifier (opamp) must have a very high slew rate. This is a consequence of the S/H topology since the output does not track the input at all times and must quickly transition from approximately 0V to the desired result within a sampling time.

Due to the system requirements of very high impedance and speed, older designs such as [8] were not useful. Thus a new opamp design was implemented. This allows for fast slew rates and high bandwidths. The schematic for the opamp is shown in Figure 6.

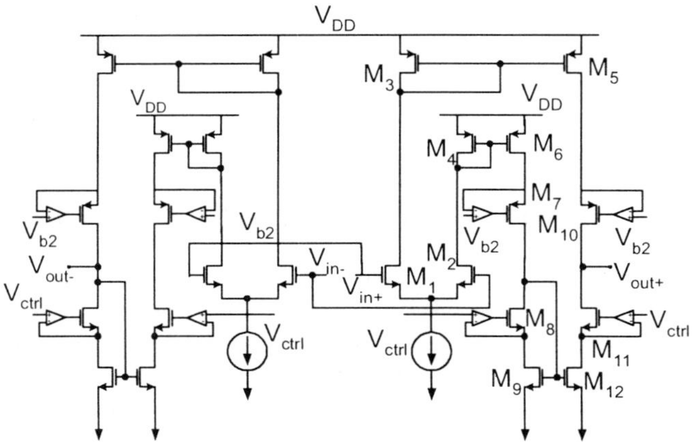

Figure 6 Opamp Schematic

The control voltage, V_{ctrl}, represents the voltage from the common mode feedback (CMFB) system. Using this system it was possible to achieve good results at 250MHz sampling speeds and reasonable results at 500MHz. It should be noted that all bias voltages are achieved within the IC. Since output impedance was increased through separate gain amplifiers, the dynamic range of the output would be reduced. However this is not a concern since the ideal output voltage range of the image sensor is between 1.1 and 1.9V.

The resistors that comprise the resistive mesh are formed using MOSFETs biased in the linear region. A simplified schematic is show in Figure 7(a). This approach was used in order to achieve a symmetric I-V characteristic in contrast to typical one-FET design.

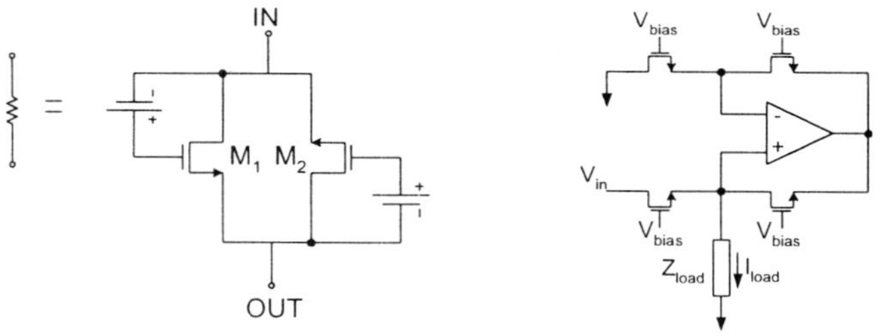

(a) Equivalent Resistor

(b) Voltage to Current Convertor

Figure 7 Equivalent Resistor and Voltage to Current Convertor

External biasing is not feasible in this design since the number of transistors is very large. Also, since the values of the resistors need to be variable (post fabrication tuning), biasing schemes separate to other subsystems were made.

Once the design of the opamp is finalized then other subsystems such as the linear voltage to current converter are fairly straightforward. Other methods described in [9], are not useful due to the large transistor count. It should be noted that minor modifications were required on the amplifier since the output is single ended and a higher dynamic range is required. Figure 7(b) shows the simplified schematic of the linear voltage to current converter.

4. RESULTS

As a demonstration of the concept of Gaussian Filter, a 5x5 kernel array was used. Care was taken to ensure scalability to a larger array size.

The results were collected and plotted as "pictures" in MATLAB. Each test first started with the middle pixel only being turned on. This way, the smoothing effect of the filter is apparent. If the smoothing effect were not present then one would know an error existed immediately. The first "pictures" were taken at a 100MHz. In Figure 9(a), the pre-filtered image is shown. Figure 9(b) and 9(c) show ideally filtered and measured filter pictures respectively. Another test case is shown in figure 10.

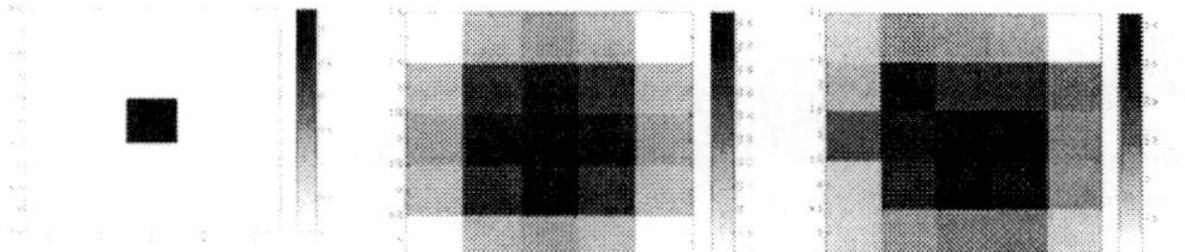

(a) Pre-Filtered (b) Ideally Filtered (c) Measured Filtered

Figure 8 Measured Results: Test Case 1

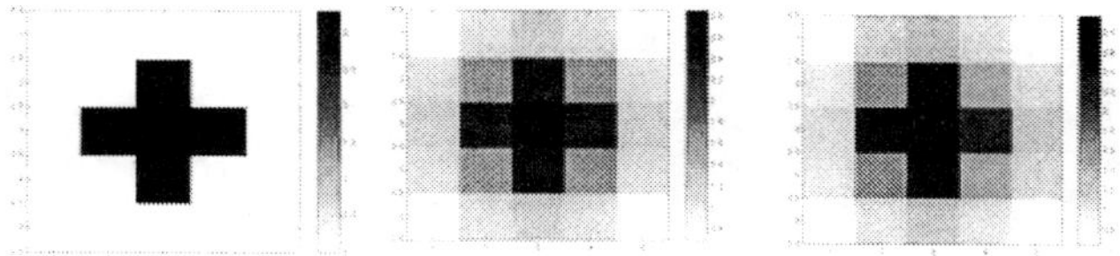

(a) Pre-Filtered (b) Ideally Filtered (c) Measured Filtered

Figure 9 Measured Results: Test Case 2

One final set of test pictures is shown in Figure 10

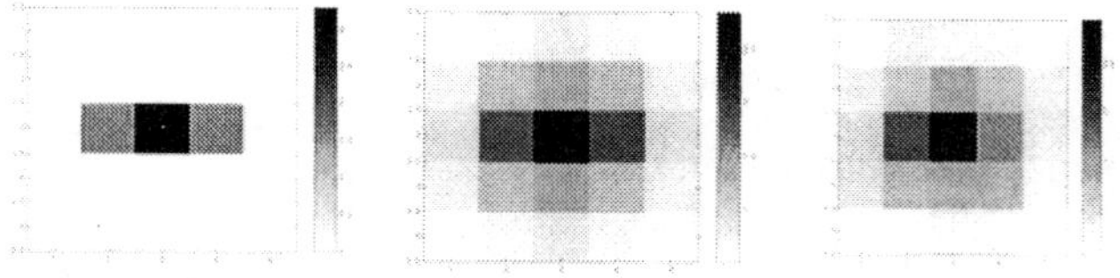

(a) Pre-Filtered (b) Ideally Filtered (c) Measured Filtered

Figure 10 Measured Results: Test Case 3

Since the edge extraction operation is a simple subtraction, that image is not shown.

As it clearly seen, the measured filtered image is only marginally different than the ideally filtered image. To quantify this, several more test pictures were filtered, and the tabulated results are shown in Table 1.

Test Case	Pre-Filtered Test Case Description	Average Error(%)
1	Figure 9	4.2
2	Figure 10	5.0
3	Figure 11	5.5
4	"Triangular Shape"	6.2
5	"L" Shape	6.7

Table 1 Average Error between Ideal Filtered Image and Measured Filtered Image at 100MHz

Since the average error is approximately 5% at the sampling frequency of 100MHz, the correct working function of the filter is proved.

To show that this filter functions at faster frequencies (250MHz), the test images were filtered again. The tabulated results are shown in Table 2.

Test Case	Pre-Filtered Test Case Description	Average Error(%)
1	Figure 9	8.6
2	Figure 10	8.2
3	Figure 11	10.1
4	"Triangular Shape"	9.6
5	"L" Shape	9.3

Table 2 Average Error between Ideal Filtered Image and Measured Filtered Image at 250MHz

Though the final case is close to the acceptable 10% error, further tuning of the resistors could further reduce this error.

5. SUMMARY

From the various measurements, the following conclusions can be made:

- A improved LOG function was implemented through a square resistive mesh. In both simulation and testing the designs have performed within acceptable parameters.
- New circuits were designed to allow for very fast sampling rates and also allowing capacitor sizes to be reduced. Thus layout size and power consumption were reduced over traditional ICs.
- The limitation of using a 5x5 kernel window is because of the required manual tuning. However, increasing the size to accommodate for more columns should not be difficult nor will the resulting system be more prone to error.

Further considerations for this system are to incorporate a automatic post fabrication tuning scheme such as those used in [10]. Furthermore, a method by which the window can be dynamically changed similar to that of many software algorithms can be implemented [11].

6. ACKNOWLEDGMENTS

Acknowledgments: The authors are grateful to the Canadian Microelectronics Corporation for access to IC design and fabrication. Financial support from NSERC Canada is gratefully acknowledged. The authors thank Gary Cheng for his helpful initial work on rectangular Gaussian filters.

7. REFERENCES

[1] A. Moini, *Vision Chips or Seeing Silicon*, University of Adelaide, Massachusetts, 1997.

[2] D. Marr, *Vision: A Computational Investigation into the Human Representation and Processing of Visual Information*, W.H. Freeman, San Francisco, 1982.

[3] H. Kobayashi et al., *An Active Resistor Network for Gaussian Filtering of Images*, *IEEE Journal of Solid-State Circuits*, Vol. 26, No.5, May 1991.

[4] C. Mead, *Analog VLSI and Neural System*, Addison-Wesley, Australia, 1989.

[5] M.A. Mahowald, Silicon Retina with Adaptive Photoreceptors, SPIE, Vol. 1473, 1991.

[6] B. Razavi, *Design of Analog CMOS Integrated Circuits*, McGraw Hill, New York, 2000.

[7] D. A. Johns and K. Martin, *Analog Integrated Circuit Design*, John Wiley and Sons, Toronto, Canada, 1996.

[8] K. Bult and G. J. G. M. Geelen, *A fast-settling CMOS Op Amp for SC circuits with 90-dB DC gain, IEEE Journal of Solid-State Circuits, Vol. 25, No.6, December*, 1990.

[9] P.P. Vervoort and R.F. Wassenaar., *A CMOS rail-to-rail linear VI-Converter, IEEE International Symposium on Circuits and Systems*, Vol. 2, 1995.

[10] J.M. Khoury, *Design of a 15-MHz CMOS Continuous-Time Filter with On-Chip Tuning, IEEE Journal of Solid-State Circuits, Vol. SC 19, No.6, December 1991.*

[11] I. Paterson-Stephens and A. Bateman, *The DSP Handbook Algorithms, Applications and Design Techniques*, Prentice Hall, New Jersey, 2002.

Normal Optical Flow Chip

Swati Mehta and Ralph Etienne-Cummings

Electrical and Computer Engineering Dept.
Johns Hopkins University
Baltimore, Maryland 21218, USA

ABSTRACT

This paper introduces a 2-D dense Normal Optical Flow estimation algorithm being implemented in VLSI which combines imaging and processing on the same chip efficiently. The algorithm computes partial derivatives with respect to time and space and uses their ratio to compute normal flow velocity. The chip implemented in 0.5µm CMOS process has a 92x52 array of APS pixels. The chip occupies an area of 4.5 mm^2.

1. INTRODUCTION

We can acquire a lot of visual information from a sequence of time-varying images. The determination of the 2-D motion of objects in the environment is an example. Optical flow, which is defined as estimating the apparent motion of the image brightness pattern, ideally corresponds to the spatial motion of objects in the scene, but not necessarily so. The existence of strong enough gradients in the images is required for computing optical flow.

VLSI systems have been developed where velocity computation and imaging have been integrated on a single chip. In previous works, either processing elements were present in the pixel itself or the wiring complexity between different blocks limited the scaling to larger arrays, thus imaging resolution suffered whenever processing was combined with imaging [3,6,7]. This paper presents an algorithm for an aVLSI chip that can extract normal optical flow, but in doing so does not interfere with the imaging process. We propose a design which has a low-power high-resolution CMOS imager at the core, computes normal optical flow at every pixel, and outputs the normal flow at the read-out frame-rate with no penalty to the imaging process.

2. MOTION DETECTION THEORY

There are 2 strategies to solve the optical flow problem [1, 2]:

1. **Matching methods**: Matching and tracking efficiently sparse image features over time. This method leads to sparse measures, i.e. computed only at a subset of image points. The key problem associated with these schemes is the correspondence problem, i.e. matching the features in frame i with the same features in frame i+1. The difficulty associated with the correspondence problem decreases with the sampling interval Δt between frames and with the number of features [6,7].
2. **Intensity based methods** use the intensity or a linear function of intensity at every location to compute the optical flow field throughout the image. These methods lead to dense measurements, i.e. computed at each image pixel. The two main approaches proposed for these types of methods are:

- Second order correlation or Spatio-temporal energy methods: The intensity I(x,y,t) is passed through a linear spatiotemporal filter and multiplied by a delayed version of the filtered intensity from a neighboring receptor. The output of these methods is a quadratic functional from which velocity or speed has to be extracted. These methods are computationally very intensive [8,9].
- Differential methods: Exploit the relationship between velocity and the ratio of temporal to spatial derivatives for constant illumination. They use estimates of time derivatives, and therefore require closely sampled image sequences. These methods yield a direct estimate of the optical flow by using the image brightness constancy equation [1, 3, 10].

$$\frac{dI(x,y,t)}{dt} = \frac{\delta I}{\delta x}\frac{\delta x}{\delta t} + \frac{\delta I}{\delta y}\frac{\delta y}{\delta t} + \frac{\delta I}{\delta t} \tag{1}$$

$$F = (\nabla I)^T V + I_t \tag{2}$$

This equation by itself is not sufficient to determine the optical flow. Only the component in the direction of the spatial image gradient V_n can be determined by this equation. This ambiguity is known as the *aperture problem*. The optical flow thus computed is called normal flow.

3. OUR DESIGN

The aim of our design is to have a high-density photosensitive array that gives us the normal motion field in two-dimensional space and a clean image. We present a technique to compute normal flow in 2-D by using the gradient method to compute velocity. In this method, the velocity is given by the ratio of temporal to spatial derivatives i.e.

$$V_x = \frac{\delta I}{\delta t}\Big/\frac{\delta I}{\delta x}, \quad V_y = \frac{\delta I}{\delta t}\Big/\frac{\delta I}{\delta y} \tag{3}$$

The algorithmic layout of our design is presented in Fig. 1 and the layout of the chip in Fig. 2. The main components of our design are discussed separately as follows.

3.1 Imaging Array

The imaging array is composed of a 92x52 array of APS pixels with local analog memories in each pixel for temporal processing. The size of each pixel is 20.3µm x 20.3µm (in a 0.5µm CMOS process). We use PMOS reset transistors and

0-7803-7761-3/03/$17.00 ©2003 IEEE

NMOS followers to eliminate reset errors and to maximize the output voltage swing respectively. Each APS pixel in the photosensitive array outputs the present photodetector output and the time-delayed output from the last frame. By using 1 row and 3 column scanners and switches as shown in Fig. 3, we have 2 buses running down the array which give us the *V(m,n,t)* and *V(m,n,t-Δt)*. All other computations are performed on these 2 buses at the edge of the photodetector array.

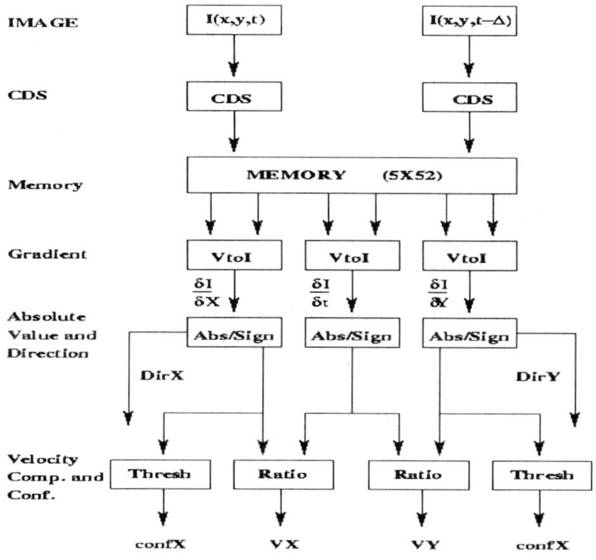

Figure 1: The flow-diagram of the algorithm where different steps for the computation of velocity are outlined.

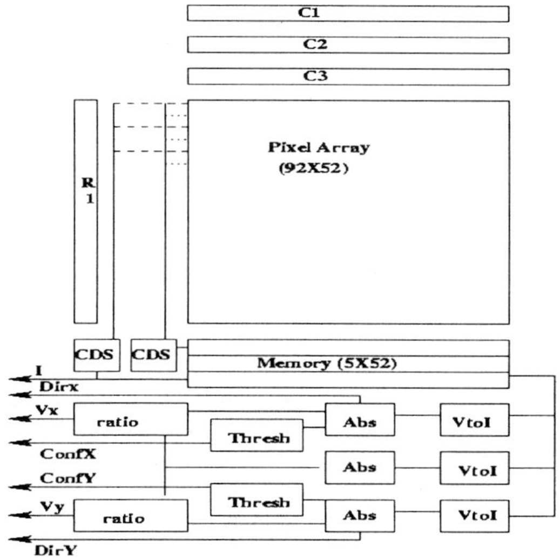

Figure 2: The different blocks comprising the chip. The imaging block is at the core of the chip while the processing is done at the bottom of the array.

3.2 Correlated double sampling

We use correlated double sampling (CDS) to reduce the fixed pattern noise (FPN) present in the image. The CDS circuits are basically subtraction circuits that compute 'cleaner' signals for *V(m,n,t)* and the delayed version *V(m,n,t-Δt)* by subtracting reset voltages from the sampled voltages for each pixel. Because our pixels use PMOS switches for reset, the reset value is always V_{dd}. Thus, we are actually removing the follower error by correlated double sampling. We use the same value for the present as well as the delayed version as we do not store the reset value from the previous frame.

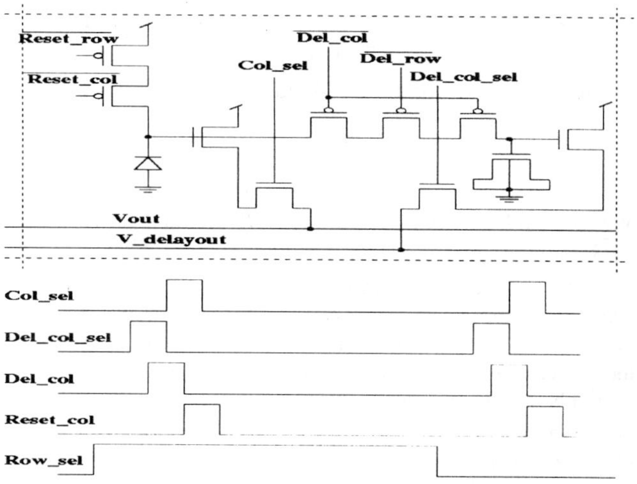

Figure 3: Pixel schematic and timing diagram. Col_sel and Del_col_sel are readout signals for the present and delayed value respectively. Del_col and Del_row sample the delayed value. Reset_col and Reset_row reset the photodiode to V_{dd}.

3.3 Analog Memory

The (5x52) memory block gets its input from the 2 CDS circuits and stores 3 rows of present image output and 2 rows of delayed image output. This block outputs 6 signals which correspond to *V(m,n,t)*, *V(m,n,t-Δt)*, *V(m-1,n,t)*, *V(m+1,n,t)*, *V(m,n-1,t)*, *V(m,n+1,t)*. We write all columns for a particular row and then move on to the next row. The row of the memory block on which we write the next row from the pixel array is determined in a cyclic fashion. While we are writing *V(m+1,n+1,t)* and *V(m+1,n+2,t-Δt)* into the analog memory, we read-out the 6 above signals. This block also adjusts the timing so that we output the pixel, its 4 neighbors and its delayed value at the same time for ease of computation of gradients.

3.4 Voltage-to-Current Converters

The 6 outputs from above are fed into 3 voltage-to-current converters (V-to-I) that convert the difference of a pair of these voltages to compute spatial and temporal derivatives. Since it is easier and more accurate to calculate ratios of currents, we decided to convert the voltages to currents and in the process compute derivatives as well. The V-to-I circuit is shown in Fig. 4 [4]. We use n and p modules in parallel for rail-to-rail operation. Two maximum current selecting circuits and a subtraction circuit are used in the output stage to generate the output currents. The output current follows the following relationship:

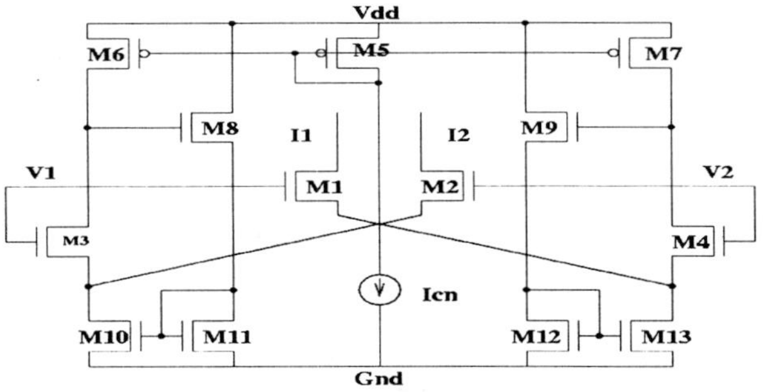

Figure 4: n module of the V-to-I block.

$$I_{out} = \max(I_{n1}, I_{p2}) - \max(I_{n2}, I_{p1}) = \sqrt{8k_n I_{cn}}(V_1 - V_2) = \sqrt{8k_p I_{cp}}(V_1 - V_2) \tag{4}$$

We use a calibration scheme to make the multiplication factors ($\sqrt{8k_n I_{cn}}$ and $\sqrt{8k_p I_{cp}}$) equal for both modules without relying on size adjustments. This step is crucial to make a linear V-to-I converter over a wide range of bipolar inputs. For the self-calibration, we have a 'n' and a 'p' module and we feed them a constant voltage of ± 0.5V; the two currents I_n and I_p should be equal for perfect matching. The difference of the two is sent back to adjust I_{cn} and I_{cp} accordingly through a high-gain feedback loop as shown in Fig. 5.

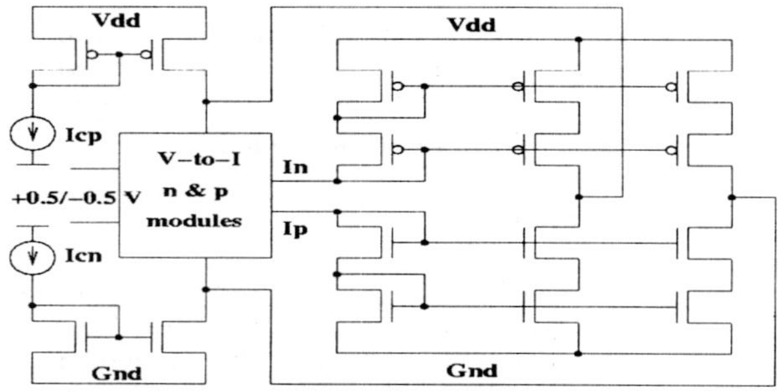

Figure 5: Block diagram for the calibration scheme of the voltage to current converter. The reference currents for the n and p blocks are adjusted through a high-gain feedback loop to make the multiplication factor equal for both modules.

3.5 Absolute Value Circuits

The 3 current-mode absolute value circuits compute the absolute value and the sign of the spatial and temporal derivatives. Depending on the sign of I_{in}, it either flows into the branch M1-M2 or M3-M4. The current through M4 is either directly from M3 or through the current mirror M1-M5. This current mirrored out through M10 is the absolute value of I_{in}. The transistors M6-M9 output the sign.

The correlation of the sign of the x and y direction derivatives and the time derivative gives us the direction of motion. Only the absolute values of these derivatives are sent to the ratio computation circuit to compute x and y direction speeds.

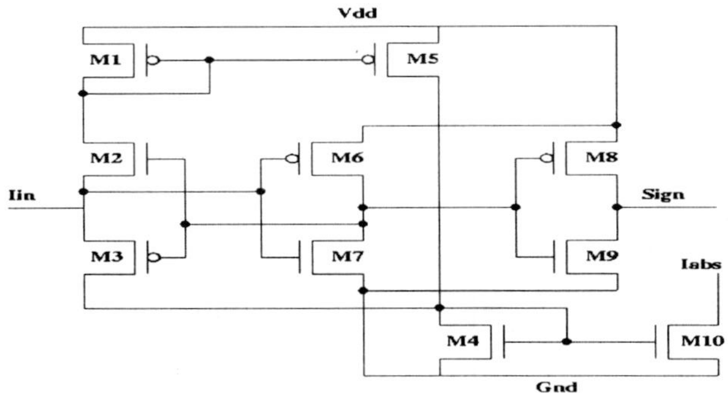

Figure 6: Absolute Value Circuit. This circuit outputs the absolute value of the current as well as the sign.

3.6 Threshold Circuits

The 2 current-mode threshold circuits compute confidence measures for the spatial derivatives. This step is useful in eliminating erroneous measurements in low contrast scenes, where the ratio will be singular.

The threshold circuit is a 7-transistor circuit. The transistors M2-M3, M6-M9 are identical to the ones in the absolute value circuit, but we have an additional transistor at the input so that I_{in} is the difference of I_{thresh} and I_{abs}.

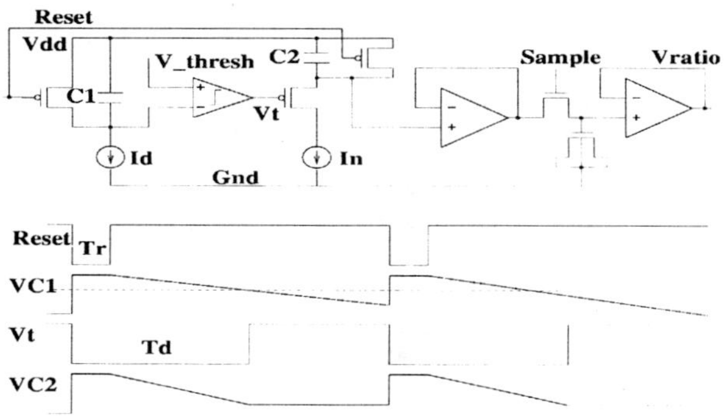

Figure 7: Ratio circuit to compute velocities.

3.7 Ratio Circuits

The two ratio circuits compute V_x and V_y by taking ratios of temporal and spatial derivatives. The circuit is given in [5]. The signal 'reset' is used to initialize the circuit using a small pulse width T_r. When reset is low, V_{C1} and V_{C2} reset to V_{dd}. The comparator output V_T is low. As soon as reset goes high, both C_1 and C_2 start discharging due to I_d and I_n respectively. When V_{C1} becomes less than V_{thresh}, the comparator output switches, ending the discharging of C_2.

The comparator output remains low for a time T_d, but the actual discharge time for C_1 and C_2 is T_d - T_r. Thus,

$$T_d - T_r = \frac{C_1(V_{dd} - V_{thresh})}{I_d} \tag{5}$$

Also, $$T_d - T_r = \frac{C_2(V_{dd} - V_{C2})}{I_n} \quad (6)$$

Thus, $$V_{dd} - V_{C2} = \frac{C_1}{C_2}(V_{dd} - V_{thresh})\frac{I_n}{I_d} = k\frac{I_n}{I_d} \quad (7)$$

From which we obtain the ratio of the two input currents.

The image is also read out from the CDS circuit. Thus, this algorithm gives us a high-resolution image plus normal optical flow at each pixel.

4. RESULTS

We present here the results from MATLAB on using this algorithm and also the same results from SPICE simulations of actual circuits for the same stimulus in Figures 8(a) and 8(b), respectively. The input stimulus is a diagonal block moving towards the bottom right of the pixel array. We simulated a 10x10 array of pixels in SPICE and MATLAB to reduce the computational load.

As can be seen from the plots, the algorithm computes velocity quite correctly. The diagonal motion of the block is detected in the simulations. We used the confidence measures from the simulations to remove the velocity measures wherever the spatial derivatives were less than a particular threshold. This helped reduce quite a few erroneous measurements. The size of the array simulated was limited by the time taken for each simulation.

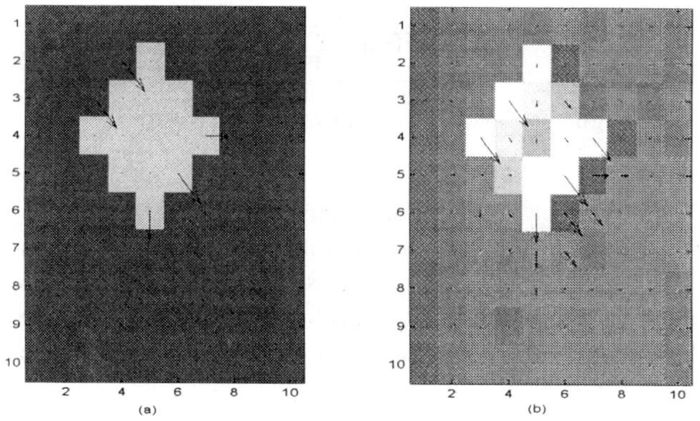

Figure 8: (a) MATLAB simulation results: Plot shows a diagonal block moving down and right. The velocity information is superimposed on the image. (b) SPICE simulation results: Plot shows the image and the superimposed velocity information. Confidence measures were used to exclude singular results. The plots clearly show a diagonal movement across the array.

5. FUTURE WORK

This chip is being fabricated in 0.5μm technology through MOSIS. Presently, we use first order smoothing to compute spatial derivatives. Smoother derivatives will imply just increasing the number of memory rows. Finally, we would like to add the smoothness constraint to the image brightness constancy equation to compute smooth optical flow, thus solving the aperture problem. The chip can be easily scaled to larger arrays without affecting the processing circuitry.

6. CONCLUSIONS

A new design is proposed for computation of high-resolution image and normal flow at each pixel at high scanning speeds. It is also possible to detect edges by using the spatial derivatives. Once normal flow is obtained, the chip can be used to compute focus of expansion, time to contact and many other motion properties of images that can be used to control robots, for example. Tracking systems can use the velocity and segmentation of moving objects can be realized using motion discontinuity. The availability of a confidence measure gives us the added option of discarding incorrect results.

7. ACKNOWLEDGEMENT

This work is supported by ONR YIP award # N140010562.

8. REFERENCES

[1] Horn B. K. and Schunck B. G., "Determining Optical Flow," Artificial Intelligence, 1981,Vol. 17, pp. 185-203.

[2] Koch, C., Moore, A., Bair, W., Horiuchi, T., Bishofberger, B. and Lazzaro, J., "Computing motion using analog VLSI vision chips: an experimental comparison among four approaches," Proceedings of the IEEE Workshop on Visual Motion, 1991, pp. 312 -324.

[3] A. Stocker and R. Douglas, "Computation of smooth optical flow in a feedback connected analog network," Advances in Neural Information Processing Systems II, 1999, MIT Press, pp. 706-712.

[4] Chung-Chih Hung, K. Halonen, V. Porra and M. Ismail, "Low-voltage CMOS GM-C filter with rail-to-rail common-mode voltage," IEEE 39th Midwest symposium on Circuits and Systems, 1996, vol. 2, pp. 921-924.

[5] C. Taillefer, Chunyan Wang and F. Devos, "Current division circuit implemented using CMOS technology," Electronics Letters, vol. 35, Issue 20, 30 Sept. 1999, pp. 1697-1698.

[6] J. Kramer, R. Sarpeshkar and C. Koch, "Pulse-based analog VLSI velocity sensors," IEEE Transactions on Circuits and Systems : Analog and Digital Signal Processing, vol. 44, pp. 86-101, 1997.

[7] R. Etienne-Cummings, S. Fernando, N. Takahashi, V. Shtonov, J. V. D. Spiegel and P. Mueller, " A new temporal domain optical flow measurement technique for focal plane VLSI implementation," Proceedings of Computer Architectures for Machine Perception, pp. 241-250, 1993.

[8] E. Adelson and J. Bergen, "Spatiotemporal Energy models for the perception of motion," J. Optical Society of America, A2, pp. 284-299, 1985.

[9] R. Etienne-Cummings, J. V. D. Spiegel and P. Mueller, "Hardware Implementation of a visual-motion pixel using oriented Spatio-temporal Neural Filters," IEEE Transactions on Circuits and Systems-II: Analog and Digital Signal Processing, vol. 46, no. 9, 1999.

[10] J. Tanner and C. A. Mead, "An integrated Analog Optical Motion Sensor," VLSI Signal Processing II, vol. 21, pp. 59-76, 1981.

HIGH-SPEED POSITION DETECTOR USING NEW ROW-PARALLEL ARCHITECTURE FOR FAST COLLISION PREVENTION SYSTEM

Y. Oike†, *M. Ikeda*†‡, *and K. Asada*†‡

†Dept. of Electronic Engineering, University of Tokyo
‡VLSI Design and Education Center, University of Tokyo
7-3-1 Hongo, Bunkyo-ku, Tokyo 113-8656, Japan

ABSTRACT

A high-speed position detector has wide variety of application fields such as real-time range finding and high-speed visual feedback in robot vision. In this paper, a row-parallel sensor architecture for high-speed position detection is presented. The edge of activated pixels is quickly detected by a row-parallel search circuit and its encoding cycles of N-pixel horizontal resolution are O(log N). The architecture keeps high-speed position detection in high pixel resolution. We have designed and fabricated the prototype position detector with a 128 × 16 pixel array in 0.35 μm CMOS process. The measurement results show it achieves high-speed detection of 450 ns. The high-speed position detection of the scanning sheet beam is demonstrated.

1. INTRODUCTION

A high-speed smart position sensor has wide variety of application fields such as real-time range finding, high-speed visual feedback in robot vision and so on. These applications require much higher speed of frame rate than the conventional imagers using serial readout and transmission. Therefore some smart sensors were proposed for high-speed position detection [1]–[6].

Fig.1 shows a fast collision prevention system with simple calculation, which is one of applications to require high-speed position detection. The scanning sheet beam activates pixels from the right to the left on the sensor plane. Two position detectors detect the edge of the activated pixels. The difference between x_R and x_L means the distance from the position detectors when the edge address of the left position detector is x_L and that of the right one is x_R. High-speed position detection realizes a fast and high-resolution collision prevention system. One forward scan with N-pixel horizontal resolution requires N frames of the scanning sheet beam. For example, 30k fps is required for real-time range finding with 1k-pixel horizontal resolution. The smart sensors [3]–[6] are useful for applications of high-speed position detection. These frame rates, however, are not enough to realize real-time or more high-speed range finding with high pixel resolution.

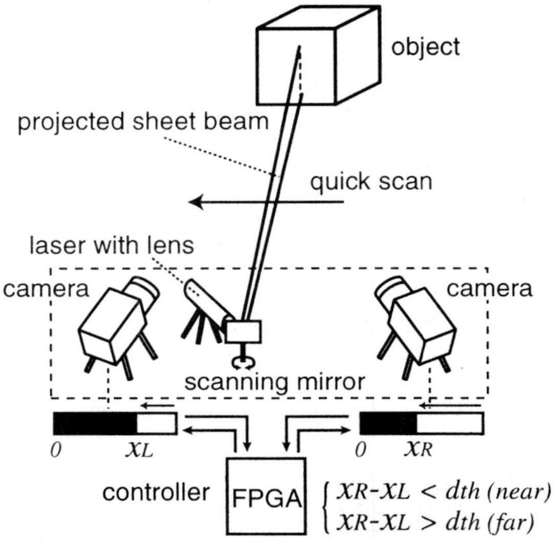

Figure 1: Fast Collision Prevention System.

In this paper, a row-parallel sensor architecture for high-speed position detection is presented. It achieves enough speed for real-time or more high-speed range finding with high pixel resolution. In this architecture, the edge of the activated pixels is quickly detected by a row-parallel search circuit and its encoding cycles of N-pixel horizontal resolution are O(log N). We have designed and successfully tested the prototype position detector with a 128 × 16 pixel array in 0.35 μm CMOS process.

2. ROW-PARALLEL ARCHITECTURE

In the position detection of a projected sheet beam, a sensor recognizes the pixels with strong incident intensity as the history of the scanning sheet beam as shown in Fig.2. Therefore it is important to quickly detect the position of the activated pixels in each row. The frontier line of the projected light provides enough information for triangulation-based range finding. Our architecture has a row-parallel search circuit for the edges of the activated pixels in each row, a row-parallel address encoder of O(log N) and a row-parallel processor to reduce data transmission.

0-7803-7761-3/03/$17.00 ©2003 IEEE

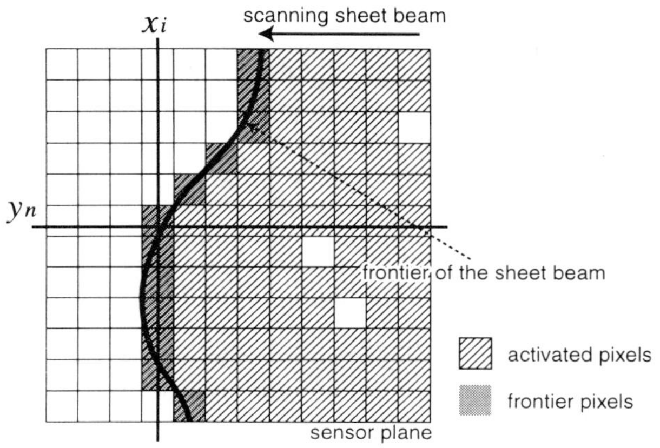

Figure 2: Captured image example of a sheet beam.

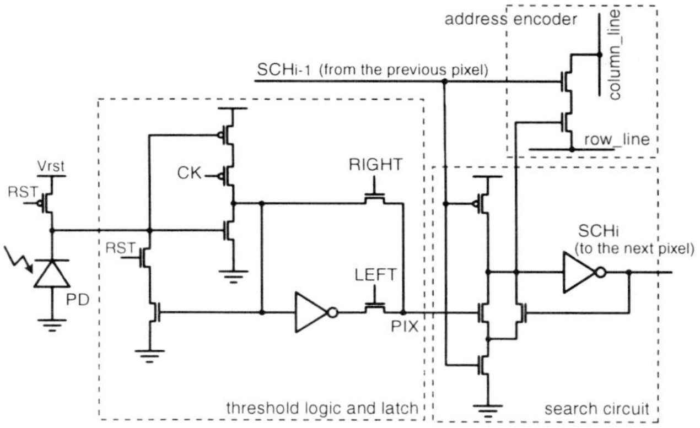

Figure 3: Pixel circuit.

2.1. Row-Parallel Search Circuit

Fig.3 shows a schematic of a pixel. It has a photodiode with a reset transistor, a threshold logic and latch circuit, a search circuit, and an address encoder. The latch circuit has an XOR circuit and can invert a pixel value PIX. At the search circuit, the search signal SCH_{i-1} from the previous pixel passes to the next pixel when the pixel value PIX is '1'. On the other hand, it stops when the pixel value PIX is '0'. Fig.4 shows a procedure of the row-parallel position search. Some pixels are activated by a strong incident light and they have a pixel value PIX of '0' as shown in Fig.4(a). The search signal SCH_0 is inputted to each row. It passes to the next pixel when the pixel value PIX is '1'. Thus the search signal stops at the left edge x_i of the activated pixels as shown in Fig.4(b). After a row-parallel encoding mentioned later, the pixel values PIX are inverted and the search signal starts again. It stops again at the next of the right edge x_j. The positions of the second and more activated pixels are detectable by the iteration of PIX inversions. It means that it is applicable to applications with a complex-shaped target object and/or multiple projected lights.

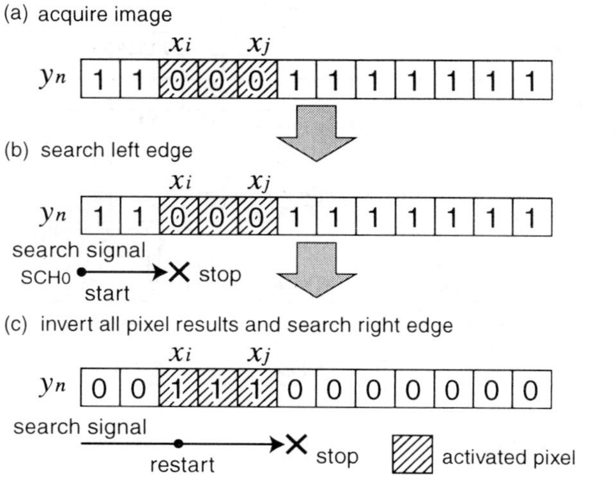

Figure 4: Procedure of row-parallel position search.

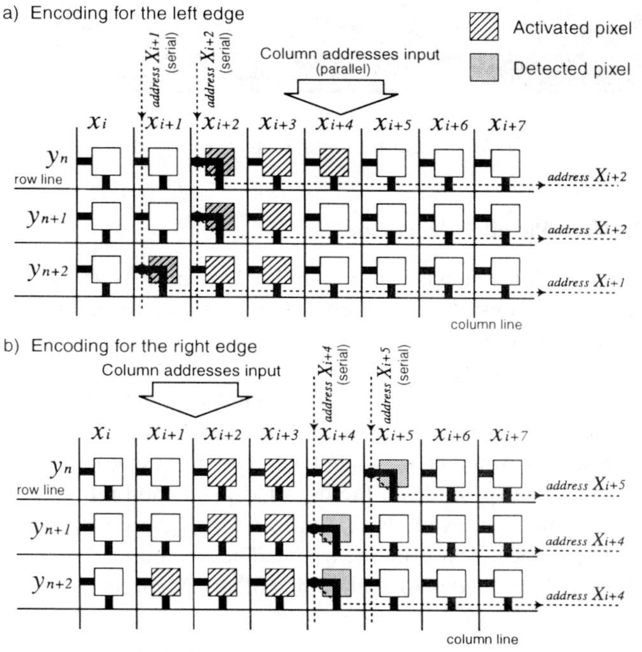

Figure 5: Method of row-parallel address encoding.

2.2. Row-Parallel Address Encoding

The address encoder of the pixel circuit consists of only 2 pass transistors as shown in Fig.3. At the detected pixel of each row, the column line is connected to the row line through the pass transistors as shown in Fig.5. Then, the serial-bit-streamed column address is inputted to each column line in parallel. Therefore the encoding cycles are O(log N) at N-pixel horizontal resolution. The compact circuit implementation and the high-speed row-parallel encoding realize high-speed position detection in high pixel resolution.

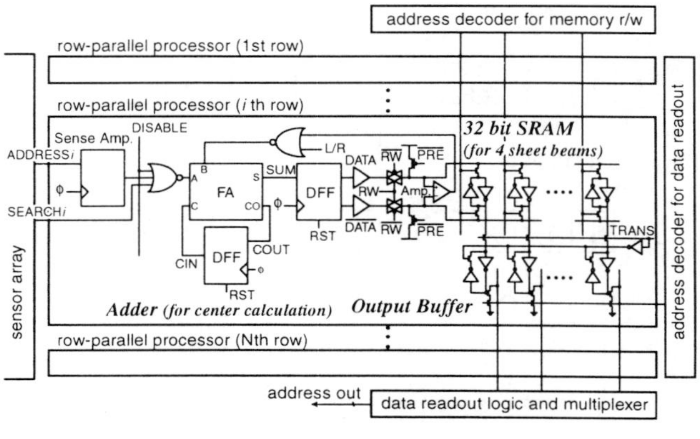

Figure 6: Structure of row-parallel processor.

Figure 7: Chip microphotograph.

2.3. Row-Parallel Processor

The photo detector has a row-parallel processor as shown in Fig.6. It consists of a latch sense amplifier to get the address data, a full adder, random access memories with a read/write circuit, output buffers for pipe-lined data readout, and some control logics. The row-parallel address encoding can acquire the addresses of x_i and $x_j + 1$ when the edges of the activated pixels are x_i and x_j. The processor calculates the center position of the detected pixels and reduces data transmission. And also it realizes to get the positions of multiple sheet beams in one frame. The row-parallel processor can be extended to deal with another data processing. For example, multiple samplings per frame can be realized for high sub-pixel accuracy when a timing memory and its control logics are implemented.

3. CHIP IMPLEMENTATION

We designed and fabricated a prototype position detector in 0.35 μm CMOS process[1]. Fig.7 shows a microphotograph of the fabricated chip. It consists of a 128 × 16 pixel array, a column address generator, row-parallel processors with 32 bit SRAM per row and a memory controller. The pixel circuit has 1 photo diode and 18 transistors in 16.25 μm × 16.25 μm pixel area with 20.15 % fill factor.

[1]The VLSI chip in this study has been fabricated through VLSI Design and Education Center (VDEC) in collaboration with Rohm Corp. and Toppan Printing Corp.

Table 1: Specifications of the prototype chip.

Process	0.35 μm CMOS 3-metal 1-poly-Si
Sensor size	2.5 mm × 0.3 mm
# pixels	128 × 16 pixels
Pixel size	16.25 μm × 16.25 μm
# trans. / pixel	18 transistors
Fill factor	20.15 %

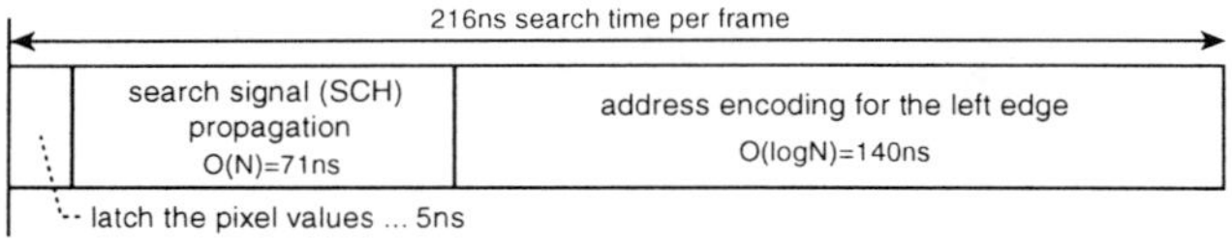

Figure 8: Simulated search time per frame for position detection of the fabricated chip.

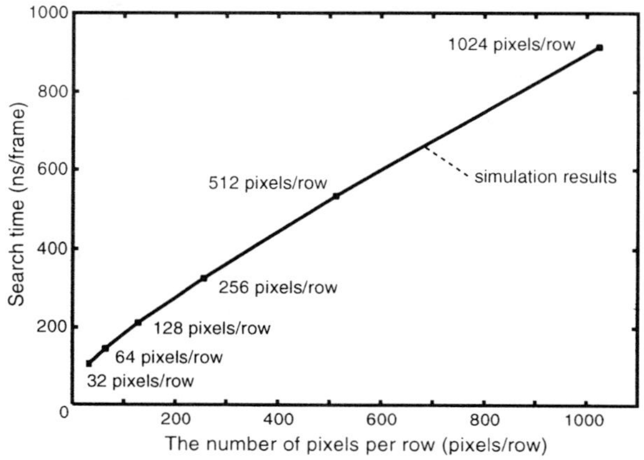

Figure 9: Search time in high pixel resolution.

4. PERFORMANCE EVALUATION

Fig.8 shows search time per frame for position detection simulated by HSPICE. The maximum propagation delay of the search signal is 71 ns and the 7-bit address encoding for 128 columns takes 140 ns. The total search time to get the position of the left edge of the activated pixels is 216 ns per frame. In a single sampling per frame, the frame interval is a total of the integration time of incident light and the search time for the left and right edges. The frame interval, however, is only the search time for the left edge in multi samplings per frame. That is, the sensor detects the frontier positions of the scanning sheet beam at each time during the integration of incident light. The present architecture achieves 918 ns search time per frame at 1024-pixel horizontal resolution. Generally real-time range finding of 1024 × 1024 pixels requires 32.5 μs search time per frame. The present architecture realizes enough speed not only for real-time range finding but also for beyond-real-time range finding and visual feedback.

Table 2: Performance comparisons.

	frame rate	# pixels
Our detector (single-sampling)	32.2k fps	128 × 16
	(31.4k fps)	1024 × 1024
Our detector (multi-sampling)	2.22M fps	128 × 16
	(1.09M fps)	1024 × 1024
Brajovic et al. [4]	6.4k fps	32 × 64
Sugiyama et al. [6]	3.3k fps	320 × 240
Required fps for real time	30.7k fps	1024 × 1024

Frame rates in parentheses are simulation results.

5. MEASUREMENT RESULTS

The measurement system of the fabricated chip has been developed as shown in Fig.1. In the system, the position detector and a scanning mirror are controlled by FPGA and the acquired position data are transferred to a PC. The FPGA was operated at 80 MHz due to the limitation of the testing equipment. The search time was 450 ns per frame and the integration time of incident light was 30 μs at V_{rst} = 1.4 V. Fig.10 shows the measurement results of the present position detector. Fig.10(a) shows the position of the left and right edges of the activated pixels. That is, the projected sheet beam is located between these edges on the sensor plane. The position detector has the processor to calculate the center position on the chip. Only the center address can be acquired to reduce data transmission as shown in Fig.10(b). Fig.10(b) shows sequentially captured positions of the scanning sheet beam of 2 kHz by single sampling per frame. It takes 30.9 μs per frame. It has 256 sub-pixel resolution due to the center calculation. Fig.10(c) shows the frontier positions of the scanning sheet beam. It takes 0.45 μs per frame. The results show the frame rate is 32.2k fps and 2.22M fps in a single sampling and multi samplings per frame respectively. The performance comparisons are shown in Table 2.

6. CONCLUSIONS

A row-parallel sensor architecture for high-speed position detection has been proposed. It has been designed and fabricated in 0.35 μm CMOS process and successfully tested. The prototype position detector has 128 × 16 pixels and it achieves 450 ns search time per frame. In the measurement system using multi samplings per frame, the high-speed position detection of a scanning sheet beam is realized at 2.22M fps. It is enough speed not only for real-time range finding but also for beyond-real-time range finding and visual feedback such as a fast collision prevention system. We have also shown its applicability to higher pixel resolution such as 1024 × 1024 pixels.

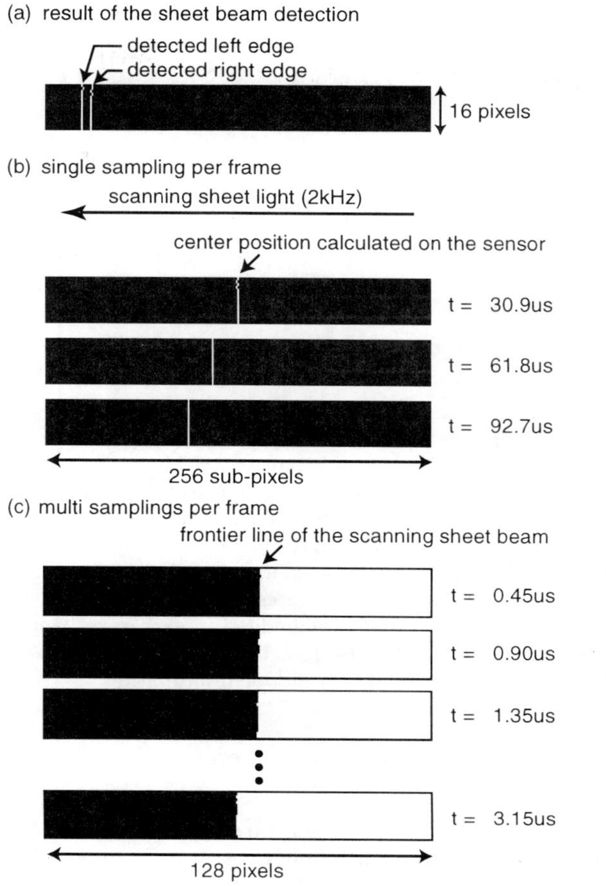

Figure 10: Measurement results.

7. REFERENCES

[1] V. Brajovic and T. Kanade, "Computational Sensor for Visual Tracking with Attention," *IEEE Journal of Solid-State Circuit*, vol. 33, pp. 1199 – 1207, 1998.

[2] T. Nezuka, M. Hoshino, M. Ikeda and K. Asada, "A Position Detection Sensor for 3-D Measurement," *in Proc. European Solid-State Circuits Conference*, pp. 412 – 415, 2000.

[3] M. de Bakker, P. W. Verbeek, E. Nieuwkoop and G. K. Steenvoorden, "A Smart Range Image Sensor," *in Proc. European Solid-State Circuits Conference*, pp. 208 – 211, 1998.

[4] V. Brajovic, K. Mori and N. Jankovic, "100 frames/s CMOS Range Image Sensor," *ISSCC Dig. of Tech. Papers*, pp. 256 – 257, 2001.

[5] S. Yoshimura, T. Sugiyama, K. Yonemoto and K. Ueda, "A 48k frame/s CMOS Image Sensor for Real-time 3-D Sensing and Motion Detection," *ISSCC Dig. of Tech. Papers*, pp. 94 – 95, 2001.

[6] T. Sugiyama, S. Yoshimura, R. Suzuki and H. Sumi, "A 1/4-inch QVGA Color Imaging and 3-D Sensing CMOS Sensor with Analog Frame Memory," *ISSCC Dig. of Tech. Papers*, pp. 434 – 435, 2002.

A SILICON RETINA SYSTEM THAT CALCULATES DIRECTION OF MOTION

S.Kameda, T.Yagi

Osaka University, 2-1 Yamadaoka, Suita, Osaka 565-0871, Japan
{kameda,yagi}@ele.eng.osaka-u.ac.jp

ABSTRACT

A silicon retina was fabricated to emulate two fundamental types of response in the vertebrate retinal circuit, namely the sustained and transient responses. The sustained response exhibits a Laplacian-Gaussian-like receptive field. The transient response is obtained by subtracting consecutive image frames. The outputs of the chip are offset-suppressed analog voltages since uncontrollable mismatches of transistor characteristics are compensated for with the aid of sample/hold circuits embedded in each pixel circuit. The chip was applied to extract direction of motion using FPGA in real-time under an indoor illumination.

1. INTRODUCTION

The retina is a part of the central nervous system in the vertebrate and plays important roles in early stage of visual information processing. Namely, images projected on the two dimensional photoreceptor arrays are transduced into electrical signals and important information to be used for the brain to express higher order visual functions is extracted by the retinal circuit. The computations carried out in the retina are quite complex as suggested by a variety of light-induced response types of neurons. However, one can find two distinct types of response. One is a sustained type in which cells respond continuously during illuminations and another is a transient type in which cells respond transiently at either illumination on or off. The sustained response is thought to be relevant to percept static images and the transient response is thought to be relevant to percept moving objects.

Inspired by the unique algorithm and the architecture of the retinal circuit, analog CMOS VLSI circuits, the silicon retinas, have been fabricated ([4, 6, 7, 9] for outlines). However, these silicon retinas have not yet reached practical engineering applications because of a low accuracy of outputs due to a low sensitivity to light and uncontrollable mismatches of transistor characteristics. Recently, we have fabricated a silicon retina having a one-dimensional Laplacian-Gaussion-like (∇^2G-like) receptive field[1, 14]. This silicon retina can provide offset-suppressed outputs.

In the present study, a two-dimensional analog silicon retina was fabricated using the similar design of the one-dimensional chip. The chip emulates the two fundamental response types of the retinal circuit. The edge of moving object and the direction of motion were extracted in real time using the outputs of the chip.

2. SILICON RETINA

The 2-dimensional silicon retina was developed using a similar design of a pixel circuit embedded in the one-dimensional chip fabricated previously. The silicon retina fabricated in the present study emulates the two fundamental types of response in the vertebrate retina: the sustained type response and the transient type response. The output of the chip emulating the sustained response has a Laplacian-Gaussian-like (∇^2G-like) receptive field and therefore carries out a smoothing and a contrast-enhancement on input images. The output of the chip emulating the transient response is obtained by subtracting consecutive image frames that are smoothed in advance by a resistive network.

Fig.1 shows the circuit design depicting a single pixel. The circuit was designed after the model of the vertebrate retina constructed with detailed physiological observations[12, 13]. The chip consists of two layers of resistive networks, which have different tightness of electrical couplings between neighboring pixels. A subtraction of outputs of these two resistive networks generates a ∇^2G-like receptive field[13, 14]. The photo sensor is an Active Pixel Sensor (APS) which consists of a photo-diode and a source-follower circuit[5, 8]. The APS has a high sensitivity to light, at the same level as general CMOS imagers, by accumulating the photoelectron in the parasitic capacitor of the photo-diode. Furthermore, the dynamic range can be controlled by changing the accumulation time. Neighboring pixels are connected with MOS resistors at the first and the second resistive networks. The resistances of MOS resisters are controllable by external bias voltages (V_{bm1},V_{bs1},V_{bm2} and V_{bs2})[7, 14]. Nbuf and DSB are sample/hold circuits[10]. The sample/hold circuits are embedded to compensate for the circuit offsets, which are the amplifier offset and the fixed-pattern noise in each pixel circuit due to the statistic mismatches of transistor characteristics[1, 14]. Two fundamental types of image processing can be performed with the chip by controlling the external signals in different timings[2]. The silicon retina was controlled by FPGA (Field Programmable Gate Array). Fig.2 shows a block diagram of the pixel arrangement. The photo-sensors (shadowed squares in Fig.2) are arranged on a hexagonal grid for a better circular symmetry as well as for a better spatial sampling efficiency compared with a square grid[3, 7]. The chip has 40 x 46 pixels. The chip

0-7803-7761-3/03/$17.00 ©2003 IEEE

was implemented with a 0.6 um, double-poly, three metals, standard analog CMOS technology and the die size was 8.9 x 8.9 mm^2. The pixel area was 178.650 x 154.725 um^2 and the fill factor was 3.14 %. The minimum readout time for all pixels was about 1.0ms.

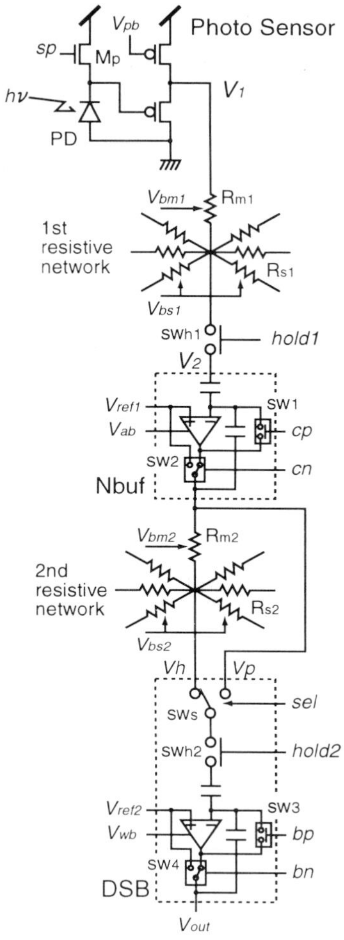

Figure 1: Circuit design of the single pixel.

2.1. Outputs of the sustained response

Retinal neurons with the sustained response often exhibit the ∇^2G-like receptive field. The ∇^2G-like receptive field possesses a band-pass filtering properties in spatial domain and can carry out a smoothing and a contrast-enhancement on input images.

Fig.3 shows the outputs of the silicon retina responding to a hand obtained by the sustained type response mode. The hand on a black and white background was presented in front of the lens. The experiment was carried out under indoor illumination (0.36 W/m^2). The accumulating time of the photo sensors, which is equal to a read out time of one frame, was 33.3 ms. A and B show the outputs

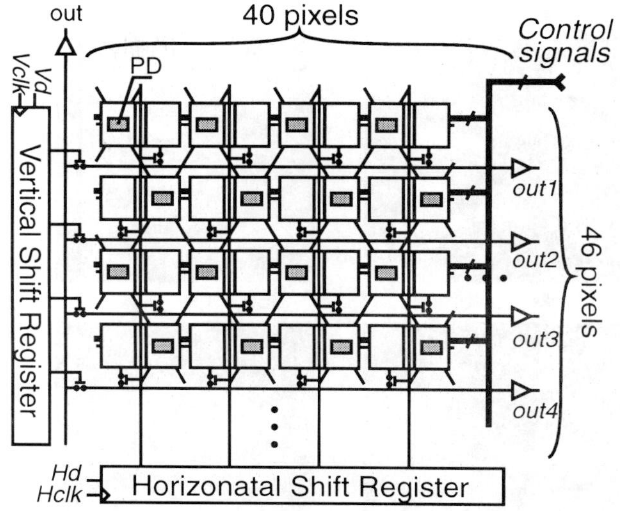

Figure 2: Block diagram of the chip

when the resistance in the second resistive network, R_{s2}, is set high and low, respectively. As shown in the Fig, the contour of the hand as well as the border between black and white was enhanced. The contrast in the image was enhanced more in A. And the response gain was larger in B. These outputs have extremely low noise because of the sample/hold circuits.

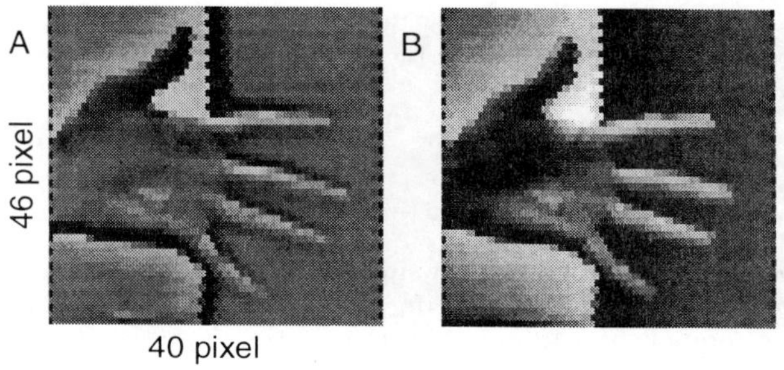

Figure 3: Response obtained by the sustained mode operation to a hand.

2.2. Output of the transient response

In the vertebrate retina, the transient type cell is sensitive to a temporal change of the intensity of light and thus thought to be relevant to detect moving objects. The transient response can be mimicked by subtracting an image obtained at a time t from the one obtained at a time $t + \Delta t$. The silicon retina fabricated in this study is able to memorize an image using the sample/hold circuit. Therefore the subtraction can be carried out on chip with external signals.

Motion detection chips using frame subtraction operations have been fabricated previously [5, 11]. However, the frame

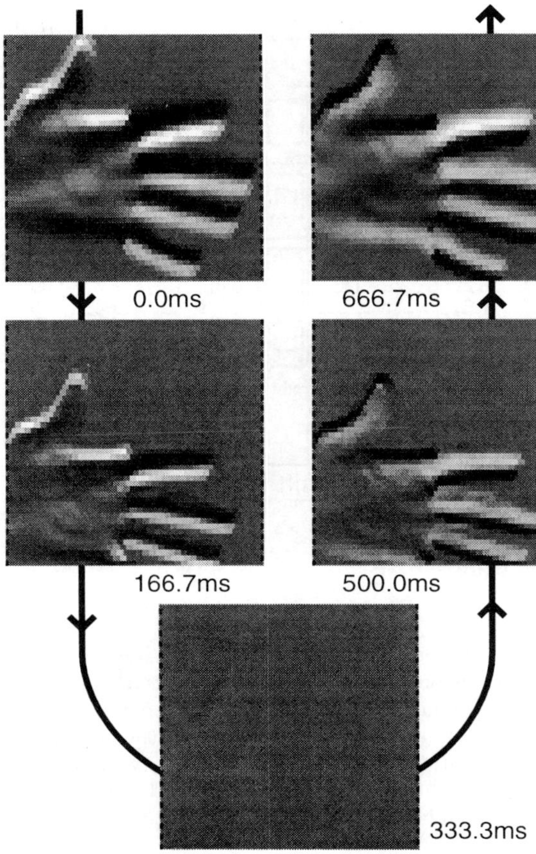

Figure 4: Outputs obtained by the transient mode operation to a moving hand.

subtractions are made on raw images in these chips, which degrades the output image due to high-frequency spatial noise components. In our chip, the subtraction is made on images that were smoothed out in advance by the resistive network to suppress such noise.

Fig.4 shows the response to a hand obtained with the frame subtraction made. The experiment was carried out under the same condition to Fig3. The hand was moved up and down in front of the lens. Each frame represents an output image of the silicon retina picked up once every five frames at a time indicated below. The hand was moved from up to down between 0.0 ms and 166.7 ms and from down to up between 500.0 ms and 666.7 ms. When the hand stopped at 333.3 ms, the image of the hand disappeared. And the black and white background was removed from the scene. It was notable that the response of an upper left side of the hand was different from the response of an upper right side because of the black and white background. The outputs had an extremely low noise because the circuit offsets were compensated by the sample/hold circuits and the high-frequency spatial noise components were suppressed by using the smoothed images.

3. EXTRACTION OF DIRECTION OF MOTION

As described above, our silicon retina outputted sustained and transient responses with a high degree of accuracy in real time. Then, by binding these two responses, we executed extraction of motion direction. Here is the algorithm of motion direction extraction. Fig.5 illustrates the responses of the sustained and transient mode to moving edges with different contrast, i.e. black-white or white-black. The two kinds of edges are moved to left or right. The response pattern can be classified 4 ways according to a combination of sustained and transient modes.

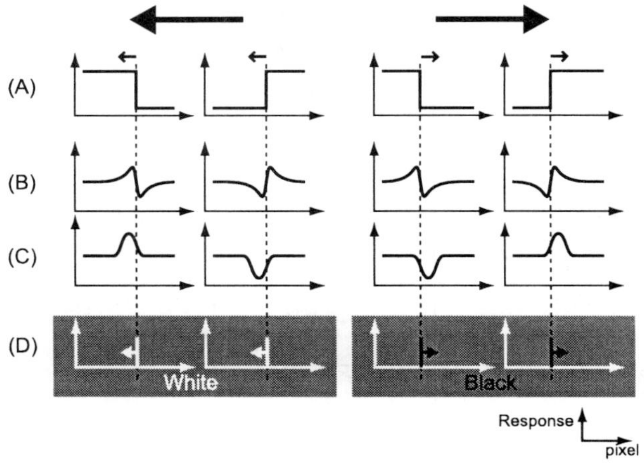

Figure 5: Algorithm sheet of extraction of moving direction. (A) Input image moving to left and right, (B) Output variations obtained with the ∇^2G filtering mode, (C) Output variations obtained with the frame subtraction mode, (D) Direction of moving edge.

We fabricated the direction of motion extraction system using FPGA to execute the classification. The silicon retina's analog responses were fed into the FPGA through an high speed A/D converter. Fig.6 shows the response of the FPGA system to a hand moving in front of a black and white background. The experiment was carried out under indoor illumination(0.36 W/m^2). The accumulating time of the chip was 33.3 ms, which corresponds to a time to obtain one frame in video rate. The hand was moved left and right in front of lens. A and B show the responses obtained with the ∇^2G filtering mode and the frame subtraction mode at 0.0ms, respectively. C shows the extracting images of motion direction on horizontal axis. Each frame represents an output image of the FPGA system picked up once every ten frames at time indicated below. The hand was moved from right to left between 0.0ms and 333.3ms and from left to right between 1000.0ms and 1333.3ms. As shown in the Fig.C, the direction of the hand is extracted in time with the hand moving independent of the black and white background. When the hand stopped at 666.7ms, the edges of the hand disappeared. And the black and white background was removed from the scene.

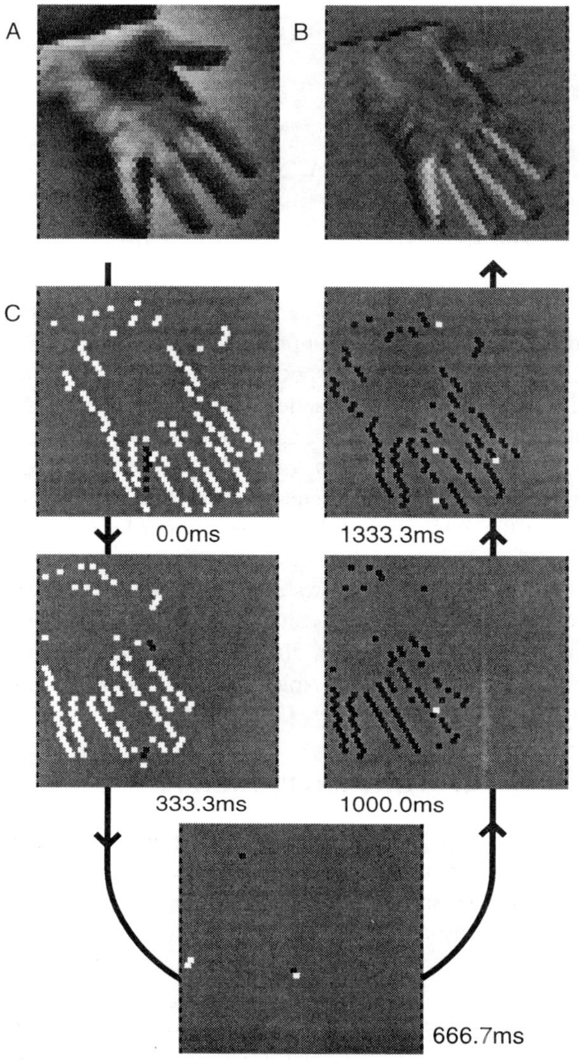

Figure 6: Outputs obtained from the FPGA system. A is the response of the sustained mode, B is the response of the transient mode, C is the direction of motion outputs.

4. CONCLUSIONS

In the present study, a silicon retina that can provide two types of output mimicking the sustained response and the transient response of the vertebrate retina was fabricated. These two filtered images i.e. sustained and transient responses were obtained in real time, (time delays not exceeding a few tens of millisecond). The chip was applied to the real time zero-crossing detection and direction of motion extraction under an indoor illumination. The silicon retina developed in the present is useful for various engineering applications, e.g. a robot vision, an automatic car navigation and etc.

5. ACKNOWLEDGEMENT

The VLSI chip in this study has been fabricated in the chip fabrication program of VLSI Design and Education Center(VDEC), the University of Tokyo with the collaboration by Rohm Corporation and Toppan Printing Corporation. This work was partially supported by the Japan Society for the Promotion of Science, grant-in-aid for Research for the Future Program, JSPS-RFTF 97I00101.

6. REFERENCES

[1] S.Kameda, A.Honda and T.Yagi, "Real Time Image Processing with an Analog Vision Chip System", *Int. J.Neural Systems*, 9(5):423-428, 1999.

[2] S.Kameda and T.Yagi, "A silicon retina calculating high-precision spatial and temporal derivatives", Proc. IJCNN'01, pp.201-205, 2001.

[3] H.Kobayashi,T.Matsumoto,T.Yagi and T.Shimmi, "Image processing regularization filters on layered architecture", *Neural Networks*, 6:327-350, 1993.

[4] C.Koch and H.Li eds., *Vision chips: implementing vision algorithms with analog VLSI circuits*, IEEE computer society press, 1995.

[5] S.-Y.Ma, L.-G.Chen, "A single-chip CMOS APS camera with direct frame difference output", *IEEE J solid-state circuits*, 34(10):1415-1418, 1999.

[6] M.Mahowald, *An analog VLSI system for stereoscopic vision*, Kluwer Academic Publishers, 1994.

[7] C. Mead, *Analog VLSI and Neural Systems*, Addison-Wesley, 1989.

[8] S.K.Mendis, E.Sabrina, R.C.Gee, B.Pain, C.O. Staller, Q.Kim and E.R.Fossum, "CMOS Active Pixel Image Sensors for Highly Integrated Imaging Systems", *IEEE J. Solid-State Circuits*, 32(2):187-197, 1997.

[9] A.Moini, *Vision Chips*, Kluwer Academic Publishers, 2000.

[10] T.Shibano, K.Iizuka, M.Miyamoto, M.Osaka, R.Miyama and A.Kito, "Matched Filter for DS-CDMA of up to 50MChip/s Based on Sampled Analog Signal Processing," *ISSCC Digest of Tech. Papers*, 100-101, 1997.

[11] A.Simoni, G.Torelli, F.Maloberiti, A.Sartori, S.Plevridis and A.Birbas, "A single-chip optical sensor with analog memory for motion detection", *IEEE J solid-state circuits*, 30(7):800-806, 1995.

[12] T.Yagi,F.Ariki and Y.Funahashi, "Dynamic model of dual layer neural network for vertebrate retina", it Proc. of Int'l Joint Conf. on Neural Networks, 1:787-789, 1989.

[13] T.Yagi, S.Ohshima and Y.Funahashi, "The role of retinal bipolar cell in early vision: an implication with analogue networks and regularization theory", *Biol. Cybern.*, 77:163-171, 1997.

[14] T.Yagi, S.Kameda and K.Iizuka, "A Parallel Analog Intelligent Vision Sensor with a Variable Receptive Field," *Systems and Computers in Japan*, 30(1):60-69, 1999, Translated from *Denshi Joho Tsushin Gakkai Ronbunshi*, J81-D-I(2):104-113, 1998.

A FAST AND LOW POWER CMOS SENSOR FOR OPTICAL TRACKING

N. Viarani, N. Massari, L. Gonzo, M. Gottardi, D. Stoppa and A. Simoni

ITC-Irst, via Sommarive 16, I-38050 Trento, Italy, Tel. (+39) 0461 314535, e-mail: nviarani@itc.it

ABSTRACT

A new CMOS optical sensor for active optical tracking is presented. The tracking function is carried out pointing the target with a collimated light beam and estimating the position of the back-reflected beam portion impinging on the device. The core of the device, fabricated in standard $0.8\mu m$ CMOS technology, is an array of 20×20 pixels where mixed analogue - digital image pre-processing, at pixel and array level, is performed. Image contour is extracted by means of distributed peak-detection, implemented at pixel level, followed by a digital extraction of the beam centroid position executed at row and column level. The image position can be estimated in $120\mu s$ with an accuracy of $0.9\mu m$. The chip exhibits a worst case power consumption of $15mW$ @ $5V$ and $3000 frames/s$.

1. INTRODUCTION

Tracking of moving objects is used in all the applications (e.g. industry robotics, virtual reality, augmented reality) where the position of a real physical object is required. Optical tracking is one of the most diffused methods for detecting the motion of a target. Within the optical tracking, different techniques can be used, depending on the performance required in terms of speed and accuracy and on the features of the target. Most diffused tracking systems use passive techniques and consist of a standard cameras together with an external digital image processing unit, which allows for maximum flexibility and computing power needed for the extraction of complex features. Alternatively tracking of objects within a volume can be accomplished by means of active techniques like those based on the active optical triangulation. In this latter case a collimated beam points to the target and a portion of the back-reflected light is focused on a position sensor. The spot position on the sensor is related to the target position by simple geometrical relations [1]. Research progresses in integrated optical sensors and technological advances in CMOS processes have recently promoted the use of integrated optical sensor arrays with on-chip image processing even for position sensing applications requiring utmost accuracy and sensitivity. With regard to this, some research activities are addressed to the development of integrated optical sensors for single and multiple spot detection [2-5]. The paper presents a novel architecture of a 2D pixel array for single spot detection and tracking [6], based on image outline extraction. The prototype device, designed in a standard $0.8\mu m$ CMOS technology, consists of an array of 20×20 pixels where the photosensitive detector is combined with analogue processing circuitry which allows for the extraction of the image contour by means of distributed peak-detection along the *x* and *y* directions. Two computational blocks, for rows and columns, read-out the information from the bit-lines and perform a digital extraction of the light spot position. High accuracy measurement is obtained by means of a multiple threshold binaryzation of the analogue image outline.

2. WORKING PRINCIPLE

Figure 1 shows a graphic representation (reduced to only one dimension for the sake of simplicity) of the sensor operation. The light spot outline is first generated along the x and y axes at sensor array level and then a corresponding binary image (*biI*) is produced by comparing the image contour with a proper threshold ($TH1$). The light spot centroid is computed by averaging the position of the two most external edges of the binary image with a precision of $\pm\frac{1}{2}$ pixel:

$$X_1 = \frac{X_{up_1} + X_{down_1}}{2}. \quad (1)$$

If higher sub-pixel resolution is required, multiple threshold modality must be adopted, i.e. the operation described above is repeated using different threshold values ($TH1, TH2, ..., THN$). The final centroid position is obtained through the average of the N results:

$$X = \frac{1}{N}\sum_{i=1}^{N} X_i. \quad (2)$$

The theoretical image centroid position accuracy is limited by the sensor pixel pitch and by the number of threshold values. The proposed algorithm works both on x and y directions adopting the superposition principle to reconstruct the final 2D result. Nevertheless, it requires intensive analogue and digital computation at pixel and array levels to be efficiently implemented.

0-7803-7761-3/03/$17.00

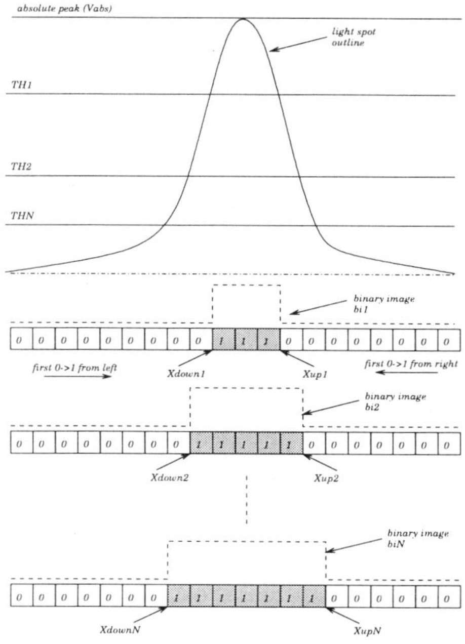

Fig. 1. Centroid extraction principle on x axis.

3. CHIP ARCHITECTURE

Overview

The sensor consists of an array of 20 × 20 pixels, each one of them is connected to one row, column and *global* line, as shown in Figure 2. Each bit-line is driven by the most illuminated pixel connected to it. This means that, after image acquisition, the rows and the columns provide the analogue sampling representation of the light spot image contour. Each bit-line is the input of a clocked comparator which compares this signal with the selected voltage threshold. The two comparators arrays generate the x and y axis binary images, which are elaborated by the related x and y Priority Nets in order to extract their two most external edges. Every Priority Net looks along the binary image for the pixel position corresponding to the first bit value transition from left to right (recall Fig. 1). Then it works backward finding the first transition from right to left. Lastly the results are written in sequence onto the data bus to be processed outside the chip. This operation is exectuted as many times as the number of thresholds used within a single image acquisition. The whole architecture is organized so that the x and y axis computational blocks (Comparators plus Priority Net) are totally independent except for the light-sensitive area, in order to guarantee the same input signal conditions. Every pixel connects its output to the bit-line x (bl_i), bit-line y (bl_j) and *global* line in order to implement three types of distributed image peak detectors, thus extracting both the 2D spot contour and the absolute image light intensity. All the connections of the pixels towards their bit-lines are made by means of bipolar transistors working as distributed low-peak detectors (Figure 3). The pixel, receiving the highest light intensity, takes the bit-line control and disconnects the others from it. The *global* line connects all the pixels of the array providing a voltage which corresponds to the absolute light peak intensity. At the end of the integration time, the 2D image outline is available on rows and columns of the array.

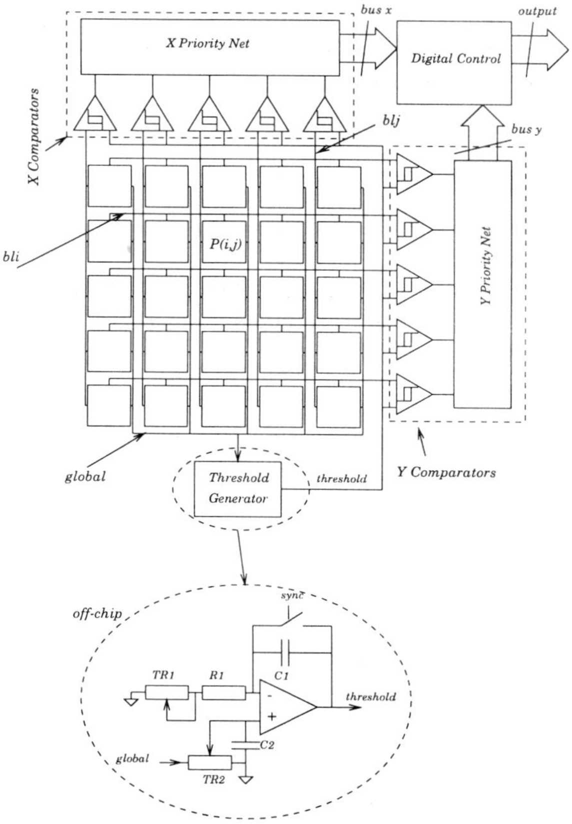

Fig. 2. Schematic block of the chip.

Pixel

Referring to the pixel schematic of Figure 3, the photodiode is precharged to V_r as well as rows, columns and *global* line. The sensing node *a*, buffered by the source-follower *M2*, controls the bipolar base *b* which linearly follows the photodiode voltage *a*. The pixel, receiving the highest light intensity, will have the lowest base voltage with respect to the other pixels connected to the same bit-line. Thus, it will take the bit-line control by forcing it to ($V_b + |V_{be}|$), being V_{be} the base-collector junction voltage drop of the vertical bipolar transistor (*VPNP*) driven by the bit-line current source I_0. All the other pixels are thus disconnected from

the bit-line, having a lower $|V_{be}|$ with respect to the winner. At the end of the integration time, each bit-line provides a voltage level corresponding to the local light peak intensity. To reduce pixel dimensions, bl_i is used both as photodiode pre-charge line, during sensor reset and as bit-line, during image outline extraction. This requires some precaution in the pixel topology. Under reset, row, column and global line are forced to V_r allowing for photodiode precharge. After transistors M3, M4, M5 are turned off, each bit-line is dynamically set to the value of the current most illuminated pixel.
During the entire operation, the power consumption is driven by the bit-lines current sources (I_0). This means that the static power consumption increases linearly with the number of pixels of the array:

$$P_{ARRAY} = (n + m + 1)I_0V_{DD}; \quad (3)$$

where n and m are the number of rows and columns of the array respectively.

The drain of *M2* is connected to Φ_{RES}, allowing the base of the three *VPNPs* to be correctly precharged under the reset phase.

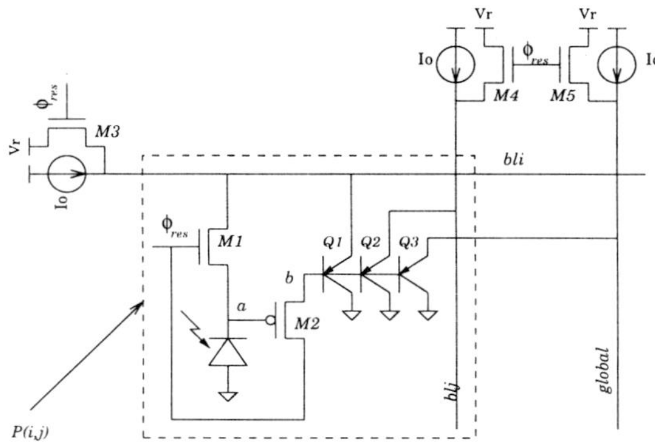

Fig. 3. Schematic of the pixel.

The reason of using vertical bipolar transistors as peak detectors is because they can be implemented with a minimum size junction and their *nwell* common base is shared with the bulk of M2. Moreover, they exhibit a logarithmic characteristic which is more selective with respect to the square root law of a MOS. The pixel pitch is $70.25\mu m$ on both directions with a fill factor of 12%.

Comparators and Priority Net

After the integration phase, when the analogue image contour is available on the sensor bit-lines, voltage comparators [7] are clocked, providing a binary image which contains the desired information in its outermost transitions. With regard to just one axis, this information is extracted in two steps. The first binary edge ($0 \rightarrow 1$) is detected searching from row 1 up to row N and its position is written onto a precharged bus to be read-out outside the chip. Then, the same operation is executed backward from row N down to row 1. After the two data are read-out, the related 1D light spot centroid is estimated by computing the average value between the two data, representing the positions of the two most external transitions of the binary image. Thus, for each cycle, the spot position is estimated with a resolution of $\pm\frac{1}{2}$ pixel. Priority Nets on x and y axes operate in parallel and the results are multiplexed on the output of the chip.

The sensor can also be operated in a multiple threshold mode, thus increasing the measurement accuracy. For each image acquisition, the light spot contour is digitized with N different threshold levels extracting N binary images. For each binary image, the centroid position is computed with a resolution of $\pm\frac{1}{2}$ pixel pitch. The final position is obtained by simply averaging the partial results, as described in eq. (2). From calculations, it results that, for a given pixel size D and a light spot diameter of σ, the minimum number of thresholds to be used, for achieving a position detection accuracy x_m, is expressed:

$$N > \frac{\sigma\sqrt{e}}{Dx_m}. \quad (4)$$

The multiple threshold mode operation is implemented by means of a voltage ramp sweeping over the full signals range, from the global value up to the dark signal level. Synchronized with the ramp, Comparators and Priority Net are clocked with a proper frequency ($1/T_{clk}$) extracting N centroid position values. As shown in Fig. 4, where the light spot involves bit-lines from *b1* up to *b5*, the first bit-line to be detected is only *b3*, corresponding to the first centroid position. In the second clock cycle, *b2, b3, b4* are detected but only *b2, b4* are selected by the priority net, corresponding to the two outermost bits, set to 1, within the current binary image. *b1, b5* is the last bit pair to be detected. The three partial estimated centroid positions will be: *b3*, *b24=(b2+b4)/2* and *b15=(b1+b5)/2*, giving a final result *bf=(b3+b24+b15)/3*.

4. EXPERIMENTAL RESULTS

Functionality tests were carried out using a $10mW$ ($645nm$) pulsed laser diode directed onto a white plane surface placed at a distance of $50cm$. The laser was synchronized with the sensor generating pulses of $8ms$, corresponding to the integration time, at a repetition rate of $120Hz$. A standard C-mount optical objective was used to focalize the spot image onto the photosensitive area down to a diameter of $350\mu m$. Consequently, only groups of 5 pixels were involved in every single axis computational process.

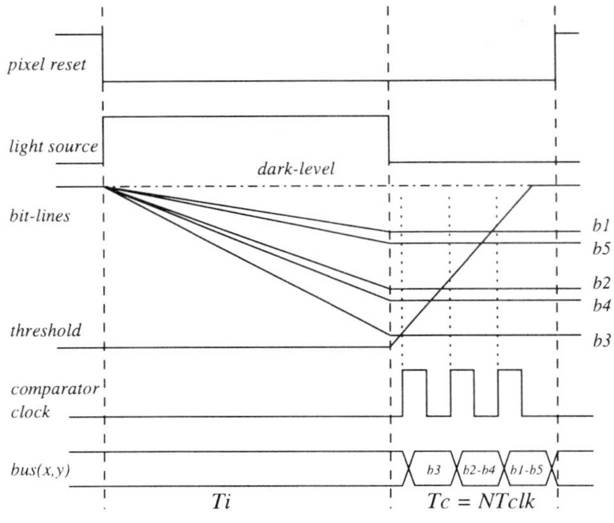

Fig. 4. Sensor multiple threshold mode operation.

Using the multiple threshold mode (over 200 different thresholds) the sensor showed an accuracy of $0.9\mu m$, shown in Fig. 5, with maximum signal range of $1.5V$. Each processing cycle of centroid computation, from image outline binarization up to data transfer, takes less than $600ns$. So the complete processing cycle takes $120\mu s$, which is much less than the integration time.

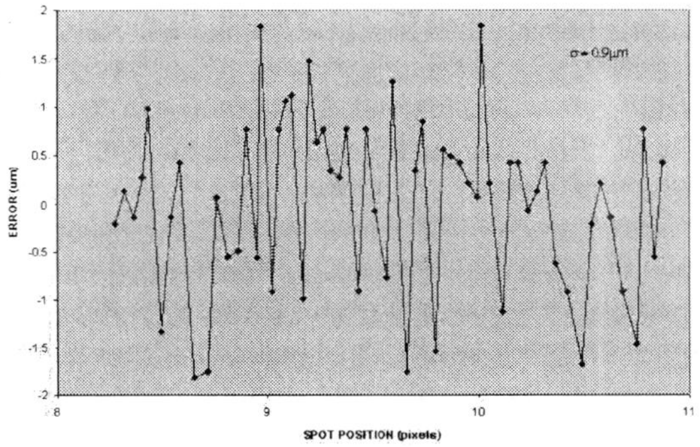

Fig. 5. Sensor differential linearity along Y axis.

5. CONCLUSIONS

A fast and low power CMOS sensor for optical tracking was presented and fully tested. The chip has been integrated in a $0.8\mu m$ CMOS process, features a 20×20 pixel array with a pitch of $70.25\mu m$ and a fill factor of 12%. The chip power consumption is $15mW@5V$ at $3000 frames/s$ operating in the multiple thresholds mode with 212 thresholds. The position sensing accuracy of $0.9\mu m$ was achieved using a $350\mu m$ spot diameter with a pulse width of $8ms$ and operating the sensor in the multiple thresold modality. The die size is $8mm^2$, pads included (Fig. 6).

Fig. 6. Microphotograph of the chip.

6. REFERENCES

[1] F. Blais, et al., "Real-time Geomtrical Tracking and Pose Estimation using Laser Triangulation and Photogrammetry", *3DIM2001, Third International Conference on 3D Digital Imaging and Modeling*, May 28 - June 1 2001, Quebec City, Canada.

[2] A. Makynen, T. Rahkonon, and J. Kostamovaara, "A CMOS Binary Position-Sensitive Photodetector (PSD) Array", *Custom Integrated Circuits Conf.*, Oct.1997, pp. 279-282.

[3] A. Makynen, H. Benten, T. Rahkonon and J. Kostamovaara, "CMOS Position-Sensitive Photodetector (PSD) for Integrated Sensor Systems", *SPIE* vol. 3100, 1997, pp. 89-100.

[4] D. Standley, "An Object Position and Orientation IC with embedded Imager", *IEEE J. Solid-State Circuits*, vol. SC-26, pp. 1853-1859, Dec. 1991.

[5] M. Tartagni and P. Perona, "Computing centroids in current-mode technique", *Electronics Letters*, vol. SC-29, pp. 1811-1813, Oct. 1993.

[6] N. Massari, L. Gonzo, M. Gottardi and A. Simoni, "High Speed Digital CMOS 2D Optical Position Sensitive Detector", *Proc. ESSCIRC 2002*, Florence, 2002

[7] Y.-T. Wang and B. Razavi, "An 8-Bit 150-MHz CMOS A/D Converter", *IEEE J. Solid-State Circuits*, vol. SC-35, pp. 308-317, March. 2000

An Orientation Selective 2D AER Transceiver

Thomas Yu Wing CHOI[1], Bertram E. SHI[1], and Kwabena BOAHEN[2]

[1]Department of Electrical and Electronic Engineering, Hong Kong University of Science and Technology, Clear Water Bay, Kowloon, Hong Kong, {eethomas, eebert}@ee.ust.hk

[2]Department of Bioengineering, University of Pennsylvania, Philadelphia, PA 19104-6392, kwabena@neuroengineering.upenn.edu

ABSTRACT

This paper describes an address event representation (AER) transceiver chip that accepts 2D images and produces 2D output images equal to the input filtered by even and odd symmetric orientation selective spatial filters. Both input and output are encoded as spike trains using a differential ON/OFF representation, conserving energy and AER bandwidth. The spatial filtering is performed by symmetric analog circuits that operate on input currents obtained by integrating the input spike trains, and which preserve the ON/OFF representation. This chip is a key component of a multi-chip system we are constructing that is inspired by the visual cortex. We present measured results from a 32 x 64 pixel prototype, which was fabricated in the TSMC0.25μm process on a 3.84mm by 2.54mm die. Quiescent power dissipation was 3mW.

1. INTRODUCTION

Moving from the retina to higher levels of visual processing in the cortex, neurons become progressively more selective to more complex stimuli. Cells in the retina are sensitive along stimulus dimensions of position, spatial frequency (size), temporal frequency and color. In the primary visual cortex, additional selectivity along the dimensions of orientation, direction of motion and binocular disparity emerges. Subsequent areas are selective to higher order dimensions such as curvature and illusory contours. Concurrently, there is a progressive increase in the size of the receptive field along stimulus dimensions established earlier, e.g. spatial position. Thus, neurons in V2 respond to visual stimuli in a much larger spatial area than ganglion cells in the retina.

A functional model that seems to account for the responses of a large proportion of cells in the primary visual cortex consists of a linear spatio-temporal filtering stage and three nonlinear mechanisms: contrast normalization, half-wave rectification and expansive exponentiation[1][2][3]. Linear spatio-temporal filtering determines the neural selectivity along different stimulus dimensions. Contrast normalization accounts for the observed saturation of the neural response with increasing contrast. The saturation occurs at a fixed contrast, independent of the response level, enabling neurons to retain selectivity over a wide input contrast range. Half-wave rectification conserves metabolic energy by mapping mean levels to a low quiescent spike rate. Signals above and below the mean are carried by complementary channels. The expansive exponent enhances stimulus selectivity.

In this work, we describe a silicon chip that implements two components of this model: linear orientation selective spatial filtering and half-wave rectification. The impulse response of the spatial filters approximate even and odd symmetric Gabor functions, which are commonly used to model the spatial receptive field profiles of visual cortical neurons[4]. Complementary ON/OFF channels carry positive and negative parts of all input, internal and output signals, which are processed using analog continuous time circuits. As in biological systems, this representation improves energy efficiency.

This chip is intended to serve as one component of a multi-chip system that takes input from a silicon retina and implements more complex visual information processing inspired by that found in the visual cortex. To facilitate construction of this system, input and output signals are encoded as spike trains, which are communicated on and off chip using the asynchronous Address-Event Representation (AER) communication protocol[7]. The AER protocol is more efficient than scanning when the spike activity within the array is sparse, as we expect here since only a few image locations will contain edges near the orientation selected by each chip and the quiescent spike activity in the array is low due to the ON/OFF signal representation. This combination of continuous time analog processing and digital communication circuits, which directly allocates power to salient areas in the image, results in better power efficiency than a conventional DSP approach.

We focus upon orientation as a first step in constructing this a system, as this dimension seems to be a fundamental primitive from which selectivity along other stimulus dimensions can be constructed. For example, linear direction-selective spatio-temporal filters can be obtained by cascading these orientation selective filters with bandpass temporal filters and combining their outputs[5]. Filters tuned to binocular disparity can be obtained by combining orientation selective filtered images from the left and right eye[6].

0-7803-7761-3/03/$17.00 ©2003 IEEE

This paper describes the architecture of this chip, as well as measurement results from a prototype.

2. CHIP ARCHITECTURE

2.1 AER Interface

The chip is a transceiver, containing both an AER transmitter and an AER receiver (see Chips A and B of Fig. 1a). The AER protocol was developed to communicate continuous time spike activity from an array of silicon neurons in one chip to another chip over a digital bus. The transmitter signals a spike occurrence in the array by placing the location (address) of the spiking neuron onto the bus. The receiver takes the address that appears on the bus and feeds a spike to the corresponding neurons in its array. The protocol is asynchronous, with the time that the address appears on the bus encoding the spike time directly. Collisions between simultaneous spikes from two neurons on the array are handled by arbitration.

The AER interface includes routing circuitry to facilitate the construction of a multi-chip network. Fig. 1a illustrates a three chip network. The split circuit enables fan out by splitting the incoming AER address stream into two: one sent to the pixel array and the other sent to an output, which can be fed into the input of another chip. A merge circuit enables fan in by combining the output of the array with an AER stream provided through a second input into a single serial output stream. In Fig. 1a, the merge output of Chip B encodes spikes from both Chips A and B.

Addresses are placed onto the bus in "bursts," where each burst encodes all of the simultaneous spikes from neurons within a given row and a given chip. We use a word serial format, where each burst is a sequence of addresses. As shown in Fig. 1b, the transmitter signals the start of a burst by placing an address identifying the source chip onto the address lines (Addr) and taking the request signal ReqY high. Subsequent addresses are signalled by taking _ReqX low. The second address identifies the row. Each of the remaining addresses identifies one of the columns containing a neuron that spiked. The transmitter signals the end of the burst by taking ReqY low. The receiver acknowledges receipt of each address by a transition on the Ack line.

We use absolute addressing to identify rows and columns within a chip, but relative addressing to identify each chip. Each chip signals its own activity with bursts whose chip addresses are set to zero. Every time a chip relays a burst from its split or a merge input, it increments the chip address. For example, a chip address of 1 at the merge output of Chip B in Fig. 1a indicates the spikes in the burst come from Chip A.

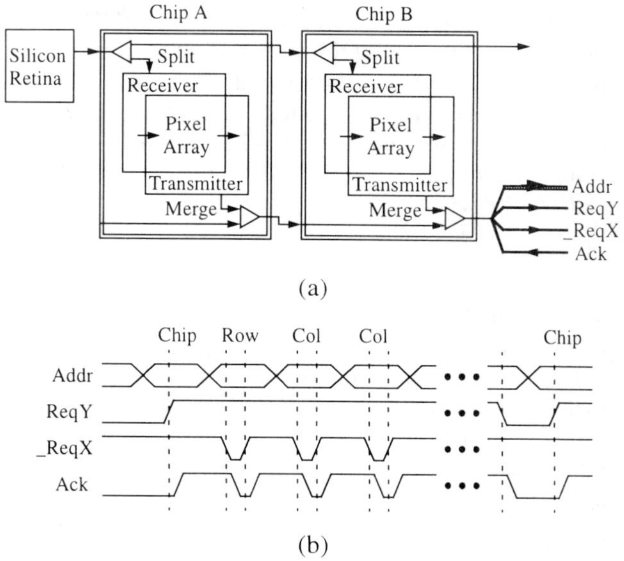

Fig. 1: (a) A three chip system where the output of a silicon retina[8] is fanned out to two orientation selective chips (Chip A and Chip B) tuned to different orientations. (b) A signal diagram of the merge output of chip B showing the addressing scheme.

2.2 Pixel level processing

The pixel processing array filters the incoming image with the transfer function:

$$H(e^{j\omega_x}, e^{j\omega_y}) = \frac{H_\Omega}{1 + \frac{(2 - 2\cos(\omega_x - \Omega_x))}{(\Delta\Omega_x)^2} + \frac{(2 - 2\cos(\omega_y - \Omega_y))}{(\Delta\Omega_y)^2}}$$

where H_Ω is the gain at resonance, (Ω_x, Ω_y) is the center spatial frequency, and $\Delta\Omega_x$ and $\Delta\Omega_y$ are the 6dB half bandwidth in the x and y directions. Pixel values at input and output are in general complex valued. For a real valued input image, the real and imaginary parts of the output equal the input image convolved with even and odd symmetric filters. Similar to Gabor functions, the impulse responses are cosine or sine waves modulated by an envelope that decays with distance from the origin. However, the envelope of these filters decays more sharply at the origin and slower at the tails than the Gaussian envelope of a Gabor function.

Each pixel within the array receives four spike trains, corresponding to the ON and OFF components of the real and imaginary parts of the input. Current mode integrators[9] convert the spike trains into currents that are approximately proportional to the incoming spike rates. In the addressing scheme, the real and imaginary parts are encoded by the least significant bit of the row address and the ON and OFF components by the least significant bit of the column address.

The four input currents are then processed by an analog neural network that produces four output currents, corresponding to the ON and OFF components of the real and imaginary parts of the filter output. The ON and OFF components of the output are interconnected in opponency, so that they mutually inhibit each other. At any time, only one component of the output is positive, the other being close to zero.

This network is a two dimensional extension of that described in [10], except that the "ON/OFF block" is placed at the output of the diffuser networks, rather than at the input. This improves the ON/OFF representation at the output by reducing the common-mode activity in the complementary channels. This change required that the diffuser or pseudo-resistor networks be implemented with NMOS rather than PMOS transistors. By adjusting analog bias voltages controlling pseudo-conductance ratios and current gains within the array, we can tune the array to arbitrary spatial frequencies and orientations between 0 and 90 degrees. Other orientations can be obtained by remapping input and output addresses to flip the array horizontally and/or vertically.

Each of the four output currents is passed to a spiking neuron circuit similar to that described in [11], which encodes each current by a spike train whose rate is proportional to the current amplitude. As activity is sparse, most pixels' input or output is zero, which can be encoded by the lack of activity in either channel, conserving power dissipated by spiking and preserving bandwidth on the AER bus.

3. EXPERIMENTAL RESULTS

We designed and fabricated an array of 32 x 64 pixels in the TSMC0.25um mixed signal/RF process available through MOSIS. This process contains 5 metal layers and 1 poly layer, uses non-epitaxial wafers, and is intended for 2.5V applications.

The array layout was generated by tiling metapixels whose layout is shown in Fig. 2. Each metapixel contains the circuits required for two pixels stacked vertically. Each metapixel is 103μm by 49μm (860λ by 390λ for $\lambda = 0.12\mu m$). The die size was 3.84mm by 2.54mm.

We laid out the metapixels to minimize interference between the analog spatial filtering circuits and the digital communication circuits (the spiking neurons, current mode integrators and the AER interface). The analog filtering circuits lie in the middle of the metapixel, sandwiched between digital circuits on the top and bottom. Within the digital parts, the integrators lie next to the analog circuits. The spiking neurons, which generate the most switching noise, lie at the top and bottom, farthest from the analog processing circuits. Guard rings are also inserted in between the Gabor cells, the integrators and the spiking neurons. The digital and analog circuits use separate power and ground lines. Bias lines connected to source voltages controlling current mirror gains run wide on the top metal layer to reduce impedance.

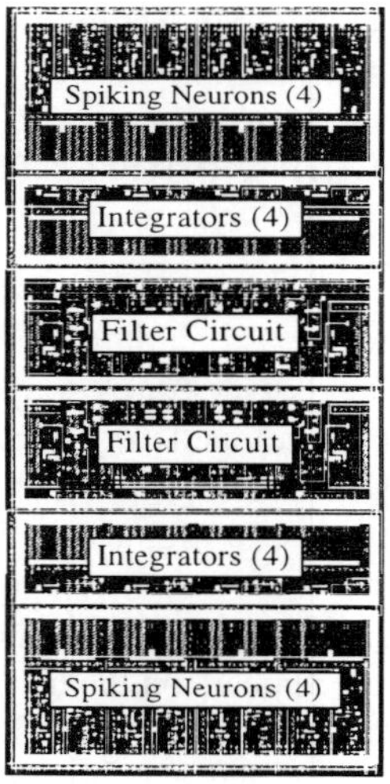

Fig. 2: Layout of one metapixel, containing the circuits necessary to process two vertically stacked pixels.

To test the response of the array, we excited pixel (16,32) with a 20kHz spike train from a pattern generator. All other inputs were silent. A logic analyzer collected the spike train at the merge output, which is digitally processed for analysis. Fig. 3(a,b,c,d) shows the average spike rates of the ON and OFF components of the real and imaginary parts of the output when the array is tuned to vertical orientations. The difference between the ON and OFF spike rates are shown in Fig. 3(e,f). Fig. 3(g,h) shows similar data when the array is tuned to horizontal orientations.

The power consumption increases with the total spike activity at the input and output. We measured the power dissipation of the chip while stimulating pixel (16,32) with spike trains ranging in frequency from 0Hz to 100kHz and plot the results in Fig. 4 as a function of average output activity per neuron, which is much higher than the input activity. The power increases linearly with the output activity. The quiescent power consumption with no input, but an average output activity of 14Hz, is about 3mW. The pads account for around 75% of the total power consumption. The digital communication circuits account for around 24%. The analog circuits consume less than 1%.

4. CONCLUSION

We have successful designed, fabricated and tested a 2D AER transceiver chip that performs orientation selective image filtering. Our initial characterizations of the chip indi-

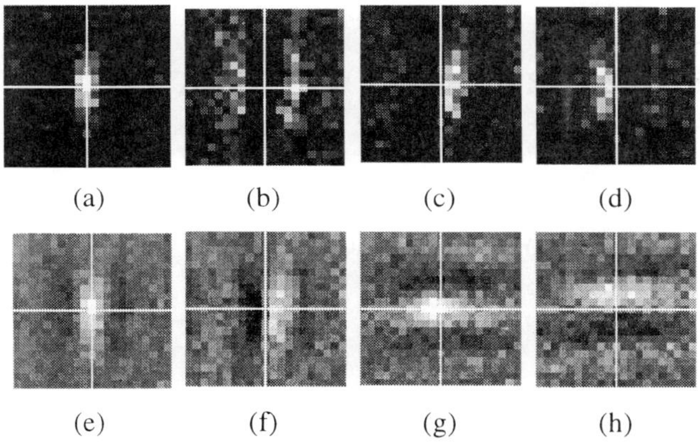

Fig. 3: Measured responses from a 21 by 21 pixel window to a spike train applied at pixel (16,32). (a) ON component of the real part of the output for the array tuned to vertical orientations. (b) OFF component of the real part. (c,d) ON and OFF components of the imaginary part. (e,f) The difference between the ON and OFF components of the real part and imaginary parts. (g,h) Similar data for the array tuned to horizontal orientations. White/black corresponds to a spike rates of (a) 267/0, (b) 98/0, (c) 142/0, (d) 149/0, (e) 267/-267, (f) 149/-149, (g) 173/-173 and (h) 134/134 Hertz. Crosshairs indicate pixel (16,32).

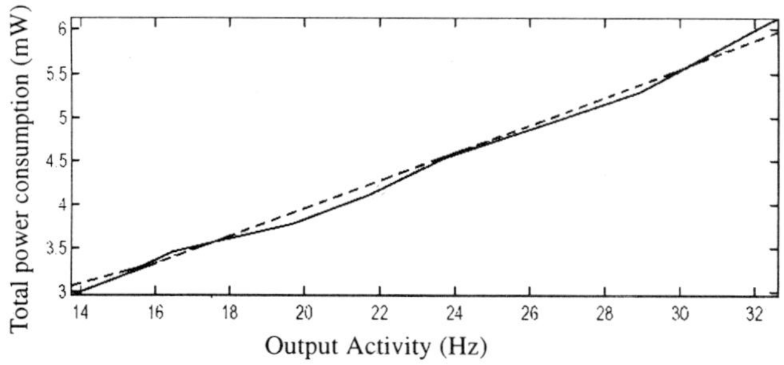

Fig. 4: The solid line plots total power consumption versus the average output activity per neuron. The dotted line is a linear least squares fit to the data, which has slope 0.16mW/Hz and vertical offset 0.77mW.

cate that it functions as expected. Our ongoing work seeks to integrate this chip into a multi-chip architecture for visual information processing.

ACKNOWLEDGEMENTS

This work was supported by the Hong Kong Research Grants Council.

REFERENCES

[1] D. G. Albrecht and W. S. Geisler, "Motion selectivity and the contrast response function of simple cells in the visual cortex," *Visual Neuroscience*, vol. 7, pp. 531-546, 1991.

[2] D. J. Heeger, "Normalization of cell responses in cat striate cortex," *Visual Neuroscience*, vol. 9, pp. 181-197, 1992.

[3] D. J. Heeger, "Half-squaring in responses of cat striate cells," *Visual Neuroscience*, vol. 9, pp. 427-443, 1992.

[4] J. P. Jones and L.A. Palmer, "An evaluation of the two-dimensional Gabor filter model of simple receptive fields in cat striate cortex," *Journal of Neuroscience*, vol. 58, no.6, pp. 1233-1258, Dec. 1987.

[5] A. B. Watson and J. A. J. Ahumada, "Model of human visual-motion sensing," *J. Optical Society of America A*, vol. 2, pp. 322-342, Feb. 1985.

[6] I. Ohzawa, G. C. DeAngelis and R. D. Freeman, "Stereoscopic depth discrimination in the visual cortex: Neurons ideally suited as disparity detectors," *Science*, vol. 249, pp. 1037-1041, 31 Aug. 1990.

[7] K. A. Boahen, "Point-to-point connectivity between neuromorphic chips using address events," *IEEE Transactions on Circuits and Systems-II: Analog and Digital Signal Processing*, vol. 47, no. 5, pp. 416-434, May 2000.

[8] K. A. Zaghloul, *A silicon implementation of a novel model for retinal processing*, Ph.D. thesis,. University of Pennsylvania, 2001.

[9] K. A. Boahen, "The retinomorphic approach: Pixel-parallel adaptive amplification, filtering and quantization," *Analog Integrated Circuits and Signal Processing*, vol. 13, pp. 53-68, 1997.

[10]B. E. Shi, T. Y. W. Choi and K. Boahen, "On-off differential current mode circuits for Gabor-type spatial filtering," *Proc. IEEE Intl. Symp. On Circuits and Systems*, Phoenix, AZ, vol. II, pp. 724-727, May 2002.

[11]E Culurciello, R Etienne-Cummings, and K. Boahen, "Arbitrated Address Event Representation Digital Image Sensor", *IEEE International Solid-State Circuits Conference*, pp 92-93, San Francisco CA, February 2001.

A SILICON RETINA WITH CONTROLLABLE WINNER-TAKE-ALL PROPERTIES

Shih-Chii Liu

Institute of Neuroinformatics, University of Zurich and ETH Zurich
Winterthurerstrasse 190, CH-8057 Zurich, Switzerland
e-mail: shih@ini.phys.ethz.ch

ABSTRACT

The winner-take-all (WTA) circuit is a useful computational circuit for signal processing and learning tasks. By adding spatial coupling between pixels, local regions of competition can be delineated. Recently, we described a normalising circuit which enhances Lazzaro's WTA circuit (1989) only with the addition of a transistor and a global bias to each pixel. This new circuit allows the network to transition between a soft-max and a WTA function through the global bias. The WTA network of Lazzaro together with spatial coupling forms the current-mode silicon retina of Boahen and Andreou (1992). This retina models the center-surround processing performed in biological retinas to enhance responses to spatial contrasts. Here, we show how our normalising network together with spatial coupling performs as a silicon retina. Results are presented from a fabricated circuit in a 2μm CMOS process.

1. INTRODUCTION

The winner-take-all (WTA) function is a useful computation in self-organizing neural networks and signal processing applications. It selects a single winner out of multiple inputs. It has been used in various aVLSI systems for computing stereo,object tracking,and image compression.Lazzaro and colleagues [1] were the first to implement a hardware model of a winner-take-all (WTA) network. Their network consists of N excitatory neurons that are inhibited by a global inhibitory neuron. It computes a single winner, the identity of which is indicated by the outputs of the excitatory cells. Localized winners can be obtained by coupling neurons together through lateral resistive connections. The properties of this network have been enriched with the addition of lateral connections and positive feedback mechanisms [2, 3, 4, 5]. This network is also the basis of the current-mode silicon retina of [7] where the inputs come from photocurrents. This retina models the center-surround processing found in biological retinas to amplify high spatial contrasts.

Recently, we described a network that performs either a soft-max or a winner-take-all function depending on a global parameter [6]. This circuit is similar to Lazzaro's WTA circuit except for the addition of a diode-connected transistor and a global bias. In this work, we describe how our network together with spatial coupling performs with normal input currents or with photocurrents. In the latter case, the network is a silicon retina with a similar architecture to the silicon retina of [7]. Results both from analysis and from the fabricated chip show that there is a smaller dependence of the smoothing space constant of the network on background intensity when compared to the space constant of the retina from [7]. The chip was fabricated in a 2μm CMOS technology process.

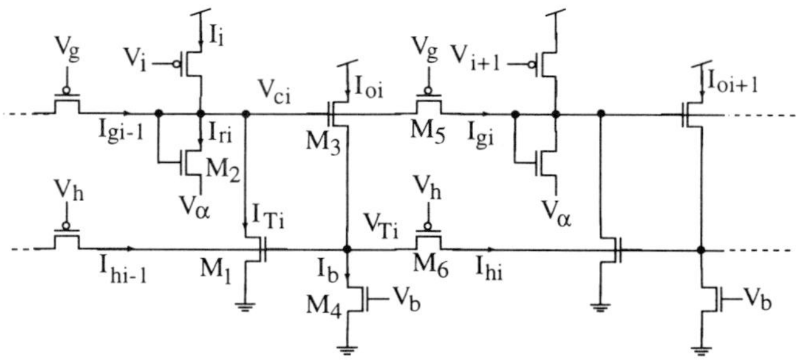

Fig. 1. Circuitry showing coupling between two excitatory neurons in an array of N excitatory neurons (in this paper, N=20). The inhibitory circuit is local to each pixel. The circuit in each excitatory neuron consists of consists of an input current source, I_i, and transistors, M_1 to M_3. The inhibitory transistor is a fixed current source, I_b through M_4. The input to the inhibitory transistor, I_{oi} is normalized with respect to NI_b.

2. CIRCUIT DESCRIPTION

The circuitry for two of the N excitatory neurons and their local inhibitory circuit is shown in Fig. 1. Each excitatory neuron is a linear threshold unit and consists of a pFET that supplies the input current, I_i, and transistors, M_1 to M_4. We ignore the spatial coupling transistors, M_5 and M_6 in each pixel for the moment and short all V_{Ti} nodes such that $V_{Ti} = V_T$. The circuit in Lazzaro's pixel is similar to this circuit but without the transistor, M_2. This transistor, M_2, introduces a rectifying nonlinearity into the system since I_{ri} cannot be negative. The inhibition current, I_{Ti}, to each neuron is determined by the gate voltage, V_{Ti}, which in turn is determined by the input current, I_i through the transistor M_3. In the hard WTA regime, the neuron with the largest I_i sets I_{Ti} for all neurons. Therefore, only the corresponding transistor M_3 in the winning neuron is not in cutoff, and its output current, I_{oi} is equal to the total bias current, NI_b.

The parameter, V_α, determines whether the circuit in Fig. 1 computes the soft-max or WTA function. In the soft-max regime, more than one I_{oi} can be positive and the relative sizes of the I_{oi} is dependent on I_i, V_α and NI_b. We can compute the "active inputs" that set the common voltage, V_T in each neuron and solve for the output current, I_{oi} in terms of I_i:

$$I_{oi} = \frac{I_i}{\sum_j^N I_j}(I_B + I_\alpha N) - I_\alpha \qquad (1)$$

where $I_\alpha = I_0 \, e^{kV_\alpha/U_T}$, U_T is the thermal voltage, and I_0 is the pre-exponential constant of the subthreshold current equation. In deriving this equation, we assume that the transistors operate in subthreshold and κ (the coupling efficiency of the gate on the channel of a transistor in subthreshold) is equal to 1.

Noting that I_{oi} cannot be non negative for "active" inputs, we get the condition:

$$I_i \geq \frac{I_\alpha \sum_j^N I_j}{I_B + NI_\alpha}. \qquad (2)$$

Equation 2 describes the condition under which a neuron stays "active" or its I_{oi} is positive.

2.1. Center-surround property

As previously shown [7], this two-layered network in which the pixels are coupled by diffusors M_5 and M_6, performs a center-surround computation. This computation is analogous to a difference of Gaussians operation to extract local high spatial contrasts. The top layer in Fig. 1 which receives the current inputs performs spatial smoothing on the inputs through the diffusors which are controlled by a global voltage V_g. The bottom layer performs spatial smoothing on the outputs, I_{oi} of the top layer through the diffusors which are controlled by a global voltage V_h. The output of this layer in return inhibits the top network. By setting the diffusor biases V_g and V_h such that the bottom network has a bigger spatial spreading constant than the top network, we can approximate the function of the two-layered network to that of the difference of Gaussians. The equations for the currents

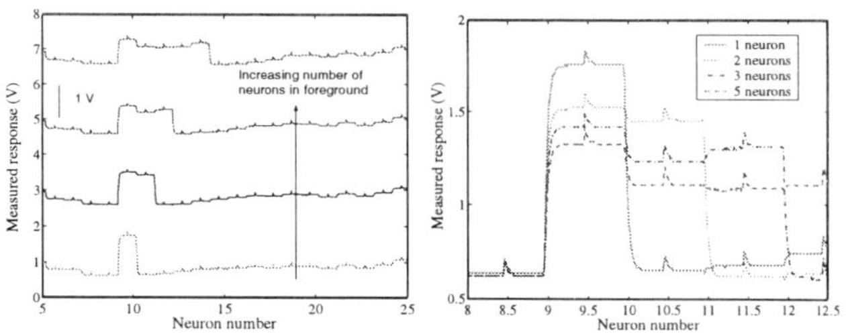

Fig. 2. Response of the network in Fig. 1 for an increasing number of neurons/pixels that receive a higher input current (considered the foreground) than the remaining neurons/pixels. The parameter V_α was set to 0.6V so that the network operated in the soft-max regime. (a) Responses to an increasing number of neurons in the foreground. The traces have been shifted relative to one another for ease of comparison. The lowermost curve is the network response for one neuron that received a larger input current (V_{in}=3.6V) than the remaining neurons (V_{in}=3.7V). The remaining three curves are obtained with an increasing number of neurons in the foreground that received the larger input current. The topmost curve is the network response for five neurons in the foreground. (b) Magnified responses of the foreground neurons. Notice the reduction in the response of the initial sole foreground neuron (the solid curve) as more neurons were added to the foreground. The figures have been adapted from Fig. 5 in [6] with permission.

in both layers of this network are

$$\begin{aligned} I_{Ti} + I_{ri} &= I_i + M(\nabla^2 I_{ri}^{1/\kappa}) \qquad (3) \\ I_{oi} &= I_b - N(\nabla^2 I_{Ti}^{1/\kappa}). \qquad (4) \end{aligned}$$

where $M \propto e^{V_\alpha} e^{(\kappa_p - 1)V_{dd} - \kappa_p V_g)}$. The spatial constant N is proportional to $e^{(\kappa_p - 1)V_{dd} - \kappa_p V_h)}$. Instead of a biharmonic operator as in Boahen and Andreou's network, the smoothing function in the top network is a Laplacian operator. The spatial smoothing constant, M and N, does not depend on the input magnitude. Ideally, the space constant should not increase with the input current. However, in reality, the current through the diffusors depend on the κ of the corresponding transistor M_5 in each pixel. (The gate coupling efficiency of the transistor, κ, changes with the magnitude of the current through the transistor.) When the input increases, the larger voltages at the nodes on the top layer lead to an increase in κ (and hence the lateral current). Hence in the network of [7], the spatial smoothing of the photocurrent inputs increases as the background intensity increases. This dependence is unlike that in biological retinas where the spatial smoothing of the inputs increases under low background intensity. In our network, because of the low impedance of the corresponding node, V_{ci}, in each pixel, the change in the voltage at V_{ci} is smaller for

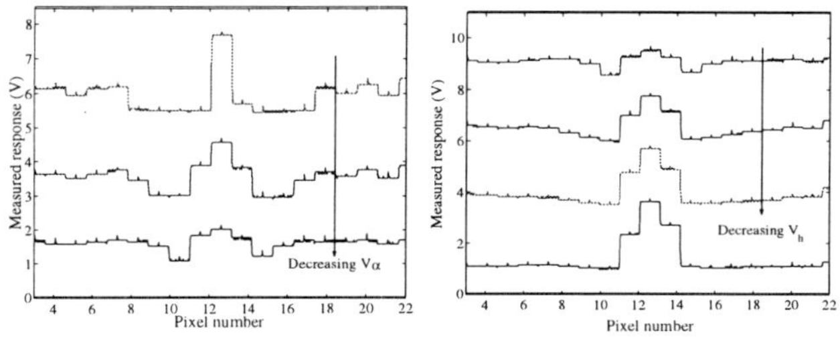

Fig. 3. Response of the network with input photocurrents. The traces have been shifted relative to one another for ease of comparison. A stimulus with a single bright strip was placed over the chip such that only one pixel was stimulated. (a) Network response of the silicon retina for V_α = 0.8V, 0.937V, and 1.2V. The lowestmost trace corresponds to the response of the network when operating in the soft-max regime while the uppermost trace corresponds to the network operation in the WTA regime. (b) Network response for different values of the diffusor voltage (V_h=0V, 0.046V, 0.081V, and 1.53V). V_g=1.28V and V_α = 0.8V.

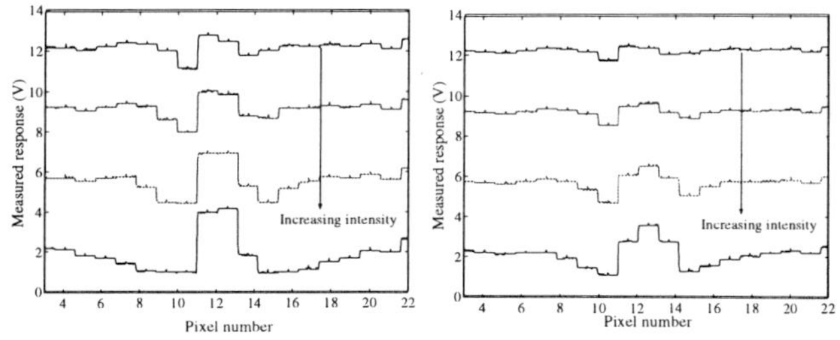

Fig. 4. Response of the network for different light intensities. The traces have been shifted relative to one another for ease of comparison. (a) Response of the silicon retina in the WTA regime (V_α=0.9V). The uppermost curve was obtained at the lowest background intensity. Successive curves were obtained at intensities which are a decade apart. We can see that the spatial smoothing of the network increased with increasing background intensity. (b) Response of retina in the soft-max regime (V_α=0.814V). The space constant was almost invariant over the four decades of background intensity.

the same increase in photocurrent input. Thus, the gain and the smoothing constant of the top layer does not increase as much with the background intensity as in the network of [7].

3. MEASURED RESULTS FROM CHIP

A network consisting of 20 pixels as shown in Fig. 1 was fabricated in a 2μm CMOS process. We describe results from the chip using current source inputs to illustrate how the network can transition between a soft-max function and a winner-take-all function. We then describe results from the same network with photocurrents as the inputs.

3.1. Results from current source inputs

We eliminate the center-surround properties of the network by setting $V_h = 0V$ and $V_g = 5V$. The output currents, I_{oi}, of the neurons were read through an on-chip scanner.These currents were converted to a voltage using an off-chip current sense amplifier and a 22 MΩ resistor.

We describe experiments that show the soft-max property of the network. The details of the other regimes of operation are given in [6]. We set the parameter V_α so that the network operated in the soft-max regime (V_α=0.6V). The input voltages, V_{in} of all the neurons (we called them the background neurons) except for one were set to 3.7V. The input voltage of the remaining neuron (which we call the foreground neuron) was set such that the neuron received a larger input current. The output response of the network for the sole foreground neuron is shown in the lowermost trace in Fig. 2(a). The network response for an increasing number of neurons in the foreground are shown by the remaining traces in Fig. 2(a). As more neurons were added to the foreground, the output current of the initial neuron in the foreground decreased as shown by the magnified responses in Fig. 2(b). The response of the network with a sole neuron (9th neuron) in the foreground is given by the solid curve. The response of this initial neuron decreased with the number of increasing foreground neurons. The responses in Fig. 2 show that the soft-max function of the network. The output currents, I_{oi}, depend on the relative magnitude of the input currents. There is no single winner as in a winner-take-all network.

3.2. Results from silicon retina

The inputs to the network now come from photodiodes (the pFET driven by V_i in each pixel is now switched off) and the network acts as a silicon retina with center-surround properties. By increasing V_α, we can change the operation of the network from the soft-max regime to the WTA regime as shown in Fig. 3(a). In these traces, a stimulus with a single bright strip was placed over the chip such that only one pixel was stimulated. In the uppermost trace where the network operated in the WTA regime, this pixel had the largest response. In the lower traces, the output of this pixel decreased dependent on the relative magnitudes of the photocurrents. The shape of the center-surround kernel of the network can be altered through the relative voltage difference between V_g and V_h as described in Sect. 2.1. By decreasing V_h (this corresponds to an increase in the spatial smoothing in the bottom layer), the inhibitory surround increases as shown in the lowermost trace of Fig. 3(b) and the

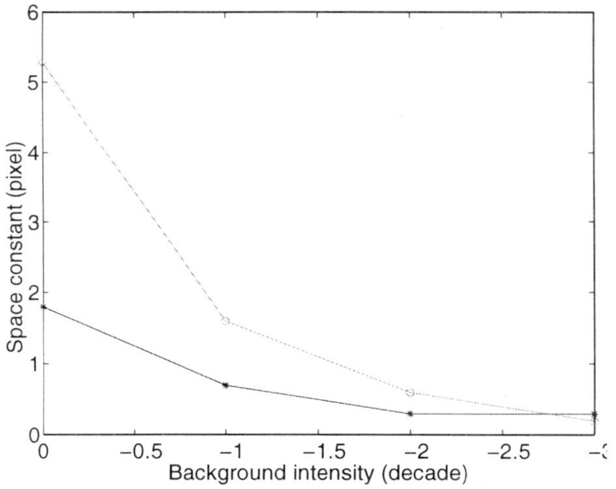

Fig. 5. The network space constant plotted against four decades of background intensity. The ordinate "0" is the brightest intensity. The remaining data points correspond to decreasing background intensity in decades. The space constant was obtained from Fig. 4. The line marked with circles is derived from the winner-take-all data and the line marked with asterisks is derived from the soft-max data. The network space constant is the interpolated pixel value where the retina response drops by 1/e from the peak response.

pixel which saw the bright strip had the largest response.

As we have seen in Fig. 3(a), changing V_α changes the function of the network. When we set V_α so that the network acts as a WTA, it is equivalent to the network of [7]. As pointed out in Sect. 2.1, the spatial spread of the impulse response of the network increases with increasing background intensity. This behavior is shown in Fig. 4(a). The opposite type of dependence is observed in the biological retina. The biological retina has a larger spatial spread under low background intensities because it needs to collect photons from a larger spatial area to get a good S/N ratio response. However, the silicon retina shows the largest spatial spread for the highest background intensity.

By decreasing V_α so that the network operates in the soft-max regime, the spatial space constant is less invariant over four decades of intensity as shown in Fig. 4(b) because of the low impedance nature of V_{ci} in Fig. 1. This smaller invariance is also depicted in Fig. 5 which shows the network space constant over the 4 decades of intensity. This data was derived from Fig. 4.

4. CONCLUSION

We described the response of a silicon retina that displays different spatial smoothing characteristics when tuned for a soft-max computation or a winner-take-all computation. When the network is tuned for a winner-take-all computation, the space constant of the smoothing increases with higher background intensities. This characteristic is undesirable for modelling the smoothing properties of the biological retina. However, if the network is tuned for a soft-max computation, the space constant of the smoothing is almost invariant to background intensity.

5. ACKNOWLEDGMENTS

I thank Rodney Douglas for supporting this work. I also thank Tobias Delbrück for critical reading of the article. This work was supported in part by the Swiss National Foundation Research SPP grant and the U.S. Office of Naval Research.

6. REFERENCES

[1] J. Lazzaro, S. Ryckebusch, M. A. Mahowald, and C. A. Mead, "Winner-take-all networks of O(n) complexity," in *Advances in Neural Information Processing Systems*, D. Touretzky, Ed., vol. 1, pp. 703–711. Morgan Kaufmann, San Mateo, CA, 1989.

[2] T.G. Morris, T. Horiuchi, and E. Niebur, "Object-based selection within an analog VLSI visual attention system," *IEEE Trans. Circuits and Systems II*, vol. 45, no. 12, pp. 1564–1572, 1998.

[3] J. Choi and B. Sheu, "A high-precision VLSI winner-take-all circuit for self-organizing neural networks," *IEEE Trans. Circuits and Systems II*, vol. 28, no. 5, pp. 576–583, 1993.

[4] J.A. Starzyk and X. Fang, "CMOS current mode winner-take-all circuit with both excitatory and inhibitory feedback," *Electronics Letters*, vol. 29, no. 10, pp. 908–910, 1993.

[5] G. Indiveri, "Winner-take-all networks with lateral excitation," *Analog Integrated Circuits and Signal Processing*, vol. 13, no. 1/2, pp. 185–193, 1997.

[6] S.-C. Liu, "A normalizing aVLSI network with controllable winner-take-all properties," *Analog Integrated Circuits and Signal Processing*, vol. 31, no. 1, pp. 47–53, 2002.

[7] K. A. Boahen and A. G. Andreou, "A contrast sensitive silicon retina with reciprocal synapses," *Advances in Neural Information Processing Systems*, vol. 4, pp. 764–772, 1992.

SINGLE CHIP STEREO IMAGER

Ralf M. Philipp, Ralph Etienne-Cummings

Dept. of Electrical and Computer Engineering, Johns Hopkins University, Baltimore, MD 21218, USA
rphilipp@jhu.edu, retienne@jhu.edu

ABSTRACT

A stereo vision chip, incorporating two 128 x 128 pixel current-mode imagers and analog disparity computation circuitry, is presented. A modified version of block matching is used to compute the disparity between the two images at each (*u*, *v*) coordinate, with the sum-of-absolute-difference value for each possible disparity being computed in parallel. The chip has been tested at computation rates up to 11.2 million checked disparities per second, while consuming only 35mW from a 5V supply, including imagers and computation circuits.

1. INTRODUCTION

This paper describes a novel stereo vision system, the Single-chip Stereo Imager (SSI). Implemented in a standard CMOS process, the SSI combines two imagers and analog computation circuitry on a single chip. Previous work includes imagers designed for use in stereo vision systems [1,2], specialized analog and digital processors [3-6], correlation processors [7]. A 1-D stereo correspondence processor is presented in [8].

The SSI is based on an algorithm designed for operation in current-mode analog VLSI [9]. The only computational blocks used are addition, subtraction, rectification (absolute value), and loser-take-all (finding the minimum).

The combination of two imagers and disparity computation circuitry on a single chip eliminates the need for separate imagers and power-hungry DSPs or MPUs. Potential applications of the SSI include obstacle detection for autonomous robots, computer user-interface devices, and vehicle navigation.

2. ARCHITECTURAL OVERVIEW

2.1 Algorithm

The SSI chip is based on the algorithm described in [9]. This algorithm, a modified version of block matching, was optimized for implementation in current-mode analog VLSI. The algorithm takes advantage of the fact that the disparity in a stereo pair is entirely horizontal. Realizing that vertical edges are the salient features when finding the disparity, one can average a block of pixels along the vertical before matching. This averaging, implemented as a sum of currents, reduces the algorithm's computational complexity by an order of magnitude, in addition to reducing sensitivity to image noise.

The algorithm solves for a disparity Δv at a point (*u*, *v*) as described by equations 1-4.

$$r_{sum}(u,v) = \sum_{i=u}^{u+U-1} r(i,v) \quad (1)$$

$$l_{sum}(u,v+d) = \sum_{i=u}^{u+U-1} l(i,v+d) \quad (2)$$

$$SAD(u,v,d) = \sum_{j=v}^{v+V-1} \left| r_{sum}(u,v) - l_{sum}(u,v+d) \right| \quad (3)$$

$$\Delta v = \arg\min_{d \in D} SAD(u,v,d) \quad (4)$$

In equations (1) and (2) a block's pixel values are summed (averaged) along the columns in order to produce row vectors r_{sum} and l_{sum}. The sum of absolute difference (SAD) metric is computed at each potential disparity *d*. The disparity yielding the lowest SAD is then selected.

Not all regions of all images provide sufficient information to find a reliable match. The most important characteristic for finding good matches is the presence of vertical edges. To test for this, one can simply verify that the contrast in the row vector r_{sum} is larger than a given threshold value *T*.

$$\sum_{j=v}^{v+V-1} \left| \nabla^2 r_{sum}(u,j) \right| > T \quad (5)$$

2.2 System Overview

The SSI consists of five main components: two 128 by 128 pixel current-mode imagers, a 15 by 114 matrix of absolute difference-of-currents elements, a 128 input loser-take-all (LTA), and a spatial highpass filter. Current-mode analog circuits are used for all of the blocks in Fig. 1, excluding the registers and other digital support circuitry (not shown).

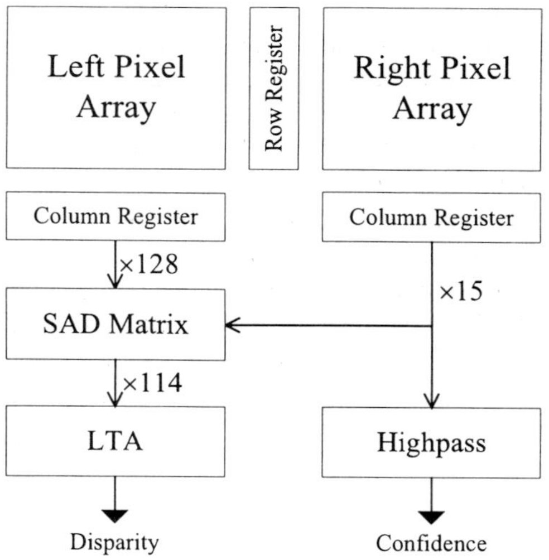

Figure 1: Block diagram of the chip

0-7803-7761-3/03/$17.00 ©2003 IEEE

Figure 2: Chip micrograph (4.2 x 4.2 mm^2 with pads)

Fig. 2 shows a micrograph of the SSI; the two imagers, SAD matrix, and loser-take-all ("LTA") circuit are highlighted.

2.3 Hardware Implementation

The current-mode APS pixel shown in Fig. 3 is composed of a photodiode and a nonlinear current amplification circuit. Current gain can be varied by adjusting the bias on the NMOS load transistor M1. An analysis of this pixel can be found in [10]. The pixels are connected to a column bus, thereby realizing column summation of all rows selected by the row registers. The two 128x128 pixel arrays are each surrounded by a ring of dummy pixels to improve matching at the edges.

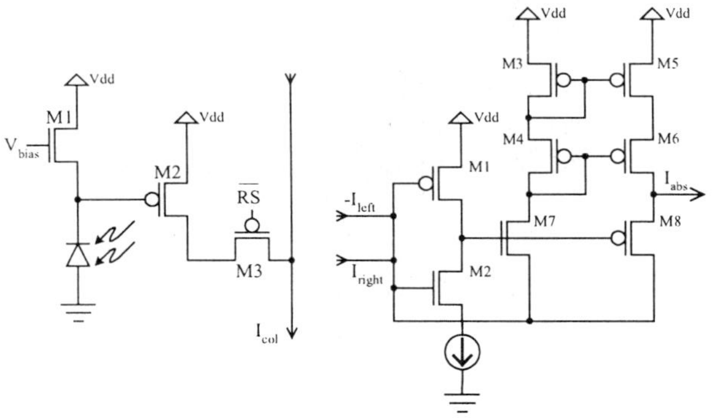

Figure 3: APS pixel and current rectifier schematics

As can be seen in the chip micrograph (Fig. 2), the two pixel arrays are located 190µm apart. This presents the challenge of presenting a stereo pair to the two imagers. Initial testing was done by mounting a standard C-mount lens over the chip and placing suitable images in the two imagers' fields of view. In the future, the SSI will be equipped with a lens arrangement that will project a stereo pair on the arrays. Two possible lens configurations are accurately aligned mirrors [11] and a biprism [12].

The two pixel arrays are controlled by a common row scanning register, capable of selecting multiple rows, such that the number of rows selected is the height of the block being matched. The right imager's column scanning register selects a 15-column window, constituting the width of the matching block. The left imager's column scanner windows the search area by turning off columns where no matches are expected. This results in a large computed sum of absolute difference (SAD) for the corresponding disparity, meaning that it will not be selected as a likely match. Use of this scanning register, while not necessary, improves the quality of the matching results by removing from consideration those disparities known *a priori* to be unlikely.

The 15 by 114 matrix of absolute difference-of-currents elements is composed of 1710 individual current rectification elements (see Fig. 3.) This circuit is composed of a current comparator (M1 & M2), a current mirror (M3-M6), and two pass transistors (M7 & M8). The comparator, a current-starved inverter, detects the input's direction. An incoming current is passed through M8, while a negative current enables the current mirror through M7. The rows of SAD matrix correspond to the columns of the block from the right image, and the columns of the matrix correspond to the columns in the left image. The rectifier output currents are summed along the columns of the matrix, producing 114 currents equal to the SAD at each location in the left image. A section of the SAD matrix is shown in Fig. 4. Each diagonal line represents a current $-l_{sum}$, each horizontal line represents an r_{sum} current, and each vertical line represents a SAD output for a disparity. At each intersection (black dot), copies of the $-l_{sum}$ and r_{sum} currents are added and rectified, placing the rectified sum on the SAD output column.

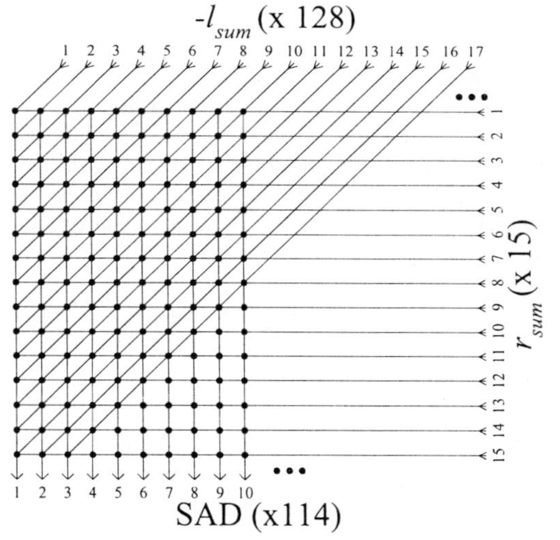

Figure 4: SAD matrix section

The 114 SAD currents are fed into a 128-input loser-take-all; the 14 remaining inputs are externally controlled current sources. The LTA is composed of 127 two-input synchronous LTA elements arranged in a binary tree configuration. A detailed description of this circuit can be found in [13]. The nodes (see Fig. 5) on each level of the LTA tree pass the smaller of two currents, while boosting the voltage at the higher current's input node to V_{dd}. The currents feed into a cross-coupled NMOS pair (M8 and

M9). The larger current raises the gate voltage of the transistor passing the smaller current. Once the circuit has latched, the booster circuit (M1-M3 and M4-M6) quickly pulls the larger current's input node to V_{dd}. The seven levels of the tree are run in sequence, starting with the input level, until all but one of the input nodes is at V_{dd}. These voltages are then encoded into a 7-bit binary output corresponding to the location of the disparity match.

The 14 externally controlled LTA inputs can be used as a confidence test by setting a maximum allowable SAD current. If the LTA disparity output is a value corresponding to the externally controlled inputs, no computed SAD value is sufficiently small, and can thus be rejected.

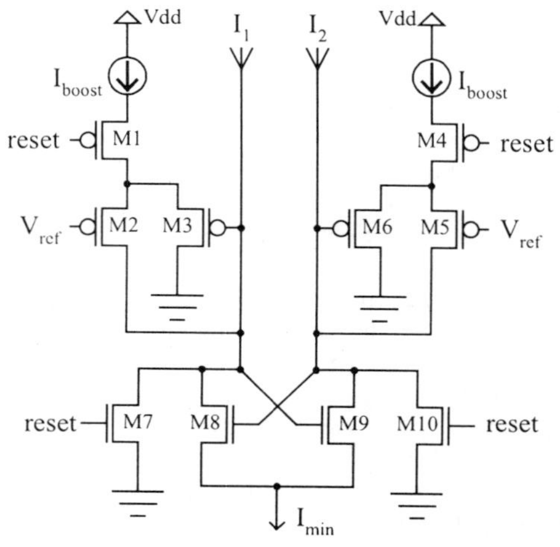

Figure 5: 2-input loser-take-all cell

The fifteen r_{sum} currents from the right imager's 15-column window are also fed into a spatial highpass filter, constructed out of current mirrors. The magnitude of the highpass filter's output corresponds to the level of confidence that can be assigned to any disparity matches from that block. Blocks with low horizontal contrast levels (the lack of vertical edges) present insufficient information to find a reliable match. The use of the confidence current is discussed in Section 3.

3. RESULTS

3.1 Imagers

Images from the left and right pixel arrays on the SSI are shown in Fig. 6. These images were obtained from a 1 by 4 pixel block. The left image exhibits significant fixed-pattern noise (FPN), about 3.5% column-to-column and 4.7% overall. The column FPN is an effect of the readout architecture; the internal computations do not see this difference in column FPN levels between the images. The FPN in the right image, about 3.3%, is typical of uncorrected current-mode APS imagers. Note that FPN measurements were obtained using a 1x1 pixel block (a 1x4 block slightly reduces noise levels) at saturation brightness, which was limited to 0.83μA by the external readout architecture.

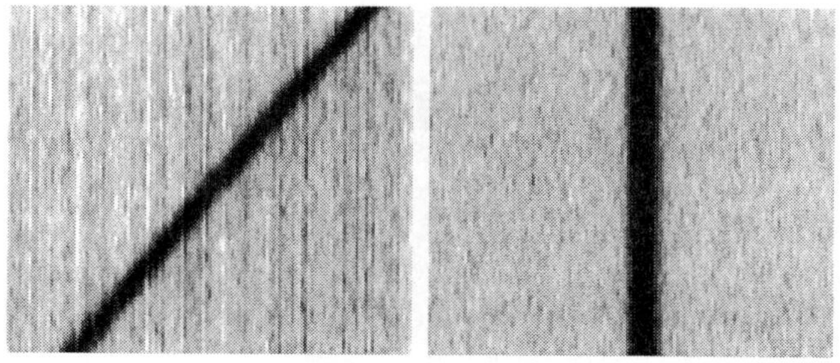

Figure 6: Left and right images

3.2 Stereo

The confidence values from the right image (Fig. 6) are shown in Fig. 7. The confidence value output is a current proportional to the level of horizontal contrast (amount of vertical 'edginess') in the image. The 15-column wide filter has the effect of creating a wider area of high confidence than the width of the white stripe in the image. The first 15 columns of the confidence image have been set to zero, as that region is cut off by windowing.

The confidence metric can be used in many ways. See [9] for a brief discussion of the possible uses. The most interesting are to improve the accuracy of computed disparity values, increase output frame rate, and decrease power consumption. Accuracy can be improved by rejecting computed disparities in regions of low confidence. Limiting the disparity calculation to regions of high confidence allows for an increased frame rate (fewer computed disparities per frame), lower power consumption (the clock speed can be lowered), or a combination of the two.

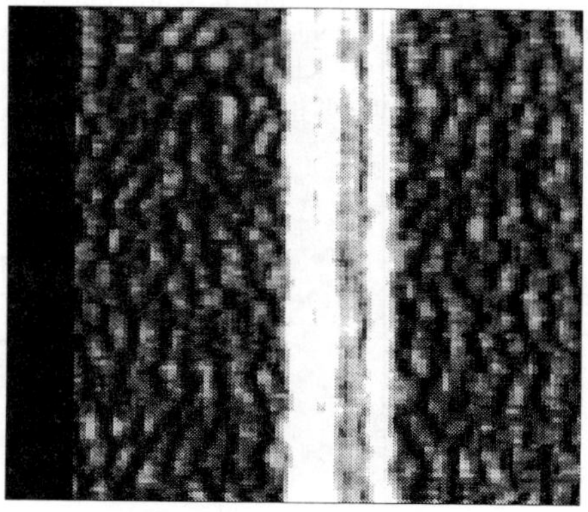

Figure 7: Confidence values (brighter is better)

The computed disparity for the images in Fig. 6 can be seen in Fig. 8. Disparity values for regions not passing the confidence test are shown as zero disparity (gray). In this example, a location passed the confidence test if it had confidence current within the top 20% of confidence currents over the image. The horizontal disparity of the diagonal stripe against the vertical stripe in the right image starts as a negative value (about -47 pixels) in the upper part of the image and slowly changes to a positive value (about 47 pixels) in the lower part of the image. Fig. 9 shows the computed disparities along columns 65-85. The standard deviation of the computed values from the best-fit line is 0.97 pixels.

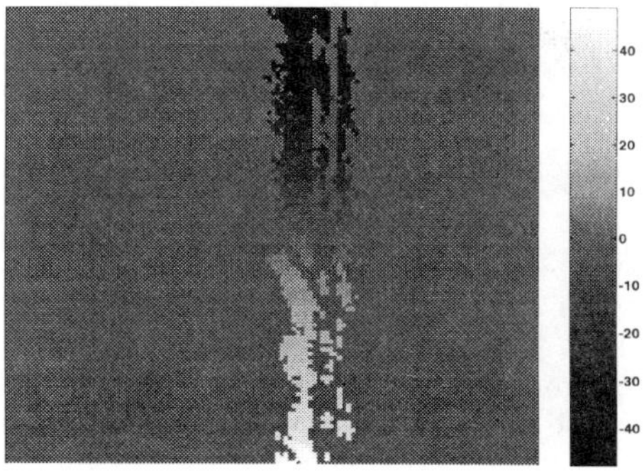

Figure 8: Calculated disparity (number of pixels from right to left image)

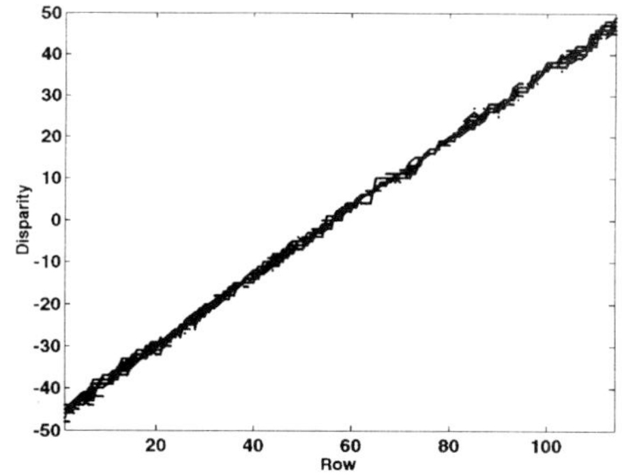

Figure 9: Disparities along columns 65-85

The chip has been tested at frame rates of up to 6fps when computing disparities. Speed is currently limited by the external test configuration; measurements indicate that speeds of at least 20fps should be possible. At this frame rate, 11.2 million possible disparities are checked per second, a figure that translates to 168 million subtraction and rectification operations per second. These figures do not include any frame rate increases that can be achieved by rejecting areas with low confidence values, which would greatly increase the potential frame rate. Table I shows the SSI's general characteristics.

Technology	0.5μm, 3M, 5V, Nwell CMOS
Chip Area	4.23 mm x 4.23mm
Transistor Count	~150,000
Pixel Size	14μm x 14μm
Pixel Fill Factor	25.4%
Imager Array Size	2 x 128 x 128
Imager FPN (right, left)	3.3%, 4.7% (current-mode imaging)
Calculation Speed	11.2M disparity comparisons/s
Energy Consumption	< 4nJ / checked disparity
Power Consumption	~35mW at 6fps (V_{dd}=5V)
Achieved Frame Rate	6fps (disparity output)

Table I: Single-chip Stereo Imager characteristics

4. SUMMARY

The combination of two 128x128 pixel imagers and parallel current-mode analog disparity-processing circuitry on a single chip enabled the creation of a compact, low-power stereo vision solution. Initial results demonstrating the functionality of the imagers and disparity computation have been shown. Computation speeds of up to 11.2 million disparity checks per second have been achieved. By taking advantage of the confidence output, the SSI chip is capable of high frame rates and low power consumption. The SSI's low power use and compact form factor make it ideally suited for embedded robotics, human interface, and navigation applications.

5. ACKNOWLEDGMENT

This work is supported by ONR YIP award #N140010562.

6. REFERENCES

[1] Y. Ni and J. Guan, "A 256x256 pixel smart CMOS image sensor for line based stereo vision applications," *IEEE J. Solid-State Circuits*, vol. 35, pp. 1055-1061, 2000.

[2] Y. Ni, F. Devos, and B. Arion, "Analog retina based real-time vision system," *Int. Semiconductor Conf. 1996*, vol. 1, pp. 275-284, 1996.

[3] D. A. Martin, H.S. Lee, I. Masaki, "A mixed-signal array processor with early vision applications," *IEEE J. Solid-State Circuits*, vol. 33, pp. 497-502, 1998.

[4] J. M. Hakkarainen and H. S. Lee, "A 40x40 CCD/CMOS absolute-value-of-difference processor for use in a stereo vision system," *IEEE J. Solid-State Circuits*, vol. 28, pp. 799-807, 1993.

[5] N. Giaquinto, M. Savino, and S. Taraglio, "A CNN-based passive optical range finder for real-time robotic applications," *IEEE Trans. Instrum. and Meas.*, vol. 51, pp. 314-319, 2002.

[6] R. A. Lane, et al., "A stereo vision processor," *Proc. IEEE 1995 Custom Integrated Circuits Conf.*, pp. 169-172, 1995.

[7] L. G. McIlrath, "A low-power analog correlation processor for real-time camera alignment and motion computation," *IEEE Trans. Circuits Syst. II*, vol. 47, pp. 1353-1364, 2000.

[8] G. Erten and R. M. Goodman, "Analog VLSI Implementation for Stereo Correspondence Between 2-D Images," *IEEE Trans. Neural Networks*, vol. 7, pp. 266-277, 1996.

[9] R. M. Philipp, et al., "Architecture for an aVLSI stereo vision system," *Proc. IEEE Int. Sym. Circ. Sys. 2002*, vol. 3, pp. 691-694.

[10] V. Gruev and R. Etienne-Cummings, "Implementation of steerable spatiotemporal image filters on the focal plane," *IEEE Trans. Circuits Syst. II*, vol. 49, pp. 233-244, 2002.

[11] W. Teoh and X. D. Zhang, "An inexpensive stereoscopic vision system for robots," *Proc. Intl. Conf. Robotics* 1984, pp. 186-189.

[12] D. H. Lee and I. S. Kweon, "A novel stereo camera system by a biprism," *IEEE Trans. Robot. Automat.*, vol. 16, pp. 528-541, 2000.

[13] B. M. Wilamowski, D. L. Jordan and O. Kaynak, "Low power current mode loser-take-all circuit for image compression," *9th NASA Symposium on VLSI Design*, 7.6, 2000.

A Scanning Thermal Microscopy System with a Temperature Dithering, Servo-Controlled Interface Circuit

Joohyung Lee[1] and Yogesh B. Gianchandani[1,2]

[1]ECE Department, University of Wisconsin, Madison, WI 53706, USA
[2]EECS Department, University of Michigan, Ann Arbor, MI 48109, USA

ABSTRACT

This paper describes a thermal imaging system which includes a customized micromachined thermal probe and circuit interface for a scanning microscopy instrument. The probe shank is made from polyimide for mechanical compliance and high thermal isolation, and has a thin-film metal tip of ≈50 nm in diameter. The circuit provides closed-loop control of the tip temperature and also permits it to be dithered, facilitating scanning microcalorimetry applications. This paper explains system design and optimization including both electrical and thermal analyses. Sample scans of patterned photoresist demonstrate noise-limited resolution of 29 pW/K in thermal conductance. Applications of the thermal imager extend from ULSI lithography research to biological diagnostics.

I. INTRODUCTION

In the past decade, scanning microscopy using thermally-sensitive probes has been applied to a variety of applications, ranging from ULSI lithography research to cellular diagnostics in biochemistry [Oc96, Li02]. Thermal probes have also been employed for data storage and other applications [Ve00, Le00, Ma99]. These are generally made from dielectric thin films on a silicon substrate, and use a metal or semiconductor film bolometer for sensing the tip temperature. Other approaches that use more involved micromachining methods have also been reported [Gi97]. A commercially available probe uses a narrow gauge wire bent into a V-shape to form a self-supporting resistor. However, for many applications, thermal probes must have very low mechanical spring constants to prevent damage to soft samples. In addition, for many applications they must have very high thermal isolation to minimize the thermal load presented to the sample. Both of these needs can be met by the use of a polymer for the probe shank. Furthermore, thermal and mechanical design challenges must be considered in conjunction with the interface circuit for best performance of the overall system.

In a frequently used microscopy technique, the scanning tip is mechanically dithered so that the sample spacing is modulated at a known frequency. This is akin to chopping the signal, which permits the detection to be phase locked to the dither, improving the overall signal-to-noise ratio. In the context of thermal microscopy, this also permits thermal capacitance measurement. However, with ultracompliant probes, the mechanical spring constant is far too low to permit physical dithering. The alternative is to dither the bias current in the bolometer, thereby placing the burden on the interface circuit. Furthermore, the very high thermal resistance of the probe shank, designed to minimize thermal loading of the sample, has the impact of reducing the thermal bandwidth of the probe below 1 KHz. This increases the susceptibility of the pick-off to flicker noise and further raises the burden on the circuit.

This paper describes a thermal imaging system (Fig. 1) which uses a customized micromachined bolometer probe and circuit interface to a commercial scanning microscopy instrument (TopoMetrix™ SPMLab v.3.06). The bias current in the bolometer can be controlled to operate the scan at a fixed temperature. The interface also provides electronic dithering of the tip temperature. Its design includes consideration of thermal and electrical interactions between the probe and circuit components based on MatLab™ modeling of the overall system. The functionality of the system is demonstrated with both microcalorimetric and imaging applications of patterned photoresist and calibration materials. To further evaluate the operation of the circuit, nodal measurements taken during a practical scan (not in a test mode), and are presented along with the scanned image obtained.

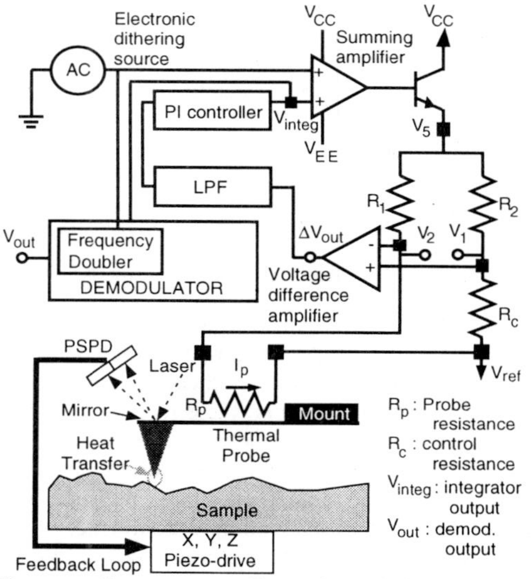

Fig. 1: The overall system configuration of the custom probe and circuit which interface with a commercial instrument.

II. SYSTEM DESCRIPTION

A. Sensor Element

The scanning thermal probe is fabricated on a Si substrate using a 7 mask process similar to those described in [Li00, Li01]. A metal thin film bolometer is sandwiched between two layers of polyimide that form a cantilever (Fig. 2). At one end of the cantilever the Ni thin film protrudes through an opening in lower polyimide layer, where it is molded into a pyramidal tip by a notch that was anisotropically wet-etched into the substrate. A tip diameter of ≈50 nm is achieved by sharpening the notch by anisotropic thermal oxidation. The tip and a portion of the probe shank are then released from the substrate by etching an underlying sacrificial layer. The released length is then folded over to extend past the die edge for clearance, and held in place by a thermo-compression bond across a thin film of Au which is deposited as the final layer on top of the polyimide. This film also serves as a mirror to permit use of the probe for AFM. The entire fabrication process is performed below 350°C, and is compatible with post-CMOS processing to accommodate the possible integration of an interface circuit. Typical dimensions of the probes after assembly are 250 μm

0-7803-7761-3/03/$17.00 ©2003 IEEE

length, 50 μm width, and 3 μm thickness, which result in a mechanical spring constant of 0.08 N/m, which is upto 100× below commercial probes. The bolometer, which has Cr/Ni at the tip and Cr/Au leads, is ≈45 Ω.

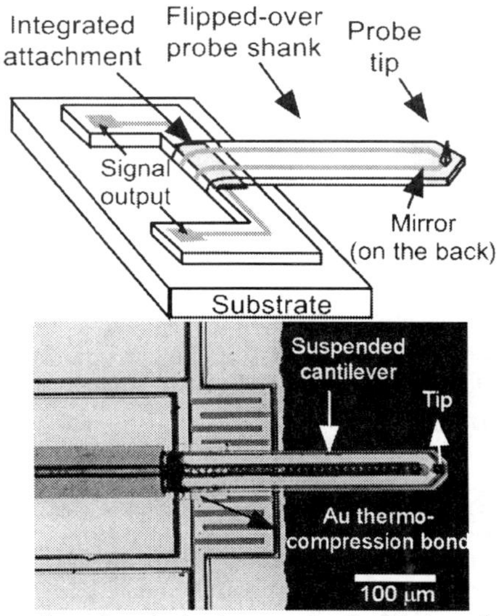

Fig. 2: Schematic and optical micrograph of a fabricated probe.

B. Interface Circuit

The bolometer readout is through a Wheatstone bridge, which is commonly used for piezoresistive pressure sensors, strain gauges, etc. It is well suited for microfabrication and allows a differential measurement that offers a higher common-mode noise rejection than a single-element measurement. Historically, the conversion of bridge resistance to current or voltage for readout has suffered from non-linearity and restricted dynamic range [Yo00]. Additionally, in DC mode the signal is subject not only to thermal noise from the resistor bridge, but also 1/f flicker noise from the electronics. To overcome these challenges, many efforts have been made to convert resistance variation to frequency [Mo95, Hu87, Gi76], to duty cycle/time [Ci90, Go93], and to both of them [Fe97]. Some require components such as a pulsed bridge supply current, or an input amplifier with very low offset and drift [Gi76]. Furthermore, these approaches are constrained by switching delays causing non-linearity between frequency (or pulse width) and resistance change, are expensive to implement, and most importantly cannot be applied directly to operating the microbolometer or anemometer in a constant temperature mode.

The system used in this effort (Fig. 1) utilizes two separate feedback loops: electrical and optomechanical. As the probe (Fig. 2) scans the sample surface, topography is mapped by detecting the laser signal reflected off a mirror located near the tip and using this in a mechanical feedback loop to maintain constant contact force. Since variations in heat loss through the probe tip cause variations in the probe resistance, this quantity maps the temperature or thermal conductance of the sample.

When both a DC and an AC signal (at ω_0) are applied to the bridge (Fig. 1), the bolometer is modulated by the square of $V_{DC}+V_{AC}\cos(\omega_0 t+\theta)$, and its resistance changes proportional to:

$$\Delta R_p \propto V_{DC}^2+2V_{DC}V_{AC}\cos(\omega_0 t+\theta) + V_{AC}^2\cos^2(\omega_0 t+\theta) \quad (1)$$

Therefore, bolometer resistance is approximately represented as:

$$R_p \approx R_{pDC} + R_{pAC}\cos(\omega_0 t+\theta) \text{ } if \text{ } V_{AC}^2 << 2\cdot V_{DC}\cdot V_{AC} \quad (2)$$

making ω_0 the dominant resistance-modulation frequency. The output of the bridge voltage difference amplifier is:

$$\Delta V_{out}=0.5\cdot I_{AC}R_{pAC}\cos(2\omega_0 t+\theta)+ I_{DC}R_{pAC}\cos(\omega_0 t+\theta) \quad (3)$$

If the $2\omega_0$ term (second harmonic) of the voltage-modulation frequency is selected, the impact of 1/f flicker noise can be reduced. In addition, better signal-to-noise ratio is expected as I_{AC} becomes high to a certain extent. In the selected implementation, V_{DC} was 5 V and V_{AC} was 0.8 V.

The interface circuit includes a PI controller (which is comprised of an integrator and an inverting amplifier), and a simple homodyne demodulator (Fig. 1), in which the input signal is multiplied by in-phase local oscillator and then low pass filtered (Fig. 3). The PI controller has integral gain of 10^4 and proportional gain of 1, showing settling time <10 msec. This demodulation technique (Method A) is applicable when phase change in the input signal is negligibly small compared to change in its magnitude. The Method A is simple to implement and does not have mismatch problems which are faced in quadrature homodyne demodulation (Method B). In Method B, in-phase (I signal) and quadrature (Q signal) signals are generated, low pass filtered, and root mean squared. Problems are caused by mismatches between the amplitude of I and Q signals and errors in the nominally 90° phase shift. Method A is consequently preferred. According to our previous investigations [Li01], the -3 dB frequency of thermal response of the probe is about 0.5 kHz with an open-loop interface circuit. It is somewhat higher with a closed loop interface circuit because external power is used to increase effective thermal conductance of the probe [Sa93]. Consequently, for this project a 1 kHz dither is selected, and scan speeds are set to provide a measured data bandwidth <50 Hz. The bridge output voltage is band pass filtered at the second harmonic 2 kHz, and multiplied by the frequency doubled output of the dither oscillator (Fig. 3). The phase of the local oscillator is synchronized with that of input carrier signal to avoid signal distortion. The final output is obtained by low pass filtering. The Q factor of band pass filter and -3 dB frequency of low pass filter are based on the dither frequency and data bandwidth, but adjusted for low frequency noise near the band edge of scan data. Bi-quad band pass filters and bi-quad low pass filters are used because of their excellent tuning features and good stability. The gain, quality factor, and salient frequencies of the filters can be independently controlled.

III. SYSTEM MODELING AND SIMULATION

The simulation of the whole sensing subsystem provides an understanding of the interaction between thermal behavior of the probe and electrical behavior of the interface circuit. It is accomplished using electrical parameters of the Simulink tool within MatLab™. Figure 4 shows the state diagram for the combined subsystem. Using this, it is demonstrated how the demodulator achieves noise reduction compared to a non-dithered DC closed loop interface circuit.

A challenge in modeling the subsystem is how to transform a thermal probe into electrical parameters. The dotted block in bottom left of Fig. 4 represents the thermal probe model. The three inputs shown are used to mimic the time variation of thermal conductance encountered during a scan of photoresist lines on a Si substrate. Thermal conductance changes smoothly in a real scan, but the variation should have a non-zero and finite bandwidth to test the circuit for signal distortion. The sum of these inputs is multiplied by the temperature bias of the tip to calculate the power variation in the probe. This variation, which would otherwise modulate the bolometer, is instantly compensated by the interface circuit which keeps the probe temperature constant.

An important optimization parameter for simulations is the ratio of V_{DC} to V_{AC} in eqn. (1). As V_{AC} increases, modulation of probe resistance by the second harmonic of applied power cannot be ignored. Additionally, simulations show that the PI controller loses its feedback control, even though the signal-to-noise at the output of demodulator becomes better in a certain range of V_{AC}. The probe temperature is supposed to be almost constant despite small AC temperature variations introduced for dither operation by the PI controller. However, as V_{AC} increases, power supplied by the AC component becomes comparable to DC power, causing the tip temperature to fluctuate significantly. Now the PI controller receives a significant AC signal in addition to the DC signal that is the differential output from the resistor bridge. The output of PI controller thus contains not only DC compensation power but also a significant amount of unnecessary AC power, which derails the PI feedback control. A low pass filter can be placed between bridge circuit and PI controller to avoid this problem. However, it is only useful when the dither frequency is much higher than bandwidth of the scan signal from the bridge. In the simulated system the mimicked signal at the input of the system contains frequency components at higher frequencies than the dithering signal.

Figure 4(a) represents a noiseless input to the system. When low frequency noise exists at 100 Hz, with a 20% variation in bolometer resistance the bridge output is deteriorated in the absence of electrical dithering (Fig. 4(b)). In contrast, the output of demodulator (Fig. 4(c)), which is used with electrical dithering, shows a much better signal-to-noise ratio. However, the output of demodulator can be distorted because high-frequency components of the input signal can be inadvertently screened by band pass filters with high quality factor. This motivates the use of the highest dither frequency (and thus a fast thermal response) to secure the maximum signal bandwidth.

IV. MEASUREMENT

Insets in Fig. 3 show frequency spectra at various circuit nodes taken while scanning a photoresist sample at a tip temperature of 45°C. Figure 3(a) shows that at the output of the bridge circuit, where the second harmonic contains the pursued power-modulated thermal signal, the amplitude ratio of the first harmonic to the second is 24.6, which is very close to the theoretical value of 25 obtained from eqn. (3). This demonstrates that the bandwidth of the thermal probe can be wider than 2 kHz and the 2 kHz-dithered signal is not distorted by thermal delay. Figure 3(b) shows that the band pass filtered signal has a dominant second harmonic. Filters with higher Q-factor can be used to suppress other harmonics, but could cause signal distortion due to reduced bandwidth. Figure 3(c) shows the output of the frequency doubling circuit, which serves as the local oscillator in demodulation. The dominant 2 kHz harmonic is obtained using a high Q-factor band pass filter. Figure 3(d) shows the multiplier output, where the DC component contains demodulated thermal signal. The output of the low pass filter shows that other harmonics can be effectively removed (Fig. 3(e)).

Figure 5 is a comparison of the thermal image with the topographic image obtained using closed loop interface circuit during measurements shown in Fig. 3. The sample was 350 nm thick, developed Shipley UV6™ photoresist, with a 1 μm pitch. The similarity between the two images is self-evident. The somewhat flatter top seen for the ridges in the thermal image is as expected because of the thermal diffusivity of the sample. According to a line scan across the photoresist patterns of Fig.5, the noise-limited minimum detectable thermal conductance change is ≈29 pW/K.

V. CONCLUSION

A scanning thermal imager with micromachined bolometer type probes and a custom interface circuit was described. Unified simulation of the transducer and circuit permits the components to be optimized together. The probe temperature can be precisely controlled by a PI controller while electrical dithering provides relative immunity to thermal bridge noise even for sub-μV low-frequency signals. Scanning thermal images obtained showed a high signal-to-noise ratio of 6 for 350nm UV photoresist in which the minimum detectable thermal conductance change was ≈29 pW/K.

ACKNOWLEDGEMENTS

This work was funded in part by the Semiconductor Research Corporation contract # 98-LP-452.005. Valuable discussions with Profs. F. Cerrina and P. Nealey, and Drs. M.-H. Li and R. Tate, all of UW-Madison, and Dr. L. Ocola of Argonne National Laboratory are gratefully acknowledged.

REFERENCES

[Ci90] A.Cichocki, R.Unbehauen, "Application of switched-capacitor self-oscillating circuits to the conversion of RLC parameters into a frequency or digital signal," *Sensors and Actuators A*, vol. 24, pp. 129-137, 1990

[Fe97] V.Ferrari, C.Ghidini, D.Marioli, A.Taroni, "Conditioning circuit for resistive sensors combining frequency and duty-cycle modulation on the same output signal," *Measurement Science & Technol.*, 8(7), pp. 827-9, '97

[Gi76] B.Gilbert, "A versatile monolithic voltage-to-frequency converter," *IEEE J. Solid-State Circuits*, SC-11(6), pp. 852-64, 1976

[Gi97] Y. Gianchandani, K. Najafi, "Scanning Thermal Profilers with Integrated Sensing and Actuation," *Transactions on Electron Devices*, 44 (11), Nov. 1997, pp. 1857-68

[Go93] F.M.L. van der Goes, P.C.De Jong, G.C.M.Meijer, "Concepts for accurate A/D converters for transducers," *Solid-State Sensors and Actuators (Transducers '93)*, Yokohama, Japan, 1993, pp. 331-4

[Hu87] J.H.Huijsing, G.A.Van Rossum, M.van der Lee, "Two-wire bridge-to-frequency converter," *J. Solid-State Circuits*, SC-22(3), pp. 343-9, 1987

[Le00] J. Lerchner, D. Caspary, G.Wolf, "Calorimetric detection of volatile organic compounds," *Sensors and Actuators B*, 70, pp. 57-66, 2000

[Li00] M.-H.Li, Y.B.Gianchandani, "Microcalorimetry applications of a surface micromachined bolometer-type thermal probe," *J. Vac. Sci. Technol. B*, 18(6), pp. 3600-3, 2000

[Li01] M.-H.Li, J.J.Wu, Y.B.Gianchandani, "Surface micromachined polyimide scanning thermocouple probes," *J. Microelectromech. Sys.*, 10(1), pp. 3-9, 2001

[Li02] M.-H.Li, J.-H.Lee, F.Cerrina, A.K.Menon, Y.B.Gianchandani, "Chemical and biological diagnostics using fully insulated ultracompliant thermal probes," *Proceedings on Solid-State Sensor, Actuator, and Microsystems Conference*, Hilton Head Island, SC, 2002, pp. 235-8

[Ma99] A.Majumdar, "Scanning thermal microscopy," *Annu. Rev. Mater. Sci.*, 29, pp. 505-85, 1999

[Mo95] K.Mochizuki, K.Watanabe, "A linear resistance-to-frequency converter," *Proc., Instrumentation and Measurement Technology Conf. (IMTC)*, Waltham, MA, 1995, pp. 339-43.

[Oc96] L.E.Ocola, D.Fryer, P.Nealey, J.dePablo, F.Cerrina, S. Kämmer, "Latent image formation: Nanoscale topography and calorimetric measurements in chemically amplified resists," *J. Vac. Sci. Technol. B*, 14(6), pp. 3974-9, 1996

[Sa93] G.R.Sarma, "Analysis of a constant voltage anemometer circuit," *Proc., Instrumentation and Measurement Technology Conf. (IMTC)*, Waltham, MA, 1993, pp. 731-6.

[Ve00] P.Vettiger, et al., "The "Millipede"-More than one thousand tips for future AFM data storage," *IBM J. Res. Develop.*, 44(3), pp. 323-40, 2000

[Yo00] D.J.Yonce, P.P.Bey Jr., T.L.J.Fare, "A DC autonulling bridge for real-time resistance measurement," *Trans. Circuits & Systems*, 47(3), pp. 273-7, 2000

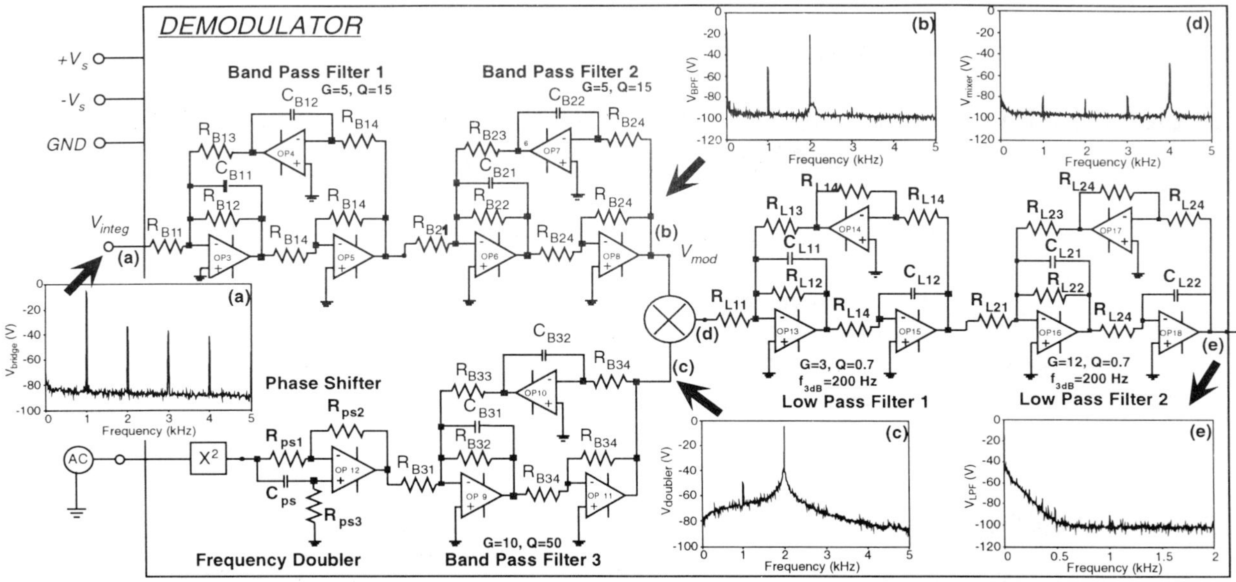

Fig. 3: Demodulator section of the interface circuit in Fig. 1. Embedded frequency spectra were obtained while scanning a real sample, not in a test mode. The scan results are present in Fig. 5.

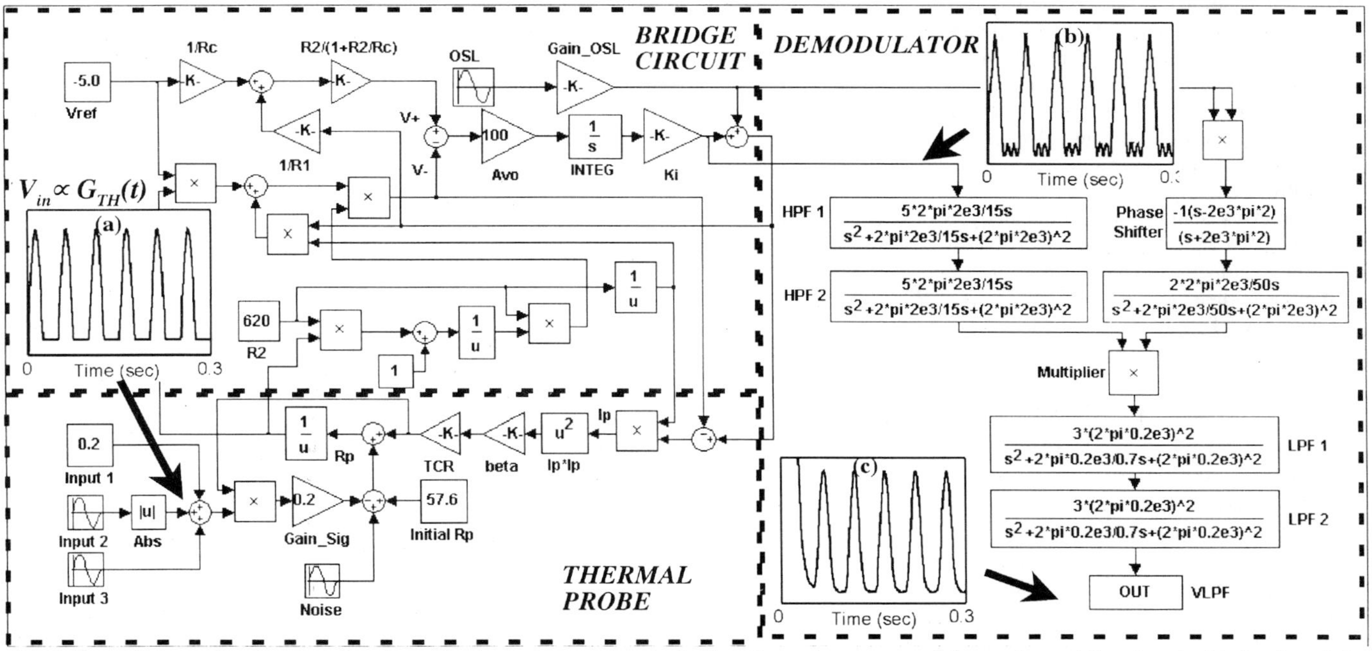

Fig. 4: State diagram for the scanning thermal microscopy system including thermal response of the probe and the electrical behavior of the circuit. The MatLab™ Simulink tool is used to optimize the circuit and evaluate overall performance, including noise immunity.

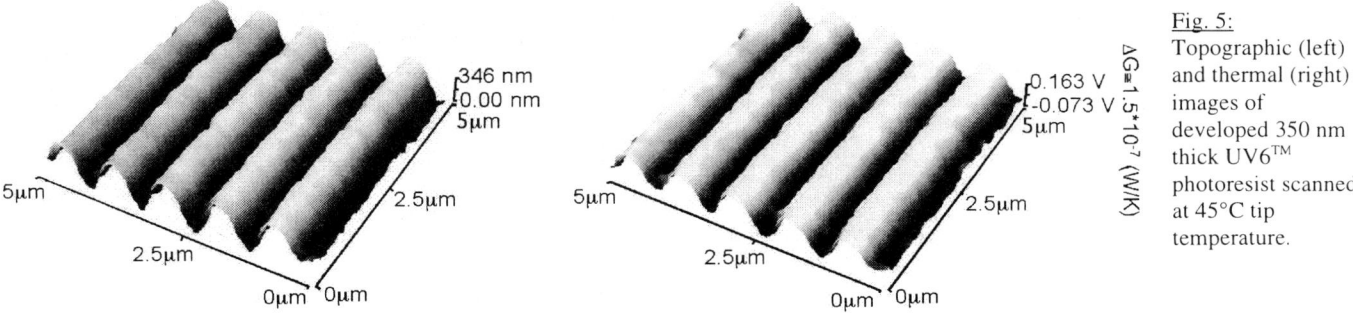

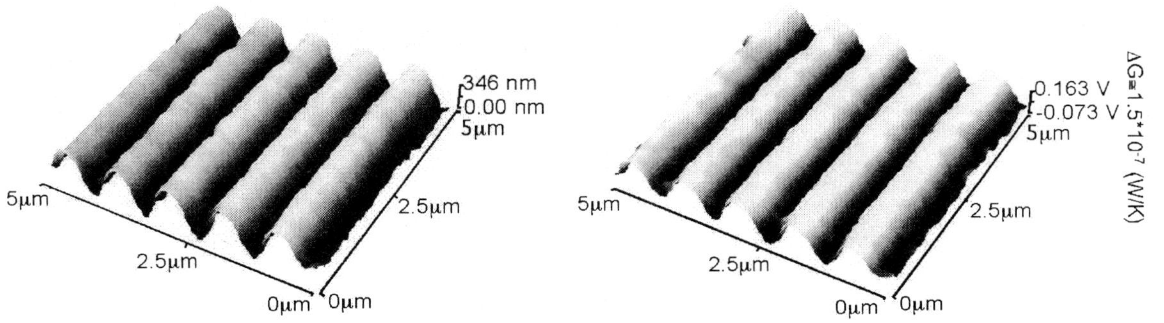

Fig. 5: Topographic (left) and thermal (right) images of developed 350 nm thick UV6™ photoresist scanned at 45°C tip temperature.

CHARACTERIZATION OF IPMC STRIP SENSORIAL PROPERTIES: PRELIMINARY RESULTS

C. Bonomo[1], C. Del Negro[2], L. Fortuna[1], S. Graziani[1]

[1]Università degli Studi di Catania
Dipartimento Elettrico, Elettronico e Sistemistico
Viale A. Doria, 6 - 95125 Catania - Italy. E-mail: cbonomo@dees.unict.it
[2]Istituto Nazionale di Geologia e Vulcanologia di Catania

ABSTRACT

Ionic Polymer Metal Composites (IPMCs) are innovative materials obtained by deposition of a metal on a ionic polymer membrane. IPMCs bend, when an electric field is applied along their thickness, and vice versa generate a voltage when are mechanically deformed. Hence, it makes sense to investigate the opportunity to use IPMCs to build either actuators or sensors.
IPMCs can be of interest because they require few volts to bend. Such an activation voltage is much lower than the one required by other moving actuators (e.g. piezoelectric materials require hundreds of volts).
In this work some results of the analysis of sensing properties of an IPMC strip are presented. Experimental data suggest that the output voltage is roughly linear in deformation. Moreover, a nonlinear behavior seems to occur in particular working conditions. To authors' knowledge such a behavior never before has been reported in literature.

1. INTRODUCTION

Ionic Polymer Metal Composites (IPMCs) are innovative materials obtained by deposition of a metal on a ionic polymer membrane [1]. Some details on their structure are necessary in order to understand their sensing properties.
Ionic polymers have inner ionizable groups. These groups dissociate in a fixed part and in a movable one in a variety of solvent media. Usually, the fixed groups have negative charge while cations can freely move. By mechanically bending the material it is possible to change the distribution of the charges with respect to the membrane neutral axis (see Fig. 1): the applied stress will contract one side of the membrane while will spread the other, the mobile ions will move consequently toward the region characterized by a lower charge density parasitically carrying the solvent molecules (i.e. deionized water).

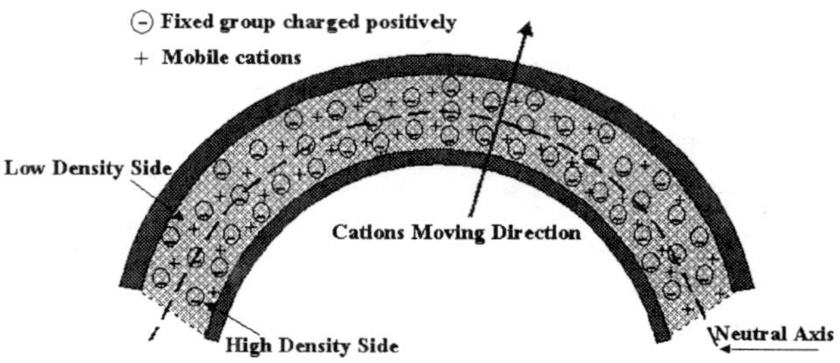

Figure 1. Effect of stress applied to the IPMC strip on charge distribution.

A deficit of negative charges and an excess of the positive ones will therefore result in the expanded side. In the contracted side the opposite will occur. This phenomenon produces a voltage gradient collected at the metal electrodes. It is intuitive as this property results in a sensing capability [2].
In the following a system for the characterization of the sensing properties of an IPMC strip is described. It allows to investigate the dependence of the voltage generated within the strip on the imposed deformation.

2. EXPERIMENTAL SETUP

The experimental analysis was performed on a strip built by using the ionic polymer Flemion^{TM} (by Asahi Glass) with gold deposed on both sides [3], as shown in Fig. 2. The sample was 38 mm long by 6 mm large, its thickness was about 200 μm.
Before each measuring survey, the strip was opportunely hydrated by immersion in deionized water. In fact a dependence of the sensing properties on the solvent contents was noticed and will be described in the following. The hydrated sample presented a weak curvature probably caused by a non uniform initial distribution of the ions.

0-7803-7761-3/03/$17.00 ©2003 IEEE

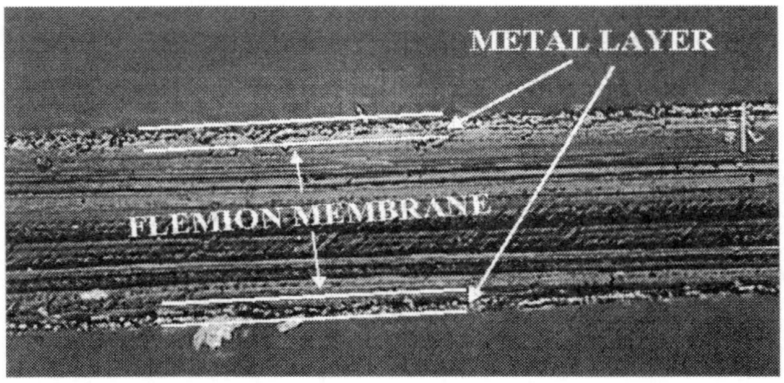

Figure 2. An optical microscopy (20x) of the cross section of the IPMC used for the analysis described in the paper. Gold distribution after deposition on the Flemion™ membrane surface is pointed out.

The experimental setup is composed of a system to impose a displacement to the membrane tip and a circuit to amplify resulting output voltage. The schematic of the experimental setup is shown in Fig. 3.

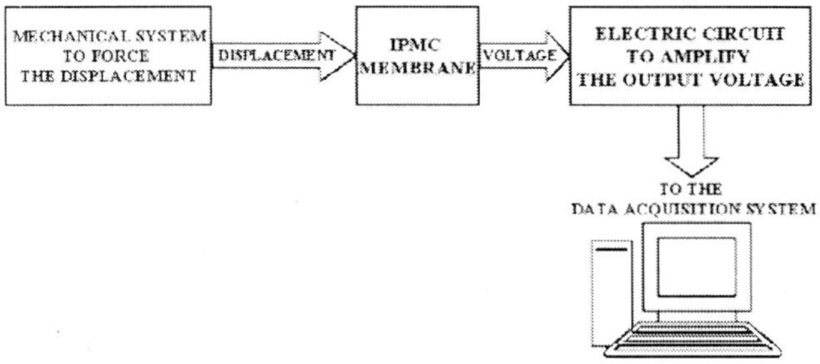

Figure 3. Schematic of the experimental setup used to characterize the sensing properties of an IPMC strip.

The strip was fixed from one end in a cantilever configuration by a clip with cupper contacts, wired to the electronic circuit. The other end was pinned to the wiper of a linear potentiometer. The wiper was moved by a rod connected to a DC motor working at different frequencies, as shown in Fig. 4.

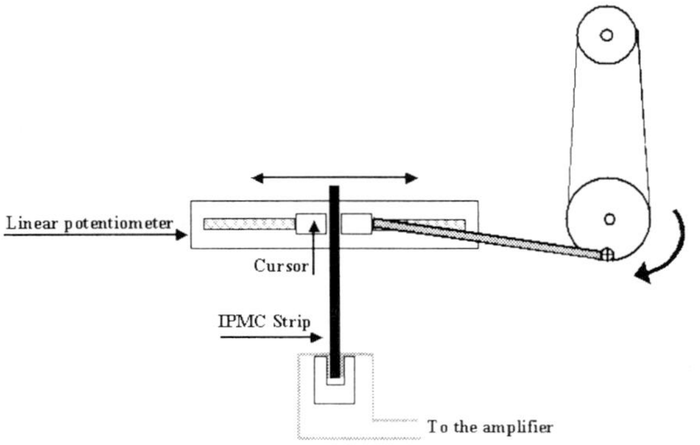

Figure 4. Schematic of the mechanical system used to force the displacement on the IPMC strip.

The strip was inserted at all times in such a way that its natural curvature was always in the positive direction of the x' axis (Fig. 5). While the motor turns, it moves the rod and the rotational movement is converted in a translatory one by the wiper of the potentiometer and consequently the membrane is bent on both direction.

This causes both a production of a voltage signal V_{IPMC} across the membrane thickness and a variation of the voltage V_l across the potentiometer. The membrane output signal was opportunely amplified. The gain was fixed to 270, in order to raise the voltage magnitude from less than 1 mV to about 300 mV. Then the amplifier output was acquired by the data acquisition board AT-MIO 16E-10 Series (National Instruments) from a Personal Computer. The potentiometer was used both to move the strip and to transduce its position into a voltage signal, as shown in Fig. 5. It must be noticed that, based on the chosen reference directions, a positive strip displacement corresponds to a negative voltage across it.

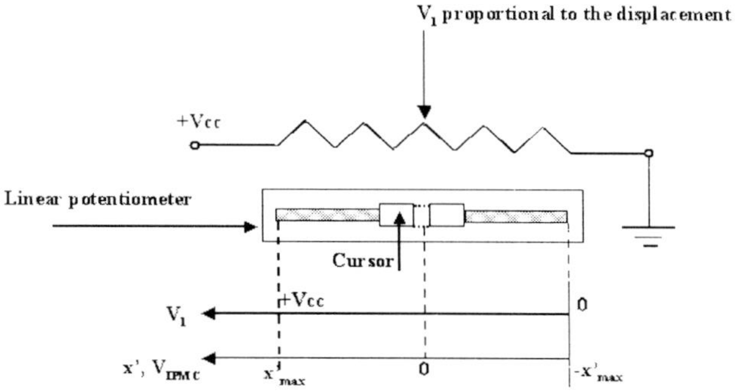

Figure 5. Electric circuit to convert the IPMC displacement into a voltage signal to be related to the voltage generated across the strip. Based on the chosen reference directions, a positive strip displacement corresponds to a negative voltage across it.

A number of data acquisition sessions were performed, tuning the DC voltage applied to the motor and hence changing the frequency of the IPMC strip excitation signal.

Actually, the excitation point on the strip varies while the rod moves the cursor of the potentiometer. However it is possible to neglect this phenomenon being the distance δ between the two farthest excitation points, that is a measure of this variation, much smaller than the distance between the fixed end and the excited one (see Fig. 6). The wiper displacement, referred to the central position, was $\pm$1.2 cm.

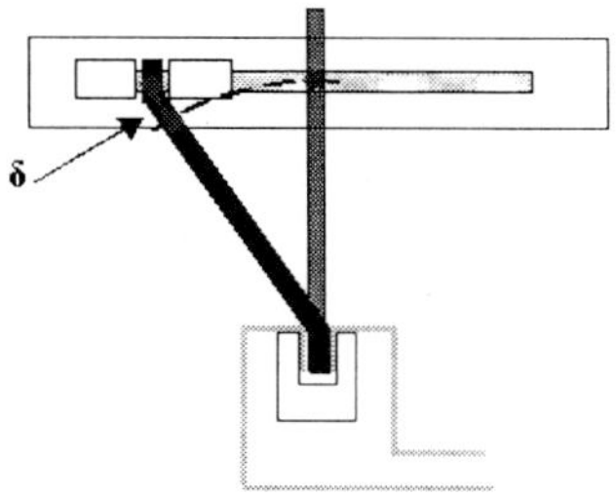

Figure 6. Variation of the excitation point due to the mechanical system. It is negligible referred to the entire length of the IPMC sensor.

3. DATA ANALYSIS

One thousand samples per second were acquired for each frequency and each acquisition lasted ten seconds, i.e. ten thousand samples for each acquisition sessions were collected. The range of the tested frequencies spanned from 1.0 Hz to 10.0 Hz, with steps of 0.5 Hz. These values were fixed because of the inertial constraints imposed by the DC motor. Recorded data were analyzed by using LabVIEW™ software.

The voltage signal V_1, picked up by the wiper of the potentiometer, is a wave oscillating approximately in the range 2.0 ÷ 6.0 V. This signal was translated into a symmetric voltage V_2 and then attenuated to the range −1.2 ÷ 1.2 V in order to give the position of the membrane tip x' directly in centimeters.

A second manipulation block was developed to estimate the displacement of a fixed point of the strip. Referring to Fig. 7, it corresponds to the estimation of the displacement x based on x'. The transfer function of such a block can be obtained by some geometric consideration:

i) $x / x' = A / A'$;

ii) $A' = \sqrt{A^2 + x'^2}$

then

$$x = \frac{2.2 * x'}{\sqrt{4.84 + x'^2}} \quad (1)$$

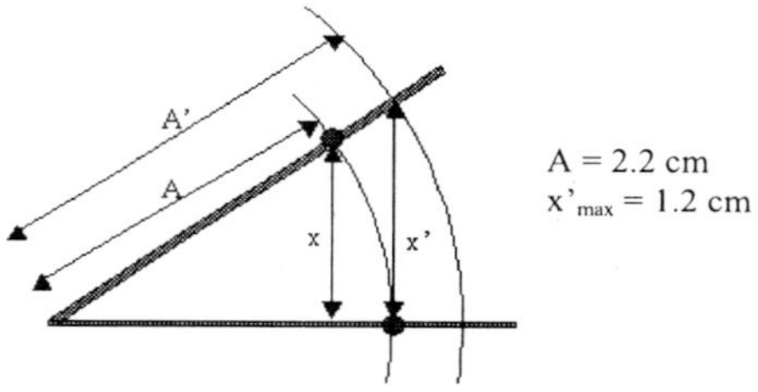

Figure 7. Dependence of the displacement x of a fixed point of the strip on the displacement x' forced by the potentiometer wiper position.

4. RESULTS

Some typical examples of collected data sets are described in the following. In particular the influence of the input signal frequency is shown in Fig. 8. Both the time plots of the input x and of the output V_{IPMC} signals and the corresponding input-output plot are shown. In Fig 8a the input signal frequency is 3.69 Hz, in Fig. 8b it is 4.9 Hz, and in Fig. 8c it is 5.6 Hz.
It can be noticed that the potentiometer output shows a threshold phenomenon at its maximum value. This effect is due to the mechanical resistance met from the rod to invert the slider's motion direction and it is more evident at the lower frequencies. The presence of a nonlinearity has to be pointed out on the sensor output signal each time the wiper crosses the zero position superimposed to a ringing signal. Also, the non linearity effect looks more evident at lower frequencies. The phenomenon could be attributed to the initial deformation of the strip.

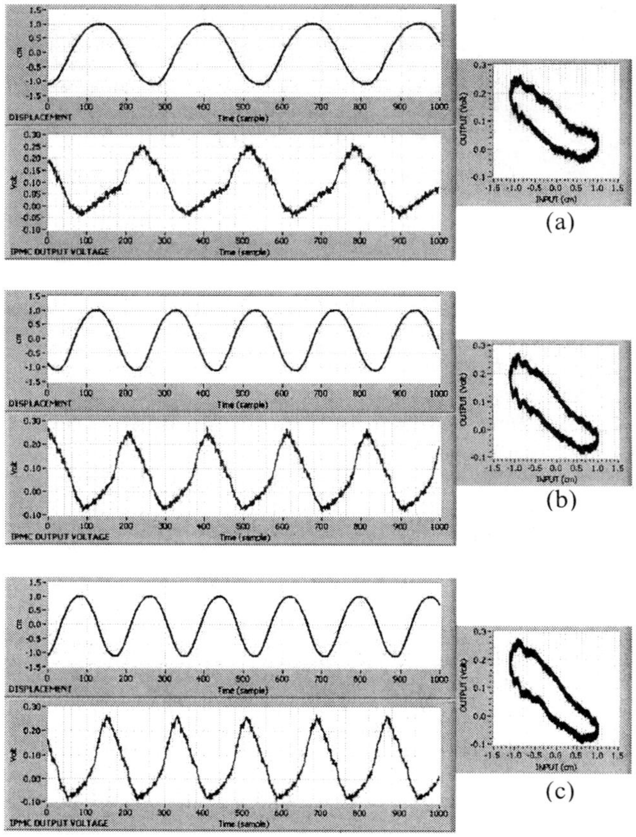

Figure 8. Time plots of the input x and of the output V_{IPMC} signals and the corresponding input-output plot: (a) at 3.69 Hz, (b) at 4.9 Hz, and (c) at 5.6 Hz.

The analysis of acquired data suggests that the input-output peak-to-peak amplitude ratio is constant in the considered range of frequencies.
Also, it is possible to observe the presence of a time delay between the imposed displacement and the voltage response of the membrane. The time delay t_d between the stimulus and the sensor response is frequency dependent as it can be noticed in the data reported in Fig. 8. It seems to decrease as the excitation frequency increase. Indeed as regards the case (a) t_d is 220 ms, in the case (b) it is 170 ms, and in the case (c) t_d is 160 ms.
The influence of the hydration level on the strip behavior was also addressed. For the sake of comparison the same set of input signal frequencies was considered also in the case of this analysis.
In Fig. 9 the time plots of the input x and of the output V_{IPMC} signals and the relative input-output plot when the strip is partially dehydrated and the input frequency is 5.6 Hz are shown (see Fig. 8c).

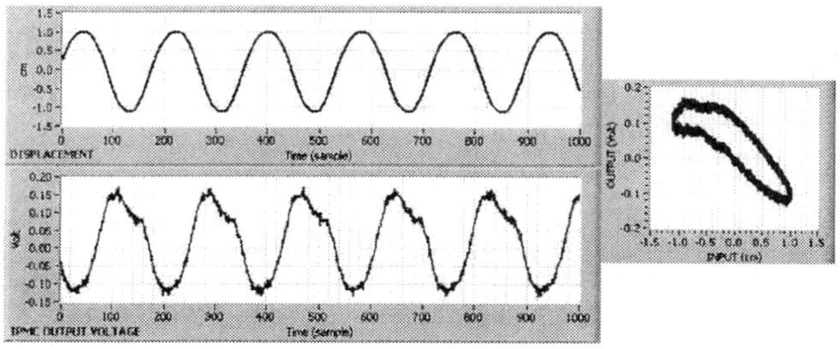

Figure 9. Time plots of the input x and of the output V_{IPMC} signals and the relative input-output plot when the strip is partially dehydrated and the input frequency is 5.6 Hz are shown (see Fig. 8c).

In Fig. 10 a photo of the discussed experimental setup is shown.

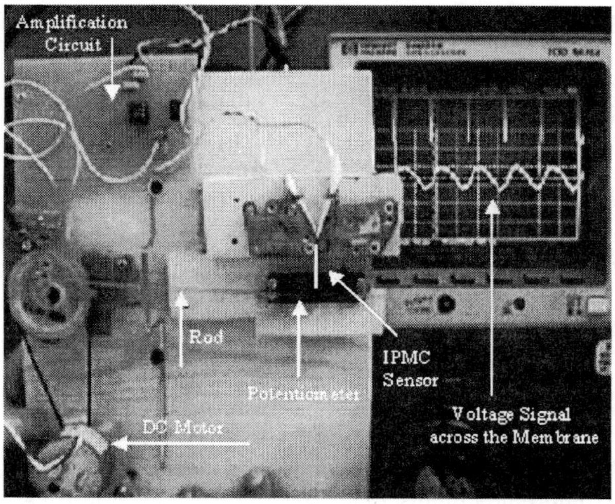

Figure 10. Photo of the experimental setup. The reader can see the mechanical system that forced the displacement and the amplification circuit for the sensor output.

5. CONCLUSIONS

In this paper some preliminary results about the sensing properties of an IPMC strip have been presented. A system to gather the required data has been built.
The dependence of a strip of IPMC on both the input signal frequency and on the hydration level of the strip have been addressed.
Further work to extend the frequency range of the acquired data in order to better characterize the sensor is in progress.
A full comprehension of the behavior of IPMCs sensing capability and a corresponding metrological characterization could be useful for their application to build sensing elements in the geophysical field, e.g. to record earthquakes and study their nature, especially because these materials are able to survey also very small displacement, at very low frequencies.

REFERENCES

[1] P. Arena, C. Bonomo, L. Fortuna, M. Frasca, "Electro-Active Polymers as CNN actuator for locomotion control", *Proceedings of ISCAS 2002, Phoenix, May 26-29.*

[2] M. Shahinpoor, Y. Bar-Cohen, T. Xue, J. O. Simpson, and J. Smith: "Ionic Polymer-Metal Composites (IPMC) as Biomimetic Sensors and Actuators", *Proceedings of SPIE's 5th Annual International Symposium on Smart Structures and Materials, San Diego, CA, March 1-5, 1998.*

[3] Yoseph Bar-Cohen1, Xiaoqi Bao, Stewart Sherrit and Shyh-Shiuh Lih: "Characterization of the Electromechanical Properties of Ionomeric Polymer-Metal Composite (IPMC)", *Proceedings of the SPIE Smart Structures and Materials Symposium, EAPAD Conference, San Diego, CA, March 18-21, 2002*

A LOW-POWER ADAPTIVE INTEGRATE-AND-FIRE NEURON CIRCUIT

Giacomo Indiveri

Institute of Neuroinformatics, University of Zurich and ETH Zurich
Winterthurerstrasse 190, CH-8057 Zurich, Switzerland
giacomo@ini.phys.ethz.ch

ABSTRACT

We present a low-power analog circuit for implementing a model of a leaky integrate and fire neuron. Next to being optimized for low-power consumption, the proposed circuit includes elements for implementing *spike frequency adaptation*, for setting an arbitrary *refractory period* and for modulating the neuron's threshold voltage. We present experimental data from a prototype chip, implemented using a standard 1.5μm CMOS process.

1. INTRODUCTION

Integrate and fire (I&F) circuits typically integrate small currents onto a capacitor until a threshold is reached. As the voltage on the capacitor exceeds the threshold a fast digital pulse is generated to signal the occurrence of a spike, or event, and the capacitor is reset. These circuits are generally integrated into large arrays, on *neuromorphic* devices that implement networks of spiking neurons, or that use spiking elements to transmit sensory signals to other neuromorphic processing elements.

The asynchronous communication protocol used to interface neuromorphic devices containing spiking elements is based on the *Address-Event Representation* (AER) [1]. In this representation input and output signals (address-events) are sent from/to VLSI devices using stereotyped non-clocked pulse-frequency modulated signals that encode the address of the sending node. Analog information is carried in the temporal structure of the inter-pulse intervals, very much as it is believed to be conveyed in spike trains of biological neural systems.

As the AER communication protocol is being embraced by a growing number of research groups [2]-[7], hardware implementations of AER pulse-based neural networks [8] and of vision sensors containing photoreceptors interfaced to spiking elements [9, 10] are becoming more and more widespread. Furthermore, continuous improvements in VLSI technology allow for the fabrication of AER devices containing thousands of elements, operating in parallel. These devices will be practically realizable only if the pulse generating elements have minimal power consumption (locally) and implement pulse-frequency saturation and adaptation mechanisms that limit and reduce the power consumption globally, and that optimize communication bandwidth for the transmission of address-events. To this end, we propose a compact, low power model of a leaky integrate and fire circuit, with spike-frequency adaptation, refractory period and voltage threshold modulation properties. Several circuits for implementing models of I&F neurons have already been proposed [9],[11]-[14]. In the following Section we briefly describe the main differences between previous circuit models of I&F neurons with the one presented here. In Section 3 we describe in detail the characteristics of the circuit and present experimental data measured from a prototype chip fabricated using a standard 1.5μm CMOS technology. Finally in Section 4 we draw the concluding remarks.

2. RELATION TO PREVIOUS WORK

Neuromorphic circuits implementing models of I&F neurons are inherently mixed-mode circuits, as they generate fast digital pulses from slowly changing analog voltages. The most direct way to implement a mixed-mode circuit that generates a pulse when an analog voltage exceeds a threshold is to connect the integrating capacitor directly to the input of an inverter. This is precisely what happens in the "Axon-Hillock" circuit, described in [11]. This circuit is very compact, comprising only six transistors and two capacitors, but it has a major drawback: it dissipates non-negligible amounts of power. This is is due to the fact that the input to the inverter (the voltage on the capacitor) changes typically with time constants of the order of milliseconds, and the inverter spends a large amount of time in the region in which both transistors conduct a short-circuit current. Next to its large-power consumption drawbacks, the Axon-Hillock circuit has a spiking threshold that depends only on CMOS process parameters (the switching threshold of the inverter), and lacks spike-frequency adaptation properties. An alternative design that has both the possibility to set an explicit threshold voltage and that implements spike-frequency adaptation is proposed in [12].

0-7803-7761-3/03/$17.00 ©2003 IEEE

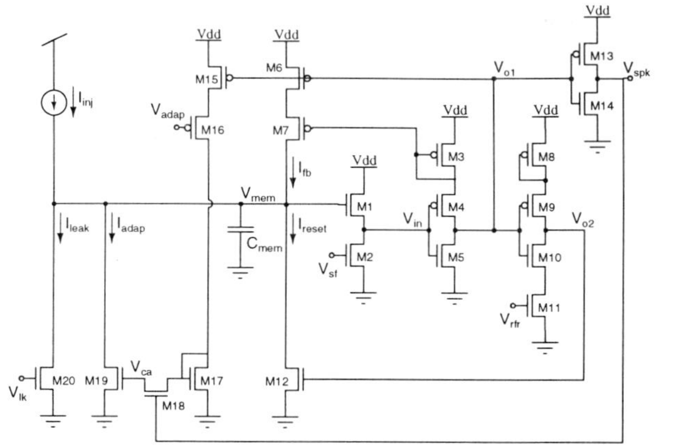

Figure 1: Circuit diagram of the integrate-and-fire neuron.

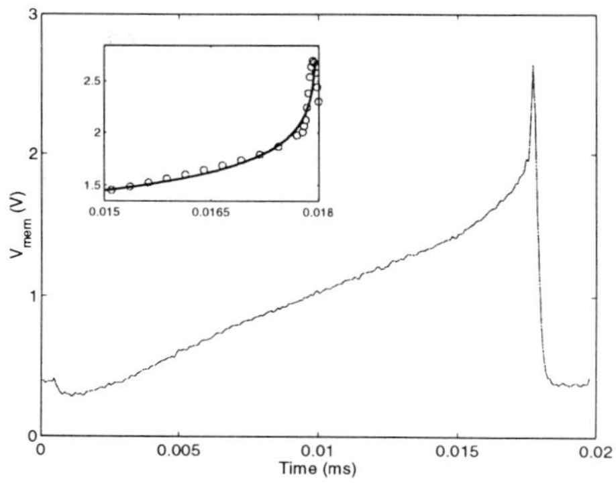

Figure 2: Measurement data showing a typical shape of a spike. The inset shows a fit of the measured data with equation (1).

This design, however, is also undermined by large power consumption, due to the very same problems present in the Axon-Hillock circuit. In [14] van Schaik proposed a circuit with an amplifier at the input, to compare the voltage on the capacitor with a desired spiking threshold voltage. As the input exceeds the spiking threshold, the amplifier drives the inverter, making it switch very rapidly. This circuit consumes less power than previously proposed ones, but has not been optimized explicitly for power consumption, and still lacks a spike-frequency adaptation mechanism. In [13] Boahen demonstrates how it is possible to implement spike-frequency adaptation by connecting a four transistor "current-mirror integrator" in negative-feedback mode to any I&F circuit. And in [9] the authors specifically address the problem of power consumption in I&F circuits (following preliminary results obtained by some of the participants of the 1999 Neuromorphic Engineering Workshop). The I&F circuit proposed in [9] did not include spike-frequency adaptation mechanisms, nor voltage threshold modulation ones, nor refractory periods, nor ways to implement explicit leak currents. The circuit we propose uses the same design tricks proposed in [9] to minimize power consumption, but also includes all of the properties mentioned above.

3. CIRCUIT IMPLEMENTATION AND MEASUREMENTS

The low-power circuit that implements the model of a leaky I&F neuron is shown in Figure 1. It comprises twenty transistors and one (explicit) capacitor. Two additional parasitic (implicit) capacitors are exploited at nodes V_{o2} and V_{ca} (see text below). The circuit can be subdivided in six main blocks: a source follower M1-M2, for increasing the linear integration range and for modulating the neuron's threshold voltage; an inverter with positive feedback M3-M7, for reducing the switching short-circuit currents at the input; an inverter with controllable slew-rate M8-M11, for setting arbitrary refractory periods; a digital inverter M13-M14, for generating the fast digital pulse that signals the occurrence of a spike; a transient current-mirror integrator M15-M19, for implementing the spike-frequency adaptation mechanism, and a minimum size transistor M20 for implementing a constant current leak.

3.1. Voltage threshold modulation and positive feedback

If V_{mem} is sufficiently low, the input current I_{inj} is integrated linearly by C_{mem}. The source-follower M1-M2, driven by V_{mem} generates the signal $V_{in} = \kappa(V_{mem} - V_{sf})$, where V_{sf} is a constant sub-threshold bias voltage and κ is the sub-threshold slope coefficient [15]. By changing V_{sf} one can change the neuron's threshold voltage and use this property to model long-term adaptation effects in cortical cells, or to reproduce traveling waves or global oscillations in the whole population of I&F neurons. As V_{mem} increases and V_{in} approaches the threshold voltage of M5, the current through M3 starts to increase, and V_{o1} starts to decrease. Consequently the feedback current I_{fb} starts to increase V_{mem} and V_{in} more rapidly. As V_{in} (and V_{o1}) approach $V_{dd}/2$, the feedback current increases, reaching a maximum value and decreasing again as V_{in} crosses $V_{dd}/2$ while approaching V_{dd}. The positive feedback has the effect of making the inverter M3-M5 switch very rapidly, reducing dramatically its power dissipation. It can be shown that to a first order approximation the positive feedback has the effect of changing the profile of $V_{mem}(t)$ from linear (for a constant I_{inj}) into a profile of the type

$$c_1 t + c_2 \ln\left(\frac{c_3}{e^{(t-t_0)} - 1}\right) \tag{1}$$

when t is close to the spike emission time t_0. The parameters c_1, c_2, and c_3 are proportionality constants. Figure 2

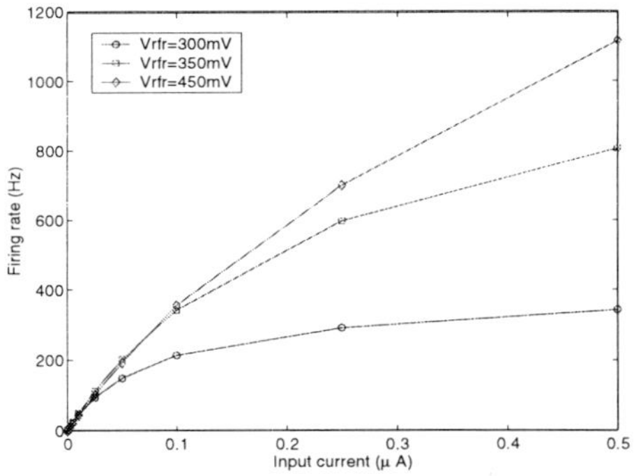

Figure 3: Firing rate (frequency of emitted spikes) versus input current I_{inj} for different refractory period settings.

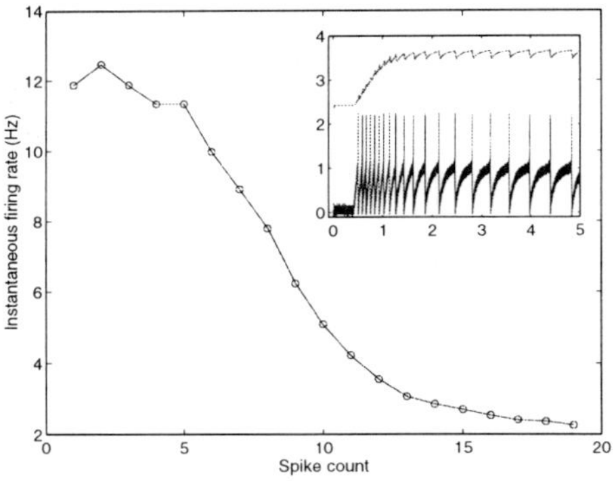

Figure 4: Spike frequency adaptation.

shows a spike trace, measured from the circuit. The inset of the figure shows the experimental data fitted with equation (1).

3.2. Refractory period

The spike is emitted when V_{mem} drives V_{o1} sufficiently low to make V_{spk} and V_{o2} switch to V_{dd}. When V_{o2} is high, the transistor M12 is fully open and C_{mem} is discharged, bringing V_{mem} rapidly to Gnd (see the falling edge of the trace in Figure 2). As V_{mem} (and V_{in}) go to ground, V_{o1} goes back to V_{dd}. The voltage V_{o2} is then discharged through the path M10-M11, at a rate set by V_{rfr} (and by the parasitic capacitance on node V_{o2}). As long as V_{o2} is high, V_{mem} is clamped to ground. During this "refractory" period, the neuron cannot spike, as all the input current I_{inj} is absorbed by M12. Figure 3 shows how different refractory period settings (V_{rfr}) saturate the maximum firing rate of the circuit at different levels.

3.3. Spike frequency adaptation

During the spike emission period (for as long as V_{spk} is high), a current with maximum amplitude set by V_{adap} is sourced into the parasitic capacitance of node V_{ca} of Figure 1 (see M15-M19). Thus, the voltage V_{ca} increases with every spike, and leaks to zero through dark (leakage) currents when there is no spiking activity. As V_{ca} increases, a negative adaptation current I_{adap} exponentially proportional to V_{ca} is subtracted from the input, and the spiking frequency of the neuron is reduced over time. The main difference between the spike-frequency adaptation mechanism described here and the one proposed in [13] is the presence of the transistor M18. In [13] transistors M17 and M19 are connected directly, and the capacitance on node V_{ca} is discharged by the diode-connected transistor M17. The presence of the switch M18 increases the effective time constant of the integrator, and thus allows for a more compact implementation, using smaller transistor geometries and corresponding parasitic capacitances.

3.4. Power consumption

The main sources of power dissipation are in the short-circuit currents that flow through the inverters during the switching time and in the DC current that flows through the source-follower M1-M2 during the periods in which $V_{mem} > (V_{sf} + 4U_T)$, where U_T is the thermal voltage and V_{sf} is the source follower's bias voltage (Fig. 1). When no input current is applied (in resting conditions) the leakage current brings V_{mem} to zero, transistor M1 of the source follower does not conduct, and the inverters do not switch. In this condition the power dissipation is null. When input current is applied, the power dissipation is a function of firing rate, of V_{sf}, and of the power supply voltage V_{dd}. In Figure 5 we set V_{sf} to 0.65V, V_{dd} to 5V and injected input currents of increasing amplitudes. The plot shows how the current flowing through the V_{dd} node increases with the neuron's firing rate (switching frequency). To evidence how the power dissipation depends on V_{sf}, we adjusted the injection current in a way to maintain the neuron's firing rate constant

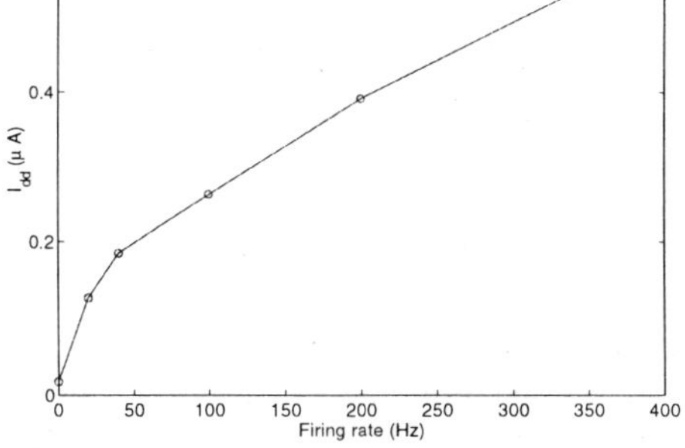

Figure 5: Dissipated current as a function of firing rate.

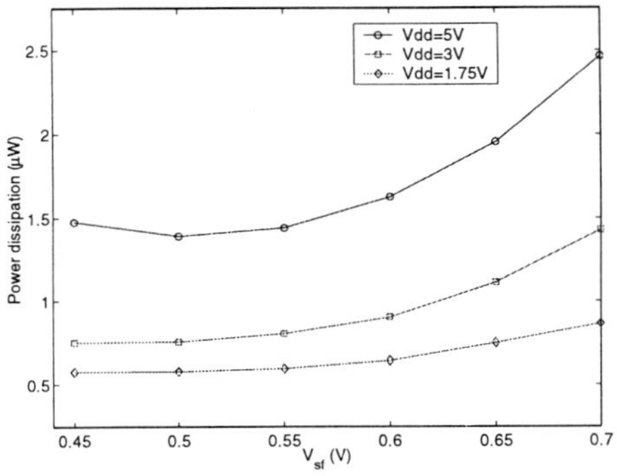

Figure 6: Power dissipation of the neuron firing at a constant rate of 200Hz, as a function of V_{sf}.

at 200Hz, while systematically increasing V_{sf} from 0.45V to 0.7V. In Figure 6 we show the data measured from the circuit in this experimental setup, for three different values of V_{dd}. The circuit was able to operate correctly with power supply voltages as low as 1.75V. As expected, power dissipation increases for high values of V_{sf}, given a fixed firing rate. But for very low values of V_{sf} the dissipation can increase again (top trace of Fig. 6). This is due to the fact that when V_{sf} assumes very low values, the source follower acts as a low-pass filter on the falling edge of the spike and the switching time of the first inverter is increased. For low values of V_{sf} and low values of V_{dd} the power dissipation is dominated by the short-circuit switching currents, and depends mainly on the neuron's firing rate (constant in Fig. 6). In typical applications the refractory period is used to limit the *maximum* firing rate of the neuron to approximately 200Hz, and the spike-frequency adaptation mechanism is used to reduce the *average* firing rate of the neuron below 10Hz. In these conditions, with V_{dd} set to 5V and V_{sf} to 0.55V, the average power consumption of the neuron is 300nW, and can be at maximum 1.5μW. For comparison, simulations of the Axon-Hillock circuit in comparable operating conditions give average power-dissipation figures greater than one order of magnitude; similarly, extrapolation of the power dissipation of the circuit presented in [9], from the simulated values that the authors provide, shows how the circuit we propose is at least 40 times more efficient.

4. CONCLUSIONS

We presented a low-power circuit for implementing a model of a leaky integrate and fire neuron with the capability of setting a refractory period and with a spike-frequency adaptation mechanism. Both refractory period settings and spike-frequency adaptation are useful tools for limiting and further reducing the power consumption of the circuit. We described the circuit schematics and presented experimental data characterizing all aspects of the circuit proposed. We plan to integrate the I&F neuron circuit in chips containing large arrays of neurons and synapses, and AER interfacing circuits, for constructing neuromorphic multi-chip systems.

5. REFERENCES

[1] M. Mahowald, *An Analog VLSI System for Stereoscopic Vision*, Kluwer, Boston, 1994.

[2] K.A. Boahen, "Communicating neuronal ensembles between neuromorphic chips," in *Neuromorphic Systems Engineering*, T. S. Lande, Ed., pp. 229–259. Kluwer Academic, Norwell, MA, 1998.

[3] A. Mortara, "A pulsed communication/computation framework for analog VLSI perceptive systems," in *Neuromorphic Systems Engineering*, T. S. Lande, Ed., pp. 217–228. Kluwer Academic, Norwell, MA, 1998.

[4] S. R. Deiss, R. J. Douglas, and A. M. Whatley, "A pulse-coded communications infrastructure for neuromorphic systems," in *Pulsed Neural Networks*, W. Maass and C. M. Bishop, Eds., chapter 6, pp. 157–178. MIT Press, 1998.

[5] C. Higgins and C. Koch, "Multi-chip motion processing," in *Proc. Conf. Advanced Research in VLSI*, Atlanta, GA, 1998.

[6] T. Serrano-Gotarredona, A. G. Andreou, and B. Linares-Barranco, "AER image filtering architecture for vision-processing systems," *IEEE Trans. Circ. Sys. I*, vol. 46, no. 9, pp. 1064–1071, Sep 1999.

[7] G. Indiveri, "Modeling selective attention using a neuromorphic analog VLSI device," *Neural Computation*, vol. 12, no. 12, pp. 2857–2880, December 2000.

[8] D.H. Goldberg, G. Cauwenberghs, and A.G. Andreou, "Probabilistic synaptic weighting in a reconfigurable network of VLSI integrate-and-fire neurons," *Neural Networks*, vol. 14, no. 6–7, pp. 781–793, Sep 2001.

[9] E. Culurciello, R. Etienne-Cummings, and K. Boahen, "Arbitrated address-event representation digital image sensor," *Electronics Letters*, vol. 37, no. 24, pp. 1443–1445, Nov 2001.

[10] J. Kramer, "An ON/OFF transient imager with event-driven, asynchronous readout," in *2002 IEEE International Symposium on Circuits and Systems*, May 2002.

[11] C.A. Mead, *Analog VLSI and Neural Systems*, Addison-Wesley, Reading, MA, 1989.

[12] S. R. Schultz and M. A. Jabri, "Analogue VLSI 'integrate-and-fire' neuron with frequency adaptation," *Electronic Letters*, vol. 31, no. 16, pp. 1357–1358, Aug 1995.

[13] K.A. Boahen, *Retinomorphic Vision Systems: Reverse Engineering the Vertebrate Retina*, Ph.D. thesis, California Institute of Technology, Pasadena CA, 1997.

[14] A. van Schaik, "Building blocks for electronic spiking neural networks," *Neural Networks*, vol. 14, no. 6–7, pp. 617–628, Jul–Sep 2001.

[15] S.-C. Liu, J. Kramer, G. Indiveri, T. Delbruck, and R. Douglas, *Analog VLSI: Circuits and Principles*, MIT Press, 2002.

EVIDENCE UPDATING IN A HETEROGENEOUS SENSOR ENVIRONMENT

K. Premaratne and D.A. Dewasurendra

University of Miami
Coral Gables, Florida USA
kamal@miami.edu

P.H. Bauer

University of Notre Dame
Notre Dame, Indiana USA
pbauer@mars.ee.nd.edu

ABSTRACT

Nodes in a distributed sensor network (DSN) gather and fuse information generated by heterogeneous sources to arrive at local decisions targeted at achieving a given global mission objective. These sources typically possess very diverse scopes of 'expertise' or frames of discernment (FoDs) that renders updating a knowledge base from the evidence received a challenging task. In this paper, we present a Dempster-Shafer (DS) theory based evidence updating strategy that accommodates such non-identical FoDs. It is composed of a linear combination of the available evidence and incoming evidence conditioned to the source it is being generated from. The linear combination weights can be used to accommodate differences in source reliability and 'inertia' of the existing knowledge base. Strategies to choose these weights are also proposed.

1. INTRODUCTION

DSNs are complex systems of information processing that show a great promise in a wide range of applications. DSN nodes gather information from a wide variety of heterogeneous sources (e.g., sensors, expert opinions, large databases, etc.), fuse this information, extract and share knowledge to arrive at local decisions targeted at some global mission objective. Development of effective methodologies for information fusion and decision making at each DSN node is hence a priority.

Most available evidence fusion methods are broadly based on probability, possibility and fuzzy theory based techniques. The theory of belief functions, or DS evidential reasoning theory [1], fall into the second group. It has been successfully applied in a wide variety of a applications [2]-[6]. The main advantage of DS theory lies in its ability to numerically quantify the lack of knowledge in an effective manner. Most of these applications however assume that each source possess identical scopes of 'expertise' or FoDs, or has prior knowledge of the global FoD. This constraint, a prerequisite of the evidence combination function typically utilized in DS theory, is often unrealistic especially in DSNs where no single node can be expected to have access to the complete body of evidence. Moreover, maintaining awareness of the global FoD is difficult, or even impossible, especially when the DSN has to keep abreast of highly dynamic global mission objectives [7].

In this paper, we present a DS theory based evidence updating strategy that accommodates sensors possessing non-identical FoDs. It can also account for both the differences in source reliability and/or the 'inertia' of the available knowledge base. How 'flexible' the decision node is for updating depends on its perception of the 'reliability' of the source providing the new information. In situations where the original knowledge base is the result of perhaps a vast amount of evidence, its inertia should not be ignored while updating. The evidence updating strategy we propose is a linear combination of the available evidence (within its own FoD) and incoming evidence (conditional to the source FoD). The flexibility of the available evidence to updating and/or its inertia can then be accounted for via these linear combination weights. Various strategies to choose these weights are also proposed. The updating strategy we propose constitutes a generalization of a result that appears in [8] which possesses an interesting probabilistic interpretation.

Section 2 has a basic review of DS theory; Section 3 presents the proposed updating strategy; Section 4 suggests various strategies for selecting the linear combination weights; Section 5 discusses how the proposed method may be applied in a DSN environment together with a simulation result; finally, Section 6 provides concluding remarks.

This work was supported by NSF Grant #s ANI-9726253 (at University of Miami) and ANI-9726247 (at University of Notre Dame).

2. PRELIMINARIES

In DS theory, support or mass assignments to non-singleton propositions from the sample space generate a notion of uncertainty. Objects for which there is no information are not assigned an *a-priori* mass. Committing support for an event does not necessarily imply that the remaining support is committed to its negation; the lack of support for any particular event simply implies support for all other events, viz., the additivity axiom in the probability formalism is relaxed in DS theory.

2.1. DS Theory: A Primer

We denote the total set of mutually exclusive and exhaustive propositions via $\Theta = \{\theta_1, \ldots, \theta_n\}$, viz., Θ is the corresponding FoD signifying the scope of 'expertise'. A singleton θ_i represents the lowest level of discernible information. Given[1] $|\Theta| = n$, 2^Θ denotes its power set. We use $A - B$, $A, B \subseteq \Theta$, to denote all propositions in A after removal of those propositions that may imply B, i.e., $A - B = \{\theta : \theta \in A,\ \theta \notin B\}$; $\overline{A}$ denotes $\Theta - A$. Elements in 2^Θ form all the propositions of interest in DS theory; the mass assigned to a proposition is free to move into the individual singleton objects that constitute the composite proposition thus generating the notion of *ignorance.* The support for any such proposition A is provided via a *basic belief assignment (bba):*

Definition 1 (BBA) *The mapping* $m : 2^\Theta \mapsto [0,1]$ *is a* bba *for the FoD* Θ *if: (i)* $m(\emptyset) = 0$*; and (ii)* $\sum_{A \subseteq \Theta} m(A) = 1$.

[1] $|\Theta|$ denotes the cardinality of set Θ.

0-7803-7761-3/03/$17.00

The set of propositions $\mathcal{F}(\Theta)$ that possesses nonzero bba s forms the *focal elements* of Θ; the triple $\{\Theta, \mathcal{F}, m\}$ is the corresponding *body of evidence (BoE)*. The quantity $m(A)$ measures the support assigned to proposition A *only;* the belief assigned to A on the other hand must take into account the supports for all proper subsets of A as well, i.e., $\text{Bel} : 2^\Theta \mapsto [0,1] : A \mapsto \text{Bel}(A) = \sum_{B \subseteq A} m(B)$. It represents the total support that can move into A without any ambiguity. The extent to which one doubts a proposition can be quantified via $\text{Dou} : 2^\Theta \mapsto [0,1] : A \mapsto \text{Dou}(A) = \text{Bel}(\overline{A})$. Then, the extent to which one finds a proposition plausible may be quantified via $\text{Pl} : 2^\Theta \mapsto [0,1] : A \mapsto \text{Pl}(A) = 1 - \text{Dou}(A) = 1 - \text{Bel}(\overline{A})$. Note that $\text{Pl}(A) = \sum_{B \cap A \neq \emptyset} m(B)$ and $\text{Pl}(A) \geq \text{Bel}(A)$.

When each focal set contains only one element, i.e., $m(A) = 0,\ \forall |A| \neq 1$, belief functions become probability functions. In such a case, the bba, belief and plausibility all reduce to probability, i.e., $m(A) = \text{Bel}(A) = \text{Pl}(A) = \text{Prob}(A)$.

Dempster's combination function allows one to combine evidence generated from two BoEs. However, its most serious limitation is its requirement that they span the *same* FoD [1].

2.2. Conditional Notions in DS Theory

Conditional notions within the context of DS theory appear in [9]. Let $\hat{\mathcal{F}}(\Theta) \doteq \{A \subseteq \Theta : \text{Bel}(A) > 0\}$. Than we have

Theorem 1 *[9] Given a BoE $\{\Theta, \mathcal{F}, m\}$ and $A \in \hat{\mathcal{F}}(\Theta)$, the conditional belief $Bel(B|A) : 2^\Theta \mapsto [0,1]$ and conditional plausibility $Pl(B|A) : 2^\Theta \mapsto [0,1]$ assigned to $B \subseteq \Theta$ are*

$$Bel(B|A) = \frac{Bel(A \cap B)}{Bel(A \cap B) + Pl(A - B)};$$
$$Pl(B|A) = \frac{Pl(A \cap B)}{Pl(A \cap B) + Bel(A - B)}. \quad (1)$$

Hence $\text{Bel}(B|A) = \text{Bel}(A \cap B|A)$ and $\text{Pl}(B|A) = \text{Pl}(A \cap B|A)$, i.e., conditioning of B with respect to A actually applies to the propositions that are in common to both A and B. In other words, in evaluating the evidence we have to support B when our view is restricted to only A, *it only makes sense to consider the propositions both A and B have access to!*

We use $m(\cdot|A) : 2^\Theta \mapsto [0,1]$ with $m(\emptyset|A) = 0$ to denote the *conditional bba given A* corresponding to these conditional notions; its existence is demonstrated in [9].

3. EVIDENCE UPDATING STRATEGY

3.1. Single BoE Case

Definition 2 (Updated Belief and Plausibility) *[8] Consider the BoE $\{\Theta, \mathcal{F}, m\}$ and a given $A \subseteq \hat{\mathcal{F}}(\Theta)$. Then, for an arbitrary $B \subseteq \Theta$, the* updated belief of B given A *$Bel_{k+1} : 2^\Theta \mapsto [0,1]$ and the corresponding* updated plausibility of B given A *$Pl_{k+1} : 2^\Theta \mapsto [0,1]$ are*

$$Bel_{k+1}(B) = \alpha_k(A)\, Bel_k(B) + \beta_k(A)\, Bel_k(B|A);$$
$$Pl_{k+1}(B) = \alpha_k(A)\, Pl_k(B) + \beta_k(A)\, Pl_k(B|A). \quad (2)$$

Here $\{\alpha_k, \beta_k\}$ are non-negative parameters dependent on the conditioning proposition A such that $\alpha_k + \beta_k = 1$; subscripts k and $k+1$ denote notions before and after updating respectively.

Numerous intuitively appealing properties of this 'symmetric' set of updating equations appear in [8]; several strategies for the selection of the weights $\{\alpha_k, \beta_k\}$ are also provided. In particular, it is argued that enforcing $\text{Bel}_{k+1}(A) = \text{Pl}_k(A)$ is perhaps the most reasonable strategy if one is unwilling to compromise the originally cast evidence. This strategy, referred to as *original plausibility based (OP-based) updating,* can be further justified via an interesting probabilistic interpretation [8]. It is these attractive properties that provide the motivation for its generalization.

3.2. Two BoEs Case

Consider two BoEs $\{\Theta_i, \mathcal{F}_i, m_i\}, i = 1, 2$, where Θ_1 and Θ_2 are not necessarily identical. For $B_i \subseteq \Theta_i,\ i = 1, 2$, consider

$$\text{Bel}_{k+1}(B_1)_1 = \alpha_k \text{Bel}_k(B_1)_1 + \beta_k \text{Bel}_k(B_1|B_2)_2, \quad (3)$$

where subscripts 1 and 2 denote the BoE where the corresponding quantities are to be computed. When both BoEs are identical, (3) reduces to (2). However, the plausibility updating equation that corresponds to (3) (which may be obtained by using the fact that, when $B_1 = \Theta_1$, we must have $1 = \alpha_k + \beta_k \text{Bel}_k(\Theta_1|B_2)_2$), is 'non-symmetric'. A 'symmetric' pair for belief and plausibility updating may be developed if $\text{Bel}_{k+1}(B_1)_1$ is taken to be

$$\alpha_k \text{Bel}_k(B_1)_1 + \frac{\beta_k}{2}\left[\text{Bel}_k(B_1|B_2)_2 + \text{Bel}_k(D_{B_1}|B_2)_2 - \text{Bel}_k(\Theta_2 - \Theta_1|B_2)_2\right], \quad (4)$$

where $D_{B_1} \equiv (\Theta_2 - \Theta_1) \cup B_1$. To arrive at the corresponding plausibility update, let $B_1 = \Theta_1$. Then we must have

$$1 = \alpha_k + \frac{\beta_k}{2}\left[\text{Bel}_k(\Theta_1|B_2)_2 + \text{Pl}_k(\Theta_1|B_2)_2\right]. \quad (5)$$

Simple manipulations on $\text{Pl}_{k+1}(B_1)_1 = 1 - \text{Bel}_{k+1}(\Theta_1 - B_1)_1$ then yield the 'symmetric' plausibility update we are seeking:

Definition 3 (Updated Belief and Plausibility) *Given the BoEs $\{\Theta_i, \mathcal{F}_i, m_i\},\ i = 1, 2$, the* updated belief and plausibility *assigned to $B_1 \subseteq \Theta_1$ conditional to $B_2 \subseteq \Theta_2$ are $Bel_{k+1} : 2^{\Theta_1} \mapsto [0,1]$ and $Pl_{k+1} : 2^{\Theta_1} \mapsto [0,1]$ where*

$$Bel_{k+1}(B_1)_1 = \alpha_k Bel_k(B_1)_1 + \frac{\beta_k}{2}\left[Bel_k(B_1|B_2)_2 + Bel_k(D_{B_1}|B_2)_2 - Bel_k(\Theta_2 - \Theta_1|B_2)_2\right];$$

$$Pl_{k+1}(B_1)_1 = \alpha_k Pl_k(B_1)_1 + \frac{\beta_k}{2}\left[Pl_k(B_1|B_2)_2 + Pl_k(D_{B_1}|B_2)_2 - Pl_k(\Theta_2 - \Theta_1|B_2)_2\right], \quad (6)$$

where $D_{B_1} = (\Theta_2 - \Theta_1) \cup B_1$, and the parameters $\{\alpha_k, \beta_k\}$ are non-negative and satisfy

$$1 = \alpha_k + \frac{\beta_k}{2}\left[Bel_k(\Theta_1|B_2)_2 + Pl_k(\Theta_1|B_2)_2\right]. \quad (7)$$

It is easy to demonstrate that $\text{Bel}_{k+1}(B_1)_1$ and $\text{Pl}_{k+1}(B_1)_1$ as defined above are indeed valid belief and plausibility functions respectively. Note that $\{\alpha_k, \beta_k\}$ are independent of B_1 and depends only on the conditioning propositions B_2. Also $\alpha_k = 1$ iff $\beta_k = 0$ and/or $\text{Pl}_k(\Theta_1|B_2)_2 = 0$.

3.3. Properties of the Updating Strategy

Due to the symmetry of the updating strategies, we concentrate only on the belief notions.

Trivial Cases. $\mathrm{Bel}_{k+1}(\emptyset)_1 = 0$ and $\mathrm{Bel}_{k+1}(\Theta_1)_1 = 1$ for arbitrary Θ_2.

$B_2 \subseteq \Theta_1 \cap \Theta_2$ Case. In this case, $1 = \alpha_k + \beta_k$ and

$$\mathrm{Bel}_{k+1}(B_1)_1 = \alpha_k \mathrm{Bel}_k(B_1)_1 + \beta_k \mathrm{Bel}_k(B_1|B_2)_2. \quad (8)$$

When both BoEs are identical, this yields the updating strategy for the single BoE case in Definition 2. When $B_2 \subseteq B_1$, observe that $\mathrm{Bel}_{k+1}(B_1)_1 \geq \mathrm{Bel}_k(B_1)_1$, i.e., belief cannot decrease.

$B_1 \cap B_2 = \emptyset$ Case. In this case, $\mathrm{Bel}_{k+1}(B_1)_1 \leq \mathrm{Bel}_k(B_1)_1$, i.e., belief cannot increase. Consider the special case $\Theta_1 \cap B_2 = \emptyset$ or $B_1 \cap \Theta_2 = \emptyset$: we get $\{\alpha_k, \beta_k\} = \{1, 0\}$ and $\mathrm{Bel}_{k+1}(B_1)_1 = \mathrm{Bel}_k(B_1)_1$, i.e., belief remain unchanged.

Updated Conditionals. When both FoDs are identical, some tedious manipulations yield $\mathrm{Bel}_{k+1}(B_1|A_1)_1 = \mathrm{Bel}_k(B_1|A_1)_1$, i.e., the conditional is invariant with updating; this agrees with [8].

Convergence Properties. Suppose the BoE $\{\Theta_1, \mathcal{F}_1, m_1\}$ repeatedly encounters the same BoE $\{\Theta_2, \mathcal{F}_2, m_2\}$ and updates its knowledge conditional to $B_2 \subseteq \Theta_2$. Noting that the belief values (computed in Θ_2) remain invariant because the new FoD Θ_2 is associated with the same BoE during each step, we may repeatedly apply Definition 3 to get

$$\lim_{n\to\infty} \mathrm{Bel}_{k+n}(B_1)_1 = \begin{cases} \frac{\mathrm{Bel}_k(B_1|B_2)_2 + \mathrm{Bel}_k(D_{B_1}|B_2)_2 - \mathrm{Bel}_k(\Theta_2 - \Theta_1|B_2)_2}{\mathrm{Bel}_k(\Theta_1|B_2)_2 + \mathrm{Pl}_k(\Theta_1|B_2)_2}, & \alpha_k \neq 1; \\ \mathrm{Bel}_k(B_1)_1, & \alpha_k = 1. \end{cases} \quad (9)$$

Hence the limiting belief values are dependent upon the incoming evidence only; the impact of the original evidence is completely eroded thus agreeing with one's intuition. The following special cases are interesting:

(a) When $B_2 \subseteq \Theta_1 \cap \Theta_2$:

$$\lim_{n\to\infty} \mathrm{Bel}_{k+n}(B_1)_1 = \begin{cases} \mathrm{Bel}_k(B_1|B_2)_2, & \text{for } \alpha_k \neq 1; \\ \mathrm{Bel}_k(B_1)_1, & \text{for } \alpha_k = 1. \end{cases} \quad (10)$$

(b) When $B_1 \cap B_2 = \emptyset$:

$$\lim_{n\to\infty} \mathrm{Bel}_{k+n}(B_1)_1 = \begin{cases} 0, & \text{for } \alpha_k \neq 1; \\ \mathrm{Bel}_k(B_1)_1, & \text{for } \alpha_k = 1. \end{cases} \quad (11)$$

Updated Incrementals.

$$\mathrm{Bel}_{k+1}(B_1)_1 - \mathrm{Bel}_{k+1}(B_1 \cap B_2)_1 = \alpha_k \left[\mathrm{Bel}_k(B_1)_1 - \mathrm{Bel}_k(B_1 \cap B_2)_1\right], \quad (12)$$

i.e., these incrementals cannot increase with updating. As was argued previously, suppose the BoE $\{\Theta_1, \mathcal{F}_1, m_1\}$ repeatedly encounters the same BoE $\{\Theta_2, \mathcal{F}_2, m_2\}$ and updates its knowledge conditional to $B_2 \subseteq \Theta_2$. Use (12) repeatedly:

$$\lim_{n\to\infty} \left[\mathrm{Bel}_{k+n}(B_1)_1 - \mathrm{Bel}_{k+n}(B_1 \cap B_2)_1\right] = 0. \quad (13)$$

This agrees with one's intuition that, when encountered with the same new information repeatedly, the belief notions converge to values supported by this incoming information only.

4. CHOICE OF WEIGHTS

4.1. Inertia of Available Evidence

The weights $\{\alpha_k, \beta_k\}$ can be interpreted as measures that indicate the flexibility or inertia of the originally cast evidence.

Definition 4 (Inertia Based Updating) *Consider the evidence updating strategy in Definition 3.*
(i) Infinite inertia based (II-based) updating *refers to* $\alpha_k = 1$.
(ii) Zero inertia based (ZI-based) updating *refers to* $\alpha_k = 0$.
(iii) Proportional inertia based (PI-based) updating *refers to* $\alpha_k = N/(N+1)$, *where* N *is to the number of 'pieces of evidence' on which the available evidence is based upon.*

Note that II-based updating indicates complete inflexibility or infinite inertia of the available evidence towards changes; this can also account for the original system's perception of complete 'unreliability' of the incoming evidence. ZI-based updating is the opposite. PI-based updating proportionally (with respect to N) accounts for the inertia of the available evidence.

4.2. Integrity of Available Evidence

To select weights that actually capture the integrity of the available evidence, we proceed as follows: Noting that the conditioning proposition $B_2 \subseteq \Theta_2$ has occurred, and we are in the process of updating the support for *all* the propositions in Θ_1, *including* $\Theta_1 \cap B_2$, consider the updates of $B_1 = \Theta_1 \cap B_2$. Clearly, we must first ensure that $\mathrm{Bel}_{k+1}(\Theta_1 \cap B_2)_1 = \mathrm{Bel}_{k+1}(B_2|\Theta_1)_1 \leq \mathrm{Pl}_k(\Theta_1 \cap B_2)_1 = \mathrm{Pl}_k(B_2|\Theta_1)_1$; second, we must also ensure that $\mathrm{Pl}_{k+1}(\Theta_1 \cap B_2)_1 = \mathrm{Pl}_{k+1}(B_2|\Theta_1)_1 \geq \mathrm{Bel}_k(\Theta_1 \cap B_2)_1 = \mathrm{Bel}_k(B_2|\Theta_1)_2$. When these are used in (6), we get

Definition 5 (Integrity Based Updating) *Consider the evidence updating strategy in Definition 3. The* original plausibility based (OP-based) updating *strategy refers to*

$$\frac{1 - Pl_k(B_2|\Theta_1)_1}{1 - Bel_k(B_2|\Theta_1)_1} \leq \alpha_k \leq 1, \textit{ for } Bel_k(B_2|\Theta_1)_1 < 1; \quad (14)$$

for $Bel_k(B_2|\Theta_1)_1 = 1$, α_k *can be arbitrarily selected in* $[0, 1]$.

Note that, when $\mathrm{Bel}_{k+1}(B_2|\Theta_1)_1 = \mathrm{Pl}_k(B_2|\Theta_1)_1$, the left side equality is achieved; it indicates that Θ_1 is most flexible to the incoming evidence *to the extent that its own evidence is not compromised.* The right side equality corresponds to current values held constant indicating that Θ_1 is least flexible. See Fig. 1.

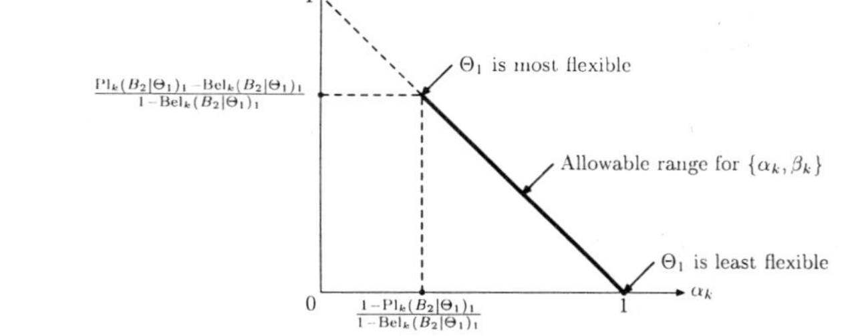

Figure 1: Choice of weights in OP-based updating.

5. EVIDENCE UPDATING IN A DSN ENVIRONMENT

When a decision node gathers evidence from two sensors having FoDs Θ_i, $i = 1, 2$, one may then use the following strategy for evidence updating. Set the FoD of the decision node as $\Theta_0 = \Theta_1 \cup \Theta_2$. Use Definition 3: for $i = 1, 2$, and $B \subseteq \Theta_0$,

$$\mathrm{Bel}_{k+1}(B)_{0\leftarrow i} = \alpha_{i,k}\mathrm{Bel}_k(B)_{0\leftarrow i} + \beta_{i,k}\mathrm{Bel}_k(B|\Theta_i)_i, \quad (15)$$

where $1 = \alpha_{i,k} + \beta_{i,k}$.

'Initialization' of Θ_0. Consider the $k = 0$ case. If the FoD Θ_0 is 'empty' and contains no prior evidence, we must have

$$\mathrm{Bel}_0(B)_{0\leftarrow i} = \begin{cases} 1, \text{for } B = \Theta_0; \\ 0, \text{for } B \subset \Theta_0. \end{cases} \quad (16)$$

Use (15):

$$\mathrm{Bel}_1(B)_{0\leftarrow i} = \begin{cases} 1, & \text{for } B = \Theta_0; \\ \beta_{i,0}\mathrm{Bel}_0(B|\Theta_i)_i, & \text{for } B \subset \Theta_0. \end{cases} \quad (17)$$

A linear combination of $\mathrm{Bel}_1(B)_{0\leftarrow 1}$ and $\mathrm{Bel}_1(B)_{0\leftarrow 2}$ may now be used to get $\mathrm{Bel}_1(B)_0$; the weights can be used to reflect, say, the source reliabilities.

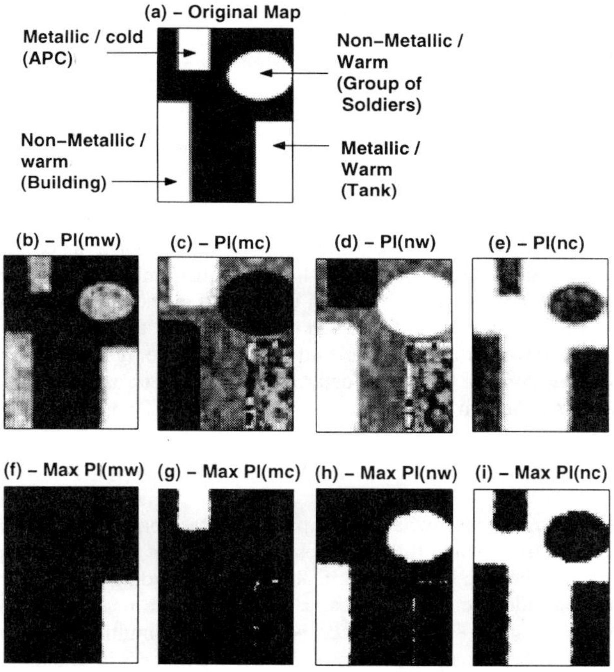

Figure 2: Object identification in a simulated battlefield.

5.1. A 'Simulated' Battlefield Example

We use the simulated 'battlefield' in Fig. 2 to illustrate the notions presented. It contains various objects and features of interest.

Evidence Updating. Suppose a DSN node fuses information generated by two mobile robots each mounted with one of two types of sensors as they move around the battlefield which, in this simulation, is divided into a 150×200 grid. The two sensor types are: (a) magnetometer (FoD $\Theta_{MG} = \{m, n\}$); and (b) IR detector (FoD $\Theta_{IR} = \{w, c\}$). Here $\{m, n, w, c\}$ denote metallic/non-metallic objects and 'warm'/'cold' objects respectively. Each sensor allocates a bba that accounts for its ambiguity within its sensing range [10]. Evidence updating was implemented using the approach in (15-17); the two sensors were equally weighted. This simulation contains 1000 pieces of evidence randomly generated throughout the battlefield. Resulting plausibility maps of $\{m, w\}$, $\{m, c\}$, $\{n, w\}$ and $\{n, c\}$ appear in Figs 2(b-e) respectively. Note that, lighter shades indicate higher plausibility values.

Decision Making. For 'hard' decision making purposes, we propose to use the 2^k combinations of plausibilities where k is the number of disparate sensors providing evidence. In this simulation, from the 2^2 plausibility maps, the maximum plausibility at each pixel is searched and displayed in Figs 2(f-i). These maps provide the best representation for decision making. For example, all metal objects can be identified by viewing the maps containing the preposition $\{m\}$ in its plausibility; the map of $\{m, w\}$ would identify objects that are *both* metal and warm.

6. CONCLUSION

A novel evidence updating strategy applicable within a heterogeneous sensor environment is proposed. It consists of a linear combination of the available evidence and the incoming evidence conditional to the FoD of the source generating it. In spite of its inability to accommodate sources possessing non-identical FoDs, its commutativity has been touted as a significant advantage of the Dempster's combination function. The proposed updating strategy does not possess this property; however, we believe that commutativity may not in fact be a desirable property in a DSN environment. Indeed, it is the lack of commutativity that enables the proposed strategy to account for knowledge base inertia. Of course, one must then pay attention to how best to 'order' the sources for updating. This is an interesting future research problem.

7. REFERENCES

[1] Shafer G (1976). *A Mathematical Theory of Evidence* (Princeton Univ. Press).

[2] Murphy RR and Hershberger D (1999). Handling sensing failures in autonomous mobile robots. *Int. J. Robotics Research* **18** 382:400.

[3] Cai D, McTear MF and McClean SI (2000). Knowledge discovery in distributed databases using evidence theory. *Int. J. Intelligent Syst.* **15** 745:761.

[4] Mukai T and Ohnishi N (2000). Object shape and camera motion recovery using sensor fusion of a video camera and gyro sensor. *Inf. Fusion* **1** 45:53.

[5] Fabre S, Appriou A and Briottet X (2001). Presentation and description of two classification methods using data fusion on sensor management. *Inf. Fusion* **2** 49:71.

[6] Smets P (2002). Decision making in a context where uncertainty is represented by belief functions. In *Belief Functions in Business Decisions* (Srivastava RP and Mock TJ, Eds.), 17:61.

[7] Zhang J, Kulasekere EC, Premaratne K and Bauer PH (2001). Resource management of task oriented distributed sensor networks. *ISCAS* (Sydney, Australia) **III** 513:516.

[8] Kulasekere EC, Premaratne K, Dewasurendra DA, Shyu M-L and Bauer PH (2002). Conditioning and updating evidence. *Int. J. Approx. Reasoning,* in review.

[9] Fagin R and Halpern J (1991). A new approach to updating beliefs. In *Uncertainty in Artificial Intelligence 6* (Bonissone PP, Henrion M, Kanal LN and Lemmer JF, Eds.), 347:374.

[10] HoseinNezhad R, Moshiri B and Asharif MR (2002). Sensor fusion for ultrasonic and laser arrays in mobile robotics: A comparative study of fuzzy, Dempster and Bayesian approaches. *IEEE Sensors* (Orlando, FL) 1682:1689.

ENERGY-BALANCING STRATEGIES FOR WIRELESS SENSOR NETWORKS

Martin Haenggi

Dept. of Electrical Engineering
University of Notre Dame
Notre Dame, IN 46556, USA
mhaenggi@nd.edu

ABSTRACT

The lifetime of wireless sensor network is crucial, since autonomous operation must be guaranteed over an extended period. As all the sensor data has to be forwarded to an observer via multi-hop routing, the traffic pattern is highly nonuniform, putting a high burden on the sensor nodes close to the observer. We propose and analyze four strategies that balance the energy consumption of the nodes to increase the lifetime of the network substantially. The analyses are based on a Rayleigh fading link model. An important result is that the energy benefits of routing over many short hops in fading environments are insignificant; especially for smaller path loss exponents, it is sensible to use fewer but longer hops.

1. INTRODUCTION

Large-scale networks of integrated wireless sensors become increasingly tractable, as advances in hardware technology and engineering design have led to dramatic reductions in size, power consumption, and cost for digital circuitry, wireless communications, and MEMS. This enables very compact and autonomous nodes, each containing one or more sensors, computation and communication capabilities, and a power supply. Multi-hop routing is typically used to reduce the transmit power and, consequently, increase the battery lifetime and decrease the interference between the nodes, thereby allowing spatial reuse of the communication channel.

Wireless sensor networks [1, 2] differ from other types of multi-hop wireless networks by the fact that, in most cases, the sensor data has to be delivered to a single sink, the *observer* or *base station*. Clearly, one of the primary concerns is the lifetime of the network. Although different definitions of lifetime exist [3], a sensor network certainly has to be considered "dead" whenever it is no longer able to forward any data to the base station. We assume that every sensor node in the network has an equal probability of generating data packets that have to be forwarded to the base station via multi-hop routing using other sensor nodes as relays. Apparently, the burden on the nodes close to the base station is considerably higher than on the nodes that are far away. Figure 1 depicts a possible arrangement of sensor nodes and identifies the most critical nodes in the network. Without appropriate measures, they will die quickly, rendering the network useless. In this paper, we propose and discuss strategies to ensure maximum lifetime of the network by balancing the energy load as equally as possible.

The partial support of the DARPA/IXO-NEST Program (AF-F30602-01-2-0526) is gratefully acknowledged.

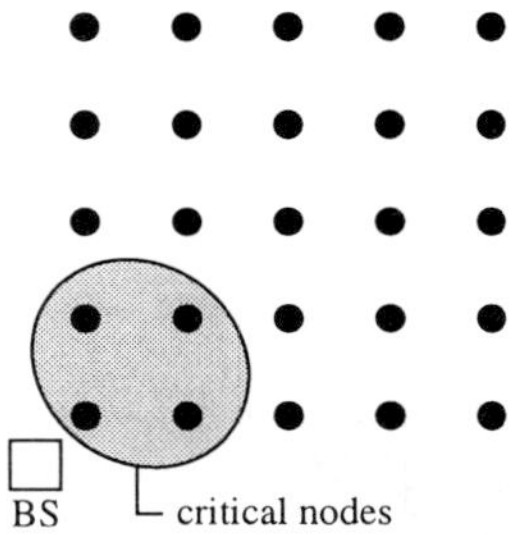

Figure 1: A sensor network with a base station.

The analysis is based on a Rayleigh fading link model, which models the wireless link more accurately than the "disk model" that is often used, where it is assumed that the radius for a successful transmission of a packet has a fixed and deterministic value R, irrespective of the condition and realization of the wireless channel [4, 5]. Such a simplified link model ignores the probabilistic nature of the wireless channel and the fact that the *signal-to-noise-and-interference ratio*, that determines the success of a transmission, is a random variable.

2. THE RAYLEIGH FADING LINK MODEL

We assume a narrowband multipath wireless channel with a coherence time longer than the packet transmission time. The channel can then be modeled as a flat Rayleigh fading channel [6] with an additive noise process z. Therefore the received signal is $y_k = a_k x_k + z_k$, where a_k is the path loss multiplied by the fading coefficient. The variance of the noise process is denoted by σ_Z^2.

The transmission from node i to node j is successful if the signal-to-noise-and-interference ratio (SINR) γ is above a certain threshold Θ that is determined by the communication hardware and the modulation and coding scheme (normally between 1 and 100 or 0dB and 20dB). With the assumptions above, γ is a discrete random process with exponential distribution $p_\gamma(x) = 1/\bar{\gamma}\, e^{-x/\bar{\gamma}}$ with mean

$$\bar{\gamma} = \frac{\bar{P}}{\sigma_Z^2 + \sigma_I^2}\,. \quad (1)$$

$\bar{P}$ denotes the average received signal power over a distance $d = \|x_i - x_j\|_2$: $\bar{P} = P_0 d^{-\alpha}$, where P_0 is proportional to the transmit power[1], and the path loss exponent is $2 \leqslant \alpha \leqslant 5$.

[1]This equation does not hold for very small distances. So, a more ac-

σ_I^2 is the interference power affecting the transmission. It is the sum of the received power of all the undesired transmitters.
In [5,7], the SINR is defined in a similar way. However, the transmission is considered to be successful whenever $\bar{\gamma}$ is bigger than some threshold. Hence, only the large-scale path loss is considered, while the probabilistic nature of the fading channel is ignored.
The following theorem proves useful for the analysis:

Theorem:
In a Rayleigh fading network, the reception probability $\mathbb{P}[\gamma \geqslant \Theta]$ *can be factorized into the reception probability of a zero-noise network and the reception probability of a zero-interference network.*

Proof: The probability that the SINR is bigger than a given threshold Θ follows from the cumulative distribution $f_\gamma(x) = 1 - e^{-x/\bar{\gamma}}$:

$$\begin{aligned}\mathbb{P}[\gamma \geqslant \Theta] &= e^{-\Theta/\bar{\gamma}} = e^{-\frac{\Theta}{\bar{P}}(\sigma_Z^2+\sigma_I^2)} \\ &= e^{-\frac{\Theta\sigma_Z^2}{\bar{P}}} \cdot e^{-\frac{\Theta\sigma_I^2}{\bar{P}}} = \mathbb{P}[\gamma_Z \geqslant \Theta] \cdot \mathbb{P}[\gamma_I \geqslant \Theta], \quad (2)\end{aligned}$$

where $\gamma_Z := \bar{P}/\sigma_Z^2$ denotes the signal-to-noise ratio (SNR) and $\gamma_I := \bar{P}/\sigma_I^2$ denotes the signal-to-interference ratio (SIR). The first factor is the reception probability in a zero-interference network as it depends only on the noise, and the second factor is the reception probability in a zero-noise network, as it depends only on the interference. Both the SNR and the SIR are exponentially distributed, and it also follows from (2) that $\bar{\gamma} = (\bar{\gamma}_Z\bar{\gamma}_I)/(\bar{\gamma}_Z + \bar{\gamma}_I)$. ∎

This allows an independent analysis of the effect caused by noise and the effect caused by interference. The focus of this paper is put on the noise, *i.e.*, on the first factor in (2). If the load is light (low interference probability), then SIR$\gg$SNR, and the noise analysis alone provides accurate results. For high load, a separate interference analysis has to be carried out [8][2].
In a zero-interference network, the reception probability over a link of distance d at a transmit power P_0, is given by

$$p_r := \mathbb{P}[\gamma_Z \geqslant \Theta] = e^{-\frac{\Theta\sigma_Z^2}{P_0 d^{-\alpha}}} . \quad (3)$$

Solving for P_0, we get for the necessary transmit power to achieve a link reliability (or reception probability) P_L:

$$P_0 = \frac{d^\alpha \Theta \sigma_Z^2}{-\ln P_L} . \quad (4)$$

3. ENERGY-BALANCING STRATEGIES

We assume that every sensor node generates an equal amount of traffic of arrival rate λ that is relayed to the base station along the shortest route. Traffic may be bursty or periodic. Since optimum routes approximately follow a straight line, the analyses of the four strategies proposed in this Section can be restricted

curate model would be $\bar{P} = P_0' \cdot (d/d_0)^{-\alpha}$, valid for $d \geqslant d_0$, with P_0' as the average value at the reference point d_0, which should be in the far field of the transmit antenna. At 916MHz, for example, the near field may extend up to 3-4ft (several wavelengths).

[2]Note that *power scaling, i.e.*, scaling the transmit powers of all the nodes by the same factor, does not change the SIR, but (slightly) increases the SINR.

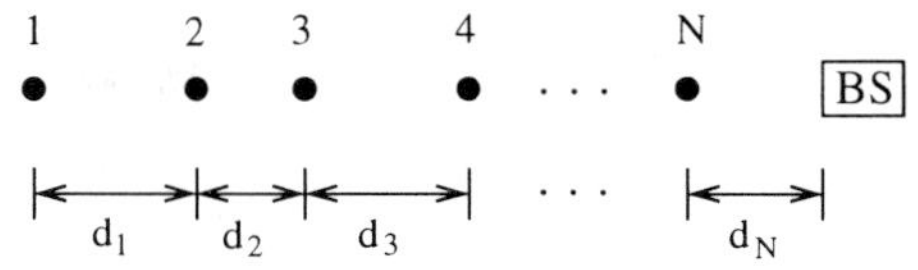

Figure 2: A one-dimensional chain of sensor nodes.

to one-dimensional chains of N nodes, as shown in Fig. 2. The strategies are compared with the simple scheme that has equal node distances d, equal link reception probabilities P_L, and employs nearest-neighbor routing (node i transmits to node $i+1$ and so on). To compare the total energy consumption, it is sufficient to calculate the energy requirements to forward one packet from each of the nodes to the base station. For the simple strategy, the total energy consumption is

$$E_{\text{tot}} = (1 + 2 + \ldots + N)\frac{d^\alpha \Theta \sigma_Z^2}{-\ln P_L} = \frac{N(N+1)}{2}\frac{d^\alpha \Theta \sigma_Z^2}{-\ln P_L} . \quad (5)$$

For the lifetime, the critical node's energy consumption has to be determined. For the simple strategy, the critical node is node N, and we get

$$E_{\max} = E_N = N\frac{d^\alpha \Theta \sigma_Z^2}{-\ln P_L} . \quad (6)$$

3.1. Distance variation

We assume nearest-neighbor routing. The idea is, given a link reliability P_L, to ensure energy-balancing by adjusting the distances d_i between the nodes. If every node generates one packet, node i has to forward a total of i packets using a total energy of

$$E_i = i \cdot \frac{d_i^\alpha \Theta \sigma_Z^2}{-\ln P_L} . \quad (7)$$

The goal $E_1 = E_2 = \ldots = E_N$ requires that all the factors id_i^α are identical. In addition, the sum of all the distances must correspond to the desired length d_{tot} of the chain. We find $d_i = d_1 i^{-1/\alpha}$ and

$$d_i = \frac{d_{\text{tot}} \cdot i^{-1/\alpha}}{\sum_{i=1}^N i^{-1/\alpha}} . \quad (8)$$

This strategy clearly leads to a non-uniform distribution of the sensor nodes, which may not be desirable. However, the distribution is not far from uniform, as manifested by the small variance of d_i. For $d_{\text{tot}} = 10$, $N = 10$ and $\alpha = 2$, the variance is about 0.16. For $\alpha = 3, 4, 5$, the variances are 0.066, 0.035, 0.022. The gain in total energy consumption varies between 22% ($\alpha = 5$) and 28% ($\alpha = 2$). The gain in lifetime is considerably higher and ranges between a factor of 2.3 ($\alpha = 5$) and 2.5 ($\alpha = 2$).

3.2. Balanced data compression

If the internode distances d_i are all equal, the incoming data flows may be compressed to ensure that every node in the path has to transmit the same number of packets. The justification for this approach is the correlation between the sensor readings of neighboring nodes. Hence, data fusion may be applied to reduce the amount of data to be transmitted. The goal is to ensure that every packet experiences the same compression factor, irrespective of its origin. At each node i, the incoming data is compressed by a factor of a_i, while the locally generated data is compressed by $b_i = 1 - a_i$. Equal compression is achieved when $b_i = 1/i$, since the total compression factor for a packet generated at node i is

$$\beta_i = b_i \cdot \prod_{k=i+1}^{N} a_i = \frac{1}{i} \cdot \prod_{k=i+1}^{N} \left(1 - \frac{1}{k}\right) = \frac{1}{N} \quad \forall i . \tag{9}$$

This way, every node transmits only one packet, so the total energy consumption is $N\frac{d^\alpha \Theta \sigma_Z^2}{-\ln P_L}$, and the gain in lifetime is a factor of N.

3.3. Routing

We again assume equal distances d between the nodes, but no longer restrict the network to strict nearest-neighbor routing. Instead, we assume that node i transmits the locally generated traffic to the next neighbor with probability a_i and directly to the sink with probability $b_i = 1 - a_i$. Incoming traffic will always be forwarded to the next node. The goal is to choose a_i to achieve energy balancing[3].

All energies in the following derivation are normalized by $d^\alpha \Theta \sigma_Z^2 / (-\ln P_L)$. The energy consumption at node i is then

$$E_i = (N - i + 1)^\alpha b_i + \sum_{k=1}^{i} a_k \tag{10}$$

$$= i + ((N - i + 1)^\alpha - 1)\, b_i - \sum_{k=1}^{i-1} b_k . \tag{11}$$

$b_N = 0$, as node N always transmits directly to the sink. $E_N = N - b_1 - b_2 - \ldots - b_{N-1} = a_1 + a_2 + \ldots + a_N$. The $N - 1$ unknowns can thus be determined by solving

$$\begin{bmatrix} N^\alpha & 1 & \ldots & 1 \\ 0 & (N-1)^\alpha & \ldots & 1 \\ \vdots & \vdots & \ddots & \vdots \\ 0 & 0 & \ldots & 2^\alpha \end{bmatrix} \begin{bmatrix} b_1 \\ b_2 \\ \vdots \\ b_{N-1} \end{bmatrix} = \begin{bmatrix} N-1 \\ N-2 \\ \vdots \\ 1 \end{bmatrix}$$

to equalize the energy consumption at every node to E_N. For a network with 5 nodes, the values for $b_1, \ldots, b_5$ are 0.0301, 0.0438, 0.0694, 0.1250, 0 for $\alpha = 3$. Since some packets are routed to the base station in a single hop, the total energy consumption is bigger than in the simple strategy. For $N = 10$, the additional energy consumption is between 60% ($\alpha = 2$) and 80% for ($\alpha = 5$). On the other hand, there is a slight gain in network lifetime, as the direct routing of some of the packets reduces the burden on node N. Therefore the sum of the a_i's is smaller than N. For 10 nodes, the increase in lifetime is 0.5% for $\alpha = 5$ and 14% for $\alpha = 2$. For $N \to \infty$, the total additional energy consumption reaches 100% and the gain in lifetime vanishes[4]. Hence this strategy is useful for smaller N, or if there are a some high-priority packets that have to be delivered with minimum delay. The average delay for packets generated at node i is $a_i(N - i + 1) + (1 - a_i) = a_i(N - i) + 1$.

[3] In [9], a similar strategy is discussed. However, the analysis is based on the "disk model" that neglects the stochastic nature of the channel.

[4] It is easily established that $b_k \leqslant (N-k) \cdot (N-k+1)^{-\alpha}$. With $E_N > \sum_{k=1}^{N-1}(1 - b_k)$, we get $E_N > \sum_{k=1}^{N-1} 1 - k \cdot (k+1)^{-\alpha}$. For $\alpha \geqslant 2$, the terms in the sum approach 1 for large k, thus $E_N \to N$ for large N, which is the same as in the simple strategy.

3.4. Equalization of the end-to-end reliability

So far, we ignored the fact that the end-to-end reliability of a multi-hop path is the product of the reception probabilities of the links. With constant link reception probabilities P_L, a packet traveling over k hops only arrives at the sink with a probability P_L^k.

We investigate a strategy where every packet, irrespective of how far away from the base station it is generated, has the same end-to-end probability P_{EE} to arrive at the base station. This *equal-end-to-end-probability* strategy is henceforth referred to as strategy A, whereas the simple *equal-power* strategy is denoted as strategy B.

Analysis of strategy A. If the desired end-to-end reliability is P_{EE}, the link probability in a k–hop connection is $P_{L_k} = P_{\mathrm{EE}}^{1/k}$. Accordingly, the transmit power at each hop in a k–hop connection with equal distances d is[5]

$$P_k^{\mathrm{A}} = \frac{d^\alpha \Theta \sigma_Z^2}{-\ln(P_{\mathrm{EE}}^{1/k})} = k \cdot \frac{d^\alpha \Theta \sigma_Z^2}{-\ln P_{\mathrm{EE}}} . \tag{12}$$

Clearly, a transmission over k hops requires a k times higher transmit power level (at each hop) than a transmission over one hop with the same probability. Thus the total energy needed for a packet to travel from node $N - k + 1$ to the base station is proportional to k^2. Note that a single large hop of length k would require an energy proportional to k^α. Let E_0^{A} denote the energy required to transmit one packet over one hop of distance d with a reliability of P_{EE}, *i.e.*, $E_0^{\mathrm{A}} := d^\alpha \Theta \sigma_Z^2 / (-\ln P_{\mathrm{EE}})$. Using nearest-neighbor routing, the energy consumption is:

Node	Energy consumption		(in units of E_0^{A})
1	N	$=$	N
2	$N + (N-1)$	$=$	$2N - 1$
3	$(2N-1) + (N-2)$	$=$	$3N - 3$
4			$4N - 6$
$\vdots$	$\vdots$		$\vdots$
i			$iN - \frac{i(i-1)}{2}$

The total energy consumption of the network (assuming one packet is generated at every node) is

$$\begin{aligned} \frac{E_{\mathrm{tot}}^{\mathrm{A}}}{E_0^{\mathrm{A}}} &= \sum_{i=1}^{N} \left(iN - \frac{i(i-1)}{2}\right) \\ &= N \cdot \frac{N(N+1)}{2} - \left(0 + 1 + 3 + 6 + \ldots + \frac{N(N-1)}{2}\right) \\ &= N \cdot \frac{N(N+1)}{2} - \left(\frac{N(N-1)}{2} + \frac{N(N-1)(N-2)}{6}\right) \\ &= \frac{N^3}{3} + \frac{N^2}{2} + \frac{N}{6} , \end{aligned} \tag{13}$$

where we have exploited the fact $0 + 1 + 3 + 6 + \ldots$ (second line) is an arithmetic series of order 2 with $q_0 = 0$, $\Delta q_0 = \Delta^2 q_0 = 1$.

Analysis of strategy B. Here, all nodes transmit at a fixed power level, corresponding to a fixed link reception probability P_L. For a fair comparison, it is assumed that the application dictates a *minimum* end-to-end reliability of P_{EE}. Packets generated at node i

[5] Note that this implies that a single node uses different power levels that depend on the origin of a packet. The power levels are assigned to flows, not to nodes.

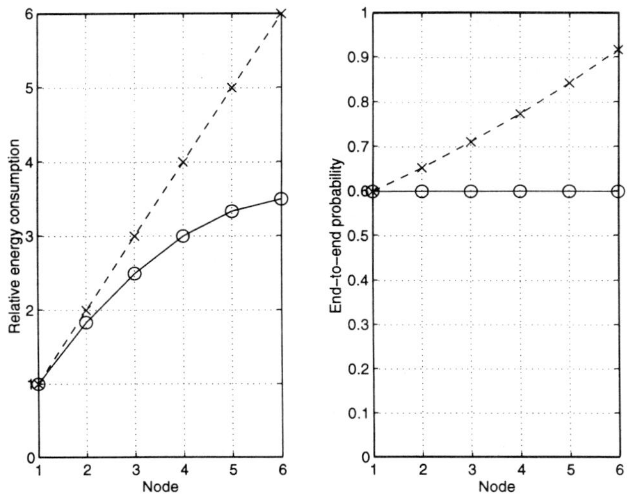

Figure 3: Comparison of the equal-power strategy B (dashed) and the equal-end-to-end-probability strategy A (solid) for $P_{\mathrm{EE}} = 0.6$. The left plot shows the relative energy consumption when every node generates one packet, the right plot the end-to-end probabilities for traffic generated at node i.

have to travel over $k := N - i + 1$ hops, resulting in an end-to-end reliability of P_L^k. To ensure that the packets from the farthest node (node 1) arrive with probability P_{EE}, a link reception probability of $P_L = P_{\mathrm{EE}}^{1/N}$ is required. The energy per hop is in this case is (cf. (5))

$$E_0^{\mathsf{B}} = d^\alpha \Theta \sigma_Z^2 / (-\ln P_L) = N \cdot d^\alpha \Theta \sigma_Z^2 / (-\ln P_{\mathrm{EE}}) = N E_0^{\mathsf{A}} .$$

The total energy consumption of the network (assuming one packet is generated at every node) is

$$E_{\mathrm{tot}}^{\mathsf{B}} = \frac{N(N+1)}{2} E_0^{\mathsf{B}} = \Big(\frac{N^3}{2} + \frac{N^2}{2}\Big) E_0^{\mathsf{A}} . \tag{14}$$

Comparison. The ratio between energy consumption of the two strategies is

$$\frac{E_{\mathrm{tot}}^{\mathsf{A}}}{E_{\mathrm{tot}}^{\mathsf{B}}} = \frac{\frac{N^2}{3} + \frac{N}{2} + \frac{1}{6}}{\frac{N^2}{2} + \frac{N}{2}} . \tag{15}$$

For large N, this ratio approximates $2/3$, hence the gain in total energy consumption for the reliability balancing strategy is 33%. More important and more significant is the gain in network lifetime, which is determined by the lifetime of the critical node N. In strategy A, the energy consumption at node N is $E_N^{\mathsf{A}} = E_0^{\mathsf{A}}(N^2 + N)/2$, whereas in case B, it is $E_N^{\mathsf{B}} = E_0^{\mathsf{B}} N = E_0^{\mathsf{A}} N^2$. The ratio is $2N/(N+1)$, thus the gain in network lifetime approaches 2 for large N. Figure 3 compares the two strategies. For strategy B, the energy consumption increases linearly with the node number, and the end-to-end probability for a packet generated at node i increases monotonically. For scheme A, the energy consumption is more balanced, and the end-to-end probability is constant.

4. CONCLUDING REMARKS

For sensor networks, where the destination of all the information gathered at the sensor nodes is a single base station, the traffic pattern is highly non-uniform, since the nodes close to the base station have to relay all the data packets. Consequently, those nodes are the first to run out of battery, thereby restricting the lifetime of the network. Four strategies have been proposed to balance the energy consumption, each of them having their application-dependent strengths and weaknesses. Since they are not mutually exclusive, several may be combined into hybrid schemes.

The analyses are based on a probabilistic link model that is derived from Rayleigh fading channels. It is shown that under Rayleigh fading, noise issues and interference issues can be analyzed separately, and that the reception probability is an exponential function of the transmit power. This non-zero probability of a packet loss even if nodes are within "transmission range" is often ignored. It entails that a transmission over k hops requires a k times higher transmit power at each hop to guarantee the same end-to-end probability as a single-hop transmission. An important consequence is that the energy benefit of multi-hop routing become much less significant. Indeed, since the energy consumption is proportional to k^2 in the hop-by-hop transmission and proportional to k^α in a single-hop scheme (see Section 3.4), the benefit vanishes for $\alpha = 2$ and, if a nonlinear power amplifier characteristics is taken into account, also for higher path loss exponents. If, in addition, the end-to-end delay is considered, minimum-hop[6] routing clearly outperforms maximum-hop (or shortest-hop) routing.

5. REFERENCES

[1] G. J. Pottie and W. J. Kaiser, "Wireless integrated network sensors," *Communications of the ACM*, vol. 43, no. 5, pp. 551–558, 2000.

[2] I. F. Akyildiz, W. Su, Y. Sankarasubramaniam, and E. Cayirci, "Wireless sensor networks: a survey," *Computer Networks*, vol. 38, pp. 393–422, Mar. 2002.

[3] A. Ephremides, "Energy Concerns in Wireless Networks," *IEEE Wireless Communications*, vol. 9, pp. 48–59, Aug. 2002.

[4] H. Takagi and L. Kleinrock, "Optimal Transmission Ranges for Randomly Distributed Packet Radio Terminals," *IEEE Transactions on Communications*, vol. COM-32, pp. 246–257, Mar. 1984.

[5] P. Gupta and P. R. Kumar, "The Capacity of Wireless Networks," *IEEE Transactions on Information Theory*, vol. 46, pp. 388–404, Mar. 2000.

[6] T. S. Rappaport, *Wireless Communications – Principles and Practice*. Prentice Hall, 1996. ISBN 0-13-375536-3.

[7] M. Grossglauser and D. Tse, "Mobility Increases the Capacity of Ad-hoc Wireless Networks," in *IEEE INFOCOM*, (Anchorage, Alaska), 2001.

[8] M. Haenggi, "Probabilistic Analysis of a Simple MAC Scheme for Ad Hoc Wireless Networks," in *IEEE CAS Workshop on Wireless Communications and Networking*, (Pasadena, CA), Sept. 2002.

[9] R. Min, M. Bhardwaj, S.-H. Cho, N. Ickes, E. Shih, A. Sinha, A. Wang, and A. Chandrakasan, "Energy-Centric Enabling Technologies for Wireless Sensor Networks," *IEEE Wireless Communications*, vol. 9, pp. 28–39, Aug. 2002.

[6] I.e., transmitting over hops that are as long as the maximum transmit power allows, given the reliability requirement.

POWER DISSIPATION LIMITS AND LARGE MARGIN IN WIRELESS SENSORS

Shantanu Chakrabartty and Gert Cauwenberghs

ECE Department, Johns Hopkins University, Baltimore, MD 21218, USA
E-mail: {*shantanu,gert*}@*jhu.edu*

ABSTRACT

Wireless smart sensors impose severe power constraints that call for power budget optimization at all levels in the design hierarchy. We elucidate a connection between statistical learning theory and rate distortion theory that allows to operate a wireless sensor array at fundamental limits of power dissipation. $Gini$SVM, a support vector machine kernel-based classifier based on quadratic entropy, is shown to encode the sensor data with maximum fidelity for a given constraint on transmission budget. The transmission power is minimized by $Gini$SVM in the form of a quadratic cost function under linear constraints. A classifier architecture that implements these principles is presented.

1. INTRODUCTION

Emerging wireless embedded sensors impose serious restrictions on power consumption [1]. These constraints derive from the vision of operating the sensors off ambient (*e.g.*, solar or thermal) power, and make it necessary to allocate power resources efficiently to the task of sensing, computation and communication.

As physical size decreases so does energy capacity. Because communication is often the single largest energy consumer the optimization of wireless communication protocols is key in meeting energy constraints. For a typical bluetooth application [2] operating at $700kbps$ and dispensing $115nJ/bit$, the total power consumption of a stand-alone communication module is approximately $800mW$ which exceeds the power budget for most miniature stand-alone sensors. To achieve sub-microwatt operation, power budget optimization has to be performed rigorously at several levels [3] as shown in increasing level in the design hierarchy:

1. **Technology Level**— Supply and threshold voltage reduction;
2. **Circuit and Logic Level**— Logic style, current starvation, switching behavior, supply switching and subthreshold design;
3. **Architectural Level**— Parallelism and pipelining;
4. **Algorithmic and System Level**— Utilization of signal statistics, floor planning, data encoding, sleep modes and reference localization.

This research is supported by a grant from The Catalyst Foundation, New York.

It has been argued [3] that higher reduction in power consumption can be obtained by optimizing at the highest level. Such a hierarchical optimization methodology is a two-way procedure where each level imposes design constraints on the subsequent higher level. For wireless smart sensor arrays, we approach the problem by addressing power-efficient classification and encoding of sensor data transmitted over a communication channel. Considerations from the perspective of low-power sensor design are:

1. **Reduction of Supply Voltages** leading to reduction in noise margins. Noise in the circuit plays a dominant role in the classifier performance, and large margin machine learning techniques may be employed during training to aid in discriminating between classes in run-time.
2. **Reduction of Transmission Rate**, an effective way to reduce power dissipation in the transmission module. Effective reduction of transmission rate amounts to efficient data encoding such that only relevant events are detected and transmitted. For a sensor the primary aim is detection, therefore a classifier is required to register detected events as illustrated in Figure 1. For a multi-event (*e.g.,* olfactory) sensor array architecture it is often necessary to transmit confidence values of events rather than indicator flags to enhance resolution at the receiver, using a more a complex decoding algorithm. With a reduced supply voltage a multi-event/class classifier has to deal with reduced noise margins and increased distortion. In designing a classifier for a low-power wireless sensor array, it is therefore crucial to encode events efficiently according to confidence values to achieve maximum fidelity given the constraint on the number of bits for encoding.

Rate Distortion Theory provides lower bounds on distortion for a fixed transmission rate, determined by the

0-7803-7761-3/03/$17.00 ©2003 IEEE

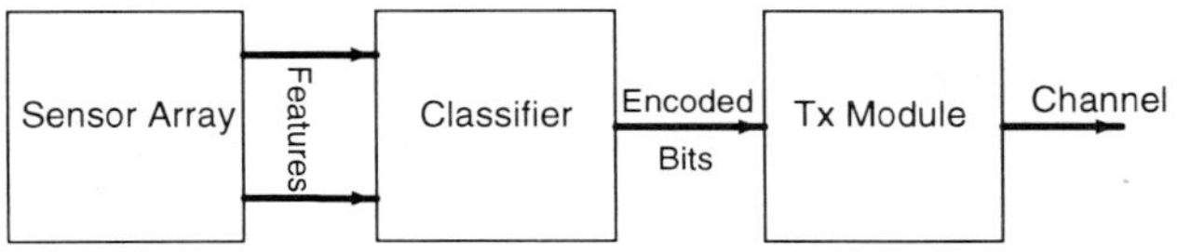

Fig. 1. *Wireless smart sensor architecture*

transmitter power budget. *Statistical Learning Theory* on the other hand provides tools to design classifiers that model signal statistics for efficient discrimination between classes. In this work we elucidate a connection between the two that provides a unified framework to a sensor and classification architecture that operates at fundamental limits of power dissipation.

The paper is organized as follows. Section II briefly reviews principles of rate distortion functions and fundamental limits of distortion given a fixed transmission rate. Section III reviews principles of large margin classifiers and statistical learning theory and discusses how the large margin concept is directly related to designing classifier with larger noise margins for low voltage operation. Section IV presents *Gini*SVM, a large margin classifier that conforms to principles of rate distortion theory to operate at fundamental limits of power dissipation. Section V discusses the practical implications and architectural implementation of *Gini*SVM. Section VI provides comments and conclusions.

2. RATE DISTORTION THEORY

Classical Rate Distortion Theory [4] specifies a lower bound on the number of bits required to encode a signal within a specified amount of distortion. Given a set of independent signals $X_1, X_2, .., X_N$, with power $S_1, S_2, .., S_N$, the optimal allocation of number of bits $R_1, R_2, .., R_N$ to represent $X_1, X_2, .., X_N$ such as to keep the total distortion $D_1 + D_2 + .. + D_N$ below an upper-bound D, is given by

$$\begin{aligned} R &= \sum_i R_i \\ &= \sum_i \frac{1}{2}(\log S_i - \log D_i) \qquad (1) \\ &= \sum_i \frac{1}{2}[\log S_i - \log D_{\min}]_+ \end{aligned}$$

where $[x]_+$ represents positive part of x. Distortion for each of the signals corresponding to this optimum bit allocation has the form

$$D_i = \begin{cases} D_{\min} & ; \quad S_i \geq D_{\min} \\ S_i & ; \quad S_i < D_{\min} \end{cases} \qquad (2)$$

where $D_{\min}$ is computed by solving $\sum_i D_i = D$ through the reverse water-filling procedure illustrated in Figure 2 [9].

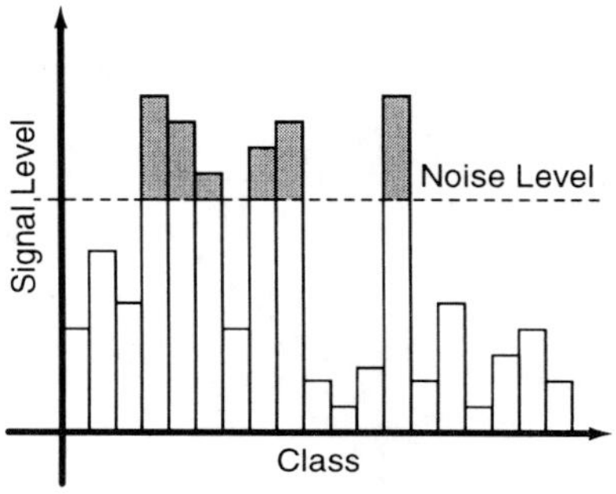

Fig. 2. *Reverse water-filling procedure: signal classes above the noise floor are encoded for transmission.*

For designing integrated sensors an alternative equivalent problem formulation is of more importance. Given the total number of bits R, specified by the transmitted power budget, the alternative seeks to optimize bit allocation amongst various signal to achieve minimum distortion in a mean square sense. The sensor utilizes the optimal allocation obtained to encode the confidence/probability measure for each of the events/classes. Current techniques use dynamic programming to solve the problem of resource allocation. We will show in the next section that by using a large margin classification architecture based on quadratic entropy, *Gini*SVM, an optimal encoding is obtained under the constraint of utilizing a fixed number of bits.

3. LARGE MARGIN CLASSIFICATION

Large margin (LM) classifiers like *Support Vector Machines* (SVM) [5] have several attractive properties from a practical implementation perspective:

1. They generalize well even with relatively few data points in the training set, and bounds on the generalization error can be directly estimated from the training data. This ensures shorter time for real-time system training.

2. The only parameter that needs to be chosen is a penalty term for mis-classification which acts as a regularizer [6] and determines a trade-off between resolution and generalization performance, to control learning ability.

3. The algorithm finds, under general conditions, a unique separating decision surface that provides best out-of-sample performance. This property is unlike neural network classifier implementation where the solution obtained is not unique and hence cannot be quantified.

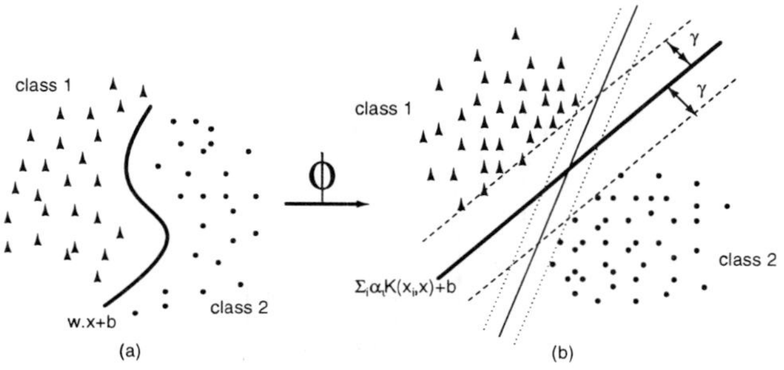

Fig. 3. *Large margin kernel machines.* (a) *Training data in data space.* (b) *Nonlinear map* $\Phi()$ *projects data space onto feature space where they are linearly separated with maximum margin.*

4. They provide a framework to model non-linear classification boundaries by projecting the input data point into higher dimensional space and then computing the distances with the aid of a kernel. This is illustrated in figure 3 where the training samples that are not linearly separable in *(a)* are projected through a non-linear mapping $\Phi()$ onto higher dimensional feature space *(b)* where they become linearly separable.

5. The learning algorithm performs model selection based on some optimization criterion, by which only the data points (support vectors) which are relevant to the classification problem are used for computation. This feature leads to optimal utilization of hardware resources (memory to store support vectors) which is critical for a power conscious design.

One attractive property of large margin classifiers from an implementation perspective, is the direct correspondence of classifier margin maximization during training to optimizing circuit noise margin. This is illustrated in figure 3(b) which shows a maximal margin hyper-plane obtained after training. The maximum margin hyper-plane is more robust to small perturbations in training samples than any other separating hyperplane, and hence is more immune to sensor noise inherent in circuit design. Traditional implementation of non-linear classifiers using (unregularized) neural networks primarily aims at separating the data without direct consideration of margin leading to less optimal noise immunity from a circuit designer perspective. Another attractive property of a large margin classifier is the relative insensitivity of classification performance to small perturbations in the non-linear mapping Φ. This enables use of simple computational elements in an array configured so it can be characterized via a kernel and optimally trained to maximize its classification performance and power consumption.

*Gini*SVM [7] is a sparse multi-class probability regression technique based on large margin principles that generates conditional probability estimates for class k based on input feature vector $\mathbf{x}$, $P_k(\mathbf{x}) \propto exp(f_i(\mathbf{x}))$. The functions $f_i(\mathbf{x})$ are estimated using empirical data consisting of training examples $\mathbf{x}[m], m = 1, .., M$ and their corresponding labels indicating prior probabilities values $y_i[m], i = 1, .., K$. As with SVMs, dot products in the expression for $f_i(\mathbf{x})$ convert into kernel expansions over the training data $\mathbf{x}[m]$ by transforming the data to feature space [8]

$$\begin{aligned} f_i(\mathbf{x}) &= \mathbf{w}_i.\mathbf{x} + b_i \\ &= \sum_m \lambda_i^m \, \mathbf{x}[m].\mathbf{x} + b_i \\ &\xrightarrow{\Phi(\cdot)} \sum_m \lambda_i^m \, K(\mathbf{x}[m], \mathbf{x}) + b_i \end{aligned} \quad (3)$$

where $K(\cdot, \cdot)$ denotes any symmetric positive-definite kernel[1] that satisfies the Mercer condition, such as a Gaussian radial basis function or a polynomial spline [6].

The parameters λ_i^m in (3) are determined by minimizing a dual formulation which for *Gini*SVM takes the form of a quadratic-entropy based potential function in the parameters [7]

$$H_g = \sum_i^M [\frac{1}{2} \sum_l^N \sum_m^N \lambda_i^l Q_{lm} \lambda_i^m + \gamma C \sum_m^N (y_i[m] - \lambda_i^m / C)^2] \quad (4)$$

subject to constraints

$$\sum_m \lambda_i^m = 0 \quad (5)$$

$$\sum_i \lambda_i^m = 0 \quad (6)$$

$$\lambda_i^m \leq C y_i[m] \quad (7)$$

where $Q_{lm} = K(\mathbf{x}[l], \mathbf{x}[m])$.

Using the first order conditions for optimizing (4) the following necessary and sufficient condition is obtained

$$\sum_i^M [f_i(x) - z]_+ = 2\gamma \quad (8)$$

which conforms to the rate distortion criterion (2). The procedure finds the optimal number of bits to encode the signal $exp(f_i(x))$ under the assumption that the log-signal power $\log S_i$ is linearly proportional to the signal strength $f_i(x)$. The parameter 2γ determines the total bit budget obtained from power consumption requirement of the transmitter module. The parameter z is adaptively computed for each classification based on the *reverse water filling*

[1] $K(\mathbf{x}, \mathbf{y}) = \Phi(\mathbf{x}).\Phi(\mathbf{y})$. The map $\Phi(\cdot)$ need not be computed explicitly, as it only appears in inner-product form.

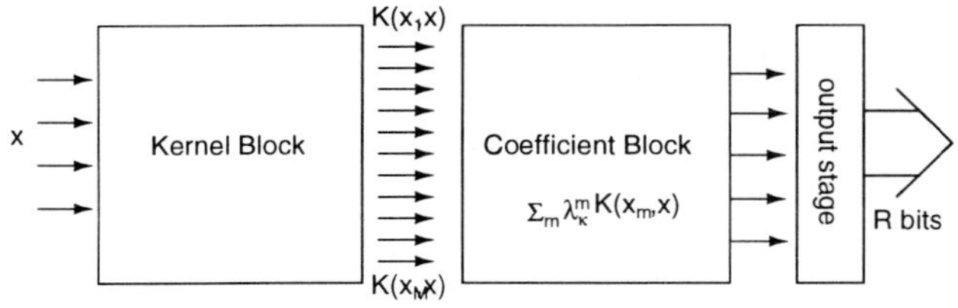

Fig. 4. *GiniSVM Architecture*

principle illustrated in figure 2. Only signals with power higher than z or the noise level are encoded for transmission and the rest are discarded. The total bit budget 2γ is then proportionately distributed according the class bit budget $R_i = [f_i(x) - z]_+$. The validity of signal independence $f_i(x)$ between K classes is obtained through training of $Gini$SVM which makes $f_i(x)$ depend only on class discriminatory features.

4. SYSTEM ARCHITECTURE

The architecture of $Gini$SVM is shown in figure 4. It consists of the following blocks:

- **Kernel Block**— stores support vectors $\mathbf{x}[m]$ and computes the kernel $K(\mathbf{x}[m], \mathbf{x})$ between support vectors and the input feature vector $\mathbf{x}$. Floating gate transistor array [10] could be a possible implementation ensuring a minimal standby power consumption.

- **Coefficient Block**— computes the inner-product between the kernel $K(\mathbf{x}[m], \mathbf{x})$ and the coefficients λ_i^m to obtain values of $f_k(\mathbf{x}) = \sum_m \lambda_i^m K(\mathbf{x}[m], \mathbf{x})$.

- **Encoder Output Block**— encodes confidence values of $P_i(\mathbf{x}) = exp(f_i(\mathbf{x})), i = 1,.., K$ using a total fixed number of bits R.

An efficient analog implementation of the kernel and coefficient block can be obtained using subthreshold CMOS circuits as described in [10].

5. CONCLUSIONS

In this paper we proposed a classifier architecture for a wireless sensor operating at fundamental limits of power dissipation by utilizing the $Gini$SVM large-margin kernel machine. The design methodology is flexible enough to incorporate user specified bit budget constraints into the classifier architecture. Circuit design parameters are then directly obtained by optimizing a classifier cost function which also ensures that the system operates with a near optimal fidelity.

6. REFERENCES

[1] Chandrakasan, A. et. al, "Power Aware Wireless Microsensor Systems," *Keynote Paper ESSCIRC*, Florence, Italy, 2002.

[2] The Official Bluetooth Wireless Info Site, *http://www.bluetooth.com*

[3] Shanbhag, N.R, "A Mathematical Basis for Power-Reduction in Digital VLSI Systems," *IEEE Transactions on Circuits and Systems-II: Analog and Digital Signal Processing,* vol. **44**, pp. 935-951, 1997.

[4] Shannon C.E, "A Mathematical Theory of Communication," *The Bell System Technical Journal,* vol. **27**, pp. 379-423,623-656, 1948.

[5] Vapnik, V. *The Nature of Statistical Learning Theory,* New York: Springer-Verlag, 1995.

[6] Girosi, F., Jones, M. and Poggio, T. "Regularization Theory and Neural Networks Architectures," *Neural Computation,* vol. **7**, pp 219-269, 1995.

[7] Chakrabartty, S. and Cauwenberghs, G. "Forward Decoding Kernel Machines: A hybrid HMM/SVM Approach to Sequence Recognition," *IEEE Int. Conf. of Pattern Recognition: SVM workshop. (ICPR'2002),* Niagara Falls, 2002.

[8] Schölkopf, B., Burges, C. and Smola, A., Eds., *Advances in Kernel Methods-Support Vector Learning,* MIT Press, Cambridge, 1998.

[9] Cover T.A, Thomas J.A, *Elements of Information Theory,* John Wiley and Sons, 1991.

[10] Chakrabartty, S., Singh, G. and Cauwenberghs, G. "Hybrid Support vector Machine/Hidden Markov Model Approach for Continuous Speech recognition," *Proc. IEEE Midwest Symp. Circuits and Systems (MWSCAS'2000),* Lansing, MI, Aug. 2000.

GNOMES: A TESTBED FOR LOW POWER HETEROGENEOUS WIRELESS SENSOR NETWORKS

Erik Welsh, Walt Fish, J. Patrick Frantz
{welsh, waltfish, jpfrantz}@rice.edu

Texas Instruments Inc.
Stafford, TX 77477

Rice University
Houston, TX 77005

ABSTRACT

Continuing trends in sensor, semiconductor and communication systems technology (smaller, faster, cheaper) make feasible very dense networks of fixed and mobile wireless devices for use in many different sensing and decision-making systems. In this paper we present the design and development of GNOMES, a low-cost hardware and software testbed. This testbed was designed to explore the properties of heterogeneous wireless sensor networks, to test theory in sensor networks architecture, and be deployed in practical application environments. We also present an overview of architectures for extending the lifetime of individual nodes in the network, along with the design tradeoffs that this presents.

1. INTRODUCTION

This paper describes the development, initial testing and deployment of a low-power and inexpensive testbed for wireless sensor networks - GNOMES. In the past few years, as the costs and sizes of sensors and wireless communications devices has rapidly decreased, several such testbeds have been created to explore and verify the properties of these systems. These hardware testbeds - such as the SmartDusts project at UC Berkeley [1] and the μAmps project at MIT [2] - were designed to test theories of power aware computing, distributed sensor data gathering, data analysis, ad hoc routing and communication. Our GNOMES testbed has many similar features to these, but it also adds some new elements that are being used to boost new theoretical research at Rice University. In addition, the testbed is also being deployed in practical application scenarios.

Our design goal was to create a system of low-cost ($\leq$ \$25 each, in quantity) sensor nodes that are self-organizing and can work in a collaborative manner to solve a given problem. Cost and power consumption were the overriding design constraints for this development effort. These constraints were based on the following assumption: traditional macro sensor devices are bulky and expensive but extremely accurate. Because of the high cost per sensor, they are also not generally fault tolerant. Therefore, by flooding an application area with many small cheap devices, one can overcome the weaknesses that are inherent in this kind of system. Data collected from multiple sensors can be combined to improve the accuracy through the use of advanced DSP algorithms (e.g. creating a high-resolution image by combining multiple low-resolution images). The system is inherently fault tolerant because of the many nodes that are in use. If a few nodes become inactive, overall system performance should not suffer much.

It is important to note that in our system, nodes are not homogeneous, but rather heterogeneous. That is, each node in the network may not be sensing the same thing. This has benefits in keeping down the cost of each node since each one will not need a full complement of sensors. Sensors can be distributed in such a manner that adequate coverage is maintained for all measured system properties, while keeping the cost of each node relatively low. Organization of similar types of heterogeneous sensor networks have been previously discussed [3, 4], and this approach to system design opens up new research opportunities which are currently being explored at Rice [5].

Financial support for this project provided in part by Raytheon Co. Erik Welsh is an employee of Texas Instruments Inc. Erik Welsh and Patrick Frantz are with the Department of Electrical and Computer Engineering. Walt Fish is with the Department of Civil Engineering. **http://cmclab.rice.edu/projects/gnomes**

2. POWER CONSIDERATIONS

In almost all applications of sensor networks, there is a desire to achieve maximum node lifetime (e.g 1 year or more). Many of these nodes will be located in areas where battery replacement may not be an option. This problem of node lifetime and accessibility is often cited in the literature [6], and many researchers tend to combat this problem by concentrating on low-power node design or through energy-aware routing algorithms [7, 8]. We extend this body of work by exploring supplemental and alternate power technologies, such as solar power, in an attempt to extend the life of each node indefinitely.

In addressing the longevity problem, the system designer can consider several approaches:

Make use of exotic, cutting-edge battery technologies. These technologies - e.g. Lithium Ion or Lithium Polymer - have extremely high energy densities. Using batteries alone however, even exotics, has several disadvantages. First, these new technologies tend to be very expensive, which defeats the purpose of a low-cost system with potentially hundreds or thousands of nodes. Second, as is shown in [9] battery technology does not adhere to Moore's law, which is a given in the semiconductor industry. There are clearly diminishing returns in each new generation of battery technologies. Additionally, it is shown in [10, 11] that current advanced battery technologies such will not even meet the standby power requirements for one year of operation for most systems. In these systems, batteries alone will not be adequate.

Use a combination of power sources. This solution uses a battery technology that is rechargeable combined with an alternate power source that can be used to charge the battery at times when it is not needed. This secondary power source could be one of several possibilities and highly application dependent. For environ-

0-7803-7761-3/03/\$17.00 ©2003 IEEE

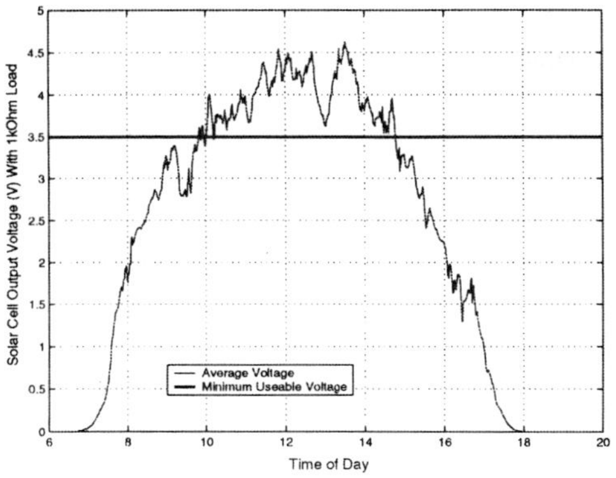

Figure 1: Average Output Voltage of Solar Cell across a 1kOhm Load

ments where outdoor operation is predominant, a solar cell could be used to recharge the battery as needed, provided that there is enough sunlight to do so. Outdoor structures such as bridges and skyscrapers could could also use small and inexpensive wind turbines to harness energy. In addition, other forms of energy could be converted to electrical potential. For example, the Piezoelectric effect allows mechanical stress to be converted into electricity, and it would be possible to store this charge in a battery.

Make use of "non-battery" technologies. In extremely embedded and low-power systems, the designer may have to forgo batteries completely. Such an example scenario may be the embedding of sensors into the concrete foundation of a building for long term measurement. In such a scenario, it may be necessary to rely completely on batteryless operation, such as the Piezoelectric effect or some other exotic power technology [12]. One could also envision powering medical sensors with body heat as the sole power supply [13]. Exploring such systems is a goal of ours at Rice, and the GNOMES project is merely a step in that direction.

Our system employs a simple solar cell as a proof of concept device to show that mixed-power nodes are a viable solution in large scale sensor networks. There are clearly some design consequences with this approach, however. A node could possibly spend much of its time in an inactive/recharging state, with little or no participation in the network. This will affect system performance as nodes become active and inactive at varying time intervals. To combat this, efficient routing and distributed computation algorithms will be necessary, which are also being studied at Rice. The system we have designed will allow the node to switch between one of two on board batteries. This will allow the node to use one battery for power to perform computations, collect sensor data and communicate with other nodes while the other battery recharges.

To test the feasibility of our architecture, we ran an experiment with the 5 Volt, 30mA solar cell used in our design. With solar cells, the output voltage is proportional to the intensity of the sunlight received. For our experiment, we placed the solar cell on the end of a rod, facing skyward in our engineering quad. In this unobstructed position, the cell received direct sunlight for most of the daylight hours. We then measured the output voltage across a 1kOhm load at 1 min. intervals over a 1 week period in early January. Figure 1 shows these readings.

To determine the useable voltage level, we have to take into

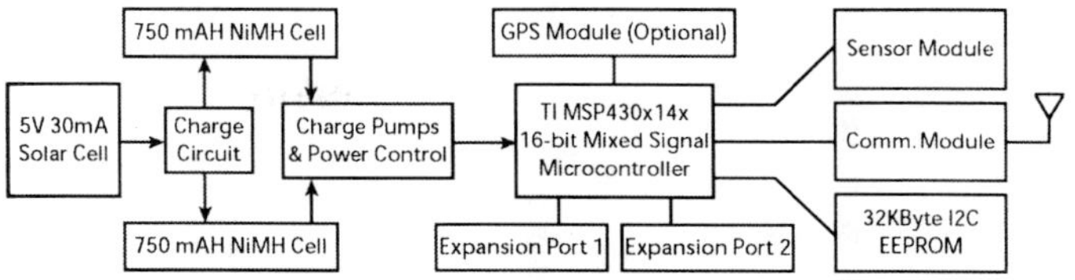

Figure 2: GNOMES Block Diagram

consideration that our charging circuit requires a minimum of 3.5 Volts to provide enough current to charge a battery. The data show that in this location we could expect about 5 hours of useful sunlight each day. Of course, the amount of sunlight received has many dependencies: time of day, weather, season, geographical location, etc. Because of this, each node's exposure to sunlight will vary. According to official records, the total amount of daylight during the experiment was approximately 10 hours and 39 minutes, with two sunny days and three cloudy, rainy days. To combat the fact that there will not always be enough time to sufficently charge the battery, we are developing intelligent adaptive algorithms to regulate a nodes activity in the network. The effect that this will have on the sensor network architecture can be dramatic. Nodes that have little chance to recharge their batteries will not be very active participants in the network. Perhaps they will be forced to sense less often and transmit fewer bits of data in order to conserve power.

In addition to solar power, we are also exploring other possibilities, such as wind, hydro and electromechanical power, and have recently obtained some very promising preliminary results using wind power.

3. SYSTEM HARDWARE

A block diagram of the node we have designed is shown in Figure 2. Our GNOMES node resembles the COTS Dust node designed as part of the SmartDusts project at UC Berkeley [14] occupying approximately 55 cm^3 (3.8 cm by 5.7 cm by 2.5 cm). While our design was inspired by this project, we have added several unique features that improve upon the original design and allow for greater system flexibility and performance.

The heart of the GNOMES node is a 16-bit MPS430 microcontroller from Texas Instruments. The particular chip that we chose, the MPS430F149, offers an impressive array of features while consuming very little power (about 3.5 MIPS/mA).

- 16-bit Operation, with Hardware Multiplier
- Up to 7.3728MHz Operation, Dynamically Controllable
- 60KBytes of Code Memory, 2KBytes of Data Memory
- 8 Integrated Channels of AD Conversion with 12 Bit Precision
- 2 UARTs for Serial Communications

In addition, our GNOMES node offers the following capabilities, as shown in Figure 2.

- An 32KByte I^2C EEPROM for Persistent Data Storage
- Support for Multiple Power Supplies (e.g. Battery and Solar)
- Support for an Optional Integrated GPS Module
- Modular Communications Header - Bluetooth or 900MHz Radio
- Modular Sensor Header
- Two Expansion ports (14 GPIO total; 5 interuptable & I^2C)

The following features are unique to GNOMES. First, as we previously mentioned, we designed this node to explore the longevity of sensor nodes. Our device draws its main power from one

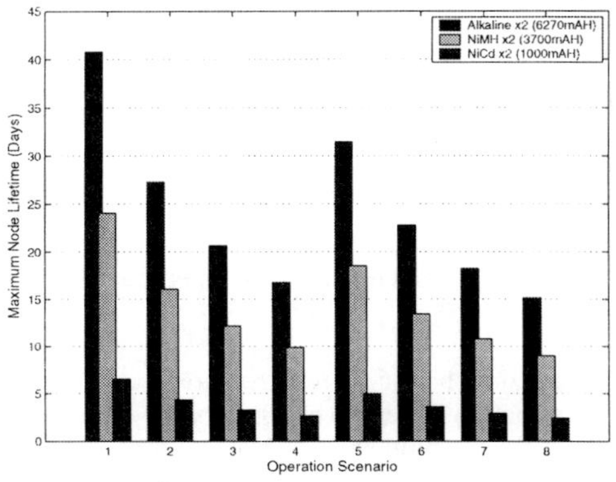

Figure 3: Maximum Lifetime of a GNOMES Node
(Accounting for Power Supply Inefficiency)

Scenario #	1	2	3	4	5	6	7	8
Sense & Process	50%	50%	100%	98%	50%	50%	98%	96%
Communication	0%	2%	0%	2%	0%	2%	0%	2%
GPS	0%	0%	0%	0%	2%	2%	2%	2%
Sleep	50%	48%	0%	0%	48%	46%	0%	0%
Avg. Current (mA)	2	3	4	4.9	2.6	3.6	4.5	5.4
Avg Power (mW)	6.7	10	13.2	16.2	8.7	12	14.9	18

Table 1: Operation Scenarios for a GNOMES Node
(Irrespective of Power Supply)

of two AA NiMH cells. In addition, auxiliary power can be supplied via an external device such as a solar cell in order to charge the battery that is not being used for main power. An adaptive software algorithm monitors system performance and battery voltages to determine when to switch sources, such that the longevity of a node can be extended indefinitely. This dual battery approach imposes little restriction on system operation. However, the node must curtail its operation if it cannot recharge for long periods of time and both batteries become low. This places some constraints on the design and operation of the sensor network, and we are currently exploring the issues that this raises in both theory and in practice.

Second, we have designed this node to include a modular sensor header, so that changing the functionality of a node is as simple as plugging in a new sensor module. In addition, this modularity allows us to design and test heterogeneous wireless sensor networks, in which each node may sense a different property of the environment than its immediate neighbor. This design approach also lowers the cost of each node since not every one of them will need to have many sensors. Our first sensor module includes accelerometers that measure two axes of acceleration from -5g to +5g. It also includes temperature sensors to aid in calibration of the accelerometers. Other sensor modules are currently in design.

Third, we designed the communications subsystem to be modular as well. Currently, it can function with either a 2.4GHz Bluetooth radio or a 900MHz radio module. For our first communication module, we chose to design around a Class 2 Bluetooth module [15]. We made this design decision for several reason. Primarily, Bluetooth is a relatively low-power radio device that offers moderately high data rates (up to 721kbps for user applications). In addition, many low-level communications functions are already built into the hardware link and physical layers, such as power efficient modulation, device discovery, smart packet detection and intelligent sleep scheduling [16]. We have also found in our previous work that typical Bluetooth radios far outperform the specifications for device range without adversely affecting data throughput. For example, we have observed Class 2 Bluetooth modules maintain connections at 5 to 10 times the specified distance of 10m [17]. We have also developed a 900MHz radio module. However, we are still running tests on the module, after which it will be ready for production.

Using data gathered from lab tests, we performed an analysis and placed some upper bounds on the lifetime of a node. These data are shown in Figure 3 and Table 1. Although our nodes recharge a battery through solar power, these tables present baseline data without any recharging of the batteries. Figure 3 shows the maximum lifetime in days for a node for three different modern battery technologies and eight different scenarios. The scenarios are described in Table 1 and show percentage of time devoted to certain activities. Average current and power consumption are also given for each scenario. We picked percentages that seemed reasonable for some of our applications.[1] These percentages are being verified in field tests.

Even in the absolute worst case scenario (#8 with the NiCd battery), we predict that a node could achieve an uptime of 25% and operate indefinitely. That is, with using the solar cell to charge the system battery, a node would have to spend 3 days recharging the battery for every 1 full day it spent sensing processing and communicating.

Our cost goal for GNOMES was to keep each node under \$25 in large quantities (> 1000). Cost depends largely on the sensors, and we expect the nodes to become cheaper as the semiconductor industry continues its pace of device integration and miniaturization. Currently a GNOMES node with 2-axes of acceleration sensing costs around \$50 without GPS and \$80 with GPS.

4. SYSTEM SOFTWARE

In order to easily create networking and other application specific software for the GNOMES testbed, we have also begun development of an embedded real time operating system and low-level software drivers to control each of GNOMES' subsystems. Modelled after TinyOS [18] from UC Berkeley and ADEOS, written by Michael Barr [19], the GNOMES operating system will offer process management and scheduling. This allows software developers to write multiple applications to be run on the node concurrently as well as allowing the operating system to take care of power management issues in the background.

We have also developed software drivers and common APIs to provide easy process-level control for each of the GNOMES subsystems. This includes: a partial implementation of a Bluetooth [15] stack to allow for device discovery and connection establishment; NMEA string parsing to better manage data from the GPS module for user applications; I^2C driver to communicate with the on-board EEPROM; an intuitive interface to control the AD converter. We have also implemented a power management pro-

[1] 1% time devoted to communication yields a maximum data transfer rate of $\approx$1MBytes/hour, well more than we expect most of our applications to use. 2% times devoted to GPS is roughly equivalent to the time needed to sample GPS data once per hour.

cess that will run in the background and look for ways to conserve power on the node. Both the drivers and the operating system were developed in a mixture of C and assembly on the "IAR Embedded Workbench" from IAR systems, a Texas Instruments third-party company, which offers both and embedded development as well as debugging environments.

5. APPLICATIONS

The low cost and long lifespan of GNOMES make them ideal for investigating various approaches to traditional sensing applications. In addition, their on-board processing power, wireless capabilities and modular design allow us to explore novel and innovative uses:

Seismic monitoring of civil structures. GNOMES allows for the rapid deployment of dense sensing networks in structures otherwise unsuited for seismic monitoring. This gives valuable insight into the nature of building behavior under various excitations. Supplementary nodes equipped with strain gauges and force transducers will be able to monitor the local performance of individual members and connections. As this data is collected, the use of on-board computing power permits pre-processing of data, thereby decreasing the necessary bandwidth for the centralized collection of data. Furthermore, structures such as bridge towers and high-rise buildings are well suited to the employment of alternate power-sources, such as solar and wind power.

Tracking and routing of personnel and machinery. GNOMES, fitted with the optional GPS module, can help track and optimize the progress heavy machinery such as graders and excavators. In indoor environments, the radio modules can act as beacons, which in conjunction with signal strength maps [20] would permit the localization of personnel operating on site. This opens the possibility of providing context-aware services to the jobsite workforce. As the cost of these devices decreases, they could be attached to large pieces of equipment as identification badges, allowing their efficient tracking and routing on the way to installation.

Examination of contaminant levels and flow. Environmental regulations and awareness lead to a reduction of the allowable concentrations of pollutants released into air, soil, and water. Thus, to detect and identify these levels of pollutants, it is increasingly important to advance the capabilities of field monitoring methods and systems. The needs for fast and cost-effective environmental assessment have been addressed with a variety of chemical sensors [21]. With the addition of chemical sensing modules, GNOMES could be employed to monitor the levels and flow patterns of environmental contaminants.

Corrosion detection and cure-rate monitoring. We are currently investigating the use of GNOMES in embedded environments such as concrete members and asphalt roadways. Through the examination of moisture content and temperature, it is possible to estimate the maturity of concrete in structural components, providing a non-destructive complement to existing strength tests. The embedding of sensors in roadways is already being investigated and could detect corrosive subsurface conditions before serious damage is done. This application could be extended to include underground pipes and drainage tiles.

6. CONCLUSION AND FUTURE DIRECTIONS

In this paper, we have presented the complete design of and applications for GNOMES, a testbed for wireless heterogeneous sensor networks. This system was built to explore the various properties of and to test and develop theories for wireless sensor networks, with a initial focus on increasing the longevity of sensor nodes and characterizing the performance this has on a sensor network. We are actively developing our testbed by creating new sensor and communication modules and by manufacturing a large quantity of the GNOMES nodes. In addition, we are considering designing a new node with either multiple MSP processors or a more powerful DSP. We realize that this will introduce more power consumption, but we also recognize that this will allow us to implement more sophisticated and interesting algorithms on each node. This will allow us to better explore the tradeoffs between local and remote data processing within a sensor network.

7. ACKNOWLEDGMENT

The authors would like to than the following Rice University undergraduate students who have helped to contribute to the success of this project: Randy Holman, Rob Gaddi, CJ Ganier, Ricky Hardy, Julie Rosser, Kileen Cheng and Patrick MacAlpine. We would also like to thank Mark Whitt, Sara Heaton and Gene Frantz of Texas Instruments for their help in the success of this project.

8. REFERENCES

[1] J. Khan, R. Katz and K. Pister, "Next Century Challenges: Mobile Networking for Smart Dust," in *Proceedings of the 5th Annual ACM/IEEE International Conference on Mobile Computing and Networking*, pp. 271–278, 1999.

[2] R. Min, M. Bhardwaj, S. Cho et al, "An Architecture for a Power-Aware Distributed Microsensor Node," in *IEEE Workshop on Signal Proc. Systems*, pp. 581–590, Oct 2000.

[3] E. Brewer, R. Katz, E. Amir et al, "A Network Architecture for Heterogeneous Mobile Computing," in *IEEE Personal Comm Magazine*, pp. 8–24, Oct 1998.

[4] L. Subramanian and R. Katz, "An Architecture for Building Self-Configurable Systems," in *1st Ann Wkshp on Mobile and Ad Hoc Networking and Comp*, pp. 63–78, 2000.

[5] A. Sabharwal, "On Capacity of Relay-Assisted Communication," in *IEEE Globecom*, November 2002.

[6] M. Bhardwaj, T. Garnett, and A. Chandrakasan, "Upper Bounds on the Lifetime of Sensor Networks," in *IEEE Intl Conf on Comm*, vol. 3, pp. 785–90, 2001.

[7] W. Heinzelman, A. Chandrakasan and H. Balakrishnan, "Energy-Efficient Routing Protocols for Wireless Microsensor Networks," in *Proc Hawaii Intl Conf on System Sciences*, 2000.

[8] F. Koushanfar, V. Prabhu, M. Potkonjak, and J. Rabaey, "Processors for Mobile Applications," in *Proc Intl Conf on Computer Design*, pp. 603–8, 2000.

[9] H. Hahn and H. Reichl, "Batteries and Power Supplies for Wearable Ubiquitous Computing," in *3rd Intl Symp on Werable Computers*, pp. 168–169, 1999.

[10] A. Wang and A. Chandrakasan, "Energy-Efficient DSPs for Wireless Sensor Networks," in *IEEE Sig Proc Magazine*, vol. 43(5), pp. 68–78, July 2002.

[11] R. Powers, "Advances and Trends in Primary and Small Secondary Batteries," in *IEEE Aerospace Electronics Systems Magazine*, vol. 9, pp. 32–36, April 1994.

[12] H. Li, A. Lala et al, "Self-Reciprocating Radioisotope-Powered Cantilever," *Journal of Applied Physics*, vol. 92, no. 2, 2002.

[13] "Prototype: Body Power." MIT Technology Review, September 2002.

[14] S. Hollar, "COTS Dust," Master's thesis, Univ of Cal, Berkeley, 2000.

[15] Bluetooth Specification version 1.1, "http://www.bluetooth.com/."

[16] J. Haartsen and S. Mattisson, "Bluetooth - A New Low-Power Radio Interface Providing Short-Range Connectivity," in *IEEE Proc*, pp. 1651–61, Oct 2000.

[17] P. Murphy, E. Welsh, and P. Frantz, "Using Bluetooth for Short-Term Ad Hoc Connections Between Moving Vehicles: A Feasibility Study," in *Proc. 55th IEEE Vehicular Technology Conf*, vol. 1, pp. 414–418, 2002.

[18] UC Berkeley TinyOS, "http://webs.cs.berkeley.edu/tos/."

[19] M. Barr, *Programming Embedded Systems in C and C++*. O'Reilly, 1999.

[20] P. Bahl and V. Padmanabhan, "RADAR: An In-Building RF-Based User Location and Tracking System," in *Proc of IEEE Infocom*, 2000.

[21] R. Potyrailo, "Environmental Sensors: What's New Since Tomas Hirschfeld?" GE Co, Global Research Center, Characterization and Combinatorial Chemistry Technologies.

ANALYSIS OF SHORT DISTANCE OPTOELECTRONIC LINK ARCHITECTURES

A. Apsel

Cornell University
Ithaca, NY, 14853

A. G. Andreou

Johns Hopkins University
Baltimore, MD, 21218

ABSTRACT

In this paper we present analysis of two types of receivers, transimpedance and low impedance, for chip-to-chip short distance interconnects. We investigate the combined transmitter and receiver power of links containing each type of receiver as part of a hard-wired interconnect. We find that the power consumption of a complete optoelectronic link as a function of bit rate for low impedance receiver links is often better than for transimpedance amplifier short distance links.

1. INTRODUCTION

It is generally accepted that optical high speed signaling over long distances outperforms electrical signaling via cables at the same distances and bit rates [1, 2]. Electrical signals suffer degradation due to parasitic resistances, capacitances, and inductances that induce losses, latency, and cross-talk into electrical signals. Optical signals suffer virtually no cross-talk, reduced latency, and can be designed with lower losses at room temperature. Although trading wires for optical interconnects is less accepted over distances of less than one meter, the same sources of signal degradation exist on these smaller scales. In fact, as CMOS feature sizes decrease further into the sub-micron regime, electrical signaling and interconnect problems are predicted to worsen and become the limit of high performance systems at both the board and chip levels [3]. Yayla et. al. have shown that when both power and speed are considered, optical signal paths are a good alternative to electrical paths at distances as short as 1 cm [4].

In the past, analysis of opto-electronic interconnects has focused on improvement in the local design of either the transmitter or receiver. Little has been said about the design of an entire opto-electronic link other than studies of link power, speed, and feasibility in comparison to purely electronic interconnects [4, 3]. In a wide area network requiring switch-able hardware and versatility, it is sensible to simply use the highest sensitivity receiver at a given bit rate and to generate as much power as possible from the transmitter. At inter and intra-chip communication scales, however, this is not appropriate. When networks are small and hard-wired, power minimization of the interconnect as a whole will dictate the number of interconnections and computations possible. Kibar et. al. investigate power consumption of an entire opto-electronic link in this manner to determine which combination of VCSEL's, MQW (multiple quantum well) structures, bipolar drivers, and CMOS drivers yield the highest performance link [5]. This analysis does not provide any insight about how link power and performance are effected by receiver design.

In this paper we examine the entire optical interconnect as a single design to determine which type of receiver architecture will produce high bit rate signals with low power consumption in a hardwired link. We will examine two types of interconnect architectures and determine whether it is better to use low impedance or simple transimpedance receivers at various bit rates.

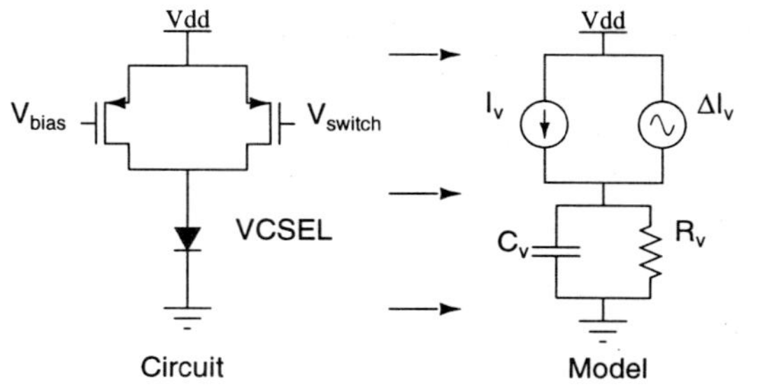

Figure 1: Simplified Driver Circuit and Model

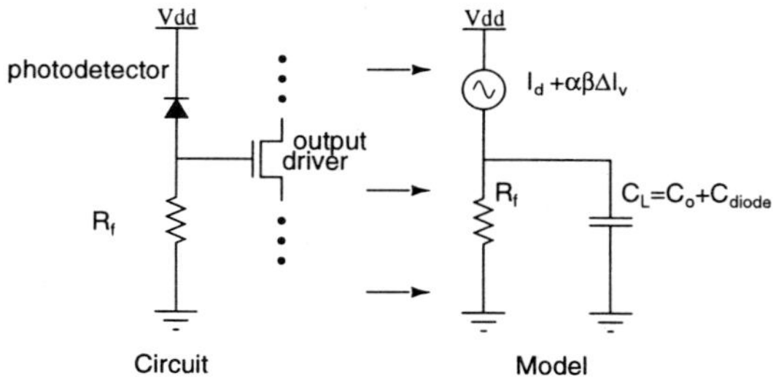

Figure 2: Simplified RC Receiver Circuit and Model

2. LINK POWER CONSUMPTION

2.1. Model Variables

We begin our analysis with a list of model variables.

R_v - VCSEL resistance
ΔI_v - VCSEL input signal current swing
I_v - VCSEL dc current, typically threshold current
γ - activity factor, represents the percentage of time on average that the signal is high. This is dependant on both duty cycle and coding method.
f - signaling frequency
α - detector optical efficiency in A/W
β - VCSEL optical efficiency in W/A
ΔV_o - receiver output signal voltage swing
R_f - Input impedance of receiver

0-7803-7761-3/03/$17.00 ©2003 IEEE

C_L - the load capacitance on the RC receiver composed of the photodetector capacitance and the capacitance of the next stage
I_d - the dc current from the detector. We assume that it is dominated by dark current. It may be dominated by ambient light photocurrent instead.
V_{DD} - supply voltage
$A(w)$ - Gain of internal amplifier as a function of frequency
V_e - Early Voltage of load transistor in internal amplifier
I_b - Bias current of internal amplifier
C_o - output capacitance, defined by capacitance of next stage
C_{in} - input capacitance, defined by photodetector capacitance and input capacitance of the internal amplifier

2.2. Low Impedance Resistive Receiver Link

Figures 1 and 2 show the low impedance (RC) link driver and receiver. In order to determine the total power consumption of an optoelectronic link we begin by determining the power consumption of the transmitter based upon the bit rate of operation and the impedance gain of the receiver. We then calculate the power consumption of the receiver as a function of bit rate. By summing these expressions, we achieve an expression for the power consumption of the entire link. The model that we use for these calculations of power is noiseless.

The power consumption of the driver is defined by the needs of the receiver. We require the VCSEL to produce enough light such that the receiver will produce an output signal of amplitude ΔV_o. The value of ΔV_o will vary depending upon the configuration of a receiver circuit and whether a limiting amplifier is used. We use a value of ΔV_o=100mV in this analysis. Considering the VCSEL efficiencies of α and β respectively, $\Delta I_v = \frac{\Delta V_o}{\alpha\beta R_f}$. The driver consumes an average power of

$$P_{TX} = V_{DD}(I_v + \gamma\Delta I_v) = V_{DD}(I_v + \frac{\gamma\Delta V_o}{\alpha\beta R_f}) \quad (1)$$

where γ defines the portion of time when the VCSEL is "on". In this analysis the operation of the VCSEL driver is assumed to be limited by the RC bandwidth of the VCSEL. We assume that the primary limitation on interconnect bandwidth is the receiver bandwidth.

We now consider the power expended by the receiver. The average power consumption of the RC receiver in this link is given by the average of the "on" and "off" power over one cycle

$$P_{RX} = V_{DD} I_{avg} \quad (2)$$

$$= f\left[V_{DD}\int_0^{\frac{1-\gamma}{f}} I_d dt + V_{DD}\int_0^{\frac{\gamma}{f}} (I_d + \alpha\beta\Delta I_v)dt\right]$$

which integrates over an average cycle to

$$P_{RX} = V_{DD}\left[I_d + \gamma\alpha\beta\Delta I_v\right] = V_{DD}\left[I_d + \gamma\frac{\Delta V_o}{R_f}\right] \quad (3)$$

since $\alpha\beta\Delta I_v$ is the current switched through the photo detector. R_f determines the 3dB bandwidth of the receiver, $\omega_{3db} = \frac{1}{R_f C_L}$ Note that the maximum bit rate of a receiver is typically defined at $(.7/\pi)\omega_{3db}$ for non-return to zero (NRZ) codes. (This standard has been determined based upon allowing only a small phase shift and amplitude reduction at signaling speed to approach flat band response.) R_f can be defined in terms of the maximum flat-band frequency, f_{max}.

$$R_f = \frac{.7}{2\pi f_{max} C_L} \quad (4)$$

Using these terms we can rewrite the power consumption of the link in terms of given parameters and frequency.

$$P_{RX} = V_{DD} I_d + \frac{\gamma V_{DD}\Delta V_o \pi C_L f_{max}}{.7} \quad (5)$$

$$P_{TX} = I_v V_{DD} + \frac{\gamma V_{DD}\Delta V_o \pi C_L f_{max}}{\alpha\beta .7} \quad (6)$$

Figure 3 shows the division of power consumption between the driver and receiver when used in a link to produce an output of 100mV at a given maximum bit rate. The total power consumption of the link is also shown, and is heavily dominated by the power consumption of the optical driver as expected for a relatively low sensitivity receiver configuration. The expression for total link power is shown below.

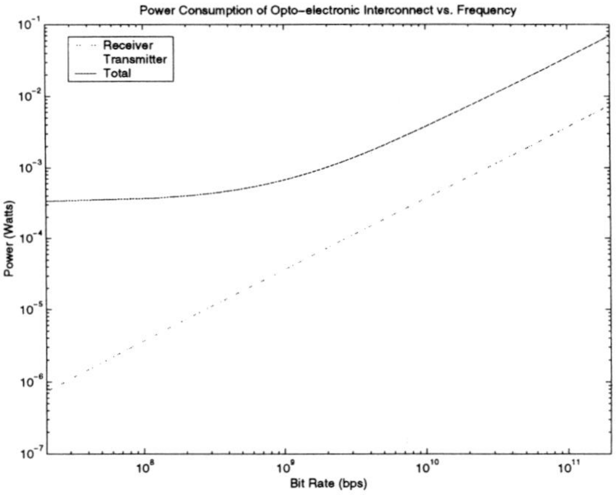

Figure 3: Power consumption of RC link. In this calculation we assume a 50 percent duty cycle, a VCSEL threshold current of 0.1mA, load capacitance of .1pF, and optical efficiencies of .4A/W and .3W/A. The values for threshold currents capacitances and the optical efficiencies are based upon detectors and VCSELs fabricated by Honeywell and Emcore.

$$P_{RC} = V_{DD}(I_v + I_d) + \gamma V_{DD}\Delta V_o C_L \frac{\pi}{.7}(1 + \frac{1}{\alpha\beta}) f_{max} \quad (7)$$

2.3. Transimpedance Receiver Link

The transimpedance link is comprised of the transmitter shown in figure 1 and the receiver shown in figure 4. We compute the power consumption of this link using the same method as we used in the RC link. This model has some added complexity due to the consideration of the internal amplifier as a power dependent and band limited element. In this analysis we consider the internal amplifier to be a simple, single ended common source amplifier as shown in figure 5.

Again, the power in the driver is defined by the requirements of the receiver as in the RC case so $P_{TX} = V_{DD}(I_v + \Delta I_v \gamma) = V_{DD}(I_v + \frac{\gamma\Delta V_o}{\alpha\beta R_f})$ as in equation 1. In this case as well, the relationship between the driver and receiver is defined primarily by ΔI_v.

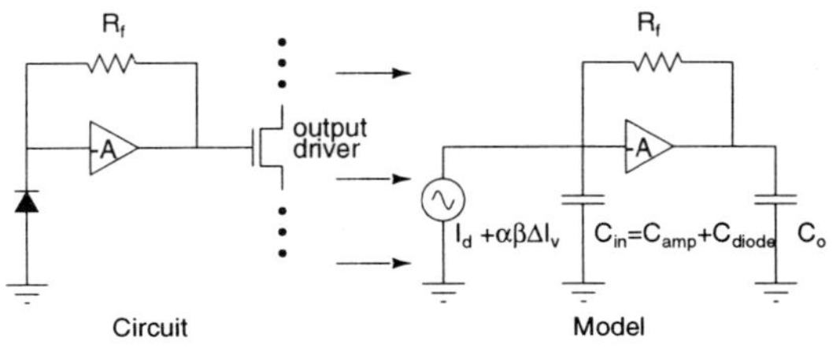

Figure 4: Simplified Transimpedance Amplifier Receiver Circuit and Model

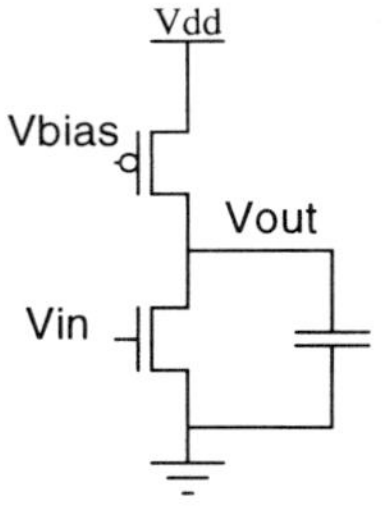

Figure 5: Common Source Amplifier

On the receiver side we begin by considering the internal amplifier. We assume a constant gain bandwidth product, GBW such that $GBW = A(\omega)\omega_{3dbint} = \frac{g_m}{C_o}$, which limits the possible operating bandwidth of the full TIA. Also, in this analysis we assume operation in a region where A is large (at least better than 2). This lets us use the approximation for the TIA that $\omega_{3dbtia} = \frac{A(\omega)}{R_f C_{in}} = \frac{\omega}{.7}$, where $\omega = 2\pi f_{max}$ and $2f_{max}$ is the maximum sustainable bitrate through the TIA. In a robust TIA design with limited peaking $BW_{TIA} < .5BW_{internalamplifier}$. For this analysis we will assume that $BW_{TIA} = .5BW_{internalamplifier}$. This condition insures that the lowest frequency pole of our amplifier is given by the feedback resistance and the input capacitance, rather than by the output conductance of our internal amplifier. The relationship between the internal amplifier bandwidth and the maximum bit rate frequency is given by $\omega = .35\omega_{3dbint}$. From here we can look at the power consumption of the receiver alone as

$$P_{RX} = I_b V_{DD} + I_d^2 R_f + \gamma(\alpha\beta\Delta I_v)^2 R_f \tag{8}$$

where I_b is the bias current of the internal amplifier and is related to the bandwidth of the amplifier as follows.

$$\omega_{3dbint} = \frac{1}{r_o C_o} = \frac{I_b}{V_e C_o} \tag{9}$$

Note that in this first order analysis we have not yet considered the capacitance of the input gate of our common source amplifiers. It may be the case that in order to maintain high gain with this amplifier given a small photodiode capacitance that the capacitance of our input gate will be large enough to dominate C_{in}. In this case, $C_{in} = C_D + C_g$, where the gate capacitance, $C_g = C_{ox}WL$ is defined in terms of gate width and length and oxide capacitance per square micron. We can find C_g in terms of L, the length of the input gate, using the expression for GBW.

$$GBW = \frac{g_m}{C_o} = \frac{\sqrt{2\frac{W}{L}\mu_n C_{ox} I_B}}{C_o} \tag{10}$$

From here we can solve for W in terms of L to find C_g.

$$C_g = WLC_{ox} = \frac{(GBWC_oL)^2}{2\mu_n I_B} = \frac{.35}{4\pi}\frac{(GBWL)^2 C_o}{\mu_n V_e f_{max}} \tag{11}$$

We see from the new expression for C_{in} that when A is large, at low frequencies, $C_{in} > C_D$. At high frequencies gain and sensitivity are sacrificed for bandwidth, and $C_{in} \approx C_D$.

The receiver power consumption can be expressed as

$$P_{RX} = \frac{.7A(\omega)I_d^2}{2\pi C_{in} f_{max}} + \left(\frac{2\pi}{.35}V_e C_o V_{DD} + \frac{2\pi\gamma\Delta V_o^2 C_{in}}{.7A(\omega)}\right) f_{max} \tag{12}$$

Noting that $\Delta V_o = k\Delta I_v R_f$ for both links, and $R_f = \frac{.7A(\omega)}{\omega C_{in}}$ we can determine the power consumption of the driver.

$$P_{TX} = I_v V_{DD} + \frac{\gamma 2\pi C_{in} V_{DD}\Delta V_o}{.7\alpha\beta A(\omega)} f_{max} \tag{13}$$

The total power consumption of the TIA link is the sum of these quantities.

$$P_{TIA} = \left(\frac{2\pi}{.35}V_e V_{DD} C_o + \frac{2\pi\gamma\Delta V_o C_{in}}{.7A}(\Delta V_o + \frac{V_{DD}}{\alpha\beta})\right) f_{max} \tag{14}$$
$$+ \frac{.7A(\omega)I_d^2}{2\pi C_{in} f_{max}} + V_{DD} I_v$$

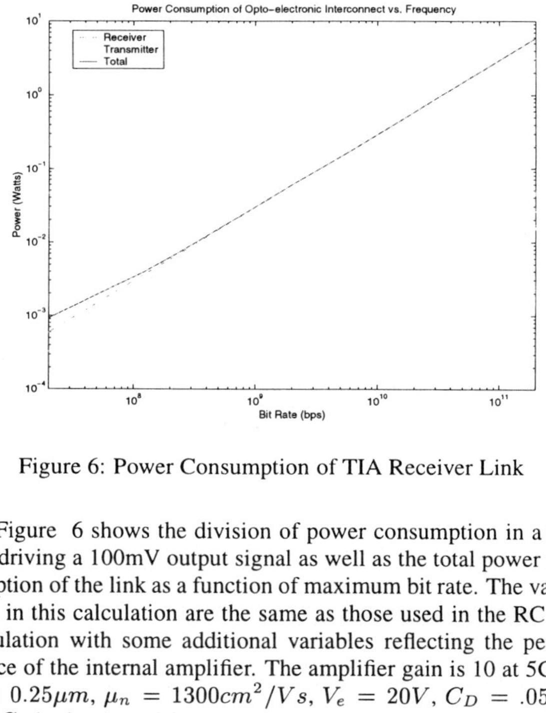

Figure 6: Power Consumption of TIA Receiver Link

Figure 6 shows the division of power consumption in a TIA link driving a 100mV output signal as well as the total power consumption of the link as a function of maximum bit rate. The values used in this calculation are the same as those used in the RC link calculation with some additional variables reflecting the performance of the internal amplifier. The amplifier gain is 10 at 5GHz, $L = 0.25\mu m$, $\mu_n = 1300cm^2/Vs$, $V_e = 20V$, $C_D = .05pF$, and C_o is dominated by the capacitance of the next input stage taken as .05pF as well. These capacitance values are plausible if we assume that C_L in the RC case is the photo-diode capacitance plus the input capacitance of the next input stage.

Unlike the previous case of the RC link, power consumption in the TIA link is not strictly dominated by the driver. In fact, through most of the frequency range of interest the receiver consumes the majority of link power. Even through the higher sensitivity transimpedance amplifier requires a lower power signal from the driver than the RC receiver requires, the power requirements of the TIA are orders of magnitude greater than those of a passive receiver.

2.4. Comparisons

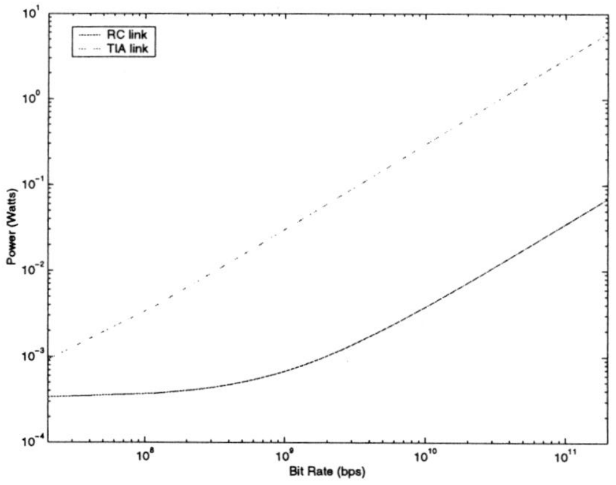

Figure 7: Power Consumption of TIA and RC Receiver Links

Figure 7 shows the relative power consumptions of the RC and TIA link architectures over the frequency range of interest. Although the power consumption of the more sensitive transimpedance receiver dominates power consumption over this range, it seems that power consumption of the two links approach each other at low bit rates. It is also possible that for some condition with very high power consumption at the driver, the transimpedance design may be preferred. Intuitively we know that the TIA receiver enables the optical power of a link to be reduced by a factor of A while maintaining the bandwidth of the link equal to the RC link. If the power consumption of the system is dominated by the driver, then a TIA system will perform better. However, the TIA link has additional power consumption in the receiver due to active internal stages. If the receiver dominates power consumption, then the RC link may be better.

Neglecting the noise limitation at this time, we can find the approximate conditions under which we prefer to use the RC link. We begin by comparing the full expressions for link power as a function of frequency such that $P_{RC} \leq P_{TIA}$.

$$V_{DD}(I_r+I_d)+\gamma V_{DD}\Delta V_o C_L \frac{\pi}{.7}(1+\frac{1}{\alpha\beta})f_{max} \leq V_{DD}I_v \quad (15)$$

$$+\frac{.7I_d^2}{2\pi C_{in} f_{max}}+\left(\frac{2\pi}{.35}V_e V_{DD} C_o + \frac{2\pi\gamma\Delta V_o C_{in}}{.7A}(\Delta V_o + \frac{V_{DD}}{\alpha\beta})\right) f_{max}$$

This expression can be reduced to a polynomial in f. We cancel the threshold current dominated terms and consider I_d to be small, and rewrite in terms of capacitances as follows.

$$\left(\frac{1}{\alpha\beta}+1\right) C_D-\left(\frac{\Delta V_o}{V_{DD}}+\frac{1}{\alpha\beta}\right)\frac{2C_{in}}{A} \leq \left(\frac{4V_e}{\gamma\Delta V_o}-\frac{1}{\alpha\beta}-1\right) C_o \quad (16)$$

Using this expression we can gain some insight about the power consumption of the two links using some general case approximations. In general, these receivers are used as first stage inputs with a gain stage to follow such that $\frac{\Delta V_o}{V_{DD}} < 1$. We also know that in real cases $\alpha\beta < 1$. If we estimate that the gate capacitances of our TIA designs are small compared to the photodiode capacitance over our frequency range of interest, we can further simplify this expression.

$$\frac{C_D}{C_o} \leq \frac{\frac{4V_e}{\gamma\Delta V_o}-1-\frac{1}{\alpha\beta}}{\frac{1}{\alpha\beta}(1-\frac{2}{A})+1-\frac{2\Delta V_o}{AV_{DD}}} \approx \frac{\frac{4V_e}{\gamma\Delta V_o}-(1+\frac{1}{\alpha\beta})}{1+\frac{1}{\alpha\beta}} \quad (17)$$

The approximation is valid for $10 < A < 100$. For $10 > A > 2$, the denominator will always be positive but will be very small as A becomes close to 2, making the inequality easier to satisfy. If $A >> 100$ the expression for C_{in} must be evaluated.

In the case that $10 < A < 100$ and the equation is most difficult to satisfy, we can examine the condition for the power consumption of the TIA link to be lower than for a passive RC link. If we define $R = \frac{C_D}{C_o}$, we can write

$$R+1 \leq \frac{4V_e}{\gamma\Delta V_o(1+\frac{1}{\alpha\beta})} \approx \frac{4V_e\alpha\beta}{\gamma\Delta V_o} \quad (18)$$

which indicates that this condition will not be satisfied in general if R is large or $\alpha\beta$ is small. These are logical results. $\frac{C_D}{C_o}$ large implies a large photodiode capacitance which enables and requires the design of a high gain TIA since an RC link needs a very small resistance to compensate for large C_D. The RC link with a small R_f would require a very large optical input signal and large amounts of transmitter power.

Similarly a small $\alpha\beta$ indicates poor optical efficiencies of the photonic elements. Clearly when the optical efficiencies are small total link power consumption can be reduced by using a high sensitivity receiver. A passive receiver will require large amounts of power from a transmitter to produce a signal. If we consider C_D =60fF, C_o =20fF and reasonable values of circuit parameters (ΔV_o =100mV, γ =0.5, V_e =5V) this condition holds for $\alpha\beta$ >0.01 or as long as the optics and channel are at least 1% efficient.

It is notable that as technology improves the speed of devices and the quality of photonic elements, a better design for minimizing link power consumption is a passive receiver.

3. REFERENCES

[1] M. C. Gupta, *Handbook of Photonics*, CRC press, Boca Raton, 1997.

[2] G. Agrawal, *Fiber-Optic Communication Systems*, John Wiley and Sons, INC., New York, 1992.

[3] Jeffrey Davis, Raguraman Venkatesan, Alain Kaloyeros, Michael Beylansky, Shukri Souri, Kaustav Banerjee, Krishna Saraswat, Arifur Rahman, Rafael Reif, and James Meindl, "Interconnect limits on gigascale integration in the 21st century," *Proceedings of the IEEE*, vol. 89, no. 3, pp. 305+, March 2001.

[4] Gokce I. Yayla, Philippe J. Marchand, and Sadik C. Esener, "Speed and energy analysis of digital interconnections: comparison of on-chip, off-chip, and free-space technologies," *Applied Optics*, vol. 37, no. 2, pp. 205+, Jan 1998.

[5] Osman Kibar, Daniel A. Van Blerkom, Chi Fan, and Sadik C. Esener, "Power minimization and technology comparisons for digital free-space optoelectronic interconnections," *Journal of Lightwave Technology*, vol. 17, no. 4, pp. 546+, Apr 1999.

BULK CARBON NANOTUBE AS THERMAL SENSING AND ELECTRONIC CIRCUIT ELEMENTS

*Victor T.S. Wong and Wen J. Li**

Centre for Micro and Nano Systems, The Chinese University of Hong Kong, HKSAR

ABSTRACT

Bulk multi-walled carbon nanotube (MWNT) were successfully and repeatably manipulated by AC electrophoresis to form resistive elements between Au microelectrodes and were demonstrated to potentially serve as novel temperature sensor and simple electronic circuit elements. We have measured the temperature coefficient of resistance (TCR) of these MWNT bundles and also integrated them into constant current configuration for dynamic characterizations. The I-V measurements of the resulting devices revealed that their power consumption were in μW range. Besides, the frequency responses of the tested devices were generally over 100 kHz in constant current mode operation. Using the same technique, bulk MWNT was manipulated between three-terminal microelectrodes to form a simple potential dividing device. The tested device was capable of dividing the input potential into 2.7:1 ratio. Our demonstrations showed that carbon nanotube is a promising material for fabricating ultra low power consumption devices for future sensing and electronic applications.

1. INTRODUCTION

Power consumption is one of the most important engineering considerations in designing electrical circuit and systems. Huge amount of efforts have been placed to minimize the power consumption of electrical systems, since high power consumption implies high heat dissipation which is undesirable in many applications. A typical example is the wall shear stress measurement in aerodynamic applications [1]. Excessive heat dissipation from a hot wire anemometrical sensor will disturb the minute fluidic motion, crippling its ability to sense true fluidic parameters. With our preliminary experimental findings on bulk MWNT, we found that bulk MWNT can be operated at μW range, which is ultra low power consumption for applications such as the shear stress and thermal sensing (e.g., in the order of mW range for typically MEMS polysilicon devices [2]). Carbon nanotubes (CNT), since discovered in 1991 by Sumio Iijima [3], have been extensively studied for their electrical [4] and mechanical properties [5]. In order to build a CNT based device, technique to manipulate the CNT has to be developed. Typical manipulation technique is by atomic force microscopy [6]. However, this conventional pick-and-place technique is time consuming, though the technique has very high positioning accuracy. Past demonstrations by K. Yamamoto et al. showed that carbon nanotube can be manipulated by AC and DC electric field [7,8]. Also, a recent report from L.A. Nagahara et al. demonstrated the individual single-walled carbon nanotube (SWNT) manipulation on nano-electrodes by AC bias voltage [9]. By using similar technique, we have successfully and efficiently manipulated bulk carbon nanotube to form resistive elements between Au microelectrodes for sensing and electronic circuits. This paper reports the technique to form bulk MWNT resistive elements between Au electrodes and our preliminary experimental findings on the electrical characterizations such as frequency response and I-V characteristics of the bulk MWNT devices. The results indicated that the carbon nanotube is promising to be used as high performance and low power consumption devices for future electronic and sensing applications.

*Contacting Author: wen@acae.cuhk.edu.hk
The Chinese University of Hong Kong
Dept. of ACAE, MMW 413,
Shatin, N.T., Hong Kong SAR
Tel: +852 2609 8475 Fax: +852 2603 6002

2. FORMATION OF CARBON NANOTUBE RESISTIVE ELEMENTS BY AC ELECTROPHORESIS

2.1 Fabrication of Microelectrodes

Array of Au microelectrodes with different geometrical shapes (see Figure 1) were fabricated on a 1.8 X 1.8 cm^2 glass substrate with standard photolithography process and wet etching process for carbon nanotube manipulation. Detailed parameters for the photolithography procedures can be found in [10].

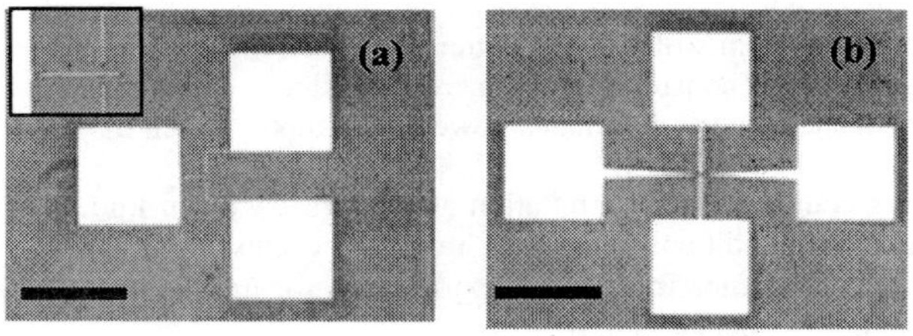

Figure 1. Optical microscopic image showing different microelectrode geometries for MWNT manipulations, a) three-terminal microelectrodes (inset showing the gap (~ 5 μm) between the microelectrodes), b) four-terminal microelectrodes. (Scale Bar = 200 μm)

0-7803-7761-3/03/$17.00 ©2003 IEEE

2.2 Carbon Nanotube Manipulations by AC Electrophoresis

AC electrophoresis (or dielectrophoresis) is a phenomenon where neutral particles undergoing mechanical motion inside non-uniform AC electric field [11]. The technology has been widely employed to manipulate micro or nano entities such as virus and latex spheres in the past few years [12]. In this project, we employed the technique of AC electrophoresis to form bulk carbon nanotubes across microfabricated electrodes as resistive elements.

The MWNT we used in the experiments was ordered commercially from [13]. According to the specifications provided, the MWNT was prepared by chemical vapour deposition. The axial dimensions and diameters of the MWNT was 1 – 10 μm and 10 – 30 nm, respectively. Prior the MWNT manipulation, 50 mg of the sample was ultrasonically dispersed in 500 ml ethanol solution and the resulting solution was diluted to 0.01 mg/ml for later usage.

After the Au microelectrodes were wire-bonded to the external circuits, about 10 μL of the MWNT/ethanol solution was transferred to the substrate by 6 mL gas syringe. Then the Au microelectrodes were excited by AC voltage source by applying 16 V peak-to-peak at 1 MHz typically (see Figure 2). The ethanol was evaporated away (within 20 seconds to 1 minute) leaving the MWNT to reside between the gap of the microelectrodes (see Figure 3).

We experimentally discovered that the resistances of bulk MWNT resistive elements were sample dependent (i.e. different MWNT samples have different room temperature resistances) and the two probe room temperature resistances of the samples were typically ranging from several kOhm to several hundred kOhm. The reason behind for this dependency was due to the complex MWNT network formation during the process of AC electrophoresis. Since the conductivity of CNT was dependent on its lattice geometry formation during the CNT growing process, therefore, the conductivities of individual CNTs cannot be well controlled and resulting in varying conductivities in individual CNTs. As a result, it is logically followed that different bulk MWNT samples exhibited different conductivities.

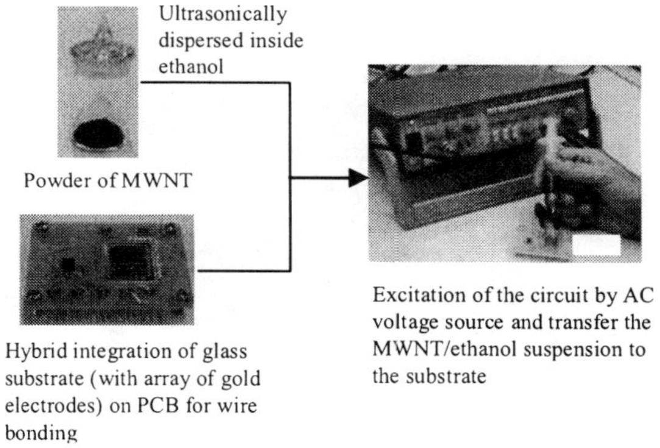

Figure 2. Experimental process flow showing the fabrication of MWNT based circuit elements.

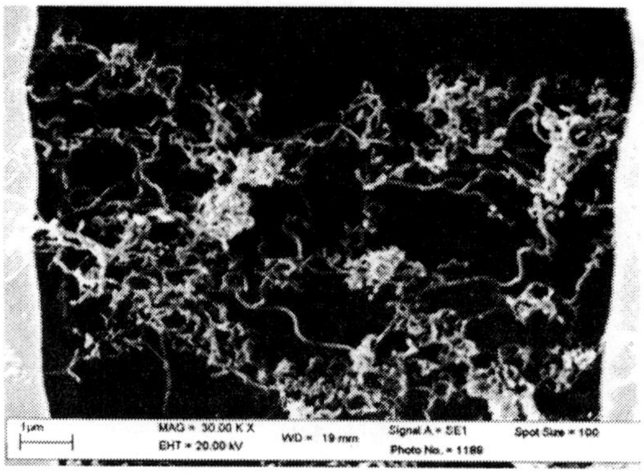

Figure 3. Scanning electron microscope (SEM) image showing the bundled MWNT connections between Au microelectrodes. (Scale Bar = 1 μm)

3. CARBON NANOTUBE AS THERMAL SENSING ELEMENT

3.1 Thermal Sensitivity

In order to measure the temperature-resistance relationship of the bulk MWNT device, the hybridly integrated circuit was put inside an oven (Lab-Line® L-C Oven) and the temperature was kept monitored by the Fluke type K thermocouple attached on the surface of the circuit board. The resistance of the bulk MWNT was then measured by Fluke 73III multi-meter. The temperature coefficient of resistance (TCR) was obtained by measuring the resistance of the bulk MWNT with the corresponding temperature. From the experimental measurements on a typical bulk MWNT device, the resistance dropped with increasing temperature, which is in agreement with [14] (i.e. negative TCR). Interestingly, the TCR measurements of all of our testing devices did not converge but the ranges were generally within -0.1 to -0.2 %/°C . Considerably drifting in room temperature resistances of the sensors were found during measurements (see Figure 4). We suspect the variations were contributed by the mismatch in thermal coefficient of expansion (TCE) between the Au electrodes and the bulk MWNT, causing some of the MWNT linkages to detach from the Au electrodes during measurements inside the oven. Another possible reason was due to contaminations to the sample such as moisture during measurements. In order to form a more robust protection to the bulk MWNT, we are currently developing a process to embed the MWNT inside parylene C diaphragms (see Figure 5). The effectiveness of the proposed method will be published elsewhere later. Nevertheless, the temperature-resistance dependency of bulk MWNT implied its thermal sensing capability. Besides, from the I-V measurement of the bulk MWNT device, the current required to induce the self heating of the device was in μA range at several volts, which suggested the power consumption of the device was in μW range (see Figure 6).

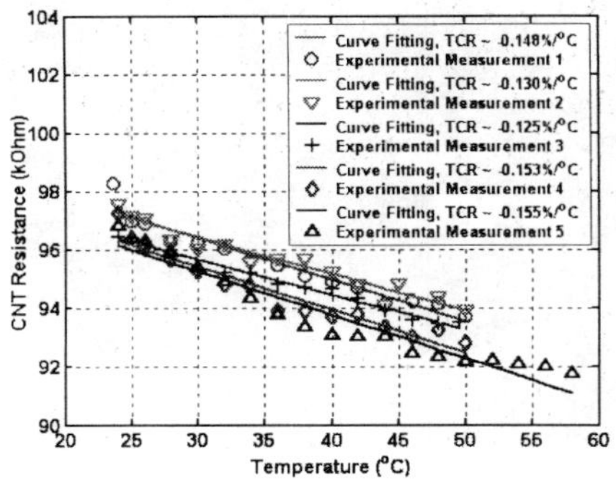

Figure 4. TCR for a typical bulk MWNT device in different measurements. Five repeated measurements were carried out for repeatability test. Considerably room temperature resistance drifting was observed in these measurements.

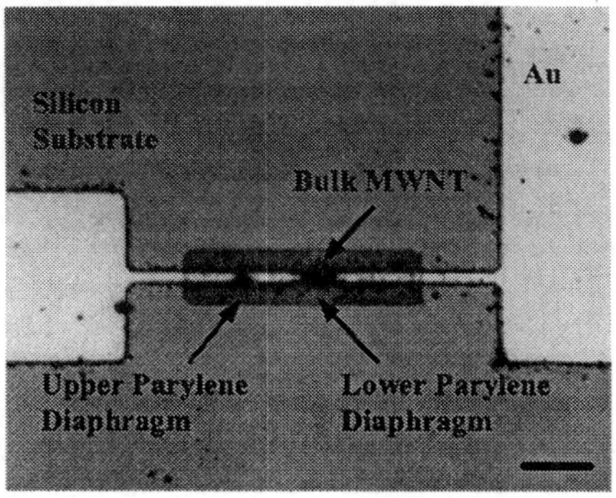

Figure 5. Optical microscopic image showing the prototyping device and the bulk MWNT was protected inside parylene C diaphragms. (Scale Bar = 20 μm)

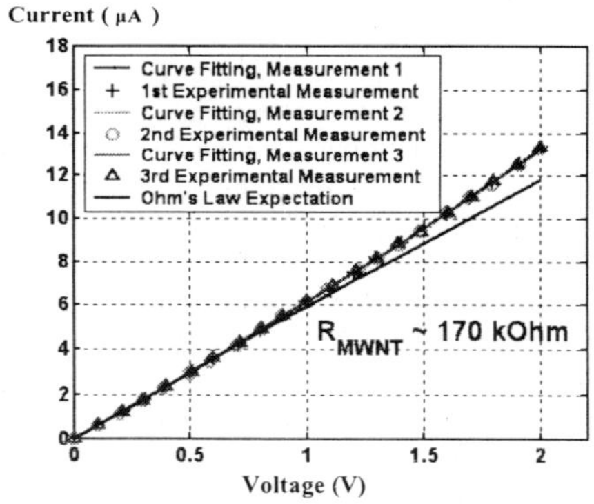

Figure 6. I-V characteristics of a typical bulk MWNT device. Three repeated measurements were performed to validate its repeatability. Non-linearity began when the applied voltage reached 0.6 V at 3.6 μA . Experimental measurements were fitted into second order curve using least square method by MatLAB® software. The straight line is the theoretical expectation using Ohm's Law and the resistance of bulk MWNT resistive elements in our sample was about 170 kΩ .

3.2 Frequency Response

In order to pick up small variations of the sensing environment, sensors with fast frequency response are highly desirable. To test the frequency response of the bulk MWNT device, input square wave of 2 V peak-to-peak at 10 kHz was fed into the hybridly integrated circuit and the output response was determined (see Figure 7). From our experimental measurements, bulk MWNT devices exhibited very fast frequency response. Using the approximation between the time constant and cutoff frequency [2],

$$f_c = 1/1.5t_c \qquad (1)$$

where f_c is the cutoff frequency, t_c is the time constant of the response, the estimated cutoff frequency of the device was about 177 kHz (see Figure 7). As a comparison, typical cutoff frequency of MEMS polysilicon sensors in constant current mode configuration without frequency compensation is around several hundred Hz to several kHz [1, 2, 15].

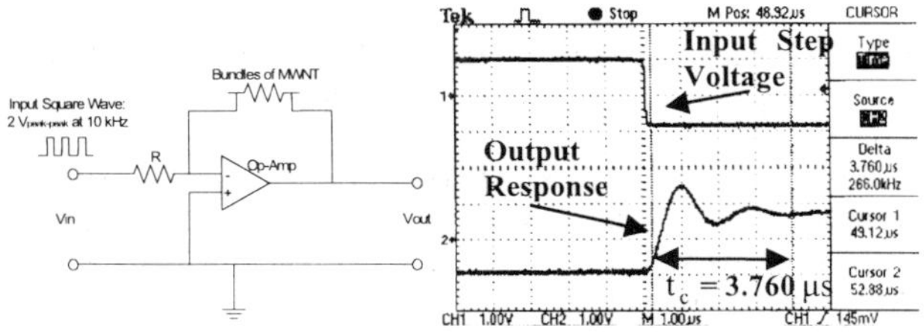

Figure 7. Output response obtained by feeding an input square wave at 2V peak-to-peak at 10 kHz into the circuit (Left). The output response was inverted due to the property of inverted amplifier configuration (Right).

4. CARBON NANOTUBE AS POTENTIAL DIVIDING DEVICE

Using the technique reported in Section 2, bulk MWNT were manipulated by a three-terminal Au microelectrodes (in Figure 1a) to form a simple potential dividing circuit (see Figure 8). Two probe resistivity measurements at room temperature for the terminals 1-2, 1-3 and 3-2 were about 1141 kΩ , 307 kΩ and 833 kΩ respectively.

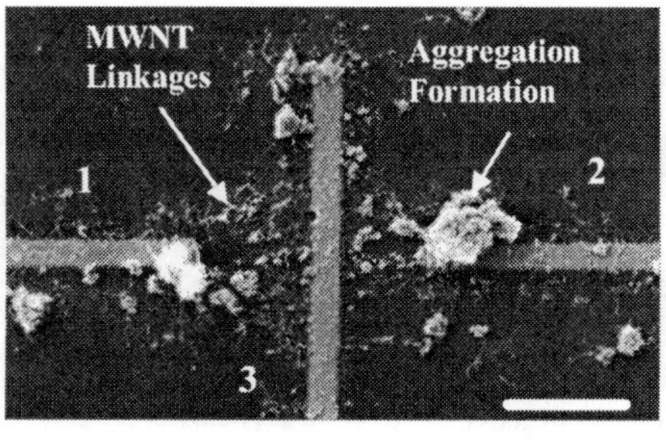

Figure 8. . SEM image showing the formation of MWNT linkages with three-terminal microelectrodes. Aggregation was observed and this was due to the instability of MWNT in ethanol medium. Terminals 1, 2 and 3 are indicated in the figure. (Scale Bar = 6 μm)

We have calculated the theoretical results based on the potential dividing formula of the resistive circuit,

$$V_{13} = \frac{R_{13}}{R_{12}} \cdot V_{12}, \quad V_{32} = \frac{R_{32}}{R_{12}} \cdot V_{12} \qquad (2)$$

where R_{12}, R_{13} and R_{32} are the resistances across the terminal 1-2 ,terminal 1-3 and terminal 3-2 respectively. V_{12} is the voltage applied across terminal 1-2. The MWNT linkages incorporated with the three-terminal microelectrodes can be used as a potential dividing circuit. The device was capable to switch the input potential into a ratio about 2.7:1(see Figure 9). The I-V measurements on terminal 1-3 and terminal 3-2 revealed the power consumption of the bulk MWNT device was in μW range (see Figure 10).

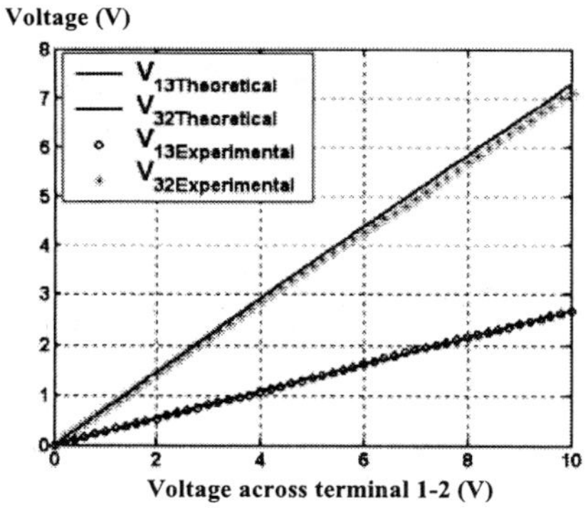

Figure 9. Comparison between the theoretical calculations and experimental results of the MWNT potential dividing circuit. The theoretical calculations are based on the potential dividing formula of a general resistive circuit.

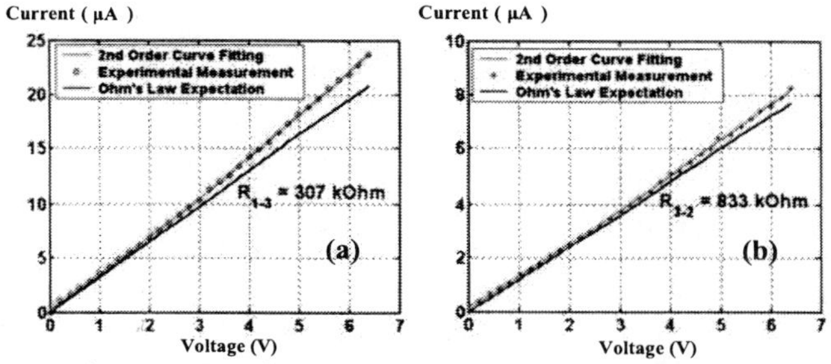

Figure 10. I-V characteristics for terminals a) 1-3 and b) 3-2. Due to the electrical conductivity dependency on different combinations of MWNT bundles, the non-linearity begun at different applied voltages in terminal 1-3 and 3-2 respectively.

5. SUMMARY

A technique to form bulk MWNT resistive elements between Au electrodes was presented. The TCR measurements and the frequency response measurement of the bulk MWNT based device showed that bulk MWNT can be used as sensing element for thermal sensing applications. Besides, the operating power of the resulting device was in μW range which is ultra low power consumption for applications such as shear stress sensing. Apart from this, a bulk MWNT based potential dividing circuit was fabricated with potential switching capability of 2.7:1 ratio and power consumption in μW range. From these demonstrations, we believe that bulk MWNT can be used for ultra low power sensing and electronic circuit applications.

6. REFERENCES

[1] J.B. Huang, C. Liu, F. Jiang, S. Tung, Y.C. Tai, C.M. Ho, "Fluidic Shear Stress Measurement Using Surface-Micromachined Sensors", Proceedings of IEEE Region 10 International Conference on Microelectronics and VLSI, (TENCON '95), pp. 16 – 19 (1995).

[2] C. Liu, J.B. Huang, Z. Zhu, F. Jiang, S. Tung, Y.C. Tai, C.M. Ho, "A Micromachined Flow Shear-Stress Sensor based on Thermal Transfer Principle", Journal of Microelectromechanical Systems, Vol. 8, No. 1, pp. 90 – 99 (1999).

[3] S. Iijima, "Helical Microtubules of Graphitic Carbon", Nature, Vol. 354, pp. 56 – 58 (1991).

[4] S. Frank, P. Poncharal, Z.L. Wang, W.A. de Heer, "Carbon Nanotube Quantum Resistors", Science, Vol. 280, pp. 1744 – 1746 (1998).

[5] E.W. Wong, P.E. Sheehan, C.M. Lieber, "Nanobeam Mechanics: Elasticity, Strength, and Toughness of Nanorods and Nanotubes", Science, Vol. 277, pp.1971 – 1975 (1997).

[6] T. Shiokawa, K. Tsukagoshi, K. Ishibashi, Y. Aoyagi, "Nanostructure Construction in Single-walled Carbon Nanotubes by AFM Manipulation", Proceedings of Microprocesses and Nanotechnology Conference 2001, pp. 164 – 165 (2001).

[7] K. Yamamoto, S. Akita, Y. Nakayama, "Orientation of Carbon Nanotubes Using Electrophoresis", Japanese Journal of Applied Physics Vol. 35, pp. L917-L918 (1996).

[8] K. Yamamoto, S. Akita, Y. Nakayama, "Orientation and Purification of Carbon Nanotubes Using AC Electrophoresis", Journal of Physics D: Applied Physics. 31, L34-L36 (1998).

[9] L.A. Nagahara, I. Amlani, J. Lewenstein and R.K. Tsui, "Directed Placement of Suspended Carbon Nanotubes for Nanometers-scale Assembly", Applied Physics Letters, Vol. 80, No. 20, pp. 3826 – 3828 (2002).

[10] V.T.S. Wong and W.J. Li, "Dependence of AC Electrophoresis Carbon Nanotube Manipulation on Microelectrode Geometry", International Journal of Non-linear Sciences and Numerical Simulation, Vol. 3, Nos. 3 – 4, pp. 769 -774 (2002).

[11] H.A. Pohl, "Dielectrophoresis: The Behaviour of Neutral Matter in Nonuniform Electric Fields", Cambridge University Press (1978).

[12] M.P. Huges, "AC Electrokinetics: Applications for Nanotechnology", Nanotechnology, Vol. 11, pp. 124 – 132 (2000).

[13] Sun Nanotech Co Ltd, Beijing, P.R. China.

[14] T.W. Ebbesen, H.J. Lezec, H. Hiura, J.W. Bennett, H.F. Ghaemi, T. Thio, "Electrical Conductivity of Individual Carbon Nanotubes", Nature, Vol. 382, pp. 54 – 56 (1996).

[15] J.B. Huang, F.K. Jiang, Y.C. Tai, C.M. Ho, "MEMS-based Thermal Shear-stress Sensor with Self-frequency Compensation", Measurement Science and Technology, Vol. 10, No. 8, pp. 687 – 696 (1999).

CMOS Integrated Gas Sensor Chip Using SAW Technology

Shahrokh Ahmadi, Can Korman, Mona Zaghloul, and Kuan-Hsun Huang
Department of Electrical and Computer Engineering,
The George Washington University, Washington DC. 20052

Abstract

The development of inexpensive and miniaturized SAW gas sensors that are highly selective and sensitive is introduced. These sensors are implemented with micro-electro-mechanical systems (MEMS) in CMOS technology. Since the sensors are fabricated on a silicon substrate, additional signal processing circuitry can easily be integrated into the chip thereby readily providing functions such as multiplexing and analog-to-digital conversion that are needed for integration into a network.

1. Introduction

For the purposes of public safety, health and wellness, it is often critical to have available smart sensors that can detect and identify toxic and poisonous gases in the air. The development of inexpensive and miniaturized sensors that are highly selective and sensitive and for which network interfacing is present all on one chip is very desirable. These types of sensors can be implemented with micro-electro-mechanical systems (MEMS) in CMOS technology. Since these sensors are fabricated on a semiconductor substrate, additional signal processing circuitry can easily be integrated into the chip thereby readily providing functions such as multiplexing and analog-to-digital conversion. In numerous other areas one could find similar uses for a smart multi-sensor array from which easy measurements can be made with a small portable device. These are the types of systems on a chip (SOC) that this research addresses.

2. System Architecture

Surface Acoustic Wave (SAW) sensors are small miniature sensors that can be used to detect various gases and other chemical compounds in the air [1]. A SAW sensor consists of an input interdigital transducer (IDT), a chemical absorbent film, and output interdigital transducer on a piezoelectric substrate. The input IDT launches acoustic waves that travel through a piezoelectric substrate coated with a gas absorbent polymer film. The primary sensing is the result of a shift in the operating frequency due to loading of the polymer film by the absorbed gas molecules. The SAW sensor devices are typically designed to run at RF (radio frequencies, in the range of hundreds of MHz) due to the fact that the sensitivity of the sensor is proportional to the square of the operating frequency. Fig. 1, shows the structure of typical SAW device.

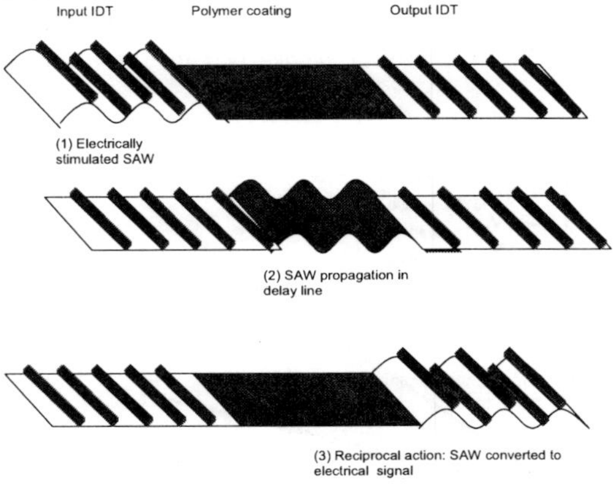

Fig. 1: Typical SAW device with IDT and delay line.

In this work, we will present the design, fabrication, and preliminary testing of the proposed SAW gas sensors based on commercial CMOS technology and novel post-fabrication steps. Due to the incorporation of the sensing element and associated RF signal processing on a common substrate, the system is expected to provide strong noise immunity and temperature sensitivity.

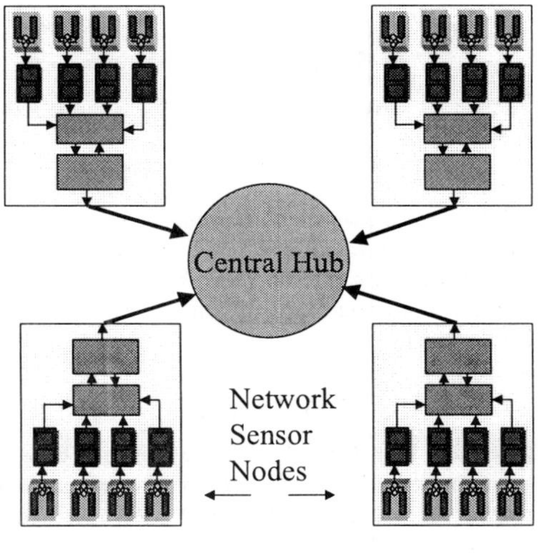

Fig. 2: A star sensor network configuration.

Fig. 2, shows the proposed architecture of network of smart nodes of chemical sensors in a star configuration. As shown in Fig. 3, each node is

This work was supported in part by NSF Grant 0225431.

0-7803-7761-3/03/$17.00 ©2003 IEEE

composed of an array of SAW sensor chips that are individually sensitive to different gases due to different absorbent chemical coatings. These nodes with associated multiplexer, control circuitry and network interfaces will transmit their signal to a central hub for further signal processing.

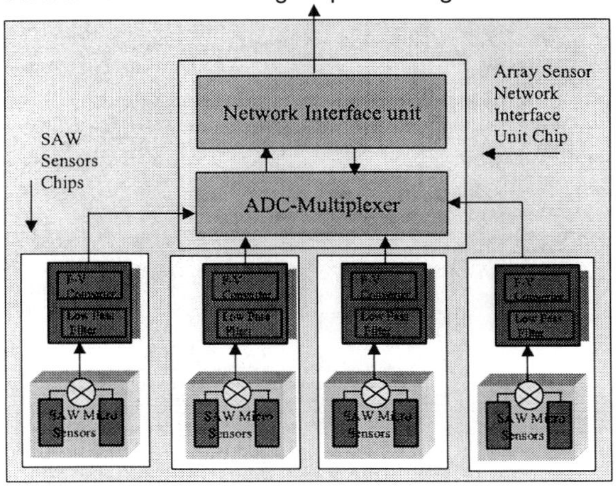

Fig. 3: Architecture for a network node.

The block diagram of the SAW sensor chip is shown in Fig. 4. Here, the SAW sensor chip is composed of two SAW devices fabricated on the same substrate in CMOS technology, where one device is coated with a gas absorbent polymer and the second one is sealed to serve as a reference frequency source. The associated RF signal processing circuitry is also integrated on the same SAW sensor chip that includes a mixer, low-pass filter, and frequency to voltage converter.

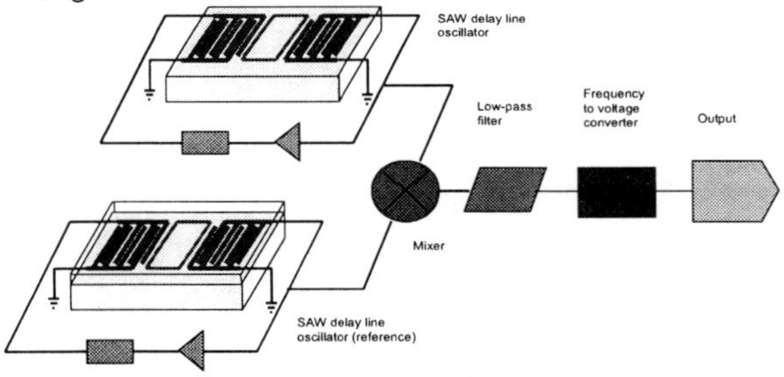

Fig. 4: System block diagram of a SAW sensor chip.

3- CMOS SAW Fabrication:

CMOS process is mainly optimized for digital/analog circuit operation. However, it has been shown that by simple micro-machining methods a variety of novel devices can be fabricated on CMOS wafers, [2], [3], [4]. The simplest CMOS micromachining method involves partial removal of silicon substrate by wet or dry etching. Since this etching and other micromachining techniques are performed after a standard CMOS process is completed, they are frequently referred to as post-CMOS processing or post-processing. Maskless post-processing of CMOS has been developed by Baltes et al. in [3] and later improved by Tea et al. [4]. Most of the thin films available in CMOS have the essential properties needed in sensor and actuator applications, such as piezoresistance in polysilicon films, but they have to be controlled very closely since they exhibit many nonlinear effects. With the help of integrated electronics, it is possible to overcome the shortcomings of the CMOS thin films and build sensor systems based on them [3].

For many MEMS devices, one or more of electrical, thermal, mechanical and optical problems must be solved consistently to obtain correct description of the overall system. Therefore, beside the electrical properties of thin films used in CMOS process, their thermal, mechanical, and optical properties must be known. Particularly for high frequency applications in CMOS, the products must be optimized to meet the quality, reliability and cost requirements.

CMOS membranes are composed of metal thin films isolated with silicon dioxide thin films. Thickness of each film layer varies between 0.01 – 4 μm. When membranes formed by such thin layers are released by etching the substrate underneath, they become very fragile and susceptible to mechanical failures. Micro-machining techniques, such as bulk micro-machining and electroplating, offer excellent opportunity to minimize these losses. These techniques are rapidly growing, and are beginning to be implemented in a variety of commercial products [5].

In SAW devices when the substrate thickness is much larger than the acoustic wavelength, two types of surface waves are often utilized: Rayleigh wave and surface Transverse Wave (STW). The type of wave propagation depends on the crystal properties, cut and direction of propagation [6]. The mechanical displacement normal to the surface is much smaller than the acoustic wavelength. The frequency independent phase velocities range from 10^3 to 10^4 m/s. Thus, depending on the IDT structure available in CMOS processes, the operating frequencies range from approximately 10MHz to above 10GHz.

For gas sensor applications a layer of adsorbent material is deposited on top of the piezoelectric crystal to absorb the gas and, hence, causing loading to the piezoelectric material. Usually polymers with different types of dopants or catalyst are used to produce a sensing thin film for individual species, such as, CH_4, CO, NO_2,O_2, etc [7].

Experiments have shown that there is a strong influence of temperature on SAW behavior [8]. The temperature dependent drift decreases the resolution of the measurement especially at small signal. Using CMOS technology the sensor devices can be heated by an underlying integrated polysilicon layer. This together with associated electronic circuits could be used to control the temperature on the surface of SAW sensors, thus, potentially minimizing the effect of temperature drift. Therefore, the combination of a sensing SAW element with polymer coating and one reference SAW element on the same substrate with accompanying temperature control offers benefits that are not possible with most discrete SAW sensors.

To apply temperature, the sensor must be isolated from silicon, as silicon is a good heat dissipation material. Thus, we have designed the CMOS sensor to be suspended on top of the silicon substrate to isolate the structure from silicon and to better control the temperature. These types of suspended structures were realized successfully in CMOS technology for gas sensors using semiconductor oxide as sensing material [9], [10].

4. CMOS Design Layers and Post Processing Steps:

Attaining dual delay line CMOS SAW sensor has two major fabrication processes. These are: CMOS fabrication using MOSIS services and post processing using MEMS Exchange services. Initially in the CMOS process, layers of silicon dioxide, polysilicon (heater), and metal (thermal plate) are formed on top of the silicon substrate. The fabricated CMOS chip will then go through the post processing. Initially, under the silicon dioxide-- through the silicon substrate-- an air gap cavity is etched away to form a suspended island on its top. The base of this island is made of the silicon dioxide and it is connected to the rest of the CMOS die through several silicon dioxide bridges (handles) all around it to secure the island structure form collapsing or deformation. The silicon dioxide island on top supports a polysilicon layer that acts as a heater and a metal layer that acts as a thermal plate. After the formation of the air gap cavity, a thin layer of piezoelectric material-- zinc oxide, ZnO--is deposited on top of the island. The piezoelectric ZnO layer is sputtered employing a PVD process, using MEMS-Exchange. [11]. A thickness between 1-5 um is required. Different thicknesses for the ZnO layers will be tested to achieve the optimal thickness. Following the ZnO layer, a metal layer is deposited to form the Interdigital Transducer (IDT). IDT's are formed using several processing steps that involve masking and etching away the metal. As a final step, to make the CMOS SAW sensors, a gas sensitive polymer is deposited on the delay line, between the transmitting and receiving IDT's.
The layers needed to realize SAW sensor in CMOS technology are shown in Fig. 5a-d
The above structure will be suspended on top of silicon substrate using the open layer defined in [9].

The top view of the structure, designed using Cadence, is shown in Fig. 6.

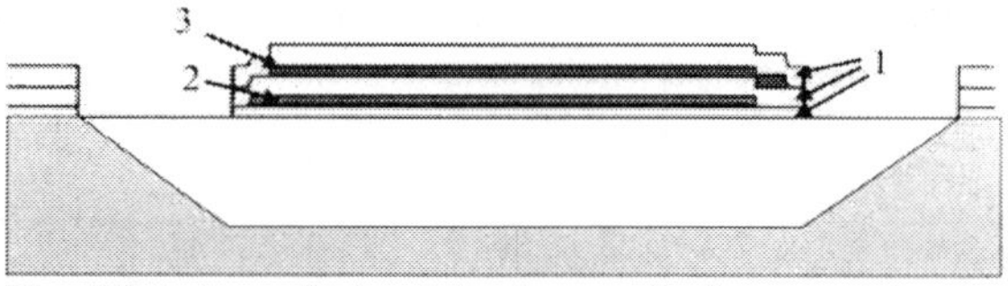

Fig. 5(a): A cavity is etched away to form an island.

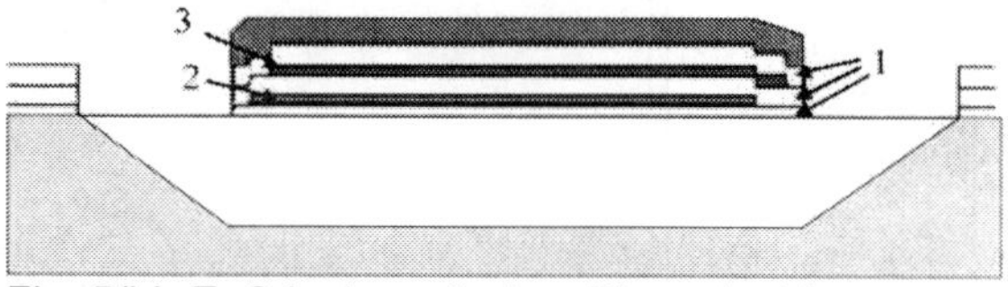

Fig. 5(b): ZnO is deposited as Piezoelectric. Materials.

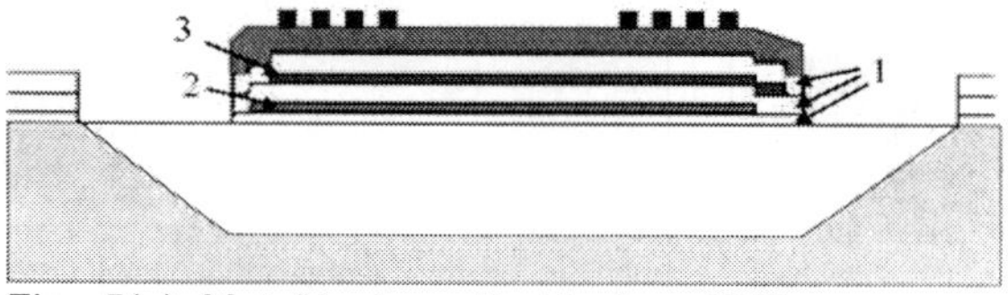

Fig. 5(c): Metal is deposited to form IDT's.

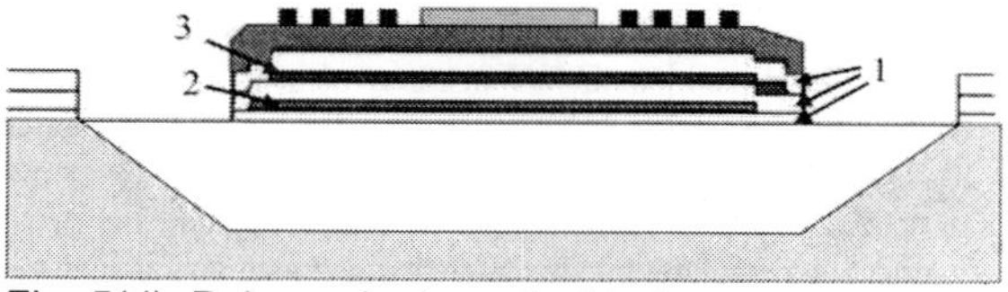

Fig. 5(d): Polymer is deposited to form gas sensor.

Fig. 5a-d: Showing SAW gas sensor Layers and its post processing steps after CMOS chip is made. Layer 1 is SiO2, layer 2 is polysilicon (heater) and layer 3 is metal (thermal plate)

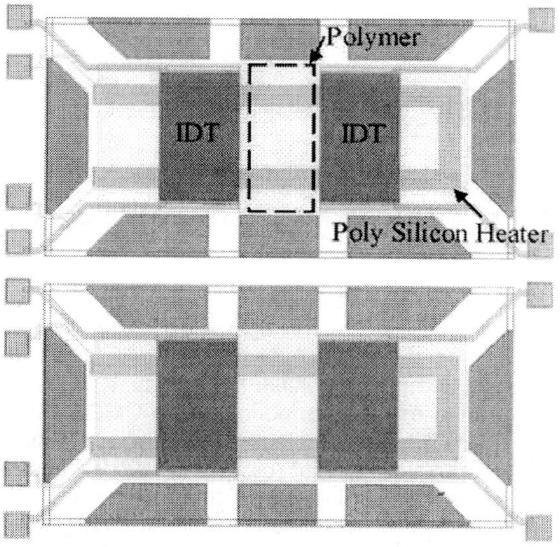

Fig. 6: Top view of duel delay line suspended SAW gas sensor structure.

5. Network Node: Delay Line Oscillator

The two IDT's that are fabricated on top of a piezoelectric material layer are going to act as a transmitter and receiver of the acoustic signal. The transmitting, [1], [12], [13]. IDT generates the surface acoustic waves. The generated surface acoustic wave travels towards the second IDT on the surface of the piezoelectric material until it reaches the receiving IDT. The received acoustic wave at the output is then amplified and fed back to the input through a phase shifter, Fig. 7. The constructed circuitry, called a delay line resonant oscillator, will oscillate at a resonant frequency f_r, provided that the total phase shift imposed by the loop is a multiple of 2π radians. The resonant frequency given in Eq. (1) is a function of the surface acoustic wave velocity v and the wavelength λ:

$$f_r = v/\lambda = 3400(m/s)/9.6\ um = 354\ MHz \quad (Eq.\ 1)$$

where $\lambda = 2d$ and $d=2.4um$ is the spacing between adjacent fingers of the IDT.

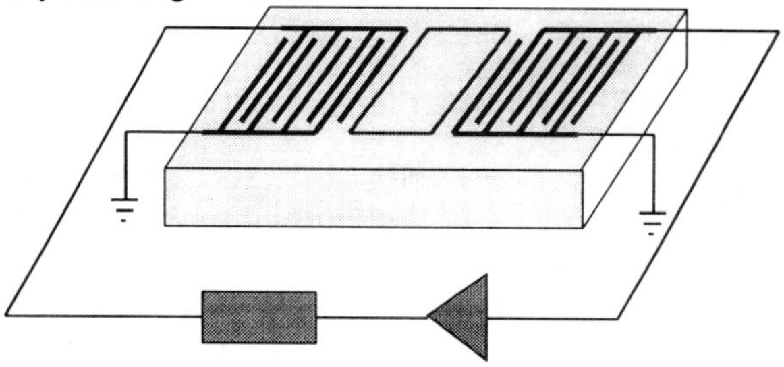

Fig. 7: A single SAW delay line oscillator.

The overall system block diagram of the SAW sensor chip is presented in Fig. 4. The system is made of two identical delay line oscillators. One is an active delay line oscillator which is exposed to the sensing chemical vapors via a polymer coating and the other one is a passive delay line oscillator which is blocked from being exposed to the sensed chemicals. Once exposed to the external chemical vapor, the active delay line oscillator resonant frequency, f_r, will be shifted, due to the change of the property of the delay line, to a new frequency of f_r'. However, due to the lack of chemical exposure, the passive delay line oscillator resonant frequency will remain the same. This constant frequency will be used a reference to measure the amount of the frequency shift of the active delay line oscillator. In addition, due to the sensitivity of surface acoustic waves to temperature fluctuations, this dual SAW delay line configuration also minimizes the effects of temperature drift.

As shown in the block diagram in Fig. 4, the shifted and un-shifted signals are modulated through a mixer device. The output of the mixer contains both the sum of frequencies, $f_r + f_r'$, and the difference of the frequencies, $f_r - f_r'$. The subsequent low pass filter in the output will reject the high frequency component of the sum, $f_r + f_r'$, while maintaining the low frequency differences of frequencies, $f_r - f_r'$. The shift in the active delay line resonant frequency, $\Delta f = f_r - f_r'$, in which the signal is the frequency, is then converted to a signal where the voltage is proportional to the input frequency. This is achieved by employing a frequency to voltage converter.

6. References

[1] D. S. Ballantine, R. M. White, S. J. Martin, A. J. Ricco, E. T. Zellers, G. C. Frye, H. Wohltjen, *Acoustic Wave Sensors: Theory, Design, and Physico-Chemical Applications,* Academic Press, New York, 1997.

[2] G. K. Fedder, S. Santhanam, M. L. Reed, S. C. Eagle, D. F. Guillou, M. S.-C. Lu, and L. R. Carley, "Laminated high-aspect ratio microstructures in a conventional CMOS process," Proc. of IEEE MEMS Workshop (MEMS'96), pp. 3-18, 1996.

[3] H. Baltes, T. Boltshauser, O. Brand, R. Lenggenhager, and D. Jaeggi, "Silicon microsensors and microstructures," *Proc. of IEEE Int. Symp. on Circuits and Systems*, pp 1820-1823, 1992.

[4] N. Tea, V. Milanovic, C. A. Zincke, J. S. Suehle, M. Gaitan, M. E. Zaghloul, and J. Geist, "Hybrid postprocessing etching for CMOS-compatible MEMS," *J. of MEMS*, vol. 6, no. 4, pp. 363-371, Dec. 1997.

[5] Special issue on integrated sensors, microactuators, and Microsystems (MEMS), *Proc. of the IEEE*, vol. 86, no. 8, Aug. 1998.

[6] G.Kovacs, Micromachined Transducers Source Book, New-York, Ny:McGraw-Hill, 1998.

[8] Clifford k.Ho, Michael T. Itamura, Michael Kelley, and Robert C. Hughes, Review of Chemical Sensors for In- Situ Monitoring of Voltile Contaminants, Sandia Report, SAND 2001-0643, March 2001.

[9] M Afridi, J. Suehle, M. Zaghloul, J. Tiffany, R. Cavicchi, "Implementation of CMOS Compatible Conductance-Based Micro-Gas-Sensor System," European Conference on Circuit Theory and Design, pp. (III) 381-384, August 2001.

[10] M.Afridi, D. Berning, A. Hefner, J. Suehle, M.Zaghloul, E.Kelly, Z. Parilla, C. Ellenwood, "Transient Heating Study of Microhotplates by Using High- Speed Thermal Imaging System", the Proceeding of Semi-Therm XVIII, March 12-14, 2002.

[11] http://www.MEMS EXCHANGE.com.

[12] M. Feldmann and J. Henaff, *Surface Acoustic Waves for Signal Processing,* Artech House, Norwood, MA, 1989.

[13] M. S. Nieuwenhuizen and A. Venema, "Surface Acoustic Wave Chemical Sensors," *Sensors and Materials,* Vol. 5, pp. 261-300, MY, Tokyo, 1989.

A MICRO-HOTPLATE-BASED MONOLITHIC CMOS GAS SENSOR ARRAY

D. Barrettino, M. Graf, M. Zimmermann, C. Hagleitner, A. Hierlemann and H. Baltes

Physical Electronics Laboratory, ETH Zurich, Switzerland

ABSTRACT

A monolithic gas sensor array fabricated in industrial CMOS-technology combined with post-CMOS micromachining is presented, which comprises three metal-oxide-covered micro-hotplates and the necessary driving and signal-conditioning circuitry. The array approach enables effective detection and discrimination of several analyte gases by using different gas-sensitive materials. The operating temperature of the metal oxide resistors varies between 200 and 350°C. Individual temperature regulation is performed by three on-chip temperature controllers with a resolution of 0.5 °C. The phase margin of the temperature controllers was improved by adding degeneration resistors in the sources of the power transistors. An implemented sequential operation mode reduces the current consumption by 15%. A ring counter provides the clock signal for sequential operation. A novel circular membrane design was developed to improve the power efficiency of the micro-hotplates.

1. INTRODUCTION

There is a strong interest in micro-hotplate-based gas sensors [1] since miniaturization and monolithic integration of transducers and circuitry offer significant advantages such as low power consumption, cost reduction and the possibility of applying new dynamic sensor operation modes. First versions of monolithic micro-hotplate-based gas sensors systems were demonstrated in refs. [2-6].

However, the selectivity of an individual sensor is still a problem. Integrated gas sensor arrays and the use of multi-component analysis algorithms, such as principal component regression (PCR) or artificial neural networks (ANN), can overcome the problems associated with poor selectivity and drift of individual gas sensors [7].

Here, three micro-hotplates were integrated with low-voltage circuitry using a standard double-poly, double-metal, 0.8μm CMOS-process as provided by austriamicrosystems (Unterpremstaetten, Austria). The membranes were covered with different types of tin dioxide (SnO_2). The SnO_2 drops were doped with different concentrations of palladium (Pd) and thus rendered selective to carbon monoxide (CO), nitrogen dioxide (NO_2), or methane (CH_4). Additionally, a sequential operation mode and a novel circular membrane architecture (high power efficiency) were implemented to reduce the overall power consumption. Key design features included stability of the temperature controllers and low power consumption. A simplified block diagram of the gas sensor array is shown in Figure 1.

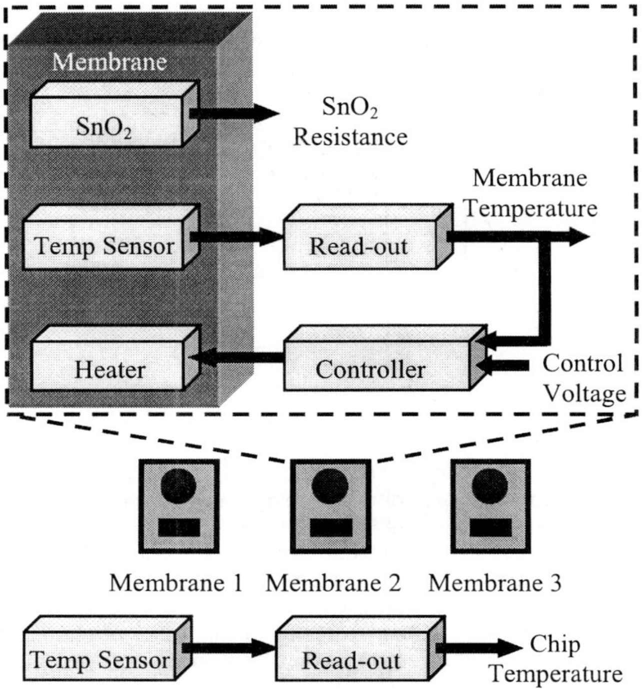

Figure 1 – Block Diagram

2. SYSTEM DESCRIPTION

2.1. Membranes

The chip hosts three circular-shape micro-hotplates (Figure 2, close-up of one of the micro-hotplates), which are thermally isolated from the rest of the chip. The membranes consist of the dielectric layers (silicon oxide/nitride), which remain after applying a potassium-hydroxide (KOH) wet-etch. The membranes are 500 by 500 μm in size and the circular heated areas (micro-hotplates) have a diameter of 300 μm and exhibit a 5.5μm-thick n-well silicon island underneath for homogeneous heat distribution and

0-7803-7761-3/03/$17.00 ©2003 IEEE

membrane stiffening. The n-well island is fabricated by using an electrochemical etch stop technique. The experimentally assessed maximum temperature variation across the membrane is 1-2% up to 350°C. The power efficiency of the circular membranes is approximately 6.0°C/mW. The distance between heated area and the edge of the bulk silicon is increased in a circular design. The metal connections along the diagonals (Figure 2) are longer, and less heat is hence dissipated to the bulk. At the same time the heated area is reduced, so that the heat dissipation through the air also decreases. The circular shape additionally reproduces the natural shape of the SnO_2-drops, which are deposited as nanocrystalline thick films from water-based slurries. More information about the preparation and deposition of the SnO_2-drops can be found in refs. [8-9]. The aluminum electrodes are first covered with 50nm TiW and then with 100nm Pt for establishing a good contact with the sensitive layer. A micrograph of the integrated gas sensor array is shown in Figure 3.

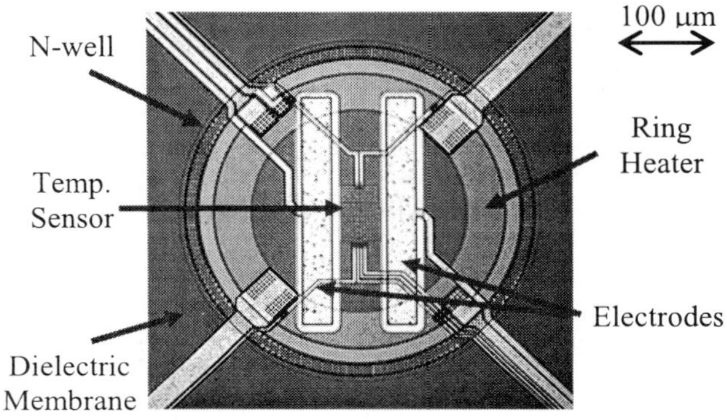

Figure 2 – Close-up of the microhotplate

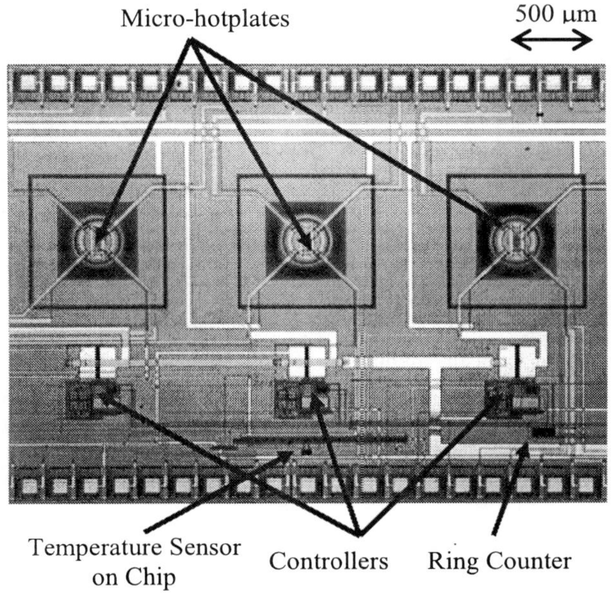

Figure 3 - Micrograph of the gas sensor array

2.2. Circuitry

The main circuitry blocks (Figure 1) include a temperature sensor on the chip, the circuitry for sequential mode operation and three analog proportional temperature controllers.

The on-chip temperature is measured via the voltage difference between a pair of diode-connected PNP transistors working at different current densities.

In the sequential mode, the membranes are heated in a defined sequence using a clock signal provided by the ring counter (Figure 4). Only one power transistor is switched on during every clock period, which reduces the average power consumption of the power transistors. When a power transistor is switched off, the thermal capacity keeps the membrane temperature until the power transistor is switched on again. The thermal time constant during cooling is 20ms. The chip operation in sequential mode reduces the current consumption by 15%.

Each proportional temperature controller (Figure 5) is implemented with an operational amplifier and an internal stabilization capacitor of 20pF. The temperature on the membrane is controlled between room temperature and 350°C. The operational amplifier drives a power transistor, which provides the current to the polysilicon heater (R_{HEATER}). The power transistor has a degeneration resistor (R_{DEG}) in the source, which limits the current. This configuration improves the phase margin of the temperature controller and enhances the long-term reliability. The transmission gates (TG1...TG4) at the input of the temperature controller are used for the sequential mode and the continuous operation mode (all 3 membranes are heated permanently). They are controlled by the ring counter and an external signal (ALL_ON).

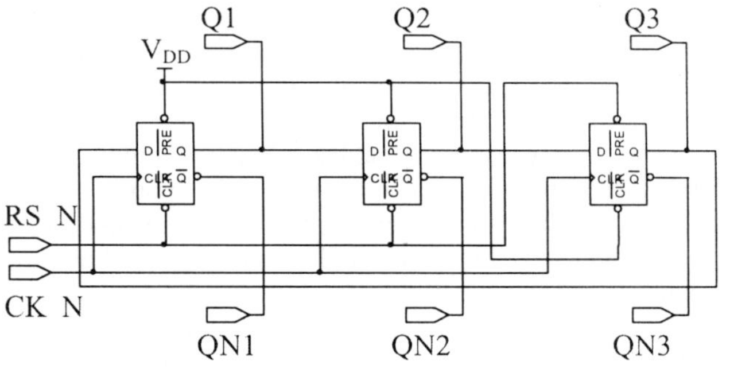

Figure 4 – Ring counter

A polysilicon resistor as temperature sensor (R_{SENSOR}) on the membrane provides the feedback signal for the temperature controller. This polysilicon resistor is biased with a temperature-independent current (I_{BIAS}).

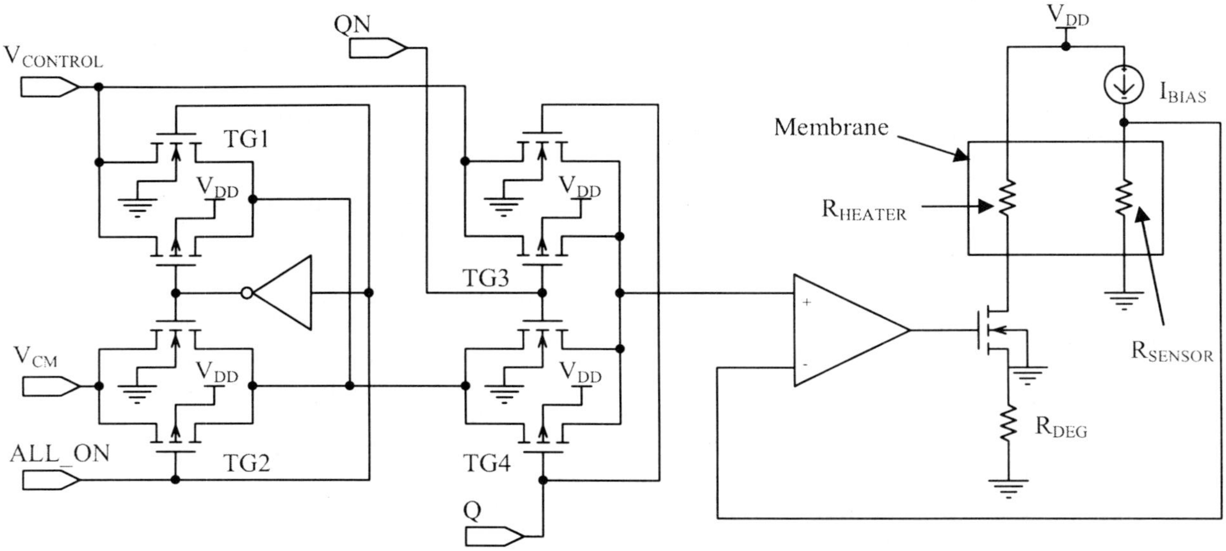

Figure 5 - Proportional temperature controller

3. RESULTS

3.1. Electrical Measurements

We measured the performance of the temperature controller in the tracking mode (Figure 6). The measurement was done at room temperature and the control voltage ($V_{CONTROL}$) of membrane 1 was increased in steps of 10mV. For example, a control voltage of 1.55V produced a membrane temperature of approximately 350°C. Membrane 2 was kept at 200°C and membrane 3 was kept at 350°C. The tracking error due to noise was less than ±0.3°C. The controller showed an excellent performance with a resolution of 0.5°C.
The behavior of the chip in the sequential mode was assessed by sweeping the clock frequency from 50Hz to 50kHz and monitoring the current consumption. A 15% reduction of the current consumption was measured using a clock frequency between 500Hz and 2kHz. For lower frequencies, the temperature fluctuations on the membrane are higher than 1% of the membrane temperature, and for higher frequencies the current consumption increases.
The performance of the temperature sensor on the chip was also measured. The ambient temperature was swept from -40 to 120°C in steps of 5°C and the temperature controllers were switched off. A 2-point calibration at -20 and 85°C was performed. The measured sensitivity was about 128μV/°C and the error due to noise was less than ±1.5°C.

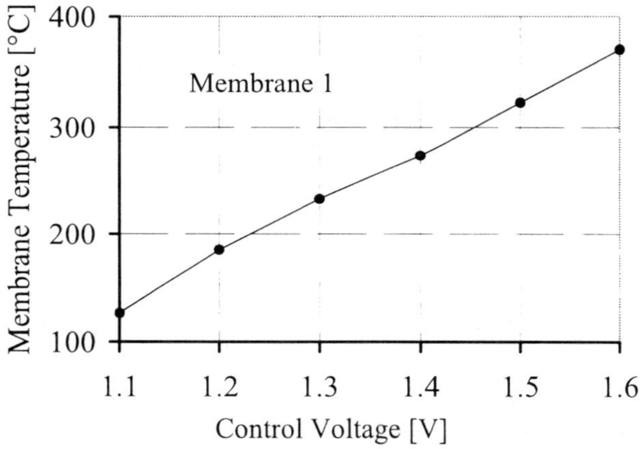

Figure 6 - Temperature controller in tracking mode

3.2. Chemical Measurements

Gas test measurements were carried out with carbon monoxide (CO). CO concentrations ranged between 10 and 50ppm. The measurements were done at a relative humidity of 40%.
Membrane 1 (doped for CO detection) was operated at 270°C which provides high CO-sensitivity. Membrane 2 (doped for NO_2 detection) was operated at 300°C and membrane 3 (doped for CH_4 detection) was operated at 350°C.

Relative SnO_2-resistance values R_0/R were measured for various CO concentrations (Figure 7), with R_0 denoting the SnO_2-resistance upon exposure to synthetic air (nitrogen-oxygen mixture), and R denoting the SnO_2-resistance upon exposure to the analyte gas. Membrane 1 showed pronounced selectivity to CO (high signal).

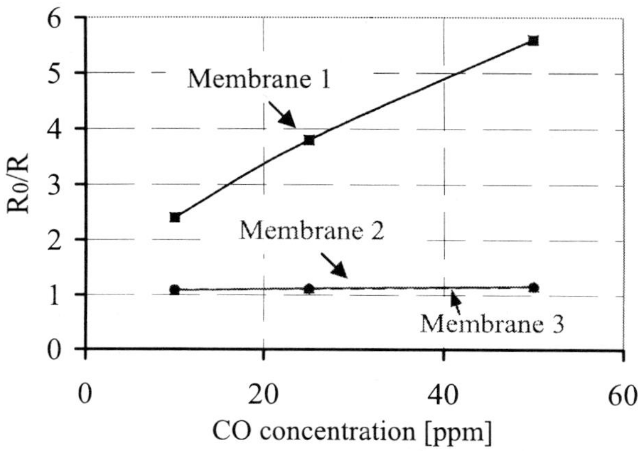

Figure 7 – Gas exposure to CO

4. CONCLUSIONS

A monolithic gas sensor array fabricated in standard 0.8μm CMOS technology combined with post-CMOS micromachining was presented. It comprises three metal-oxide-covered micro-hotplates and the necessary driving and signal-conditioning circuitry on the same chip.

The on-chip temperature controllers can accurately adjust the membrane temperature up to 350°C with a resolution of 0.5°C.

The sequential operation mode reduces the current consumption by 15%. The stability (phase margin) of the temperature controllers was improved by adding degeneration resistors in the sources of the power transistors.

The SnO_2 drops were doped with different concentrations of palladium (Pd) and thus rendered selective to carbon monoxide (CO), nitrogen dioxide (NO_2), or methane (CH_4). Gas test results upon exposure to CO at 40% relative humidity demonstrate the effectivity of the doping and the discrimination capability.

5. REFERENCES

[1] S. Semancik, R. E. Cavicchi, M. C. Wheeler, J. E. Tiffany, G. E. Poirier, R. M. Walton, J. S. Suehle, B. Panchapakesan and D. L. DeVoe, "Microhotplate platforms for chemical sensor research", *Sensors and Actuators B 77*, pp. 579-591, 2001.

[2] V. Demarne and A. Grisel, "An integrated low-power thin-film CO gas sensor on silicon", *Sensors and Actuators, 13,* pp. 301-313, 1988.

[3] D. Briand, A. Krauss, B. van der Schoot, U. Weimar, N. Bârsan, W. Göpel and N.F.de Rooij, "Design and fabrication of high-temperature micro-hotplates for drop-coated gas sensors", *Sensors and Actuators, B 68,* pp. 223-233, 2000.

[4] D. Barrettino, M. Graf, M. Zimmermann, A. Hierlemann, H. Baltes, S. Hahn, N. Bârsan and U. Weimar, "A smart single-chip micro-hotplate-based chemical sensor system in CMOS-technology", *Proceedings of the IEEE International Symposium on Circuits and Systems, Vol. 2,* pp. 157-160, 2002.

[5] M. Y. Afridi, J. S. Suehle, M. E. Zaghloul, D. W. Berning, A. R. Hefner, S. Semancik and R. E. Cavicchi, "A Monolithic Implementation of Interface Circuitry for CMOS Compatible Gas-Sensor System", *Proceedings of the IEEE International Symposium on Circuits and Systems, Vol. 2,* pp. 732-735, 2002.

[6] Y. Mo, Y. Okawa, K. Inoue and K. Natukawa, "Low-voltage and low-power optimization of micro-heater and its on-chip drive circuitry for gas sensor array", *Sensors and Actuators, A 100*, pp. 94–101, 2002.

[7] A. Hierlemann, M. Schweizer-Berberich, U. Weimar, G. Kraus, A. Pfau and W. Göpel, "Pattern Recognition and Multicomponent Analysis", *Sensors Update, Vol. 2,* VCH, pp. 119-180, 1996.

[8] A. Dieguez, A. Romano-Rodriguez, J.R. Morante, U. Weimar, M. Schweizer-Berberich and W. Göpel, "Morphological analysis of nanocrystalline SnO_2 for gas sensor applications", *Sensors and Actuators, B 31*, pp. 1–8, 1996.

[9] J. Kappler, N. Bârsan, U. Weimar, A. Dieguez, J.L. Alay, A. Romano-Rodriguez, J.R. Morante and W. Göpel, "Correlation between XPS, Raman and TEM measurements and the gas sensitivity of Pt and Pd doped SnO_2 based gas sensors", *Fresenius' Journal of Analytical Chemistry 361*, pp. 110–114, 1998.

6. ACKNOWLEDGMENTS

The authors would like to thank Advanced Sensing Devices (ASD) for coating the micro-hotplates, and the Bundesamt für Bildung und Wissenschaft (contract number 99-0135) and the European Union (contract number IST-1999-10579) for financial support

MICROMACHINED PIEZORESISTIVE TACTILE SENSOR ARRAY FABRICATED BY BULK-ETCHED MUMPS PROCESS

Tanom Lomas, Adisorn Tuantranont, and Fusak Cheevasuvit[*]

National Electronics and Computer Technology Center (NECTEC)
112 Thailand Science Park, Pahol Yothin Rd., Klong Laung, Pathumthani 12120 Thailand
Tel: +662-564-6900, Fax: +662-564-6771, Email: adisorn_t@notes.nectec.or.th
[*]King Mongkut's Institute of Technology Ladkrabang, Bangkok 10520, Thailand. Tel: +662-326-4204

ABSTRACT

This paper discusses the design, fabrication and testing of a 5×5 micromachined tactile sensor array for the detection of an extremely small force (micrometer-Newton range). Central contacting pads that are trampoline-shape suspended structures and sensor beams are formed using an anisotropic etching of silicon substrate of a MUMPs process chip. A piezoresistive layer of polysilicon embedded in sensor beams is used to detect the displacement of the suspended contacting pad. Each square tactile has dimension of 200 μm × 200 μm with 250 μm center-to-center spacing. The entire sensor area is 1.25 mm × 1.25 mm. The device was characterized under various normal force loads using weight microneedles. The individual sensor element shows the linear response to normal force with good repeatability.

1. INTRODUCTION

One of mostly sensitive human perceptions is tactile perception. Good example of tactile perception is how we feel when our finger tip contact to the object. Micro-Electro-Mechanical Systems (MEMS) technologies have found the broad field in intelligent solid-state microsensors, the sensing devices that are batch-fabricated by micromachining techniques and integrated with electrical circuits on the same chip. A micromachined tactile sensor is a promising area in the field of physical MEMS sensors. It has a function similar to the surface of a human fingertip. The measurement and processing of a contact stress found useful in robotic dexterous manipulation applications. When a robot grasps object, information on contact, shear force and torque determination are needed for feedback control of robot. Other potential applications of this sensor would be such as sensing of organic tissue on a small scale at the end of catheter or on the fingers of an endoscopic-surgery telemanipulator [1]. Recently, various micromachined tactile sensors capable of measuring normal and shear stress are individually developed by many research groups [2,3] but none of them are foundry-fabricated for potential low cost devices. In this paper, we present a novel piezoresistive tactile sensor array designed for measurement in medium-density sub-millimeter tactile sensing and fabricated by a commercial available Multi-Users MEMS Process (MUMPs) with bulk etching in post processing step.

2. SENSOR DESIGN

2.1 Sensor Configuration

The individual sensing element in the array is a trampoline-shape structure composed of a central contacting plate suspended by four beams over an etched pit. Embedded in each of beams is a polysilicon piezoresistor (2.8 μm wide and 168 μm long). Each tactile sensor element has dimension of 200 μm × 200 μm with 250 μm center-to-center spacing. Each of four sensor beams has dimension of 90 μm long and 10 μm wide. The central contacting plate is a square plate of 40μm × 40 μm. The entire sensor area is 1.25 mm × 1.25 mm. Figure 1(a) shows the scanning electron micrograph (SEM) of the micromachined piezoresistive tactile sensor array and Figure 1(b) shows the configuration of an individual sensor element.

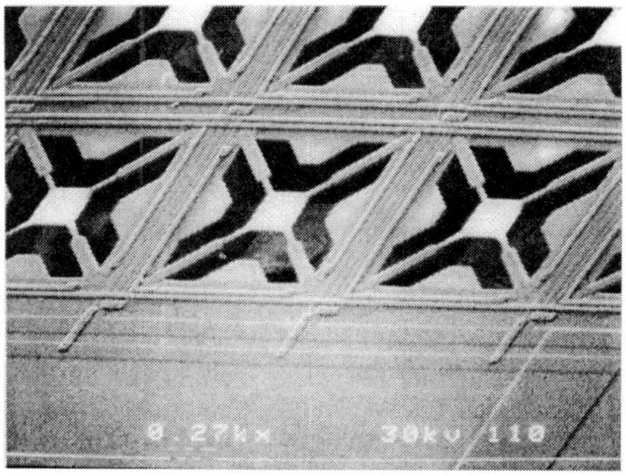

(a)

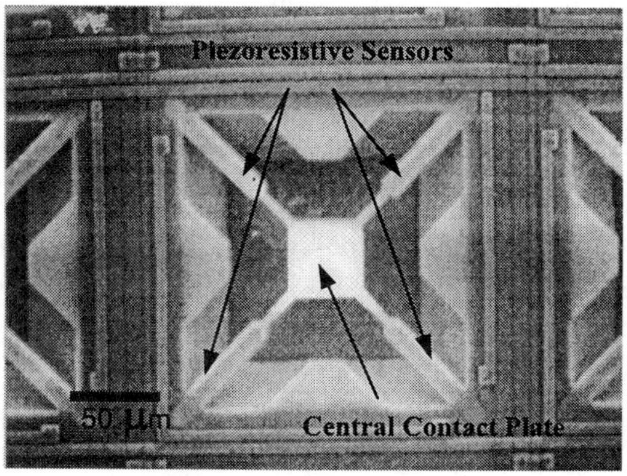

(b)

Figure 1. Scanning electron micrograph of the micromachined piezoresistive tactile sensor array (a) and configuration of a sensor element (b).

0-7803-7761-3/03/$17.00 ©2003 IEEE

2.2 Piezoresistive Sensor

The response of the sensor structure to pure normal stress loading was evaluated. An applied normal force causes the deflection of the central plate in a direction normal to the plane of substrate and induces equal tension in each of the four beam elements as axial elongation occurs. For a small deflection of the central plate, the beams can be modeled as solid bars with pin joints at both points where the beam joins the central plate and the substrate [3]. We assume that the applied normal force is uniformly distributed on the central plate. With this simplified model, the strain induced in each beam due to applied normal force is characterized as:

$$\varepsilon_n = \frac{L}{2(EA)d} F \qquad (1)$$

where
- ε_n strain induced in a beam due to normal force
- L length of the sensor beam
- F the applied normal force
- d the normal deflection of the central plate
- E Young's modulus
- A cross-section area of the sensor beam

The working principle of the piezoresistive tactile sensor is piezoresistivity, which is the property of some materials to change their resistivity under strain. The sensitivity to strain of a certain material is referred to as the gauge factor (GF). The gauge factor is defined as the ratio between the fractional change in resistance ($\Delta R/R$) and the strain (ε_n) induced in the resistor by an applied stress [4]. Longitudinal gauge factors (GF_l) in which the direction of an electrical current flow is parallel to the applied strain is of interest here. The change of resistance due to strain in the beams is given by:

$$\frac{\Delta R}{R} = \varepsilon_n \times (GF_l) \qquad (2)$$

From the mechanical beam theory, the deflection d at the tip of the cantilever beam is given by:

$$d = \frac{1}{2} kL^2 \qquad (3)$$

where k is the beam's spring constant and L is beam length. Thus, combining Eq.(1)-(3), the change of resistance as a function of the applied normal force is given by:

$$\frac{\Delta R}{R} = \left[\frac{(GF_l)}{k(AE)L}\right] \cdot F \qquad (4)$$

A commercially available finite element analysis tool (MARC) was used to determine the deflection of the structure under applied normal force [5]. When normal force is applied on the central plate, the sensor beams bend toward the substrate as shown in Fig. 2 and the maximum stress is at base of the beams where piezoresistors are embedded.

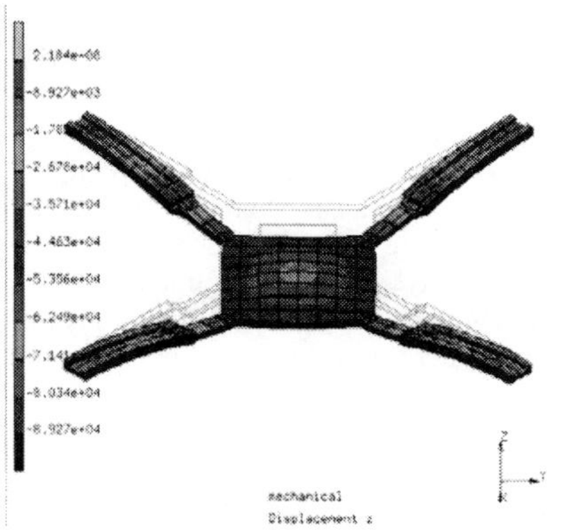

Figure 2. FEM result showing the deflection of the structure under normal force load.

3. DEVICE FABRICATION

3.1 Bulk-Etched MUMPs

Piezoresistive tactile sensors have been fabricated through the commercially available Multi-Users MEMS Process (MUMPs) [6]. This is a three-layer surface micromachining polycrystalline silicon (polysilicon) process. The lowest polysilicon layer, POLY0, is non-releasable. The upper two polysilicon layers, POLY1 and POLY2, can be released to form the movable mechanical structures. The two phosphosilicate glass (PSG) sacrificial layers, OXIDE1 and OXIDE2, are deposited between the two polysilicon layers and are etched away to release the mechanical layers that are used to form the flexures and suspended structure. Silicon nitride layer is used as electrical isolation between the polysilicon and the substrate. The gold is deposited on the top of POLY2 layer and used to create wire-bonding pads. To suspend trampoline-shape structures over cavities, the bare single-crystal silicon substrate is exposed to chemical etchant using the superimposition of "Anchor1", "Anchor2", Poly1-poly2 via", "Dimple", "Hole1" and "Hole2" layers on top of each other as shown in Fig. 3 to create "cut through substrate" mask.

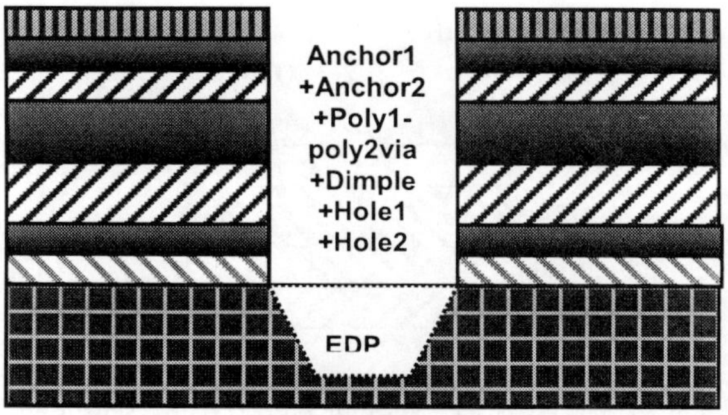

Figure 3. Schematic diagram of MUMPs and "cut through

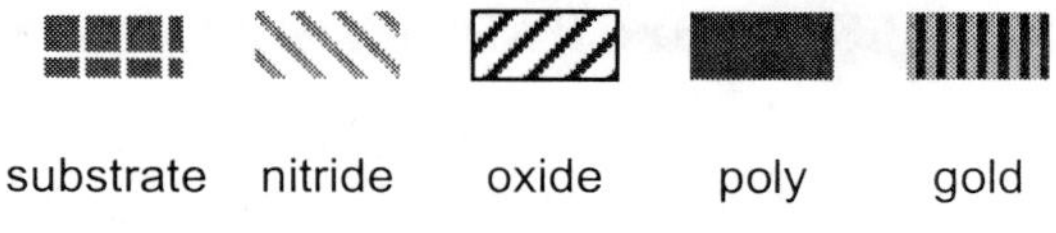

substrate" mask for bulk etching post-processing step with EDP.

By this method, the nitride and all the oxide layers above nitride are opened through during regular MUMPs fabrication, leaving exposed bare silicon while the other area of the chip has covered oxide and nitride layers on [7]. The silicon substrate is selectively etched in-house using the method of anisotropic etching by EDP (ethylene diamine pyrochatechol). The polysilicon structures are not effected by EDP etching because the etching rate of polysilicon in this chemical etchant is much slower than single-crystal silicon and can be negligible. By this method, the bulk micromachining can be done on the surface micromachining chip (MUMPs) using only a single maskless post-processing etching step.

3.2 Tactile Sensor Array Fabrication

The sensor array was fabricated using a bulk-etched MUMPs process as discussed in 3.1. Silicon nitride layer comprised the base layer of the free-standing sensor structures. The thinnest polysilicon, POLY0, is used to construct the piezoresistive structures and embedded in silicon dioxide layer of OXIDE1 and OXIDE2. The piezoresistor is located at the base of all four beams where the induced strain from beam bending is at maximum. On the top layer, polysilicon, POLY2, was used to encapsulate silicon dioxide and increase the total thickness of the beams (total beam thickness = 5.35 μm). Figure 4 shows the cross-section of the piezoresistive sensor beam. The central plate was constructed with the same structure as the beam except there was no piezoresistors embedded. The sensing structures were then released from the underlying silicon substrate using bulk silicon wet etching with an EDP solution at temperature of 90 °C for 30 minutes. The single crystal silicon is anisotropically etched in the area opened by the combination of "cut through substrate " layer (as discussed in section 3.1) until the central plate was undercut. The pits are formed underneath structures and their depths were controlled by etching time. The device was packaged in a ceramic DIP package and the read-out signal is electrically connected by gold wire bonding. A thin protective layer of polyvinyl chloride (PVC) or elastomer rubber can be adhered to the polysilicon surface of the sensor to provide interpolation of normal load between elements.

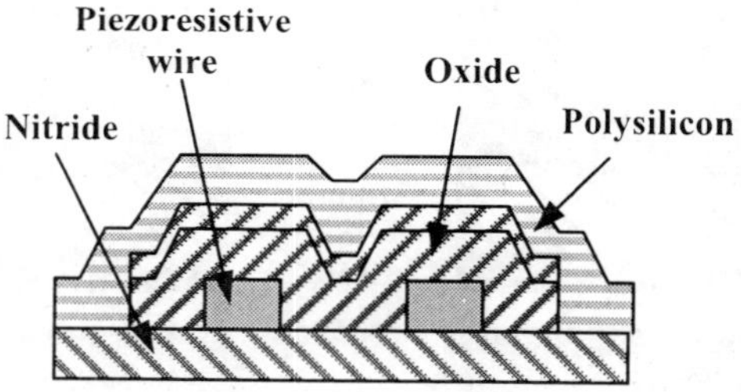

Figure 4. The cross section of the piezoresistive sensor beam.

4. TEST AND CHARACTERIZATION

First, the device has been electrically tested to determine the continuity of interconnects within the array. High-resolution X-Y-Z micromanipulator is used to position the pre-known weight microneedle on the bare sensor (without protective layer) directly for force testing. The entire setup was placed under a microscope for visual inspection. The tactile sensor array was connected to the external circuit on a board to measure the resistance and voltage change when force is applied. Figure 5 shows the experimental setup of piezoresistive sensitivity measurement with the applied normal force.

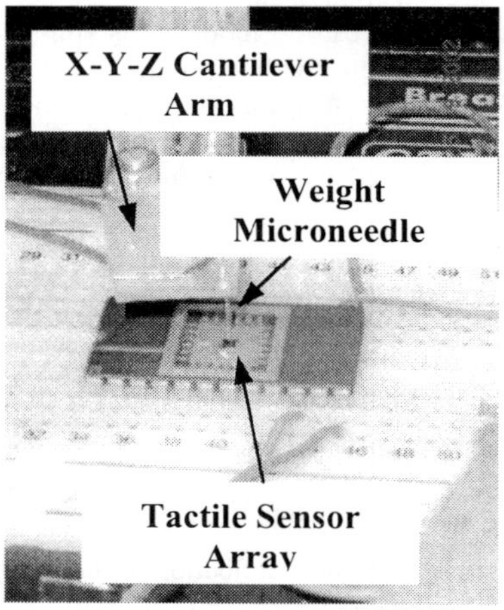

Figure 5. Experimental setup of piezoresistive sensitivity measurement with applied normal force by using pre-known weight microneedle.

A wheatstone bridge circuit as shown in Figure 6 is used to measure the extremely small change in the resistivity of piezoresistors. The resistivities of R_1, R_2, R_3 are adjusted evenly until equal to resistivity of the sensor, R_M, of 9.4kΩ. Therefore read-out voltage, V_0 , is equal to zero before normal force is applied. Plot of the change in read-out voltage correspond to applied normal force is shown in Figure 7. Sensor response shows linearity behavior with 0.02 mV/μN gain.

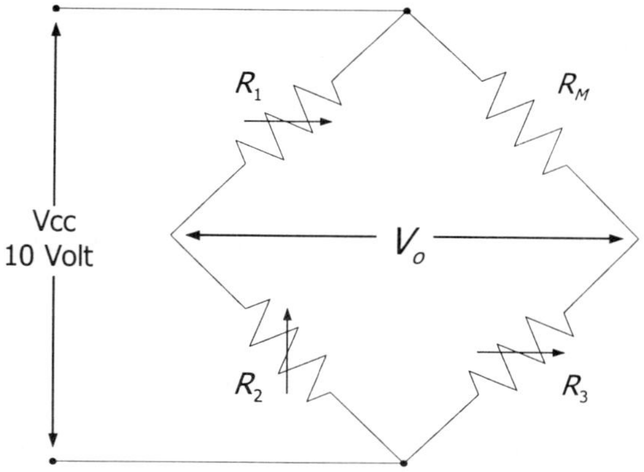

Figure 6. Wheatstone bridge for small change in resistivity measurement.

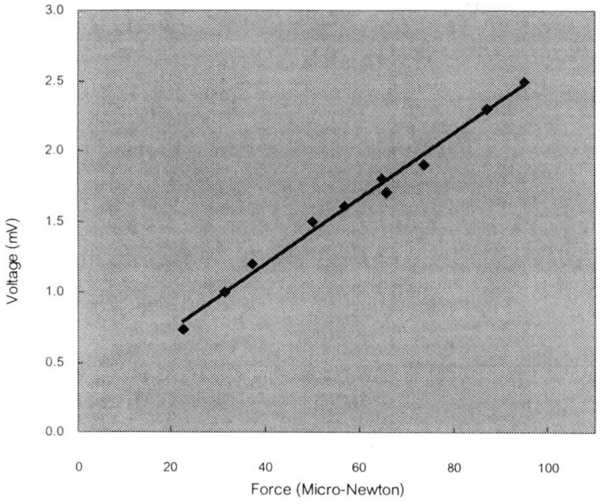

Figure 7. Plot of read-out voltage versus applied normal force.

5. SUMMARY

The micromachined piezoresistive tactile sensor array has been successfully fabricated by the commercial available surface micromachining foundry (MUMPs) with bulk-etching post-processing technique. The bulk etching of silicon substrate through "cut through substrate" mask is successfully done with a complete undercut of a central contacting plate and create a suspended structures. The response of the sensor structure to pure normal stress loading was experimentally evaluated. Linear sensitivity of the sensors with normal force load was approximately 0.02 mV/μN as tested with pre-known weight microneedle. Through preliminary test and characterization studies, the sensor array was shown to meet the design objective. More tests on operating temperature dependence and hysteresis behavior when the various protective layers are applied need to be studied further. Moreover, the packaging issues of the sensor must be addressed before the sensor can be considered useful for practical applications.

6. REFERENCES

[1] B. L. Gray and R. S. Fearing "A surface micromachined microtactile sensor array". *Proceedings of IEEE International Conference on Robotics and Automation*, Minneapolis, Minnesota, pp. 1-6, 1996.

[2] R. H. Liu, L. Wang, and D. J. Beebe "Progress towards a smart skin: fabrication and preliminary testing". *Proceedings of International Conference of the IEEE Engineering in Medicine and Biology Society*, Vol. 20, No. 4, pp. 1841-1844, 1998.

[3] B. J. Kane, M. R. Cutkosky and G. T. A. Kovacs "A traction stress sensor array for use in high-resolution robotic tactile imaging". *Journal of Microelectromechanical Systems*, Vol. 9, No. 4, pp. 425-434, 2000.

[4] G. T. A. Kovacs, *Micromachined Transducer Sourcebook*, WCB McGraw-Hill, New York, 1998.

[5] L. A. Liew, A. Tuantranont, and V. M. Bright "Modeling of thermal actuation in a bulk-micromachined CMOS micromirror". *Microelectronics Journal*, 31, pp. 791-801, 2000.

[6] D. A. Koester, R. Mahadevan, and K. W. Markus "Multi-user MEMS process (MUMPs): introduction and design rules, rev.4" MCNC MEMS Technical Application Center, Research Triangle Park, NC, 1996.

[7] A. Tuantranont, L. A. Liew, V. M. Bright, J. L. Zhang, W. Zhang, and Y. C. Lee "Bulk-etched surface micromachined and flip chip integrated micromirror array for infrared applications". *Proceedings of IEEE/LEOS International Conference on Optical MEMS*, Kauai, Hawaii, pp. 71-72, 2000.

FABRICATION PROCESS FOR A MICROFLUIDIC VALVE

Antonio Luque, José M. Quero

Dpto. Ingeniería Electrónica
Universidad de Sevilla
Av. Descubrimientos, s/n 41092 Sevilla, Spain
aluque@gte.esi.us.es

Cyrille Hibert, Philippe Flückiger

Center of Micro-Nano-Technology (CMI)
Swiss Federal Institute of Technology
CH-1015 Lausanne, Switzerland
cyrille.hibert@epfl.ch

ABSTRACT

In this paper, necessity for a high pressure valve is discussed, and a design for such a valve that has been previously presented is described. This valve can be built using simple fabrication techniques available in microsystem foundries. Its fabrication process is also shown. Later, a brief description of expected behaviour is presented.

1. INTRODUCTION

In today's application fields of microsystem technologies, microfluidics play an important and ever increasing role. A large number of applications in medical, biological and industrial fields would benefit from having an efficient microfluidic system that could treat fluids at a micrometer scale (with the improved efficiency obtained from that) like traditional processing machines have done at a larger scale.

In macroscopic world, improving efficiency in a fluidic system is achieved by using greater flows, and this is usually done by applying a bigger pressure difference over the fluid. If a high flow could be combined with the precision in processing typical of microfluidics, a much useful system would be built.

To accomplish this, a valve capable of controlling routing of fluid trough fluidic circuit, and at the same time able to drive high pressures must be designed. At the moment, no valve can open and close itself when a pressure as high as several atmospheres is applied on it.

A number of approximations to this problem has been tried (e.g. based on thermal expansion [1]), but electrostatic valves have proven their versatility and ease of use [2]. A way of making an electrostatic valve work with higher pressures is desirable.

2. AIR BALANCED PRESSURE

It has been proposed before [3] to use fluid pressure itself to compensate its own high pressure. Such designs use the fact that fluid can press valve seat both in opening and closing way, thereby compensating the pressure and allowing the valve to open or close by applying only a relatively small actuation force.

These designs are good at moderate pressures, but are not enough at high pressures. In this range, another kind of compensation must be used. The design here presented try to compensate fluid pressure with air (or another gas).

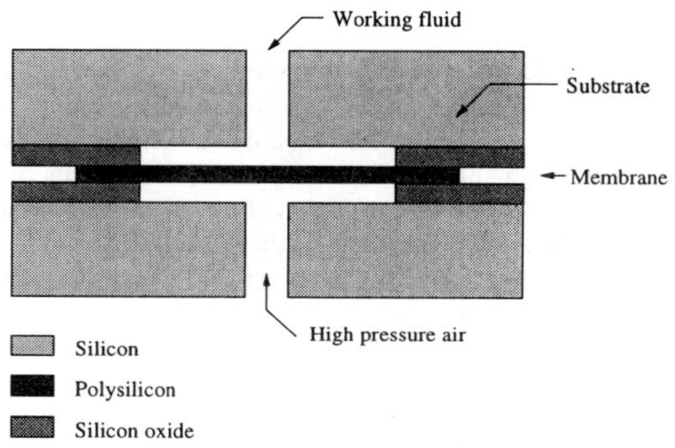

Figure 1: Cross-section view of proposed design

This objective of pressure compensation can be accomplished with the structure shown in fig. 1. In it, two different fluids apply pressure over a membrane, located in the middle of the drawing. In equilibrium, the membrane is in a half way between two orifices. A voltage can be applied between the membrane and any one of the substrates, creating an electrostatic attractive force that will displace membrane until it reaches the substrate. There is a thin dielectric layer that avoids electrical contact in this case.

In the drawing, only a orifice is shown for each fluid, but there are two of them, trough which liquid or air flows from one to another. When membrane is totally deflected touching top body, working fluid does not flow, and the valve is closed.

Membrane cannot be deflected in its entirety because of the limit imposed by upper and lower bodies. This has an effect over stability of the system, given that the elastic recuperation of membrane is much smaller than the one that

0-7803-7761-3/03/$17.00 ©2003 IEEE

would exist if the membrane would be totally deflected.

The valve operation has been successfully simulated using finite element software and has proven to be useful for the desired objective ([3], [4]). The next step is to design a fabrication process that satisfies the demands of the application.

3. FABRICATION PROCESS

In this section, we are going to describe each step in the fabrication process of the previously shown microvalve. This process is displayed in fig. 2.

The process starts with a standard double-side polished p-doped silicon wafer. The polishing will help with the bonding in the last step (see below). The low resistivity (about 0.1 - 0.5 Ω·cm) is needed because the substrate will be used as a electrode in electrostatic actuation. Additional doping or ion implantation can be used in order to decrease resistivity. The substrate will be used as a mechanical material in working system, so it shouldn't be too thin, but a standard 380-μm will suffice.

The wafer should be cleaned before starting depositions steps. The first of these depositions is a LPCVD of 2.5 μm SiO_2 that will be used as a sacrificial layer. Mechanical properties of this oxide are not extremely important, and a low temperature oxide is preferred, as this is less likely to degrade properties of underlying silicon.

Next, a very thin layer of nitride (Si_3N_4) is deposited, also by LPCVD. The only function of this layer is to isolate valve membrane from body and to avoid electrical breakdown between them. As nitride has a very high dielectric strength, only 200 nm are needed to achieve this. A thicker layer could influence mechanical properties of membrane, while a thinner one could not be enough.

The last deposition step consists of LPCVD of 2.5 μm of polysilicon that will build up the membrane. This membrane will be itself a working electrode, so a low resistivity in it is even more important that in valve substrate. A doping of polysilicon, followed by a deglazing to remove oxide, is performed. After the doping, measured resistivity of polysilicon is about $5.5 \cdot 10^{-4}$ Ω·cm, low enough for our application.

Next step is polysilicon patterning with dry etching preceded by photolithography. No special care must be taken in this photolithography, as all device dimensions are large enough to not have to care about exposure distance, and soft contact can be used.

Plasma etching is then performed at an etch rate of about 1 μm/min [5], [6] . It is fundamental to completely etch all polysilicon outside the membrane to avoid electrical contact between neighbour devices. One of the most important characteristics of this device is its ability to operate in arrays that can be made from the same wafer. If an electrical contact would exist between two or more adjacent devices, they couldn't be operated independently and this advantage would be lost. The etching process is a pulsed room temperature process that uses SF_6 and C_4F_8 plasma.

Once top-side polysilicon has been micromachined, bottom-side polysilicon and nitride are removed using plasma etching. Dry etching is preferred over wet etching because it doesn't affect recently formed structure on top side. Bottom polysilicon and nitride can be removed in about 4.5 min, employing two different plasma recipes. It is not convenient to remove bottom-side oxide, as it can serve as a mask for the next deep etching step. After said step, all oxide will go away and silicon substrate will be electrically contactable to act as an electrode.

The inlet and outlet orifices, through which the fluid will flow will be opened next. They are etched in substrate. Oxide will act as a mask, as said before, and it needs to be patterned. As above, photolithography is very easy; both holes are big enough (100 μm) so soft contact and thick photorresist (about 2.7 μm) can be used. The most important thing is to carefully align bottom photolitography mask with top one, not for positioning orifices under polysilicon membrane, which is easy given their sizes, but for later aligning of two pieces that will be bonded together to form design shown in fig. 1. Oxide etching will also strip photorresist, and oxide will be the only mask for remaining bottom-side etching.

Next step, shown in fig. 2 f) is Deep Reactive Ion Etching (DRIE) [7], [8]. It is a very anisotropic etching that will form channels with almost vertical walls. The etching is performed using a Bosch process on an Alcatel 601E equipment, with an etch rate of about 6 μm/min [9]. This step opens holes trough the whole wafer.

Last steps (shown in fig. 2 g and h) consist in removing the sacrificial oxide layer by etching oxide in HF, and then bonding two identical pieces to form a valve.

In figure 3, resulting membrane can be seen on top of one of the holes etched trough the substrate.

It should be noted that last step (releasing membrane in HF) also etches silicon nitride and thus removes the electrical isolation between membrane and substrate that this nitride provided. We are now considering add on more polysilicon layer to the fabrication process. This layer would protect nitride from being removed by HF and electrical isolation would be preserved.

4. OPERATIONAL BEHAVIOUR

Working of a electrostatic parallel-plate actuator, like the one included in the valve, has been extensively described in the literature ([10], [11]). In the present case, applying a constant voltage of 200 V between both electrodes gives a displacement like the one shown in fig. 4. It can be seen

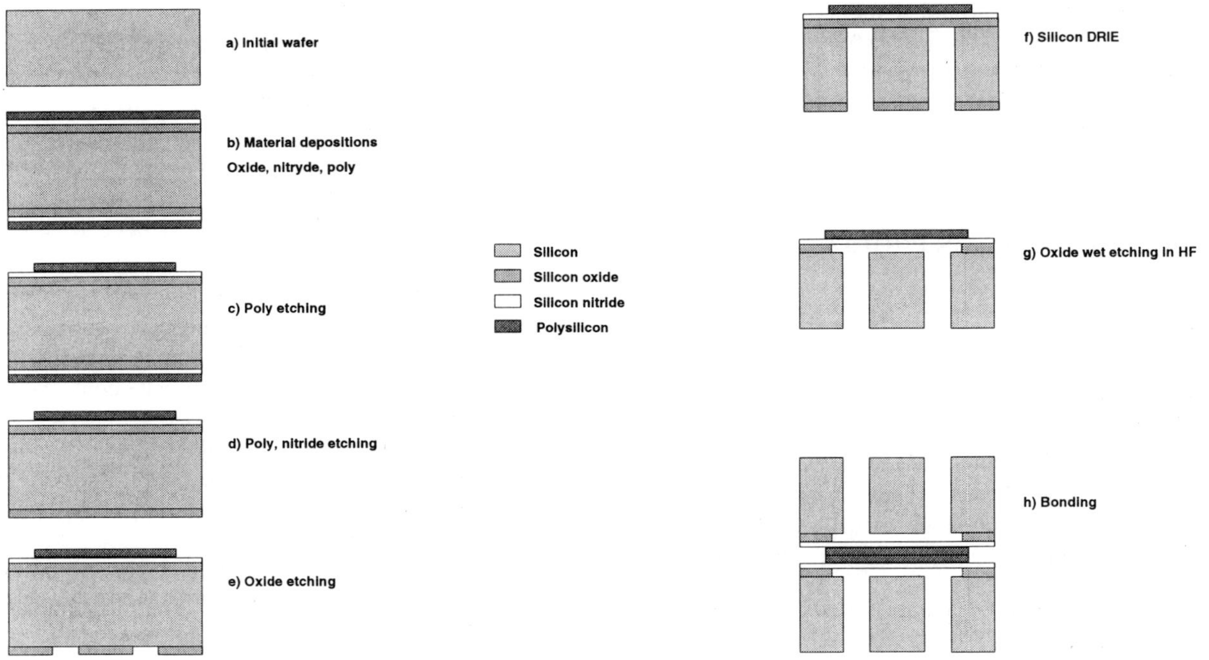

Figure 2: Fabrication process

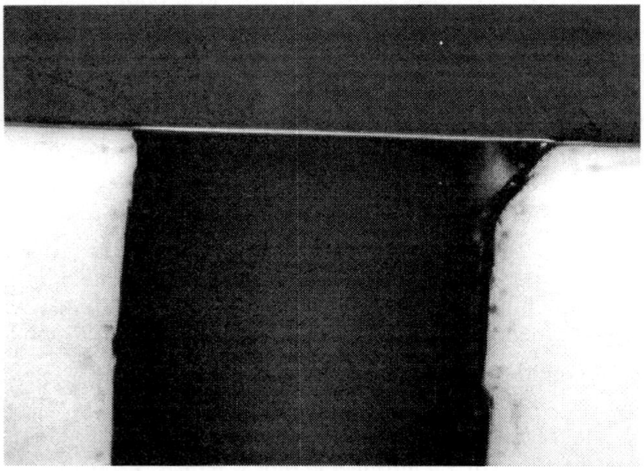

Figure 3: Detail view of polysilicon membrane and fluidic orifice

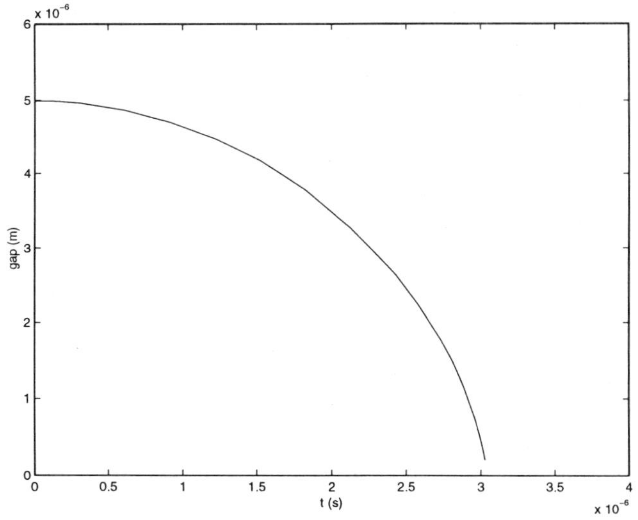

Figure 4: Dynamic evolution of gap between electrodes

that valve fully closes (or opens) in a time of about 3 microseconds [12].

In figure 5, evolution of current during closing is shown. To obtain these figures, values used are: gap between electrodes 5 μm, membrane size 300 μm, membrane thickness 2.5 μm.

These figures are obtained by numerically solving the differential equations giving displacement of and current through a movable capacitor, and that are widely known.

5. CONCLUSIONS

A valve capable of driving pressures as high as several atmospheres and its fabrication process have been presented. The device has been successfully fabricated using the described process and it is currently under test, in order to obtain working curves in the fluidic domain.

The valve here described is protected under Spanish patent

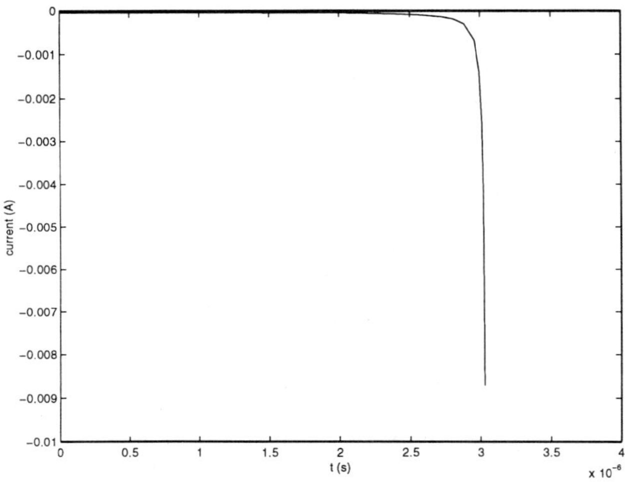

Figure 5: Dynamic evolution of current through capacitor

P200102361.

6. REFERENCES

[1] E. T. Carlen and C. H. Mastrangelo, "Paraffin actuated surface micromachined valve," in *13th IEEE Intl. Micro Electro Mechanical Systems Conference*, 2000.

[2] J. Braneberg and P. Gravesen, "A new electrostatic actuator providing improved stroke length and force," in *Proc. of Micro Electro Mechanical Systems*, Feb 1992, pp. 6–10.

[3] J. M. Quero, A. Luque, and L. G. Franquelo, "A novel pressure balanced microfluidic valve," in *IEEE Intl. Symposium on Circuits and Systems (ISCAS)*, 2002, vol. 2, pp. 588–591.

[4] A. Luque, F. R. Palomo, and J. M. Quero, "Analysis and simulation of a microfluidic valve," in *XVII Conference on Design of Circuits and Integrated Systems (DCIS)*, Nov 2002.

[5] Cyrille Hibert, Willy Dufour, and Philippe Flückiger, "Deep anisotropic etching of silicon using low pressure high density plasma. presentation of complementary techniques and their applications in microtechnology," in *Intl. Symposium on Plasma Chemistry*, Jul 2001.

[6] T. Syau, B. Baliga, and R. Hamaker, "Reactive ion etching of silicon trenches using SF_6/O_2 gas mixtures," *J. Electrochem. Soc*, vol. 138, no. 10, pp. 3076–3081, 1991.

[7] A. Schilp, M. Hausner, M. Puech, N. Launay, H. Karagoezoglu, and F. Laermer, "Advanced etch tool for high etch rate deep reactive ion etching in silicon micromachining production environment," in *Proceedings MST*, 2001.

[8] M. Takinami, K. Minami, and M. Esashi, "High-speed directional low-temperature dry etching for bulk silicon micromachining," in *Digests of the 11th Sensor Symposium*, 1992, pp. 15–28.

[9] J. Jorne, Y. J. Lii, K. C. Cadien, and J. E. Schoenholtz, "Plasma etching of silicon in SF_6: experimental and reactor modeling studies," *J. Electrochem. Soc*, vol. 137, no. 11, pp. 3633–3638, Nov. 1990.

[10] J. L. Seeger and B. E. Boser, "Dynamics and control of parallel-plate actuators beyond the electrostatic instability," in *Proc. Intl. Conf. on Solid-State Sensors and Actuators (Transducers'99)*, Jun 1999, pp. 474–477.

[11] E. K. Chan and R. W. Dutton, "Electrostatic micromechanical actuator with extended range of travel," *J. Microelectromech. Syst.*, vol. 9, pp. 321–328, Sept 2000.

[12] M. A. Huff, J. R. Gilbert, and M. A. Schmidt, "Flow characteristics of a pressure-balanced microvalve," in *Tech. Dig. of Transducers'93*, 1993, pp. 98–101.

A Neuromorphic Sound Localizer for a Smart MEMS System

André van Schaik and Shihab Shamma

School of Electrical and Information Engineering
University of Sydney
Sydney, NSW 2006, Australia

Institute for Systems Research
University of Maryland
College Park, MD 20742, USA

ABSTRACT

In this paper we present an analog circuit that determines the direction of incoming sound using two microphones. The circuit is inspired by biology and uses two silicon cochlea to determine the azimuthal angle of the sound source with respect to the axis of the two microphones using the time difference between the two microphone signals. A new algorithm, adapted to an analog VLSI implementation, is presented together with simulation and measurement results.

1. INTRODUCTION

Air-coupled acoustic MEMS offer exciting opportunities for a wide range of applications for robust sound detection, analysis, and recognition in noisy environments. The most important advance these sensors offer is the potential for fabricating and utilizing miniature, low-power, and intelligent sensor elements and arrays. In particular, MEMS make it possible for the first time to conceive of applications which employ arrays of interacting micro-sensors, creating in effect spatially distributed sensory fields. To achieve this potential, however, it is essential that these sensors be coupled to signal conditioning and processing circuitry that can tolerate their inherent noise and environmental sensitivity, without sacrificing the unique advantages of compactness and efficiency.

The authors, together with several colleagues, are currently focusing their efforts on developing a smart microphone, suitable for outdoor acoustic surveillance on robotic vehicles. This smart microphone will incorporate MEMS sensors for acoustic sensing and adaptive noise-reduction circuitry. These intelligent and noise robust interface capabilities will enable a new class of small, effective air-coupled surveillance sensors, which will be small enough to be mounted on future robots and will consume less power than current systems. By including silicon cochlea based detection, classification, and localization processing, these sensors can perform end-to-end acoustic surveillance. The resulting smart microphone technology will be very power efficient, enabling a networked array of autonomous sensors that can be deployed in the field.

We envision such a sensory processing system to be fully integrated with sophisticated capabilities beyond the passive sound reception of typical microphones. Smart MEMS sensors may possess a wide range of intelligent capabilities depending on the specific application, e.g., they may simply extract and transmit elementary acoustic features (sound loudness, pitch, or location), or learn and perform high-level decisions and recognition. To achieve these goals, we aim to develop and utilize novel technologies that can perform these functions robustly, inexpensively, and at extremely low power. An equally important issue is the formulation of algorithms that are intrinsically matched to the characteristic strengths and weaknesses of this technology. In this paper we present an implementation of one such algorithm, which is inspired by biology, but adapted to the strengths and weaknesses of analog VLSI, for localizing sounds in the horizontal plane using two MEMS microphones.

2. THE ALGORITHM

Humans rely heavily on the Interaural Time Difference (ITD) for localization of sounds in the horizontal plane. When a sound source is in-line with the axis through both ears, sound will reach the furthest ear with a certain delay after reaching the closest ear. To a first approximation, ignoring diffraction effects around the head, this time delay is equal to the distance between the ears divided by the speed of sound. On the other hand, if the sound source is straight ahead or behind the listener, it will reach both ears at the same time. In between, the ITD varies as the sine of the angle of incidence of the sound.

The most common method for determining the time difference between two signals in engineering is to look for the delay at which there is a peak in the cross-correlation of the two signals. In biology, a similar strategy is used, known as Jeffress' coincidence model [1]. In the ear sound captured by the ear-drum is transmitted to the cochlea via the middle-ear bones. The fluid-filled cochlea is divided into two parts with a flexible membrane, the basilar membrane, which has mechanical properties such that high-frequency sound makes the start of the membrane vibrate most, whereas low-frequency sound vibrates the end most. Inner Hair Cells on the basilar membrane transduce this vibration into a neural signal. For frequencies below 1-2kHz, the spikes generated on the auditory nerve are phase-locked to the vibration of the basilar membrane and therefore to the input signal. In Jeffress' model, this phase-locking is used together with neural delay-lines to extract the interaural time difference in each frequency band. Delayed spikes from one ear are compared with the spikes from the other ear and coincidences are detected. The position along the delay-line where the spikes coincide is a measure of the ITD.

A hardware implementation of this model has been developed by Lazarro [2]. Such a hardware implementation needs the creation of delay lines with a maximum delay value of the maximum time difference expected and a minimum delay value equal to the

0-7803-7761-3/03/$17.00 ©2003 IEEE

resolution needed. This will have to be done at each cochlear output, which makes the model rather large.

An alternative approach uses the fact that a silicon cochlea itself not only functions as a cascade of filters, but also as a delay line, since each filter adds a certain delay. Cross-correlation of the output of two cochleae, one for each ear, will thus give us information about the ITD [3]. However, the delays are proportional to the inverse of the cut-off frequency of the filters and are therefore scaled exponentially. This makes obtaining an actual ITD estimate from a silicon implementation of this algorithm rather tricky [4].

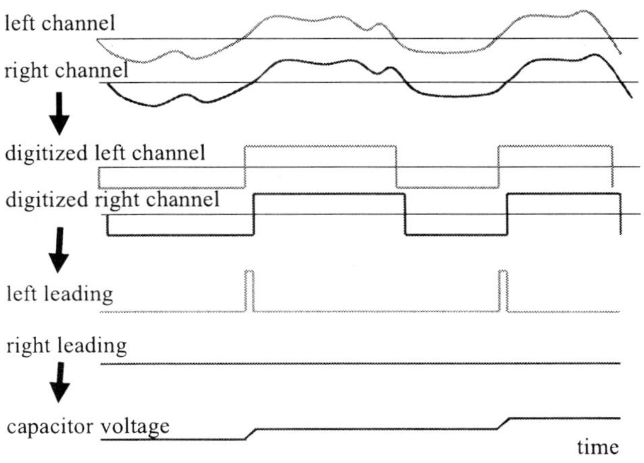

Figure 1. Signals in the sound localizer algorithm.

Instead of the algorithms discussed above, we have developed an algorithm that is adapted for aVLSI implementation. The algorithm is illustrated in Figure 1. First, the left and right signals are digitized by detecting if they are above or below zero. Next, the delay between the positive zero-crossing in both signals is detected and a pulse is created with a width equal to this delay. Finally, a known constant current is integrated on a capacitor for the duration of the pulse, so that the change in voltage is equal to the pulse width. A voltage proportional to the average pulse width can be obtained by integrating over a fixed number of pulses. In our implementation, separate pulses are created for a left-leading signal and for a right-leading signal. The left-leading pulses increase the capacitor voltage, whereas the right-leading pulses decrease the capacitor voltage. Once a fixed number of pulses has been counted, the capacitor voltage is read and reset to its initial value.

This algorithm was simulated in Matlab using sound files that were recorded in the field with the MEMS microphones that we intend to use in the final system. The sounds to localize can best be described as low-frequency noise, with most of the energy between 50Hz and 300Hz. Furthermore, as the microphones are only a few centimeters apart, the ITD for a sound played from 1 degree of angle in front is about 4μs, which is much less than 0.1% of phase shift for the lowest frequency components. The results of these tests are shown in Figure 2. It can be seen that the average value of the ITD estimate (thick line) corresponds well with the theoretical curve (dashed line), but that the standard deviation of the responses is rather high.

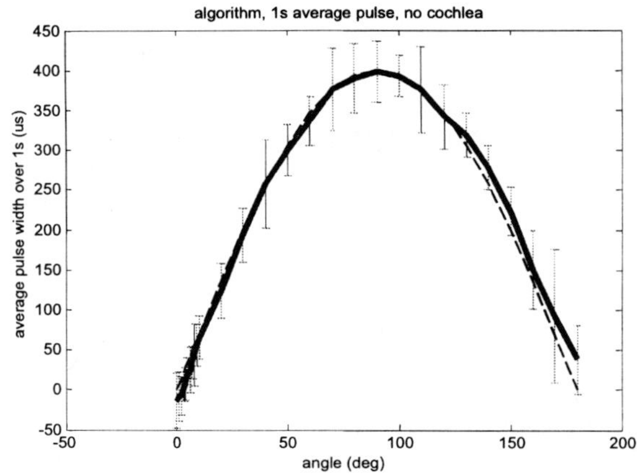

Figure 2. Results of simulation of the algorithm. The thick line shows the average value of the time difference estimate. Each estimate is the average value for one second. As the sound files are 30s long, a total of 30 estimates is obtained The error bars indicate one standard deviation and the dashed line indicates a theoretical fit proportional to sin(angle).

In reality the algorithm is not applied directly to the input signal, but pair-wise to the output of two cochlear models containing 32 sections. Each cochlear section has a band-pass filter characteristic and the best-frequencies of the 32 filters are scaled exponentially between 300Hz and 60Hz. Band-pass filtering increases the periodicity of signals that the algorithm is applied to, which improves its performance. The results of the Matlab simulation of the algorithm including Lyon's cochlear model [5] is shown in Figure 3. Using the cochlea and averaging the ITD estimate over all sections reduces the standard deviation of the response somewhat, but at the same time, the mean is slightly less close to the ideal curve.

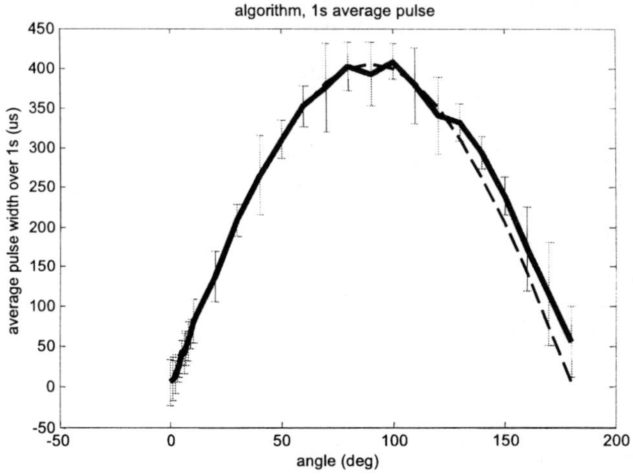

Figure 3. Same as Figure 2 but using two silicon cochleae with 32 sections each. The average is obtained by averaging over one second and over all 32 sections.

A problem with both versions of the Matlab simulation is the very fine time resolution needed in order to extract delays in the order of microseconds.

3. THE IMPLEMENTATION

The hardware implementation of our algorithm uses two identical silicon cochleae with 32 sections each. At each of the 32 sections the output of both cochleae is used to create digital pulses that are as wide as the time delay between the two signals. This time delay is measured within each section and averaged over all active sections in order to obtain a global estimate. Inactive sections are sections that do not contain enough signal during the period over which the ITD is estimated and are therefore not included in the global estimate. The total size of this implementation is $5mm^2$ in a 0.5µm process, with 75% of the circuit area devoted to the implementation of the capacitors. If the circuit were to operate at sound frequencies that humans use for ITD detection, the capacitor sizes could easily be reduced by a factor 3, cutting the total circuit size in half.

3.1 THE SILICON COCHLEA

The silicon cochlea used is similar to the one we have presented in [6], which has already proven its use in other neuromorphic sound processing systems [7, 8]. The basic building block for the filters in this cochlear model is the transconductance amplifier, operated in weak inversion. For input voltages smaller than about 60 mV_{pp}, the amplifier can be approximated as a linear transconductance:

$$I_{out} = g_m(V_{in+} - V_{in-}) \tag{1}$$

with transconductance g_m given by:

$$g_m = \frac{I_0}{2nU_T} \tag{2}$$

where I_0 is the bias current, n is the slope factor, and the thermal voltage $U_T = kT/q = 25.6$ mV at room temperature.

It has been shown that if all three amplifiers in the circuit are identical, the second-order section may be stable for small signals, but will exhibit large signal instability due to slew-rate limitations [9]. This can be solved by using a transconductance amplifier with a wider linear input range in the forward path [9]. This also allows larger input signals to be used, up to about 140 mVpp.

Our silicon cochlea is implemented by cascading 32 of these second-order sections with exponentially decreasing cut-off frequencies. The exponential decrease is obtained by creating the bias currents of the second-order section with CMOS Compatible Lateral Bipolar Transistors, as proposed in [6].

3.2 THE ITD DETECTION

The circuit of Figure 4 implements a second order low-pass filter, where the output voltage, V_{C2}, of each section is a low-pass filtered version of the section's input signal and serves as the input for the next section. A normalized band-pass output is obtained by subtracting V_{C2} from V_{C1}. This signal is not created explicitly on chip. Instead, the signal is digitized using a comparator to detect when $V_{C1} > V_{C2}$. The same is done for the second cochlea. Both comparators are, however, controlled by a latch which will first need to be set to allow the comparison. This is done when $g_m(V_{C2} - V_{C1})$ is larger than a certain threshold current I_{th} in both cochleae at the same time, where g_m is the transconductance of the differential pair in the comparator. This ensures that zero-crossings are only detected on signals for which $V_{C1} - V_{C2}$ is well below zero in both cochleae just before the zero-crossing, which prevents the generation of numerous zero-crossings that would be created by a small, noisy signal around zero and improves the performance of the algorithm.

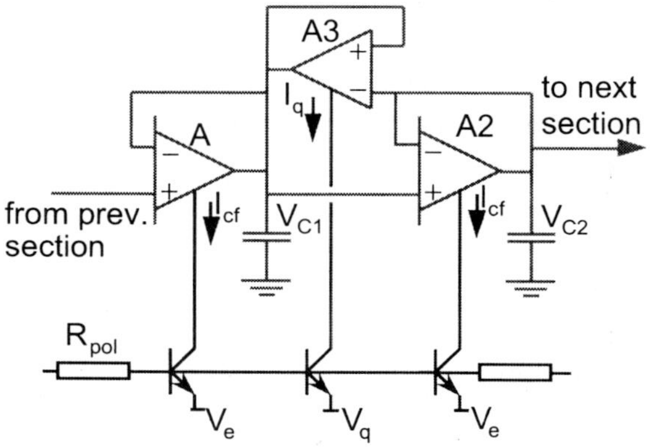

Figure 4. The second order section used in the cochleae.

If the comparators are enabled and V_{C1} becomes larger than V_{C2} in the "left ear" cochlea, we detect a zero-crossing in left ear leading channel, which we will call ZCDL=1. The same is done for the right ear to detect ZCDR=1. However, one of these two will occur first, depending on which ear is leading. Furthermore, when both ZCDR and ZCDL are high, the latch that enables the comparators is reset which in turn forces both ZCDL and ZCDR low. In reality the feedback is so fast that whichever happens second – ZCDL=1 or ZCDR=1 – never really goes high. Therefore, the zero-crossing is only detected in the leading channel and is only high for the duration of the time difference between the zero-crossings in the two channels.

Finally, the number of pulses created is counted, while for the duration of each pulse a fixed reference current is integrated onto a unit capacitor at each section. ZCDL pulses source the reference current onto the capacitor, whereas ZCDR pulses sink the current from the capacitor. When 64 pulses are counted in a particular section, the capacitor in that section is connected to a common bus and pulse generation in that section is disabled until a reset signal occurs. As the capacitors in all sections that reach 64 pulses are connected to the bus, the total charge in all the active sections is shared across all unit capacitors in those sections, leading to an averaging of the capacitor voltage. This averaged voltage will be an estimate of the ITD of the input sound. After this value is read out, a reset signal is given and a new estimate begins.

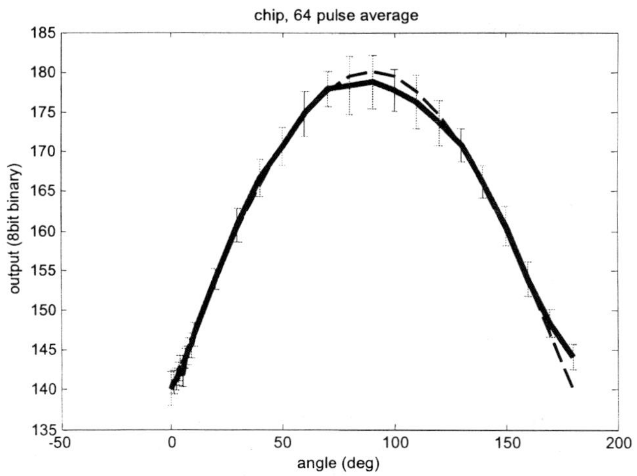

Figure 5. Measured digital (8 bit) representation of the average capacitor voltage after 64 pulses.

4. MEASUREMENT RESULTS

The silicon implementation has been tested using the same recorded sounds played using a PC soundcard. Each second of sound an estimate was obtained and a reset signal given. The estimates were recorded using an off-chip A/D converter interfaced with the PC. The mean and standard deviation of the 30 estimates per sound were calculated using Matlab and are plotted in Figure 5.

The aVLSI sound localizer actually has less standard deviation of the ITD estimate than the Matlab simulation. This is a result of the noise suppression – the detection of $g_m(V_{C2} - V_{C1}) < I_{th}$ in both channels – that was implemented on-chip, but not used in the Matlab simulation, because of the excessive simulation time that this would result in. For the extreme values of the ITD, around an angle of 90 degrees, the circuit systematically underestimates the delay. This is due to saturation of the integrator capacitor voltage in some of the sections.

The power consumption of the circuit depends on the ITD measured, as more current is integrated onto the capacitors for larger ITDs. For an ITD generated using a 200Hz square wave delayed by 100μs in one of the cochleae with respect to the other, the power consumption of the circuit was 370μA at a 5V supply voltage, i.e., less than 2mW. When I_{th} was increased to disable all pulse generation so that the power consumption is mainly due to the cochlear filters, the power consumption dropped to 400μW.

5. CONCLUSIONS

We have successfully realized a low-power aVLSI sound localizer that is capable of detecting phase differences of less than 0.1% in low-frequency sounds captured by two MEMS microphones. The standard deviation of the estimate is less than a few degrees of angle, except for the most lateral positions, where the ITD varies little with angle.

6. ACKNOWLEDGEMENTS

The work in this paper has been supported by DARPA through the "Air-coupled acoustic microsensor technology" program (BAA 00-08).

7. REFERENCES

[1] L. A. Jeffress, "A place theory of sound localization," *Journal of Comparative and Physiological Psychology*, vol. 41, pp. 35-39, 1948.

[2] J. Lazzaro and C. A. Mead, "A silicon model of auditory localization," *Neural Computation*, vol. 1, pp. 47-57, 1989.

[3] S. A. Shamma, S. Naiming, and P. Gopalaswamy, "Stereausis: binaural processing without neural delays," *Journal of the Acoustical Society of America*, vol. 86, pp. 989-1006, 1989.

[4] C. A. Mead, X. Arreguit, and J. Lazzaro, "Analog VLSI model of binaural hearing," *IEEE Transactions on Neural Networks*, vol. 2, pp. 230-6, 1991.

[5] M. Slaney, "Auditory Toolbox: Version 2," Interval Research Corporation 1998-010, 1998.

[6] A. van Schaik, E. Fragnière, and E. Vittoz, "Improved Silicon Cochlea using Compatible Lateral Bipolar Transistors," in Advances in Neural Information Processing Systems 8, Cambridge MA, 1996.

[7] A. van Schaik, "An analog VLSI model of periodicity extraction in the human auditory system," *Analog Integrated Circuits & Signal Processing*, vol. 26, pp. 157-77, 2001.

[8] A. van Schaik and R. Meddis, "Analog very large-scale integrated (VLSI) implementation of a model of amplitude-modulation sensitivity in the auditory brainstem," *Journal of the Acoustical Society of America*, vol. 105, pp. 811-821, 1999.

[9] L. Watts, D. A. Kerns, R. F. Lyon, and C. A. Mead, "Improved implementation of the silicon cochlea," *IEEE Journal of Solid-State Circuits*, vol. 27, pp. 692-700, 1992.

A Time-Series Processor for Sonar Mapping and Novelty Detection

T. Horiuchi[1,2,3] *and R. Etienne-Cummings*[1,3]

[1]Electrical and Computer Engineering Dept.,
[2]Neuroscience and Cognitive Sciences Program
[3]Institute for Systems Research
University of Maryland
College Park, Maryland 90742, USA

ABSTRACT

A time-series processor chip, intended for sonar mapping and novelty detection applications, has been designed, fabricated and tested. The chip, when coupled with a sonar bearing and range estimation unit, receives an image of the environment as a voltage waveform. The bearing of an object is given by the magnitude of the signal, while its range is given by its time following the transmission of the sonar 'ping'. The chip stores this voltage waveform in a bank of sample-and-hold (S/H) elements and compares it to a previously stored trace. Objects that move in either azimuth or range are immediately detected and reported. The chip contains 54 S/H elements that are triggered by an on-board timer. The change detector can detect motion in bearing and range of 10 degrees/s and 11 cm/s, respectively. The maximum range is approximately 5m. Operating in the CMOS subthreshold region of operation, the novelty detection chip is designed for ultra-low power and micro-footprint smart surveillance systems. Implemented in a CMOS 0.5um process, it consumes less than 20 uW @ 8 Hz repetition rate (and less than 6 uW quiescent) with a 3V supply and occupies less than 0.3 sq. mm.

1. INTRODUCTION

The real-time processing of time domain, analog or digital, signals is required in many fields. This is particularly true for sonar data conditioning and understanding, where the received data has a pseudo-transient nature and is episodically distributed in time. Furthermore, active sensing with sonar is typically used to control behaving systems, e.g. robot navigation, or make real-time decisions, e.g. movement detection, which emphasizes the need for real-time, on-line and continuous time signal processing. Hence, working with time domain signals, analysis can be performed during signal acquisition, which leads to decisions and control with minimum latency.

Time series processing VLSI chips have been implemented primarily for two applications. The most common application is speech or transient acoustic signal classification [1],[2]. The second application is for sonar signal analysis [3]-[6]. In the former, the processor performs signal conditioning, storing and classification/identification based on well-known algorithms, such as Vector Quantization [1] and other correlation-based template matching schemes. In the case of the sonar signal analysis, the time-series processors are typically used to precondition sonar returns, i.e. filter, and digitize for output, such that standard digital computation hardware, e.g. DSPs, can be used to implement mapping and tracking algorithms [3],[4]. In rare cases, the chip may include some form of classification, e.g. coded "ping" identification, and output digitization [5]. This allows the system to distinguish between its own returns and those from other sources (on the same frequency).

On the other hand, we are working towards a self-contained chip that includes bearing and range estimation, mapping, storage and change/novelty detection. All computation will be performed with custom mixed-signal VLSI circuits. The architecture for the bearing and range estimation chip is presented in [6]. This paper focuses on the circuits and measurement results for mapping, storage and change/novelty detection. The two systems will eventually be merged into a single chip.

2. SYSTEM ARCHITECTURE

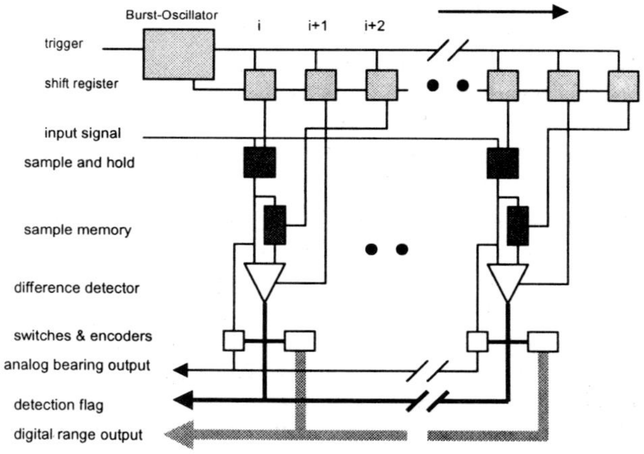

Figure 1. A block diagram of the time-series novelty detection chip. Upon receiving a trigger signal, the burst-oscillator begins driving a single bit down the shift register. The sample-and-hold unit *i* is activated by shift register (*SR*) unit *i*, the difference-detector unit *i* is activated by *SR(i+1)*, and the sample-memory transfer circuit *i* is activated by *SR(i+2)*. Likewise, the detection unit at position *i+1* (not shown) receives signals from SR units *i+1*, *i+2*, and *i+3*. The chip contains 54 S/H units

0-7803-7761-3/03/$17.00 ©2003 IEEE

and 56 shift register units. The chip completely powers down once the bit leaves the shift register.

2.1 Block Diagram

Figure 1 shows the block diagram of the system. For normal operation, there are two inputs: the trigger and input signals. In our sonar application, an external source initiates the sonar ping and triggers the chip to begin sampling the returning echoes. The intended input to the chip is a real-time computed bearing angle, represented as a voltage, and range, encoded the time delay of the return relative to the sonar "ping". Any signal based on the returning sonar echo, however, can be used.

Following a trigger, the analog input signal is sampled (and recorded) at approximately 1.5 KHz for about 38 msec to an accuracy of about 16 mV. Any change in the sampled values since the last scan (above a threshold) produces output signals that indicate at what range and bearing it occurred.

2.2 Circuits

The primary goals in the design process were to minimize power consumption yet retain the precision needed for the change detection process. As a result, the design centered around using a digital shift register that would power the analog circuits up and down in a wave as they were needed in sequence. Each location in the array analyzes a specific interval of time following the trigger. Since the system only operates with a low duty cycle, (In our example, ~25% or less.) the on-board clock shuts down after the sampling period to save power.

Sample and Hold Circuit

The sample and hold circuit (Figure 2) is a transmission gate that connects the input line to a 210fF inter-poly capacitor using small transistor sizes (W/L=1.5um / 0.75um) to minimize offset problems from gate-capacitance-induced charge injection. The sampling capacitor was kept small to minimize the time and current required to charge it.

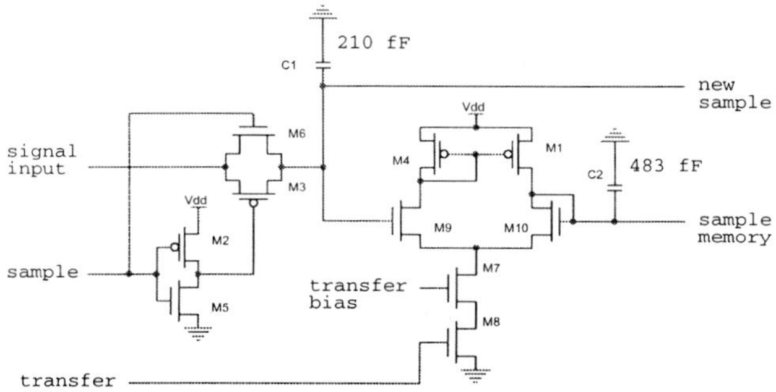

Figure 2. Sample-and-hold and memory transfer circuits. The digital sample signal connects the signal input to the 'new sample' capacitor. After the comparison (2 clock cycles later), the digital transfer signal copies the value into the sample memory capacitor.

Burst Oscillator and Single Bit Shift Register

The burst oscillator (Figure 3) is designed to begin producing a 50% duty-cycle square-wave output upon receiving a trigger signal that will continue oscillating until the bit traveling down the shift register 'pops' out the end. The oscillator is based on two inverters coupled by a simple RC-like circuit to set the oscillation frequency. When the burstrun signal is high, the transconductance amplifier-follower acts like a resistance, its output current saturating for large voltage differences. The effective resistance is controlled by the bias voltage oscbias. When 'burstrun' is low, the follower is powered down and the oscillator is clamped into a known state, reducing power.

There are two concerns about this circuit, high power consumption and frequency stability. High power consumption in the first inverter is an issue because the input voltage moves up and down around the threshold voltage where the current is highest. Frequency stability is a concern due to the strong dependence of the bias current on the oscbias voltage and on temperature. On the other hand, by modifying oscbias in a controlled way, the sampling rate of the system can be changed so that data can be stored/processed with variable temporal acuity. For the sonar application, this feature allows us to interrogate particular locations with higher range resolution.

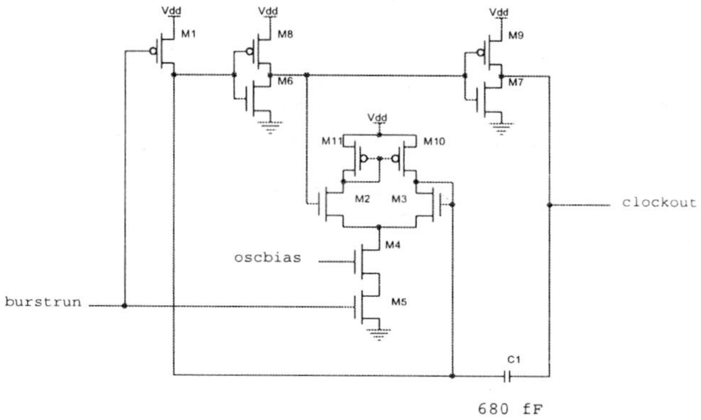

Figure 3. The on-board oscillator circuit.

Difference Detection Circuit

The difference detection circuit (Figure 4) is based on a transconductance amplifier followed by a current-mode full-wave rectifier. The resulting current is compared against a threshold current and the result is a digital voltage signal. The transconductance amplifier is activated by the output from shift register unit *i+1*, in the clock cycle following sampling.

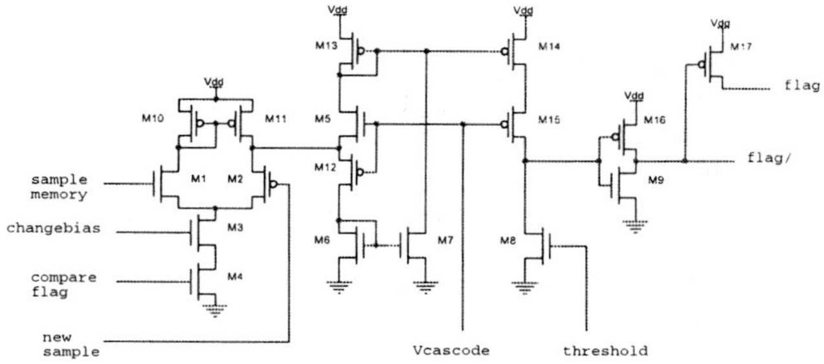

Figure 4. The difference detection circuit.

The transconductance amplifier operating in subthreshold has an input dynamic range of about +/- 100 mV. The sensitivity of this circuit to differential input voltage is dependent on the bias current, the threshold current, and (at the highest sensitivity settings,) the drain conductance of the current comparison transistors (M15 and M8). For this reason, the pFET mirror has been cascoded. For power conservation, when 'compare flag' is low, there is no quiescent current in this circuit.

Bearing and Range Outputs

Upon detecting a change, the flag/ signal (Figure 4) is pulled low, activating a wired-OR pull-up transistor to indicate the detection off-chip. This signal also locally activates two other circuits, the analog 'bearing output' and a digital 'range address' bus. The flag/ signal turns on a voltage-follower circuit to drive the newly sampled value off-chip, signaling the ('bearing') value that triggered the detection event. The flag/ signal also drives a 6-bit binary encoder to indicate which element is signaling the event. These circuits are not shown.

Memory Transfer Circuit

Once the comparison between old and new values has been performed, the output from shift register unit $i+2$ is used to turn on a follower circuit (Figure 2) driving the memory capacitor voltage to the sampled voltage level. The sample memory capacitor needed to be large enough to minimize voltage loss, but kept small to minimize the current needed to charge it. A 483fF capacitor was chosen based on estimates from simulation.

3. CHIP TESTING

The Triggered Sample-and-Hold Subsystem

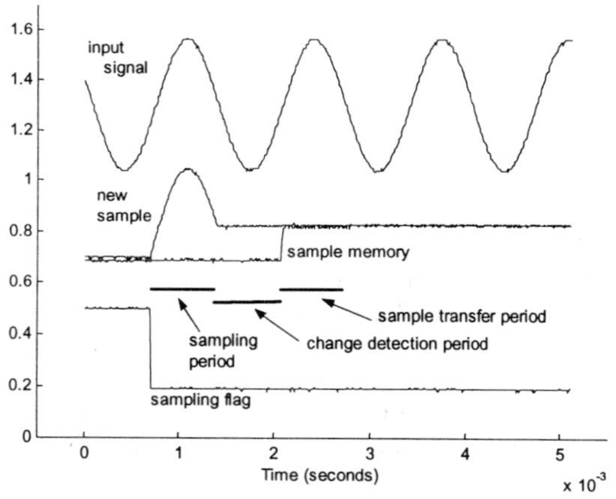

Figure 5. A test of the sample-and-hold circuit (tap #1) shows the 'new sample' and 'sample memory' voltages. Processing at each pixel occurs in three stages: sampling of the input signal, comparison to the stored previous value and transfer of the new value to the memory location. The input signal trace has been voltage shifted and the sampling flag voltage has been rescaled to fit in the figure.

The sample and hold circuit for tap #1 is shown operating in Figure 5. A 500 mV p-p sinewave is the input signal, showing the S/H capacitor following the input during the sampling period. The input value at the close of the unit sample period is the stored value. At the end of the change detection period, the new sample is transferred to the sample memory capacitor.

The memory transfer circuit produces a slight systematic offset of about -10 mV due to the parasitic gate-source capacitance between the sampling capacitor and the common node of the differential pair. The 483fF storage capacitor has been tested and shown to lose voltage approximately linearly at 6.1mV / second in the dark (at room temperature).

The Difference Detection and Bearing Reporting Subsystem

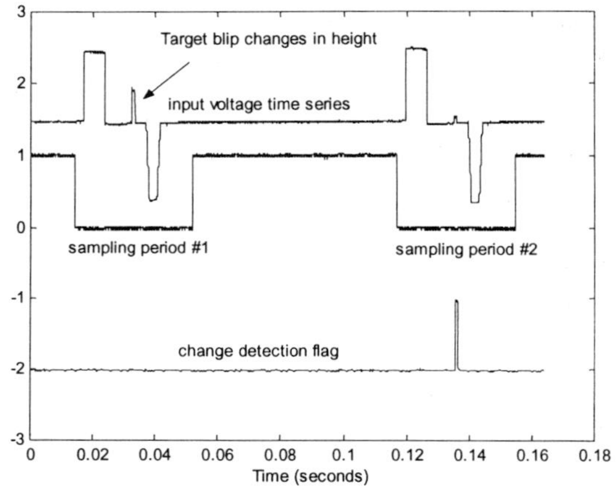

Figure 6. Demonstration of amplitude change detection. In this example, a moveable "blip" (middle feature) is changed in amplitude (-60 mV). The change detection flag indicates the change one internal clock cycle following the new sample.

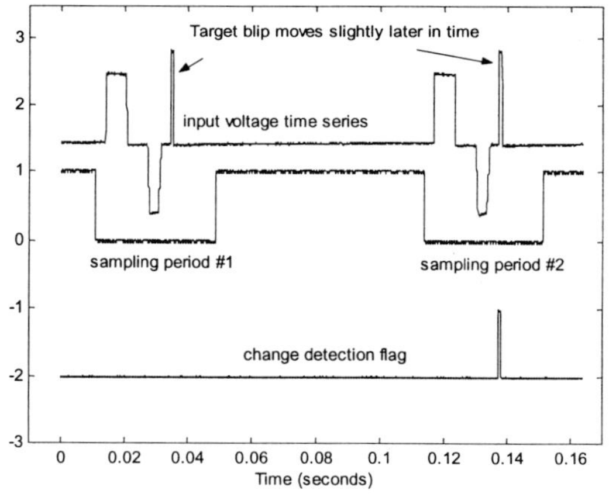

Figure 7. Demonstration of range change detection. In this example, a static background pattern is played to the chip (first two features) and a moveable "blip" is moved slightly later in time between the two sampling periods. In this case, only 40 usec was needed for the blip to cross into a new detection unit (change detected).

The difference detection circuits were tested by repeatedly presenting an input waveform and modifying only one part of the stimulus (narrow 'blip') in either amplitude (Figure 6) or occurrence in time (which produces changes in amplitude) (Figure 7). The change detection threshold was set to prevent detection in the static input case.

Random offsets in the S/H, memory transfer, and change detection circuits produced shifts in the detection threshold of individual circuits. Systematic (e.g., memory leakage) and random offsets (e.g., comparator input offsets) produce a direction asymmetry in the detection threshold; For example, a positive offset will reduce the minimum rise and increase the minimum drop needed to trigger the unit.

Change-Triggered Bearing/Range Output

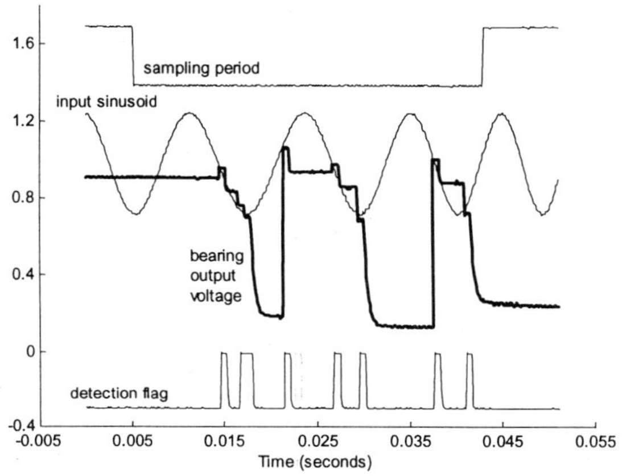

Figure 8. Demonstration of the bearing output signal.

Upon detecting a difference between the newly received sample and the previously stored value (in our sonar example, this will represent the 'target bearing'), the detection signal turns on a follower circuit that drives the newly sample voltage off-chip on a common analog line. Figure 8 shows an example where an unsynchronized FM sweep triggers many change detection events, and the bearing output voltage jumps to report the latest sample at the location of the change..

Power Consumption

There are several modes in which the power consumption is relevant. Reading the analog bearing voltage (and other test points in the circuit) requires biasing large follower circuits to drive the pads quickly. In this mode, quiescent power consumption of the circuit (no triggering of the sample-and-hold system) is at about (1.8uA x 3v) 5.4uW. When triggered at an 8 Hz repetition rate (with 32 ms sampling time) the power consumption rises to (6.2uA x 3v) 18.6uW.

If, however, the detection flag and the digital range output are sufficient as a wake-up signal for other systems, at an 8 Hz repetition rate and with no change detection events, a power of (2uA x 3v) 6uW is measured. In the quiescent mode (no bearing output, and no sampling), the system now draws (335nA x 3v) 1uW. Note that these measurements do not include the power required to generate the seven bias voltages needed to operate the chip.

3.1 Sonar System Integration and the Future

This chip was designed for sonar-derived target *bearing* inputs (as part of a larger research project [6]), however, the circuit operates on any repetitive event-triggered voltage time-series. We have integrated this chip with an acoustic sonar front-end [7] using echo amplitude signals (instead of bearing) to drive the input to the chip.

A number of improvements are planned, particularly in the area of power consumption, offset errors, and operational flexibility. A new, low-power oscillator circuit is planned, an even lower offset S/H circuit (smaller input transistors for the memory transfer circuit), and lower-offset difference detection circuits (larger transistors) are planned.

4. SUMMARY

We have designed, fabricated, and tested a low power time-series novelty detector chip intended for use in a sonar application, but easily applicable to other problems. Following an input trigger signal this chip records samples from the incoming signal and compares them to previously recorded values at the same relative time from the trigger. We have integrated this chip into a functioning acoustic sonar system to investigate its detection capabilities. The architecture has been chosen to reduce power consumption by asynchronously activating circuits only when they are needed.

5. ACKNOWLEDGEMENTS

This work was supported by a DARPA Air-Coupled Microsensors Grant (N0001400C0315).

6. REFERENCES

[1] Edwards, T. and G. Cauwenberghs, "Mixed-Mode Correlator for Micropower Acoustic Transient Classification," *IJSSC*, Vol. 24, No. 10, pp. 1367, 1999.

[2] Vlassis, S., et al., "Analog CMOS Design of the Incremental Credit Assignment (ICRA) Scheme for Time Series Classification," *ISCAS*, pp. II-324, 1998.

[3] Heale, A., and L. Kleeman, "A Real-Time DSP Sonar Echo Processor," *ICIRS*, pp. 1261, 2000.

[4] Fisher, G., and A. Davis, "VLSI Implementation of a Wide-Band Sonar Receiver," *ISCAS*, pp. V-673, 2000.

[5] Sabatini, A., and A. Rocchi, "DSP Techinques for the Design of Coded Excitation Sonar Ranging Systems," *ICRA*, pp. 335, 1996.

[6] Clapp, M., and R. Etienne-Cummings, "Ultrasonic Bearing Estimation Using a MEMS Microphone Array and Spatiotemporal Filters," *ISCAS*, pp. I-661, 2002.

[7] Horiuchi T. and Hynna, K. M. "Spike-based Modeling of the ILD System in the Echolocating Bat", *Neural Networks* 14:755-762, 2001.

A VLSI MODEL OF RANGE-TUNED NEURONS IN THE BAT ECHOLOCATION SYSTEM

M. Cheely[1] *and T. Horiuchi*[1,2,3]

[1]Neuroscience and Cognitive Science Program
[2]Electrical and Computer Engineering
[3]Institute for Systems Research
University of Maryland
College Park, MD 20742

ABSTRACT

The neural computations that support bat echolocation are of great interest to both neuroscientists and engineers, due to the complex and extremely time-constrained nature of the problem and its potential for application to engineered systems. In various areas of the bat's brain, there exist neural circuits that are sensitive to the specific difference in time between the outgoing sonar vocalization and the returning echo, or time-of-flight. While some of the details of the neural mechanisms are known to be species-specific, a basic model of re-afference triggered, post-inhibitory rebound timing is reasonably well supported by the available data. We have designed low-power neuromorphic VLSI circuits to mimic this mechanism and have demonstrated range-dependent outputs for use in a real time sonar system. These circuits are being used to implement range-dependent vocalization rates and amplitudes.

1. INTRODUCTION

Information about target range has many uses for bats during both prey-capture and navigation tasks. Beyond the extraction of target distance and velocity, it may be important for less obvious tasks, such as optimizing the parameters of the echolocation process. As a bat approaches a target, it alters many parameters of its vocalization, including pulse repetition rate, pulse duration, spectral content, and amplitude [1].

Neurons have been found in bats that show a selective response to paired sounds (simulated vocalization and echo) presented at particular delays. The cells' responses to the delayed sounds are much greater than the sum of the responses to the individual sounds presented alone. These neurons are referred to as delay-tuned cells and are found in many levels of the bat brain [2][3][4]. Disruption of cortical delay-tuned cells has been shown to impair a bat's ability to discriminate between artificial pulse-echo pairs with different delays [5].

The largest amount of information related to the mechanisms underlying delay-tuning comes from the mustached bat, *Pteronotus parnellii* [2][6][7][8][9][10][11][12]. In this species, delay-tuned cells respond specifically to the first harmonic of the echolocation call (FM_1) paired with a delayed higher harmonic ($FM_{2\text{-}4}$) [13]. In contrast, delay-tuned cells in the big brown bat, *Eptesicus fuscus* respond preferentially to an initial loud sound followed by a delayed softer sound [14]. These patterns of response are thought to relate to the discrimination between the outgoing vocalization and the returning echo. In the case of *Pteronotus*, the first harmonic of the vocalization is weak, and is probably too attenuated in returning echoes to impact the ranging system. For *Eptesicus*, the outgoing vocalization is obviously much louder than the returning echoes. Note, however, that in both cases the timing is started by the sound of the outgoing pulse, and not an internal trigger.

The earliest point in the auditory pathway where delay-tuned cells have been found is in the inferior colliculus (IC) of *Pteronotus* [2], and it is hypothesized that this is where the delay-tuned responses are formed. When presented with FM_1 tones, IC delay-tuned cells either do not respond at all, or respond weakly at latencies as long as 30ms [10]. In response to FM_n tones, the cells respond weakly with consistent short latencies [10]. In response to paired FM_1-FM_n sounds, IC delay-tuned cells have tuning curves similar to figure 1. The delay eliciting the maximum response in delay-tuned cells is called the best delay (BD) and is highly correlated with the latency of the response to FM_1 tones [10]. This leads to a latency-coincidence hypothesis for delay tuning, in which the cell's facilitated response at long delays is due to the coincidence of the long latency FM_1 response and the short latency FM_n response [3][10].

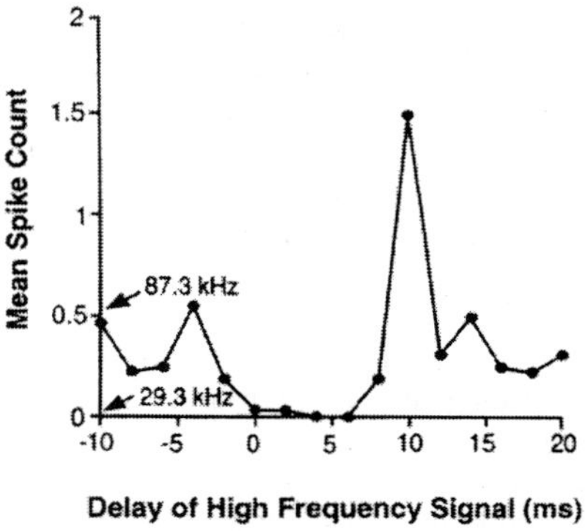

Figure 1. From Portfors and Wenstrup (1999). Tuning curve of a delay-tuned neuron in mustached bat IC. Average number of spikes is plotted vs. the time interval between the FM_1 (29.3 kHz) and FM_n (87.3 kHz) presentations. Arrows indicate the average number of spikes in response to each stimulus presented alone

An interesting question is how delay-tuned cells can have such long latency responses to the FM_1 sounds. It is unlikely that the short neural pathways from the cochlear nucleus to IC could create

0-7803-7761-3/03/$17.00 ©2003 IEEE

such long latencies. In Figure 1, there is a clear period of suppressed response beginning at 0ms delay. This suppression is consistent with a period of inhibition triggered by the FM_1 harmonic and lasting until the time of facilitation. This has led to the proposal of a post-inhibitory-rebound (PIR) mechanism for generating long-latency responses and facilitation [15].

In the PIR model for delay-tuning, the outgoing vocalization triggers a long lasting inhibition in delay-tuned cells. At the offset of inhibition, an instability in the membrane dynamics of the delay-tuned neuron leads to a brief excitatory jump above resting potential. If excitatory input from a returning echo coincides with this rebound event, the membrane potential will cross threshold, and the cell will spike. This model predicts that blocking of inhibition in the IC should reduce or eliminate the latency-coincident facilitation of delay-tuned cells. Results from a study by Wenstrup and Leroy [12] indicate that blocking the inhibitory transmitter glycine has such an effect. VNLLc, a lower brainstem area, which contains neurons that respond with short and remarkably precise latencies, has been shown to contain glycenergic cells which project to IC. [16]

2. VLSI CIRCUIT IMPLEMENTATION

2.1 Neuron Circuit

The design of our delay-tuned neuron is intended to capture the fundamental functional aspects of biological delay-tuned neurons while maintaining simplicity of use and maximum control over neuron behavior.

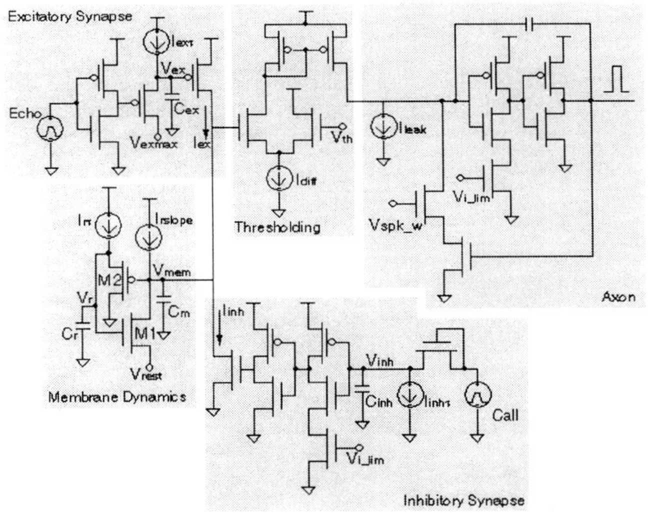

Figure 2. Circuit schematic for the rebound neuron. Thirteen neurons per chip were fabricated in an AMI 1.5μm process by MOSIS.

The analog processing components in the neuron are implemented using transistors operating in the subthreshold regime. For analysis, we assume transistors are operating in saturation, and the drain current can be computed as:

for the nFET, $$I_{dn} = I_0 e^{\frac{\kappa_n V_{GB} - V_{SB}}{V_T}} \tag{1.1}$$

for the pFET $$I_{dp} = I_0 e^{\frac{\kappa_p V_{BG} - V_{BS}}{V_T}} \tag{1.2}$$

The delay-tuned neuron circuit is constructed from component circuits designed to mimic the basic elements of a PIR delay-tuned neuron. The circuit diagram for the neuron is shown in figure 2.

2.2 Inhibitory Synapse

Input spikes based on the outgoing vocalization trigger the inhibitory synapse, charging C_{inh} and driving V_{inh} to a high voltage, V_{inhH}. This signal is passed through a pair of inverters, activating an nFET that draws current from the membrane dynamics circuit, pulling V_{mem} to ground. $I_{inh\tau}$ discharges C_{inh} and the output nFET shuts down when V_{inh} drops below the switching point of the first inverter. The length of inhibition, which sets the best delay of the cell, is then given by:

$$t_{inh} = \frac{C_{inh}(V_{inhH} - V_{hl})}{I_{inh\tau}} \tag{1.3}$$

2.3 Membrane Dynamics

The membrane dynamics circuit generates post-inhibitory rebound in the cell, opening a facilitation window that leads to delay-tuning. In the absence of external input from the excitatory or inhibitory synapses, we can derive the equilibrium state of the membrane circuit. Applying the drain current equation to transistor M1, and solving for V_r, we obtain the equation:

$$V_{r\_eq} = \frac{1}{\kappa_n}\left(V_{rest} + V_T \ln\left(\frac{I_{rslope}}{I_0}\right)\right) \tag{1.4}$$

Following the same process for M2, and solving for V_{mem}, we obtain:

$$V_{mem\_eq} = \frac{1}{\kappa_p}\left(V_{r\_eq} - V_T \ln\left(\frac{I_{r\tau}}{I_0}\right)\right) + V_{dd}\left(1 - \frac{1}{\kappa_p}\right) \tag{1.5}$$

Substituting equation (1.4) into equation (1.5), and assuming the ideal case where $\kappa_n = \kappa_p = 1$:

$$V_{mem\_eq} = V_{rest} + V_T \ln\left(\frac{I_{rslope}}{I_{r\tau}}\right) \tag{1.6}$$

In the case where $I_{rslope} \le I_{r\tau}$, Transistor M1 will leave saturation and V_{mem} will sit slightly above V_{rest}.

The dynamic behavior of the circuit is best described in a stepwise manner. When an outgoing call triggers the inhibitory synapse, V_{mem} is pulled to ground, and V_r, which is connected to V_{mem} through a source follower, will also drop to some minimum voltage level, V_{r_min} which depends on $I_{r\tau}$.

On the release of inhibition, V_{mem} is less than V_{rest}, and the source and drain of M1 are reversed from the equilibrium state. Current flowing through M1 combined with I_{rslope} acts to drive V_{mem} very quickly toward V_{rest}.

It is reasonable to assume that with the appropriate biasing, V_{mem} will rise faster than V_r such that during the rebound, the current through M2 will be negligible, and V_r will rise at a rate of:

$$\dot{V}_r = \frac{I_{r\tau}}{C_r} \tag{1.7}$$

Under these conditions, once V_{mem} reaches V_{rest}, the current through M1 will be negligible, and V_{mem} will rise at a rate of:

$$\dot{V}_{mem} = \frac{I_{rslope}}{C_m} \tag{1.8}$$

V_{mem} will continue to rise at a linear rate, exceeding its equilibrium value, until V_r reaches V_{rest} and transistor M1 begins to turn on. The rebound will peak $\left(\dot{V}_{mem} = 0\right)$ when the drain current in M1 equals I_{rslope}, which occurs at the equilibrium voltage for V_r, given in equation (1.4). We define the duration of the rebound, t_{reb}, as the time interval after inhibition during which V_{mem} is rising. This is easily computed, provided that V_{mem} rises faster than V_r and the assumption that V_r rises linearly holds. The duration of the rebound is then:

$$t_{reb} = \frac{C_r\left(V_{r\_eq} - V_{r\_min}\right)}{I_{r\tau}} \tag{1.9}$$

An estimate for peak voltage of the rebound, V_{m_peak} can be obtained by assuming V_{mem} rises to V_{rest} nearly instantaneously, then applying the value of $\dot{V}_{mem}$ given in equation (1.8) and t_{reb}:

$$V_{m\_peak} = V_{rest} + \frac{I_{rslope}}{C_m} t_{reb} \tag{1.10}$$

The parameters of the circuit are adjusted so that V_{m_peak} is less than the neuron's voltage threshold.

The neuron will be most responsive to an excitatory current during the rising portion of the rebound. Normally, the active membrane properties dampen the response of the cell to excitatory currents. During the rebound period however, the active processes of the membrane have not recovered sufficiently to compensate for excitatory input. During this time window, I_{ex} sums with I_{rslope} to drive V_{mem}. $\dot{V}_{mem}$ can be computed as:

$$\dot{V}_{mem} = \frac{I_{ex} + I_{rslope}}{C_m} \tag{1.11}$$

In this condition, the current through M1 must compensate for I_{ex} + I_{rslope} before the membrane voltage peaks. If we assume a constant excitatory current, the rising time of the system, t_r, would be:

$$t_r = \frac{C_r\left(V_{rest} + V_T \ln\left(\frac{I_{ex} + I_{rslope}}{I_{r\tau}}\right) + V_{dd}(\kappa_p - 1)\right)}{\kappa_n I_{r\tau}} \tag{1.12}$$

As in equation (1.10), we can form an estimate of the peak voltage that the membrane reaches:

$$V_{m\_peak} = V_{rest} + \frac{I_{ex} + I_{rslope}}{C_m} t_r \tag{1.13}$$

We can see that V_{mem} now rises at a faster rate for a longer period of time than in equation (1.10). Under these conditions, V_{mem} easily exceeds threshold before V_r increases enough to compensate. Regardless of the actual time course of I_{ex}, the response of the neuron to excitatory inputs during this time window is facilitated over the normal condition.

Once the rising portion of the rebound ends, V_{mem} begins to fall and V_r continues to rise until $I_{r\tau}$ is balanced by the current in M2. Once this point is reached, V_r begins to fall, following V_m quickly. Both voltages return to their equilibrium value with no ringing, due to the fast downward action of the source follower, and the fact that M1 cannot drive V_{mem} below V_{rest}.

2.4 Excitatory Synapse

In response to spikes from returning echoes, the excitatory synapse drives and exponentially decaying current into the membrane dynamics circuit, which is given by:

$$I_{ex}(t) = I_{max} e^{\frac{-\kappa_p I_{ext}}{C_{ex}} t} \tag{1.14}$$

Where I_{max} is set by the bias voltage, V_{exmax}.

2.5 Chip Performance

Membrane voltage traces in response different pulse-echo delays are shown in Figure 3. Tuning curves for the 13-neuron array are shown in Figure 4. Long-delay neurons are biased to have wider tuning curves than short-delay neurons using I_{ext}.

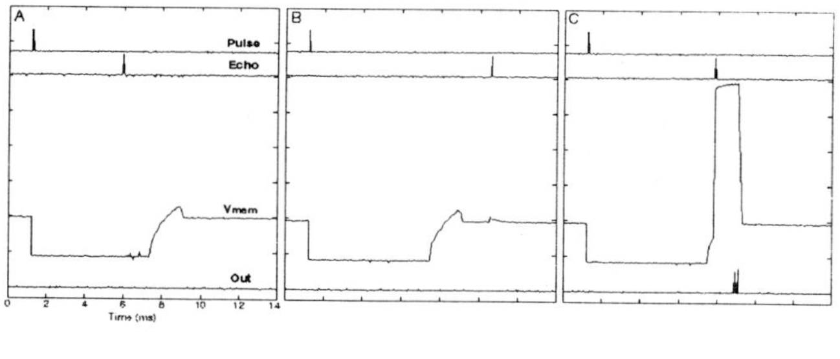

Figure 3. Response of a silicon delay-tuned neuron to artificial stimuli at different delays. Top trace is the spike representing an outgoing pulse. Second trace is a spike representing an echo. Third trace is V_{mem} of the neuron. Bottom trace shows output spikes from the neuron.

The entire 13-neuron array with biasing circuitry measures 927x390 μm in the AMI 1.5 μm process. A large portion of the layout size is due to capacitors that will be significantly reduced in future designs. Direct measurement of power consumption was not possible, due to additional testing circuits on chip with shared biasing, but simulations indicate that quiescent power consumption is on the order of 33μW (66nA over 5V power supply) and for echolocation at a rate of 50Hz with a single target echo, average power consumption is about 550μW (110υA over 5V).

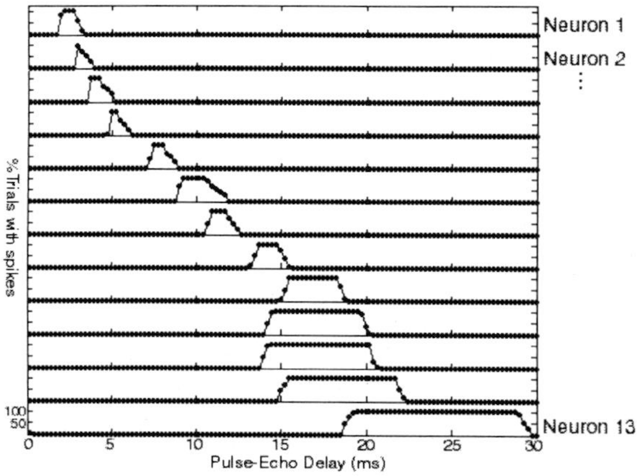

Figure 4. Example tuning curves for the 13-neuron array. Each neuron was presented with 100 pulse-echo spike pairs at delays ranging from 0 to 30ms in .25 ms intervals. Percent of trials on which the neuron spiked are plotted vs. time. (The peak in each plot reaches 100%)

3. SUMMARY

Neuromorphic VLSI design strives to capture the essential elements of any specific instance of neural computation and produce a circuit that will not only reproduce behavior in normal ecological conditions, but will react in qualitatively similar ways to damage and extreme stimulus conditions. The primary purpose is to test these neural algorithms in closed-loop, real-world conditions.

In this paper we present the design of an analog VLSI circuit that mimics the behavior of delay-tuned neurons in the bat midbrain. The circuit produces the delay-tuned responses by implementing the PIR model supported by numerous neurophysiological and anatomical studies. We have incorporated this chip into an artificial bat echolocation system to test these neurons in the closed-loop behavior of reporting target range and modifying parameters in response to a moving target.

This silicon implementation of the delay-tuned neurons of the bat provides a biologically realistic input layer for more detailed neural processing of target range such as attentional tracking and feature binding. While only a small piece of the sophisticated bat echolocation system, these circuits are a critical gateway for processing of range-related information.

Acknowledgements:

The authors would like to acknowledge the help and advice of Cynthia Moss, Shihab Shamma and P.S. Krishnaprasad for their discussions about the basic philosophy and design of auditory computation, and Chris Diorio (Univ. Washington) and his group for their assistance with the chip padframes and fabrication.

This work was supported by a DARPA Air-Coupled Microsensors Grant (N0001400C0315) and an AFOSR Cooperative Control Grant (F496200110415)

4. REFERENCES

[1] Fenton M.B. "Natural History and Biosonar Signals," in Popper A.N. and Fay R.R. (Ed.) *Hearing by Bats*. Springer-Verlag, New York, 1995

[2] Mittman D.H. and Wenstrup J.J. "Combination-sensitive neurons in the inferior colliculus". *Hear. Res.*, 90:185-191, 1995

[3] Olsen J.F. and Suga N. "Combination-sensitive neurons in the medial geniculate body of the mustached bat: encoding of target range information". *J. Neurophysiol.*, 65(6):1275-1296, 1991

[4] Feng A.S., Simmons J.A., and Kick S.A. "Echo detection and target-ranging neurons in the auditory system of the bat *Eptesicus fuscus*". *Science*, 202(4368):645-648, 1978

[5] Riquimaroux H., Gaioni S.J., and Suga N. "Cortical computational maps control auditory perception". *Science*, 251(4993):565-568, 1991

[6] Saitoh I. and Suga N. "Long delay lines for ranging are created by inhibiton in the inferior colliculus of the mustached bat". *J. Neurophysiol.*, 74(1):1-11, 1995

[7] Hattori T. and Suga N. "The inferior colliculus of the mustached bat has the frequency-vs-latency coordinates". *J. Comp. Physiol. A*, 180:271-284, 1997

[8] Yan J. and Suga N. "The midbrain creates and the thalamus sharpens echo-delay tuning for the cortical representation of target distance information in the mustached bat". *Hear. Res.*, 93:102-110, 1996

[9] Wenstrup, J.J. Mittmann, D.H. and Grose C.D. "Inputs to the combination-sensitive neurons of the inferior colliculus" *J. Comp. Neurol.*, 409:509-528, 1999

[10] Portfors C.V. and Wenstrup J.J. "Delay-tuned neurons in the inferior colliculus of the mustached bat: implications for analyses of target distance". *J. Neurophysiol.*, 82:1326-1338, 1999

[11] Portfors C.V. and Wenstrup J.J. "Topographical distribution of delay-tuned responses in the mustached bat inferior colliculus". *Hear. Res.*, 151:95-105, 2001

[12] Wenstrup J.J. and Leroy S.A. "Spectral integration in the inferior colliculus: role of glycinergic inhibition in response facilitation" *J. Neurosci.*, 21(RC124):1-6, 2001

[13] O'Neill W.E. and Suga N. "Encoding of target range and its representation in the auditory cortex of the mustached bat" *J. Neurosci.*, 2(1):17-31, 1982

[14] Dear S.P. and Suga N. "Delay-tuned neurons in the midbrain of the big brown bat" *J. Neurophysiol.*, 73(3):1084-1100, 1995

[15] Sullivan W.E. "Possible neural mechanisms of target distance coding in the auditory system of the echolocating bat *Myotis lucifugus*". *J. Neurophysiol.*, 48(4):1033-1047, 1982

[16] Vater M., Covey E., and Casseday, J.J. "The columnar region of the ventral nucleus of the lateral lemniscus in the big brown bat (*Eptesicus fuscus*): synaptic arrangements and structural correlates of feedforward inhibitory function". *Cell Tissue Res.* 289:223-233, 1997.

Development of an AA Size Energy Transducer with Micro Resonators

Johnny M.H.Lee[1,2], *Steve C.L.Yuen*[3], *Wen J.Li*[1,2,*] *and Philip H.W. Leong*[3]

[1]Centre for Micro and Nano Systems
[2]Dept. of Automation and Computer-Aided Engineering
[3]Dept. of Computer Science and Engineering
The Chinese University of Hong Kong,
Shatin, N.T., Hong Kong SAR

ABSTRACT

This paper presents the preliminary design and experimental results of a standard AA size vibration-induced micro energy transducer which is integrated with a power-management circuit. The generator is a spring mass system which uses laser-micromachined copper springs to convert mechanical energy into electrical power by Faraday's Law of Induction. A power-management circuit is used to step up the AC output and act as a reservoir to store the electrical energy generated. Our goal is for the generator to provide 3V DC output with low input mechanical frequencies, and to produce enough power for low-power wireless applications. Potential applications for this micro power generator to serve as a power supply for infrared transmission and radio frequency transmission was proved to be possible with input frequencies below 100Hz and amplitudes be 250microns..

Keywords: micro power generator, micro energy transducer, power-management circuit.

1. INTRODUCTION

Traditional alkaline battery has being used for almost a century, and has brought dramatic revolutions to human life. However, shelf life, replacement accessibility and potential hazards of chemical are some of the problems when chemical batteries are used. Our ongoing work is to develop a brand new power supply with unlimited shelf life and is environmentally safe. Three main advancements in engineering technology in the last 20 years allow possible applications for magnetic-induction based micro energy transducers: 1) increase in magnetic flux density of rare-earth-magnets; 2) continual reduction of power consumption of low-power circuits and sensor; 3) MEMS fabrication technology that allows precise and low cost production of spring-mass system. Thus far, we have successfully developed a vibration-induced power transducer with total volume of ~1 cm^3 [1] and demonstrated it to be useful for IR and RF wireless transmissions. When input vibration frequencies ranging from 60 to 110Hz with ~200μm amplitude is provided, the generator is capable of producing up to 4.4V peak-to-peak, which have a maximum rms power of ~680μW with loading resistance of 1500Ω. We are now targeting to develop a micro power generator integrated with a power-management circuit with total dimension equal to an AA size battery. The development of this AA size micro power generator is presented in this paper.

Research on micro power generator have been done by various groups throughout the world. Williams and Yates developed an electromagnetic micro generator to produce 0.3μW in 1997 [2], Amirtharajah & Chandra-Kasan used a vibration-based power generator to drive a signal processing circuitry in 1998 [3]. Nevertheless, neither of them has fabricated a micro power generator which integrated with a power-management circuit that have enough power to drive an off-the-shelf circuit.

2. GENERATOR PRINCIPLE AND DESIGN

The prototype micro power generator is consist of five main components: 1) inner and outer housing which is used to carry the resonating structure and the power generating system, respectively, 2) a laser-micromachined resonating spring with spring constant k, 3) a N45 grading rare earth permanent magnet of mass m and magnetic field strength B, 4) copper coil of length l, and 5) a power-management circuit for output voltage step up and energy storing purpose. The resonating spring is attached to the magnet and packed by the inner housing. The orientation of inner housing, magnet and the resonating spring is shown in Figure 1a, and the illustrative drawing of AA size's micro power generator is shown in Figure 1b.

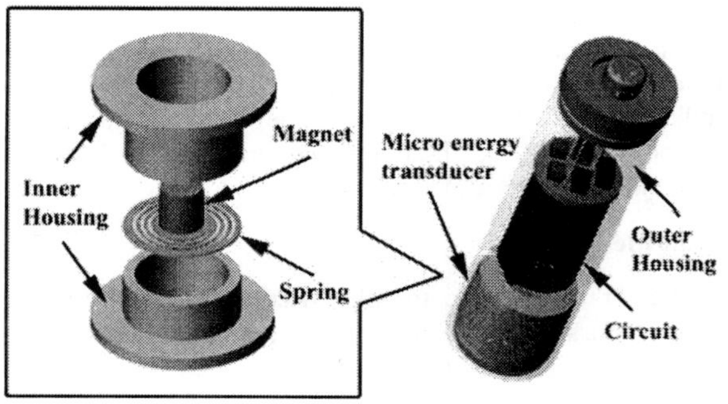

(a) (b)

Figure 1. *Illustrations of: (a) Inner structure of the micro power generator; (b) the AA size micro power generator which is integrated with a power-management circuit.*

**Contacting Author :* wen@acae.cuhk.edu.hk; CMNS, The Chinese University of Hong Kong, MMW Bldg., Room 422 Shatin, NT, Hong Kong SAR; phone: +852 2609-8059; fax: +852 2603-6002.

This project is funded by a grant from the Hong Kong Innovation and Technology Commission (ITF/185/01).

When the generator housing is vibrated with an amplitude of $y(t)$, the magnet will vibrate with a relative amplitude of $z(t)$. This relative movement of the magnet results in the varying amount of magnetic flux density cutting through the coil. According to Faraday's law of electromagnetic induction, a voltage is induced in the loop of coil. The average power output of the vibration-induced power generating system can be derived as [4]:

$$P = m\xi_e Y_0^2 (\omega/\omega_n)^3 \omega^3 \Big/ \left(\left[1-(\omega/\omega_n)^2\right]^2 + (2\xi\omega/\omega_n)^2 \right) \qquad \textbf{Eq.1}$$

where ξ_e is the electrical damping factor, Y_0 is the input vibration amplitude, ω is the input vibration frequency (angular), ω_n is the resonance frequency of the spring-mass system and ξ is the sum of electrical damping factor and mechanical damping factor of the system. From the above equation, at resonance, the average power and voltage output is maximized:

$$P = m\xi_e Y_0^2 \omega_n^3 / 4\xi^2 \qquad \textbf{Eq.2}$$

$$V = BlY_0\omega_n / 2\xi \qquad \textbf{Eq.3}$$

Based on the above equations, the power generator will have maximum power and voltage output when vibrating in resonance frequency with maximum amplitude and electrical damping factor. Using a spring constant k of ~ 40 N/m, damping ratio of ~0.01, magnet weight of 140mg, magnetic field strength of 3600 Guass (experimentally measured) and input vibration amplitude ~150μm, we performed Matlab simulations and compared with experimental results. The results were close to experimental measurements and the comparisons are shown in Table1.

***Table 1**: Comparison between experimental and simulation results.*

	Experimental results	Simulation results
$V_L(R_L = 0\Omega)$	4 V p-p	4 V p-p
$V_L (R_L=1000\Omega)$	2.58 V p-p	2.71 V p-p
$P (R_L=1000\Omega)$	830 μW	919 μW

3. DESIGN OF RESONATING STRUCTURE

The resonance frequency of the spring-mass system depends on the materials used for the resonating structure, and hence, the choice of spring material will affect the performance of power generator. Copper was chosen to be the material for the resonating structure because of its relatively low Young's modulus and high yield stress where compared to Silicon (See [6]). Some other materials such as brass, titanium and 55-Ni-45-Ti can also be considered, depending on the operation environment. For instance, titanium should be used if the power generator is designed to vibrate in extremely large displacement, as its yield stress is higher than copper. We have experimentally verified that brass and 55-Ni-45-Ti resonating structures could obtain a lower resonance frequency than copper due to their lower Young's modulus. The material properties of some potential metals which may be suitable to fabricate the resonating spring are compared in Table 2.

***Table 2**: Potential materials for the resonating spring [5].*

	Young's modulus (GPa)	Yield Stress (MPa)	Ultimate Stress (MPa)	Fatigue Limits (MPa)	Fatigue Ratio
Aluminum	70	270	310	21	0.30
Brass	96 – 110	70 – 550	200 – 620	98 - 147	0.31
Copper	130	55 – 760	230 – 830	63	0.29
Nickel	200	100 – 620	310 – 760	109	0.35
Titanium	120	760 – 1000	900 – 1200	364	0.59
55-Ni-45-Ti	83	195 – 690	895	---	---
Silicon	160(ave)	---	---	---	---

Using ANSYS to simulate the resonating structures, it was found that springs with spiral geometry have lower spring constant and stress concentration than other designs, such that a larger displacement can be obtained [6]. We have used a Q-switch Nd:YAG (1.06μm wavelength) laser to micromachine the spiral resonating spring as shown in Figure 2a and b. A copper spring with diameter of 8mm and 0.1mm thickness will be used for the AA size micro power generator.

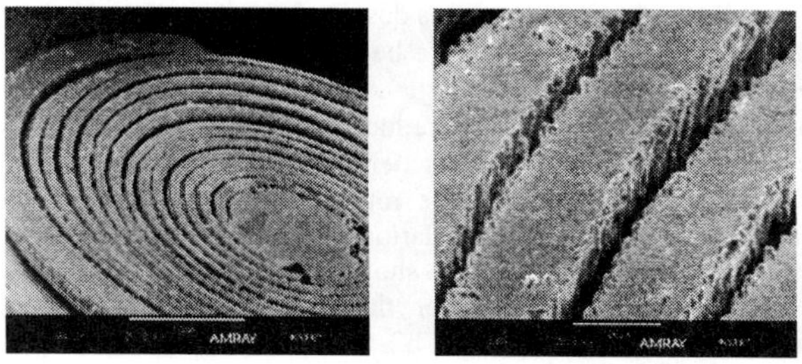

(a) (b)

***Figure 2**. SEM pictures of : (a) a laser-micromachined copper spring with diameter of 5mm;. (b) close up of the copper spring; width of the spring is ~100μm.*

The micro power generator was experimentally found to have different motion of resonance vibration in different frequencies. It gave relatively high voltage at higher frequencies even though the vibration amplitude was almost negligible in the vertical direction. The 3 different modes of vibration captured by a strobe light is shown in Figure 3a to 3c. It was observed that the spring was vibrated vertically in 1st mode, but appeared to cyclically rotate about an axis parallel to the plane of the coil in 2nd and 3rd mode.

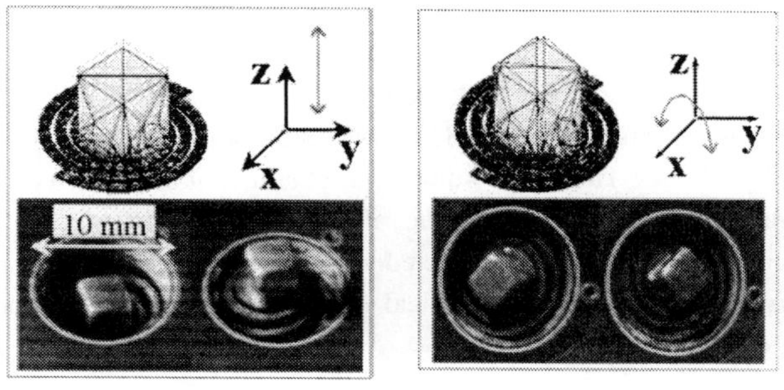

(a) (b)

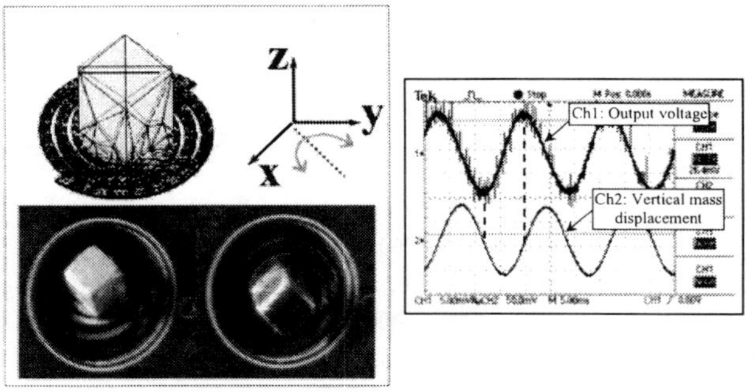

(c) (d)

***Figure 3**. Simulation and experimental results for 3 different resonance vibration modes were matched: (a) 1st mode vibration (vertical); (b) 2nd mode vibration (rotation about x axis); (c) 3rd mode vibration (rotation about an axis between x and y axis); (d) The phase difference between the output voltage of the generator and vertical displacement of the magnet at the 3rd mode vibration.*

Based on the experimental was designed to vibrate in a horizontal plane with rotation results, we believe if the a spring instead of vibrating vertically, the voltage output can be increased and the stress on the spring can be reduced. This can be explained by Faraday's Law of induction which stated that the voltage output should be proportional to the rate of changing magnetic flux. Therefore the faster the translation and rotation of the mass, the greater current induction. As shown in Figure 3d, at 3rd mode resonant vibration, the faster the rate of change of vertical displacement (i.e., slope of Ch2, which represents the "angular velocity" of the magnet), greater is the voltage induced.

Using laser-micromachining to fabricate the copper spring is direct, fast, but the cutting resolution is not ideal (see Figure 2b). We are now developing another process which will involve high-aspect-ratio electroplating of copper using lithographic techniques.

4. POWER – MANAGEMENT CIRCUIT DESIGN

A quadrupler circuit was integrated with the micro power generator to step up and rectify the AC output to DC voltage. A schematic diagram of the circuit, a photograph of the prototype and the output voltage verses load is shown in Figure4. The prototype circuit was built using KEMET type T491 10μF, 10V capacitors and Toshiba 1SS374 Silicon Epitaxial Schottky Barrier Type diodes. The 1SS374 diode was chosen as it has a low voltage drop (0.23V), increasing the efficiency of the quadrupler. A capacitor of 1.2mF is connected with the quadrupler and acts as a reservoir to store the electrical energy generated by the micro power generator.

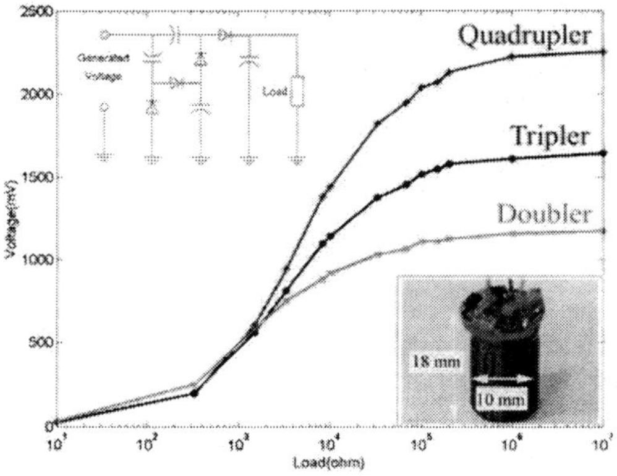

***Figure 4**. Schematic diagram, picture of quadrupler and comparison of the output voltage for doubler, tripler and quadrupler.*

5. APPLICATIONS

5.1 Infrared Transmission

An Infrared (IR) transmitter was built using a commercial SM5021 encoder chip. Experimental result shows that this circuit could operate properly with a voltage as low as 1.8V. An IR signal would be sent to a receiver every time a key was pressed. The signal was a 140.8ms long IR pulse train. For a 2.0V power supply, the current drawn during a key press was measured to be 1.5mA and, in standby mode, 2.4μA. We have used a ~1cm^3 volume micro energy transducer to successfully drive the above IR transmitter (as reported in [1]). The schematic diagram for this system is shown in Figure 5.

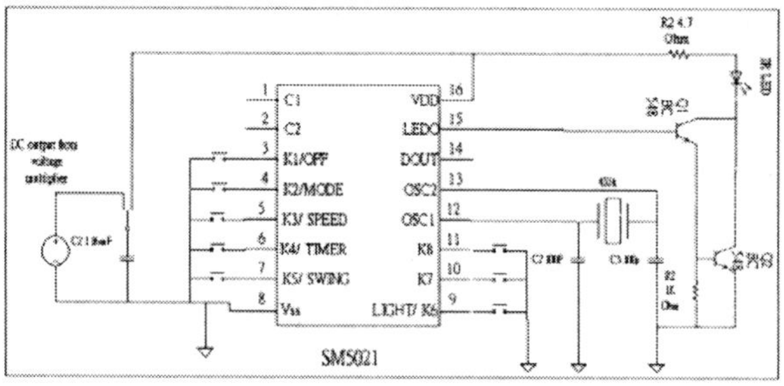

***Figure 5.** Diagram of the commercial SM5021 encoder chip*

5.2 RF Temperature Transmission

We have also demonstrated that the ~1cm^3 volume micro energy transducer is able to drive a 914.8MHz FM wireless temperature sensing system [7]. The schematic diagram of this system is shown in Figure 6. A microcontroller is used to instruct the temperature sensor to convert the ambient temperature to digital format and read the converted temperature through 3-wire communication protocol. The controller then sends the data serially to the *TXD* pin of the transmitter, which will modulate the digital temperature data into FM signal at 914.5MHz to be received by the receiver module at a distance of 25m. The transmitted data is readily available in digital form from *RXD* pin of the receiver.

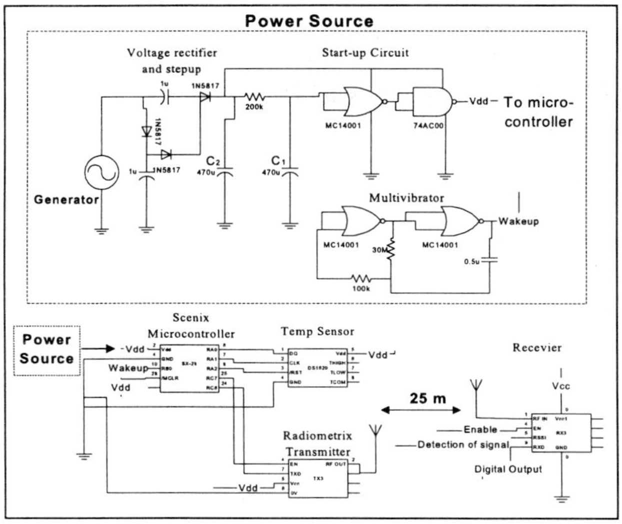

Figure 6. Schematic diagram of the wireless temperature sensing system.

5.3 FM Transmission

Transmission using frequency modulation (FM) has advantages over the infrared system described in Section 5.1 if line of sight to the transmitter is not available. A prototype implementation which employed the same temperature sensing circuit as described in the previous sections, but using frequency shift keying (FSK) was developed and is shown in Figure 7.

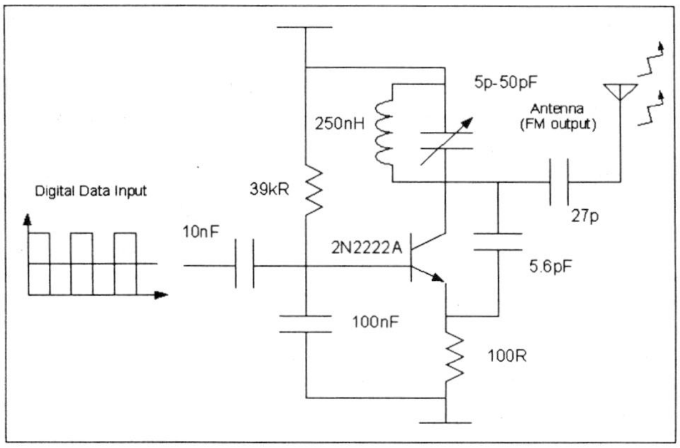

Figure 7. Circuit diagram for the FM transmitter.

A commercial FM radio was used for the receiver. This circuit consumes 7.6mA at 3V in continuous. In real operation, a very low duty cycle can be used, reducing the power requirements proportionally. Thus at a 1% duty cycle, power consumption is expected to be approximately 75μA. We are currently building the AA size power source to generate sufficient power to drive the transmission system described above. The results will be reported at the conference.

6. CONCLUSION

We have designed a magnet-based AA size micro power transducer that converts mechanical energy into electrical power by Faraday's Law of Induction. Potential applications using the micro power source for wireless (IR and RF) transmission systems were shown to be possible. We have also found that if the resonating spring is designed to vibrate horizontally with a vertical input vibration, significantly higher output voltage for the generator could be obtained.

Future work for this project include 1) using ANSYS modeling to aid the design of low frequency resonating springs and springs which give significant horizontal vibrations; 2) development of a MEMS-based process to fabricate the spring-mass and coil system; 3) integration of the generator system with low-power consumption devices. With the growing interests and the advantages of technologies to minimize power consumption of portable systems, we believe our micro power generator will find many applications in the future.

7. ACKNOWLEDGMENT

We would like to thank the Hong Kong Innovation and Technology Commission, Brilliant System Limited, DAKA Development Limited, and Varitronix International Limited for funding this project (ITF/185/01). We also deeply appreciate Magtech Industrial Company (Hong Kong) for donating the permanent magnets needed for this project. Special thanks are due to Thomas K. F. Lei and Gordon M. H. Chan for their contributions to this project.

8. REFERENCES

[1] Neil N. H. Ching, H. Y. Wong, Wen J. Li, Philip H. W. Leong, and Zhiyu Wen, "A laser-micromachined multi-modal resonating power transducer for wireless sensing systems", Sensors and Actuators, A: Physical, 2002, pp. 685-690.

[2] C. B. Williams, and R. B. Yates, "Analysis of a micro-electric generator for microsystems", Sensors and Actuators, A 52, 1996, pp. 8-11.

[3] R. Amirtharajah, and A.P. Chandrakasan, "Self-powered signal processing using vibration-based power generator", IEEE J. of Solid-State Circuits, vol. 33, May 1998, pp. 687-695.

[4] W. J. Li, Z. Y. Wen, P. K. Wong, G. M. H. Chan, and P. H. W. Leong, "A micromachined vibration-induced power generator for low power sensors of robotic systems", Proc. Of the World Automation Congress, Hawaii, USA, June 11-14, 2000.

[5] F. Cardarelli, *Materials Handbook : A Concise Desktop Reference*, Springer-Verlag London Limited, 2000.

[6] W. J. Li, G. M. H. Chan, N. N. H. Ching, P. H. W. Leong, and H. Y. Wong, "Dynamical modeling and simulation of a laser-micromachined vibration-based micro power generator", International Journal of Non-linear Sciences and Simulation, vol. 1, 2000, pp. 345-353.

[7] Neil N. H. Ching, Hiu Yung Wong, Wen J. Li, and Philip H. W. Leong, "A laser-micromachined vibrational to electrical power transducer for wireless sensing systems", 11th International Conference on Solid-State Sensors and Actuators, Munich, Germany, June 2001.

A DISPLACEMENT-TO-VOLTAGE CONVERTER CIRCUIT USING A SWITCHED-CAPACITOR TECHNIQUE

W. L. Lien, B. S. Quek, R. Walia

Institute of Microelectronics, Singapore
11, Science Park Road
Singapore Science Park 2, Singapore 117685

ABSTRACT

A technique for precise measurement of changes in displacement of capacitor gap is presented. In order to obtain a linear relationship between the output voltage and changes in capacitor gap over a wide dynamic range, a novel switched-capacitor circuit has been applied. This circuit has been implemented in 0.8-μm CMOS process. Results showed that the theoretical values match that of measurement results with accuracy up to 1%.

1. INTRODUCTION

Improvements in precision silicon micromachining technology have allowed the development of a wide range of improved sensing structures. Nowadays, it is common to see micro-sensors interfacing with signal conditioning circuits on the same chip. Capacitive sensors are typically used to convert variations in measured variables (e.g. pressure, acceleration) to variations in capacitance. This variation is then converted to variations in voltage, which is observable. Sensor electronics are precision circuit that attempts to derive the variables to be measured accurately, while ensuring that non-idealities are taken into design considerations. Various circuit techniques have been reported to measure the changes in the sensor capacitance [1-3]. Some of them are accurate up first order approximation, while others use complex closed-loop feedback system.

In this circuit to be presented, the output voltage is directly proportional to the change in capacitor gap, and at the same time a wide dynamic range can be achieved through a unique way of connecting and biasing of various nodes within the switched-capacitor circuit.

2. CIRCUIT CONFIGURATION

In switched-capacitor circuits, the signal is processed in the form of charges, converted to voltages and back into the charges again. The required conversion between charges and voltages is achieved by the capacitors used in the circuits.

In this sensor interface circuitry, one of the capacitors used in the switched-capacitor design will be the sensor itself. This capacitor is placed in the feedback path of the switched-capacitor circuit. A generic configuration of the interface circuitry is shown in Fig. 1.

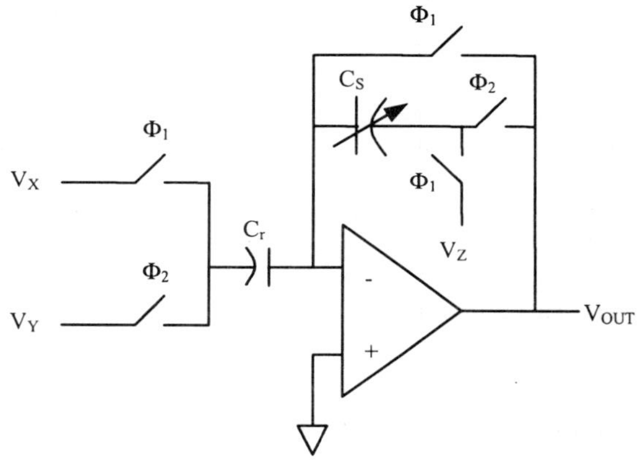

Figure 1. General diagram of sensor interface circuitry.

Φ_1 and Φ_2 are outputs from two-phase non-overlapping clock circuitry. Nodal voltages V_X, V_Y and V_Z are fixed DC voltages that have a unique mathematical relationship to be derived shortly. The capacitive sensor (C_S) is connected between the inverting input of an operational amplifier and voltage level V_Z during phase 1 (Φ_1) and between the inverting input and output of the operational amplifier during phase 2 (Φ_2). C_r can be a fixed on-chip reference capacitor or external reference sensor capacitor.

During Φ_1, the corresponding switches are turned on. The circuit's behavior will be analyzed from a charge perspective. Assuming ideal opamp with infinite gain, the charge stored at the inverting node of the opamp is

$$Q_1 = C_r(0 - V_X) + C_S(0 - V_Z) \tag{1}$$

During Φ_2, the capacitive sensor is connected between the inverting input and output of the opamp. Similarly, the charge stored at the inverting node of the opamp is

$$Q_2 = C_r(0 - V_Y) + C_S(0 - V_{OUT}) \tag{2}$$

Due to charge conservation at this node, Q_1=Q_2. Equating both equations together and rearranging, the following relationship is observed:

$$V_{out} = \frac{C_r}{C_S}(V_X - V_Y) + \frac{C_S}{C_r}V_Z \tag{3}$$

Let

$$V_X = xV_{DD}$$
$$V_Y = yV_{DD}$$
$$V_Z = zV_{DD} \tag{4}$$
$$C_{S0} = \frac{K}{d}$$
$$C_S = \frac{K}{d-\Delta d}$$
$$C_r = \frac{1}{w}C_{S0}$$

where w, x, y and z are fixed values, C_{S0} is the capacitance value of the sensor during quiescent state, K is a capacitance proportionality constant, d is the capacitor gap of the sensor during quiescent state and Δd is the change in capacitor gap to be measured. Substituting the equations from (4) into (3) and rearranging, the following general equation can be derived:

$$V_{out} = \frac{V_{DD}}{w\times d}\{[(x-y)+wz]d+(y-x)\Delta d\} \tag{5}$$

An important observation is that if the variables in the first term are selected such that the term (x-y)+wz becomes zero, then the output voltage is directly proportional to the change in capacitor gap Δd. In this case,

$$V_{out} = \frac{(y-x)V_{DD}}{w\times d}\Delta d \tag{6}$$

Equation (6) shows that the output is ratio-metric to the supply voltage V_{DD}, which is required in some applications.

To simplify the equation even further, since x-y+wz=0, then

$$x = y - wz \tag{7}$$

Substituting (7) into (6) and simplifying, we get

$$V_{out} = zV_{DD}\frac{\Delta d}{d} \tag{8}$$

It can be observed that the relative change in capacitor gap is a function of V_{DD} and voltage at node V_Z only.

3. IMPACT OF NON-IDEALITIES ON CIRCUIT PERFORMANCE

3.1 Parasitic Capacitances

Parasitic capacitance exists at both terminals of the on-chip and sensor capacitors that are implemented. It is common for parasitic capacitance values to be 1 to 20% of the actual capacitor at both terminals depending on the process technology. While it is not easy to reduce the existence of these parasitic capacitors, circuit techniques are available to reduce their effects on circuit performance [4], [5]. Parasitic-insensitive switched-capacitor circuit implementation is a critical requirement for high-accuracy precision sensor electronics.

The same switched-capacitor sensor interface circuit with their associated parasitic capacitance is illustrated in Fig. 2.

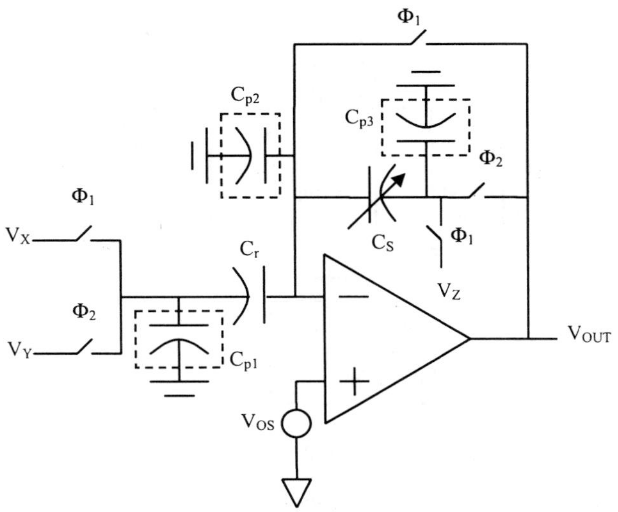

Figure 2. sensor interface circuitry including non-idealities.

C_{p1} represents the associated parasitic capacitance due to the bottom plate of capacitor C_r. C_{p2} represents the parallel combination of parasitic capacitance due to the top plates of C_r and of C_s. Finally, C_{p3} represents the parasitic capacitance formed by the bottom plate of C_s.

From observation, it can be seen that during either phase transitions (Φ_1 or Φ_2), the nodes with the parasitic capacitances are either driven nodes or virtual ground. As such, the effects of parasitic capacitances on the output voltage V_{OUT} are greatly reduced.

3.2 Offset and 1/f noise of Amplifier

Especially in CMOS technology, opamps are known to have offset voltages of around 5-20mV. Flicker noise, or better known as 1/f noise, is also present at low frequency range. Correlated double sampling techniques (CDS) is a well-known technique to minimize errors due to finite offset voltages and 1/f noise from amplifier [6].

In this current architecture, the CDS technique is also employed. The errors due to offset and 1/f noise is stored in the sensor capacitor (C_s) during Φ_1 and cancelled out in Φ_2.

3.3 Capacitance value of Sensor

Capacitor sensor values C_{s0} can vary due to process variations. In (4), there is a requirement for C_{s0}=wC_r, a fixed value. In general, C_{s0} can be of any value as long as (7) can be satisfied. In other words, V_x can be used to calibrate for variations in C_{s0}.

4. OVERALL SIGNAL CONDITIONING

The interface circuit for the capacitive sensor has been designed using 0.8-μm CMOS process. A block diagram description of the entire ASIC is illustrated in Fig. 3.

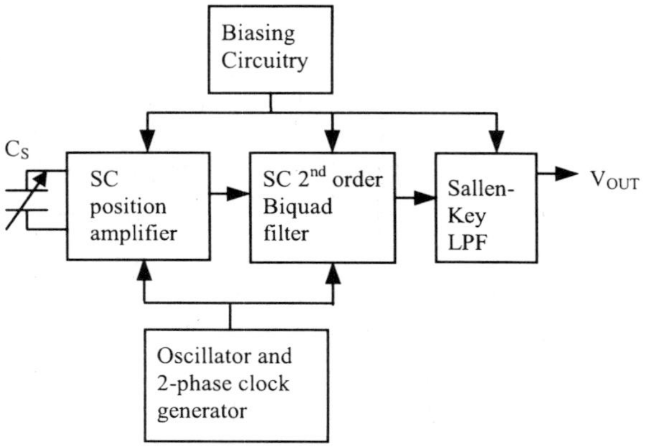

Figure 3. General diagram of sensor interface circuitry.

The front-end stage is the sensor electronics described earlier. The variables used are w=2, x=½ and y=z= -½ (Fig. 4).

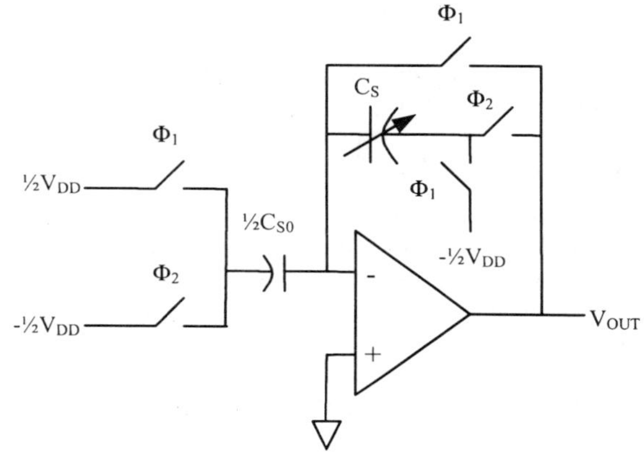

Figure 4. switched-capacitor position amplifier

It can be shown that by applying these constants to equation (5), the following is obtained

$$V_{out} = \frac{V_{DD}}{2 \times d}\left\{\left[\left(\frac{1}{2}+\frac{1}{2}\right)-2\times\frac{1}{2}\right]d+\left(-\frac{1}{2}-\frac{1}{2}\right)\Delta d\right\} \quad (9)$$

Simplifying,

$$V_{out} = -\frac{V_{DD}}{2 \times d}\Delta d \quad (10)$$

which displays the concept that output is now directly proportional to changes in sensor capacitor gap.

The values are chosen such that the voltages are achievable in a 5V CMOS process. In this case, V_X is simply the chip supply (5V) and V_Y and V_Z are the chip ground (0V). The analog ground connected to the non-inverting input of the opamp will be biased at 2.5V. The advantage of selecting these values is that no external voltage bias circuitry is needed to generate V_X through V_Z.

In the current implementation, V_X is applied externally for trimming purposes, whereas V_Y and V_Z are internal chip ground and $Cr=\frac{1}{2}C_{s0}=8pF$. A folded-cascode opamp is used in the position sensing stage.

The stage following the position sensing amplifier is a second-order switched-capacitor biquad stage that serves as a low-pass filter to the sensor capacitor mechanical variations. The final stage is a continuous time Sallen-Key low pass filter that eliminates the high frequency components in the prior switched-capacitor stages.

The chip micrograph of the sensor ASIC is shown in Fig. 5. The chip measures 2.5 × 2.1 mm^2 in 0.8-μm double-poly double-metal (DPDM) CMOS, and is packaged in 24-pin DIP.

Figure 5. microphotograh of the sensor ASIC.

5. EXPERIMENTAL RESULTS

The ASIC was tested to correlate theoretical results described above with actual implementation. A prototype interface was implemented using capacitor components, both discrete and internal. The on-chip oscillator is running at 800kHz and divided by 2 internally.

The calibration process is as follows. The ASIC is connected to a capacitor with value C_{s0}. The output V_{OUT} is then trimmed to the default DC value at AGND. This determines the reference output voltage for Δd=0. To confirm its principles of operation, this interface was applied to capacitive measurement with various capacitance values and V_{OUT} measured.

The ratio of change in capacitor gap can be re-written in terms of the actual capacitance values,

$$\frac{\Delta d}{d} = 1 - \frac{wC_r}{C_S} \quad (11)$$

Substituting (11) into (10), referring analog ground as $V_{DD}/2$ and re-arranging,

$$V_{OUT} = \frac{w \times V_{DD}}{2}\left(\frac{C_r}{C_S}\right) \quad (12)$$

A graph of output voltage versus the reciprocal values of the various sensor capacitor values $1/C_s$ is plotted in Fig. 6.

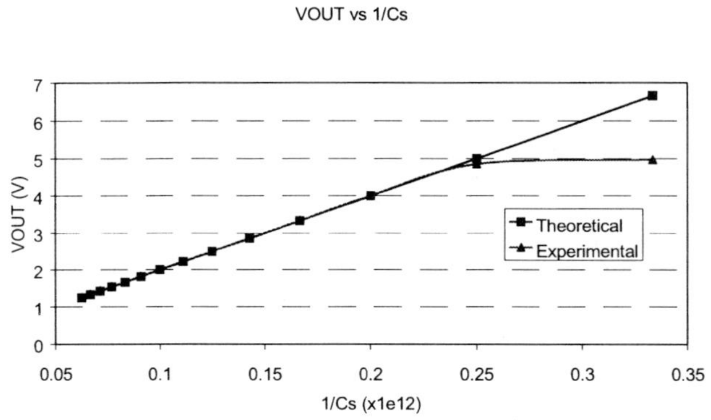

Figure 6. Output voltage as a function of changes in sensor capacitance values.

It can be observed that the experimental values agree with the theoretical results for a wide range of capacitance values. It is also obvious that the experimental results cannot exceed 5V since this is the chip supply for the ASIC.

Fig. 7 shows the output noise spectrum from the ASIC. The noise floor is around $460nV_{rms}/\sqrt{Hz}$ in the frequency band of interest.

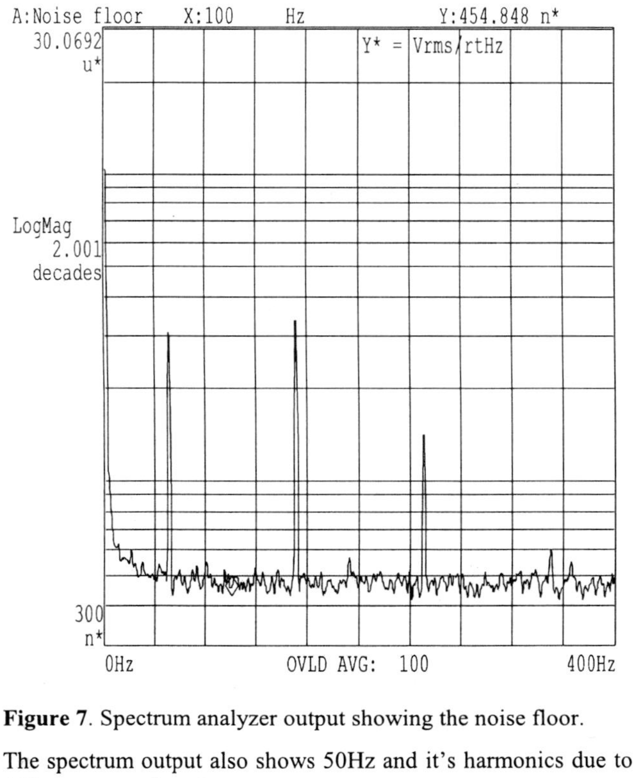

Figure 7. Spectrum analyzer output showing the noise floor.

The spectrum output also shows 50Hz and it's harmonics due to AC power supply noise.

Table 1 summarizes the performances of the overall ASIC.

supply voltage V_{DD} (V)	5 ± 0.25
sensor capacitance values C_{s0} (pF)	4 - 15
non-linearity error (%)	±1
capacitance gap variation Δd/d	±0.4
Conversion rate (ksamples/sec)	400
current consumption @ V_{DD}=5V (mA)	3.4

Table 1. Measurement performances summary.

6. CONCLUSIONS

A capacitive sensor interface that provides an output that is proportional to the change in sensor capacitor gap has been presented.

A novel switched-capacitor technique has been used to eliminate errors due to opamp limitations and parasitic capacitances, while at the same time obtain highly linear and wide dynamic range.

7. REFERENCES

[1] Y. E. Park and K. D. Wise. "An MOS Switched-Capacitor Readout Amplifier for Capacitive Pressure Sensors," *IEEE Custom IC Conf*, pp. 380-384, 1983.

[2] H. Leuthold and F. Rudolf, "An ASIC for High-Resolution Capacitive Microaccelerometers," *Sensors and Actuators*, A21-A23, pp. 278-281, 1990.

[3] T. Smith e.a. "A 15b Electromechanical Sigma-Delta Converter for Acceleration Measurements," *ISSCC'94*, San Francisco, USA, February 1994.

[4] G. M. Jacobs, D. J. Allstot, R.W. Brodersen, and P. R. Gray., "MOS Switched-Capacitor Ladder Filters," *IEEE Trans. Circuits and Systems*, Vol. CAS-25, pp. 1014-1021, December 1978.

[5] K. Martin. "Improved Circuits for the Realization of Switched-Capacitor Filters," *IEEE Trans. Circuits and Systems*, Vol. CAS-27, no. 4, pp. 237-244, April 1980.

[6] C. Enz and G. C. Temes, "Circuit Techniques for Reducing the Effects of Op-Amp Imperfections: Autozeroing, Correlated Double Sampling, and Chopper Stabilization," *IEEE Proceedings*, Vol 84, no. 11, November 1996.

HIGH-SPEED PROCESSOR FOR QUANTUM-COMPUTING EMULATION AND ITS APPLICATIONS

M. Fujishima[1], *K. Saito*[1], *M. Onouchi*[1] *and H. Hoh*[1, 2]

[1]School of Frontier Sciences, University of Tokyo, [2]CREST
7-3-1-703 Kashiwanoha
Kashiwa, Chiba 277-8561, Japan

ABSTRACT

A high-speed and large-scale processor dedicated to quantum computing is proposed, which has the minimum operation function needed for execution of a quantum algorithm. In this processor, the probability amplitudes determined by the states of the quantum bits in the quantum computer are described using only one bit, and we named it a logic quantum processor. It is found that for this processor an execution speed 275 times higher than that of a software simulation is obtained when the processor executing the quantum algorithm for 16 qubit has been implemented in a FPGA. As an example application, it is shown that the processor can solve the satisfiability problem and that it can also simulate the decoherence in the quantum computer at high speed.

1. INTRODUCTION

The quantum computer has attracted much attention since it solves integer problems such as factorization [1] and database search [2] at high speed better than conventional computers. Research of hardware using quantum mechanics is progressing and the quantum computer with seven quantum bits (qubits) is also realized [3]. However, there are several problems in programming a practical problem; such as the scale of current hardware is too small to deal with practical problems, a complicated interface is required to operate the quantum computer, and the period of maintaining the quantum state is so short that the number of available operating steps is restricted. Although the simulation of the quantum algorithm is, on the other hand, also possible using a conventional computer, there will be a problem of calculation time increasing exponentially with an increase in the number of qubits.

We are studying high-speed quantum processors executing the parallel operation realized in the quantum computer by operating hardware in parallel [4,5]. In this paper, a quantum processor with improved computing speed and scale is reported, which executes the minimum function required for the quantum algorithm. This processor enables the development of the quantum algorithm and the emulation of the error generated in quantum computing with quantum mechanics. In the following sections, after the new quantum processor is explained, application using the proposed processor is described.

2. LOGIC QUANTUM PROCESSOR

2.1 Minimum Required Operations

2^n bases exist in a quantum computer with n qubits. Each base has the probability amplitude of a complex number. In quantum computing, executing quantum operation for a certain qubit results in applying unitary transformations to the probability amplitude vector of a base, as shown in Fig. 1. In the unitary transformations corresponding to the quantum operation, since derivation of one probability amplitude requires information on two bases, two sets of sum-of-product operations with a complex number are needed per base, as shown in Fig. 2. In order to complete the quantum operation in one step, 2^n processing elements (PEs) are necessary. At each quantum operation, calculations are executed at the bases whose Humming distance is one. Consequently, all PEs make a hypercube network with 2^n communication wires, and each PE is connected to n PEs, as shown in Fig. 3.

The general quantum algorithm consists of the distribution of probability amplitude, the main operation, extraction of a solution, and observation, as shown in Fig. 4. Until the main operation, only two values of 0 and $1/\sqrt{m}$ exist in the probability amplitudes. Here, m is the number of the candidates of a solution. Therefore, it is possible to substitute 0 and 1 for these two values. Since numerical operation is unnecessary if binary numbers are substituted for probability amplitude, a PE can consist of only simple logic circuits, as shown in Fig. 5. Such a processor is called a logic quantum processor (LQP). The number of PE gates in a logic quantum processor becomes 1/16 of that in the case in which the sum-of-product operation of a complex number is executed using the 8-bit fixed point. In a control command, the cases of both 0 and 1 can be used as a control qubit.

2.2 Commands of Logic Quantum Processor

The LQP has two reversible commands called the Walsh-Hadamard transform for distributing probability amplitude, and the NOT for exchange of probability amplitude, as shown in Fig. 6. Bit masks corresponding to control 0 and 1 are used for specification of a control bit.

The LQP cannot execute the quantum Fourier transform used by extraction of a solution and the quantum observation. Therefore, the logic observation command INQUIRY is added instead of quantum observation. The INQUIRY command investigates multiple qubits simultaneously and returns the existent possibility of the specified state to the specified answer register. The number of answer registers is the same as the number of qubits. In an INQUIRY command, a bit mask is used for the specification of the quantum state. It is set to *false*, when the bit mask of control 1 is 1 and the qubit specified is never 1. It is also set to *false*, when the bit mask of control 0 is 1 and the qubit specified is never 0. Otherwise, it is set to *true*. When both bit masks are 0, the quantum state is specified according to the answer register. INQUIRY0 returning negative logic and

0-7803-7761-3/03/$17.00 ©2003 IEEE

INQUIRY1 returning positive logic are prepared for the observation. A binary search utilizing the INQUIRY0 and INQUIRY1 commands gives the minimum and maximum values, respectively, as shown in Fig. 7. To observe the maximum, the MSBs of the bit masks of control 1 and 0 are set to 1 and 0, respectively, and all the remaining bit masks are set to 1. As a result, the possibility of 1 for the MSB is returned. If INQUIRY1 is used, the MSB of the answer register is set to 1 when 1 exists in the MSB, and it is set to 0 when 1 does not exist. When executing the next INQUIRY1 command, both the MSBs of the bit masks of control; 0 and 1 are set to 0, and the bit mask for the second qubit is set so that 1 may be observed. When 1 exists in the MSB, an observation bit is set to 1 if 11* exists, and it is set to 0 if 11* does not exist. On the other hand, when 1 does not exist in the MSB, the answer register is set to 1 if 01* exists, and it will be set to 0 if 01* does not exist. If this process is repeated in the LSB, the maximum number is saved in the answer register. The minimum number can be obtained by repeating the observation for 0 for the MSB using INQUIRY0 by a similar method.

2.3 Implementation of the LQP

The LQP is implemented using FPGA and its block diagram is shown in Fig. 8. Parallel operations are executed using 2^{10} PEs and sequential operations are executed using 2^{6} memories per PE. Consequently, the logic quantum operation with 16 qubits has been realized. Six-stage pipelines are used in an operation circuit. Since part of the unitary transformation turns into the identical transformation in the control command, skipping calculation accelerates the operation speed. The clock frequency of the fabricated processor is 40MHz, and the specification of the processor is summarized in Table I.

3. APPLICATIONS OF THE LQP

3.1 Fast Solver for the Satisfiability Problem

The satisfiability (SAT) problem is a problem which verifies whether the group of the input corresponding to the output of a logic circuit exists. The existence of the solution of the formula called product-of-sum form is investigated for a SAT problem. A SAT problem can be solved using the LQP. Two parts of state selection and state elimination in the LQP constitute a SAT problem, as shown in Fig. 9. In the state selection, the case where there is no value corresponding to each product term of the given logic equation is chosen. The selected value is eliminated in the state elimination. The answer is obtained by repeating this procedure for all product terms. The solution for a SAT problem is obtained by observing whether or not a solution exists. In the case of the quantum computer, a counter is used for state elimination. For a common quantum computer, the Grover's algorithm is used for observation of a solution. Thereby, a SAT problem can be solved in $2^{N/2}$ steps. On the other hand, the LQP sets 0 to the qubit for a flag, when not fulfilling the conditions. Finally, an INQUIRY command is used and the state where a flag is 1 is observed. Consequently, regardless of the number of variables, a SAT problem can be solved with the number of steps of product terms. When the same calculation was carried out in a PC simulation, it was found that the LPQ operation speed became 275 times that of the PC, as summarized in Table II.

3.2 Simulation of Stochastic Bit Errors

In the quantum calculation, decoherence by which the quantum state is lost with time is a problem. The bit error, which is one of the reasons for causing decoherence, can be simulated using the LQP. A bit error is an error which reverses the bit information stochastically with thermal excitation. In the quantum state, the probability amplitudes of 0 and 1 invert, which is equivalent to a NOT command. In the bit error simulation, the NOT command between the quantum operations is inserted stochastically, as shown in Fig. 10. Assuming that relaxation time is τ, the stochastic NOT command is generated by the Poisson process with an average generating interval τ. When τ_0 is given for the interval between quantum operations, the quantum operation is simulated inserting NOT stochastically. By executing simulations many times, the distribution of the result depending on decoherence can be obtained as shown in Fig. 11.

4. CONCLUSION

We proposed a logic quantum processor, improving both operation speed and scale by expressing probability amplitude by 1 bit. This processor has two reversible commands, NOT and Walsh-Hadamard, which are needed for execution of the quantum algorithm, and also has INQUIRY as a substitute for the quantum observation command. As a result of implementing this processor on FPGA, it was shown that an operation speed 275 times higher than that of the software simulation is possible. Moreover, a high-speed SAT solver and a high-speed bit-error simulation were shown as applications of the LQP. By combining the assembler and a monitor on a PC, coding the quantum algorithm and executing the program are easily realized. Development of the quantum algorithm utilizing the LQP will play a significant role in utilization of the quantum computer.

5. REFERENCES

[1] P. Shor, "Algorithms for quantum computation: discrete logarithms and factoring," *Proceedings 35th Annual Symposium on Foundations of Computer Science*, 1994, pp. 124-134.

[2] L. K. Grover, "A fast quantum mechanical algorithm for database search," *Proceedings 28th Annual ACM Symposium on Theory of Computing*, 1996, pp. 212-219.

[3] L. Vandersypen, M. Steffen, G. Breyta, C. Yannoni, M. Sherwood, and I. Chung, "Experimental realization of Shor's quantum factoring algorithm using nuclear magnetic resonance," *Nature*, 2001, pp. 883-887

[4] S. O'uchi, M. Fujishima, and K. Hoh, "Emulation of Quantum Computing by Finite Impulse Reponses," *Ext. Abs. of 1999 Int. Conf. on Solid State Devices and Materials*, 1999, pp. 96-97.

[5] S. O'uchi, M. Fujishima, and K. Hoh, "An 8-Qubit Quantum-Circuit Processor"*IEEE Symp. on Circuits and Systems*, 2002, pp. V-209-212.

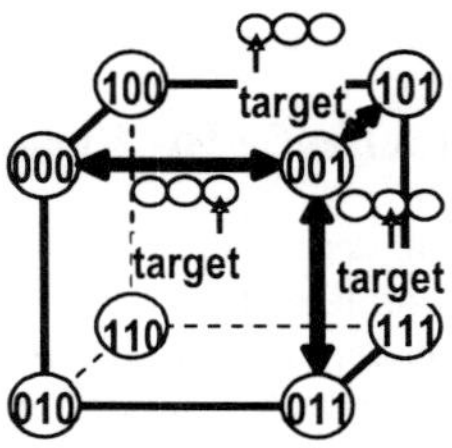

Fig. 1. Conceptual connection view of each base determined by qubits. One of the *n* connections is selected according to the target qubit.

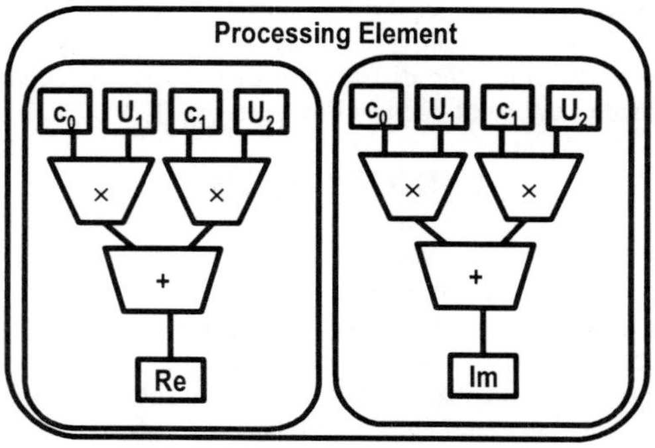

Fig. 2. Block diagram in the processing element for the calculation of quantum operations. Two sum-of-product operations are required for the unitary transformation.

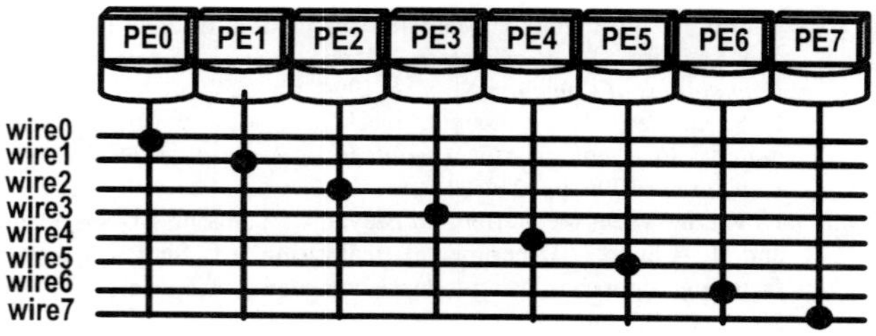

Fig. 3. Hypercube network formed by processing elements (PEs) in a quantum processor. Each PE has *n* I/Os and the number of communication wires is 2^n.

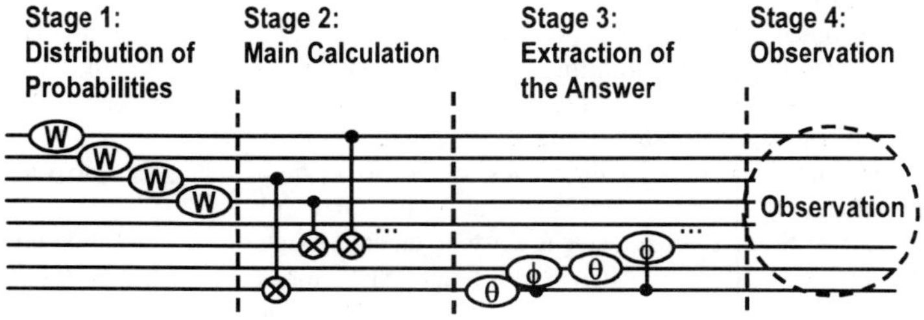

Fig. 4. Quantum circuit for the general quantum algoritum. The first stage is the distribution of probability amplitudes, the second stage is the main calculation, the third stage is the extraction of the answer, and the fourth stage is quantum observation. Until the second stage, only two types of probability amplitude exist.

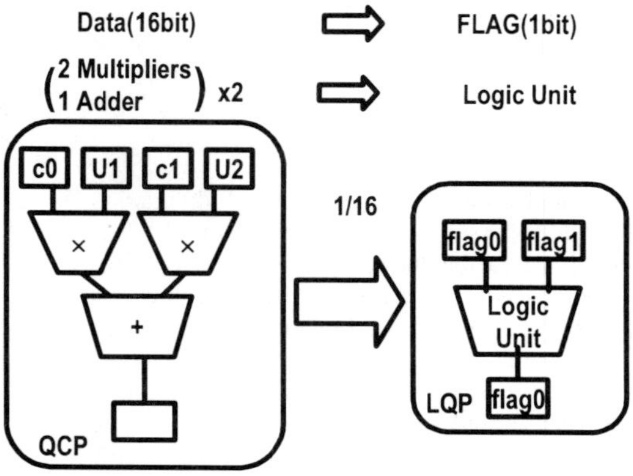

Fig. 5. The processing elements in the logic quantum processor require only simple logic elements. The hardware scale of the PE of the LQP is about 1/16 of the PE of the conventional QCP.

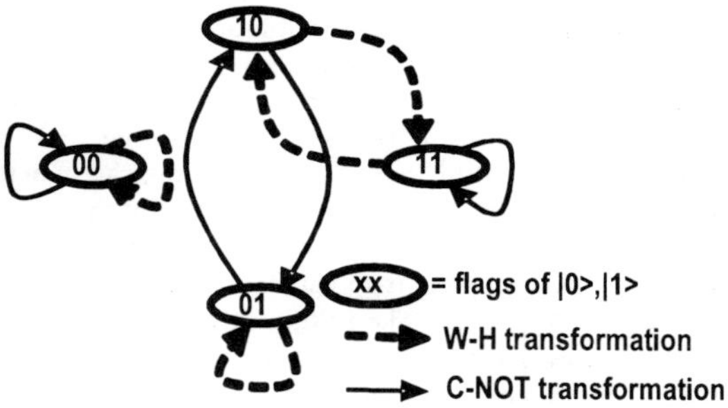

Fig. 6. Reversible operations executed in the LQP. One is the Walsh-Hadamard (W-H) transformation and the other is NOT. W-H in the LQP is slightly different from that in the original quantum computing since the LQP cannot handle negative numbers.

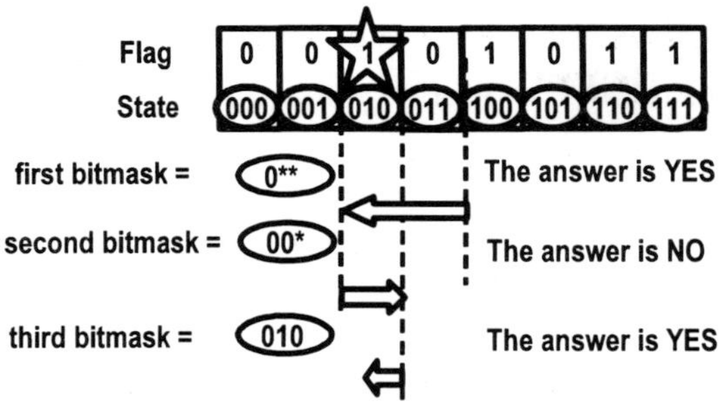

Fig. 7. Binary search executed by the LQP utilizing INQUIRY command. The bitmask and the answer register determine the qubit(s) to be inquired.

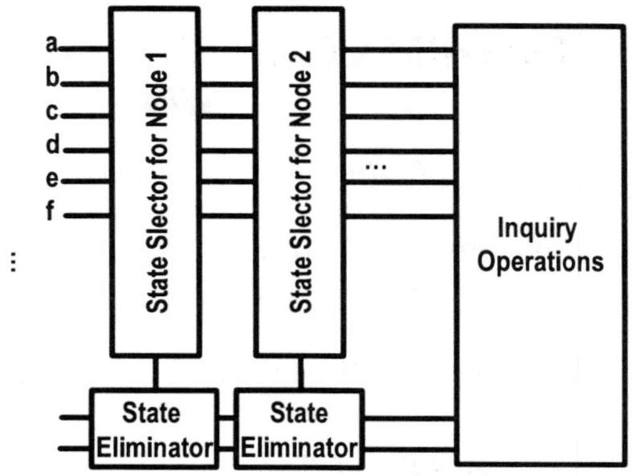

Fig. 9. Quantum circuit for the SAT problem. Each state selector and eliminator search candidates for the answer of the given logic equation.

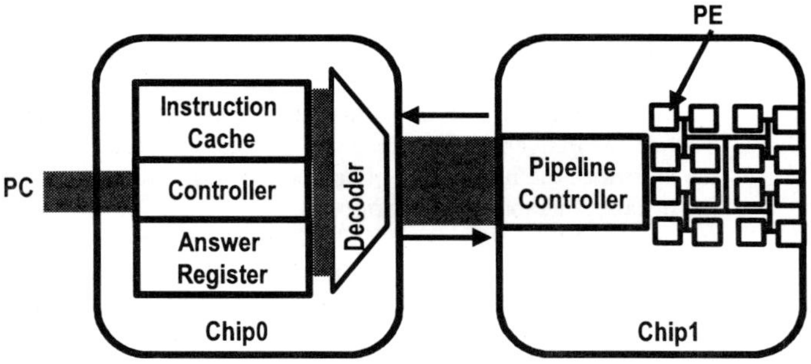

Fig. 8. Schematic block diagram of the LQP implemented in a FPGA. Two FPGAs are used for the LQP. The main calculations for the quantum algorithm are executed in the chip 1 and the interface between chip 1 and a PC is tuned in chip 0.

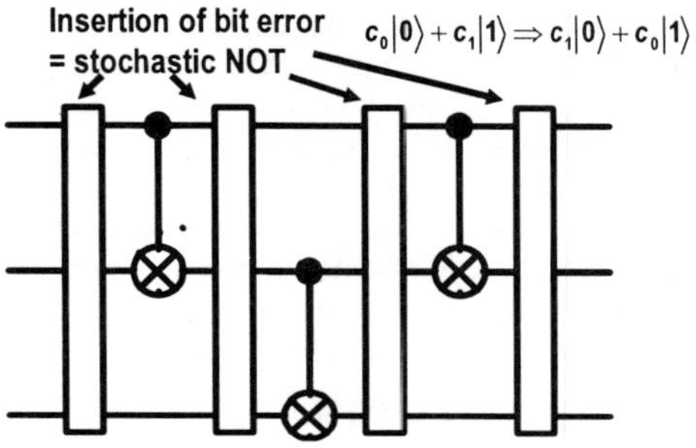

Fig. 10. Bit error simulation by the LQP. Stochastic NOT operations are inserted between original quantum operations.

Table I. Summary of the specification of the 16-qubit LQP.

Chip	APEX EP20K1500E ×2
Size	10+6 qubits
Logic Element (Used #/ Total #)	Chip0:960/51840(1%), Chip1:38846/51840(74%)
Memory (Used #/ Total #)	Chip0:196608/442368(44%), Chip1:65536/442368(14%)
Clock Frequency	40MHz < 70clk/QOperation

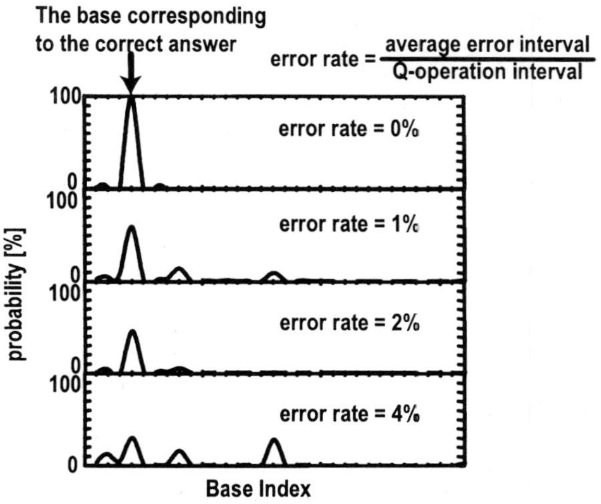

Fig. 11. Distribution of the state probabilities. The result is obtained after the same quantum circuit with stochastic errors is simulated 1000 times.

Table II. Summary of the performance comparison between the LQP and PC simulation.

	PC simulation	LQP
System	Intel Xeon 2.2GHz	16qubit 1024 parallel 40MHz
SAT problem	15 variables 182 nodes	
Result (problems/sec)	24	6600

INFORMATION EXCHANGES IN QUANTUM ARRAYS DUE TO SPATIAL DIVERSITY

M. Bucolo[1], L. Fortuna[1], M. La Rosa [1,2], D. Nicolosi[2], D. Porto[2]

[1]Dipartimento Elettrico Elettronico e Sistemistico, Università degli Studi di Catania, Viale A. Doria 6, 95100 Catania, Italy, e-mail: mlarosa@dees.unict.it

[2]Soft Computing Group, ST Microelectronics, Stradale Primosole 50, 95121 Catania, Italy

ABSTRACT

The effects of parameter spatial disorder are investigated in quantum arrays focusing on collective behaviors and communication between connected units. The amount of information exchanged has been correlated to the global dynamics and the parameters of the complex systems. A two-cell Quantum Cellular Neural Networks (QCNNs) oscillator is chosen as fundamental unit; chaotic dynamics characterizes the oscillators coupled through Coulomb interaction. Depending on the diffusion coefficient, two opposite behaviors have been recognized: strong communication between the cells and break of the information exchanges among the systems. Spatial parametric dissymmetry, random and chaotic, has been introduced increasing the number of communicating cells. The application of the deterministic disorder gives better results than the random one. Spatial stochastic resonance effects are recognizable in the chaotic approach.

1. INTRODUCTION

High-developed processing methods are continuously required in order to overcome the technological limits of the semiconductor industry products. Submicron and nanometric devices, realized using innovative approaches, offer an alternative solution for computation. In the last decade quantum-dot cells have been proposed as an innovative nanoelectronic paradigm to process binary information or to carry analog information. The dynamical features of quantum-dot cells and the possibility to couple them gave birth to interesting structures like Quantum Cellular Neural Networks (QCNN). QCNNs represent an innovative architecture for massive signal processing and exhibit interesting spatio-temporal phenomena [1]. The processing capability evaluation can be carried out using extended nonlinear systems approaches. In this paper, collective dynamics of quantum arrays are investigated focusing the attention on information exchange attitudes. The chaotic oscillator, generated by the connection of two QCNN cells, is chosen as basic brick of the lattices [2]. The behavior of the chain is tested by varying the coupling parameter and by weighting the reciprocal influence among oscillators. The cell dynamics are either collective, in case of efficient exchange of information, or saturated if the cell time evolution oscillates around a limit value. Higher values of diffusion coefficient increase the number of systems able to transfer information and decrease the percentage of oscillators persisting in a saturated behavior.

A parameter spatial disorder approach is proposed to improve the percentage of communicating cells. Previous studies have demonstrated that the introduction of noise in a network of nonlinear systems gives better regularization performances [3]. Moreover, the technological limits existing in the realization of the nano-structures imply that spatial physical differences cannot be disregarded in the evaluation of quantum systems.

Introducing both random and chaotic variation of oscillator parameters, the percentage of not-saturated cells decreases. Comparing the two approaches, the deterministic generation of spatial diversity exhibits systems with higher information transfer attitude. Moreover, the existence of a suitable value of the noise demonstrates the hypothesis of spatial stochastic resonance effects [4] in case of chaotic perturbation.

In Section 2 the structure of QCNN arrays are briefly introduced showing the equations of the two-cells QCNN oscillator and their dynamics with some examples. In Section 3, the spatio-temporal behaviors of the arrays are shown in case of spatial diversity. Section 4 is dedicated to conclusions.

2. QUANTUM ARRAY DYNAMICS

2.1 Two-cell QCNN

A Quantum dot is an "artificial atom" obtained in a semiconductor nanostructure by including a small quantity of material in a substrate. The definition "artificial atom" for a quantum dot is intended in an electrical sense: the transition of a single electron (or a single hole) is allowed among a couple of quantum dots.

Four dots realized on the same layer constitute a Quantum Dots Cellular Automata (QCA) cell [5]. In each cell two extra electrons can assume different locations by tunneling between the dots, providing a certain polarization *P*. Polarization varies continuously between -1 and +1 and constitutes the macroscopic degree of freedom of the cell. In the description of the system dynamics, *P* can be assumed as a state variable. In Fig.1 different values of *P* are shown for a QCA cell.

Every structure of interacting quantum cells can be considered a Quantum Cellular Automata.

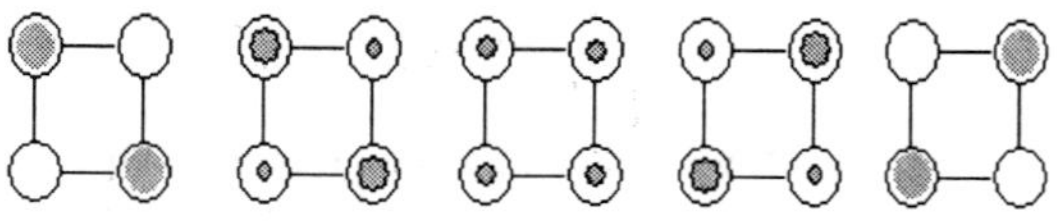

Figure 1. Different cases of cell polarization. Variable *P* is treated as continuous.

Following Schrödinger equation it is possible to transform the quantum mechanical description into a CNN-style description having the polarization P as macroscopic degree of freedom and a quantum phase displacement φ as microscopic degree of freedom [2]. Complex trajectories can be observed considering a two-cell QCNN. Denoting by P_i the ith cell polarization and by φ_i the phase displacement, this system is characterized by the presence of two quantum variables φ_1 and φ_2 in addition to the macroscopic degrees of freedom P_1 and P_2.

In its general form, the model can be described by the following equations [2]:

$$\begin{aligned}
\frac{\partial}{\partial t} P_1 &= -2a_1\sqrt{1-P_1^2}\sin\varphi_1 \\
\frac{\partial}{\partial t} \varphi_1 &= -w_1(P_1-P_2) + 2a_1\frac{P_1}{\sqrt{1-P_1^2}}\cos\varphi_1 \\
\frac{\partial}{\partial t} P_2 &= -2a_2\sqrt{1-P_2^2}\sin\varphi_2 \\
\frac{\partial}{\partial t} \varphi_2 &= -w_2(P_2-P_1) + 2a_2\frac{P_2}{\sqrt{1-P_2^2}}\cos\varphi_2
\end{aligned} \qquad (1)$$

where the parameters a_1, a_2, w_1 and w_2 are defined in [1,2]. In particular a_1 and a_2 are proportional to the interdot energy inside each cell (for identical cells $a_1 = a_2$) while w_1 and w_2 weight the effects of the difference of neighboring cell polarization and assume the same role of cloning templates in traditional CNNs.

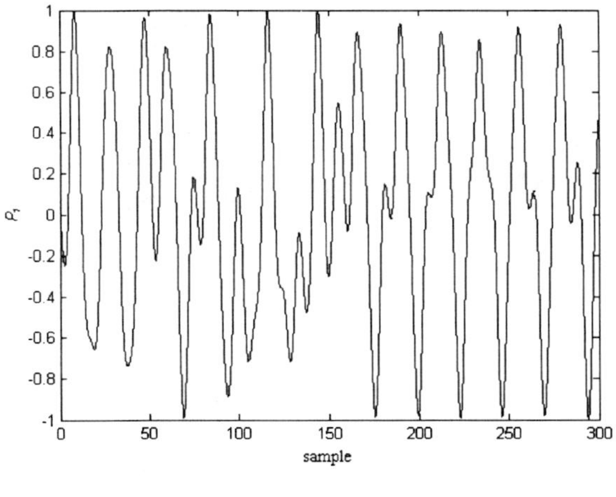

Figure 2. Time evolution of P_1 state variable for $a_1=a_2=0.25$, $w_1=0.3$ and $w_2=0.8$.

Chaotic dynamics are obtained in case of unbalanced cells with different initial polarizations and different coupling parameters. It has been demonstrated in [2], by evaluating the Lyapunov exponents, that for parameter values $a_1=a_2=0.25$, $w_1=0.3$ and $w_2=0.8$, the state variables reproduce a chaotic oscillation. The obtained chaotic dynamics is depicted in Figg. 2 and 3.

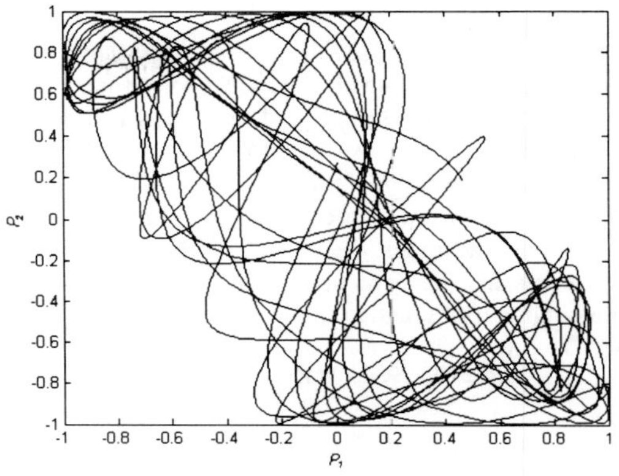

Figure 3. State trajectories of a two-cell QCNN with $a_1=a_2=0.25$, $w_1=0.3$ and $w_2=0.8$.

2.2 QCNN Arrays

Spatial distribution and local coupling characterize real quantum dot structures. The connection properties are fundamental in the investigation of QCNN lattices as suitable structures for computation and signal processing. Starting from the dynamics features of the single QCNN cell, extended lattices are built.

In this paper, a key example is studied in order to correlate the dynamic evolution of the systems with the coupling coefficient. The lattice considered is an array of N chaotic oscillators regulated by the differential equations reported above. The equations of the single two-cell QCNN oscillator are reformulated considering the number of oscillators N and the diffusion coefficient D.

The dynamics of the ith cell is described by the following equations:

$$\begin{aligned}
\frac{\partial}{\partial t} P_{i,1} &= -2a_1\sqrt{1-P_{i,1}^2}\sin\varphi_{i,1} + D\left(P_{i-1,1} - 2P_{i,1} + P_{i+1,1}\right) \\
\frac{\partial}{\partial t} \varphi_{i,1} &= -w_1\left(P_{i,1}-P_{i,2}\right) + 2a_1\frac{P_{i,1}}{\sqrt{1-P_{i,1}^2}}\cos\varphi_{i,1} \\
\frac{\partial}{\partial t} P_{i,2} &= -2a_2\sqrt{1-P_{i,2}^2}\sin\varphi_{i,2} \\
\frac{\partial}{\partial t} \varphi_{i,2} &= -w_2\left(P_{i,2}-P_{i,1}\right) + 2a_2\frac{P_{i,2}}{\sqrt{1-P_{i,2}^2}}\cos\varphi_{i,2} \\
i &= 1,\cdots,N
\end{aligned} \qquad (2)$$

The influence of the neighborhood oscillators is taken into account in the first state variable equation. The resistive coupling between two cells is obtained by multiplying the diffusion coefficient D by the difference between their first variables P_1. All the connections are assumed to be bi-directional depending on the coupling features.

For a fixed diffusion coefficient D, each unit shows one of two different behaviors: while a consistent number of oscillators perform a collective communication, several others persist in a saturated state. The second dynamic performs an oscillation around the limit values of the state variable range and hinders the information transmission.

Fig. 4 depicts the first variable evolutions in a *100*-cells array by using a 64-colors map. The spread number of solid lines related to the saturated behavior is visible in Fig. 4 (a). Fig. 4 from (b) to (d) shows how the density of these stripes decreases by increasing the diffusion coefficient from $D=0.075$ to $D=0.15$.

An index g is defined in order to compare the effects of D on the information exchange attitudes.

The index g represents the percentage of not-saturated oscillators over the whole number of units:

$$g = \frac{N_{notsaturated}}{N} * 100 \qquad (3)$$

The values of g in function of D are reported in Tab. 1. The percentage of not-saturated systems increases with D underlining the influence of diffusion over the number of not-saturated oscillators.

	g
D= 0.05	57%
D= 0.075	70%
D= 0.1	78%
D= 0.15	94%

Table 1. Percentage g of units performing a collective behavior for $N=100$.

3. PARAMETRIC SPATIAL DISORDER IN QUANTUM ARRAYS

Random spatial disorder has been applied to improve autosyncronization and pattern formation in arrays of nonlinear systems [6]. Parameters characterizing each fundamental cell are perturbed around the nominal value introducing dissymmetry in the structure of the lattices. The parametric disorder is similar to noise that is introduced on spatial instead of temporal dimension. This approach underlines a new methodology to control lattice dynamics by using spatial stochastic resonance features.

In recent works, the introduction of a deterministic dissymmetry has been proposed instead of the random one [3]. The parameters are modified by using a chaotic generator.

The nominal values are perturbed, by sampling the time evolution of a chaotic system variable.

In this work, both random and chaotic spatial dissymmetry is considered. The improvements obtained in information exchanges are evident in Fig. 5. The number of not-saturated units increases even for a small value of D. Moreover, the number of saturated oscillators is furthermore decreased if the parameter w_1 is chaotically varied.

It can be noticed, from Tab. 2, that introducing a random dissymmetry, the entity of perturbation does not modify the percentage g of communicating cells.

Using a chaotic parameter perturbation, instead, optimal noise intensity can be detected giving evidence to a spatial stochastic resonance effect.

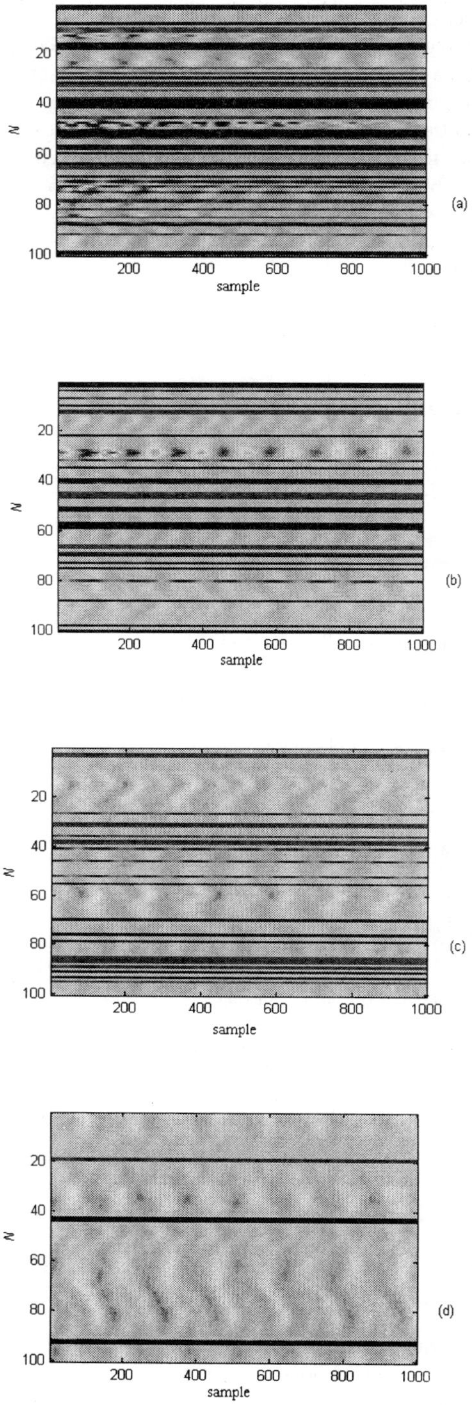

Figure 4. Spatio-temporal maps of two-cell QCNN arrays with: $D=0.05$ (a), $D=0.075$ (b), $D=0.1$ (c), $D=0.15$ (d).

	g
Random 10%	61%
Random 20%	61%
Chaotic 10%	68%
Chaotic 20%	66%

Table 2. Percentage g of not-saturated cells in case of spatial diversity of parameter w_1 for $N=100$ and $D=0.05$.

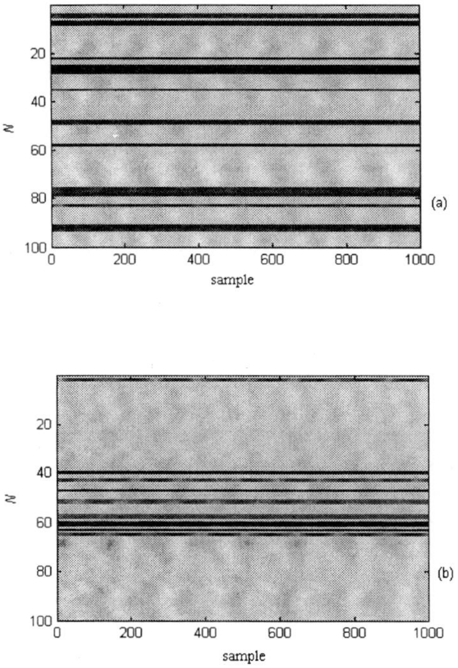

Figure 5. Spatio-temporal maps of two-cell QCNN array for $D=0.05$ with 10% spatial diversity of parameter w_1: random (a), chaotic (b).

4. CONCLUSIONS

QCNNs represent an innovative architecture for massive signal processing. The dynamics of an array of QCNN oscillators has been investigated, with particular interest to information exchanges enhancing. The number of oscillators exhibiting communication attitudes increases with the weight D of the connection. The introduction of spatial diversity in the cells parameters improves the information exchange between them and better results have been obtained with a deterministic diversity in respect to the random one. A spatial stochastic resonance effect can be underlined in the chaotic approach: comparing the intensities of perturbations investigated, an optimal noise value can be evaluated.

5. REFERENCES

[1] Porod W. "Quantum dot devices and Quantum-dot cellular automata", in *Visions of Nonlinear Science in the 21st Century,* Huertas J. L., Chen W-K., and Madan R. N. Editors, World Scientific Series on Nonlinear Science, Series A – 26:495-527, July 1999.

[2] Fortuna L., Porto D. "Chaotic Phenomena in Quantum Cellular Neural Networks". *Proceedings of the 7th IEEE International Workshop on Cellular Neural Networks and their Applications (CNNA'02),* Frankfurt, Germany, May 2002, pages 369-377.

[3] Arena P., Caponetto R., Fortuna L. and Rizzo A. "Nonorganized deterministic dissymetries induce regularity in spatiotemporal dynamics". *International Journal of Bifurcation and Chaos*, 10(1):73-85, 2000.

[4] Gammaitoni L., Hanggi P., Jung P. and Marchesoni F. "Stochastic Resonance". *Rev. Mod. Phys.* 70(1), 223-287, January 1998.

[5] Amlani I., Orlov A., Tóth G., Lent C. S., Bernstein G. H. and Snider G. L. "Digital Logic Gate Using Quantum-dot Cellular Automata", *Science*, 284:289-291, 9 April 1999.

[6] Braiman Y., Lidner J. F. and Ditto W. L. "Taming spatiotemporal chaos with disorder". *Nature,* 378:465-467, November 1995.

A Global Wire Planning Scheme for Network-on-Chip

J. Liu, L-R Zheng, D. Pamunuwa, H. Tenhunen

Laboratory of Electronics & Computer Systems (LECS)
Royal Institute of Technology (KTH)
Electrum 229, SE-164 40 Kista, Sweden
{jianliu, lrzheng, dinesh, hannu}@imit.kth.se

Abstract

As technology scales down, the interconnect for on-chip global communication becomes the delay bottleneck. In order to provide well-controlled global wire delay and efficient global communication, a packet switched Network-on-Chip (NoC) architecture was proposed by different authors [1][2]. In this paper, the NoC system parameters constrained by the interconnections are studied. Predictions on scaled system parameters such as clock frequency, resource size, global communication bandwidth and inter-resource delay are made for future technologies. Based on these parameters, a global wire planning scheme is proposed.

1. Introduction

Interconnect has been the major design constraint in deep sub-micron circuits. The downscaled wire size, increased aspect ratio, combined with higher signal speed cause many signal integrity challenges and time closure problems. Traditionally, these issues are tackled mainly from an electrical design point of view. Recent studies show that the problem also can be coped with interconnect-centric system architectures [1][2]. One such emerging architecture is the Network-on-Chip (NoC). The NoC architecture is a packet switched network on a single chip [1][2]. It scales from a few dozens to several hundreds or even thousands of resources. A resource may be a processor core, a DSP core, an FPGA block, a dedicated HW block, or a memory block. Any kind of inter-resource information is sent in packets over the network. The structured network wiring gives well-controlled electrical parameters and enables reusing of building blocks. Clearly, any topology that fully connects the resources can be used for the network. However, a two-dimensional mesh topology turns out to be simple and effective [2][3]. Thus, the following study will be based on this specific topology.

The NoC uses a backbone to provide a reliable and efficient communication platform for user-specified resources. The NoC backbone consists of resources and switches organized in a two-dimensional mesh, as shown in **Figure 1**. A data packet from one resource is first passed to the switch attached to the resource. The switch then routes the packet onto the appropriate link.

As the NoC is targeted to future DSM and nanometer technologies, the following questions are interesting: what is the appropriate size of each synchronous resource; how many resources can be integrated in one chip in future technologies; how fast can signals travel from one resource to another through the on-chip communication network and how to plan the wires to get an optimal data bandwidth with limited wire resource. In this paper, we study the NoC system parameters constrained by the interconnections and answer the above questions. In section 2, we use empirical rules to derive the gate delays for future DSM technologies, which is followed by an estimation of the maximum clock frequency and the corresponding resource size. In section 3, the inter-resource delay is studied and a global wire planning scheme providing maximum bandwidth is proposed.

The NoC is a typical interconnect-centric architecture, which means that the wire planning is the first design step. In this early planning stage, detailed system parameters for the wires are often unknown, making it impractical to consider layout-related properties such as 3D multiplayer interconnections. Therefore, a simpler wire model is used below. When the planning is done and various requirements on the wires, such as delay and noise level, are determined, a dynamic interconnect model can be used to generate a wire structure meeting these requirements in later design phases. One dynamic interconnect model using 3D capacitance, resistance and inductance is described in [4]. Similar CAD tools like Magma's FixedTiming [www.magma-da.com] are also emerging commercially.

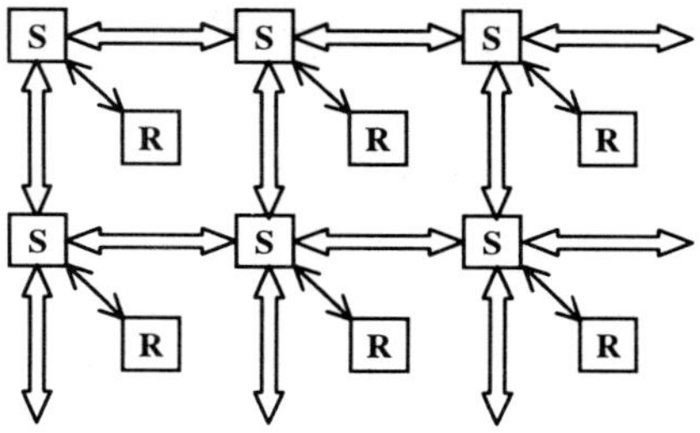

Figure 1. The 2D-mesh backbone of the NoC, with switches (S) and resources (R).

2. NoC Interconnect Fabric Optimization

The performance of interconnections is a major concern in scaled technologies. Under scaling, the gate delay decreases. However, the global wires do not scale in length since they communicate signals across the chip. For these wires, the delay per unit length can be kept constant if optimal repeaters are used [5]. In NoC, we assume that all global wires are reserved for global communications and semi-global/local wires are used within a resource.

2.1 Technology Scaling and Gate Delay

Since four is the typical average gate connectivity, "fan-out-of-four inverter delay", or simply FO4 is a reasonable parameter to

0-7803-7761-3/03/$17.00 ©2003 IEEE

be used for measuring gate delays. As the name suggests, an FO4 is the delay through an inverter driving four identical copies. *Ron Ho* [5] pointed out that, historically, gates have scaled linearly with technology, and an accurate model of recent FO4 delays has been $360 \cdot L_{gate}$ ps at typical and $500 \cdot L_{gate}$ ps under worst-case environmental conditions. After studying today's existing nanometer scale devices, he also predicts that this trend will continue for future generations of transistors, which means $500 \cdot L_{gate}$ ps is a lower limit for future FO4 delays. This model of gate delay will be used later when estimating clock cycle time and comparing with wiring delays.

2.2 Clock Cycle Analysis

A resource in a NoC can run at different speed. To study how the clock cycle within a NoC resource scales with the gate delay, we first examine the relationship between clock cycle and FO4 delay. Recent Pentium4 micro architecture and the aggressive Compaq/DEC alpha chips have 14 to 16 FO4s per clock cycle. Older processors, for example PentiumPro/II, run at 20 to 40 FO4s per clock cycle. It shows that the number of FO4s required in a clock cycle decreases as the technology scales down. Extrapolating historical data would lead to 6-8 FO4s per clock cycle within a few generations [5]. However, such fast-cycling machines pose many difficulties. With 6-8 FO4s per clock cycle, clock skew of a few FO4s would be extreme hard to manage. Furthermore, generating a clock of 8 FO4s per clock cycle is a difficult task since the rise and fall time of a clock wave take more than 2 FO4s to fully transition. With these difficulties in consideration, a clock cycle of 20 FO4s is projected for a cost-performance NoC resource and 10 FO4s for a high-performance one. Thus, with 0.05-µm technology, the clock cycle becomes $20 \cdot 500 \cdot 0.05 = 500$ ps for a cost-performance NoC resource, giving a clock frequency of 2 GHz. **Table 1** shows projected clock frequencies for some different technologies.

	0.18-µm	0.13-µm	0.10-µm	0.07-µm	0.05-µm
Cost Perf. (GHz)	0.56	0.77	1.0	1.4	2.0
High Perf. (GHz)	1.1	1.5	2.0	2.9	4.0

Table 1. Projected clock frequencies for NoC resources under worse-case FO4 delays.

2.3 Synchronous NoC Resource Size Estimation

Knowing the projected clock cycle, the maximum size of a synchronous NoC resource is limited by the wiring delays since the clock signal must be able to traverse 2 resource edges within a clock cycle (assuming the resource is quadratic) in the worst case, see **Figure 2**.

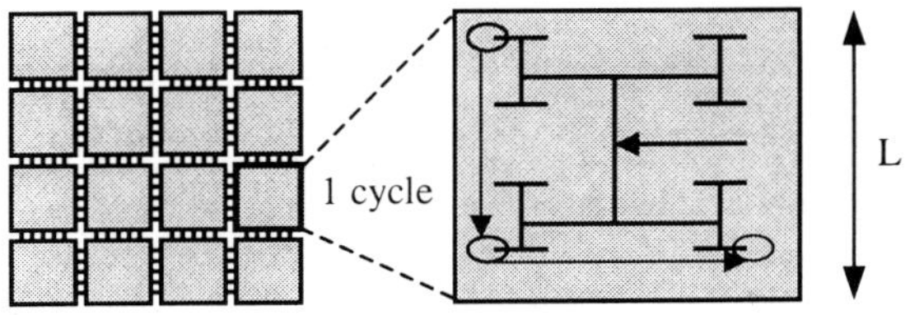

Figure 2. The worst-case delay in a NoC resource.

The wiring delay of a distributed *RC* line can be modeled as:

$$T_{wire} = 0.4rcl^2$$

Here T_{wire} is the wiring delay, l is the wire length, r is the resistance per unit length and c is the capacitance per unit length. This is a very good approximation and is reported to be accurate to within 4% for a very wide range of r and c [6]. Knowing the clock cycle time and *RC* delay model, the maximum resource size satisfies:

$$\max\_wiring\_delay < clock\_cycle$$

$$\Rightarrow 0.4rc(2L)^2 < clock\_cycle$$

Here, L is the maximum resource edge length. The clock cycle estimation is described in previous section and qualified predictions on wire resistance and capacitance for future technologies are available in a number of different papers.

The *RC*-model given above shows that the wiring delay grows quadratically with wire length. To reduce the delay for semi-global and global wires, a long line can be broken into shorter sections, with a repeater (an inverter) driving each section, see **Figure 3**. This makes the total wire delay equal to the number of repeated sections multiplied by the individual section delay:

$$T_{total} = k \cdot (T_{drv} + 0.4 \cdot rc(l/k)^2)$$

Now, a first order model of the driver (repeater), with lumped output resistance and input capacitance, gives the driver delay as:

$$T_{drv} = 0.7\frac{R}{h}(hC_0 + hC_g + c\frac{l}{k}) + 0.7r\frac{l}{k}hC_g$$

Here, R is the minimum sized inverter resistance, C_0 and C_g are diffusion and gate capacitances of a minimum sized inverter and r and c are wire resistance and capacitance per unit length.

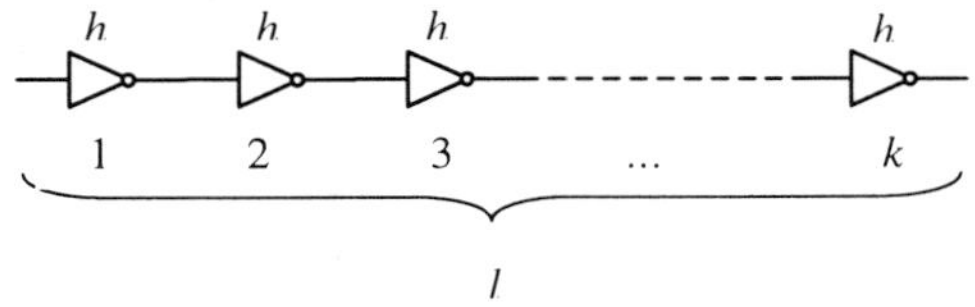

Figure 3. A long wire with k repeaters, each with a size of h times the minimum sized inverter.

The expression above for the total delay can be minimized and the minimum delay per unit length can be shown to be $2.13\sqrt{rcFO1}$ ps/mm [5][7]. Here, FO1 stands for fan-out-of-one delay and $1FO4 \approx 3FO1$. The time for a signal to traverse 2 resource edge lengths should be less than a clock cycle, suggesting the inequality $4.26 \cdot L \cdot \sqrt{rcFO1} < 1\ clock\_cycle$. Using the predicted future semi-global wire parameters provided in [7], as shown in **Table 2**, the maximum synchronous resource size and the number of resources on a single chip are calculated and listed in **Table 3**.

Wire Type	Parameter	0.18-µm	0.13-µm	0.10-µm	0.07-µm	0.05-µm
Semi-Global	R (ohm/mm)	107	185	317	611	1196
	c (fF/mm)	331	268	208	170	155

Table 2. Wire parameters for different technologies.

	Technology	0.18-μm	0.13-μm	0.10-μm	0.07-μm	0.05-μm
	Chip Size (mm)	20	21	23	25	28
High Performance	Max Resource Size	6.5	4.7	3.5	2.4	1.5
	Nr of Resources	9	20	42	112	350
Cost Performance	Max Resource Size	13	9.3	7.1	4.7	3.0
	Nr of Resources	2	5	10	28	87

Table 3. Maximum resource size and number of resources on a single chip, with different technologies.

The resistance and capacitance used to calculate **Table 3** are for semi-global wire, since the semi-global wire is normally used within a resource. Routing with global wires within a resource would allow larger resource size, since global wires, in general, have lower resistance and therefore also smaller delay per unit length than semi-global wires. From the table, we have that the maximum size of a synchronous high performance resource is 1.5 mm using 0.05 μm technology. For a cost performance resource with a cycle time of 20 FO4s, twice as long as the high performance resource cycle time, the maximum resource size is also twice as large.

It should be noticed that the analysis made above is valid for single wires. Crosstalk effects are not taken into consideration. If many wires are in parallel and switch simultaneously, the delay will be higher for unfavorable switch patterns, requiring smaller resource size. Therefore, the derived maximum resource size above should be seen as an upper bound.

3. Inter-Resource Delay and Bandwidth

3.1 Inter-Resource Delay

The inter-resource communication link will most likely consist of a large number of parallel wires, with uniform coupling over most of the wire length. For such closely coupled parallel wire structures, the crosstalk effects are considerable and cannot be neglected. Hence, the single wire model used in previous section is not valid here. Instead, the model shown in **Figure 4** is used. Each wire is modeled as a distributed *RC* line with total resistance *R*, total self-capacitance C_s, and total coupling capacitance C_c uniformly distributed over the whole line.

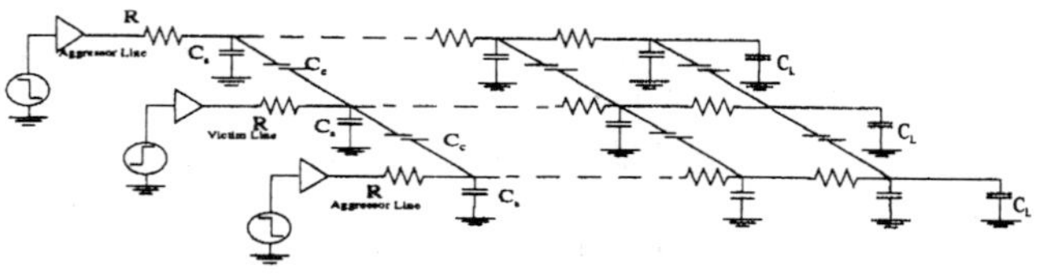

Figure 4. Distributed *RC* lines with uniform coupling.

The effect of crosstalk on the delay depends on the switching pattern of the aggressor (adjacent) lines. Most often, static timing models that take crosstalk into account are based on a *switch factor.* To model the crosstalk effects, the coupling capacitance is multiplied by this switch factor, which takes the value between 0 and 2 for the best and worst case respectively. In **Figure 4**, suppose that the victim line in the middle switches up from zero to one, the switching pattern that gives rise to the worst case delay on the victim line is when the two aggressor lines switch down from one to zero (almost) simultaneously [6]. The worst-case delay is then given by:

$$t_{0.5} = 0.7R_{drv}(C_s + 4.4C_c + C_{drv}) + R(0.4C_s + 1.5C_c + 0.7C_{drv})$$

Here, $t_{0.5}$ is the delay for step response to reach 50% point, R_{drv} is the driver (minimum sized inverter) output resistance and C_{drv} is the driver capacitance. Similar to the single wire case, the second term in this expression grows quadratically with the wire length. Inserting repeaters reduces the total wire delay. As shown in **Figure 5**, a long wire is broken into *k* sections, with an *h*-sized repeater driving each section. For each section, the driver has a lumped resistance of R_{drv}/h and capacitance of $h \cdot C_{drv}$, the wire has a distributed resistance of R/k and self-capacitance C_s/k, the mutual capacitance becomes C_c/k between two adjacent lines.

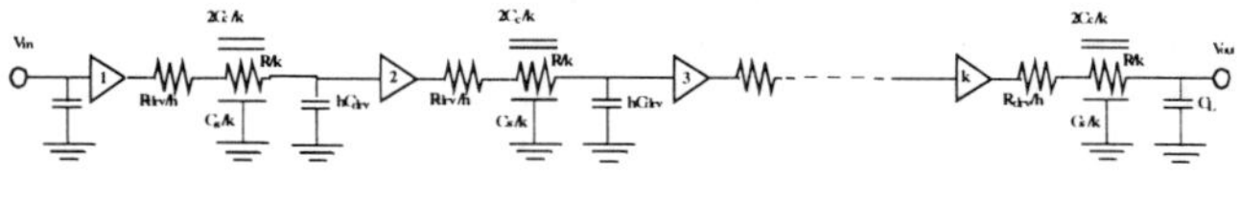

Figure 5. Insertion of repeaters in a long uniformly coupled *RC* line.

Applying the formula for worst-case delay for each section, the total wire delay becomes:

$$t_{0.5} = k\left[0.7\frac{R_{drv}}{h}\left(\frac{C_s}{k} + hC_{drv} + 4.4\frac{C_c}{k}\right) + \frac{R}{k}\left(0.4\frac{C_s}{k} + 1.5\frac{C_c}{k} + 0.7hC_{drv}\right)\right]$$

To obtain the optimal *k* and *h* value, the partial derivatives are equaled to zero, giving:

$$\frac{\partial t_{0.5}}{\partial k} = 0 \Rightarrow k_{opt} = \sqrt{\frac{0.4RC_s + 1.5RC_c}{0.7R_{drv}C_{drv}}}$$

$$\frac{\partial t_{0.5}}{\partial h} = 0 \Rightarrow h_{opt} = \sqrt{\frac{0.7R_{drv}C_s + 3.1R_{drv}C_c}{0.7RC_{drv}}}$$

Now, the optimal value of *k* must be a positive integer. Using the minimum sized inverter resistance and capacitance from [8], as shown in **Table 4**, the optimal *k* and *h* values are calculated and listed in **Table 5**. If the optimal *k* is not an integer, both of the two closest integers are used and corresponding delays are compared to each other in order to find the smallest delay.

	0.18-μm	0.13-μm	0.10-μm	0.07-μm	0.05-μm
Inv. Resistance (ohm)	9020	10560	11370	13710	15080
Inv. Capacitance (fF)	1.795	1.267	0.996	0.709	0.532

Table 4. Resistance and capacitance of minimum sized inverter for different technologies.

From **Table 5**, we see that the optimal size of the repeaters is large and the number of sections does not seem to be very significant for the delay. The increased number of repeaters only gives marginal improvement in delay. This means that the trade-off between the number of repeaters and the delay should be considered.

Technology	0.18-μm	0.13-μm	0.10-μm	0.07-μm	0.05-μm
Optimal h	322	296	226	187	154
Optimal k (1/mm)	0.99	1.30	1.66	2.28	3.33
Integer k (1/mm)	1	1	1	2	3
Total Delay (ps/mm)	65.5	73.2	83.7	91.8	110
Integer k (1/mm)	1	2	2	3	4
Total Delay (ps/mm)	65.5	71.3	76.0	90.1	108

Table 5. Optimal size of the repeaters, h, optimal number of sections, k, closest integer values to k and corresponding delay per unit length.

3.2 Inter-Resource Bandwidth Estimation

We have seen that repeater insertion can reduce the wire delay. However, the repeaters tend to be area- and power hungry and repeaters for global wires require many via cuts from the upper-layer wires all the way down to the substrate, introducing considerable via-resistances. Therefore, it is preferable to avoid repeaters in inter-resource communication.

The wire delay makes demand on the inter-resource bandwidth and distance. To see how these quantities are related, we first assume that a good signal has duration of at least $3t_r$, where t_r is the time for a rising signal to rise from 10% to 90% of its final value. Usually, for RC delays, 0-50% time $t_{0.5} = 0.69\tau$ and $t_r = 2.2\tau$ [5], where τ is the RC time constant. Thus, the bandwidth of a single wire is limited by $\frac{1}{9t_{0.5}}$. **Figure 6** shows the allowed maximum length of a global wire at different bandwidths, with and without repeaters. Clearly, for same technology and wire length, wires with repeaters can have higher bandwidth due to their low propagation delay. For an inter-resource distance of 1.5 mm with 0.05-μm technology (assuming that the resources are close to each other and the inter-resource distance is therefore equal to the resource size), the bandwidth between two adjacent resources is estimated to 0.6 Gbps per global wire without repeaters.

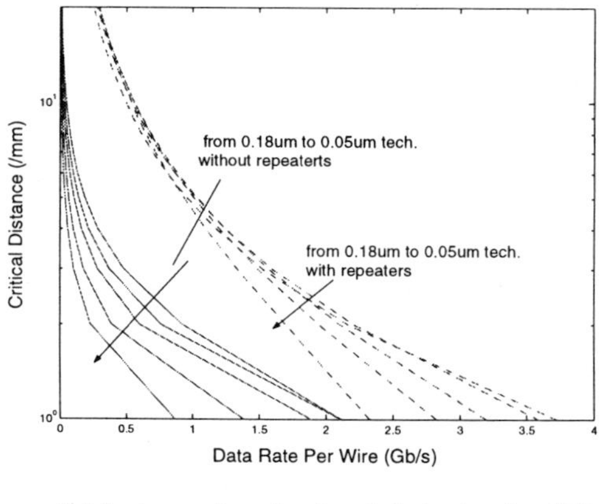

Figure 6. Maximum length of a global wire for different bandwidths and technologies, with and without repeaters.

4. Summary and Future Works

In this paper, we study the NoC system parameters constrained by the interconnections. Predictions on future technology feature size, clock speed in a synchronous resource, maximum NoC resource size, optimal global communication bandwidth and inter-resource distance, are made. These quantities are closely related to each other. The technology determines the gate delay, which in turn determines the maximum clock frequency. The maximum resource size can then be derived from the obtained clock frequency and the semi-global wire delay. At last, the global communication bandwidth is limited by the distance between resources and the global wire delay. Based on these estimated quantities, this paper provides a global wire planning scheme for NoC and can be used as a guideline for NoC system architecture definition. This can be demonstrated in a numerical example: for a NoC in 50-nm technology, the clock frequency is estimated to be 4 GHz for a high-performance synchronous resource with an edge length of 1.5 mm. With an inter-resource distance of 1.5 mm, there is room for about 350 such resources on a single chip of 28×28 mm. The bandwidth between two adjacent resources is estimated to be 0.6 Gbps per global wire without using repeaters.

Future work involves global communication bandwidth optimization strategies under different constraints such as area, power consumption, etc. In addition, the role of multilayer interconnection and real-world application integration in NoC are important and should be studied more closer.

5. References

[1] A. Hemani, A. Jantsch, S. Kumar, A. Postula, J. Öberg, M. Millberg, and D. Lindqvist. "Network on Chip: An Architecture for Billion Transistor Era", *Proceeding of the IEEE NorChip Conference*, November 2000.

[2] W. J. Dally and B. Towles, "Route Packets, Not Wires: On-Chip Interconnection Networks", *Design Automation Conference, Proceedings*, 684-689, 2001.

[3] E. Nilsson, "Design and Implementation of a Hot-potato Switch in Network on Chip", Master of Science thesis, Laboratory of Electronics and Computer Systems, Royal Institute of Technology (KTH), Sweden, June 2002.

[4] L-R Zheng, H. Tenhunen, "Design and Analysis of Power Integrity in Deep Submicron System-on-Chip Circuits", *Analog Integrated Circuits and Signal Processing, 30*, 15-29, 2002.

[5] R. Ho, K. W. Mai and M Horowitz, "The Future of Wires", *Proceedings of The IEEE, vol. 89, no. 4*, April 2001.

[6] D. Pamunuwa, L-R. Zheng and H. Tenhunen, "Maximizing Throughput over Parallel Wire Structures in the Deep Sub-micro Regime", in manuscript, Laboratory of Electronics and Computer Systems, Royal Institute of Technology (KTH), Sweden.

[7] H. Tenhunen, workshop "Systems on Chip, Systems in Package", *ESSCIRC 2001*, Villach Austria, Sep 2001.

[8] A. Maheshwari, S. Srinivasaraghavan and W. Burleson, "Quantifying the Impact of Current-Sensing on Interconnect Delay Trends", *ASIC/SOC Conference, 15th Annual IEEE International*, 461-465, 2002.

QUANTUM DOT NETWORKS WITH WEIGHTED COUPLING

Koray Karahaliloğlu, Sina Balkır

Department of Electrical Engineering, 209N WSEC,
University of Nebraska-Lincoln, Lincoln, NE 68588-0511, USA
E-mail: koray@destroyer.unl.edu

ABSTRACT

The capabilities of a particular quantum dot network with weighted dot coupling is investigated. A network configuration method is introduced for large scale applications. The method is based on using template images in order to assign weighted coupling between the dot pairs. A symmetrical update rule is defined for the coupling conductances as a function of the template image data. The described method is incorporated into a dedicated simulator, and the configured networks are simulated at the circuit-system level. It is shown that additional image processing capabilities like the line and motion detection are possible with such configurations of the quantum dot network connectivity.

1. INTRODUCTION

Recently, the quantum dot self-assembled nanostructures have appeared as attractive candidates for certain large scale parallel processing tasks. They have relatively simple implementation steps while sustaining a much higher degree of integration compared to their silicon counterparts. The quantum dot nanostructure considered in this work is a massively parallel array of dots formed by self-assembled metallic islands on the surface of a double barrier resonant tunneling diode (RTD) structure [1],[2]. The lumped circuit models can be assumed for both cases if quantum mechanical effects are negligible. A dot diameter within the range of 20 nm to 100 nm is considered as appropriate for classical approaches to be utilized [1]. The processing cells may involve one or more of these physical dots and essentially the same circuit model can be assumed for such clusters. This is because the dots in the cluster will receive the same inputs and initial conditions. Hence we use the term *dot* as a generic term referring to the processing cells.

Certain image processing capabilities of the RTD based quantum dot network are mentioned in [3]. The basic network structure resembles cellular neural networks (CNN) and is able to realize the smoothing or edge detection and enhancement processes. In this work we demonstrate that a pre-defined weighted configuration of the coupling conductances results in further possible applications for this device. A network configuration scheme at the circuit-system level is proposed which effectively defines the weighted coupling.

The outline of the paper is as follows: Section 2 briefly introduces the quantum dot network. Certain steady state characteristics related to the dot array system are given. In Section 3 the template image method is introduced which can be used to assign any weighted connectivity for the dot network. This method is used in the numerical analysis to obtain different network configurations. In Section 4 configured quantum dot networks are simulated with the resistive-capacitive coupling model. The novel processing capabilities of the quantum dot networks with weighted dot coupling are demonstrated. Section 5 concludes the paper.

2. QUANTUM DOT NETWORK

If one considers the circuit model in [1], a current equation for each dot i can be cast as

$$\begin{aligned}(C_{si} + \sum_{j\neq i} C_{ij})\dot{v}_i &= \\ I_{Bi}(t) - J_{si}(v_i) &+ \sum_{j\neq i} [G_{ij}(v_j - v_i) + C_{ij}\dot{v}_j]\end{aligned} \quad (1)$$

where $I_{Bi}(t)$ is the bias current to control the dot state, v_i is the potential and $\dot{v}_i$ denotes the derivative with respect to time. C_{si} is the substrate capacitance of the dot structure. The tunnel diode current J_{si} can be modeled with piecewise linear approximation as in Fig. 1. The coupling between the dots are represented by a linear capacitive-resistive model. The assumed network connectivity and the dot pair circuit model are also shown in Fig. 1. Since the dots have capacitive coupling, the system is tightly coupled for a general set of network parameters. For the general case one has to apply the numerical integration methods to the entire system. For this purpose, a dedicated circuit level simulator has been developed for the simulation of large networks including capacitive dot coupling.

This work was supported by the Office of Naval Research under grant N00014-01-1-0742.

0-7803-7761-3/03/$17.00 ©2003 IEEE

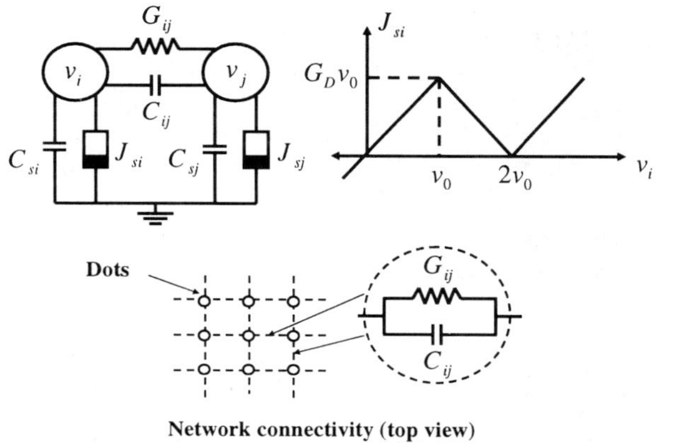

Figure 1: Circuit and connectivity model for the dot pair.

Under certain assumptions it is possible to obtain steady-state equilibrium points for the dot voltages. If one assumes the same steady-state dot voltages and biasing conditions for a particular region of the quantum dot array, then these voltages can be defined in terms of the model parameters and biasing conditions. The spatial differential term becomes zero in these regions. Using the steady-state conditions and the RTD model, the stable states for v_i in such regions follow as

$$V_{Low} = \frac{I_B}{G_D} \tag{2}$$

for the low equilibrium state and

$$V_{High} = 2v_0 + V_{Low} \tag{3}$$

for the high equilibrium state. However, note that these are not the only possible steady-state values for v_i. At the borders of such regions intermediate steady-state values are possible for the dot voltages.

3. TEMPLATE IMAGE METHOD

Compared to general CNN architecture [4], the quantum dot network structure is relatively simple. However, it is possible to extend the number of its applications by making suitable changes in the connectivity structure. For instance, the coupling conductances denoted as G_{ij} can be arranged to change the system response. In principle, this is achievable with the variation of the coupling between the dot clusters at the implementation level. Therefore, a pre-configured network can be obtained. For the numerical analysis, we give a convenient template image method to assign weighted coupling to any dot pair. Due to the connection model, the update rule is considered as symmetrical in indices. An initial template image is applied to the network, and the template image data is used to update the dot coupling resistances. For this purpose, we define differential and additive update terms for the interconnection resistance of a dot pair i and j as

$$R_{ij} = R_{ij}^{(0)} + T_D|v_j - v_i| + T_A|v_j + v_i| \tag{4}$$

Here $R_{ij}^{(0)}$ denotes the default coupling resistance before configuration. The linear update relation includes constants T_D and T_A as differential and additive template coefficients. In terms of conductances the relation in (4) is

$$G_{ij} = \frac{G_{ij}^{(0)}}{1 + G_{ij}^{(0)}[T_D|v_j - v_i| + T_A|v_j + v_i|]} \tag{5}$$

The update rule given in Equation (5) refers to a single symmetrical modification of the conductances in terms of the voltage states for any dot pair i and j. Since the rule is a function of the states, it is possible to set the couplings with a representative image as template. For this purpose the image intensity information is linearly mapped to the voltage range given by V_{Low} (black) and V_{High} (white). As an example, assume that the template coefficient $T_A = 0$ and a suitable differential coefficient T_D is chosen. Then in a given template image, the coupling of the identical pixel states will remain the same. However, the connections at the contrasting pixel state borders are changed according to Equation (5). The order of this change can be manipulated by the value chosen for the coefficient T_D. The described update scheme is illustrated across a 2×2 pixel region as shown in Fig. 2.

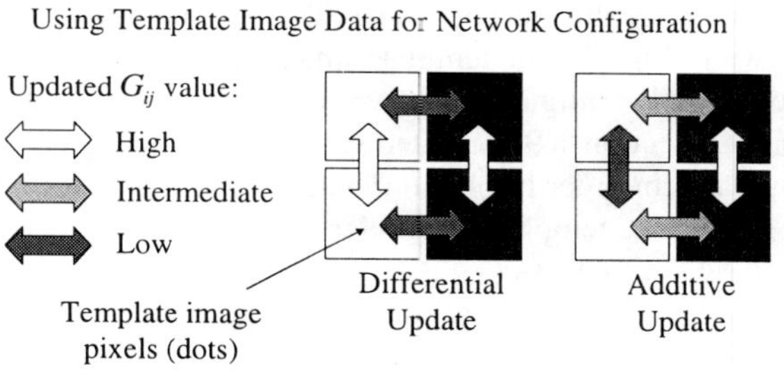

Figure 2: Template image method.

The method described is very convenient in the numerical analysis if one deals with the interconnections of large size arrays. Any image can be used as a template to obtain different network configurations. It can be recognized that the differential component in the update rule will be sensitive to the contrasts in the template image. The additive term however produces a different effect which depends on the similarity of the states within a region of the image. Therefore it can be used to construct isolated regions in the dot network. By using this template image method we show that the configured quantum dot network realizes certain interesting tasks.

4. SYSTEM SIMULATIONS

The system state is represented by the dot states or voltages v_i. These state variables can be regarded as the image pixel intensities and the intensity information can be mapped to the system as initial dot voltages. Then the whole system is left to evolve dynamically to its final state which corresponds to the processed output image. The input images are considered as black and white and the maximum intensity information of an input image is used to normalize and convert the pixel intensity data for these initial dot voltages. The physical properties of the device are given in [1] and [2]. An estimation for the order of circuit parameter values can be made from those results. The typical peak value current density is between $10^4 - 10^5 A/cm^2$ for a dot density of $10^{12} cm^{-2}$. Hence the peak current for the tunnel diode model is in the order of $\sim 0.1 \mu A$. This is used as the peak value for the diode current. For the peak current voltage, $v_0 = 0.2\ V$ is chosen. The input image dot voltage range is taken from $v_i = 0$ to $0.5\ V$. The coupling and substrate capacitances are $C_{ij} = C_{si} = 10^{-2}\ fF$. The parameter values above are assumed to be identical throughout the network in the examples. In order to obtain appropriate state space conditions, a constant bias current is assumed for all the dots as $I_{Bi} = G_D v_0 / 2$.

Example 1: Line detection. One useful configuration of the system is obtained if different coupling conductances G_{ij} for the horizontal and vertical directions are assumed. Such a configuration can be exploited to perform line detections for an input image. We make use of the template image method presented for this purpose. The 80×80 pixel input image is shown in Fig. 3. The template image in Fig. 3 is applied for the network configuration of vertical detection. For the horizontal detection, a $90°$ rotated version of this image is used as template. In order to obtain the desired stable states one has to choose the template coefficients accordingly. For this example the initial conductances are chosen as $G_{ij}^{(0)} = G^{(0)} = 0.1\ \mu S$ throughout the network. An optimum value is chosen for T_D by numerical experimentation as $T_D = 60\ \mu A^{-1}$ and the additive coefficient is set to $T_A = 0$. Then the coupling conductances are updated using the template images. The result is two network configurations as $G_{hor} = G_{ver}/4$ ($G_{ver} = G^{(0)}$) for the vertical detection and $G_{ver} = G_{hor}/4$ ($G_{hor} = G^{(0)}$) for the horizontal detection. After updating the coupling conductances, the input image is applied to the both of these configured systems. Steady-state responses from these systems are shown in Fig. 3 for the vertical and horizontal configurations respectively. It is observed that the system evolves in favor of the direction where the coupling is enhanced in both cases. With an appropriate value of conductances, the system output is stabilized in that direction. Therefore using such arrangements of coupling conductances, the system is able to perform line detection tasks.

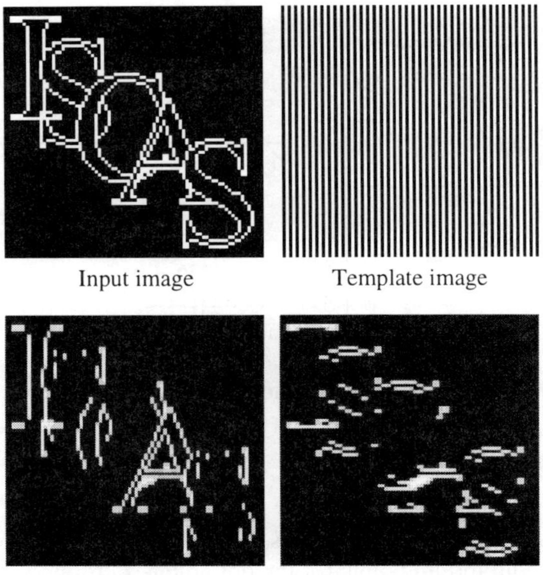

Figure 3: Vertical and horizontal line detection.

Example 2: Motion detection. It is possible to make the system learn a particular input pattern and respond to the successive input images accordingly. Hence, if the interconnections can be modified in real-time, this feature can be employed in applications like the detection and tracking of motion. However, the quantum dot network response is not equivalent to the subtraction of the successive images. Furthermore, the behavior is not independent of the input image. For example, for particular white pixel shapes moving in front of a dark background the intersected regions of two successive images can be extracted by the system. The related system response is examined with an example. The 80×80 primary image shown in Fig. 4 is used as the template and the network connections are configured according to this image. However,

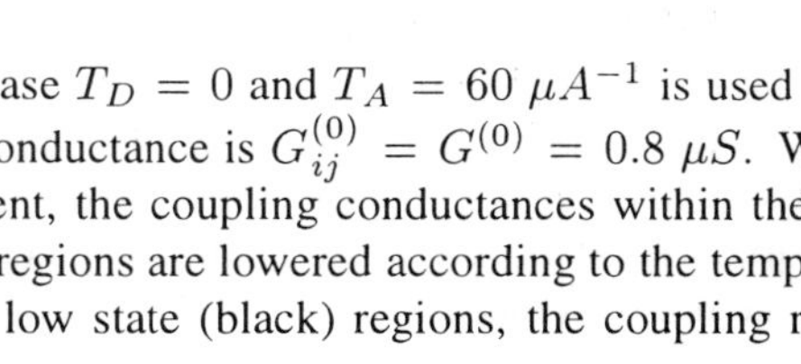

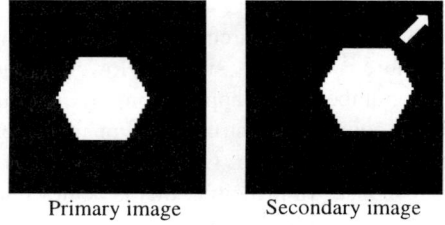

Figure 4: Successive images for motion detection.

in this case $T_D = 0$ and $T_A = 60\ \mu A^{-1}$ is used in (5). The initial conductance is $G_{ij}^{(0)} = G^{(0)} = 0.8\ \mu S$. With this arrangement, the coupling conductances within the high state (white) regions are lowered according to the template image. For the low state (black) regions, the coupling remains the same since additive update term is small. At the borders of these regions, the conductances will assume intermediate values depending on the template image. After this configuration, the secondary image in Fig. 4 is introduced to the configured system where the same hexagonal shape is shifted in the upper right direction. The related response is shown in Fig. 5 with snapshots from the dynamical state changes. It is observed that the states in the weakened coupling region rapidly stabilize with the new image data. On the other hand, the non-intersecting portion of the secondary image eventually dies away since the conductances are high at that region and the dot states move to the low equilibrium point. As a result, the system extracts the intersecting portion of the initially memorized primary image and the successive secondary image. In addition, the transient behavior of the network can be used to estimate the direction of the motion. These features can be employed in applications like motion detection and tracking. However, this operational characteristic also depends on the image data. This is because the stable state of the system does not only depend on the configuration, but also the initial dot states established.

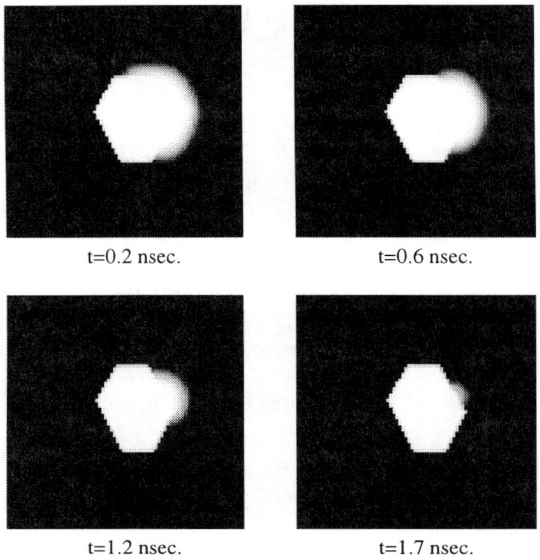

Figure 5: System response to secondary image.

5. CONCLUSIONS

The simulations depict that the quantum dot network with weighted connections is able to perform additional tasks combined with the main characteristic behavior. The simple network structure is capable of image smoothing and edge detection with the cooperative dynamical behavior of the dots. If one configures the network by arranging the coupling conductances, the application area can be extended. We have introduced two such applications as line and motion detection for configured quantum dot networks.

The numerical results are obtained with the proposed template image method where the dot connections are configured according to a particular image information acting as a template. An update rule for the resistive dot coupling in the network is presented. According to this rule, the connections G_{ij} of each dot i and j is updated according to its state v_i and its neighbor state v_j defined by the template image. The related dependency is kept linear and symmetrical in indices with the differential and additive coefficients. By variation of these coefficients, the amount of change in conductances can be manipulated. The method is especially useful to configure a large number of interconnections and it is also more instructive when the image applications are considered. This method is applicable only for numerical analysis purposes at the present time. However, it can be used in the configuration of the physical devices if the coupling connections of the device can be modified in real-time.

The desired response of the network will not be independent of the input image, when a general set of inputs is considered. Therefore pre-processing may be necessary depending on the intended application. Even when this is the case, such pre-processing will include relatively simple tasks like scaling and inversion of the image intensity data. Hence, the main processing advantages of the quantum dot network remain intact. The implementation issues of such configured quantum dot devices still involve substantial challenges. However, the numerical analysis shows that the quantum dot nanodevice is a promising candidate for a broader range of applications.

6. REFERENCES

[1] V.P. Roychowdhury, D.B. Janes, and S. Bandyopadhyay, "Collective Computational Activity in Self-Assembled Arrays of Quantum Dots: A Novel Neuromorphic Architecture for Nanoelectronics," *IEEE Trans. on Electron Devices*, vol. 43, no.10, pp. 1688-1699, 1996.

[2] V.P. Roychowdhury, D.B. Janes, and S. Bandyopadhyay, "Nanoelectronic Architecture for Boolean Logic," *Proceedings of IEEE*, vol. 85, no. 4, pp. 574-587, 1997.

[3] K. Karahaliloğlu, and S. Balkır, "Image Processing with Quantum Dot Nanostructures," *In Proceedings of the 2002 IEEE Int. Symposium on Circuits and Systems*, vol. 5, pp. 217-220, Phoenix, AZ, 2002.

[4] L.O. Chua, and L. Yang, "Cellular Neural Networks: Theory," *IEEE Transactions on Circuits and Systems*, vol. 35, no. 10, pp. 1257-1272, 1988.

PERFORMANCE MODELING OF RESONANT TUNNELING BASED RAMS

Hui Zhang, Pinaki Mazumder, Li Ding

Dept. of EECS, University of Michigan,
Ann Arbor, MI 48109,USA

Kyounghoon Yang

Dept. of EECS, KAIST,
Republic of Korea

ABSTRACT

Tunneling based random-access memories (TRAM's) have recently garnered a great amount of interests among the memory designers due to their intrinsic merits such as reduced power consumption by elimination of refreshing operation, faster read and write cycles, and improved reliability in comparison to conventional silicon DRAM's. In order to understand the precise principle of operation of TRAM's, an in-depth circuit analysis has been attempted in this paper and analytical models for memory cycle time, soft error rate, and power consumption have been derived. The analytical results are then validated by simulation experiments performed with HSPICE. These results are then compared with conventional DRAM's to establish the claim of superiority of TRAM performance to DRAM performance.

1. INTRODUCTION

Silicon dynamic random-access memories (DRAM's) are currently dominant commercial commodity in the semiconductor memories market due to their lowest cost per bit as well as gargantuan integration scale that allows DRAM manufacturers to monolithically fabricate over 256 million cells per chip. However, these mega-size DRAM chips are encountering several formidable problems due to a host of reasons, some of which are listed below.

First, DRAM's are becoming increasingly prone to soft errors. Soft error is caused by extra charge collection in the storage node of memories, generally induced by external charged particles and neutrons. The chances of loss of a stored bit depend on the amount of critical charge of the storage node. Technology scaling that achieves lowered capacitance, reduced power supply voltage, tinier transistor geometries, is generally deployed to increase the density and performance of the DRAM's; however, the scaling also concomitantly reduces the critical charge of the DRAM cell, thus increasing the Soft Error Rate (SER).

Second, DRAM's power consumption largely depends on periodic refreshing of memory cells deemed necessary due to excessive leakage currents. The continuous down scaling of the transistor threshold voltage as well as packing of memory cells more densely sharply aggravates leakage currents, thereby significantly increasing the power consumption of DRAM chips.

With the objective to solving these problems, memory manufacturers are continuously pursuing circuit and technology innovations. Tunneling based RAM's (TRAM's) proposed in [1]-[2] are of interest because of their great potentials in increasing critical charge, while reducing power consumption due to dispensing with mandatory refreshing of cells in DRAM's. A TRAM cell is composed of a conventional DRAM cell being augmented by co-integrating along the cell capacitor a pair of series connected resonant tunneling diodes (RTD), as illustrated in Fig. 1.

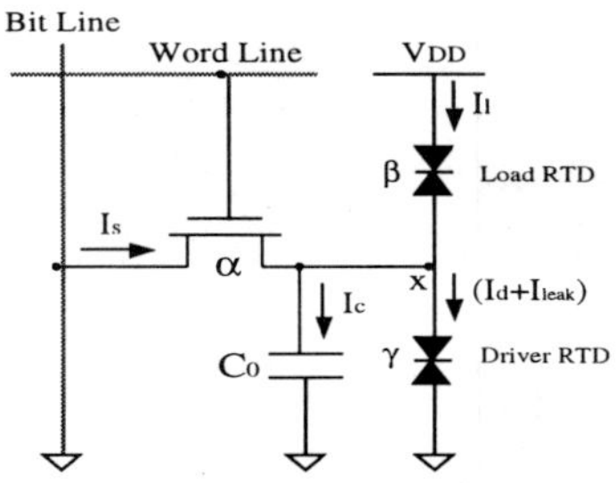

Figure 1: Schematic of 1T RTD-based RAM. α, β, and γ are sizing parameters.

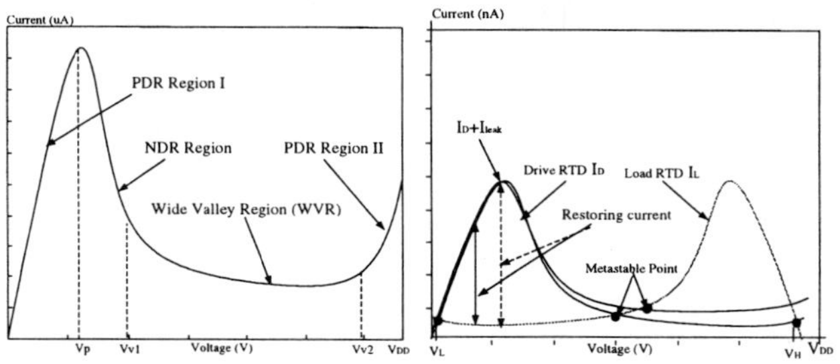

Figure 2: (a) RTD I-V characteristic. (b) Bistable property of RTD-pair.

RTD has a novel I-V characteristic as illustrated in Fig. 2(a). Instead of having a monotonic I-V characteristic, RTD has two positive differential resistance (PDR) regions interspersed by a negative differential resistance (NDR) region. This nonlinear tunneling characteristic renders RTD into a very promising device for a wide class of circuit applications, namely, multi-valued logic, high-speed and low-power circuits, and radiation-hardened reliable circuits. Two series connected RTDs have the self-latching or bistable property as is shown in Fig. 2(b). The RTD-pair can latch at either V_H or V_L, corresponding to logic '1' or logic '0', respectively. This bistable property of the RTD-pair in TRAM can be exploited to improve the soft error immunity, the standby power consumption and the speed of memories. From circuit design point of view, however, a detailed analytical study of the impact of augmentation of conventional DRAM cell by a RTD-pair is necessary.

In this paper, an exact analysis of speed, soft errors and power consumption in a TRAM is presented. The organization of the paper is as follows. In Section II, an analytical study of speed is

0-7803-7761-3/03/$17.00 ©2003 IEEE

given and the formulas are validated by HSPICE simulation. In Section III, the critical charge, one of the most important parameters in SER analysis, is derived. A comparison between the critical charge of TRAM and that of DRAM is presented. Finally, the power consumption of TRAM is analyzed in Section IV.

2. READ AND WRITE OPERATIONS ANALYSIS

In conventional DRAM, the READ operation is destructive. The read access time is therefore increased by the decreasing drive ability. In the case of TRAM, as shown in Fig. 2(b), the restoring current generated by the RTD-pair will help to drive the bit-line capacitance. Therefore, the TRAM read access time is potentially smaller than that of the conventional DRAM. For the WRITE operation, the restoring current plays two conflicting roles. Initially, it opposes the transition between V_L and V_H. But once the voltage of the storage node reaches the meta-stable point, the restoring current begins to help the switch over. Therefore, for the WRITE operation, the RTD-pair should be sized very carefully to obtain a reasonable fast speed. In this section, we first analytically study the READ and WRITE operations. Then, the results are validated with HSPICE and compared with conventional DRAM.

2.1. Analytical study

TRAM is stable at either V_L or V_H in the standby mode. These stable values are not exactly V_{SS} or V_{DD} and are determined by the I-V characteristic of the RTD-pair. For simplicity, we use the piece-wise linear model [3] for the resonant tunneling diodes:

$$I = \begin{cases} \frac{1}{R_1}V & 0 \le V < V_p \quad (PDR-I) \\ I_p + (V - V_p)\frac{1}{R_n} & V_p \le V < V_{v1} \quad (NDR) \\ I_{v1} & V_{v1} \le V < V_{v2} \quad (WVR) \\ I_q + (V - V_{DD})\frac{1}{R_2} & V_{v2} \le V \le V_{DD} \quad (PDR-II), \end{cases} \tag{1}$$

$$R_1 = \frac{V_p}{I_p}, \quad R_n = \frac{V_p - V_{v1}}{I_p - I_{v1}}, \quad R_2 = \frac{V_{DD} - V_{v2}}{I_q - I_{v1}},$$

where I_p and V_p represent the peak current and peak voltage; I_{v1} and V_{v1} are the valley current and valley voltage; and I_q is used to model the second peak current at $V = V_{DD}$. R_1 is the resistance of the PDR-I region; R_n is the resistance of the NDR region; and R_2 is the resistance of PDR-II region.

Let x denote the voltage of the storage node of the TRAM. We use x_0 and x_1 to respresent the voltages of logic '0' and logic '1', respectively. To balance the performance of READ '0' and READ '1', we choose β equals γ. At stable point, the driver RTD current (I_d) equals to the load RTD current (I_l). We derive x_0 and x_1 as:

$$x_0 = I_q R_{12}, \quad x_1 = V_{DD} - I_q R_{12}, \quad R_{12} = (R_1 R_2)/(R_1 + R_2).$$

Therefore, in TRAM, the voltage of the storage node is determined by RTD's characteristic instead of V_{SS} and V_{DD}.

2.1.1. Analysis of READ *operation*

In DRAM, READ operation is always destructive. In TRAM, READ operation can also be destructive if $\alpha|I_{sr}| > \beta(I_p - I_v)$, where I_{sr} is the read access current. We use κ to denote the size ratio β/α. The minimal κ value κ_{min} that makes the READ operation non-destructive is $\kappa_{min} = \alpha|I_{sr}|/(I_p - I_v)$. In READ operation, the access transistor operates within a small region. Therefore, we use a linear model for the access transistor: $I_{sr} = A - Bx$. Let the charge (discharge) current of the storage node be I_c. Using KCL at the storage node, we obtain the waveform expression for the bit-line voltage as follows:

$$\Delta V(t) = \frac{\alpha}{C_{bit}}\left\{\frac{B}{\lambda^*}(x_0 - \frac{A^*}{B^*})(e^{-\lambda^* t} - 1) + (A - B\frac{A^*}{B^*})t\right\}.$$

For READ '0' operation, we have $A^* = A + \Delta A_0$, $B^* = B + \Delta B_0$, and $\lambda^* = \alpha B^*/C_0$. The values of ΔA_0 and ΔB_0 are derived as:

$$\Delta A_0 = \kappa I_q, \quad \Delta B_0 = \frac{\kappa}{R_{12}} \quad (0 \le x < V_{DD} - V_{v2}),$$
$$\Delta A_0 = \kappa I_v, \quad \Delta B_0 = \frac{\kappa}{R_1} \quad (V_{DD} - V_{v2} \le x < V_p).$$

For READ '1' operation, $A^* = \Delta A_1 - A$, $B^* = \Delta B_1 - B$, and

$$\Delta A_1 = \kappa\frac{V_{DD}}{R_{12}} - \kappa I_q, \quad \Delta B_1 = \frac{\kappa}{R_{12}} \quad (V_{v2} \le x < V_{DD}),$$
$$\Delta A_1 = \kappa\frac{V_{DD}}{R_1} - \kappa I_v, \quad \Delta B_1 = \frac{\kappa}{R_1} \quad (V_{DD} - V_p \le x < V_{v2}).$$

2.1.2. Analysis of WRITE *operation*

For WRITE operation, the restoring current of the RTD-pair is against the state transition at first and then it helps to flip the state. Therefore, the write operation is only successful when $\alpha|I_{sw}| > \beta(I_p - I_v)$, where I_{sw} is the write access current. And κ should be smaller than $\kappa_{max} = \alpha|I_{sw}|/(I_p - I_v)$. The microstate diagram of WRITE operation is shown in Fig. 3.

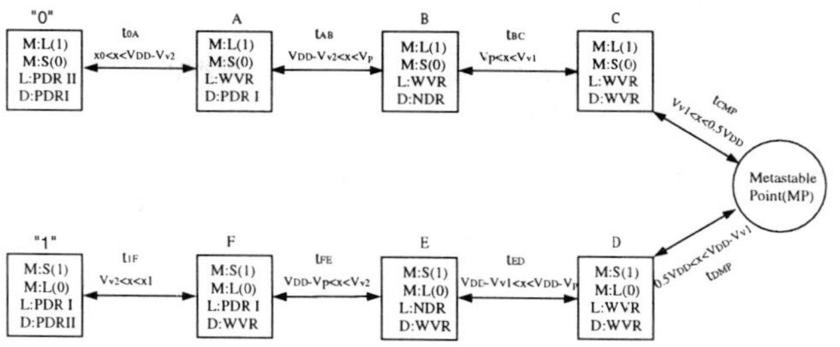

Figure 3: TRAM WRITE operation microstate diagram. $M: S(1)$, $M: S(0)$, $M: L(1)$ and $M: L(0)$ represent the saturation (S) and linear (L) operation region of the access transistor. (1) denotes WRITE '1' operation and (0) denotes WRITE '0' operation.

The write access time is the time when the voltage of the storage node reaches the meta-stable point. It can be obtained according to the microstate diagram as follows:

$$T_1 = t_{0A} + t_{AB} + t_{BC} + t_{CM},$$

$$T_0 = t_{1F} + t_{FE} + t_{ED} + t_{DM},$$

where T_1 and T_0 represent the write '1' access time and write '0' access time, respectively. By using KCL at storage node, we find that for each transition region, the transition time (each term in T_1 and T_0) can be computed by using the identical expression as follows:

$$t = \frac{1}{\hat{\lambda}} \ln \frac{a^* - b^* x_{t0}}{a^* - b^* x} + t_0,$$

where, $\hat{\lambda} = \alpha b^*/C_0$. The parameters a^* and b^* are different for each transition region. They are determined by the access transistor parameters A, B and the RTD electrical parameters. Table 1 shows a^* and b^* for all different transition regions.

Table 1: Parameters for WRITE operations.

	State Transition (WRITE '1')			
	'0'→A	A→B	B→C	C→M
a^*	$A+\kappa I_q$	$A+\kappa I_v$	$A+\kappa(I_v-I_p+\frac{V_p}{R_n})$	A
b^*	$B+\frac{\kappa}{R_{12}}$	$B+\frac{\kappa}{R_1}$	$B+\frac{\kappa}{R_n}$	B
	State Transition (WRITE '0')			
	'1'→F	F→E	E→D	D→M
a^*	$-A-\kappa Iq$ $+\kappa\frac{V_{DD}}{R_{12}}$	$-A-\kappa I_v$ $+\kappa\frac{V_{DD}}{R_1}$	$-\kappa(I_v-I_p+\frac{V_p}{R_n})$ $-A+\frac{\kappa V_{DD}}{R_n}$	$-A$
b^*	$-B+\frac{\kappa}{R_{12}}$	$-B+\frac{\kappa}{R_1}$	$-B+\frac{\kappa}{R_n}$	$-B$

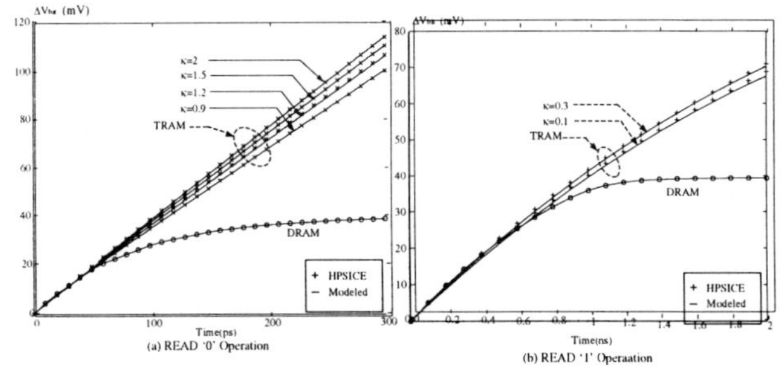

Figure 4: Bit-Line waveforms comparison during READ operation.

2.2. Validation and comparison

Figure 4 shows the bit-line waveforms during READ operation for TRAM and DRAM. We have used $V_{DD}=1.6V$, $C_0=20fF$, $C_{bit}=400fF$ and $\alpha=1$. The κ_{min} is 0.8 for READ '0' operation and a much smaller value 0.1 for READ '1' due to small I_{sr} caused by the body effect of the access transistor. Since increasing κ does not help the READ '1' access speed which is limited by the access transistor current, we have used κ to be 0.1 and 0.3 to compare with conventional DRAM as shown in Fig. 4(b). We finally choose the κ value of 0.9 in the effort to optimize both READ '0' and READ '1' operations in the comparison. The speed improvements of READ '0' and READ '1' operations compared with DRAM are 27.8% and 9.2% at $\Delta V_{bit}=30mV$, respectively. As is shown, the derived result ΔV_t matches the result of HSPICE simulation very well in a large ΔV_{bit} region and the relative value keeps increasing when ΔV_{bit} of DRAM saturates. Therefore, TRAM does not require stringent sensitivity of sense amplifier and can get by using simple sense amplifier.

Further research shows that the derived WRITE results also agree with the experimental results obtained by HSPICE simulation. For the WRITE operation, with $\alpha=1$, $V_{DD}=1.6V$, the κ_{max} in our case is 1.4. The WRITE '0' speed improvement compared with DRAM is 37.8% at $\kappa=0.9$. For WRITE '1' operation, due to the body effect and V_T drop, the write time is determined by the access transistor. However, TRAM also shows comparable speed as conventional DRAM.

3. CRITICAL CHARGE ANALYSIS

Soft error in memories is becoming a critical issue as technology continues to shrink. Soft error occurs at the storage node of memory cells when the induced external charge is larger than the critical charge (Q_c). Therefore, the critical charge is one of the most important parameters for estimating the soft error rates in memories. In this section, we analytically study the critical charge in TRAM and compare it with conventional DRAM technology.

In both DRAM and TRAM, the worst case for charge collection happens during the READ operation. The critical charge for DRAM cell can be expressed as [4]:

$$Q_c=\frac{1}{2}C_0V_{node}-(C_{bit}+C_s)\Delta V_{sen},$$

where C_0, C_{bit}, V_{node} and ΔV_{sen} are the storage capacitance, bit-line capacitance, storage node voltage and sense margin voltage, respectively.

The TRAM will flip when the meta-stable point is met. Let x_m represent the meta-stable point voltage. When the memory cell is in the standby mode, the critical charge for logic '0' and logic '1' are $Q_{sc0}=C_0(x_m-x_0)$ and $Q_{sc1}=C_0(x_1-x_m)$, respectively. Since the two RTDs are identical, in the standby mode, we have

$$Q_{sc0}=Q_{sc1}=C_0(\frac{1}{2}V_{DD}-I_qR_{12}).$$

In the worst case scenario, the critical charge Q_{r1} for READ '1' operation is obtained as:

$$Q_{r1}=Q_{sc1}-C_0(x_1-x), \qquad (2)$$

$$x=\begin{cases}(\kappa(\frac{V_{DD}}{R_1}-I_v)-A)/(\frac{\kappa}{R_1}-B) & (V_{DD}-V_p\leq x<V_{v2})\\ (\kappa(\frac{V_{DD}}{R_{12}}-I_q)-A)/(\frac{\kappa}{R_{12}}-B) & (V_{v2}\leq x<x_1).\end{cases}$$

Similarly, the critical charge Q_{r0} for READ '0' operation is obtained as:

$$Q_{r0}=Q_{sc1}-C_0(x-x_0), \qquad (3)$$

$$x=\begin{cases}(\kappa I_v+A)/(\frac{\kappa}{R_1}+B) & (V_{DD}-V_{v2}\leq x<V_{vp})\\ (\kappa I_q+A)/(\frac{\kappa}{R_{12}}+B) & (x_0\leq x<V_{DD}-V_{v2}).\end{cases}$$

Equation (2) and (3) show that critical charge is determined by the electrical characteristic of the RTD device, the supply voltage, as well as the size ratio κ. As shown in Fig. 5. Q_c initially increases very quickly and then tends to be saturated with the increase of κ. In order to get large critical charge for both READ '1' and READ '0' operations with small area penalty, κ should be choose properly. Table 2 shows the comparison of the critical charge in TRAM and DRAM. We have used $V_{DD}=1.6V$, $\kappa=0.9$. A and B are extracted from a 0.18 micron process technology by curve fitting the I-V characteristic generated by the HSPICE Level-49 model. The result shows that even in the worst case, TRAM still has considerably larger critical charge than DRAM.

4. POWER CONSUMPTION ANALYSIS

TRAM has lower power consumption than conventional DRAM. First of all, the bistable property eliminates the requirement of refreshing operation. Leakage currents are replenished by restoring current of the RTD-pair. Second, in DRAM, due to the cell leakage current variation, the worst-case leakage current has to be accounted for at each cell when performing the refreshing of cells which means that power consumption for refreshing operation is

Table 2: Critical charge comparison of TRAM and conventional DRAM

$C_{bit}(fF)$	DRAM Q_C (fC) 250		300		350		TRAM Q_C (fC)		
C_0 (fF)	$\Delta V_{sen}(mV)$ 40	50	$\Delta V_{sen}(mV)$ 40	50	$\Delta V_{sen}(mV)$ 40	50	Q_{sc} (fC)	Q_{r1} (fC)	Q_{r0} (fC)
40	20.4	17.5	18.4	15.0	16.4	12.5	28.0	27.2	22.1
35	16.6	13.8	14.6	11.3	12.6	8.80	24.5	23.8	19.4
30	12.8	10.0	10.8	7.50	8.80	5.00	21.0	20.4	16.6
25	9.00	6.25	7.00	3.80	5.00	1.25	17.5	17.0	13.8

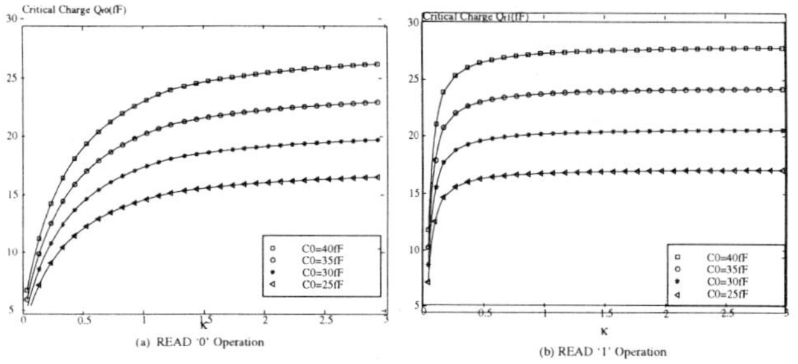

Figure 5: Critical charge of READ operation w.r.t. κ.

much more than what is actually required. In case of TRAM, the restoring current always equals to the actual leakage current of each single cell. Therefore, the power consumption is reduced further. In this section, we derive the power consumption of TRAM and then compare it with conventional DRAM.

Static power consumption of TRAM is determined by the leakage current and the dc current of the RTD-pair [5]. Assuming the average leakage current of the cell is I_{leak}, in order to guarantee the bistability of the TRAM, the following condition should be satisfied: $I_P > I_{leak} + I_v$. Let δ be the ratio of the maximum leakage current I_{leak}^{max} and the average leakage current I_{leak} in a DRAM chip. For all cells, the valley current I_v should be larger than a minimum valley current I_v^{min}:

$$I_v^{min} = \delta \frac{I_{leak}}{PVCR - 1}.$$

Therefore, for the worst case that the maximal leakage current is assumed for all the cells, the standby power is obtained as:

$$P_{TRAM} = I_{leak} V_{DD} \left(1 + \frac{\delta}{PVCR - 1}\right).$$

Because of the dynamic compensation of the leakage current, the standby power is then given by

$$P_{TRAM}^{Avg} = I_{leak} V_{DD} \left(\frac{1}{\delta} + \frac{1}{PVCR - 1}\right).$$

The power consumption of the DRAM due to the refresh operation is given by [5]:

$$P_{DRAM} = \delta I_{leak} V_{DD} \left(1 + \frac{C_{bit}}{C_0}\right) \frac{1}{1 - \frac{C_{bit}}{C_0} \frac{1}{\frac{V_{DD}}{2V_r} - 1}},$$

where V_r is the sense margin of the sense amplifier. Therefore, the power consumption of TRAM versus that of DRAM is derived as:

$$\eta = \frac{P_{TRAM}^{Avg}}{P_{DRAM}} = \frac{\left(\frac{1}{\delta} + \frac{1}{PVCR - 1}\right)}{\delta \left(1 + \frac{C_{bit}}{C_0}\right) \frac{1}{1 - \frac{C_{bit}}{C_0} \frac{2V_r}{V_{DD} - 2V_r}}}$$

Using the typical value $\delta = 50$, $PVCR = 10$, $C_{bit}/C_0 = 10$, the power consumption ratio is estimated to be $\eta \simeq 10^{-2}$ for an ideal sense amplifier with $V_r = 0$. With increase of the DRAM density, the power consumption ratio η will continue to decrease.

5. CONCLUSIONS

In this paper, we analytically model the READ and WRITE operations, critical charge for soft errors, and power consumption of single-transistor tunneling based RAM. The results are validated by HSPICE simulation. The size ratio of the RTD pair to the access transistor plays an important role in determining the access speed and the critical charge of TRAM. We show that critical charge is not as sensitive to ΔV_{sen} as in DRAM due to its self-latching property. The analytical study of power consumption shows that the dynamic compensation to the leakage current by the restoring current of RTD-pair will reduce the power consumption by one or two orders of magnitude. In a nut shell, TRAM has a great potential in future high-density, low-power, fast and highly reliable memory design. This paper analytically establishes the above claim.

6. REFERENCES

[1] J. P. A. van der Wagt, "Tunneling-Based SRAM," *Proceedings of the IEEE*, vol. 87, no. 4, pp. 571-595, Apr. 1999.

[2] E. Goto, *et al.*, "Esaki Diode High-speed logical circuits," *IRE Trans. Electron. Comput.*, vol. 9, pp. 25-29, Mar. 1960.

[3] S. Mohan, J. P. Sun, P. Mazumder, and G. I. Haddad, "Device and Circuit Simulation of Quantum Electronic Devices," *IEEE Trans. CAD* , vol. 14, no. 6, pp. 653-662, June 1995.

[4] S. Hyungsoon, "Modeling of Alpha-Particle-Induced Soft Error Rate in DRAM," *IEEE Trans. Elec. Dev.*, vol. 46, no. 9, pp. 1850-1857, Sept. 1999.

[5] T. Uemura and P. Mazumder, "Design and Analysis of Resonant Tunneling Diode (RTD) Based High Performance Memory System," *IEICE Trans. on Electrons*, vol. E82-C, no. 9, pp. 1630-1637, Sept. 1999.

THE CMOS/NANO INTERFACE FROM A CIRCUITS PERSPECTIVE

Matthew M. Ziegler and Mircea R. Stan
University of Virginia, ECE Department, Charlottesville, VA 22904
{ziegler, mircea}@virginia.edu

Abstract—We consider a circuit paradigm that combines conventional silicon microelectronics with emerging self-assembled nanoelectronics. Peripheral CMOS circuitry is used to drive the input signals and restore the output signals of nanoscale crossbar structures. We address a number of issues dealing with interfacing CMOS and nanoelectronics. Furthermore, we consider important metrics, such as, delay, area, and energy for a full-adder implemented in the mixed circuit paradigm.

I. Introduction

The past few years have put forth remarkable demonstrations of novel nanoscale and molecular devices assembled into small circuits [1]. In addition, rapid advancements in conventional silicon electronics have spawned predictions of an accelerated path towards the silicon "brick wall" [2]. When considered separately, conventional silicon electronics and nanoelectronics present different views for the future of integrated circuits. Silicon has a proven track record and has already revolutionized our society, but it's ability to continue upon Moore's Law will eventually cease. On the other hand, nanoscale and molecular circuits have not yet been assembled into complex systems, yet they may hold the potential for reaching new levels of computation.

However, rather than considering a future based on either silicon or nanoelectronics, we believe hybrid solutions composed of both technologies will be the key for achieving new levels of performance, power, and density [3]. Although, the maturity of nanoscale and molecular circuits has not yet reached the VLSI level of fabrication, we are able to explore nanoscale circuits via simulation.

II. A CMOS/nano Circuit Paradigm

Developing an integrated circuit composed of conventional silicon electronics and novel nanoelectronics will encounter a variety of issues. Fabrication challenges for achieving acceptable yields will no doubt be numerous. However, in this paper we address some of the circuit design issues that arise.

A. Crossbar-Based Nanoelectronics

As conventional silicon devices push the 100nm barrier traditional conceptions of nanoelectronics begin to blur. Therefore, we narrow our definition of nanoelectronics in this paper to nanoscale technologies that employ a "bottom-up" fabrication approach, such as self-assembly. We will refer to these technologies as *nano*. In turn, we will use the term *CMOS* refer to conventional solid-state technologies that rely on a "top-down" fabrication process, such as lithography. The key distinctions between these two fundamentally different fabrication approaches are the attainable features sizes and the freedom in determining arbitrary patterns. While a bottom-up process should allow smaller feature sizes, the resulting structures will typically be restricted to regular or periodic patterns. On the hand, top-down processes will be limited to larger feature sizes [4], but can realize arbitrary patterns. Thus, we believe the challenge of optimal CMOS/nano integration will be based on balancing the appropriate mixtures of smaller regular circuits (nano) and larger (CMOS) arbitrary circuits.

The most commonly targeted regular structure for self-assembled nanoelectronics is a crossbar. Fig. 1 shows a typical nanoscale crossbar structure that consists of a lower plane of parallel nanowires crossed perpendicularly by an upper plane of nanowires. A bistable junction exists between the wire crossings. The bistable junctions can be electrically altered to switch between a low resistance state (on-state) and a high resistance state (off-state). Typically the junction also has a diode-like behavior that allows currents flow under forward bias, but inhibits reverse bias current. This crossbar structure is a target for a variety physical implementations, such as, self-assembled nanowires, nanotubes, and nanoimprinted wires. Likwise, MRAM also consists conceptually of a similar crossbar structure with hysteretic junctions.

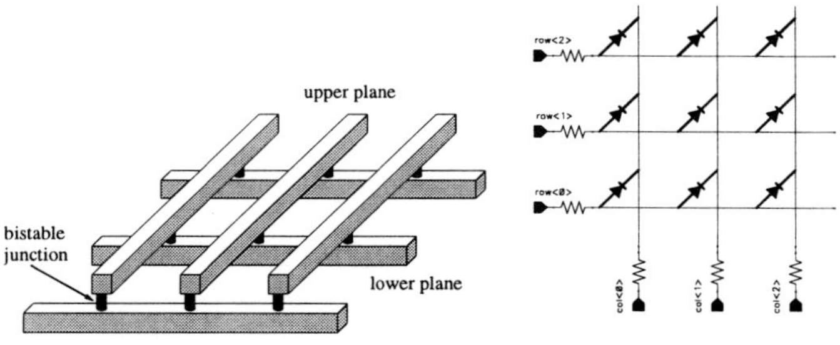

Fig. 1. A typical nanoscale crossbar structure consists of two planes of crossed nanowires with bistable devices at the junctions.

In particular, Hewlett-Packard and UCLA have pioneered the crossbar for molecular electronics [5], [6], [7], [8], [9]. Harvard has also demonstrated nanowire circuits that could be assembled in crossbar structures. [10]. The promise of crossbar-based nanoelectronics has also spawned a number of architectural ideas [11], [12], [13]. These works have motivated us to study circuit design issues for crossbars [3], [14].

B. The CMOS/nano Interface

In this paper, we consider interaction between on-chip CMOS circuits and nanoscale crossbar circuits. Unlike conventional CMOS, the nanoscale crossbars we consider do not have a fundamental dependency on the substrate, since the active devices are formed between the wire junctions. Thus, as in [3], we adopt the paradigm where the nano circuitry is fabricated on top of a conventional CMOS IC.

The physical properties of the CMOS/nano interface creates a number of fabrication challenges. However, there are higher-level issues when considering the CMOS/nano interface as well. Previous nano interface suggestions include nano decoder designs that are stochastic [7] or require nanoscale patterning resolution [13]. These CMOS/nano interface designs attempt to combine the following two goals:

Pitch Reduction - Communicating signals from the wire pitch in the CMOS technology, i.e., the microwire pitch (P_{micro}), to the nanoscale pitch in the nano crossbar, i.e., the nanowire pitch (P_{nano}).

Decoding - The ability to address a large address space in the nano crossbar with a smaller number of CMOS wires.

While we believe both pitch reduction and decoding are necessary, a solution we briefly outlined in [3] decouples the two goals. This approach relies on alignment precision for pitch reduction and a decoder

0-7803-7761-3/03/$17.00 ©2003 IEEE

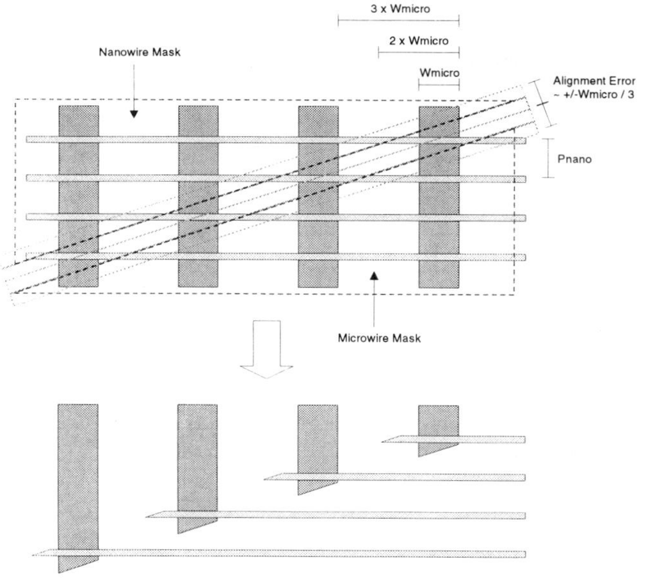

Fig. 2. Pitch reduction can be achieved by employing an interface scheme that relies on alignment accuracy.

programmed into the crossbar. As shown in Fig. 2, this approach begins with parallel nanowires perpendicularly overlapping an equal number of parallel microwires. Two masks are then used to remove the portions of the nanowires and microwires along a diagonal. This process assumes that the nanowires and microwires can be etched separately. The reliance on nanoscale pattern resolution is shifted to mask alignment precision. Using the rule of thumb that alignment will be equal to approximately one-third of the line width, this approach should be able reduce P_{nano} to at least the microwire line width (W_{micro}), as depicted in Fig. 2. Additional mask alignment accuracy as well as a number of other variations on this scheme should allow P_{nano} to be even smaller.

We can create nanowire crossbars by employing two or more of the interface structures described above, as shown in Fig. 3. We can then program the crossbar junctions by addressing individual crosspoints with the vertical programming microwires (connected to the horizontal nanowires) and horizontal address and data wires (connected to vertical nanowires). This scheme provides a mechanism to program a decoder into the crossbar. The decoder addresses a second crossbar we will refer to as the data array.

The decoder and data array pair can be used for logic or memory. The decoder is equivalent to a plane of AND gates, while the data array is essentially a plane of OR gates. Thus, two-level logic representations can be easily mapped to the crossbar. Fig. 4 shows a schematic of a full-adder realized in a crossbar along with the truth table.

Memory can also be implemented in these structures. However, memory requires the ability to program the data array via the decoder. Thus, to avoid inadvertently programming the decoder when writing to the data array, it may be necessary for the decoder to consist of junctions that are programmed at higher voltages than the data array.

To further reduce the area overhead associated with the CMOS/nano interface structure, it may also be desirable for multiple decoder and data array pairs to share the horizontal nanowires during the initial crossbar programming. This can be accomplished via a structure shown in Fig. 3. A method of post progamming isolation can then be used to allow each decoder and data array pair to function independently.

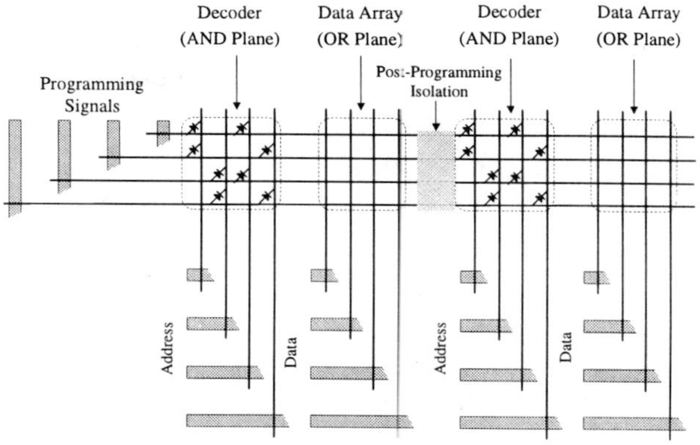

Fig. 3. A platform for logic and memory can be assembled by combining a number of diagonal interface structures.

Cin	A	B	Sum	Cout
1	1	1	1	1
1	1	0	0	1
1	0	1	0	1
1	0	0	1	0
0	1	1	0	1
0	1	0	1	0
0	0	1	1	0
0	0	0	0	0

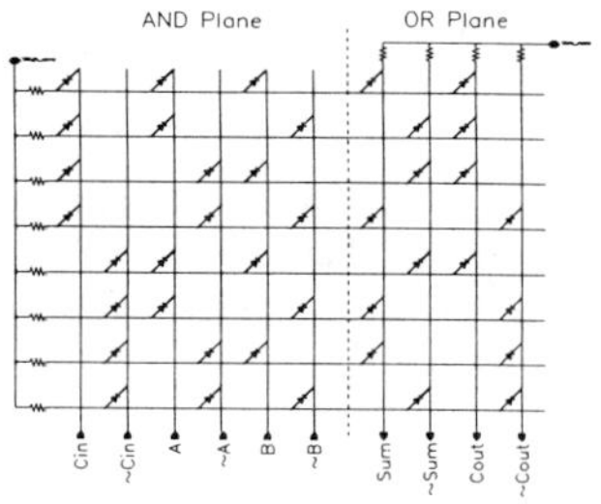

Fig. 4. The truth table for a full-adder and the corresponding mapping to a nano crossbar technology employing bistable diode junctions.

C. CMOS Interface Circuitry

Although the nano crossbar circuits hold the potential for high device densities, there are a number of inherent characteristics that pose problems for this technology alone. One of the most significant is the lack of signal gain in a diode-resistor circuit paradigm. Thus, signals on the crossbar must be periodically restored to the appropriate logic levels. In addition, the crossbar suffers from an inability to implement analog circuitry, a lack of an inversion function, and difficulty in implementing sequential logic. We suggested mixing nano crossbar circuits and CMOS on the same chip to achieve optimal device densities, performance, and power in [3]. In this paper we expand upon that notion by considering CMOS circuitry that can drive the nano crossbar inputs and sense the outputs. Fig. 5 shows a schematic of the mixed circuit. We have found that to meet certain design goals it may be advantageous to drive the nano crossbar with a different signal swing than the CMOS operating voltage levels. Thus, we use CMOS level shifters to drive the crossbar input lines. Furthermore, sense amplfiers are needed to restore the crossbar output signal to CMOS voltage levels. We use standard CMOS circuits for the level shifters and sense amplifiers, as shown in Fig. 6 a) and Fig. 6 b), respectively.

III. Circuit Simulation Results

For simulation we use the 70nm CMOS Predictive Technology Models (PTM) [15]. At the nano crossbar junctions we use models of the rotaxane molecules as described in [5] and [6]. The crossbar junctions are modeled with the Universal Device Model (UDM) [16] and implemented in Verilog-A. Our simulations use the Spectre circuit simulator from Cadence. Two additional system parameters of high importance are the contact resistance between the microwires and the nanowires

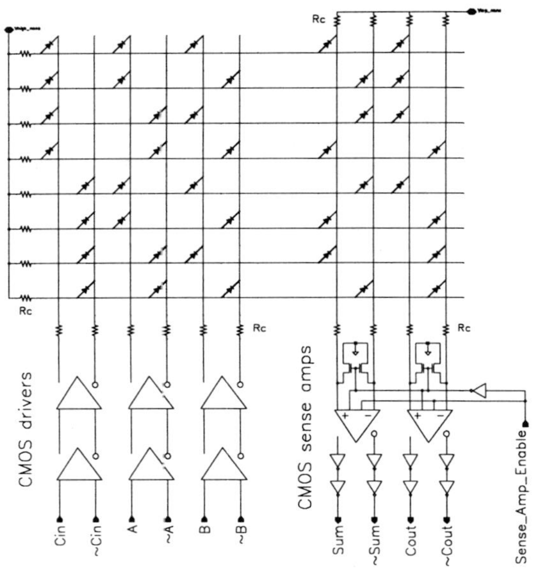

Fig. 5. The system we consider involves CMOS level shifters to drive the nano crossbar and CMOS sense amplifiers to restore the crossbar output to CMOS voltage levels.

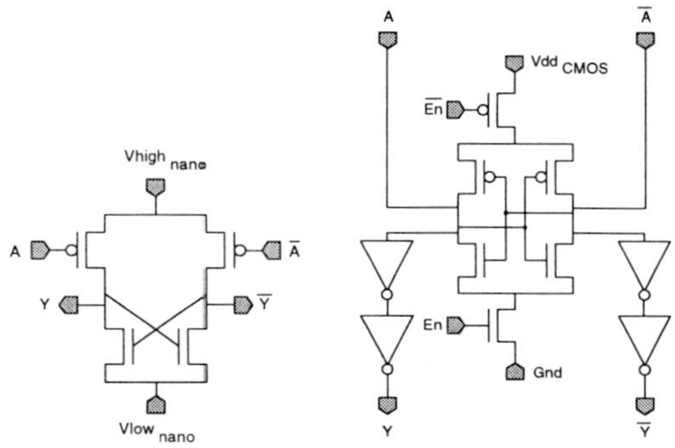

a) b)

Fig. 6. We use standard CMOS circuits for the CMOS interface, a) level shifters driving the nano crossbar inputs, b) cross-coupled inverter latches to sense the outputs.

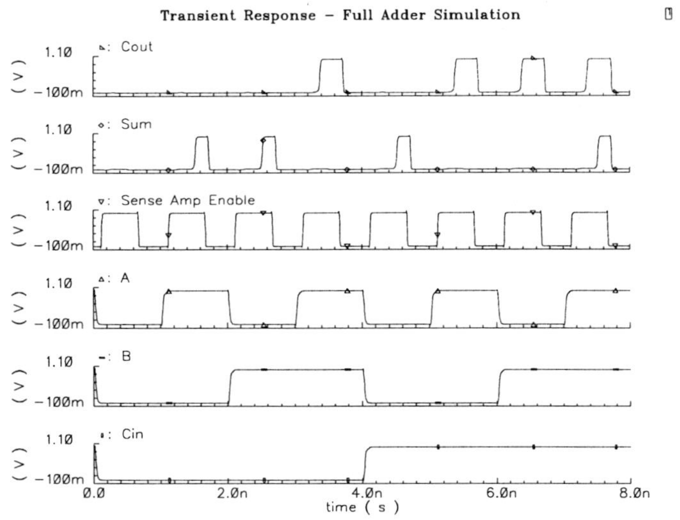

Fig. 7. Simulation waveforms for the nano crossbar and peripheral CMOS show the full-adder crossbar functioning correctly.

Performance	
Delay	491 ps
Cycletime	1 ns
Area	
Nano Crossbar	3 μm^2
Drivers	30 μm^2
Sense Amps	40 μm^2
Static Power	
Nano Crossbar	549 nW
Drivers	1.01 μW
Sense Amps	67.71 μW
Average Energy / Cycle	
Nano Crossbar	5.78e-16 J
Drivers	4.73e-14 J
Sense Amps	6.77e-14 J
Energy-Delay Product	5.67e-23 J-S

TABLE I
SIMULATED FIGURES OF MERIT FOR THE CROSSBAR FULL-ADDER.

(R_c) as well as the junction capacitance between the nanowires in the crossbar (C_j). The value of R_c is particularly important for the system we consider because it acts as the pull-up resistor for the AND plane and the pull-down resistor for OR plane. Likewise, the value of C_j is important because it determines the amount of capacitance that needs charged/discarged during switching. We set R_c to 1MΩ and C_j to 1aF, which is consistent with values from [8]. We assume the wire resistance of the nanowires, capacitance between the nanowires and substrate, and crosstalk capacitance between parallel wires are all negligible.

We use a 1 volt supply (Vdd_{CMOS}) for the 70nm CMOS circuitry; however, we have found that driving the nano crossbar with a 1 volt ($Vhigh_{nano}$) to -0.25 volt ($Vlow_{nano}$) signal swing provides a reasonable balance between performance and power consumption. Furthermore, we use 3 volts and -1.5 volts for the AND plane supply and OR plane supply of the crossbar, respectively. Finally, we use a differential sensing scheme that requires both the output signal and its complement to switch the sense amp. Despite the very small output voltage changes caused by the high resistance of the diode junction in the on-state in comparison to the contact resistance, the differential amplifiers are able to deterimine the correct output values. The simulated waveforms are shown in Fig. 7. Table I shows other figures of merits for the nano crossbar full-adder and peripheral CMOS.

Although the nano crossbar full-adder and peripheral CMOS we have simulated would be slower, larger, and consume more power in comparison to a full-adder implemented in 70nm CMOS, there are still a number of reasons to pursue mixed CMOS/nano circuits in the paradigm we have described above. More complex CMOS/nano systems will most likely target networks of crossbars, which may provide opportunities for reducing the amount of CMOS circuitry. Secondly, only about 0.5% of the average energy in Table I is consumed by the nano crossbar, while the rest is consumed by the peripheral CMOS. Further optimization of the peripheral CMOS may achieve additional reductions in energy, delay, and area.

In terms of area, we assume the nanowire pitch to be 70nm. To achieve the 70nm nanowire pitch, we estimate the microwire pitch to be 210nm. Table I shows our area estimates for the nano crossbar, which includes the diagonal interface structures and the crossbar itself. Area estimates for all the drivers and the sense amplifiers are also shown. Since the nano crossbar is on top of the peripheral CMOS, the total planar area will be determined only by nano crossbar area or the peripheral CMOS area, depending on which is greater. It is clear that the CMOS peripheral circuitry dominates the total area, while the nano crossbar occupies less than 4% of the total area. In comparison with a 70nm process standard library full-adder cell, we estimate the overall mixed CMOS/nano full-adder to be over twice the size of the library cell. However, considering only the nano crossbar and interface structures, the area would be only 11% of the full-adder library cell. From this we see that to take advantage of the small nano crossbar area, we need to reduce the amount of peripheral CMOS circuitry. A smaller amount of CMOS peripheral circuitry will also reduce energy. New nanoscale devices that can achieve signal gain and more complex crossbar structures are two possiblities for reducing the amount of CMOS peripherial circuitry.

Improving the present characteristics of the crossbar, such as, increasing the on-state current through the junctions and reducing the contact resistance, will enhance the performance of the crossbar. This would be particularly beneficial in the system we consider in this paper. Seeing that the resistance of the diode junction in the on-state is very large in comparison to the contact resistance, the output signals are inherently hampered.

However, entirely new devices will create new circuit paradigms. For example, devices that can provide gain without leaving the nano crossbar will allow extended levels of logic before interfacing to CMOS and allow a reduction in the amount peripheral CMOS. More elaborate crossbar structures may also make the CMOS/nano paradigm more advantageous. The addition of a third plane of parallel wires that cross the output nanowires provides an opportunity for a column decoder. Fig. 8 shows a crossbar system that allows multiple columns to share a sense amplifier. Likewise, this scheme allows parallel access to multiple data arrays, which reduces the number of drivers. Vertically stacking three or more nanowire planes may also allow additional density.

IV. Conclusion

While self-assembled nanoelectronics are still at an early stage of development, it is not too early to evaluate the potentials of these technologies. In this paper, we considered a crossbar-based nanoscale paradigm with peripheral CMOS circuitry. The nano crossbar paradigm contains inherent properties that will be advantageous for future electronics systems. However, for the paradigm we have considered here, it is clear that the overhead of the CMOS peripheral circuitry will need to be reduced to take advantage of the properties of the nanoscale technology. Further optimization of CMOS drivers and sense amplifiers specifically for nano integration may reduce the delay and energy of the mixed CMOS/nano circuits. In addition, new nanoscale devices that exhibit gain could allow extended computation without interfacing to CMOS, in turn reducing the amount of peripherial CMOS. Furthermore, more complex crossbar structures will allow more functionality to be shifted from the CMOS circuits onto the nano crossbar. Thus, there are a number of device, circuit, and architectural avenues that may lead to the raw density potentials inherent with self-assembled nanoelectronics.

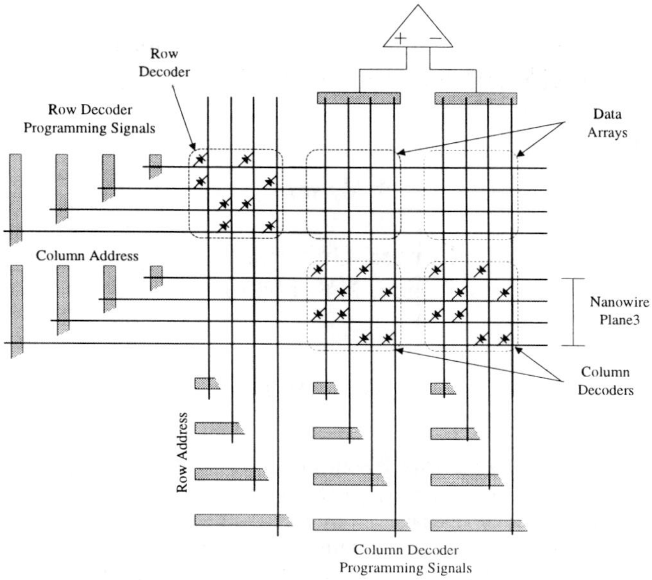

Fig. 8. The addition of a third plane of nanowires allows a column decoder to included in the nano crossbar structure.

V. Acknowledgments

Thanks to Dr. James Ellenbogen from the MITRE Corporation as well as Garrett Rose and Prof. Lloyd Harriott from the University of Virginia for interesting discussions on this topic. This work was supported by NSF grant No. 0210585 and by a UVa FEST award.

References

[1] R. F. Service, "Molecules get wired," *Science*, vol. 294, pp. 2442–2443, December 2001.

[2] Semiconductor Industry Association, "International technology roadmap for semiconductors," 2001.

[3] M. M. Ziegler and M. R. Stan, "A case for cmos/nano co-design," in *International Conference on Computer Aided Design*, November 2002.

[4] L. R. Harriott, "Limits of lithography," *Proceedings of the IEEE*, vol. 89, no. 3, pp. 366–374, March 2001.

[5] C. P. Collier, E. W. Wong, M. Belohradsky, F. M. Raymo, J. F. Stoddart, P. J. Kuekes, R. S. Williams, and J. R. Heath, "Electronically configurable molecular-based logic gates," *Science*, vol. 285, pp. 391–394, July 1999.

[6] P. J. Kuekes, J. R. Heath, and R. S. Williams, "Molecular wire crossbar memory," US Patent, number 6128214 (Hewlett-Packard), October 2000.

[7] P. J. Kuekes and R. S. Williams, "Demultimplexer for a molecular wire crossbar network (MWCN DEMUX)," US Patent, number 6256767 (Hewlett-Packard), July 2001.

[8] P. J. Kuekes, J. R. Heath, and R. S. Williams, "Molecular-wire crossbar interconnect (MWCI) for signal routing and communications," US Patent, number 6314019 (Hewlett-Packard), November 2001.

[9] Y. Luo, C. P. Collier, J. O. Jeppesen, K. A. Nielson, E. Delonno, G. Ho, J. Perkins, H. Tseng, T. Yamamoto, J. F. Stoddart, and J. R. Heath, "Two-dimensional molecular electronics circuits," *ChemPhysChem*, vol. 3, pp. 519–525, 2002.

[10] Y. Huang, X. Duan, Y. Cui, L. J. Lauhon, K. Kim, and C. M. Lieber, "Logic gates and computation from assembled nanowire building blocks," *Science*, vol. 294, pp. 1313–1316, November 2001.

[11] J. R. Heath, P. J. Kuekes, G. S. Snider, and R. S. Williams, "A defect-tolerant computer architecture: Opportunities for nanotechnology," *Science*, vol. 280, pp. 1716–1721, June 1998.

[12] S. C. Goldstein and M. Budiu, "Nanofabrics: Spatial computing using molecular nanoelectronics," in *28th International Symposium on Computer Architecture*, June 2001.

[13] A. DeHon, "Array-based architecture for molecular electronics," in *1st Workshop on Non-Silicon Computation (NSC-1)*, 2002.

[14] M. M. Ziegler and M. R. Stan, "Design and analysis of crossbar circuits for molecular nanoelectronics," in *The IEEE Nanotechnology Conference*, August 2002.

[15] Device Group at UC Berkeley, "Berkeley predictive technology model (BPTM)," http://www-device.eecs.berkeley.edu/˜ptm/.

[16] M. M. Ziegler, G. S. Rose, and M. R. Stan, "A universal device model for nanoelectronic circuit simulation," in *The IEEE Nanotechnology Conference*, August 2002.

THIN FILM PIN PHOTODIODES FOR OPTOELECTRONIC SILICON ON SAPPHIRE CMOS

A. Apsel

Cornell University
Ithaca, NY, 14853

E. Culurciello, A. G. Andreou, K. Aliberti

Johns Hopkins University
Baltimore, MD, 21218

ABSTRACT

In this paper, we consider both the utility of SOS substrates as a vehicle for optoelectronic packaging and the high speed silicon photodiodes available on a commercial SOS process. We show optical responses for six configurations of PIN photodiodes designed in this process. Our results indicate that photodiodes native to this process will operate at better than gigabit rates and produce signals over a range of visible wavelengths.

1. INTRODUCTION

Silicon on Insulator (SOI) CMOS technology is an emergent technology, poised to replace bulk CMOS for deep sub-micron VLSI [1]. The strength of traditional SOI as a candidate technology comes purely from its electrical properties. Devices produced in SOI are electrically isolated from one another, reducing parasitic capacitances, eliminating latch-up paths, and decreasing cross-talk. As a result, the delay and power consumption of these circuits can have up to a 300 percent performance gain over bulk CMOS circuitry [2].

Silicon on Sapphire (SOS) CMOS is a subset of SOI technology. It has long been considered a low performance alternative to standard SOI due to poor lattice matching of epitaxially grown silicon on sapphire. Recent developments in production of SOS CMOS using strained silicon have virtually eliminated these problems. Peregrine's ultra-thin silicon on sapphire (UTSi^{TM}) CMOS process now produces low leakage currents and f_{max}s in excess of 50 GHz for a .5μm process.

Unlike oxide based SOI CMOS technology, SOS CMOS offers advantages beyond its electrical properties. In addition to its usefulness as a high speed, low power CMOS platform, SOS wafers have desirable optical properties. In this paper we discuss the characteristics of sapphire substrates and thin film PIN photodiodes fabricated in a 0.5 μm SOS process. This technology is unique in that the sapphire substrate is transparent to wavelengths that may be absorbed by native photodetectors. While optical transmission through bulk substrates of 1330nm and 1550nm light has been demonstrated [3], native detectors cannot be utilized in these bulk CMOS or conventional oxide based SOI routing schemes as they can in SOS systems. We contend that thin film SOS photodiodes along with the characteristics of sapphire substrates allow the construction of opto-electronic systems with high-speed native photodetectors.

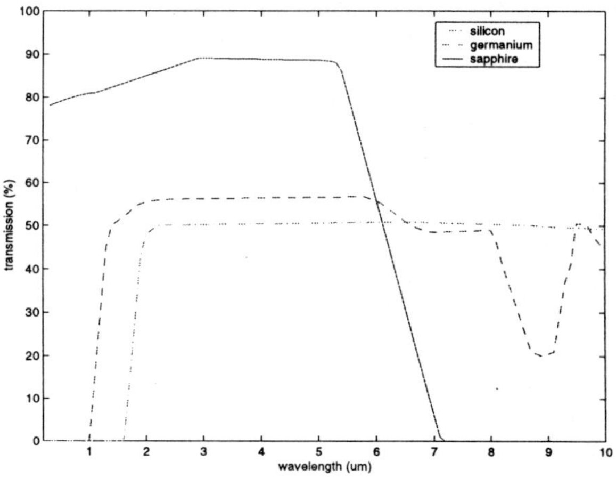

Figure 1: Transmission curves for sapphire, silicon, and germanium. [4]

2. OPTICAL PROPERTIES OF SAPPHIRE SUBSTRATES

Packaging of optoelectronic VLSI systems can be greatly simplified when signals transmit directly through electronic substrates [5, 3]. This also simplifies flip bonding of any front side contact optical elements such as VCSEL's or passive nano-photonics. Unfortunately, standard bulk and SOI CMOS wafers do not have good optical properties for our purposes, as shown in figure 1. The data indicates that over the wavelengths of commercially available VCSEL sources (850nm, 980nm) both silicon and germanium are opaque. Even in upper wavelength ranges of 1550nm, transmission in silicon is well below 100 percent and development of through wafer interconnects requires significant thinning of the substrate. Working systems have been produced at 1310nm and 1550nm using this technique, [3], however the complexity of this approach as well as the optical loss incurred adds cost and reduces yield significantly. Furthermore, working systems have not been constructed which include native photodetectors, crucial to high speed monolithic receiver design.

As opposed to bulk CMOS and SiGe wafers, SOS wafers allow through substrate transmission of optical signals from 300nm up to 6μm as shown in figure 1. We see that the transmission through sapphire is typically better than 80 percent over this wavelength range, reducing the need for thinned substrates when compared to bulk CMOS wafers. While internal reflections at interfaces of silicon nitride, silicon dioxide and sapphire do create addi-

0-7803-7761-3/03/$17.00 ©2003 IEEE

tional losses if post-processing is not performed, these losses will account for no more than a 12 percent loss of the transmitted light in the 400nm to 6μm wavelength range. We also note that SOS wafers combine two materials, silicon and sapphire that allow both transmission and absorption over a range of wavelengths.

3. NATIVE PHOTODIODES

In this section we present the characteristics of PIN photodiodes constructed in a commercial silicon on sapphire process. All of the photodiodes tested were fabricated in the Peregrine .5 μm UTSi^{TM} SOS process. The silicon layer in this process is epitaxially grown on a sapphire substrate and oxidized down to a thickness of 100nm. We present the results of tests performed on six PIN photodiodes of various geometries. We constructed all devices from abutting regions of low threshold "intrinsic" n and p silicon. We blocked the source drain implant doping in the junction region. The width of the source/drain block varies for different device geometries.

Figure 2: SOS PIN photodiode with interdigitated fingers.

Figure 3: SOS PIN photodiode with concentric ring.

We tested six devices of two geometries shown in figures 2 and 3. All devices are in either a parallel finger PIN photodiode geometry as in figure 2 or in a ring geometry as shown in figure 3. Three devices are 65μm$\times$65μm in size. Of these there is one ring photodiode with metal contacts to the source/drain implanted regions, one finger photodiode with metal contacts to the source/drain implanted regions, and one finger photodiode with metal contacts placed only at the edges of the device. The other three devices are 40μm$\times$40μm rings with variations in the lengths of designated junction regions where source/drain implants are blocked. The source/drain implant blocks are 2μm, 3μm, and 4.5μm respectively. Since these three devices have metal contacts to their source/drain implanted regions, the width of the blocked regions effects the spacing of metal contacts.

3.1. Spectral Response

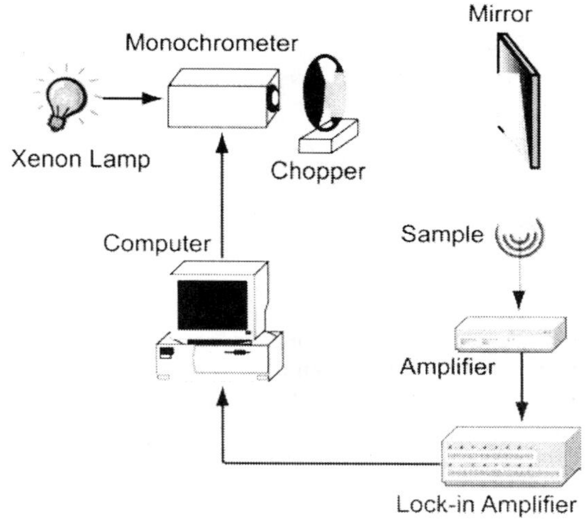

Figure 4: Experimental setup used to measure the spectral response of photo-diodes.

We measured the spectral response of the 6 geometries of thin film silicon PIN photodiodes using a Xenon lamp and the experimental set-up shown in figure 4. We measured the sensitivity of these geometries from 350nm to 850nm wavelengths with a reverse bias of 2.5V.

The resulting spectral response curves are shown in figure 5. These curves show the relative sensitivity of all photodiodes as a function of wavelength. There are several notable characteristics that are revealed in this data. We are not surprised to see that the shape of all 6 responses is very similar given that the material and the material thickness is the same in all cases. As a result the absorption of each photodiode should be similar. It is also not surprising to find that the most responsive photodiode is the device described above constructed from interdigitated fingers of p-type and n-type silicon with metal contacts only at the ends of the photodiode and not on each finger. We expect that since this device has the most open active area that it should respond most to incident light.

Next we examine the spectral response of two photodiode geometries covering the same total area. Both diodes, one ring configuration and one constructed of interdigitated fingers are 65μm$\times$65μm and have metal contacts to p and n silicon throughout the diode but distant from the junctions. Figure 5 indicates that the response of the ring photo-diode of the same size with the same active area between contacts of 1.2μm as the finger photo-diode, has higher responsivity over the wavelength range.

Finally, we examine the response of the three 40μm$\times$40μm ring photodiodes. The lengths of the active regions of these photodiodes are larger than the minimum 1.2μm width at 2μm, 3μm,

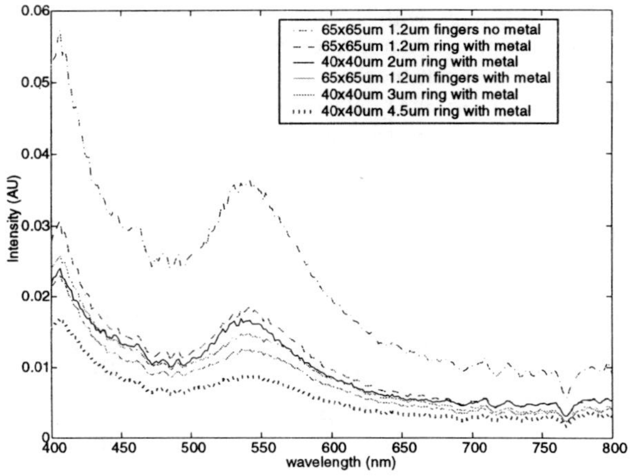

Figure 5: Spectral response of 6 geometries of thin film photodiodes in an SOS process.

and 4.5μ respectively. The intensity vs. wavelength curve indicates that the response of the photo-diodes drops with an increase in the length of the active region. We can conclude from this plot that the optimum width of an active region is below 2μm for photodiodes in this process. It is relatively simple to calculate a theoretical bound on the performance of PIN photodiodes in this process. We compare the measured responsivity of the most responsive photo-diode measure to this theoretical bound. From a series of sensitivity measurements using a source of known power output, we find that 1V on the intensity curve corresponds to .27 A/W of responsivity. We normalize this to the active area of the photodiode to determine the actual responsivity of this photo-diode, shown in figure 6.

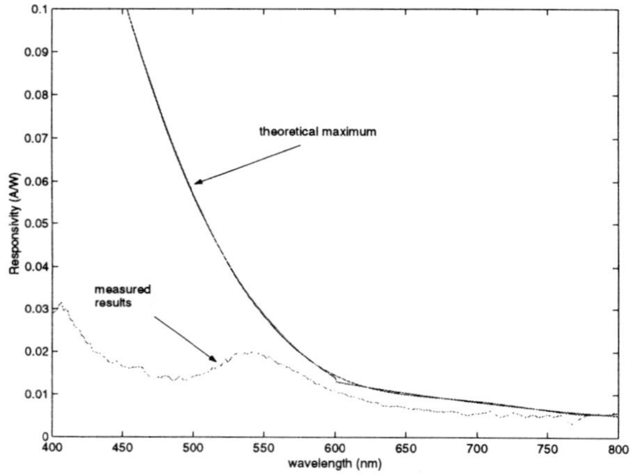

Figure 6: Responsivity of a 65$\times$65μm thin film SOS photodiode with minimum width 1.2μm junction vs. theoretical responsivity.

We begin calculating the maximum theoretical responsivity of PIN detectors in this process by calculating the theoretical maximum quantum efficiency, η, of a thin film of silicon. The following equation gives the quantum efficiency for a film of thickness L, given a surface reflectivity of r, and absorption as a function of wavelength for silicon, $\alpha(\lambda)$, that can be easily found in most optics texts. ζ is a fraction reflecting the percentage of electron hole pairs that contribute to photo-current.

$$\eta = (1-r)\zeta[1-e^{-\alpha L}] \quad (1)$$

ζ is a function of the quality of material used for the photodiode. For most commercial processes material is very good and this can be assumed to be unity.

Using this equation and the knowledge that L=100nm we can find a function for η by calculating r and using the absorption curve for silicon. In the peregrine UTSi^{TM} process, there are a series of three interfaces through which light must be transmitted before it is incident upon the silicon sample. The first interface is from air (n=1.03) to silicon nitride (n=2). The second is from nitride to silicon dioxide (n=1.45). The third interface is between silicon dioxide and silicon (n=3.38). We can determine the total transmission to the silicon sample by assuming that reflections from back surfaces have negligible retransmission and using the following transmission equation.

$$(1-r) = T = \frac{2n_1 n_2}{(n_1+n_2)^2} \quad (2)$$

The total transmission of 0.75 is the product of the three individual surface transmissions, equal to $1-r$. This gives us a complete expression for quantum efficiency as a function of wavelength. However, we want to examine the responsivity, not quantum efficiency in our comparison. Since responsivity is the current output for a given optical power input, we use the power of incident photons as a function of wavelength to determine the theoretical responsivity as follows.

$$R = \frac{\eta q}{h\nu} = \eta\frac{\lambda(\mu m)}{1.24} = \frac{0.75\lambda}{1.24}[1-e^{-\alpha L}] \quad (3)$$

Figure 6 shows the measured responsivity of the interdigitated finger PIN photo-diode compared to the calculated maximum responsivity of a thin film silicon PIN detector. We can see from this figure that the shape of the measured and theoretical responsivity match well from 570nm to 800nm wavelengths. Below 570nm there is a dip in the responsivity of the measured curve at approximately 490nm. This dip relates to an absorption phenomena at this wavelength. This may be due to a half wave reflectance set up by a 240nm oxide layer in the layers covering the silicon. Etching and removal of these reflective layers on the native SOS photodiodes will certainly improve the responsivity of these photodiodes and perhaps bring them to the theoretical response limit.

3.2. Frequency Response

In addition to the spectral response discussed in the previous section, we also measured the frequency response of the aforementioned six photodetectors. We probed the samples with 40 GHz 100 μm pitch ground-signal-ground probes, focused the 785nm beam onto the sample, modulated the beam with a signal generator, and measured the output of the photodiode with a spectrum analyzer. The bias across the photodiodes is 2.5V in all cases. Figures 7 and 8 show the resulting frequency responses for the photodiodes. All responses are compensated for the bandwidth of the laser diode and normalized to show the measured bandwidth of the devices, denoted by a 3dB drop in response.

For an intrinsic region of silicon less than 5μm wide the transit time of carriers limits the speed of our detectors at approximately

10 GHz. Below this the bandwidth is limited by the resistance and capacitance of the detector itself and the circuit configuration of the experimental set-up. Figure 7 shows the frequency response

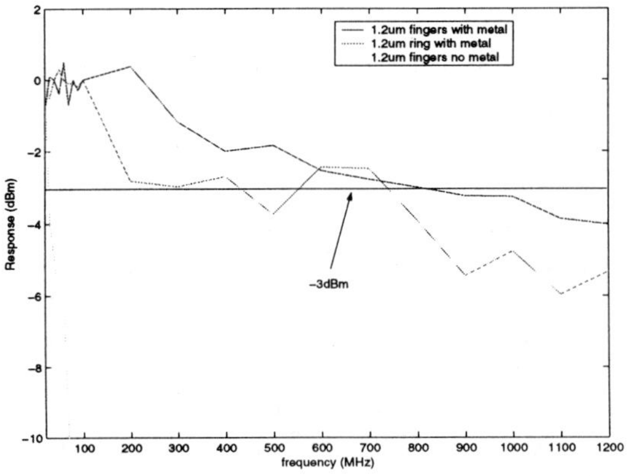

Figure 7: Frequency response of different geometries of 65×65μm SOS photodiodes.

for the three 65μm×65μm photodiodes. When we examined the spectral response of these above we noted that the most responsive device was the device which had metal contacts only on it's edges. While the responsivity of this device was nearly twice that of the equally sized devices with metal contacts, we note that the increased responsivity of this device comes with a dramatically reduced bandwidth due to increased resistance. The 3dB drop in frequency response occurs below 25 MHz, while the same geometry device with metal contacts throughout has a 3dB drop at approximately 850 MHz. The ring device has a somewhat reduced bandwidth over this geometry with a 3dB drop at approximately 450 MHz. These speeds are achievable with approximately half of the responsivity of the device without metal, making them more desirable for high speed designs where sensitivity can often be increased with only a linear reduction in bandwidth.

We note that these bandwidths are slower than expected for our junction widths. Transit times are limited to 10GHz for silicon PIN diodes less than 5μm wide. Considering the lack of substrate capacitance we approximate junction capacitances for these geometries to be much less than 100pF. If we calculate resistances for the two finger geometry devices with and without metal fingers, we find device resistances of approximately 4 Ω and 7 kΩ respectively. In the case of the 4 Ω detector, the 50 Ω input capacitance of the spectrum analyzer will dominate the intrinsic resistance of this device in measurement. When we consider the cases of these two detectors, we find that the measured bandwidths indicate a capacitance between 2pF and 3pF in both devices. This is clearly not a reasonable result for 65μm×65μm thin film silicon photo-detectors. We conclude that this capacitance must be induced by the experimental set-up. If we calculate the capacitance of the 40A-100-C GSG GGB picoprobes, we find that the total capacitance to ground of these probes is approximately 2.3pF. We attribute the bandwidth limitation of these measurements to this parasitic capacitance. More accurate and aggressive photodiode bandwidths can be measured by the addition of an output buffer or amplifier stage with a small gate load on these photodiodes. These measurements still provide useful information about the photodiodes. We note that the ring configuration must be more resistive

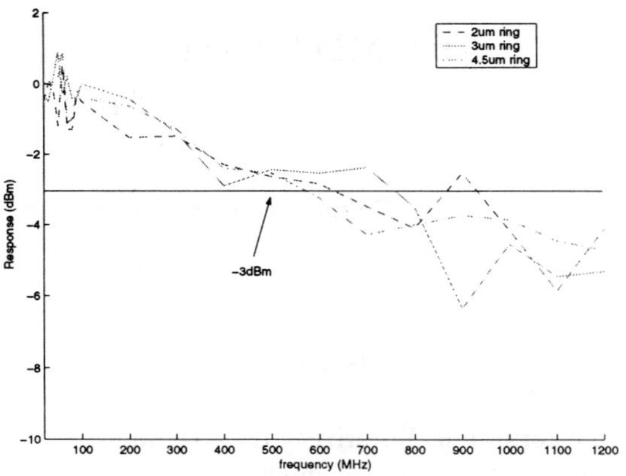

Figure 8: Frequency response of 40×40μm SOS ring photodiodes with varying junction widths.

than the finger configuration, for instance, and we see by our measured bandwidths that we can easily achieve GHz signaling rates with 65μm×65μm devices by limiting the load capacitance on the output node (now 2.3pF of probe capacitance) below 1.5pF.

While these speeds seem slow in comparison to GaAs and other direct bandgap detectors, they are fast compared to most commercial silicon PIN detectors. Honeywell's highest speed PIN detectors, for instance, have minimum rise times of 5ns under normal bias conditions. Furthermore, the SOS detectors are monolithic in a CMOS process, allowing easy integration of detectors and receiver circuitry for inexpensive packaging of opto-electronic systems at gigabit rates.

The next set of devices we tested were 40μm×40μm ring devices described in the previous section. Since the variation in these devices is the length of the region around the junction which does not have a highly doped source/drain implantation, the distance between metal contact rings is also varied. We expect that this varies the capacitance across the diode somewhat and would change the frequency response of these devices, favoring more distant contacts. However the results shown in figure 8 indicate no conclusive evidence of this due to RC bandwidth limitations of our set-up.

4. REFERENCES

[1] J. B. Kuo and Ker-Wei Su, CMOS VLSI *Engineering Silicon-on-Insulator*, Kluwer Accademic Publishers, 2000.

[2] S. Cristoloveanu, "SOI a metamorphosis of silicon," *IEEE Circuits and Devices*, vol. 15, no. 1, pp. 26–32, 1999.

[3] O. Vendier, S. Bond, M. Lee, S. Jung, M. Brooke, N. M. Jokerst, and R. P. Leavitt, "Stacked si CMOS circuits with a 40-Mb/s through-silicon optical interconnect," *IEEE Photonics Technology Letters*, vol. 10, no. 4, pp. 606–608, Apr. 1998.

[4] J. Jamieson, R. McFee, G. Plass, R. Grube, and R. Richards, *Infrared Physics and Engineering*, McGraw-Hill, 1963.

[5] A. Andreou, Z. Kalayjian, A. Apsel, P. O. Pouliquen, R. A. Athale, G. Simonis, and R. Reedy, "Hybrid integration of surface emitting vcsel's with ultra-thin silicon on sapphire (sos) cmos vlsi circuits," *Circuits and Systems Magazine*, Aug. 2001.

The IMD Cancellation Characteristics of Predistortion Linearizer in Microwave Transistor

Ung Hee Park, Kyung Hee Lee

Satellite Communications Antenna Research Team, Radio & Broadcasting Technology Lab., ETRI, Korea

Tel: +82-42-860-6987, E-mail: uhpark@etri.re.kr

Abstract

A predistortortion linearizer, the method that IMD signals are combined at the input port of the amplifier, because of its small size and good efficiency is frequently used in High Power Amplifier linearizer system. The amount of IMD signal cancellation by adjusting the amplitude and phase of predistorter is investigated by new experiment method. In the combining method of predistorter type, the magnitude and phase of combining signals cannot be easily expected due to different magnitude and phase of incoming signals. By experiment, it is measured that a predistortortion linearizer has lower amounts of IMD signal cancellation than those of the parallel cancellation method at the same condition (amplitude and phase).

I. Introduction

To obtain the maximum efficiency in High power amplifier, its operation point must be located near the saturation region where the highly nonlinear characteristic is inevitable. The nonlinear characteristics of high power amplifier are amplitude distortion, phase distortion, AM to AM conversion, AM to PM conversion and IMD, etc. The IMD (Intermodulation Distortion) is generated when two or more RF input signals are propagated through an amplifier[1][2]. It may be added to the received noise level, producing a new carrier to noise ratio level. The IMD specifications of base station in a mobile communication are very tight. The unwanted products should be at least 60dB lower than the main signal output. In order to meet the requirement, HPA in base station needs the linearization system. Currently the most popular linearization systems are feedforward method and predistortion method. A feedforward linearizer has the advantage of good IMD suppression and wide bandwidth, but it has low efficiency and large size. However, a Predistortion linearizer has the advantages of small size, low cost and good efficiency comparing with a feedforward linearizer.

The amount of IMD cancellation using a feedforward linearizer can be easily acquired with the mathematical method or the exercise using the parallel cancellation method. But, it is difficult to predict the amount of IMD cancellation using a predistortion linearizer exactly. This paper investigated the amount of IMD cancellation with a predistortion linearizer by adjusting the amplitude variation and phase variation experimentally. The results are analyzed and then compared with the amount of IMD cancellation using the parallel cancellation method.

II. The signal cancellation of the parallel Cancellation method [3][4]

The parallel combining method is that the power of combined signal by the signal phase and magnitude of the different paths is increased or decreased at the combining circuit. The example of the combining circuit is shown in Fig 1.

The amount of power cancellation in the coupler according to amplitude variance and phase variance of two input ports is shown in Fig2. Fig.2(b) is vector diagram of the coupler. A(A-port input power), D(B-port input power) and C(C-port output power) are vector. Output voltage C is given as eq.(1). Here, θ is the phase difference between A-port signal and B-port signal. Eq.(2) is voltage expression of vector C and Eq.(3) is power expression of vector C. If V_A, V_B have the same magnitude and θ has 180^0 degree in eq.(2), V_{OUT} is zero. Fig.3 is Cancellation chart of the different signal paths combining by phase and magnitude using ideal mathematical method.

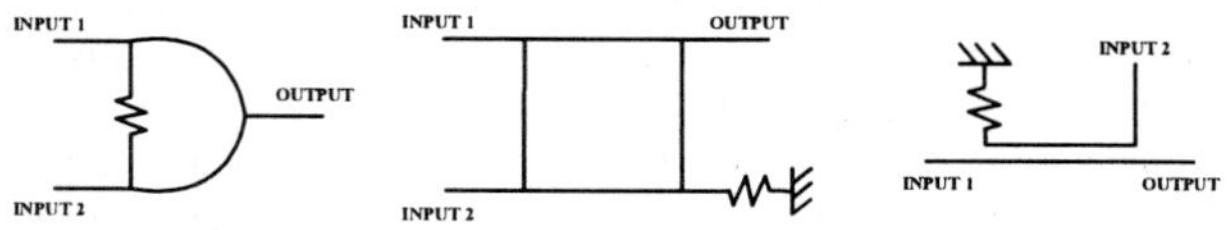

(a) Wilkinson (b) Branch-line (c) Coupler

Fig 1. Parallel cancellation circuit

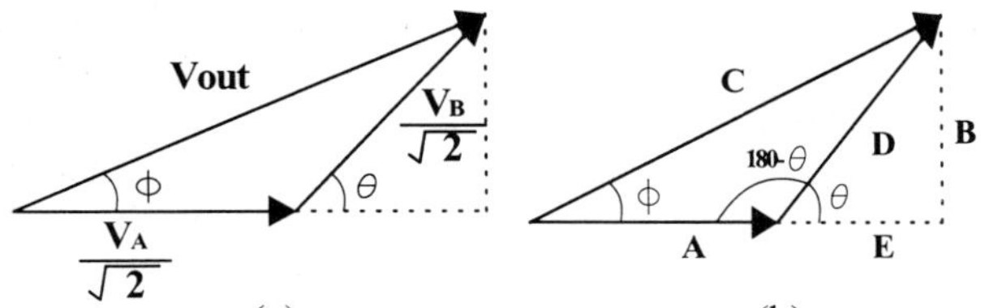

(a) (b)

Fig 2. Signal combining of the parallel cancellation method (a) signal combining (b) vector

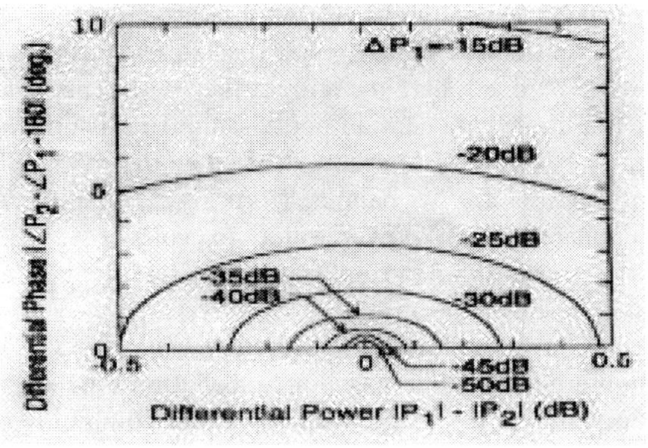

Fig 3. Cancellation chart of the parallel cancellation method by phase and magnitude

$$C^2 = A^2 + 2AD\,\mathrm{Cos}\,\theta + D^2 \qquad (1)$$

$$V_{out}^2 = \frac{V_A^2}{2} + \frac{V_B^2}{2} + 2\frac{V_A V_B}{2}\mathrm{Cos}\,\theta \qquad (2)$$

$$P_{out} = \frac{\left(\frac{V_{out}}{\sqrt{2}}\right)^2}{R_L} = \frac{V_{out}^2}{2R_L} \qquad (3)$$

The signal phase of V_{OUT} can be calculated with V_A and V_B. Eq.(4) and Eq.(5) are the signal phase of V_{OUT}.

$$\mathrm{Tan}\,\Phi = \frac{B}{A+E} \qquad (4)$$

$$\mathrm{Tan}\,\Phi = \frac{\frac{V_B}{\sqrt{2}}\mathrm{Sin}\,\theta}{\frac{V_A}{\sqrt{2}} + \frac{V_B}{\sqrt{2}}\mathrm{Cos}\,\theta} = \frac{V_B\,\mathrm{Sin}\,\theta}{V_A + V_B\,\mathrm{Cos}\,\theta} \qquad (5)$$

III. The IMD cancellation of a predistortion linearizer [3][4]

A. The estimation method

Predistortion linearizer has the circuit producing a distortion characteristic that is precisely complementary to the distortion characteristic of the high power amplifier and cascading the two paths in order to ensure that the linearized system has little intermodulation signals. The output power $v_o(t)$ of the high power amplifier can be expressed power series form as in eq.(6). If input signal($v_{i(t)}$) is A($\cos w_1 t + \cos w_2 t$), the output signal($v_{o(t)}$) has input frequency signal (ω_1, ω_2), second intermodulation frequency signal (dc, $2\omega_1$, $2\omega_2$, $\omega_1 \pm \omega_2$), third intermodulation frequency signal ($3\omega_1$, $3\omega_2$, $2\omega_1 \pm \omega_2$, $2\omega_2 \pm \omega_1$), etc. Except for third intermodulation frequency $2\omega_1 \pm \omega_2$, $2\omega_2 \pm \omega_1$, other intermodulation frequencies are out of band in communication system.

$$v_o = k_1 v_i + k_2 v_i^2 + k_3 v_i^3 + \cdots \qquad (6)$$

$$\begin{aligned} v_o &= k_1 A(\cos\omega_1 t + \cos\omega_2 t) + k_2 A^2(\cos\omega_1 t + \cos\omega_2 t)^2 + k_3 A^3(\cos\omega_1 t + \cos\omega_2 t)^3 \\ &= k_2 A^2 + k_2 A^2 \cos(\omega_1 - \omega_2)t + (k_1 A + \tfrac{4}{9}k_3 A^3)\cos\omega_1 t + (k_1 A + \tfrac{4}{9}k_3 A^3)\cos\omega_2 t \\ &+ \tfrac{3}{4}k_3 A^3 \cos(2\omega_1 - \omega_2)t + \tfrac{3}{4}k_3 A^3 \cos(2\omega_2 - \omega_1)t \\ &+ k_2 A^2 \cos(\omega_1 + \omega_2)t + \tfrac{1}{2}k_2 A^2 \cos 2\omega_1 t + \tfrac{1}{2}k_2 A^2 \cos 2\omega_2 t \\ &+ \tfrac{3}{4}k_3 A^3 \cos(2\omega_1 + \omega_2)t + \tfrac{3}{4}k_3 A^3 \cos(2\omega_2 + \omega_1)t \\ &+ \tfrac{1}{4}k_3 A^3 \cos 3\omega_1 t + \tfrac{1}{4}k_3 A^3 \cos 3\omega_2 t + \ldots \end{aligned}$$

.... (7)

This paper describes the amount of IMD suppression in test amplifier by adjusting IMD amplitude and phase of predistorter which is comprised of IM(Intermodulation) generator, voltage variable phase shifter and voltage variable attenuator as in Fig.4. In Fig.4, V_{S1} is Vin signal and V_{S3} is third IMD of Vin. If Vin is equal to eq.(8), V_{S1} is given as eq.(9) and V_{S3} is given as eq.(10), here, α means delay phase of the element. When V_{S1} is put into the amplifier having nonlinear characteristic, the amplifier output Vout,s1 has intermodulation signals as in eq.(11). In eq.(11) β is the phase delay of the amplifier. When V_{S3} is put into the amplifier,

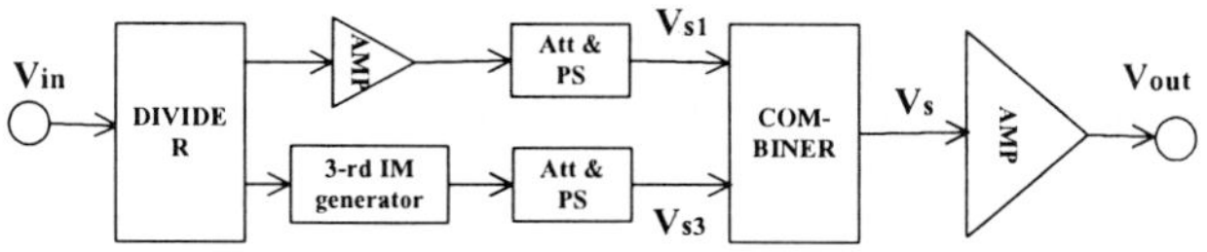

Fig 4. Test circuit block diagram between input IMD signal and output IMD signal of the amplifier

$V_{out,S3}$ is obtained eq.(12). If Vs1 and Vs3 in put at the input of the amplifier simultaneously, the total output is given as in eq.(13). If $K_3 K^3_{S1} A'_3 = K_1 K_{S3} A_3$, $(\alpha_1+\beta_3) = (\alpha_3+\beta_1+180^o)$ in eq.(13), the output of the amplifier has no IMD signal. This paper also shows the amount of IMD cancellation with the variation of K_{S3} and α_3.

$$V_i = A\cos(\omega_1 t + \phi_1) + A\cos(\omega_2 t + \phi_2) \quad (8)$$

$$V_{S1} = K_{S1} A\cos(\omega_1 t + \phi_1 + \alpha_1) + K_{S1} A\cos(\omega_2 t + \phi_2 + \alpha_1) \quad (9)$$

$$V_{S3} = K_{S3} A_3 \cos((2\omega_1 - \omega_2)t + 2\phi_1 - \phi_2 + \alpha_3) + K_{S3} A_3 \cos((2\omega_2 - \omega_2)t + 2\phi_2 - \phi_1 + \alpha_3) \quad (10)$$

$$\begin{aligned} V_{out,S1} &= K_1 V_{S1} + K_3 V^3_{S3} + K_5 V^5_{S5} \\ &= K_1 K_{S1} A\cos(\omega_1 t + \phi_1 + \alpha_1 + \beta_1) \\ &\quad + K_1 K_{S1} A\cos(\omega_2 t + \phi_2 + \alpha_1 + \beta_1) \\ &\quad + K_3 K^3_{S1} A'_3 \cos((2\omega_1 - \omega_2)t + 2\phi_1 - \phi_2 + \alpha_1 + \beta_3) \\ &\quad + K_3 K^3_{S1} A'_3 \cos((2\omega_2 - \omega_1)t + 2\phi_2 - \phi_1 + \alpha_1 + \beta_3) \end{aligned} \quad (11)$$

$$\begin{aligned} V_{out,S3} &= K_3 V_{S3} = K_1 K_{S3} A_3 \cos((2\omega_1 - \omega_2)t \\ &\quad + 2\phi_1 - \phi_2 + \alpha_3 + \beta_1) \\ &\quad + K_1 K_{S3} A_3 \cos((2\omega_2 - \omega_1)t \\ &\quad + 2\phi_2 - \phi_1 + \alpha_3 + \beta_1) \end{aligned} \quad (12)$$

$$\begin{aligned} V_{out} &= V_{out,S1} + V_{out,S3} \\ &= K_1 K_{S1} A\cos(\omega_1 t + \phi_1 + \alpha_1 + \beta_1) + \\ &\quad K_1 K_{S1} A\cos(\omega_2 t + \phi_2 + \alpha_1 + \beta_1) \\ &\quad + [K_3 K^3_{S1} A'_3 \cos((2\omega_1 - \omega_2)t + \\ &\quad 2\phi_1 - \phi_2 + \alpha_1 + \beta_3) \\ &\quad + K_1 K_{S3} A_3 \cos((2\omega_1 - \omega_2)t + \\ &\quad 2\phi_1 - \phi_2 + \alpha_3 + \beta_1)] \\ &\quad + [K_3 K^3_{S1} A'_3 \cos((2\omega_2 - \omega_1)t + \\ &\quad 2\phi_2 - \phi_1 + \alpha_1 + \beta_3) \\ &\quad + K_1 K_{S3} A_3 \cos((2\omega_2 - \omega_1)t + \\ &\quad 2\phi_2 - \phi_1 + \alpha_3 + \beta_1)] \end{aligned} \quad (13)$$

B. The results

For the experiment, an amplifier which has 1.85GHz centered amplifier was implemented using Motorola MRF-6401 with AB-class (bias condition: 26V, 80mA). When input power is 18.4dBm/t, Output power is 23.5dBm/t and 3-rd IMD power is 2.9dBm/t. In Fig.4, the combiner is replaced by the 27.3dB coupler. Table.1 shows the IMD cancellation value according to the variation of the amplitude and the phase of the predistorter in MRF-6401 amplifier system. The reference value of the 3-rd IMD amplitude is –24.4 dBm(=2.9 dBm-27.3 dB) and the reference value of the 3-rd IMD phase is 24.0° (= –156.0° + 180.0°) in predistorter block. Fig.5 shows the amount of IMD cancellation by amplitude error and phase error for the optimum amplitude and phase (-24.4dBm, 24.0°) figures. It is realized that the amplitude ± variance of the optimum amplitude cancellation point,–24.4dBm, is unbalance. However, the phase ± variance of the optimum phase cancellation point,24.0°, is balance from Fig.5. In spite of the amplitude change from power value to voltage value, the amplitude ± variance of the optimum amplitude cancellation point is unbalance.

In Table.2, it compares the difference value between the cancellation amount by the parallel cancellation type with the cancellation amount by the predistorter power combining type. Though 25dB of IMD cancellation value was obtained by the predistortion power combine method, the IMD cancellation value by the parallel cancellation method was from 31dB to 35dB in the same amplitude error and phase error.

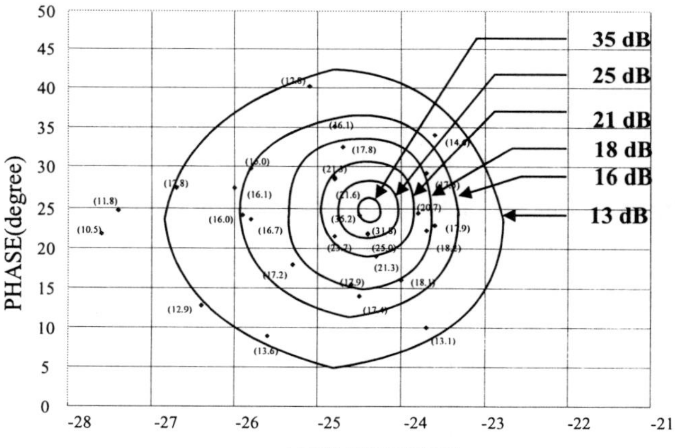

Fig 5. Amplifier IMD cancellation characteristic by the input IMD variation of predistorter

Table 1. Output IMD Cancellation value by the input IMD variation of predistorter

Input power (dBm)	Input phase (degree)	Output IM (dBm)	Cancellation (dBc)
-25.3	17.9	-14.3	17.2
-24.8	21.5	-20.8	23.7
-24.8	56.9	-5.1	8
-24.6	15.4	-15	17.9
-25.6	8.9	-10.7	13.6
-26.4	12.8	-10	12.9
-24.3	19	-18.4	21.3
-27.4	24.7	-8.9	11.8
-23.8	24.4	-17.8	20.7
-23.7	10	-10.2	13.1
-23.6	22.9	-15	17.9
-24.5	14	-14.5	17.4
-24	16	-15.2	18.1
-24.7	32.5	-14.9	17.8
-24.8	28.7	-18.4	21.3

(Tr. : MRF-6401, Input power: 18.4dBm/t, Output power: 23.5dBm/t, Output 3-rd IMD power: 2.9dBm)

Table 2. Cancellation comparison between parallel Cancellation and predistorter power combine (Reference point: Power –24.4dBm. Phase –156.0°)

Cancellation (dB)	Parallel Cancellation (dBm,degree)	Predistorter Combine type (dBm,degree)	The difference value of cancellation
35dB	(-24.4, 26)	(-24.4, 25)	6dB
	(-24.4, 22)	(-24.4, 23)	6dB
	(-24.7, 24)	(-24.5, 24)	10dB
	(-24.1, 24)	(-24.3, 24)	10dB
25dB	(-24.4,30.4)	(-24.4, 27)	6dB
	(-24.4,17.6)	(-24.4, 21)	6dB
	(-25.4, 24)	(-24.75, 24)	9dB
	(-23.4, 24)	(-24.1, 24)	10dB
21dB	(-24.4,34.2)	(-24.4, 29)	6dB
	(-24.4 13.8)	(-24.4, 19)	6dB
	(-25.55, 24)	(-25.1, 24)	7dB
	(-22.85, 24)	(-23.8, 24)	8dB
18dB	(-24.4, 38)	(-24.4,32.5)	5dB
	(-24.4, 10)	(-24.4,15.6)	5dB
	(-26.6, 24)	(-25.5, 24)	6dB
	(-22.2, 24)	(-23.6, 24)	9dB

IV. Conclusion

This paper dealt with the IMD cancellation condition of predistortion linearizer by varying 3-rd IMD amplitude and phase of predistorter and the IMD cancellation difference between parallel cancellation type and predistorter type by the experiment. As a result, the IMD cancellation condition of predistorter type showed that the amplitude (±) variance is unbalance and the phase (±) variance is balance at the optimum cancellation point and the IMD cancellation value of predistorter type has lower value than the IMD cancellation value of parallel cancellation at the same amplitude difference and phase difference. It is ,also, figured out that the lower value of the optimum amplitude cancellation point has more good cancellation amount than the higher value of the optimum cancellation point in the predistortion linearizer.

Reference

[1] Stephen A. Maas, Nonlinear Microwave Circuits, Artech House, 1988, pp.155-207

[2] N.Suematsu, Y.Iyama, O.Ishida, "Transfer Characteristic of IM3 Relative Phase for a GaAs FET Amplifier",*IEEE Trans. Microwave Theory and Tech.*, Vol.45, No.12,pp.2509-2513, Dec. 1997

[3] Tri T. Ha, Solid-State Microwave Amplifier Design, (John Wiely and Sons, 1981), pp.202-283.

[4] P.B.Kenington, High-Linearity RF Amplifier Design, Artech House, 2000

A

B

AUTHOR INDEX ISCAS 2003

Baki, Rola I-657
Balasa, Florin V-489
Balasuriya, Arjuna III-558
Balkir, Sina IV-776
Balkir, Sina IV-896
Balsi, Marco IV-636
Baltes, Henry III-542
Baltes, Henry IV-852
Balya, David III-838
Balya, David III-774
Bampi, Sergio I-773
Bampi, Sergio V-421
Bandi Rachaiah, Sripriya IV-596
Bandyopadhyay, Abhishek V-13
Banerjee, Ansuman V-249
Banerjee, Soumitro III-92
Bansilal, P H V-661
Bao, Feng II-596
Barcella, Marco V-429
Barrera, Paola V-777
Barrettino, Diego III-542
Barrettino, Diego IV-852
Barry, John II-13
Bartoloni, Stefano IV-624
Baruqui, Fernando I-477
Baschirotto, Andrea I-493
Basedau, Philipp I-1049
Bashirullah, Rizwan V-169
Bashirullah, Rizwan V-5
Basile, Adriano III-842
Basu, Anup II-812
Bauer, Peter H. IV-568
Bauer, Peter H. IV-824
Baxter, Chris III-786
Bayoumi, Magdy V-481
Patch Beadle II-1
Beck, Douglas I-853
Beck, Robert IV-289
Beck, Sungho I-285
Becker, Rolf I-1049
Becker, Rolf III-602
Bednara, Marcus II-268
Beerel, Peter V-361
Begum, Momotaz II-428
Ben Jebara, Sofia IV-432
Benaissa, Mohammed II-264
Benjamin Premkumar, Annamalai IV-229
Benjapolakul, Watit V-785
Berg, Yngvar I-345
Berg, Yngvar I-125
Berg, Yngvar V-193
Bergen, Stuart IV-169
Berkeman, Anders V-333
Berkovich, Yefim III-435
Berkovich, Yefim III-296
Bermak, Amine IV-772
Bermak, Amine V-685
Bermak, Amine V-833
Bertoni, Guido V-145
Besbes, Hichem IV-432
Bettarelli, Ferruccio IV-624
Bhasaputra, Pornrapeepat III-375
Bhattacharya, Mrinmoy IV-193
Bhattacharya, Mrinmoy IV-237
Bhattacharya, Mrinmoy IV-249
Bhattacharya, Sambuddha IV-660
Bhattacharya, Sambuddha IV-704
Biernacki, Janusz I-97
Biey, Mario III-590
Bilavarn, Sebastien V-589
Bircan, Aril V-145
Bistritz, Yuval III-682
Biswas, Mainak II-604
Bjoernsen, Johnny III-906
Bjork, Christian II-161
Blionas, S. II-73
Blaauw, David IV-668
Boahen, Kwabena IV-800
Boashash, Boualem II-212
Boashash, Boualem V-29
Boashash, Boualem V-33
Bode, Peter III-614
Bofill-i-Petit, Adria V-817
Bonomo, Claudia IV-816
Boole, E. IV-516
Boonchu, Boonchai I-405
Boonprasert, Udomsak III-371
Boonyaroonate, Itsda III-284
Boonyaroonate, Itsda III-355
Boonyaroonate, Itsda III-359
Bora, Prabin K. II-396
Borghetti, Fausto III-526
Bose, Nirmal III-662
Bose, Tamal II-388
Bose, Tamal IV-85
Boser, Bernhard III-930
Bossuet, Lilian V-589
Bouguezel, Saad III-698
Bouguezel, Saad IV-65
Bouguezel, Saad IV-97
Bouzerdoum, Abdesselam II-464
Brajovic, Vladimir V-825
Brand, Oliver III-542
Bregovic, Robert IV-141
Brennan, Paul I-961
Breveglieri, Luca V-145
Brooke, Martin I-1009
Brookes, Mike V-613
Brouse, Keith I-693
Brown, Andrew III-894
Brown, Richard I-665
Bruguera, Javier IV-121
Bruton, Leonard III-702
Bruton, Leonard IV-476
Bruun, Erik I-1069
Bruun, Erik I-1065
Bryant, Brad III-292
Buchaillot, Lionel III-530
Bucolo, Maide IV-888
Bucolo, Maide V-793
Bull, David II-959
Bulsara, Adi IV-768
Bumrungkeeree, Surachet III-359
Bunnjaweht, Sawat I-713
Buonomo, Antonio III-144
Burdenski, Ralf III-602
Burg, Andreas V-113
Burian, Adrian IV-488
Burian, Adrian V-729
Burton, Phil I-593
Byun, Hyun-Geun V-93
Byun, Hyunil II-780

C

Caballero Gaudes, Cesar II-85
Caccetta, Lou III-886
Cacopardi, Saverio II-41
Cafagna, Donato III-60
Caffarena, Gabriel IV-309
Cagdaser, Baris III-930
Cai, Yici IV-708
Cai, Yici IV-720
Cai, Yici V-493
Calabro, Antonino V-777
Calderon-Martinez, Jose V-749
CalôBa, Luiz V-701
Calvo, Belen I-557
Campi, Fabio V-133
Campos, Marcello L. R. IV-13
Campoy-Cervera, Pascual V-749
Canagarajah, Nishan II-959
Canagarajah, Nishan III-44
Cannas, Barbara III-156
Cantoni, Antonio II-516
Cao, Bin IV-520
Cao, Bin IV-536
Carcenac, Franck III-462
Cardarilli, Gian Carlo II-37
Cardarilli, Gian Carlo V-413
Cardarilli, Giancarlo V-649
Cardells-Tormo, Francisco II-260
Cardozo, Rodrigo V-421
Carlosena, Alfonso I-13
Carmona, Ricardo III-522
Carrara, Francesco I-289
Carreras, Carlos IV-309
Carrillo, Juan I-165
Carrillo, Juan I-277
Carro, Luigi I-773
Carro, Luigi IV-61
Carro, Luigi V-421
Carusone, Anthony V-449
Carvajal, Ramon I-589
Carvajal, Ramon I-237
Carvajal, Ramon I-269
Carvajal, Ramon I-677
Carvajal, Ramon I-805
Carvajal, Ramon I-817
Carvajal, Ramón I-157
Carvajal, Ramon G. I-781
Carvajal, Ramon G. I-801
Carvajal, Ramon G. I-813
Cassia, Marco I-1065
Castorina, Salvatore III-510
Castorina, Salvatore IV-760
Castro-López, Rafael IV-632
Cauwenberghs, Gert I-769
Cauwenberghs, Gert I-777
Cauwenberghs, Gert IV-832
Cauwenberghs, Gert IV-508
Cauwenberghs, Gert V-13
Cavin, Ralph V-169
Celinski, Peter V-233
Celma, Santiago I-5
Celma, Santiago I-557
Centurelli, Francesco IV-636
Cetin, Ediz IV-69
Chai, Douglas II-464
Chaji, G. IV-137
Chakrabarti, Partha V-249
Chakrabartty, Shantanu IV-832
Chakraborty, Mrityunjoy IV-452
Chakraborty, Satyabrata III-140
Chambers, Jonathon III-40
Chamnongthai, Kosin III-355
Chamnongthai, Kosin V-809
Chan, Cheong Fat II-185
Chan, Cheong Fat II-744
Chan, Cheong Fat V-337
Chan, Francis H.Y II-1
Chan, Mansun I-705

D

E

F

G

H

I

J

K

L

M

N

O

P

Q

R

T

U

V

W

X

Y

Z